T4-AQL-646

COMPREHENSIVE ORGANIC SYNTHESIS

IN 9 VOLUMES

COMPREHENSIVE ORGANIC SYNTHESIS

Selectivity, Strategy & Efficiency in Modern Organic Chemistry

Editor-in-Chief
BARRY M. TROST
Stanford University, CA, USA

Deputy Editor-in-Chief
IAN FLEMING
University of Cambridge, UK

Volume 8
Reduction

Volume Editor
IAN FLEMING
University of Cambridge, UK

PERGAMON PRESS
OXFORD • NEW YORK • SEOUL • TOKYO

ELSEVIER SCIENCE Ltd
The Boulevard, Langford Lane
Kidlington, Oxford OX5 1GB, UK

First edition 1991
Second impression 1993
Third impression 1999

Library of Congress Cataloging in Publication Data

Comprehensive organic synthesis: selectivity, strategy and efficiency in modern organic chemistry/editor[s] Barry M. Trost, Ian Fleming.
p. cm.
Includes indexes.
Contents: Vol. 1.–2. Additions to C-X[pi]-Bonds — v. 3. Carbon–carbon sigma-Bond formation — v. 4. Additions to and substitutions at C-C[pi]-Bonds — v. 5. Combining C-C[pi]-Bonds — v. 6. Heteroatom manipulation — v. 7. Oxidation — v. 8. Reduction — v. 9. Cumulative indexes.
8. Organic Compounds — Synthesis I. Trost, Barry M. 1941–
II. Fleming, Ian, 1935–
QD262.C535 1991
547.2—dc20 90-26621

British Library Cataloguing in Publication Data

Comprehensive organic synthesis
8. Organic compounds. Synthesis
I. Trost, Barry M. (Barry Martin) 1941–
547.2

ISBN 0-08-040599-1 (Vol. 8)
ISBN 0-08-035929-9 (set)

♾ ™ The paper used in this publication meets the minimum requirements of American National Standard for Information Sciences — Permanence of Paper for Printed Library Materials, ANSI Z39.48-1984.

Contents

Preface

The emergence of organic chemistry as a scientific discipline heralded a new era in human development. Applications of organic chemistry contributed significantly to satisfying the basic needs for food, clothing and shelter. While expanding our ability to cope with our basic needs remained an important goal, we could, for the first time, worry about the quality of life. Indeed, there appears to be an excellent correlation between investment in research and applications of organic chemistry and the standard of living. Such advances arise from the creation of compounds and materials. Continuation of these contributions requires a vigorous effort in research and development, for which information such as that provided by the *Comprehensive* series of Pergamon Press is a valuable resource.

Since the publication in 1979 of *Comprehensive Organic Chemistry*, it has become an important first source of information. However, considering the pace of advancements and the ever-shrinking timeframe in which initial discoveries are rapidly assimilated into the basic fabric of the science, it is clear that a new treatment is needed. It was tempting simply to update a series that had been so successful. However, this new series took a totally different approach. In deciding to embark upon *Comprehensive Organic Synthesis*, the Editors and Publisher recognized that synthesis stands at the heart of organic chemistry.

The construction of molecules and molecular systems transcends many fields of science. Needs in electronics, agriculture, medicine and textiles, to name but a few, provide a powerful driving force for more effective ways to make known materials and for routes to new materials. Physical and theoretical studies, extrapolations from current knowledge, and serendipity all help to identify the direction in which research should be moving. All of these forces help the synthetic chemist in translating vague notions to specific structures, in executing complex multistep sequences, and in seeking new knowledge to develop new reactions and reagents. The increasing degree of sophistication of the types of problems that need to be addressed require increasingly complex molecular architecture to target better the function of the resulting substances. The ability to make such substances available depends upon the sharpening of our sculptors' tools: the reactions and reagents of synthesis.

The Volume Editors have spent great time and effort in considering the format of the work. The intention is to focus on transformations in the way that synthetic chemists think about their problems. In terms of organic molecules, the work divides into the formation of carbon–carbon bonds, the introduction of heteroatoms, and heteroatom interconversions. Thus, Volumes 1–5 focus mainly on carbon–carbon bond formation, but also include many aspects of the introduction of heteroatoms. Volumes 6–8 focus on interconversion of heteroatoms, but also deal with exchange of carbon–carbon bonds for carbon–heteroatom bonds.

The Editors recognize that the assignment of subjects to any particular volume may be arbitrary in part. For example, reactions of enolates can be considered to be additions to C—C π-bonds. However, the vastness of the field leads it to be subdivided into components based upon the nature of the bond-forming process. Some subjects will undoubtedly appear in more than one place.

In attacking a synthetic target, the critical question about the suitability of any method involves selectivity: chemo-, regio-, diastereo- and enantio-selectivity. Both from an educational point-of-view for the reader who wants to learn about a new field, and an experimental viewpoint for the practitioner who seeks a reference source for practical information, an organization of the chapters along the theme of selectivity becomes most informative.

The Editors believe this organization will help emphasize the common threads that underlie many seemingly disparate areas of organic chemistry. The relationships among various transformations becomes clearer and the applicability of transformations across a large number of compound classes becomes apparent. Thus, it is intended that an integration of many specialized areas such as terpenoid, heterocyclic, carbohydrate, nucleic acid chemistry, *etc*. within the more general transformation class will provide an impetus to the consideration of methods to solve problems outside the traditional ones for any specialist.

In general, presentation of topics concentrates on work of the last decade. Reference to earlier work, as necessary and relevant, is made by citing key reviews. All topics in organic synthesis cannot be treated with equal depth within the constraints of any single series. Decisions as to which aspects of a topic require greater depth are guided by the topics covered in other recent *Comprehensive* series. This new treatise focuses on being comprehensive in the context of synthetically useful concepts.

The Editors and Publisher believe that *Comprehensive Organic Synthesis* will serve all those who must face the problem of preparing organic compounds. We intend it to be an essential reference work for the experienced practitioner who seeks information to solve a particular problem. At the same time, we must also serve the chemist whose major interest lies outside organic synthesis and therefore is only an occasional practitioner. In addition, the series has an educational role. We hope to instruct experienced investigators who want to learn the essential facts and concepts of an area new to them. We also hope to teach the novice student by providing an authoritative account of an area and by conveying the excitement of the field.

The need for this series was evident from the enthusiastic response from the scientific community in the most meaningful way — their willingness to devote their time to the task. I am deeply indebted to an exceptional board of editors, beginning with my deputy editor-in-chief Ian Fleming, and extending to the entire board — Clayton H. Heathcock, Ryoji Noyori, Steven V. Ley, Leo A. Paquette, Gerald Pattenden, Martin F. Semmelhack, Stuart L. Schreiber and Ekkehard Winterfeldt.

The substance of the work was created by over 250 authors from 15 countries, illustrating the truly international nature of the effort. I thank each and every one for the magnificent effort put forth. Finally, such a work is impossible without a publisher. The continuing commitment of Pergamon Press to serve the scientific community by providing this *Comprehensive* series is commendable. Specific credit goes to Colin Drayton for the critical role he played in allowing us to realize this work and also to Helen McPherson for guiding it through the publishing maze.

A work of this kind, which obviously summarizes accomplishments, may engender in some the feeling that there is little more to achieve. Quite the opposite is the case. In looking back and seeing how far we have come, it becomes only more obvious how very much more we have yet to achieve. The vastness of the problems and opportunities ensures that research in organic synthesis will be vibrant for a very long time to come.

BARRY M. TROST
Palo Alto, California

Contributors to Volume 8

Professor A. G. M. Barrett
Department of Chemistry, Colorado State University, Fort Collins, CO 80523, USA

Professor J. H. Brewster
Department of Chemistry, Purdue University, West Lafayette, IN 47906, USA

Professor P. Caubère
Université de Nancy, Faculté des Sciences, Laboratoire de Chimie Organique I, BP 239, F-54506 Vandoeuvre Les Nancy, France

Professor P. Coutrot
Université de Nancy, Faculté des Sciences, Laboratoire de Chimie Organique I, BP 239, F-54506 Vandoeuvre Les Nancy, France

Professor A. R. Chamberlin
Department of Chemistry, University of California at Irvine, Irvine, CA 92717, USA

Dr A. P. Davis
Department of Chemistry, Trinity College, Dublin 2, Republic of Ireland

Professor J. J. Eisch
Department of Chemistry, State University of New York at Binghamton, PO Box 6000, Binghamton, NY 13901, USA

Dr I. D. Entwistle
Shell Research Ltd, Sittingbourne Research Centre, Kent ME9 8AG, UK

Professor A. J. Fry
Department of Chemistry, Wesleyan University, Middletown, CT 06457, USA

Dr T. L. Gilchrist
Department of Organic Chemistry, University of Liverpool, PO Box 147, Liverpool L69 3BX, UK

Dr N. Greenspoon
Department of Organic Chemistry, The Weizmann Institute of Science, Rehovot, Israel

Dr N. Greeves
Department of Organic Chemistry, University of Liverpool, PO Box 147, Liverpool L69 3BX, UK

Professor G. W. Gribble
Department of Chemistry, Dartmouth College, Hanover, NH 03755, USA

Professor K. Harada
Department of Chemistry, Tsukuba University, Tsukuba, Ibaraki 305, Japan

Dr T. Hiyama
Sagami Chemical Research Center, 4-4-1 Nishiohnuma, Sagamihara, Kanagawa 229, Japan

Professor M. Hudlicky
Department of Chemistry, Virginia Polytechnic Institute & State University, Blacksburg, VA 24061-0212, USA

Professor J. W. Huffman
223 Howard L Hunter Chemistry Laboratory, Clemson University, Clemson, SC 29634-1905, USA

Dr M. K. Hutchins
ICI Americas, Advanced Materials Division, Exton, PA 19341, USA

Professor R. O. Hutchins
Department of Chemistry, College of Science, Drexel University, Philadelphia, PA 19104, USA

Dr T. Imamoto
Department of Chemistry, Faculty of Science, Chiba University, Yayoi-cho, Chiba 260, Japan

Dr R. A. W. Johnstone
Department of Organic Chemistry, University of Liverpool, PO Box 147, Liverpool L69 3BX, UK

Professor J. B. Jones
Lash Miller Chemical Laboratories, University of Toronto, 80 St George Street, Toronto, Ontario M5S 1A1, Canada

Professor G. W. Kabalka
Department of Chemistry, University of Tennessee, 575 Buehler Hall, Knoxville, TN 37996-1600, USA

Dr S. Kato
Department of Chemistry, Faculty of Engineering, Gifu University, Yanagido, Gifu 501-11, Japan

Dr J. G. Keay
Reilly Industries Inc, 1500 South Tibbs Avenue, PO Box 41076, Indianapolis, IN 46241, USA

Professor E. Keinan
Technion — Israel Institute of Technology, Technion City, Haifa 32 000, Israel

Professor R. M. Kellogg
Department of Organic Chemistry, University of Groningen, Nijenborgen 16, 9747 AG Groningen, The Netherlands

Dr T. Kusumoto
Sagami Chemical Research Center, 4-4-1 Nishiohnuma, Sagamihara, Kanagawa 229, Japan

Professor J. A. Labinger
Division of Chemistry & Chemical Engineering 127-72, California Institute of Technology, Pasadena, CA 91125, USA

Professor L. N. Mander
Research School of Chemistry, Australian National University, GPO Box 4, Canberra, ACT 2601, Australia

Dr S. W. McCombie
Schering Research, Schering-Plough Corporation, 60 Orange Street, Bloomfield, NJ 07003, USA

Dr L. Moëns
Solar Energy Research Institute, Solar Fuels Research Division, 1617 Cole Blvd, Golden, CO 80401, USA

Professor T. Munegumi
Department of Chemistry, Tsukuba University, Tsukuba, Ibaraki 305, Japan

Professor S. Murai
Department of Applied Chemistry, Faculty of Engineering, Osaka University, Suita, Osaka 565, Japan

Dr T. Murai
Department of Chemistry, Faculty of Engineering, Gifu University, Yanagido, Gifu 501-11, Japan

Professor S. Nishiyama
Department of Chemistry, Faculty of Science & Technology, Keio University, 14-1 Hiyoshi, 3-chome, Kohokuku, Yokohama 223, Japan

Professor M. Nishizawa
Faculty of Pharmaceutical Sciences, Tokushima Bunri University, Yamashiro-Cho, Tokushima 770, Japan

Professor R. Noyori
Department of Chemistry, Faculty of Science, Nagoya University, Chikusa, Nagoya 464, Japan

Professor D. J. Pasto
Department of Chemistry, College of Science, University of Notre Dame, Notre Dame, IN 46556, USA

Professor A. Pelter
Department of Chemistry, University College Swansea, Singleton Park, Swansea SA2 8PP, UK

Dr M. Sainsbury
School of Chemistry, University of Bath, Claverton Down, Bath BA2 7AY, UK

Professor S. Siegel
Department of Chemistry & Biochemistry, University of Arkansas, Fayetteville, AR 72701, USA

Professor K. Smith
Department of Chemistry, University College Swansea, Singleton Park, Swansea SA2 8PP, UK

Professor I. W. J. Still
Department of Chemistry, University of Toronto, Mississauga, Ontario L5L 1C6, Canada

Professor H. Takaya
Department of Industrial Chemistry, Faculty of Engineering, Kyoto University, Yoshida, Sakyo-ku, Kyoto 606, Japan

Dr R. S. Varma
Houston Biotechnology Incorporated and Center for Biotechnology, Baylor College of Medicine, 4000 Research Forest Drive, The Woodlands, TX 77381, USA

Dr W. W. Wood
Shell Research Ltd, Sittingbourne Research Centre, Kent ME9 8AG, UK

Professor S. Yamamura
Department of Chemistry, Faculty of Science & Technology, Keio University, 14-1 Hiyoshi, 3-chome, Kohokuku, Yokohama 223, Japan

Abbreviations

The following abbreviations have been used where relevant. All other abbreviations have been defined the first time they occur in a chapter.

Techniques

CD	circular dichroism
CIDNP	chemically induced dynamic nuclear polarization
CNDO	complete neglect of differential overlap
CT	charge transfer
GLC	gas–liquid chromatography
HOMO	highest occupied molecular orbital
HPLC	high-performance liquid chromatography
ICR	ion cyclotron resonance
INDO	incomplete neglect of differential overlap
IR	infrared
LCAO	linear combination of atomic orbitals
LUMO	lowest unoccupied molecular orbital
MS	mass spectrometry
NMR	nuclear magnetic resonance
ORD	optical rotatory dispersion
PE	photoelectron
SCF	self-consistent field
TLC	thin layer chromatography
UV	ultraviolet

***Reagents, solvents,* etc.**

Ac	acetyl
acac	acetylacetonate
AIBN	2,2′-azobisisobutyronitrile
Ar	aryl
ATP	adenosine triphosphate
9-BBN	9-borabicyclo[3.3.1]nonyl
9-BBN-H	9-borabicyclo[3.3.1]nonane
BHT	2,6-di-*t*-butyl-4-methylphenol (butylated hydroxytoluene)
bipy	2,2′-bipyridyl
Bn	benzyl
t-BOC	*t*-butoxycarbonyl
BSA	*N,O*-bis(trimethylsilyl)acetamide
BSTFA	*N,O*-bis(trimethylsilyl)trifluoroacetamide
BTAF	benzyltrimethylammonium fluoride
Bz	benzoyl
CAN	ceric ammonium nitrate
COD	1,5-cyclooctadiene
COT	cyclooctatetraene
Cp	cyclopentadienyl
Cp*	pentamethylcyclopentadienyl
18-crown-6	1,4,7,10,13,16-hexaoxacyclooctadecane
CSA	camphorsulfonic acid
CSI	chlorosulfonyl isocyanate
DABCO	1,4-diazabicyclo[2.2.2]octane
DBA	dibenzylideneacetone
DBN	1,5-diazabicyclo[4.3.0]non-5-ene
DBU	1,8-diazabicyclo[5.4.0]undec-7-ene

DCC	dicyclohexylcarbodiimide
DDQ	2,3-dichloro-5,6-dicyano-1,4-benzoquinone
DEAC	diethylaluminum chloride
DEAD	diethyl azodicarboxylate
DET	diethyl tartrate (+ or –)
DHP	dihydropyran
DIBAL-H	diisobutylaluminum hydride
diglyme	diethylene glycol dimethyl ether
dimsyl Na	sodium methylsulfinylmethide
DIOP	2,3-*O*-isopropylidene-2,3-dihydroxy-1,4-bis(diphenylphosphino)butane
DIPT	diisopropyl tartrate (+ or –)
DMA	dimethylacetamide
DMAC	dimethylaluminum chloride
DMAD	dimethyl acetylenedicarboxylate
DMAP	4-dimethylaminopyridine
DME	dimethoxyethane
DMF	dimethylformamide
DMI	*N,N'*-dimethylimidazolone
DMSO	dimethyl sulfoxide
DMTSF	dimethyl(methylthio)sulfonium fluoroborate
DPPB	1,4-bis(diphenylphosphino)butane
DPPE	1,2-bis(diphenylphosphino)ethane
DPPF	1,1'-bis(diphenylphosphino)ferrocene
DPPP	1,3-bis(diphenylphosphino)propane
E^+	electrophile
EADC	ethylaluminum dichloride
EDG	electron-donating group
EDTA	ethylenediaminetetraacetic acid
EEDQ	*N*-ethoxycarbonyl-2-ethoxy-1,2-dihydroquinoline
EWG	electron-withdrawing group
HMPA	hexamethylphosphoric triamide
HOBT	hydroxybenzotriazole
$IpcBH_2$	isopinocampheylborane
Ipc_2BH	diisopinocampheylborane
KAPA	potassium 3-aminopropylamide
K-selectride	potassium tri-*s*-butylborohydride
LAH	lithium aluminum hydride
LDA	lithium diisopropylamide
LICA	lithium isopropylcyclohexylamide
LITMP	lithium tetramethylpiperidide
L-selectride	lithium tri-*s*-butylborohydride
LTA	lead tetraacetate
MCPBA	*m*-chloroperbenzoic acid
MEM	methoxyethoxymethyl
MEM-Cl	β-methoxyethoxymethyl chloride
MMA	methyl methacrylate
MMC	methylmagnesium carbonate
MOM	methoxymethyl
Ms	methanesulfonyl
MSA	methanesulfonic acid
MsCl	methanesulfonyl chloride
MVK	methyl vinyl ketone
NBS	*N*-bromosuccinimide
NCS	*N*-chlorosuccinimide

NMO	*N*-methylmorpholine *N*-oxide
NMP	*N*-methyl-2-pyrrolidone
Nu^-	nucleophile
PPA	polyphosphoric acid
PCC	pyridinium chlorochromate
PDC	pyridinium dichromate
phen	1,10-phenanthroline
Phth	phthaloyl
PPE	polyphosphate ester
PPTS	pyridinium *p*-toluenesulfonate
Red-Al	sodium bis(methoxyethoxy)aluminum dihydride
SEM	β-trimethylsilylethoxymethyl
Sia_2BH	disiamylborane
TAS	tris(diethylamino)sulfonium
TBAF	tetra-*n*-butylammonium fluoride
TBDMS	*t*-butyldimethylsilyl
TBDMS-Cl	*t*-butyldimethylsilyl chloride
TBHP	*t*-butyl hydroperoxide
TCE	2,2,2-trichloroethanol
TCNE	tetracyanoethylene
TES	triethylsilyl
Tf	triflyl (trifluoromethanesulfonyl)
TFA	trifluoroacetic acid
TFAA	trifluoroacetic anhydride
THF	tetrahydrofuran
THP	tetrahydropyranyl
TIPBS-Cl	2,4,6-triisopropylbenzenesulfonyl chloride
TIPS-Cl	1,3-dichloro-1,1,3,3-tetraisopropyldisiloxane
TMEDA	tetramethylethylenediamine [1,2-bis(dimethylamino)ethane]
TMS	trimethylsilyl
TMS-Cl	trimethylsilyl chloride
TMS-CN	trimethylsilyl cyanide
Tol	tolyl
TosMIC	tosylmethyl isocyanide
TPP	*meso*-tetraphenylporphyrin
Tr	trityl (triphenylmethyl)
Ts	tosyl (*p*-toluenesulfonyl)
TTFA	thallium trifluoroacetate
TTN	thallium(III) nitrate

Contents of All Volumes

Volume 5 Combining C—C π-Bonds

Volume 6 Heteroatom Manipulation

Volume 7 Oxidation

1.1
Reduction of C═O to CHOH by Metal Hydrides

NICHOLAS GREEVES
University of Liverpool, UK

1.1.1 INTRODUCTION

The reduction of the carbonyl group of aldehydes and ketones to the corresponding alcohol is ubiquitous in organic synthesis. Since the first report of reduction by diborane more than half a century ago,[1] metal hydride reagents have achieved preeminence as the reagents of choice for performing this synthetic transformation. The opportunities for variation in the metal, ligands, counterion and reaction conditions have enabled most problems of stereo-, regio- or chemo-selectivity in synthesis to be overcome satisfactorily. The majority of the complex metal hydrides described in this chapter that exhibit useful reducing properties are readily available from commercial sources, which has contributed enormously to their widespread acceptance and application.

The state of the art of reductions with metal hydrides a decade ago was the subject of comprehensive reviews. A detailed survey of reductions of carbonyl compounds with alkali and alkaline earth metal hydrides, borane and derivatives, alane and derivatives, metal borohydrides, metal aluminohydrides, silanes, stannanes and transition metal hydrides was compiled.[2] The properties, preparation and applications of each reagent were discussed together with methods for their determination, handling techniques and mechanistic aspects. The functional group selectivity of selected reagents was assessed,[3] and Brown described a 40-year odyssey in the development of a range of complex nucleophilic alumino- and borohydrides from their electrophilic counterparts, borane and alane.[4] Reductions by metal alkoxyaluminum hydrides were the subject of an exhaustive *Organic Reactions* review which emphasized the mechanism, scope, limitations and synthetic utility of the reagents.[5]

It is not the intention of this chapter to reiterate the conclusions of these compilations, but rather to concentrate on important synthetic developments that have appeared subsequently.

1.1.1.1 Kinetics and Mechanism

In spite of four decades of investigation, the full details of the mechanism of reduction of ketones by sodium borohydride and lithium aluminum hydride remain to be established.[6] Sodium borohydride reacts with first-order kinetics involving a rate-determining hydride attack on the carbonyl carbon, but the nature of the interaction of the metal cation or boron atom with the carbonyl oxygen is unknown. An *ab initio* theoretical study of borohydride addition to formaldehyde concluded that the traditional [2 + 2] four-center transition state, with simultaneous CH and BO bond formation and BH bond breaking, was not the optimum representation. The preferred alternative was a single-step mechanism with a nonsynchronous four-center transition state with a product-like geometry involving transfer of BH_3 from the hydride, already bonded to carbon, to the carbonyl oxygen atom.[7] The energy of this pathway was further reduced by the incorporation of a water molecule, an OH substituent on boron or a metal counterion. The importance of a metal counterion was demonstrated experimentally in a series of ketone reductions with lithium and sodium alumino- and boro-hydrides. In ethereal solvents reduction proceeded uneventfully, but, when macrocyclic cryptands, specific to either metal ion, were added, no reduction was observed.[8] The details of the effect of the metal counterion and reaction medium on the reduction of carbonyl compounds with alkali metal alumino- or boro-hydrides has been reviewed recently.[9]

The reduction of a series of chiral acyclic ketones, lacking polar functional groups, with a range of lithium, sodium and potassium alumino- and boro-hydrides was investigated in various solvents under different reaction conditions. The changes in steric bulk of the reagents and reaction medium enabled a semiempirical scale for the effective size of the reagent to be applied to the stereochemical analysis of the observed diastereoselectivity.[10]

The reduction kinetics of mesityl phenyl ketone, selected for its convenient rate, with lithium and sodium aluminum hydrides in THF were studied in detail.[11] In the presence of excess hydride, the reaction was first-order in hydride and ketone. A lithium counterion was worth a 10-fold increase in rate over sodium, indicating the influence of the cation on the mechanism. The observed deuterium kinetic isotope effect (k_H/k_D = 1.27) was consistent with a rate-determining transfer of hydride to the carbonyl carbon. Further evidence for the association of the counterion with the carbonyl oxygen during reduction was obtained from comparison of entropies of activation of both reactions. The results were rationalized by considering the metal hydrides as solvent-separated ion pairs and/or free ions, prior to coordination of the metal cation to the carbonyl oxygen and hydride delivery *via* a cyclic transition state. Analogous experiments on the reduction of camphor led to the same kinetic conclusions. Significantly, lithium *t*-butoxy- and methoxy-aluminohydrides were observed to follow kinetics consistent with disproportionation of the alkoxide species to give LAH which was the predominant reducing agent.[12]

The kinetics of reduction of 15 cyclohexanones by $Li(Bu^tO)_3AlH$ were investigated and found to follow a simple second-order process.[13] The rate constants were determined at various temperatures and were found to exceed the corresponding values for sodium borohydride. Activation parameters were derived for the reduction, which was nearly isoenthalpic, and rate differences were attributed to the entropic contribution. The results were consistent with a simple four-centred transition state, but additional work is required for a definitive conclusion.

First-order kinetic behavior was observed in the reduction of aldehydes and reactive ketones with 9-BBN-H dimer, but with hindered ketones intermediate or three-halves order kinetics were observed.[14] Monomeric 9-BBN-H was the active intermediate, as in the case of the hydroboration of alkenes and alkynes by this reagent. 9-BBN-H was less sensitive to steric effects than sodium borohydride; hindrance on one side of the ketone caused a moderate rate decrease, while a significant drop was observed with bulky groups on either side of the carbonyl group. The rate was also sensitive to electronic factors; electron-releasing substituents increased the rate of reduction, and electron-withdrawing groups retarded it. These data strongly suggested that the boron atom was coordinated to the carbonyl oxygen during reduction, but it was not established if this was in advance of hydride transfer or simply in the transition state.

Although a polar mechanism is generally accepted for hydride reduction, evidence for an electron transfer mechanism in the reduction of hindered aromatic ketones by main group hydrides has been presented.[15] Dimesityl ketone formed deeply colored, electron paramagnetic resonance active solutions with neutral hydrides of aluminum, boron and magnesium in THF. The kinetic analysis for alane reduction, which produced a long-lived paramagnetic intermediate, suggested a radical cation–radical anion pair as a relatively stable intermediate. The contribution of this pathway in the reduction of dialkyl ketones remains to be established.

1.1.2 THEORY OF STEREOSELECTIVITY

1.1.2.1 Acyclic Carbonyl Compounds

The origin and magnitude of the stereoselectivity observed in the reduction of chiral carbonyl compounds has long been an area of intense theoretical and practical study. Efforts have concentrated largely on the 1,2-asymmetric induction that occurs in the hydride addition to a carbonyl group flanked by an asymmetric center (Scheme 1). Cram's rule was formulated to rationalize the results of nucleophilic addition to aldehydes and ketones containing nonpolar groups.[16] The most stable conformation (**1**) was assumed to arise by minimization of the interaction between the largest group (R^l) and the carbonyl group, which was coordinated to the incoming reagent. Addition then occurred preferentially on the side of the smallest substituent (R^s) rather than the larger medium sized group (R^m). The outcome of Cram reduction of ketone (**2**) to alcohol (**3**) is illustrated. This rule enabled a large body of experimental results to be correlated, but its theoretical basis was subsequently shown to be flawed.

The results for reduction of α-halo aldehydes and ketones were anomalous and led to Cornforth's dipolar model (**4**), in which the dipoles due to the carbonyl group and the carbon–halogen bond were in an antiperiplanar arrangement. Reduction then proceeded from the less-hindered side of ketone (**5**), leading to alcohol (**6**).[17] The possibility of chelation when the α-substituent was an alcohol, alkoxy or amino group (X) was covered by Cram's cyclic or chelate model (**7**).[18] The chelating group (X) and the carbonyl group were eclipsed, coordinating to the metal (M), and reduction occurred from the less-hindered side. This model has been widely used to rationalize diastereoselectivity in reduction of ketones (**8**) when chelation is important, depending on the nature of the substituent (X) and the metal ion.

Information about the ground state conformations of carbonyl compounds that demonstrated that conformations with one bond eclipsing the carbonyl group were energetically favored led Karabatsos to propose an alternative model.[19] Calculations suggested that the most favored conformation (**9**) would have the medium-sized group (R^m) of ketone (**10**) eclipsing the carbonyl group, and addition of hydride would occur from the side of the less bulky substituent (R^s) to give alcohol (**11**). Comparison of the calculated energies of the other possible transition states with this conformation allowed the magnitude of diastereoselectivity to be correlated with experimental results.

The most influential contribution to the interpretation of 1,2-asymmetric induction in carbonyl group reduction was that of Felkin.[20] Attention was directed for the first time to the structure of the transition state, which was assumed to be very similar to that of the substrate. Torsional strain caused by interactions between the partially formed hydride–carbonyl bond and the full bonds at the adjacent center was of paramount importance. The nucleophile was assumed to attack perpendicular to the carbonyl group plane and staggered to the largest or most electronegative group (R^l). This angle of approach was later revised to the Bürgi–Dunitz trajectory,[21] derived from crystallographic studies, which placed the incoming hydride much closer to the substituent (R^s). This interaction was the decisive influence on the selection of the transition state illustrated (**12**) over the alternative which would have the positions of the small (R^s) and medium (R^m) groups exchanged. This model was successfully applied to acyclic and cyclic ketones and allowed crude quantitative rationalization of experimental diastereoisomeric ratios.[22]

Support for the Felkin model was provided by *ab initio* calculations of a range of transition state geometries for reductions of carbonyl compounds with and without an adjacent polar substituent.[23] The transition state energies were found to be minimized in the Felkin conformation as a consequence of the *anti* disposition of the bond forming to the incoming hydride and the bond between the adjacent carbon and the largest or most electronegative group (R^l). Other models that correctly predicted the stereochemical outcome required transition states of significantly higher energy.

Further refinement of the model by calculated trajectory analysis enabled the steric influence of the group (R) attached directly to the carbonyl to be assessed. Increasing the size of R was known to enhance the reduction diastereoselectivity.[22] This may be understood by the perturbation of the trajectory of the

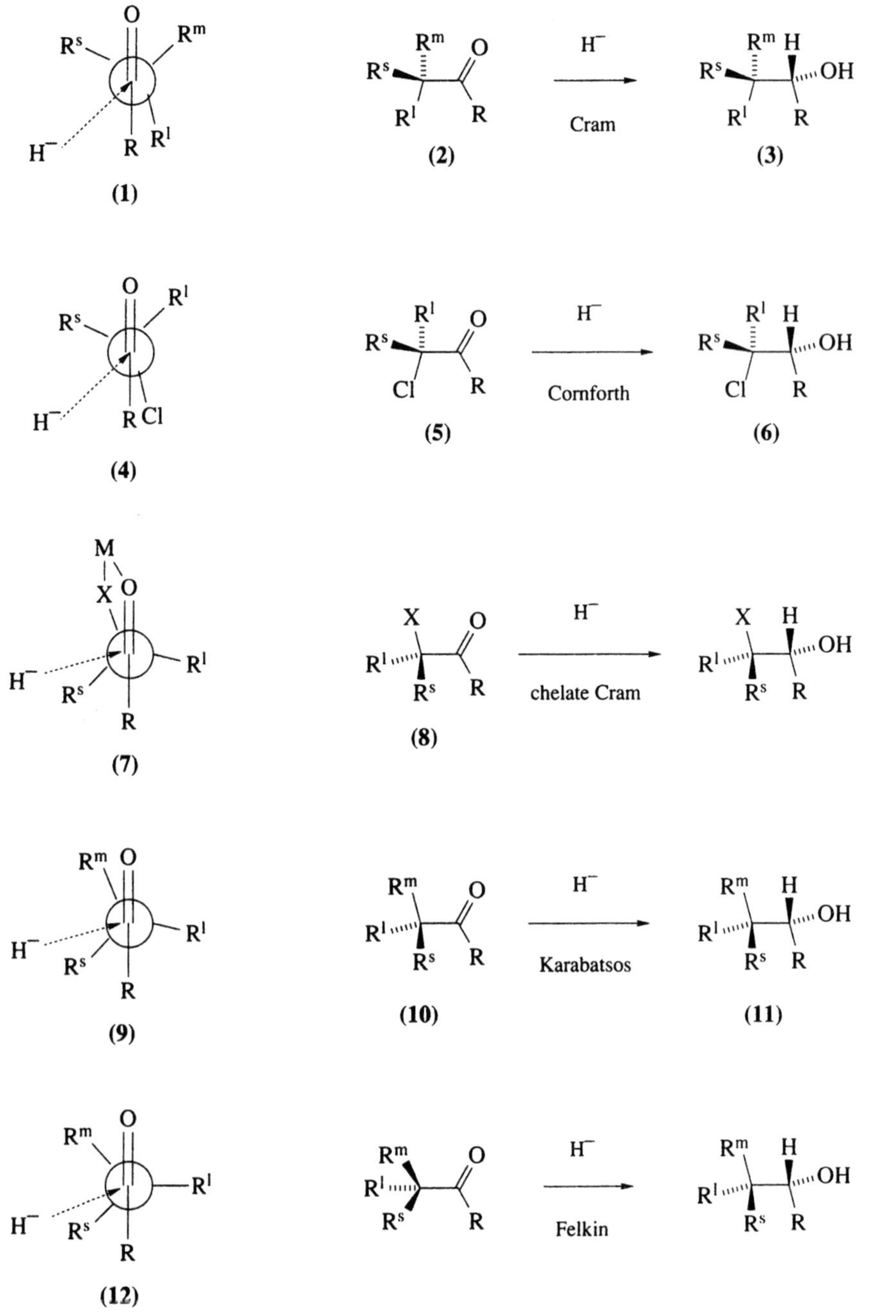

Scheme 1

nucleophile from perpendicular to the carbonyl plane away from the bulky group; the interactions of the nucleophile with the α-center would thus become more decisive. Quantitative support for such deviations from normal approach was provided by calculations on hydride addition to pivaldehyde.[24]

Ab initio calculations of transition state geometries for additions of lithium and sodium hydrides to formaldehyde, acetaldehyde, propionaldehyde and acetone indicated that nucleophilic attack *anti* to a methyl group was disfavored over attack *anti* to a hydrogen by 1–2 kcal mol^{-1} (1 kcal = 4.18 kJ). The combination of the torsional effects identified by Felkin, steric influences and the tendency to avoid attack *anti* to an alkyl group controlled the observed stereoselectivities. Importantly, in agreement with the Felkin model, when two alkyl substituents were present, the most stable conformation had the larger group perpendicular to the carbonyl group and the remaining alkyl group away from the incoming nucleophile. The alternative conformation with an *anti* carbon–hydrogen bond was disfavored because the alkyl groups were unable to attain their preferred dihedral angles. The incorporation of these effects into an MM2 force field enabled good qualitative rationalization of $LiAlH_4$ reductions of a wide range of alkyl ketones.[25]

The computational support for Felkin's torsional strain model and its success in interpretation of experimental diastereoselectivities has led to its widespread adoption. It appears to be the preeminent open transition state involved in reductions when chelation is not important. Complementary selectivity observed in reductions that do involve chelation may be understood in terms of Cram's cyclic model.

1.1.2.2 Cyclic Carbonyl Compounds

The stereochemistry and mechanism of reduction of cyclic ketones by metal hydride reagents provided a unique opportunity for comparison of experimental results with theoretical expectation. The models proposed by Cram, Cornforth and Karabatsos described above were inadequate to explain the stereochemical outcome, and so a wide range of models was developed to explain the dichotomy between cyclic and acyclic results.[26] The theoretical basis, applications and limitations of these models have been critically reviewed.[6] The effect of steric influences, torsional and electronic factors, and the nature of the cation on the rate of reduction, stereochemical outcome and position of the transition state have also been surveyed.[27]

The stereochemical characteristics of lithium trimethoxyaluminohydride and lithium aluminum hydride in the reduction of cyclic ketones were analyzed by a linear combination of steric strain and product stability control. Qualitative and quantitative explanation of the experimental observations was possible using this approach.[28]

Consideration of the stabilizing interaction between the low-lying σ^*-orbital associated with the bond forming between the incoming hydride and the carbonyl carbon, and remote electron-donor σ-orbitals led Cieplak to an explanation for many kinetic and stereochemical effects in cyclohexanones that were previously unexplained.[29] The normal preference for axial attack in simple cyclohexanones was attributed to the improved electron-donor ability of carbon–hydrogen bonds over carbon–carbon bonds that would be antiperiplanar to the incoming nucleophile in the transition state.

The electronic contribution to reduction stereoselectivity was assessed with a series of substituted 9-benzonorbornenones (**13**) with a range of reducing agents.[30] The observed selectivity, increasing *anti* attack with electron-rich benzene rings, paralleled the homoconjugation sequence. Analogous results were observed in the reduction of a series of 5-substituted adamantanones (**14**) with electronically varied substituents (X), providing strong support for the importance of σ-participation in diastereoselectivity.[31] The unexpected preference for axial reduction of 2-methoxy-4-pyranones (equation 1) with an equatorial methoxy group, even with L-selectride, was explained by the electronic effects of the two conformationally defined carbon–oxygen bonds.[32]

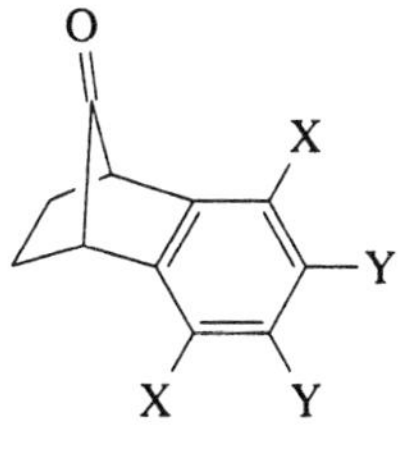

X = F, Cl, H, OMe
Y = F, Cl, H, OMe

(**13**)

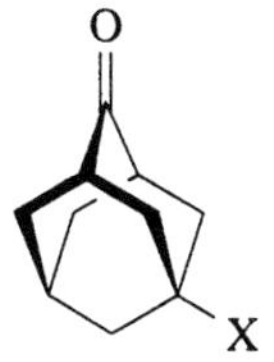

X = Ar, Bu^t, F, Cl, OH, CF_3

(**14**)

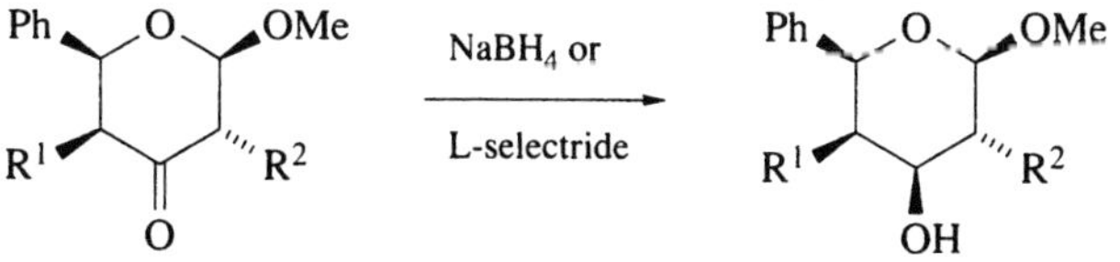

Ph, O, OMe, R^1, R^2, O —— $NaBH_4$ or L-selectride ——→ Ph, O, OMe, R^1, R^2, OH (1)

Felkin identified torsional effects in cyclohexanone reductions that accounted for the observed stereoselectivity. Minimization of these torsional effects, in the absence of steric hindrance, led to the predominance of axial attack (equation 2).[33] Recently, a computational approach has provided quantitative support for this model.[25] The eclipsing interactions between the incoming nucleophile and the bonds α to the carbonyl group are more serious, for any given trajectory, from the equatorial direction. The differences are clearly illustrated in the Newman projection of cyclohexanone (**15**). Differences in bond lengths and thus torsional bond energies were used to explain observed stereoselectivity in hetero-substituted cyclohexanones.

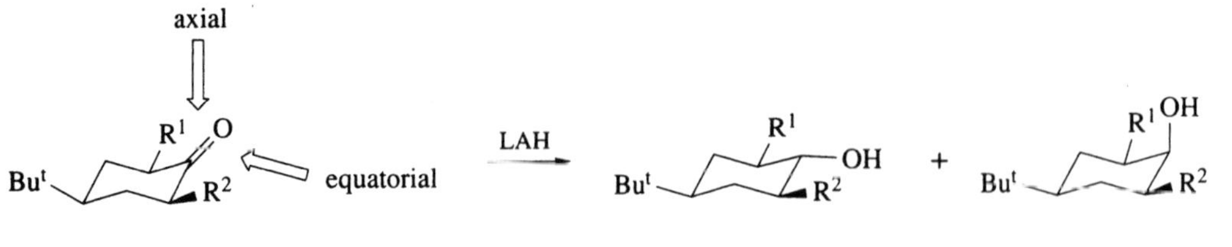

$R^1 = R^2 = H$	92	:	8
R^1 = Me, $R^2 = H$	83	:	17
$R^1 = R^2$ = Me	53	:	47

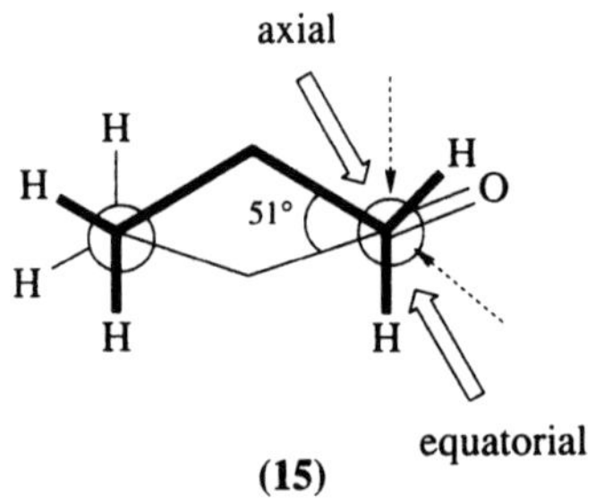

(15)

The dramatically enhanced axial selectivity demonstrated in the addition of sterically undemanding nucleophiles, such as lithium aluminum hydride, to conjugated cyclohexenones was explained by the application of the modified MM2 model.[34] The important difference between cyclohexanones and their unsaturated counterparts was the flattening of the ring. The internal dihedral angle was reduced from 51° in cyclohexanone to 22° in cyclohexenone. This in turn produced a dramatic change in the torsional interactions corresponding to axial and equatorial attack, illustrated on the Newman projection (**16**). Calculated energies of transition states for axial and equatorial addition were used to calculate diastereoisomeric ratios that showed good agreement with experimental results.

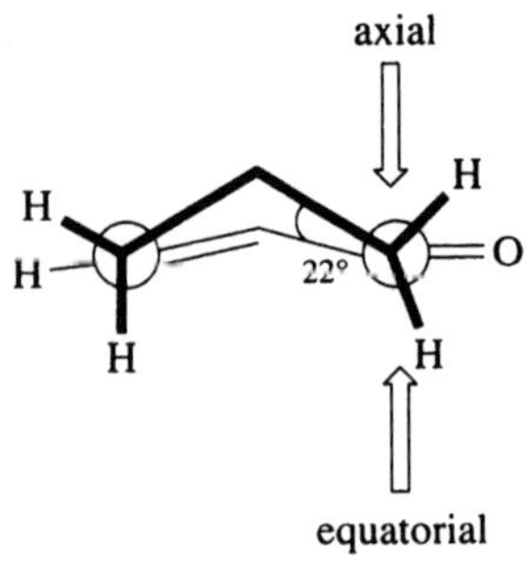

(16)

The accuracy and applicability of the model was tested further on the lithium aluminum hydride reduction of a series of benzocycloheptenones (**17**; equation 3). Dynamic NMR studies and MM2 calculations demonstrated that the chair conformation shown was the most stable. Since the substrates were sterically unhindered most models predicted that axial attack would be favored. Conversely, MM2 calculations indicated that the transition state for equatorial approach was of lower energy than its axial counterpart. In fact equatorial reduction to give the axial alcohol predominated (60:40) with a single methyl substituent (**17**; R^1 = Me), and was the exclusive outcome in the more hindered case with two methyl groups (**17**; R^1

= R^2 = Me).[35] Felkin's torsional strain approach was considered to be the only model consistent with these results, but an alternative interpretation based on participation of the aromatic ring, consistent with Cieplak's model, has recently been advanced.[36] The significance of torsional interactions was related to the degree of puckering in the cyclic ketone according to the 'flattening rule'.[37]

axial

equatorial

LAH

(3)

(17)

R^1 = Me, R^2 = H	40	:	60
R^1 = R^2 = Me	0	:	100

The success of this method for calculating quantitative stereoselectivities of reductions of substituted five-, six- and seven-membered ring and bicyclic ketones was impressive, but additional examples are required to demonstrate its generality.

1.1.3 DIASTEREOSELECTIVITY

1.1.3.1 Acyclic Carbonyl Compounds

The development of reliable methods for the diastereoselective reduction of carbonyl compounds in a wide range of acyclic systems has been an area of explosive growth in recent years.[38] This was prompted by the requirements of modern total synthesis in which redundant diastereoisomers are avoided,[39] together with enhanced theoretical understanding of stereoselectivity which allows rationalization of the results.

An exhaustive compilation of examples of reduction of acyclic ketones with an adjacent chiral center appeared recently and is not reproduced here.[40] Those methods that give excellent levels of asymmetric induction likely to be useful in synthesis are highlighted here. The results are collected according to the substrate being reduced rather than the reducing agent. Most of the examples can be rationalized by consideration of the Felkin transition state or, where appropriate, the chelated transition state (Section 1.1.2.1).

α-Hydroxy ketones (**18**) were reduced to the corresponding *anti*-diols (**19**) by zinc borohydride with good selectivity (77:23–99:1) *via* a chelated transition state (equation 4). Silylation with the bulky TBDMS group gave the protected derivatives (**20**) which were reduced by Red-Al at –78 °C preferentially to the *syn* products (**21**) according to a Felkin transition state with the silyl ether in the perpendicular position (equation 5).[41] Selectivity was good (76:24–98:2), except when R^1 was a branched or long chain alkyl group, presumably due to an unfavorable interaction with the silyl ether. These complementary methods were used to synthesize various diastereoisomers of a possible fragment of polyoxygenated antibiotics.[42] Analogous results were observed when a range of reducing agents were screened for reduction of α-benzyloxy alkynic ketone (**22**; Scheme 2).[43] The *anti* isomer (**23**) was produced by zinc borohydride (95:5) *via* chelation control, and K-selectride gave the *syn* Felkin product (**24**; 90:10). The generality of these results was demonstrated on five additional substrates. Significantly, Red-Al was *anti* selective, due to coordination to the benzyloxy group, which was not as effective as TBDMS at suppressing chelation. The dramatic difference between these two protecting groups was again manifested in the

$Zn(BH_4)_2$, ether, 0 °C

(4)

(18) (19)

LAH reduction of α-alkoxy enone (**25**; equation 6). Chelation to the benzyloxy group favored formation of the *anti* product (**26**; 98:2) while the TBDMS ether prevented chelation so that reduction occurred *via* an open transition state to give the *syn*-alcohol (**27**; 95:5).[44]

OSiBu^{t}Me$_2$; i, Red-Al/toluene, −78 °C; ii, TBAF/THF (5)

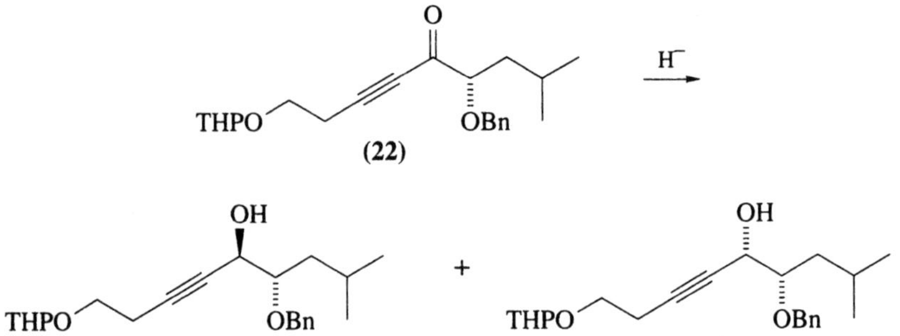

(**20**) (**21**)

(**22**) H$^-$

(**23**) + (**24**)

Zn(BH$_4$)$_2$, ether, −30 °C 95 : 5
K-selectride, THF, −95 °C 10 : 90

Scheme 2

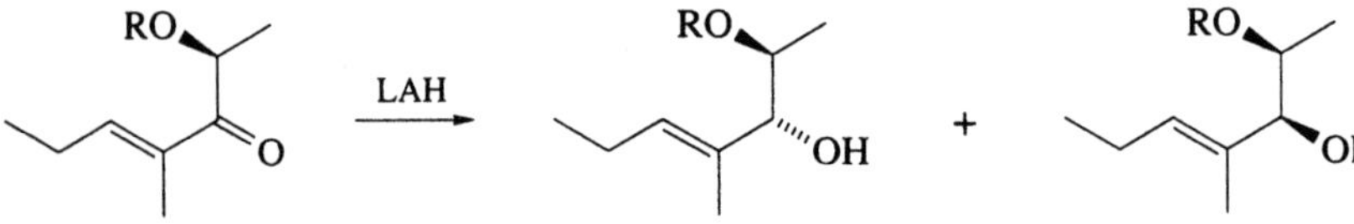

LAH (6)

(**25**) (**26**) (**27**)

R = Bn, ether, −10 °C 98:2
R = Bu^{t}Me$_2$Si, THF, −20 °C 5:95

Hydrosilylation provided a novel alternative reduction of α-oxy ketones (**29**) with tunable diastereoselectivity (Scheme 3). Fluoride-catalyzed reduction with phenyldimethylsilane in HMPA provided the *syn*-alcohols (**28**) with high selectivity (87:13–96:4).[45] The absence of a coordinating cation and the bulkiness of the reducing species combined to favor the Felkin model for these reductions. Conversely, reduction in trifluoroacetic acid proceeded *via* a proton-bridged cyclic transition state to give the *anti* products (**30**; 84:16–99:1).[46] These complementary methods constitute a powerful tool in stereoselective synthesis.

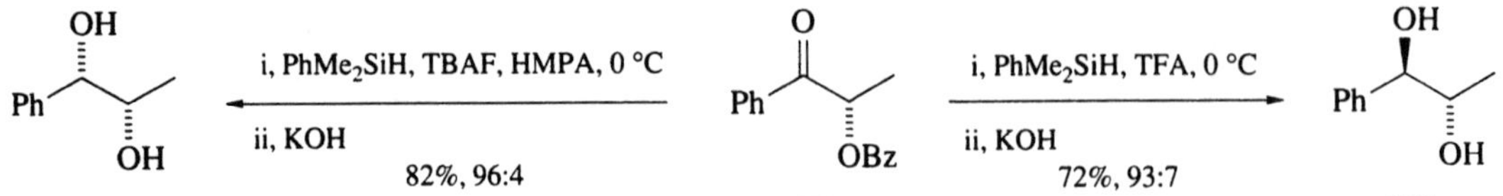

i, PhMe$_2$SiH, TBAF, HMPA, 0 °C; ii, KOH; 82%, 96:4 — i, PhMe$_2$SiH, TFA, 0 °C; ii, KOH; 72%, 93:7

(**28**) (**29**) (**30**)

Scheme 3

The 1,3-diol unit is an important constituent of many highly oxygenated natural products. Consequently, highly diastereoselective methods for reducing β-hydroxy ketones to either of the possible diastereoisomers have been developed. The *syn* isomer was produced by attack of an external hydride reagent on a six-membered chelate such as (**31**), the conformation of which was governed by the substituent at the alcohol center. Conversely, intramolecular delivery of hydride from a group bound to the alcohol *via* a chair transition state (**32**) gave the *anti* diastereoisomer.

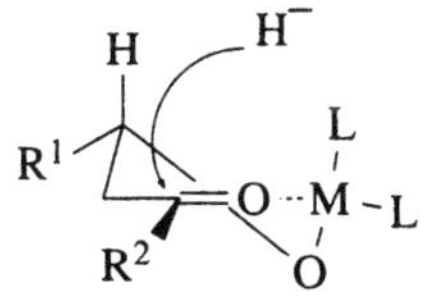

(**31**)
M = B; L = Bu, Et
M = Li; L = THF

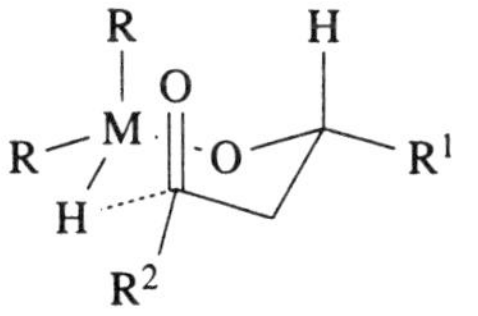

(**32**)
M = Si; R = Pr^i
M = B; R = OAc

The combination of trialkylboranes and sodium borohydride in THF at –100 °C was the first highly *syn* selective method of this type.[47] This was later refined by using alkoxydialkylboranes to chelate the β-hydroxy ketones (**33**) and conducting the reduction at –70 °C in THF–methanol (4:1).[48] The *syn*-1,3-diols (**34**) predominated by at least 98:2 (equation 7). Reduction of β,δ-diketo esters to the corresponding *syn*-β,δ-dihydroxy esters was accomplished with excellent diastereoselectivity.[49] A comparative study of hydrogenation with iron(II) chloride to enhance carbonyl reduction, alkylborane-mediated $NaBH_4$ reduction and zinc borohydride alone, demonstrated the generality of useful *syn* selectivity.[50] DIBAL-H has also been used to reduce β-hydroxy ketones to *syn*-1,3-diols with high diastereofacial selectivity (≥92:8) at –78 °C in THF.[51] Tributyltin hydride with both Lewis acid and radical catalysis exhibited lower levels of selectivity.

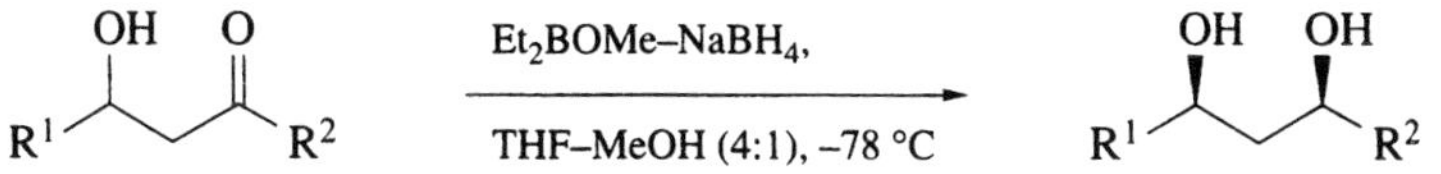

$Et_2BOMe–NaBH_4$, THF–MeOH (4:1), –78 °C: (**33**) → (**34**) (7)

Lithium iodide apparently was chelated by β-alkoxy ketones to form a similar six-membered transition state, allowing LAH to reduce the carbonyl group at –100 °C in ether with high *syn* selectivity (≥95:5) (Scheme 4).[52]

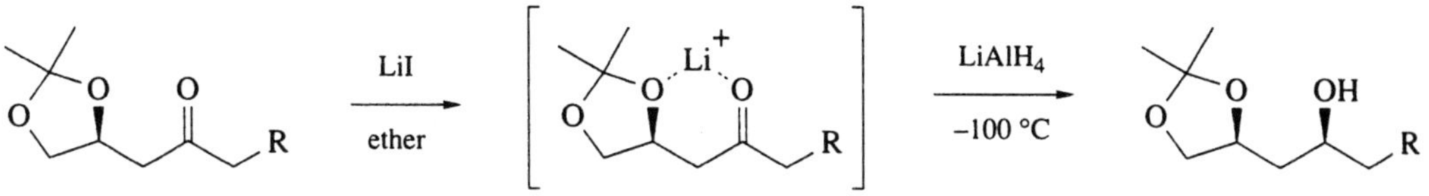

LiI, ether; $LiAlH_4$, –100 °C

Scheme 4

Intramolecular hydrosilylation of the carbonyl group of β-silyloxy ketones (**35**) induced by tin(IV) chloride catalysis at –80 °C in dichloromethane generated the *anti*-1,3-diols (**36**) after desilylation with excellent stereoselectivity (>95%).[53] Unfortunately, the overall yield (60–69%) of the process was reduced by inefficient initial silylation (Scheme 5). Tetramethylammonium triacetoxyborohydride in acetic acid–acetonitrile, a mild reducing agent, reduced acyclic β-hydroxy ketones to the *anti*-1,3-diols with high diastereoselectivity (>92:8), independent of α-alkyl group stereochemistry.[54] The reducing agent requires coordination to an adjacent hydroxy group to allow intramolecular delivery of the hydride and so is highly chemoselective. Internal stereopropagation allowed sequential diastereoselective reductions of hydroxy diketone (**37**) to give *anti,anti*-triol (**38**) in 50% isolated yield (equation 8). Most of the β-hydroxy ketones screened had a bulky isopropyl group at the β-position; it is not clear if this is a requirement for high diastereoselectivity.

Reduction of TBDMS ethers of α-substituted-β-hydroxy ketones (**39** and **41**) with LAH in ether at – 78 °C proceeded with high 1,2-*anti* diastereoselectivity (>96:4) to give the corresponding *syn,anti*- (**40**) or *anti,anti*-1,3-diols (**42**) after acidic hydrolysis.[55] The stereochemistry results from a chelation-free Felkin transition state with the bulky alkylsilyloxy group in the perpendicular position. The result was independent of the configuration of the silyloxy substituent (equations 9 and 10).

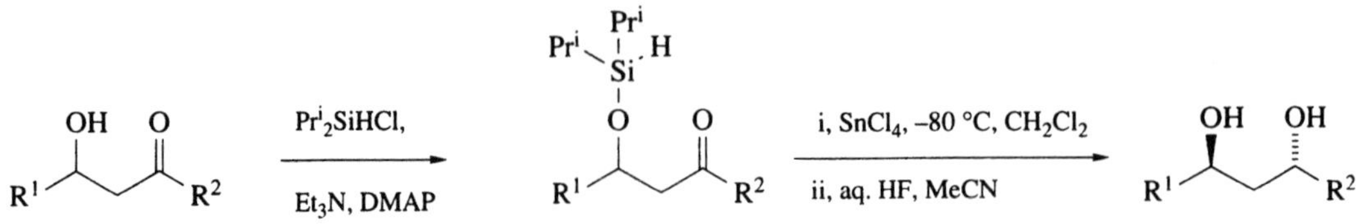

(33) Pr^i_2SiHCl, Et_3N, DMAP (35) i, $SnCl_4$, –80 °C, CH_2Cl_2; ii, aq. HF, MeCN (36)

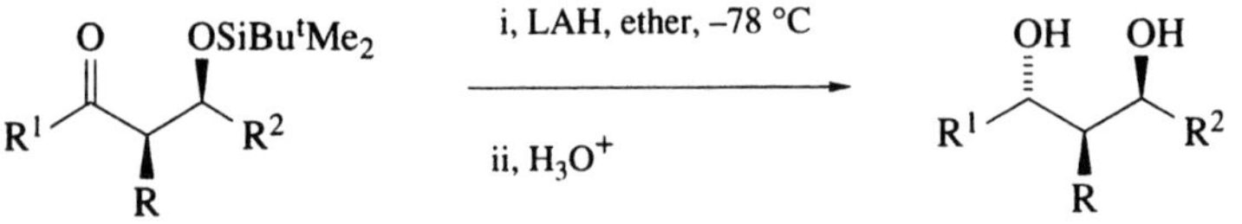

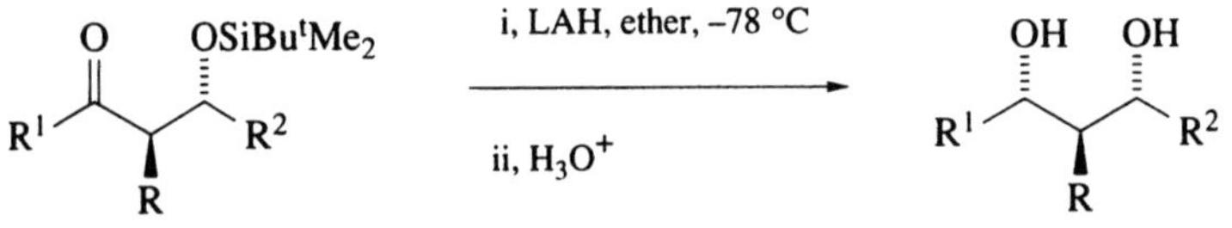

Scheme 5

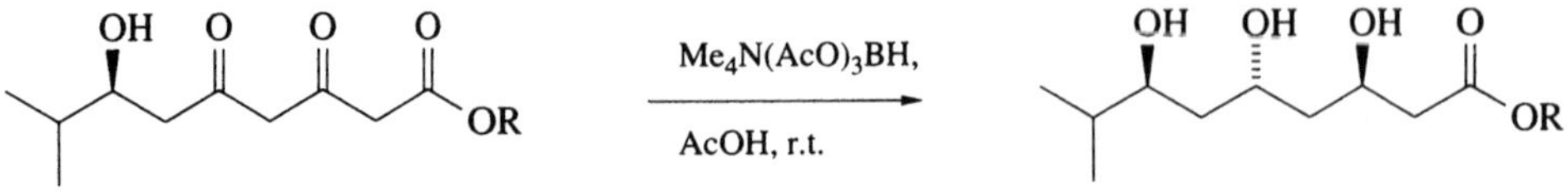

(37) $Me_4N(AcO)_3BH$, AcOH, r.t. (38) 50% isolated yield (8)

(39) i, LAH, ether, –78 °C; ii, H_3O^+ (40) (9)

(41) i, LAH, ether, –78 °C; ii, H_3O^+ (42) (10)

Examples of asymmetric induction from more remote hydroxy groups in the reduction of hydroxy ketones are rare. However the correct choice of reducing agent allowed complete diastereoselectivity to be achieved in the reduction of γ-hydroxy ketone (**43**) in a synthesis of ancistrofuran.[56] 2 equiv. of $LiEt_3BH$ in THF, one presumably coordinating the γ-hydroxy ketone in a fixed conformation, gave the diol (**44**) as the sole product in 88% yield (equation 11).

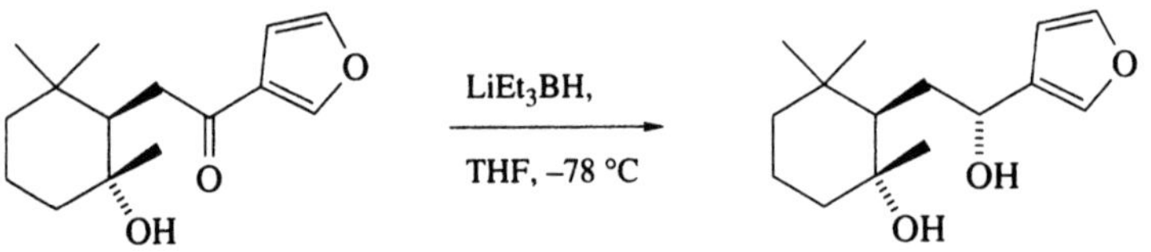

(43) $LiEt_3BH$, THF, –78 °C (44) (11)

The unsaturated linkage in enantiomerically pure α-methyl β,γ-unsaturated ketones (**45**) exerted a powerful stereochemical influence on their reduction with L-selectride, particularly when R^1 is a trimethylsilyl group.[57] The *anti* homoallylic alcohols (**46**) were produced with uniformly excellent stereoselectivity (>93:7) *via* a Felkin transition state in which the double bond occupied the perpendicular position (equation 12). This Felkin selectivity was sufficient to overcome any chelation-mediated contribution in the reduction of α-vinyl-β-hydroxy ketones (**47**) to the 1,2-*syn* diols (**48**) with $LiEt_3BH$ in THF at –78 °C (equation 13).[58]

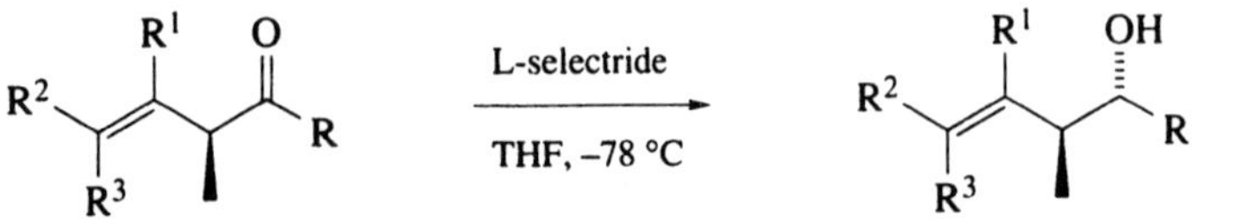

(45) L-selectride, THF, –78 °C (46) (12)

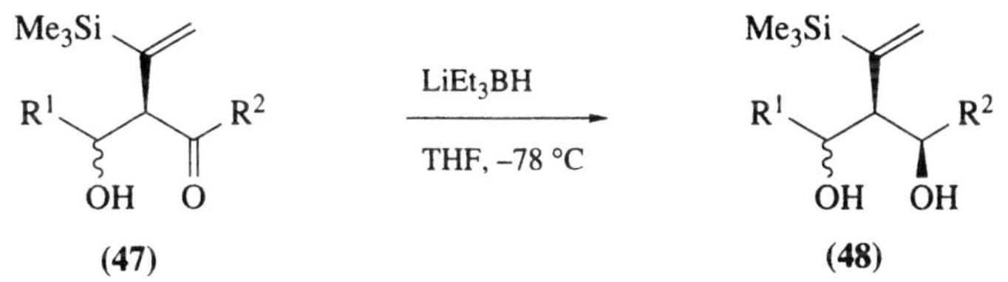

$$\text{(47)} \xrightarrow[\text{THF, }-78\ ^\circ\text{C}]{\text{LiEt}_3\text{BH}} \text{(48)} \quad (13)$$

Zinc borohydride was effective for the reduction of α,β-epoxy ketones (**49**) to the corresponding *anti*-α,β-epoxy alcohols (**50**) in ether at 0 °C irrespective of the substituents on the epoxide (equation 14).[59] The selectivity was rationalized by intramolecular hydride delivery from a five-membered zinc chelate avoiding the epoxide ring. In a limited study of the stereoselective reduction of γ,δ-epoxy ketones (**51**), LAH and di-2-(*o*-toluidinomethyl)pyrrolidine in ether at –78 °C gave the desired *cis*-epoxy alcohols (**52**) required for ionophore synthesis with good selectivity (>10:1) (equation 15).[60]

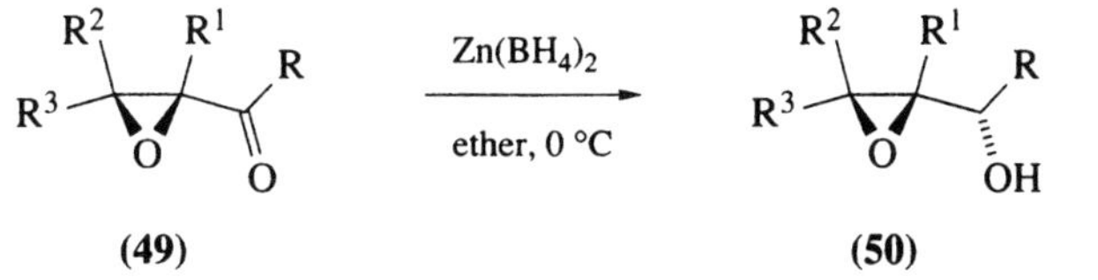

$$\text{(49)} \xrightarrow[\text{ether, }0\ ^\circ\text{C}]{\text{Zn(BH}_4)_2} \text{(50)} \quad (14)$$

$$\text{(51)} \xrightarrow[\text{ether, }-78\ ^\circ\text{C}]{\text{LAH, diamine}} \text{(52)} \quad (15)$$

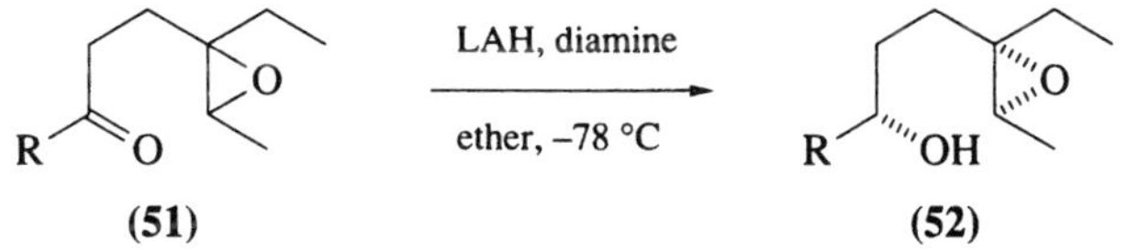

Stereoselective reduction of α-alkyl-β-keto acid derivatives represents an attractive alternative to stereoselective aldol condensation. Complementary methods for production of either diastereoisomer of α-alkyl-β-hydroxy amides from the corresponding α-alkyl-β-keto amides (**53**) have been developed. Zinc borohydride in ether at –78 °C gave the *syn* isomer (**54**) with excellent selectivity (≥97:3) in high yield *via* a chelated transition state.[61] A Felkin transition state with the amide in the perpendicular position accounted for reduction with potassium triethylborohydride in ether at 0 °C to give the stereochemically pure *anti* diastereoisomer (**55**). The combination of these methods with asymmetric acylation provided an effective solution to the asymmetric aldol problem (Scheme 6).[62] In contrast, the reduction of α-methyl-β-keto esters with zinc borohydride was highly *syn* selective when the ketone was aromatic or α,β-unsaturated, but less reliable in aliphatic cases.[38] Hydrosilylation also provided complete diastereocontrol (Scheme 7). The fluoride-mediated reaction was *anti* selective (≥98:2) while reduction in trifluoroacetic acid favored production of the *syn* isomer (≥98:2).[63] No loss of optical purity was observed under these mild conditions.

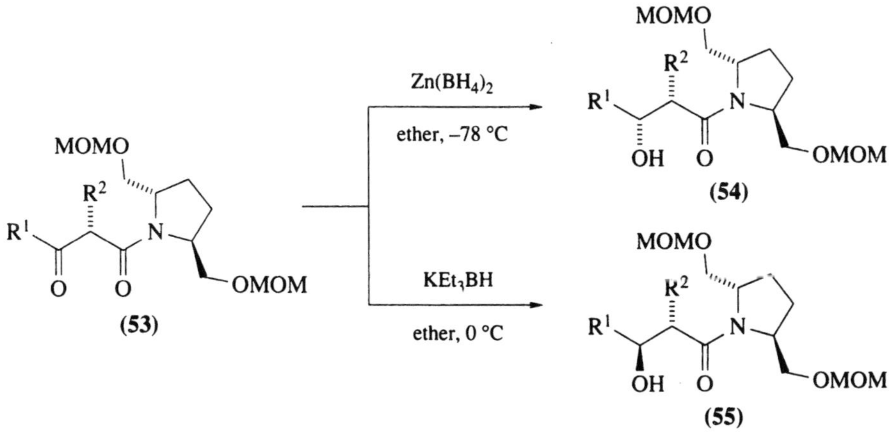

$$\text{(53)} \xrightarrow[\text{ether, }-78\ ^\circ\text{C}]{\text{Zn(BH}_4)_2} \text{(54)}$$

$$\text{(53)} \xrightarrow[\text{ether, }0\ ^\circ\text{C}]{\text{KEt}_3\text{BH}} \text{(55)}$$

Scheme 6

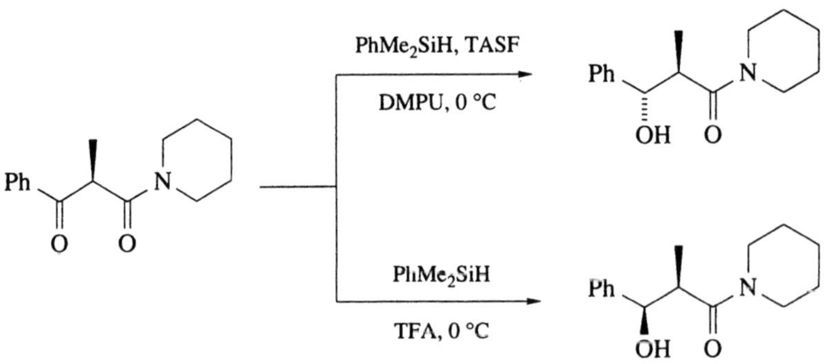

PhMe$_2$SiH, TASF
DMPU, 0 °C
PhMe$_2$SiH
TFA, 0 °C

Scheme 7

Diastereoisomerically pure β-hydroxy sulfides are useful synthetic intermediates for the preparation of geometrically pure alkenes or single diastereoisomers of epoxides. Reduction of β-keto sulfides (**56**) with L-selectride in THF gave the *syn*-β-hydroxy sulfides (**57**) *via* a Felkin transition state.[64] The selectivity was good (≥92:8) unless the acyl group was branched. Chelation-controlled reduction with zinc borohydride was successful in favorable situations, but a more general solution to the production of the *anti* diastereoisomers (**59**) and ultimately *trans*-epoxides was reduction of the sulfonium salts (**58**), prepared by methylation of the β-keto sulfides (**56**), with sodium borohydride in dichloromethane (Scheme 8). The selectivity was excellent (≥95:5) when R^2 was a branched alkyl group, but declined when it was not. This is consistent with a cyclic charge-controlled transition state analogous to a chelated conformation without a metal ion.[65]

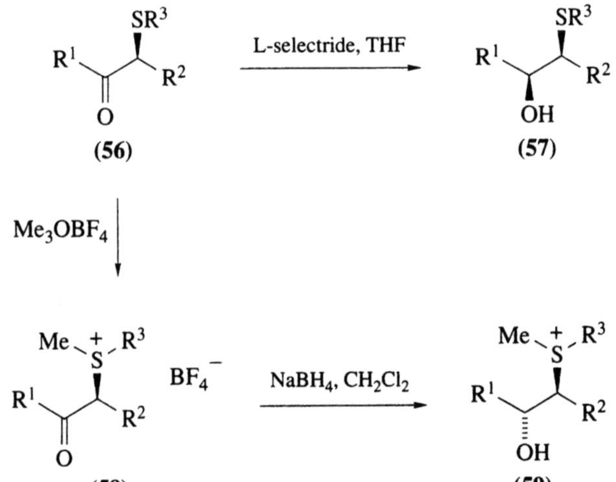

(**56**) L-selectride, THF → (**57**)

(**56**) Me_3OBF_4 → (**58**)

(**58**) BF_4^- NaBH$_4$, CH$_2$Cl$_2$ → (**59**)

Scheme 8

The stereocontrolled reduction of optically pure β-keto sulfoxides (**60**) with DIBAL-H (*anti* selective, ≥93:7) or DIBAL-H in the presence of zinc chloride (*syn* selective, >95:5) provided an entry to enantiomerically pure alcohols after desulfurization (Scheme 9).[66] The stereoselectivity may be rationalized by consideration of transition states analogous to those described for β-hydroxy ketone reduction (**31** and **32**), cyclic chelation by zinc chloride and external hydride delivery giving the *syn* isomer, and coordination of the DIBAL-H to the sulfoxide and internal hydride delivery giving the *anti* product.

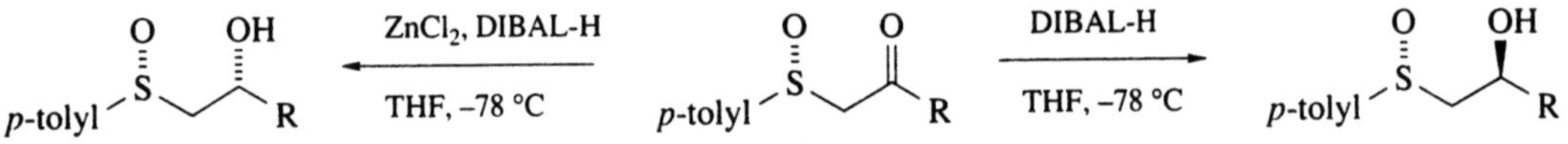

ZnCl$_2$, DIBAL-H, THF, –78 °C ← (**60**) → DIBAL-H, THF, –78 °C

Scheme 9

Geometrically pure alkenes were generated by Horner–Wittig elimination of stereochemically pure β-hydroxy diphenylphosphine oxides available from reduction of the corresponding β-keto phosphine oxides.[67] Reduction with a range of reagents was generally *syn* selective according to the Felkin model, but the success of the method depended on the ease of separation of diastereoisomers rather than intrinsically high stereoselectivity. However, in the presence of cerium chloride, sodium borohydride reduced sterically hindered α′-diphenylphosphinoyl ketones and enones (**61**) to the *anti*-alcohols (**62**) exclusively, presumably *via* a chelated transition state (equation 16).[68]

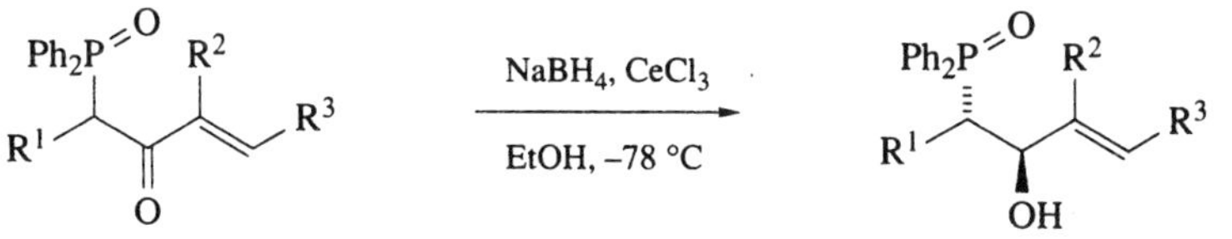

$NaBH_4$, $CeCl_3$; EtOH, –78 °C (16)

(**61**) (**62**)

Diastereocontrolled reduction of amino ketones represents an attractive route to amino alcohols, many of which are pharmacologically important, and has been exhaustively reviewed.[69] Even in the case of α-amino ketones, examples of high stereoselectivity were rare[70] with conventional metal hydride reagents,[71] and mixtures were common as the amino group became more distant.[72] In contrast, α-triazolyl ketones (**64**) were reduced with high stereoselectivity by tetraalkylammonium borohydrides to the *syn*-alcohols (**63**) in dichloromethane or to the *anti* isomers (**65**) when titanium tetrachloride was added (Scheme 10).[73]

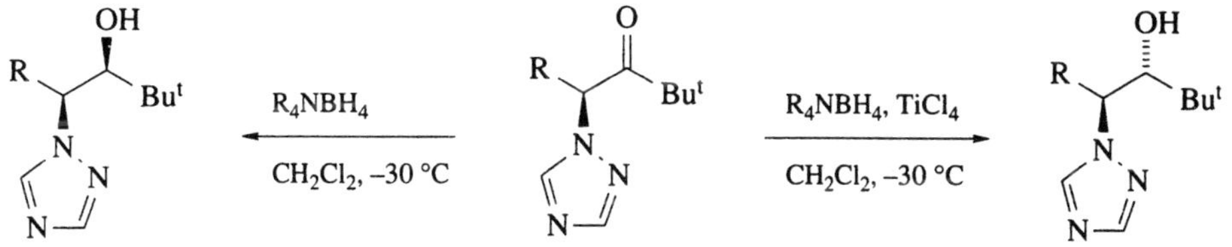

(**63**) ← R_4NBH_4, CH_2Cl_2, –30 °C — (**64**) — R_4NBH_4, $TiCl_4$, CH_2Cl_2, –30 °C → (**65**)

Scheme 10

Hydrosilylation of α-amino ketones (**66** and **68**) exhibited extremely high levels of selectivity in either direction depending on the reaction conditions (equations 17 and 18). The *syn*-alcohol (**67**) was the exclusive product of the fluoride-catalyzed reduction,[45] while trifluoroacetic acid catalysis generated the *anti* isomer (**69**).[46] Both these examples of highly diastereocontrolled reduction depended on the selection of either an open Felkin or a proton- or metal-bridged transition state.

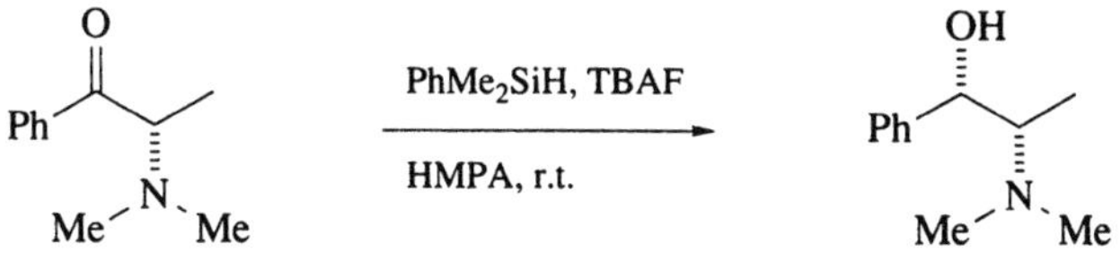

(**66**) — $PhMe_2SiH$, TBAF; HMPA, r.t. → (**67**) (17)

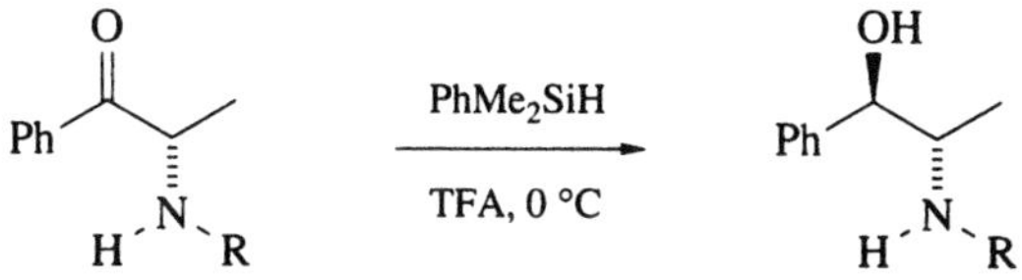

(**68**) — $PhMe_2SiH$; TFA, 0 °C → (**69**) (18)

The diastereoselectivity of reduction of a series of symmetrical diketones to the corresponding diols revealed an intriguing dependance on the separation of the carbonyl groups, but the selectivity was not generally useful.[74] However, in the case of 1,3-diketone (**71**) lithium borohydride alone produced the *anti* isomer (**70**) with 91% diastereoselectivity, but prior addition of titanium tetrachloride gave the *syn*-diol (**72**) with 96% diastereoselectivity *via* a chelated intermediate analogous to the crystalline complex between the diketone and $TiCl_4$ (Scheme 11).[75]

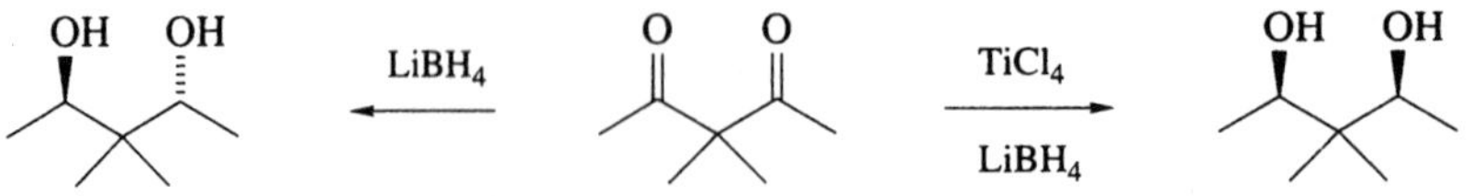

(70)	← $LiBH_4$	(71)	$TiCl_4$, $LiBH_4$ →	(72)

Scheme 11

1.1.3.2 Cyclic Carbonyl Compounds

The stereoselective reduction of cyclic ketones is profoundly influenced by conformational effects and is an area of intense theoretical study (Section 1.1.2.2). From a practical perspective, impressive diastereoselectivity in the required direction is possible by the correct choice of reagent (Table 1). Conformational homogeneity has been recognized as an important factor in achieving high diastereoselectivity. Temperature control is extremely important as at –78 °C a methyl group has almost the same conformational biasing effect as a *t*-butyl group at 0 °C.[76]

Table 1 Comparison of Selectivity for Equatorial Attack in Reductions of Alkyl Substituted Cyclic Ketones

Substituent(s) / *Reagent*	*2-Me* C_6[a]	*3-Me* C_6	*4-Me* C_6	*4-Bu*^{t} C_6	*3,3,5-Me*$_3$ C_6	*2-Me* C_5[b]	*Ref.*
$LiBu^s{}_3BH$	99.3	95	90	96.5	99.8	99.3	27
$LiMes_2BH_2$	99	99	94	94	—	98	80
$Li(MeCp)_3BH$	99.5	99	98	99.4	—	—	79
$LiSia_3BH$	99.7	99.6	99	99.4	—	99.5	79
K9-OThx-9-BBN-H	98.5	90	85.5	87	99.9	—	84
K9-Bu^t-9-BBN-H	99.5	98	94	98.5	99	—	83
$LiBuBH_3$	—	8	6	2	—	—	77
$LiMeBH_3$	13	5.7	4.5	1.7	66	—	77
NiCRA	7	—	5	6	1	—	86
$LiAlH_4$THF	24	16	—	10	80	23	27
$NaBH_4$	30	14	—	14	52	26	27

[a]C_6, cyclohexanone. [b]C_5, cyclopentanone.

Unhindered cyclohexanones exhibited an intrinsic preference for axial attack by small hydride reagents ($NaBH_4$, $LiAlH_4$) leading to the equatorial alcohol. Monoalkylborohydrides, prepared from borane–dimethyl sulfide and the corresponding alkyllithium, proved to be superior to the traditional reagents, but the selectivity declined as the ketone became more hindered.[77] Axial alcohols, *via* equatorial attack, were produced from both hindered and unhindered ketones with bulky hydride reducing agents. The most successful were highly branched alkylborohydrides generally with a lithium counterion.[78] Lithium trisiamylborohydride[79] demonstrated exceptional stereoselectivity, better than the tris(*trans*-methylcyclopentyl)borohydride ($Li(MeCp)_3BH$), L-selectride or lithium dimesitylborohydride ($LiMes_2BH_2$).[80] In contrast, the analogous ate complex formed from *t*-butyllithium and DIBAL-H, which was expected to be highly hindered and hence stereoselective, reduced 4-*t*-butylcyclohexanone without any preference.[81] Potassium trisiamylborohydride, prepared from potassium hydride and trisiamylborane with triisopropoxyborane catalysis, exhibited stereoselectivity comparable with the lithium analog.[82]

Two new classes of bulky reducing agents generated from potassium hydride and *B*-alkyl-[83] or *B*-alkoxy-9-boratabicyclo[3.3.1]nonanes[84] have been investigated. The optimum stereoselectivity was obtained with maximum branching in the alkyl group and *t*-butyl was significantly better than 2,4-dimethyl-2-pentyloxy (OThx). Various reagents based on magnesium hydride with a variety of ligands on magnesium (hydridomagnesium alkoxides, dialkylaminomagnesium hydrides, and lithium alkoxy- and dialkoxy-magnesium hydrides) have been prepared and some bulky examples provided useful selectivity in ketone reduction.[85]

Complex reducing agents containing sodium hydride, a metallic salt and an alcohol were screened against six representative cyclic ketones for stereoselectivity. Cadmium, cobalt and manganese chlorides gave rise to reagents that favored formation of the axial alkoxides, but the selectivity was not outstanding. Those derived from zinc chloride and particularly nickel acetate 2,5-dimethylhexanediol (NiCRA) provided the more stable equatorial alkoxides in very good yield with high selectivity.[86] The proposed equilibration to the more stable product parallels the classical solution to the production of equatorial alcohols from hindered cyclohexanones based on equilibration with Raney nickel.[87]

Correct reagent selection allowed reduction of steroidal enone (**74**) to either diastereoisomeric allylic alcohol, uncontaminated by its isomer. Sodium borohydride/cerium chloride in methanol–THF gave the equatorial alcohol (**73**), while L-selectride produced the axial isomer (**75**) *via* equatorial attack (Scheme 12).[88] Unexpected axial attack on diketone (**76**) to give equatorial alcohol (**77**; equation 19) led to the proposition that for hydride additions to decalones two 1,3-diaxial interactions override one *peri* interaction which in turn takes precedence over a single 1,3-diaxial interaction.[89]

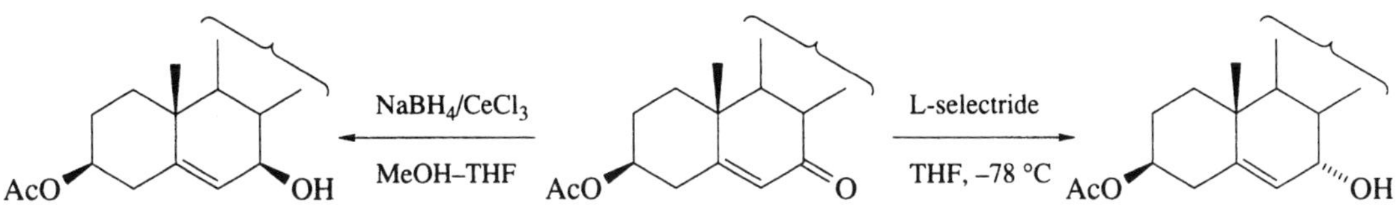

(**73**) ← $NaBH_4/CeCl_3$, MeOH–THF — (**74**) — L-selectride, THF, –78 °C → (**75**)

Scheme 12

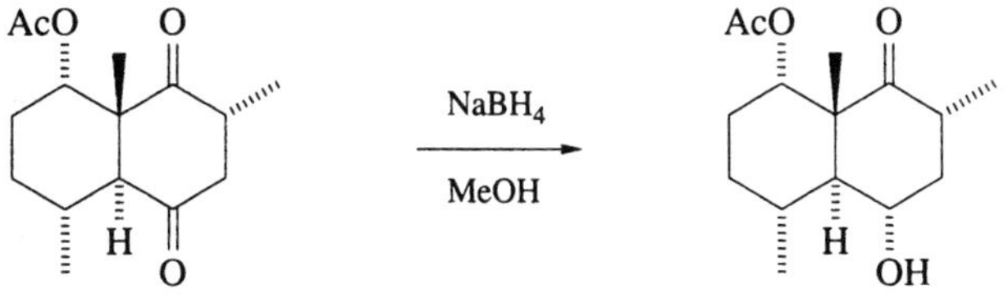

(**76**) — $NaBH_4$, MeOH → (**77**) (19)

The influence of polar substituents on the stereoselectivity of reduction of cyclic ketones has not been widely studied. A series of cyclohexanones with hydroxy groups in differing positions on the ring and on alkyl chains attached to the ring was reduced with a selection of reducing agents to investigate the selectivity, but, although some examples of high stereoselectivity were observed, no clear pattern emerged.[90] Conformationally flexible and conformationally locked 2-methylthio- and 2-methylsulfonyl-cyclohexanones were studied with a range of reducing agents. Exclusive axial approach, even with bulky hydrides, was observed when the sulfur adopted the axial disposition, but diastereoisomeric mixtures resulted when the sulfur was equatorial.[91] The selectivity was rationalized by a combination of steric and stereoelectronic factors. The effect of ring size on reduction selectivity for a series of α-phenylthio ketones was investigated.[92] Small rings (five- and six-membered) exhibited reasonable selectivity (>7:1), consistent with the preceding results, but medium sized (seven- and eight-membered) and large rings produced low selectivity (<2:1). The reversal was explained by a conformational change in the ring position in the transition state.

1.1.4 CHEMOSELECTIVITY

1.1.4.1 Reduction of Enones

Reduction of α,β-unsaturated ketones by metal hydrides can follow two pathways; addition to the carbonyl group (1,2-addition, charge control) to give the allylic alcohol, or addition to the conjugated double bond (1,4-addition, frontier orbital control) to give the saturated ketone. The concept of hard and soft acids and bases has been used to explain observed selectivities.[93] The hard metal hydrides attack the carbonyl group while the softer metal hydrides, with a more covalent M—H bond, favor conjugate addition. Borohydrides are softer than their aluminum counterparts due to the more covalent hydride bond and both can be hardened by incorporating alkoxy groups on the metal. The counterion is also important, sequestration of lithium by a [2.1.1]cryptand causes conjugate reduction of cyclohexenone by $LiAlH_4$ and $LiBH_4$ to predominate[94] while added hard lanthanide salts enhance reduction to the allylic alcohols (*vide infra*).

In spite of substantial evidence, the tendency for sodium borohydride to reduce enones in a conjugate sense is often ignored,[95] but the need for reduction to the corresponding allylic alcohols has led to the development of several new specific reagent combinations. The most widely accepted of these involves sodium borohydride in the presence of cerium chloride, which has been optimized to give excellent selectivity under mild and experimentally convenient conditions.[96] The success of this reagent may be explained by considering the active species as an alkoxy borohydride in combination with a hard cerium cation which coordinates to the carbonyl to enhance 1,2-selectivity. Other lanthanoid-mediated processes include sodium borohydride with $LnCpCl_2(THF)_3$ (Ln = Sm and Er) in methanol[97] and LAH with a lanthanoid salt,[98] particularly cerium chloride, in THF.

DIBAL-H has often proved effective in 1,2-reduction of enones[99] and the ate complex generated with *n*-butyllithium is a powerful and selective reducing agent, which has been used in toluene–hexane to reduce cyclic enones and in toluene–THF to reduce acyclic examples.[100] 1-Pyrrolylborane–THF complex, which may be prepared from pyrrole and borane–THF, is a reactive aminoborane and has been used to reduce enones and enals to the corresponding allylic alcohols, but its scope remains to be established.[101]

Complex reducing agents prepared from sodium hydride, in combination with sodium alkoxides and metal salts to moderate its basicity, have proved effective for highly regioselective 1,4- and 1,2-reductions of enones depending on the choice of metal salt.[102] The reagent based on nickel acetate was selective for conjugate reduction while that derived from zinc chloride produced allylic alcohols both in good yields.[103] This method has been extended to hydrosilylation of the carbonyl group of enones by incorporation of trimethylsilyl chloride into the reagent.

Molecular orbital calculations have suggested that cyclopentenone is intrinsically more prone to conjugate reduction than cyclohexenone[104] and thus is a good substrate on which to test new 1,2-selective reagents. The selectivity of reduction of both these enones with the best of the new reagents together with the results for 9-BBN-H, the previous reagent of choice,[105] are tabulated for comparison (Table 2).

Table 2 Comparison of Selective Reagents for Allylic Alcohol Formation over Saturated Ketone or Alcohol in Cyclic Ketone Reduction

Reagent	*Cyclopentenone*		*Cyclohexenone*		*Ref.*
	Yield (%)	*Selectivity*	*Yield (%)*	*Selectivity*	
$LiAlH_4$	100	85:15	97	94:6	95
$NaBH_4$	100	0:100	90	59:41	95
$NaBH_4/CeCl_3$	100	97:3	100	>99:1	96
$NaBH_4/SmCpCl_2$	95	98.8:1.2	99	97.9:2.1	97
$NaBH_4/ErCpCl_2$	83	100:0	99	97.8:2.2	97
$LiAlH_4/CeCl_3$	99	64:36	100	98:2	98
DIBAL-H	81	98:2	94	98:2	100
DIBAL-H/BuLi	83	99:1	96	94:6	100
9-BBN-H	85	>99:1	85	>99:1	105

1.1.4.2 Aldehydes *versus* Ketones

The selective reduction of the more reactive and sterically less hindered aldehyde functionality in the presence of a ketone is a problem often encountered in synthesis. Moderation of the reactivity of the reagent by changes in its structure or to the nature of the reaction medium have been the aims of recent solutions to this problem. An exhaustive listing of 30 modern methods has appeared recently and will not be duplicated here.[106] Those methods that are likely to prove successful in synthesis have been selected for consideration (Table 3).

The reactivity of sodium borohydride has been controlled by employing a mixed solvent system (ethanol/CH_2Cl_2 3:7) and performing the reduction at –78 °C. Under these conditions most aldehydes were reduced in the presence of ketones with >95% selectivity.[106] The comparison between benzaldehyde and 4-methylcyclohexanone revealed the limitation of this method, as substantial reduction of the ketone occurred during the long reaction time that was required. Borohydride exchange resin, prepared from anionic exchange resin and sodium borohydride, reduced aldehydes in preference to ketones in ethanol at 25 °C.[107] The polymeric nature of this reagent facilitates isolation of the products and regeneration of the polymer is possible. Remarkably, at –10 °C in methanol, selectivity between aldehydes was observed. Aromatic aldehydes were reduced in the presence of aliphatic compounds and the reaction was sensitive to the electronic effects of aromatic substituents. Cyclohexanone was reduced much more rapidly than acyclic or bicyclic ketones.

Tetra-*n*-butylammonium triacetoxyborohydride in refluxing benzene reduced aldehydes but not acyclic ketones, the selectivity was demonstrated in competition experiments and keto aldehyde reductions.[108] The more reactive cyclohexanones were reduced only slowly under the same conditions. The limitation of this convenient method is that proximal hydroxy groups activated the reagent enabling the

Table 3 Comparison of Chemoselectivity in Reductions of Aldehydes in the Presence of Reactive Ketones

Reagent	*Aldehyde/Ketone*	*Yield of Alcohol (%)*	*Ref.*
$NaBH_4EtOH/CH_2Cl_2$	Hexanal	97	106
	4-Me-Cyclohexanone	5	
Borohydride exchange resin	Benzaldehyde	99	107
	Acetophenone	1.0	
$Bu_4NBH(OAc)_3$	Benzaldehyde	95	108
	Acetophenone	<4	
9-BBN-H–py	Benzaldehyde	93	109
	Cyclohexanone	1.5	
B-Siamyl-9-BBN	Benzaldehyde	>95	110
	Acetophenone	0	
8-Oxyquinoline dihydroboronite/BF_3	Nonanal	90	111
	4-Me-Cyclohexanone	28	
$Li(Et_3CO)_3AlH$	Hexanal	99.6	112
	Cyclohexanone	0.4	
$Et_4N[\mu\text{-}HMo_2(CO)_{10}]/HOAc$	Hydrocinnamaldehyde	100	113
	4-Bu^t-Cyclohexanone	0	
Bu_3SnH/SiO_2	*n*-Octanal	99	114
	Cyclohexanone	13	
$NaBH_4/CeCl_3/EtOH/H_2O$	Hexanal	2	116
	Cyclohexanone	100	

stereoselective reduction of ketones (Section 1.1.3.1). Thus a β-keto aldehyde was reduced to the corresponding diol in high yield.

The complex formed from 9-BBN-H and pyridine is a readily prepared stable crystalline solid, which is a mild highly selective reducing agent.[109] Even unhindered ketones were not significantly reduced under conditions that led to complete reduction of aldehydes to the corresponding alcohols. Various representative functional groups were inert to this reagent. 9-BBN-H has also been modified by hydroborating 2-methyl-2-butene to give *B*-siamyl-9-BBN which reduced a wide range of aldehydes at rates at least 100–200 times faster than all ketones tested.[110] 8-Hydroxyquinoline has been used to moderate the reactivity of borane–dimethyl sulfide, the resulting complex could be stored for several weeks under nitrogen at low temperature. In the presence of catalytic boron trifluoride etherate the 8-oxyquinoline dihydroboronite selectively reduced aldehydes rather than ketones.[111] The selectivity was good for alkyl ketones over alkyl aldehydes and two ketoaldehydes were reduced exclusively to the corresponding hydroxy ketones. Cyclohexanones proved to be rather reactive and competed with the aldehydes. As a result 4-methylcyclohexanone could be selectively reduced in preference to 3-pentanone.

A comparison of four tri-*t*-alkoxyaluminum hydrides revealed that lithium tris[(3-ethyl-3-pentyl)oxy]aluminum hydride, prepared from LAH and 3-ethyl-3-pentanol, was the most selective for reduction of aldehydes over ketones of all types.[112] Even the less reactive benzaldehyde was reduced in THF at –78 °C faster than cyclohexanone (97.7:2.3). A good correlation between the steric demands of the reducing agent and the observed chemoselectivity was observed.

Highly selective reduction of aldehydes rather than ketones, even 4-*t*-butylcyclohexanone, has been achieved by a binuclear hydrido molybdenum anion ($Et_4N[\mu\text{-}HMo_2(CO)_{10}]$) and acetic acid in refluxing THF.[113] The stable complex is readily prepared from molybdenum hexacarbonyl, which can be recovered after reduction. The full value of this reagent in synthesis remains to be established. Tributyltin hydride has been shown to be effective for the selective reduction of aldehydes in the presence of dried silica gel which acts as a mild acid catalyst.[114] The reaction proceeded at room temperature in cyclohexane, many common functional groups were inert under these conditions.

Investigations into the selective hydration and ketalization of various classes of aldehydes and ketones with lanthanoid chlorides have revealed interesting opportunities for selective reductions of less reactive or more hindered carbonyl groups. Ketalization with methanol catalyzed by cerium, erbium or neodymium chlorides protected the aldehydes *in situ*, while sodium borohydride reduced the unreacted ketones; on work-up the aldehydes were released.[115] The selectivity of this method declined with easily ketalized ketones, but conducting the reaction in aqueous ethanol with $NaBH_4/CeCl_3$ provided a more useful alternative.[116] Selective hydration discriminated between ketones and aldehydes and between conjugated and nonconjugated aldehydes.[117] Similarly, selective imine formation with *t*-butylamine and 4Å sieves, reduction with $Li(Bu^tO)_3AlH$, and mild acidic hydrolysis allowed aldehydes to remain unchanged while ketones were reduced.[118]

1.1.4.3 Ketones *versus* other Carbonyl Groups

The requirement for a powerful nucleophilic hydride reagent able to reduce even very hindered ketones rapidly and quantitatively has been satisfied by lithium triethylborohydride which was screened against selected representative functional groups.[119] The quest for increased chemoselectivity in total synthesis has led to the development of reducing agents that will discriminate between various classes of ketone or between ketones and other carbonyl groups.

Potassium triisopropoxyborohydride, a mild selective reducing agent, rapidly converted ketones and aldehydes to the corresponding alcohols, while many common functional groups were inert.[120] The reaction of potassium hydride with triphenylborane produced the triphenylborohydride, which is highly hindered and which exhibited excellent chemoselectivity between ketones.[121] Cyclohexanone was reduced in preference to cyclopentanone (97:3) and 4-heptanone (99.4:0.6), while methyl ketones were more reactive than 4-heptanone (2-heptanone, 94:6; acetophenone, 97.8:2.2).

Zinc-modified cyanoborohydride, prepared from anhydrous zinc chloride and sodium cyanoborohydride in the ratio 1:2 in ether, selectively reduced aldehydes and ketones but not acids, anhydrides, esters and tertiary amides.[122] In methanol the reactivity paralleled the unmodified reagent. Zinc and cadmium borohydrides form solid complexes with DMF, which may prove to be convenient sources of the reducing agents.[123] Aromatic and α,β-unsaturated ketones were reduced much more slowly than saturated ketones, so chemoselective reduction should be possible.

Sodium borohydride in methanol/dichloromethane (1:1) at –78 °C reduced ketones in the presence of conjugated enones with excellent selectivity.[124] Competition experiments were complemented by reductions of diones to illustrate the synthetic value of the method. The Wieland–Miescher ketone was reduced to the 1-β-alcohol contaminated by less than 3% diol in quantitative yield. The same reagent in acetic acid was shown to be effective at reducing aromatic ketones with *ortho* hydroxy or *ortho* amino substituents rapidly.[125] Other aromatic ketones were relatively inert under the same conditions. Presumably the success of this procedure relies on activation and intramolecular delivery of the acetoxyborohydride by the *ortho* substituent.

Borane was modified by reaction with half a molar equivalent of 2-aminoethanol to give a reagent that reduced aldehydes and ketones rapidly at room temperature in THF, but did not reduce esters, oxime esters, tetiary amides, nitriles, halides and acid chlorides.[126] Complete chemoselectivity between acetophenone and ethyl benzoate was demonstrated. The combination of polymer-bound amino alcohols and borane was also investigated to take advantage of simplification in work-up and the possibility of enhanced selectivity. Various amines were combined with borane to produce reagents of moderated reactivity. The most promising was *t*-butylamine borane, which exhibited dramatic rate differences between different types of ketone, suggesting that chemoselective reduction of certain diketones would be possible with this reagent.[127] The reducing power of LAH has been tempered by pretreatment with dry silica gel in ether to enable reductions of keto esters to hydroxy esters in nonpolar solvents.[128] Diphenylstibine has been used, for the first time, in combination with Lewis acids, particularly aluminum trichloride, to reduce aldehydes and ketones in THF at room temperature in excellent yields.[129] Esters, acid chlorides, alkyl halides and alkenes were unreactive under these conditions, so this new reagent holds promise as a mild chemoselective reducing agent.

The concept of *in situ* protection of the less hindered or more Lewis basic of two ketones to enable selective reduction of the usually less reactive groups has been successfully developed.[130] The sterically hindered Lewis acid MAD (**78**) derived from BHT and trimethylaluminum was used to coordinate preferentially to the less hindered ketone and DIBAL-H reduced the more hindered ketone that remained uncomplexed. An approximate order of comparative reactivity for various classes of ketones has been established. The selectivity was improved by using the more hindered Lewis acid MAB (**79**) and/or dibromoalane as the reducing agent. The discrimination between aromatic ketones is good but less successful between two dialkyl ketones. The chemoselectivity was demonstrated in the reduction of diketone (**80**) to keto alcohol (**81**) in 87% yield and excellent selectivity (equation 20).

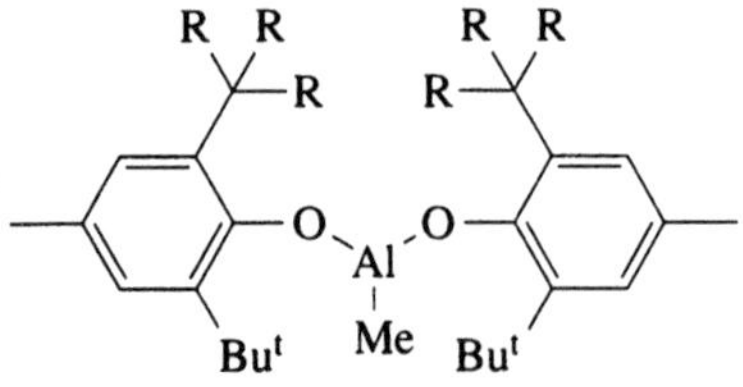

(**78**) MAD, R = Me
(**79**) MAB, R = Et

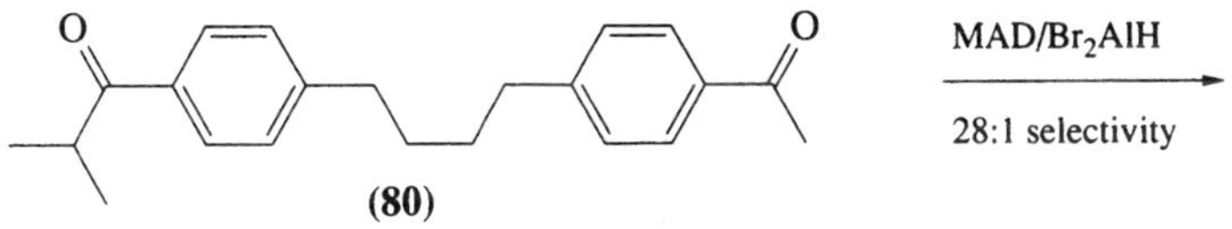

MAD/Br_2AlH

28:1 selectivity

(80)

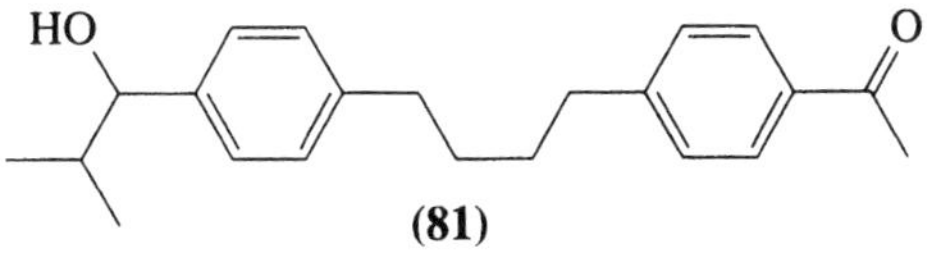

(81)

(20)

1.1.4.4 Quinones to Hydroquinones

The reduction of quinones to the corresponding hydroquinones is an important synthetic transformation that has been accomplished with various conventional reducing agents.[131] Complex metal hydride reagents have not tended to be the method of choice, as competitive formation of the dihydrodiols is frequently a problem. These two possible outcomes are illustrated for anthraquinone (**83**) giving the hydroquinone (**82**) and dihydrodiol (**84**), respectively (Scheme 13). Route 1, to the hydroquinone, diverges from route 2 after the first hydride equivalent has been added, as the remaining carbonyl group is enolized rather than reduced; hydrogen is liberated as a result.

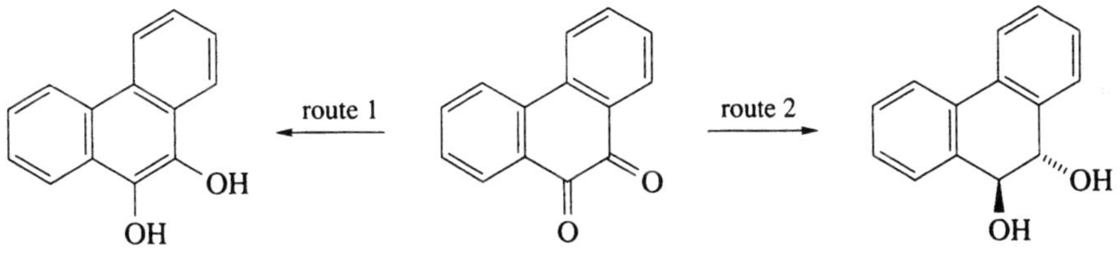

route 1 route 2

(82) (83) (84)

Scheme 13

Alkoxyaluminum hydrides including Red-Al, $Li(Bu^tO)_3AlH$ and $Li(MeO)_3AlH$ have been used extensively, particularly for the reduction of *o*-quinones to the corresponding *o*-hydroquinones.[5] It appears that the use of more soluble reagents and powerful ethereal solvents also favors reduction to the hydroquinone.[132]

Hydrosilylation to give bis(trimethylsilyloxy)benzenes has proved an attractive alternative.[133] The most convenient and high yielding method used 1,1,3,3-tetramethyldisiloxane (TMDS) and a catalytic quantity of iodine or trimethylsilyl iodide[134] in refluxing dichloromethane. A range of *p*-quinones (**85**), including 1,4-naphthoquinone, was reduced and the resulting silyl ethers were hydrolyzed to give uniformly good yields (85–98%) of the corresponding hydroquinones (**86**; equation 21).

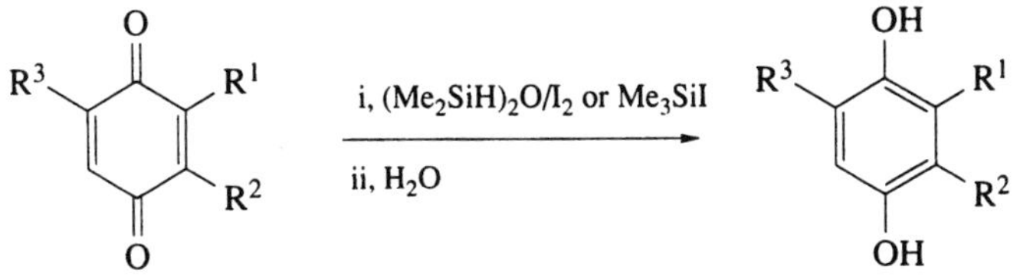

i, $(Me_2SiH)_2O/I_2$ or Me_3SiI

ii, H_2O

(85) (86)

(21)

Hydrides of tin and germanium also reduce quinones, ultimately to the hydroquinones. The detailed mechanism of hydrogen transfer between four stannanes and four quinones has been studied which suggested a radical mechanism in which hydrogen transfer was the first step.[135]

1.1.4.5 α-Halo Ketones

α-Halocarbonyl compounds can be reduced at the carbonyl group to give the chlorohydrins or at the carbon–halogen bond to give the saturated ketones or alcohols. Hydrogenolysis of the carbon–halogen bond is often a severe limitation on the value of this route to chlorohydrins, particularly when nucleophilic reducing reagents are used.

A detailed study of a series of α,α-halo ketones demonstrated that electrophilic reducing agents, notably borane, DIBAL-H and BH_3·DMS, provided excellent yields of the alcohols without reduction of the halogen bond.[136] In contrast, nucleophilic reagents ($NaBH_4$, Li($Bu^tO)_3AlH$ and K-selectride) were unsatisfactory. The results for a representative α,α-dibromo ketone (**87**) being reduced to the mono- (**89**) and di-bromohydrin (**88**; equation 22) are shown in Table 4.

(**87**) —reagent / solvent→ (**88**) + (**89**) (22)

Table 4 Comparison of Reducing Agents for the Reduction of α,α-Dibromoketone (**87**)

Reagent	*Solvent*	*Yield of (**88**) (%)*	*Yield of (**89**) (%)*
$NaBH_4$	MeOH	50	50
$Li(Bu^tO)_3AlH$	THF	55	45
K-Selectride	Ether	25	75
DIBAL-H	Hexane	90	0
BH_3·DMS	Ether	87	0
Borane	THF	90	0

Under radical conditions tributyltin hydride reduced α-chlorocarbonyl compounds to the corresponding saturated carbonyl derivatives but in the presence of phosphine oxides (HMPA or Bu_3PO) chemoselective reduction of the carbonyl group was observed. α-Chloro ketones and aldehydes were reduced with this reagent combination to give good yields of the chlorohydrins.[137] The choice of ligand was important as neither phosphines nor amines were effective substitutes for the phosphine oxides.

1.1.5 OTHER METAL HYDRIDES

1.1.5.1 Silicon

Silicon hydrides are an interesting class of carbonyl reducing agents as they are reasonably stable under normal conditions requiring activation with a transition metal complex,[138] fluoride ion[45] or Lewis acid.[46] The correct choice of reaction conditions allows highly chemo- and stereo-selective reduction of particular classes of carbonyl compounds with these convenient reagents.

Tris(triphenylphosphine)chlororhodium promoted hydrosilylation of α,β-unsaturated carbonyl compounds and the resulting adducts gave the corresponding alcohols on methanolysis.[139] The regioselectivity of addition depended on the silane; monohydrosilanes (Et_3SiH,$EtMe_2SiH$) favor conjugate reduction to give the silyl enol ether, but dihydrosilanes (Ph_2SiH_2, Et_2SiH_2) reduced the carbonyl group directly, generating allylic alcohol derivatives. The yields were uniformly high and the regioselectivity in the substrates screened was impressive.

Aldehydes and ketones were reduced chemoselectively under mild conditions by hydrosilanes (generally $PhMe_2SiH$) in dipolar aprotic solvents with TBAF or tris(diethylamino)sulfonium difluorotrimethylsilicate (TASF) as catalytic fluoride source. The reduction of α-substituted ketones is highly *syn* selective (Section 1.1.3.1). The mechanism of this reaction has been investigated and involves rate-determining hydride transfer from a hexavalent silicate $[HSiR_3F(HMPA)]^-$ (**90**) and not prior coordination of the fluorosilane with the carbonyl group (Scheme 14).[45] Minor amounts of radical-derived reduction products due to single-electron transfer processes, together with predominant hydride transfer, have been observed in certain mechanistic probe systems.[140]

An analogous pentacoordinate hydridosilicate reagent (**91**), prepared from trichlorosilane and catechol or 2,2′-dihydroxybiphenyl with butyllithium (equation 23), reduced aldehydes and ketones in THF under mild conditions in high yield to the corresponding alcohols without any added catalyst (equation 24).[141]

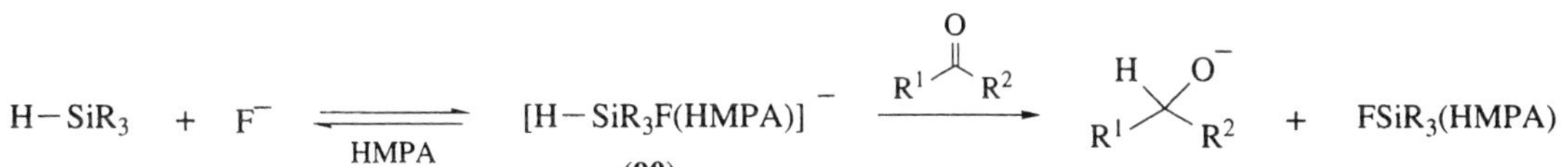

$$H{-}SiR_3 + F^- \underset{HMPA}{\rightleftharpoons} [H{-}SiR_3F(HMPA)]^- \ \textbf{(90)} \xrightarrow{R^1C(O)R^2} R^1R^2C(H)O^- + FSiR_3(HMPA)$$

Scheme 14

The chemo- and stereo-selectivity of this reagent was briefly examined, as was the mechanism indicating that rate-determining hydride transfer was involved.

$$2\ C_6H_4(OLi)_2 + HSiCl_3 \xrightarrow{THF} [(C_6H_4O_2)_2SiH]^- \ \textbf{(91)} \quad (23)$$

$$R^1C(O)R^2 \xrightarrow[\text{ii, } H_2O]{\text{i, } \textbf{(91)}} R^1R^2C(H)OH \quad (24)$$

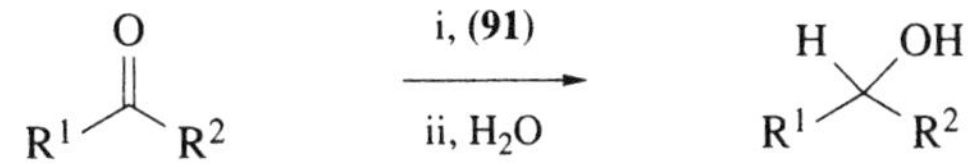

Although hydrosilanes reduce ketones, in trifluoroacetic acid, to the corresponding methylene compounds or dimeric ethers *via* ionic hydrogenation, the reduction of α-amino and α-oxy ketones and β-keto acid derivatives with hydrosilanes, particularly $PhMe_2SiH$, under these conditions proceeded with high *anti* selectivity to the alcohols.[46] No racemization was observed at the carbon α to the carbonyl group. Intramolecular hydrosilylation catalyzed by Lewis acids provided a highly stereoselective route to *anti*-1,3-diols from β-hydroxy ketones (Section 1.1.3.1).[53]

1.1.5.2 Tin

Hydrostannation of carbonyl compounds with tributyltin hydride is promoted by radical initiation and Lewis or protic acid catalysis.[142] The activation of the carbonyl group by the acidic species allows the weakly nucleophilic tin hydride to react *via* a polar mechanism. Silica gel was a suitable catalyst allowing chemoselective reduction of carbonyl groups under conditions that left many functional groups unchanged.[114] Tributyltin triflate generated *in situ* from the tin hydride and triflic acid was a particularly efficient catalyst for the reduction of aldehydes and ketones with tributyltin hydride in benzene or 1,2-dichloromethane at room temperature.[143] Esters and ketals were not affected under these conditions and certain aldehydes were reduced selectively in preference to ketones.

Sterically hindered ketones were more difficult to reduce with tin hydrides but reduction at high pressure (1 GPa) without radical or Lewis acid catalyst in methanol at 55 °C was effective.[144] Reduction of the very hindered ketone (**92**) did not occur at atmospheric pressure but proceeded usefully at high pressure (equation 25). The absence of radical intermediates allowed even α,β-epoxy and cyclopropyl ketones such as (**93**) to be reduced in high yield, largely without cleavage of the strained ring (equation 26). Under conventional AIBN-initiated conditions radical-mediated processes predominated.

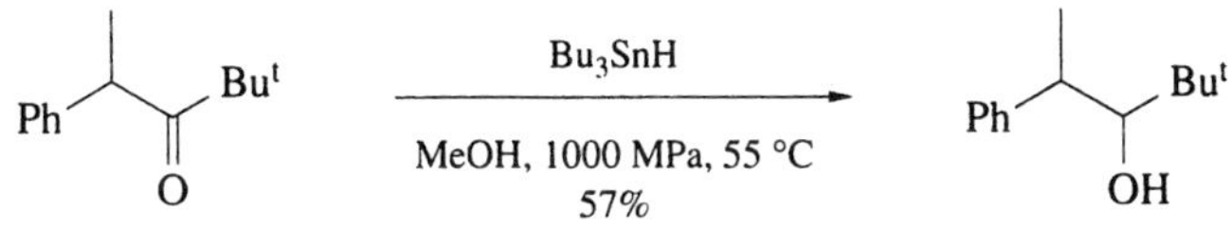

$$\textbf{(92)} \xrightarrow[\text{MeOH, 1000 MPa, 55 °C; 57\%}]{Bu_3SnH} \text{alcohol} \quad anti/syn = 0.14 \quad (25)$$

Ph, Bu^t, O; Ph, Bu^t, OH

$$\textbf{(93)} \xrightarrow[\text{MeOH, 1400 MPa, 65 °C; 99\%}]{Bu_3SnH} \text{cyclopropyl alcohol} \quad (26)$$

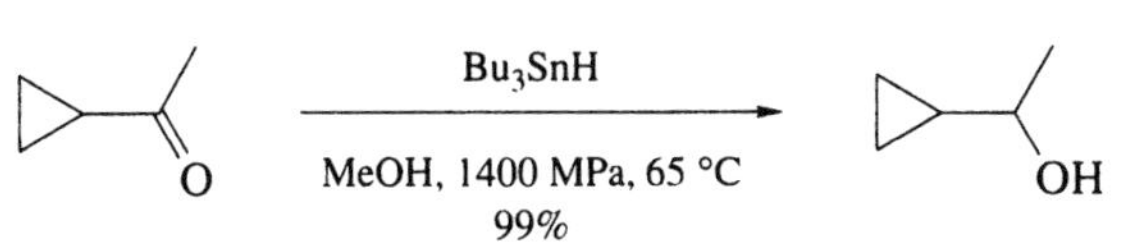

1.1.5.3 Transition Metals

Transition metal hydrides have also been investigated as potential reducing agents for aldehydes and ketones. A preliminary study of the reaction of neutral metal hydrides with acetone and trifluoroacetone indicated that complexes of metals on the left of the transition series were more likely to reduce the ketones.[145] Anionic Group VI transition metal hydrides [$HM(CO)_5^-$ and *cis*-$HM(CO)_4P(OMe)_3^-$, M = Cr, W] were screened against a range of aldehydes and ketones.[146] The former complex required acid activation for all aldehydes except formaldehyde, but the latter was more nucleophilic and reduced all the aldehydes to the corresponding alcohols in good yield. Ketone reduction was possible in the presence of acetic acid, but competitive destruction of the metal complex was a problem. Useful chemoselectivity allowing reduction of aldehydes but not ketones was noted with an anionic molybdenum hydride complex.[113]

Lithium hydride, prepared by hydrogenolysis of *t*-butyllithium under pressure, was observed to reduce aldehydes and cyclohexanones to the corresponding alcohols in the presence of transition metal halides.[147] An equimolar quantity of vanadium trichloride was found to be the most effective catalyst, but the identity of the active reducing species was not established.

1.1.6 REFERENCES

1. H. C. Brown, H. I. Schlesinger and A. B. Burg, *J. Am. Chem. Soc.*, 1939, **61**, 673.
2. A. Hajós, 'Complex Hydrides', Elsevier, Amsterdam, 1979.
3. E. R. H. Walker, *Chem. Soc. Rev.*, 1976, **5**, 23.
4. H. C. Brown and S. Krishnamurthy, *Tetrahedron*, 1979, **35**, 567; H. C. Brown and S. Krishnamurthy, *Aldrichimica Acta*, 1979, **12**, 3.
5. J. Málek, *Org. React. (N.Y.)*, 1985, **34**, 1.
6. D. C. Wigfield, *Tetrahedron*, 1979, **35**, 449.
7. O. Eisenstein, H. B. Schlegel and M. M. Kayser, *J. Org. Chem.*, 1982, **47**, 2886.
8. H. Handel and J.-L. Pierre, *Tetrahedron Lett.*, 1976, 741.
9. A. A. Solov'yanov and I. P. Beletskaya, *Usp. Khim.*, 1987, **56**, 793 (*Chem. Abstr.*, 1987, **107**, 216 792k).
10. C. A. Ibarra, R. Pérez-Ossorio, M. L. Quiroga, M. S. A. Pérez and M. J. F. Dominguez, *J. Chem. Soc., Perkin Trans. 2*, 1988, 101.
11. E. C. Ashby and J. R. Boone, *J. Am. Chem. Soc.*, 1976, **98**, 5524.
12. K. E. Wiegers and S. G. Smith, *J. Org. Chem.*, 1978, **43**, 1126.
13. D. C. Wigfield and F. W. Gowland, *J. Org. Chem.*, 1980, **45**, 653.
14. H. C. Brown, K. K. Wang and J. Chandrasekharan, *J. Am. Chem. Soc.*, 1983, **105**, 2340.
15. E. C. Ashby, A. B. Goel and R. N. DePriest, *J. Am. Chem. Soc.*, 1980, **102**, 7779.
16. D. J. Cram and F. A. Abd Elhafez, *J. Am. Chem. Soc.*, 1952, **74**, 5828.
17. J. W. Cornforth, R. H. Cornforth and K. K. Mathew, *J. Chem. Soc.*, 1959, 112.
18. D. J. Cram and K. R. Kopecky, *J. Am. Chem. Soc.*, 1959, **81**, 2748.
19. G. J. Karabatsos, *J. Am. Chem. Soc.*, 1967, **89**, 1367.
20. M. Chérest, H. Felkin and N. Prudent, *Tetrahedron Lett.*, 1968, 2199.
21. H. B. Bürgi, J. D. Dunitz, J.-M. Lehn and G. Wipff, *Tetrahedron*, 1974, **30**, 1563; H. B. Bürgi and J. D. Dunitz, *Acc. Chem. Res.*, 1983, **16**, 153.
22. M. Chérest and N. Prudent, *Tetrahedron*, 1980, **36**, 1599.
23. N. T. Anh, *Top. Curr. Chem.*, 1980, **88**, 145.
24. E. P. Lodge and C. H. Heathcock, *J. Am. Chem. Soc.*, 1987, **109**, 2819.
25. K. N. Houk and Y.-D. Wu, in 'Stereochemistry of Organic and Bioorganic Transformations', ed. W. Bartmann and K. B. Sharpless, VCH, Weinheim, 1987, p. 247; Y.-D. Wu and K. N. Houk, *J. Am. Chem. Soc.*, 1987, **109**, 908.
26. D. Kruger, A. E. Sopchik and C. A. Kingsbury, *J. Org. Chem.*, 1984, **49**, 778.
27. B. Caro, B. Boyer, G. Lamaty and G. Jaouen, *Bull. Soc. Chim. Fr.*, 1983, 281.
28. M.-H. Rei, *J. Org. Chem.*, 1983, **48**, 5386.
29. A. S. Cieplak, *J. Am. Chem. Soc.*, 1981, **103**, 4540.
30. K. Okada, S. Tomita and M. Oda, *Bull. Chem. Soc. Jpn.*, 1989, **62**, 459.
31. C. K. Cheung, L. T. Tseng, M.-H. Lin, S. Srivastava and W. J. le Noble, *J. Am. Chem. Soc.*, 1986, **108**, 1598.
32. S. J. Danishefsky and M. E. Langer, *J. Org. Chem.*, 1985, **50**, 3672.
33. M. Chérest and H. Felkin, *Tetrahedron Lett.*, 1968, 2205.
34. Y.-D. Wu, K. N. Houk and B. M. Trost, *J. Am. Chem. Soc.*, 1987, **109**, 5560.
35. D. Mukherjee, Y.-D. Wu, F. R. Fronczek and K. N. Houk, *J. Am. Chem. Soc.*, 1988, **110**, 3328.
36. M.-H. Lin, J. E. Silver and W. J. le Noble, *J. Org. Chem.*, 1988, **53**, 5155.
37. J. Huet, Y. Maroni-Barnaud, N. T. Anh and J. Seyden-Penne, *Tetrahedron Lett.*, 1976, 159.
38. T. Oishi and T. Nakata, *Acc. Chem. Res.*, 1984, **17**, 338.
39. P. A. Bartlett, *Tetrahedron*, 1980, **36**, 3.
40. M. Nógrádi, 'Stereoselective Synthesis', VCH, Weinheim, 1986.
41. T. Nakata, T. Tanaka and T. Oishi, *Tetrahedron Lett.*, 1983, **24**, 2653.
42. T. Nakata, M. Fukui, H. Ohtsuka and T. Oishi, *Tetrahedron Lett.*, 1983, **24**, 2661.
43. T. Takahashi, M. Miyazawa and J. Tsuji, *Tetrahedron Lett.*, 1985, **26**, 5139.
44. L. E. Overman and R. J. McCready, *Tetrahedron Lett.*, 1982, **23**, 2355.

45. M. Fujita and T. Hiyama, *J. Org. Chem.*, 1988, **53**, 5405.
46. M. Fujita and T. Hiyama, *J. Org. Chem.*, 1988, **53**, 5415.
47. K. Narasaka and F.-C. Pai, *Tetrahedron*, 1984, **40**, 2233.
48. K.-M. Chen, G. E. Hardtmann, K. Prasad, O. Repič and M. J. Shapiro, *Tetrahedron Lett.*, 1987, **28**, 155.
49. T. Hanamoto and T. Hiyama, *Tetrahedron Lett.*, 1988, **29**, 6467.
50. F. G. Kathawala, B. Prager, K. Prasad, O. Repič, M. J. Shapiro, R. S. Stabler and L. Widler, *Helv. Chim. Acta*, 1986, **69**, 803.
51. S. Kiyooka, H. Kuroda and Y. Shimasaki, *Tetrahedron Lett.*, 1986, **27**, 3009.
52. Y. Mori, M. Kuhara, A. Takeuchi and M. Suzuki, *Tetrahedron Lett.*, 1988, **29**, 5419.
53. S. Anwar and A. P. Davis, *Tetrahedron*, 1988, **44**, 3761.
54. D. A. Evans, K. T. Chapman and E. M. Carreira, *J. Am. Chem. Soc.*, 1988, **110**, 3560.
55. R. Bloch, L. Gilbert and C. Girard, *Tetrahedron Lett.*, 1988, **29**, 1021.
56. R. Baker, P. D. Ravenscroft and C. J. Swain, *J. Chem. Soc., Chem. Commun.*, 1984, 74.
57. K. Suzuki, E. Katayama and G. Tsuchihashi, *Tetrahedron Lett.*, 1984, **25**, 2479.
58. K. Suzuki, M. Shimazaki and G. Tsuchihashi, *Tetrahedron Lett.*, 1986, **27**, 6233.
59. T. Nakata, T. Tanaka and T. Oishi, *Tetrahedron Lett.*, 1981, **22**, 4723.
60. T. Fukuyama, B. Vranesic, D. P. Negri and Y. Kishi, *Tetrahedron Lett.*, 1978, 2741.
61. Y. Ito and M. Yamaguchi, *Tetrahedron Lett.*, 1983, **24**, 5385; R. M. DiPardo and M. G. Bock, *Tetrahedron Lett.*, 1983, **24**, 4805; D. A. Evans, M. D. Ennis, T. Le, N. S. Mandel and G. S. Mandel, *J. Am. Chem. Soc.*, 1984, **106**, 1154.
62. Y. Ito, T. Katsuki and M. Yamaguchi, *Tetrahedron Lett.*, 1985, **26**, 4643.
63. M. Fujita and T. Hiyama, *J. Am. Chem. Soc.*, 1985, **107**, 8294.
64. M. Shimagaki, T. Maeda, Y. Matsuzaki, I. Hori, T. Nakata and T. Oishi, *Tetrahedron Lett.*, 1984, **25**, 4775.
65. M. Shimagaki, Y. Matsuzaki, I. Hori, T. Nakata and T. Oishi, *Tetrahedron Lett.*, 1984, **25**, 4779.
66. G. Solladié, C. Fréchou, G. Demailly and C. Greck, *J. Org. Chem.*, 1986, **51**, 1912; G. Solladié, C. Fréchou and G. Demailly, *Tetrahedron Lett.*, 1986, **27**, 2867; H. Kosugi, H. Konta and H. Uda, *J. Chem. Soc., Chem. Commun.*, 1985, 211.
67. A. D. Buss, N. Greeves, R. Mason and S. Warren, *J. Chem. Soc., Perkin Trans. 1*, 1987, 2569.
68. J. Elliott, D. Hall and S. Warren, *Tetrahedron Lett.*, 1989, **30**, 601.
69. M. Tramontini, *Synthesis*, 1982, 605.
70. A. Ookawa and K. Soai, *J. Chem. Soc., Perkin Trans. 1*, 1987, 1465.
71. M.-N. Dufour, P. Jouin, J. Poncet, A. Pantaloni and B. Castro, *J. Chem. Soc., Perkin Trans. 1*, 1986, 1895.
72. J. Barluenga, B. Olano and S. Fustero, *J. Org. Chem.*, 1985, **50**, 4052.
73. P. C. Thieme, H. Sauter and G. Reissenweber, *Chem. Ber.*, 1988, **121**, 1059.
74. G. Maier, C. Roth and R. K. Schmitt, *Chem. Ber.*, 1985, **118**, 704.
75. G. Maier, U. Seipp and R. Boese, *Tetrahedron Lett.*, 1987, **28**, 4515.
76. R. O. Hutchins, *J. Org. Chem.*, 1977, **42**, 920.
77. S. Kim, S. J. Lee and H. J. Kang, *Synth. Commun.*, 1982, **12**, 723; S. Kim, Y. C. Moon and K. H. Ahn, *J. Org. Chem.*, 1982, **47**, 3311.
78. H. C. Brown, J. L. Hubbard and B. Singaram, *Tetrahedron*, 1981, **37**, 2359.
79. S. Krishnamurthy and H. C. Brown, *J. Am. Chem. Soc.*, 1976, **98**, 3383.
80. J. Hooz, S. Akiyama, F. J. Cedar, M. J. Bennett and R. M. Tuggle, *J. Am. Chem. Soc.*, 1974, **96**, 274.
81. S. Kim, K. H. Ahn and Y. W. Chung, *J. Org. Chem.*, 1982, **47**, 4581.
82. C. A. Brown and S. Krishnamurthy, *J. Org. Chem.*, 1986, **51**, 238.
83. J. S. Cha, M. S. Yoon, Y. S. Kim and K. W. Lee, *Tetrahedron Lett.*, 1988, **29**, 1069.
84. H. C. Brown, J. S. Cha, B. Nazer and C. A. Brown, *J. Org. Chem.*, 1985, **50**, 549.
85. E. C. Ashby, S. A. Noding and A. B. Goel, *J. Org. Chem.*, 1980, **45**, 1028.
86. A. Feghouli, Y. Fort, R. Vanderesse and P. Caubère, *Tetrahedron Lett.*, 1988, **29**, 1379.
87. E. L. Eliel and S. H. Schroeter, *J. Am. Chem. Soc.*, 1965, **87**, 5031.
88. A. Amann, G. Ourisson and B. Luu, *Synthesis*, 1987, 1002.
89. C. Schmidt and T. Breining, *Tetrahedron*, 1983, **39**, 3929.
90. H. C. Brown and F. G. M. Vogel, *Justus Liebigs Ann. Chem.*, 1978, 695.
91. M. C. Carreño, E. Domínguez, J. L. García-Ruano and A. Rubio, *J. Org. Chem.*, 1987, **52**, 3619.
92. M. Hannaby and S. Warren, *Tetrahedron Lett.*, 1986, **27**, 765.
93. T.-L. Ho, *Tetrahedron*, 1985, **41**, 3.
94. A. Loupy and J. Seyden-Penne, *Tetrahedron*, 1980, **36**, 1937.
95. G. R. Meyer, *J. Chem. Educ.*, 1981, **58**, 628.
96. A. L. Gemal and J.-L. Luche, *J. Am. Chem. Soc.*, 1981, **103**, 5454; J.-L. Luche, *J. Am. Chem. Soc.*, 1978, **100**, 2226.
97. S. Komiya and O. Tsutsumi, *Bull. Chem. Soc. Jpn.*, 1987, **60**, 3423.
98. S. Fukuzawa, T. Fujinami, S. Yamaguchi and S. Sakai, *J. Chem. Soc., Perkin Trans. 1*, 1986, 1929.
99. K. E. Wilson, R. T. Seidner and S. Masamune, *J. Chem. Soc., Chem. Commun.*, 1970, 213; N. M. Yoon and Y. S. Gyoung, *J. Org. Chem.*, 1985, **50**, 2443.
100. S. Kim and K. H. Ahn, *J. Org. Chem.*, 1984, **49**, 1717.
101. M. Añez, G. Uribe, L. Mendoza and R. Contreras, *Synthesis*, 1981, 214.
102. P. Caubère, *Angew. Chem., Int. Ed. Engl.*, 1983, **22**, 599.
103. L. Mordenti, J.-J. Brunet and P. Caubère, *J. Org. Chem.*, 1979, **44**, 2203.
104. J. Durand, N. T. Anh and J. Huet, *Tetrahedron Lett.*, 1974, 2397.
105. S. Krishnamurthy and H. C. Brown, *J. Org. Chem.*, 1977, **42**, 1197.
106. D. E. Ward and C. K. Rhee, *Synth. Commun.*, 1988, **18**, 1927.
107. N. M. Yoon, K. B. Park and Y. S. Gyoung, *Tetrahedron Lett.*, 1983, **24**, 5367.
108. C. F. Nutaitis and G. W. Gribble, *Tetrahedron Lett.*, 1983, **24**, 4287; G. W. Gribble and C. F. Nutaitis, *Org. Prep. Proced. Int.*, 1985, **17**, 317.

109. H. C. Brown and S. U. Kulkarni, *J. Org. Chem.*, 1977, **42**, 4169.
110. M. M. Midland and A. Tramontano, *J. Org. Chem.*, 1978, **43**, 1470.
111. S. Kim, H. J. Kang and S. Yang, *Tetrahedron Lett.*, 1984, **25**, 2985.
112. S. Krishnamurthy, *J. Org. Chem.*, 1981, **46**, 4628.
113. D. H. Gibson and Y. S. El-Omrani, *Organometallics*, 1985, **4**, 1473.
114. N. Y. M. Fung, P. de Mayo, J. H. Schauble and A. C. Weedon, *J. Org. Chem.*, 1978, **43**, 3977.
115. A. L. Gemal and J.-L. Luche, *J. Org. Chem.*, 1979, **44**, 4187.
116. J.-L. Luche and A. L. Gemal, *J. Am. Chem. Soc.*, 1979, **101**, 5848.
117. A. L. Gemal and J.-L. Luche, *Tetrahedron Lett.*, 1981, **22**, 4077.
118. M. P. Paradisi, G. P. Zecchini and G. Ortar, *Tetrahedron Lett.*, 1980, **21**, 5085.
119. H. C. Brown, S.-C. Kim and S. Krishnamurthy, *J. Org. Chem.*, 1980, **45**, 1.
120. H. C. Brown, J. S. Cha, B. Nazer, S.-C. Kim, S. Krishnamurthy and C. A. Brown, *J. Org. Chem.*, 1984, **49**, 885.
121. N. M. Yoon, K. E. Kim and J. Kang, *J. Org. Chem.*, 1986, **51**, 226.
122. S. Kim, C. H. Oh, J. S. Ko, K. H. Ahn and Y. J. Kim, *J. Org. Chem.*, 1985, **50**, 1927.
123. B. J. Hussey, R. A. W. Johnstone, P. Boehm and I. D. Entwistle, *Tetrahedron*, 1982, **38**, 3769.
124. D. E. Ward, C. K. Rhee and W. M. Zoghaib, *Tetrahedron Lett.*, 1988, **29**, 517.
125. T. E. A. Nieminen and T. A. Hase, *Tetrahedron Lett.*, 1987, **28**, 4725.
126. S. Itsuno, T. Wakasugi, K. Ito, A. Hirao and S. Nakahama, *Bull. Chem. Soc. Jpn.*, 1985, **58**, 1669.
127. G. C. Andrews, *Tetrahedron Lett.*, 1980, **21**, 697.
128. Y. Kamitori, M. Hojo, R. Masuda, T. Izumi and T. Inoue, *Synthesis*, 1983, 387.
129. Y. Z. Huang, Y. Shen and C. Chen, *Tetrahedron Lett.*, 1985, **26**, 5171.
130. K. Maruoka, Y. Araki and H. Yamamoto, *J. Am. Chem. Soc.*, 1988, **110**, 2650.
131. S. Patai (ed.), 'The Chemistry of Quinoid Compounds', Wiley, New York, 1974.
132. R. G. Harvey, S. H. Goh and C. Cortez, *J. Am. Chem. Soc.*, 1975, **97**, 3468.
133. B. Lecea, J. M. Aizpurua and C. Palomo, *Tetrahedron*, 1985, **41**, 4657.
134. J. M. Aizpurua, B. Lecea and C. Palomo, *Can. J. Chem.*, 1986, **64**, 2342.
135. R. J. Klingler, K. Mochida and J. K. Kochi, *J. Am. Chem. Soc.*, 1979, **101**, 6626.
136. B. L. Jensen, J. Jewett-Bronson, S. B. Hadley and L. G. French, *Synthesis*, 1982, 732.
137. I. Shibata, T. Suzuki, A. Baba and H. Matsuda, *J. Chem. Soc., Chem. Commun.*, 1988, 882.
138. M. F. Semmelhack and R. N. Misra, *J. Org. Chem.*, 1982, **47**, 2469.
139. I. Ojima and T. Kogure, *Organometallics*, 1982, **1**, 1390.
140. D. Yang and D. D. Tanner, *J. Org. Chem.*, 1986, **51**, 2267.
141. M. Kira, K. Sato and H. Sakurai, *J. Org. Chem.*, 1987, **52**, 948.
142. W. P. Neumann, *Synthesis*, 1987, 665.
143. Y. T. Xian, P. Four, F. Guibe and G. Balavoine, *Nouv. J. Chim.*, 1984, **8**, 611.
144. M. Degueil-Castaing, A. Rahm and N. Dahan, *J. Org. Chem.*, 1986, **51**, 1672.
145. J. A. Labinger and K. H. Komadina, *J. Organomet. Chem.*, 1978, **155**, C25.
146. P. L. Gaus, S. C. Kao, K. Youngdahl and M. Y. Darensbourg, *J. Am. Chem. Soc.*, 1985, **107**, 2428.
147. E. C. Ashby and S. A. Noding, *J. Org. Chem.*, 1980, **45**, 1041.

1.2 Reduction of C=N to CHNH by Metal Hydrides

ROBERT O. HUTCHINS
Drexel University, Philadelphia, PA, USA

and

MARYGAIL K. HUTCHINS
ICI Americas, Exton, PA, USA

1.2.1 INTRODUCTION

The reduction of various carbon–nitrogen π-systems to saturated derivatives with metal hydrides, principally aluminum- and boron-based reagents, provides highly useful processes for the preparation of amines and related functionalities.

Since their introduction in the 1940s, lithium aluminum hydride ($LiAlH_4$, LAH)[1] and sodium borohydride ($NaBH_4$)[2] have, of course, enjoyed enormous popularity for reduction of organic groups including nitrogen-containing double bonds. In addition, systematic alterations of these reagents, usually by replacement of one or more hydrogens with other functionalities or by combination with other reagents, have provided a great number of useful, modified hydrides, many of which have found significant utility for the manipulation of carbon–nitrogen π-systems, as well as other organic functional types.

The dominance of various metal hydrides for reductions of organic functionalities has led to several reviews, monographs and books devoted to discussions of their utility (including this volume), embedded in which are often included applications to carbon–nitrogen π-systems. Thus, LAH and modified derivatives,[3–10] $NaBH_4$ (in alcohol, ether,[4–13] and carboxylic acid[10–12,14] solvents) and altered derivatives,[7–9,11–13,17] sodium cyanoborohydride ($NaBH_3CN$),[7,9,11,12,15,16] borane (BH_3)/THF,[4,5,7,11–13,17,18] BH_3SMe_2[19] and amine·BH_3[20] complexes, as well as less-utilized hydride reagents,[7,9–11] have received considerable coverage, and much general information concerning carbon–nitrogen π-reductions may be extracted from the above, as well as from most recent advanced texts and discussions of imine chemistry.[21,22] However, no extensive reviews have appeared which focus primarily on imino π-reductions, offer comparisons between hydride reagents and present an overview of synthetic possibilities. Thus, coverage in this chapter will survey metal hydride reductions of the various carbon–nitrogen π-systems with an eye toward identifying superior choices for a given functionality and to augment, not duplicate, the above discussions, although some overlap is inevitable. Other, complementary, reducing systems, not involving metal hydrides, as well as asymmetric conversions are discussed elsewhere in this volume (*i.e.* Chapters 1.3–1.8). Likewise, reductions of nitriles (Chapters 1.10–1.12), most heterocyclic C=N systems (Chapters 3.6–3.8) and C=N reductions which proceed further to products devoid of nitrogen (*e.g.* sulfonylhydrazone reductions to hydrocarbons, see Chapter 1.14) are not included.

1.2.2 REDUCTION OF IMINES AND IMINIUM SALTS OF AMMONIA AND AMINES

1.2.2.1 General Considerations and Comparisons to Carbonyl Reductions

In a general sense, reductions of imine-type π-bonds with metal hydrides broadly parallel the corresponding reductions of carbonyls to alcohols, and many of the same reagents are routinely utilized (*e.g.* LAH, $NaBH_4$, $NaBH_3CN$, BH_3, *etc.*, see also Section 1.2.2.2). However, as expected, the type of *N*-substitution present in imino π-systems does alter, often substantially, the electrophilicity of the carbon under attack, and this in turn provides wide variations in reduction rates or, in some cases, prevention of reduction altogether. Thus, for example, while trialkylborohydrides rapidly reduce most carbonyls (even at –78 °C),[23] oximes and oxime ethers are essentially inert at 65 °C.[24,25] As with carbonyls, protonation or complexation of imino nitrogens enhances the electrophilicity of the resulting iminium salts and increases reduction rates. This has led to the development of highly successful reducing systems conducted under mildly acidic conditions (pH *ca.* 3–6) using reagents (*e.g.* $NaBH_3CN$,[15,16,26] $NaBH_4/RCO_2H$,[14] amineboranes[20]) which tolerate acidic media and a substantial portion of imino group reductions presently utilize such systems. Furthermore, the relatively enhanced reduction rates of imines compared to carbonyls under acidic conditions, presumably because of the higher basicity of the former, permits the *in situ* generation of the requisite iminium ions from carbonyls and amines, and eliminates the need to preform the imino derivatives.[26] This has led to direct reductive amination processes, particularly with $NaBH_3CN$,[26] which are presently the most successful and often used procedures for this important transformation.

The stereochemical outcomes of the reductions of cyclic imino derivatives often display surprisingly great differences from those obtained with cycloalkanones[7,27] and, in fact, attenuations or even reversals in stereoselectivity are common.[28]

Mechanistic details involved in imine and carbonyl[27,29] reductions are undoubtedly similar, although thorough investigations of the former are lacking. Certainly, hydride transfer to the electrophilic carbon, with or without prior activation by protonation or complexation is essential for both types of π-systems (Scheme 1). Whether or not alcohol solvents participate in imine reductions by borohydride (in the absence of added acid) to furnish the amine proton (as is the case with carbonyls)[29] is not known and must await detailed kinetic study and analysis of the initial intermediates formed before hydrolysis. Direct, *in situ*, reductive amination with $NaBH_3CN$ has been attributed to initial, reversible formation (*via* an intermediate hydroxyamine, (**1**) of an iminium ion (**2**) from carbonyls and amines followed by rapid attack by hydride (Scheme 2).[26] However, the inertness of an imine (partial structure **3**) to the usual reductive amination conditions ($NaBH_3CN$, MeOH, H^+) coupled with successful reductive amination of the corresponding ketone of (**3**) suggests that, at least in some cases, direct reduction of the hydroxy amine (**1**) occurs.[30] This possibility is supported by the successful reduction of bis(methoxymethyl)amines to dimethylamines[31] and methylols (*e.g.* **4**) to *N*-methylamides[32] with $NaBH_3CN$ in acidic media. Further investigation of this question appears warranted.

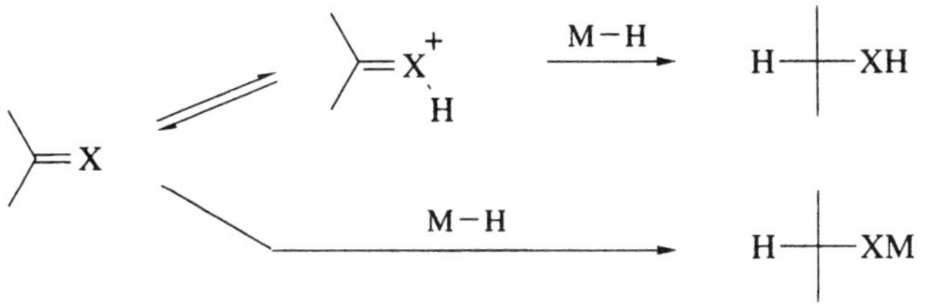

Scheme 1

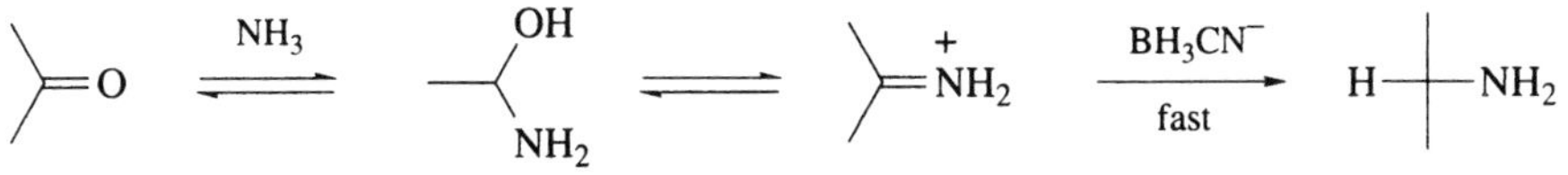

Scheme 2

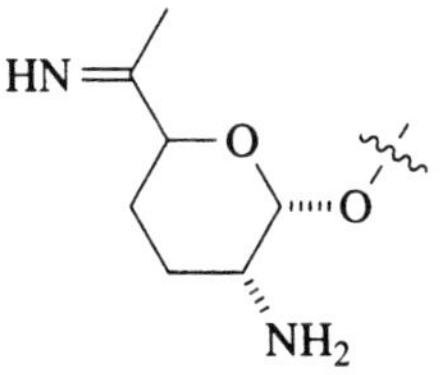

The sections which follow present an overview of of the most effective and convenient reagent(s), and the chemo- and stereo-selectivities (where known) expected with them. Of course, the choice of reagents and conditions is often dictated by a combination of these considerations, so that structural types and functional group complexity play key roles in reagent selection. A wide selection of examples are presented, chosen to illustrate the range of successful structural types and potential pitfalls. Further examples may be obtained from reviews and monographs.[3–20]

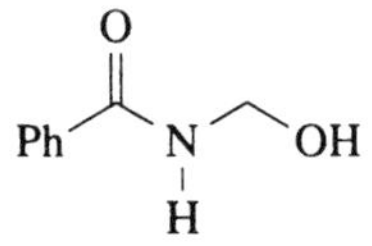

1.2.2.2 Reduction of Preformed Imines and Iminium Salts of Ammonia, Primary and Secondary Amines

1.2.2.2.1 Choice of hydride reagent and solvent

As mentioned above, a variety of metal hydrides have been employed for successful imine and iminium ion reductions, particularly LAH and its derivatives[3–10] and $NaBH_4$ and derived reagents.[4–20] For discussion here, iminium ions imply those derived from secondary amines, although protonated ammonia derivatives and primary imines are actually the intermediates involved in reductions in acidic media. The most popular reagents for imine reductions, especially in recent years and with complex synthetic targets, include $NaBH_4$ in alcohol or carboxylic acid [usually acetic acid in which the actual reducing species is $NaBH(OAc)_3$][14] solvents and $NaBH_3CN$ in acidic media (usually methanol/HCl or acetic acid). To illustrate the versatility of $NaBH_4$ and $NaBH_3CN$ for imine reductions, collections of successful conversions, chosen to represent varieties of structural types, are presented in Tables 1 and 2.

Table 1 demonstrates that $NaBH_4$ may be used in the presence of heterocyclic rings (entries 8–10, 13–17), esters (entries 8–10, 17), amides (entries 1, 9), conjugated double bonds (entry 7), alkynes (entry 5) and acetals (entry 4). The process can also be used to methylate an amine with formaldehyde as the carbonyl (entry 13). With certain structures, further reactions may occur subsequent to reduction (*e.g.* entry 6). Entry 15 illustrates a synthetically useful amine alkylation reaction which may occur in acetic acid. This remarkable reaction is attributed to self-reduction of an acyloxyborohydride to an aldehyde

Table 1 Reduction of Imines with $NaBH_4$

Entry	*Imine*	*Solvent*	*Temp.* (°C)	*Product*	*Yield* (%)	*Ref.*
1	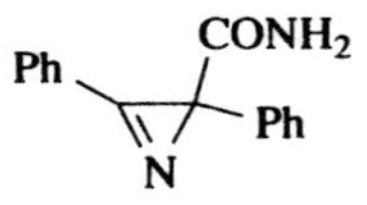 Ph, $CONH_2$, Ph, N	MeOH	65	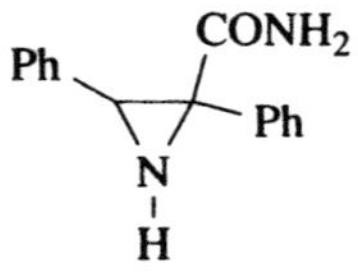 Ph, $CONH_2$, Ph, N, H	73	33
2	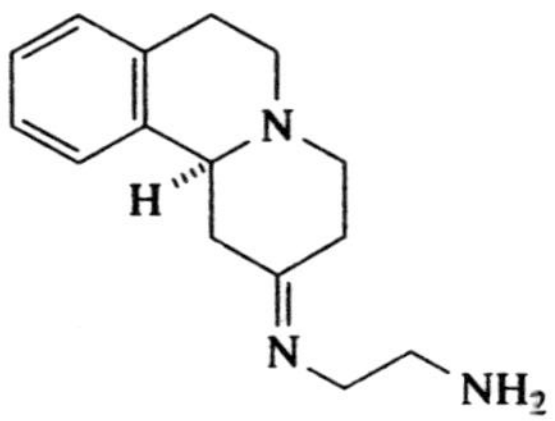 N, H, N, NH_2	MeOH	35	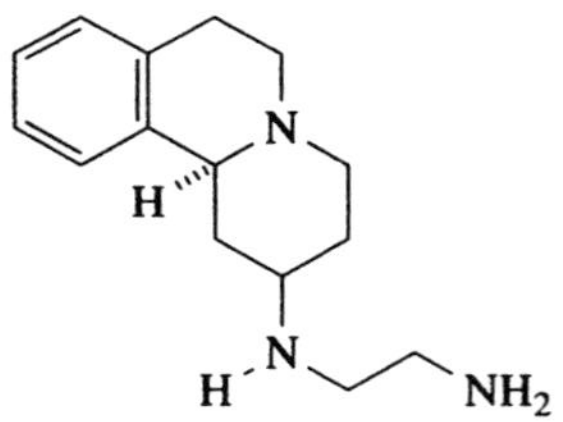 N, H, H, N, NH_2	86	34
3	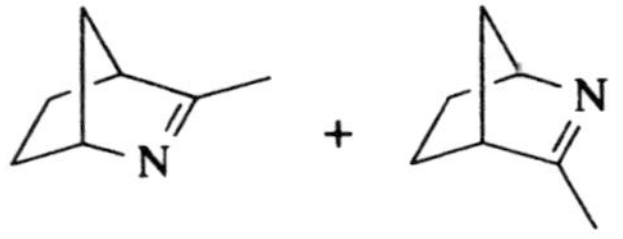 N + N	MeOH	—	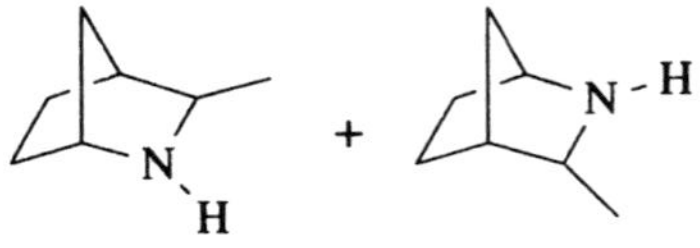 N, H + N-H	—	35
4	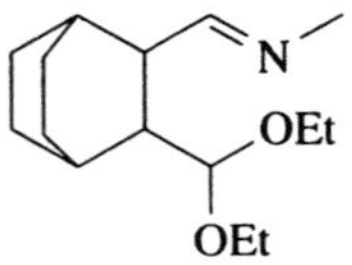 N, OEt, OEt	MeOH	0	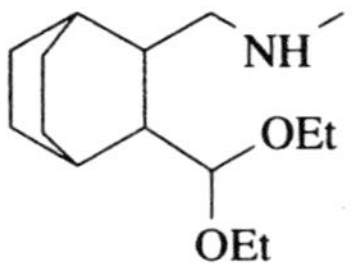 NH, OEt, OEt	—	36
5	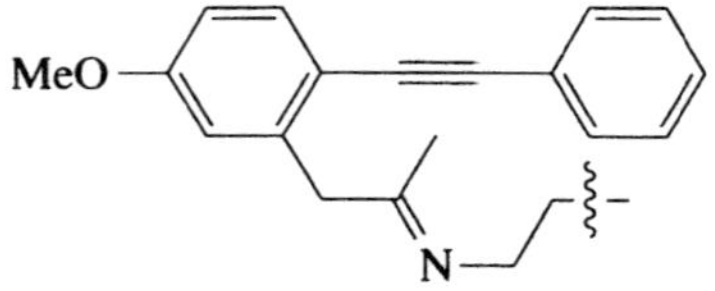 MeO, N	MeOH	25	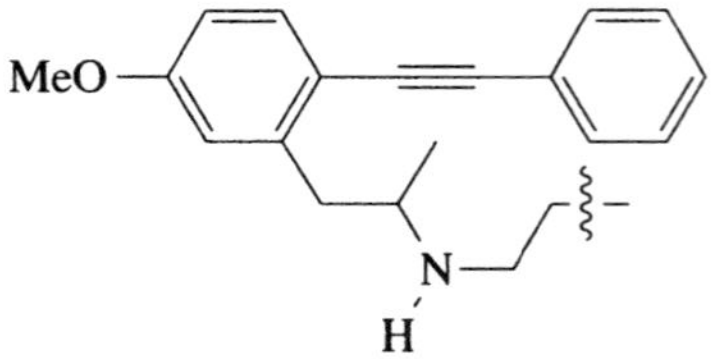 MeO, N, H	56	37

Table 1 *(continued)*

Entry	*Imine*	*Solvent*	*Temp.* (°C)	*Product*	*Yield* (%)	*Ref.*
6	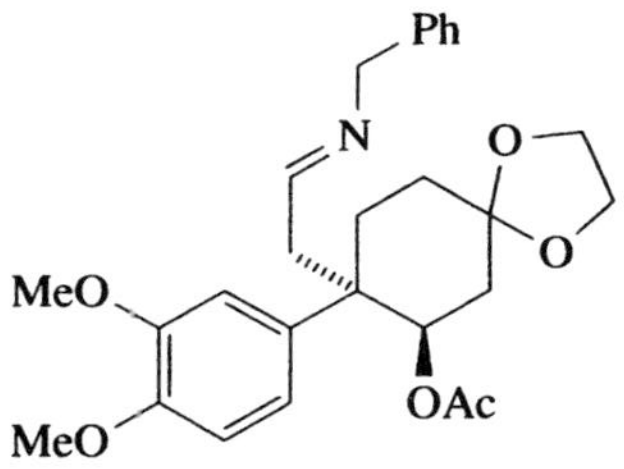	MeOH	25	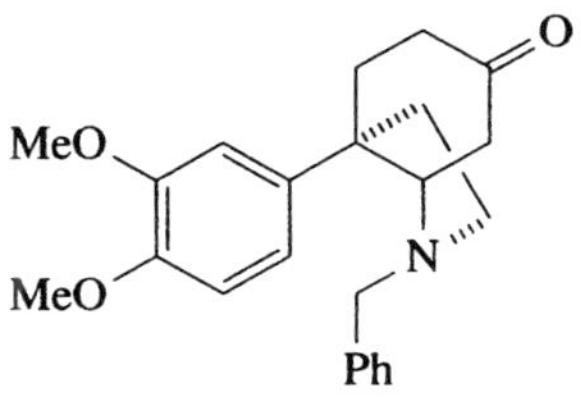	75	38
7	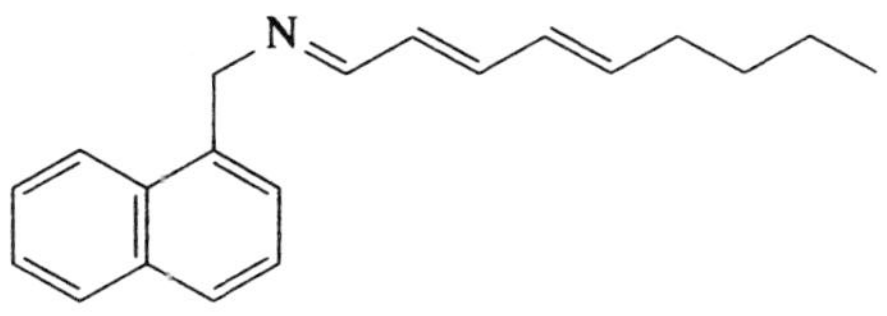	MeOH[a]	40	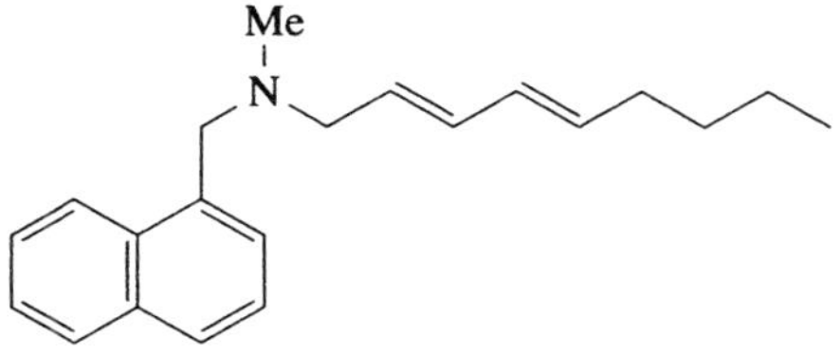	67	39
8	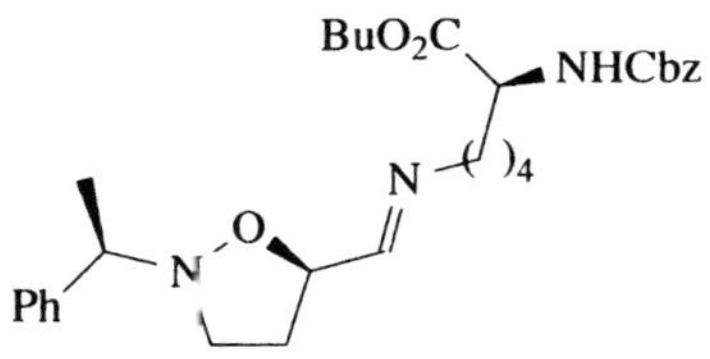	MeOH	5	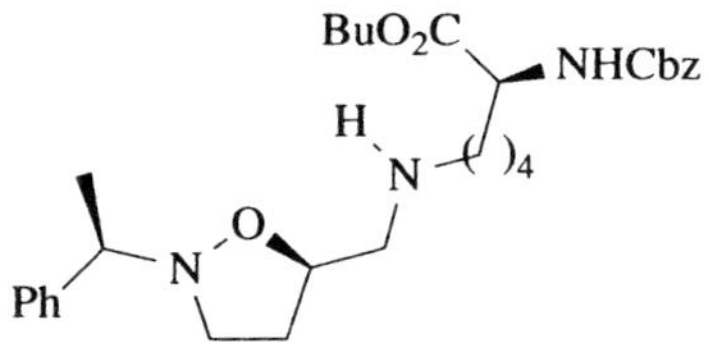	68	40
9	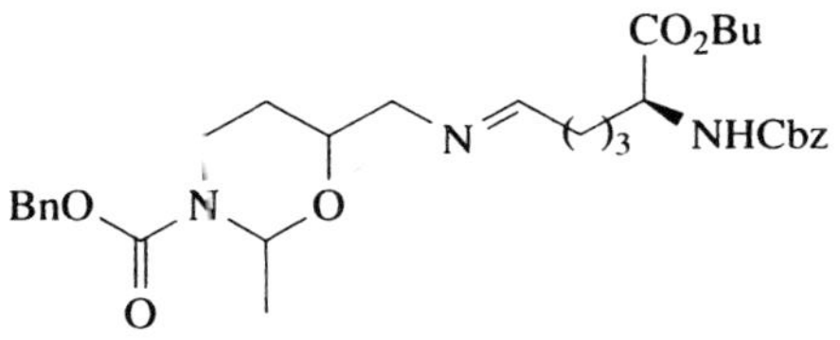	THF/MeOH	5	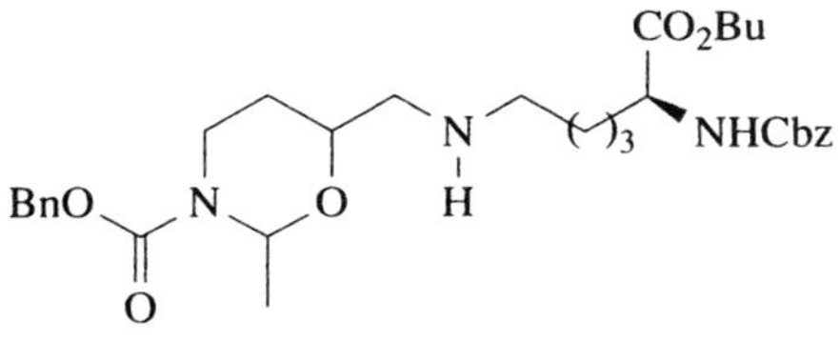	46	41

Table 1 *(continued)*

Entry	*Imine*	*Solvent*	*Temp.* (°C)	*Product*	*Yield* (%)	*Ref.*
10	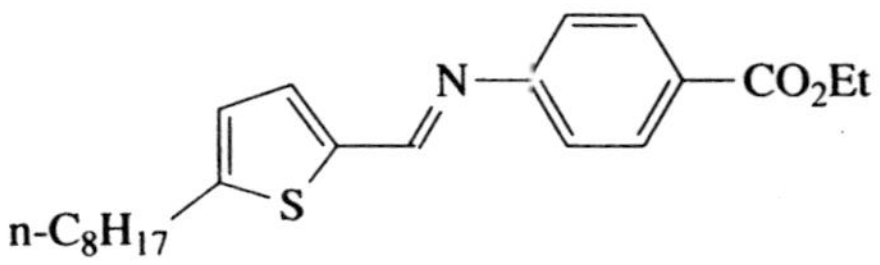	EtOH	78	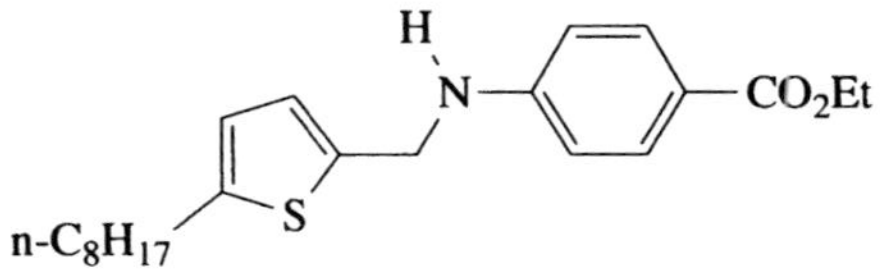	75	42
11	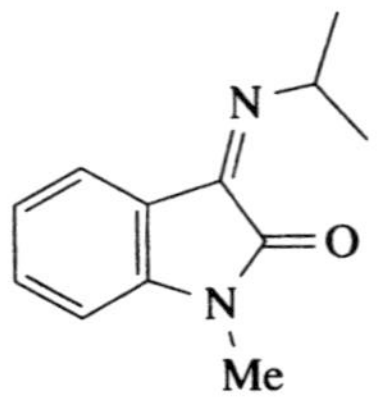	MeOH	—	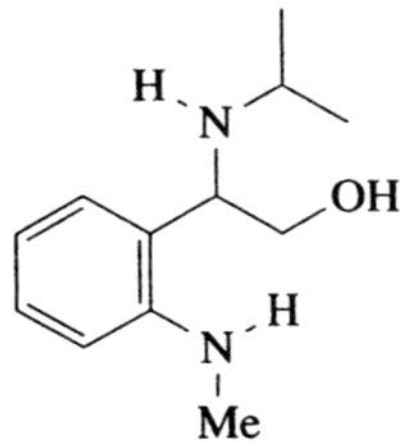	—	43
12	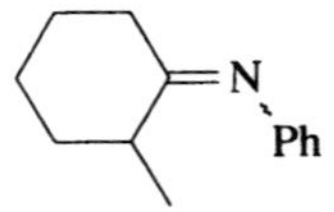	Pr^iOH	25	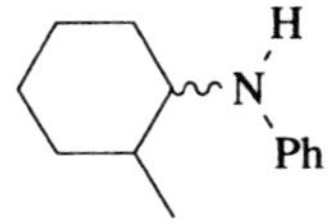	93[b]	28
13	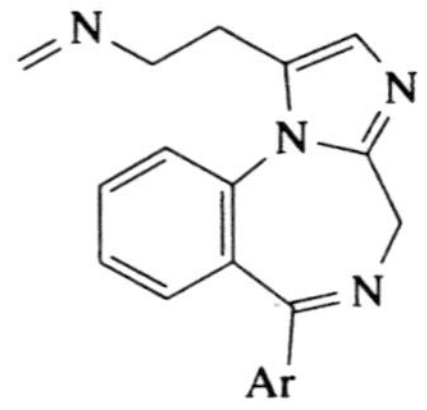	MeOH	25	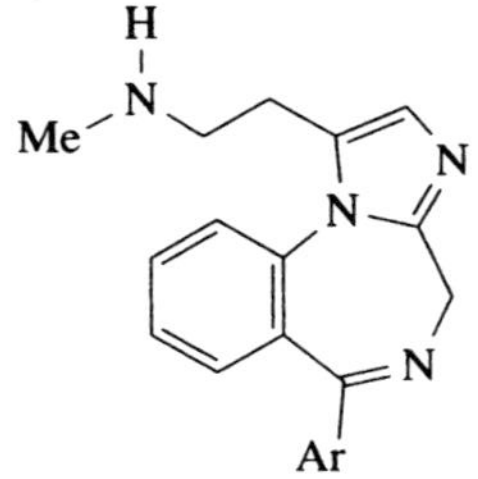	43	44

Table 1 *(continued)*

Entry	*Imine*	*Solvent*	*Temp.* (°C)	*Product*	*Yield* (%)	*Ref.*
14	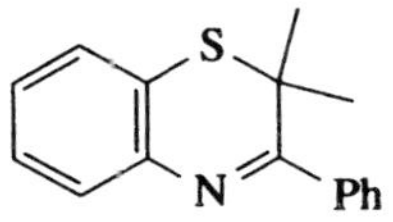	AcOH	80	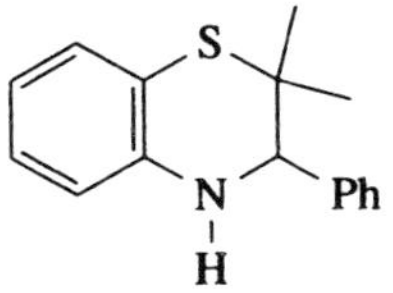	60	45
15	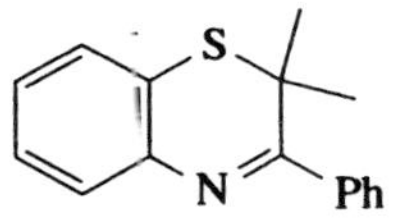	AcOH	80	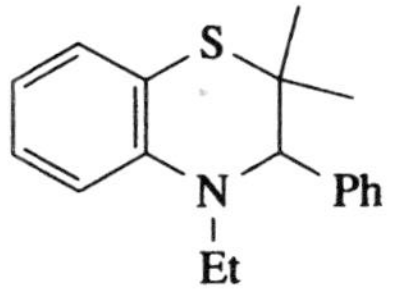	75	45
16	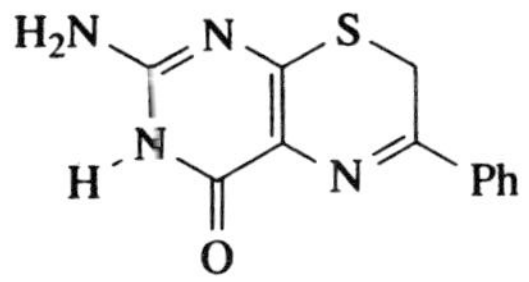	CF_3CO_2H	80	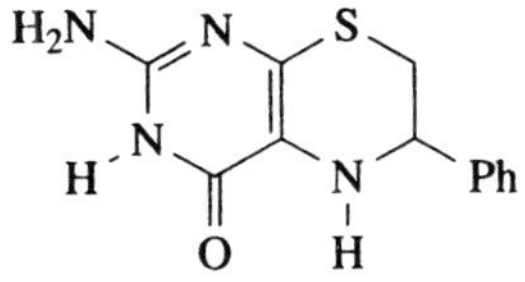	51	46
17	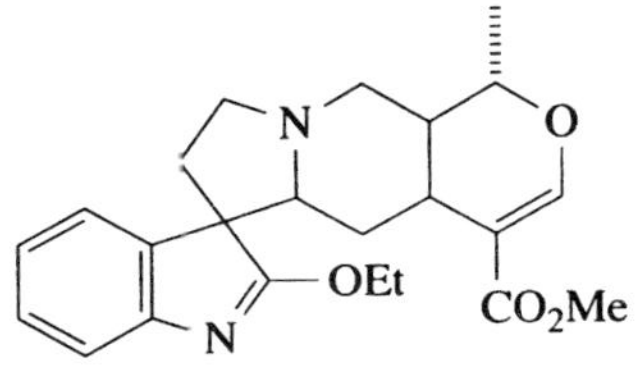	AcOH	25	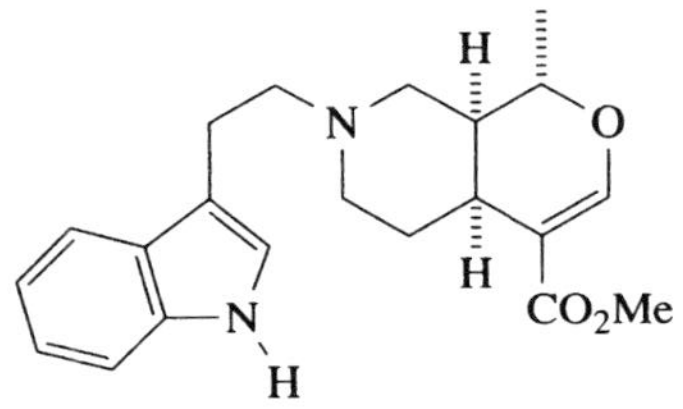	55	47

[a] Reduction was followed by reductive *N*-methylation with $CH_2O/NaBH_4$. [b] *Cis/trans* ratio 75:25.

Table 2 Reduction of Imines with $NaBH_3CN$

Entry	*Imine*	*Solvent*	*Temp.* (°C)	*Product*	*Yield* (%)	*Ref.*
1	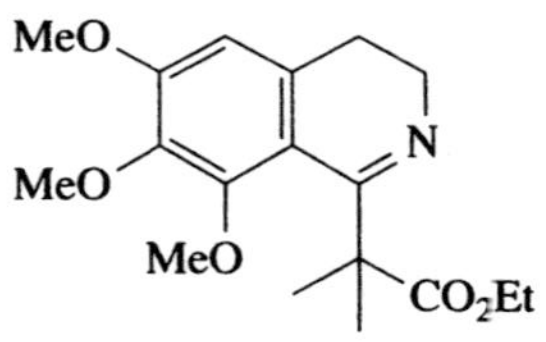	AcOH	25–30	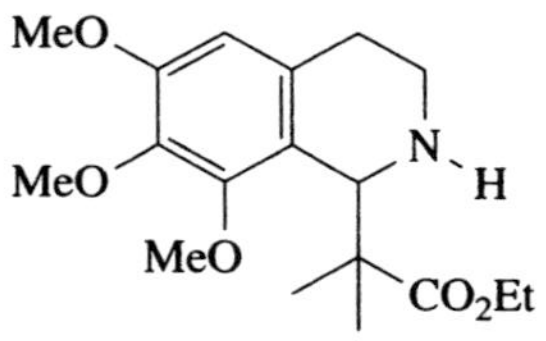	94	53
2	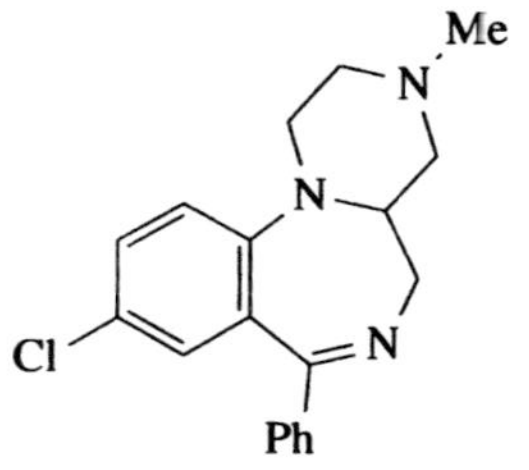	CF_3CO_2H	25	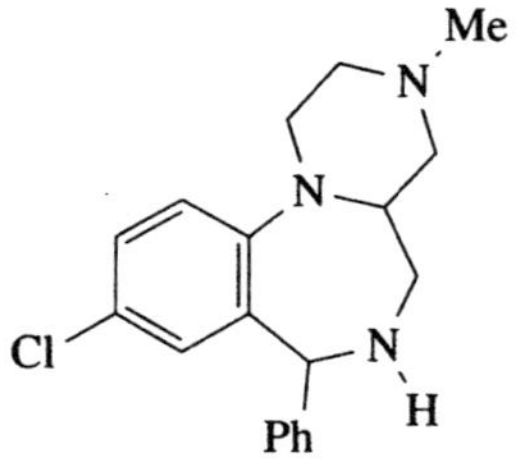	95	54
3	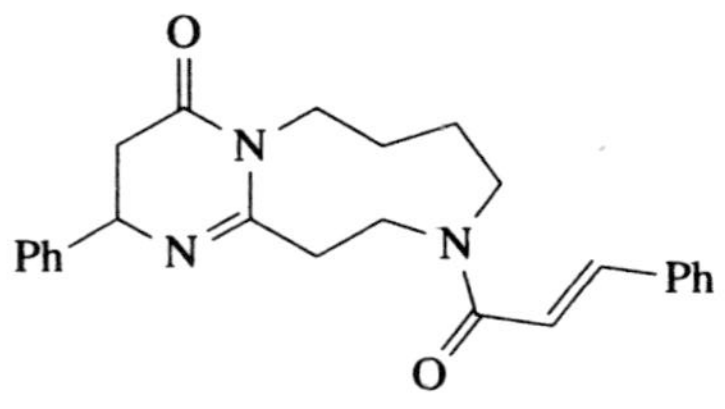	AcOH	25	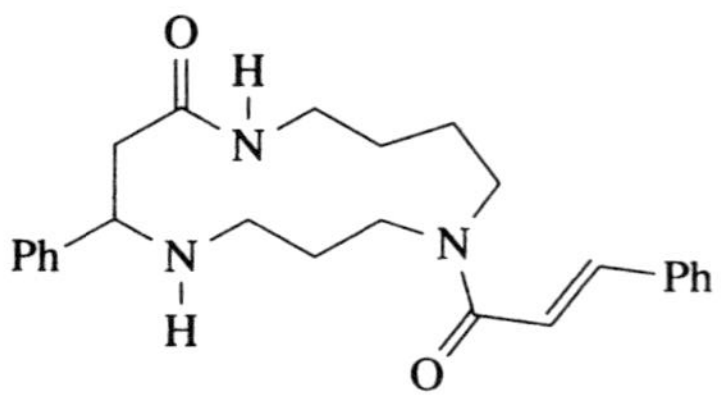	93	55

Table 2 *(continued)*

Entry	*Imine*	*Solvent*	*Temp.* (°C)	*Product*	*Yield* (%)	*Ref.*
4	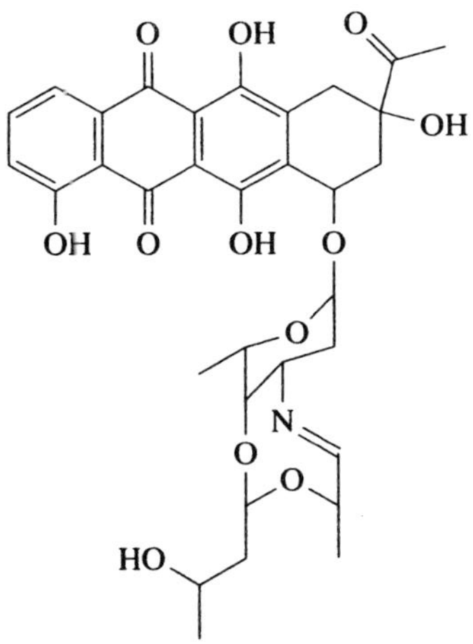	AcOH/MeOH	0	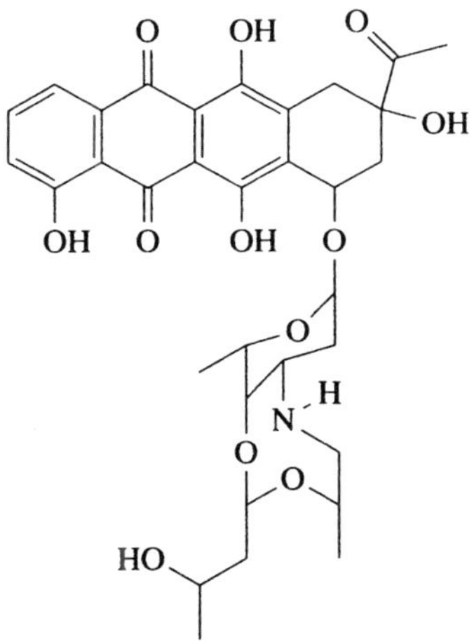	—	56
5	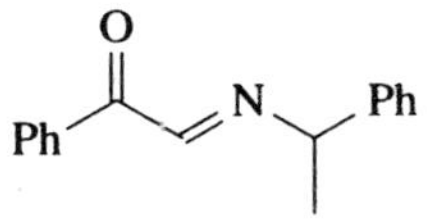	AcOH/MeCN	−10	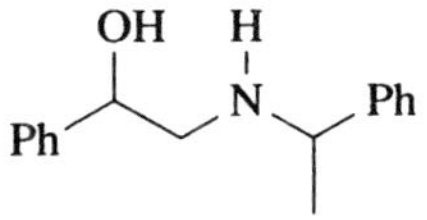	62	57
6	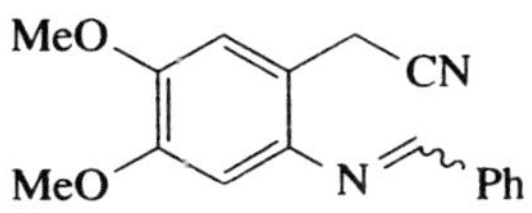	MeOH/HCl	25	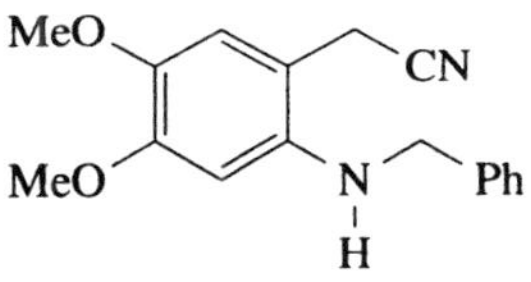	76	58
7	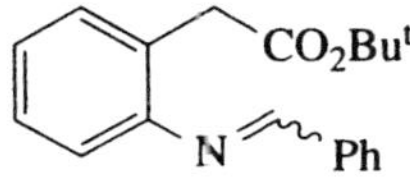	MeOH/HCl	25	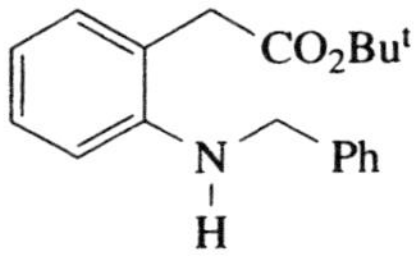	75	58

Table 2 *(continued)*

Entry	*Imine*	*Solvent*	*Temp.* (°C)	*Product*	*Yield* (%)	*Ref.*
8	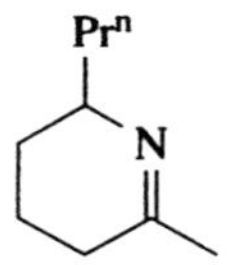	THF/H_2O	—	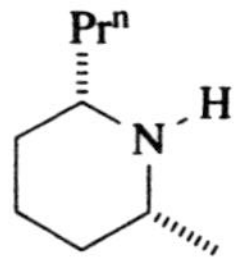	81	59
9	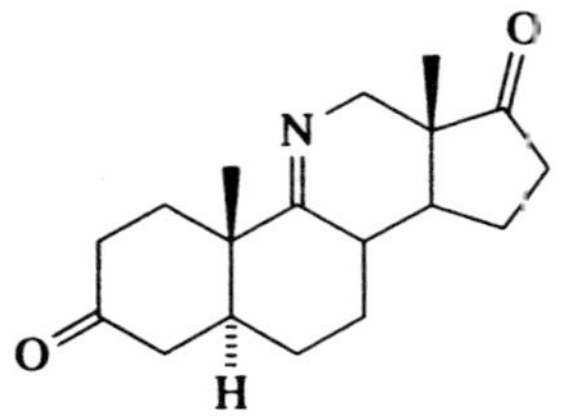	MeOH/HCl	25	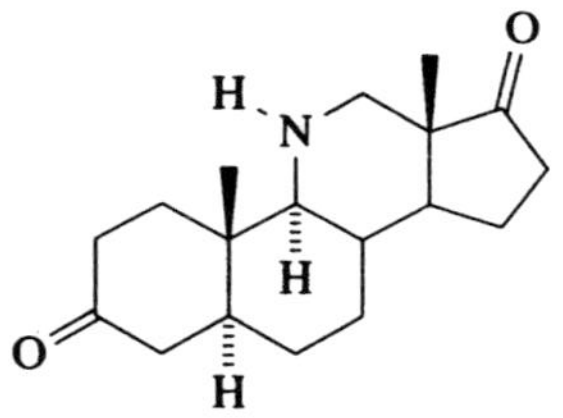	65[a]	60
10	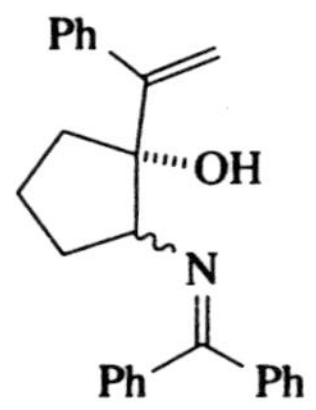	—	—	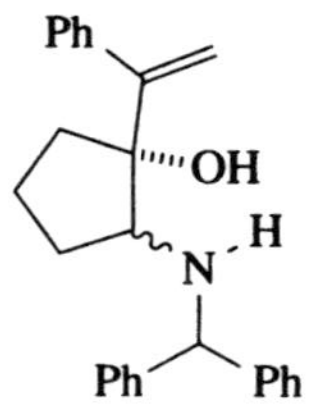	63	61
11	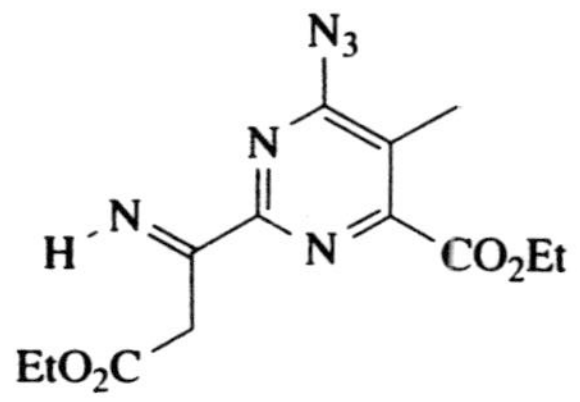	MeOH/HCl	25	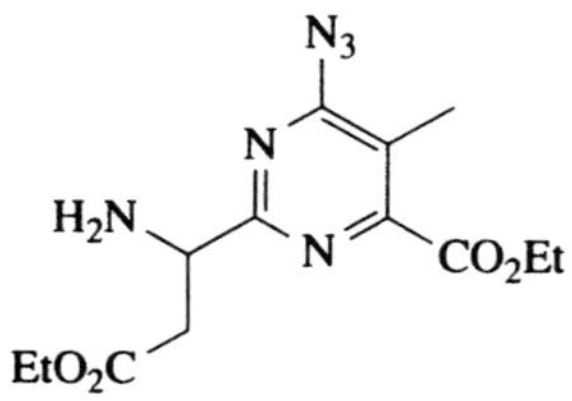	80	62

Table 2 *(continued)*

Entry	*Imine*	*Solvent*	*Temp.* (°C)	*Product*	*Yield* (%)	*Ref.*
12	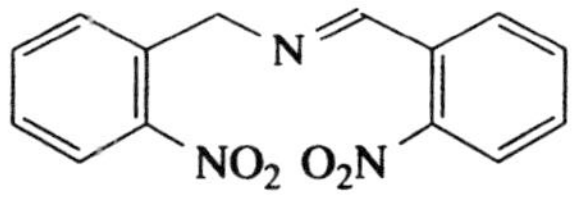	MeOH/THF/HCl	25	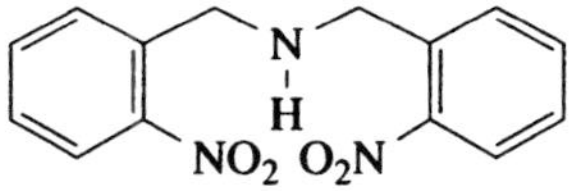	92	63
13	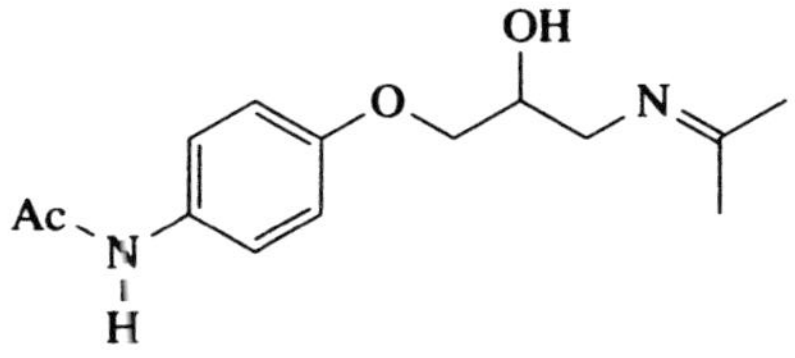	MeOH	—	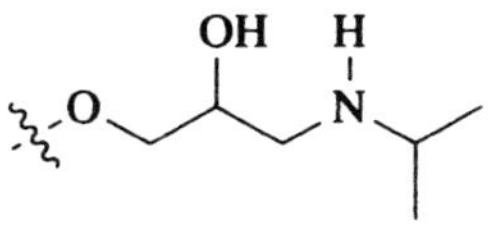	—	64
14	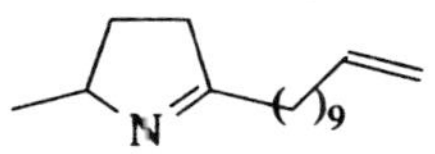	MeOH/HCl	—	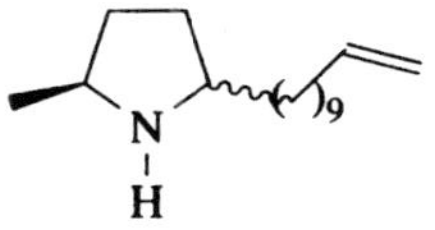	>93	65
15	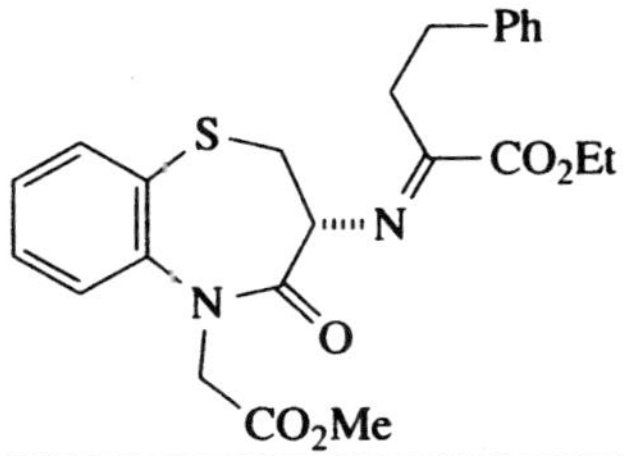	MeOH/AcOH	25	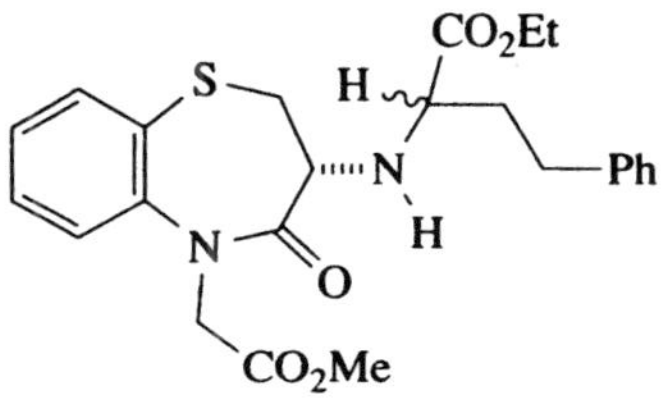	48	66

[a] After reoxidation with Jones reagent.

followed by condensation with an amine and subsequent reduction of the resulting imine.[11,14,45,48] Imine reductions with $NaBH_4$ have also been employed in the biochemical arena, for example, to couple drugs (*e.g.* digoxin) to polydimethylsiloxanes (in methanol),[49] to attach immunoglobulins to liposomes *via* glycosphingolipids (in water, pH 9.5),[50] to attach spin labels (for ESR studies) to adenylic acid and tRNA,[51] and to affinity label (and introduce tritium *via* $NaBH_3T$) diphtheria toxin with ADP-ribose (in aqueous phosphate buffer, pH 8).[52] It should be noted that the even milder reagent $NaBH_3CN$ is more commonly used in biological applications, as will be addressed below.

A variety of imine reductions using $NaBH_3CN$ in acidic media (where the actual species reduced are protonated iminium ion intermediates) are presented in Table 2 and illustrate the versatility of this mild reagent for synthetic applications. Thus, esters (entries 1, 7, 11, 15), amides (entries 3, 13, 15), cyano (entry 6), nitro (entry 12) and even such normally sensitive moieties as ketones (entry 4; however, see entry 5) and azide (entry 11) remain immune to this exceptionally tolerant reagent.

As mentioned, the mildness of $NaBH_3CN$ (coupled with its effectiveness and stability in aqueous media) has attracted considerable interest for applications in biochemical areas. Examples include the trapping of suspected imine intermediates produced in enzyme (mitochondrial monoamine oxidase) inactivation by amines,[67] the establishment by reduction of the positions of imine-forming amines in 2-keto-3-deoxy-6-phosphogluconate aldolase,[68] and the transfer labeling of methionyl-tRNA synthetase[69] and methionyl-tRNA transformalase[70] by treatment with periodate-treated tRNA. In fact, most biochemical applications of $NaBH_3CN$ have utilized *in situ* imine formation–reduction (*i.e.* reductive amination) conditions and will be further discussed in Section 1.2.2.3.1.

Although $NaBH_4$ and $NaBH_3CN$ usually attract the most attention, many other reagents reduce imines to amines and have been utilized in synthetic applications. These include the previously mentioned LAH,[3–10,71–74] borane[18,19,75] and amineborane complexes,[20,24,28] as well as various substituted derivatives such as KBH_4,[33] $NaAlH_4$,[76] $NaAlH_2(OCH_2CH_2OMe)_2$,[28,33] diisobutylaluminum hydride (DIBAL-H),[77] lithium triethyl-[24,78] and tri-*s*-butyl-borohydride[24,28,72,75,78] ('Superhydride' and 'L-Selectride,' respectively), alkyl- and dialkyl-cyanoborohydrides,[78] sodium dithionite,[79] hydridoferrates [*e.g.* $KFeH(CO)_4$,[80,81] and $Fe_3H(CO)_{11}$[82]], silanes [*e.g.* Et_3SiH/CF_3CO_2H,[83] $Et_3SiH/(Ph_3P)RhCl$[84]] and selenophenol (PhSeH).[85] A brief selection of representative examples of imine reductions with various hydride reagents listed above is presented in Table 3. Entry 2 illustrates a side reaction with LAH, in which initial reduction is followed by an internal displacement to afford an aziridine.[73] Furthermore, reduction of imines with trimethylamineborane in acetic acid gives *N*-acylation concomitant with C=N reduction (*e.g.* entry 5).[87] This result is probably distantly related to the previously described alkylation of amines with $NaBH_4/AcOH$[14,45,48] except that in this case the amine is acylated by attack on an acetoxyborane. As with ketones, bulky reagents (*e.g.* $LiBHEt_3$ and, particularly, $LiBHBu^s{}_3$) find utility in imine reductions where stereocontrol of cyclic systems is desired (*e.g.* entries 6–8, see also Section 1.2.2.2.3).

With certain structural types, the reduction course and subsequent reactions exhibited with hydride reagents may depend on the relative reduction rates of groups present. Thus, for example, reduction of the keto imine (**5**) with LAH (THF, reflux) afforded a mixture of the amino alcohol (**6**; 29%) and the *O*-bridged derivative (**7**; 10%), while $NaBH_4$ (THF/MeOH) gave (**7**; 49%) and $NaBH_3CN$ (EtOH/AcOH) afforded only the *N*-bridged compound (**8**). This suggests that LAH reduced the imine and ketone at comparable rates, while $NaBH_4$ and $NaBH_3CN$ selected the ketone and the imine (*via* the iminium ion) groups, respectively.[88]

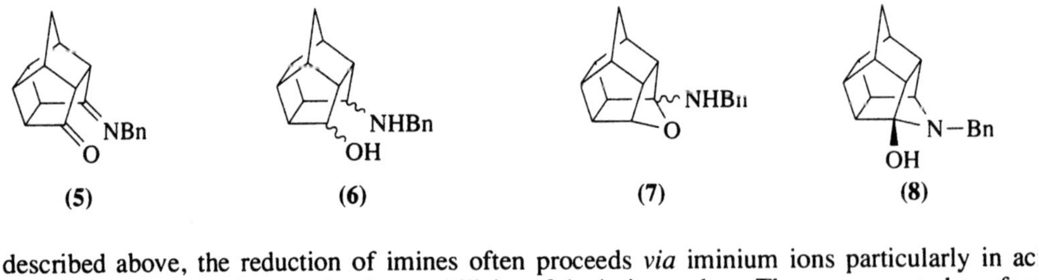

NBn O (**5**) | NHBn OH (**6**) | NHBn O (**7**) | N−Bn OH (**8**)

As described above, the reduction of imines often proceeds *via* iminium ions particularly in acidic media, since protonation enhances the electrophilicity of the imine carbon. Thus, as expected, preformed iminium salts generated from carbonyls and secondary amines are also readily reduced by most hydride reagents. Several examples of synthetic applications with a variety of reagents are illustrated in Table 4. Entries 9–12 illustrate the use of iminium intermediates for the reductive removal of amide carbonyls. Thus, treatment of amides with $POCl_3$ affords the iminium derivatives (*e.g.* **9**; Scheme 3), which are reduced by $NaBH_4$ to the corresponding amines (Table 4, entries 9, 10).[96–98] Likewise, reaction of amides with trialkyloxonium salts to give imidic esters (entry 12) or thioamides with methyl iodide to give thioiminium salts (*e.g.* entry 11) followed by reduction with $NaBH_4/SnCl_4$[100] (or $NaBH_4$)[101] or $NaBH_4$ (also $NaBH_3CN$),[99] respectively, affords amines.

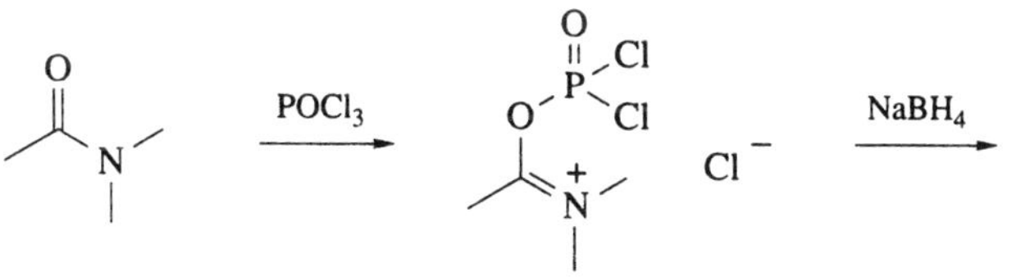

POCl₃ NaBH₄ Cl⁻

(9)

Scheme 3

1.2.2.2.2 Chemoselectivity of hydride reagents

An important consideration in the selection of a metal hydride reagent for C=N reductions is, of course, that other functionalities present are not affected. Summaries and tables of the chemoselectivities of various hydride reagents are available,[8,102] (*e.g.* also ref. 9, p. 177; ref. 11, p. 129) and can be consulted for particular situations. However, in general, the principal reason that $NaBH_4$ and, particularly, $NaBH_3CN$ have enjoyed extensive popularity is directly related to the selectivity available with these reagents, since both are essentially inert toward most common functional groups except aldehydes and ketones, and even these latter are resistant toward $NaBH_3CN$ in neutral or basic conditions.[16,26] As previously mentioned, this tolerance has been exploited to allow imine reductions in the presence of esters, amides, alkenes, ethers, nitriles, nitro groups, ketones, a variety of heterocyclic rings (see Tables 1 and 2), and with sensitive biological systems. Other functional groups such as aryl and alkyl halides (*e.g.* entry 8, Table 4), sulfoxides, sulfones and even some epoxides also remain intact under conditions used for imine reductions.[103]

1.2.2.2.3 Stereoselectivity with cyclic derivatives

Another consideration which often must be addressed in reductions of cycloalkanone imines involves the control of stereochemistry, so that proper diastereorelationships can be installed in complex molecules. Although considerably less investigation of this topic has occurred compared to the immense efforts devoted to corresponding ketone reductions,[27,29,104] (see also ref. 7, chap. 12; ref. 11, p. 141 and references cited therein) several features have emerged which allow some degree of prediction and control of stereochemical outcomes.[24,28,72,75,78,91] Thus, in general, and in analogy to ketones, 'small' unsubstituted metal hydrides (*e.g.* LAH,[72,91] $NaBH_4$,[28] $NaBH_3CN$[28,72,78]) prefer to approach from the axial direction of unhindered cyclohexyl imines and iminium ions (*i.e.* **10**; equation 1) to afford predominantly the corresponding equatorial amines (*i.e.* **11**; equation 1). Likewise, results with very bulky reagents (*e.g.* $LiBHBu^s_3$) also correspond with those obtained with ketones[22] in that equatorial attack of (**10**) to give **12**) is highly favored.[24,28,72,78] On the other hand, major differences between ketones and imines are observed with even moderately bulky reagents [*e.g.* $BH(OAc)_3$,[28] $NaAlH_2(OCH_2CH_2OMe)_2$,[28] $(alkyl)_2BHCN$[78]]. Thus, such reagents normally afford principally axial attack with ketones but equatorial approach with imines and, in many cases, the difference is substantial.[28] Thus, for example, the imine (**10a**; R = Bu^t) afforded a 16:84 *trans:cis* mixture of amines (**11a**; *trans*) and (**12a**; *cis*) with $Bu^tNH_2{\cdot}BH(OAc)_2$, while the analogous ketone, 4-*t*-butylcyclohexanone, gave an 84:16 *trans:cis* mixture of the corresponding alcohols.[28] This divergent behavior has been attributed to increased steric impedance toward axial approach induced by substituents on nitrogen (*i.e.* as indicated in **13** and **14**).[28] Table 5 presents stereochemical results for several 4-substituted cycloalkyl imines with a variety of hydride reagents, along with comparisons to the analogous ketones (see also entries 3 and 5, Table 4), and illustrates the similarities and differences described above.

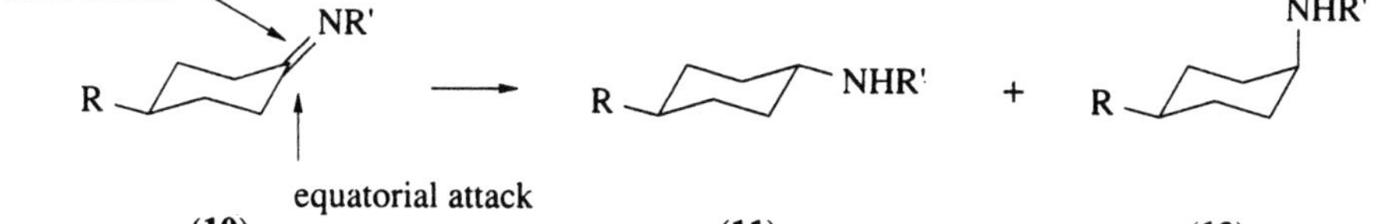

axial attack NR' equatorial attack → NHR' + NHR' (1)

(10) **(11)** **(12)**

Table 3 Reduction of Imines with Various Hydride Reagents

Entry	*Imine*	*Reagent*	*Solvent*	*Product*	*Yield (%)*	*Ref.*
1		$LiAlH_4$	THF		a	72
2		$LiAlH_4$	Et_2O		66	73
3		BH_3•THF	PhMe/THF		99[b]	75
4		Me_2NHBH_3	AcOH		71	86
5		Me_3NBH_3	AcOH		61	87

Table 3 *(continued)*

Entry	*Imine*	*Reagent*	*Solvent*	*Product*	*Yield (%)*	*Ref.*
6	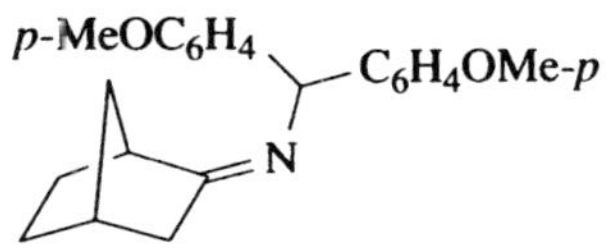 p-MeOC$_6$H$_4$, C$_6$H$_4$OMe-p, N	$LiBHEt_3$	THF	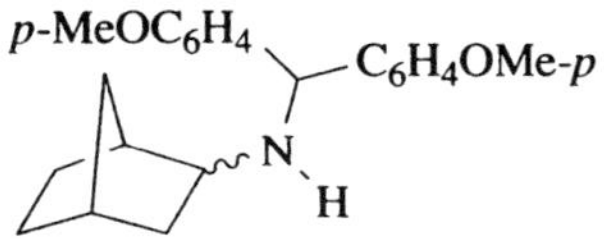 p-MeOC$_6$H$_4$, C$_6$H$_4$OMe-p, N, H	72[c]	24
7	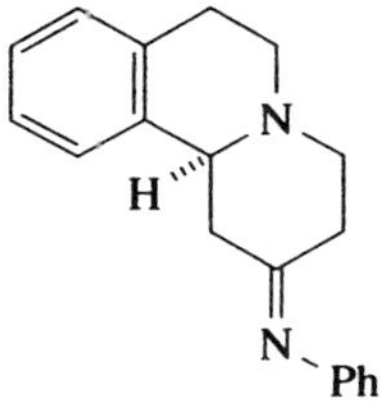 N, H, N, Ph	$LiBHBu^s_3$	THF/PhMe	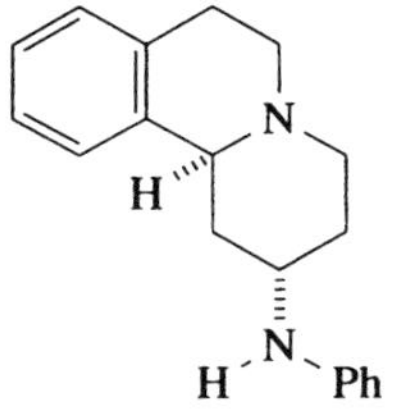 N, H, H, N, Ph	86	75
8	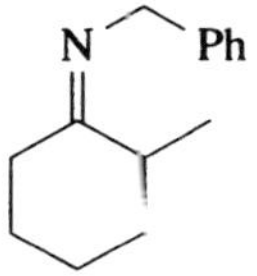 N, Ph	Na 9-BBN	MeOH/THF	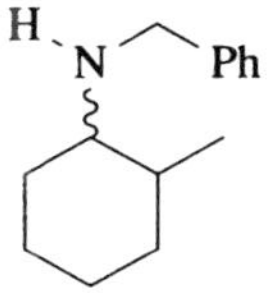 H, N, Ph	80[d]	78
9	 Ph, N, Ph	$Fe_3H(CO)_{11}$	—	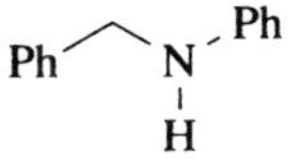 Ph, N, Ph, H	88	82
10	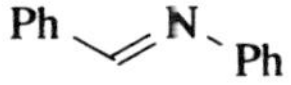 Ph, N, Ph	$Et_3SiH/(Ph_3P)_3RhCl$	PhH	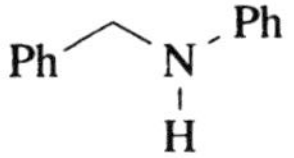 Ph, N, Ph, H	91	84
11	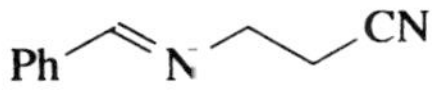 Ph, N, CN	$(Ph_3P)SeH$	$CHCl_3$	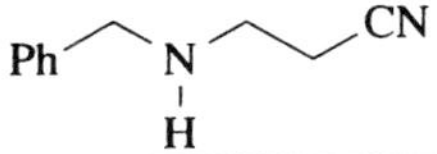 Ph, N, CN, H	80	85

[a] 1:1 mixture of diastereomers. [b] Mixture of diastereomers. [c] *Endo/exo* ratio 94:6. [d] *Cis/trans* ratio 84:16.

Table 4 Reduction of Iminium Salts with Various Hydride Reagents

Entry	*Iminium Salt*	*Reagent*	*Solvent*	*Product*	*Yield (%)*	*Ref.*
1	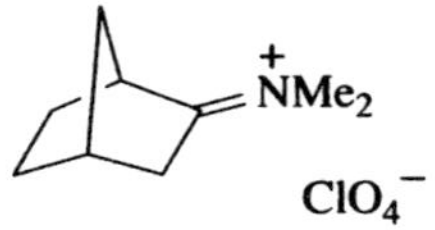 $\overset{+}{N}Me_2$ ClO_4^-	$LiAlH_4$	—	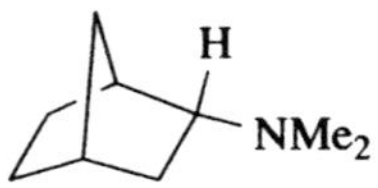 H, NMe_2	85[a]	89
2	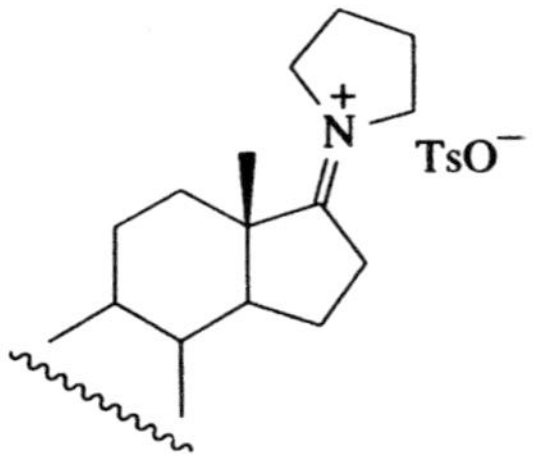 $\overset{+}{N}$ TsO^-	$LiAlH_4$	Et_2O	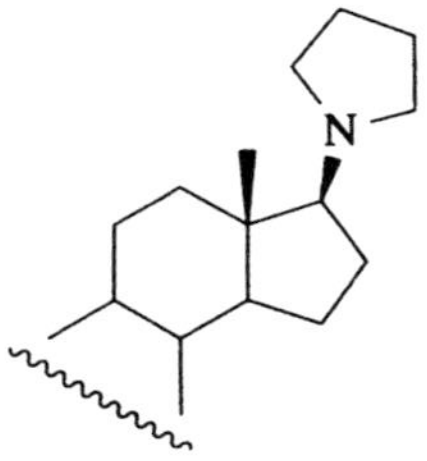 N	93[a]	90
3	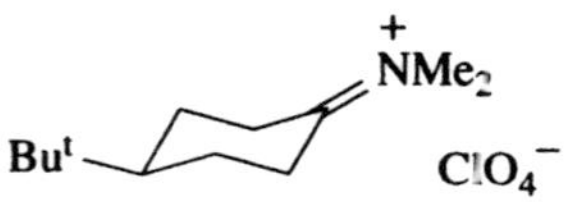 Bu^t $\overset{+}{N}Me_2$ ClO_4^-	$LiAlH_4$	Et_2O	 Bu^t NMe_2	b	91
4	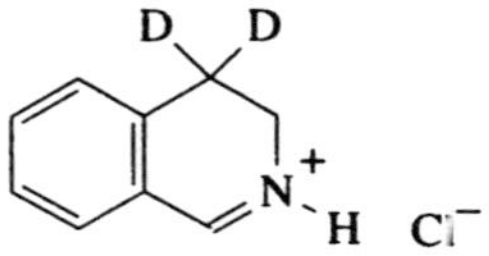 D D $\overset{+}{N}$ H Cl^-	$NaBH_4$	MeOH	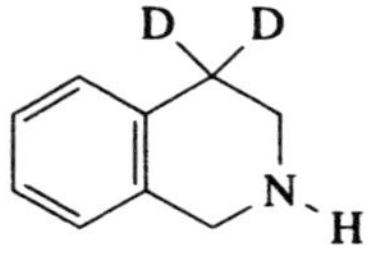 D D N H	88	92
5	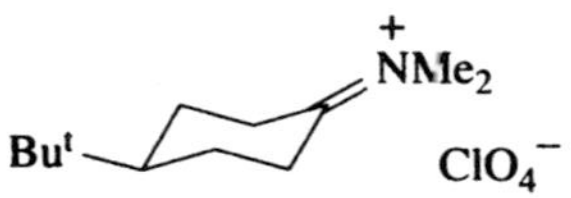 Bu^t $\overset{+}{N}Me_2$ ClO_4^-	$NaBH_4$	Pr^iOH	 Bu^t NMe_2	97[c]	28

Table 4 *(continued)*

Entry	*Iminium Salt*	*Reagent*	*Solvent*	*Product*	*Yield (%)*	*Ref.*
6	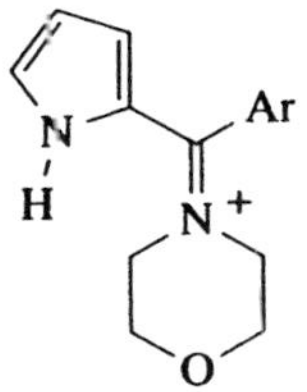	$LiBH_4$	THF	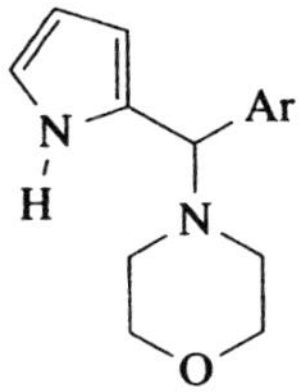	99	93
7	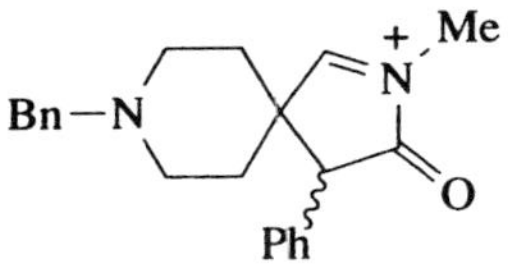	$NaBH_3CN$	HMPA	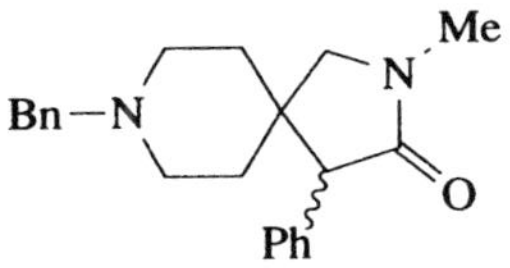	90	94
8	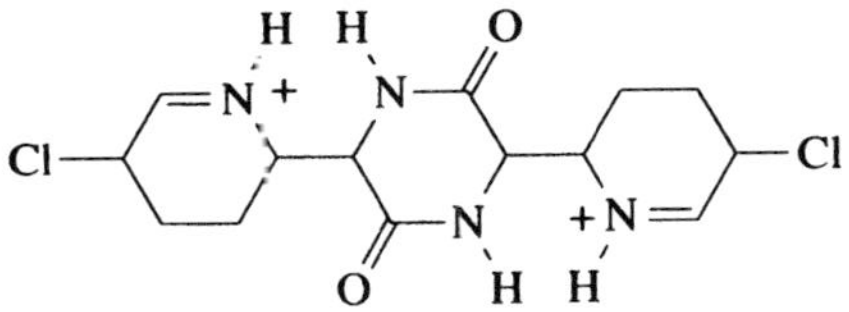	$NaBH_3CN$	MeOH/AcOH	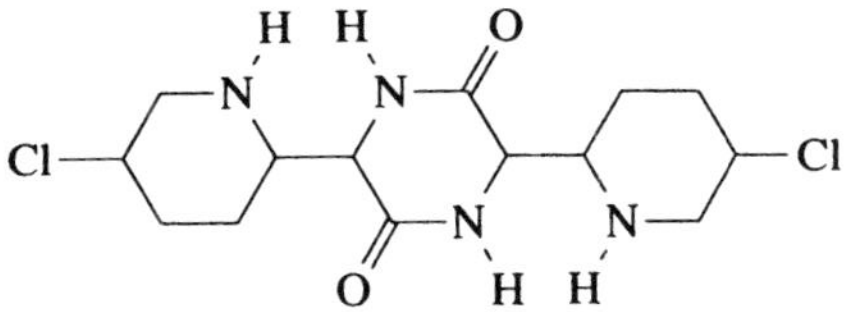	62	95
9	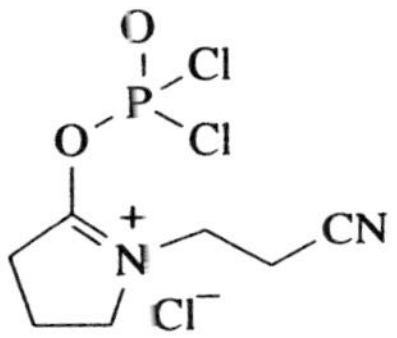	$NaBH_4$	DME		66	96

Table 4 *(Continued)*

Entry	*Iminium Salt*	*Reagent*	*Solvent*	*Product*	*Yield (%)*	*Ref.*
10		$NaBH_4$	DME/EtOH		57–66	97, 98
11		$NaBH_4$	MeOH		96	99
12		$NaBH_4/SnCl_4$	DME		51	100

[a] Only product detected. [b] *Cis/trans* ratio 36:64. [c] *Cis/trans* ratio 14:86.

(13)

(14)

The reduction of 2-alkylcyclohexyl imines is even more dramatic with regard to differences compared to the corresponding 2-alkylcyclohexanones. As illustrated in Table 6, all studied reagents prefer equatorial approach to give the *cis*-amines (**12**). This is strikingly divergent from results with 2-methylcyclohexanone, and even from other substituted cyclohexyl imines (*e.g.* entries 1, 2, 9, 10, Table 5). These enhanced differences are attributed to a combination of the decreased axial attack on the equatorial methyl conformer (**15**), coupled with an augmentation of the axial methyl conformer (**16**) induced by 1,3-allylic strain[106] in (**15**; Scheme 4). In this latter conformer (**16**), the axial methyl apparently retards equatorial approach thus favoring axial attack. The combination of favored equatorial approach to (**15**) and axial approach to (**16**) results in the observed high proportion of *cis* isomers.[28] The reluctance of equatorial approach to axial 2-methyl systems was demonstrated by the nondiscriminatory attack of 2-methylcyclohexyl iminium derivatives (*e.g.* entry 11, Table 6) by $LiBHBu^s{}_3$[28] in contrast to the usual extreme stereoselectivity for equatorial approach observed with ketones[23] and unhindered imines (see Tables 5 and 6).

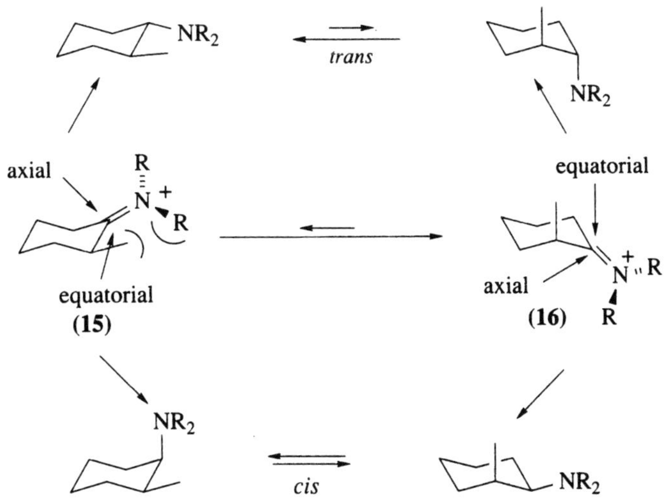

Scheme 4

Other cyclic imines also show enhanced stereoselectivities compared to ketones. Thus, 3,3,5-trimethylcyclohexyl imine (**17**) afforded a 94:6 *trans:cis* ratio with $NaBH_4$[28] compared to 58:42 with 3,3,5-trimethylcyclohexanone,[107] again suggesting increased steric interference from *N*-substituents.

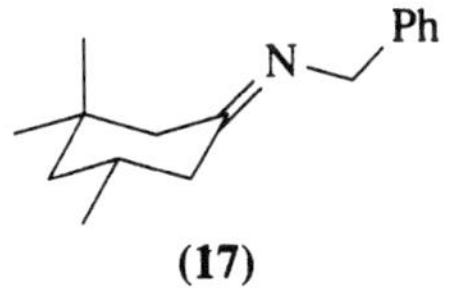

(17)

Reductions of substituted rings containing internal imines with hydrides display varying stereoselectivity dependent on the reagent. Thus, the 2,6-dialkylpiperideine (**18a**; equation 2, R^1 = Me, R^2 = *n*-undecyl, *n* = 2) afforded predominantly (>99%) the *cis*-piperidine (**19a**) with DIBAL-H or LAH/NaOMe, but the *trans* isomer (**20**) with LAH/Me_3Al.[108] The difference in stereoselectivity has been attributed to a change in conformation stemming from Me_3Al complexation with nitrogen forcing the 2-methyl group into an axial orientation, which reverses the normal direction of attack caused by preferred

Table 5 Stereochemistry of Reduction of 4-*t*-Butylcyclohexyliminium Salts and other Cyclohexyl Imines with Hydride Reagents

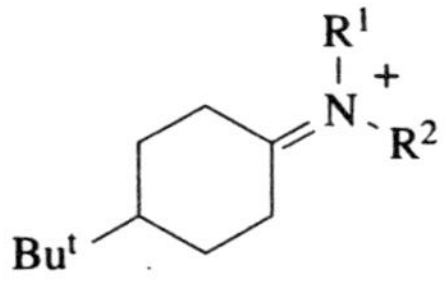

Entry	R^1	R^2	*Hydride*	*Solvent*	*Yield (%)*	*Product cis:trans ratio*	*Ketone reduction cis:trans ratio (ref.)*	*Ref.*
1	Bn	—	$NaBH_4$	Pr^iOH	82	30:70	17:83 (21)	28
2	Bn	—	$NaBH_3CN$	MeOH/HCl	87–97	35:65–46:54	16:84 (27)	28, 78
3	Bn	—	Me_2NHBH_3	$MeCO_2H$	74	66:34	—	28
4	Bn	—	$NaAlH_2[O(CH_2)_2OMe]_2$	THF	94	64:36	8:92 (28)	28
5	Bn	—	$LiBHEt_3$	THF	89	67:33	—	78
6	Bn	—	$LiBHBu^s_3$	THF	77–95	97:3–98:2	93:7 (23)	28, 78
7	Me	—	$LiAlH_4$	THF	—	25:75	10:90 (27)	72
8	Me	—	$NaBH_3CN$	MeOH, H^+	—	40:60	16:84 (28)	72
9	Me	Me	$NaBH_4$	Pr^iOH	70	16:84	17:83 (23)	28
10	Me	Me	$LiAlH_4$	—	—	35:65	10:90 (27)	91
11	Me	Me	$NaAlH_2[O(CH_2)_2OMe]_2$	—	74	72:28	8:92 (28)	28
12	$CH(C_6H_4\text{-}p\text{-}OMe)_2$	—	$LiBHBu^s_3$	—	59	99:1	93:7 (23)	24
13	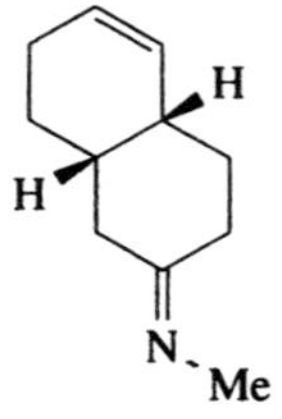 (N-methyl imine, bicyclic structure)		$LiBHBu^s_3$	THF		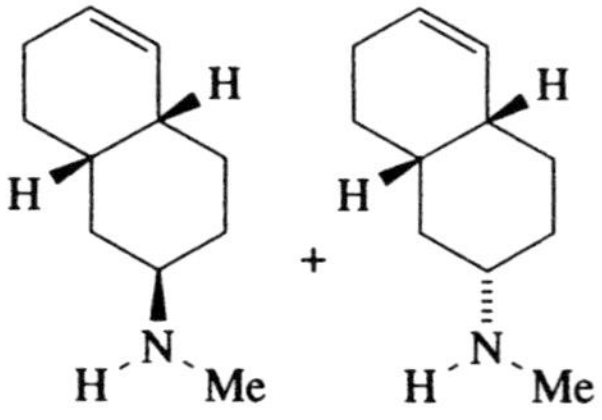 92:8	—	105

Table 5 *(continued)*

Entry	R^1	R^2	*Hydride*	*Solvent*	*Yield (%)*	*Product cis:trans ratio*	*Ketone reduction cis:trans ratio (ref.)*	*Ref.*
14			$LiBHBu^s{}_3$	THF	86	0:100	—	75

Table 6 Stereochemistry of Reduction of 2-Methylcyclohexyl Imines and Iminium Salts with Hydride Reagents

Entry	R^1	R^2	*Hydride*	*Solvent*	*Yield (%)*	*Product cis:trans ratio*	*Ketone reduction cis:trans ratio (Ref.)*	*Ref.*
1	Bn	—	$NaBH_4$	Pr^iOH	90	85:15	25:75 (27)	28
2	Bn	—	$NaBH_3CN$	MeOH/HCl	82–92	64:36–76:24	28:72 (28)	28,78
3	Bn	—	Na 9-BBNCN	MeOH/THF/HCl	80	84:16	—	78
4	Bn	—	Na disiamylBHCN	MeOH/THF/HCl	73	81:19	—	78
5	Bn	—	9-BBN	MeOH/THF/HCl	70	83:17	—	78
6	Bn	—	$Bu^tNH_2BH_3$	THF	78	90:10	—	78
7	Bn	—	$LiBHEt_3$	THF	84	97:3	—	78
8	Bn	—	$LiBHBu^s_3$	THF	78	99:1	99.3:0.7 (23)	28
9	Bn	—	$NaAlH_2[O(CH_2)_2OMe]$	THF	81	78:22	—	28
10	$CH(C_6H_4$-*p*-$OMe)_2$	—	$LiBHBu^s_3$	THF	<10	—	99.3:0.7 (23)	24
11	—$(CH_2)_4$—		$LiBHBu^s_3$	THF	60	48:52	99.3:0.7 (23)	28

orbital overlap stabilization.[108] Likewise, 2,5-dialkylpyrrolines (*e.g.* **18b**; $n = 1$, R^1 = Me, R^2 = *n*-nonyl) gave the *cis*-pyrrolidine (**19b**) with DIBAL-H (CH_2Cl_2), while (**18c**; $n = 1$, R^1 = Me, R^2 = *n*-undecyl) provided a 95:5 *trans:cis* (**20c:19c**) mixture with LAH/Me_3Al (THF).[65] Other hydride reagents (*e.g.* $NaBH_4$, $NaBH_3CN$) afforded much less stereoselectivity with analogs of (**18**; *e.g.* entry 14, Table 2).[65]

(18) → (19) + (20) (2)

1.2.2.3 Reduction of *In Situ* Generated Imines and Iminium Salts of Ammonia, Primary and Secondary Amines; Reductive Amination

1.2.2.3.1 Reductive aminations with sodium cyanoborohydride

As described in the sections above, imines and iminium salts are effectively reduced by a number of reagents. However, most hydride reducing agents require preformation of the carbon–nitrogen π-system, since reduction of aldehydes and ketones is very facile with most of them. As mentioned previously, the introduction of $NaBH_3CN$ into the arsenal of reducing agents altered this situation, primarily because of the exceptional functional group tolerance exhibited by this reagent. Thus, under neutral or slightly acidic (*i.e.* pH > 5) conditions, reductions of nearly all common functionalities, including aldehydes and ketones, are slow.[15,16,26] The principal exceptions to this are iminium ions, which are rapidly attacked by cyanoborohydride. These properties, then, allow the sequence depicted in Scheme 2 to be accomplished *in situ* without prior formation and isolation of the intermediate imines. Thus, reductive amination of carbonyls may be accomplished directly by reaction of an aldehyde or ketone with ammonia, primary or secondary amines at pH *ca.* 5–7 in the presence of BH_3CN^- (usually in MeOH, but occasionally in other solvents, *e.g.* H_2O, EtOH, MeCN, THF, DMF) at room temperature, leading directly to primary, secondary or tertiary amines. Since the process relies on successful iminium ion generation (*i.e.* **1**; Scheme 2), the use of molecular sieves (and presumably other drying agents) to absorb water produced in the initial equilibrium (Scheme 2) is beneficial, especially with carbonyls which form imines slowly (*e.g.* norbornanone).[26] As expected, hindered ketones (*e.g.* phenyl *t*-butyl ketone) which resist imine formation are not successfully reduced using the standard procedure,[26] and require more forcing conditions to preform the imines. Nevertheless, the overall reliability and convenience of reductive amination with $NaBH_3CN$, coupled with the reagent's stability in most solvents, including acidic media, and its compatibility with other functional groups, have combined to provide enormous popularity for the process. Indeed, several hundred reports have appeared describing diverse applications to synthetic endeavors requiring the insertion of amino groups, and this popularity continues unabated. At present, then, this methodology must be regarded as the protocol of choice for reductive amination.

The utility of reductive amination with $NaBH_3CN$ in synthesis is contained in reviews[15,16] and successful applications have been compiled through 1978.[16] Table 7 provides a variety of examples taken from more recent accounts and chosen to illustrate the versatility and compatibility of the process with diverse structural types and chemoselectivity demands. Thus, esters (entries 2–4, 8–12), amides (entries 3, 6–9, 12), nitro groups (entry 13), alkenes (entry 2), cyclopropyl groups (entry 2), organometallics (entry 5), amine oxides (entry 14) and various heterocyclic rings (entries 1, 3, 5–10) all survive intact. Entry 6 illustrates that deuterium can be conveniently inserted *via* the readily available $NaBD_3CN$,[15,16] and entry 15 demonstrates that double reductive amination with diones can be utilized to afford cyclic amines.

The functional group tolerance possible with $NaBH_3CN$ is even more amply illustrated by the successful reductive amination of the C-20 aldehyde in tylosin (**21**) with several amines,[124] of the C-3a aldehyde of the chlorophyll derivative Chl b (**22**) with NH_3 (80% yield),[125] and the C-11 aldehyde of the modified tetrodotoxin derivative (**23**) with several amino acids.[126]

The usefulness of reductive amination is augmented by the facile methylation of amines with formaldehyde (usually in MeCN),[127] which provides a convenient, mild alternative to Clark–Eschweiler and other methylation procedures. Table 8 presents a selection of successful methylation applications with various amines, and further illustrates the chemoselectivity and versatility of the process.

As described previously, $NaBH_3CN$ has found wide usage in the biochemical arena where mildness and compatibility with aqueous media are essential, and reductive amination has been utilized especially

Table 7 Reductive Amination of Carbonyls with $NaBH_3CN$

Entry	*Carbonyl*	*Amine*	*Product*	*Yield (%)*	*Ref.*
1	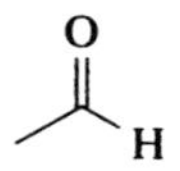	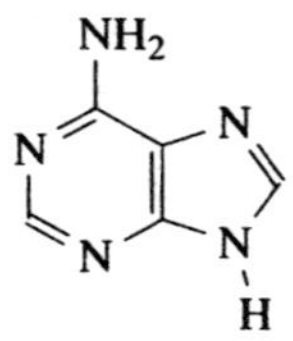	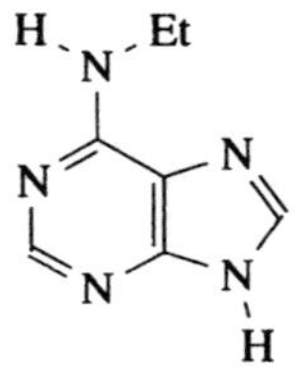	80	109
2	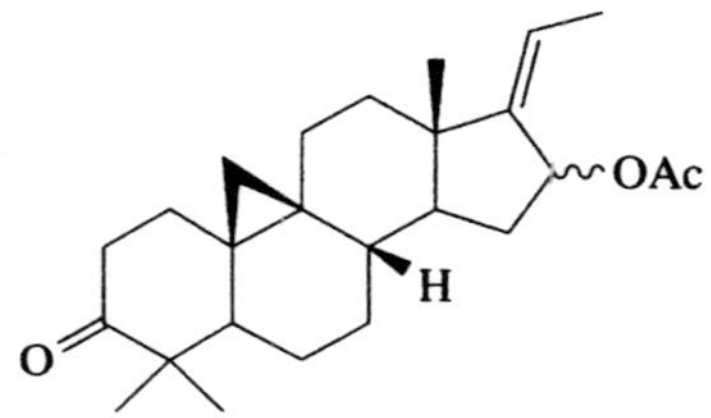	$MeNH_2$	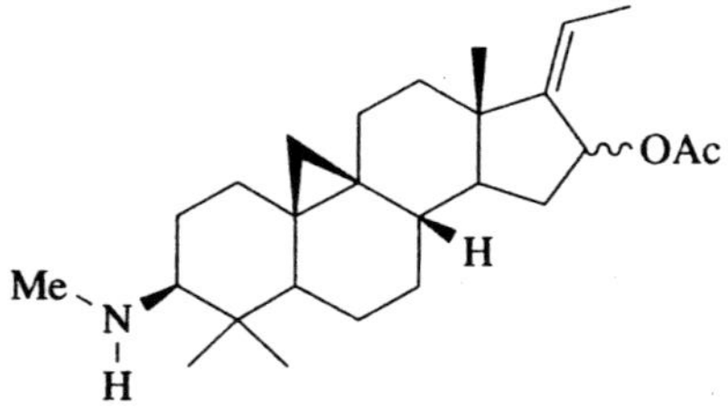	67	110
3	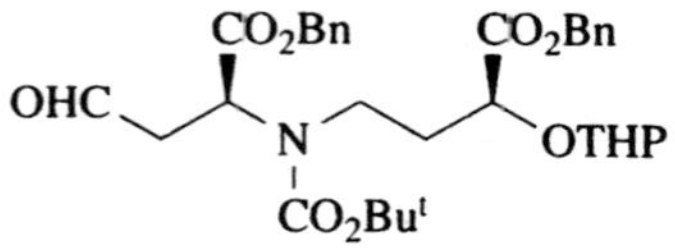	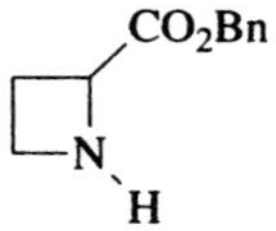	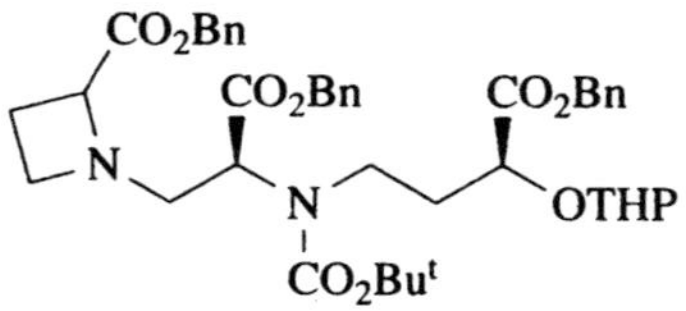	59	111
4	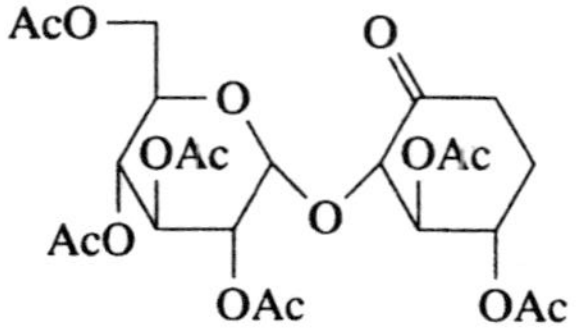	NH_3	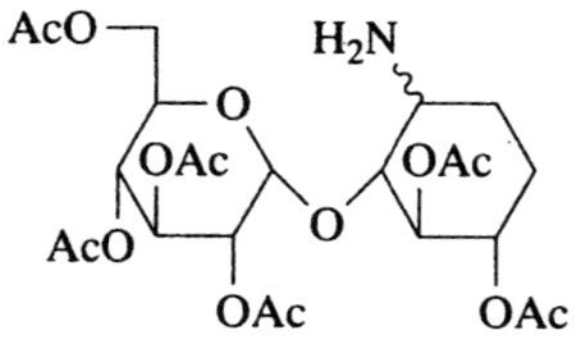	41	112

Table 7 *(continued)*

Entry	*Carbonyl*	*Amine*	*Product*	*Yield (%)*	*Ref.*
5	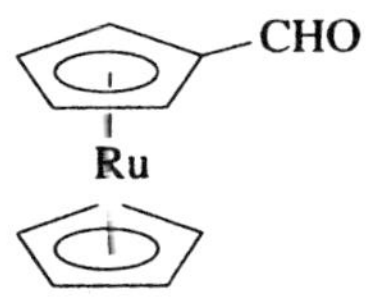	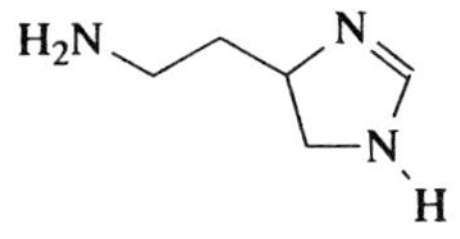	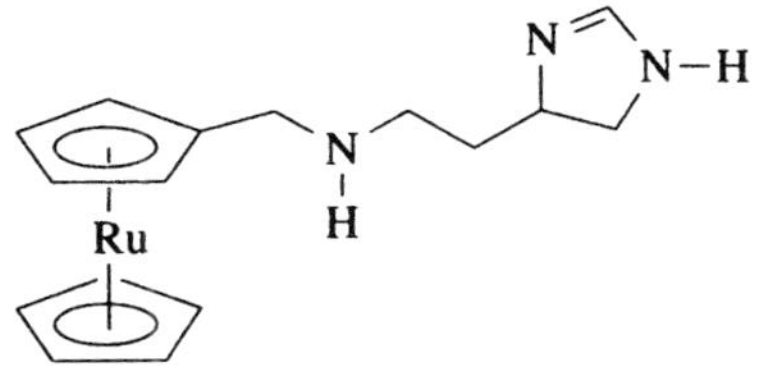	55	113
6	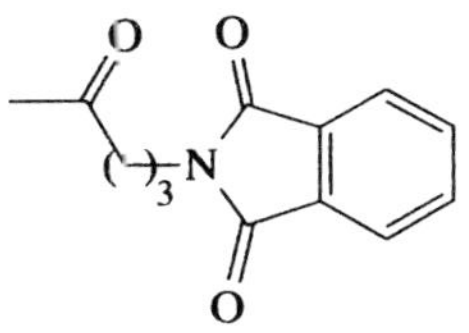	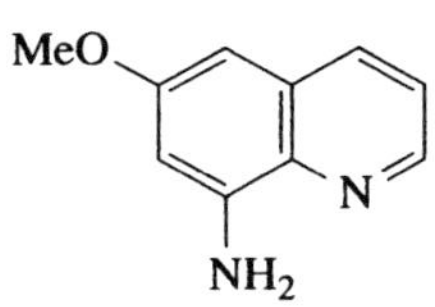	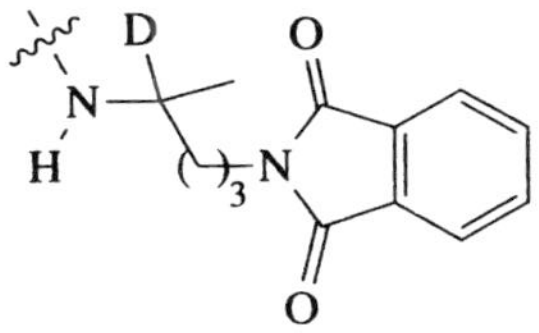	46[a]	114
7	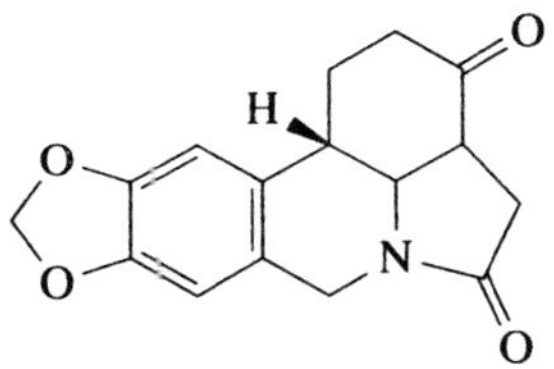	$HNMe_2$	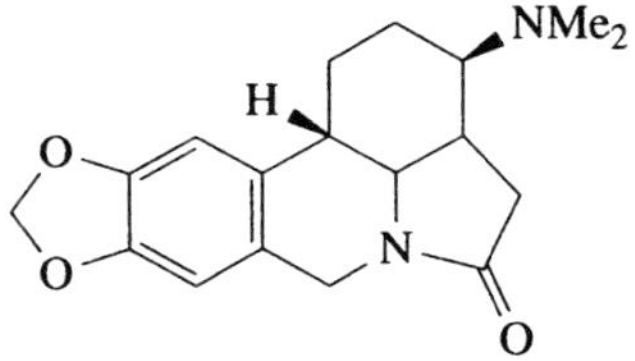	43	115
8	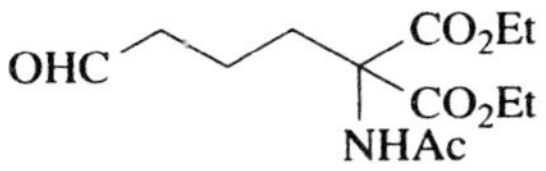	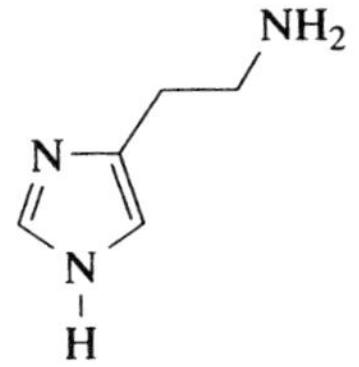	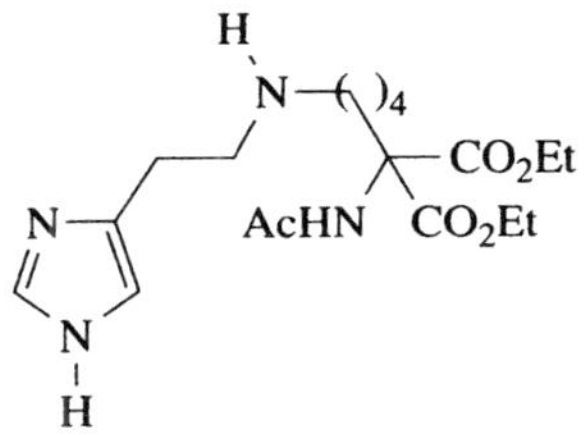	87	116

Table 7 *(continued)*

Entry	*Carbonyl*	*Amine*	*Product*	*Yield (%)*	*Ref.*
9	O, EtO_2C, Ph	NH_2, N, O, CO_2Bn	H, N, CO_2Et, Ph, N, O, CO_2Bn	68[b,c]	117
10	O, H	MeO, NH_2, CO_2Et, N, H	MeO, NEt_2, CO_2Et, N, H	76[d]	118
11	CHO, O, O, Ar, AcO	$HNMe_2$	Me, Me–N, O, O, Ar, AcO	84	119
12	CbzHN, CO_2Me, Ph, CHO	Ph, H_2N, CO_2Me	CbzHN, CO_2Me, Ph, Ph, N, H, CO_2Me	78	120

Table 7 *(continued)*

Entry	*Carbonyl*	*Amine*	*Product*	*Yield (%)*	*Ref.*
13	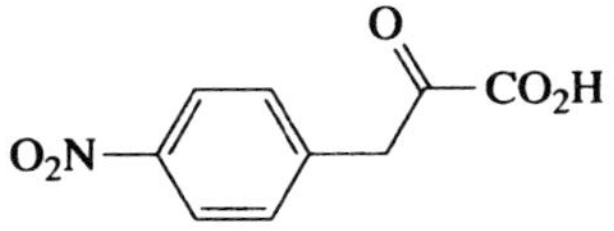 O_2N, O, CO_2H	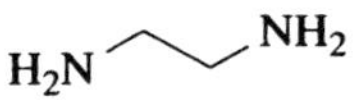 H_2N, NH_2	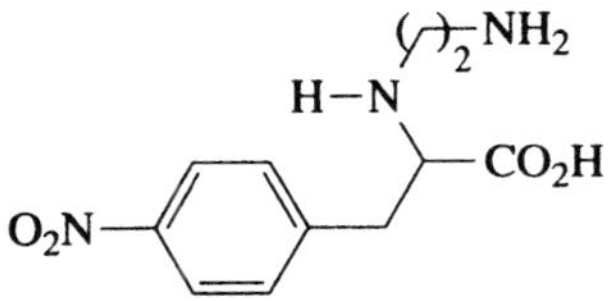 NH_2, H–N, CO_2H, O_2N	23	121
14	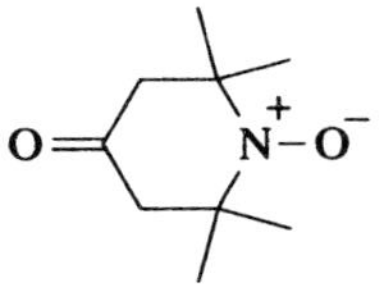 O, N–O (+, −)	H_2NMe	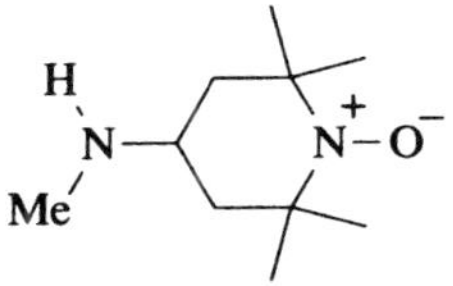 H, N, Me, N–O (+, −)	73	122
15	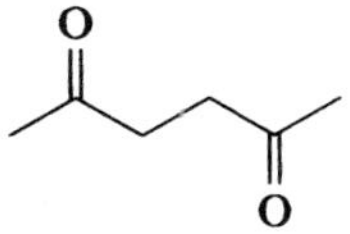 O, O	NH_3	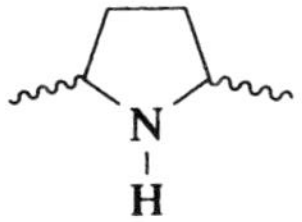 N, H	62	123

[a] $NaBD_3CN$ used. [b] Ethanol solvent. [c] Mixture of diastereomers. [d] MeCN solvent.

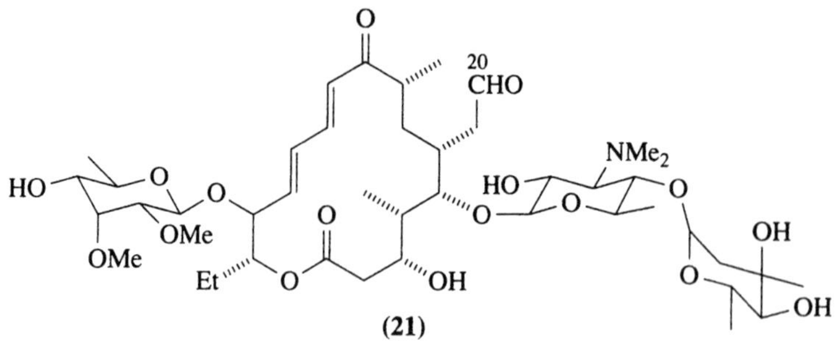

(21)

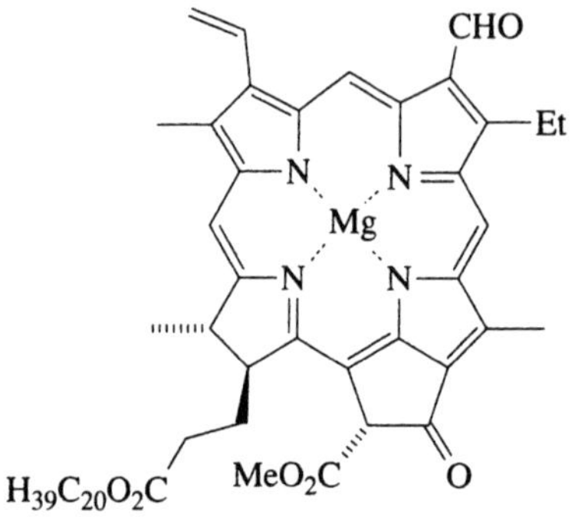

(22)

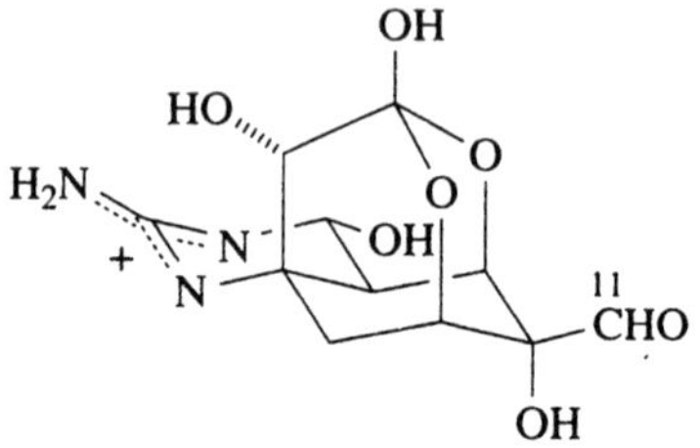

(23)

in linking proteins and related polypeptides to various supports. Examples include the coupling of aldehyde-bearing glycosides to amino groups of proteins (*e.g.* bovine serum albumin,[134,135] human fibrinogen,[135] bovine pancreatic ribonuclease-A,[135] low density lipoprotein,[136] invertase,[137] collagen,[138] and polylysine,[139] to name a few), the binding of the globular protein concanavalin A to silica,[140] of a number of enzymes to alumina (*via* pyridoxyl 5′-phosphate),[141] and the reductive alkylation of the opioid peptide methionine-enkephalin.[142] Other applications encompass the labeling of oligosaccharides with UV-absorbing aromatic amines,[143] specific methylation of amino acid residues in enzymes and proteins (*e.g.* aspartate aminotransferase,[144] concanavalin A,[145] and the L-subparticles of rabbit cytoplasmic ribosomes[146]) and the spin labeling of AMP and NaD^+.[147] It must be noted that the above represent only a small portion of the extensive applications of reductive amination to biochemical problems, further discussion of which is beyond the scope of this chapter.

Although reductive amination with $NaBH_3CN$ has been successfully utilized for numerous synthetic applications, side reactions and failures have been noted. Thus, reductive amination (with ammonia) of the keto sugar (**24**) afforded a mixture of products including two isomeric cyanohydrins and alcohols (partial structures **25** and **26**, respectively), in addition to the expected diaminated product.[148] Also, the diketone corresponding to (**5**) gave a diol (resulting from reduction of both carbonyls) and the cyclic amino alcohol (**8**), in addition to the diamine resulting from reductive amination with ammonia, all in low yields.[149] Likewise, carbohydrate derivatives (**27**)[150] and retinal (**28**)[151] failed to give satisfactory yields of amine products upon reductive amination.

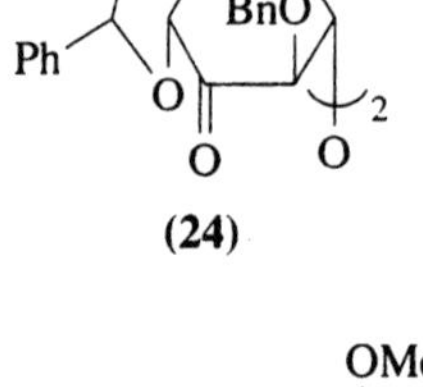

(24) 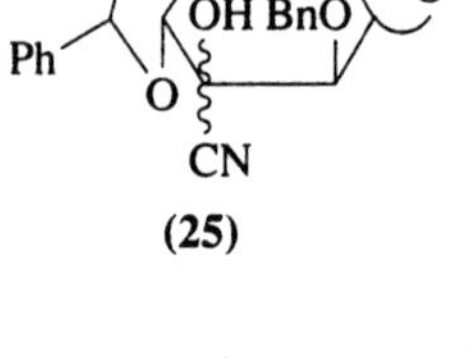 (25) 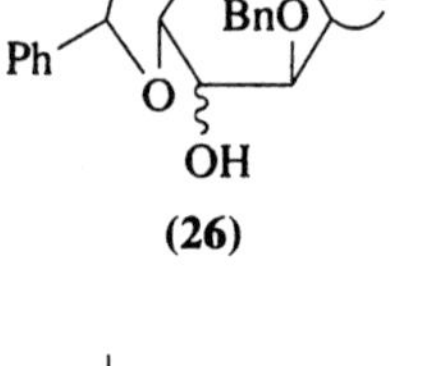 (26)

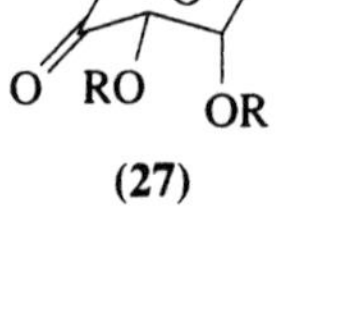

(27) 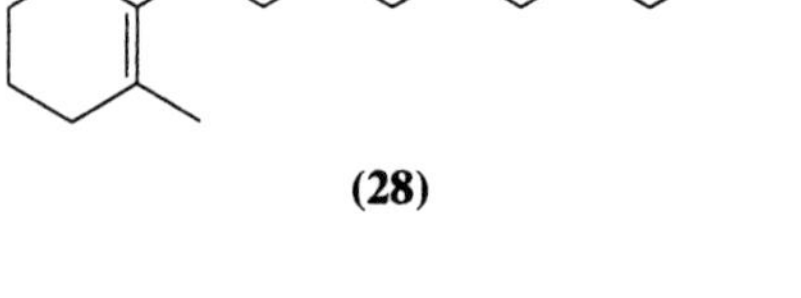 (28)

Table 8 Reductive Methylation with $NaBH_3CN$/Formaldehyde

Entry	*Amine*	*Product*	*Yield (%)*	*Ref.*
1	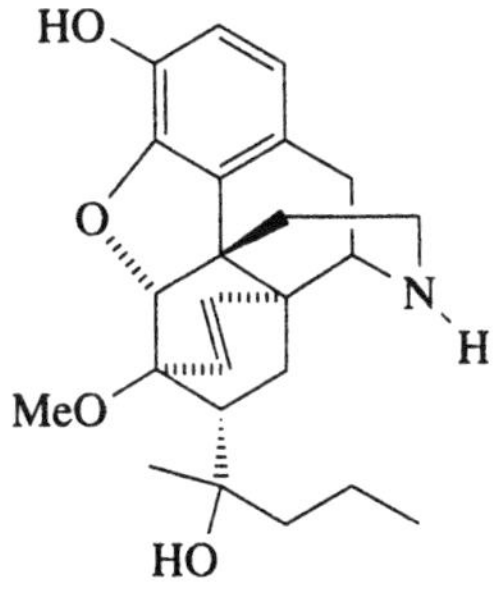	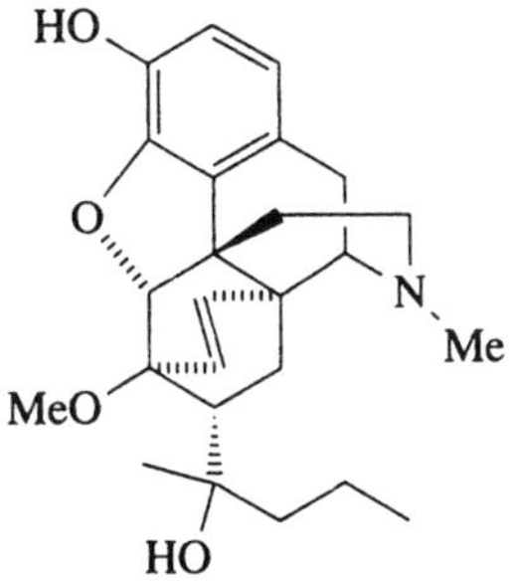	—	128
2	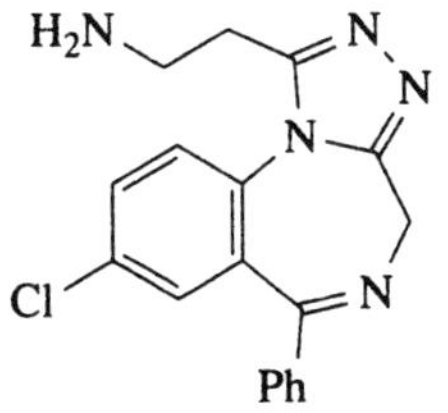	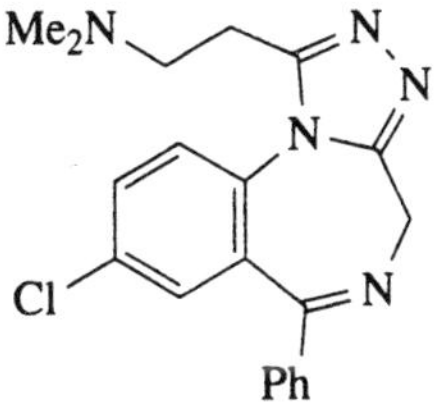	68	129
3	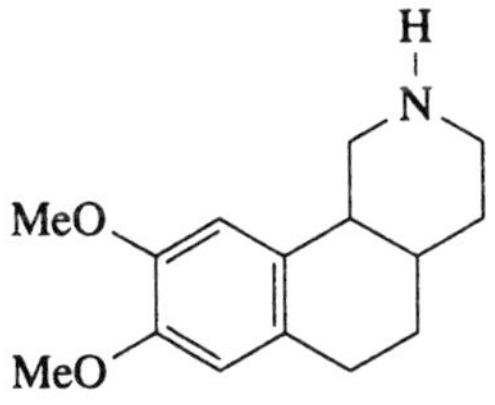	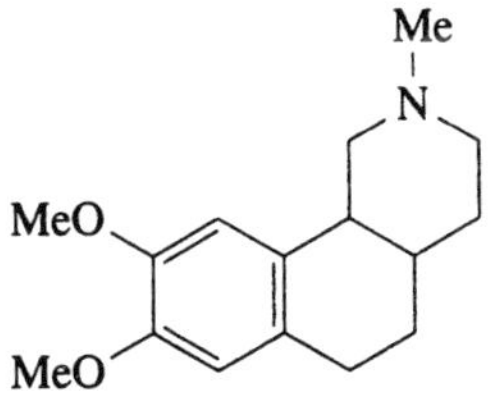	83	130
4	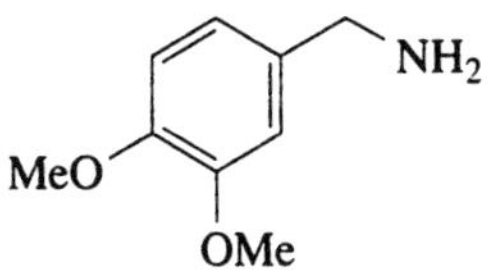	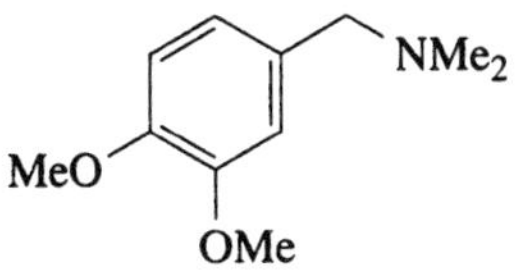	—	131
5	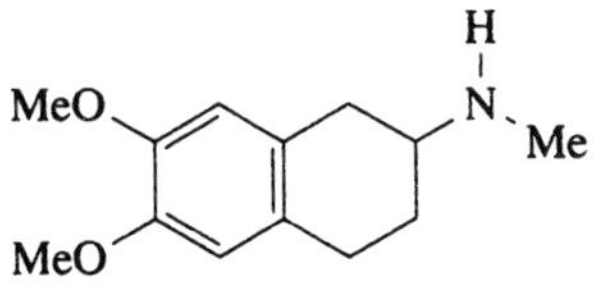	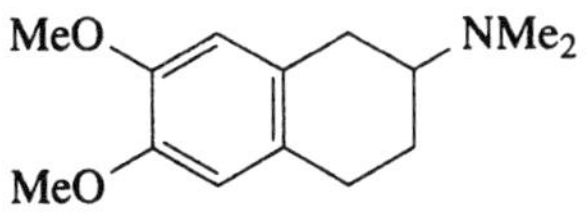	—	132
6	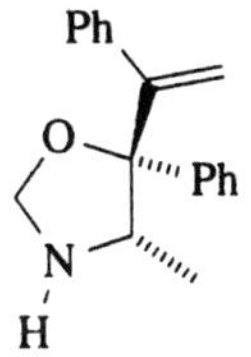	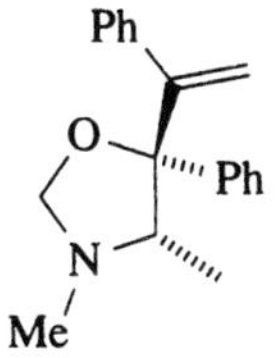	84	133

1.2.2.3.2 Reductive aminations with other hydride reagents

Although the use of $NaBH_3CN$ alone for reductive amination has thus far dominated applications in synthesis, modified cyanoborohydride derivatives and other reagents, which offer viable alternatives and potential improvements, have recently been introduced. Thus, $Zn(BH_3CN)_2$ (methanol solvent) is apparently equally effective for reductive aminations and amine methylations and, in most cases, requires short reaction times (*ca.* 1–6 h).[152] In addition, the combination of $NaBH_3CN$ and $Ti(OPr^i)_4$ is reported to afford superior results with difficult carbonyls and those sensitive to acidic conditions,[153] and other Lewis acids (*i.e.* $MgCl_2$) have also been employed with $NaBH_3CN$.[154] Alkylammonium cyanoborohydrides allow reductive aminations to be conducted in nonpolar solvents.[155] One study reports the use of Na-9-BBNCN-H for the reductive amination of 2-methylcyclohexanone with ammonia, but the yield was low (31%).[78] Other boron-based mild reducing reagents also provide alternative reductive amination processes. These include $BH_3{\cdot}THF$,[156] pyridine·BH_3,[20,157,158] $NaBH_4$ (or $NaBD_4$) in liquid or aqueous ammonia,[159] and $NaBH(OAc)_3$ in THF or $ClCH_2CH_2Cl$.[160] The previously mentioned $KFeH(CO)_4$[80,81] reductively aminates carbonyls in ethanol with best results obtained under a CO atmosphere. The reagent also provides methodology for the methylation of amines using formaldehyde as the carbonyl partner.[161]

1.2.2.3.3 Stereoselectivity with cyclic derivatives

Investigations of stereochemical outcomes of *in situ* reductive aminations have been sparse and the few reports generally indicate relatively poor diastereoselection with cyclohexyl systems. Table 9 presents a selection of available stereochemical results with the reagents which successfully reductively aminate ketones and indicate that selectivities generally parallel those obtained with preformed imines where comparisons are available (see Table 5, entries 2, 3, 8; Table 6, entries 2–4). Thus, 4-substituted cyclohexanones display no or low selectivity with 'small' reagents (*i.e.* BH_3CN^-, entries 1, 7) while moderately bulky reagents [*i.e.* $NaBH(OAc)_3$, entries 2, 3] afford a predominance of *cis* (*i.e.* axial amine) diastereomers. Likewise, 2-methylcyclohexanone gives principally *cis* isomers with $NaBH_3CN$, as well as with the more hindered Na-9-BBNCN-H. In contrast to this, electrolytic reductive amination (Hg cathode, aqueous ethanol, pH 10–11) of 2-methylcyclohexanone (with $MeNH_2$) provided a 2:1 *trans:cis* mixture of amines, while 4-*t*-butylcyclohexanone gave high *trans* diastereoselectivities with $MeNH_2$ and Me_2CHNH_2 (*trans:cis* 14:1 and 20:1, respectively).[163]

Table 9 Stereochemistry of Reductive Amination of Cyclohexanones with Amines and Hydride Reagents

Entry	*Cyclohexanone*	*Amine*	*Reagent*	*Yield (%)*	*Product ratio cis:trans*	*Ref.*
1	4-*t*-butyl	pyrrolidine (N–H)	Bu_4NBH_3CN	73	33:67	155
2	4-*t*-butyl	pyrrolidine (N–H)	$NaBH(OAc)_3$	96	71:29	160
3	4-*t*-butyl	Me_2CHNH_2	$NaBH(OAc)_3$	91	79:21	160
4	2-methyl	$PhCH(Me)NH_2$	$NaBH_3CN$	63	65:35	162
5	2-methyl	NH_3	Na-9-BBNCN-H	31	70:30	78
6	2-carboethoxy	Me_2NH	$NaBH_3CN$	53	40:60	26
7	[structure: Ph, N, H, O]	$PhNH_2$	$NaBH_3CN$	96	50:50	75

1.2.2.4 Reduction of Enamines *via* Iminium Ion Intermediates

1.2.2.4.1 Choice of hydride reagent and solvent

Although enamines are normally resistant to reduction by hydride reagents, rapid and reversible protonation on carbon in acidic media generates iminium ions, which, as presented above, are rapidly attacked by hydride reagents (Scheme 5). Thus, reagents which tolerate acidic conditions [*e.g.* $NaBH_3CN$,[15,16,26] $NaBH(OAc)_3$,[14] $BH(OAc)_2$[20,28], $BH(O_2CCF_3)_2$[164,165]] provide effective systems for enamine reductions to amines. Normally the latter three reagents are generated *in situ* from $NaBH_4$, BH_3 or amine·BH_3 complexes and the corresponding carboxylic acid as solvent, followed by addition of the enamine substrate. For this reason, the exact reducing species are not accurately known and therefore may be mixes of variously substituted acyloxyboranes. The reviews and reports listed above for acid-stable reagents provide collections of successful enamine reductions. Table 10 presents further examples with a variety of reducing systems and illustrates the versatility of the methodology and, again, the chemoselectivity possible with such mild reagents as $NaBH_3CN$ and acetoxyboranes. Thus, esters (entries 5, 6, 9, 10, 12), lactones (entry 2), certain heterocycles (entry 4), phosphites (entry 8), and even sensitive ketones (entry 7) and epoxides (entry 10) remain intact. Other types of enamines including vinylogous amides (entries 5–7, 9, 10, 12, 13) and indoles (entries 14, 15) are also effectively reduced. Occasionally, elimination of appropriately placed groups has been observed (*e.g.* a β-amino ketone, entry 13).

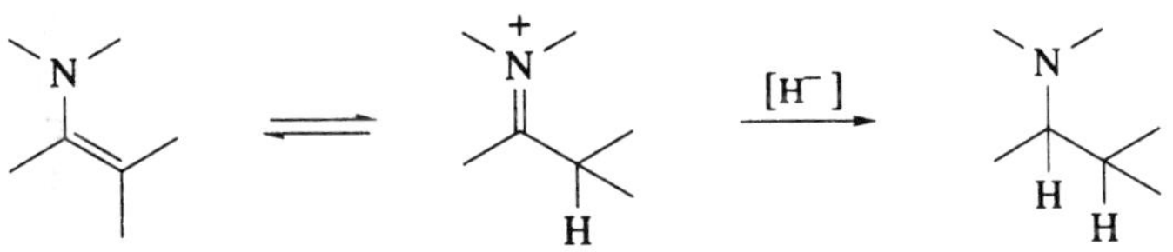

N N⁺ [H⁻] N H H H

Scheme 5

Other reagent systems not listed in Table 10 which successfully reduce enamines (*i.e.* indoles) include BH_3·pyridine in AcOH[179] or TFA,[164] catecholborane/TFA,[164] $NaBH_3CN$/TFA,[164] and $KFeH(CO)_4$/CO/ethanol.[180]

1.2.2.4.2 Stereoselectivity with cyclic derivatives

Since enamine reductions proceed through iminium ion intermediates, stereochemical features with cyclic examples should parallel results with preformed derivatives, and this is generally the case, at least with the acid-stable systems described above.[28] Furthermore, since the successful reagents are usually the moderately bulky acetoxyborane or acetoxyborohydride derivatives, a preponderance of *cis* diastereomers (from equatorial attack on cyclohexyl systems, equation 1 and Scheme 4) is obtained with 2- and 4-substituted derivatives.[28] In addition, 2-methylcyclopentenylamines are reduced in high stereoselectivity to the corresponding *cis*-2-methylcyclopentylamines with borane and borohydride reagents in AcOH (Scheme 6). This result is attributed to 1,3-allylic strain in the iminium ion: the methyl group is pushed into a pseudoaxial orientation which blocks the approaching reagent and forces attack from the opposite side.[28]

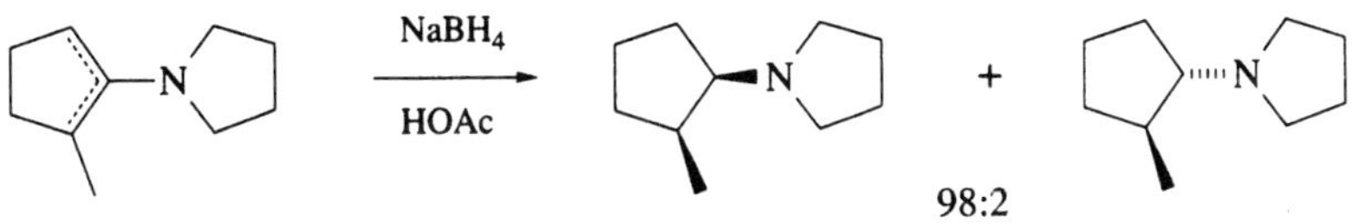

N $NaBH_4$ HOAc N + N 98:2

Scheme 6

Table 10 Reduction of Enamines with Hydride Reagents in Acidic Media

Entry	*Enamine*	*Reagent*	*Solvent*	*Product*	*Yield (%)*	*Ref.*
1	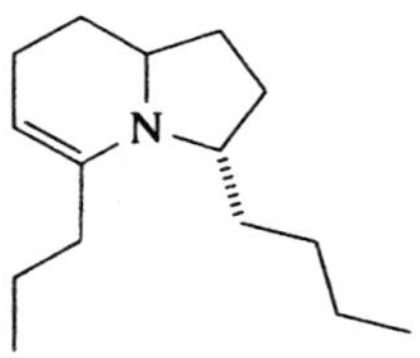	$NaBH_3CN/HCl$	MeOH	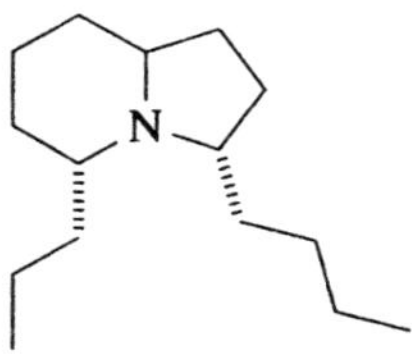	62	166
2	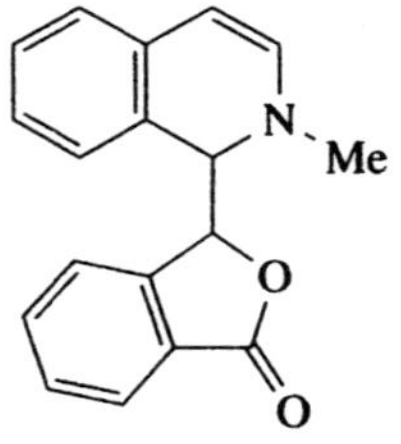	$NaBH_3CN/HCl$	MeOH	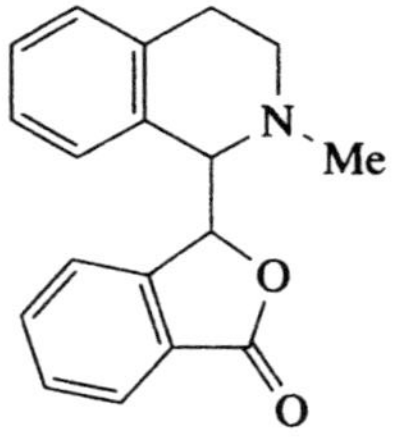	45	167
3	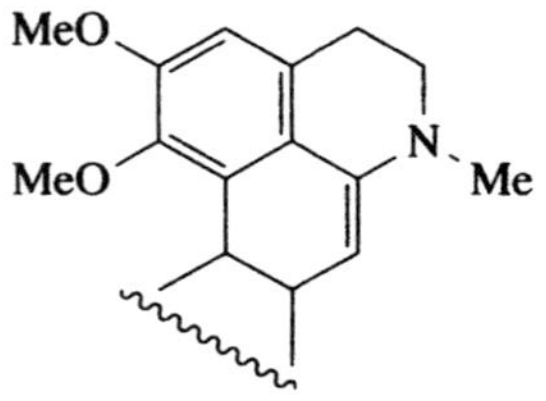	$NaBH_3CN/HCl$	EtOH	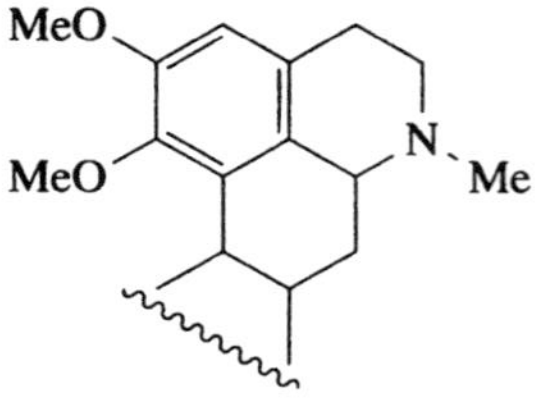	—	168
4	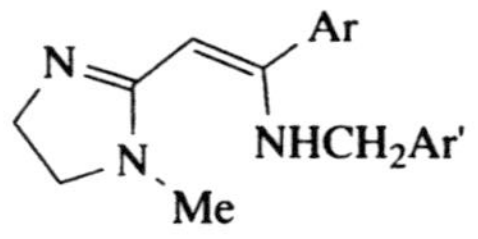	$NaBH_3CN/HCl$	THF/MeOH	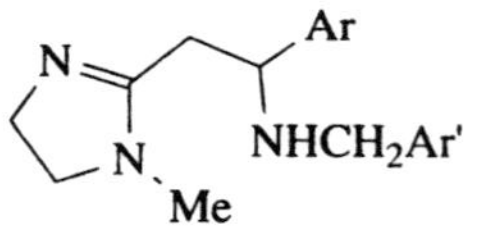	70–77	169
5	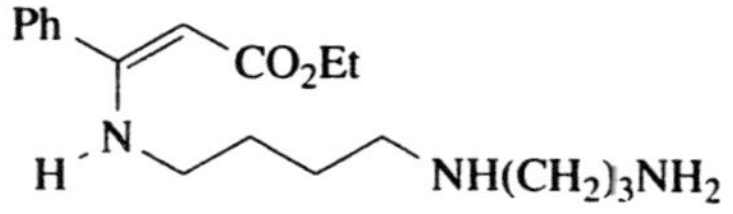	$NaBH_3CN$	—	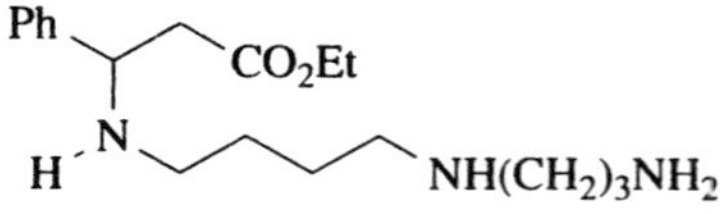	43	170

Table 10 *(continued)*

Entry	*Enamine*	*Reagent*	*Solvent*	*Product*	*Yield (%)*	*Ref.*
6	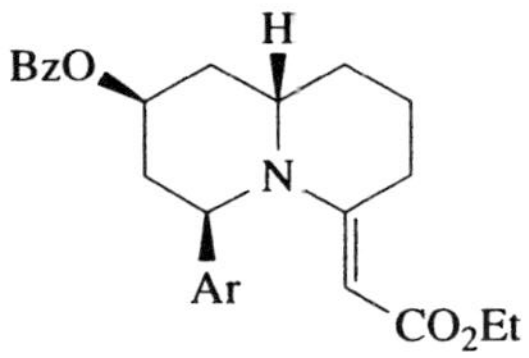 BzO, H, N, Ar, CO_2Et	$NaBH_3CN/HCl$	THF/MeOH	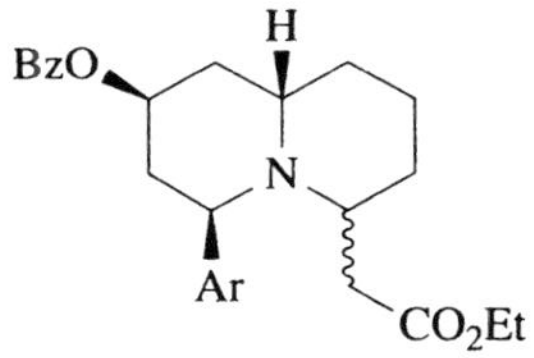 BzO, H, N, Ar, CO_2Et	98[a]	171
7	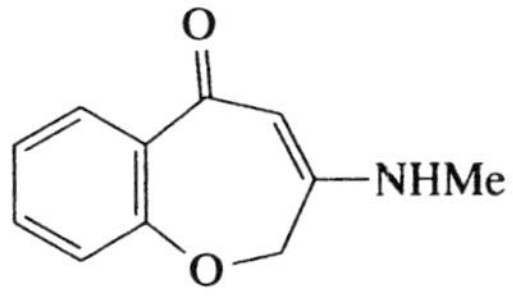 O, O, NHMe	$NaBH_3CN/AcOH$	*p*-Dioxane	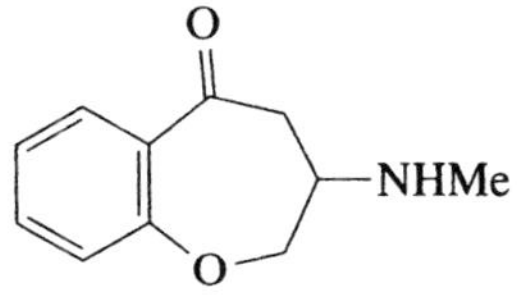 O, O, NHMe	79	172
8	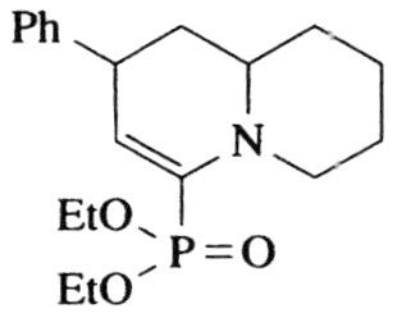 Ph, N, EtO, EtO, P=O	$NaBH_3CN/HCl$	—	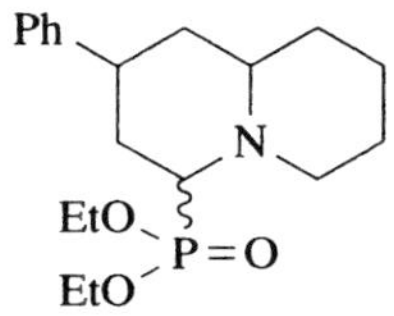 Ph, N, EtO, EtO, P=O	50	173
9	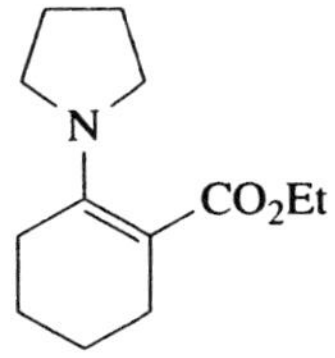 N, CO_2Et	$NaBH_3CN/H^+$	MeOH	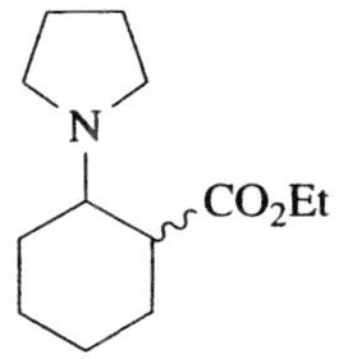 N, CO_2Et	50	26

Table 10 *(continued)*

Entry	*Enamine*	*Reagent*	*Solvent*	*Product*	*Yield (%)*	*Ref.*
10	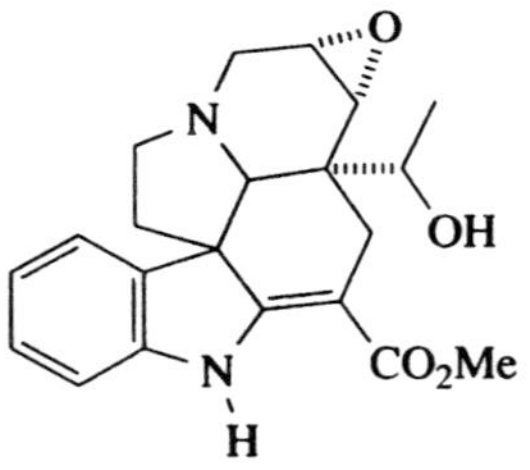	$NaBH_3CN$	AcOH	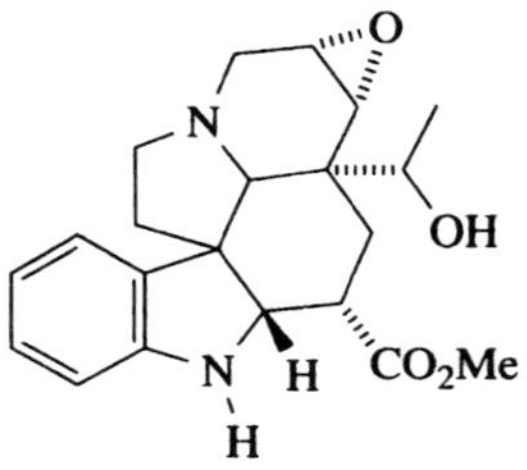	43[b]	174
11	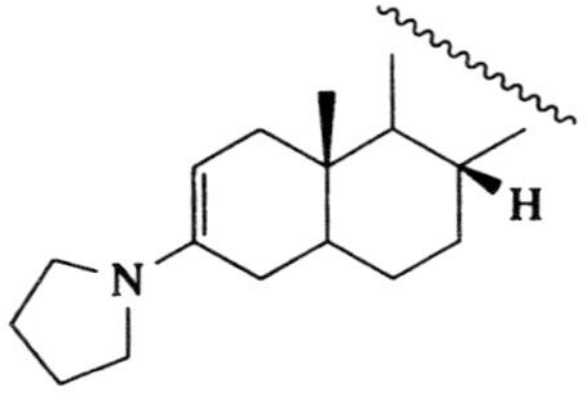	$NaBH_4$/AcOH	THF	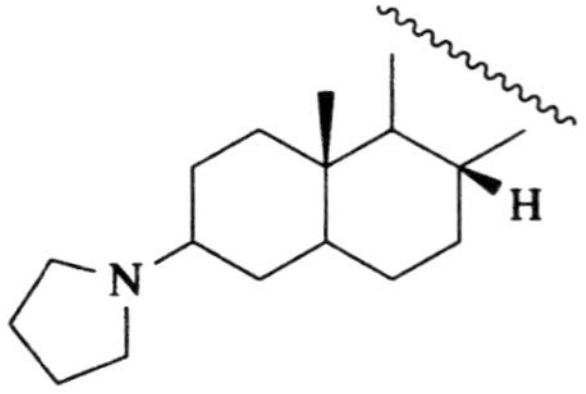	70	175
12	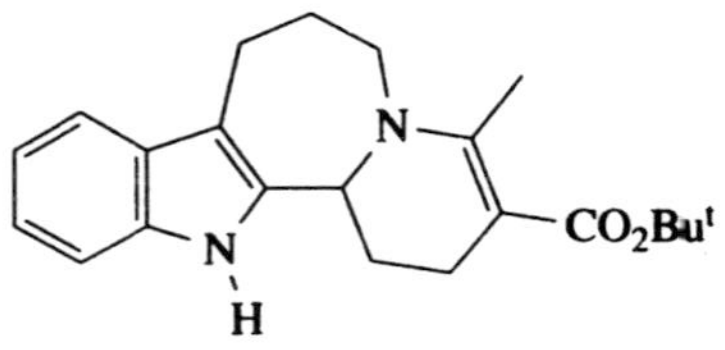	$NaBH_4$	AcOH	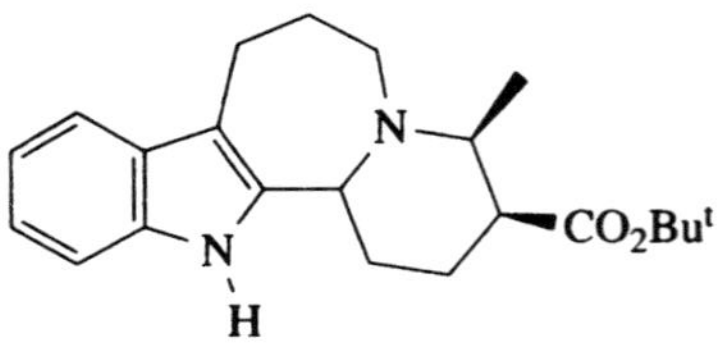	80	176
13	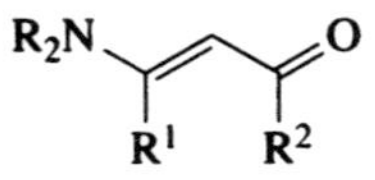	$NaBH_4$/AcOH	Pr^iOH	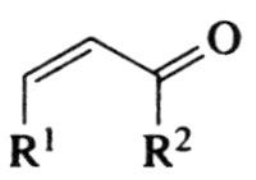	60–80	177

Table 10 *(continued)*

Entry	*Enamine*	*Reagent*	*Solvent*	*Product*	*Yield (%)*	*Ref.*
14	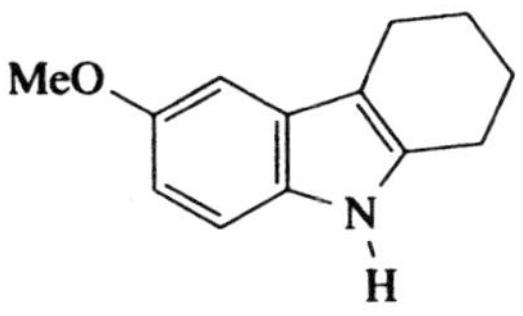 MeO, N, H	BH_3	CF_3CO_2H	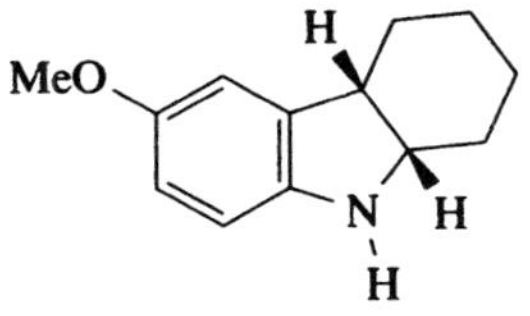 MeO, H, N, H, H	90	164
15	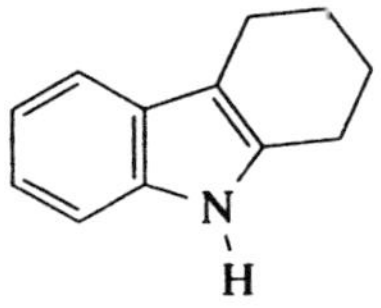 N, H	$Me_3N{\bullet}BH_3/HCl$	*p*-Dioxane	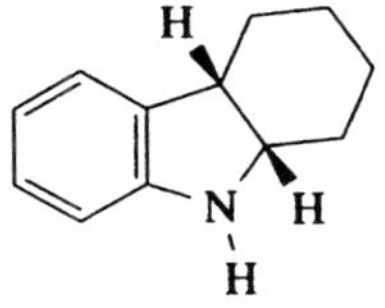 H, N, H, H	87	178

[a] *Cis/trans* = 9:1. [b] Plus 28% lactone side product.

1.2.3 REDUCTIONS OF *N*-HETEROATOM-SUBSTITUTED IMINES

1.2.3.1 Reduction of Oximes, Oxime Ethers and Oxime Esters to Hydroxylamines and Derivatives

1.2.3.1.1 Reduction of oximes to hydroxylamines

Aldoximes and ketoximes are reduced to hydroxylamines with an assortment of mild reducing reagents including BH_3/THF (at moderate temperatures),[181] BH_3·pyridine/HCl,[20,182,183] $BH_3·NMe_3$/7 M HCl,[20,183] $LiBH_4$,[184] $NaBH_4$ in basic[185] or acidic (*e.g.* carboxylic acid)[14] media and $NaBH_3CN$ with HCl[15,16,26] or AcOH.[20,186] The products obtained with the latter two reagents depend somewhat upon the reaction conditions. Thus, $NaBH_4$/AcOH at 25°C affords *N*-monoalkylhydroxylamines while *N,N*-dialkylhydroxylamines are obtained at higher temperatures *via* reduction of imines generated *in situ* from aldehydes (Scheme 7).[20,186] In addition, aldoximes may also give *N,N*-dialkylhydroxylamines *via* condensation of an initially produced hydroxylamine (**29**) with the aldoxime and subsequent reduction.[20,186] This latter side reaction is also observed in reductions with $NaBH_3CN/H^+$ at pH *ca.* 4 but not at pH 3 (Scheme 8).[16,26] Table 11 presents a selection of successful oxime reductions with various reagents, and illustrates the chemoselectivity toward acetals (entry 1), alkenes (entries 2, 3), esters (entries 4, 7) and nitro groups (entry 6) Reduction may be followed by cyclization to *N*-hydroxyamides, as is the case in entry 4.

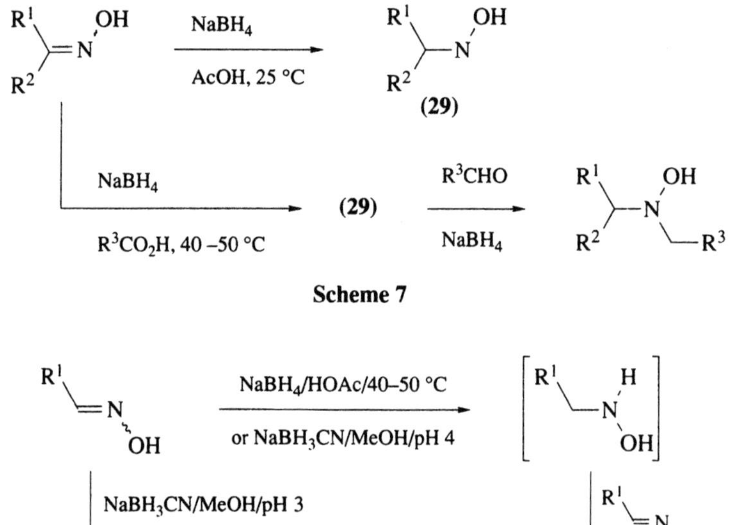

$NaBH_4$, AcOH, 25 °C; (**29**); $NaBH_4$, R^3CO_2H, 40–50 °C; (**29**); R^3CHO, $NaBH_4$

Scheme 7

$NaBH_4$/HOAc/40–50 °C or $NaBH_3CN$/MeOH/pH 4; $NaBH_3CN$/MeOH/pH 3; (**29**) R^2 = H; red

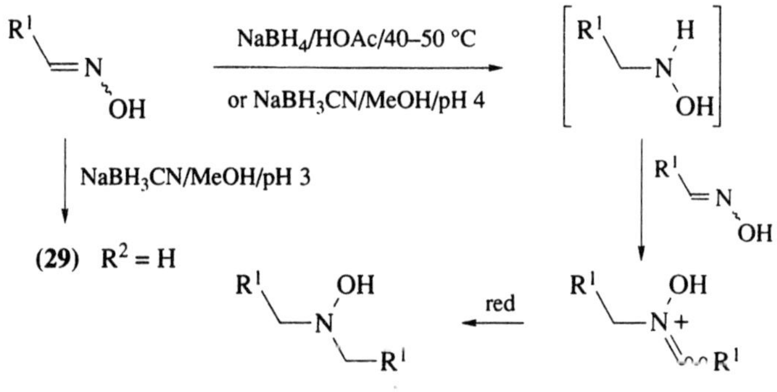

Scheme 8

1.2.3.1.2 Reduction of oxime ethers and esters to hydroxylamine ethers and esters

Many of the same reagents which reduce oximes to hydroxylamines are also effective for the conversion of *O*-alkyl, *O*-aryl and/or *O*-acyl oximes to the corresponding *O*-alkyl-, *O*-aryl- and *O*-acyl-hydroxylamines, respectively. Thus, BH_3,[181] BH_3·pyridine/HCl,[20,182,183] $BH_3·NMe_3$/HCl,[20,183] $NaBH_3CN$ with HCl[192,193] or AcOH,[194] all reduce *O*-alkyl (or *O*-aryl)[193] derivatives, a selection of which are presented in Table 12. Again, the chemoselectivity available with the reagents is displayed by the inertness of cyano, carbohydrate, amide and ester groups (entries 1, 2, 5 and 7, respectively). An interesting synthesis of *N*-alkoxyazetidines is illustrated (entry 4) in which the initially produced *O*-alkylhydroxylamine underwent an intramolecular substitution to give the four-membered ring.[194] Reduction of the oxime corre-

Table 11 Reduction of Oximes to Hydroxylamines

Entry	*Oxime*	*Reagent*	*Product*	*Yield (%)*	*Ref.*
1	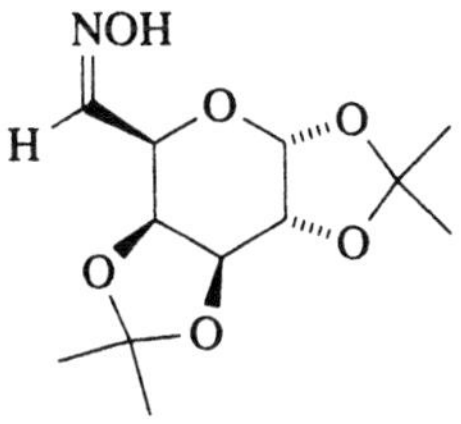	$NaBH_3CN/HCl$	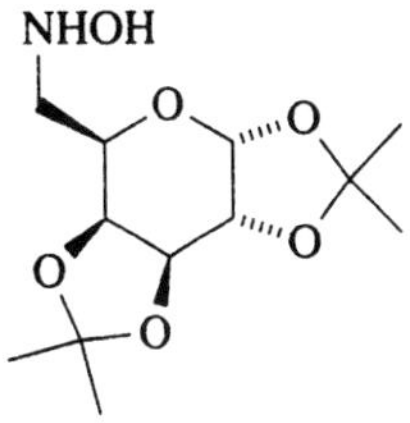	73	187
2	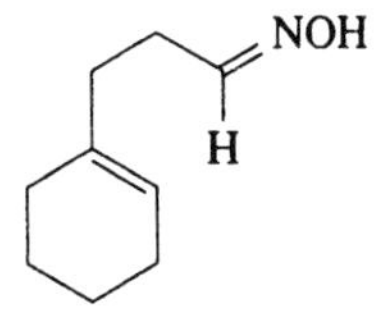	$NaBH_3CN/HCl$	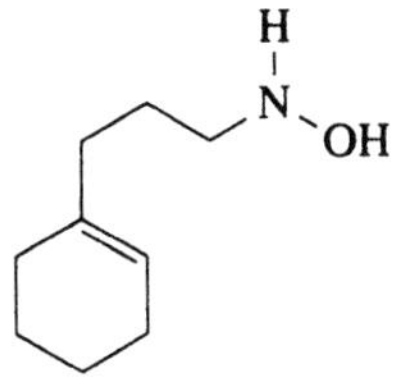	95	188
3	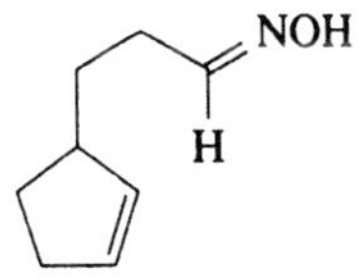	$NaBH_3CN/HCl$	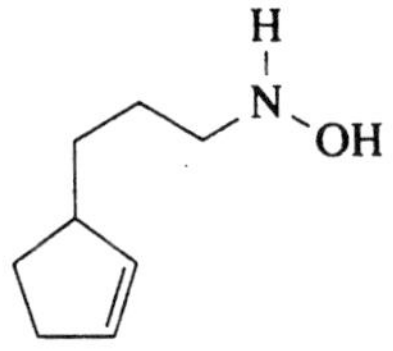	95	189
4	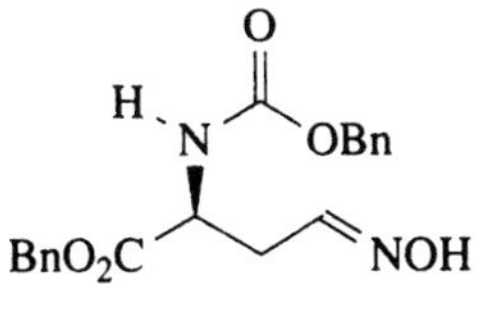	$NaBH_3CN/HCl$	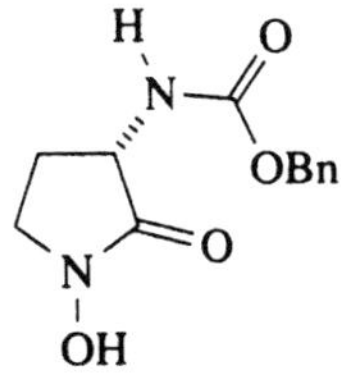	29	190
5	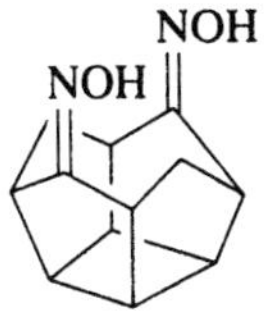	$NaBH_3CN/AcOH$	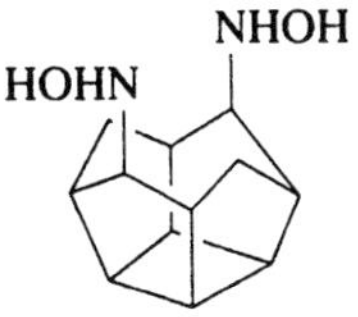	85	191
6	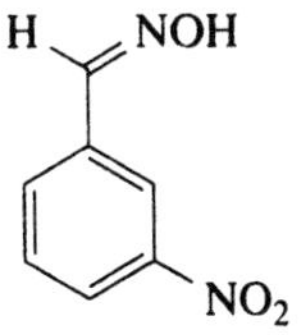	$Py•BH_3/HCl$	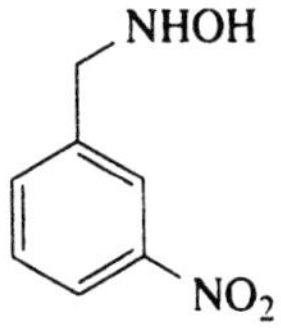	91	182
7	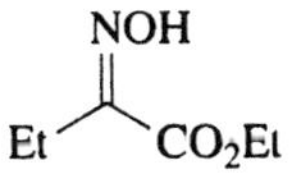	$Et_3N•BH_3/HCl$	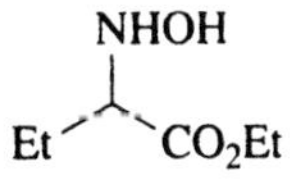	92	183
8	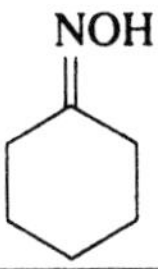	$LiBH_4$	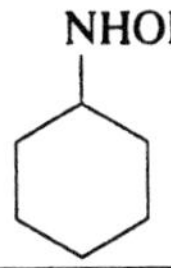	93	184[a]

[a] Ref. 186 reported 81% yield with $NaBH_3CN/AcOH$.

Table 12 Reduction of *O*-Alkyl, *O*-Aryl and *O*-Acyl Oximes to *O*-Substituted Hydroxylamines

Entry	*Oxime*	*Reagent*	*Product*	*Yield (%)*	*Ref.*
1	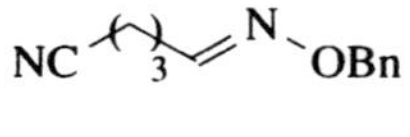 NC, 3, N, OBn	$NaBH_3CN/HCl$	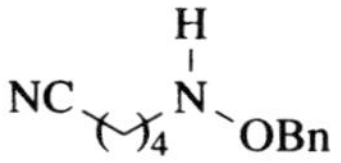 NC, 4, H, N, OBn	84	195
2	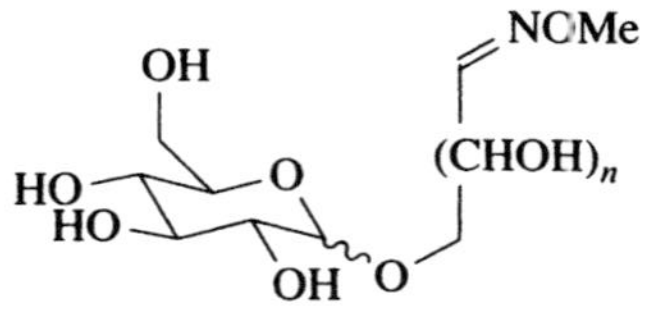 HO, HO, OH, O, OH, O, $(CHOH)_n$, NOMe	$NaBH_3CN/HCl$	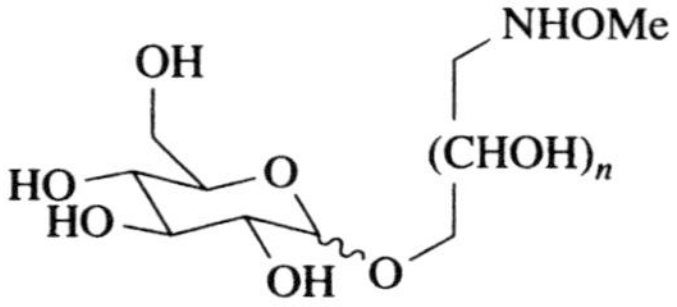 HO, HO, OH, O, OH, O, $(CHOH)_n$, NHOMe	—	196
3	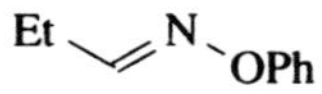 Et, N, OPh	$NaBH_3CN/HCl$	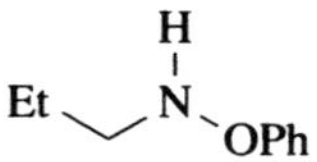 Et, H, N, OPh	—	193
4	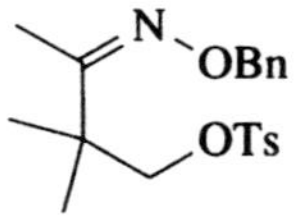 N, OBn, OTs	$NaBH_3CN/AcOH$	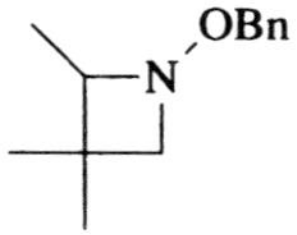 N, OBn	63	194
5	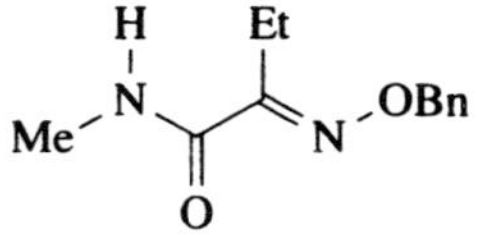 Me, H, N, O, Et, N, OBn	$Py{\bullet}BH_3/HCl$	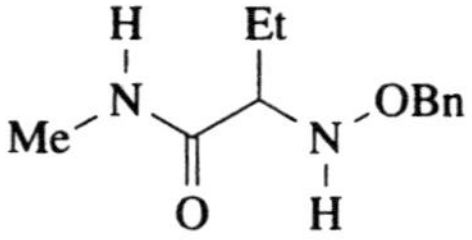 Me, H, N, O, Et, N, H, OBn	70	183
6	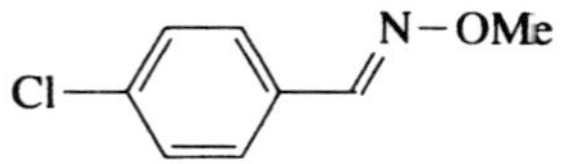 Cl, N–OMe	$Py{\bullet}BH_3/HCl$	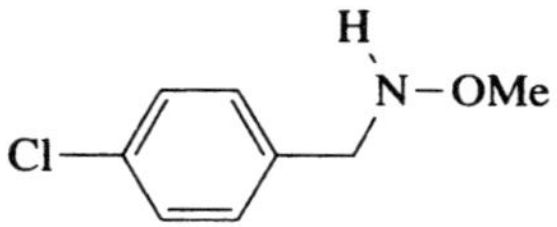 Cl, H, N–OMe	92	182
7	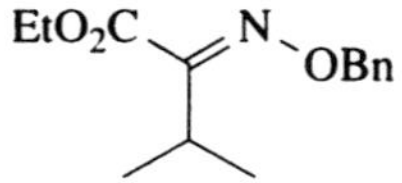 EtO_2C, N, OBn	$Me_3N{\bullet}BH_3/HCl$	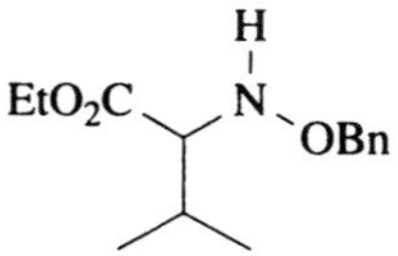 EtO_2C, H, N, OBn	82	183

Table 12 *(continued)*

Entry	*Oxime*	*Reagent*	*Product*	*Yield (%)*	*Ref.*
8	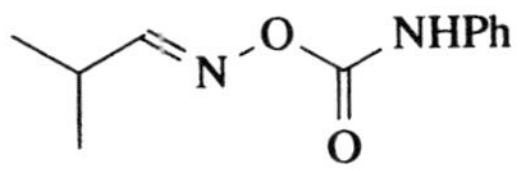 N O NHPh O	$NaBH_3CN/HCl$	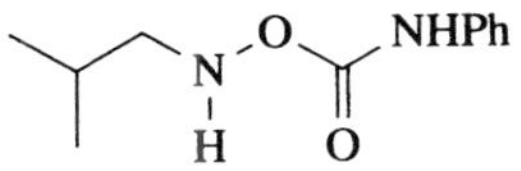 N O NHPh H O	—	197
9	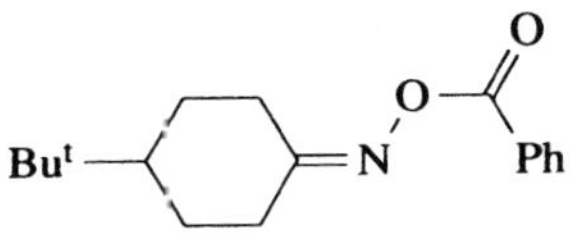 Bu[t] N O O Ph	$NaBH_3CN/AcOH$	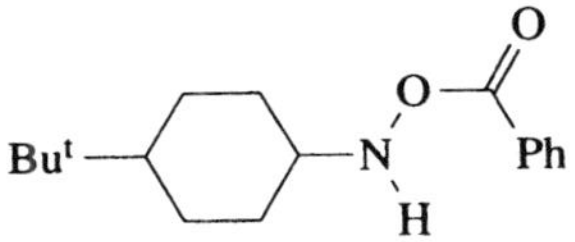 Bu[t] N O O Ph H	90[a]	198
10	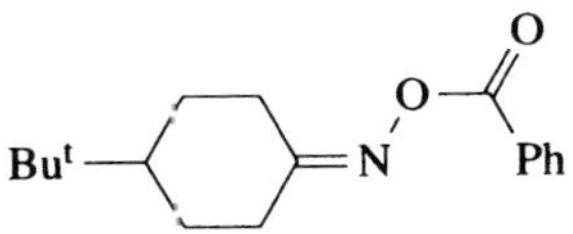 Bu[t] N O O Ph	Et_3SiH/TFA	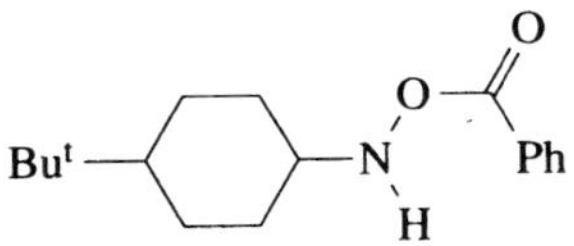 Bu[t] N O O Ph H	85[b]	198

[a] *Cis:trans* = 1:4. [b] *Cis:trans* = 5:1.

sponding to entry 4 afforded the five-membered isoxazolidine resulting from displacement of tosyl by the hydroxylamine oxygen.[194]

The reduction of *O*-acyl oximes to *O*-acylhydroxylamines also is accomplished under mild conditions with pyridine·BH_3/HCl,[182] $NaBH_3CN$/HCl,[197] and $NaBH_3CN$/AcOH[198] in a similar fashion as with other oxime derivatives. In addition, BH_3 (1 equiv.)[199] and the combination of Et_3SiH/TFA[198] also effects efficient reduction. With the former reagent, rearrangement to *N*-acylhydroxylamines has been observed (equation 3).[199] Representative examples of *O*-acyl oxime reductions are presented in Table 12. The related derivatives nitrones are also reduced by LAH,[200,201] or $NaBH_3CN/H^+$[202] to hydroxylamines (equation 4).[202]

$$\text{C}_6\text{H}_{13}\text{-oxime O-acetate} \xrightarrow[60\%]{BH_3} \text{N-acetyl-N-hydroxylamine (C}_6\text{H}_{13}\text{)} \quad (3)$$

$$\text{nitrone (MeO, MeO-aryl; N}^+\text{-O}^-\text{; Ph)} \xrightarrow[60\%]{NaBH_3CN/H^+} \text{N-OH hydroxylamine (Ph)} \quad (4)$$

1.2.3.2 Reduction of Oximes, Oxime Ethers and Oxime Esters to Amines

The reduction of oximes and subsequent cleavage of the N—O bond to afford primary amines occurs with a variety of potent hydride reagents including LAH,[7,9,201,203] $NaAlH_2(OCH_2CH_2OMe)_2$,[7,204] BH_3 (at 105–110 °C),[205] $NaBH_2S_3$,[206] and $NaBH_4$ in combination with a variety of transition metal salts including $TiCl_3$,[207a,b] (also with $NaBH_3CN$[207c]), $ZrCl_4$,[208] $NiCl_2$,[209] and MoO_3.[209] Several representative examples of oxime reductions to amines with a variety of reagents are presented in Table 13. Although thorough chemoselectivity studies with the transition metal salts/$NaBH_4$ systems have not been conducted, several functionalities do survive, including NO_2 (entry 3), cyclopropyl (entry 5), ketal, amide and certain heterocycles (entry 8). With $NiCl_2/NaBH_4$, conjugated oximes afford saturated amines (entry 6) while $MoO_3/NaBH_4$ gives mostly the allylic amine (entry 7). Alternate pathways are obtained with DIBAL-H which efficiently affords rearranged secondary amines as the only product (entry 9), presumably *via* Beckmann-type rearrangement and subsequent reduction of the resulting imine. This result also sometimes occurs as a side reaction with LAH. Ketoximes and aldoximines bearing a β-phenyl and ketoximes with an α-phenyl generally afford aziridines with LAH in ether solvents although yields are usually moderate (entry 10). Likewise, $NaAlH_2(OCH_2CH_2OMe)_2$ affords mixtures of amines and aziridines with oximes.[204]

Treatment of oximes with LAH in HMPA does not afford the expected amine products, but rather gives ketones from ketoximes and nitriles and/or aldehydes from aldoximes. These unusual results are attributed to the formation of imine intermediates which are hydrolyzed to carbonyls or undergo elimination to nitriles (which may be further reduced and hydrolyzed to aldehydes).[215]

Oxime ethers yield amines from reduction with BH_3 (25 °C),[205] $NaBH_4$ in combination with $ZrCl_4$[208] or CF_3CO_2H ($NaBH_3O_2CCF_3$),[216,217] LAH/NaOMe,[218] and $NaAlH_2(OCH_2CH_2OMe)_2$.[204] This latter reagent affords aziridines in addition to primary amines.[204] Likewise, *O*-acyl oximes afford amines upon reduction with BH_3,[205,219] or $NaBH_4$ in combination with $NiCl_2$ or MoO_3,[212] while oxime sulfonate esters undergo Beckmann rearrangement and reduction to secondary amines with DIBAL-H[220] in a similar fashion as observed with oximes (Table 13, entry 9).[213] A selection of reductions of *O*-substituted oximes is presented in Table 14.

1.2.3.3 Stereoselectivity of Reductions of Cyclic Oximes and Oxime Derivatives

Systematic investigations of the stereochemical outcomes of reductions of cyclic oximes and their derivatives have been sparse; the available information suggests that results sometimes parallel those obtained in other imine reductions (Sections 1.2.2.2.3, 1.2.2.3.3 and 1.2.2.4.2), but differences are also noted. A collection of examples in which the stereochemistry was determined is presented in Table 15. Thus, with cyclohexyl systems enhanced equatorial attack (compared to ketones) leading to axial amines

Table 13 Reduction of Oximes to Amines

Entry	*Oxime*	*Reagent*	*Product*	*Yield (%)*	*Ref.*
1	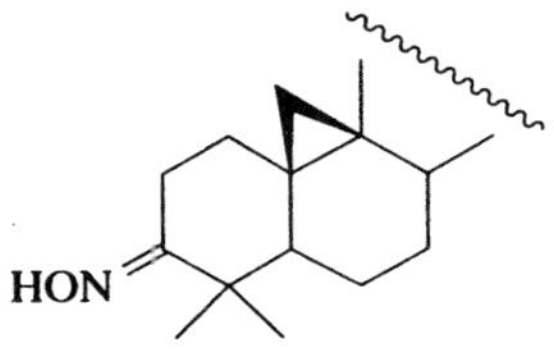 HON	$LiAlH_4$	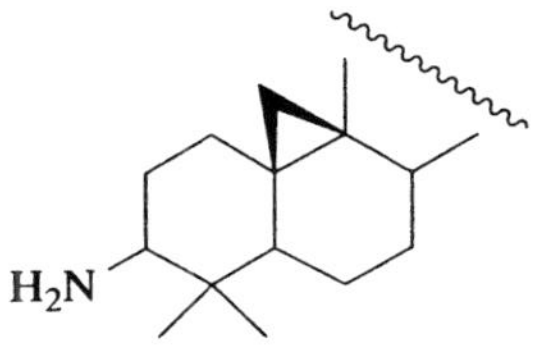 H_2N	—	210
2	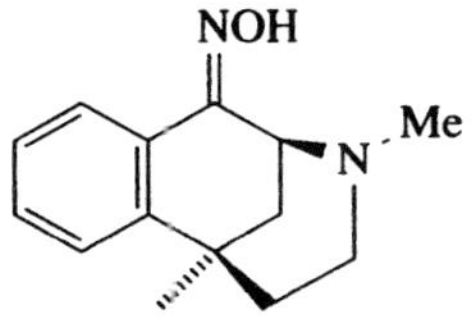 NOH, N, Me	$LiAlH_4$	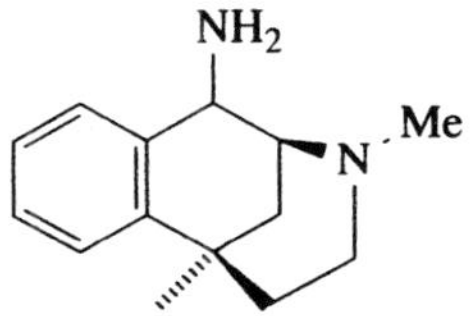 NH_2, N, Me	80	211
3	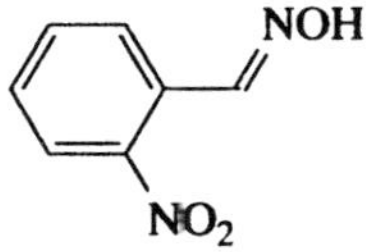 NOH, NO_2	$NaBH_4/TiCl_4$	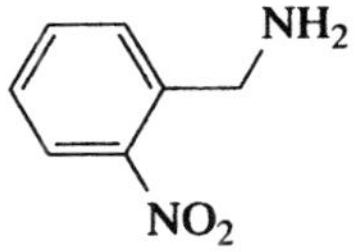 NH_2, NO_2	82	207
4	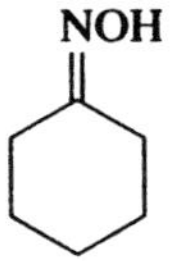 NOH	$NaBH_4/ZrCl_4$	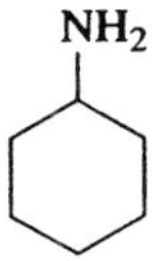 NH_2	85	208
5	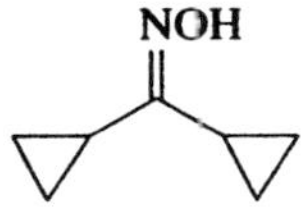 NOH	$NaBH_4/NiCl_2$	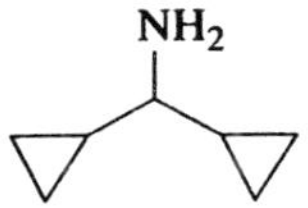 NH_2	70	209
6	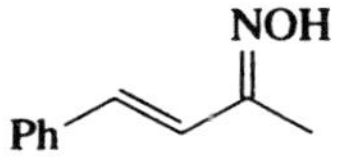 NOH, Ph	$NaBH_4/NiCl_2$	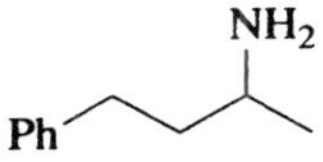 NH_2, Ph	95	209

Table 13 *(continued)*

Entry	*Oxime*	*Reagent*	*Product*	*Yield (%)*	*Ref.*
7		$NaBH_4/MoO_3$		66[a]	209
8		$NaBH_4/MoO_3$		60	212
9		Bu^i_2AlH		92	213
10		$LiAlH_4$		34	214

[a] *ca.* 29% of the saturated amine also obtained.

Table 14 Reduction of Oxime Derivatives to Amines

Entry	*Oxime derivative*	*Reagent*	*Product*	*Yield (%)*	*Ref.*
1	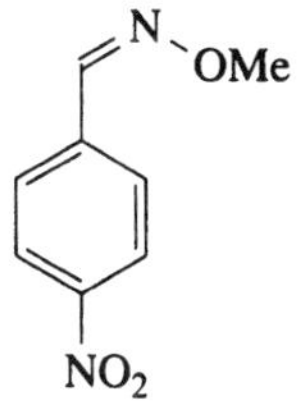 N, OMe, NO_2	BH_3/THF	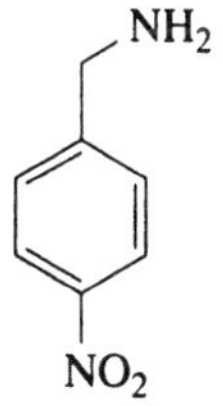 NH_2, NO_2	82	205
2	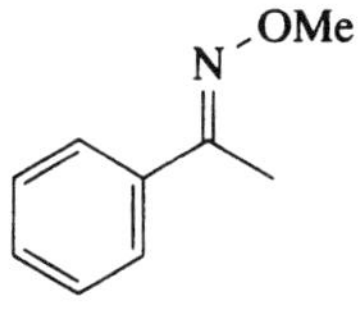 N, OMe	$NaBH_4/ZrCl_4$	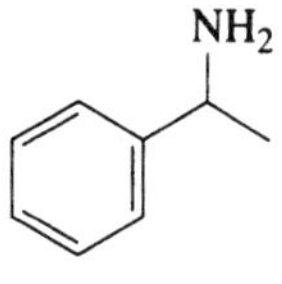 NH_2	95	208
3	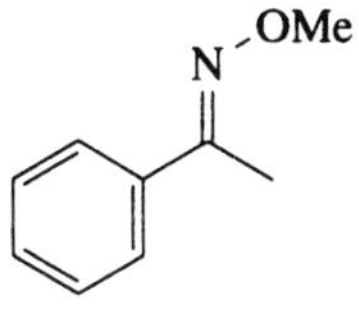 N, OMe	$NaBH_3O_2CCF_3$	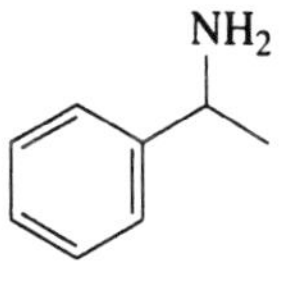 NH_2	90	216
4	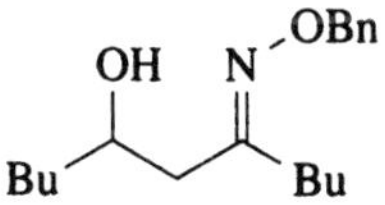 OH, N, OBn, Bu, Bu	$LiAlH_4$/NaOMe	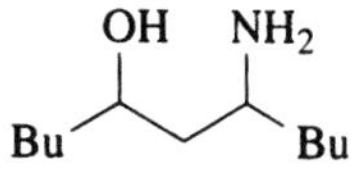 OH, NH_2, Bu, Bu	92[a]	218
5	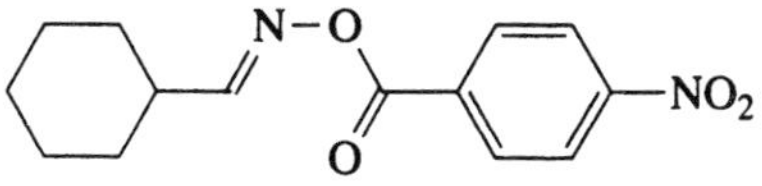 N–O, O, NO_2	BH_3/THF	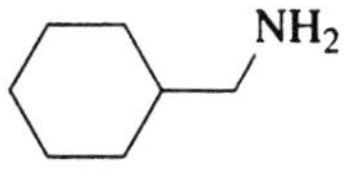 NH_2	68[b]	205
6	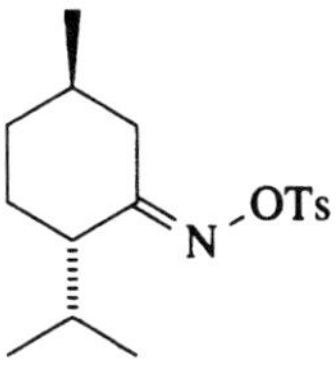 N, OTs	DIBAL-H/Et_2O	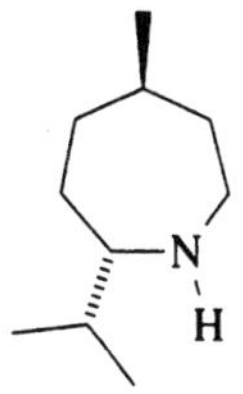 N, H	82	220
7	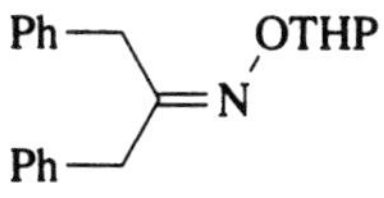 Ph, Ph, N, OTHP	$NaAlH_2[O(CH_2)_2OMe]$	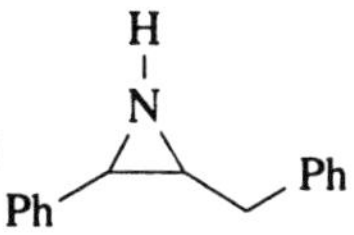 H, N, Ph, Ph	88[c]	204

[a] *Syn*/*anti* = 96:4. [b] Also, 81% *p*-nitrobenzyl alcohol. [c] Also, 12% primary amine.

(or derivatives) is observed with $NaBH_4/NiCl_2$ (entries 2–4)[209] while $NaBH_4/MoO_3$ gives predominatly axial attack (compare entries 4 and 5).[209] However, with more substituted rings this latter reagent gives almost entirely equatorial approach (*e.g.* entry 6; also entry 8, Table 13),[212] possibly induced by the *syn*-axial alkoxy. Since the actual reducing species with these transition metal modified reagents is not known, speculation as to the origin of these differences is not warranted without additional study. The combination of Et_3SiH/TFA also gives predominately axial amine derivatives, while $NaBH_3CN$/HOAc affords mostly (80%) axial attack to give the equatorial amine (entries 7, 8).[198] Unfortunately, as mentioned previously, the use of bulky trialkylborohydrides to obtain highly stereoselective reductions of oximes and oxime ethers (as occurs with imines, Section 1.2.2.2.3) is prevented by the inertness of these derivatives toward the reagents.[24,25]

Table 15 Stereochemistry of Reduction of Oximes and their Derivatives

Entry	*Oxime derivative*	*Reagent*	*Product*	*Ratio of isomers*	*Yield (%)*	*Ref.*
1	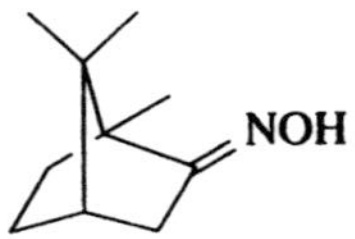 NOH	$NaBH_4/NiCl_2$	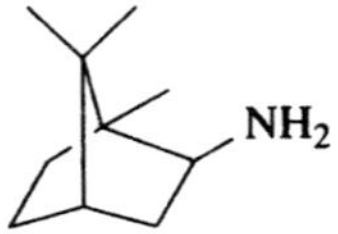 NH_2		92	209
2	 HON	$NaBH_4/NiCl_2$	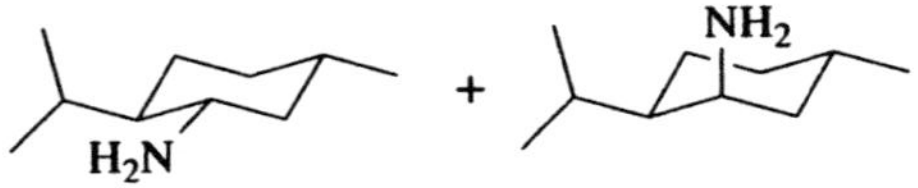 H_2N + NH_2	33:67	92	209
3	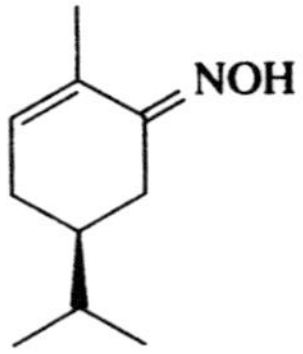 NOH	$NaBH_4/NiCl_2$	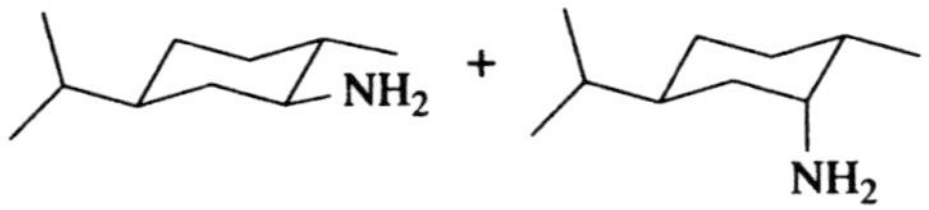 NH_2 + NH_2	50:50	90	209
4	 Bu^t NOH	$NaBH_4/NiCl_2$	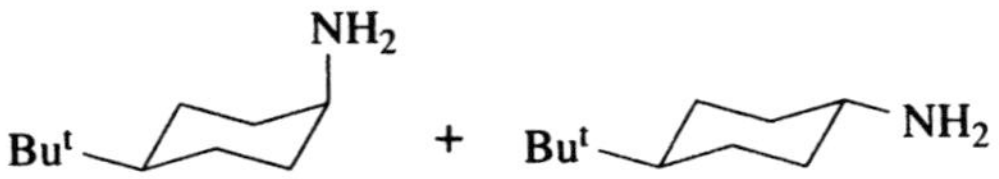 Bu^t NH_2 + Bu^t NH_2	62:38	95	209
5	 Bu^t NOH	$NaBH_4/MoO_3$	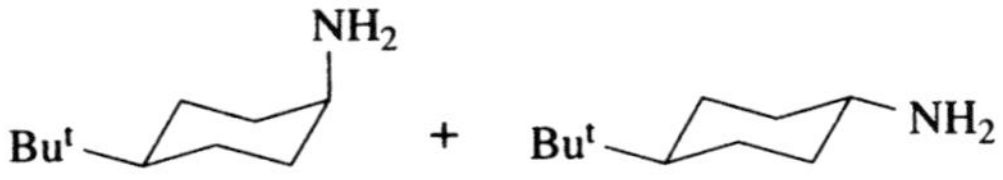 Bu^t NH_2 + Bu^t NH_2	25:75	87	209
6	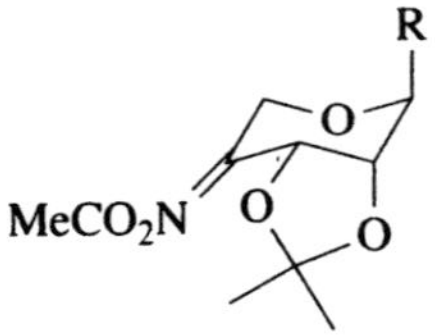 R O $MeCO_2N$ O O	$NaBH_4/MoO_3$	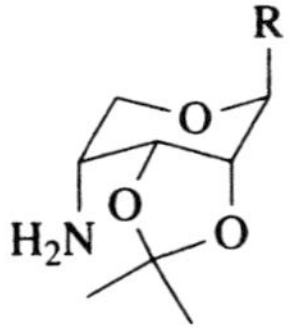 R O H_2N O O		70	212

Table 15 *(continued)*

Entry	*Oxime derivative*	*Reagent*	*Product*	*Ratio of isomers*	*Yield (%)*	*Ref.*
7	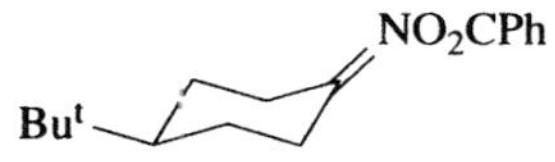 Bu^t … NO_2CPh	$NaBH_3CN/AcOH$	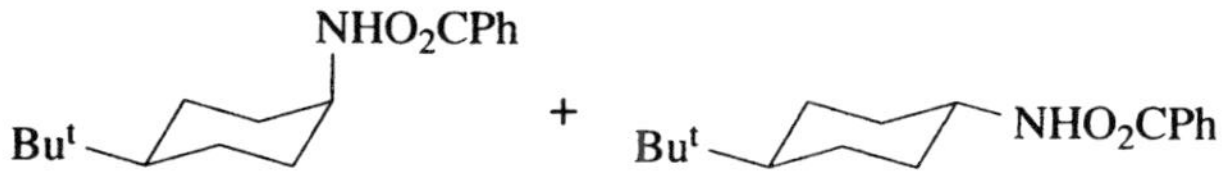 Bu^t … NHO_2CPh + Bu^t … NHO_2CPh	20:80	90	198
8	 Bu^t … NO_2CPh	Et_3SiH/TFA	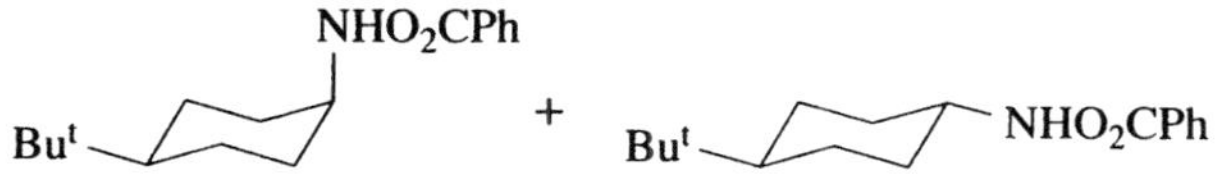 Bu^t … NHO_2CPh + Bu^t … NHO_2CPh	83:17	85	198
9	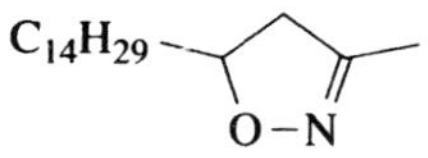 $C_{14}H_{29}$ … O–N	$LiAlH_4$	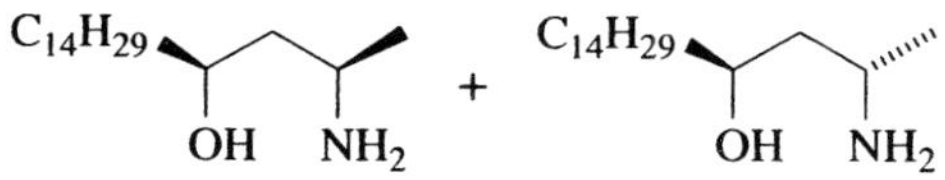 $C_{14}H_{29}$ … OH NH_2 + $C_{14}H_{29}$ … OH NH_2	88:12	95	221

Entry 9 in Table 15 illustrates another synthetically useful stereocontrolled reduction of cyclic oxime ethers (isoxazolines) to alicyclic amino alcohols using LAH. The stereoselectivity obtained is further enhanced by the incorporation of a 4α-hydroxy group which, upon reduction, affords almost entirely the *erythro* isomer (equation 5).[221–223] High diastereoselectivity in the reductive cleavage of isoxazolines has also been obtained using $LiBH_4$ (equation 6).[224]

LAH, 95%; *erythro* (5)

$LiBH_4$, 69%; + *erythro*, 95:5 (6)

1.2.3.4 Reduction of Hydrazone Derivatives to Hydrazine Derivatives

As expected, many of the same hydride reagents which convert other imine derivatives to the saturated analogs also reduce hydrazones, although extensive surveys have not been conducted. Successful reagents with varying types of hydrazones include LAH,[4,7,9,225] BH_3,[7,226] $NaBH_4$,[227,228] catechol borane,[229] pyridine·BH_3/HCl,[20,230] and $NaBH_3CN$/HCl.[16,231–233] Representative examples of hydrazone reductions are presented in Table 16 along with conversions of other *N*-heteroatom-substituted derivatives.

Probably the most synthetically important conversion in the hydrazone category involves reduction of *N*-arylsulfonyl hydrazones to arylsulfonyl hydrazine intermediates which undergo further decomposition to give hydrocarbons, a topic covered in Chapter 1.14, this volume. Under mild conditions (*e.g.* pyridine·BH_3/HCl,[20,230] $NaBH_3CN$/HCl[16,231,234]), the arylsulfonylhydrazines survive, providing preparative methods for these derivatives (entries 3, 4).

Aldehydes and ketones undergo *in situ* reductive amination with hydrazine (*via* hydrazone intermediates) using $NaBH_3CN$/HCl in an analogous fashion, as previously described for amines (Section 1.2.2.3), to afford alkyl-substituted hydrazines, and the process can be utilized to synthesize nitrogen heterocyclic rings with dialdehydes (entries 5, 6).[235] Reductions of hydrazones with $NaBH_3CN$ have also been used in the biochemical area to attach fluorescent chromophores to gangliosides.[233]

Reductions of hydrazones with LAH may, as expected, proceed further if carbonyls or other susceptible groups are present. Thus, *N*-acyl hydrazones (*e.g.* **30**) give *N*-alkylhydrazines along with the *N*-acyl derivative (equation 7),[236] while the phenylosazone (**31**) gave further rearrangements of the intermediate dihydrazine to afford (**32**; equation 8).[237]

(**30**) LAH, THF; 13–70% + 48–12% (7)

(**31**) $LiAlH_4$, dioxane, 50%; (**32**) (8)

Table 16 Reduction of Various *N*-Heterosubstituted Imines to *N*-Heterosubstituted Amines

Entry	*N-Heteroimine*	*Reagent*	*Product*	*Yield (%)*	*Ref.*
1	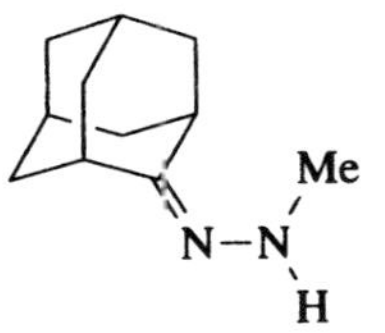 Me, N–N, H	$LiAlH_4/Et_2O$	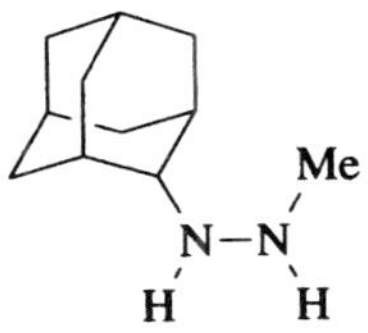 Me, N–N, H, H	80	225
2	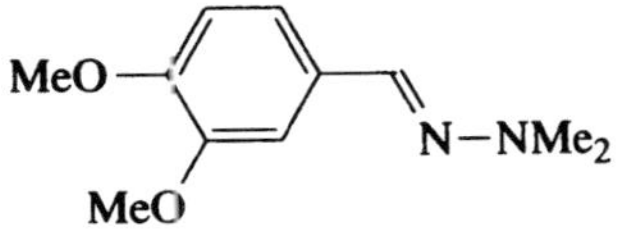 MeO, MeO, N–NMe_2	$NaBH_4$/MeOH	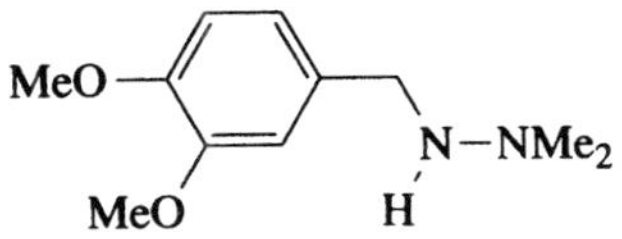 MeO, MeO, N–NMe_2, H	56	227
3	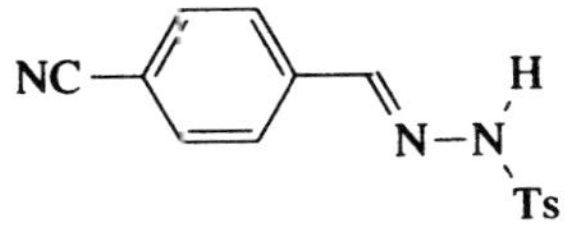 NC, H, N–N, Ts	Py•BH_3	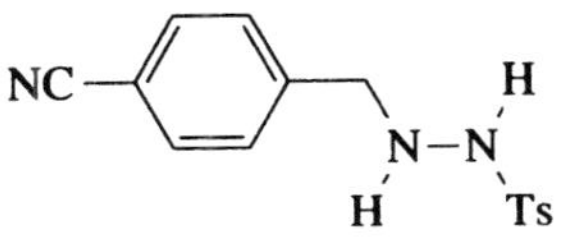 NC, H, N–N, H, Ts	94	230
4	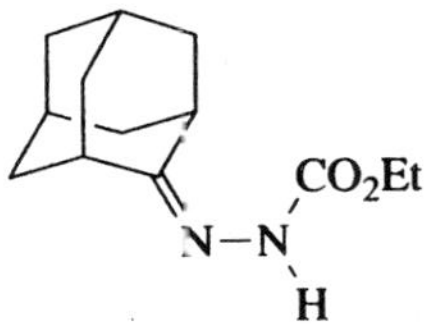 CO_2Et, N–N, H	$NaBH_3CN$/HCl	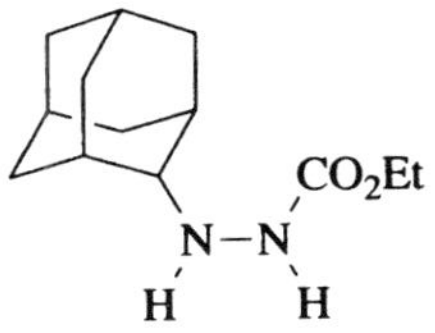 CO_2Et, N–N, H, H	83	225
5	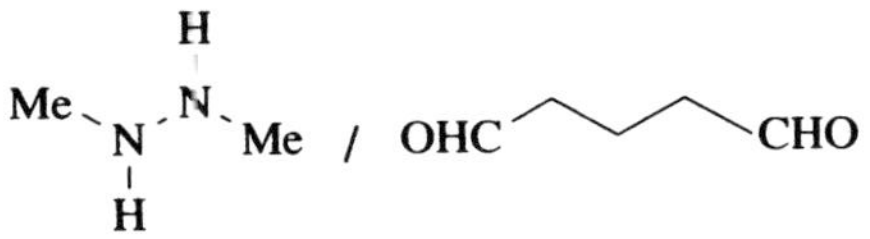 H, Me, N, N, Me, H / OHC, CHO	$NaBH_3CN$/HCl	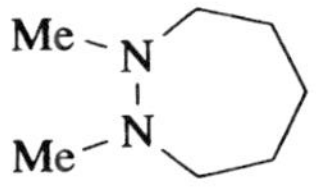 Me, N, N, Me	27	235
6	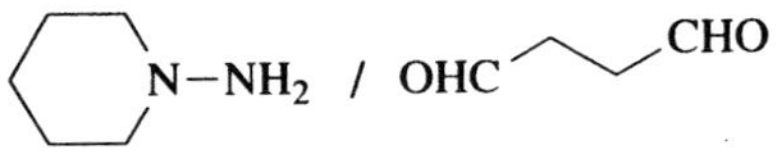 N–NH_2 / OHC, CHO	$NaBH_3CN$/HCl	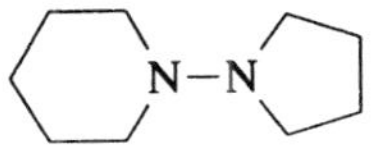 N–N	36	235

Table 16 *(continued)*

Entry	*N-Heteroimine*	*Reagent*	*Product*	*Yield (%)*	*Ref.*
7		$NaBH_3CN/HCl$		40	232
8		$NaBH_3CN/TFA$		91–96	238
9		$NaBH_4/EtOH$		90–100	239
10		$NaBH_4/MeOH$		82	240
11		$LiAlH_4$		—	241
12		$NaBH_4/THF$		65	242
13		$LiBHBu^s_3/THF$		93[a]	243

Table 16 *(continued)*

Entry	*N-Heteroimine*	*Reagent*	*Product*	*Yield (%)*	*Ref.*
14	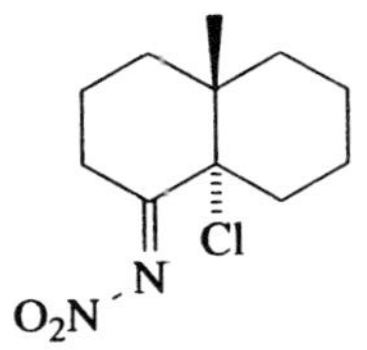 O_2N–N═, Cl	$NaBH_4$/AcOH	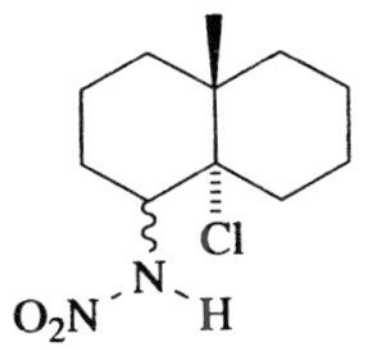 O_2N–N–H, Cl	70	247
15	 $(\,)_6$ NO_2^- K^+	$NaBH_4$/dioxane/EtOH/AcOH	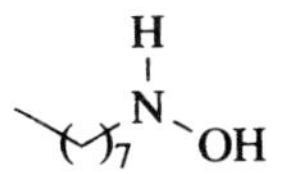 $(\,)_7$ N(H)–OH	32	248

[a] 99% *cis*.

1.2.3.5 Reduction of *N*-Sulfur-substituted Imines to *N*-Sulfur-substituted Amines

Relatively little investigation of the reduction of sulfur-substituted imines has appeared. Tritylsulfenimines are reduced with $NaBH_3CN/CF_3CO_2H$ to the corresponding tritylsulfenamides, but are resistant toward $NaBH_4$/THF/EtOH.[238] The more highly activated sulfinyl imines (i.e. *N*-alkylidenesulfinamides) are reduced by LAH/Et_2O (and various alkoxy-modified reagents) and $NaBH_4$/EtOH to sulfinylamides (*e.g.* Table 16, entry 9).[239] With optically active derivatives, significant diastereoselectivities (20–80%) are obtained in the reductions and are enhanced by reduction of racemic derivatives with optically active LAH-derived reagents.[239] More thorough examination of such asymmetric reductions are found in Chapter 1.7, this volume. The product sulfinamides may be cleaved with TFA to the corresponding primary amine or oxidized to sulfonamides.[239] Thc activated imine linkage in *N*-sulfonyl imines is reduced under mild conditions with $NaBH_4$/MeOH (entry 10)[240] or LAH (entry 11)[241] and probably many other reagents, although the range has not been explored. Entry 11 illustrates an unusual reductive removal of a sulfonyl imine with LAH.[241]

1.2.3.6 Reduction of *N*-Phosphorus-substituted Imines to *N*-Phosphorus-substituted Amines

The attachment of the strongly electronegative phosphinyl (P—O) group to an imine, usually *via* reaction of an oxime with a chlorophosphine,[242] also gives highly electrophilic imines which are reduced by $NaBH_4$/THF[242,243] and various modified borohydride and LAH derivatives[243–245] under mild conditions. The product *N*-phosphinylamines are protected forms of primary amines since removal of the phosphorus substituent is accomplished under mild acidic conditions.[242,246] Entries 12 and 13 (Table 16) present representative reductions and illustrate (entry 13) that highly diastereoselective reductions of cyclic systems to axial amine derivatives are accomplished with $LiBHBu^s_3$.[243] Enantioselective reductions of *N*-diphenylphosphinyl imines to optically active amine derivatives have also been reported (Chapter 1.7, this volume).[244,245]

1.2.3.7 Reduction of other *N*-Heteroatom-substituted Imines to *N*-Heteroatom-substituted Amines

A few other *N*-substituted imines have occasionally been reduced to amine derivatives. Thus, *N*-nitro imines (nitrimines) are reduced by $NaBH_4$/dioxane/EtOH/AcOH[247] to *N*-nitroamines (Table 16, entry 14), and nitronate salts are reduced by BH_3/THF to hydroxylamines (entry 15) *via* oximes.[248] Isocyanates are reduced by $LiBHEt_3$,[25] $LiAlH(OBu^t)_3$ (at low temperature),[249] or Ph_3SnH[250] to formamides, which may be hydrolyzed in acidic workup to amines,[25] while reduction with LAH affords *N*-methylamines.[251]

1.2.4 REFERENCES

1. A. E. Finholt, A. C. Bond, Jr. and H. I. Schlesinger, *J. Am. Chem. Soc.*, 1947, **69**, 1199.
2. H. I. Schlesinger, H. C. Brown and A. E. Finholt, *J. Am. Chem. Soc.*, 1953, **75**, 205.
3. W. G. Brown, *Org. React. (N.Y.)*, 1951, **6**, 469.
4. N. G. Gaylord, 'Reduction with Complex Metal Hydrides', Interscience, New York, 1956.
5. K. Harada, in 'The Chemistry of the Carbon–Nitrogen Double Bond,' ed. S. Patai, Wiley, Chichester, 1970, p. 255.
6. J. S. Pizey, 'Lithium Aluminum Hydride', Horwood, Chichester, 1977.
7. A. Hajos, 'Complex Hydrides', Elsevier, New York, 1979.
8. H. C. Brown, in 'Boranes in Organic Chemistry', Cornell University Press, Ithaca, 1972, p. 209.
9. M. Hudlicky, in 'Reductions in Organic Chemistry', Wiley, New York, 1984, p. 132.
10. L. F. Fieser and M. Fieser, 'Reagents for Organic Synthesis', Wiley-Interscience, New York. 1967–1986, vols. 1–13.
11. A. Pelter, K. Smith and H. C. Brown, in 'Borane Reagents', Academic Press, New York, 1988, pp. 45, 125.
12. A. Pelter and K. Smith, in 'Comprehensive Organic Chemistry', ed. D. H. R. Barton and W. D. Ollis, Pergamon Press, Oxford, 1979, vol 3, p. 689.
13. H. C. Brown, in 'Hydroboration', Benjamin, New York, 1963, p. 238.
14. G. W. Gribble and C. F. Nutaitis, *Org. Prep. Proced. Int.*, 1985, **17**, 317.
15. C. F. Lane, *Synthesis*, 1975, 135.
16. R. O. Hutchins and N. R. Natale, *Org. Prep. Proced. Int.*, 1979, **11**, 201.
17. C. F. Lane, *Chem. Rev.*, 1976, **76**, 773.
18. H. C. Brown and S. Krishnamurthy, *Tetrahedron*, 1979, **35**, 567.
19. R. O. Hutchins and F. Cistone, *Org. Prep. Proced. Int.*, 1981, **13**, 225.
20. R. O. Hutchins, K. Learn, B. Nazer, D. Pytlewski and A. Pelter, *Org. Prep. Proced. Int.*, 1984, **16**, 335.

21. R. W. Layer, *Chem. Rev.*, 1963, **63**, 489.
22. K. Harada, in 'The Chemistry of the Carbon–Nitrogen Double Bond,' ed. S. Patai, Wiley, Chichester, 1970, p. 255.
23. H. C. Brown and S. Krishnamurthy, *J. Am. Chem. Soc.*, 1972, **94**, 7159.
24. R. O. Hutchins and W.-Y. Su, *Tetrahedron Lett.*, 1984, **25**, 695.
25. H. C. Brown, S.-C. Kim and S. Krishnamurthy, *J. Org. Chem.*, 1980, **45**, 1.
26. R. F. Borch, M. D. Bernstein and H. D. Durst, *J. Am. Chem. Soc.*, 1971, **93**, 2897.
27. J. R. Boone and E. C. Ashby, *Top. Stereochem.*, 1979, **11**, 53, and cited references.
28. R. O. Hutchins, W.-Y. Su, R. Sivakumar, F. Cistone and Y. P. Stercho, *J. Org. Chem.*, 1983, **48**, 3412.
29. D. C. Wigfield, *Tetrahedron*, 1979, **35**, 449, and cited references.
30. J. Tadanier, R. Hallas, J. R. Martin and R. S. Stanaszek, *Tetrahedron*, 1981, **37**, 1309.
31. H. Kapnang, G. Charles, B. L. Sondengam and J. H. Hemo, *Tetrahedron Lett.*, 1977, 3469.
32. A. Basha, J. Orlando and S. M. Weinreb, *Synth. Commun.*, 1977, **7**, 549.
33. T. Nishiwaki and F. Fujiyama, *Synthesis*, 1972, 569.
34. T. J. Ward, G. J. Warrellow, J. A. Stirrup, N. Lattimer and K. F. Rhodes, *J. Med. Chem.*, 1989, **32**, 179.
35. R. V. Hoffman and G. A. Buntain, *J. Org. Chem.*, 1988, **53**, 3316.
36. D. Stephan, A. Gorgues and A. Le Coq, *Tetrahedron Lett.*, 1988, **29**, 1025.
37. J. R. Carson, H. R. Almond, M. D. Brannan, R. J. Carmosin, S. F. Flaim, A. Gill, M. M. Gleason, S. L. Keely, D. W. Ludovici, P. M. Pitis, M. C. Rebarchak and F. J. Villani, *J. Med. Chem.*, 1988, **31**, 630.
38. O. Hoshino, S. Sawaki, N. Shimamura, A. Onodera and B. Umezawa, *Chem. Pharm. Bull.*, 1987, **35**, 2734.
39. A. Stütz and G. Petranyi, *J. Med. Chem.*, 1984, **27**, 1539.
40. C. M. Tice and B. Ganem, *J. Org. Chem.*, 1983, **48**, 5048.
41. C. M. Tice and B. Ganem, *J. Org. Chem.*, 1983, **48**, 5043.
42. J. D. Albright, V. G. DeVries, E. E. Largis, T. G. Miner, M. F. Reich, S. A. Schaffer, R. G. Shepherd and J. Upeslacis, *J. Med. Chem.*, 1983, **26**, 1378.
43. J. Ashby and E. M. Ramage, *J. Heterocycl. Chem.*, 1978, **15**, 1501.
44. M. Gall and B. V. Kamdar, *J. Org. Chem.*, 1981, **46**, 1575.
45. P. Marchini, G. Liso, A. Reho, F. Liberatore and F. M. Moracci, *J. Org. Chem.*, 1975, **40**, 3453.
46. R. N. Henrie, II, R. A. Lazarus and S. J. Benkovic, *J. Med. Chem.*, 1983, **26**, 559.
47. N. Aimi, E. Yamanaka, J. Endo, S. Sakai and J. Haginiwa, *Tetrahedron*, 1973, **29**, 2015.
48. G. W. Gribble, P. D. Lord, J. Skotnicki, S. E. Dietz, J. T. Eaton and J. L. Johnson, *J. Am. Chem. Soc.*, 1974, **96**, 7812.
49. A. Bachrach and A. Zilkha, *Eur. Polym. J.*, 1984, **20**, 493.
50. T. D. Heath, B. A. Macher and D. Papahadjopoulos, *Biochim. Biophys. Acta*, 1981, **640**, 66.
51. R. H. Pscheidt and B. D. Wells, *J. Biol. Chem.*, 1986, **261**, 7253.
52. R. L. Proia, S. K. Wray, D. A. Hart and L. Eidels, *J. Biol. Chem.*, 1980, **255**, 12 025.
53. J. Bosch, A. Domingo and A. Linares, *J. Org. Chem.*, 1983, **48**, 1075.
54. R. G. Smith, R. A. Lucas and J. W. F. Wasley, *J. Med. Chem.*, 1980, **23**, 952.
55. H. H. Wasserman and H. Matsuyama, *J. Am. Chem. Soc.*, 1981, **103**, 461.
56. K. Kimura, S. Nakayama, N. Miyata, Y. Takeshita and G. Kawanishi, *J. Antibiot.*, 1988, **41**, 411.
57. S. W. Garry and D. G. Neilson, *J. Chem. Soc., Perkin Trans. 1*, 1987, 601.
58. E. O. M. Orlemans, A. H. Schreuder, P. G. M. Conti, W. Verboom and D. N. Reinhoudt, *Tetrahedron*, 1987, **43**, 3817.
59. D. M. Ryckman and R. V. Stevens, *J. Org. Chem.*, 1987, **52**, 4274.
60. M. Gumulka, I. H. Ibrahim, Z. Bończa-Tomaszewski and C. R. Engel, *Can J. Chem.*, 1985, **63**, 766.
61. L. E. Overman, L. T. Mendelson and E. J. Jacobsen, *J. Am. Chem. Soc.*, 1983, **105**, 6629.
62. W. K. Hagmann, F. Z. Basha, M. Hashimoto, R. B. Frye, S. Kojo and S. M. Hecht, *J. Org. Chem.*, 1981, **46**, 1413.
63. M. A. Smith, B. Weinstein and F. D. Greene, *J. Org. Chem.*, 1980, **45**, 4597.
64. G. Berger, C. Prenant, J. Sastre, A. Syrota and D. Comar, *Int. J. Appl. Radiat. Isot.*, 1983, **34**, 1556.
65. D. Bacos, J. P. Célérier, E. Marx, C. Saliou and G. Lhommet, *Tetrahedron Lett.*, 1989, **30**, 1081.
66. J. Slade, J. L. Stanton, D. Ben-David and G. C. Mazzenga, *J. Med. Chem.*, 1985, **28**, 1517.
67. R. B. Silverman and R. B. Yamasaki, *Biochemistry*, 1984, **23**, 1322.
68. N. Suzuki and W. A. Wood, *J. Biol. Chem.*, 1980, **255**, 3427.
69. C. Hountondji, G. Fayat and S. Blanquet, *Eur. J. Biochem.*, 1979, **102**, 247.
70. C. Hountondji, G. Fayat and S. Blanquet, *Eur. J. Biochem.*, 1980, **107**, 403.
71. J. R. Charles, H. Christol and G. Solladie, *Bull. Soc. Chim. Fr.*, 1970, 4439
72. D. A. Evans, A. M. Golob, N. S. Mandel and G. S. Mandel, *J. Am. Chem. Soc.*, 1978, **100**, 8170.
73. N. De Kimpe, R. Verhé, L. De Buyck and N. Schamp, *J. Org. Chem.*, 1981, **46**, 2079.
74. B. Alcaide, F. de la Pradilla, C. López-Mardomingo, R. Pérez-Ossorio and J. Plumet, *J. Org. Chem.*, 1981, **46**, 3234.
75. B. E. Maryanoff, D. F. McComsey, R. J. Taylor, Jr. and J. F. Gardocki, *J. Med. Chem.*, 1981, **24**, 79.
76. C. L. Bumgardner, E. L. Lawton and J. G. Carver, *J. Org. Chem.*, 1972, **37**, 407.
77. W. P. Neuman, *Justus Liebigs Ann. Chem.*, 1963, **667**, 1963, and cited references.
78. J. E. Wrobel and B. Ganem, *Tetrahedron Lett.*, 1981, **22**, 3447.
79. P. M. Pojer, *Aust. J. Chem.*, 1979, **32**, 201.
80. G. P. Boldrini, M. Panunzio and A. Umani-Ronchi, *Synthesis*, 1974, 733.
81. Y. Watanabe, M. Yamashita, T. Mitsudo, M. Tanaka and Y. Takegami, *Tetrahedron Lett.*, 1974, 1879.
82. H. Alper, *J. Org. Chem.*, 1972, **37**, 3972.
83. N. M. Loim, *Izv. Akad. Nauk SSSR, Ser. Khim.*, 1968, 1418 (*Chem. Abstr.*, 1968, **69**, 96 127v).
84. I. Ojima, T. Kogure and Y. Nagai, *Tetrahedron Lett.*, 1973, 2475.
85. K. Fujimori, H. Yoshimoto and S. Oae, *Tetrahedron Lett.*, 1980, **21**, 3385.
86. J. H. Billman and J. W. McDowell, *J. Org. Chem.*, 1961, **26**, 1437.

87. J. H. Billman and J. W. McDowell, *J. Org. Chem.*, 1962, **27**, 2640.
88. A. P. Marchand, B. E. Arney, Jr., P. R. Dave, N. Satyanarayana, W. H. Watson and A. Nagl, *J. Org. Chem.*, 1988, **53**, 2644.
89. A. G. Cook and C. R. Schulz, *J. Org. Chem.*, 1967, **32**, 473.
90. R. M. Scribner, *J. Org. Chem.*, 1965, **30**, 3203.
91. D. Cabaret, G. Chauvière and Z. Welvart, *Tetrahedron Lett.*, 1966, 4109.
92. C. O. Meese and T. Ebner, *J. Labelled Comp. Radiopharm.*, 1988, **25**, 335.
93. G. McGillivray and E. Smal, *S. Afr. J. Chem.*, 1983, **36**, 93.
94. M. Van Parys and M. Vandewalle, *Bull. Soc. Chim. Belg.*, 1981, **90**, 757.
95. T. Fukuyama, R. K. Frank and C. F. Jewell, Jr., *J. Am. Chem. Soc.*, 1980, **102**, 2122.
96. M. E. Kuehne and P. J. Shannon, *J. Org. Chem.*, 1977, **42**, 2082.
97. M. Cushman and W. C. Wong, *J. Org. Chem.*, 1984, **49**, 1278.
98. R. D. Clark and Jahangir, *J. Org. Chem.*, 1989, **54**, 1174.
99. R. J. Sundberg, C. P. Walters and J. D. Bloom, *J. Org. Chem.*, 1981, **46**, 3730.
100. Y. Tsuda, T. Sano and H. Watanabe, *Synthesis*, 1977, 652.
101. R. F. Borch, *Tetrahedron Lett.*, 1968, 61.
102. E. R. H. Walker, *Chem. Soc. Rev.*, 1976, **5**, 23.
103. R. O. Hutchins, C. A. Milewski and B. E. Maryanoff, *J. Am. Chem. Soc.*, 1973, **95**, 3662.
104. E. L. Eliel and Y. Senda, *Tetrahedron*, 1970, **26**, 2411.
105. A. P. Kozikowski and R. J. Schmiesing, *J. Org. Chem.*, 1983, **48**, 1000.
106. F. Johnson, *Chem. Rev.*, 1968, **68**, 375.
107. P. T. Lansbury and R. E. MacLeay, *J. Org. Chem.*, 1963, **28**, 1940.
108. K. Maruoka, T. Miyazaki, M. Ando, Y. Matsumura, S. Sakane, K. Hattori and H. Yamamoto, *J. Am. Chem. Soc.*, 1983, **105**, 2831.
109. P. D. Sattsangi, J. R. Barrio and N. J. Leonard, *J. Am. Chem. Soc.*, 1980, **102**, 770.
110. M. C. Desai, J. Singh, H. P. S. Chawla and S. Dev, *Tetrahedron*, 1981, **37**, 2935.
111. Y. Ohfune, M. Tomita and K. Nomoto, *J. Am. Chem. Soc.*, 1981, **103**, 2409.
112. F. Sugawara and H. Kuzuhara, *Agric. Biol. Chem.*, 1981, **45**, 301.
113. M. Wenzel, P. Asindraza and G. Schachschneider, *J. Labelled Comp. Radiopharm.*, 1983, **20**, 1061.
114. J. D. McChesney and S. Sarangan, *J. Labelled Comp. Radiopharm.*, 1984, **21**, 293.
115. B. Umezawa, O. Hoshino, S. Sawaki, H. Sashida, K. Mori, Y. Hamada, K. Kotera and Y. Iitaka, *Tetrahedron*, 1984, **40**, 1783.
116. K. Mori, T. Sugai, Y. Maeda, T. Okazaki, T. Noguchi and H. Naito, *Tetrahedron*, 1985, **41**, 5307.
117. E. D. Thorsett, E. E. Harris, S. D. Aster, E. R. Peterson, J. P. Snyder, J. P. Springer, J. Hirshfield, E. W. Tristram, A. A. Patchett, E. H. Ulm and T. C. Vassil, *J. Med. Chem.*, 1986, **29**, 251.
118. P. C. Unangst, D. T. Conner and S. R. Stabler, *J. Heterocycl. Chem.*, 1987, **24**, 817.
119. O. Hoshino, M. Ishizaki, S. Sawaki, M. Yuasa and B. Umezawa, *Chem. Pharm. Bull.*, 1988, **36**, 3373.
120. T. M. Zydowsky, J. F. Dellaria, Jr. and H. N. Nellans, *J. Org. Chem.*, 1988, **53**, 5607.
121. D. A. Westerberg, P. L. Carney, P. E. Rogers, S. J. Kline and D. K. Johnson, *J. Med. Chem.*, 1989, **32**, 236.
122. G. Sosnovsky and B. D. Paul, *J. Med. Chem.*, 1984, **27**, 782.
123. T. H. Jones, J. B. Franko, M. S. Blum and H. M. Fales, *Tetrahedron Lett.*, 1980, **21**, 789.
124. H. A. Kirst, J. E. Toth, M. Debono, K. E. Willard, B. A. Truedell, J. L. Ott, F. T. Counter, A. M. Felty-Duckworth and R. S. Pekarek, *J. Med. Chem.*, 1988, **31**, 1631.
125. R. C. Davis, S. L. Ditson, A. F. Fentiman and R. M. Pearlstein, *J. Am. Chem. Soc.*, 1981, **103**, 6823.
126. R. Chicheportiche, M. Balerna, A. Lombet, G. Romey and M. Lazdunski, *Eur. J. Biochem.*, 1980, **104**, 617.
127. R. F. Borch and A. I. Hassid, *J. Org. Chem.*, 1972, **37**, 1673.
128. M. Maziere, J. M. Godot, G. Berger, C. Prenant and D. Comar, *J. Radioanal. Chem.*, 1981, **62**, 279.
129. J. B. Hester, Jr., A. D. Rudzik and P. F. VonVoigtlander, *J. Med. Chem.*, 1980, **23**, 392.
130. J. G. Cannon, T. Lee, F. Hsu, J. P. Long and J. R. Flynn, *J. Med. Chem.*, 1980, **23**, 502.
131. T. V. Hung, B. A. Mooney, R. H. Prager and J. M. Tippett, *Aust. J. Chem.*, 1981, **34**, 383.
132. J. F. Van Der Werf, W. Vaalburg, J. Korf, T. Kuilman and T. Wiegman, *Int. J. Appl. Radiat. Isot.*, 1984, **35**, 377.
133. E. J. Jacobsen, J. Levin and L. E. Overman, *J. Am. Chem. Soc.*, 1988, **110**, 4329.
134. M. A. Bernstein and L. D. Hall, *Carbohydr. Res.*, 1980, **78**, C1.
135. J. R. Shainoff, *Biochem. Biophys. Res. Commun.*, 1980, **95**, 690.
136. G. J. Murray and D. M. Neville, Jr., *J. Biol. Chem.*, 1980, **255**, 11 942.
137. P. Monsan, D. Combes and I. Alemzadeh, *Biotechnol. Bioeng.*, 1984, **26**, 658.
138. S. Rogozinski, O. O. Blumenfeld and S. Seifter, *Arch. Biochem. Biophys.*, 1983, **221**, 428.
139. S. J. Danielson and G. R. Gray, *Glycoconjugate*, 1986, **3**, 363.
140. A. J. Muller and P. W. Carr, *J. Chromatogr.*, 1984, **284**, 33.
141. M. Pugniere, C. San Juan, M.-A. Coletti-Previero and A. Previero, *Biosci. Rep.*, 1988, **8**, 263.
142. M. C. Summers and R. J. Hayes, *J. Biol. Chem.*, 1981, **256**, 4951.
143. F. Matsuura and A. Imaoka, *Glycoconjugate*, 1988, **5**, 13.
144. W. J. Roberts, E. Hubert, A. Iriarte and M. Martinez-Carrion, *J. Biol. Chem.*, 1988, **263**, 7196.
145. G. Berger, M. Maziere, C. Prenant, J. Sastre and D. Comar, *Int. J. Appl. Radiat. Isot.*, 1984, **35**, 81.
146. R. A. Cox and P. Greenwell, *Biochem. J.*, 1980, **186**, 861.
147. K. G. Gloggler, K. Balasubramanian, A. Beth, T. M. Fritzche, J. H. Park, D. E. Pearson, W. E. Trommer and S. D. Venkataramu, *Biochim. Biophys. Acta*, 1982, **701**, 224.
148. H. H. Baer and B. Radatus, *Carbohydr. Res.*, 1986, **157**, 65.
149. A. P. Marchand, P. R. Dave, N. Satyanarayana and B. E. Arney, Jr., *J. Org. Chem.*, 1988, **53**, 1088.
150. R. M. Giuliano and S. Kasperowicz, *Carbohydr. Res.*, 1988, **183**, 277.
151. J. M. Pisano and R. A. Firestone, *Synth. Commun.*, 1981, **11**, 375.
152. S. Kim, C. H. Oh, J. S. Ko, K. H. Ahn and Y. J. Kim, *J. Org. Chem.*, 1985, **50**, 1927.

153. R. J. Mattson, K. M. Pham, D. J. Leuck and K. A. Cowen, *J. Org. Chem.*, 1990, **55**, 2552.
154. M. Renaud, F. Fasiolo, M. Baltzinger, Y. Boulanger and P. Remy, *Eur. J. Biochem.*, 1982, **123**, 267.
155. R. O. Hutchins and M. Markowitz, *J. Org. Chem.*, 1981, **46**, 3571.
156. H. R. Morales, M. Pérez-Juárez, L. Cuéllar, L. Mendoza, H. Fernández and R. Contreras, *Synth. Commun.*, 1984, **14**, 1213.
157. A. Pelter, R. M. Rosser and S. Mills, *J. Chem. Soc., Perkin Trans. 1*, 1984, 717.
158. W. S. D. Wong, D. T. Osuga and R. E. Feeney, *Anal. Biochem.*, 1984, **139**, 58.
159. U.-H. Dolling, A. W. Douglas, E. J. J. Grabowski, E. F. Schoenewaldt, P. Sohar and M. Sletzinger, *J. Org. Chem.*, 1978, **43**, 1634.
160. A. F. Abdel-Magid, C. A. Maryanoff and K. G. Carson, *Tetrahedron Lett.*, 1990, **31**, 5595; A.F. Abdel-Magid and C. A. Maryanoff, *Synth. Lett.*, 1990, 537.
161. Y. Watanabe, S. C. Shim, T. Mitsudo, M. Yamashita and Y. Takegami, *Bull. Chem. Soc. Jpn.*, 1976, **49**, 1378.
162. G. Knupp and A. W. Frahm, *Arch. Pharm. (Weinheim, Ger.)*, 1985, **318**, 250.
163. T. Pienemann and H. J. Schaefer, *Synthesis*, 1987, 1005.
164. B. E. Maryanoff and D. F. McComsey, *J. Org. Chem.*, 1978, **43**, 2733.
165. B. E. Maryanoff, D. F. McComsey and S. O. Nortey, *J. Org. Chem.*, 1981, **46**, 355.
166. Y. Nakagawa and R. V. Stevens, *J. Org. Chem.*, 1988, **53**, 1871.
167. R. H. Prager, J. M. Tippett and A. D. Ward, *Aust. J. Chem.*, 1981, **34**, 1085.
168. P. J. Davis, D. Wiese and J. P. Rosazza, *J. Chem. Soc., Perkins Trans. 1*, 1977, 1.
169. E. F. Godefroi, J. J. H. Geenen, B. van Klingeren and L. J. van Wijngaarden, *J. Med. Chem.*, 1975, **18**, 530.
170. J. S. McManis and B. Ganem, *J. Org. Chem.*, 1980, **45**, 2041.
171. D. J. Hart, W.-P. Hong and L.-Y. Hsu, *J. Org. Chem.*, 1987, **52**, 4665.
172. W. Kaupmann, H. W. Ohlendorf and K. U. Wolf, *Eur. J. Med. Chem.*, 1985, **20**, 207.
173. M. A. Whitesell and E. P. Kyba, *Tetrahedron Lett.*, 1984, **25**, 2119.
174. A. Chiaroni, C. Riche, L. Diatta, R. Z. Andriamialisoa, N. Langlois and P. Potier, *Tetrahedron*, 1976, **32**, 1899.
175. J. A. Marshall and W. S. Johnson, *J. Org. Chem.*, 1963, **28**, 421.
176. D. Thielke, J. Wegener and E. Winterfeldt, *Chem. Ber.*, 1975, **108**, 1791.
177. L. Nilsson, *Acta Chem. Scand., Ser. B*, 1979, **33**, 547.
178. J. G. Berger, *Synthesis*, 1974, 508.
179. Y. Kikugawa, K. Saito and S. Yamada, *Synthesis*, 1978, 447.
180. T. Mitsudo, Y. Watanabe, M. Tanaka, S. Atsuta, K. Yamamoto and Y. Takegami, *Bull. Chem. Soc. Jpn.*, 1975, **48**, 1506.
181. H. Feuer, B. F. Vincent, Jr. and R. S. Bartlett, *J. Org. Chem.*, 1965, **30**, 2877.
182. M. Kawase and Y. Kikugawa, *J. Chem. Soc., Perkin Trans. 1*, 1979, 643.
183. M. W. Tijhuis, J. D. M. Herscheid and H. C. J. Ottenheijm, *Synthesis*, 1980, 890.
184. B. T. Cho and S. Y. Seong, *Bull. Korean Chem. Soc.*, 1988, **9**, 322.
185. K. H. Bell, *Aust. J. Chem.*, 1970, **23**, 1415.
186. G. W. Gribble, R. W. Leiby and M. N. Sheehan, *Synthesis*, 1977, 856.
187. J. M. J. Reonchet, G. Zosimo-Landolfo, N. Bizzozero, D. Cabrini, F. Habaschi, E. Jean and M. Geoffroy, *J. Carbohydr. Chem.*, 1988, **7**, 169.
188. B. B. Snider and C. P. Cartaya-Marin, *J. Org. Chem.*, 1984, **49**, 1688.
189. W. Oppolzer, S. Siles, R. L. Snowden, B. H. Bakker and M. Petrzilka, *Tetrahedron*, 1985, **41**, 3497.
190. S. Fushiya, K. Maeda, T. Funayama and S. Nozoe, *J. Med. Chem.*, 1988, **31**, 480.
191. L. M. Waykole, C. C. Shen and L. A. Paquette, *J. Org. Chem.*, 1988, **53**, 4969.
192. C. Bernhart and C.-G. Wermuth, *Tetrahedron Lett.*, 1974, 2493.
193. T. Sheradsky and E. Nov, *J. Chem. Soc., Perkin Trans. 1*, 1980, 2781.
194. M. L. M. Pennings and D. N. Reinhoudt, *J. Org. Chem.*, 1983, **48**, 4043.
195. R. J. Bergeron and J. J. Pegram, *J. Org. Chem.*, 1988, **53**, 3131.
196. H. J. Chaves Das Neves and A. M. V. Riscado, *J. Chromatogr.*, 1986, **367**, 135.
197. G. Zinner, H. Grigat and M. Psiorz, *Arch. Pharm. (Weinheim, Ger.)*, 1986, **319**, 787.
198. D. D. Sternbach and W. C. L. Jamison, *Tetrahedron Lett.*, 1981, **22**, 3331.
199. B. Ganem, *Tetrahedron Lett.*, 1976, 1951.
200. T. S. Sulkowski and S. J. Childress, *J. Org. Chem.*, 1963, **28**, 2150.
201. H. C. Brown, P. M. Weissman and N. M. Yoon, *J. Am. Chem. Soc.*, 1966, **88**, 1458.
202. P. H. Morgan and A. H. Beckett, *Tetrahedron*, 1975, **31**, 2595.
203. D. R. Smith, M. Maienthal and J. Tipton, *J. Org. Chem.*, 1952, **17**, 294.
204. S. R. Landor, O. O. Sonola and A. R. Tatchell, *J. Chem. Soc., Perkin Trans. 1*, 1974, 1294.
205. H. Feuer and D. M. Braunstein, *J. Org. Chem.*, 1969, **34**, 1817.
206. J. M. Lalancette and J. R. Brindle, *Can. J. Chem.*, 1970, **48**, 735.
207. (a) S. Kano, Y. Tanaka, E. Sugino and S. Hibino, *Synthesis*, 1980, 695; (b) C. Hoffman, R. S. Tanke and M. J. Miller, *J. Org. Chem.*, 1989, **54**, 3750; (c) J. P. Leeds and H. A. Kirst, *Synth. Commun.*, 1988, **18**, 777.
208. S. Itsuno, Y. Sakurai and K. Ito, *Synthesis*, 1988, 995.
209. J. Ipaktschi, *Chem. Ber.*, 1984, **117**, 856.
210. S. Dev, *Proc. Indian Acad. Sci.*, 1984, **93**, 1015.
211. D. Carr, B. Iddon, H. Suschitzky and R. T. Parfitt, *J. Chem. Soc., Perkin Trans. 1*, 1980, 2374.
212. J. Herscovici, M.-J. Egron and K. Antonakis, *J. Chem. Soc., Perkin Trans. 1*, 1988, 1219.
213. S. Sasatani, T. Miyazaki, K. Maruoka and H. Yamamoto, *Tetrahedron Lett.*, 1983, **24**, 4711.
214. K. Kotera and K. Kitahonoki, *Org. Synth.*, 1968, **48**, 20.
215. N. Balachander, S.- S. Wang and C. N. Sukenik, *Tetrahedron Lett.*, 1986, **27**, 4849.
216. N. Umino, T. Iwakuma, M. Ikezaki and N. Itoh, *Chem. Pharm. Bull.*, 1978, **26**, 2897.
217. M. Numazawa and Y. Osawa, *Steroids*, 1981, **38**, 149.
218. K. Narasaka, Y. Ukaji and S. Yamazaki, *Bull. Chem. Soc. Jpn.*, 1986, **59**, 525.

219. A. Hassner and P. Catsoulacos, *J. Chem. Soc., Chem. Commun.*, 1967, 590.
220. K. Maruoka, T. Miyazaki, M. Ando, Y. Matsumura, S. Sakane, K. Hattori and H. Yamamoto, *J. Am. Chem. Soc.*, 1983, **105**, 2831.
221. V. Jäger, W. Schwab and V. Buss, *Angew. Chem., Int. Ed. Engl.*, 1981, **20**, 601.
222. W. Schwab and V. Jäger, *Angew. Chem., Int. Ed. Engl.*, 1981, **20**, 603.
223. V. Jäger, I. Müller, R. Schohe, M. Frey, R. Ehrler, B. Häfele and D. Schröter, *D. Lect. Heterocycl. Chem.*, 1985, **8**, 79.
224. P. A. Wade and D. T. Price, *Tetrahedron Lett.*, 1989, **30**, 1185.
225. G. Zinner, H. Blass, W. Kilwing and B. Geister, *Arch. Pharm. (Weimheim, Ger.)*, 1984, **317**, 1024.
226. J. A. Blair and R. J. Gardner, *J. Chem. Soc. C*, 1970, 1714.
227. G. N. Walker, M. A. Moore and B. N. Weaver, *J. Org. Chem.*, 1961, **26**, 2740.
228. K. C. Ingham and S. A. Brew, *Biochim. Biophys. Acta*, 1981, **670**, 181.
229. G. W. Kabalka, *Org. Prep. Proced. Int.*, 1977, **9**, 133.
230. Y. Kikugawa and M. Kawase, *Synth. Commun.*, 1979, **9**, 49.
231. V. Nair and A. K. Sinhababu, *J. Org. Chem.*, 1978, **43**, 5013.
232. S. N. Ege, M. L. C. Carter, D. F. Ortwine, S.-S. P. Chou and J. F. Richman, *J. Chem. Soc., Perkin Trans. 1*, 1977, 1252.
233. A. P. Winiski, M. Eisenberg, M. Langner and S. McLaughlin, *Biochemistry*, 1988, **27**, 386.
234. G. Rosini, A. Medici and M. Soverini, *Synthesis*, 1979, 789.
235. S. F. Nelsen and G. R. Weisman, *Tetrahedron Lett.*, 1973, 2321.
236. R. L. Hinman, *J. Am. Chem. Soc.*, 1957, **79**, 414.
237. H. El Khadem, Z. M. El-Shafei and M. El Sekeili, *J. Org. Chem.*, 1972, **37**, 3523.
238. B. P. Branchaud, *J. Org. Chem.*, 1983, **48**, 3531.
239. R. Annunziata, M. Cinquini and F. Cozzi, *J. Chem. Soc., Perkin Trans. 1*, 1982, 339.
240. W. R. McKay and G. R. Proctor, *J. Chem. Soc., Perkin Trans. 1*, 1981, 2443.
241. F. Clerici, A. Di Mare, M. L. Gelmi and D. Pocar, *Synthesis*, 1987, 719.
242. B. Krzyzanowska and W. J. Stec, *Synthesis*, 1978, 521.
243. R. O. Hutchins and M. C. Rutledge, *Tetrahedron Lett.*, 1987, **28**, 5619.
244. B. Krzyzanowska and W. J. Stec, *Synthesis*, 1982, 270.
245. R. O. Hutchins, A. F. Abdel-Magid, Y. P. Stercho and A. Wambsgans, *J. Org. Chem.*, 1987, **52**, 702.
246. T. W. Greene, in 'Protective Groups in Organic Synthesis', Wiley, New York, 1981, p. 281.
247. M. J. Haire, *J. Org. Chem.*, 1977, **42**, 3446.
248. H. Feuer, R. S. Bartlett, B. F. Vincent, Jr. and R. S. Anderson, *J. Org. Chem.*, 1965, **30**, 2880.
249. H. M. Walborsky and G. E. Niznik, *J. Org. Chem.*, 1972, **37**, 187.
250. D. H. Lorenz and E. I. Becker, *J. Org. Chem.*, 1963, **28**, 1707.
251. H. C. Brown, P. M. Weissman and N. M. Yoon, *J. Am. Chem. Soc.*, 1966, **88**, 1458.

1.3
Reduction of C=X to CHXH by Hydride Delivery from Carbon

RICHARD M. KELLOGG
University of Groningen, The Netherlands

1.3.1 INTRODUCTION

Unanticipated developments help to put known facts into place. Results from biochemistry drove home to organic chemists the message that it was not a chemical rarity for carbon–hydrogen bonds to be sources of hydride equivalents. Westheimer, Vennesland *et al.*[1] established beyond doubt that in a redox reaction mediated by the coenzyme couple $NAD(P)^+$/NAD(P)H the carbon–hydrogen bond of ethanol could serve directly as a hydride donor to an electron-deficient carbon of a pyridinium ion, and that this hydride equivalent could in turn be donated directly to the electropositive carbon of a carbonyl group. Thus the hydride donor capacities of carbon are also part and parcel of life. All this can occur under physiological conditions with the help of an enzyme, which somehow activates these reactants. The sequence is illustrated schematically in equation (1). In either direction hydride is transferred from carbon to carbon.

The organic ground on which these observations fell was fertile. The reducing power of aldehydes was well known, for example the use of formaldehyde (together with zinc dithionite) in the dye industry.[2]

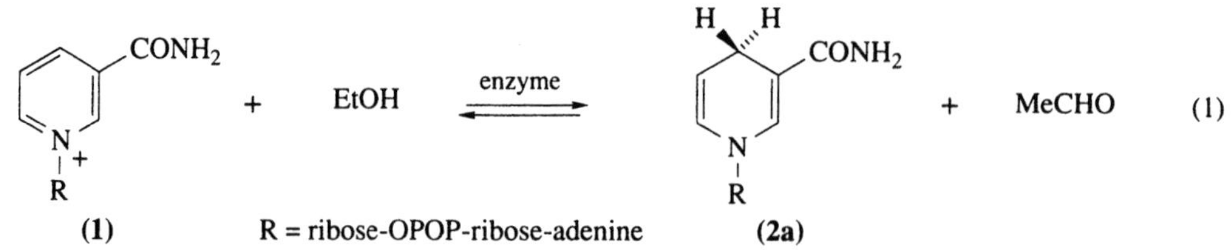

(1) + EtOH ⇌ (enzyme) (2a) + MeCHO (1)

R = ribose-OPOP-ribose-adenine

The Cannizzaro reaction (Section 1.3.3.2), the Leuckart reaction (Section 1.3.3.1), wherein formic acid is used as a reductant, and the Meerwein–Ponndorf–Verley (MPV) reaction (Section 1.3.3.3), wherein alcohols serve as reducing agents, had been used and studied widely by the organic community. The results from biochemistry underscored the kindredness of these different processes.

Another development also had an effect, but in this case a delaying one, on the development of carbon–hydrogen bonds as hydride donors; the advent of highly reactive but safe metal hydrides as reducing agents tempered interest for a time in the less potent organic reducing agents. In recent years, however, interest in some new synthetic and mechanistic aspects of 'hydride transfer' from carbon has been rekindled.

The quotation marks of the foregoing sentence refer to a convention that will be adopted here, namely that the descriptors 'hydride' and 'hydride transfer' will represent an end result, not a mechanistic implication. The mechanistic aspects will be discussed when relevant to the material of this chapter. Indeed the majority of more recent work is more mechanistically than synthetically oriented.

1.3.2 STRUCTURE AND REACTIVITY

1.3.2.1 Structural Types that may donate Hydride

Some representative examples of hydride donor couples are triarylmethanes (**3a,b**), cycloheptatrienes (**4a,b**), α-heterosubstituted methanes (**5a,b**), and heterocycles such as 1,4-dihydropyridines (**6a,b**).

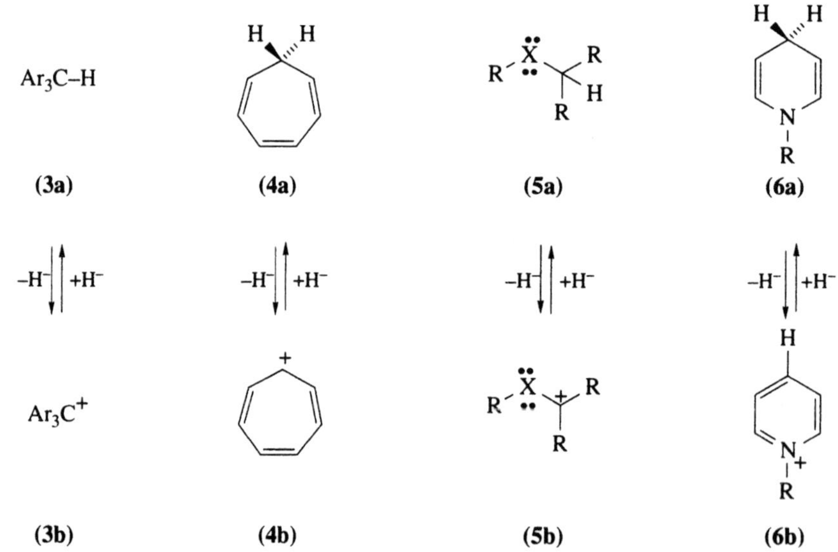

Ar_3C–H (**3a**) ⇌ Ar_3C^+ (**3b**) (−H⁻ / +H⁻)

(**4a**) ⇌ (**4b**)

(**5a**) ⇌ (**5b**)

(**6a**) ⇌ (**6b**)

Primary and secondary alcohols or alkoxides may serve as hydride sources; the generation of a strong carbonyl bond serves as a driving force (equation 2). Other atoms can replace oxygen; an analogous process whereby a carbon–carbon double bond is formed *via* loss of hydride from a carbanion has much synthetic potential (equation 3). In some cases the metal remains bonded to the alkene (Section 1.3.3.7). Some relatively isolated examples wherein bonds other than double bonds are formed will be discussed later. Most of these reactions will be seen to proceed often (but not obligatorily) through six-membered ring transition states.

Examples of transpositions of hydride *via* sigmatropic reactions (equation 4)[3] will be restricted to one special example, which will be discussed in Section 1.3.3.5.3.

$$X{-}O{-}CR_2H \longrightarrow X^+ + O{=}CR_2 + \text{'}H^-\text{'} \qquad (2)$$

X = H, metal ion

$$X_nM{-}CR_2{-}CR_2{-}H \longrightarrow X_nM^+ + R_2C{=}CR_2 + \text{'}H^-\text{'} \qquad (3)$$

X = ligand; M = B, Al , Zn , Mg, Mo, *etc.*

$$(4)$$

$n = 1, 3, 5,...$

Some cases of hydride donation from carbon will not be treated here, including various industrial catalyzed cracking processes.[4]

1.3.2.2 Some Mechanisms for Hydride Donation

Proper kinetic and thermodynamic meshing of the reactants is necessary for hydride transfer from a donor to an acceptor. There are at least three obviously different mechanisms by which hydride equivalents can be transferred. These are: (i) concerted transfer of the proton and two electrons (equation 5); (ii) homolytic cleavage of the carbon–hydrogen bond followed by subsequent transfer of an electron (equation 6); and (iii) initial loss of a proton followed by transfer of two electrons, either together or stepwise (equation 7). Fusion of pathways is imaginable, dependent on the structures of the participating molecules and possibilities for catalysis.

$$\text{>C–H} + X{=}Y \longrightarrow \text{>C}^+ + H{-}X{-}Y^- \qquad (5)$$

$$\text{>C–H} + X{=}Y \longrightarrow \text{>C}^\bullet + H{-}X{-}Y^\bullet \longrightarrow \text{>C}^+ + H{-}X{-}Y^- \qquad (6)$$

$$\text{>C–H} \rightleftharpoons \text{>C}^- + H^+ \xrightarrow{X=Y} \text{>C}^+ + H{-}X{-}Y^-$$
$$\searrow^{X=Y}\ \text{>C}^\bullet + H{-}X{-}Y^\bullet \uparrow \qquad (7)$$

The concerted mechanism (equation 5) is often encountered (or experimental techniques are not sensitive enough to detect very rapid consecutive steps). Unless otherwise stated a concerted mechanism will be presumed. In a number of examples to be cited, however, the final word is doubtless not yet in.

1.3.2.3 Catalysis

Catalysis and activation are not the same. In Friedel–Crafts acylations of aromatics, aluminum chloride is used in more than stoichiometric amounts as an activator and coreagent. A similar approach is often used on a stoichiometric basis to activate carbon–hydrogen bonds toward hydride donation. This chemical approach stands in sharp contrast to the enzyme alcohol dehydrogenase, which works catalytically *via* a complex mechanism wherein activation is achieved but not at the expense of irreversible consumption of a coreagent.[1b]

A brief discussion of some aspects of alcohol dehydrogenase will be used to illustrate the potential for catalysis. This system is chosen for illustration because it has been studied so extensively. Lessons drawn can be applied in a broader context. The 1,4-dihydropyridine (**2a**) is the reductant and this affords a nicotinium ion (**1**) on transfer of hydride, as illustrated in equation (1). This process is mimicked in many abiotic systems by derivatives of (**2**; R = alkyl or benzyl), by Hantzsch esters (**7**), which are synthetically readily accessible, and 1,4-dihydro derivatives (**8**) of pyridine-3,5-dicarboxylic acid. A typical abiotic reaction is the reduction of the activated carbonyl group of an alkyl phenylglyoxylate (**9**), activated by a stoichiometric amount of the powerful electrophile $Mg(ClO_4)_2$, by, for example, (**2b**; equation 8). After acrimonious debate the consensus seems to be that such reactions involve a one-step mechanism (*i.e.* equation 5), unless the reaction partner strongly demands a radical intermediate, as in the reduction of iron(II) to iron(III).

(**7**) R^1 = alkyl, aryl; X = OR, NR_2, CR^2; R^2 = alkyl, aryl

(**8**) R^1 = alkyl, benzyl; X = OR, NR^2

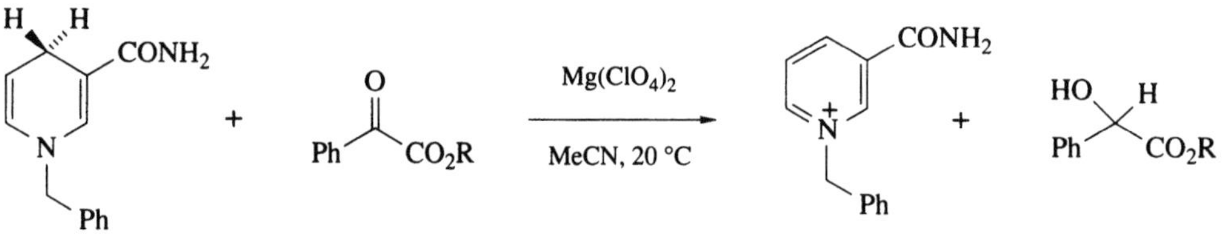

(**2b**) + (**9**) Ph–CO–CO_2R → [$Mg(ClO_4)_2$, MeCN, 20 °C] (**1b**) + (**10**) Ph–CH(OH)–CO_2R (8)

How may the kinetic barriers to hydride transfer be lowered? The following arguments pertain to both biotic and abiotic systems. Activation is possible. Most likely magnesium ion activates the carbonyl group *via* complexation as depicted in equation (9). The hydride-donating potential can be modified by structural change. Using (**7**) and (**8**) as examples, it is known that introduction of carboxamide groups (X = NR_2) increases the redox potential relative to X = OR(see **8a** and **8b**), and introduction of substituents at the 2,6-positions of the 1,4-dihydropyridine ring also has the same effect (see **7a** and **8a**). As a marker NADH (**2a**) has E^0 = −361 mV.[5]

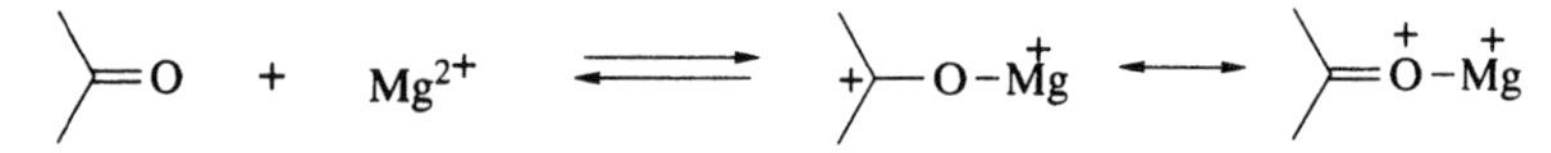

$$\text{>C=O} + Mg^{2+} \rightleftharpoons \text{>}\overset{+}{C}\text{–O–}\bar{Mg} \longleftrightarrow \text{>C=}\overset{+}{O}\text{–}\overset{+}{Mg} \quad (9)$$

Hard electrophiles like $Mg(ClO_4)_2$ are used to activate abiotic systems. In the enzyme liver alcohol dehydrogenase (LAD) a considerably different catalytic apparatus is present; a zinc ion coordinated to two cysteines and a histidine serves as a coordinating site for the carbonyl compound/alcoholate, as illustrated in equation (10). This zinc ion has amphoteric properties consistent with the capacity to activate the reaction in both directions without being consumed, in other words to act as a catalyst. Synthetic models of this catalytically active zinc have been shown to possess some catalytic activity in analogy to the enzyme (see Section 1.3.3.5.1iii).

Various experimental and theoretical approaches suggest that in the transition state only slight, if any, negative charge is present on the hydrogen atom being transferred; B in Scheme 1 rather than A.[6] In fact argument on the basis of precedent suggests that the transition state may actually involve a positively charged hydrogen atom (C in Scheme 1). For example, the bridged 1,5-hydridocyclodecyl carbocation

(2c) $E^\circ = -434\text{mV}$ **(7a)** $E^\circ = -363$ mV

(8a) $E^\circ = -260$ mV **(8b)** $E^\circ = -365$ mV

$$\xrightleftharpoons[-H^+]{+H^+} \qquad (10)$$

(**A**) is a model for the transition state for hydride transfer.[7] The situation in (**A**) is a particularly good model for hydride transfer from an sp^3 carbon to an sp^2 carbonium ion center. NMR investigations of (**A**) indicate that the hydrogen undergoing transfer has indeed only about 0.1 electron.

A $\quad \text{C–H} + \text{C}^+ \longrightarrow \left[\overset{+}{\text{C}}\cdots\overset{-}{\text{H}}\cdots\overset{+}{\text{C}}\right]^{\ddagger}$

B $\quad \text{C–H} + \text{C}^+ \longrightarrow \left[\overset{1/2+}{\text{C}}\cdots\text{H}\cdots\overset{1/2+}{\text{C}}\right]^{\ddagger}$

C $\quad \text{C–H} + \text{C}^+ \longrightarrow \left[\text{C}\cdots\overset{+}{\text{H}}\cdots\text{C}\right]^{\ddagger}$

Scheme 1

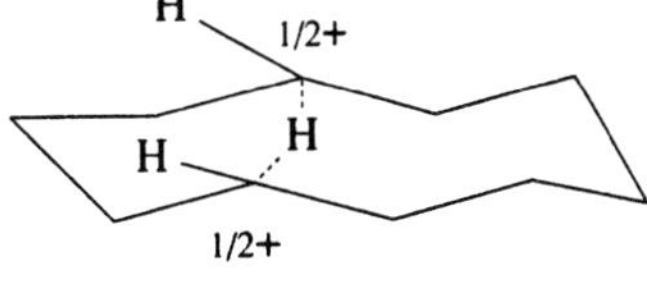

(A)

Simple frontier orbital arguments have been made on how a transition state analogous to C (Scheme 1) could be stabilized by the carboxamide of nicotinamide or amide units from the peptide chain. This is, of course, the essence of complementarity in an enzyme.[6] More rigorous calculations (Section 1.3.3.5.1i) indeed indicate participation of the carboxamide in stabilization of the transition state for hydride transfer. These simple schemes can be put in even more primitive acid/base terms. The hydrogen being transferred has the character of an acid and as such will be stabilized by interaction with bases like carbonyl oxygen. One may expect acceleration of hydride transfer when such opportunities for stabilization are present.

1.3.3 COMPOUND CLASSES

1.3.3.1 Hydride Transfer from Formic Acid

Formic acid is an exception among organic acids in having reducing capacity because of its combined acid/aldehyde character and the possibility for exothermic release of carbon dioxide on hydride donation. Ammonium formate, NH_4CO_2H, has long been used for reductive alkylation of amines. This is the Leuckart reaction, which has been reviewed.[8] In this process aldehydes or ketones are used to alkylate ammonia and form a primary amine (**11**), as depicted in equation (11). The reductive step is the reaction of formate with an iminium ion or its equivalent (equation 12).[9]

$$R^1C(O)R^2 + 2\ NH_4CO_2H \longrightarrow R^1CH(NH_2)R^2\ \textbf{(11)} + 2H_2O + NH_3 + CO_2 \qquad (11)$$

$$H_2N^+{=}C(R^1)R^2 + HCO_2^- \longrightarrow \textbf{(11)} + CO_2 \qquad (12)$$

Formic acid also reduces stable iminium functionalities (but not in general carbonyl groups). Acridinium ions are also readily reduced. This process bears some outward resemblance to the enzymatic transformation of formate to carbon dioxide by formate dehydrogenase, although a molybdenum/sulfur cluster is probably the hydride acceptor in the enzyme.[10]

Probably the most use has been made of the Eschweiler–Clarke modification wherein formaldehyde is used together with formic acid to methylate amines.[11] This procedure is especially useful for the clean preparation of *N,N*-dimethylamines, a particularly useful example being the *N,N*-dimethylation of amino acids.[12]

1.3.3.1.1 Catalysis

The synthetic potential of reductions by formate has been extended considerably by the use of ammonium formate with transition metal catalysts like palladium and rhodium. This forms a safe alternative to use of hydrogen. In this fashion it is possible to reduce hydrazones to hydrazines, azides and nitro groups to amines, to dehalogenate chloro-substituted aromatics, and to carry out various reductive removals of functional groups. For example, phenol triflates are selectively deoxygenated to the aromatic derivatives using triethylammonium formate as reductant and a palladium catalyst.[13] These recent applications have been reviewed.[14]

Catalyzed reductions with ammonium formate proceed with retention of configuration, at least for the case of nitro compounds. Treatment of steroid derivative (**12**) with ammonium formate in methanolic solution and a catalytic amount of palladium on charcoal gave exclusively amino steroid (**13**), as shown in equation (13).[15] A similar technique has been used to produce members of the ephedrine family.

Hydrogenation of itaconic acid (**14**) with $Rh(COD)Cl_2$ catalyst and commercially available triethylammonium formate as hydrogen source delivers (*S*)-(**15**) in good enantiomeric excess (equation 14); with hydrogen as reductant instead of ammonium formate a 94% *ee* is obtained.[16]

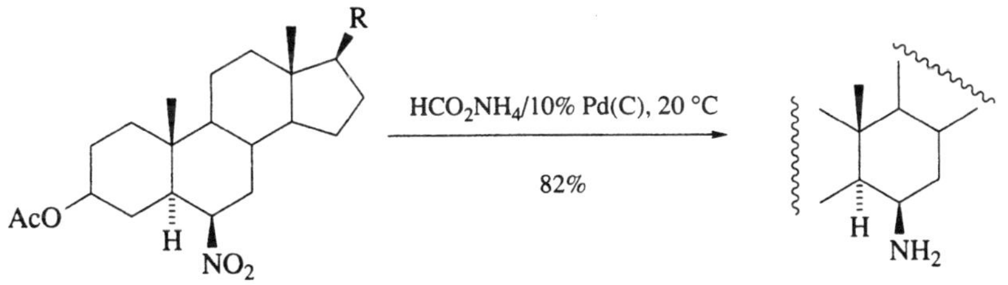

HCO$_2$NH$_4$/10% Pd(C), 20 °C, 82% (13)

(12) (13)

Et$_3$N/HO$_2$CH/Rh(COD)Cl$_2$, 45 °C, 100% (14)

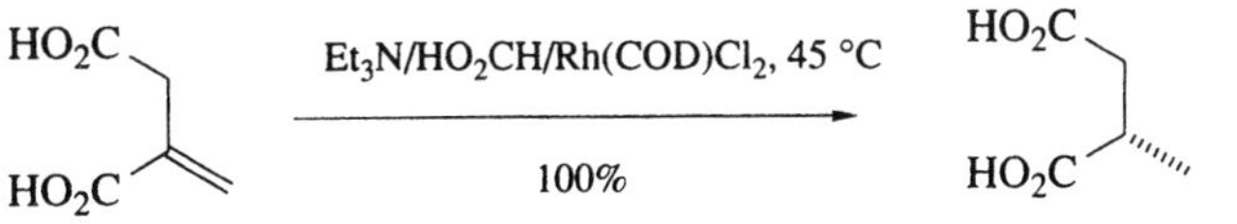

(14) (15) 84% *ee*

An attractive method for the regeneration of NAD(P)H (see also Section 1.3.3.5.1) is *via* regioselective reduction of NAD(P)$^+$ by formate ion, catalyzed by a rhodium derivative.[17]

Ortho esters or amides or mixed ortho derivatives of formic acid might also be expected to be fairly good hydride donors. The absence of proton acidity is attractive for potential application in catalyzed reactions. This type of reaction is, however, rather uncommon, owing to ready loss of a heteroatom and formation of a stabilized carbocation.[18] Compound (**16**) represents an interesting case that has been examined by Wuest *et al.*[19] In the indicated conformation the lone pairs on the nitrogen atoms are antiperiplanar to the carbon–hydrogen bond; at first sight good stereoelectronic activation of the hydride-donating potential would be expected. Although (**16**) will generate molecular hydrogen in the presence of a strong acid like HBF$_4$ (equation 15) with generation of (**17**), reduction of ethyl phenylglyoxylate (**9a**) to ethyl mandelate (**10a**) in the presence of Mg^{2+} occurs with hydride donation from a methylene carbon (Scheme 2), as evidenced by the formation of (**19**). On the other hand (**20**) does provide under analogous conditions (**10a**) and ion (**21**; equation 16).

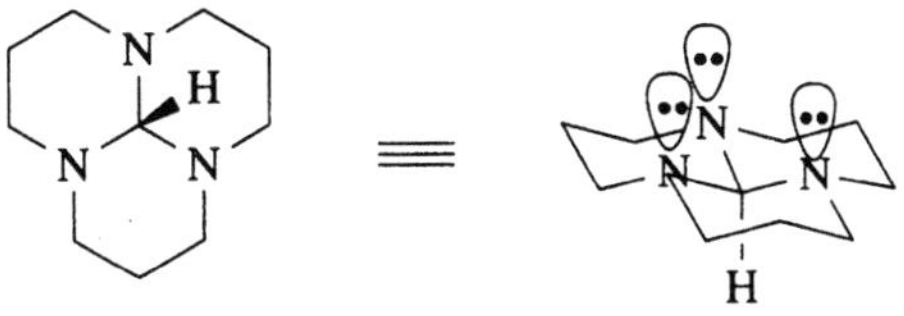

(16) ≡ (16)

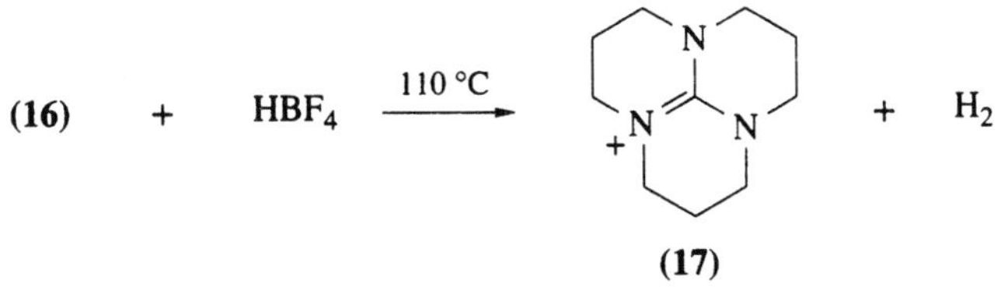

(16) + HBF$_4$ —110 °C→ (17) + H$_2$ (15)

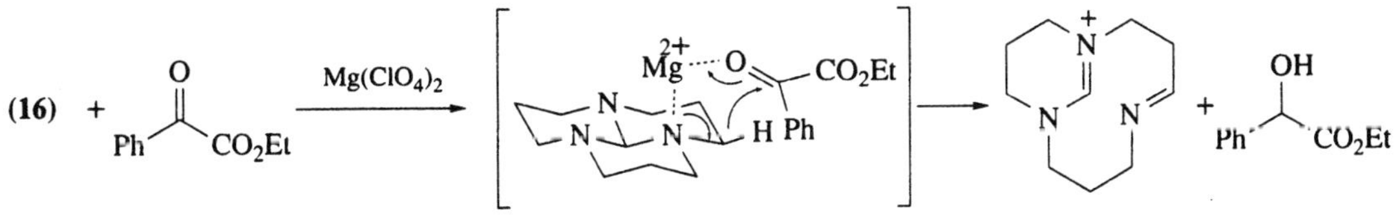

(16) + (9a) —Mg(ClO$_4$)$_2$→ [(18)] → (19) + (10a)

Scheme 2

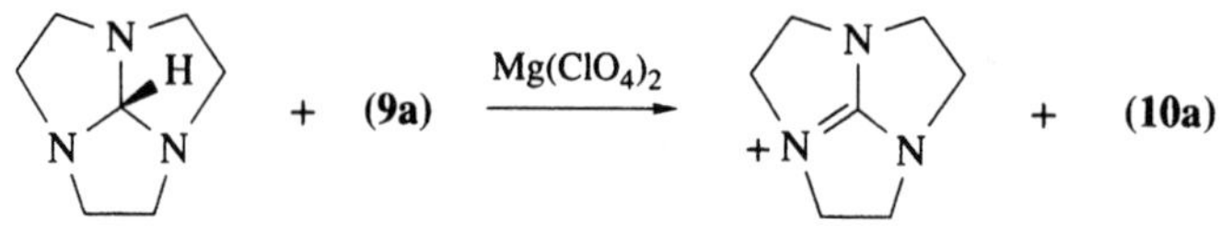

(20) + (9a) $\xrightarrow{Mg(ClO_4)_2}$ (21) + (10a) (16)

1.3.3.2 Hydride Transfer from Aldehydes

The Cannizzaro reaction (Scheme 3), in which a nonenolizable aldehyde (**22**) disproportionates in the presence of strong base, usually NaOH, to an acid (**25**) and alcohol (**26**), is one of the longest known organic reactions. The crossed version wherein excess formaldehyde is used as the reductant is especially popular. Geissman has adequately reviewed these variations.[20] Current mechanistic thinking entails hydride transfer from a tetracoordinate intermediate (**23**) formed on addition of hydroxide to the aldehyde. The most straightforward course of reaction is rate-determining transfer of hydride *via* a linear or bent transition state (**24**) to a second molecule of aldehyde.[21] MNDO-SCF calculations support a linear geometry for transition state (**24**) with only 0.11 electron on the hydrogen atom.[22]

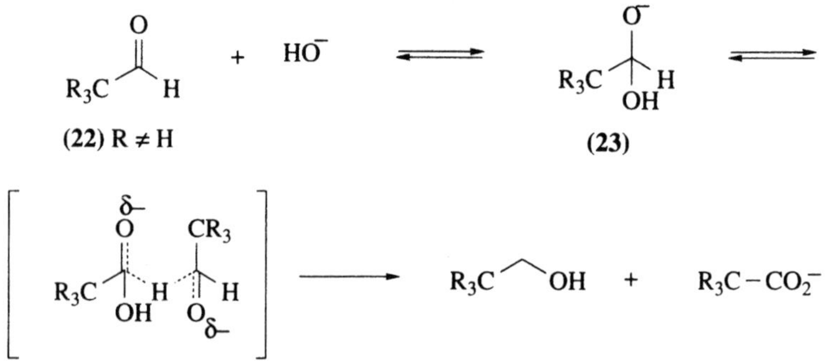

(22) R ≠ H + HO^- ⇌ (23) ⇌ [(24)] ⟶ R_3C–CH_2–OH (26) + $R_3C-CO_2^-$ (25)

Scheme 3

Intermediate (**23**) could also react *via* a ketyl radical (**27**) and hydrogen atom (equation 17). Ketyl radical intermediates have been detected, isotopic exchange experiments suggest radical pathways, and calculations indicate that a radical pathway could well be followed, particularly in the case of aromatic aldehydes.[23,24]

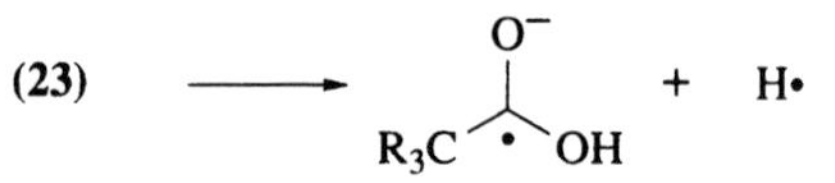

(23) ⟶ (27) + H• (17)

An additional complication in mechanistic interpretation is the fact that methoxide (reactions are often run in methanol) serves also as an effective hydride donor towards aldehydes and the derived formaldehyde may participate in a crossed Cannizzaro reaction.[21] The conditions of the Cannizzaro reaction are fairly extreme (strong base, reasonably high temperatures) and complex mechanistic behavior is therefore not surprising.

1.3.3.2.1 Catalysis

Various methods, many with an eye towards industrial application, have been examined to accelerate Cannizzaro reactions. For example copper–silica catalysts,[25] Na_2S, $Na_2S_2O_3$ or Na_2SO_3 supported on Al_2O_3, and ultrasound all have been reported to accelerate certain Cannizzaro processes.[26] An extremely promising development is the use of transition metal catalysts, which may be employed under neutral conditions in the absence of strong base. This permits the use of enolizable aldehydes like (**28**), which under basic conditions would immediately be consumed *via* aldol reactions. For example, in the presence of $H_4Ru(CO)_8(PBu_3)_3$ or $Ru_2(CO)_4$ $(MeCO_2)_2(PBu_3)_2$, the disproportionation of (**28**) to (**29**)/(**30**) indicated in equation (18) occurs readily. Small enantiomeric excesses of acid have been obtained when an optically active ruthenium complex is used as catalyst.[27]

2 (**28**) CHO —cat. Ru→ (**29**) CO_2H + (**30**) OH (18)

In view of the hydride-donating potential of (**23**), one might anticipate similar reactivity for adducts derived from, for example, amines and thiols. Intermolecular demonstrations of such reactivity are scarce; on the other hand intramolecular examples are well known, although they may be mechanistically ambiguous. The prototype reaction is the intramolecular disproportionation of glyoxal (**31**) to glycolic acid (**32**), shown in equation (19). This reaction may be induced by hydroxide.[28] With the aid of *ab initio* and MNDO-SCF-MO calculations a highly bent transition state has been calculated for hydride transfer.[29]

$$(CHO)_2 + HO^- \rightleftharpoons HOCH_2-CO_2^- \qquad (19)$$

(**31**) (**32**)

Chemically analogous transformations occur enzymatically. For example, in the glycolase system glutathione (**34**; GSH, Glu-Cys-Gly) catalyzes the conversion of methylglyoxal (**33**) to lactic acid (**35**), as shown in equation (20). This transformation is initiated not by hydroxide, but by the addition of thiolate from GSH to the aldehyde carbonyl group.[30]

(**33**) + GSH (**34**) $\underset{}{\overset{H_2O}{\rightleftharpoons}}$ (**35**) OH, CO_2H (20)

The enzyme systems are complex; two reactions are catalyzed, namely formation of a thioester and its subsequent hydrolysis to free acid. Because little or no tritium or deuterium from solution is incorporated into the rearranged product, a hydride shift was assumed.[31] With optically pure thiols like (**36**) the rearrangement of (**37**) to (**10b**) can be catalyzed (equation 21). The methyl mandelate formed has a very modest enantiomeric excess.[32]

SH, NMe_2

(**36**)

Ph(CO)CHO (**37**) + MeOH —(**36**), 85%→ Ph CH(OH) CO_2Me (**10b**) (21)

Evidence later became available that indicated that the lack of isotope incorporation from solvent in enzymic experiments is misleading.[33] Trapping experiments have established quite conclusively that an enediol (**38**) is a kinetically competent intermediate. Hydride transfer occurring through a transition state like (**39**; equation 22) is therefore not required.[34]

The glyoxalase system bound on a solid phase has been used for the asymmetric synthesis of α-hydroxy acids (equation 23). Aldehyde (**40**; $R = Pr^n$) provides the acid (**42**) in 79% yield and in >99% *ee*.[35] The enzyme system GX-I/GX-II is not especially stable when isolated on a matrix and scale-up is difficult because large quantities of GSH are required.

(22)

(38) (39)

(23)

(40) R = p-ClC$_6$H$_4$, Ph, p-MeOC$_6$H$_4$, Me, Pr^n

(41) (42)

Intramolecular redox reactions *via* enediol chemistry are, of course, well known. Ascorbic acid and the multitude of known related structures which readily form enediols have even been coined 'reductones' by von Euler and Eistert.[36]

1.3.3.3 Hydride Transfer from Alcohols and Amines

The classical Meerwein–Ponndorf–Verley (MPV) process, named after the independent originators, can be illustrated by the reduction of crotonaldehyde (**43**) by aluminum isopropoxide (**44**) in isopropyl alcohol (equation 24). Aluminum isopropoxide transfers hydride reversibly to a carbonyl acceptor. Acetone is formed as a volatile side product, which can be removed during reaction. The reaction of equation (24) is forced even further to the right by the use of excess isopropyl alcohol. MPV reactions have been reviewed.[37] In the Oppenauer variant of this reaction an alcohol is oxidized to a ketone, and acetone is used as hydride acceptor in the presence of a strong base like t-butoxide. This reaction was originally developed for the selective oxidation of sterols. The synthetic aspects of this procedure have also been reviewed.[38]

(43) + (Pr^{i}O)$_3$Al (44) —Pr^{i}OH, 60%→ (45) + Me$_2$CO (24)

Examples of amines serving as hydride donors are scarce (for a special case, see Section 1.3.3.5.3). That this reaction is possible is exemplified by a side reaction that occurs when simple aldehydes are treated with lithium diisopropylamide (**46**); reduction of the aldehyde with formation of an imine (**47**) occurs (equation 25).[39]

(46) + RCHO → (47) + R OH (25)

A mass of experimental evidence on MPV reactions, much of which has been summarized by Morrison and Mosher,[40,41] points to hydride transfer through a six-membered Lewis salt (**50**; equation 26). No account of aggregation is taken in this model.[42] The observation (equation 26) that the major enantiomer obtained from reductions with optically pure aluminum alkoxides is correctly predicted by (**50**) provides additional support for the six-membered intermediate postulate. Examples of reductions by alkoxides wherein this geometry is unattainable will shortly be discussed, however.

Calculational results also support a transition state formed from the six-membered Lewis salt (**50**). As calculated at the 3-21G level the transition state for the reaction of lithium methoxide with formaldehyde

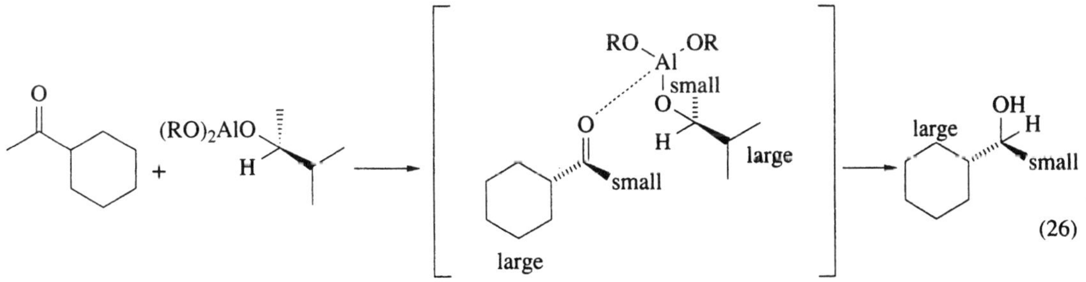

(48) (49) (50) (S)-(51) 22% *ee* (26)

involves a bent C—H—C bond with rather low bending energies.[43] Moreover, the hydrogen atom being transferred is calculated to bear little negative charge.

Assessment of the role of the cation in MPV reduction is difficult. Conversion of alcohol to alkoxide certainly enhances the reactivity towards hydride donation, and aluminum ions aid *via* chelation in arranging alkoxide and carbonyl compound properly for reaction. However, alkoxides bearing cations other than aluminum may also exhibit good hydride-donating tendencies. Lithium isopropoxide reduces steroidal ketones efficiently[44] and magnesium alkoxides derived from chiral alcohols have been used extensively in chiral syntheses. Isobornyloxy magnesium bromide (**52**) has been used widely for this purpose (equation 27).[41]

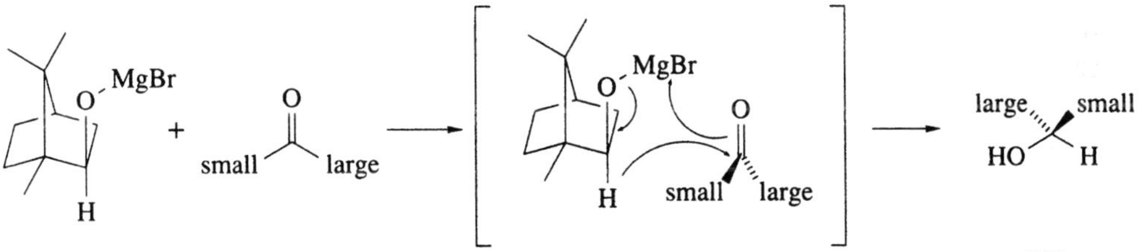

(52) (53) (27)

Proper orientation plays a large role. For example (**54**) rearranges cleanly to (**55**) merely with sodium sulfite or dilute sodium hydroxide (pH 10.5), an equilibrium mixture of 75% (**55**) and 25% (**54**) being formed (equation 28).[45]

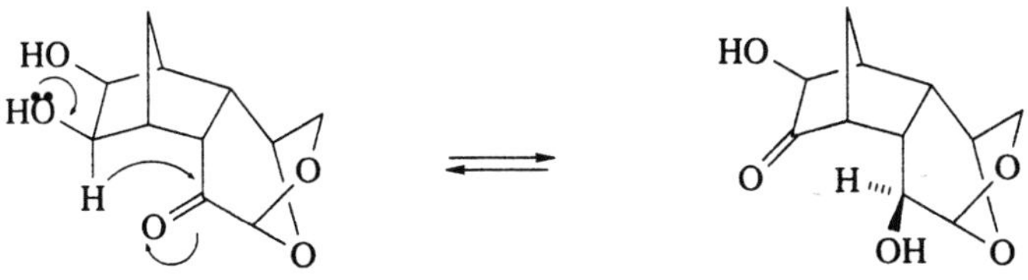

(54) (55) (28)

The rather complex conversion of (**56**) to (**58**) depicted in Scheme 4 doubtless involves an intramolecular hydride shift through a boat conformation (**57**). Benzaldehyde generated by this process undergoes

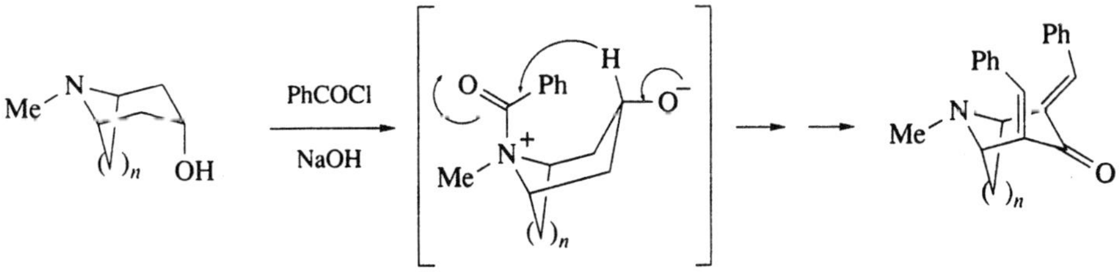

(56) (57) (58)

Scheme 4

subsequent aldol condensation to give (**58**). This entire sequence occurs under mildly basic Schotten–Baumann conditions.[46]

Thus intramolecular hydride transfers occur quite readily if the conformation is suitable for the efficient interaction of acceptor, usually carbonyl, and hydride donor carbon atom. In the transformation of (**59**) to (**62**; Scheme 5) intramolecular hydride transfer occurs from an amine (**59**) rather than an alcohol.[47] This reaction is probably mechanistically akin to the chemistry discussed in Section 1.3.3.5.3.

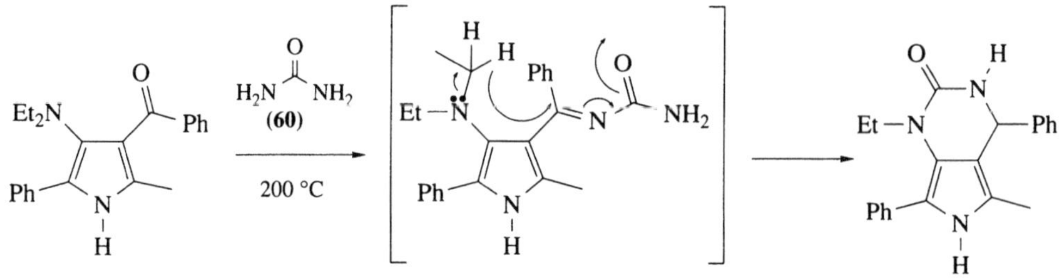

(**59**) (**61**) (**62**)

Scheme 5

In the presence of a metal isopropoxide (**63**) undergoes either intermolecular reduction to form (**66**), or reacts intramolecularly to form (**65**). The latter appears to be the product expected from transition state (**64**), derived from a boat conformation (Scheme 6). The intramolecular reaction is much faster than its intermolecular competitor with Ba^{2+} or K^+ as cation but when Al^{3+} is the cation, *i.e.* classical MPV conditions, then only intermolecular reduction occurs.[48]

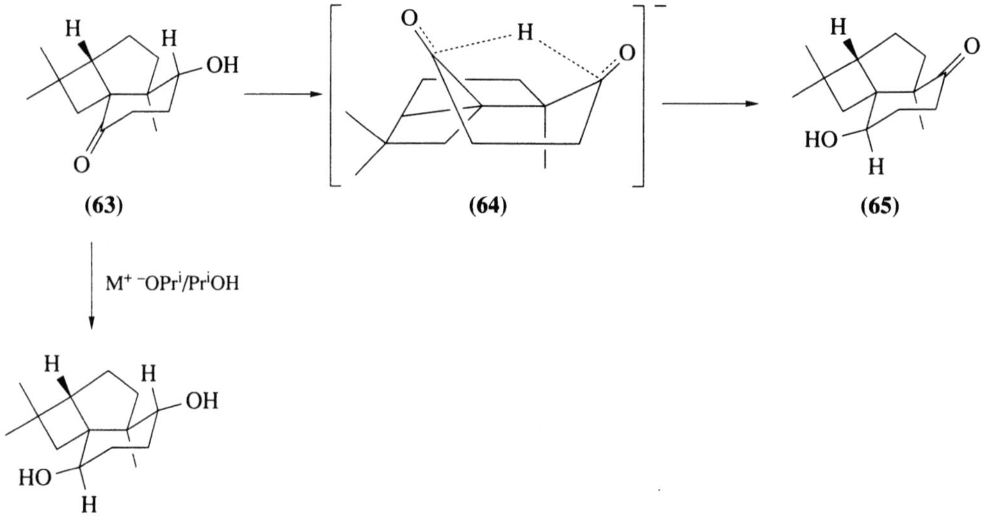

(**63**) (**64**) (**65**)

$M^+ {}^-OPr^i/Pr^iOH$

(**66**)

Scheme 6

Hydride is transferred intramolecularly in (**67**; equation 29) in a degenerate process.[49] Clearly the six-membered transition state characteristic of the MPV reaction cannot be achieved here because the orientation of hydroxy and carbonyl is incompatible with simultaneous coordination to a metal ion. The rates depend on the cation; if the alkoxide is a solvent-separated ion pair the hydride-donating potential in the intramolecular reaction is increased, consistent with the greater charge localized on alkoxide oxygen.

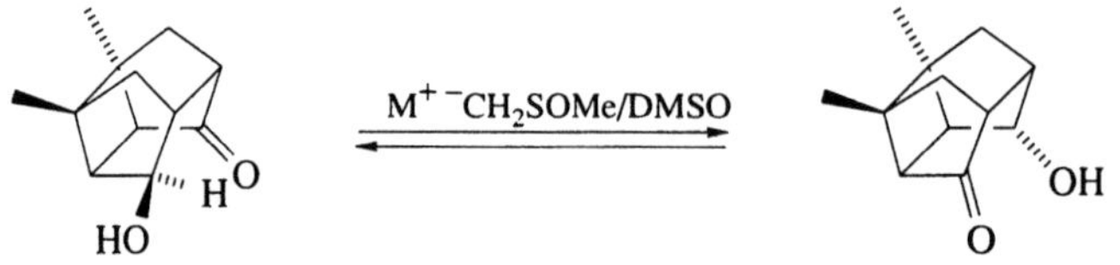

$M^+ {}^-CH_2SOMe/DMSO$ (29)

(R)-(**67**) *(S)*-(**68**)

Alcohols also transfer hydride to carbonium ions.[50] In seminal experiments Bartlett[51] demonstrated that isopropyl alcohol as well as other alcohols readily transferred hydride to various triaryl carbonium ions; oxidation of the alcohol and reduction of the carbonium ion is the result.

In a similar fashion ethers, acetals[52] and formic acid may serve as efficient hydride donors towards carbonium ions, aryldiazonium salts (reduction with loss of nitrogen occurs) and pyrylium salts. A review of this chemistry is available.[53]

1.3.3.3.1 Catalysis

Promise is held in MPV reactions carried out under catalytic conditions. Instead of, for example, stoichiometric amounts of aluminum as the metal ion activator, catalytic quantities of complexes of rhodium and iridium can sometimes be used to bring about the same reactions. Although the catalytic mechanisms have not been established, postulation of the usual six-membered transition state in the critical step of hydride transfer appears reasonable. The strongly basic conditions of the MPV reaction are avoided. Reductions of aryl ketones (**69**; equation 30) using (excess) isopropyl alcohol as hydrogen donor and at partial conversions have led to the formation of alcohol (**70**) in modest enantiomeric excesses with various chiral ligands.[54–57]

COR^2 … R^1 + $\rangle$—OH —Rh/ligand→ HO, R^2 … R^1 + $\rangle$=O (30)

(69) **(70)**

Metal carbonyls like $Mo(CO)_6$ or $Mn_2(CO)_{10}$ will catalyze the reduction of carbon tetrachloride to chloroform by isopropyl alcohol. Apparently these are radical reactions initiated by abstraction of a chloride from carbon tetrachloride by the metal complex.[58]

Homogeneous (or heterogeneous) catalyst systems offer the exciting perspective of high efficiency, optimal use of (chiral) ligands in enantioselective reactions, and the possibility of taking many different compounds as 'hydride' sources. The use of isopropyl alcohol has been illustrated in this regard, but dioxane, dihydrofuran, aldehydes, formic acid, cyclohexene or *N*-benzylaniline are some of the other compounds that already have been employed as alternatives.[4]

1.3.3.3.2 Photochemical reactions

Alcohols (as well as amines, sulfides and many hydrocarbons) may act as overall hydride donors towards excited nonbonding–π^* states of carbonyl compounds and heterocycles. The synthetic as well as the photophysical aspects of these processes have been discussed extensively.[59] These reactions will not be dealt with further here. Singlet oxygen also accepts hydride from alkoxides in the gas phase; the mechanisms of such reactions have received considerable study.[60]

1.3.3.4 Hydride Transfer from Hydrocarbons

Cycloheptatriene and derivatives thereof donate hydride readily to a variety of carbonium ion acceptors. The position of the end equilibrium depends on the thermodynamics of the exchange.[61–63] These reactions are prototypes of a broad area of carbonium ion chemistry wherein carbonium ions equilibrate *via* intra- and inter-molecular hydride shifts between a donor C—H bond, usually sp^3 hybridized, and a carbonium ion acceptor. This chemistry is often achieved with heterogeneous catalysts and is of great industrial significance; it lies outside the emphasis of this review, however. Excellent treatises are available,[4,64] and a review has appeared on the use of carriers like adamantane to promote hydride transfer in hydrocarbons under strongly acidic conditions.[65]

Although not many examples are known, properly constrained organic compounds can transfer the elements of hydrogen in what appears to be a concerted process akin to reductions by diimide, HN=NH. The reduction of 1,2-dimethylcyclohexene (**72**) by *cis*-9,10-dihydronaphthalene (**71**) is an example (equation 31). Yields are only moderate, but the stereospecificity is consistent with a concerted process.[66]

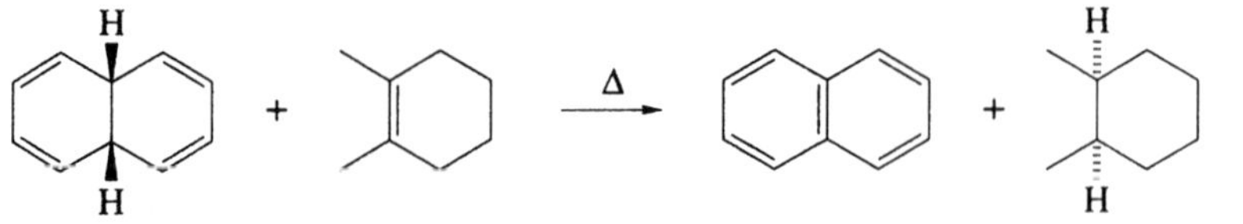

(**71**) + (**72**) —Δ→ (**73**) + (**74**) (31)

1.3.3.5 Hydride Transfer from Heterocycles

Heterocyclic redox couples do not differ in any profound chemical sense from other hydride donors discussed here. Practical aspects such as reasonable stability of both partners of the redox pair as well as synthetic accessibility, do play a major role, however. In the majority of cases a (hetero)aromatic cation is formed, the developing aromaticity being the driving force for the loss of hydride. By far the most used couple is the dihydropyridine/pyridinium salt (or pyridine) combination already discussed.

1.3.3.5.1 1,4-Dihydropyridines

The hydride-donating potential of 1,4-dihydropyridines has already been discussed in a mechanistic context in Section 1.3.2.3. Commonly available 1,4-dihydropyridines are the 'Hantzsch esters' (**7**; X = OR), obtained from condensation of a β-keto ester with ammonia and an aldehyde.[67] By far the simplest syntheses of Hantzsch esters involve condensation of ammonia (substituted amines react in general poorly) and an aldehyde if a hydride equivalent at the 4-position is desired. The Hantzsch synthesis has been used in many guises, also to produce nonsymmetrical systems. The interest in these compounds as calcium antagonists has doubtless stimulated the extensive synthetic effort.[68]

Hantzsch esters like (**7b**) react with appropriate acceptors to provide reduced product and pyridine (**75**), as shown in equation (32). The elements of hydrogen are transferred, but there are no effective ways to transform (**75**) cleanly back to (**7b**) so that a catalytic cycle may be established.

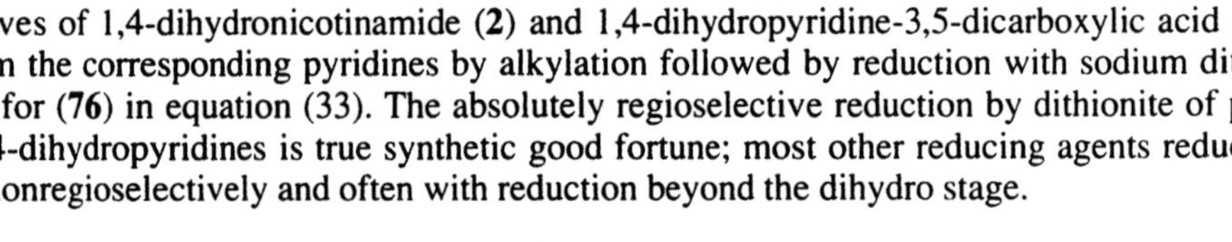

(**7b**) + X=Y ⟶ (**75**) + HXYH (32)

Derivatives of 1,4-dihydronicotinamide (**2**) and 1,4-dihydropyridine-3,5-dicarboxylic acid (**8**) are obtained from the corresponding pyridines by alkylation followed by reduction with sodium dithionite, as illustrated for (**76**) in equation (33). The absolutely regioselective reduction by dithionite of pyridinium salts to 1,4-dihydropyridines is true synthetic good fortune; most other reducing agents reduce pyridinium salts nonregioselectively and often with reduction beyond the dihydro stage.

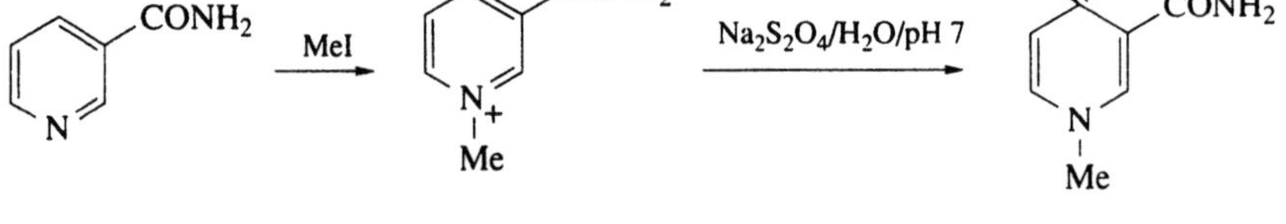

(**76**) —MeI→ (**1c**) —$Na_2S_2O_4/H_2O$/pH 7→ (**2c**) (33)

In general only very electron-deficient double bonds will react spontaneously with 1,4-dihydropyridines. (This statement applies to nonenzymic chemistry; enzymatically, as in LAD, NAD(P)H, with a lesser reduction potential than many of the 1,4-dihydropyridines used in abiotic reactions, reduces unactivated carbonyl groups.) Thiobenzophenone, quinone, maleic acid, and hexachloroacetone react spontaneously with simple dihydropyridines and undergo reduction (the carbon–carbon double bond of maleic anhydride is reduced).[69] Trifluoroacetophenone will also often react spontaneously with 1,4-dihydropyridines.

Reduction of iminium salt (**77**) is illustrated in equation (34). On hydrolysis (**78**) is obtained in good yield.[70]

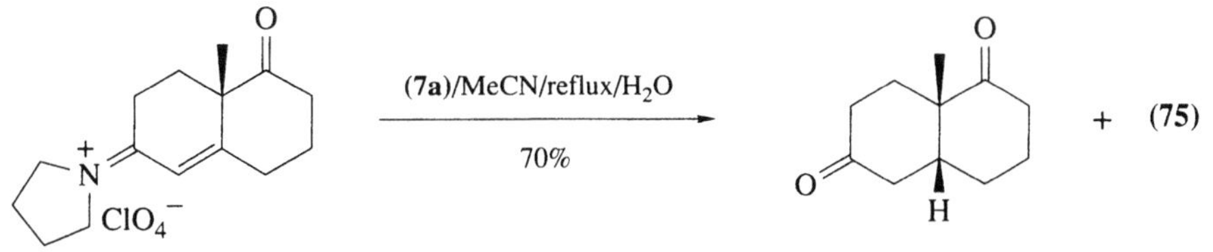

(**7a**)/MeCN/reflux/H_2O, 70%; (**77**) → (**78**) + (**75**) (34)

More complex pathways can be followed dependent on the structure of the hydride acceptor. For example, dihydroacridine reacts cleanly with 2,3,5,6-tetracyanobenzoquinone to generate a pyridinium/radical anion pair.[71]

The acceptors thus far discussed react spontaneously. By far the greatest amount of work has been done by use of an electrophilic activator, present in stoichiometric amounts, to activate the receptor, usually carbonyl. Dry $Mg(ClO_4)_2$ is particularly effective;[72] zinc ions are sometimes used although their effectiveness is often lower. Presumably the magnesium ion polarizes the carbonyl group making it more electrophilic and a better hydride acceptor (equation 9). Ethyl phenylglyoxylate (**9a**) is not reactive enough to react spontaneously with (**2b**); however, on addition of a stoichiometric amount of $Mg(ClO_4)_2$ spontaneous reduction occurs at room temperature in the dark (equation 8).

(i) Enantioselective reductions

In a timely experiment Ohno demonstrated that 1,4-dihydropyridine (*R*)-(**79**) in the presence of $Mg(ClO_4)_2$ reduced (**9a**) to (**10a**) with modest (18%) enantiomeric excess.[73] This observation gave rise to an enormous amount of work on the design, synthesis and reactions of optically active 1,4-dihydropyridines.

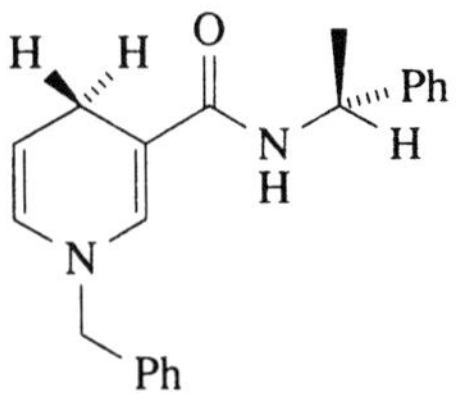

(*R*)-(**79**)

Before further discussion some consideration of the structure and reactivity of 1,4-dihydropyridines is in order. A pyridinium salt is almost certainly capable of accepting a hydride from a 1,4-dihydropyridine. As long as hydride transfer occurs between the 4-positions of structurally analogous 1,4-dihydropyridine and pyridinium salts, the reaction is thermoneutral and leads to no net change. An example is given in equation (35).[74] At 25 °C only 'blind' exchange between the 4-positions occurs and this can only be made visible by isotopic labeling. This '4,4-exchange' reaction can be used to establish the relative redox potentials of 1,4-dihydropyridines (Section 1.3.2.3). But at 67 °C irreversible competitive hydride transfer to the 2(6)-position occurs to form the 1,2-dihydropyridine; all the dihydropyridine present eventually falls into this thermodynamic pit. In any reaction whereby pyridinium salt is generated during the reaction in the presence of unconsumed 1,4-dihydropyridine, exchange and isomerization may be complicating factors especially in the case that the substrates to be reduced are only sluggishly reactive. Additional mechanistic and synthetic aspects of these types of hydride transfers have been considered.[75–77]

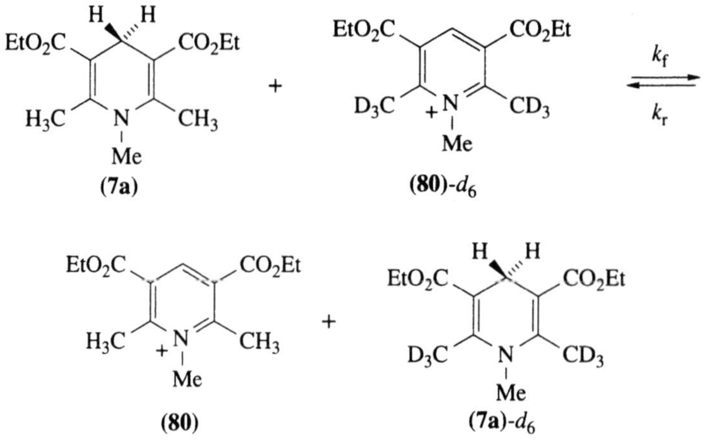

(7a) + (80)-d_6 $\underset{k_r}{\overset{k_f}{\rightleftharpoons}}$ (80) + (7a)-d_6 (35)

K_{eq}(25 °C) = 0.84

An additional complication peculiar to derivatives of 1,4-dihydronicotinamide (**2**) is the pronounced sensitivity of the enaminic 5-position to electrophilic attack.[77,78] The (reversible) addition of water to 1,4-dihydronicotinamides (protonation at the 5-position and addition of hydroxide to the 6-position) has caused difficulties in kinetic measurements[79] and the instability of NADH towards acid is doubtless also the result of chemistry initiated at the enaminic 5-carbon.[69] This problem is not, or only barely, present in 1,4-dihydropyridines like (**7**) and (**8**) which bear electron-withdrawing substituents at the 5- as well as 3-positions.

With this knowledge in mind, enantioselective reductions by chiral 1,4-dihydropyridines can be considered in more detail. Two types will be used to illustrate cogent points: 1,4-dihydropyridines wherein C-4 of the heterocyclic ring is stereogenic, and 1,4-dihydropyridines incorporated into a macrocyclic framework that contains stereogenic centers.

It has been found that (4*R*,9*R*)-(**81**) in the presence of a stoichiometric amount of $Mg(ClO_4)_2$ affords (**10b**) in virtually enantiomerically pure (*R*)-form (equation 36).[80] Similar reductions of 2-benzoylpyridine, 2-acetylpyridine, and 4-chlorotrifluoromethylacetophenone proceeded with enantiomeric excesses ranging from 60–100%.

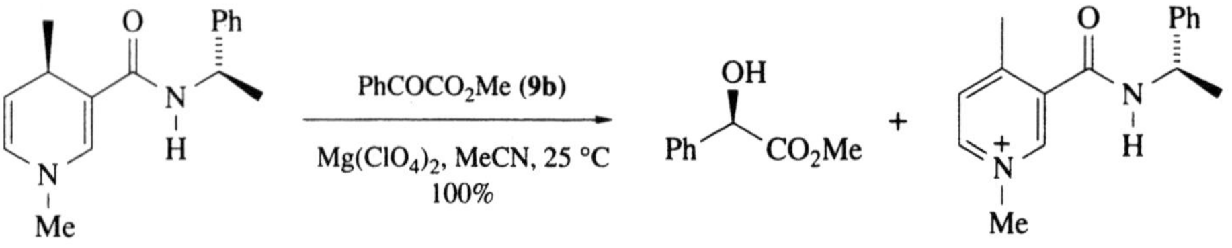

(4*R*,9*R*)-(**81**) $\xrightarrow[\text{Mg(ClO}_4)_2\text{, MeCN, 25 °C; 100\%}]{\text{PhCOCO}_2\text{Me (9b)}}$ (**10b**) 97.6% *ee* + (9*R*)-(**82**) (36)

The stereochemical situation in these experiments may be more involved than one would anticipate at first sight. Buck *et al.*[81] have succeeded in resolving (**83**). These enantiomers can only be the result of axial chirality caused by a torsional barrier to rotation about the aromatic–carbonyl bond being high enough to freeze the amide groups out of plane in the pyridinium salt.

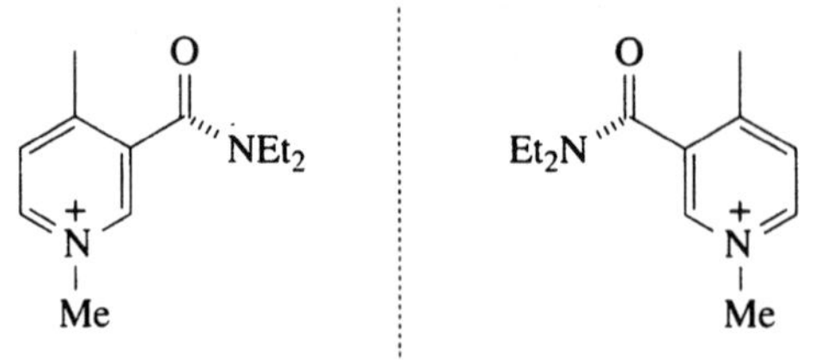

(**83**)

Buck and Donkersloot[82] had suggested on the basis of MINDO/3 calculations that the carbonyl dipole of the amide group will be oriented in the direction of a hydride approaching C-4 of (**83**) (or departing from the corresponding 1,4-dihydropyridine). This hypothesis is supported by several lines of experimental evidence. In the crystal structure the carboxamide group of (4*R*,9*R*)-(**81**) is 65° out of plane and the oxygen is indeed oriented along the line of the carbon–hydrogen bond at C-4.[83] Circumstantial evidence has also been accumulated that nicotinamidium salts with locked conformations about the carboxamide group are reduced on the oxygen side of the carboxamide.[84]

A model for reduction (**84**) proposed by Ohno[85] entails folding the carbonyl back over the dihydropyridine ring; magnesium is complexed simultaneously to both the heterocyclic nitrogen and carbonyl oxygen. It is not clear how axial chirality should be incorporated in this model.

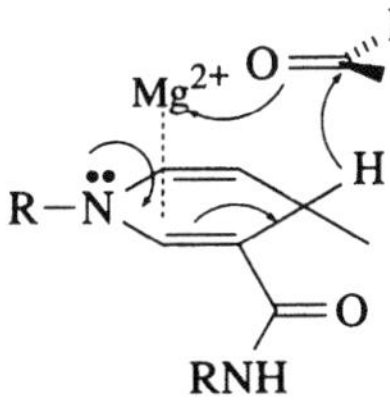

(**84**)

More potent electrophilic activators can be used with 1,4-dihydropyridines more stable to acid conditions. Considerable stabilization is achieved by fusion of an aromatic ring to the 5,6-positions. For example (**85**) will reduce benzaldehyde to benzyl alcohol in moderate yield in the presence of benzenesulfonic acid (equation 37).[86] An optically active variant analogous to (**81**) has also been studied.[87] In addition to proton sources, electrophiles like $AlCl_3$ and $TiCl_4$ may also be used.[88,89]

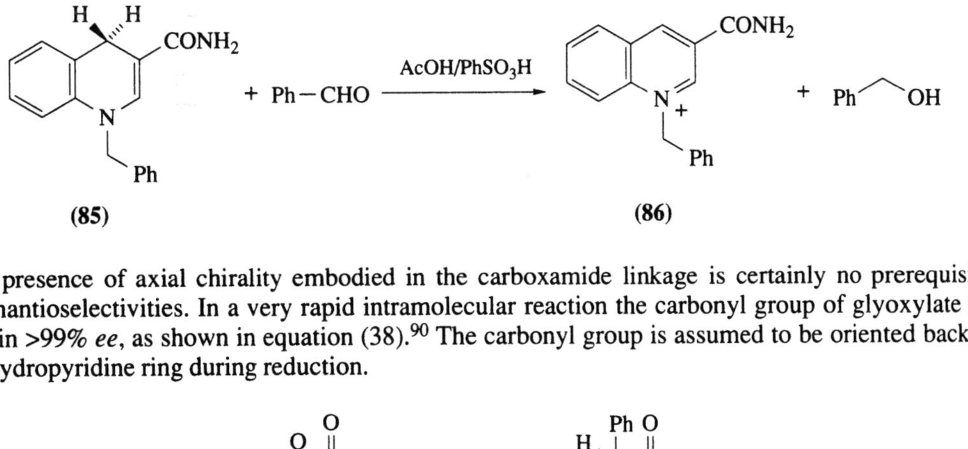

(**85**) + Ph–CHO —AcOH/$PhSO_3H$→ (**86**) + Ph⌒OH (37)

The presence of axial chirality embodied in the carboxamide linkage is certainly no prerequisite to high enantioselectivities. In a very rapid intramolecular reaction the carbonyl group of glyoxylate is reduced in >99% *ee*, as shown in equation (38).[90] The carbonyl group is assumed to be oriented back over the dihydropyridine ring during reduction.

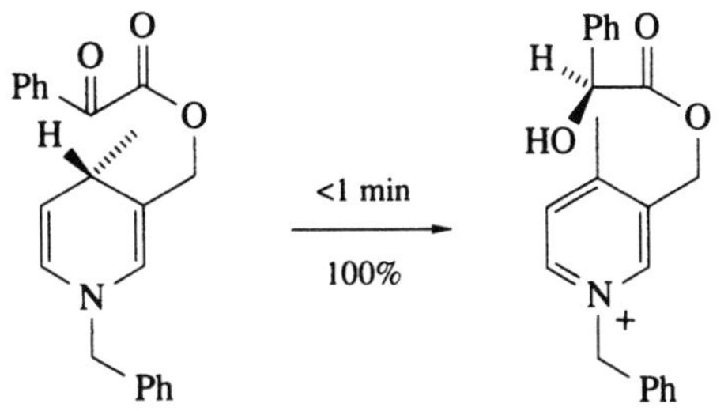

(**87**) —<1 min, 100%→ (**88**) >99% *ee* (38)

A considerably different approach to enantioselective reductions is represented by chiral 1,4-dihydropyridine (**89**).[91] The anticipation was that a structured ternary complex could be formed (equation 39).

As in other experiments with 1,4-dihydropyridines only activated carbonyl compounds like (**9**) could be reduced; (**9a**) is reduced to (*S*)-(**10a**) in 80% yield and 86% *ee* at room temperature in the presence of a stoichiometric amount of $Mg(ClO_4)_2$ (equation 40). This agrees entirely with the prediction embodied in equation (39) if the phenyl group is regarded as the larger and the ester as the smaller substituent.

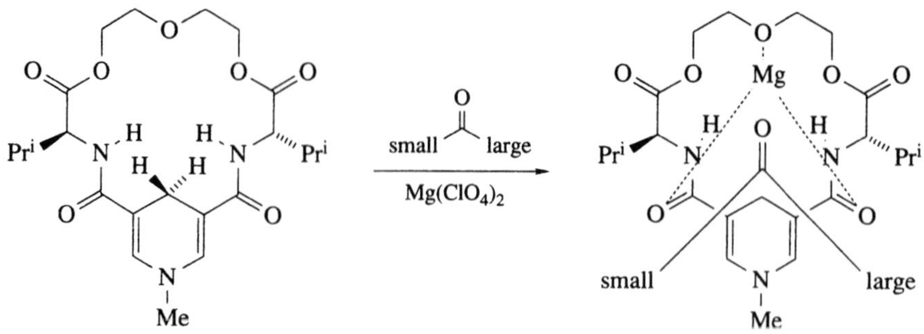

(89) + small/large ketone, $Mg(ClO_4)_2$ → (89)-complex (39)

(89)

(89)-complex

(9a) + (89) $\xrightarrow[80\%]{Mg(ClO_4)_2/MeCN/20\ °C}$ (10a) 86% *ee* (40)

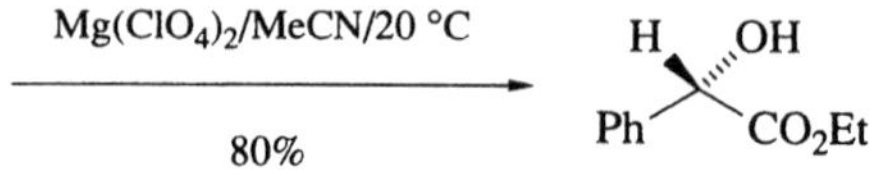

(10a) 86% *ee*

Macrocycles related to (**89**) but derived from alanine, phenylalanine, *t*-leucine,[92] proline and phenylglycine have been prepared. The enantioselectivities for the reduction of (**9a**) increase with the size of the amino acid substituent (proline 0% *ee*, methyl 65% *ee*, benzyl 87% *ee*, isopropyl 86% *ee*). In all cases the natural amino acids of (*S*)-configuration always deliver an excess of the (*S*)-enantiomer of the alcohol.

The bridge that holds the two amino acid carboxylate groups can be varied in shape and length. For bridges five atoms long, *i.e.*—$(CH_2)_2O(CH_2)_2$— or —$(CH_2)_5$—, up to eight atoms long the enantioselectivities remain high. For longer bridges considerable loss of enantioselection occurs, apparently because of the greater conformational flexibility.

From the body of experimental data available a model fairly close to that initially postulated (equation 39) has been developed, except that the carbonyl group is skewed relative to this model about 120°.

It is conceivable that all the various geometries suggested for arrangement of dihydropyridine and carbonyl during the transition state for hydride transfer may be correct. Hydride transfer could occur *via* various orientations analogous to alkoxides; the classical six-membered transition state of the MPV reductions is not mandatory (Section 1.3.3.3). MINDO calculations on the transition state for thermoneutral hydride transfer between a 1,4-dihydropyridine and a pyridinium salt indicate virtually no energy barrier to rotation of the participants, *i.e.* (**90**) is as stable as (**91**).[93]

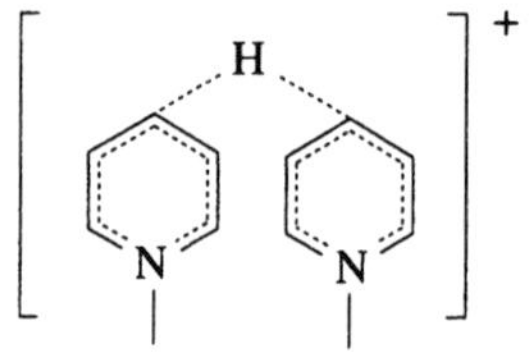

(90) 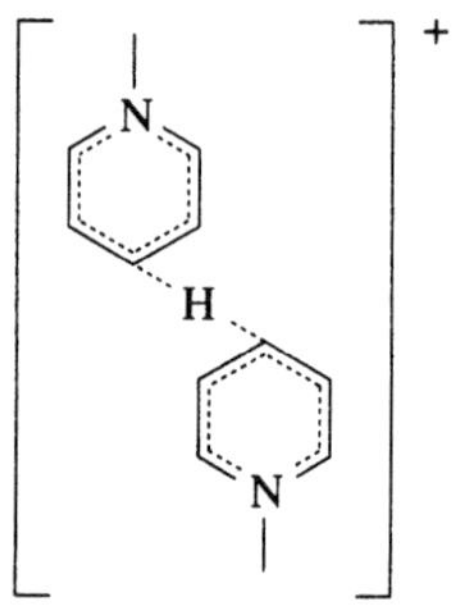 (91)

In other words the relative orientations of hydride donor and double bond acceptor may not be as crucial as once thought.

(ii) Oxidative reactions

The oxidation of primary and secondary alcohols by pyridinium salts, *i.e.* the reverse of the reactions just discussed, is seldom seen. An experimental difficulty is that MPV conditions, that is formation of metal alkoxides, must be used to obtain any reaction at all. Under these strongly basic conditions most pyridinium salts are unstable. There is a dearth of processes of obvious synthetic usefulness based on this approach.[94]

A notable exception is catalytic oxidation wherein a pyridodipyrimidine is used as a shuttle in the oxidation of alcohols by molecular oxygen. Pyridodipyrimidine (**92**) oxidizes cyclohexanol to cyclohexanone and the dihydro form (**93**) is oxidized *in situ* back to (**92**) by oxygen (equation 41). The yield of cyclohexanone is >22 000% based on (**92**).[95]

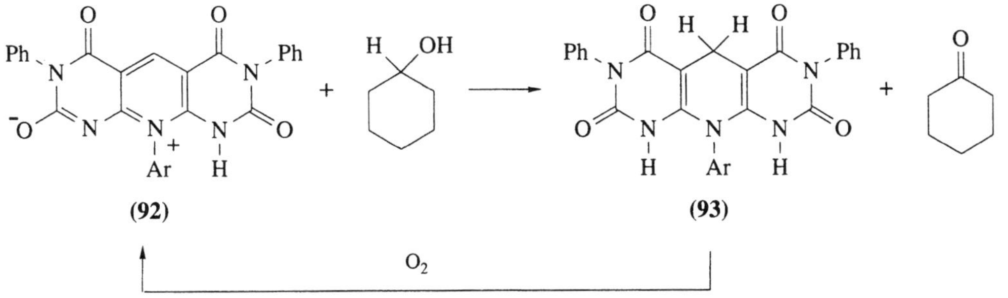

(**92**) + cyclohexanol → (**93**) + cyclohexanone (41)

(**93**) —O_2→ (**92**)

(iii) Catalyzed hydride transfer

Few truly catalytic systems have been reported. An *N*-alkylated 1,4-dihydropyridine forms a pyridinium salt on hydride loss. In principle the pyridinium salt may be recovered and reduced regioselectively with sodium dithionite (equation 33). This reaction is very general and fails only when the 1,4-dihydropyridines are excessively strained or too reactive to be isolated.

Attempts to design systems wherein $Na_2S_2O_4$ is used as a continuous electron source *via* an *in situ* reduction of the pyridinium salt formed have met with little success. In principle one would expect a two-phase system consisting of $Na_2S_2O_4$ in the aqueous layer and 1,4-dihydropyridine/substrate in the organic layer to be effective. The charged pyridinium salt will have some solubility in water so that it can be reduced. Unfortunately the electrophilic activators tend to remain in the aqueous rather than organic layer and $Na_2S_2O_4$ has a pronounced tendency to reduce directly most potential ketonic substrates. These problems can be circumvented, at least on a small scale, by using a water/toluene two-phase system, zinc metal as electron source, the bis-*N*-methylpyridinium salt of 4,4′-bipyridine as electron carrier, and a 1,4-dihydronicotinamide/nicotinamidium salt couple as reductor. Reasonable rates of reduction of (**9a**) to (**10a**) can be obtained.[96] Attachment of the 1,4-dihydropyridine to a polymer or other solid matrix is another potential solution to the problem of regeneration.

Certain transition metal complexes exhibit activating properties and act with turnover on the metal center analogously to the catalytically active zinc ion in the active center of liver alcohol dehydrogenase. Various chiral europium shift reagents, for example $Eu(hfc)_3$, induce reduction of (**9b**) by 1,4-dihydronicotinamides. Turnovers of about 100 are obtained on the metal complexes and methyl mandelate is formed with enantiomeric excesses of 25–44%.[97]

Complex (**94**) (the corresponding Zn^{2+} complex is insoluble) will induce the same reduction with turnovers of about 100 on the complex. This complex approximates the coordination about zinc in the active site of liver alcohol dehydrogenase.[98]

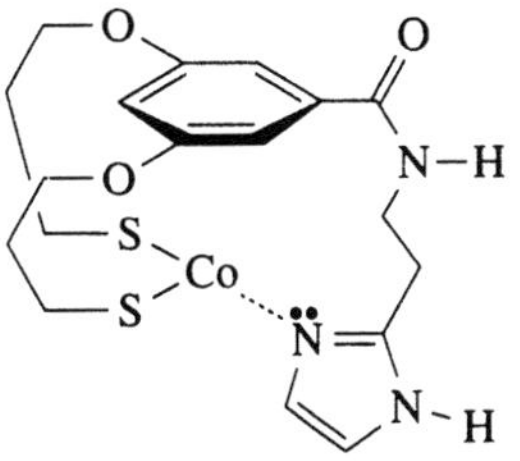

(**94**)

1.3.3.5.2 Other heterocyclic hydride donors

The 1,4-dihydropyridine/pyridinium salt couple is by far the most common six-membered heterocyclic ring redox couple. Other combinations are possible, for example pyrimidines and pyrazines. However, the dihydro forms tend often to be poorly defined, consist of more than one regioisomer, and/or are too reactive to be readily handled.

Five-membered heterocyclic ring systems can also act as redox shuttles. An example using an uncharged form of the reduced heterocycle is shown in equation (42).[99,100] The *N*-alkylated derivatives of (**95**), which provide positively charged ions on hydride loss, seem to be somewhat more reactive.[101,102]

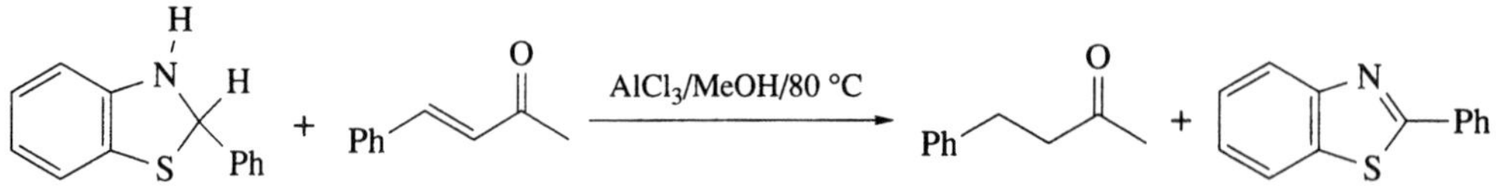

$$\text{(95)} + \text{(96)} \xrightarrow{\text{AlCl}_3\text{/MeOH/80 °C}} \text{(97) 87\%} + \text{(98) 98\%} \quad (42)$$

1.3.3.5.3 *Tertiary anilines as hydride donors*

Although also a special case of a MPV reduction (Section 1.3.3.3) some unusual reactions of practical import will be considered here because of the heterocyclic characteristics. Donation in an intramolecular reaction of a hydride from a tertiary aniline derivative to generate an iminium ion initiates the reaction. An example is given in Scheme 7. Tertiary aniline derivative (**99**) rearranges thermally apparently *via* intermediate (**100**) to pyrrolo[1,2-*a*]quinoline (**101**).[103] The highly electron-deficient double bond is clearly the hydride acceptor; in this case exclusively the most substituted iminium salt is formed. Subsequent ring closure provides the end product (**101**). Formation of the more substituted iminium ion is not always the case.[104,105]

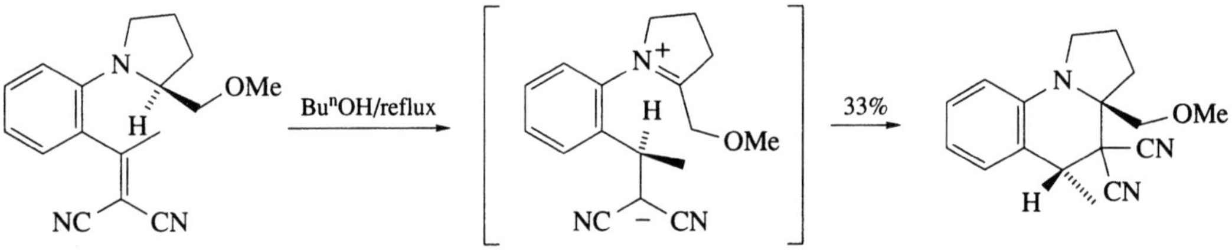

(**99**) —Bu^{n}OH/reflux→ [(**100**)] —33%→ (**101**)

Scheme 7

Extensive use of this process for synthesis of various heterocycles has been made. The cited examples are probably the best studied of a broad set of unusual reactions of tertiary aniline derivatives wherein the aromatic ring bears an *ortho* substituent. These examples have been compiled in a remarkably prescient review by Meth-Cohn and Suschitzky.[106]

1.3.3.6 Hydride Transfer from Organometallics

Hydrogens located on a carbon adjacent to a negatively charged atom (*i.e.* β) are always potential sources of hydride; the possibility of formation of a double bond (equation 43) is an effective driving force. As seen with MPV reductions (X = O) the success of the hydride transfer depends also on the thermodynamics of the reaction, the possibility of finding a proper orientation of the reacting partners, the cation and the nature of ion pairing. The possibilities for carbanions are in general greater since the carbanionic character is profoundly affected by the choice of metal counterions. Covalency, back donation possibilities from the metal, as well as ligation of the carbanion acceptor to the metal with concomitant increase of effective molarity are some of the additional factors to be considered.

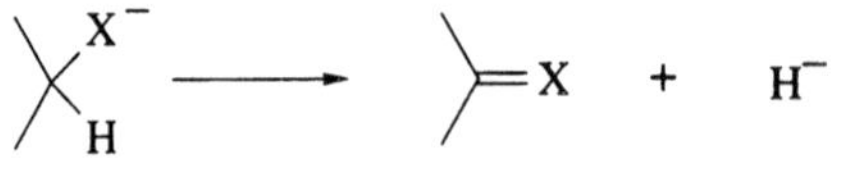

$$\text{>C(H)–X}^- \longrightarrow \text{>C=X} + \text{H}^- \quad (43)$$

The additional stereochemical control available in carbanions relative to alkoxides arises both from the extra ligation about carbon and the contributions of the metal. The stereochemistry of hydride transfer from organostannanes has been particularly well investigated. Coordinatively saturated metals like tin function less well as Lewis acids in a cyclic mechanism, and tend to induce hydride loss through an antiperiplanar conformation.[107] The hydride transfer reactions of organometallics have long been known and are in many cases chiefly an annoying complication to C—C bond formation, especially in the reactions of Grignard reagents.

1.3.3.6.1 Divalent metals: Grignard reagents and organozinc compounds

An example of reduction of a ketone by a Grignard reagent is given in equation (44).[108] The combination of a hindered ketone (**103**) and branched Grignard reagent (**102**) inhibits carbanionic addition to the carbonyl group (<3% occurs) and promotes reduction. On the basis of extensive work ably reviewed by Morrison and Mosher,[109] it appears that a MPV-like transition state (**106**) is involved. Not an antiperiplanar but a *syn* arrangement of metal and departing hydride is involved in this case. Stereochemical analyses are complicated by the complex nature of Grignard reagents and the complexation of ether molecules to the magnesium. The scope and limitations and the possibilities for enantioselective synthesis have been discussed in detail.[109] The use of chiral Grignard reagents for enantioselective syntheses has been largely supplanted by boron reagents (Section 1.3.3.6.2ii).

Et_2O/35 °C, 97% (44)

(*S*)-(**102**) (**103**) (*R*)-(**104**) 36% *ee* (**105**)

(**106**)

Dialkylzincs are on the whole less reactive towards addition to carbonyl groups.[110] Diethylzinc reduces benzophenone without significant formation of a tertiary alcohol by addition.[111] Reaction of phenyl isopropyl ketone (**107**) with (+)-bis[(*S*)-2-methylbutyl]zinc (**108**) affords a modest enantiomeric excess of alcohol (**109**) and alkene (**110**; equation 45).[112] The kinetics of the reaction are second order; a six-membered transition state analogous to the Grignard reduction seems probable. Reaction is clearly sluggish compared to the corresponding Grignard reagents. The enantiomeric excesses obtained are not high and the factors that lead to enantioselection are not readily identified. As in the case of the Grignard reagents, the coordination about zinc is uncertain. This latter point is no triviality; the reactivity and potential for enantiomeric recognition are profoundly affected by the nature of the ligation.[113]

no solvent/24 h/86.5 °C, 79% (45)

(**107**) (**108**) (**109**) 15.2% *ee* (**110**)

1.3.3.6.2 Trivalent metals: alkyl-aluminum and -boron compounds

(i) Alkylaluminums

Trialkylaluminum compounds and trialkylboranes seem to reduce carbonyls in the same fashion as Grignard reagents and dialkylzincs. The reactivity towards carbonyl groups is greater than that of dialkylzincs, despite the third substituent, which does provide, however, additional opportunities for recognition. The compounds are well defined and can encompass a fairly wide range of alkyl substituents. This is especially true of the trialkylboranes.

Triisobutylaluminum (racemic) is commercially available in toluene solution. Triethylaluminum and related compounds are used, of course, commercially in Ziegler–Natta polymerization.[114] These solutions can be handled safely in contrast to the pure materials, which are violently reactive. The applications of triisobutylaluminum have been reviewed.[115] Its use is difficult to divorce from its chemical relative, diisobutylaluminum hydride, which is probably more often used for reductions of carbonyl groups. This latter reagent reduces, of course, *via* the reactive aluminum–hydride bond. The thought that the dialkylaluminum is less bulky than the trimer is misleading; there is a greater tendency of the former towards aggregation.[116]

Triisobutylaluminum (**111**), of the trialkylaluminums, is probably used most often for reductions. The reducing capacity of only one of the isobutyl groups is used. Reduction of carbonyl groups apparently occurs by the now familar MPV six-membered ring mechanism found with Grignard reagents and dialkylzincs (equation 46). Alkynes and nitriles (benzonitrile has been studied) are also reduced by triisobutylaluminum *via* analogous six-membered ring transition states. The very bulky reagent (**115**), prepared from diisobutylaluminum hydride and 2,6-di-*t*-butyl-4-methylphenol is useful for the stereoselective reduction of the α,β-unsaturated carbonyl functionality of prostaglandins; reduction of (**114**) affords a ratio of (15*S*):(15*R*)-alcohols of 92:8 (equation 47).[117] The (15*S*)-diastereomer is the desired product and also the more difficult to obtain.

Bu^i_3Al (**111**) → [n-C_5H_{11}, OH, H, O; −78 °C; toluene] → [Bu^i_2Al … H … O=C–R] (**112**) R = (*S*)-C(OH)(n-C_5H_{11})H → [52% (2% *threo*)] → n-C_5H_{11}, OH, H, OH, H (**113**) (46)

(**114**) + Bu^i_2AlO–Ar (Bu^t, Bu^t) (**115**) → [toluene/−70 °C; 95%] → (**116**) 92:8 *de* (47)

Quite extensive studies of the reduction of (chiefly phenyl) ketones have been carried out with optically active trialkylaluminum (**117**). Enantiomeric excesses are at best only moderate. Again it is difficult to develop a model that rationalizes the interactions between groups that determine the enantiomeric recognition capacities of the reagent. Comparisons of reactivity and recognition capacities of RMgCl, R_3Al, and R_2Be reagents with the same R-groups have been made.[118]

(*S*)-[ ]$_3$Al

(**117**)

Much better results in enantioselective reductions have been obtained with reagent (**118**) derived from β-pinene.[118e] For example, reduction of (**119**) affords the corresponding alcohol (**120**) in excellent enantiomeric excess (equation 48). Good results have also been obtained in the reductions of a number of other carbonyls.

(48)

(118) **(119)** **(120)** 78% *ee*

(ii) Reductions by trialkylborons

Quite practical reagents, although of relatively high molecular weight per hydride equivalent, have been developed from sterically crowded trialkylboranes, derived from addition of a dialkylborane to an alkene (equation 49).[119] That trialkylboranes can react with carbonyl groups *via* donation of hydride and freeing of an alkene was apparent from the reaction of tributylborane (**121**) with formaldehyde (equation 50).[120]

(49)

$$Bu_3B + CH_2O \longrightarrow Bu_2BOMe + \text{propene} \qquad (50)$$

(121) **(122)**

The trialkylborane (**123**) derived from addition of 9-borobicyclo[3.3.1]nonane to α-pinene reduces benzaldehyde-α-*d* (**124**) in nearly 100% *ee* (equation 51) after correction for optical purity of (**123**) (commercial α-pinene is only about 92% enantiomerically pure).[121] Only a single hydride equivalent, the tertiary hydrogen of pinene, is transferred. Under analogous conditions simple aldehydes are reduced to α-*d*-alcohols in 81–100% *ee*.[122]

i, THF/reflux; ii, $HOCH_2CH_2NH_2$; 82% (51)

(123) **(124)** **(125)** 100% *ee* **(126)**

Again rationalization of the stereochemical course of reaction is not easy. The Lewis salt (**127**), formed by complexation of oxygen with boron (equation 52), may be the actual reactive species. To explain the stereochemistry of reduction observed in equation (51), (**128**) has been suggested, with the R-group oriented over the pinane ring in what seems at first glance an abuse of stereochemical logic. The explanation may lie in the formation of a 'pocket' in which the alkyl group can fit in hydrophobic surroundings.

$$R_3B + RCHO \rightleftharpoons R_3\bar{B}-\overset{+}{O}=CHR \qquad (52)$$

(127)

The rates of hydride transfer with (**123**) are only moderate with aldehydes in THF solution. However if no solvent is used and reaction times of several days are tolerated both aliphatic and aryl ketones can be reduced to the corresponding secondary alcohols (equation 53).[123] Although reactions are sluggish,

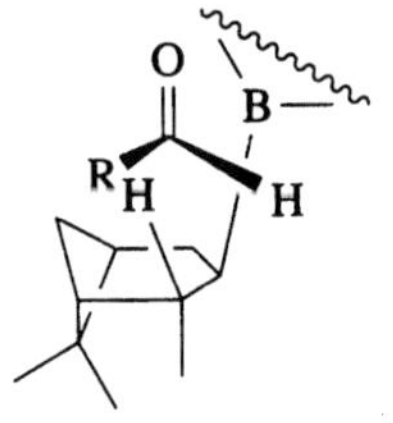

(128)

several hours in THF solution at room temperature being required for reaction, alkynic ketones (**131**) are reduced by (**123**) in quite reasonable enantiomeric excesses (equation 54).[124]

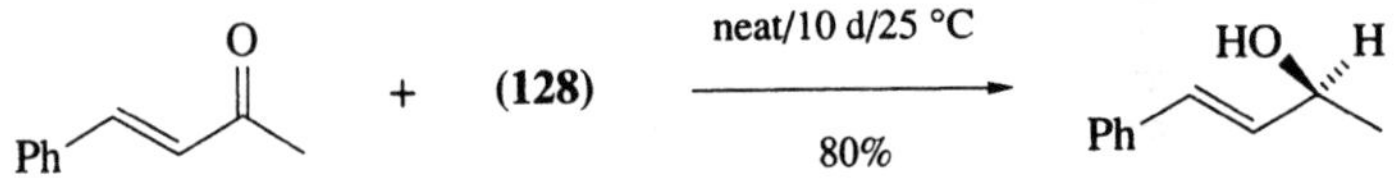

(129) + (128) —neat/10 d/25 °C, 80%→ (130) 97% *ee* (53)

(131) + (123) —i, THF/35 °C; ii, hydrolysis→ (132) (54)

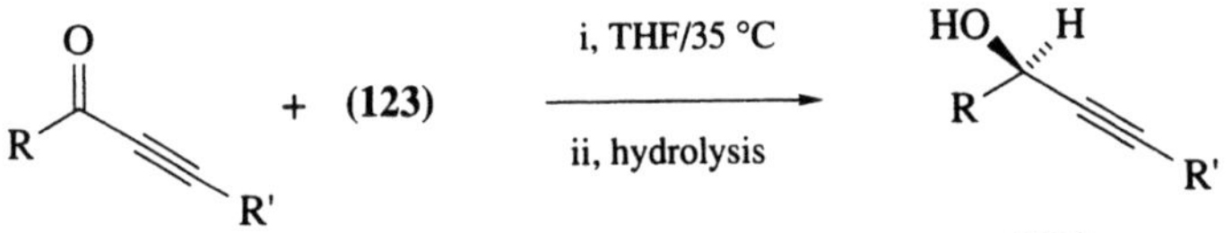

R	*R'*	*yield* (%)	*ee* (%)
Ph	Bu^n	72	89
Pr^i	H	78	91
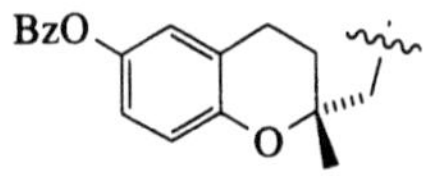 BzO-substituted chromanyl	Me	77	85:15 *de*
Ph	CO_2Et	64	92

Synthetic applications of this methodology have been made[125] and mechanistic investigations have been carried out.[126] Reagents (**133**) and (**134**), derived, respectively, from addition of 9-borobicyclo[3.3.1]nonane to β-pinene and nopol benzyl ether, are also capable of delivering a single hydride equivalent with good enantioselectivity.[127,128]

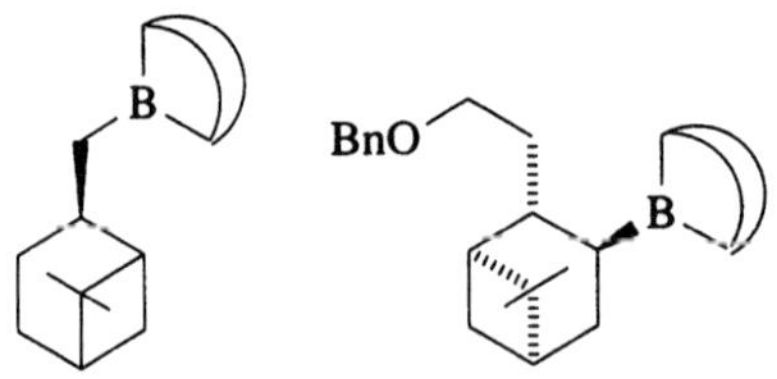

(133) (134)

The boroadamantane (**135**), formed by addition of methyllithium to the corresponding trialkylborane, transfers a bridgehead hydride to acetyl chloride to form (**136**), as shown in equation (55).[129]

(135) + MeCOCl ——→ (136) + MeCHO (55)

1.3.3.7 Hydride Transfer from Transition Metal Alkyls

Considerable mechanistic examination of hydride transfer from transition metal alkyls where M is, for example, $C_5H_5Fe(CO)_2$ or $C_5H_5Re(NO)PPh_3$ has been carried out. Hydride transfer may occur either from an α- or β-position on the alkyl framework, dependent on structural features and the identity of the transition metal. Stereoelectronic effects, kinetic factors, and electronegativity considerations have been considered in recent review articles.[130,131] Because of the excellent coverage already available, no detailed discussion will be given here.

1.3.4 REFERENCES

1. (a) F. H. Westheimer, H. F. Fisher, E. E. Conn and B. Vennesland, *J. Am. Chem. Soc.*, 1951, **73**, 2403; (b) F. H. Westheimer, in 'Coenzymes and Cofactors. Pyridine Nucleotide Coenzymes, Part A', ed. D. Dolphin, R. Poulson and O. Avramovíc, Wiley, New York, 1987, p. 253.
2. C. R. Noller, 'Chemistry of Organic Compounds', 3rd edn. Saunders, Philadelphia, 1965, p. 245.
3. R. B. Woodward and R. Hoffmann, in 'The Conservation of Orbital Symmetry', Verlag Chemie, Weinheim, 1970, p. 114.
4. R. A. W. Johnstone, A. H. Wilby and I. D. Entwistle, *Chem. Rev.*, 1985, **85**, 129.
5. R. M. Kellogg and O. Piepers, *J. Chem. Soc., Chem. Commun.*, 1982, 402.
6. C. M. Visser and R. M. Kellogg, 'Redesigning The Molecules of Life', ed. S. Benner, Springer-Verlag, Berlin, 1988.
7. (a) R. P. Kirchen, T. S. Sorensen, K. Wagstaff and A. M. Walker, *Tetrahedron*, 1986, **42**, 1063; (b) R. P. Kirchen, T. S. Sorensen and K. Wagstaff, *J. Am. Chem. Soc.*, 1978, **100**, 6761.
8. M. L. Moore, *Org. React. (N.Y.)*, 1941, **5**, 301.
9. W. S. Emerson, *Org. React. (N.Y.)*, 1940, **4**, 174.
10. J. E. C. Hutchins, D. A. Binder and M. M. Kreevoy, *Tetrahedron*, 1986, **42**, 993.
11. R. E. Bouwman and H. H. Stroud, *J. Chem. Soc.*, 1950, 1342.
12. (a) T. Hayashi, M. Konishi, M. Fukushima, K. Kanehira, T. Hioki and M. Kumada, *J. Org. Chem.*, 1983, **48**, 2195; (b) J. H. Griffin and R. M. Kellogg, *J. Org. Chem.*, 1985, **50**, 3261; (c) B. K. Vriesema, M. Lemaire, J. Buter and R. M. Kellogg, *J. Org. Chem.*, 1986, **51**, 5169.
13. S. Cacchi, P. G. Ciattini, E. Morera and G. Ortar, *Tetrahedron Lett.*, 1986, **27**, 5541.
14. S. Ram and R. E. Ehrenkaufer, *Synthesis*, 1988, 91.
15. A. G. M. Barrett and C. D. Spilling, *Tetrahedron Lett.*, 1988, **29**, 5733.
16. H. Brunner and W. Leitner, *Angew. Chem., Int. Ed. Engl.*, 1988, **27**, 1180.
17. R. Ruppert, S. Herrmann and E. Steckhan, *J. Chem. Soc., Chem. Commun.*, 1988, 1150.
18. W. Kantlehner, 'The Chemistry of Functional Groups, Supplement B: The Chemistry of Acid Derivatives', ed. S. Patai, Interscience, New York, 1979, p. 533.
19. (a) J. M. Erhardt and J. D. Wuest, *J. Am. Chem. Soc.*, 1980, **102**, 6363; (b) J. M. Erhardt, E. R. Grover and J. D. Wuest, *J. Am. Chem. Soc.*, 1980, **102**, 6365; (c) see also T. J. Atkins, *J. Am. Chem. Soc.*, 1980, **102**, 6364.
20. T. A. Geissman, *Org. React. (N.Y.)*, 1944, **2**, 94.
21. For various mechanistic suggestions see: (a) C. G. Swain, A. L. Powell, W. A. Sheppard and C. R. Morgan, *J. Am. Chem. Soc.*, 1979, **101**, 3576; (b) C. G. Swain, A. L. Powell, T. J. Lynch, S. R. Alpha and R. P. Dunlap, *J. Am. Chem. Soc.*, 1979, **101**, 3584.
22. (a) H. S. Rzepa and J. Miller, *J. Chem. Soc., Perkin Trans. 2*, 1985, 717.
23. (a) E. C. Ashby, D. T. Coleman, III and M. P. Gamasa, *Tetrahedron Lett.*, 1983, **24**, 851; *J. Org. Chem.*, 1987, **52**, 4079.
24. S.-K. Chung, *J. Chem. Soc., Chem. Commun.*, 1982, 480.
25. K. Takahashi, N. Tekezawa and H. Kobayashi, *Chem. Lett.*, 1983, 1061.
26. (a) S. Toma and V. Kaliska, *Acta Fac. Rerum Nat. Univ. Cumenianae, Chim.*, 1984, **32**, 147 (*Chem. Abstr.*, 1984, **101**, 210 666b); (b) A. Fuentes and J. V. Sinisterra, *Tetrahedron Lett.*, 1986, **27**, 2967; (c) T. Ando, J. H. Clark, D. G. Cork and T. Kimura, *Chem. Lett.*, 1987, 1315.
27. M. Bianchi, U. Matteoli, G. Menchi, P. Frediani and F. Piacenti, *J. Organomet. Chem.*, 1982, **240**, 65.
28. H. Fredenhagen and K. F. Bonhoeffer, *Z. Phys. Chem., Abt. A.*, 1938, **181**, 379.
29. I. H. Rajyaguru and H. S. Rzepa, *J. Chem. Soc., Chem. Commun.*, 1987, 998.
30. (a) L. G. Együd, L. A. McLaughlin and A. Szent-Gyogyi, *Proc. Natl. Acad. Sci. USA*, 1967, **67**, 1422; (b) R. Vince and S. Daluge, *J. Med. Chem.*, 1971, **14**, 35.
31. (a) I. A. Rose, *Biochim. Biophys. Acta*, 1957, **25**, 214; (b) V. Franzen, *Chem. Ber.*, 1956, **89**, 1020; (c) V. Franzen, *Chem. Ber.*, 1955, **88**, 1361; (d) J. Hine and C. D. Fischer, Jr., *J. Am. Chem. Soc.*, 1975, **97**, 6513; (e) C. Walsh, in 'Enzymatic Reaction Mechanisms', Freeman, San Francisco, 1979, p. 319.
32. J. D. Morrison and H. S. Mosher, 'Asymmetric Organic Reactions', Prentice-Hall, Englewood Cliffs, NJ, 1971, p. 154.
33. (a) S. S. Hall, A. M. Doweyko and F. Jordan, *J. Am. Chem. Soc.*, 1978, **100**, 5934, and references cited therein; (b) T. Okuyama, K. Kimura and T. Fueno, *Bull. Chem. Soc. Jpn.*, 1982, **55**, 1493.
34. S. Shinkai, T. Yamashita, Y. Kusano and O. Manabe, *J. Am. Chem. Soc.*, 1981, **103**, 2070.
35. M. A. K. Patterson, R. P. Szajewski and G. M. Whitesides, *J. Org. Chem.*, 1981, **46**, 4682.
36. (a) H. von Euler and B. Eistert, 'Chemie und Biochemie der Reduktone und Reduktonate', Enke Verlag, Stuttgart, 1957; (b) K. Schank, *Synthesis*, 1972, 176.
37. A. L. Wilds, *Org. React. (N.Y.)*, 1944, **2**, 178.
38. C. Djerassi, *Org. React. (N.Y.)*, 1951, **6**, 207.
39. M. Majewski, *Tetrahedron Lett.*, 1988, **29**, 4057.

40. J. D. Morrison and H. S. Mosher, 'Asymmetric Organic Reactions', Prentice-Hall, Englewood Cliffs, NJ, 1971, p. 160.
41. M. P. Doyle and C. T. West, 'Stereoselective Reductions', Dowden, Hutchinson and Ross, Stroudsburg, PA, 1976.
42. W. von E. Doering and R. W. Young, *J. Am. Chem. Soc.*, 1950, **72**, 631.
43. (a) Y.-D. Wu and K. N. Houk, *J. Am. Chem. Soc.*, 1987, **109**, 906; (b) see also B. G. Hutley, A. E. Mountain, I. H. Williams, G. M. Maggiora and R. J. Schowen, *J. Chem. Soc., Chem. Commun.*, 1986, 267; (c) for earlier work see H. B. Bürgi, J. D. Dunitz, J.-M. Lehn and G. Wipff, *Tetrahedron*, 1974, **30**, 1563; (d) S. Scheiner, W. N. Lipscomb and D. A. Kleier, *J. Am. Chem. Soc.*, 1976, **98**, 4770; (e) A. J. Stone and R. W. Erskine, *J. Am. Chem. Soc.*, 1980, **102**, 7185; (f) S. Ingemann, J. C. Kleingeld and M. N. Nibbering, *J. Chem. Soc., Chem. Commun.*, 1982, 1009.
44. D. N. Kirk and A. Mudd, *J. Chem. Soc. C*, 1969, 804.
45. M. G. Essig, T. T. Stevenson, F. Shafizadeh, R. E. Stenkamp and L. H. Jensen, *J. Org. Chem.*, 1984, **49**, 3652.
46. B. J. Calvert and J. D. Hobson, *J. Chem. Soc.*, 1965, 2723.
47. G. Tarzia and G. Panzone, *Gazz. Chim. Ital.*, 1978, **108**, 591 (*Chem. Abstr.*, 1979, **91**, 39 425w).
48. E. W. Warnhoff, P. Reynolds-Warnhoff and M. Y. H. Wong, *J. Am. Chem. Soc.*, 1980, **102**, 5956, and references cited therein.
49. I. Watt, S. N. Whittleton and S. M. Whitworth, *Tetrahedron*, 1986, **42**, 1047, and references cited therein.
50. O. R. Martin, S. P. Rao, H. A. El-Shenawy, K. G. Kurz and A. B. Cutler, *J. Org. Chem.*, 1988, **53**, 3287.
51. P. D. Bartlett and J. D. McCollum, *J. Am. Chem. Soc.*, 1956, **78**, 1441.
52. Kabir-ud-Din and P. H. Plesch, *J. Chem. Soc., Perkin Trans. 2*, 1978, 937.
53. C. D. Nenitzescu, in 'Carbonium Ions', ed. G. A. Olah and P. von R. Schleyer, Wiley-Interscience, New York, 1970, vol. 2, p. 482.
54. S. Gladiali, G. Chelucci, G. Chessa, G. Delogu and F. Soccolini, *J. Organomet. Chem.*, 1987, **327**, C15.
55. P. Kvintovics, B. R. James and B. Heil, *J. Chem. Soc., Chem. Commun.*, 1986, 1810.
56. G. Zassinovich and G. Mestroni, *J. Mol. Catal.*, 1987, **42**, 86.
57. H. Brunner, *Synthesis*, 1988, 645.
58. E. C. Chukovskaya, R. K. Friedlina and N. A. Kuz'mina, *Synthesis*, 1983, 773.
59. N. J. Turro, 'Modern Molecular Photochemistry', Benjamin/Cummings, Menlo Park, CA, 1978.
60. R. J. Schmitt, V. M. Bierbaum and C. H. DePuy, *J. Am. Chem. Soc.*, 1979, **101**, 6443.
61. C. D. Nenitzescu, in 'Carbonium Ions', ed. G. A. Olah and P. von R. Schleyer, Interscience, New York, 1970, vol. 2, p. 463, and references cited therein.
62. H. H. Freedman and A. M. Frantz, Jr., *J. Am. Chem. Soc.*, 1962, **84**, 4165.
63. (a) B. Föhlisch and R. Braun, *Tetrahedron Lett.*, 1978, 2735; (b) K. Conrow, *J. Am. Chem. Soc.*, 1961, **83**, 2343.
64. G. A. Olah and P. von R. Schleyer (eds.), 'Carbonium Ions', Wiley-Interscience, New York, 1970; ref. 4, p. 129.
65. G. M. Kramer, *Tetrahedron*, 1986, **42**, 1071.
66. W. von E. Doering and J. W. Rosenthal, *J. Am. Chem. Soc.*, 1967, **89**, 4534.
67. (a) U. Eisner and J. Kuthan, *Chem. Rev.*, 1972, **72**, 1.
68. (a) B. Loev, M. M. Goodman, K. M. Snader, R. Tedeschi and E. Macko, *J. Med. Chem.*, 1974, **17**, 956; (b) L. Dagnino, M. C. Li-Kwon-Ken, M. W. Wolowyk, H. Wynn, C. R. Triggle and E. E. Knaus, *J. Med. Chem.*, 1986, **29**, 2524, (c) D. M. Stout and A. I. Meyers, *Chem. Rev.*, 1982, **82**, 223.
69. T. C. Bruice and S. J. Benkovic, 'Bioorganic Mechanisms', Benjamin, New York, 1966, vol. 2, p. 301.
70. (a) R. A. Gase and U. K. Pandit, *J. Am. Chem. Soc.*, 1979, **101**, 7059; (b) M. J. de Nie-Sarink and U. K. Pandit, *Tetrahedron Lett.*, 1978, 1335.
71. G. Saito and A. K. Colter, *Tetrahedron Lett.*, 1977, 3325.
72. (a) Y. Ohnishi, M. Numakunai and A. Ohno, *Tetrahedron Lett.*, 1975, 3813; (b) A. Ohno, M. Ikeguchi, T. Kimura and S. Oka, *J. Chem. Soc., Chem. Commun.*, 1978, 328; (c) Y. Ohnishi, M. Kagami and A. Ohno, *J. Am. Chem. Soc.*, 1975, **97**, 4766; (d) R. A. Gase and U. K. Pandit, *J. Am. Chem. Soc.*, 1979, **101**, 7059.
73. Y. Ohnishi, M. Kagami and A. Ohno, *J. Am. Chem. Soc.*, 1975, **97**, 4766.
74. T. J. van Bergen, T. Mulder, R. A. van der Veen and R. M. Kellogg, *Tetrahedron*, 1978, **34**, 2377.
75. J. W. Bunting and J. L. Bolton, *Tetrahedron*, 1986, **42**, 1007.
76. (a) W. W. Epstein and H. C. Rilling, *J. Biol. Chem.*, 1970, **245**, 4597; (b) J. W. Cornforth, *Chem. Soc. Rev.*, 1973, **2**, 1; (c) B. M. Trost, P. Conway and J. Stanton, *J. Chem. Soc., Chem. Commun.*, 1971, 1639.
77. (a) T. J. van Bergen and R. M. Kellogg, *J. Am. Chem. Soc.*, 1976, **98**, 1962; (b) T. J. van Bergen, D. M. Hedstrand, W. H. Kruizinga and R. M. Kellogg, *J. Org. Chem.*, 1979, **44**, 4953, and references cited therein.
78. U. K. Pandit, R. A. Gase, F. R. Mas Cabré and M. J. de Nie-Sarink, *J. Chem. Soc., Chem. Commun.*, 1975, 211.
79. (a) J. J. Steffens and D. M. Chipman, *J. Am. Chem. Soc.*, 1971, **93**, 6694; (b) D. M. Chipman, R. Yaniv and P. van Eikeren, *J. Am. Chem. Soc.*, 1980, **102**, 3244; (c) P. van Eikeren, D. L. Grier and J. Eliason, *J. Am. Chem. Soc.*, 1979, **101**, 7406.
80. A. Ohno, M. Ikeguchi, T. Kimura and S. Oka, *J. Am. Chem. Soc.*, 1979, **101**, 7036.
81. H. J. G. van Hooff, P. M. van Lier, L. A. M. Bastiaansen and H. M. Buck, *Recl. Trav. Chim. Pays-Bas*, 1982, **101**, 191.
82. (a) M. C. A. Donkersloot and H. M. Buck, *J. Am. Chem. Soc.*, 1981, **103**, 6549; (b) M. C. A. Donkersloot and H. M. Buck, *J. Am. Chem. Soc.*, 1981, **103**, 6554; (c) R. H. A. M. Brounts and H. M. Buck, *J. Am. Chem. Soc.*, 1983, **105**, 1284; (d) see also P. M. T. de Kok, M. C. A. Donkersloot, P. M. van Lier, G. H. W. M. Meulendijks, L. A. M. Bastiaansen, H. J. G. van Hooff, J. A. Kanters and H. M. Buck, *Tetrahedron*, 1986, **42**, 941.
83. P. M. van Lier, M. C. A. Donkersloot, A. S. Koster, H. J. G. van Hooff and H. M. Buck, *Recl. Trav. Chim. Pays-Bas*, 1982, **101**, 119.
84. P. M. T. de Kok and H. M. Buck, *J. Chem. Soc., Chem. Commun.*, 1985, 1009.

85. A. Ohno, T. Kimura, H. Yamamoto, S. G. Kim, S. Oka and Y. Ohnishi, *Bull. Chem. Soc. Jpn.*, 1977, **50**, 1535.
86. S. Fukuzumi, M. Ishikawa and T. Tanaka, *Tetrahedron*, 1986, **42**, 1021.
87. A. Ohno, M. Kashiwagi, Y. Ishihara, S. Ushida and S. Oka, *Tetrahedron*, 1986, **42**, 961.
88. S. Shinkai, H. Hamada and O. Manabe, *Tetrahedron Lett.*, 1979, 1397.
89. A. Ohno, Y. Ishihara, S. Ushida and S. Oka, *Tetrahedron Lett.*, 1982, **23**, 3185.
90. (a) A. I. Meyers and T. Oppenlaender, *J. Am. Chem. Soc.*, 1986, **108**, 1989; (b) A. I. Meyers and J. D. Brown, *J. Am. Chem. Soc.*, 1987, **109**, 3155; (c) A. I. Meyers and J. D. Brown, *Tetrahedron Lett.*, 1988, **29**, 5617; (d) see also S. G. Davies, R. T. Skerlj and M. Whittaker, *Tetrahedron Lett.*, 1990, **31**, 3213; (e) S. G. Davies, R. T. Skerlj and M. Whittaker, *Tetrahedron Asym.*, 1990, **1**, 725.
91. A. G. Talma, P. Jouin, J. G. de Vries, C. B. Troostwijk, G. H. Werumeus Buning, J. K. Waninge, J. Visscher and R. M. Kellogg, *J. Am. Chem. Soc.*, 1985, **107**, 3981.
92. J. Speelman, A. G. Talma, R. M. Kellogg, A. Meetsma, J. L. de Boer, P. T. Beurskens and W. P. Bosman, *J. Org. Chem.*, 1989, **54**, 1055.
93. (a) J. W. Verhoeven, W. van Gerresheim, F. M. Martens and S. M. van der Kerk, *Tetrahedron*, 1986, **42**, 975; (b) see also M. C. A. Donkersloot and H. M. Buck, *J. Am. Chem. Soc.*, 1981, **103**, 6549.
94. (a) A. Shirra and C. J. Suckling, *Tetrahedron Lett.*, 1975, 3332; (b) A. Shirra and C. J. Suckling, *J. Chem. Soc., Perkin Trans. 2*, 1977, 759; (c) Y. Ohnishi and M. Kitami, *Tetrahedron Lett.*, 1978, 4035; (d) D. C. Dittmer and B. B. Blidner, *J. Org. Chem.*, 1973, **38**, 2873; (e) L. E. Overman, *J. Org. Chem.*, 1972, **37**, 4214.
95. (a) F. Yoneda, H. Yamato and M. Ono, *J. Am. Chem. Soc.*, 1981, **103**, 5943; (b) F. Yoneda, Y. Sakuma, Y. Kadokawa and A. Koshiro, *Chem. Lett.*, 1979, 1467; (c) F. Yoneda, K. Tanaka, H. Yamato, K. Moriyama and I. Nagamatsu, *J. Am. Chem. Soc.*, 1989, **111**, 9199
96. (a) P. Dan and I. Willner, *Nouv. J. Chim.*, 1984, **8**, 719; (b) D. Mandler and I. Willner, *J. Am. Chem. Soc.*, 1984, **106**, 5352; (c) for an example of photochemically induced regeneration of NADH see J. Kiwi, *J. Photochem.*, 1981, **16**, 193.
97. S. Zehani and G. Gelbard, *J. Chem. Soc., Chem. Commun.*, 1985, 1162.
98. B. Kaptein, L. Wang-Griffin, G. Barf and R. M. Kellogg, *J. Chem. Soc., Chem. Commun.*, 1987, 1457.
99. H. Chikashita, M. Miyazaki and K. Itoh, *Synthesis*, 1984, 308.
100. A. V. Elltsov, N. Y. Pavlish and V. A. Ketlinskii, *Zh. Org. Khim.*, 1978, **14**, 1751 (*Chem. Abstr.*, 1978, **89**, 216 812b).
101. S. H. Mashraqui and R. M. Kellogg, *Tetrahedron Lett.*, 1985, **26**, 1453.
102. (a) S. M. Ramos, M. Tarazi and J. D. Wuest, *J. Org. Chem.*, 1987, **52**, 5437; (b) B. Bachand, S. M. Ramos and J. D. Wuest, *J. Org. Chem.*, 1987, **52**, 5443; (c) F. Montgrain, S. M. Ramos and J. D. Wuest, *J. Org. Chem.*, 1988, **53**, 1489.
103. W. H. N. Nijhuis, W. Verboom, D. N. Reinhoudt and S. Harkema, *J. Am. Chem. Soc.*, 1987, **109**, 3136.
104. L. C. Groenen, W. Verboom, W. H. N. Nijhuis, D. N. Reinhoudt, G. J. van Hummel and D. Feil, *Tetrahedron*, 1988, **44**, 4637.
105. (a) W. H. N. Nijhuis, W. Verboom, A. A. El-Fadl, S. Harkema and D. N. Reinhoudt, *J. Org. Chem.*, 1989, **54**, 199; (b) W. H. N. Nijhuis, W. Verboom, A. A. El-Fadl, G. J. van Hummel and D. N. Reinhoudt, *J. Org. Chem.*, 1989, **54**, 209.
106. O. Meth-Cohn and H. Suschitzky, *Adv. Heterocycl. Chem.*, 1972, **14**, 211.
107. (a) S. J. Hannon and T. G. Traylor, *J. Org. Chem.*, 1981, **46**, 3645, and references cited therein; (b) T. G. Traylor and G. S. Koermer, *J. Org. Chem.*, 1981, **46**, 3651; (c) E. V. Uglova, T. G. Brodskaya, Y. K. Grishin and O. A. Reutov, *J. Org. Chem. USSR (Engl. Transl.)*, 1977, **13**, 217; (d) C. E. Coates, M. L. H. Green and K. Wade, in 'Organometallic Chemistry', Methuen, London, 1968, vol. 2, p. 211; (e) T. G. Traylor, H. J. Berwin, J. Jerkunica and M. L. Hall, *Pure Appl. Chem.*, 1972, **30**, 599; (f) E. Murayama, M. Uematsu, M. Nishi and T. Sato, *Tetrahedron Lett.*, 1984, **25**, 313; (g) I. Fleming and M. Rowley, *Tetrahedron*, 1986, **42**, 3181.
108. G. Giacomelli, L. Lardicci and A. M. Caporusso, *J. Chem. Soc., Perkin Trans. 1*, 1975, 1795.
109. Ref. 41, p. 177.
110. J. Boersma, in 'Comprehensive Organometallic Chemistry', ed. G. Wilkinson, F. G. A. Stone and E. W. Abel, Pergamon Press, Oxford, 1982, vol. 2, p. 824.
111. G. A. Coates and D. D. Ridley, *J. Chem. Soc. A*, 1966, 1064.
112. G. Giacomelli, L. Lardicci and R. Santi, *J. Org. Chem.*, 1974, **39**, 2736.
113. With respect to carbanionic additions to the carbonyl groups of aldehydes: (a) N. Oguni, Y. Matsuda and T. Kaneko, *J. Am. Chem. Soc.*, 1988, **110**, 7877; (b) also H. Wynberg and B. L. Feringa, *Tetrahedron*, 1976, **32**, 2831.
114. J. R. Zietz, Jr., G. C. Robinson and K. L. Lindsay, in 'Comprehensive Organometallic Chemistry', ed. G. Wilkinson, F. G. A. Stone and E. W. Abel, Pergamon Press, Oxford, 1982, vol. 7, p. 366.
115. E. Winterfeldt, *Synthesis*, 1975, 617.
116. (a) E. G. Hoffmann, *Justus Liebigs Ann. Chem.*, 1960, **629**, 104; (b) K. Ziegler, W. R. Kroll, W. Larbig and O. W. Steudle, *Justus Liebigs Ann. Chem.*, 1960, **629**, 53; (c) H. W. Schroetter and E. G. Hoffmann, *Ber. Bunsenges. Phys. Chem.*, 1964, **68**, 627.
117. S. Iguchi, H. Nakai, M. Hayashi, H. Yamamoto and K. Maruoka, *Bull. Chem. Soc. Jpn.*, 1981, **54**, 3033.
118. (a) G. Giacomelli, R. Menicagli and L. Lardicci, *J. Org. Chem.*, 1973, **38**, 2370; (b) G. Giacomelli, R. Menicagli and L. Lardicci, *J. Org. Chem.*, 1974, **39**, 1757; (c) G. Giacomelli, R. Menicagli and L. Lardicci, *J. Am. Chem. Soc.*, 1975, **97**, 4009; (d) G. Giacomelli, R. Menicagli, A. M. Caporusso and L. Lardicci, *J. Org. Chem.*, 1978, **43**, 1790; (e) G. Giacomelli, L. Lardicci and F. Palla, *J. Org. Chem.*, 1984, **49**, 310.
119. (a) D. S. Matteson, *Synthesis*, 1986, 973; (b) M. M. Midland, in 'Asymmetric Synthesis', ed. J. D. Morrison; Academic Press, New York, 1983, vol. 2, p. 45.
120. N. Miyaura, M. Itoh, A. Suzuki, H. C. Brown, M. M. Midland and P. Jacob, III, *J. Am. Chem. Soc.*, 1972, **94**, 6549.
121. M. M. Midland, A. Tramontano and S. A. Zderic, *J. Am. Chem. Soc.*, 1977, **99**, 5211.
122. M. M. Midland, S. Green, A. Tramontano and S. A. Zderic, *J. Am. Chem. Soc.*, 1979, **101**, 2352.

123. H. C. Brown and G. G. Pai, *J. Org. Chem.*, 1982, **47**, 1606.
124. M. M. Midland, D. C. McDowell, R. L. Hatch and A. Tramontano, *J. Am. Chem. Soc.*, 1980, **102**, 867.
125. (a) M. M. Midland and P. E. Lee, *J. Org. Chem.*, 1981, **46**, 3933; (b) M. M. Midland and N. H. Nguyen, *J. Org. Chem.*, 1981, **46**, 4107.
126. (a) M. M. Midland and S. A. Zderic, *J. Am. Chem. Soc.*, 1982, **104**, 525; (b) M. M. Midland, J. E. Petre, S. A. Zderic and A. Kazubski, *J. Am. Chem. Soc.*, 1982, **104**, 528.
127. M. M. Midland and J. I. McLoughlin, *J. Org. Chem.*, 1984, **49**, 4101.
128. M. M. Midland and A. Kazubski, *J. Org. Chem.*, 1982, **47**, 2814.
129. Yu. N. Bubnov, M. E. Gurskii, A. I. Grandberg and D. G. Pershin, *Tetrahedron*, 1986, **42**, 1079.
130. R. S. Bly, R. K. Bly, A. M. M. Hossain, G. S. Silverman and E. Wallace, *Tetrahedron*, 1986, **42**, 1093.
131. S. Georgiou and J. A. Gladysz, *Tetrahedron*, 1986, **42**, 1109.

1.4 Reduction of C=X to CHXH by Dissolving Metals and Related Methods

JOHN W. HUFFMAN
Clemson University, SC, USA

1.4.1 INTRODUCTION

Dissolving metals have been used as reducing agents in organic synthesis for well over a century and for many years metal–alcohol or metal amalgam–water systems were the principal methods employed for the reduction of aldehydes and ketones to primary and secondary alcohols, respectively. The same reagents were employed for the reduction of imines and oximes to the corresponding amines. Catalytic hydrogenation was the only viable alternative to dissolving metal reductions until the development of the Meerwein–Pondorff–Verley reduction in the mid 1920s and the introduction of complex metal hydrides in the years following World War II.

Although complex metal hydrides have, to a large extent, replaced dissolving metal systems for the reduction of carbonyl compounds and their derivatives, dissolving metal reductions frequently offer advantages for the control of stereochemistry. In particular, the dissolving metal reduction of hindered cyclic ketones provides a highly stereoselective method for the preparation of equatorial alcohols, which are difficult to prepare by other methods. The prototype for this synthetic methodology was the nearly simultaneous observation that Li–NH_3–alcohol[1] or Na–propanol[2] reduction of 11-keto steroids gives exclusively the equatorial 11α-alcohol, while metal hydride reduction of these hindered ketones gives the axial 11β-alcohol.

The stereochemical course of these, and other similar reductions, led Barton to suggest that dissolving metal reductions of ketones and oximes to secondary alcohols and primary amines would lead to mixtures of products rich in the thermodynamically more stable product.[3] However, in the early 1960s a number of reports appeared in which the reduction of ketones gave primarily the thermodynamically less stable epimeric alcohol. These observations have prompted a continuing series of investigations into the mechanism of these reductions.

In contemporary organic synthesis, dissolving metal reductions still provide the most effective method for reducing hindered cyclic ketones steroselectively to the more stable secondary alcohol, and in many cases are more stereoselective than complex metal hydrides for the reduction of unhindered cyclic ketones. These procedures also continue to be useful for the reduction of oximes and imines to the corresponding amines. Although many of the dissolving metal reductions employed in organic synthesis are carried out using either active metals (Li, Na, K) in liquid NH_3, or the same metals in alcohols, a number of variations on these precedures have been developed, and several low-valent metal ions, most notably Sm^{2+}, have been found to give improved chemoselectively. However, all of these reactions are mechanistically similar in that they proceed by initial addition of a single electron to the substrate to form a radical anion, the fate of which is determined by a complex set of variables. Among the more important of these variables are the structure of the substrate, the metal or metal ion used as reducing agent, and the solvent system.

1.4.2 MECHANISM OF CARBONYL REDUCTION

1.4.2.1 Early Mechanistic Studies

The overall transformation effected in the reduction of an aldehyde or a ketone to the corresponding alcohol by a dissolving metal is conceptually simple; however, the reaction is actually quite complex, and an evolving series of mechanisms has been suggested to explain the experimental observations. Historically, the reducing properties of dissolving metals were attributed to 'nascent hydrogen', but it is now recognized that these reactions proceed *via* single electron transfer (SET) reaction paths.

It was suggested in the 1950s that the reduction of aliphatic ketones by dissolving metals proceeded by two sequential one-electron additions to provide a dianion (equation 1). This mechanism was based on the observation that benzophenone affords a dianion on reaction with excess Na in liquid NH_3, and it was inferred that aliphatic ketones would behave similarly. A number of workers presented mechanistic rationalizations for the stereochemical course of the dissolving metal reductions of cyclic aliphatic ketones based on this dianion concept. However, in a 1972 review, it was noted that the reduction potentials of alkali metals were not sufficient to effect the addition of two electrons to an aliphatic carbonyl group, and an alternative mechanism was suggested which with some modification is now generally accepted.[4]

$$R_2C{=}O \xrightarrow{e^-} R_2\dot{C}{-}O^- \xrightarrow{e^-} R_2\bar{C}{-}O^- \quad (1)$$

The contemporary view of these reductions recognizes that the reactions of carbonyl compounds with dissolving metals follow one of two general reaction paths. One of these prevails in reductions carried out in the absence of proton donors, the other in reductions in the presence of an alcohol or other proton source, frequently NH_4Cl. Two recent reviews present rather different mechanistic explanations for these reactions, particularly those in liquid NH_3 in the absence of added proton donors.[5,6]

1.4.2.2 Reduction in the Absence of Proton Donors

For reductions carried out in the absence of an added proton donor the usual solvents are liquid NH_3, usually with an ethereal cosolvent, or less commonly an ether (THF, DME or diethyl ether). The reducing agent is usually one of the common alkali metals (Li, Na, K), although Rb, Cs and alkaline earth metals have also been used.[4–6] At least in the case of camphor both types of solvent systems give similar ratios of epimeric alcohols; however, product ratios may vary as a function of the metal used as reducing agent.[4–7] In reductions carried out in ethereal solvents the use of ultrasound increases the rate of the reaction, but does not affect the product distribution.[7]

A number of experimental details have contributed to an understanding of the mechanism of reductions carried out under these conditions. Among the more important observations are the facts that ketones react with one and only one equivalent of alkali metal in NH_3; enolizable ketones afford equal amounts of enolate and alcohol, while nonenolizable ketones give metal ketyls which are stable at low temperature.[5,7–9] Also, pinacol formation is a major reaction path with Li, but K affords little or no pinacol.[5,10] Finally, α-deuterio ketones afford product alcohols in which deuterium has been transferred to the carbinol carbon of the product alcohol or alcohols.[9,10]

The substrate ketone which has been investigated most extensively is camphor (**1**; equation 2), both as one optical antipode (usually the (+)-enantiomer) and as the racemate. In addition to borneol (**2**) and isoborneol (**3**), reduction using Li–NH_3 or Li–THF gives up to 70% yield of dimeric reduction products, while Na–NH_3 gives 20%.[11] Pinacols are not usually obtained when K is used as the reducing agent. Although (+)-camphor can, in theory, afford three pinacols, only two, which correspond to *exo–endo* and *endo–endo* coupling have been isolated.[10,12,13] A third dimeric reduction product which apparently arises by β-cleavage of an intermediate alkoxyl radical has also been observed.[13] Although the relative proportions of isomeric pinacols varies as a function of reaction conditions, the ratios of epimeric alcohols are apparently independent of the composition of the pinacol mixtures.[13] For a detailed discussion of pinacol formation see Volume 3, Chapter 2.6.

(**1**) $\xrightarrow{\text{M–NH}_3}$ (**2**) OH + (**3**) OH + dimeric products (2)

It has been long known that different ratios of product alcohols (**2** and **3**) are obtained in the reduction of (+)-camphor with various metals in liquid ammonia. On the other hand, reduction of (±)-camphor with Li, Na or K in liquid NH_3 affords, within experimental error, the same 82:18 ratio of (**2**):(**3**). An explanation in terms of diastereomeric bimolecular reactions has been presented.[14]

On the basis of these observations and a variety of other experimental data, a mechanism has been suggested which in modified form is presented in equations (3) to (5).[5] The first step (equation 3) is the reversible transfer of an electron to a carbonyl group, generating a metal ketyl, which may be in equilibrium with a dimer or higher aggregate, although not necessarily of the structure depicted. The position of the equilibrium for the addition of an electron to a carbonyl group is a function of the difference in reduction potential of the carbonyl group and the oxidation potential of the metal. Unfortunately the reduction potentials of aldehydes and ketones in liquid NH_3 are not known, but for the reaction of camphor with an alkali metal K_{eq} has been estimated, using reduction potentials in other solvents, to be 3.4×10^2 based on $\Delta V = 0.15$ V.[7] The intervention of ketyl aggregates in these reactions was inferred from studies carried out on aromatic and nonenolizable aliphatic ketones.[5] Regardless of the exact nature of the intermediate at this stage, it must be formed at a considerably greater rate than its subsequent irreversible conversion to products, because it has been observed that in reductions of (+)-camphor carried out with mixtures of alkali metals and the salt of a second metal, the products reflect a weighted average of that predicted assuming a mixture of metals.[11]

$$R_2C{=}O \underset{}{\overset{e^-}{\rightleftharpoons}} R_2\dot{C}{-}O^- \rightleftharpoons R_2\dot{C}{-}O^- \cdots M^+ \cdots M^+ \cdots {}^-O{-}\dot{C}R_2 \quad (3)$$

$$\text{R}_2\dot{\text{C}}\text{–O}^-\cdots\text{M}^+\cdots\text{O}^-\text{–}\dot{\text{C}}\text{R}_2 \ (\text{M}^+\ \text{bridged}) \longrightarrow \text{pinacol} \quad (4)$$

$$\text{R}_2\dot{\text{C}}\text{–O}^-\cdots\text{M}^+\cdots\text{O}^-\text{–}\dot{\text{C}}\text{R}_2 \ (\text{M}^+\ \text{bridged}) \longrightarrow \text{R}_2\text{CH–O}^- + \text{enolate} \quad (5)$$

Pinacol formation (equation 4) is considered to occur *via* coupling of two ketyl units, either within a higher aggregate as depicted or by direct coupling of two ketyls. Production of alcohols by hydrogen transfer (equation 5) probably occurs either within a ketyl aggregate or by the direct interaction of two metal ketyls.[5,9,10]

In a recent review it was once again suggested that these reductions proceed by way of the vicinal dianion path.[6] This argument has again been refuted on the basis of reduction potentials and stoichiometry[7] and most convincingly by a consideration of the relative acidities of the various species present in the reaction medium.[15]

1.4.2.3 Reduction in the Presence of Proton Donors

In terms of synthetic utility, the reduction of carbonyl compounds by a dissolving metal in liquid NH_3 in the presence of an alcohol, water or NH_4Cl is far more common and usually far more efficient than reduction in the absence of a proton donor. Historically these reductions were carried out using active metals, usually Na, in alcohols and the experimental results are similar in both systems.[4–6]

Although reductions under these conditions are more common than those carried out in the absence of a proton donor, relatively few mechanistic studies have been carried out. It is known that in the presence of NH_4Cl the reduction of (+)–[3,3–$^2H_2$]camphor by Li, Na or K in liquid NH_3 gives very little product resulting from deuterium transfer.[10] In the presence of water or ethanol–water mixtures some deuterium transfer occurs, but to a lesser extent than is observed in the absence of an added proton donor, which indicates that the disproportionation path presented in equation (5) is attenuated in the presence of strongly acidic proton donors.[10] The generally accepted alternative mechanism is that suggested by House,[4] in which a ketyl is protonated to give a carbon radical which is subsequently reduced to a carbanion (equation 6) which on protonation provides the product alcohol. Although the details of the mechanism are not certain, it appears that the stereochemical course of the reaction is relatively insensitive to the nature of the metal used for reduction since (+)-camphor gives a borneol:isoborneol ratio of 9.0–9.5:1 using Li, Na or K in NH_3 with NH_4Cl present.[10] With less acidic proton donors there is competition by reduction *via* protonation and hydrogen transfer.[5,10]

$$\text{R}_2\dot{\text{C}}\text{–O}^- \underset{-\text{H}^+}{\overset{\text{H}^+}{\rightleftharpoons}} \text{R}_2\dot{\text{C}}\text{–OH} \xrightarrow{e^-} \text{R}_2\bar{\text{C}}\text{–OH} \quad (6)$$

Although it was once believed that reductions by dissolving metals in alcohols or liquid NH_3 would invariably provide the thermodynamically more stable alcohol as the major product,[3] it has been known for many years that the thermodynamically less stable epimeric alcohol is sometimes the major product on either metal–alcohol or metal–NH_3–proton donor reduction.[4–6] A detailed explanation for the stereochemistry of these reductions based on a combination of steric and frontier molecular orbital interactions has been presented.[6]

It must be emphasized that reduction of carbonyl compounds by dissolving metals, either in the presence or absence of an added proton donor is a kinetically controlled process. This was tacitly stated in 1972,[4] and has been repeated or implied in more recent reviews of this topic.[5,6] A recent study of the reduction of several bicyclo[2.2.1]heptanones using alkali metal–NH_3–NH_4Cl systems emphasizes that these reductions are kinetically controlled.[16]

A variety of other less common reaction conditions, which employ a variety of either dissolving metals or low-valence metal cations, have been used to effect the reduction of carbonyl groups to primary or secondary alcohols. Although the mechanisms of these reactions have not been explored in detail, they almost certainly proceed by mechanisms similar to those outlined above.

1.4.3 SYNTHETIC UTILITY OF CARBONYL REDUCTION

1.4.3.1 Reaction Medium

1.4.3.1.1 Alcohols and water

The reduction of various substrates by dissolving metals in alcoholic and aqueous media is a very old procedure in synthetic organic chemistry. In addition to aldehydes, ketones, imines and other unsaturated nitrogen compounds, many other functional groups are reduced under these conditions. Historically, the most common reduction conditions were Na in ethanol, and the reductions were carried out by adding the metal to a solution of the substrate in alcohol and the reaction mixture was heated at reflux for varying periods of time. Other reduction systems included Na–Hg amalgam in water or alcohols and, for easily reduced compounds such as aldehydes and aromatic ketones, Zn–NaOH or Fe–acetic acid have been used.[4]

Although reductions of ketones by active metals in alcohols have been largely supplanted by other procedures in modern synthetic chemistry, these methods still find occasional use. A modification employing K in *t*-butyl or *t*-pentyl alcohol has been used for the stereoselective reduction of 7-keto steroids in high yield,[17,18] and the Li–ethanol and Na–ethanol reductions of 16-keto steroids have been investigated.[19] Both traditional Na–ethanol[20–22] reductions and a variation using Na–propan-2-ol in toluene[4,23] have also been used recently in selected systems.

In terms of product distribution, reductions in these protic media are similar to those obtained using metal–NH_3–proton donor systems and it is generally accepted that they proceed by similar mechanisms.[4–6] However, in terms of synthetic utility, metal/alcohol systems suffer from two inherent liabilities. The first, which has been commented upon in various reviews, is the possibility of equilibration of epimeric product alcohols as their alkoxide *via* a Meerwein–Pondorff–Verley-type mechanism.[4,6] Thus, although the initial product distribution is the result of a kinetically controlled reduction, prolonged heating of the reaction mixture will lead to mixtures approaching the composition predicted on the basis of thermodynamic control. It has been suggested that equilibration *via* hydrogen transfer from the solvent alcohol may be alleviated by carrying out metal–alcohol reductions in *t*-butyl alcohol,[6] other tertiary alcohols appear to be equally effective.[18]

The second inherent problem is encountered in the reduction of ketones with a proton-bearing chiral center adjacent to the carbonyl group. In this case, enolate formation may lead to equilibration before reduction. For example, reduction of a 98.6:1.4 mixture of menthone (**4**) and its epimer isomenthone (**5**) with Na or K in ethanol gave approximately 15% of a mixture of alcohols (**8**) and (**9**), in addition to 82% of menthol (**6**) and a small amount of neomenthol (**7**),[22] because of partial isomerization.

Enolization is an inherent problem with any metal–alcohol reducing system, and cannot be avoided due to the inevitable formation of metal alkoxides. This side reaction also occurs, although to a somewhat lesser extent, in the reduction of (**4**) with Na in moist ether.[22]

(**4**) → (**6**) + (**7**) (7)

(**5**) → (**8**) + (**9**)

1.4.3.1.2 Ammonia, added proton source

In terms of mechanism and stereochemical consequences, reductions by dissolving metals in liquid NH_3 are very similar to reductions by the same metals in alcoholic media.[4–6] However, reductions carried out in liquid ammonia do not suffer from the same inherent problems as those by metals in alcohols. There is no evidence for equilibration of the product alcohols, and ketones which undergo epimerization prior to reduction with metals in alcohols are reduced cleanly by metals in NH_3. For example, menthone (**4**) on reduction with Li–NH_3–ethanol gives a mixture of alcohols (**6**) and (**7**), with no trace of alcohols (**8**) and (**9**).[22]

The conventional method for carrying out reductions under these conditions consists of adding the appropriate metal to a solution of the compound to be reduced in a mixture of liquid NH_3 and an ethereal cosolvent, usually ether or THF, plus the proton donor, commonly NH_4Cl or an alcohol. The use of NH_4Cl has been advocated due to the observation that reduction by hydrogen transfer (Section 1.4.2.2) is effectively suppressed under these conditions.[10] However, NH_4Cl is a very strong acid which is only sparingly soluble in NH_3–ether mixtures and in preparative scale work a large excess of metal must be used to ensure complete reduction.[24] A suitable alternative is the use of an excess of a primary alcohol (methanol or ethanol) which affords little of the product obtained by hydrogen transfer.[25] This procedure results in a homogeneous reaction mixture, and complete reduction is obtained with only a modest excess of metal.[22] Tertiary alcohols and relatively small amounts of primary alcohols lead to product mixtures resulting from reduction by a combination of hydrogen transfer and protonation.[5,26]

Reductions carried out under these procedures do not lead to bimolecular products and usually afford excellent yields of reduction products. Dissolving metal reductions in liquid NH_3 in the presence of a proton donor are usually the method of choice for effecting the reduction of a substrate by a dissolving metal.

1.4.3.1.3 Ammonia, no added proton source

As noted above (Section 1.4.2.2) reduction of carbonyl compounds under these conditions proceeds with hydrogen transfer to afford an equimolar mixture of alkoxide and enolate, plus varying quantities of dimeric reduction products. As a consequence, at least in theory, this procedure should afford an equimolar mixture of recovered ketone and reduction product. This appears to be the case if less than one equivalent of metal is used; however, with excess metal, camphor,[10,11] some 12-keto steroids[25] and several 1-decalones[24] afforded 70–99% yields of secondary alcohols. The explanation which has been offered is that the product enolate is protonated by NH_3 to regenerate the starting ketone, which is recycled through the reduction process.[10]

The usual experimental procedure for carrying out reductions in the absence of an added proton donor entails the addition of a metal (Li, Na, K) to a solution of the substrate ketone in a mixture of NH_3 and an ethereal cosolvent, usually ether or THF at the reflux temperature of liquid NH_3. The metal is added until a permanent blue color is obtained and the reaction is stirred for 10 to 30 min. The excess Li is decomposed, traditionally with NH_4Cl or an alcohol; however, sodium benzoate is probably superior.[6]

Reductions by dissolving metals in liquid NH_3 in the absence of proton donors, are, in general, inferior in terms of general synthetic utility to reductions carried out under other conditions. In many cases, particularly when Li is used as a reducing agent, bimolecular reduction is difficult to predict and the ratios of epimeric alcohols may vary as a function of the metal used as the reducing agent.[4–6]

1.4.3.1.4 Other solvents

The vast majority of the dissolving metal reductions of carbonyl compounds which have been carried out synthetically have used either alcohols or liquid NH_3 as the solvent.[4–6] However, a variety of other solvents have been employed, frequently in connection with studies of the mechanism of the reductions or in exploratory synthetic studies.

Ethereal solvents, principally THF, either with or without sonication, have been reported to give results similar to those obtained on reductions in NH_3 with no added proton donor, and pinacol formation as a major reaction path.[7,27] A potentially useful selective reduction of unhindered cyclohexanones in the presence of other ketones using Al amalgam in aqueous THF has been described and will be discussed in detail subsequently (Section 1.4.3.3.2).[28] In this procedure aliphatic ketones give no pinacols; however, aromatic ketones give only the corresponding pinacol.[28]

A number of steroidal and triterpenoid ketones have been reduced with Li–ethylenediamine to give results consistent with those anticipated by analogy with NH_3.[29–31] In addition to NH_3 and aliphatic amines, a number of other solvents provide relatively stable solutions of solvated electrons, usually using Na as the metal. These solvents, which include HMPA, *N,N*-diethylacetamide and *N*-ethylpyrrolidone, provide a medium for producing Na ketyls, which have been reacted with alkenes both inter- and intramolecularly.[32] In these reactions, secondary alcohols are by-products, and presumably conditions could be found which would provide a viable synthetic method for the reduction of ketones to secondary alcohols. These solvents, however, would appear to offer no advantages compared to NH_3.

The reduction of a number of ketones by alkali metals in HMPA has been carried out, both in the presence and absence of added proton donors.[27,33] The results are qualitatively similar to those obtained using the corresponding metal in NH_3 under similar conditions. If deuterated *t*-butyl alcohol is used as the proton donor, high levels of deuterium incorporation at the carbinol position are obtained.[27] Due to the carcinogenicity of HMPA it cannot be considered as a viable solvent for preparative chemistry and it also offers no advantage over other solvents.

1.4.3.2 Metal

In reductions carried out by dissolving metals in alcohols, Na in primary alcohols or K in tertiary alcohols are the most common systems. Somewhat different epimeric mixtures are occasionally obtained with these systems[18] and it has been reported that Li–ethanol does not reduce a menthone/isomenthone mixture.[22]

All five Group I metals (Li, Na, K, Rb and Cs) and three Group II metals (Ca, Sr, Ba) have been used in NH_3 to effect the reduction of ketones to secondary alcohols.[4–6,11] In addition, it has been reported that Yb–NH_3 reduction of α,β-unsaturated ketones affords the saturated alcohol as the major product, which presumably arises *via* reduction of the intermediate saturated ketone.[34] Reduction of (+)-camphor with Yb–NH_3 both in the absence and presence of NH_4Cl affords the same 86:14 ratio of borneol (**2**) to isoborneol (**3**). In the presence of NH_4Cl, the ketone is completely consumed and dimeric reduction products are not observed.[35] Excess Yb–THF–HMPA effects bimolecular reduction of aromatic ketones,[36] but aliphatic ketones are apparently inert to Yb–THF.[35,36]

A variety of other electron transfer reagents have been employed in reactions which appear to be mechanistically similar to the more common metal–NH_3 or metal–alcohol systems. These include K–graphite,[37,38] Zn–KOH–DMSO[39] and both Li[40] and Al[28] amalgams. The amalgams from Zn, Mg, Ni, Cu, Sn and Pb have been found not to be effective in the reduction of cyclohexanone in aqueous THF.[41] Also, several low-valent metal cations have been employed in the reduction of carbonyl compounds to alcohols. Among these reagents are low-valence salts of Ti,[42–46] Ce[47] and Sm.[48,49]

Some of these procedures have considerable potential as chemoselective reagents for the reduction of different types of carbonyl groups as will be discussed below. However, others, in particular CeI_2[45] and low-valence Ti salts,[42–45] are primarily of interest as reagents for dimeric reduction and are of little synthetic utility in reductions to form alcohols (see Volume 3, Chapter 2.6).

1.4.3.3 Chemoselectivity

1.4.3.3.1 Metal–ammonia and metal–alcohol systems

Under the usual reaction conditions alkali metal–NH_3 and alkali metal–alcohol systems are very powerful reducing systems. As noted above (Section 1.4.2.2) the reduction potentials of various functional groups in liquid NH_3 are not known. However, if it is assumed that the same relative order of reduction potentials is maintained in aqueous and nonaqueous systems, then it is apparent that in polyfunctional molecules the functional group with the least negative reduction potential should be reduced most readily. The reduction potentials (*versus* saturated calomel electrode) for aliphatic ketones range from −2.3 to −2.7 V and aromatic carbonyl compounds are considerably less negative (approximately −1.5 V).[50] Many common functional groups, including conjugated dienes, aromatic hydrocarbons, unsaturated carbonyl, halo, nitro and nitroso compounds all have reduction potentials less negative than those of aliphatic ketones and do not survive dissolving metal reductions using alkali metals in NH_3 or alcohols. Although carboxylic acids are stable to metal–NH_3 reductions,[4,5] they are reduced to aldehydes by Li in methylamine (see Chapter 1.12, this volume).[4] The reduction of esters to primary alcohols with Na in alcohol is the classical Bouveault–Blanc reduction (see Chapter 1.10, this volume).[4]

Both disubstituted alkynes (Chapter 3.3, this volume) and isolated terminal double bonds may be reduced by alkali metals in NH_3, but isolated double bonds are usually stable to these conditions. However, 16,17-secopregnanes (**10**; equation 8) afford mixtures of cyclization products (**11**) and (**12**) in 61% to 80% yield with Na naphthalenide–THF, Na–NH_3–THF, Na–THF or Li–NH_3–THF. With Na–NH_3–THF–*t*-butyl alcohol, a 91% yield of a 72:28 mixture of (**11**):(**12**) (R = Me) is obtained.[51] This type of radical cyclization of alkenes and alkynes under dissolving metal reduction conditions to form cyclopentanols in the absence of added proton donors is a general reaction, and in other cases it competes with reduction of the carbonyl group.[6,51,52] Under the conditions of these reactions which involve brief reaction times, neither competitive reduction of a terminal double bond nor an alkyne was observed.[51,52] However, allenic aldehydes and ketones (**13**) with Li–NH_3–*t*-butyl alcohol afford no reduction products in which the diene system survives.[53]

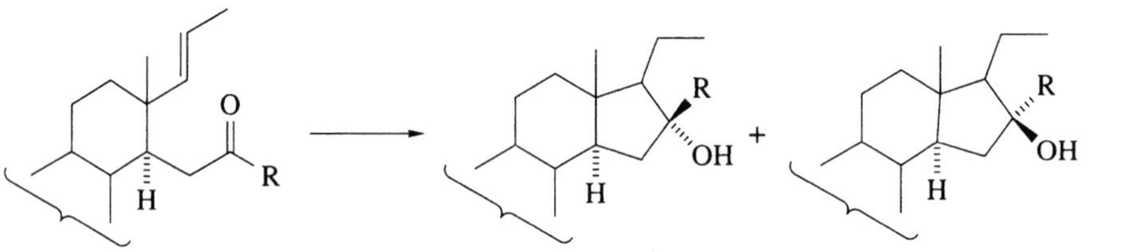

(8)

(**10**) R = H, Me (**11**) (**12**)

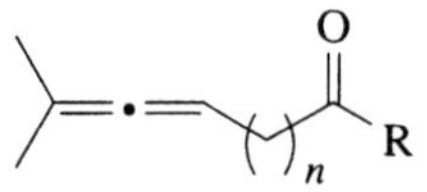

(**13**) n = 2, 3; R = H, Me

In addition to the inherent problems associated with easily reduced functional groups in dissolving metal reductions, ketones with an α-hetero substituent suffer removal of the α-substituent rather than reduction of the carbonyl group (Chapter 4.8, this volume). Cyclopropyl ketones constitute a special example of reduction of α-substituted ketones in which the ketyl formed by addition of one electron to the carbonyl group behaves as a cyclopropyl carbinyl radical, providing ketonic products arising from cleavage of the cyclopropane ring.[54–56]

The metal–NH_3 reductions of carbonyl groups are exceedingly fast reactions; for the reaction of acetone with an ammoniated electron the rate is 9×10^7 M^{-1} s^{-1}.[57] Although many, particularly older, published experimental procedures for the metal–NH_3 reduction of ketones employ prolonged reaction times with excess metal, these conditions are unnecessarily harsh. The reactions of carbonyl compounds with metals in NH_3 are effectively instantaneous and by using short reaction times it appears that reduction of terminal alkenes and disubstituted alkynes can be avoided.[52] In addition to the functional groups mentioned above, alcohols, amines and ethers, other than epoxides, are usually stable to reductions of aldehydes and ketones by dissolving metals.[4–6]

Aromatic ketones represent a rather special case in dissolving metal reductions. Under many conditions pinacol formation is the predominent reaction path (see Volume 3, Chapter 2.6). Also, the reduction potentials of aromatic carbonyl compounds are approximately 1 V less negative than their aliphatic counterparts.[50] The reductions of aromatic ketones by metals in ammonia are further complicated by the fact that hydrogenolysis of the carbon–oxygen bond can take place (Chapter 1.13, this volume) and Birch reduction may intervene (Chapter 3.4, this volume).

The course of reductions of aromatic ketones is far more dependent on the exact reaction conditions than those of aliphatic ketones. A series of aromatic ketones, of which 1-tetralone (**14**; Scheme 1) is typical, on reduction with excess Li–NH_3 in the presence of a trace of Co or Al and quenching with NH_4Cl gave almost exclusively tetralin (**15**).[58,59] However, quenching with sodium benzoate[58] or inverse quenching into aqueous NH_4Cl gave 1-tetralol (**16**) as the major product.[59] Reduction at –78 °C with Li gave significant amounts of dimeric reduction products. Up to 31% of a Birch reduction product was obtained when Li or Na was added to a solution of (**14**) in THF–NH_3.[59]

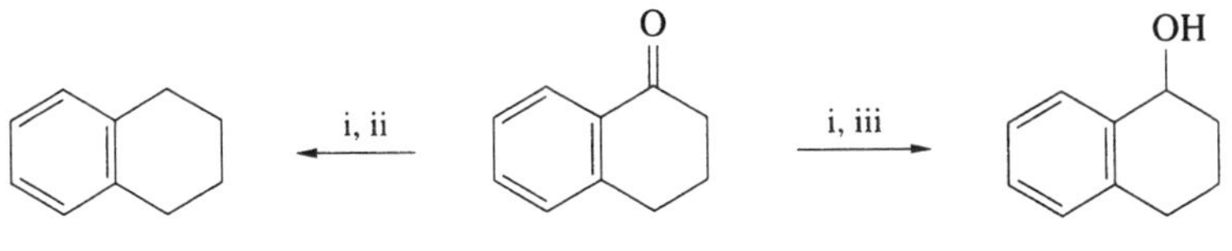

(15) ← i, ii — (14) — i, iii → (16)

i, Li–NH_3–THF, –33 °C; ii, NH_4Cl; iii, NaOBz

Scheme 1

The metal–NH_3 and metal–amine reductions of acetophenone and acetyl derivatives of polycyclic aromatics are complex and afford primarily mixtures of Birch reduction products. In some cases a ketonic carbonyl survives the reduction, and in some cases it is reduced to the corresponding alcohol.[60–63]

1.4.3.3.2 Other systems

Reduction of aromatic ketones using dissolving metal systems other than alkali metals–NH_3 in the absence of proton donors usually provide good yields of pinacols (see Volume 3, Chapter 2.6) with only small amounts of the secondary alcohol as by-product. A detailed study of the reduction of acetophenone by Li amalgam in several solvents using a variety of proton donors has been carried out.[40] In benzene the ratio of carbinol to pinacol was found not to vary in a regular way with the pK_a of the proton donor. Using ethanol as a proton donor and a variety of solvents the carbinol:pinacol ratio was greatest in solvents which coordinate poorly with Li (cyclohexane, benzene, ether). In polar solvents which coordinate well with Li ions (propan-2-ol, acetonitrile, THF) the pinacol was the major product. In the presence of optically active quaternary ammonium salts some asymmetric induction was observed, but the optical purity of the 1-phenylethanol did not exceed 8.9%.[40]

A potentially useful chemoselective dissolving metal reagent for the reduction of aromatic ketones in the presence of other functional groups is the combination Zn–DMSO and aqueous potassium hydroxide.[39] In three examples (benzophenone, fluorenone and 4-benzoylpyridine), the yields of secondary alcohols were over 90%. Two other ketones (xanthone and thioxanthone) gave mixtures of alcohol and the hydrocarbon obtained by hydrogenolysis of a carbon–oxygen bond.[39]

In general, the reduction of aromatic carbonyl compounds to the corresponding alcohols by dissolving metals is not a particularly valuable synthetic procedure. Better yields and chemoselectivity are usually obtained using complex metal hydrides.

In contrast to traditional dissolving metal systems, low-valence metal cations, particularly Sm^{2+}, show considerable chemoselectivity in carbonyl reductions. In some cases these reagents can effect transformations which are difficult using metal hydrides or other traditional reducing agents. It has been suggested, and is almost certain, that these reductions are mechanistically similar to alkali metal reductions.[49]

Samarium diiodide is usually prepared *in situ* by the reaction of Sm metal with 1,2-diiodoethane in THF,[48,64] but a more convenient method appears to be the direct interaction of a slight excess of the metal with iodine in THF.[65] Although this reagent has found its principal utility as a reagent for effecting Barbier-type coupling reactions, in the presence of a proton donor it is an effective reagent for the reduction of aldehydes and ketones to the corresponding alcohols.[48,49] With methanol as a proton donor, aldehydes and aromatic ketones are reduced efficiently to the corresponding alcohol. Aliphatic ketones are, however, only partially reduced at room temperatures in THF solution. This provides an effective and experimentally simple method of effecting the selective reduction of an aldehyde in the presence of an aliphatic ketone.[48] This procedure appears to be at least as efficient as other procedures for the selective reduction of aldehydes.[66] With water as the proton donor, 2-octanone is reduced to the corresponding alcohol in 64% yield; in the absence of proton donors pinacols are the major product in the reaction of carbonyl compounds with SmI_2.[64] Ytterbium diiodide behaves similarly, but is not as effective a reducing agent.[48]

Although SmI_2 is more chemoselective than traditional dissolving metal reagents, it does react with sulfoxides, epoxides, the conjugated double bonds of unsaturated ketones, aldehydes and esters, alkyl bromides, iodides and *p*-toluenesulfonates.[48] It does not, however, reduce carboxylic acids, esters, phosphine oxides or alkyl chlorides.[48] In common with most dissolving metal systems, ketones with an α-hetero substituent suffer loss of the substituent rather than reduction of the carbonyl group.[67]

Other low-valence metal ions which can effect selective reductions are VCl_2 and $TiCl_3$, both in aqueous THF, which selectively reduce benzils (**17**; equation 9) to the corresponding benzoin (**18**).[46] A similar chemo- and enantio-selective reduction of benzil employs the combination SmI_2–HMPA–quinidine, which affords the (*R*)-enantiomer of benzoin (56% *ee*). Other chiral reagents are not as effective.[68] In basic media (methanol–NH_4OH or NaOH), $TiCl_3$ reduces both benzil (**17**) and benzoin (**18**) to the corresponding diol in 98% yield, with a (±):*meso* ratio of approximately 3:1.[43] Acetophenones and substituted acetophenones with alkaline $TiCl_3$ usually give pinacols; however, *p*-amino- and *p*-hydroxyacetophenone afford the corresponding alcohols in good yield.[43]

(**17**) $\xrightarrow[\text{H}_2\text{O–THF, 82–100\%}]{\text{VCl}_2 \text{ or TiCl}_3}$ (**18**) (9)

R = H, Me, OMe, Ph

Aluminum amalgam in aqueous THF is a reagent which shows some promise for the selective reduction of unhindered cyclohexanones in the presence of a variety of other aliphatic ketones.[28,41] Reduction of cyclohexanone and 3- and 4-methylcyclohexanone gave, respectively, 94%, 71% and 70% yields (GLC) of the corresponding secondary alcohols; however, cyclooctanone, cyclononanone, cyclododecanone, norcamphor, menthone, and 2-, 3- and 4-heptanone gave at most traces of reduction products. Cyclobutanone, cyclopentanone and cycloheptanone gave 12–33% yields of the corresponding alcohols.[28] Aromatic ketones gave good yields of pinacol under these conditions.[28] Based on preliminary competitive rate studies cyclohexanones are reduced approximately 10 times faster than cyclopentanones by Al amalgam–aqueous THF.[41]

1.4.3.4 Stereoselectivity

1.4.3.4.1 Generalizations

It was originally believed that the dissolving metal reduction of cyclic ketones would invariably afford the more stable of a pair of epimeric ketones as the major product.[3] Although it has since been established beyond reasonable doubt that these reactions are kinetically controlled and that the less stable epimeric alcohol frequently predominates,[4–6,16] the belief persists that these reductions are under thermodynamic control.[29,69]

Reductions of cyclic ketones by dissolving metals are frequently highly stereoselective and these reductions have been used to obtain secondary alcohols which are difficult or impossible to prepare by metal hydride reduction. In terms of yield, the best results are usually obtained either by reductions with alkali metals (commonly Li) in liquid NH_3 in the presence of proton donors[4–6,10] or with active metals in an alcohol.[4,6] Although a number of explanations have been advanced for the stereoselectivity of these reductions, they are all rationalizations with dubious predictive value.[4–6] There are, however, a number of empirical generalizations which are based on a considerable body of experimental data, specifically: (i) cyclohexanones with one or no α-alkyl substituents afford almost exclusively the equatorial alcohol; (ii) very hindered cyclic ketones afford almost exclusively the thermodynamically more stable alcohol; (iii) 2,2- or 2,6-disubstituted cyclohexanones may afford the less stable epimeric alcohol as the major product; and (iv) bicyclo[2.2.1]heptanones afford the *endo*-alcohol without regard to the relative stability of the *exo*- and *endo*-alcohols.

In reductions carried out in the absence of an added proton donor, it is difficult to predict the stereochemical consequences of a given reduction. The details of reductions under various conditions are discussed in Sections 1.4.3.4.2 through 1.4.3.4.6.

1.4.3.4.2 Cyclohexanones with one or no α-substituents

Reduction of a series of cyclohexanones (2- and 4-methyl, 4-isopropyl, 4-*t*-butyl and 3,3,5-trimethyl), both in the presence and absence of proton donors gave a 99:1 ratio of equatorial to axial alcohol. For 3-methylcyclohexanone the ratio was 94:6.[70] Also, reduction of menthone (**4**) affords a 97:3 ratio of equatorial alcohol to axial alcohol at –30 °C[22] These are somewhat larger ratios of equatorial to axial alcohol than can be obtained by equilibration and considerably greater than can be obtained by conventional complex metal hydride reduction. Although LAH reduction of 4-*t*-butylcyclohexanone affords an equatorial:axial ratio (90:10) which is comparable to that obtained using dissolving metals, reduction of 3,3,5-trimethylcyclohexanone by either LAH or sodium borohydride gives the axial alcohol as the major product.[71]

Although the metal–NH_3 reduction of unhindered cyclohexanones usually affords equatorial alcohols with greater stereoselectivity then metal hydride reductions, this method has not been used frequently in synthesis. An exception is the highly stereoselective reduction of ketoxime (**19**; equation 10), an intermediate in the synthesis of (±)-perhydrohistrionicotoxin, which gave an excellent yield of equatorial alcohol (**20**) on reduction with Na–NH_3. This reduction is noteworthy in that the oxime survives the reduction, which was carried out below the boiling point of liquid NH_3 for a relatively short period of time (30 min).[72]

(10)

(19) **(20)**

An unhindered keto triterpene, friedelan-3-one (**21**), with a large excess of Li in refluxing ethylenediamine has been reported to give a good (85–90%) yield of equatorial alcohol (**22**).[29,30] Reduction of a second 3-ketofriedelane which also contained a sterically hindered lactone afforded the 3α-ol with concomitant hydrogenolysis of the lactone.[73] The hydrogenolysis of esters by alkali metals in amines is a general method for the conversion of alcohols to the corresponding alkane.[74]

(21) R, R' = O
(22) R = H, R' = OH

Although metal–alcohol reductions of ketones suffer from possible epimerization of the product alcohol as well as isomerization of the substrate ketone (see Section 1.4.3.1.1), it was found possible to employ the latter type of epimerization to advantage in the preparation of an alcohol used as a chiral director in asymmetric synthesis.[23] In this preparation an 85:15 mixture of ketones (**23**) and (**24**; equation 11) was reduced with Na–isopropyl alcohol in toluene to afford alcohol (**25**) as the only isolated product in good yield.[23]

(11)

(23) **(24)** **(25)**

Metal–alcohol reductions have also been used in the stereoselective reduction of 3α-hydroxy-7-ketocholanic acid (**26**) to the commercially important equatorial 7β-ol (**27**).[17,18] These reductions have been carried out with several alkali metals in secondary and tertiary alcohols, where reduction with K in tertiary alcohols is more stereoselective than Na–isopropyl alcohol.[17,18] These reductions afford the equatorial alcohol as the major product in good yield, in contrast to sodium borohydride reduction, which provides the axial alcohol (**28**) almost exclusively.

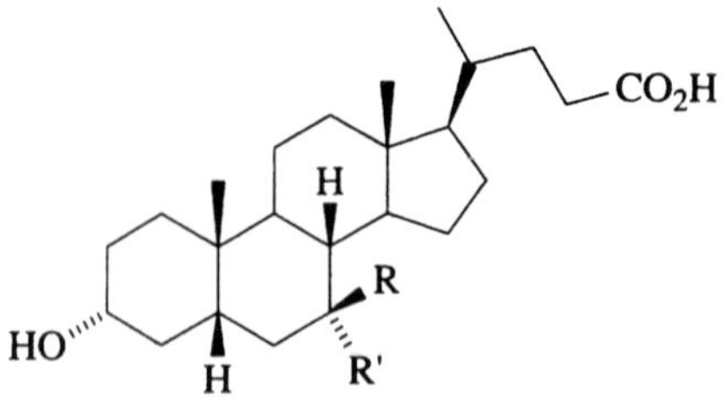

(**26**) R, R' = O
(**27**) R = OH, R' = H
(**28**) R = H, R' = OH

1.4.3.4.3 Sterically hindered cyclohexanones

Dissolving metal reductions remain the method of choice, and are frequently the only viable method, for the reduction of sterically hindered cyclohexanones to equatorial alcohols. In the early 1950s it was found that reduction of 11-keto steroids using either Na–propan-1-ol[2] or Li–NH_3–dioxane–ethanol[1] gave good yields of the equatorial 11α-alcohol. 11-Keto steroids, such as androstan-11-one (**29**; equation 12) have two axial methyl groups in a 1,3-relationship to the carbonyl group and afford exclusively the axial 11β-ol (**30**) on reduction with metal hydrides.

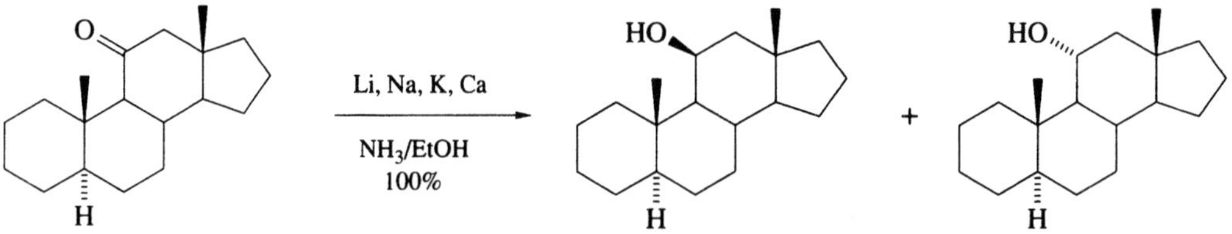

Reduction of the ketone (**29**) in NH_3–ethanol gave a quantitative yield of a mixture in which the equatorial 11α-ol (**31**) predominated.[75] With Li–NH_3–NH_4Cl a quantitative yield of alcohols was also obtained, but quite surprisingly the ratio of (**31**):(**30**) was 55:45, which is similar to that obtained using alkali and alkaline earth metals (Li, Na, K, Cs, Sr) in the absence of a proton donor. Under these conditions Ca and Ba gave 79:21 and 65:35 ratios of (**31**):(**30**), respectively. In none of these reductions is there any evidence for pinacol formation.[75] Reduction of 19-keto-α-agarofuran (**32**; equation 13) using excess Li–NH_3–ethanol gave almost exclusively the equatorial 9α-ol (**34**), while LAH gave the axial 9β-ol (**33**) as the only isolable product.[26] As expected with a severely hindered ketone, no pinacols were obtained with any alkali metal–proton donor combination.[26]

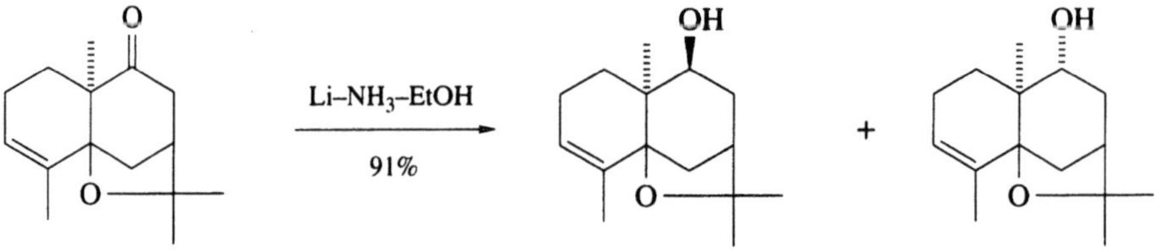

Although alkali metal–NH_3 reductions are usually preferable to Na–alcohol reductions in terms of yield and convenience, the stereochemical consequences of both systems are similar, at least for all sterically hindered systems studied thus far. Illustrative examples are the reductions of bicyclo[3.3.1]nonanones (**35**; equation 14) and (**36**; equation 15).[20] In both cases the equatorial alcohols (**37**) and (**38**) were obtained in excellent yield and with complete stereoselectivity. LAH reduction of ketone (**35**) gave only the epimeric axial alcohol, while (**36**) gave a mixture of (**38**) and its epimer.[20]

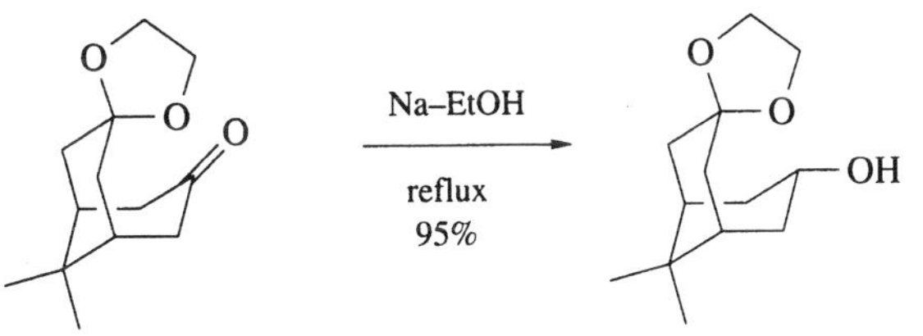

Na–EtOH

reflux
95%

(14)

(35) **(37)**

Na–EtOH

reflux
95%

(15)

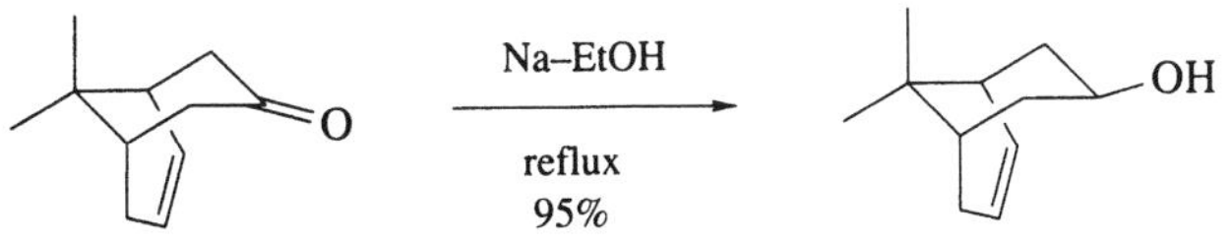

(36) **(38)**

Several sterically hindered keto triterpenes, including friedelan-7-one (**39**; equation 16), were reduced with a large excess of Li in refluxing ethylenediamine.[29] This exceedingly hindered ketone (three axial methyl groups with a 1,3-diaxial relationship to the carbonyl) affords only the equatorial 7α-ol (**40**) in good yield. Other ketones which were reduced under these conditions are considerably less hindered than ketone (**39**), but all gave the equatorial alcohol in good yield.[29]

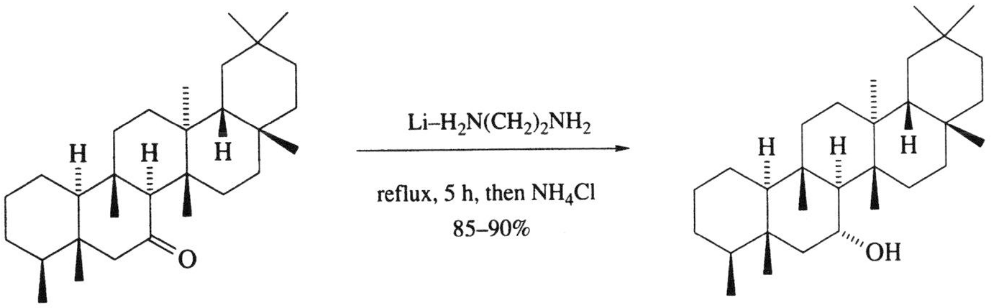

$Li–H_2N(CH_2)_2NH_2$

reflux, 5 h, then NH_4Cl
85–90%

(16)

(39) **(40)**

1.4.3.4.4 Moderately hindered cyclohexanones

A group of moderately hindered cyclohexanones which has been thoroughly studied is the 12-keto steroids, in which the stereochemical course of the reduction is governed by the structure of the C-17 alkyl group.[5,6,25] Reduction of 12-keto steroids which have a secondary substituent on C-17, (**41**), (**42**) and (**43**), gave mixtures of alcohols with $Li–NH_3$–methanol in which the axial 12α-ol, (**45**), (**46**) and (**47**), was the major product (70–80%).[5,25] However, reduction of 5β-pregnan-12-one (**44**) under the same conditions gave a mixture containing only 11% of the axial alcohol. Similar results were obtained using Na–propan-1-ol.[6] Reduction of ketones (**41**) and (**42**) with $Li–NH_3$ in the absence of an added proton donor gave mixtures of epimeric alcohols containing from 10–20% of axial alcohols (**45**) and (**46**). Considerable quantities of pinacol were obtained under these conditions and pinacol was the major product from the reduction of ketone (**44**) in the absence of a proton donor.[25] Sodium borohydride reductions of ketones (**41**), (**42**) and (**43**) were comparable in terms of yield and stereoselectivity to reduction by $Li–NH_3$–alcohol. However, borohydride reduction of ketone (**44**) gave an approximately 1:1 mixture of alcohol (**48**) and the equatorial 12β-ol.[25]

Although they have apparently not been studied in detail, dissolving metal reductions of 1-keto steroids appear similar to those of the 12-keto steroids. In one report, cholestan-1-one (**49**; equation 17) on reduction with Na in either ethanol or 1-pentanol gave the axial alcohol, cholestan-1α-ol (**50**), as the major reduction product in unspecified yield.[76] The equatorial 1β-ol was detected by TLC, but could not be isolated.

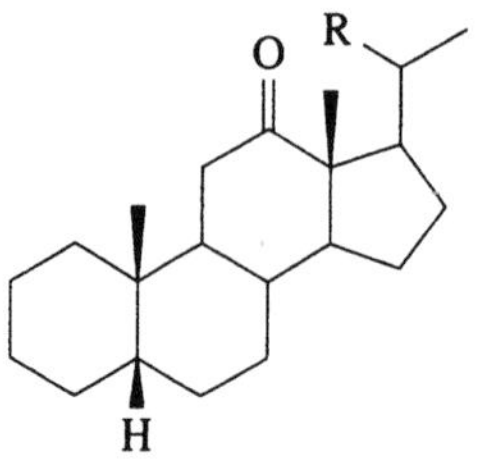

(41) R = Pr
(42) R = Et
(43) R = Me
(44) R = H

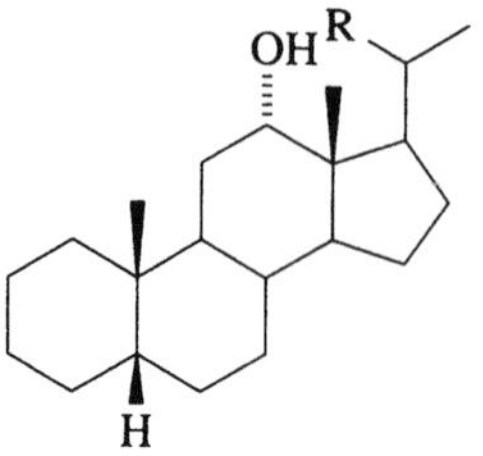

(45) R = Pr
(46) R = Et
(47) R = Me
(48) R = H

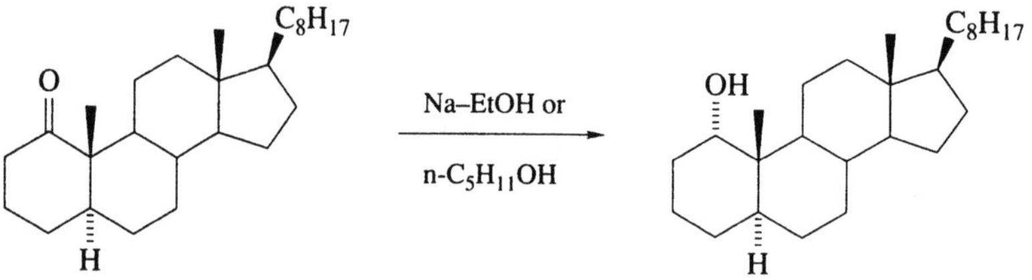

Na–EtOH or n-C_5H_{11}OH (17)

(49) (50)

The other group of moderately hindered cyclohexanones which has been examined in detailed is a group of substituted 1-decalones, which gave results similar to those of the 12-keto steroids on Li–NH_3 reduction.[24] These decalones were all derived from protected diketone (**51**) and included two different 2-monosubstituted derivatives (**52**) and (**53**). These decalones were reduced both in the presence of a proton donor (NH_4Cl) and under anhydrous conditions in yields of 69–99%. The axial alcohol was the major product in the presence of a proton donor, while the equatorial alcohol was predominant in its absence.[24]

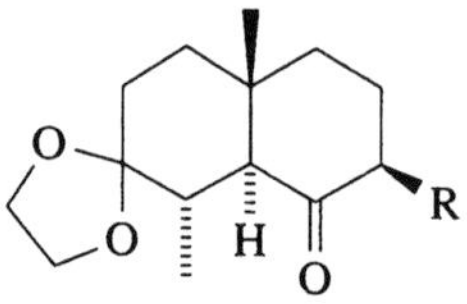

(**51**) R = H
(**52**) R = $CH_2CH{=}CMe_2$
(**53**) R = CH_2CO_2H

The reductions of 1- and 12-keto steroids and their 1-decalone derivatives graphically illustrate the fact that dissolving metal reductions of ketones do not necessarily afford the more stable of a pair of epimeric alcohols. As a corollary, while the reduction of cyclic ketones is a synthetically useful procedure for the stereoselective preparation of secondary alcohols, it cannot be assumed that the thermodynamically stable alcohol will be the product which is obtained stereoselectively.

1.4.3.4.5 Bicyclo[2.2.1]heptan-2-ones

The dissolving metal reductions of bicyclo[2.2.1]heptanones have been studied extensively, and it has been established that both metal–alcohol and metal–NH_3–proton donor systems provide the *endo*-alcohol regardless of its stability relative to the *exo* isomer.[4–6,10,16,70] In the case of camphor (**1**) which has been studied in the most detail, the ratio of *endo*-alcohol (borneol; **2**) to *exo*-alcohol (isoborneol; **3**) is very close to 90:10 for all metal–NH_3 conditions employed. The variables include temperature (–33 and –78 °C),[16] cosolvent (ether and THF),[16] metal (Li, Na, K, Rb)[16,70] and proton donor (NH_4Cl and ethanol).[16,70] The same results are obtained with both (+)- and (±)-camphor.[16] These results are, coincidentally, almost identical to the equilibrium ratio for alcohols (**2**) and (**3**).[16]

In addition to camphor (**1**), the metal–NH_3–proton donor reduction of four other bicyclo[2.2.1]heptan-2-ones and bicyclo[2.2.1]hept-5-en-2-one (**54**) have been studied systematically.[16,70] The other ketones include the parent bicyclo[2.2.1]heptan-2-one (**55**), fenchone (**56**), 1-methylbicyclo[2.2.1]heptan-2-one (**57**) and 7,7-dimethylbicycloheptan-2-one (**58**). The results of these reductions and, where known, the equilibrium ratio of the alcohols are summarized in Table 1. In all cases and under the conditions noted, the *endo*-alcohol is produced stereoselectively.

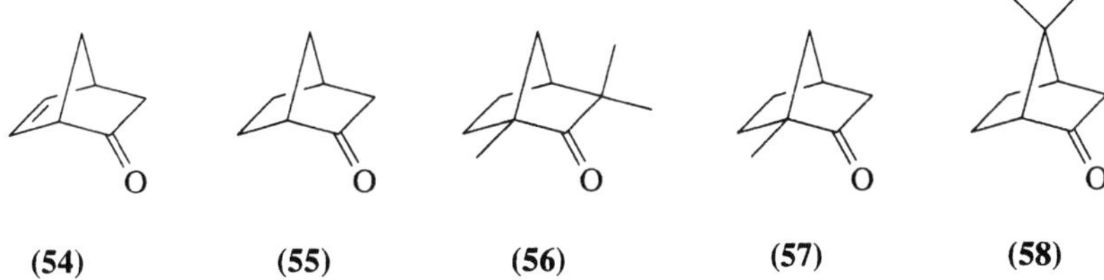

(54) **(55)** **(56)** **(57)** **(58)**

Table 1 Metal–Ammonia Reductions of Bicyclo[2.2.1]heptanones and Equilibrium Ratios of the Corresponding Alcohols

Ketone	*Metal*	*Cosolvent*	*Proton donor*	*Endo:exo (reduction)*	*Endo:exo (equilibrium)*	*Ref.*
(**54**)	Li	THF	NH_4Cl	99:1	50:50	16
(**55**)	Li	Et_2O	NH_4Cl	89:11	5:95	16, 70
(**55**)	Li	Et_2O	EtOH	85:15	—	69
(**55**)	Na	Et_2O	NH_4Cl	89:11	—	16, 70
(**55**)	Na	Et_2O	EtOH	85:15	—	69
(**55**)	K	Et_2O	NH_4Cl	90:10	—	16, 70
(**55**)	K	Et_2O	EtOH	90:10	—	70
(**56**)	Li	THF	NH_4Cl	>99:1	72:28	16
(**57**)	Li	Et_2O	NH_4Cl	91:9	40:60	16
(**58**)	Li	Et_2O	NH_4Cl	83:17	Unknown	16

The reduction of bicycloheptanones, particularly camphor (**1**) in the absence of added proton donors has been studied extensively in connection with the mechanism of the dissolving metal reductions and are discussed in Section 1.4.2.2. As previously noted, reductions under these conditions are of considerably less utility in synthesis than those carried out in the presence of a relatively acidic proton donor (NH_4Cl or ethanol).

The stereoselective dissolving metal reduction of bicycloheptanones has been applied to the synthesis of two tricyclic sesquiterpenes, longiborneol (**59**; equation 18) and ylangoborneol (**60**; equation 19). Alcohol (**59**) was originally prepared by metal–alcohol reduction of longicamphor (**61**)[77] and in improved yield using a metal–NH_3–alcohol procedure.[78] Using a similar procedure, ylangocamphor (**62**) gave alcohol (**60**) in excellent yield.[79] Reduction of ketone (**62**) using LAH afforded exclusively the *exo* isomer of alcohol (**60**).

(**61**) —[Na–Pr^nOH, reflux, 58% or Ca–NH_3–Pr^nOH, 97%]→ (**59**) (18)

(**62**) —[Ca–NH_3, –33 °C then EtOH]→ (**60**) (19)

1.4.3.4.6 Cyclopentanones

The stereochemistry of the reduction of substituted cyclopentanones has not been studied in great detail and it is not possible to make detailed generalizations regarding their course. Based on limited experimental data, it appears that metal–alcohol and metal–NH_3–proton donor reductions of alkyl-substituted cyclopentanones will usually afford a greater than equilibrium ratio of the thermodynamically more stable alcohol. In the absence of an added proton donor, pinacol formation may be a problem and different metals may give different ratios of epimeric alcohols.

The reductions of two steroidal ketones, androstan-17-one (**63**)[75] and androst-5-en-16-one (**64**)[19] under various conditions have been studied in some detail. In the case of 17-ketone (**63**) the β-ol (**65**) is the stable epimer and for the 16-ol (**66**), the α-isomer is more stable. Dissolving metal reductions of both ketones in the presence of proton donors gave the more stable alcohol as the major product; however, reduction of 17-keto steroid (**63**) is considerably more stereoselective as noted in Table 2. Although pinacols are not usually obtained in dissolving metal reductions carried out in the presence of proton donors, ketone (**63**) gave from 6 to 34% of dimeric products under these conditions (Li, 6%; Na, 34%; K, 13%).[75]

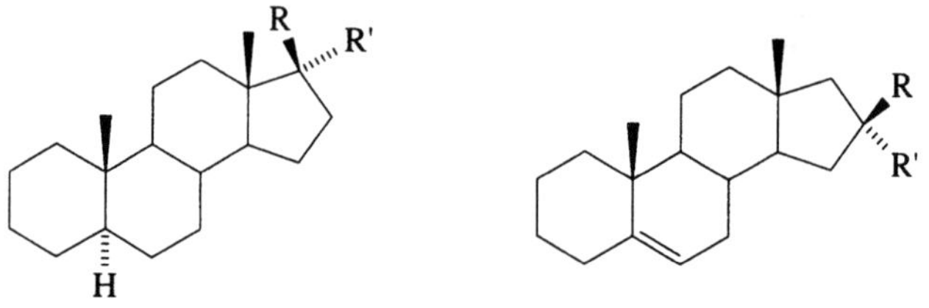

(**63**) R, R' = O
(**65**) R = OH, R' = H

(**64**) R, R' = O
(**66**) R = H, R' = OH

Table 2 Dissolving Metal Reductions of Androstan-17-one (**63**) and Androst-5-en-16-one (**64**)

Ketone	*Reduction conditions*	*α-ol:β-ol (Reduction)*	*α-ol:β-ol (Equilibrium)*	*Ref.*
(**63**)	Li–NH_3–EtOH	3:97	42:58	75
(**63**)	Na–NH_3–EtOH	0:100	—	75
(**63**)	K–NH_3–EtOH	0:100	—	75
(**64**)	Li–EtOH	63:37	82:18	19
(**64**)	Na–EtOH	85:15	—	19
(**64**)	Li–NH_3–THF	55:45	—	19
(**64**)	Na–NH_3–THF	42:58	—	19
(**64**)	K–NH_3–THF	22:78	—	19

In the absence of an added proton donor, the 17-keto steroid again gave the 17β-ol (**65**) as the predominant or exclusive product with seven metals (Li, Na, K, Cs, Ca, Ba, Sr) in NH_3. All the alkali metals gave the 17β-ol (**65**) exclusively and the alkaline earth metals gave only 2–5% of the α-ol. With the exception of Ca, all of the metals gave pinacols in yields of 3% (Cs) to 28% (Li).[75] Reduction of 16-keto steroid (**64**) with alkali metals–NH_3 in the absence of an added proton donor gave less 16α-ol with Na than with Li and still less with K (Table 2).[19] Pinacols were not reported in the reductions of ketone (**64**).

A series of dialkylcyclopentenones (**67**) has been reduced with Li–NH_3 in the presence of various proton donors (phenol and methyl, ethyl, isopropyl and *t*-butyl alcohols).[80] These reductions gave complex mixtures of products, most of which were cyclopentanols (**68**), arising from sequential conjugate reduction to the cyclopentanone followed by reduction to the saturated secondary alcohol. In all cases, and regardless of the relative stereochemistry of R, R′ and X, the major product was a secondary alcohol in which the hydroxy group was *trans* to the adjacent alkyl group (**68**).[80]

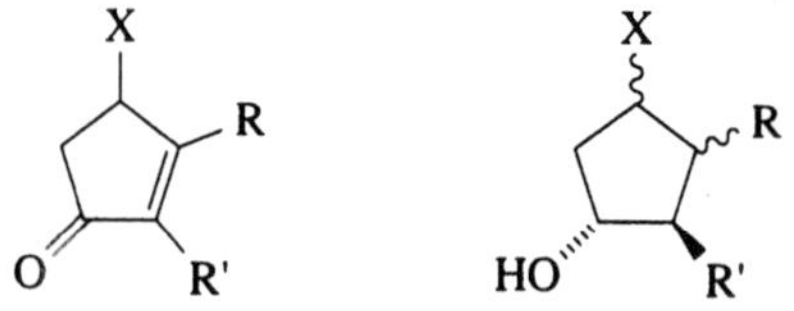

(**67**) (**68**)

X = H or OH; R = R' = Me; R = Me, R' = Bu^t; R = Me, R' = Pr^i

A stereoselective Li–NH_3 reduction of a cyclopentenone has been employed in two different syntheses of the cytotoxic sesquiterpene coriolin (**69**; Scheme 2). In one synthesis, tricyclic ketone (**70**) was reduced stereoselectively to alcohol (**71**) using Li–NH_3–methanol.[81] In the second synthesis, tetracyclic enone (**72**) was converted in a single step to (**71**).[82] This reduction proceeds by initial cleavage of the cyclopropyl ketone unit of (**72**) to give ketone (**70**), which is then reduced to (**71**).

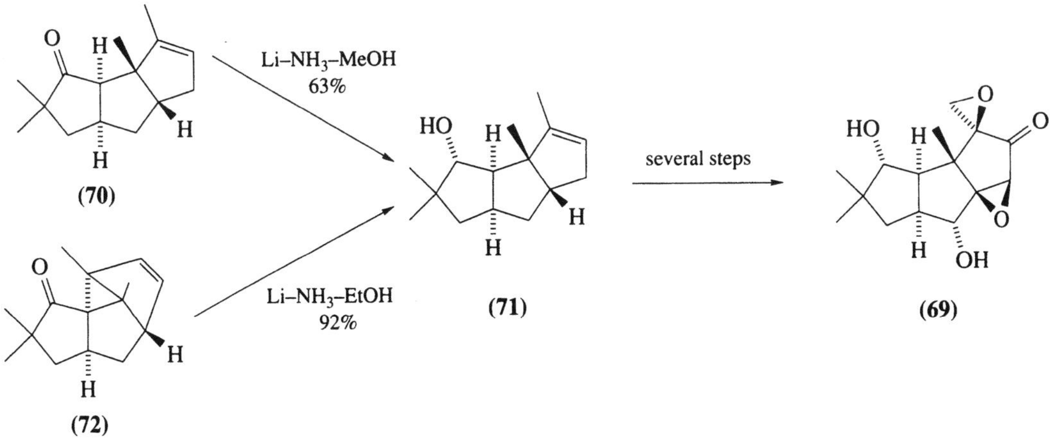

Li–NH_3–MeOH 63%

several steps

Li–NH_3–EtOH 92%

(**70**) (**71**) (**69**) (**72**)

Scheme 2

Metal–alcohol reductions of substituted indanone (**73**; equation 20) which afford mixtures of alcohols (**74**) and (**75**) have been described.[21] Although alcohol (**75**) is the more stable isomer, Li and Na both afford alcohol (**74**) as the major reduction product (**74**:**75** = 60:40). Aluminum gives a (**74**):(**75**) ratio of 90:10, but with K it is 30:70. Prolonged (12 h) heating of the Al and K reactions changes the ratios to (**70**):(**30**) and (**10**):(**90**), respectively.[21] It is apparent that some equilibration is occurring under these conditions, but it is not clear if the results after 3 h represent partial equilibration or if they are truly the result of kinetically controlled reductions.

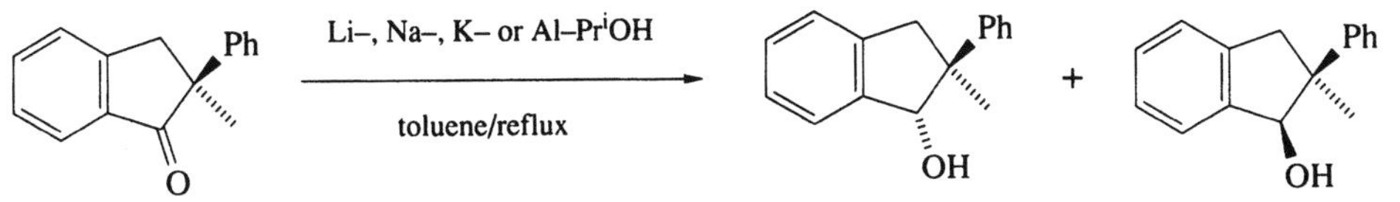

Li–, Na–, K– or Al–Pr^iOH, toluene/reflux (20)

(**73**) (**74**) (**75**)

1.4.4 REDUCTION OF C=N AND C=S

1.4.4.1 Reduction of Imines and Oximes

1.4.4.1.1 Reduction of imines

The reduction of imines to amines (equation 21) by dissolving metals is usually carried out using active metals in a protic solvent, typically Na–alcohol, Zn–NaOH and Al or Mg in alcohols.[83–85] Although the mechanism of these reductions has not been investigated in detail it is almost certainly analogous to that of the reduction of ketones (Section 1.4.2). It has been established that radical anions are intermediates in these reductions and in the absence of a proton donor reductive dimerization is the principal reaction path.[86]

$$R_2C{=}NR \longrightarrow R_2CH{-}NHR \qquad (21)$$

The reduction of a series of 1,3-diimines (**76**) has been investigated in which the dissolving metal reductions have been carried out both in the presence[87] and absence[88] of proton donors. These compounds exist as a tautomeric mixture of two enaminoimines (**77**) and (**78**) and the diimine (**76**). Reduction by Na–propan-2-ol of five examples (Ar = Ph, β-tolyl; R = Me; R′ = Ph, β-tolyl, cyclohexyl) gave excellent (90–99%) yields of diastereomeric mixtures of three of the four possible fully reduced diamines in which no isomer comprised more than 50% of the reaction mixture. Diamines (**79**) were not observed, and their absence was attributed to steric factors.[87] On reduction with Li–THF for several hours, followed by quenching with methanol, ethanol or water a number of these tautomeric mixtures (**76**, Ar = Ph, β-tolyl; R = H, Me, $PhCH_2$, $CH_2{=}CH{-}CH_2$; R′ = Ph, β-tolyl, Me, cyclohexyl) gave saturated ketones, (**80**) in 70 to 93% yield.[88]

The metal–alcohol reduction of imines usually produces the reduced amine in good yield; however, the reduction of some allylic imines (**81**, R = Pr^n,Bu^n) with Na–alcohol proceeds with hydrogenolysis to give alkene (**82**).[89]

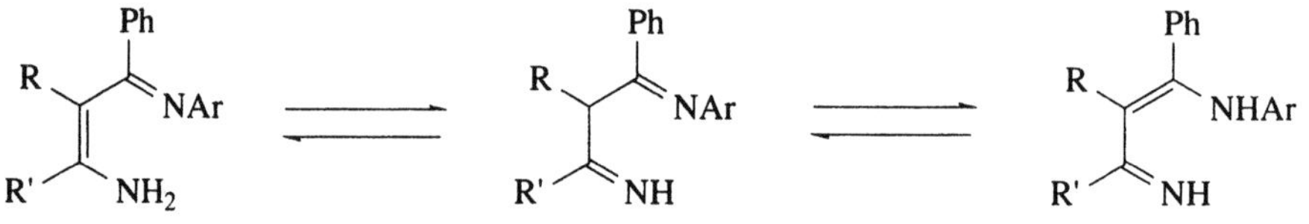

(**78**) (**76**) (**77**)

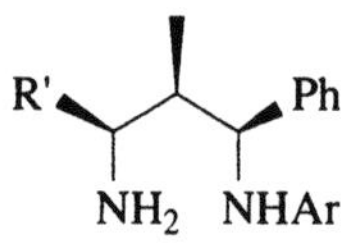

(**79**) 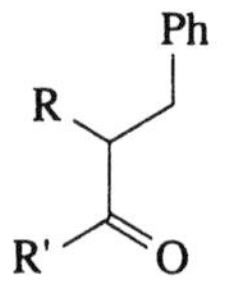 (**80**) 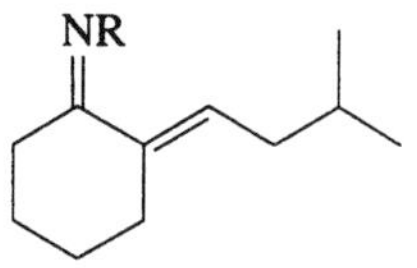 (**81**) 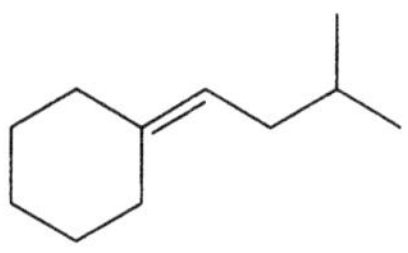 (**82**)

Although by far the most usual method for the dissolving metal reduction of imines is metal–alcohols, a few other systems have been explored with mixed results. In one of these, the reduction of three imines (**83**)–(**85**) with potassium graphite (C_8K) in anhydrous THF gave the corresponding amines in 83–90% yields at room temperature in 30 min.[90] Much less favorable results were found in the reduction of benzalaniline (**86**) with SmI_2 in methanolic THF which gave *N*-benzylaniline in 50% yield, with much recovered starting material. An attempted Barbier-type coupling reaction employing (**86**), l-iodobutane and SmI_2–THF gave a mixture of products of which *N*-benzylaniline constituted 60%.[64]

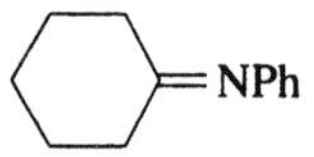

(**83**) 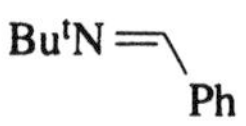 (**84**) 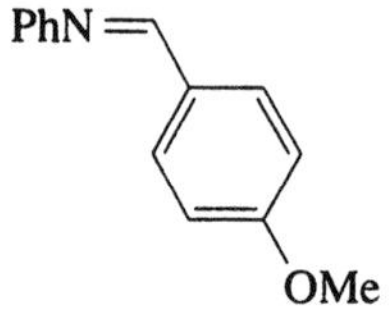 (**85**) 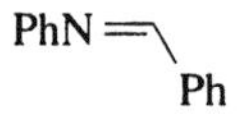 (**86**)

1.4.4.1.2 Reduction of oximes

The reduction of oximes to primary amines by dissolving metals, usually Na–alcohol, is an established synthetic procedure which has been employed for many years.[83,84] Although in some cases LAH reduction is superior, the reduction of many oximes with LAH leads either to aziridines or Beckman-type rearrangements.[91]

The reduction of a series of bicyclic and tricyclic ketoximes using both Na–ethanol and LAH has been carried out and the product mixtures analyzed in detail.[92] In this investigation it was found that reduction of the oximes derived from two tricyclo[2.2.1.$0^{2,6}$]heptan-3-ones (**87**) and (**88**) with either LAH–THF or Na–ethanol gave mixtures of *exo* (**89**) and (**90**) and *endo* (**91**) and (**92**) amines. Neither method of reduction was particularly stereoselective and the yields were somewhat better with LAH (**87**; 76% *versus* 67% and **88**; 67% *versus* 44%). In contrast, reduction of the oximes of three bicyclo[2.2.1]heptan-2-ones (**93**)–(**95**) with LAH–THF gave significant quantities of azabicyclo[3.2.1]octanes (**96**)–(**98**) in addition to mixtures of *exo* and *endo* primary amines. Reduction of oximes (**93**)–(**95**) with Na–alcohol gave only mixtures of epimeric amines.[92] These results, outlined in Table 3, show that the *endo*-amine is the major stereoisomer, although the unsubstituted oxime (**93**) had previously been reported to give an *exo:endo* ratio of 75:25 on Na–alcohol reduction and exclusively the *endo*-amine on LAH reduction.[93]

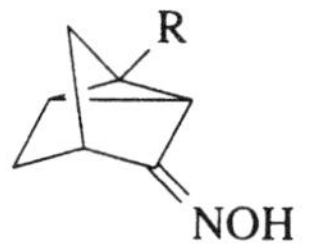

R NOH R NH_2 R NH_2

(**87**) R = Me
(**88**) R = Et

(**89**) R = Me
(**90**) R = Et

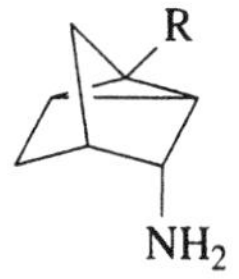

(**91**) R = Me
(**92**) R = Et

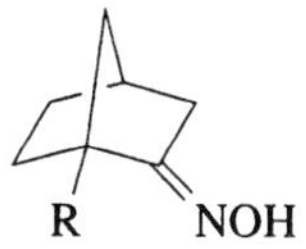

R NOH R N-H

(**93**) R = H
(**94**) R = Me
(**95**) R = Et

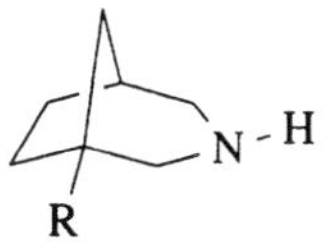

(**96**) R = H
(**97**) R = Me
(**98**) R = Et

Table 3 Reduction of Oximes of Bicyclo[2.2.1]heptan-2-ones[91]

Oxime	*Conditions*	*Yield (%)*	*Rearrangement (rel. %)*	*Amine (rel. %)*	*Exo:endo*
(**93**)	Na–EtOH, 80 °C	67	0	100	27:73
(**93**)	LAH–THF, 65 °C	79	36	64	36:64
(**94**)	Na–EtOH, 80 °C	52	0	100	32:68
(**94**)	LAH–THF, 65 °C	87	60	40	10:90
(**95**)	Na–EtOH, 80 °C	50	0	100	14:86
(**95**)	LAH–THF, 65 °C	59	52	48	10:90

The chemoselective reduction of a series of 2-oximo-1-tetralones (**99**–**103**; equation 22) to the corresponding aminotetralones has been carried out using Zn–HOAc–Ac_2O in 55–65% yield. The amino group is acetylated under these conditions and the 2-acetamidotetralone is the product which is isolated.[94]

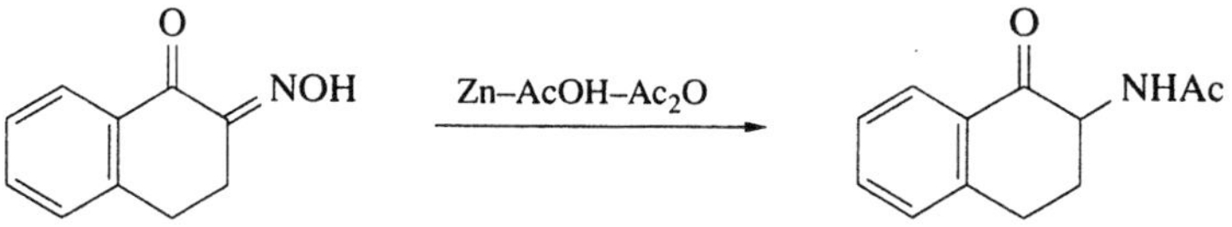

O NOH $\xrightarrow{\text{Zn–AcOH–Ac}_2\text{O}}$ O NHAc (22)

(**99**) unsubstituted; 65%
(**100**) 5-OMe; 62%
(**101**) 6-OBn; 65%
(**102**) 7-OBn; 60%
(**103**) 5,8-di-OMe; 55%

The reduction of oximes by SmI_2 has been attempted, but complex mixtures of products are obtained.[64] In the case of benzaldehyde oxime (**104**; equation 23) two dimeric reduction products (**105**) and (**106**) were isolated, apparently in good yield.[64]

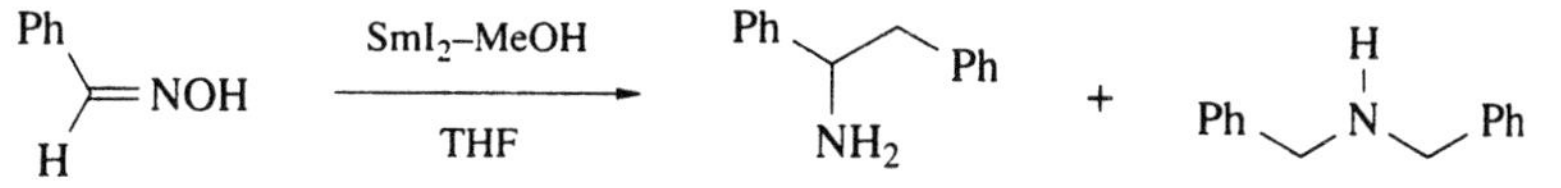

Ph(H)C=NOH $\xrightarrow[\text{THF}]{\text{SmI}_2\text{–MeOH}}$ PhCH(NH_2)CH$_2$Ph + PhCH$_2$N(H)CH$_2$Ph (23)

(**104**) (**105**) (**106**)

1.4.4.2 Reduction of Thioketones

There is apparently only one report of the dissolving metal reduction of thioketones; thiobenzophenone (**107**) has been reduced with excess Na–THF to give dianion (**108**), which on acidification gave thiol (**109**). The thiol was not isolated, but was oxidized with iodine to give the corresponding disulfide in 65% overall yield.[95] The mechanism of the reduction is suggested to be the sequential addition of two electrons to the thiocarbonyl, which was confirmed by electron spin resonance spectroscopy studies of the intermediate thioketyl, and trapping dianion (**108**) with a variety of electrophiles.[95]

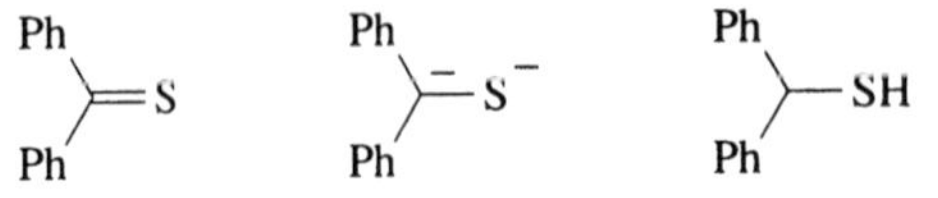

(**107**) (**108**) (**109**)

1.4.5 REFERENCES

1. F. Sondheimer, R. Yashin, G. Rosenkranz and C. Djerassi, *J. Am. Chem. Soc.*, 1952, **74**, 2696.
2. H. L. Heusser, R. Anliker and O. Jeger, *Helv. Chim. Acta*, 1952, **35**, 1537.
3. D. H. R. Barton, *J. Chem. Soc.*, 1953, 1027.
4. H. O. House, in 'Modern Synthetic Reactions', 2nd edn., Benjamin, Menlo Park, CA, 1972, p. 152.
5. J. W. Huffman, *Acc. Chem. Res.*, 1983, **16**, 399.
6. S. K. Pradhan, *Tetrahedron*, 1986, **42**, 6351.
7. J. W. Huffman, W.-P. Liao and R. H. Wallace, *Tetrahedron Lett.*, 1987, **28**, 3315.
8. V. Rautenstrauch and M. Geoffroy, *J. Am. Chem. Soc.*, 1976, **98**, 5035.
9. V. Rautenstrauch and M. Geoffroy, *J. Am. Chem. Soc.*, 1977, **99**, 6280.
10. V. Rautenstrauch, B. Willhalm, W. Thommen and U. Burger, *Helv. Chim. Acta*, 1981, **64**, 2109.
11. W. S. Murphy and D. F. Sullivan, *J. Chem. Soc., Perkin Trans. 1*, 1972, 999.
12. S. K. Pradhan, K. R. Thakker and A. T. McPhail, *Tetrahedron Lett.*, 1987, **28**, 1813.
13. J. W. Huffman, R. H. Wallace and W. T. Pennington, *Tetrahedron Lett.*, 1988, **29**, 2527.
14. V. Rautenstrauch, *Helv. Chim. Acta*, 1982, **65**, 402.
15. V. Rautenstrauch, *Tetrahedron*, 1988, **44**, 1613.
16. V. Rautenstrauch, *J. Chem. Soc., Chem. Commun.*, 1986, 1558.
17. G. Castaldi, G. Perdoncin and C. Giordano, *Tetrahedron Lett.*, 1983, **24**, 2487.
18. C. Giordano, G. Perdoncin and G. Castaldi, *Angew. Chem., Int. Ed. Engl.*, 1985, **24**, 499.
19. S. K. Pradhan and S. V. Sohani, *Tetrahedron Lett.*, 1981, **22**, 4133.
20. G. Aranda, J.-M. Bernassau, M. Fétizon and I. Hanna, *J. Org. Chem.*, 1985, **50**, 1156.
21. J. Berlan, P. Sztajnbok, Y. Besace and P. Cresson, *C. R. Hebd. Seances Acad. Sci., Ser. B*, 1985, **301**, 693.
22. J. Solodar, *J. Org. Chem.*, 1976, **41**, 3461.
23. H. E. Ensley, C. A. Parnell and E. J. Corey, *J. Org. Chem.*, 1978, **43**, 1610.
24. P. A. Grieco, S. Burke, W. Metz and M. Nishizawa, *J. Org. Chem.*, 1979, **44**, 152.
25. J. W. Huffman and D. J. Copley, *J. Org. Chem.*, 1977, **42**, 3811.
26. J. W. Huffman, R. C. Desai and J. E. LaPrade, *J. Org. Chem.*, 1983, **48**, 1474.
27. J. W. Huffman and W.-P. Liao, unpublished results.
28. M. Hulce and T. LaVaute, *Tetrahedron Lett.*, 1988, **29**, 525.
29. P. Sengupta, S. Das and K. Das, *Indian J. Chem., Sect. B*, 1984, **23**, 1113.
30. B. P. Pradhan, D. K. Chakrabarti and S. Chakraborty, *Indian J. Chem., Sect. B*, 1984, **23**, 1115.
31. P. Sengupta, M. Sen, A. Sarkar and S. Das, *Indian J. Chem., Sect. B*, 1986, **25**, 975.
32. A. F. Sowinski and G. M. Whitesides, *J. Org. Chem.*, 1979, **44**, 2369.
33. M. Larcheveque and T. Cuvigny, *Bull. Soc. Chim. Fr.*, 1973, 1445.
34. J. D. White and G. L. Larson, *J. Org. Chem.*, 1978, **43**, 4555.
35. J. W. Huffman and R. H. Wallace, unpublished results.
36. Z. Hou, K. Takamine, Y. Fujiwara and H. Taniguchi, *Chem. Lett.*, 1987, 2061.
37. J. M. Lalancette, G. Rollin and P. Dumas, *Can. J. Chem.*, 1972, **50**, 3058.
38. D. E. Bergbreiter and J. M. Killough, *J. Am. Chem. Soc.*, 1978, **100**, 2126.
39. K. L. Handoo and K. Gadru, *Tetrahedron Lett.*, 1986, **27**, 1371.
40. L. Horner and K. Dickerhof, *Liebigs Ann. Chem.*, 1984, 1240.
41. M. Hulce, private communication.
42. G. R. Newkome and J. M. Roper, *J. Org. Chem.*, 1979, **44**, 502.
43. A. Clerici and O. Porta, *J. Org. Chem.*, 1985, **50**, 76.
44. A. Clerici, O. Porta and P. Zago, *Tetrahedron*, 1986, **42**, 561.
45. A. Clerici and O. Porta *J. Org. Chem.*, 1987, **52**, 5099.
46. T.-L. Ho and G. A. Olah, *Synthesis*, 1976, 815.
47. T. Imamoto, T. Kusumoto, Y. Hatanaku and M. Yokoyama, *Tetrahedron. Lett.*, 1982, **23**, 1353.
48. P. Girard, J. L. Namy and H. B. Kagan, *J. Am. Chem. Soc.*, 1980, **102**, 2693.
49. H. B. Kagan, J. L. Namy and P. Girard, *Tetrahedron Suppl.*, 1981, **9**, 269.
50. J. A. Dean, 'Handbook of Organic Chemistry', McGraw-Hill, New York, 1987, p. 8-76.
51. S. K. Pradhan, S. R. Kadam, J. N. Kolhe, T. V. Radhakrishnan, S. V. Sohani and V. B. Thaker, *J. Org. Chem.*, 1981, **46**, 2622.
52. S. K. Pradhan, S. R. Kadam and J. N. Kolhe, *J. Org. Chem.*, 1981, **46**, 2633.

53. J. K. Crandall and M. Mualla, *Tetrahedron Lett.*, 1986, **27**, 2243.
54. A. J. Bellamy, E. A. Campbell and I. R. Hall, *J. Chem. Soc., Perkin Trans. 2*, 1974, 1347.
55. W. S. Murphy and S. Wattanasin, *J. Chem. Soc., Perkin Trans. 1*, 1986, 1445.
56. S. K. Attah-Poku, S. J. Alward and A. G. Fallis, *Tetrahedron Lett.*, 1983, **24**, 681.
57. U. Schindewolf and P. Wünschel, *Can. J. Chem.*, 1977, **55**, 2159.
58. S. S. Hall, S. D. Lipsky, F. J. McEnroe and A. P. Bartels, *J. Org. Chem.*, 1971, **36**, 2588.
59. Z. Marcinow and P. W. Rabideau, *J. Org. Chem.*, 1988, **53**, 2117.
60. S. Mejer and R. Pacut, *Pol. J. Chem.*, 1980, **54**, 453.
61. P. W. Rabideau, C. A. Husted and D. M. Young, *J. Org. Chem.*, 1983, **48**, 4149.
62. S. Mejer and R. Pacut, *Pol. J. Chem.*, 1978, **52**, 529.
63. G. S. R. Subba Rao and N. Shyama Sundar, *J. Chem. Soc., Perkin Trans. 1*, 1982, 875.
64. J. Souppe, L. Danon, J. L. Namy and H. B. Kagan, *J. Organomet. Chem.*, 1983, **250**, 227.
65. T. Imamoto and M. Ono, *Chem. Lett.*, 1987, 501.
66. N. Y. M. Fung, P. de Mayo, J. H. Schauble and A. C. Weedon, *J. Org. Chem.*, 1978, **43**, 3977.
67. G. A. Molander and G. Hahn, *J. Org. Chem.*, 1986, **51**, 1135.
68. S. Takeuchi and Y. Ohgo, *Chem. Lett.*, 1988, 403.
69. S. Chandrasekhar, *Chem. Soc. Rev.*, 1987, **16**, 313.
70. J. W. Huffman and J. T. Charles, *J. Am. Chem. Soc.*, 1968, **90**, 6486.
71. H. O. House, in 'Modern Synthetic Reactions', 2nd edn., Benjamin, Menlo Park, CA, 1972, p. 60.
72. E. J. Corey, M. Petrzilka and Y. Ueda, *Tetrahedron Lett.*, 1975, 4343.
73. B. P. Pradhan, A. Hassan and J. N. Shoolery, *Tetrahedron Lett.*, 1984, **25**, 865.
74. A. G. M. Barrett, C. R. A. Godfrey, D. M. Hollinshead, P. A. Prokopiou, D. H. R. Barton, R. B. Boar, L. Joukhadar, J. F. McGhie and S. C. Misra, *J. Chem. Soc., Perkin Trans.*, **1**, 1981, 1501.
75. A. M. Giroud and A. Rassat, *Bull. Soc. Chim. Fr.*, 1976, 1881.
76. C. W. Shoppee, R. E. Lack, S. C. Sharma and L. R. Smith, *J. Chem. Soc. C*, 1967, 1155.
77. D. H. R. Barton and W. H. Werstiuk, *J. Chem. Soc. C*, 1968, 148.
78. S. C. Welch and R. L. Walters, *Synth. Commun.*, 1973, **3**, 419.
79. E. Piers, R. W. Britton, M. B. Geraghty, R. J. Keziere and F. Kido, *Can. J. Chem.*, 1975, **53**, 2838.
80. M. Samson, P. DeClercq and M. Vandewalle, *Tetrahedron*, 1977, **33**, 249.
81. G. Mehta, A. V. Reddy, A. N. Murthy and D. S. Reddy, *J. Chem. Soc., Chem. Commun.*, 1982, 540.
82. P. A. Wender and J. J. Howbert, *Tetrahedron Lett.*, 1983, **24**, 5325.
83. H. O. House, in 'Modern Synthetic Reactions', 2nd edn., Benjamin, Menlo Park, CA, 1972, p. 210.
84. K. Harada and S. Patai, in 'The Chemistry of the Carbon Nitrogen Double Bond', Wiley-Interscience, London, 1970, p. 279.
85. M. Tramontini, *Synthesis*, 1982, 605.
86. J. G. Smith and I. Ho, *J. Org. Chem.*, 1972, **37**, 653.
87. J. Barluenga, B. Olano and S. Fustero, *J. Org. Chem.*, 1983, **48**, 2255.
88. J. Barluenga, E. Aguilar, B. Olano and S. Fustero, *J. Org. Chem.*, 1988, **53**, 1741.
89. N. Barbulescu, O. Cuza, E. Barbulescu, S. Moya-Gheorghe and D. Zavoranu, *Rev. Chim. (Bucharest)*, 1985, **36**, 295 (*Chem. Abstr.*, 1985, **103**, 123 042t).
90. M. Contento, D. Savoia, C. Trombini and A. Umani-Ronchi, *Synthesis*, 1979, 30.
91. S.-C. Chen, *Synthesis*, 1974, 691.
92. M. D. Ordubadi, T. I. Pekhk, N. A. Belikova, T. M. Rakhmanchik and A. F. Plate, *J. Org. Chem. USSR (Engl. Transl.)*, 1984, **20**, 678.
93. A. J. Fry and J. H. Newberg, *J. Am. Chem. Soc.*, 1967, **89**, 6374.
94. A. Delgado, J. M. Garcia, D. Mauléon, C. Minguillon, J. R. Subirato, M. Felix, F. Lopez and D. Velasco, *Can. J. Chem.*, 1988, **66**, 517.
95. Y. Minoura and S. Tsuboi, *J. Org. Chem.*, 1972, **37**, 2064.

1.5
Reduction of C═X to CHXH Electrolytically

LUC MOËNS*
Solar Energy Research Institute, Golden, CO, USA

1.5.1 INTRODUCTION

Electrochemical methodology has proven to be a useful tool for carrying out certain redox reactions which normally require stoichiometric amounts of (organo)metallic reagents.[1] The reagent in this case is an electrode which, when negatively charged (cathode), can transfer an electron onto a substrate molecule in solution (*i.e.* reduction), or, when positively charged (anode), can abstract an electron from it (*i.e.* oxidation). In this chapter a review will be presented dealing with the cathodic reduction of C═O and C═N groups. In order to make the text more intelligible for the synthetic chemist who may not be familiar with electrochemical jargon, a short review of the basic principles of electrosynthesis is presented here.

In a cathodic reduction process electrons are transferred from the electrode surface onto the substrate molecules only if the former possess enough potential energy, *i.e.* if the electrode potential is negative enough. Each organic compound has its own reduction potential (expressed in volts) at which it is reduced but the value depends greatly on the solvent, supporting electrolyte and the presence of other compounds that may interact with the substrate or electrode surface. In anodic oxidations the term 'oxidation potential' is used, referring to the electrode potential at which electrons will be transferred from the sub-

*This chapter is dedicated to the memory of Dr. Manuel M. Baizer

strate to the positively charged anode surface. Thus, the higher the oxidation potential of a compound, the more difficult it is to oxidize it. In an electrochemical reaction electrons will only flow in a closed circuit (Figure 1) which includes a 'counter electrode' (sometimes called 'auxiliary electrode'). The cathode receives electrons from the counter electrode, which becomes positive (anode) with the help of a power supply (potentiostat), which serves as an 'electron pump'. As a result an oxidation process needs to take place near the anode surface to maintain the current flow. The electrode at which the desired reaction occurs is called the 'working electrode'. The term 'electrode potential' is in reality a potential difference measured relative to a reference electrode, such as the 'saturated calomel electrode' (SCE) or the silver–silver halide electrode. Whenever a potential value is reported for a compound, the type of reference electrode must be indicated. When the reduction is carried out at a constant potential (see Figure 1), the potentiostat uses the reference electrode as a zero point against which it will maintain the negative potential value required at the cathode surface. In order to maintain electroneutrality in the solution during the electrolysis it is necessary to add a so-called 'supporting electrolyte' which makes the medium electrically conductive. The most popular are the tetraalkylammonium salts (*e.g.* Bu_4NBr) and alkali metal salts (*e.g.* $LiClO_4$), although virtually any soluble salt can be used. The solvent required for dissolving these salts must be very polar (high dielectric constant) and, therefore, electrosynthetic reactions are usually carried out in dimethylformamide, acetonitrile, dimethyl sulfoxide, alcohols, water and sometimes even ethers (*e.g.* tetrahydrofuran).

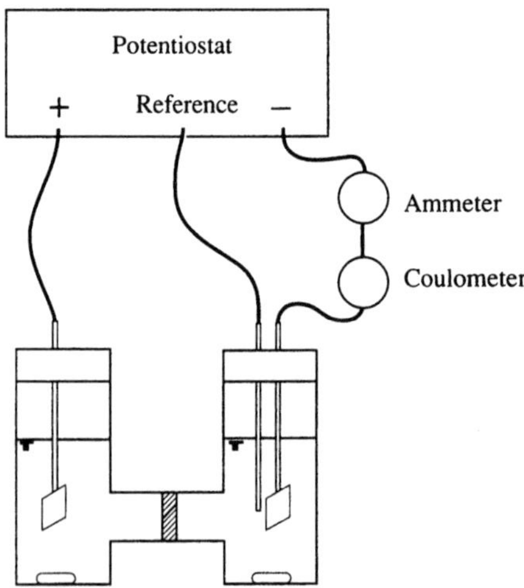

Figure 1 Controlled potential electroreduction in H–cell

The cell shown in Figure 1 (also called an H-cell because of its shape) is a divided cell, wherein a porous glass frit or polymer (*e.g.* Nafion) prevents the anolyte and catholyte solutions from mixing with each other. In some cases, however, a simple beaker can be used (undivided cell).

Once the potential is dialed on the potentiostat and the electrodes are activated, a current flows through the cell but decreases gradually until all the substrate at the working electrode is converted into products. The current is monitored with an ammeter (Figure 1) and it is also common to include a coulometer so that at the end of the electrolysis the total charge that has passed through the cell can be measured. For instance, in a 'one-electron' process a total charge of 96 487 C will be consumed by 1 mol of substrate knowing that 1 mol of electrons (each carrying a charge of 1.6×10^{-19} C) is required. That total charge of 1 mol of electrons is commonly referred to as one Faraday (1 F = 96 487 C) and the electron consumption is expressed in Faraday per mol of substrate (F mol^{-1}). Suppose that one deals with a well-established one-electron process and at the end of the experiment the coulometer indicates the consumption of 2 F mol^{-1}, then it is obvious that 1 F mol^{-1} of electricity must have been consumed by impurities or secondary electron transfer reactions. While it is theoretically possible to obtain 100% chemical yield in such an electrolysis, the 'current efficiency' or 'current yield' is 50%: only 50% of the current is used for the desired reaction.

In the electrochemical literature one often encounters reports dealing with electrolyses carried out under a constant current (*i.e.* galvanostatic), expressed as 'current density' (*e.g.* A cm^{-2}), instead of a constant potential. Such experiments do not require the use of a reference electrode and may also take less time, but a major drawback is that the potential value of the working electrode increases gradually as the substrate in the solution is consumed. For a reduction it means that the cathode potential becomes more negative and consequently the product selectivity may suffer if the composition of the reaction mixture is not carefully monitored during the electrolysis.

There are many electrochemical reactions known that involve the use of so-called 'mediators', which are basically redox catalysts. The active form acts as an electron transfer agent to convert the substrate into products. The catalyst is then regenerated at the electrode surface and the cycle can be repeated until all the substrate is consumed. Such an 'indirect' electrode reaction is schematically shown in Figure 2 for the case of an electroreduction. The mediator is reduced to its active form at potentials that can be up to 600 mV more positive than the potential required for direct reduction of the substrate.

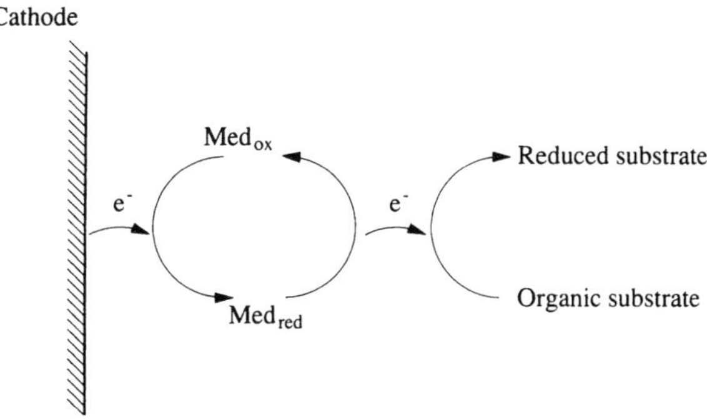

Figure 2 Indirect electroreduction

Before embarking on an electrosynthesis it is desirable to know the potential at which the desired reaction is expected to occur. That information can be obtained from electroanalytical techniques, such as polarography and cyclic voltammetry.[2,3]

These basic principles will hopefully clarify some data presented in the following review that covers the literature from 1979 until the present, dealing with the electroreduction of a carbonyl group to an alcohol, and of an imino group to the corresponding amine. Earlier literature on this topic has been reviewed elsewhere.[1]

1.5.2 KETONES AND ALDEHYDES

1.5.2.1 Nonalkenic Ketones and Aldehydes

1.5.2.1.1 General mechanism and products

The product distribution obtained in the cathodic reduction of ketones and aldehydes depends strongly on the right choice of the solvent (aqueous or nonaqueous), the supporting electrolyte, the electrode material and additives. Scheme 1 shows a simplified overview of the reduction pathways that lead to alcohols (pinacols **2** and methanols **3**) or deoxygenated products (**4**). In solutions containing aqueous acid the carbonyl group exists partly in the protonated form, which is more electrophilic and thus easier to reduce to the neutral radical (**1**). In some cases complete reduction of the carbonyl group to a methylene group (as in **4**) can be observed.

Van Tilborg and coworkers have reported synthetically useful data that support this mechanistic scheme.[4,5] They found that the reduction of acetophenones often leads to nearly quantitative yields of pinacols in aprotic media, as shown in equation (1).

Presumably a layer of tetraalkylammonium ions is formed at the mercury surface from which the radical anion of acetophenone is quickly expelled into the bulk solution, where pinacolization can take place. An increased bulk of the alkyl chain of the ketone caused a lower selectivity: reduction of isobutyrophenone afforded, besides 68% pinacol, 17% of the corresponding alcohol. Reduction of isobutyrophenone under protic conditions gave predominantly alcohol and only 13% of the pinacol. Aryl alkyl ketones also undergo pinacolization in excellent yields when lithium ions are added to the aprotic medium.[6] However,

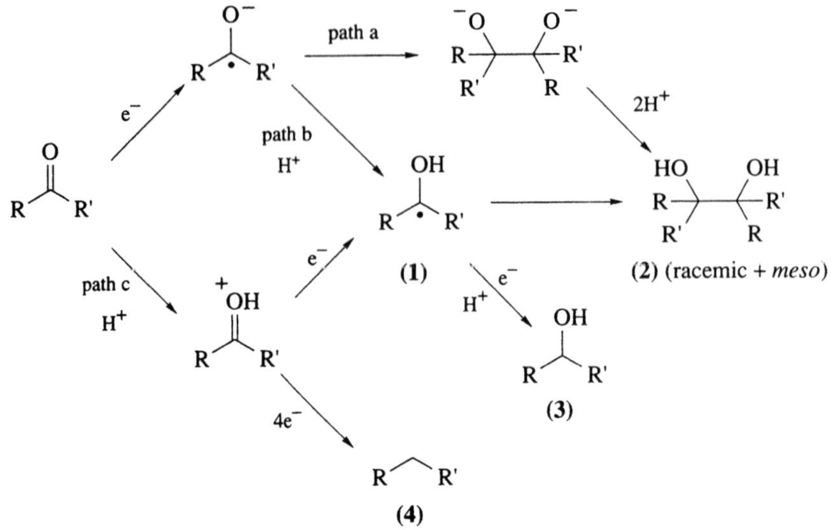

Scheme 1

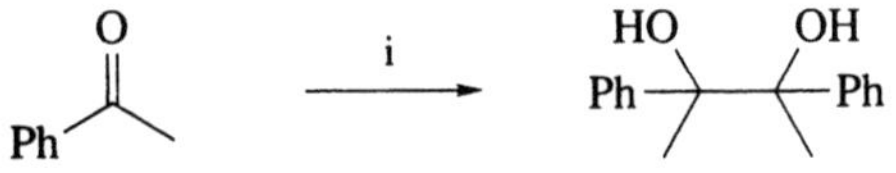

(racemic:*meso* = 6:1) (1)

i, Hg, –2.0 V (SCE), 1.3 F mol^{-1}, Et_4NClO_4, MeCN, 98%

in the case of diaryl ketones a disproportionation of the radical anion has been observed in the presence of lithium ions, affording the corresponding alcohols.[7] Dialkyl ketones have received much less attention in such reduction studies, mainly because they can undergo aldol-type reactions when enolizable protons are present within the molecule.[4]

Pletcher and Razaq studied the four-electron reduction of acetophenone to ethylbenzene in strongly acidic media.[8] Such an electroreduction is analogous to the classical Clemmensen reaction, which is carried out with amalgamated zinc in concentrated hydrochloric acid.

The pathway for the cathodic reduction of acetophenone can be profoundly altered when the aromatic ring is complexed inside the hydrophobic cavity of β-cyclodextrin.[9] Products resulting from *ortho* and *para* coupling to the carbonyl carbon are isolated in excellent yield, as shown in equation (2).

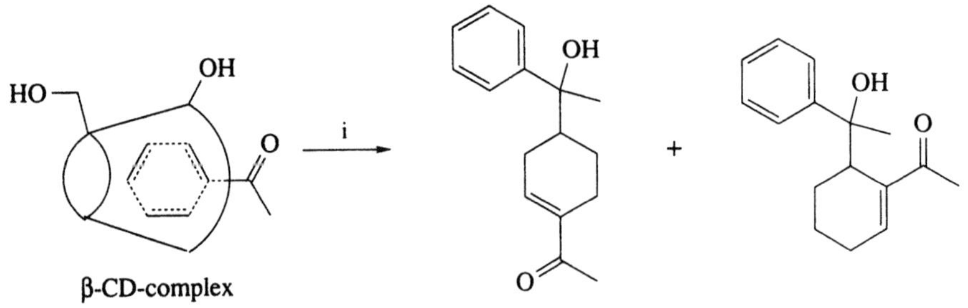

β-CD-complex (1:1) (2)

i, Hg, –1.34 V(Ag/AgI), 1.5 F mol^{-1}, Bu_4NBF_4, DMF, 90%

1.5.2.1.2 Indirect electroreduction

The reduction potential of an unactivated carbonyl group in aprotic media lies so close to the reduction potential of the supporting electrolyte that the latter is often reduced first. Kariv-Miller and coworkers recently exploited this seemingly undesirable effect in the development of extremely powerful electroreductions.[10–14] They showed that the tetraalkylammonium cation can act as a mediator, which accepts an electron from the cathode and transfers it onto the carbonyl group. The reduction of cyclohexanone (**5**) shown in Scheme 2 serves as an example of the high yields and high selectivities that can be obtained by this method. Addition of dimethylpyrrolidinium ions (DMP^+) causes the reduction to occur at less negative potentials (compared to Bu_4N^+ ions) and at the same time favors pinacolization (*i.e.* the one-electron process). The DMP^+ ion is reduced by one electron to generate a radical species that reacts with the mercury cathode, forming a black solid material characterized as $DMPHg_5$. The latter is believed to be the active reducing agent in these indirect electroreductions.

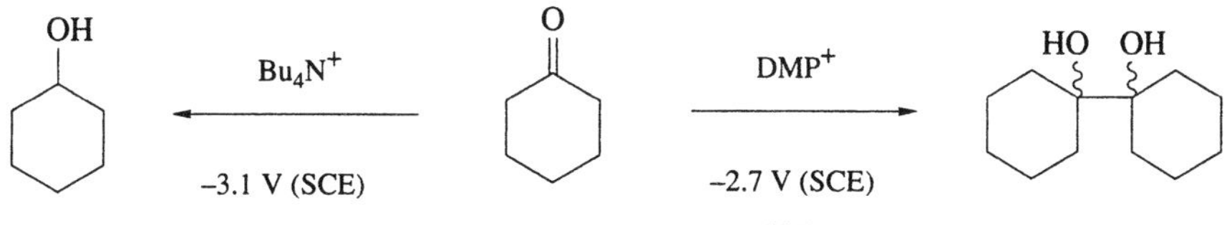

OH ← Bu_4N^+, –3.1 V (SCE), 75% — (**5**) — DMP^+, –2.7 V (SCE), 95% → HO OH

Scheme 2

1.5.2.1.3 Electroreduction in the presence of transition metal ions

Several studies have been done on the product distribution resulting from electroreduction of ketones and aldehydes in the presence of transition metal ions.[15–22] The complexation of such ions with the carbonyl group causes the latter to be reduced at less negative potentials and also determines the regio- and stereo-chemical outcome of the reduction. Sopher and Utley showed that $CrCl_3$ favors the pinacolization of ketones and aldehydes.[15,16] Controlled potential electrolysis of benzophenone (Hg cathode, DMF, $NaClO_4$) required a potential of –1.3 V (*versus* Ag/AgI) to afford 80% alcohol, while addition of 1 equiv. of $CrCl_3$ resulted in the formation of the pinacol (70%) and alcohol (10%) at only –1.0 V. The cathodic reduction of sterically hindered aromatic ketones and aldehydes in the presence of Cr^{3+}, Mn^{2+}, Fe^{2+}, Co^{2+}, Ni^{2+} and Zn^{2+} also gave good yields of the corresponding pinacols with predominantly the racemic diastereoisomer.[19–21] A stereoselective pinacolization (100% racemic diastereoisomer) of acetophenone (**6**) was observed when the electroreduction was carried out in the presence of Eu^{3+} ions (equation 3).[22]

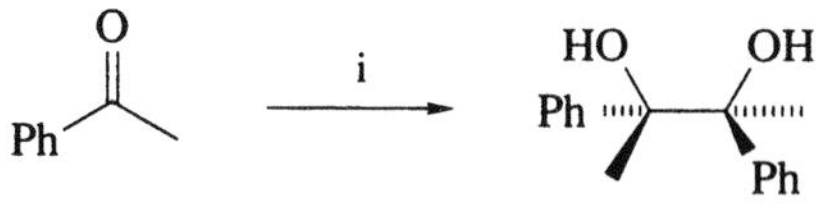

(**6**) —i→ 100% racemic isomer (3)

i, $EuCl_3 \cdot 6H_2O$, Hg, –1.4 V (Ag/AgI), Bu_4NClO_4, DMF, 97%

1.5.2.1.4 Stereocontrol in the electroreduction of ketones

An important issue in the reduction of a carbonyl group is the degree of stereocontrol that can be obtained by using a particular set of conditions. A few systematic studies dealing with stereocontrol in cathodic reductions of cyclic ketones were reported independently by Utley[23–25] and Shono[26] in the early 1970s. More recently, Le Guillanton reported a study on the electroreduction of 2-ethoxycarbonyl cycloalkanones (**7**) in aqueous media, as shown in equation (4).[27] Both *cis*- and *trans*-2-hydroxy esters (**8a**) and (**8b**) were isolated in a ratio depending strongly on the nature of the supporting electrolyte and the cathode material.

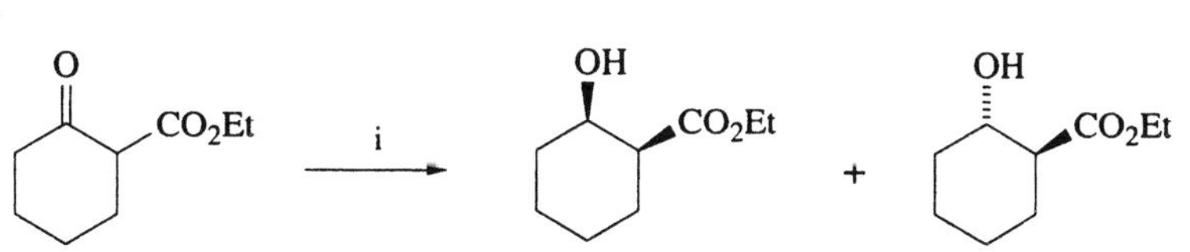

(**7**) —i→ (**8a**) + (**8b**) (4)

i, Hg, –2.2 V (SCE), Et_4NI, H_2O, EtOH, 90%

The most important parameter was the temperature: at –6 °C *cis* isomer (**8a**) predominated (*cis:trans* = 65:35), while at 80 °C the thermodynamically more stable *trans* isomer (**8b**) was the major compound (*cis:trans* = 15:85), suggesting that the isomers result from an equilibrium of an intermediate radical or anionic species. The electroreduction of acyclic ketones with a chiral center (optically inactive) at the α- and β-positions has been studied by Nonaka and coworkers.[28,29] Unfortunately the yields were generally very low for aliphatic ketones, probably because the reactions were carried out under extreme pH conditions. Tetraols are formed stereoselectively in the coupling of 2-substituted indan-1-ones.[30,31]

1.5.2.1.5 Asymmetric electroreduction of a carbonyl group

The development of asymmetric electroreductions remains a major challenge in spite of the many different approaches that have been tested during the last two decades. The literature on this topic prior to 1979 has been reviewed.[1,4,32] Chirality can be induced during electropinacolization of acetophenones by using chiral supporting electrolyte salts.[4] The enantiomeric excess (*ee*) in the (*RS,RS*)-pinacols varied between 3 and 25%, depending on the nature of the chiral supporting electrolyte, the applied potential and the solvent. Anhydrous media gave the highest optical yield.[32] Chemically modified electrodes, such as poly-L-valine coated graphite electrodes have been tested but they gave less than 7% optical yield in the reduction of phenylglyoxylic acid.[33] A very intriguing method reported by Takahashi deals with the electrolytic reduction of phenylglyoxylic acid in a magnetic field of 1680 G (1 G = 10^{-4} T), affording dextrorotatory mandelic acid with 21% *ee*.[34]

1.5.2.2 Alkenic and Alkynic Ketones and Aldehydes

1.5.2.2.1 Nonconjugated alkenic ketones

Kariv-Miller and coworkers have developed indirect electroreductive cyclizations with the dimethylpyrrolidinium ion (DMP^+) as a mediator. Preparative electrolysis of 6-hepten-2-one (**9**) at a graphite cathode afforded *cis*-dimethylcyclopentanol (**10**) in 90% yield (equation 5).[13] The reduction is believed to occur *via* the ketyl radical anion, which cyclizes onto the alkenic bond.[10] In the absence of DMP^+ simple reduction to 6-hepten-2-ol takes place.[13] Very recently it was shown that instead of DMP^+ several aromatic hydrocarbons can be used as mediators to initiate the cyclization reaction.[10] The carbonyl group can also be cyclized onto an alkynic bond[12] and even an aromatic ring.[12,35]

O OH i (5)

(**9**) (**10**)

i, C, –2.75 V (SCE), DMF, $DMP^+ BF_4^-$, Bu_4NBF_4, 90%

1.5.2.2.2 α,β-Unsaturated ketones and aldehydes

When an alkenic bond is conjugated with the carbonyl group, the carbonyl carbon and the β-carbon become the two reactive centers and a variety of products can be obtained depending on the medium.[1] Presence of water in the reduction of 4-methyl-2-cyclohexenones results in a mixture of products.[36] However, when the β-position is substituted such as in retinal (**11**), pinacolization takes place to form the pinacol (**12**) in 89% yield, provided that the electroreduction is carried out in an aprotic medium in the presence of a mild proton donor, such as diethyl malonate (equation 6).[37]

2 CHO i OH 2 (6)

(**11**) (**12**)

i, Hg, 1.09 F mol^{-1}, Bu_4NClO_4, MeCN, $CH_2(CO_2Et)_2$, 89%

Similarly, the pinacols of β-ionone could be obtained in 75% yield by carrying out the electrolysis in the presence of chromium(III) chloride.[16] Touboul and Dana observed a unique case of enantiomeric recognition when they reduced a series of racemic 1,9,10,10a-tetrahydro-3(2*H*)-phenanthrones (**13**). Pinacolization occurred with formation of a new carbon–carbon bond preferentially between identical enantiomers, as shown in equation (7).[38].

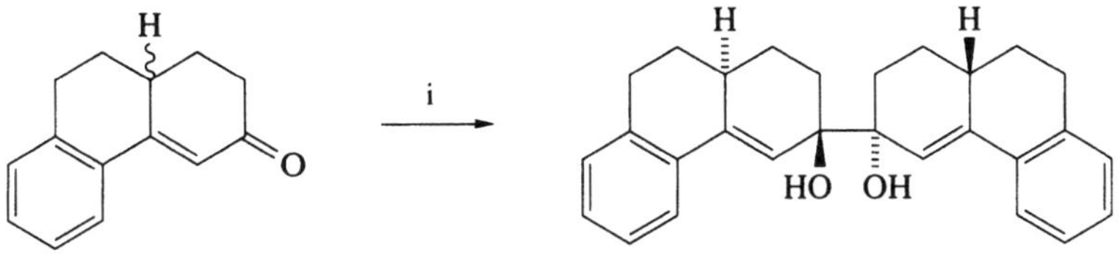

(7)

(13)

i, Hg, –1.4 V (SCE), 0.96 F mol^{-1}, EtOH, AcOH, 79%

1.5.2.3 Electrocatalytic Hydrogenation

A different type of electroreduction is the catalytic hydrogenation of organic compounds at an activated electrode surface in an aqueous medium. The water is cathodically reduced to hydrogen gas, which reacts immediately with the dissolved organic substrate. Lessard and coworkers found that benzophenone can be reduced to benzhydrol in nearly quantitative yields using a Raney nickel cathode.[39] Other catalytic electrode materials, such as palladium black[40] and platinized platinum,[41] have also been used. Chiba and coworkers showed that a series of aromatic and aliphatic ketones and aldehydes can be reduced to the corresponding alcohols with Raney nickel electrodes in a methanolic medium.[42] Under certain conditions the ketone can be reduced directly to the hydrocarbon.[43,44]

1.5.3 IMINES AND OXIMES

1.5.3.1 Electroreduction of Acyclic and Cyclic Compounds with a C═N Bond

1.5.3.1.1 General mechanism and products

The mechanism of the cathodic reduction of carbon–nitrogen double bonds is not very different from that of a carbonyl group. As shown in Scheme 3, an initial one-electron reduction leads to a radical anion (**14**), which in most cases is quickly protonated due to its high basicity.[45] The resulting neutral radical (**15**) can dimerize to form vicinal diamines (**16**)[46,47] or can be further reduced to lead to the amine (**17**) after a second protonation. Reversibility has been observed with several imines for the first and second electron transfer step.[48,49]

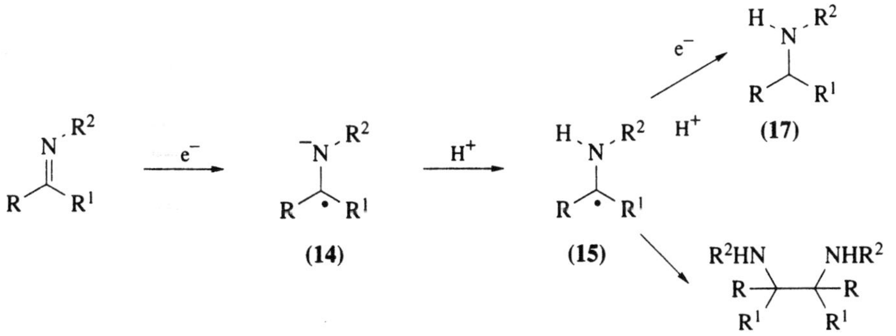

(14) **(15)** **(17)** **(16)** (racemic + *meso*)

Scheme 3

The equilibrium existing in an aqueous medium between a Schiff base and its corresponding aldehyde or ketone can be exploited in reductive aminations, as shown recently by Pienemann and Schäfer.[50] Because of its slightly more positive reduction potential compared to a carbonyl group, the Schiff base can be selectively reduced in a controlled potential electrolysis, to yield an amine in high yields, as shown in equation (8). This method is applicable to a variety of ketones and aldehydes.

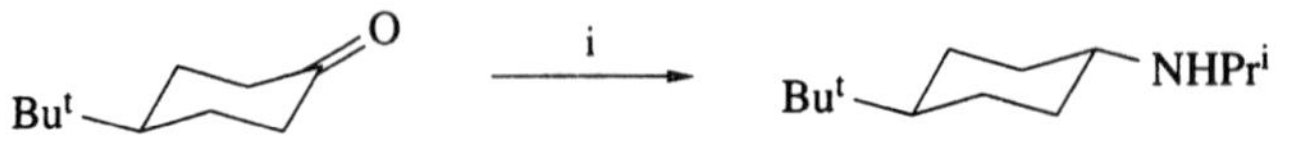

(8)

i, Pr^iNH_2, Hg, −1.68 V (SCE), H_2O, EtOH, 81%

A nice example of regioselectivity that can be obtained with controlled potential electrolyses has been reported by Tallec.[47] Electroreduction of the 2-phenyl-6*H*-1,3-thiazine (**18**) in aqueous ethanol with an acetate buffer and a mercury pool cathode afforded the dimer (**19**; 80%) at a working potential of −0.80 V (SCE) or the 2,3-dihydro compound (**20**: 75%) at −1.40 V (SCE) (Scheme 4).

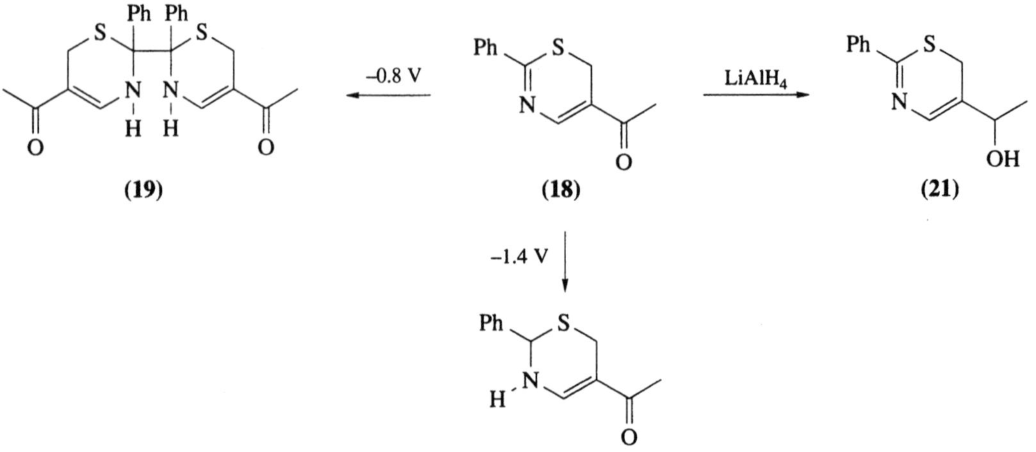

(**19**) ← −0.8 V — (**18**) — $LiAlH_4$ → (**21**)

(**18**) — −1.4 V → (**20**)

Scheme 4

In contrast, sodium borohydride or lithium aluminum hydride reduced the carbonyl group to the alcohol (**21**) leaving the C=N double bond intact.

Imino lactones, such as (**22**), can be reduced cleanly to the morpholinone (**23**) in an aprotic medium using diethyl malonate as a mild proton donor (equation 9).[51] The electroreductive behavior of this class of activated imines is presently under study.

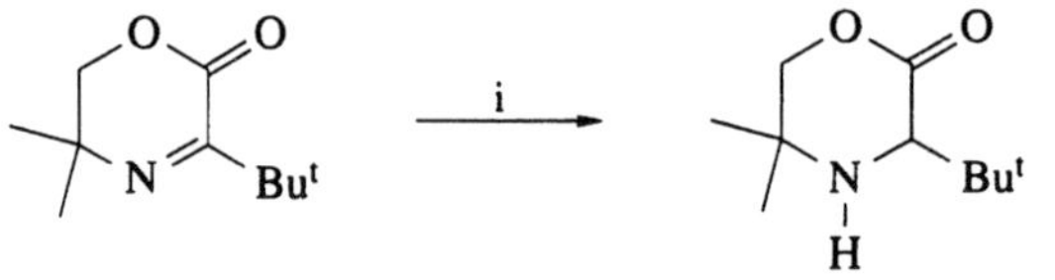

(**22**) → (**23**) (9)

i, Hg, −1.83 V (SCE), 1.9 F mol^{-1}, Bu_4NBr, DMF, $CH_2(CO_2Et)_2$, 81%

1.5.3.1.2 Electroreductive cyclization

Schiff bases have been used in electroreductive cyclizations, leading to heterocyclic compounds. The bisimine (**24**), when reduced in glyme, affords the piperazine (**25**) in 42% yield after quenching with methyl iodide (equation 10).[52] When the solvent is changed to dimethylformamide, saturation of the imino bonds takes place without ring closure. Transannular ring closure of diazocine (**26**) led to indoloindole (**27**) with a *cis* ring junction in 94% yield, as shown in equation (11). The radical anion generated from an imino bond can also act as a nitrogen base and consequently intramolecular deprotonation reactions may compete with the cyclization process if acidic protons are present in the chain.[53]

(**24**) —i, 2e⁻ (Hg), Bu_4NClO_4, glyme; ii, MeI→ (**25**) (10)

(**26**) —2e⁻, 94%→ (**27**) (11)

1.5.3.2 Electroreduction of Oximes

Another method for reductive amination of carbonyl compounds is based on the electroreduction of oximes in aqueous media. Thus, sugar oximes were reduced to their corresponding glycamines in good yields (Hg, KCl, acetate buffer).[54] Similar reductions of oximes derived from furfural, salicylaldehyde, benzophenone and cyclohexanone have also been described,[55–58] but reports of preparative electrolyses in strictly aprotic media seem to be absent in the literature.

1.5.3.3 Asymmetric Electroreduction of a C=N Bond

The asymmetric electroreduction of an imino group has also been the subject of several studies. The use of modified electrodes,[33] or chiral additives that supposedly create a chiral environment close to the electrode surface, give only low optical yields.[32,59,60] Other strategies will have to be explored in order to obtain high optical purities useful for synthetic work.

1.5.4 REFERENCES

1. M. M. Baizer and H. Lund (ed.), 'Organic Electrochemistry, an Introduction and a Guide', Dekker, New York, 1983.
2. 'State of the Art Symposium: Electrochemistry', *J. Chem. Educ.*, 1983, **60**, 258.
3. J. Heinze, *Angew. Chem., Int. Ed. Engl.*, 1984, **23**, 831.
4. W. J. M. van Tilborg and C. J. Smit, *Recl. Trav. Chim. Pays-Bas*, 1979, **98**, 532.
5. W. J. M. van Tilborg, C. J. Smit and R. A. van Santen, *Recl. Trav. Chim. Pays-Bas*, 1979, **98**, 526.
6. N. Egashira, T. Minami, T. Kondo and F. Hori, *Electrochim. Acta*, 1986, **31**, 463.
7. N. Egashira, Y. Takita and F. Hori, *Bull. Chem. Soc. Jpn.*, 1982, **55**, 3331.
8. D. Pletcher and M. Razaq, *Electrochim. Acta*, 1981, **26**, 819.
9. C. Z. Smith and J. H. P. Utley, *J. Chem. Soc., Chem. Commun.*, 1981, 492.
10. J. E. Swartz, E. Kariv-Miller and S. J. Harrold, *J. Am. Chem. Soc.*, 1989, **111**, 1211.
11. P. B. Lawin, A. C. Hutson and E. Kariv-Miller, *J. Org. Chem.*, 1989, **54**, 526.
12. J. E. Swartz, T. J. Mahachi and E. Kariv-Miller, *J. Am. Chem. Soc.*, 1988, **110**, 3622.
13. E. Kariv-Miller and T. J. Mahachi, *J. Org. Chem.*, 1986, **51**, 1041.
14. E. Kariv-Miller, K. E. Swenson, G. K. Lehman and R. Andruzzi, *J. Org. Chem.*, 1985, **50**, 556.
15. D. W. Sopher and J. H. P. Utley, *J. Chem. Soc., Chem. Commun.*, 1979, 1087.
16. D. W. Sopher and J. H. P. Utley, *J. Chem. Soc., Perkin Trans. 2*, 1984, 1361.
17. M. Perrin, P. Pouillen, G. Mousset and P. Martinet, *Tetrahedron*, 1980, **36**, 221.
18. M. T. Escot, P. Pouillen and P. Martinet, *Electrochim. Acta*, 1983, **28**, 1697.
19. F. Fournier, J. Berthelot and Y.-L. Pascal, *Can. J. Chem.*, 1983, **61**, 2121.
20. F. Fournier, J. Berthelot and Y.-L. Pascal, *Tetrahedron*, 1984, **40**, 339.

21. F. Fournier and M. Fournier, *Can. J. Chem.*, 1986, **64**, 881.
22. J. Douch and G. Mousset, *Can. J. Chem.*, 1987, **65**, 549.
23. J. P. Coleman, R. J. Holman and J. H. P. Utley, *J. Chem. Soc., Perkin Trans. 2*, 1976, 879.
24. R. J. Holman and J. H. P. Utley, *J. Chem. Soc., Perkin Trans. 2*, 1976, 884.
25. J. P. Coleman, R. J. Kobylecki and J. H. P. Utley, *J. Chem. Soc., Chem. Commun.*, 1971, 104.
26. T. Shono and M. Mitani, *Tetrahedron*, 1972, **28**, 47.
27. G. Le Guillanton, *Electrochim. Acta*, 1980, **25**, 1351.
28. T. Nonaka, Y. Kusayanagi and T. Fuchigami, *Electrochim. Acta*, 1980, **25**, 1679.
29. T. Nonaka and Y. Kusayanagi, *Electrochim. Acta*, 1981, **26**, 893.
30. A. Orliac-Le Moing, J. Delannay and J. Simonet, *Electrochim. Acta*, 1987, **32**, 1769.
31. J. Delaunay, A. Orliac-Le Moing, A. Lebouc, J. Simonet and L. Toupet, *J. Chem. Res. (S)*, 1986, 418.
32. M. Jubault, E. Raoult and D. Peltier, *Electrochim. Acta*, 1980, **25**, 1359.
33. S. Abe, T. Fuchigami and T. Nonaka, *Chem. Lett.*, 1983, 1033.
34. F. Takahashi, K. Tomii and H. Takahashi, *Electrochim. Acta*, 1986, **31**, 127.
35. T. Shono, N. Kise, T. Suzumoto and T. Morimoto, *J. Am. Chem. Soc.*, 1986, **108**, 4676.
36. P. Tissot, J.-P. Surbeck, F. O. Gülaçar and P. Margaretha, *Helv. Chim. Acta*, 1981, **64**, 1570.
37. L. A. Powell and R. M. Wightman, *J. Am. Chem. Soc.*, 1979, **101**, 4412.
38. E. Touboul and G. Dana, *J. Org. Chem.*, 1979, **44**, 1397.
39. G. Belot, S. Desjardins and J. Lessard, *Tetrahedron Lett.*, 1984, **25**, 5347.
40. T. Nonaka, M. Takahashi and T. Fuchigami, *Bull. Chem. Soc. Jpn.*, 1983, **56**, 2584.
41. M. A. Casadei and D. Pletcher, *Electrochim. Acta*, 1988, **33**, 117.
42. T. Chiba, M. Okimoto, H. Nakai and Y. Takata, *Bull. Chem. Soc. Jpn.*, 1983, **56**, 719.
43. M. J. Lain and D. Pletcher, *Electrochim. Acta*, 1987, **32**, 109.
44. D. Pletcher and M. Razaq, *Electrochim. Acta*, 1981, **26**, 819.
45. A. J. Fry and R. G. Reed, *J. Am. Chem. Soc.*, 1969, **91**, 6448.
46. L. Horner and D. H. Skaletz, *Justus Liebigs Ann. Chem.*, 1977, 1365.
47. B. Bujoli, M. Jubault, J.-C. Rozé and A. Tallec, *Tetrahedron*, 1987, **43**, 2709.
48. J. H. Barnes, F. M. Triebe and M. D. Hawley, *J. Electroanal. Chem. Interfacial Electrochem.*, 1982, **139**, 395.
49. C. P. Andrieux and J. M. Savéant, *J. Electroanal. Chem. Interfacial Electrochem.*, 1971, **33**, 453.
50. T. Pienemann and H.-J. Schäfer, *Synthesis*, 1987, 1005.
51. L. Moëns, F. Beaulieu and J. Roby, unpublished results.
52. R. W. Koch and R. E. Dessy, *J. Org. Chem.*, 1982, **47**, 4452.
53. C. Degrand, P.-L. Compagnon, G. Belot and D. Jacquin, *J. Org. Chem.*, 1980, **45**, 1189.
54. G. Ryan and J. H. P. Utley, *Tetrahedron Lett.*, 1988, **29**, 3699.
55. E. Ramanathan, P. S. M. Kannan, S. Thangavelu and K. S. Udupa, *Bull. Electrochem.*, 1988, **4**, 365.
56. N. Ayyaswami and V. Krishnan, *Indian J. Chem., Sect. A*, 1983, **22**, 555.
57. N. Ayyaswami and V. Krishnan, *J. Appl. Electrochem.*, 1983, **13**, 731.
58. N. Ayyaswami and V. Krishnan, *J. Appl. Electrochem.*, 1984, **14**, 557.
59. M. Jubault, A. Lebouc and A. Tallec, *Electrochim. Acta*, 1982, **27**, 1339.
60. E. Raoult, M. Jubault and D. Peltier, *Electrochim. Acta*, 1979, **24**, 1219.

1.6
Reduction of C=X to CHXH by Catalytic Hydrogenation

KAORU HARADA and TORATANE MUNEGUMI
University of Tsukuba, Japan

1.6.1 INTRODUCTION

This chapter deals with heterogeneous and homogeneous catalytic hydrogenation of the C=X bond, mainly compounds containing carbon–oxygen and carbon–nitrogen double bonds.[1–8] Various catalysts are employed for the hydrogenation reaction, the most widely used for heterogeneous catalytic hydrogenation being based on Fe, Ni, Co, Ru, Rh, Pd, Os, Ir and Pt, with the most popular catalysts based on Pt, Pd and Ni. The so-called copper chromite has also been used for hydrogenation of C=X double bonds.

Heterogeneous catalytic hydrogenation is a classical method of hydrogenation for preparative organic chemistry. Recently various reducing reagents of the metal hydride type have become popular in synthetic organic chemistry; however, classical heterogeneous hydrogenation is still an important and convenient method for the preparation of organic compounds both on the macro and the micro scale.

Some metallic elements such as rhodium have been used for homogeneous catalytic hydrogenation in a properly designed complex. However, most homogeneous catalytic hydrogenations, including asymmetric reactions, have been applied to substrates containing a carbon–carbon double bond, and few have been applied to the reduction of carbon–oxygen double bonds.

Heterogeneous catalytic hydrogenations using chirally modified metal catalysts have been carried out successfully. This method is interesting chemically and could be a useful production method for chiral organic compounds. However, the application of this type of modified catalyst is, for the present, rather limited.

1.6.2 HETEROGENEOUS CATALYTIC HYDROGENATION OF THE C=O BOND

1.6.2.1 Hydrogenation of Aldehydes

Saturated aliphatic aldehydes are easily reduced to the corresponding alcohols in high yield by heterogeneous catalytic hydrogenation. Saturated aliphatic aldehydes are converted to the corresponding alcohols quantitatively using Raney nickel,[9–11] and hydrogenation of glucose over nickel to produce sorbitol has been used on an industrial scale.[12] Platinum oxide is also an efficient catalyst.[13]

Unsaturated aliphatic aldehydes can be selectively hydrogenated in three ways: reduction of (a) the aldehyde group, (b) the carbon–carbon double bond, or (c) both the aldehyde and carbon–carbon double bond. A rhenium catalyst was suitable for selective hydrogenation of the carbonyl group;[14] crotonaldehyde (**1**) was hydrogenated, using osmium on charcoal, to crotyl alcohol in high yield.[15] In contrast, crotonaldehyde was converted to butyraldehyde using a palladium catalyst prepared by reduction of Pd salts with sodium borohydride.[16] Finally, using a nickel catalyst, crotonaldehyde was reduced to *n*-butanol (Scheme 1), indicating that the palladium or osmium catalyst is more selective than the nickel catalyst. Similar catalytic hydrogenations of cinnamaldehyde were also carried out. The substrate was reduced to cinnamyl alcohol by osmium on charcoal,[15] to hydrocinnamaldehyde using a palladium catalyst[16] and to hydrocinnamyl alcohol by Raney nickel.[9] Palladium[17] and platinum[18] catalysts, modified with iron(II) ion and zinc acetate, were also effective for selective hydrogenation of the carbonyl group of unsaturated aldehydes.[14]

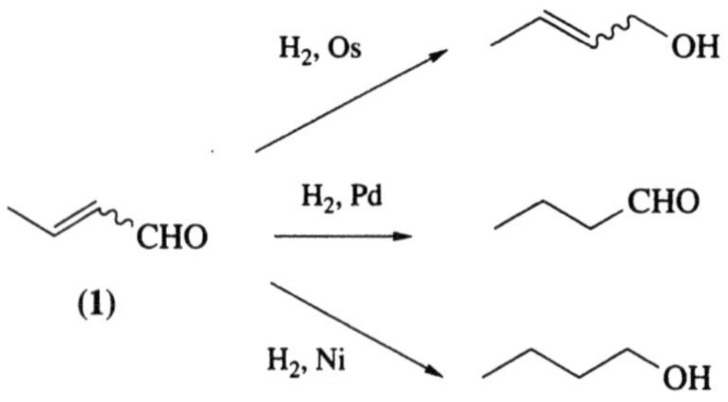

CHO (**1**) — H_2, Os → OH; H_2, Pd → CHO; H_2, Ni → OH

Scheme 1

2,2-Dimethyl-4-pentenal was hydrogenated to form 2,2-dimethylpentanal using 5% palladium on alumina.[16] On the other hand, the substrate was converted to 2,2-dimethylpentanol using Raney nickel under stronger conditions.[19] Similar hydrogenation of citral (**2**) to geraniol (**3**) has been carried out using a modified platinum oxide catalyst (equation 1).[20]

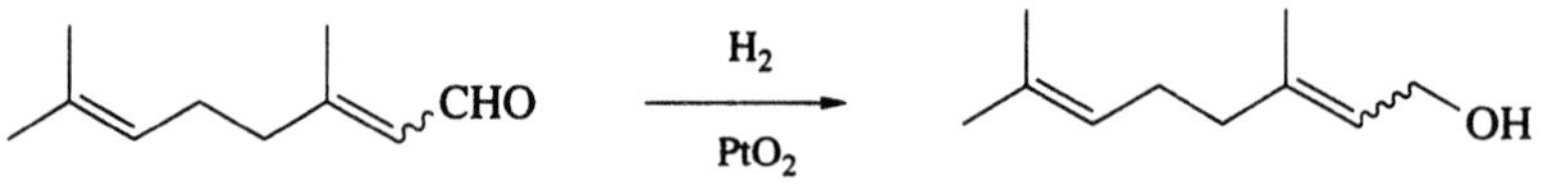

$$\text{CHO } (\mathbf{2}) \xrightarrow[\text{PtO}_2]{\text{H}_2} \text{OH } (\mathbf{3}) \qquad (1)$$

The catalytic hydrogenation of benzaldehyde to benzyl alcohol has been carried out using various catalysts. Prolonged hydrogenation results in the formation of toluene, and hydrogenation must be stopped when the uptake of hydrogen reaches the theoretical value.[13,21–23] Further hydrogenolysis of the resulting benzyl alcohol tends to proceed to a larger extent when a platinum catalyst is employed as compared to palladium. Nickel,[10,24] Raney nickel[9] and copper chromite[25] are suitable for hydrogenation of benzaldehyde to benzyl alcohol practically without any hydrogenolysis. The use of platinum oxide modified by the addition of iron(II) chloride is also a good catalyst for hydrogenation, preventing further hydrogenolysis.[13] Benzaldehyde is hydrogenated to benzyl alcohol using a platinum catalyst in ethanol,[22] but in acetic acid it is further hydrogenolyzed to toluene.[22] Benzaldehyde is similarly converted to toluene with a palladium catalyst in acetic acid,[26] and fluorene-1-carbaldehyde and furfural are converted to 1-methylfluorene and methylfuran, respectively.[26,27]

1.6.2.2 Hydrogenation of Ketones

Heterogeneous catalytic hydrogenation of saturated ketones proceeds easily to form secondary alcohols using various metal catalysts such as platinum, palladium, nickel, copper chromite and rhodium. However, the rate of hydrogenation of ketones is generally lower than that of the corresponding aldehydes. Catalytic hydrogenation over platinum oxide,[28] rhodium, ruthenium[28–30] and Raney nickel[10] gives the corresponding alcohols under mild conditions. On the other hand, hydrogenation over nickel[24] and copper chromite[25] requires stronger reaction conditions, although it still proceeds almost quantitatively. For example, hydrogenation of acetone over Raney nickel (25 °C, 1 atm) or copper chromite (150 °C, 100 atm) gives isopropyl alcohol in almost 100% yield.

Heterogeneous catalytic hydrogenation of aromatic ketones takes place easily to form the corresponding alcohols using various metal catalysts. Special care should be taken to avoid further hydrogenolysis if the products are of the benzyl alcohol type. Acetophenone was hydrogenated over palladium oxide in ethanol to form the secondary alcohol in good yield without hydrogenolysis. However, the same reaction in acetic acid gave the hydrogenated and hydrogenolyzed product ethylcyclohexane,[28,29] and nitroacetophenone was converted to ethylaniline over a palladium catalyst at room temperature in acetic acid.[31] Noble metal catalysts tend to hydrogenate the aromatic ring and also hydrogenolyze the resulting benzylic alcohol. The hydrogenation of acetophenone over rhodium–platinum oxide in ethanol gave 1-cyclohexylethanol in good yield, whereas hydrogenation of the same substrate over palladium or platinum oxide in acetic acid in the presence of hydrochloric acid gave ethylbenzene.[28] The same hydrogenation reaction using nickel and copper chromite gave 1-phenylethanol without hydrogenolysis.[9,10,24,25]

Hydrogenations of acetylpyridines were carried out similarly, and yielded various products depending on the reaction conditions and especially on the catalyst used.[32] Thus 4-acetylpyridine was hydrogenated over palladium oxide to form the corresponding alcohol, 4-(1-hydroxyethyl)pyridine, with a small amount of the pinacol (**4**), but it was converted mainly to the pinacol using palladium on charcoal or rhodium on alumina. However, under different conditions, the pyridyl ring of 3-acetylpyridine was reduced to a mixture of 3-acetyl-1,4,5,6-tetrahydropyridine and 3-acetylpiperidine.

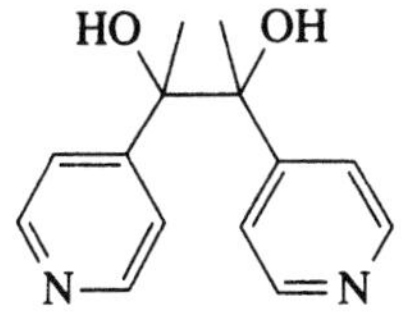

HO OH

N N

(4)

Hydrogenation of cyclic ketones yields secondary alcohols. In these reactions, the stereochemistry of the products depends on the reaction conditions.[33]

Heterogeneous catalytic hydrogenation generally takes place with *syn* addition to the double bond. An empirical rule for heterogeneous hydrogenation of cyclohexanones over Pt or Ni catalysts has been proposed by von Auwers and Skita.[34–36] According to this rule, when Pt is employed as a catalyst, *cis* compounds are obtained under acidic conditions, while the reaction tends to favor the *trans* isomer in neutral or basic conditions. Barton has also expressed the rule of hydrogenation in terms of the conformations of the product; the catalytic hydrogenation of a derivative of cyclohexanone should yield the axial alcohol predominantly in acidic conditions, while in neutral conditions the equatorial alcohol is obtained predominantly.[37] Table 1 shows the results of catalytic hydrogenation of cyclohexanone derivatives under various conditions.[38–43]

In the reactions (a) and (b), the percentage of the *cis* products depended on the catalysts and the solvent, hydrogenation over platinum or platinum oxide giving the *cis* isomer, but giving the *trans* isomer with a basic catalyst (Raney Ni).[44] Hydrogenation in acidic media favors the formation of axial hydroxy products, and alkaline conditions gave alcohols with an equatorial hydroxy group.[38] In reaction (c), a higher proportion of the *cis* product was obtained.

Several factors such as the structure of the substrate, the catalyst, the solvent, the reaction temperature, the pressure of hydrogen and other reaction conditions determine the stereochemistry of the catalytic hydrogenation of cyclic ketones,[36] and it is sometimes difficult to predict the major product of catalytic hydrogenation. One reason for the complexity of the stereochemistry of the hydrogenation of cyclic ketones, at least in part, is related to the isomerization of the products under the reaction conditions. Some cyclohexanols were isomerized in the presence of platinum or nickel catalysts at room temperature or at higher temperature under a hydrogen atmosphere, and the isomerization reached a *cis–trans* equilibrium.[45] For example, *trans*-3,3,5-trimethylcyclohexanol isomerized in the presence of a nickel catalyst,

Table 1 Catalytic Hydrogenation of Cyclic Ketones

Substrate	Solvent	Catalyst	Percentage of cis-product
(a)	AcOH, HCl	PtO_2	80
	AcOH	PtO_2	47
	MeOH	PtO_2	26
(b)	AcOH	Pt	73
	AcOH	PtO_2	38
	EtOH	R, Ni	23
(c)	AcOH, HCl	Pt	93
	AcOH	PtO_2	70
	MeOH	R, Ni	70

and reached equilibrium (*cis:trans* = 73:27). The presence of alkali accelerates the isomerization[39] and, therefore, isomerization of cyclohexanol takes place in the presence of basic catalysts such as Raney nickel or Adams' platinum oxide.[46] However, isomerization occurred to a much lower extent when an acid-washed platinum catalysts was used.[47] The mechanism of the isomerization has been considered in two ways as shown in equation (2).[39,48] Reaction (a) is an equilibrium between the *cis* and *trans* isomers by dehydrogenation and subsequent hydrogenation.[39] A kinetic study of the isomerization suggests that the racemization reaction takes place directly following adsorption of the alcohol on the catalyst (reaction b).[47,48]

hexanone

(a) (a)

cis-hexanol ⇄ *trans*-hexanol (2)

(b)

Unsaturated ketones can be hydrogenated to saturated alcohols, or to saturated ketones by controlled catalytic hydrogenation. The latter is not the subject of this article and hydrogenation to saturated alcohols will be discussed. Hydrogenation of a ketone group to a secondary alcohol was achieved for chalcone (benzylideneacetophenone) and phenylbutan-2-one using platinum oxide[49] or copper chromite,[25] and a nickel catalyst was also used for the hydrogenation.[10] Chalcone was converted to the unsaturated alcohol using a modified colloidal palladium.[50] A temperature effect in the catalytic hydrogenation of unsaturated aromatic ketones was observed using nickel catalyst.[51] The carbon–carbon double bond was hydrogenated at room temperature, while the carbonyl group was reduced to an alcohol at 120 °C; at even higher temperatures (260 °C) the phenyl group was hydrogenated to a cyclohexyl group.

α-Keto alcohols (acyloins) were hydrogenated to the mixture of glycols (**5**) over copper chromite under relatively strong conditions (150 °C, 100 atm; equation 3).[52] Benzoquinone was converted to hydroquinone easily over a palladium catalyst under mild conditions,[53] and 1,4-cyclohexanedione was hydrogenated to 4-hydroxycyclohexanone using ruthenium on silica.[54]

$$(CH_2)_8[CO, CHOH] \xrightarrow[CuCr_2O_4]{H_2} (CH_2)_8[\overset{*}{C}HOH, \underset{*}{C}HOH] \quad (3)$$

(**5**)

1.6.3 HETEROGENEOUS CATALYTIC HYDROGENATION OF THE C═N BOND

1.6.3.1 Hydrogenation of Oximes and Hydrazones

The carbon–nitrogen double bonds of oximes and hydrazones are hydrogenated to form the corresponding amines by heterogeneous catalytic hydrogenation.

The oximes of cyclohexanone and cycloheptanone were converted to cyclohexylamine[9] and cycloheptylamine[55] in good yield over Raney nickel or rhodium on alumina, respectively. Benzaldehyde oxime was converted to benzylamine over a nickel catalyst in good yield.[56] The stereochemistry of the hydrogenation is different depending on the catalyst and the reaction conditions. The catalytic hydrogenation of 2-alkylcyclohexanone oxime (**6**) over platinum yielded *cis*-2-alkylcyclohexylamine as the major product,[57] and hydrogenation over Raney nickel gave the *trans* product (Scheme 2).[58] Other stereochemically interesting results are obtained by heterogeneous catalytic hydrogenation of the monooximes of diketones such as compounds (**7**)[59,60] and (**8**).[61] These were both hydrogenated to form *erythro*-amino alcohols over platinum oxide and palladium catalysts.[62,63] The stereochemistry of the catalytic hydrogenation will be discussed later along with studies on asymmetric syntheses using heterogeneous catalytic hydrogenation (Section 1.6.4).

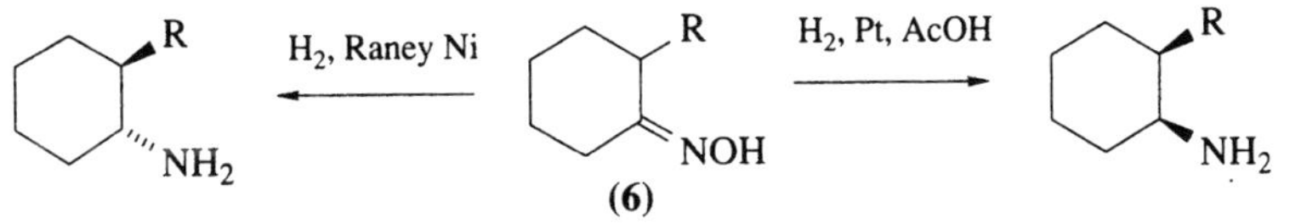

Scheme 2

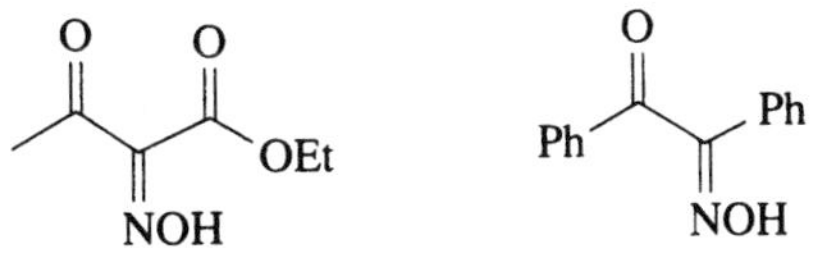

The phenylhydrazone of acetone was hydrogenated to form *N*-isopropyl-*N*-phenylhydrazine over colloidal platinum.[64] However, a similar hydrazone, 1-(*N*-acetyl-2-piperidyl)acetone phenylhydrazone, was hydrogenated and hydrogenolyzed to form 1-(*N*-acetyl-2-piperidyl)-2-aminopropane in good yield.[65]

1.6.3.2 Hydrogenation of Schiff's Bases (Aldimines and Ketimines)

Hydrogenation of a mixture of amines and aldehydes or ketones results in the formation of unsymmetrical secondary amines. The hydrogenation of the intermediate aldimine or ketimine can be carried out by heterogeneous catalytic hydrogenation over metal catalysts such as platinum, palladium, nickel or Raney nickel. The overall reaction can be regarded as an alkylation of the amino compounds, and also as a reductive amination of the aldehyde or ketone (Scheme 3).

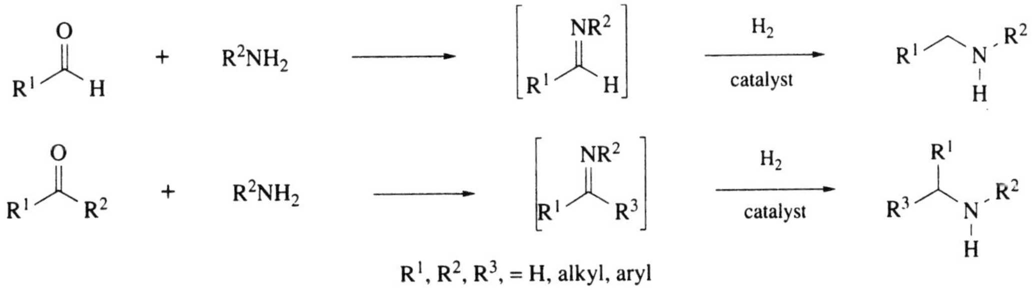

R^1, R^2, R^3, = H, alkyl, aryl

Scheme 3

The intermediate aldimine or ketimine is prepared from the amine and the aldehyde or ketone in an organic solvent, and is usually hydrogenated without isolation. However, in some cases, the intermediate azomethine compounds are isolated and then hydrogenated. Extensive studies on this two-step hydrogenation have been carried out, and the reactions have been reviewed by Emerson.[1] Therefore, the catalytic hydrogenation of aldimines and ketimines are not included in this review.

Catalytic hydrogenation of aldehydes and ketones in the presence of ammonia is used for the preparation of primary amines. The reaction product usually contains the secondary amine and an excess of ammonia should be used to minimize its formation. On the other hand, symmetrical secondary amines can be synthesized by catalytic hydrogenation using an excess of aldehyde.[66]

The synthesis of α-amino acids has been carried out from α-keto acids by reductive amination using a platinum or palladium catalyst, mimicking their biosynthesis. Thus alanine,[67] leucine,[68] phenylalanine,[67–69] aspartic acid[67] and glutamic acid[67,70] have been synthesized by reductive amination. When an optically active primary amine is used, asymmetric induction can proceed in the course of the reductive amination.

1.6.4 DIASTEREOSELECTIVE ASYMMETRIC HYDROGENATION

1.6.4.1 Asymmetric Hydrogenation of the C=O Bond

Catalytic hydrogenation of esters of α-keto acids with optically active alcohols using metal catalysts has been carried out.[71–73] The alcohols used were (–)-menthol and (+)-borneol, and the catalysts used were PtO_2, Pd/C, Pd/$CaCO_3$, Pd/$BaSO_4$ and Raney nickel. The optical purities of the resulting α-hydroxy acid esters were not high. When (–)-menthyl benzoylformate (**9**) and (+)-bornyl benzoylformate were hydrogenated over Raney nickel, PtO_2 or alkali-treated Pd/C, the major products were (–)-and (+)-mandelate, respectively.[72] However, when acid-treated catalysts were used, the hydrogenation products were (+)- and (–)-mandelate, respectively (Scheme 4). Thus the configuration of the resulting mandelic acid was not determined by the chiral source but by the manner in which the catalyst was modified.[71,72] Similar inversion of configuration was observed in the catalytic hydrogenation of (*S*)-(–)-α-phenethyl benzoylformate[73] and menthyl α-naphthylglyoxylate.[74] However, in the catalytic hydrogenation of menthyl pyruvate, the change in configuration of the product using a basic or an acidic catalyst was not observed.[75] The change was observed in the hydrogenation of aromatic α-keto acid esters, but not when aliphatic α-keto acid esters were used.[72–76] In order to explain the changeover in configuration, the substrate was presumed to prefer an *s-trans* conformation when an alkali catalyst was used, and an *s-cis* conformation with an acidic catalyst.[72–75] The interconversion was explained by the formation of a hemiketal (equation 4),[77] so that the *s-trans* conformer was converted to an *s-cis* conformer in acidic media and then hydrogenated to form the α-hydroxy acid ester that has a configuration opposite to that obtained using the basic catalyst.

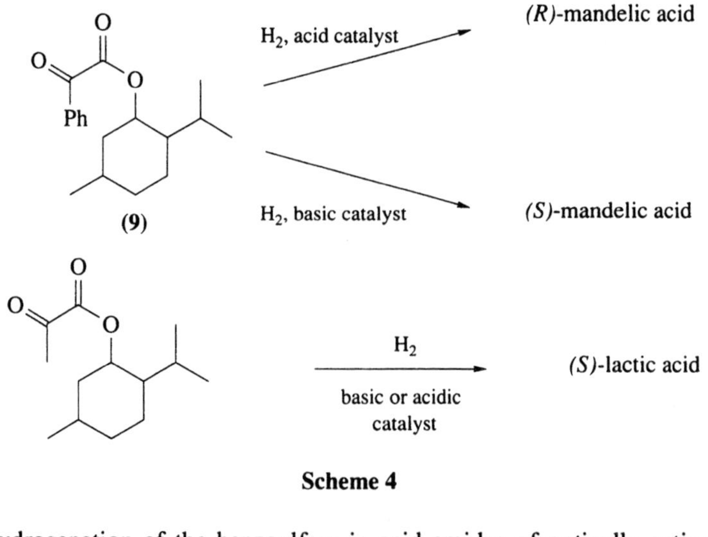

(**9**) —H_2, acid catalyst→ (*R*)-mandelic acid

(**9**) —H_2, basic catalyst→ (*S*)-mandelic acid

menthyl pyruvate —H_2, basic or acidic catalyst→ (*S*)-lactic acid

Scheme 4

The catalytic hydrogenation of the benzoylformic acid amides of optically active amino acid esters was carried out.[78] When the (*S*)-amino acid ester was used, the resulting mandelic acid had the (*R*)-configuration. When pyruvic acid amides of optically active benzylic amines were hydrogenated over palladium, optically active lactic acid was obtained in relatively high enantiomeric excess (*ee* 60%).[79] The

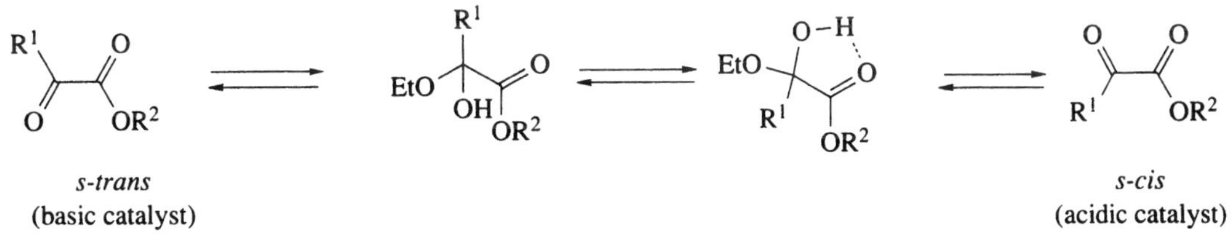

(4)

s-trans (basic catalyst) ⇌ s-cis (acidic catalyst)

asymmetric hydrogenation shows a solvent and also a temperature effect.[80] The catalytic hydrogenation of pyruvic acid amides of optically active amino acid esters also showed a solvent effect: with a polar solvent, the configuration of the lactic acid was (*R*), but with a less polar solvent, the percentage of the (*S*)-isomer increased.[81] The solvent effect was explained by assuming competing substrate–catalyst complexes (**10**) and (**11**), as shown in Scheme 5 (chelation mechanism). In the less polar solvent, the interaction between the ester carbonyl and the catalyst was greater, and (**11**) could be the major conformation resulting in (*S*)-lactic acid. Asymmetric catalytic hydrogenation of *N*-pyruvoyl-(*S*)-proline esters was carried out under various conditions to afford *N*-[(*S*)-lactoyl]-(*S*)-proline esters with a diastereomeric excess (*de*) of up to 60%.[82] The effects of temperature and the bulkiness of the ester groups were studied, and the steric course explained by the chelation mechanism which will be described later. Catalytic hydrogenation of benzoylformic acid amide with optically active *trans* 2,5-disubstituted pyrrolidine was carried out over Pd/C, and the *de* of the product was 56%.[83]

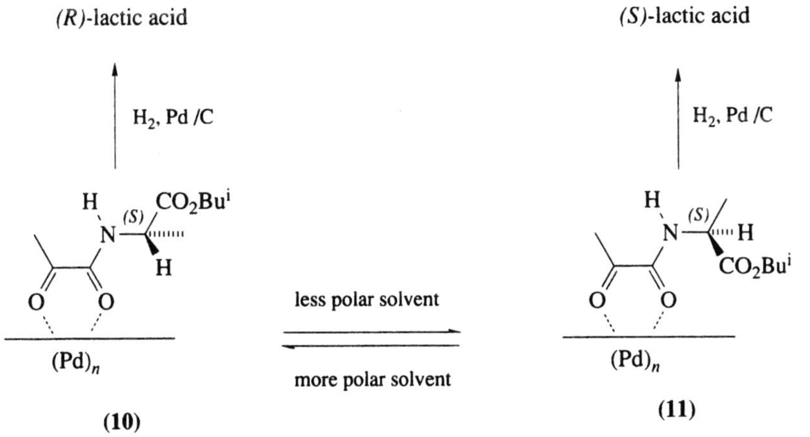

(*R*)-lactic acid ← H_2, Pd/C ← (**10**) [on $(Pd)_n$] ⇌ (less polar solvent / more polar solvent) (**11**) [on $(Pd)_n$] → H_2, Pd/C → (*S*)-lactic acid

Scheme 5

1.6.4.2 Asymmetric Hydrogenation of the C═N Bond

Numerous asymmetric catalytic hydrogenations of carbon–nitrogen double bonds have been carried out.[84,85] Some of the substrates used are oximes and hydrazones, but most of the reactions were carried out using Schiff's bases of ketones. α-Keto acids are precursors of α-amino acids in biosynthesis, and therefore α-keto acids have been used for the asymmetric syntheses of α-amino acids.[84,85]

An aqueous alkaline solution of a mixture of (*S*)-arginine and pyruvic acid was hydrogenated over platinum oxide to yield (+)-isooctopine (**12**),[86,87] a diastereomer of natural octopine, with 70% *de* (equation 5). Detailed syntheses and properties of the diastereomers of octopine have been reported.[88]

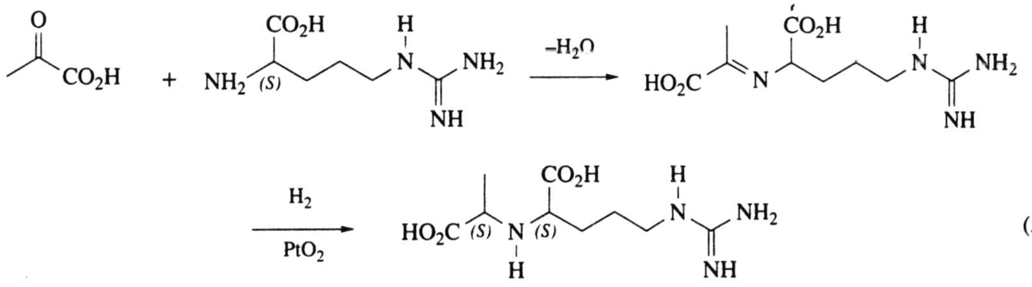

pyruvic acid + (*S*)-arginine —($-H_2O$)→ imine —(H_2, PtO_2)→ (**12**) (5)

A systematic study of the catalytic hydrogenation of the Schiff's bases prepared from α-keto acids and optically active benzylic amines in organic solvents has been carried out, and some of the results summarized.[84,85] The resulting imino acids were hydrogenolyzed to obtain α-amino acids (equation 6). The reaction is actually a hydrogenolytic asymmetric transamination between amino compounds and α-keto acids. The chemical yields were in the range 20–80% and the enantiomeric excesses of the resulting amino acids were 12–81%. The sense of the asymmetric hydrogenation was based on the Prelog rule.[89,90] Asymmetric syntheses of alanine, α-aminobutyric acid, phenylglycine, phenylalanine and glutamic acid were carried out starting with the Schiff's bases prepared from the corresponding α-keto acids and optically active α-(*S*)-methylbenzylamine [Me(–)], α-(*S*)-ethylbenzylamine [Et(–)] and α-(*R*)-(1-naphthyl)ethylamine [Naph(+)].[91,92] The results obtained using α-methylbenzylamine were similar to those obtained earlier.[89] However, the results with the other amines were quite different from those expected from the steric course proposed.[90] The results, shown in Table 2,[91,92] suggest that the most probable conformation of the substrate in the catalytic hydrogenation could be structure (**13**). Structure (**13**) results in the formation of (*S*)-amino acid when (*S*)-amine is used, and the substituent effect of the α-keto acid can be explained by assuming conformation (**13**). The substrate initially interacts with the catalyst to form a substrate–catalyst complex, as shown in structure (**14**), which is adsorbed onto the catalyst from the less bulky face of the molecule, where the catalytic hydrogenation takes place.

$$^{-}O_2C\text{–}C(R)\text{=N–}CH(CH_3)Ph\ (S) \xrightarrow[\text{Pd /C}]{H_2} {}^{-}O_2C\text{–}CH(R)\text{–NH–}CH(CH_3)Ph \xrightarrow[\text{Pd(OH)}_2\text{ /C}]{H_2} R\text{–}CH(NH_2)\text{–}CO_2H\ (S) + PhCH_2CH_3 \quad (6)$$

Table 2 Asymmetric Synthesis of Amino Acids by Catalytic Hydrogenation[92]

α-Keto acid	*Amine*	*Amino acid*	*ee (%)*
$MeCOCO_2H$	Me(–)	(*S*)-Alanine	67
	Et(–)	(*S*)-Alanine	52
	Naph(+)	(*R*)-Alanine	83
$EtCOCO_2H$	Me(–)	(*S*)-Butyrine	44
	Et(–)	(*S*)-Butyrine	33
$PhCOCO_2H$	Me(–)	(*S*)-Phenylglycine	30
	Et(–)	(*S*)-Phenylglycine	24
$PhCH_2COCO_2H$	Me(–)	(*S*)-Phenylglycine	14
	Et(–)	(*S*)-Phenylglycine	10
$HO_2C(CH_2)_2COCO_2H$	Me(–)	(*S*)-Glutamic acid	12
	Et(–)	(*S*)-Glutamic acid	6

H R (*S*) Ph N CO_2^- — (**13**)

R O H N O Ph $(Pd)_n$ — (**14**)

The 'chelation mechanism' has been tested in several hydrogenation reactions of the Schiff's bases of α-keto acids and their derivatives in organic solvents. A solvent effect in the hydrogenolytic asymmetric transamination between benzylic amines and α-keto acids[91] and their esters[93,94] was observed (Table 3). The effect is explained by the chelation mechanism as shown in Scheme 6 using the unchelated structure (**15**). Ethyl phenylglycinate can also be used as a chiral source for the asymmetric synthesis of amino acids, due to the ability of phenylglycine to be hydrogenolyzed to phenylacetic acid and ammonia in alkaline conditions.[95,96] Similarly an amino alcohol, 2-amino-2-phenylethanol,[94] amphetamine[97] and methylbenzylamine[98] have all been used as chiral sources which can also be hydrogenolyzed.

Studies on the transamination reaction between *t*-butyl esters of optically active amino acids and methyl pyruvate were carried out, as shown in Scheme 7.[99] The resulting iminodicarboxylic acid (**16**) was partially hydrolyzed and then oxidized with *t*-butyl hypochlorite to form alanine. The oxidation is a generally applicable one, and the optical purity of alanine is high (50–70%). Similar asymmetric transamination between an (*S*)-amino acid and ketones was carried out.[100] Catalytic hydrogenation of the Schiff's bases prepared from α-keto acid esters and amino acid esters was carried out, and a substituent and temperature effect observed (*de* 40–70%).[101]

Table 3 Solvent Effect in the Asymmetric Synthesis of Alanine Using Me (–)[91]

Solvent	*Yield* (%)	*Configuration*	*ee* (%)
Hexane	75	(*S*)	75
AcOEt	49	(*S*)	60
DMFA	47	(*S*)	50
Pr^iOH	56	(*S*)	46
MeOH	61	(*S*)	38
MeOH:H_2O (1:2)	75	(*S*)	35
MeOH:H_2O (1:4)	76	(*S*)	29

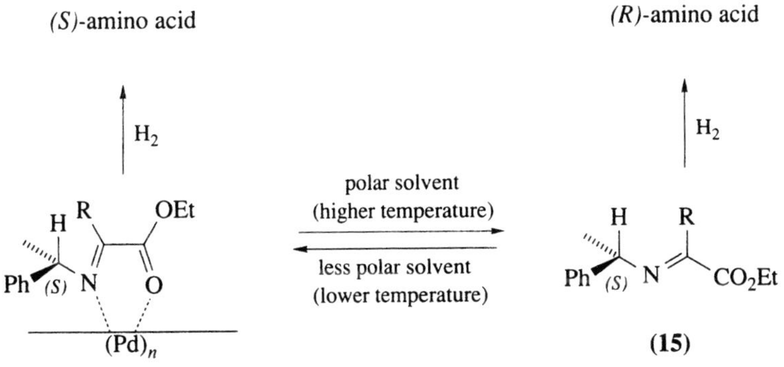

(*S*)-amino acid

H_2

(*R*)-amino acid

H_2

polar solvent (higher temperature)

less polar solvent (lower temperature)

H R OEt Ph (*S*) N O $(Pd)_n$

H R Ph (*S*) N CO_2Et

(15)

Scheme 6

O CO_2Me + NH_2 R (*S*) CO_2Bu^t $-H_2O$ MeO_2C CO_2Bu^t N R H_2 Pd/C MeO_2C CO_2Bu^t N H R

(16)

MeO_2C CO_2H N H R Bu^tOCl [MeO_2C O O N Cl R $-CO_2$ MeO_2C N R]

CO_2H NH_2

(*S*)-Alanine

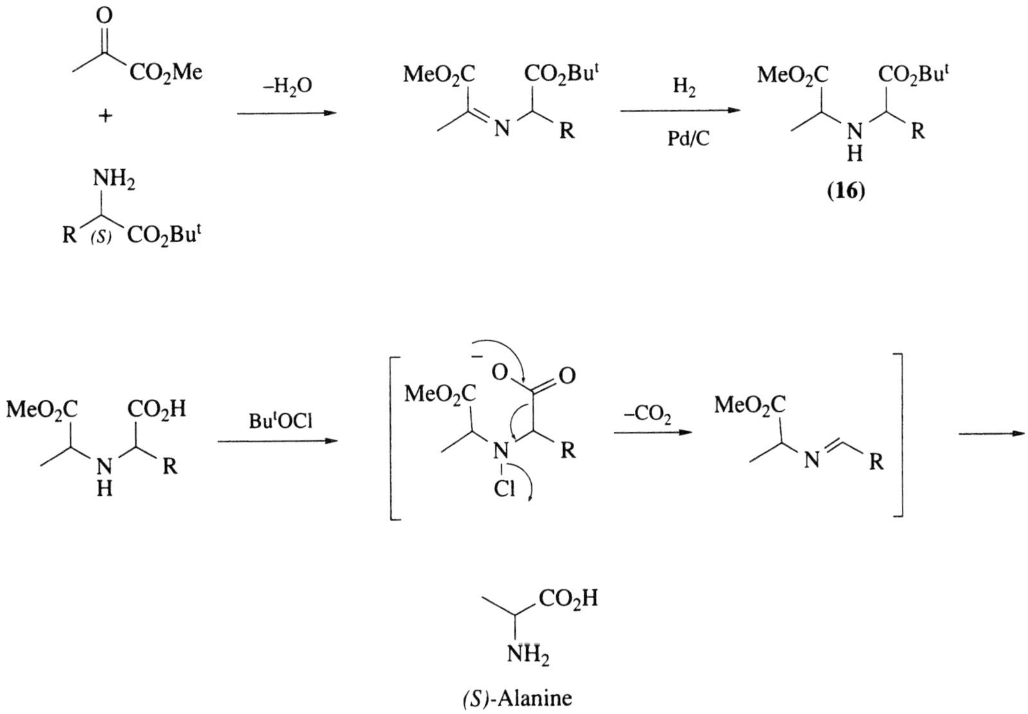

R = Pr^i, $PhCH_2$, Bu^i, $Bu^tO_2CCH_2$, $Bu^tO_2CCH_2CH_2$

Scheme 7

In the hydrogenolytic asymmetric transamination between a benzylic amine and oxalacetic acid[102] the Schiff's base was hydrogenated in ethanol over palladium, but alanine rather than aspartic acid was the product (*ee* 52–69%; equation 7). A fast decarboxylation of the Schiff's base must have taken place. A study of the time course of the reaction indicated that after half an hour at room temperature, the Schiff's base of oxalacetic acid (**17**) was converted to that of pyruvic acid (**18**). Thus the reaction using oxalacetic acid is actually the same as using pyruvic acid in ethanol. The reaction is interesting in connection with aspartic acid β-decarboxylase.

HO_2C CO_2H O + NH_2 Ph (S) → HO_2C N Ph CO_2H (**17**) $\xrightarrow{-CO_2}$

N Ph CO_2H (**18**) $\xrightarrow[Pd/C]{H_2}$ $\xrightarrow[Pd(OH)_2/C]{H_2}$ (*S*)-Alanine (7)

A clear temperature effect in the hydrogenolytic asymmetric transamination between optically active amines and ethyl pyruvate was observed (–20 to 65 °C).[103,104] At relatively low temperatures, when the optically active amine had the (*S*)-configuration, the resulting alanine also had the (*S*)-configuration. The *ee* of the (*S*)-alanine (80% *ee* at –20 °C) decreased as the reaction temperature increased, and at higher temperature (50 °C) the configuration of the alanine inverted to the (*R*)-configuration.

The temperature effect could be explained by the chelation mechanism.[91,92] The Schiff's base composed of α-keto acid and optically active amine interacted with the catalyst to form a substrate–catalyst complex (**14**) at lower temperature; at higher temperatures, the population of the unchelated structure (**15**) would increase as shown in Scheme 6. Asymmetric hydrogenation involving (*R*)-α-phenylglycinate and ethyl pyruvate has also been studied.[95,96]

An asymmetric catalytic hydrogenation of the Schiff's base of α-ketoacylamide derivatives $R^1C(=NR^3)CONHR^2$ (**19**) was carried out.[105] In the synthesis of phenylglycine, when the chiral center was changed from (*S*)-α-methylbenzylamine to (*S*)-α-ethylbenzylamine, the configuration of the product was inverted.[92] In order to clarify this inversion in the sense of the induction, the Schiff's bases of pyruvylamino acid isobutyl esters were hydrogenated over palladium on charcoal.[106] When the chiral moiety was (*S*)-alanine, the configuration of the newly formed alanyl residue was predominantly (*R*) (ratio of diastereomers: 82:18). However, when (*S*)-α-aminobutyric acid was used, the configuration of the resulting alanine was inverted, and the ratio of diastereomers was now 29:71. The change of configuration in the catalytic hydrogenation was also explained by the chelation mechanism.[106]

Finally, some historical background for the 'chelation mechanism' in heterogeneous catalytic hydrogenation is given. It is known that the catalytic hydrogenation of benzil monooxime yields diphenylethanolamine. If the conformation of the substrate were planar and *anti*, as expected from steric and electronic properties, the resulting product would be the *threo* isomer as shown in equation (8). However, the resulting ethanolamine was predominantly the *erythro* isomer (equation 9).[107] Benzoin oxime also gave *erythro*-diphenylethanolamine.[108,110] Similar results have been reported in the synthesis of threonine[59,60,62,63] and phenylserine[109,111] by catalytic hydrogenation (Scheme 8). The catalytic hydrogenation is so specific that optically active *erythro*-diphenylethanolamine was converted to benzoin, and the oxime could be converted back to the original optically active *erythro*-diphenylethanolamine.[110] There-

NOH O $\xrightarrow{H_2}$ OH NH_2 (8)

(*threo*)

fore, there must be some interaction between the substrate and the catalyst leading to an *s-cis* conformation. Chang and Hartung[111,112] proposed an interesting mechanism for the heterogeneous catalytic hydrogenation of α-oximino ketones that explains the stereoselectivity of the reaction whereby a single racemic modification is produced. This explanation can be regarded as a kind of chelation mechanism. Thus the predominant formation of the *erythro* isomer in the catalytic hydrogenation of benzil monooxime, benzoin oxime, ethyl acetoacetate oxime and ethylbenzoylacetate oxime can be explained by the formation of a substrate–catalyst complex prior to hydrogenation. Physicochemical evidence for the formation of a substrate–catalyst complex has also been obtained by IR absorption spectra of several substances on nickel or palladium surfaces.[113–115]

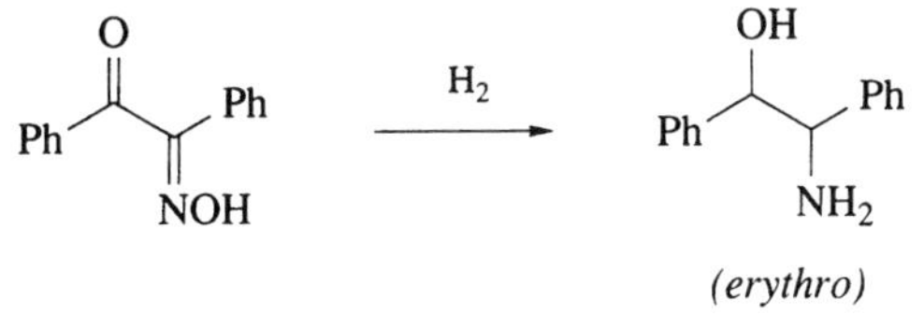

O, Ph, Ph, NOH $\xrightarrow{H_2}$ OH, Ph, Ph, NH_2 *(erythro)* (9)

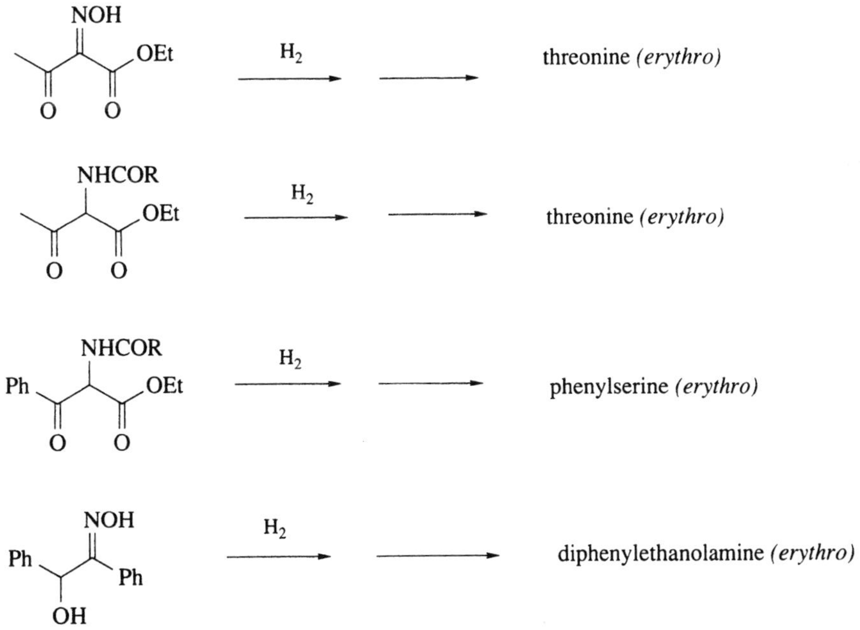

NOH, OEt, O, O $\xrightarrow{H_2}$ ⟶ threonine *(erythro)*

NHCOR, OEt, O, O $\xrightarrow{H_2}$ ⟶ threonine *(erythro)*

NHCOR, Ph, OEt, O, O $\xrightarrow{H_2}$ ⟶ phenylserine *(erythro)*

NOH, Ph, Ph, OH $\xrightarrow{H_2}$ ⟶ diphenylethanolamine *(erythro)*

Scheme 8

1.6.5 HETEROGENEOUS CATALYTIC HYDROGENATION USING A CHIRAL CATALYST

Heterogeneous catalytic hydrogenation of a carbon–nitrogen double bond over a metal catalyst in the presence of a chiral compound was first carried out by Nakamura.[116] Acetophenone oxime was hydrogenated over platinum black with ethyl menthoxyacetate or tartaric acid to give α-methylbenzylamine with 15–18% *ee*.

Enantioselective heterogeneous catalytic hydrogenation using a chiral catalyst was pioneered by Akabori and Izumi, who prepared a palladium catalyst supported on silk fibroin. The oxime acetates of diethyl α-ketoglutarate or of ethyl phenylpyruvate were hydrogenated to form glutamic acid (7–15% *ee*) and phenylalanine (30% *ee*).[117–119] Similarly, a palladium–poly-L-leucine catalyst was used for the asymmetric synthesis of phenylalanine.[120]

1.6.5.1 Hydrogenation Using a Chirally Modified Catalyst

Systematic studies on the enantioselective heterogeneous catalytic hydrogenation of carbonyl compounds were carried out by Izumi using Raney nickel modified with various chiral reagents. Hydroxy acids or amino acids were used for the modification of the nickel catalyst, and (+)-tartaric acid (2*R*,3*R*) was found to be the best chiral modifying reagent. The substrates used were β-keto esters,[121,122] β-keto alcohols[123,124] and ketones.[125] Among these substrates, methyl acetoacetate is the best, and the hydrogenation of this compound has been investigated extensively using the chirally modified nickel catalyst (equation 10). The heterogeneous catalytic hydrogenation of carbonyl compounds using chirally modified nickel catalysts has been reviewed by Izumi,[126] Klabunovskii,[127] Fish[128] and Tai.[129]

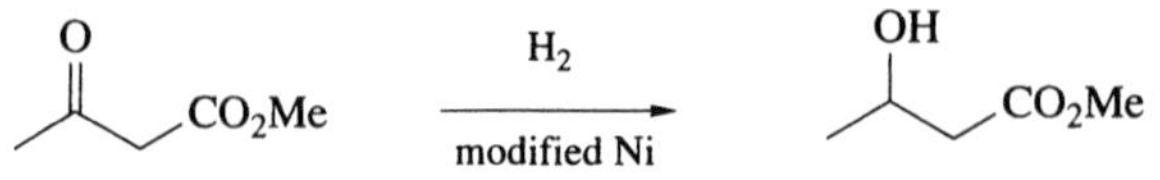

O, CO_2Me —H_2, modified Ni→ OH, CO_2Me (10)

More than a hundred modifying reagents, which include amino acids and their derivatives, peptides, amino alcohols and hydroxy acids have been tested.[126] α-Amino acids or α-hydroxy acids were found to be effective modifying reagents in asymmetric hydrogenation. The order of the efficiency of the modifying reagents using Raney nickel in the hydrogenation of methyl acetoacetate is as follows: tartaric acid > malic acid > glutamic acid > aspartic acid > valine. The (*S*)-α-amino acid modified nickel gave methyl (*R*)-β-hydroxybutyrate in excess and (S)-α-hydroxy acid modified nickel gave (*S*)-β-hydroxybutyrate in excess (Scheme 9).[130] The pH and the temperature strongly affect the *ee* of the product. Sodium hydroxide is the best alkali hydroxide in the modification process, and when LiOH, KOH, RbOH and aqueous ammonia are used for the modification, the catalyst gives a low *ee* in the product.[131]

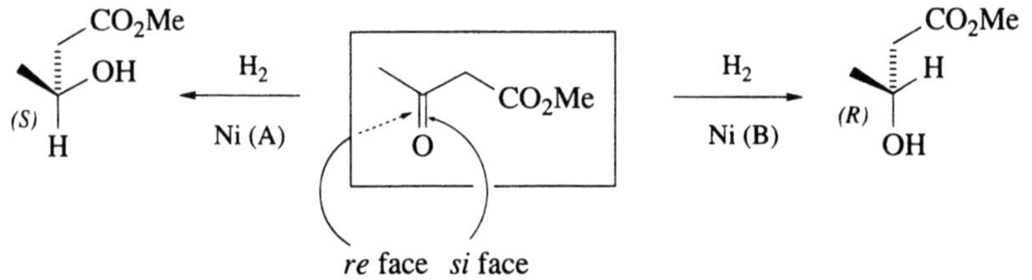

(*S*) CO_2Me, OH, H ←H_2, Ni (A)— CO_2Me, O —H_2, Ni (B)→ (*R*) CO_2Me, H, OH

re face *si* face

Ni (A): Modified with *(S,S)*-tartaric acid, *(S)*-malic acid, *(R)*-glutamic acid and *(R)*-valine
Ni (B): Modified with *(R,R)*-tartaric acid, *(R)*-malic acid, *(S)*-glutamic acid and *(S)*-valine

Scheme 9

Several other metal catalysts (Raney Co, Fe, Cu and Ru) have been investigated, but nickel catalysts show the highest activity in the asymmetric hydrogenation of ethyl acetoacetate.[132] Various nickel catalysts such as Raney nickel,[133–136] nickel acetate,[137] nickel carbonyl,[133] nickel oxide,[133,135] a nickel catalyst on Kieselguhr,[133,138] nickel on alumina[133] or on silica,[138,139] and a palladium-containing nickel catalyst on Kieselguhr[140] have been tested under various modifying and hydrogenation conditions.

The optical purity of the product is dependent not only on the preparation method of the catalyst, but also on the reaction conditions. Variables include the pressure of hydrogen, the solvent, the reaction temperature and others. The optical purity of the product is not affected by pressure at high pressure (30–120 kg cm^{-2}; 0.3–1.2 × 10^6 Pa), but it is affected in a complicated way at lower pressure.[142] Similarly, a change in the reaction temperature does not affect the optical purity of the product at high pressure, but it is affected by the reaction temperature at 1 atm of hydrogen.[142]

The solvent has a large effect on the reaction. It was first thought that a polar solvent such as methanol or ethanol was best, but later studies indicated that aprotic solvents like tetrahydrofuran and ethyl acetate are better.[143] The best solvent examined to date is methyl propionate.[121]

Additives such as water, carboxylic and other organic acids, and some inorganic salts enhance the optical purity of the product.[144] Table 4[133] shows the effect of supplementary reagents in the modification

Table 4 Hydrogenation of Methyl Acetoacetate over Raney Nickel Modified with Tartaric Acid–Inorganic Salt

Inorganic salt	$[\alpha]_D^{20}$ *of product*	*ee (%)*
None	– 8.99	39
NaF	–13.95	61
NaCl	–16.55	72
NaBr	–19.07	83
NaI	–11.74	51
Na_2SO_4	–12.95	56
$LiBr.H_2O$	–14.25	62
$NiBr_2$	–14.36	63

of Raney nickel with tartaric acid.[145] The optical purity increased considerably with the addition of inorganic salts, especially of sodium bromide. Thus Raney nickel, modified in a solution of tartaric acid with sodium bromide, gave a higher level of optical purity without any loss of activity.[121,133,135]

In the following few paragraphs, some typical examples of the catalytic hydrogenation of alkyl acetoacetate with Raney nickel modified by (+)-tartaric acid and sodium bromide[146] will be described. The *ee* of the product is close to 88% (*R*) for all of the higher alkyl esters shown in Table 5.[146–148] Higher homologs of acetoacetate also give products of high optical purity (83–88% *ee*).[147,148]

Table 5 Hydrogenation of Alkyl Acetoacetate over Raney Nickel Modified with (+)-Tartaric Acid and NaBr

β-Keto ester $R^1COCH_2COOR^2$ R^1	R^2	*ee (%)*	*Configuration*
Me	Me	83.0–85.0	(*R*)
Me	Et	87.9	(*R*)
Me	Pr^n	87.6	(*R*)
Me	Pr^i	87.7	(*R*)
Me	Bu^n	88.0	(*R*)
Me	Bu^i	88.1	(*R*)
Me	n-C_8H_{17}	87.7	(*R*)
Et	Me	86.0	(*R*)
Pr^n	Me	87.0	(*R*)
Pr^i	Me	87.5	(*R*)
Bu^n	Me	86.7	(*R*)
Bu^i	Me	85.5	(*R*)
n-C_8H_{17}	Me	87.0	(*R*)

Enantioselective hydrogenations of acetylacetone[149] and other β-diketones[150] were carried out using the same modified nickel catalyst. The hydrogenation of acetylacetone proceeds stepwise to form 4-hydroxy-2-pentanone, which is further hydrogenated to 2,4-pentanediol.[149] The first step of the hydrogenation is the usual enantioselective reaction yielding preferentially the (*R*)-keto alcohol (**20**; 87%). In the second step, both diastereoselective and enantioselective effects operate simultaneously. The (*R*)-keto alcohol is converted mainly to the (*R,R*)-diol (**21**; 78.3%), and the ratio of the minor diastereoisomers is shown in Scheme 10. The composition of the four diastereoisomers in the reaction products indicates that the stereoselective hydrogenation of the (*R*)-keto alcohol (**20**) to (*R,R*)-diol (**21**) is highly selective (90%). The minor product, (*S*)-keto alcohol, was converted to (*S,R*)- and (*S,S*)-diol in almost equal amounts. Thus, the enantioselective and diastereoselective effects function in the same direction to form the (*R,R*)-diol but oppose each other in the hydrogenation of the (*S*)-keto alcohol. When the hydrogenation of racemic keto alcohol was stopped halfway, the remaining keto alcohol was (*S*)-rich and the hydrogenated diol was (*R,R*)-rich, which indicates that kinetic resolution is taking place. The resulting (*R,R*)-diol is easily purified by recrystallization.[151]

In addition to alkyl acetoacetates and acetylacetone, several other carbonyl compounds were used for hydrogenation using the same modified nickel catalyst. The enantioselective hydrogenation of methyl alkyl ketones in the presence of pivalic acid gave the corresponding chiral secondary alcohols [(*S*)-isomer, *ee* 49–74%].[152] Similarly, the hydrogenation of 4-hydroxy-2-butanone and its methyl ether in the presence of acetic acid gave secondary alcohols [(*R*)-isomer, *ee* 68–70%].[153] The hydrogenation of β-keto sulfone proceeded smoothly to form the corresponding secondary alcohol [(*R*)-isomer, *ee* 67–71%] in the presence of acetic acid.[154]

Heterogeneous catalytic hydrogenation of the methyl esters of α-keto acids over modified metal catalysts other than nickel have been studied using a cinchonidine-modified platinum catalyst.[155–157] Methyl pyruvate and methyl benzoylformate were hydrogenated to form methyl (*R*)-lactate and (*R*)-mandelate with high *ee* (81–84%).

These studies on catalytic hydrogenation using chirally modified catalysts suggest that enantioselective hydrogenation is still an open area for the study of asymmetric synthesis, especially as a practical synthetic method for chiral nonracemic organic compounds. However, catalytic hydrogenation is affected by so many variables that it is not easy to optimize the stereoselectivity. Many physicochemical studies have been made to elucidate the reaction mechanism, and these have been reviewed by Tai.[129]

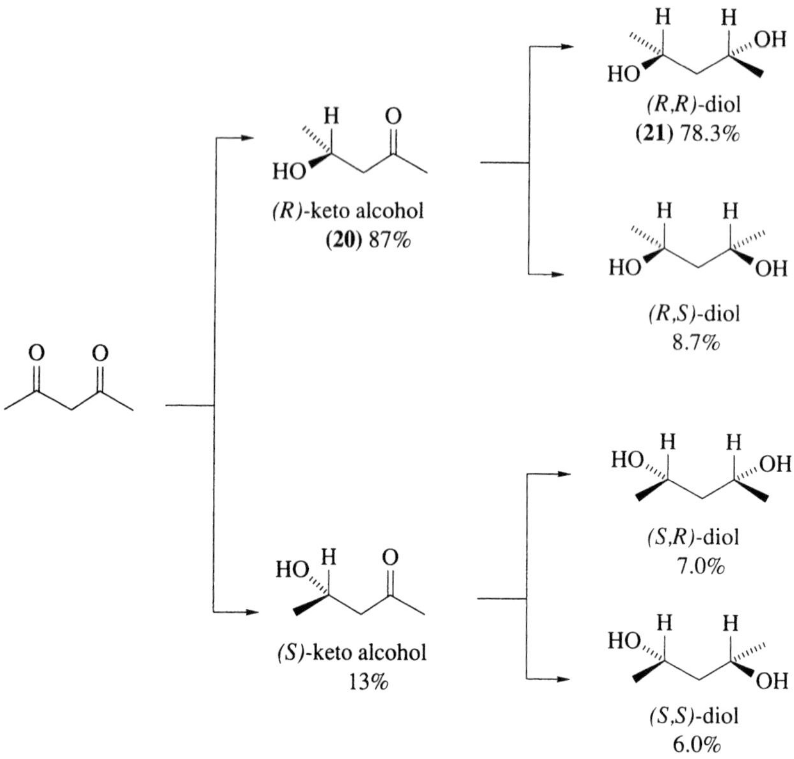

(*R*)-keto alcohol
(**20**) 87%

(*R*,*R*)-diol
(**21**) 78.3%

(*R*,*S*)-diol
8.7%

(*S*)-keto alcohol
13%

(*S*,*R*)-diol
7.0%

(*S*,*S*)-diol
6.0%

Scheme 10

1.6.6 HOMOGENEOUS CATALYTIC HYDROGENATION OF C=O AND C=N DOUBLE BONDS

In 1938, Calvin reported that *p*-benzoquinone could be reduced by hydrogen in the presence of copper salts [$Cu(OAc)_2$ and bis(salicylaldehydato)copper][158,159] and this is one of the first examples of a homogeneous catalytic hydrogenation. Hydrogenation of quinone in the presence of a palladium complex of ethylenediamine ($[Pd(en)]Cl_2$)[160] and of the rhodium complexes ($[Rh(NH_3)_5(OH)_2]Cl_3$, $[Rh(NH_3)_4Cl_2]$)[161,162] has also been reported. In the middle 1960s, rhodium–phosphine complexes, the Wilkinson catalysts, were synthesized[163,164] and their usefulness in homogeneous catalytic hydrogenation was demonstrated.[165–167]

1.6.6.1 Homogeneous Hydrogenation by Rhodium or Other Complexes[2,168–172]

Early work on the hydrogenation of carbonyl compounds using Wilkinson's rhodium complex were unsuccesful, due to the low reactivity of the carbonyl group and the high susceptibility for decarbonylation.[173] The first successful reduction of a carbonyl compound was achieved for ketones using soluble rhodium complexes. Cationic rhodium complexes ($[RhH_2P_2S_2]^+ClO_4^-$ or $[RhH_2P_2S_2]^+PF_6^-$; S = solvent, P = phosphine ligand) possessing ligands (PPh_2Me, $PPhMe_2$, PMe_3) more basic than the conventional triphenylphosphine were found to be effective (equation 11).[174] The first homogeneous asymmetric catalytic hydrogenations of ketones were performed in the presence of a rhodium complex with (*R*)-(+)-benzylmethylphenylphosphine as a ligand (equation 12). Several chiral phosphine ligands have been employed for the hydrogenation of carbonyl compounds,[175–180] and a number of asymmetric hydrogenations are described below. Asymmetric hydrogenations of acetophenone and related compounds were carried out to give *ee* values of up to 95% using (aminoalkylferrocenyl)phosphine as a chiral ligand (equation 13).[179,180] A rhodium phosphine complex with a neutral phosphine ligand (BPPM) was found to be effective for the hydrogenation of α-ketocarboxylic esters[181] and ketopantolactone (86.7% *ee*; equation 14).[182] Catalysts with a phosphine ligand (BPPM or DIOP;**22**) were tested for the asymmetric hydrogenation of *N*-(α-ketoacyl)amino acid esters, but this chiral ligand was not effective.[183]

H_2 / $[RhH_2(PPhMe_2)_2L_2]^+$ PF_6^- or ClO_4^-

25 °C, 1% water (v/v) (11)

R_2 = Me, Me; Me, Et; Ph, Me

H_2 / $[Rh(NBD)L_2]^+$ ClO_4^-

ethanol (12)

R = Ph, *ee* = 8.6% *(S)*; R = Et, *ee* = 1.9% *(R)*; L = *(R)*-benzylmethylphenylphosphine; NBD = norborna-2,5-diene

HO HO O H N Me•HCl H_2 [Rh*] HO HO OH H N Me•HCl (13)

conversion 100%; 95% *ee*

[Rh*] = $[Rh\{(R,S)\text{-BPPFOH}\}(NBD)]^+$ ClO_4^-

(R)-(S)-BPPFOH : H OH Fe PPh_2 PPh_2

H_2 (5×10^5 Pa)

[Rh(BPPM)] /benzene

30 °C, 48 h (14)

conversion 100%; *ee* 86.7%

BPPM = Ph_2P N PPh_2 CO_2Bu^t

(–)-DIOP (**22**)

(–)-CyDIOP (**23**)

Cy = cyclohexyl

Cy_2P N R PPh_2

(**24**)

BCPM: R = CO_2Bu^t

MCPM: R = CO_2Me

BCCPM: R = $CONHBu^t$

MCCPM: R = CONHMe

On the other hand, rhodium complexes with fully alkylated phosphine ligands[184,185] were used for the hydrogenation of carbonyl compounds,[184] α-ketoamides[184–187] and ketopantolactone (66% *ee*).[187] When a rhodium complex with CyDIOP (**23**) was used, a slightly higher asymmetric yield (71% *ee*) was observed in the hydrogenation of α-ketoamides.[188] Hydrogenation of *N*-(α-ketoacyl)amino acid esters was also performed by using a complex containing CyDIOP to give *N*-mandeloylamino acid esters with 72% *de* (equation 15).[189] Asymmetric hydrogenations of ketopantolactone (92% *ee*; equation 16),[190] α-keto esters (87% *ee*) and α-keto acetals[191] were carried out with the rhodium catalysts containing the ligands (**24**).

H_2 (1 atm), 20 °C; (–)-CyDIOP–Rh^N (15)

R^1 = Ph, R^2 = CH_2Ph; conversion 100%; *de* 72%

(–)-CyDIOP–Rh^N = (–)-CyDIOP + 1/2 $[Rh(C_2H_4)_2Cl]_2$

H_2 (50 atm); $[Rh(COD)Cl]_2$, L*/THF (16)

L* = BCPM: conversion 100%; *ee* 92% *(R)*

Several studies on the hydrogenation of carbonyl compounds using cobalt complexes have been reported,[192–195] including the hydrogenation of benzil in the presence of bis(dimethylglyoximato)cobalt(II)–chiral amine complexes.[193] The highest *ee* (78%) was obtained when quinine was used as the chiral base (equation 17).[194] Cyano cobalt complexes have also been reported to be effective for the hydrogenations of ketones.[196,197]

H_2; $Co(DMG)_2$, B–Q*, –10 °C (17)

ee 78%

DMG = dimethylglyoxime; Q = quinine; B = benzylamine

Ruthenium complexes have been used for the hydrogenation of alkenes[2] but they had not been used for hydrogenations of carbonyl compounds until recently, when several highly selective catalytic hydrogenations of carbonyl compounds (β-keto esters, α-diketones, α-amino ketones *etc*., *ee* > 95%) were reported using BINAP–Ru^{II} complexes (equation 18).[198–202]

H_2 (73–100 atm); Ru–*(S)*-BINAP; 93–99.5% (18)

ee 98–100%

R^1 = Me, Et, Bu, Pr^i; R^2 = Me, Et, Pr^i, Bu^t

(S)-BINAP = [structure with PPh_2, PPh_2]

Ru–BINAP = $RuCl_2$\{*(S)*- or *(R)*-BINAP\}

There are some reports on the asymmetric catalytic hydrogenation of C═N double bonds.[176,188,197] Rhodium phosphine complexes do not have high activity, and the stereoselectivity is low.[176,188] Homogeneous hydrogenation of oximes,[188] Schiff's bases[176] and cyclic imines[203] have been reported. A cyano cobalt complex has been used for the homogeneous catalytic hydrogenation of oximes and Schiff's bases[197] but the degree of asymmetric induction is unknown.

1.6.7 REFERENCES

1. W. S. Emerson, *Org. React. (N.Y.)*, 1948, **4**, 174.
2. A. J. Birch and D. H. Williamson, *Org. React. (N.Y.)*, 1976, **24**, 1.
3. R. L. Augustine, 'Catalytic Hydrogenation', Dekker, New York, 1965.
4. P. N. Rylander, 'Catalytic Hydrogenation over Platinum Metals', Academic Press, New York, 1967.
5. M. Freifelder, 'Practical Catalytic Hydrogenation', Wiley-Interscience, New York, 1971.
6. M. Freifelder, 'Catalytic Hydrogenation in Organic Synthesis. Procedures and Commentary', Wiley-Interscience, New York, 1978.
7. P. N. Rylander, 'Catalytic Hydrogenations in Organic Syntheses', Academic Press, New York, 1979.
8. M. Hudlicky, 'Reductions in Organic Chemistry', Ellis Horwood, New York, 1984.
9. H. Adkins and H. R. Billica, *J. Am. Chem. Soc.*, 1948, **70**, 695.
10. J.-J. Brunet, P. Gallois and P. Caubère, *J. Org. Chem.*, 1980, **45**, 1937, 1946.
11. W. Wenner, *J. Org. Chem.*, 1950, **15**, 301.
12. V. I. Scharkov, *Chem.-Ing.-Tech.*, 1963, **35**, 494.
13. W. H. Carothers and R. Adams, *J. Am. Chem. Soc.*, 1924, **46**, 1675.
14. H. S. Broadbent, G. C. Campbell, W. J. Bartley and J. H. Johnson, *J. Org. Chem.*, 1959, **24**, 1847.
15. P. N. Rylander and D. R. Steele, *Tetrahedron Lett.*, 1969, 1579.
16. T. W. Russell, D. M. Duncan and S. C. Hansen, *J. Org. Chem.* 1977, **42**, 551.
17. Z. Csuros, K. Zech and I. Géczy, *Hung. Acta Chim.*, 1946, **1**, 1 (*Chem. Abstr.*, 1947, **41**, 109i).
18. P. N. Rylander, N. Himelstein and M. Kilroy, *Engelhard Ind. Tech. Bull.*, 1963, **4**, 49, 131.
19. K. C. Brannock, *J. Am. Chem. Soc.*, 1959, **81**, 3379.
20. R. Adams and B. S. Garvey, *J. Am. Chem. Soc.*, 1926, **48**, 477.
21. R. L. Shriner and R. Adams, *J. Am. Chem. Soc.*, 1924, **46**, 1683.
22. A. Skita, *Chem. Ber.*, 1915, **48**, 1486.
23. R. Adams, S. McKenzie, Jr. and S. Loewe, *J. Am. Chem. Soc.*, 1948, **70**, 664.
24. H. Adkins and H. I. Cramer, *J. Am. Chem. Soc.*, 1930, **52**, 4349.
25. H. Adkins and R. Connor, *J. Am. Chem. Soc.*, 1931, **53**, 1091.
26. T. P. C. Mulholland and G. Ward, *J. Chem. Soc.*, 1954, 4676.
27. L. W. Burnette, I. B. Johns, R. F. Holdren and R. M. Hixon, *Ind. Eng. Chem.*, 1948, **40**, 502.
28. S. Nishimura, *Bull. Chem. Soc. Jpn.*, 1961, **34**, 32.
29. E. Breitner, E. Roginski and P. N. Rylander, *J. Org. Chem.*, 1959, **24**, 1855.
30. R. H. Hasek, E. U. Elam, J. C. Martin and R. G. Nations, *J. Org. Chem.*, 1961, **26**, 700.
31. H. Oelschläger, *Chem. Ber.*, 1956, **89**, 2025.
32. M. Freifelder, *J. Org. Chem.*, 1964, **29**, 2895.
33. E. Hardegger, P. A. Blank and F. Blank, *Helv. Chim. Acta*, 1944, **27**, 793.
34. K. von Auwers, *Justus Liebigs Ann. Chem.*, 1920, **420**, 84.
35. A. Skita, *Justus Liebigs Ann. Chem.*, 1923, **431**, 1.
36. A. Skita and W. Faust, *Chem. Ber.*, 1931, **64**, 2878.
37. D. H. R. Barton, *J. Chem. Soc.*, 1953, 1027.
38. E. L. Eliel and R. S. Ro, *J. Am. Chem. Soc.*, 1957, **79**, 5992.
39. R. J. Wicker, *J. Chem. Soc.*, 1956, 2165.
40. W. Hückel and A. Hubele, *Justus Liebigs Ann. Chem.*, 1958, **613**, 27.
41. W. Hückel and J. Kurz, *Chem. Ber.*, 1958, **91**, 1290.
42. A. K. Macbeth and J. A. Mills, *J. Chem. Soc.*, 1945, 709.
43. W. Hückel, M. Maier, E. Jordan and W. Seeger, *Justus Liebigs Ann. Chem.*, 1958, **616**, 46.
44. R. Helg and H. Schinz, *Helv. Chim. Acta*, 1952, **35**, 2406.
45. E. L. Eliel and S. H. Schroeter, *J. Am. Chem. Soc.*, 1965, **87**, 5031.
46. A. Farkas and L. Farkas, *J. Am. Chem. Soc.*, 1939, **61**, 1336.
47. M. Faillebin, *Ann. Chim. (Paris)*, 1925, **4**, 156, 410.
48. D. Cornet and F. G. Gault, *J. Catal.*, 1967, **7**, 140.
49. R. Adams, J. W. Kern and R. L. Shriner, *Org. Synth., Coll. Vol.*, 1932, **1**, 101.
50. M. Kawai, T. Imanaka and S. Teranishi, *Nippon Kagaku Kaishi*, 1969, **90**, 42 (*Chem. Abstr.*, 1969, **70**, 77 012h).
51. Z. Zafiriadis, *C. R. Hebd. Seances Acad. Sci.*, 1948, **226**, 731.
52. A. T. Blomquist and A. Goldstein, *Org. Synth., Coll. Vol.*, 1963, **4**, 216.
53. E. F. Rosenblatt, *J. Am. Chem. Soc.*, 1940, **62**, 1092.
54. M. Bonnet, P. Geneste and M. Rodriguez, *J. Org. Chem.*, 1980, **45**, 40.
55. M. Freifelder, W. D. Smart and G. R. Stone, *J. Org. Chem.*, 1962, **27**, 2209.
56. C. F. Winans and H. Adkins, *J. Am. Chem. Soc.*, 1932, **54**, 306.
57. P. Anziani and R. Cornubert, *Bull. Soc. Chim. Fr.*, 1948, 857.
58. F. E. King, J. A. Barltrop and R. J. Walley, *J. Chem. Soc.*, 1945, 277.
59. H. Adkins and E. W. Reeve, *J. Am. Chem. Soc.*, 1938, **60**, 1328.
60. N. F. Alberstron, B. F. Tullar, J. A. King, B. B. Fishburn and S. Archer, *J. Am. Chem. Soc.*, 1948, **70**, 1150.

61. T. Ishimaru, *Nippon Kagaku Zasshi*, 1960, **81**, 643.
62. K. Pfister, III, C. A. Robinson, A. C. Shabica and M. Tishler, *J. Am. Chem. Soc.*, 1948, **70**, 2297.
63. K. Pfister, III, C. A. Robinson, A. C. Shabica and M. Tishler, *J. Am. Chem. Soc.*, 1949, **71**, 1101.
64. R. C. Goodwin and J. R. Bailey, *J. Am. Chem. Soc.*, 1925, **47**, 167.
65. T. R. Norton, A. A. Benson, R. A. Seibert and F. W. Bergstrom, *J. Am. Chem. Soc.*, 1946, **68**, 1330.
66. C. F. Winans, *J. Am. Chem. Soc.*, 1939, **61**, 3566.
67. F. Knoop and H. Oesterlin, *Hoppe-Seyler's Z. Physiol. Chem.*, 1925, **148**, 294.
68. A. Darapsky, *J. Prakt. Chem.*, 1936, **146**, 219.
69. F. Knoop and H. Oesterlin, *Hoppe-Seyler's Z. Physiol. Chem.*, 1927, **170**, 186.
70. F. Kögl, J. Halberstadt and T. J. Basendregt, *Recl. Trav. Chim. Pays-Bas*, 1949, **68**, 387.
71. S. Mitsui, T. Kamaishi, T. Imaizumi and I. Takamura, *Nippon Kagaku Zasshi*, 1962, **83**, 1115 (*Chem. Abstr.*, 1963, **59**, 11 569e).
72. T. Kamaishi and S. Mitsui, *Nippon Kagaku Zasshi*, 1965, **86**, 623 (*Chem. Abstr.*, 1966, **65**, 2326e).
73. S. Mitsui and M. Kanai, *Nippon Kagaku Zasshi*, 1965, **86**, 627 (*Chem. Abstr.*, 1966, **65**, 2101g).
74. S. Mitsui and Y. Imai, *Nippon Kagaku Zasshi*, 1967, **88**, 86 (*Chem. Abstr.*, 1968, **67**, 43 934h).
75. S. Mitsui and M. Kanai, *Nippon Kagaku Zasshi*, 1966, **87**, 179 (*Chem. Abstr.*, 1966, **65**, 17 006f).
76. M. Kawana and S. Emoto, *Bull. Chem. Soc. Jpn.*, 1968, **41**, 259.
77. M. M. Jamison and E. E. Turner, *J. Chem. Soc.*, 1941, 538.
78. K. Harada and K. Matsumoto, in 'Peptides: Chemistry and Biochemistry', ed. B. Weinstein and S. Lande, H. Dekker, New York, 1970, p. 451.
79. K. Harada, T. Munegumi and S. Nomoto, *Tetrahedron Lett.*, 1981, **22**, 111.
80. K. Harada and T. Munegumi, *Bull. Chem. Soc. Jpn.*, 1984, **57**, 3203.
81. K. Harada and T. Munegumi, *Bull. Chem. Soc. Jpn.*, 1983, **56**, 2774.
82. T. Munegumi, M. Fujita, T. Maruyama, S. Shiono, M. Takasaki and K. Harada, *Bull. Chem. Soc. Jpn.*, 1987, **60**, 249.
83. Y. Kawanami, I. Fujita, Y. Taniguchi, T. Katsuki and M. Yamaguchi, *Chem. Lett.*, 1987, 2021.
84. J. D. Morrison and H. S. Mosher, 'Asymmetric Organic Reactions', Prentice-Hall, Englewood Cliffs, 1971, p. 303.
85. K. Harada, in 'Asymmetric Synthesis', ed. J. D. Morrison, Academic Press, New York, 1985, vol. 5, p. 345.
86. F. Knoop and C. Martius, *Hoppe-Seyler's Z. Physiol. Chem.*, 1939, **258**, 242.
87. R. M. Herbst and E. A. Swart, *J. Org. Chem.*, 1946, **11**, 368.
88. N. Izumiya, R. Wada, M. Winitz, M. C. Otey, S. M. Birnbaum, R. J. Koegel and J. P. Greenstein, *J. Am. Chem. Soc.*, 1957, **79**, 652.
89. R. G. Hiskey and R. C. Northrop, Jr., *J. Am. Chem. Soc.*, 1961, **83**, 4798.
90. A. Kanai and S. Mitsui, *Nippon Kagaku Zasshi*, 1966, **87**, 183 (*Chem. Abstr.*, 1966, **65**, 16 835h).
91. K. Harada and K. Matsumoto, *J. Org. Chem.*, 1968, **33**, 4467.
92. K. Harada and K. Matsumoto, *J. Org. Chem.*, 1967, **32**, 1794.
93. K. Harada and T. Yoshida, *Bull. Chem. Soc. Jpn.*, 1970, **43**, 921.
94. K. Harada and M. Tamura, *Bull. Chem. Soc. Jpn.*, 1979, **52**, 1227.
95. K. Harada and Y. Kataoka, *Tetrahedron Lett.*, 1978, 2103.
96. K. Harada and Y. Kataoka, *Chem. Lett.*, 1978, 791.
97. D. E. Nichols, C. F. Barfknecht, D. B. Rusterholz, F. Benington and R. D. Morin, *J. Med. Chem.*, 1973, **16**, 480.
98. W. H. Pirkle and J. R. Hauske, *J. Org. Chem.*, 1977, **42**, 2436.
99. S. Yamada and S. Hashimoto, *Tetrahedron Lett.*, 1976, 997.
100. S. Yamada, N. Ikota and K. Achiwa, *Tetrahedron Lett.*, 1976, 1001.
101. K. Harada and S. Shiono, *Bull. Chem. Soc. Jpn.*, 1984, **57**, 1367.
102. K. Matsumoto and K. Harada, *J. Org. Chem.*, 1968, **33**, 4526.
103. K. Harada and T. Yoshida, *J. Chem. Soc., Chem. Commun.*, 1970, 1071.
104. K. Harada and T. Yoshida, *J. Org. Chem.*, 1972, **37**, 4366.
105. R. G. Hiskey and R. C. Northrop, Jr., *J. Am. Chem. Soc.*, 1965, **87**, 1753.
106. K. Harada and K. Matsumoto, *Bull. Chem. Soc. Jpn.*, 1971, **44**, 1068.
107. J. Weijland, K. Pfister, III, E. F. Swanezy, C. A. Robinson and M. Tishler, *J. Am. Chem. Soc.*, 1951, **73**, 1216.
108. T. Ishimaru, *Nippon Kagaku Zasshi*, 1960, **81**, 643 (*Chem. Abstr.*, 1962, **56**, 368b).
109. W. A. Bolhofer, *J. Am. Chem. Soc.*, 1952, **74**, 5459.
110. K. Harada and S. Shiono, *Bull. Chem. Soc. Jpn.*, 1984, **57**, 1040.
111. Y.-T. Chang and W. H. Hartung, *J. Am. Chem. Soc.*, 1952, **74**, 5927.
112. Y.-T. Chang and W. H. Hartung, *J. Am. Chem. Soc.*, 1953, **75**, 89.
113. A. Hatta and W. Suëtaka, *Bull. Chem. Soc. Jpn.*, 1975, **48**, 2428.
114. A. Hatta, Y. Moriya and W. Suëtaka, *Bull. Chem. Soc. Jpn.*, 1975, **48**, 3441.
115. M. Osawa, A. Hatta, K. Harada and W. Suëtaka, *Bull. Chem. Soc. Jpn.*, 1976, **49**, 1512.
116. Y. Nakamura, *Nippon Kagaku Zasshi*, 1940, **61**, 1051 (*Chem. Abstr.*, 1943, **37**, 377).
117. S. Akabori, S. Sakurai, Y. Izumi and Y. Fujii, *Nature (London)*, 1956, **178**, 323.
118. S. Akabori, Y. Izumi, Y. Fujii and S. Sakurai, *Nippon Kagaku Zasshi*, 1956, **77**, 1374 (*Chem. Abstr.*, 1959, **53**, 5149b).
119. S. Akabori, Y. Izumi and Y. Fujii, *Nippon Kagaku Zasshi*, 1957, **78**, 886 (*Chem. Abstr.*, 1960, **54**, 9889e).
120. R. L. Beamer, C. S. Fickling and J. H. Ewing, *J. Pharm. Sci.*, 1967, **56**, 1029.
121. T. Harada and Y. Izumi, *Chem. Lett.*, 1978, 1195.
122. A. Tai, M. Nakahata, T. Harada, Y. Izumi, S. Kusumoto, M. Inage and T. Shiba, *Chem. Lett.*, 1980, 1125.
123. A. Tai, T. Harada, Y. Hiraki and S. Murakami, *Bull. Chem. Soc. Jpn.*, 1983, **56**, 1414.
124. S. Murakami, T. Harada and A. Tai, *Bull. Chem. Soc. Jpn.*, 1980, **53**, 1356.
125. T. Osawa and T. Harada, *Bull. Chem. Soc. Jpn.*, 1984, **57**, 1518.
126. Y. Izumi, *Angew. Chem., Int. Ed. Engl.*, 1971, **10**, 871.

127. E. I. Klabunovskii and A. A. Vedenyapin, 'Asymmetricheskiy Kataliz, Hidrogenizatsya no metalakh', Nauka, Moscow, 1980.
128. M. J. Fish and D. F. Ollis, *Catal. Rev.*, 1978, **18**, 259.
129. A. Tai and T. Harada, 'Tailored Metal Catalysts', ed. Y. Iwasawa, Reidel, Dordrecht, 1986, p. 265.
130. H. Ozaki, A. Tai and Y. Izumi, *Chem. Lett.*, 1974, 935.
131. T. Tanabe, K. Okuda and Y. Izumi, *Bull. Chem. Soc. Jpn.*, 1973, **46**, 514.
132. E. I. Klabunovskii, V. I. Neupokoev and Yu. I. Petrov, *Izv. Akad. Nauk SSSR, Ser. Khim.*, 1970, 2829 (*Chem. Abstr.*, 1971, **74**, 99 415b); E. I. Klabunovskii, N. P. Sokolova, A. A. Vendenyapin and Yu. M. Talanov, *Izv. Akad. Nauk SSSR., Ser. Khim.*, 1971, 1803 (*Chem. Abstr.*, 1972, **76**, 3317p); E. I. Klabunovskii, N. P. Sokolova, A. A. Vendenyapin, Yu. M. Talanov, N. D. Zubareva, V. P. Polyakova and N. V. Gorina, *Izv. Akad. Nauk SSSR, Ser. Khim.*, 1972, 2361 (*Chem. Abstr.*, 1973, **78**, 71 346p).
133. T. Harada, M. Yamamoto, S. Onaka, M. Imaida, H. Ozaki, A. Tai and Y. Izumi, *Bull. Chem. Soc. Jpn.*, 1981, **54**, 2323.
134. S. Tatsumi, *Bull. Chem. Soc. Jpn.*, 1968, **41**, 408.
135. T. Harada, A. Tai, M. Yamamoto, H. Ozaki and Y. Izumi, 'Proceedings of the 7th. International Congress on Catalysis, Tokyo, 1980', Elsevier/Kodansha, 1981, p. 364.
136. G. V. Smith and M. Musoiu, *J. Catal.*, 1979, **60**, 184.
137. Y. Nitta, F. Sekine, T. Imanaka and S. Teranishi, *Bull. Chem. Soc. Jpn.*, 1981, **54**, 980.
138. Y. Nitta, F. Sekine, T. Imanaka and S. Teranishi, *J. Catal.*, 1982, **74**, 382.
139. A. Hoek, H. W. Woerde and W. M. H. Sachtler, 'Proceedings of the 7th International Congress on Catalysis, Tokyo, 1980', Elsevier/Kodansha, 1981, p. 376.
140. Y. Orito, S. Imai and S. Niwa, *Yuki Gosei Kagaku Kyokai Shi*, 1976, **34**, 236, 672 (*Chem. Abstr.*, 1976, **85**, 123 295n; 1977, **86**, 43 148x); Y. Orito, S. Imai and S. Niwa, *Yuki Gosei Kagaku Kyokai Shi*, 1977, **35**, 672, 753 (*Chem. Abstr.*, 1977, **87**, 183 934w; 1978, **88**, 37 219w).
141. E. N. Lipgart, Yu. L. Petrov and E. I. Klabunovskii, *Kinet. Katal.*, 1971, **12**, 1491 (*Chem. Abstr.*, 1972, **76**, 98 985s).
142. H. Ozaki, *Bull. Chem. Soc. Jpn.*, 1978, **51**, 257.
143. Y. Orito, S. Niwa and S. Imai, *Yuki Gosei Kagaku Kyokai Shi*, 1976, **34**, 236 (*Chem. Abstr.*, 1976, **85**, 123 295n).
144. F. Higashi, T. Ninomiya and Y. Izumi, *Bull. Chem. Soc. Jpn.*, 1971, **44**, 1333.
145. T. Harada, M. Yamamoto, S. Onaka, M. Imaida, H. Ozaki, A. Tai and Y. Izumi, *Bull. Chem. Soc. Jpn.*, 1981, **54**, 2323.
146. A. Tai, T. Harada, Y. Hiraki and S. Murakami, *Bull. Chem. Soc. Jpn.*, 1983, **56**, 1414.
147. T. Aoki and M. Munemori, *Bull. Chem. Soc. Jpn.*, 1982, **55**, 730.
148. A. Tai, M. Nakahata, T. Harada, Y. Izumi, S. Kusumoto, M. Inage and T. Shiba, *Chem. Lett.*, 1980, 1125.
149. A. Tai, K. Ito and T. Harada, *Bull. Chem. Soc. Jpn.*, 1981, **54**, 223.
150. K. Ito, T. Harada and A. Tai, *Bull. Chem. Soc. Jpn.*, 1980, **53**, 3367.
151. K. Ito, T. Harada, A. Tai and Y. Izumi, *Chem. Lett.*, 1979, 1049.
152. T. Osawa and T. Harada, *Chem. Lett.*, 1982, 315.
153. S. Murakami, T. Harada and A. Tai, *Bull. Chem. Soc. Jpn.*, 1980, **53**, 1356.
154. Y. Hiraki, K. Ito, T. Harada and A. Tai, *Chem. Lett.*, 1981, 131.
155. Y. Orito, S. Imai and S. Niwa, *Nippon Kagaku Kaishi*, 1979, 1118 (*Chem. Abstr.*, 1979, **91**, 192 483h).
156. Y. Orito, S. Imai and S. Niwa, *Nippon Kagaku Kaishi*, 1980, 670 (*Chem. Abstr.*, 1980, **93**, 113 912n).
157. S. Niwa, S. Imai and Y. Orito, *Nippon Kagaku Kaishi*, 1982, 137 (*Chem. Abstr.*, 1982, **96**, 68 117f).
158. M. M. Jones, 'Ligand Reactivity and Catalysis', Academic Press, New York, 1968.
159. M. Calvin, *Trans. Faraday Soc.*, 1938, **34**, 1181.
160. Y. Shibata and E. Matsumoto, *Nippon Kagaku Zasshi*, 1939, **60**, 1173.
161. M. Iguchi, *Nippon Kagaku Zasshi*, 1939, **60**, 1787.
162. M. Iguchi, *Nippon Kagaku Zasshi*, 1942, **63**, 634.
163. J. A. Osborn, G. Wilkinson and J. F. Young, *J. Chem. Soc., Chem. Commun.*, 1965, 17.
164. M. A. Bennett and P. A. Longstaff, *Chem. Ind. (London)*, 1965, 846.
165. J. A. Osborn, F. H. Jardine, J. F. Young and G. Wilkinson, *J. Chem. Soc. A*, 1966, 1711.
166. J. F. Young, J. A. Osborn, F. H. Jardine and G. Wilkinson, *J. Chem. Soc., Chem. Commun.*, 1965, 131.
167. F. H. Jardine, J. A. Osborn, G. Wilkinson and J. F. Young, *Chem. Ind. (London)*, 1965, 560.
168. D. Valentine, Jr. and J. W. Scott, *Synthesis*, 1978, 329.
169. V. Čapler, G. Comisso and V. Šunjič, *Synthesis*, 1981, 85.
170. H. Brunner, *Synthesis*, 1988, 645.
171. B. R. James, 'Homogeneous Hydrogenation', Wiley-Interscience, New York, 1973.
172. K. E. Koenig, in 'Asymmetric Synthesis', ed. J. D. Morrison, Academic Press, New York, 1983, vol. 1, p. 71.
173. A. J. Birch and K. A. M. Walker, *J. Chem. Soc. C*, 1966, 1894.
174. R. R. Schrock and J. A. Osborn, *J. Chem. Soc., Chem. Commun.*, 1970, 567.
175. P. Bonvicini, A. Levi, G. Modena and G. Scorrano, *J. Chem. Soc., Chem. Commun.*, 1972, 1188.
176. A. Levi, G. Modena and G. Scorrano, *J. Chem. Soc., Chem. Commun.*, 1975, 6.
177. B. Heil, S. Törös, S. Vastag and L. Markó, *J. Organomet. Chem.* 1975, **94**, C47.
178. T. Hayashi, M. Tanaka and I. Ogata, *Tetrahedron Lett.*, 1977, 295.
179. T. Hayashi, T. Mise and M. Kumada, *Tetrahedron Lett.*, 1976, 4351.
180. T. Hayashi, A. Katsumura, M. Konishi and M. Kumada, *Tetrahedron Lett.*, 1979, 425.
181. I. Ojima, T. Kogure and K. Achiwa, *J. Chem. Soc., Chem. Commun.*, 1977, 428.
182. I. Ojima, T. Kogure and T. Terasaki, *J. Org. Chem.*, 1978, **43**, 3444.
183. I. Ojima, T. Tanaka and T. Kogure, *Chem. Lett.*, 1981, 823.
184. K. Tani, K. Suwa, E. Tanigawa, T. Yoshida, T. Okano and S. Otsuka, *Chem. Lett.*, 1982, 261.
185. K. Tani, E. Tanigawa, Y. Tatsuno and S. Otsuka, *J. Organomet. Chem.*, 1985, **279**, 87.
186. K. Tani, K. Suwa, Y. Yamagata and S. Otsuka, *Chem. Lett.*, 1982, 265.

187. K. Tani, T. Ise, Y. Tatsuno and T. Saito, *J. Chem. Soc., Chem. Commun.*, 1984, 1641.
188. K. Yamamoto and Saeed-ur-Reiman, *Chem. Lett.*, 1984, 1603.
189. K. Tani, E. Tanigawa, Y. Tatsuno and S. Otsuka, *Chem. Lett.*, 1986, 737.
190. H. Takahashi, M. Hattori, M. Chiba, T. Morimoto and K. Achiwa, *Tetrahedron Lett.*, 1986, **27**, 4477.
191. H. Takahashi, T. Morimoto and K. Achiwa, *Chem. Lett.*, 1987, 855.
192. Y. Ohgo, S. Takeuchi and J. Yoshimura, *Bull. Chem. Soc. Jpn.*, 1971, **44**, 283.
193. Y. Ohgo, S. Takeuchi and J. Yoshimura, *Bull. Chem. Soc. Jpn.*, 1971, **44**, 583.
194. Y. Ohgo, Y. Natori, S. Takeuchi and J. Yoshimura, *Chem. Lett.*, 1974. 1327.
195. S. Takeuchi and Y. Ohgo, *Bull. Chem. Soc. Jpn.*, 1981, **54**, 2136.
196. J. Kwiatek, *Catal. Rev.*, 1967, **1**, 37.
197. M. Murakami and J.-W. Kang, *Bull. Chem. Soc. Jpn.*, 1963, **36**, 763.
198. R. Noyori, T. Ohkuma and M. Kitamura, *J. Am. Chem. Soc.*, 1987, **109**, 5856.
199. M. Kitamura, T. Ohkuma, S. Inoue, N. Sayo, H. Kumabayashi, S. Akutagawa, T. Ohta, H. Takaya and R. Noyori, *J. Am. Chem. Soc.*, 1988, **110**, 629.
200. M. Kitamura, T. Ohkuma, H. Takaya and R. Noyori, *Tetrahedron Lett.*, 1988, **29**, 1555.
201. T. Ohta, H. Takaya and R. Noyori, *Inorg. Chem.*, 1988, *27*, 566.
202. T. Nishi, M. Kitamura, T. Ohkuma and R. Noyori, *Tetrahedron Lett.*, 1988, **29**, 6327.
203. S. Murata, T. Sugimoto and S. Matsuura, *Heterocycles*, 1987, **26**, 763.

1.7
Reduction of C=X to CHXH by Chirally Modified Hydride Reagents

MUGIO NISHIZAWA
Tokushima Bunri University, Japan

and

RYOJI NOYORI
Nagoya University, Japan

1.7.1 INTRODUCTION

The modification of aluminum or boron hydrides with chiral protic substances, such as R*OH or RR*NH, generates useful reagents for the asymmetric reduction of prochiral ketones or imines leading to optically active secondary alcohols and amines, respectively. Some reviews have appearered in the literature.[1–11]

1.7.2 ASYMMETRIC REDUCTIONS OF THE CARBONYL GROUP

In 1951 Bothner-By first attempted asymmetric reductions based on the conversion of lithium aluminum hydride (LAH) into a chiral alkoxy derivative by reaction with (+)-camphor.[12] Since this pioneering work, the use of chirally modified LAH reagents has been the focus of much attention. In 1979, the first virtually complete enantiofacial recognition of prochiral carbonyl compounds was accomplished by using LAH modified with optically pure 2,2′-dihydroxy-1,1′-binaphthyl and a simple alcohol (BINAL-H).[34] Asymmetric reduction with chiral 2,5-dimethylborolane also gave alcohols in high optical yields.[108] Recently, excellent results have been obtained using a chirally modified sodium borohydride reagent (K-glucoride),[90] and chirally modified borane reagents.[100,102] The latter is particularly useful since chirality is introduced using a catalytic quantity of a chiral source.

Although *B*-3-pinanyl-9-borabicyclo[3.3.1]nonane and related substances have also been developed as efficient asymmetric reducing agents for carbonyl compounds (Volume 8, Chapter 1.3), we discuss here only asymmetric reductions using chirally modified metal hydride reagents. The asymmetric hydrosilylation of a carbonyl group catalyzed by a chirally modified transition metal is mentioned briefly.

1.7.2.1 Chirally Modified Lithium Aluminum Hydride Reagents

Most reactions have been conducted using readily available, naturally occurring, chiral modifiers. The auxiliaries employed for the modification of LAH are classified into three types: (i) alcohol modifiers; (ii) dialkylamino alcohol modifiers; and (iii) primary or secondary amines and amino or monoalkylamino alcohol modifiers. Most asymmetric reductions have been investigated with acetophenone (**1**) as the substrate. Structures (**3**) to (**24**) summarize the chiral modifiers, enantiomeric excesses (*ee*) and absolute configurations of the 1-phenyl-1-ethanol (**2**) produced.

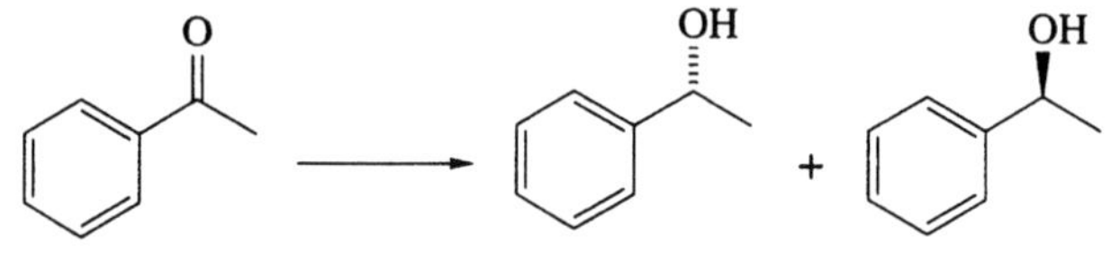

(**1**) (*R*)-(**2**) (*S*)-(**2**)

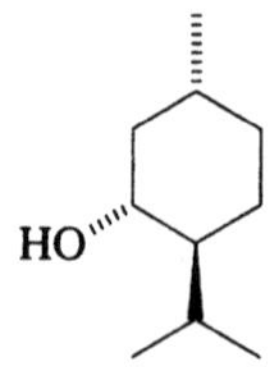

(**3**) 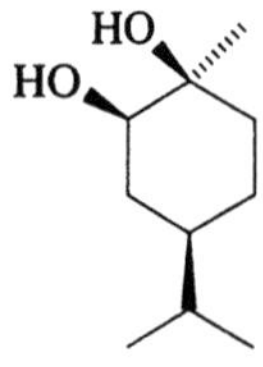 (**4**) 30% *ee*, (*R*) 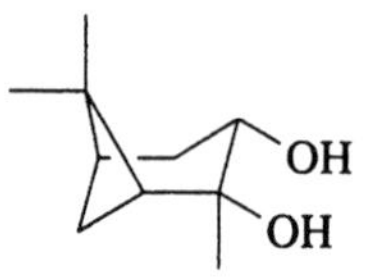 (**5**) 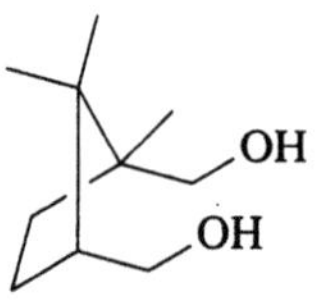 (**6**) 18% *ee*, (*R*)

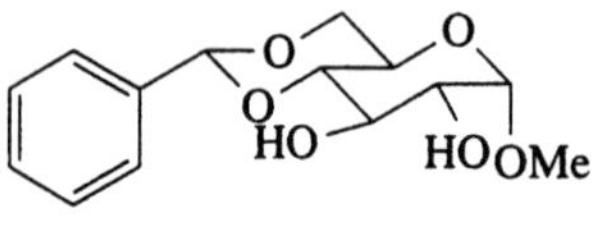

(**7**) 7.7% *ee*, (*S*) 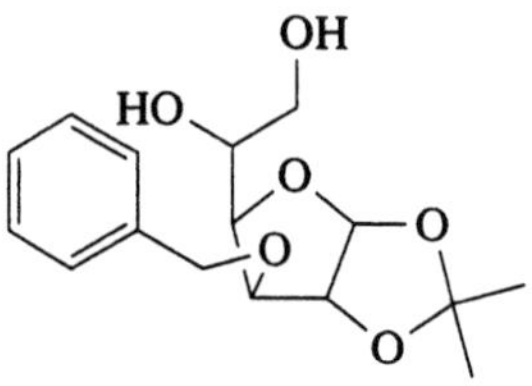 (**8**) 6% *ee* (*S*) 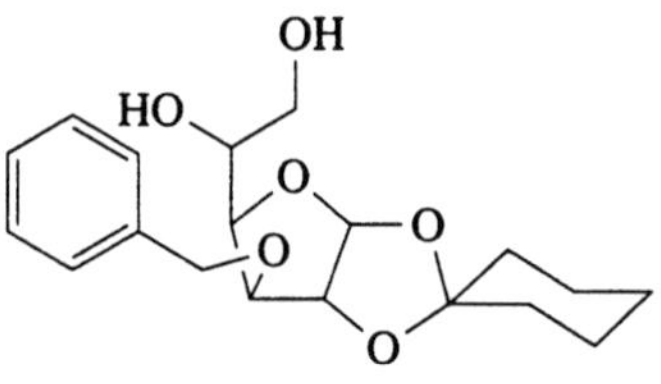 (**9**) 71% *ee*, (*R*)

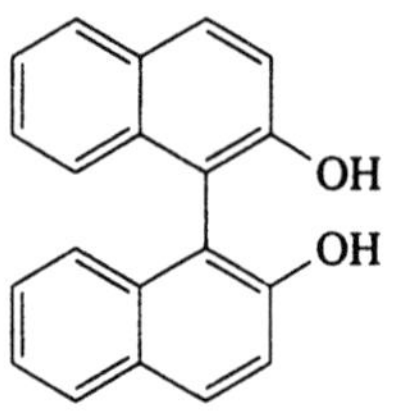

(**10**) 95% *ee*, (*R*)→(*R*) and (*S*)→(*S*) 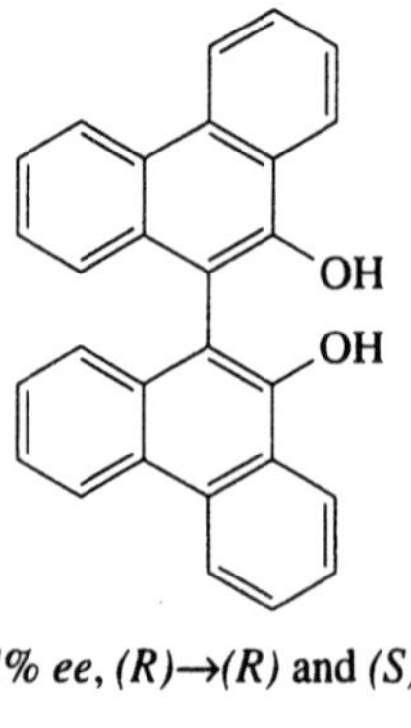 (**11**) 97% *ee*, (*R*)→(*R*) and (*S*)→(*S*)

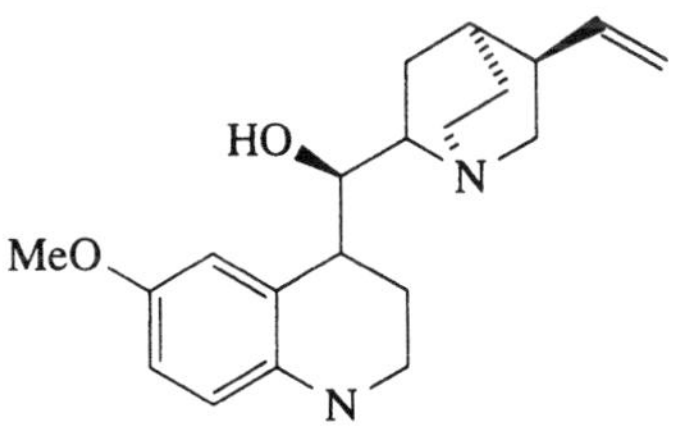

(12) 48% *ee*, *(R)* 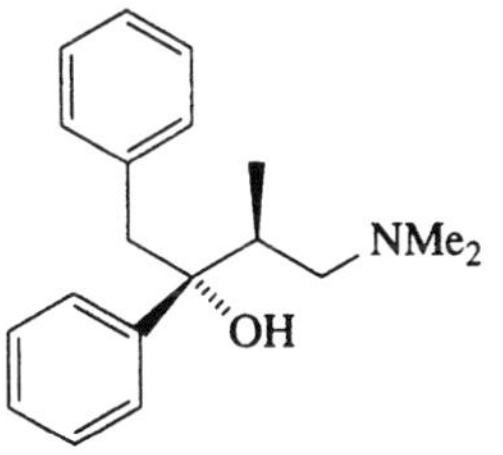 **(13)** 75% *ee*, *(R)* 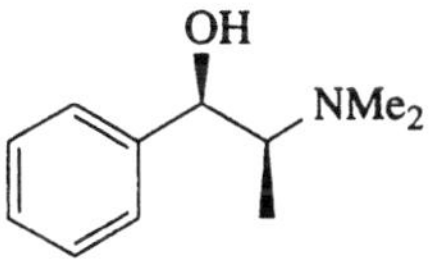 **(14)** 83% *ee*, *(R)*

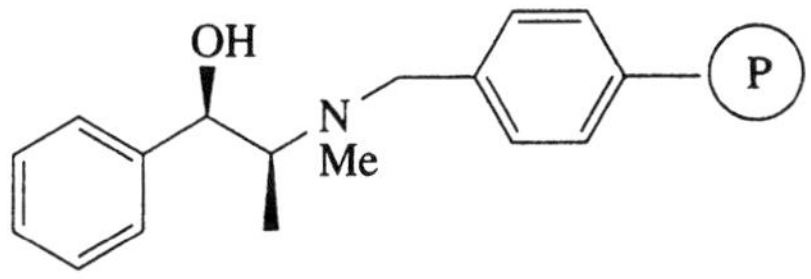

(15) 79% *ee*, *(R)* 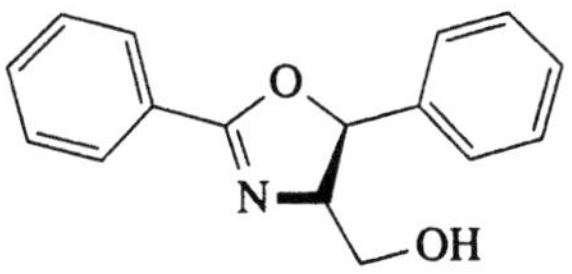 **(16)** 65% *ee*, *(R)* 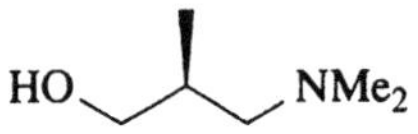 **(17)** 60% *ee*, *(S)*

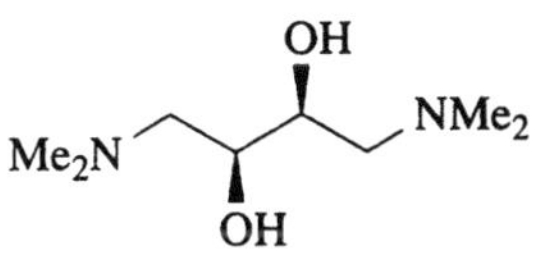

(18) 42% *ee*, *(S)*→*(S)* and *(R)*→*(R)* 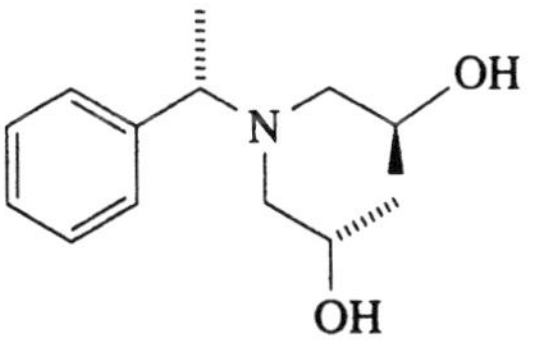 **(19)** 82% *ee*, *(R)*

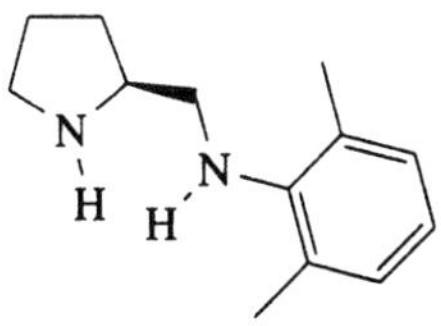

(20) 95% *ee*, *(S)* 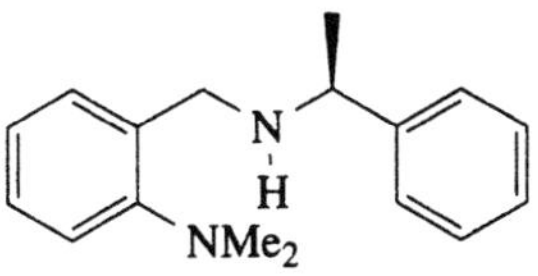 **(21)** 43% *ee*, *(R)* 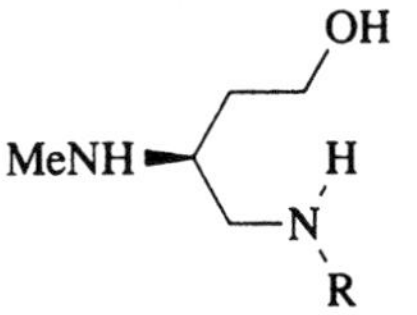 **(22)** 51% *ee*, *(S)*

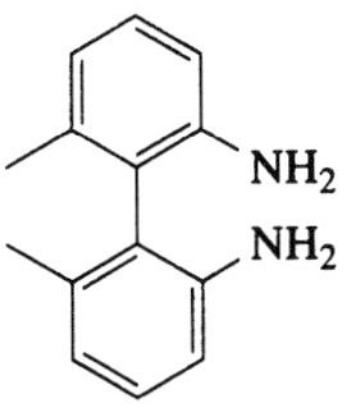

(23) 46% *ee*, *(R)*→*(R)* and *(S)*→*(S)* 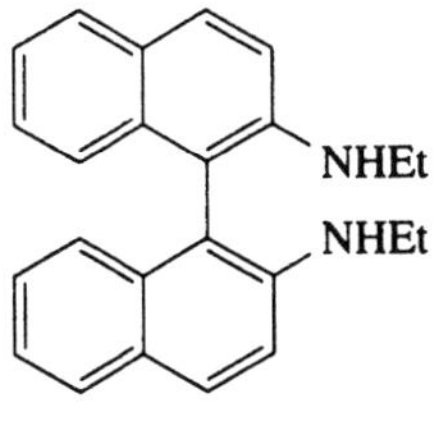 **(24)** 43% *ee*, *(R)*→*(R)* and *(S)*→*(S)*

1.7.2.1.1 Alcohol modifiers

A large number of asymmetric reductions of simple prochiral ketones were carried out using LAH reagents modified with monoterpene alcohols, such as **(3)** to **(6)**. However, in most cases only moderate optical yields were observed.[13–21] The failure to obtain a high level of stereoselectivity is partly due to the instability of these complex hydride reagents, which disproportionate to generate achiral reducing species. Exceptionally, a LAH–(–)-menthol reagent **(3)** reduces α- and β-amino ketones, **(25)** and **(26)**, to give (*R*)-carbinols with rather good optical yields, 75% *ee* and 37% *ee*, respectively.[22,23]

When diols (**7**) to (**9**), derived from D-glucose, are employed for modification of LAH, optical yields for the reduction of acetophenone do not exceed 20% *ee*.[24] Double modification of LAH by **(9)** and etha-

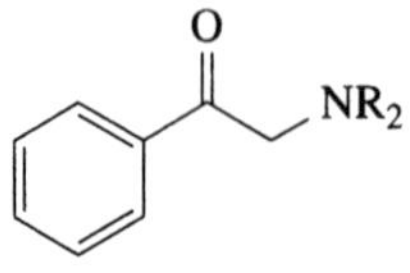

(25) 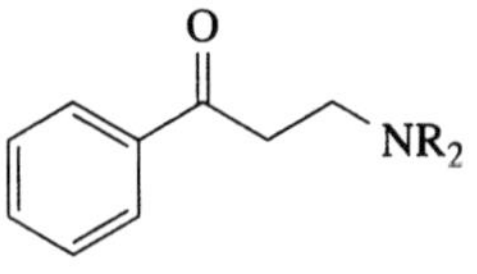 (26)

nol, giving empirical formula (**27**), increases the optical yields for the reduction of simple prochiral ketones, giving (*R*)-carbinols in up to 71% *ee* (Scheme 1).[25–28]

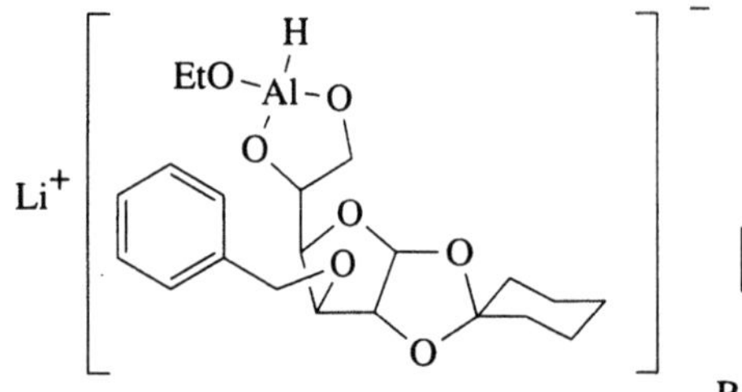

(27)

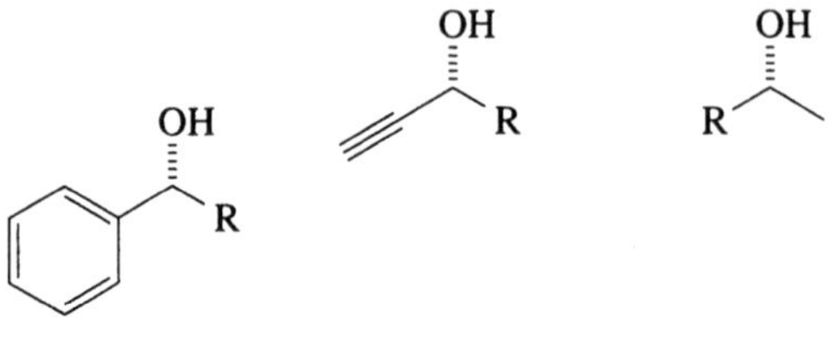

R = Me, 71% *ee*
R = Et, 45.8% *ee*

R = Pr^n, 3.5% *ee*
R = Pr^i, 5.9% *ee*

R = Bu^i, 16.6% *ee*
R = $CH{=}CMe_2$, 31.4% *ee*
R = $C{\equiv}CBu^n$, 7% *ee*

Scheme 1

Minimization of the number of reactive hydride species is crucial for obtaining a high level of enantiofacial differentiation. Noyori introduced, as a chiral auxiliary to LAH, an axially dissymmetric and bifunctional (*S*)-(–)- and (*R*)-(+)-2,2′-dihydroxy-1,1′-binaphthyl (**10**; binaphthol),[29] easily available *via* optical resolution[30,31] or asymmetric synthesis,[32] and achieved virtually complete enantiofacial differentiation of the carbonyl group.[33,34] This binaphthol-modified aluminum hydride reagent (hereafter referred to as BINAL-H), (*S*)- or (*R*)-(**28**), is prepared *in situ* by mixing LAH with equimolar amounts of optically pure (*S*)- or (*R*)-binaphthol (**10**) and a second alcoholic component, ethanol. The optical yield is generally enhanced by lowering the reaction temperature. For example, 95% *ee* is obtained by carrying out the reaction of acetophenone with (**28**) at –100 to –78 °C in THF. The asymmetric reduction exhibits high selectivity, and, in some cases, optically pure alcohols are prepared. Although the degree of stereoselectivity depends on the nature of the alkyl group attached to the carbonyl function, the reduction using (*R*)-(**28**) generally gives (*R*)-carbinols, while (*S*)-(**28**) produces the (*S*)-enantiomers (Scheme 2).

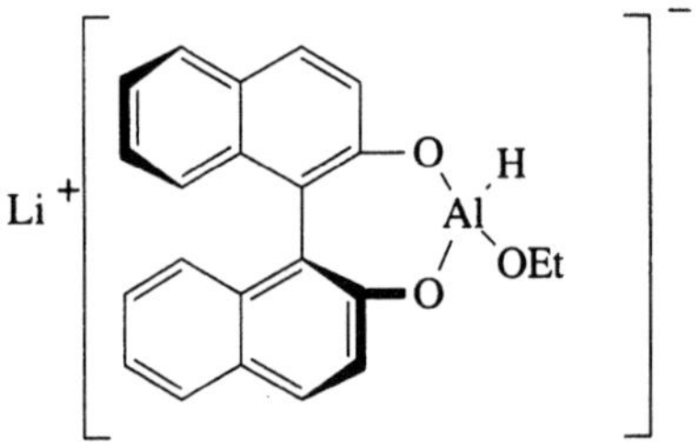

(*S*)-BINAL-H [(S)-(**28**)]

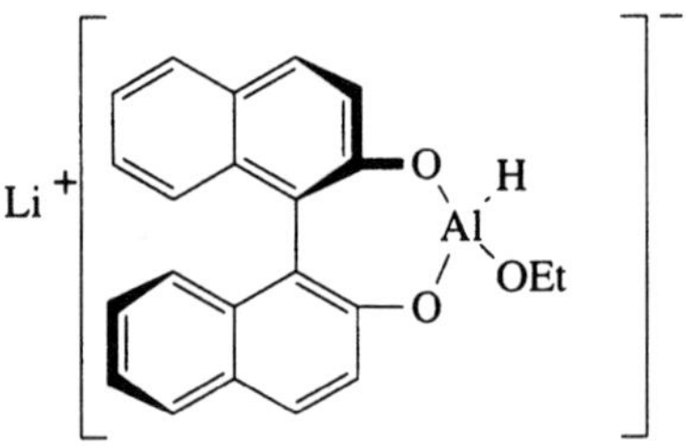

(*R*)-BINAL-H [(*R*)-(**28**)]

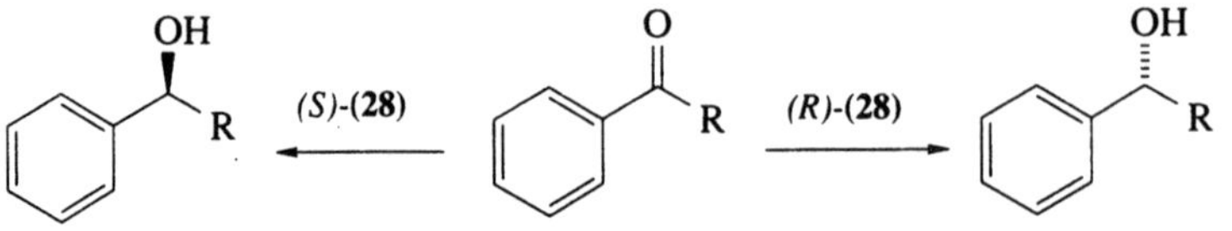

(*S*)-(**28**) ← → (*R*)-(**28**)

R = Me, 95% *ee*
R = Et, 98% *ee*
R = Pr^n, 100% *ee*

R = Bu^n, 100% *ee*
R = Pr^i, 71% *ee*
R = Bu^t, 44% *ee*

Scheme 2

The BINAL-H reduction works well with alkenic[35,36] and alkynic ketones affording alcohols with satisfactory optical purities (Scheme 3).[37,38] It is noteworthy that (*S*)-BINAL-H [(*S*)-(**28**)] reduces 1-halo-1-octen-3-one and 2-cyclopentene-1,4-dione to give the corresponding carbinols, which are important chiral building blocks for the synthesis of prostaglandins by the conjugate addition approach.[39]

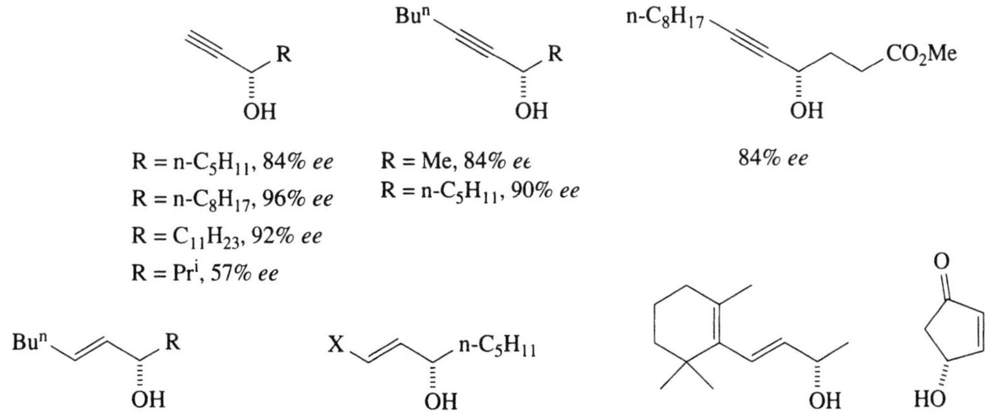

R = n-C_5H_{11}, 84% *ee*
R = n-C_8H_{17}, 96% *ee*
R = $C_{11}H_{23}$, 92% *ee*
R = Pr^i, 57% *ee*

Bu^n ... R, OH
R = Me, 84% *ee*
R = n-C_5H_{11}, 90% *ee*

n-C_8H_{17} ... CO_2Me, OH
84% *ee*

Bu^n ... R, OH
R = Me, 79% *ee*
R = n-C_5H_{11}, 91% *ee*

X ... n-C_5H_{11}, OH
X = I, 97% *ee*
X = Br, 96% *ee*
X = c-C_5H_9, 92% ee

100% *ee*

94% *ee*

Scheme 3

BINAL-H reduction can also be used to control the requisite (15*S*)-configuration of prostaglandins in the Corey route.[40] When the α,β-unsaturated ketones (**37**) to (**39**) are treated with (*S*)-(**28**), the corresponding allylic alcohols possessing the (15*S*)-absolute stereochemistry are produced with >99:1 stereoselectivity.[35,37] In contrast, when (**37**) is treated with (*R*)-BINAL-H, the (15*R*)-alcohol is obtained as the major isomer but the selectivity is only 68:32 (15*R*:15*S*).[37] These results can be rationalized in terms of double stereodifferentiation. Reagent control differentiation of the hypothetical enantiofaces is calculated to be 20.6:1, and substrate control, namely the efficiency of intramolecular chirality transfer from the bicyclo[3.3.0] system to the carbonyl center attained with a hypothetical achiral agent, is evaluated to be 9.7:1 favoring the (*S*)-isomer. In a similar manner, the reduction of the monocyclic substrate (**40**) affords the prostaglandin $F_{2\alpha}$ derivative as a single stereoisomer.[35] Diastereoselective reduction of enones (**41**) with (*S*)-BINAL-H gives (15*S*)-alcohols exclusively.[41,42] A kinetic differentiation is seen in the reduction of enone (**42**) with BINAL-H. (*R*)-BINAL-H matches well with the chiral environment of (**42**) to form the (9*S*)-carbinol (9α:9β = 99:1), whereas the enantiomeric (*S*)-BINAL-H fails to react with (**42**) under identical conditions.[39]

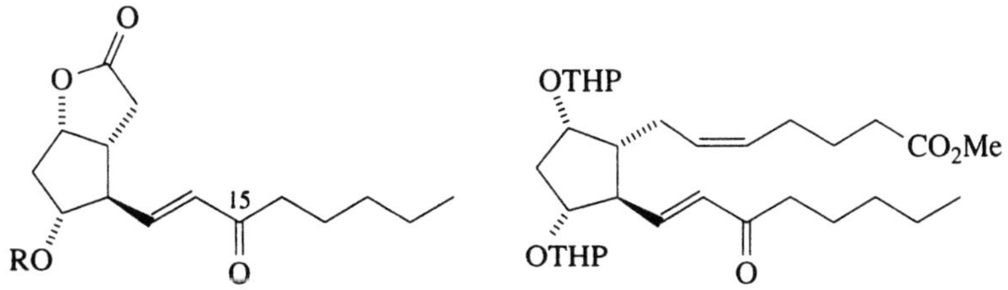

(**37**) R = COMe, (15*S*):(15*R*) = 99.4:0.6
(**38**) R = THP, 15(*S*):(15*R*) = 99.5:0.5
(**39**) R = OH, (15*S*):(15*R*) = 100:0

OTHP ... CO_2Me, OTHP

(**40**) (15*S*):(15*R*) = 100:0

When terpenic 1-deuterioaldehydes are subjected to the reduction with (*R*)- or (*S*)-BINAL-H, optically active, deuterium-labeled primary alcohols are produced (Scheme 4).[43]

The most important feature of the BINAL-H asymmetric reduction is the empirical rule for the orientation observed with simple prochiral carbonyl substrates of type (**43**) (Un = phenyl, alkenyl, alkynyl; R =

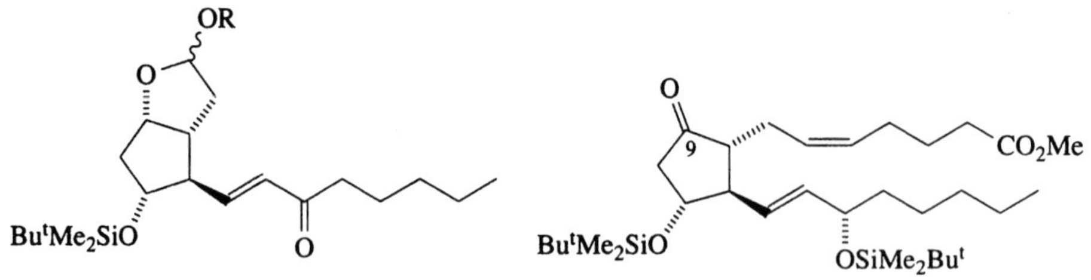

(41) R = Me or Et

(42)

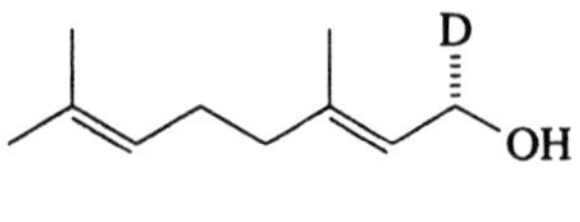

84% *ee* from (*S*)-BINAL-H

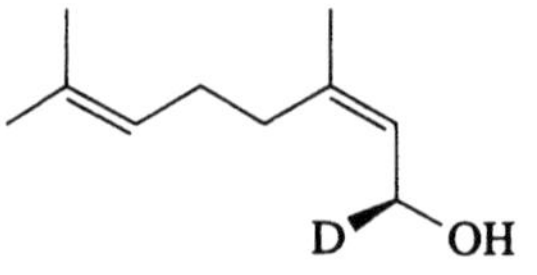

72% *ee* from (*S*)-BINAL-H

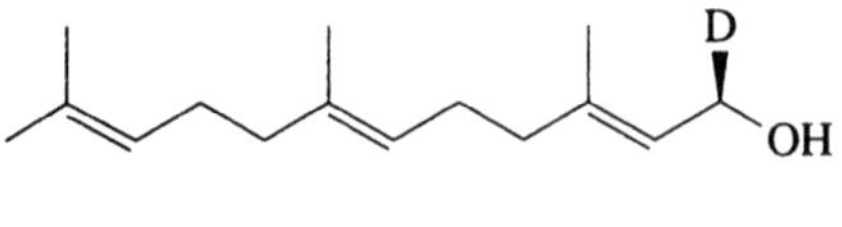

88% *ee* from (*R*)-BINAL-H

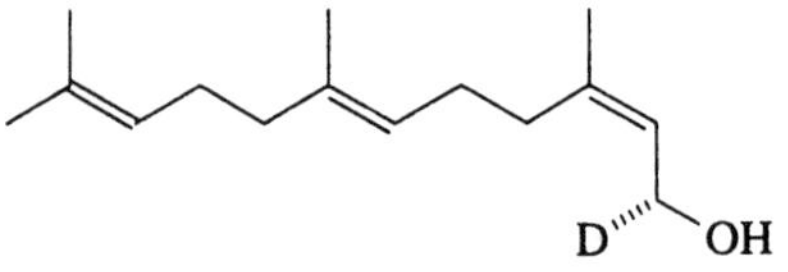

82% *ee* from (*R*)-BINAL-H

Scheme 4

alkyl, hydrogen). (*S*)-BINAL-H [(*S*)-(**28**)] generally affords the carbinol (**44**), in which the (*S*)-enantiomer predominates, whereas (*R*)-(**28**) forms the (*R*)-antipode. Electronically, phenyl, alkenyl and alkynyl groups exert qualitatively the same directing influence in the creation of the new asymmetric centers, but steric factors are also of some significance.[34]

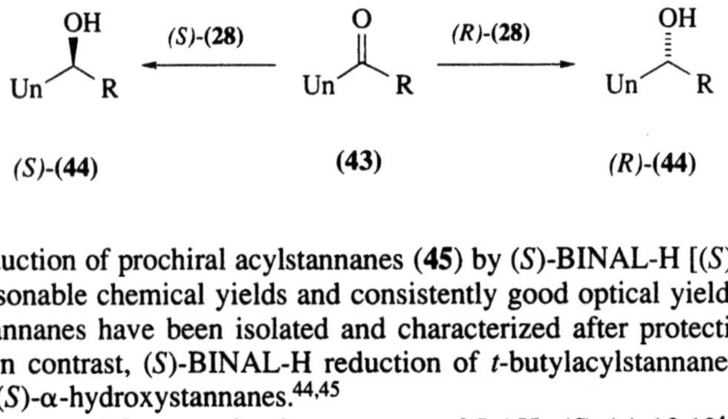

(*S*)-(**44**) ← (*S*)-(**28**) — (**43**) — (*R*)-(**28**) → (*R*)-(**44**)

Enantioselective reduction of prochiral acylstannanes (**45**) by (*S*)-BINAL-H [(*S*)-(**28**)] gives (*R*)-α-hydroxystannanes in reasonable chemical yields and consistently good optical yields (up to 96% *ee*). The unstable α-hydroxystannanes have been isolated and characterized after protection as methoxymethyl (MOM) ethers (**46**). In contrast, (*S*)-BINAL-H reduction of *t*-butylacylstannanes and α,β-unsaturated acylstannanes affords (*S*)-α-hydroxystannanes.[44,45]

The complex (**47**), prepared from equimolar amounts of LAH, (*S*)-(–)-10,10′-dihydroxy-9,9′-biphenanthryl (**11**) and ethanol, reduces aryl alkyl ketones to give the (*S*)-alcohols in 97–98% *ee* (Scheme 5). As with (**28**), the relationship between the chirality of the reagent and the product is (*S*) to (*S*) and (*R*) to (*R*).[46]

1.7.2.1.2 Dialkylamino alcohol modifiers

The reduction of acetophenone with a 1:1 complex of LAH and quinine (**12**), a β-dialkylamino alcohol, affords the (*R*)-carbinol in 48% *ee*.[47] Asymmetric induction in this chirally modified LAH system is enhanced by the presence of a suitably positioned tertiary nitrogen atom along with a chiral carbinol center in the ligand. A γ-dialkylamino alcohol (**13**), from which the analgesic Darvon is made, is an effective chiral modifier for LAH. In this system an aging effect is observed with respect to the sense and degree of asymmetric induction. A reagent freshly prepared from LAH and Darvon alcohol (**13**) in a 1:2.3 ratio reduces acetophenone to give the (*R*)-alcohol in 75% *ee*, while the reagent aged by refluxing in ether for 3 min gives rise to the (*S*)-alcohol in 75% *ee*. The structure of the fresh reagent was postulated to be (**48**). Scheme 6 illustrates the enantiomeric excesses and the absolute configurations of the

(45) **(46)**

R = Me, 94% *ee*
R = Et, 96% *ee*
R = n-C_5H_{11}, 91% *ee*
R = Pr^i, 96% *ee*
R = Bu^t, 80% *ee (S)*

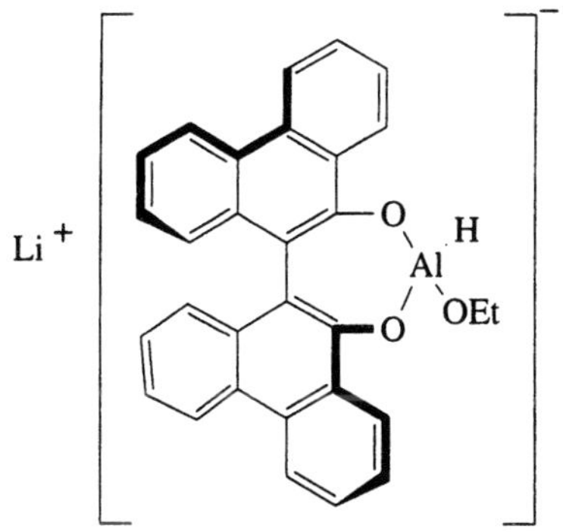

(S)-**(47)**

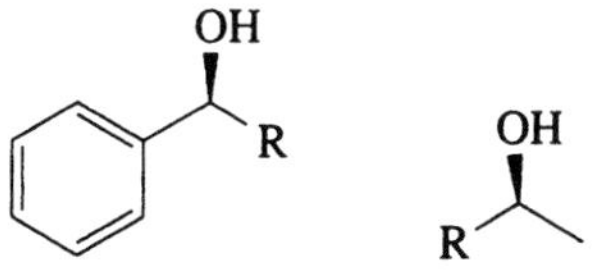

R = Me, 97% *ee*
R = Et, 98% *ee*
R = CH_2Ph, 98% *ee*
R = D, 87% *ee*

R = CH_2Ph, 33% *ee*
R = Bu^i, 21% *ee*

Scheme 5

major alcohols obtained with this reagent for a series of aryl and alkynyl ketones,[48,49] and a single example of an enone.[50]

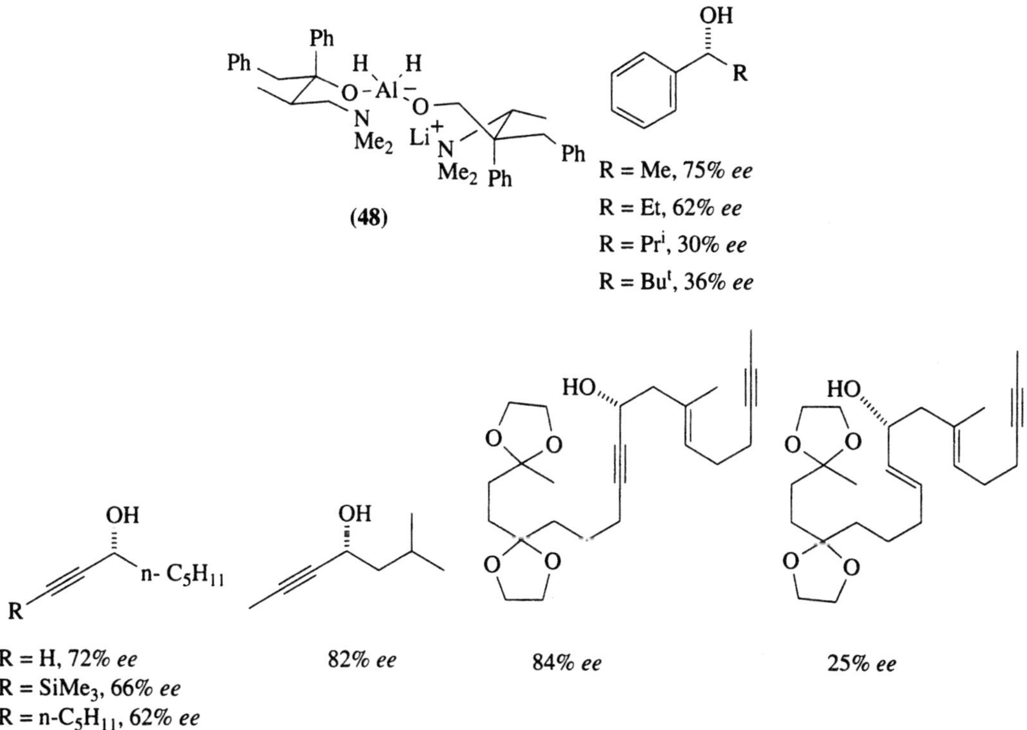

Scheme 6

An asymmetric reducing agent (**49**), prepared from LAH, (–)-*N*-methylephedrine (**14**) and 3,5-dimethylphenol in a 1:1:2 ratio, reduces acetophenone with 83.8% *ee*.[51,52] A series of aromatic and alkynyl ketones are reduced by (**49**) to the corresponding (*R*)-alcohols (Scheme 7).[53] Some of the products are useful intermediates for the synthesis of γ-lactone insect pheromones.[54]

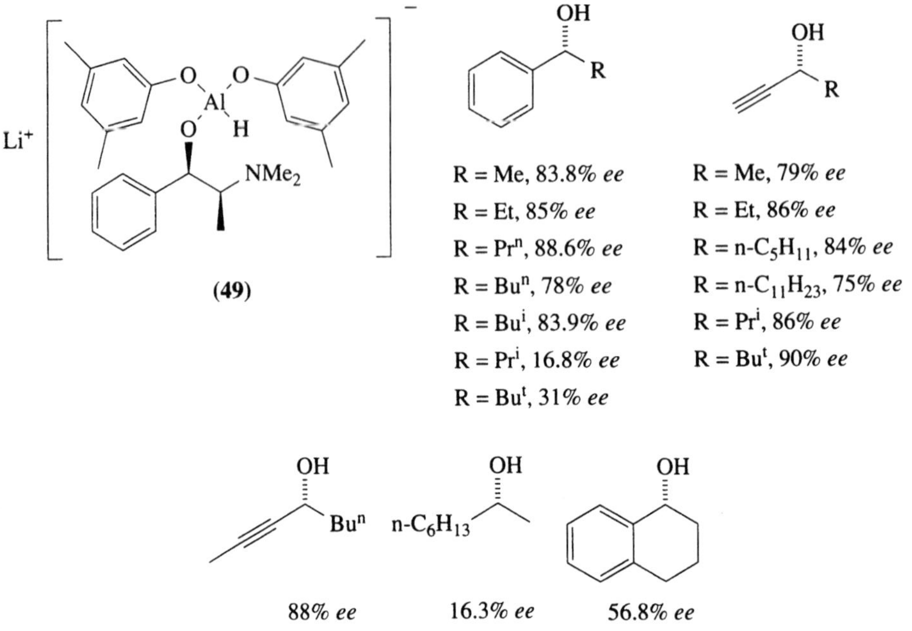

Li^+ (**49**)

R = Me, 83.8% *ee*
R = Et, 85% *ee*
R = Pr^n, 88.6% *ee*
R = Bu^n, 78% *ee*
R = Bu^i, 83.9% *ee*
R = Pr^i, 16.8% *ee*
R = Bu^t, 31% *ee*

R = Me, 79% *ee*
R = Et, 86% *ee*
R = n-C_5H_{11}, 84% *ee*
R = n-$C_{11}H_{23}$, 75% *ee*
R = Pr^i, 86% *ee*
R = Bu^t, 90% *ee*

88% *ee* 16.3% *ee* 56.8% *ee*

Scheme 7

Modification of LAH with (–)-*N*-methylephedrine (**14**) and *N*-ethylaniline affords another chiral reducing agent (**50**).[55–61] A variety of aromatic ketones are reduced by (**50**) to (*S*)-carbinols in high optical yields (Scheme 8). Acyclic α,β-unsaturated ketones are also reduced to (*S*)-carbinols with high selectivity (76–92% *ee*), but cyclic enones are reduced with only moderate selectivity.

The pyridyl analog (**51**) leads to the opposite asymmetric orientation in most cases. Asymmetric reduction of 2-cyclohexenone affords (*R*)-2-cyclohexenol with 98% *ee*. Other cyclic ketones are also converted to (*R*)-carbinols in high optical yields (Scheme 9), but the reduction of acyclic ketones is only moderately stereoselective.[56]

Generally chiral auxiliaries are liberated unchanged during or after the reaction, but attachment to an insoluble polymer facilitates recovery and reuse. This 'purification advantage', a main asset of polymer-assisted reactions, as developed by Merrifield in the early 1960s,[62] was first introduced to the chiral modification of LAH using ephedrine and following a procedure reported by Vigneron.[52] (–)-Ephedrine is immobilized on 1% cross-linked polystyrene, mixed with LAH, (**15**) and 2 equiv. of 3,5-dimethylphenol to afford the polymer-supported chiral reducing agent (**52**). The degree of functionalization is an important factor in achieving effective chiral reduction, and using 0.7 mmol g^{-1} of polymer capacity 78.8% *ee* was obtained in the reduction of acetophenone at –15 °C.[63] The direction of asymmetric induction and optical yields are comparable with those of the original solution phase reaction.[52]

Asymmetric reductions with LAH modified with (**16**) and (**17**) are known, but the optical yields of the 1-phenylethyl alcohol (**2**) were only moderate. A C_2 dissymmetrical bis(dialkylamino) diol (DBD, **18**) and its enantiomer are available from diethyl tartrates. The (*S*)-chiral complex (**53**) of this diol reduces a variety of aryl alkyl ketones to the corresponding (*S*)-secondary alcohols in moderate optical purity in each case (Scheme 10).[64]

A LAH complex (**54**) prepared from 2.28 equiv. of the amino diol (**19**) reduces acetophenone and propiophenone to the corresponding (*R*)-alcohols with 82% *ee* and 77% *ee*, respectively.[65] The key feature of this carbinolamine-modified LAH reagent is a lithium ion chelate.

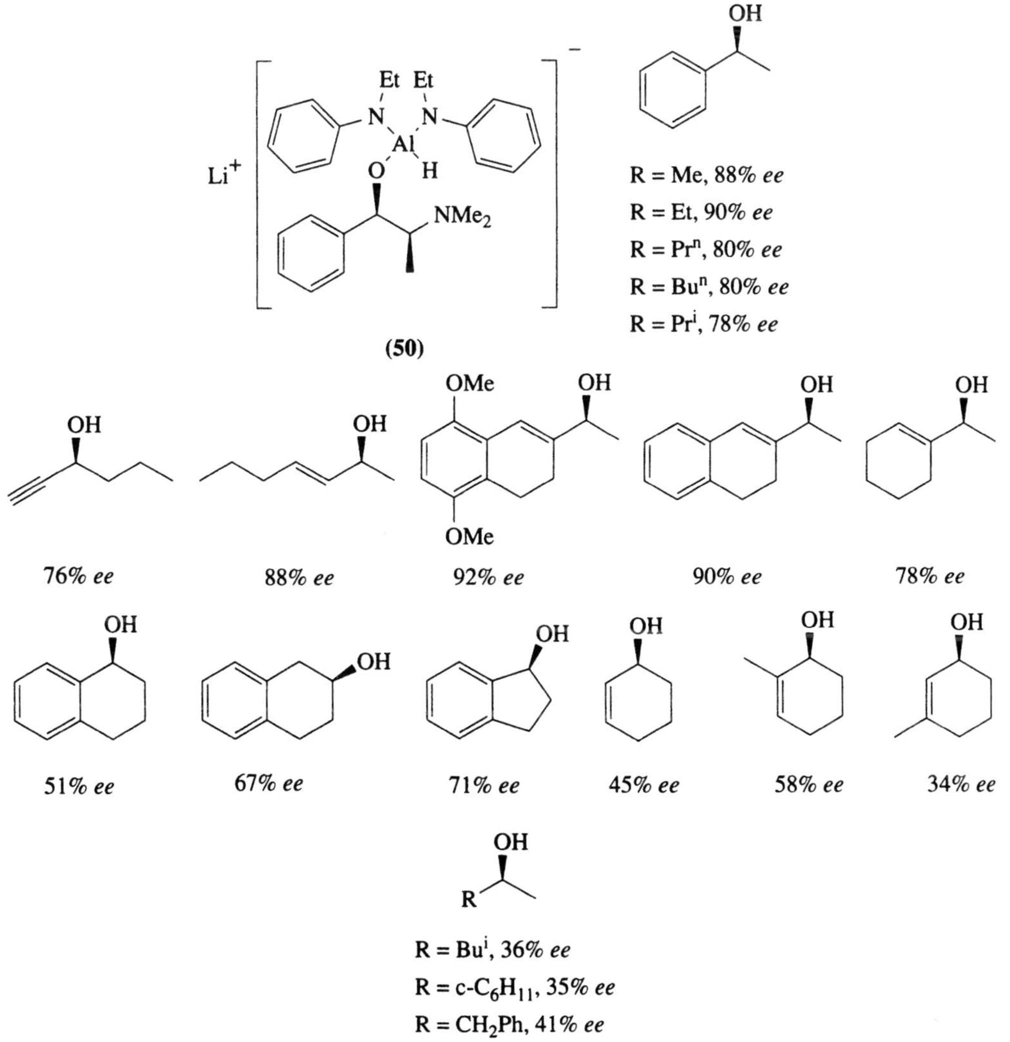

(50)

R = Me, 88% *ee*
R = Et, 90% *ee*
R = Pr^n, 80% *ee*
R = Bu^n, 80% *ee*
R = Pr^i, 78% *ee*

76% *ee* 88% *ee* 92% *ee* 90% *ee* 78% *ee*

51% *ee* 67% *ee* 71% *ee* 45% *ee* 58% *ee* 34% *ee*

R = Bu^i, 36% *ee*
R = c-C_6H_{11}, 35% *ee*
R = CH_2Ph, 41% *ee*

Scheme 8

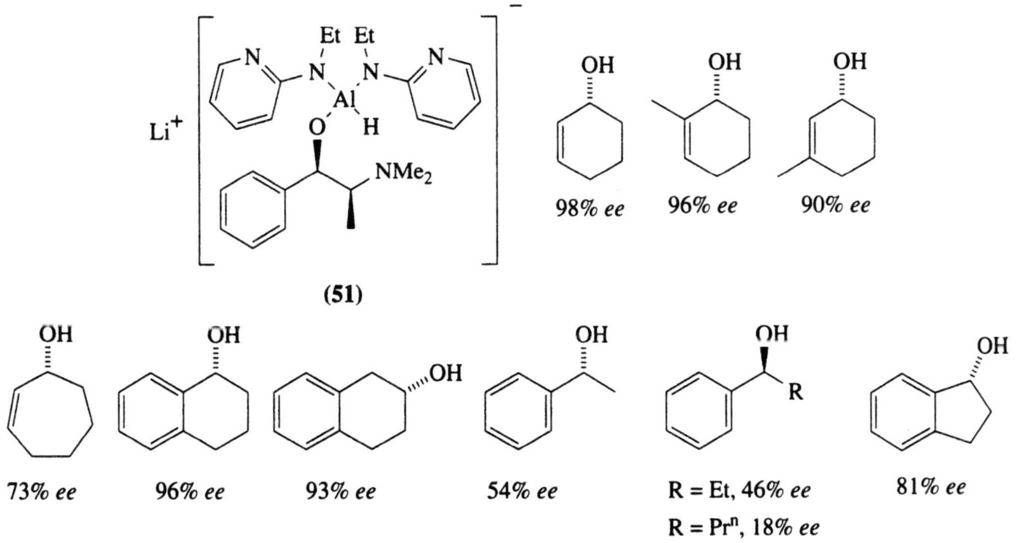

(51)

98% *ee* 96% *ee* 90% *ee*

73% *ee* 96% *ee* 93% *ee* 54% *ee*

R = Et, 46% *ee*
R = Pr^n, 18% *ee*

81% *ee*

Scheme 9

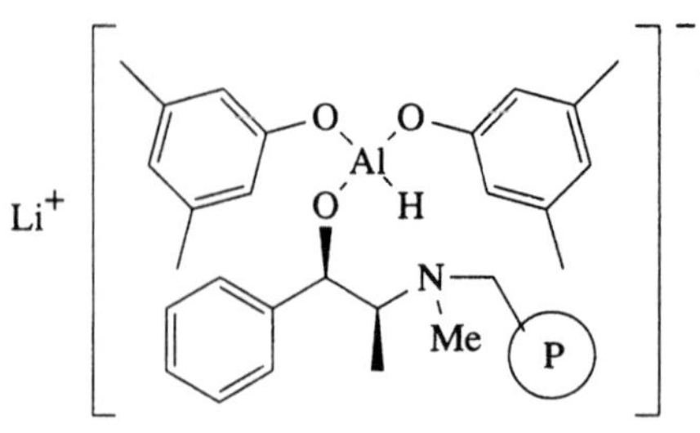

Li$^+$

(52)

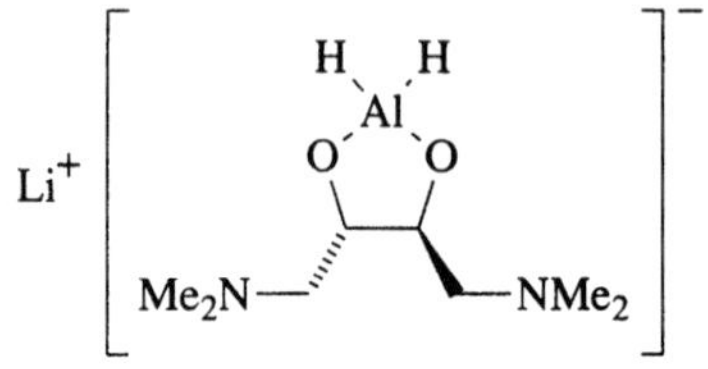

Li$^+$

(53)

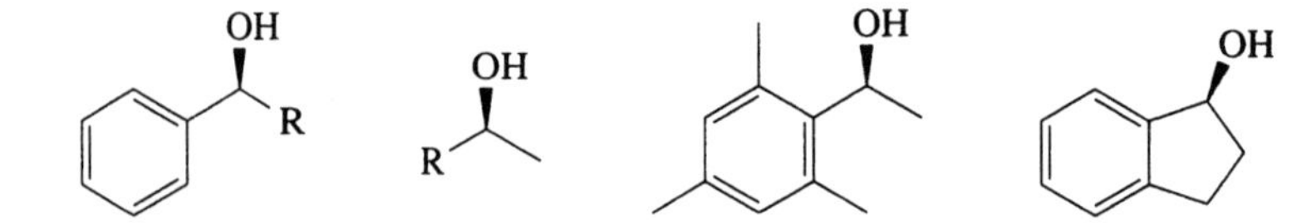

R = Me, 42% *ee*
R = Et, 44% *ee*
R = Pr^n, 47% *ee*
R = Pr^i, 27% *ee*
R = c-C_6H_{11}, 53% *ee*
R = Bu^t, 21% *ee*

R = Et, 2.4% *ee*
R = Bu^i, 16% *ee*

75% *ee*

15% *ee*

Scheme 10

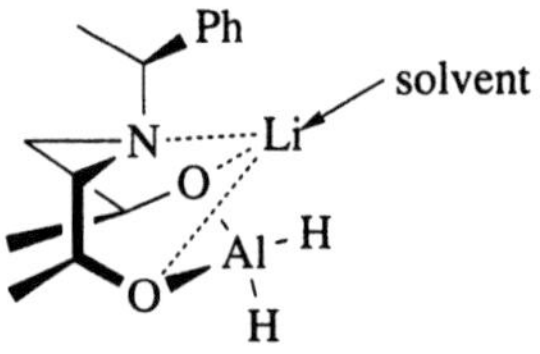

(54)

1.7.2.1.3 Primary or secondary amine and amino or monoalkylamino alcohol modifiers

A highly efficient asymmetric reducing agent (**55**) is prepared by reaction of LAH and the β-diamine (**20**) derived from (*S*)-proline in five steps. Acetophenone is reduced by (**55**) at –100 °C to give the (*S*)-alcohol in 95% *ee*. Some other aromatic ketones also give high optical yields (Scheme 11), but this procedure is not effective for prochiral dialkyl ketones. The high stereodifferentiation has been interpreted in terms of the bicyclo[3.3.0] rigid structure of (**55**), where only one hydrogen is responsible for the reduction. The presence of additives such as tetramethylethylenediamine, DME or $MgBr_2$ reduces the optical yield, indicating that the lithium cation plays an important role.[66–68]

(*S*)-4-Anilino-3-methylamino-1-butanol (**22a**) and (*S*)-4-(2,6-xylidino)-3-methylamino-1-butanol (**22b**), tridentate chiral auxiliaries, are easily derived from (*S*)-aspartic acid. A complex (**56a**) prepared from equimolar amounts of LAH and (**22a**) in THF reduces aromatic ketones to (*S*)-carbinols in 51–88% *ee*. This complex is also effective in reducing several enones to (*S*)-allylic alcohols (Scheme 12), whereas the complex (**56b**) affords the (*R*)-enantiomers. Particularly noteworthy is the virtually complete enantiofacial differentiation of cyclohexenone giving (*S*)-2-cyclohexenol.[69,70]

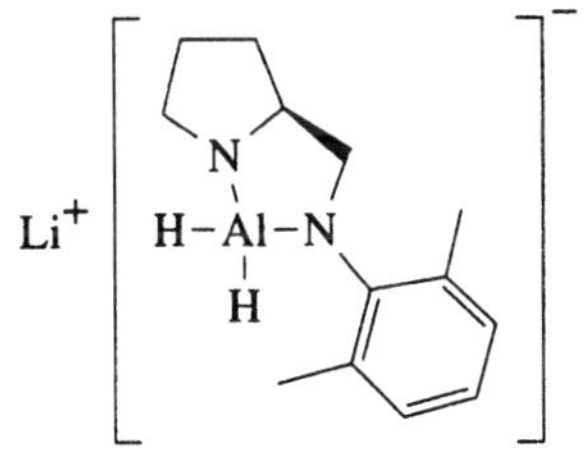

Li$^+$ H–Al–N H N

(55)

OH R OH R OH

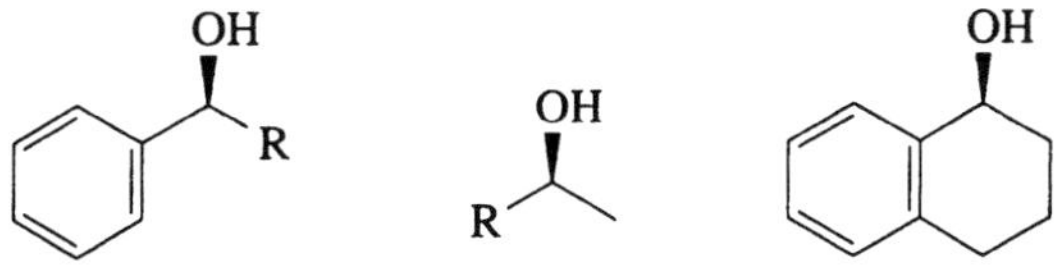

R = Me, 95% *ee*
R = Et, 96% *ee*
R = Pr^i, 89% *ee*

R = n-C_6H_{13}, 26% *ee*
R = CH_2Ph, 42% *ee*

86% *ee*

Scheme 11

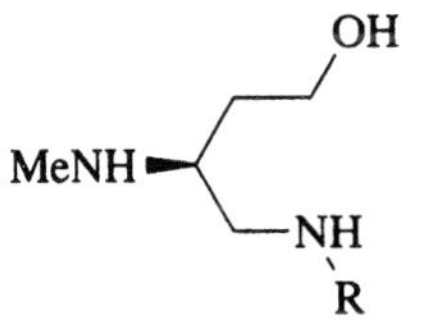

OH MeNH NH R

(22a) R = Ph
(22b) R = 2,6-$Me_2C_6H_3$

Li$^+$ N O Me Al H N R

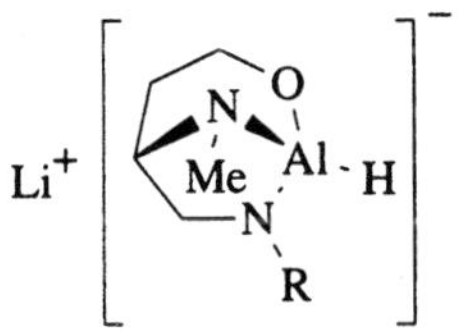

(56a) R = Ph
(56b) 2,6-$Me_2C_6H_3$

From reduction with **(56a)**

OH R OH

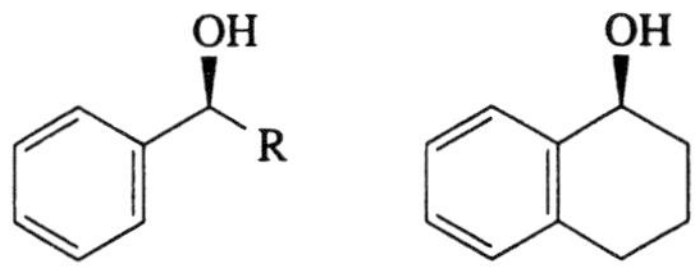

R = Me, 51% *ee*
R = Et, 68% *ee*
R = Pr^i, 77% *ee*
R = Bu^t, 86% *ee*

88% *ee*

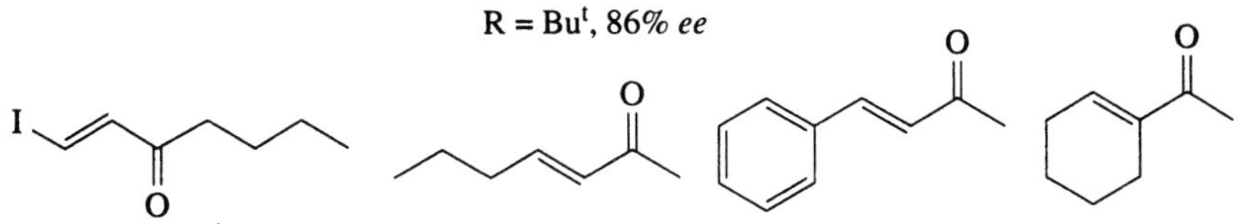

I O O O O

51% ee, (S)
21% ee, (R)

88% ee, (S)
84% ee, (R)

72% ee, (S)
42% ee, (R)

5% ee, (S)
61% ee, (R)

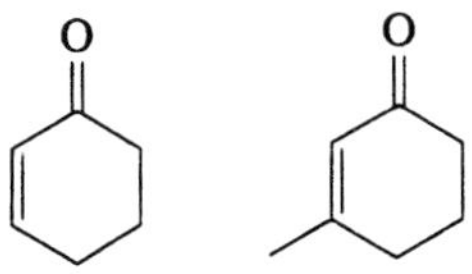

O O

100% ee, (S)
13% ee, (R)

28% ee, (S)
2% ee, (S)

(S)-products from reduction with **(56a)**
(R)-products from reduction with **(56b)**

Scheme 12

Reduction of acetophenone with the LAH complexes of nitrogen analogs of binaphthol, (*R*)-(**23**)[71] and (*S*)-(**24**),[72] affords (*R*)-1-phenylethyl alcohol in 43% *ee* and 46% *ee*, respectively. Optical yields for the reduction of other prochiral ketones are similarly moderate.

1.7.2.2 Asymmetric Reduction with Borohydride Reagents

When borohydride reductions are carried out in the presence of either a chiral phase transfer catalyst or a chiral crown ether, asymmetric reduction of ketones occurs but optical yields are low. In the reduction of acetophenone with $NaBH_4$ aided with a phase transfer catalyst (**57**), 10% *ee* was obtained.[73–78] Similarly, reduction of acetophenone with $NaBH_4$ in the presence of the chiral crown ether (**58**) was ineffective (6% *ee*).[79,80] Sodium borohydride reduction of aryl alkyl ketones in the presence of a protein, bovine serum albumin, in 0.01 M borax buffer at pH 9.2 affords (*R*)-carbinols in maximum 78% *ee*.[81]

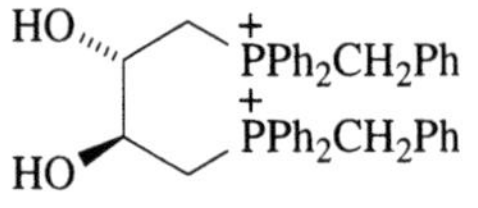

(**57**) 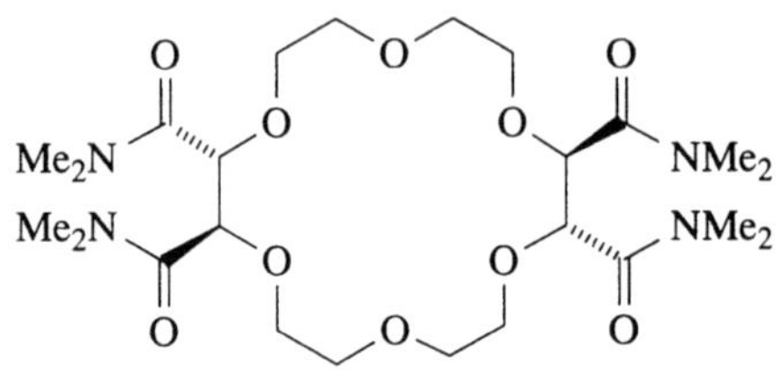 (**58**)

Partial decomposition of $NaBH_4$ or $LiBH_4$ by chiral protic substances such as R*OH or RR*NH generates modified borohydride reagents. When aromatic ketones are reduced by $NaBH_4$ modified with the sugar-derived monoalcohols (**59**) or (**60**), moderate optical yields (6–35% *ee*) result.[82,83] Modification of $NaBH_4$ with the sugar (**60**) and 2-methylpropionic acid permits the reduction of propiophenone to give the (*R*)-alcohol in 63% *ee*.[84] Racemic 2-phenylbutyric acid is also an effective comodifier affording 1-phenylpropanol in 51% *ee*.[85]

Lithium borohydride decomposed by *N*-benzoylcysteine (**61**) or *N,N'*-dibenzoylcystine (**62**), a sulfur-containing modifier, is a highly efficient chiral reducing agent. A complex prepared from (**61**), *t*-butyl alcohol and $LiBH_4$ affords carbinols in maximum 92% *ee* by the reduction of aryl alkyl ketones in THF at –78 °C (Scheme 13). A $LiBH_4$ complex with (**62**) and *t*-butyl alcohol is useful for the reduction of β-keto esters to give (*R*)-β-hydroxy esters in up to 91% *ee*. In both cases the use of *t*-butyl alcohol is essential in order to achieve efficient enantiofacial differentiation.[86–88]

A highly promising chiral reducing agent, K-glucoride (**64**), is prepared from 9-borabicyclo[3.3.1]nonane (9-BBN) and a D-glucose derivative (**59**) *via* the stable borinic ester (**63**). Reduction of aryl alkyl ketones with (**64**) in THF at –78 °C affords (*R*)-carbinols in high optical yields (78–97% *ee*). Although the enantioselectivity of the reduction of acyclic saturated dialkyl ketones is not high, prochiral cyclic ketones are converted to (*R*)-alcohols with high optical purity (Scheme 14). This procedure is also applicable to the reduction of α-keto esters to (*S*)-α-hydroxy esters.[89,90]

1.7.2.3 Asymmetric Reduction with Chirally Modified Boranes and Alanes

Chiral modification is not limited to boronate and aluminate complexes. Boranes or alanes are partially decomposed with protic substances such as chiral amines, alcohols or amino alcohols to form useful reagents for enantioselective reduction of carbonyl compounds. For example, reduction of acetophenone with borane modified with the amines (**65**) to (**67**) gives (*S*)-1-phenylethyl alcohol with 3.5–20% *ee*.[91–93] A 1:1 mixture of sodium L-prolinate (**68**) and borane gives rise to carbinols with as much as 62% *ee*,[94] and the complex prepared by mixing borane, *m*-nitroaniline and (*S*)-(**69**) in a 1:1:1 ratio reduces acetophenone in 84% *ee*.[95]

Itsuno's amino alcohol (**70**), prepared from L-valine, is an extremely efficient auxiliary for enantioselective reduction of aryl alkyl ketones using BH_3. The corresponding alcohols are obtained in up to 100% *ee* using BH_3 and 0.5 equiv. of (**70**) in THF at 30 °C. Reduction of dialkyl ketones affords (*R*)-carbinols in 55–73% *ee*.[96–99] Halomethyl *t*-butyl ketones are also converted to the corresponding (*S*)-carbinols in high optical purity (Scheme 15). Immobilized amino alcohol (**70**) permits reduction in a continuous flow system. 1-Phenylpentanol of 90% *ee* was prepared by this catalytic process in almost 1000% chemical yield based on the quantity of chiral auxiliary used.[100,101]

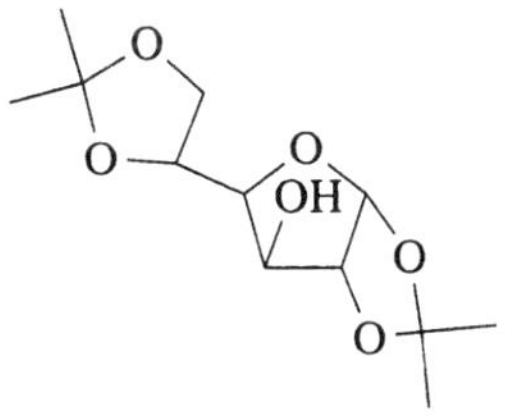

(59) 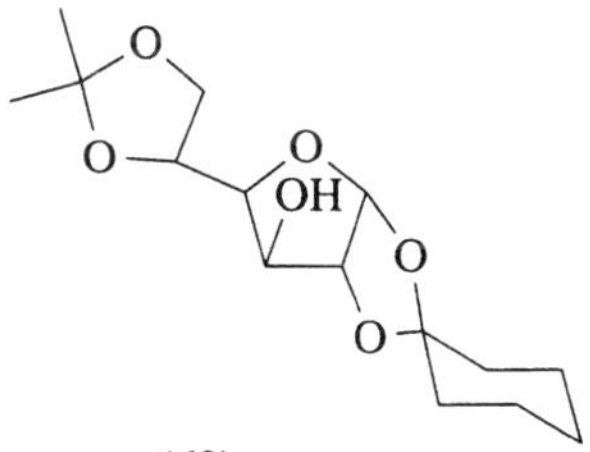 **(60)**

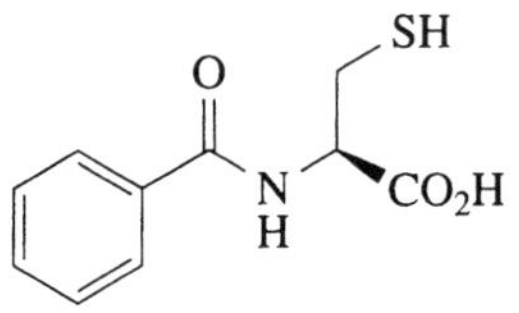

(61) 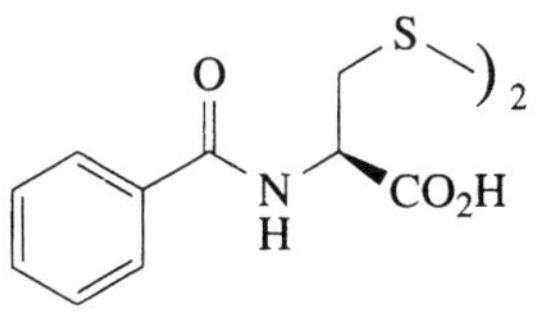 **(62)**

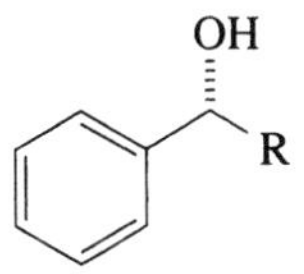

R = Me, 87% *ee*
R = Et, 89% *ee*
R = Pr^n, 92% *ee*
R = Bu^n, 88% *ee*
R = Pr^i, 57% *ee*

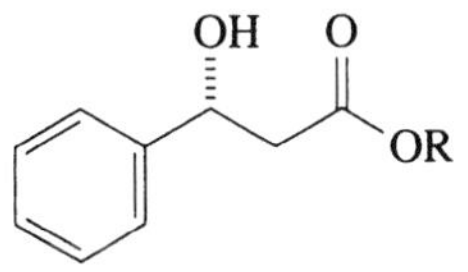

R = Me, 84% *ee*
R = Et, 87% *ee*
R = Bu^n, 80% *ee*
R = n-C_6H_{13}, 86% *ee*
R = Pr^i, 91% *ee*
R = Bu^t, 90% *ee*

Scheme 13

Borane complex catalyzed reductions have been greatly extended by Corey and coworkers,[102] who found that oxazaborolidine (**73**) could be prepared as crystals from (*S*)-2-(diphenylhydroxymethyl)pyrrolidine by heating with BH_3·THF followed by sublimation, while *B*-methylated oxazaborolidine (**74**), prepared from (**72**) and methylboronic acid, is stable enough to store in closed containers at room temperature and can be weighed or transferred in air.[104] Borane reduction of prochiral ketones in the presence of 0.05 to 0.1 equiv. of (**73**) or (**74**) generated both unsaturated and saturated carbinols with high enantioselectivity. The utility of these reagents is shown by some simple examples in Scheme 16 and by the effective chiral syntheses of compounds (**78**) to (**80**), which are intermediates in the syntheses of ginkgolide B,[106] forskolin[105] and prostaglandins,[103] respectively. Moreover, ^{11}B NMR and IR analysis of (**73**) revealed that it is a mixture of monomer and dimer and has double bond character between the B and N atoms. A 1:1 borane/oxazaboridine (**75**) is well set up to serve as an effective reagent for carbonyl reduction. Coordination of the ketone oxygen to the electrophilic ring boron atom promotes hydride transfer from the NBH_3^- unit *via* a six-membered cyclic transition state (**76**). The resulting borate (**77**) reacts with BH_3, regenerating the active species (**75**), thus completing a catalytic cycle.[102]

Masamune synthesized the chiral borane (**81**), possessing a chiral organic backbone, by the reaction of dihydridoborate (**82**) and methanesulfonic acid.[107] The former is obtained by reaction of (diethylamino)dichloroborane and the Grignard reagent derived from 2,5-dibromohexane, followed by resolution. Enantioselective reduction of prochiral aliphatic ketones has been accomplished using (*R,R*)-(**81**), generating (*R*)-carbinols in 79–98.4% *ee* (Scheme 17).[108,109] A borolanyl mesylate (**83**), generated *in situ* along with (**81**), plays a catalytic role by coordination with the ketone oxygen.

The sole example of a chiral alane reagent is dialkylaminoalane (**85**), prepared by the reaction of LAH and (*S*)-*N*-methyl-*N*-phenylethylamine hydrochloride (**84**). The reduction of acetophenone gives the (*S*)-alcohol in 84.5% *ee*.[110]

K+

(63) **(64)**

OH, R

R = Me, 78% *ee*
R = Et, 92% *ee*
R = Pr[n], 87% *ee*
R = Bu[n], 85.4% *ee*
R = Pr[i], 87% *ee*
R = Bu[t], 97% *ee*

OH, R

R = Et, 3% *ee*
R = n-C_6H_{13}, 27% *ee*
R = Pr[i], 39% *ee*
R = c-C_6H_{11}, 23% *ee*
R = Bu[t], 70% *ee*

84% *ee* 82% *ee* 64% *ee*

70% *ee* 42% *ee* 42% *ee*

RO_2C, OH

R = Me, 86% *ee*
R = Et, 86% *ee*
R = Pr[i], 87% *ee*
R = Bu[t], 81% *ee*

EtO_2C, OH, R

R = Et, 92% *ee*
R = Pr[n], 94% *ee*
R = Bu[i], 93% *ee*
R = Pr[i], 97% *ee*
R = Bu[t], 98% *ee*
R = Ph, 94% *ee*

Scheme 14

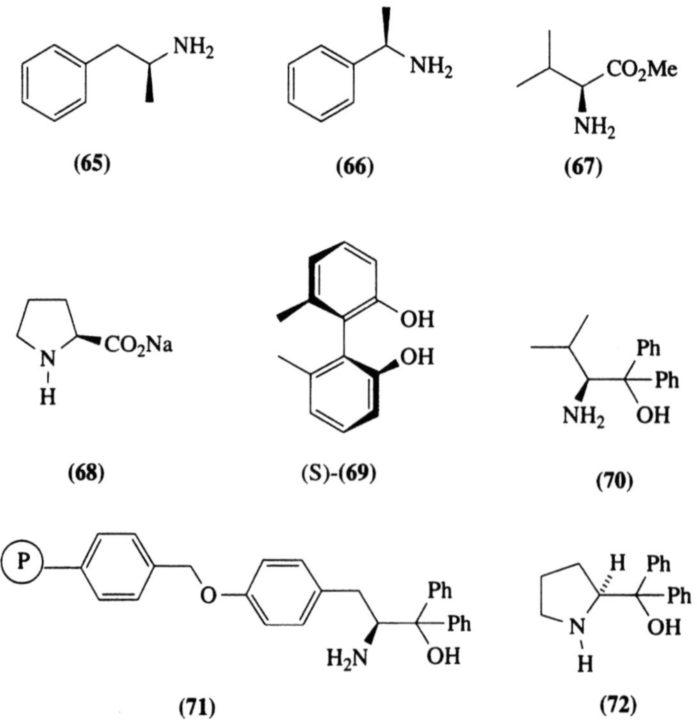

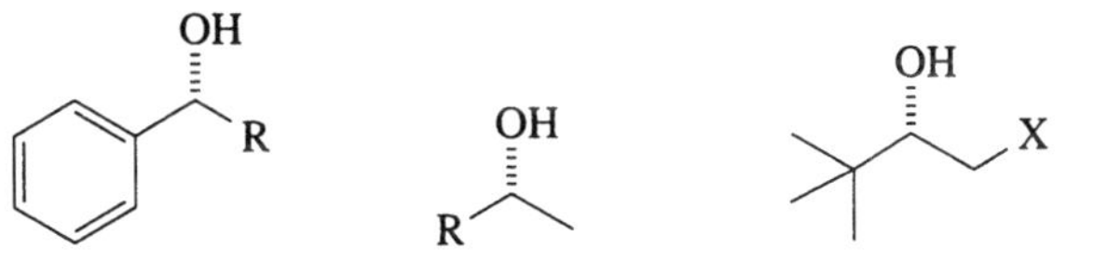

R = Me, 94% *ee*
R = Et, 94% *ee*
R = Pr^n, 96%*ee*
R = Bu^n, 100% *ee*
R = CH_2Cl, 96% *ee*
R = CH_2Br, 83% *ee*

R = Bu^n, 55% *ee*
R = n-C_5H_{11}, 56% *ee*
R = n-C_6H_{13}, 58% *ee*
R = Bu^i, 61% *ee*
R = Pr^i, 60% *ee*
R = Bu^t, 73% *ee*

X = Cl, 90% *ee*
X = Br, 93% *ee*

Scheme 15

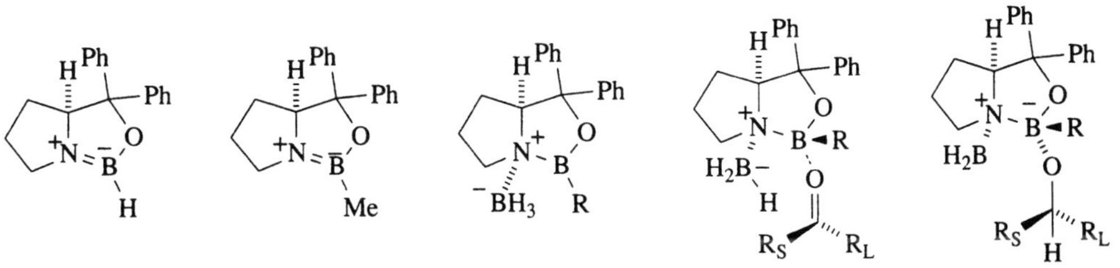

(73) **(74)** **(75)** **(76)** **(77)**

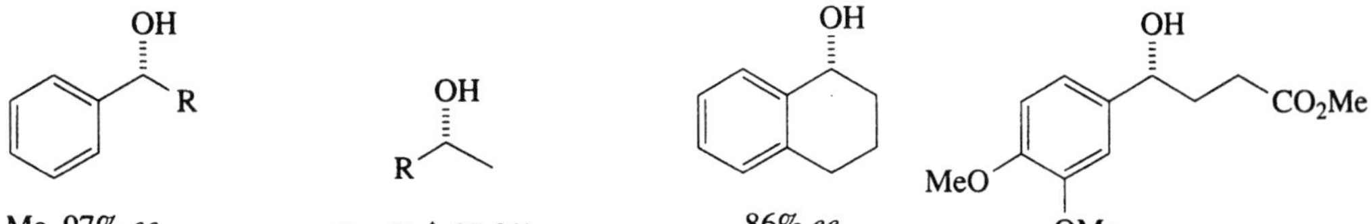

R = Me, 97% *ee*
R = Et, 96.7% *ee*
R = CH_2Cl, 95.3% *ee*

R = Bu^t, 97.3% *ee*
R = c-C_6H_{11}, 84% *ee*

86% *ee*

95% *ee*

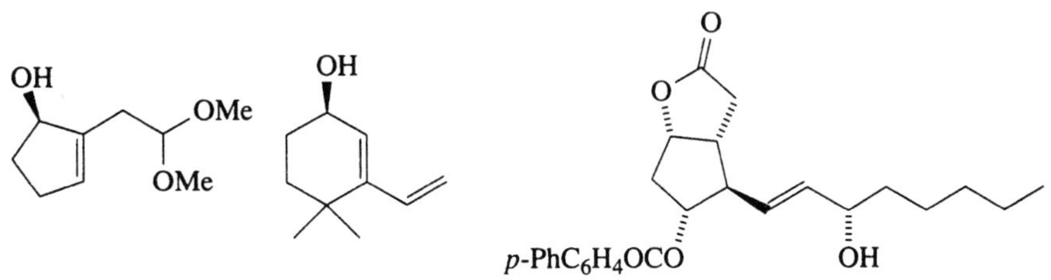

(78) 93% *ee* **(79)** 90% *ee*

(80) (15*S*):(15*R*) = 91:9

Scheme 16

1.7.2.4 Asymmetric Hydrosilylation

Addition of the elements of Si—H to a carbonyl group produces silyl ethers which are synthetically equivalent to chiral secondary alcohols since the silyl groups are easily hydrolyzed. Hydrosilylation can be catalyzed by acids or transition metal complexes. Enantioselective hydrosilylation of prochiral ketones has been extensively studied using platinum or rhodium complexes possessing chiral ligands such as BMPP (**86**), DIOP (**87**), NORPHOS (**88**), PYTHIA (**89**) and PYBOX (**90**).[111]

Hydrosilylation of aryl alkyl ketones with trimethylsilane, catalyzed by a [$PtCl_2${(*R*)-BMPP}] complex, gave (*R*)-silyl ethers (**91**) with low enantiomeric excesses,[112,113] but a combination of dimethylphenylsilane and a Rh^0/(*S*)-BMPP complex gave (*R*)-silyl ethers (**92**) in 44–56% *ee*.[114–117] Use of a [{Rh(COD)Cl}$_2$]/(–)-DIOP complex and α-naphthylphenylsilane[118] is effective for the reaction of α-keto esters, giving (*R*)-α-hydroxy esters (**93**).[119,120]

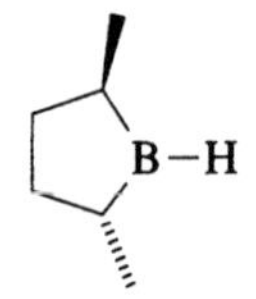

(R,R)-(**81**) 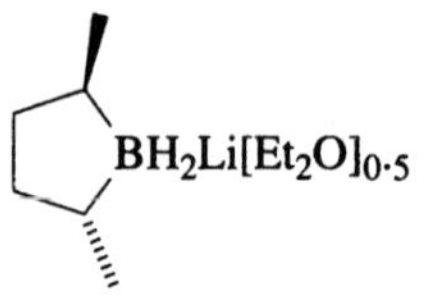 (**82**) 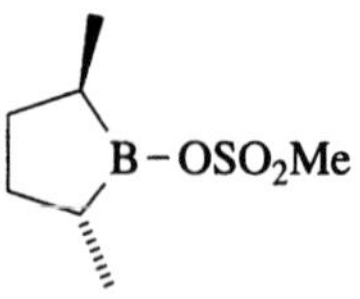 (**83**)

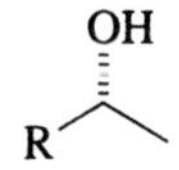

R = Et, 79% *ee*
R = n-C_6H_{13}, 80% *ee*
R = CH_2Ph, 97.3% *ee*
R = CH_2Bu^t, 98.4% *ee*
R = Bu^i, 97% *ee*
R = Pr^i, 98.4% *ee*
R = c-C_6H_{11}, 98% *ee*
R = Bu^t, 98.1% *ee*

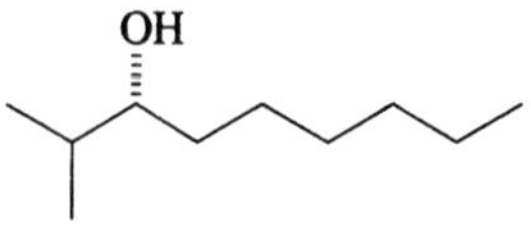

91% *ee*

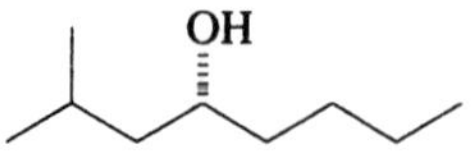

95.2% *ee*

Scheme 17

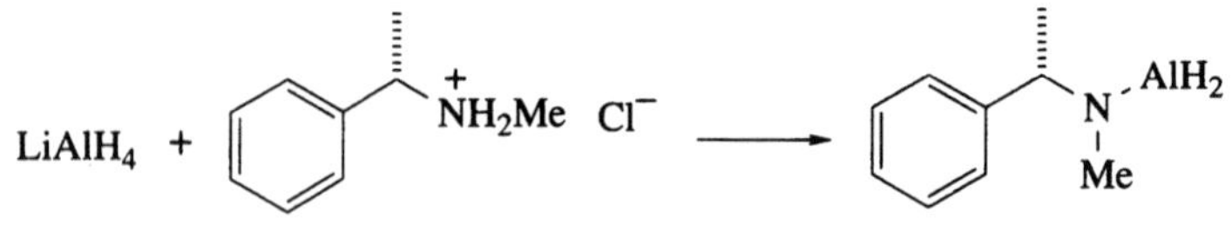

$LiAlH_4$ + (**84**) → (**85**)

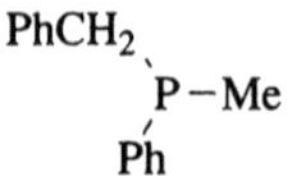

BMPP (**86**) 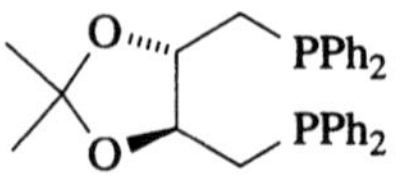 DIOP (**87**)

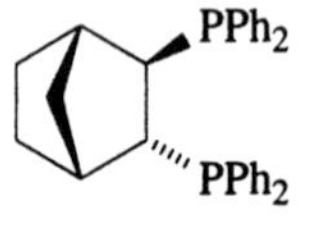

NORPHOS (**88**) 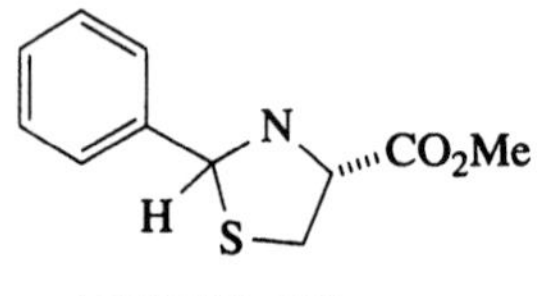 PYTHIA (**89**) 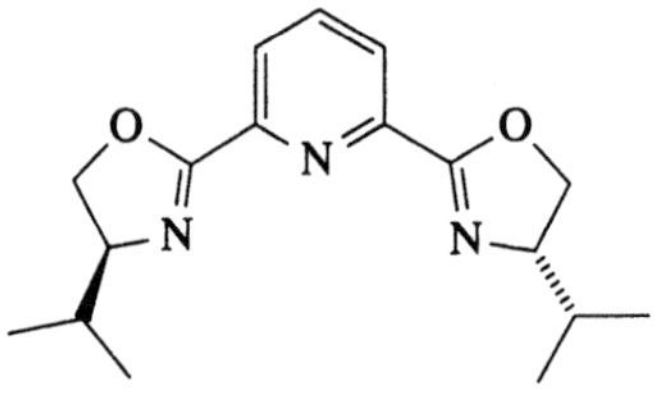 PYBOX (**90**)

Hydrosilylation of acetophenone to give silyl ether (**94**; R = Me) can also be achieved using copper(I) complexes with the chiral phosphine ligands (–)-DIOP (**87**) or (+)-NORPHOS (**88**). The enantioselectivity is rather low,[121] but a nonphosphine auxiliary, PYTHIA (**89**) with a $Rh(COD)Cl_2$ catalyst using neat diphenylsilane reduces aryl ketones to (*R*)-1-phenylethyl alcohol silyl ethers in high yield and with high enantiomeric excess (Scheme 18).[122–125]

The C_2 symmetrical bis(oxazolinylpyridine), PYBOX (**90**), also serves as an excellent ligand. Treatment of (**90**) with $[RhCl_3(H_2O)_3]$ in ethanol affords the complex (**95**), which has been characterized by single crystal X-ray analysis. Hydrosilylation of ketones with diphenylsilane (1.5 equiv.) in the presence of $AgBF_4$ (0.2 equiv.), (**90**) (0.06 equiv.) and (**95**) (0.01 equiv.) at 0 °C gives (*S*)-alcohols of high enantiomeric excess.[126]

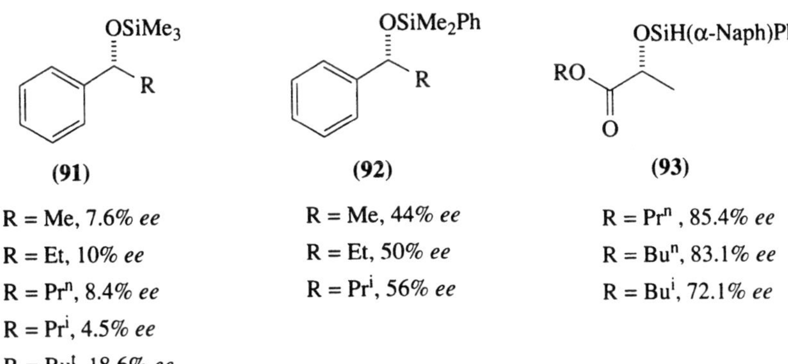

(91)

R = Me, 7.6% *ee*
R = Et, 10% *ee*
R = Pr^n, 8.4% *ee*
R = Pr^i, 4.5% *ee*
R = Bu^t, 18.6% *ee*

(92)

R = Me, 44% *ee*
R = Et, 50% *ee*
R = Pr^i, 56% *ee*

(93)

R = Pr^n, 85.4% *ee*
R = Bu^n, 83.1% *ee*
R = Bu^i, 72.1% *ee*

OSiHPh2 R OSiHPh2 R

(94)

R = Me, 86.7% *ee*
R = Et, 76.7% *ee*
R = Pr^n, 81.3% *ee*

R = 4-MeC_6H_4, 86.2% *ee*
R = 2-ClC_6H_4, 83.1% *ee*
R = 3-$MeOC_6H_4$, 91.1% *ee*
R = 4-$O_2NC_6H_4$, 61.2% *ee*

Ph_2HSiO — 83.3% *ee*

OSiHPh2 — 77.9% *ee*

OSiHPh2 — 82.6% *ee*

OSiHPh2 (N) — 89.6% *ee*

OSiHPh2 — 52.2% *ee*

OSiHPh2 — 55.8% *ee*

Scheme 18

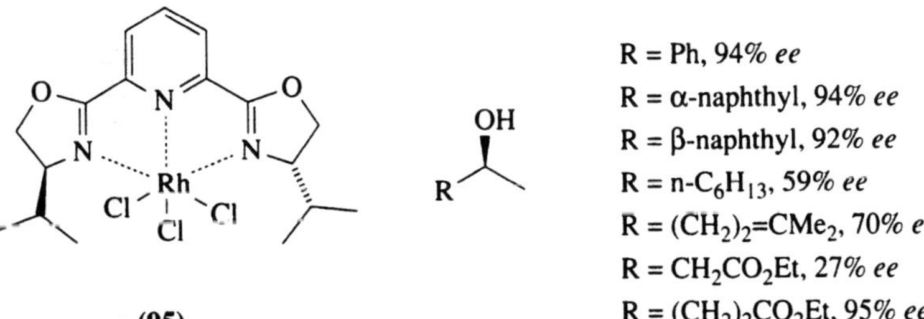

(95)

R = Ph, 94% *ee*
R = α-naphthyl, 94% *ee*
R = β-naphthyl, 92% *ee*
R = n-C_6H_{13}, 59% *ee*
R = $(CH_2)_2$=CMe_2, 70% *ee*
R = CH_2CO_2Et, 27% *ee*
R = $(CH_2)_2CO_2Et$, 95% *ee*

Scheme 19

1.7.2.5 Asymmetric Reduction of the C═N Double Bond

A general way to reduce a C═N bond to an optically active amine is still elusive. Attempted reduction, using a LAH/(–)-menthol reagent, of 2-substituted iminium salts (**96**, R = methyl, ethyl, propyl or benzyl)[127] or (**97**, R = methyl, ethyl, propyl or benzyl)[128] and imine (**98**)[129] gives the corresponding amines with low optical purities.

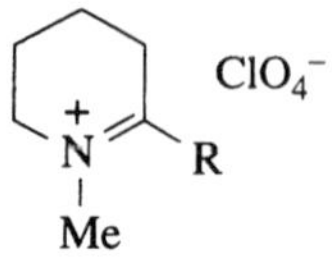

ClO_4^- ClO_4^- NH

(**96**) 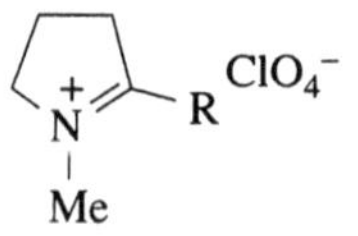 (**97**) 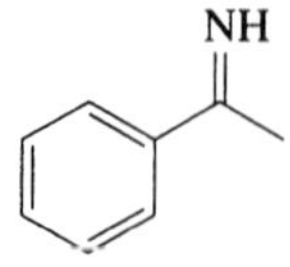 (**98**)

Reduction of oximes (**100**) and (**102**) with a LAH complex modified by a D-glucose derivative (**99**) affords (*S*)-primary amines (**101**) and (**103**) directly with 32% *ee* and 49% *ee*, respectively.[130,131]

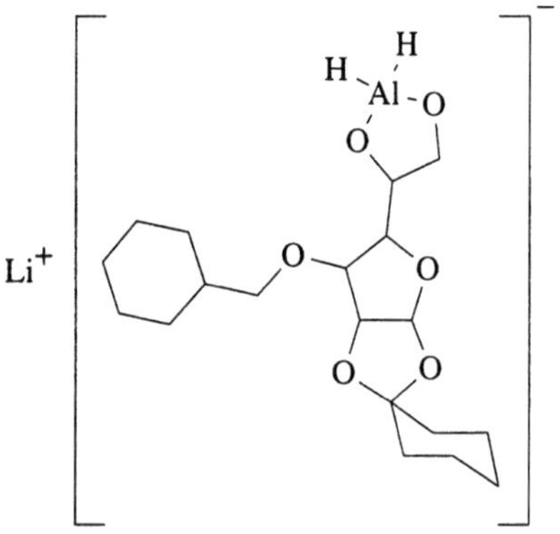

(**99**) 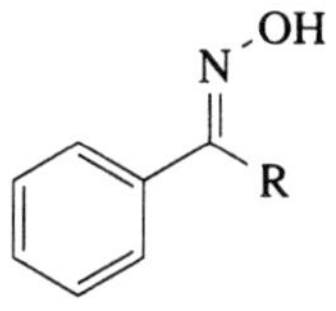 (**100**) 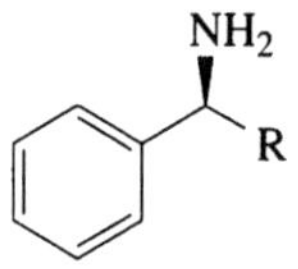 (**101**)

R = Me, 39.3% *ee*
R = Et, 45.7% *ee*

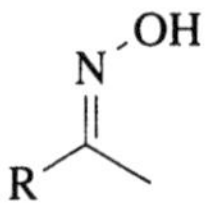

(**102**) 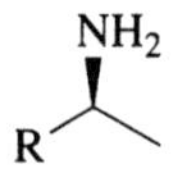 (**103**)

R = Et, 32.5% *ee*
R = n-C_6H_{13}, 49.3% *ee*
R = CH_2Ph, 43.6% *ee*
R = Pr^i, 44.8% *ee*

Sodium (*S*)-prolinate–borane complex (**104**) reduces cyclic imine (**105**) to give the (*S*)-tetrahydroisoquinoline derivative (**106**) in 87% *ee*. Related amines (**107**) and (**108**) are also prepared in a like manner in 86% *ee* and 71% *ee*, respectively.[132–134]

A high degree of asymmetric induction has been realized by reduction of *O*-methyloximes (**109**) to give the corresponding amines (**110**) using a borane reagent modified with (**70**). Acetophenone *O*-methyl oxime is converted to (*S*)-1-phenylethylamine in 99% *ee* after hydrolysis of the resulting methoxylamine derivative. Curiously, the asymmetric reduction of acetophenone with the same reagent induces the opposite chirality. The reduction of oxime *O*-methyl ethers derived from α-naphthyl methyl ketone occurs in only moderate optical yield.[135]

Asymmetric reduction of prochiral diphenylphosphinylimines (**111**) by (*S*)-BINAL-H (**28**), LAH–Darvon alcohol complex (**48**) or K-glucoride (**64**) affords the (*S*)-amines (**112**; Scheme 20). The protecting group of (**112**) is readily cleaved under acidic conditions to give the free amine. When the alkyl group is methyl or ethyl, (*S*)-BINAL-H shows high enantioselectivity.[136]

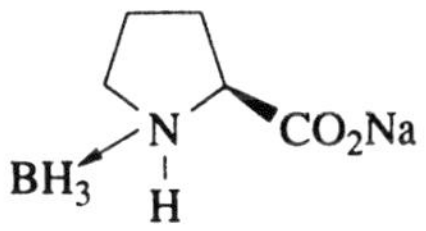

(104) 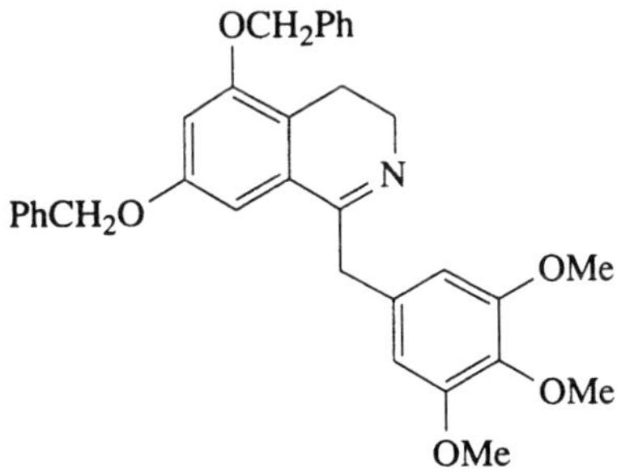 **(105)** 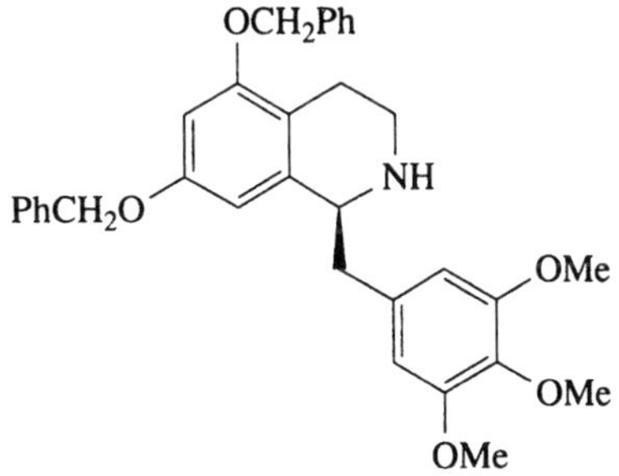 **(106)**

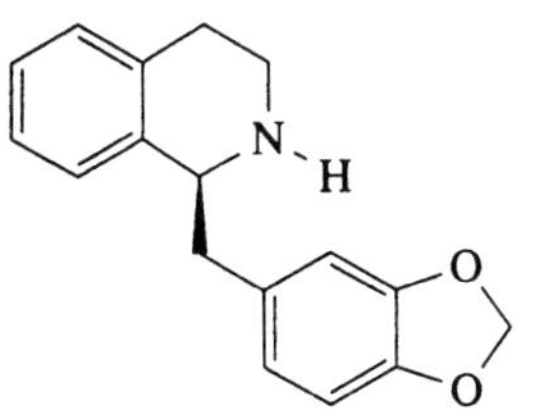

(107) 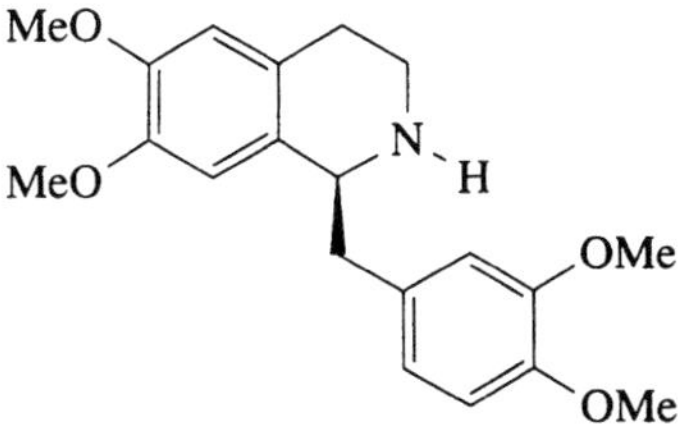 **(108)**

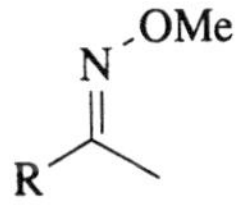

(109) 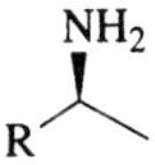 **(110)**

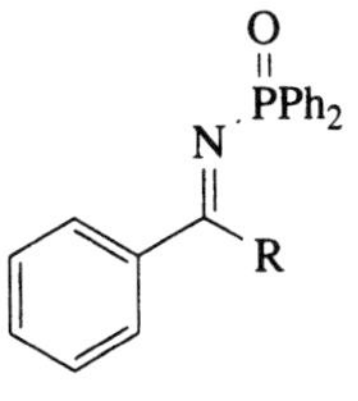

(111) 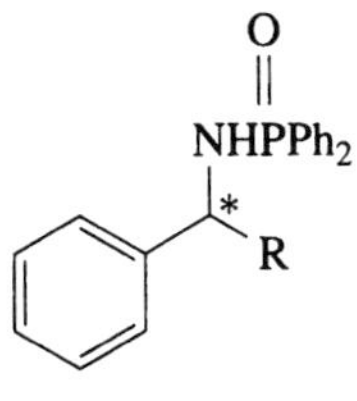 **(112)**

Reducing agent

R	(*S*)-**(28)**	**(48)**	**(64)**
Me	100% *ee*, (*S*)	41% *ee*, (*S*)	15% *ee*, (*R*)
Et	80% *ee*, (*S*)	17% *ee*, (*S*)	28% *ee*, (*S*)
Pr[i]	2% *ee*, –[a]	13% *ee*, –[a]	57% *ee*, –[a]

[a] Absolute configuration was not determined

Scheme 20

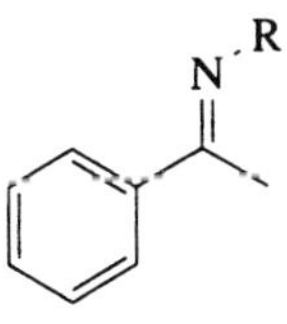

(113) 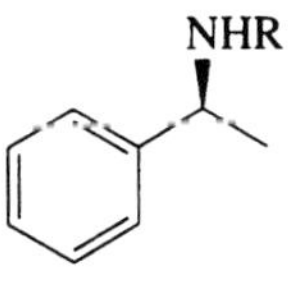 **(114)**

R = CH_2Ph, 65% *ee*
R = Ph, 40% *ee*

Table 1 Optical Yields for Asymmetric Reduction of Typical C=X Compounds with Selected Chiral Hydride Reagents

Reducing agent or method	*Substrate* $PhCOR$: *Me*	*Et*	*Pr*^{n}	*Bu*^{n}	*Pr*^{i}	*Bu*^{t}	*CO$_2$Me*	*CH$_2$Cl*	α-tetralone (1-oxo-1,2,3,4-tetrahydronaphthalene)	$HC{\equiv}C{-}CO{-}C_5H_{11}$-n	$PhC{\equiv}C{-}CO{-}CH_3$	$I{-}CH{=}CH{-}CO{-}C_5H_{11}$-n	*Ref.*
(28)	95	98	100	100	71	44	24	95	62	84	89	97	34
(47)	97	98	98	—	—	—	—	—	—	—	—	—	46
(48)	75	—	62	—	48	36	—	—	—	72	—	—	49
(49)	83.8	85	88.6	78.1	16.8	31	—	—	56.8	84	—	—	52
(50)	88	90	80	80	78	—	—	—	51	76	—	—	56
(51)	46	18	—	—	—	—	—	96	—	—	—	—	60
(53)	42	44	47	27	21	21	—	—	—	—	—	—	64
(54)	82	77	—	—	—	—	—	—	—	—	—	—	65
(55)	95	96	—	—	89	—	—	—	86	—	—	—	68
(56a)	51	68	—	—	77	86	—	—	89	—	—	51	69
$NaBH_4$/**(61)** or **(62)**	87	89	92	88	57	—	87	—	48	—	—	—	86
(64)	78	92	87	85.4	87	97	92	77	—	—	61	—	90
BH_3/**(70)**	94	96	100	—	—	25	96	—	—	7	—	—	100
(75)	97	90	—	—	—	—	—	97	89	—	—	—	102
(81)	—	—	—	—	—	—	—	—	—	—	—	—	108
$H_2SiPh_2/RhCl$/**(89)**	86.7	76.7	81.3	—	9	—	—	—	82.6	—	—	—	125
$H_2SiPh_2/RhCl_3$/**(90)**	94	—	—	—	—	—	—	—	—	—	—	—	126

Table 1 *(Continued)*

Reducing agent or method	Substrate: H	Me		CH=CHPh	CH=CHBu^n	Bu^i	n-C_6H_{13}	CH_2Ph	Pr^i	Bu^t	$EtC(O)SnBu^n_3$	$PhC(=NOPPh_2)Me$	Ref.
(28)	—	—	79	70	79	—	24	13	78	—	96	100	34
(47)	—	—	—	—	—	21	—	33	—	—	—	—	46
(48)	—	—	—	—	—	—	—	—	—	28	—	41	49
(49)	—	—	—	—	—	20.2	18.6	45.5	41.2	20.9	—	—	52
(50)	45	34	88	98	88	36	0	41	—	15	—	—	56
(51)	98	90	—	98	90	—	—	—	—	—	—	—	60
(53)	—	—	—	—	—	2.4	—	8	16	12	—	—	64
(53)	—	—	—	—	—	—	—	—	—	—	—	—	65
(55)	—	—	—	—	—	—	26	42	—	—	75	—	68
(56a)	100	28	—	72	88	—	33	—	—	—	—	—	69
$NaBH_4$/**(61)** or **(62)**	—	—	—	76	—	—	—	—	—	—	—	—	86
(64)	—	—	84	60	—	—	27	—	39	70	—	15	90
BH_3/**(70)**	35	—	96	6	—	61	58	—	60	79	68	—	100
(75)	—	—	—	—	—	—	—	—	—	92	—	—	102
(81)	—	—	—	—	—	98.6	81.3	98.9	—	99.3	—	—	108
$H_2SiPh_2/RhCl$/**(89)**	—	—	—	13.8	—	38.5	—	—	—	55.8	—	—	125
$H_2SiPh_2/RhCl_3$/**(90)**	—	—	—	—	—	—	59	—	—	—	—	—	126

Hydrosilylation by diphenylsilane, catalyzed by [{Rh(COD)Cl}$_2$]/(+)-DIOP, has been applied to the reduction of the C=N double bond of imines (**113**) to give amines (**114**), but only in moderate optical yield.[137]

1.7.3 SUMMARY

Brown and coworkers have reported a critical examination of the relative effectiveness of various reducing agents for the asymmetric reduction of different classes of ketones.[10] In Table 1 we present an expanded list of 24 typical C—X compounds and 17 selected asymmetric reducing agents or methods.

1.7.4 REFERENCES

1. J. D. Morrison and H. S. Mosher, 'Asymmetric Organic Reactions', Prentice–Hall, Englewood Cliffs, NJ, 1971; corrected paperback edn. American Chemical Society, Washington DC, 1979.
2. D. Valentine, Jr. and J. W. Scott, *Synthesis*, 1978, 329.
3. H. B. Kagan and J.-C. Fiaud, *Top. Stereochem.*, 1978, **10**, 175.
4. J. W. ApSimon and R. P. Seguin, *Tetrahedron*, 1979, **35**, 2797.
5. M. Nishizawa and R. Noyori, *Kagaku Zokan (Kyoto)*, 1981, **91**, 181.
6. H. Haubenstock, *Top. Stereochem.*, 1983, **14**, 231.
7. E. R. Grandbois, S. I. Howard and J. D. Morrison, in 'Asymmetric Synthesis', ed J. D. Morrison, Academic Press, New York, 1983, Vol. 2, part A, p. 71.
8. T. Mukaiyama and M. Asami, *Top. Curr. Chem.*, 1985, **127**, 133.
9. J. W. ApSimon and T. L. Collier, *Tetrahedron*, 1986, **42**, 5157.
10. H. C. Brown, W. S. Park, B. T. Cho and P. V. Ramachandran, *J. Org. Chem.*, 1987, **52**, 5406.
11. M. Nogradi, 'Stereoselective Synthesis', VCH, Weinheim, 1987.
12. A. A. Bothner-By, *J. Am. Chem. Soc.*, 1951, **73**, 846.
13. P. S. Portoghese, *J. Org. Chem.*, 1962, **27**, 3359.
14. S. R. Landor, B. J. Miller and A. R. Tatchell, *Proc. Chem. Soc., London*, 1964, 227.
15. S. R. Landor, B. J. Miller and A. R. Tatchell, *J. Chem. Soc. C*, 1966, 1822.
16. Y. Minoura and H. Yamaguchi, *J. Polym. Sci., Polym. Chem. Ed.*, 1968, **6**, 2013.
17. O. Červinka, V. Suchan, O. Kotýnek and V. Dudek, *Collect. Czech. Chem. Commun.*, 1965, **30**, 2484.
18. A. Horeau, H. B. Kagan and J.-P. Vigneron, *Bull. Soc. Chim. Fr.*, 1968, 3795.
19. R. Haller and H. J. Schneider, *Chem. Ber.*, 1973, **106**, 1312.
20. E. D. Lund and P. E. Shaw, *J. Org. Chem.*, 1977, **42**, 2073.
21. T. H. Johnson and K. C. Klein, *J. Org. Chem.*, 1979, **44**, 461.
22. R. Andrisano, A. S. Angeloni and S. Marzocchi, *Tetrahedron*, 1973, **29**, 913.
23. S. Yamaguchi and K. Kabuto, *Bull. Chem. Soc. Jpn.*, 1977, **50**, 3033.
24. S. R. Landor, B. J. Miller and A. R. Tatchell, *Proc. Chem. Soc., London*, 1964, 227.
25. S. R. Landor, B. J. Miller and A. R. Tatchell, *J. Chem. Soc. C*, 1966, 1822.
26. S. R. Landor, B. J. Miller and A. R. Tatchell, *J. Chem. Soc. C*, 1967, 197.
27. S. R. Landor, B. J. Miller, and A. R. Tatchell, *J. Chem. Soc. C*, 1971, 2339.
28. N. Baggett and P. Stribblehill, *J. Chem. Soc., Perkin Trans. 1*, 1977, 1123.
29. R. Noyori, *Pure Appl. Chem.*, 1981, **53**, 2315.
30. F. Toda and K. Tanaka, *J. Org. Chem.*, 1988, **53**, 3607.
31. J. Jacques and C. Fouquey, *Org. Synth.*, 1988, **67**, 1.
32. A. I. Meyers and K. A. Lutomski, *J. Am. Chem. Soc.*, 1982, **104**, 879.
33. R. Noyori, I. Tomino and Y. Tanimoto, *J. Am. Chem. Soc.*, 1979, **101**, 3129.
34. R. Noyori, I. Tomino, Y. Tanimoto and M. Nishizawa, *J. Am. Chem. Soc.*, 1984, **106**, 6709.
35. R. Noyori, I. Tomino and M. Nishizawa, *J. Am. Chem. Soc.*, 1979, **101**, 5843.
36. P. Baeckström, F. Björkling, H.-E. Högberg and T. Norin, *Acta Chem. Scand., Ser. B*, 1983, **37**, 1.
37. R. Noyori, I. Tomino, M. Yamada and M. Nishizawa, *J. Am. Chem. Soc.*, 1984, **106**, 6717.
38. M. Nishizawa, M. Yamada and R. Noyori, *Tetrahedron Lett.*, 1981, **22**, 247.
39. M. Suzuki, A. Yanagisawa and R. Noyori, *J. Am. Chem. Soc.*, 1988, **110**, 4718; Y. Morita, M. Suzuki and R. Noyori, *J. Org. Chem.*, 1989, **54**, 1785.
40. E. J. Corey, N. M. Weinshenker, T. K. Schaaf and W. Huber, *J. Am. Chem. Soc.*, 1969, **91**, 5675.
41. G. Stork, P. M. Sher and H. L. Chen, *J. Am. Chem. Soc.*, 1986, **108**, 6384.
42. B. Achmatowicz, S. Marczak and J. Wicha, *J. Chem. Soc., Chem. Commun.*, 1987, 1226.
43. M. Nishizawa and R. Noyori, *Tetrahedron Lett.*, 1980, **21**, 2821.
44. P. C.-M. Chan and J. M. Chong, *J. Org. Chem.*, 1988, **53**, 5586.
45. J. A. Marshall and W. Y. Gung, *Tetrahedron*, 1989, **45**, 1043.
46. K. Yamamoto, H. Fukushima and M. Nakazaki, *J. Chem. Soc., Chem. Commun.*, 1984, 1490.
47. O. Červinka and O. Bělovský, *Collect. Czech. Chem. Commun.*, 1967, **32**, 3897.
48. S. Yamaguchi, H. S. Mosher and A. Pohland, *J. Am. Chem. Soc.*, 1972, **94**, 9254.
49. S. Yamaguchi and H. S. Mosher, *J. Org. Chem.*, 1973, **38**, 1870.
50. R. S. Brinkmeyer and V. M. Kapoor, *J. Am. Chem. Soc.*, 1977, **99**, 8339.
51. I. Jacquet and J.-P. Vigneron, *Tetrahedron Lett.*, 1974, 2065.
52. J.-P. Vigneron and I. Jacquet, *Tetrahedron*, 1976, **32**, 939.
53. J.-P. Vigneron and V. Bloy, *Tetrahedron Lett.*, 1979, 2683.

54. J.-P. Vigneron and V. Bloy, *Tetrahedron Lett.*, 1980, **21**, 1735.
55. S. Terashima, N. Tanno and K. Koga, *J. Chem. Soc., Chem. Commun.*, 1980, 1026.
56. S. Terashima, N. Tanno and K. Koga, *Tetrahedron Lett.*, 1980, **21**, 2753.
57. S. Terashima, N. Tanno and K. Koga, *Chem. Lett.*, 1980, 981.
58. N. Tanno and S. Terashima, *Chem. Pharm. Bull.*, 1983, **31**, 811.
59. N. Tanno and S. Terashima, *Chem. Pharm. Bull.*, 1983, **31**, 821.
60. N. Tanno and S. Terashima, *Chem. Pharm. Bull.*, 1983, **31**, 837.
61. M. Kawasaki, Y. Suzuki and S. Terashima, *Chem. Pharm. Bull.*, 1985, **33**, 52.
62. R. B. Merrifield, *J. Am. Chem. Soc.*, 1963, **85**, 2149.
63. J. M. J. Fréchet, E. Bald and P. Lecavalier, *J. Org. Chem.*, 1986, **51**, 3462.
64. D. Seebach and H. Daum, *Chem. Ber.*, 1974, **107**, 1748.
65. J. D. Morrison, E. R. Grandbois, S. I. Howard and G. R. Weisman, *Tetrahedron Lett.*, 1981, **22**, 2619.
66. T. Mukaiyama, M. Asami, J. Hanna and S. Kobayashi, *Chem. Lett.*, 1977, 783.
67. M. Asami, H. Ohno, S. Kobayashi and T. Mukaiyama, *Bull. Chem. Soc. Jpn.*, 1978, **51**, 1869.
68. M. Asami and T. Mukaiyama, *Heterocycles*, 1979, **12**, 499.
69. T. Sato, Y. Goto and T. Fujisawa, *Tetrahedron Lett.*, 1982, **23**, 4111.
70. T. Sato, Y. Gotoh, Y. Wakabayashi and T. Fujisawa, *Tetrahedron Lett.*, 1983, **24**, 4123.
71. H. Suda, M. Motoi, M. Fujii, S. Kanoh and H. Yoshida, *Tetrahedron Lett.*, 1979, 4565.
72. K. Kabuto, T. Yoshida, S. Yamaguchi, S. Miyano and H. Hashimoto, *J. Org. Chem.*, 1985, **50**, 3013.
73. J. Balcells, S. Colonna and R. Fornasier, *Synthesis*, 1976, 266.
74. S. I. Goldberg, N. Baba, R. L. Green, R. Pandian, J. Stowers and R. B. Dunlap, *J. Am. Chem. Soc.*, 1978, **100**, 6768.
75. S. Colonna and R. Fornasier, *J. Chem. Soc., Perkin Trans. 1*, 1978, 371.
76. S. Colonna, R. Fornasier and U. Pfeiffer, *J. Chem. Soc., Perkin Trans. 1*, 1978, 8.
77. R. Kinishi, Y. Nakajima, J. Oda and Y. Inouye, *Agric. Biol. Chem.*, 1978, **42**, 869.
78. R. Kinishi, N. Uchida, Y. Yamamoto, J. Oda and Y. Inouye, *Agric. Biol. Chem.*, 1980, **44**, 643.
79. Y. Shida, N. Ando, Y. Yamamoto, J. Oda and Y. Inouye, *Agric. Biol. Chem.*, 1979, **43**, 1797.
80. I. Takahashi, K. Odashima and K. Koga, *Chem. Pharm. Bull.*, 1985, **33**, 3571.
81. T. Sugimoto, Y. Matsumura, S. Tanimoto and M. Okano, *J. Chem. Soc., Chem. Commun.*, 1978, 926.
82. A. Hirao, S. Nakahama, D. Mochizuki, S. Itsuno, M. Ohowa and N. Yamazaki, *J. Chem. Soc., Chem. Commun.*, 1979, 807.
83. A. Hirao, H. Mochizuki, S. Nakahama and N. Yamazaki, *J. Org. Chem.*, 1979, **44**, 1720.
84. A. Hirao, S. Nakahama, H. Mochizuki, S. Itsuno and N. Yamazaki, *J. Org. Chem.*, 1980, **45**, 4231.
85. J. D. Morrison, E. R. Grandbois and S. I. Howard, *J. Org. Chem.*, 1980, **45**, 4229.
86. K. Soai, T. Yamanoi and H. Oyamada, *Chem. Lett.*, 1984, 251.
87. K. Soai, H. Oyamada and T. Yamanoi, *J. Chem. Soc., Chem. Commun.*, 1984, 413.
88. K. Soai, T. Yamanoi, H. Hikima and H. Oyamada, *J. Chem. Soc., Chem. Commun.*, 1985, 138.
89. H. C. Brown, W. S. Park and B. T. Cho, *J. Org. Chem.*, 1986, **51**, 1936.
90. H. C. Brown, B. T. Cho and W. S. Park, *J. Org. Chem.*, 1988, **53**, 1231.
91. J.-C. Fiaud and H. B. Kagan, *Bull. Soc. Chim. Fr.*, 1969, 2742.
92. R. F. Borch and S. R. Levitan, *J. Org. Chem.*, 1972, **37**, 2347.
93. M. F. Grundon, D. G. McCleery and J. W. Wilson, *J. Chem. Soc., Perkin Trans. 1*, 1981, 231.
94. N. Umino, T. Iwakuma and N. Itoh, *Chem. Pharm. Bull.*, 1979, **27**, 1479.
95. H. Suda, S. Kanoh, N. Umeda, T. Nakajo and M. Motoi, *Tetrahedron Lett.*, 1983, **24**, 1513.
96. A. Hirao, S. Itsuno, S. Nakahama and N. Yamazaki, *J. Chem. Soc., Chem. Commun.*, 1981, 315.
97. S. Itsuno, A. Hirao, S. Nakahama and N. Yamazaki, *J. Chem. Soc., Perkin Trans. 1*, 1983, 1673.
98. S. Itsuno, K. Ito, A. Hirao and S. Nakahama, *J. Chem. Soc., Chem. Commun.*, 1983, 469.
99. S. Itsuno, K. Ito, A. Hirao and S. Nakahama, *J. Org. Chem.*, 1984, **49**, 555.
100. S. Itsuno, M. Nakano, K. Miyazaki, H. Masuda, K. Ito, A. Hirao and S. Nakahama, *J. Chem. Soc., Perkin Trans. 1*, 1985, 2039.
101. S. Itsuno, K. Ito, T. Maruyama, N. Kanda, A. Hirao and S. Nakahama, *Bull. Chem. Soc. Jpn.*, 1986, **59**, 3329.
102. E. J. Corey, R. K. Bakshi and S. Shibata, *J. Am. Chem. Soc.*, 1987, **109**, 5551.
103. E. J. Corey, R. K. Bakshi, S. Shibata, C.-P. Chen and V. K. Singh, *J. Am. Chem. Soc.*, 1987, **109**, 7925.
104. E. J. Corey, S. Shibata and R. K. Bakshi, *J. Org. Chem.*, 1988, **53**, 2861.
105. E. J. Corey, P. D. S. Jardine and J. C. Rohloff, *J. Am. Chem. Soc.*, 1988, **110**, 3672.
106. E. J. Corey and A. V. Gavai, *Tetrahedron Lett.*, 1988, **29**, 3201.
107. S. Masamune, B. M. Kim, J. S. Petersen, T. Sato and S. J. Veenstra, *J. Am. Chem. Soc.*, 1985, **107**, 4549.
108. T. Imai, T. Tamura, A. Yamamura, T. Sato, T. A. Wollmann, R. M. Kennedy and S. Masamune, *J. Am. Chem. Soc.*, 1986, **108**, 7402.
109. S. Masamune, R. M. Kennedy, J. S. Petersen, K. N. Houk and Y.-D. Wu, *J. Am. Chem. Soc.*, 1986, **108**, 7404.
110. G. M. Giongo, F. Di Gregorio, N. Palladino and W. Marconi, *Tetrahedron Lett.*, 1973, 3195.
111. I. Ojima and K. Hirai, in 'Asymmetric Synthesis', ed J. D. Morrison, Academic Press, New York, 1985, vol. 5, p. 103.
112. K. Yamamoto, T. Hayashi and M. Kumada, *J. Organomet. Chem.*, 1972, **46**, C65.
113. T. Hayashi, K. Yamamoto, K. Kasuga, H. Omizu and M. Kumada, *J. Organomet. Chem.*, 1976, **113**, 127.
114. I. Ojima, T. Kogure and Y. Nagai, *Chem. Lett.*, 1973, 541.
115. I. Ojima and Y. Nagai, *Chem. Lett.*, 1974, 223.
116. I. Ojima, T. Kogure and Y. Nagai, *Tetrahedron Lett.*, 1974, 1889.
117. I. Ojima, T. Kogure, M. Kumagai, S. Horiuchi and T. Sato, *J. Organomet. Chem.*, 1976, **122**, 83.
118. W. Dumont, J.-C. Poulin, T.-P. Dang and H. B. Kagan, *J. Am. Chem. Soc.*, 1973, **95**, 8295.
119. I. Ojima, T. Kogure and M. Kumagai, *J. Org. Chem.*, 1977, **42**, 1671.
120. I. Kogure and I. Ojima, *J. Organomet. Chem.*, 1982, **234**, 249.
121. H. Brunner and W. Miehling, *J. Organomet. Chem.*, 1984, **275**, C17.

122. H. Brunner and G. Riepl, *Angew. Chem., Int. Ed. Engl.*, 1982, **21**, 377.
123. H. Brunner, B. Reiter and G. Riepl, *Chem Ber.*, 1984, **117**, 1330.
124. H. Brunner, G. Riepl and H. Weitzer, *Angew. Chem., Int. Ed. Engl.*, 1983, **22**, 331.
125. H. Brunner and A. Kürzinger, *J. Organomet. Chem.*, 1988, **346**, 413.
126. H. Nishiyama, H. Sakaguchi, T. Nakamura, M. Horihata, M. Kondo and K. Ito, *Organometallics*, 1989, **8**, 846.
127. O. Červinka, *Collect. Czech. Chem. Commun.*, 1961, **26**, 673.
128. O. Červinka, *Collect. Czech. Chem. Commun.*, 1965, **30**, 2403.
129. O. Červinka, V. Suchan, O. Kotýnek, and V. Dudek, *Collect. Czech. Chem. Commun.*, 1965, **30**, 2484.
130. S. R. Landor, O. O. Sonola and A. R. Tatchell, *J. Chem. Soc., Perkin Trans. 1*, 1974, 1902.
131. S. R. Landor, Y. M. Chan, O. O. Sonola and A. R. Tatchell, *J. Chem. Soc., Perkin Trans. 1*, 1984, 493.
132. K. Yamada, M. Takeda and T. Iwakuma, *Tetrahedron Lett.*, 1981, **22**, 3869.
133. K. Yamada, M. Takeda and T. Iwakuma, *J. Chem. Soc., Perkin Trans. 1*, 1983, 265.
134. K. Yamada, M. Takeda, H. Ohtsuka and T. Iwakuma, *Chem. Pharm. Bull.*, 1983, **31**, 70.
135. S. Itsuno, M. Nakano, K. Miyazaki, H. Masuda, K. Ito, A. Hirao and S. Nakahama, *J. Chem. Soc., Perkin Trans. 1*, 1985, 2039.
136. R. O. Hutchins, A. F. Abdel-Magid, Y. P. Stercho and A. Wambsgans, *J. Org. Chem.*, 1987, **52**, 702.
137. N. Langlois, T.-P. Dang and H. B. Kagan, *Tetrahedron Lett.*, 1973, 4865.

1.8
Reduction of C=X to CHXH Using Enzymes and Microorganisms

J. BRYAN JONES
University of Toronto, Canada

1.8.1 INTRODUCTION

Applications of enzymes as specific and chiral catalysts in synthetic organic chemistry are now well established, with the utilization of oxidoreductases for C=X → CHXH conversions being one of the best documented areas of their exploitation.[1–3]

Enzymic methods of reduction take place under mild conditions, often at room temperature and at neutral pH, minimizing such problems as isomerization, racemization, epimerization and rearrangement that can plague more traditional methodology. Enzyme specificity permits the catalysis of C=X reductions to be chemospecific in the presence of other, unprotected, functional groups and with regio- and stereospecific control that is difficult, or currently impossible, to achieve nonenzymically in single-step reactions.

The following enzymes and coenzymes are abbreviated: HLADH, horse liver alcohol dehydrogenase; YADH, yeast alcohol dehydrogenase; PTADH, *Pseudomonas testosteronii* alcohol dehydrogenase; NAD(P) and NAD(P)H, oxidized and reduced forms of nicotinamide adenine diphosphate (or its phosphate) respectively; BY, baker's yeast; TBADH, *Thermoanaerobium brockii* alcohol dehydrogenase;

GDH, glycerol dehydrogenase; HSDH, hydroxysteroid dehydrogenase; MJADH, *Mucor javanicus* alcohol dehydrogenase; CFADH, *Curvularia falcata* alcohol dehydrogenase; LDH, lactate dehydrogenase; HSDH, hydroxysteroid dehydrogenase; G6PDH, glucose-6-phosphate dehydrogenase; 3HBDH, 3-hydroxybutyrate dehydrogenase; FDH, formate dehydrogenase; LeuDH, leucine dehydrogenase; and PLADH, pig liver alcohol dehydrogenase.

It is not necessary to use a different enzyme for each new substrate encountered. Some enzymes, such as HLADH have very broad specificities and can accommodate wide structural variations in their substrates. Also, a broad structural range of substrates is often accessible using a very limited number of enzymes of overlapping specificities (Scheme 1).

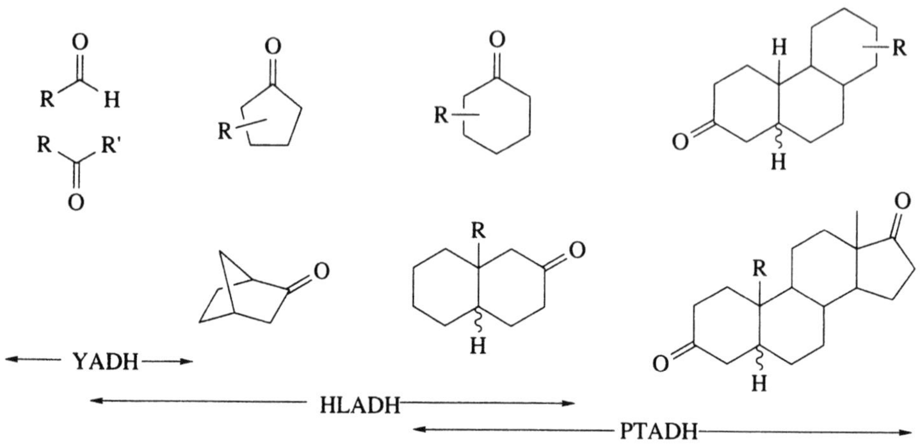

← YADH →
← HLADH →
← PTADH →

Scheme 1

1.8.2 SCOPE OF THE METHOD

In selecting the examples for this chapter, preference was given to the most recent literature. However, it must be remembered that research on this topic has been performed by many groups for over 70 years, beginning with the original, and often still definitive, work of the early 1900s.[4]

All biological C=X → CHXH reductions are dependent on the nicotinamide coenzymes NAD(P)H, as illustrated for ketone reduction in equation (1), which shows only the reactive dihydronicotinamide moiety of the NAD(P)H cofactor.

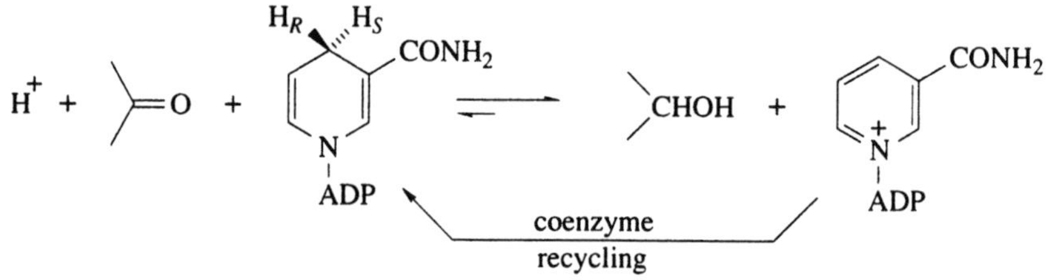

$$H^+ + {>}C{=}O + \text{(dihydronicotinamide: } H_R, H_S, CONH_2, N\text{–ADP)} \rightleftharpoons {>}CHOH + \text{(nicotinamide: } CONH_2, N^+\text{–ADP)} \quad (1)$$

coenzyme recycling

1.8.3 SOURCES OF ENZYMES AND MICROORGANISMS

The oxidoreductases of synthetic value can be grouped into four main supply categories. These are: (i) fully or partially purified enzymes, (ii) yeasts, (iii) other microorganisms, and (iv) mammalian sources. Mammalian enzymes usually have broader specificities than those of microorganisms, but this is compensated for by the much larger selection of microorganisms available.

Biochemical catalogs list several hundred enzymes, including most of the oxidoreductases that have been used to date in preparative-scale reductions. These catalogs also list various yeasts, but the most commonly used of these, BY, is easily obtained at the local supermarket. More specialized yeasts, and the complete range of other microorganisms available, are listed in the catalogs of the national culture collections in the USA, Japan and Europe.[5] In the future, the cloning and site-directed mutagenesis techniques of molecular biology will permit natural enzymes from any source, and mutant enzymes tailored for particular synthetic applications, to be provided in plentiful supply.

Identifying a suitable enzyme or microorganism for reduction of a new substrate is best done by perusal of reviews[1,2,6,7] and by literature search for an analogous transformation of a closely related structure. Chemical Abstracts Registry Numbers for enzymes are obtained from the Index Guide, using the enzyme commission (EC) number for the enzyme of interest. The EC numbers of all known enzymes are listed in the Enzyme Nomenclature handbook.[8] For yeasts and microorganisms, the Chemical Abstract search protocol is based on the name of the organism, again using the Index Guide for clarification, if necessary.

1.8.4 PRACTICAL CONSIDERATIONS

The experimental methodology for exploiting enzymes is very straightforward and can be performed with common organic equipment, and with routine experimental and work-up procedures. A range of reaction conditions can be applied, including the use of immobilized enzymes[9] and organic solvents.[10–14] For yeast, often all that is required is to suspend it, together with the substrate, in water at pH 7, and to add some glucose if a growing culture is needed. The procedures for fermentations with other microorganisms are somewhat more sophisticated, but still easily performed in the organic laboratory,[7] perhaps with some minimal initial guidance from a microbiological colleague.

The metabolisms of yeasts and microorganisms ensure a sufficient supply of the NAD(P)H coenzyme during fermentative reductions. Since the NAD(P)H needed for *in vitro* reductions catalyzed by isolated enzymes is expensive, it is used in catalytic amounts only, and continuously regenerated from the inactive pyridinium form of NAD or NADP by an economical recycling system.[15]

Unless specified otherwise, all reductions included in this chapter gave good yields of >90% enantiomeric excess (*ee*) products. Not all products of enzyme-catalyzed reactions meet the minimum % *ee* levels normally required for asymmetric synthetic applications. However, protocols exist for improving *ee*'s of imperfectly specific enzyme-mediated transformations.[14,16–19]

1.8.5 ORGANIZATION OF TOPICS

Preparative-scale enzyme-catalyzed reactions are conveniently considered in terms of the different aspects of specificity involved and this approach is followed here. Reactions involving a single aspect of specificity are considered first, followed by examples of different, and increasingly complex, combinations of specificity in a single-step catalysis.

Reductions that do not capitalize on the abilities of enzymes to make stereochemical distinctions are not covered (in this regard, it should be noted that the term 'stereospecificity' is used throughout in its broad, biochemical, connotation) nor are mechanistic aspects, details of which can be found in several excellent texts.[20–23] Although space limitations preclude discussion of the synthetic motivations for the reductions cited, the majority were used to prepare chiral synthons needed in total synthesis.

1.8.6 REDUCTION OF C=O TO CHOH

1.8.6.1 Enantiomeric Distinctions

Most resolutions of racemates in reductions with oxidoreductases are accompanied by stereoheterotopic face specificity and are covered in a subsequent section. Enantiomeric discrimination alone is exemplified (Scheme 2) by the partial resolutions of binaphthyls (±)-(**1**),[24] of the formyl derivatives (±)-(**2**)[25] and (±)-(**3**),[26] and of the organometallic aldehyde (±)-(**4**)[27] with BY. The racemic 1-indanone complex analog of (±)-(**4**) is also a substrate, but its BY reduction proceeds with low enantioselectivity and to give a mixture of (mainly) *endo*- and *exo*-alcohol products.[28] For the Scheme 2 reactions, the product *ee*'s can be influenced by the nature of the substituents. In the case of the binaphthyl reduction, replacement of —OMe by —Me gave the same *ee* levels, but afforded product alcohol of the opposite absolute configuration. Replacement of the 2-methyl substituent of (±)-(**2**), by —CH_2Ph for example, can also result in reversals of the absolute configurations of the products.[29] Of the various thiocarbamates of (±)-(**3**) evaluated, that with R = CH_2Bu^t gave the highest *ee*.[25] HLADH can also be used for resolution of racemic aldehydes, as shown in Scheme 3.[10,30]

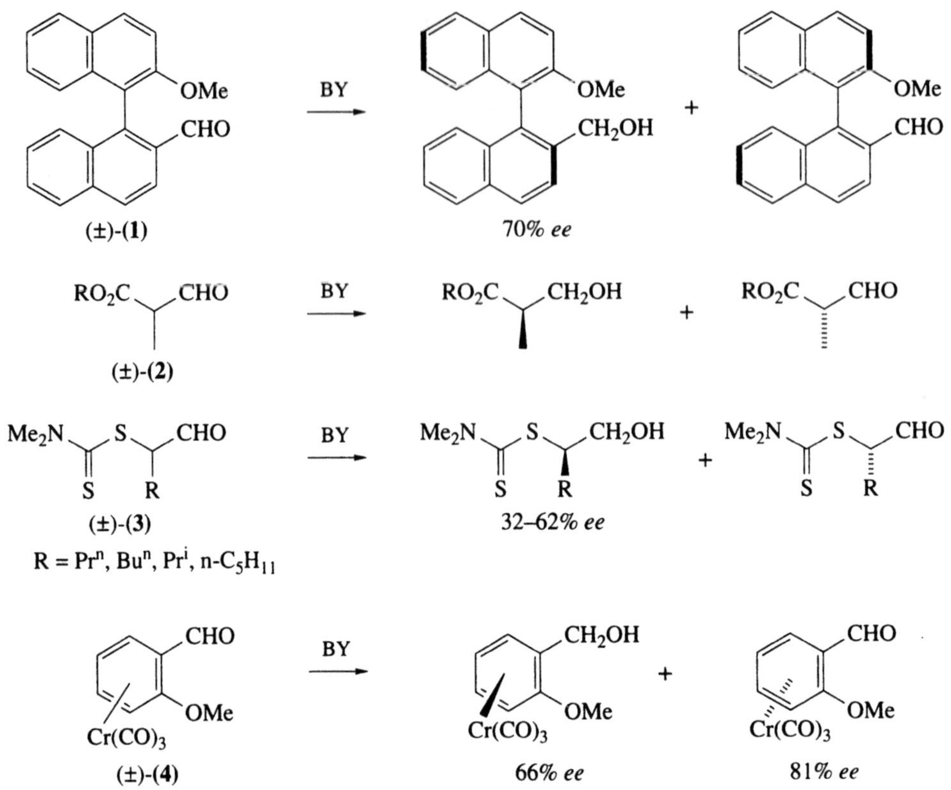

OMe CHO (±)-(1) BY OMe CH_2OH 70% ee + OMe CHO

RO_2C CHO (±)-(2) BY RO_2C CH_2OH + RO_2C CHO

Me_2N S CHO S R (±)-(3) BY Me_2N S CH_2OH S R 32–62% ee + Me_2N S CHO S R

R = Pr^n, Bu^n, Pr^i, n-C_5H_{11}

CHO OMe $Cr(CO)_3$ (±)-(4) BY CH_2OH OMe $Cr(CO)_3$ 66% ee + CHO OMe $Cr(CO)_3$ 81% ee

Scheme 2

CHO HLADH CH_2OH + CHO

CHO R HLADH CH_2OH R 52–93% ee + CHO R 42–90% ee

R = H, Me, Ph, $CH=CH_2$,

CHO R HLADH CH_2OH R 21–84% ee + CHO R 22–62% ee

R =

CHO HLADH CH_2OH 73% ee + CHO unstable

CHO CO_2Et HLADH CH_2OH CO_2Et 41% ee + CHO CO_2Et 48% ee

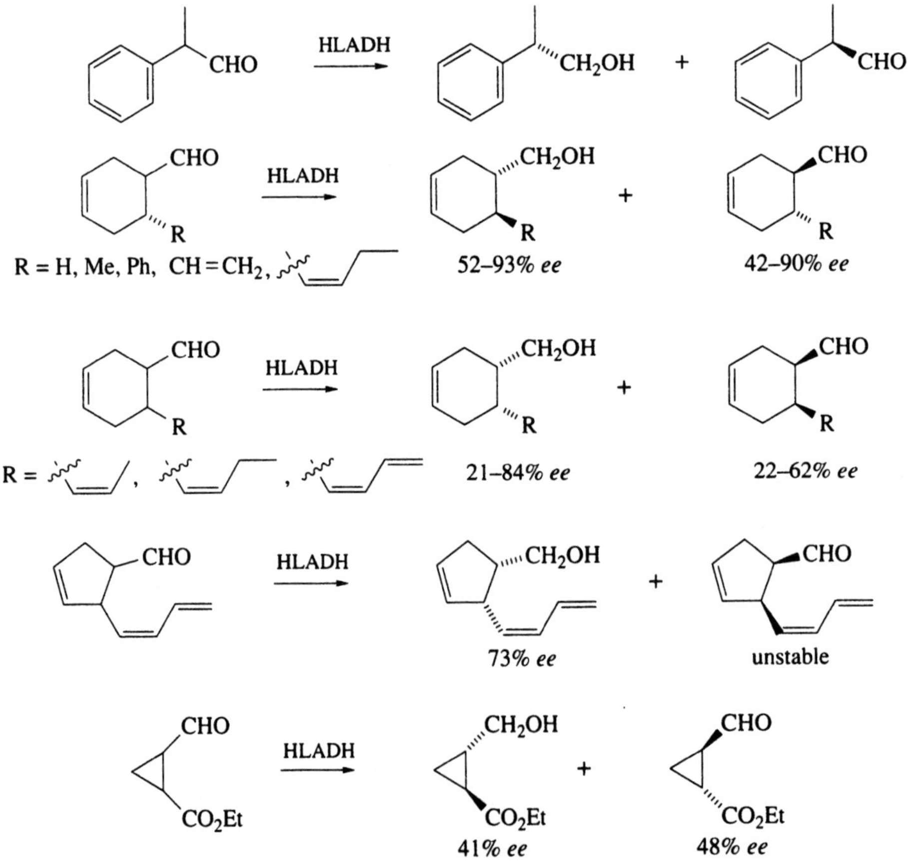

Scheme 3

1.8.6.2 Enantiotopic Face Distinctions

1.8.6.2.1 Monocarbonyl substrates

A broad spectrum of stereospecific reductions of prochiral acyclic ketones has been recorded (Scheme 4). These arise by enzyme-controlled delivery of the hydride equivalent to only one of the two enantiotopic faces of a C=O group. Such reductions are achievable with many different oxidoreductases, as illustrated for the conversions of (**5a**) → (**6a**) with GDH,[31] of (**5b–d**) → (**6b–d**) with TBADH,[32–34] of (**5e–h**) → (**6e–h**) with BY,[35–41] of (**5i**) → (**6i**) with *Kloeckera magna*[42] and of (**5j**) → (**6j**) with *Acinetobacter calcoaceticus*.[43]

The yeast-induced reduction of (**5e**) → (**6e**), (R = substituted phenyls, R′ = Me) is one of the very first preparative-scale alcohol dehydrogenase catalyzed reactions to have been reported.[35] While the transformations of (**5a–j**) demonstrate that broad structural tolerances are possible in the substrate ketones, the enzymes are sometimes very discriminating. For example, while the BY-catalyzed reductions of the 2-, 3- and 4-substituted pyridyl ketones (**5g**) proceed smoothly with high stereospecificity,[39,41] the analogous furanyl and thiophenyl ketones give virtually racemic product alcohols, and 2-acetylpyrrole is not a substrate at all.[39]

Diketones or keto aldehydes in which one carbonyl group is protected, as in (**7**) and (**10**) of Scheme 5, may also be reduced stereospecifically to the corresponding alcohols, (**8**) and (**11**). These can then be hydrolyzed chemically to give the hydroxy aldehydes (**9**) or hydroxy ketones (**12**), respectively.[44,45]

The absolute configurations of the product alcohols of acyclic ketone reductions can be reliably predicted for many oxidoreductases by using the Prelog rule (Scheme 6).[1a,46] More complex models are required for cyclic ketones, as will be seen later. For a given substrate, oxidoreductases are usually available that deliver hydride equivalents from opposite enantiotopic faces. This can be exploited to produce either alcohol enantiomer at will, as noted in Scheme 7 by the reduction of (**13**) to either (*R*)- or (*S*)-(**14**).[47] Organometallic ketones can also be reduced with enantiotopic specificity. This is demonstrated by the conversions (Scheme 8) of the ferrocenyl[48] and chromium carbonyl ketones (**15**) → (**16**) and (**17**) → (**18**), respectively.[28]

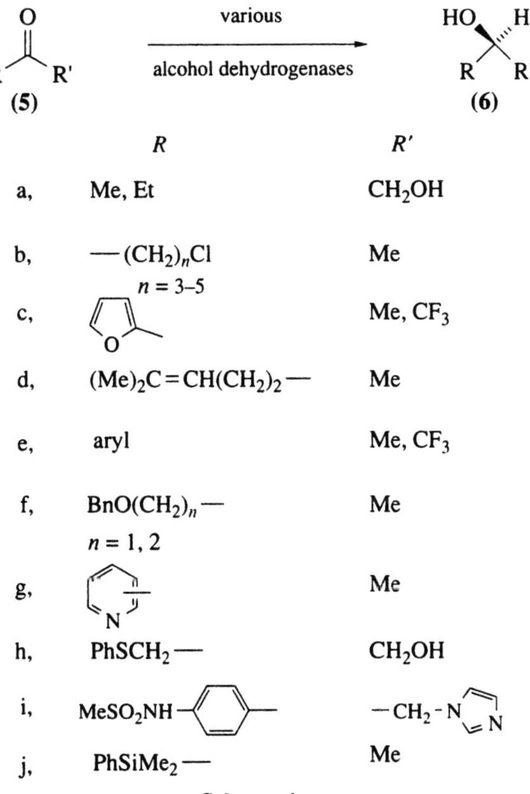

Scheme 4

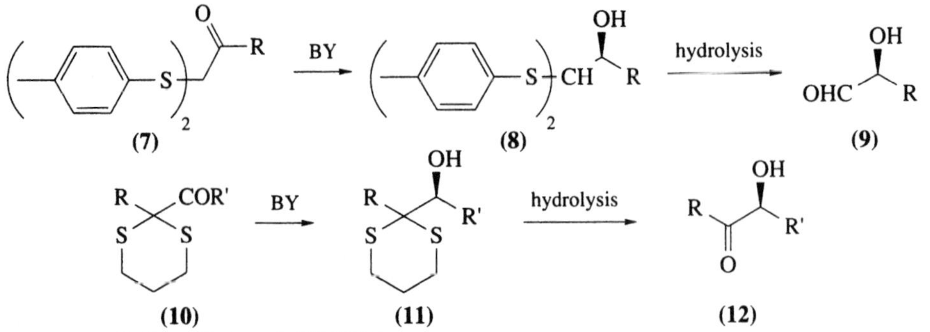

BY
hydrolysis
(7) (8) (9)
R, R' = alkyl, aryl, $HO(CH_2)_2-$
BY
hydrolysis
(10) (11) (12)

Scheme 5

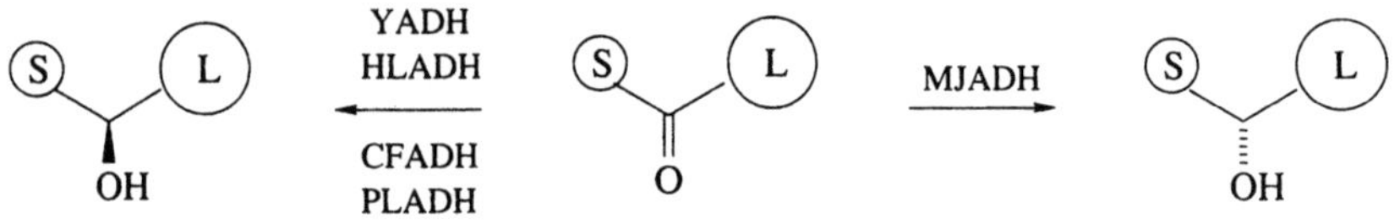

YADH
HLADH
CFADH
PLADH
MJADH

Scheme 6

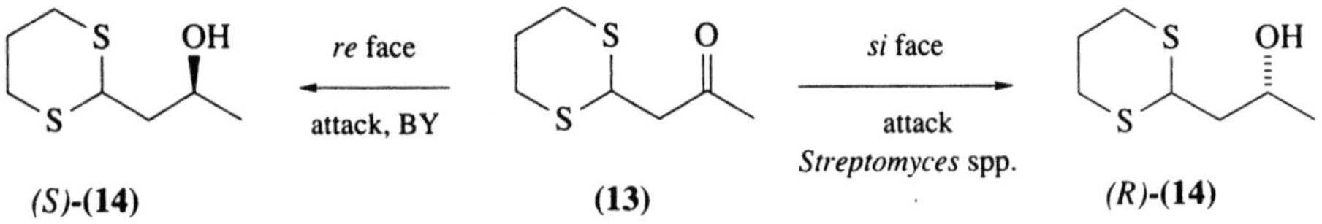

re face attack, BY
si face attack *Streptomyces* spp.
(*S*)-(14) (13) (*R*)-(14)

Scheme 7

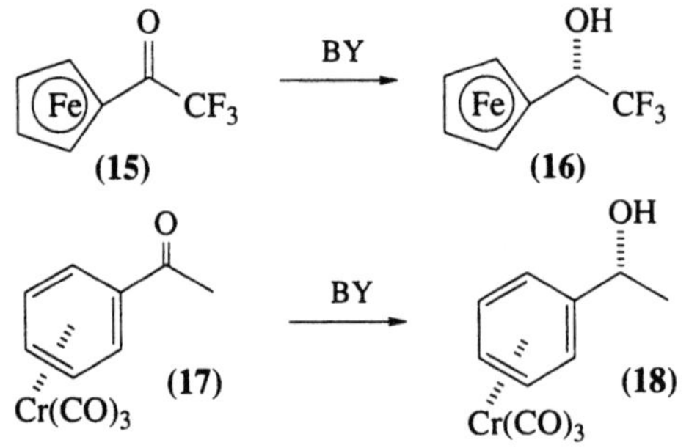

BY
(15) (16)
BY
(17) (18)

Scheme 8

1.8.6.2.2 Diketone substrates

The two carbonyl groups of symmetrical diketones are distinguishable, with the carbonyl group undergoing reduction doing so with enantiotopic specificity. Some acyclic and monocyclic examples are shown in Scheme 9.[32,44,49,50] Once more, enantiomeric products can be selected by the use of organisms with opposite enantiotopic face specificities, as shown for the reduction of (**19**) to (*R*)- or (*S*)-(**20**) (Scheme 10).[51]

The (**21**) → (**22**) transformation of Scheme 11 is an interesting variation of this category in that, while reduction of both homotopic carbonyl groups of the racemic substrate (**21**) occurs with diastereotopic *re* face specificity, the product (**22**) is still racemic.[52] The (**23**) → (**24**) reduction is analogous, with only one of the two homotopic carbonyl groups reduced, again with diastereotopic face specificity.[53]

Reductions of symmetrical bicyclic diketones may also be effected selectively, as illustrated in Scheme 12. The stereospecificities of the HLADH-catalyzed transformations of the unsaturated decalindiones (**25**) and (**27**) to the corresponding hydroxy ketones (**26**) and (**28**),[54] and in fact all specificity aspects of this enzyme, are fully predictable using a simple to use, cubic-space model of the enzyme's active site.[55]

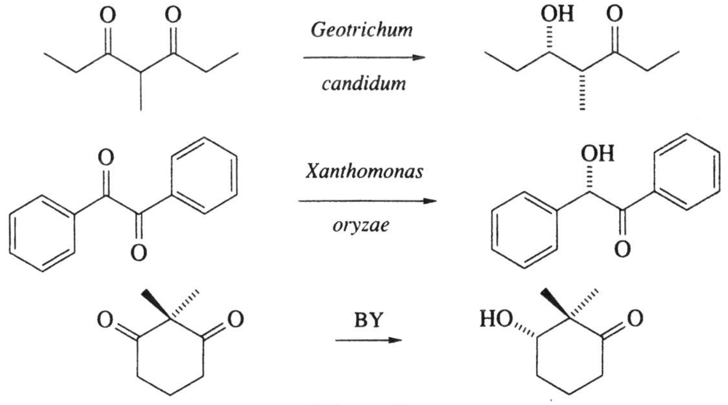

Geotrichum candidum

Xanthomonas oryzae

BY

Scheme 9

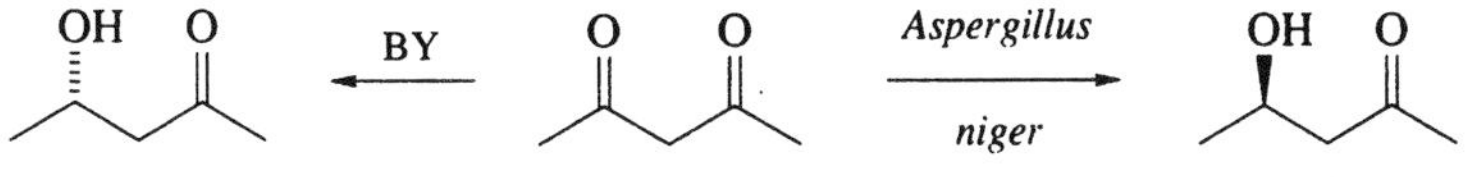

(S)-(20) BY (19) Aspergillus niger (R)-(20)

Scheme 10

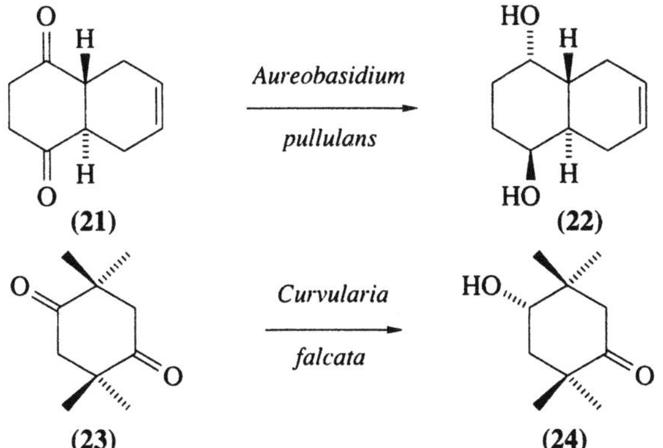

(21) Aureobasidium pullulans (22)

(23) Curvularia falcata (24)

Scheme 11

(25) HLADH (26)

(27) HLADH (28)

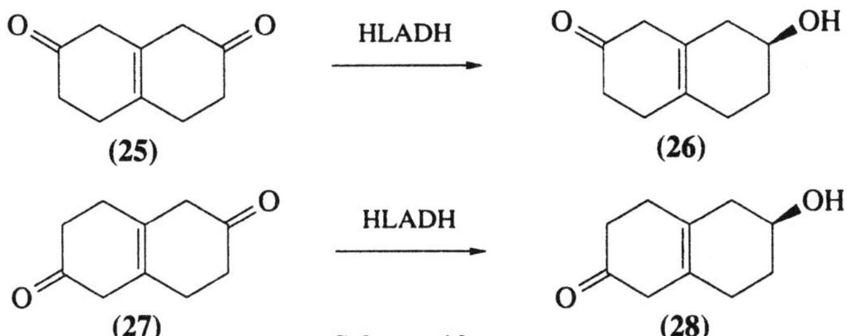

Scheme 12

1.8.6.2.3 Keto acid and ester substrates

Stereospecific reductions of α-keto acids are well documented.[56–60] L- and D-lactate dehydrogenases from common mammalian or bacterial sources are available for this purpose. By using a preselected L- or D-LDH, the preparation of a (2*S*)- or (2*R*)-hydroxy acid of >99% *ee* can be virtually assured. The structural range of α-keto acids (**29**) that has been subjected to preparative-scale reductions to hydroxy acids (**30**) is summarized in Scheme 13 for both L-[58,60,61] and D-LDH.[59,61,62] Plentiful supplies of such enzymes are guaranteed by their cloning and overexpression.[60]

Stereospecific reduction of α-keto esters is also readily achieved (Scheme 14). BY is effective for the (**31**) → (**32**) reduction,[63] but this is not always the case. Different enzyme sources are more effective for

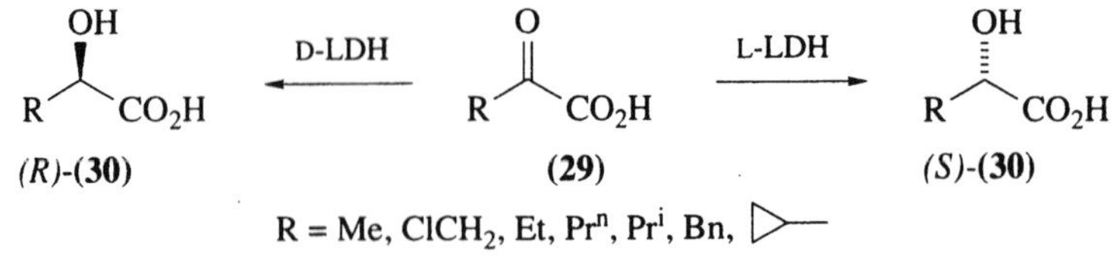

(R)-(30) ← D-LDH — (29) — L-LDH → (S)-(30)

R = Me, $ClCH_2$, Et, Pr^n, Pr^i, Bn, cyclopropyl

Scheme 13

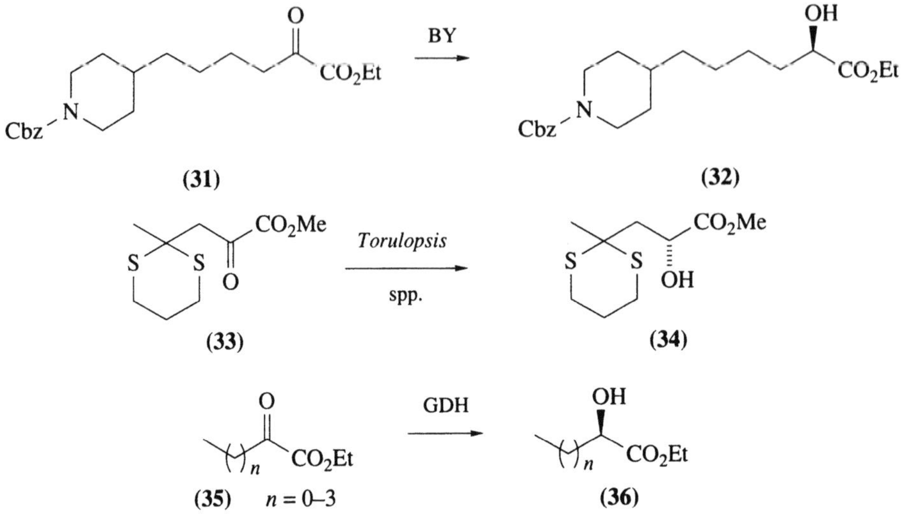

(31) —BY→ (32)

(33) —*Torulopsis* spp.→ (34)

(35) n = 0–3 —GDH→ (36)

Scheme 14

L-(38) ← *Mucor racemosus* — (37) — *Candida parapsilosis* → D-(38)

(S)-(40) ← BY — (39) R = alkyl — GDH → (R)-(40)

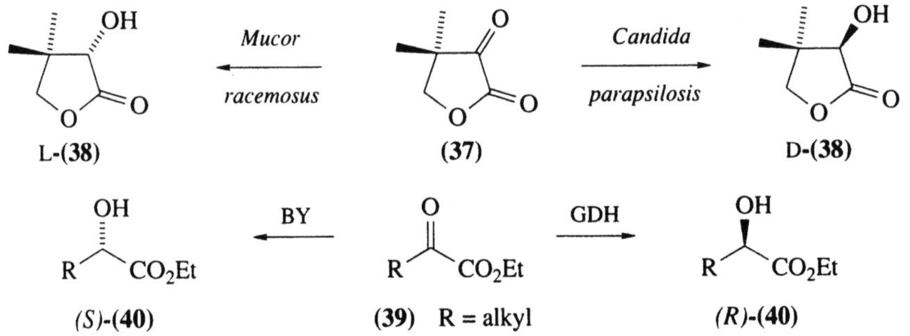

Scheme 15

the (**33**) → (**34**)[64] and (**35**) → (**36**)[65] conversions. Using different organisms with opposite enantiotopic face specificities permits both enantiomers of a product to be produced at will from the same substrate. This is demonstrated in Scheme 15 by the reduction of (**37**) to either natural[66] D- or unnatural L-pantoyllactone (**38**)[67] and of simple α-keto esters (**39**) to either the (*R*)-[65] or (*S*)-enantiomer[68] of (**40**).

A broad structural range of β-keto esters (**41**) can be reduced with enantiotopic specificity to give either the (*R*)- or (*S*)-β-hydroxy esters (**42**; Scheme 16).[26,69–81] Whether the configurations of the stereogenic centers of (**42**) will be (*R*) or (*S*) is determined by the nature of the organisms and of the substituents. However, with the appropriate combination of organisms, hydroxy esters of either (*R*)- or (*S*)-configuration can be produced from the same substrate, with either *si* or *re* face attack of the carbonyl group being selected. This is shown in Scheme 17, where the reduction of (**43**) to either (*R*)- or (*S*)-(**44**) can be achieved.[82] In this latter case, the enzymes exploited were located in different organisms. However, it is possible for 'enantiotopic' enzymes to be contained in the same organism. For example, yeast contains two fatty acid synthetases[83,84] with alcohol dehydrogenase like capabilities. These have opposite enantiotopic specificities and also distinct substrate specificities. It is possible to select between these 'L' and 'D' enzymes by varying the steric size of the β-keto ester substituents, as exemplified in Scheme 18.[19] Controlled formation of an (*R*)- or (*S*)-β-hydroxy ester from a common β-keto ester precursor can also be preordained by using enzymes with opposite enantiotopic face specificities.[85] Additional control of enantiotopic selectivity in a multiple-enzyme organism can be exerted by restricting substrate or product diffusion, as between the free and immobilized yeasts.[18] This selectivity is again controllable by vari-

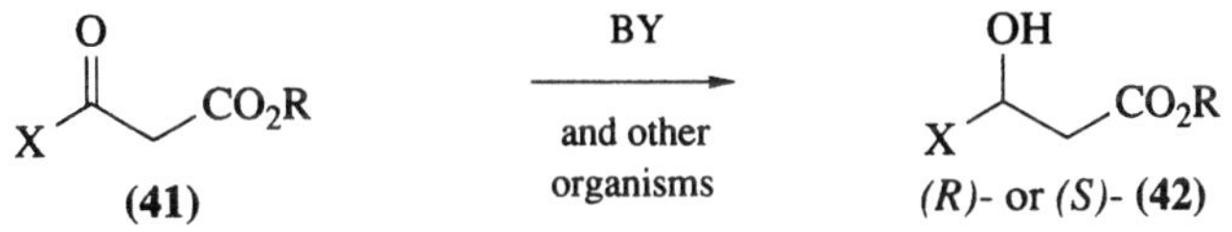

X = C_1–C_{15} alkyl, aryl, CH_2Oalkyl/aryl, CH_2Cl, OH, S-alkyl/aryl, NHCOR, CH_2N_3, CCl_3, CF_3, CO_2Me, CH_2CO_2H; R = C_1–C_8 alkyl, aryl; CO_2R = CONHR, CSSMe, SO_2Ph, $SCSNMe_2$

Scheme 16

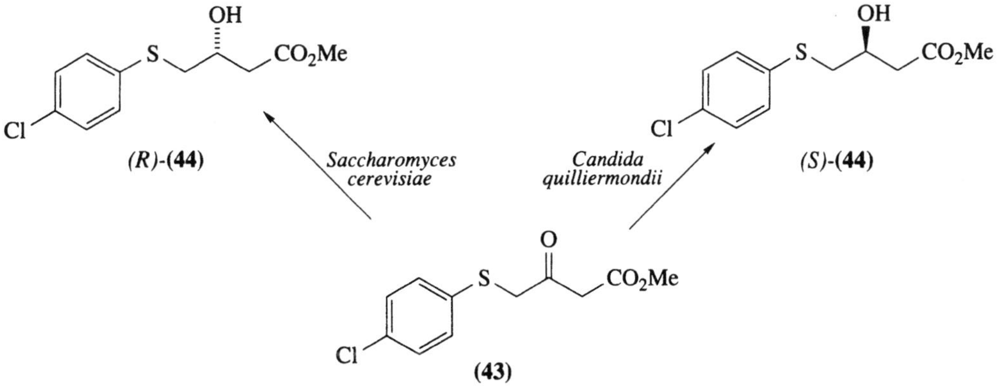

Scheme 17

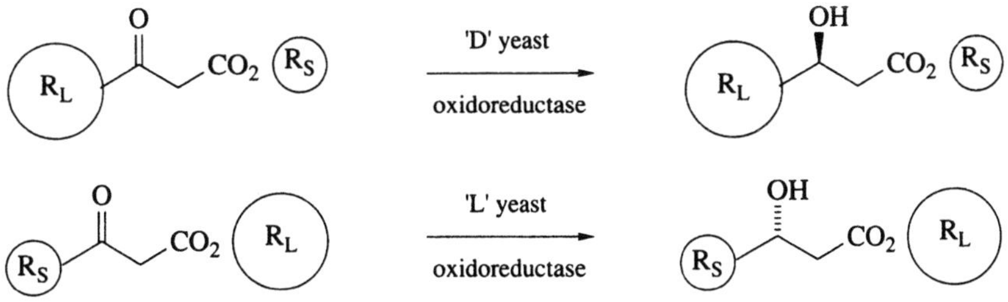

Scheme 18

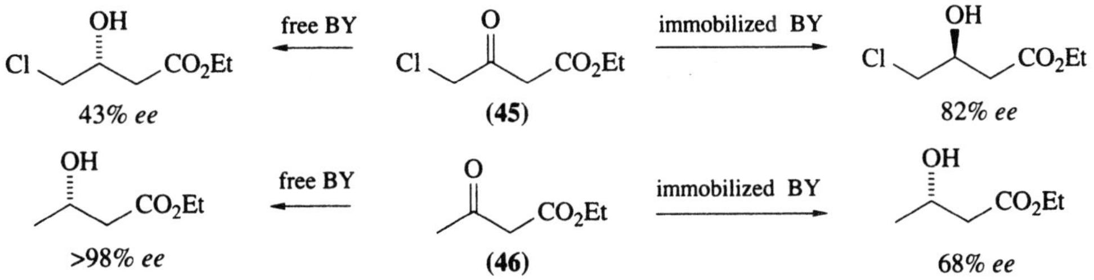

Scheme 19

ations in substrate structure, as illustrated by the different stereospecificity responses towards (**45**) and (**46**; Scheme 19). Stereoselectivity can also be influenced by growth conditions[73,86] and by added solvents.[73,87]

Although reductions of β-keto esters are by far the most common, yeast-catalyzed reductions of β-keto acids are also viable.[88,89] Reductions of other keto acids and esters, where the carbonyl group is γ, or further, separated from the acid or ester function, also proceed stereoselectively, as demonstrated by the Scheme 20 conversions.[73,90–94] Reduction of an α,β-unsaturated ketone is somewhat unusual, but examples are known (Scheme 21).[95]

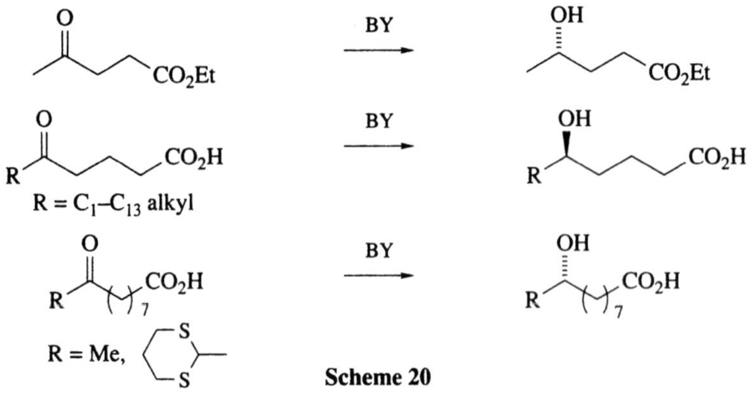

BY

CO_2Et

BY

CO_2H

R = C_1–C_{13} alkyl

BY

CO_2H

R = Me,

Scheme 20

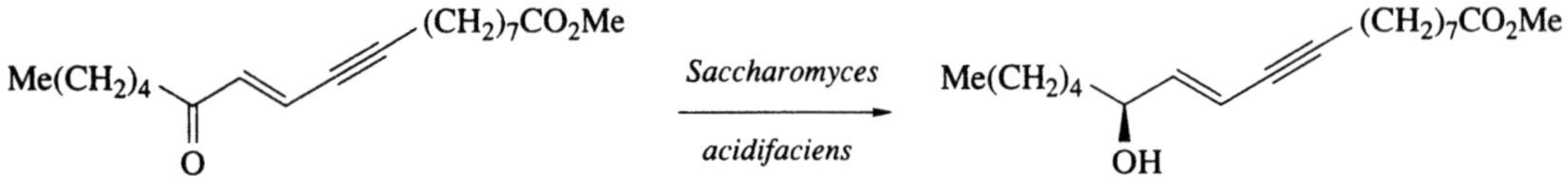

$Me(CH_2)_4$ $(CH_2)_7CO_2Me$

Saccharomyces acidifaciens

Scheme 21

1.8.6.3 Diastereotopic Face Distinctions

In contrast to the multiplicity of examples of diastereotopic face discriminations in combination with other enzyme specificity discussed later, there are relatively few alcohol dehydrogenase catalyzed reductions that involve diastereotopic face selectivity alone. However, whenever a carbonyl group is present in a single chiral stereoisomer, diastereotopically face-selective reductions are possible. The reductions of the L-homocysteine derivative (**47**) → (**48**),[96] and of the (–)-oxocineole (1*S*,4*R*)-(**49**) → (**50**),[97] are two such examples (Scheme 22). A single stereoisomer can also be converted into epimeric alcohol products by employing organisms with opposite diastereotopic face specificities (equation 2 and Scheme 23).[97–99] The conversions of (1*R*,4*S*)-(**49**) to (**51**) and (**52**)[97] complement that of the reduction of its enantiomer (1*S*,4*R*)-(**49**) to (**50**) in Scheme 22. The facility with which reduction of quadrone (**53**) to either (**54**) or (**55**)[99] can be accomplished demonstrates very clearly that structural complexity is not a barrier to enzyme-catalyzed reductions.

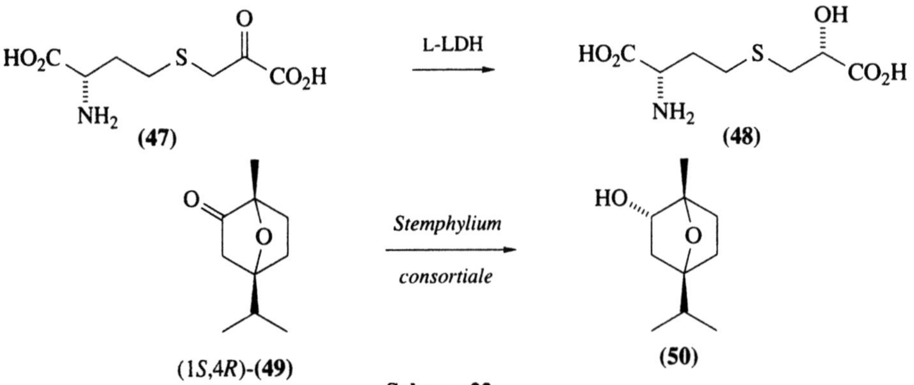

(**47**) L-LDH (**48**)

(1*S*,4*R*)-(**49**) *Stemphylium consortiale* (**50**)

Scheme 22

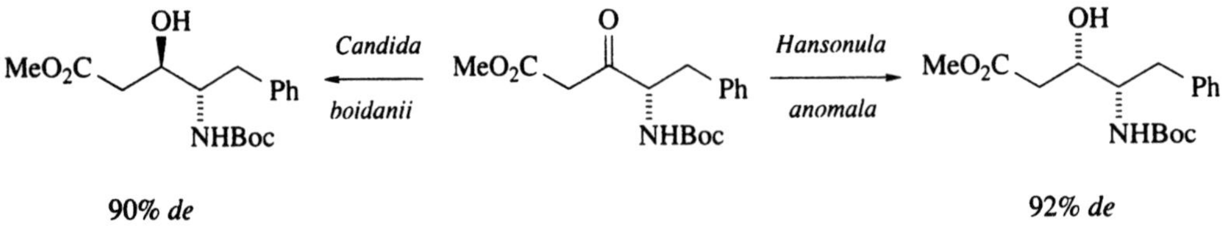

Candida boidanii ← MeO_2C ... NHBoc ... Ph → *Hansonula anomala*

90% *de* 92% *de* (2)

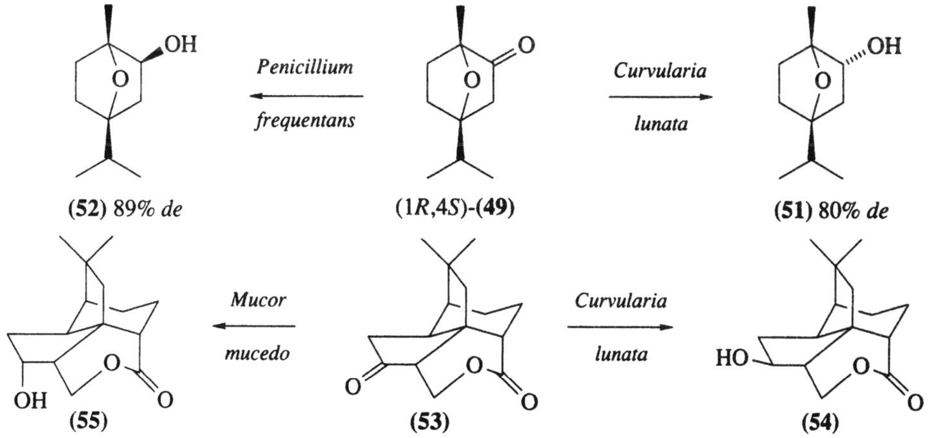

(52) 89% *de* | *Penicillium frequentans* | (1*R*,4*S*)-(49) | *Curvularia lunata* | (51) 80% *de*

(55) | *Mucor mucedo* | (53) | *Curvularia lunata* | (54)

Scheme 23

1.8.6.4 Combinations of Specificity

Despite the advances that have been made using nonenzymic methods in asymmetric synthesis, the degrees of control that are possible with enzymes able to combine two or more specificities in a single-step reaction remain unchallenged. Several different chemo- or regio-specific and enantiotopic face specific combinations are illustrated in Scheme 24 by the reduction of α-[100] or β-diketones[100–102] (see also Scheme 27). The control of regiospecificity that can be exerted by small structural changes is exemplified by the effect of either an *o*- or *p*-methyl group on the reductions of **(56)** and **(58)** to **(57)** and **(59)** respectively.[49] Opposite enantiotopic face specificities can be selected by using appropriate organisms (Scheme 25).

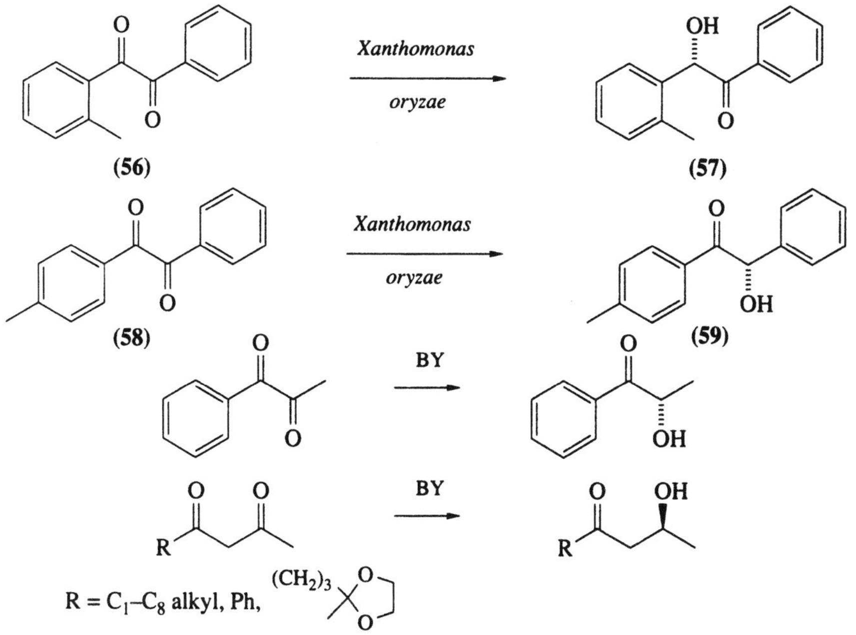

(56) → *Xanthomonas oryzae* → (57)

(58) → *Xanthomonas oryzae* → (59)

BY

BY

R = C_1–C_8 alkyl, Ph, $(CH_2)_3$

Scheme 24

BY ← | → *Aspergillus niger*

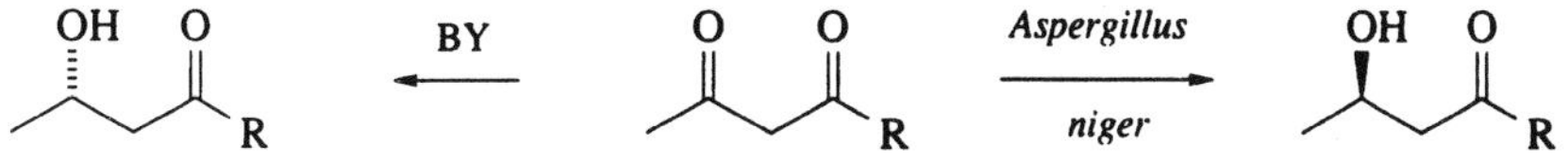

R = Et, $(CH_2)_3CH{=}CH_2$, Ph

Scheme 25

Regiospecificity combined with diastereotopic face specificity is seen in the reduction of the trione (**60**) to (**61**) (Scheme 26).[103] In Scheme 27 the reduction is quite regiospecific, but both diastereotopic faces of the exocyclic ketone function of (**62**) are attacked to give the *erythro*-(**63**) and *threo*-(**64**) hydroxy ketone products.[104a] The degree to which regio specificity and diastereotopic face specificity can be controlled with different enzymes is shown in the bile acid reductions of Scheme 28.[105]

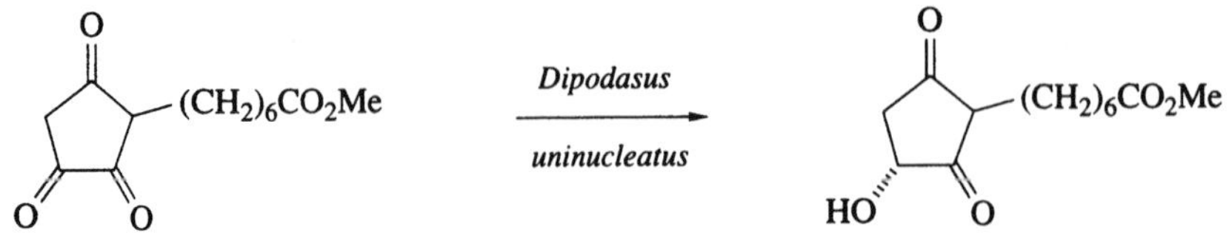

(CH$_2$)$_6$CO$_2$Me (**60**) —*Dipodasus uninucleatus*→ (CH$_2$)$_6$CO$_2$Me (**61**)

Scheme 26

CO$_2$Et (±)-(**62**) —BY→ CO$_2$Et (**63**) + CO$_2$Et (**64**)

Scheme 27

3α-HSDH, 3β-HSDH, 7α-HSDH, 7β-HSDH, 12α-HSDH

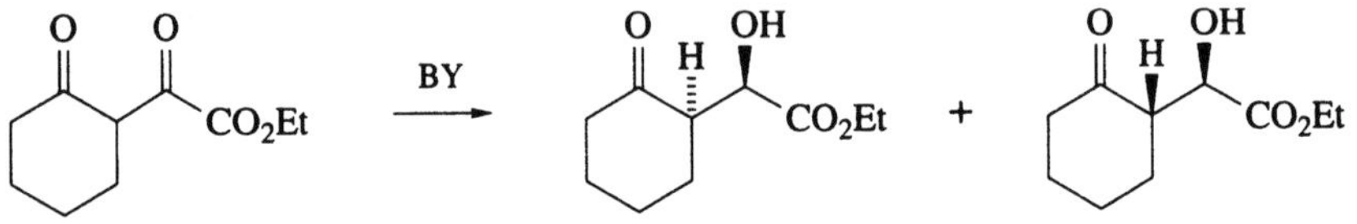

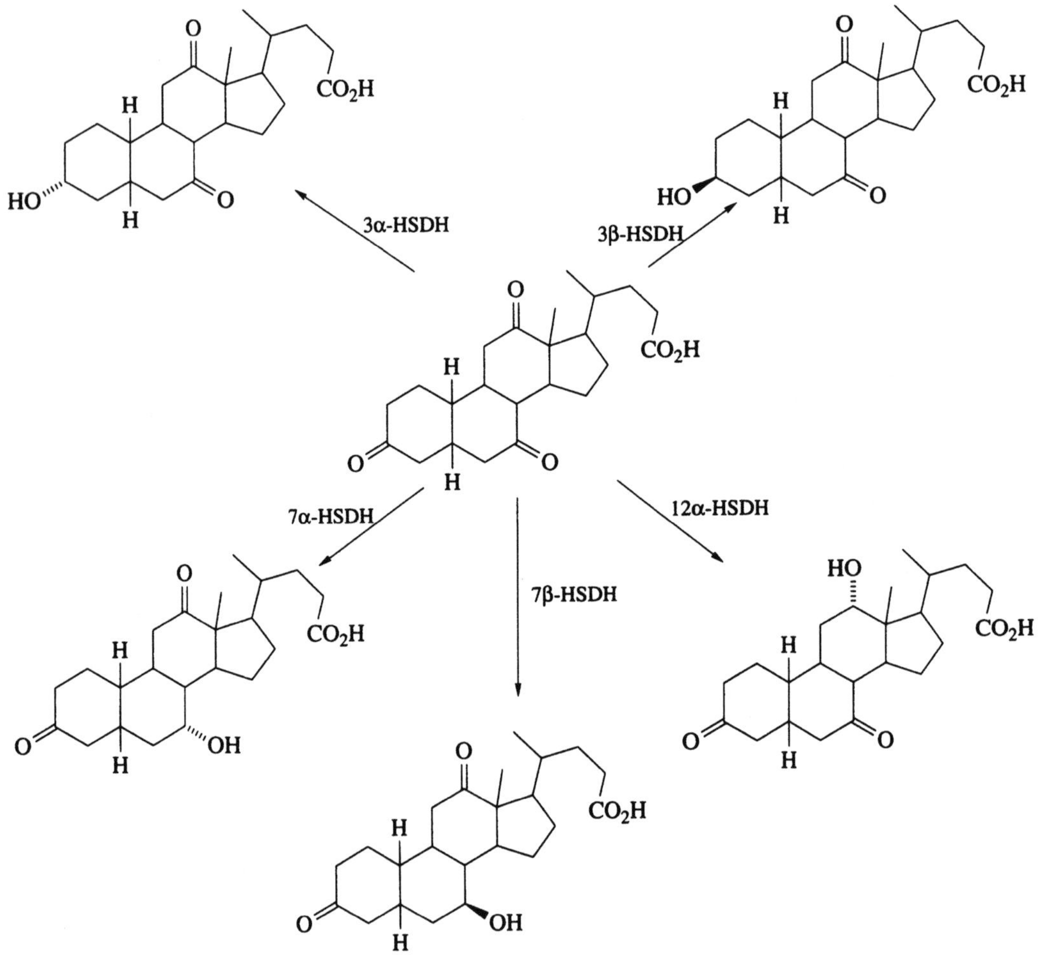

Scheme 28

The largest group of substrates reduced with combined specificity is that involving concurrent enantiomeric and diastereotopic face discrimination. Examples for acyclic substrates are given in Schemes 29 and 30. As Scheme 29 shows, variously functionalized racemic ketones (**65**),[31] α-keto esters (**66**)[104b] and β-keto esters (**67**),[106] (**68**)[107,108] and (**69**)[109] are transformed into the corresponding *threo*- and *erythro*-hydroxy diastereomers. The transformations of Scheme 30 are of particular interest since they lead to carbohydrate chirons of unusual or unnatural configurations.[110–112] The combination of diastereomeric and diastereotopic face selectivity is also feasible. This is shown in Scheme 31, again for carbohydrate synthon production.[113] For the BY-catalyzed reduction of (±)-(**70**), the configuration at the asterisked carbon is critical. Only diastereomers that have the (*S*)-configuration at this center are substrates. This keystone selection factor, together with the yeast preferences at the other stereogenic centers, results in (**71**) being the only product.[113]

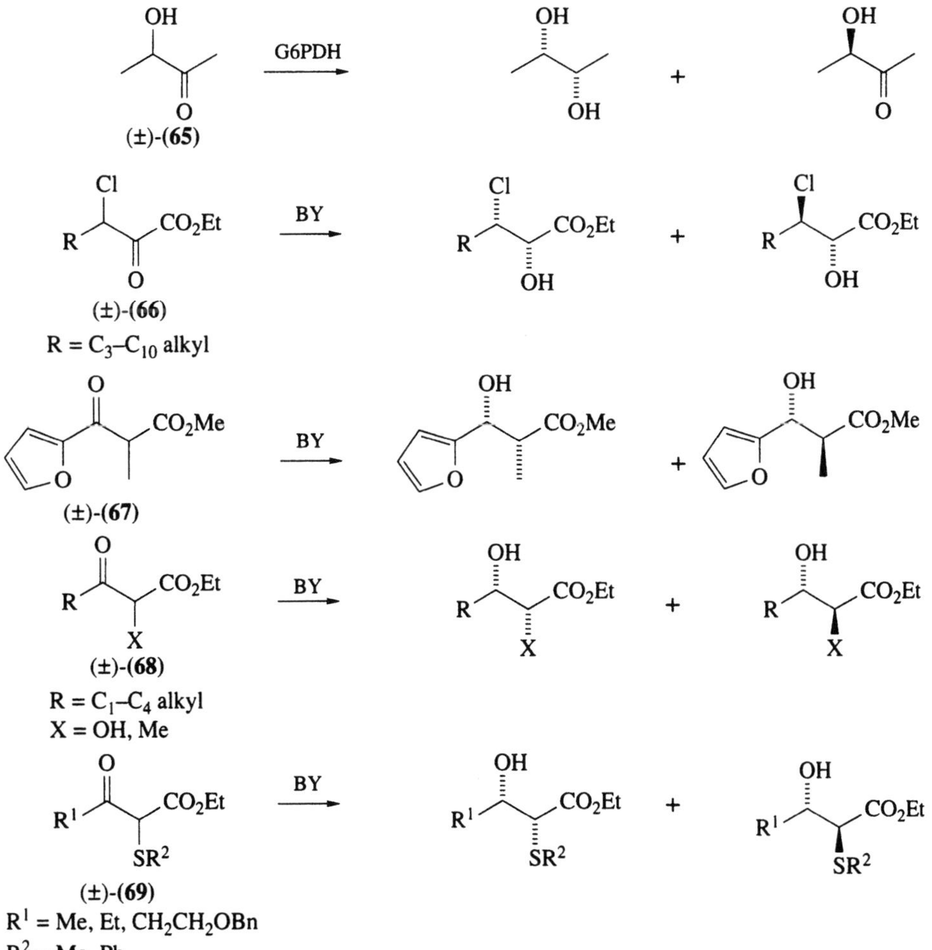

(±)-(**65**)

(±)-(**66**)
R = C_3–C_{10} alkyl

(±)-(**67**)

(±)-(**68**)
R = C_1–C_4 alkyl
X = OH, Me

(±)-(**69**)
R^1 = Me, Et, CH_2CH_2OBn
R^2 = Me, Ph

Scheme 29

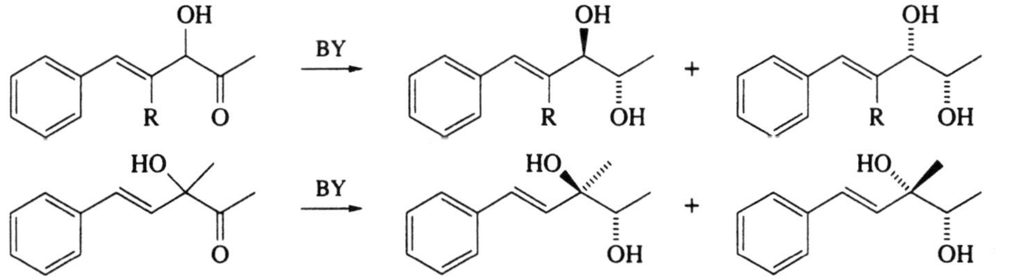

Scheme 30

When the stereogenic center of the racemate can undergo epimerization under the reduction conditions, the unreactive enantiomer will racemize continuously, thereby enabling complete conversion of

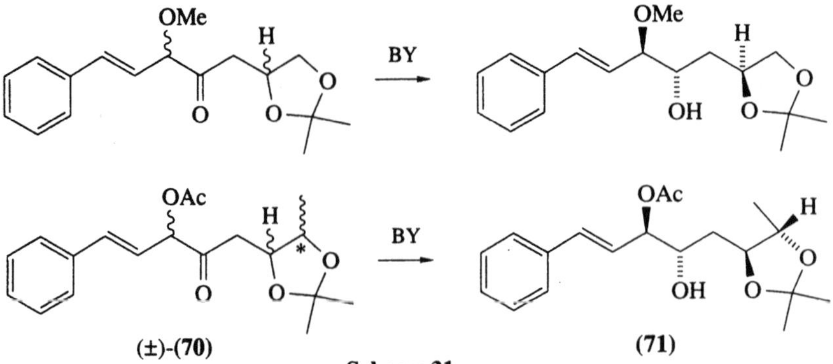

Scheme 31

substrate to a single product stereoisomer to be achieved. This is illustrated in Scheme 32 for the reduction of racemic β-keto esters.[77] By using the appropriate organism, different enantiomeric and diastereotopic face specificities can be preselected (Scheme 33).[114]

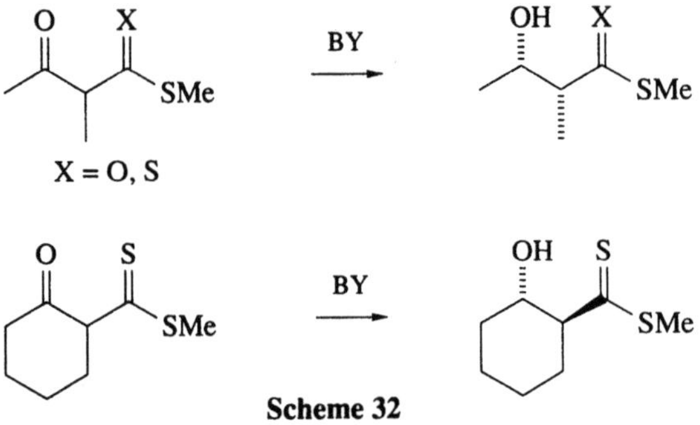

Scheme 32

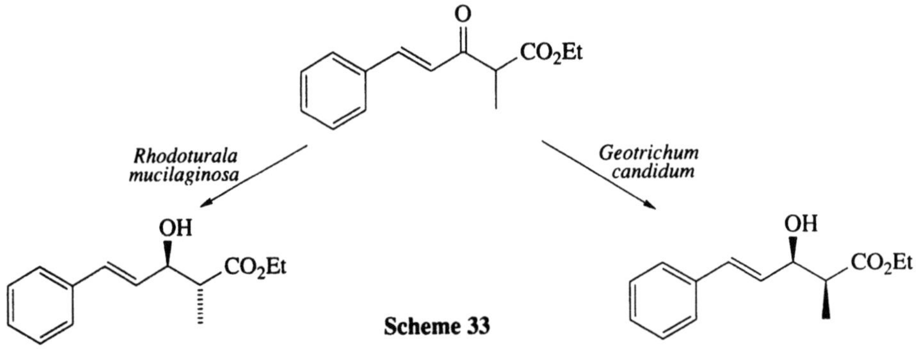

Scheme 33

The results with monocyclic ketones are similar. In Scheme 34, enzymic control of enantiomeric plus diastereotopic face specificity enables stereoisomerically pure product diols (**73**) and (**75**) and unreactive enantiomers (*R*)-(**72**) and (*R*)-(**74**) to be obtained from the racemates (±)-(**72**) and (±)-(**74**).[1a,31,115] In the case of the pyranones (±)-(**76**),[116] both enantiomers are reduced. When enantiomeric specificity is lacking in this way, the initial reaction is often that no resolution is possible by the enzymic procedure. However, this is not true when each enantiomer is reduced stereospecifically to a single diastereomeric product, such as (**77**) or (**78**); these are then easily separated by chromatography. In such situations, which are readily identified for HLADH using the cubic-space active-site model,[55] the reactions are allowed to proceed to completion rather than terminating at the 50% point that would apply for enantiomerically specific reductions. Using this approach, quantitative yields of each pure diastereomer may be isolated.[116] When the stereocenter is adjacent to a carbonyl group, epimerization of the unreactive enantiomer can occur under the reaction conditions. Then only the single stereoisomer product derived from the reactive enantiomer is formed. This is outlined in Scheme 35 for the BY-mediated reductions of racemic carbo- and hetero-cyclic β-keto esters and β-keto-thio esters.[117–122] By appropriate selection of the enzyme or microorganism, reductions in which the unreactive enantiomer racemizes under the reaction conditions can be directed towards either enantiomer of the starting racemate. Scheme 36 shows controlled reductions of this type, of 2-oxocyclohexanecarboxylic acid[123] or ester[117] (**79**) that give only the β-hydroxy products (**80**) of the (1*R*,2*S*) or (1*S*,2*R*) series respectively. With the right microorganisms, combinations of enantiomeric and diastereotopic face specificities can be exploited with racemic substrates to give all four possible product stereoisomers (Scheme 37).[106]

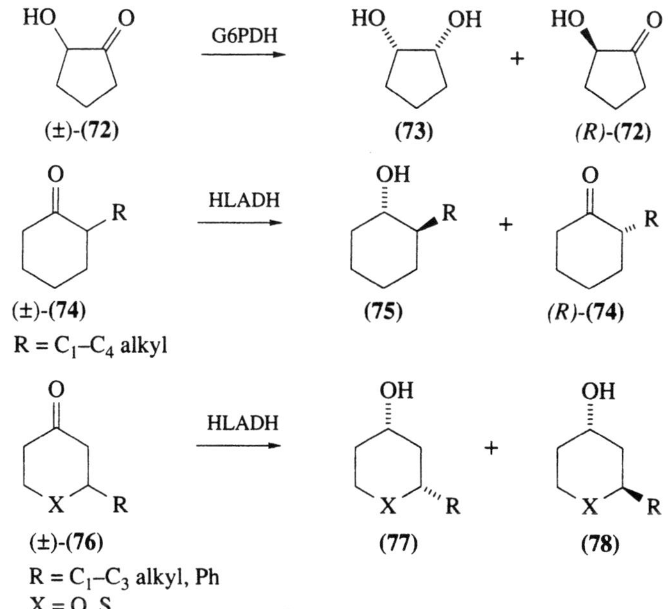

(±)-(**72**) —G6PDH→ (**73**) + (*R*)-(**72**)

(±)-(**74**), R = C_1–C_4 alkyl —HLADH→ (**75**) + (*R*)-(**74**)

(±)-(**76**), R = C_1–C_3 alkyl, Ph; X = O, S —HLADH→ (**77**) + (**78**)

Scheme 34

BY (CO_2Et; R)

BY (CO_2Et; $(CH_2)_n$, S; *n* = 1, 2)

BY (COX; X = O-*n*-octyl, SEt)

BY (CO_2Et; N, Boc)

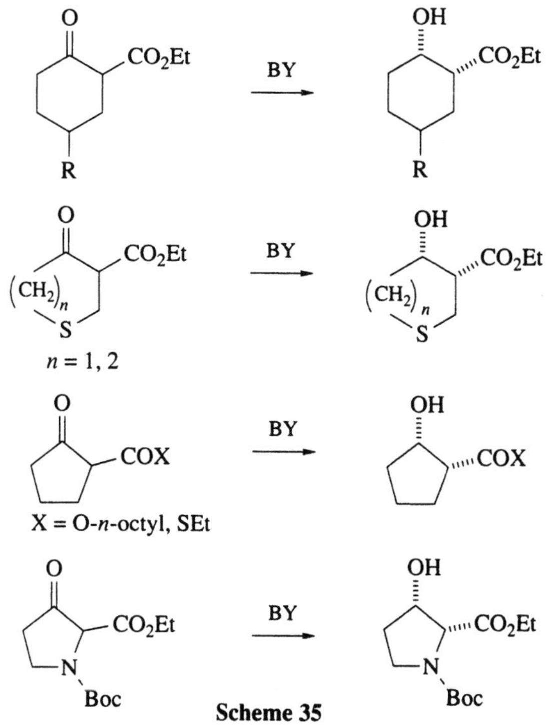

Scheme 35

Enantiomeric and diastereotopic face specific reductions are also readily effected on racemic bicyclic ketones. An illustration of the broad structural range that is amenable to enzyme-catalyzed transformation in this way is given in Scheme 38. While 2-decalones, such as (±)-(**81**)–(**83**),[124,125] and the related heterocyclic analogs (±)-(**85**)[126] are good substrates for HLADH, the 1-decalone (±)-(**84**) is not. However, by changing enzymes to MJADH, (±)-(**84**) becomes a good substrate.[127–129] Similarly, TBADH is a highly satisfactory catalyst for stereospecific reduction of (±)-(**86**), but will not accept its dimethyl

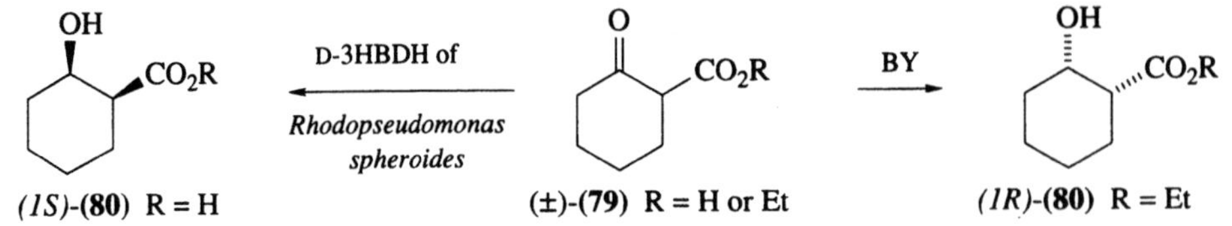

(1S)-(80) R = H ← D-3HBDH of *Rhodopseudomonas spheroides* — (±)-(79) R = H or Et — BY → (1R)-(80) R = Et

Scheme 36

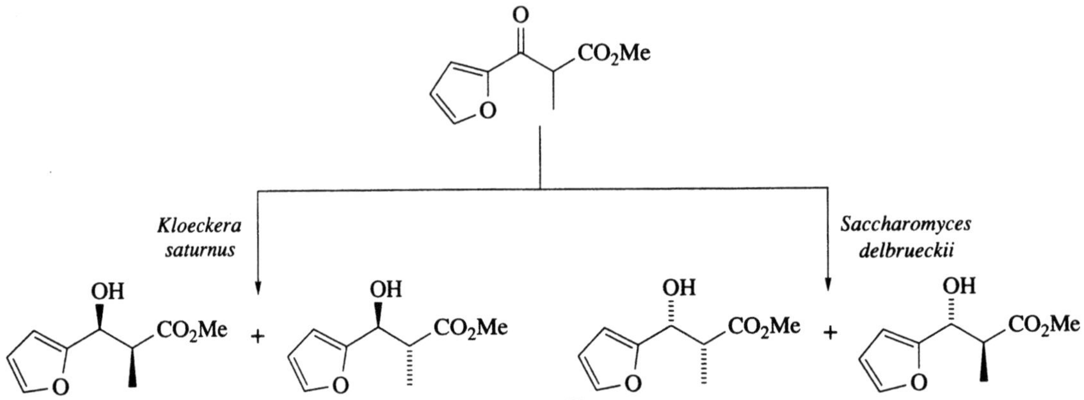

Kloeckera saturnus / *Saccharomyces delbrueckii*

Scheme 37

analog (±)-(**87**).[130] Here again, a change of enzyme to 3α,20β-HSDH solves the problem,[130] while BY is the recommended catalyst for conversions of (±)-(**88**).[131] As would be expected for similar structures, the same stereochemical pattern is followed by BY-catalyzed reductions of (±)-(**89**),[132] (±)-(**90**)[133] and (±)-(**91**).[134]

As with some of the racemic ketones discussed previously (Schemes 27, 29, 30, 31 and 37), reductions of bicyclic substrates to all possible stereoisomeric products can be achieved. The BY-mediated reduction of (±)-(**86**) to (**92**) and (**93**)[135] establishes this point (Scheme 39) and contrasts the more specific TBADH-promoted reaction of Scheme 38.[130]

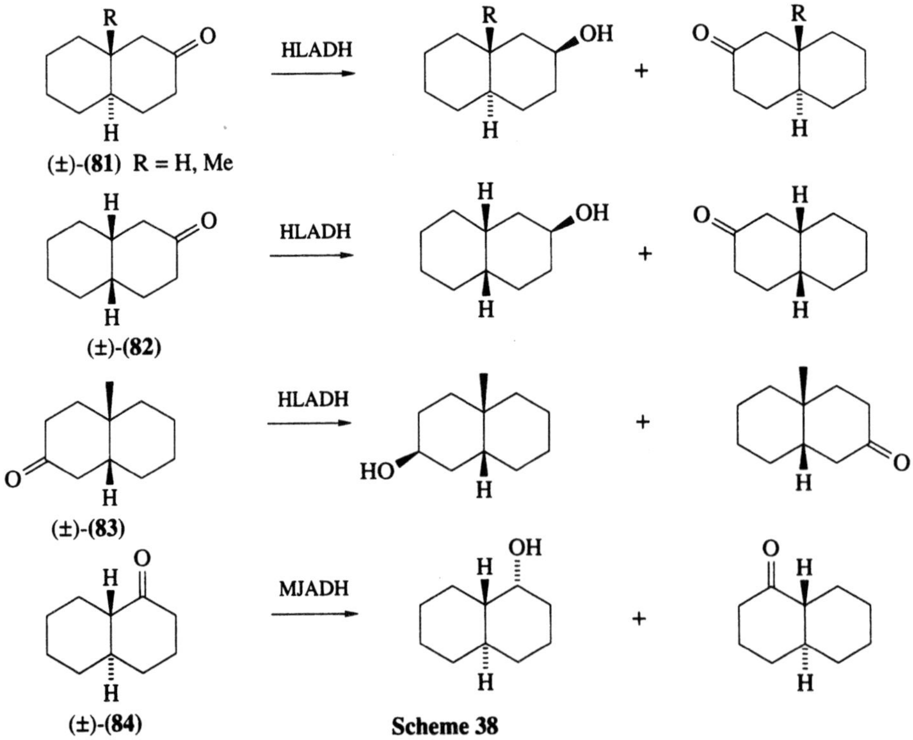

(±)-(**81**) R = H, Me — HLADH

(±)-(**82**) — HLADH

(±)-(**83**) — HLADH

(±)-(**84**) — MJADH

Scheme 38

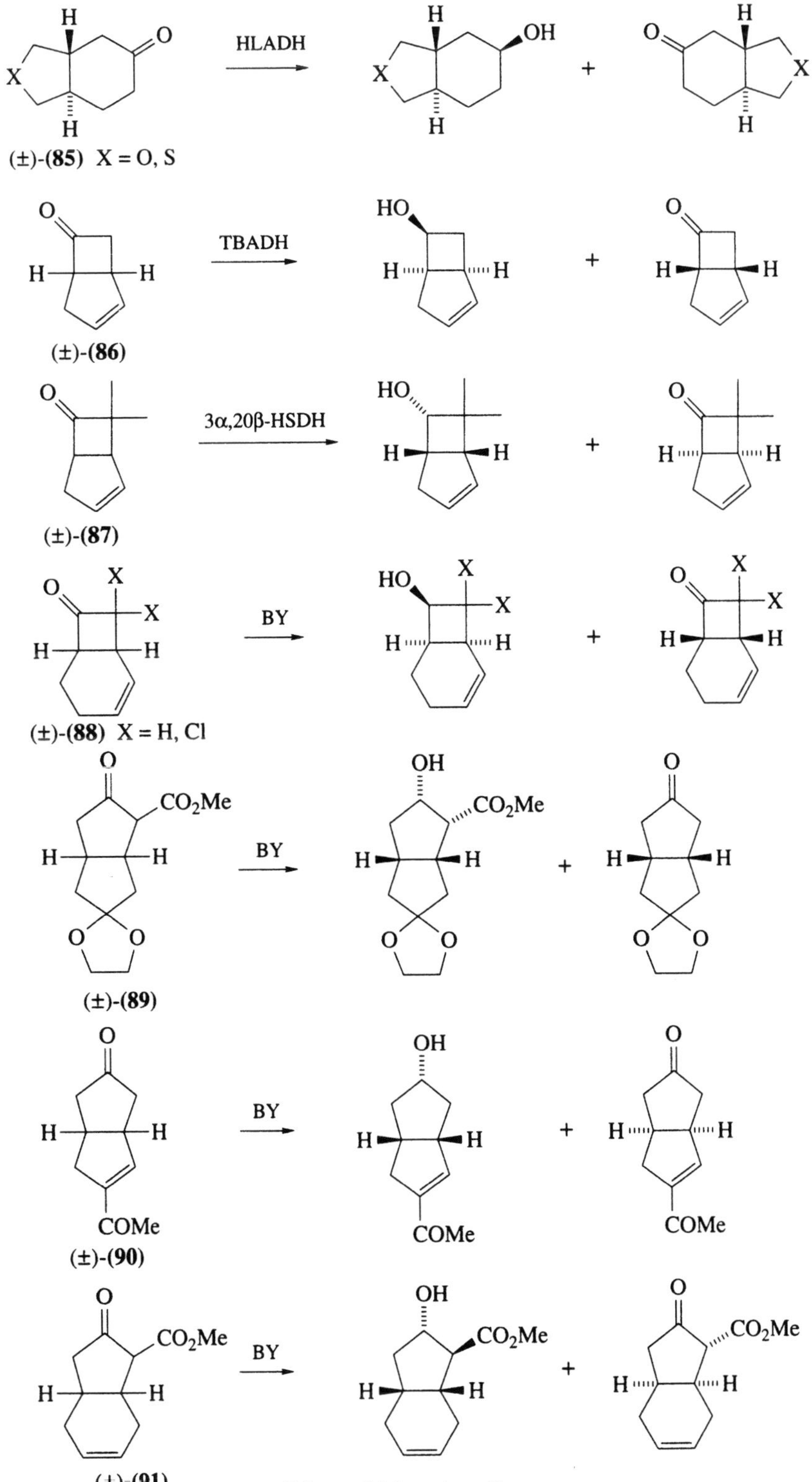

(±)-(85) X = O, S

HLADH

(±)-(86)

TBADH

(±)-(87)

3α,20β-HSDH

(±)-(88) X = H, Cl

BY

(±)-(89)

BY

(±)-(90)

BY

(±)-(91)

BY

Scheme 38 ***(continued)***

Exploitation of the complementary specificities of enzymes from different sources towards the same racemic substrate permits very precise control of the product stereochemistry. For example, any one of the three diastereomeric 2-decalols (**94**)–(**96**) can be obtained at will from (±)-*trans*-2-decalone (**81**; R = H) using the alcohol dehydrogenases HLADH, MJADH or CFADH, respectively (Scheme 40). The stereospecificities of these three enzymes are well documented and a simple active site model of predictive value is available for each.[1a,55] Racemic bridged bicyclic ketones are similarly discriminated, either

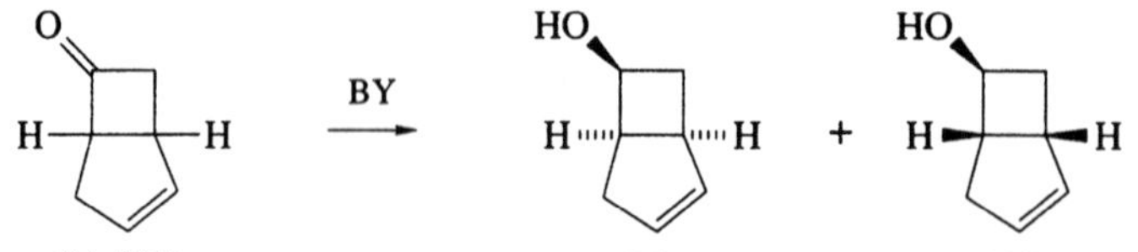

(±)-(**86**) BY (**92**) + (**93**)

Scheme 39

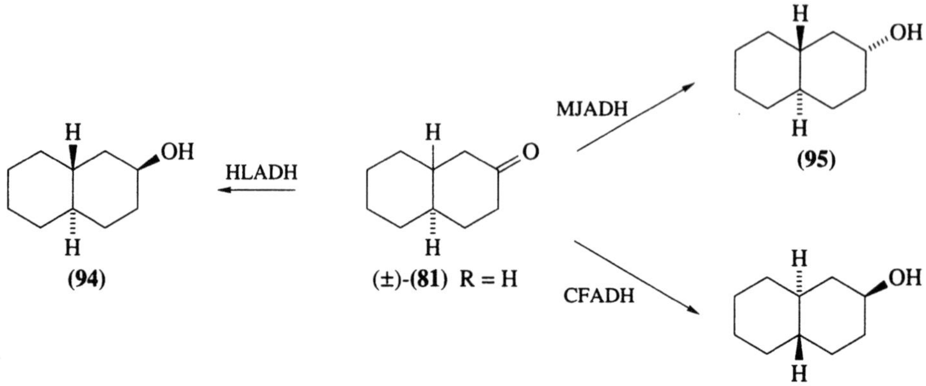

(**94**) HLADH (±)-(**81**) R = H MJADH (**95**) CFADH (**96**)

Scheme 40

to give the reduction products of one enantiomer, with recovery of the unreactive ketone enantiomer (Scheme 41)[136–138] or to give two diastereomerically pure alcohols from stereospecific reduction of each enantiomer (Scheme 42).[138]

The HLADH-catalyzed reduction of (±)-(**97**) is of interest since the alcohol product (**98**) is the thermodynamically less-preferred *exo* epimer. This emphasizes the fact that the geometry of an initial, kinetically controlled, product of an enzyme-catalyzed reduction reflects only the direction of attack of the hydride equivalent that is imposed by the orientation of substrate binding in the ES-complex. In this case, the reduction occurs on the *re* face of (1*S*)-(**97**),[136] in accord with the predictions of the active site

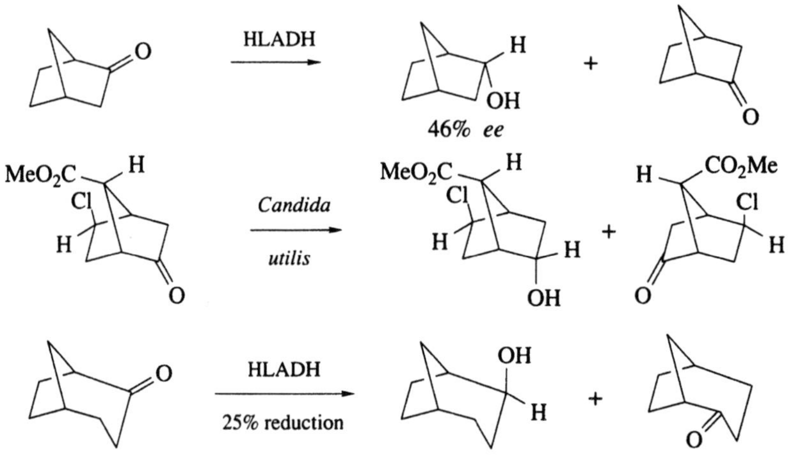

HLADH 46% *ee*

Candida utilis

(±)-(**97**) HLADH 25% reduction (**98**) 83% *ee* 18% *ee*

Scheme 41

BY

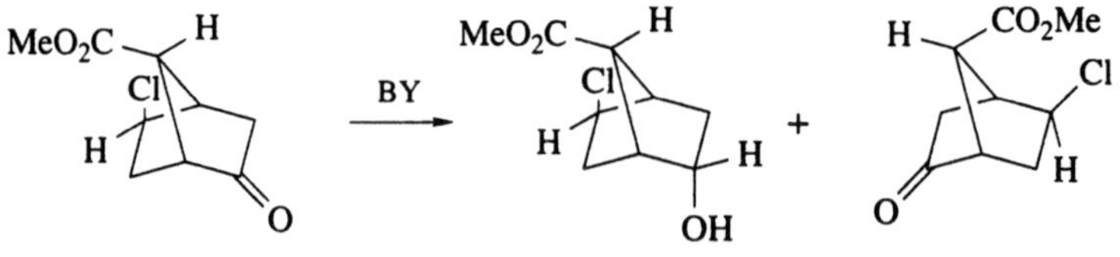

Scheme 42

model.[55] The feasibility of such discriminations for bicyclic diketones, such as (**99**), is demonstrated in Scheme 43. The hydroxy ketone (**100**) is the only alcohol product.[54] When α,β-unsaturated ketone functions are present in substrates, such as (±)-(**101**), chemospecific control can be exerted. In the example shown in Scheme 44, each enantiomer of (**101**) is reduced stereospecifically to the diastereomeric products (**102**) and (**103**), in each of which the new alcohol center has the (*S*) configuration.[139]

The examples discussed so far have been for ketones of chirality types commonly encountered. However, similar discriminations are possible with molecules of unusual chiralities, as represented in Scheme 45.

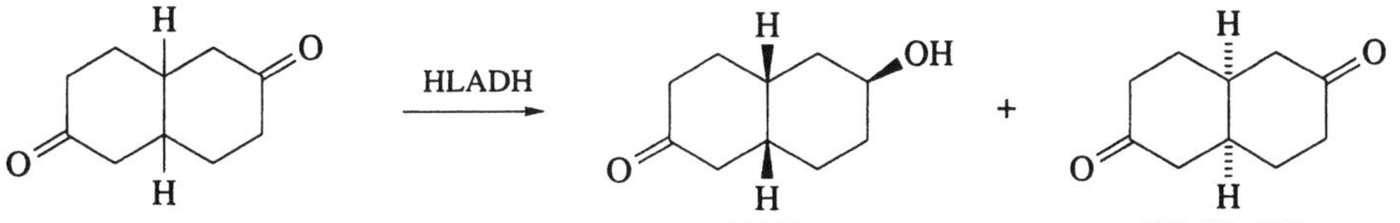

(±)-(**99**) HLADH (**100**) + *(1S,6S)*-(**99**)

Scheme 43

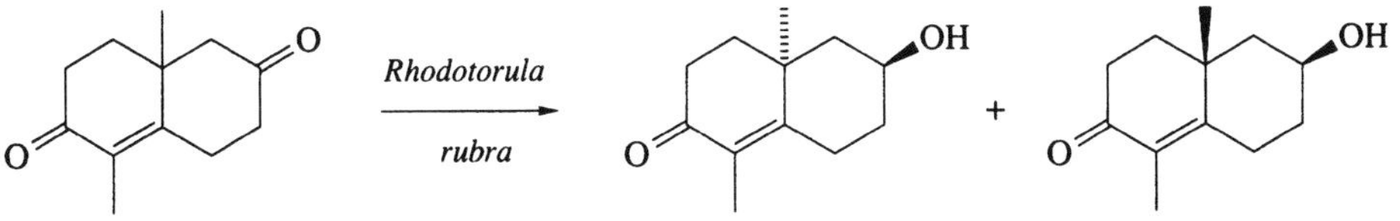

(±)-(**101**) *Rhodotorula rubra* (**102**) + (**103**)

Scheme 44

Rhodotorula rubra

HLADH

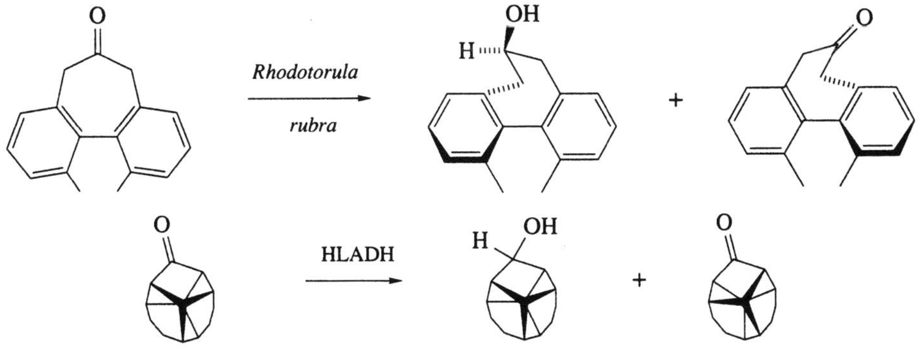

Scheme 45

Another important specificity combination is that of concurrent enantiotopic group and diastereotopic face selection. This is exemplified for mono-, bi- and poly-cyclic diketones in Scheme 46. Reduction of variously substituted cyclopentanediones (**104**) with BY gives products of the diastereomeric series (**105**).[140] Interestingly, *Pichia terricola* catalyzed reduction[141] of (**106**) → (**107**) proceeds in the same stereochemical senses as for the BY-mediated reductions of (**104**),[142] although (**106**) itself is not a substrate for BY.[141] For such compounds, the BY specificity selection parameters are controlled by the nature of the alkyl substituents and the ring size. The configuration of the alcohol center produced is (2*S*) in every case, but BY-mediated reduction of the 3-substituted cyclopentane- and cyclohexane-1,3-diones proceed in opposite diastereomeric senses with respect to the configuration of the 3-position.[140] The reductions of the bicyclic diketones (**108**), (**109**)[54] and (**110**),[1a,125] and the polycyclic substrate (**111**)[143] are stereospecific. Once again, different organisms can be employed to give different stereoisomers from the same substrate. The diketone (**112**) of Scheme 47 can be converted stereospecifically into the natural (**113**) or unnatural (**114**) steroid precursors.[144,145] An example of simultaneous diastereotopic group and face combination is given in Scheme 48.[54]

Even higher multiplicities of specificity combinations are possible. In Scheme 49, conversion of (**115**) to (**116**)[144] involves regiospecific reduction of an enantiotopic carbonyl group concurrently with enantiotopic face specificity. This transformation can also be achieved using *Saccharomyces* species, with the level of (**116**) produced being enhanced by the addition of unsaturated carbonyl compounds, such as acrolein.[145]

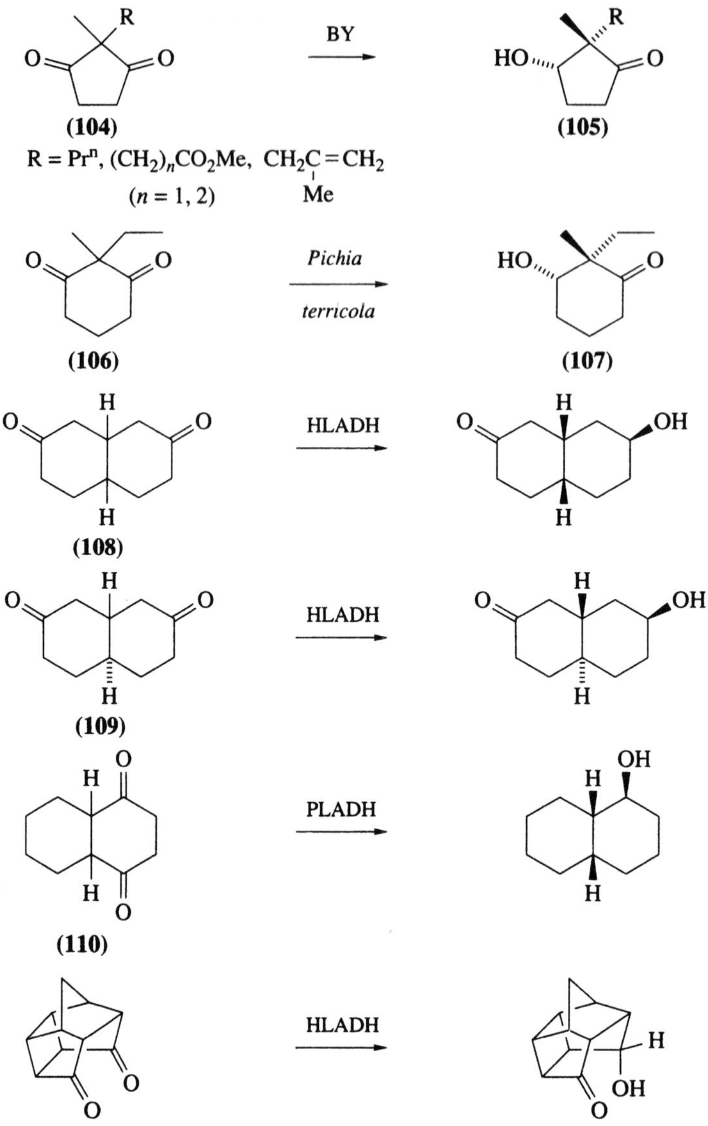

BY
(104)
(105)
R = Pr^n, $(CH_2)_nCO_2Me$, $CH_2C(Me){=}CH_2$
(n = 1, 2)
Pichia terricola
(106)
(107)
HLADH
(108)
HLADH
(109)
PLADH
(110)
HLADH
(111)

Scheme 46

Saccharomyces uvarum
Bacillus thuringiensis
MeO
(113)
(112)
(114)

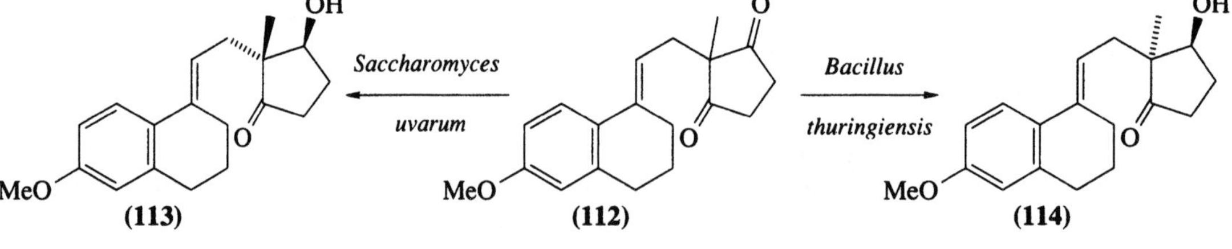

Scheme 47

HLADH

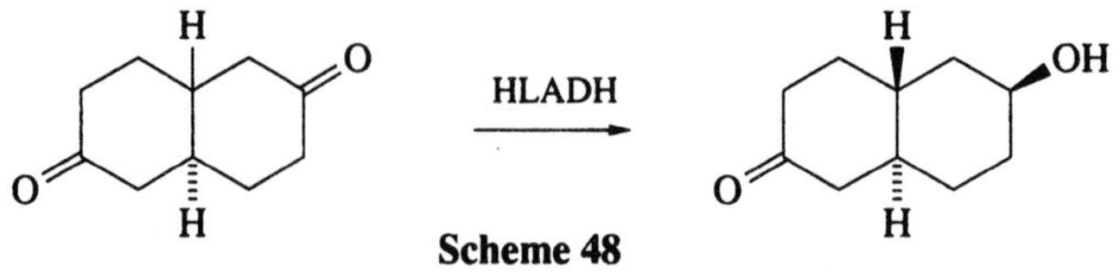

Scheme 48

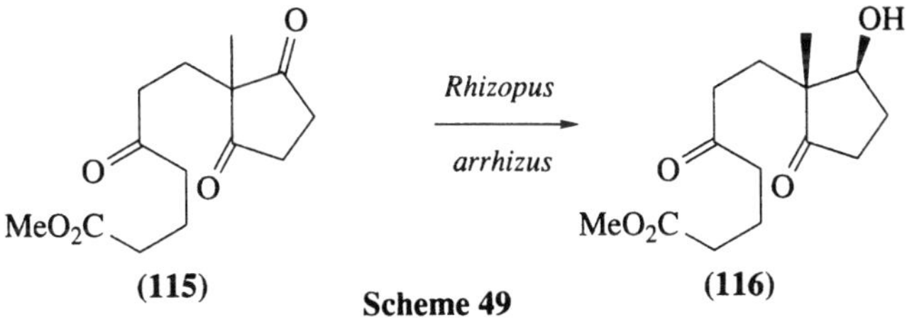
O
OH
Rhizopus
arrhizus
O
O
O
O
MeO2C
MeO2C
(115)
(116)

Scheme 49

Sequential reductions are possible for diketones. That of Scheme 50 involves straightforward initial enantiotopic face specificity followed by diastereotopic face control.[146] In contrast, the initial alcohol (*S*)-(**118**) produced from (**117**; Scheme 51) must epimerize to (*R*)-(**118**) before the second reduction to (*R*,*R*)-(**119**) can take place.[147]

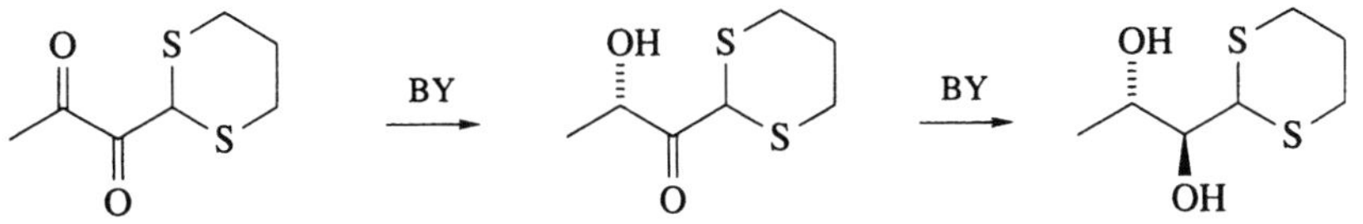
O
S
S
O
BY
OH
S
S
O
BY
OH
S
S
OH

Scheme 50

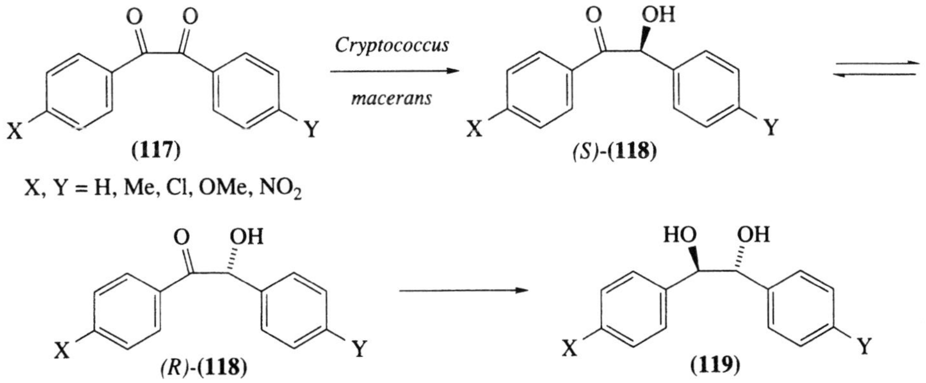
O O
X
Y
(117)
X, Y = H, Me, Cl, OMe, NO2
Cryptococcus
macerans
O OH
X
Y
(S)-(118)
O OH
X
Y
(R)-(118)
HO OH
X
Y
(119)

Scheme 51

Yeast-mediated reductions can provide benefits in addition to their control of the stereospecificity of hydride delivery. For example, the acetaldehyde produced in the fermentation process can participate in acyloin condensations prior to the reduction steps (Scheme 52).[111,148–150]

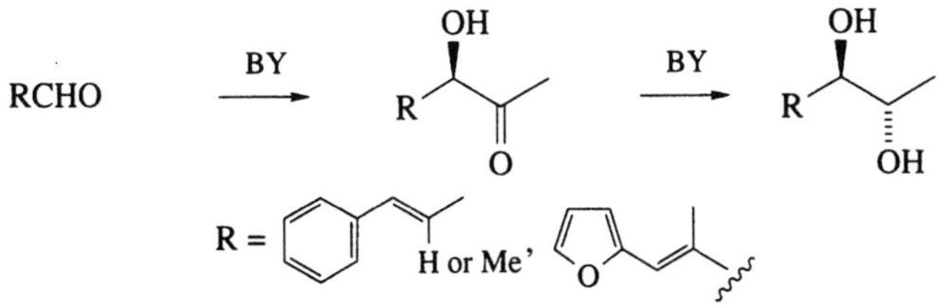
RCHO
BY
OH
R
O
BY
OH
R
OH
R =
H or Me'
O

Scheme 52

Enzyme-mediated carbonyl reductions are not restricted to aldehydes and ketones. With some organisms, carboxy groups can be reduced to primary alcohols. The example in Scheme 53 proceeds with enantiomeric specificity.[151]

1.8.6.5 Preparations of Deuterated or Tritiated Alcohols

From the broad range of stereospecific reductions listed above, it is evident that enzymes can provide attractive, and often unique, opportunities for the stereospecific introduction of isotopes. Introduction of

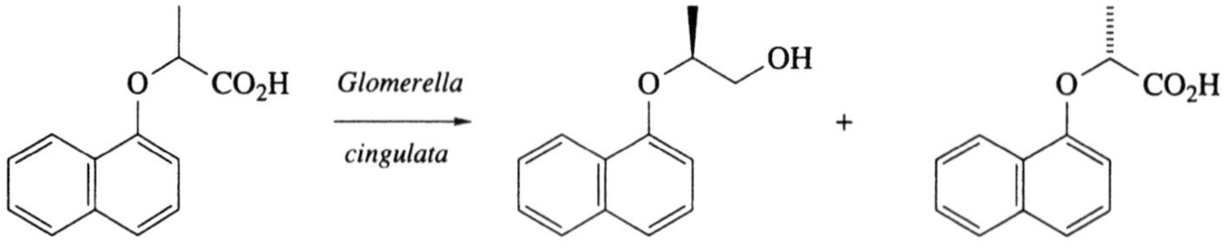

Glomerella cingulata

36% *ee*

Scheme 53

deuterium or tritium can be accomplished in several ways. The predictable stereospecificity of reductions of aldehydes can be exploited to prepare either (*R*) or (*S*) deuterated or tritiated primary alcohols. This is manifest in Scheme 54 for YADH-catalyzed reactions,[1] which occur with *re* face specificity. The opposite configurations can be produced by swapping the locations of protium and deuterium/tritium in the substrate or coenzyme respectively. If BY is employed, the label must be located in the substrate aldehyde, as represented in Scheme 55.[152] Alternatively, enzymes of opposite enantiotopic face specificities can be employed (Scheme 56).[153]

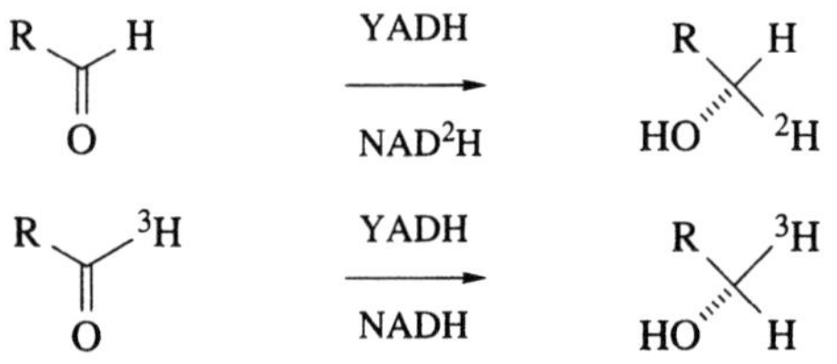

YADH, NAD^2H

YADH, NADH

Scheme 54

BY

Scheme 55

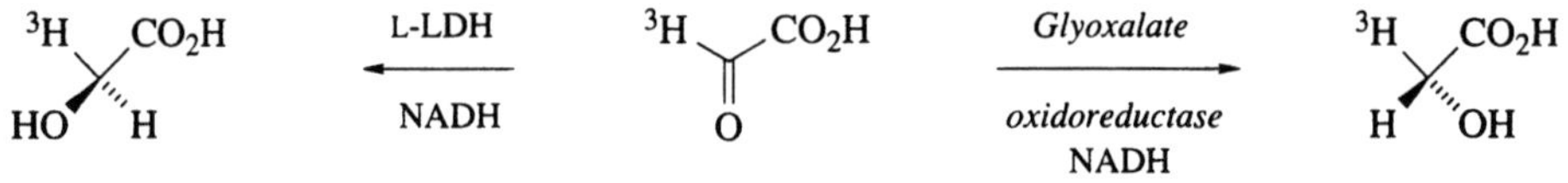

L-LDH, NADH

Glyoxalate oxidoreductase, NADH

Scheme 56

The source of the label may be the solvent or a coupled substrate. It can be introduced by a single enzyme (Scheme 57)[1] or using a coupled enzyme system (Scheme 58).[31]

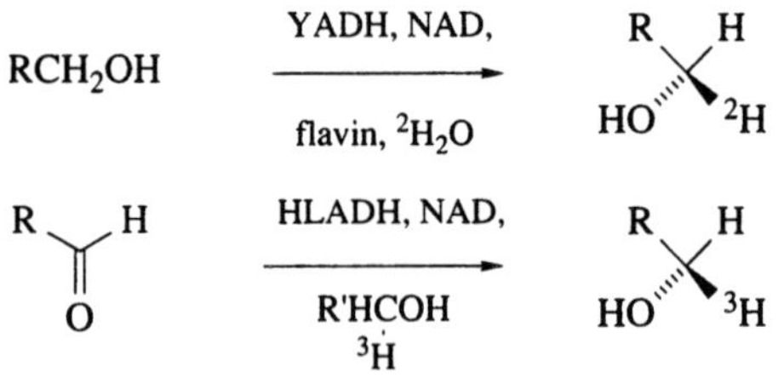

RCH_2OH → YADH, NAD, flavin, ^2H_2O

HLADH, NAD, R'HCOH (^3H)

Scheme 57

1.8.7 REDUCTION OF C=N TO CHNH

The only preparative-scale reactions of synthetic value in this category are those catalyzed by the amino acid dehydrogenases. These enzymes catalyze the reductions of α-imino acids to α-amino acids. This can be done on a very large scale, as demonstrated by the LeuDH-catalyzed reduction of (**120**) to *t*-butyl leucine (**121**) shown in Scheme 59.[154] Another enzyme of this group with preparative promise is

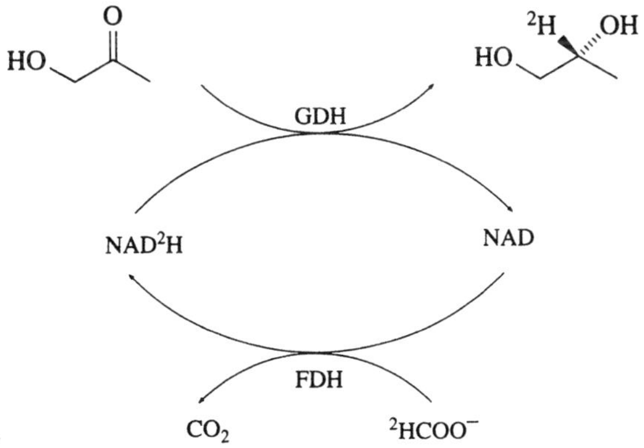

Scheme 58

pimelate dehydrogenase.[155] Enzyme-catalyzed reductions of imines have also been documented in alkaloid biosynthetic pathways,[156] but as yet are of limited synthetic practicality.

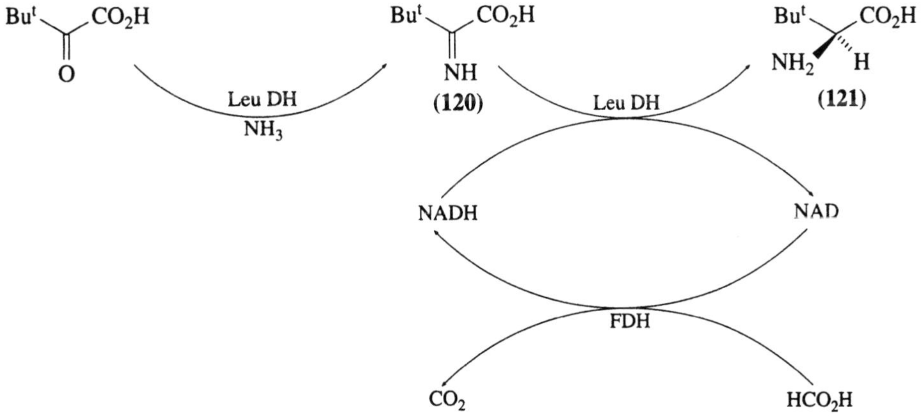

Scheme 59

1.8.8 REDUCTION OF C=C TO CHCH

The alkene reduction reactions most frequently observed are of α,β-unsaturated aldehydes, ketones, acids and esters. Examples of stereospecific reductions of acyclic substrates are given in Scheme 60.[148,157–159] In the formation of (**123**), the double bond of (**122**) is reduced prior to the aldehyde function.[148] The conversion of (**124**) to (**125**) involves oxidation of the intermediate alcohol to the carboxylic acid by bubbling air into the fermentation medium.[157] Stereospecific reductions of α,β-unsaturated ketones may be similarly effected (Scheme 61).[160–163] The reduction of the chloro ketone (**126**) gives (**127**) initially. This epimerizes under the reaction conditions, and each enantiomer is then reduced further to (**128**) and (**129**), with the predominance of the (**128**) stereoisomer increasing with the size of the R-group.[166] Reduction of (±)-(**130**) leads to (**131**) and (**132**).[163]

Reductions of allylic alcohols can also go stereospecifically (Scheme 62). However, they are not always substrates. For example, the allylic alcohol arising from reduction of the formyl group of the α,β-unsaturated aldehyde (**122**) is not reduced to (**123**) by *B. sulfurescens*.[148]

For the racemic allenic acid (**133**), each enantiomer is reduced stereospecifically to give the alkenoic acid diastereomers (**134**) and (**135**) (Scheme 63).[164] In contrast, the levels of enantiomeric discrimination observed in microbial reductions of allenic alcohols have so far been very low.[165]

1.8.9 FUTURE DEVELOPMENTS

The examples of enzyme-catalyzed C=X → CHXH transformations documented in this chapter show that the approach is already a well developed and versatile one. We may look forward to advances in the

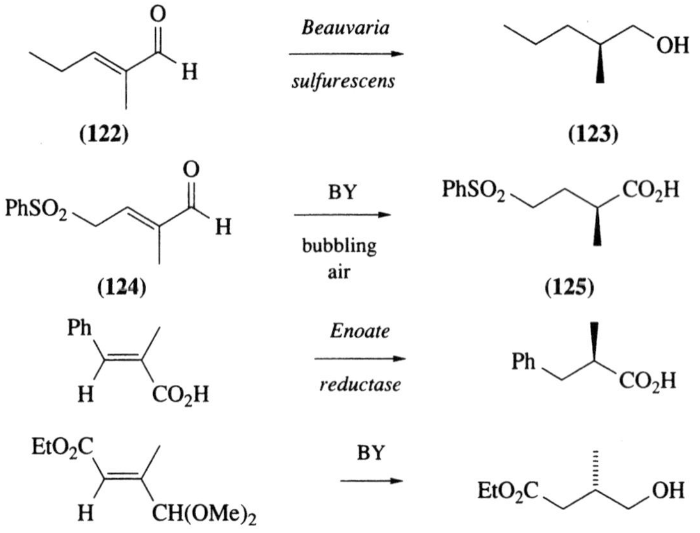

Beauvaria sulfurescens

(122) (123)

BY bubbling air

(124) (125)

Enoate reductase

BY

Scheme 60

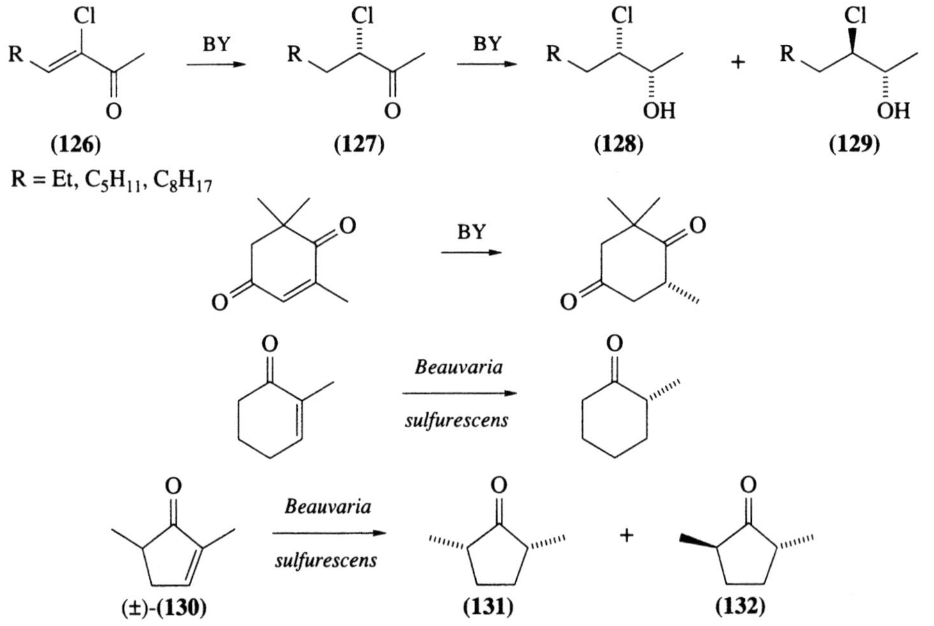

BY BY

(126) (127) (128) (129)

R = Et, C_5H_{11}, C_8H_{17}

BY

Beauvaria sulfurescens

Beauvaria sulfurescens

(±)-(130) (131) (132)

Scheme 61

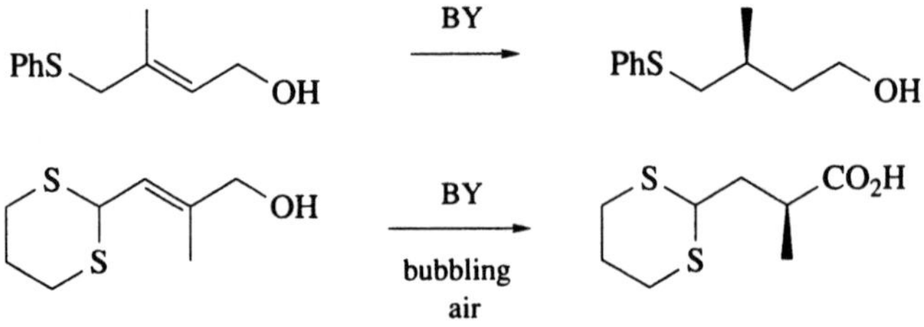

BY

BY bubbling air

Scheme 62

use of organic solvents[13,14] and unnatural coenzymes.[166] Further ahead, there are even more exciting possibilities in the genetic manipulation of microorganisms, in the antibody-based 'abzyme' approach to the design of enzyme specificity,[167,168] and in tailoring enzyme specificities,[169,170] including those of oxidoreductases,[171–173] *via* the site-directed techniques of molecular biology.

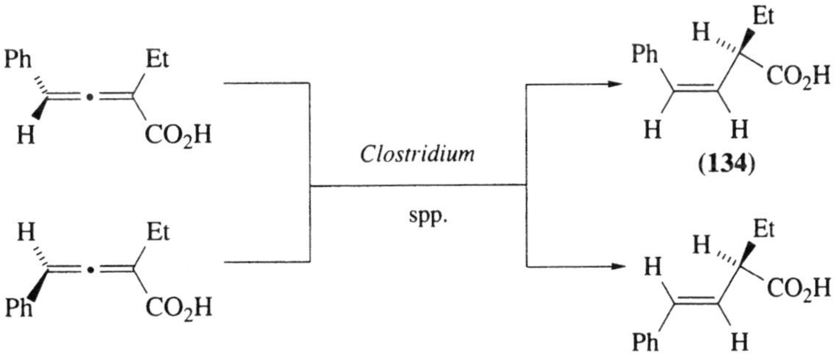

(±)-(133)

Clostridium spp.

(134)

(135)

Scheme 63

1.8.10 REFERENCES

1. (a) J. B. Jones and J. F. Beck, *Tech. Chem. (N.Y.)*, 1976, **10**, 107; (b) J. B. Jones, *Tetrahedron*, 1986, **42**, 3351.
2. C.-H. Wong and G. M. Whitesides, *Angew. Chem., Int. Ed. Engl.*, 1985, **24**, 617.
3. C.-H. Wong, *Science (Washington, D.C.)*, 1989, **244**, 1145.
4. P. D. Ritchie 'Asymmetric Synthesis and Asymmetric Induction', Oxford University Press, London, 1933.
5. Ref. 1a, 491.
6. K. Kieslich, 'Microbial Transformations of Non-steroid Cyclic Compounds', Thieme, Stuttgart, 1976; K. Kieslich, in 'Biotechnology', ed. H. J. Rehm and G. Reed, Verlag Chemie, Weinheim, 1984, vol. 6A.
7. D. Perlman, *Tech. Chem. (N.Y.)*, 1976, **10**, 47.
8. 'Enzyme Nomenclature', Academic Press, New York, 1979.
9. K. Mosbach, *Methods Enzymol.*, 1987, **136** (c), 1.
10. J. Grunwald, B. Wirz, M. P. Scollar and A. M. Klibanov, *J. Am. Chem. Soc.*, 1986, **108**, 6732.
11. J. B. Jones and H. M. Schwartz, *Can. J. Chem.*, 1982, **60**, 335.
12. D. G. Drueckhammer, S. K. Sadozai and C.-H. Wong, *Enzyme Microb. Technol.*, 1987, **9**, 564.
13. A. Zaks and A. M. Klibanov, *J. Biol. Chem.*, 1988, **263**, 8017.
14. C. Laane, J. Tramper and M. D. Lilly (eds.), 'Biocatalysis in Organic Media', Elsevier, Amsterdam, 1987.
15. H. K. Chenault and G. M. Whitesides, *Appl. Biochem. Biotechnol.*, 1987, **14**, 147; H. K. Chenault, E. S. Simon and G. M. Whitesides, in 'Biotechnology and Genetic Engineering Reviews', ed. G. E. Russell, Intercept, Wimbourne, UK, 1988, vol. 6, p. 221.
16. L. K. P. Lam, R. A. H. F. Hui and J. B. Jones, *J. Org. Chem.*, 1986, **51**, 2047.
17. R. Chênevert and L. D'Astous, *Can. J. Chem.*, 1988, **66**, 1219.
18. K. Nakamura, M. Higaki, K. Ushio, S. Oka and A. Ohno, *Tetrahedron Lett.*, 1985, **26**, 4213.
19. C. J. Sih and C.-S. Chen, *Angew. Chem., Int. Ed. Engl.*, 1984, **23**, 570.
20. D. E. Metzler, 'Biochemistry', Academic Press, New York, 1977.
21. L. Stryer, 'Biochemistry', 3rd edn., Freeman, New York, 1988.
22. J. D. Rawn, 'Biochemistry', Patterson, Burlington, NC, 1989.
23. C. Walsh, 'Enzymatic Reaction Mechanisms', Freeman, New York, 1979.
24. K. Kawahara, M. Matsumoto, H. Hashimoto and S. Miyano, *Chem. Lett.*, 1988, 1163.
25. K. Nakamura, T. Miyai, K. Ushio, S. Oka and A. Ohno, *Bull. Chem. Soc. Jpn.*, 1988, **61**, 2089.
26. S. Tsuboi, N. Kohara, K. Doi, M. Utaka and A. Takeda, *Bull. Soc. Chem. Jpn.*, 1988, **61**, 3205.
27. S. Top, G. Jaouen, J. Gillois, C. Baldoli and S. Maiorana, *J. Chem. Soc., Chem. Commun.*, 1988, 1284.
28. J. Gillois, D. Buisson, R. Azerad and G. Jaouen, *J. Chem. Soc., Chem. Commun.*, 1988, 1224.
29. M. F. Zueger, F. Giovannini and D. Seebach, *Angew. Chem., Int. Ed. Engl.*, 1983, **22**, 1012.
30. W. Boland and U. Niedermeyer, *Synthesis*, 1987, 28.
31. L. G. Lee and G. M. Whitesides, *J. Org. Chem.*, 1986, **51**, 25.
32. A. Belan, J. Bolte, A. Fauve, J. G. Gourcy and H. Veschambre, *J. Org. Chem.*, 1987, **52**, 256; A. Fauve and H. Veschambre, *Tetrahedron Lett.*, 1987, **28**, 5037.
33. E. Keinan, K. K. Seth and R. Lamed, *J. Am. Chem. Soc.*, 1986, **108**, 3474.
34. D. G. Drueckhammer, C. F. Barbas, III, K. Nozaki, C.-H. Wong, C. Y. Wood and M. A. Ciufolini, *J. Org. Chem.*, 1988, **53**, 1607.
35. C. Neuberg and A. Lewite, *Biochem. Z.*, 1918, **91**, 257; C. Neuberg and F. F. Nord, *Chem. Ber.*, 1919, **52**, 2237, 2248.
36. D. R. Boyd and D. C. Neill, *J. Chem. Soc., Chem. Commun.*, 1977, 31.
37. M. Bucciarelli, A. Forni, I. Moretti and G. Torre, *J. Chem. Soc., Chem. Commun.*, 1978, 456; A. Forni, I. Moretti and G. Torre, *J. Chem. Soc., Chem. Commun.*, 1977, 731.
38. A. Manzocchi, A. Fiecchi and E. Santaniello, *Synthesis*, 1987, 1007.
39. M. Takeshita, K. Terada, N. Akutsu, S. Yoshida and T. Sato, *Heterocycles*, 1987, **26**, 3051.
40. T. Fujisawa, T. Itoh, M. Nakai and T. Sato, *Tetrahedron Lett.*, 1985, **26**, 771.
41. M. Imuta and H. Ziffer, *J. Org. Chem.*, 1978, **43**, 3530.
42. R. Lis, W. B. Caldwell, G. A. Hoyer and K. Petzoldt, *Tetrahedron Lett.*, 1987, **28**, 1487.
43. C. Syldatk, A. Stoffregen, F. Wuttke and R. Tacke, *Biotechnol. Lett.*, 1988, **10**, 731.
44. T. Fujisawa, E. Kojima, T. Itoh and T. Sato, *Chem. Lett.*, 1985, 1751.
45. G. Guanti, L. Banfi and E. Narisano, *Tetrahedron Lett.*, 1986, **27**, 3547.
46. V. Prelog, *Pure Appl. Chem.*, 1964, **9**, 119.

47. R. Bernardi, R. Cardillo, D. Ghiringhelli and O. Vajna de Pava, *J. Chem. Soc., Perkin Trans. 1*, 1987, 1607.
48. V. I. Solokov, L. L. Troitskaya and T. I. Rozhkova, *Gazz. Chim. Ital.*, 1987, **117**, 525.
49. H. Ohta, J. Konishi, Y. Kato and G. Tsuchihashi, *Agric. Biol. Chem.*, 1987, **51**, 2421.
50. K. Mori and H. Watanabe, *Tetrahedron*, 1986, **42**, 273.
51. A. Fauve and H. Veschambre, *J. Org. Chem.*, 1988, **53**, 5215.
52. N.-Y. Wang, C.-T. Hsu and C. J. Sih, *J. Am. Chem. Soc.*, 1981, **103**, 6538.
53. J. d'Angelo, G. Revial, R. Azerad and D. Buisson, *J. Org. Chem.*, 1986, **51**, 40.
54. D. R. Dodds and J. B. Jones, *J. Am. Chem. Soc.*, 1988, **110**, 577.
55. J. B. Jones and I. J. Jakovac, *Can. J. Chem.*, 1982, **60**, 19; I. J. Jakovac, H. B. Goodbrand, K. P. Lok and J. B. Jones, *J. Am. Chem. Soc.*, 1982, **104**, 4659.
56. H. Schuette, W. Hummel and M. R. Kula, *Appl. Microbiol. Biotechnol.*, 1984, **19**, 167.
57. W. Hummel, H. Schuette and M. R. Kula, *Appl. Microbiol. Biotechnol.*, 1985, **21**, 7.
58. M.-J. Kim and G. M. Whitesides, *J. Am. Chem. Soc.*, 1988, **110**, 2959.
59. E. S. Simon, R. Plante and G. M. Whitesides, *Appl. Biochem. Biotech.*, 1989, **22**, 169.
60. D. Bur, M. A. Luyten, H. Wynn, L. R. Provencher and J. B. Jones, *Can. J. Chem.*, 1989, **67**, 1065.
61. B. L. Hirschbein and G. M. Whitesides, *J. Am. Chem. Soc.*, 1982, **104**, 4458.
62. Y. Yamazaki and H. Maeda, *Agric. Biol. Chem.*, 1986, **50**, 2621.
63. M. Kori, K. Itoh and H. Sugihara, *Chem. Pharm. Bull.*, 1987, **35**, 2319.
64. H. Suemune, Y. Mizuhara, H. Akita, T. Oishi and K. Sakai, *Chem. Pharm. Bull.*, 1987, **35**, 3112.
65. K. Nakamura, T. Yoneda, T. Miyai, K. Ushio, S. Oka and A. Ohno, *Tetrahedron Lett.*, 1988, **29**, 2453.
66. S. Shimizu, H. Hata and H. Yamada, *Agric. Biol. Chem.*, 1984, **48**, 2285.
67. H. Hata, S. Shimizu and H. Yamada, *Agric. Biol. Chem.*, 1987, **51**, 3011.
68. K. Nakamura, K. Inoue, K. Ushio, S. Oka and A. Ohno, *J. Org. Chem.*, 1988, **53**, 2589.
69. D. Seebach, F. Giovannini and B. Lamatsch, *Helv. Chim. Acta*, 1985, **68**, 958.
70. K. Mori and M. Ikunaka, *Tetrahedron*, 1987, **43**, 45.
71. A. Manzocchi, A. Fiecchi and E. Santaniello, *J. Org. Chem.*, 1988, **53**, 4405.
72. D. Seebach and M. Eberle, *Synthesis*, 1986, 37.
73. A. Manzocchi, P. Casati, A. Fiecchi and E. Santaniello, *J. Chem. Soc., Perkin Trans. 1*, 1987, 2753.
74. H. Yamada, H. Sugiyama and M. Kajiwara, *Heterocycles*, 1987, **26**, 2841.
75. D. W. Brooks, R. P. Kellogg and C. S. Cooper, *J. Org. Chem.*, 1987, **52**, 192.
76. A. P. Kozikowski, B. B. Mugrage, C. S. Li and L. Felder, *Tetrahedron Lett.*, 1986, **27**, 4817.
77. T. Itoh, Y. Yonekawa, T. Sato and T. Fujisawa, *Tetrahedron Lett.*, 1986, **27**, 5405.
78. D. Seebach, P. Renaud, W. B. Schweizer and M. F. Zueger, *Helv. Chim. Acta*, 1984, **67**, 1843.
79. C. Fuganti, P. Grasselli, P. F. Seneci and P. Casati, *Tetrahedron Lett.*, 1986, **27**, 5275; C. Fuganti and P. Grasselli, *Tetrahedron Lett.*, 1985, **26**, 101.
80. M. Kawai, K. Tajima, S. Mizuno, K. Niimi, H. Sugioka, Y. Butsugan, A. Kozawa, T. Asano and Y. Imai, *Bull. Chem. Soc. Jpn.*, 1988, **61**, 3014.
81. F. Moriuchi, H. Muroi and H. Aibe, *Chem. Lett.*, 1987, 1141.
82. M. Christen and D. H. G. Crout, *J. Chem. Soc., Chem. Commun.*, 1988, 264.
83. W.-R. Shieh, A. S. Gopalan and C. J. Sih, *J. Am. Chem. Soc.*, 1985, **107**, 2993.
84. J. Heidlas, K.-H. Engel and R. Tressl, *Eur. J. Biochem.*, 1988, **172**, 633.
85. C.-S. Chen, B. N. Zhou, G. Girdaukas, W.-R. Shieh, F. Van Middlesworth, A. S. Gopalan and C. J. Sih, *Bioorg. Chem.*, 1984, **12**, 98.
86. K. Ushio, K. Inouye, K. Nakamura, S. Oka and A. Ohno, *Tetrahedron Lett.*, 1986, **27**, 2657.
87. K. Nakamura, K. Inoue, K. Ushio, S. Oka and A. Ohno, *Chem. Lett.*, 1987, 679.
88. M. Hirama, T. Noda and S. Ito, *J. Org. Chem.*, 1985, **50**, 127.
89. M. Utaka, H. Higashi and A. Takeda, *J. Chem. Soc., Chem. Commun.*, 1987, 1368.
90. M. Utaka, H. Watabu and A. Takeda, *J. Org. Chem.*, 1987, **52**, 4363.
91. Y. Naoshima, H. Hasegawa and T. Saeki, *Agric. Biol. Chem.*, 1987, **51**, 3417.
92. M. Gessner, C. Guenther and A. Mosandl, *Z. Naturforsch., Teil C*, 1987, **42**, 1159.
93. Y. Naoshima and H. Hasegawa, *Chem. Lett.*, 1987, 2379.
94. C.-Q. Han, D. Di'Tullio, Y.-F. Wang and C. J. Sih, *J. Org. Chem.*, 1986, **51**, 1253.
95. H. Suemune, N. Hayashi, K. Funakoshi, H. Akita, T. Oishi and T. Sakai, *Chem. Pharm. Bull.*, 1985, **33**, 2168.
96. M. Nardini, H. Kodama, G. Ricci, G. Federici and D. Cavallini, *Biochem. Int.*, 1985, **11**, 789.
97. A. Goswami, R. P. Steffek, W. G. Liu, J. P. N. Rosazza and J. J. Steffens, *Enzyme Microb. Technol.*, 1987, **9**, 521.
98. P. Raddatz, H. E. Radunz, G. Schneider and H. M. Schwartz, *Angew. Chem., Int. Ed. Engl.*, 1988, **27**, 426.
99. J. M. Beale, Jr., J. M. Hewitt and J. P. N. Rosazza, *Enzyme Microb. Technol.*, 1984, **6**, 543.
100. R. Chênevert and S. Thiboutot, *Chem. Lett.*, 1988, 1191; *Can. J. Chem.*, 1986, **64**, 1599.
101. H. Ohta, K. Ozaki and G. Tsuchihashi, *Agric. Biol. Chem.*, 1986, **50**, 2499; *Chem. Lett.*, 1987, 2225.
102. S. Tsuboi, E. Nishiyama, H. Furutani, M. Utaka and A. Takeda, *J. Org. Chem.*, 1987, **52**, 1359.
103. C. J. Sih, J. B. Heather, G. P. Peruzzotti, P. Price, R. Sood and L.-F. Hsu Lee, *J. Am. Chem. Soc.*, 1973, **95**, 1676.
104. (a) S. Tsuboi, E. Nishiyama, M. Utaka and A. Takeda, *Tetrahedron Lett.*, 1986, **27**, 1915; (b) S. Tsuboi, H. Furutani, M. Utaka and A. Takeda, *Tetrahedron Lett.*, 1987, **28**, 2709.
105. S. Riva, R. Bovara, P. Pasta and G. Carrea, *J. Org. Chem.*, 1986, **51**, 2902.
106. H. Akita, A. Furuichi, H. Koshiji, K. Horikoshi and T. Oishi, *Tetrahedron Lett.*, 1982, **23**, 4051; K. Horikoshi, A. Furuichi, H. Koshiji, H. Akita and T. Oishi, *Agric. Biol. Chem.*, 1983, **47**, 435.
107. T. Sato, M. Tsurumaki and T. Fujisawa, *Chem. Lett.*, 1986, 1367.
108. K. Nakamura, T. Miyai, K. Nozaki, K. Ushio, S. Oka and A. Ohno, *Tetrahedron Lett.*, 1986, **27**, 3155.
109. T. Fujisawa, T. Itoh and T. Sato, *Tetrahedron Lett.*, 1984, **25**, 5083.
110. G. Fronza, C. Fuganti, P. Grasselli and S. Servi, *Tetrahedron Lett.*, 1985, **26**, 4961.
111. C. Fuganti, P. Grasselli, S. Servi, F. Spreafico, C. Zirotti and P. Casati, *J. Org. Chem.*, 1984, **49**, 4087.

112. C. Fronza, C. Fuganti, G. Pedrocchi-Fantoni and S. Servi, *J. Org. Chem.*, 1987, **52**, 1141.
113. G. Fronza, C. Fuganti, P. Grasselli and S. Servi, *J. Org. Chem.*, 1987, **52**, 2086.
114. D. Buisson, S. Henrot, M. Larcheveque and R. Azerad, *Tetrahedron Lett.*, 1987, **28**, 5033.
115. R. E. Helmchen-Zeier, Ph. D. Thesis, ETH, Zürich, 1973.
116. J. Davies and J. B. Jones, *J. Am. Chem. Soc.*, 1979, **101**, 5405; J. A. Haslegrave and J. B. Jones, *J. Am. Chem. Soc.*, 1982, **104**, 4666.
117. B. S. Deol, D. D. Ridley and G. W. Simpson, *Aust. J. Chem.*, 1976, **29**, 2459.
118. G. Fráter, *Helv. Chim. Acta*, 1980, **63**, 1383.
119. T. Kitahara and K. Mori, *Tetrahedron Lett.*, 1985, **26**, 451; T. Kitahara, H. Kurata and K. Mori, *Tetrahedron*, 1988, **44**, 4339.
120. T. Sato, H. Maeno, T. Noro and T. Fujisawa, *Chem. Lett.*, 1988, 1739.
121. R. W. Hoffmann, W. Helbig and W. Ladner, *Tetrahedron Lett.*, 1982, **23**, 3479.
122. J. Cooper, P. T. Gallagher and D. W. Knight, *J. Chem. Soc., Chem. Commun.*, 1988, 509.
123. S. Benner and T. H. Morton, *J. Am. Chem. Soc.*, 1981, **103**, 991.
124. Ref. 1a, 303.
125. R. Mislin, Ph. D. Thesis, ETH, Zürich, 1968.
126. L. K. P. Lam, I. A. Gair and J. B. Jones, *J. Org. Chem.*, 1988, **53**, 1611.
127. Ref. 1a, 339.
128. E. Hochuli, Ph. D. Thesis, ETH, Zürich, 1974.
129. Z. Kis, Ph. D. Thesis, ETH, Zürich, 1965.
130. S. Butt, H. G. Davies, M. J. Dawson, G. C. Lawrence, J. Leaver, S. M. Roberts, M. K. Turner, B. J. Wakefield, W. F. Wall and J. A. Winders, *Tetrahedron Lett.*, 1985, **26**, 5077; *J. Chem. Soc., Perkin Trans. 1*, 1987, 903.
131. D. J. Kertesz and A. F. Kluge, *J. Org. Chem.*, 1988, **53**, 4962.
132. D. W. Brooks, M. Wilson and M. Webb, *J. Org. Chem.*, 1987, **52**, 2244.
133. Z.-F. Xie, H. Suemune, I. Nakamura and K. Sakai, *Chem. Pharm. Bull.*, 1987, **35**, 4454.
134. Z.-F. Xie, K. Funakoshi, H. Suemune, T. Oishi, H. Akita and K. Sakai, *Chem. Pharm. Bull.*, 1986, **34**, 3058.
135. R. F. Newton, J. Paton, D. P. Reynolds, S. Young and S. M. Roberts, *J. Chem. Soc., Chem. Commun.*, 1979, 908.
136. A. J. Irwin and J. B. Jones, *J. Am. Chem. Soc.*, 1976, **98**, 8476.
137. O. Abril and G. M. Whitesides, *J. Am. Chem. Soc.*, 1982, **104**, 1552.
138. I. Stibor, I. Veselý, J. Paleček and J. Mostecký, *Synthesis*, 1986, 640.
139. S. Inayama, N. Shimuzu, T. Ohkura, H. Akita, T. Oishi and Y. Iitaka, *Chem. Pharm. Bull.*, 1987, **35**, 429.
140. D. W. Brooks, H. Mazdiyasni and P. G. Grothaus, *J. Org. Chem.*, 1987, **52**, 3223.
141. K. Mori and M. Fujiwara, *Tetrahedron*, 1988, **44**, 343.
142. D. W. Brooks, P. G. Grothaus and W. L. Irwin, *J. Org. Chem.*, 1982, **47**, 2820.
143. K. Naemura, T. Fujii and H. Chikamatsu, *Chem. Lett.*, 1986, 923.
144. L. Velluz, G. Valls and G. Nominé, *Angew. Chem., Int. Ed. Engl.*, 1965, **4**, 181; P. Bellet, G. Nominé and J. Mathieu, *C. R. Hebd. Seances Acad. Sci., Ser. C*, 1966, **263**, 88.
145. R. P. Lanzilotta, D. G. Bradley and C. C. Beard, *Appl. Microbiol.*, 1975, **29**, 427.
146. T. Fujisawa, E. Kojima, T. Itoh and T. Sato, *Tetrahedron Lett.*, 1985, **26**, 6089.
147. M. Imuta and H. Ziffer, *J. Org. Chem.*, 1978, **43**, 3319.
148. M. Bostmembrun-Desrut, G. Dauphin, A. Kergomard, M. F. Renard and H. Veschambre, *Tetrahedron*, 1985, **41**, 3679.
149. C. Fuganti, P. Grasselli, F. Spreafico, C. Zirotti and P. Casati, *J. Org. Chem.*, 1984, **49**, 543.
150. C. Fuganti and P. Grasselli, *Chem. Ind. (London)*, 1977, 983; *J. Chem. Soc., Chem. Commun.*, 1982, 205.
151. Y. Tsuda, *Chem. Pharm. Bull.*, 1987, **35**, 2554.
152. H. Yamada, K. Kurumaya, T. Eguchi and M. Kajiwara, *J. Labelled Comp. Radiopharm.*, 1986, **24**, 561.
153. J. Lüthy, J. Rétey and D. Arigoni, *Nature (London)*, 1969, **221**, 1213.
154. R. Wichmann, C. Wandrey, A. F. Bueckmann and M. R. Kula, *Biotechnol. Bioeng.*, 1981, **23**, 2789.
155. L. K. P. Lam, L. D. Arnold, T. H. Kalantar, J. G. Kelland, P. M. Lane-Bell, M. M. Palcic, M. A. Pickard and J. C. Vederas, *J. Biol. Chem.*, 1988, **263**, 11 814.
156. J. P. Kutney, C. A. Boulet, L. S. L. Choi, W. Gustowski, M. McHugh, J. Nakano, T. Nikaido, H. Tsukamoto, G. M. Hewitt and R. Suen, *Heterocycles*, 1988, **27**, 621.
157. T. Sato, K. Hanayama and T. Fujisawa, *Tetrahedron Lett.*, 1988, **29**, 2197.
158. H. Simon, H. Günther, J. Bader and W. Tischler, *Angew. Chem., Int. Ed. Engl.*, 1981, **20**, 861.
159. H. G. W. Leuenberger, W. Boguth, R. Barner, M. Schmid and R. Zell, *Helv. Chim. Acta*, 1979, **62**, 455.
160. M. Utaka, S. Konishi and A. Takeda, *Tetrahedron Lett.*, 1986, **27**, 4737.
161. H. G. W. Leuenberger, W. Boguth, E. Widmer and R. Zell, *Helv. Chim. Acta*, 1976, **59**, 1832.
162. A. Kergomard, M. F. Renard and H. Veschambre, *J. Org. Chem.*, 1982, **47**, 792.
163. J.-C. Gramain, A. Kergomard, M. F. Renard and H. Veschambre, *J. Org. Chem.*, 1985, **50**, 120.
164. H. Günther, J. Bader, H. Simon and W. Tischler, *Angew. Chem., Int. Ed. Engl.*, 1981, **20**, 861.
165. G. Gil, E. Ferre, M. Barre and J. LePetit, *Tetrahedron Lett.*, 1988, **29**, 3797.
166. R. Kazlauskas, *J. Org. Chem.*, 1988, **53**, 4633.
167. P. G. Schultz, *Science (Washington, D.C.)*, 1988, **240**, 426; S. J. Pollack, J. W. Jacobs and P. G. Schultz, *Science (Washington, D. C.)*, 1986, **234**, 1570.
168. A. Napper, S. J. Benkovic, A. Tramontano and R. A. Lerner, *Science (Washington, D. C.)*, 1987, **237**, 1041.
169. R. J. Leatherbarrow and A. R. Fersht, *Protein Eng.*, 1986, **1**, 7.
170. J. A. Gerlt, *Chem. Rev.*, 1987, **87**, 1079.
171. C. Murali and E. H. Creaser, *Protein Eng.*, 1986, **1**, 55.
172. A. R. Clarke, C. J. Smith, K. W. Hart, H. M. Wilks, W. N. Chia, T. V. Lee, J. J. Birktoft, L. J. Banaszak, D. A. Barstow, T. Atkinson and J. J. Holbrook, *Biochem. Biophys. Res. Commun.*, 1987, **148**, 15.
173. M. A. Luyten, M. Gold, J. D. Friesen and J. B. Jones, *Biochemistry*, 1989, **28**, 6605.

1.9
Reduction of Acetals, Azaacetals and Thioacetals to Ethers

JAMES H. BREWSTER
Purdue University, West Lafayette, IN, USA

1.9.1 INTRODUCTION

Acetals and ketals are passive under neutral and basic conditions; the ether groups resist alkylation and acylation, and the masked carbonyl group is protected against attack by nucleophiles and against enolization by bases,[1] although nearby unsaturation may open the system to reaction. When, however, one of the oxygen atoms acquires a Lewis or protic acid as a ligand it becomes a better leaving group, and the carbonyl carbon atom can suffer S_N2 and, especially, S_N1 reactions. These reactions can be highly regio- and stereo-selective when the two oxygen atoms are in different chemical environments. Neither the 'anomeric effect',[2] which is so important in setting the ground state conformation, nor the 'stereoelectronic effect',[3] which has come under serious attack,[4] have an overriding effect, but differences in the stability of the two possible oxocarbonium ions clearly do.

With cyclic acetals and ketals, selective reductions allow the blocked hydroxy groups of the diol to be deprotected one at a time, a matter of some importance in carbohydrate chemistry.[5] Although there have been a few studies of stereoselective reductions at the masked carbonium center of chiral ketals, more has been done with the formally related reactions in which C—C bonds are formed stereoselectively.[6]

1.9.2 MAJOR METHODS

1.9.2.1 Catalytic Hydrogenation

Aliphatic ethers are stable to ordinary catalytic hydrogenation conditions, but benzyl ethers can be cleaved in good yields and under mild conditions, making them useful as protecting groups.[1] Palladium, promoted by small amounts of acid, is the best catalyst. Thus, benzylidene acetals are cleaved to methylbenzenes in high yield,[7–9] and ketals derived from benzophenone behave similarly.[10] Allyl ethers may also be cleaved over palladium, but with other catalysts a competing hydrogenation of the double bond becomes more important.

Benzyl ethers are obtained from benzylidene acetals over nickel at 180 °C.[11] The yields are moderate, however, because the ethers are cleaved about as readily as the acetals;[12] at lower temperatures ring reduction of the hydrocarbon product is also competitive with hydrogenolysis. The best catalyst for the preparation of benzyl ethers from benzylidene acetals would appear to be cobalt carbonyl, under synthesis gas at temperatures from 128–210 °C.[13] The active reagent is probably $HCo(CO)_4$.

Better yields of ethers are obtained from aliphatic acetals;[11,14] vinyl ethers, which can be prepared from acetals over nickel (alone) at high temperatures,[15] may be intermediates. With rhodium on alumina in the presence of acid, ketals of acetone (**1**) are hydrogenolyzed to ethers (**3**) at about the same rate that the corresponding vinyl ethers (**4**) are hydrogenated,[16] indicating that the latter may be intermediates (Scheme 1). Vinyl ethers have also been suggested as intermediates in the direct formation of ethers from ketones during catalytic hydrogenation over platinum in MeOH–HCl,[17] but the oxocarbonium ion (**2**), which is a common intermediate for the formation of both the acetal and the vinyl ether, might be reduced directly.

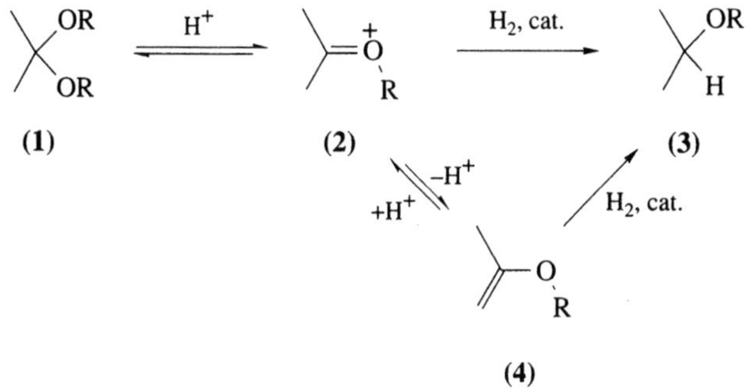

OR OR (**1**) H^+ $\overset{+}{O}$ R (**2**) H_2, cat. OR H (**3**) $-H^+$ $+H^+$ H_2, cat. O R (**4**)

Scheme 1

1.9.2.2 Reduction with Dissolving Metals

Saturated alcohols and ethers are inert to the usual methods of reduction by dissolving metals, but benzylic alcohols, ethers, esters and acetals are completely hydrogenolyzed under the conditions of the Birch reduction.[18–20] The fission process probably occurs after two electrons have been added to the ring;[18] proton donors are not required and may, when *p*-methoxy groups are present, promote reduction of the ring instead.[21] The corresponding 1,3-dioxolanes, however, give complete hydrogenolysis.[22] The furan nucleus also promotes 'benzylic' hydrogenolysis with lithium in ammonia in the absence of proton donors.[23] The location of the side chain makes a difference. In the 3-furanyl system (equation 1)[24] single cleavage of a dioxolane predominates, but in the 2-furanyl system[25] only double cleavage, to remove both acetal oxygen atoms, occurs.

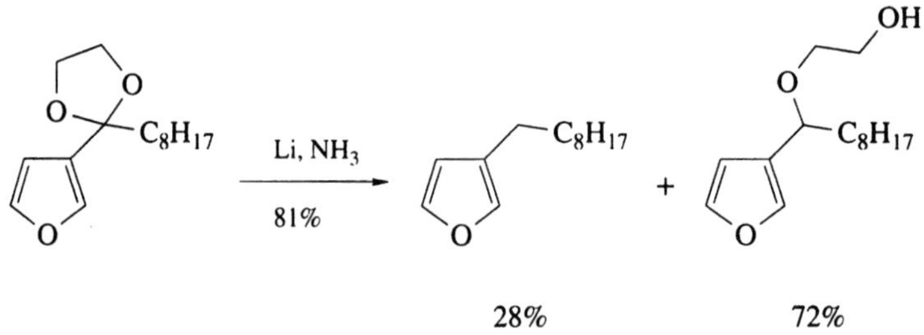

O O C_8H_{17} O —Li, NH_3, 81%→ C_8H_{17} O (28%) + OH O C_8H_{17} O (72%) (1)

1.9.2.3 Reduction with Metal Hydrides

The tetrasubstituted hydrides of Group III and IV metals act only as hydride donors; for the reduction of acetals a Lewis acid partner, one that will not react faster with the hydride donor, is required and several such silicon and boron hydride systems have been developed. The trisubstituted hydrides are, of themselves, both Lewis acids and hydride donors; the halo- and alkyl-aluminum hydrides have proven to be valuable reagents of this type. In all of them modulation of reducing power can be achieved by changing the kind and numbers of substituents. Halogen atoms increase Lewis acidity in both series. Alkoxy groups enhance the reducing power of boranes but diminish that of alanes; they also permit variation in steric requirements and might allow the development of chiral reagents.

1.9.2.3.1 Derivatives of alane

Lithium aluminum hydride (LAH) does not, as a rule, react with ethers,[26,27] but does reduce ortho esters to acetals (equation 2).[28] 2-Vinyl-1,3-dioxolane (**5**) is reduced to a vinyl ether with double-bond rearrangement (equation 3), but the cinnamyl analog (**6**) is only reduced at the double bond (equation 4).

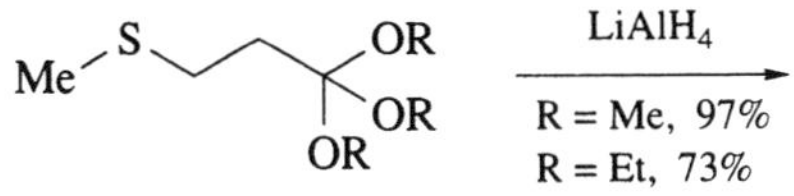

Me–S–CH$_2$CH$_2$–C(OR)$_3$ —[$LiAlH_4$; R = Me, 97%; R = Et, 73%]→ Me–S–CH$_2$CH$_2$–CH(OR)$_2$ (2)

(**5**) —[$LiAlH_4$, 76%]→ vinyl ether –O–CH$_2$CH$_2$–OH (3)

(**6**) —[$LiAlH_4$]→ Ph–CH$_2$CH$_2$–(1,3-dioxolan-2-yl) (4)

Although some propargylic alcohols are reduced to allenes,[29] such behavior is not seen with the simple open chain propargylic acetal (**7**).[30] Studies with LAD and D_2O show that the triple bond is reduced first, and that the alumino derivative then undergoes acetal cleavage without rearrangement (Scheme 2). The resulting crotyl acetal then undergoes S_N2 displacement of one ethoxy group without reduction of the double bond.

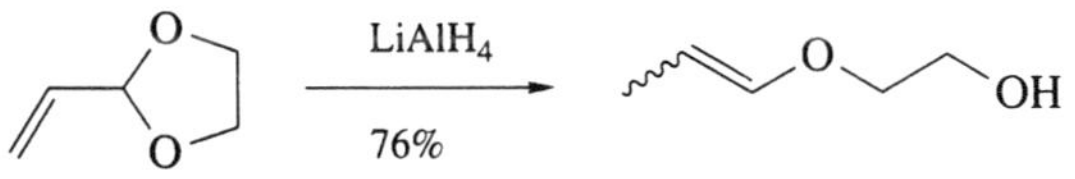

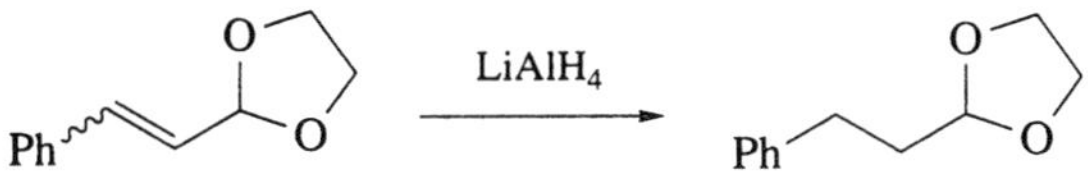

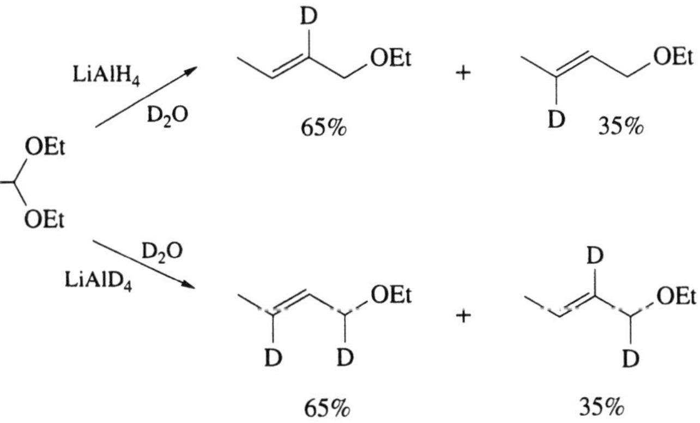

(**7**) —[$LiAlH_4$; D_2O]→ D-substituted OEt product 65% + D-substituted OEt product 35%

(**7**) —[$LiAlD_4$; D_2O]→ D, D-substituted OEt product 65% + D, D-substituted OEt product 35%

Scheme 2

Alane is best prepared by reaction of LAH with H_2SO_4; prepared so, it is stable in THF for several days and is a potentially useful selective reducing agent.[31] It can also be prepared by reaction of LAH with $AlCl_3$ (3:1); in other proportions, these reagents form the chloroalanes.[32] Alkoxyalanes and chloroalkoxyalanes can be made from these reagents by adding appropriate amounts of alcohols.[33] Of these, the haloalanes ('mixed hydride' reagents) have been most widely used, particularly where cleavage of C—O bonds is desired.[34,35] Dichloroalane, with an excess of $AlCl_3$ (1LAH:4$AlCl_3$) and used in 100% molar excess, gives good preparative cleavages of simple acetals (equation 5),[36,37] but the milder reagent H_2AlCl (1LAH:1$AlCl_3$) has been used extensively in mechanistic studies.[38] Alane is a still weaker hydrogenolyzing agent.[33,38] Alkoxyalanes are even less active but offer the possibility of variation in steric requirements; dialkoxyalanes are too unreactive to be useful.

$LiAlH_4$ (1 mol) + $AlCl_3$ (4.1 mol), Et_2O, 83–94% (5)

Diisobutylaluminum hydride (DIBAL-H) hydrogenolyzes simple ortho esters to acetals at room temperature and reduces acetals and ketals to ethers at 70–80 °C (equation 6); benzyl ethers are cleaved at still higher temperatures.[39] This reagent shows good selectivity and considerable versatility. It has been used to reduce acetals of formaldehyde (equation 7),[40] which few other reagents can accomplish. With catechol ketals a single reductive cleavage occurs at room temperature (equation 8).[41]

OEt, OEt, OEt —DIBAL-H, 30 °C, 90%→ OEt, OEt —DIBAL-H, 70–80 °C, 80%→ OEt (6)

R^1, R^2, O, O —DIBAL-H, 90–120 °C, 30–90%→ HO, R^1, R^2, OMe (7)

R^1 = H, R^2 = H, Ph; R^1 = Me, R^2 = Me, Ph

O, O, n-$C_{11}H_{23}$, n-$C_{11}H_{23}$ —DIBAL-H, r.t., 80%→ O, n-$C_{11}H_{23}$, OH, n-$C_{11}H_{23}$ (8)

1.9.2.3.2 Derivatives of borane

A mixture of LAH and BF_3 generates diborane, but other products are formed first when BF_3 is added to LAH.[34] This system reduces lactones to cyclic ethers and gives some cleavage of steroidal spiroketals.[42] It gives slightly better yields of dihydroxy ethers from acetals of pentaerythritol than does chloroalane (equation 9).[43] Borane in THF gives good cleavage of acetals and ketals, usually at the less-hindered oxygen atom.[44,45] Propargylic acetals undergo reductive cleavage *via* an elimination mechanism during hydroboration (Scheme 3; *cf.* Scheme 2).[46]

Hydroboration is faster than cleavage with borane,[47] but H_2BCl etherate cleaves unsaturated acetals without giving hydroboration (equation 10).[48] Lower yields are obtained with the less reactive Me_2S complex, but that reagent gives good yields of dihydroxy diethers from chiral bisacetals (equation 11).[49] The diethyl acetal of benzaldehyde can be reduced to ethyl benzyl ether in moderate yield by decaborane in toluene at 120–130 °C (equation 12).[50]

The borohydride ion is a good trap for spontaneously generated carbonium ions (or ion pairs), such as those formed by secondary or tertiary benzylic or allylic halides in ionizing solvents,[51] but would be expected to be inactive when the leaving group is alkoxy unless an acidic partner were also present.

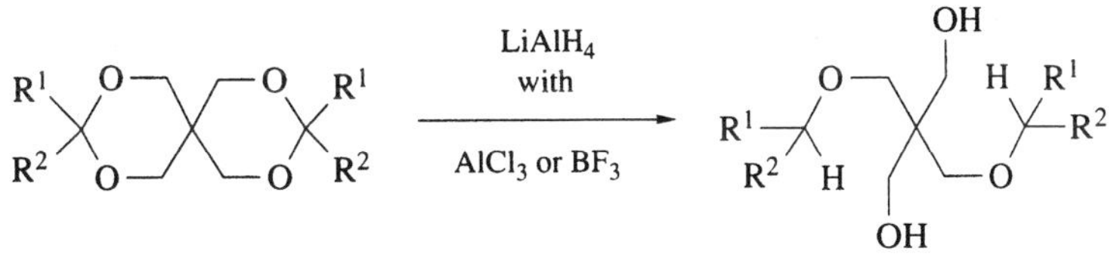

LiAlH$_4$ with AlCl$_3$ or BF$_3$ (9)

R^1 = Ph, R^2 = H; AlCl$_3$, 80%; BF$_3$, 82%

R^1, R^2 = (cyclohexane); AlCl$_3$, 85%; BF$_3$, 91%

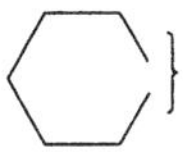

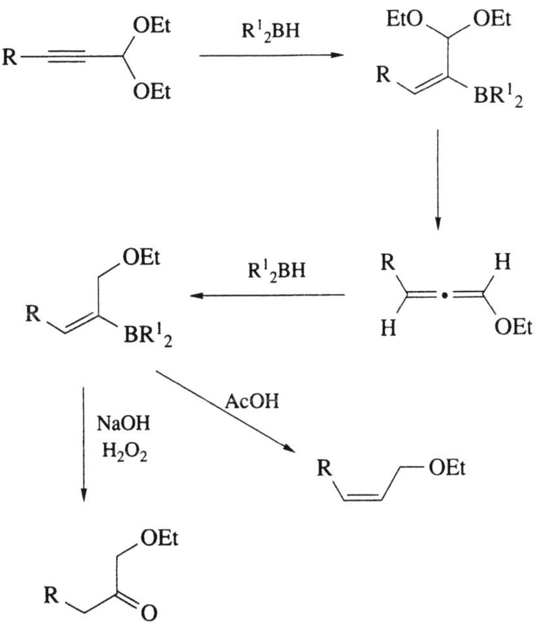

R^1_2BH; R^1_2BH; NaOH, H_2O_2; AcOH

Scheme 3

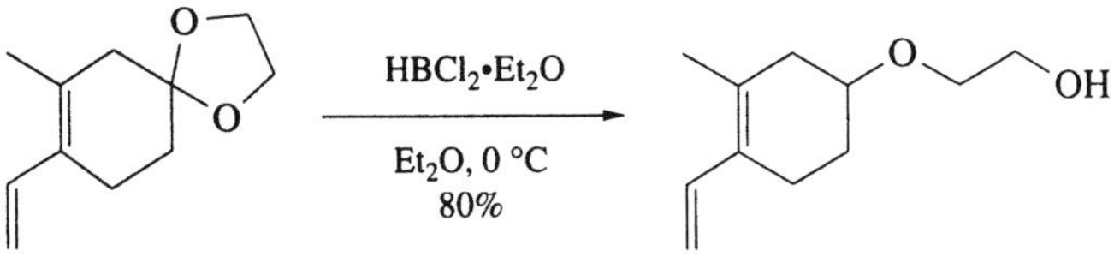

HBCl$_2$•Et$_2$O; Et$_2$O, 0 °C; 80% (10)

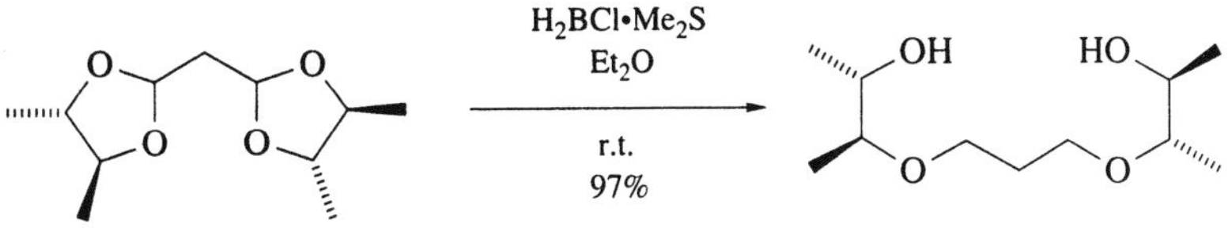

H$_2$BCl•Me$_2$S; Et$_2$O; r.t.; 97% (11)

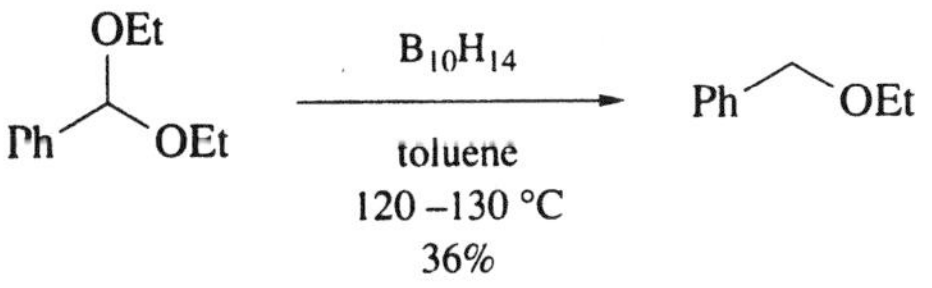

B$_{10}$H$_{14}$; toluene; 120–130 °C; 36% (12)

Sodium borohydride and trifluoroacetic acid (TFA) reduce 2-aryl-1,3-dioxolanes to hydroxy ethers.[52] Zinc borohydride, with TMS-Cl as its acidic partner, has proven to be effective in the reductive cleavage of a variety of acetals.[53] MOM ethers are reduced to methyl ethers in good yield (equation 13). Cyclohexene is hydroborated under these conditions, indicating that diborane may be present.

$$\text{R}\diagup\!\!\diagdown\text{O}\diagup\!\!\diagdown\text{OMe} \xrightarrow[\text{Me}_3\text{SiCl}]{\text{Zn(BH}_4)_2} \text{R}\diagup\!\!\diagdown\text{OMe} \quad (13)$$

Methyl acetals and ketals are rapidly reduced to methyl ethers by sodium cyanoborohydride in methanol with dry HCl at ice temperatures.[54] A dioxolane is completely cleaved to a methyl ether, showing intervention by the solvent at some stage (equation 14), but when an inert solvent such as THF is used only single cleavage occurs; this reagent shows interesting selectivity in the reduction of benzylidene acetals in the carbohydrate series (see Section 1.9.3.4).

$$\text{(2-methyl-2-phenyl-1,3-dioxolane)} \xrightarrow[\text{HCl, MeOH, } 0\ ^\circ\text{C, } 76\%]{\text{Na}^+\ \text{N}\equiv\text{C-BH}_3^-} \text{Ph-CH(Me)-OMe} \quad (14)$$

1.9.2.3.3 Derivatives of silane

'Ionic hydrogenation' occurs when alkyl- or aryl-silicon hydrides reduce carbonium ions.[55–57] Thus, acetals and ketals can be reduced by trialkylsilanes in the presence of 5–10% of $ZnCl_2$.[58] Yields of 50–85% of ether are obtained from acyclic acetals (equation 15), but cyclic acetals give appreciable amounts of a by-product (**8**; equation 16), which is believed not to be the result of further reduction.

$$\text{MeCH(OEt)}_2 \xrightarrow[5\%\ \text{ZnCl}_2,\ 25\ \text{h},\ 85\%]{\text{Et}_3\text{SiH}} \text{EtOEt} + \text{EtOSiEt}_3 \quad (15)$$

$$\text{(2-methyl-1,3-dioxane)} \xrightarrow[120\ ^\circ\text{C, } 120\ \text{h}]{\text{Et}_3\text{SiH (ZnCl}_2)} \underset{50\%}{\text{EtO(CH}_2)_3\text{OSiEt}_3} + \underset{(\mathbf{8})\ 16.5\%}{\text{Et}_3\text{SiO(CH}_2)_3\text{OSiEt}_3} \quad (16)$$

Trimethylsilyl triflate (TMS-OTf; 1%) catalyzes the reduction of acetals by TMS-H in CH_2Cl_2 at 0 °C,[59] with the TMS moiety acting as a chain-carrying species (equation 17).[59]

$$\text{PhCH}_2\text{CH(OMe)}_2 \xrightarrow{\text{Me}_3\text{SiOTf}} \text{PhCH}_2\text{CH}{=}\overset{+}{\text{O}}\text{Me}\ \ \text{OTf}^- \xrightarrow[96\%]{\text{Me}_3\text{SiH}} \text{PhCH}_2\text{CH}_2\text{OMe} + \text{Me}_3\text{SiOTf} \quad (17)$$

Carbonyl compounds are reduced to symmetrical ethers, probably by way of reduction of some of the starting material to a silyl ether (**9**), reaction to form the mixed ketal (**10**) and then reductive replacement of the silyloxy group. Some hydrocarbon may be obtained as a by-product by reduction of (**9**; Scheme 4). Among the acid partners that have been used are trifluoroacetic acid,[55,56] trityl perchlorate[60] (with aldehydes) and electrogenerated protons.[61] With Nafion resin[57] symmetrical ethers are obtained from aldehydes, but silyl ethers are obtained from ketones.[62,63]

Ketones in alcohol solutions give mixed ethers with triethylsilane and trifluoroacetic acid,[64] presumably *via* the hemiketal (or ketal). Similar results were obtained with ketones and silylated alcohols.[60] Studies of the acid-catalyzed reduction of alcohols with silanes provide strong support for the postulate that carbonium ions are intermediates.[65,66]

1.9.3 SELECTIVE REDUCTIONS

The rate-controlling step in the acid-mediated reductive cleavage of an acetal or ketal involves the breaking of one of the C—O bonds. This step is product-controlling when the two oxygen atoms are dif-

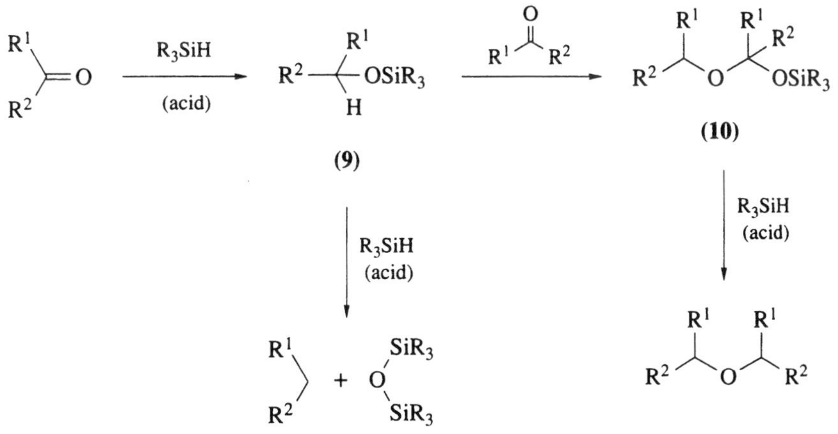

Scheme 4

ferent. Selectivity at this stage depends in part on the relative basicity of the two oxygen atoms and their relative accessibility to the Lewis acid. For S_N1-like reactions, the degree to which the incipient carbonium ion, which contains the other oxygen atom, can be stabilized by polar effects or destabilized by steric crowding may be more important. The ease with which the hydride donor can approach the reacting center will control the stereochemistry of the product in these reactions. In S_N2-like reactions, of course, the accessibility of the reacting center to the hydride donor determines both the structure and the stereochemistry of the product.

1.9.3.1 Acyclic Ketals

In hydrogenolyses with $HAlCl_2$, the dimethyl acetals of cyclobutanone and cyclohexanone are cleaved more slowly than that of 3-pentanone, while those of cyclopentanone and cycloheptanone are cleaved more rapidly (Table 1),[67] as would be expected for a carbonium ion process.[68] The differences in rate are small, suggesting that carbonium ion character is not strongly developed in the transition state. With the dimethyl acetal of 4-*t*-butylcyclohexanone, the hydride addition step occurs with strongly predominating axial addition when $HAlCl_2$ is used;[36] $Zn(BH_4)_2$ with TMS-Cl,[53] and TMS-H with TMSO-Tf[59] are less selective (Table 2). Equatorial attack predominates, however, in the reduction of the ketone itself with TBDMS-H and TBDMS-OTf.[63]

Table 1 Relative Rates of Reduction of Dimethyl Ketals of Acyclic and Cyclic Ketones with $HAlCl_2$[67]

Ketone	*Relative rate*
3-Pentanone	1.0
Cyclobutanone	0.06
Cyclopentanone	1.3
Cyclohexanone	0.5
Cycloheptanone	7.0

Table 2 Diastereoselectivity in Reductions of the Dimethyl Ketal of 4-*t*-Butylcyclohexanone

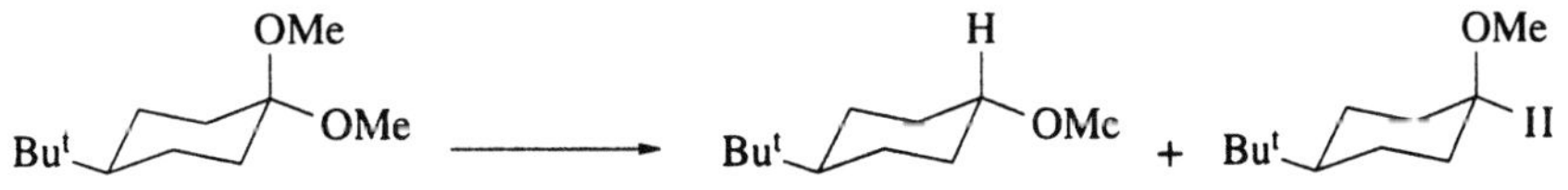

Reducing system	*Yield (%)*	*Trans (%)*	*Cis (%)*	*Ref.*
$HAlCl_2$	92	82	18	36
$Zn(BH_4)_2$:TMS-Cl	95	59	41	53
TMS-H:TMS-OTf	89	56	44	59

In reductions with $HAlCl_2$, mixed ketals of norcamphor lose the less-branched alkoxy group preferentially, regardless of its orientation, but give only *endo*-ether as product.[69] This indicates that the oxocarbonium ion becomes essentially free and is reduced exclusively on the less-hindered *exo* face (Table 3). Some evidence for preferential cleavage of the more exposed *exo*-alkoxy group is seen in the case where the alkoxy groups differ only isotopically, but this effect is swamped in the other cases, suggesting that stabilization of the oxocarbonium ion is the main product-controlling effect.

Table 3 Hydrogenolysis of Mixed Ketals of Norcamphor by $HAlCl_2$ in Ether at Room Temperature[69]

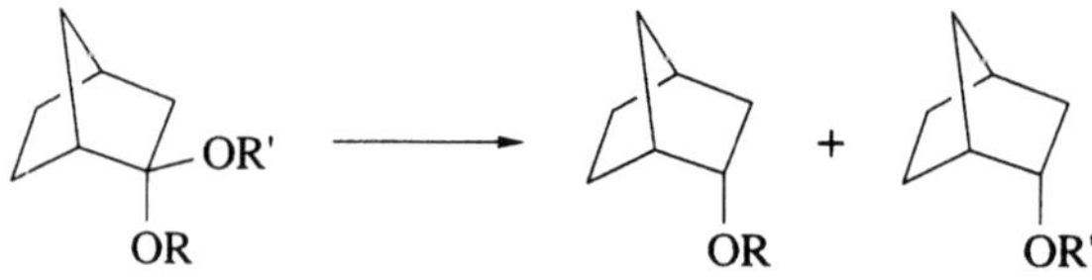

R	*R′*	*Retained alkoxy groups* *O—R* Yield (%)	*O—R′* Yield (%)
Me	CD_3	67	33
Me	Et	20	80
Et	Me	86	14
Me	Pr^i	2	98
Pr^i	Me	99	1
Me	Bu^t	8	83
		(plus 9% *endo*-norborneol)	

1.9.3.2 Furanosides and Pyranosides

For synthetic purposes there is probably more interest in the reductive ring cleavage of tetrahydrofuranyl and tetrahydropyranyl ethers than in *exo* cleavage of their alkoxy side chains. With $HAlCl_2$[70] and H_2AlCl,[71,72] comparable furanosides react more rapidly than pyranosides and tend to give more ring cleavage (Tables 4 and 5), consistent with the greater basicity of THF.[73] Increased branching in the alkoxy group OR also promotes ring cleavage, but substituents on the other side of the acetal system (R^1) promote side chain cleavage. In contrast, aryloxy compounds (R = Ar)[71,74] give only side chain cleavage, while aryl substitution at R^1 promotes ring cleavage. These effects have been ascribed[71,72] to the electron-releasing effect of branched alkyl groups, which, located at the oxygen end of the oxocarbonium ion, would tend to stabilize it, thereby promoting cleavage at the other oxygen atom. The steric effects of branching would work in the same way. On the other hand, aryl groups can act only as electron-withdrawing groups when they are at the oxygen end of the oxocarbonium system; in such cases the polar effects would seem to outweigh the steric. A comparison of the action of different alanes on THFOBn shows that the weaker Lewis acid (stronger nucleophile) gives more ring opening (Table 6).[72] The LAH–BF_3 reagent resembles the less-halogenated alanes (Table 5).[70]

Table 4 Reductive Cleavage of Tetrahydrofuranyl Ethers with H_2AlCl in Ether at Room Temperature

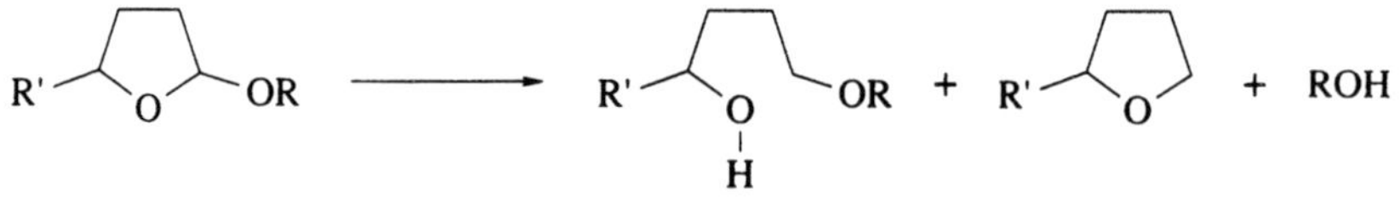

R′	*R*	*Cleavage (%)* *Ring*	*Exo*	*Ref.*
H	Me, Et, Pr^i, Bu^t	100	0	72
Me	Me	37.5	62.5	72
Pr^i	Me	14	86	72
Bu^t	Me	0	100	72
Ph	Me	84.3	15.7	72
H	Ph	0	100	74
H	Bn	51.5	48.5	72

Table 5 Cleavage of Tetrahydrofuranyl and Tetrahydropyranyl Ethers with $HAlCl_2$ and with $LiAlH_4$:BF_3 Reagents[70]

R	Cleavage (%) Ring	Exo	Reagent	Cleavage (%) Ring	Exo
			$HAlCl_2$		
n-C_6H_{13}	27	40		11	*ca.* 90
c-C_6H_{11}	63	*ca.* 15		45	38
Bu^t	58	—		72	—
Bn	4	83		—	90
			$LiAlH_4$–BF_3		
n-C_6H_{13}	66			41	
c-C_6H_{11}	69			69	
Bu^t	16			46	
Bn	48			25	

Table 6 Cleavage of Tetrahydrofuranyl Benzyl Ether with Alanes in Ether at Room Temperature (2 h)

Reagent	Reacted (%)	Recovered (%)	Cleavage in recovered product Ring (%)	Exo (%)	Ref.
AlH_3	67.5	84	79.5	20.5	72
H_2AlCl	100	95	51.5	48.5	72
$HAlCl_2$	100	90	9.5	90.5	72
$HAlCl_2$–$AlCl_3$	100	91	8.7	91.3	72
$LiAlH_4$–BF_3	70			48 (isolated)	70

Side chain cleavage is of interest as a tool for locating links in oligosaccharides (equation 18).[75] Triethylsilane, catalyzed by BF_3 alone or by a 5:1 mixture of TMS-OTf and BF_3,[76] has proven to be useful for this purpose with both pyranosides and furanosides. In these reactions the hydride atom is delivered axially at C-1.[77]

$$\xrightarrow[\text{Me}_3\text{SiOTf, BF}_3]{\text{Et}_3\text{SiH}} \quad (18)$$

2,3-Unsaturated methylpyranosides give 3-deoxyglycals with chloride-free LAH, best in THF or dioxane.[78–81] It was shown, by use of LAD, that hydride usually enters the molecule stereospecifically on the same side as the departing oxygen (equations 19 and 20). Ease of reduction appears to be related to ease of formation of the rather tight transition state (**11**). These reactions have been used to synthesize stereospecifically 2-deuterated 2-deoxyriboses,[79] which were then used to establish the stereochemistry of nucleotide deoxygenation.[82] An axial 4-hydroxy or -alkoxy group may also serve to direct the aluminohydride to one face of the ring; in that case the oxygen atom at C-4 may be eliminated instead. This process predominates when the C-4 oxygen atom is part of a rigid ether system.

$$\xrightarrow[\text{fast}]{\text{LiAlD}_4} \quad + \quad (19)$$

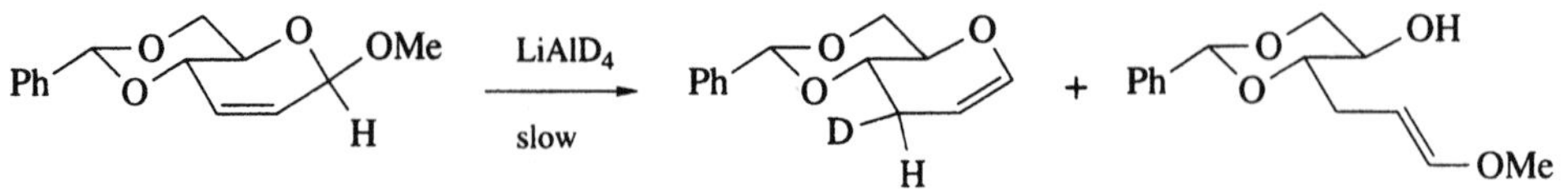

(20)

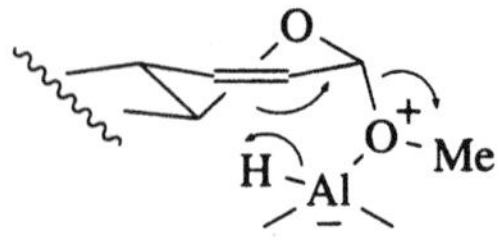

(11)

1.9.3.3 Spiroketals

The 1,6-dioxaspiro [4.5]decane nucleus (**12**; Scheme 5) might be expected, under acidic conditions, to give preferential opening of the five-membered ring (path a) in analogy with the easier opening of furanosides (see Section 1.9.3.2). On the other hand, since cyclopentyl carbonium ions form more easily (see Section 1.9.3.1), a more ready opening of the six-membered ring (path b) might be anticipated. In fact, path a is followed in $HAlCl_2$ reduction of the relatively simple ketal (**13**; equation 21),[83] while path b is followed when steroidal sapogenins are reduced (equation 22). The latter result probably reflects the higher degree of alkyl substitution in the five-membered ring.

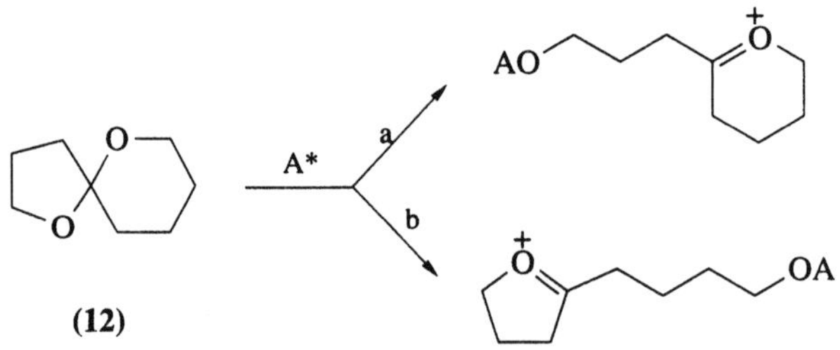

(12)

Scheme 5

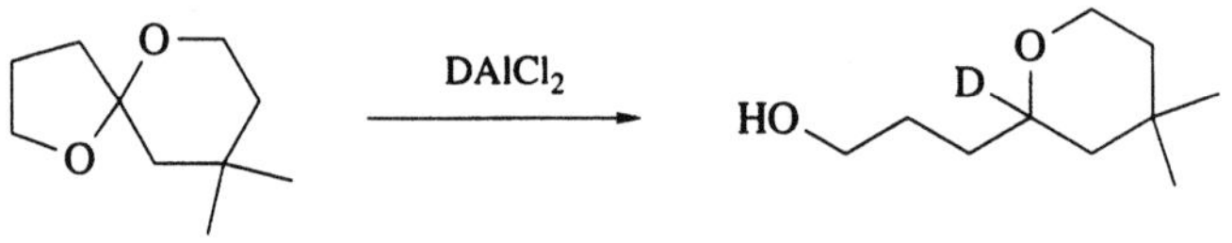

(21)

(13)

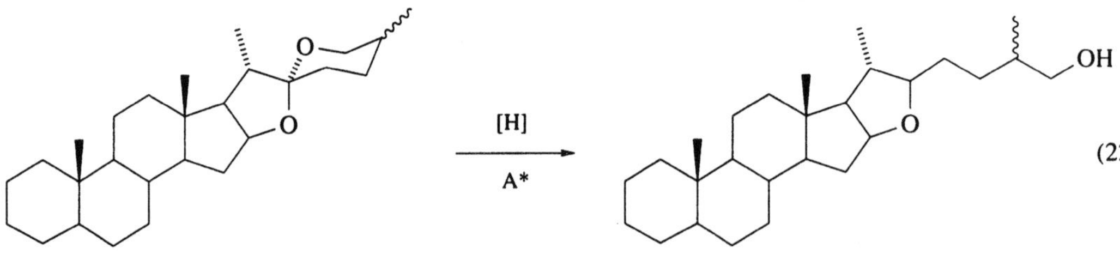

(22)

The sapogenin 'side chain' has a splendid repertoire of acid-mediated reactions[84] which could be understood only when its structure was worked out, a development not brought to fruition until it was realized that catalytic reduction in acidic medium was cleaving a spiroketal.[85] The same cleavage occurs with LAH and dry HCl,[86] and with $HAlCl_2$.[87] The LAH–BF_3 reagent gives some reduction of the spiroketal nucleus of several steroidal sapogenins, but hydroboration and the reduction of esters to ethers are major competing reactions.[42]

1.9.3.4 Dioxolanes and Dioxanes

Acetals derived from 1,2-ethanediol (1,3-dioxolanes; **14**; R^1 = H) are more readily hydrolyzed in acidic media than those from 1,3-propanediol (1,3-dioxanes; **15**; R^1 = H), but the reverse is true for the ketals (**14**, **15**; R = Me, Ph).[88] A parallel difference, suggesting a parallel mechanism for cleavage *via* an oxocarbonium ion (**18**), is seen in reductions with H_2AlCl (Table 7).[89,90] These effects may be understood as arising from: (i) the greater basicity of five-membered cyclic ethers;[73] (ii) the relief of eclipsing strains in five-membered rings; and (iii) the relief of 1,3-diaxial crowding in the six-membered ring ketals. Electron-releasing substituents at C-2 (Me, Ph) would stabilize a transition state leading to (**18**), formation of which is thought to be rate controlling.[89–91] Electron-withdrawing groups would, then, retard the rection (Tables 7 and 8).

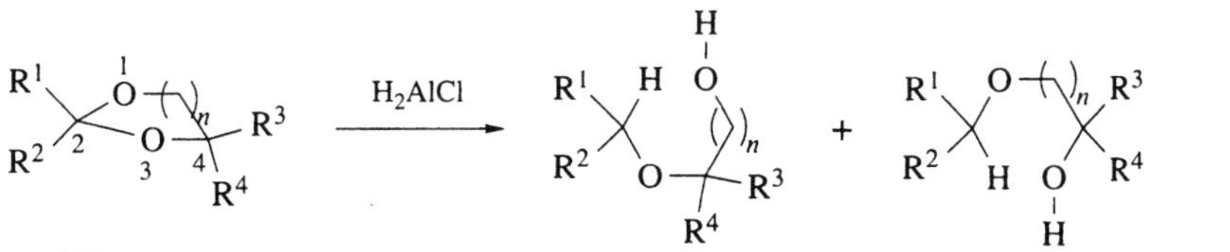

(**14**) $n = 1$
(**15**) $n = 2$

H_2AlCl → (**16**) + (**17**) (23)

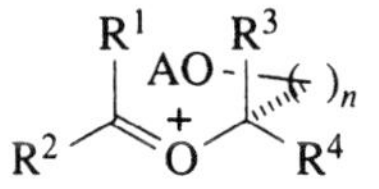

(**18**)

Table 7 Comparative and Competitive Reductions of 1,3-Dioxolanes (**14**) and 1,3-Dioxanes (**15**) by $HAlCl_2$[89,90]

Substituents		*Time*	*Total recovery*	*Reduced (%)*	
At C-2	*At C-4*	(h)	(%)	(**14**)	(**15**)
H, H	H, H	24	55	60[a]	
			75		28[a]
Me, H	H, H	2	70	66	0
H, H	Me, H	2	80	40	0
Ph, H	H, H	0.33	89	100	30
Me, Me	Me, H	0.33	73	41	92

[a]Comparative reactions; all others competitive.

Table 8 Hydrogenolysis of 2-Substituted-1,3-dioxolanes (**14**) by H_2AlCl in Ether at Room Temperature[91]

Substituents R^1	R^2	*Time* (h)	*Total recovery* (%)	*Reduced* (%)
H	H	24	55	60
Me	H	1	70	100
Me	Me	1.5	71	100
CH_2Cl	H	48	80	14
$CHCl_2$	H	48	85	0
Ph	H	1	80	100

When there is no more than one substituent at C-4, cleavage tends to occur at the oxygen atom remote from the larger or more electron-releasing groups, favoring the formation of (**16**; equation 23). When the substituent is electron withdrawing the reverse occurs and (**17**) is favored (Table 9). Steric hindrance to approach by the Lewis acid (A) and inductive electron release toward the positive (but not electron-deficient) oxygen atom of (**18**) appear to be important. Steric crowding between R^1 and R^3 in the transition state leading to (**18**) has a strong effect; thus 2,2,4,4-tetramethyl-1,3-dioxolane cleaves to give (**17**), contrary to initial expectations on steric and electronic grounds. That this is due to a steric retardation is shown by the fact that 2,2,4,4,5,5-hexamethyl-1,3-dioxolane is untouched after 1 h and only 30% reduced after 48 h.[92] Consistent with this, *cis*-2,4-dimethyl-1,3-dioxolane (**14**; R^2, R^4 = Me) cleaves more selectively at O-1 than does the *trans* isomer (**14**; R^1, R^4 = Me) or any of the other materials shown in Table 9; it reacts faster than, and during reaction isomerizes to some extent to, the *trans* isomer.[92] It hydrolyzes faster, too.[93] Borane in THF gives comparable results (Table 9).[44] Norcamphor ethylene ketal (**19**) gives almost exclusive *exo* reduction (equation 24) but camphor ketal (**20**) gives predominant *endo* reduction (equation 25),[94] consistent with the behavior of the parent ketones with $HAlCl_2$.

Table 9 Hydrogenolysis of 2- and 4-Substituted-1,3-Dioxolanes (**14**) by H_2AlCl[a] and by BH_3–THF[b] (equation 23)

Reagent	*Substituents*				*Total recovery*	*Recovered as:*		
	R^1	R^2	R^3	R^4	(%)	(**16**) (%)	(**17**) (%)	*Ref.*
H_2AlCl	Me	Me	Me	H	86	82	18	9
BH_3–THF					75	80	20	44
H_2AlCl	Me	Me	Ph	H	81	52	48	91
BH_3–THF					88	58	42	44
H_2AlCl	Me	Me	CH_2OH	H	79	19	81	91
H_2AlCl	Me	Me	CH_2Cl	H	96	5	95	91
BH_3–THF					88	15	85	44
H_2AlCl	Me	Me	Me	Me	87	6	94	91
H_2AlCl	Me	H	Me	H (*cis*)		96	4	92
H_2AlCl	Me	H	H	Me (*trans*)		35	65	92
H_2AlCl	Ph	H	Me	Me	81	90	10	91
BH_3–THF					89	97	3	91

[a]Ether, r.t.; acetal:$AlCl_3$:LAH = 1:1:1. [b]Mix at –10 °C, let warm to r.t., 24 h, acetal:BH_3–THF = 1:1.5.

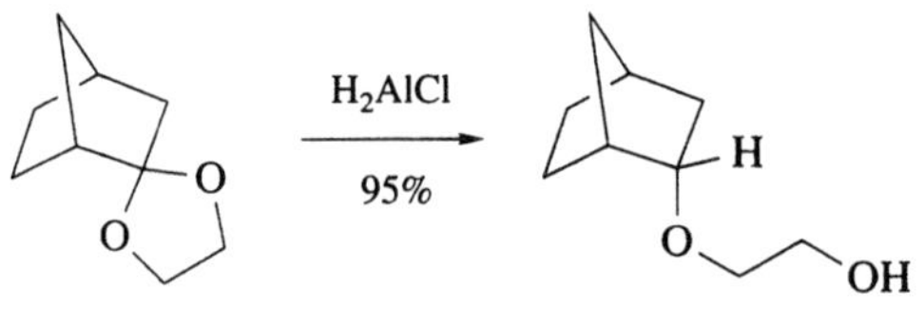

(**19**) —H_2AlCl, 95%→ (24)

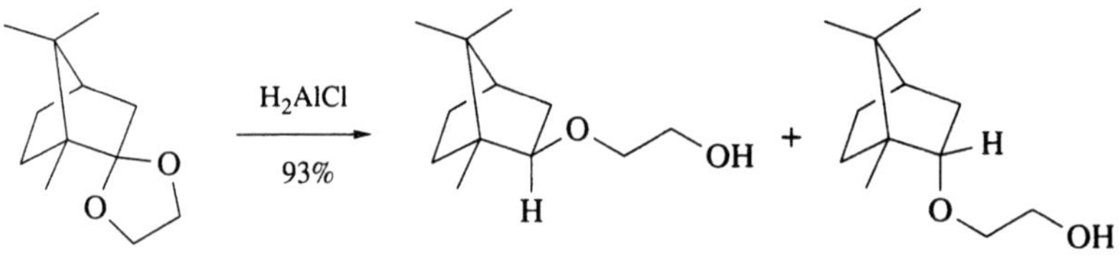

(**20**) —H_2AlCl, 93%→ 78% + 22% (25)

Norcamphor isobutylene ketal appears (NMR) to be a nearly equal mixture of the isomers (**21**) and (**22**) (Scheme 6),[95] but it is reduced as though it were only (**21**), reacting *via* a retention mechanism. The reduction is very slow (168 h) and it is this author's view that one cannot exclude the possibility that (**22**) is not reduced at all but is isomerized to replace (**21**) as it is consumed (a few such isomerizations have been observed).

The retention mechanism was first clearly enunciated to account for results in the steroid series,[96] where the simple ketals (**23**) and (**24**) are reduced exclusively from the less-hindered underside and where the 1,2-propylenedioxy ketal of cholest-5-en-3-one (**25**) is both opened and reduced from that side. This mechanism may be visualized by imagining that A in (**18**) carries the hydride and attacks the carbonium center as soon as it forms.[96]

With a variety of substrates, DIBAL-H in toluene at 0 °C is effective in selectively cleaving benzylidene dioxolanes and dioxanes in the same way as H_2AlCl.[97]

Acetals or ketals (R^1, R^2 different) of chiral 2,3-butanediol (**26**) or 2,4-pentanediol (**27**) have diastereotopic oxygen atoms which can be distinguished by Lewis acids. Stereoselective reduction will occur if the subsequent step must follow an obligatory steric course. It could also take place at the carbonium center, under the influence of the chiral ligand, but this would not be expected to be very efficient. In either case the chiral adjuvant, by remaining attached, would serve as a resolving agent and allow separation of the reduction products as diastereomers. Efficient removal of the adjuvant can be achieved by way of mild oxidation of the free hydroxy group and reduction with metals or cleavage with base.[6]

The oxygen atoms of the dioxolane (**26**) are not very different and the degree of asymmetric reduction with ketals[98] is not much by modern standards (Table 10), although some acetals do show good stereose-

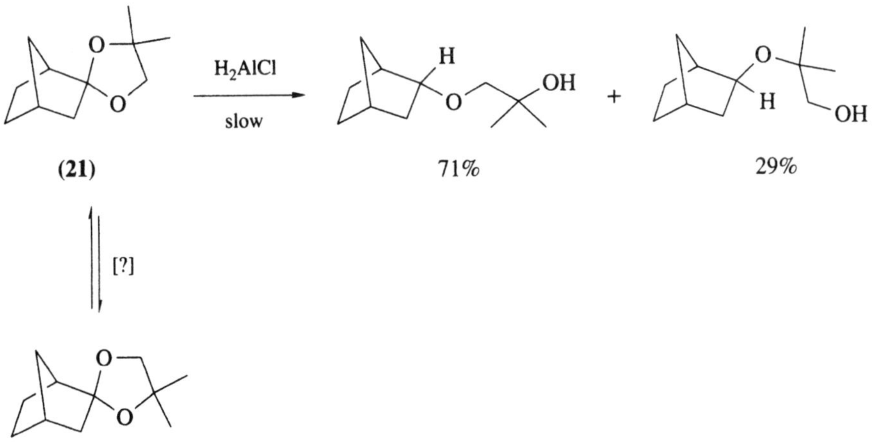

(21)

H_2AlCl

slow

71%

29%

[?]

(22)

Scheme 6

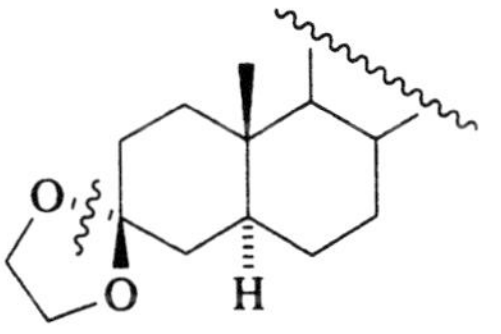

(23)

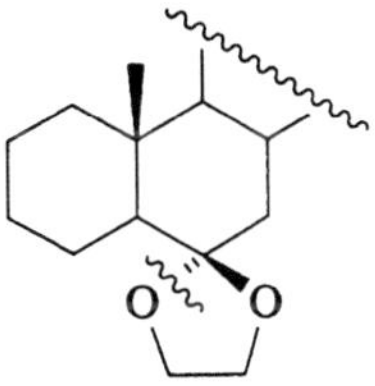

(24)

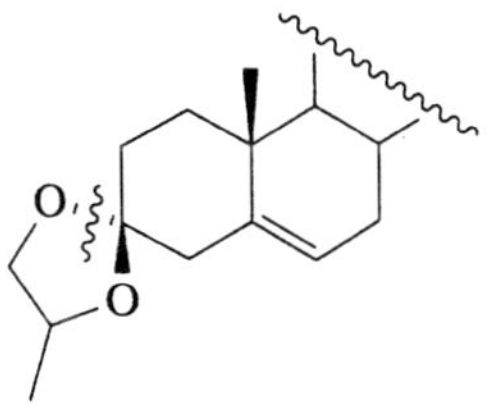

(25)

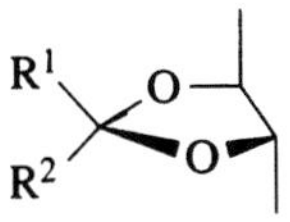

(26)

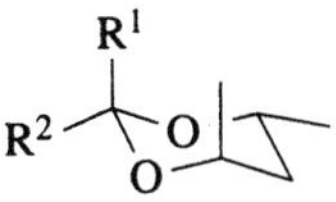

(27)

lectivity with $DAlCl_2$. Racemic diol was used in this work, so the sense of the asymmetric reduction is not known.

The oxygen atoms of (**27**) are, however, quite different. One is next to an axial methyl group and is quite open to attack, the other is hindered by an equatorial methyl group. Reactions of compounds of this series with various aluminum hydrides ($HAlBr_2$, $HAlCl_2$ and DIBAL-H are best) at low temperatures give good yields of the products expected from the retention mechanism (equation 26; Table 11).[99,100] Inversion (equation 27) is obtained with triethylsilane and various Lewis acid partners.

(26)

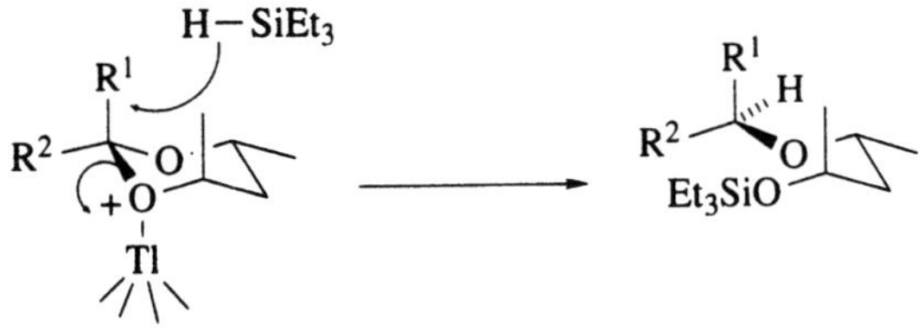

H—$SiEt_3$

Et_3SiO

(27)

Table 10 Reduction of 2-Substituted-2,4,5-trimethyl-1,3-dioxolanes (**26**) with $HAlCl_2$ in Ether at Room Temperature (2 h)[98]

R^1	R^2	Isomer ratio (configurations unspecified)
Me	Et	75.5:24.5
	Pr^i	79:21
	Bu^t	86.6:13.4
	c-C_6H_{11}	75:25
	1-Adamantyl	88.5:11.5
	Pr^n	69.3:30.7
	Bu^i	75.8:24.2
	CH_2Bu^t	64:36
	$CH{=}CH_2$	54.5:45.6
	$C{\equiv}CH$	51.7:48.3
	Ph	52.5:47.5
	Bn	59.5:40.5
	(with $DAlCl_2$)	
H	Me	98:2
	Et	99:1
	c-C_6H_{11}	90:10
	Bu^t	78:22
	Ph	86:14
	Bn	90.6:9.4

Table 11 Stereoselective Reductions of 2-Substituted-2,4,6-trimethyl-1,3-dioxanes (**27**)[99,100]

Substituents at C-2 R^1 (Small)	R^2 (Large)	Reagent (conditions)	Yield (%)	Configuration at C-2	Diastereomer ratio	Ref.
Me	c-C_6H_{11}	DIBAL-H (CH_2Cl_2)	82	(*S*)	13:1	99
		$HAlCl_2$ (Et_2O)	98	(*S*)	19:1	99
		$HAlBr_2$ (Et_2O, –20 °C)	99	(*S*)	23:1	99
		Et_3SiH ($TiCl_4$, –78 °C)	85	(*R*)	49:1	100
	n-C_6H_{13}	DIBAL-H (CH_2Cl_2)	58	(*S*)	3.5:1	99
		$HAlCl_2$ (Et_2O)	73	(*S*)	2:1	99
		$HAlBr_2$ (Et_2O, –40 °C)	87	(*S*)	4:1	99
		Et_3SiH ($TiCl_4$, –78 °C)	97	(*R*)	7.5:1	100
	Ph	DIBAL-H (CH_2Cl_2)	88	(*S*)	28:1	99
		$HAlBr_2$ (Et_2O, –78 °C)	94	(*S*)	57:1	99
		Et_3SiH ($SnCl_4$, –78 °C)	24	(*R*)	4:1	100
$C{\equiv}C$—Bu	Me	$HAlBr_2$ (Et_2O, –20 °C)	100	(*R*)	17:1	100
		Et_3SiH ($TiCl_4$, –78 °C)	58	(*S*)	16:1	100

This principle has been used to convert a substance with a chirotopic, but not stereogenic, central carbon atom (**28**) to one in which that atom has, selectively, become chiral (**29**; Scheme 7).[101]

It has been stated:[102] 'It is probably true that no sugar derivative has been manipulated as successfully as methyl-4,6-*O*-benzylidene-α-D-glucopyranoside.' Reduction of derivatives of (**30**) with H_2AlCl or H_2AlBr in ether solvents gives increasing amounts of cleavage at O-6 to form derivatives of (**31**) as crowding by substituents at C-3 increases (Scheme 8; Table 12).[103] DIBAL-H is much less selective but gives clean cleavage at O-6 at –30 °C.[104] Borane in THF is ineffective and reacts only slowly at higher temperatures.[103] Sodium cyanoborohydride,[54] with dry HCl in THF[105] or trifluoroacetic acid in DMF,[106] reacts cleanly in the opposite sense with *p*-methoxybenzylidene acetals to form analogs of the series (**32**). It seems likely that the smaller Lewis acid can coordinate more readily at O-4 (perhaps assisted by the neighboring oxygen atom, O-3) while the hydride donor attacks from the other, less-hindered, side of the reacting carbon atom. With a more bulky reagent, Me_3SiCl, as acidic partner, cleavage occurs again at O-6.[106] A striking solvent effect is observed in reductions with borane:trimethylamine and $AlCl_3$;[107] with toluene as solvent cleavage occurs at O-6, but with THF it occurs at O-4. Reduction of the galactose analogs (**33**) with H_2AlCl gives cleavage at O-4 (Scheme 9) when the group at C-3 is small,[108,109] but at O-6 when it is large.[109]

Acetopheneone forms a pair of 4,6-ketals with glucose.[110] The kinetic product has an equatorial phenyl group (**36**) and reacts with H_2AlCl by cleavage at O-6 to give (**38**), in analogy with the conversion of (**30**) to (**31**). The more stable isomer (**37**), however, has an axial phenyl group, as found in simpler systems.[111] It reacts by cleavage at O-4. Models indicate that only an equatorial phenyl group can stabilize the incipient oxocarbonium ion by resonance; this would seem to require a boat-like transition state (**39**)

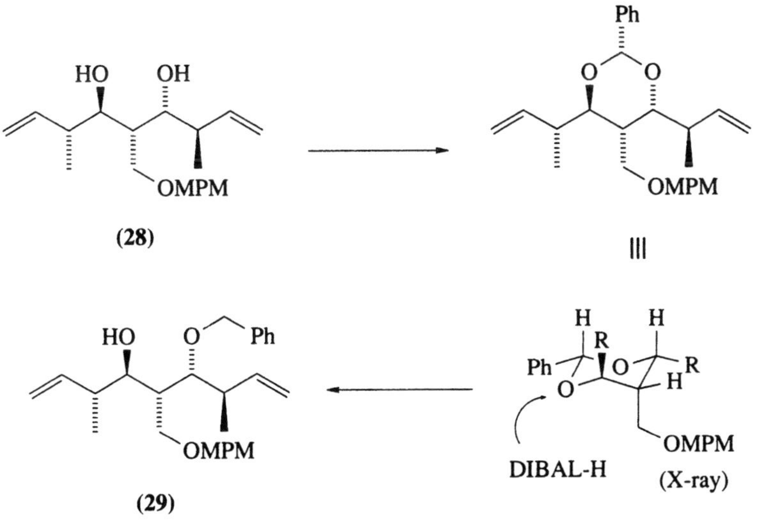

(28)

III

(29)

DIBAL-H (X-ray)

Scheme 7

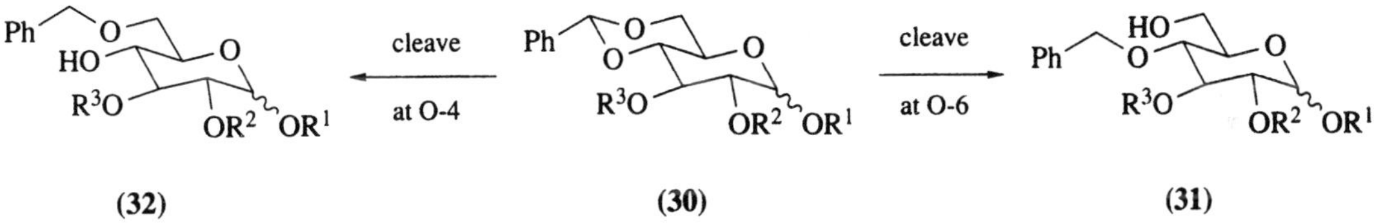

(32) cleave at O-4 (30) cleave at O-6 (31)

Scheme 8

Table 12 Regioselectivity in the Reduction of 4,6-*O*-Benzylidene-D-glucopyranosides (**30**) with Aluminum Hydride Reagents[103]

Reagent (Solvent) (Temperature)	*Substituent at C-2*	*Isolated yield (%)* (**31**)[a]	(**32**)[a]	*Product ratio* (**31**):(**32**)
H_2AlCl (2:1 = Et_2O:CH_2Cl_2) (45 °C)	—OMe	68	14	77:23
	—OEt	75	7	91:9
	—OPr^n	91	3	94:6
	—OBn	91	4	93:7
	—H (R^2 = Me)	43	41	53:47
H_2AlBr (as above)	—OMe	73	12	84:16
DIBAL-H (9:1 = Et_2O:CH_2Cl_2)	—OMe	68	25	71:29
(C_6H_6, 0 °C)		34	43	46:54

[a](**30**); R^1 = Bn (β); R^2 = R^3.

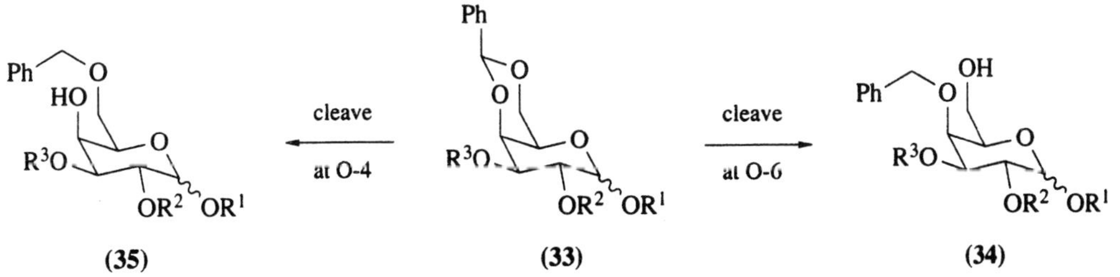

(35) cleave at O-4 (33) cleave at O-6 (34)

Scheme 9

for this case (Scheme 10). This would give the observed cleavage at O-4. The corresponding galactose derivatives behave in the same way.[110]

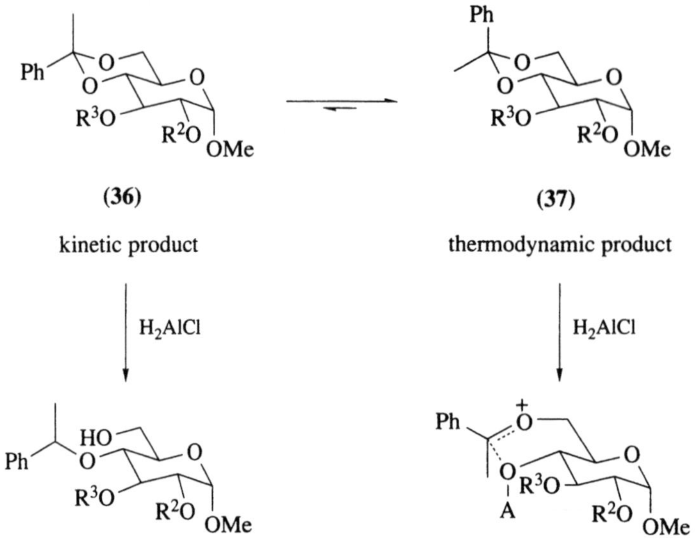

(36) kinetic product

(37) thermodynamic product

H_2AlCl

(38)

(39)

Scheme 10

A dioxolane fused to a six-membered ring is subject to major constraints in forming a transition state resembling (**18**), where C-4, O-3, C-2, R^1 and R^2 must all be coplanar. The oxygen atom that becomes a part of the oxocarbonium system is puckered out of the plane of the other four atoms. The substituents on the carbon atom of the developing oxocarbonium ion then adopt quasi-axial (R^1) or quasi-equatorial (R^2) orientations, the latter being required of a phenyl group if it, too, is to become coplanar and contribute to stabilization of the transition state. This now means that when the phenyl group has an *endo* orientation relative to the six-membered ring (**40**) the oxygen atom that is breaking away will be equatorial, while the one that becomes an ether oxygen is axial. The reverse is true when the phenyl group is *exo* to the six-membered ring (**41**).

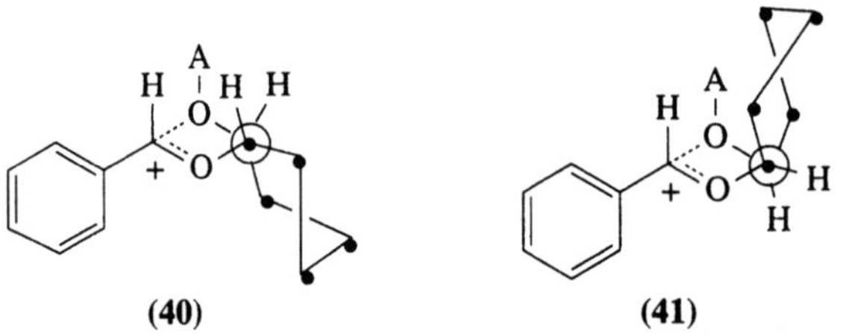

(40)

(41)

The *exo* and *endo* isomers of benzylidene acetals of the carbohydrate series can be obtained separately. Their behavior on reduction with H_2AlCl accords with the analysis outlined above; frequently only one product is obtained. Thus, the 1,2,4,6-di-*O*-benzylidene derivatives of glucose (**42**) and (**43**; R = H or Bn) cleave rapidly with 1 equiv. of H_2AlCl at 0 °C (30 min for **42** and 10 min for **43**) at the marked bonds.[112] Here, at least, it is clear that steric effects outweigh polar ones. The 2,3-*endo* and -*exo* isomers of the α-D-mannoside series (**44**) and (**45**) likewise cleave mainly equatorial (78:22) and axial (96:4), respectively, when treated with 1 equiv. of H_2AlCl at room temperature.[113] The same behavior is seen with the methyl 2,3-benzylidene-α-L-rhamnosides,[114] the methyl 3,4-*O*-benzylidene-β-L-arabinosides[114] and

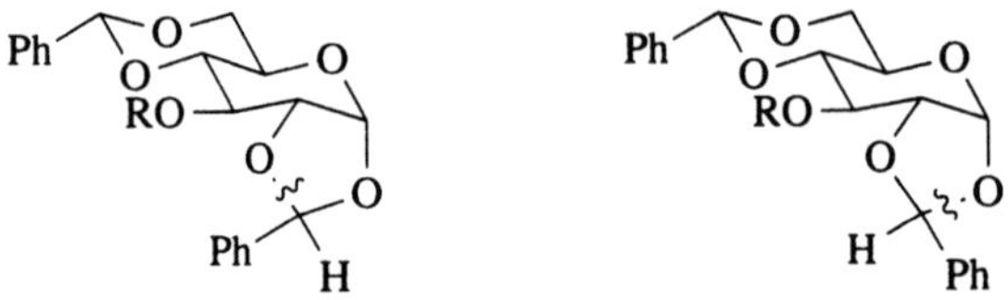

(42) *endo*

(43) *exo*

the methyl 3,4-*O*-benzylidene-α- and methyl-3,4-*O*-benzylidene-β-D-galactopyranosides.[115] Only with the benzylidene acetals of 1,6-anhydro-β-D-galactopyranose (**46**) and (**47**) do both acetals give the same product,[116] both cleaving at O-4. This is the normal mode of cleavage for the *endo* isomer (**46**) and it occurs readily. The normal reaction for the *exo* isomer would require puckering at O-4; this would produce a very close approach of the incipient carbonium carbon atom to C-6, perhaps producing prohibitive strain. The more difficult cleavage *via* a transition state with a quasi-axial phenyl group would then be required; consistent with this, the *exo* isomer (**47**) is distinctly more difficult to reduce.

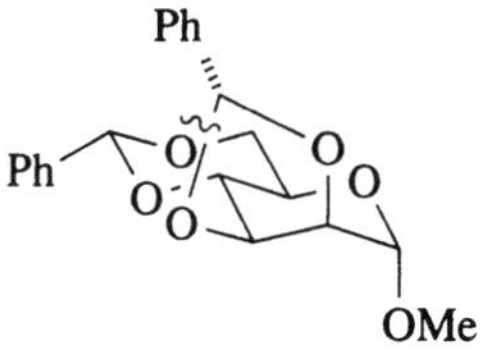

(**44**) *endo* 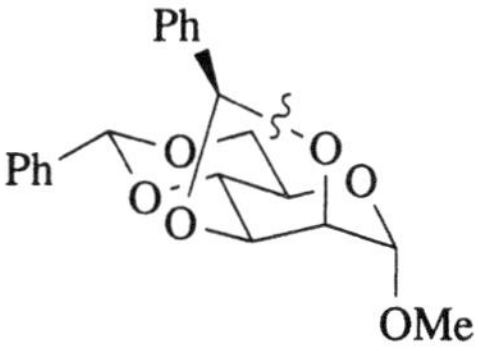 (**45**) *exo*

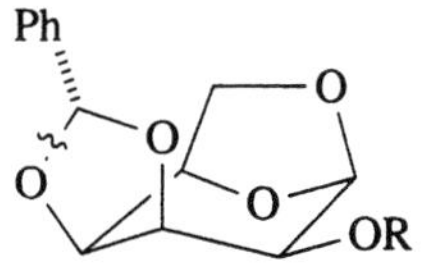

(**46**) *endo* 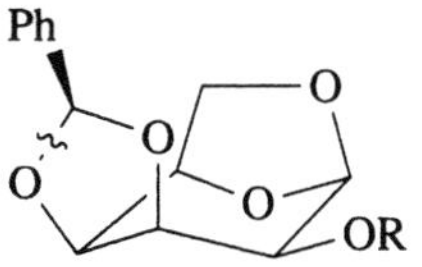 (**47**) *exo*

1.9.3.5 Bicyclic Acetals and Ketals

Ring strain appears to be important with bicyclic acetals—they tend to cleave under reducing conditions so as to leave six-membered rings or to avoid leaving five-membered ones. Thus, hydrogenolysis of 6,8-dioxabicyclo[3.2.1]octane (**48**; R = H)[117,118] and the 3,6,8-trioxa analog (**49**; R = H)[119] with H_2AlCl occurs with cleavage of the oxygen atom that is, essentially, axial to the six-membered ring. Ketals (R = Me) give mixtures of stereoisomers,[117,119] indicating that the carbonium ion intermediate is long lived enough to be attacked from either side.

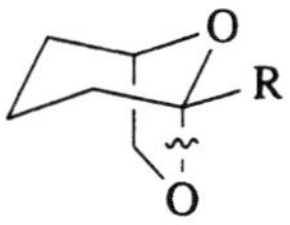

(**48**) 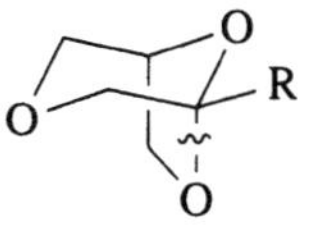 (**49**)

Good stereoselectivity has been obtained with other reagents, however. Thus, the dioxabicyclo[3.3.1]nonane (**50**) opens with retention using DIBAL-H in CCl_4 but with inversion using diphenylsilane and $TiCl_4$ (Scheme 11).[120] Similar stereoselectivity is shown with both reagents in reductions of dioxa[3.2.1]octanes and dioxa[2.2.1]heptanes (**51**); in the latter case cleavage occurs nearly equally at both oxygen atoms (Scheme 12).[120] In like fashion, the dioxa[3.2.1]octane (**52**) gives mainly *trans* product with DIBAL-H in CH_2Cl_2 at –78 °C, but *cis* product with the $Zn(BH_4)_2$:$TiCl_4$ pair or with Et_3SiH:$TiCl_4$, both at –78 °C, (Scheme 13).[121] These reactions are nearly quantitative and give only traces of the other isomer.

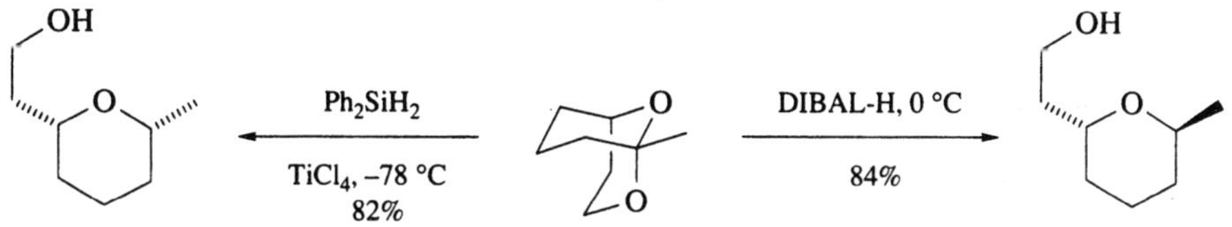

Ph₂SiH₂, TiCl₄, –78 °C, 82% (only) ← (**50**) → DIBAL-H, 0 °C, 84% (*trans*:*cis* = 92:8)

Scheme 11

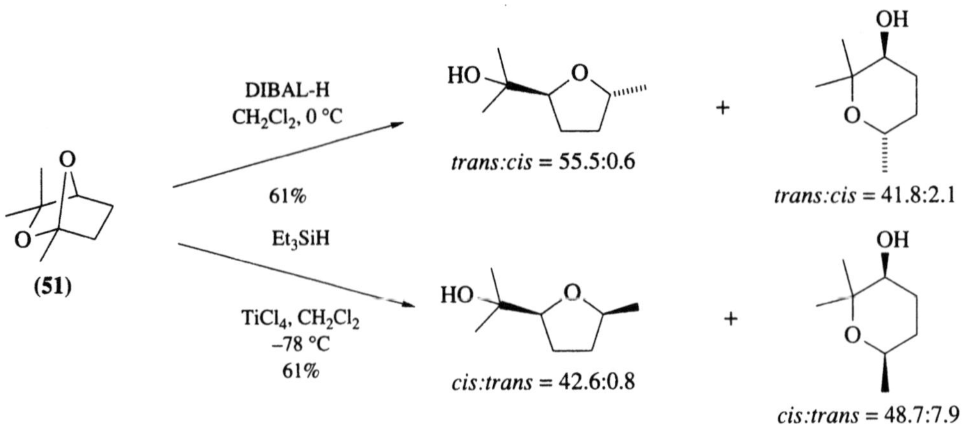

DIBAL-H
CH_2Cl_2, 0 °C
61%

trans:cis = 55.5:0.6

trans:cis = 41.8:2.1

(51)

Et_3SiH
$TiCl_4$, CH_2Cl_2
−78 °C
61%

cis:trans = 42.6:0.8

cis:trans = 48.7:7.9

Scheme 12

$Zn(BH_4)_2$, $TiCl_4$
or
Et_3SiH, $TiCl_4$
−78 °C

(52)

DIBAL-H
CH_2Cl_2, −78 °C

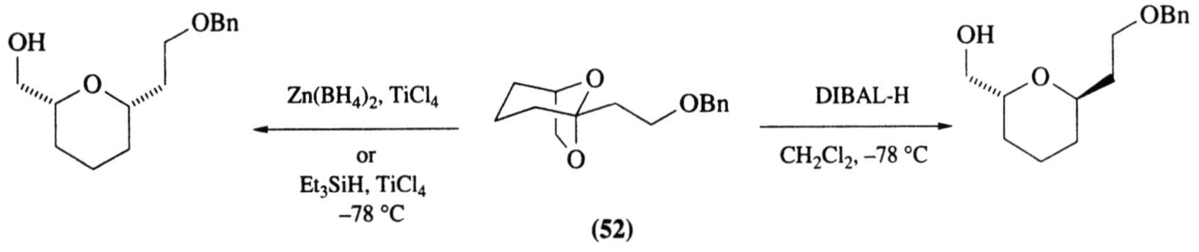

Scheme 13

1.9.3.6 Azaacetals and Azaketals

Oxazolidines (**53**) are readily formed from aldehydes or ketones and ethanolamines; they can be hydrolyzed with ease and show reactions that might be expected of the imino alcohol intermediate (**54**). Among these are the addition of Grignard reagents[122] and catalytic hydrogenolysis of the C—O bond (equation 28).[122,123] This reaction is exothermic over Adam's catalyst in methanol but slower in acetic acid.[123] Nickel and copper chromite are also effective but at higher temperatures and pressures,[123] as is the case with palladium.[122] The same cleavage occurs with LAH (unassisted)[124,125] and with the borane–THF complex.[126]

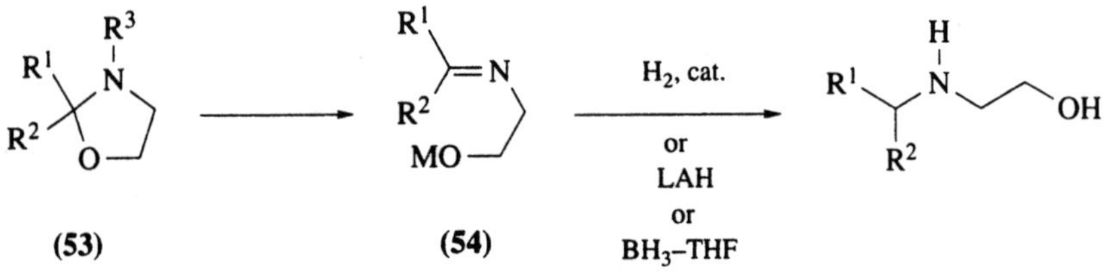

(53) → (54) —H_2, cat. or LAH or BH_3–THF→ (28)

This pattern of cleavage is also followed when the azaacetal is acyclic (equation 29)[127] and even when the nitrogen atom is part of a three-membered ring (equation 30).[128] Tetrahydropyranylamines are reduced to ε-hydroxypentylamines by hydrogenation over Ni[129] or reduction with LAH (equation 31).[130]

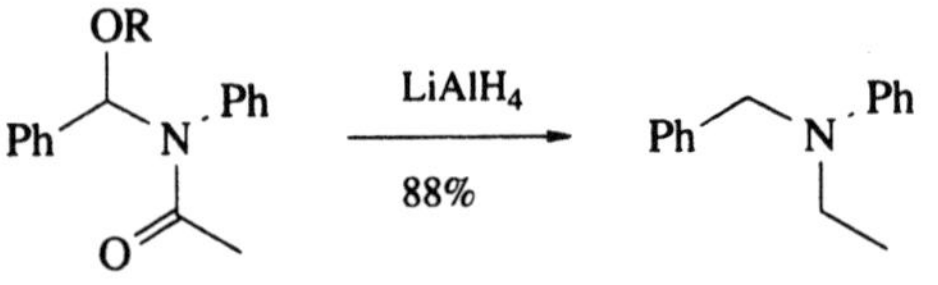

$LiAlH_4$, 88% (29)

The sapogenin-like side chain of steroidal alkaloids such as solasodine[131] and tomatidine (**55**),[132,133] undergoes cleavage of the five-membered oxa ring with hydrogen over Pt[131,133] or with LAH[131,132] (equation 32; *cf.* Section 1.9.3.3). When, however, the electron-donating power of the nitrogen atom is diminished by acylation the pattern is reversed and C—N cleavage takes place (equation 33).[134]

Diazaspiroalkanes (**56**) give nearly quantitative hydrogenolysis of one ring over platinum or nickel (equation 34).[135]

1.9.3.7 Thioacetals and Thioketals

The isomeric 1,3-oxathiolanes (**57**) and (**58**) are readily interconverted by Lewis acids such as $AlCl_3$ or BF_3 (Scheme 14).[136] They are reduced with previously prepared $HAlCl_2$ (but not with BF_3–LAH)[136] to the same mixture of products,[137] with exclusive C—O cleavage. Addition of BF_3 to a mixture of ketal with LAH or, better, of LAH to a mixture of ketal and BF_3 does, however, give reduction.[138] The latter results provided a firm basis for the accepted carbonium ion mechanism and serve to emphasize that BF_3–LAH mixtures can generate several reducing agents (*i.e.* $LiBH_4$ and B_2H_6) depending on how they are made and how quickly they are used (see Section 1.9.2.3.2). It is suggested that the Lewis acid coordinates more readily with the oxygen atom than with the sulfur atom and that this determines the direction of cleavage. With longer reaction times, the oxygen atom is removed from the product (Scheme 15); studies with deuterium-labeled reagents show that this reaction proceeds *via* the sulfonium ion (**59**).[136]

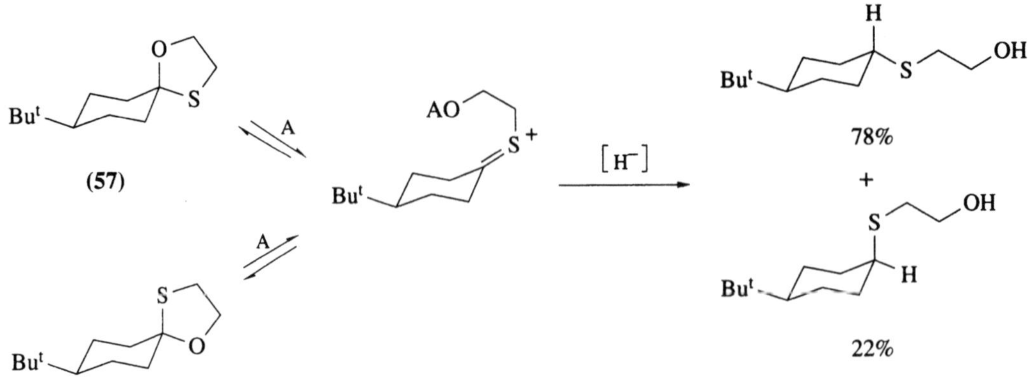

Scheme 14

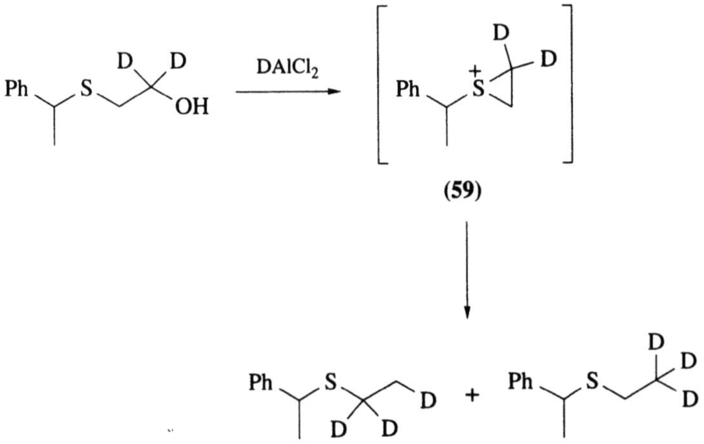

Scheme 15

1,3-Thioxanes (**60**) give similar ring-opening reactions but the subsequent reaction to lose oxygen is not observed.[137] 1,3-Dithiolanes such as the *cis* and *trans* forms of (**61**) can be isomerized by Lewis acids[138] but fail to undergo reduction with H_2AlCl; this may be a consequence of solubility problems.

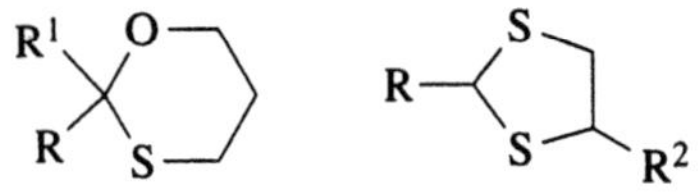

(**60**) (**61**)

Tetrahydrofuranyl sulfides (equation 35; $n = 1$) and the tetrahydropyranyl analogs (equation 35; $n = 2$) give ring opening with $HAlCl_2$;[139] the subsequent removal of oxygen does not occur here.

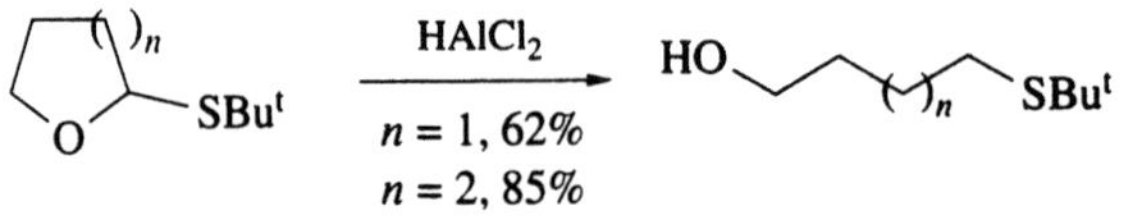

$HAlCl_2$; $n = 1$, 62%; $n = 2$, 85% (35)

The methyl *endo*- and *exo*-benzylidenegalactopyranosides, (**62**) and (**63**), respectively, give the same product (**64**) on reduction with H_2AlCl, *via* the expected C—O bond cleavage (Scheme 16).[140] In conformity with the analysis presented above (Section 1.9.3.4) the *endo* isomer reacts rapidly, with its normal equatorial cleavage, while the *exo* isomer, for which equatorial cleavage is disfavored, reacts slowly at reflux.

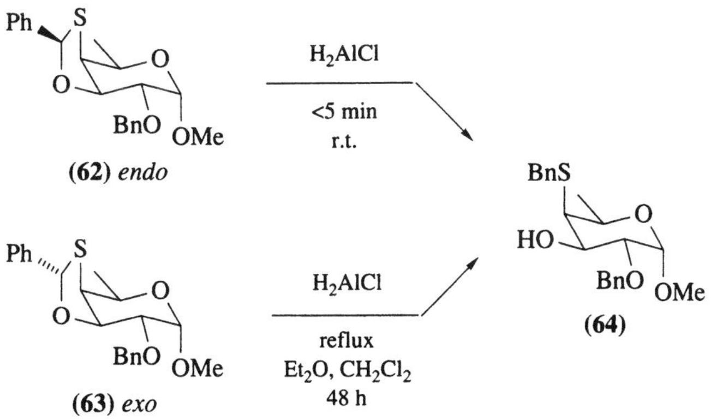

(**62**) *endo* —H_2AlCl, <5 min, r.t.→ (**64**)

(**63**) *exo* —H_2AlCl, reflux, Et_2O, CH_2Cl_2, 48 h→ (**64**)

Scheme 16

Thiazolidinones (**65**) can be reduced, with C—S bond cleavage, by LAH (equation 36);[141] thiazolidines (**66**; which may be intermediates) are reduced in the same way (equation 37) with LAH or $NaBH_4$.[142] An aromatic group at the reacting center promotes cleavage; thus the reduction of 5,6-dihydro-1,3-thiazines (**67**) with $NaBH_4$[143] can be stopped at the thiazane stage (**68**; equation 38), except when R is phenyl, when only the ring-opened product (**69**) is obtained. Thiazanes in general can be reduced with LAH.[144]

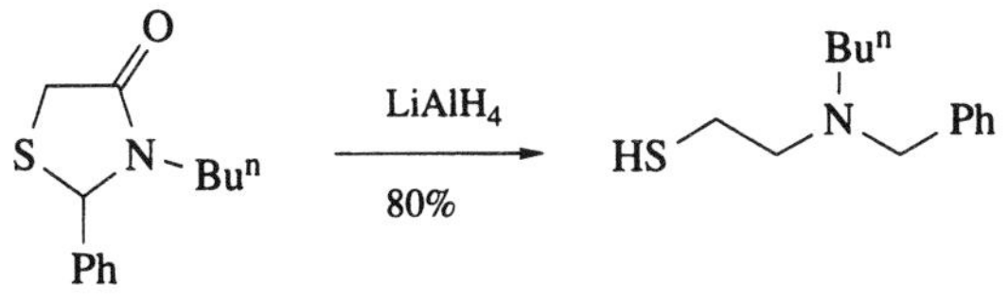

(**65**) —$LiAlH_4$, 80%→ $HS(CH_2)_2N(Bu^n)CH_2Ph$ (36)

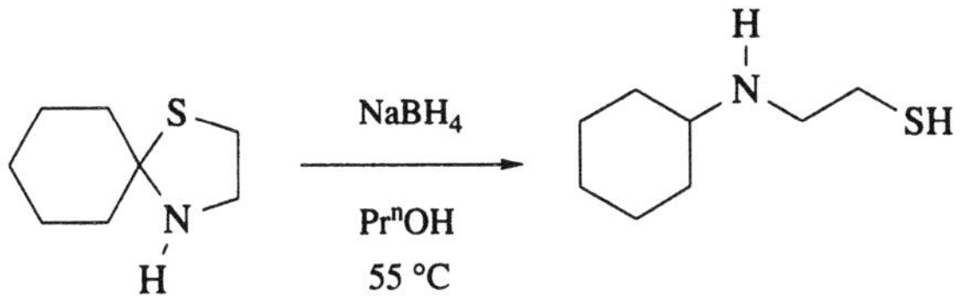

(**66**) —$NaBH_4$, Pr^nOH, 55 °C, 60%→ cyclohexyl-NH-$(CH_2)_2SH$ (37)

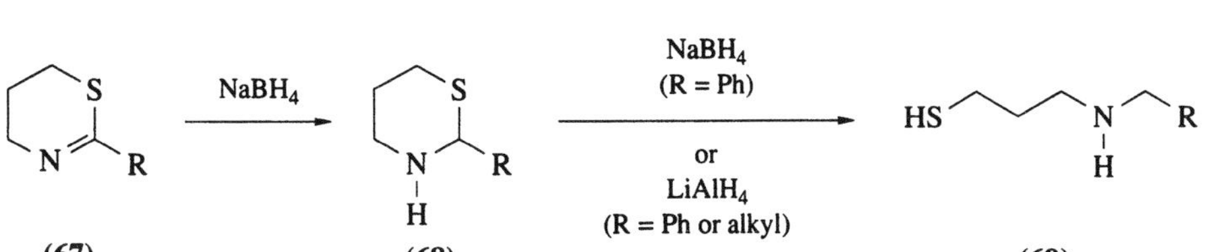

(**67**) —$NaBH_4$→ (**68**) —$NaBH_4$ (R = Ph) or $LiAlH_4$ (R = Ph or alkyl)→ (**69**) $HS(CH_2)_3NHCH_2R$ (38)

C—S bonds can be cleaved reductively by dissolving metals, by processes which depend on the ability of sulfur to accept electrons.[145] 1,3-Dithiolanes tend strongly to suffer cleavage of both C—S bonds with sodium in liquid ammonia when C-2 is unsubstituted or carries a phenyl group;[146] isopropylidene derivatives (**70**), however, give single cleavage (equation 39). Bond rupture requires formation of a relatively stable carbanion.

Calcium in ammonia offers some advantages[147,148] in that most cases of overreduction can be prevented by limiting the amount of active metal, although derivatives of benzaldehyde lose both C—S bonds.[147] Oxathiolanes and oxathianes give clean cleavage of the C—S bond.[146,148] In the *t*-butylcyclohexylidene series, the isomeric pairs (**71**) and (**72**) give almost convergent results, with the *trans*

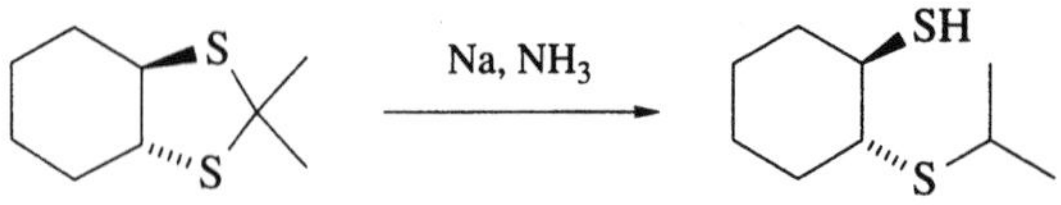

(70) (39)

product (**74**) predominating strongly over the *cis* product (**75**; Table 13), probably *via* protonation of the carbanion (**73**), which must have an appreciable lifetime (Scheme 17).

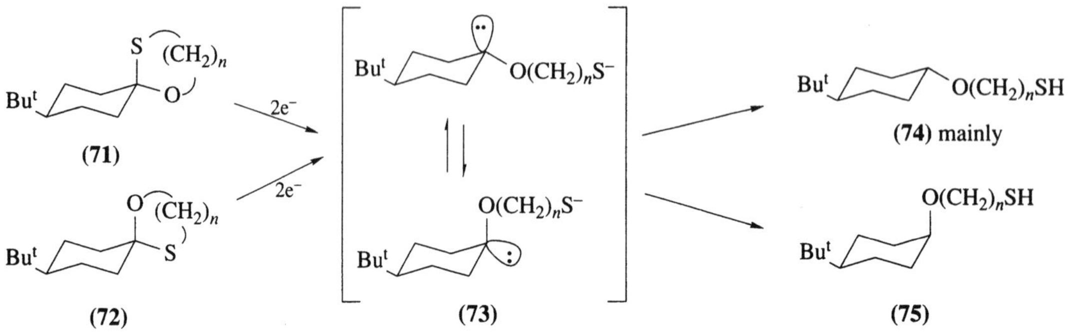

(71) (72) (73) (74) mainly (75)

Scheme 17

Table 13 Reductions of Oxathiolanes and Oxathianes with Calcium in Liquid Ammonia[148]

Thioketal	*n*	*Yield (%)*	*Product (%)* (**74**)	(**75**)
(**71**)	2	83–89	97–98	3–2
(**72**)	2	47–61	90–97	10–3
(**71**)	3	84–90	99.5	0.5
(**72**)	3	24–52	84–93	16–7

1.9.4 REFERENCES

1. T. W. Greene, 'Protective Groups in Organic Synthesis', Wiley, New York, 1981.
2. A. J. Kirby, 'The Anomeric Effect and Related Stereoelectronic Effects at Oxygen', Springer-Verlag, Berlin, 1983.
3. P. Deslongchamps, 'Stereoelectronic Effects in Organic Chemistry', Pergamon Press, Oxford, 1983.
4. M. L. Sinnott, *Adv. Phys. Org. Chem.*, 1988, **24**, 113.
5. J. Gelas, *Adv. Carbohydr. Chem. Biochem.*, 1981, **39**, 71.
6. H.-J. Altenbach, *Nachr. Chem., Tech. Lab.*, 1988, **36**, 1212 (*Chem. Abstr.*, 1989, **110**, 38 282w).
7. T. Bieg and W. Szeja, *Carbohydr. Res.*, 1986, **154**, 296.
8. R. L. Augustine, 'Catalytic Hydrogenation', Dekker, New York, 1965, p. 135.
9. T. Kariyone, T. Kajiura, A. Ueno and N. Suzuki, *Yakugaku Zasshi*, 1953, **73**, 493 (*Chem. Abstr.*, 1954, **48**, 3296a).
10. P. E. Papadakis, *J. Am. Chem. Soc.*, 1936, **58**, 665.
11. F. Sigmund and G. Marchart, *Monatsh. Chem.*, 1927, **48**, 267 (*Chem. Abstr.*, 1927, **21**, 3608).
12. L. W. Covert, R. Connor and H. Adkins, *J. Am. Chem. Soc.*, 1932, **54**, 1651.
13. B. I. Fleming and H. I. Bolker, *Can. J. Chem.*, 1976, **54**, 685.
14. M. Cabanac, *C. R. Hebd. Seances Acad. Sci., Ser. B*, 1929, **188**, 1257.
15. F. Sigmund and R. Uchann, *Monatsh. Chem.*, 1929, **51**, 234 (*Chem. Abstr.*, 1929, **23**, 2 416).
16. W. L. Howard and J. H. Brown, Jr., *J. Org. Chem.*, 1961, **26**, 1026.
17. M. Acke and M. Anteunis, *Bull. Soc. Chim. Belg.*, 1965, **74**, 41.
18. A. J. Birch and G. S. R. Subba Rao, *Adv. Org. Chem.*, 1972, **8**, 1.
19. M. Smith, in 'Reduction', ed. R. L. Augustine, Dekker, New York, 1968, p. 95.
20. J. M. Hook and L. N. Mander, *Nat. Prod. Rep.*, 1986, **3**, 35.
21. K. S. J. Stapleford, *Synth. Commun.*, 1982, **12**, 651.
22. A. R. Pinder and H. Smith, *J. Chem. Soc.*, 1954, 113.
23. E. Dimitriadis and R. A. Massy-Westropp, *Aust. J. Chem.*, 1984, **37**, 619.
24. R. A. Massy-Westropp and R. F. O. Warren, *Aust. J. Chem.*, 1984, **37**, 1023.
25. R. A. Massy-Westropp and R. F. O. Warren, *Aust. J. Chem.*, 1984, **37**, 1303.
26. N. G. Gaylord, 'Reductions with Complex Metal Hydrides', Interscience, New York, 1956.
27. S. S. Pizey, 'Synthetic Reagents', Wiley, New York, 1974, vol. 1.

28. C. J. Claus and J. L. Morgenthau, Jr., *J. Am. Chem. Soc.*, 1951, **73**, 5005.
29. J. W. Blunt, M. P. Hartshorn, L. T. Soong and M. H. G. Munro, *Aust. J. Chem.*, 1982, **35**, 2519.
30. W. W. Zajac, Jr. and K. J. Byrne, *J. Org. Chem.*, 1975, **40**, 530.
31. N. M. Yoon and H. C. Brown, *J. Am. Chem. Soc.*, 1968, **90**, 2927.
32. E. C. Ashby and J. Prather, *J. Am. Chem. Soc.*, 1966, **88**, 729.
33. W. W. Zajac, Jr. and K. J. Byrne, *J. Org. Chem.*, 1973, **38**, 384.
34. M. N. Rerick, in 'Reduction', ed. R. L. Augustine, Dekker, New York, 1968, p. 1.
35. J. Jonas, *Chem. Listy*, 1977, **71**, 695, and equations and references cited therein.
36. E. L. Eliel, V. G. Badding and M. N. Rerick, *J. Am. Chem. Soc.*, 1962, **84**, 2371.
37. R. A. Daignault and E. L. Eliel, *Org. Synth., Coll. Vol.*, 1973, **5**, 303.
38. U. E. Diner, H. A. Davis and R. K. Brown, *Can. J. Chem.*, 1967, **45**, 207.
39. L. I. Zakharkin and I. M. Khorlina, *Izv. Akad. Nauk SSSR, Otd. Khim. Nauk*, 1959, 2255 (*Chem. Abstr.*, 1960, **54**, 10 837h).
40. T. K. Kiladze, I. A. Melnitskii, O. F. Glukhova, E. A. Kantor, D. L. Rakhmankulov and J. M. Paushkin, *Dokl. Akad. Nauk SSSR*, 1987, **293**, 1133 (*Chem. Abstr.*, 1988, **108**, 112 410p).
41. G. Schill, G. Doerjer, E. Logemann and W. Vetter, *Chem. Ber.*, 1980, **113**, 3697.
42. G. R. Pettit and T. R. Kasturi, *J. Org. Chem.*, 1961, **26**, 4553, 4557.
43. A. R. Abdun-nur and C. H. Issidorides, *J. Org. Chem.*, 1962, **27**, 67.
44. B. I. Fleming and H. I. Bolker, *Can. J. Chem.*, 1974, **52**, 888.
45. H. I. Bolker and B. I. Fleming, *Can. J. Chem.*, 1975, **53**, 2818.
46. G. Zweifel, A. Horng and J. E. Plamondon, *J. Am. Chem. Soc.*, 1974, **96**, 316.
47. A. Pelter, *Chem. Ind. (London)*, 1976, 888.
48. R. J. Borders and T. A. Bryson, *Chem. Lett.*, 1984, 9.
49. P. P. Castro, S. Tihomirov and C. G. Gutierrez, *J. Org. Chem.*, 1988, **53**, 5179.
50. L. I. Zakharkin, V. I. Stanko and Y. A. Chapovskii, *Izv. Akad. Nauk SSSR, Otd. Khim. Nauk*, 1962, 1118 (*Chem. Abstr.*, 1962, **57**, 14 981b).
51. H. M. Bell and H. C. Brown, *J. Am. Chem. Soc.*, 1966, **88**, 1473.
52. C. F. Nutaitis and G. W. Gribble, *Org. Prep. Proced. Int.*, 1985, **17**, 11.
53. H. Kotsuki, Y. Ushio, N. Yoshimura and M. Ochi, *J. Org. Chem.*, 1987, **52**, 2594.
54. D. A. Horne and A. Jordan, *Tetrahedron Lett.*, 1978, 1357.
55. L. M. Loim, Z. N. Parnes, S. P. Vasil'eva and D. N. Kursanov, *Zh. Org. Khim.*, 1972, **8**, 896 (*Chem. Abstr.*, 1972, **77**, 125 841w).
56. D. N. Kursanov, Z. N. Parnes and N. M. Loim, *Synthesis*, 1974, 636.
57. G. A. Olah, T. Yamato, P. S. Iyer and G. K. Surya Prakash, *J. Org. Chem.*, 1986, **51**, 2826.
58. E. Frainnet and C. Esclamadon, *C. R. Hebd. Seances Acad. Sci., Ser. B*, 1962, **254**, 1814.
59. T. Tsunoda, M. Suzuki and R. Noyori, *Tetrahedron Lett.*, 1979, 4679.
60. J. Kato, N. Iwasawa and T. Mukaiyama, *Chem. Lett.*, 1985, **6**, 743.
61. S. Torii, S. Takagishi, T. Inokuchi and H. Okumoto, *Bull. Chem. Soc. Jpn.*, 1987, **60**, 775.
62. M. B. Sassaman, K. D. Kotian, G. K. Surya Prakash and G. A. Olah, *J. Org. Chem.*, 1987, **44**, 4314.
63. M. B. Sassaman, G. K. Surya Prakash and G. A. Olah, *Tetrahedron*, 1988, **44**, 3771.
64. M. P. Doyle, D. J. De Bruyn and D. A. Kooistra, *J. Am. Chem. Soc.*, 1972, **94**, 3659.
65. F. A. Carey and H. S. Tremper, *J. Org. Chem.*, 1971, **36**, 758.
66. I. Smonou and M. Orfanopoulos, *Tetrahedron Lett.*, 1988, **29**, 5793.
67. W. W. Zajac, Jr. and K. J. Byrne, *J. Org. Chem.*, 1970, **35**, 3375.
68. H. C. Brown, R. S. Fletcher and R. B. Johannesen, *J. Am. Chem. Soc.*, 1951, **73**, 212.
69. W. W. Zajac, Jr. and K. J. Byrne, *J. Org. Chem.*, 1972, **37**, 521.
70. E. L. Eliel, B. E. Nowak, R. A. Daignault and V. G. Badding, *J. Org. Chem.*, 1965, **30**, 2441.
71. U. E. Diner and R. K. Brown, *Can. J. Chem.*, 1967, **45**, 2547.
72. P. C. Loewen, L. P. Makhubu and R. K. Brown, *Can. J. Chem.*, 1972, **50**, 1502.
73. D. E. McLaughlin, M. Tamres, S. Searles, Jr. and S. Nukina, *J. Inorg. Nucl. Chem.*, 1961, **17**, 112.
74. P. C. Loewen and R. K. Brown, *Can. J. Chem.*, 1972, **50**, 3639.
75. D. Rolf and G. R. Gray, *J. Am. Chem. Soc.*, 1982, **104**, 3539.
76. J.-G. Jun and G. R. Gray, *Carbohydr. Res.*, 1987, **163**, 247.
77. D. Rolf, J. A. Bennek and G. R. Gray, *J. Carbohydr. Chem.*, 1983, **2**, 373.
78. B. Fraser-Reid and B. Radatus, *J. Am. Chem. Soc.*, 1970, **92**, 6661.
79. B. Radatus, M. Yunker and B. Fraser-Reid, *J. Am. Chem. Soc.*, 1971, **93**, 3086.
80. S. Y.-K. Tam and B. Fraser-Reid, *Tetrahedron Lett.*, 1973, 4897.
81. O. Achmatowicz, Jr. and B. Szechner, *Tetrahedron Lett.*, 1972, 1205.
82. B. Fraser-Reid and B. Radatus, *J. Am. Chem. Soc.*, 1971, **93**, 6342.
83. G. R. Pettit, A. H. Albert and P. Brown, *J. Am. Chem. Soc.*, 1972, **94**, 8095.
84. L. F. Fieser and M. Fieser, 'Steroids', Reinhold, New York, 1959, p. 810.
85. R. E. Marker and E. Rohrmann, *J. Am. Chem. Soc.*, 1939, **61**, 846.
86. H. M. Doukas and T. D. Fontaine, *J. Am. Chem. Soc.*, 1951, **73**, 5917; 1953, **75**, 5355.
87. G. R. Pettit and W. J. Bowyer, *J. Org. Chem.*, 1960, **25**, 84.
88. O. Ceder, *Ark. Kemi*, 1954, **6**, 523; *Chem. Abstr.*, 1954, **48**, 7 406b.
89. B. E. Leggetter, U. E. Diner and R. K. Brown, *Can. J. Chem.*, 1964, **42**, 2113.
90. U. E. Diner and R. K. Brown, *Can. J. Chem.*, 1967, **45**, 1297.
91. B. E. Leggetter and R. K. Brown, *Can. J. Chem.*, 1964, **42**, 990, 1005.
92. B. E. Leggetter and R. K. Brown, *Can. J. Chem.*, 1965, **43**, 1030.
93. P. Salomaa, *Turun Yliopiston Julk., Sar. A*, 1961, **46**, 15 (*Chem. Abstr.*, 1961, **55**, 12 387).
94. W. W. Zajak, Jr., B. Rhee and R. K. Brown, *Can. J. Chem.*, 1966, **43**, 1547.
95. P. C. Loewen, W. W. Zajac, Jr. and R. K. Brown, *Can. J. Chem.*, 1969, **47**, 4059.
96. M. S. Ahmad and S. C. Logani, *Aust. J. Chem.*, 1971, **24**, 143.

97. S. Takano, M. Akiyama, S. Sato and K. Ogasawara, *Chem. Lett.*, 1983, 1593.
98. W. J. Richter, *J. Org. Chem.*, 1981, **46**, 5119.
99. A. Mori, J. Fujiwara, K. Maruoka and H. Yamamoto, *J. Organomet. Chem.*, 1985, **285**, 83.
100. A. Mori, K. Ishihara, I. Arai and H. Yamamoto, *Tetrahedron*, 1987, **43**, 755.
101. S. L. Schreiber, Z. Wang and G. Schulte, *Tetrahedron Lett.*, 1988, **29**, 4085.
102. T. D. Inch, *Tetrahedron*, 1984, **40**, 3161.
103. P. Fügedi, A. Lipták, P. Nánási and J. Szejtli, *Carbohydr. Res.*, 1982, **104**, 55.
104. T. Mikami, H. Asano and O. Mitsunobu, *Chem. Lett.*, 1987, 2033.
105. P. J. Garegg, H. Hultberg and S. Wallin, *Carbohydr. Res.*, 1982, **108**, 97.
106. R. Johansson and B. Samuelsson, *J. Chem. Soc., Perkin Trans. 1*, 1984, 2371.
107. M. Ek, P. J. Garegg, H. Hultberg and S. Oscarson, *J. Carbohydr. Chem.*, 1983, **2**, 305.
108. S. S. Bhatacharjee and P. A. J. Gorin, *Can. J. Chem.*, 1969, **47**, 1195.
109. A. Lipták, I. Jodal and P. Nánási, *Carbohydr. Res.*, 1975, **44**, 1.
110. A. Lipták and P. Fügedi, *Angew. Chem., Int. Ed. Engl.*, 1983, **22**, 255.
111. W. F. Bailey, H. Connon, E. L. Eliel and K. B. Wiberg, *J. Am. Chem. Soc.*, 1978, **100**, 2202.
112. A. Lipták, J. Imre, J. Harangi and P. Nánási, *Carbohydr. Res.*, 1983, **116**, 217.
113. A. Lipták, I. Czegeny, J. Harangi and P. Nánási, *Carbohydr. Res.*, 1979, **73**, 327.
114. J. Harangi, A. Lipták, V. A. Oláh and P. Nánási, *Carbohydr. Res.*, 1981, **98**, 165.
115. A. Lipták, L. Jánossy, J. Imre and P. Nánási, *Acta Chim. Acad. Sci. Hung.*, 1979, **101**, 81.
116. C. Subero, L. Fillol and M. Martin-Lomas, *Carbohydr. Res.*, 1980, **86**, 27.
117. J. Colonge, J. Buendia and H. Guignard, *Bull. Soc. Chim. Fr.*, 1969, 956.
118. P. Clasper and R. K. Brown, *J. Org. Chem.*, 1972, **37**, 3346.
119. J. Gelas and S. Veyssières-Rambaud, *Carbohydr. Res.*, 1974, **37**, 293.
120. K. Ishihara, A. Mori and H. Yamamoto, *Tetrahedron Lett.*, 1987, **28**, 6613.
121. H. Kotsuki, Y. Ushio, I. Kadota and M. Ochi, *Chem. Lett.*, 1988, 927.
122. M. Senkus, *J. Am. Chem. Soc.*, 1945, **67**, 1515.
123. A. C. Cope and E. M. Hancock, *J. Am. Chem. Soc.*, 1942, **64**, 1503.
124. H. Heusser, P. T. Herzig, A. Fürst and P. A. Plattner, *Helv. Chim. Acta*, 1950, **33**, 1093.
125. E. D. Bergmann, D. Lavie and S. Pinchas, *J. Am. Chem. Soc.*, 1951, **73**, 5662.
126. R. C. Northrop, Jr. and P. L. Russ, *J. Org. Chem.*, 1975, **40**, 558.
127. A. W. Burgstahler, *J. Am. Chem. Soc.*, 1951, **73**, 3021.
128. M. J. Hatch and D. J. Cram, *J. Am. Chem. Soc.*, 1953, **75**, 38.
129. C. Glacet and J.-C. Couturier, *C. R. Hebd. Seances Acad. Sci., Ser. B*, 1961, **252**, 1624.
130. E. L. Eliel and R. A. Daignault, *J. Org. Chem.*, 1965, **30**, 2450.
131. L. H. Briggs and R. H. Locker, *J. Chem. Soc.*, 1950, 3020.
132. T. D. Fontaine, J. S. Ard and R. M. Ma, *J. Am. Chem. Soc.*, 1951, **73**, 878.
133. Y. Sato and H. G. Latham, Jr., *J. Am. Chem. Soc.*, 1956, **78**, 3146.
134. Y. Sato and H. G. Latham, Jr., *J. Am. Chem. Soc.*, 1956, **78**, 3150.
135. F. Korte, A. K. Bocz and K. H. Büchel, *Chem. Ber.*, 1966, **99**, 737.
136. E. L. Eliel, L. A. Pilato and V. G. Badding, *J. Am. Chem. Soc.*, 1962, **84**, 2377.
137. E. L. Eliel, E. W. Della and M. M. Rogić, *J. Org. Chem.*, 1965, **30**, 855.
138. B. E. Leggetter and R. K. Brown, *Can. J. Chem.*, 1963, **41**, 2671.
139. E. L. Eliel, B. E. Nowak and R. A. Daignault, *J. Org. Chem.*, 1965, **30**, 2448.
140. P. Fügedi and A. Lipták, *J. Chem. Soc., Chem. Commun.*, 1980, 1234.
141. I. R. Schmolka and P. E. Spoerri, *J. Am. Chem. Soc.*, 1957, **79**, 4716.
142. E. L. Eliel, E. W. Della and M. M. Rogić, *J. Org. Chem.*, 1962, **27**, 4712.
143. J. C. Getson, J. M. Greene and A. I. Meyers, *J. Heterocycl. Chem.*, 1964, **1**, 300.
144. E. L. Eliel and J. Roy, *J. Org. Chem.*, 1965, **30**, 3092.
145. R. Gerdil and E. A. C. Lucken, *J. Chem. Soc.*, 1964, 3916.
146. E. D. Brown, S. M. Iqbal and L. N. Owen, *J. Chem. Soc. C*, 1966, 415.
147. B. C. Newman and E. L. Eliel, *J. Org. Chem.*, 1970, **35**, 3641.
148. E. L. Eliel and T. W. Doyle, *J. Org. Chem.*, 1970, **35**, 2716.

1.10
Reduction of Carboxylic Acid Derivatives to Alcohols, Ethers and Amines

ANTHONY G. M. BARRETT
Colorado State University, Fort Collins, CO, USA

1.10.1 REDUCTIONS OF CARBOXYLIC ACIDS

1.10.1.1 Hydrogenation Reactions

Carboxylic acids may be reduced to produce primary alcohols by hydrogenation over heterogeneous catalysts at elevated temperatures and pressures. Copper chromite, barium oxide promoted copper chromite, Raney nickel, ruthenium dioxide and rhenium oxides are the most commonly employed catalysts.[1,2] For example, glyoxalic acid was hydrogenated over ruthenium dioxide at 145–150 °C and 700–775 atm (1 atm = 101 kPa) to produce 1,2-ethanediol (83%). Such ruthenium-mediated hydrogenation reactions are frequently carried out in the presence of water to suppress ester formation. Rhenium oxides are useful for the reduction of aromatic carboxylic acids without concomitant arene hydrogenation. Representative hydrogenation reactions include the preparations of 1-decanol (100%; Re_2O_7, H_2, 137 °C, 173 atm, H_2O, dioxane), 2-methyl-1-propanol (90%; ReO_3, H_2, 144 °C, 205 atm, cyclohexane), 1-hexanol (93%; Cu–Ba chromite, H_2, 200 °C, 188 atm, H_2O) and 1,4-butanediol (59%, RuO_2, H_2, 150–190 °C, 720–950 atm) from the corresponding carboxylic acids or succinic acid. In general, the preparation of primary alcohols from carboxylic acids *via* such hydrogenation reactions is inconvenient. Specialist equipment and techniques are required.

There are few reports on the use of homogeneous catalysts. $H_4Ru_4(CO)_8(PBu_3)_4$ has been used to catalyze the hydrogenation of aliphatic monocarboxylic acids to produce primary alcohols at 180–200 °C and 130 atm.[3] Generally, yields of the primary alcohols were low due to poor conversions (1–44%) and concomitant formation of esters.

1.10.1.2 Electrochemical and Dissolving Metal Reductions

The electrochemical reduction of aromatic carboxylic acids may be used to prepare aldehydes, primary alcohols, methylarenes or dihydroarene carboxylic acids, depending on the substrate and reaction conditions employed.[2,4,5] Primary alcohol formation is observed when aromatic carboxylic acids are reduced in strongly acidic media using cathode material of high hydrogen overpotential (typically mercury or lead). These reductions proceed *via* the protonated carboxylic acid and the intermediacy of the geminal diol and the aldehyde. In general, complete reduction to the primary alcohol is observed unless the intermediate aldehyde is trapped, for example as its bisulfite adduct. Cathodic reductions of aliphatic carboxylic acids are invariably inefficient and limited. Frequently such reductions produce copious quantities of hydrogen and little primary alcohol. Applications of carboxylic acid electrochemistry in synthesis are exemplified by the transformations in Scheme 1. The dissolving metal reduction of carboxylic acids is not useful for the direct preparation of primary alcohols. Reduction of aliphatic carboxylic acids may be used to prepare aldehydes using excess Li in $MeNH_2$ or $EtNH_2$[6] at reflux, or to prepare alkanes using Mg and water at 420–450 °C and 2–5 atm pressure.[7]

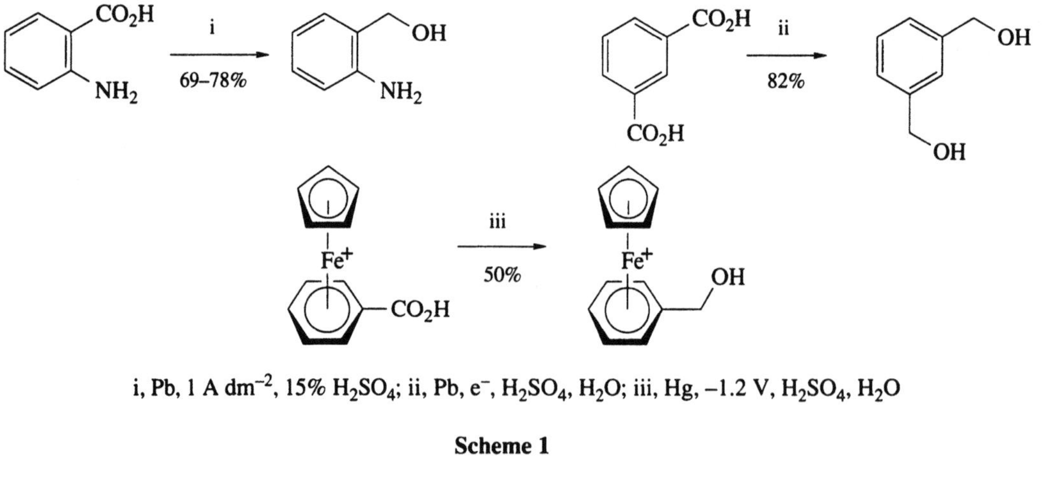

CO₂H NH₂ —i, 69–78%→ OH NH₂; CO₂H CO₂H —ii, 82%→ OH OH; Fe⁺ CO₂H —iii, 50%→ Fe⁺ OH

i, Pb, 1 A dm^{-2}, 15% H_2SO_4; ii, Pb, e^-, H_2SO_4, H_2O; iii, Hg, −1.2 V, H_2SO_4, H_2O

Scheme 1

1.10.1.3 Reductions using Metal Hydride Reagents

Metal hydrides are unquestionably the most useful reagents for the reduction of carboxylic acids to produce primary alcohols. A diverse variety of derivatives of B_2H_6 and LAH have been applied for the chemoselective reduction of polyfunctional organic molecules. B_2H_6 is an excellent reagent for the rapid and efficient reduction of aliphatic and aromatic carboxylic acids to produce borate esters, which are hydrolyzed on work-up to produce primary alcohols.[8] B_2H_6 is frequently used in THF solution, in which it exists as the complex $BH_3{\cdot}THF$. $BH_3{\cdot}SMe_2$ in THF is an alternative very soluble, more stable reagent for carboxylic acid reductions. $BH_3{\cdot}THF$ and $BH_3{\cdot}SMe_2$ rapidly reduce aldehydes, ketones, lactones, carboxylic acids and tertiary amides at 0 °C in THF. In contrast, acid chlorides, epoxides, esters, carboxylic acid salts and nitro compounds are only very slowly reduced, if at all. Alkenes may be hydroborated under the reduction conditions.[2,8] Thus $BH_3{\cdot}THF$ and $BH_3{\cdot}SMe_2$ are very useful for the selective reduction of carboxylic acids in the presence of esters *etc*. Such selective transformations are exemplified by the conversions in Scheme 2. Chemoselectivities in the reduction of carboxylic acids bearing ester, nitro, lactone, halide, amide, nitrile or (diene)tricarbonyliron functionality are noteworthy.[2,8–11] Although lactones and nitriles are reduced by $BH_3{\cdot}THF$, they do not react as rapidly as carboxylic acids. $BH_3{\cdot}THF$ is a very convenient reagent for the reduction of α-amino acids and α-acylamino acids (including peptides) to produce the corresponding β-amino (acylamino) alcohols.[2,12] $BH_3{\cdot}SMe_2$ is less reactive than $BH_3{\cdot}THF$, and thus aliphatic carboxylic acids are rapidly reduced at 25 °C in THF, while aromatic carboxylic acids react slowly. However, reduction may be accelerated by the addition of trimethyl borate.[2,8] $BH_3{\cdot}SMe_2$ is also particularly efficient for the reduction of amino acids to β-amino alcohols when used in the presence of $BF_3{\cdot}OEt_2$. Although borane in its various forms is invarably the reagent of choice for carboxylic acid reduction, occasionally problems arise. Alkenes may be hydroborated during the reaction. Additionally, deoxygenation may be observed, particularly with electron-rich heteroaromatic ring systems; one example of this is given in Scheme 2.

Various mono- (BH_2X) and di-substituted boranes (BHX_2) or borane complexes ($BH_3{\cdot}X$) have been used for the reduction of carboxylic acids to produce primary alcohols or aldehydes. Catecholborane in $CHCl_3$ at 25 °C rapidly reduces aldehydes, ketones, hydrazones, carboxylic acid salts, sulfoxides, anhydrides, epoxides, alkynes, acetals and ketals. Ketones, acid chlorides, alkenes, nitriles, carboxylic acids and tertiary amides are reduced at a much slower rate. Esters, lactones and anhydrides are not reduced.[13] $BH_3{\cdot}NHPh_2$ is a crystalline reducing agent that in THF rapidly reduces ketones, aldehydes, carboxylic acids and aliphatic esters.[14] Anhydrides and aromatic esters are inert. Both catecholborane and $BH_3{\cdot}NHPh_2$ reduce carboxylic acids to primary alcohols. 9-BBN-H in THF will slowly reduce aliphatic but not aromatic carboxylic acids.[15] Ketones, esters, lactones and tertiary amides are reduced more rapidly. The thexylborane·diethylaniline complex reduces aldehydes, ketones, carboxylic acids, anhydrides and tertiary amides. Esters, acid chlorides and nitriles are only slowly reduced. It is interesting to note that reductions of carboxylic acids with this reagent give primary alcohols. This is in contrast to thexylborane, thexylchloroborane or thexylbromoborane, which are particularly useful for the reduction of carboxylic acids to produce aldehydes.[16]

$NaBH_4$ and related reagents do not reduce carboxylic acids when used in alcohol or ether solvents. Borohydride salts in combination with powerful Lewis acids show reactivities comparable to diborane.[2,8,17] Thus, carboxylic acids may be reduced using $NaBH_4$ in the presence of $AlCl_3$, $BF_3{\cdot}OEt_2$, $ZnCl_2$, $TiCl_4$ or $MeSO_3H$. $NaBH_4$ and 1,2-ethanedithiol in THF at reflux has been used to reduce esters and carboxylic acids to primary alcohols and amides to primary amines.[18] Nitriles do not react with the reagent. For example, reduction of benzoic acid gave benzyl alcohol (53%). Finally, Cho and Yoon have reported that various sodium acyloxyborohydrides slowly reduce carboxylic acids on reflux in THF. The method is useful for the selective reduction of aliphatic carboxylic acids in the presence of aromatic carboxylic acids.[19]

LAH is a powerful reducing agent that rapidly converts carboxylic acids or their salts into primary alcohols at 0 °C in Et_2O or THF solution. A plethora of other groups are reduced by this reagent, including aldehydes, ketones, acid chlorides, esters, lactones, amides, nitriles, nitro compounds, aromatic halides and epoxides.[2,9,20] In consequence of its high reactivity, LAH is less useful than borane derivatives for chemoselective reduction of polyfunctional carboxylic acids. Nonetheless, unlike B_2H_6, it does not generally react with alkenes and is therefore useful for the selective reduction of unsaturated carboxylic acids to provide the corresponding primary alcohols. Low carboxylic acid or salt solubility in ethereal solvents may result in slow reduction and, in these cases, reduction of derived esters or acid chlorides are more convenient. A vast array of both aliphatic and aromatic carboxylic acids have been reduced using LAH, and representative examples are provided in Scheme 3. α,β-Unsaturated acids may be reduced to produce allylic or saturated alcohols, depending on the stoichiometry and temperature of reaction. Overreduction is the result of metallocycle (**1**) formation, and such intermediates may give rise to arylcyclo-

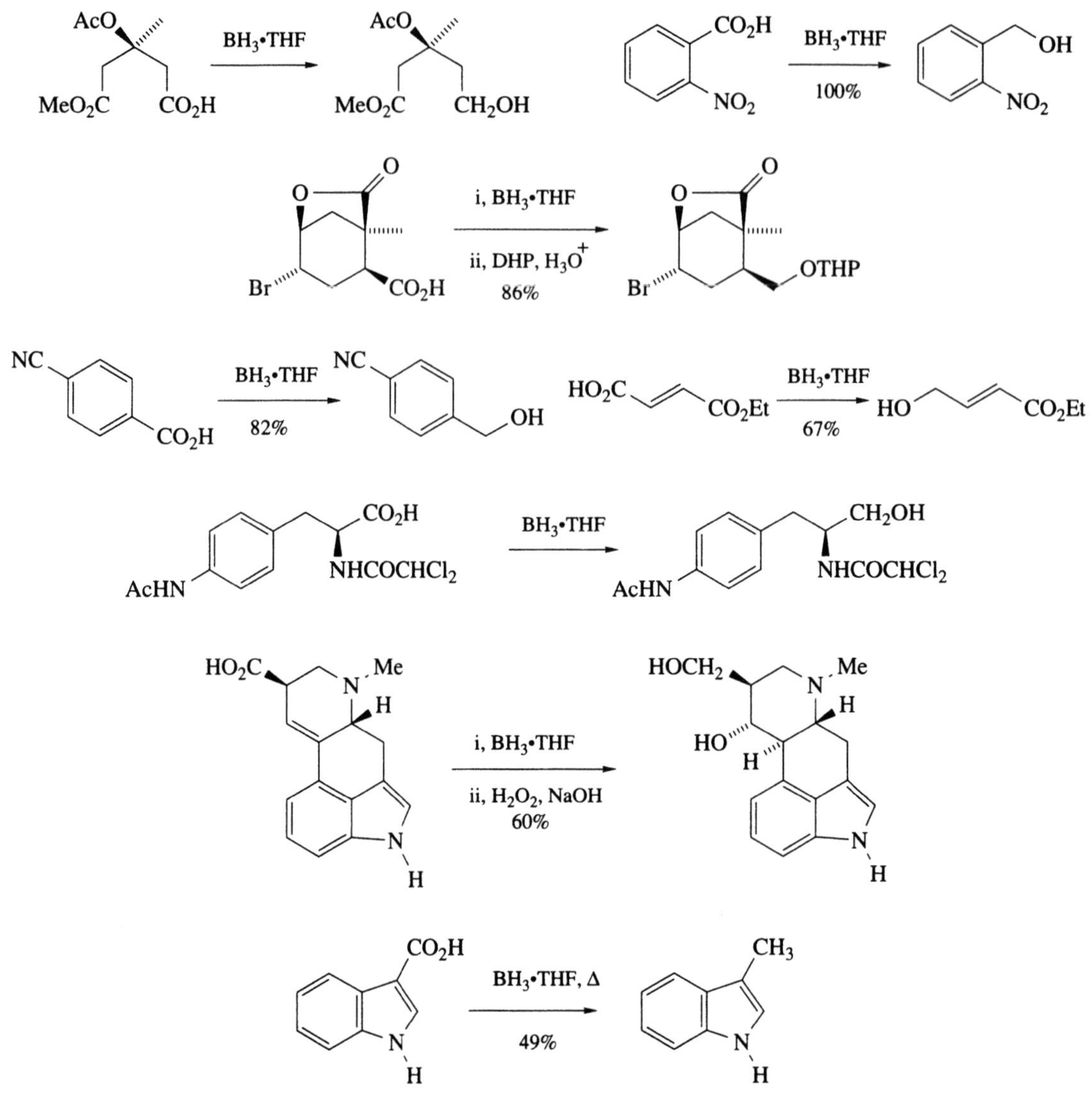

Scheme 2

propanes on the prolonged reduction of cinnamic acid derivatives.[20] A variety of metal alkoxyaluminum hydride reagents including $LiAlH(OMe)_3$, $NaAlH(OEt)_3$, Red-Al, $Ca(AlH_2(OCH_2CHEtBu)_2)_2{\cdot}THF$, $Ca(AlH_2(OBu^i)_2)_2{\cdot}THF$ and $Ca(AlH_2(OPr^i)_2)_2{\cdot}THF$ have been used for the reduction of carboxylic acids or their salts, to produce primary alcohols.[2,9,21] Red-Al is a particularly useful reagent that is readily soluble in aromatic hydrocarbon solvents. It is convenient for the reduction of carboxylic acids of low solubility, including α-amino acids.

Alane (AlH_3) and its derivatives have also been utilized in the reduction of carboxylic acids to primary alcohols.[2,9] It rapidly reduces aldehydes, ketones, acid chlorides, lactones, esters, carboxylic acids and salts, tertiary amides, nitriles and epoxides. In contrast, nitro compounds and alkenes are slow to react. AlH_3 is particularly useful for the chemoselective reduction of carboxylic acids containing halogen or nitro substituents, to produce the corresponding primary alcohols. DIBAL-H reduces aliphatic or aromatic carboxylic acids to produce either aldehydes (–75 °C) or primary alcohols (25 °C).[22] Aminoaluminum hydrides are less reactive reagents and are superior for aldehyde synthesis.[23]

Aromatic carboxylic acids have been reduced using $HSiCl_3$ and tertiary amines. Generally, this reaction leads to deoxygenation and the formation of (trichlorosilylmethyl)arenes. However, this reaction was found to produce ether (**3**) on the reduction of diacid (**2**; Scheme 3). Reduction by $HSiCl_3$ is selective for aromatic carboxylic acids in the presence of esters.[24] A representative reduction using $HSiCl_3$ is also given in Scheme 3.

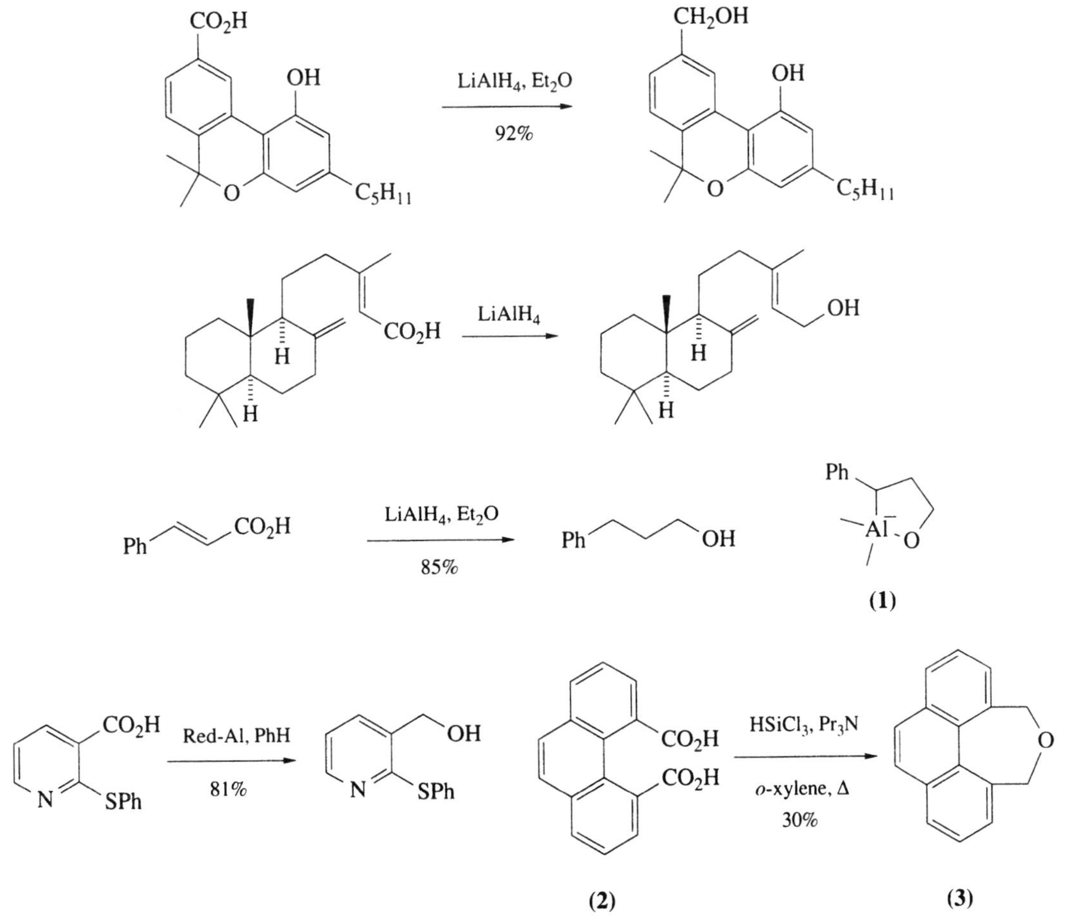

(1)

(2)

(3)

Scheme 3

1.10.2 REDUCTIONS OF ACYL HALIDES, ANHYDRIDES AND RELATED SPECIES

1.10.2.1 Hydrogenation Reactions

The conversion of carboxylic acid chlorides into aldehydes *via* low temperature and low pressure hydrogenation over poisoned $Pd/BaSO_4$ is the classical Rosenmund reduction. Further hydrogenation, particularly with more active catalysts, gives rise to the corresponding primary alcohols.[25] The hydrogenation of anhydrides over heterogeneous catalysts may be used to prepare alcohols, lactones or ethers depending on the substrate and reaction conditions.[1,26] Examples of these reactions include the reduction of succinic anhydride to give γ-butyrolactone (80–94%; Pd/C, H_2, 35–100 °C, 16–75 atm) and of phthalic anhydride to give phthalide (90%; Raney Ni, H_2, 30 °C, 100 atm, dioxane). Several homogenous catalysts have been employed to reduce cyclic anhydrides to produce lactones.[3,27,28] Lyons reported that $(Ph_3P)_3RuCl_2$ efficiently catalyzed the hydrogenation of succinic anhydride at 100 °C and 5 atm to produce γ-butyrolactone (50%). Reduction of acetic anhydride using the same catalyst gave ethyl acetate (36%) and acetic acid (50%). This homogeneous hydrogenation catalyst is particularly useful for the regioselective hydrogenation of unsymmetrical anhydrides. Thus, anhydride (**4**) was converted into lactones (**5**) and (**6**) on hydrogenation at 100 °C and 10 atm in toluene solution (Scheme 4). The regioselectivity of reduction was the reverse of LAH reduction, in that the less-substituted carbonyl was preferentially reduced (see Section 1.10.2.3). Yoshikawa and coworkers have reported that cyclic anhydrides can be hydrogenated to produce optically enriched (*ee* 5–20%) γ- and δ-lactones using the ruthenium complex (**7**).[28] The authors postulated that the reaction took place *via* the insertion of a ruthenium hydride into the acyl oxygen bond. The process is exemplified by conversion of anhydride (**8**) into lactone (**9**; 20% *ee*). Clearly there is considerable opportunity for optimization in this enantioselective approach to lactones.

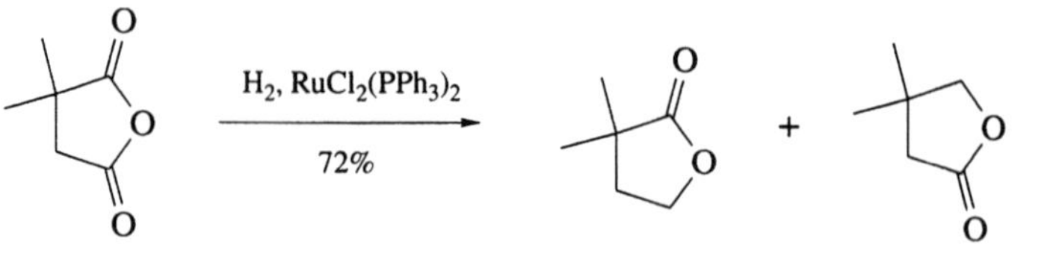

H_2, $RuCl_2(PPh_3)_2$

72%

(4) **(5)** 90% **(6)** 10%

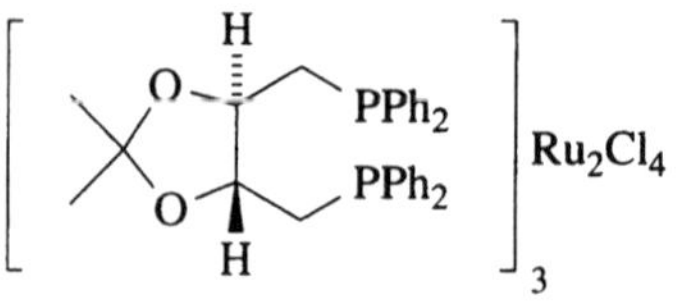

PPh_2 PPh_2 Ru_2Cl_4

(7)

H_2, PhMe, Et_3N

(7) (0.5 mol %), 100 °C

56%

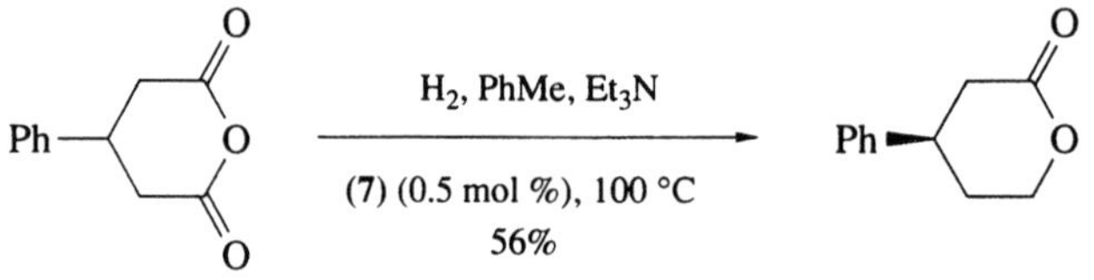

(8) **(9)**

Scheme 4

1.10.2.2 Electrochemical and Dissolving Metal Reductions

In general, the dissolving metal or electrochemical reductions of acyl halides or acyclic anhydrides are not useful for the preparation of primary alcohols. Such reductions invariably provide acyloin esters, enediolate diesters or related species.[29] Cyclic anhydrides may be reduced to give lactones.[5,27] For example, the reduction of phthalic anhydride at a mercury cathode has been used in the synthesis of phthalide (90%). In general, however, such reduction are not widely employed in synthesis.

1.10.2.3 Reductions using Metal Hydride Reagents

Carboxylic acid chlorides are readily reduced to provide primary alcohols by reaction with nucleophilic hydride reagents. In contrast, electrophilic reagents including B_2H_6 and its derivatives react only at a very slow rate.[25] However, derivatives of B_2H_6 of reduced Lewis acidity will reduce acyl chlorides to produce primary alcohols. For example, 9-BBN-H in THF has been reported to reduce hexanoyl chloride to hexanol (81%).[15] Acid anhydrides, being less electrophilic than acid chlorides, are reduced by B_2H_6 to give diols or primary alcohols (2 mol).[2,8] Examples include the reductions of hexanoic anhydride to produce hexanol ($BH_3{\cdot}THF$; 94%) and of naphthalene-1,2-dicarboxylic acid anhydride to give 1,2-naphthalenedimethanol ($NaBH_4$, $BF_3{\cdot}Et_2O$, diglyme; 63%). In contrast, reduction of the 1,8-naphthoic anhydride (**10**) gave the cyclic ether (**11**; Scheme 5). Clearly in this unusual case, cyclization is favored by the proximity and activation of the *peri* substituents. Derivatives of B_2H_6 including $BH_3{\cdot}NHPh_2$[14] and catecholborane[13] are also useful for the reduction of acyclic anhydrides.

Reductions of carboxylic acid chlorides using $NaBH_4$ in ethereal solvents (dioxane, DME, THF, poly(ethylene glycol) *etc.*)[2,9,25] rapidly provide primary alcohols. Reductions in poly(ethylene glycol)[30] are noteworthy in that esters, acyl chlorides, alkyl halides and sulfonate esters are reduced, whereas amides and nitriles are inert. Acid chlorides may alternatively be reduced using $NaBH_4$ on alumina (Et_2O),[31] $NaBH_4$–$TiCl_4$ (DME),[17] $NaBH_4$–$CeCl_3$ (MeCN), $NaBH(OMe)_3$ (THF), $Zn(BH_4)_2{\cdot}TMEDA$ (THF),[32] Bu_4NBH_4 or $Bu_4NB_3H_8$ (CH_2Cl_2),[33] Li-9-BBN-H_2[34] and $KBH(OPr^i)_3{\cdot}(THF)$[35] *etc.*[2,9] Since acyl chlorides are reduced very rapidly by nucleophilic borohydride reagents, chemoselectivity in the reduction of polyfunctional molecules is generally not a problem providing that temperature and stoichiometry are controlled. The choice of reagent is frequently based upon substrate solubility and ease of work-up considerations. However there is one point of note amongst this plethora of choice. Raber has reported that Bu_4NBH_4 in CH_2Cl_2 reduced benzoyl chloride to benzyl alcohol faster than benzaldehyde to benzyl alcohol.

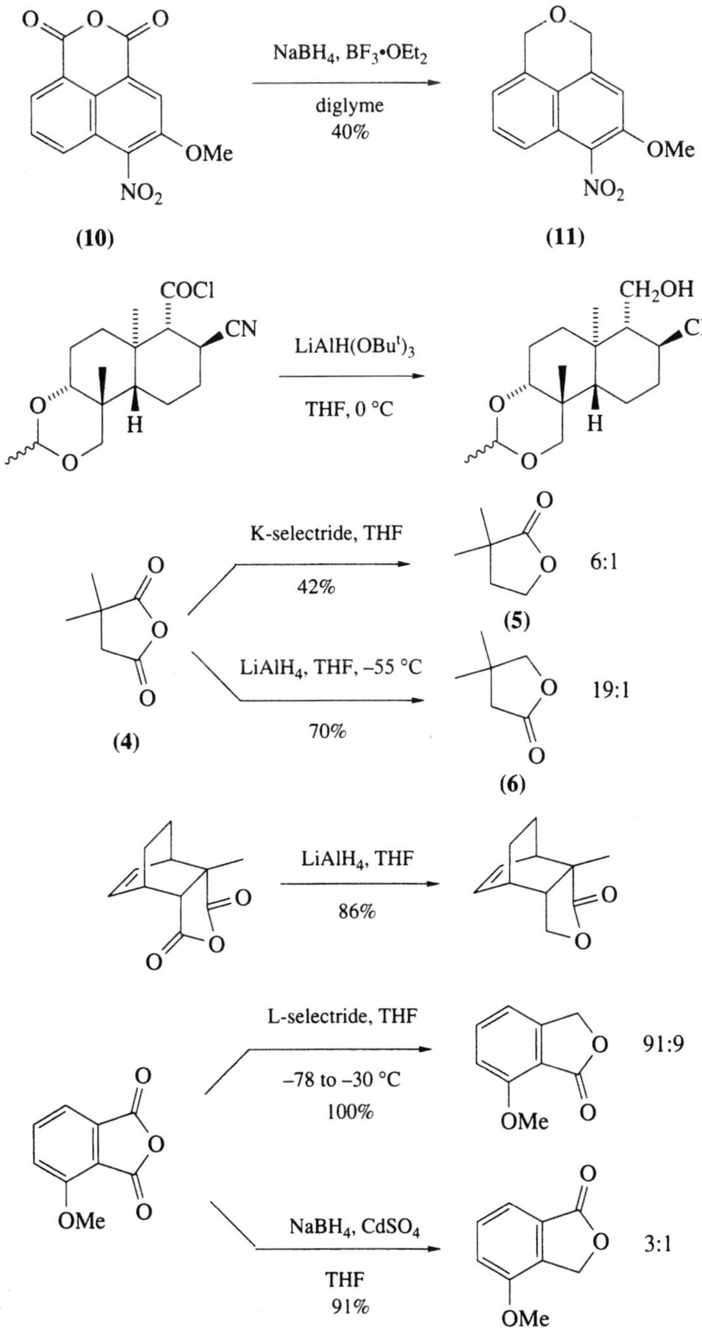

Scheme 5

Acyclic acid anhydrides are reduced to produce primary alcohols using $NaBH_4$ and its many derivatives.[2,9,36] In most cases, the reaction gives both the primary alcohol and the carboxylate salt (1:1). Whilst such monoreduction may be desirable in cyclic anhydride chemistry (*vide infra*), it is inefficient when acyclic anhydrides are reduced. Thus, B_2H_6 or LAH reductions are the preferred synthetic methods. The reductions of mixed anhydrides, for example carboxylic/diphenylphosphoric or carboxylic/carbonic anhydrides, produce the primary alcohol from the acyl component.[2,9,37]

LAH rapidly reduces aliphatic and aromatic acid chlorides or anhydrides to produce primary alcohols.[2,9,20,25] Activated esters, electrophilic amides and mixed anhydrides are alternative substrates for reduction to provide primary alcohols.[38] Numerous related reducing agents including $LiAlH(OMe)_3$, $NaAlH(OEt)_3$, Red-Al, $Ca(AlH_2(OCH_2CHMe_2)_2)_2$,[21] DIBAL-H,[22] DIBAL-H/Bu^nLi[39] and AlH_3[9] are also used. In general, reductions of acyl chlorides are used when the reduction of the corresponding carboxylic acid is slow for solubility reasons, or overreduction is a problem.

The reductions of cyclic anhydrides may be used to prepare hydroxylactones, lactones, lactols, cyclic ethers or diols depending on the substrate, reagents and conditions of reduction. Regioselective reductions of unsymmetrical anhydrides to produce γ- or δ-lactones have been the subject of much investigation.[27,40,41] $NaBH_4$ or LAH reductions at low temperatures are both synthetically useful for the preparation of lactones. Alternative reagents include $LiBH_4$, various lithium trialkylborohydrides, $LiAlH(OBu^t)_3$, $LiAlH(OMe)_3$, AlH_3 and Red-Al.[21,22] The reduction of cyclic anhydrides may alternatively be stopped at the hydroxylactone stage using $LiAlH(OBu^t)_3$ at low temperature. Alternatively, cyclic anhydrides may be completely reduced to diols using an excess of LAH or an equivalent reagent. Synthetically, the most useful reactions are reductions to produce γ- or δ-lactones. In general, unsymmetrical anhydrides are reduced using LAH or $NaBH_4$ to produce the lactone in which the more substituted carbonyl has been reduced. There are many exceptions, however. The reductions of rigid bridged ring anhydrides show the reverse selectivity. Additionally, the regiochemistry of reaction may be changed on using hindered reductants such as K-selectride. Regioselectivity is controlled by a balance between the relative electrophilicities of the two carbonyls, steric approach control (Bürgi–Dunitz vector approach), the influence of neighboring bonds coplanar with the carbonyl π-system ('the antiperiplanar effect') and the opportunity for chelation. A detailed analysis of regiocontrol is presented in the series of papers by Kayser and coworkers.[40] Representative acyl chloride and anhydride reductions are given in Scheme 5.

1.10.3 REDUCTIONS OF CARBOXYLIC ESTERS

1.10.3.1 Hydrogenation Reactions

Carboxylic esters may be hydrogenated more readily than the corresponding acids to produce primary alcohols.[1] Nonetheless, such reactions using heterogeneous catalysts require elevated temperatures and pressures. Although yields of the primary alcohols, corresponding to the acyl fragment, are frequently excellent, such reductions are synthetically inconvenient and now not widely used in small-scale synthesis.[42] Hydrogenation using copper chromite at 60–200 atm and 150–300 °C are the most frequently employed conditions. The ester is either hydrogenated alone or in a solvent such as MeOH, EtOH or dioxane. Alternatively, esters may be converted into primary alcohols by high pressure and high temperature hydrogenation over zinc chromate, copper oxide, Raney nickel, nickel/copper/aluminum/rhenium oxides or related species.[2] Hydrogenation of α-amino esters over W-6 Raney nickel is particularly useful for the preparation of amino alcohols with minimal racemization.[2,42] Specific examples include the conversions of ethyl pentanoate into 1-pentanol (94%; Cu chromite, H_2, 250 °C, 100 atm), diethyl tetradecanedioate into 1,14-tetradecanediol (95%; Cu chromite, H_2, 250 °C, 116 atm), ethyl benzoate into benzyl alcohol (65%; Cu chromite, H_2, 160 °C, 156 atm), *meso*-diethyl tartrate into erythritol (95%; Cu chromite, H_2 165 °C, 340 atm) and ethyl alaninate into 2-amino-1-propanol (80%; W-5 Raney Ni, H_2, 100 °C, 156 atm). It is germane to note that esters of both aliphatic and aromatic carboxylic acids may be hydrogenated to produce primary alcohols. However, benzylic C—O hydrogenolysis is a frequent problem with esters of aromatic carboxylic acids. Additionally, aromatic ring hydrogenation takes place on phenol ester reduction. Hydrogenation may be used to prepare diols from keto acids, hydroxy acids or diesters. Esters of perfluorocarboxylic acids may be conveniently converted into the corresponding primary alcohols by hydrogenation over ruthenium catalysts. Thiol esters are readily converted into primary alcohols by reduction over W-4 Raney nickel.[43] This method is useful for the reduction of steroidal, carbohydrate and α-amino carboxylic acids. The intermediate aldehyde may be produced if deactivated Raney nickel is employed.

Grey *et al.* have reported that activated carboxylic acids may be hydrogenated under mild conditions using the homogeneous catalysts $K^+[(Ph_3P)_2Ph_2\overline{PC_6H_4}RuH_2]^-\cdot C_{10}H_8\cdot Et_2O$ or $2K^+[(Ph_3P)_3(Ph_2P)Ru_2H_4]^{2-}\cdot 2$ diglyme.[44] For example, methyl trifluoroacetate was reduced with the second catalyst at 90 °C and 62 atm to produce 2,2,2-trifluoroethanol and methanol (88% conversion). These catalysts are not efficient for the hydrogenation of nonactivated esters.

1.10.3.2 Electrochemical and Dissolving Metal Reductions

Aromatic carboxylic esters may be reduced to produce the corresponding primary alcohol by electrolysis at a mercury, lead or cadmium cathode.[2,4,5,45] For example, methyl benzoate is readily reduced to benzyl alcohol (91%) at a mercury cathode in MeOH containing Me_4NCl. Ring substituents in the benzoyl group including methyl, methoxy or chloro substituents (*ortho-*, *meta-* and *para-*) are tolerated. In acidic media, benzylic alcohol formation may be suppressed and alkyl benzyl ether formation observed. The electrochemical reduction of ethyl benzoate (Pb, H_2SO_4, H_2O, EtOH) has been reported to provide benzyl ethyl ether (40%) and benzyl alcohol. Clearly the ether is formed *via* late benzylic cation formation. Aromatic 1,2-diesters may be selectively reduced to produce γ-lactones. For example diethyl phthalate has been reduced (Hg, –1.7 V, EtOH, H_2O, Me_4NCl) to provide phthalide. In contrast, diethyl 1,3- and 1,4-benzenedicarboxylates are electrochemically reduced (Hg, 10–24 F mol^{-1}, AcOH, EtOH, Me_4NCl) to produce benzene-1,3- (72%) and -1,4-dimethanol (75%), respectively. The intermediate monoesters may be isolated if the reductions are interrupted after 4.5–5 F mol^{-1}. Aliphatic carboxylic esters may be electrochemically reduced to produce primary alcohols. For example, reduction of methyl octanoate (Hg, 6 F mol^{-1}) gave 1-octanol (49%). Aliphatic esters of phenols are electrochemically reduced to produce the aliphatic primary alcohols in 30–40% yield.

The dissolving metal reduction of esters has found widespread use in synthesis. Reduction of a carboxylic ester (**12**) using an alkali metal in an alcohol solvent provides two alcohols (**13**) and (**14**; Scheme 6). This reaction is the classical Bouveault–Blanc reduction and it was the method of choice for preparing primary alcohols from aliphatic esters prior to the development of hydride reducing agents. Typically, Bouveault–Blanc reductions are carried out by adding sodium to a solution of the ester in EtOH.[46] The sodium may be replaced by other alkali or alkali earth metals or sodium amalgam and the EtOH by alternative alcohols, ethanolic ammonia or by phenol. Representative examples of Bouveault–Blanc reductions are provided in Scheme 6. The method is useful for the preparation of fatty alcohols or diols. Ketones or aldehydes may be protected from reduction *via* ketal formation. Aromatic esters undergo competitive arene reduction (Birch reduction). Esters may be reduced without concomitant carboxylic acid reduction[2,10] and such transformations are complementary to B_2H_6 reductions (Section 1.10.1.3). Bouveault–Blanc reductions in $EtOH/NH_3$ are frequently more efficient than reductions in EtOH alone due to the suppression of competitive ester hydrolysis or Claisen condensations at the lower temperature of reaction. The Bouveault–Blanc reduction takes place *via* two sequential one-electron transfers, protonation, aldehyde formation and further reduction.

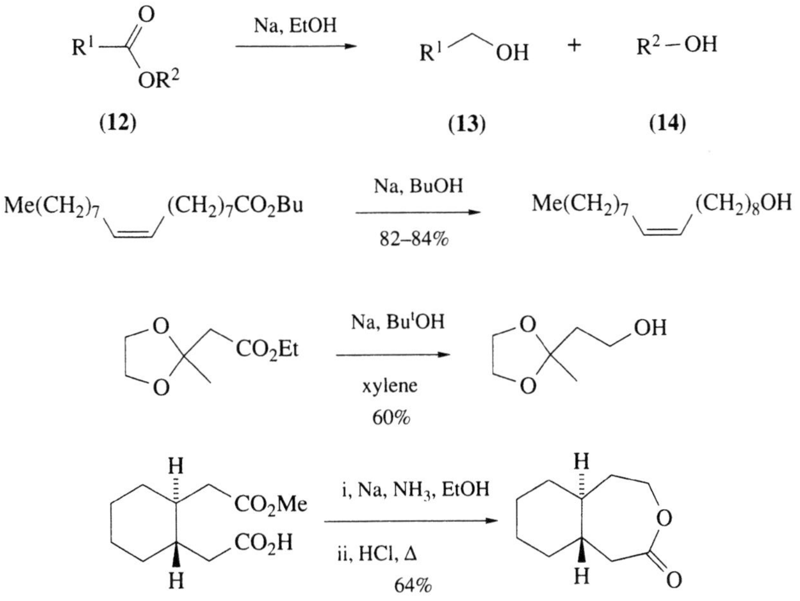

$R^1CO_2R^2$ (**12**) —(Na, EtOH)→ R^1CH_2OH (**13**) + R^2–OH (**14**)

$Me(CH_2)_7CH=CH(CH_2)_7CO_2Bu$ —(Na, BuOH; 82–84%)→ $Me(CH_2)_7CH=CH(CH_2)_8OH$

(1,3-dioxolane ketal)CH_2CO_2Et —(Na, Bu^tOH, xylene; 60%)→ (1,3-dioxolane ketal)CH_2CH_2OH

(cyclohexane with CH_2CO_2Me and CH_2CO_2H) —(i, Na, NH_3, EtOH; ii, HCl, Δ; 64%)→ (fused bicyclic lactone)

Scheme 6

The acyloin condensation[29] and deoxygenation[47] are two alternative dissolving metal reduction reactions of carboxylic esters. Since neither reaction results in primary alcohol formation from the acyl fragment, both are beyond the scope of this chapter. The acyloin condensation is carried out using molten Na in PhMe or in PhMe and TMS-Cl, or Na in NH_3. Clearly, this condensation differs from the Bouveault–Blanc reduction in that a protic solvent/cosolvent is not used, and therefore radical anion and/or dianion intermediates are not protonated and thereby undergo reactions with C—C bond formation. The author and others have reported that the reduction of alkyl carboxylates using, for example K and 18-crown-6 in Bu^tNH_2 or Na in HMPA, gave carboxylic acids and alkanes. These reactions again are of only indirect importance in this chapter since the ester acyl fragment is not converted to a primary alcohol. Deoxygenation *via* ester radical anion cleavage is the major reaction pathway when esters are reduced in homogeneous solution in the absence of proton sources ($pK_a < 20$) or nucleophiles.

1.10.3.3 Reductions using Metal Hydride Reagents

The reduction of carboxylic esters to produce primary alcohols is most conveniently carried out using metal hydride reagents. Indeed, such reductions are unquestionably the most important methods for effecting these transformations. B_2H_6, $NaBH_4$, LAH and legions of derivatives have found widespread use in ester reductions. BH_3·THF in THF solution at 0 °C only slowly reduces carboxylic esters. Thus, this reagent is frequently used to reduce aldehydes, ketones, amides or carboxylic acids chemoselectively, without accompanying ester reduction.[2,8] Prolonged reaction may lead to the reduction of aliphatic carboxylic esters to produce primary alcohols. Esters of aromatic carboxylic acids are essentially unreactive at 0 °C but may be slowly reduced at elevated temperatures. Brown and coworkers have found that BH_3·SMe_2 in THF at reflux may efficiently be used to reduce both aliphatic and aromatic carboxylic esters rapidly, by removing the Me_2S liberated during the reaction.[48] For example, ethyl benzoate and hexanoate were reduced to produce benzyl alcohol (90%) and 1-hexanol (89%), respectively. Rodrigo and coworkers have observed that diester monoreduction may be used in 1-arylnaphthalide lignan synthesis.[49] Reductions in THF may be catalyzed by sodium borohydride (5 mol %).[50] Various borane derivatives have been employed in ester reduction. BH_3·$NHPh_2$ has been used to reduce aliphatic esters, although aromatic esters are not readily reduced.[14] Additionally, 9-BBN-H reduces aliphatic esters much more rapidly than aromatic esters, to produce primary alcohols.[15] More substituted boranes, including thexylborane,[51] thexylchloroborane·SMe_2,[16] catecholborane[13] and Sia_2BH, react only very slowly with esters.[2,9,15] A representative borane reduction is given in Scheme 7.

$NaBH_4$, in general, is slow to reduce carboxylic esters to produce primary alcohols. Borohydride salts of more Lewis acidic cations, including $LiBH_4$,[52,53] $Mg(BH_4)_2$[2] and $Ca(BH_4)_2$,[53,54] are effective reductants. $LiBH_4$ is particularly useful for the chemoselective reduction of carboxylic esters in the presence of carboxylic acids, amides or nitriles. $NaBH_4$ will slowly reduce esters when used in alcoholic solvents at room temperature. However, reductions are dramatically accelerated in Bu^tOH and MeOH[55] at reflux or in poly(ethylene glycol) at 65–80 °C.[30,56] The reduction of esters may also be carried out using Bu_4NBH_4 in CH_2Cl_2 solution.[57] It is noteworthy that carboxylic esters bearing an α-oxygen substituent are rapidly reduced by $NaBH_4$ at room temperature. Presumably, chelation and/or inductive activation accelerate the reaction. Thus, for example, glycidic esters are readily reduced to the corresponding primary alcohols.[58] Soai and Ookawa have reported that chemoselectivity in $LiBH_4$ reductions is strongly solvent dependent. In Et_2O at reflux, $LiBH_4$ and MeOH (1 equiv.) selectively reduce esters, lactones and epoxides in the presence of carboxylic acids, carbamates, chlorides, nitro compounds and secondary amides. In THF or diglyme containing MeOH (4 equiv.) $LiBH_4$ is more reactive and nitroalkanes, nitriles, carboxylic acids and amides are also efficiently reduced.[59] $NaBH_4$ or $LiBH_4$ reductions of carboxylic esters may be accelerated by Lewis acids, including $AlCl_3$,[60] $(MeO)_3B$, 9-BBN-OMe,[61] Li-9-BBN-H_2 or $LiBHEt_3$.[62] These methods are also useful for the selective reductions of esters in the presence of amides. Under powerfully Lewis acidic conditions, ester reductions can be diverted to produce ethers. For example, esters may be converted into ethers using $NaBH_4$ and BF_3·OEt_2 at elevated temperatures.[63] Clearly this reaction involves a BF_3-catalyzed B_2H_6 reduction.

There are many substituted borohydride reagents known, and several of these have been employed in ester reductions. Such reagents include $NaBH_3(OH)$,[64] $NaBH_2(SCH_2CH_2S)$,[18,65] $NaBH(OMe)_3$,[66] $NaBH_3(NMe_2)$, $NaBH_3(NHBu^t)$ and $NaBH_3(N(Me)Ac)$.[67] In general, these offer little advantage over $LiBH_4$. However, the more-substituted reagents selectively reduce nonhindered carboxylic esters in the presence of hindered ester functionality. Alkyl-substituted borohydrides, including $LiBH_3Bu$ and $LiBHEt_3$, are potent reductants, and these readily reduce esters.[68] However, again, these reagents offer little advantage over $LiBH_4$ other than higher rates of reaction. Ireland and Thompson have reported that di-MEM esters may be selectively monoreduced using $LiBHEt_3$.[69] Finally, borohydride reductions of thiol esters and activated esters, including *N*-acyloxysuccinimide derivatives, may be used to prepare primary alcohols.[70] Representative reductions of esters using borohydride reagents are given in Scheme 7. Again, these examples are chosen to illustrate selectivities.

AlH_3 and alkyl derivatives are powerful Lewis acidic reducing agents that rapidly convert carboxylic esters into primary alcohols.[2,9,22,71] In many cases, the reduction may be stopped at the intermediate aldehyde stage, providing that the stoichiometry and temperature are carefully controlled. However, complete reduction is experimentally easier to carry out. DIBAL-H is the most useful and widely used alane.

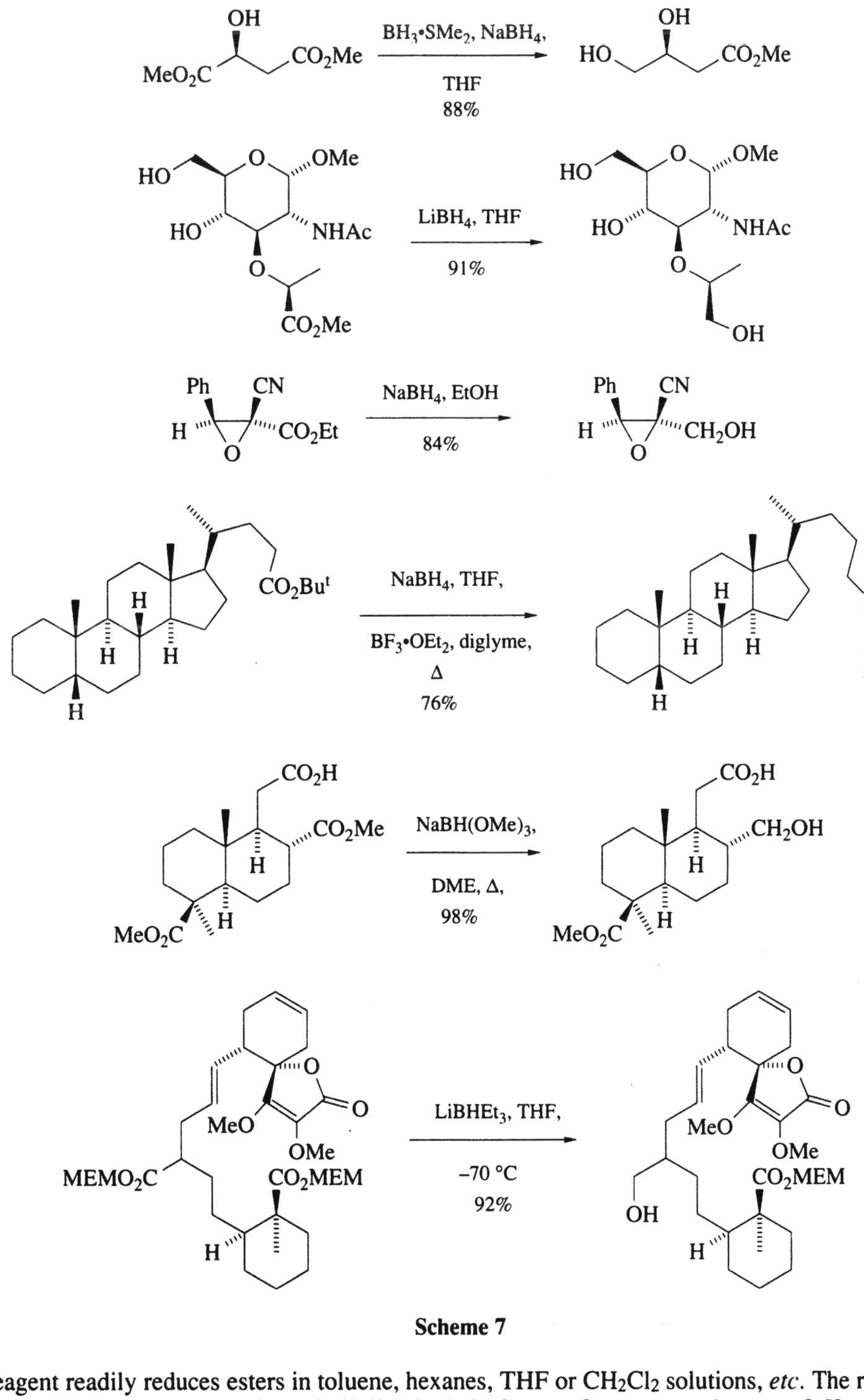

Scheme 7

This reagent readily reduces esters in toluene, hexanes, THF or CH_2Cl_2 solutions, *etc.* The reaction is especially useful for the preparation of allylic alcohols from α,β-unsaturated esters. β-Keto esters have been selectively reduced to β-hydroxy ketones *via* ketone enolate formation and AlH_3 reduction.[72]

LAH[2,9,20,73,74] and numerous related reagents, including LAH-silica,[75] $LiAlH(OMe)_3$, $NaAlH(OEt)_3$, Red-Al, $Ca(AlH_2(OC_6H_{11})_2)_2 \cdot THF$, $Ca(AlH_2(OBu^i)_2)_2 \cdot THF$, $NaAlH_2(OCH_2CH_2OPr^i)_2$, $NaAlH_2(OCH_2CH_2NMe_2)_2$, $LiAlH_2(OBu^t)_2$, $NaAlH_2Bu^i{}_2$ and $LiAlH_2Bu^i{}_2$, *etc.*,[21] readily reduce aliphatic and aromatic carboxylic esters to produce primary alcohols. More sterically hindered reagents such as $LiAlH(OBu^t)_3$ are slow to reduce esters. The most commonly used reagents are LAH or $LiAlH(OMe)_3$ in ether solvents or Red-Al in aromatic hydrocarbons or ether solvents. Dicarboxylic esters are rapidly reduced to diols although, in some cases, monoreduction of the less-hindered ester may be carried out at low temperature and with careful stoichiometric control. It should be noted that LAH also reduces aldehydes, ketones, acid chlorides, epoxides, carboxylic acids, amides, nitriles and nitro compounds. Thus it is recommended for global rather than chemoselective reduction. However, the reactivity of the alkoxy-substituted reagents decrease with increase in substitution and steric congestion. Therefore, they are easier to apply for chemo- or regio-selective control. Esters of α,β-unsaturated carboxylic acids are reduced by LAH to produce the saturated primary alcohol or the allylic alcohol depending on reaction stoichiometry and temperature. There is a myriad of ester reductions in the literature. Representative examples are given in Scheme 8. The LAH reduction of carboxylic esters may be diverted to ether formation when carried out in the presence of $BF_3{\cdot}OEt_2$.[63,76] In general, yields are only modest. Alternatively, sulfides (37–80%) may be prepared from thiol esters in the same way.[77]

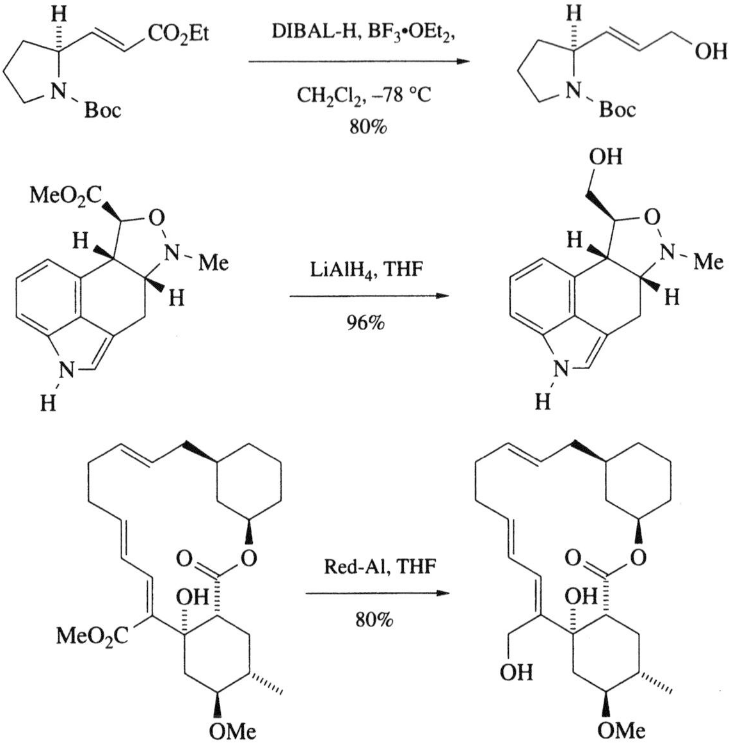

DIBAL-H, $BF_3{\bullet}OEt_2$, CH_2Cl_2, −78 °C, 80%

$LiAlH_4$, THF, 96%

Red-Al, THF, 80%

Scheme 8

Aliphatic carboxylic esters have been reduced to produce simple primary alcohols (84–92%) *via* Cp_2TiCl_2-catalyzed reaction with Bu^iMgBr.[78] Additionally, hydrosilylation of aliphatic and aromatic carboxylic esters using Ph_2SiH_2, $(EtO)_2Si(Me)H$, $(EtO)_3SiH$ or $MeSiO[Si(Me)H—O]_nSiMe_3$ has been used to produce simple primary alcohols (50–100%). In contrast, the photolytic reactions of esters with $HSiCl_3$ produces ethers. For example, 1-dodecyl acetate was converted into 1-dodecyl ethyl ether in this way. Competitive deoxygenation complicates ether formation particularly with secondary and tertiary alkyl esters.[79] All these exotic methods have little synthetic merit.

1.10.4 REDUCTIONS OF LACTONES

1.10.4.1 Hydrogenation Reactions

Lactones may be hydrogenated to produce diols under comparable conditions to ester hydrogenation reactions (see Section 1.10.3.1). Alternatively, the hydrogenation reaction may be used to prepare cyclic ethers. Hydrogenation (250 °C, 100 atm) of γ-butyrolactone over Cu chromite (89%) or Ni/Co/Th oxides (98%) has been used to prepare 1,4-butanediol efficiently.[1,42] Alternatively, hydrogenation (240 °C, 150 atm) of γ-butyrolactone over a Co/Re catalyst gave THF (92%). Substitution of the lactone at the γ-position may radically alter the course of reduction. For example, hydrogenation of γ,γ-dialkyl-γ-butyrolactones over Cu chromite results in C—O hydrogenolysis and formation of 4,4-dialkyl-1-butanol. Hydrogenation of γ-valerolactone over Cu chromite gave 1,4-pentanediol, 1-pentanol and/or 2-methyltetrahydrofuran. Diol formation was favored at lower temperatures (250 °C and 200–300 atm), whereas hydrogenation at 270–290°C gave mostly 1-pentanol and 2-methyltetrahydrofuran. δ-Lactones may be converted into cyclic ethers by hydrogenation (25 °C, 1 atm) over Adams' catalyst in HOAc containing $HClO_4$.[80] For example, 4-oxa-5α-cholestan-3-one was converted into 4-oxa-5α-cholestane (92%) in this way. The procedure is notable since it is particularly mild and γ-lactones are not reduced under these conditions. In general, the synthesis of diols or ethers from the catalytic hydrogenation of lactones is not widely used since it is experimentally inconvenient relative to other methods.

1.10.4.2 Electrochemical and Dissolving Metal Reductions

Although lactones may be reduced electrochemically or *via* Bouveault–Blanc reactions to produce diols, such reactions are more frequently used to prepare lactols.[4] Both cathodic (Hg or Pb) and Na/Hg reduction are useful in the preparation of alditols from aldonic acid γ-lactones. The reductions may be easily stopped at the intermediate aldose stage.[81]

1.10.4.3 Reductions using Metal Hydride Reagents

Lactones are reduced rapidly by LAH,[20] $LiAlH(OMe)_3$, $LiAlH(OEt)_3$,[21] AlH_3, DIBAl-H,[22] $LiBH_4$,[53] $NaBH_4$–$AlCl_3$, $LiBHEt_3$,[68] $LiAlHBu^i{}_2Bu$,[39] Li-9-BBN-H_2[34] and $LiBH_3Bu$ *etc*.[2,9] The reductions may also be carried out with $NaBH_4$, $BH_3{\cdot}SMe_2$, $BH_3{\cdot}THF$, Red-Al, $LiAlH(OBu^t)_3$, thexylborane, Sia_2BH and 9-BBN-H, *etc*., but at much slower rates.[2,8,9,21,48] These reductions may be stopped at the intermediate lactol stage in many cases with temperature and stoichiometric control, particularly with the less reactive reagents. Reductions using DIBAL-H, Sia_2BH or $LiAlH(OBu^t)_3$ are especially useful in lactol synthesis.[21,71] In general, lactones are more rapidly reduced than esters and thus chemoselectivity is fre-

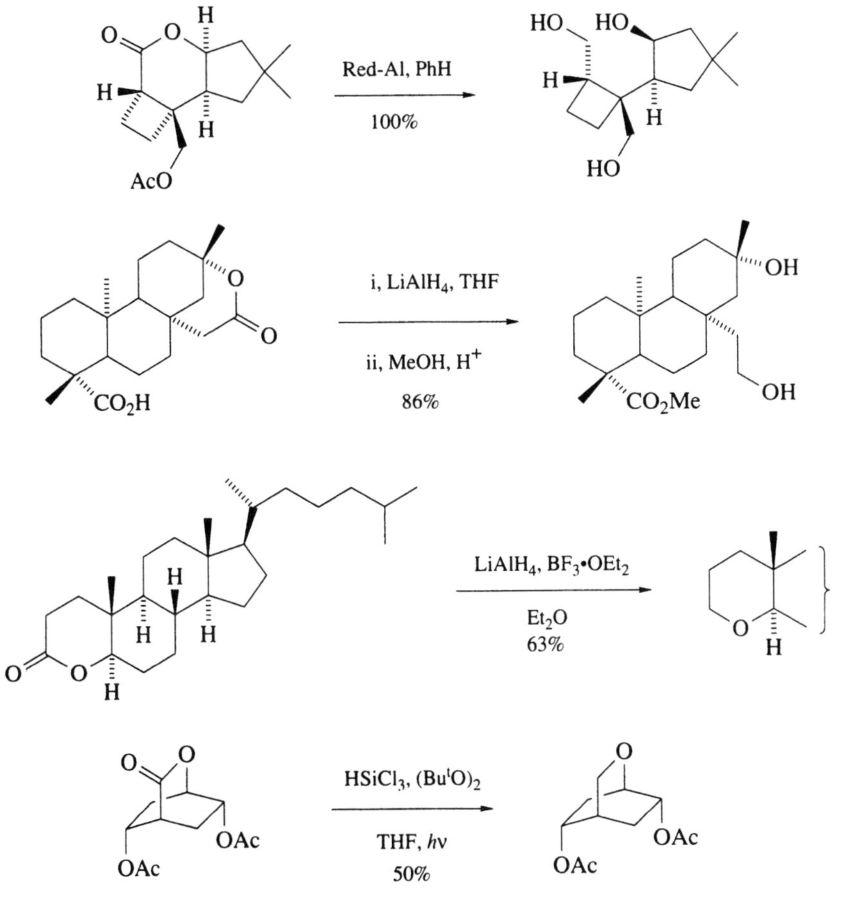

Red-Al, PhH; 100%

i, $LiAlH_4$, THF; ii, MeOH, H^+; 86%

$LiAlH_4$, $BF_3{\bullet}OEt_2$; Et_2O; 63%

$HSiCl_3$, $(Bu^tO)_2$; THF, *hν*; 50%

Scheme 9

quently possible in reduction. Additionally, carboxylic acids and amides are reduced more slowly using nucleophilic reagents; again, this provides opportunities for selectivity. Lactones are only slowly reduced by $NaBH_4$ in alcohol solvents at 25 °C unless the carbonyl is flanked by an α-heteroatom functionality.[58,82] Alternatively reductions of simple lactones using $NaBH_4$ may be efficiently carried out using refluxing Bu^tOH or THF as solvent with the slow addition of methanol.[55]

Pettit and others have reduced lactones to produce cyclic ethers using $LAH/BF_3{\cdot}OEt_2$, $LAH/AlCl_3$, $BH_3{\cdot}THF/BF_3{\cdot}OEt_2$[63,76,83] or even with excess B_2H_6.[84] Under these Lewis acidic conditions, reaction takes place *via* lactol C—O fission and further 2-tetrahydrofuranyl or 2-tetrahydropyranyl cation reduction. These methods are not useful for preparing strained bicyclic ethers due to preferential ring C—O bond cleavage at the intermediate lactol stage. Additionally, the method is not satisfactory for the synthesis of hindered oxepanes, although branching α to the ring oxygen facilitates deoxygenation in these cases. Ager and Sutherland have reported that crown lactones may be reduced with LAH to produce crown ethers.[85] It is clear that lithium cation complexation prevents ring scission during these unusual reductions. γ- and δ-lactones have been reduced to provide cyclic ethers *via* reaction with DIBAL-H followed by Et_3SiH and $BF_3{\cdot}OEt_2$ (50–88%), or directly using $HSiCl_3$ and di-*t*-butyl peroxide under irradiation (γ or UV) (55–100%).[86,87] The $HSiCl_3$ method is useful for the chemoselective reduction of strained lactones bearing ester groups.[87] Representative examples of lactone reductions to produce diols or cyclic ethers are given in Scheme 9.

1.10.5 REDUCTIONS OF AMIDES AND LACTAMS

1.10.5.1 Hydrogenation Reactions

Primary, secondary and tertiary carboxamides may be reduced to prepare primary, secondary or tertiary amines by heterogeneous catalytic hydrogenation at elevated temperatures and pressures.[1,88] Such reactions frequently require more drastic conditions than are necessary for the hydrogenation of carboxylic acids or esters. Additionally, the preparation of amines from amides in this way requires specialist equipment and reaction conditions. It is, therefore, only recommended for the desperate. Most hydrogenation reactions have been carried out over Cu chromite or Ba/Cu chromite. Raney nickel, Raney cobalt, ruthenium on carbon and rhenium(VII) oxide[89] are alternative catalysts. The hydrogenation of primary amides frequently produces secondary amines. These are presumably formed *via* aminolysis of an intermediate imine. Primary amide hydrogenations may be directed to primary amine formation by the use of ammonia as the (co)solvent. For example, decanamide has been converted into decylamine (90%; Ba–Cu chromite, H_2, 350 °C, 411 atm, NH_3) or *N,N*-didecylamine (73%; Ba–Cu chromite, H_2, 350 °C, 411 atm) depending on the reaction solvent. Hydrogenation of amides in alcoholic solvents may be used to prepare primary alcohols *via* amide transacylation and ester reduction. Alternatively, *N*-alkylation may be observed when primary or secondary amides are hydrogenated in alcoholic media. Both secondary and tertiary amides may be hydrogenated to produce the corresponding amines. Tertiary amide reductions can be carried out under relatively mild conditions (260 °C and ≥15 atm). Diamides may also be reduced to produce α,ω-diamines. However, reductions of 1,4-, 1,5- and 1,6- primary or secondary dicarboxamides produce pyrrolidine, piperidine or hexahydroazepine derivatives. Such systems are most probably formed *via* aminolysis at the imine stage. Hydrogenation of amides derived from aromatic carboxylic acids generally do not provide good yields of benzylic amines. Under the drastic reaction conditions, benzylic C—N cleavage is a dominant pathway. *N*-Benzoylpiperidine has been reported to produce mostly toluene (79%) on hydrogenation (Cu chromite, 250 °C, 200–300 atm). However, rhenium(VII) oxide minimizes C—N cleavage: benzamide has been reduced to yield benzylamine (69%) using this catalyst (220 °C, 205 atm, EtOH). Lactams may also be hydrogenated to produce cyclic amines; again, drastic reaction conditions are needed. The recalcitrance of amides to undergo reduction by catalytic hydrogenation may be used to advantage in synthesis. Esters may be selectively hydrogenated in the presence of amide groups.

1.10.5.2 Electrochemical and Dissolving Metal Reductions

The electrochemical reduction of carboxamides ($R^1CONR^2R^3$) has been used to prepare aldehydes (R^1CHO), amines ($R^1CH_2NR^2R^3$) or primary alcohols (R^1CH_2OH) depending on the structure of the substrate and reaction conditions.[4,5] Aldehyde formation may be favored by increasing the lifetime of the carbinolamine intermediate ($R^1CH(OH)NR^2R^3$; 5- and 6-ring cyclic carbinolamines are more stable than acyclic systems) or by trapping the aldehyde (R^1CHO) as it is formed. Aliphatic amides have been reduced to produce primary alcohols by electrolysis under neutral or basic conditions. Typically, the primary, secondary or tertiary amide is electrolyzed in an alcohol or amine solvent using an alkali metal or tetraalkylammonium salt as the supporting electrolyte.[2,4,5,88,90] Formation of the primary alcohol is thus favored by rapid carbinolamine ($R^1CH(OH)NR^2R^3$) fragmentation and subsequent aldehyde (R^1CHO) reduction. Amides may be electrochemically reduced to produce amines using cathode material of high hydrogen overpotential (Pb, Hg, Cd) under strongly acidic conditions. In the presence of strong acids, the carbinolamine intermediate readily loses water to produce the iminium salt $[(R^1CH{=}NR^2R^3)^+]$, which is reduced further to produce the amine. The electrolysis of amides in sulfuric acid at a lead cathode is a classical method for amine synthesis. The process is general for tertiary and secondary amines. Primary amides are reduced to amines only with considerable difficulty due to the lower stability of the intermediate iminium salt ($[R^1CH{=}NH_2]^+$). Representative electrochemical reductions include the conversion of dodecanamide into 1-dodecanol (92%; Pt, $MeNH_2$, LiCl), of *N,N*-dimethylhexanamide into 1-hexanol (97%; Pt, $MeNH_2$, LiCl), of *N,N*-dimethylbenzamide into *N,N*-dimethylbenzylamine (63%; Pb, H_2O, H_2SO_4) and of *N,N*-dibenzyl-1,4-piperazine-2,5-dione into *N,N*-dibenzyl-1,4-piperazine (70%; Pb, AcOH, H_2SO_4, H_2O).

The dissolving metal reduction of amides has been used to prepare aldehydes, primary alcohols or amines.[88,91] Tertiary but not primary or secondary amides have been reduced to provide aldehydes using Li in $MeNH_2$ or Na in NH_3 in the presence of EtOH, AcOH, or H_2O, *etc*. The production of aldehydes require that the carbinolamine fragmentation is slow on the reduction time scale. Reductions using excess alkali metal in NH_3 or HMPA containing a proton source have been used to prepare primary alcohols. Under these conditions, the excess metal is available to reduce the aldehyde formed from the carbinolamine fragmentation. Amides may be reduced by dissolving metals to produce amines when carbinolamine fragmentation to form imine intermediates is favored. For example, γ- or δ-lactams are converted in high yield into cyclic amines on dissolving alkali metal reduction in alcoholic solution. Finally, amides may be reduced indirectly to yield amines *via* dissolving metal reductions of imino chlorides, thioamides or related species.[88,92] Representative dissolving metal reductions of amides include the conversion of phenylalaninamide into 3-(2,5-dihydrophenyl)-2-amino-1-propanol (77%; Na, NH_3 MeOH, NH_4Cl, –78 °C) and of 5-phenyl-2-piperidinone into 5-phenylpiperidine (57%; Na, BuOH, Δ).

1.10.5.3 Reductions using Metal Hydride Reagents

Amides may be reduced using hydride reagents to produce amines, primary alcohols or aldehydes. Aliphatic and aromatic primary, secondary or tertiary amides are rapidly reduced to provide the corresponding amines using $BH_3{\cdot}THF$ or $BH_3{\cdot}SMe_2$ in ether solvents. Lactams are reduced to cyclic amines. The reaction may be chemoselective in that less reactive groups including alkyl halides, carbamates, epoxides, esters and nitro compounds may be easily tolerated.[2,8,48,93] Formamides may be selectively reduced in the presence of other amide functionality, providing that the reaction stoichiometry is carefully controlled.[94] Representative chemoselective reductions are given in Scheme 10. B_2H_6 is the reagent of choice for amide reductions, unless the molecule contains alkene units. In these cases, hydroboration is also observed. The B_2H_6 reduction of β-lactams produces β-amino alcohols and not azetidines. Clearly ring fragmentation is favored by the release of strain. Alkylboranes react more slowly with amides; both thexylborane and Sia_2BH slowly reduce tertiary amides to produce aldehydes.[51,95] 9-BBN-H reduces tertiary amides to produce mostly primary alcohols.[15] For example, reduction of $PhCONEt_2$ gave $PhCH_2OH$ (80%) and $PhCH_2NEt_2$ (20%). The thexylborane·$PhNEt_2$ complex reduces aliphatic and aromatic tertiary amides to produce tertiary amines. Clearly alkyl substitution of borane alters the rate of formation and fragmentation of the intermediate carbinolamine borate esters. None of the alkylborane reagents are particularly useful for amide reductions.

Neither $NaBH_4$ in alcohol solvents nor $LiBH_4$ in ethers readily reduce amides. Under more forcing conditions, amide recalcitrance can be overwhelmed.[17,18,96] Thus, tertiary but not secondary or primary amides may be reduced using $NaBH_4$ in pyridine at reflux. Although secondary amides are recovered unchanged, primary amides are dehydrated to produce nitriles. Primary, secondary and tertiary amides are reduced to the corresponding amines with $NaBH_4$ and $MeSO_3H$ in DMSO[17] or Bu_4NBH_4 in CH_2Cl_2.[97] Primary, secondary and tertiary amides are reduced to amines (75–96%) using $NaBH_4$–$TiCl_4$ in DME or with $NaBH_4$–$CoCl_2$ or $NiCl_2$ in methanol.[17,98] $NaBH_4$–$AlCl_3$ reduces tertiary amides to amines whereas primary amides form only salts, which do not undergo reduction.[60] γ- and δ-lactams have been reduced to the corresponding cyclic amines using excess $NaBH_4$ in refluxing MeOH and Bu^tOH. Primary and

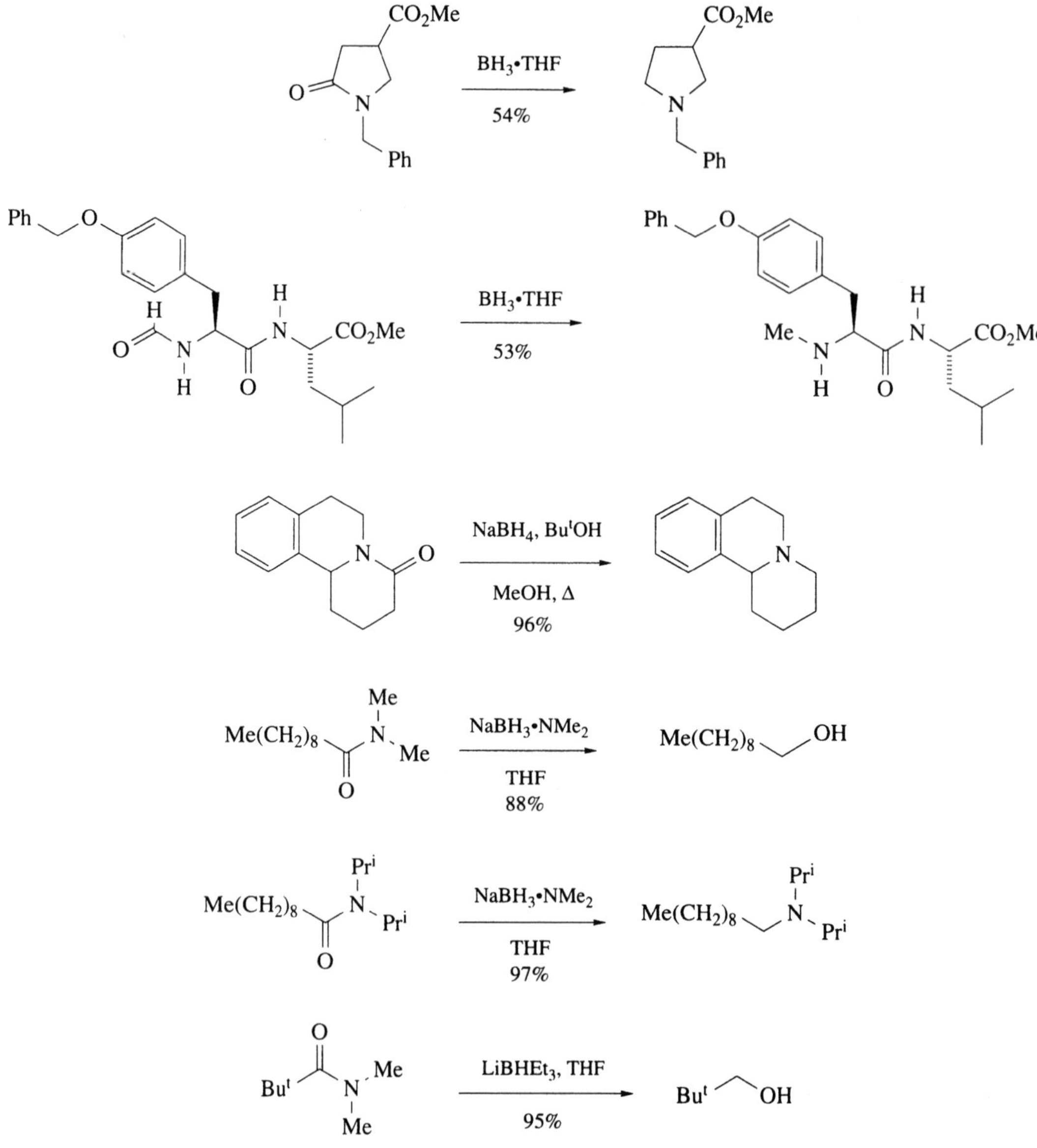

Scheme 10

secondary amides are reduced to the corresponding amines using $LiBH_4$ in THF and MeOH at reflux. Under these conditions, tertiary amides are reduced to primary alcohols.[59] Sulfur-, oxygen- and nitrogen-substituted borohydride reagents, which are more reactive than $NaBH_4$, have been used in amide reductions.[18,67,99–101] $NaBH_2(SCH_2CH_2S)$, $NaBH_2S_3$, $NaBH_4$/EtSH, $NaBH_4$/PhSH, $NaBH_3NMe_2$, $NaBH_3(NHBu^t)$, $NaBH_3(OAc)$, $NaBH_3(O_2CPh)$, and $NaBH_3(O_2CCl_3)$ have all been used in amide reductions to produce amines. Reductions of tertiary amides using $NaBH_3(NMe_2)$ and $NaBH_3(NHBu^t)$ are especially noteworthy in that primary alcohols are usually formed, unless the amide nitrogen is substituted by bulky groups when tertiary amines are the products. Presumably, N—C cleavage at the intermediate carbinolamine stage requires prior complexation by a boron Lewis acid and clearly this is disfavored by bulky alkyl groups. Primary amides are reduced by $NaBH_3(NMe_2)$ or $NaBH_3(NHBu^t)$ to give amines (21–77%), whereas secondary amines are recovered unchanged. Alkyl-substituted borohydride reagents reduce amides to amines or primary alcohols. $LiBHEt_3$ in THF is a powerful reducing agent that rapidly converts tertiary amides into primary alcohols[68] and it is the reagent of choice for such reductions. Li-9-BBN-H_2 reduces tertiary amides to amines[34] and $LiBH_3Bu$ converts tertiary amides into amine/primary alcohol mixtures.[68] None of the alkyl borohydrides are useful for the reduction of primary or secondary amides since the initially formed amide salts are insufficiently electrophilic. Amides may be rapidly reduced to produce amines by activation prior to borohydride reduction. Thus, tertiary amides are readily reduced *via* Vilsmeier salt formation or *O*-alkylation, followed by $NaBH_4$ reduction.[102] *N*-Nitrosoamides are reduced by $NaBH_4$ to produce primary alcohols (41–84%).[103] Representative examples of borohydride reductions are given in Scheme 10.

Amides and lactams are reduced by $Al_2H_3(OCH_2CH_2OMe)_3$, AlH_3, DIBAL-H or AlH_2Cl to produce the corresponding amine. Tertiary amides are reduced more rapidly than secondary or primary amides.[22,71,104] Reductions of tertiary amides may be stopped at the intermediate carbinolamine stage, thereby providing aldehydes or enamines from lactams. In general, AlH_3 or DIBAL-H reductions are useful for the reduction of unsaturated amides when hydroboration or overreduction are problems. Aromatic and aliphatic amides and lactams are reduced by LAH or $NaAlH_4$ to produce the corresponding amines.[20,88,105] The reaction is general for the preparation of primary, secondary or tertiary acyclic amines and cyclic secondary or tertiary amines. The reduction of primary amides may involve the intermediacy of the nitrile, and these have been isolated in several cases. Amides and lactams have also been reduced to the corresponding amines using alkoxyaluminum hydride reagents including $LiAlH(OMe)_3$, $LiAlH(OEt)_3$, Red-Al and $Ca[AlH_2(OBu^i)_2]_2 \cdot THF$ *etc*., although they are not reduced by hindered reagents including $LiAlH(OBu^t)_3$. All these reductions may be stopped at the intermediate aldehyde stage, although reactions using a slight excess of reagents readily provide the corresponding amines.[21] As with LAH, tertiary amides are reduced more readily than primary or secondary amides. The alkoxy-substituted reagents are less reactive and therefore more selective. Representative amide reductions using AlH_3, LAH and derivatives are given in Scheme 11.

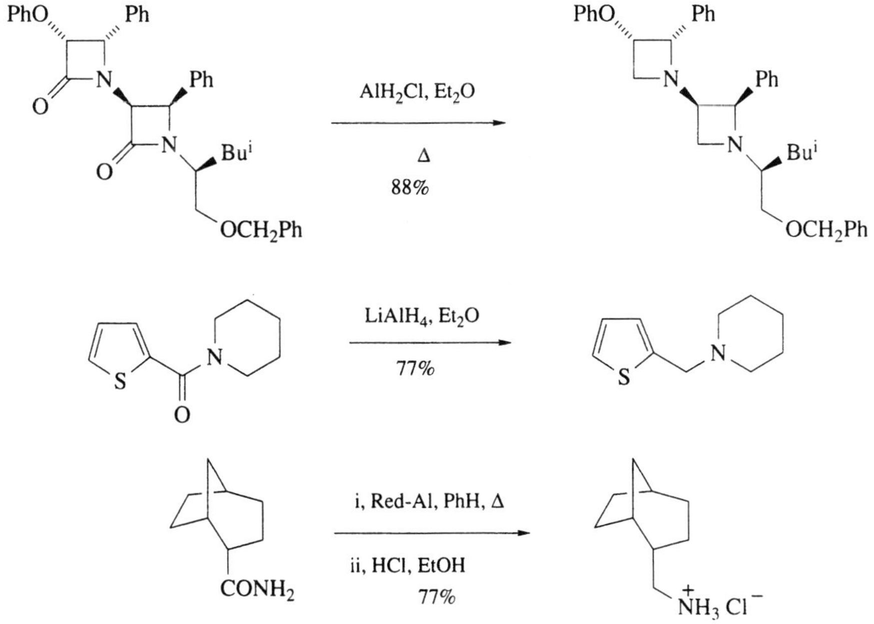

AlH_2Cl, Et_2O, Δ, 88%

$LiAlH_4$, Et_2O, 77%

i, Red-Al, PhH, Δ; ii, HCl, EtOH; 77%

Scheme 11

Finally, both aliphatic and aromatic tertiary amides have been slowly reduced using $HSiCl_3$ at elevated temperatures (90–110 °C) to produce the corresponding amines (33–90%).[106] There is little reason to recommend these hydrosilylation methods for synthesis.

1.10.6 REDUCTIONS OF NITRILES

1.10.6.1 Hydrogenation Reactions

Nitriles have been hydrogenated at low temperatures and pressures over heterogeneous and homogeneous catalysts to produce amines, aldehydes, primary alcohols or alkanes.[107,108] The reduction to produce amines is by far the most widely used transformation. The most commonly used catalysts are Raney nickel, Raney cobalt, nickel boride, cobalt boride, rhodium, palladium or platinum on various supports. Products formed in the hydrogenation of a nitrile (RCN) are determined by the fate of the intermediate imine (RCH=NH). Hydrogenation may be stopped at the intermediate imine (RCH=NH) stage under carefully controlled conditions (*vide infra*). Rapid hydrolysis of the imine and further hydrogenation may be used to produce the primary alcohol. The imine may be hydrogenated further to produce the primary amine. However, unless the primary amine is removed as it is formed, transimination of RCH=NH readily takes place. Subsequent hydrogenation of $RCH{=}NCH_2R$ provides the secondary amine $(RCH_2)_2NH$. Transimination with RCH=NH and further hydrogenation may occasionally produce the tertiary amine $(RCH_2)_3N$. In spite of all these complications, nitrile hydrogenation reactions may be used to selectively prepare primary or secondary amines, primary alcohols or aldehydes. Complete reduction to the alkanes (RCH_3) is only observed on prolonged hydrogenation with aromatic cyanides where C—N or C—O benzylic hydrogenolysis is facile. Primary amines may be easily prepared from nitriles by hydrogenation using Raney nickel in anhydrous MeOH or EtOH containing NH_3 (≥6 equiv.) at 100 °C and 100 atm. Lower temperatures and pressures may be used with larger quantities of highly active catalyst. Addition of ammonia favors primary amine formation by displacement of the RCH=NH and $RCH{=}NCH_2R$ equilibrium in favor of the former. Alternatively, nitriles may be reduced over Rh/Al_2O_3 or Rh/C under mild conditions. These procedures are particularly useful for minimizing C—Cl or benzylic C—N or C—O hydrogenolyses, and additionally has the advantage of experimental convenience. Secondary amine formation may also be suppressed by hydrogenation in the presence of an acid, an anhydride or $CHCl_3$ (which generates HCl *in situ*). The $CHCl_3$ method is particularly useful for acid-sensitive nitriles. Hydrogenation of nitriles in alcoholic solvents in the absence of ammonia or primary amine traps may be used to prepare symmetrical secondary amines in good yields. In general, efficient formation of tertiary amines is not observed, since the necessary intermediate $RCH{=}N^+(CH_2R)_2$ is not produced in appreciable quantities. Secondary amine formation is suppressed at elevated pressures and with sterically hindered substrates. Hydrogenation of nitriles in the presence of excess amines may be used to prepare unsymmetrical diamines. These are again formed *via* transimination. Transimination may also be used to prepare cyclic amines (5-, 6- and 7-membered) from the hydrogenation of amino nitriles or dinitriles. Alternatively, heterocyclic ring systems may be prepared from the hydrogenations of keto nitriles, keto amides, *etc*. Aldehydes may be produced from the transfer hydrogenation of nitriles in aqueous media using Raney nickel and formic acid or sodium hypophosphite. Alternatively, production of the aldehyde may be favored by nitrile hydrogenation in the presence of aldehyde traps such as hydrazine. Both aromatic and aliphatic nitriles have been converted into primary alcohols by hydrogenation over Raney nickel in the presence of acid resins.[109] Under these conditions, the imine (RCH=NH) and aldehyde (RCHO) equilibrium is displaced by removal of ammonia. Representative nitrile hydrogenations include the conversion of phenylacetonitrile into 2-phenylethylamine (>90%; Raney Ni, H_2, MeOH, NH_3, 100 °C, 100 atm) or (93%; Rh/Al_2O_3, H_2, EtOH, NH_3, 25 °C, 3 atm), of dodecanenitrile into *N*-dodecylacetamide (100%, Raney Ni, H_2, Ac_2O, NaOAc, 50 °C, 2 atm), of pentanenitrile into *N*-butylpentanamine (93%; Pd/C, H_2, $BuNH_2$, 25 °C, 2 atm) and of 4-methylbenzonitrile into 4-methylbenzyl alcohol (72%; Raney Ni, H_2, H_2O, Amberlyst-15, 25 °C, 1 atm).

Nitriles have been hydrogenated using homogeneous catalysts to produce primary amines.[110] For example, $(Pr^i_3P)_3RhH$ smoothly catalyzes the hydrogenation of aliphatic and aromatic nitriles to produce primary amines (44–100%) in THF solution at 20 °C and 1 atm. Alternatively, $(K^+[(Ph_3P)_2Ph_2PC_6H_4RuH_2]^-{\cdot}C_{10}H_8{\cdot}Et_2O$ and $2K^+[(Ph_3P)_3(Ph_2P)Ru_2H_4]^{2-}{\cdot}(diglyme)_2$ catalyze the hydrogenation of nitriles at 90 °C and 62 atm in toluene containing 18-crown-6 to produce primary amines (18–100%).[44]

In conclusion, it is germane to mention selectivity in the hydrogenation of polyfunctional molecules containing nitrile groups. Alkynes, alkenes, acid chlorides and nitro compounds are generally reduced more rapidly than nitriles. Nitriles, however, may be selectively reduced in the presence of carboxylic acids, esters, amides, aromatic ring systems, ketones and hydrogenolyzable groups.

1.10.6.2 Electrochemical and Dissolving Metal Reductions

The conversions of nitriles into primary amines *via* cathodic or dissolving metal reduction are treacherous procedures. Nitrile radical anions frequently undergo C—CN homolysis, thereby resulting in reductive decyanation. Electrolysis of nitriles in $EtNH_2$ containing LiCl or dissolving metal reductions using alkali metals in Bu^tOH or HMPA, or on alumina, in NH_3, or NH_3 and Bu^tOH, K and crown ethers in PhMe or Na and $Fe(acac)_3$ result in significant decyanation.[107,111] Since the decyanation reaction takes place *via* radical anion fragmentation to produce an alkyl radical and cyanide anion, it is particularly favored with tertiary or allylic nitriles. Under certain conditions, decyanation may be observed even with acetonitrile (Na/$Fe(acac)_3$, 98%). However, the conversion of primary and secondary nitriles into the corresponding amines may be carried out in reasonable to excellent yields. Tridecanenitrile has been reduced using Na in BuOH to produce tridecanamine (90–93%).[88] Ca in NH_3 generally provides primary amines in better yields than Li or Na in NH_3 or $MeNH_2$.[112] Indeed, it is possible to prepare 1-amino-2,2-dimethylpropane (42%) from the corresponding tertiary nitrile using this reagent. Several nitriles have been reduced to provide primary amines using $Cr_2(OAc)_4$, Raney Ni alloy in alkali solution, Ni in water at reflux, and electrochemically at a lead cathode in aqueous mineral acids, *etc.*[107,113] Additionally, attention is directed to the recently described efficient reductions of nitriles to produce amines *via* electrolysis using Raney nickel[114] or *via* cobalamin-catalyzed zinc reductions.[115] These methods are exemplified by the conversions of phenylacetonitrile into 2-phenylethylamine (71%; Ni cathode, 0.75 A, NaOMe, MeOH) and of cyanocyclododecane into di(cyclododecyl)methylamine (81%; Zn, HOAc, cobalt(I) cobalamin catalyst, 25 °C).

In conclusion, nitriles may be reduced electrochemically or with dissolving metal to produce amines. Nonetheless, such reactions must be used with caution. In addition to competitive decyanation, aldehyde or 2,4,6-trialkylhexahydro-1,3,5-triazine formation and reductive dimerization are waiting to overwhelm the unwary.

1.10.6.3 Reductions using Metal Hydride Reagents

Aliphatic and aromatic nitriles are readily reduced by $BH_3{\cdot}THF$ to produce primary amines. Excellent yields are obtained when the intermediate borazine salts are hydrolyzed with HCl in EtOH.[2,8,48,107,116] $BH_3{\cdot}SMe_2$ in THF is an alternative reagent, that is most efficiently used at reflux with continuous removal of Me_2S. Nitriles may also be reduced by catecholborane and thexylborane, although rates are slow.[13,16] Nitriles are not readily reduced by more hindered dialkylboranes such as Sia_2BH or 9-BBN-H nor are they rapidly reduced by $NaBH_4$, although the electrophilic polyfluoroalkyl nitriles may be converted into the corresponding primary amines.[107] Aromatic and aliphatic nitriles may be reduced to the corresponding amines (53–87%) using Bu_4NBH_4 in CH_2Cl_2.[97] Additionally, more reactive oxygen- and sulfur-substituted borohydrides are useful for converting nitriles into primary amines. $NaBH_3(OH)$,[64] $NaBH_2S_3$[100] and $NaBH_3(OCOCF_3)$[117] are all useful reagents, although thioamide formation is a complication with $NaBH_2S_3$. $NaBH_3(OCOCF_3)$ is particularly efficient for the reduction of nitriles in the presence of esters, nitroarenes or carbamates.

$NaBH_4/AlCl_3$ in diglyme, a more powerful reducing agent than $NaBH_4$, rapidly reduces aliphatic and aromatic nitriles to the corresponding primary amines (60–90%).[60,107] Alternatively, Raney nickel has been employed to catalyze the reductions of aromatic nitriles to benzylamines.[118] However, the most useful and general $NaBH_4$ reductions are those catalyzed by $CoCl_2$.[98,119] Nickel(II), osmium(IV), iridium(III), platinum(II), copper(II) and rhodium(III) salts are alternative less frequently employed catalysts. Reductions using $NaBH_4/CoCl_2$ involve the formation of cobalt boride, and this insoluble species functions as a heterogeneous catalyst for the $NaBH_4$ reduction of amides, nitriles and nitro compounds. Secondary amines have been prepared from nitriles *via* *N*-alkylation and subsequent $NaBH_4$ reduction of the resultant nitrilium salts.[120]

Nitriles may be reduced by AlH_3 in THF or DIBAL-H in PhMe or hexane to produce the corresponding primary amine in excellent yield.[22,71,104] However, DIBAL-H is much more frequently used to prepare aldehydes by reduction at low temperature. Under these conditions, the *N*-alumino imine is only slowly reduced further. AlH_3 and DIBAL-H are superior reagents to LAH in many cases. In particular α-deprotonation of acidic nitriles is less pronounced with these Lewis acidic reagents. For example, Ph_2CHCN is reduced to give $Ph_2CHCH_2NH_2$ in superior yield (91%) using AlH_3 (46% with LAH). The basicity of LAH is underscored by the reduction of 2-aryl-1,4-butanedinitriles to produce 3-arylpropanamines.[121] Clearly, this reaction takes place *via* the cinnamonitrile and complete reduction. However, in general, simple nitriles are reduced by excess LAH in ether solvents to produce primary amines in good yields.[20,88,107] Since LAH is a very reactive species, the reduction of nitriles containing other groups generally proceeds with global rather than selective reduction. Alkoxy-substituted reagents are less reactive than LAH and may be used for selective reductions.[21] Aromatic nitriles are rapidly reduced by $LiAlH(OMe)_3$ and Red-Al to produce primary amines in high yields. Aliphatic nitriles are not efficiently reduced by Red-Al due to facile α-deprotonation and thus the starting nitrile is recovered on work-up. Nitriles are readily transformed into aldehydes on reduction with $NaAlH(OEt)_3$, $LiAlH_2(OEt)_2$, $LiAlH(OEt)_3$ or Red-Al using inverse addition techniques and/or careful temperature and stoichiometric control. Murai *et al.* have reported that $Co_2(CO)_8$ catalyzes the reduction of aromatic nitriles using Me_3SiH to produce *N*,*N*-bis(trimethylsilyl)benzylamines (11–91%).[122]

In conclusion, aromatic nitriles are most conveniently and efficiently reduced by $BH_3 \cdot SMe_2$, $NaBH_4$–$CoCl_2$, AlH_3 or LAH. Aliphatic systems should be reduced with $BH_3 \cdot SMe_2$, $NaBH_4$–$CoCl_2$ or AlH_3. Representative nitrile reductions are given in Scheme 12.

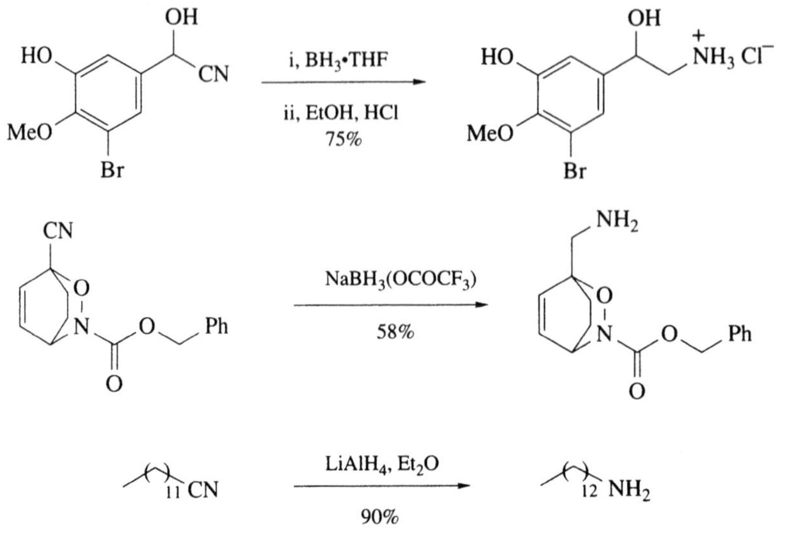

Scheme 12

1.10.7 REDUCTIONS OF IMIDES, ISOCYANATES AND CARBAMATES

Cyclic imides may be reduced to give hydroxy lactams, lactones, lactams, hydroxy amides or cyclic amines. The most useful synthetic transformations are reductions ($NaBH_4$, MeOH) to produce hydroxy lactams since these are precursors to α-acyliminium salts. However, such reductions are beyond the scope of this chapter. *N*-Alkylphthalimides may be reduced using $NaBH_4$ in 2-propanol to produce phthalide and the corresponding amine (70–97%) and this method provides a convenient procedure for releasing amines from phthalimides.[123] Phthalimide may alternatively be reduced using Zn/Cu to provide phthalide (71%).[124] Cyclic imides (5- and 6-membered) have been reduced using LAH, lithium alkoxyaluminum hydrides or $NaBH_4/BF_3 \cdot OEt_2$ to provide pyrrolidine or piperidine derivatives.[21,125] For example, *N*-methylsuccinimide was reduced to produce *N*-methylpyrrolidine (92%) using Red-Al.

Aliphatic or aromatic isocyanates may be reduced to produce *N*-methylalkyl- (or *N*-methylaryl-) amines using LAH,[88] $LiAlH(OMe)_3$, Red-Al[21] or $NaBH_2S_3$.[126] Isothiocyanates show similar reactivities. Alternatively, carbamates may be reduced to yield *N*-methylalkyl- (or *N*-methylaryl-) amines using LAH or Red-Al although the rates of reduction are slow.[21,127] Representative reductions include the conversion of phenyl isocyanate into *N*-methylaniline (75–89%; LAH, Et_2O) and the preparation of *N*-methyl-2-(3,5-dimethoxyphenyl)ethylamine from the corresponding ethyl carbamate (83%; Red-Al, Et_2O, PhH). In general reduction of isocyanates or urethanes have found little use in synthesis.

1.10.8 REFERENCES

1. P. N. Rylander, 'Hydrogenation Methods', Academic Press, New York, 1985; P. N. Rylander, 'Catalytic Hydrogenation in Organic Syntheses', Academic Press, New York, 1979.
2. H. Kropf, *Methoden Org. Chem. (Houben-Weyl)*, 1984, **VI/1b**, 1.
3. M. Bianchi, G. Menchi, F. Francalanci, F. Piacenti, U. Matteoli, P. Frediani and C. Botteghi, *J. Organomet. Chem.*, 1980, **188**, 109.
4. S. Wawzonek, *Synthesis*, 1971, 285; F. D. Popp and H. P. Schultz, *Chem. Rev.*, 1962, **62**, 19.
5. J. P. Coleman, in 'The Chemistry of Acid Derivatives', ed. S. Patai, Interscience, New York, 1979, part 2, p. 781.
6. A. O. Bedenbaugh, J. H. Bedenbaugh, W. A. Bergin and J. D. Adkins, *J. Am. Chem. Soc.*, 1970, **92**, 5774.
7. I. M. Cheklovskaya and M. M. Aleksankin, *Ukr. Khim. Zh. (Russ. Ed.)*, 1979, **45**, 1086 (*Chem. Abstr.*, 1980, **92**, 93 856w).
8. C. F. Lane, *Chem. Rev.*, 1976, **76**, 773; C. F. Lane, in 'Synthetic Reagents'; ed. J. S. Pizey, Horwood, Chichester, 1977, vol. 3; H. C. Brown, 'Organic Syntheses *via* Boranes', Wiley-Interscience, New York, 1975.

9. H. C. Brown and S. Krishnamurthy, *Tetrahedron*, 1979, **35**, 567; E. R. H. Walker, *Chem. Soc. Rev.*, 1976, **5**, 23.
10. F.-C. Huang, L. F. H. Lee, R. S. D. Mittal, P. R. Ravikumar, J. A. Chan, C. J. Sih, E. Caspi and C. R. Eck, *J. Am. Chem. Soc.*, 1975, **97**, 4144.
11. N. Cohen, R. J. Lopresti and G. Saucy, *J. Am. Chem. Soc.*, 1979, **101**, 6710; A. S. Kende and P. Fludzinski, *Org. Synth.*, 1985, **64**, 104.
12. M.-L. Anhoury, M. Arickx, P. Crooy, R. DeNeys, and J. Eliaers, *J. Chem. Soc., Perkin Trans. 1*, 1974, 191.
13. G. W. Kabalka, J. D. Baker, Jr. and G. W. Neal, *J. Org. Chem.*, 1977, **42**, 512.
14. C. Camacho, G. Uribe and R. Contreras, *Synthesis*, 1982, 1027.
15. H. C. Brown, S. Krishnamurthy and N. M. Yoon, *J. Org. Chem.*, 1976, **41**, 1778.
16. H. C. Brown, B. Nazer, J. S. Cha and J. A. Sikorski, *J. Org. Chem.*, 1986, **51**; 5264; H. C. Brown, J. S. Cha, N. M. Yoon and B. Nazer, *J. Org. Chem.*, 1987, **52**, 5400; J. S. Cha, J. E. Kim and K. W. Lee, *J. Org. Chem.*, 1987, **52**, 5030; A. Pelter, D. J. Ryder and J. H. Sheppard, *Tetrahedron Lett.*, 1978, 4715.
17. S. R. Wann, P. T. Thorsen and M. M. Kreevoy, *J. Org. Chem.*, 1981, **46**, 2579; S. Kano, Y. Tanaka, E. Sugino and S. Hibino, *Synthesis*, 1980, 695.
18. W. C. Guida, E. E. Entreken and A. R. Guida, *J. Org. Chem.*, 1984, **49**, 3024.
19. B. T. Cho and N. M. Yoon, *Synth. Commun.*, 1985, **15**, 917.
20. W. G. Brown, *Org. React. (N.Y.)*, 1951, **6**, 469; J. S. Pizey, in 'Synthetic Reagents', Horwood, Chichester, 1974, vol. 1.
21. J. Málek, *Org. React. (N. Y.)*, 1988, **36**, 249.
22. N. M. Yoon and Y. S. Gyoung, *J. Org. Chem.*, 1985, **50**, 2443.
23. T. D. Hubert, D. P. Eyman and D. F. Wiemer, *J. Org. Chem.*, 1984, **49**, 2279; M. Muraki and T. Mukaiyama, *Chem. Lett.*, 1974, 1447.
24. G. S. Li, D. F. Ehler and R. A. Benkeser, *Org. Synth.*, 1977, **56**, 83; K. Ramakrishnan and P. Bickart, *J. Chem. Soc., Chem. Commun.*, 1972, 1338.
25. O. H. Wheeler, in 'The Chemistry of Acyl Halides', ed. S. Patai, Interscience, New York, 1972, p. 231.
26. P. N. Rylander, in 'Catalytic Hydrogenation over Platinum Metals', Academic Press, New York, 1967, p. 235.
27. J. E. Lyons, *J. Chem. Soc., Chem. Commun.*, 1975, 412; P. Morand and M. M. Kayser, *J. Chem. Soc., Chem. Commun.*, 1976, 314.
28. K. Osakada, M. Obana, T. Ikariya, M. Saburi and S. Yoshikawa, *Tetrahedron Lett.*, 1981, **22**, 4297.
29. J. J. Bloomfield, D. C. Owsley and J. M. Nelke, *Org. React. (N.Y.)* 1976, **23**, 259.
30. E. Santaniello, A. Fiecchi, A. Manzocchi and P. Ferraboschi, *J. Org. Chem.*, 1983, **48**, 3074.
31. E. Santaniello, C. Farachi and A. Manzocchi, *Synthesis*, 1979, 912.
32. H. Kotsuki, Y. Ushio, N. Yoshimura and M. Ochi, *Tetrahedron Lett.*, 1986, **27**, 4213.
33. D. J. Raber and W. C. Guida, *J. Org. Chem.*, 1976, **41**, 690; W. B. Tamblyn, R. E. Aquadro, O. D. DeLuca, D. H. Weingold and T. V. Dao, *Tetrahedron Lett.*, 1983, **24**, 4955.
34. H. C. Brown, C. P. Mathew, C. Pyun, J. C. Son and N. M. Yoon, *J. Org. Chem.*, 1984, **49**, 3091.
35. H. C. Brown, J. S. Cha, B. Nazer, S.-C. Kim, S. Krishnamurthy and C. A. Brown, *J. Org. Chem.*, 1984, **49**, 885.
36. K. Soai, S. Yokoyama and D. Mochida, *Synthesis*, 1987, 647.
37. K. Ishizumi, K. Koga and S.-I. Yamada, *Chem. Pharm. Bull.*, 1968, **16**, 492; T. Koizumi, N. Yamamoto and E. Yoshii, *Chem. Pharm. Bull.*, 1973, **21**, 312; N. Minami and S. Kijima, *Chem. Pharm. Bull.*, 1979, **27**, 816.
38. P. L. Hall and R. B. Perfetti, *J. Org. Chem.*, 1974, **39**, 111; Y. Nagao, K. Kawabata, K. Seno and E. Fujita, *J. Chem. Soc., Perkin Trans. 1*, 1980, 2470; T. Fujisawa, T. Mori and T. Sato, *Chem. Lett.*, 1983, 835.
39. S. Kim and K. H. Ahn, *J. Org. Chem.*, 1984, **49**, 1717.
40. C. Soucy, D. Favreau and M. M. Kayser, *J. Org. Chem.*, 1987, **52**, 129; M. M. Kayser, L. Breau, S. Eliev, P. Morand and H. S. Ip, *Can. J. Chem.*, 1986, **64**, 104; M. M. Kayser, J. Salvador and P. Morand, *Can. J. Chem.*, 1983, **61**, 439; M. M. Kayser, J. Salvador, P. Morand and H. G. Krishnamurty, *Can. J. Chem.*, 1982, **60**, 1199; M. M. Kayser and P. Morand, *Can. J. Chem.*, 1978, **56**, 1524; P. Stanetty, H. Fröhlich and F. Sauter, *Monatsh. Chem.*, 1986, **117**, 69.
41. J. J. Bloomfield and S. L. Lee, *J. Org. Chem.*, 1967, **32**, 3919; M. E. Birckelbaw, P. W. LeQuesne and C. K. Wocholski, *J. Org. Chem.*, 1970, **35**, 558; B. E. Cross and J. C. Stewart, *Tetrahedron Lett.*, 1968, 3589.
42. H. Adkins, *Org. React. (N.Y.)*, 1954, **8**, 1.
43. A. V. McIntosh, Jr., E. M. Meinzer and R. H. Levin, *J. Am. Chem. Soc.*, 1948, **70**, 2955.
44. R. A. Grey, G. P. Pez and A. Wallo, *J. Am. Chem. Soc.*, 1981, **103**, 7536.
45. L. Horner and H. Höenl, *Justus Liebigs Ann. Chem.*, 1977, 2036.
46. R. H. Manske, *Org. Synth., Coll. Vol.*, 1943, **2**, 154; E. E. Reid, F. O. Cockerille, J. D. Meyer, W. M. Cox, Jr. and J. R. Ruhoff, *Org. Synth., Coll. Vol.*, 1943, **2**, 468.
47. A. G. M. Barrett, C. R. A. Godfrey, D. M. Hollinshead, P. A. Prokopiou, D. H. R. Barton, R. B. Boar, L. Joukhadar, J. F. McGhie and S. C. Misra, *J. Chem. Soc., Perkin Trans. 1*, 1981, 1501.
48. H. C. Brown, Y. M. Choi and S. Narasimhan, *J. Org. Chem.*, 1982, **47**, 3153.
49. H. P. Plaumann, J. G. Smith and R. Rodrigo, *J. Chem. Soc., Chem. Commun.*, 1980, 354.
50. S. Saito, T. Hasegawa, M. Inaba, R. Nishida, T. Fujii, S. Nomizu and T. Moriwake, *Chem. Lett.*, 1984, 1389.
51. H. C. Brown, P. Heim and N. M. Yoon, *J. Org. Chem.*, 1972, **37**, 2942.
52. R. W. Jeanloz and E. Walker, *Carbohydr. Res.*, 1967, **4**, 504.
53. H. C. Brown, S. Narasimhan and Y. M. Choi, *J. Org. Chem.*, 1982, **47**, 4702.
54. R. Dohmori, R. Yoshimura, S.-I. Kitahara, Y. Tanaka and T. Naito, *Chem. Pharm. Bull.*, 1970, **18**, 1908.
55. K. Soai, H. Oyamada, M. Takase and A. Ookawa, *Bull. Chem. Soc. Jpn.*, 1984, **57**, 1948.
56. E. Santaniello, P. Ferraboschi, A. Fiecchi, P. Grisenti and A. Manzocchi, *J. Org. Chem.*, 1987, **52**, 671.
57. T. Wakamatsu, H. Inaki, A. Ogawa, M. Watanabe and Y. Ban, *Heterocycles*, 1980, **14**, 1441.
58. J. Mauger and A. Robert, *J. Chem. Soc., Chem. Commun.*, 1986, 395; S. Corsano and G. Piancatelli, *J. Chem. Soc., Chem. Commun.*, 1971, 1106.
59. K. Soai and A. Ookawa, *J. Org. Chem.*, 1986, **51**, 4000.

60. H. C. Brown and B. C. Subba Rao, *J. Am. Chem. Soc.*, 1956, **78**, 2582.
61. H. C. Brown and S. Narasimhan, *J. Org. Chem.*, 1984, **49**, 3891.
62. H. C. Brown and S. Narasimhan, *J. Org. Chem.*, 1982, **47**, 1604.
63. G. R. Pettit and D. M. Piatak, *J. Org. Chem.*, 1962, **27**, 2127.
64. J. W. Reed and W. L. Jolly, *J. Org. Chem.*, 1977, **42**, 3963.
65. Y. Maki, K. Kikuchi, H. Sugiyama and S. Seto, *Tetrahedron Lett.*, 1975, 3295.
66. R. A. Bell and M. B. Gravestock, *Can. J. Chem.*, 1969, **47**, 2099.
67. R. O. Hutchins, K. Learn, F. El-Telbany and Y. P. Stercho, *J. Org. Chem.*, 1984, **49**, 2438; Y. Kikugawa, *Chem. Lett.*, 1975, 1029.
68. H. C. Brown, S.-C. Kim and S. Krishnamurthy, *J. Org. Chem.*, 1980, **45**, 1; S. Kim, Y. C. Moon and K. H. Ahn, *J. Org. Chem.*, 1982, **47**, 3311.
69. R. E. Ireland and W. J. Thompson, *Tetrahedron Lett.*, 1979, 4705.
70. J.-I. Nikawa and T. Shiba, *Chem. Lett.*, 1979, 981; H. Kotsuki, N. Yoshimura, Y. Ushio, T. Ohtsuka and M. Ochi, *Chem. Lett.*, 1986, 1003, H.-J. Liu, R. R. Bukownik and P. R. Pednekar, *Synth. Commun.*, 1981, **11**, 599.
71. E. Winterfeldt, *Synthesis*, 1975, 617; T. Moriwake, S.-I. Hamano, D. Miki, S. Saito and S. Torii, *Chem. Lett.*, 1986, 815.
72. K. Isobe, K. Mohri, H. Sano, J.-I. Taga and Y. Tsuda, *Chem. Pharm. Bull.*, 1986, **34**, 3029.
73. K. Barlos, D. Papaioannou, S. Patrianakou and T. Tsegenidis, *Liebigs Ann. Chem.*, 1986, 952; S. Hanessian, P. J. Murray and S. P. Sahoo, *Tetrahedron Lett.*, 1985, **26**, 5623, 5627.
74. G. A. Kraus and K. Frazier, *J. Org. Chem.*, 1980, **45**, 4262.
75. Y. Kamitori, M. Hojo, R. Masuda, T. Inone and T. Izumi, *Tetrahedron Lett.*, 1983, **24**, 2575; Y. Kamitori, M. Hojo, R. Masuda, T. Izumi and T. Inoue, *Synthesis*, 1983, 387.
76. G. R. Pettit and T. R. Kasturi, *J. Org. Chem.*, 1960, **25**, 875.
77. E. L. Eliel and R. A. Daignault, *J. Org. Chem.*, 1964, **29**, 1630.
78. F. Sato, T. Jinbo and M. Sato, *Tetrahedron Lett.*, 1980, **21**, 2175.
79. S. W. Baldwin and S. A. Haut, *J. Org. Chem.*, 1975, **40**, 3885; J. Boyer, R. J. P. Corriu, R. Perz, M. Poirier and C. Reyé, *Synthesis*, 1981, 558; C. Chuit, R. J. P. Corriu, R. Perz and C. Reyé, *Synthesis*, 1982, 981.
80. J. T. Edward and J. M. Ferland, *Chem. Ind. (London)*, 1964, 975.
81. L. Hough and A. C. Richardson, in 'Comprehensive Organic Chemistry', ed. D. H. R. Barton and W. D. Ollis, Pergamon Press, Oxford, 1979, vol. 5, p. 687.
82. K. Hanaya, Y. Koga, A. Yamaguchi, H. Kudo and Y. L. Chow, *Nouv. J. Chim.*, 1982, **6**, 149; S. V. Attwood and A. G. M. Barrett, *J. Chem. Soc., Perkin Trans. 1*, 1984, 1315.
83. S. Wolff, A. B. Smith, III and W. C. Agosta, *J. Org. Chem.*, 1974, **39**, 1607; G. R. Pettit, B. Green, T. R. Kasturi and U. R. Ghatak, *Tetrahedron*, 1962, **18**, 953.
84. J. R. Dias and G. R. Pettit, *J. Org. Chem.*, 1971, **36**, 3485.
85. D. J. Ager and I. O. Sutherland, *J. Chem. Soc., Chem. Commun.*, 1982, 248.
86. R. Nakao, T. Fukumoto and J. Tsurugi, *J. Org. Chem.*, 1972, **37**, 76; G. A. Kraus, K. Frazier, B. D. Roth, M. J. Taschner and K. Neuenschwander, *J. Org. Chem.*, 1981, **46**, 2417.
87. S. W. Baldwin, R. J. Doll and S. A. Haut, *J. Org. Chem.*, 1974, **39**, 2470.
88. E. Müller, *Methoden Org. Chem. (Houben-Weyl)*, 1957, **XI/1**, 1.
89. H. S. Broadbent and W. J. Bartley, *J. Org. Chem.*, 1963, **28**, 2345; A. Guyer, A. Bieler and G. Gerliczy, *Helv. Chim. Acta*, 1955, **38**, 1649.
90. L. Horner and H. Neumann, *Chem. Ber.*, 1965, **98**, 3462.
91. R. A. Benkeser, H. Watanabe, S. J. Mels and M. A. Sabol, *J. Org. Chem.*, 1970, **35**, 1210; A. O. Bedenbaugh, A. L. Payton and J. H. Bedenbaugh, *J. Org. Chem.*, 1979, **44**, 4703; A. J. Birch, J. Cymerman-Craig and M. Slaytor, *Aust. J. Chem.*, 1955, **8**, 512; M. Larchevêque and T. Cuvigny, *C. R. Hebd. Seances Acad. Sci., Ser. C*, 1973, **276**, 209.
92. A. Basha and A.-U. Rahman, *Experientia*, 1977, **33**, 101.
93. H. C. Brown, S. Narasimhan and Y. M. Choi, *Synthesis*, 1981, 996; P. G. Sammes and S. Smith, *J. Chem. Soc., Chem. Commun.*, 1982, 1143.
94. R. C. Northrop, Jr. and P. L. Russ, *J. Org. Chem.*, 1977, **42**, 4148; S. Krishnamurthy, *Tetrahedron Lett.*, 1982, **23**, 3315.
95. H. C. Brown, D. B. Bigley, S. K. Arora and N. M. Yoon, *J. Am. Chem. Soc.*, 1970, **92**, 7161.
96. I. Saito, Y. Kikugawa and S.-I. Yamada, *Chem. Pharm. Bull.*, 1970, **18**, 1731; S. B. Mandal, V. S. Giri, M. S. Sabeena and S. C. Pakrashi, *J. Org. Chem.*, 1988, **53**, 4236.
97. T. Wakamatsu, H. Inaki, A. Ogawa, M. Watanabe and Y. Ban, *Heterocycles*, 1980, **14**, 1437.
98. T. Satoh, S. Suzuki, Y. Suzuki, Y. Miyaji and Z. Imai, *Tetrahedron Lett.*, 1969, 4555.
99. Y. Maki, K. Kikuchi, H. Sugiyama and S. Seto, *Chem. Ind. (London)*, 1976, 322.
100. J. M. Lalancette and J. R. Brindle, *Can. J. Chem.*, 1971, **49**, 2990.
101. N. Umino, T. Iwakuma and N. Itoh, *Tetrahedron Lett.*, 1976, 763.
102. M. E. Kuehne and P. J. Shannon, *J. Org. Chem.*, 1977, **42**, 2082; R. F. Borch, *Tetrahedron Lett.*, 1968, 61.
103. J. E. Saavedra, *J. Org. Chem.*, 1979, **44**, 860.
104. N. M. Yoon and H. C. Brown, *J. Am. Chem. Soc.*, 1968, **90**, 2927; I. Ojima, T. Yamato and K. Nakahashi, *Tetrahedron Lett.*, 1985, **26**, 2035.
105. C. V. Wilson and J. F. Stenberg, *Org. Synth. Coll. Vol.*, 1963, **4**, 564; A. C. Cope and E. Ciganek, *Org. Synth. Coll. Vol.*, 1963, **4**, 339.
106. R. A. Benkeser, G. S. Li and E. C. Mozdzen, *J. Organomet. Chem.*, 1979, **178**, 21.
107. M. Rabinovitz, in 'The Chemistry of the Cyano Group', ed. Z. Rappoport, Interscience, New York, 1970, p. 307.
108. M. Freifelder, in 'Catalytic Hydrogenation in Organic Synthesis. Procedures and Commentary', Wiley, New York, 1978, p. 43; M. Freifelder, in 'Practical Catalytic Hydrogenation', Wiley, New York, 1971, p. 238.
109. A. Gauvreau, A. Lattes and J. Perie, *Bull. Soc. Chim. Fr.*, 1969, 126.

110. T. Yoshida, T. Okano and S. Otsuka, *J. Chem. Soc., Chem. Commun.*, 1979, 870.
111. E. E. van Tamelen, H. Rudler and C. Bjorklund, *J. Am. Chem. Soc.*, 1971, **93**, 7113; T. Ohsawa, T. Kobayashi, Y. Mizuguchi, T. Saitoh and T. Oishi, *Tetrahedron Lett.*, 1985, **26**, 6103; J. A. Marshall and R. Bierenbaum, *J. Org. Chem.*, 1977, **42**, 3309; D. Savoia, E. Tagliavini, C. Trombini and A. Umani-Ronchi, *J. Org. Chem.*, 1980, **45**, 3227.
112. A. R. Doumaux, Jr., *J. Org. Chem.*, 1972, **38**, 508.
113. K. Sakai and K.-I. Watanabe, *Bull. Chem. Soc. Jpn.*, 1967, **40**, 1548; Y. D. Smirnov, A. P. Tomilov and S. K. Smirnov, *Z. Org. Khim.*, 1975, **11**, 522.
114. T. Chiba, M. Okimoto, H. Nagai and Y. Takata, *Bull. Chem. Soc. Jpn.*, 1983, **56**, 719.
115. A. Fischli, *Helv. Chim. Acta*, 1978, **61**, 3028.
116. H. C. Brown, Y. M. Choi and S. Narasimhan, *Synthesis*, 1981, 605.
117. N. Umino, T. Iwakuma and N. Itoh, *Tetrahedron Lett.*, 1976, 2875; J. E. Baldwin, M. Otsuki and P. M. Wallace, *J. Chem. Soc., Chem. Commun.*, 1985, 1549.
118. R. A. Egli, *Helv. Chim. Acta*, 1970, **53**, 47.
119. J. O. Osby, S. W. Heinzman and B. Ganem, *J. Am. Chem. Soc.*, 1986, **108**, 67.
120. R. F. Borch, *J. Org. Chem.*, 1969, **34**, 627.
121. J. S. New and J. P. Yevich, *Synthesis*, 1983, 388.
122. T. Murai, T. Sakane and S. Kato, *Tetrahedron Lett.*, 1985, **26**, 5145.
123. J. O. Osby, M. G. Martin and B. Ganem, *Tetrahedron Lett.*, 1984, **25**, 2093.
124. J. H. Gardner and C. A. Naylor, Jr., *Org. Synth., Coll. Vol.*, 1943, **2**, 526.
125. D. W. Henry, *J. Heterocycl. Chem.*, 1966, **3**, 503; A. P. Gray and D. E. Heitmeier, *J. Org. Chem.*, 1969, **34**, 3253; W. Merkel, D. Mania and D. Bormann, *Liebigs Ann. Chem.*, 1979, 461.
126. J. R. Brindle, J.-L. Liard and N. Bérubé, *Can. J. Chem.*, 1976, **54**, 871.
127. H. L. Holland and G. B. Johnson, *Tetrahedron Lett.*, 1979, 3395.

1.11
Reduction of Carboxylic Acids to Aldehydes by Metal Hydrides

ROBERT A. W. JOHNSTONE
University of Liverpool, UK

1.11.1 INTRODUCTION

Apart from their intrinsic importance, aldehydes are useful functional groups in organic synthesis because they can be transformed readily into a wide variety of other functions. Since aldehydes are easily protected and deprotected, such transformations can be effected at various stages of a synthesis. It is unfortunate that building in an aldehyde function by 'direct' methods is generally not easy, so that methods have been sought for many years for the conversion of carboxylic acids and a variety of their derivatives into the corresponding aldehydes by direct reduction (similarly, many oxidative methods for the conversion of alcohols into aldehydes have been sought). However, aldehydes are very reactive to a large number of reagents, so much so that their synthesis by reduction of carboxylic acids or their derivatives is difficult because the aldehydes are normally more readily reduced than the carboxylic acid or its derivative, and reduction through to the alcohol stage is easy and common.

Until about 1950, reduction of carboxylic acids and their derivatives to aldehydes was not straightforward, and even one of the best methods, the Rosenmund hydrogenation of acid chlorides, required very careful control of both the reaction conditions and preparation of catalyst. The advent of aluminum and boron hydrides and their ready commercial availability transformed the situation to such an extent that the formation of aldehydes from carboxylic acids, acid chlorides, esters, amides, nitriles and similar groups in the presence of other reducible functional groups has become a relatively easy operation on both small and large scale.

Nowadays, there is a wide variety of hydride reagents but this section will concern itself mostly with those derived from boron, aluminum, tin and silicon. Even within this restricted group there is a wide range of hydride types, many of which are far too vigorous for controlled reduction of carboxylic acid derivatives to aldehydes. Accordingly, rather than select the hydrides one by one and discuss their ability to effect the desired reduction, this section will categorize carboxylic acids and their derivatives and discuss which hydrides can be used profitably for their selective reduction. Also, this review will concentrate on developments of methods rather than their application in actual synthesis in order to restrict the number of references. Suffice it to say that there are many examples of elegant, skilled use of hydride reagents to reduce carboxylic acids or their derivatives to aldehydes or their derivatives as parts of syntheses.

The ideal hydride to effect reduction of acids to aldehydes does not exist, nor is it axiomatic that the ideal reagent should react only with carboxylic acids because, in synthesis, the carboxylic acid is usually already protected in some way and it may be more convenient to reduce an amide rather than the free acid. Probably, for most organic chemists, the ideal reagent would be stable, nontoxic, cheap, easy to obtain, not reactive to air or water, would reduce only the group it was needed for and would present no difficulties on work-up: a formidable ideal.

There have been two major approaches towards achieving selective reduction of carboxylic acid derivatives to aldehydes (or derivatives) by hydrides. Firstly, hydride reagents themselves have been modified as, for example, sodium borohydride and sodium cyanoborohydride. Sometimes these modifications have led to 'finely tuned' reducing agents, as with the dimethyl sulfide adduct of thexylbromoborane (see later). Secondly, the type of carboxylic acid derivative has been modified for the optimum yield of aldehyde. For example, amides have been made from an assortment of amines in efforts to maximize yields of aldehyde. Best yields of aldehydes are obtained usually by a combination of modified hydride reagent with a modified carboxylic acid derivative.

The various functional groups will be considered in the following order: carboxylic acids, acyl chlorides, esters (including ortho esters), lactones, amides, lactams and nitriles (including imidoyl halides). Nomenclature for aluminum and boron hydrides is in a state of flux. There are the 'authorized' versions and those used by most practising organic chemists. Thus, lithium tetrahydridoaluminate is the correct name for lithium aluminum hydride ($LiAlH_4$), although the latter is more easily and widely recognized. In writing this article, the view has been taken that, for now and the near future, the more widely used nomenclature will be more readily recognized by readers and so common everyday names will be used. To avoid confusion and encourage the use of the more systematic, recommended nomenclature, this will be given in parentheses at the first mention of any one hydride; for example, tri-*n*-butyltin hydride (tri-*n*-butylstannane). Similarly, trivial but widely used names will be given, as with sodium bis(2-methoxyethoxy)aluminum hydride (sodium bis(2-methoxyethoxy)dihydridoaluminate; Red-Al).

1.11.2 CARBOXYLIC ACIDS

Direct reduction of carboxylic acids to aldehydes has been a long-sought transformation since acids are usually relatively easy to obtain. The number of hydride reagents which effect this transformation is limited.

Although it had been reported in 1959 that diisobutylaluminum hydride (DIBAH or DIBAL-H; diisobutylalane) reduced acids and esters to alcohols[1] at 70 °C, it was shown later that, at –75 to –70 °C, good yields of aldehydes could be obtained.[2] At higher temperatures, yields dropped dramatically. For example, caproic acid gave a 70% yield of caproaldehyde at –75 °C but, at –60 °C, the yield fell to 30%. This reaction with an alane appears to be the first reported direct reduction of acids to aldehydes. The very low temperatures required, the ease of overreduction and the possibility of reduction of other functional groups makes this reaction less attractive on a large scale. Shortly after this work, it was shown that the parent aluminum hydride itself (alane) was generally too vigorous a reducing agent, giving alcohols with acids and esters and amines with amides and nitriles; a 6% yield of caproaldehyde from caproic acid was reported.[3]

Attempts to modify the alanes led to lithium trimethoxyaluminum hydride (lithium trimethoxyhydridoaluminate)[4] and lithium tri-*t*-butoxyaluminum hydride (lithium tri-*t*-butoxyhydridoaluminate),[3] but the trimethoxy compound, with carboxylic acids, gave alcohols and the tributoxy hydride would effect no reduction at all. However, other workers[5] showed that alane could be modified with secondary amines to give compounds of the type (**1**; $HAlX_2$) and that these hydrides reduced acids directly to aldehydes in yields of 60–80% on refluxing them in tetrahydrofuran for 3–20 h. These same workers confirmed that alane itself gave almost no aldehyde and even H_2AlX gave inferior yields. To prepare the reagent, an alane solution must be calibrated and reacted with exact amounts of amine; two hydride ions are lost in the preparation and the reduction requires 2–4 equiv. of reagent so that the method is neither easy nor efficient in usage of hydride. A more convenient preparation has been proposed for the modified alanes, $HAlX_2$ (**1**), starting from lithium aluminum hydride;[6] this last publication confirmed the excellent yields of aldehydes which can be obtained.

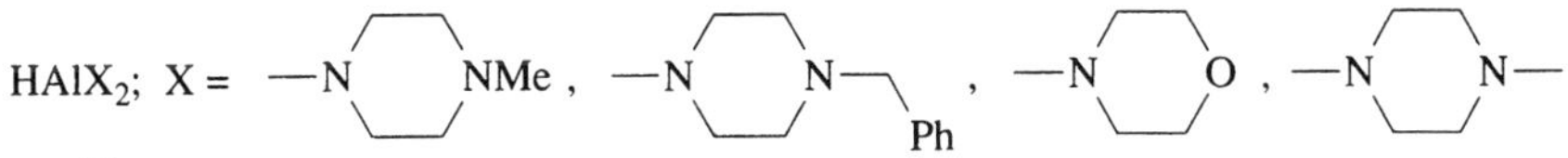

$HAlX_2$; X = —N(piperazine)NMe , —N(piperazine)N—CH₂Ph , —N(morpholine)O , —N(piperazine)N—

(**1**)

Analogous boranes have been investigated as suitable modified reducing agents for carboxylic acids. Diborane itself reacts vigorously with acids but does not give aldehydes.[7] In a comparative review of diborane and thexylborane (2,3-dimethyl-2-butylborane) it was pointed out that the latter reacts with acids to form hydrogen and a 1:1 complex, but with an excess of hydride reagent at –20 °C in tetrahydrofuran a sluggish reaction affords aldehydes.[8] It was emphasized later that thexylborane reacts with other functional groups so that care is needed for selective reduction of acids to aldehydes.[9] Furthermore, thexylborane adds to double or triple bonds much faster than it reduces carbonyl functional groups, thereby diminishing its usefulness for selective reduction to aldehydes.

The reducing power of diborane has been blunted by forming its adduct with dimethyl sulfide. This adduct, $Me_2S{\cdot}BH_3$, is stable and commercially available and therefore more attractive as a hydride reagent than diborane itself.[10] Nevertheless, the adduct still reduces carboxylic acids to alcohols which are isolated as cyclic boroxins (**2**; Scheme 1). In a 'one-pot' reaction, carboxylic acids can be reduced to boroxins and then oxidized with pyridinium chlorochromate to the required aldehyde (Scheme 1).[11]

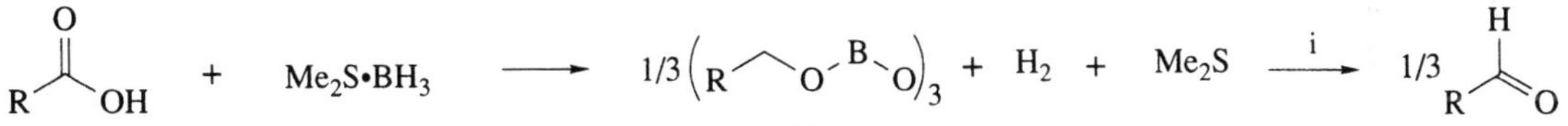

$$RCO_2H + Me_2S{\bullet}BH_3 \longrightarrow 1/3\,(RCH_2OBO)_3 + H_2 + Me_2S \xrightarrow{\text{i}} 1/3\,RCHO$$

(**2**)

i, pyridinium chlorochromate

Scheme 1

In much the same way, thexylborane has been modified by reaction with HCl and dimethyl sulfide to give thexylchloroborane–dimethyl sulfide ($Me_2CHCMe_2BHCl{\cdot}SMe_2$).[12] This stable reagent is much more reactive to acids than is thexylborane itself; reduction takes place at room temperature and an excess of reagent can be tolerated. No reaction is observed with esters, epoxides, aromatic nitriles and nitro compounds. Interestingly, aliphatic acids are reduced to aldehydes in about 15 min, but aromatic acids require about 24 h so that selective reactions can be effected on aliphatic and aromatic carboxylic acids. An earlier survey of the reduction of carboxylic acids with the thexylchloroborane–dimethyl sulfide adduct showed that 46–92% yields of aldehydes could be obtained and that diacids gave high yields of dialdehydes.[13] One practical difficulty with their use lies in the removal of the resulting thexylboronic acid during work-up. Although aromatic acids require a long reaction time, it was pointed out that aromatic acid chlorides are more favorably reduced to aldehydes by lithium tri-*t*-butoxyaluminum hydride, which is therefore a complementary reagent (see Section 1.11.3).

A slight but significant development of thexylchloroborane has given a reagent which reacts equally well with carboxylic acids but does not add to double or triple bonds. Thus, the adduct thexylbromoborane–dimethyl sulfide gives approximately 95% yields of aldehydes of all types in CS_2/CH_2Cl_2 at room temperature;[14] the reagent is somewhat laborious to prepare but is stable and can be stored. The high molecular weight of the reagent may make it unsuitable for large scale use.

Further adaptations of the boranes has led to reagents which reduce carboxylic acids but afford thioacetals rather than the aldehyde itself.[15] Thus, with the thioborane (**3**; equation 1), aliphatic acids give 80–87% yields of thioacetals but aromatic acids respond less well in giving significant quantities of sulfides as well. Carboxylic acid esters are inert to this reagent but give sulfides if a Lewis acid is included. Similarly, the 1,3,2-dithiaborinane–dimethyl sulfide adduct (**4**; equation 2) affords cyclic dithioacetals in 70–90% isolated yields in the presence of $SnCl_2$.[16] Aliphatic acids react in about 6 h at room temperature but aromatic acids need about 20 h and yields are somewhat poorer. This area has been reviewed.[17] From a practical viewpoint, it should be noted that the dithiaborinane (**4**) requires a week for its preparation.

A completely different approach has been taken in the use of hypervalent silicon hydrides. It has been shown that pentacoordinated silanes possess enhanced reducing power. Thus, in the presence of a nucleophile, which may be external (reaction (a) in Scheme 2) or internal (reaction (b) in Scheme 2), hy-

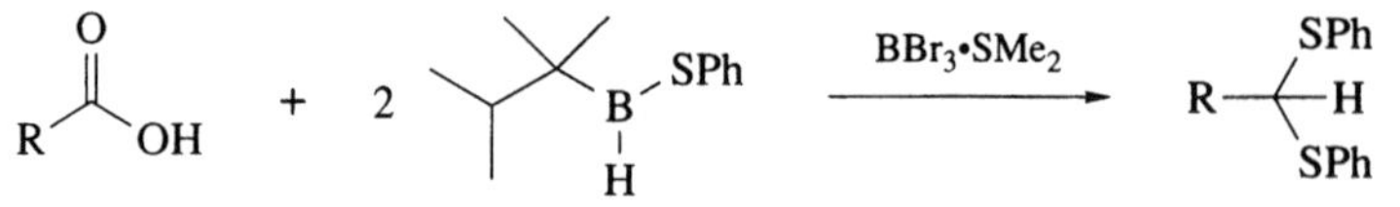

(1)

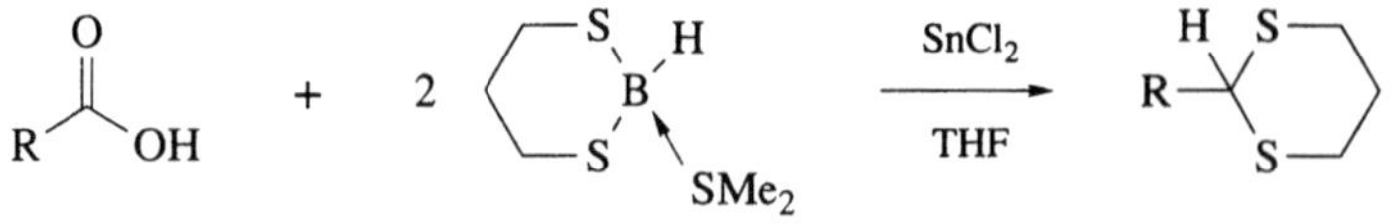

(2)

drosilanes can reduce carbonyl compounds or can react with acids because of the activation of the hydrogen in the hypervalent state.[18] In pursuing this work, it was found that the hydrosilyl ester (**5**) on heating to 110–160 °C gave aldehydes in 50–90% yields (reaction (c) in Scheme 2). It was not necessary to isolate the ester (**5**).[19] From examination of similar dihydrosilanes, the best conversion was given by compound (**6**); the importance of internal coordination of nitrogen was demonstrated by the failure of compound (**7**) to provide more than a trace of aldehyde under similar reaction conditions. α,β-Unsaturated acids can be reduced by the dihydrosilane (**6**) to the aldehyde without reduction of the double bond. The molecular weight and cost of the dihydrosilane (**6**) would appear to make it unsuitable for large scale work, whilst the need for relatively high temperatures is not suitable for relatively unstable compounds. This last drawback is obviated through the use of an acid chloride (see Section 1.11.3).

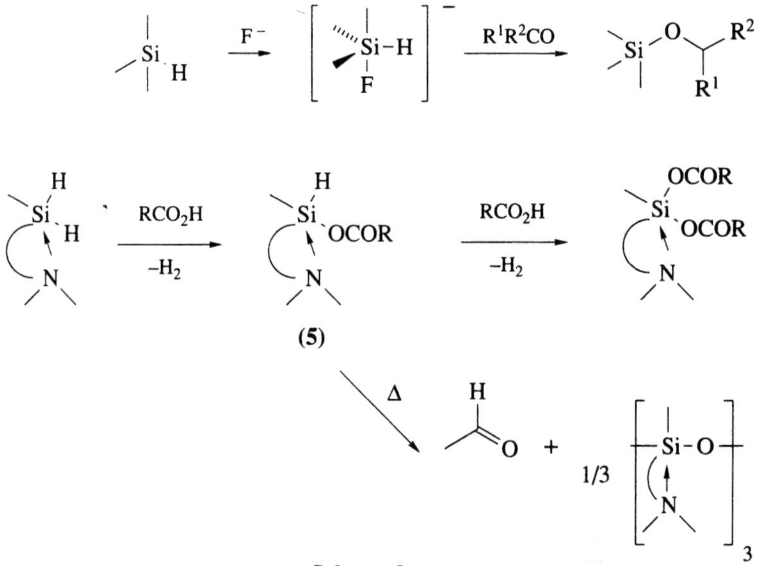

Scheme 2

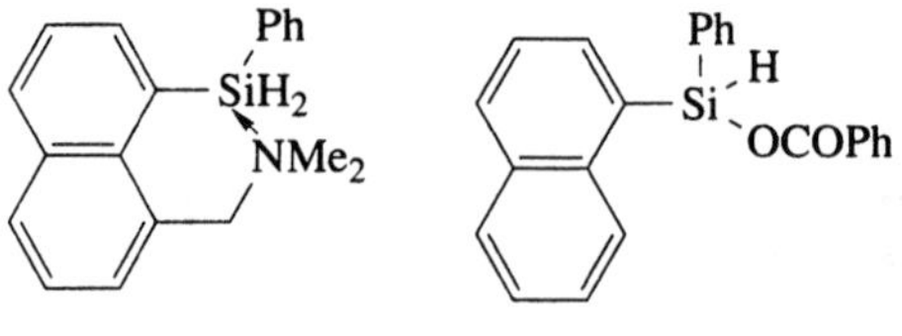

1.11.3 ACYL HALIDES

Direct displacement of chloride in acyl chlorides by hydride has been used to prepare aldehydes in 25–50% yield through the use of lithium hydride.[20] It was necessary to reflux the acid chloride in benzene or toluene with freshly prepared lithium hydride for 4–20 h. Under these conditions, there is the distinct likelihood that, when formed, aldehydes with α-hydrogen atoms would undergo aldol condensation. Sodium and calcium hydrides afforded only traces of aldehyde.

Lithium aluminum hydride is much too powerful a reducing agent to stop the reduction of acyl halides at the aldehyde stage and further reduction to alcohol is the usual result. In a preliminary communication, it was shown that even lithium trimethoxyaluminum hydride, in which the sole hydride is much less reactive than the hydrides in lithium aluminum hydride, was too powerful a reducing agent for acyl halides but lithium tri-*t*-butoxyaluminum hydride could give aldehydes.[21] This report was confirmed in two subsequent publications in which the reduction of a variety of acid chlorides in diglyme at –75 to –80 °C was described.[22,23] Aromatic acyl halides gave 60–80% yields of aldehyde but aliphatic compounds afforded only 40–60% yields. Very slow inverse addition of the hydride to the acid chloride is recommended at the very low temperatures involved. The method is convenient, particularly for aromatic aldehydes, since the reagent can be prepared *in situ* from lithium aluminum hydride and *t*-butyl alcohol, albeit with loss of three hydride equivalents. Lithium tri-*t*-amyloxyaluminum hydride gave much inferior conversions to aldehyde. The usefulness of lithium tri-*t*-butoxyaluminum hydride was confirmed in the formation of 3,5-dinitrobenzaldehyde in 60–63% yield from the acid chloride. The authors report that this hydride, with proper control, does not affect oxides, esters, acetals, nitriles and lactones.[25]

In the boron series, an analog of the alkoxyaluminum hydrides, *viz.* sodium trimethoxyborohydride (sodium trimethoxyhydroborate), has been shown to give aldehydes from acid chlorides but only in poor yield.[24] As an exception, other workers found that the same reducing agent at –80 °C afforded the aldehyde (**8**) in 85% yield from the corresponding acid chloride.[26]

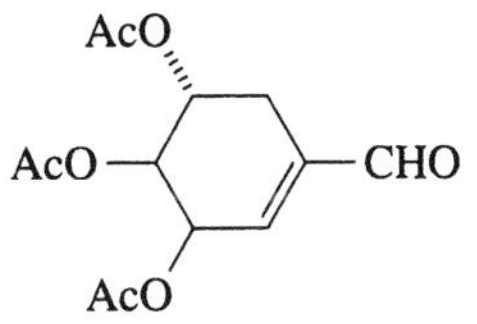

AcO, AcO, AcO, CHO

(8)

O_2N, Cl, SMe_2, H, $O-BH_2$

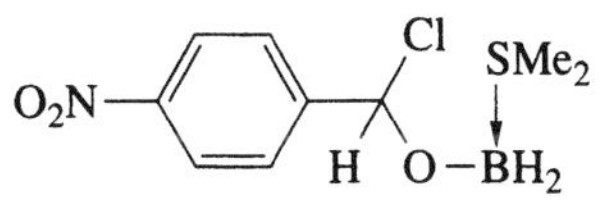

(9)

Diborane reduces acyl halides to alcohols as does thexylborane, but very slowly,[8,9] and disiamylborane has no effect.[27] Complexing diborane with dimethyl sulfide has a marked effect on its reactivity, but it has been found that acyl halides are still reduced to the alcohol stage, with the one exception of 4-nitrobenzoyl chloride which gave a 60% yield of 4-nitrobenzaldehyde.[28] This effect of an electron-withdrawing substituent is observed with other reducing agents described later and is probably caused by the substituent counteracting the loss of the leaving group (Cl^-), thereby stabilizing the intermediate (**9**) and preventing its rapid collapse to aldehyde.

Sodium borohydride and its derivatives have been examined extensively as convenient means of converting acyl halides to aldehydes. The borohydride itself or in polyethylene glycol[29] reduces the halides to alcohols, but, if care is taken to utilize only one of its hydrides, successful reduction to the aldehyde can be achieved. In one of the earliest modifications of sodium borohydride, it was mixed with cadmium chloride and dimethylformamide to give a reagent which reduced acid chlorides to aldehydes in good yield in 5 min at 0 °C.[30] The reaction is believed to involve the bidentate borohydride (**10**) and it was suggested that it proceeded through an intermediate iminium species (**11**; equation 3).[31] This suggestion was followed up by other workers who synthesized the iminium species directly from the carboxylic acid (equation 4), and, in a one-pot reaction, reduced compounds (**11**) to aldehydes in 70–90% yield using lithium tri-*t*-butoxyaluminum hydride in the presence of copper(I) iodide at –78 °C (see Section 1.11.4).[32] The reaction in equation (3) aroused the attention of other workers who demonstrated that selective reduction could be effected in dimethylformamide in the absence of the cadmium salt if the reaction was carried out at low temperatures and then quenched with acid and ethyl vinyl ether. It was surmised that liberation of diborane on quenching the reaction mixture was the cause of overreduction to the alcohol.[33] The method of quenching left difficulties with the work-up of reaction products and, in a second publication,[34] it was shown that the addition of pyridine as a diborane scavenger allowed sodium borohydride to be used at 0 °C to reduce acyl chlorides in only a few minutes. Small amounts of alcohol (5–10%) were observed with very good yields of aldehyde.

H_2B, H, H, Cd, H, H, BH_2

(10)

$$\mathrm{RCOCl} + \mathrm{HCONMe_2} \longrightarrow \left[\mathrm{RC(O)OCH{=}\overset{+}{N}Me_2}\ \mathrm{Cl^-}\right]\ \textbf{(11)} \xrightarrow[-\mathrm{Cl^-}]{+\mathrm{BH_4^-}} \mathrm{RCHO} + \mathrm{HCONMe_2} + \mathrm{BH_3} \quad (3)$$

$$\mathrm{RCO_2H} + \mathrm{ClCH{=}\overset{+}{N}Me_2\ Cl^-} \xrightarrow{\text{i}} \left[\mathrm{RC(O)OCH{=}\overset{+}{N}Me_2}\ \mathrm{Cl^-}\right]\ \textbf{(11)} \xrightarrow{\text{ii}} \mathrm{RCHO} + \mathrm{HCONMe_2} \quad (4)$$

i, pyridine, –30 °C; ii, $LiAlH(OBu^t)_3$, CuI, –78 °C

Contemporary with the use of $CdCl_2$ as a moderator of borohydride reducing activity, other reports appeared on the use of copper salts. Bis(triphenylphosphine)copper borohydride (bis(triphenylphosphine)tetrahydroboratocopper) was used in acetone to reduce acid chlorides to aldehydes in 2–15 min at 25 °C (equation 5).[35] Recovery of tris(triphenylphosphine)copper chloride is relatively straightforward but extra triphenylphosphine must be added to the reaction mixture, making large scale use inconvenient. This drawback was recognized in a further publication,[36] in which was reported the same reaction, except that trimethoxyphosphine or triisopropoxyphosphine ligands were used in place of triphenylphosphine in an attempt to reduce the molecular weight of the reagent. Reduction of acid chlorides was fast, but the extra, more soluble ligand made work-up to isolate the aldehyde difficult. An advantage of this copper reagent is its stated inertness to all functional groups except acid halides and iminium salts. At the same time that this work appeared, other workers reported the same reaction with bis(triphenylphosphine)copper borohydride to give 60–85% yields of a range of aldehydes at room temperature;[37] a further publication described a simple method for recycling the copper salt.[38] Later, it was found that, whereas the moderated hydride, sodium cyanoborohydride, was still active enough to reduce acid chlorides to the alcohol stage, μ-bis(cyanotrihydroborato)-tetrakis(triphenylphosphine)dicopper reduced them to aldehydes in 50–90% yields.[39] Acidity in the reaction medium was important because, at pH 3, this same copper reagent caused overreduction to the alcohol.

$$\mathrm{(Ph_3P)_2CuBH_4} + \mathrm{RCOCl} \xrightarrow{\mathrm{2PPh_3}} \mathrm{(Ph_3P)_3CuCl} + \mathrm{RCHO} + \mathrm{Ph_3PBH_3} \quad (5)$$

$$\mathrm{R^1COX} + \mathrm{R^2{}_3SnH\ (in\ excess)} \longrightarrow x\ \mathrm{R^1CHO} + \frac{(1-x)}{2}\ \mathrm{R^1CO_2CH_2R^1}\ \textbf{(12)} + \mathrm{R^2{}_3SnX} \quad (6)$$

X = F, Cl, Br; R^2 = Ph, Bu^n

Tin hydrides or stannanes have been known to form aldehydes from acyl halides for many years. Acid chlorides are reduced to aldehydes in poor to modest yields by triphenyltin hydride (triphenylstannane),[40] but it has been found that tri-*n*-butyltin hydride (tri-*n*-butylstannane) was a better, more convenient reagent.[41] A second product from these reductions is the ester (**12**; equation 6), notionally formed by acylation of the alcohol produced by overreduction. In some reactions the ester forms the major product, particularly with triphenyltin hydride.[42] The ratio of aldehyde to ester appears to depend on the type of acid halide, on the experimental conditions (neat liquids or in solution) and on which stannane is used.[43–45] Thus, acid bromides with tri-*n*-butyltin hydride give good to high yields of aldehyde but acid chlorides with trimethyltin hydride afford mainly esters.[45] Acid fluorides with either hydride give esters. The yield of aldehyde is increased as the group attached to the acid chloride becomes larger or branched and solvent exerts a profound effect.[43] The reduction has been considered to be a radical process, with the initial formation of acyl radicals followed by hydrogen abstraction or coupling and hydride extraction (Scheme 3).[44] This view of the reaction has been accepted generally,[46] but more recent evidence has indicated that radicals are not involved.[47] This reassessment of the reaction revealed that the product distributions from carefully chosen acid chlorides were not those expected (from other work) if acyl radicals were important. Further, intentional addition of a radical initiator altered the product distributions so that they became in keeping with those expected for a radical reaction. It was thought that an addition compound (**13**; Scheme 4) is formed first and then undergoes further reaction, as shown.

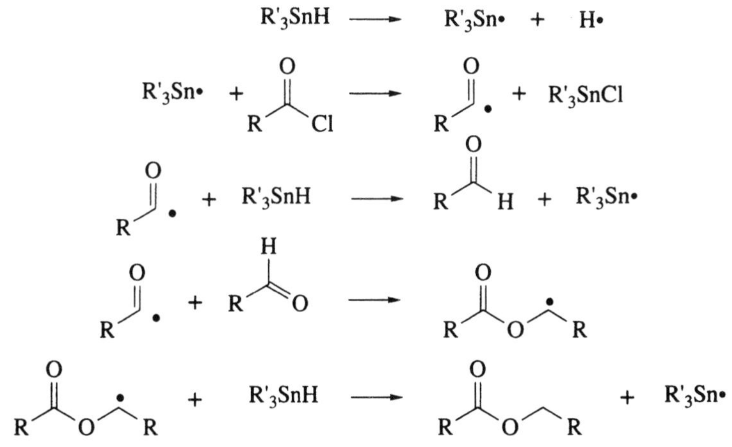

Scheme 3

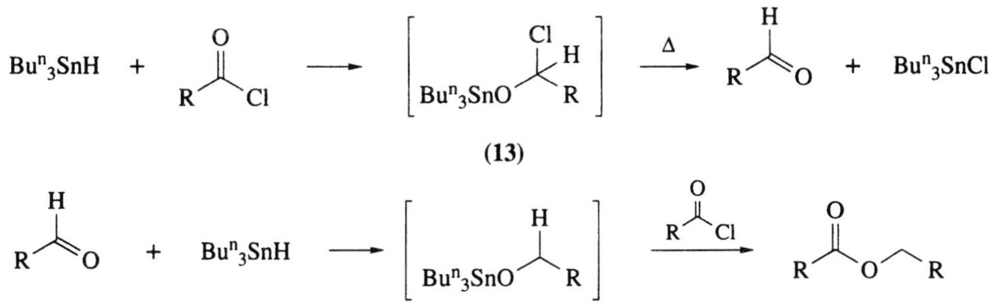

(**13**)

Scheme 4

A significant improvement in this reaction was the discovery that it could be catalyzed by palladium to give aldehydes in 75–95% yield; in the absence of catalyst yields of only about 10% were observed.[48,49] Tri-*n*-butyltin hydride with tetrakis(triphenylphosphine)palladium were the preferred reagents. It was noted that work-up can be troublesome with regard to removal of tri-*n*-butyltin halide. The reaction is claimed to be better than the closely analogous reaction of acyl chlorides with trialkylsilicon hydrides (see below). This metal-catalyzed reduction is also generally considered not to be a radical process.[50]

Development of a comparable reduction of acyl chlorides by silicon hydrides (silanes) began even earlier than that of the tin hydrides. It was found that acid chlorides could be reduced to aldehydes in modest yields with a variety of silanes but best results were obtained with triethylsilicon hydride (triethylsilane).[51] This reaction was much improved by use of a 10% palladium-on-charcoal catalyst which provided 30–70% yields of aldehydes.[52] Unbranched aliphatic acyl chlorides were reduced satisfactorily but, unlike the later-developed stannane reaction described above,[48,49] α-branched chlorides gave poor or no yields of aldehydes; aromatic acyl halides gave poorer results than did the aliphatic analogs. The palladium-catalyzed reduction with silanes, carried out at ambient or higher temperatures, was considered to be a better process than reduction with lithium tri-*t*-butoxyaluminum hydride[21–23] for aliphatic compounds but not for aromatics. There were difficulties in work-up of the silane reaction in that the product aldehydes could not be distilled directly from the reaction mixture because of their destruction if silanes or silyl chlorides were present. Other workers have used *cis*-$[PtCl_2(PPh_3)_2]$ and various Rh complexes as catalysts in this reaction but were able to isolate only poor quantities of aldehydes. Indeed, ketones formed significant reaction products under these conditions.[53]

In an extension of the high temperature reaction of hypervalent silanes with carboxylic acids to produce aldehydes (see Section 1.11.2),[19] it was found that acyl chlorides reacted at room temperature with compound (**14**) to give 85–95% yields of aldehydes.[54] Halogen, methoxy, nitroaryl and heteroaryl groups and carbon–carbon double bonds were unaffected; dialdehydes could be produced as readily as monoaldehydes. During work-up, it was found necessary to trap the resulting silyl chloride on wet silica, or in other ways, because its presence severely reduced yields of the aldehyde during distillation, as found with the silane and stannane reductions described above.

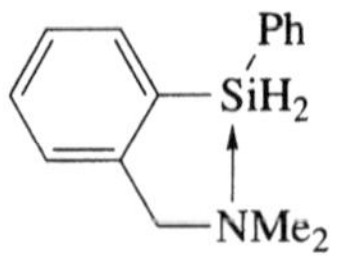

Ph SiH$_2$ NMe$_2$

(**14**)

1.11.4 ESTERS OF CARBOXYLIC ACIDS (INCLUDING ORTHO ESTERS)

Hydrides based on aluminum have proved most versatile and selective for the reduction of esters (RCO_2R^1) to aldehydes (RCHO), but it should be noted that the nature of R and R^1 (aliphatic, aromatic, branched, electron-withdrawing or electron-donating, steric size) is important. Most of the reactions described here have been carried out on simple methyl or ethyl esters. It is frequently found that aliphatic esters give much better yields than do aromatic esters. Reductions are carried out at very low temperatures such that esters are still reactive to the hydride but any aldehyde produced is not. Alternatively, the intermediate formed by addition of metal hydride to the ester (**15**; Scheme 5) may be sufficiently stable that no significant amount of aldehyde is formed. Thus, if the rate constant $k_1 > k_2$ then little aldehyde will be released, and no aldehyde formed even if $k_3 > k_1$. It also seems likely that the difference in the Lewis basicity of esters and aldehydes affects the complex of rate constants shown as k_1, k_2 and k_3, and hence affects the relative reactivity of these two functional groups towards the different hydride reagents.

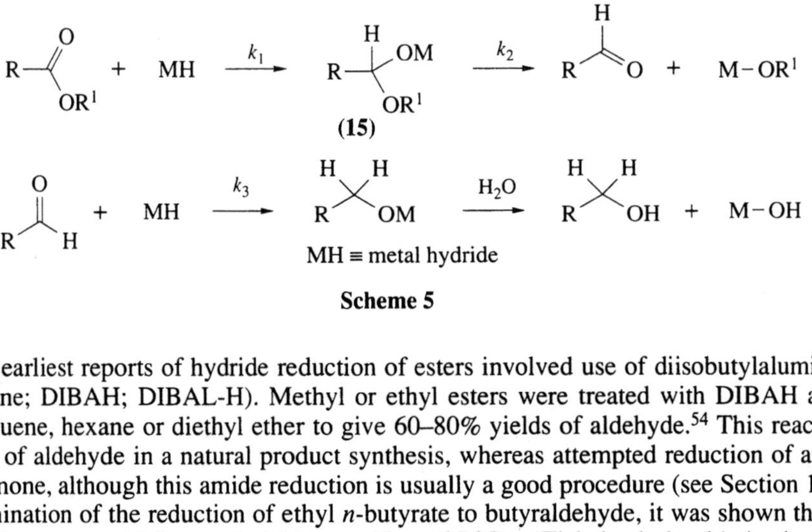

$$RC(O)OR^1 + MH \xrightarrow{k_1} RCH(OM)OR^1 \ (\mathbf{15}) \xrightarrow{k_2} RCHO + M{-}OR^1$$

$$RCHO + MH \xrightarrow{k_3} RCH_2OM \xrightarrow{H_2O} RCH_2OH + M{-}OH$$

MH ≡ metal hydride

Scheme 5

One of the earliest reports of hydride reduction of esters involved use of diisobutylaluminum hydride (diisobutylalane; DIBAH; DIBAL-H). Methyl or ethyl esters were treated with DIBAH at –70 °C for 0.5–1 h in toluene, hexane or diethyl ether to give 60–80% yields of aldehyde.[54] This reaction afforded an 83% yield of aldehyde in a natural product synthesis, whereas attempted reduction of an amide with $LiAlH_4$ gave none, although this amide reduction is usually a good procedure (see Section 1.11.6).[55] In a detailed examination of the reduction of ethyl *n*-butyrate to butyraldehyde, it was shown that a temperature of –70 °C was essential for good yields; at –45 to –35 °C, a 'Tishchenko' oxidation/reduction led to reduced yields of aldehyde and the formation of isobutyl esters.[56] In more recent work, benzyloxycarbonyl derivatives of α-amino acid methyl or ethyl esters were reacted with DIBAH in toluene or tetrahydrofuran at –50 °C to give modest to good yields of α-amino aldehydes.[57] Importantly, these easily racemized aldehydes were isolated in high optical purity. In another useful application, DIBAH was used to reduce methyl esters to aldehydes selectively in the presence of *t*-butyl esters, an illustration of the part played by steric effects.[58] It was noted that no other hydride reducing agent could achieve the same selectivity. The use of DIBAH (and triisobutylaluminum) has been reviewed in detail.[59]

To increase the selectivity of DIBAH, it has been converted into amine derivatives, $HAlX_2$ (**1**), with X = *N*-methylpiperidino being the preferred reagent.[60] Yields of aldehydes of 60–80% from both aliphatic and aromatic esters were obtained on refluxing them with bis(*N*-methylpiperidino)aluminum hydride (bis(*N*-methylpiperidino)alane; BMPA) in tetrahydrofuran for 6–20 h; an excess of reagent can be tolerated. The higher temperatures are a distinct advantage because difficulties with the solubility of substrate, as occur at the very low temperatures required in the use of DIBAH itself, do not usually arise. It should be noted that the authors believe that the reaction proceeds through the sequence shown in Scheme 6 so that, if the reaction is stopped too early, amides and not aldehydes are isolated.

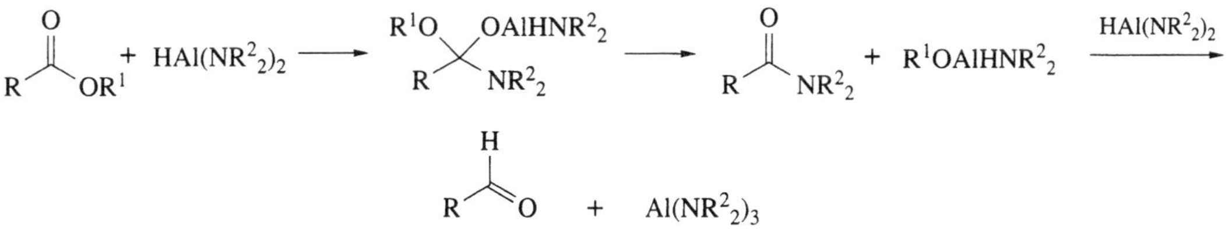

$$RC(O)OR^1 + HAl(NR^2_2)_2 \longrightarrow R(R^1O)C(OAlHNR^2_2)(NR^2_2) \longrightarrow RC(O)NR^2_2 + R^1OAlHNR^2_2 \xrightarrow{HAl(NR^2_2)_2} RCHO + Al(NR^2_2)_3$$

Scheme 6

At about the same time that DIBAH was developed as a selective reducing agent for the production of aldehydes from esters, the same workers[54] examined the use of sodium aluminum hydride (sodium tetrahydridoaluminate) for this purpose.[61] In tetrahydrofuran or pyridine at –65 to –45 °C, aliphatic and aromatic methyl or ethyl esters gave aldehydes in 50–80% yield. Better yields for aliphatics were found and aromatics required much longer reaction times. Control of temperature was crucial, and, if not controlled, led to sharp decreases in yields of aldehyde. A major difficulty with sodium aluminum hydride has been the low solubility of both it and the reduction intermediates at the low temperatures necessary for selective reduction to the aldehyde. In a development of this reagent, it was shown that sodium bis(2-methoxyethoxy)aluminum hydride (sodium bis(2-methoxyethoxy)dihydridoaluminate; Vitride; Red-Al) was more convenient and provided yields of about 40–90% of the aldehyde for a range of methyl esters at –50 to –70 °C; below –50 °C, aldehydes are reported to be almost unreactive towards this hydride reagent.[62] Its further modification by addition of one equivalent of *N*-methylpiperidine gives sodium bis(2-methoxyethoxy)-*N*-methylpiperidinoaluminum hydride, which is suitable for reducing aliphatic, aromatic and α,β-unsaturated esters in yields of 50–85%.[63]

Modified lithium aluminum hydride has been used successfully for the reduction of esters at temperatures of about 0 °C. Thus, lithium tri-*t*-butoxyaluminum hydride readily reduces phenyl esters of carboxylic acids to aldehydes in 33–77% yields; other esters are reported to be unreactive, as are many other functional groups (acyl chlorides react with the same reagent at –70 °C, however).[64,65] Phenylbenzoate and phenyl cyclopropanecarboxylate do not give the aldehyde. Iminium salt esters (**11**; equation 4) can be reduced with lithium tri-*t*-butoxyaluminum hydride (see Section 1.11.3).[32]

Reduction of carboxylic acid esters and ortho esters to acetals or ketals can be effected in high yield even with vigorous reducing agents. The ortho ester (**16**; Scheme 7), formed from 2-cyanoethanethiol, was reduced in refluxing benzene by using 0.25 molar equiv. of lithium aluminum hydride to give the acetal (**17**) in 97% yield, which could be converted into the aldehyde, also in excellent yield.[66] The triethyl ortho ester of benzoic acid was reduced to benzaldehyde diethylacetal with DIBAH at 30 °C over a period of 1 h. By raising the temperature to 70–80 °C, further reduction to isobutyl ethyl ether occurred.[67]

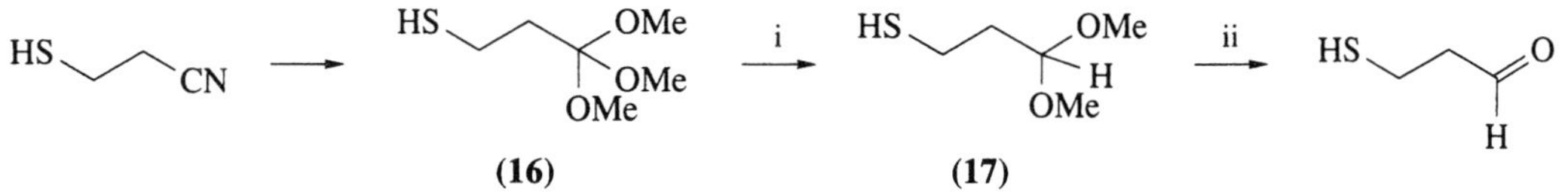

$$HSCH_2CH_2CN \longrightarrow HSCH_2CH_2C(OMe)_3\ \textbf{(16)} \xrightarrow{i} HSCH_2CH_2CH(OMe)_2\ \textbf{(17)} \xrightarrow{ii} HSCH_2CH_2CHO$$

i, 1/4 $LiAlH_4$; ii, Rochelle salt

Scheme 7

In a novel reaction, the ester (**18**; Scheme 8) was heated with $NaBH_4$ in pyridine to 115 °C. The reaction was stopped at the ketal stage (**19**) so that overreduction to the alcohol was not possible.[68] Overall yields of about 50% were reported but it might be noted that the other product of the reaction was the pyridine/borane adduct $C_5H_5N{\cdot}BH_3$, so that only a quarter of the available hydride was used in the reduction.

With the strongly electron-withdrawing group in ester (**20**; equation 7), reaction with sodium borohydride afforded the hemiacetal (**21**), although normally this hydride does not react with esters.[69] In some cases, electron-withdrawing groups allow isolation of the aldehydes after reaction with lithium aluminum hydride, such as with trichloropentadienoic esters[70] and chloroepoxy-α,β-unsaturated esters.[71]

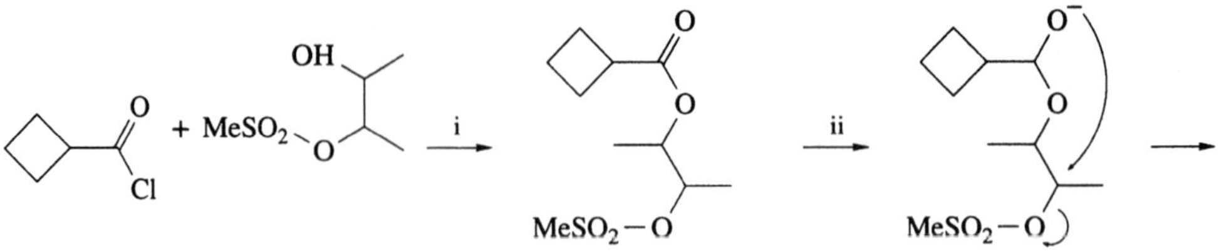

i, pyridine; ii, $NaBH_4$/pyridine/115 °C; iii, aq. acid

Scheme 8

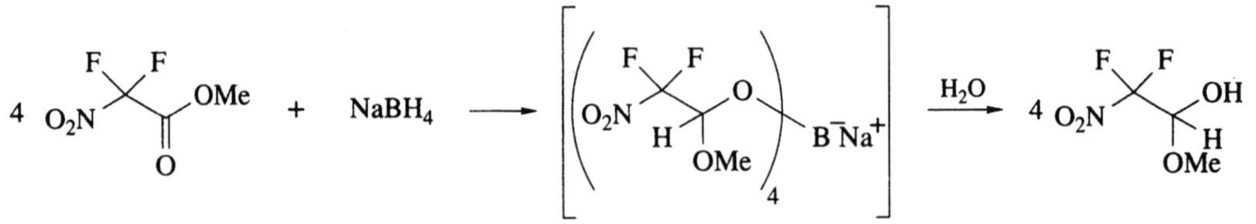

(20) (21) (7)

Finally, thione thiolates (**22**) have been reduced to the dithioacetals (**23**) in almost quantitative yield with tri-*n*-butyltin hydride under the action of ultraviolet light.[72]

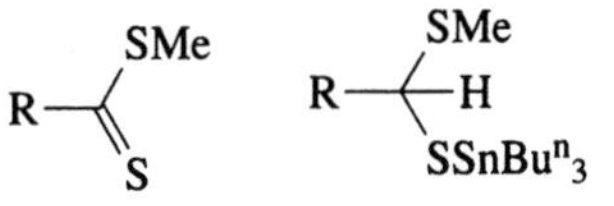

(22) (23)

1.11.5 LACTONES

Normally, lithium aluminum hydride is too vigorous a reagent for the reduction of lactones to the hemiacetal (hydroxyaldehyde; lactol) stage, but, in certain circumstances under carefully controlled conditions, it can be used successfully. Usually, lactones give diols with $LiAlH_4$ but, with 0.25 equiv., hemiacetals have been prepared from γ-lactones in high yield (Scheme 9).[73,74] Sodium bis(2-methoxyethoxy)aluminum dihydride (Red-Al) has been modified further by addition of one molar equivalent of ethanol to give sodium bis(2-methoxyethoxy)ethoxyaluminum hydride, a reagent which is reported to reduce γ-lactones to lactols in 84–94% yield in 25–60 min at 0 °C; one δ-lactone required a temperature of –60 °C and gave only 64% of the lactol.[63] Without the addition of ethanol, poor yields of lactol were obtained in uncontrollable reactions.

hemiacetal (lactol) hydroxy aldehyde diol

i, 1/4 $LiAlH_4$; ii, excess of $LiAlH_4$

Scheme 9

In uncontrolled reductions, lactones are reduced by lithium trialkyoxyaluminum hydrides to diols but the reduction with $LiAlH(OBu^t)_3$ is very slow and advantage can be taken of this to stop the reaction at the lactol stage. The first reported examples using this reagent showed that δ-lactones could be converted into lactols in high yield at about –60 °C.[75–77] A later publication suggested that even better conversions may be possible through the use of preformed, pure $LiAlH(OBu^t)_3$ rather than by its preparation first *in situ*.[78] This area has been reviewed.[79]

Reduction of lactones to lactols with diisobutylaluminum hydride has become an important, high-yielding reaction in the wake of many syntheses of prostaglandins.[59] First introduced in this role in 1969,[80] DIBAH can give 98% yields of lactols from γ-lactones by reaction at –60 °C for 20 min in toluene.[81,82]

Diborane has been used to reduce lactones to cyclic hemiacetals and, when boron trifluoride etherate was added, ethoxy compounds (**24**; Scheme 10) were isolated.[83] It was later shown that yields of hemiacetals in excess of 80% were obtainable with short reaction times.[84]

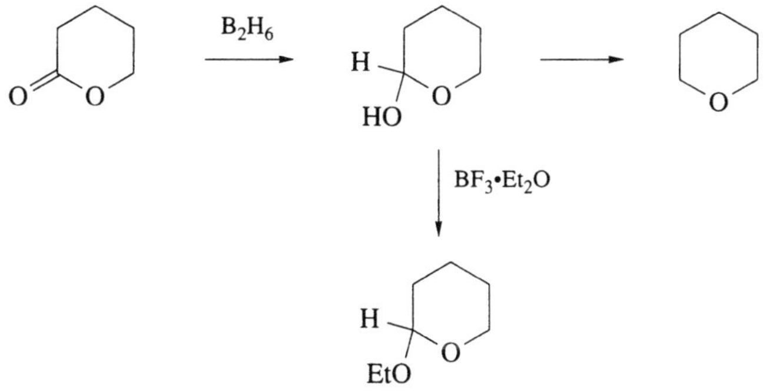

B_2H_6

H HO O

BF_3•Et_2O

H EtO O

(24)

Scheme 10

Sodium borohydride in water at pH 3 has been effective in reducing aldonic lactones to aldoses in approximately 65% yield.[85] Thus, at 0 °C, D-gluco-D-guloheptano-γ-lactone gave D-gluco-D-guloheptose on simultaneous dropwise additions of aqueous $NaBH_4$ and dilute H_2SO_4. Similar reductions of other lactones confirmed this convenient method for the production of aldoses.[86]

Although thexylborane reduces lactones slowly,[9,87] it appears to have been little used for the purpose of synthesizing lactols; however, its close analog disiamylborane (bis(3-methyl-2-butyl)borane) has been used successfully. Simple reduction of a variety of lactones to hydroxy aldehydes or hemiacetals (Scheme 9) appears to be quite general for disiamylborane.[27] At –10 °C, this reagent was found to reduce tetraacylhexano-γ-lactones to tetraacylhexanofuranoses in 100% yield.[88] Similarly, unprotected sugars could be reduced, as with the conversion of D-erythronolactone into D-erythrose in good yield.[89] In a comparison of the rates of reduction of lactones to lactols and their yields, it has been shown that disiamylborane is superior to other analogous boranes either more or less sterically demanding. For example, disiamylborane reduces acylated aldonolactones to acylated aldoses in 100% yield at 45 °C in 6 h, but bis-2-(2,3-dimethylbutyl)borane only gives a poor yield over a long period at the same temperature.[90] Confirmation of the excellent yields afforded by disiamylborane came in the quantitative reduction of a γ-lactone to a lactol.[91] Whilst this reagent gives such good yields, it is worth bearing in mind that it is also an excellent hydroborating reagent, reacting readily with double and triple bonds. It also reacts with alcohols, phenols, thiols and amines to liberate hydrogen rapidly, and such groups would need to be protected if the use of excessive amounts of disiamylborane is to be avoided.

1.11.6 AMIDES OF CARBOXYLIC ACIDS

The use of lithium aluminum hydride for the reduction of carboxylic acid derivatives to aldehydes goes back to some of the earliest-reported applications of this reagent.[92] An example of this type of reduction appeared when it was reported[93] that the piperidide of cyclopropanecarboxylic acid could be reduced to cyclopropanecarbaldehyde in moderate to good yield. The following few years saw considerable interest in the reduction of amides to aldehydes and many types of amide were investigated with a view to optimizing yields. One of the difficulties with this reaction lies in there being two other possible products of reduction, *viz.* amines and alcohols (see later, Scheme 11). In an attempt to reduce the dipiperidide of phthalic acid with $LiAlH_4$, only 20% yield of the dialdehyde was obtained, but, on changing to the dimethylamide, the yield of phthalaldehyde increased to 70%.[94] However, it was later demonstrated that the piperidide of cyclohexanecarboxylic acid could be reduced to the aldehyde in good yield if the lithium aluminum hydride was added to the amide (inverse addition), instead of the usual addition of substrate to reducing agent.[95] Although inverse addition became the standard procedure in this reaction for the best yields, interest in piperidides gave way to other tertiary amides. In a preliminary communication, it was reported that *N*-methylanilides of carboxylic acids in tetrahydrofuran with inverse addition of solid $LiAlH_4$ afforded 54–75% yields of aldehydes, including dialdehydes.[96] In following up this work, a much wider range of aldehydes was synthesized, thereby demonstrating the general usefulness of the reaction in the absence of other functional groups reducible by $LiAlH_4$.[97]

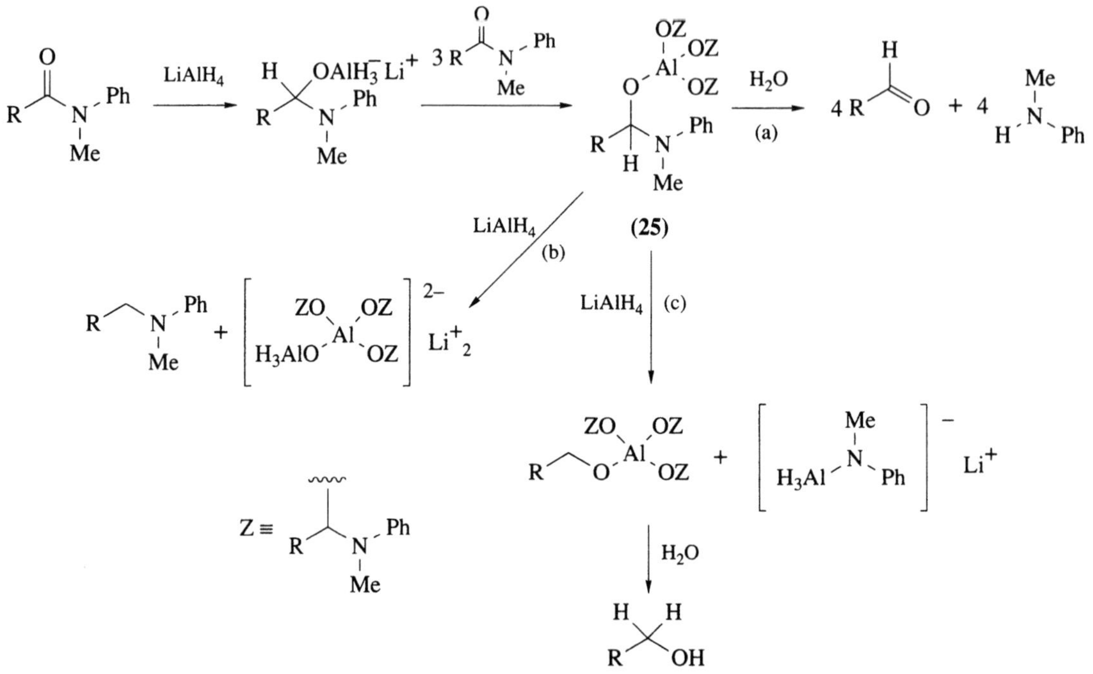

Scheme 11

The overall mechanism of the reaction has been discussed (Scheme 11).[97] For successful production of the aldehyde, the reaction needs to stop at the stage represented by structure (**25**; Scheme 11) and inverse addition of exactly the required amount of $LiAlH_4$ helps achieve this step; subsequent hydrolysis (step a) affords the aldehyde. If an excess of $LiAlH_4$ is used, or if the rates for steps b and c are greater than the rate of formation of structure (**25**), then alcohols and/or amines are formed (Scheme 11). The type of amine used to prepare the amide plays a considerable role in determining the stability of structure (**25**), *viz.* in determining whether steps b and c are faster or slower than the rate of formation of (**25**). The effects of the amine are apparently both steric and electronic. These factors were examined extensively when it was shown that bulky amines favored aldehyde formation (or stability in structure **25**), as did delocalization of the amide nitrogen lone pair electron density away from the amide and into an aromatic structure.[98] Especially good yields of aldehydes were obtained with acylated pyrroles, indoles or carbazoles. A fuller mechanistic study of the reduction of *N*-methylanilides with $LiAlH_4$ indicated that the reaction sequence shown in Scheme 11 is probably correct. Whereas electron-withdrawing substituents in the phenyl ring of the *N*-methylanilide decreased the rates for steps b and c, similar substituents in the phenyl ring of the benzoic acid from which the amides were made had little or no effect.[99] This reaction has been exploited as a general method for the production of aromatic aldehydes. Reaction of arenes with chloroformyl-*N*-methylanilide (**26**; Scheme 12) gave the corresponding arylcarboxy-*N*-methylanilides which were reduced to arene aldehydes with $LiAlH_4$.[100] Dialdehydes could be synthesized and some overreduction to the alcohol was observed.

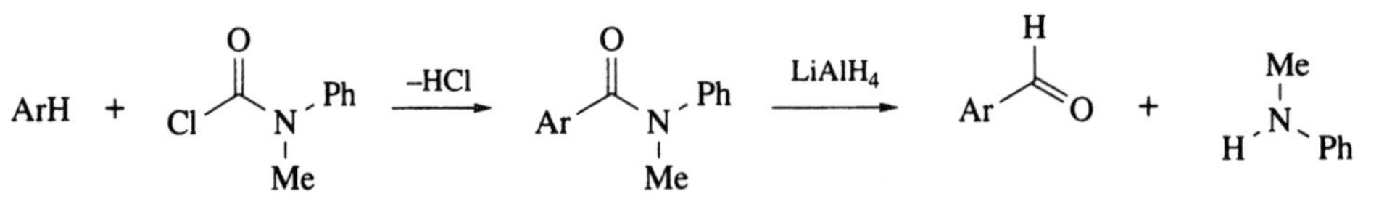

Scheme 12

The ideas for delocalization of nitrogen lone pair electron density into an aromatic or heteroaromatic system were pursued through reduction of acylated pyrazoles and imidazoles to aldehydes in high yield. 3,5-Dimethyl-*N*-acylpyrazoles are easy to prepare and afford 77–96% yields of aldehydes with $LiAlH_4$ in diethyl ether at 0 °C.[101] Further examples of this reaction have appeared.[102,103] Although these later publications commented unfavorably on the ability of $LiAlH_4$ to reduce acyl imidazoles to aldehydes (low yields), other workers have demonstrated that yields of 60–80% could be attained at temperatures of –20 to +20 °C in diethyl ether.[104] It was considered that the earlier failure may have been caused by the presence of impurities in the acyl imidazoles. The latter are easy to prepare from the parent carboxylic acid and *N,N′*-carbonyldiimidazole.

Unexpectedly, aziridine was found to be particularly good for the preparation of amides which could be reduced with $LiAlH_4$ in high yield to aldehydes. With butyric acid, the yields of butyraldehyde from various amides were: dimethylamide 25%, diethylamide 22%, diisopropylamide no reaction, *N*-methylphenylamide 58%, piperidide 33%, pyrrolidide 16% and aziridide 88%.[105] A later report indicated that yields of 40–80% of the aldehyde could be attained even with a 100% excess of $LiAlH_4$ if aziridides were used.[106]

With primary or secondary amides, reaction with $LiAlH_4$ is wasteful in that hydrogen gas is evolved and reduction usually proceeds through to the amine. Nevertheless, it has been shown that *t*-butylamides, refluxed with an excess of $LiAlD_4$ for 15 h in diethyl ether, give very good yields of deuteriated aldehydes, RCDO.[107]

Sodium aluminum hydride and its alkoxy derivatives have been used in place of lithium aluminum hydride.[108] In a survey of various types of amide it was found that, with $NaAlH_4$, dimethylamides were most convenient, giving 72–92% yields of aldehydes, with aromatic amides giving better results than aliphatics. Although sodium tri-*t*-butoxyaluminum hydride was more selective for the reduction of amides (being very slow acting), it was pointed out that the use of $NaAlH_4$ was more convenient and less expensive. Sodium bis(2-methoxyethoxy)aluminum dihydride (Red-Al) is readily available and has been used to reduce acyl saccharides to aldehydes (Scheme 13). Chlorosaccharin (**27**; Scheme 13) reacts with acids to give the amide (**28**).[109] Alternatively, the same amide can be prepared simply by melting together an acid chloride and sodiosaccharin (**29**).[110] The amide (**28**) has been reduced at 0–5 °C over 5–10 min using Red-Al in benzene; yields of aldehydes reached 65–80%.[110] By utilizing the melt reaction to form the amide (**28**), the conversion of acid chlorides to aldehydes *via* the amide can be made a convenient one-pot reaction.

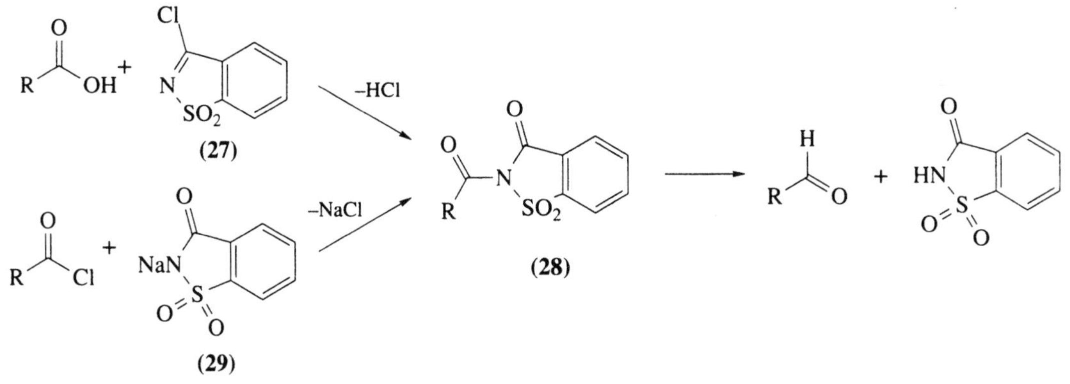

(**27**) (**28**) (**29**)

Scheme 13

In attempts to moderate $LiAlH_4$, it was reacted with ethanol or ethyl acetate to give either lithium diethoxyaluminum dihydride (lithium diethoxydihydridoaluminate) or lithium triethoxyaluminum hydride (lithium triethoxyhydridoaluminate).[111] Either of these hydrides when reacted with suitable amides afforded aldehydes in 70–90% yield.[112] It was pointed out that, whereas dimethylamides were not satisfactory with $LiAlH_4$ (see above), the ethoxyhydridoaluminates gave the best results with dimethylamides, although aziridides are better still. Since the latter are not so convenient and are more expensive than dimethylamides, it was concluded that the small increase in yields when using aziridides was not generally worth their use.[112]

It is well known that substitution of hydrogen in $LiAlH_4$ by alkoxy groups moderates its reducing activity and that substitution by alkyl groups increases its reducing capability. Thus, most emphasis on the formation of aldehydes from amides has concentrated on either the type of amide or on decreasing the reducing capacity of hydride reagents. Unusually, one high-yielding synthesis of aldehydes from amides utilizes a trialkyl-substituted $LiAlH_4$. By reacting diisobutylaluminum hydride with *n*-butyllithium (equation 8), lithium *n*-butyldiisobutylhydridoaluminate can be readily prepared.[113] At 0 °C for 2 h, or even at room temperature, this aluminate with diethylamides, pyrrolidides and piperidides gave approximately 95–98% yields of a variety of aldehydes.[113] It is probable that severe steric crowding around the aluminum effectively decreases the availability of the hydride ion so that the analog of the intermediate (**25**; Scheme 11) is not further attacked by more aluminate. It was noted that the reagent is not as selective as might be preferred in that acid chlorides, anhydrides, esters, lactones and thiol esters were all reduced to the alcohol stage; nitriles were reduced to aldehydes very slowly at room temperature.

$$\mathrm{HAlBu^i_2} + \mathrm{Bu^nLi} \longrightarrow \mathrm{LiAlHBu^nBu^i_2} \quad (8)$$

Just as considerable effort has been exerted to utilize $LiAlH_4$ or its derivatives for the selective reduction of amides to aldehydes, so too has much exertion gone into the use of diisobutylaluminum hydride. This area has been reviewed.[79,114] With DIBAH itself in diethyl ether at 0 °C, *N*-methylanilides furnish aldehydes in yields of 25–70%, the remainder of the reaction products being mostly amines.[115] The yield of aldehyde drops severely above 0 °C. DIBAH, tempered by the addition of various amines, has been discussed in Section 1.11.2 (ref. 5) and Section 1.11.4 (ref. 60). One of these reagents, bis(4-methyl-1-piperazinyl)aluminum hydride has been recommended for the reduction of dimethylamides (and other amides) to aldehydes in 65–80% yield.[116] The hydride is mild enough for it to be necessary to reflux the amide in diethyl ether with the reagent for several hours. It should be noted that under such conditions acids and esters will be reduced also.

Almost simultaneously, two groups of workers have reported the reduction of 3-acylthiazolidine-2-ones (**30**) to aldehydes with DIBAH (equation 9).[117–120] The acylthiazolidine compounds are made easily from an acyl chloride and thiazolidine-2-thione and then reacted with DIBAH in toluene or $LiAlH(OBu^t)_3$ in tetrahydrofuran at low temperatures; the end of the reduction is indicated by the disappearance of a yellow color. Hydrolysis of the resulting complex (**31**) with acid gives the aldehyde.[117,118] An almost identical reaction used DIBAH in hexane,[119] although initially the wrong structure was assigned to the starting materials.[120] Yields of aldehyde reported by both sets of workers vary between 50% and 90%. Amides can be reduced in the presence of ester functions but low reaction temperatures are necessary to prevent overreduction to the alcohol stage. The reaction appears suitable for aliphatic, aromatic and α,β-unsaturated amides.

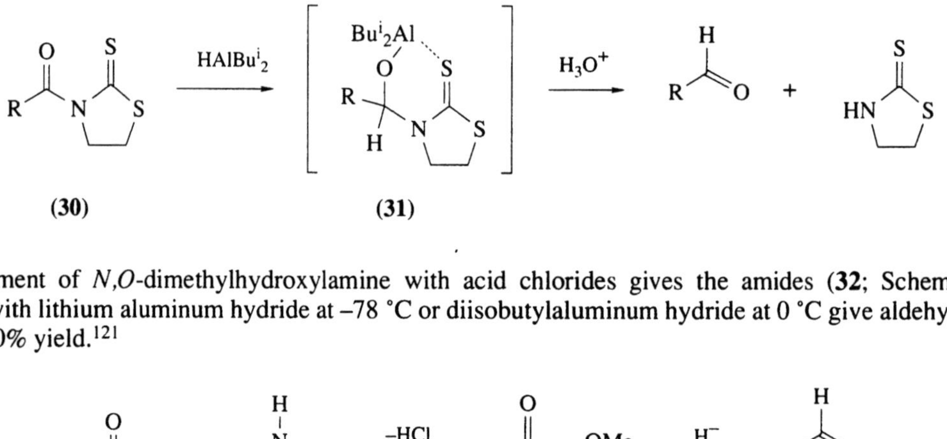

HAlBu$^i_2$; H_3O^+ ; R–CHO + HN(thiazolidine-2-thione) (9)

(**30**) (**31**)

Treatment of *N*,*O*-dimethylhydroxylamine with acid chlorides gives the amides (**32**; Scheme 14) which with lithium aluminum hydride at –78 °C or diisobutylaluminum hydride at 0 °C give aldehydes in about 70% yield.[121]

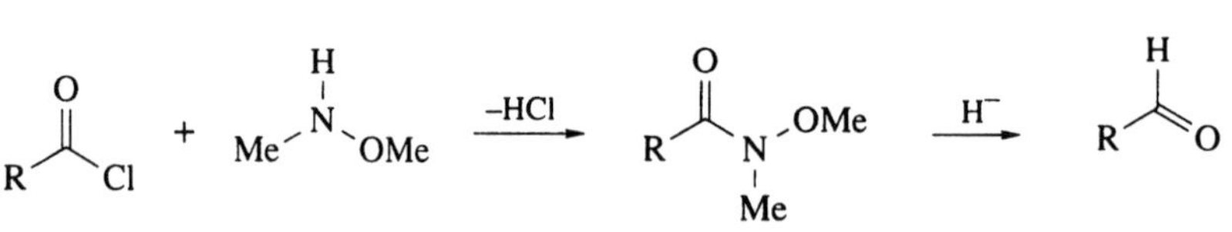

RCOCl + MeNHOMe —(–HCl)→ RCON(Me)OMe —(H^-)→ RCHO

(**32**)

Scheme 14

Amides have been converted into imidoyl chlorides and then reduced to aldimines with $LiAlH(OBu^t)_3$, as in Scheme 15.[122] Although not claimed as a synthesis of aldehydes, the aldimines can be hydrolyzed to aldehydes quite readily. Interestingly, the authors say that an excess of the reducing agent can be used because further reduction to the amine requires 24 h, whereas the first stage to the aldimine requires only 30 min at –78 °C.

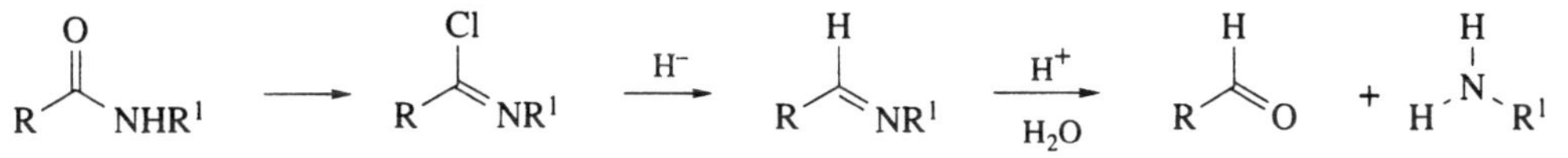

Scheme 15

In an unusual examination of the reducing capacities of several metal hydrides, the yields of aldehydes from carbazole amides were found to change markedly.[123] For one particular carbazide, *N*-9-phenylnona-2,4,6,8-tetraenoylcarbazole, yields of 9-phenylnona-2,4,6,8-tetraenaldehyde with various reducing agents were: $LiAlH_4$ 81%, $LiBH_4$ 69%, $LiBH(OBu^t)_3$ 62%, $LiZnHPh_2$ 45% and $LiBeHPh_2$ 37%. It was concluded that general use of the more readily available $LiAlH_4$ with *N*-acylcarbazoles was a convenient and simple reaction for the preparation of aldehydes, rather than the use of more esoteric hydrides.

Reference has been made already to reviews of the reactions of thexylborane[8] and disiamylborane.[27] Thexylborane reacts slowly with primary amides; tertiary amides are reduced slowly, but although some aldehyde could be isolated from the reaction mixtures there was no obvious break in the uptake of hydride at the aldehyde stage and the reaction proceeded smoothly to alcohol, viz. reduction to aldehyde and to alcohol took place simultaneously. Disiamylborane reacts with primary amides with the release of two molar equivalents of hydrogen gas but no reduction is observed. However, with dimethylamides one hydride equivalent is taken up to yield the aldehyde.

1.11.7 LACTAMS AND CYCLIC IMIDES

Comparatively little research has been carried out in this area. The sequence of reductions which might be expected is shown in Scheme 16. Initial reduction of the lactam should give the cyclic amino alcohol or carbinolamine (**33**), which can be expected to exist as a tautomer of the open chain amino aldehyde (**34**). Either of these tautomers can be reduced, the first to the saturated cyclic amine and the second to an open chain amino alcohol. All of these stages have been observed in hydride reductions of lactams. As might be expected, compounds (**33**) and (**34**) are easily polymerized and may not be isolable. Early in the use of $LiAlH_4$, it was found that 1-methyl-2-pyrrolidine and 1-methyl-2-piperidone could be reduced to the corresponding cyclic amino alcohols (amino aldehydes; **33** and **34**; R = Me, n = 1,2; Scheme 16) if stoichiometric quantities of hydride reagent were used;[124] the amino alcohols were used in syntheses of hygrine and pelletierine.[125] With an excess of $LiAlH_4$, oxygen free bases were isolated by overreduction of the intermediate amino alcohol/amino aldehyde. Other, similar reductions have been recorded.[126] The oxygenated lactam (**35**) was reduced to the corresponding cyclic amino alcohol but other structurally related amides did not give similar results.[127]

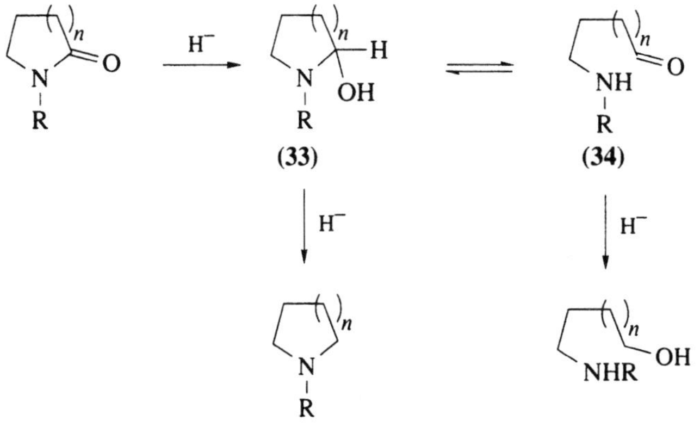

Scheme 16

Reduction of one carbonyl group to a hydroxy group in cyclic imides has been achieved to afford useful synthetic intermediates. The reaction (Scheme 17) is analogous to that shown for lactams (Scheme 16). Thus, reduction of cyclic imides (**36**) with $NaBH_4$ in acidulated water has given good yields of the hydroxy lactam (**37**).[128] Many similar examples by the same research group have appeared.[129] With so-

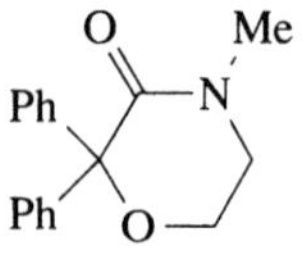

(35)

dium borohydride in methanol, phthalimides have been reduced in moderate to good yields to 3-hydroxyphthalimidines, with considerable overreduction to 2-hydroxymethylbenzamide.[130] This overreduction was found with diborane and 1,3,3-trisubstituted succinimides, which gave only lactams.[131] Unexpectedly, only the imide carbonyl group between the nitrogen and the 3-position (more sterically crowded) was found to be affected.

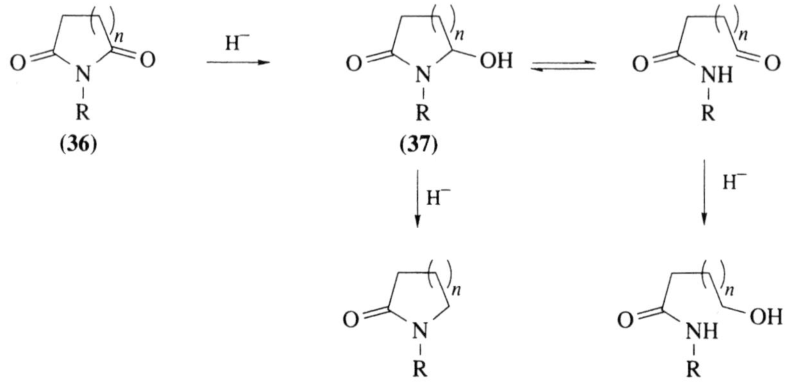

(36) (37)

Scheme 17

1.11.8 NITRILES

Reduction of nitriles to aldehydes with $LiAlH_4$ at low temperatures has been described,[92] but the reaction does not seem to be easy to control and it does not work satisfactorily in many cases. By moderating the activity of $LiAlH_4$ through introduction of alkoxy groups, reduction of nitriles becomes more readily controllable. Reaction of $LiAlH_4$ with three equivalents of ethanol gives $LiAlH(OEt)_3$, which is found to be more reactive than the corresponding $LiAlH(OBu^t)_3$ but less so than $LiAlH_4$ itself.[132] Whereas $LiAlH_4$ reduces benzonitrile to benzaldehyde at 0 °C in 92% yield, with butyronitrile overreduction is observed. $LiAlH(OBu^t)_3$ did not react with nitriles and $LiAlH(OMe)_3$ was too reactive, giving only 15–30% yields of aldehyde. However, in tetrahydrofuran $LiAlH(OEt)_3$ afforded a 59% yield of butyraldehyde from butyronitrile. In a follow-up to this work,[132] it was found that the reaction of $LiAlH(OEt)_3$ with nitriles to give aldehydes was quite general for both aromatic and aliphatic types, but that the best solvent was diethyl ether.[133] In solvents like tetrahydrofuran or diglyme, inferior yields of aldehyde were obtained.

In other work, $NaAlH_4$ was found to be no milder than $LiAlH_4$.[134] Inverse addition of the hydride reagent at 0 °C to benzonitrile gave a 70% yield of benzaldehyde but addition of the same nitrile to the hydride reagent gave 92% of benzylamine. Similarly, it has been found that $NaAlH_4$ reduces aromatic and heterocyclic nitriles to aldehydes in 65–85% yields at 0–30 °C, but that aliphatic compounds give only 15–28% yields.[135]

Again, as with $LiAlH_4$, moderation of the activity of $NaAlH_4$ by incorporation of alkoxy groups has given hydride reagents which are less active in the reduction of nitriles to aldehydes. In view of the readier availability of $LiAlH_4$, there does not seem to be a major advantage in the use of $NaAlH_4$. Sodium triethoxyaluminum hydride has been shown to reduce aromatic nitriles at room temperature to give excellent yields of aldehydes.[136] However, phthalonitrile, 9-cyanofluorene and aliphatic nitriles do not give aldehydes; esters and acid chlorides are reduced to alcohols. Sodium bis(2-methoxyethoxy)aluminum dihydride is soluble in hydrocarbon solvents and more stable in air than many other complex hydrides but reduction of aromatic nitriles was shown to give amines.[137] Later research revealed that the solvent played a major role in this reduction and heterocyclic nitriles could be reduced to aldehydes in about 50% yield.[138] Use of sodium diisobutylaluminum dihydride at –70 °C gave only moderate yields of aldehydes.[139]

DIBAH has long been known to partially reduce nitriles. In diethyl ether, heptane or benzene, reaction of nitriles with DIBAH gives aldimines directly bonded to aluminum (**38**; Scheme 18).[140] These aldiminoaluminum compounds can be isolated by direct distillation from the reaction mixture or can be stirred with acidified water for 30 min to give typical yields of 50–90% of aldehydes. Later, it was shown that a 2:1 ratio of hydride to benzonitrile gave 82% of benzaldehyde but a 1:1 ratio gave 90%; the other major product of reduction was benzalbenzylamine.[141] In a deft adaptation of the intermediate formation of aldiminoaluminum compounds, nitriles have been reduced to this stage with DIBAH (Scheme 18) and then reacted with lithium diethylamide or *n*-butyllithium prior to reaction with an alkyl halide to give ketones (**39**).[142]

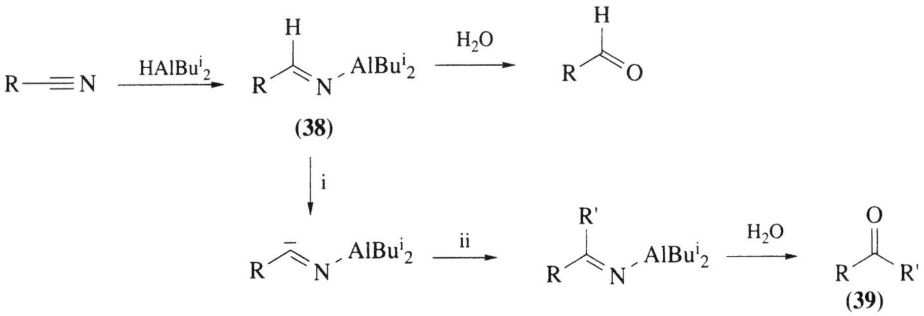

$$R{-}C{\equiv}N \xrightarrow{HAlBu^i_2} \underset{\textbf{(38)}}{RCH{=}N{-}AlBu^i_2} \xrightarrow{H_2O} RCHO$$

$$\textbf{(38)} \xrightarrow{i} R\bar{C}{=}N{-}AlBu^i_2 \xrightarrow{ii} RR'C{=}N{-}AlBu^i_2 \xrightarrow{H_2O} \underset{\textbf{(39)}}{RCOR'}$$

i, $LiNEt_2$ or Bu^nLi; ii, R'X, X = halide, sulfonate

Scheme 18

Nitrilium salts (**40**; equation 10) can be prepared from nitriles and triethyloxonium fluoroborate. Early workers concluded that the nitrilium salts could be reduced to amines but that the reduction could not be stopped at the aldehyde (aldimine) stage.[143] Later, it was demonstrated that reduction of the nitrilium salts with $NaBH_4$ was rapid and did give the corresponding amine, but with triethylsilane yields of 60–90% of both aliphatic and aromatic aldehydes were obtained.[144] Triethylsilane is a mild reagent which reduces carbocations but neutral compounds are generally not reduced and so overreduction is not a problem with this reagent. Formation of the nitrilium salts with triethyloxonium fluoroborate is slow but can be speeded up by the use of iron(III) chloride (Scheme 19). There were some notable exceptions to aldehyde formation with some aromatic nitriles.

$$R{-}C{\equiv}N \;+\; Et_3\overset{+}{O}\,BF_4^- \longrightarrow \underset{\textbf{(40)}}{R{-}C{\equiv}\overset{+}{N}Et\;BF_4^-} \;+\; Et_2O \qquad (10)$$

$$R{-}C{\equiv}N \;+\; FeCl_3 \longrightarrow R{-}C{\equiv}\overset{+}{N}\overset{-}{Fe}Cl_3 \xrightarrow{Pr^iCl} R{-}C{\equiv}\overset{+}{N}Pr^i\;FeCl_4^- \xrightarrow{Et_3SiH} RCHO$$

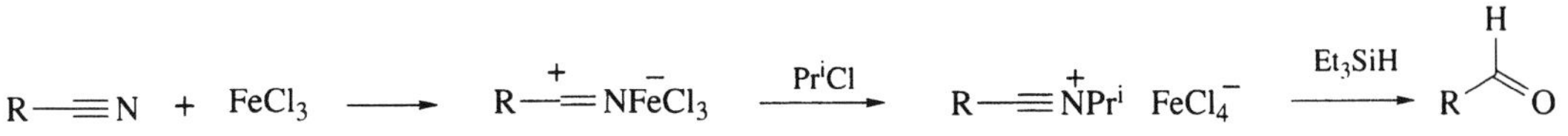

Scheme 19

Disiamylborane and thexylborane react very slowly with nitriles and give only low yields of aldehydes.[27,9]

1.11.9 MISCELLANEOUS REDUCTIONS TO ALDEHYDES

Several related reactions involve reduction of cyclic carboxylic acid derivatives to masked aldehydes which resist further reduction but can be converted into the required aldehydes by acid hydrolysis. In a series of papers, it was established that carboxylic acids could be converted into dihydro-1,3-thiazines or dihydro-1,3-oxazines which could be reduced by $NaBH_4$ in weakly acidic ethanol. Thus, as shown in Scheme 20, dihydro-1,3-thiazines (**41**) were reduced to tetrahydro-1,3-thiazines (**42**) in yields of 66–84%. The resulting tetrahydro compounds could be hydrolyzed to aldehydes by aqueous acid.[145] In a later publication, these workers showed that there was little evidence for ring opening during reduction and that other methods of reduction (*e.g.* hydrogenation over Pt, Pd or Rh or use of dissolving metals such as Zn, Sn or Na) were totally unsuccessful.[146] In closely similar work, reduction of 5,6-dihydro-4*H*-

1,3-oxazines (**43**; Scheme 21) with $NaBH_4$ at –30 °C and pH 7 gave the corresponding tetrahydrooxazines (**44**) which could be hydrolyzed in aqueous acetic acid to provide the aldehydes:[147] overall yields were moderate to good. Without pH control or at higher temperatures ring-opened products (amino alcohols) were obtained, sometimes predominantly. Several other papers[148] describe the use of this method which was held to be useful on a large scale, unlike many other conversions of carboxylic acids into aldehydes. In a novel development of this aldehyde synthesis, the authors showed that the oxazines could be used as two-carbon synthons.[149] For example, the dihydro-1,3-oxazines (**45**; Scheme 22) were lithiated with *n*-butyllithium and reacted with an epoxide to give the oxazines (**46**). Reduction with $NaBH_4$ and hydrolysis afforded γ-hydroxy aldehydes (**47**) in good overall yields.[150]

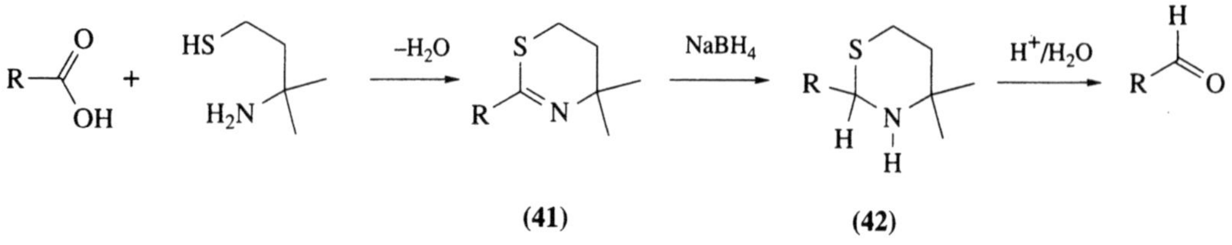

(**41**) (**42**)

Scheme 20

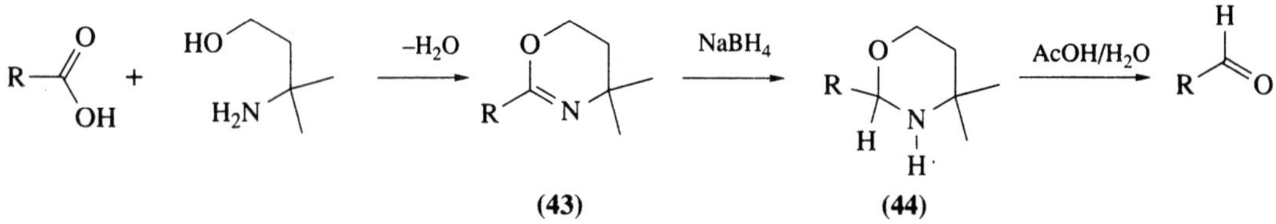

(**43**) (**44**)

Scheme 21

(**45**) (**46**) (**47**)

i, Bu^nLi; ii, R^1-epoxide $/H^+$; iii, $NaBH_4/H_3O^+$

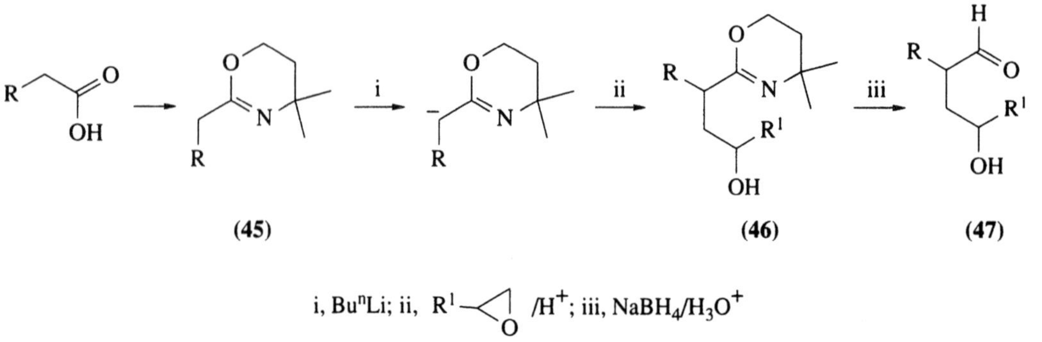

Scheme 22

(*S*)-Triazolium salts (**49**; Scheme 23), prepared from an acid chloride and the thiohydrazo compound (**48**), have been reduced to the hydro compound (**50**) and then hydrolyzed in dilute sulfuric acid to the aldehyde.[151] Overall yields of aldehyde are reported to be 46–90% and a variety of other functional groups can be supported such as alkene, halogen, nitro, ketone and ester; there can be one or two side chains on the α-carbon or it can be part of an alicyclic system.

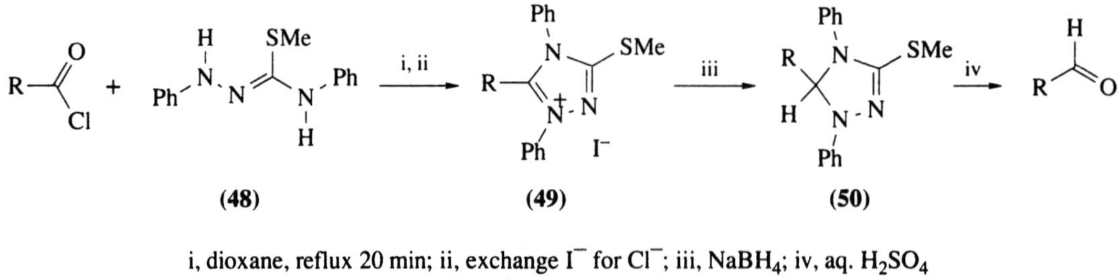

(**48**) (**49**) (**50**)

i, dioxane, reflux 20 min; ii, exchange I^- for Cl^-; iii, $NaBH_4$; iv, aq. H_2SO_4

Scheme 23

In conceptually much the same sort of reaction, carboxylic acids have been treated with *o*-phenylenediamine and then methyl iodide to yield benzimidazolium salts (**51**; Scheme 24). Reduction of the latter with $NaBH_4$ gives the hydro compound (**52**), which can be hydrolyzed by dilute acid to give the aldehyde.[152] For the few aldehydes described, yields were about 70%.

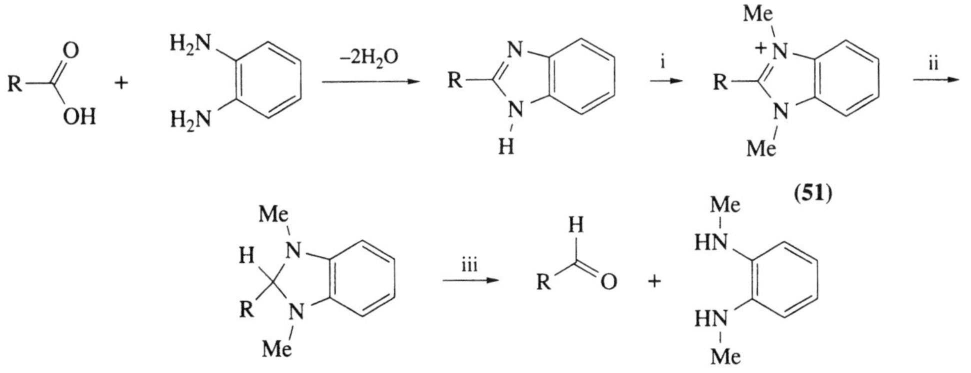

i, NaOMe/MeOH/MeI; ii, $NaBH_4$; iii, 4% aq. HCl

Scheme 24

2-Substituted-1,3-benzoxathiolium tetrafluoroborates (**53**; Scheme 25) can be prepared readily from acids, anhydrides, esters and trihalomethyl compounds.[153,154] The thiolium salts can be reduced easily with $NaBH_4$ in acetonitrile at 0–20 °C to give aldehydes in 71–91% overall yields.

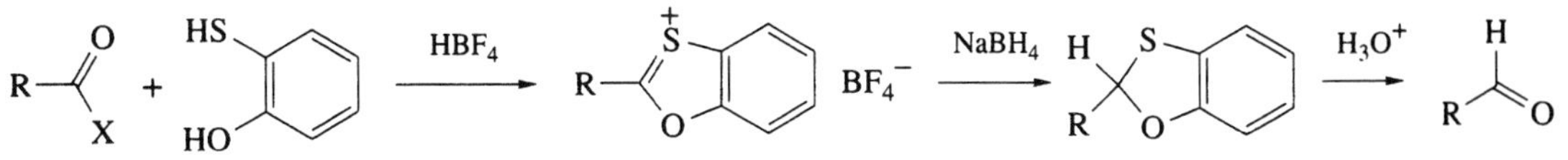

X = OH, OR^1, $OCOR^1$, CCl_3

Scheme 25

Two methods for the formation of aldehydes embody C—C bond formation and then its cleavage by a reverse aldol reaction after reduction, and a third involves similar formation and breakage of a C—P bond. Reaction of esters with the sulfur compound (**55**; Scheme 26) affords ketones (**56**) which can be reduced with $NaBH_4$ at 0 °C to the alcohols (**57**). Simply warming this alcohol, or better, treating it with aqueous potassium carbonate, gives the aldehyde in about 95% yield for aromatic compounds but somewhat less for aliphatic; dialdehydes can be obtained from diacids.[155] Acylmalonates (**58**; Scheme 27)

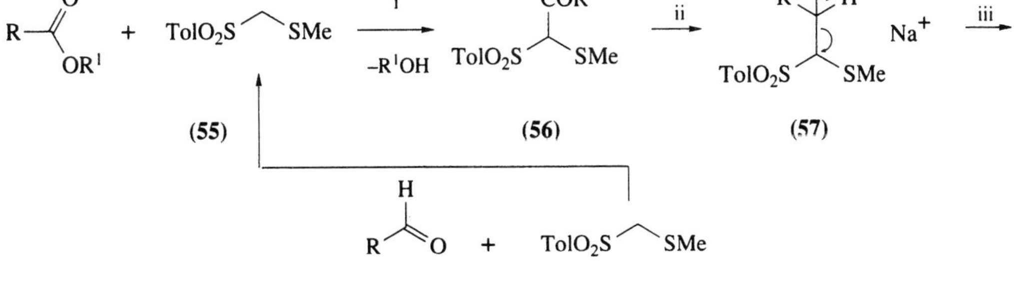

i NaH; ii, $NaBH_4$; iii, base/H_2O

Scheme 26

have been reduced with $NaBH_4$ at 0 °C to give alcohols (**59**) which, under the alkaline conditions, eliminate malonate to give aldehydes in moderate to good overall yields.[156] It is worth noting that malonates can give small yields of aldehydes directly during their reduction with $LiAlH_4$.[157] Finally, reaction of acyl halides with trialkyl phosphites (Scheme 28) gives acyl phosphites (**60**) which can be reduced with buffered $NaBH_4$ to the alcohols (**61**); under alkaline conditions, these alcohols eliminate phosphite to give aldehydes in high yields.[158]

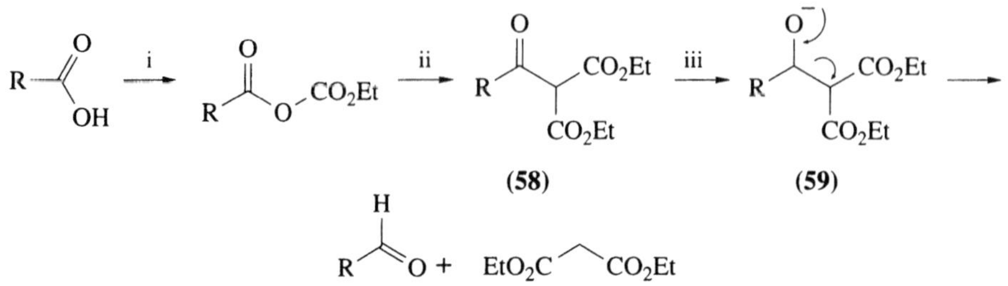

i, $ClCO_2Et$; ii, $Mg[CH(CO_2Et)_2]_2$; iii, $NaBH_4$

Scheme 27

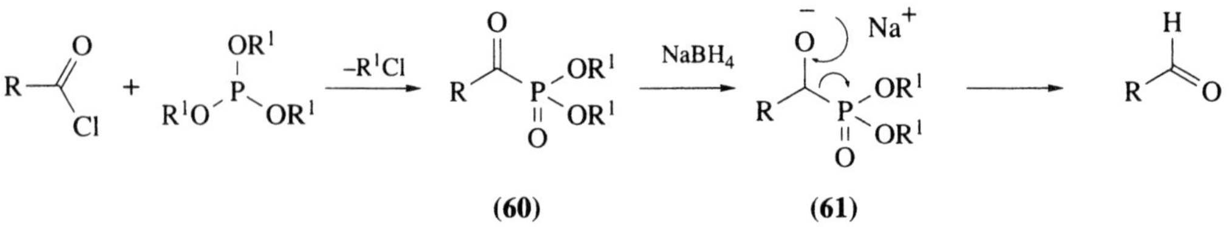

Scheme 28

1.11.10 REVIEWS AND BOOKS

The following articles are recommended because they contain specific discussions of one or more of the above sections. This list is not meant to be exhaustive and many good books and reviews which touch on reduction methods generally have been omitted simply because of their generality. The position of articles in the following list is not intended to be in order of priority or preference.

(a) E. Mosetigg, *Org. React. (N.Y.)*, 1954, **8**, 218: discusses the then-existing major methods for the conversion of carboxylic acids to aldehydes and leads up to first uses of $LiAlH_4$.
(b) N. C. Gaylord, 'Reduction with Complex Metal Hydrides', Interscience, New York, 1956: includes much of the early chemistry of hydrides including some reactions leading to aldehydes.
(c) E. R. H. Walker, *Chem. Soc. Rev.*, 1976, 23: lists all aluminum and boron hydrides used for the reduction of organic compounds. A table of functional group selectivity is given.
(d) H. C. Brown and S. Krishnamurthy, *Tetrahedron*, 1979, **35**, 567: a useful, readable review of hydride reductions with $NaBH_4$ and $LiAlH_4$ and their derivatives.
(e) M. Hudlicky, 'Reductions in Organic Chemistry', Horwood, Chichester, 1984: this gives a good account of many reduction methods with useful, practical details
(f) M. Pereyre, J. Quintard and A. Rahm, 'Tin in Organic Synthesis', Butterworths, London, 1987, pp. 76, 95: covers reductions with stannanes in organic synthesis.
(g) W. P. Neumann, *Synthesis*, 1987, 665: discusses the reduction of acyl halides to aldehydes with tri-*n*-butyltin hydride.
(h) 'Compendium of Organic Synthetic Methods', Wiley, New York, I. T. Harrison and S. Harrison, Vol. 1, 1971; vol. 2, 1974; L. S. Hegedus and L. G. Wade, vol. 3, 1977; L. G. Wade, vol. 4, 1980; L. G. Wade, vol. 5, 1984. Excellent tabulations of interconversions of functional groups, including carboxylic acids and esters to aldehydes, are contained in these volumes.
(i) Sections on oxidation and reduction by many different authors in 'Organic Reaction Mechanisms', Wiley, Chichester. This series stretching over many years includes most of the mechanistic work on hydride reductions from 1966.
(j) A. Hajos, *Methoden Org. Chem. (Houben-Weyl)*, 1981, **IV/1d**, 1. This is an excellent comprehensive review, entitled 'Metal Hydrides or Complex Hydrides as Reducing Agents', with many leading references and selected practical details (in German).

(k) A. Hajos, 'Complex Hydrides', Elsevier, Amsterdam, 1979: a comprehensive examination of the reducing properties of a whole range of mostly aluminum and boron hydrides. Functional group selectivity is considered.

(l) E. Negishi, 'Organometallics in Organic Synthesis', Wiley, New York, 1980, vol. 1, pp. 343, 377: some reductions with alanes and boranes are covered.

(m) 'Comprehensive Organometallic Chemistry', ed. G. Wilkinson, Pergamon, Oxford, 1982: contains many specific but scattered references to reduction listed under hydride types and substrate groups.

(n) E. Negishi and H. C. Brown, *Synthesis*, 1974, 77: a review of reactions of thexylborane, including reduction of carboxylic acids and derivatives.

(o) H. C. Brown, D. B. Bigley, S. K. Arora and N. M. Yoon, *J. Am. Chem. Soc.*, 1970, **92**, 7161: a useful comparison of functional group reactivity towards disiamylborane.

(p) J. Malek and M. Cerny, *Synthesis*, 1972, 217: a general review of uses of alkoxyaluminum hydrides as reducing agents for a wide range of functional groups.

(q) E. Winterfeldt, *Synthesis*, 1975, 617: a discussion of reductions in which diisobutylaluminum hydride or triisobutylaluminum are the preferred reagents in synthesis with emphasis on functional group selectivity.

(r) 'Comprehensive Organic Chemistry', ed. D. H. R. Barton and W. D. Ollis, Pergamon, Oxford, 1979: reduction discussed under a wide range of headings but mostly by functional group with a variety of reagents, including hydrides.

1.11.11 REFERENCES

1. K. Ziegler, K. Schneider and J. Schneider, *Justus Liebigs Ann. Chem.*, 1959, **623**, 9.
2. L. I. Zakharkin and I. M. Khorlina, *Zh. Obshch. Khim.*, 1964, **34**, 1029.
3. H. C. Brown and N. M. Yoon, *J. Am. Chem. Soc.*, 1966, **88**, 1464.
4. H. C. Brown and P. M. Weissman, *J. Am. Chem. Soc.*, 1965, **87**, 5614.
5. M. Muraki and T. Mukaiyama, *Chem. Lett.*, 1974, 1447.
6. T. D. Hubert, D. P. Eyman and D. F. Wiemer, *J. Org. Chem.*, 1984, **49**, 2279.
7. H. C. Brown, P. Heim and N. M. Yoon, *J. Am. Chem. Soc.*, 1970, **92**, 1637.
8. H. C. Brown, P. Heim and N. M. Yoon, *J. Org. Chem.*, 1972, **37**, 2942.
9. E. Negishi and H. C. Brown, *Synthesis*, 1974, 77.
10. S. Krishnamurthy and K. L. Thompson, *J. Chem. Educ.*, 1977, **54**, 778.
11. H. C. Brown, C. G. Rao and S. U. Kulkarni, *Synthesis*, 1979, 704.
12. H. C. Brown, J. S. Cha, N. M. Yoon and B. Nazer, *J. Org. Chem.*, 1987, **52**, 5400.
13. H. C. Brown, J. S. Cha, B. Nazer and N. M. Yoon, *J. Am. Chem. Soc.*, 1984, **106**, 8001.
14. J. S. Cha, J. E. Kim and K. W. Lee, *J. Org. Chem.*, 1987, **52**, 5030.
15. S. Kim and S. S. Kim, *Tetrahedron Lett.*, 1987, **28**, 1913.
16. S. Kim, S. S. Kim, S. T. Lim and S. C. Shim, *J. Org. Chem.*, 1987, **52**, 2114.
17. S. Kim, *Pure Appl. Chem.*, 1987, **59**, 1005.
18. J. Boyer, C. Breliere, R. J. P. Corriu, A. Kpoton, M. Poitier and G. Royo, *J. Organomet. Chem.*, 1986, **311**, C39.
19. R. J. P. Corriu, G. F. Lanneau and M. Perrot, *Tetrahedron Lett.*, 1987, **28**, 3941.
20. P. Brandt, *Acta Chem. Scand.*, 1949, **3**, 1050.
21. H. C. Brown and R. F. McFarlin, *J. Am. Chem. Soc.*, 1956, **78**, 252.
22. H. C. Brown and R. F. McFarlin, *J. Am. Chem. Soc.*, 1958, **80**, 5372.
23. H. C. Brown and B. C. Subba Rao, *J. Am. Chem. Soc.*, 1958, **80**, 5377.
24. H. C. Brown and E. J. Mead, *J. Am. Chem. Soc.*, 1953, **75**, 6263.
25. J. E. Siggins, A. A. Larsen, J. H. Ackerman and C. D. Carabateas, *Org. Synth.*, 1973, **53**, 52.
26. R. Grewe and H. Büttner, *Chem. Ber.*, 1958, **91**, 2452.
27. H. C. Brown, D. B. Bigley, S. K. Arora and N. M. Yoon, *J. Am. Chem. Soc.*, 1970, **92**, 7161.
28. E. Mincione, *J. Org. Chem.*, 1978, **43**, 1829.
29. E. Santaniello, A. Fiecchi, A. Manzocchi and P. Ferraboschi, *J. Org. Chem.*, 1983, **48**, 3074.
30. R. A. W. Johnstone and R. P. Telford, *J. Chem. Soc., Chem. Commun.*, 1978, 354.
31. I. D. Entwistle, P. Boehm, R. A. W. Johnstone and R. P. Telford, *J. Chem. Soc., Perkin Trans. 1*, 1980, 27.
32. T. Fujisawa, T. Mori, S. Tsuge and T. Sato, *Tetrahedron Lett.*, 1983, **24**, 1543.
33. J. H. Babler and B. J. Invergo, *Tetrahedron Lett.*, 1981, **22**, 11.
34. J. H. Babler, *Synth. Commun.*, 1982, **12**, 839.
35. T. N. Sorrell and R. J. Spillane, *Tetrahedron Lett.*, 1978, 2473.
36. T. N. Sorrell and P. S. Pearlman, *J. Org. Chem.*, 1980, **45**, 3449.
37. G. W. J. Fleet, C. J. Fuller and P. J. C. Harding, *Tetrahedron Lett.*, 1978, 1437.
38. G. W. J. Fleet and P. J. C. Harding, *Tetrahedron Lett.*, 1979, 975.
39. R. O. Hutchins and M. Markowitz, *Tetrahedron Lett.*, 1980, **21**, 813.
40. G. J. M. Van Der Kerk, J. G. Noltes and J. G. A. Luijten, *J. Appl. Chem.*, 1957, **7**, 356.
41. H. G. Kuivila, *J. Org. Chem.*, 1960, **25**, 284.
42. E. J. Kupchik and R. J. Kiesel, *J. Org. Chem.*, 1966, **31**, 456.
43. H. G. Kuivila and E. J. Walsh, Jr., *J. Am. Chem. Soc.*, 1966, **88**, 571.
44. E. J. Walsh, Jr. and H. G. Kuivila, *J. Am. Chem. Soc.*, 1966, **88**, 576.
45. E. J. Walsh, Jr., R. L. Stoneberg, M. Yorke and H. G. Kuivila, *J. Org. Chem.*, 1969, **34**, 1156.
46. W. P. Neumann, *Synthesis*, 1987, 665.
47. J. Lusztyk, E. Lusztyk, B. Maillard and K. U. Ingold, *J. Am. Chem. Soc.*, 1984, **106**, 2923.
48. F. Guibe, P. Four and H. Rivière, *J. Chem. Soc., Chem. Commun.*, 1980, 432.
49. P. Four and F. Guibe, *J. Org. Chem.*, 1981, **46**, 4439.

50. W. P. Neumann, *Synthesis*, 1987, 665.
51. J. W. Jenkins and H. W. Post, *J. Org. Chem.*, 1950, **15**, 556.
52. J. D. Citron, *J. Org. Chem.*, 1969, **34**, 1977.
53. S. P. Dent, C. Eaborn and A. Pidcock, *J. Chem. Soc., Chem. Commun.*, 1970, 1703.
54. L. I. Zakharkin and I. M. Khorlina, *Tetrahedron Lett.*, 1962, 619.
55. C. Szántay, L. Tökés and P. Kolonits, *J. Org. Chem.*, 1966, **31**, 1447.
56. L. I. Zakharkin and L. P. Sorokina, *Zh. Obshch. Khim.*, 1967, **37**, 561.
57. A. Ito, R. Takahashi and Y. Baba, *Chem. Pharm. Bull.*, 1975, **23**, 3081.
58. M. Boch, T. Korth, J. M. Nelke, D. Pike, H. E. Radunz and E. Winterfeldt, *Chem. Ber.*, 1972, **105**, 2126.
59. E. Winterfeldt, *Synthesis*, 1975, 617.
60. M. Muraki and T. Mukaiyama, *Chem. Lett.*, 1975, 215.
61. L. I. Zakharkin, V. V. Gavrilenko, D. N. Maslin and I. M. Khorlina, *Tetrahedron Lett.*, 1963, 2087.
62. J. Vit, *Eastman Org. Chem. Bull.*, 1970, **42** (3), 1.
63. R. Kanazawa and T. Tokoroyama, *Synthesis*, 1976, 526.
64. H. C. Brown and P. M. Weissman, *Isr. J. Chem.*, 1963, **1**, 430.
65. P. M. Weissman and H. C. Brown, *J. Org. Chem.*, 1966, **31**, 283.
66. C. J. Claus and J. L. Morgenthau, Jr., *J. Am. Chem. Soc.*, 1951, **73**, 5005.
67. L. I. Zakharkin and I. M. Khorlina, *Izv. Akad. Nauk SSSR, Ser. Khim.*, 1959, 2255 (*Chem. Abstr.*, 1960, **54**, 10 837).
68. M. R. Johnson and B. Rickborn, *Org. Synth.*, 1971, **51**, 11.
69. M. E. Hill and L. O. Ross, *J. Org. Chem.*, 1967, **32**, 2595.
70. A. Roedig, H.-A. Renk, V. Schaal and D. Scheutzow, *Chem. Ber.*, 1974, **107**, 1136.
71. P. Coutrot, J. Villieras and J. C. Combret, *C. R. Hebd. Seances Acad. Sci., Ser. C*, 1972, **274**, 1531.
72. L. Barre, M. Saquet and A. Thuillier, *Sulfur Lett.*, 1984, **2**, 81.
73. G. E. Arth, *J. Am. Chem. Soc.*, 1953, **75**, 2413.
74. J. Schmidlin and A. Wettstein, *Helv. Chim. Acta*, 1963, **46**, 2799.
75. W. E. Parham and L. D. Huestis, *J. Am. Chem. Soc.*, 1962, **84**, 813.
76. F. J. McQuillin and R. B. Yeats, *J. Chem. Soc.*, 1965, 4273.
77. J. Martin, W. Parker and R. A. Raphael, *J. Chem. Soc.*, 1964, 289.
78. W. Carruthers and M. I. Qureshi, *J. Chem. Soc. C*, 1970, 2238.
79. J. Málek and M. Černý, *Synthesis*, 1972, 217.
80. E. J. Corey, N. M. Weinshenker, T. K. Schaaf and W. Huber, *J. Am. Chem. Soc.*, 1969, **91**, 5675.
81. E. J. Corey, H. Shirahama, H. Yamamoto, S. Terashima, A. Venkateswarlu and T. K. Schaaf, *J. Am. Chem. Soc.*, 1971, **93**, 1490.
82. J. J. Partridge, N. K. Chadha and M. Usković, *J. Am. Chem. Soc.*, 1973, **95**, 7171.
83. G. R. Pettit, T. R. Kasturi, B. Green and J. C. Knight, *J. Org. Chem.*, 1961, **26**, 4773.
84. J. R. Dias and G. R. Pettit, *J. Org. Chem.*, 1971, **36**, 3485.
85. M. L. Wolfrom and H. B. Wood, *J. Am. Chem. Soc.*, 1951, **73**, 2933.
86. M. L. Wolfrom and K. Anno, *J. Am. Chem. Soc.*, 1952, **74**, 5583.
87. H. C. Brown and D. B. Bigley, *J. Am. Chem. Soc.*, 1961, **83**, 486.
88. P. Kohn, R. H. Samaritano and L. M. Lerner, *J. Am. Chem. Soc.*, 1965, **87**, 5475.
89. T. A. Giudici and A. L. Fluharty, *J. Org. Chem.*, 1967, **32**, 2043.
90. P. Kohn, L. M. Lerner, A. Chan, S. D. Ginocchio and C. A. Zitrin, *Carbohydr. Res.*, 1968, **7**, 21.
91. R. E. Ireland, D. A. Evans, D. Glover, G. M. Rubottom and H. Young, *J. Org. Chem.*, 1969, **34**, 3717.
92. L. Friedman, in 'Abstracts of 116th ACS Meeting, Atlantic City, NJ', ACS, Washington, DC, September 1949, p. 5M.
93. L. I. Smith and E. R. Rogier, *J. Am. Chem. Soc.*, 1951, **73**, 4047.
94. F. Weygand and D. Tietjen, *Chem. Ber.*, 1951, **84**, 625.
95. M. Mouseron, R. Jacquier, M. Mousseron-Canet and R. Zagdoun, *Bull. Soc. Chim. Fr.*, 1952, **19**, 1042.
96. F. Weygand and G. Eberhardt, *Angew. Chem.*, 1952, **64**, 458.
97. F. Weygand, G. Eberhardt, H. Linden, F. Schäfer and I. Eigen, *Angew. Chem.*, 1953, **65**, 525.
98. V. M. Micovic and M. L. Mihailović, *J. Org. Chem.*, 1953, **18**, 1190.
99. B. L. Fox and R. J. Doll, *J. Org. Chem.*, 1973, **38**, 1136.
100. F. Weygand and R. Mitgau, *Chem. Ber.*, 1955, **88**, 301.
101. W. Ried and F. J. Königstein, *Angew. Chem.*, 1958, **70**, 165.
102. W. Ried, G. Deuschel and A. Kitélko, *Justus Liebigs Ann. Chem.*, 1961, **642**, 121.
103. W. Ried and F. J. Königstein, *Justus Liebigs Ann. Chem.*, 1959, **622**, 37.
104. H. A. Staab and H. Bräunling, *Justus Liebigs Ann. Chem.*, 1962, **654**, 119.
105. H. C. Brown and A. Tsukomoto, *J. Am. Chem. Soc.*, 1961, **83**, 2016.
106. H. C. Brown and A. Tsukomoto, *J. Am. Chem. Soc.*, 1961, **83**, 4549.
107. T. Axenrod, L. Loew and P. S. Pregosin, *J. Org. Chem.*, 1968, **33**, 1274.
108. L. I. Zakharkin, D. N. Maslin and V. V. Gavrilenko, *Tetrahedron*, 1969, **25**, 5555.
109. E. Stephen and H. Stephen, *J. Chem. Soc.*, 1957, 490.
110. N. S. Ramegowda, M. N. Modi, A. K. Koul, J. M. Bora, C. K. Narang and N. K. Mathur, *Tetrahedron*, 1973, **29**, 3985.
111. H. C. Brown, C. J. Shoaf and C. P. Garg, *Tetrahedron Lett.*, 1959, 9.
112. H. C. Brown and A. Tsukomoto, *J. Am. Chem. Soc.*, 1964, **86**, 1089.
113. S. Kim and K. H. Ahn, *J. Org. Chem.*, 1984, **49**, 1717.
114. N. M. Yoon and Y. S. Gyoung, *J. Org. Chem.*, 1985, **50**, 2443.
115. L. I. Zakharkin and I. M. Khorlina, *Izv. Akad. Nauk SSSR, Ser. Khim.*, 1959, 2146.
116. M. Muraki and T. Mukaiyama, *Chem. Lett.*, 1975, 875.
117. T. Izawa and T. Mukaiyama, *Chem. Lett.*, 1977, 1443.
118. T. Izawa and T. Mukaiyama, *Bull. Chem. Soc. Jpn.*, 1979, **52**, 555.

119. Y. Nagao, K. Kawabata and E. Fujita, *J. Chem. Soc., Chem. Commun.*, 1978, 330.
120. Y. Nagao, K. Kawabata, K. Seno and E. Fujita, *J. Chem. Soc., Perkin Trans. 1*, 1980, 2470.
121. S. Nahm and S. M. Weinreb, *Tetrahedron Lett.*, 1981, **22**, 3815.
122. S. Karady, J. S. Amato, L. M. Weinstock and M. Sletzinger, *Tetrahedron Lett.*, 1978, 403.
123. G. Wittig and P. Hornberger, *Justus Liebigs Ann. Chem.*, 1952, **577**, 11.
124. F. Galinovsky and R. Weiser, *Experientia*, 1950, **6**, 377.
125. F. Galinovsky, A. Wagner and R. Weiser, *Monatsh. Chem.*, 1951, **82**, 551.
126. G. A. Swan and J. D. Wilcock, *J. Chem. Soc., Perkin Trans. 1*, 1974, 885.
127. A. L. Morrison, R. F. Long and M. Königstein, *J. Chem. Soc.*, 1951, 952.
128. J. C. Hubert, J. B. P. A. Wijnberg and W. N. Speckamp, *Tetrahedron*, 1975, **31**, 1437.
129. J. Dijkink and W. N. Speckamp, *Tetrahedron Lett.*, 1975, 4047 and refs. therein.
130. Z. Horri, C. Iwata and Y. Tamura, *J. Org. Chem.*, 1961, **26**, 2273.
131. R. Suess, *Helv. Chim. Acta*, 1977, **60**, 1650.
132. H. C. Brown and C. J. Shoaf, *J. Am. Chem. Soc.*, 1964, **86**, 1079.
133. H. C. Brown and C. P. Garg, *J. Am. Chem. Soc.*, 1964, **86**, 1085.
134. A. E. Finholt, E. C. Jacobson, A. E. Ogard and P. Thompson, *J. Am. Chem. Soc.*, 1955, **77**, 4163.
135. L. I. Zakharkin, D. N. Maslin and V. V. Gavrilenko, *Izv. Akad. Nauk SSSR, Ser. Khim.*, 1964, 1511.
136. G. Hesse and R. Schrödel, *Justus Liebigs Ann. Chem.*, 1957, **607**, 24.
137. V. Bažant, M. Čapka, M. Černý, Y. Chvalovský, K. Kochloefl, M. Kraus and J. Málek, *Tetrahedron Lett.*, 1968, **29**, 3303.
138. I. Stibor, M. Janda and J. Srogl, *Z. Chem.*, 1970, **10**, 342.
139. L. I. Zakharkin and I. M. Khorlina, *Izv. Akad. Nauk SSSR, Ser. Khim.*, 1964, 465 (*Chem. Abstr.*, 1964, **60**, 15 765d).
140. L. I. Zakharkin and I. M. Khorlina, *Dokl. Akad. Nauk SSSR*, 1957, **116**, 422 (*Chem. Abstr.*, 1958, **52**, 8 040f).
141. A. E. G. Miller, J. W. Biss and L. H. Schwartzman, *J. Org. Chem.*, 1959, **24**, 627.
142. H. L. Goering and C. C. Tseng, *J. Org. Chem.*, 1981, **46**, 5250.
143. R. F. Borch, *J. Chem. Soc., Chem. Commun.*, 1968, 442; *J. Org. Chem.*, 1969, **34**, 627.
144. J. L. Fry, *J. Chem. Soc., Chem. Commun.*, 1974, 45; J. L. Fry and R. A. Ott, *J. Org. Chem.*, 1981, **46**, 602.
145. J. C. Getson, J. M. Greene and A. I. Meyers, *J. Heterocycl. Chem.*, 1964, **1**, 300.
146. A. I. Meyers and J. M. Greene, *J. Org. Chem.*, 1966, **31**, 556.
147. A. I. Meyers and A. Nabeya, *J. Chem. Soc., Chem. Commun.*, 1967, 1163.
148. A. I. Meyers, A. Nabeya, H. W. Adickes and I. R. Politzer, *J. Am. Chem. Soc.*, 1969, **91**, 763; I. R. Politzer and A. I. Meyers, *Org. Synth.*, 1971, **51**, 24.
149. A. I. Meyers, A. Nabeya, H. W. Adickes, I. R. Politzer, G. R. Malone, A. Kovelesky, R. L. Nolan and R. C. Portnoy, *J. Org. Chem.*, 1973, **38**, 36.
150. H. W. Adickes, I. R. Politzer and A. I. Meyers, *J. Am. Chem. Soc.*, 1969, **91**, 2155.
151. G. Doleschall, *Tetrahedron*, 1976, **32**, 2549; *Tetrahedron Lett.*, 1974, 2649; 1975, 681, 1889.
152. J. C. Craig, N. N. Ekwuribe, C. C. Fu and K. A. M. Walker, *Synthesis*, 1981, 303.
153. M. Barbero, S. Cadamuro, I. Degani, R. Fochi, A. Gatti and V. Regondi, *Synthesis*, 1986, 1074.
154. I. Degani and R. Fochi, *J. Chem. Soc., Perkin Trans. 1*, 1976, 323.
155. K. Ogura, N. Yahata, K. Takahashi and H. Iida, *Tetrahedron Lett.*, 1983, **24**, 5761.
156. H. Muxfeldt, W. Rogalski and G. Klauenberg, *Chem. Ber.*, 1965, **98**, 3040.
157. J. A. Marshall, N. H. Anderson and A. R. Hochstetler, *J. Org. Chem.*, 1967, **32**, 113.
158. L. Horner and H. Röder, *Chem. Ber.*, 1970, **103**, 2984.

1.12
Reduction of Carboxylic Acids to Aldehydes by Other Methods

ANTHONY P. DAVIS

Trinity College, Dublin, Republic of Ireland

1.12.1 INTRODUCTION

Because so many of the recent developments in reduction methodology have involved metal hydride reducing agents, this chapter inevitably has an old-fashioned flavor. Most of the methods described below originated many years ago, and many of them do not figure prominently in the current literature. However, this is not to say that they have no value. In many cases they are economical and suitable for large scale use, and it is quite likely that the best way to make substantial amounts of a simple aldehyde (especially an aromatic one) will be found in this chapter. Moreover, there are some cases where recent improvements, which may not be widely appreciated, have given a classical method a new lease of life. The reader is directed particularly to the 'amine-modified Rosenmund' described in Section 1.12.3.

The question of demarcation between this and the previous chapter deserves some comment. It is not always clear whether a reduction method involves a metal hydride or not. Catalytic hydrogenation, for example, presumably involves species with metal–hydrogen bonds, although it is not normally regarded as a 'metal hydride method'. However, a clear distinction can generally be made on the basis of the source of the stoichiometric reducing equivalents, and this has been the guideline used. Thus, the methods discussed in this chapter are those in which the reducing equivalents are supplied by hydrogen gas, organic molecules, metals, low valent metal salts or (nonhydride) complexes, or an electrode. Note, however, that it has proved convenient to depart from this principle at one point. In Section 1.12.3, a number of low valent metal complex reducing agents are discussed together, in spite of the fact that some of them do behave as hydride donors.

As in the previous chapter, the material will be organized according to substrate category, starting with the carboxylic acids themselves and then moving on to the various derivatives. The methods have been placed generally in sections corresponding to the immediate precursors of the aldehydes. Thus, the hydrolysis of Reissert compounds appears under amides as opposed to acyl chlorides, and the Sonn–Müller appears under imidoyl chlorides and not amides.

1.12.2 CARBOXYLIC ACIDS

The central problem in the reduction of carboxylic acids and their derivatives to aldehydes is the avoidance of overreduction to primary alcohols. The difficulty is particularly acute when, as in the case of the carboxylic acids themselves, the substrates are not especially reactive towards reducing agents. Thus there are few nonhydride methods of reducing carboxylic acids to aldehydes, and none which are routinely used. Those which are available generally have some specific feature which results in the trapping or protection of the product aldehydes.

Probably the most generally useful methods are those involving: (i) Bu^iMgBr with $[Cp_2TiCl_2]$ as a catalyst;[1] and (ii) $Li/MeNH_2$.[2] The former is thought to occur by the mechanism shown in Scheme 1, in which the aldehyde product is protected as adduct (**1**), and only released on hydrolysis. As shown in Table 1, the method has been applied with reasonable success to a number of aliphatic acids and to benzoic acid (although in the latter case, some overreduction did occur). However, it failed completely with cinnamic acid.

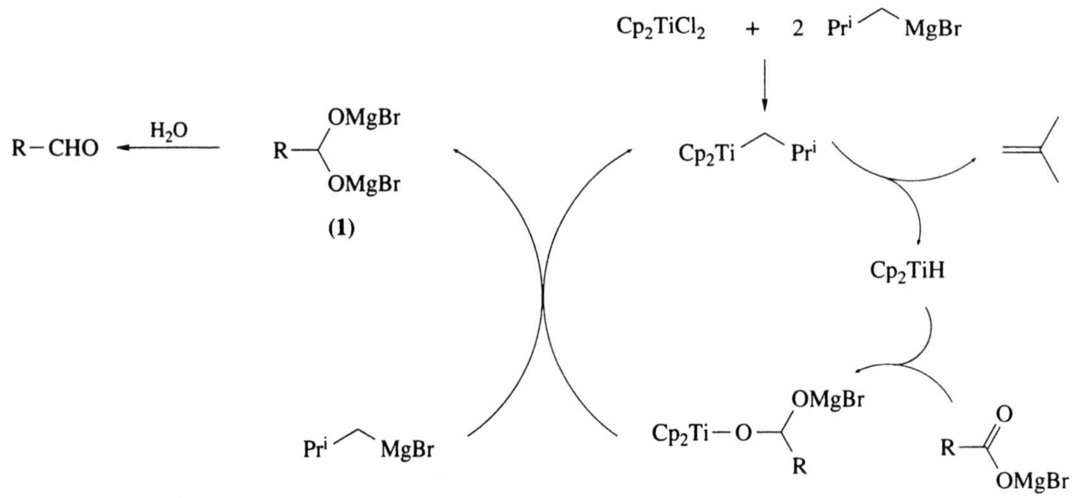

Cp_2TiCl_2 + 2 Pr^i⌒MgBr → Cp_2Ti⌒Pr^i → Cp_2TiH + alkene

R–CHO ←(H_2O)— R–CH(OMgBr)(OMgBr) (**1**)

Pr^i⌒MgBr; $Cp_2Ti–O–CH(OMgBr)R$; R–C(=O)OMgBr

Scheme 1

Table 1 Yields of Aldehydes RCHO from RCO_2H using $Bu^iMgBr/[Cp_2TiCl_2]$[1]

R	*Yield (%)*
Hexyl	73
PrCHMe	65
Cyclohexyl	59
$PhCH_2$	48
Phenyl	55
PhCH=CH	0

The success of the method employing $Li/MeNH_2$ again arises from the formation of the aldehyde in a protected form. In this case it is the amino alcohol salt (**2**; Scheme 2). As shown in Scheme 2, the reaction mixture is quenched with saturated aqueous ammonium chloride and may then be extracted with pentane, to yield an imine, or with ether and washed with acid to give the aldehyde. The imine is clearly not the protected form of the aldehyde, as it is reduced to the corresponding secondary amine under the conditions of the reaction. The reaction succeeds for aliphatic saturated acids, and also for one example containing an isolated C=C double bond (Table 2). However, additional reduction occurs with substrates containing aromatic rings or conjugated double bonds.

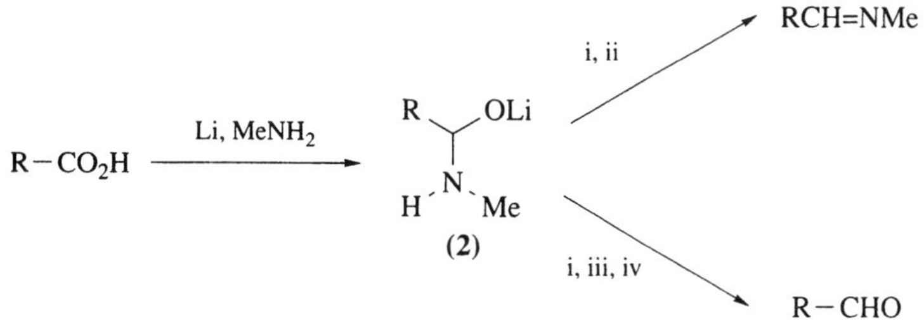

R−CO_2H —(Li, $MeNH_2$)→ (2) [R–CH(OLi)–N(H)Me] —(i, ii)→ RCH=NMe; —(i, iii, iv)→ R−CHO

i, sat. aq. NH_4Cl; ii, pentane extraction; iii, ether extraction; iv, wash with aq. HCl

Scheme 2

Table 2 Yields of Aldehydes RCHO from RCO_2H using Li/$MeNH_2$[2]

R	*Yield (%)*
Butyl	66
Heptyl	59
Nonyl	61
Undecyl	61
Tridecyl	84
$Bu^t(CH_2)_2$	62
$Me(CH_2)_8CH{=}CH(CH_2)_6$	60

There has been considerable research into the electrolytic reduction of aromatic carboxylic acids to the corresponding aldehydes. A general procedure has been described[3] in which key elements are the use of the ammonium salt of the acid, careful control of the pH and the presence of an organic phase (benzene) to extract the aldehyde and thus minimize overreduction. The method appears to work best for relatively acidic substrates; for example, salicylaldehyde was obtained in 80% yield. Danish workers have shown that, under acidic conditions, controlled electrolytic reductions are possible for certain pyridine-, imidazole- and thiazole-carboxylic acids.[4] In these cases, it is thought that the product aldehydes are protected by geminal diol formation. A chemical method which is closely related to electrolysis is the use of sodium amalgam as reductant. Although not widely used, it was successfully employed in the synthesis of a fluorinated salicylaldehyde.[5]

A classical method for preparing aldehydes from carboxylic acids is treatment with formic acid in the vapour phase over a metal oxide catalyst at temperatures of 300–400 °C. For example, aldehyde (**3**) was synthesized from the corresponding acid in 78% yield (but with only 40% conversion) using manganese(II) oxide as catalyst.[6] Closely related methods using hydrogen as reductant have been reported in patents from a number of industrial laboratories. Good selectivities and conversions are claimed, examples being the synthesis of (**4**) over ZrO_2 (99% selectivity, 81% conversion),[7] of benzaldehyde over MnO_2 (95% selectivity, 91% conversion)[8] and of pivalaldehyde (**5**) over a complex mixture of metal oxides supported on alumina (100% selectivity, 73% conversion).[9]

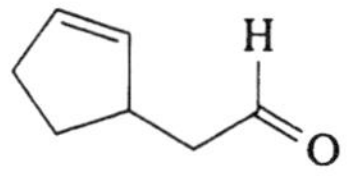

(**3**) 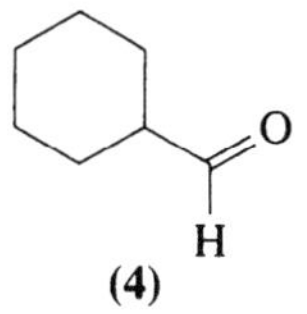 (**4**) 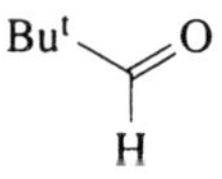 (**5**)

Finally, it may be useful to note a method which, although not actually a reduction, has the same overall effect of transforming RCO_2H to RCHO. As shown in Scheme 3,[10] the acid (**6**) is doubly deprotonated then treated with a formiminium salt (**7**). Carbon–carbon bond formation is followed by loss of CO_2 and methoxide to give enamine (**8**), which can be isolated or transformed by acid hydrolysis to the corresponding aldehyde. The scope and efficiency of the method are difficult to judge accurately, as the authors were principally interested in the synthesis of particular enamines, but it seems likely to be quite useful for acids (**6**) where R^1, R^2 = alkyl or R^1 = H; R^2 = aryl, secondary alkyl or tertiary alkyl.

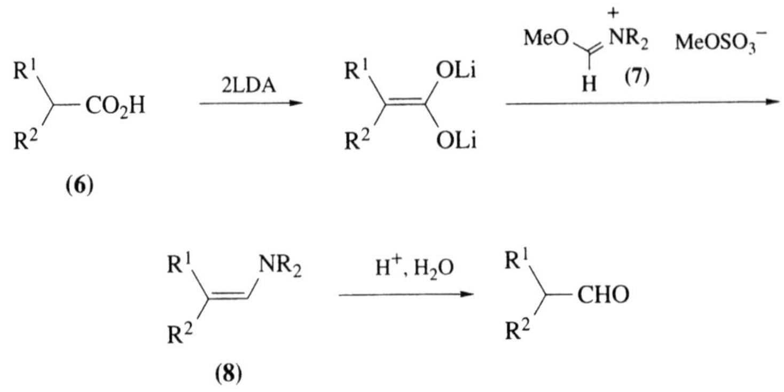

$$R^1R^2CHCO_2H\ (\mathbf{6}) \xrightarrow{2LDA} R^1R^2C{=}C(OLi)_2 \xrightarrow{MeOCH{=}\overset{+}{N}R_2\ MeOSO_3^-\ (\mathbf{7})} R^1R^2C{=}CHNR_2\ (\mathbf{8}) \xrightarrow{H^+,\ H_2O} R^1R^2CHCHO$$

Scheme 3

1.12.3 ACYL CHLORIDES

As acyl chlorides are the most reactive of the common acid derivatives, it is not surprising that they are the easiest to reduce. All the reducing methods described in the rest of this chapter are, in a sense, methods for the reduction of acyl chlorides, as they all involve derivatives which can be made from acid chlorides. Moreover, it is possible to find reducing systems sufficiently mild to convert the acyl chlorides directly to aldehydes without then reducing the aldehydes to alcohols.

In the area of nonhydride reducing agents, the principal means of accomplishing this task is the Rosenmund reduction,[11] the catalytic hydrogenation of acyl chlorides. Before the advent of metal hydride reducing agents, this was probably the principal reductive method of aldehyde synthesis. Although it may seem outdated, modern developments have greatly improved it, so that it still merits serious consideration.

The classical Rosenmund reaction is discussed in depth in an article by Mosettig and Mozingo.[12] It was most usually accomplished by passing hydrogen gas through a refluxing solution of the substrate in toluene or xylene, in the presence of the catalyst. The hydrogen chloride formed was collected by passage through water, and was titrated occasionally in order to follow the progress of the reaction. The usual catalyst was palladium on barium sulphate, with a 'regulator' to moderate the catalytic activity and prevent overreduction (although the regulator was often omitted, especially for nonaromatic substrates). The most common regulator was quinoline-S, a crude preparation of thioquinanthrene obtained by heating together quinoline and elemental sulfur. Other sulfur-containing compounds could also be used, and later workers recommended tetramethylthiourea.[13] Table 3 shows a selection of yields and conditions, including two examples (the first two entries) which are recorded in *Organic Syntheses*.[14,15]

Table 3 Yields and Conditions for Selected 'Classical' Rosenmund Reductions of Acyl Chlorides RCOCl with Pd–$BaSO_4$ as Catalyst

R	*Solvent*	*Temperature* (°C)	*Regulator*	*Yield* (%)
β-Naphthyl	Xylene	150	Quinoline-S	84
2,4,6-Trimethylphenyl	Xylene	Reflux	None	70–80
p-Nitrophenyl	Xylene	150	Quinoline-S	91
3,5-Dimethoxy-4-(ethoxycarbonyl)phenyl	Toluene	120–125	None	49
3-Furyl	Xylene	Reflux	None	55
$MeO_2C(CH_2)_3$	Xylene	140–150	Quinoline-S	52
$Me(CH_2)_{16}$	Xylene	150	None	17–21
PhCH=CH	Xylene	122	Thioquinanthrene, quinoline-S	50–60

The classical Rosenmund reduction is quite broad in scope, and can be applied to fairly complex molecules. Not included in Table 3 are several examples giving terpenoid aldehydes which occur in yields of up to 88%.[12] Nevertheless, there are serious limitations. The yields are variable, being affected by side reactions such as overreduction and decarbonylation (Scheme 4). Moreover, the high temperatures required and the production of free hydrogen chloride are obvious disadvantages.

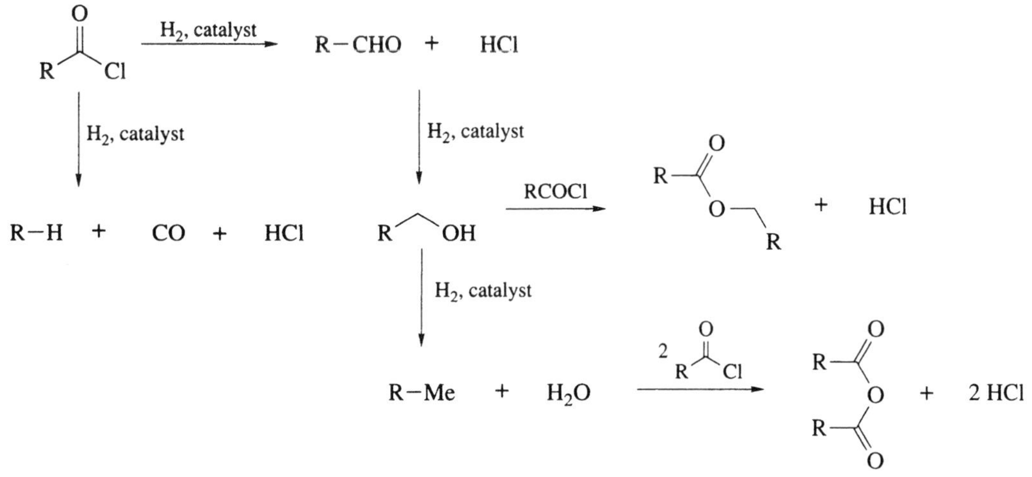

RCOCl $\xrightarrow{H_2,\ catalyst}$ R–CHO + HCl

RCOCl $\xrightarrow{H_2,\ catalyst}$ R–H + CO + HCl

R–CHO $\xrightarrow{H_2,\ catalyst}$ RCH_2OH $\xrightarrow{RCOCl}$ RCO_2CH_2R + HCl

RCH_2OH $\xrightarrow{H_2,\ catalyst}$ R–Me + H_2O $\xrightarrow{2\ RCOCl}$ $(RCO)_2O$ + 2 HCl

Scheme 4

However, in 1944 it was reported by Sakurai and Tanabe that with acetone or ethyl acetate as solvent, Pd–$BaSO_4$ as catalyst (no regulator) and in the presence of *N*,*N*-dimethylaniline as an HCl acceptor, the Rosenmund reduction occurs cleanly and rapidly at room temperature.[16] More recently, three other groups have reported mild and effective variants of the Rosenmund. In each case the key element appears to be the use of a basic additive to remove the HCl produced by the reduction and possibly also to regulate the catalyst activity. All the methods involve the use of closed hydrogenation apparatus and thus have a clear safety advantage over the classical procedure.

In the method of Rachlin *et al.*,[17] the basic additive is sodium acetate, the catalyst 10% palladium on carbon, the solvent toluene and the temperature 35–40 °C. The reaction is carried out in a medium pressure hydrogenation apparatus at 50 p.s.i. (1 p.s.i = *ca.* 6.9 kPa) of hydrogen. Several aromatic aldehydes have been produced in good yields, including examples containing nitro and benzyloxy groups.

Peters and van Bekkum[18,19] noted that the original method of Sakurai and Tanabe suffered from problems due to the competitive reduction of the *N*,*N*-dimethylaniline. In their modified procedure, diisopropylethylamine is used as HCl acceptor/catalyst regulator, and 10% palladium on carbon as catalyst. As for the Sukarai–Tanabe procedure, the reaction is carried out at room temperature and atmospheric pressure in acetone[18] or ethyl acetate.[19] In the latter solvent, which is said to be marginally preferable, the reduction is complete within minutes for aliphatic acyl chlorides, although aromatic substrates may take several days.

In the third procedure, due to Burgstahler *et al.*,[20] the basic additive is 2,6-dimethylpyridine. For aliphatic acyl chlorides, THF is used as solvent and 10% palladium on carbon or 5% Pd–$BaSO_4$ as catalyst. Aromatic acyl chlorides require benzene as solvent, 10% palladium on carbon/quinoline-S as catalyst and a raised temperature.

Table 4 lists most of the results reported in two papers by Peters and van Bekkum[19] and Burgstahler *et al.*[20] The reactions were carried out at room temperature except where otherwise specified, and the yields, where given, are of purified product (with the exception of entry 23). It can be seen that the 'amine-modified Rosenmund' is quite broad in scope, giving clean conversions and good to excellent yields. Ester and keto groups are tolerated. Acyl chlorides containing either C═C double bonds or aromatic nitro groups can be reduced selectively to aldehydes, provided appropriate conditions are chosen; only the method of Peters and van Bekkum (Table 4, entries 1–15) is suitable for nitroaromatics, and the conditions employing 10% palladium on carbon, 2,6-dimethylpyridine and THF (entries 16–26) were found to hydrogenate C═C double bonds. Notable are the successful syntheses of hindered aldehydes such as (**9**) (entry 26) and (**10**).[18] Tertiary acyl chlorides are particularly inclined to decarbonylation during the Rosenmund reduction, and, indeed, application of the classical conditions to the precursor of (**10**) gave only *t*-butylcyclohexane.[18]

The mechanism of the Rosenmund reduction has attracted occasional attention. The discovery of Tsuji *et al.*[21] that palladium metal can catalyze the conversion of acyl chlorides (**11**) into alkenes (**13**), carbon monoxide and HCl, presumably *via* intermediates of the form (**12**; Scheme 5), suggests that (**12**) may also be an intermediate in the Rosenmund reaction. Consistent with this suggestion is the observation that the unsaturated acyl chloride (**14**) is converted to phenol by a palladium catalyst.[22]

Affrossman and Thomson[13] used a ^{14}C-labeling experiment to confirm that overreduction in the Rosenmund is a result of further reduction of the aldehyde product, and not a competing, direct reduction

Table 4 Aldehydes Prepared by Modified Rosenmund Reduction

Entry	*R*	*Catalyst*	*HCl acceptor*	*Solvent*	*Yield (%)*	*Ref.*
1	Heptanal	10% Pd–C	Pr^i_2NEt	EtOAc	a	19
2	c-$C_6H_{12}CHO$	10% Pd–C	Pr^i_2NEt	EtOAc	a	19
3	Benzaldehyde	10% Pd–C	Pr^i_2NEt	EtOAc	a	19
4	4-Methylbenzaldehyde	10% Pd–C	Pr^i_2NEt	EtOAc	a	19
5	4-*t*-Butylbenzaldehyde	10% Pd–C	Pr^i_2NEt	EtOAc	a	19
6	4-Methoxybenzaldehyde	10% Pd–C	Pr^i_2NEt	EtOAc	a	19
7	4-Chlorobenzaldehyde	10% Pd–C	Pr^i_2NEt	EtOAc	a	19
8	3-Chlorobenzaldehyde	10% Pd–C	Pr^i_2NEt	EtOAc	a	19
9	4-Nitrobenzaldehyde	10% Pd–C	Pr^i_2NEt	EtOAc	a	19
10	2,4,6-Trimethylbenzaldehyde	10% Pd–C	Pr^i_2NEt	EtOAc	a	19
11	3,5-Di-*t*-butylbenzaldehyde	10% Pd–C	Pr^i_2NEt	EtOAc	a	19
12	3,4,5-Trimethoxybenzaldehyde	10% Pd–C	Pr^i_2NEt	EtOAc	a	19
13	Ferrocenylcarbaldehyde	10% Pd–C	Pr^i_2NEt	EtOAc	a	19
14	$Ph(CH_2)_2CHO$	10% Pd–C	Pr^i_2NEt	EtOAc	a	19
15	(*E*)-PhCH=CHCHO	10% Pd–C	Pr^i_2NEt	EtOAc	a	19
16	Dodecanal	10%Pd–C	2,6-Dimethylpyridine	THF	95	20
17	Octadecanal	10%Pd–C	2,6-Dimethylpyridine	THF	96	20
18	$OCH(CH_2)_4CHO$	10%Pd–C	2,6-Dimethylpyridine	THF	74	20
19	$OCH(CH_2)_8CHO$	10% Pd–C	2,6-Dimethylpyridine	THF	77	20
20	$MeCO(CH_2)_6CHO$	10% Pd–C	2,6-Dimethylpyridine	THF	85	20
21	$Pr^iO_2C(CH_2)_2CHO$	10% Pd–C	2,6-Dimethylpyridine	THF	89	20
22	$Pr^iO_2C(CH_2)_3CHO$	10% Pd–C	2,6-Dimethylpyridine	THF	89	20
23	3-Oxo-5-octanolide	10% Pd–C	2,6-Dimethylpyridine	THF	94	20
24	2-Ethylhexanal	10% Pd–C	2,6-Dimethylpyridine	THF	87	20
25	c-C_5H_9CHO	10% Pd–C	2,6-Dimethylpyridine	THF	78	20
26	(**9**)[b]	10% Pd–C	2,6-Dimethylpyridine	THF	92	20
27	(*Z*)-n-$C_8H_{17}CH=CH(CH_2)_7CHO$	5% Pd–$BaSO_4$	2,6-Dimethylpyridine	THF	96	20
28	(*E*)-PhCH=CHCHO	5% Pd–$BaSO_4$	2,6-Dimethylpyridine	THF	84	20
29	Benzaldehyde[c]	10% Pd–C/quinoline-S	2,6-Dimethylpyridine	C_6H_6	93	20
30	4-Chlorobenzaldehyde[c]	10% Pd–C/quinoline-S	2,6-Dimethylpyridine	C_6H_6	77	20

[a]Reaction followed to completion by g.c. Negligible overreduction and decarbonylation.[19] [b]Carried out at 0 °C. [c]Carried out at 40–50 °C.

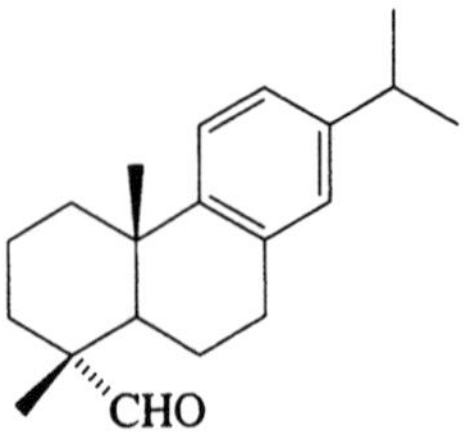

CHO

(9)

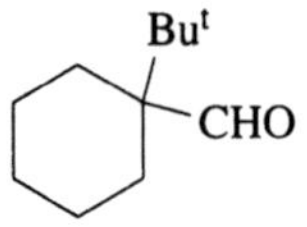

Bu^t CHO

(10)

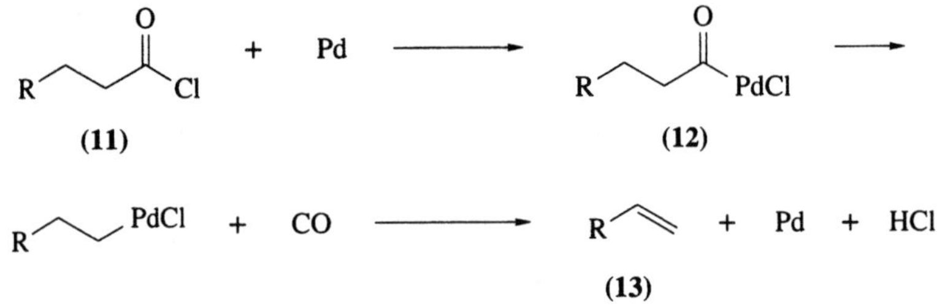

R–C(O)Cl + Pd → R–C(O)PdCl →

(11) **(12)**

R–PdCl + CO → R–CH=CH₂ + Pd + HCl

(13)

Scheme 5

Cl

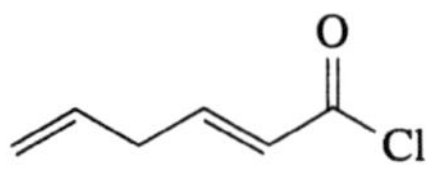

(14)

of the acid chloride. They also investigated the nature of the poisoning of the catalyst in the classical Rosenmund, without being able to draw any firm conclusions. More recently, McEwen *et al.*[23] argued that, considering the wide range of catalyst regulators employed in the Rosenmund, the role of the regulators might not be to form specific chemical bonds with the palladium but rather to alter its physical structure. They found that: (i) treatment of palladium on silica with indole gave an effective Rosenmund catalyst which did not, however, contain any indole; (ii) treatment of palladium black with triethylamine at 150 °C for 20 h causes rearrangement into metallic palladium with a much lower surface area; and (iii) ultrapure palladium powder is a catalyst (though somewhat inactive) for the conversion of benzoyl chloride to benzaldehyde at 0 °C.

The next group of methods to be considered represents a far more recent development than the Rosenmund: the use of the anionic metal complexes $[Fe(CO)_4]^{2-}$, $[HFe(CO)_4]^-$, $[HCr(CO)_5]^-$, $[HW(CO)_5]^-$ and $[CpV(CO)_3H]^-$. Although it is only the first of these reagents which is clearly a nonhydride reducing agent, they are all considered here because of their close structural relationship and because of possible mechanistic similarities (*vide infra*).

The method involving $[Fe(CO)_4]^{2-}$ was reported first by Watanabe *et al.*[24] The actual reagent used is the disodium salt of the anion, formed as a solution in THF by the reduction of $[Fe(CO)_5]$ by sodium amalgam. As shown in Scheme 6, the anion reacts with acyl chlorides to give acyltetracarbonylferrates (**15**), which can be decomposed with acetic acid to yield the corresponding aldehydes. This method has been applied to give simple aldehydes, typical examples and yields (by gas chromatography) being benzaldehyde 95%, *o*-chlorobenzaldehyde 65% and isobutyraldehyde 71%. Subsequently, an essentially similar procedure was reported by Siegl and Collman.[25] These authors placed more emphasis on the structural organometallic chemistry involved, and were able to isolate the intermediate anions (**15**) as salts of $[Ph_3P{=}N{=}PPh_3]^+$.

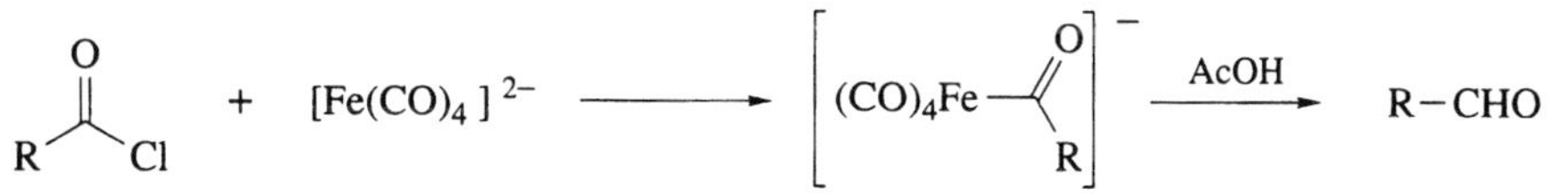

$$\mathrm{RC(O)Cl} + [\mathrm{Fe(CO)_4}]^{2-} \longrightarrow \left[(\mathrm{CO})_4\mathrm{Fe}{-}\mathrm{C(O)R}\right]^- \xrightarrow{\mathrm{AcOH}} \mathrm{R{-}CHO}$$

(**15**)

Scheme 6

The remainder of the anionic metal complexes, all of which bear a hydrogen, have been employed subsequently. The most convenient experimentally appears to be the $[HFe(CO)_4]^-$ anion, employed either as the Me_4N^+ salt[26] or in immobilized form on an anion exchange resin.[27] The former reagent is readily synthesized from $[Fe(CO)_5]$, Me_4NBr and aqueous potassium hydroxide. In aprotic solvents such as dichloromethane, it reacts with acid chlorides giving direct formation of the corresponding aldehydes in 1–4 h at room temperature. The stoichiometry of the reaction is shown in equation (1). As some of the reagent is consumed by formation of the inert by-product (**16**), at least 1.5 equiv. are required. The method has been applied to a number of simple aldehydes, as shown in Table 5. It can be seen that it is incompatible with an aromatic nitro group, and unsatisfactory with cinnamoyl chloride. The alternative, polymer bound version of the reagent is also easy to prepare, and has the usual advantage of being easily separated from the product. It has been used to prepare several aldehydes in good isolated yields, including $OCH(CH_2)_8CHO$ (85%) and $MeO_2C(CH_2)_7CHO$ (90%).[27]

$$2\,\mathrm{RCOCl} + 3\,[\mathrm{NMe_4}][\mathrm{HFe(CO)_4}] \longrightarrow 2\,\mathrm{RCHO} + 2\,\mathrm{NMe_4Cl} + \underset{(\mathbf{16})}{[\mathrm{NMe_4}][\mathrm{HFe_3(CO)_{11}}]} + \mathrm{CO} \quad (1)$$

The remaining complex anions, $[HCr(CO)_5]^-$,[28] $[HW(CO)_5]^-$,[28] and $[CpV(CO)_3H]^-$,[29] have been used as their salts with the bulky cation $[Ph_3P{=}N{=}PPh_3]^+$. They are capable of reducing acyl halides to aldehydes in good yield and at a much faster rate than $[HFe(CO)_4]^-$. These reagents are, however, somewhat less convenient to prepare. Although they are also capable of reducing alkyl halides, they are far more reactive towards acyl chlorides. Unlike $[HFe(CO)_4]^-$, $[HCr(CO)_5]^-$ has been shown to be fairly unreactive towards nitrobenzene. A useful feature is that it undergoes rapid H/D exchange with MeOD, giving a reagent for the conversion of RCOCl to RCDO.

The mechanism of action of these metal hydrides is somewhat uncertain. In the original report on $[HFe(CO)_4]^-$, it was suggested that, by analogy with $[Fe(CO)_4]^{2-}$, the anion displaces a chloride ion from the acyl chloride to give an intermediate (**17**) which then collapses to give the aldehyde (Scheme 7).[26] However, there are alternative possibilities, particularly that the reagents act as H-atom donors (like

Table 5 Aldehydes RCHO from RCOCl by Treatment with $NMe_4{\cdot}HFe(CO)_4$[26]

R	*Yield (%)* *G.c.*	*Isolated*
Phenyl	91	
2-Furyl	90	
Hexyl	99	
Cyclohexyl	95	
p-Bromophenyl	80	75
Isopropyl	99	82[a]
t-Butyl		80
Methyl	100	
PhCH=CH	22	
p-Nitrophenyl	0	

[a]As 2,4-dinitrophenylhydrazone.

Bu_3SnH). Although the question was subjected to a detailed investigation in the case of $[CpV(CO)_3H]^-$, it did not prove possible to give a definitive answer.[29]

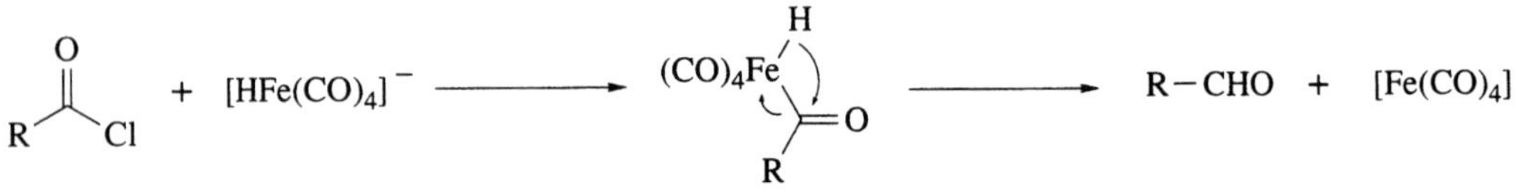

$$\text{RCOCl} + [\text{HFe(CO)}_4]^- \longrightarrow (\text{CO})_4\text{Fe(H)(COR)}\ \mathbf{(17)} \longrightarrow \text{R–CHO} + [\text{Fe(CO)}_4]$$

(17)

Scheme 7

Closely related to the anionic complexes described above are the neutral species $[Cp_2M(CO)H]$ (M = Nb or Ta), also capable of reducing acyl chlorides to aldehydes.[30] The reactions are reported to be virtually instantaneous, and the reagents appear to have useful selectivity. For example, *p*-nitrobenzoyl chloride and cinnamoyl chloride are reduced to their corresponding aldehydes (although yields are not recorded). However, as with the Cr, W and V anionic hydrides, these neutral complexes may prove too inaccessible to be generally useful.

A reduction method which is conceptually somewhat similar to the use of $[Fe(CO)_4]^{2-}$ is shown in Scheme 8.[31] Treatment of an aroyl chloride with phospholen (**18**) and triethylamine gives a salt (**19**) which can be decomposed with water to give the corresponding aldehyde and phospholen oxide (**20**). It appears that ring strain in (**19**) is crucial in directing attack of water to the phosphorus, as opposed to the carbonyl carbon. The method gives good yields for a number of aroyl chlorides (*e.g.* benzoyl chloride, converted to benzaldehyde in 80% yield) but is limited in scope, being inapplicable to aliphatic acyl chlorides and to aroyl chlorides containing $—NO_2$ and —OH groups.

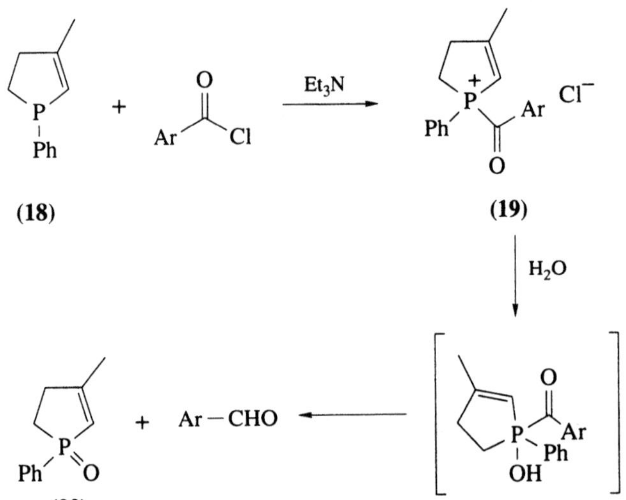

(18) + ArCOCl —(Et_3N)→ (19) [P⁺(Ph)(COAr) phospholenium, Cl^-] —(H_2O)→ [P(OH)(Ph)(COAr) intermediate] → (20) + Ar—CHO

(18) **(19)** **(20)**

Scheme 8

Recently it has been shown by Chikashita *et al.* that hydride transfer is possible from the benzimidazoline (**21**) to an acyl chloride, giving the corresponding aldehyde and benzimidazolium salt (**22**; equation 2).[32] The reaction is most effective in the presence of 1 mol equiv. of acetic acid. Although it has only been used for a few aldehydes, it is successful with aromatic and aliphatic examples (*e.g.* *p*-nitrobenzaldehyde, 82% g.c. yield; cyclohexanecarbaldehyde, 80% g.c. yield) and may have substantial potential. Another new method is the treatment of aroyl chlorides with dialkylzinc reagents in the presence of catalytic palladium complexes.[33] However, the applicability would seem to be limited by the rather low yields.

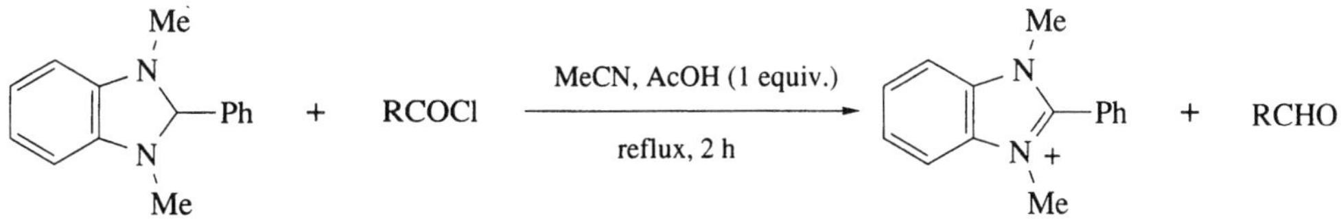

(**21**) + RCOCl —[MeCN, AcOH (1 equiv.), reflux, 2 h]→ (**22**) + RCHO (2)

Finally, mention should be made of the Grundmann method which, though long winded and unlikely to be of use today, was once of some importance.[34] It is summarized in Scheme 9. In favorable cases the sequence could be carried through in over 50% overall yield, and it was valued for its ability to synthesize aldehydes inaccessible *via* the classical Rosenmund.

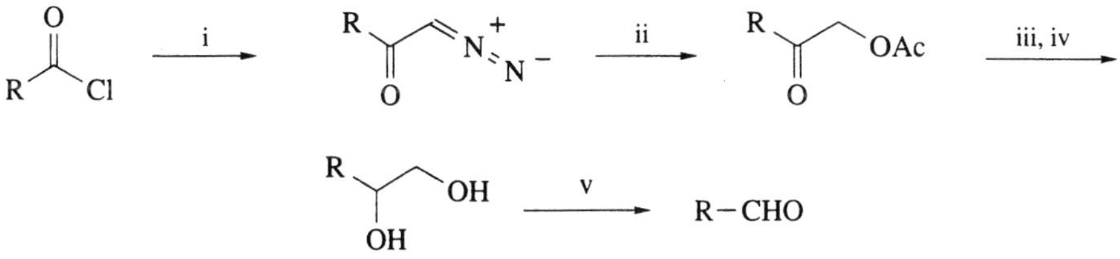

RCOCl —i→ $RCOCH=N^+=N^-$ —ii→ $RCOCH_2OAc$ —iii, iv→ $RCH(OH)CH_2OH$ —v→ R−CHO

i, CH_2N_2; ii, AcOH; iii, $Al(OPr^i)_3$; iv, aq. HCl; v, $Pb(OAc)_4$

Scheme 9

1.12.4 CARBOXYLIC ACID ANHYDRIDES

The reduction of carboxylic acid anhydrides to aldehydes does not appear to have attracted much attention, despite the intrinsic reactivity of anhydrides. The only well-developed method is the use of $Na_2[Fe(CO)_4]$ (Scheme 10),[35,36] closely analogous to the use of the same reagent to reduce acyl chlorides (*vide supra*). Again, the reaction proceeds *via* acyltetracarbonylferrates, the formation of which can be monitored by IR. The method has been used to produce simple aldehydes from the corresponding symmetrical anhydrides in fair to good g.c. yields (*e.g.* propanal, 90%; butanal, 61%; benzaldehyde, 73%).[36] It is also successful with cyclic anhydrides, giving the corresponding aldehydic acids in quite good isolated yields, as shown in Table 6. In practical terms, this is probably the most valuable application.

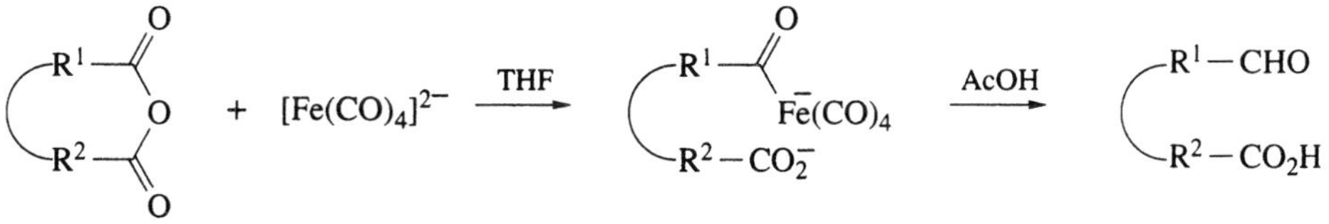

cyclic anhydride (R^1, R^2) + $[Fe(CO)_4]^{2-}$ —THF→ $R^1-C(O)\bar{F}e(CO)_4$ / $R^2-CO_2^-$ —AcOH→ R^1-CHO / R^2-CO_2H

Scheme 10

With mixed diaryl or alkyl aryl anhydrides, the regioselectivity of attack by the tetracarbonylferrate is, not surprisingly, too poor to be synthetically useful.[36] However, the reagent can be usefully applied to the mixed carboxylic alkylcarbonic anhydrides (**23**), giving aldehydes after treatment of the intermediate acylferrates with acetic acid (equation 3).[37] The substrates (**23**) are readily available from carboxylic acids by treatment with ethylchloroformate and triethylamine. Unfortunately the quoted g.c. yields are variable (*e.g.* benzaldehyde, 41%; *p*-tolualdehyde, 76%; nonanal, 51%), suggesting that further development may be required.

Table 6 Cyclic Carboxylic Acid Anhydrides Reduced to the Corresponding Aldehydic Acids by $Na_2[Fe(CO)_4]$[36]

Anhydride	*Isolated yield of aldehydic acid (%)*
Succinic	81[a]
Glutaric	60
Phthalic	61
Tetrachlorophthalic	83
1,8-Naphthalic	75

[a]As 2,4-dinitrophenylhydrazone

$$\mathrm{RC(O)OC(O)OEt}\ \textbf{(23)}\ \xrightarrow[\text{ii, AcOH}]{\text{i, Na}_2[\text{Fe(CO)}_4],\ \text{THF}}\ \text{R–CHO} \qquad (3)$$

It is also possible to reduce anhydrides to aldehydes by catalytic hydrogenation, although there is no method which has been shown to be generally useful. Acetic and propionic anhydrides have been hydrogenated with Pd–$BaSO_4$ as catalyst, giving the corresponding aldehydes with reasonable selectivities (acetaldehyde, 78%; propanal, 90%) but low conversions.[38] Another method employing catalytic $[Co_2(CO)_8]$ and high pressures of H_2 and CO gives yields of 57% or less with simple acyclic and cyclic anhydrides.[39] High temperature hydrogenation over a metal oxide–alumina catalyst (see Section 1.12.2) has also been used with anhydrides.[9]

1.12.5 CARBOXYLIC ACID ESTERS

There is no generally useful nonhydride method for the direct reduction of carboxylic acid esters to aldehydes. There are, however, procedures which are valuable under particular circumstances. An important example is the one-electron reduction of aldonolactones to aldoses. Two factors presumably contribute to the success of these reactions; firstly the presence of electron-withdrawing substituents in the substrates, raising the reactivity of the carbonyl group, and secondly the ability of the products to form cyclic hemiacetals stable to further reduction.

The use of sodium amalgam originates with E. Fischer.[40] The method was a cornerstone of his aldose homologation (cyanohydrin formation, hydrolysis, lactone formation and reduction) which was so important to the development of carbohydrate chemistry. Although the yields obtained by Fischer were moderate (*ca.* 20–50%), more recent work by Sperber *et al.* has resulted in significant improvements.[41] In particular, they discovered that control of the pH of the reaction mixture was very important. At pH 3–3.5, yields in the range 52–82% were obtained with a variety of aldonolactones. As an example, the preparation of arabinose is shown in equation (4). If the pH was allowed to rise, yields were lower due to overreduction. Methyl esters of aldonic acids could also be used as substrates.

$$\xrightarrow[\text{pH 3–3.5, 9 °C, 8 min; 75\%}]{\text{2.5\% Na/Hg, aq. H}_2\text{SO}_4} \qquad (4)$$

An alternative which is attractive for large scale work is the electrochemical reduction of aldonolactones.[42,43] Particular attention has been paid to the electroreductive synthesis of ribose from ribonolactone because of the importance of the former in the synthesis of riboflavin (vitamin B_2).[43] Processes generally involve a mercury cathode and maintenance of an acidic pH, often with the assistance of a phosphate or borate buffer. It has been reported that alkali metal ions are also necessary, suggesting that the reduction occurs *via* metal amalgam formation.[42] However, other accounts make no mention of metal ions.[44]

Early work showed that the reduction of aldonolactones to aldoses may also be carried out by catalytic hydrogenation over PtO_2 in aqueous solution.[45] Using this method, D-glucono-1,5-lactone was reduced to glucose in up to 80% yield, the remainder of the product being D-gluconic acid. However, careful optimization was required to achieve this result, and there appears to have been little further development of the procedure.

Whereas acyl chlorides and carboxylic acid anhydrides can be reduced to aldehydes using $Na_2[Fe(CO)_4]$ (*e.g.* Scheme 10) most esters cannot, because of their low reactivity as acylating agents. However, $Na_2[Fe(CO)_4]$ does give low yields of aldehydes (<40%) with esters of aromatic alcohols.[46]

Finally, some of the methods for carboxylic acid reduction (Section 1.12.2) have also been applied to esters. Thus, the ethyl ester of thiazole-2-carboxylic acid has been reduced electrochemically to the corresponding aldehyde,[4] and some of the high temperature metal oxide catalyzed hydrogenations have been used on esters.[8,9]

1.12.6 CARBOXYLIC ACID THIOL ESTERS

One of the classical reductive syntheses of aldehydes is the treatment of carboxylic acid thiol esters with Raney nickel.[34] The first report of this reaction, due to Wolfrom and Karabinos,[47] appeared around the same time as a report from Jeger *et al.* that Raney nickel under similar conditions reduced thiol esters to alcohols.[48] Later investigations by a group at the Upjohn Company[49–51] confirmed the latter result, but showed that if the Raney nickel was first deactivated by boiling in acetone (for *ca.* 1 h) it could indeed be used to convert thiol esters to aldehydes (equation 5). Some of the results obtained using the method are summarized in Table 7. It is notable that carbon–carbon double bonds in positions 5 and 11 of the steroid nucleus were unaffected. Thiol esters derived from a variety of thiols (R^1SH) could be used as substrates, the results being marginally better for R^1 = Et or CH_2Ph than for Ph or Pr^i. The thiol esters were prepared from the corresponding acid chlorides using lead sulfides, or thiols in the presence of pyridine.

$$R{-}C(=O){-}SR^1 \xrightarrow[\text{reflux } ca.\text{ 1 h}]{\text{Raney Ni (deactivated), acetone}} R{-}C(=O){-}H \qquad (5)$$

Table 7 The Reduction of Thiol Esters $RCOSR^1$ to Aldehydes RCHO by Deactivated Raney Nickel

RCHO	R^1	*Yield (%)*	*Ref.*
PhCHO	Et	62[a]	47
EtCHO	Et	73[a]	47
3β-Acetoxy-5-cholen-24-al	Et	63, 67[b]	50
3β-Acetoxy-5-cholen-24-al	$PhCH_2$	68[b]	50
3β-Acetoxy-5-cholen-24-al	Ph	62[b]	50
3β-Acetoxy-5-cholen-24-al	Pr^i	49[b]	50
3α-Formoxy-5-cholen-24-al	Et	78	51
3α-Acetoxy-11-cholen-24-al	Et	53	51

[a]As $NaHSO_3$ adduct. [b]As 2,4-dinitrophenylhydrazone.

As thiol esters are fairly good acylating agents, they can act as substrates for the $[Fe(CO)_4]^{2-}$ procedure.[46] Thus, treatment of four simple *p*-chlorophenylthiol esters with $Na_2[Fe(CO)_4]$/THF followed by acetic acid leads to the corresponding aldehydes in low to reasonable yields (nonanal, 45%; benzaldehyde, 49%; *p*-chlorobenzaldehyde, 56%; *p*-tolualdehyde, 70%).

1.12.7 CARBOXYLIC ACID AMIDES

Although amides are the least reactive of the common acid derivatives under most circumstances, they are involved in several reductive syntheses of aldehydes. As described below the conversion can be accomplished by one-electron reducing agents (though relatively powerful ones), and there are specific types of amide which can be caused to decompose to aldehydes. Moreover, amides are precursors of imidoyl chlorides, Vilsmeier complexes and thioamides, which can also be reduced to aldehydes as discussed in later sections.

A number of groups have reported one-electron reductions of amides to aldehydes. Most involve the use of ammonia or methylamine as solvent, with electrons provided electrochemically or by an alkali metal. It must be said, however, that none of the methods is sufficiently effective or general to be routinely useful. In one of the earlier procedures, due to Birch *et al.*,[52] the tertiary amides (**24**) to (**26**) were reduced to the corresponding aldehydes in 80%, 53% and 47% yields, respectively, using sodium in liquid ammonia containing acetic acid. The use of acetic acid as a proton source was crucial to success; with ethanol yields were far lower, the acyl portion of the molecule being converted to nitrogen-free products. It was postulated that the reaction occurred as shown in Scheme 11. The intermediate hemiaminal (**27**) would be stable toward further reduction, and if it survived would yield the aldehyde RCHO in the work-up. However, if the proton donor BH were too weak an acid, base-catalyzed decomposition of (**27**) would occur within the reaction mixture, resulting in overreduction.

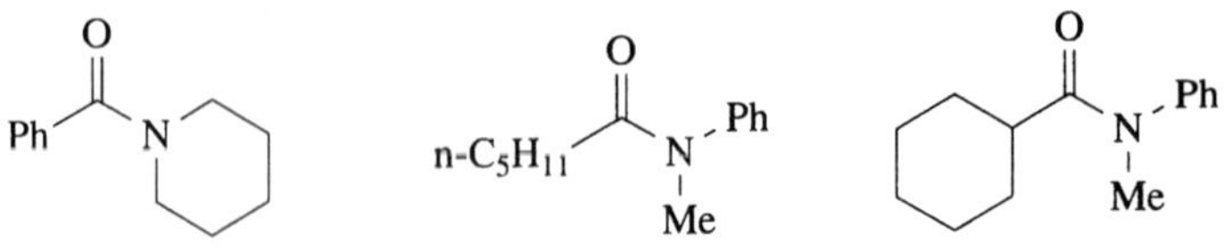

(**24**) (**25**) (**26**)

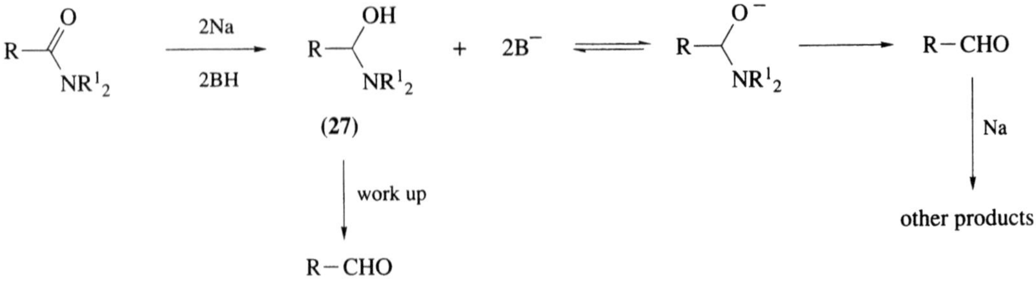

$RCONR^1_2 \xrightarrow[2BH]{2Na} RCH(OH)NR^1_2$ (**27**) $+ 2B^- \rightleftharpoons RCH(O^-)NR^1_2 \longrightarrow R-CHO \xrightarrow{Na}$ other products

(**27**) $\xrightarrow{\text{work up}}$ R−CHO

Scheme 11

In a later development by Bedenbaugh *et al.*,[53] methylamine was used as solvent and lithium as electron donor. No proton donor was required, suggesting that the lithium salt (**28**) of hemiaminal (**27**) is stable under the reaction conditions (both aldehydes and aldimines are reduced by the reagent; *cf.* the analogous reduction of carboxylic acids, Section 1.12.2 and Scheme 2). Yields of aldehydes produced by this method are shown in Table 8. It is notable that only tertiary amides are reduced satisfactorily. A major limitation of the reaction is the substantial formation of side products resulting from transamidation by the methylamine solvent (*i.e.* RCONHMe from $RCONR^1_2$).

Table 8 Amides Reduced to the Corresponding Aldehydes by (A) $Li/MeNH_2$[53] or (B) $e^-/MeNH_2/LiCl/EtOH$[54]

Amide	*Method*	*Yield of Aldehyde (%)*
$Bu^nCONHMe$	A	4.4
n-$C_7H_{15}CONHMe$	A[a]	13.2
n-$C_7H_{15}CONMe_2$	A	51.7
n-$C_9H_{19}CONMe_2$	A	
n-$C_9H_{19}CONH_2$	B	28
n-$C_9H_{19}CONHMe$	B	58
n-$C_9H_{19}CONMe_2$	B	45

[a]Large excess of Li used.

Amides may be reduced to aldehydes electrochemically in methylamine, with lithium chloride as the supporting electrolyte.[54] Despite the apparent similarity between this and the foregoing method, the results are rather different. With no added proton donor, overreduction occurs leading to the corresponding alcohol. It is only in the presence of ethanol that aldehydes are obtained in reasonable yields. Another difference is that secondary amides give satisfactory yields, as shown in Table 8 where this and the foregoing method can be compared.

It has been mentioned in earlier Sections that certain electron deficient heterocyclic carboxylic acids and esters can be reduced electrochemically to aldehydes under acidic conditions, the aldehydes being protected from overreduction by geminal diol formation. This method is also applicable to the corresponding amides, allowing, for example, the conversion of amide (**29**) into aldehyde (**30**) in 93% yield.[4]

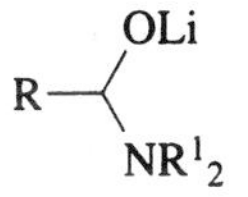

(28) 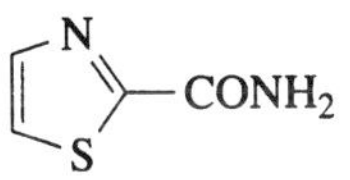 (29) 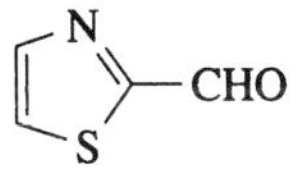 (30)

As mentioned above, there are two particular types of amide which can be induced to decompose to aldehydes without requiring reducing agents. They are the Reissert compounds, to be discussed first, and the sulfonylhydrazides involved in the McFadyen–Stevens method, to be discussed second.

Formally speaking, a Reissert compound[55] is an adduct of an acylium cation and cyanide anion across the C=N double bond of a heterocyclic amine. The process is illustrated in equation (6), taking quinoline as the amine. Isoquinoline and phenanthridine may also be employed as amines (giving (**32**) and (**33**), respectively) as well as a variety of more complex derivatives and analogs. The chemistry of these compounds has been investigated quite intensively over the years, and has been reviewed regularly.[34,56–58] Broadly speaking, they are of interest for two reasons. Currently, the more important is that they are valuable intermediates in heterocyclic synthesis. However, the original focus of attention was the fact that, on treatment with acid, the acyl groups RCO are cleaved from the compounds to give the corresponding aldehydes RCHO.[34,56,57]

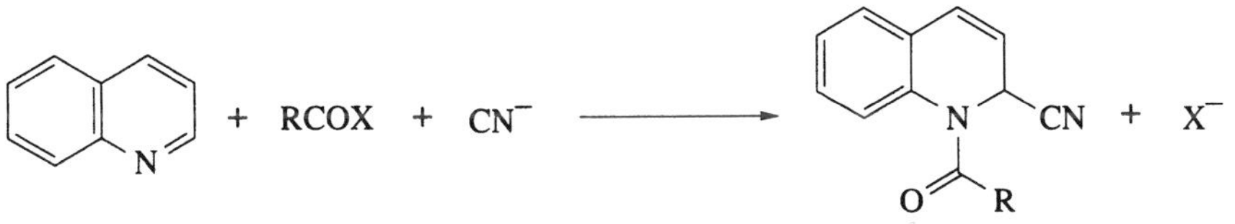

$$+ \ RCOX \ + \ CN^- \longrightarrow \ (31) \ + \ X^- \qquad (6)$$

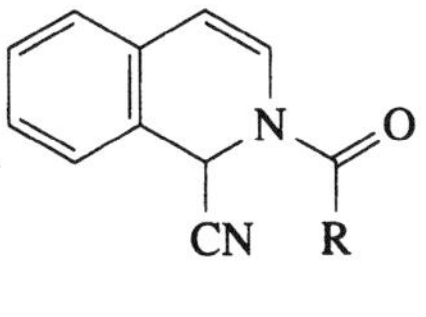

(32) 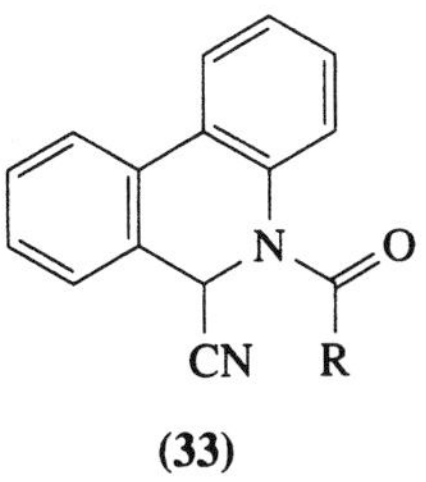 (33)

A mechanism for the latter transformation, suggested by Cobb and McEwan,[59] is shown in Scheme 12. This proposal was supported by the observation that the cleavage was invariably accompanied by hydrolysis of the cyano group giving amide (**34**) as well as the derived carboxylic acid (**35**). The nitrile (**36**) was never isolated. Further support for the participation of the carbonyl oxygen in the reaction was provided by the discovery that isoquinoline-derived Reissert compounds (**32**) could be treated with anhydrous acids to give salts which could be decomposed to aldehydes RCHO with aqueous acid.[60] However, although originally assigned the structure (**37**), these compounds were later reformulated as the amino

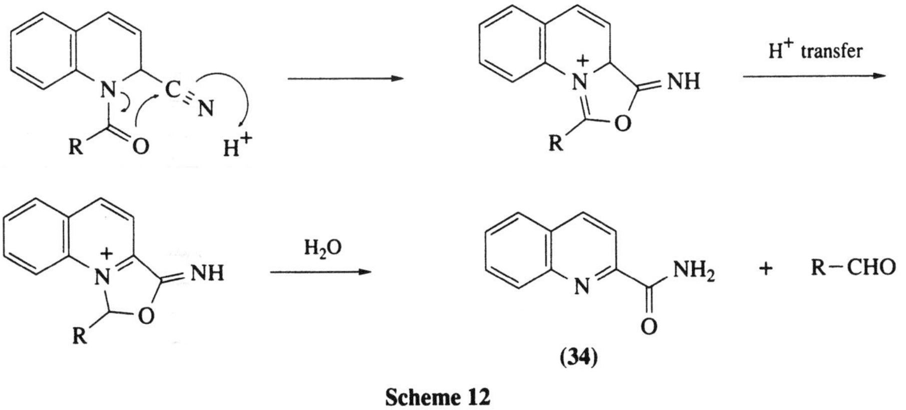

H^+ transfer

H_2O

(34) + R–CHO

Scheme 12

tautomers (**38**).[61] This suggests that (**39**), the analogous tautomer in the quinoline series, should perhaps play a central role in Scheme 12.

The synthesis of the original Reissert's compound (**31**; R = Ph) was accomplished by mixing quinoline and benzoyl chloride in aqueous potassium cyanide.[55] A later development, which broadened the applicability of the reaction, involved the addition of dichloromethane as a solvent for the organic phase.[62] Alternatively, the synthesis can be accomplished by treating the acid chloride with the amine and anhydrous HCN in benzene.[63] The scope and limitations of these two methods are illustrated by the selection of results in Table 9. The following points may be noted: (i) although the yields listed for aldehyde formation from the Reissert compounds are generally excellent (especially for Method B, where aqueous sulfuric acid is used), most refer to the derived 2,4-dinitrophenylhydrazones. For aromatic R, the isolation of the aldehyde itself does not generally pose a problem, but it is not clear that aliphatic aldehydes can be obtained directly; and (ii) Method B, employing anhydrous HCN in benzene, is quite general for aromatic aldehydes and is also successful for unhindered aliphatic R groups. It is not, however, attractive experimentally. Method A, though more convenient, is less general, especially in that it gives lower yields with the more reactive halides. In these cases it is often more successful when isoquinoline is used as the base.

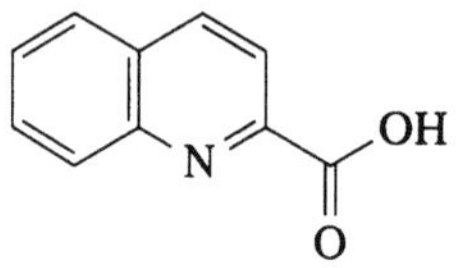

(**35**) 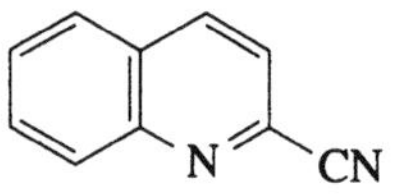 (**36**)

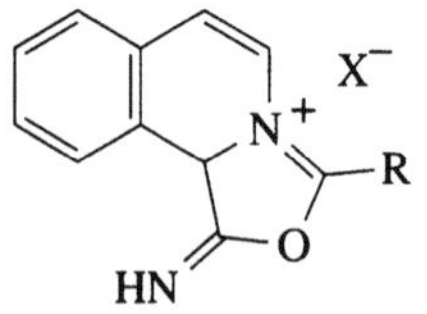

(**37**) 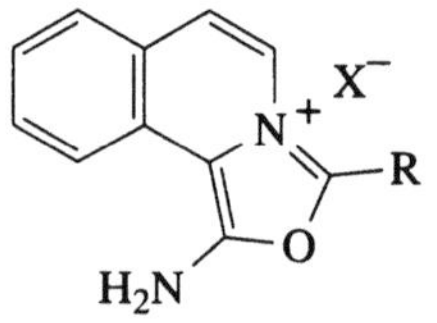 (**38**) 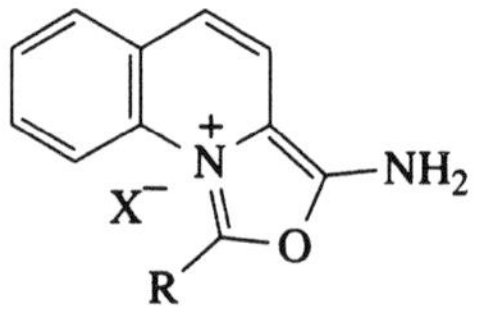 (**39**)

Table 9 Yields of Reissert Compounds (**31**) and (**32**), and Derived Aldehydes RCHO, Prepared by RCOCl/Amine/CH_2Cl_2/aq.KCN (Method A)[62] or RCOCl/Amine/HCN/Benzene (Method B)[63]

Method	*Reissert compound*	*R*	*Yield of Reissert compound (%)*	*Yield of aldehyde (%)*[a]
A	(**32**)	Ph	69	
A	(**31**)	Ph	70	
B	(**31**)	Ph	96	98[b]
A	(**32**)	*p*-Methoxyphenyl	72	90[b]
A	(**31**)	*p*-Methoxyphenyl	80	96[b]
B	(**31**)	*p*-Methoxyphenyl	88	98
A	(**32**)	*p*-Chlorophenyl	30	92[b]
A	(**31**)	*p*-Chlorophenyl	20	
B	(**31**)	*p*-Chlorophenyl	77	92
A	(**32**)	Pr^n	64	57[b]
A	(**31**)	Pr^n	25	
B	(**31**)	Pr^n	64	97[b]
A	(**32**)	Pr^i	11	
A	(**31**)	Pr^i	18	80[b]
B	(**31**)	Pr^i	28	98[b]
A	(**32**)	PhCH═CH	64	
A	(**31**)	PhCH═CH	43	
B	(**31**)	PhCH═CH	91	82

[a]Method A, hydrolysis by conc. HCl or conc. HCl/H_2O 6:5, in presence of 2,4-dinitrophenylhydrazine; method B, hydrolysis by 2.5–5 M aq. H_2SO_4. [b]As 2,4-dinitrophenylhydrazone.

In the light of the foregoing, it is perhaps not surprising that interest has lapsed in the hydrolysis of Reissert compounds as a method of aldehyde synthesis. However, the value of the compounds in heterocyclic chemistry has led to continued developments in their synthesis,[58] and it is not impossible that, for those aldehydes stable to the hydrolysis conditions, the method may at some stage regain its usefulness.

The McFadyen–Stevens aldehyde synthesis[34,64] is based on the reaction shown in Scheme 13, in which arenesulfonylhydrazides (**40**) are decomposed to aldehydes by treatment with base. It is generally presumed that the acylimides (**41**) are intermediates in the procedure. The arenesulfonylhydrazides can be synthesized either from acid hydrazides and arenesulfonyl chlorides[34] or from acyl chlorides and arenesulfonylhydrazines. The latter method is somewhat more common in recent work[65,66] and is reported to give superior yields.[65]

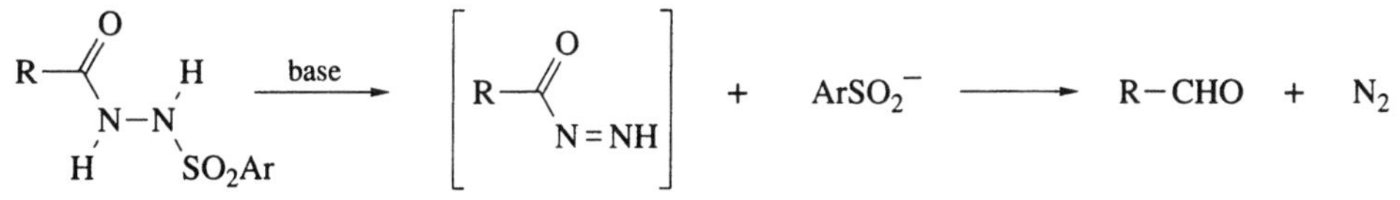

$$\underset{\textbf{(40)}}{RC(=O)NH{-}NHSO_2Ar} \xrightarrow{\text{base}} \underset{\textbf{(41)}}{\left[RC(=O)N{=}NH\right]} + ArSO_2^- \longrightarrow R{-}CHO + N_2$$

Scheme 13

In the classical McFadyen–Stevens method, the Ar group in (**40**) is generally phenyl or *p*-tolyl and the conditions for the decomposition are sodium or potassium carbonate in ethylene glycol at 160 °C for *ca.* 2 min.[34] This procedure is suitable for the synthesis of aromatic aldehydes, and appears to have been quite widely used. Table 10 shows a few examples; many more are tabulated elsewhere.[34] It can be seen that the best yields are obtained with electron rich aromatic nuclei.

Table 10 Yields of Aldehydes Prepared from Arenesulfonylhydrazides (**40**; Ar = Ph or *p*-tolyl) by Treatment with Alkali Metal Carbonate in Hot Ethylene Glycol[34]

Aldehyde	*Yield (%)*
Benzaldehyde	70
p-Methoxybenzaldehyde	77
3,4-Methylenedioxybenzaldehyde	87
m-Fluorobenzaldehyde	50
m-Nitrobenzaldehyde	42
3-Quinolinecarbaldehyde	33

Curiously, a study of the mechanism of the reaction revealed that the presence of a finely divided solid is necessary for the decomposition.[67] A variety of solids such as sodium carbonate, charcoal, zinc powder and powdered glass were all effective, and addition of the latter to the classical reaction conditions improved the yields slightly (*e.g.* PhCHO, 81%).

Quite recently, an attempt was made to improve the method by increasing the bulk of the Ar group in (**40**), employing 2,4,6-triisopropylphenyl as opposed to phenyl or *p*-tolyl.[66] As expected, the new derivatives were found to decompose under far milder conditions (K_2CO_3, methanol, reflux, 15–60 min), presumably due to steric acceleration of the S—N bond cleavage. Unfortunately, for reasons which are unclear, the yields of the resulting aldehydes were poorer than those from the classical method. Nonetheless, if the reaction was carried out in the presence of hydrazine, good yields of the corresponding aldehyde hydrazones could be obtained, and the method could be useful under some circumstances.

None of the above conditions are suitable for the production of aliphatic aldehydes, and recent work has been aimed at a remedy for this deficiency. Sprecher *et al.* demonstrated that aliphatic aldehydes with no α-H could be produced in moderate yields if the reaction time was reduced to 30 s.[68] Babad *et al.* then showed that volatile aldehydes could be isolated (again in moderate yields) if they were distilled from the reaction mixture as they were formed.[65] Both results imply that the difficulty with aliphatic aldehydes is their instability in the reaction mixture.

Finally, Nair and Schechter developed a version of the McFadyen–Stevens which does appear to be genuinely useful for aliphatic aldehydes, provided that they are reasonably volatile.[69] The salts (**42**) are produced from the corresponding tosylhydrazides by treatment with NaOMe/MeOH followed by evaporation, and are then heated to 140–155 °C under a vacuum. The product aldehydes RCHO are collected in a cold trap. The method was used for butanal (68% yield), 2-methylpropanal (71% yield), 3-phenylpropanal (85% yield) and 10-undecenal (60% yield). Yields were also good for simple aromatic aldehydes.

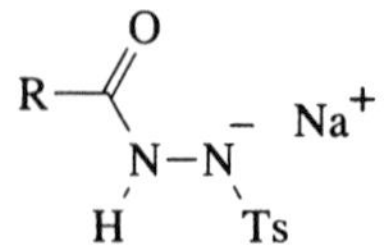

R–C(=O)–NH–N⁻(Ts) Na⁺

(42)

1.12.8 NITRILES

Nitriles appear to be quite useful as precursors of aldehydes, and there are a number of 'nonhydride' methods for accomplishing the conversion.[70] Classically the best known is probably the Stephen reduction.[34,70,71] It involves a reagent system prepared from anhydrous tin(II) chloride and hydrogen chloride gas in ether. The nitrile is added to this mixture, and, as shown in Scheme 14, adds HCl to give the imidoyl chloride salt (**43**), which is then reduced by the tin(II) chloride. The initial product is the complex (**44**) (often crystalline), which is then hydrolyzed to the aldehyde. Although the reaction conditions are generally thought of as anhydrous, it has in fact been found that a small amount of water is beneficial.[34]

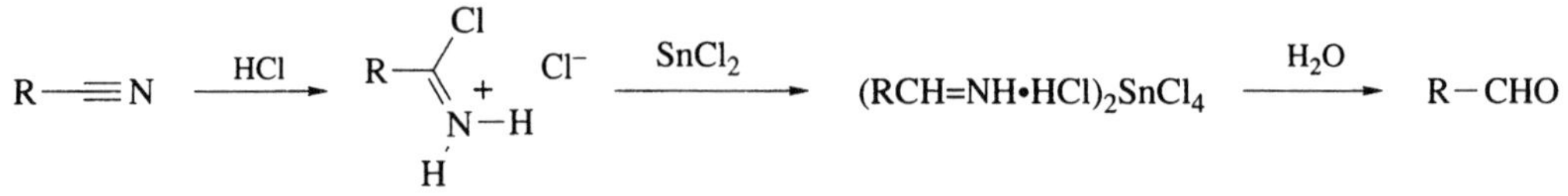

$$R{-}C{\equiv}N \xrightarrow{HCl} R{-}C(Cl){=}N^{+}H_2\ Cl^- \xrightarrow{SnCl_2} (RCH{=}NH{\bullet}HCl)_2SnCl_4 \xrightarrow{H_2O} R{-}CHO$$

(43) **(44)**

Scheme 14

The Stephen reduction has been applied to a variety of nitriles, but is genuinely useful only for aromatic substrates. Table 11 gives a number of representative examples, and more are available elsewhere.[34] The failure with *o*-toluonitrile is untypical, and was attributed to steric hindrance. A variant employing tin(II) bromide was successful in synthesizing cinnamaldehyde and its vinylog (**45**) in 65% and 50% yields, respectively.[72]

Table 11 Yields of Aldehydes Prepared by the Stephen Reduction of Aromatic Nitriles[34]

Aldehyde	*Yield (%)*
Benzaldehyde	97
o-Tolualdehyde	9
p-Iodobenzaldehyde	70
4-(Methoxycarbonyl)benzaldehyde	90
β-Naphthaldehyde	75
3-Pyridinecarbaldehyde	83

With aliphatic aldehydes the yields are poor at best. In a relatively recent study, Zil'berman and Pyryalova used a modified Stephen procedure to reduce a series of aliphatic nitriles RCN.[73] As the size of R increased, the yield decreased from a maximum of 51% (R = Et). A major cause of the low yields was found to be the formation of trimeric side products (**46**).

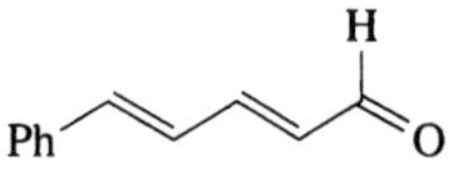

Ph–CH=CH–CH=CH–CHO   RC(=O)–NH–CH(R)–NH–C(=O)R

(45) 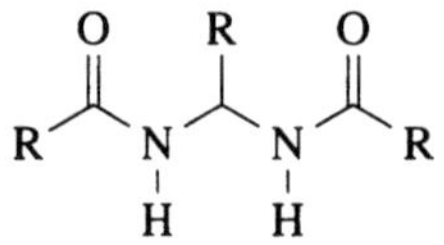 **(46)**

Aldehydes may also be synthesized by catalytic hydrogenation of the corresponding nitriles.[70] One method, applied only to aromatic aldehydes, employs Raney nickel, an atmosphere of hydrogen and a solvent system of THF/water (10:1) containing one mole equivalent of sulfuric acid.[74,75] Examples of its use are shown in Table 12.[75] Convenient and effective alternatives involve the use of phosphinic acid[76] or formic acid[77,78] as both solvent and hydrogen source. Good to excellent yields of aromatic aldehydes were obtained using these methods, although in some cases careful optimization was required. In general the best results appear to have come from using 75% aqueous formic acid as the solvent system.[78] Examples are given in Table 12, and a procedure is described in the literature.[79]

Table 12 Aromatic Aldehydes Synthesized by the Catalytic Reduction of Nitriles using Raney Nickel in THF/H_2O with 1 mol equiv. H_2SO_4 (Method A)[75] or 75% Aqueous HCO_2H (Method B)[78]

Aldehyde	*Method*	*Yield (%)*
3-Formylacetophenone	A	75
Ethyl 4-formylbenzoate	A	78
4-Formylbenzenesulfonamide	A	76
4-Formylbenzenesulfonamide	B	70[a]
Benzaldehyde	B	97[a]
p-Chlorobenzaldehyde	B	100[a]
p-Methoxybenzaldehyde	B	93[a]

[a]As 2,4-dinitrophenylhydrazone.

The catalytic hydrogenation of aliphatic nitriles to aldehydes appears to be intrinsically more difficult. The Raney nickel/formic acid method was applied to certain aliphatic aldehydes, but only with limited success (*e.g.* *n*-hexanal produced in 40% yield).[78] The situation can be improved by undertaking the hydrogenation in the presence of a reagent which can trap the aldehyde (or imine) as it is formed and prevent overreduction.[70] For example, hydrogenation of various aliphatic nitriles with Raney nickel in the presence of semicarbazide hydrochloride gave reasonable yields of the corresponding aldehyde semicarbazones (*e.g.* phenylacetaldehyde semicarbazone, 70%; 3-cyanopropanal semicarbazone, 60%).[80] Phenylhydrazine has also been used as a trapping agent.[81] A disadvantage of these methods is, of course, that the derivatives must be hydrolyzed in separate steps to free the aldehydes. An alternative additive which minimizes this problem is *N,N'*-diphenylethylenediamine. This forms tetrahydroimidazole derivatives which are quite easily hydrolyzed to the corresponding aldehydes.[80,82] Although not developed into a general method, this procedure proved valuable in a specific case. Thus, hydrogenation of (3,5-dimethoxyphenyl)acetonitrile gave the derivative (**47**) in 60% yield.[82] The corresponding aldehyde, obtained by acid hydrolysis of (**47**), was used for synthetic studies on yohimbane derivatives.

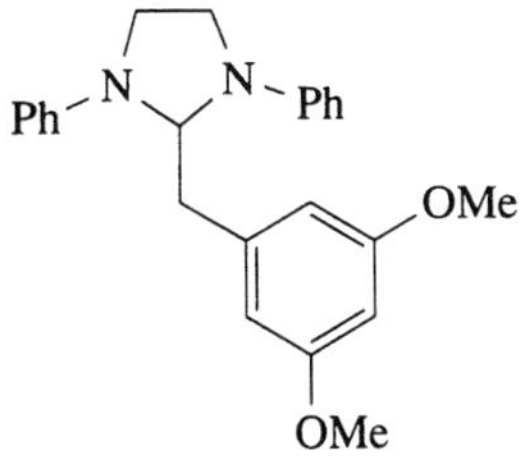

Ph N N Ph OMe OMe

(**47**)

For the direct conversion of aliphatic nitriles to aldehydes, the best 'nonhydride' method appears to be that of Fischli.[83] The nitrile is treated with a large excess of activated zinc and a catalytic amount (0.1 equiv.) of aquocobalamine in acetic acid/water (4:1) at room temperature for several hours. Using this method, a variety of aliphatic aldehydes were obtained in yields which were generally very satisfactory. As examples, the aldehydes (**48**), (**49**) and (**50**) were produced in yields of 63%, 82% and 90%, respectively. Aldehyde (**50**) was also formed from the unsaturated nitrile (**51**) under the same conditions. The proposed mechanism is shown in Scheme 15, the key steps being the attack of cob(I)alamine on the nitrile to give intermediate (**52**) and the reductive cleavage of the latter by the zinc. It was suggested that the product aldehyde was not reduced because the adduct between itself and cob(I)alamine is, for electronic reasons, less stable than (**52**).

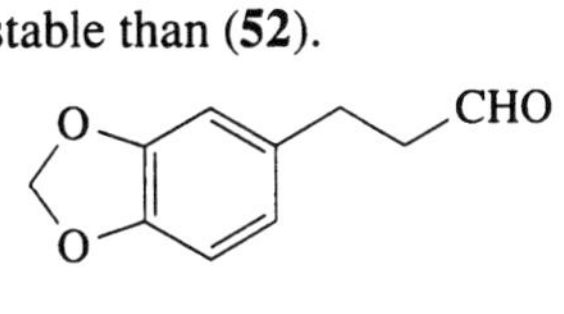

O O CHO

(**48**)

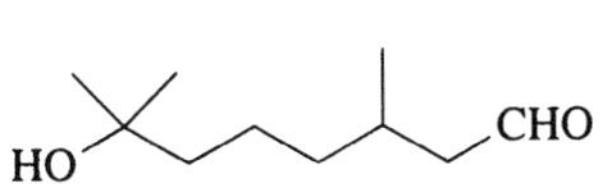

HO CHO

(**49**)

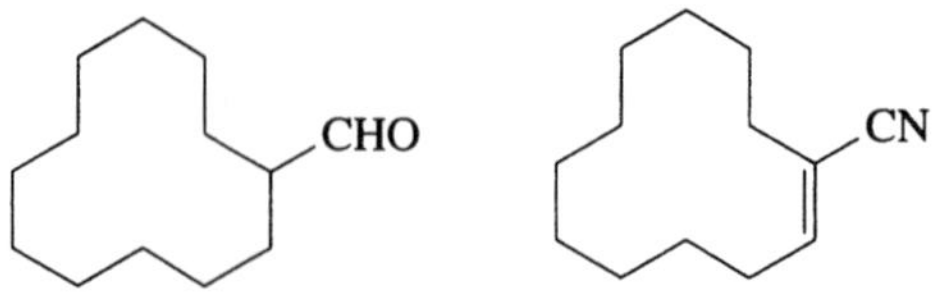

(50) (51)

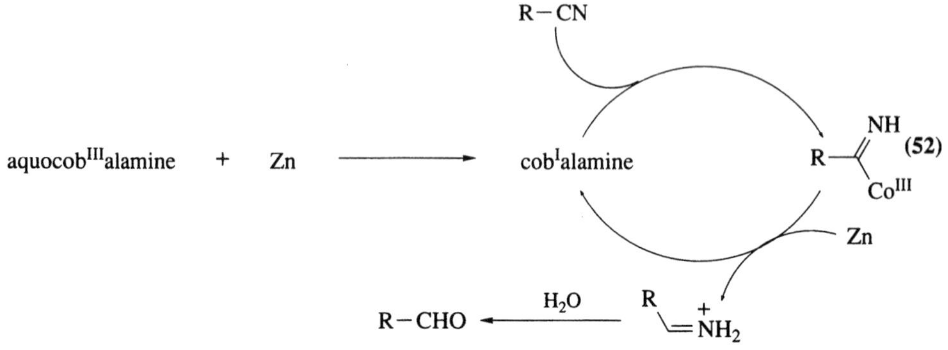

Scheme 15

A method which is apparently of greater theoretical than synthetic interest is the use of photochemically generated solvated electrons to reduce aromatic nitriles to aldehydes. If 3- or 4-cyanophenolate ions are irradiated at 254 nm, the corresponding aldehydes are obtained in 67% and 62% yields, respectively.[84] It appears that the absorption of a photon by the phenolate ion results in the generation of a solvated electron, which then reduces the nitrile (although the ultimate source of the reducing equivalents is not clear). The solvated electrons can also be generated from added iodide ions. Unfortunately, neither variant of the reaction gives useful yields with other substrates.

Finally, there have been several investigations into hydride transfer from Grignard reagents to nitriles, giving iminomagnesium complexes, and thus (after hydrolysis) aldehydes. Initially this type of method appeared to be of little use, being restricted to hindered tertiary nitriles and not especially successful even then.[85] However, it has been reported recently by Chinese workers that the addition of titanocene dichloride improves matters considerably, allowing the synthesis of aldehydes from nitriles and isopropylmagnesium bromide in yields of 52–68%.[86]

1.12.9 IMIDOYL CHLORIDES

One of the classical reductive aldehyde syntheses is the Sonn–Müller reduction of imidoyl chlorides.[34,87] The method is closely related to the Stephen reduction described in the previous section. As shown in Scheme 16, substrates (**53**) may be prepared either by the treatment of anilides with phosphorus pentachloride or sulfinyl chloride, or less usually *via* the Beckmann rearrangement.[88] They are

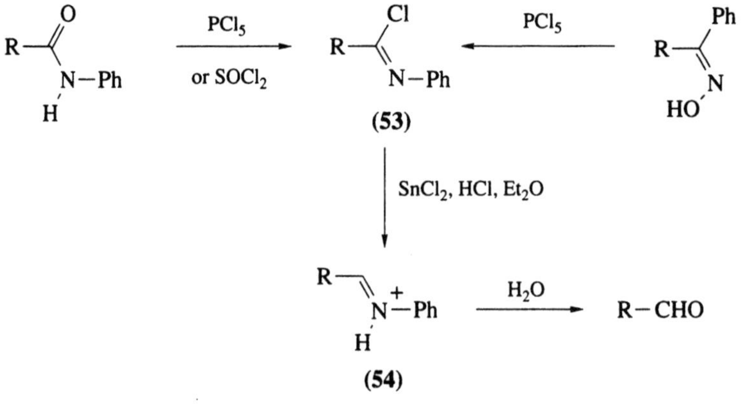

(53)

(54)

Scheme 16

then reduced with the same reagent as used in the Stephen reduction (*i.e.* tin(II) chloride/HCl/ether) to give iminium ions (**54**) which are hydrolyzed to aldehydes.

The method cannot be applied generally to the synthesis of aliphatic aldehydes, but for aromatic aldehydes and cinnamaldehyde the yields are good (Table 13). An example, the synthesis of *o*-tolualdehyde, is given in the literature.[89] A variant which has proved successful for α,β-unsaturated aliphatic aldehydes is the use of chromium(II) chloride.[90] However, as exemplified in Table 13, the yields are only fair. Moreover, there appears to be some question as to exactly how the reagent should be prepared.[34]

Table 13 Aldehydes Synthesized by the Reduction of Imidoyl Chlorides using $SnCl_2$ (Sonn–Müller) or $CrCl_2$[34]

Aldehyde	*Reducing agent*	*Yield (%)*
o-Tolualdehyde	$SnCl_2$	80
3,5-Dimethoxybenzaldehyde	$SnCl_2$	86
p-(Phenylsulfonyl)benzaldehyde	$SnCl_2$	57
Cinnamaldehyde	$SnCl_2$	92
2-Hexenal	$CrCl_2$	50
6-Methyl-1-cyclohexenal	$CrCl_2$	60
5,9-Dimethyl-2,4,8-decatrienal	$CrCl_2$	30

Two alternatives to the Sonn–Müller have been published in recent years. In the first, due to Alper and Tanaka, an aromatic imidoyl chloride (**55**) is reduced with the $[HFe(CO)_4]^-$ ion (equation 7) to give a mixture of imine (**56**) and amine (**57**).[91] For most of the substrates tried R was aromatic, and overreduction to (**57**) was the predominant outcome. However, in one case an aliphatic R group was used (R = Bu^n, Ar = Ph), and this gave a clean reduction to the corresponding imine (**56**) in 71% yield. More recently it was shown by Tanaka and Kobayashi that imidoyl chlorides could be reduced to imines using catalytic hydrogenation, with a soluble palladium complex as catalyst precursor (equation 8).[92] Although the variation of R^1 in their substrates (**58**) was somewhat limited, fair to good yields were obtained for R^1 = Ph, Bu^t and 2-thiophenyl, and the method appears potentially useful.

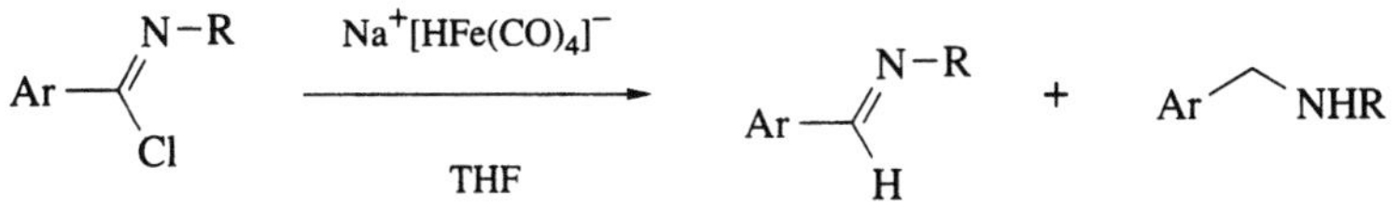

$$\text{ArC(Cl)=NR}\ (\mathbf{55}) \xrightarrow[\text{THF}]{\text{Na}^+[\text{HFe(CO)}_4]^-} \text{ArCH=NR}\ (\mathbf{56}) + \text{ArCH}_2\text{NHR}\ (\mathbf{57}) \quad (7)$$

$$\text{R}^1\text{C(Cl)=NR}^2\ (\mathbf{58}) \xrightarrow[\text{benzene, 120 °C, 2–9 h}]{\text{H}_2,\ [\text{PdCl}_2(\text{PPh}_3)_2],\ \text{Et}_3\text{N}} \text{R}^1\text{CH=NR}^2 \quad (8)$$

Finally, treatment of amides with phosphorus oxychloride gives Vilsmeier complexes, which are often formulated as *N*-alkylated imidoyl chlorides (**59**). Irrespective of their precise nature,[93] it has been found that they can be converted very efficiently to aldehydes by reduction with zinc followed by aqueous work-up (Scheme 17). Although the method has only been used for benzaldehyde and a number of chlorinated and brominated analogs, the yields reported are consistently high (87–97%).[94]

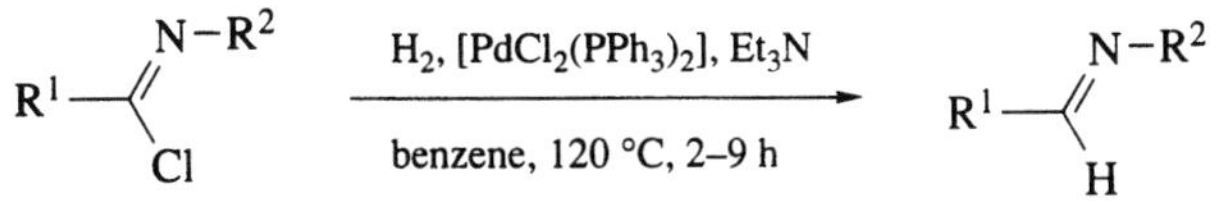

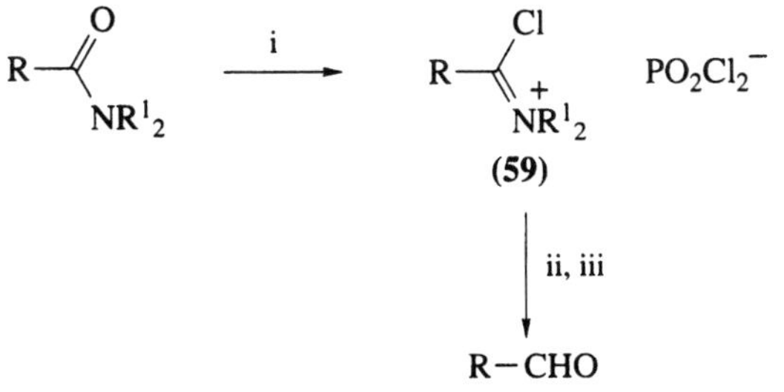

$$\text{RC(=O)NR}^1_2 \xrightarrow{\text{i}} \text{RC(Cl)=}\overset{+}{\text{N}}\text{R}^1_2\ \text{PO}_2\text{Cl}_2^-\ (\mathbf{59}) \xrightarrow{\text{ii, iii}} \text{R–CHO}$$

i, $POCl_3$; ii, Zn dust; iii, H_2O

Scheme 17

1.12.10 AMIDINES, IMIDATES AND THIOIMIDATES

Amidines are apparently rather good substrates for selective one-electron reduction to aldehydes. As early as 1908, Merling obtained good yields of some fairly complex aliphatic aldehydes by reduction of the corresponding *N*,*N*-diphenylamidines using sodium in ethanol.[95] As an example, amidine (**60**) was converted into aldehyde (**61**) in 60–70% yield.

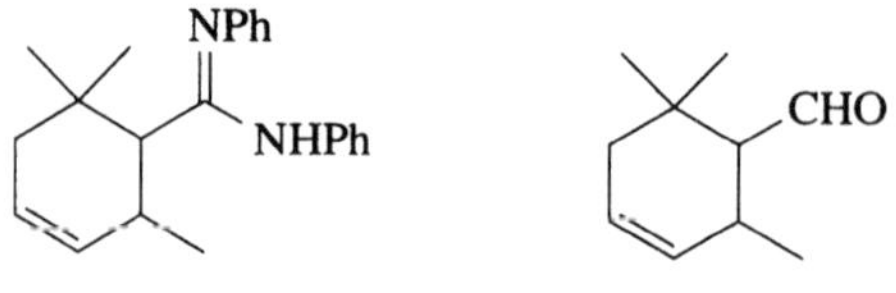

NPh NHPh CHO

(60) **(61)**

The method was not generalized, however, and was largely ignored until reinvestigated by Birch and coworkers,[52] who got rather better results using liquid ammonia as the solvent (retaining some ethanol as a proton source), provided the substrate was soluble therein. *N*-Unsubstituted amidines were thus preferred. As can be seen in Table 14, good yields could be obtained with both aromatic and aliphatic amidines. Presumably the geminal diamines (**62**) are intermediates (Scheme 18), and are stable under the reaction conditions, in contrast to the hemiaminal intermediates (**27**).

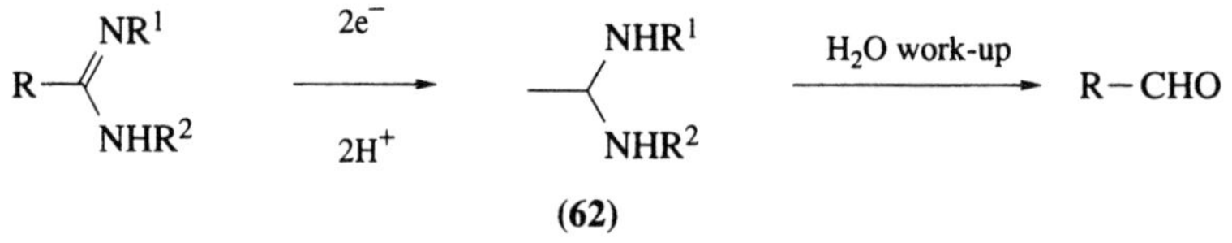

$$\mathrm{R{-}C(=NR^1)NHR^2} \xrightarrow[2H^+]{2e^-} \mathrm{R{-}CH(NHR^1)NHR^2}\ (\mathbf{62}) \xrightarrow{H_2O\ \text{work-up}} \mathrm{R{-}CHO}$$

(62)

Scheme 18

Table 14 Amidines Reduced to the Corresponding Aldehydes by Sodium and Ethanol in Liquid Ammonia[52]

Amidine	*Yield of aldehyde (%)*
Benzamidine	100
Acetamidine	53
Hexanoamidine	91
Phenylacetamidine	91

The selective reduction of cyclic imidates (**63**) to the aldehyde oxidation level has been demonstrated by Shono *et al.*[96] As shown in Scheme 19, the imidate is first alkylated on nitrogen and then reduced electrolytically in DMF in the presence of methanesulfonic acid. Unfortunately the scope of the method is unclear, as the main purpose of the work was to generate intermediate (**64**) in the presence of alkylating agents, leading to 2,2-disubstituted imidazolidines. Nonetheless, it was reported that decanal and dodecanal could be obtained in 82% and 70% yields, respectively.

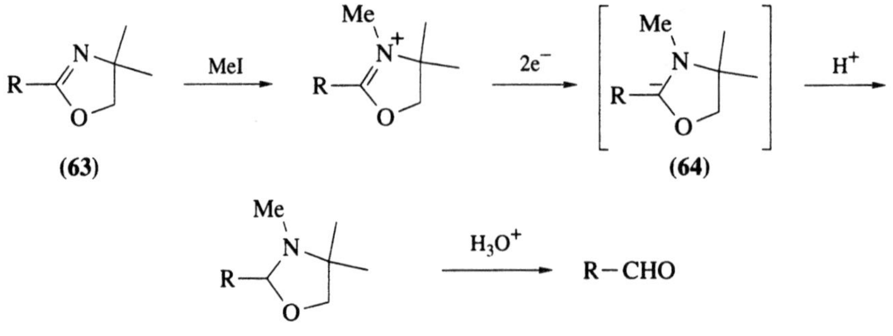

R N O —MeI→ Me N+ R O —2e⁻→ [Me N R O] —H+→

(63) **(64)**

Me N R O —H_3O^+→ R−CHO

Scheme 19

The reduction of thioimidates to the aldehyde oxidation level can be accomplished with aluminum amalgam. The reaction has not been investigated as a general method, but was developed to solve a specific problem. Thus, amide (**65**) was converted to thioimidate (**66**), which was reduced to (**67**) in 38% yield (Scheme 20).[97] Although the yield was moderate, the method succeeded in a molecule with a fairly high level of functionality.

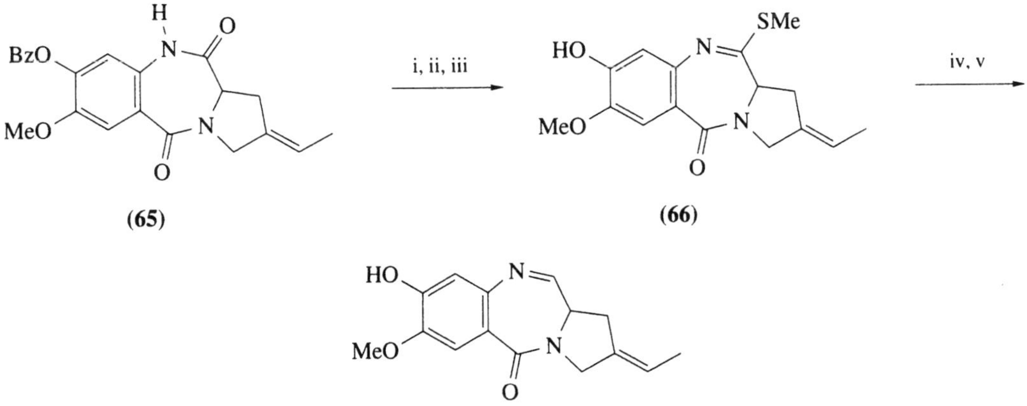

(**65**) —i, ii, iii→ (**66**) —iv, v→ (**67**)

i, (p-$MeOC_6H_4PS_2)_2$, benzene; ii, MeI; iii, K_2CO_3, MeOH; iv, Al/Hg, aq. THF; v, $HgCl_2$, MeOH

Scheme 20

1.12.11 THIOAMIDES, THIOESTERS AND DITHIOESTERS

It has been reported from two laboratories that thioamides can be reduced to aldehydes by Raney nickel. Brovet found that thiobenzamide and thio-p-toluamide gave the corresponding aldehydes in yields of 77% and 42%, respectively (as 2,4-dinitrophenylhydrazones).[98] In contrast, Cronyn and Goodrich found that thiobenzamide gave poor yields, but that thiobenzanilide, p-methoxythiobenzanilide and p-hydroxythiobenzanilide gave yields in the range 78–96% (as 2,4-dinitrophenylhydrazones)[99] using Raney nickel which had previously been deactivated by boiling in acetone. Application of the method to aliphatic thioanilides was less successful, the yields being reduced to *ca.* 30%.

Recent work by Voss and coworkers has shown that tertiary thioamides,[100] thioesters[101] and dithioesters[101,102] can all be reduced electrochemically to the aldehyde level in the presence of alkylating agents. The process is summarized in equation (9). The cleanest and most efficient of the reactions appears to be that of the thioamides (**68**; X = $NR^2{}_2$), using anhydrous acetonitrile as solvent and simple alkyl bromides (*e.g.* EtBr, Bu^iBr) as alkylating agents.[100] Although the method was only demonstrated for (**68**; R = Ph and o-ClC_6H_4), the yield of α-amino sulfides was 95% in both cases.

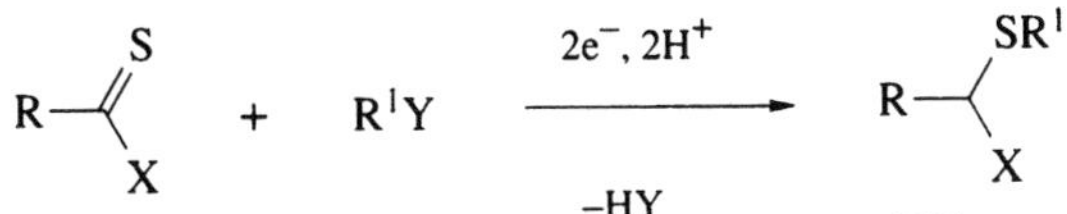

$$R\text{–}C(=S)X\ (\mathbf{68}) + R^1Y \xrightarrow[-HY]{2e^-,\ 2H^+} R\text{–}CH(SR^1)X\ (\mathbf{69}) \qquad (9)$$

$X = NR^2{}_2, OR^2, SR^2$; Y = leaving group

Dithioesters can be reduced in anhydrous acetonitrile[101,102] or in methanol,[102] the best results being obtained with dimethyl sulfate as alkylating agent. The most convenient procedure involves a simple electrolysis cell with a lead cathode and methanol as solvent.[102] Eight substrates (**68**; X = SMe) were tested, with R equal to phenyl, chloro- and methoxy-substituted phenyls, and t-butyl. Unlike the corresponding reductions of thioamides, the reactions were not entirely clean and gave substantial amounts of side products such as (**70**). The product dithioacetals (**69**; X = SMe, R^1 = Me) were formed in yields of 40–60%. On two of the substrates, better yields (*ca.* 70%) could be obtained in anhydrous acetonitrile, but a more sophisticated apparatus was required.[102] One enolizable substrate (**71**) was tested, but the yield was only 30%.[101]

Finally, the thiobenzoate (**72**) was also subjected to electrochemical reduction–alkylation.[101] This reaction was still less successful, possibly because of the strongly negative reduction potential of the substrate. The expected monothioacetal product was formed, but only in 25% yield.

Ar(MeS)C=C(SMe)Ar Et–C(=S)SMe Ph–C(=S)OMe

(70) **(71)** **(72)**

1.12.12 REFERENCES

1. F. Sato, T. Jinbo and M. Sato, *Synthesis*, 1981, 871.
2. A. O. Bedenbaugh, J. H. Bedenbaugh, W. A. Bergin and J. D. Adkins, *J. Am. Chem. Soc.*, 1970, **92**, 5774.
3. J. H. Wagenknecht, *J. Org. Chem.*, 1972, **37**, 1513.
4. H. Lund, *Acta Chem. Scand.*, 1963, **17**, 972, 1077 and 2325; P. E. Iversen and H. Lund, *Acta Chem. Scand.*, 1967, **21**, 279; P. E. Iversen, *Acta Chem. Scand.*, 1970, **24**, 2459.
5. L. N. Ferguson, J. C. Reid and M. Calvin, *J. Am. Chem. Soc.*, 1946, **68**, 2502.
6. H. Moureau, P. Chovin and M. Brunet, *Bull. Soc. Chim. Fr.*, 1948, **15**, 96; P. Sabatier and A. Mailhe, *C. R. Hebd. Seances Acad. Sci.*, 1914, **158**, 986.
7. T. Maki, M. Nakajima, T. Yokoyama and T. Setyoama (Mitsubishi), *Jpn. Pat.* (1987) 62 108 832 [87 108 832] (*Chem. Abstr.*, 1987, **107**, 197 540h).
8. A. P. Gelbein and R. Hansen (Chem. Systems Inc.), *US Pat.* 4 585 899 (1986) (*Chem. Abstr.*, 1986, **105**, 42 332b).
9. F. Wattimena and H. J. Heijman (Shell), *Eur. Pat. Appl.* 101 111 (1984) (*Chem. Abstr.*, 1984, **100**, 191 377u).
10. R. Knorr, P. Löw and P. Hassel, *Synthesis*, 1983, 785.
11. K. W. Rosenmund and F. Zetzsche, *Chem. Ber.*, 1921, **54**, 425.
12. E. Mosettig and R. Mozingo, *Org. React. (N.Y.)*, 1948, **4**, 362.
13. S. Affrossman and S. J. Thomson, *J. Chem. Soc.*, 1962, 2024.
14. E. B. Hershberg and J. Cason, *Org. Synth.*, 1941, **21**, 84.
15. R. P. Barnes, *Org. Synth.*, 1941, **21**, 110.
16. Y. Sakurai and Y. Tanabe, *Yakugaku Zasshi*, 1944, **64**, 25 (*Chem. Abstr.*, 1951, **45**, 5613e).
17. A. I. Rachlin, H. Gurien and D. P. Wagner, *Org. Synth.*, 1971, **51**, 8.
18. J. A. Peters and H. Van Bekkum, *Recl. Trav. Chim. Pays-Bas*, 1971, **90**, 1323.
19. J. A. Peters and H. Van Bekkum, *Recl. Trav. Chim. Pays-Bas*, 1981, **100**, 21.
20. A. W. Burgstahler, L. O. Weigel and C. G. Shaefer, *Synthesis*, 1976, 767.
21. J. Tsuji, K. Ohno and T. Kajimoto, *Tetrahedron Lett.*, 1965, 4565.
22. G. P. Chiusoli and G. Agnès, *Chim. Ind. (Milan)*, 1964, **46**, 548.
23. A. B. McEwen, M. J. Guttieri, W. F. Maier, R. M. Laine and Y. Shvo, *J. Org. Chem.*, 1983, **48**, 4436.
24. Y. Watanabe, T. Mitsudo, M. Tanaka, K. Yamamoto, T. Okajima and Y. Takegami, *Bull. Chem. Soc. Jpn.*, 1971, **44**, 2569.
25. W. O. Siegl and J. P. Collman, *J. Am. Chem. Soc.*, 1972, **94**, 2516.
26. T. E. Cole and R. Pettit, *Tetrahedron Lett.*, 1977, 781.
27. G. F. Cainelli, F. Manescalchi and A. Umani-Ronchi, *J. Organomet. Chem.*, 1984, **276**, 205.
28. S. C. Kao, P. L. Gaus, K. Youngdahl and M. Y. Darensbourg, *Organometallics*, 1984, **3**, 1601.
29. R. J. Kinney, W. D. Jones and R. G. Bergman, *J. Am. Chem. Soc.*, 1978, **100**, 7902.
30. J. C. LeBlanc, C. Moise and J. Tirouflet, *J. Organomet. Chem.*, 1985, **292**, 225.
31. D. G. Smith and D. J. H. Smith, *J. Chem. Soc., Chem. Commun.*, 1975, 459.
32. H. Chikashita, H. Ide and K. Itoh, *J. Org. Chem.*, 1986, **51**, 5400.
33. R. A. Grey (Atlantic Richfield), *US Pat.*, 4 533 754 (1985) (*Chem. Abstr.*, 1986, **104**, 33 859f).
34. E. Mosettig, *Org. React. (N.Y.)*, 1954, **8**, 218.
35. Y. Watanabe, M. Yamashita, T. Mitsudo, M. Tanaka and Y. Takegami, *Tetrahedron Lett.*, 1973, 3535.
36. Y. Watanabe, M. Yamashita, T. Mitsudo, M. Igami and Y. Takegami, *Bull. Chem. Soc. Jpn.*, 1975, **48**, 2490.
37. Y. Watanabe, M. Yamashita, T. Mitsudo, M. Igami, K. Tomii and Y. Takegami, *Tetrahedron Lett.*, 1975, 1063.
38. H. Musso and K. Figge, *Chem. Ber.*, 1962, **95**, 1844.
39. H. Wakamatsu, J. Furukawa and N. Yamakami, *Bull. Chem. Soc. Jpn.*, 1971, **44**, 288.
40. E. Fischer, *Chem. Ber.*, 1889, **22**, 2204; E. Fischer, *Chem. Ber.*, 1890, **23**, 930.
41. N. Sperber, H. E. Zaugg and W. M. Sandstrom, *J. Am. Chem. Soc.*, 1947, **69**, 915.
42. M. Fedoronko, *Adv. Carbohydr. Chem. Biochem.*, 1974, **29**, 107.
43. J. P. Coleman, in 'The Chemistry of Functional Groups', ed. S. Patai, Wiley Interscience, Chichester, 1979, suppl. B, p. 804.
44. S. Sugasawa and M. Matsumoto (Tanabe Seiyaku Co.), *Jpn. Pat.* 64 613 (1964) (*Chem. Abstr.*, 1964, **60**, 12 096h); A. Sturc, *Czech. Pat.* 222 096 (1985) (*Chem. Abstr.*, 1987, **106**, 18 983q).
45. J. W. E. Glattfeld and G. W. Schimpff, *J. Am. Chem. Soc.*, 1935, **57**, 2204.
46. Y. Watanabe, M. Yamashita, M. Igami, T. Mitsudo and Y. Takegami, *Bull. Chem. Soc. Jpn.*, 1976, **49**, 2824.
47. M. L. Wolfrom and J. V. Karabinos, *J. Am. Chem. Soc.*, 1946, **68**, 1455.
48. O. Jeger, J. Norymberski, S. Szpilfogel and V. Prelog, *Helv. Chim. Acta*, 1946, **29**, 684.
49. G. B. Spero, A. V. McIntosh, Jr. and R. H. Levin, *J. Am. Chem. Soc.*, 1948, **70**, 1907.

50. A. V. McIntosh, Jr., E. M. Meinzer and R. H. Levin, *J. Am. Chem. Soc.*, 1948, **70**, 2955.
51. A. V. McIntosh, Jr., A. M. Searcy, E. M. Meinzer and R. H. Levin, *J. Am. Chem. Soc.*, 1949, **71**, 3317.
52. A. J. Birch, J. Cymerman-Craig and M. Slaytor, *Aust. J. Chem.*, 1955, **8**, 512.
53. A. O. Bedenbaugh, A. L. Payton and J. L. Bedenbaugh, *J. Org. Chem.*, 1979, **44**, 4703.
54. R. A. Benkeser, H. Watanabe, S. J. Mels and M. A. Sabol, *J. Org. Chem.*, 1970, **35**, 1210.
55. A. Reissert, *Chem. Ber.*, 1905, **38**, 1603.
56. W. E. McEwen and R. L. Cobb, *Chem. Rev.*, 1955, **55**, 511.
57. F. D. Popp, *Adv. Heterocycl. Chem.*, 1968, **9**, 1.
58. F. D. Popp, *Adv. Heterocycl. Chem.*, 1979, **24**, 187.
59. R. L. Cobb and W. E. McEwen, *J. Am. Chem. Soc.*, 1955, **77**, 5042.
60. J. W. Davis, Jr., *J. Org. Chem.*, 1960, **25**, 376; W. E. McEwen, M. A. Calabro, I. C. Mineo and I. C. Wang, *J. Am. Chem. Soc.*, 1973, **95**, 2392.
61. M. J. Cook, A. R. Katritzky and A. D. Page, *J. Am. Chem. Soc.*, 1977, **99**, 165.
62. F. D. Popp and A. Soto, *J. Chem. Soc.*, 1963, 1760.
63. J. M. Grosheintz and H. O. L. Fischer, *J. Am. Chem. Soc.*, 1941, **63**, 2021.
64. J. S. McFadyen and T. S. Stevens, *J. Chem. Soc.*, 1936, 584.
65. H. Babad, W. Herbert and A. W. Stiles, *Tetrahedron Lett.*, 1966, 2927.
66. C. C. Dudman, P. Grice and C. B. Reese, *Tetrahedron Lett.*, 1980, **21**, 4645.
67. M. S. Newman and E. G. Caflisch, Jr., *J. Am. Chem. Soc.*, 1958, **80**, 862.
68. M. Sprecher, M. Feldkimel and M. Wilchek, *J. Org. Chem.*, 1961, **26**, 3664.
69. M. Nair and H. Shechter, *J. Chem. Soc., Chem. Commun.*, 1978, 793.
70. M. Rabinovitz, in 'The Chemistry of the Cyano Group', ed. Z. Rappoport, Wiley Interscience, New York, 1970, p. 307.
71. H. Stephen, *J. Chem. Soc.*, 1925, **127**, 1874.
72. G. Wittig and H. Hartmann, *Chem. Ber.*, 1939, **72**, 1387.
73. E. N. Zil'berman and P. S. Pyryalova, *Zh. Obshch. Khim.*, 1963, **33**, 3420 (*Chem. Abstr.*, 1964, **60**, 4005d).
74. P. Tinapp, *Chem. Ber.*, 1969, **102**, 2770.
75. P. Tinapp and E. Möltgen, *Arch. Pharm. (Weinheim, Ger.)*, 1976, **309**, 766.
76. O. G. Backeberg and B. Staskun, *J. Chem. Soc.*, 1962, 3961.
77. B. Staskun and O. G. Backeberg, *J. Chem. Soc.*, 1964, 5880.
78. T. van Es and B. Staskun, *J. Chem. Soc.*, 1965, 5775.
79. T. van Es and B. Staskun, *Org. Synth.*, 1971, **51**, 20.
80. H. Plieninger and G. Werst, *Chem. Ber.*, 1955, **88**, 1956.
81. A. Gaiffe, *Chim. Ind. (Paris)*, 1965, **93**, 259.
82. H. Plieninger and B. Kiefer, *Chem. Ber.*, 1957, **90**, 617.
83. A. Fischli, *Helv. Chim. Acta*, 1978, **61**, 2560.
84. J. P. Ferris and F. R. Antonucci, *J. Am. Chem. Soc.*, 1972, **94**, 8091.
85. N. Rabjohn and E. L. Crow, *J. Org. Chem.*, 1963, **28**, 2907.
86. X. Huang and Y. He, *Youji Huaxue*, 1986, 204 (*Chem. Abstr.*, 1987, **106**, 4287d).
87. A. Sonn and E. Müller, *Chem. Ber.*, 1919, **52**, 1927.
88. G. H. Coleman and R. E. Pyle, *J. Am. Chem. Soc.*, 1946, **68**, 2007.
89. J. W. Williams, C. H. Witten and J. A. Krynitsky, *Org. Synth.*, 1946, **26**, 97.
90. J. von Braun and W. Rudolph, *Chem. Ber.*, 1934, **67**, 269.
91. H. Alper and M. Tanaka, *Synthesis*, 1978, 781.
92. M. Tanaka and T. Kobayashi, *Synthesis*, 1985, 967.
93. B. C. Challis and J. A. Challis, in 'Comprehensive Organic Chemistry', ed. D. H. R. Barton and W. D. Ollis, Pergamon Press, Oxford, 1979, vol. 2, p. 1023.
94. A. Rahman and A. Basha, *J. Chem. Soc., Chem. Commun.*, 1976, 594.
95. G. Merling, *Chem. Ber.*, 1908, **41**, 1519.
96. T. Shono, S. Kashimura, Y. Yamaguchi, O. Ishige, H. Uyama and F. Kuwata, *Chem. Lett.*, 1987, 1511.
97. T. Kaneko, H. Wong and T. W. Doyle, *Tetrahedron Lett.*, 1983, **24**, 5165.
98. D. Brovet, *Arh. Kem.*, 1948, **20**, 70 (*Chem. Abstr.*, 1950, **44**, 6829i)
99. M. W. Cronyn and J. E. Goodrich, *J. Am. Chem. Soc.*, 1952, **74**, 3936.
100. J. Voss, G. H. Wiegand and K. Hülsmeyer, *Chem. Ber.*, 1985, **118**, 4806.
101. L. Kistenbrügger and J. Voss, *Liebigs Ann. Chem.*, 1980, 472.
102. G. Drosten, P. Mischke and J. Voss, *Chem. Ber.*, 1987, **120**, 1757.

1.13
Reduction of C═X to CH_2 by Dissolving Metals and Related Methods

SHOSUKE YAMAMURA and SHIGERU NISHIYAMA
Keio University, Yokohama, Japan

1.13.1 INTRODUCTION

Of a variety of reduction methods,[1-3] Clemmensen reduction has long been known to convert a carbonyl function to the corresponding methylene group in strongly acidic conditions. This method cannot be utilized for an acid-labile compound, but is complemented by Wolff–Kishner reduction and related reactions which are carried out in basic media (Chapter 1.14, this volume). Electrochemical reduction is similar to Clemmensen reduction in its reaction mechanism, wherein two electrons are transferred in a stepwise manner to the substrate from the surface of a cathode instead of from the zinc metal. Both methods are able to reduce both conjugated and nonconjugated C═O functions to the corresponding methylene groups. Except for some special cases, the utility of the remaining reduction methods is limited to aromatic aldehydes and ketones, where a rather stable carbocation conjugated to the aromatic ring is presumably formed as an intermediate.

1.13.2 DISSOLVING METAL REDUCTION AND RELATED REACTIONS OF NONCONJUGATED AND CONJUGATED ALDEHYDES AND KETONES

1.13.2.1 Reduction with Alkali Metals in Liquid Ammonia

It has been known since the early part of this century that reduction of benzophenone with sodium in liquid ammonia affords diphenylmethanol *via* a ketyl intermediate,[4] and Hall *et al.* have shown more recently that aromatic aldehydes and ketones can be reduced with lithium in liquid ammonia to the corresponding methylene derivatives (equations 1–3).[5,6]

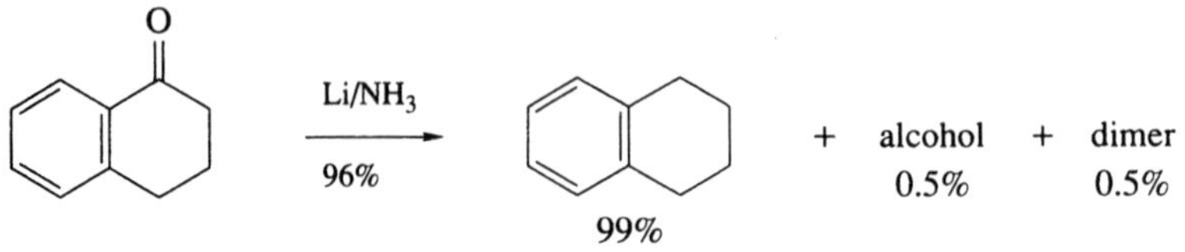

Li/NH_3, 96% → 99% + alcohol 0.5% + dimer 0.5% (1)

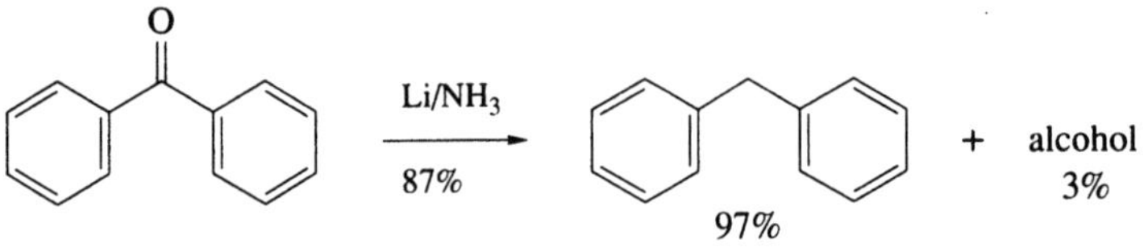

Li/NH_3, 87% → 97% + alcohol 3% (2)

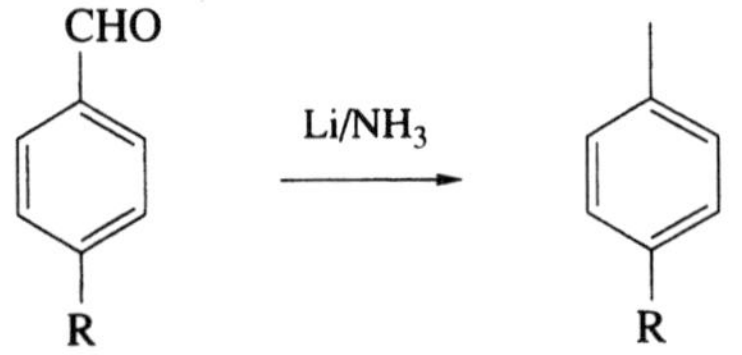

CHO, R → Li/NH_3 → R (3)

R = Bu^t; 89%, toluene:dimer = 94:2
R = Pr^i; 90%, toluene:dimer = 98:2

As indicated in Scheme 1, the reaction mechanism is similar to that of Birch reduction, and an alcohol intermediate has to be passed through. Therefore, while activated alcohols such as benzyl alcohols and their derivatives can be reduced under these conditions, reduction of aliphatic alcohols carrying no activating group does not take place.

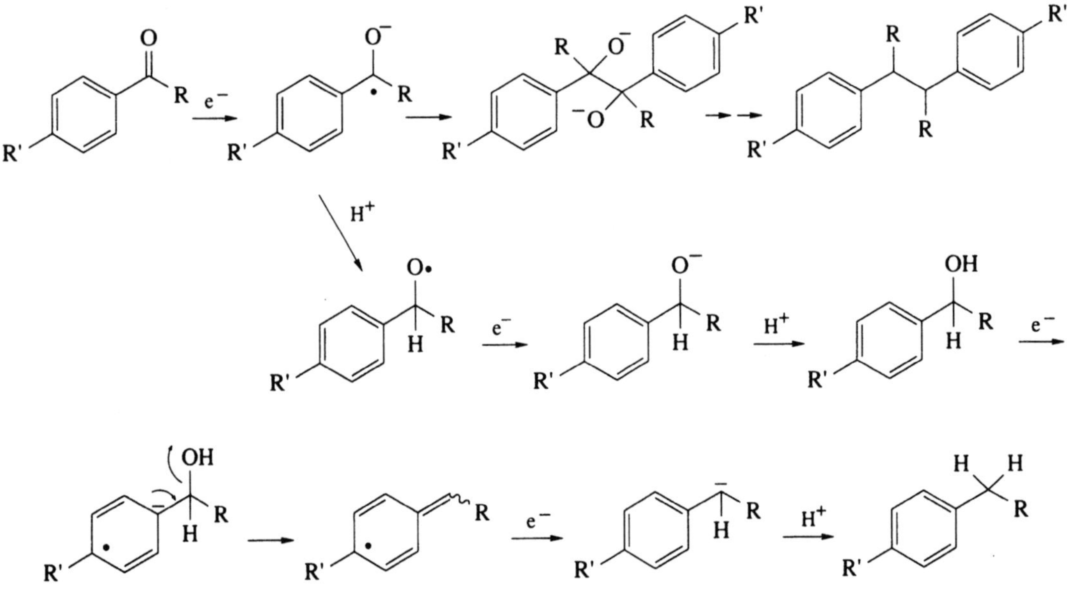

Scheme 1

In the case of cyclopropyl derivatives, although cyclopropyl methyl ketone gave a mixture of pentan-2-one and pentan-2-ol, cyclopropyl phenyl ketone afforded cyclopropylphenylmethane or cyclopropylphenylmethanol depending on the method of quenching (Scheme 2).[7]

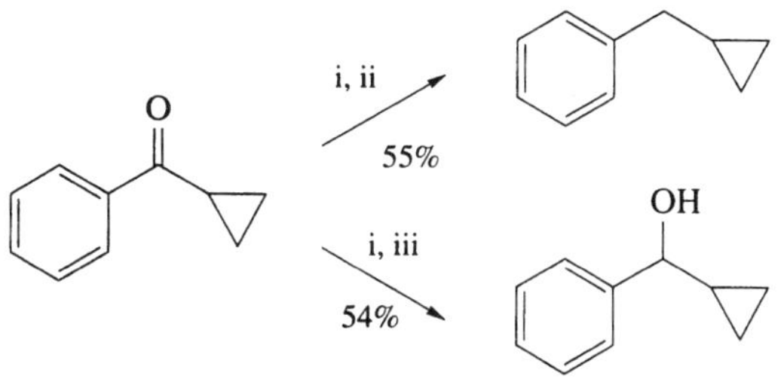

i, Li/NH_3; ii, NH_4Cl; iii, NaOBz

Scheme 2

In comparing aromatic aldehydes with aromatic ketones, the former are reduced faster than the latter. Thus, the presence of a trace metal such as cobalt or aluminum affected the reaction rate in the case of ketone derivatives, but this was not appreciable in the case of aldehyde derivatives, which react too fast. Also, the formation of dimeric by-products is much easier in the aldehyde case than the ketone. To prevent this dimerization, the addition of a proton source such as *t*-butyl alcohol is effective.[5,6]

1.13.2.2 Clemmensen Reduction

Clemmensen reduction is one of the most powerful reduction methods in organic synthesis, and many reviews have been published.[3,8–10] Generally, zinc amalgam and highly concentrated hydrochloric acid under reflux are employed to suppress the formation of by-products such as alcohols, dimerization products including pinacols, and related compounds. However, among several modifications of this synthetic procedure, the method using zinc powder in acetic anhydride or ether saturated with hydrogen chloride[11–13] is recommended for its mild conditions (0 °C, 1–2 h), and is available for reductions of aldehydes and ketones in molecules carrying such functional groups as cyano, acetoxy, phenol ether and alkoxycarbonyl, as described later.

The reaction mechanism of Clemmensen reduction has not completely been clarified, but it is well known that the alcohol is not an intermediate. As summarized in Scheme 3,[8,14] the reduction is thought to occur on the zinc metal surface, and involves protonation of the carbonyl function and a concomitant

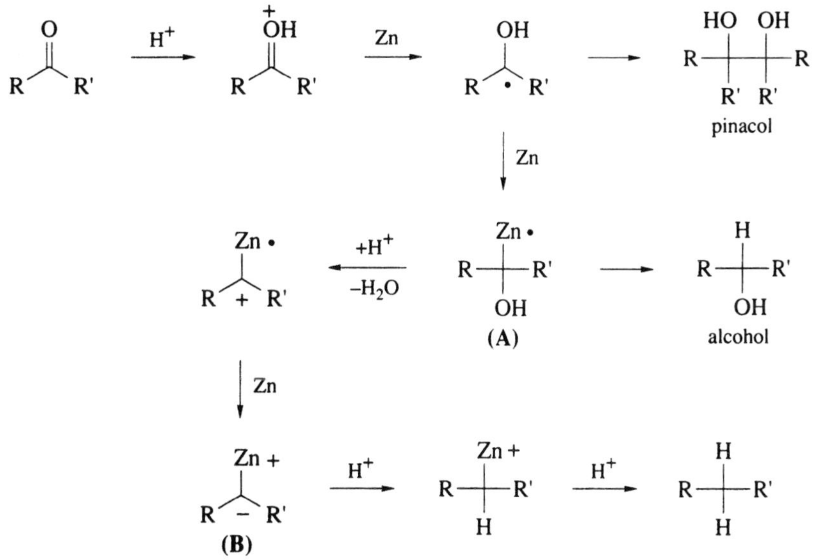

Scheme 3

electron transfer process to give an organozinc intermediate (**A**). Further protonation of (**A**) followed by abstraction of water and stepwise electron transfer yield a carbanion (**B**), which traps a proton, and in the final stage the corresponding methylene group is formed by exchange of the zinc with another proton.

One possibility in the case of benzaldehyde (Scheme 4) is that reduction proceeds *via* a carbene or carbenoid intermediate, which can be trapped by an alkene to give a cyclopropane (**1**) under aprotic conditions.[15]

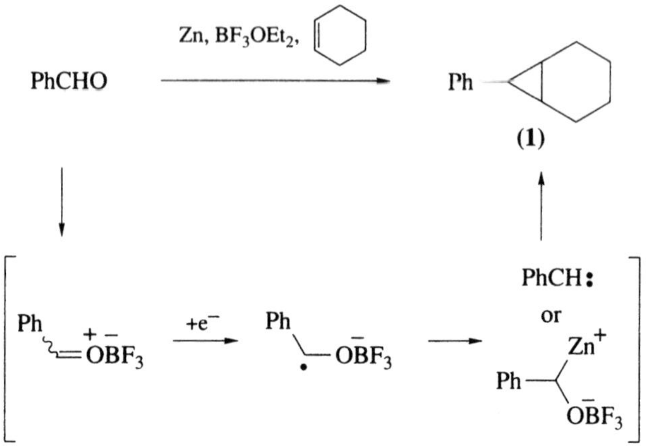

Zn, BF_3OEt_2,

PhCHO → Ph (**1**)

Ph $\overset{+}{=}\overset{-}{O}BF_3$ $\xrightarrow{+e^-}$ Ph $\overset{-}{O}BF_3$ → PhCH: or Ph Zn^+ $\overset{-}{O}BF_3$

Scheme 4

Because of carbonium ion generation, aromatic aldehydes and ketones can usually be reduced more easily than the corresponding aliphatic compounds. However, a modified Clemmensen reduction is an effective method to reduce isolated aliphatic carbonyl groups directly to methylene groups, and typical examples are shown in equations (4)–(6).[11,13,16]

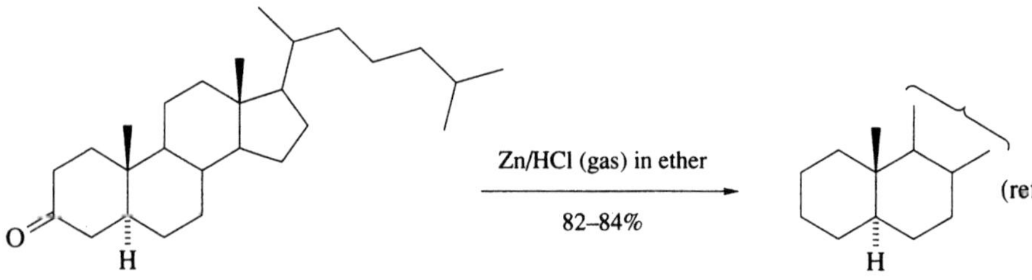

Zn/HCl (gas) in ether, 82–84% (ref. 11) (4)

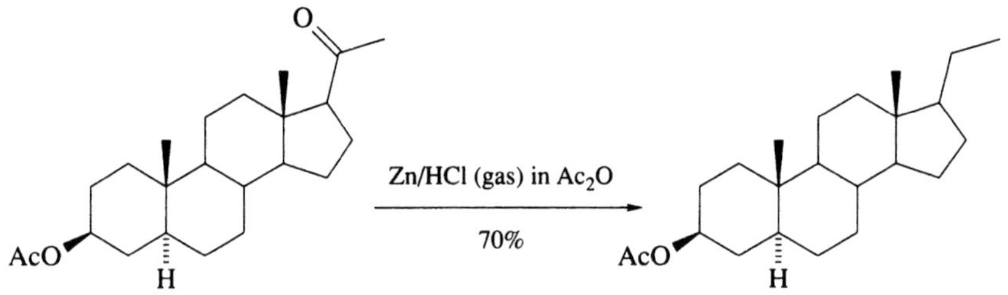

Zn/HCl (gas) in Ac_2O, 70%; AcO, H (ref. 13) (5)

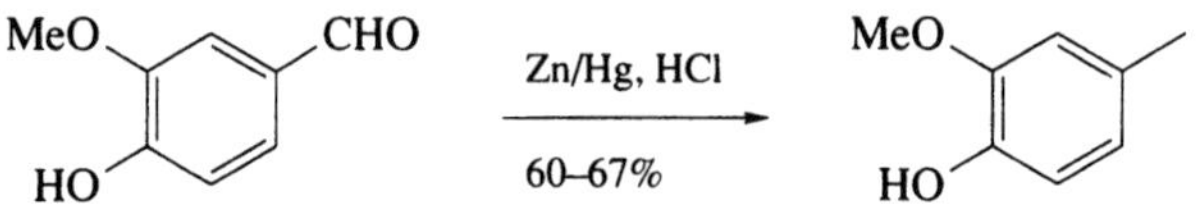

MeO, CHO, HO → Zn/Hg, HCl, 60–67% → MeO, HO (ref 16) (6)

In deuterium-labeling experiments, 4,4-diphenylcyclohexanone (**2**) reacted with zinc dust/deuterium chloride (prepared from trimethylchlorosilane and deuterium oxide), and gave 4,4-dideuterio-1,1-diphenylcyclohexane (**3**) as the major product (isotopic purity of 81% d_2, 10% d, 6% d_3 and 3% d_4; equation 7),[8] while the ketone (**4**) was converted to (**5**) in 83% isotopic purity, along with the trideuterio compound (17%), under rather vigorous conditions (equation 8).[17]

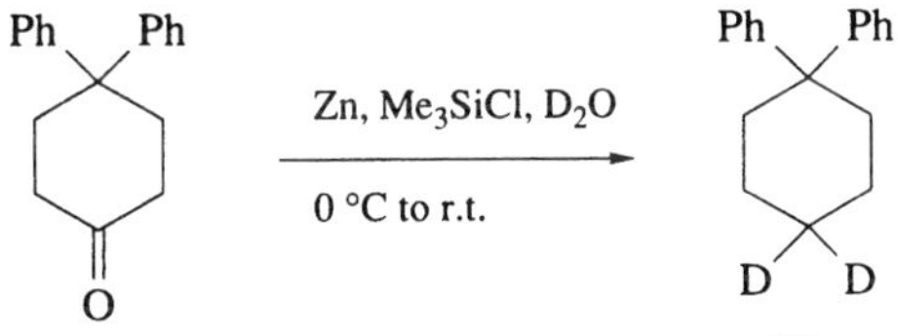

Ph Ph
Zn, Me_3SiCl, D_2O
0 °C to r.t.
Ph Ph
D D
(2) (3)

(7)

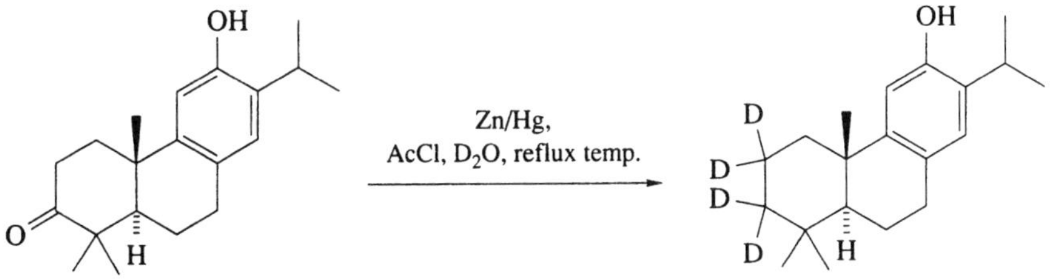

OH
Zn/Hg,
AcCl, D_2O, reflux temp.
OH
D
D
D
D
O H
H
(4) (5)

(8)

In conjugated carbonyl systems, the usual Clemmensen conditions may give rise to reduction and isomerization of the C═C double bond. When treated with zinc powder in acetic anhydride saturated with hydrogen chloride, cholest-1-en-3-one (**6**; Scheme 5) is converted to a mixture of cholestane (**7**; 30–32%), 3-acetoxycholest-2-ene (**8**; 10–24%) and cholestan-3-one (**9**; 30–40%).[12]

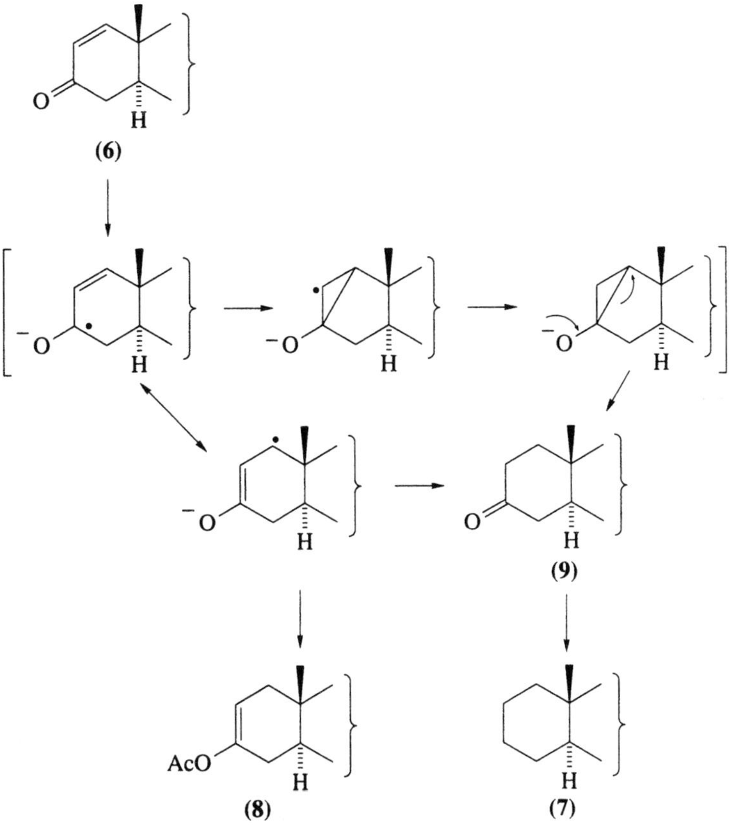

O
H
(6)

$^-$O H
$^-$O H
$^-$O H
$^-$O H
O H
(9)

AcO H
(8)
H
(7)

Scheme 5

The reaction may include a cyclopropanol intermediate derived from an anion radical, as seen in the reduction of cyclohexenones under Clemmensen conditions to afford ring-contracted cyclopentanones along with cyclohexanone derivatives.[10] Thus the diastereomeric cyclopropanol acetates (**12**) and (**13**) can be obtained in different ratios from both (**10**) and (**11**) (**12**/**13** = 3 from **10**, **12**/**13** > 100 from **11**; Scheme 6).[18]

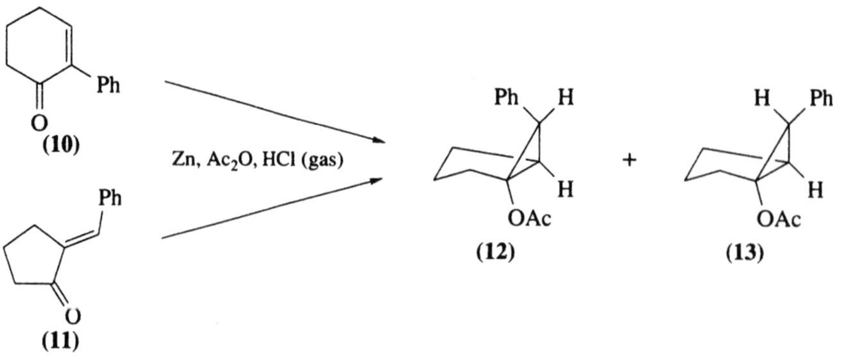

Zn, Ac_2O, HCl (gas)

(10) (11) (12) (13)

Scheme 6

A typical dimerization reaction in a conjugated system has been found on reduction of the octalone (**14**) with zinc amalgam in concentrated hydrochloric acid under reflux, which gave a geometrical mixture of the dehydro dimers (30–40%) along with octalin (**15**; equation 9).[19]

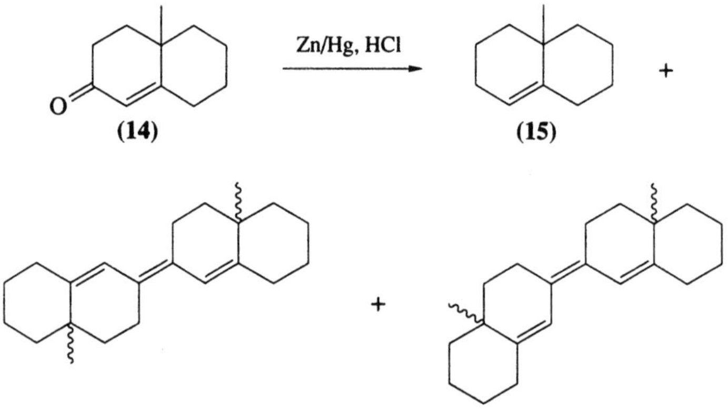

Zn/Hg, HCl

(14) (15)

(9)

Reduction of 4,4-diphenylcyclohex-2-en-1-one with zinc dust in hydrogen chloride in aprotic solvents affords different reduction products, depending on the acid concentration. Thus, on using excess hydrogen chloride the reaction proceeds exhaustively to give 1,1-diphenylcyclohexane, while a dimeric product is obtained with only 3 equiv. of hydrogen chloride, which permits organozinc intermediates enough time to undergo the coupling reaction (Scheme 7).[9]

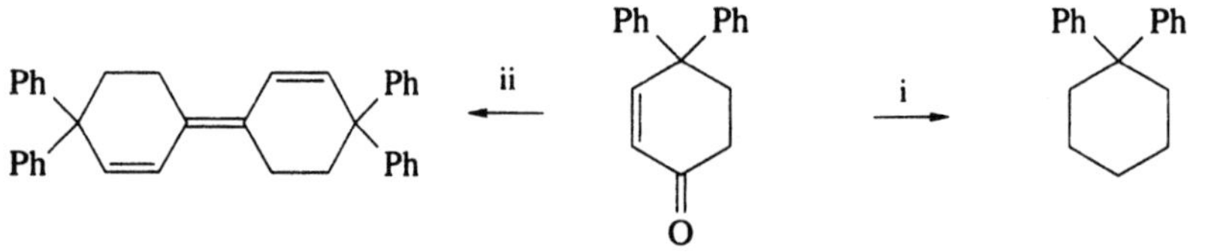

i, excess HCl (gas), Zn in Et_2O, ii, 3 equiv. HCl (gas), Zn in THF

Scheme 7

In the case of derivatives carrying two carbonyl functions stereochemically close to each other, Clemmensen reduction affords an interesting product distribution, passing through characteristic intermediates.[10] In the case of 1,3-diketones, a cyclopropanediol is involved, similar to the cyclopropanes formed from conjugated carbonyl systems.[20] For instance, 5,5-dimethylcyclohexane-1,3-dione (**16**) is converted into 2,4,4-trimethylcyclopentanone (**17**) as the major product (equation 10).[21]

Similar rearranged products can be seen in the Clemmensen reduction of many cyclic and acyclic compounds carrying the 1,3-diketone system.[9,10] The intermediacy of a cyclopropanediol derivative was established by Curphey *et al.*, who showed that the cyclopropane diacetate (**19**) can be obtained in high yield by reaction of (**18**) with zinc dust in acetic anhydride and hydrogen chloride, as well as by electrolysis.[22] The diketone (**18**) gives a mixture of rearranged products (**20** and **21**) under normal Clemmensen

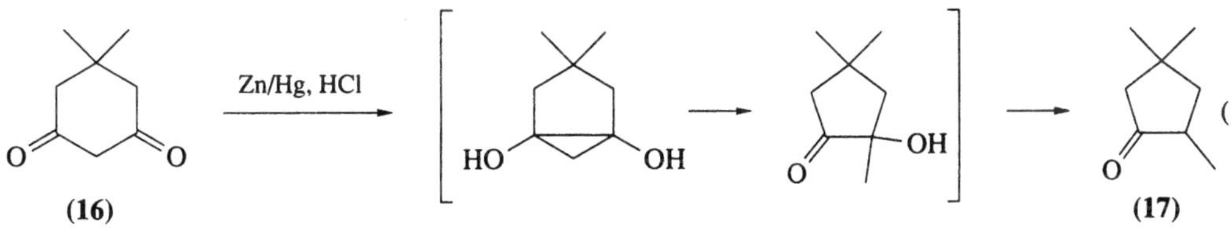

Zn/Hg, HCl

(16) (17) (10)

conditions using hydrochloric acid.[23] Reduction of cyclopentane-1,3-dione derivatives did not afford any ring contraction product because of the highly strained structure of the bicyclo[2.1.0]pentane-1,4-diol intermediate.[10]

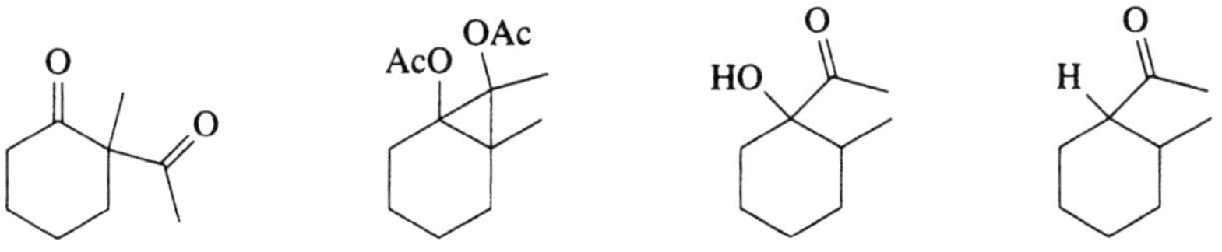

(18) (19) (20) (21)

With 1,4-diketones the distribution of the reduction products is dependent on the stereochemical situation of the two carbonyl functions.[10] In acyclic derivatives, with no stereochemical interaction between the two carbonyl functions, the ketone groups are independently reduced to give methylene products in the usual manner. On the other hand, cyclic 1,4-diketones react differently. For example, cyclohexane-1,4-dione (**22**) suffers ring opening to give hexane-2,5-dione and hex-5-en-2-one derivatives, and products of further reduction are also detectable (equation 11). A 1,4-diketone (**23**) in which the two carbonyls are stereochemically close, gave the diketone (**24**) under relatively mild conditions (Zn/AcOH, 25 °C), formed by the same C—C bond cleavage as seen in cyclohexane-1,4-dione. Under Clemmensen conditions this derivative was then converted to cyclobutane-1,4-diol (**25**; equation 12) in 98% yield, which is closely related to the aforementioned cyclopropanediol intermediate.[24]

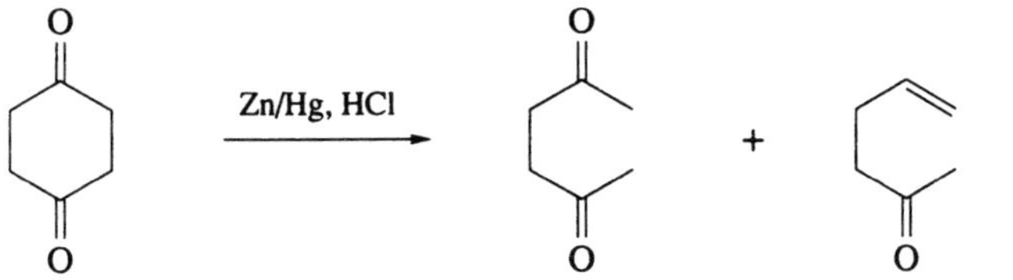

Zn/Hg, HCl (11)

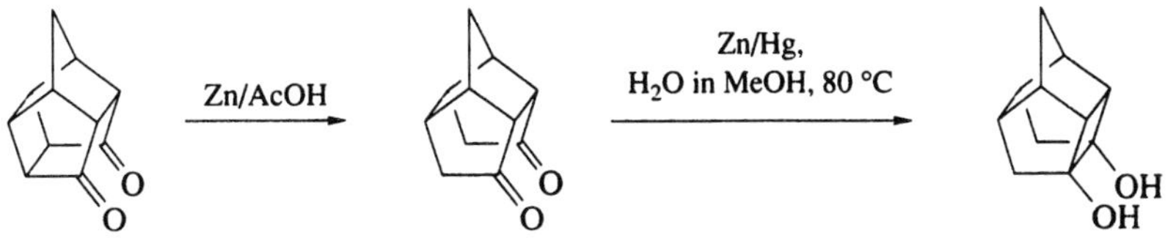

(22)

Zn/AcOH; Zn/Hg, H_2O in MeOH, 80 °C (12)

(23) (24) (25)

1.13.3 METAL HYDRIDE REDUCTIONS OF AROMATIC ALDEHYDES AND KETONES

1.13.3.1 Reduction with $NaBH_4$, $LiAlH_4$ and Related Reagents

Generally, on using metal hydride reagents, reduction of the carbonyl group to the corresponding methylene involves two steps: at the first stage, metal hydride addition permits the formation of an alkoxide intermediate, which on quenching gives an alcohol product as shown in alkyl ketones and aldehydes. In aryl systems, however, abstraction of the oxygen can occur in a second stage to generate a carbonium ion stabilized by delocalization. Further hydride addition to the carbonium ion then gives the methylene derivative. For this reason, reduction of carbonyls with metal hydrides to the corresponding methylene compounds is limited to aryl aldehydes and ketones (Scheme 8).

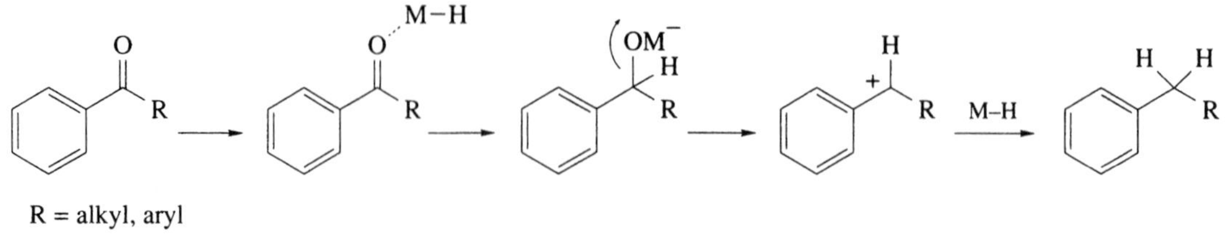

R = alkyl, aryl

Scheme 8

Hydrogenolysis using lithium aluminum hydride under reflux without any additives requires the presence of strongly electron-donating groups such as alkoxy, hydroxy, and amino groups at the *ortho* or *para* position to the carbonyl substituent in the aromatic system (equation 13).[25,26]

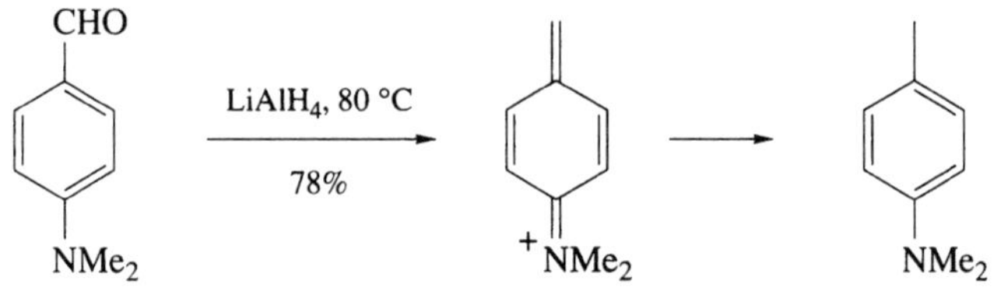

LiAlH$_4$, 80 °C; 78% (13)

Similarly reduction of *o*- or *p*-hydroxy-substituted aryl ketones with $NaBH_4$ in refluxing aqueous NaOH solution[27] gives rise to an intermediate methylene quinone (**26**), which is further reduced to the desired methylene product (equation 14). When the hydroxy group is situated only at the *meta* position, there is no possibility to make such a methylene quinone intermediate, and the corresponding benzyl alcohol (**27**) is formed (equation 15).

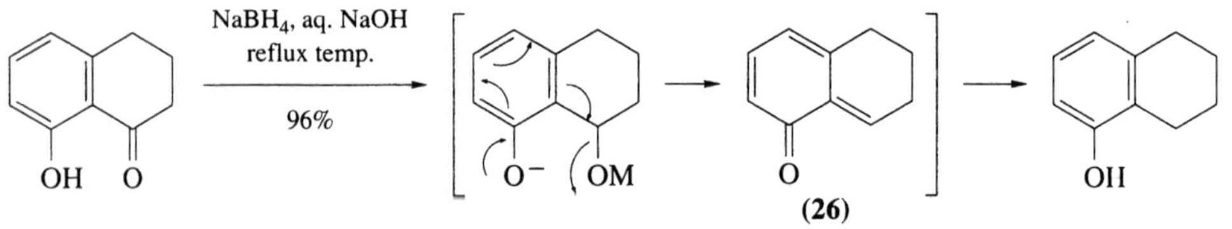

NaBH$_4$, aq. NaOH, reflux temp.; 96% (14)

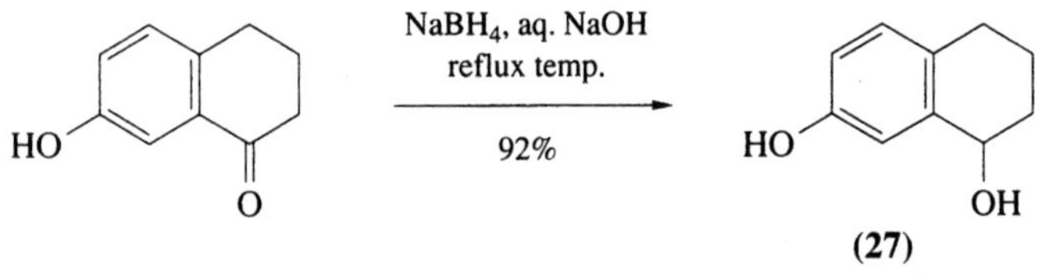

NaBH$_4$, aq. NaOH, reflux temp.; 92% (15)

There is evidence that acyl protection of an *o*-phenolic group improves the yield of the corresponding methylene compound, and this method has been utilized for the reduction of flavanones to flavans.[28,29]

Methods using sodium cyanoborohydride in acidic solvents[30,31] are also applicable for ketone reduction. Černý and his coworkers postulated that sodium bis(methoxyethoxy)aluminum dihydride in xylene at *ca.* 140 °C[32–36] has a nearly equal effect compared with lithium aluminum hydride. The rather drastic conditions allow the cleavage of phenyl benzyl ether, and this alternative deprotection method was used in the synthesis of pentazocine (**28**; equation 16).[37]

Lithium aluminum hydride/Lewis acid reduction of aryl aldehydes carrying electron-donating groups at the *ortho* and/or *para* position proceeds even in ether under reflux within a short period, to give the corresponding methylene derivatives. Both diaryl and alkyl aryl ketones, with or without such substituents, can be reduced in the same way.[38–42] It should be noted that the active species in this system may be AlH_3 or $AlCl_xH_y$ generated from the 1:1–1:3 mixture of $LiAlH_4$ and $AlCl_3$. A related reduction with $NaBH_4/AlCl_3$ has also been reported.[43]

Brewster *et al.* have investigated reductions of α,β-unsaturated carbonyls with a 1:3 mixture of $LiAlH_4/AlCl_3$. The reduction is closely similar to the results mentioned above, with an allyl cation intermediate trapped by hydride attack to afford an alkenic product. As seen in equations (17) and (18),

BnO … $NaAlH_2(OC_2H_4OMe)_2$ in xylene, reflux temp. → HO … **(28)** (16)

benzylidenecyclohexanone (**29**) gave benzylidenecyclohexane (**30**) and benzylcyclohex-1-ene (**31**) in a ratio of 7:3, and optically active carvone (**32**) afforded racemic limonene (**33**) through a symmetrical allyl cation intermediate.[41,42] This reagent was also used for reduction of steroidal unsaturated ketones, where the proportions of the various reduction products varied with the stability of the intermediate carbonium ions.[44,45]

(29) $\xrightarrow{LiAlH_4/AlCl_3,\ <35\ °C}$ 88% → **(30)** 70% + **(31)** 30% (17)

(32) $\xrightarrow{LiAlH_4/AlCl_3}$ 70% → [+] → **(33)** (18)

Although the mechanism is still uncertain, alkyl aryl ketones with a variety of substituents can be reduced with $LiAlH_4/P_2I_4$ in refluxing benzene.[46] Similarly, under very mild and rather neutral conditions ($NaBH_4/PdCl_2$, r.t.),[47] alkyl aryl or diaryl ketones are converted to the corresponding methylene derivatives without affecting an ester function. Similarly, aryl aldehydes in indole derivatives can be reduced to a methyl group using $NaBH_4$/10% Pd–C in boiling isopropyl alcohol.[48]

Reduction of diaryl ketones carrying functional groups at the *para* position can also be performed with $NaBH_4$/TFA even in the presence of an electron-withdrawing group such as an alkoxycarbonyl, nitro, or cyano group.[49,50] A similar reagent system, $NaBH_4/BF_3{\cdot}OEt_2$, was used for the reduction of indole derivatives;[51] however, the active species might be an electrophilic borane derivative, as discussed in Section 1.13.3.2.

Sodium cyanoborohydride/zinc iodide[52] can also reduce aryl ketones, including indoles and flavanones, to the corresponding methylenes in moderate yields, which depend on the aromatic substituents. For example, electron-withdrawing groups at the *ortho* and/or *para* positions depress the yields, whereas electron-donating groups give high yields. Lau and coworkers have proposed that this reaction may proceed by single-electron transfer, as shown in Scheme 9.[53]

1.13.3.2 Reduction with BH_3, DIBAL-H and Related Reagents

Diborane has been known to be an effective reagent for reductions of aldehydes and ketones to borate esters, which give alcohols on quenching. In the case of aryl carbonyl functions, reduction with this electrophilic reagent is closely related to those mentioned in Section 1.13.3.1. For further reduction of the borate ester to the corresponding hydrocarbon, elimination of the borate anion demands acceleration by electron-donating substituents at the *ortho* or *para* positions in the aromatic system, or else $BF_3{\cdot}OEt_2$ is needed as an additive, except for highly activated cases.[54,55] For this reason, *in situ* generated diborane,

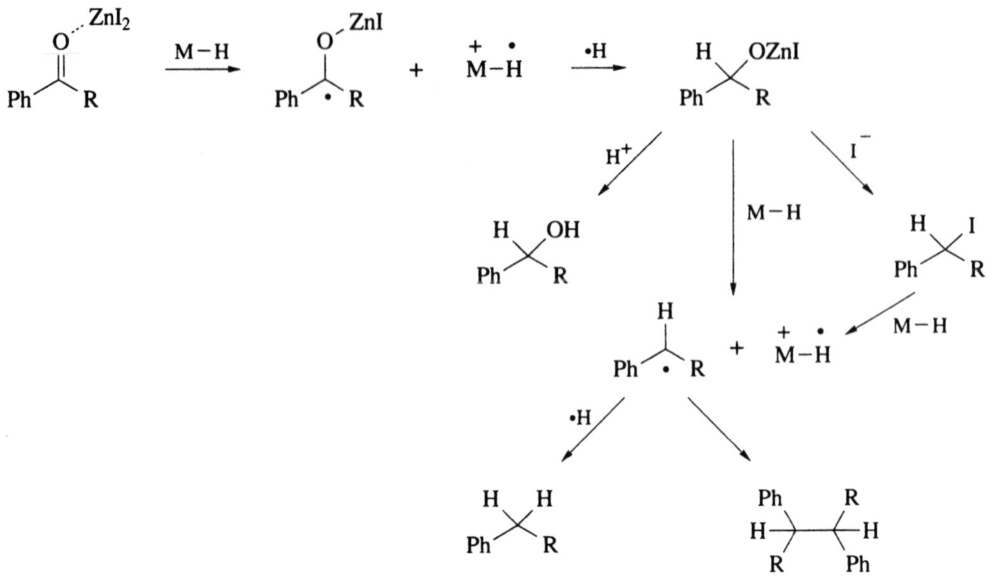

Scheme 9

still containing $BF_3{\cdot}OEt_2$ and $NaBH_4$, may be the method of choice. Thus, substituted benzophenone derivatives were cleanly reduced to the corresponding hydrocarbons in high yields (equation 19), and 4-dimethylaminobenzaldehyde afforded 4-dimethylaminotoluene in 95% yield (equation 20).

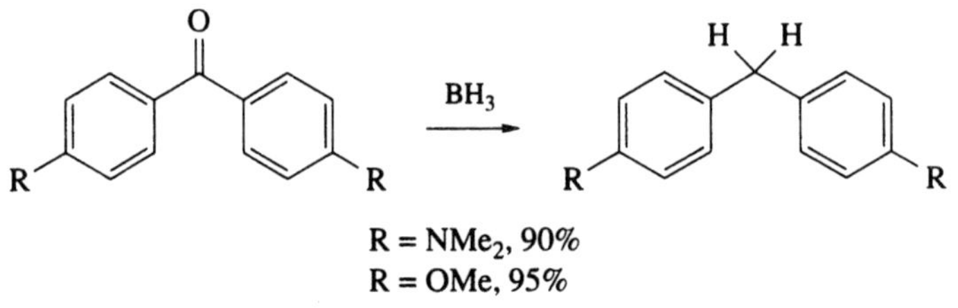

R = NMe$_2$, 90%
R = OMe, 95% (19)

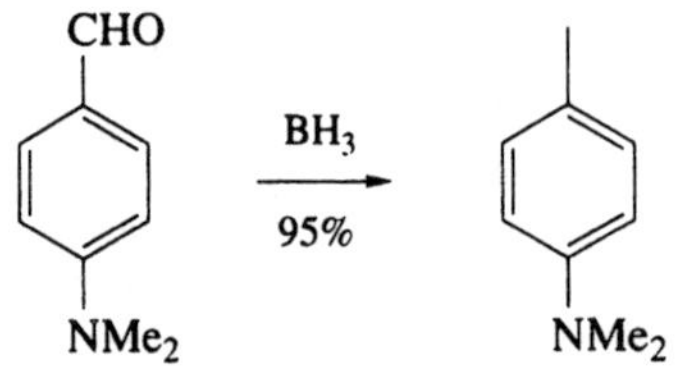

(20)

In the case of anisaldehyde, however, different results were obtained: on treatment with *in situ* generated diborane, the aldehyde gave a complex mixture including polymerized products derived from the cation (**A**), and *p*-methoxytoluene was not obtained (Scheme 10).[55]

Xanthone (**34**) and thioxanthone (**35**) are smoothly reduced with diborane in tetrahydrofuran at 0 °C to give the corresponding xanthenes (**36** and **37**). The ease of this reaction is explained by the formation of a stable xanthonium ion (**38**), generated by elimination of the borate ion (Scheme 11).[56]

The $BH_3/BF_3{\cdot}OEt_2$ reagent reduces not only aryl ketones and aldehydes, but also cyclopropyl aryl or dicyclopropyl ketones to the methylene compounds without opening or rearrangement in the cyclopropyl unit (equation 21).[54]

t-Butylamine/borane/aluminum chloride has been reported to reduce aryl ketones and aldehydes to the corresponding methylenes without any effect on aliphatic and aromatic chloro or bromo substituents (equations 22 and 23). However, this reagent cannot be used for carbonyl compounds carrying such functional groups as acetoxy, alkoxycarbonyl and nitro groups, because of their easy reduction.[57]

During alkylation studies on 3,6-diethoxyxanthen-9-one (**39**), stepwise addition of diisobutylaluminum hydride and triisobutylaluminum in hexane yielded 3,6-diethoxyxanthene (**40**) in 75% yield, although on using these reagents separately the yield of (**40**) was much less, as shown in Scheme 12.[58]

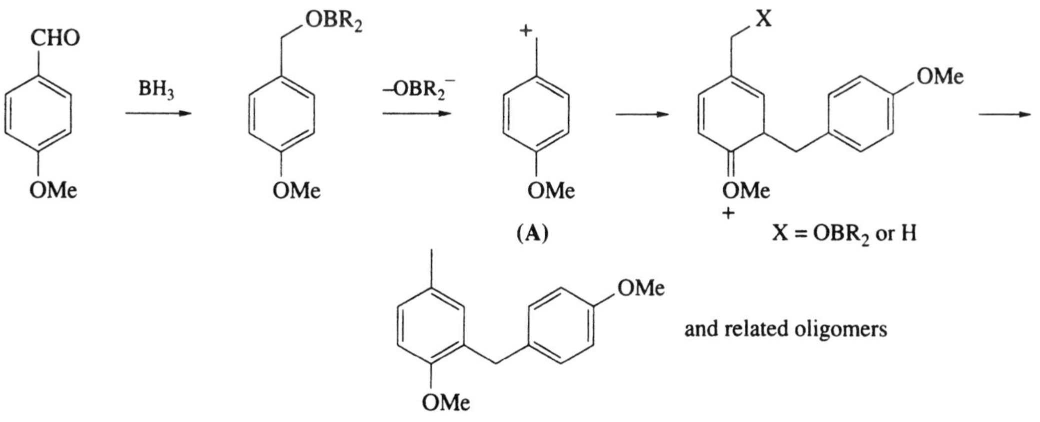

CHO OBR$_2$ + X OMe BH$_3$ –OBR$_2^-$ OMe OMe OMe OMe (A) + X = OBR$_2$ or H

OMe and related oligomers OMe

Scheme 10

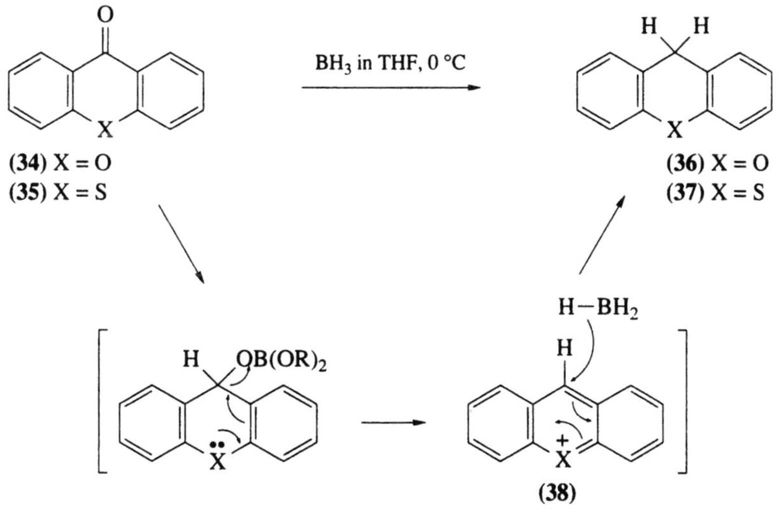

O BH$_3$ in THF, 0 °C H H X X **(34)** X = O **(35)** X = S **(36)** X = O **(37)** X = S

H–BH$_2$ H OB(OR)$_2$ H X + X **(38)**

Scheme 11

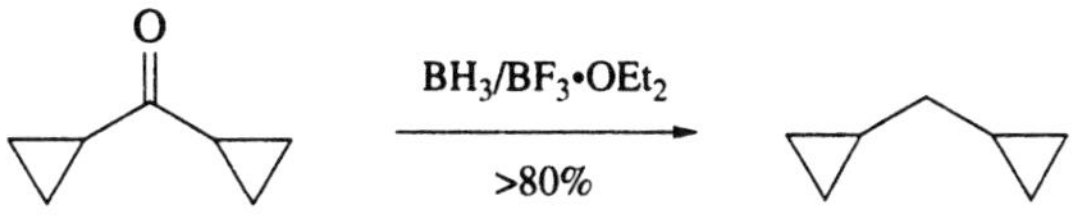

O BH$_3$/BF$_3$•OEt$_2$ >80% (21)

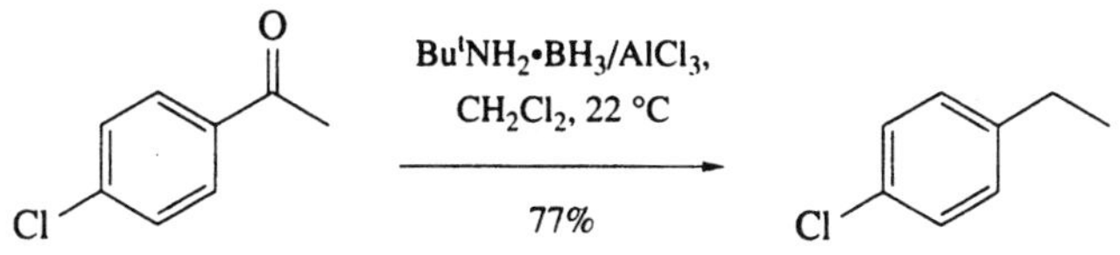

O Bu^{t}NH$_2$•BH$_3$/AlCl$_3$, CH$_2$Cl$_2$, 22 °C 77% Cl Cl (22)

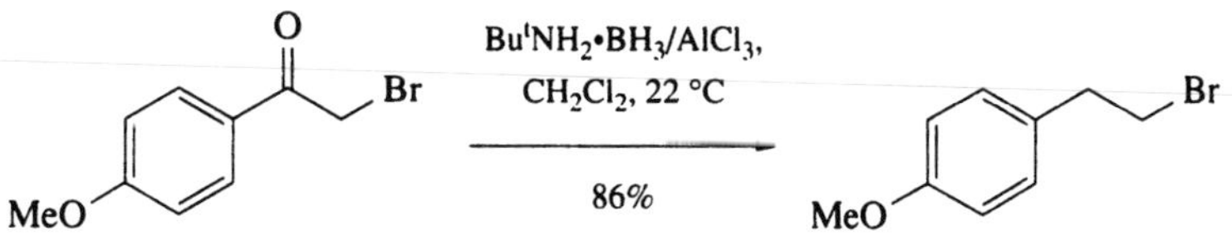

O Br Bu^{t}NH$_2$•BH$_3$/AlCl$_3$, CH$_2$Cl$_2$, 22 °C 86% Br MeO MeO (23)

1.13.3.3 Ionic Hydrogenation

Ionic hydrogenation is the noncatalytic addition of hydrogen to carbon–carbon or carbon–heteroatom double bonds, and the hydrogenolysis of carbon–oxygen or carbon–halogen single bonds. Thus, the conversion of aldehydes and ketones to the corresponding hydrocarbon is one of its many manifesta-

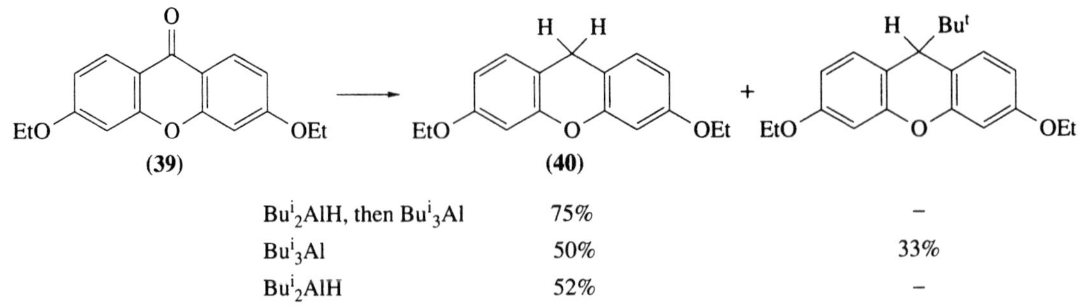

	(40)	
Bu^i_2AlH, then Bu^i_3Al	75%	–
Bu^i_3Al	50%	33%
Bu^i_2AlH	52%	–

Scheme 12

tions.[59,60] The reduction of the carbonyl system (Scheme 13) involves repeated protonation and hydride addition, and proceeds by way of an alcohol or its derivatives, which with some alkyl aldehydes and ketones can be isolated. From this viewpoint, ionic hydrogenation shows the same features as the other metal hydride methods, in which the ease of oxygen abstraction by electron-donating *ortho* or *para* substituents in the aromatic ring facilitates the reduction of aryl aldehydes and ketones. In this noncatalytic hydrogenation dimeric ether products are obtained as by-products, when the alcohol intermediate reacts with the starting material to form a hemiketal, which undergoes further protonation, dehydration, and hydride addition. Thus, benzaldehyde on ionic hydrogenation yielded dibenzyl ether in 80% yield.[61] This reduction is comparable with the conversion of ketals to ethers, which involves the formation of an alkoxy carboxonium ion followed by hydride addition.[59,60]

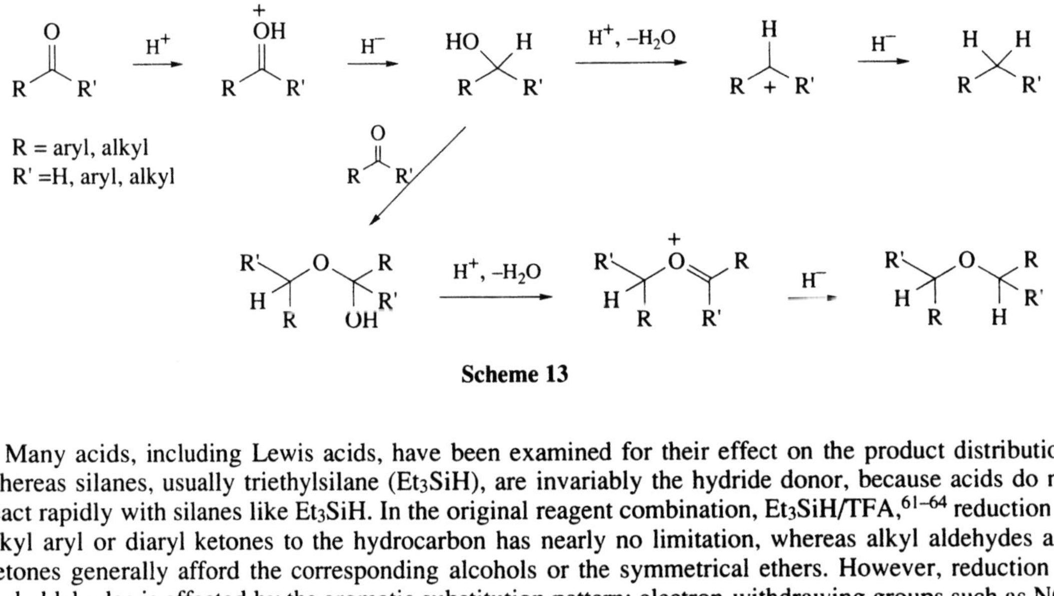

R = aryl, alkyl
R' =H, aryl, alkyl

Scheme 13

Many acids, including Lewis acids, have been examined for their effect on the product distribution, whereas silanes, usually triethylsilane (Et_3SiH), are invariably the hydride donor, because acids do not react rapidly with silanes like Et_3SiH. In the original reagent combination, Et_3SiH/TFA,[61–64] reduction of alkyl aryl or diaryl ketones to the hydrocarbon has nearly no limitation, whereas alkyl aldehydes and ketones generally afford the corresponding alcohols or the symmetrical ethers. However, reduction of aryl aldehydes is affected by the aromatic substitution pattern: electron-withdrawing groups such as NO_2 and CN impede the second step, and yield the corresponding alcohols or ethers. Cyclopropyl phenyl ketone (**41**) can be reduced to cyclopropylphenylmethane (**42**) under these conditions, but ring opening also occurs, affording phenylbutane (**43**) as a by-product. Cyclobutyl phenyl ketone (**44**) gave cyclobutylphenylmethane (**45**), coupled with cyclopentane derivatives (**46** and **47**) derived from ring expansion of the carbenium ion intermediate (Scheme 14).[64]

Aryl vinyl ketones undergo reduction exhaustively to give arylpropanes in high yields.[65,66] Anthraquinone derivatives also react with Et_3SiH/TFA to afford the corresponding hydrocarbons, while *p*-quinone was converted to hydroquinone in 98% yield.[67]

The system consisting of gaseous BF_3/Et_3SiH has been reported[68,69] to reduce aldehydes and ketones; thus benzaldehyde was reduced to toluene in 52% yield. In alkyl carbonyl compounds, aldehydes afford the corresponding alcohol products, whereas ketones, such as undecan-2-one and cyclohexanone, are readily converted to the hydrocarbons in high yields. On reduction with limited amounts of both reagents these alkyl ketones were reduced to the alcohols, indicating the intermediacy of the corresponding borate ester. In the case of aromatic compounds, aryl aldehydes and diaryl and alkyl aryl ketones are reduced to

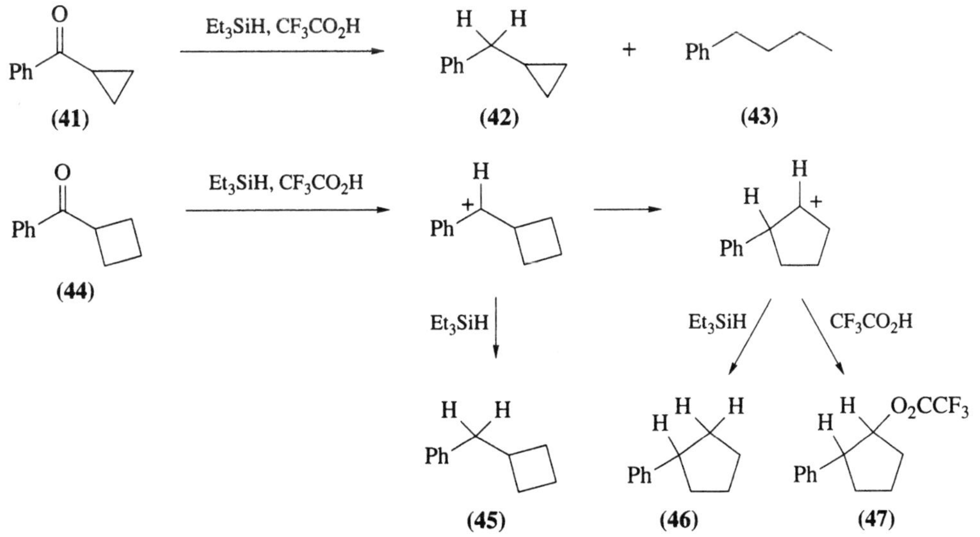

Scheme 14

the corresponding hydrocarbons in high yields, except for those carrying electron-withdrawing groups such as NO_2 and CN.

When $BF_3{\cdot}OEt_2$ was used in place of BF_3, the products in many cases were alcohols or symmetrical ethers. However, benzophenone and acetophenone were reduced to diphenylmethane and ethylbenzene in quantitative yields.[70] During attempts to isolate a borate ester, cyclohexene was obtained in 68% yield by direct distillation of a mixture of cyclohexanone, Et_3SiH and $BF_3{\cdot}OEt_2$.[70] An analogous method using $BF_3{\cdot}OH_2$, the acidity of which is comparable to anhydrous sulfuric and hydrofluoric acids,[71] permits reductions of adamantanone, 1-tetralone and 2-acetonaphthalene to their methylene derivatives in 78, 67 and 70% yields, respectively, though the scope and limitation of this method have not yet been established.[72]

An alternative reduction method using CF_3SO_3H and Et_3SiH has been developed.[73] Diaryl ketones with *ortho*- or *para*-substituted groups including OH, Me, Br, F, CF_3 and NO_2, and acetophenone-type ketones undergo reduction to the corresponding methylenes. However, 4,4′-dimethoxybenzophenone did not react with this reagent, which evidently did not attack the highly stabilized protonated carbonyl group. Steric effects also interfere with the reduction. Dimesityl and mesityl xylyl ketones, for example, were not reduced, though mesityl phenyl ketone could be reduced to the hydrocarbon. The alkyl ketones, cyclohexanone and 2-adamantanone afforded the corresponding hydrocarbons in quite low yields; the major products were their alcohols.

1.13.4 HYDROGENOLYSIS OF AROMATIC ALDEHYDES AND KETONES

1.13.4.1 Catalytic Hydrogenation

Catalytic hydrogenolysis of carbonyl compounds to the hydrocarbons is a convenient method, in which the reaction proceeds *via* the corresponding alcohol. Although a variety of catalysts are known, this reduction is generally effective only with aromatic aldehydes and ketones. Platinum oxide is often used for reductions of aryl,[74] and α,β-unsaturated carbonyl compounds.[75] Depending on conditions, this catalyst can reduce even an aromatic system exhaustively; acetophenone for instance was converted to ethylcyclohexane in 84% yield.[76] A benzofluorenone derivative could be hydrogenolized to the benzofluorene (89%) using 10% Pd–C in ethanol, but on quenching after consumption of 1 mol equiv. of H_2, the corresponding benzyl alcohol was obtained in 93% yield.[77] Deuteriation of aryl carbonyl compounds with dideuterium in acetic *d*-acid in the presence of catalytic 10% Pd–C has been carried out. Under these conditions, the hydrocarbon products are labeled at the carbonyl carbon atom, and if present, at the benzylic methylene carbon atom.[78] Hydrogenolysis of relatively unhindered alkyl ketones at 190 °C in the presence of Ni/Al_2O_3 afforded the corresponding hydrocarbons;[79] however, substrates for this reduction are limited to relatively simple molecules, because other functional groups are sensitive towards this reagent.[80] As alternative catalysts, osmium on carbon,[81] molybdenum trioxide,[82] and iridium or rhodium derivatives[83] are also applicable for reduction of aromatic aldehydes and ketones.

Early work by Papa *et al.* indicated that reduction of carbonyl compounds with Raney nickel in alkaline solution gave the corresponding hydrocarbon or alcohol products,[84] and formation of the hydrocarbon was only feasible in the case of aromatic carbonyl compounds at 80–90 °C.[85] Mitchell *et al.* reported an improved method: under neutral conditions using W-7 Raney nickel in 50% aqueous ethanol, aryl aldehydes, alkyl aryl and diaryl ketones can be reduced to the methylene products in high yields.[86] Aromatic substituents such as nitro, cyano and halogen also suffer reduction under these conditions.

1.13.4.2 Hydrogen Transfer Reduction

Catalytic hydrogen transfer reduction is well known as a powerful method, not only for deprotection of benzyl ethers,[87] but also for the reduction of functional groups.[88] Reduction of carbonyl compounds provides either the corresponding alcohols or, particularly in the case of aromatic derivatives, the hydrocarbons. Brieger *et al.* found that reduction of aryl aldehydes and ketones to the aryl methylene products can be realized in good yields using cyclohexene or limonene as the hydrogen donor in the presence of a Pd–C catalyst, coupled with a promoter such as iron(III) chloride, aluminum chloride or even water.[89] The intermediacy of the benzyl alcohol under these conditions was proved by trapping its derivatives; for example, the reduction of anisaldehyde in the presence of acetic anhydride provided *p*-methoxybenzyl acetate in 83% yield (equation 24).[90] In the case of *o*-carboxybenzaldehyde, the benzyl alcohol intermediate had a chance to be trapped by the neighboring carboxyl group to form a γ-lactone which was stable under the reaction conditions (equation 25). Consequently, a mixture of (**48**) and (**49**) was obtained in 45 and 35% yields, respectively.[89]

CHO, OMe — cyclohexene/Ac_2O, 10% Pd–C, $FeCl_3$, 83% → OAc, OMe (24)

CHO, CO_2H — cyclohexene, 10% Pd–C, $FeCl_3$ → O, O (**48**) + CO_2H (**49**) (25)

Cyclopropyl phenyl ketone afforded quantitatively *n*-butylbenzene, instead of cyclopropylphenylmethane, because of the reactivity of the cyclopropyl group.[90] In the case of 6-methoxytetralone, a dehydrogenation product (**50**) was obtained in 20% yield together with the desired methylene compound (**51**) in 33% yield (equation 26). Because the tetraline structure of (**51**) can be a hydrogen donor, the newly formed product (**51**) may be consumed during the reaction; hence the formation of the naphthalene (**50**).[90]

O, MeO — limonene, 10% Pd–C, $FeCl_3$ → MeO (**50**) + MeO (**51**) (26)

The method using ammonium formate as a hydrogen donor has been reported to accelerate the reaction, but the selectivity of the reduction excludes nitro, halo or alkenic groups, which are reduced faster than the carbonyl group.[91]

1.13.5 ELECTROLYSIS OF NONCONJUGATED AND CONJUGATED ALDEHYDES AND KETONES

Electrolysis of carbonyl compounds provides pinacols, alcohols or hydrocarbons, depending on the conditions, such as pH, the nature of the electrode, and its potential.[92] Fundamental studies have been carried out on the mechanisms of hydrocarbon formation using acetone as a substrate.[93–95] Although several electrodes, such as Cd, Pt, Pb or Zn, are recommended, carbonyl compounds, including aryl and alkyl derivatives, require strong aqueous acidic media for reduction to the hydrocarbons. The mechanism of the electrolytic reduction is probably similar to that of Clemmensen reduction, which starts from anion radical formation by one-electron transfer, as indicated in Scheme 3. The difference is that electrolytic reduction takes place in an electric double layer, rather than on the surface of the zinc metal.

In the case of aryl aldehydes and ketones, benzaldehyde afforded benzyl alcohol as the major product, but acetophenone and its *para*-substituted derivatives carrying such groups as OMe, Cl or OH provided ethylbenzene derivatives in good yields. As with Clemmensen reduction, the alcohol produced in this reduction cannot be further reduced, and the alcohol is not therefore an intermediate. Still uncertain in the reaction mechanism of electrolytic reduction, however, is the role of adsorbed hydrogen.[96]

Electrolytic reduction of alkyl ketones, such as the conversion of menthone to menthane using a cadmium cathode in dilute sulfuric acid, can be seen in early work,[97,98] and electrolytic reduction of alkyl ketones has also been applied to such steroidal ketones as 7-oxo-,[99] 12-oxo-,[100] 3-oxo- (equation 27), 19-oxo- and 20-oxo-derivatives[101,102] using Pb,[99,102] Cd,[100] or Hg[101] electrodes in solvents containing sulfuric acid. One purpose of these deoxygenations was deuterium labeling of the carbonyl carbon atom, and products with high isotopic purity have been obtained, except with the 3-one derivative which afforded a significant amount of d_3 and d_4 isotopic impurities, because of enolizable protons α to the carbonyl group.[99,101]

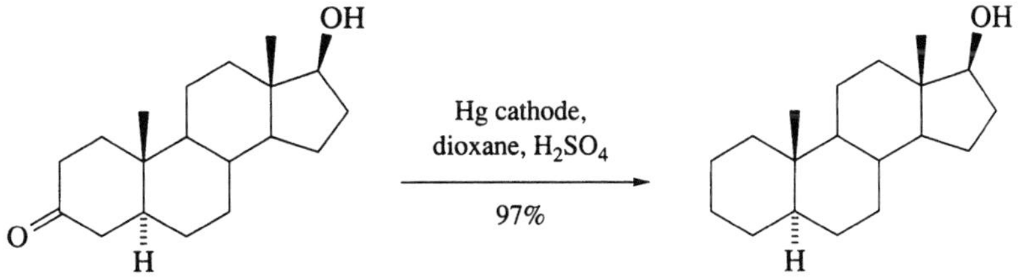

Hg cathode, dioxane, H_2SO_4, 97% (27)

As shown in equation (28) the 17-hydroxy group at the α position to the carbonyl group at C-20 was easily removed, whereas the isolated 3-hydroxy group was unaffected. This fact supports the carbanion intermediate in the mechanism. During these studies, it was found that alkyl aldehydes can also be reduced to the corresponding methyl derivatives.[102,103]

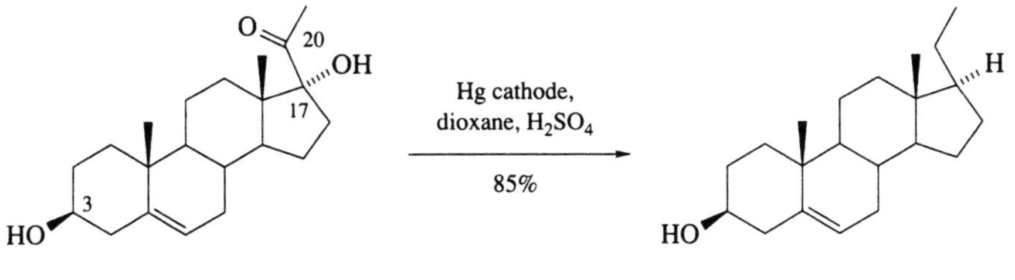

Hg cathode, dioxane, H_2SO_4, 85% (28)

As in the electrolysis of α-ketols, the α-amino ketone (**52**) was reduced using a Pb cathode in 30% sulfuric acid to give (**53**) *via* C(α)—N bond cleavage.[104] This reduction was extended to medium-size ring formation, in which the product distribution between the ring-expanded product (**54**) and the ring-rearranged product (**55**), depends on the ring size of the starting material (Scheme 15).[105–107]

As mentioned in Section 1.13.2.2 (Clemmensen reduction), 1,3-diketone derivatives give rise to ring contraction products *via* cyclopropanediols, which can be trapped as acetates. On electrolysis using a Hg cathode in the presence of acetic anhydride, the 1,3-diketone (**18**) afforded a mixture of stereoisomers (**19**) in 33% yield, and a similar electrolysis of (**56**) gave (**57**) in 71% yield (equation 29).[22]

1.13.6 MISCELLANEOUS

In addition to the aforementioned methods and to those discussed in Chapter 1.14 (this volume), there are several other methods for the reduction of carbonyl functions to the corresponding methylenes.

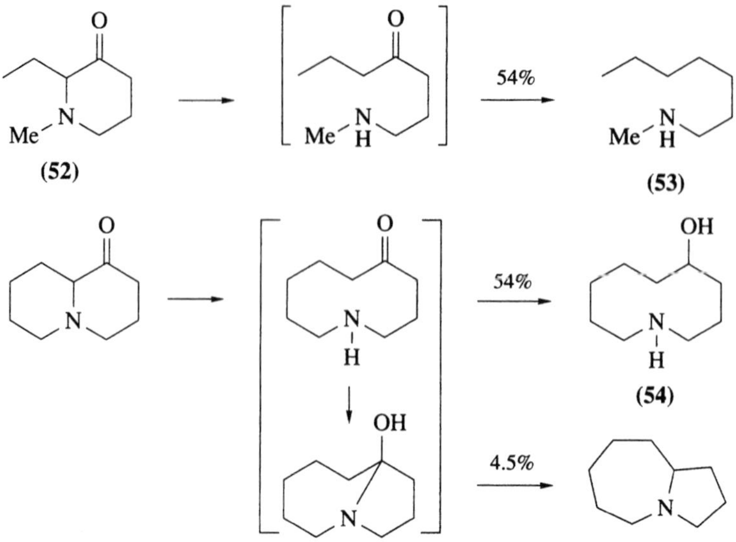

Scheme 15

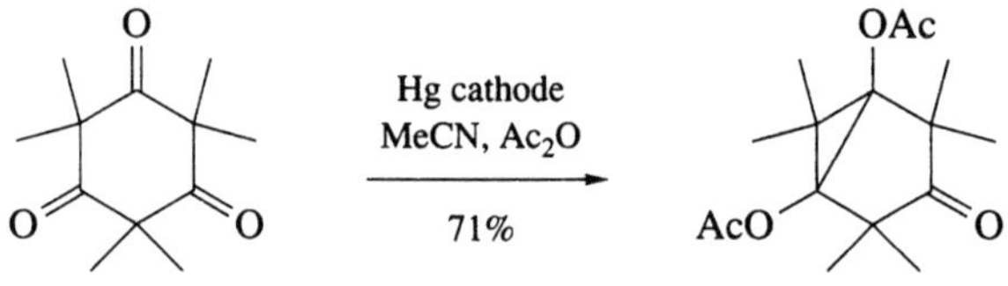

(29)

The reducing ability of silicon hydride has long been known, and in Section 1.13.3.3 we discussed ionic hydrogenation using the triethylsilane/acid combination. In addition, diphenylsilane at high temperature without any additives converted several diaryl ketones to the corresponding hydrocarbons. For example, 10-thiaxanthenone could be reduced with Ph_2SiH_2 at 260 °C to 10-thiaxanthene in 64% yield. A more detailed study has been carried out on benzophenone, which was reduced at *ca.* 260 °C to diphenylmethane in 37% yield. Benzhydryloxydiphenylsilane (**58**) was formed at 230 °C, and this compound could be converted at 270 °C to diphenylmethane, showing that (**58**; Scheme 16) is an intermediate in this reduction.[108,109]

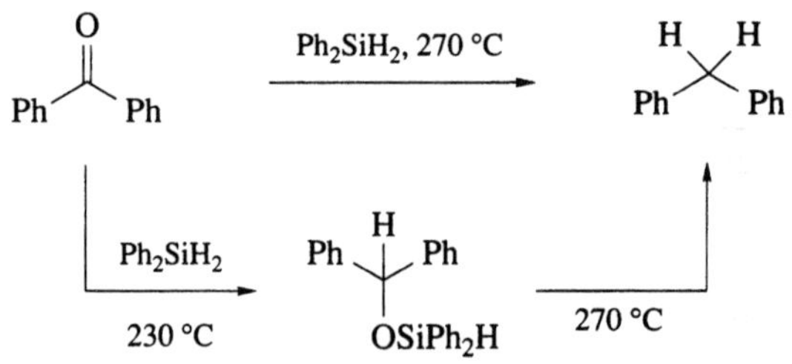

Scheme 16

A related reduction utilizing other silicon hydrides has been investigated by Benkeser.[59,110] Aryl aldehydes and ketones reacted with trichlorosilane–tri-*n*-propylamine to give an organosilicon derivative (**59**; Scheme 17), which on alkaline hydrolysis underwent silicon–hydrogen exchange, and consequently yielded the corresponding hydrocarbons.

Reduction of aryl ketones using Ph_3SnH and acetyl chloride has been reported.[111] The stepwise process is indicated in Scheme 18. Depending on the nature of the carbonyl carbon, chloride and/or acetate intermediates are formed, of which the latter is trapped, but the former is further reduced by the tin hydride.

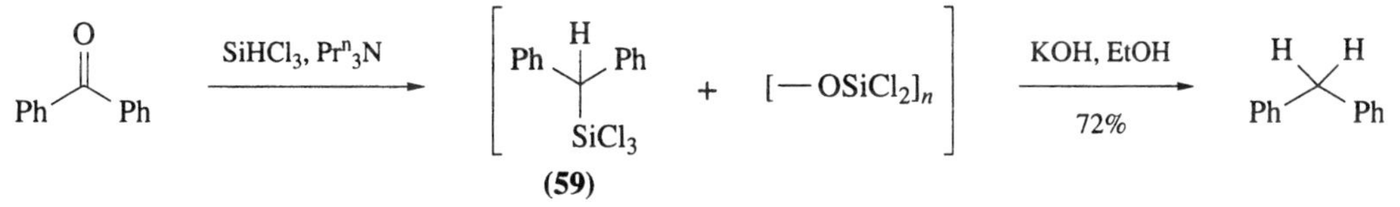

Scheme 17

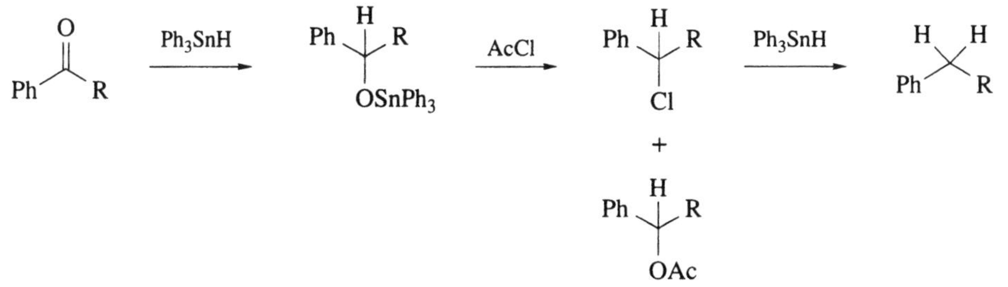

Scheme 18

As an application of selenium chemistry, reduction of aryl ketones with CO and H_2O in the presences of selenium and DBU has been carried out. The reduction, indicated in Scheme 19, can be considered as a redox reaction by CO and H_2O. This reaction proceeds in high yield with only a catalytic amount of selenium. Although the reaction mechanism has not been clarified, this reaction probably involves an organoselenium hydride intermediate (**60**), which is known to give the corresponding hydrocarbon under the reaction conditions.[112]

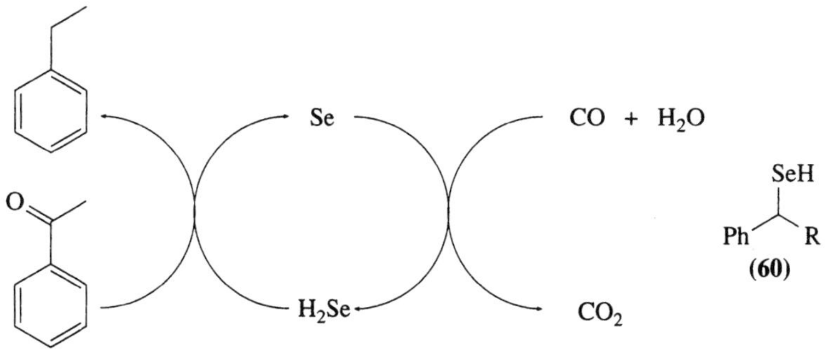

Scheme 19

As a variation of the hydrogen transfer method, reduction of diaryl ketones with H_2S generated from hydrogen and sulfur in the presence of catalytic MoS_3 and *p*-TsOH afforded the corresponding diarylmethane under high pressure and at high temperature. In this case, clean reduction is limited to ketones, such as diarylketones, carrying no proton α to the carbonyl group (Scheme 20).[113]

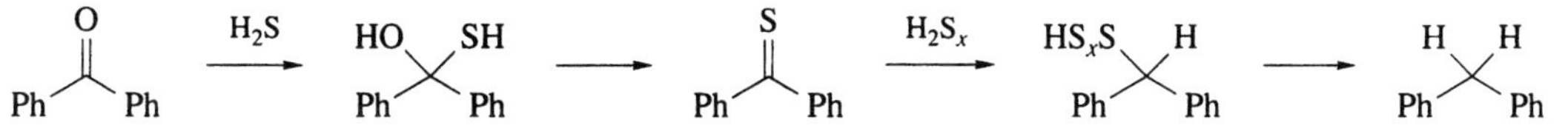

Scheme 20

Diaryl ketones, such as *o*-bromobenzophenone[114] and fluorenone carboxylic acids,[115,116] are reduced by HI and red phosphorus in acetic acid or propionic acid under reflux, giving *o*-bromophenylphenylmethane and fluorene carboxylic acid in high yields.

As a rather special case, alkyl aldehydes are reduced with titanocen dichloride to the hydrocarbons through a titanium-bonded alkenic intermediate as shown by a deuterium-labeling experiment. Thus, dodecanal was converted to dodecane in 71% yield, along with dodecan-1-ol in 15–20% yield; alkyl ketones such as adamantan-2-one and dodecan-6-one afforded alcohols as the major products. No reduction occurred in the case of aryl aldehydes where alkene formation is impossible.[117]

1.13.7 REFERENCES

1. M. Hudlicky, 'Reductions in Organic Chemistry', Wiley, New York, 1984.
2. A. Hajós, *Stud. Org. Chem. (Amsterdam)*, 1979, **1**, 1.
3. R. N. Augustine (ed.), 'Reduction', Dekker, New York, 1968.
4. H. Schulbach, *Ber. Dtsch. Chem. Ges.*, 1915, **48**, 12; W. E. Bachmann, *J. Am. Chem. Soc.*, 1933, **55**, 1179; C. B. Wooster, *J. Am. Chem. Soc.*, 1937, **59**, 377.
5. S. S. Hall, S. D. Lipsky, F. J. McEnroe and A. P. Bartels, *J. Org. Chem.*, 1971, **36**, 2588.
6. S. S. Hall, A. P. Bartels and A. M. Engman, *J. Org. Chem.*, 1972, **37**, 760.
7. S. S. Hall and C.-K. Sha, *Chem. Ind. (London)*, 1976, 216.
8. E. L. Martin, *Org. React. (N.Y.)*, 1942, **1**, 155.
9. E. Vedejs. *Org. React. (N.Y.)*, 1975, **22**, 401.
10. J. G. St C. Buchanan and P. D. Woodgate, *Q. Rev., Chem. Soc.*, 1969, **23**, 522.
11. S. Yamamura, M. Toda and Y. Hirata, *Org. Synth., Coll. Vol.*, 1988, **6**, 289.
12. M. Toda, M. Hayashi, Y. Hirata and S. Yamamura, *Bull. Chem. Soc. Jpn.*, 1972, **45**, 264.
13. S. Yamamura and Y. Hirata, *J. Chem. Soc. C*, 1968, 2887.
14. H. O. House, in 'Modern Synthetic Reactions', Benjamin, Menlo Park, CA, 1972, p. 166.
15. I. Elphimoff-Felkin and P. Sarda, *J. Chem. Soc., D*, 1969, 1065.
16. R. Schwarz and H. Hering, *Org. Synth., Coll. Vol.*, 1963, **4**, 203.
17. C. R. Enzell, *Tetrahedron Lett.*, 1966, 1285.
18. I. Elphimoff-Felkin and P. Sarda, *Tetrahedron Lett.*, 1969, 3045.
19. A. K. Banerjee, J. Alvárez, M. Santana and M. C. Carrasco, *Tetrahedron*, 1986, **42**, 6615.
20. N. J. Cusack and B. R. Davis, *Chem. Ind. (London)*, 1964, 1426; *J. Org. Chem.*, 1965, **30**, 2062.
21. A. N. Dey and R. P. Linstead, *J. Chem. Soc.*, 1935, 1063.
22. T. J. Curphey, C. W. Amelotti, T. P. Layloff, R. L. McCartney and J. H. Williams, *J. Am. Chem. Soc.*, 1969, **91**, 2817.
23. E. Wenkert and E. Kariv, *Chem. Commun.*, 1965, 570.
24. E. Wenkert and J. E. Yoder, *J. Org. Chem.*, 1970, **35**, 2986.
25. L. H. Conover and D. S. Tarbell, *J. Am. Chem. Soc.*, 1950, **72**, 3586.
26. H. E. Zieger and J. A. Dixon, *J. Am. Chem. Soc.*, 1960, **82**, 3702.
27. K. H. Bell, *Aust. J. Chem.*, 1969, **22**, 601.
28. J. G. Sweeny and G. A. Iacobucci, *Tetrahedron*, 1977, **33**, 2927.
29. J. G. Sweeny and G. A. Iacobucci, *Tetrahedron*, 1977, **33**, 2923.
30. J. M. Saá and M. P. Cava, *J. Org. Chem.*, 1977, **42**, 347.
31. C. A. Elliger, *Synth. Commun.*, 1985, **15**, 1315.
32. M. Černý and J. Málek, *Collect. Czech. Chem. Commun.*, 1970, **35**, 1216.
33. M. Černý and J. Málek, *Collect. Czech. Chem. Commun.*, 1970, **35**, 2030.
34. M. Černý and J. Málek, *Collect. Czech. Chem. Commun.*, 1970, **35**, 3079.
35. D. J. Humphreys and G. R. Proctor, *J. Chem. Soc., Perkin Trans. 1*, 1972, 722.
36. M. Černý and J. Málek, *Tetrahedron Lett.*, 1969, 1739.
37. T. Kametani, S.-P. Huang, M. Ihara and K. Fukumoto, *J. Org. Chem.*, 1976, **41**, 2545.
38. B. R. Brown and A. M. S. White, *J. Chem. Soc.*, 1957, 3755.
39. J. Blackwell and W. J. Hickinbottom, *J. Chem. Soc.*, 1961, 1405.
40. R. F. Nystrom and C. R. A. Berger, *J. Am. Chem. Soc.*, 1958, **80**, 2896.
41. J. H. Brewster and H. O. Bayer, *J. Org. Chem.*, 1964, **29**, 116.
42. J. H. Brewster and J. E. Privett, *J. Am. Chem. Soc.*, 1966, **88**, 1419.
43. A. Ono, N. Suzuki and J. Kamimura, *Synthesis*, 1987, 736.
44. J. Broome, B. R. Brown, A. Roberts and A. M. S. White, *J. Chem. Soc.*, 1960, 1406.
45. R. Albrecht and C. Tamm, *Helv. Chim. Acta*, 1957, **40**, 2216.
46. H. Suzuki, R. Masuda, H. Kubota and A. Osuka, *Chem. Lett.*, 1983, 909.
47. T. Satoh, N. Mitsuo, M. Nishiki, K. Nanba and S. Suzuki, *Chem. Lett.*, 1981, 1029.
48. R. A. Heacock and O. Hutzinger, *Can. J. Chem.*, 1964, **42**, 514.
49. G. W. Gribble, R. M. Leese and B. E. Evans, *Synthesis*, 1977, 172.
50. G. W. Gribble, W. J. Kelly and S. E. Emery, *Synthesis*, 1978, 763.
51. A. H. Jackson, B. Naidoo and P. Smith, *Tetrahedron*, 1968, **24**, 6119.
52. C. K. Lau, C. Dufresne, P. C. Bélanger, S. Piétré and J. Scheigetz, *J. Org. Chem.*, 1986, **51**, 3038.
53. E. C. Ashby, A. B. Goel and R. N. DePriest, *J. Am. Chem. Soc.*, 1980, **102**, 7779; E. C. Ashby and A. B. Goel, *Tetrahedron Lett.*, 1981, **22**, 1879; E. C. Ashby, R. N. DePriest, A. B. Goel, B. Wenderoth and T. N. Pham, *J. Org. Chem.*, 1984, **49**, 3545.
54. E. Breuer, *Tetrahedron Lett.*, 1967, 1849.
55. K. M. Biswas, L. E. Houghton and A. H. Jackson, *Tetrahedron Suppl.*, 1966, **7**, 261.
56. W. J. Wechter, *J. Org. Chem.*, 1963, **28**, 2935.
57. C. K. Lau, S. Tardif, C. Dufresne and J. Scheigetz, *J. Org. Chem.*, 1989, **54**, 491.
58. J. J. Evans, L. Bagnell, R. Eibl, G. Holan and T. Mole, *Aust. J. Chem.*, 1975, **28**, 519.
59. W. P. Weber, in 'Reactivity and Structure Concepts in Organic Chemistry', ed. K. Hafner, J.-M. Lehn, C. W. Rees, P. von R. Schleyer, B. M. Trost and R. Zahradnik, Springer-Verlag, New York, 1983, vol. 14, p. 273.
60. D. N. Kursanov, Z. N. Parnes and N. M. Loim, *Synthesis*, 1974, 633.
61. D. N. Kursanov, Z. N. Parnes, N. M. Loim and G. V. Bakalova, *Dokl. Akad. Nauk SSSR*, 1968, **179**, 1106 (*Chem. Abstr.*, 1968, **69**, 43 570c).
62. D. N. Kursanov, Z. N. Parnes and N. M. Loim, *Izv. Akad. Nauk SSSR, Ser. Khim.*, 1966, 1289 (*Chem. Abstr.*, 1966, **65**, 15 183c).
63. D. N. Kursanov, Z. N. Parnes, G. I. Bassova, N. M. Loim and V. I. Zdanovich, *Tetrahedron*, 1967, **23**, 2235.
64. C. T. West, S. J. Donnelly, D. A. Kooistra and M. P. Doyle, *J. Org. Chem.*, 1973, **38**, 2675.

65. Z. N. Parnes, N. M. Loim, V. A. Baranova and D. N. Kursanov, *Zh. Org. Khim.*, 1971, **7**, 2066 (*Chem. Abstr.*, 1972, **76**, 13 495j).
66. D. N. Kursanov, N. M. Loim, V. A. Baranova, L. V. Moiseeva, L. P. Zalukaev and Z. N. Parnes, *Synthesis*, 1973, 420.
67. N. M. Loim, Z. N. Parnes, I. I. Brunovlenskaya and D. N. Kursanov, *Dokl. Akad. Nauk SSSR*, 1971, **196**, 1361 (*Chem. Abstr.*, 1971, **74**, 141 369s).
68. J. L. Fry, M. Orfanopoulos, M. G. Adlington, W. R. Dittman, Jr. and S. B. Silverman, *J. Org. Chem.*, 1978, **43**, 374.
69. J. L. Fry and S. B. Silverman, *Org. Synth.*, 1981, **60**, 108.
70. M. P. Doyle, C. T. West, S. J. Donnelly and C. C. McOsker, *J. Organomet. Chem.*, 1976, **117**, 129.
71. C. H. Rochester, in 'Acidity Functions', Academic Press, New York, 1970, p. 52.
72. J. W. Larsen and L. W. Chang, *J. Org. Chem.*, 1979, **44**, 1168.
73. G. A. Olah, M. Arvanaghi and L. Ohannesian, *Synthesis*, 1986, 770.
74. E. J. Corey and E. W. Cantrall, *J. Am. Chem. Soc.*, 1959, **81**, 1745.
75. H. Rapoport and G. Smolinsky, *J. Am. Chem. Soc.*, 1960, **82**, 1171.
76. S. Nishimura, *Bull. Chem. Soc. Jpn.*, 1961, **34**, 32.
77. R. L. Letsinger and J. D. Jamison, *J. Am. Chem. Soc.*, 1961, **83**, 193.
78. K. Ofusu-Asante and L. M. Stock, *J. Org. Chem.*, 1987, **52**, 2938.
79. W. F. Maier, K. Bergmann, W. Bleicher and P. von R. Schleyer, *Tetrahedron Lett.*, 1981, **22**, 4227.
80. W. F. Maier, P. Grumüller, I. Thies, P. M. Stein, M. A. McKervey and P. von R. Schleyer, *Angew. Chem., Int. Ed. Engl.*, 1979, **18**, 939.
81. K. Teranishi, T. Shimizu, Y. Zenitani and T. Yoshioka (Mitsui Petro. Chem. Co.), *Jpn. Pat.* 14 601 (1975) (*Chem. Abstr.*, 1976, **84**, 4732x).
82. K. Teranishi, T. Shimizu, T. Nagaoka and T. Yoshioka (Mitsui Petro. Chem. Co.), *Jpn. Pat.* 134 637 (1974) (*Chem. Abstr.*, 1975, **82**, 155 961f).
83. G. M. Intille (Monsato Co.), *US Pat.* 4 067 900 (1978) (*Chem. Abstr.*, 1978, **89**, 23 930e).
84. D. Papa, E. Schwenk and B. Whitman, *J. Org. Chem.*, 1942, **7**, 587.
85. P. L. Cook, *J. Org. Chem.*, 1962, **27**, 3873.
86. R. H. Mitchell and Y.-H. Lai, *Tetrahedron Lett.*, 1980, **21**, 2637.
87. T. W. Greene, 'Protective Groups in Organic Synthesis', Wiley, New York, 1981.
88. G. Brieger and T. J. Nestrick, *Chem. Rev.*, 1974, **74**, 567.
89. G. Brieger and T.-H. Fu, *J. Chem. Soc., Chem. Commun.*, 1976, 757.
90. G. Brieger, T. J. Nestrick and T.-H. Fu, *J. Org. Chem.*, 1979, **44**, 1876.
91. S. Ram and L. D. Spicer, *Tetrahedron Lett.*, 1988, **29**, 3741.
92. M. R. Rifi, *Tech. Chem. (New York)*, 1975, **5** (II), 83.
93. T. Sekine, A. Yamura and K. Sugino, *J. Electrochem. Soc.*, 1965, **112**, 439.
94. X. De Hemptinne and K. Schunck, *Trans. Faraday Soc.*, 1969, **65**, 591.
95. X. De Hemptinne and K. Schunck, *Ann. Soc. Sci. Bruxelles, Ser. 1*, 1966, **80**, 289.
96. D. Pletcher and M. Razaq, *Electrochim. Acta*, 1981, **26**, 819.
97. G. H. Keats, *J. Chem. Soc.*, 1937, 2003.
98. C. Schall and W. Kirst, *Z. Elektrochem.*, 1923, **29**, 537.
99. G. Phillipou, C. J. Seaborn and I. A. Blair, *Aust. J. Chem.*, 1979, **32**, 2767.
100. Showa Denko K. K., *Jpn. Pat.* 32 900 (1983) (*Chem. Abstr.*, 1983, **99**, 88 476n).
101. L. J. Throop and L. Tökés, *J. Am. Chem. Soc.*, 1967, **89**, 4789.
102. L. J. Throop and A. D. Cross (Syntex Corp.), *US Pat.* 3 506 549 (1970) (*Chem. Abstr.*, 1970, **73**, 136 807h).
103. I. A. Blair, R. G. Frith, G. Phillipou and C. J. Seaborn, *Aust. J. Chem.*, 1979, **32**, 2327.
104. N. J. Leonard, S. Swann, Jr. and H. L. Dryden, Jr., *J. Am. Chem. Soc.*, 1952, **74**, 2871.
105. N. J. Leonard, S. Swann, Jr. and J. Figueras, Jr., *J. Am. Chem. Soc.*, 1952, **74**, 4620.
106. N. J. Leonard, S. Swann, Jr., and G. Fuller, *J. Am. Chem. Soc.*, 1954, **76**, 3193.
107. N. J. Leonard, S. Swann, Jr. and E. H. Mottus, *J. Am. Chem. Soc.*, 1952, **74**, 6251.
108. J. W. Diehl and H. Gilman, *Chem. Ind. (London)*, 1959, 1095.
109. H. Gilman and J. W. Diehl, *J. Org. Chem.*, 1961, **26**, 4817.
110. R. A. Benkeser, *Acc. Chem. Res.*, 1971, **4**, 94.
111. H. Patin, L. Roullier, and R. Dabard, *C. R. Hebd. Seances Acad. Sci., Ser. C*, 1971, **272**, 675.
112. Y. Nishiyama, S. Hamanaka, A. Ogawa, N. Kambe and N. Sonoda, *J. Org. Chem.*, 1988, **53**, 1326.
113. T. Takido, Y. Yamane and K. Itabayashi, *Yuki Gosei Kagaku Kyokaishi*, 1975, **33**, 694 (*Chem. Abstr.*, 1976, **84**, 105 145p).
114. C. K. Bradsher and F. A. Vingiello, *J. Org. Chem.*, 1948, **13**, 786.
115. D. C. Morrison, *J. Org. Chem.*, 1958, **23**, 1772.
116. D. C. Morrison, *J. Org. Chem.*, 1958, **23**, 1371.
117. E. E. van Tamelen and J. A. Gladysz, *J. Am. Chem. Soc.*, 1974, **96**, 5290.

1.14

Reduction of C=X to CH_2 by Wolff–Kishner and Other Hydrazone Methods

ROBERT O. HUTCHINS
Drexel University, Philadelphia, PA, USA

and

MARYGAIL K. HUTCHINS
ICI Americas, Exton, PA, USA

1.14.1 INTRODUCTION

Since its introduction early in this century,[1,2] the deoxygenation of aldehydes and ketones to methyl or methylene derivatives, respectively, *via* base treatment of hydrazone intermediates (equation 1) has proven to be one of the most convenient and synthetically useful processes available for this important type of transformation. The reaction is termed the Wolff–Kishner reduction in recognition of the two original independent discoverers.[1,2] However, the initial recipes introduced proved tedious and unreliable with many structural, especially hindered, examples. This led to substantial efforts devoted over the years to developing more convenient and successful experimental procedures, resulting in a number of improved and more reliable modifications which are most often utilized at present. More recently, modified procedures have been provided which utilize hydride reductions of *p*-toluenesulfonylhydrazone (tosylhydrazone) derivatives and subsequent decomposition to release the hydrocarbon products under much milder and less basic conditions than those normally required for Wolff–Kishner reductions (equation 2).

$$R_2C{=}N{-}NH_2 \xrightarrow[\text{sol-H}]{\text{base}} R_2CH_2 \qquad (1)$$

$$R_2C{=}N{-}NH{-}SO_2R \xrightarrow{\text{reduction}} R_2CH{-}NH{-}NH{-}SO_2R \xrightarrow{\text{base}} R_2CH_2 + N_2 \qquad (2)$$

The synthetic utility of the Wolff–Kishner reduction has been the subject of several reviews.[3–6] However, the latest of these is over 20 years old[6] and fails to incorporate adequately many of the important advancements which have occurred in the past 25 years. This is particularly evident regarding the emergence over the past two decades of the above-mentioned reductions of sulfonylhydrazones with hydride donors as alternatives to standard Wolff–Kishner conditions. Consequently, this discussion will focus primarily on the period since the last major review (1968),[6] although some overlap with earlier work is essential for continuity and completeness. For further information and references to the older literature, the excellent aforementioned treatments[3–6] should be consulted. Reference 3 contains extensive tables of Wolff–Kishner reductions up to about 1947.

1.14.2 THE WOLFF–KISHNER REDUCTION OF ALDEHYDE AND KETONE HYDRAZONES

1.14.2.1 General Procedures and Mechanistic Aspects

The original protocols introduced for conversion of carbonyls to hydrocarbons by treatment with hydrazine and base involved the addition of the preformed hydrazone directly to hot potassium hydroxide (with[7] or without[1,8–10] added crushed platinized porous plate or palladium/barium sulfate[11]) at atmospheric pressure, or heating the hydrazone or semicarbazone with sodium ethoxide in a sealed tube at 160–200 °C.[2] In the ensuing years many variations of these procedures were introduced,[3] the most important of which was the use of high boiling point solvents to alleviate the inconvenience of sealed tubes or solid hydroxide and still achieve the high temperature (180–190 °C) necessary for decomposition of the hydrazones. Thus, Soffer[12,13] and, independently, Whitmore[14] (extending initial investigations by Ruzicka[15]) introduced the high boiling point solvents ethylene and triethylene glycols with which direct reductions of carbonyls were achieved *via* refluxing with hydrazine and dissolved sodium metal or sodium methoxide. These studies set the stage for the process known as the Huang-Minlon modification[16,17] and other improvements as discussed in the subsections to follow.

Mechanistic investigations[18–21] of Wolff–Kishner reductions implicate a hydrazone (**4**) as the initial intermediate generated from the carbonyl (**1**) and hydrazine or from decomposition of a preformed semicarbazone (**2**) or from an azine (**3**; Scheme 1). Subsequent rapid formation of the *trans*-hydrazone anion (**5**) by base-induced proton abstraction is followed by a rate-determining step that is viewed as an essentially concerted proton capture at the carbon coupled with a solvent-induced proton abstraction at the terminal nitrogen (*i.e.* **5**) to give a diimide anion (**6**). The process is completed by a rapid expulsion of nitrogen and protonation of the resulting carbanion (Scheme 1). An alkali metal hydroxide or alkoxide is normally employed as the basic catalyst, although with some systems (*i.e.* fluorenones[22]), excess hydrazine is also effective. Use of polar, aprotic dimethyl sulfoxide as solvent[23] or cosolvent with protic media[20] has been observed to increase the rate of, and greatly decrease the temperature required (from *ca.* 190 °C to as low as 25 °C)[23] for, Wolff–Kishner reductions, properties attributed to less efficient solvation of anions and the enhancement of the rate of proton removal by DMSO.[18,20,23,24] The greatest rate enhancement occurs with DMSO containing about 0.1–0.2 mol fraction of 2-(*t*-butoxyethoxy)ethanol[20] or *t*-butyl alcohol[24] as proton sources; greater concentrations of alcohol cause rapid decreases in the rates.[20,24]

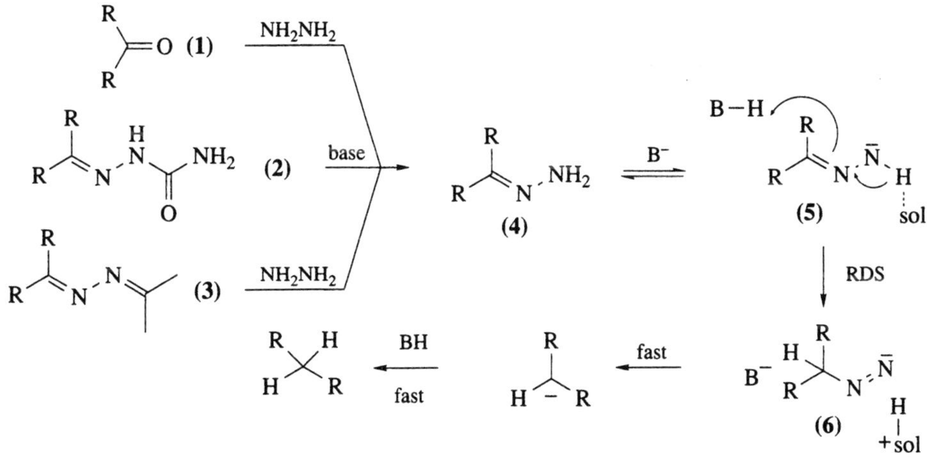

Scheme 1

1.14.2.2 Modified Procedures

The success of Wolff–Kishner reductions relies on two key processes, namely the reversible reaction of hydrazine with the carbonyl compound, which must be driven toward hydrazone formation, followed by base-induced decomposition of the hydrazone to generate a diimide anion and, subsequently, the hydrocarbon as depicted in Scheme 1. As expected, generation of the required hydrazones is highly dependent on the steric environment flanking the carbonyl and, consequently, much of the efforts devoted to improvements have focused on shifting the equilibrium toward the hydrazone by removal of water and/or the use of high concentrations of hydrazine. Coupled with this is the requirement to provide sufficient rates for the hydrazone decomposition step, which has generally meant modifications to provide increased reaction temperatures.

The most successful and synthetically useful of these modifications are discussed separately in the subsections which follow. It should be kept in mind, however, that like most synthetic methodologies, individual molecules often require further, usually slight, modifications so that within each category practitioners have utilized further variations as the need dictated. The most synthetically important of these will also be discussed.

1.14.2.2.1 The Huang-Minlon and related modifications

As mentioned, in 1945, Soffer *et al.*[12,13] and Whitmore and coworkers[14] introduced the use of high boiling point solvents, such as ethylene or triethylene glycol, for Wolff–Kishner reductions in order to attain the high reaction temperatures necessary for the decomposition of the hydrazones. However, disadvantages remained, since large excesses of sodium metal and solvent were required, coupled with the need for 100% hydrazine and long reaction times (50–100 h). This was primarily due to the temperature-lowering effect of water produced in hydrazone formation, resulting in the necessity of vigorous conditions and long reaction times. In 1946, Huang-Minlon[16] provided adjusted reaction conditions (serendipitously)[25] in which excess hydrazine and water were removed by distillation subsequent to hydrazone formation. This raised the reaction temperature to the required 190–200 °C and permitted the use of the more convenient 85% hydrazine hydrate, much less solvent and sodium or potassium hydroxide as the base. Reaction times were also cut dramatically to 3–6 h. Thus, in the original disclosure,[16] the keto acid (**7**) was deoxygenated to (**8**) in 95% isolated yield (equation 3) compared to 48%[25] with the Soffer procedure and 54%[16] under Clemmensen conditions (Volume 8, Chapter 1.13).

(**7**) $\xrightarrow[\text{KOH, (HOCH}_2\text{CH}_2\text{)O, 195 °C, 4 h, 95\%}]{\text{85\% NH}_2\text{NH}_2}$ (**8**) (3)

This efficient modification has become the standard general protocol for Wolff–Kishner reductions and has provided the basis for most further improvements introduced over the years. For instance, Lock[26] demonstrated that, for compounds sensitive to strong base, better results are obtained if the hydrazone is preformed by refluxing the carbonyl compound and hydrazine hydrate, while distilling out low boiling point components, followed by cooling, addition of the base and further refluxing (180–200 °C). On the other hand, the high refluxing temperatures usually associated with the Huang-Minlon procedure are not always necessary and several examples (*i.e.* veratraldehyde[27]) have been found which proceed well in the 135–160 °C range.[28,29]

The Huang-Minlon modification has been employed in an enormous number of synthetic endeavors, full coverage of which is beyond the scope of this discussion. However, Table 1 presents a collection of successful conversions, mostly from recent applications, chosen to illustrate a variety of structural types.

1.14.2.2.2 The Barton and related modifications

In the course of applying his modification to steroid ketones, Huang-Minlon[17] observed that while carbonyls at positions C-3, C-7, C-12, C-17 and C-20 were smoothly deoxygenated, the highly hindered C(11)-keto site remained unaffected (see also entry 2, Table 1). To alleviate this problem, Barton and coworkers[45,46] introduced a modification, more vigorous than the Huang-Minlon process, by scrupulously avoiding moisture, and raising the temperature and time for the hydrazone decomposition. Thus, sodium was dissolved in dry diethylene glycol to generate an alkoxide base, hydrazine hydrate was replaced with anhydrous hydrazine, the apparatus was protected from outside moisture, excess hydrazine distilled until the temperature reached 210 °C, and refluxing extended for 12 h. Using this methodology, both carbonyls of 7,11-dioxolanostanyl acetate (**9**) were successfully removed[45,46] to afford (after reacetylation) lanosterol in 69% yield (equation 4). The requirement of anhydrous conditions is illustrated by the removal of only the 7-keto group of (**9**) in low yield (41%) using hydrazine hydrate and KOH.[47]

(**9**) $\xrightarrow[\text{DEG, 210 °C, 12 h, 69\%}]{\text{anhydrous NH}_2\text{NH}_2\text{, Na}}$ (4)

AcO, C_8H_{15}

Since its introduction in 1954, the Barton modification[45] has enjoyed considerable utility in Wolff–Kishner reductions, especially with sterically hindered or otherwise relatively inaccessable ketones. A selection of structural examples is presented in Table 2.

Although the Barton procedure is usually successful with hindered carbonyls, failures or low yields have been noted. For example, ketone (**10**) was recovered unaltered, a result attributed to the extremely hindered environment of the carbonyl (**10** was also unaffected by $NaBH_4$ and failed to give a semicarbazone or 2,4-dinitrophenylhydrazone).[48] Likewise, Barton conditions failed to remove the highly hindered ketone in a triterpenoid (partial structure **11**).[49]

Table 1 Wolff–Kishner Reduction of Carbonyls Using the Huang–Minlon Modification

Entry	*Carbonyl*	*Product*	*Yield (%)*	*Ref.*
1	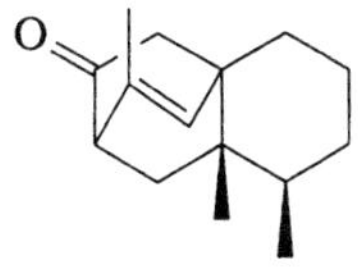		74	30
2	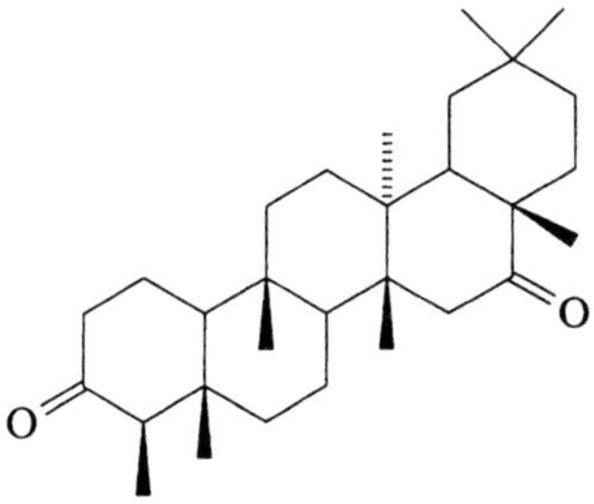	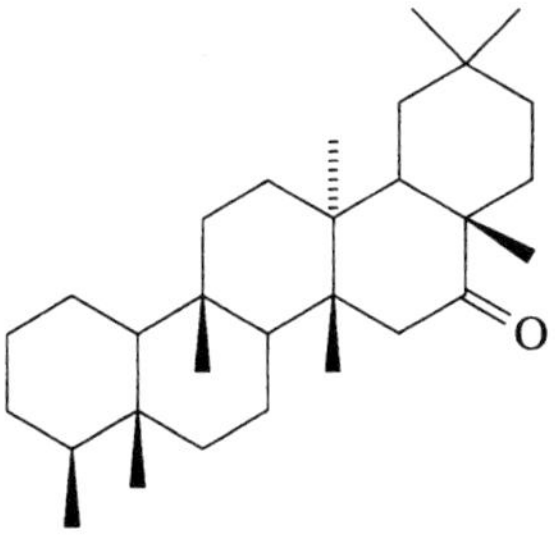	62	31
3	Mixture 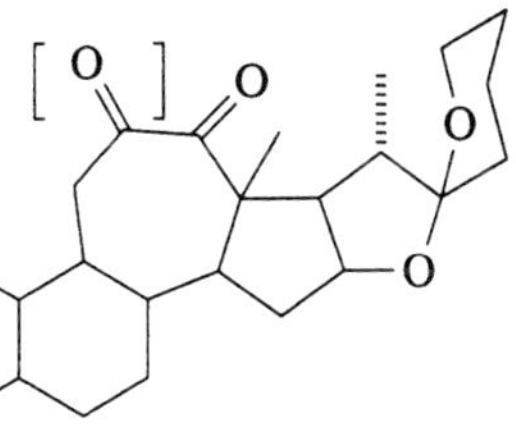 [O] HO	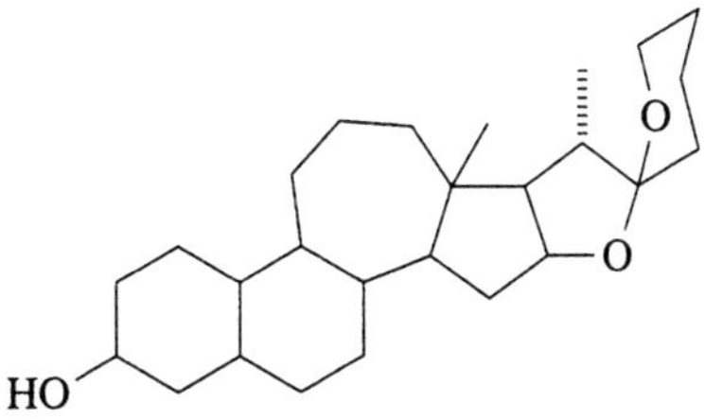 HO	66	32
4	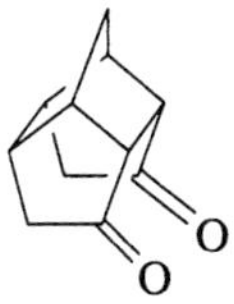		75	33
5	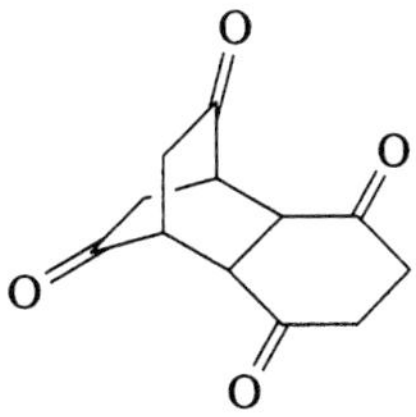	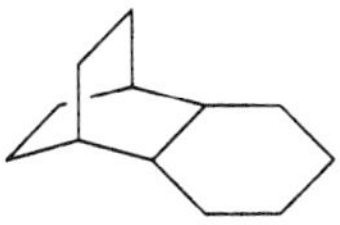	—	34
6	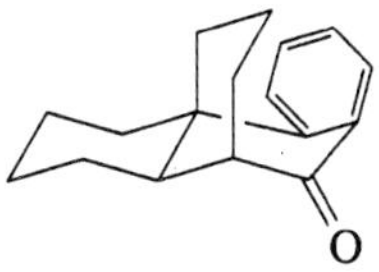	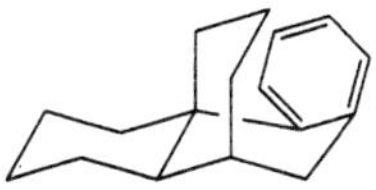	85	35
7	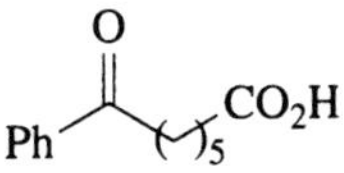 $PhCO(CH_2)_5CO_2H$	 $Ph(CH_2)_6CO_2H$	75	36
8	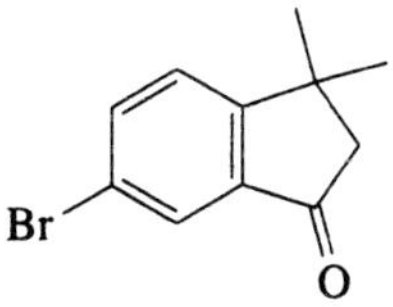 Br	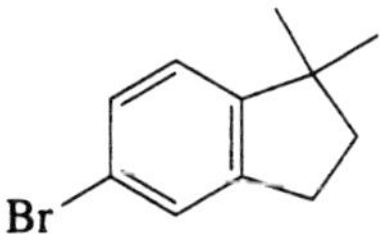 Br	88	37

Table 1 *(continued)*

Entry	*Carbonyl*	*Product*	*Yield (%)*	*Ref.*
9	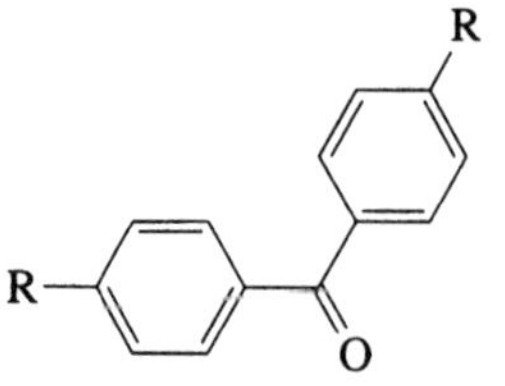 R = F, Cl, SO_2Me, CN, NMe_2	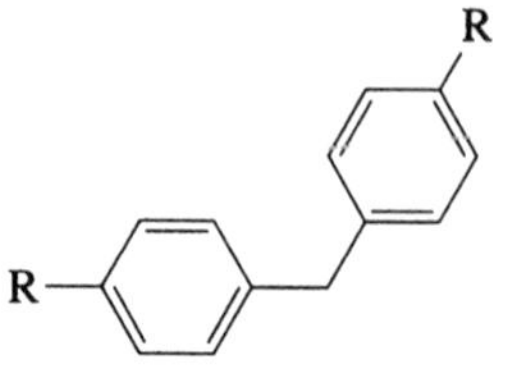	62–94	38
10	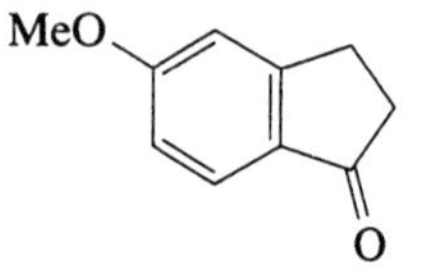	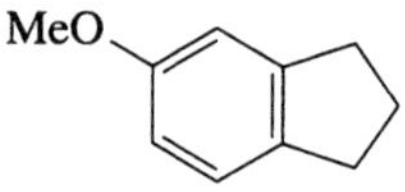	56	39
11	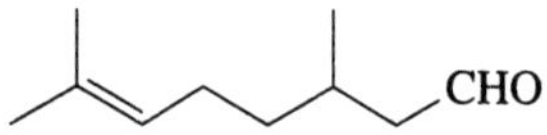	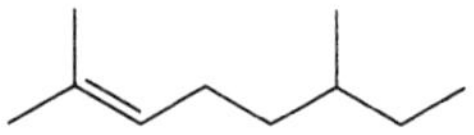	—	40
12	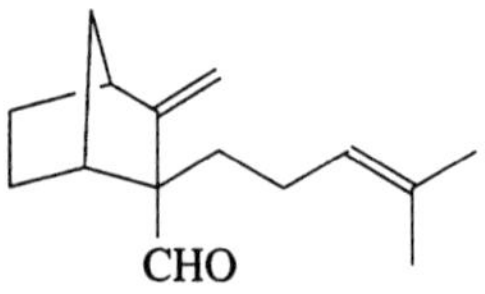	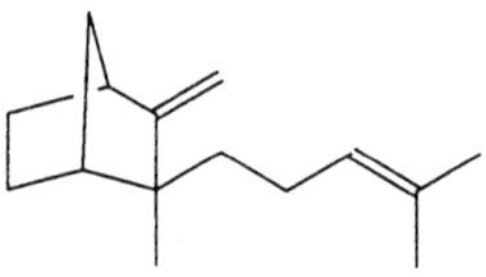	85	41
13	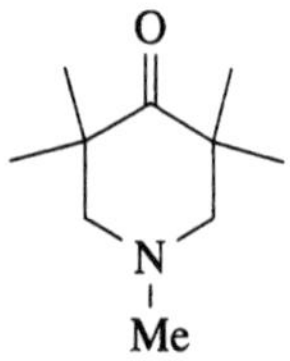	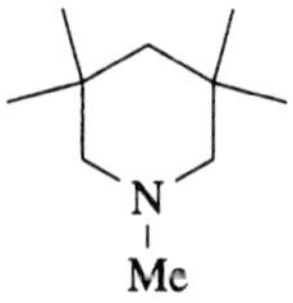	59[a]	42
14	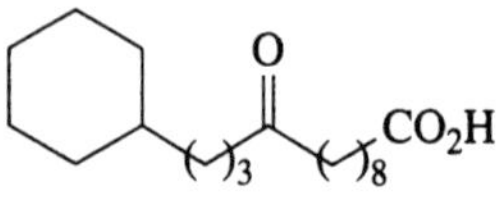	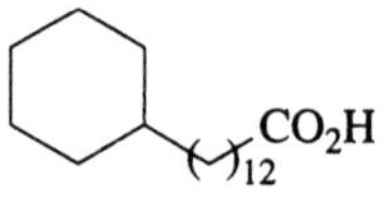	59	43
15	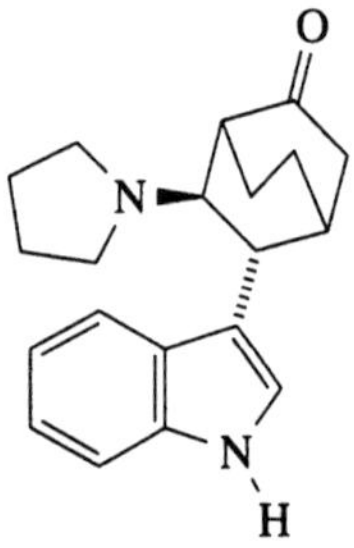	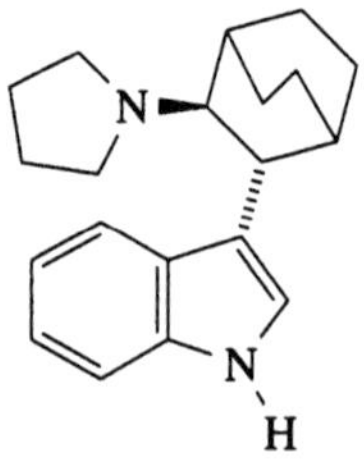	45	44

[a] *Via* the preformed hydrazone.

A modification reportedly even more vigorous than the Barton procedure was introduced by Nagata and Itazaki[50] in which large excesses of hydrazine hydrate (66–400 mol %), hydrazine hydrochloride (8–30 mol %) and KOH (20–70 mol %) in triethylene glycol are heated (130 °C, 2.5–7 h) and then distilled until the temperature reaches 210–220 °C and heating continued for 3–5 h. Treatment of a hindered 11-keto steroid (partial structure **12**) under these conditions[50] afforded a 52% yield of the deoxygenated product, but only 36% using Barton conditions. However, reaction of the spiroketone (**13**) under the Nagata conditions afforded 100% recovery of starting material, while use of the Barton modification

Table 2 Wolff–Kishner Reduction of Carbonyls Using the Barton Modification

Entry	*Carbonyl*	*Product*	*Yield (%)*	*Ref.*
1	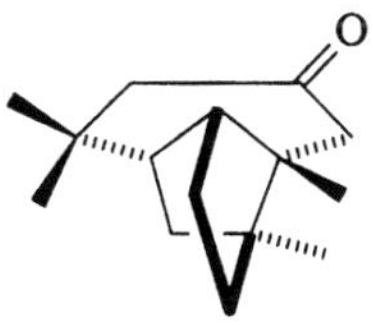 O	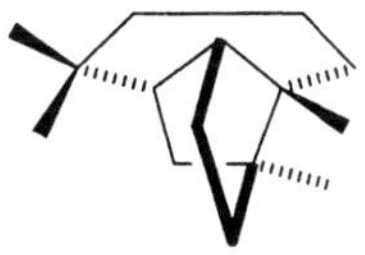	89	52
2	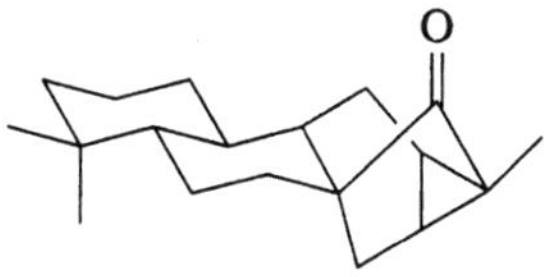 O	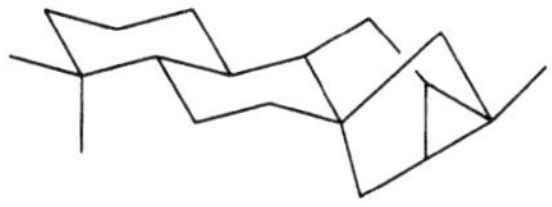	56	53
3	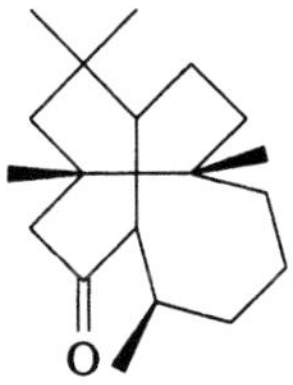 O	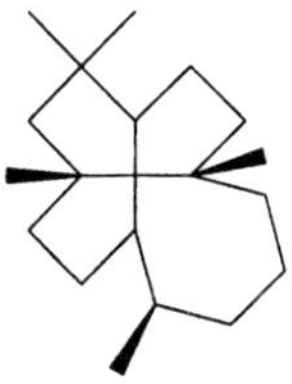	50	54
4	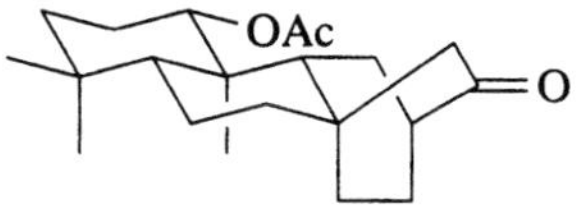 OAc, O	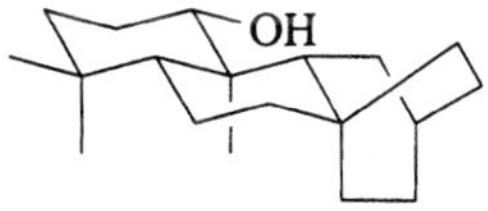 OH	62	55
5	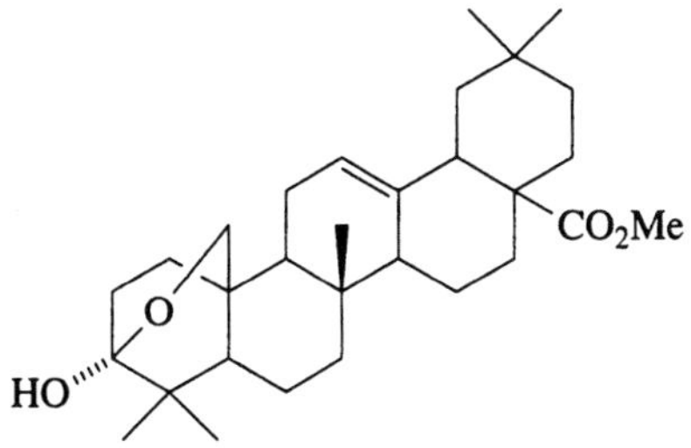 CO_2Me, O, HO	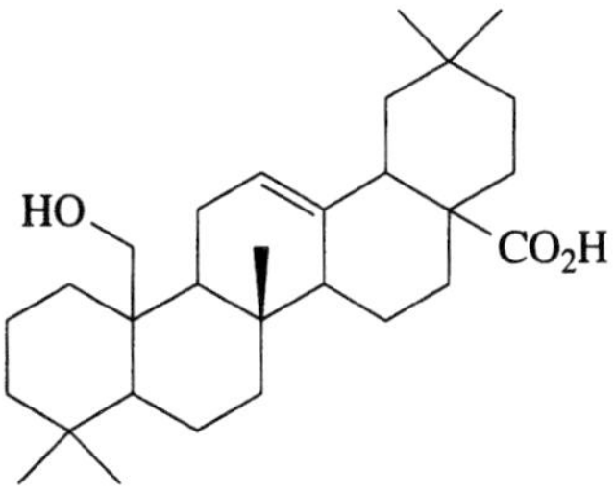 HO, CO_2H	—	56
6	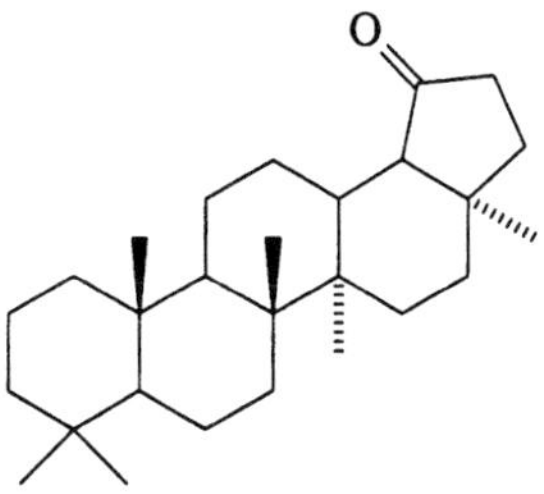 O	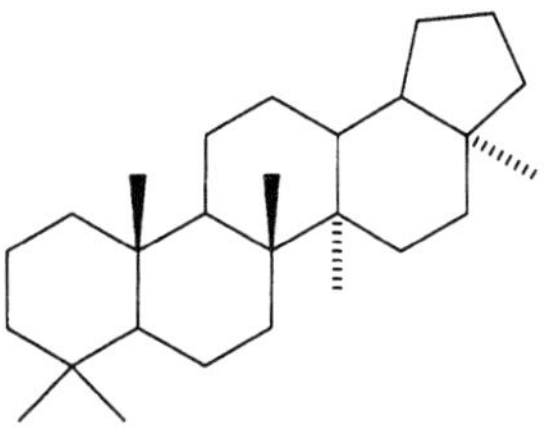	59	57
7	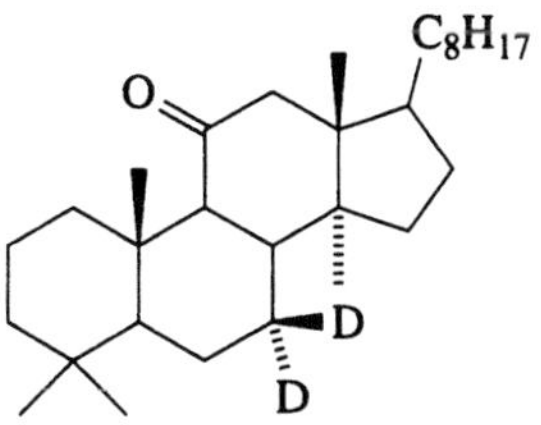 C_8H_{17}, O, D, D	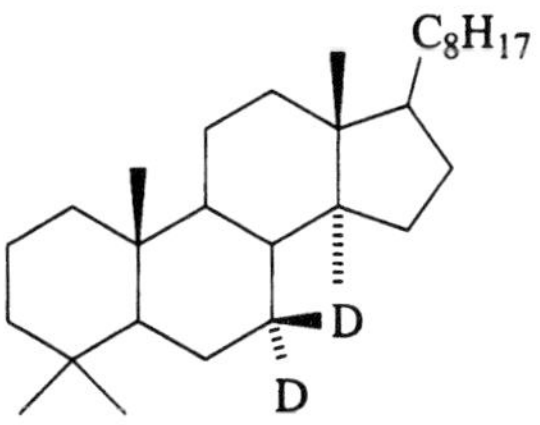 C_8H_{17}, D, D	54	58

Table 2 *(continued)*

Entry	*Carbonyl*	*Product*	*Yield (%)*	*Ref.*
8	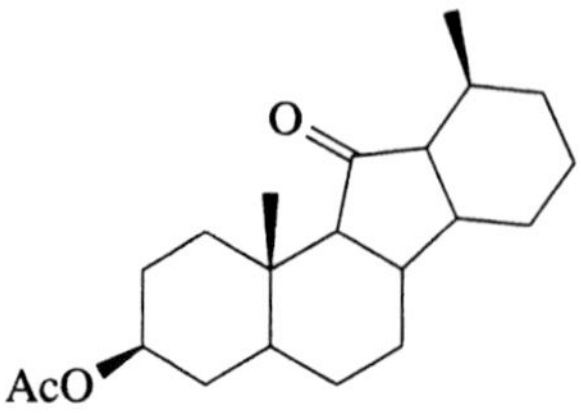 AcO, O	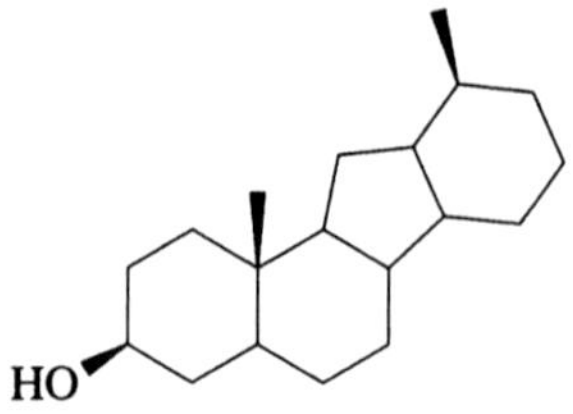 HO	62	59
9	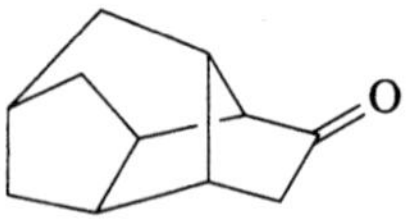 O	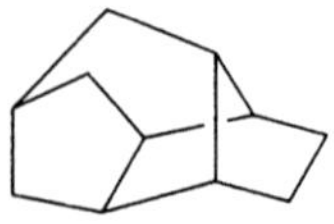	46	60
10	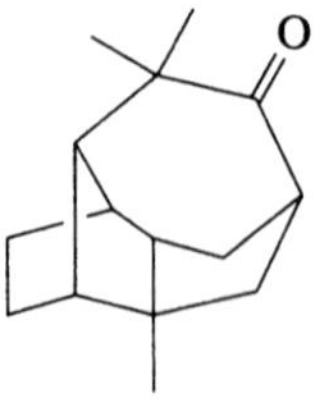 O	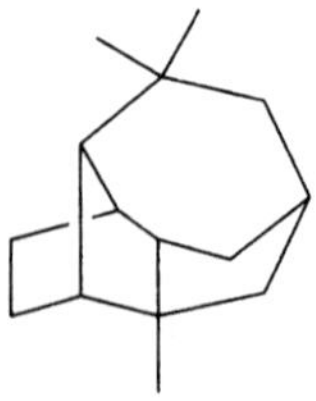	52	61
11	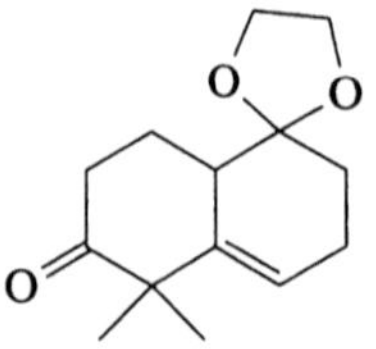 O, O, O	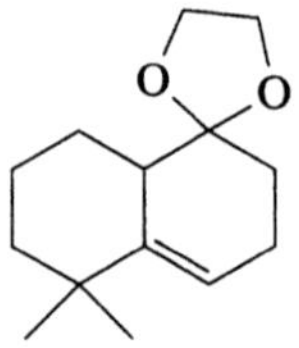 O, O	—	62
12	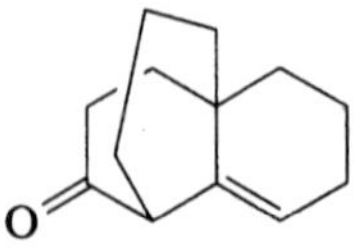 O	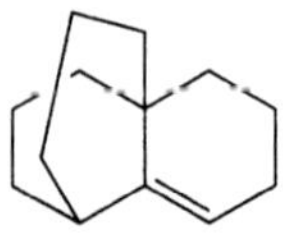	90	63
13	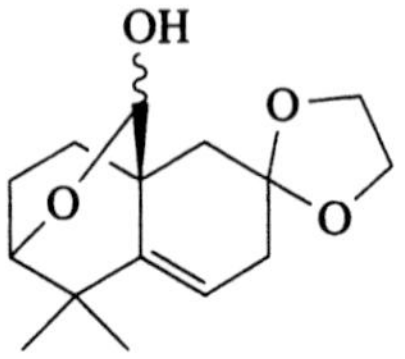 OH, O, O, O	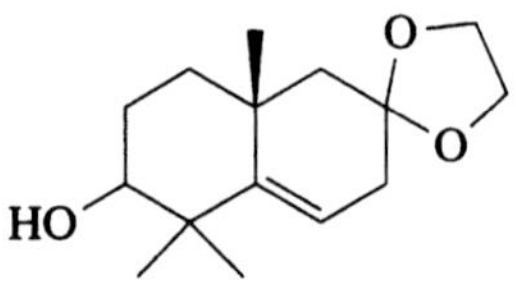 HO, O, O	65	64
14	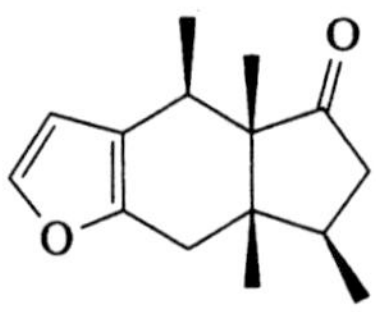 O, O	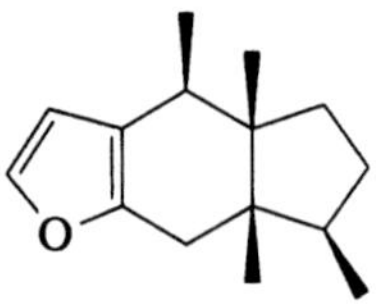 O	15	65

Table 2 *(continued)*

Entry	*Carbonyl*	*Product*	*Yield (%)*	*Ref.*
15	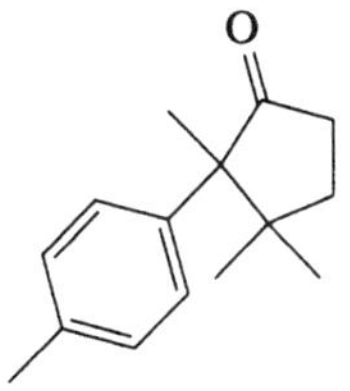	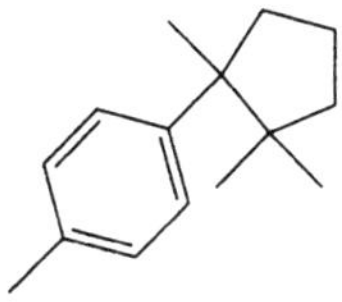	65	66
16	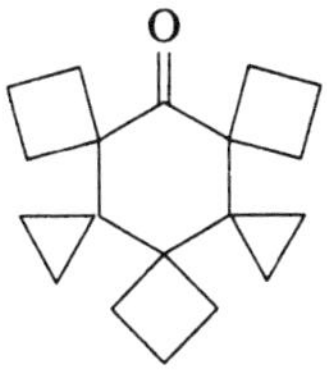	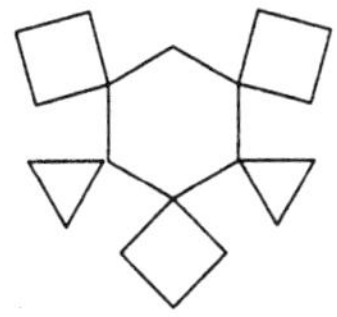	93	67

gave 85% of the desired hydrocarbon (**14**).[51] Thus, the efficiencies of these modifications appear to be structure dependent, and the proper choice is not always evident.

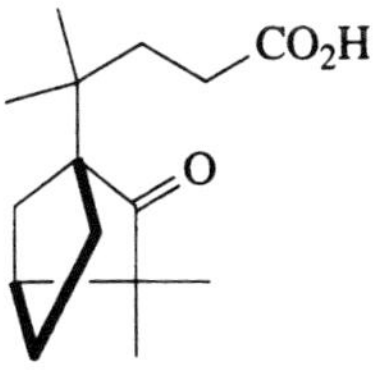

(10) 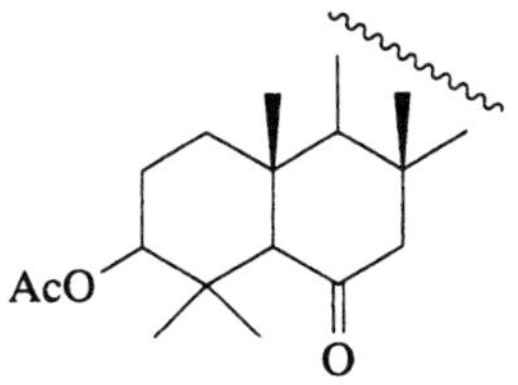 (11) 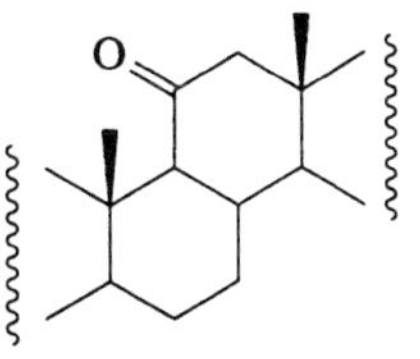 (12)

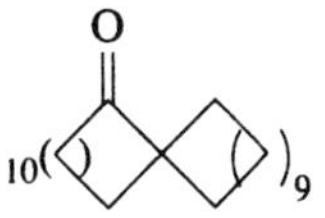

(13) 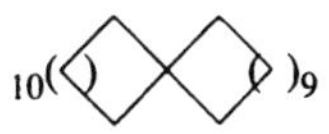 (14)

1.14.2.2.3 The Cram modification

As mentioned previously, DMSO as the reaction medium provides significant enhancement of Wolff–Kishner reaction rates and this allows the use of much lower temperatures to effect reductions. In 1962 Cram *et al.*[23] introduced the use of *t*-butoxide in dry DMSO for the successful reduction of preformed hydrazones at room temperature. Using this process, benzophenone hydrazone (**15**) afforded an 88% yield of diphenylmethane (**16**), along with 11% of benzophenone azine (**17**) as side product (equation 5). However, maximum success requires very slow addition (*i.e.* over 8 h) of the hydrazone to the reaction solution, otherwise yields of reduced products are decreased and azine formation augmented. Thus, addition of (**15**) over 0.5 h in the above reaction lowered the yield of (**16**) to 72%, while the yield of (**17**) was increased to 22%.[23] Other successful reductions reported[23] include hydrazones of benzaldehyde (67%), camphor (64%) and cyclohexanone (80%).

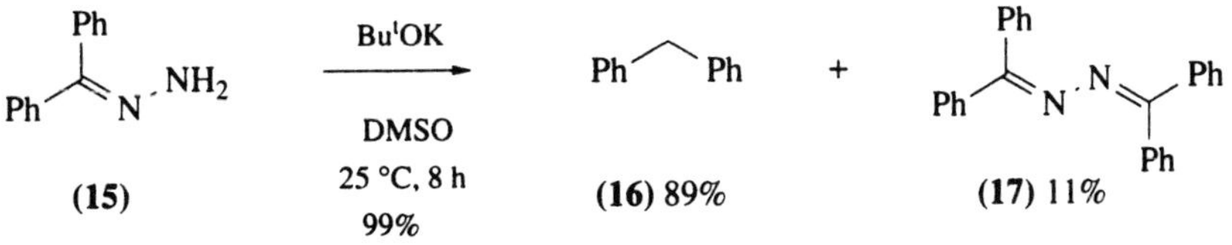

(15) —(Bu^tOK, DMSO, 25 °C, 8 h, 99%)→ (16) 89% + (17) 11% (5)

The Cram modification has not been greatly exploited in multistep synthetic endeavors, possibly because of the above-mentioned need to preform and isolate the hydrazone intermediate coupled with very slow addition of the hydrazone. This latter requirement is particularly inconvenient when milligram quantities are involved. The aldehyde in a triterpenoid (partial structure **18**) was successfully reduced to the corresponding methyl group using the Cram method, albeit in low yield (32%).[68] Notably, the hydrazone (10 mg) was not added slowly. The Cram process was not successful for the reduction of the cyclobutanone (**19**)[69] (the Huang-Minlon process succeeded), or the pyrrole aldehydes (**20**)[70] and (**21**).[71]

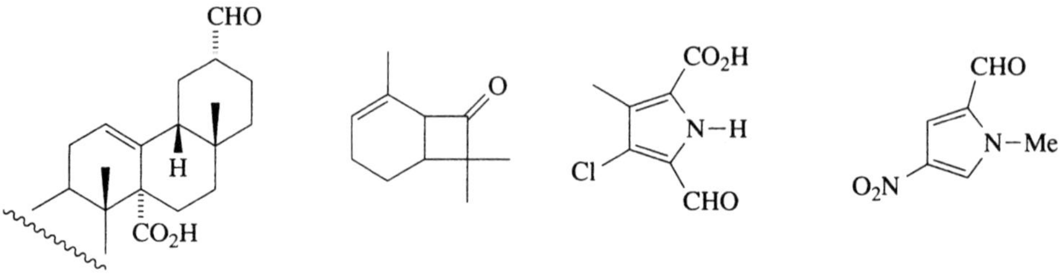

(**18**) (**19**) (**20**) (**21**)

In view of the observations of Szmant and coworkers[20,24] that combinations of DMSO and protic solvents [*i.e.* *t*-butyl alcohol, 2-(*t*-butoxyethoxy)ethanol] provide more effective solvent systems for enhancement of Wolff–Kishner reduction rates than DMSO alone, the relative paucity of possible synthetic applications in this area is somewhat surprising. Indeed, Ramuz[72] successfully reduced the hydrazone of aldehyde (**22**) in 90% yield using the combination of potassium *t*-butoxide in DMSO containing ethylene glycol (at 100 °C). Further investigations concerning the synthetic utility of DMSO and protic solvent mixtures appears to offer significant potential.

1.14.2.2.4 The Henbest modification

Another lower temperature modification of Wolff–Kishner reductions was introduced by Henbest and coworkers[73] in which carbonyl hydrazones and *t*-butoxide are refluxed in dry toluene (110 °C). Using this procedure, benzophenone hydrazone (**15**) afforded an 85% yield of diphenylmethane (**16**) in 4 h. Other conditions, including the use of DMSO at 40 and 100 °C, were explored, but with less success. The less vigorous conditions and lower temperatures appear particularly suited for carbonyl compounds prone to base-induced side reactions, primarily double bond migrations in α,β-enones[74] and functional group eliminations of certain α-substituted ketones.[75] Thus, the α-amino ketone (**23**) afforded an 83% yield of the reduced product (**24**),[73] while standard Wolff–Kishner conditions afforded (**24**) in only 44% yield, along with the elimination products 3,3-dimethylbutene and 2-methylpiperidine in 38% yield.[75] The Henbest modification has been used successfully to deoxygenate a variety of hydrazones, examples of which are illustrated in Table 3. Note that in entries 2 and 5 (Table 3), reductions were effected without concomitant migration of the α,β-double bonds and the lactam functionalities in entries 7 and 8 survived. Likewise, the dibenzocyclooctanone hydrazone (entry 4) was reduced, but elimination of the very labile α-hydroxy group was also observed.[79] Failures with the Henbest modification have been noted with the cycloheptanone (**25**)[84] and the enone (**26**).[85]

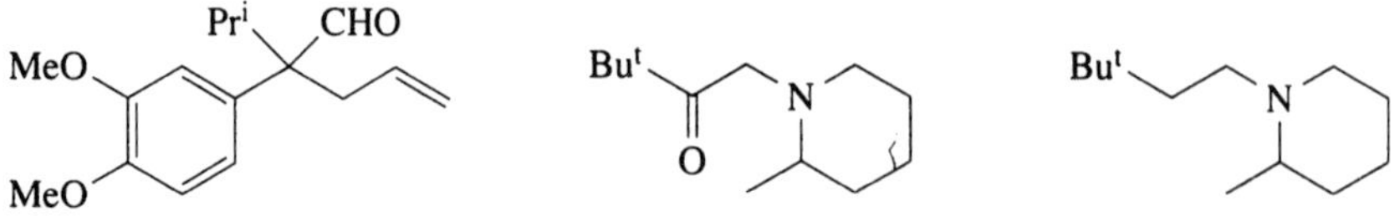

(**22**) (**23**) (**24**)

1.14.2.2.5 Reductions of semicarbazones

Carbonyl semicarbazones often provide suitable alternative derivatives for Wolff–Kishner reductions *via* generation *in situ* of the requisite hydrazone intermediates under basic conditions (Scheme 1).[2] Although the process is successful under standard Wolff–Kishner conditions (*i.e.* sealed tube, 205 °C[86] and

Table 3 Wolff–Kishner Reduction of Carbonyl Hydrazones Using the Henbest Modification

Entry	*Carbonyl*	*Product*	*Yield (%)*	*Ref.*
1	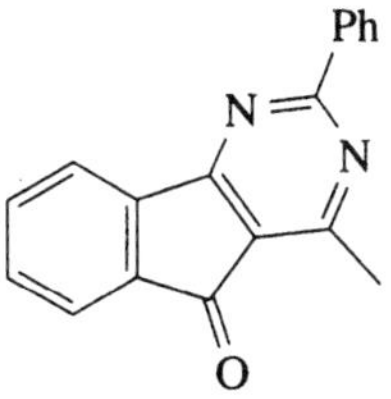	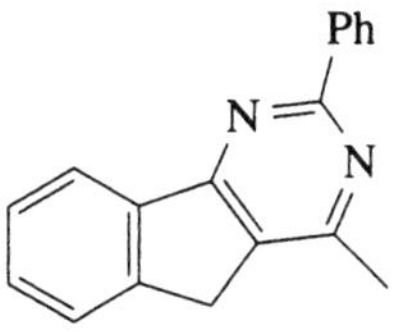	37	76
2	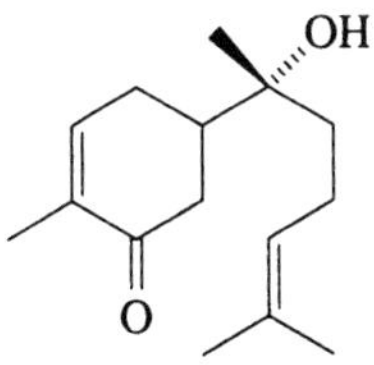	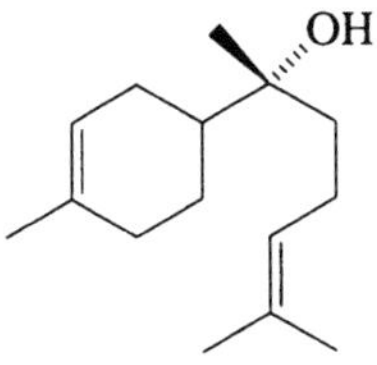	—	77
3	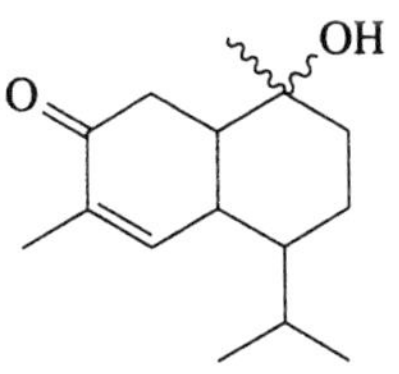	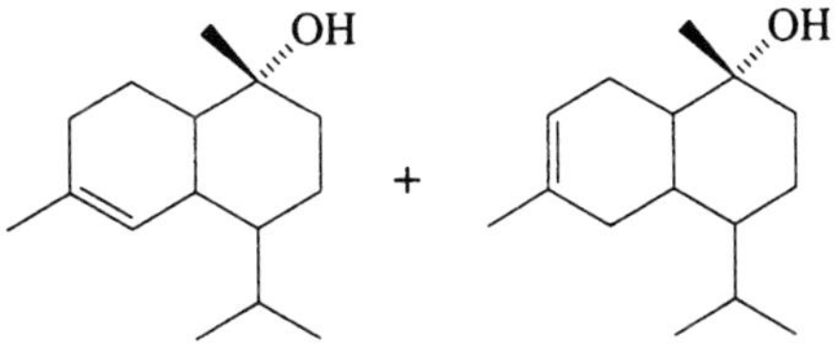 2:1	94	78
4	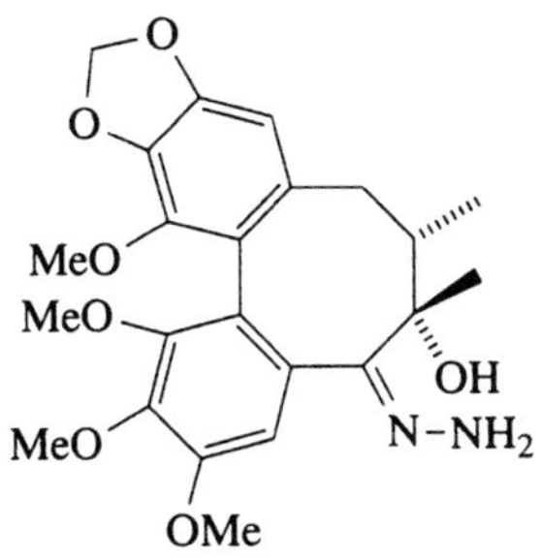	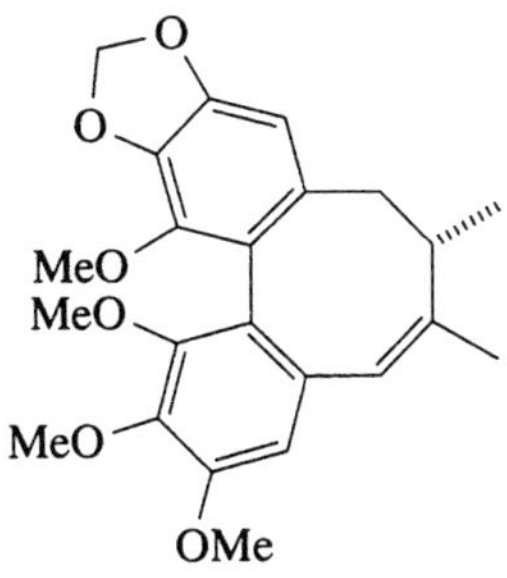	60	79
5	Cholest-4-en-3-one	4-Cholestene	65	73
6	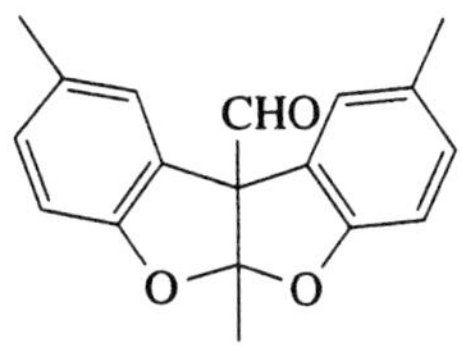	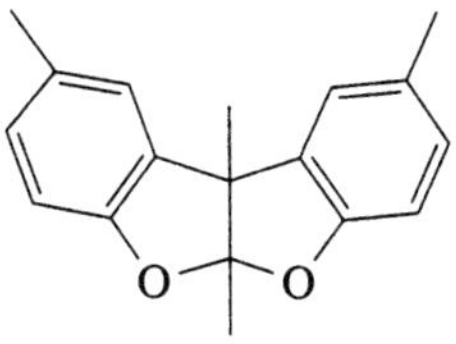	50	80
7	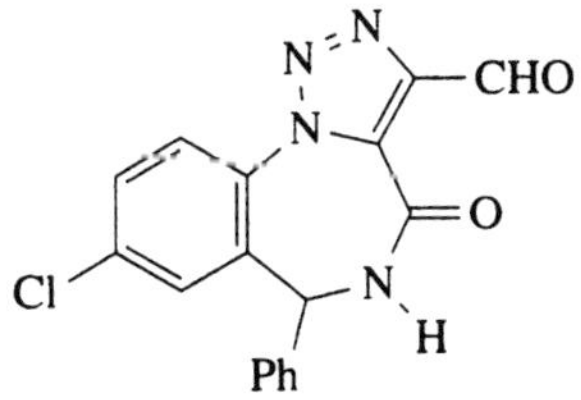	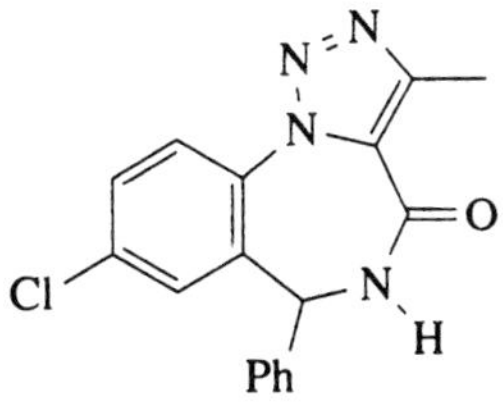	91	81

Table 3 *(continued)*

Entry	Carbonyl	Product	Yield (%)	Ref.
8	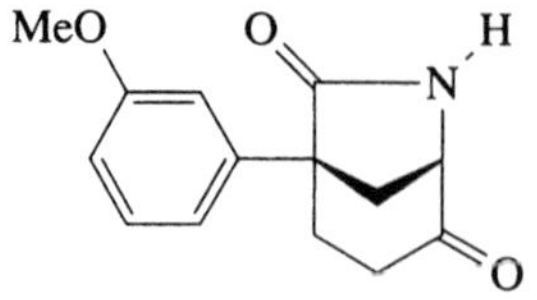 MeO, O, H, N, O	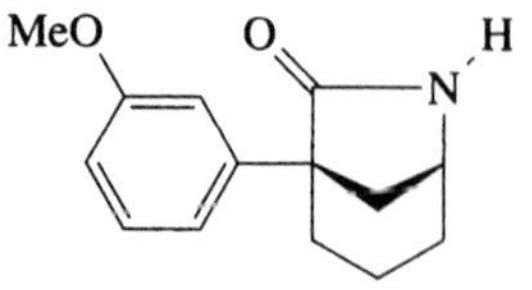 MeO, O, H, N	91	82
9	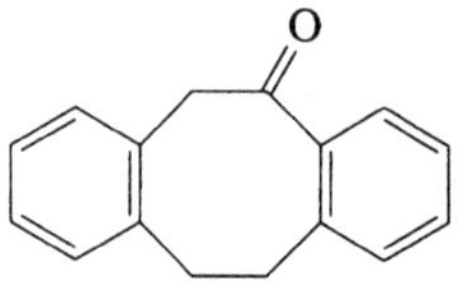 O	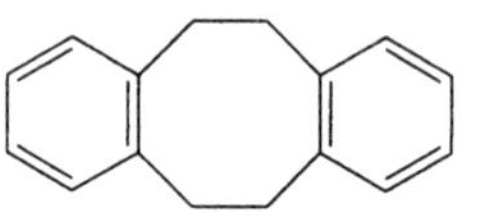	75	83

the Huang-Minlon modification[87,88]), utilization of the Henbest modification,[73] described in the previous section for hydrazones, appears to offer the most synthetic potential, especially for α,β-unsaturated carbonyls. Examples of successful conversions using the Henbest procedure are presented in Table 4. Entries 1–5 illustrate the absence of migration of double bonds adjacent to the carbonyl under these conditions. Noteworthy in this regard, the α,β-unsaturated enone in entry 5 afforded only the rearranged product (**27**) when reduced using Huang-Minlon conditions.[92] The Henbest procedure was unsuccessful in reducing the semicarbazone of 3-ketonorlongifolane (**28**).[94]

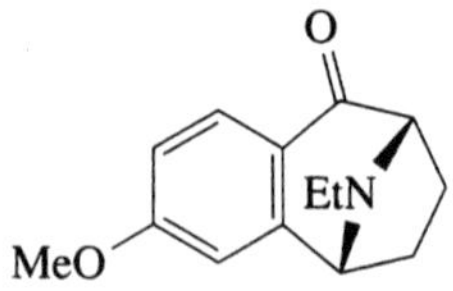

O, EtN, MeO — Ar, N, Me, O — OH, Pr^i — O

(**25**) 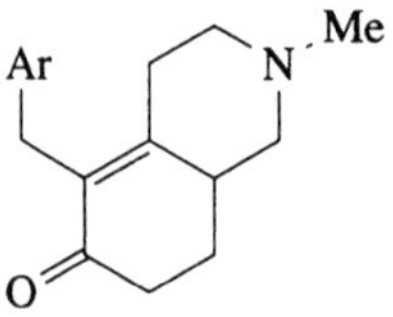 (**26**) 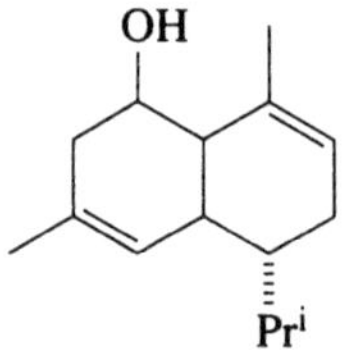 (**27**) 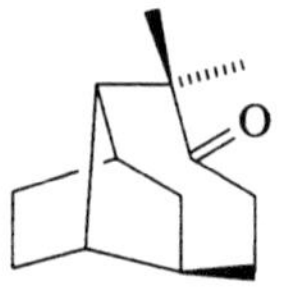 (**28**)

1.14.2.2.6 *Other modifications*

As noted above, certain structural types of carbonyls have been observed to be reduced to the hydrocarbons by excess hydrazine without the use of strong bases such as hydroxide or alkoxide. Thus, fluorenones,[22] benzoylnaphthalenes, benzophenone, benzaldehyde and pyrene-3-carbaldehyde are reduced in good yields by heating with hydrazine hydrate.[3] In the absence of strong base, these reductions may proceed in an alternative manner to that depicted in Scheme 1 *via* isomerization of the hydrazones to the corresponding diimides (**29**), which are known to degrade with expulsion of nitrogen to give hydrocarbons (equation 6).[95]

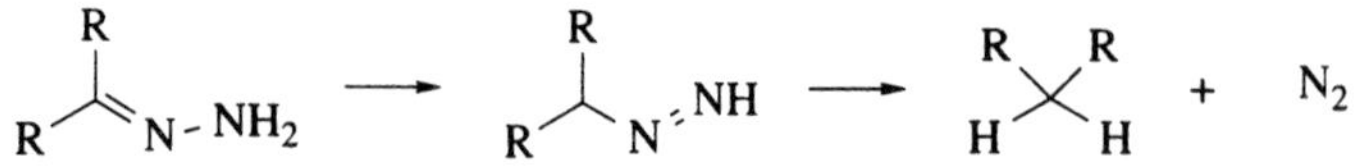

$$R_2C{=}N{-}NH_2 \;(\mathbf{4}) \longrightarrow R_2CH{-}N{=}NH \;(\mathbf{29}) \longrightarrow R_2CH_2 + N_2 \qquad (6)$$

1.14.2.3 Scope and Limitations

1.14.2.3.1 *Chemoselectivity*

As expected, functional groups which are sensitive to the vigorous basic conditions of most Wolff–Kishner reduction conditions do not survive intact. Thus, esters and lactones are normally hydrolyzed to the corresponding acids and alcohols[6,17] (see, for examples, entries 4, 5 and 8, Table 2) and require

Table 4 Wolff–Kishner Reductions of Carbonyl Semicarbazones Using the Henbest Modification

Entry	*Carbonyl*	*Product*	*Yield (%)*	*Ref.*
1	4-Cholesten-3-one	4-Cholestene	81	73
2	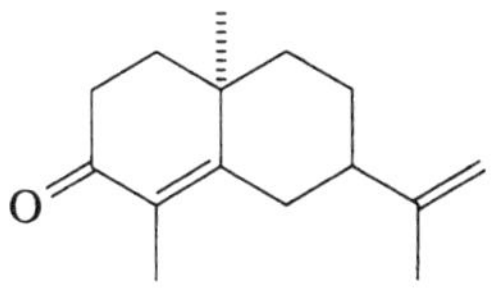 O	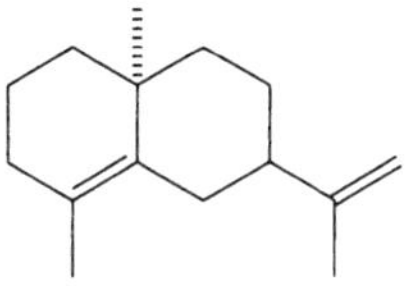	—	73
3	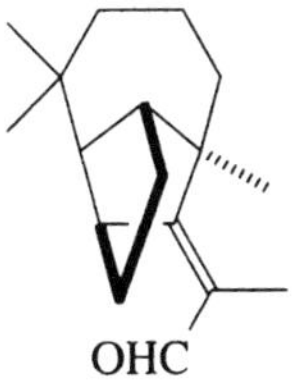 OHC		85	90
4	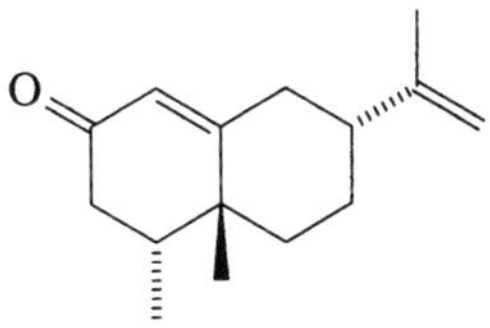 O	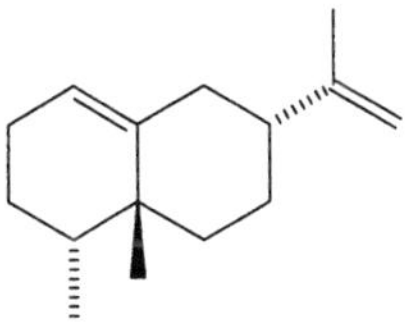	—	91
5	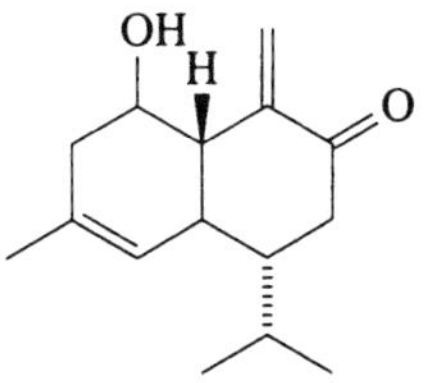 OH H O	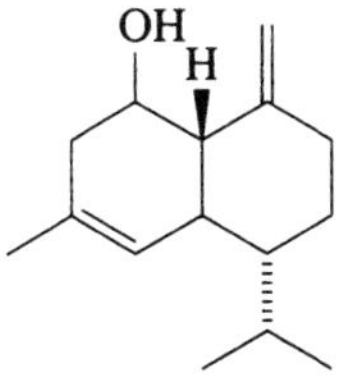 OH H	—	92
6	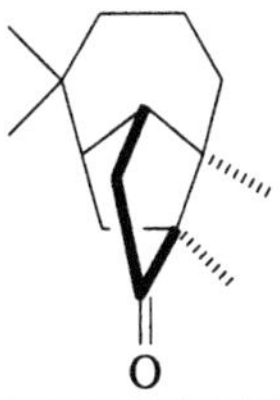 O	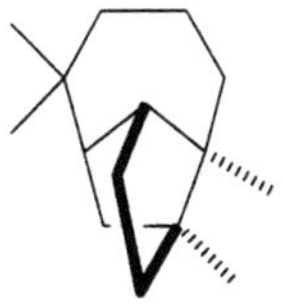	—	93

reesterification. Likewise, amides and lactam moieties are susceptible to attack. For instance, the lactam substrate of entry 8, Table 3 gave a low yield (36%) of the Wolff–Kishner product along with hydrolysis of the lactam. As indicated, this problem may be at least partly avoided by use of the Henbest modification (see also entry 7, Table 3).[82] It might be expected that nitriles would also be affected, but this does not always appear to be the case, at least with aryl derivatives (entry 9, Table 1).[38] Occasionally, reduction of ketones to alcohols is observed[3,6] if the Wolff–Kishner reduction is not successful. Thus, compound (**25**) afforded a mixture of the corresponding alcohol along with starting material under Henbest conditions,[84] and a triterpene ketone (partial structure **30**) gave only the corresponding alcohol, even under the forcing Nagata modification.[96]

Simple aliphatic halogenated compounds containing β-hydrogens are, of course, not expected to survive since Wolff–Kishner conditions simulate typical β-elimination and substitution conditions (hot hydroxide, alkoxide). An 11,20-diketo-12-bromo steroid (partial structure **31**) was debrominated along with deoxygenation of the 20-keto group to afford the 11-keto derivative (partial structure **32**) along with an unidentified isomer of (**32**).[17] The product possibly arises from attack on bromine by hydrazine to afford the corresponding enol, which survives and returns (**32**) upon protonation. Reduction of the chlorotetracyclo derivative (**33**) afforded the pentacyclo compound (**34**; 49%) resulting from internal substitution by the intermediate carbanion (Scheme 1).[97]

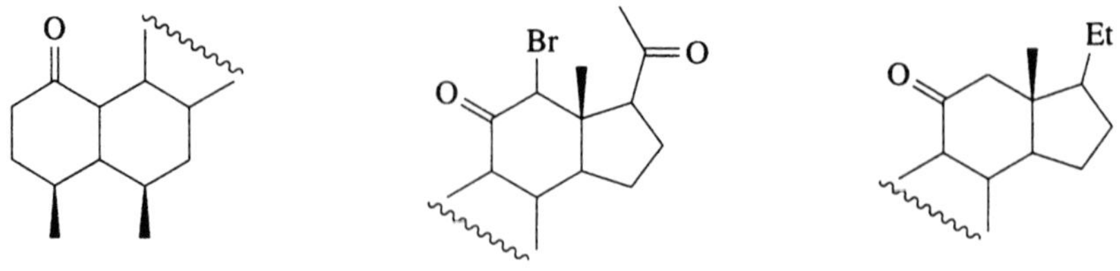

(30) **(31)** **(32)**

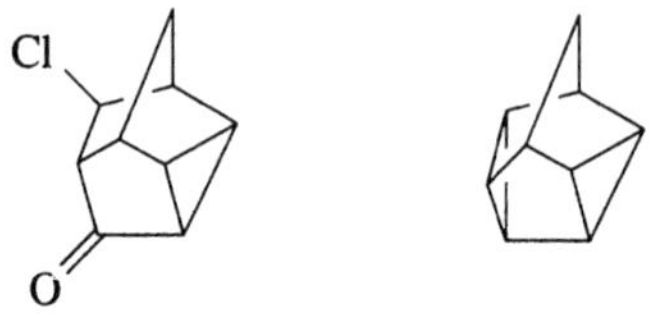

(33) **(34)**

Although reductions of aromatic iodides and bromides have been noted,[4,6] such is not always the case, at least with bromides (see entry 8, Table 1; the corresponding 6-bromo derivative also survived Wolff–Kishner conditions[37]). Aromatic chlorides are normally not affected (entries 9, Table 1 and 7, Table 2).

Aromatic nitro compounds are reduced to amines by hydrazine in hot diethylene glycol with or without base.[98] In the presence of alkali *p*-nitrotoluene affords the dimeric *p,p'*-diaminostilbene, while *p*-toluidine is obtained without the base.[98]

1.14.2.3.2 Steric effects

As previously discussed, Wolff–Kishner reductions are hampered by steric hindrance surrounding the carbonyl primarily due to difficulty in the requisite hydrazone formation. Forcing conditions such as the Barton[45] and Nagata[50] modifications have alleviated many of the problems with hindered ketones (*i.e.* 11-keto steroids, equation 4), but unyielding problems still remain (*i.e.* structures **10** and **11**) requiring alternative strategies for oxygen removal.

1.14.2.3.3 Isomerization of conjugated and other double bonds

Subjection of α,β-unsaturated carbonyls to Wolff–Kishner conditions often leads to at least partial migration of the double bond to give mixtures of alkene products.[6,74] Such migrations are not unexpected since an allylic carbanion is produced, which may be protonated at two positions. Thus, isophorone affords a mixture of the alkene (**35**; 30%) along with the unrearranged isomer (**36**; 70%; equation 7).[99] This problem is at least partly alleviated by application of the Henbest modification as illustrated in Tables 3 and 4. Thus, while the enone semicarbazone in entry 2, Table 4 afforded only the unrearranged alkene using *t*-butoxide in toluene, conventional Wolff–Kishner conditions gave 50% of the rearranged isomers (**37**).[89]

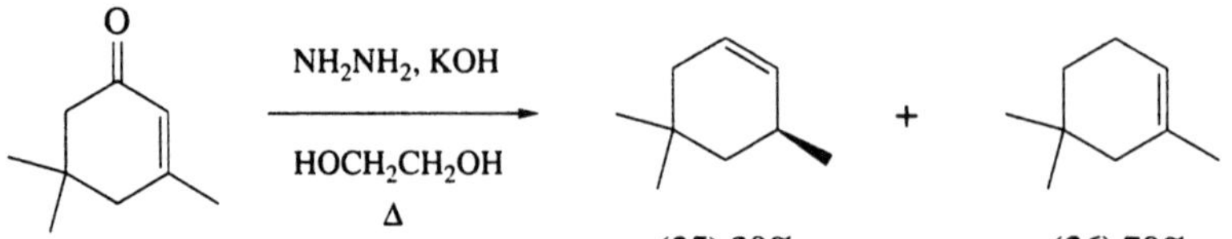

NH_2NH_2, KOH; $HOCH_2CH_2OH$, Δ (7)

(35) 30% **(36)** 70%

Rearrangements of nonconjugated alkenes have also been observed. Thus, a triterpenoid (partial structure **38**) afforded the alkene (**39**; 32% yield) upon reduction (Barton modification), a result attributed to initial base-induced isomerization of (**38**) to the corresponding α,β-unsaturated ketone followed by rearrangement during reduction.[100]

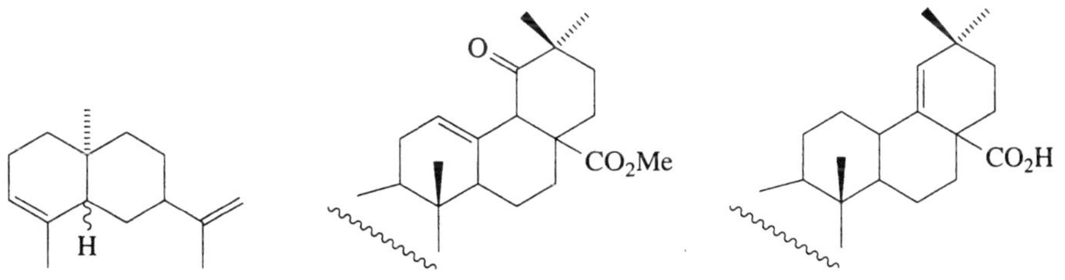

(37) (38) (39)

1.14.2.3.4 Elimination of adjacent substituents (Kishner–Leonard eliminations)

Aldehydes and ketones bearing heteroatom substituents on the carbon adjacent to the carbonyl often afford alkene side products when subjected to Wolff–Kishner conditions,[1,6,75,101] the amount depending on the leaving ability and steric environment of the substituent.[75] Thus, a series of substituted pinacolones Bu^tCOCH_2X (**40**) with X = *N*-piperidinyl, trimethylacetoxy, thiophenoxy and phenoxy afforded the elimination product 3,3-dimethylbutene in 29, 52, 76 and 80% yields, respectively, in approximate order of leaving ability. The reaction (Kishner–Leonard elimination) is depicted as proceeding either through the diimide anion (**6a**), which fragments *via* elimination to an alkene (equation 8) instead of continuing to the reduction product,[75] or *via* a vinyl diimide (**29a**) as pictured in equation (9).[102] As noted previously, application of the Henbest modification (*t*-butoxide, refluxing toluene)[73] lowers the tendency for elimination, at least for amino substituents.

(**6a**) → N_2 + alkene (8)

(**29a**) (9)

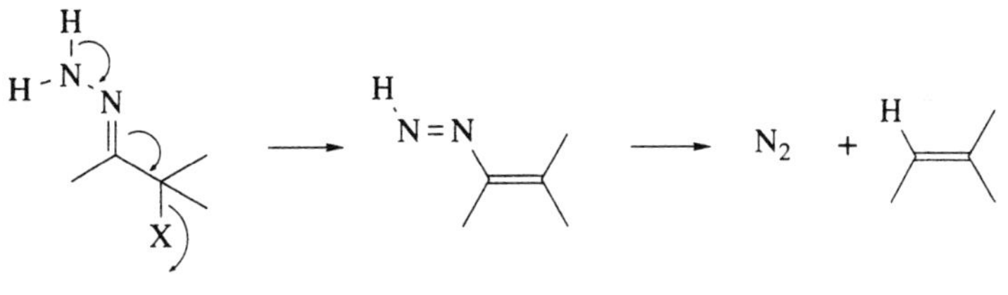

Kishner–Leonard type eliminations also occur with certain α-substituted carbonyls upon treatment with hydrazine alone, the most synthetically important of which is the fragmentation of α,β-epoxy ketones to allylic alcohols (the Wharton reaction, equation 10).[6,102] The Wharton process has found application in a significant number of synthetic schemes, two of which are illustrated in equations (11)[103] and (12).[104] The latter provides the pivotal step in an efficient protocol for enone transpositions. Further reduction of the initial alkene products by diimide to afford saturated derivatives is often obtained,[6,105,106] but this can usually be prevented by exclusion of oxygen and/or adding an alkene to absorb any diimide generated.[105,106] Other side reactions,[105] including pyrazole formation, have also been noted.[107,108] Similar eliminations of α-halocyclohexanones (including α-fluoro derivatives) also appear synthetically useful, while α-sulfonyloxy and corresponding aliphatic derivatives give inferior yields.[106]

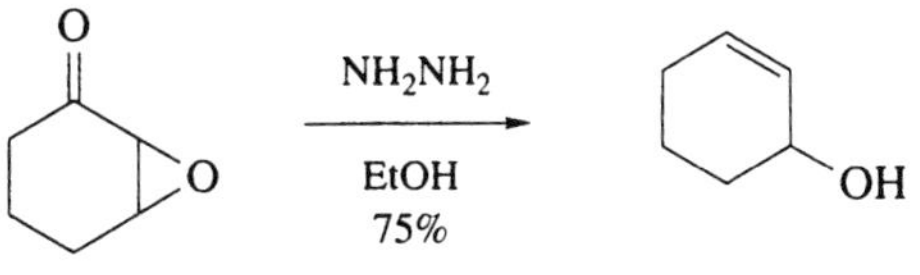

NH_2NH_2, EtOH, 75% (10)

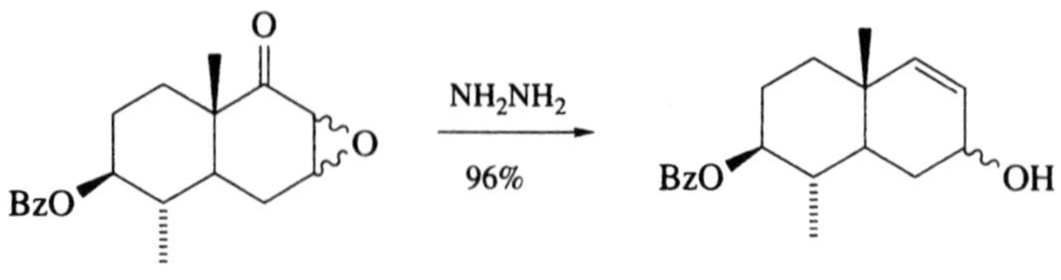

(11)

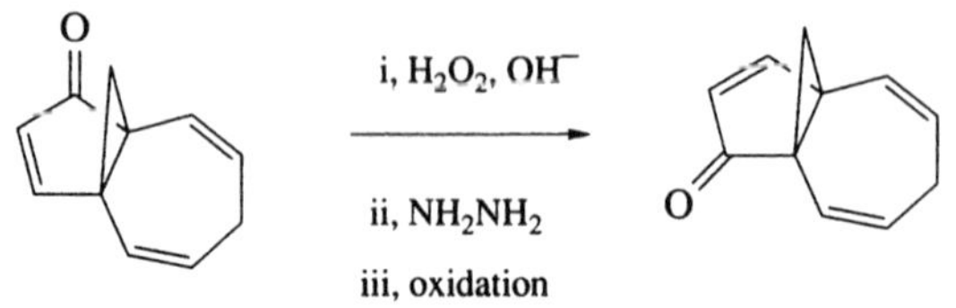

(12)

1.14.2.3.5 Other side and/or unusual reactions

A number of side reactions have been reported to occur during Wolff–Kishner reductions in addition to those discussed above. In particular, cleavage of strained rings located adjacent to the carbonyl may accompany reduction to afford saturated[109] and/or alkene products.[109,110] For example, the pentacyclic diketone (**41**) afforded the unsaturated tetracyclic compound (**42**) and saturated derivative (**43**) in addition to the normal Wolff–Kishner product[46] (**44**; equation 13).[109] Likewise, the cyclopropyl ring in alkaloid (**45**) suffered cleavage during reduction to give the alkene (**46**; 65%; equation 14).[110]

(13)

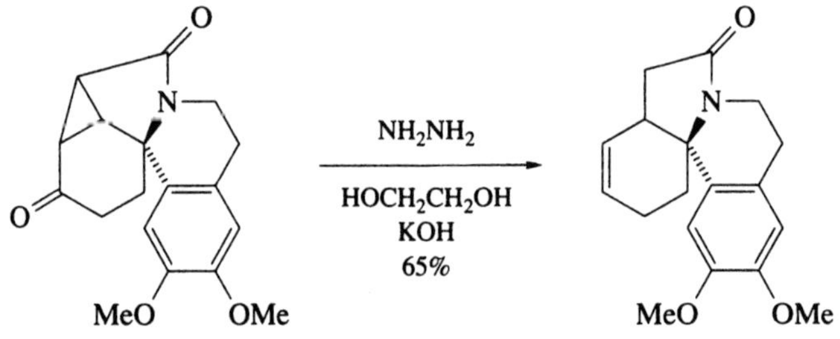

(14)

An unusual disproportionation reaction was observed when the amino ketone hydrazone (**47**) was reduced at temperatures above 200 °C to afford mixtures of (**48**), (**49**), and (**50**) (equation 15).[111] Lower temperature reductions gave almost entirely the expected Wolff–Kishner product (**51**; 52%), which, upon refluxing with KOH in triethylene glycol, returned a mixture of (**48**), (**49**) and (**50**).[111]

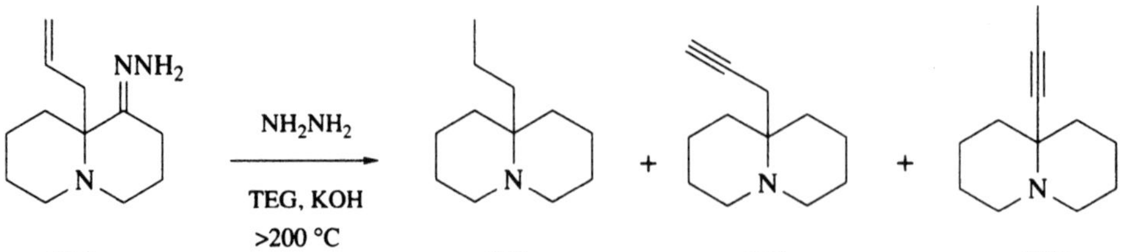

(15)

The presence of a second carbonyl β or γ to an aldehyde or ketone may afford cyclic derivatives upon treatment with hydrazine.[6] The intermediates from β-diketones (*e.g.* **52**) usually resist further reduction,

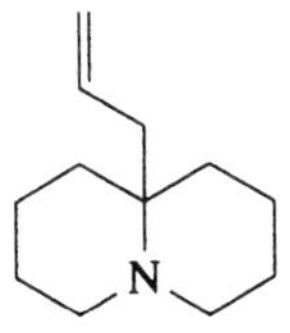

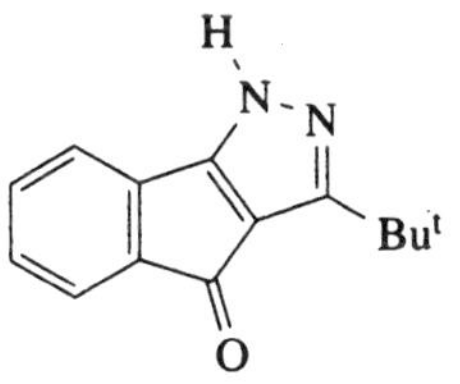

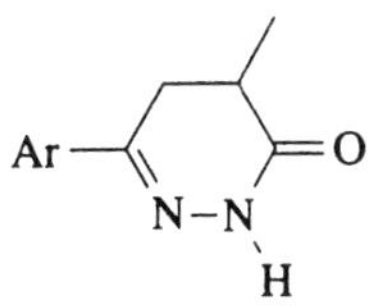

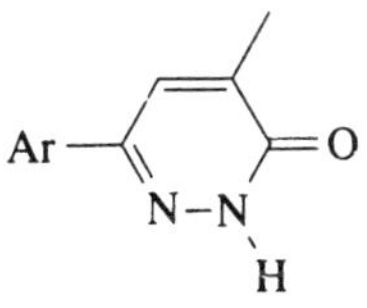

(51) (52) (53) (54)

while (2*H*)-pyridazinones from γ-keto acids (*e.g.* **53**) are further reduced to the normal Wolff–Kishner products.[112] However, α,β-unsaturated-γ-keto acids afford cyclic derivatives (*e.g.* **54**) which do not further react,[112] while pyrazoline intermediates from α,β-unsaturated carbonyls may degrade to cyclopropyl products (equation 16).[3,6]

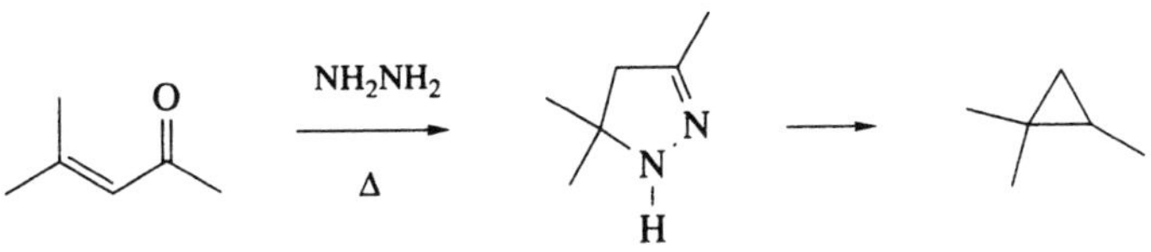

NH_2NH_2, Δ (16)

Treatment of the semicarbazone (**55**) under Wolff–Kishner conditions (KOH, NaOMe, KOBu^t, no solvent) has been utilized to synthesize the substituted cyclopentenes *cis*- and *trans*-(**56**; equation 17).[113] In DEG at 225 °C, the normal Wolff–Kishner product was obtained.[113]

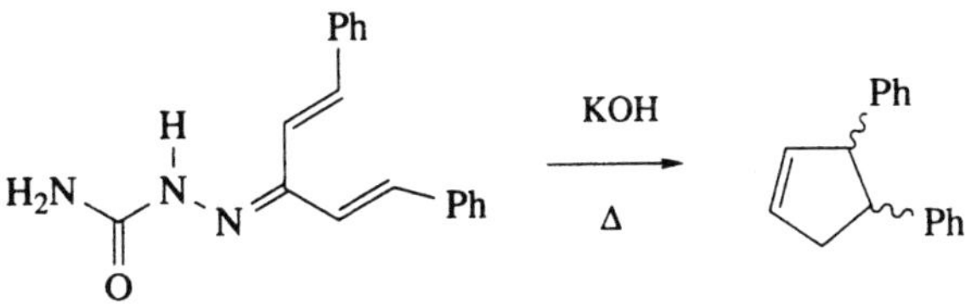

KOH, Δ (17)

(55) (56)

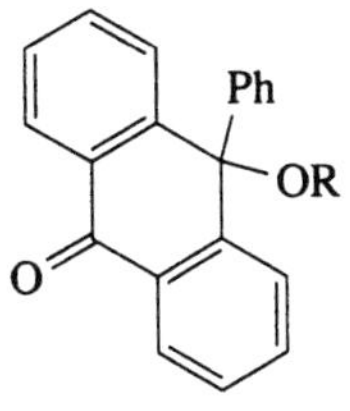

(57)

Tritylone ethers (**57**), used as acid-stable protecting groups for alcohols, are cleaved under Wolff–Kishner (Huang-Minlon) conditions as the unblocking step.[114]

1.14.3 REDUCTION OF ALDEHYDE AND KETONE ARYLSULFONYLHYDRAZONES WITH HYDRIDE REAGENTS

In the early 1960s, Caglioti and Magi[115] reported that the reduction of aldehyde and ketone *p*-toluenesulfonylhydrazones (tosylhydrazones) with $LiAlH_4$ in THF affords hydrocarbon products (equation 2). This was soon followed by the disclosure[116] that the milder reducing reagent $NaBH_4$ in refluxing methanol or dioxane was also effective for such reductions and this ushered in processes (Caglioti reactions) as alternatives to Wolff–Kishner reductions for carbonyl deoxygenations. Following these pioneering investigations, other reduction systems, particularly $NaBH_3CN$[117] and catechol borane,[118] have been introduced over the ensuing years to provide even milder and more selective methodology to effect reductions. These reduction systems have found considerable synthetic utility, especially in synthetic situations which would not tolerate the relatively harsh basic conditions of standard Wolff–Kishner reductions or the highly acidic requirements dictated in Clemmensen reductions (see Volume 8, Chapter 1.13). The availability of the reagents in isotopically labeled form (*i.e.* $LiAlD_4$, $NaBD_4$, $NaBD_3CN$, catechol borane-*d*) permits incorporation of deuterium (or tritium) into synthetic targets, which augments the utility of the methodology. Each of the major categories of reducing systems will be discussed separately in the subsections below.

Caglioti-type reductions are mechanistically related to Wolff–Kishner processes in that diimide intermediates are implicated as depicted in Scheme 2.[115,117–120] In the absence of strong base, diimide (**29a**) represents the key intermediate rather than the diimide anion (**6**) found in the Wolff–Kishner reduction (Scheme 1). Reduction with $LiAlH_4$ is probably an exception to this since this highly basic reagent should readily abstract the very acidic *N*-proton to afford (**59**), followed by hydride delivery and loss of tosyl anion to give the diimide anion (**6**).[119,120] Recent studies[121] indicate that reductions of certain structural types (*e.g.* some sugar derivatives[121] and aryl examples bearing electron deficient groups[121b]) with $NaBH_3CN$ under highly acidic conditions (HCl, pH *ca.* 3.8) in methanol proceed *via* an alternate route involving initial acid-catalyzed isomerization to the azo derivative (**60**), followed by reduction to the tosylhydrazine (**61**; equation 18). Subsequent elimination of tosylic acid leads to the hydrocarbon *via* diimide (**29**) as in Scheme 2. This process appears to be important only when tautomerization to (**60**) is facilitated by electron-withdrawing inductive effects and/or conformational restrictions.[121]

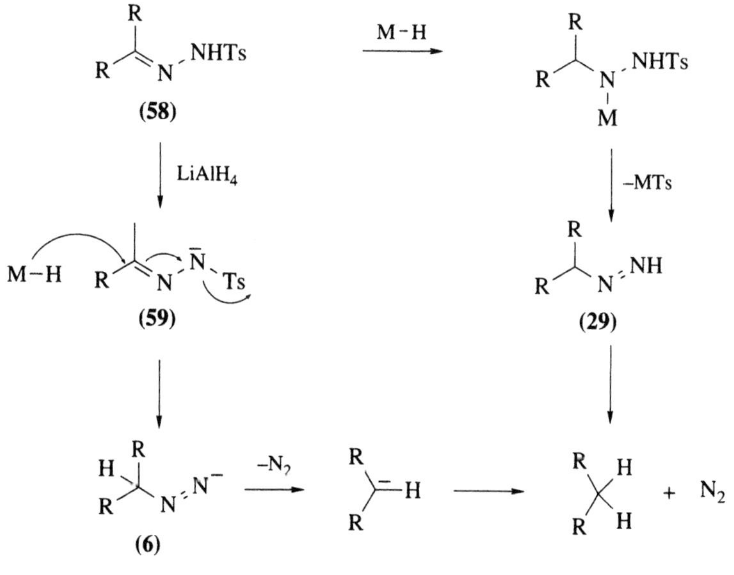

(58) M–H → (N–M intermediate) ; (58) LiAlH4 → (59) ; –MTs → (29) ; (59) → (6) ; (6) –N2 → → R2CH2 + N2

Scheme 2

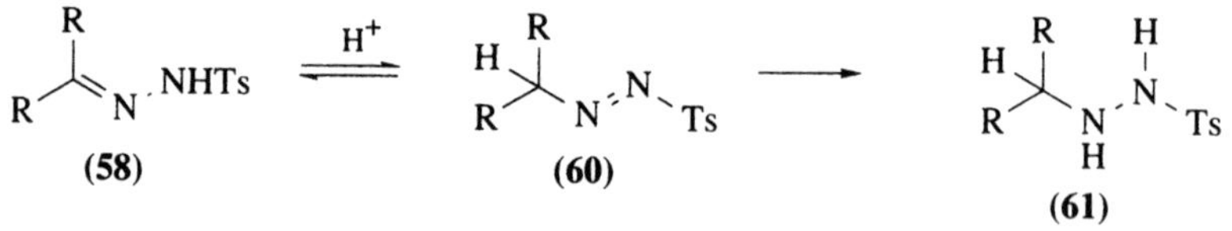

(58) $\xrightleftharpoons{H^+}$ (60) → (61) (18)

As is the case with Wolff–Kishner reactions, the application of tosylhydrazone reductions relies on the successful conversion of carbonyls to the requisite hydrazone derivatives (**58**). This usually presents no problem except with relatively hindered or deactivated ketones and, in fact, several failures of the methodology are traced to unsuccessful tosylhydrazone formation (*i.e.* for examples with **62**,[122] **63**,[123] diketone (**9**) in equation 4, and the substrate of entry 3, Table 2).

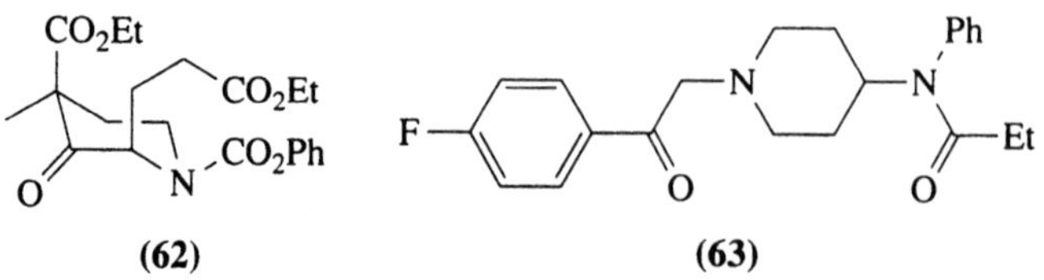

(62) (63)

1.14.3.1 Reductions with Lithium Aluminum Hydride

1.14.3.1.1 Scope and limitations

As mentioned, $LiAlH_4$ in refluxing THF was the initial system introduced to reduce preformed tosylhydrazones to hydrocarbons[115] and a number of successful conversions have been reported, representative examples of which are presented in Table 5. Alkene side products often accompany the hydrocarbon products,[115,119] a result attributed to proton abstraction from the α-carbon of intermediate (**59**), leading to a vinyldiimide anion (**64**), followed by N_2 expulsion and protonation during work-up (Scheme 3).[119] With certain ketones, including 17-keto steroids, alkenes are the major[128] or sole product[115,127] (entries 7–9, Table 5). This side reaction mimics the elimination obtained upon treatment of tosylhydrazones with other strong bases (*i.e.* alkyllithiums, the Shapiro reaction[129]). Note that use of $LiAlD_4$ introduces one deuterium (with H_2O work-up) or two deuteriums (with D_2O work-up; entries 5 and 6, Table 5, respectively).

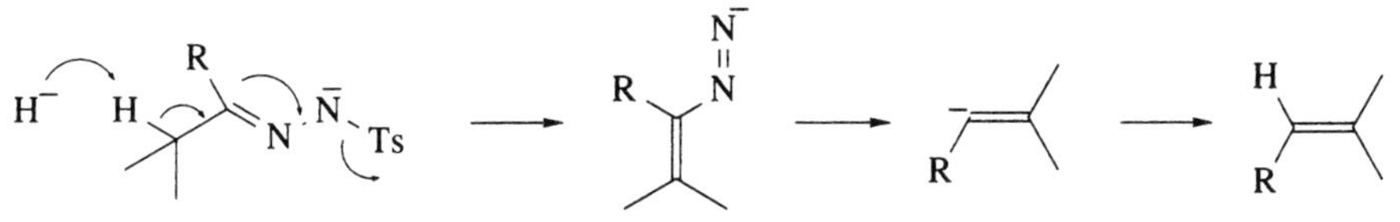

(59) **(64)**

Scheme 3

1.14.3.1.2 Chemoselectivity

The powerful reducing ability of $LiAlH_4$ precludes the presence of many other functional groups, *i.e.* CO_2H, CO_2R (entries 2 and 4, Table 5), CN, CON— (entry 6, Table 5), NO_2, halides, *etc.*, and this limits the synthetic applications to synthetic targets where such groups are absent or their reduction desired or unimportant.

1.14.3.1.3 Reduction of conjugated derivatives

Reductions of α,β-unsaturated carbonyl tosylhydrazones with $LiAlH_4$ often give mixtures of unrearranged and rearranged alkene products similar to the situation in Wolff–Kishner reductions. Thus, 3-keto-cholest-1-ene gave significant quantities of both the 1- and 2-alkene products while 3-keto-cholest-4-ene afforded 3- and 4-alkene products along with the product of elimination, 2,4-cholestadiene.[115]

1.14.3.2 Reductions with Sodium Borohydride

1.14.3.2.1 Scope and limitations

The introduction of the milder reagent sodium borohydride in refluxing methanol or dioxane for reduction of tosylhydrazones[116,120] represented an improvement over $LiAlH_4$ since many more functional groups are tolerated and the lessened basicity decreases alkene side products observed with the latter reagent. Reductions with $NaBH_4$ have found wide usage in synthesis mostly with dioxane or methanol as the solvent (where the reducing species is probably $NaBH(OMe)_3$), although other solvents, such as isopropanol and DMF, have occasionally been employed. Table 6 presents a selection of successful applications chosen to illustrate various structural types. Several features of the reductions are evident from the table. Thus, the 17-keto group in 3β-acetoxy-5α-androstane (entry 3, Table 6) was effectively removed, while reduction with $LiAlH_4$ gave only elimination (entry 7, Table 5). However, the 17-keto group in a related derivative (entry 4, Table 6) gave a mixture of the 16-alkene along with the unusual cyclopropyl derivative shown, indicating that seemingly slight structural alterations may greatly influence the course

Table 5 Wolff–Kishner Reductions of Carbonyl Tosylhydrazones with Lithium Aluminum Hydride

Entry	*Carbonyl*	*Product*	*Yield (%)*	*Ref.*
1	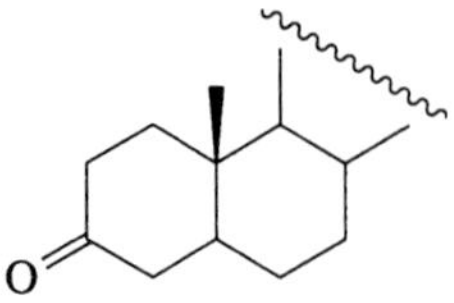 O	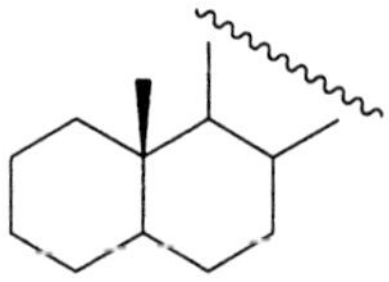	70	115
2	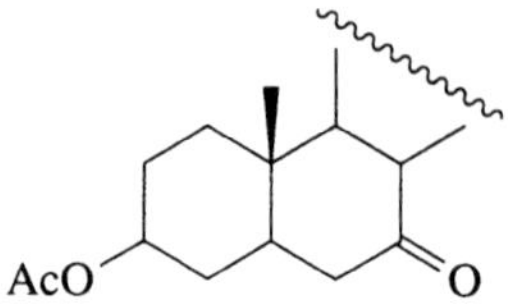 AcO, O	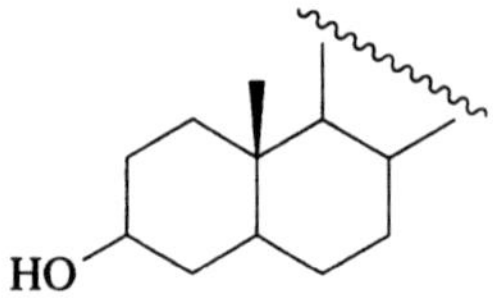 HO	65–75	115
3	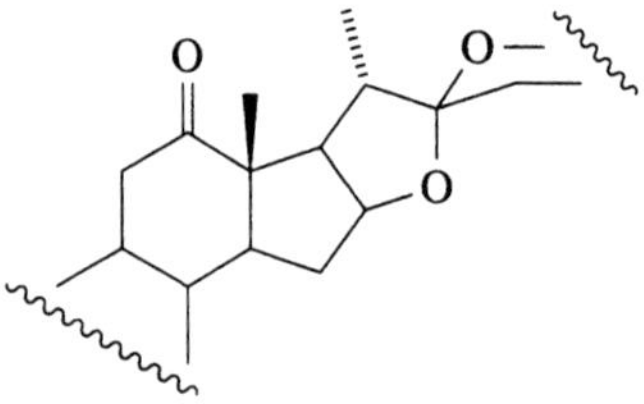 O, O, O	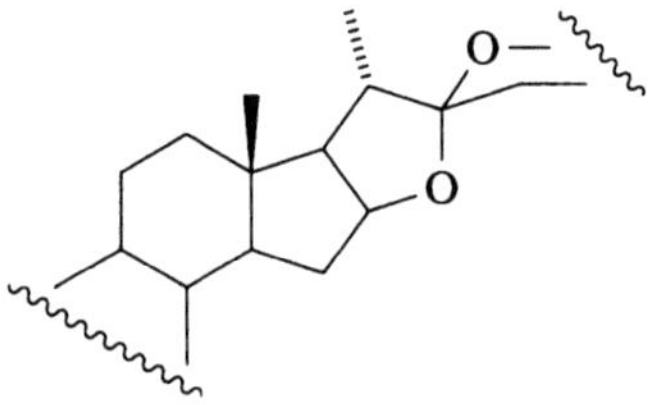 O, O	50	115
4	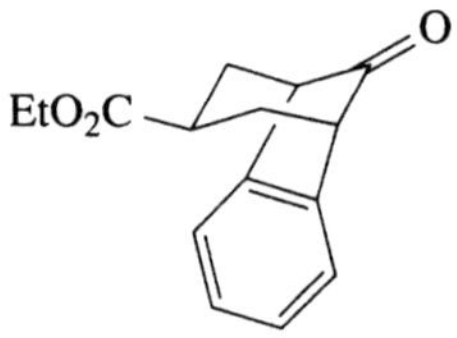 EtO_2C, O	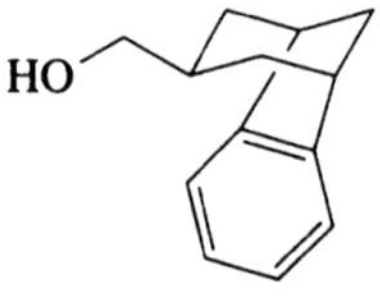 HO	90	124
5	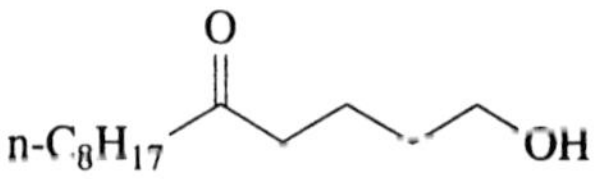 n-C_8H_{17}, O, OH	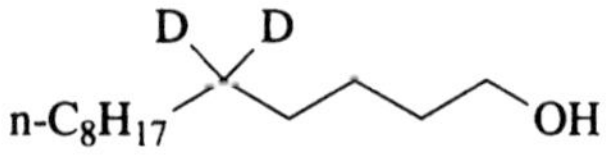 n-C_8H_{17}, D, D, OH	27[a]	125
6	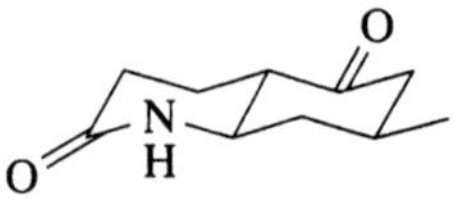 O, N H, O	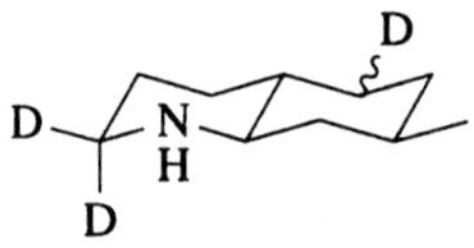 D, D, N H, D	—[b]	126
7	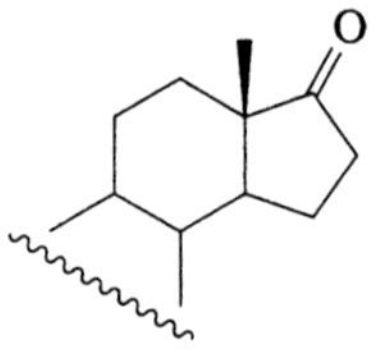 O	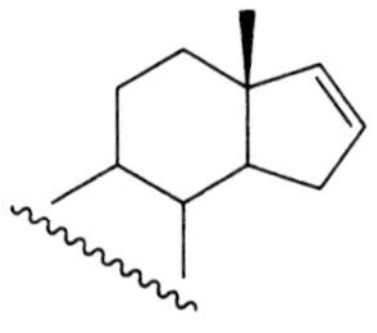	60–70	115
8	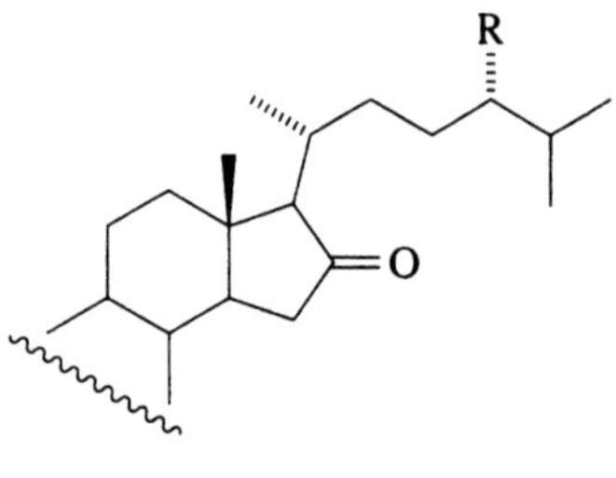 R, O	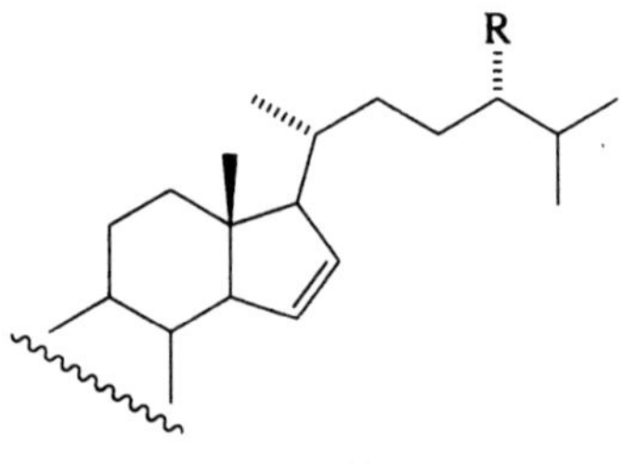 R	—	127

Table 5 *(continued)*

Entry	Carbonyl	Product	Yield (%)	Ref.
9	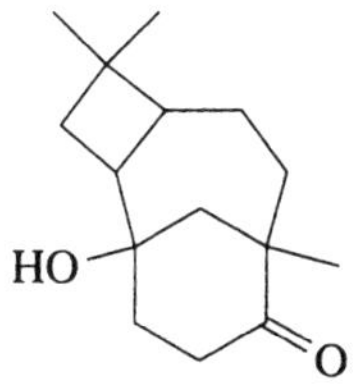 HO, O	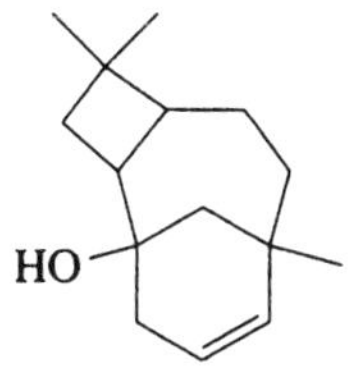 HO	40	128

[a] $LiAlD_4$; D_2O. [b] $LiAlD_4$; H_2O.

of reductions. Also, configurations adjacent to carbonyls are not altered by base-catalyzed enolization as indicated in entries 5 and 6. Deuterium is incorporated *via* use of $NaBD_4$ for the reducing agent as presented in entry 7, Table 6. The introduction of only one deuterium, even though D_2O was used in the work-up,[133] provides evidence for the intermediacy of a diimide (**29a**) in the mechanism (Scheme 2) since (**29a**) represents the only source of hydrogen for the second incorporation.

The attempted conversion of the aldehyde in compound (**65**) to a methyl group by reduction of the tosylhydrazone with $NaBH_4$ (ethanol solvent) led instead to the furan derivative (**66**), resulting from an unusual cyclization with elimination of a $MeOCH_2$ group.[138]

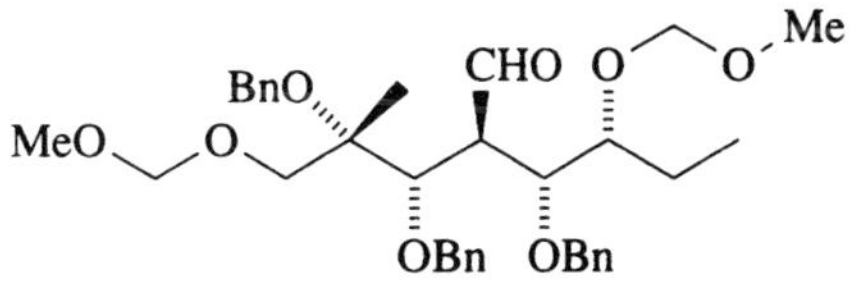

MeO O BnO CHO O O Me OBn OBn

(65)

MeO O BnO O OBn OBn

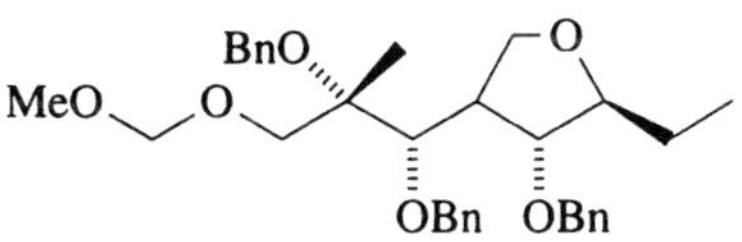

(66)

The combination of $NaBH_4$ in acetic acid has also proven to be a very effective system for the reduction of carbonyl tosylhydrazones.[139] The actual reducing species is probably $NaBH(OAc)_3$[140] and the acidic medium enhances the electrophilicity of the π-bonded carbon by *N*-protonation and allows the reductions to occur at mild temperatures (25–70 °C) with most aldehydes and ketones. This reducing system has been successfully employed in a number of synthetic applications, representative examples of which are displayed in Table 7. The use of $NaBD_4$ in $MeCO_2H$ or $MeCO_2D$ allows the regioselective introduction of one or two deuteriums, respectively (entries 6 and 7, Table 7) thus augmenting the versatility of the methodology.[139]

The transition metal complex bis(triphenylphosphine)copper(I) borohydride, $(Ph_3P)_2CuBH_4$, has also been shown to be effective for the reduction of tosylhydrazones to hydrocarbons under mild conditions (refluxing chloroform).[144] Yields from unhindered aliphatic aldehyde and ketone tosylhydrazones are generally in the range 48–84%. Reductions of hindered ketones (*e.g.* camphor) and aromatic aldehydes were less successful giving 0–20% of reduced products.[144]

1.14.3.2.2 Chemoselectivity

The mild reducing ability of $NaBH_4$ (and $NaBH(OMe)_3$ or $NaBH(OAc)_3$) permits the presence of many more functional groups compared to $LiAlH_4$. Thus, CON, CO_2H, CN (entry 1, Table 7), and halides are expected to survive. Esters remain intact in HOAc (entries 4 and 5, Table 7) but not in MeOH (entry 2, Table 6), probably because of transesterification. Even relatively hindered or deactivated ketones (*e.g.* aryl) have been recovered untouched (entry 8, Table 6 and entry 3, Table 7, respectively). Surprisingly, the tertiary hydroxyl group in entry 3, Table 7 apparently did not give elimination, although no yield for the product was provided.[141] A complex mixture was reported[139] with *p*-nitrobenzaldehyde, indicating that the nitro group was probably attacked.

Table 6 Reduction of Carbonyl Tosylhydrazones with Sodium Borohydride

Entry	*Carbonyl*	*Solvent*	*Product*	*Yield (%)*	*Ref.*
1	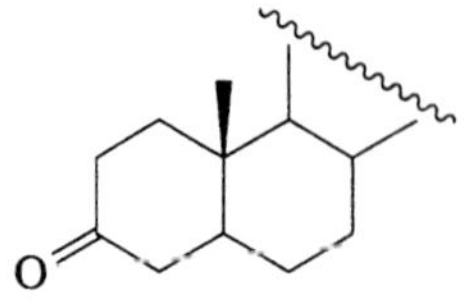 O	MeOH	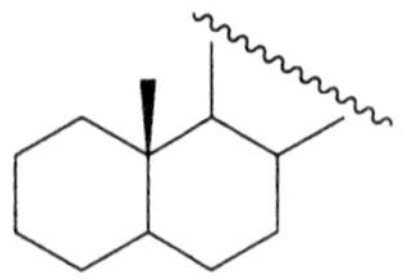	70–80	120
2	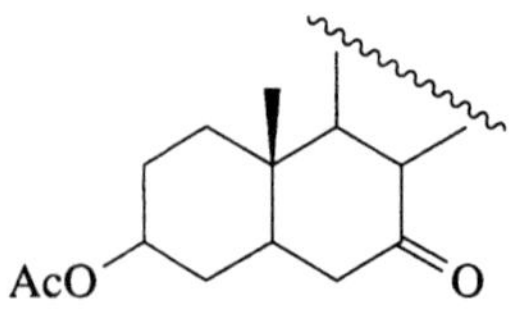 AcO, O	MeOH	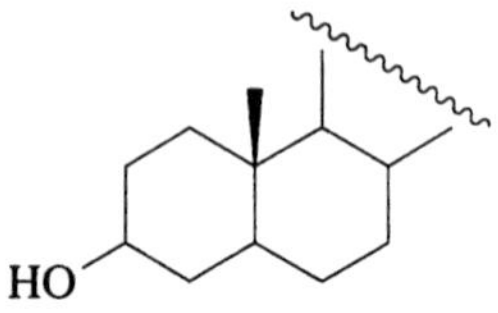 HO	70–80	120
3	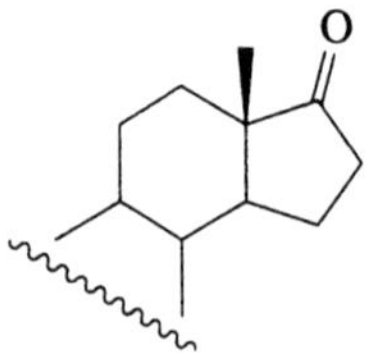 O	Dioxane	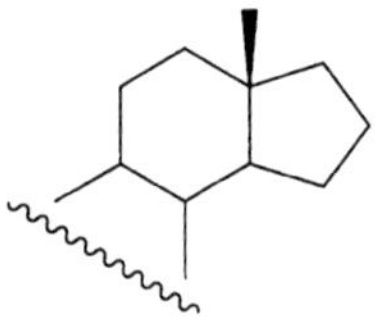	60–70	120
4	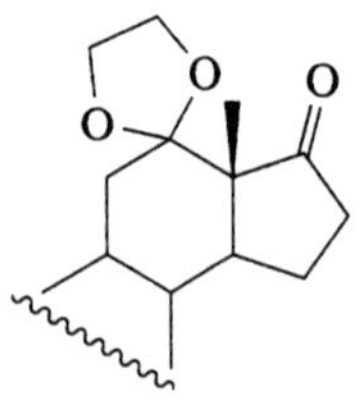 O, O, O	Dioxane	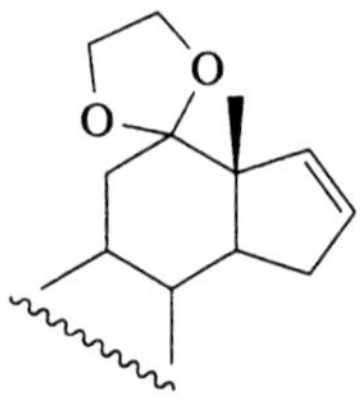 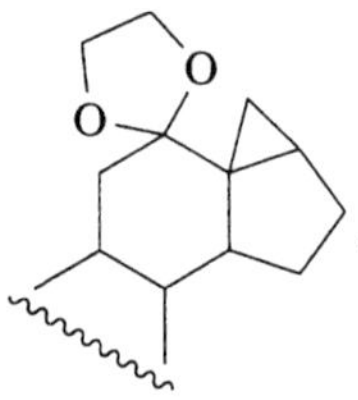 O, O + O, O	71	130
5	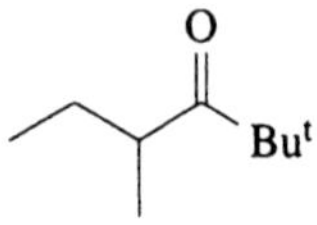 O, Bu[t]; optically active	Dioxane	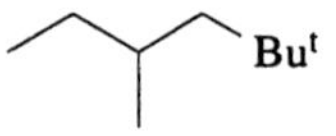 Bu[t]; optically active	37	131
6	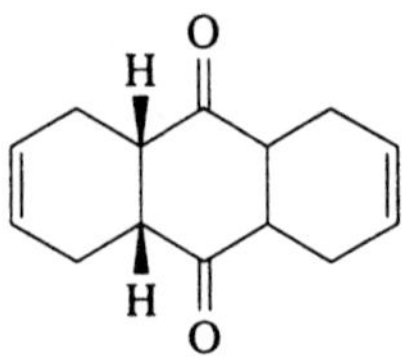 H, O, H, O	DMF	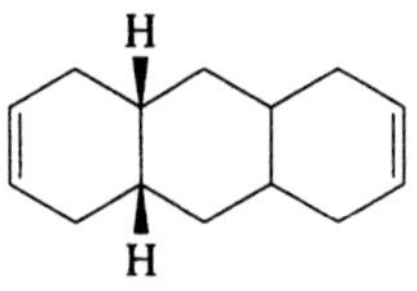 H, H	31	132
7	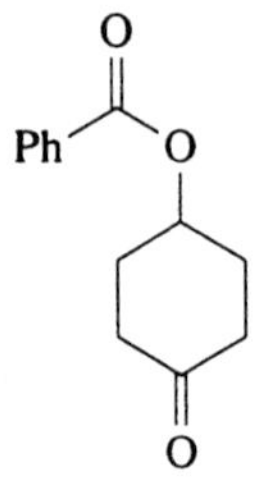 O, Ph, O, O	—	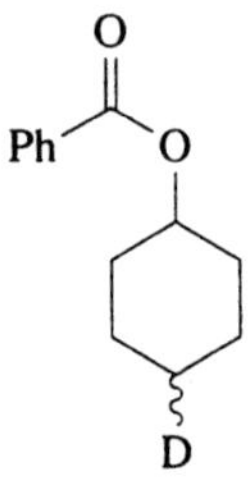 O, Ph, O, D	—[a]	133

Table 6 *(continued)*

Entry	*Carbonyl*	*Solvent*	*Product*	*Yield (%)*	*Ref.*
8	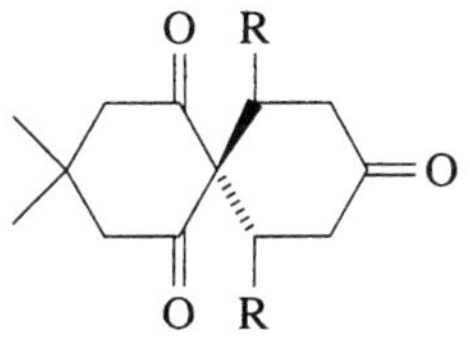 O R / =O / O R	MeOH	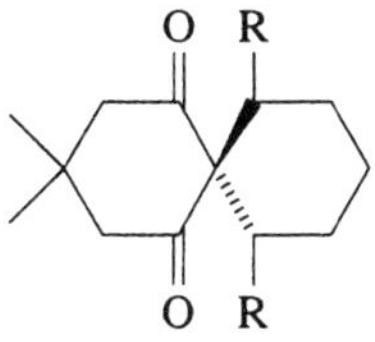 O R / O R	75	134
9	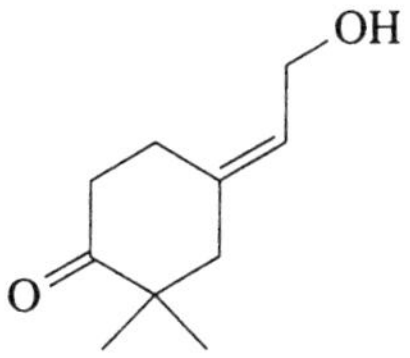 OH / O	MeOH	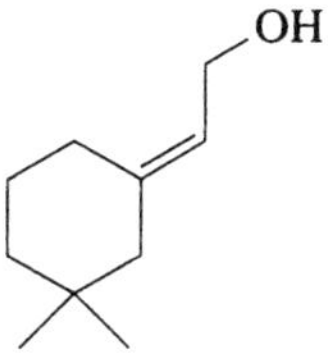 OH	65	135
10	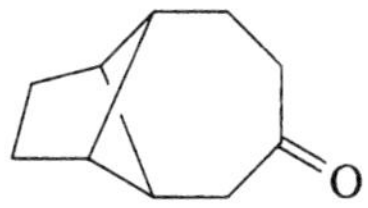 O	MeOH	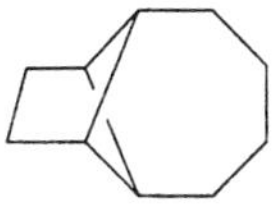	47	136
11	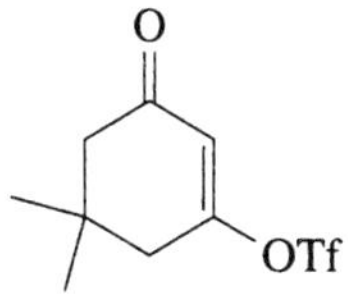 O / OTf	MeOH	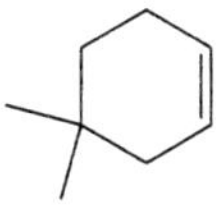	49	137
12	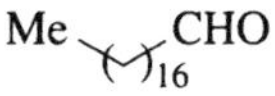 $Me(\sim)_{16}CHO$	MeOH	 $Me(\sim)_{16}Me$	60–70	116

[a] Worked up with D_2O.

1.14.3.2.3 Reduction of conjugated derivatives

Treatment of α,β-unsaturated tosylhydrazones with $NaBH_4$ in MeOH affords principally allylic ethers from cyclic derivatives[145] and pyrazoles with most noncyclic examples.[146] This divergent behavior compared to saturated tosylhydrazones has been attributed[145,146] to a lessening of the electrophilicity of conjugated imine π-bonds, which allows initial abstraction of the acidic N—H proton by BH_4^- to compete with reduction, and gives alternative reactions related to the Bamford–Stevens process as depicted in Scheme 4. An exception to this may be the deoxygenation of conjugated vinyl triflates (entry 11, Table 6).[137] The cyclopropanation and elimination products produced in entry 4, Table 6 also probably arise from similar, alternative reaction paths.[130]

With acetic acid as solvent, the above paths are circumvented, since protonation of the imine nitrogen increases nucleophilic attack and prevents formation of the offending tosylhydrazone anion. The result is that deoxygenation is obtained, but with concomitant rearrangement of the double bond to the site formerly occupied by the carbonyl.[139] This unusual reaction apparently proceeds through a diimide intermediate, which deposits a hydride *via* a 1,5-sigmatropic migration as indicated in Scheme 5. The mechanistic requirement embodied in equation (20) generates the double bond migration even when the resulting alkene is thermodynamically less stable. A selection of representative reductions is presented in Table 8 and illustrates synthetically useful applications as deconjugating alkenes from double bonds (entries 1–3) or aromatic rings (entries 6, 9), producing exocyclic alkenes (entry 4) and regioselectively introducing one (entry 3) or two (entry 2) deuteriums. A limitation was noted in that isophorone tosylhydrazone afforded only an 18% yield of the rearranged alkene (entry 7, Table 8). Apparently, situating a pseudoaxial diimide group over the ring required for migration is opposed by the axial 5-methyl and other side reactions (*i.e.* Michael-type addition of hydride) compete favorably.[139]

Table 7 Reduction of Carbonyl Tosylhydrazones with Sodium Borohydride in Acetic Acid

Entry	*Carbonyl*	*Product*	*Yield (%)*	*Ref.*
1	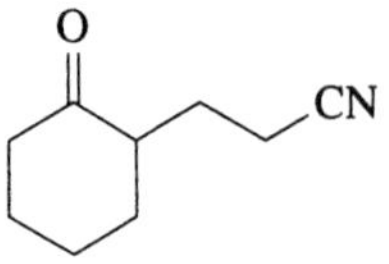	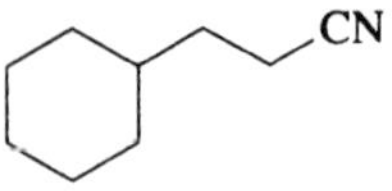	70	139
2	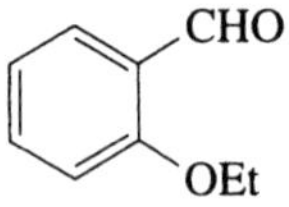	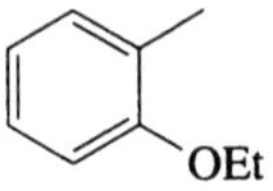	80	139
3	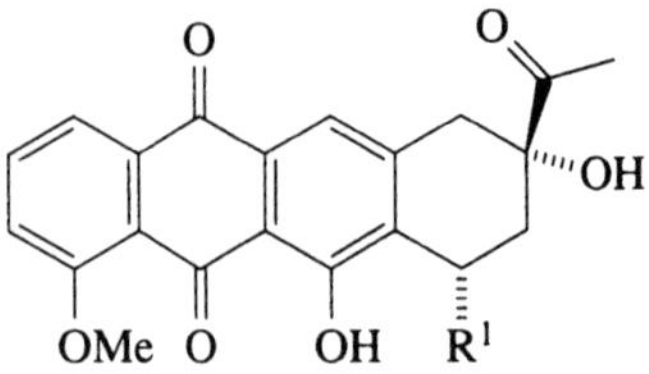	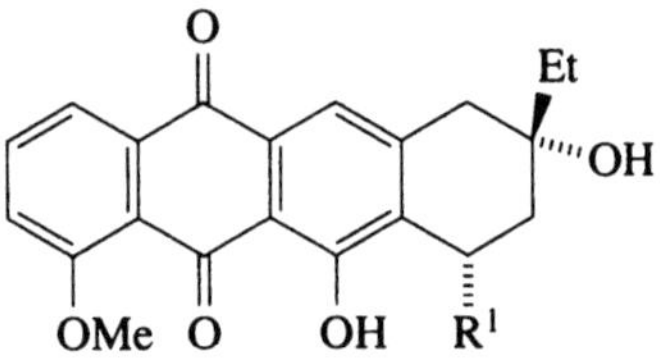	—	141
4	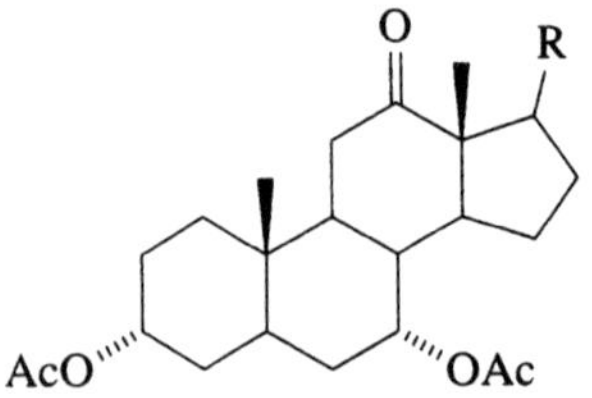	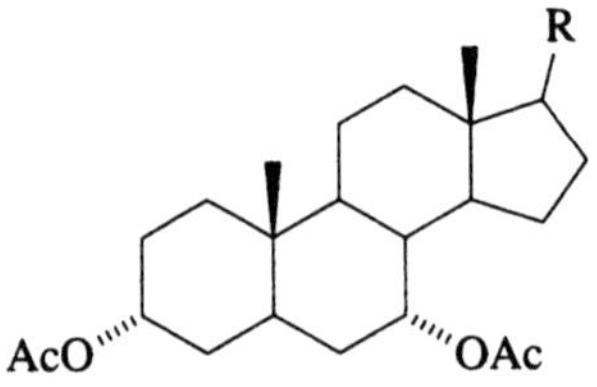	53	142
5	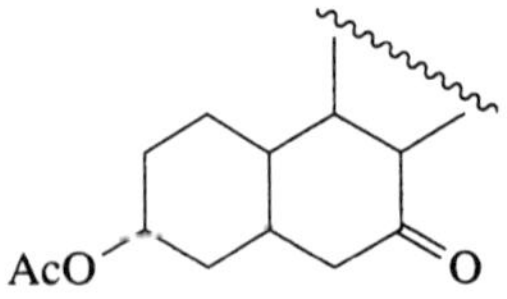	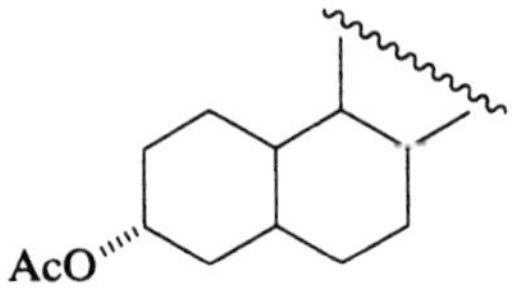	76	143
6	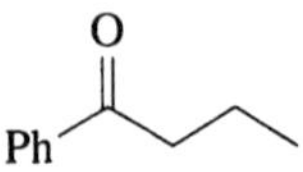	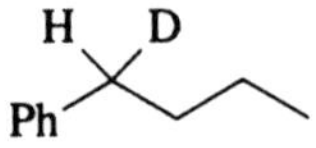	60[a]	139
7	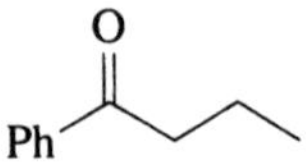	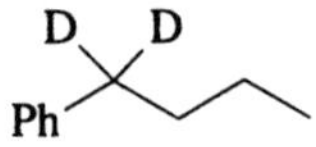	72[b]	139

[a] $NaBD_4$, AcOH. [b] $NaBD_4$, AcOD.

1.14.3.3 Reductions with Sodium Cyanoborohydride

1.14.3.3.1 Scope and limitations

Substitution for one of the hydrogens of borohydride with the strong electron-withdrawing cyano group greatly increases the reluctance of the resulting cyanoborohydride to deliver a hydride. The result is a moderated reducing ability (and an enhanced stability to acid) compared to borohydride. This, coupled with a high dependency of reducing capability on pH, allows remarkable, often pH controllable, selectivity among functional groups.[149]

One successful application of cyanoborohydride reductions is the reduction of tosylhydrazones to hydrocarbons which, since its introduction in 1971,[150] has been employed extensively for such conversions. The reductions are most often conducted in 1:1 DMF:sulfolane containing a small amount of acid at

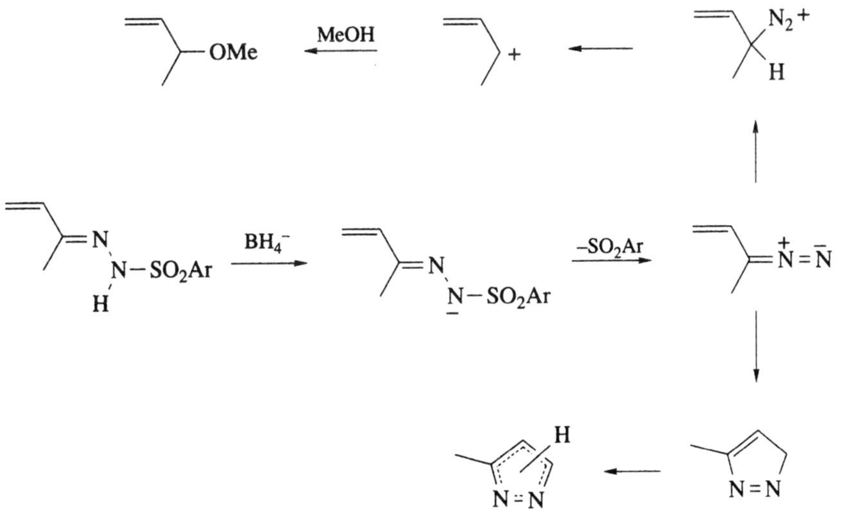

Scheme 4

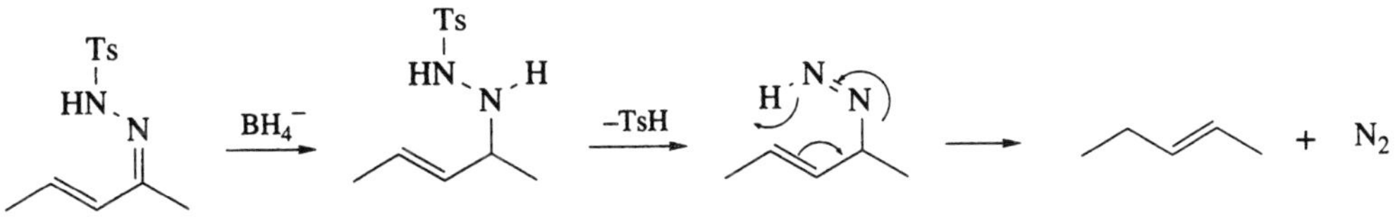

Scheme 5

about 100–110 °C.[117,149] The relatively slow rate of carbonyl attack compared to iminium ions[151] under these reaction conditions permits the *in situ* generation of tosylhydrazones and subsequent reduction of unhindered carbonyls to be conducted in one step (Scheme 6). Hindered ketones (*e.g.* α-phenyl, α-isopropyl or α-*t*-butyl or more than one alkyl group flanking the carbonyl) generally require preformation of the tosylhydrazone and fairly large excesses of cyanoborohydride.[117] Likewise, aryl ketone tosylhydrazones are not readily deoxygenated.[117]

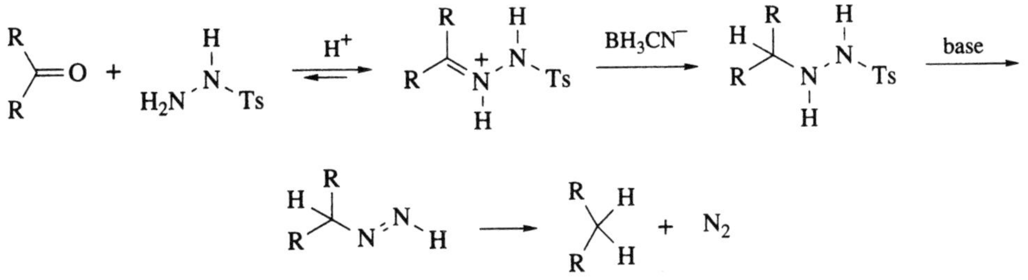

Scheme 6

Synthetic applications of the reducing methodology through 1978 have been reviewed.[149] Table 9 presents further examples chosen to illustrate the range of structural types, many of which contain other functionalities, in which carbonyls have been successfully removed with cyanoborohydride. Thus, relatively hindered types are effectively deoxygenated (entries 2, 3, 5, 13) as well as systems containing strained rings (entry 7). Deuterium (entry 12) and tritium (entry 13) may be incorporated using the appropriate isotopically labeled reagent. Sites adjacent to the carbonyl are normally not isomerized (entry 8), although exceptions have been uncovered.[165,166]

Milder modifications have been developed for systems in which the above methodology was ineffective. Thus, the tosylhydrazone of the carbohydrate (**67**) afforded mixtures of products with $NaBH_3CN$ in DMF at 110 °C, with or without acid, giving a maximum of 20% of the desired product (**68**). However, treatment of (**67**) tosylhydrazone with $NaBH_3CN$ in acidic (HCl) methanol gave the corresponding tosylhydrazine, which underwent smooth conversion to (**68**) upon treatment with NaOAc in refluxing ethanol. This procedure appears to be superior with carbohydrate[167] and other substrates,[168] although treatment

Table 8 Reduction of Conjugated Carbonyl Tosylhydrazones with Sodium Borohydride in Acetic Acid

Entry	*Carbonyl*	*Product*	*Yield (%)*	*Ref.*
1	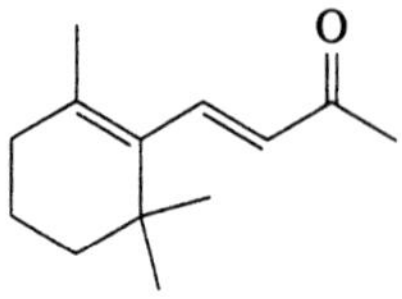	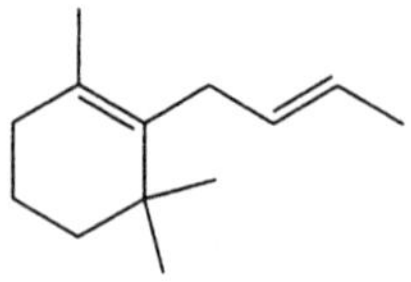	81–89	139
2	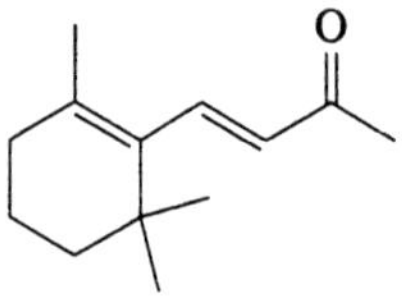	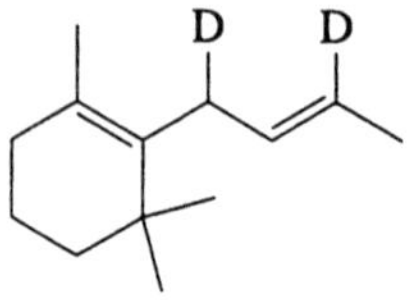	81[a]	139
3	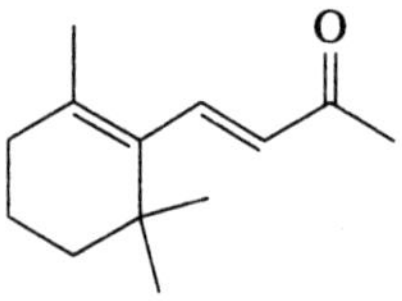	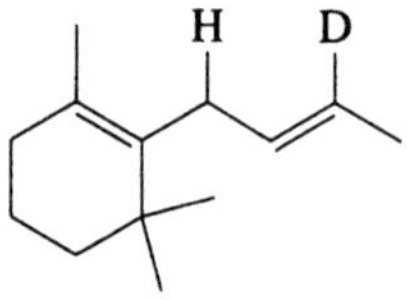	61–72[b]	139
4	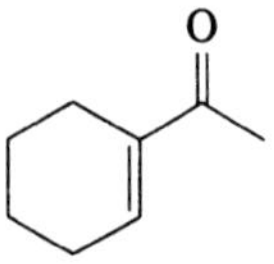	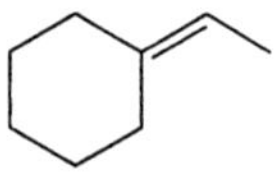	53	142
5			67	139
6	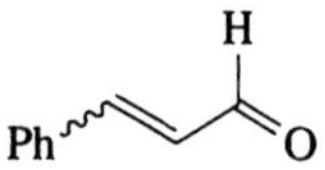	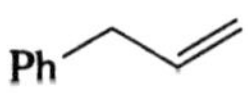	42–56	139
7	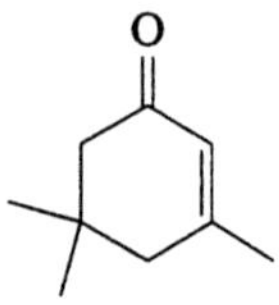	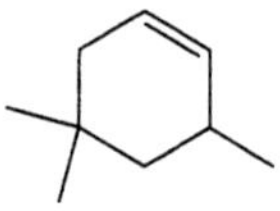	18	139
8	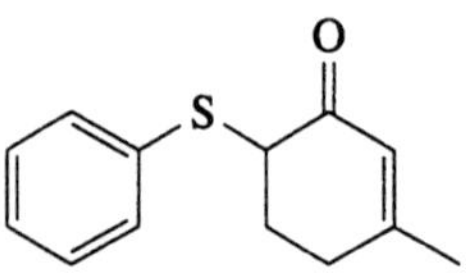	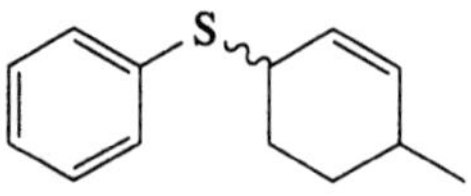	78	147
9	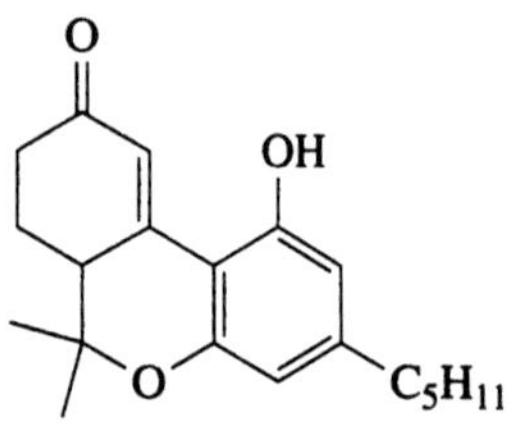	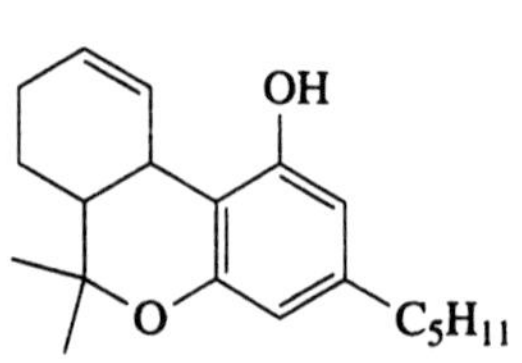	44	148

[a] $NaBD_4$, $MeCO_2D$. [b] $NaBD_4$, $MeCO_2H$.

Table 9 Reduction of Carbonyl Tosylhydrazones with Sodium Cyanoborohydride in DMF–Sulfolane

Entry	*Carbonyl*	*Product*	*Yield (%)*	*Ref.*
1	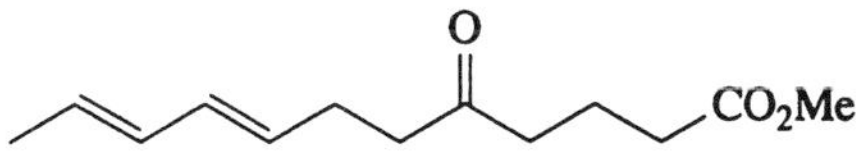 O; CO_2Me	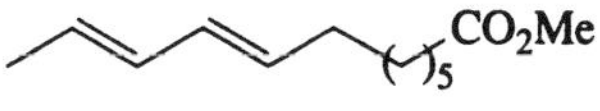 CO_2Me; ( )$_5$	38	152
2	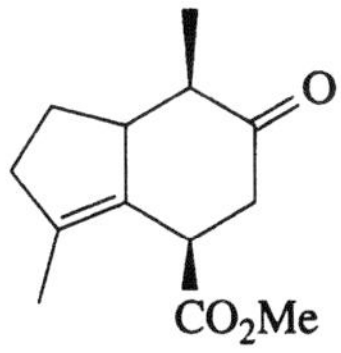 O; CO_2Me	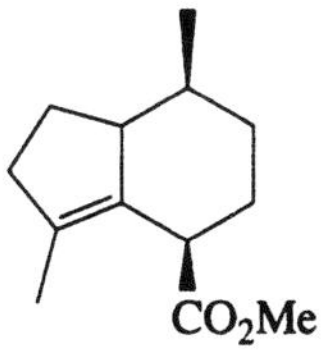 CO_2Me	68	153
3	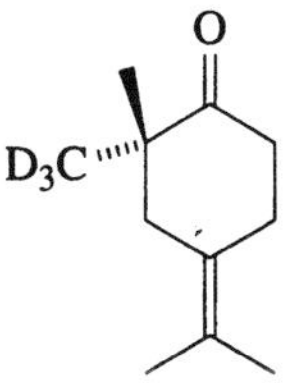 O; D_3C	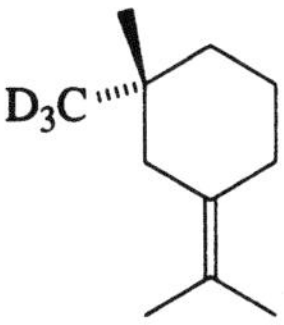 D_3C	75	154
4	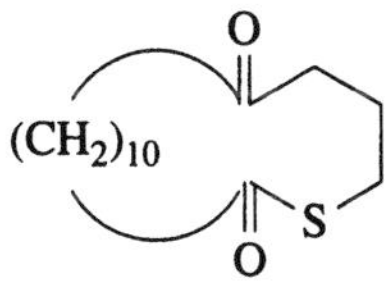 O; $(CH_2)_{10}$; S; O	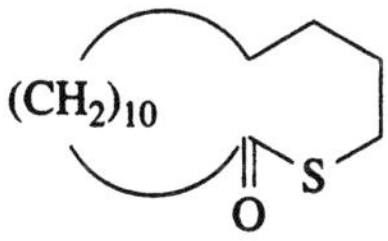 $(CH_2)_{10}$; S; O	44	155
5	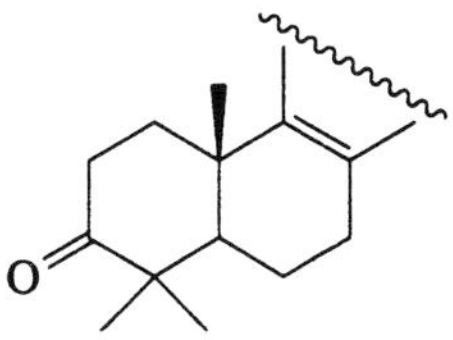 O	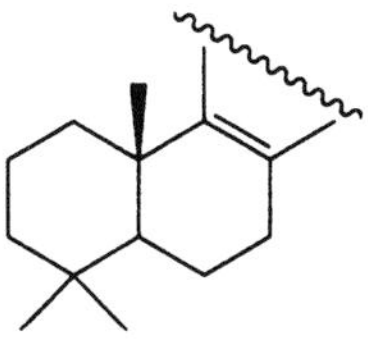	74	156
6	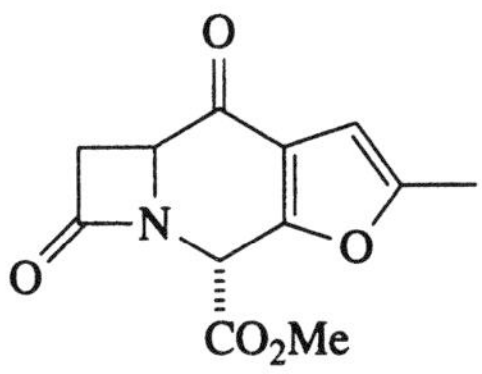 O; N; O; O; CO_2Me	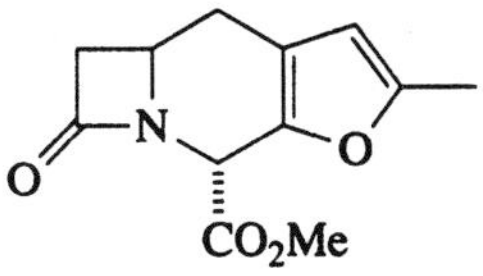 N; O; O; CO_2Me	53	157
7	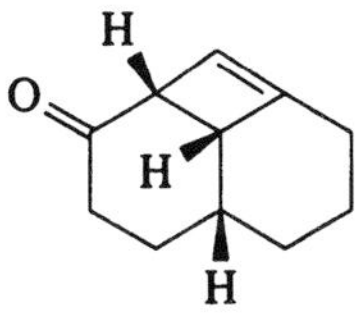 H; O; H; H	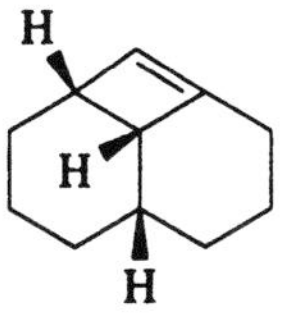 H; H; H	40	158
8	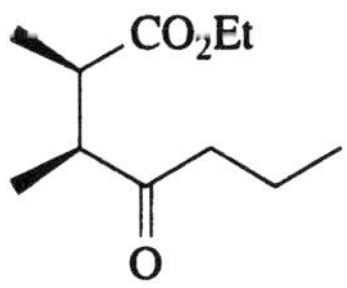 CO_2Et; O	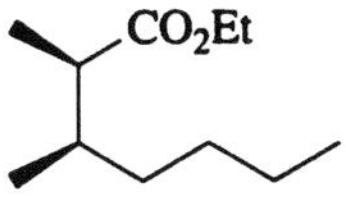 CO_2Et	77	159
9	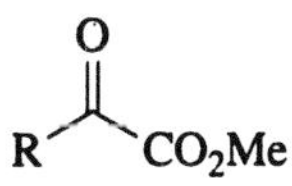 O; R; CO_2Me	 R; CO_2Me	—	160

Table 9 *(continued)*

Entry	Carbonyl	Product	Yield (%)	Ref.
10	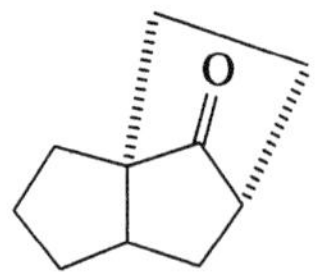 O	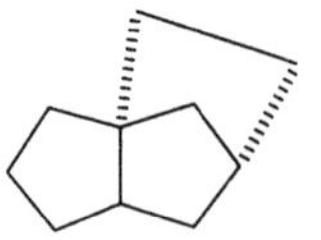	*ca.* 100	161
11	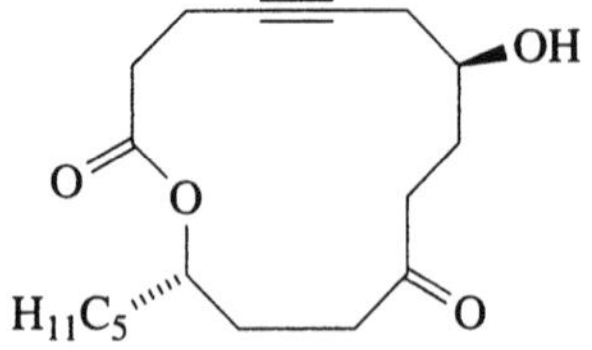 OH, O, O, $H_{11}C_5$, O	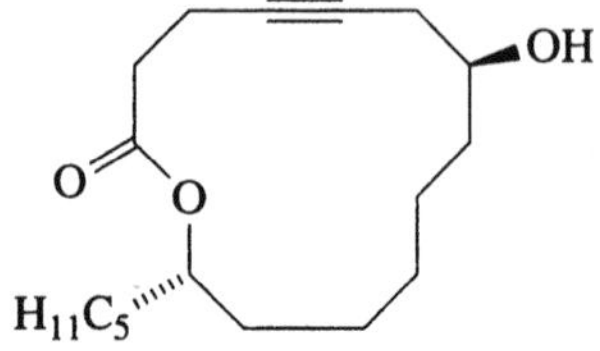 OH, O, O, $H_{11}C_5$	35	162
12	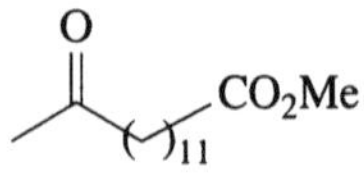 O, CO_2Me, 11	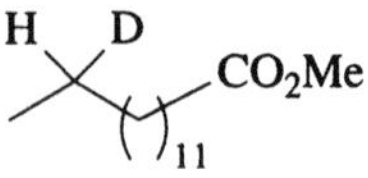 H D, CO_2Me, 11	54[a]	163
13	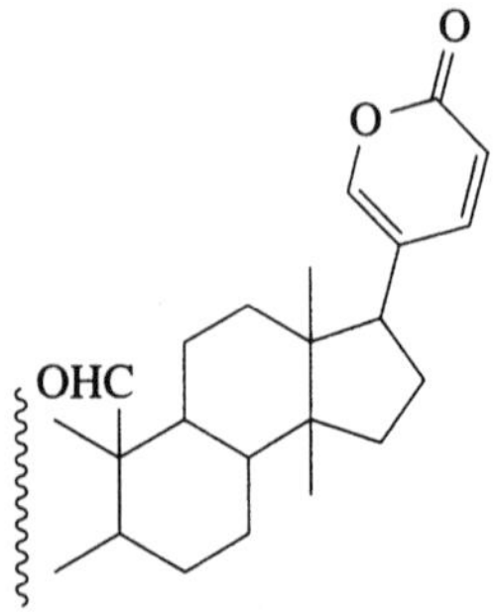 O, O, OHC	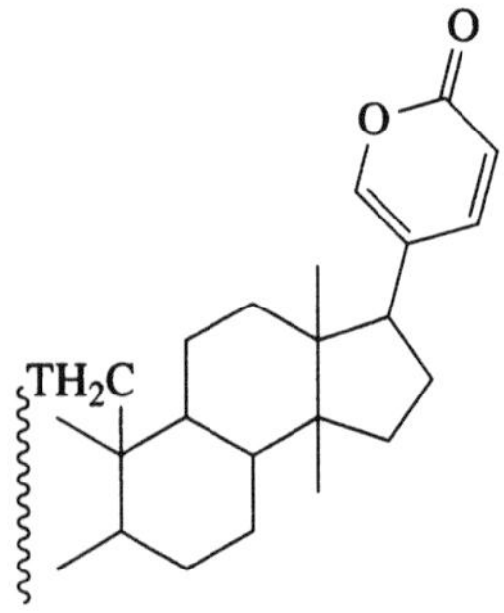 O, O, TH_2C	—[b]	164

[a] $NaBD_3CN$ used. [b] $NaBH_2TCN$ used.

with NaOAc is not always required, especially in refluxing methanol[169–171] or ethanol.[172] Similar mild reductions of tosylhydrazones to isolable tosylhydrazines by $NaBH_3CN$ occurs in acidic THF at room temperature,[173] and reductions in refluxing acidic THF[174] (*e.g.* with **69**) and in THF/HOAc (room temperature) have been successful.[175] Further modifications involve the combination of $NaBH_3CN$ with zinc chloride as the reducing system,[176,177] and the use of mercury complexes with tosylhydrazones as a method of activating reluctant aryl derivatives for reduction with cyanoborohydride (in acidic THF and followed by treatment with KOH/MeOH).[178]

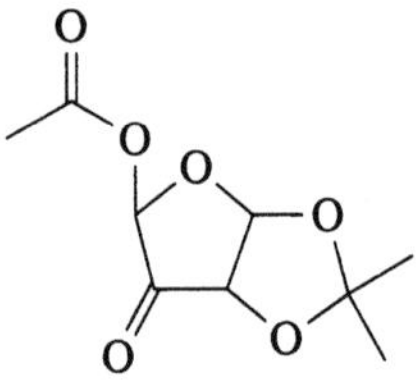

O O O O O O O O O O MeO_2C N O

(67) 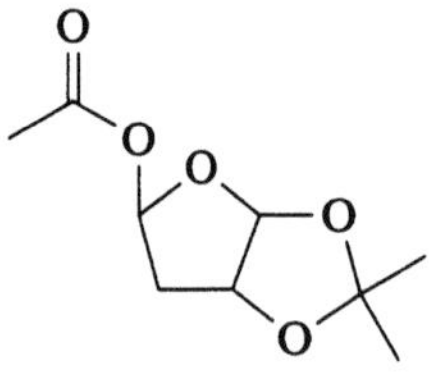 **(68)** 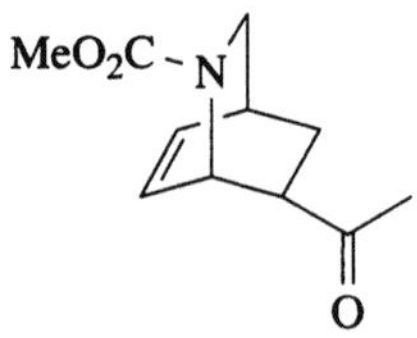 **(69)**

1.14.3.3.2 Chemoselectivity

One of the major advantages of cyanoborohydride for deoxygenations is the high tolerance exhibited for other functionalities, which allows for highly discriminate carbonyl removal.[117,149] Thus, most common groups may be present, including CO_2H, CON— (including β-lactams; entry 6, Table 9), CO_2R (entries 1, 2, 6, 8, 9, 13, Table 9), lactones (entries 11, 13, Table 9), thiolactones (entry 4, Table 9), furans (entry 6, Table 9), NO_2,[117] sulfoxides, aryl halides, and alkyl chlorides.[117] Indeed, the success of the one-step deoxygenation process relies on the slow reduction of even aldehydes and ketones under neutral or mildly acidic (pH > *ca.* 4.5) conditions. Alkyl bromides and iodides are reductively removed in the polar aprotic solvents DMF and sulfolane,[117] but these would probably survive with the acidic methanol methodology.[167]

1.14.3.3.3 Reduction of conjugated derivatives

As with $NaBH_4$ in $MeCO_2H$, reduction of α,β-unsaturated carbonyls with $NaBH_3CN$ affords alkenes in which the double bond has migrated to the position formally occupied by the carbonyl (Scheme 5).[117,149,179,180] Compilations of successful synthetic applications are found in refs. 117 and 149. As with $NaBH_4$, isophorone tosylhydrazone affords only a low yield (36%) of the alkene product contaminated with 1,1,3-trimethylcyclohexane.[179] Other ketones which cannot readily assume a cisoid orientation for intramolecular migration (Scheme 5) also give mixtures of alkene and alkane products resulting from competing Michael-type delivery of hydride (Scheme 7).[180] This problem may be alleviated by use of $NaBH_3CN$ in $MeCO_2H$. Thus, a tetracyclic derivative (partial structure **70**) gave the single rearranged alkene (**71**; 55%).[181]

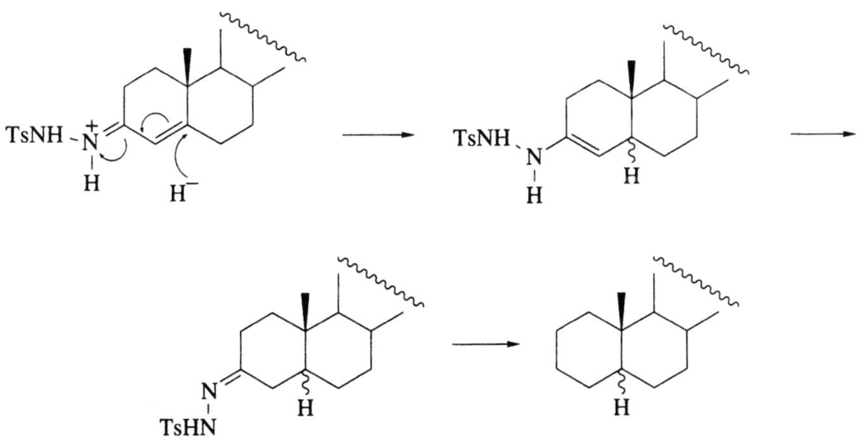

TsNH–N⁺(H)= ... H⁻ → TsNH–N(H)– ... H → N(–NHTs)= ... H → ... H

Scheme 7

An interesting application of the reduction–migration methodology involves the conversion of α,β-alkynic ketones to allenes.[182] Thus, reduction of the tosylhydrazone of (**72**) under slightly modified conditions (lower temperature and pH, greater excess of $NaBH_3CN$) afforded the trimethylsilylallene (**73**) in 69% yield, while the original recipe[117] gave the pyrazole (**74**; 81%).[182]

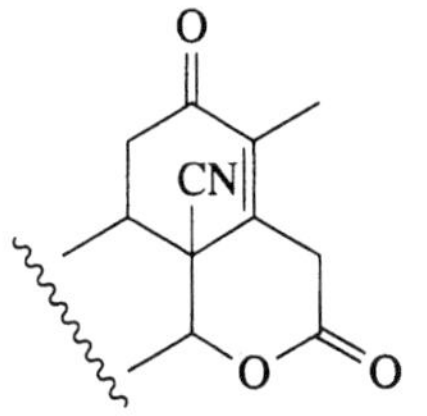

O CN O O **(70)**

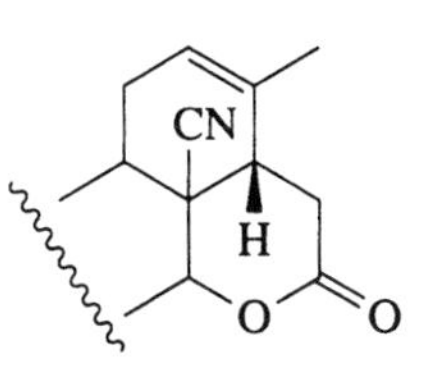

CN H O O **(71)**

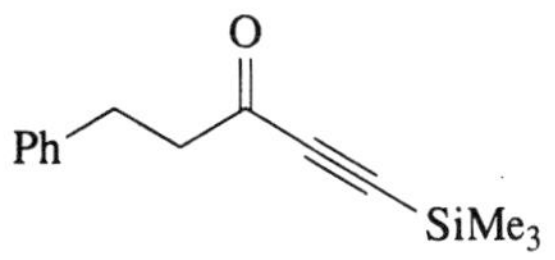

O Ph $SiMe_3$ **(72)**

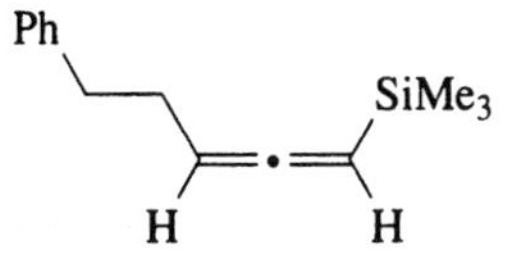

Ph $SiMe_3$ H H **(73)**

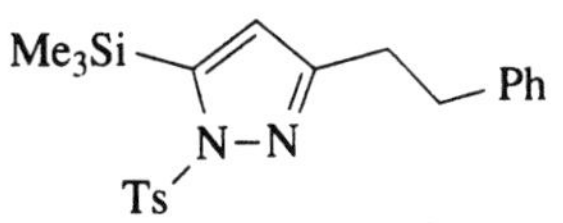

Me_3Si Ph N–N Ts **(74)**

1.14.3.4 Reductions with Catechol Borane and Related Reagents

1.14.3.4.1 Scope and limitations

The reduction of tosylhydrazones with catechol borane under very mild conditions ($CHCl_3$, 25 °C followed by NaOAc[118] or, preferably, Bu_4NOAc[183]) is also a highly effective deoxygenation process, which has been successfully utilized in a number of synthetic conversions. A 1977 review of catechol borane chemistry, including reductive deoxygenations, is available.[184] Table 10 contains a selection of successful conversions from the literature and illustrates the effectiveness of the reagent as well as the ability to incorporate a deuterium regioselectively by utilizing NaOAc·D_2O (entry 4) or two deuteriums using catechol borane-*d* and NaOAc·D_2O (entry 5, Table 10).[187] Reduction of a hindered cyclobutanone (entry 6) afforded a low yield of reduction product (20%). In this case, the Cram modification of the Wolff–Kishner reduction was more successful (56% yield).[188]

Table 10 Reduction of Carbonyl Tosylhydrazones with Catechol Borane and Bis(benzoyloxy)borane

Entry	*Carbonyl*	*Product*	*Yield (%)*	*Ref.*
1			94	183
2			—	185
3			68	186
4			97[a]	187
5			95[b]	187
6			20	188
7	Me$(\sim)_8$CHO	Me$(\sim)_8$Me	91[c]	189
8			70–80[d]	190

[a] Using NaOAc•3D_2O in workup. [b] Using catechol borane-*d* and NaOAc•3D_2O in work-up. [c] $(PhCO_2)_2BH$ as reducing agent. [d] $(PhCO_2)_2BH$ as reducing agent: NaOD, D_2O used in work-up.

The related borane bis(benzoyloxy)borane[189] has also been found to be effective for tosylhydrazone reductions, examples of which are presented in Table 10 (entries 7 and 8). The latter case required the use of $NaOD/D_2O$ for efficient deuterium incorporation[190] (instead of $NaOAc/D_2O$).[189] A pyridine–borane complex likewise reduces tosylhydrazones in acidic ethanol/dioxane[191] to tosylhydrazines (91–98% yields), which may be converted to hydrocarbons by treatment with KOH/MeOH[178] or $NaOAc{\cdot}3H_2O/CHCl_3$.[191]

1.14.3.4.2 Chemoselectivity

As with cyanoborohydride, very few functional groups are affected by catechol borane under the tosylhydrazone reduction conditions (25 °C), allowing highly selective conversions in the presence of most moieties, including alkenes and alkynes which are hydroborated at more elevated temperatures (70–100 °C).[192] The only exceptions to this appear to be aldehydes, carboxylates, sulfoxides, amine oxides and anhydrides, which are reduced at rates comparable to tosylhydrazones.[192]

1.14.3.4.3 Reduction of conjugated derivatives

As with other hydride reagents, reductions of α,β-unsaturated carbonyl tosylhydrazones with catechol borane provides the alkene product resulting from migration of the double bond (Scheme 5).[118,193,194] Furthermore, even certain cyclic systems which afford low yields and alkane side products with $NaBH_4$ and $NaBH_3CN$ (*e.g.* isophorone, see entry 7, Table 8) normally provide only the rearranged alkene and no alkane side products.[184,193–195] Apparently, the low nucleophilicity of the reagent prevents competing Michael addition to the double bond (Scheme 5), which afflicts the reductions with anionic hydrides. Table 11 provides a collection of successful conversions of conjugated systems and illustrates the versatility (and chemoselectivity) of the method. With β-substituted rigid cyclohexenone systems, highly stereoselective introduction of hydrogen is observed. Thus, the initial attack by the reagent is apparently favored (as expected)[204] from the axial (α) direction leading to the equatorial (β) orientated diimide derivative which, in turn, leads to delivery of hydride to the top (β) face of the double bond as observed in entries 1, 4 and 7, Table 11. The geometric constraints imposed by deliverance of a hydride by an equatorial diimide to the β-position implicates a boat-type conformation (Scheme 8) or a bimolecular mechanism.[95] In one instance, (entry 4, Table 11) the principal rearranged alkene was accompanied by a small amount of the deoxygenated product in which the alkene had remained in place.[198] Other stereoselective conversions also appear to be controlled by initial approach of the reagent from the less-hindered face of the hydrazone (entry 6, Table 11). When the faces are comparable in accessibility, mixtures of isomers are obtained (*e.g.* entry 8, Table 11). The methodology is also useful for the conversion of α,β-alkynic ketones to allenes (Table 11, entry 9).[203]

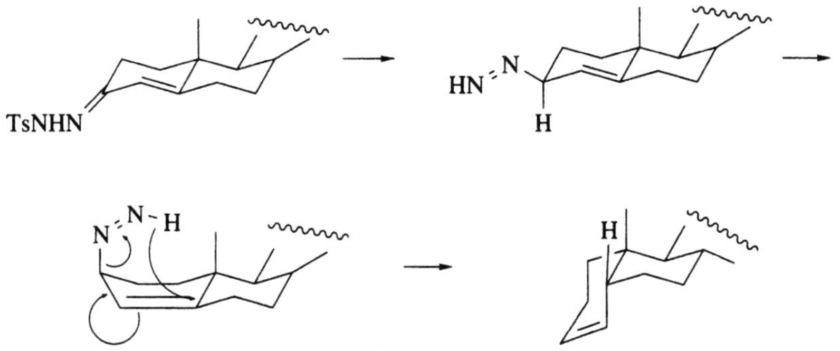

TsNHN

HN=N H

N=N–H

H

Scheme 8

An unusual reduction course is illustrated by results with the crossed dienone in Table 11, entry 7, in which reductive migration was accompanied by removal of the second double bond, the mechanism of which is not obvious.[201]

Table 11 Reduction of Conjugated Carbonyls with Catechol Borane

Entry	*Carbonyl*	*Product*	*Yield (%)*	*Ref.*
1	4-Cholesten-3-one	3-Cholestene	83–88	195
2	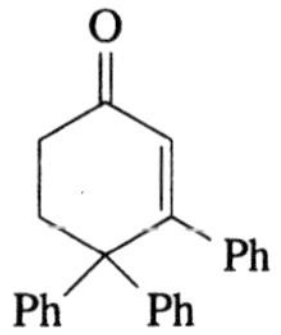	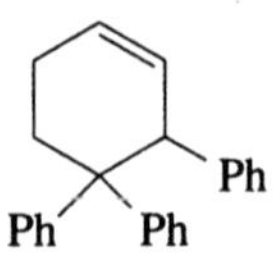	77	196
3	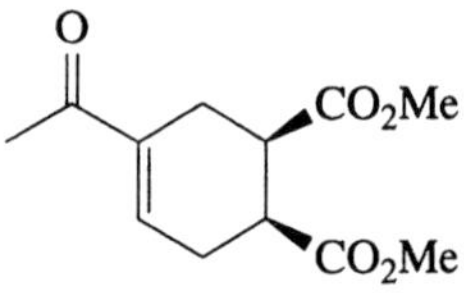	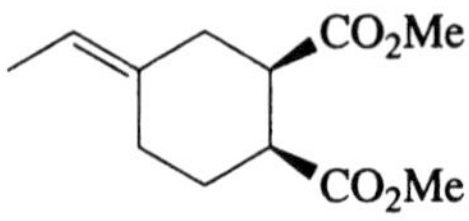	89	197
4	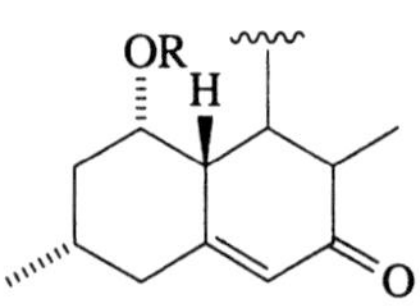	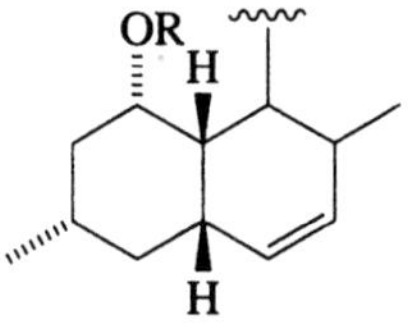 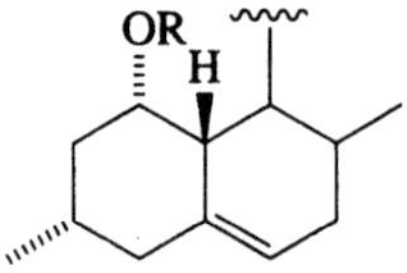 84:16	72	198
5	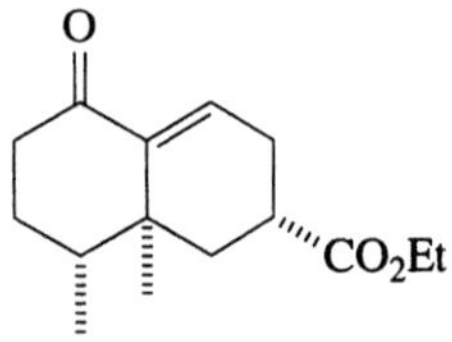	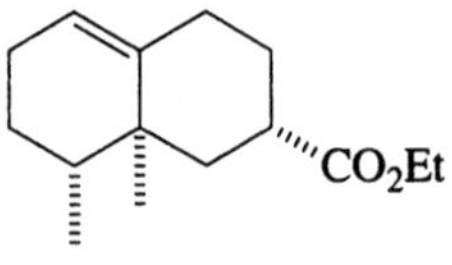	55	199
6	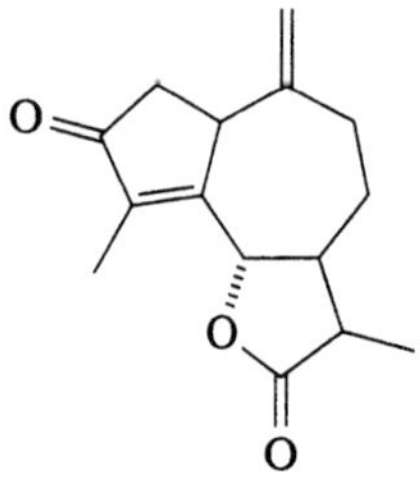	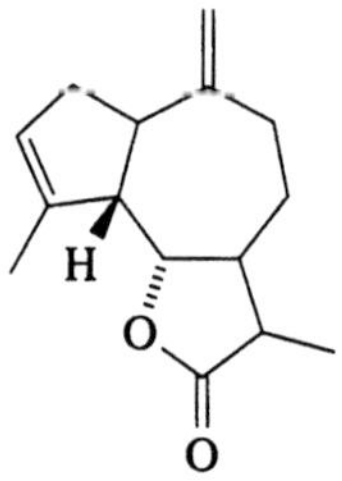	55	200
7	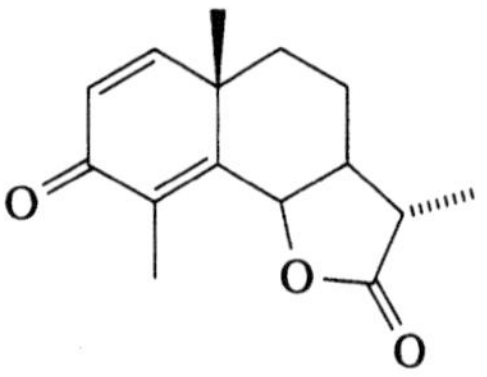	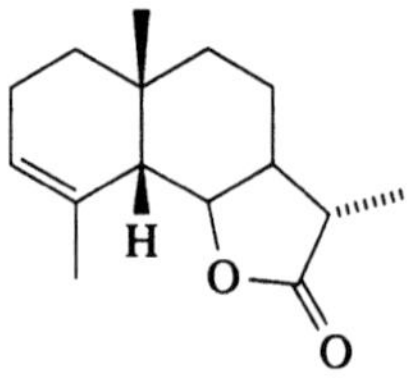	55	200, 201
8	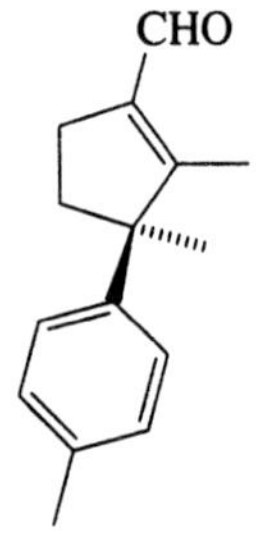	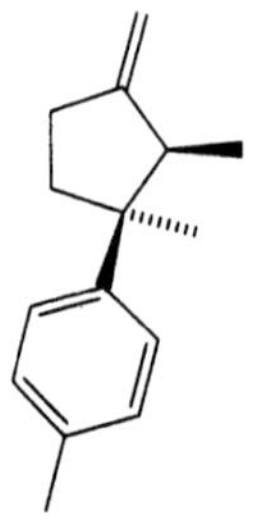 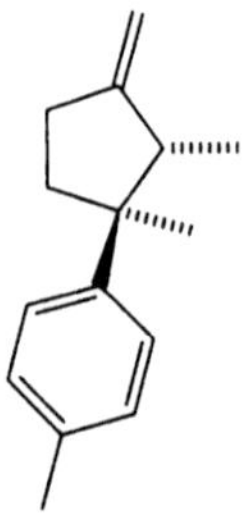 65:35	93	202

Table 11 *(continued)*

Entry	Carbonyl	Product	Yield (%)	Ref.
9	Ph–C≡C–C(=O)–CH3	PhCH=C=CH–CH3	75	203

1.14.4 REFERENCES

1. N. Kishner, *Zh. Russ. Fiz.-Khim. O-va., Chast. Khim.*, 1911, **43**, 582.
2. L. Wolff, *Justus Liebigs Ann. Chem.*, 1912, **394**, 86.
3. D. Todd, *Org. React. (N.Y.)*, 1948, **4**, 378.
4. N. P. Buu-Hoi, N. Hoán and N. D. Xuong, *Recl. Trav. Chim. Pays-Bas*, 1952, **71**, 285.
5. Huang-Minlon, *Sci. Sinica*, 1961, **10**, 711.
6. W. Reusch, in 'Reduction', ed. R. L. Augustine, Dekker, New York, 1968, p. 171.
7. N. Kishner, *Zh. Russ. Fiz.-Khim. O-va., Chast. Khim.*, 1912, **44**, 1754.
8. A. H. Cook and R. P. Linstead, *J. Chem. Soc.*, 1934, 946.
9. J. W. Barrett and R. P. Linstead, *J. Chem. Soc.*, 1935, 436.
10. Y. Asahina and H. Nogami, *Chem. Ber.*, 1935, **68**, 1500.
11. R. P. Linstead and E. M. Meade, *J. Chem. Soc.*, 1934, 935.
12. M. D. Soffer, M. B. Soffer and K. W. Sherk, *J. Am. Chem. Soc.*, 1945, **67**, 1435.
13. K. W. Sherk, M. V. Augur and M. D. Soffer, *J. Am. Chem. Soc.*, 1945, **67**, 2239.
14. C. H. Herr, F. C. Whitmore and R. W. Schiessler, *J. Am. Chem. Soc.*, 1945, **67**, 2061.
15. L. Ruzicka and M. W. Goldberg, *Helv. Chim. Acta*, 1935, **18**, 668.
16. Huang-Minlon, *J. Am. Chem. Soc.*, 1946, **68**, 2487.
17. Huang-Minlon, *J. Am. Chem. Soc.*, 1949, **71**, 3301.
18. H. H. Szmant, *Angew. Chem., Int. Ed. Engl.*, 1968, **7**, 120.
19. H. H. Szmant and C. M. Harmuth, *J. Am. Chem. Soc.*, 1964, **86**, 2909.
20. H. H. Szmant and M. N. Román, *J. Am. Chem. Soc.*, 1966, **88**, 4034.
21. E. M. Kaiser, F. E. Henoch and C. R. Hauser, *J. Am. Chem. Soc.*, 1968, **90**, 7287.
22. A. S. Harris, E. N. White and D. McNeil, *J. Chem. Soc.*, 1955, 4216.
23. D. J. Cram, M. R. V. Sahyun and G. R. Knox, *J. Am. Chem. Soc.*, 1962, **84**, 1734.
24. H. H. Szmant, A. Birke and M. P. Lau, *J. Am. Chem. Soc.*, 1977, **99**, 1863.
25. L. F. Fieser and M. Fieser, 'Topics in Organic Chemistry', Reinhold, New York, 1963, p. 258.
26. G. Lock, *Monatsh. Chem.*, 1954, **85**, 802.
27. J. M. Bruce and F. Sudcliffe, *J. Chem. Soc.*, 1956, 3824.
28. M. Gates and G. Tschudi, *J. Am. Chem. Soc.*, 1956, **78**, 1380.
29. M. Gates and W. G. Webb, *J. Am. Chem. Soc.*, 1958, **80**, 1186.
30. R. B. Kelly, J. Zamecnik and B. A. Beckett, *Can. J. Chem.*, 1972, **50**, 3455.
31. T. Kikuchi, M. Takayama, T. Toyoda, M. Arimoto and M. Niwa, *Chem. Pharm. Bull.*, 1973, **21**, 2243.
32. G. Haffer, U. Ender, G. Neef, G. Sauer and R. Wiechert, *Liebigs Ann. Chem.*, 1981, 425.
33. S. A. Godleski, P. von R. Schleyer, E. Ōsawa, Y. Inamoto and Y. Fujikura, *J. Org. Chem.*, 1976, **41**, 2596.
34. C. H. Foster and D. A. Payne, *J. Am. Chem. Soc.*, 1978, **100**, 2834.
35. A. K. Chakraborti, S. K. Alam, P. C. Chakraborti, R. Dasgupta, J. Chakravarty, U. R. Ghatak, A. Kabiraj and S. G. Biswas, *J. Chem. Soc., Perkin Trans. 1*, 1986, 1243.
36. W. A. Denny and B. F. Cain, *J. Med. Chem.*, 1978, **21**, 430.
37. J. G. Smith and M. P. Massicotte, *Org. Prep. Proced. Int.*, 1978, **10**, 123.
38. A. Streitwieser, Jr., E. R. Vorpagel and C.-C. Chen, *J. Am. Chem. Soc.*, 1985, **107**, 6970.
39. B. Raju and G. S. Krishna Rao, *Indian J. Chem., Sect. B*, 1987, **26**, 1185.
40. T. L. Schaeffer, S. G. Cantwell, J. L. Brown, D. S. Watt and R. R. Fall, *Appl. Environ. Microbiol.*, 1979, **38**, 742.
41. M. Bertrand, H. Monti and K. C. Huong, *Tetrahedron Lett.*, 1979, 15.
42. H. Mohrle and M. Class, *Pharmazie*, 1986, **41**, 553.
43. T. Endo, K. Inoue, S. Nojima, S. Terashima and T. Oshima, *Chem. Phys. Lipids*, 1982, **31**, 61.
44. M. F. Schlecht and S. Giandinoto, *Heterocycles*, 1987, **25**, 485.
45. D. H. R. Barton, D. A. J. Ives and B. R. Thomas, *J. Chem. Soc.*, 1954, 903.
46. D. H. R. Barton, D. A. J. Ives and B. R. Thomas, *J. Chem. Soc.*, 1955, 2056.
47. M. Yamashita, M. Naora, T. Murae, T. Tsuyuki and T. Takahashi, *Bull. Chem. Soc. Jpn.*, 1987, **60**, 1383.
48. R. Ranganathan, U. R. Nayak, T. S. Santhanakrishnan and S. Dev, *Tetrahedron*, 1970, **26**, 621.
49. A. G. Gonzalez, J. Amaro, B. M. Fraga and J. G. Luis, *Phytochemistry*, 1983, **22**, 1828.
50. W. Nagata and H. Itazaki, *Chem. Ind. (London)*, 1964, 1194.
51. A. Nickon and P. S. J. Zurer, *J. Org. Chem.*, 1981, **46**, 4685.
52. J. R. Prahlad, U. R. Nayak and S. Dev, *Tetrahedron*, 1970, **26**, 663.
53. R. B. Kelly, J. Eber and H. K. Hung, *Can. J. Chem.*, 1973, **51**, 2534.
54. P. J. Eaton, D. R. Lauren, A. W. O'Conner and R. T. Weavers, *Aust. J. Chem.*, 1981, **34**, 1303.

55. W. A. Ayer, J.-A. H. Ball, B. Rodríguez and S. Valverde, *Can. J. Chem.*, 1974, **52**, 2792.
56. A. K. Barua, P. Chakrabarti, S. P. Dutta, D. K. Mukherjee and B. C. Das, *Tetrahedron*, 1971, **27**, 1141.
57. N. K. Dunlap, M. R. Sabol, P. E. Bauer, D. S. Watt, J. H. Reibenspies, O. P. Anderson, W. K. Seifert and J. M. Moldowan, *J. Org. Chem.*, 1985, **50**, 1826.
58. R. R. Muccino and C. Djerassi, *J. Am. Chem. Soc.*, 1974, **96**, 556.
59. A. Murai, T. Nishimura and T. Masamune, *Bull. Chem. Soc. Jpn.*, 1976, **49**, 1612.
60. S. A. Godleski, P. von R. Schleyer and E. Ōsawa, *J. Chem. Soc., Chem. Commun.*, 1976, 38.
61. P. K. Jadhav and U. R. Nayak, *Indian J. Chem., Sect. B*, 1978, **16**, 1047.
62. D. L. Snitman, M.-Y. Tsai and D. S. Watt, *Synth. Commun.*, 1978, **8**, 195.
63. E. Piers, M. Zbozny and D. C. Wigfield, *Can. J. Chem.*, 1979, **57**, 1064.
64. Y. Tsuda, K. Yoshimoto, T. Yamashita and M. Kaneda, *Chem. Pharm. Bull.*, 1981, **29**, 3238.
65. S. M. Krutov, Z. Samek, V. Benesova and V. Herout, *Phytochemistry*, 1973, **12**, 1405.
66. P. de Mayo and R. Suau, *J. Chem. Soc., Perkin Trans. 1*, 1974, 2559.
67. L. Fitjer, U. Klages, D. Wehle, M. Giersig, N. Schormann, W. Clegg, D. S. Stephenson and G. Binsch, *Tetrahedron*, 1988, **44**, 405.
68. T. K. Chen, D. C. Ales, N. C. Baenziger and D. F. Wiemer, *J. Org. Chem.*, 1983, **48**, 3525.
69. K. J. Crowley, K. L. Erickson, A. Eckell and J. Meinwald, *J. Chem. Soc., Chem. Commun.*, 1973, 2671.
70. G. Angelini and G. Sleiter, *Gazz. Chim. Ital.*, 1975, **105**, 961.
71. L. Grehn, *Chem. Scr.*, 1979, **13**, 67.
72. H. Ramuz, *Helv. Chim. Acta*, 1975, **58**, 2050.
73. M. F. Grundon, H. B. Henbest and M. D. Scott, *J. Chem. Soc.*, 1963, 1855.
74. G. Lardelli and O. Jeger, *Helv. Chim. Acta*, 1949, **32**, 1817.
75. N. J. Leonard and S. Gelfand, *J. Am. Chem. Soc.*, 1955, **77**, 3272.
76. T. L. Gilchrist, C. J. Moody and C. W. Rees, *J. Chem. Soc., Perkin Trans. 1*, 1979, 1871.
77. L. M. Harwood and M. Julia, *Tetrahedron Lett.*, 1980, **21**, 1743.
78. D. Caine and A. S. Frobese, *Tetrahedron Lett.*, 1977, 3107.
79. T. Rui, L. Lian-niang and F. Qicheng, *Planta Med.*, 1986, **52**, 49.
80. M. Ramah and B. Laude, *Bull. Soc. Chim. Fr.*, 1975, 2649.
81. D. L. Coffen, R. I. Fryer, D. A. Katonak and F. Wong, *J. Org. Chem.*, 1975, **40**, 894.
82. M. Takeda, H. Inoue, K. Noguchi, Y. Honma, M. Kawamori, G. Tsukamoto and S. Saito, *Chem. Pharm. Bull.*, 1976, **24**, 1002.
83. J. A. Moore and T. D. Mitchell, *Org. Prep. Proced. Int.*, 1984, **16**, 411.
84. M. Curphey, M. Lennon, E. Murphy and G. R. Proctor, *J. Chem. Soc., Perkin Trans. 1*, 1979, 2013.
85. J. R. L. Smith, R. O. C. Norman, M. E. Rose and A. C. W. Curran, *J. Chem. Soc., Perkin Trans. 1*, 1979, 2863.
86. N. J. Leonard and W. J. Musliner, *J. Org. Chem.*, 1966, **31**, 639.
87. F. Sondheimer and D. Rosenthal, *J. Am. Chem. Soc.*, 1958, **80**, 3995.
88. A. Nickon, H. R. Kwasnik, C. T. Mathew, T. D. Swartz, R. O. Williams and J. B. DiGiorgio, *J. Org. Chem.*, 1978, **43**, 3904.
89. L. H. Zalkow, M. Smith, G. L. Chetty, A. W. Shaligram and P. Ingwalson, *J. Org. Chem.*, 1976, **41**, 3710.
90. S. N. Suryawanshi and U. R. Nayak, *Indian J. Chem., Sect. B*, 1979, **18**, 102.
91. J. Jakupovic, T. Teetz and F. Bohlmann, *Phytochemistry*, 1987, **26**, 1841.
92. P. S. Kalsi and K. K. Talwar, *Tetrahedron*, 1987, **43**, 2985.
93. S. N. Suryawanshi and U. R. Nayak, *Indian J. Chem., Sect. B*, 1980, **19**, 1.
94. P. K. Jadhav and U. R. Nayak, *Indian J. Chem., Sect. B*, 1978, **16**, 947.
95. T. Tsuji and E. M. Kosower, *J. Am. Chem. Soc.*, 1971, **93**, 1992.
96. B. S. Joshi, V. N. Kamat and N. Viswanathan, *Tetrahedron*, 1973, **29**, 1365.
97. P. K. Freeman, B. K. Stevenson, D. M. Balls and D. H. Jones, *J. Org. Chem.*, 1974, **39**, 546.
98. Huang-Minlon, *J. Am. Chem. Soc.*, 1948, **70**, 2802.
99. R. A. Sneen and N. P. Matheny, *J. Am. Chem. Soc.*, 1964, **86**, 5503.
100. T. Honda, T. Murae, T. Tsuyuki and T. Takahashi, *Bull. Chem. Soc. Jpn.*, 1978, **51**, 884.
101. N. J. Leonard and S. Gelfand, *J. Am. Chem. Soc.*, 1955, **77**, 3269.
102. P. S. Wharton and D. H. Bohlen, *J. Org. Chem.*, 1961, **26**, 3615.
103. F. E. Ziegler, K.-J. Hwang, J. F. Kadow, S. I. Klein, U. K. Pati and T.-F. Wang, *J. Org. Chem.*, 1986, **51**, 4573.
104. L. T. Scott and M. Oda, *Tetrahedron Lett.*, 1986, **27**, 779.
105. L. K. G. Wickremesinghe and K. N. Slessor, *Can. J. Chem.*, 1980, **58**, 2628.
106. P. S. Wharton, S. Dunny and L. S. Krebs, *J. Org. Chem.*, 1964, **29**, 958.
107. S. V. Kessar and A. L. Rampal, *Tetrahedron*, 1968, **24**, 887.
108. Y. Sonoda and Y. Sato, *Chem. Pharm. Bull.*, 1983, **31**, 907.
109. G. J. Kent, S. A. Godleski, E. Ōsawa and P. von R. Schleyer, *J. Org. Chem.*, 1977, **42**, 3852.
110. A. Mondon, H. G. Vilhuber, C. Fischer, M. Epe, B. Epe and C. Wolff, *Chem. Ber.*, 1979, **112**, 1110.
111. J. M. McIntosh, *Can. J. Chem.*, 1979, **57**, 2114.
112. H. Cousse, G. Mouzin, J. P. Rieu, A. Delhon, F. Bruniquel and F. Fauran, *Eur. J. Med. Chem.*, 1987, **22**, 45.
113. C. W. Shoppee and G. N. Henderson, *J. Chem. Soc., Perkin Trans. 1*, 1975, 765.
114. W. E. Barnette, L. L. Needham and R. W. Powell, *Tetrahedron*, 1972, **28**, 419.
115. L. Caglioti and M. Magi, *Tetrahedron*, 1963, **19**, 1127.
116. L. Caglioti and P. Grasselli, *Chem. Ind. (London)*, 1964, 153.
117. R. O. Hutchins, C. A. Milewski and B. E. Maryanoff, *J. Am. Chem. Soc.*, 1973, **95**, 3662.
118. G. W. Kabalka and J. D. Baker, Jr., *J. Org. Chem.*, 1975, **40**, 1834.
119. M. Fischer, Z. Pelah, D. H. Williams and C. Djerassi, *Chem. Ber.*, 1965, **98**, 3236.
120. L. Caglioti, *Tetrahedron*, 1966, **22**, 487.

121. (a) O. Han, Y. Shih, L. Liu and H. Liu, *J. Org. Chem.*, 1988, **53**, 2105; (b) V. P. Miller, D. Yang, T. M. Weigel, O. Han and H. Liu, *J. Org. Chem.*, 1989, **54**, 4175.
122. J. Bosch and J. Bonjoch, *J. Org. Chem.*, 1981, **46**, 1538.
123. D. R. Hwang, A. L. Feliu, A. P. Wolf, R. R. MacGregor, J. S. Fowler, C. D. Arnett, M. J. Holland, K. Carr and E. J. Simon, *J. Labelled Comp. Radiopharm.*, 1986, **23**, 277.
124. M. Haslanger, S. Zawacky and R. G. Lawton, *J. Org. Chem.*, 1976, **41**, 1807.
125. G. J. Bukovits and H. Budzikiewicz, *Org. Mass Spectrom.*, 1984, **19**, 23.
126. A. Leniewski, D. B. MacLean and J. K. Saunders, *Can. J. Chem.*, 1981, **59**, 2695.
127. P. E. Bauer, D. A. Nelson, D. S. Watt, J. H. Reibenspies, O. P. Anderson, W. K. Seifert and J. M. Moldowan, *J. Org. Chem.*, 1985, **50**, 5460.
128. F. Y. Edamura and A. Nickon, *J. Org. Chem.*, 1970, **35**, 1509.
129. R. H. Shapiro, *Tetrahedron Lett.*, 1966, 3401.
130. M. Biesemans, G. V. de Woude and L. von Hove, *Bull. Soc. Chim. Belg.*, 1983, **92**, 151.
131. A. M. Caporusso, G. Giacomelli and L. Lardicci, *J. Chem. Soc., Perkin Trans. 1*, 1979, 3139.
132. D. Tavernier, N. Hosten and M. Anteunis, *Synthesis*, 1979, 613.
133. J. B. Lambert, *J. Am. Chem. Soc.*, 1967, **89**, 1836.
134. W. T. Hoeve and H. Wynberg, *J. Org. Chem.*, 1980, **45**, 2925.
135. Y. Masaki, K. Hashimoto, K. Sakuma and K. Kaji, *Bull. Chem. Soc. Jpn.*, 1984, **57**, 3466.
136. R. Gleiter, W. Sander and I. Butler-Ransohoff, *Helv. Chim. Acta*, 1986, **69**, 1872.
137. A. G. Martínez, R. M. Alvarez, M. M. Casado, L. R. Subramanian and M. Hanack, *Tetrahedron*, 1987, **43**, 275.
138. D. R. Mootoo and B. Fraser-Reid, *Carbohydr. Res.*, 1988, **174**, 99.
139. R. O. Hutchins and N. R. Natale, *J. Org. Chem.*, 1978, **43**, 2299.
140. G. W. Gribble, P. D. Lord, J. Skotnicki, S. E. Dietz, J. T. Eaton and J. L. Johnson, *J. Am. Chem. Soc.*, 1974, **96**, 7812.
141. F. Arcamone, G. Cassinelli, F. DiMatteo, S. Forenza, M. C. Ripamonti, G. Rivola, A. Vigevani, J. Clardy and T. McCabe, *J. Am. Chem. Soc.*, 1980, **102**, 1462.
142. T. Iida and F. C. Chang, *J. Org. Chem.*, 1982, **47**, 2966.
143. T. Iida, T. Tamura, T. Matsumoto and F. C. Chang, *Synthesis*, 1984, 957.
144. G. W. J. Fleet, P. J. C. Harding and M. J. Whitcombe, *Tetrahedron Lett.*, 1980, **21**, 4031.
145. R. Grandi, A. Marchesini, U. M. Pagnoni and R. Trave, *J. Org. Chem.*, 1976, **41**, 1755.
146. R. Grandi, W. Messerotti, U. M. Pagnoni and R. Trave, *J. Org. Chem.*, 1977, **42**, 1352.
147. T. Mimura and T. Nakai, *Chem. Lett.*, 1981, 1579.
148. M. Srebnik, N. Lander, A. Breuer and R. Mechoulam, *J. Chem. Soc., Perkin Trans. 1*, 1984, 2881.
149. R. O. Hutchins and N. R. Natale, *Org. Prep. Proced. Int.*, 1979, **11**, 201.
150. R. O. Hutchins, B. E. Maryanoff and C. A. Milewski, *J. Am. Chem. Soc.*, 1971, **93**, 1793.
151. R. F. Borch, M. D. Bernstein and H. D. Durst, *J. Am. Chem. Soc.*, 1971, **93**, 2897.
152. Y. Naoshima, E. Nakagawa, S. Wakabayashi and S. Hayashi, *Agric. Biol. Chem.*, 1980, **44**, 1419.
153. F. Bohlmann and M. Lonitz, *Chem. Ber.*, 1980, **113**, 2410.
154. S.-F. Lee, M. Edgar, C. S. Pak, G. Barth and C. Djerassi, *J. Am. Chem. Soc.*, 1980, **102**, 4784.
155. J. R. Mahajan and H. C. de Araujo, *Synthesis*, 1980, 64.
156. W. S. Johnson, A. B. Shenvi and S. G. Boots, *Tetrahedron*, 1982, **38**, 1397.
157. A. Sato, T. Hirata and N. Nakamizo, *Agric. Biol. Chem.*, 1983, **47**, 799.
158. E. R. Koft and A. B. Smith, III, *J. Am. Chem. Soc.*, 1984, **106**, 2115.
159. Y. Yamamoto and K. Maruyama, *J. Chem. Soc., Chem. Commun.*, 1984, 904.
160. R. N. A. H. Lewis and R. N. McElhaney, *Biochemistry*, 1985, **24**, 2431.
161. K. Avasthi and R. G. Salomon, *J. Org. Chem.*, 1986, **51**, 2556.
162. G. Adam, R. Zibuck and D. Seebach, *J. Am. Chem. Soc.*, 1987, **109**, 6176.
163. P. W. Westerman and N. Ghrayeb, *Chem. Phys. Lipids*, 1982, **30**, 381.
164. G. Schenk, H. P. Albrecht and H. Lietz, *Arzneim.-Forsch.*, 1978, **28**, 518.
165. L. Anzalone and J. A. Hirsch, *J. Org. Chem.*, 1985, **50**, 2607.
166. J. A. Hirsch and V. C. Truc, *J. Org. Chem.*, 1986, **51**, 2218.
167. V. Nair and A. K. Sinhababu, *J. Org. Chem.*, 1978, **43**, 5013.
168. J. R. Mahajan and H. C. de Araujo, *Synthesis*, 1981, 49.
169. O. E. Edwards and Z. Paryzek, *Can. J. Chem.*, 1983, **61**, 1973.
170. W. Tochtermann, A. Pahl, E.-M. Peters, K. Peters and H. G. von Schnering, *Chem. Ber.*, 1988, **121**, 493.
171. T. H. Smith, A. N. Fujiwara and D. W. Henry, *J. Med. Chem.*, 1978, **21**, 280.
172. J. G. Henkel and W. C. Faith, *J. Org. Chem.*, 1981, **46**, 4953.
173. G. Rosini, A. Medici and M. Soverini, *Synthesis*, 1979, 789.
174. G. R. Krow, D. A. Shaw, B. Lynch, W. Lester, S. W. Szczepanski, R. Raghavachari and A. E. Derome, *J. Org. Chem.*, 1988, **53**, 2258.
175. Y. Masaki, K. Sakuma, K. Hashimoto and K. Kaji, *Chem. Lett.*, 1981, 1283.
176. S. Kim, C. H. Oh, J. S. Ko, K. H. Ahn and Y. J. Kim, *J. Org. Chem.*, 1985, **50**, 1927.
177. R. C. Bansal, C. E. Browne, E. J. Eisenbraun and J. S. Thomson, *J. Org. Chem.*, 1988, **53**, 452.
178. G. Rosini and A. Medici, *Synthesis*, 1976, 530.
179. R. O. Hutchins, M. Kacher and L. Rua, *J. Org. Chem.*, 1975, **40**, 923.
180. E. J. Taylor and C. Djerassi, *J. Am. Chem. Soc.*, 1976, **98**, 2275.
181. D. G. Batt, N. Takamura and B. Ganem, *J. Am. Chem. Soc.*, 1984, **106**, 3353.
182. R. L. Danheiser, D. J. Carini, D. M. Fink and A. Basak, *Tetrahedron*, 1983, **39**, 935.
183. G. W. Kabalka and J. H. Chandler, *Synth. Commun.*, 1979, **9**, 275.
184. G. W. Kabalka, *Org. Prep. Proced. Int.*, 1977, **9**, 131.
185. R. Noyori, M. Nishizawa, F. Shimizu, Y. Hayakawa, K. Maruoka, S. Hashimoto, H. Yamamoto and H. Nozaki, *J. Am. Chem. Soc.*, 1979, **101**, 220.

186. J. M. Hornback and R. D. Barrows, *J. Org. Chem.*, 1983, **48**, 90.
187. G. W. Kabalka, D. T. C. Yang, J. H. Chandler and J. D. Baker, Jr., *Synthesis*, 1977, 124.
188. J. D. Burkey, P. A. Leber and L. S. Silverman, *Synth. Commun.*, 1986, **16**, 1363.
189. G. W. Kabalka and S. T. Summers, *J. Org. Chem.*, 1981, **46**, 1217.
190. Z. Majerski, M. Žuanić and B. Metelko, *J. Am. Chem. Soc.*, 1985, **107**, 1721.
191. Y. Kikugawa and M. Kawase, *Synth. Commun.*, 1979, **9**, 49.
192. G. W. Kabalka, J. D. Baker, Jr. and G. W. Neal, *J. Org. Chem.*, 1977, **42**, 512.
193. G. W. Kabalka, D. T. C. Yang and J. D. Baker, Jr., *J. Org. Chem.*, 1976, **41**, 574.
194. D. T. C. Yang and G. W. Kabalka, *Org. Prep. Proced. Int.*, 1977, **9**, 85.
195. G. W. Kabalka, R. O. Hutchins, N. R. Natale, D. T. C. Yang and V. Broach, *Org. Synth.*, 1979, **59**, 42.
196. H. E. Zimmerman and R. D. Solomon, *J. Am. Chem. Soc.*, 1986, **108**, 6276.
197. S. Penco, F. Angelucci, M. Ballabio, G. Barchielli, A. Suarato, E. Vanotti, A. Vigevani and F. Arcamone, *Tetrahedron*, 1984, **40**, 4677.
198. C. H. Kuo, A. A. Patchett and N. L. Wendler, *J. Org. Chem.*, 1983, **48**, 1991.
199. F. Naf, R. Decorzant and W. Thommen, *Helv. Chim. Acta*, 1982, **65**, 2212.
200. A. E. Greene, *J. Am. Chem. Soc.*, 1980, **102**, 5337.
201. A. E. Greene, *Tetrahedron Lett.*, 1979, **20**, 63.
202. D. F. Taber and J. M. Anthony, *Tetrahedron Lett.*, 1980, **21**, 2779.
203. G. W. Kabalka, R. J. Newton, Jr. and J. H. Chandler, *J. Chem. Soc., Chem. Commun.*, 1978, 726.
204. R. O. Hutchins, W.-Y. Su, R. Sivakumar, F. Cistone and Y. P. Stercho, *J. Org. Chem.*, 1983, **48**, 3412.

2.1
Reduction of Nitro and Nitroso Compounds

GEORGE W. KABALKA
University of Tennessee, Knoxville, TN, USA
and
RAJENDER S. VARMA
Baylor College of Medicine, The Woodlands, TX, USA

2.1.1 INTRODUCTION

Nitro compounds are versatile synthetic intermediates which have found widespread utility in industrial applications. Aromatic nitro compounds are the usual starting materials for commercial applications, but aliphatic compounds exhibit a greater diversity of chemical behavior under reducing conditions.[1–4] Nitroso compounds, hydroxylamines, oximes, amines, nitrones, ketones and silyl nitronates are frequently encountered during the reduction of nitro compounds. Several specialized reviews have appeared which highlight the versatility of the nitro group in organic chemistry.[1–8]

2.1.2 REDUCTION OF AROMATIC NITRO COMPOUNDS TO NITROSO COMPOUNDS

The reduction of nitro compounds should initially produce nitroso compounds. This area has not been systematically explored because the nitroso group can be more easily introduced by alternative methods such as direct nitrosation, condensation and oxidative procedures.[9–11] In fact, there have been few instances in which nitroso compounds have been isolated as intermediates in reductions of nitro compounds. For example, it was initially believed that *m*-trifluoromethylnitrobenzene produced the corresponding nitroso compound upon reduction,[12] but subsequently the product was shown to be *m*-trifluoromethylazoxybenzene.[13] Low yields of an intramolecular dimeric, nitroso compound, benzo[*c*]cinnoline dioxide (**1**), can be obtained by reducing 2,2′-dinitrobiphenyl with zinc or sodium sulfide (equation 1).[14,15]

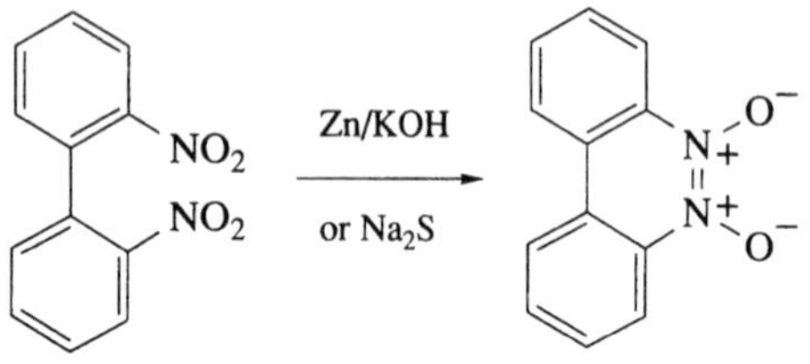

NO_2 NO_2 Zn/KOH or Na_2S N^+ O^- N^+ O^- (1)

(1)

2.1.3 REDUCTION OF AROMATIC NITRO AND NITROSO COMPOUNDS TO AZO AND AZOXY COMPOUNDS

Reductions of aromatic nitro compounds often proceed to generate mixtures of nitroso and hydroxylamine products which then condense to form azoxy and, eventually, azo compounds. This bimolecular reduction is practical only for the generation of symmetrically substituted azo compounds. The situation can be further complicated if the reduction continues such that aromatic amines are formed; the amines may then condense with the intermediate nitroso compounds to generate hydrazo compounds which can then undergo a benzidine rearrangement.

A variety of reducing agents have been used to reduce nitroarenes to azo compounds. However, a mixture of zinc and sodium hydroxide is used most frequently.[16,17] Reduction under these conditions produces hydrazo compounds, which are then oxidized to azo compounds by dissolved atmospheric oxygen;[18,19] alternatively, air can be drawn through the product solution to achieve the conversion.[20,21] Occasionally, activation of the zinc dust is required prior to its use,[22] and, since metal ions form chelated complexes with azo compounds, a vigorous post-treatment with acids is recommended.[23] The zinc–sodium hydroxide reduction conditions are sufficiently mild that *p*-nitrostyrene (**2**; equation 2), can be reduced without reduction of the vinyl groups.[24]

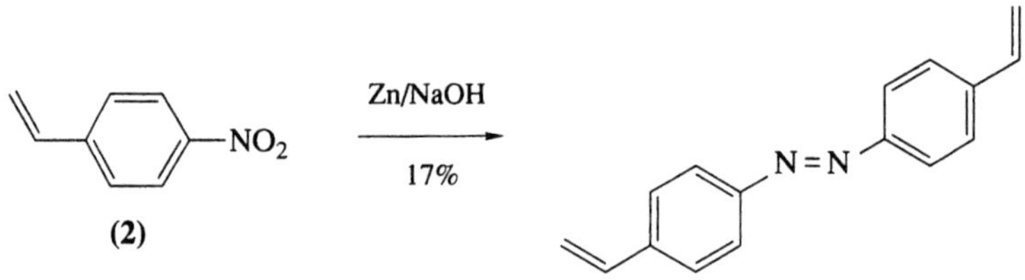

NO_2 Zn/NaOH 17% N=N (2)

(2)

The reduction of (**3**) is an interesting example involving the formation of an intramolecular azo bridge (equation 3).[25]

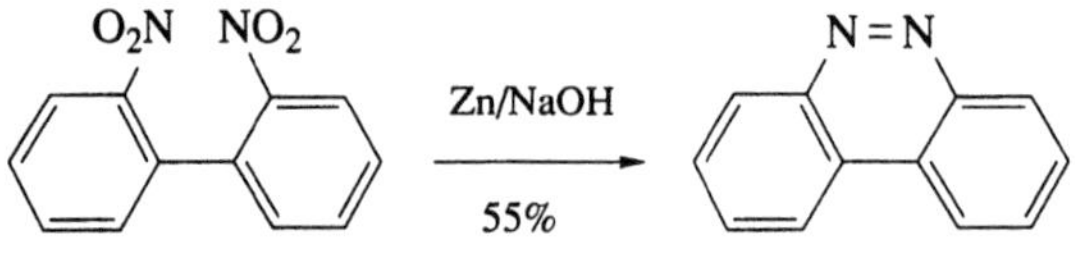

O_2N NO_2 Zn/NaOH 55% N=N (3)

(3) **(4)**

As noted, the bimolecular reduction of aromatic nitro compounds may produce azoxy compounds, azo compounds, hydrazo compounds (1,2-diarylhydrazines), benzidines or amines (Scheme 1) depending on the reaction conditions. Zinc reduction under basic conditions generates azo compounds, whereas the use of acetic anhydride/acetic acid as the solvent system affords symmetrical azoxy compounds.[26] Although unsymmetrical azoxy compounds are accessible in the aliphatic series, aromatic reagents yield only symmetrical products. Azoxy compounds are of current interest because of their physiological activities and their utilization in liquid crystal systems.[27] The chemistry of aromatic azoxy compounds has been reviewed.[28,29]

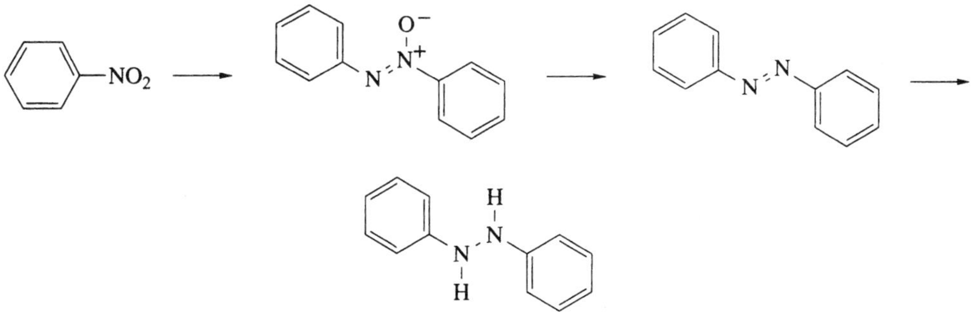

Scheme 1

The preparation of *o,o′*-dicyanoazoxybenzene (**6**) is representative of the reduction of aromatic nitro compounds using zinc under acidic conditions (equation 4).[26] The reduction of aromatic nitro compounds to azoxy derivatives has also been accomplished with tin(II) chloride in a basic medium.[30]

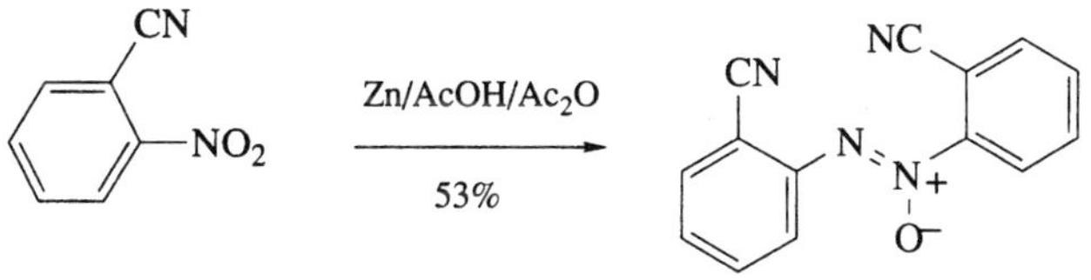

Zn/AcOH/Ac_2O, 53% (4)

(**5**) (**6**)

Magnesium is another metallic reducing agent which produces a mixture of azoxy and azo compounds (~75%).[31] An alloy of sodium and lead (Drynap[32]) results in a similar mixture of products. A simple procedure involving the reaction of an aromatic nitro compound with thallium metal affords good yields of the corresponding azoxy compounds (equation 5).[33] Although high yields of azoxy compounds are obtainable from nitroarenes with alkyl or ether substituents, numerous other functionalities inhibit the reaction; these include CHO, COR, CO_2H, CO_2R, CN, OH and amino groups.

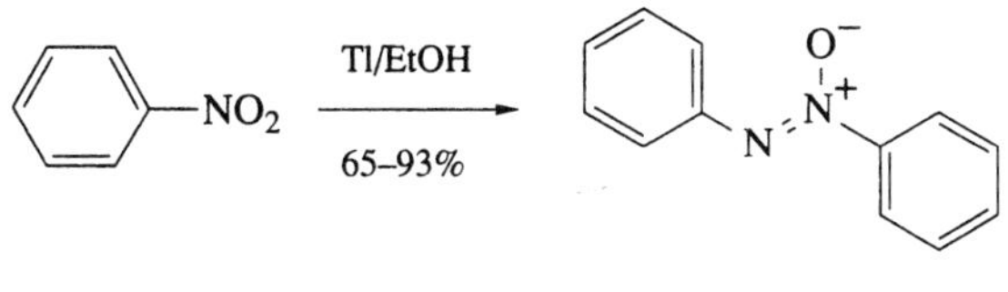

Tl/EtOH, 65–93% (5)

Alcoholic potassium hydroxide and sodium alcoholates have also been used as reducing agents for aromatic nitro compounds; the free alcohol or the alcoholate ion presumably acts as the reducing agent in these cases as illustrated for *m,m′*-diiodoazoxybenzene (**8**; equation 6).[17] In a few instances, where benzyl alcohol was used to prepare sodium benzylate, pure azoxy products were obtained without the formation of amino by-products.

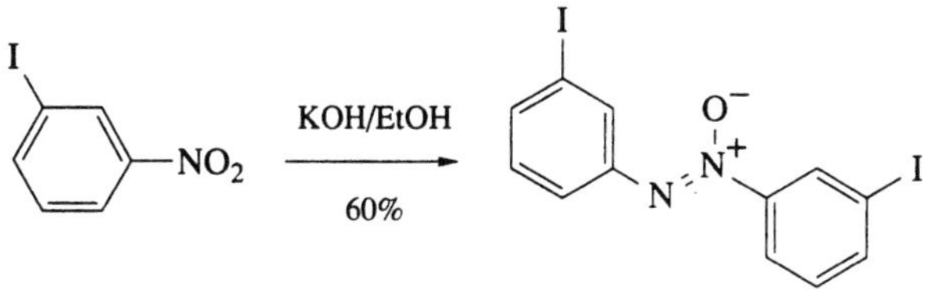

KOH/EtOH, 60% (6)

(**7**) (**8**)

An interesting reduction of aromatic nitro compounds which uses glucose in an alkaline medium (equation 7) has received little attention. The advantages of this reaction include high yields, rapid rate and ease of product isolation from oxidation by-products.[34] Other reagents which bring about the reduction of nitroarenes to azoxy compounds include potassium borohydride,[35] sodium arsenate,[36] phosphine[37] and yellow phosphorus.[38] Electrolytic methods have also been utilized.[39,40]

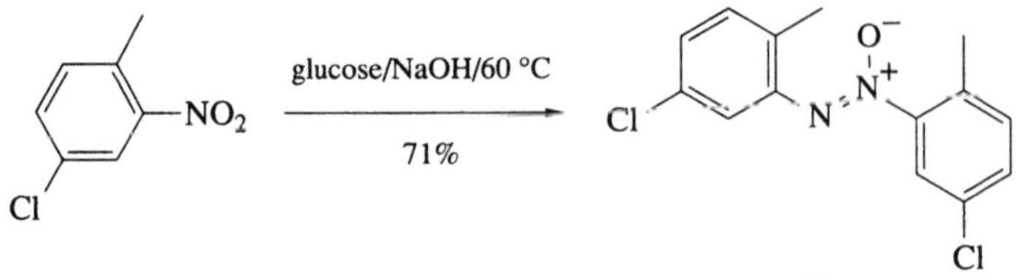

glucose/NaOH/60 °C, 71% (7)

(9)

2.1.4 REDUCTION OF AROMATIC NITRO AND NITROSO COMPOUNDS TO HYDROXYLAMINES

Reduction of nitro compounds proceeds through intermediate stages involving nitroso compounds and hydroxylamines prior to amine formation (equation 8).

$$RNO_2 \longrightarrow RNO \longrightarrow RNHOH \longrightarrow RNH_2 \qquad (8)$$

Selective reduction to hydroxylamine can be achieved in a variety of ways; the most widely applicable systems utilize zinc and ammonium chloride in an aqueous or alcoholic medium.[41] The overreduction to amines can be prevented by using a two-phase solvent system. Hydroxylamines have also been obtained from nitro compounds using molecular hydrogen and iridium catalysts.[42] A rapid metal-catalyzed transfer reduction of aromatic nitroarenes to *N*-substituted hydroxylamines has also been developed; the method employs palladium and rhodium on charcoal as catalyst and a variety of hydrogen donors such as cyclohexene, hydrazine, formic acid and phosphinic acid.[43] The reduction of nitroarenes to arylhydroxylamines can also be achieved using hydrazine in the presence of Raney nickel[44] or iron(III) oxide.[45]

The preparation of *N*-phenylhydroxylamine in high yields from nitrobenzene under catalytic transfer hydrogenation conditions is also possible utilizing wet 5% rhodium on carbon and hydrazine hydrate.[46] Unfortunately, the transition metal catalysts tend to be expensive and the high temperatures required can be detrimental, particularly when the resulting hydroxylamines are explosive in nature.[47]

Tellurides (H_2Te, NaTeH, PhTeH and Na_2Te) are inexpensive and effective agents for reducing aromatic nitro compounds to hydroxylamines. Catalytic quantities of tellurium, in the presence of sodium borohydride, reduce *p*-substituted nitrobenzenes to *N*-arylhydroxylamines (equation 9).[48] Mild reaction conditions, absence of side reactions and experimental simplicity are the main features of this reduction sequence.

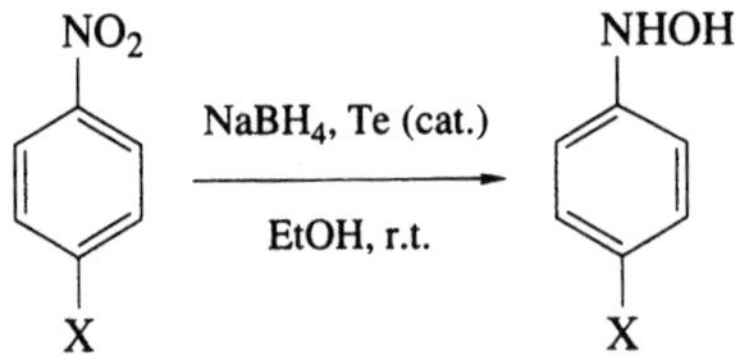

$NaBH_4$, Te (cat.), EtOH, r.t. (9)

X = H, Me, Cl, CN, NO_2, CO_2Et

Selenium also catalyzes the sodium borohydride reduction of nitroarenes to the corresponding *N*-arylhydroxylamines (equation 10).[49] The active species in this selenium-catalyzed reduction was confirmed to be the hydrogen selenide anion.[50] The reaction is accelerated by electron-withdrawing groups and hindered by electron-donating substituents; nitro compounds bearing strongly electron-donating groups are completely inert under the normal reaction conditions.

The iron sulfide complex (**10**) resembles the active sites of oxidized rubredoxins: nonheme iron–sulfur proteins.[51] The complex catalyzes the reductions of aromatic nitro compounds to *N*-arylhydroxylamines by thiol (equation 11).[52] The method offers a facile, high yield approach to *N*-arylhydroxylamines. For example, *p*-dinitrobenzene was reduced to *p*-nitrophenylhydroxylamine in 92% yield.

$$\text{4-X-C}_6\text{H}_4\text{NO}_2 \xrightarrow[\text{37–88\%}]{\text{NaBH}_4\text{/Se, EtOH, r.t.}} \text{4-X-C}_6\text{H}_4\text{NHOH} \tag{10}$$

$X = H, Me, Cl, CN, NO_2, CO_2Et$

$[Fe(S_2\text{-}o\text{-xylyl})_2]^- \, [Et_4N]^+$

(10)

$$\text{X-C}_6\text{H}_4\text{NO}_2 \xrightarrow[\text{o-C}_6\text{H}_4(\text{CH}_2\text{SH})_2]{\textbf{(10)}} \text{X-C}_6\text{H}_4\text{NHOH} \tag{11}$$

$X = NO_2, CN, CO_2Me$

2.1.5 REDUCTION OF AROMATIC NITRO AND NITROSO COMPOUNDS TO ANILINES

The synthetic, and industrially important, reduction of aromatic nitro compounds to anilines has been extensively investigated.[53–56]

2.1.5.1 Transfer Hydrogenation

The general area of catalytic transfer hydrogenation, as applied to nitroarenes, has also been reviewed.[57] The reduction of nitroarenes to aminoarenes using hydrazine as the hydrogen donor takes place readily in the presence of a wide variety of catalysts, including Cu, Fe, Ni, Rh, Ru and Pd.[58,59] It has been observed that controlling the reduction rates is difficult with active catalysts such as Pd but the utilization of less active catalysts results in better control, although substitution and dehalogenation reactions are quite common. In the presence of an excess of hydrazine hydrate, a number of nitrobenzenes are readily reduced to amines in high yields using iron(III) chloride on active carbon,[60] including the reduction of 5-chloro-2,4-dimethoxynitrobenzene without concomitant loss of the chlorine. The procedure is equally effective for the partial reduction of dinitrobenzene derivatives to the corresponding aminonitroarenes.[61] The superiority of catalytic transfer hydrogenation for the reduction of poly- and di-nitrobenzenes is noteworthy. Since the reduction of aminonitroarenes is considerably slower than their rate of formation, selective reduction to the half-reduced stage is the method of choice for the preparation of aminonitroarenes.

The reductive cyclization of the 2-nitrodihydrocinnamoyl group has found application for the protection of alcohols and amines using sodium phosphinate or cyclohexene as the hydrogen donor.[62] The hydrogen transfer reduction of both esters and amides regenerates the alcohols or amines under mild conditions (equation 12).

$$\underset{\textbf{(11)}}{o\text{-O}_2\text{NC}_6\text{H}_4\text{CH}_2\text{CH}_2\text{CO}_2\text{R}} \xrightarrow[\text{80–95\%}]{\text{10\% Pd/C, H-donor}} o\text{-H}_2\text{NC}_6\text{H}_4\text{CH}_2\text{CH}_2\text{C(O)OR} \xrightarrow{\text{H}^+} \underset{\textbf{(12)}}{\text{3,4-dihydroquinolin-2(1H)-one}} + \text{ROH} \tag{12}$$

The search for more active donors for transfer hydrogenation has led to the identification of economical and low cost reagents such as formic, phosphinic (hypophosphorous) and phosphorous acids.[63] Although reductions with formic acid are facile, anions of strong acids (such as Cl^-) quickly terminate the reduction. Thus nitroarenes containing a halogen substituent other than fluorine cannot be reduced using formic acid because of the catalyst-poisoning effect of the halide generated *in situ*. Arenes containing heterocyclic sulfur atoms are also unreactive. The use of formate salts[64] or $Pd/AlPO_4/SiO_2$ catalysts[65] obviates the difficulties encountered with termination of reductions by halogens.

The selective and rapid reduction of nitro compounds is an active area of research, particularly when other potentially reducible moieties are present in the molecules. Anhydrous ammonium formate has been developed as a catalytic hydrogen transfer agent for selectively reducing nitro groups in the presence of acid, ester, amide, halogen and nitrile groups (equation 13).[66]

$$Ar-NO_2 \xrightarrow{HCO_2NH_4,\ 10\%\ Pd/C,\ r.t.} Ar-NH_2 \qquad (13)$$

The selective reduction of nitroarenes containing benzyl-protected phenolic groups without concomitant hydrogenolysis has also been achieved using hydrazine hydrate and Raney nickel.[67] This procedure avoids strongly acidic conditions. The reaction is selective and a variety of *N*-benzyl- and chloro-substituted nitroarenes are reduced to the corresponding anilines without dehalogenation or debenzylation.

A detailed investigation of selectivity in the catalytic reduction of (nitroaryl)alkylnitriles with hydrazine and metal catalysis has been carried out.[68] On a small scale, selective reduction of (nitroaryl)alkylnitrile (**13**) to either (aminoaryl)alkylnitrile (**14**) or (aminoaryl)alkylamine (**15**) is readily achieved using a catalytic quantity of Raney nickel with hydrazine hydrate (2–5 equiv.) if adequate control of temperature is maintained. A lower temperature range (20–25 °C) affords (**14**), whereas warming (50–55 °C) with excess hydrazine hydrate (4–5 equiv.) produces (**15**) in nearly quantitative yields (Scheme 2). The advantages of the procedure include mild reaction conditions, the absence of high pressure equipment and the possibility of a simple scale-up to multikilogram quantities. Furthermore, aromatic rings are not reduced, a reaction which frequently occurs when pyridine derivatives are reduced.

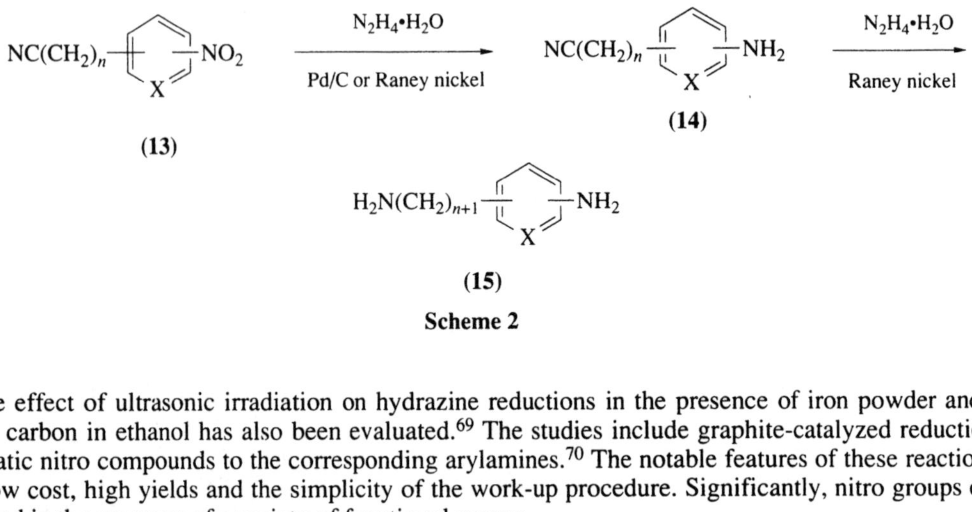

$NC(CH_2)_n$–C$_5$H$_3$X–NO_2 (**13**) $\xrightarrow[\text{Pd/C or Raney nickel}]{N_2H_4\bullet H_2O}$ $NC(CH_2)_n$–C$_5$H$_3$X–NH_2 (**14**) $\xrightarrow[\text{Raney nickel}]{N_2H_4\bullet H_2O}$ $H_2N(CH_2)_{n+1}$–C$_5$H$_3$X–NH_2 (**15**)

Scheme 2

The effect of ultrasonic irradiation on hydrazine reductions in the presence of iron powder and activated carbon in ethanol has also been evaluated.[69] The studies include graphite-catalyzed reductions of aromatic nitro compounds to the corresponding arylamines.[70] The notable features of these reactions are the low cost, high yields and the simplicity of the work-up procedure. Significantly, nitro groups can be reduced in the presence of a variety of functional groups.

Raney nickel and hydrazine hydrate have also been used to prepare 4-(benzyloxy)indoles (**16**) *via* reductive cyclization.[71] An improvement can be achieved by using nickel boride instead of Raney nickel;[72] the advantages of this method include ease of preparation of the catalyst and its nonpyrophoric nature (equation 14).

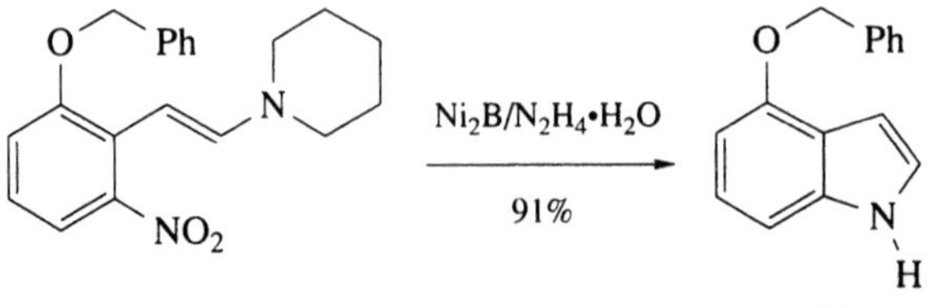

$$\xrightarrow[91\%]{Ni_2B/N_2H_4\bullet H_2O} \qquad (14)$$

(**16**)

The observation that formic acid can be decomposed to hydrogen and carbon dioxide at room temperature using $[RuCl_2(PPh_3)_3]$ in the presence of triethylamine led to the development of the $[RuCl_2(PPh_3)_3]/HCO_2H/Et_3N$ system as a hydrogen source. Pd/C increases the reactivity of this system, which is selective in reducing nitro groups in the presence of alkenes, esters and cyano groups.[73] Surprisingly, the hydrogenolysis of aromatic halides is slower than the reduction of the nitro group.

2.1.5.2 Borohydride and Borane Reductions

Sodium borohydride is an effective and widely used reducing agent. The relatively low cost and the ease with which it can be handled contribute to its popularity. However, borohydride does not generally reduce aromatic nitro compounds to the corresponding aniline derivatives in the absence of a catalyst[74] such as cobalt(II) chloride,[75] palladium,[76] titanium(IV) chloride[77] or tin(II) chloride.[78] Relatively inexpensive systems such as $NaBH_4$–iron(II) chloride[79] and $NaBH_4$–copper(I) chloride[80] also reduce nitroarenes containing electron-donating groups in the *ortho* or *para* positions.

Sodium borohydride reduces disulfides to thiols, which can then be used to reduce nitro groups. Based on the redox properties of 1,2-dithiolane, lipoamide (**17**) was used for the selective reduction of monosubstituted nitrobenzenes to the corresponding anilines.[81] Lipoamide (**17**) can also be immobilized on hydrophilic polymers such as polyvinylamine, polyethyleneimine and chitosan.[82] These polymeric reducing catalysts can be recycled and are easy to separate from the reaction mixture. The system has been used to reduce nitroarenes to anilines.[83]

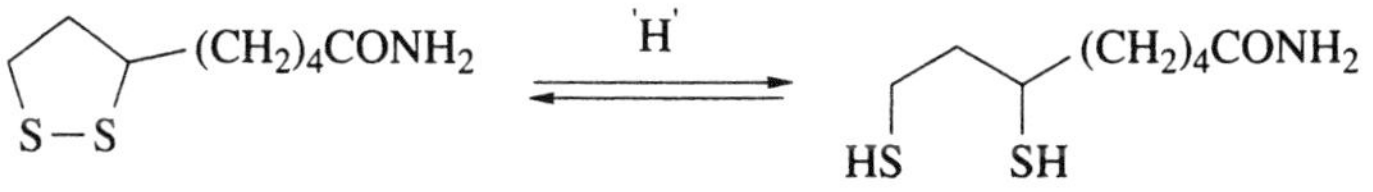

(CH$_2$)$_4$CONH$_2$ / S–S ⇌ 'H' ⇌ (CH$_2$)$_4$CONH$_2$ / HS SH (15)

(**17**) LAm (**18**) DHLAm

Sodium borohydride can also be used to couple dinitrobenzenes.[84] 1,3-Dinitrobenzene (**19**) and 1,3,5-trinitrobenzene (**21**) produce diphenylamine derivatives in reactions that are of preparative value when the nitro compound is present in excess (Scheme 3). The method involves the addition of a suspension of

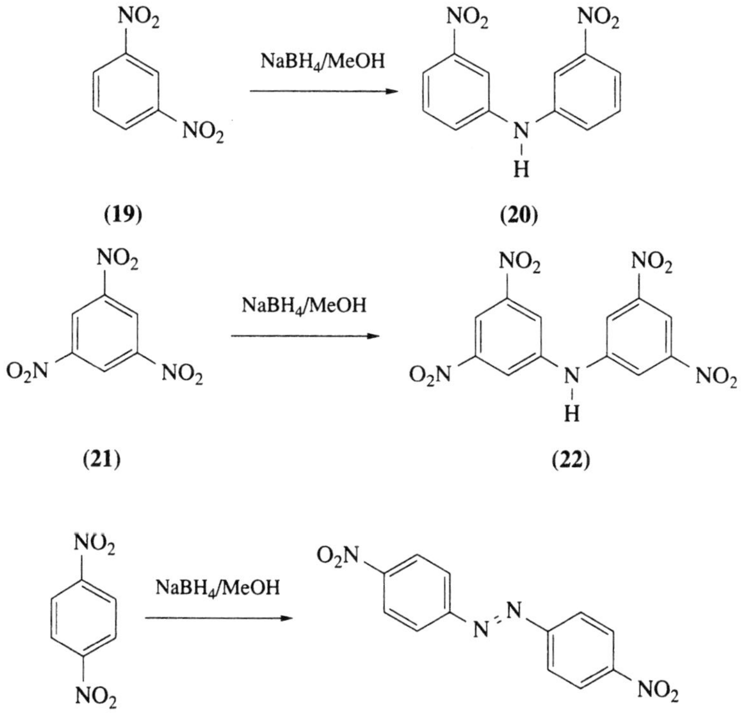

NO$_2$, NO$_2$ (**19**) —$NaBH_4$/MeOH→ NO$_2$, NO$_2$, N–H (**20**)

NO$_2$, O$_2$N, NO$_2$ (**21**) —$NaBH_4$/MeOH→ NO$_2$, NO$_2$, O$_2$N, N–H, NO$_2$ (**22**)

NO$_2$, NO$_2$ (**23**) —$NaBH_4$/MeOH→ O$_2$N, N=N, NO$_2$ (**24**)

Scheme 3

sodium borohydride in methanolic sodium hydroxide to a solution of the nitro compound in refluxing methanol. Under these conditions, no reduction to dinitrocyclohexenes is observed.[85,86] Coupling products are not obtained from 1,2,4-trinitrobenzene, even though two of the nitro groups have a *meta* relationship. 1,4-Dinitrobenzene (**23**) undergoes reductive coupling to yield 4,4′-dinitroazobenzene (**24**).

Curiously, 1,2-dinitrobenzene (**25**) couples with the loss of one nitro group; however, the product is a diphenylhydroxylamine derivative (**26**) rather than a diphenylamine (equation 16). Nitroso compounds can also be reduced to amines in good yields using diborane (equation 17).[87]

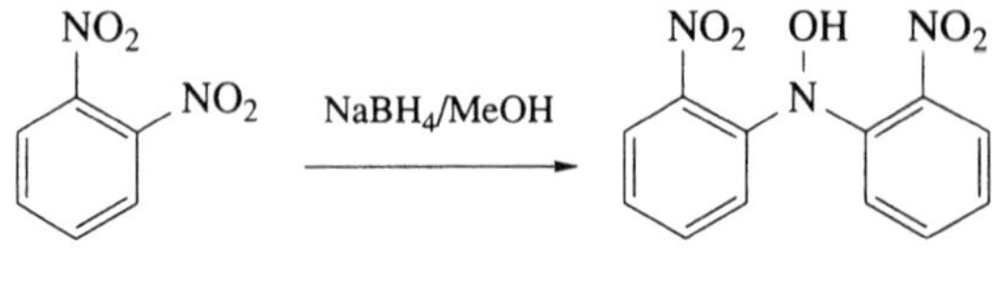

NaBH$_4$/MeOH; (**25**) → (**26**) (16)

BH$_3$•THF, 25 °C; 62–90% (17)

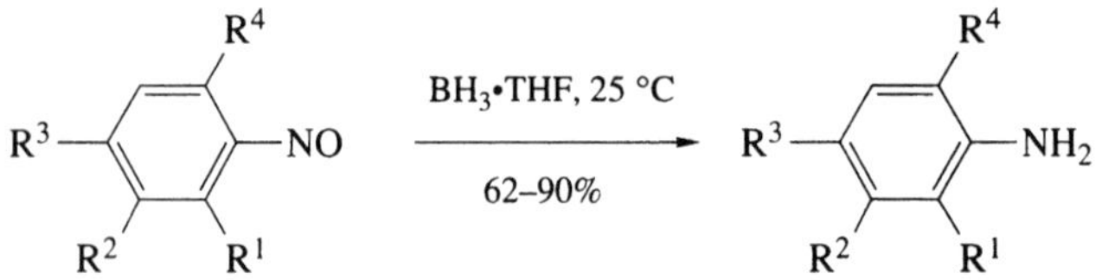

2.1.5.3 Reductions Involving Tellurium, Selenium and Sulfur

Selenol,[88] a very soft nucleophile, reduces nitroarenes.[89] Hydrogen telluride (H_2Te),[90] sodium hydrogen telluride (NaTeH)[91] and related organotellurols are also effective reducing agents. The ready autooxidation of organotellurols to ditellurides prevents their isolation; consequently these reagents are generated and used *in situ*.[92] Benzenetellurol (**27**) is prepared[93] by methanolysis of phenyl trimethylsilyl telluride or by reduction of diphenyl ditelluride with phosphinic acid or sodium borohydride and used to reduce aromatic nitro compounds to amines (Scheme 4).

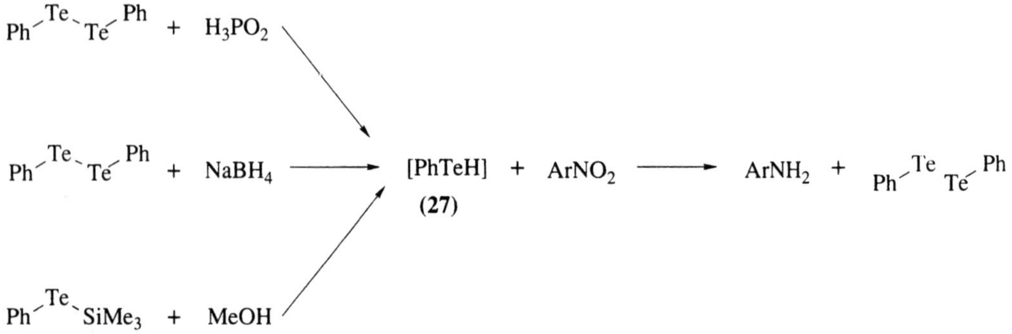

PhTeTePh + H_3PO_2 → [PhTeH] (**27**)
PhTeTePh + $NaBH_4$ → [PhTeH] (**27**)
PhTeSiMe$_3$ + MeOH → [PhTeH] (**27**)
[PhTeH] + $ArNO_2$ → $ArNH_2$ + PhTeTePh

Scheme 4

Heating tellurium with an excess of Rongalite ($HOCH_2SO_2Na{\cdot}2H_2O$) in aqueous sodium hydroxide under an inert atmosphere affords sodium telluride (Na_2Te),[94] which reduces nitroarenes to anilines. The reductions are carried out using a catalytic amount of tellurium. The tellurium-mediated reduction of nitroarenes with Rongalite proceeds directly to anilines, and is the method of choice for compounds containing acid sensitive functionalities.

Interest in an old reducing agent, sodium sulfide,[95] was renewed when it was discovered that its reactivity toward nitro groups was sensitive to water.[96] In the absence of water, the aromatic nitro group in (**28**) can be reduced selectively to (**29**), without reduction of the aliphatic nitro group, but compounds such as (**30**), which contain a tertiary aliphatic nitro group eliminate the nitro moiety to yield styrene derivatives (**31**; Scheme 5). A unique application of the sulfide reduction involves the preparation of isomerically pure substituted aromatic nitro compounds and anilines.[97]

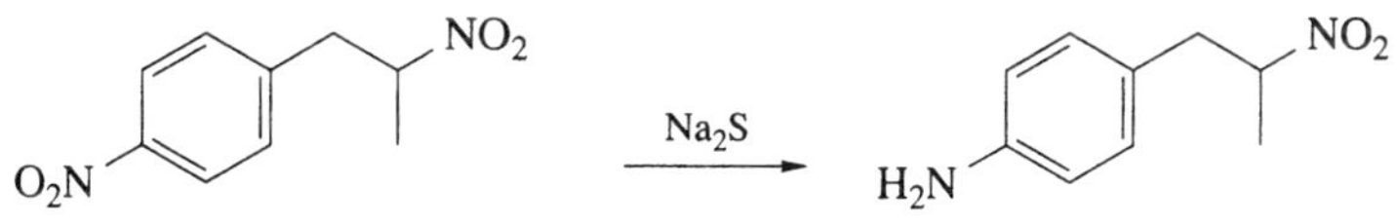

Na_2S

(28) → **(29)**

Na_2S

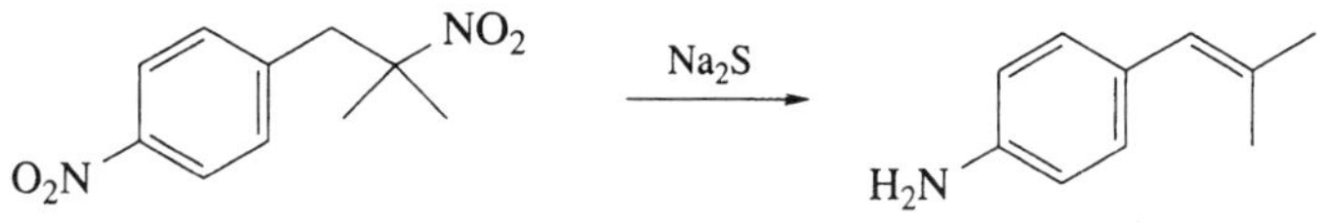

(30) → **(31)**

Scheme 5

2.1.5.4 Transition Metal Reductions

Low valent species of transition metals (Groups IVB, VB and VIB) have found widespread use in the reduction of nitro compounds.[98] Chromium(II) chloride,[99] titanium(III) chloride[100] and vanadium(II) sulfate[101] readily reduce aromatic nitro compounds to arylamines. Chromium(II) chloride[102] also reduces nitrobenzene and nitroquinoline to the corresponding amines in excellent yields.[103]

Oxime deoxygenation[104] by titanium(III) chloride spurred investigation of this agent for reducing nitro aliphatics;[105] the method is applicable to other nitro compounds.[106,107] The active species are thought to be $ArNO_2H^+$ and a hydrolyzed form of titanium(II).[108] The reaction conditions have been optimized for the reduction of aromatic and even heteroaromatic nitro compounds.[109]

A combination of titanium(IV) chloride and dialkyl telluride in relatively inert solvent systems reduces nitroarenes to arylamines in moderate yields.[110] Presumably, titanium(IV) chloride is reduced to a low valent titanium species which subsequently reacts with aromatic nitro compounds to afford the corresponding amines (equation 18).

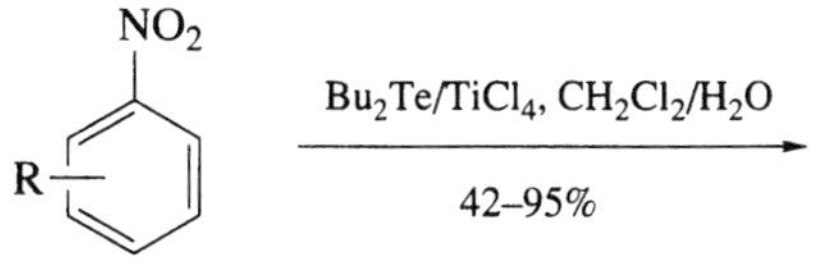

$Bu_2Te/TiCl_4$, CH_2Cl_2/H_2O; 42–95% (18)

Tin(II) chloride is an effective reducing agent under acidic conditions.[111] This inexpensive reagent selectively reduces aromatic nitro compounds in nonacidic and nonaqueous media.[112] Nearly quantitative yields of arylamines are obtained using tin(II) chloride dihydrate in alcohol or ethyl acetate (equation 19). Under these conditions other reducible or acid sensitive groups such as aldehyde, ketone, ester, cyano, halogen and benzyl ethers are not affected.

(19)

R = H, Cl, OH, CO_2H, CHO, OMe, CO_2Me, OAc

The iron cluster $[Fe_4S_4(SPh)_4]^{2-}$ catalyzes the reduction of nitroarenes to arylamines.[113] A less hydridic nucleophile $[HFe(CO)_4]^-$ has also found application as a selective reducing agent for nitroarenes.[114] Although $[HFe(CO)_4]^-$] is known to reduce aldehydes, ketones and acid halides,[115] in THF solvent with trifluoroacetic acid, it selectively reduces nitrobenzenes to anilines in the presence of aldehyde and acid halide groups.

2.1.5.5 Hydrogenation

The catalytic reduction of aromatic nitroso compounds to amines is well documented.[116] Selective reduction of nitro aromatic compounds can be achieved at room temperature and atmospheric pressure using an interlamellar montmorillonitesilylpalladium(II) complex (equation 20).[117]

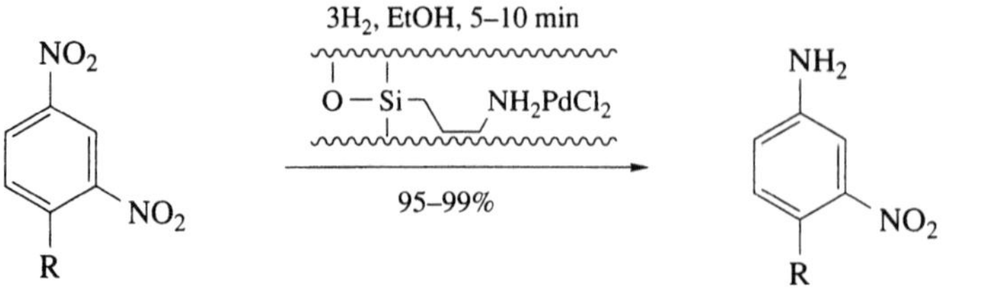

$$\text{1-R-2,4-}(NO_2)_2C_6H_3 \xrightarrow[\text{95–99\%}]{3H_2,\ EtOH,\ 5\text{–}10\ min;\ \text{O–Si–}\diagdown\text{–}NH_2PdCl_2} \text{4-}NH_2\text{-2-}NO_2\text{-1-R-}C_6H_3 \qquad (20)$$

R = H, Me

There has also been continuous interest in the preparation of thermally stable polymers which can be utilized as heterogeneous catalyst supports. Polybenzimidazole (PBI) (**32**) is a thermally stable heterocyclic polymer.[118] Polybenzimidazole forms a complex with $PdCl_2$ (**33**) which can be reduced to PBI–Pd^0 (**34**). This catalyst is extremely stable yet it can be used to reduce a variety of nitro compounds (Scheme 6). The catalyst may be recycled simply by washing with methanol to remove the reduced substrate from the polymer; the recycled catalyst shows no loss of activity or of Pd content after more than 20 cycles.

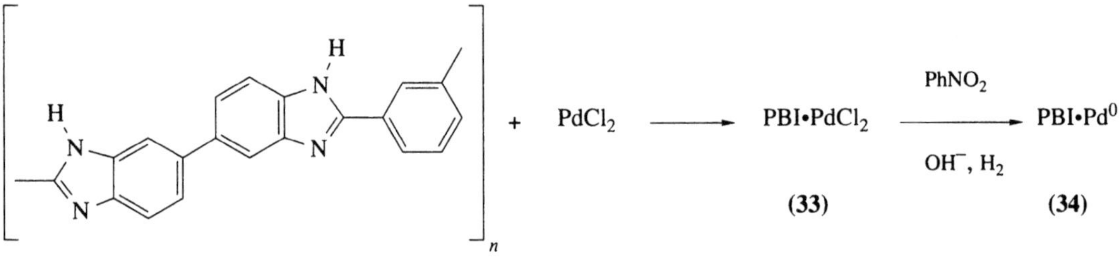

$$[PBI]_n\ (\mathbf{32}) + PdCl_2 \longrightarrow PBI{\bullet}PdCl_2\ (\mathbf{33}) \xrightarrow[OH^-,\ H_2]{PhNO_2} PBI{\bullet}Pd^0\ (\mathbf{34})$$

$$Ar{-}NO_2 \xrightarrow[H_2,\ r.t.]{PBI{\bullet}Pd^0} Ar{-}NH_2$$

Scheme 6

2.1.5.6 Carbon Monoxide

Carbon monoxide reduces aromatic nitro compounds when iron pentacarbonyl is used as catalyst.[119] A direct homogeneous catalytic reduction of nitro derivatives with water under moderate carbon monoxide pressure also occurs when rhodium carbonyl derivatives in aqueous organic bases are used as catalysts (equation 21).[120] Presumably hydridorhodium carbonyl species[121] are the active agents whose preferred formation in aqueous organic base may be analogous to that of iron carbonyl hydrides.[122]

$$Ar{-}NO_2 + 3CO + H_2O \xrightarrow{\text{catalyst}} Ar{-}NH_2 + 3CO_2 \qquad (21)$$

Several other polynuclear metal carbonyls were tested as catalysts for the reduction of nitrobenzene to aniline using carbon monoxide and water as the reducing agent. Rhodium, iridium and osmium clusters were found to be very effective and they were far less susceptible to oxidative degradation than $[Fe(CO)_5]$.[123]

The search for a novel, low cost replacement for elemental hydrogen resulted in the discovery of a mixed gas consisting of hydrogen sulfide and carbon monoxide.[124] This gas, when used with Fe on Al_2O_3, reduces dinitroarenes to diamino aromatic compounds in high yield (equation 22).

$$Ar(NO_2)_2 \xrightarrow[\text{catalyst}]{H_2S/CO} Ar(NH_2)_2 \qquad (22)$$

Amines are also obtained in excellent yields, at room temperature and atmospheric pressure, using carbon monoxide, ruthenium carbonyl and benzyltriethylammonium chloride in an aqueous base-organic solvent system.[125] The method is significantly milder than previously described water gas shift reaction conditions.[123]

2.1.5.7 Miscellaneous Reduction Methods

An atmospheric pressure hydrogenation of mononitro aromatic compounds to anilines can be achieved using hydrogen sulfide as reducing agent and titanium dioxide as catalyst.[126] The reduction of arylnitroso compounds by NADH and N_1-(2,6-dichlorobenzyl)-1,4-dihydronicotinamide (DBDN-4H_2) has been investigated in model systems.[127] Arylnitroso compounds such as (**35**) are readily reduced by NADH and DBDN-4H_2, but the reaction conditions require the absence of oxygen, or neutral and weakly alkaline aqueous buffers (Scheme 7).

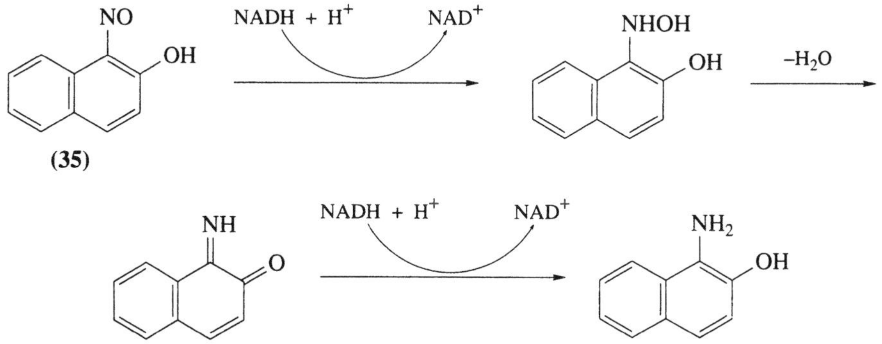

NO, OH (**35**) — $NADH + H^+$ / NAD^+ → NHOH, OH — $-H_2O$ →

NH, O — $NADH + H^+$ / NAD^+ → NH_2, OH

Scheme 7

Copper salt–amine complexes can also be used for the reduction of aromatic compounds to the corresponding amine.[128] A more general and convenient method for reducing nitroso compounds to amines involves the use of a nickel/aluminum alloy.[129] The low cost and ready commercial availability of nickel/aluminum alloy are important features of this reduction procedure which may find wide acceptance as a preparative method.

2.1.6 REDUCTION OF ALIPHATIC NITRO COMPOUNDS TO HYDROXYLAMINES

The reduction of nitroalkanes to *N*-monosubstituted hydroxylamines has not been extensively explored. The classical approach involves electrolytic reduction of primary and secondary nitroalkanes.[130–132] Catalytic hydrogenation[133] and hydride reductions[134] of nitroalkenes also yield hydroxylamine derivatives.

Diborane, which does not reduce nitro compounds,[135] readily reduces the salts of primary and secondary nitro compounds to the corresponding *N*-monosubstituted hydroxylamines (equation 23).[136]

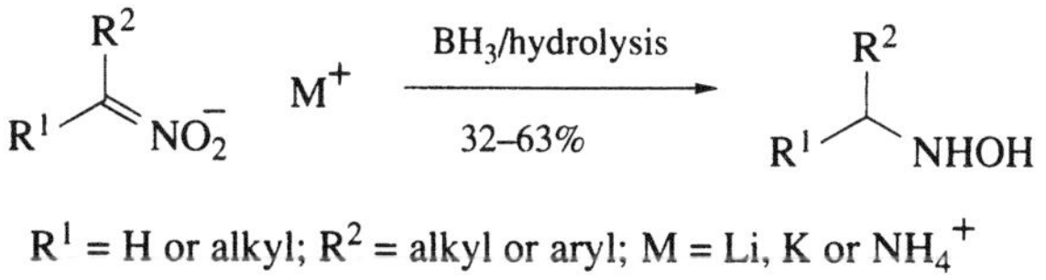

$$R^1R^2C{=}NO_2^- \; M^+ \xrightarrow[32\text{–}63\%]{BH_3/\text{hydrolysis}} R^1R^2CH\text{–}NHOH \qquad (23)$$

R^1 = H or alkyl; R^2 = alkyl or aryl; M = Li, K or NH_4^+

Trialkylboron hydrides react with conjugated nitroalkenes to yield nitronate intermediates *via* a 1,4-addition of the hydride. The corresponding nitroalkanes are easily obtained by using silica gel to protonate the nitronate salt.[137] These reactions presumably occur through a common intermediate (**36**), which is then hydrolyzed to the nitroalkane or reduced by a borane complex to yield hydroxylamine, after hydrolysis.

Even sodium borohydride catalyzes the reaction of borane complexes with conjugated nitroalkenes (equation 24).[138] This straightforward approach affords pure hydroxylamines in high yields. As an

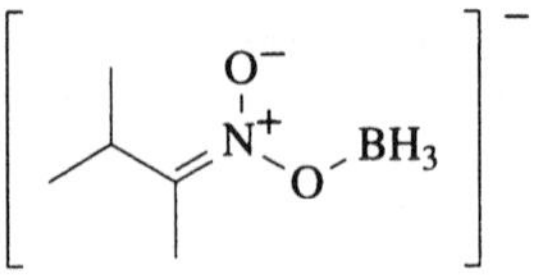

(36)

example, the reduction of 3-nitrochromenes (**37**) by a borane–borohydride system results in novel hydroxylaminochroman derivatives (**38**; equation 25).[139] A modification of these procedures involving the use of *in situ* generated borane (from $NaBH_4$ and $BF_3{\cdot}Et_2O$) obviates the need for $BH_3{\cdot}THF$, which is not universally available.[140]

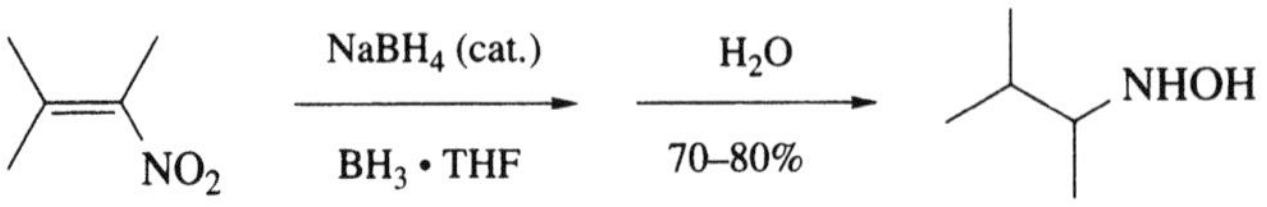

NaBH$_4$ (cat.), BH$_3$ • THF; H$_2$O, 70–80% (24)

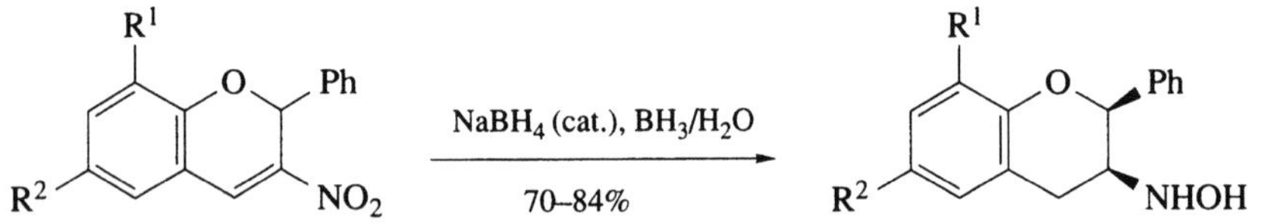

NaBH$_4$ (cat.), BH$_3$/H$_2$O, 70–84% (25)

(**37**) (**38**)

2.1.7 REDUCTION OF ALIPHATIC NITROALKANES TO AMINES

While aromatic nitro compounds are reduced to anilines by a variety of methods,[1,141] comparatively few reagents reduce aliphatic nitro compounds to the corresponding amines. Traditionally, nitro aliphatic materials have been reduced using high pressure hydrogenation,[142,143] lithium aluminum hydride[144,145] or aluminum amalgam.[146–150]

2.1.7.1 Nitroalkane Reductions

The value of the catalytic transfer hydrogenation route is demonstrated by the selective, high yield and rapid reduction of nitro aliphatic compounds to their corresponding amine derivatives using anhydrous ammonium formate (equation 26).[67] A wide variety of nitro compounds are reduced in the presence of other functional groups including acids, esters and nitriles. Furthermore, the method is stereospecific and proceeds with retention of configuration:[151] pure racemic *syn*-nitro alcohols (**39a**) and (**39b**) were converted to the *syn*-amino alcohols (**40a**) and (**40b**) and the axial nitrosteroid (**41a**) afforded the 6β-amine (**41b**).

$$\text{Alkyl–NO}_2 \xrightarrow[\text{r.t., MeOH; 31–98\%}]{\text{HCO}_2\text{NH}_4\text{, 10\% Pd–C}} \text{Alkyl–NH}_2 \quad (26)$$

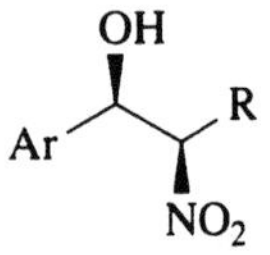

(**39a**) R = Me
(**39b**) R = CH_2OTHP

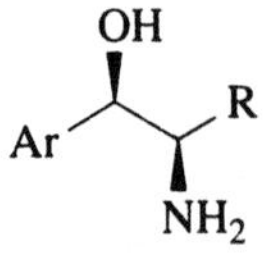

(**40a**) R = Me
(**40b**) R = CH_2OTHP

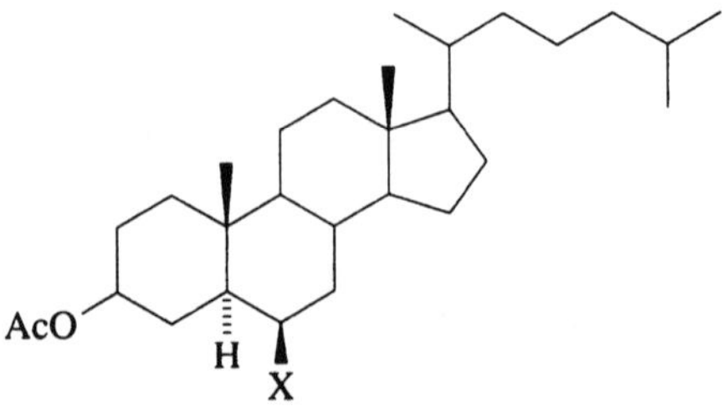

(**41a**) X = NO_2
(**41b**) X = NH_2

In view of its industrial importance, a number of homogeneous catalysts have been examined for use in selective hydrogenations of nitroalkanes to amines.[152] Tris(triphenylphosphine)ruthenium(II) chloride is particularly effective.[153] The mechanism of this hydrogenation is known to involve nitronate anion formation.[154]

Hydrazine hydrate can also be used to reduce nitroalkanes in the presence of a graphite catalyst.[70] The low cost of the catalyst serves to make this a particularly attractive reduction method. Electrochemical reduction of tertiary nitroalkanes such as the 2-substituted-1,1-dimethyl-1-nitroethanes (**42a–42c**) yields amines (**44a–44c**) as well as hydroxylamines (**43a–43c**; equation 27).[155]

R^2, R^1, NO_2 —$+e^-$→ R^2, R^1, NHOH + R^2, R^1, NH_2 (27)

(**42a**) R^1 = Ar, R^2 = H
(**42b**) R^1 = Ar, R^2 = OH
(**42c**) R^1 = H, R^2 = OH

(**43a–43c**) (**44a–44c**)

Titanium(II) reagents have also been used to reduce aliphatic nitro compounds to amines; halo, cyano and ester groups are not reduced.[156] Sodium borohydride, in the presence of catalytic amounts of nickel(II) chloride, reduces a variety of aliphatic nitro compounds to amines.[157] Nickel boride (Ni_2B) is an active catalyst for reductions of primary, secondary and tertiary nitro aliphatic compounds to amines.[158] The reduction of nitrocyclohexane (**45**) yields cyclohexylamine (**47**) as well as small amounts of dicyclohexylamine (**49**), the latter being formed *via* reaction of intermediates (**46**) and (**48**; equation 28).

NO_2 → [NH] + NH_2 →

(**45**) (**46**) (**47**)

N → NH, 2 (28)

(**48**) (**49**)

1-(Indol-3-yl)-2-nitroalkanes (**50**) are reduced to the corresponding α-alkyltryptamines (**51**) using nickel boride and hydrazine hydrate (equation 29).[72]

NO_2, R^1, R^2, N, H —$Ni_2B/N_2H_4 \cdot H_2O$→ NH_2, R^1, R^2, N, H (29)

(**50**) (**51**)

2.1.7.2 Nitroalkene Reductions

The reduction of α,β-unsaturated nitroalkenes provides a simple access to a vast array of functionalities, including amines.[4] Early studies focused on the catalytic hydrogenation of the nitropropene derivative (**52**); 1-(2,5-dimethoxy-4-methylphenyl)-2-aminopropane (**53**) is obtained as a minor product in a complex reaction.[133] The reverse addition of lithium aluminum hydride to β-methyl-β-nitrostyrene affords a mixture of β-phenylisopropylamine and *N*-(β-phenylisopropyl)hydroxylamine.[134]

NO_2 OMe MeO **(52)** NH_2 OMe MeO **(53)**

Alkylamines are generally accessible *via* the reduction of nitroalkenes with lithium aluminum hydride.[159–161] Substituted thienylethylamines[162] can also be obtained using this methodology (equation 30).

X, S, R, NO_2 —$LiAlH_4$, 50–70%→ X, S, R, NH_2 (30)

A general synthesis of β-(2- or 3-pyrrolyl)alkylamines (**55**) containing an unsubstituted nitrogen atom is available *via* the reduction of the corresponding nitroalkene (**54**; equation 31).[163]

N–SO_2Ph, R, NO_2 —$LiAlH_4$→ N–H, R, NH_2 (31)

	Position	*R*		
(54a)	3	H	**(55a)**	69%
(54b)	3	Me	**(55b)**	68%
(54c)	2	H	**(55c)**	70%
(54d)	2	Me	**(55d)**	72%

Since hydroxylamines, as well as their precursor oxime derivatives, are reduced by diborane to amines,[164] the reaction has been extended by reducing the initially formed hydroxylamine intermediates (**57**) to amines (**58**).[165] Thus excess borane reduces nitroalkenes (**56**) to amines (**58**) in the presence of a catalytic amount of sodium borohydride (equation 32).

NO_2 **(56)** —$M^+BH_4^-$, $BH_3 \cdot THF$→ $NHOBH_2$ **(57)** —Δ, $BH_3 \cdot THF$, 85–90%→ NH_2 **(58)** (32)

The reduction can also be achieved by utilizing *in situ* generated $BH_3 \cdot THF$ (from sodium borohydride and boron trifluoride etherate).[166] The scope of this reaction includes the synthesis of novel 3-chromanamine derivatives (**60**; equation 33).[140] This stereoselective reaction proceeds *via* the hydroxylamine intermediate; only *cis*-2-aryl-3-amino derivatives are obtained.

R^1, R^2, O, Ph, NO_2 **(59)** —$NaBH_4/BH_3$, 74–94%→ R^1, R^2, O, Ph, NH_2 **(60)** (33)

In the course of investigations involving the reaction of lithium triethylhydroborate with conjugated nitroalkenes (**61**),[137,167] *N*-ethylamines (**62**) were consistently observed as by-products (equation 34).[168] The intermediacy of nitroso compounds has been confirmed in this reaction which provides a useful method for the preparation of *N*-ethylated amine derivatives. These *N*-alkylated products are not produced when sterically demanding reagents, such as potassium tri-*s*-butylborohydride, are used. Apparently, the competition between reduction and alkylation of the nitroso group is sensitive to steric effects.

NO_2 ... R —$LiEt_3BH/BH_3$, 62–72%→ NH_2 ... R + N(H)Et ... R (34)

(**61**) R = H or Me (**62**)

2.1.8 REFERENCES

1. D. Seebach, E. W. Colvin, F. Lehr and T. Weller, *Chimia*, 1979, **33**, 1.
2. A. G. M. Barrett and G. G. Graboski, *Chem. Rev.*, 1986, **86**, 751.
3. R. S. Varma and G. W. Kabalka, *Heterocycles*, 1986, **24**, 2645.
4. G. W. Kabalka and R. S. Varma, *Org. Prep. Proced. Int.*, 1987, **19**, 283; G. W. Kabalka, L. H. M. Guindi and R. S. Varma, *Tetrahedron*, 1990, **46**, 7443.
5. T. Urbański, *Synthesis*, 1974, 613.
6. N. Kornblum, *Angew. Chem., Int. Ed. Engl.*, 1975, **14**, 734; *Org. React. (N.Y.)*, 1962, **12**, 101.
7. H. Feuer, in 'The Chemistry of the Nitro and Nitroso Groups,' ed. S. Patai, Wiley-Interscience, New York, parts 1 and 2, 1969 and 1970.
8. P. A. S. Smith, 'The Chemistry of Open-chain Organic Nitrogen Compounds', Benjamin, New York, 1966, vol. II.
9. B. G. Gowenlock and W. Lüttke, *Q. Rev., Chem. Soc.*, 1958, **12**, 321.
10. T. L. Gilchrist, *Chem. Soc. Rev.*, 1983, **12**, 53.
11. S. R. Sandler and W. Karo, 'Organic Functional Group Preparation', 2nd edn., Academic Press, New York, 1986, vol. II, p. 456.
12. L. A. Errede and H. R. Davis, *J. Org. Chem.*, 1963, **28**, 1430.
13. R. R. Holmes, R. P. Bayer, L. A. Errede, H. R. Davis, A. W. Wiesenfeld, P. M. Bergman and D. L. Nicholas, *J. Org. Chem.*, 1965, **30**, 3837.
14. S. D. Ross and I. Kuntz, *J. Am. Chem. Soc.*, 1952, **74**, 1297.
15. S. D. Ross, G. J. Kahan and W. A. Leach, *J. Am. Chem. Soc.*, 1952, **74**, 4122.
16. J. Meisenheimer and K. Witte, *Chem. Ber.*, 1903, **36**, 4153.
17. B. T. Newbold, *J. Chem. Soc.*, 1961, 4260.
18. D. A. Blackadder and C. Hinshelwood, *J. Chem. Soc.*, 1957, 2898.
19. H. E. Bigelow and D. B. Robinson, *Org. Synth., Coll. Vol.*, 1955, **3**, 103.
20. S. Shinkai, T. Nakaji, T. Ogawa, K. Shigematsu and O. Manabe, *J. Am. Chem. Soc.*, 1981, **103**, 111.
21. G. M. Badger and G. E. Lewis, *J. Chem. Soc.*, 1953, 2147.
22. R. Adams and J. R. Johnson, 'Laboratory Experiments in Organic Chemistry', 4th edn., Macmillan, New York, 1949, p. 359.
23. S. R. Sandler and W. Karo, 'Organic Functional Group Preparation', 2nd edn., Academic Press, New York, 1986, vol. II, p. 353.
24. H. J. Shine and J. T. Chamness, *J. Org. Chem.*, 1963, **28**, 1232.
25. J. Radell, L. Spialter and J. Hollander, *J. Org. Chem.*, 1956, **21**, 1051.
26. E. Cullen and Ph. L'Ecuyer, *Can. J. Chem.*, 1961, **39**, 862.
27. J. P. Snyder, V. T. Bandurco, F. Darack and H. Olsen, *J. Am. Chem. Soc.*, 1974, **96**, 5158 and refs. cited therein.
28. H. E. Bigelow, *Chem. Rev.*, 1931, **9**, 117.
29. S. Patai (ed.), 'The Chemistry of Hydrazo, Azo and Azoxy Groups', Wiley, New York, 1975.
30. S. R. Sandler and W. Karo, 'Organic Functional Group Preparation', 2nd edn., Academic Press, New York, 1986, vol. II, p. 442.
31. L. Zechmeister and P. Rom, *Justus Liebigs Ann. Chem.*, 1929, **468**, 117.
32. K. Tabei and M. Yamaguchi, *Bull. Chem. Soc. Jpn.*, 1967, **40**, 1538.
33. A. McKillop, R. A. Raphael and E. C. Taylor, *J. Org. Chem.*, 1970, **35**, 1670.
34. H. W. Galbraith, E. F. Degering and E. F. Hitch, *J. Am. Chem. Soc.*, 1951, **73**, 1323.
35. H. J. Shine and H. E. Mallory, *J. Org. Chem.*, 1962, **27**, 2390.
36. H. E. Bigelow and A. Palmer, *Org. Synth., Coll. Vol.*, 1943, **2**, 57.
37. S. A. Buckler, L. Doll, F. K. Lind and M. Epstein, *J. Org. Chem.*, 1962, **27**, 794.
38. N. S. Kozlov and V. A. Soshin, *Uch. Zap. Permsk. Gos. Pedagog. Inst.*, 1965, **32**, 84 (*Chem. Abstr.*, 1967, **66**, 2290k).
39. M. V. King, *J. Org. Chem.*, 1961, **26**, 3323.
40. S. Swann, Jr., in 'Techniques of Organic Chemistry', ed. A. Weissberger, Wiley, New York, 1956, vol. II, p. 478.

41. T. B. Patrick, J. A. Schield and D. G. Kirchner, *J. Org. Chem.*, 1974, **39**, 1758.
42. K. Taya, *J. Chem. Soc., Chem. Commun.*, 1966, 464.
43. I. D. Entwistle, T. Gilkerson, R. A. W. Johnstone and R. P. Telford, *Tetrahedron*, 1978, **34**, 213.
44. N. R. Ayyangar, K. C. Brahme, U. R. Kalkote and K. V. Srinivasan, *Synthesis*, 1984, 938.
45. T. Miyata, Y. Ishino and T. Hirashima, *Synthesis*, 1978, 834.
46. P. W. Oxley, B. M. Adgar, M. J. Sasse and M. A. Forth, *Org. Synth.*, 1988, **67**, 187.
47. C. S. Rondestvedt, Jr. and T. A. Johnson, *Synthesis*, 1977, 851.
48. S. Uchida, K. Yanada, H. Yamaguchi and H. Meguri, *Chem. Lett.*, 1986, 1069.
49. K. Yanada, H. Yamaguchi, H. Meguri and S. Uchida, *J. Chem. Soc., Chem. Commun.*, 1986, 1655.
50. D. L. Klayman and T. S. Griffin, *J. Am. Chem. Soc.*, 1973, **95**, 197.
51. R. W. Lane, J. A. Ibers, R. B. Frankel and R. H. Holm, *Proc. Natl. Acad. Sci. USA*, 1975, **72**, 2868; R. W. Lane, J. A. Ibers, R. B. Frankel, G. C. Papaefthymiou and R. H. Holm, *J. Am. Chem. Soc.*, 1977, **99**, 84.
52. K. Yanada, T. Nagano and M. Hirobe, *Tetrahedron Lett.*, 1986, **27**, 5113.
53. R. Schröter and F. Müller, *Methoden Org. Chem. (Houben–Weyl)*, 1957, **XI/I**, 360.
54. M. S. Gibson, in 'The Chemistry of the Amino Group', ed. S. Patai, Interscience, New York, 1968, p. 66.
55. P. N. Rylander, 'Catalytic Hydrogenation over Platinum Metal', Academic Press, New York, 1967, pp. 22, 181.
56. M. Hudlicky, 'Reductions in Organic Chemistry', Ellis Horwood, Chichester, 1984, p. 13.
57. R. A. W. Johnstone, A. H. Wilby and I. D. Entwistle, *Chem. Rev.*, 1985, **85**, 129.
58. A. Fürst, R. G. Berlo and S. Hooton, *Chem. Rev.*, 1965, **65**, 51.
59. D. J. Pasto, *J. Am. Chem. Soc.*, 1979, **101**, 6852.
60. T. Hirashima and O. Manabe, *Chem. Lett.*, 1975, 259.
61. N. R. Ayyangar, A. G. Lugade, P. V. Nikrad and V. K. Sharma, *Synthesis*, 1981, 640.
62. I. D. Entwistle, *Tetrahedron Lett.*, 1979, 555.
63. I. D. Entwistle, A. E. Jackson, R. A. W. Johnstone and R. P. Telford, *J. Chem. Soc., Perkin Trans. 1*, 1977, 443.
64. N. A. Cortese and R. F. Heck, *J. Org. Chem.*, 1977, **42**, 3491.
65. M. A. Aramendia, M. S. Climent, C. Jiménez and J. M. Marinas, *React. Kinet. Catal. Lett.*, 1975, **31**, 311.
66. S. Ram and R. E. Ehrenkaufer, *Tetrahedron Lett.*, 1984, **25**, 3415.
67. F. Yuste, M. Saldaña and F. Walls, *Tetrahedron Lett.*, 1982, **23**, 147.
68. B. M. Adgar and R. G. Young, *Tetrahedron Lett.*, 1984, **25**, 5219.
69. B.-H. Han, D. M. Shin and S. Y. Cho, *Bull. Korean Chem. Soc.*, 1985, **6**, 320 (*Chem. Abstr.*, 1986, **104**, 129 304x).
70. B.-H. Han, D. M. Shin and S. Y. Cho, *Tetrahedron Lett.*, 1985, **26**, 6233.
71. A. D. Batcho and W. Leimgruber, *Org. Synth.*, 1984, **63**, 214.
72. D. H. Lloyd and D. E. Nichols, *J. Org. Chem.*, 1986, **51**, 4294.
73. B. T. Khai and A. Arcelli, *J. Organomet. Chem.*, 1986, **309**, C63.
74. K. Hanaya, N. Fujita and H. Kudo, *Chem. Ind. (London)*, 1973, 794.
75. T. Satoh, S. Suzuki, Y. Suzuki, Y. Miyaji and Z. Imai, *Tetrahedron Lett.*, 1969, 4555.
76. T. Neilson, H. C. S. Wood and A. G. Wylie, *J. Chem. Soc.*, 1962, 371.
77. S. Kanno, Y. Tanaka, E. Sugino and S. Hibino, *Synthesis*, 1980, 695.
78. T. Satoh, N. Mitsuo, M. Nishiki, Y. Inoue and Y. Ooi, *Chem. Pharm. Bull.*, 1981, **29**, 1443.
79. A. Ono, H. Sasaki and F. Yaginuma, *Chem. Ind. (London)*, 1983, 480.
80. A. Ono, M. Hiroi and K. Shimazaki, *Chem. Ind. (London)*, 1986, 75.
81. M. Kijima, Y. Nambu, T. Endo and M. Okawara, *J. Org. Chem.*, 1984, **49**, 1434.
82. M. Kijima, Y. Nambu, T. Endo and M. Okawara, *J. Polym. Sci., Polym. Chem. Ed.*, 1984, **22**, 821.
83. M. Kijima, Y. Nambu and T. Endo, *J. Polym. Sci., Polym. Chem. Ed.*, 1985, **23**, 1723.
84. J. Dale and L. Vickersveen, *Acta Chem. Scand., Ser. B*, 1988, **42**, 354.
85. T. Severin and R. Schmitz, *Chem. Ber.*, 1962, **95**, 1417.
86. T. Severin and M. Adam, *Chem. Ber.*, 1963, **96**, 448.
87. H. Feuer and D. M. Braunstein, *J. Org. Chem.*, 1969, **34**, 2024.
88. W. H. H. Günther, *J. Org. Chem.*, 1966, **31**, 1202.
89. K. Fujimori, H. Yoshimoto and S. Oae, *Tetrahedron Lett.*, 1979, 4397.
90. N. Kambe, K. Kondo, S. Morita, S. Murai and N. Sonoda, *Angew. Chem., Int. Ed. Engl.*, 1980, **19**, 1009.
91. D. H. R. Barton and S. W. McCombie, *J. Chem. Soc., Perkin Trans. 1*, 1975, 1574; A. Osuka, H. Shimizu and H. Suzuki, *Chem. Lett.*, 1983, 1373.
92. M. Akiba and M. P. Cava, *Synth. Commun.*, 1984, **14**, 1119.
93. N. Ohira, Y. Aso, T. Otsubo and F. Ogura, *Chem. Lett.*, 1984, 853.
94. H. Suzuki, H. Manabe and M. Inouye, *Chem. Lett.*, 1985, 1671.
95. H. H. Hodgson and H. S. Turner, *J. Chem. Soc.*, 1948, 318.
96. D. Huber, G. Andermann and G. Leclerc, *Tetrahedron Lett.*, 1988, **29**, 635.
97. T. E. Nickson, *J. Org. Chem.*, 1986, **51**, 3903.
98. T.-L. Ho, *Synthesis*, 1979, 1.
99. K. Someya, *Z. Anorg. Allg. Chem.*, 1928, **169**, 293.
100. E. Knecht, *Ber. Dtsch. Chem. Ges.*, 1903, **36**, 166.
101. P. C. Banerjee, *J. Indian Chem. Soc.*, 1942, **19**, 30.
102. J. R. Hanson, *Synthesis*, 1974, 1.
103. Y. Akita, M. Inaba, H. Uchida and A. Ohta, *Synthesis*, 1977, 792.
104. G. H. Timms and E. Wildsmith, *Tetrahedron Lett.*, 1971, 195.
105. J. E. McMurry and J. Melton, *J. Org. Chem.*, 1973, **38**, 4367.
106. J. E. McMurry and M. G. Silvestri, *J. Org. Chem.*, 1975, **40**, 1502.
107. T.-L. Ho and C. M. Wong, *Synthesis*, 1974, 45.
108. S. A. Newton, Jr., F. J. Stubbs and C. Hinshelwood, *J. Chem. Soc.*, 1953, 3384.

109. M. Somei, K. Kato and S. Inoue, *Chem. Pharm. Bull.*, 1980, **28**, 2515.
110. H. Suzuki and Y. Hanazaki, *Chem. Lett.*, 1986, 549.
111. W. K. Xing and Y. Ogata, *J. Org. Chem.*, 1982, **47**, 3577 and refs. cited therein.
112. F. D. Bellamy and K. Ou, *Tetrahedron Lett.*, 1984, **25**, 839.
113. H. Inoue, M. Shirai and E. Haruki, *J. Chem. Soc., Chem. Commun.*, 1987, 674.
114. P. L. Gaus, S. W. Gerritz and P. M. Jeffries, *Tetrahedron Lett.*, 1988, **29**, 5083.
115. T. E. Cole and R. Pettit, *Tetrahedron Lett.*, 1977, 781.
116. P. A. S. Smith, 'The Chemistry of Open Chain Organic Nitrogen Compounds', Benjamin, New York, 1966, vol. II, p. 375.
117. K. Mukkanti, Y. V. Subba Rao and B. M. Choudary, *Tetrahedron Lett.*, 1989, **30**, 251.
118. N.-H. Li and J. M. J. Fréchet, *J. Chem. Soc., Chem. Commun.*, 1985, 1100.
119. J. E. Kmiecik, *J. Org. Chem.*, 1965, **30**, 2014.
120. A. F. M. Iqbal, *Tetrahedron Lett.*, 1971, 3385.
121. P. Chini and S. Martinengo, *J. Chem. Soc., Chem. Commun.*, 1969, 1092.
122. I. Wender and P. Pino, 'Organic Synthesis *via* Metal Carbonyls', Interscience, New York, 1968, p. 100.
123. K. Cann, T. E. Cole, W. Slegeir and R. Pettit, *J. Am. Chem. Soc.*, 1978, **100**, 3969.
124. C. T. Ratcliffe and G. Pap, *J. Chem. Soc., Chem. Commun.*, 1980, 260.
125. H. Alper and S. Amaratunga, *Tetrahedron Lett.*, 1980, **21**, 2603.
126. C. T. Ratcliffe, S. L. Soled, A. J. Signorelli and I. L. Mador (Allied Corp.), *US Pat.* 4 326 081 (1982) (*Chem. Abstr.*, 1982, **97**, 38 660f).
127. V. Leskovac and S. Trivić, *J. Org. Chem.*, 1988, **53**, 6123.
128. H. R. Appell, *J. Org. Chem.*, 1967, **32**, 2021.
129. G. Lunn, E. B. Sansone and L. K. Keefer, *Synthesis*, 1985, 1104.
130. P. Pierron, *J. Am. Chem. Soc.*, 1899, **21**, 780.
131. M. W. Leeds and G. B. L. Smith, *J. Electrochem. Soc.*, 1951, **98**, 129.
132. V. L. Kornienko, M. A. Kedrinskii, G. V. Kornienko, N. V. Kalinichenko and Yu. G. Chirkov (Siberian Technological Institute), *USSR Pat.* SU 677 277 A1 (1984) (*Chem. Abstr.*, 1985, **101**, 151 395n).
133. R. T. Coutts and J. L. Malicky, *Can. J. Chem.*, 1974, **52**, 395.
134. R. T. Gilsdorf and F. F. Nord, *J. Am. Chem. Soc.*, 1952, **74**, 1837.
135. H. C. Brown, 'Hydroboration', Benjamin, New York, 1962, p. 249.
136. H. Feuer, R. S. Bartlett, B. F. Vincent, Jr. and R. S. Anderson, *J. Org. Chem.*, 1965, **30**, 2880.
137. R. S. Varma and G. W. Kabalka, *Synth. Commun.*, 1984, **14**, 1093.
138. M. S. Mourad, R. S. Varma and G. W. Kabalka, *J. Org. Chem.*, 1985, **50**, 133.
139. R. S. Varma, Y.-Z. Gai and G. W. Kabalka, *J. Heterocycl. Chem.*, 1987, **24**, 767.
140. R. S. Varma and G. W. Kabalka, *Org. Prep. Proced. Int.*, 1985, **17**, 254.
141. R. C. Wade, *J. Mol. Catal.*, 1983, **18**, 273.
142. H. L. Finkbeiner and M. Stiles, *J. Am. Chem. Soc.*, 1959, **81**, 505.
143. H. L. Finkbeiner and M. Stiles, *J. Am. Chem. Soc.*, 1963, **85**, 616.
144. H. H. Wasserman, M. J. Hearn, B. Haveaux and M. Thyes, *J. Org. Chem.*, 1976, **41**, 153.
145. E. W. Colvin and D. Seebach, *J. Chem. Soc., Chem. Commun.*, 1978, 689.
146. W. Hückel and M. L. Blohm, *Justus Liebigs Ann. Chem.*, 1933, **502**, 114.
147. E. J. Corey, N. H. Andersen, R. M. Carlson, J. Paust, E. Vedejs, I. Vlattas and R. E. K. Winter, *J. Am. Chem. Soc.*, 1968, **90**, 3245.
148. E. J. Corey, I. Vlattas, N. H. Anderson and K. Harding, *J. Am. Chem. Soc.*, 1968, **90**, 3247, 5947.
149. E. McDonald and R. T. Martin, *Tetrahedron Lett.*, 1977, 1317.
150. C. Shin, Y. Kosuge, M. Yamaura and J. Yoshimura, *Bull. Chem. Soc. Jpn.*, 1978, **51**, 1137.
151. A. G. M. Barrett and C. D. Spilling, *Tetrahedron Lett.*, 1988, **29**, 5733.
152. J. F. Knifton, *J. Org. Chem.*, 1975, **40**, 519 and refs. cited therein.
153. P. S. Hallman, B. R. McGarvey and G. Wilkinson, *J. Chem. Soc. A*, 1968, 3143.
154. J. F. Knifton, *J. Org. Chem.*, 1973, **38**, 3296.
155. H. Ohmori, S. Furusako, M. Kashu, C. Ueda and M. Masui, *Chem. Pharm. Bull.*, 1984, **32**, 3345.
156. J. George and S. Chandrasekaran, *Synth. Commun.*, 1983, **13**, 495.
157. J. O. Osby and B. Ganem, *Tetrahedron Lett.*, 1985, **26**, 6413.
158. C. A. Brown and V. K. Ahuja, *J. Org. Chem.*, 1973, **38**, 2226.
159. F. A. Ramirez and A. Burger, *J. Am. Chem. Soc.*, 1950, **72**, 2781.
160. A. T. Shulgin, *J. Med. Chem.*, 1966, **9**, 445.
161. B. T. Ho, W. M. McIsaac, R. An, L. W. Tansey, K. E. Walker, L. F. Englert, Jr. and M. B. Noel, *J. Med. Chem.*, 1970, **13**, 26.
162. R. T. Gilsdorf and F. F. Nord, *J. Org. Chem.*, 1950, **15**, 807.
163. A. Hamdan and J. W. F. Wasley, *Synth. Commun.*, 1985, **15**, 71.
164. H. Feuer and D. M. Braunstein, *J. Org. Chem.*, 1969, **34**, 1817.
165. M. S. Mourad, R. S. Varma and G. W. Kabalka, *Synth. Commun.*, 1984, **14**, 1099.
166. R. S. Varma and G. W. Kabalka, *Synth. Commun.*, 1985, **15**, 843.
167. M. S. Mourad, R. S. Varma and G. W. Kabalka, *Synthesis*, 1985, 654.
168. G. W. Kabalka, Y.-Z. Gai, N. M. Goudgaon, R. S. Varma and E. E. Gooch, *Organometallics*, 1988, **7**, 493.

2.2
Reduction of N=N, N—N, N—O and O—O Bonds

THOMAS L. GILCHRIST
University of Liverpool, UK

2.2.1 INTRODUCTION

This chapter covers methods of reduction of nitrogen–nitrogen multiple bonds and methods for the reductive cleavage of N—N, N—O and O—O bonds. Perhaps the most important use of reactions of these types is that they can allow the introduction of amino and hydroxy groups with considerable control of stereochemistry. Amino groups are often introduced by the use of azide ion and occasionally by hydroxylamines or hydrazines as nucleophiles, followed by reductive cleavage. Hydroxy groups can similarly be introduced by way of peroxides. Activated azo and nitroso compounds and singlet oxygen are excellent dienophiles in the Diels–Alder reaction and the cleavage of the heteroatom to heteroatom bonds permits two functional groups to be introduced in a controlled manner. Alkenes and alkynes can also be functionalized at both termini by the cycloaddition of 1,3-dipoles such as nitrones, nitrile imides and nitrile oxides, followed by reductive cleavage.

Inevitably there is some overlap with other chapters, particularly those dealing with reduction of C=N bonded groups (Chapters 1.2–1.8, this volume) and that on the reduction of heterocycles (Chapter 3.8, this volume). For completeness, most of the important reduction methods for the appropriate groups are also referred to here. An attempt has been made to cover the literature to late 1989; a few later references are also included.

2.2.2 REDUCTION OF GROUPS WITH NITROGEN–NITROGEN MULTIPLE BONDS

2.2.2.1 Reduction to N—N Bonded Groups

2.2.2.1.1 Azo and azoxy to hydrazo groups

Standard methods which existed in the mid-1970s for the reduction of azo and azoxy compounds to hydrazo compounds have been reviewed.[1] The reduction of aliphatic azo compounds is not of much use for the synthesis of aliphatic hydrazo compounds because the azo compounds are usually prepared from them. Methods for the reduction of aliphatic azoxy compounds directly to the corresponding hydrazo compounds involve the use of tin(II) chloride in hydrochloric acid and catalytic hydrogenation over platinum oxide; electrochemical reduction can also be used.

The reduction is an important method for the preparation of aromatic hydrazo compounds. Useful reagents for the conversion of azobenzenes to hydrazobenzenes in good yield include diimide, sodium borohydride in the presence of metal catalysts, lithium aluminum hydride[2] and cobalt boride with hydrazine hydrate.[3] Catalytic hydrogenation has also been used,[4] although further reductive cleavage of the N—N bond can occur. Zinc and other metals can be used for the reduction, but strongly acidic conditions have to be avoided in order to prevent the benzidine rearrangement of the hydrazobenzenes. For the reduction of azoxybenzenes to hydrazobenzenes, commonly used reagents are zinc dust in acetic acid or in alkali and aqueous sodium disulfide.

There are several recent methods for the reduction of azobenzene to hydrazobenzene in near-quantitative yield.[5] Samarium(II) iodide reduces azobenzene to hydrazobenzene rapidly at room temperature.[6] Hydrogen telluride, generated *in situ* from aluminum telluride and water, reduces both azobenzene and azoxybenzene to hydrazobenzene;[7] a mixture of phenyllithium and tellurium powder has been used to reduce azobenzene.[8] A complex of the coenzyme dihydrolipoamide and iron(II) is also effective for the reduction of azo- and azoxy-benzene to hydrazobenzene; the reduction probably involves coordination of the azobenzene to iron(II) as shown in structure (**1**).[9] Electrochemical reduction has been used to prepare a number of hydrazobenzenes from the corresponding azobenzenes. In the presence of an acylating agent a diacylhydrazine (*e.g.* the pyridazinedione derivative **2**) can be isolated from the electrochemical reduction of azobenzene.[10]

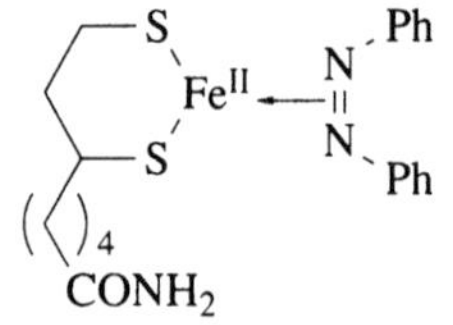

S, Fe^{II}, S, N=N, Ph, Ph, $(\ )_4$, $CONH_2$

(1)

Ph, N, N, Ph, O, O

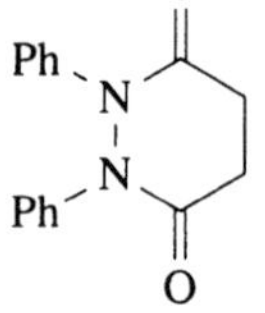

(2)

$ArNHNH_3^+\ BF_4^-$

(3)

2.2.2.1.2 Diazo compounds and diazonium salts to hydrazines

A few examples of the reduction of aliphatic diazo compounds to hydrazines exist,[11] but this is not a generally applicable method for the synthesis of alkylhydrazines. On the other hand, arylhydrazines can be prepared by reduction of aromatic diazonium salts.[12] The most commonly used reagents for this conversion are sulfur dioxide (or sodium sulfite) and tin(II) chloride, these being used to reduce arenediazonium chlorides in aqueous solution. Several other reagents, including sodium amalgam and triphenylphosphine, have been used for specific reductions of this type.[12] Arenediazonium tetrafluoroborates have been reduced to the corresponding hydrazinium salts (**3**) by benzeneselenol in dichloromethane solution.[13] Partial reduction of the benzenediazonium cation leads to the formation of the corresponding radical[14] or of benzenediimide, PhN═NH. Benzenediimide is an unstable (although detectable)[15] compound which readily loses nitrogen; the usual products of such partial reduction are, therefore, benzene and nitrogen.

2.2.2.2 Reductive Cleavage

2.2.2.2.1 Azo, azoxy, diazo and diazonium compounds to amines

Standard methods for the reductive cleavage of azo and azoxy groups are described in several reviews.[1,2,4,16] Catalytic hydrogenation of azo compounds over platinum, or with Raney nickel, often leads to reductive cleavage; an example is the cleavage of the azo compound (**4**) to 2-methylindol-3-amine (85%) by catalytic reduction over platinum.[17] Catalytic transfer hydrogenation over palladium (with cyclohexene as coreagent) is also an efficient method for the reductive cleavage of azobenzenes to anilines.[18] A reagent which appears to have wide applications for reductions of this type is nickel–aluminum alloy.[19,20] When added to the substrate in aqueous methanolic potassium hydroxide at room temperature, azobenzene was reduced to aniline (79%) and the azoxy compound MeN═N$^+$O$^-$Ph gave methylamine quantitatively. Older methods include metal and acid combinations (for example, tin(II) chloride and hydrochloric acid), hydriodic acid, sodium dithionite and diborane.

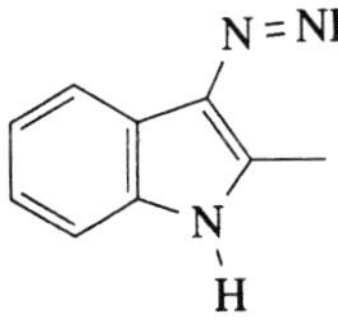

N=NPh N$_2$ NH$_2$ Ph CO$_2$Me Ph CO$_2$Me N H O OH

(**4**) 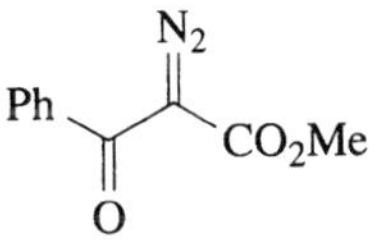 (**5**) 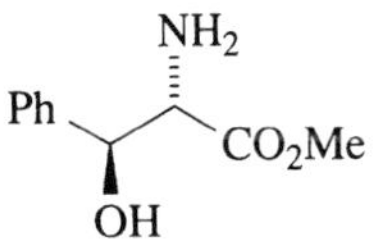 (**6**)

The reductive cleavage of diazo compounds to amines is not a reaction which has found wide use.[4,11] Catalytic hydrogenolysis of some diazo esters in acidic media can lead to the formation of amines in fair to good yield,[4,21] although other functional groups present may also be reduced; for example, reduction of the diazo compound (**5**) over palladium in 70% acetic acid gave the amine (**6**; 82%).[4] The conversion of aromatic diazonium salts to the corresponding anilines is also a reaction with little general synthetic applicability;[12] examples of the reaction have been described using zinc or tin in acidic media and ammonia (this last reagent probably forming an unstable triazene by nucleophilic attack on the diazonium ion).

2.2.2.2.2 Azides and triazenes to amines

Since the azide ion is an excellent nucleophile it is often the reagent of choice for introducing a nitrogen functional group. The reduction of the azido group to a primary amino group is consequently often an important step in a reaction sequence. The attack of azide ion on a primary or secondary carbon center bearing a good leaving group proceeds with inversion of configuration and the subsequent reduction of the azido to an amino group proceeds with retention. The amino group can, therefore, be introduced with control of stereochemistry. An example of a practical application, in which these reactions are combined with enantioselective enzymic ester hydrolysis to produce optically active amino alcohols from cyclohexene oxide, is shown in Scheme 1.[22]

A wide range of reduction methods is available, including several which are claimed to permit selective reduction of the azido group in the presence of other functional groups. Many of the methods fall into one of three broad categories: (i) involving the use of hydrogen and metal catalysts, (ii) involving low-valent metals, and (iii) involving the use of nucleophiles which initially attack the terminal nitrogen of the azide with the formation of a triazene (**7**). These triazenes can sometimes be isolated, but they are usually reduced *in situ*, by the same reagent or by a different one, to the primary amine. The ease of reduction of the azido group by reagents of the third type, therefore, depends upon the susceptibility of the functional group to nucleophilic attack: aromatic azides, especially those with electron-withdrawing substituents, are generally easier to reduce than simple aliphatic azides.

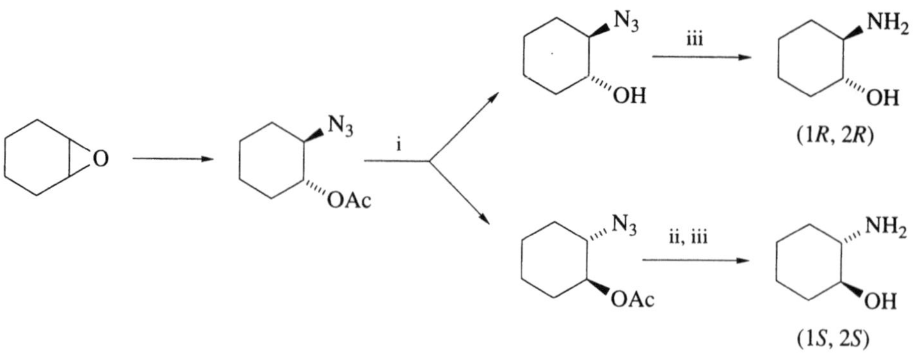

i, lipase from *Candida cyclindracea*; ii, NaOMe, MeOH; iii, H_2, Pd/C

Scheme 1

RN = NNHNu

(7) Nu = nucleophile

Reviews published in 1971 and 1988 cover the methods known at the time for the reduction of azides.[23] There are more selective reviews of catalytic methods[4] and of reduction by metal hydrides.[2]

Catalytic hydrogenation over platinum or palladium is generally an efficient method of reduction of azides when there are no other sensitive functional groups present. Nitrogen is evolved in the reaction so it is not possible, with normal hydrogenation apparatus, to follow the course of the hydrogenation by the uptake of gas. The mechanism of hydrogenation has not been established, but the triazenes RN=NNH_2 have been suggested as intermediates. Selective hydrogenolysis of azido groups in the presence of double bonds and of benzyloxy groups has been achieved using Lindlar catalyst in ethanol;[24] for example, the azide $PhCH_2O(CH_2)_3N_3$ was reduced to the corresponding amine (96%) by this method. There are also examples of the reduction of azides over palladium on carbon without affecting other sensitive functional groups, such as the benzyloxycarbonyl group.[25] Raney nickel catalyzed reduction has similarly been used for selective attack on the azido group in the presence of a double bond and a benzyloxy group.[26] Other catalytic methods include catalytic transfer hydrogenation[27] and reduction over rhodium(III) chloride in the presence of carbon monoxide.[28]

Several alkyl and aryl azides have been reduced to the corresponding amines in good yield by tin(II) chloride at room temperature; the less reactive azides require a catalytic amount of aluminum trichloride to be added.[29] Aqueous vanadium(II) chloride is a useful reducing agent for aryl azides;[30] heteroaryl and arenesulfonyl azides have been reduced with aqueous titanium(III) chloride[31] and with a molybdenum(III) catalyst generated from molybdenum(V) chloride and zinc.[32]

There are many nucleophilic reagents which can effect the conversion of azides into amines and some are very selective. Lithium aluminum hydride is a good reagent in cases where high chemoselectivity is not required. Sodium borohydride has also been used to reduce aryl and other azides, either under phase transfer conditions[33] or by adding the reagent in methanol dropwise to a solution of the azide in THF.[34] 2-Nitrophenyl azide was reduced to 2-nitroaniline (94%) by this reagent,[34] although alkoxycarbonyl groups were reduced. The complex hydride $(Ph_3P)_2CuBH_4$ has also been used to reduce 2-nitrophenyl azide; aromatic azides bearing azo and ethoxycarbonyl groups were also selectively reduced.[35]

Hydrogen sulfide has long been known as a reagent for the reduction of azides and it is selective for the azido group in the presence of many other functional groups, reaction taking place by nucleophilic attack on the terminal nitrogen of the azide. Its uses include the selective reduction of unsaturated nucleoside azides,[36] the reduction of azidothiophenes bearing formyl and cyano groups[37] and the reduction of vinyl azides to carbonyl compounds (Scheme 2).[38] A better, and highly selective, reagent is 1,3-propanedithiol (Scheme 3).[39] The dithiol can reduce aryl and alkyl azides in the presence of double and triple bonds, nitro, cyano, carbonyl and other groups. Its sensitivity to electronic and steric effects can also allow for selective reduction of different azides; for example, 4-nitrophenyl azide is reduced in 1 min at room temperature, whereas 1-azidoadamantane requires 120 h at 60 °C. Thioacetic acid has also been used for selective reduction.[40] This reagent permits reduction in the presence of nitro, *t*-butoxycarbonyl and benzyl groups at room temperature. The reduction is accompanied by acetylation so that the products are acetamides rather than free amines.

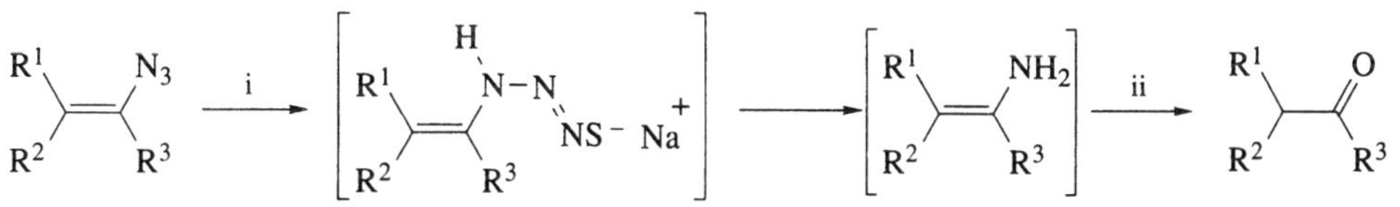

i, Na_2S; ii, H_2O

Scheme 2

$RN_3 + HS\text{(CH}_2)_3\text{SH} \longrightarrow RNH_2 + \text{1,2-dithiolane} + N_2$

Scheme 3

These sulfur-based compounds are mild reducing agents. Two tin-based reagents have been described which are claimed to combine high reducing power with good selectivity.[41] The more reactive of these, $Et_3NH^+Sn(SPh)_3^-$, rapidly reduces a variety of aliphatic azides in organic solution at room temperature and without attack on a carbonyl group. The reagent Bu_2SnH_2 can be used to carry out the same reductions but is slower.

Phosphines are known to react with azides to give phosphine imides with elimination of nitrogen. These imides, which are often isolable, can then be converted into amines by hydrolysis (Scheme 4). The method has effectively been made into a one-pot procedure by carrying out the reaction of triphenylphosphine with azides in THF at room temperature in the presence of a slight excess of water.[42] This is a highly chemoselective method of reduction and many other functional groups, including nitro, alkoxycarbonyl and epoxide, are unaffected. The method has proved to be a useful alternative to catalytic hydrogenolysis in oligonucleotide synthesis.[43] Similar reactions have been carried out with trialkylphosphines[41] and with triethyl phosphite.[44] Diphosphorus tetraiodide has been used to reduce aryl and acyl azides directly.[45]

$$RN_3 + PPh_3 \longrightarrow RN{=}PPh_3 \xrightarrow{H_2O} RNH_2 + Ph_3PO$$

Scheme 4

A useful method of electrophilic amination of aromatic substrates is provided by the reaction of organomagnesium or organolithium derivatives with *p*-toluenesulfonyl azide, followed by reductive cleavage of the resulting triazene (Scheme 5).[46] The same reducing agents which cleave azides (for example, lithium aluminum hydride, sodium borohydride in phase transfer conditions and hydrogen sulfide) are used. With the increasing importance of selective lithiation of aromatic and heteroaromatic compounds, the method has become a very useful one for the regioselective introduction of an amino group. An example is provided by the selective amination of diethylamides (**8**) *ortho* to the carboxamido group (Scheme 6).[47] The same type of process has been described between vinyl azides and aryllithium reagents[48] and between (phenylthio)methyl azide and Grignard reagents.[49] In both cases the intermediate triazenes were cleaved by hydrolysis.

$$ArMgBr + TsN_3 \longrightarrow ArN{=}N\bar{N}Ts\ ^{+}MgBr \longrightarrow ArNH_2$$

Scheme 5

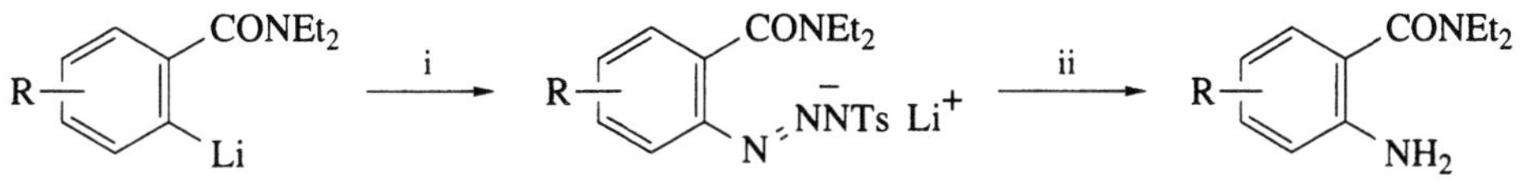

(**8**)

i, TsN_3; ii, $NaBH_4$, phase transfer catalyst

Scheme 6

All these reduction methods lead to the formation of primary amines (or to acetamides derived from them). It is also possible to convert azides reductively into secondary amines. One special case of this type is the conversion of β-iodoalkyl azides and related compounds into aziridines by reduction, usually with lithium aluminum hydride (Scheme 7).[50] A more general method, which was introduced by Brown and coworkers, is the reductive alkylation of azides by heating with trialkylboranes (Scheme 8).[51] For example, triethylborane and azidocyclohexane gave cyclohexylethylamine (73%) by heating in xylene followed by hydrolysis with methanol. (If hydrazoic acid or azidotrimethylsilane is used in this reaction the product is a primary amine.[52]) An intramolecular version of the reaction was used as the cyclization step in a synthesis of the cyclic hexapeptide echinocandin D.[53] The azide (**9**) gave the amino ester (**10**; 72%) as a single stereoisomer when treated with dicyclohexylborane at room temperature. The reaction (Scheme 9) probably involves selective migration of the methylene group of the trialkylborane by way of the cyclic intermediate shown.

N_3 LiAlH$_4$ I H N

Scheme 7

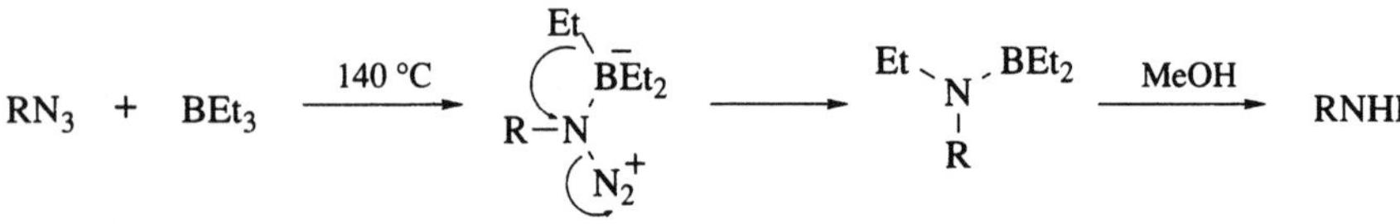

RN_3 + BEt_3 140 °C Et $\bar{B}Et_2$ R–N N_2^+ Et N BEt_2 R MeOH RNHEt

Scheme 8

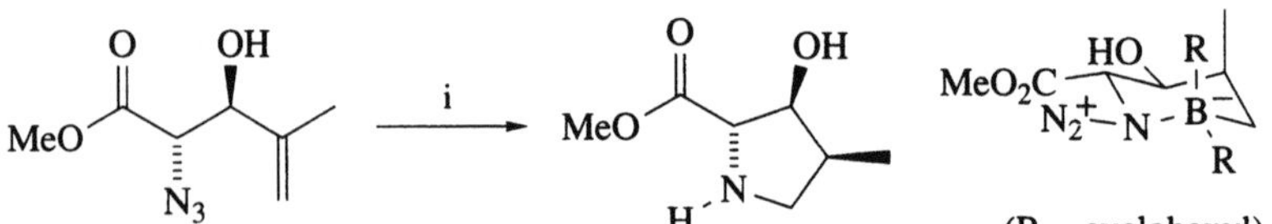

O OH MeO N_3 **(9)** i O OH MeO H N **(10)** HO R MeO_2C N_2^+ N B R (R = cyclohexyl)

i, Dicyclohexylborane, r.t.

Scheme 9

Intermolecular reactions of this type can be slow if either of the reaction partners has a bulky substituent. Reactions with chloroboranes R_2BCl or $RBCl_2$ go under much milder conditions.[51,54] The stereochemistry of the carbon–boron bond is retained in the product; for example, the borane (**11**) gave the amine (**12**; 90%) on reaction with azidocyclohexane.

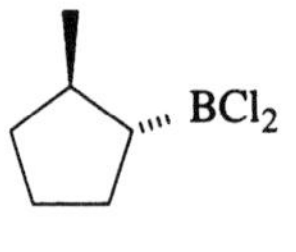

BCl_2 **(11)** H N **(12)** O N–H N H **(13)**

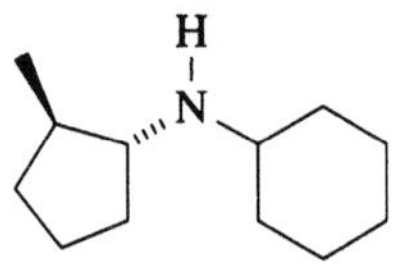

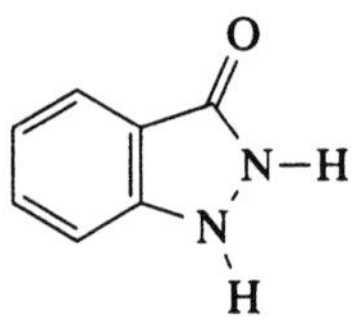

Reaction of aryl azides with hydrazine hydrate can result in reductive cleavage of the azido group. Azidobenzene gives a mixture of benzene and aniline when heated with an excess of hydrazine hydrate in ethanol.[55] If there is an acyl or an alkoxycarbonyl group at the *ortho* position, cyclization products are formed instead; for example, methyl 2-azidobenzoate gave indazolinone (**13**; 90%).

2.2.3 REDUCTION OF GROUPS WITH NITROGEN–NITROGEN SINGLE BONDS

2.2.3.1 Cleavage of Hydrazones

There are several examples of useful syntheses of amines which are based on the reduction and N—N bond cleavage of hydrazones. Some of these undoubtedly involve reduction of the C=N bond as the first step; in a few cases the N—N bond is probably cleaved first and in others the order in which the reactions take place has not been established. Reactions which involve reduction of the C=N bond are covered in the appropriate sections of Chapters 1.2–1.8, this volume; methods for cleavage of the resulting hydrazines are discussed in the following section (Section 2.2.3.2).

The most commonly used method for the cleavage of hydrazones is catalytic hydrogenation in the presence of acid. An example is provided by the synthesis of secondary and tertiary amines by reduction of benzophenone hydrazones (**14**), derived as shown in Scheme 10.[56] Nickel–aluminium alloy can also be used to reduce hydrazones to amines.[20]

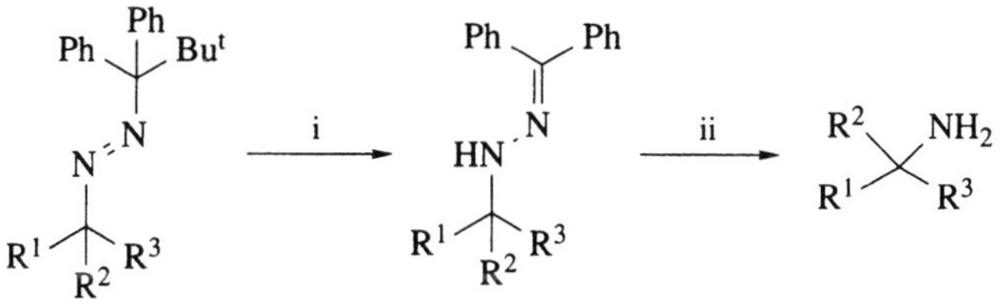

(14)

i, CF_3CO_2H; ii, H_2, Pd/C, HCl/EtOH, 1 atm, 50 °C

Scheme 10

α-Keto acids can be reduced to α-amino acids by way of their phenylhydrazones.[57] The reaction can be carried out by reduction of the hydrazone with hydrogen over palladium or platinum, or with zinc and mercury(II) chloride. As illustrated for the synthesis of alanine in Scheme 11, it is suggested that the N—N bond is cleaved first. Carbonyl compounds can be regenerated from their (2,4-dinitrophenyl)hydrazones by titanium(III) chloride; in this reaction the nitro groups are reduced and the N—N bond is then cleaved to give benzene-1,2,4-triamine, which can be isolated, and the imine of the carbonyl compound, which is hydrolyzed *in situ*.[58] Hydrazonium salts (*e.g.* **15**) derived from diaryl ketones are cleaved by Grignard reagents to give imines (Scheme 12); these reductions probably involve two successive one-electron transfers from the Grignard reagent.[59]

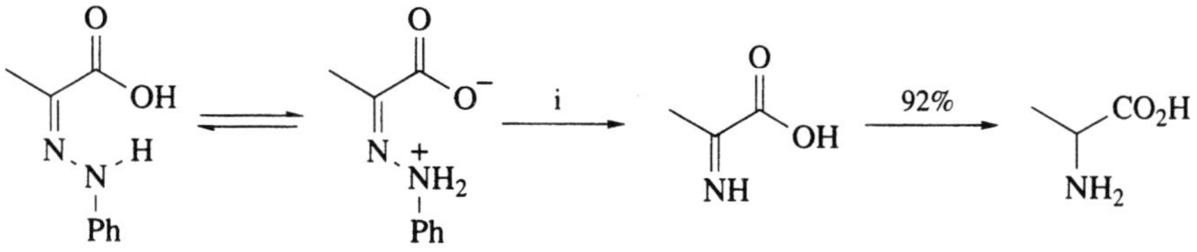

i, H_2, Pd/C, H_2O

Scheme 11

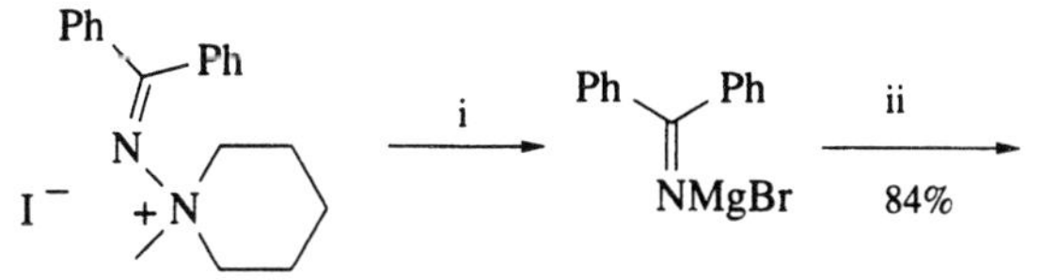

(15)

i, $Me(CH_2)_9MgBr$; ii, HCl

Scheme 12

2.2.3.2 Cleavage of Hydrazines, Hydrazides and *N*-Nitrosamines

There are few general methods available for the reductive cleavage of the N—N bond and the conditions required to cleave this bond can depend markedly on the substituents attached to it. The most widely used method for the cleavage of alkylhydrazines is catalytic hydrogenolysis over Raney nickel, platinum or (less commonly) palladium.[4] These reactions often require the use of hydrogen under pressure and, with platinum and palladium catalysts, an acidic medium. The cleavage reaction is a key step in a number of asymmetric syntheses of amines and amino acids. An approach which has been used by several research groups is outlined in Scheme 13; a chiral hydrazone is reduced or is converted by addition of an organometallic reagent into a hydrazine, the N—N bond of which is then cleaved catalytically.[60] A related reaction in which the choice of reduction conditions proved crucial is the synthesis of amino alcohols from aldehyde *N*,*N*-dimethylhydrazones illustrated in Scheme 14; it was important to use platinum oxide in the presence of acetic acid, rather than Raney nickel, to avoid epimerization.[61] These catalytic methods have also proved useful for the cleavage of cyclic hydrazines to diamines;[62,63] an example is provided by the reduction of the hydrazine (**16**) to the diamine (**17**) in high yield by hydrogenation over Raney nickel or platinum in the presence of hydrochloric acid.[63]

Scheme 13

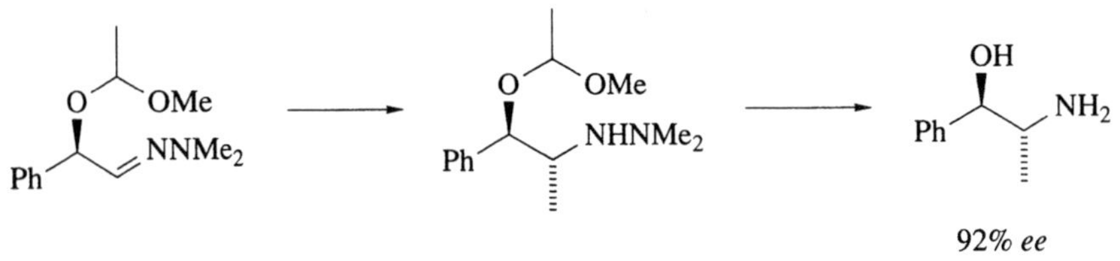

92% *ee*

Scheme 14

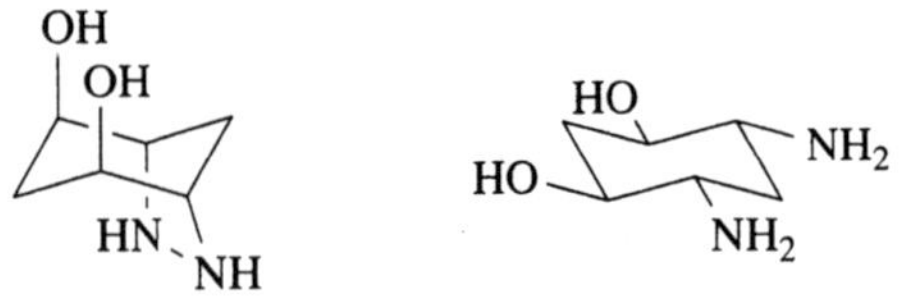

(**16**) (**17**)

The structure of the substrate determines whether other reducing agents can be used. Arylhydrazines and hydrazinium salts are more easily cleaved than simple alkylhydrazines; for example, catalytic transfer hydrogenation over palladium has been used to cleave hydrazobenzene to aniline.[18] A substance in which the bond is cleaved very easily is the cyclic dication (**18**); the N—N bridge is opened with a wide range of reducing agents, including iron in aqueous acid.[64] Hydrazides can be cleaved electroreductively or by electron transfer from metals, sodium in liquid ammonia being the most widely used reagent. Two very useful surveys have been carried out, one on electroreduction[65] and the other on metal reduction,[66] which relate the ease of reduction of hydrazides to their structure. Electrochemical reduction of cyclic hydrazides is facilitated by relief of ring strain; the presence of more than one acyl substituent in a hydrazide also makes reduction easier. Mellor and Smith investigated the use of a range of metallic reagents, including zinc in acetic acid, aluminum amalgam and sodium in liquid ammonia, for the cleavage of acyclic and cyclic hydrazides.[66] The mild reducing agent aluminum amalgam efficiently cleaved 1,2-bis(*p*-toluenesulfonyl)-1,2-dimethylhydrazine and the bis(trifluoroacetyl)hydrazine (**19a**), but failed to react with the bis(ethoxycarbonyl) derivative (**19b**). On the other hand sodium in liquid ammonia proved to be a good reducing agent for compound (**19b**) but not for (**19a**). The N—N bond of a related hydrazide was selectively cleaved by sodium in liquid ammonia in the presence of an epoxide function.

Azodicarboxylic esters are excellent dienophiles and they can also add to electron-rich double bonds as heterodienes. Adducts of both types have been used as intermediates in the synthesis of amino sugars.

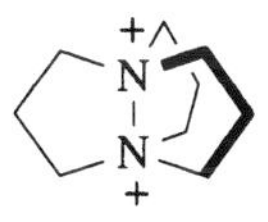

(18) 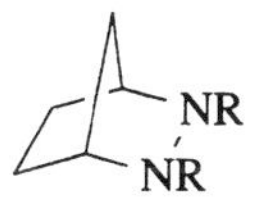 (19) a: R = $COCF_3$
b: R = CO_2Et (20)

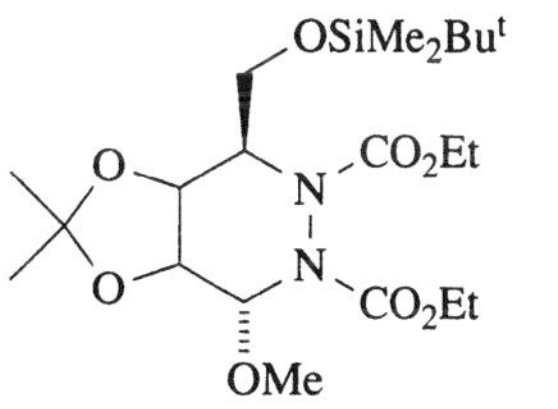

Schmidt and his coworkers have cleaved the N—N bond of tetrahydropyridazines (**20**) with sodium in liquid ammonia as a step in the synthesis of 4-aminolyxose derivatives.[67] Leblanc and coworkers have used carbohydrate-derived glycals as two-electron components in cycloadditions to dibenzyl azodicarboxylate; the adducts (*e.g.* **21** in Scheme 15) were subjected to methanolysis and the N—N bond was then cleaved using Raney nickel.[68]

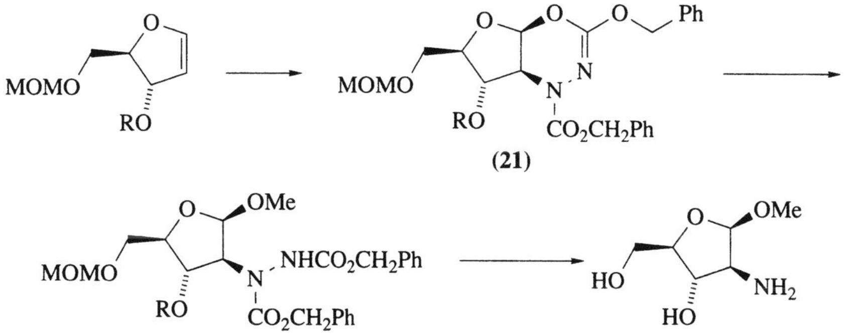

(21)

Scheme 15

Bicyclic hydrazides in which the nitrogen atoms are at the ring junctions have been cleaved by sodium in liquid ammonia and the reaction provides a good method of creating medium-ring amides.[69,70] The cleavage of the hydrazide (**22**) provides an example; this gave compound (**23**), which was a key intermediate in a synthesis of the spermidine alkaloid celacinnine.[70]

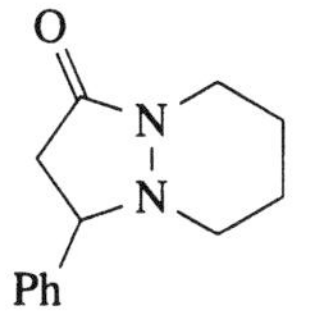

(22) 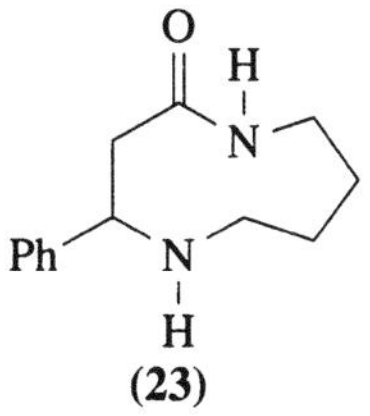 (23)

Other reagents which have occasionally been used to cleave hydrazides include diborane (which also reduces the carbonyl groups),[71] sodium naphthalenide,[69,72] *O,O*-diethyldithiophosphoric acid, $(EtO)_2PS_2H$,[73] and sulfur monochloride.[74] Nickel–aluminum alloy in aqueous methanolic potassium hydroxide is a good reagent for reductively cleaving a number of N—N bonded compounds, such as *N*-methyl-*N*-phenylhydrazine and *N,N*-dimethylnitrosamine.[19,20] Nitrosamines have also been cleaved with titanium(IV) chloride–sodium borohydride[75] and lithium aluminium hydride.

2.2.4 REDUCTION OF N—O BONDED GROUPS

2.2.4.1 Deoxygenation of Azoxy Compounds

This section covers the deoxygenation of compounds containing the structural units (**24**) and (**25**). Compounds of the first type include not only aliphatic and aromatic azoxy compounds, but also heterocyclic compounds, such as pyridazine *N*-oxides; those of the second type include dimers of nitroso compounds and heterocycles, such as pyridazine 1,2-dioxides.

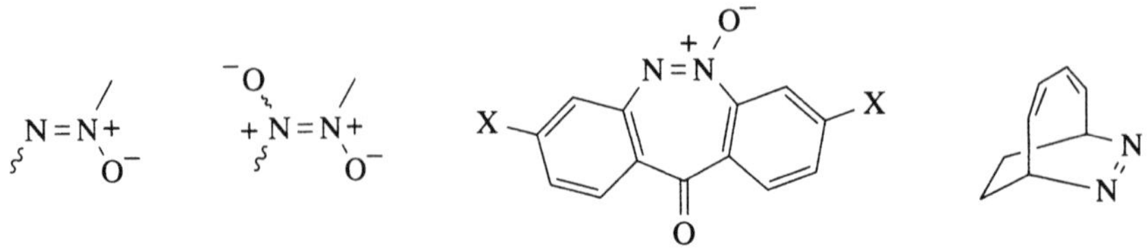

(**24**) (**25**) (**26**) (**27**)

There are several well-established methods for the deoxygenation of azoxy compounds, the most common of which are catalytic hydrogenation, reduction with lithium aluminum hydride, and the use of zinc in an alkaline medium.[1] These methods can be rather unselective, either because they cause over-reduction of the azoxy group or because other functional groups can be reduced. Phosphorus trichloride and trialkyl phosphites are more selective;[1] phosphorus trichloride has been used to deoxygenate pyridazine *N*-oxides containing azido or nitro groups.[76] Several metal carbonyls convert azoxybenzene into azobenzene;[77,78] iron pentacarbonyl gives azobenzene in 77% yield, probably by the route shown in Scheme 16.[78] Aliphatic[1] and aromatic[79] azoxy compounds are deoxygenated with magnesium in methanol or ethanol; thus, compounds (**26**; X = Br or Cl) were cleanly deoxygenated with magnesium in ethanol. The corresponding di-*N*-oxides were also deoxygenated by this reagent. Hexachlorodisilane is a mild and selective reagent for the deoxygenation of *N*-oxides of various types;[80] it was used in a synthesis of the bicyclic azo compound (**27**) from the corresponding azoxy compound.[81]

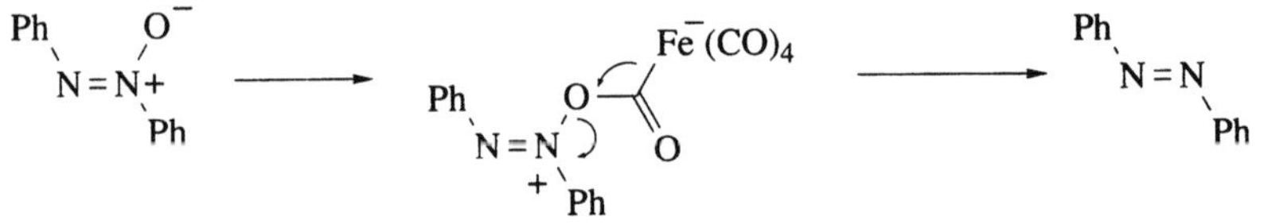

Scheme 16

Stepwise reduction of di-*N*-oxides has been achieved by using catalytic hydrogenation over palladium to remove one oxygen atom.[82] An electrochemical reduction method also allows controlled deoxygenation of these compounds.[83]

2.2.4.2 Deoxygenation of Nitrones, Nitrile Oxides and Tertiary Amine Oxides

Methods of deoxygenation of nitrones (**28**), nitrile oxides (**29**), heteroaromatic *N*-oxides (**30**) and tertiary amine oxides (**31**) are described in this section. There are some reagents, such as trialkyl phosphites, which can deoxygenate compounds of all these types as well as those in the preceding section, whereas others are more limited in scope. Oae and coworkers have outlined three distinct mechanistic types of deoxygenation process, which are illustrated in Scheme 17.[84] Clearly, a mechanism of type C will not apply to tertiary amine oxides (**31**); on the other hand, these compounds are more easily deoxygenated than heteroaromatic *N*-oxides, such as (**30**), by some reagents because the aromatic *N*-oxides are inherently more stable.

A comprehensive account of methods of deoxygenation of heteroaromatic *N*-oxides, including descriptions of experimental procedures, has been given by Ochiai.[85] The most widely used methods are catalytic reduction, reaction with phosphorus trichloride and reduction by metals or metal salts. Catalytic hydrogenation over palladium in strong acid or in a mixture of acetic acid and acetic anhydride is a good but unselective method of deoxygenation. Raney nickel in alcoholic solution can be used to deoxygenate

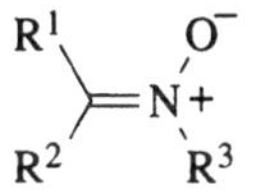

(28) 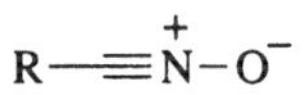 (29) 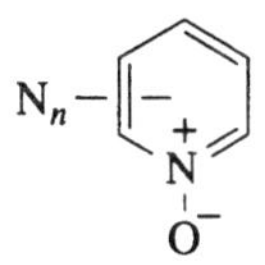 (30) 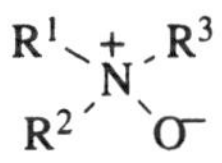 (31)

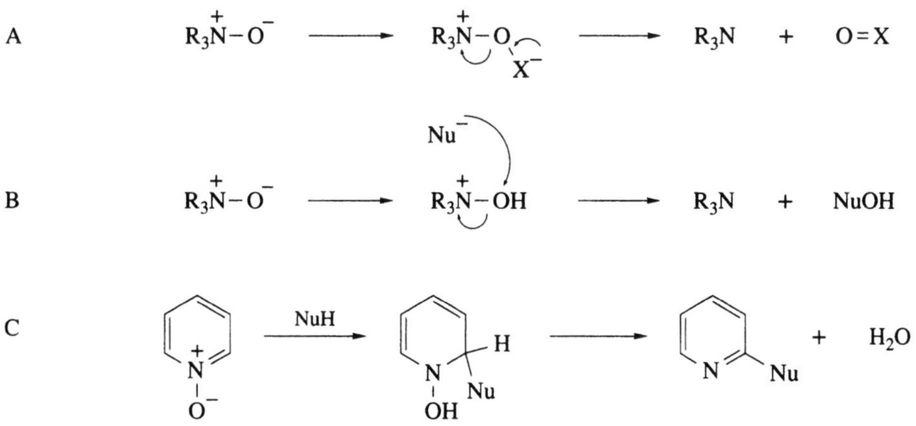

Scheme 17

N-oxides of virtually all types and it shows some selectivity; halogen substituents are unaffected and benzyloxy groups are reduced relatively slowly. Phosphorus trichloride has also been widely used; with the exception of hydroxy groups and others bearing an acidic hydrogen atom, most common functional groups are tolerated in the deoxygenation. The apparently related deoxygenation of pyridine *N*-oxides by triethyl phosphite was shown by Emerson and Rees to require the presence of oxygen and a peroxide; a radical mechanism was suggested for the deoxygenation.[86] Trimethyl phosphite has proved useful as a reagent for selective monodeoxygenation of quinoxaline 1,4-dioxides, the reaction taking place preferentially at the nitrogen atom adjacent to a carbon bearing an electron-withdrawing substituent. For example, the dioxide (**32**) gave the *N*-oxide (**33**; 81%).[87]

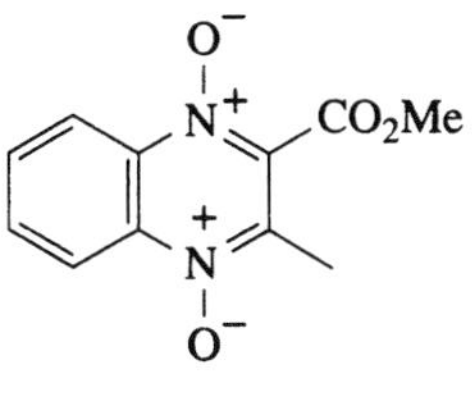

(32) 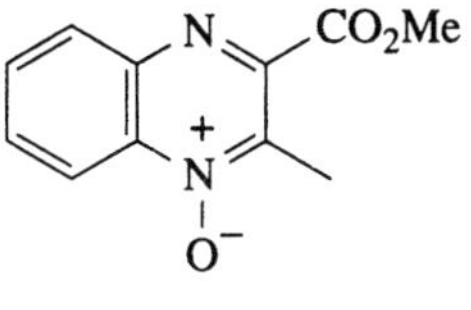 (33) 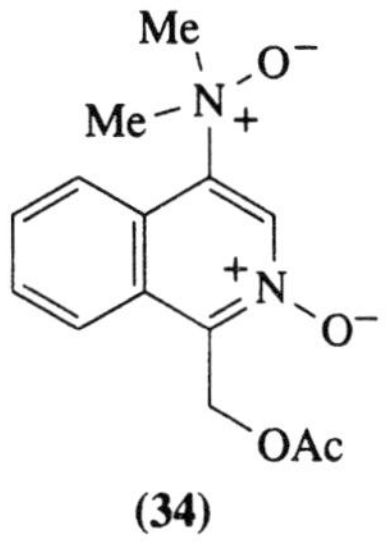 (34)

There are many other literature methods for the deoxygenation reaction, some of which are discussed in Ochiai's review.[85] Pyridine and quinoline *N*-oxides are deoxygenated when heated with iron powder or zinc in acetic acid and this was an early method for the one-pot synthesis of 4-aminopyridines from 4-nitropyridine *N*-oxides. Nitrogen monoxide or nitrosylsulfuric acid have also been used; it is possible to carry out a one-pot nitration and deoxygenation of pyridine *N*-oxides by carrying out the nitration with concentrated sulfuric and fuming nitric acids at elevated temperatures.[88] Sulfur dioxide can be used at room temperature for the selective deoxygenation of tertiary amine oxides of type (**31**), most heteroaromatic oxides being unaffected. Under more vigorous conditions pyridine and quinoline *N*-oxides are deoxygenated; reactions can conveniently be carried out using triethylamine–sulfur dioxide complex in dioxane[89] or by using sulfolene as a source of sulfur dioxide.[90] Another reagent which is selective for alkyl-substituted tertiary amine oxides under mild conditions is carbon disulfide (it will also reduce nitrones).[84] For example, the dioxide (**34**) was selectively reduced to its 2-oxide (70%) by carbon disulfide in methanol.[91] Isoquinoline 2-oxide is, however, deoxygenated by carbon disulfide at 110 °C.[92] Other reagents which have been reported to reduce amine oxides of type (**31**) under very mild conditions include nickel–aluminum alloy,[19] acetic formic anhydride,[93] tetraphenylporphyrinatoiron(II)[94] and sodium hydrogentelluride.[95] This last reagent reduces amine oxides but not sulfoxides; at pH 10 to 11 it can also be used to deoxygenate nitrones.

Some deoxygenation reagents are useful for particular types of aromatic *N*-oxides. Alkyl- and halo-pyridine oxides are cleanly reduced at 0 °C by a titanium(0) reagent prepared from titanium(IV) chloride and lithium aluminum hydride or magnesium.[96] Tributyltin hydride can deoxygenate *N*-oxides of pyridine and quinoline at 80 °C in the presence of AIBN: a radical mechanism is indicated.[97] Hexabutylditin and dichlorotetrabutylditin, $ClBu_2SnSnBu_2Cl$, are claimed to be superior reagents and will tolerate nitro, cyano, chloro and hydroxy groups.[98] Electrochemical reduction of pyridine *N*-oxide has also been described.[10] Simple pyridine and quinoline *N*-oxides have been deoxygenated under very mild conditions by dipropyl sulfoxylate, $(PrO)_2S$,[99] by sodium hypophosphite, NaH_2PO_2,[100] and by chlorotrimethylsilane in the presence of sodium iodide and zinc.[101] Hexamethyldisilane has also been used in the presence of a catalytic amount of TBAF;[102] the mechanism proposed is illustrated in Scheme 18. This is an example of type C in Scheme 17, in which deoxygenation follows nucleophilic attack at the α-carbon atom. Other reactions of this type, involving attack by organometallic reagents and by acetic anhydride, are well known and lead to the formation of a variety of 2-substituted pyridines from pyridine *N*-oxide.[103]

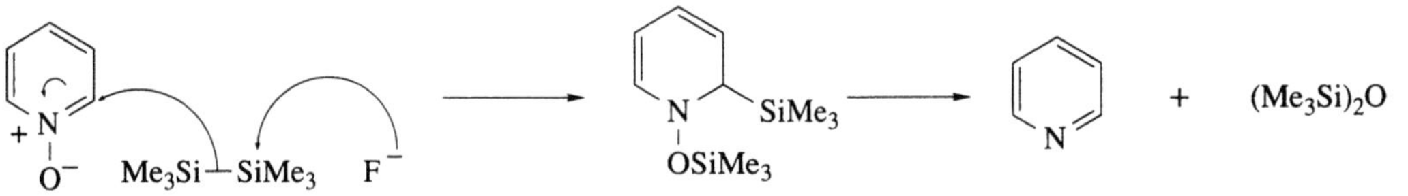

Me_3Si–$SiMe_3$ F^- → N-OSiMe₃ / SiMe₃ → pyridine + $(Me_3Si)_2O$

Scheme 18

The deoxygenation of nitrile oxides is rarely an important reaction, but some of the standard methods referred to above have been used, including reaction with trimethyl phosphite[104] and with iron pentacarbonyl.[105]

2.2.4.3 Reduction of Oximes to Imines

The reductive cleavage of the N—O bond of oximes is a reaction which has been widely used by synthetic chemists. When accompanied by reduction of the C═N bond the reaction leads to the formation of primary amines. Reactions of this type are described in Chapters 1.2–1.8, this volume. The reductive cleavage of cyclic oximes, particularly isoxazoles and 4,5-dihydroisoxazoles, has been used as a key step in several target syntheses of natural products. These cleavage reactions are covered in Chapter 3.8, this volume; however, some of the methods which clearly bring about cleavage of the N—O bond before reduction of the C═N bond are also described in this section. The products of these reactions are *N*-unsubstituted imines. Unless the imines are sterically protected or unless special precautions are employed in the work-up, the isolated products are usually carbonyl compounds formed by hydrolysis (Scheme 19). Reductive hydrolysis reactions of this type are also included here.

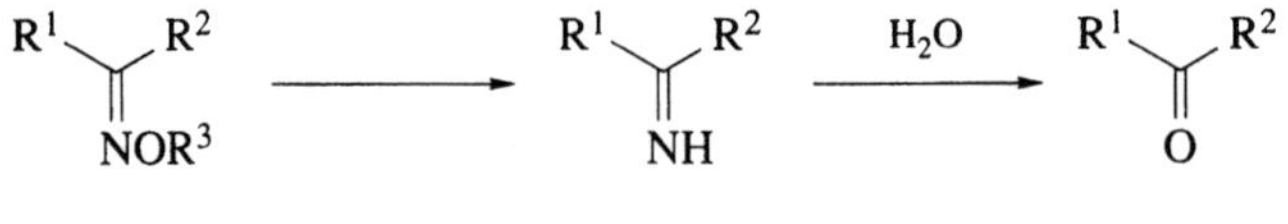

$R^1R^2C{=}NOR^3 \longrightarrow R^1R^2C{=}NH \xrightarrow{H_2O} R^1R^2C{=}O$

Scheme 19

A general method for the cleavage of 4,5-dihydroisoxazoles is hydrogenation over Raney nickel in the presence of an acid or a Lewis acid, such as boric acid[106] or boron trichloride.[107] The products are normally β-hydroxy ketones but in the case of compound (**35**), which was prepared as an intermediate in the construction of the AB-ring system of forskolin, reduction over Raney nickel gave an isolable imine (**36**).[108]

Barton *et al.* have described a particularly mild method for the conversion of ketoximes into ketimines (aldoximes give nitriles instead).[109] A mixture of tributylphosphine and diphenyl disulfide reacts under anhydrous conditions with ketoximes, probably by way of the phosphorane intermediates shown in Scheme 20, to give ketimines. The reagent is effectively a 'self-drying' one since it reacts irreversibly with water. Chlorodiphenylphosphine is also capable of deoxygenating ketoximes; with this reagent

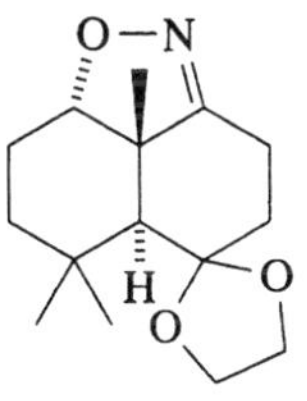

(35) 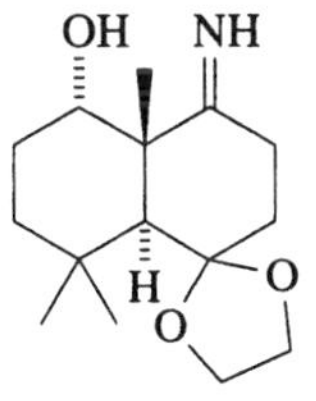 **(36)**

intermediate phosphinylimines can be isolated (Scheme 21).[110] These compounds can then be further reduced or they can be hydrolyzed to ketones.

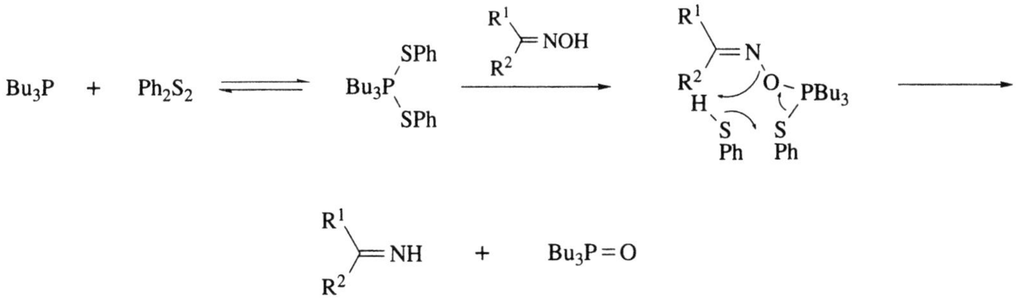

Scheme 20

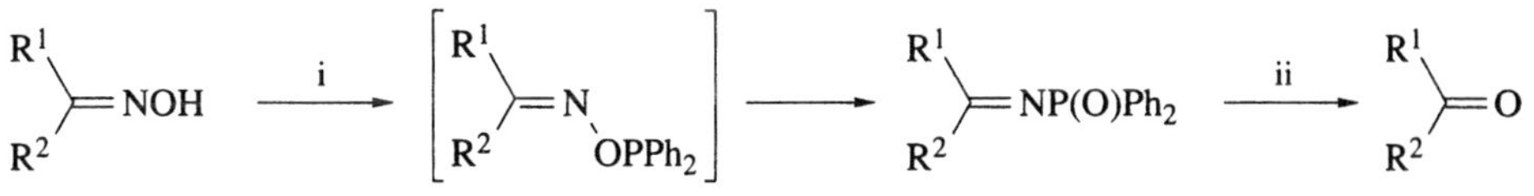

i, Ph_2PCl; ii, H_2O

Scheme 21

Low-valent metal salts have been used to bring about reductive cleavage of oximes. Corey and Richman used chromium(II) acetate to convert *O*-acetyl ketoximes into imines, which were hydrolyzed to ketones.[111] Aqueous titanium(III) chloride and vanadium(II) salts also reduce oximes; again, the imines are usually hydrolyzed *in situ*, but some hindered imines, such as compound (**37**), are isolable.[112] A method of preventing hydrolysis is to carry out the reduction in anhydrous conditions in the presence of an acylating agent. The products of such reactions, when applied to oximes of enolizable ketones, are enamides. For example, these ketoximes are converted into *N*-formylenamines when heated in acetonitrile with anhydrous titanium(III) acetate and acetic formic anhydride; cyclohexanone oxime gives the enamide (**38**; 97%; Scheme 22).[113] This type of reduction has been used by Barton and coworkers to prepare enamides from steroidal oximes. They reported that the reaction could be performed by acetic

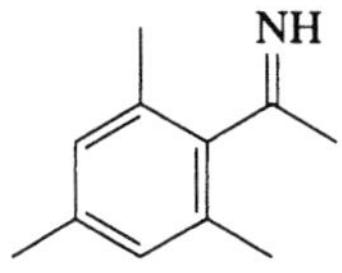

(37)

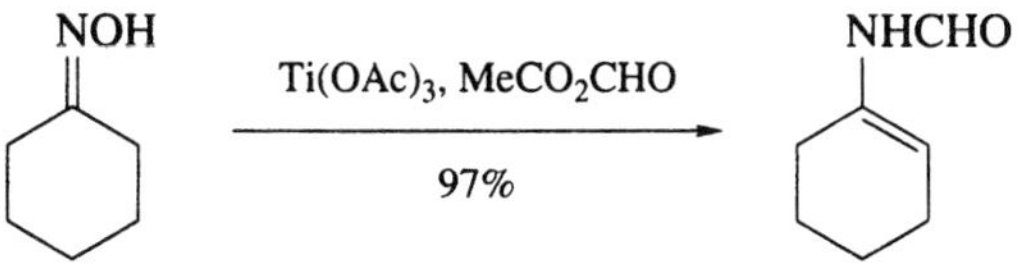

$Ti(OAc)_3$, $MeCO_2CHO$

97%

(38)

Scheme 22

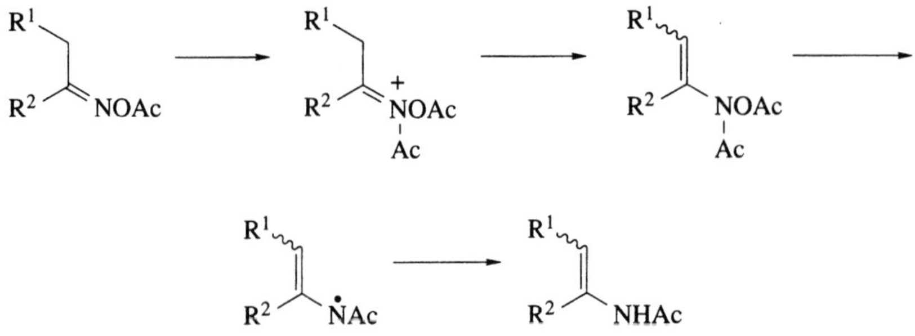

Scheme 23

anhydride alone and recommended acetic anhydride in pyridine under reflux as the reagent of choice.[114] A homolytic cleavage of the N—O bond, as shown in Scheme 23, is suggested to account for the formal reduction by acetic anhydride.

Samarium(II) iodide,[115] molybdenum hexacarbonyl[116] and iron pentacarbonyl[117] can also cleave the N—O bond of oximes. Amidines (*e.g.* **39**) can be prepared in good yield by reduction of the corresponding amidoximes with iron pentacarbonyl. The same type of reduction can be achieved electrochemically.[10] Reactions of nucleophilic reducing agents normally result in reduction of the C=N bond, but there are a few exceptions; ketoximes are reported to be cleaved to ketones, by way of the corresponding imines, with lithium aluminum hydride and HMPA.[118] Organolithium, organomagnesium and organoaluminum reagents can react with *O*-substituted oximes to give imines, which incorporate a substituent from the reagent;[119] for example, *O*-methyl oximes (**40**) of aldehydes react with organolithium reagents to give lithioketimines (**41**; Scheme 24).

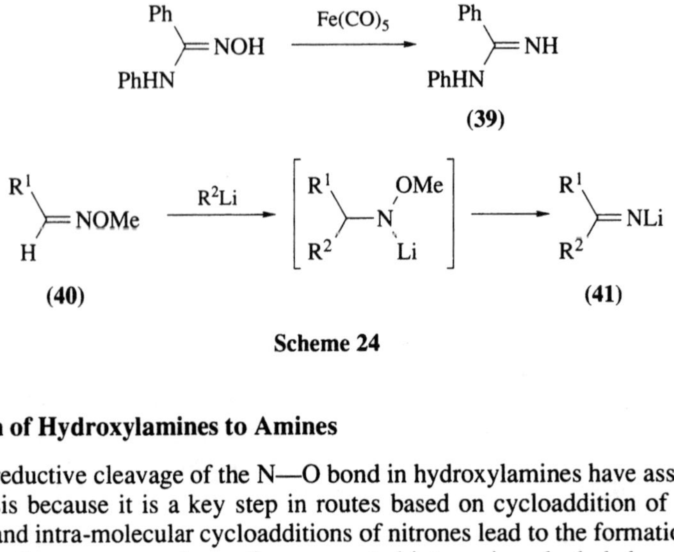

Scheme 24

2.2.4.4 Reduction of Hydroxylamines to Amines

Methods for the reductive cleavage of the N—O bond in hydroxylamines have assumed increasing importance in synthesis because it is a key step in routes based on cycloaddition of nitrones and nitroso compounds. Inter- and intra-molecular cycloadditions of nitrones lead to the formation of tetrahydroisoxazoles (Scheme 25); these compounds are then converted into amino alcohols by reductive cleavage of the N—O bonds. This type of reaction sequence has been exploited in the synthesis of a number of alkaloids.[120] Nitroso compounds, particularly those activated by electron-withdrawing substituents, act as dienophiles in the Diels–Alder reaction (Scheme 26)[121] and the cycloadducts can be used as synthetic intermediates by cleaving the N—O bonds.

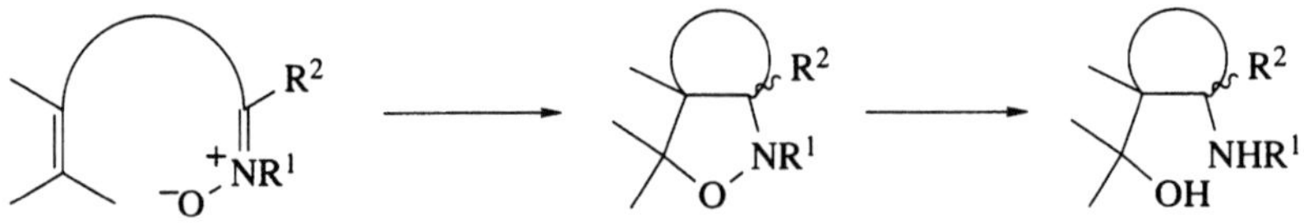

Scheme 25

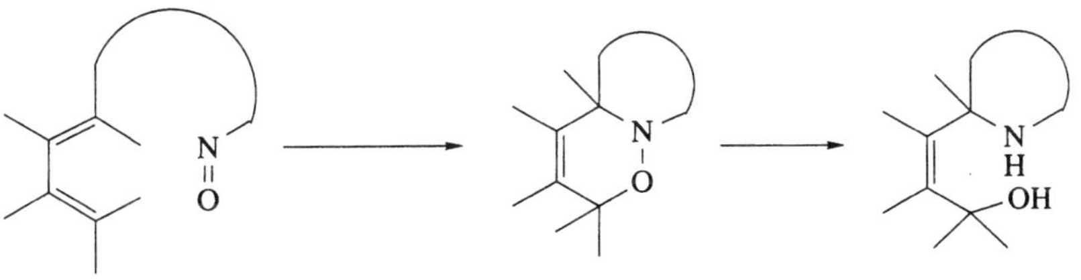

Scheme 26

N,N-Dialkylhydroxylamines can be cleaved to secondary amines by titanium(III) chloride in aqueous methanol[122] and by zinc powder in aqueous hydrochloric acid.[123] Titanium(III) chloride also reduces *N*-hydroxyimidazoles[124] and hydroxamic acids; thus the protected hydroxamic acid (**42**) was reduced to the β-lactam (**43**; 67%) with no racemization or loss of the protecting group.[125] Tetrahydro-1,2-oxazines are cleanly reduced by zinc in acetic acid. This method was used to cleave a tetrahydrooxazine (85%), generated by intramolecular Diels–Alder cycloaddition, in a synthesis of the alkaloid gephyrotoxin (Scheme 27).[126] Several tetrahydroisoxazoles have been reduced by anhydrous nickel(II) chloride and lithium aluminum hydride in THF.[127] Sodium naphthalenide proved to be the best reagent for the reduction of tri-*t*-butylhydroxylamine to di-*t*-butylamine.[128] The N—O bond of hydroxylamines can be cleaved catalytically over palladium or nickel, although any double bonds present may be preferentially reduced.[129] Diiron nonacarbonyl has also been used to cleave the bond.[130]

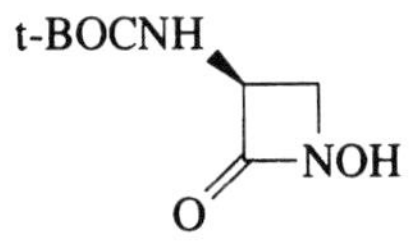

t-BOCNH NOH O

(**42**)

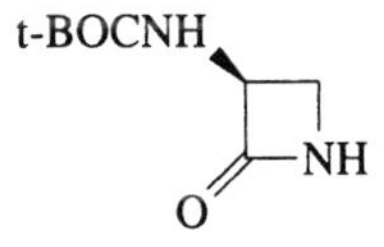

t-BOCNH NH O

(**43**)

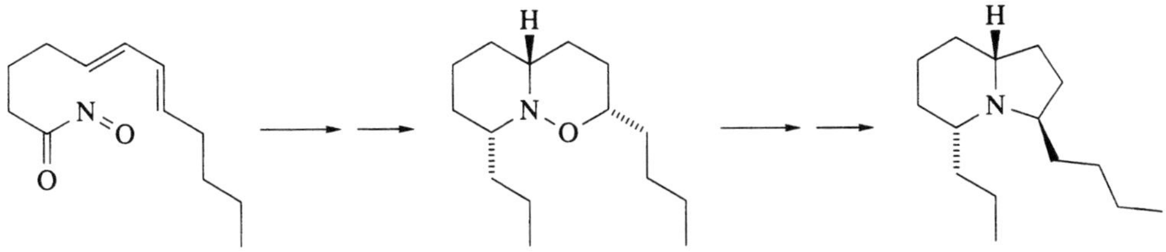

Scheme 27

When the nitrogen of a 1,2-oxazine is activated by acylation aluminum amalgam is a suitable reagent for reductive cleavage; sodium amalgam has also been used in cases where this reagent is too mild.[131] Two other very mild methods of reduction apply to hydroxylamines of specific types. Several hydroxylamines of relatively high basicity have been cleaved by dihydrolipoamide (**44**), which is formed from the coenzyme lipoamide by reduction with sodium borohydride.[132] Cyclic hydroxylammonium chlorides of the general structure (**45**) have been reduced by hydrogen in the presence of various microorganisms, the carbon–carbon double bond being unaffected.[133]

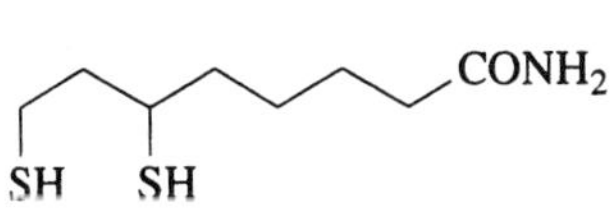

SH SH $CONH_2$

(**44**)

NH_2^+ Cl^- O R^2 R^1

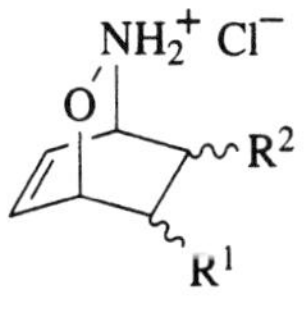

(**45**)

Oxaziridines can be regarded as a special category of cyclic hydroxylamines. Activated compounds of this type, such as the *N*-benzenesulfonyl derivative (**46**)[134] and the salt (**47**),[135] are very easily reduced. They are proving to be useful in synthesis as oxygen-transfer reagents to alkenes, thiols and other nucleophiles.

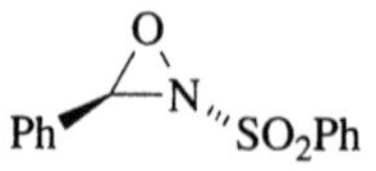

(46) 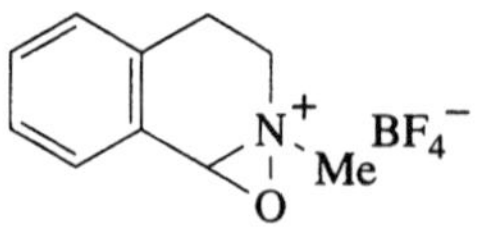 (47)

2.2.5 REDUCTION OF O—O BONDED GROUPS

2.2.5.1 Introduction

The oxygen–oxygen bond of peroxides is easily reduced and many standard reducing agents are capable of cleaving the bond efficiently. Catalytic[4] and other[136] methods have been reviewed. Whereas the reduction of hydroperoxides leads to the formation of alcohols, considerable selectivity is possible in the products derived from disubstituted peroxides. Hydroperoxides and disubstituted peroxides are, therefore, discussed separately below, even though some of the reduction methods are identical. Reductive ozonolysis of alkenes has also been included as a separate third category.

2.2.5.2 Reduction of Hydroperoxides to Alcohols

Alkyl hydroperoxides (**48**) are available by radical oxidation of alkanes, by nucleophilic displacement using peroxide anion, and by other methods. The reductive cleavage of the O—O bond, therefore, represents a useful method for the synthesis of alcohols. The reduction of hydroperoxides to alcohols can be brought about by many reagents. Catalytic methods have been widely used;[4] although the O—O bond is very susceptible to reduction, not all catalysts allow the bond to be cleaved selectively. It has been suggested that the best catalysts for selective reduction are the types which are used to reduce alkynes to alkenes; one such catalyst, palladium and lead(II) acetate, was used to reduce the hydroperoxide (**49**) to cyclohexenol (89%).[137] Salts of tin(II), iron(II) and other low-valent metals can cleave the bond. Many nucleophilic reducing agents have also been used, the most common being sodium hydrogensulfite, lithium aluminum hydride, sodium iodide, sodium thiosulfate, dimethyl sulfide and triphenylphosphine. The mechanism most often put forward for these reductions is nucleophilic attack on oxygen (Scheme 28).[138] These methods are usually more selective than standard catalytic reduction methods.

ROOH

(48) (49)

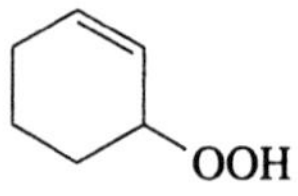

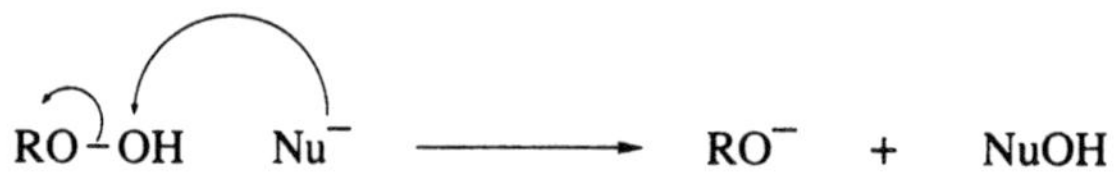

$RO{-}OH \quad Nu^- \longrightarrow RO^- + NuOH$

Scheme 28

2.2.5.3 Reduction of Disubstituted Peroxides

Disubstituted peroxides are cleaved reductively by the same reagents as are used for hydroperoxides. The usual products are alcohols derived from the two substituents, but considerable selectivity is possible with some substrates. This is illustrated by the reduction of cyclic endoperoxides of the general formula (**50**); these compounds are available from the cycloaddition of singlet oxygen to cyclic conjugated dienes.[139] The reduction of such compounds provides a good method of synthesis of *cis*-1,4-diols, which can be formed with retention of the double bond or with reduction of it (Scheme 29). It is also possible to prepare unsaturated epoxides by reduction with triphenylphosphine or other phosphorus(III) reagents.

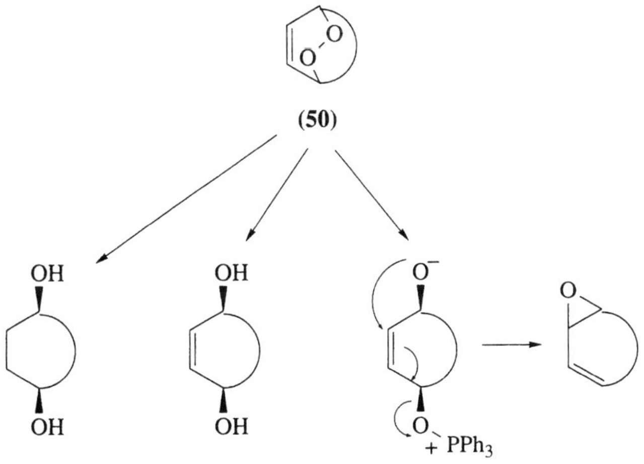

(50)

Scheme 29

Of the methods for reductive cleavage in the presence of a carbon–carbon double bond, reduction with lithium aluminum hydride and with thiourea have been most commonly used. Thiourea is very selective and has been used to reduce unstable cyclic endoperoxides, such as compound (**51**), *in situ* when they are generated by dye-sensitized photooxygenation of dienes (Scheme 30).[140] Even when reduction of a double bond is also required, it may be advantageous to use a selective method of cleavage of the peroxide. For example, in the synthesis of the sesquiterpene cybullol shown in Scheme 31, the final product was most efficiently formed in two steps from the endoperoxide, by reductive cleavage (77%) with aluminum amalgam, followed by hydrogenation over platinum (100%).[141]

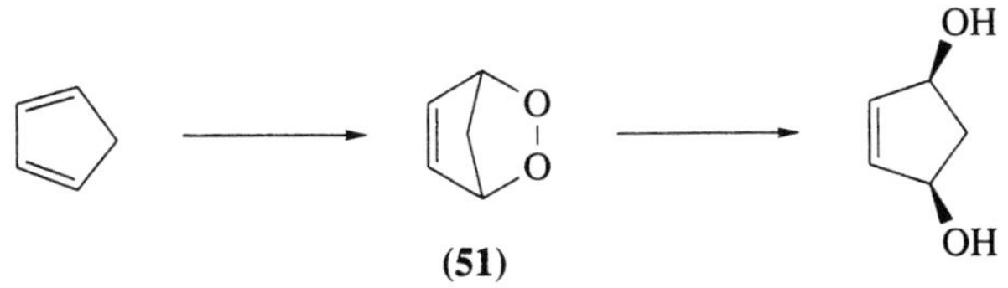

(51)

Scheme 30

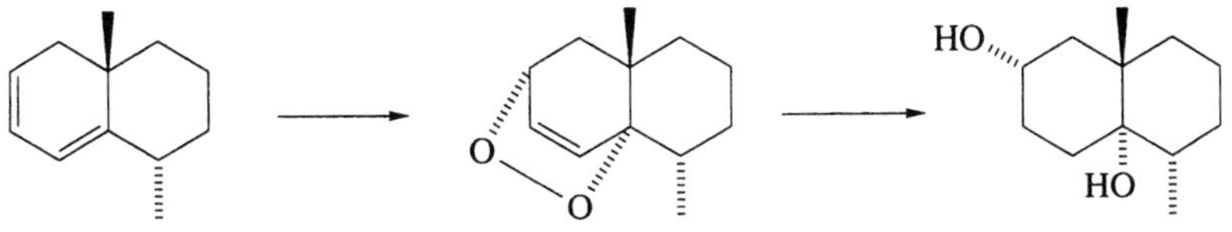

Scheme 31

Reactions of peroxides with triphenylphosphine usually result in monodeoxygenation.[142] Dialkyl peroxides react relatively slowly with triphenylphosphine and give ethers as the major products; for example, di-*t*-butyl peroxide gave di-*t*-butyl ether (81%) after being heated with triphenylphosphine at 110–120 °C for 30 h. The mechanism suggested for these reactions is an ionic one, triphenylphosphine acting as a nucleophile. The reaction illustrated in Scheme 29 can then be rationalized by the loss of triphenylphosphine oxide and intramolecular attack on the double bond. The reaction has found several useful applications in synthesis; an example is a preparation of naphthalene 1,2-oxide (Scheme 32).[143]

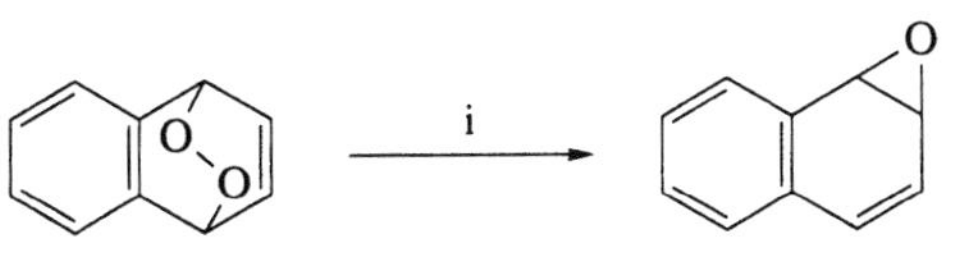

i, PPh_3, –78 °C

Scheme 32

Several thiols and sulfides (particularly dimethyl sulfide) have also been used as reducing agents. An example of this reaction which may well have biological significance is the reduction of 1,2-dioxetanes. These compounds are capable of causing damage to DNA, but they can be efficiently reduced to diols by glutathione and several other thiols. Glutathione, a tripeptide, probably protects living cells by deactivating dioxetanes and other peroxides in this way.[144] Although dimethyl sulfide can also reduce dioxetanes, the products are different, as illustrated in the example in Scheme 33. It is suggested that dimethyl sulfide acts as a reducing agent through electron transfer to the peroxide; CIDNP experiments and other evidence support this view. Ascorbic acid has been reported to cause cleavage of dilauroyl peroxide; an electron transfer mechanism has also been suggested to account for this reaction.[145]

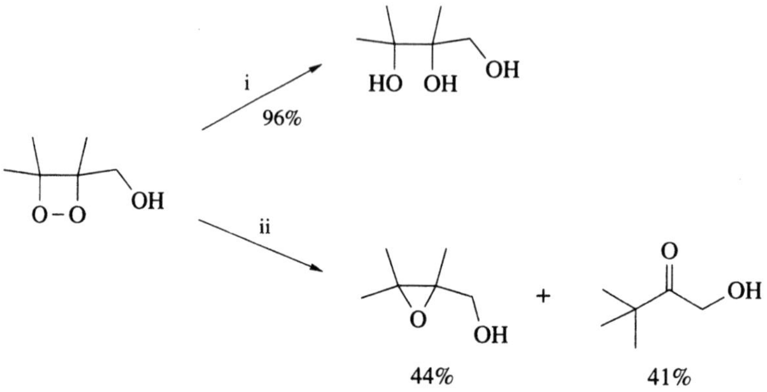

i, L-glutathione, H_2O, 5 °C; ii, Me_2S, $CHCl_3$, 5 °C

Scheme 33

Dimethyldioxirane (**52**) and other methyldioxiranes are easily generated peroxides which have proved to be very useful oxygen atom transfer agents; alkenes, sulfides and amines can all be oxidized and dimethyldioxirane is reduced to acetone.[146]

(**52**)

2.2.5.4 Reductive Ozonolysis of Alkenes

The cleavage of alkenes by ozone, usually to give carbonyl compounds as products, is a reaction which has been widely exploited in synthesis and on which a great deal of mechanistic work has been carried out. Reviews of this chemistry are available.[138,147] The ozonolysis usually leads to the formation of one of two distinct types of peroxidic product (Scheme 34); the cyclic peroxides (**53**), which are formed in nonnucleophilic solvents, and acyclic hydroperoxides, such as (**54**), which are formed in the

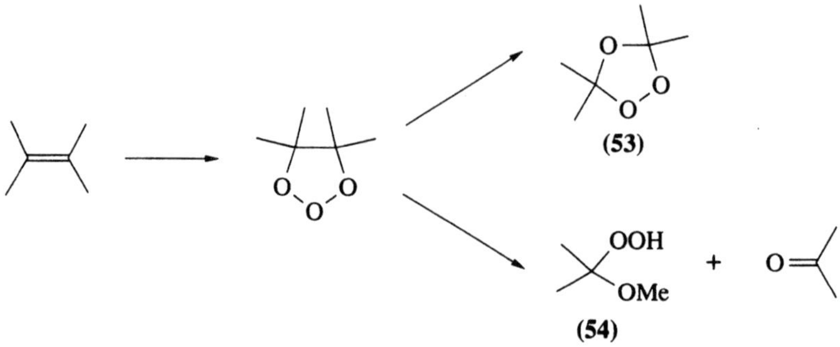

(**53**)

(**54**)

Scheme 34

presence of nucleophiles. These can then be reductively cleaved by the methods described in the preceding sections. With nucleophilic reducing agents, as shown in Scheme 28, the products can be rationalized by attack on the peroxide function. Further reduction is obviously possible with some reagents, such as lithium aluminum hydride. Dimethyl sulfide, sodium hydrogensulfite and triphenylphosphine have been commonly used when reduction to the carbonyl compounds is required. Hydrogenation over modified catalysts allows the formation of unsaturated carbonyl compounds; thus, the dialdehyde (**55**) was prepared from 1,5-cyclooctadiene by ozonolysis of one double bond followed by catalytic reduction of the ozonide. Ozonides can be reductively cleaved directly to alcohols; the best reagent for this purpose is borane–dimethyl sulfide complex.[148]

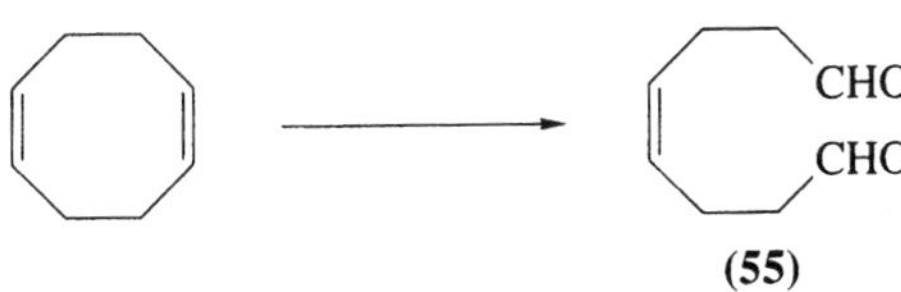

(**55**)

2.2.6 REFERENCES

1. B. T. Newbold, in 'The Chemistry of the Hydrazo, Azo, and Azoxy Groups', ed. S. Patai, Wiley, New York, 1975, p. 599.
2. A. Hajós, *Methoden Org. Chem. (Houben-Weyl)*, 1981, **4** (1d), 1.
3. J. M. Pratt and G. Swinden, *Chem. Commun.*, 1969, 1321.
4. J. Lehmann, *Methoden Org. Chem. (Houben-Weyl)*, 1980, **4** (1c), 482; P. N. Rylander, in 'Hydrogenation Methods', Academic Press, London, 1985, p. 168.
5. S. R. Sandler and W. Karo, in 'Organic Functional Group Preparations', 2nd edn., Academic Press, Orlando, 1983, vol. 1, p. 452.
6. Y. Zhang and R. Lin, *Synth. Commun.*, 1987, **17**, 329.
7. N. Kambe, K. Kondo and N. Sonoda, *Angew. Chem., Int. Ed. Engl.*, 1980, **19**, 1009.
8. M. Akiba and M. P. Cava, *Synth. Commun.*, 1984, **14**, 1119.
9. M. Kijima, Y. Nambu, T. Endo and M. Okawara, *J. Org. Chem.*, 1983, **48**, 2407.
10. J. Bracht, *Methoden Org. Chem. (Houben-Weyl)*, 1981, **4** (1d), 575.
11. B. Eistert, M. Regitz, G. Heck and H. Schwall, *Methoden Org. Chem. (Houben-Weyl)*, 1968, **10** (4), 473.
12. E. Enders, *Methoden Org. Chem. (Houben-Weyl)*, 1967, **10** (2), 169; K. H. Saunders and R. L. M. Allen, in 'Aromatic Diazo Compounds', 3rd edn., Arnold, London, 1985, p. 200.
13. F. G. James, M. J. Perkins, O. Porta and B. V. Smith, *J. Chem. Soc., Chem. Commun.*, 1977, 131.
14. R. O. Hutchins and M. G. K. Hutchins, in 'The Chemistry of Functional Groups; Supplement C: The Chemistry of Triply Bonded Functional Groups', ed. S. Patai and Z. Rappoport, Wiley, Chichester, 1983, p. 571; K. C. Brown and M. P. Doyle, *J. Org. Chem.*, 1988, **53**, 3255.
15. M. R. Smith, III and G. L. Hillhouse, *J. Am. Chem. Soc.*, 1988, **110**, 4066.
16. H. Muth and M. Sauerbier, *Methoden Org. Chem. (Houben-Weyl)*, 1980, **4** (1c), 563.
17. J. Schmitt, M. Langlois, C. Perrin and G. Callet, *Bull. Soc. Chim. Fr.*, 1969, 2004.
18. T.-L. Ho and G. A. Olah, *Synthesis*, 1977, 169.
19. G. Lunn, E. B. Sansone and L. K. Keefer, *Synthesis*, 1985, 1104.
20. L. K. Keefer and G. Lunn, *Chem. Rev.*, 1989, **89**, 459.
21. C. Shiraki, H. Saito, K. Takahashi, C. Urakawa and T. Hirata, *Synthesis*, 1988, 399.
22. K. Faber, H. Hönig and P. Seufer-Wasserthal, *Tetrahedron Lett.*, 1988, **29**, 1903.
23. T. Sheradsky, in 'The Chemistry of the Azido Group', ed. S. Patai, Interscience, London, 1971, p. 333; see also E. F. V. Scriven and K. Turnbull, *Chem. Rev.*, 1988, **88**, 297.
24. E. J. Corey, K. C. Nicolaou, R. D. Balanson and Y. Machida, *Synthesis*, 1975, 590; G. Kunesch, *Tetrahedron Lett.*, 1983, **24**, 5211.
25. L. D. Cama, W. J. Leanza, T. R. Beattie and B. G. Christensen, *J. Am. Chem. Soc.*, 1972, **94**, 1408; Y. Gaoni, *Tetrahedron Lett.*, 1988, **29**, 1591.
26. D. Balderman and A. Kalir, *Synthesis*, 1978, 24.
27. T. Gartiser, C. Selve and J.-J. Delpuech, *Tetrahedron Lett.*, 1983, **24**, 1609; A. A. Malik, S. B. Preston, T. G. Archibald, M. P. Cohen and K. Baum, *Synthesis*, 1989, 450.
28. S. C. Shim, K. N. Choi and Y. K. Yeo, *Chem. Lett.*, 1986, 1149.
29. S. N. Maiti, M. P. Singh and R. G. Micetich, *Tetrahedron Lett.*, 1986, **27**, 1423; see also S. N. Maiti, P. Spevak and A. V. Reddy, *Synth. Commun.*, 1988, **18**, 1201.
30. T.-L. Ho, M. Henninger and G. A. Olah, *Synthesis*, 1976, 815.
31. B. Stanovnik, M. Tišler, S. Polanc and M. Gračner, *Synthesis*, 1978, 65.
32. S. Polanc, B. Stanovnik and M. Tišler, *Synthesis*, 1980, 830.
33. F. Rolla, *J. Org. Chem.*, 1982, **47**, 4327.
34. K. Soai, S. Yokoyama and A. Ookawa, *Synthesis*, 1987, 48.
35. S. J. Clarke, G. W. J. Fleet and E. M. Irving, *J. Chem. Res. (S)*, 1981, 17; for other hydride reductions see G. Eberle, I. Lagerlund, I. Ugi and R. Urban, *Tetrahedron*, 1978, **34**, 977; S. C. Shim and K. N. Choi, *Tetrahedron Lett.*, 1985, **26**, 3277.
36. T. Adachi, Y. Yamada, I. Inoue and M. Saneyoshi, *Synthesis*, 1977, 45.
37. S. Gronowitz, C. Westerlund and A.-B. Hörnfeldt, *Acta Chem. Scand., Ser. B*, 1975, **29**, 224.

38. B. A. Belinka, Jr. and A. Hassner, *J. Org. Chem.*, 1979, **44**, 4712.
39. H. Bayley, D. N. Standring and J. R. Knowles, *Tetrahedron Lett.*, 1978, 3633.
40. S. S. Nikam, R. K. Shastri, A. D. Sahasrabudhe, V. A. Mudumbai and S. Ramanathan, *Chem. Ind. (London)*, 1982, 720; T. Rosen, I. M. Lico and D. T. W. Chu, *J. Org. Chem.*, 1988, **53**, 1580.
41. M. Bartra, F. Urpí and J. Vilarrasa, *Tetrahedron Lett.*, 1987, **28**, 5941.
42. N. Knouzi, M. Vaultier and R. Carrié, *Bull. Soc. Chim. Fr.*, 1985, 815.
43. W. S. Mungall, G. L. Greene, G. A. Heavner and R. L. Letsinger, *J. Org. Chem.*, 1975, **40**, 1659.
44. A. Koziara, K. Osowska-Pacewicka, S. Zawadzki and A. Zwierzak, *Synthesis*, 1985, 202.
45. H. Suzuki, H. Tani and S. Ishida, *Bull. Chem. Soc. Jpn.*, 1985, **58**, 1861.
46. P. A. S. Smith, C. D. Rowe and L. B. Bruner, *J. Org. Chem.*, 1969, **34**, 3430; P. Spagnolo and P. Zanirato, *J. Org. Chem.*, 1982, **47**, 3177.
47. J. N. Reed and V. Snieckus, *Tetrahedron Lett.*, 1983, **24**, 3795; see also N. S. Narasimhan and R. Ammanamanchi, *Tetrahedron Lett.*, 1983, **24**, 4733.
48. A. Hassner, P. Munger and B. A. Belinka, Jr., *Tetrahedron Lett.*, 1982, **23**, 699.
49. B. M. Trost and W. H. Pearson, *J. Am. Chem. Soc.*, 1983, **105**, 1054.
50. A. Hassner, G. J. Matthews and F. W. Fowler, *J. Am. Chem. Soc.*, 1969, **91**, 5046; J. A. Deyrup, in 'Small Ring Heterocycles', ed. A. Hassner, Wiley, New York, 1983, part 1, p. 1.
51. A. Suzuki, S. Sono, M. Itoh, H. C. Brown and M. M. Midland, *J. Am. Chem. Soc.*, 1971, **93**, 4329; H. C. Brown, M. M. Midland and A. B. Levy, *J. Am. Chem. Soc.*, 1972, **94**, 2114; H. C. Brown, M. M. Midland, A. B. Levy, A. Suzuki, S. Sono and M. Itoh, *Tetrahedron*, 1987, **43**, 4079.
52. G. W. Kabalka, N. M. Goudgaon and Y. Liang, *Synth. Commun.*, 1988, **18**, 1363.
53. D. A. Evans and A. E. Weber, *J. Am. Chem. Soc.*, 1987, **109**, 7151.
54. B. Carboni, M. Vaultier and R. Carrié, *Tetrahedron*, 1987, **43**, 1799; *Tetrahedron Lett.*, 1988, **29**, 1279.
55. T. M. Paterson, R. K. Smalley and H. Suschitzky, *Tetrahedron Lett.*, 1977, 3973.
56. J. E. Baldwin, R. M. Adlington and I. M. Newington, *J. Chem. Soc., Chem. Commun.*, 1986, 176.
57. N. H. Khan and A. R. Kidwai, *J. Org. Chem.*, 1973, **38**, 822.
58. J. E. McMurry and M. G. Silvestri, *J. Org. Chem.*, 1975, **40**, 1502.
59. P. A. S. Smith and C. R. Messing, *J. Org. Chem.*, 1988, **53**, 2959.
60. E. J. Corey, R. J. McCaully and H. S. Sachdev, *J. Am. Chem. Soc.*, 1970, **92**, 2476; S. Kiyooka, K. Takeshima, H. Yamamoto and K. Suzuki, *Bull. Chem. Soc. Jpn.*, 1976, **49**, 1897; H. Takahashi and Y. Suzuki, *Chem. Pharm. Bull.*, 1983, **31**, 4295; D. Enders, H. Schubert and C. Nübling, *Angew. Chem., Int. Ed. Engl.*, 1986, **25**, 1109; S. E. Denmark, O. Nicaise and J. P. Edwards, *J. Org. Chem.*, 1990, **55**, 6219.
61. D. A. Claremon, P. K. Lumma and B. T. Phillips, *J. Am. Chem. Soc.*, 1986, **108**, 8265.
62. H. Stetter and H. Spangenberger, *Chem. Ber.*, 1958, **91**, 1982; S. F. Nelsen and M. R. Willi, *J. Org. Chem.*, 1984, **49**, 1; for a review of methods of N—N bond cleavage see M. F. Marshalkin and L. N. Yakhontov, *Russ. Chem. Rev.*, 1986, **55**, 1016.
63. T. Suami, S. Ogawa, H. Uchino and Y. Funaki, *J. Org. Chem.*, 1975, **40**, 456.
64. R. W. Alder, R. B. Sessions, A. J. Bennet and R. E. Moss, *J. Chem. Soc., Perkin Trans. 1*, 1982, 603.
65. L. Horner and M. Jordan, *Justus Liebigs Ann. Chem.*, 1978, 1505.
66. J. M. Mellor and N. M. Smith, *J. Chem. Soc., Perkin Trans. 1*, 1984, 2927.
67. A. K. Forrest, R. R. Schmidt, III, G. Huttner and I. Jibril, *J. Chem. Soc., Perkin Trans. 1*, 1984, 1981.
68. B. J. Fitzsimmons, Y. Leblanc and J. Rokach, *J. Am. Chem. Soc.*, 1987, **109**, 285; B. J. Fitzsimmons, Y. Leblanc, N. Chan and J. Rokach, *J. Am. Chem. Soc.*, 1988, **110**, 5229.
69. D. S. Kemp, M. D. Sidell and T. J. Shortridge, *J. Org. Chem.*, 1979, **44**, 4473.
70. H. H. Wasserman, R. P. Robinson and H. Matsuyama, *Tetrahedron Lett.*, 1980, 3493; H. Matsuyama, M. Kobayashi and H. H. Wasserman, *Heterocycles*, 1987, **26**, 85.
71. H. Feuer and F. Brown, Jr., *J. Org. Chem.*, 1970, **35**, 1468; D. S. Kemp, J. C. Chabala and S. A. Marson, *Tetrahedron Lett.*, 1978, 543; G. Lawton, C. J. Moody and C. J. Pearson, *J. Chem. Soc., Perkin Trans. 1*, 1987, 877.
72. A. J. Fry, W. E. Britton, R. Wilson, F. D. Greene and J. G. Pacifici, *J. Org. Chem.*, 1973, **38**, 2620.
73. S. Oae, N. Tsujimoto and A. Nakanishi, *Bull. Chem. Soc. Jpn.*, 1973, **46**, 535.
74. F. Bellesia, U. M. Pagnoni and A. Pinetti, *J. Chem. Res. (S)*, 1982, 222.
75. S. Kano, Y. Tanaka, E. Sugino, S. Shibuya and S. Hibino, *Synthesis*, 1980, 741.
76. T. Itai, in 'Pyridazines', ed. R. N. Castle, Wiley, New York, 1973, p. 675; J. P. Freeman and R. C. Grabiak, *J. Org. Chem.*, 1976, **41**, 3970.
77. R. B. King and C. A. Harmon, *J. Organomet. Chem.*, 1975, **86**, 139; H. Alper and M. Gopal, *J. Org. Chem.*, 1981, **46**, 2593.
78. H. Alper and J. T. Edward, *Can. J. Chem.*, 1970, **48**, 1543.
79. C. P. Joshua and P. K. Ramdas, *Synthesis*, 1974, 873; see also R. Sanchez, G. Vest, W. Scott and P. S. Engel, *J. Org Chem.*, 1989, **54**, 4026.
80. A. G. Hortmann, J. Koo and C.-C. Yu, *J. Org. Chem.*, 1978, **43**, 2289.
81. H. Olsen and J. P. Snyder, *J. Am. Chem. Soc.*, 1978, **100**, 285.
82. W. Lüttke and V. Schabacker, *Justus Liebigs Ann. Chem.*, 1965, **687**, 236; M. Nakadate, S. Sueyoshi and I. Suzuki, *Chem. Pharm. Bull.*, 1970, **18**, 1211.
83. G. Belot and C. Degrand, *Tetrahedron Lett.*, 1976, 153.
84. T. Yoshimura, K. Asada and S. Oae, *Bull. Chem. Soc. Jpn.*, 1982, **55**, 3000.
85. E. Ochiai, in 'Aromatic Amine Oxides', Elsevier, Amsterdam, 1967, p. 184; see also R. Balicki, *Synthesis*, 1989, 645; R. Balicki, L. Kaczmarek and M. Malinowski, *Synth. Commun.*, 1989, **19**, 897.
86. T. R. Emerson and C. W. Rees, *J. Chem. Soc.*, 1962, 1917.
87. J. P. Dirlam and J. W. McFarland, *J. Org. Chem.*, 1977, **42**, 1360.
88. F. Kröhnke and H. Schäfer, *Chem. Ber.*, 1962, **95**, 1098.
89. G. A. Olah, M. Arvanaghi and Y. D. Vankar, *Synthesis*, 1980, 660.
90. C. Kaneko, R. Hayashi, H. Fujii and A. Yamamoto, *Chem. Pharm. Bull.*, 1978, **26**, 3582.

91. S. Nishimura and M. Saneyoshi, *Chem. Pharm. Bull.*, 1980, **28**, 1695; J. S. Wieczorek and E. Plażek, *Recl. Trav. Chim. Pays-Bas*, 1965, **84**, 785.
92. H. Seidl, R. Huisgen and R. Grashey, *Chem. Ber.*, 1969, **102**, 926.
93. N. Tokitoh and R. Okazaki, *Chem. Lett.*, 1985, 1517.
94. N. Miyata, T. Santa and M. Hirobe, *Chem. Pharm. Bull.*, 1984, **32**, 377.
95. D. H. R. Barton, A. Fekih and X. Lusinchi, *Tetrahedron Lett.*, 1985, **26**, 4603.
96. M. Malinowski, *Synthesis*, 1987, 732; M. Malinowski and L. Kaczmarek, *Synthesis*, 1987, 1013; see also J. M. McCall and R. E. TenBrink, *Synthesis*, 1975, 335.
97. S. Kozuka, T. Akasaka, S. Furumai and S. Oae, *Chem. Ind. (London)*, 1974, 452.
98. B. Jousseaume and E. Chanson, *Synthesis*, 1987, 55.
99. H. Kagami and S. Motoki, *J. Org. Chem.*, 1978, **43**, 1267.
100. S. K. Boyer, J. Bach, J. McKenna and E. Jagdmann, Jr., *J. Org. Chem.*, 1985, **50**, 3408.
101. T. Morita, K. Kuroda, Y. Okamoto and H. Sakurai, *Chem. Lett.*, 1981, 921.
102. H. Vorbrüggen and K. Krolikiewicz, *Tetrahedron Lett.*, 1983, **24**, 5337.
103. B. C. Uff, in 'Comprehensive Heterocyclic Chemistry' ed. A. R. Katritzky and C. W. Rees, Pergamon Press, Oxford, 1984, vol. 2, p. 315.
104. C. Grundmann and H.-D. Frommeld, *J. Org. Chem.*, 1965, **30**, 2077.
105. N. A. Genco, R. A. Partis and H. Alper, *J. Org. Chem.*, 1973, **38**, 4365.
106. D. P. Curran, *J. Am. Chem. Soc.*, 1983, **105**, 5826; D. P. Curran and P. B. Jacobs, *Tetrahedron Lett.*, 1985, **26**, 2031.
107. A. P. Kozikowski and J. G. Scripko, *J. Am. Chem. Soc.*, 1984, **106**, 353.
108. P. G. Baraldi, A. Barco, S. Benetti, G. P. Pollini, E. Polo and D. Simoni, *J. Chem. Soc., Chem. Commun.*, 1986, 757.
109. D. H. R. Barton, W. B. Motherwell, E. S. Simon and S. Z. Zard, *J. Chem. Soc., Perkin Trans. 1*, 1986, 2243.
110. B. Krzyzanowska and W. J. Stec, *Synthesis*, 1978, 521; R. O. Hutchins, A. F. Abdel-Magid, Y. P. Stercho and A. Wambsgans, *J. Org. Chem.*, 1987, **52**, 702; R. O. Hutchins and M. C. Rutledge, *Tetrahedron Lett.*, 1987, **28**, 5619.
111. E. J. Corey and J. E. Richman, *J. Am. Chem. Soc.*, 1970, **92**, 5276.
112. G. H. Timms and E. Wildsmith, *Tetrahedron Lett.*, 1971, 195.
113. D. H. R. Barton, T. Bowles, S. Husinec, J. E. Forbes, A. Llobera, A. E. A. Porter and S. Z. Zard, *Tetrahedron Lett.*, 1988, **29**, 3343.
114. R. B. Boar, J. F. McGhie, M. Robinson, D. H. R. Barton, D. C. Horwell and R. V. Stick, *J. Chem. Soc., Perkin Trans. 1*, 1975, 1237.
115. N. R. Natale, *Tetrahedron Lett.*, 1982, **23**, 5009.
116. M. Nitta and T. Kobayashi, *J. Chem. Soc., Chem. Commun.*, 1982, 877; P. G. Baraldi, A. Barco, S. Benetti, S. Manfredini and D. Simoni, *Synthesis*, 1987, 276.
117. A. Dondoni and G. Barbaro, *J. Chem. Soc., Chem. Commun.*, 1975, 761; S. Nakanishi, Y. Shirai, K. Takahashi and Y. Otsuji, *Chem. Lett.*, 1981, 869; H. Alper and J. T. Edward, *J. Org. Chem.*, 1967, **32**, 2938.
118. N. Balachander, S.-S. Wang and C. N. Sukenik, *Tetrahedron Lett.*, 1986, **27**, 4849.
119. S. Itsuno, K. Miyazaki and K. Ito, *Tetrahedron Lett.*, 1986, **27**, 3033; K. Maruoka, T. Miyazaki, M. Ando, Y. Matsumura, S. Sakane, K. Hattori and H. Yamamoto, *J. Am. Chem. Soc.*, 1983, **105**, 2831.
120. J. J. Tufariello, in '1,3-Dipolar Cycloaddition Chemistry', ed. A. Padwa, Wiley, New York, 1984, vol. 2, p. 83.
121. D. L. Boger and S. M. Weinreb, 'Hetero Diels–Alder Methodology in Organic Synthesis', Academic Press, San Diego, 1987, p. 71.
122. S.-I. Murahashi and Y. Kodera, *Tetrahedron Lett.*, 1985, **26**, 4633.
123. S.-I. Murahashi, Y. Imada, Y. Taniguchi and Y. Kodera, *Tetrahedron Lett.*, 1988, **29**, 2973.
124. B. H. Lipshutz and M. C. Morey, *Tetrahedron Lett.*, 1984, **25**, 1319.
125. P. G. Mattingly and M. J. Miller, *J. Org. Chem.*, 1980, **45**, 410.
126. H. Iida, Y. Watanabe and C. Kibayashi, *J. Am. Chem. Soc.*, 1985, **107**, 5534; *Tetrahedron Lett.*, 1986, **27**, 5513.
127. J. J. Tufariello, H. Meckler and K. P. A. Senaratne, *Tetrahedron*, 1985, **41**, 3447.
128. E. J. Corey and A. W. Gross, *Tetrahedron Lett.*, 1984, **25**, 491.
129. E. C. Taylor, K. McDaniel and J. Skotnicki, *J. Org. Chem.*, 1984, **49**, 2500; G. Augelmann, J. Streith and H. Fritz, *Helv. Chim. Acta*, 1985, **68**, 95; J. E. Baldwin, M. Otsuka and P. M. Wallace, *J. Chem. Soc., Chem. Commun.*, 1985, 1549; Y. Takeuchi and F. Furusaki, *Adv. Heterocycl. Chem.*, 1977, **21**, 207.
130. Y. Becker, A. Eisenstadt and Y. Shvo, *Tetrahedron*, 1978, **34**, 799.
131. G. E. Keck and S. A. Fleming, *Tetrahedron Lett.*, 1978, 4763; G. E. Keck, S. A. Fleming, D. Nickell and P. Weider, *Synth. Commun.*, 1979, **9**, 281; M. Saburi, G. Kresze and H. Braun, *Tetrahedron Lett.*, 1984, **25**, 5377; T. P. Burkholder and P. L. Fuchs, *J. Am. Chem. Soc.*, 1988, **110**, 2341.
132. M. Kijima, Y. Nambu and T. Endo, *J. Org. Chem.*, 1985, **50**, 1140.
133. K. Klier, G. Kresze, D. Werbitzy and H. Simon, *Tetrahedron Lett.*, 1987, **28**, 2677.
134. L. C. Vishwakarma, O. D. Stringer and F. A. Davis, *Org. Synth.*, 1987, **66**, 203.
135. C. Hanquet, X. Lusinchi and P. Milliet, *Tetrahedron Lett.*, 1988, **29**, 3941.
136. R. A. Sheldon, in 'The Chemistry of Peroxides', ed. S. Patai, Wiley, Chichester, 1983, p. 161.
137. M. Rebeller and G. Clément, *Bull. Soc. Chim. Fr.*, 1964, 1302.
138. V. N. Odinokov and G. A. Tolstikov, *Russ. Chem. Rev. (Engl. Transl.)*, 1981, **50**, 636.
139. M. Balci, *Chem. Rev.*, 1981, **81**, 91; I. Saito and S. S. Nittala, in 'The Chemistry of Peroxides', ed. S. Patai, Wiley, New York, 1983, p. 311.
140. C. Kaneko, A. Sugimoto and S. Tanaka, *Synthesis*, 1974, 876.
141. W. A. Ayer, L. M. Browne and S. Fung, *Can. J. Chem.*, 1976, **54**, 3276.
142. L. Horner and W. Jurgeleit, *Justus Liebigs Ann. Chem.*, 1955, **591**, 138.
143. M. Schäfer-Ridder, U. Brocker and E. Vogel, *Angew. Chem., Int. Ed. Engl.*, 1976, **15**, 228.

144. W. Adam, B. Epe, D. Schiffmann, F. Vargas and D. Wild, *Angew. Chem., Int. Ed. Engl.*, 1988, **27**, 429.
145. K. F. Jenkins, S. A. Hershberger, J. W. Hershberger and R. Marshall, *J. Org. Chem.*, 1988, **53**, 3393.
146. R. W. Murray, *Chem. Rev.*, 1989, **89**, 1187.
147. P. S. Bailey, in 'Ozonation in Organic Chemistry', Academic Press, New York, 1978, vol. 1.
148. L. A. Flippin, D. W. Gallagher and K. Jalali-Araghi, *J. Org. Chem.*, 1989, **54**, 1430.

2.3

Reduction of S=O and SO_2 to S, P=O to P, and of S—X to S—H

IAN W. J. STILL

University of Toronto, Mississauga, Canada

2.3.1 INTRODUCTION

The reduction of the important sulfoxide functional group has attracted considerable attention during the past 20 years. This topic was extensively reviewed in the late 1970s,[1,2] and there have been several excellent reviews of the area very recently.[3–5] The reviews by Grossert[4] and Madesclaire[5] present a particularly thorough coverage and, in each case, include references as recent as 1987. Effective methods for the reduction of the related sulfone functionality have proved to be more elusive, in large part because of the much stronger S=O bonds present, but this topic is also given comprehensive treatment in the review by Grossert[4] and in a much earlier one by Durst.[6]

The reduction of P=O is clearly a more daunting task than the reduction of the S=O bond in a sulfone and the dearth of literature references to this topic reflects this fact. In contrast, much work has been done on the reduction of disulfides and other systems with S—X bonds,[7] including their selenium and tellurium counterparts.[8]

A number of reviews which relate to reduction in general and only in part to the types of reduction presently being considered may also be of interest to the reader. These include a review of hydride reductions in general by Brown and Krishnamurthy[9] and of metal alkoxyaluminum hydrides in particular by Malek.[10] The important role of electrophilic catalysis in achieving reductive S—O bond cleavage is summarized in a review of the applications of iodotrimethylsilane by Olah and Narang.[11]

2.3.2 REDUCTION OF S=O AND RELATED FUNCTIONAL GROUPS

2.3.2.1 Reduction (Deoxygenation) of Sulfoxides

This section will deal only with those processes which involve a straightforward reduction or deoxygenation of the sulfoxide and not, for example, with processes such as the Pummerer or sila-Pummerer reactions, which involve reduction of the sulfoxide with concomitant oxidation at one of the α-carbon atoms. Also not included will be those reactions such as desulfinylation, which involve reductive cleavage of one of the C—S bonds. As indicated in the Introduction, there have been several reviews of sulfoxide deoxygenation to the corresponding sulfide.[1-5] While many different procedures for achieving this type of reduction have been reported, it is possible to identify three major approaches, which will be discussed in turn. Although most of the literature cited is from the last 10 to 15 years, the trends already emerging in the mid 1970s have become more clearly established, in particular the need for reagents selective enough to reduce, for example, a sulfoxide in the presence of a ketone or other carbonyl functionality. Thus, while vigorous methods such as lithium aluminum hydride or dissolving metal reductions may still be used in simple cases, they have now been largely supplanted by more sophisticated procedures.

The first broad class of reagents to find general application to the reduction or deoxygenation of sulfoxides was that of the 'low valent' transition metals. These methods rely on reducing agents such as molybdenum(II),[12,13] molybdenum(III),[12,14] titanium(II),[15] vanadium(II)[14] or tungsten(III),[12] which are often generated *in situ* by zinc reduction (equation 1). These reagents give excellent yields of sulfides of all major types and can be employed in the range 25–100 °C. Other functional groups, such as ketone, ester, nitrile, (aromatic) nitro, phosphine oxide or sulfone, were shown by San Filippo and coworkers[12] to be unaffected and the reducing capabilities of such low valent metal systems have been reviewed by Ho.[16] More recently, the use of cobalt(II) in combination with $NaBH_4$,[17] and the use of titanium(III),[18] have been reported for the reduction of sulfoxides to sulfides.

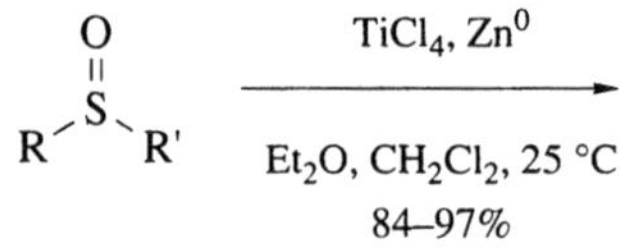

$$R-S(=O)-R' \xrightarrow[\text{Et}_2\text{O, CH}_2\text{Cl}_2\text{, 25 °C; 84–97\%}]{\text{TiCl}_4\text{, Zn}^0} R-S-R' \quad (1)$$

A second, single reagent procedure relies upon the use of pentavalent phosphorus compounds, in which the formation of a very strong P=O bond provides a thermodynamic driving force for the deoxygenation to proceed, often at 25 °C or lower. Thus, Still *et al.*[19] used phosphorus pentasulfide (tetraphosphorus decasulfide) for the reduction of several types of sulfoxide to sulfide, generally in yields of 60% or higher, although lower yields were recorded for sterically congested sulfoxides and for heterocyclic sulfoxides. Although P_4S_{10} is well known to be capable of effecting carbonyl–thiocarbonyl exchange, the example in equation (2) illustrates that ketones (as well as ester, amide, nitro or halide compounds) are not affected under the mild conditions used. Phosphorus pentasulfide is ineffective in reducing sulfones, sulfinates or sulfites but was successfully applied in the reduction of sulfimides (sulfilimines) and selenoxides (see Sections 2.3.2.4 and 2.3.2.5). A plausible mechanistic suggestion for the P_4S_{10} reduction of sulfoxides, which involves a four-center transition state and initial formation of an unisolable thiosulfoxide intermediate, has been made independently by Still *et al.*[19] and by Baechler *et al.*[20] (Scheme 1).

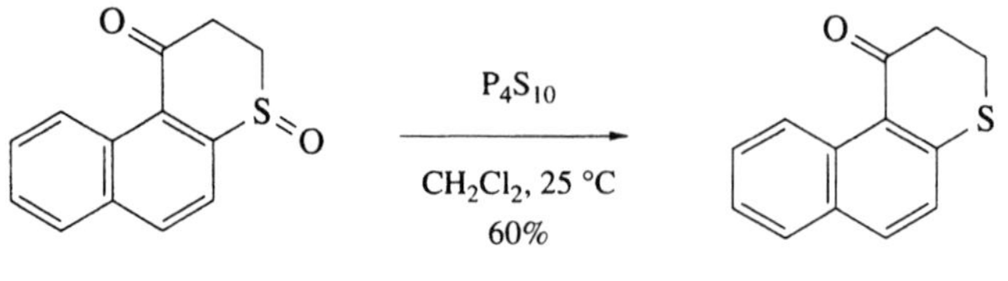

$$\xrightarrow[\text{CH}_2\text{Cl}_2\text{, 25 °C; 60\%}]{\text{P}_4\text{S}_{10}} \quad (2)$$

$$S_3P{=}S \; + \; R-S(=O)-R' \longrightarrow [\text{four-center TS}] \longrightarrow S_3P{=}O \; + \; R-S(=S)-R' \longrightarrow R-S-R' \; + \; 1/8\; S_8$$

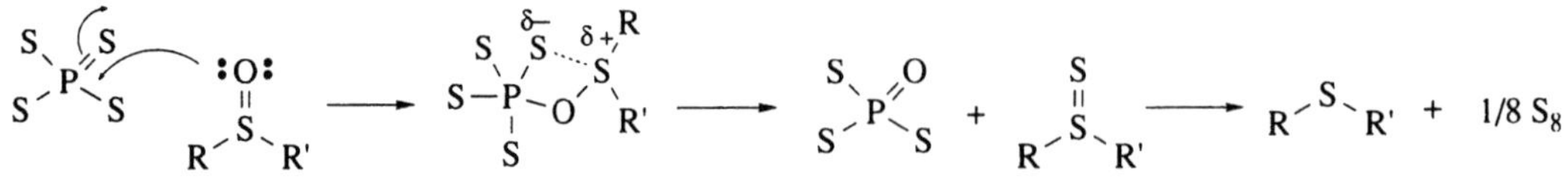

Scheme 1

Still *et al.*[21] have also developed thiophosphoryl bromide ($PSBr_3$) as an alternative to P_4S_{10} which does not suffer from the drawback of insolubility in organic solvents. This reagent, readily prepared from P_4S_{10}, red phosphorus and bromine, gives very high yields of the sulfides in reaction times that vary from a few minutes to a few hours at 25 °C. These features make $PSBr_3$ generally superior for the sulfoxide–sulfide interconversion but the reverse was found to be true for the related conversion of sulfines (thione *S*-oxides) to thiones[22] shown in equation (3), where the requirement for aqueous work-up in the $PSBr_3$ reactions created serious difficulties in isolating water-sensitive thiones. In a promising development, Yokoyama *et al.*[23] have reported that an analog (**1**) of the organic phase soluble Lawesson reagent reduces dibenzyl sulfoxide to the sulfide in 83% yield.

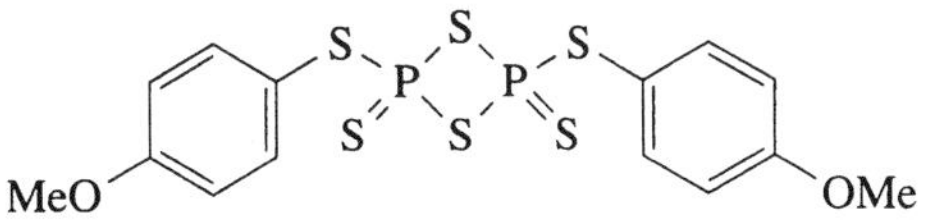

MeO–C_6H_4–S–P(=S)(μ-S)$_2$P(=S)–S–C_6H_4–OMe

(1)

$C_6H_4O_2$P–Cl

(2)

$$RR'C{=}S{=}O \xrightarrow[\text{CH}_2\text{Cl}_2,\ 25\ ^\circ\text{C}\quad 58\text{–}95\%]{\text{P}_4\text{S}_{10}\text{ or PSBr}_3} RR'C{=}S \tag{3}$$

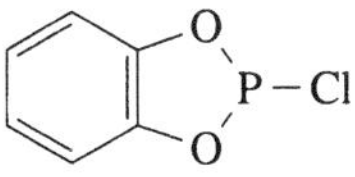

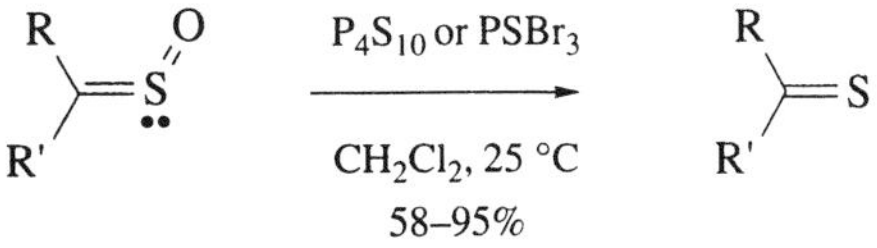

Other pentavalent phosphorus reagents which have been used for the deoxygenation of sulfoxides include phosphorus pentachloride[24] and various combinations of phosphines[25] or phosphites[26] with I_2 which also appear to involve pentavalent phosphorus species. In the former case, the addition of a simple enamine or tertiary aromatic amine was found to be essential to prevent formation of Pummerer-type products. $POCl_3$ or $SiCl_4$ could be used instead of PCl_5. Olah *et al.*[25] have suggested that a similar mechanism is operative in the reduction of sulfoxides with the reagent tris(dimethylamino)phosphine–iodine–sodium iodide (equation 4), which they find to be superior to an earlier version of this system using triphenylphosphine. The iodine appears to play a catalytic role, *via* initial formation of $(Me_2N)_3PI_2$. While not involving pentavalent phosphorus analogs, the application of other phosphorus iodides such as P_2I_4 or PI_3 to the reduction of sulfoxides and selenoxides has been reported by Denis and Krief[27,28] and, independently, by Suzuki[29] in the case of the former reagent. Finally, in an elegant application of a sulfoxide-deoxygenating agent first reported by Chasar and Pratt,[30] Ottenheijm and coworkers[31] have successfully used 1,2-phenylene phosphorochloridite (**2**) as the reducing agent in a novel reduction–oxidation sequence for the selective epimerization (at sulfur) of a series of sulfoxides, in overall yields of 26–37%.

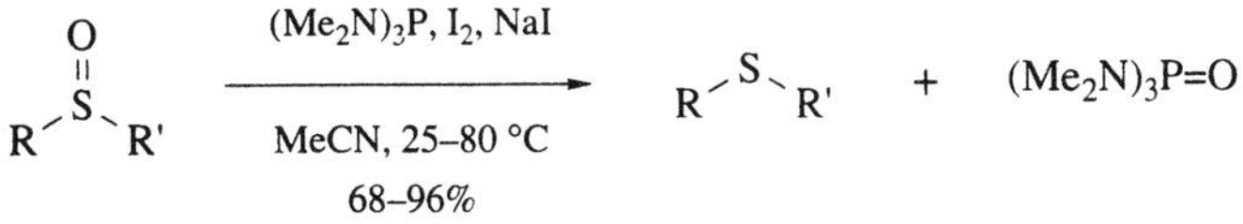

$$R{-}S(=O){-}R' \xrightarrow[\text{MeCN, 25–80 °C}\quad 68\text{–}96\%]{(\text{Me}_2\text{N})_3\text{P, I}_2\text{, NaI}} R{-}S{-}R' + (\text{Me}_2\text{N})_3\text{P=O} \tag{4}$$

The third and final subclass of methodology in the reduction of sulfoxides to sulfides relies upon what is basically a dual reagent or 'push–pull' strategy. The nucleophilicity of the sulfoxide oxygen is utilized in an initial reaction with an electrophilic species, usually a halosilane, followed by displacement of the silanolate leaving group at the sulfonium center by various reducing species, such as I^-,[32] Br^-,[33] or PhSH.[34] The reaction is completed (Scheme 2) by a final redox step involving the generation of I_2, Br_2, or PhSSPh, respectively. In several cases, two distinct reagents for the initial two steps were employed but in other instances a single reagent could be used, such as the use by Oae and coworkers[34] of the re-

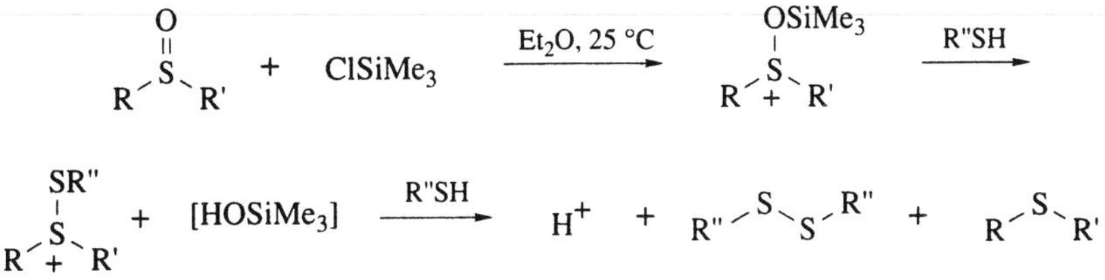

$$R{-}S(=O){-}R' + \text{ClSiMe}_3 \xrightarrow{\text{Et}_2\text{O, 25 °C}} R{-}\overset{+}{S}(\text{OSiMe}_3){-}R' \xrightarrow{\text{R''SH}}$$

$$R{-}\overset{+}{S}(\text{SR''}){-}R' + [\text{HOSiMe}_3] \xrightarrow{\text{R''SH}} \text{H}^+ + \text{R''SSR''} + R{-}S{-}R'$$

Scheme 2

agent $PhSSiMe_3$ or of $ISiMe_3$ by Olah *et al.*[32] In the former case, the advantage of a single reagent was offset by lowered yields and much longer reaction times.

Much ingenuity has been displayed in modifying the reagent system used in the above type of reduction. The basic thermodynamic driving force in all such cases is the formation of a strong Si—O bond, accompanied by the sacrifice of a relatively weak X—Si bond. Thus, reagents such as $(Me_3Si)_2S$[35] and the analogous selenides and tellurides[36] have been used to reduce sulfoxides, as well as several selenoxides and telluroxides (see Section 2.3.2.5), in high yields at 25 °C. Other variations include the use of various chlorosilanes in combination with zinc dust (equation 5).[37,38] These procedures give good to excellent yields even for heterocyclic sulfoxides, with little or no Pummerer product formation, and can be carried out in the presence of hydroxy or carboxy groups or double bonds. Olah *et al.*[39] have used trichloromethylsilane–sodium iodide as an alternative to the rather unstable iodotrimethylsilane.[32] The sulfides are produced, in yields of over 85%, in minutes at 25 °C using acetonitrile as the solvent, by this technique. While not directly relevant to the present topic, it is of interest to note that Miller *et al.*[40] have recently developed a general route to triarylsulfonium salts from diaryl sulfoxides, using similar activation by electrophilic silicon reagents, accompanied by nucleophilic displacement of silanolate by aryllithium or aryl Grignard reagents (equation 6).

$$\mathrm{RS(O)R'} + \mathrm{Cl_2SiMe_2} \xrightarrow[\text{0–5 °C, 70–95\%}]{\mathrm{Zn^0,\ acetone}} \mathrm{RSR'} + \mathrm{ZnCl_2} + [\mathrm{Me_2Si{=}O}] \qquad (5)$$

$$\mathrm{Ar_2S{=}O} + \mathrm{Ar'MgBr} \xrightarrow{\mathrm{TfOSiR_3}} \mathrm{Ar_2S^{+}{-}Ar'} + \mathrm{TfO^-} \qquad (6)$$

In addition to the above variations, the nature of the Lewis acid electrophilic activator has also been examined, from the standpoint of elements other than silicon, and several recent papers have utilized boron electrophiles in this way.[41–43] The yields of sulfide obtained are very high and the reductions can often be carried out at 0 °C or lower. The reagents used comprise either combined electrophile–nucleophile systems such as BF_3–NaI[41] or thexylchloroborane–dimethyl sulfide,[42] or the bromodimethylborane or bromoborane (**3**) reagents used by Guindon *et al.*[43] These workers also used BBr_3 but, unlike the other boron bromides, it reacted only sluggishly with diaryl sulfoxides and had the additional disadvantage of reacting with any free hydroxy groups present. Other electrophile–nucleophile combinations of a related type which have been examined include the strong Lewis acid AlI_3[44] and the weak Lewis acid $FeCl_3$ used in combination with $NaBH_4$ in aqueous ethanol.[45]

(**3**) (**4**) (**5**)

Other procedures which presumably operate by a very similar type of mechanism include the use of alternative electrophilic activators such as acid anhydrides[46] and acid chlorides,[47] while Olah *et al.*[48] have also used amine complexes of SO_2 or SO_3 as electrophiles, with sodium iodide in each case as the nucleophilic reductant. Among the simplest systems of this type are those reported by Chasar and Shockcor[49] using HCl–KI in chloroform, which gives excellent yields even with the keto sulfoxide (**4**), and, interestingly, even *t*-butyl bromide in refluxing chloroform, reported by Marchelli and coworkers,[50] which gives sulfides almost quantitatively. An unusual procedure reported by Olah *et al.*[51] was the use of the fluorinated triazine (**5**) or the corresponding trichloro analog (cyanuric chloride).

A number of procedures have been reported for the reduction of sulfoxides to sulfides which do not appear to fit the above broad categories but which are also versatile procedures, employing relatively low temperatures and short reaction times. For example, Drabowicz and Mikolajczyk[52] used the system thiourea *S,S*-dioxide–iodine in refluxing acetonitrile to reduce several sulfoxides to sulfides in yields of 89–95%. The sulfinic acid (**6**) is quite well known as a reducing agent but the catalytic role played by iodine in the present reduction is unclear. Amos[53] has successfully used the poly(styryl)diphenylphosphine–CCl_4 system to produce sulfides in yields of 81–99%. The method is selective for sulfoxides in the presence of aromatic nitro groups and, while very similar mechanistically to the reagent systems used by Olah *et al.*,[25] is also useful in that it avoids the separation and purification difficulties associated with the use of triphenylphosphine. Yoshihara and coworkers[54] have used the thioformamizium salt (**7**), readily prepared from *N,N*-dimethylthioformamide, in the reduction in dry chloroform of a series of sulfoxides, with yields in the range 53–82%.

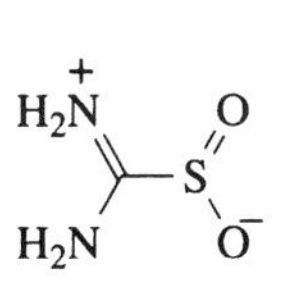

(**6**) 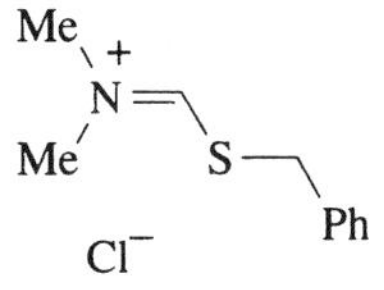 (**7**)

2.3.2.2 Reduction of Sulfones

The reduction of the S—O bond in sulfones is much more difficult, partly because this bond is stronger than the S—O bond in the corresponding sulfoxides.[55] In addition, powerful nucleophilic hydride reducing agents such as $LiAlH_4$[56] meet with limited success as a result of the normal steric hindrance to attack at tetracoordinate sulfur atoms and competing α- or α′-deprotonation by these strongly basic species. The use of electrophilic hydride reagents such as diisobutylaluminum hydride (DIBAL-H)[57] has been shown to be somewhat effective but only under severe conditions, presumably due to the weakly nucleophilic nature of the sulfone oxygen atoms. Grossert[4] has discussed these and related mechanistic problems in a recent comprehensive review. A further complication not found with the sulfoxide reduction is the possibility of obtaining the intermediate sulfoxide, as well as the sulfide.[58]

In spite of these obstacles, Still and Szilagyi[59] were able to achieve the so far unique reduction of sulfones to sulfoxides using a variant of the 'push–pull' strategy which has proved to be so successful in the sulfoxide reduction. After many unsuccessful attempts, it was found that electrophilic activation of one of the S—O bonds in a series of aliphatic and aromatic sulfones could be achieved using the *p*-chlorophenyl carbonium ion, derived from the corresponding diazonium tetrafluoroborate by thermal decomposition. Reduction of the aryloxysulfoxonium salt (**8**) was achieved by the addition of a $NaBH_4$–Al_2O_3 reagent to yield the corresponding sulfoxide (Scheme 3). No further reduction takes place under these conditions. The proposed mechanism, involving nucleophilic attack at the sulfoxonium center by hydride to give the *S*-protonated sulfoxide (**9**), was confirmed as the major pathway by deuterium-labeling experiments.[60] The yields are moderate and the nature of the electrophile restricts the reaction to simple, monofunctional sulfones, but this remains the only general method of converting sulfones into the corresponding sulfoxides.

p-$ClC_6H_4N_2^+ BF_4^-$, Δ; (**8**) BF_4^-; $NaBH_4$–Al_2O_3; (**9**) + ArO^-; $-H^+$; 25–55%

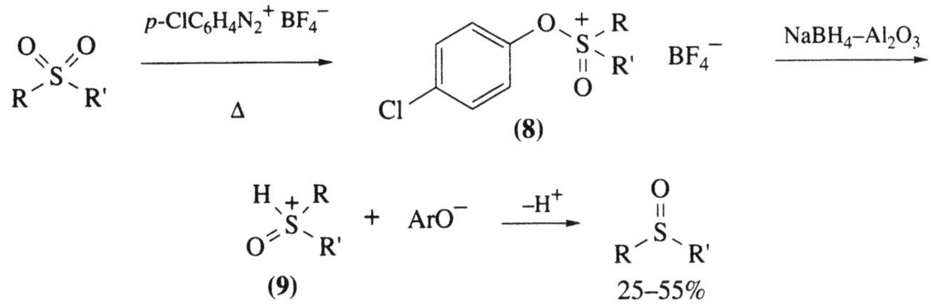

Scheme 3

Because of the importance of the sulfone moiety in organic synthesis as an 'activating' group and the need, therefore, for its facile subsequent removal, it is no surprise that much more attention has been paid to methods of reductively cleaving the C—S bond in sulfones. Many such methods are available and they have been thoroughly examined in the recent review by Grossert.[4] They will not be discussed further here.

2.3.2.3 Reduction of Sulfonic and Sulfinic Acids and Their Derivatives

The pronounced affinity of sulfur for the higher (+4, +6) oxidation states means that other subclasses of compound exist which contain S=O units. Many of these are little known but a number of successful reductions have been reported, especially of sulfonic acids and their ester, acid halide, *etc.* derivatives. These will be discussed in this section.

Although a common functional group in organic chemistry, the sulfonic acid (RSO_3H) unit is remarkably inert to reduction to RSH or any intermediate oxidation level. In contrast to carboxylic acids, for example, it is claimed that sulfonic acids resist refluxing with $LiAlH_4$ in dibutyl ether for three days.[61] On the other hand, it has long been known that sulfonyl chlorides can be reduced to the corresponding thiols by using amalgamated zinc–sulfuric acid[62] or tin(II) chloride–hydrochloric acid.[63] This works very well in the aromatic series where chlorosulfonyl groups can be introduced by chlorosulfonation and then reduced to the corresponding thiols. Lithium aluminum hydride can also be used to achieve this reduction but at –70 °C the intermediate sulfinic acid can be isolated in over 60% yield.[64] Brown and Rao[65] also achieved the sulfonic acid to thiol reduction using a $NaBH_4$–$AlCl_3$ system. Much more recently Nose and Kudo[66] have reported the use of $NaBH_4$ alone in THF at 0 °C for the conversion of aromatic sulfonyl chlorides to the corresponding sulfinic acids in good yields. The same reagent at higher temperatures produced mixtures of the disulfide and arenethiol.

In recent times, several groups have attempted to employ the indirect method of (electrophilic) activation of one of the S—O bonds, followed by reduction, often with a mild reducing agent. One of the earliest such attempts was reported by Chan *et al.*,[67] who found that sulfonyl chlorides and sulfonate esters, as well as the analogous sulfenyl and sulfinyl compounds, could be reduced to the corresponding disulfides (equation 7) using an $HSiCl_3$–NPr_3 system. Oae and coworkers[68] were the first to develop a successful reduction of arene- and alkane-sulfonic acids to thiols, using a system, $(CF_3CO)_2O$–$Bu_4N^+I^-$ at 25 °C, very similar to that used earlier[46] for the reduction of sulfoxides. Yields, however, were moderate and the thiol was present predominantly as the trifluoroacetate ester (equation 8). Oae and Togo[69] found Ph_3P–I_2 successful in reducing arenesulfonic acids quantitatively to arenethiols in refluxing benzene. While the reduction also proceeds with alkanesulfonic acids, the thiols so formed are transformed into alkyl iodides under the reaction conditions (equation 9). In the same way presumably, arenesulfonic acids are reduced to thiols with Ph_3P–Ar_2S_2 but the alkanesulfonic acids give high yields of the mixed sulfides (RSAr) in this reaction.[70] Interestingly, Oae and Togo[71] also found that both arene- and alkanesulfonic acids could be reduced to the corresponding disulfides in good yields, using P_4O_{10} or other polyphosphoric acid derivatives with KI. The reaction is presumed to proceed *via* the sulfonyl iodide — it has been known for many years that sulfonyl halides are quantitatively reduced to the disulfides with HI.[72] Olah *et al.*[73] used iodotrimethylsilane in dichloromethane at 25 °C to convert sulfonyl, sulfinyl and sulfenyl halides into the analogous disulfides. Here iodotrimethylsilane can be viewed as a milder, more convenient form of the HI reagent.

$$2\ PhSO_2Cl \xrightarrow[\text{PhH, 80 °C}\ 85\%]{HSiCl_3,\ Pr_3N} PhS{-}SPh \quad (7)$$

$$RSO_2OH \xrightarrow[CH_2Cl_2,\ 25\ °C]{(CF_3CO)_2O,\ Bu_4N^+\ I^-} R{-}SH + RS{-}C(=O)CF_3 \quad (8)$$

$$RSO_2OH \xrightarrow[\text{PhH, 25 °C}]{Ph_3P,\ I_2} R{-}SH \xrightarrow{Ph_3P,\ I_2} RS{-}P^+Ph_3\ I^- \longrightarrow RI \quad (9)$$

Other mild procedures for achieving the conversion of sulfonic acids to disulfides include BBr_3–KI in refluxing 1,2-dichloroethane[74] and P_2I_4 in refluxing acetonitrile.[75] Interestingly, P_4S_{10} in sulfolane at 100 °C converted sulfonic acids into polysulfides (ArS_nAr, $n > 2$).[76] The polysulfides were subsequently converted into the thiols by $LiAlH_4$ or $NaBH_4$ reduction.[76] Cipris and Pouli[77] found that a large excess of the corresponding thiol converted aromatic sulfonyl chlorides into disulfides according to equation (10), while AlI_3 alone has been shown recently to convert sulfonyl chlorides into disulfides in yields of over 80%.[44] Two other reductive transformations of sulfonyl compounds of a less simple type have been noted. Arenesulfonyl azides were converted into aryl alkyl sulfides in moderate yield on treatment with trialkylboranes, followed by alkaline H_2O_2 (equation 11).[78] Kagabu *et al.*[79] have reported the successful conversion of a series of arenesulfonyl chlorides with trimethylsilyl cyanide into the corresponding thiocyanates. The reaction proceeds *via* the (sodium) sulfinate as intermediate and requires a reducing system (Na_2SO_3–K_2CO_3) for successful completion (equation 12).

$$ArSO_2Cl + 5\ Ar{-}SH \longrightarrow 3\ ArSSAr + HCl + 2\ H_2O \quad (10)$$

$$ArSO_2N_3 \xrightarrow{R_3B} \xrightarrow[\text{aq. NaOH}]{H_2O_2} ArSR \ \ (45\text{–}70\%) \quad (11)$$

$$ArSO_2Cl \xrightarrow{Na_2SO_3} ArS(O)ONa \xrightarrow[\text{reflux, MeCN}]{Me_3SiCN} ArSCN \ \ (47\text{–}87\%) \quad (12)$$

In contrast to the other types of S═O functionality already discussed, there has been very little systematic study of the reduction of sulfinic acids and their derivatives. This probably reflects nothing more than the relative paucity of such species in organic synthesis, since sulfinates have often been identified or proposed as fleeting intermediates in the reduction of sulfonic acid derivatives, as discussed above. Pinnick *et al.*[80] described an interesting reductive coupling of aromatic sulfinate salts to disulfides, using ethyl phosphinate in DMSO. Yields are moderate and the reaction is confined to aromatic sulfinates. A mechanism has been proposed *via* a sulfoxy sulfone intermediate. In a reaction reminiscent of the work of Cipris and Pouli,[77] Oae *et al.*[81] have reduced sulfinic acids to disulfides in better yields (80–100%) than those in the earlier procedure by using $ClSiMe_3$ as coreagent, along with a large excess of the thiol reducing agent (equation 13). If a mixed disulfide is required, yields are lower due to disproportionation. Two examples of reductions of sulfinyl systems in recent syntheses of natural products were noted.[82,83] One involves the reduction of allylic sulfinamides to allylic thiols with $LiAlH_4$ (equation 14), and the other the conversion of sulfinic acids into *S*-acetyl thiols by a sequence involving $SOCl_2$, thioacetic acid and triphenylphosphine (equation 15).

$$RS(O)OH + 3\ R'{-}SH + 4\ ClSiMe_3 \longrightarrow RSSR' + R'SSR' + 2\ (Me_3Si)_2O + 4\ HCl \quad (13)$$

$$CH_2{=}CHCH_2S(O)NHSO_2Ph \xrightarrow[\substack{0\text{–}25\ ^\circ C \\ 72\text{–}93\%}]{LiAlH_4} CH_2{=}CHCH_2SH + H_2NSO_2Ph \quad (14)$$

$$RS(O)OH \xrightarrow[\text{MeCOSH}]{SOCl_2} RS{-}SC(O)CH_3 \xrightarrow[CH_2Cl_2,\ 0\ ^\circ C]{PPh_3} RSC(O)CH_3 \quad (15)$$

There has been recent controversy over the reduction of seleninic acids ($RSeO_2H$) to selenenic acids (RSeOH). Reich *et al.*[84] and Kice *et al.*[85] have independently concluded that earlier claims for the isolation of 'stable' selenenic acids are probably in error. While selenenic acids are stable in dilute solution, they show a pronounced tendency to form the corresponding anhydrides in the absence of water.

2.3.2.4 Reduction of Sulfimides (Sulfilimines) and Sulfoximides (Sulfoximines)

These nitrogen analogs of sulfoxides and sulfones are still relatively unusual types of organosulfur compound but a few methods are known for their reduction. Colonna and coworkers,[86] for example, found that *N*-tosylsulfimides could be reduced to sulfides in good yields, using formamidinesulfinic acid (thiourea *S,S*-dioxide) (**6**) under phase-transfer conditions at 70 °C (equation 16). This reagent was later adopted for the reduction of disulfides (see Section 2.3.4.1) and sulfoxides[52] and the latter were shown not to be intermediates in the reduction of sulfimides. Still and Turnbull[87] showed that phosphorus pentasulfide is as effective for the reduction of sulfimides as it had earlier proved to be for sulfoxides.[19] In addition to the *N*-tosyl derivatives, free diarylsulfimides were also reduced by this method. Oae and coworkers[34] showed that one of their procedures is capable of extension to *N*-tosylsulfimides (72–96% yields) and have used dichlorocarbene (generated from chloroform in a phase-transfer system) for the reduction of both sulfimides and sulfoximides to sulfides.[88] Oae and coworkers[89] have also used 1-benzyl-1,4-dihydronicotinamide (BNAH), catalyzed by various metalloporphins, for the reduction of both sulfimides and sulfoxides in benzene at 25–80 °C. Young and Reid[90] have conducted a thorough examination of the kinetics of the reduction of sulfimides to sulfides with the thiolate ion.

$$\mathrm{Ar{-}S({=}NTs){-}R} \xrightarrow[60\text{–}100\%]{(NH_2)_2C{=}SO_2} \mathrm{Ar{-}S{-}R} \qquad (16)$$

The sulfoximide system is of particular interest because stable chiral analogs are readily obtained (R ≠ R′). A thorough examination of a series of stereochemical reaction cycles involving sulfoximides (**10**) was carried out by Cram and coworkers.[91] While the conversion of sulfoximides to sulfoxides is of course a reduction, it is perhaps more correctly viewed as a deimination reaction and Cram and coworkers reported the conversion of sulfoximides to sulfoxides in 99% yield, with 99% retention of configuration, using either $NO^+PF_6^-$ or, better, aqueous nitrous acid (equation 17). Oae *et al.*[92] have shown that other nitrosating agents such as TsNO 4-(*S*-nitrososulfonyl)toluene and *t*-butyl thionitrate may be used in dry acetonitrile and are as effective as aqueous HNO_2 (equation 18) for the deimination of sulfoximides (and sulfimides). Racemization of the sulfoxides obtained, however, was noted. These workers showed conclusively that N_2O was evolved, as expected, from the initial *N*-nitrosation step. An interesting procedure reported by Oae *et al.*[93] for the conversion of sulfoximides to sulfoxides involves the use of elemental sulfur or diphenyl disulfide and again proceeds with essentially complete retention of stereochemical configuration.

$$\mathrm{R{-}S(O)(NH){-}R'}\ (\mathbf{10}) \xrightarrow[25\,°C]{aq.\ HNO_2} \mathrm{R{-}S(O){-}R'} \qquad (17)$$

$$\mathrm{R{-}S(O)(NH){-}R'} \xrightarrow[MeCN,\ 25\,°C]{TsNO} \mathrm{R{-}S(O)(N{-}NO){-}R'} \longrightarrow \mathrm{R{-}S(O){-}R'} + N_2O \qquad (18)$$

2.3.2.5 Reduction of Se and Te Analogs

While the literature in this area is sparse, it is useful to bear in mind that the strength of the relevant bonds is in the order S=O > Se=O > Te=O. Thus, while sulfoxides have been used as general oxidizing agents in various methods, such as the Pfitzner–Moffat or Swern procedures for the mild oxidation of primary and secondary alcohols,[94] diaryl telluroxides have been shown to be selective oxidants for the direct transformation of thiols (mercaptans) to disulfides, phosphines to phosphine oxides or hydroquinones to quinones.[95] The telluroxides could also be used in a catalytic cycle for the mild conversion of thiocarbonyl compounds to the corresponding carbonyl compounds.[95] Engman and Cava[96] have reported the first fully characterized tellurone (**11**). Not surprisingly, this is a more powerful oxidant than the telluroxide analog,[95] and will oxidize benzylic alcohols and oxidatively cleave 1,2-diols.

$$(4\text{-MeOC}_6\text{H}_4)_2\text{TeO}_2$$

(11)

Likewise, diphenyl selenoxide has been shown to convert aromatic amines to the corresponding azobenzenes,[97] while, in contrast to their sulfur analogs, benzeneseleninic acid and its anhydride are well established as versatile oxidants in organic chemistry.[98]

Several of the procedures discussed in the sulfoxide section describe the successful extension of the method to the reduction of selenoxides,[19,27,28] and there is little doubt that many of the other procedures cited earlier could be used likewise. Sakaki and Oae[99] used triphenylphosphine selenide and similar selenides to reduce selenoxides to selenides in 79–93% yield (equation 19). Using a chiral phosphine selenide, these workers showed that the phosphine oxide formed had suffered predominant inversion, with a stereospecificity of over 80%. Detty[100] has reported the application of the silane $PhSeSiMe_3$ (**12**) to the reduction of selenoxides and telluroxides. The reactions are rapid and proceed essentially quantitatively, even in the presence of a hydroxy or carbonyl group.

$$R(Se{=}O)R' + Ar_3P{=}Se \longrightarrow RSeR' + Ar_3P{=}O + Se \qquad (19)$$

Lang and Comasseto[101] have shown that formamidinesulfinic acid, or thiourea dioxide (**6**), is an efficient and inexpensive reagent for reducing selenoxides and telluroxides, as well as various selenium(IV) halide derivatives, to the corresponding selenides and tellurides.

2.3.3 REDUCTION OF P═O COMPOUNDS

Little work appears to have been done with the reduction of phosphine oxides to phosphines, in sharp contrast with the voluminous literature on the reduction of sulfoxides (see Section 2.3.2.1). Two simple explanations which suggest themselves are: (i) the thermodynamic problem of a very strong P═O bond (*cf.* the sulfones); and (ii) the fact that, in contradistinction to the vast number of stable sulfone- and sulfoxide-containing intermediates encountered in organic synthesis, the supporting role of phosphorus intermediates is usually fleeting and confined very often, for example, to the single step involved in a Wittig reaction. Changes in oxidation state of the phosphorus atom, therefore, have so far been of little importance.

Nevertheless, a number of papers have appeared on this topic, including an important one by Mislow and coworkers[102] which examined the reduction of a series of chiral acyclic phosphine oxides by Si_2Cl_6 or Si_3Cl_8. These reactions proceed in high yield (70–90%), with stereochemical inversion at phosphorus and with almost complete stereospecificity. A complex mechanistic scheme is proposed to account for the observed results. The authors found that amine oxides and sulfoxides were also deoxygenated by Si_2Cl_6, although it should be noted that Pummerer-like chlorination products were observed in significant amounts for sulfoxides with α-C—H bonds. Since optically active phosphine oxides may readily be obtained from the reaction of diastereomerically pure menthyl phosphinate esters with Grignard reagents,[103] it is of interest to note that Marsi[104] has found phenylsilane ($PhSiH_3$) to be a reliable agent for the conversion of phosphine oxides to the corresponding phosphines, in quantitative yield and with complete retention of configuration (equation 20).

$$Ph(Pr)(Me)P{=}O \xrightarrow[\text{80–100 °C}\\ 96\%]{PhSiH_3} Ph(Pr)(Me)P{:} \qquad (20)$$

If further evidence of the difficulty of reducing the P═O bond were needed, the recent findings of Yamashita *et al.*[105] provide interesting confirmation. Many α-hydroxyalkyldiphenylphosphine oxides (**13**) were converted into the corresponding alkyldiphenylphosphine oxides (**14**) in 33–73% yield, with PCl_3–KI (equation 21). No competing reduction of the P═O bond was apparently observed. Majewski[106] has reported the disproportionation of dialkylphosphine oxides into dialkylphosphines and phosphinic acids (equation 22). The reaction proceeds in yields of over 90% and the phosphines are easily isolated by distillation.

$$\underset{(\mathbf{13})}{Ph_2P(O)C(OH)RR'} \xrightarrow[\text{CHCl}_3\text{–Me}_2\text{CO}\\ 25\text{–}50\ °\text{C}]{PCl_3,\ KI} \underset{(\mathbf{14})}{Ph_2P(O)CHRR'} \qquad (21)$$

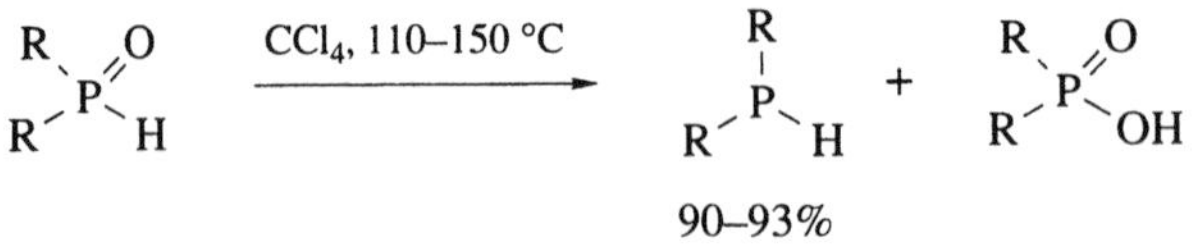

$$R_2P(=O)H \xrightarrow{CCl_4,\ 110\text{–}150\ °C} R_2P\text{–}H\ (90\text{–}93\%) + R_2P(=O)OH \qquad (22)$$

In other developments, Lecat and Devaud[107] have proposed an electrochemical reduction (Scheme 4) to recover triphenylphosphine from triphenylphosphine oxide produced, for example, in the Wittig reaction. Yields of up to 86% were obtained. Engel and Chakrabarty[108a] have reduced phosphoric and phosphonic chlorides to the corresponding monobasic phosphorous and phosphonous acid derivatives (equations 23 and 24), using $NaBH_4$ in refluxing dioxane. Triphenylphosphine oxide is reportedly[108b] reduced to triphenylphosphine in 75% yield by the electron-transfer system SmI_2–THF–HMPA, at 65 °C.

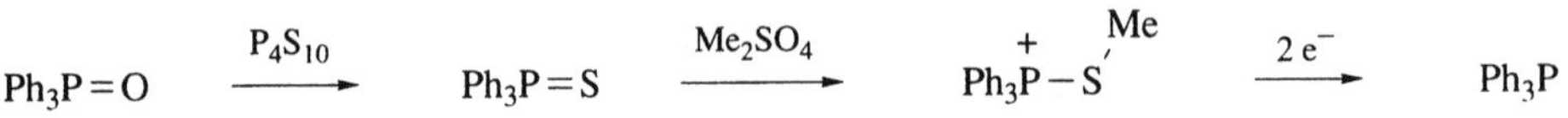

$$Ph_3P{=}O \xrightarrow{P_4S_{10}} Ph_3P{=}S \xrightarrow{Me_2SO_4} Ph_3\overset{+}{P}\text{–}S\text{–}Me \xrightarrow{2e^-} Ph_3P$$

Scheme 4

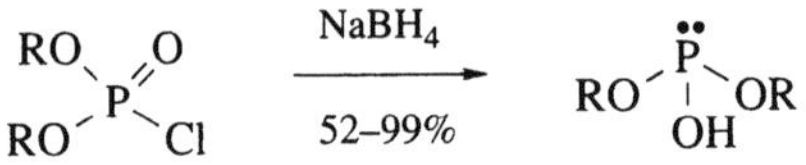

$$(RO)_2P(=O)Cl \xrightarrow[52\text{–}99\%]{NaBH_4} (RO)_2\ddot{P}\text{–}OH \qquad (23)$$

$$(RO)R'P(=O)Cl \xrightarrow{NaBH_4} (RO)R'\ddot{P}\text{–}OH \qquad (24)$$

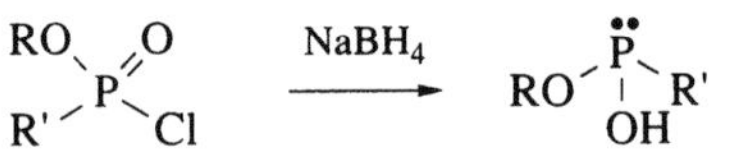

2.3.4 REDUCTION OF S—X TO S—H

2.3.4.1 Reduction of Disulfides and Diselenides

The reduction of disulfides to thiols (equation 25) is a familiar reduction process in organic chemistry, being easily effected by a variety of reducing agents. The subject has been reviewed[109] but still continues to attract attention and some recent methods will be discussed here. Various hydride reducing agents have been used, such as $NaBH_4$ (in refluxing THF–MeOH),[110] potassium triisopropoxyborohydride in THF,[111] and the mild reagents potassium triphenylborohydride[112] and lithium tri-*t*-butoxyaluminum hydride[113] in THF. In all cases, the authors noted that diaryl disulfides were reduced much faster than their dialkyl counterparts. Given the great importance of the disulfide–thiol redox relationship in living organisms, it is interesting to note that a biomimetic reducing system of 1-benzyl-1,4-dihydronicotinamide (BNAH), catalyzed by 3-methyllumiflavin, has been used by Oae and coworkers[114] for the reduction of diaryl disulfides.

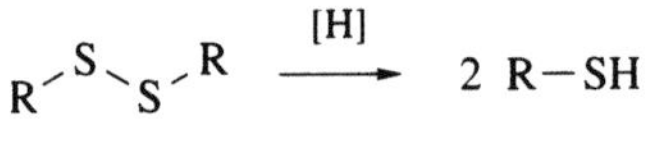

$$R\text{–}S\text{–}S\text{–}R \xrightarrow{[H]} 2\ R\text{–}SH \qquad (25)$$

Maiti *et al.*[115] reported the reduction of a series of disulfides to thiols, using hydrazines. Other approaches to this reduction have employed thiourea *S*,*S*-dioxide (**6**),[116] a reagent known to reduce sulfoxides and sulfimides,[86] as well as selenoxides and telluroxides,[101] and an interesting Birch-type reduction using aluminum in liquid ammonia.[117] Activation of the aluminum used (by halogen or halide) appeared to be essential but the role of the activating agent was not explained. In some procedures the thiols were not isolated, but trapped by reaction of thiolate ion with alkyl halides,[116] as in Amos and Fawcett's[118] reductive cleavage using a polymer-supported phosphine reagent, or with Michael acceptors, as in an interesting procedure using CO–H_2O–Se[119] in an autoclave at 50–100 °C. In a similar vein, Suzuki and Sato[120] showed that P_2I_4 treatment of dialkyl or diaryl disulfides could be used to form mixed sulfides from benzyl alcohols (equation 26).

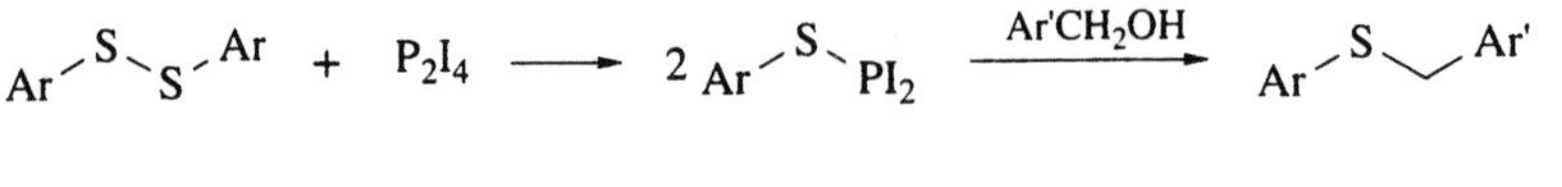

$$Ar\text{–}S\text{–}S\text{–}Ar + P_2I_4 \longrightarrow 2\ Ar\text{–}S\text{–}PI_2 \xrightarrow{Ar'CH_2OH} Ar\text{–}S\text{–}CH_2\text{–}Ar' \qquad (26)$$

Diselenides have been cleaved reductively to the selenolate ion with CO–H_2O–Se,[119] $NaBH_4$ or sodium or potassium hydrides.[121] Very recently, Kong and Zhou[122] have utilized sodium hydrogen telluride to selectively reduce a variety of disulfides to thiols in yields ranging from 65–98%. Bunte salts ($RSSO_3Na$) are also efficiently reduced to the corresponding thiols using this reagent.

The desulfurization of disulfides to sulfides is also, formally, a reduction. Harpp and Gleason[123] showed that tris(dialkylamino)phosphines readily produced the corresponding sulfides, except in the case of diaryl disulfides. For example, important natural products such as α-lipoic acid and cystine could be converted to their monosulfides in excellent yield. Harpp and Gleason showed that inversion of configuration occurred at only one of the α-carbon centers, in accord with the proposed mechanism (equation 27). Later, Middleton *et al.*[124] showed that triphenylphosphine could be used to convert even diaryl disulfides into sulfides, but only under very forcing conditions (250–300 °C) in the absence of solvent. Interestingly, when diaryl disulfides are treated with triphenylphosphine in wet dioxane, the primary product obtained is not the sulfide desulfurization product, but the thiol reduction product.[125] Trapping of the arenethiolate presumed to be formed initially with a sulfenylating species (sulfur transfer agent) was shown by these workers to produce unsymmetrical disulfides in yields of 73–83%. The reactions of disulfides with trivalent phosphorus compounds have been reviewed by Mukaiyama and Takei.[126]

$$\mathrm{RSSR'} \xrightarrow[\text{benzene, 25–80 °C}]{(\mathrm{R''_2N})_3\mathrm{P}} \mathrm{RS^-} + (\mathrm{R''_2N})_3\overset{+}{\mathrm{P}}\mathrm{-SR'} \longrightarrow \mathrm{RSR'} + (\mathrm{R''_2N})_3\mathrm{P{=}S} \quad (27)$$

2.3.4.2 Reduction of Thiocyanates and their Se and Te Analogs

The reduction of thiocyanates to the corresponding thiols (equation 28) has long been known and has been extensively reviewed.[127,128] While much less well known, the reduction of seleno- and tellurocyanates has also been reviewed recently.[129,130] Among the reducing agents which have been employed in this area are sodium dithionite ($Na_2S_2O_4$), dithiothreitol (Cleland's reagent) and phosphinic acid (H_3PO_2). While reports of the preparation of tellurocyanates exist,[131] Engman[132] has noted that very few compounds of this type have been completely characterized, perhaps because of a facile disproportionation reaction, as shown in equation (29).

$$\mathrm{RYCN} \xrightarrow{[\mathrm{H}]} \mathrm{R{-}YH} \quad (28)$$

$$\mathrm{Y = S, Se, Te}$$

$$2\,\mathrm{ArTeCN} \longrightarrow \mathrm{ArTeAr} + \mathrm{NCTeCN} \quad (29)$$

$NaBH_4$ has been found to be an effective reducing agent for the reduction of both thiocyanates (in DMF at 25 °C)[133] and selenocyanates (in pyridine at 25 °C).[134] A procedure for the reduction of thiocyanates using tributyltin hydride (equation 30) has been reported by Ueno *et al.*[135] The mechanism of the reduction is probably homolytic and the thiostannanes (**15**) formed are stable and can be readily isolated. The thiostannanes, in addition to being more pleasant to handle than the parent thiols, can be converted into typical thiol derivatives, such as thiol esters, in excellent yields.[135]

$$\mathrm{RSCN} \xrightarrow[\text{AIBN, 60 °C; 86–99\%}]{\mathrm{Bu_3SnH}} \underset{(\mathbf{15})}{\mathrm{RSSnBu_3}} \quad (30)$$

2.3.5 REFERENCES

1. J. Drabowicz, T. Numata and S. Oae, *Org. Prep. Proced. Int.*, 1977, **9**, 63.
2. T. Durst, in 'Comprehensive Organic Chemistry', ed. D. H. R. Barton and W. D. Ollis, Pergamon Press, Oxford, 1979, vol. 3, p. 131.
3. J. Drabowicz, H. Togo, M. Mikolajczyk and S. Oae, *Org. Prep. Proced. Int.*, 1984, **16**, 171.
4. J. S. Grossert, in 'The Chemistry of Sulphones and Sulphoxides', ed. S. Patai, Z. Rappoport and C. J. M. Stirling, Wiley, Chichester, 1988, chap. 20, p. 925.

5. M. Madesclaire, *Tetrahedron*, 1988, **44**, 6537.
6. T. Durst, in 'Comprehensive Organic Chemistry', ed. D. H. R. Barton and W. D. Ollis, Pergamon Press, Oxford, 1979, vol. 3, p. 181.
7. G. C. Barrett, in 'Comprehensive Organic Chemistry', ed. D. H. R. Barton and W. D. Ollis, Pergamon Press, Oxford, 1979, vol. 3, p. 7.
8. P. D. Magnus, in 'Comprehensive Organic Chemistry', ed. D. H. R. Barton and W. D. Ollis, Pergamon Press, Oxford, 1979, vol. 3, p. 506.
9. H. C. Brown and S. Krishnamurthy, *Tetrahedron*, 1979, **35**, 567.
10. J. Målek, *Org. React. (N.Y.)*, 1988, **36**, 249.
11. G. A. Olah and S. C. Narang, *Tetrahedron*, 1982, **38**, 2225.
12. R. G. Nuzzo, H. J. Simon and J. San Filippo, Jr., *J. Org. Chem.*, 1977, **42**, 568.
13. T.-L. Ho, *Synth. Commun.*, 1977, **7**, 321.
14. G. A. Olah, G. K. Surya Prakash and T.-L. Ho, *Synthesis*, 1976, 810.
15. J. Drabowicz and M. Mikolajczyk, *Synthesis*, 1978, 138.
16. T.-L. Ho, *Synthesis*, 1979, 1.
17. S.-K. Chung and G. Han, *Synth. Commun.*, 1982, **12**, 903.
18. H. Suzuki, H. Manabe, R. Enokiya and Y. Hanazaki, *Chem. Lett.*, 1986, 1339.
19. I. W. J. Still, S. K. Hasan and K. Turnbull, *Can. J. Chem.*, 1978, **56**, 1423.
20. R. D. Baechler, S. K. Daley, B. Daly and K. McGlynn, *Tetrahedron Lett.*, 1978, 105.
21. I. W. J. Still, J. N. Reed and K. Turnbull, *Tetrahedron Lett.*, 1979, 1481.
22. J. A. M. Kuipers, B. H. M. Lammerink, I. W. J. Still and B. Zwanenburg, *Synthesis*, 1981, 295.
23. M. Yokoyama, Y. Hasegawa, M. Hatanaka, Y. Kawazoe and T. Imamoto, *Synthesis*, 1984, 827.
24. M. Wakisaka, M. Hatanaka, H. Nitta, M. Hatamura and T. Ishimaru, *Synthesis*, 1980, 67.
25. G. A. Olah, B. G. Balaram Gupta and S. C. Narang, *J. Org. Chem.*, 1978, **43**, 4503.
26. S. P. Singh and R. K. Saxena, *Synth. Commun.*, 1987, **17**, 575.
27. J. N. Denis and A. Krief, *Tetrahedron Lett.*, 1979, 3995.
28. J. N. Denis and A. Krief, *J. Chem. Soc., Chem. Commun.*, 1980, 544.
29. H. Suzuki, N. Sato and A. Osuka, *Chem. Lett.*, 1980, 143.
30. D. W. Chasar and T. M. Pratt, *Synthesis*, 1976, 262.
31. L. A. G. M. van den Broek, P. J. A. in't Veld, J. H. Colstee and H. C. J. Ottenheijm, *Synth. Commun.*, 1989, **19**, 3397.
32. G. A. Olah, B. G. Balaram Gupta and S. C. Narang, *Synthesis*, 1977, 583.
33. A. H. Schmidt and M. Russ, *Chem. Ber.*, 1981, **114**, 1099.
34. T. Numata, H. Togo and S. Oae, *Chem. Lett.*, 1979, 329.
35. H. S. D. Soysa and W. P. Weber, *Tetrahedron Lett.*, 1978, 235.
36. M. R. Detty and M. D. Seidler, *J. Org. Chem.*, 1982, **47**, 1354.
37. A. H. Schmidt and M. Russ, *Chem. Ber.*, 1981, **114**, 822.
38. K. Nagasawa, A. Yoneta, T. Umezawa and K. Ito, *Heterocycles*, 1987, **26**, 2607.
39. G. A. Olah, A. Husain, B. P. Singh and A. K. Mehrotra, *J. Org. Chem.*, 1983, **48**, 3667.
40. R. D. Miller, A. F. Renaldo and H. Ito, *J. Org. Chem.*, 1988, **53**, 5571.
41. G. Palumbo, C. Ferreri and R. Caputo, *Phosphorus Sulfur*, 1983, **15**, 19.
42. J. S. Cha, J. E. Kim and J. D. Kim, *Tetrahedron Lett.*, 1985, **26**, 6453.
43. Y. Guindon, J. G. Atkinson and H. E. Morton, *J. Org. Chem.*, 1984, **49**, 4538.
44. J. Ramesh Babu and M. V. Bhatt, *Tetrahedron Lett.*, 1986, **27**, 1073.
45. R. Lin and Y. Zhang, *Synth. Commun.*, 1987, **17**, 1403.
46. J. Drabowicz and S. Oae, *Synthesis*, 1977, 404.
47. G. A. Olah, R. Malhotra and S. C. Narang, *Synthesis*, 1979, 58.
48. G. A. Olah, Y. D. Vankar and M. Arvanaghi, *Synthesis*, 1979, 584.
49. D. W. Chasar and J. P. Shockcor, *Phosphorus Sulfur*, 1980, **8**, 187.
50. C. Tenca, A. Dossena, R. Marchelli and G. Casnati, *Synthesis*, 1981, 141.
51. G. A. Olah, A. P. Fung, B. G. Balaram Gupta and S. C. Narang, *Synthesis*, 1980, 221.
52. J. Drabowicz and M. Mikolajczyk, *Synthesis*, 1978, 542.
53. R. A. Amos, *J. Org. Chem.*, 1985, **50**, 1311.
54. K. Nozaki, M. Yoshihara, K. Enomura, Y. Matsubara and T. Maeshima, *Phosphorus Sulfur*, 1985, **22**, 131.
55. S. W. Benson, *Chem. Rev.*, 1978, **78**, 23.
56. F. G. Bordwell and W. H. McKellin, *J. Am. Chem. Soc.*, 1951, **73**, 2251.
57. J. N. Gardner, S. Kaiser, A. M. Krubiner and H. Lucas, *Can. J. Chem.*, 1973, **51**, 1419.
58. A. G. Anastassiou, J. C. Wetzel and B. Y.-H. Chao, *J. Am. Chem. Soc.*, 1975, **97**, 1124.
59. I. W. J. Still and S. Szilagyi, *Synth. Commun.*, 1979, **9**, 923.
60. I. W. J. Still and F. J. Ablenas, *J. Org. Chem.*, 1983, **48**, 1617.
61. L. D. Dolby and D. R. Rosencrantz, *J. Org. Chem.*, 1963, **28**, 1888.
62. P. D. Caesar, *Org. Synth.*, 1953, **33**, 47.
63. C. G. Overberger, H. Biletch and F. W. Orttung, *J. Org. Chem.*, 1959, **24**, 289.
64. L. Field and F. A. Grunwald, *J. Org. Chem.*, 1951, **16**, 946.
65. H. C. Brown and B. C. S. Rao, *J. Am. Chem. Soc.*, 1956, **78**, 2582.
66. A. Nose and T. Kudo, *Chem. Pharm. Bull.*, 1987, **35**, 1770.
67. T. H. Chan, J. P. Montillier, W. F. Van Horn and D. N. Harpp, *J. Am. Chem. Soc.*, 1970, **92**, 7224.
68. T. Numata, H. Awano and S. Oae, *Tetrahedron Lett.*, 1980, **21**, 1235.
69. S. Oae and H. Togo, *Bull. Chem. Soc. Jpn.*, 1983, **56**, 3802.
70. S. Oae and H. Togo, *Bull. Chem. Soc. Jpn.*, 1984, **57**, 232.
71. S. Oae and H. Togo, *Synthesis*, 1982, 152.
72. W. A. Sheppard, *Org. Synth.*, 1960, **40**, 80.
73. G. A. Olah, S. C. Narang, L. Field and G. F. Salem, *J. Org. Chem.*, 1980, **45**, 4792.

74. G. A. Olah, S. C. Narang, L. Field and R. Karpeles, *J. Org. Chem.*, 1981, **46**, 2408.
75. H. Suzuki, H. Tani and A. Osuka, *Chem. Lett.*, 1984, 139.
76. S. Oae and H. Togo, *Tetrahedron Lett.*, 1982, **23**, 4701.
77. D. Cipris and D. Pouli, *Synth. Commun.*, 1979, **9**, 207.
78. M. Ortiz and G. L. Larson, *Synth. Commun.*, 1982, **12**, 43.
79. S. Kagabu, M. Maehara, K. Sawahara and K. Saito, *J. Chem. Soc., Chem. Commun.*, 1988, 1485.
80. H. W. Pinnick, M. A. Reynolds, R. T. McDonald, Jr. and W. D. Brewster, *J. Org. Chem.*, 1980, **45**, 930.
81. S. Oae, H. Togo, T. Numata and K. Fujimori, *Chem. Lett.*, 1980, 1193.
82. A. Gadras, J. Dunoguès, R. Calas and G. Déléris, *J. Org. Chem.*, 1984, **49**, 442.
83. J. R. Irving, E. Perrone and R. J. Stoodley, *Tetrahedron Lett.*, 1983, **24**, 1429.
84. H. J. Reich, W. W. Willis, Jr. and S. Wollowitz, *Tetrahedron Lett.*, 1982, **23**, 3319.
85. J. L. Kice, F. McAfee and H. Slebocka-Tilk, *Tetrahedron Lett.*, 1982, **23**, 3323.
86. G. Borgogno, S. Colonna and R. Fornasier, *Synthesis*, 1975, 529.
87. I. W. J. Still and K. Turnbull, *Synthesis*, 1978, 540.
88. N. Furukawa, K. Hoshino, T. Morishita and S. Oae, *Synthesis*, 1984, 317.
89. T. Nagata, T. Yoshimura, K. Fujimori and S. Oae, *Tetrahedron Lett.*, 1984, **25**, 341.
90. P. R. Young and K. J. Reid, *J. Org. Chem.*, 1987, **52**, 2695.
91. T. R. Williams, A. Nudelman, R. E. Booms and D. J. Cram, *J. Am. Chem. Soc.*, 1972, **94**, 4684.
92. S. Oae, K. Iida and T. Takata, *Phosphorus Sulfur*, 1981, **12**, 103.
93. S. Oae, Y. Tsuchida and N. Furukawa, *Bull. Chem. Soc. Jpn.*, 1973, **46**, 648.
94. J. March, 'Advanced Organic Chemistry', 3rd edn., Wiley, New York, 1985, p. 1081.
95. S. V. Ley, C. A. Meerholz and D. H. R. Barton, *Tetrahedron*, 1981, **37** suppl. 1, 213.
96. L. Engman and M. P. Cava, *J. Chem. Soc., Chem. Commun.*, 1982, 164.
97. V. I. Naddaka, V. P. Gar'kin and V. I. Minkin, *Zh. Org. Khim.*, 1976, **12**, 2481.
98. T. G. Back, S. Collins and R. G. Kerr, *J. Org. Chem.*, 1981, **46**, 1564.
99. K. Sakaki and S. Oae, *Chem. Lett.*, 1977, 1003.
100. M. R. Detty, *J. Org. Chem.*, 1979, **44**, 4528.
101. E. S. Lang and J. V. Comasseto, *Synth. Commun.*, 1988, **18**, 301.
102. K. Naumann, G. Zon and K. Mislow, *J. Am. Chem. Soc.*, 1969, **91**, 7012.
103. O. Korpiun, R. A. Lewis, J. Chickos and K. Mislow, *J. Am. Chem. Soc.*, 1968, **90**, 4842.
104. K. L. Marsi, *J. Org. Chem.*, 1974, **39**, 265.
105. M. Yamashita, T. Suzuki, H. Imoto, T. Oshikawa and S. Inokawa, *Synthesis*, 1986, 556.
106. P. Majewski, *Synthesis*, 1987, 554.
107. J.-L. Lecat and M. Devaud, *Tetrahedron Lett.*, 1987, **28**, 5821.
108. (a) R. Engel and S. Chakraborty, *Synth. Commun.*, 1988, **18**, 665; (b) Y. Handa, J. Inanaga and M. Yamaguchi, *J. Chem. Soc., Chem. Commun.*, 1989, 298.
109. J. L. Wardell, in 'The Chemistry of the Thiol Group', ed. S. Patai, Wiley, London, 1974, part 1, p. 220.
110. A. Ookawa, S. Yokoyama and K. Soai, *Synth. Commun.*, 1986, **16**, 819.
111. H. C. Brown, B. Nazer and J. S. Cha, *Synthesis*, 1984, 498.
112. N. M. Yoon and K. E. Kim, *J. Org. Chem.*, 1987, **52**, 5564.
113. S. Krishnamurthy and D. Aimino, *J. Org. Chem.*, 1989, **54**, 4458.
114. K. Fujimori, T. Nagata and S. Oae, *Tetrahedron Lett.*, 1983, **24**, 5231.
115. S. N. Maiti, P. Spevak, M. P. Singh and R. G. Micetich, *Synth. Commun.*, 1988, **18**, 575.
116. J. V. Comasseto, E. S. Lang, J. T. B. Ferreira, F. Simonelli and V. R. Correia, *J. Organomet. Chem.*, 1987, **334**, 329.
117. R. Sato, R. Akaishi, T. Goto and M. Saito, *Bull. Chem. Soc. Jpn.*, 1987, **60**, 773.
118. R. A. Amos and S. M. Fawcett, *J. Org. Chem.*, 1984, **49**, 2637.
119. A. Ogawa, Y. Nishiyama, N. Kambe, S. Murai and N. Sonoda, *Tetrahedron Lett.*, 1987, **28**, 3271.
120. H. Suzuki and N. Sato, *Chem. Lett.*, 1981, 267.
121. P. Dowd and P. Kennedy, *Synth. Commun.*, 1981, **11**, 935.
122. F. Kong and X. Zhou, *Synth. Commun.*, 1989, **19**, 3143.
123. D. N. Harpp and J. G. Gleason, *J. Am. Chem. Soc.*, 1971, **93**, 2437.
124. D. J. Middleton, E. G. Samsel and G. H. Wiegand, *Phosphorus Sulfur*, 1979, **7**, 339.
125. L. E. Overman, J. Smoot and J. D. Overman, *Synthesis*, 1974, 59.
126. T. Mukaiyama and H. Takei, in 'Topics in Phosphorus Chemistry', ed. E. J. Griffith and M. Grayson, Wiley, New York, 1976, vol. 8, p. 587.
127. J. L. Wardell, in 'The Chemistry of the Thiol Group', ed. S. Patai, Wiley, London, 1974, part 1, p. 232.
128. R. G. Guy, in 'The Chemistry of Cyanates and Their Thio Derivatives', ed. S. Patai, Wiley, Chichester, 1977, part 2, p. 867.
129. E. Bulka, in 'The Chemistry of Cyanates and Their Thio Derivatives', ed. S. Patai, Wiley, Chichester, 1977, part 2, p. 900.
130. A. Toshemitsu and S. Uemura, in 'The Chemistry of Organic Selenium and Tellurium Compounds', ed. S. Patai, Wiley, Chichester, 1987, vol. 2, pp. 554, 584.
131. F. Ogura, H. Yamaguchi, T. Otsubo and K. Chikamatsu, *Synth. Commun.*, 1982, **12**, 131.
132. L. Engman, *J. Org. Chem.*, 1983, **48**, 2920.
133. I. W. J. Still and V. A. Sayeed, *Synth. Commun.*, 1983, **13**, 1181.
134. C. H. Banks, J. R. Daniel and R. A. Zingaro, *J. Med. Chem.*, 1979, **22**, 572.
135. Y. Ueno, M. Nozomi and M. Okawara, *Chem. Lett.*, 1982, 1199.

3.1
Heterogeneous Catalytic Hydrogenation of C=C and C≡C

SAMUEL SIEGEL
University of Arkansas, Fayetteville, AR, USA

3.1.1 INTRODUCTION

3.1.1.1 General

Heterogeneous catalytic hydrogenation continues to be a most useful technique for the addition of H_2 to carbon–carbon double and triple bonds and to the aromatic nucleus, particularly if the objective involves chemo-, regio- or stereo-selectivity.[1–6] Generally, the conditions for a laboratory preparation are mild and the catalyst is easily separated from the product. A simple apparatus often can suffice, although commercial reactors add safety, convenience and the ability to conduct hydrogenations over a wide range of hydrogen pressures and temperatures. Reliable catalysts can be purchased and stored for years without significant loss of activity, and some active catalysts can be prepared conveniently when needed.

The addition of hydrogen is catalyzed by a large variety of materials, but synthetically useful procedures generally employ nickel or the platinum metals, whose surface provides the reaction centers. To increase the fraction of the metal which is exposed to the fluid phase, the metal is finely divided or distributed over the surface of an appropriate, presumedly inert, support. Although the properties of a catalyst are affected by the method of preparation, the characteristics of the metal are dominant.[3,4]

The choice of catalyst and conditions for optimal results sometimes can be critical, but often is not. A number of generalizations can aid in the selection of a suitable procedure; however, the best guide for an initial choice is based on the recorded experience of others.[1–6] However, in order to improve or adapt the procedure to a particular objective, some understanding of the dynamics and mechanism of heterogeneous catalytic hydrogenation can be helpful.

3.1.1.2 Catalysts

Knowledge of the source of the catalysts which are most commonly used in synthetic procedures is useful. Supported platinum metals which contain 1–10% of the metal dispersed on carbon, alumina or silica are available from commercial sources, as is Adams' platinum oxide (a catalyst much used in the past).[3] The latter is a mixture of Pt, α-$PtO_2{\cdot}xH_2O$, and $Na_xPt_3O_4$, according to Cahen and Ibers.[7] The presence of sodium is thought to be related to the promoting effect of acids on its activity.[3] If KNO_3 is used in place of $NaNO_3$ in its preparation, the product is a mixture of Pt and α-$PtO_2{\cdot}xH_2O$, which is reduced to Pt by H_2. A rhodium/platinum (7:3) oxide is prepared by a similar procedure and, when reduced by H_2, it exhibits the activity and selectivity of a pure rhodium catalyst.[8] Unsupported platinum metals have been prepared by the hydrogen reduction of the corresponding hydroxide (Pd, Pt, Rh, Ir) or oxide (Os) in water at atmospheric or high pressure.[9,10] High surface purity ruthenium powders have been described for use in studies of the effect of water vapor on the hydrogenation of benzene.[11] Water has an unusual promoting effect on ruthenium-catalyzed hydrogenations.[3]

Raney nickel can be purchased in a form which is ready for use (W-2), or can be prepared by published procedures which furnish catalysts with a range of activities.[3] The increase in selectivity of Raney Ni catalysts upon 'aging' apparently results from the formation of CO by the decomposition of the ethanol under which the catalyst is usually stored.[12] Highly regio- and stereo-selective catalysts are prepared as slurries by the reduction of the metal acetates, and are used *in situ* or stored for later use. P-2 nickel is prepared by reducing an aqueous alcoholic solution of $Ni(OAc)_2$ with $NaBH_4$.[13] More recently, highly selective Ni and Pd catalysts were prepared by the reduction of $Ni(OAc)_2$ or $Pd(OAc)_2$ with NaH and *t*-pentyl alcohol in THF.[14,15]

Chemically bonding metal complexes to an insoluble solid is a method of preparing catalysts whose active centers have well-defined structures, but the results have yet to be applied to synthesis.[16] Polymer-bound palladium and rhodium complexes appear to offer the combined advantages of the selectivities of soluble hydrogenation catalysts and the ease of separating heterogeneous catalysts from the products.[17]

3.1.2 DYNAMICS AND MECHANISM[18]

3.1.2.1 Transport Phenomena[19]

The rate and selectivity of a surface-catalyzed reaction can be affected by the existence of concentration or temperature gradients in the quiescent layer of fluid which surrounds the catalyst particle or is contained within its pores upon whose surface much of the exposed active material is distributed. The reactants in the bulk phase reach the reactive sites by diffusion through these regions of the fluid, and they affect the kinetics when the rates of the surface-catalyzed reactions are fast relative to the rate of transport of reactants to the catalytic sites. In the hydrogenation of unsaturated liquids or compounds in solution, the agitation of the liquid–catalyst mixture must be adequate to assure that the solution remains saturated with hydrogen.

The transport of either the unsaturated reactant or hydrogen may limit the rate. The diffusivity of H_2 is much greater than any unsaturated reactant, but, in a liquid phase, the transport of hydrogen may limit the rate because of its low solubility in the commonly used solvents.[20] To minimize diffusional effects, small amounts of the metal are deposited on nonporous supports, or on supports with wide pores with the metal confined to an outer shell of the supporting particle. Colloidal particles also afford short diffusion paths. Temperature gradients are unlikely to be important if the catalyst particle is immersed in a liquid and the agitation of the mixture is vigorous. Madon *et al.* have recommended procedures to establish that neither heat nor mass transfer, either interphase or intraparticle, affect the rate.[19]

3.1.2.2 Kinetics[20]

The formal kinetic expressions for the rate of surface-catalyzed reactions are like those of homogeneous catalysis, whether simplified by assuming a rapidly established equilibrium between reactants and the catalytic site, as in the Michaelis–Menten equations for enzyme catalysis, or based on less restrictive assumptions such as the steady state for the reactive intermediates. Following the simplified Langmuir–Hinshelwood method, the rate is assumed to be a function of the fraction of the active sites which are occupied by the reactants. The fraction occupied is related to the concentration of the reactant in the bulk phase through a Langmuir adsorption isotherm. One useful formulation assumes that the adsorption of the unsaturated hydrocarbon (A) is not competitive with hydrogen.[1] Under conditions in which the surface reactions are rate limiting, the expression for the rate of forming the product is equation (1) where K_H and K_A represent the equilibrium constants for adsorption of H_2 and A, respectively, and $[H_2]$ and [A] are the concentrations of the reactants in the fluid phase. If $K_A[A]$ is much greater than unity, then the rate is independent of [A]. Often this condition is met by an alkene or alkyne, giving equation (2) which holds until near the end of the reaction when the concentration of A becomes small.[21]

$$-d[A]/dt = k_A K_H K_A [H_2][A]/(1 + K_H[H_2])(1 + K_A[A]) \quad (1)$$

$$d[A]/dt = k_A K_H [H_2] \quad (2)$$

If the system is well represented by Langmuir–Hinshelwood kinetics, then the relative values of the adsorption equilibrium constants of various unsaturated compounds can be obtained from the relative rates determined individually and in competition.[22] The expression for the competitive reaction between two compounds A and B results from the division of the corresponding rate expressions for each compound (equation 3) to give equation (4). The ratio of the individual rates of hydrogenation, k_A/k_B, permits the abstraction of the ratio K_A/K_B from the competitively determined ratio k_AK_A/k_BK_B (equation 5). The assumption that the adsorption of the unsaturated compound is rapid and reversible relative to the rates of the surface-catalyzed reactions may not be correct. In that case, K_A/K_B is an 'apparent' relative adsorption equilibrium constant (see Section 3.1.3.2.2).

$$-d[A]/dt = k_A K_H [H_2] K_A [A]/(1 + K_H[H_2])(1 + K_A[A] + K_B[B]) \quad (3)$$

$$d[A]/d[B] = k_A K_A [A]/k_B K_B [B] \quad (4)$$

$$\ln([A]_o/[A]_t) = (k_A K_A/k_B K_B)\ln([B]_o/[B]_t) \quad (5)$$

$$-d[A]/dt = k_A K_H[H_2]K_A[A]/(1 + K_H[H_2])(K_B[B]) \quad (6)$$

If B represents a nonreducible compound, a reversibly adsorbed inhibitor or poison, and $K_B[B]$ is greater than $(1 + K_A[A])$, then equation (3) takes the form of equation (6), which shows that a reversibly adsorbed poison causes the rate of hydrogenation of A to become proportional to its concentration, *i.e.* kinetically first order. An irreversibly adsorbed poison will slow the rate but not affect the rate expression unless particular sites are poisoned selectively or the decrease in rate moves the reaction out of a mass transport limiting regime. Although these equations are approximations, they often reproduce the form of the physical phenomena of heterogeneous catalytic hydrogenation, and provide a useful guide for the interpretation of results.

Generally, solvents exert a greater effect upon competitive reactions than upon the individual reaction rates when corrected for the solubility of hydrogen.[19–22] Solvents which are strongly nucleophilic, such as amines, compete for active sites, and benzene slows the hydrogenation of alkenes, particularly on Pd.[19,21]

3.1.2.3 Mechanism[1,20,23]

3.1.2.3.1 Formalism of Horiuti–Polanyi

The early use of deuterium in place of hydrogen in the study of catalytic hydrogenation led to the recognition that the process was not simply the addition of H_2 to the double bond. Horiuti and Polanyi proposed that both H_2 and alkene (**1**) are bound to the catalyst surface and transformed to products by a sequence of elementary steps, which they represented as shown in Scheme 1, where an asterisk (*) represents a vacant site on the catalyst.[24] The last step, (d), is virtually irreversible under the usual hydrogenation conditions, but can be observed in the exchange reactions of D_2 with alkanes. The mechanism accounts for the isomerization of an alkene if the reversal of step (c), which involves the formation of the alkyl intermediate (**3**), involves the abstraction of a hydrogen atom other than the one first added, and is coupled with the desorption of the alkene, (**2**) → (**1**). At present, the bond between the alkene and the metal often is represented as a π-complex (**4**), as in equation (7).[1–5]

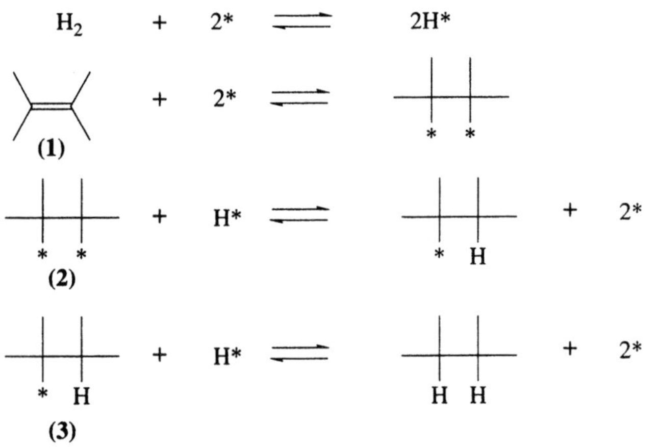

H_2 + 2* ⇌ 2H* (a)

(**1**) + 2* ⇌ (**2**) [C–C bonded to * *] (b)

(**2**) + H* ⇌ (**3**) [C–C bonded to * H] + 2* (c)

(**3**) + H* ⇌ [C–C bonded to H H] + 2* (d)

Scheme 1 The Horiuti–Polanyi mechanism of alkene hydrogenation

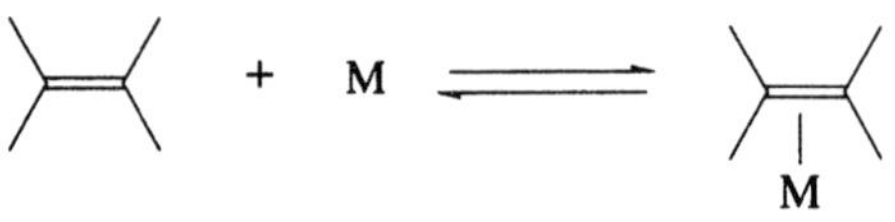

alkene + M ⇌ alkene–M (**4**) (7)

3.1.2.3.2 Organometallic analogy

The rapid development of catalysis by soluble transition metal complexes and the results of surface science studies afford useful insights into the mechanisms of heterogeneous catalysis.[16,25,26] The surface of a metallic particle presents an array of atoms differing in coordination number which depends on the atom's location on a terrace, step or kink.[27] The spectroscopy of adsorbed reactants and intermediates have properties which suggest structural relationships to organometallic compounds and, presumably, their reactions are analogous.[27,28]

Depending upon its location, and assuming octahedral coordination as a maximum, a metal atom may have one, two or three vacant coordination sites at which a monodentate group may be bound.[29] These sites correspond approximately to a location on a terrace, step or kink.[27] A convenient symbol for the exposed atom is ^{x}M where x represents the number of vacant sites before exposure to the reactants. The elementary reactions which occur are assumed to be those of an organometallic complex which contains the analogous structural unit.[28] For example, the dissociation of H_2 on a metal surface may involve a single atom, providing it has at least two vacant coordination sites (equation 8). The migration of hydrogen atoms to other parts of the surface may be represented as in equation (9) where ^{x}M may be ^{1}M, ^{2}M or ^{3}M. Indeed H_2 dissociates more rapidly on the edges of a stepped platinum surface than on atoms located on the terraces or close-packed planes.[27] A single vacant site can bind an alkene, or alkyne, as a π-complex and, if it is located on an atom which also bears a hydrogen atom, its insertion into a metal–hydrogen bond may occur (equation 10). If both the alkyl group and a hydrogen atom are bound to the same metal atom then reductive elimination can proceed to the alkane (equation 11). The reactions may involve neighboring surface atoms in ways analogous to dinuclear or polynuclear metal complexes, and can be represented by an extension of the above symbolism.

$$H_2 + {}^{2}M \rightleftharpoons {}^{2}M(H)-H \quad (8)$$

$$^{x}M\text{-}{}^{2}M(H)\text{-}H \rightleftharpoons {}^{x}M(H)-{}^{2}M\text{-}H \quad (9)$$

$$\text{(alkene)}{\cdot}{}^{2}M\text{-}H \rightleftharpoons \text{(alkyl-H)}{}^{2}M \quad (10)$$

$$\text{(alkyl-H)}{}^{2}M-H \rightleftharpoons {}^{2}M + \text{H-alkane-H} \quad (11)$$

Obviously, these later representations imply more knowledge of structure than the asterisks of Horiuti and Polanyi, but they serve to emphasize possible mechanistic relationships with homogeneous catalysis. Interestingly, as in hydrogenations catalyzed by soluble transition metal complexes,[28] the spectroscopically observable structures may not include important intermediates along the main reaction paths. For example, Yates *et al.* have shown that the spectral intensity of the IR bands associated with ethylidyne, a previously postulated intermediate in the hydrogenation of ethylene,[30] is not related to the rate of the latter reaction.[31] Presumably, reaction occurs on a small fraction of the surface, most of it being occupied by unreactive groups.

3.1.3 HYDROGENATION OF C=C

The addition of hydrogen to simple alkenes proceeds rapidly with any of the readily available forms of palladium, platinum, rhodium or nickel catalysts. However, the structure of more complex alkenes and the presence of additional functional groups, whether or not reducible, can dictate the particular catalyst and reaction conditions which should be used to ensure that the addition of hydrogen is complete, or to promote regio- or stereo-selectivity.

3.1.3.1 Double Bond Migration

The isomerization of an alkene often accompanies the addition of hydrogen but goes unnoticed unless the isomer is appreciably less reactive, results in the loss of chirality in the alkene, or differs in stereoselectivity.[3,4] The relative rate of hydrogenation to isomerization depends on the structure of the alkene, the catalyst and the conditions used.[3,4,32]

Metals differ in their tendency to promote isomerization. Nishimura *et al.* found that, on unsupported platinum metals, the order of decreasing activity is Pd >> Rh, Ru, Pt > Os > Ir.[33] The more active forms of Raney Ni exhibit a greater tendency than Rh to catalyze isomerization, but aged Raney Ni (see Section 3.1.1.2) and P-2 Ni are less active in this regard.[12,13] Of the common catalysts, platinum often is used when isomerization is to be avoided, and palladium is chosen for its selectivity and reactivity when isomerization is acceptable or desired.[3,4]

3.1.3.1.1 Mechanism

The mechanisms of isomerization which have been considered fall into two categories; 'associative', first proposed by Horiuti and Polanyi,[24] and 'dissociative', advanced by Farkas *et al.*[34] The 'associative' mechanism is a consequence of the reversibility of the formation of the alkyl intermediate shown in Scheme 1 and in equation (10), while the dissociative mechanism, in its current form, involves allylic species (**5**)–(**7**) shown in Scheme 2. The Horiuti–Polanyi mechanism implies that double bond isomerization and the addition of H_2 proceed through a common intermediate, whereas the dissociative mechanism represents an independent path.[26]

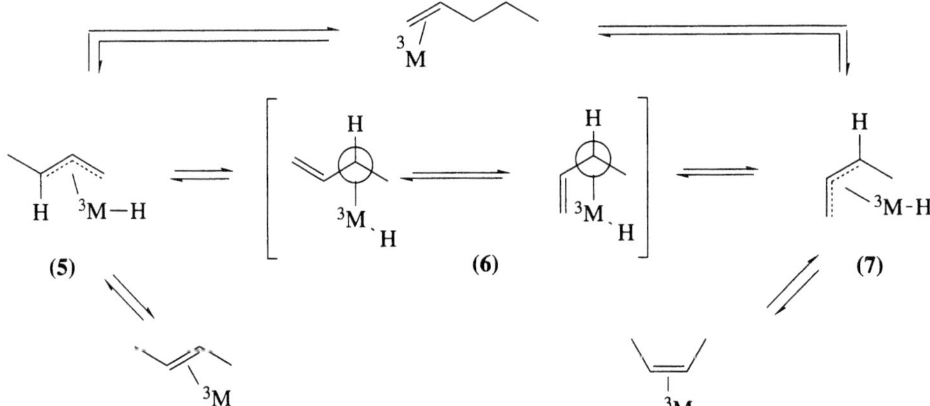

Scheme 2 Dissociative mechanism: isomerization of 1-butene to *cis*- and *trans*-2-butene

The representations in Scheme 2 and equations (7)–(11) use a formalism which indicates important elements of structure that a surface atom requires for the respective mechanisms.[29] On a ^{2}M site, the associative mechanism can take place through the reversibility of equation (9). This site provides the required hydrido group and a vacant coordination position to accept the alkene, a two-electron donor ligand.[28]

The 'dissociative' mechanism of double bond migration *via* an allylic intermediate can occur at a ^{3}M site, as indicated in Scheme 2, or at a two atom site such as ^{x}M—^{2}M which can provide the three vacant coordination positions which are required to bind both the π-allyl, a four-electron donor, and the hydrido group.[28,29] Because the surface of a metal provides sites which differ in structure, both mechanisms may operate on a given catalyst, the relative importance being a function of the metal, the alkene and the reaction conditions.[3,4,32]

Microwave spectroscopy identifies the location of the deuterium introduced by exchange into simple alkenes, and has been used with both homogeneous and heterogeneous catalysts to clarify the mechanisms of isomerization.[35] Using this method, Naito and Tanimoto showed that the addition and exchange reactions of propene over Pd/SiO_2 at 187 K take place *via* *n*-propyl and *s*-propyl adsorbed intermediates.[36] The adsorbed *s*-propyl group is not symmetric for hydrogen abstraction because of restricted rotation about the alkyl–metal bond.[26] In addition, an apparent intramolecular hydrogen migration between the 1- and 3-positions was assumed to be evidence of a suprafacial 1,3-sigmatropic shift of a hydrogen atom within the π-complexed propene. This mechanism, however, is not supported by the stereoselective exchange of D_2 with the 8(14)-cholestene (**10**; Scheme 3), or by the quantum mechanical calculations of

Anderson *et al.*, who concluded that a mechanism involving coadsorbed allyl and a hydrogen atom, as represented in Scheme 2, is allowed.[37]

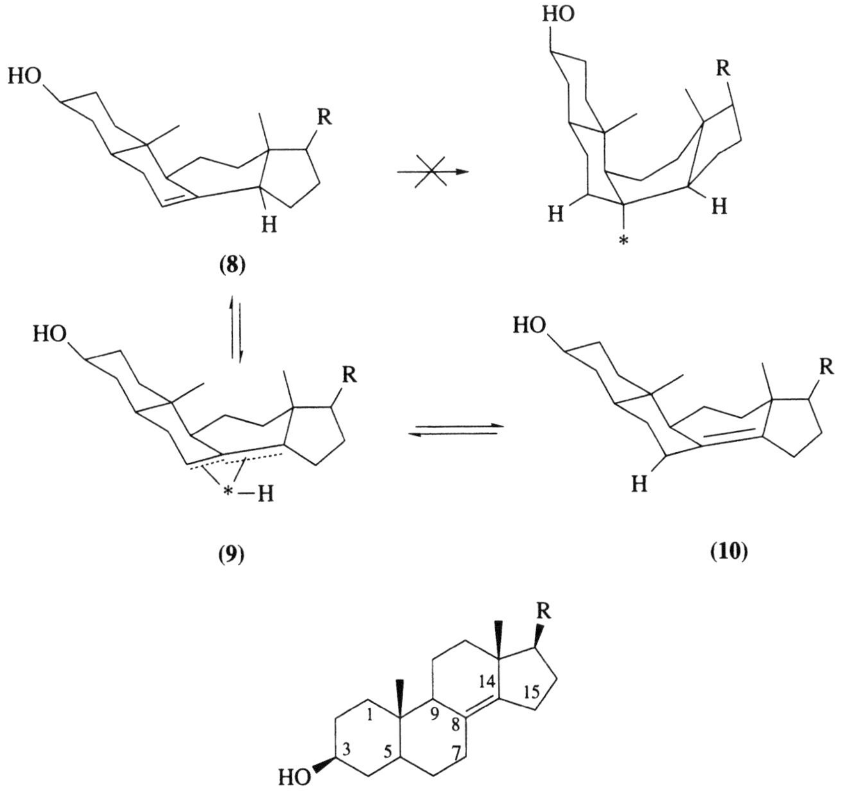

(8)

(9)

(10)

(10)

Scheme 3

An example of the exclusive operation of the dissociative mechanism is seen in the palladium-catalyzed isomerization of cholest-7-en-3β-ol (**8**) to cholest-8(14)-en-3β-ol (**10**) which does not add H_2 (D_2) in the absence of a strong acid (Scheme 3).[23,38] The formation of the alkyl intermediate is prevented by the increase in the internal nonbonded interactions which would have to be overcome to form a *cis* junction of the B/C rings (see Section 3.1.3.3.4). The conversion to the allylic intermediate (**9**) avoids these internal interactions. In the presence of D_2 and palladium, the 8(14)-ene (**10**) exchanges the α-hydrogens at carbons 7, 9 and 15 with deuterium more rapidly than at any other position, which implicates three different π-allyl complexes.[38]

3.1.3.1.2 Modifiers of isomerization

A comparison of the mechanisms of isomerization and hydrogen addition shows that an increase in the availability of hydrogen at the reaction site should favor the addition of hydrogen over isomerization, and this has been demonstrated repeatedly.[1–5] The relative rate of isomerization to the addition of hydrogen can be affected also by nucleophilic reagents which include strong bases, amines, phosphines and CO.[4,39] Apparently, these nucleophiles are adsorbed reversibly and compete for vacant coordination sites with H_2 along with the unsaturated compound. Whether isomerization proceeds *via* the associative or the dissociative mechanism, a vacant site must be provided to accept a hydride from the adsorbed alkyl intermediate or the adsorbed alkene, respectively. The adsorption of H_2 not only furnishes the hydrido groups required for addition but also decreases the number of vacant sites.

The alkene itself is a nucleophile. Thus, increasing the ratio of alkene to catalyst (Pd/C) from 4:1 to 30:1 decreases the rate of isomerization of 4-*t*-butylmethylenecyclohexane (**11**) to 4-*t*-butyl-1-methylcyclohexene (**12**; equation 12), relative to the rate of hydrogen addition.[40] With the 4:1 ratio, isomerization is about 10 times as fast as addition, and the saturated product is mainly the more stable *trans* isomer (see

Section 3.1.3.3.5). Tertiary amines, triphenylphosphine and CO are much more effective inhibitors of isomerization.

$$\text{Bu}^t\text{-C}_6\text{H}_{9}\text{=CH}_2 \; \textbf{(11)} \rightleftharpoons \text{Bu}^t\text{-C}_6\text{H}_8\text{-CH}_3 \; \textbf{(12)} \qquad (12)$$

Benzene appears to inhibit isomerization with Pt and Rh catalysts;[4] however, the effect may be due in part to differences in the solubility of hydrogen in the solvents. The rate of hydrogenation of cyclohexene catalyzed by Pt/SiO_2 is independent of the solvent, including benzene, provided that the concentration of dissolved hydrogen is used in the rate expression.[19] With Pd however, the corrected rate constants are smaller in benzene and smaller still in xylene, than in nonaromatic solvents.[21]

To compensate for the decrease in the rate of hydrogenation caused by the added nucleophile, the amount of catalyst or the hydrogen pressure may be increased.[3,4] The nucleophile shifts the kinetics from zero to first order in the alkene, which increases selectivity.

3.1.3.2 Structure–Reactivity

3.1.3.2.1 Individual rates

The rate of hydrogenation of a carbon–carbon double bond depends upon its structural environment, the catalyst and the reaction conditions. Attempts to find linear free energy relationships between structure and reactivity have had limited success in determining whether the substituent effects are mainly polar or steric, but the studies serve to assemble much of the available pertinent information on the subject.[41,42]

The rate of hydrogenation decreases with an increase in the number of alkyl substituents in the order $CH_2{=}CHR^1 > CH_2{=}CR^1R^2 > cis\text{-}R^1CH{=}CHR^2 > trans\text{-}R^1CH{=}CHR^2 > R^1CH{=}CHR^2R^3 > R^1R^2C{=}CR^3R^4$. Variation in the structure of the alkyl group has a secondary effect as does the polarity of the attached group, providing that the substituent is not conjugated with the double bond.[41] On palladium, polar effects are small or negligible.[41]

3.1.3.2.2 Competitive rates

The structure of the alkene has a greater effect on the competitive reactivities than on the individual rates. For example, Maurel and Tellier determined the individual and competitive rates in cyclohexane on a 7.5% Pt/SiO_2 catalyst (20 °C, 1 atm; 1 atm = 101.3 kPa) of 15 substituted alkenes (Table 1).[43] The relative values of the individual rates agree well with the rates of those compounds which are also in-

Table 1 Individual and Competitive Rates of Hydrogenation[a] of Alkenes on Platinum[43]

Structural type	*Compound*	k_A[b]	k_AK_A[c]
$RCH{=}CH_2$	1-Hexene	110	7.4×10^5
	4-Methyl-1-pentene	125	4.7×10^5
$R^1R^2C{=}CH_2$	2-Methyl-1-pentene	120	1.3×10^4
	2,3-Dimethyl-1-butene	160	0.9×10^4
$\overline{\text{CH=CH}}\,(CH_2)_n$ (cyclic)	Cyclopentene	100	1.3×10^5
	Cyclohexene	(100)	(1.0×10^4)
$R^1R^2C{=}CHR^3$	1-Methylcyclopentene	95	3.7×10^3
	3-Ethyl-2-pentene	65	3.5×10^2
	1-Methylcyclohexene	40	1.0×10^2
$R^1R^2C{=}CR^3R^4$	2,3-Dimethyl-2-butene	75	1.8
	1,2-Dimethylcyclohexene	20	0.6

[a]Catalyst, 7.5% Pt/SiO_2. [b]Individual rate constants relative to the rate of cyclohexene scaled to 100. [c]Competitive rates scaled relative to cyclohexene, $k_cK_c = 1.0 \times 10^4$.

cluded in the study by Hussey *et al.*, who used a 0.52% Pt/Al_2O_3 catalyst at 25 °C in cyclohexane.[44] The rate constants cover a range of less than 10, whereas the equilibrium constants have a range of 2×10^5, and furnish a linear plot of the logarithm of the equilibrium constants against the Taft polar substituent constant σ^*. However, the polar and steric substituent constants are intercorrelated and therefore the result is ambiguous.[42]

Hussey *et al.* interpret the relative competitive rates on platinum catalysts as measures of competitive rates of alkene adsorption.[45] They note that very little isomerization accompanies hydrogenation, which suggests that desorption of the alkene is slow. A rapid interconversion of adsorbed alkene and the alkyl intermediate of the Horiuti–Polanyi mechanism is indicated by the distribution of deuterium in the deuterated alkane formed when D_2 is used in place of H_2; yet little or no deuterium appears in the recovered alkene.[32]

The appropriate interpretation may depend on the range of structures considered, the catalyst and the conditions employed. The interpretations differ in that in one, the rate of desorption is assumed to be fast relative to the conversion of adsorbed alkene to the alkyl intermediate, while in the other, desorption is assumed to be relatively slow. Catalysis by palladium may conform more closely to the reversible adsorption model than does platinum.

The relative individual rates and the relative competitive rates are determined by the difference in energy of the same transition states but different ground states; in the former the alkene is bound to the catalyst, in the latter it is free. The effect of alkene structure on relative reactivity indicates that the structural change on adsorption is greater than the changes which follow adsorption on the catalyst. Interestingly, hydrogenations over P-2 Ni are apparently first order in alkene, which probably accounts for the greater sensitivity to structure shown by this catalyst in individual hydrogenations as compared to Pt.[13]

3.1.3.2.3 Alkene association constants

The relative equilibrium constants for the adsorption of alkenes on the surfaces of platinum metals are not available from static measurements. Accordingly, the kinetically derived constants have been compared with the association constants of alkenes with metal ions in solution. Of these, the measurements by Tolman are the most instructive.[46]

Tolman has shown that the equilibrium constants for the reactions of 38 substituted alkenes with Ni[P(O-*o*-tolyl)$_3$]$_3$ (**13**) in benzene, to form (ENE)bis(tri-*o*-tolylphosphite)nickel complexes (**14**), are sensitive to the structure of the alkene (equation 13). Values of K_1 at 25 °C vary from 10^{-4} to 4×10^8. The stability of the complex is enhanced by electron-withdrawing substituents such as cyano and carboxy and lowered by alkyl groups. That resonance involving unshared electrons on the oxygen of an alkoxy group overpowers the inductive effect is indicated by the relative values of K_1 for allyl methyl ether, 1-hexene and vinyl butyl ether which diminish in that order by factors of 3:1:0.006.

$$\underset{(\mathbf{13})}{NiL_3} + ENE = \underset{(\mathbf{14})}{(ENE)NiL_2} + L \quad \text{where } L = P(O\text{-}o\text{-tolyl})_3 \qquad (13)$$

The equilibrium constants for a series of cycloalkenes decrease in the order norbornene > *cis*-cyclooctene > cyclopentene > cycloheptene > cyclohexene, which correlates with the calculated strain energies as well as the kinetically determined relative adsorption constants on Pt (Table 2). Tolman states that electron donation from a filled metal *d*-orbital to an empty alkene π^*-orbital is extremely important in determining the stability of these complexes. Steric effects of substituents are relatively unimportant compared to electronic effects, and resonance is more important than inductive interactions. The ability of the metal to back bond is lowered progressively in the series $Ni^0 > Pt^0 > Rh^I > Pt^{II} > Ag^I$ which reduces the importance of resonance and decreases the selectivity of the metal for different substituted alkenes.

The diimide reduction of an alkene has been suggested as a model for the adsorption of an alkene on a metal, and has been used to help identify the intramolecular interactions which determine the stereochemistry of hydrogenating alkyl-substituted cycloalkenes.[47] Both processes involve *syn* addition; in one, two hydrogen atoms are transferred to the face of the double bond, in the other, a π-alkene complex is formed. Both represent changes in the direction of forming an eclipsed conformation of an alkane. Garbisch *et al.* found that the major factors that contribute to the observed reactivities in diimide reductions of unsaturated hydrocarbons are torsional strain, bond angle strain and α-alkyl substituent effects, as indicated by the good agreement between calculated and observed relative reactivities which covered a range of 38 000.[48] This treatment was applied also to stereoselectivities. Comparisons with catalytic hydrogenations are limited but show a parallelism, and clearly, diimide reductions are more sensitive to structure than are the individual rates of hydrogenation and, possibly, the kinetically determined equilibrium constants for adsorption on Pt (Table 2).

Table 2 Structure–Reactivity of Cycloalkenes. Comparisons of Individual and Competitive Hydrogenation Rates with Related Reactions

Compound	H_2, Pt^a k_A[44]	K_{AB}[45]	*Ni(ENE)*[46] K_{eq}	*STRAIN*[46] ΔH *(Kcal)*	*Diimide*[48] k_{rel}
Norbornene	223	—	4.4	27.2	4.5×10^{2}
cis-Cyclooctene	10	25	6.2×10^{-2}	7.4	17
Cyclopentene	121	7.5	2.6×10^{-2}	6.8	16
Cycloheptene	78	6.3	2.3×10^{-2}	6.7	12
Cyclohexene	113	(1.0)	3.5×10^{-4}	2.5	(1.0)

[a]0.52% Pt/Al_2O_3 at 25 °C, 1 atm.

3.1.3.3 Stereochemistry

3.1.3.3.1 Syn *addition*

With few exceptions, the stereochemistry of the hydrogenation of a carbon–carbon double bond is the result of *syn* addition of two atoms of hydrogen from the catalyst to the proximal face of the double bond.[23] A classical demonstration is given by the hydrogenation (Pd, AcOH, 1 atm) of the (*Z*)- and (*E*)-isomers of 2,3-diphenyl-2-butene (**15**) and (**16**) which yield respectively the *meso* (98%) and (±) (98%) isomers of 2,3-diphenylbutane (equations 14 and 15).[32] Likewise the hydrogenation of dimethyl 1,2-cyclohexenedicarboxylate yields only the *cis* saturated isomer.[23] Such stereospecificity is not always found.

Ph Ph (**15**) —Pd, H_2, 1 atm→ *meso*-2,3-diphenylbutane 98% (14)

Ph Ph (**16**) —Pd, H_2, 1 atm→ (±)-2,3-diphenylbutane 98% (15)

The hydrogenation of 1,2-dimethylcyclohexene (**17**; PtO_2, HOAc, 1 atm) yields *trans*-(**19**) as well as *cis*-1,2-dimethylcyclohexane (**18**),[23,32] but Ir and Os in *t*-butyl alcohol are much more selective (equation 16).[49] The addition of hydrogen to (**17**) competes with isomerization to 1,6-dimethylcyclohexene (**20**), which not only reacts more rapidly but yields *trans* (**19**) as well as the *cis* product (**18**) *via syn* addition of the two hydrogen atoms (equation 17). In the Pt-catalyzed reaction, the intermediate (**20**) can be observed, although its steady-state concentration remains low (0.21% of the 1,2-isomer **17**) because of its

(**17**) —[H]→ (**18**) + (**19**) (16)

	Pressure (atm)	(**18**) (%)	(**19**) (%)
Pt	1	82	18
Pt	300	96	4
Os	1	98.6	1.4
Ir	1	98.9	1.1

greater competitiveness. This explanation for the apparent antarafacial addition of hydrogen is supported by the fact that the amount of the *trans* isomer (**19**) which is formed over a particular metal correlates with its ability to catalyze the migration of a double bond.[33]

(17) ⇌ (20) —*syn* addition→ (18) + (19) (17)

3.1.3.3.2 *Apparent* anti *addition*

Anti addition has been observed in compounds which lack an allylic hydrogen that can participate in a 1,3-hydrogen shift as above. With heptane as solvent (25 °C, 1 atm), the hydrogenation of dimethyl bicyclo[2.2.2]oct-2-ene-2,3-dicarboxylate (**21**) yields 7.1% of the *anti* addition product (**24**) over a Pt/C catalyst, but only *syn* addition (**22**) occurs over Rh/C.[50] An acidic medium (small amounts of a strong acid in methanol) has little effect on catalysis by Rh, but greatly increases the fraction of *anti* addition catalyzed by Pd/C, as much as 60% in the presence of *p*-TsOH. Evidence for the participation of the solvent was assumed to indicate a proton transfer to the alkyl intermediate with inversion of configuration at the attached carbon atom.

(21) CO_2Me, CO_2Me —Rh/C→ (22) CO_2Me, CO_2Me

(21) —Pd/C→ (22) + (24)

(23) CO_2Me, H, OMe, H-M, O ——→ (22) + (24) CO_2Me, CO_2Me

However, the carbonyl of the methoxycarbonyl group may be involved in the formation of an oxallylic intermediate (**23**), which affords a path to both *syn* and *anti* isomers. A similar structure was proposed by Teratani *et al.* to explain the distribution of deuterium in the products of deuteration of 2-methylcyclohexanone catalyzed by palladium; such a structure was not required to explain the results of catalysis by rhodium.[51]

3.1.3.3.3 *Catalyst hindrance*

If *syn* addition to the opposite faces of the double bond yields diastereomeric compounds, then the orientation is subject to a number of factors. Linstead *et al.* asserted that adsorption on the catalyst occurs

X —D_2, Pd/C→ X (D, D) + X (D, D) (18)

	exo (%)	*endo* (%)
(**25**) X = OH	100	0
(**26**) X = Bu^t	75	25

so that steric hindrance with the catalyst is minimized, and that hydrogen is transferred to the proximal face of the double bond to yield the principal product.[52] Often, the application of this principle appears straightforward. For example, the addition of deuterium to norbornene yields the *exo*-2,3-dideuteronorbornane on Pd, Pt, or Rh.[32] Substituents at C-7 affect the orientation. An *anti* hydroxy group (**25**) has no effect, but the *t*-butyl group (**26**) exhibits an opposing influence (equation 18). A *syn* *t*-butyl group (**27**) directs 80% *endo* and 20% *exo*, while a *syn*-OH (**28**) directs 60–70% *exo* (equation 19).

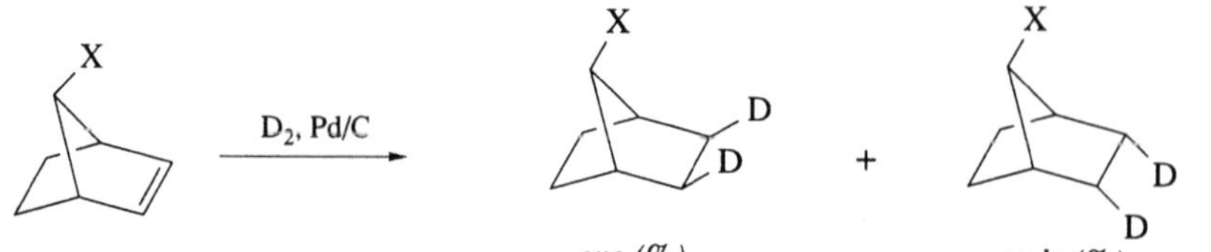

D_2, Pd/C (19)

	exo (%)	*endo* (%)
(**27**) X = Bu^t	20	80
(**28**) X = OH	100	0

3.1.3.3.4 Intramolecular nonbonding interactions

Adsorption alters the structure of an alkene, and the subsequent hydrogen transfers are accompanied by further changes in the organic moiety.[23] The transition state for adsorption undoubtedly has greater double bond character than the transition state for the formation of the alkyl intermediate, which implies that intramolecular nonbonding interactions will be greater in the latter.

Hydrogenation of the strained hexamethylbicyclo[2.2.0]hexa-2,5-diene (**29**) is highly stereospecific over the platinum metals at 25 °C, 1 atm.[32] Addition of hydrogen from the less-hindered *exo* direction yields the bicyclohexene (**30**) but further addition is very slow (Scheme 4). Raising the pressure (150 atm) results in the *syn, exo* addition of a second mol of H_2 over Ru and Rh to form (**31**; >98%). The resistance to addition can be attributed to the increase in nonbonded interactions among the four *endo* methyl groups which accompanies the formation of the alkyl intermediate. These interactions are avoided in the dissociative mechanism of isomerization to the *exo*-methylene intermediate, apparently the path taken on Pt and Ni at 50 °C, where both the *syn, exo*-(**31**) and the *anti* addition product (**32**) are formed.

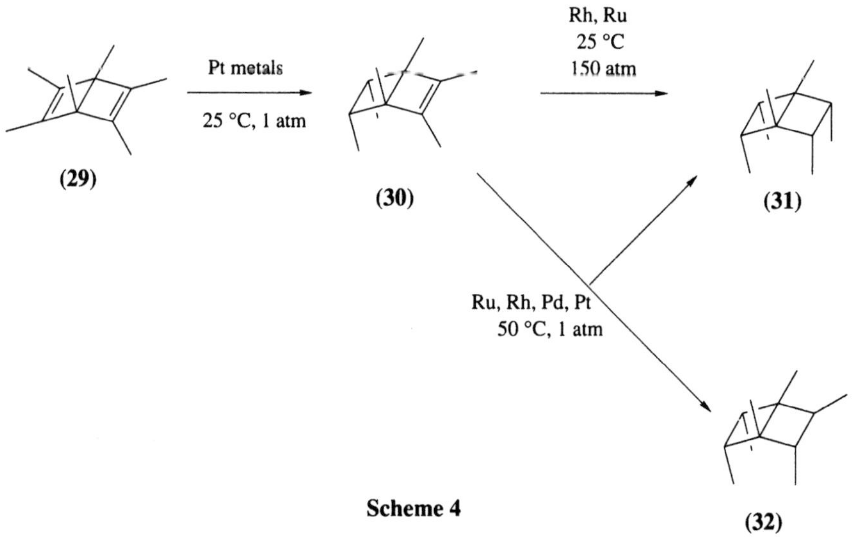

Pt metals, 25 °C, 1 atm; Rh, Ru, 25 °C, 150 atm; Ru, Rh, Pd, Pt, 50 °C, 1 atm

(**29**) (**30**) (**31**) (**32**)

Scheme 4

The inability of cholest-8(14)-en-3β-ol (**10**) to be hydrogenated except in the presence of a strong acid has been explained in a similar manner (see Scheme 3).[23,38] The transfer of the first hydrogen to the less-hindered α-face of the double bond is resisted by the large increase in the repulsive nonbonding interactions involving the methyl groups at C-10 and C-13. A Δ-7 unsaturated steroid (**33**) which contains a 9β-H is hydrogenated from the β-side of the molecule to form (**34**; equation 20).[32] Other examples are found in Molnar's review which contains data on the stereochemistry of hydrogenating a large variety of steroids.[32]

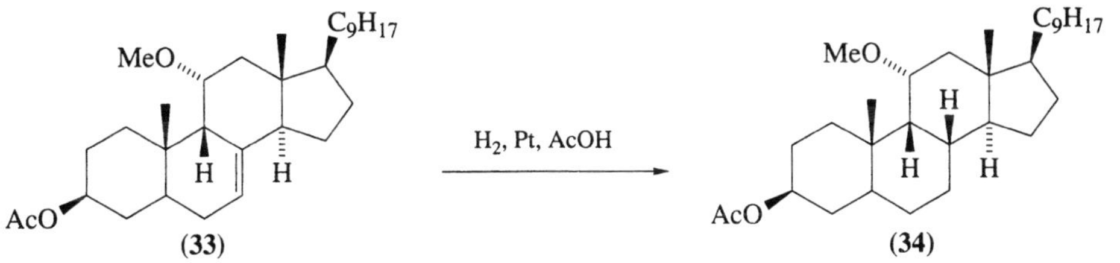

$$\xrightarrow{H_2,\ Pt,\ AcOH} \quad (20)$$

(33) (34)

3.1.3.3.5 *Application of conformational analysis*

The geometry of the alkene moiety at the transition state for adsorption is assumed to approximate to the most stable conformation of the alkene.[23] This model correctly explains the predominance of the axial–equatorial-disubstituted cyclohexanes in the products from alkyl-substituted methylenecyclohexanes under conditions in which alkene isomerization is a minor factor (see Section 3.1.3.1.2).[40] The effect of an increase in pressure (0.25–150 atm) on the ratio of isomers formed on hydrogenating 4-*t*-butylmethylenecyclohexane (**11**; PtO_2, AcOH) is explained as indicating a change in the product-controlling reaction from the formation of the alkyl intermediate to the adsorption of the alkene.[23] Similarly, the effect of increasing the hydrogen pressure on the change in stereoselectivity in the hydrogenation of methyl (*Z*)-α-acetamidocinnamate, which is catalyzed by the chiral cationic complex [(*R*,*R*)-1,2-bis(phenyl, *o*-anisylphosphino)ethane]rhodium(I), has been demonstrated to be due to a change in the rate-limiting reaction, from the oxidative addition of hydrogen to the formation of the alkene rhodium complex.[53]

If the reductive elimination of the alkane is product-controlling, and if all the preceding reactions of the Horiuti–Polanyi mechanism are relatively rapid and reversible, including the desorption of the alkene, then the principal product should be the more stable isomer with substituents in equatorial positions. Palladium appears to conform to the expectations for this mechanism because isomerization is often rapid, and the more stable isomers predominate in the products of addition (see Section 3.1.3.1.2). However, when isomerization is inhibited by added nucleophiles, the less stable stereoisomer is the main saturated product. Gonzo and Boudart concluded from the kinetics of hydrogenation of cyclohexene catalyzed by Pd/SiO_2 that the rate-controlling step is the reductive elimination of alkane.[21] If their conclusions apply to the methylenecyclohexanes, then it suggests that the preceding reactions may be comparatively rapid, but not rapid enough to obtain an equilibrium distribution of the alkyl intermediates.

The principles described above may serve as a useful guide for unsaturated hydrocarbons, but substituents other than alkyl groups may involve attractive interactions with a catalytic site. A compound in which a phenyl group is attached to a double bond in a five- or six-membered cycloalkene and is also vicinal to an alkyl substituent yields a larger fraction of the *cis* stereoisomer than is formed from the related dialkylcycloalkene.[32] *Cis* isomers are formed preferentially on Pd and Rh, and are believed to involve structures in which the phenyl group is bound to the surface through its π-electrons.[32]

The Pd-catalyzed hydrogenation of alkyl-substituted 1-alkoxycyclohexenes is stereoselective. The axial/equatorial stereoisomer predominates no matter where the alkyl substituent is located on the ring.[32] Nishimura suggested that the result indicates that the product-controlling step involves the reductive elimination of the most stable alkyl intermediate which is attached to the metal at the carbon atom that bears the alkoxy group (equation 21).

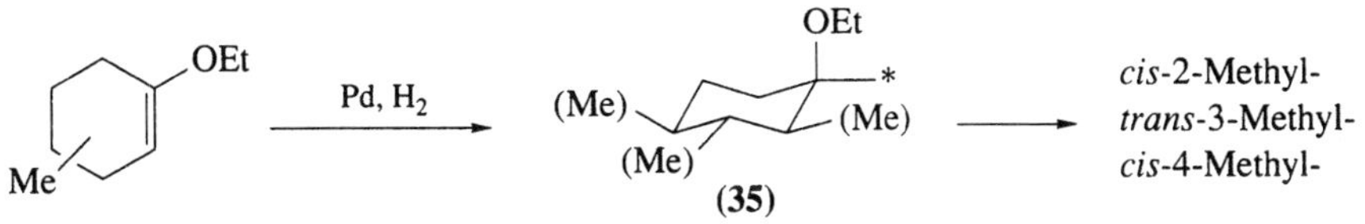

$$\xrightarrow{Pd,\ H_2} \quad \text{(35)} \longrightarrow \quad (21)$$

cis-2-Methyl-
trans-3-Methyl-
cis-4-Methyl-

3.1.3.4 Haptophilicity

The location of hydroxy groups relative to the carbon–carbon double bond appears at times to direct the addition to the side of the molecule which presents the greater catalyst hindrance.[32] This has been noted by Dart and Henbest,[54] who termed it an 'anchor effect', and later by Nishimura and Mori for the hydrogenation of several steroid structures.[55]

A hydroxy group situated on an allylic carbon of a five- or a six-membered ring can exhibit an anchoring effect on a Ni catalyst (equation 22); with other metals the group's steric effect is dominant.[32] The hydrogenation of 1-methylcyclohexenes, with an amino group (NH_2 or NHMe) in place of the hydroxy group, exhibits stereoselectivities like the corresponding alcohols on a Ni catalyst (equation 23), but the dimethylamino group exerts only a steric effect, as do the amine hydrochlorides.[32]

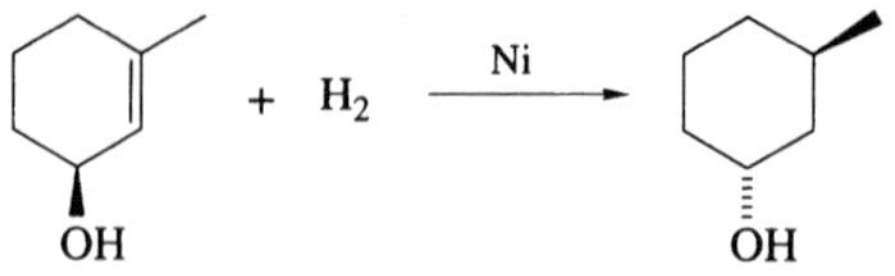

+ H_2 —Ni→ (22)

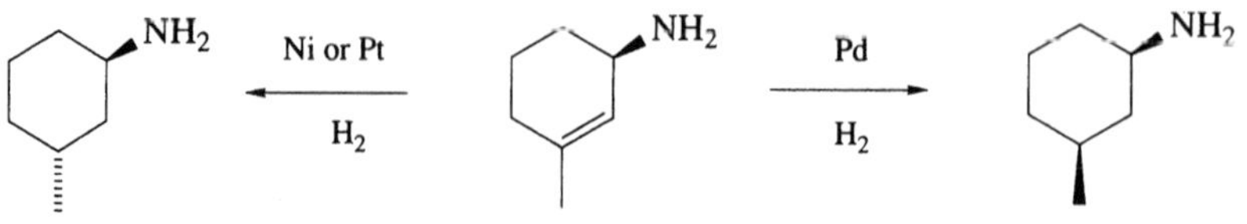

OH OH

NH_2 ←Ni or Pt, H_2— NH_2 —Pd, H_2→ NH_2 (23)

80% 60%

The name 'haptophilicity' was coined by Thompson to describe this effect of polar groups, which is presumed to arise through the ability of the group to coordinate to the surface of the catalyst.[56] A study of the Pd/C-catalyzed hydrogenation of a series of tetrahydrofluorenes (**36**) to the hexahydro derivatives (**37**) shows that the effect does not correlate well with the basicity of the group, *e.g.* $CH_2OH > CO_2Na$ in the solvent 2-methoxyethanol (equation 24). That the groups also exert a steric effect which may be dominant is recognized, and solvent effects can be important. Other structural elements in complex molecules can influence the stereoselectivity, so it is difficult to predict the relative importance of haptophilicity.[32]

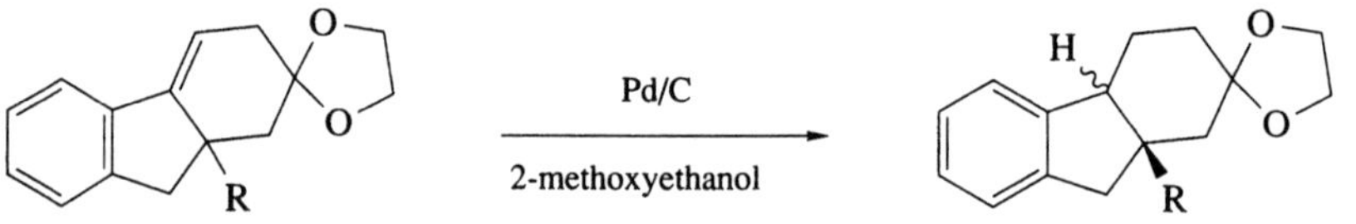

—Pd/C, 2-methoxyethanol→ (24)

(**36**) (**37**)

R	*cis (%)*	*trans (%)*
CH_2OH	95	5
CO_2Na	55	45

3.1.4 HYDROGENATION OF C≡C

3.1.4.1 Semihydrogenation

The semihydrogenation of the carbon–carbon triple bond is a particularly valuable and frequently used application of heterogeneous catalysis to synthetic chemistry, and is the subject of several recent reviews.[4,6,57–59] Catalysts prepared from palladium and nickel are most commonly used, but the form of the catalyst and the conditions of use affect the results (see Section 3.1.1.2). A polymer-bound palladium catalyst, $PdCl_2$ with poly-4-diphenylphosphinomethylstyrene, is intended to combine the selective properties of mononuclear transition metal complexes with the ease of separating the product from a solid.[17] Whether catalysts of this type will replace the more traditional heterogeneous catalysts remains to be seen.

The hydrogenation of an alkyne can be virtually stopped at the semihydrogenated stage with various preparations of Pd and Ni catalysts by adding to the reaction mixture a nitrogen base, such as quinoline, pyridine or ethylenediamine, that is reversibly adsorbed on the catalyst in competition with the alkyne and the product alkene.[3,4,58,59] Solvents affect the adsorption equilibria. Depending upon the catalyst and the structure of the alkyne, the strength of adsorption of the amine from an alkane solvent may be so great as to slow the hydrogenation to a virtual standstill. The use of a hydrogen-bonding solvent (MeOH, EtOH) will lower the surface coverage by the amine and increase that of the hydrocarbons.[20,22]

High selectivity can be achieved because the alkyne is more strongly bound than the alkene to a metal center and either displaces or blocks the readsorption of any alkene which leaves the surface of the catalyst. Selectivity suffers unless the diffusion of the reactants to the reaction sites is faster than the catalyzed reaction (see Section 3.1.2.1). To optimize selectivity, the catalytic activity can be lowered and/or the diffusion path shortened. The latter accounts for the high selectivity for the conversion of dienes to monoenes of the recently prepared and characterized colloidal palladium catalysts.[60]

Lowering the temperature, –10 to –30 °C, has improved the selectivity of the Lindlar catalyst when used in the presence of quinoline dissolved in pentane, hexane or hexane/THF.[58] However, raising the temperature increases the rate of desorption of the alkene, which appears to account largely for the resulting increase in the selectivity of the heavier Group VIII metals for the hydrogenation of the simple alkynes.[61]

Various materials have been incorporated into the preparation of the catalysts to lower the metal's intrinsic activity.[1–4,59] Lindlar's treatment of palladium deposited on $CaCO_3$ with $Pb(OAc)_2$ lowers its activity, apparently by affecting the morphology of the metal.[62] Unfortunately, Lindlar's recommendation that quinoline be used with his catalyst is sometimes overlooked, which may account for some of the poorer selectivities reported when its use is not mentioned.

The function of quinoline is illustrated by the kinetic study of Steenhoek *et al.* who reported high stereoselectivities using quinoline with $Pd/CaCO_3$ prepared according to Lindlar but omitting the treatment with lead acetate.[1,63] Quinoline is adsorbed reversibly on the reactive sites, competing with the alkyne and the product alkene, but much more effectively with the latter. Accordingly, as the concentration of alkyne diminishes, the quinoline occupies more of the surface as does the alkene but the amount of adsorbed alkene remains too low to affect the composition of the product. Quinoline is unaffected by the usual mild experimental conditions so its concentration remains constant throughout the system, including the boundary layer and pores of the catalyst where it is most needed. Not surprisingly, the stereospecificity of the P-2 nickel catalyst for the hydrogenation of alkynes is improved markedly by the use of ethylenediamine in 2–3 times the molar amount of the catalyst.[1–4,59]

Selectivity also is increased by the use of supported catalysts in which a small amount of metal (~0.1%) is highly dispersed.[4,59] However, this strategy may fail if the reagents are not sufficiently free of adventitious catalyst poisons. A catalyst containing approximately 5% of the metal can act as a sop for such materials.[3] Very small loadings of Pd on wide pore aluminas have been shown to increase the selectivity of converting acetylene to ethylene in a large excess of the latter, an important industrial objective. Carbon monoxide is usually added to the gaseous mixture to improve selectivity, the CO functioning as a reversibly adsorbed competitor, as does quinoline when the fluid phase is a liquid.[64]

3.1.4.2 Mechanism

The reaction of D_2 with simple alkynes on Group VIII metals indicates that the reaction is more complicated than is implied by equation (25). Which of these complications should be assigned to the readsorption and subsequent reactions of the first-formed alkene or to other transformations of the adsorbed intermediates is unsettled.[57–59,61]

$$R{-}C{\equiv}C{-}R + M \rightleftharpoons R{-}C{\equiv}C{-}R\ (\text{M}) \underset{}{\overset{H_2}{\rightleftharpoons}} \text{RC(M)}{=}\text{C(R)H} \cdot \text{MH} \longrightarrow \textit{cis-}\text{RHC}{=}\text{CHR} + M \quad (25)$$

Reactions of soluble metal complexes, whose mechanisms of catalysis appear to be reasonably well known, can serve as a guide to the main reaction paths followed on heterogeneous catalysts. Mononuclear complexes catalyze *syn* addition of H_2 to alkynes to yield initially only *cis* isomers, as in equation (25).[65] More recently, Muetterties and coworkers showed that the dinuclear rhodium hydride complex $\{(\mu\text{-H})Rh[P(OPr^i)_3]_2\}_2$ (**38**) converts alkynes to *trans* isomers as initial products (equation 26).[66] The alkyne addition compound (**39**) was isolated; its structure shows the vinyl group bonded to one rhodium atom by a σ-bond and to the other by a π-bond, while the substituents on the vinyl group are *trans* to one another. This structure resembles ones hypothesized earlier to explain the formation of *trans* isomers and alkanes.[59,61] Hydrogenations of alkynes which are catalyzed by the dinuclear rhodium hydride are much slower than the hydrogenation of an alkene catalyzed by the dinuclear tetrahydride (**40**), which is formed rapidly from (**38**) in the presence of H_2 (equation 27).[66]

Additional evidence that reaction paths analogous to the reactions of mononuclear and dinuclear rhodium complexes exist on palladium is shown by the effect of triphenylphosphine on the stereoselectivity of the hydrogenation of di-*t*-butylacetylene (DTBA).[67] In the absence of PPh_3 but on a highly dispersed

$$R-C\equiv C-R + L_2Rh(\mu-H)_2RhL_2\ (\mathbf{38}) \longrightarrow (\mathbf{39}) \xrightarrow{H_2} trans\text{-}RHC{=}CHR \quad (26)$$

$L = P(OPr^i)_3$

$$(\mathbf{38}) + H_2 \rightleftharpoons (\mathbf{40}) \quad (27)$$

catalyst, DTBA yields 84% *cis* and 16% *trans* to about 80% conversion when the isomerization of the *cis* to the *trans* isomer comes into play. With a ratio of PPh_3 to the exposed Pd atoms of 2.0:1 the reaction is slow but the products are the *cis* and *trans* isomers in a fixed ratio of about 1:2, which does not change during the complete semihydrogenation of the alkyne; the isomerization of *cis*-DTBA to *trans*-DTBA is rapid in the absence of PPh_3, but does not occur in this experiment. The result can be understood if the alkyne functions as a monodentate ligand at the site at which the more rapid hydrogenation to *cis*-alkene occurs but as a bidentate ligand at sites which yield *trans* as well as *cis* isomers as initial products. The phosphine, a monodentate ligand, is a less effective competitor at dinuclear sites. The less sterically encumbered 3-hexyne is a much more effective competitor than DTBA for the mononuclear sites and its *cis* stereoselectivity is improved by PPh_3, the usual consequence of adding nucleophiles.

The predominance of the *cis* isomers in the products of the Pd- or Ni-catalyzed hydrogenations of disubstituted alkynes suggests that the main reaction path involves the alkyne bound to a single metal atom as in catalysis by the mononuclear complexes $ClRhL_3$ or $HRhL_4$.[28] Surface sites which can furnish the required coordinative unsaturation correspond to the designations ^3M and $^2M—^2M$. The latter designation indicates a structural element which also can support a mechanism like that of the dinuclear rhodium complex (**38**).

An alternative mechanism to account for the formation of alkanes is that the alkene which is released from the catalytic site is immediately readsorbed unless the vacated site is filled by some other donor molecule. The process may be likened to the 'cage effect' in the formation of radical pairs in solution. Adsorption of the alkene is exothermic and, accordingly, is likely to have a low energy barrier. The lower selectivity of Os, Ir, and Pt has been attributed to the readsorption of alkene.[61]

3.1.4.3 Regio- and Stereo-selectivity

The semihydrogenation of a carbon–carbon triple bond can be accomplished in molecules containing double bonds, whether isolated or conjugated with one another.[57–59] A classical example is the final stage but one in the synthesis of β-carotene.[1] Stereospecificity is high if the triple bond is separated from the double bond by one or more methylene groups as in enyne (**41**; equation 28), but, according to Henrick, conjugated enynes such as enyne (**42**; equation 29) usually yield mixtures over Lindlar catalyst (in petroleum ether with added quinoline).[58]

$$(\mathbf{41}) \xrightarrow[\text{THF, hexane, quinoline}]{H_2,\ \text{L-Pd},\ -5\ \text{to}\ 0\ °C} \quad (28)$$

96%, 98.1% pure

OH

(42)

H_2, L-Pd, –20 °C

methanol, no quinoline

OH

(29)

92% *(7E,9Z)*, 8% *(7E,9E)*

A diyne can be semihydrogenated to the enyne if one triple bond is terminal and the other is not, the terminal bond being hydrogenated selectively.[59] If both bonds are nonterminal and unconjugated, hydrogenation to the (Z,Z)-diene is accomplished in high yield.[57–59] Interestingly, the (Z,Z)-octadeca-9,12-dienecarboxylic acid as well as the (Z)-enyne appear as initial products of the hydrogenation of octadeca-9,12-diynecarboxylic acid ($Pd/CaCO_3$, hexane, quinoline at 25 °C, 1 atm).[63]

A triple bond can be semihydrogenated in the presence of a variety of reducible or hydrogenolyzable groups, although G conjugation with the carbonyl group of aldehydes and ketones poses problems of stereoselectivity.[57,59]

3.1.5 DIENES AND POLYENES, REGIO- AND STEREO-SELECTIVITY

3.1.5.1 Isolated Dienes

Regioselectivity in hydrogenating isolated double bonds is governed mainly by the same structural effects which determine the relative competitive reactivities of monoenes. Thus terminal double bonds are reduced in preference to nonterminal and the less alkyl-substituted to the more. Strain in the double bond alters this preference, for example 5-methylenebicyclo[2.2.1]hept-2-ene is hydrogenated to 2-methylenebicyclo[2.2.1]heptane.[1,3,32] Nickel and palladium catalysts are usually most selective.

1,5-Cyclooctadiene (1,5-COD) is converted selectively to cyclooctene (97.8% on the addition of 1 mol of H_2 per mol of diene) over a colloidal Pd supported on poly(*N*-vinyl-2-pyrrolidone) in methanol.[60] The reaction is in part direct and in part proceeds through isomerization to the 1,4-COD, which is observed, and, presumably, through 1,3-COD, which is not observed. 1,3-COD is the most reactive and selective of the three dienes; at its complete conversion it yields 99.9% cyclooctene. Nishimura discovered that the reduction of 1,5-COD is highly selective over $Pd/CaCO_3$ or Pd black in the presence of phenylacetaldehyde with almost complete suppression of the reduction of cyclooctene.[68] The inhibition is probably due to the decarbonylation of the aldehyde; the adsorbed CO competes with the cyclooctene for surface sites.[69]

3.1.5.2 Conjugated Dienes

The hydrogenation of conjugated dienes to monoalkenes poses questions of both regio- and stereoselectivity and is a function of the metal and the conditions.[1–4,32] The distribution of products depends not only on the inherent selectivity of the initial addition process but also on the competition between the diene and the first-formed alkene(s) for the reactive sites. The latter may add hydrogen or be isomerized. If the structure of the diene allows the change, the diene itself may be isomerized. As for alkynes, palladium and nickel catalysts tend to be the most selective to the monoene.

3.1.5.3 Conjugated Diene Mechanisms

The monoenes formed on hydrogenating 1,3-alkadienes correspond to 1,2- and 1,4-adducts of hydrogen.[23,32] To account for the main products of the reaction of D_2 with 1,3-butadiene over Pd/Al_2O_3, 1-butene-3,4-d_2 and *trans*-2-butene-1,4-d_2, Meyer and Burwell assumed that the diene is adsorbed on the surface in the *trans* conformation.[23,32] Addition of deuterium to a terminal carbon atom forms an allylic species which is a common intermediate for the principal products. Later studies led to more detailed mechanisms designed to explain the observed regio- and stereo-selectivities.[32]

Adsorption may involve one or both double bonds. If only one double bond is involved initially the reaction would follow the sequence in equation (30). This appears to characterize additions on Cu and Au, for example.[32]

+ M + D_2 ⇌ MD_2 ⇌ MD D → D D + M (30)

The σ-allyl intermediate may be transformed to a π-allyl structure with either the *syn* or *anti* configuration (relative to the methynyl hydrogen) as in equation (31), see also Scheme 2. The union with a second hydrogen leads to adsorbed alkenes: 1-butene and either *trans*-2-butene from the *syn*-allyl intermediate or *cis*-2-butene from the *anti* form (Scheme 2). The desorption of the alkene competes with the further addition of hydrogen to form the alkane (equation 32). The reaction of a σ-allyl structure with hydrogen can yield the unbound alkene directly. The selectivity may depend upon the relative importance of these competitive reactions, which are likely to be a function of the metal as well as the reaction conditions.[32]

MD ⇌ MD ⇌ MD (31)

syn-π-allyl σ-allyl *anti*-π-allyl

M —D_2→ MD_2 → alkene-d_2 + M (32)

→ M +

3.1.5.4 Allenes and Cumulenes

Acyclic allenes can be hydrogenated with high selectivities to the monoenes over palladium catalysts.[32] The less-substituted double bond tends to be the more readily reduced. Crombie *et al.* showed that, when allowed to react competitively over $Pd/BaSO_4$, a terminal alkyne is hydrogenated in preference to a terminal allenyl isomer, which in turn competes for the surface more effectively than the nonterminal alkyne, *e.g.* $HC{\equiv}CCH_2CO_2H > H_2C{=}C{=}CHCO_2H > MeC{\equiv}CCO_2H$. They also compared the regio- and stereo-selectivities of Lindlar Pd, 10% Pt/C, 10% Rh/C and a Raney Ni, all used without quinoline, to hydrogenate these compounds and found the palladium catalyst to be most selective to the enoic acid (90–99%) and stereoselective to the *cis*-2-butenoic acids, 96% from tetrolic acid and 93% from allenecarboxylic acid. The catalysts exhibit similar selectivity for the hydrogenation of both alkynes and allenes and, accordingly, one should anticipate that catalysts and conditions which lead to highly selective semihydrogenations of alkynes should be about equally effective for the semihydrogenation of allenes. Quinoline or other bases should provide self-terminating hydrogenations.

syn-1,2-Addition appears to be the dominant process for the addition of H_2 to allenes. Passing 1,2-butadiene in a stream of H_2 over Pd/Al_2O_3 produces a mixture of products consisting of 1-butene (40%), *cis*-2-butene (53%) and *trans*-2-butene (7%). The regioselectivity is only moderate; the less-substituted double bond is favored by a ratio of 3:2; however, the ratio *cis-/trans*-2-butene is 7.5:1. This difference was attributed to adsorption, which minimizes steric hindrance with the surface.[23,32] This conclusion was supported later by Crombie *et al.* from their observation of the stereochemistry of hydogenation of a group of systematically substituted allenes under ambient conditions over 5% $Pd/BaSO_4$.[32] Vinyl intermediates were invoked to explain the results (equation 33).

R
R
MH
R
(33)
+
R
MH_2
MH
R
R
MH

The hydrogenations of allenes may proceed through allylic intermediates.[32] The distribution of products formed during the reaction of 1,2-butadiene with D_2 depends on the particular nickel catalyst used. Reactions on the more selective catalysts (Ni powder, Ni/SiO_2) are presumed to proceed mainly *via* vinyl intermediates; reactions on the less selective catalysts (Ni/Al_2O_3) proceed *via* allylic intermediates (equation 34). A similar dichotomy of mechanisms is assumed to characterize reactions on Pd.

R
R
MH
R
R
+ MH_2
(34)
HM R
R

The palladium-catalyzed hydrogenations of 1,2-cyclononadiene and 1,2-cyclodecadiene form substantial amounts of *trans*- as well the *cis*-cycloalkene.[23,32] The only unhindered approaches to either double bond of these dienes lead to a *cis* isomer. Because the *trans* is the less stable geometrical isomer in rings with fewer than 11 carbon atoms, its formation implies that an intermediate is produced which is capable of yielding both *cis*- and *trans*-cycloalkenes at comparable rates. Moore proposed that the intermediate has the *syn–anti*-π-allyl structure (**43**; equation 35).

$(CH_2)_4$
H
cis-cyclononene ⟵ ⟶ *trans*-cyclononene (35)
H M H

(43)
syn,anti-π-allyl

This hypothesis is supported by Lin's studies of the reactions of the C_9 to C_{13} 1,2-cycloalkadienes with $HRh(PPh_3)_4$ and $ClRh(PPh_3)_3$,[70] complexes classified by Collman *et al.* as monohydride and dihydride hydrogenation catalysts, respectively.[28] The monohydride complex unites with allenes to form π-allyl compounds which have characteristic NMR spectra in solution. 1,2-Cyclononadiene yields almost the same proportion of *cis*- and *trans*-cyclononene with either Pd/Al_2O_3 or $HRh(PPh_3)_4$ as catalyst; little of the *trans* isomer is formed when $ClRh(PPh_3)_3$ is used. The X-ray crystal structure of the complex formed from the reaction of $ClRh(PPh_3)_3$ with 1,2-cyclononadiene shows that only one double bond of the diene is coordinated with rhodium.[71]

The hydrogenation of *cis*-1,4-diphenyl-1,2,3-butatriene (**44**) over Lindlar's Pd yields mainly the (*Z*,*Z*)-diene (**45**), accompanied by a little of the (*E*,*E*)-diene (equation 36).[3,32] Lower stereoselectivity accompanies the hydrogenation of 1,6-diindenylene-2,3,4-hexatriene in which the cumulene unit is conjugated with double bonds.[3] Recall that the stereoselectivity of the hydrogenation of a triple bond is lowered when it is conjugated with a double bond. Possibly the intermediate formed has a π-allylic structure which can account for the various products, just as allylic intermediates can account for the products of hydrogenating conjugated dienes on Ni and Pd catalysts.[32]

Ph, Ph (**44**) —Pd/ $CaCO_3$→ Ph, Ph (**45**) (36)

3.1.6 AROMATIC HYDROCARBONS

The hydrogenation of aromatic hydrocarbons is catalyzed by various transition metal oxides and sulfides, but for a laboratory preparation the catalyst usually is selected from Ni, Ru, Rh, Pd or Pt.[3,4] Nickel catalysts require elevated temperatures and pressures, which dictate the use of a high pressure autoclave. Rhodium, platinum and ruthenium are effective under mild conditions, rhodium being the most active. Rhodium and ruthenium are used when the hydrogenolysis of carbon–oxygen or carbon–nitrogen bonds is to be avoided. Under mild conditions, the hydrogenation of mononuclear aromatic hydrocarbons proceeds slowly over palladium, which suggests its use when the aromatic ring is to be retained.

3.1.6.1 Structure–Reactivity

The rate of hydrogenation of the benzene ring is lowered by alkyl substiuents, although with two or more, the arrangement of the groups affects the rate.[41] Smith *et al.* determined the individual and competitive rates of all the methyl-substituted benzenes on Pt and Rh catalysts in the solvent acetic acid, and found that not only the individual rates (k_A) but also the derived relative equilibrium constants (K_{AB}) decrease with increasing methyl substitution.[41,72] Within groups of isomers, k_A and K_{AB} change in opposite directions. For the xylenes, the rate constants (k_A) increase but the K_{AB} decrease in the order 1,2- > 1,3- > 1,4-. A similar relationship applies to the trimethylbenzenes; the order of increasing rate is 1,2,3- < 1,2,4- < 1,3,5-, while K_{AB} decreases in the same order. The more limited group of compounds studied on Ni at 170 °C by Wauquier and Jungers shows the same relationships.[22]

That the kinetically derived relative adsorption constants, K_{AB}, decrease with the numbers of alkyl substituents is surprising because alkyl substituents increase the basicity of the benzene ring and stabilize η^6-arene transition metal complexes.[28] The directly measured adsorption coefficients of benzene, toluene, *p*-xylene and mesitylene on a cobalt catalyst at 89 °C do increase with the number of methyl groups and the rates of hydrogenation decrease in that order.[41] A consensus regarding the significance of the kinetically determined adsorption constants has not been reached.[41]

Another kinetic method of determining relative adsorption constants does show that toluene is adsorbed more strongly than benzene on the platinum metals.[73] The method determines the effect of the partial pressure of toluene on the rate of hydrogenation of benzene. With the assumption that adsorption is reversible, the ratio of adsorption coefficients, b_T/b_B, is evaluated by the use of equation (37) where V^o_B and $V_{B(T)}$ represent the rate of hydrogenation of benzene in the absence of toluene and in its presence at stated partial pressures of toluene (P_T) and benzene (P_B). This method avoids the determination of the relative individual rate constants that is required by the method of Wauquier and Jungers.[22] Xylene inhibits the hydrogenation of alkenes on Pd more effectively than does benzene, which is consistent with the effect of alkyl groups on the basicity of benzene.[21]

$$V^o_B/V_{B(T)} = 1 + (b_T/b_B)(P_T/P_B) \qquad (37)$$

The importance of steric effects on the individual rates of arene hydrogenation is not well defined, but 1,4-di-*t*-butylbenzene is more reactive than its 1,3-isomer or *t*-butylbenzene; competitive experiments show that, of the three, it has the smallest apparent adsorption constant.[74] It is not surprising that in hydrogenations of isomeric arenes, which are zero order in arene, the more strongly adsorbed arene reacts more slowly. The orientation of the substituents in the di-*t*-butylbenzenes not only affects their relative reactivity but also the proportions of the unsaturated intermediates which are formed.[74]

3.1.6.2 Intermediates and Mechanism

A priori, the mechanism of hydrogenation of benzene may be represented as a series of hydrogen transfers from the catalyst to the adsorbed benzene and the adsorbed intermediates (Scheme 5).[11,74] The often observed first order reaction in H_2 suggests that the addition of the second hydrogen atom is the more difficult step, indicating that the largest energy barrier lies between adsorbed arene and adsorbed diene. However, no cyclohexadienes have been detected as intermediates.

$$\begin{array}{ccccccccccc} C_6H_6\,(g) & & & & C_6H_8\,(g) & & & & C_6H_{10}\,(g) & & C_6H_{12}\,(g) \\ \updownarrows & & & & \updownarrows & & & & \updownarrows & & \uparrow \\ C_6H_6\,(a) & \rightleftharpoons & C_6H_7\,(a) & \rightleftharpoons & C_6H_8\,(a) & \rightleftharpoons & C_6H_9\,(a) & \rightleftharpoons & C_6H_{10}\,(a) & \rightleftharpoons & C_6H_{11}\,(a) \end{array}$$

Scheme 5

A 1,4-cyclohexadienic structure (**46**) was proposed for the octahydroparacyclophane formed in the hydrogenation (PtO_2) of paracyclophane (**47**), whose unusual properties provided a degree of plausibility to the claim (equation 38).[1,75] Actually, four isomeric dienes, none of which has the first postulated structure, are formed.[76] The possible structures have one trisubstituted double bond in each six-carbon cycle, as in formula (**48**). Dienes are more reactive than the related arenes or cyclohexenes and apparently are more tightly bound to a catalytic site.[77,78]

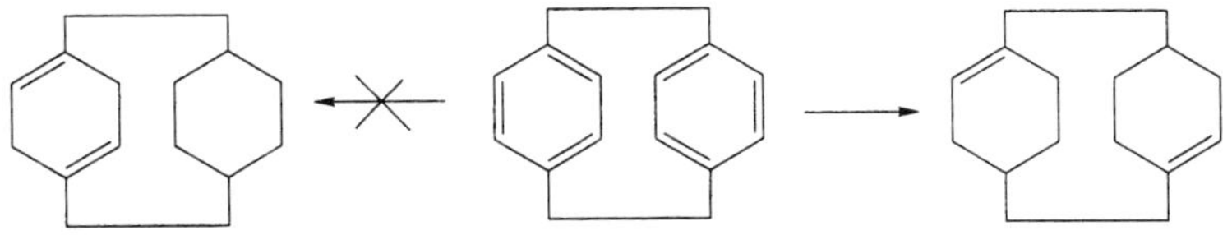

(46) ⟵✕ (47) ⟶ (48) + 3 $C_{16}H_{24}$ isomers (38)

Intermediate cyclohexene(s) can be detected in amounts which depend upon the structure of the aromatic compound, the catalyst and the reaction conditions, all being variables which are likely to affect the relative rates of the processes indicated in Scheme 5.[1,23,75] Strain in the arene, which is released in part upon adsorption or the addition of the first hydrogen atom, increases the competitiveness of the arene relative to the derived cyclohexene, as illustrated in the hydrogenation of paracyclophane.[76] The strained 1,2-di-*t*-butylbenzene yields 1,6-di-*t*-butylcyclohexene whose maximum concentration depends upon the catalyst and the solvent: 45% (Rh/C, AcOH), 30% (Rh/C, EtOH), and 6% (Pt/C, EtOH).[1] The reactivity of the derived cyclohexenes also are lowered by the bulky substituents, presumably a steric effect. Apparently this pathway competes with one that does not involve the desorbed cyclohexene, a path which is indicated by the stereochemistry of the reduction.[74]

Ruthenium was recognized early to be the catalyst which produced the largest amount of the intermediate cyclohexene from the xylenes, rhodium and platinum giving decreasing amounts.[23] The importance of water in maximizing the yield of cyclohexene using ruthenium as catalyst was demonstrated by Don and Scholten.[11] A recent patent application claims yields of 48% cyclohexene at 60% conversion of the benzene.[79]

3.1.6.3 Stereochemistry

Although *cis* isomers are the principal products of hydrogenating disubstituted benzenes under mild conditions, *trans* isomers are also formed. The reaction is kinetically controlled at near ambient conditions but *cis–trans* isomerization will occur at higher temperatures where the reversibility of the hydrogen addition becomes significant.

A comparison of the proportion of the saturated stereoisomers, formed under the same conditions (PtO_2, AcOH) from the isomeric xylenes and the derived dimethylcyclohexenes, leads to the proposal that the reaction proceeds through the desorbed cyclohexenes.[23] Low concentrations of the intermediate cyclohexenes were detected later. The effect of the metal on the stereoselectivity of hydrogenating *o*-xylene follows closely the effect of the metal on the hydrogenation of 1,2- and 1,6-dimethylcyclohexene.[49] The highest selectivity for the conversion of *o*-xylene to the *cis* isomer is given by iridium

(99.2%), the least by palladium (57.7%); the percent given by 1,2-dimethylcyclohexene on the two catalysts are 99.2 and 25.9%, the 1,6-isomer gives 89.0 and 25.6%, respectively.

That desorbed cyclohexenes give rise to the *trans*-disubstituted cyclohexanes is most clearly illustrated by van Bekkum and coworkers' hydrogenation of *t*-butyl-substituted benzenes.[1] For example, the hydrogenation of 2-*t*-butylbenzoic acid over Rh/C in ethanol at 22 °C and 1 atm yields *cis*-2-*t*-butylcyclohexanecarboxylic acid and 6-*t*-butylcyclohexenecarboxylic acid at a fixed ratio until about 80% conversion, at which point the hydrogenation of the cyclohexene begins and yields both *cis*- and *trans*-2-t-butylcyclohexanecarboxylic acid. 1,2-Di-*t*-butylbenzene and 1,3,5-tri-*t*-butylbenzene exhibit similar properties.[1]

The effect of pressure on the stereochemistry of hydrogenating 1,4-di-*t*-butylbenzene on rhodium also indicates a competition between the addition of six hydrogen atoms during a single period of residence on the catalyst and a path involving desorbed cycloalkenes.[74] Accessible high pressures, which are expected to increase the rate of addition of hydrogen, are unable to cause the xylenes or 1,4-di-*t*-butylbenzene to yield only the *cis* isomers. Whether an increase in pressure can improve the stereoselectivity of the iridium- and osmium-catalyzed hydrogenation of *m*- and *p*-xylene has not been tested. Small amounts of dimethylcyclohexenes are detected in the solutions (26 °C, 1 atm, Bu^tOH), which indicates the existence of a path through desorbed intermediates.[49]

3.1.7 POLYCYCLIC AROMATIC HYDROCARBONS

3.1.7.1 Regioselectivity

A considerable degree of control of the regioselective hydrogenation of polycyclic aromatic hydrocarbons has been achieved through the choice of catalyst and optimized conditions. The review by Nishimura and Takagi and Rylander's monographs furnish many examples and references.[3,4,6]

Of the platinum metals, palladium, which is least able to catalyze the hydrogenation of the benzene ring under mild conditions, is the most selective for converting biphenyl to phenylcyclohexane or naphthalene to tetralin. With alkyl substituents at the 2-position, the unsubstituted ring is reduced preferentially. However, the regioselectivity of reducing a 1-alkylnaphthalene depends upon the size of the alkyl group; a methyl substituent favors the unsubstituted ring, the *t*-butyl group directs to the substituted ring (equation 39).[1,75] This structural effect is attributed to the release of *peri* strain in the critical transition state, an effect which is magnified in the enhanced reactivity of the 1,8-naphthalenes.[1,4]

R R R
Pd/C
AcOH, 80 °C, 1 atm
+
(39)

R = Me 66% 34%
= Bu^t 3% 97%

Anthracene and phenanthrene can be converted to their 9,10-dihydroderivatives in high yield over copper chromite at 100–150 °C and 150 atm. The change can be accomplished under milder conditions using platinum metal catalysts, Rh/Al_2O_3 with anthracene and Pd/C with phenanthrene.[3,4,6]

Using palladium as catalyst, the region of minimal bond delocalization is reduced preferentially, whereas platinum tends to catalyze the reduction of terminal rings (equation 40).[3,4,6]

Pd/C
1–3 atm
Pt/C
1–3 atm
(40)

97% 95%

3.1.7.2 Stereoselectivity

The stereochemistry of hydrogenating naphthalene is described in Weitkamp's review.[80] Mainly *cis*-decalin is formed when the hydrogenation is catalyzed by platinum metals under relatively mild conditions; ruthenium is the most stereoselective (95–98% *cis*) while palladium is the least. The *trans*-decalin which is formed appears to result from the reduction of octalin intermediates *via* a mechanism similar to that which accounts for the formation of *trans*-dialkylcyclohexanes from dialkylbenzenes.

The *cis*-stereoselectivity and the rate of hydrogenation (Pd/C) of the substituted ring in 1,4-dialkylnaphthalenes increase with the bulk of the substituents; equal amounts of the *cis*- and *trans*-tetralins form from 1,4-diethylnaphthalene, while the di-*t*-butylnaphthalene yields almost exclusively the *cis* isomer (99%).[1,3,75] A similar affect of structure on rate and stereoselectivity is observed in the hydrogenations (Rh/Al_2O_3) of 1,4-dialkylbenzenes.[74]

The hydrogenation of phenanthrene proceeds through aromatic intermediates to the perhydrophenanthrene.[75] Under mild conditions the product has the *syn-cis-syn* configuration, but more vigorous conditions yield products of lower selectivity. Similar results can be expected for the hydrogenation of more complex polycyclic aromatic hydrocarbons.[3,6,75]

3.1.8 CHEMOSELECTIVITY

3.1.8.1 Nonconjugated Functional Groups

Carbon–carbon unsaturation can be hydrogenated selectively in compounds which also contain other reducible or hydrogenolyzeable functional groups.[1,3,4,32,59] Apparently this reflects the strong interaction of the metal with carbon–carbon double and triple bonds.

Aliphatic nonconjugated nitroalkenes can be reduced to the saturated nitro compounds without difficulty using platinum or palladium catalysts.[3,4,6] A similarly situated alkynic group too is reduced selectively using Pd.

Unconjugated unsaturated aldehydes and ketones can be reduced without affecting the carbonyl group, particularly with a Pd catalyst; Ni and Pt have been used, but the reduction must be terminated after the addition of 1 equiv. of hydrogen.[3,4,6,32] The nitrile group also can survive the reduction of carbon–carbon unsaturation in the same molecule.

The hydrogenolysis of allyl–oxygen bonds can be avoided by the use of Rh or Ru catalysts, and modified Pt catalysts have been effective.[3] Epoxides also survive conditions used to hydrogenate carbon–carbon double and triple bonds.

3.1.8.2 Conjugated Functional Groups

Generally, α,β-unsaturated carbonyl groups are more stable towards hydrogenation than their unconjugated isomers; however, the conjugated alkenic double bond can be hydrogenated in preference to an isolated one.[3–6] This reflects the effect of an electron-attracting substituent on stabilizing the association of an alkene with a transition metal.[46] Palladium catalysts are most useful, but selectivity is enhanced by the choice of solvent or the use of added base. Many examples are found in the reviews by Molnar and by Augustine.[32,81]

Vinyl and allyl groups can be saturated but hydrogenolysis and other undesired reactions may occur. The choice of metal as a catalyst and the solvent affects the results of hydrogenating substituted cyclohexenyl ethers.[3,32] Besides the saturated ether, the cyclohexanone and its diethyl acetal, the cyclohexanol and cyclohexane are formed in ethanol in proportions which vary with the metal. The relative rates of the competing reactions depend on both solvent and metal and yields exceeding 96% of the saturated ether are reported for catalysis by Pd in ethanol, and Ru or Rh in *t*-butyl alcohol (see Section 3.1.3.3.5).

The enamine (**49**) can be hydrogenated to the tertiary amine; Pd gives the highest yield (99%), the product having the *trans* configuration (equation 41).[4] The yield is lower with PtO_2 (in ethanol, 2-propanol or ethyl acetate); however, the *cis* isomer predominates (92–98%). The stereochemical result on Pd suggests that the product-controlling reaction is the union of hydrogen with the most stable alkyl intermediate, as proposed for the Pd-catalyzed hydrogenation of vinyl ethers (see Section 3.1.3.3.5).[32] The association constants measured by Tolman show that a vinyl ether is more weakly bound to a metal than is an alkene and one may infer that the vinyl amino group would affect the metal–alkene π-bond in the same way (see Section 3.1.3.2). The weakening of the π-bond facilitates the equilibration of the alkyl intermediates.

CHNMe$_2$ (49) —Pd/C, 99%→ NMe$_2$ 100% *trans* (41)

3.1.9 CATALYTIC TRANSFER HYDROGENATION

The possibility that organic molecules of relatively low oxidation potential could serve as a source of hydrogen which could be transferred catalytically to unsaturated compounds under mild conditions was proposed by Braude, Linstead *et al.* in 1952, and the scope of the reaction was subsequently explored.[82] Hydrogen transfer between like molecules, as in the disproportionation of cyclohexene on Pt and Pd to form cyclohexane and benzene, had been discovered much earlier by Zelinskii *et al.* Cyclohexadienes disproportionate more rapidly on these catalysts forming equimolar amounts of cyclohexane and benzene. Indeed cyclohexene and cyclohexadiene, or their readily available alkyl derivatives, are among the most frequently cited hydrogen donors for catalytic transfer hydrogenations.[82,83] A considerable body of work has appeared since the review by Brieger and Nestrick, and the improvements in technique, including greater catalyst loadings and different hydrogen donors, have increased the utility of the process.[83]

Palladium in the form of Pd black or Pd/C is the most effective catalyst. Although Raney nickel has been used, there is doubt that it serves only as a hydrogen transfer catalyst because it contains a considerable amount of adsorbed hydrogen. Platinum and rhodium have been found to be ineffective. Both alkenes and alkynes have been hydrogenated and *syn* addition to 1,2-diphenylacetylene has been demonstrated.[59,83]

Regioselectivity in the catalytic transfer hydrogenation of dimethyl bicyclo[2.2.1]hepta-2,5-diene-2,3-dicarboxylate is a function of the hydrogen donor.[4] Using cyclohexene or 1,3-cyclohexadiene as donor, addition to the unsubstituted double bond predominates, forming dimethyl bicyclo[2.2.1]hept-2-ene-2,3-dicarboxylate while, with 1-methylcyclohexene or α-terpinene as donor, *endo*-dimethyl[2.2.1]hepta-2-ene-5,6-dicarboxylate is the main product. Alkyl substitution in the donor slows the hydrogen transfer and allows a redistribution of the adsorbed diene species to the favored 2,3-double bond, a structure which is stabilized by the electron-withdrawing methoxycarbonyl groups (see Section 3.1.3.2.3).

Hydrogen transfer reductions are facilitated by the use of a simple column reactor containing a small amount of palladium catalyst in the form of palladium black and with formic acid as the hydrogen donor.[84] Transfer hydrogenation is accomplished by allowing a solution of the starting material, in an appropriate solvent plus 1% formic acid, to pass slowly through the column. The optimal flow rate is easily adjusted, the effluent collected and the solvent evaporated. A single pass is sufficient to complete the addition of 1 mol of hydrogen to cinnamic acid, coumarin, oleic acid, fumaric acid or cholesterol.

Discussions of the mechanisms of these reactions usually consider whether the hydrogen transfers involve a catalyst-mediated donor–acceptor complex or proceed according to mechanisms which are part of the hydrogenation–dehydrogenation reactions which are observed at low hydrogen pressures.[82,83] Both donor and acceptor molecules compete for adsorption on the catalyst and their transformations are affected by the amount of adsorbed hydrogen which is present. A recent study of the disproportionation of specifically deuterated 1,4-cyclohexadienes catalyzed by colloidal nickel shows that it is a multistep reaction.[85] The dehydrogenation of the diene is strictly *cis* selective but the hydrogen loses its molecular identity before the hydrogen addition step. The authors conclude that the direct H-transfer between two molecules of 1,4-cyclohexadiene is excluded.

3.1.10 REFERENCES

1. A. P. G. Kieboom and F. van Rantwijk, 'Hydrogenation and Hydrogenolysis in Synthetic Organic Chemistry', Delft University Press, Delft, 1977.
2. M. Freifelder, 'Catalytic Hydrogenation in Organic Synthesis. Procedures and Commentary', Wiley-Interscience, New York, 1978.
3. P. N. Rylander, 'Catalytic Hydrogenation in Organic Synthesis', Academic Press, New York, 1979.

4. P. N. Rylander, 'Hydrogenation Methods', Academic Press, London, 1985.
5. M. Bartók, J. Czombos, K. Felfoldi, L. Gera, G. Gondos, A. Molnar, F. Notheisz, I. Palinko, G. Wittman and A. G. Zsigmond, 'Stereochemistry of Heterogeneous Metal Catalysis', Wiley, New York, 1985.
6. S. Nishimura and U. Takagi, 'Catalytic Hydrogenation. Application to Organic Synthesis', Tokyo Kagaku Dojin, Tokyo, 1987. (Written in Japanese but most references, line formulas, reaction equations and schemes can be understood by a reader of English.)
7. D. Cahen and J. A. Ibers, *J. Catal.*, 1973, **31**, 369.
8. S. Nishimura, *Bull. Chem. Soc. Jpn.*, 1961, **34**, 1544.
9. S. Nishimura, M. Ishige and M. Shiota, *Chem. Lett.*, 1977, 535.
10. S. Nishimura, M. Katagiri, T. Watanabe and M. Uramoto, *Bull. Chem. Soc. Jpn.*, 1971, **44**, 166.
11. J. A. Don and J. J. F. Scholten, *Faraday Discuss. Chem. Soc.*, 1981, **72**, 145.
12. S. Imaizumi, H. Murayama, J. Ishiyama and Y. Senda, *Bull. Chem. Soc. Jpn.*, 1985, **58**, 1071.
13. C. A. Brown and V. K. Ahuja, *J. Org. Chem.*, 1973, **38**, 2226.
14. J.-J. Brunet, P. Gallois and P. Caubère, *J. Org. Chem.*, 1980, **45**, 1937.
15. J.-J. Brunet and P. Caubère, *J. Org. Chem.*, 1984, **49**, 4058.
16. Y. Iwasawa, *Adv. Catal.*, 1987, **35**, 187.
17. M. Terasawa, K. Kaneda, H. Yamamoto, T. Imanaka and S. Teranishi, *J. Catal.*, 1979, **57**, 315.
18. R. L. Burwell, Jr., *Pure Appl. Chem.*, 1976, **46**, 71.
19. R. J. Madon, J. P. O'Connell and M. Boudart, *AIChE J.*, 1978, **24**, 904.
20. M. Boudart and G. Djega-Mariadassou, 'Kinetics of Heterogeneous Catalytic Reactions', Princeton University Press, Princeton, NJ, 1984.
21. E. E. Gonzo and M. Boudart, *J. Catal.*, 1978, **52**, 462.
22. J. P. Wauquier and J. C. Jungers, *Bull. Soc. Chim. Fr.*, 1957, 1280.
23. S. Siegel, *Adv. Catal.*, 1966, **16**, 123.
24. I. Horiuti and M. Polanyi, *Trans. Faraday Soc.*, 1934, **30**, 1164.
25. S. Siegel, *J. Catal.*, 1973, **30**, 139.
26. K.-I. Tanaka, *Adv. Catal.*, 1985, **33**, 99.
27. G. A. Somorjai, *Adv. Catal.*, 1977, **26**, 1.
28. J. P. Collman, L. S. Hegedus, J. R. Norton and R. G. Finke, 'Principles and Applications of Organotransition Metal Chemistry', University Science Books, Mill Valley, CA, 1987.
29. S. Siegel, J. Outlaw, Jr. and N. Garti, *J. Catal.*, 1978, **52**, 102.
30. F. Zaera and G. A. Somorjai, *J. Am. Chem. Soc.*, 1984, **106**, 2288.
31. T. P. Beebe, Jr. and J. T. Yates, Jr., *J. Am. Chem. Soc.*, 1986, **108**, 663.
32. A. Molnar, ref. 5, chap. III.
33. S. Nishimura, H. Sakamoto and T. Ozawa, *Chem. Lett.*, 1973, 855.
34. A. Farkas, L. Farkas and E. K. Rideal, *Proc. R. Soc. London, Ser. A*, 1934, **146**, 630.
35. E. Hirota, M. Ito and T. Ueda, in *Proc. Int. Congr. Catal., 6th, 1976*, 1977, **1**, 518.
36. S. Naito and M. Tanimoto, *J. Catal.*, 1986, **102**, 377.
37. A. B. Anderson, D. B. Kang and Y. Kim, *J. Am. Chem. Soc.*, 1984, **106**, 6597.
38. (a) V. Ku, J. Palmer, S. Siegel and R. Clough, *J. Catal.*, 1976, **44**, 449; (b) S. Siegel, *J. Catal.*, 1986, **102**, 475.
39. J. A. Hawkins, *Diss. Abstr. Int. B*, 1983, **44** (6), 1827.
40. S. Mitsui, K. Gohke, H. Saito, A. Nanbu and Y. Senda, *Tetrahedron*, 1973, **29**, 1523.
41. M. Kraus, *Adv. Catal.*, 1980, **29**, 151.
42. L. Červený and V. Ružička, *Adv. Catal.*, 1981, **30**, 335.
43. R. Maurel and J. Tellier. *Bull. Soc. Chim. Fr.*, **1968**, 4650.
44. A. S. Hussey, G. W. Keulks, G. P. Nowack and R. H. Baker, *J. Org. Chem.*, 1968, **33**, 610.
45. A. S. Hussey, R. H. Baker and G. W. Keulks, *J. Catal.*, 1968, **10**, 258.
46. C. A. Tolman, *J. Am. Chem. Soc.*, 1974, **96**, 2780.
47. S. Siegel, G. M. Foreman and D. Johnson, *J. Org. Chem.*, 1975, **40**, 3589.
48. E. W. Garbisch, Jr., S. M. Schildcrout, D. B. Patterson and C. M. Sprecher, *J. Am. Chem. Soc.*, 1965, **87**, 2932.
49. S. Nishimura, F. Mochizuki and S. Kobayakawa, *Bull. Chem. Soc. Jpn.*, 1970, **43**, 1919.
50. (a) F. Van Rantwijk, A. Van Vliet and H. Van Bekkum, *J. Mol. Catal.*, 1980, **9**, 283; (b) F. Van Rantwijk, A. P. G. Kieboom and H. Van Bekkum, *J. Mol. Catal.*, 1975, **1**, 27.
51. S. Teratani, K. Tanaka, H. Ogawa and K. Taya, *J. Catal.*, 1981, **70**, 347.
52. R. P. Linstead, W. von E. Doering, S. B. Davis, P. Levine and R. B. Whetstone, *J. Am. Chem. Soc.*, 1942, **64**, 1985.
53. C. R. Landis and J. Halpern, *J. Am. Chem. Soc.*, 1987, **109**, 1746.
54. M. C. Dart and H. B. Henbest, *J. Chem. Soc.*, 1960, 3563.
55. S. Nishimura and K. Mori, *Bull. Chem. Soc. Jpn.*, 1963, **36**, 318.
56. H. W. Thompson and R. E. Naipawer, *J. Am. Chem. Soc.*, 1973, **95**, 6379.
57. E. N. Marvell and T. Li, *Synthesis*, 1973, 457.
58. C. A. Henrick, *Tetrahedron*, 1977, **33**, 1845.
59. J. Czombos, in ref. 5, chap. IV.
60. H. Hirai, H. Chawanya and N. Toshima, *Bull. Chem. Soc. Jpn.*, 1985, **58**, 682.
61. G. C. Bond and P. B. Wells, *Adv. Catal.*, 1964, **15**, 91.
62. J. G. Ulan, E. Kuo and W. F. Maier, *J. Org. Chem.*, 1987, **52**, 3126.
63. A. Steenhoek, B. H. Van Wijngaarden and H. J. J. Pabon, *Recl. Trav. Chim. Pays-Bas*, 1971, **90**, 961.
64. W. T. McGown, C. Kemball and D. A. Whan, *J. Chem. Soc., Faraday Trans. 1*, 1977, **73**, 632.
65. E. L. Muetterties, *Inorg. Chim. Acta*, 1981, **50**, 1.
66. R. R. Burch, A. J. Shusterman, E. L. Muetterties, R. G. Teller and J. M. Williams, *J. Am. Chem. Soc.*, 1983, **105**, 3546.

67. J. A. Hawkins and S. Siegel, *J. Org. Chem.*, 1986, **51**, 1638.
68. S. Nishimura, M. Ishibashi, H. Takamiya, N. Koike and T. Matsunaga, *Chem. Lett.*, 1987, 167.
69. W. A. Bonner and T. W. Greenlee, *J. Am. Chem. Soc.*, 1959, **81**, 2122.
70. S. T. Lin, *Diss. Abstr. Int. B*, 1983, **44** (6), 1829.
71. A. W. Cordes, S. Siegel, S.-T. Lin, M. C. Noble, S. Barsoum and J. Fairman, *Acta Crystallogr.*, 1989, **C45**, 1426.
72. C. P. Rader and H. A. Smith, *J. Am. Chem. Soc.*, 1962, **84**, 1443.
73. R. Szymanski, H. Charcosset, P. Gallezot, J. Massardier and L. Tournayan, *J. Catal.*, 1986, **97**, 366.
74. S. Siegel, J. F. Outlaw, Jr. and N. Garti, *J. Catal.*, 1979, **58**, 370.
75. J. Czombos, in ref. 5, chap. V.
76. S. T. Lin, D. L. Wharry, R. L. Yates, N. Garti and S. Siegel, Abstracts, 192nd American Chemical Society National Meeting, Anaheim, CA, Sept. 7–12, 1986, ORGN 014.
77. J. F. Outlaw, Jr., J. R. Cozort, N. Garti and S. Siegel, *J. Org. Chem.*, 1983, **48**, 4186.
78. J. R. Cozort, J. F. Outlaw, Jr., A. Hawkins and S. Siegel, *J. Org. Chem.*, 1983, **48**, 4190.
79. H. Nagahara and M. Konishi (Asahi Chem. Ind. Co., Ltd.), *Eur. Pat. Appl.* 220 525 A1 (1987) (*Chem. Abstr.*, 1987, **110**, 134 779).
80. A. W. Weitkamp, *J. Catal.*, 1966, **6**, 431.
81. R. L. Augustine, *Adv. Catal.*, 1976, **25**, 56.
82. G. Brieger and T. J. Nestrick, *Chem. Rev.*, 1974, **74**, 567.
83. R. A. W. Johnstone, A. H. Wilby and I. D. Entwistle, *Chem. Rev.*, 1985, **85**, 129.
84. B. Elamin, J. W. Park and G. E. Means, *Tetrahedron Lett.*, 1988, **29**, 5599.
85. H. Hintze and A. Hessing, *Chem. Ber.*, 1988, **121**, 1133.

3.2
Homogeneous Catalytic Hydrogenation of C═C and C≡C

HIDEMASA TAKAYA
Kyoto University, Japan

and

RYOJI NOYORI
Nagoya University, Japan

3.2.1 HYDROGENATION OF NONCONJUGATED C═C BONDS

Chlorotris(triphenylphosphine)rhodium(I), $[RhCl(PPh_3)_3]$, was reported independently by three groups in 1965,[2–4] and its application to catalytic homogeneous hydrogenation has been studied intensively by Wilkinson's group.[5] The Wilkinson catalyst is now the most widely used for hydrogenation of a variety of unsaturated substrates, and several extensive reviews of this catalyst have been published.[1]

With this catalyst alkenes such as 1-hexene, 2-pentene, 3-hexene, cyclohexene, 2,3-dimethyl-2-butene, styrene, allyl alcohol, allyl cyanide and acrylamide, are reduced under mild conditions. Rates of hydrogenation of some representative alkenes catalyzed by [RhCl(PPh3)3] are listed in Tables 1 and 2. The addition of triphenylphosphine decreases the rate of hydrogenation.[6] Small amounts of impurities such as oxygen and hydroperoxides in catalytic systems sometimes increase the rate of hydrogenation,[7,8] but large quantities of impurities cause deposition of rhodium metal.[9,10] The steric effect has been studied.[11] The less-substituted double bonds of unconjugated alkadienes are usually reduced selectively.

Table 1 Rate Constants for the Hydrogenation of Alkenes[5b] (1.25 mM Solution of [RhCl(PPh3)3] in Benzene)

Alkene	k'_{298} ($\times 10^2$)[a] (l mol^{-1} s^{-1})
Cyclopentene	34.3
Cyclohcxene	31.6
Cycloheptene	21.8
Hex-l-ene	29.1
Dodec-l-ene	34.3
1-Methylcyclohexene	0.6
cis-Pent-2-ene	23.2
2-Methylpent-1-ene	26.6
Styrene	93.0
cis-4-Methylpent-2-ene	9.9
cis-4-Methylpent-2-ene[b]	12.6
trans-4-Methylpent-2-ene	1.8
trans-4-Methylpent-2-ene[b]	1.9

[a]Errors in the rate constants ($k' >$ *ca.* 20 x 10^{-2} M^{-1} s^{-1}) do not exceed ±5%; errors on low k' values are somewhat greater.
[b]Deuteriation by D_2.

Table 2 Rate of Hydrogen Consumption (mmol min^{-1}) at 0.66 atm of Hydrogen Partial Pressure for Unsaturated Substrates under Standard Conditions; Catalyst [RhCl(PPh3)3], 1.25 mM in Benzene, Substrate Concentration 1.25 M[5b]

Substrate[a]	*Rate* (mmol min^{-1})	*Substrate*	*Rate* (mmol min^{-1})
Acenaphthylene	1.76 ± 0.11	3-Ethylpent-2-ene	0.02 ± (0.002)
Acrylamide	0.22 ± 0.01	Hexa-1,5-diene	0.21 ± 0.01
Allyl alcohol	3.02 ± 0.18	*trans*-Hex-3-ene	0.05 ± (0.005)
Allyl cyanide	0.45 ± 0.03	Cyclohexa-1,3-diene	0.13 ± 0.01
Octa-1,7-diene	0.65 ± 0.04	Penta-1,3-diene	0.06 ± (0.004)
Cyclohexene	0.80 ± 0.05	2,3-Dimethylbut-2-ene	0.01 ± (0.001)
Styrene	2.56 ± 0.15	4-Fluorostyrene	3.70 ± 0.26
		4-Methoxystyrene	3.22 ± 0.22

[a]The following were either not reduced or were reduced extremely slowly (<0.01 mmol min^{-1}) under the standard conditions: acrylic acid, allylamine, allyl chloride, bicyclo[2.2.1]hepta-2,5-diene, cyclohexa-1,4-diene, maleic anhydride and tetrachloroethylene.

(1) (2) $\xrightarrow{H_2, <1 \text{ atm}}$ (3) + (4)

(5) $\xrightarrow[\text{[RhCl(PPh}_3)_3]\text{, 94\%}]{H_2, 1 \text{ atm}, 8 \text{ h}}$

Both isotetralin (**1**) and 1,4-dihydrotetralin (**2**) are reduced in the presence of the catalyst to an 80:20 mixture of 9,10-octaline (**3**) and 1,9-octaline (**4**).[12] The sterically less-hindered double bond of eremophilone (**5**) is preferentially hydrogenated.[13]

Labeling experiments show that the addition of hydrogen to C═C bonds proceeds with *syn* stereochemistry.[5,14,15] Thus a *trans*-alkene is hydrogenated to give the *threo* product,[15] and deuteration has been shown to afford *cis* isomers.[14,16,17]

The variation of the halogen and the other ligands coordinated to rhodium exerts important effects on the activity and selectivity. [$RhBr(PPh_3)_3$] and [$RhI(PPh_3)_3$] show higher catalytic activities in hydrogenation of terminal alkenes than [$RhCl(PPh_3)_3$], and lower rates of isomerization.[18] Other rhodium–phosphine complexes, such as [$Rh(NO)(PPh_3)_3$],[19,20] [$Rh(OCOR)(PPh_3)_3$] (R = alkyl, aryl, substituted alkyl),[21] [$RhCl(PR_3)_3$] (R = substituted phenyl, alkyl),[22–24] and complexes having aminophosphine,[25] triphenylarsine, or triphenylstibine,[26] also catalyze hydrogenation of alkenes.

Halogen can be replaced by hydride to produce a catalyst, [$RhH(CO)(PPh_3)_3$],[27,28] which catalyzes the selective hydrogenation of 1-alkenes. 1-Hexene gives hexane under mild conditions, while cyclohexene and *cis*-4-methyl-2-pentene are not reduced. The activity of [$RhH(CO)(PPh_3)_3$] decreases with time,[28] but the catalyst can be reactivated by UV irradiation.[29] The hydridorhodium(I) complex [RhH(5-phenyl-5*H*-dibenzophosphole)$_4$], catalyzes rapid hydrogenation of terminal alkenes under mild conditions.[30,31] Internal alkenes, conjugated alkenes, and alkynes can be differentiated. The rhodium hydride complex derived from the reaction of [$RhCl(PPh_3)_3$], Cs^+(7-butenyl-7,8-carborane)$^-$, and trimethylvinylsilane in THF, has been used for the reduction of alkenes,[32] as have [$RhCl_3(pyridine)_3$],[33] [$Rh_2HCl_3(C_5Me_5)_2$],[34] and [$RhHCl_2(PR_3)_2$].[35–37] A water soluble, air stable complex, [RhCl(TPPTS)$_3$] (TPPTS = triphenylphosphine *m*-trisulfonate), has been used for hydrogenation of linear and cyclic alkenes and polyenes with or without functional groups under very mild reaction conditions. Selective hydrogenations are achieved by controlling the volume of hydrogen absorbed. For example, (**6**) is selectively converted to (**7**) in 32 h or to (**8**) in 76 h.[38] [Rh(1,5-COD)(MeCN)$_2$]BF_4 is an efficient catalyst for the hydrogenation of 1-hexene,[39] while isomerization occurs with the analogous iridium complexes.[40]

H_2, 1 atm
[RhCl(TPPTS)$_3$]
r.t., water

(**6**) → (**7**) + (**8**)

Chelating bidentate ligands affect the hydrogenation, and especially chiral ligands have been used for asymmetric hydrogenation (see Section 3.2.5). The complex [$RhCl(PPh_3)$(dppcb)] (dppcb = 1,2-bis(diphenylphosphino)-1,2-dicarbadodecaborane) is an effective catalyst only at higher temperatures.[41]

Bis(diphenylphosphino)methane (dppm) upon coordination to rhodium forms dinuclear complexes. Some of them exhibit high activity for hydrogenation of alkenes, as has been shown with [$Rh_2(\mu$-$Cl)(CO)_2(dppm)_2]^+$ in methanol.[42] Tridentate phosphine ligands also form dinuclear complexes such as [$Rh_2(COD)\{CH(PPh_2)_3\}](CF_3SO_3)_2$·acetone, which reacts with hydrogen in acetone to liberate cyclooctane, and on addition of 1-hexene catalytic hydrogenation proceeds (25 °C, 1 atm H_2).[43]

Cationic rhodium(I) complexes, such as [$RhCl(Py)_2(DMF)(BH_4)$]Cl,[44,45] [$Rh(diene)(PPh_3)_2]^+$, [$RhH_2(PPh_3)_2L_2]^+$ (L = solvent),[46–50] [$Rh(CH_2$═CHCN)(CO)$(PPh_3)_2]ClO_4$[51] and [$Rh(\eta^4$-$C_8H_{10})(\eta^5$-$C_8H_{10})]^+$,[52] also catalyze alkene hydrogenation.

Phosphine-substituted rhodium carbonyl clusters, such as [$Rh_6(CO)_{10}L_6$] (L_6 = $(PPh_3)_6$, [$P(OMe)_3]_6$, or [(–)-DIOP]$_3$), are good catalyst precursors for cyclohexene hydrogenation.[53]

The pentamethylcyclopentadienyl (Cp*) derivatives of rhodium[54] such as [$(RhCp^*)_2(OH)_3$]Cl[54d] and the catalyst system prepared from [$(RhCp^*)_2(\mu$-$Cl)_2$][54c] and pyrazole-type ligands,[55] are used for alkene hydrogenation in 2-propanol.

The complexes [$RuCl_2(PPh_3)_3$] and [$RuHCl(PPh_3)_3$], when used in the presence of methanol, ethanol or triethylamine, serve as efficient catalysts for the hydrogenation of 1-alkenes.[1g,56–59] At ambient temperatures rates of hydrogenation of internal alkenes and cycloalkenes and isomerization of alkenes are very low.[60] Thus, the selective reduction of terminal alkenes has been attained (equation 1).[61] When a solution of [$RuCl_2(PPh_3)_3$] or [$RuHCl(PPh_3)_3$] is exposed to oxygen, green complexes of triphenylphosphine oxide are formed and the activity of the catalyst is lost.[46,57] The dihydride complex [$RuH_2(PPh_3)_4$] is also used as catalyst for selective hydrogenation of 1-alkenes.[62] The alkene-coordinated complex

$[Ru(alkene)(PPh_3)_3]$ has been isolated as a catalytically active intermediate. An analogous iron dihydride complex, $[FeH_2N_2(PEtPh_2)_3]$, is usable as a catalyst for the hydrogenation of ethylene.[63] The orthometalated complex of ruthenium, $[RuCl\{P(OC_6H_4)(OPh)_2\}\{P(OPh)_3\}_3]$ is also an active catalyst, but the analog $[RuHCl\{P(OPh)_3\}_4]$, which is not orthometalated, is not active.[64]

H_2, 1 atm
$[RuCl_2(PPh_3)_3]$
EtOH–benzene
r.t., 99 min
98%
< 1%
(1)

$[RuCl_2(AsPh_3)_3]$,[65] $[RuH(NO)(PR_3)_3]$,[66] $[RuH(OAc)(PPh_3)_3]$,[67] $[Ru(OAc)_2(PPh_3)_2]$,[68] and $[Ru(PPh_3)_3(CF_3SO_3)_2]$,[68] also show catalytic activities for the hydrogenation of alkenes.

The cationic complex $[RuCl(DPPB)(MeCN)_3]^+PF_6^-$ (DPPB = 1,4-bis(diphenylphosphino)butane) is a good catalyst for the hydrogenation of terminal alkenes, acetonitrile and imines.[69]

The anionic ruthenium cluster $[Ru_3(NCO)(CO)_{10}]^-$ acts as an efficient catalyst for the reduction of terminal and unactivated alkenes under mild conditions.[70] A mechanistic study using the osmium analog has been done.[71] The complexes $[\{Ru(\eta^4\text{-}Ph_4C_4CO)(CO)_2\}_2]$ and $[\{Ru(\eta^4\text{-}Ph_2Me_2C_4CO)(CO)_2\}_2]$ have been used as catalysts for the hydrogenation of alkenes, alkynes, ketones, aldehydes and anthracene under mild conditions.[72] The complex $[Ru_2(CO)_5(PPh_3)(\mu\text{-}PPh_2)(\mu\text{-}O{=}CPh)]$, obtained by pyrolysis of $[Ru_3(CO)_9(PPh_3)_3]$, is a catalyst for the hydrogenation of cyclohexene and 2-cyclohexen-1-one, giving cyclohexane and cyclohexanone with a small amount of cyclohexanol, respectively.[73]

The complex $[OsHBr(CO)(PPh_3)_3]$ is effective for the selective hydrogenation of conjugated and nonconjugated dienes to monoenes, and this complex also catalyzes the hydrogenation of linear and cyclic alkenes. (–)-Carvone (**9**) is selectively reduced to 2-methyl-5-(1-methylethyl)cyclohex-2-en-1-one (**10**) or 2-methyl-5-(1-methylethyl)cyclohexanone (**11**), depending on the reaction conditions.[74]

(**9**)
i
ii
(**10**) 100%
(**10**) 12% (**11**) 88%

i, $[OsHBr(CO)(PPh_3)_3]$, H_2, 1 atm, r.t.; ii, $[OsHBr(CO)(PPh_3)_3]$, H_2, 5 atm, 100 °C

Vaska's complex $[IrCl(CO)(PPh_3)_2]$ and its derivatives can be used as catalysts for the hydrogenation of alkenic compounds.[75–88] The rates of hydrogenation depend on steric effects similar to those observed for $[RhCl(PPh_3)_3]$ (terminal alkene > *cis*-alkene > *trans*-alkene). Isomerization is more rapid than hydrogenation. The catalyst is activated by UV irradiation.[80] $[Ir(COD)(PR_3)_2]^+$-type complexes are inactive in aromatic and hydrocarbon solvents, but active in chlorinated solvents such as chloroform and dichloromethane.[89–91] The cationic iridium complexes of the type $[Ir(COD)(PR_3)(Py)]PF_6$ are highly efficient catalysts for hydrogenation of hindered alkenes.[92] The complex (R = C_6H_{11}) readily hydrogenates sterically congested steroid alkene groups without reducing carbonyl groups, carbon–halogen bonds and cyclopropane rings. For example, (**12**) is reduced to (**13**) in the presence of the above complex.[93]

In the presence of the cobalt hydride complexes, $[CoH_3(PPh_3)_3]$,[94] $[CoH(CO)(PPh_3)_3]$,[94] $[CoH(CO)(PBu_3)_3]$[95] and $[CoH(CO)_2(PBu_3)_2]$,[95] alkenes are reduced at elevated temperatures and pressures.

C_8H_{17} Bu^tMe_2SiO (12) $\xrightarrow[100\%]{H_2,\ 1\ atm;\ [Ir(COD)(PCy_3)(Py)]PF_6}$ C_8H_{17} Bu^tMe_2SiO H (13)

The complexes $[RCo(CO)_2\{P(OMe)_3\}_2]$ (R = Me and MeCO) show a marked catalytic activity (450 turnovers h^{-1}) for hydrogenation of terminal alkenes under ambient conditions,[96] while more-hindered alkenes are reduced slowly. Cobalt complexes such as $[CoH_2(PR_3)_3]$[97,98] and $[CoHN_2(PPh_3)_3]$[99,100] are also active catalysts for alkene hydrogenation.

Several complexes of Pd^0 and Pd^{II} containing PR_3, DMSO, *etc.* catalyze hydrogenation of monoalkenes.[101–104] Salen complexes of Pd^{II} and Ni^{II} are employed as catalysts for alkene hydrogenation.[105] The cluster complex of palladium $[Pd_5(PPh)_2]$, is a catalyst for the hydrogenation of alkenes, dienes, alkynes and many other kinds of unsaturated compound.[106]

The complexes derived from platinum complexes and $SnCl_2$ exhibit catalytic activity for hydrogenation under mild conditions.[107–116] These complexes have also been used for alkene hydroformylation. The above catalyst system and $[PtCl_2(PPh_3)_2]$ assist hydrogenation of polyenes to monoalkenes, but isomerization of double bonds has been observed. The anionic cluster complex $[\{Pt_3(CO)_6\}_5]^{2-}$ has been used as a catalyst for hydrogenation of cyclohexene.[117]

A number of Ziegler-type catalysts based on early transition metals exhibit catalytic activity for alkene hydrogenation.[1a,118] Hydrogenation of alkenes is also catalyzed by d^0-metallocene species such as $[Cp_2ZrH_2]$,[119a] $[Cp_2ZrMe_2]$,[119b] $[Cp_2MHR]$ (M = Zr, Hf; R = Me, $CHMe_2$),[120] and so on.[121] Zirconium(II) complexes catalyze the hydrogenation of 1,3- and 1,5-cyclooctadiene to give cyclooctene.[122] Titanocene derivatives[123–125] and polymer-anchored titanocenes[126] have also been used for alkene hydrogenation. Ti^{II} complexes with chiral cyclopentadienyl ligands have been used for asymmetric hydrogenation of simple alkenes (see Section 3.2.5).[127–129]

Recently, both actinide and lanthanoid complexes have been used as catalysts for the hydrogenation of alkenes and alkynes. 1-Hexene is hydrogenated rapidly at atmospheric pressure in the presence of the species prepared from UCl_4 and *t*-butyllithium.[130] Actinide–hydride complexes are also efficient catalysts for 1-hexene.[131] Hydrogenation of propene proceeds effectively with the organoactinide complexes supported on alumina.[132] The hydride complexes, $[\{(C_5Me_5)_2LnH\}_2]$[133] and $[\{Me_2Si(C_5Me_4)_2LnH\}_2]$,[134] and the species derived from $[Cp_3Ln]$ and NaH,[135] are highly active catalysts for alkene hydrogenation. The relative ordering of activities is found to be approximately inversely proportional to metal ionic radius: Lu > Sm > Nd > La. $[\{(C_5Me_5)_2LuH\}_2]$ is the most active catalyst among the complexes shown in Table 3, exceeding 120 000 turnovers h^{-1} at 25 °C and 1 atm of hydrogen.[136] Organoyttrium hydride complexes are also catalysts for alkene hydrogenation.[137]

Table 3 Hydrogenation of 1-Hexene[136]

Catalyst	*Temperature* (°C)	*Rate* (h^{-1})	*Ref.*
$[RhCl(PPh_3)_3]$	0	60	90
	25	650	90
$[RuHCl(PPh_3)_3]$	25	9000	90
$[Rh(COD)(PPh_3)_2]PF_6$	25	4000	90
$[Ir(COD)(PMePh_2)_2]PF_6$	0	5100	90
$[Ir(COD)(PCy_3)(Py)]PF_6$	0	6400	90
$[\{(C_5Me_5)_2LuH\}_2]$	25	120 000	136

Functional groups such as hydroxy, ester and amide often direct the sterochemistry of hydrogenation. Hydrogenation of allylic and homoallylic alcohols occurs with high stereoselectivity.[138] Catalysts used in the hydroxy-directed hydrogenation include: $[RhCl(PPh_3)_3]$, a cationic chelating diphosphine–rhodium complex, $[Rh(NBD)(Ph_2P(CH_2)_4PPh_2)]BF_4$ (**15**), and a cationic iridium complex, $[Ir(COD)(PCy_3)(Py)]PF_6$ (**20**).

Homogeneous hydrogenation of chiral allylic and homoallylic alcohols is accomplished with a high degree of diastereoselectivity, as shown in the hydrogenation of 3-phenylbut-3-en-2-ol (**14**) and 4-phenylpent-4-en-2-ol (**16**) in the presence of (**15**).[139]

H_2, $[Rh(NBD)(Ph_2P(CH_2)_4PPh_2)]BF_4$ (**15**), CH_2Cl_2; (**14**) → 97 : 3

(**15**), THF; (**16**) → 88 : 12

With the cationic rhodium complex (**15**) diastereoselectivity is high (>91%) at 44 atm of hydrogen,[140] but it falls off at atmospheric pressure. Hydrogenation with the cationic iridium complex (**20**) in dichloromethane (25 °C, 1 atm H_2) exhibits a low level of diastereoselection.[140,141] The rhodium complex (**15**) is also superior to iridium complex (**20**) in the hydrogenation of homoallyl alcohols,[142] as in the stereoselective synthesis of (**18**) from (**17**).[143]

(**17**) → (**18**): (**15**), H_2, 68 atm, 94%

Excellent diastereoselectivity has been observed in the directed hydrogenation of cyclic alkenic alcohols (**19**) and (**21**).[141,144–149] In all cases, *trans* products are formed predominantly.

(**19**): H_2, $[Ir(COD)(PCy_3)Py]PF_6$ (**20**), R = Me, 74%; >100 : 1

(**21**): R = H, (**15**), H_2, 95%, *trans:cis* = 290:1; R = Me, (**20**), H_2, 64%, *trans:cis* = >100:1

The directed hydrogenation of cyclic unsaturated ethers[145b] and carboxylic acids and their esters,[150] has been studied using the catalysts (**15**) and (**20**), but the selectivities are usually lower than those of allylic and homoallylic alcohols (equations 2a, 2b and 3).

H_2, 1 atm; (**15**), 96.9% yield; (**20**), 99.9% yield (2a)

H_2, 1 atm; (**15**), 90.4% yield; (**20**), 81.4% yield (2b)

H_2, 1 atm (3)

R = H, (**15**), 98:2 (**20**), 99.7:0.3
R = Me, (**15**),86:14 (**20**), 97:3

H_2, 1 atm, (**20**) 89% (4)

trans:*cis* = 130:1

The amide group is superior to the ester as a directing group (equation 4).[151,152]

The cationic rhodium complex (**15**) catalyzes the hydrogenation of 3-substituted itaconic acid esters with high diastereoselectivity. When a chiral rhodium complex is employed, effective kinetic resolution occurs.[152]

3.2.2 HYDROGENATION OF CONJUGATED ALKENES

3.2.2.1 Dienes, Trienes and Polyenes to Saturated Hydrocarbons

Sterically unhindered conjugated alkenes and chelating nonconjugated dienes (1,5-cyclooctadiene, norbornadiene) can be hydrogenated by $[RhCl(PPh_3)_3]$ under low hydrogen pressure (1 atm), though reaction proceeds slowly. At elevated hydrogen pressure (60 atm), however, the rate increases and some rather sterically hindered dienes are also rapidly hydrogenated.[5a]

Dienes and trienes are hydrogenated in aqueous–organic two-phase media using $[\{RhCl(1,5\text{-hexadiene})\}_2]$ and a phase transfer catalyst.[153]

Detailed kinetic and mechanistic studies for the homogeneous hydrogenation of polybutadiene (90% 1,2-addition; M_n = 10 000) in the presence of $[RhCl(PPh_3)_3]$ as catalyst have been carried out by monitoring the amount of hydrogen consumed. Quantitative hydrogenation of carbon–carbon unsaturation present in polybutadiene proceeds at 20–50 °C under subatmospheric pressures (<1 atm).[154]

Unsaturated polymers (*cis*-1,4-polybutadiene and butadiene–styrene copolymer) are also hydrogenated by Ziegler catalysts.[155,156]

3.2.2.2 Dienes and Trienes to Alkenes by 1,2-Reduction

The rhodium complexes $[RhH(PPh_3)_4]$ and $[Rh(CO)_2(PPh_3)_2]\cdot 2C_6H_6$, which interact strongly with 1,3-dienes, allow selective hydrogenation of 1,3-dienes to the corresponding terminal monoalkenes (equation 5). The hydrogenation proceeds smoothly in cyclohexane at 50–100 °C under 15 atm of hydrogen.[157]

H_2, 15 atm, $[RhH(PPh_3)_4]$ + PEt_3, 46 °C, 60% (5)

The selective hydrogenation of terminal double bonds of conjugated linear dienes can be attained using $[CpNiRu_3(\mu\text{-}H)_3(CO)_9]$ as catalyst.[158]

Potassium pentacyanocobaltate(II), derived form cobalt(II) chloride and KCN, catalyzes the hydrogenation of 1,3-dienes to monoalkenes. 1,3-Butadiene (**22**), isoprene[159] and 1-phenyl-1,3-butadiene (**23**),[160] have been converted to mixtures of the corresponding 1-butenes, *trans*-2-butenes and *cis*-2-butenes. The product distribution depends highly on reaction conditions such as the cyanide:Co ratio, the concentra-

(**22**) H_2, 1 atm, 20 °C

Solvent	*CN/Co*	1-butene *Yield* (%)	*trans*-2-butene *Yield* (%)	*cis*-2-butene *Yield* (%)
H_2O	4.8	16	81	2
H_2O	6.0	95	4	1
Glycerol–MeOH	5.0	9	87	4
Glycerol–MeOH	5.7	50	9	41

tion of added base and the kind of solvent. Relatively high selectivity has been attained by finely tuning these factors. The reaction is usually carried out in water, aqueous alcohols, or mixed solvent systems such as methanol and ethylene glycol. Sorbic acid sodium salt (**24**) affords sodium 2-hexenoate (**25**) in high yield.[161] Tertiary aryl- or alkyl-phosphine complexes of the type $[\{Co(CO)_3(PR_3)\}_2]$ can be used as catalysts in the absence of carbon monoxide, although high substrate:catalyst ratios cause decomposition of the catalyst.[162]

H_2, 1 atm, $K_3[Co(CN)_5]$, MeOH, r.t.

(**23**) → 95% + 4% + 1%

H_2, 1 atm, $K_3[Co(CN)_5]$, MeOH, 96%

(**24**) CO_2Na → (**25**) CO_2Na

Several other transition metal complexes, including $[RuCl_2(PPh_3)_3]$,[163] $Pd_2(DPPM)_3$[101,102] and $PdCl_2(DMSO)$,[103,104] $[Pd(OAc)_2(PPh_3)_2]$,[106] $[Ni(acac)_2]–Al_2Et_3Cl_3–PPh_3$,[164] $[MoH_2(Cp)_2]$,[165] $[IrCl(CO)(PPh_3)_2]$,[166] $[CoH_2(bipy)(PR_3)_2]$,[167] $[Co(C_3H_5)\{P(OMe)_3\}_3]$[168] and Ziegler catalysts,[169] are used for the selective reduction of dienes to monoalkenes.

Acyclic and cyclic allenes are converted to alkenes at 60 °C under atmospheric pressure of hydrogen with $[RhCl(PPh_3)_3]$. 1,2-Nonadiene (**26**), 3-ethyl-1,2-pentadiene and 1,2-cyclotridecadiene are hydrogenated to give *cis*-2-nonene (**27**), 3-ethyl-2-pentene and cyclotridecene (*cis:trans* = 85:15), respectively.[170]

H_2, 1 atm, $[RhCl(PPh_3)_3]$, 60 °C, 66%

(**26**) → (**27**)

1,3-Cyclooctadiene (**28**) can be converted to cyclooctene (**29**) almost quantitatively by $[RhH(PPh_3)_4]$ and $[Rh(CO)_2(PPh_3)_2]\cdot 2C_6H_6$.[157] On the other hand, 1,5- and 1,4-cyclooctadiene (**30** and **31**) have been hydrogenated to cyclooctene (**29**) with $[Ir_2H_2Cl_2(1,5\text{-}COD)(PPh_3)_2]$[171] or $[Ir(1,5\text{-}COD)_2]PF_6$.[48] Zirconium(III) complexes containing the chelated bis(diphenylphosphino)methane act as catalysts for the hydrogenation of both 1,3- and 1,5-cyclooctadiene to cyclooctene in quantitative yields.[123,124]

H_2, 15 atm, $[\{Rh(CO)_2(PPh_3)\}_2]\bullet 2C_6H_6 + 2PEt_3$, 100 °C, 99.7%

(**28**) → (**29**)

H_2, < 1 atm, $[Ir(1,5\text{-}COD)_2]PF_6$, 30 °C, > 98%

(**30**) → (**29**)

H_2, 1 atm, $[Ir_2H_2Cl_2(1,5\text{-}COD)(PPh_3)_2]$

(**31**) → (**29**)

In the presence of a catalytic amount of [Ru(η^4-COD)(η^6-COT)], cycloheptatriene is hydrogenated to cycloheptene under 1 atm of H_2 at room temperature.[172]

1,5,9-Cyclododecatriene (**32**) has been hydrogenated to cyclododecene (**33**) with various catalysts[173] such as [{Co(CO)$_3$(PBu$_3$)}$_2$],[162] [NiI$_2$(PPh$_3$)$_2$],[174] [Pt(SnCl$_3$)$_5$]$^{3-}$,[175] [RuCl$_2$(PPh$_3$)$_3$],[163c] [RuCl$_2$(CO)$_2$(PPh$_3$)$_2$][176] and [RhCl$_3$(Py)$_3$]/NaBH$_4$.[177]

(**32**) → (**33**): H_2, 41 atm, [Ru(CO)$_3$(PPh$_3$)$_2$], 158 °C, 1.2 h, 94.6%

Reduction of cyclic conjugated dienes such as 1,3-cyclohexadiene to cyclohexene is also attained with [Co(CN)$_2$(bipyridine)] more rapidly than with [Co(CN)$_5$]$^{3-}$.[178]

With [Cr(PhCO$_2$Me)(CO)$_3$], 1,3,5-cycloheptatriene gives 1,3-cycloheptadiene and cycloheptene in successive steps.[179]

3.2.2.3 Dienes to Alkenes by 1,4-Reduction

Several complexes having arenes, [Cr(CO)$_3$(arene)] (arene = benzene, mesitylene, methyl benzoate, *etc.*) and [Cr(CO)$_3$(cycloheptatriene)], are excellent catalysts for selective 1,4-hydrogen addition to 1,3-dienes.[179,180] 1,4-Hydrogenation of methyl sorbate (**34**), isoprene (**36**) and 1,3-pentadiene (**38**) in cyclohexane affords methyl *cis*-3-hexenoate (**35**), 2-methyl-2-butene (**37**) and *cis*-2-pentene (**39**), respectively. [Cr(CO)$_3$(MeCN)$_3$] is also an active catalyst under mild conditions (40 °C, H_2 1.4 atm),[181] as are [Cr(CO)$_6$] or [Cr(NBD)(CO)$_4$] under UV irradiation.[182]

CO$_2$Me (**34**) → CO$_2$Me (**35**)

	H_2 (atm)	*Temperature* (°C)	*Time* (h)	*Yield* (%)
[Cr(CO)$_3$(benzene)]	48	160	8	94
[Cr(CO)$_3$(methyl benzoate)]	48	150	2	99
[Cr(CO)$_3$(cycloheptatriene)]	30	120	1	98
[Cr(CO)$_3$(chlorobenzene)]	30	150	2	96
[Mo(CO)$_3$(cycloheptatriene)]	30	100	6	90
[Mo(CO)$_3$(mesitylene)]	30	150	0.5	28 (4-monoene 55)
[W(CO)$_3$(mesitylene)]	30	165	6	13 (2-monoene 43)

(**36**) → (**37**)

[Cr(CO)$_3$(MeCN)$_3$], H_2,1.4 atm, 40 °C, 90%
[CpCr(CO)$_3$], H_2, 90 atm, 70 °C, 95%

(**38**) → (**39**)

[Cr(CO)$_3$(MeCN)$_3$], H_2, 1.4 atm, 40 °C, 100%
[CpCr(CO)$_3$], H_2, 90 atm, 70 °C, 100%

Hydrogenation of methyl β-eleostearate (methyl *trans,trans,trans*-9,11,13-octadecatrienoate) with [Cr(CO)$_3$(arene)] yields the diene products from 1,4-addition (*trans*-9-*cis*-12- and *cis*-10-*trans*-13-octadecadienoates). With α-eleostearate (methyl *cis,trans,trans*-9,11,13-octadecatrienoate), stereoselective 1,4-reduction of the *trans,trans*-diene moiety yields linoleate (*cis,cis*-9,12) accompanied by *cis,trans*-1,4-dienes which are formed from the isomerization of α- to β-eleostearate.[183]

3.2.2.4 Hydrogenation of C═C Bonds Conjugated with Aromatic Rings, CN, CO, CO_2R, *etc.*

Complex [$RhCl(PPh_3)_3$] is a good catalyst for the hydrogenation of alkenes conjugated with aromatics, nitriles, aldehydes, ketones, carboxylic acids, esters, lactones, amides, nitro compounds, furans, thiophenes and other unsaturated heterocyclic compounds.[5b,13,145] With aldehydes, decarbonylations sometimes occur on hydrogenation by [$RhCl(PPh_3)_3$].[184] α,β-Unsaturated ketones resist decarbonylation, but are converted into saturated dimethyl acetals during hydrogenation in benzene–methanol. Electron-withdrawing substituents increase the rate of hydrogenation; for the alkenes H_2C═CHR, the order is R = CN > CO_2Me > Ph > alkyl. (–)-Dehydrogriseofulvin (**40**) gives griseofulvin (**41**), indicating that hydrogenation at the MeC═C(CO) group is faster than at the MeOC═C(CO) group.[13]

H_2, 1 atm
[$RhCl(PPh_3)_3$]
55–60 °C

(**40**) (**41**)

[$RhCl(PPh_3)_3$] catalyzes the hydrogenation of α,β-unsaturated nitro compounds, and the nitro group survives.[184a] A modified catalyst having a phenyldipiperidylphosphine ligand hydrogenates styrene to ethylbenzene.[25,26,185]

The air stable complex [$RhCl(Py)_2(DMF)(BH_4)$]Cl can hydrogenate α,β-unsaturated carbonyl compounds to saturated products. Methyl 3-phenylbutenoate (**42**) is converted to methyl 3-phenylbutanoate (**43**). 4-Cholesten-3-one, testosterone, 17-methyltestosterone and progesterone are also hydrogenated in good yields.[44,185]

H_2, 1 atm
[$RhCl(Py)_2(DMF)(BH_4)$]Cl

(**42**) (**43**)

Cationic Rh^I complexes [$RhL(CO)(PPh_3)_2$]ClO_4 (L = unsaturated nitrile) show catalytic activities for the hydrogenation of α,β-unsaturated nitriles under mild conditions (30 °C).[186] Selective reduction of the carbon–carbon double bonds of α,β-unsaturated carbonyls also occurs in high yield with [{Rh(1,5-hexadiene)Cl}$_2$] and a phase transfer catalyst in aqueous media.[153]

With [$RuCl_2(PPh_3)_3$], α,β-unsaturated ketones are reduced to saturated ketones.[57,58] Hydrogenation of 1,4-androstadiene-3,17-dione (**44**) gives the 4-ene-3,17-dione (**45**) in high yield.[187] [$OsHBr(CO)(PPh_3)_3$] also catalyzes hydrogenation of α,β-unsaturated ketones to give saturated ketones.[74]

H_2, 100 atm
[$RuCl_2\{P(C_6H_4OMe)_3\}_3$]
50 °C
93.7%

(**44**) (**45**)

Crabtree's catalyst, [$Ir(COD)(PCy_3)(Py)$]PF_6, is quite inert to other functional groups under hydrogenation conditions, as illustrated by the reduction of the enone (**46**), in which the *gem*-dibromocyclopropane survives.[93]

The complexes [(arene)$Cr(CO)_3$] (arene = methyl benzoate, naphthalene) are highly efficient catalysts for selective hydrogenation of acyclic enones to ketones.[188] Isolated C═C bonds remain intact.

α,β-Unsaturated carbonyl compounds such as crotonaldehyde, acrolein, methyl vinyl ketone, mesityl oxide and ethyl cinnamate can be hydrogenated in fair yields by use of [$Co_2(CO)_8$] and [$CoH(CO)_4$] under a carbon monoxide and hydrogen atmosphere.[94,95a,95b,189a,b] Alkenes and alkynes conjugated with aromatic rings are also hydrogenated.[189c] At elevated temperatures, however, overreduction of aldehydes and ketones to alcohols and hydrocarbons occurs.[190]

H_2, 1 atm
[Ir(COD)Py(PCy$_3$)]PF$_6$
94%

Br Br
(46)

Br Br

Appropriately substituted α,β-unsaturated acids or aldehydes such as α-phenylacrylic acid, α-methylacrylic acid, α-methylcrotonaldehyde, and allylic alcohols such as cinnamyl alcohol, have been reduced in the presence of $[Co(CN)_3]^{3-}$ in fair to good yields,[191] while acrylic acid and acrolein cannot be reduced. Styrene and substituted styrenes are also reduced.[192]

The complexes derived from chloroplatinic acid and tin(II) chloride can catalyze the hydrogenation of α,β-unsaturated ketones (equation 6).[111,114,115]

H_2, 34 atm
$[PtCl_2(PPh_3)_2]$, $SnCl_2{\cdot}5H_2O$
90 °C
100%
(6)

Using Vaska's complex, hydrogenation of methyl acrylate is accelerated by weak UV irradiation.[75] A related rhodium complex, $[RhH(CO)(PPh_3)_3]$, is also activated by UV irradiation in the hydrogenation of ethyl acrylate.[29]

3.2.2.5 Transfer Hydrogenation of Conjugated C═C Bonds

α,β-Unsaturated ketones are converted to saturated ketones in the presence of $[RuCl_2(PPh_3)_3]$ by means of transfer hydrogenation from organic compounds such as benzyl alcohol, ethylene glycol and formic acid, at relatively high temperatures.[193] For example, α,β-unsaturated ketones (**47**) and (**49**) are reduced to the corressponding ketones (**48**) and (**50**), respectively.

Ph (**47**) + $PhCH_2OH$ —$[RuCl_2(PPh_3)_3]$, 180 °C, 1 h, 95%→ Ph (**48**) + PhCHO

(**49**) + HCO_2H —$[RuCl_2(PPh_3)_3]$, 200 °C, 2 h, 90%→ (**50**) + CO_2

3.2.3 HYDROGENATION OF AROMATIC AND HETEROAROMATIC RINGS

The hydrogenation of isolated aromatic rings is not an easy task with homogeneous transition metal catalysts. Several complexes, however, have been successfully used for this purpose.

Benzene and substituted benzenes are hydrogenated to cyclohexanes by several catalyst systems. With $[\{(\eta^6\text{-}C_6Me_6)RuCl_2\}_2]$ and Na_2CO_3, 9 000 molecules of benzene are reduced to cyclohexane per molecule of catalyst at 50 °C under 50 atm of hydrogen for 36 h.[194] Under the same conditions, styrene, anisole, methyl benzoate, acetophenone and benzophenone are completely hydrogenated to ethylcyclohexane, methoxycyclohexane, cyclohexanecarboxylic acid methyl ester, methyl cyclohexyl ketone and dicyclohexyl ketone, respectively (equation 7). Hydrogenation of benzene-d_6 gives $C_6D_6H_6$ in 95% yield.[194]

Other ruthenium complexes also catalyze the hydrogenation of benzenes. Under high pressure of hydrogen, benzene is hydrogenated to cyclohexane by use of $[(\eta^6\text{-}C_6Me_6)RuHCl(PPh_3)]$.[195] With

H$_2$, 50 atm
[{(η^6-C_6Me_6)$RuCl_2$}$_2$] + Na_2CO_3
50 °C
100%
(7)

R = H, Me, OMe, CO_2Me, COMe

[(η^6-C_6Me_6)(η^4-C_6Me_6)Ru], *p*-xylene is hydrogenated to *cis*-1,4-dimethylcyclohexane. The rate of hydrogenation decreases with alkyl substitution.[196]

Ziegler catalysts prepared from nickel or cobalt 2-ethylhexanoate and triethylaluminum (3 or 4:1 ratio) are very efficient for the hydrogenation of aromatic compounds.[118,197,198] *o*-Xylene (**51**) gives a mixture of *cis*- (**52**) and *trans*-1,2-dimethylcyclohexane (**53**) in 65:35 ratio. Phenol (**54**), dimethyl phthalate and dimethyl terephthalate are also reduced in high yield, while nitro-substituted benzene and phenols cannot be reduced. Catalytic activity toward benzene hydrogenation decreases in the order Ni ≥ Co > Fe > Cr > Cu.[197]

H$_2$, 68 atm
Ni^{II} + $AlEt_3$
150 °C

(**51**) (**52**) 65% + (**53**) 35%

H$_2$, 68 atm
Ni^{II} + $AlEt_3$
150–160 °C

OH (**54**) O 92% + OH 5%

Arenes are hydrogenated under mild conditions using Rh^I–*N*-phenylanthranilic acid anchored to polystyrene beads. For example, benzene is reduced to cyclohexane at room temperature under 3.4 atm of hydrogen.[199]

Benzene-d_6 is converted to $C_6D_6H_6$ (nearly all *cis*) by hydrogenation with [{(η^5-C_5Me_5)$RhCl_2$}$_2$] in the presence of bases. Benzene is hydrogenated faster than toluene, and acetophenone is reduced to methyl cyclohexyl ketone (50 °C, 50 atm).[200]

The complex [{RhCl(1,5-COD)}$_2$] and a phase transfer catalyst (cetyltrimethylammonium bromide or tetrabutylammonium hydrogen sulfate) also reduce aromatic hydrocarbons under hydrogen.[201]

Benzene and substituted benzenes are reduced under mild conditions using [Co(C_3H_5){P(OMe)$_3$}$_3$] as catalyst. Benzene, anisole and alkylbenzenes are converted to cyclohexanes, while alkyl benzoates give 1-alkoxycarbonylcyclohexenes.[168,202] Interesting is the fact that benzene is more easily reduced than cyclohexene because of the special affinity of the benzene ring for cobalt.

Strongly acidic systems (TaF_5/HF or SbF_5/HF) have also been used for hydrogenation of aromatic compounds at 50 °C under 50–100 atm of H_2.[203]

Naphthalene (**55**) and anthracene (**57**) are hydrogenated to decalin (**56**) and perhydroanthracene (**58**) catalyzed by [Co(C_3H_5){P(OMe)$_3$}$_3$].[168,202] Rh^I–*N*-phenylanthranilic acid anchored to polystyrene beads can also hydrogenate naphthalene to decalin.[199]

H$_2$, 1–3 atm
[Co(C_3H_5){P(OMe)$_3$}$_3$]
r.t.
99%

(**55**) (**56**)

H$_2$, 1–3 atm
[Co(C_3H_5){P(OMe)$_3$}$_3$]
r.t.
76%

(**57**) (**58**)

With nickel(II) 2-ethylhexanoate and triethylaluminum, tetralin (**59**) is obtained by hydrogenation of naphthalene (**55**).[197] Polycyclic aromatics, such as anthracene (**57**; equation 8), 9-methylanthracene and 9-trifluoroacetylanthracene, are partially hydrogenated to 1,2,3,4-tetrahydroanthracene derivatives by use of $[Rh(DPPE)(arene)]^+$ in methanol[204] and by ruthenium hydride complexes having triphenylphosphine ligands.[91,205]

H_2, 68 atm; Ni^{II} + $AlEt_3$; 210 °C; 84%

(55) → **(59)**

H_2, 1 atm; $[Rh(DPPE)(arene)]^+$; <60 °C

(57) → (8)

Partial hydrogenation of aromatic rings can also be accomplished with catalysts such as $[Co_2(CO)_8]$ under an atmosphere of carbon monoxide and hydrogen, but the isolated benzene nucleus cannot be reduced.[206] Anthracene (**57**), naphthacene, perylene (**61**) and pyrene (**63**) are converted to 9,10-dihydroanthracene (**60**), 5,12-dihydronaphthacene, 1,2,3,10,11,12-hexahydroperylene (**62**) and 4,5-dihydropyrene (**64**), respectively.

H_2, 197 atm; $[Co_2(CO)_8]$; 135 °C; 99%

(57) → **(60)**

H_2, 204 atm; $[Co_2(CO)_8]$; 150 °C; 72%

(61) → **(62)**

H_2, 204 atm; $[Co_2(CO)_8]$; 200 °C; 69%

(63) → **(64)**

Pyridine (**65**) is converted to piperidine (**66**) in 98% yield with Ziegler catalyst, Ni^{II} + $AlEt_3$.[118,197] The phase transfer catalyst system $[\{RhCl(1,5\text{-}COD)\}_2]$ with cetyltrimethylammonium bromide or tetrabutylammonium hydrogen sulfate can also reduce heteroaromatic compounds.[201]

H_2, 68 atm; Ni^{II} + $AlEt_3$; 150–174 °C; 98%

(65) → **(66)**

Pyridine and quinoline are also reduced to piperidine and 1,2,3,4-tetrahydroquinoline by [$RhCl_3(Py)_3$] and $NaBH_4$. Isoquinoline and indole are not hydrogenated.[207]

[$RuHCl(PPh_3)_3$] and [$RuH_2(PPh_3)_2(PPh_2C_6H_4)$]$^-$ can reduce polyaromatic compounds under hydrogen. Quinoline (**67**) and phenanthridine (**69**) are converted to 1,2,3,4-tetrahydroquinoline (**68**) and 9,10-dihydrophenanthridine (**70**).[91,205a]

H_2, 21 atm, [$RuHCl(PPh_3)_3$], 85 °C: (**67**) → (**68**)

H_2, 21 atm, [$RuHCl(PPh_3)_3$], 85 °C: (**69**) → (**70**)

Sulfur-containing heteroaromatics can be smoothly hydrogenated with [$Co_2(CO)_8$] under carbon monoxide and a hydrogen atmosphere (238–272 atm), as demonstrated by the reduction of thiophene derivatives (**71**).[208]

H_2, 238–272 atm, [$Co_2(CO)_8$], 180–190 °C: (**71**) → tetrahydrothiophene

R = H, 66%; R = Me, 77%; R = Et, 82%

3.2.4 HYDROGENATION OF C≡C BONDS

3.2.4.1 Reduction of C≡C Bonds to Saturated Hydrocarbons

With [$RhCl(PPh_3)_3$] or [$Rh(OAc)(PPh_3)_3$], 1-hexyne (**72**) and other alkynes are hydrogenated to the saturated compounds by way of the alkenes.[5a,21]

H_2, 0.5 atm, 25 °C: (**72**) → (**73**)

	Relative rate
[$RhCl(PPh_3)_3$]	2.22
[$Rh(OAc)(PPh_3)_3$]	1

Under an atmosphere of hydrogen and acetylene (2:1), [$Rh_2(CO)_2(DPPM)_2$] in toluene hydrogenates acetylene at 80 °C to ethane at a rate of ~3.3 mol of ethane mol^{-1} of Rh dimer h^{-1}. No cyclotrimerization of acetylene is observed in this reaction.[209]

[$Ti(Cp)_2(CO)_2$] is a catalyst for the hydrogenation of phenylacetylene to ethylbenzene, while alkyl-substituted terminal alkynes are reduced to alkenes.[210] Electron rich titanium(II) complexes, [$Cp_2Ti(PhC≡CPh)(PMe_3)$], [$(MeCp)_2Ti(PhC≡CPh)(PMe_3)$] and [$CpCp^*Ti(PhC≡CPh)$] are also catalyst precursors for the hydrogenation of alkynes to alkanes at 20 °C under atmospheric pressure of hydrogen.[211]

3.2.4.2 Reduction of Terminal C≡C Bonds to Alkenes

In benzene and with acidic alcohols such as 2,2,2-trifluoroethanol or phenol as cosolvents, the rate of hydrogenation of alkyne by Wilkinson's catalyst, $[RhCl(PPh_3)_3]$, is accelerated, while that of C=C bonds does not change, which allows selective hydrogenation of alkynes.[212] $[Rh(COD)(PPh_3)(Py)]PF_6$ can also catalyze the selective reduction of 1-alkynes to 1-alkenes in the presence of benzoate ion. For example, phenylacetylene is slowly reduced with 1 equiv. of D_2 (1 atm) to *cis*-PhCD=CHD by $[Rh(COD)(PPh_3)(Py)]PF_6$ in the presence of an excess of $Et_3NH^+PhCO_2^-$. Thus compound **(74)** is converted selectively to **(75)**.[213]

(74) OH — H_2, 1 atm; $[Rh(COD)(PPh_3)(Py)]PF_6$; C_6H_6, $PhCO_2H$, NEt_3; 20 °C → **(75)** OH

Methanolic solutions of the dinuclear complexes $[Rh_2(\mu\text{-}X)(CO)_2(Ph_2ECH_2EPh_2)_2]^+$ (X = Cl, Br; E = P, As) and $[Ir_2Cl(CO)_3(Ph_2PCH_2PPh_2)_2]$ are catalysts for the hydrogenation of alkynes to alkenes.[42] The complex $[RuH(PMe_2Ph)_5]PF_5$ selectively converts terminal alkynes to alkenes without hydrogenating C=C bonds.[214] $[CoH(CO)(PBu^n{}_3)_3]$ catalyzes partial hydrogenation of 1-pentyne **(76)** to 1-pentene **(77)** in 95% yield without isomerization of the alkenic bond, but aldehydes are not tolerated under the reaction conditions.[95a,95b] $[CoCl\{P(OEt)_3\}_n]$ (n=3,4) is also active for hydrogenation of alkynes to alkenes.[215]

(76) — H_2, 30 atm; $[CoH(CO)(PBu^n{}_3)_3]$; 40 °C, 45 min; 95% → **(77)**

$[Co_2(CO)_6(HC{\equiv}CH)]$ has catalytic activity for hydrogenation of acetylene to ethylene. On increasing the acetylene concentration, however, cyclotrimerization to benzene becomes the dominant process.[216]

1-Pentyne and 1-hexyne are hydrogenated with $[Ti(Cp)_2(CO)_2]$ to 1-pentene and 1-hexene, respectively, but phenylacetylene gives ethylbenzene.[210] Reduced compounds of $[Ti(Cp)_2Cl_2]$ with Na, Mg, Ca, sodium naphthalenide, or butyllithium catalyze the hydrogenation of a variety of alkynes to alkenes.[122,126]

The C≡C bond in phenylacetylene can be selectively reduced, giving styrene, by $[OsHBr(CO)(PPh_3)_3]$ under H_2 at atmospheric pressure. After *ca.* 100 catalytic cycles, the selectivity for the production of styrene is still close to 100%.[74] In the absence of the alkyne, styrene is reduced to ethylbenzene at a comparable rate, but the presence of as little as 1% phenylacetylene suppresses the reduction of the double bond almost completely.

The iron cluster $[\{(\eta^5\text{-}C_5H_5)Fe(\mu_3\text{-}CO)\}_4]$ catalyzes the selective hydrogenation of alkynes to alkenes at 100–130 °C and 6.8–68 atm of hydrogen. Terminal alkynes are selectively hydrogenated to alkenes even in the presence of alkenes or internal alkynes.[217]

3.2.4.3 Reduction of C≡C Bonds to *cis*-Alkenes

Highly stereo- and chemo-selective hydrogenation of internal alkynes to *cis*-alkenes has been attained with $[(arene)Cr(CO)_3]$ complexes.[188] For example, reduction of 7-tetradecyne **(78)** gives **(79)** in quantitative yield.

$C_6H_{13}{-}{\equiv}{-}C_6H_{13}$ **(78)** — H_2, 70 atm; $[(methyl\ benzoate)Cr(CO)_3]$; 120 °C, 15 h; 100% → C_6H_{13} C_6H_{13} **(79)**

The complex $[CoH(CO)(PBu^n{}_3)_3]$ catalyzes partial hydrogenation of 2-pentyne to *cis*-2-pentene in 85% yield.[98a,98b,215] Catalysts can also be generated from $[\{Co(CO)_2(PR_3)\}_3]$ (R = Ph, Bu^n) under hydrogenation conditions.[95c,218]

The complexes of the type $[Rh(diene)L_n]PF_6$ (diene = norbornadiene, 1,5-cyclooctadiene; L = tertiary phosphine; n =2 or 3) are also used as catalysts for the reduction of internal alkynes to *cis*-alkenes (equation 9). Selectivity is greater than 95%.[48,50] At 1 atm of H_2 $[(\mu\text{-H})_2H_2Rh_2(triphos)_2]^{2+}$ in DMF at 22 °C catalyzes, although very slowly, the hydrogenation of dimethyl acetylenedicarboxylate (**80**) to dimethyl maleate (**81**).[219]

$$\text{R}-\!\!\equiv\!\!-\text{R}' \xrightarrow[\text{yield >98\%, selectivity >95\%}]{H_2\text{, 1 atm, [Rh(diene)}L_n\text{]}PF_6} \textit{cis-}\text{RCH=CHR}' \qquad (9)$$

$$\underset{(\mathbf{80})}{MeO_2C-\!\!\equiv\!\!-CO_2Me} \xrightarrow[22\ ^\circ C]{H_2\text{, 1 atm, }[(\mu\text{-H})_2H_2Rh_2(triphos)_2]^{2+}} \underset{(\mathbf{81})}{\textit{cis-}MeO_2CCH=CHCO_2Me}$$

Hydrogenation of alkynes to *cis*-alkenes with $[RuHCl(PPh_3)_3]$ proceeds at a rate roughly 0.1 times as fast as that of alkenes.[220] The complex $[RuH(PMe_2Ph)_5]PF_5$ selectively reduces internal and terminal alkynes without hydrogenating the corresponding alkenes. 2-Butyne is hydrogenated to *cis*-2-butene at 20 °C under 1 atm of H_2.[214] Disubstituted alkynes are hydrogenated stereoselectively in the presence of $[RuCl_2(PPh_3)_3]$ to give *cis*-alkenes in a 1:1 mixture of ethanol and benzene at 20 °C.[221]

$[Pd_5(PPh)_2(PPh_3)_2]$ is a catalyst for the hydrogenation of various alkynes and conjugated dienes, with alkynes selectively hydrogenated to *cis*-alkenes.[222]

In the presence of $[Cp_2Mo_2(CO)_4]$,[223b] $[Ni_2(COD)_2(\mu\text{-}\eta^2\text{-RC}{\equiv}\text{CR})]$,[223c] and $[Ni_4(RNC)_4(\mu_3\text{-}\eta^2\text{-RC}{\equiv}\text{CR})_3]$,[223d] alkynes are converted to *cis*-alkenes. The alkenes once formed no longer bind to the complex and hydrogenation does not proceed further to give alkanes or even isomerized alkenes.[223]

Cocondensation of lanthanoid metal atoms with internal alkynes generates lanthanoid complexes of alkynes, which are potential catalysts for hydrogenation. $[\{Sm(1\text{-hexyne})\}_n]$ or $[\{Er(3\text{-hexyne})\}_n]$ catalyzes hydrogenation of hex-3-yne to *cis*-hex-3-ene (97% *cis*) at room temperature and atmospheric pressure of H_2.[224]

Cyclopentadienyltitanium(II) complexes, $[Cp_2Ti(PhC{\equiv}CPh)(PMe_3)]$, $[CpCp^*Ti(PhC{\equiv}CPh)]$ and $[(MeCp)_2Ti(PhC{\equiv}CPh)(PMe_3)]$, are catalyst precursors for the hydrogenation of alkynes to *cis*-alkenes in the presence of an excess of PMe_3 under mild conditions (1 atm of H_2).[211]

The iron cluster $[\{(\eta^5\text{-}C_5H_5)Fe(\mu_3\text{-CO})\}_4]$, catalyzes the selective hydrogenation of alkynes to alkenes at 100–130 °C and 6.8–68 atm of H_2. Internal alkynes are slowly reduced to *cis*-alkenes, but nitro groups attached to aromatic rings and terminal carbon–carbon double bonds are also hydrogenated.[218]

3.2.4.4 Reduction of C≡C Bonds to *trans*-Alkenes

The hydrogenation of acetylenedicarboxylic acid dimethyl ester (**80**) and diphenylacetylene (**83**) with $[RhH_2(OC({=}O)OH)(PPr^i_3)_2]$ gives dimethyl fumarate (**82**) and *trans*-stilbene (**84**), respectively. This complex also catalyzes an isomerization of *cis*-stilbene to *trans*-stilbene, though hydrogenation of alkyne is about eight times faster than the isomerization.[225]

$$\underset{(\mathbf{80})}{MeO_2C-\!\!\equiv\!\!-CO_2Me} \xrightarrow[20\ ^\circ C]{H_2\text{, 1 atm, }[RhH_2\{OC(=O)OH\}(PPr^i_3)_2]} \underset{(\mathbf{82})}{\textit{trans-}MeO_2CCH=CHCO_2Me}$$

$$\underset{(\mathbf{83})}{Ph-\!\!\equiv\!\!-Ph} \xrightarrow[20\ ^\circ C]{H_2\text{, 1 atm, }[RhH_2\{OC(=O)OH\}(PPr^i_3)_2]} \underset{(\mathbf{84})}{\textit{trans-}PhCH=CHPh}$$

The dinuclear hydridorhodium complex $[\{(\mu\text{-H})Rh\{P(O\text{—}Pr^i)_3\}_2\}_2]$ is a catalyst for the stereoselective hydrogenation of dialkylalkynes and diarylalkynes to the corresponding *trans*-alkenes. Although the hydrogenation rates are much lower than for terminal alkenes (approximately 1 turnover min^{-1} at 20 °C), the catalytic reaction is totally stereoselective. Unfortunately, the lifetime of the catalyst is very short; after approximately 5 min under the catalytic conditions, *cis*-alkenes become significant products.[226]

With $[RhCl_3(Py)_3]$, DMF and $NaBH_4$, diphenylacetylene is hydrogenated to *trans*-stilbene as the sole product, while dimethyl acetylenedicarboxylate is converted to dimethyl maleate, and butyne-1,4-diol to *cis*-but-2-ene-1,4-diol.[227]

3.2.5 HOMOGENEOUS ASYMMETRIC HYDROGENATION OF C═C BONDS USING CHIRAL CATALYSTS

3.2.5.1 Chiral Catalysts

Attempts to prepare optically active organic compounds by enantioselective hydrogenation of prochiral alkenes with the aid of homogeneous catalysts emerged in the late 1960s and now this methodology has become one of the most attractive synthetic approaches to this class of compounds.[228] Horner and Knowles independently demonstrated, for the first time, that phosphine–rhodium complexes were promising catalysts for asymmetric hydrogenation.[229,230] Early in the 1970s Kagan[231] and Knowles[232] developed the chiral chelating diphosphines (**85a**)[231] and (**85b**),[232] respectively, and had great success in the asymmetric hydrogenation of certain alkenic substrates. Since then, a great number of chiral diphosphines have been synthesized and used as chiral ligands.[233] Most of the highly efficient diphosphines are 1,2-diphosphines and 1,4-diphosphines such as (**85c**)–(**95**). The ferrocenylphosphines (**96**) are also important chiral ligands.[245]

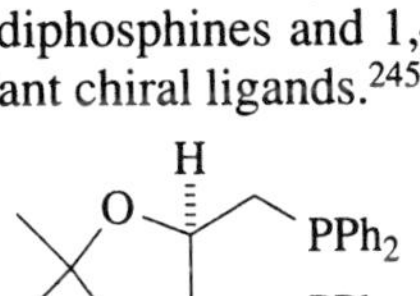

H, O, O, H, PPh_2, PPh_2

(R,R)-DIOP (**85a**)

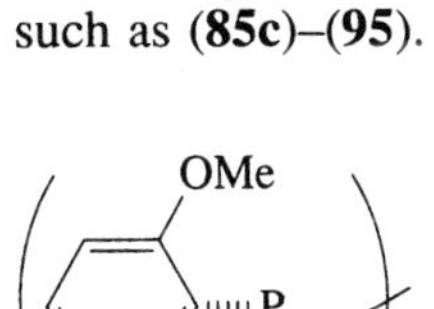

OMe, P, Ph, 2

(R,R)-DIPAMP (**85b**)

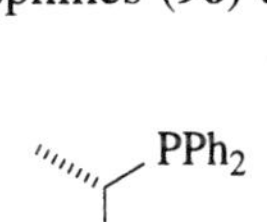

PPh_2, PPh_2

(S,S)-CHIRAPHOS (**85c**) (ref. 234)

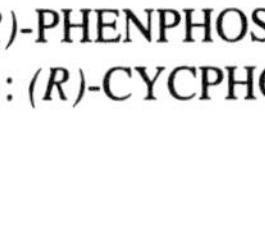

R, PPh_2, PPh_2

R = Me: *(R)*-PROPHOS (**86**) (ref. 235)
R = Ph: *(R)*-PHENPHOS (**87**) (ref. 236)
R = C_6H_{11}: *(R)*-CYCPHOS (**88**) (ref. 237)

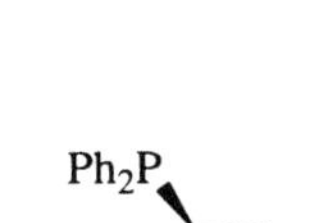

PPh_2, PPh_2

(R,R)-NORPHOS (**89**) (ref. 238)

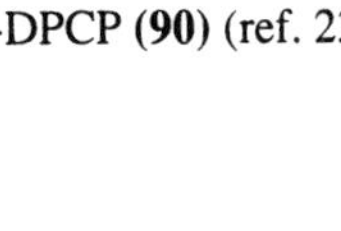

PPh_2, PPh_2

(S,S)-DPCP (**90**) (ref. 239)

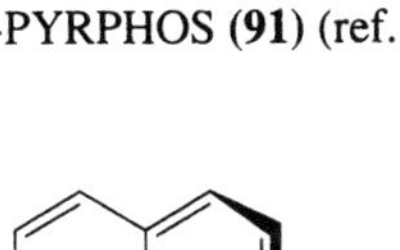

R—N, PPh_2, PPh_2

(S,S)-PYRPHOS (**91**) (ref. 240)

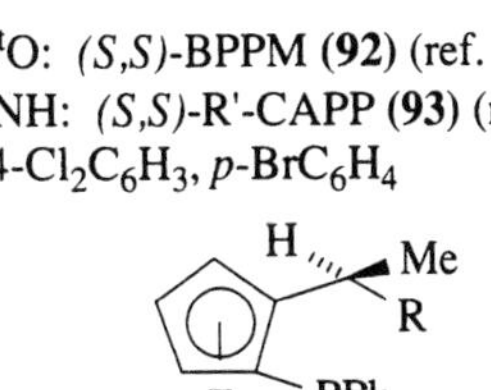

Ph_2P, N, COR, PPh_2

R = Bu^{t}O: *(S,S)*-BPPM (**92**) (ref. 241)
R = R^{1}NH: *(S,S)*-R'-CAPP (**93**) (ref. 242)
R' = 3,4-$Cl_2C_6H_3$, *p*-BrC_6H_4

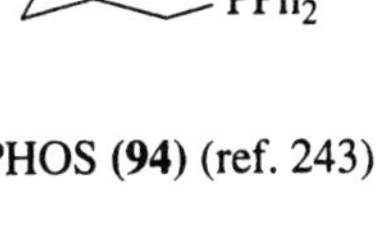

PPh_2, PPh_2

CAMPHOS (**94**) (ref. 243)

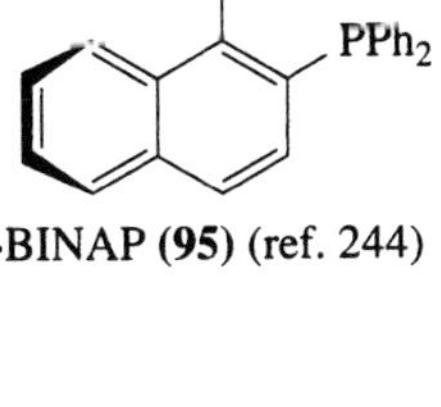

PPh_2, PPh_2

(R)-BINAP (**95**) (ref. 244)

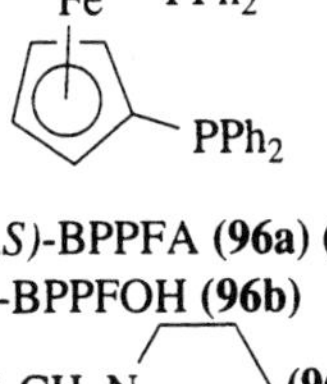

H, Me, R, Fe, PPh_2, PPh_2

R = NMe_2: *(R,S)*-BPPFA (**96a**) (ref. 245)
R = OH: *(R,S)*-BPPFOH (**96b**)
R = $N(Me)CH_2CH_2N$ (piperidine ring) (**96c**)

Chiral bidentate ligands such as diphosphinites,[228b,246] diaminophosphines[228b,247] and aminophosphine–phosphinites,[228b,248] have also been used.[228b] Although much information has been accumulated the choice of chiral catalysts and reaction conditions still remains empirical, and one must make an effort to find the most suitable catalytic system and reaction conditions for every new substrate.

3.2.5.2 Asymmetric Hydrogenation of Enamides and Related Substrates

α-Acylaminoacrylic acids (**97**) were the first successful alkenic substrates used in homogeneous asymmetric hydrogenation. Kagan reported asymmetric hydrogenation of (*Z*)-α-acetylaminocinnamic acid with DIOP–Rh^+ complexes giving *N*-acetylphenylalanine in 72% *ee*.[231] The Monsanto group also developed chiral ligands, (*R*)-*o*-anisylcyclohexylmethylphosphine [(*R*)-CAMP][230] and (**85b**),[232] and obtained very high optical yields (>90%) in the Rh^I-catalyzed asymmetric hydrogenation of (**97**). Now, a number of chiral phosphine ligand–metal complexes are known to catalyze this type of asymmetric hydrogenation, some of which are listed in Table 4. In contrast to the high *ee* obtained for (*Z*)-α-acylaminoacrylic acids and esters, the hydrogenation of the (*E*)-isomers usually proceeds very slowly and in poor *ee*, though some alkyl-substituted (*E*)-α-acylaminoacrylic acids are hydrogenated in high *ee*.[249] Using Rh–PYRPHOS complexes, (**97**) can be hydrogenated with a high substrate to catalyst ratio (up to 50 000).[250] With a water soluble catalyst, [Rh(1,5-COD){3*R*,4*R*)-3,4-bis(diphenylphosphino)-1,1-dimethylpyrrolidinium-*P*,*P′*}]·$2BF_4$, the sodium salt of α-(acetylamino)cinnamic acid has been hydrogenated in aqueous media in 90% *ee*.[251]

Extensive studies on the mechanism of the reaction have been carried out by Halpern[252] and Brown.[253]

$R^2(R^3)C{=}C(CO_2H)(NHCOR^1)$ (**97**) —[H_2, 1 atm; Rh–diphosphine*]→ $R^2(R^3)C^*H{-}C^*H(CO_2H)(NHCOR^1)$ (**98**)

Table 4 Asymmetric Hydrogenation of α-Acylaminoacrylic Acids[a]

Substrate	*DIOP (% ee)*	*DIPAMP (% ee)*	*NORPHOS (% ee)*	*BPPM (% ee)*	*BINAP (% ee)*	*BPPFA (% ee)*
a $CH_2{=}C(CO_2H)NHAc$	73	94	90	99	(98)	93
b $PhCH{=}C(CO_2H)NHCOMe$	82	96	97	91	(100)	
c 3-AcO-4-MeO-$C_6H_3CH{=}C(CO_2H)NHAc$	84	94	94	86	(79)	86

[a]Figures in parentheses are given for the hydrogenation of *N*-benzoyl derivatives.

The carboxylic functionality in α-acylaminoacrylic acids can be replaced by other electron-withdrawing groups such as alkoxycarbonyl, carbonyl, keto and cyano groups.[228b] These substrates have been reduced in high enantioselectivity (more than 90% *ee;* equation 10). Dipeptides and oligopeptides have been obtained in high *ee* by asymmetric hydrogenation of the corresponding prochiral dehydropeptides.[254] Although enamides without any other functional groups have been reduced only with difficulty in high *ee*, BINAP–ruthenium(II) complexes bring about high enantioselectivities (99% *ee*).[255,256] Thus the hydrogenation of *N*-acyl-1-alkylidene-1,2,3,4-tetrahydroisoquinolines in the presence of $[Ru(OCOR)_2(BINAP)]$ gives isoquinoline alkaloids generally in high optical purity (equation 11). This method finds a very wide applicability and has been successfully applied to asymmetric synthesis of morphines, benzomorphans and morphinans.[257]

(*R,R*)-DIPAMP–Rh (10)

R = OMe, 95% *ee*; R = NH_2, 94% *ee*

H_2, 4 atm, $[Ru(OAc)_2\{(R)\text{-BINAP}\}]$, >99.5% *ee* (11)

3.2.5.3 Asymmetric Hydrogenation of Acrylic Acids and Esters

Acrylic acids were the substrates chosen at the early stage of the asymmetric hydrogenation of alkenes. Reduction of (*E*)-3-phenyl-2-butenoic acid with a neomenthyldiphenylphosphine–rhodium complex as catalyst was achieved in 61% *ee*, which showed that asymmetric hydrogenation of acrylic acids is a promising way to produce optically active carboxylic acids.[258] However, it was soon revealed that other coordinating functionalities such as amido, carboxy, amidomethyl, alkoxycarbonylmethyl and hydroxycarbonylmethyl groups, are required for obtaining high enantioselectivity. Thus, itaconic acid (**99**) has been reduced to (*R*)-2-methylsuccinic acid (**100**) in 95% *ee* with the (*S,S*)-R-CAPP–Rh^+ complex.[259] Asymmetric reduction of itaconic acid by transfer hydrogenation with the catalyst derived from $[\{Rh(COD)Cl\}_2]$ and BPPM, and triethylammonium formate as hydrogen source, also proceeds with high enantioselectivity.[260]

H_2, 1 atm, (*S,S*)-R-CAPP–Rh, 95% *ee*

(**99**) → (**100**)

Certain alkyl- or aryl-substituted acrylic acids such as tiglic acid and 2-(6-methoxy-2-naphthyl)propenoic acid (**101**) are hydrogenated in 92% and 97% *ee*, respectively, with $[Ru(OAc)_2\{(S)\text{-BINAP}\}]$.[261] Acrylic acids having hydroxyalkyl substituents or their esters can also be hydrogenated in high enantiomeric excesses with the same catalyst system (equation 12). The sense of stereochemistry is the same as that found for the corresponding substrates without these functional groups. Trisubstituted acrylic acids have been reduced with the modified ferrocenyldiphosphine–Rh complex in 98% *ee* (equation 13).[262]

H_2, 135 atm, $[Ru(OAc)_2\{(S)\text{-BINAP}\}]$, 97% *ee*

(**101**)

H_2, 86 atm, $[Ru(OAc)_2\{(R)\text{-BINAP}\}]$ (12)

93% *ee*

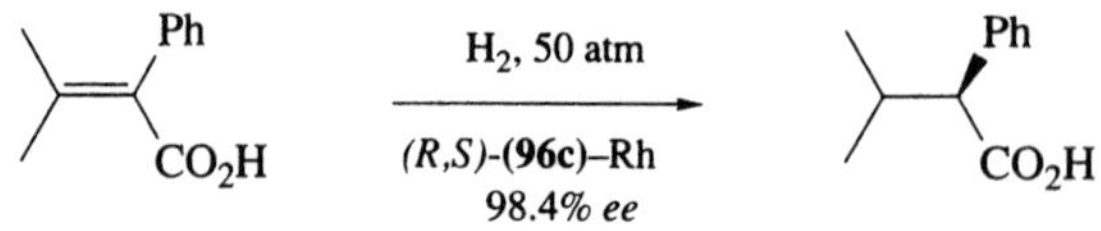

$$\text{(CH}_3)_2\text{C=C(Ph)CO}_2\text{H} \xrightarrow[\text{(}R,S\text{)-(96c)–Rh, 98.4\% }ee]{\text{H}_2\text{, 50 atm}} \text{(CH}_3)_2\text{CH–CH(Ph)CO}_2\text{H} \quad (13)$$

Hydrogenation of acrylic acid esters with high enantioselectivity has usually been accomplished with difficulty. The enantioselective reduction of α,β-unsaturated carboxylates with sodium borohydride in the presence of cobalt–semicorrin complexes has been achieved in up to 96% *ee* (equation 14). The (*E*)- and (*Z*)-isomers each afford products of opposite configuration, and the isolated double bonds remain untouched.[263]

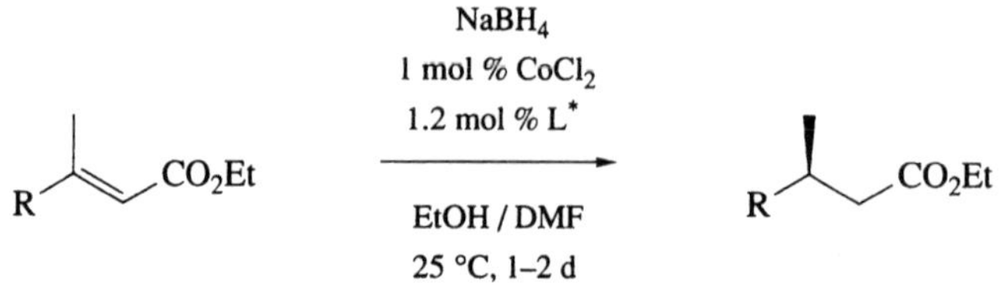

$$\text{R(CH}_3\text{)C=CHCO}_2\text{Et} \xrightarrow[\text{EtOH / DMF, 25 °C, 1–2 d}]{\text{NaBH}_4\text{, 1 mol \% CoCl}_2\text{, 1.2 mol \% L}^*} \text{R(CH}_3\text{)CH–CH}_2\text{CO}_2\text{Et} \quad (14)$$

R = $PhCH_2CH_2$, 94% *ee*; R = $Me_2C{=}CH(CH_2)_2$, 94% *ee*; R = Me_2CH, 96% *ee*; R = Ph, 81% *ee*

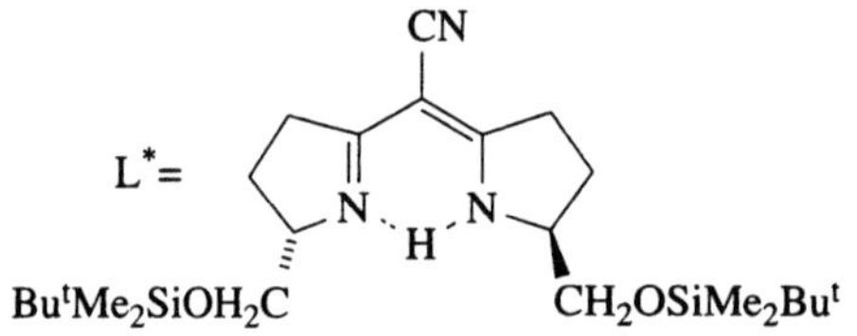

L* = (semicorrin ligand with CN, N···H···N, $Bu^tMe_2SiOH_2C$ and $CH_2OSiMe_2Bu^t$ substituents)

With the catalysts derived from (*S*,*S*)-1,2-bis(diphenylphosphinomethyl)cyclobutane and $[RhH(CO)(PPh_3)_3]$ or rhodium carbonyls, the α,β-unsaturated aldehydes, neral and geranial, are hydrogenated to (*R*)- and (*S*)-citronellal in 79% and 60% *ee*, respectively.[264] Cyclic α,β-unsaturated ketones such as isophorone and 2-methyl-2-cyclohexenone have been hydrogenated using ruthenium hydrides coordinated with chiral diphosphines in up to 62% *ee* to give chiral ketones, though conversions are not satisfactory.[265]

3.2.5.4 Asymmetric Hydrogenation of Allylic Alcohols

Some cationic Rh– and Ir–phosphine complexes have been used for diastereoselective hydrogenation of chiral allylic and homoallylic alcohols.[266] Here, the preexisting chirality of the sp^3-hybridized carbons induces new asymmetry on the neighboring alkenic diastereofaces through coordination of the hydroxy group to the transition metals (see Section 3.2.1).

Highly enantioselective hydrogenation of prochiral alkenic alcohols has been attained with BINAP–Ru^{II} catalysts.[267] (*R*)- or (*S*)-citronellol (**102**) was obtained in nearly quantitative yield and with 96–99% enantioselectivity by hydrogenation of geraniol or nerol using $[Ru(OAc)_2(BINAP)]$. The stereochemical outcome is outlined in Scheme 1. The hydrogenation is carried out in methanol at room temperature, at an initial hydrogen pressure higher than 30 atm. The substrate to catalyst mol ratio reaches more than 20 000 without difficulty. The allylic and nonallylic double bonds in the substrates can be clearly differentiated and the product was contaminated by less than 0.5% dihydrocitronellol. This procedure has

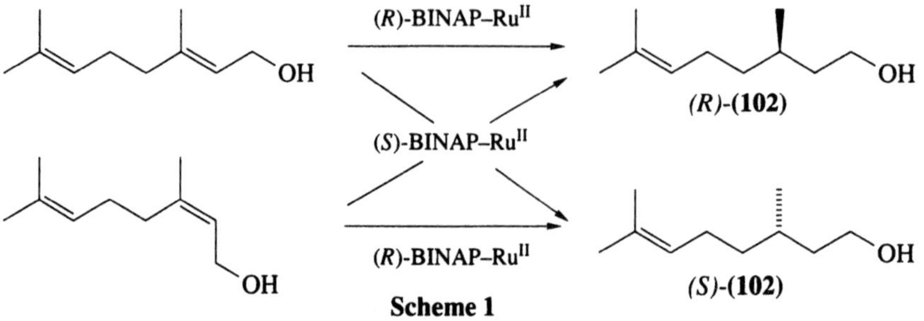

Geraniol —(*R*)-BINAP–Ru^{II}→ (*R*)-(**102**); Geraniol —(*S*)-BINAP–Ru^{II}→ (*S*)-(**102**)

Nerol —(*S*)-BINAP–Ru^{II}→ (*R*)-(**102**); Nerol —(*R*)-BINAP–Ru^{II}→ (*S*)-(**102**)

Scheme 1

been successfully applied to the asymmetric synthesis of (3*R*,7*R*)-3,7,11-trimethyldodecanol, a versatile intermediate for the synthesis of α-tocopherol. Homogeraniol can be hydrogenated in 92% *ee*.

In the hydrogenation of racemic allylic alcohols catalyzed by a chiral Rh catalyst, at most, 20:1 discrimination has been attained for some acyclic substrates.[268] BINAP–Ru complexes have been used for kinetic resolution of chiral acyclic and cyclic secondary alcohols with up to 74:1 differentiation between the enantiomeric alcohols (equation 15).[269]

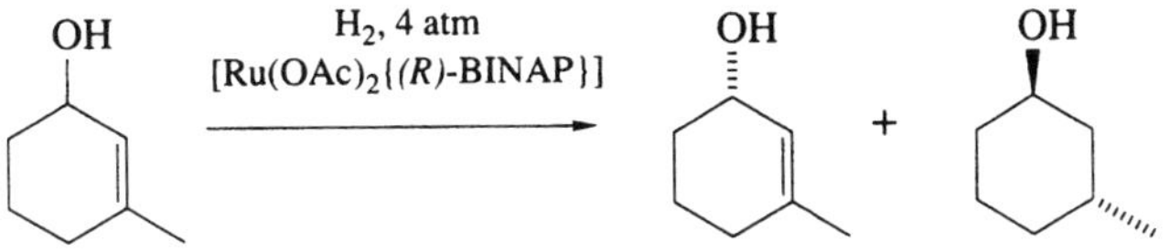

H_2, 4 atm; [Ru(OAc)$_2${(*R*)-BINAP}] (15)

3.2.5.5 Asymmetric Hydrogenation of Simple Alkenes

Hydrogenation of simple alkenes in high enantiomeric excesses has not been attained with the chiral rhodium catalysts,[270] but a fused cyclopentadienyl ligand with C_2 symmetry has been used in enantioselective titanocene-catalyzed hydrogenation of alkenes.[127–129] Thus 2-phenyl-1-butene (**103**) was hydrogenated in 96% *ee*.

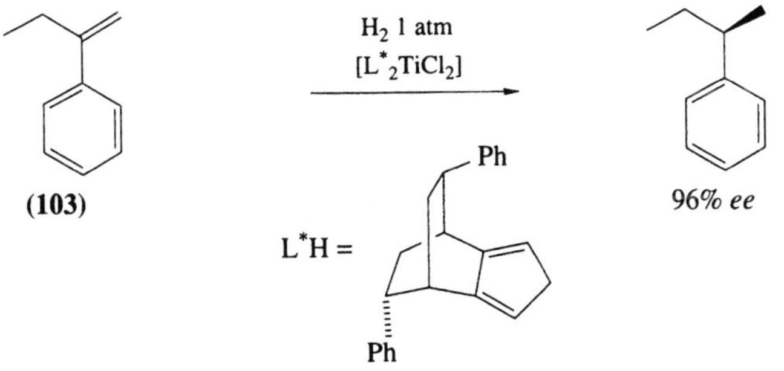

(**103**) H_2 1 atm; [$L^*{}_2TiCl_2$] 96% *ee*

L^*H = Ph Ph

3.2.6 REFERENCES

1. (a) A. J. Birch and D. H. Williamson, *Org. React. (N. Y.)*, 1976, **24**, 1; (b) B. R. James, in 'Comprehensive Organometallic Chemistry', ed. G. Wilkinson, F. G. A. Stone and E. W. Abel, Pergamon Press, Oxford, 1982, vol. 8, p. 285; (c) B. R. James, 'Homogeneous Hydrogenation', Wiley, New York, 1974; (d) B. R. James, *Adv. Organomet. Chem.*, 1979, **17**, 319; (e) J. W. Faller, in 'Homogeneous Catalysis with Metal Phosphine Complexes', ed. L. H. Pignolet, Plenum Press, New York, 1983, chap. 2; (f) F. H. Jardine, *Prog. Inorg. Chem.*, 1981, **28**, 63; (g) F. H. Jardine, *Prog. Inorg. Chem.*, 1984, **31**, 266; (h) A. Andreetta, F. Conti and G. F. Ferrari, 'Aspects Homogeneous Catalysis', ed R. Ugo, Reidel, Milan, vol. 1, p. 203.
2. (a) J. F. Young, J. A. Osborn, F. H. Jardine and G. Wilkinson, *J. Chem. Soc., Chem. Commun.*, 1965, 131; (b) F. H. Jardine, J. A. Osborn, G. Wilkinson and J. F. Young, *Chem. Ind. (London)*, 1965, 560.
3. M. A. Bennett and P. A. Longstaff, *Chem. Ind. (London)*, 1965, 846.
4. ICI Ltd., Neth. Appl., 6 602 062 (*Chem. Abstr.*, 1967, **66**, 10 556y).
5. (a) J. A. Osborn, F. H. Jardine, J. F. Young and G. Wilkinson, *J. Chem. Soc. A*, 1966, 1711; (b) F. H. Jardine, J. A. Osborn and G. Wilkinson, *J. Chem. Soc. A*, 1967, 1574.
6. Y. Ohtani, A. Yamagishi and M. Fujimoto, *Bull. Chem. Soc. Jpn.*, 1979, **52**, 69.
7. H. van Bekkum, F. van Rantwijk and T. van de Putte, *Tetrahedron Lett.*, 1969, 1.
8. A. A. Blanc, H. Arzoumanian, E. J. Vincent and J. Metzger, *Bull. Soc. Chim. Fr.*, 1974, 2175.
9. W. Strohmeier, E. Hitzel and B. Kraft, *J. Mol. Catal.*, 1977, **3**, 61.
10. C. W. Dudley, G. Read and P. J. C. Walker, *J. Chem. Soc., Dalton Trans.*, 1974, 1926.
11. M. Brown and L. W. Piszkiewicz, *J. Org. Chem.*, 1967, **32**, 2013.
12. J. J. Sims, V. K. Honwad and L. H. Selman, *Tetrahedron Lett.*, 1969, 87.
13. A. J. Birch and K. A. M. Walker, *Aust. J. Chem.*, 1971, **24**, 513.
14. A. J. Birch and K. A. M. Walker, *Tetrahedron Lett.*, 1966, 4939.
15. Y. Senda, S. Mitsui, H. Sugiyama and S. Seto, *Bull. Chem. Soc. Jpn.*, 1972, **45**, 3498.
16. C. Djerassi and J. Gutzwiller, *J. Am. Chem. Soc.*, 1966, **88**, 4537.
17. J.-F. Biellmann and H. Liesenfelt, *Bull. Soc. Chim. Fr.*, 1966, 4029.
18. G. C. Bond and R. A. Hillyard, *Faraday Discuss. Chem. Soc.*, 1968, **46**, 20.
19. G. Dolcetti, *Inorg. Nucl. Chem. Lett.*, 1973, **9**, 705.
20. W. Strohmeier and R. Endres, *Z. Naturforsch., Teil B*, 1972, **27**, 1415.
21. R. W. Mitchell, J. D. Ruddick and G. Wilkinson, *J. Chem. Soc. A*, 1971, 3224.
22. L. Horner, H. Büthe and H. Siegel, *Tetrahedron Lett.*, 1968, 4023.
23. S. Montelatici, A. van der Ent, J. A. Osborn and G. Wilkinson, *J. Chem. Soc. A*, 1968, 1054.

24. L. Horner and H. Siegel, *Justus Liebigs Ann. Chem.*, 1971, **751**, 135.
25. Y. Chevallier, R. Stern and L. Sajus, *Tetrahedron Lett.*, 1969, 1197.
26. R. Stern, Y. Chevallier and L. Sajus, *C. R. Hebd. Seances Acad. Sci., Ser. C*, 1967, **264**, 1740 (*Chem. Abstr.*, 1967, **67**, 73 672p).
27. (a) C. O'Connor and G. Wilkinson, *J. Chem. Soc. A*, 1968, 2665; (b) C. O'Connor, G. Yagupsky, D. Evans and G. Wilkinson, *J. Chem. Soc., Chem. Commun.*, 1968, 420.
28. M. Yagupsky, C. K. Brown, G. Yagupsky and G. Wilkinson, *J. Chem. Soc. A*, 1970, 937.
29. W. Strohmeier and G. Csontos, *J. Organomet. Chem.*, 1974, **67**, C27.
30. D. E. Budd, D. G. Holah, A. N. Hughes and B. C. Hui, *Can. J. Chem.*, 1974, **52**, 775.
31. D. G. Holah, I. M. Hoodless, A. N. Hughes, B. C. Hui and D. Martin, *Can. J. Chem.*, 1974, **52**, 3758.
32. M. S. Delaney, C. B. Knobler and M. F. Hawthorne, *J. Chem. Soc., Chem. Commun.*, 1980, 849.
33. R. D. Gillard, J. A. Osborn, P. B. Stockwell and G. Wilkinson, *Proc. Chem. Soc., London*, 1964, 284.
34. C. White, D. S. Gill, J.-W. Kang, H. B. Lee and P. M. Maitlis, *J. Chem. Soc., Chem. Commun.*, 1971, 734.
35. C. Masters, W. S. McDonald, G. Raper and B. L. Shaw, *J. Chem. Soc., Chem. Commun.*, 1971, 210.
36. C. Masters and B. L. Shaw, *J. Chem. Soc. A*, 1971, 3679.
37. D. G. Holah, A. N. Highes and B. C. Hui, *Can. J. Chem.*, 1972, **50**, 3714.
38. C. Larpent, R. Dabard and H. Patin, *Tetrahedron Lett.*, 1987, **28**, 2507.
39. M. Green, T. A. Kuc and S. H. Taylor, *J. Chem. Soc., Chem. Commun.*, 1970, 1553.
40. J. Solodar, *J. Org. Chem.*, 1972, **37**, 1840.
41. F. A. Hart and D. W. Owen, *Inorg. Chim. Acta*, 1985, **103**, L1.
42. A. R. Sanger, *Can. J. Chem.*, 1982, **60**, 1363.
43. H. El-Amouri, A. A. Bahsoun, J. Fischer and J. A. Osborn, *Angew. Chem., Int. Ed. Engl.*, 1987, **26**, 1169.
44. P. Abley, I. Jardine and F. J. McQuillin, *J. Chem. Soc. C*, 1971, 840.
45. I. Jardine and F. J. McQuillin, *J. Chem. Soc., Chem. Commun.*, 1969, 477.
46. R. R. Schrock and J. A. Osborn, *J. Chem. Soc., Chem. Commun.*, 1970, 567.
47. R. R. Schrock and J. A. Osborn, *J. Am. Chem. Soc.*, 1971, **93**, 2397.
48. R. R. Schrock and J. A. Osborn, *J. Am. Chem. Soc.*, 1971, **93**, 3089.
49. J. R. Shapley, R. R. Schrock and J. A. Osborn, *J. Am. Chem. Soc.*, 1969, **91**, 2816.
50. R. R. Schrock and J. A. Osborn, *J. Am. Chem. Soc.*, 1976, **98**, 2143.
51. T. Kwon, J. C. Woo and C. S. Chin, *Polyhedron*, 1983, **2**, 1225.
52. P. T. Draggett, M. Green and S. F. W. Lowrie, *J. Organomet. Chem.*, 1977, **135**, C60.
53. W. Reimann, W. Abboud, J. M. Basset, R. Mutin, G. L. Rempel and A. K. Smith, *J. Mol. Catal.*, 1980, **9**, 349.
54. (a) J. E. Hamlin, K. Hirai, V. C. Gibson and P. M. Maitlis, *J. Mol. Catal.*, 1982, **15**, 337; (b) P. M. Maitlis, *Acc. Chem. Res.*, 1978, **11**, 301; (c) D. S. Gill, C. White and P. M. Maitlis, *J. Chem. Soc., Dalton Trans.*, 1978, 617; (d) J. E. Hamlin, K. Hirai, V. C. Gibson and P. M. Maitlis, *J. Mol. Catal.*, 1982, **15**, 337.
55. L. A. Oro, M. Campo and D. Carmona, *J. Mol. Catal.*, 1987, **39**, 341.
56. D. Evans, J. A. Osborn, F. H. Jardine and G. Wilkinson, *Nature (London)*, 1965, **208**, 1203.
57. P. S. Hallman, B. R. McGarvey and G. Wilkinson, *J. Chem. Soc. A*, 1968, 3143.
58. P. S. Hallman, D. Evans, J. A. Osborn and G. Wilkinson, *J. Chem. Soc., Chem. Commun.*, 1967, 305.
59. I. Jardine and F. J. McQuillin, *Tetrahedron Lett.*, 1968, 5189.
60. J. E. Lyons, *J. Catal.*, 1973, **28**, 500.
61. J. P. Candlin, R. W. Dunning, R. S. McKenna and A. R. Oldham, *Br. Pat.* 1 141 847 (1971) (*Chem. Abstr.*, 1972, **76**, 3464).
62. (a) S. Komiya and A. Yamamoto, *J. Mol. Catal.*, 1979, **5**, 279; (b) S. Komiya, A. Yamamoto and S. Ikeda, *J. Organomet. Chem.*, 1972, **42**, C65.
63. V. D. Bianco, S. Doronzo and M. Aresta, *J. Organomet. Chem.*, 1972, **42**, C63.
64. L. N. Lewis, *J. Am. Chem. Soc.*, 1986, **108**, 743.
65. M. M. Taqui Khan, R. K. Andal and P. T. Manoharan, *J. Chem. Soc., Chem. Commun.*, 1971, 561.
66. S. T. Wilson and J. A. Osborn, *J. Am. Chem. Soc.*, 1971, **93**, 3068.
67. A. C. Skapski and F. A. Stephens, *J. Chem. Soc., Dalton Trans.*, 1974, 390.
68. R. W. Mitchell, A. Spencer and G. Wilkinson, *J. Chem. Soc., Dalton Trans.*, 1973, 846.
69. I. S. Thorburn, S. J. Rettig and B. R. James, *J. Organomet. Chem.*, 1985, **296**, 103.
70. J. L. Zuffa, M. L. Blohm and W. L. Gladfelter, *J. Am. Chem. Soc.*, 1986, **108**, 552.
71. J. L. Zuffa and W. L. Gladfelter, *J. Am. Chem. Soc.*, 1986, **108**, 4669.
72. Y. Blum, D. Czarkie, Y. Rahamim and Y. Shvo, *Organometallics*, 1985, **4**, 1459.
73. A. Basu, S. Bhaduri, H. Khwaja, P. G. Jones, T. Schroeder and G. M. Sheldrick, *J. Organomet. Chem.*, 1985, **290**, C19.
74. R. A. Sánchez-Delgado, A. Andriollo, E. González, N. Valencia, V. León and J. Espidel, *J. Chem. Soc., Dalton Trans.*, 1985, 1859.
75. W. Strohmeier, W. Rehder-Stirnweiss and R. Fleischmann, *Z. Naturforsch., Teil B*, 1970, **25**, 1481.
76. W. Strohmeier and R. Fleischmann, *J. Organomet. Chem.*, 1972, **42**, 163.
77. W. Strohmeier, R. Fleischmann and W. Rehder-Stirnweiss, *J. Organomet. Chem.*, 1973, **47**, C37.
78. J. E. Lyons, *J. Chem. Soc., Chem. Commun.*, 1969, 564.
79. J. E. Lyons, *J. Catal.*, 1973, **30**, 490.
80. W. Strohmeier and G. Csontos, *J. Organomet. Chem.*, 1974, **72**, 277.
81. M. Yamaguchi, *Kogyo Kagaku Zasshi*, 1967, **70**, 675; *Chem. Abstr.*, 1967, **67**, 99 542w.
82. L. Vaska, *Inorg. Nucl. Chem. Lett.*, 1967, **1**, 89.
83. W. Strohmeier and S. Hohmann, *Z. Naturforsch., Teil B*, 1970, **25**, 1309.
84. W. Strohmeier and W. Rehder-Stirnweiss, *Z. Naturforsch., Teil B*, 1971, **26**, 193.
85. F. Glockling and M. D. Wilbey, *J. Chem. Soc. A*, 1970, 1675.
86. M. G. Burnett and R. J. Morrison, *J. Chem. Soc. A*, 1971, 2325.
87. M. G. Burnett and R. J. Morrison, *J. Chem. Soc., Dalton Trans.*, 1973, 632.
88. M. G. Burnett, R. J. Morrison and C. J. Strugnell, *J. Chem. Soc., Dalton Trans.*, 1974, 1663.

89. R. H. Crabtree, H. Felkin and G. E. Morris, *J. Organomet. Chem.*, 1977, **141**, 205.
90. R. H. Crabtree, *Acc. Chem. Res.*, 1979, **12**, 331.
91. R. H. Crabtree, D. F. Chodosh, J. M. Quirk, H. Felkin, T. Fillebeen-Kharr and G. E. Morris, *Fundam. Res. Homogeneous Catal.*, 1979, **3**, 475.
92. R. H. Crabtree and G. E. Morris, *J. Organomet. Chem.*, 1977, **135**, 395.
93. J. W. Suggs, S. D. Cox, R. H. Crabtree and J. M. Quirk, *Tetrahedron Lett.*, 1981, **22**, 303.
94. M. Hidai, T. Kuse, T. Hikita, Y. Uchida and A. Misono, *Tetrahedron Lett.*, 1970, 1715.
95. (a) G. F. Pregaglia, A. Andreetta, G. F. Ferrari and R. Ugo, *J. Organomet. Chem.*, 1971, **30**, 387; (b) G. F. Ferrari, A. Andreetta, G. F. Pregaglia and R. Ugo, *J. Organomet. Chem.*, 1972, **43**, 213; (c) G. F. Pregaglia, A. Andreetta, G. F. Ferrari and R. Ugo, *J. Chem. Soc., Chem. Commun.*, 1969, 590.
96. (a) T. S. Janik, M. F. Pyszczek and J. L. Atwood, *J. Mol. Catal.*, 1981, **11**, 33; (b) T. S. Janik, M. F. Pyszczek, P. S. Sullivan and J. D. Atwood, *J. Organomet. Chem.*, 1984, **272**, 427.
97. A. Misono, Y. Uchida, T. Saito and K. M. Song, *J. Chem. Soc., Chem. Commun.*, 1967, 419.
98. J. L. Hendrikse and J. W. E. Coenen, *J. Catal.*, 1973, **30**, 72.
99. E. Balogh-Hergovich, G. Speier and L. Markó, *J. Organomet. Chem.*, 1974, **66**, 303.
100. S. Tyrlik, *J. Organomet. Chem.*, 1973, **50**, C46.
101. E. W. Stern and P. K. Maples, *J. Catal.*, 1972, **27**, 120.
102. E. W. Stern and P. K. Maples, *J. Catal.*, 1972, **27**, 134.
103. L. Kh. Freidlin, N. M. Nazarova and Yu. A. Kopyttsev, *Izv. Akad. Nauk SSSR, Ser. Khim.*, 1972, 201 (*Chem. Abstr.*, 1972, **77**, 4 634x).
104. N. M. Nazarova, L. Kh. Freidlin, Yu. A. Kopyttsev and T. I. Varava, *Izv. Akad. Nauk SSSR, Ser. Khim.*, 1972, 1422 (*Chem. Abstr.*, 1972, **77**, 100 943).
105. G. Henrici-Olivé and S. Olive, *J. Mol. Catal.*, 1975/76, **1**, 121.
106. A. S. Berenblyum, A. G. Knizhnik, S. L. Mund and I. I. Moiseev, *J. Organomet. Chem.*, 1982, **234**, 219.
107. J. H. Flynn and H. M. Hulburt, *J. Am. Chem. Soc.*, 1954, **76**, 3393.
108. R. D. Cramer, E. L. Jenner, R. V. Lindsey, Jr. and U. G. Stolberg, *J. Am. Chem. Soc.*, 1963, **85**, 1691.
109. R. V. Lindsey, Jr., G. W. Parshall and U. G. Stolberg, *J. Am. Chem. Soc.*, 1965, **87**, 658.
110. J. C. Bailar, Jr. and H. Itatani, *J. Am. Chem. Soc.*, 1967, **89**, 1592.
111. H. A. Tayim and J. C. Bailar, Jr., *J. Am. Chem. Soc.*, 1967, **89**, 4330.
112. J. C. Bailar, Jr. and H. Itatani, *Inorg. Chem.*, 1965, **4**, 1618.
113. J. C. Bailar, Jr. and H. Itatani, *J. Am. Oil Chem. Soc.*, 1966, **43**, 337.
114. H. van Bekkum, J. van Gogh and G. van Minnen-Pathuis, *J. Catal.*, 1967, **7**, 292.
115. H. van Bekkum, F. van Rantwijk, G. van Minnen-Pathuis, J. D. Remijnse and A. van Veen, *Recl. Trav. Chim. Pays-Bas*, 1969, **88**, 911 (*Chem. Abstr.*, 1969, **71**, 60 815v).
116. G. K. Anderson, C. Billard, H. C. Clark, J. A. Davies and C. S. Wong, *Inorg. Chem.*, 1983, **22**, 439.
117. S. Bhaduri and K. R. Sharma, *J. Chem. Soc., Dalton Trans.*, 1982, 727.
118. M. F. Sloan, A. S. Matlack and D. S. Breslow, *J. Am. Chem. Soc.*, 1963, **85**, 4014.
119. (a) P. C. Wailes, H. Weigold and A. P. Bell, *J. Organomet. Chem.*, 1972, **43**, C32; (b) P. C. Wailes, H. Weigold and A. P. Bell, *J. Organomet. Chem.*, 1972, **34**, 155.
120. S. Couturier, G. Tainturier and B. Gautheron, *J. Organomet. Chem.*, 1980, **195**, 291.
121. K. I. Gell and J. Schwartz, *J. Am. Chem. Soc.*, 1978, **100**, 3246.
122. R. Choukroun, M. Basso-Bert and D. Gervais, *J. Chem. Soc., Chem. Commun.*, 1986, 1317.
123. H. A. Martin and R. O. deJongh, *J. Chem. Soc., Chem. Commun.*, 1969, 1366.
124. H. A. Martin and R. O. deJongh, *Recl. Trav. Chim. Pays-Bas*, 1971, **90**, 713 (*Chem. Abstr.*, 1971, **75**, 87 903s).
125. G. Fachinetti and C. Floriani, *J. Chem. Soc., Chem. Commun.*, 1974, 66.
126. R. H. Grubbs, C. Gibbons, L. C. Kroll, W. D. Bonds, Jr. and C. H. Brubaker, Jr., *J. Am. Chem. Soc.*, 1973, **95**, 2373.
127. (a) R. L. Halterman and K. P. C. Vollhardt, *Organometallics*, 1988, **7**, 883; (b) R. L. Halterman, K. P. C. Vollhardt, M. E. Welker, D. Bläser and R. Boese, *J. Am. Chem. Soc.*, 1987, **109**, 8105.
128. L. A. Paquette, J. A. McKinney, M. L. McLaughlin and A. L. Rheingold, *Tetrahedron Lett.*, 1986, **27**, 5599.
129. J. C. Gallucci, B. Gautheron, M. Gugelchuk, P. Meunier and L. A. Paquette, *Organometallics*, 1987, **6**, 15.
130. W. J. Evans, D. J. Wink and D. R. Stanley, *Inorg. Chem.*, 1982, **21**, 2565.
131. P. J. Fagan, J. M. Manriquez, E. A. Maatta, A. M. Seyam and T. J. Marks, *J. Am. Chem. Soc.*, 1981, **103**, 6650.
132. R. G. Bowman, R. Nakamura, P. J. Fagan, R. L. Burwell, Jr. and T. J. Marks, *J. Chem. Soc., Chem. Commun.*, 1981, 257.
133. G. Jeske, H. Lauke, H. Mauermann, P. N. Swepston, H. Schumann and T. J. Marks, *J. Am. Chem. Soc.*, 1985, **107**, 8091.
134. G. Jeske, L. E. Schock, P. N. Swepston, H. Schumann and T. J. Marks, *J. Am. Chem. Soc.*, 1985, **107**, 8103.
135. C. Qian, Y. Ge, D. Deng, Y. Gu and C. Zhang, *J. Organomet. Chem.*, 1988, **344**, 175.
136. G. Jeske, H. Lauke, H. Mauermann, H. Schumann and T. J. Marks, *J. Am. Chem. Soc.*, 1985, **107**, 8111.
137. W. J. Evans, J. H. Meadows, W. E. Hunter and J. L. Atwood, *J. Am. Chem. Soc.*, 1984, **106**, 1291.
138. J. M. Brown, *Angew. Chem., Int. Ed. Engl.*, 1987, **26**, 190.
139. J. M. Brown and R. G. Naik, *J. Chem. Soc., Chem. Commun.*, 1982, 348.
140. D. A. Evans and M. M. Morrissey, *J. Am. Chem. Soc.*, 1984, **106**, 3866.
141. D. A. Evans and M. M. Morrissey, *Tetrahedron Lett.*, 1984, **25**, 4637.
142. D. A. Evans, M. M. Morrissey and R. L. Dow, *Tetrahedron Lett.*, 1985, **26**, 6005.
143. A. Villalobos and S. J. Danishefsky, *J. Org. Chem.*, 1989, **54**, 12.
144. H. W. Thompson and E. McPherson, *J. Am. Chem. Soc.*, 1974, **96**, 6232.
145. (a) J. M. Brown and S. A. Hall, *Tetrahedron Lett.*, 1984, **25**, 1393; (b) J. M. Brown and S. A. Hall, *Tetrahedron*, 1985, **41**, 4639.

146. (a) R. H. Crabtree and M. W. Davis, *Organometallics*, 1983, **2**, 681; (b) R. H. Crabtree and M. W. Davis, *J. Org. Chem.*, 1986, **51**, 2655.
147. G. Stork and D. E. Kahne, *J. Am. Chem. Soc.*, 1983, **105**, 1072.
148. E. J. Corey and T. A. Engler, *Tetrahedron Lett.*, 1984, **25**, 149.
149. C. A. Hoeger and W. H. Okamura, *J. Am. Chem. Soc.*, 1985, **107**, 268.
150. J. M. Brown and S. A. Hall, *J. Organomet. Chem.*, 1985, **285**, 333.
151. A. G. Schultz and P. J. McCloskey, *J. Org. Chem.*, 1985, **50**, 5905.
152. J. M. Brown and A. P. James, *J. Chem. Soc., Chem. Commun.*, 1987, 181.
153. K. R. Januszkiewicz and H. Alper, *Can. J. Chem.*, 1984, **62**, 1031.
154. N. A. Mohammadi and G. L. Rempel, *J. Mol. Catal.*, 1989, **50**, 259.
155. (a) B. I. Tikhomirov, I. A. Klopotova and A. I. Yakubchik, *Vysokomol. Soedin., Ser. B*, 1967, **9**, 427 (*Chem. Abstr.*, 1967, **67**, 82 418n); (b) B. I. Tikhomirov, I. A. Klopotova and A. I. Yakubchik, *Vestn. Leningr. Univ., Fiz., Khim.*, 1967, **22**, 147 (*Chem. Abstr.*, 1968, **68**, 59 020p); (c) A. I. Yakubchik, B. I. Tikhomirov, I. A. Klopotova and L. N. Mikhailova, *Dokl. Akad. Nauk SSSR*, 1965, **161**, 1365 (*Chem. Abstr.*, 1965, **63**, 4 412g).
156. J. C. Falk, *Makromol. Chem.*, 1972, **160**, 291.
157. G. F. Pregaglia, G. F. Ferrari, A. Andreetta, G. Capparella, F. Genoni and R. Ugo, *J. Organomet. Chem.*, 1974, **70**, 89.
158. M. Castiglioni, R. Giordano and E. Sappa, *J. Organomet. Chem.*, 1987, **319**, 167.
159. (a) J. Kwiatek, K. L. Mador and J. K. Seyler, *Adv. Chem. Ser.*, 1963, **37**, 201; (b) K. Tarama and T. Funabiki, *Bull. Chem. Soc. Jpn.*, 1968, **41**, 1744; (c) T. Funabiki and K. Tarama, *Bull. Chem. Soc. Jpn.*, 1971, **44**, 945.
160. T. Funabiki, M. Mohri and K. Tarama, *J. Chem. Soc., Dalton Trans.*, 1973, 1813.
161. A. F. Mabrouk, H. J. Dutton and J. C. Cowan, *J. Am. Oil Chem. Soc.*, 1964, **41**, 153.
162. A. Misono and I. Ogata, *Bull. Chem. Soc. Jpn.*, 1967, **40**, 2718.
163. (a) S. Nishimura and K. Tsuneda, *Bull. Chem. Soc. Jpn.*, 1969, **42**, 852; (b) E. F. Litvin, L. Kh. Freidlin and K. G. Karimov, *Neftekhimiya*, 1972, **12**, 318 (*Chem. Abstr.*, 1972, **77**, 100 613k); (c) J. Tsuji and H. Suzuki, *Chem. Lett.*, 1977, 1083.
164. M. Sakai, F. Harada, Y. Sakakibara and N. Uchino, *Bull. Chem. Soc. Jpn.*, 1982, **55**, 343.
165. (a) A. Nakamura and S. Otsuka, *Tetrahedron Lett.*, 1973, 4529; (b) A. Nakamura and S. Otsuka, *J. Am. Chem. Soc.*, 1973, **95**, 7262.
166. W. Strohmeier and L. Weigelt, *J. Organomet. Chem.*, 1974, **82**, 417.
167. A. Camus, C. Cocevar and G. Mestroni, *J. Organomet. Chem.*, 1972, **39**, 355.
168. F. J. Hirsekorn, M. C. Rakowski and E. L. Muetterties, *J. Am. Chem. Soc.*, 1975, **97**, 237.
169. (a) Y. Tajima and E. Kunioka, *J. Am. Oil Chem. Soc.*, 1968, **45**, 478; (b) Y. Tajima and E. Kunioka, *J. Org. Chem.*, 1968, **33**, 1689; (c) Y. Tajima and E. Kunioka, *J. Catal.*, 1968, **11**, 83.
170. M. M. Bhagwat and D. Devaprabhakara, *Tetrahedron Lett.*, 1972, 1391.
171. M. Gargano, P. Giannoccaro and M. Rossi, *J. Organomet. Chem.*, 1975, **84**, 389.
172. M. Airoldi, G. Deganello, G. Dia and G. Gennaro, *J. Organomet. Chem.*, 1980, **187**, 391.
173. G. W. Parshall, 'Homogeneous Catalysis', Wiley, New York, 1980, p. 41, and references cited therein.
174. L. W. Gosser, *Ger. Pat.*, 1 940 303 (*Chem. Abstr.*, 1970, **72**, P 100 161h).
175. G. W. Parshall, *J. Am. Chem. Soc.*, 1972, **94**, 8716.
176. (a) D. R. Fahey, *J. Org. Chem.*, 1973, **38**, 80; (b) D. R. Fahey, *J. Org. Chem.*, 1973, **38**, 3343.
177. A. D. Shebaldova, V. I. Bystrenina, V. N. Kravtsova and M. L. Khidekel, *Izv. Akad. Nauk SSSR, Ser. Khim.*, 1975, 2101 (*Chem. Abstr.*, 1976, **84**, 16 834v).
178. G. M. Schwab and G. Mandre, *J. Catal.*, 1968, **12**, 103.
179. E. N. Frankel, *J. Org. Chem.*, 1972, **37**, 1549.
180. (a) M. Cais, E. N. Frankel and A. Rejoan, *Tetrahedron Lett.*, 1968, 1919; (b) E. N. Frankel, E. Selke and C. A. Glass, *J. Am. Chem. Soc.*, 1968, **90**, 2446; (c) E. N. Frankel and R. O. Butterfield, *J. Org. Chem.*, 1969, **34**, 3930; (d) E. N. Frankel and F. L. Little, *J. Am. Oil Chem. Soc.*, 1969, **46**, 256; (e) A. Miyake and H. Kondo, *Angew. Chem., Int. Ed. Engl.*, 1968, **7**, 631; (f) R. Dabard, G. Jaouen, G. Simonneaux, M. Cais, D. H. Kohn, A. Lapid and D. Tatarsky, *J. Organomet. Chem.*, 1980, **184**, 91; (g) M. Cais, D. Fraenkel and K. Weidenbaum, *Coord. Chem. Rev.*, 1975, **16**, 27; (h) M. Cais and A. Rejoan, *Inorg. Chim. Acta*, 1970, **4**, 509; (i) G. Yagupsky and M. Cais, *Inorg. Chim. Acta*, 1975, **12**, L27; (j) M. Cais, M. Kaftory, D. H. Kohn and D. Tatarsky, *J. Organomet. Chem.*, 1979, **184**, 103; (k) J. R. Tucker and D. P. Riley, *J. Organomet. Chem.*, 1985, **279**, 49; (l) M. J. Mirbach, T. Nhu Phu and A. Saus, *J. Organomet. Chem.*, 1982, **236**, 309; (m) P. Le Maux, G. Jaouen and J. Y. Saillard, *J. Organomet. Chem.*, 1981, **212**, 193; (n) E. N. Frankel, in 'Catalysis in Organic Synthesis', ed. W. H. Jones, Academic Press, New York, 1980, p. 185; (o) E. N. Frankel, E. Selke and C. A. Glass, *J. Org. Chem.*, 1969, **34**, 3936; (p) E. N. Frankel, F. L. Thomas and J. C. Cowan, *J. Am. Oil Chem. Soc.*, 1970, **47**, 497.
181. M. A. Schroeder and M. S. Wrighton, *J. Organomet. Chem.*, 1974, **74**, C29.
182. (a) J. Nasielski, P. Kirsch and L. Wilputte-Steinert, *J. Organomet. Chem.*, 1971, **27**, C13; (b) M. S. Wrighton and M. A. Schroeder, *J. Am. Chem. Soc.*, 1973, **95**, 5764.
183. E. N. Frankel and F. L. Thomas, *J. Am. Oil Chem. Soc.*, 1972, **49**, 70.
184. (a) A. J. Birch and K. A. M. Walker, *J. Chem. Soc. C*, 1966, 1894; (b) F. H. Jardine and G. Wilkinson, *J. Chem. Soc. C*, 1967, 270.
185. I. Jardine and F. J. McQuillin, *J. Chem. Soc., Chem. Commun.*, 1969, 503.
186. M. K. Lee, I. B. Kim and C. S. Chin, *J. Organomet. Chem.*, 1985, **290**, 115.
187. S. Nishimura, T. Ichino, A. Akimoto and K. Tsuneda, *Bull. Chem. Soc. Jpn.*, 1973, **46**, 279.
188. M. Sodeoka and M. Shibasaki, *J. Org. Chem.*, 1985, **50**, 1147, 3246.
189. (a) H. Adkins and G. Krsek, *J. Am. Chem. Soc.*, 1949, **71**, 3051; (b) R. Ercoli and R. E. Torregrosa, *Chim. Ind. (Milan)*, 1958, **40**, 552 (*Chem. Abstr.*, 1959, **53**, 3 186g); (c) H. Greenfield, J. H. Wotiz and I. Wender, *J. Org. Chem.*, 1957, **22**, 542.

190. (a) L. Markó, *Proc. Chem. Soc., London*, 1962, 67; (b) C. L. Aldridge and H. B. Jonassen, *J. Am. Chem. Soc.*, 1963, **85**, 886; (c) I. Wender, R. Levine and M. Orchin, *J. Am. Chem. Soc.*, 1950, **72**, 4375; (d) W. Dawydoff, *Chem. Tech. (Berlin)*, 1959, **11**, 431.
191. (a) J. Kwiatek, I. L. Madov and J. K. Seyler, *Adv. Chem. Ser.*, 1963, **37**, 201; (b) T. Suzuki and T. Kwan, *J. Chem. Soc. Jpn.*, 1965, **86**, 713; (c) K. Tarama and T. Funabiki, *Bull. Chem. Soc. Jpn.*, 1968, **41**, 1744; (d) T. Funabiki, M. Matsumoto and K. Tarama, *Bull. Chem. Soc. Jpn.*, 1972, **45**, 2723; (e) T. Suzuki and T. Kwan, *Nippon Kagaku Zasshi*, 1965, **86**, 1341; *Chem. Abstr.*, 1966, **65**, 12 097d; (f) L. Simandi and F. Nagy, *Acta Chim. Acad. Sci. Hung.*, 1965, **46**, 137; (g) M. Murakami, K. Suzuki and J.-W. Kang, *Nippon Kagaku Zasshi*, 1962, **83**, 1226; *Chem. Abstr.*, 1963, **59**, 13 868a; (h) B. de Vries, *Proc. K. Ned. Akad. Wet., Ser. B: Phys. Sci.*, 1960, **63**, 443; *Chem. Abstr.*, 1961, **55**, 9 142i; (i) A. F. Mabrouk, E. Selke, W. K. Rohwedder and H. J. Dutton, *J. Am. Oil Chem. Soc.*, 1965, **42**, 432; (j) T. Takagi, *Nippon Kagaku Zasshi*, 1966, **87**, 600 (*Chem. Abstr.*, 1966, **65**, 15 217f).
192. W. Strohmeier and N. Iglauer, *Z. Phys. Chem. (Leipzig)*, 1966, **51**, 50.
193. (a) Y. Sasson and J. Blum, *Tetrahedron Lett.*, 1971, 2167; (b) Y. Sasson and J. Blum, *J. Org. Chem.*, 1975, **40**, 1887; (c) Y. Sasson, M. Cohen and J. Blum, *Synthesis*, 1973, 359; (d) J. Blum, Y. Sasson and S. Iflah, *Tetrahedron Lett.*, 1972, 1015.
194. M. A. Bennett, T.-N. Huang and T. W. Turney, *J. Chem. Soc., Chem. Commun.*, 1979, 312.
195. M. A. Bennett, T.-N. Huang, A. K. Smith and T. W. Turney, *J. Chem. Soc., Chem. Commun.*, 1978, 582.
196. (a) E. L. Muetterties and J. R. Bleeke, *Acc. Chem. Res.*, 1979, **12**, 324; (b) J. W. Johnson and E. L. Muetterties, *J. Am. Chem. Soc.*, 1977, **99**, 7395.
197. S. J. Lapporte and W. R. Schuett, *J. Org. Chem.*, 1963, **28**, 1947.
198. (a) W. R. Kroll, *J. Catal.*, 1969, **15**, 281; (b) S. J. Lapporte, *Ann. N.Y. Acad. Sci.*, 1969, **158**, 510; (c) K. A. Klinedinst and M. Boudart, *J. Catal.*, 1973, **28**, 322; (d) V. G. Lipovich, F. K. Shmidt and I. V. Kalechits, *Kinet. Katal.*, 1967, **8**, 939 (*Chem. Abstr.*, 1968, **68**, 59 185w).
199. N. L. Holy, *J. Org. Chem.*, 1979, **44**, 239.
200. M. J. Russell, C. White and P. M. Maitlis, *J. Chem. Soc., Chem. Commun.*, 1977, 427, and references cited therein.
201. K. R. Januszkiewicz and H. Alper, *Organometallics*, 1983, **2**, 1055.
202. (a) E. L. Muetterties and F. J. Hirsekorn, *J. Am. Chem. Soc.*, 1974, **96**, 4063; (b) L. S. Stuhl, M. Rakowski DuBois, F. J. Hirsekorn, J. R. Bleeke, A. E. Stevens and E. L. Muetterties, *J. Am. Chem. Soc.*, 1978, **100**, 2405; (c) J. R. Bleeke and E. L. Muetterties, *J. Am. Chem. Soc.*, 1981, **103**, 556.
203. (a) M. Siskin and J. Porcelli, *J. Am. Chem. Soc.*, 1974, **96**, 3640; (b) J. Wristers, *J. Am. Chem. Soc.*, 1975, **97**, 4312.
204. C. R. Landis and J. Halpern, *Organometallics*, 1983, **2**, 840.
205. (a) R. H. Fish, J. L. Tan and A. D. Thormodsen, *Organometallics*, 1985, **4**, 1743; (b) R. A. Grey, G. P. Pez and A. Wallo, *J. Am. Chem. Soc.*, 1980, **102**, 5948; (c) R. Wilczynski, W. A. Fordyce and J. Halpern, *J. Am. Chem. Soc.*, 1983, **105**, 2066; (d) W. A. Fordyce, R. Wilczynski and J. Halpern, *J. Organomet. Chem.*, 1985, **296**, 115; (e) D. E. Linn, Jr. and J. Halpern, *J. Am. Chem. Soc.*, 1987, **109**, 2969.
206. (a) P. D. Taylor and M. Orchin, *J. Org. Chem.*, 1972, **37**, 3913; (b) S. Friedman, S. Metlin, A. Svedi and I. Wender, *J. Org. Chem.*, 1959, **24**, 1287; (c) H. M. Feder and J. Halpern, *J. Am. Chem. Soc.*, 1975, **97**, 7186.
207. I. Jardine and F. J. McQuillin, *J. Chem. Soc., Chem. Commun.*, 1970, 626.
208. H. Greenfield, S. Metlin, M. Orchin and I. Wender, *J. Org. Chem.*, 1958, **23**, 1054.
209. C. P. Kubiak and R. Eisenberg, *J. Am. Chem. Soc.*, 1980, **102**, 3637.
210. K. Sonogashira and N. Hagihara, *Bull. Chem. Soc. Jpn.*, 1966, **39**, 1178.
211. B. Demerseman, P. Le Coupanec and P. H. Dixneuf, *J. Organomet. Chem.*, 1985, **287**, C35.
212. J. P. Candlin and A. R. Oldham, *Faraday Discuss. Chem. Soc.*, 1968, **46**, 60.
213. R. H. Crabtree, A. Gautier, G. Giordano and T. Khan, *J. Organomet. Chem.*, 1977, **141**, 113.
214. M. O. Albers, E. Singleton and M. M. Viney, *J. Mol. Catal.*, 1985, **30**, 213.
215. M. E. Vol'pin and I. S. Kolomnikov, *Katal. Reakts. Zhidk. Faze, Tr. Vses. Konf., 2nd*, 1966, 429 (*Chem. Abstr.*, 1968, **69**, 46 340p).
216. D. Osella, S. Aime, D. Boccardo, M. Castiglioni and L. Milone, *Inorg. Chim. Acta*, 1985, **100**, 97.
217. C. U. Pittman, Jr., R. C. Ryan, J. McGee and J. P. O'Connor, *J. Organomet. Chem.*, 1979, **178**, C43.
218. G. F. Pregaglia, A. Andreetta, G. F. Ferrari, G. Montrasi and R. Ugo, *J. Organomet Chem.*, 1971, **33**, 73.
219. C. Bianchini, A. Meli, F. Laschi, J. A. Ramirez, P. Zanello and A. Vacca, *Inorg. Chem.*, 1988, **27**, 4429.
220. I. Jardine and F. J. McQuillin, *Tetrahedron Lett.*, 1966, 4871.
221. E. F. Litvin, A. Kh. Freidlin and K. K. Karimov, *Izv. Akad. Nauk SSSR, Ser. Khim.*, 1972, 1853 (*Chem. Abstr.*, 1972, **77**, 151 382s).
222. A. Sisak, F. Ungváry and G. Kiss, *J. Mol. Catal.*, 1983, **18**, 223.
223. (a) E. L. Muetterties, *Inorg. Chim. Acta*, 1981, **50**, 1; (b) S. Slater and E. L. Muetterties, *Inorg. Chem.*, 1980, **19**, 3337; (c) E. L. Muetterties, W. R. Pretzer, M. G. Thomas, B. F. Beier, D. L. Thorn, V. W. Day and A. B. Anderson, *J. Am. Chem. Soc.*, 1978, **100**, 2090; (d) E. L. Muetterties, E. Band, A. Kokorin, W. R. Pretzer and M. G. Thomas, *Inorg. Chem.*, 1980, **19**, 1552.
224. (a) W. J. Evans, S. C. Engerer and K. M. Coleson, *J. Am. Chem. Soc.*, 1981, **103**, 6672; (b) W. J. Evans, S. C. Engerer, P. A. Piliero and A. L. Wayda, *J. Chem. Soc., Chem. Commun.*, 1979, 1007.
225. T. Yoshida, W. J. Youngs, T. Sakaeda, T. Ueda, S. Otsuka and J. A. Ibers, *J. Am. Chem. Soc.*, 1983, **105**, 6273.
226. (a) R. R. Burch, A. J. Shusterman, E. L. Muetterties, R. G. Teller and J. M. Williams, *J. Am. Chem. Soc.*, 1983, **105**, 3546; (b) R. R. Burch, E. L. Muetterties, R. G. Teller and J. M. Williams, *J. Am. Chem. Soc.*, 1982, **104**, 4257.
227. P. Abley and F. J. McQuillin, *J. Chem. Soc., Chem. Commun.*, 1969, 1503.
228. (a) H. B. Kagan, in 'Asymmetric Synthesis', ed. J. D. Morrison, Academic Press, New York, 1985, vol. 5, p. 1; (b) K. E. Koenig, in 'Asymmetric Synthesis', ed. J. D. Morrison, Academic Press, New York, 1985, vol. 5, p. 71; (c) J. W. Scott and D. Valentine, Jr., *Science*, 1974, **184**, 943; (d) J. D. Morrison, W. F. Melser and S.

Hathaway, 'Catalysis in Organic Chemistry', ed. P. N. Rylander and H. Greenfield, Academic Press, New York, 1976, p. 229; (e) V. Čaplar, G. Comisso and V. Šunjić, *Synthesis*, 1981, 85; (f) D. Valentine, Jr. and J. W. Scott, *Synthesis*, 1978, 329; (g) A. P. Kozikowski and H. F. Wetter, *Synthesis*, 1976, 561; (h) T. H. Maugh, *Science*, 1983, **221**, 351; (i) H. S. Mosher and J. D. Morrison, *Science*, 1983, **221**, 1013; (j) W. S. Knowles, W. C. Christopfel, K. E. Koenig and C. F. Hobbs in 'Catalytic Aspects of Metal Phosphine Complexes', ed. E. C. Alyea and D. W. Meek, Am. Chem. Soc., Washington DC, 1982, p. 325; (k) R. E. Merrill, *CHEMTECH*, 1981, 118; (l) W. S. Knowles, *Acc. Chem. Res.*, 1983, **16**, 106; (m) K. Drauz, A. Kleeman and J. Martens, *Angew. Chem., Int. Ed. Engl.*, 1982, **21**, 584; (n) H. B. Kagan, *Bull. Soc. Chim. Fr.*, 1988, 846; (o) H. Brunner, *Synthesis*, 1988, 645; (p) R. Noyori and M. Kitamura, in 'Modern Synthetic Methods', ed. R. Scheffold, Springer Verlag, Berlin, 1989, vol. 5. p. 115; (q) I. Ojima, N. Clos and C. Bastos, *Tetrahedron*, 1989, **45**, 6901; (r) S. L. Blyston, *Chem. Rev.*, 1989, **89**, 1663; (s) H. B. Kagan and M. Sasaki, in 'The chemistry of Organophosphorous Compounds', ed. F. R. Hartley, Wiley, New York, 1990, vol.1, chap. 3, p. 53.

229. L. Horner, H. Siegel and H. Büthe, *Angew. Chem., Int. Ed. Engl.*, 1968, **7**, 942.
230. W. S. Knowles and M. J. Sabacky, *J. Chem. Soc., Chem. Commun.*, 1968, 1445.
231. T.-P. Dang and H. B. Kagan, *J. Chem. Soc., Chem. Commun.*, 1971, 481.
232. (a) W. S. Knowles, M. J. Sabacky and B. D. Vineyard, *J. Chem. Soc., Chem. Commun.*, 1972, 10; (b). B. D. Vineyard, W. S. Knowles, M. J. Sabacky, G. L. Bachman and D. J. Weinkauff, *J. Am. Chem. Soc.*, 1977, **99**, 5946.
233. H. Brunner, *Top. Stereochem.*, 1988, **18**, 129.
234. M. D. Fryzuk and B. Bosnich, *J. Am. Chem. Soc.*, 1977, **99**, 6262.
235. M. D. Fryzuk and B. Bosnich, *J. Am. Chem. Soc.*, 1978, **100**, 5491.
236. R. B. King, J. Bakos, C. D. Hoff and L. Markó, *J. Org. Chem.*, 1979, **44**, 1729.
237. (a) D. P. Riley and R. E. Shumate, *J. Org. Chem.*, 1980, **45**, 5187; (b) J. D. Oliver and D. P. Riley, *Organometallics*, 1983, **2**, 1032.
238. H. Brunner, W. Pieronczyk, B. Schönhammer, K. Streng, I. Bernal and J. Korp, *Chem. Ber.*, 1981, **114**, 1137.
239. D. L. Allen, V. C. Gibson, M. L. H. Green, J. F. Skinner, J. Bashkin and P. D. Grebenik, *J. Chem. Soc., Chem. Commun.*, 1983, 895.
240. U. Nagel, E. Kinzel, J. Andrade and G. Prescher, *Chem. Ber.*, 1986, **119**, 3326.
241. K. Achiwa, *J. Am. Chem. Soc.*, 1976, **98**, 8265.
242. I. Ojima and N. Yoda, *Tetrahedron Lett.*, 1980, **21**, 1051.
243. J. D. Morrison, W. F. Masler and M. K. Neuberg, *Adv. Catal.*, 1976, **25**, 81.
244. (a) A. Miyashita, A. Yasuda, H. Takaya, K. Toriumi, T. Ito, T. Souchi and R. Noyori, *J. Am. Chem. Soc.*, 1980, **102**, 7932; (b) A. Miyashita, H. Takaya, T. Souchi and R. Noyori, *Tetrahedron*, 1984, **40**, 1245; (c) H. Takaya, K. Mashima, K. Koyano, M. Yagi, H. Kumobayashi, T. Taketomi, S. Akutagawa and R. Noyori, *J. Org. Chem.*, 1986, **51**, 629; (d) H. Takaya and R. Noyori, *Org. Synth.*, 1988, **67**, 20.
245. (a) T. Hayashi and M. Kumada, *Fundam. Res. Homogeneous Catal.*, 1978, **2**, 159; (b) T. Hayashi and M. Kumada, *Acc. Chem. Res.*, 1982, **15**, 395; (c) T. Hayashi, M. Konishi, M. Fukushima, T. Mise, M. Kagotani, M. Tajika and M. Kumada, *J. Am. Chem. Soc.*, 1982, **104**, 180.
246. (a) R. Selke and H. Pracejus, *J. Mol. Catal.*, 1986, **37**, 213; (b) I. Habus, Z. Raza and V. Šunjić, *J. Mol. Catal.*, 1987, **42**, 173.
247. (a) M. Fiorini and G. M. Giongo, *J. Mol. Catal.*, 1979, **5**, 303; 1980, **7**, 411.
248. (a) E. Cesarotti, A. Chiesa, G. Ciani and A. Sironi, *J. Organomet. Chem.*, 1983, **251**, 79; (b) A. Karim, A. Mortreux and F. Petit, *J. Organomet. Chem.*, 1986, **312**, 375.
249. J. W. Scott, D. D. Keith, G. Nix, Jr., D. R. Parrish, S. Remington, G. P. Roth, J. M. Townsend, D. Valentine, Jr. and R. Yang, *J. Org. Chem.*, 1981, **46**, 5086.
250. U. Nagel, E. Kinzel, J. Andrade and G. Prescher, *Chem. Ber.*, 1986, **119**, 3326.
251. U. Nagel and E. Kinzel, *Chem. Ber.*, 1986, **119**, 1731.
252. (a) J. Halpern, *Pure Appl. Chem.*, 1983, **55**, 99; (b) C. R. Landis and J. Halpern, *J. Am. Chem. Soc.*, 1987, **109**, 1746; (c) J. Halpern, in 'Asymmetric Catalysis', ed. J. D. Morrison, Academic Press, New York, 1985, vol. 5, p. 41.
253. (a) J. M. Brown and P. A. Chaloner, *J. Am. Chem. Soc.*, 1980, **102**, 3040; (b) J. M. Brown, P. A. Chaloner and G. A. Morris, *J. Chem. Soc., Chem. Commun.*, 1983, 664; (c) J. M. Brown, L. R. Canning, A. J. Downs and A. M. Forster, *J. Organomet. Chem.*, 1983, **255**, 103; (d) J. M. Brown and P. L. Evans, *Tetrahedron*, 1988, **44**, 4905.
254. (a) K. Onuma, T. Ito and A. Nakamura, *Chem. Lett.*, 1980, 481; (b) D. Meyer, J.-C. Poulin, H. B. Kagan, H. Levine-Pinto, J.-L. Morgat and P. Fromageot, *J. Org. Chem.*, 1980, **45**, 4680; (c) A. Kleeman, J. Martens, M. Samson and W. Bergstein, *Synthesis*, 1981, 740; (d) I. Ojima and M. Yatabe, *Chem. Lett.*, 1982, 1335; (e) I. Ojima and N. Yoda, *Tetrahedron Lett.*, 1982, **23**, 3913; (f) I. Ojima, N. Yoda and M. Yatabe, *Tetrahedron Lett.*, 1982, **23**, 3917; (g) I. Ojima, T. Kogure, N. Yoda, T. Suzuki, M. Yatabe and T. Tanaka, *J. Org. Chem.*, 1982, **47**, 1329; (h) H. Levine-Pinto, J.-L. Morgat, P. Fromageot, D. Meyer, J.-C. Poulin and H. B. Kagan, *Tetrahedron*, 1982, **38**, 119; (i) J.-C. Poulin and H. B. Kagan, *J. Chem. Soc., Chem. Commun.*, 1982, 1261; (j) I. Ojima, N. Yoda, M. Yatabe, T. Tanaka and T. Kogure, *Tetrahedron*, 1984, **40**, 1255; (k) T. Yamagishi, S. Ikeda, M. Yatagai, M. Yamaguchi and M. Hidai, *J. Chem. Soc., Perkin Trans. 1*, 1988, 1787.
255. R. Noyori, M. Ohta, Y. Hsiao, M. Kitamura, T. Ohta and H. Takaya, *J. Am. Chem. Soc.*, 1986, **108**, 7117.
256. T. Ohta, H. Takaya and R. Noyori, *Inorg. Chem.*, 1988, **27**, 566.
257. M. Kitamura, Y. Hsiao, R. Noyori and H. Takaya, *Tetrahedron Lett.*, 1987, **28**, 4829.
258. J. D. Morrison, R. E. Burnett, A. M. Aguiar, C. J. Morrow and C. Phillips, *J. Am. Chem. Soc.*, 1971, **93**, 1301.
259. I. Ojima and N. Yoda, *Tetrahedron Lett.*, 1980, **21**, 1051.
260. H. Brunner and W. Leitner, *Angew. Chem., Int. Ed. Engl.*, 1988, **27**, 1180.
261. T. Ohta, H. Takaya, M. Kitamura, K. Nagai and R. Noyori, *J. Org. Chem.*, 1987, **52**, 3174.
262. (a) T. Hayashi, N. Kawamura and Y. Ito, *J. Am. Chem. Soc.*, 1987, **109**, 7876; (b) T. Hayashi, N. Kawamura and Y. Ito, *Tetrahedron Lett.*, 1988, **29**, 5969.

263. U. Leutenegger, A. Madin and A. Pfaltz, *Angew. Chem., Int. Ed., Engl.*, 1989, **28**, 60.
264. T.-P. Dang, P. Aviron-Violet, Y. Colleuille and J. Varaggnat, *J. Mol. Catal.*, 1982, **16**, 51.
265. V. Massoneau, P. Le Maux and G. Simonneaux, *Tetrahedron Lett.*, 1986, **27**, 5497.
266. J. M. Brown, *Angew. Chem., Int. Ed. Engl.*, 1987, **26**, 190.
267. H. Takaya, T. Ohta, N. Sayo, H. Kumobayashi, S. Akutagawa, S. Inoue, I. Kasahara and R. Noyori, *J. Am. Chem. Soc.*, 1987, **109**, 1596, 4129.
268. J. M. Brown, *Angew. Chem., Int. Ed. Engl.*, 1987, **26**, 190, and references cited therein.
269. M. Kitamura, I. Kasahara, K. Manabe, R. Noyori and H. Takaya, *J. Org. Chem.*, 1988, **53**, 708.
270. M. Tanaka and I. Ogata, *J. Chem. Soc., Chem. Commun.*, 1975, 735; Y. Kawabata, M. Tanaka and I. Ogata, *Chem. Lett.*, 1976, 1213.

3.3
Reductions of C=C and C≡C by Noncatalytic Chemical Methods

DANIEL J. PASTO
University of Notre Dame, IN, USA

3.3.1 INTRODUCTION

This chapter is devoted to the discussion of the reduction of carbon–carbon double and triple bonds by noncatalytic methods. These methods include reductions by diimide, by dissolving metals in the presence or absence of proton donors, by low-valent metal ions, by metal hydride–metal halide combinations and by so-called 'ionic hydrogenation' procedures. Of these widely diverse methods of reduction of carbon–carbon double and triple bonds, the reduction by diimide appears to be the most versatile. The reduction of carbon–carbon double and triple bonds by diimide occurs with complete stereoselectivity and stereospecificity, and can be effected in the presence of a variety of other, very chemically reactive functional groups. The other methods of reduction of carbon–carbon double and triple bonds described in this chapter possess excellent synthetic utility in certain cases; however, their application depends critically on the structure of the unsaturated organic substrate and the reagent(s) used to effect the reduction.

3.3.2 REDUCTION OF C=C AND C≡C BY DIIMIDE

3.3.2.1 Introduction

The first observation of the reduction of a carbon–carbon double bond by 'diimide' appears to be the report of the reduction of glyceryl oleate to stearic hydrazide in the presence of hydrazine, although at the time it was not recognized that diimide was the active reducing agent.[1] (Diimide has also been referred to as diimine and diazene. As diimide is the name most commonly used in the current literature, diimide is the term used in this chapter.) Although there were several other reports of the reduction of double bonds in the presence of hydrazine, it was not until the early 1960s that the synthetic potential of the reaction was recognized when results of studies from several independent laboratories implicated diimide (HN=NH) as the actual reducing agent.[2–5] Following the proposal that diimide is the reactive intermediate in these reduction reactions, numerous experimental and theoretical studies were initiated to find other methods for the generation of diimide, and to determine the structure(s) of the reactive intermediate(s) and the mechanism of the reduction reaction.

3.3.2.2 Methods of Generation of Diimide

Many methods have been discovered for the generation of diimide in solution. However, not all of these methods are synthetically useful or practical. The most widely used procedures are described in the following paragraphs.

1. Diimide can be generated by the oxidation of hydrazine with oxygen (air) in the presence of a catalytic quantity of Cu^{II} and/or a carboxylic acid in a wide range of protic and aprotic solvents. The organic substrate is dissolved in a suitable solvent, hydrazine is added along with a catalytic amount of Cu^{II} and acetic or propionic acid. The solution is vigorously stirred, or air is bubbled through the solution.

2. Diimide can be generated by the oxidation of hydrazine with hydrogen peroxide, generally in the presence of a catalytic quantity of Cu^{II} and an added carboxylic acid.[2] The organic compound is dissolved in a suitable solvent, hydrazine is added along with a catalytic amount of Cu^{II} and acetic or propionic acid and 30% hydrogen peroxide is slowly added dropwise.

3. Diimide is generated by the acid-catalyzed hydrolysis of the dipotassium or disodium salt of azodiformate ($K^+\ ^-O_2CN{=}NCO_2^-\ K^+$) in protic or aprotic solvents.[4] The azodiformate is dissolved or suspended in the organic solvent containing the compound to be reduced and acetic acid is slowly added.

4. Diimide is also formed in the thermal and base-catalyzed decomposition of benzene-, 4-methylbenzene-,[4] 4-nitrobenzene-,[6] and 2,4,6-triisopropylbenzene-sulfonyl hydrazide.[7]

The protic solvents most widely used are methanol, ethanol, ethoxyethanol and methoxyethanol in which most organic molecules are soluble. Aprotic solvents that have found extensive use are diglyme, dimethoxyethane, tetrahydrofuran, methylene chloride and chloroform.

3.3.2.3 Mechanism of Reduction by Diimide

There are three isomeric structures for diimide, N_2H_2; *cis*- and *trans*-diimide (**1** and **2**), and 1,1-diimide (aminonitrene; **3**). 'Diimide' has been generated and trapped at low temperatures from the gas-phase electric discharge of hydrazine,[8] and by the thermal decomposition of metal salts of *p*-toluenesulfonyl hydrazide.[9] A careful analysis of the available spectroscopic data has indicated the formation and presence only of (**2**).[10] 1,1-Diimide (**3**) has been recently generated by the low temperature photochemi-

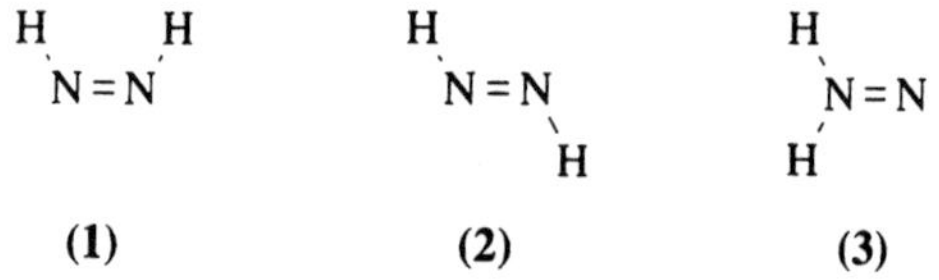

H H
N=N

(**1**)

H
N=N
H

(**2**)

H
N=N
H

(**3**)

cal decomposition of carbamoyl azide.[11] As theoretical studies have suggested, *cis*-diimide must be formed as a reactive intermediate in the reduction systems, although it has not yet been physically characterized.

The N_2H_2 system has been subjected to several theoretical studies at many different levels of theory. To summarize, *trans*-diimide is calculated to be lowest in energy, with *cis*-diimide being higher in energy by 4.7–7.3 kcal mol^{-1} (1 kcal mol^{-1} = 4.18 kJ mol^{-1}) and 1,1-diimide being 24.5–27.4 kcal mol^{-1} higher in energy. The results of the theoretical calculations also indicate that the *trans*-to-*cis* inversion and rotation barriers are very high (over 40 kcal mol^{-1}).

Stereochemical studies on the reduction of C=C and C≡C by diimide have shown that the transfer of hydrogens from 'diimide' occurs in a completely *syn* manner.[2] The reduction of (**4**) and (**5**) with dideuteriodiimide, generated by the deuterolysis of dipotassium azodiformate, resulted in the formation of the *meso*- and (±)-reduction products (**6**) and (**7**) in at least 97% stereospecificity (the lower limit of detectability with IR spectral analysis).[2] The reduction of diphenylacetylene (**8**) produces only *cis*-stilbene (**9**) as an intermediate reduction product.[2] It was considered that the reduction of multiple bonds by diimide occurred as a synchronous transport of a pair of hydrogens to a single face of the π-system *via* a transition state represented as (**10**).[2]

(**4**) + N_2D_2 → (**6**) (1)

R = CO_2H, CH_2OH, Ph

(**5**) + N_2D_2 → (**7**) (2)

R = CO_2H, CH_2OH, Ph

Ph—≡—Ph (**8**) + N_2H_2 → (**9**) (3)

HN=NH + $H_2C=CH_2$ → [(**10**)] → N≡N + CH_3CH_3 (4)

Theoretical calculations of the energy surfaces for the concerted transfer of hydrogen from (**1**), (**2**) and (**3**) to ethylene give calculated energy barriers of 26.7, 45.3 and 45.8 kcal mol^{-1}, respectively.[12] Although *cis* diimide has not been observed, the results of the theoretical studies indicate that *cis*-diimide undergoes hydrogen atom transfer to ethylene with a considerably lower energy barrier than does *trans*- or 1,1-diimide.[12] Since most of the procedures used to generate diimide in solution probably produce *trans*-diimide, isomerization of *trans*- to *cis*-diimide must occur, most probably involving a rapid protonation–deprotonation sequence catalyzed by the added carboxylic acid. In this chapter the term 'diimide' will imply *cis*-diimide as the reducing agent.

During diimide reductions considerable nitrogen gas is formed by the disproportionation of diimide. Theoretical calculations on the energy surfaces for the transfer of hydrogen from *cis*-diimide to *cis*- and *trans*-diimide give energy barriers of 19.3 and 23.8 kcal mol^{-1}, respectively, considerably smaller than that for the transfer of hydrogen from (**1**) to ethylene.[13] From a practical point of view, the occurrence of the disproportionation of diimide requires the use of a considerable excess of the diimide precursor, and places a lower limit of reactivity for diimide reduction of a double or triple bond if it is to be observed.

3.3.2.3.1 Relative reactivities

Several studies have been carried out to determine the relative reactivities of substituted alkenes and dienes toward reduction by diimide. Garbisch *et al.* have determined the relative reactivities of a large number of substituted alkenes toward reduction by diimide generated by the triethylamine-induced decomposition of *p*-toluenesulfonyl hydrazide in diglyme at 80 °C.[14] Representative results appear in Table 1. Increasing alkyl substitution on the double bond results in decreased reactivity. *Trans* double bonds are, in general, more reactive than *cis* double bonds, and strained double bonds are more reactive than nonstrained double bonds. It was concluded that the major factors contributing to the observed relative reactivity differences are torsional strain, bond angle bending strain and α-alkyl substituent effects.[14] (A more detailed description of these interactions and examples will be cited in Section 3.3.2.4.) An empirical correlation was developed to calculate the relative rates of the reduction of substituted alkenes with diimide.[14]

Table 1 Relative Reactivities of Substituted Alkenes Toward Reduction with Diimide[14]

Substrate	k_{rel}[a]	*Substrate*	k_{rel}[a]
1-Pentene	20.2	1-Methylcyclohexene	0.11
trans-2-Pentene	2.59	1,2-Dimethylcyclohexene	0.012
cis-2-Pentene	2.65	Cycloheptene	17.0
2-Methyl-1-pentene	2.04	Cyclooctene	17.0
2-Methyl-2-butene	0.28	Cyclononene	5.7
2,3-Dimethyl-2-butene	0.50	Cyclodecene	0.85
Cyclopentene	15.5	Bicyclo[2.2.1]heptene	450
Cyclohexene	1.00	Bicyclo[2.2.2]octene	29

[a]k_{rel} relative to cyclohexene.

A similar trend in relative reactivities has been observed in the reductions of unsaturated acids (Table 2).[15] The presence of strongly electron-withdrawing groups, such as the carboxylic acid function, enhances reactivity toward reduction of a double or triple bond by diimide.

Table 2 Relative Reactivities of Unsaturated Acids Toward Reduction with Diimide[15]

Acid	k_{rel}[a]	*Acid*	k_{rel}[a]
Fumaric	100	*trans*-Cinnamic	10
Maleic	10	*cis*-Cinnamic	3
Methylfumaric	3	*trans*-Methylcinnamic	1.4
Methylmaleic	0.7	*cis*-Methylcinnamic	1.4

[a]k_{rel} relative to fumaric acid.

Conjugated dienes are more reactive than monoenes (Table 3).[16] The selectivity between double bonds of an unsymmetrically substituted diene follows the substitution trend noted for monoenes, *i.e.* the less substituted double bond preferentially undergoes reduction.

Table 3 Relative Reactivities of Conjugated Dienes Toward Reduction with Diimide[16]

Diene	k_{rel}[a]	*Diene*	k_{rel}[a]
1,3-Cyclohexadiene	47	2,3-Dimethyl-1,3-butadiene	3.1
2-Methyl-1,3-butadiene	13.6[b]	2,5-Dimethyl-2,4-hexadiene	0.5

[a]k_{rel} relative to rate of reduction of cyclohexene. [b]Relative rate of reduction of the least substituted double bond.

3.3.2.3.2 Stereoselectivity of diimide reductions

The results of many stereochemical studies have indicated that diimide approaches from the less sterically hindered side of the double bond, *i.e.* steric approach control dominates the stereoselectivity of the diimide reduction reaction. This is consistent with the rather exothermic nature of the reaction,[12] which suggests that the transition state for the reaction occurs early along the reaction coordinate, in which case ground state electronic and steric effects will control the stereoselectivity of the reaction. However, when the approach to the faces of the double bond are similar in steric hindrance, alkene conformational and torsional effects play dominant roles in determining the stereoselectivity of the reduction process.[14] Examples showing the effect of steric hindrance on the approach of diimide to the two faces of the double bond on the stereoselectivity of the reduction of a double bond are illustrated in equations (5)–(9). In equation (5), the reduction of norbornadiene occurs exclusively from the *exo* direction to produce (**12**), which undergoes further reduction exclusively from the less sterically hindered *exo* direction to produce (**13**).[17] 3,4-Dimethylcyclobutene (**14**) undergoes reduction on the face of the ring opposite the two methyl groups (equation 6).[18] Reduction of the bicyclo[2.2.0]hexadiene (**16**) occurs exclusively from the *exo* direction (equation 7).[19] The reduction of α-pinene (**18**) occurs highly stereoselectively on the face of the double bond on the same side as the bridging methylene (equation 8).[20] And finally, the steroid derivative (**21**) undergoes reduction from the less hindered α side (equation 9).[21]

(**11**) + N_2D_2 → (**12**) 44% + (**13**) 26% (5)

(**14**) + N_2D_2 → (**15**) (6)

(**16**) + N_2H_2 → (**17**) (7)

(**18**) + N_2H_2 → (**19**) 99% + (**20**) 1% (8)

(**21**) + N_2H_2 → (**22**) 85–95% (9)

Interesting exceptions have been reported in which the diimide approaches the most sterically hindered double bond. The reduction of 7-hydroxy-, 7-acetoxy- and 7-*t*-butoxy-norbornadiene (**23a–c**) occur in a highly *syn,exo* manner (equation 10).[17] Two possible explanations have been forwarded to account for these results.[17] The first involves an electronic interaction of the functional groups with the *syn* double bond, which enhances its reactivity toward reduction with diimide. The second involves hydrogen bond formation between diimide and the oxygen-containing functions, which directs the diimide reduction of the *syn* double bond.

(10)

(23) **a:** R = H
b: R = COMe
c: R = OBu^t

(24) **a:** 8%
b: 10%
c: 0–3%

(25) **a:** 27%
b: 50%
c: 63–65%

In the reduction of alkylidenecyclohexanes the approach to the faces of the double bond are similarly sterically hindered, and roughly equal amounts of products derived from attack at the two faces of the double bond should be formed. This is the case with (**26a**). Replacement of the vinyl hydrogens with methyl groups should not greatly affect the approach of diimide to either face of the double bond. However, such substitution results in increases favoring formation of the *trans* isomer (**28**). This trend has been interpreted in terms of subtle changes in the conformations of the alkenes, which affect the ease of approach of diimide to the two faces of the double bonds.[14,23]

(11)

(26) **a:** $R^1 = R^2 = H$
b: $R^1 = Me$, $R^2 = H$
c: $R^1 = R^2 = Me$

(27) **a:** 49%
b: 46%
c: 30%

(28) **a:** 51%
b: 54%
c: 70%

The reduction of 1,5-dimethylcyclopentene (**29a**) and 1,6-dimethylcyclohexene (**29b**) results in the preferential formation of the *trans*-dimethyl products (**31**; equation 12).[22] In these cases the approach of the diimide occurs from the direction which will allow the methyl groups to move away from each other, *i.e.* in the transition state for hydrogen transfer from diimide the direction of the approach of diimide is dictated by torsional effects generated between the methyl groups.[23]

(12)

(29) **a:** $n = 1$
b: $n = 2$

(30) **a:** 31%
b: 29%

(31) **a:** 69%
b: 71%

3.3.2.4 Scope and Limitations

The reduction of carbon–carbon double and triple bonds with diimide can be carried out under exceptionally mild conditions, which can be selected to tolerate the presence of a number of reactive functional groups that would either be reduced or would suffer hydrogenolysis under catalytic hydrogenation conditions. Diimide reductions can be carried out in the presence of reactive functional groups including allylic halides,[23,24] esters,[23] amines[23] and disulfides.[25] Vinyl halides and ethers undergo reduction, but they react only very slowly.[23,26] Unsaturated ketones are reduced to saturated ketones by diimide generated by the hydrolysis of dipotassium azodiformate.[23,27] (The use of hydrazine as the diimide precursor results in the formation of azines with aldehydes and ketones.) Other functions that are rather sensitive toward reduction by other reducing agents, but are not reduced by diimide, are N—O[28] and O—N—N[29] containing systems. The reduction of double bonds in very thermally labile systems can be accomplished at –70 °C in methylene chloride by diimide generated by the acidolysis of dipotassium azadiformate with acetic acid.[30] The reaction shown in equation (13) illustrates the use of this procedure to reduce the highly reactive endoperoxide (**32**).[30a]

$$K^+ \, {}^-O_2CN{=}NCO_2^- \, K^+ \quad -70\,°C,\ CHCl_3,\ MeCO_2H \qquad (13)$$

(**32**) → (**33**)

Substituted allenes readily undergo reduction with diimide to produce alkenes, which undergo further slower reduction to alkanes.[31,32] The diimide preferentially approaches the allene chromophore from the less hindered side of the less substituted double bond to produce the alkene having the *cis* stereochemistry as is illustrated by the reductions of allenes (**34**) and (**37**) in equations (14)[32] and (15).[33] The reactivity of the allene chromophore towards reduction by diimide appears to be only slightly greater than that of the product alkenes; however, extensive overreduction occurs in the presence of the highly electronegative carboxylic acid functional group, as illustrated in equation (15).[33] Increasing the degree of substitution on the allene chromophore increases the reactivity of the allene to reduction. This effect must be due to the long-range electronic effect of functional groups attached to the remote end of the allene chromophore on the energy of the π MOs of the bond undergoing reduction.[34] An interesting example is the reduction of allene (**41**) involving the more highly strained exocyclic double bond from the less sterically hindered direction opposite the phenyl group as illustrated in equation (16).[35]

(**34**) + N_2H_2 → (**35**) 32% + (**36**) 1% (14)

(**37**) CO_2H + N_2H_2 → (**38**) 20% CO_2H + (**39**) 6% CO_2H + (**40**) 31% CO_2H (15)

(**41**) Ph + N_2H_2 → (**42**) Ph, H (16)

Alkynes undergo reduction to produce *cis*-alkenes, which in turn undergo further reduction to the alkane. With alkyl-substituted alkynes, the reactivity of the triple bond is comparable with that of the double bond of the product alkenes, and extensive overreduction occurs. With 1-iodoalkynes, however, the reduced reactivity of the product *cis*-1-iodoalkenes toward further reduction allows for the isolation of the *cis*-1-iodoalkenes in good yields. An example is the reduction of the iodoalkyne (**43**) shown in equation (17).[36]

(**43**) I, OTHP + N_2H_2 → (**44**) OTHP, I (17)

Highly regio- and stereo-selective deuterium and tritium labeling can be readily accomplished by diimide reduction in aprotic solvents using dipotassium azodiformate and labeled acetic acid, or by the hydrolysis of azodiformate salts in labeled protic solvents.

A limitation to the use of diimide as a reducing agent appears to be the relative rate at which diimide reacts with the unsaturated substrate. If the rate of reduction is sufficiently slower than that of the disproportionation of diimide, the latter reaction will dominate and no reduction will be accomplished. A further consideration is the reactivity of any other functional groups present in the substrate toward reaction with hydrazine, which may be used as the source of the diimide, or, if not, formed by the disproportionation of the diimide generated from other sources.

Although diimide was initially thought to be useful only for the reduction of symmetical double and triple bonds (*i.e.* C=C, C≡C, N=N, *etc.*), it was subsequently demonstrated that aromatic aldehydes can be reduced to the corresponding benzyl alcohols in excellent yield by diimide generated by the hydrolysis of dipotassium azodiformate.[37] Aromatic ketones are reduced more slowly, while aliphatic aldehydes and ketones are reduced even more slowly yet.[23] Imines of aromatic and aliphatic aldehydes and ketones undergo reduction in good yield with diimide generated from hydrazine.[38]

3.3.3 REDUCTION OF C=C AND C≡C BY DISSOLVING METALS

3.3.3.1 Introduction

The reduction of a carbon–carbon multiple bond by the use of a 'dissolving' metal was first accomplished by Campbell and Eby in 1941.[39] The reduction of disubstituted alkynes to *cis*-alkenes by catalytic hydrogenation, for example by the use of Raney nickel, provided an excellent method for the preparation of isomerically pure *cis*-alkenes. At the time, however, there were no practical synthetic methods for the preparation of pure *trans*-alkenes. All of the previously existing procedures for the formation of an alkene resulted in the formation of mixtures of the *cis*- and *trans*-alkenes, which were extremely difficult to separate with the techniques existing at that time (basically fractional distillation) into the pure components. Campbell and Eby discovered that dialkylacetylenes could be reduced to pure *trans*-alkenes with sodium in liquid ammonia in good yields and in remarkable states of isomeric purity.[39] Since that time several metal/solvent systems have been found useful for the reduction of C=C and C≡C bonds in alkenes and alkynes, including lithium/alkylamine,[40] calcium/alkylamine,[41] sodium/HMPA in the absence or presence of a proton donor,[42,43] activated zinc in the presence of a proton donor (an alcohol),[44,45] and ytterbium in liquid ammonia.[46] Although most of these reductions involve the reduction of an alkyne to an alkene, several very synthetically useful reactions involve the reduction of α,β-unsaturated ketones to saturated ketones.[47]

3.3.3.2 Mechanisms of Dissolving Metal Reductions

The mechanism of dissolving metal reductions depends on the nature of the solvent and the nature of the substrate. The proposed mechanism for the reduction of dialkylacetylenes by sodium in HMPA in the presence of a proton donor is illustrated in equation (18).[42] The addition of an electron to the triple bond of (**45**) is proposed to produce the *trans*-sodiovinyl radical (**46**), or the corresponding radical anion (**47**), which undergoes protonation by the added alcohol to produce the radical (**48**). Further reduction of (**48**) by sodium produces the *trans*-sodiovinyl compound (**49**), which on protonation produces the *trans*-alkene (**50**). In the absence of a proton donor, the reduction of (**45**) with sodium in HMPA results in the formation of a mixture of *cis*- and *trans*-2- and 3-hexenes.[42] Control studies showed that the isomerization products 2- and 3-hexene are not formed by rearrangement of the *cis*- or *trans*-3-hexenes. It was concluded that the starting alkyne (**45**) acts as a reversible proton donor reacting with an intermediate anion or radical anion to produce the delocalized anion (**51**) which is then protonated to produce the allene (**52**). Reduction of the allene (**52**), or further rearrangement to the alkyne (**53**) followed by reduction, then leads to the formation of the mixture of the *cis*- and *trans*-2- and 3-hexenes (equation 19).[42]

(**45**) —e⁻, Na⁺→ (**46**) or (**47**) —Bu^tOH→ (**48**) —Na→ (**49**) —Bu^tOH→ (**50**) (18)

Although the mechanisms of the reductions of substituted alkynes by the other dissolving metal/solvent combinations have not been studied in such detail, it would appear that similar mechanisms are operative.

$-H^+$ (19)

(45) **(51)** **(52)** **(53)**

The mechanism and stereochemistry of the lithium in ammonia reduction of α,β-unsaturated ketones has been studied by Stork and coworkers.[47] The reduction is proposed to proceed by formation of a metallated oxyallyl anion (**55**; equation 20), which undergoes protonation at the β-carbon atom.[47a] The stereochemistry of the protonation step was probed by the use of substituted octalones.[47b] The results supported the suggestion that the reduction of an octalone system with lithium in ammonia will produce the product of greatest stability (*cis* or *trans*) having the newly introduced hydrogen axial to the ketone-containing ring. This is required by the necessity of maintaining continual overlap of the π-orbital with the enolate double bond as shown in intermediate (**57**) in equation (21).[47b]

$$\underset{\textbf{(54)}}{C{=}C{-}C{=}O} \xrightarrow[NH_3]{Na} \underset{\textbf{(55)}}{C{\cdots}C{\cdots}C{-}OM} \tag{20}$$

OMe; Li; NH_3, ROH; MO; OMe; H; OMe; O; H (21)

(56) **(57)** **(58)**

The reduction of α,β-unsaturated ketones with alkali metals in HMPA in the absence of a proton donor has been proposed to proceed *via* the mechanism shown in equation (22), in which a dianion is formed.[48] Protonation during work-up results in the formation of the final reduced product (**63**).

$$\underset{\textbf{(59)}}{C{=}C{-}C{=}O} \xrightarrow{e^-} \underset{\textbf{(60)}}{\dot{C}{-}C{=}C{-}O^-} \xrightarrow{e^-} \underset{\textbf{(61)}}{\bar{C}{-}C{=}C{-}O^-} \xrightarrow{2H^+}$$

$$\underset{\textbf{(62)}}{CH{-}C{=}C{-}OH} \longrightarrow \underset{\textbf{(63)}}{CH{-}CH{-}C{=}O} \tag{22}$$

3.3.3.3 Scope and Limitations

In the initial studies by Campbell and Eby it was noted that 3- and 4-octyne, 3-hexyne and 5-decyne could be efficiently reduced to the corresponding *trans*-alkenes in good yields and with remarkably high stereoselectivity.[39] Shortly thereafter, Henne and Greenlee reported the quantitative reduction of 1-alkynes to 1-alkenes using sodium in ammonia in the presence of ammonium ion.[49] In the absence of ammonium ion, however, extensive metallation of the 1-alkyne occurs.[49] In the presence of ammonium ion dialkylacetylenes are inefficiently reduced (extensive hydrogen evolution occurs, in which sodium is consumed).[49]

Benkeser *et al.* have reported that the reduction of 3- and 5-octyne with lithium in ethylamine at –78 °C produces the corresponding *trans*-alkenes in good yield.[40] At 17 °C, however, overreduction of the alkene can occur if excess lithium is present.[40]

Although most of the reductions of carbon–carbon π-systems have involved the reduction of triple to double bonds, Whitesides and Ehmann have described the reduction of alkenes to alkanes using sodium

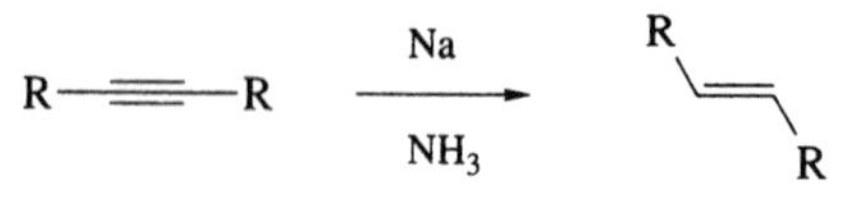

$$\text{R—≡—R} \xrightarrow[\text{NH}_3]{\text{Na}} \text{(E)-RCH=CHR} \tag{23}$$

(64) R = Et, Pr^n, Bu^n **(65)**

in HMPA in the presence of *t*-butyl alcohol.[43] These reductions appear to be fairly stereoselective and produce the alkane probably arising from the more thermodynamically stable conformation of the intermediate anion, as illustrated in the reductions of **(66)** and **(69)** shown in equations (24) and (25).[43]

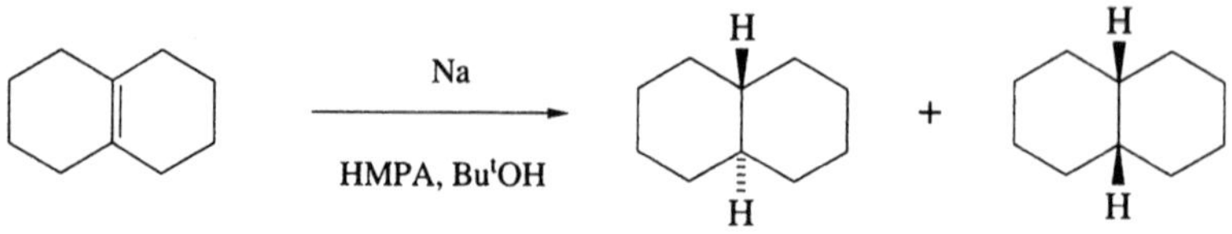

$$\textbf{(66)} \xrightarrow[\text{HMPA, Bu}^t\text{OH}]{\text{Na}} \textbf{(67)}\ 91\% + \textbf{(68)}\ 3\% \tag{24}$$

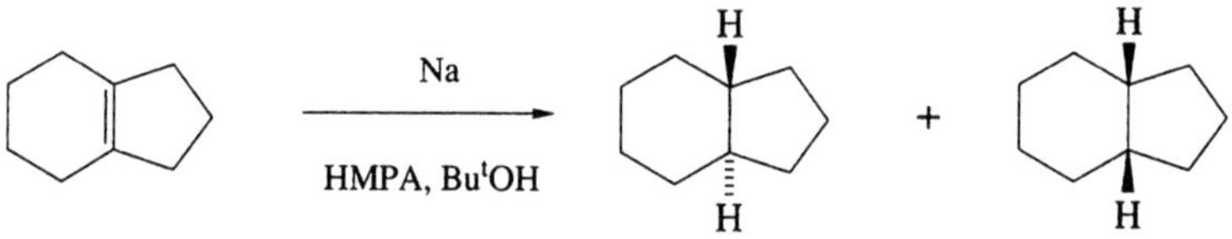

$$\textbf{(69)} \xrightarrow[\text{HMPA, Bu}^t\text{OH}]{\text{Na}} \textbf{(70)}\ 73\% + \textbf{(71)}\ 27\% \tag{25}$$

The reduction of disubstituted acetylenes with calcium in mixed solvent systems or methylamine, ethylenediamine or amine–mixed solvent systems produces *trans*-alkenes in 70–88% yields; however, extensive migration (9–28%) of the π-bond occurs, which is attributed to isomerization of the alkyne as illustrated in equation (19).[41]

Zinc, either activated as a zinc/copper couple or by reaction with 1,2-dibromoethane, in the presence of an alcohol solvent effectively reduces alkynes to alkenes. Terminal alkynes are reduced essentially quantitatively to 1-alkenes.[44] Disubstituted acetylenes, interestingly, are reduced to the corresponding *cis*-alkenes, one example being illustrated in equation (26).[44] Zinc, activated by either $LiCuBr_2$ or dibromoethane, reduces conjugated diynes to conjugated (*Z*)-eneynes.[45] The triple bond nearest to a heteroatom functional group is preferentially reduced (equation 27).[45] The triple bonds of eneynes are also preferentially reduced (equation 28).[45] Finally, the disubstituted alkyne **(78)** is smoothly reduced to the

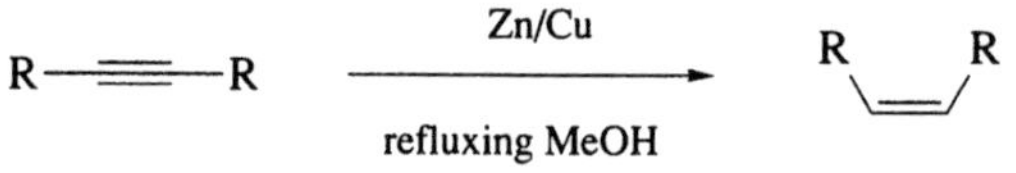

$$\text{R—≡—R} \xrightarrow[\text{refluxing MeOH}]{\text{Zn/Cu}} \text{(Z)-RCH=CHR} \tag{26}$$

(72) R = Ph, CH_2OH **(73)**

$$\textbf{(74)} \xrightarrow[\text{refluxing EtOH}]{\text{Zn/BrCH}_2\text{CH}_2\text{Br}} cis\text{-}\textbf{(75)} \tag{27}$$

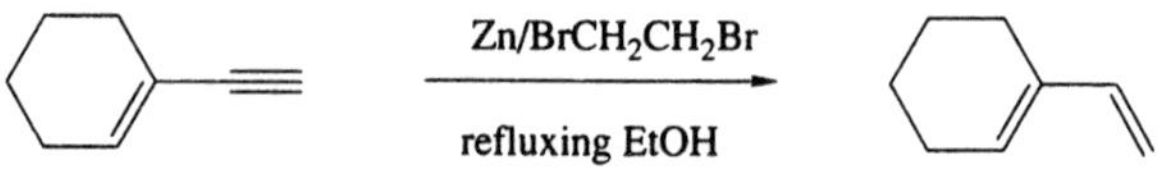

(74): diyne with $CH(OEt)_2$; cis-(75): EtO, OEt

$$\textbf{(76)} \xrightarrow[\text{refluxing EtOH}]{\text{Zn/BrCH}_2\text{CH}_2\text{Br}} \textbf{(77)} \tag{28}$$

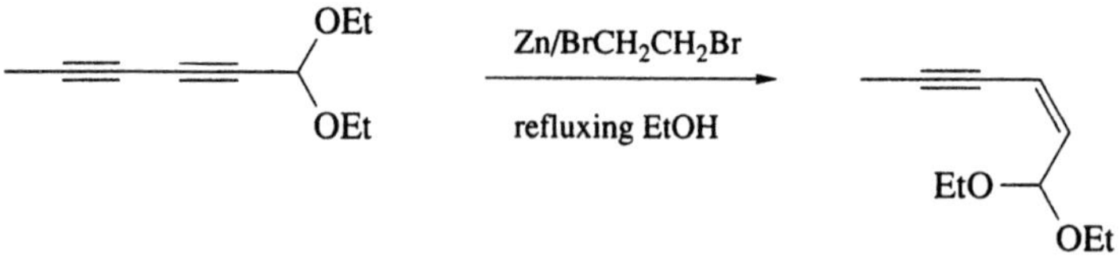

EtO—C≡C—OEt (78) ⟶ EtO—CH=CH—OEt (79) (29)

(*Z*)-alkene (**79**) without loss of the alkoxy functional groups (equation 29).[45] The mechanism of the zinc reduction of alkynes does not appear to have been studied in detail and the factors leading to the observed chemoselectivity and stereoselectivity are not readily apparent.

The reduction of disubstituted acetylenes with ytterbium in liquid ammonia also produces *trans*-alkenes in good yields.[46] This reducing system does reduce some double bonds, such as the strained double bond in norbornadiene; however, in general, carbon–carbon double bonds do not undergo reduction with this reducing system.[46] The expense of powdered metallic ytterbium does not make this a very practical reducing agent for synthetic purposes.

α,β-Unsaturated ketones undergo reduction with lithium in ammonia in the presence of an alcohol, as exemplified in equation (21).[47] A number of acyclic and cyclic α,β-unsaturated ketones have been reduced by alkali metals in anhydrous HMPA (*via* the dianion), or HMPA-containing nonprotic solvents, to the corresponding saturated ketones in 45–95% yields (equations 30 and 31).[48] The formation of the intermediate dianion apparently prevents further reduction of the carbonyl group.

(**80**) —Li, HMPA, ether→ (**81**) 55% (30)

(**82**) —K, HMPA→ (**83**) 72% + (**84**) 18% (31)

In summary, dissolving metal reductions are useful for the reduction of conjugated and nonconjugated alkynes, and the carbon–carbon double bonds of α,β-unsaturated ketones. The reduction of conjugated dienes appears not to have been extensively reported, and would probably give rise to mixtures of isomers. The reduction of conjugated dienes is a source of by-products in Birch reduction, when the intermediate anion is protonated at the terminus of the pentadienye system (see Chapters 3.4 and 3.5). Only one report on the reduction of nonconjugated alkenes has appeared. Although the results of studies on the attempted reductions of functionally substituted alkenes other than α,β-unsaturated ketones have not been reported, it should be obvious that this reduction procedure cannot be applied to systems that contain functional groups that will react with the dissolving metal, or the highly basic conditions that are generated during the course of the reaction.

3.3.4 REDUCTION OF C=C AND C≡C BY LOW-VALENT METAL SPECIES

3.3.4.1 Introduction

The reduction of alkynes[50] and electronegatively substituted alkenes[51] by Cr^{II} in aqueous DMF was first reported by Castro and coworkers in 1964, who also carried out a detailed study of the mechanism of the reactions. It does not appear that reductions by other low-valent metal ions have been studied, or if so they are not capable of reducing carbon–carbon double and triple bonds.

3.3.4.2 Mechanism of the Reduction of C═C and C≡C by Cr^{II}

Alkynes are reduced to *trans*-alkenes in excellent yields by $Cr^{II}SO_4$ in aqueous DMF at room temperature (equation 32).[50] Terminal alkynes are reduced to terminal alkenes in excellent yield (equation 33).[50] The reduction of alkynes is second order in Cr^{II}. The proposed mechanism for the reduction involves rate-determining attack by Cr^{II} on a 1:1 alkyne–Cr^{II} complex as is illustrated in equation (34).[50] In order to account for the observed stereochemistry, the reduction is proposed to occur *via* the transition state structure (**91**).

R^1—≡—R^2 (**85**) —[$CrSO_4$, DMF/H_2O]→ (**86**) >85% (32)

Ph—≡ (**87**) —[$CrSO_4$, DMF/H_2O]→ (**88**) (33)

R—≡—R (**89**) + $Cr^{II}(H_2O)_n$ ⟶ R—≡—R / $Cr^{II}(H_2O)_n$ (**90**) + $Cr^{II}(H_2O)_m$ ⟶ [(**91**)]‡ ⟶ (34)

Simple alkenes do not undergo reduction by Cr^{II}.[51] However, electronegatively substituted alkenes, or apparently those bearing functional groups that are capable of forming a complex with Cr^{II}, are readily reduced by $Cr^{II}SO_4$ in good yields in aqueous DMF at room temperature.[51] As in the reduction of the substituted alkynes, this reaction has been shown to be second order in Cr^{II} and first order in alkene. The reduction of alkenes, in general, shows little stereoselectivity or stereospecificity as is illustrated in equations (35) and (36), but in one case, equation (37), an unexplained high level of selectivity was observed. A mechanism was proposed for the reduction of electronegatively substituted alkenes involving com-

(**92**) ⟶ (**93**) 85%; *meso*:(±) = 48:52 (35)

(**94**) ⟶ (**95**) 84%; *meso*:(±) = 50:50 (36)

plexation of the substrate and reduction to an enol complex of general structure (**98**).[51] The presence of a carboxy, ester or nitrile function appears to be necessary to induce the reduction of carbon–carbon double bonds.

$$\text{MeO}_2\text{C}-\text{C(Me)}=\text{C(Me)}-\text{CO}_2\text{Me}\ (\mathbf{96}) \xrightarrow[\text{H}_2\text{O, DMF}]{\text{Cr}^{\text{II}}} \text{MeO}_2\text{C}-\text{CH(Me)}-\text{CH(Me)}-\text{CO}_2\text{Me}\ (\mathbf{97})\ 100\%;\ meso:(\pm) = 100:0 \qquad (37)$$

(**98**)

3.3.4.3 Scope and Limitations

The reduction of a carbon–carbon triple bond to a *trans*-alkene is a synthetically useful reaction. However, the synthetic utility of this reaction does not appear to have been evaluated, in particular with respect to the compatibility of other functional groups that might be present in the substrate molecule with the highly reducing capability of the Cr^{II}.

The reduction of carbon–carbon double bonds requires the presence of electronegative functional groups capable of forming a complex with Cr^{II}. Again, the problem of the compatibility of other functional groups with Cr^{II} must be condsidered. The essentially total lack of stereospecificity in the reduction of such substituted alkenes also severely limits the synthetic applicability of this reduction reaction.

3.3.5 REDUCTION OF C=C AND C≡C BY METAL HYDRIDE–TRANSITION METAL HALIDE COMBINATIONS

3.3.5.1 Introduction

Alkynes and alkenes have been found to undergo reduction to alkenes and alkanes with $LiAlH_4$ in the presence of transition metal-(II) and -(III) halides from titanium to nickel.[52,53] The extent of reduction depends on the nature of the transition metal and the reaction conditions.[52] Copper and zinc halides do not effect reduction with $LiAlH_4$.[52,53] $LiAlH_4$ in the presence of Ti^{IV} halides[54] and Zr^{IV} halides[55] also effectively reduces alkenes to alkanes. $NaBH_4$ appears to less effective, only reducing alkynes and alkenes in the presence of Co^{II}.[56] NaH in the presence of Ni^{II} acetate[57] and MgH_2 in the presence of Cu^{I} iodide[58] reduce alkynes peferentially to *cis*-alkenes, and alkenes to alkanes.

3.3.5.2 Mechanism of Metal Hydride–Transition Metal Halide Reductions

All of the proposed mechanisms for the reduction of alkynes with metal hydride–transition metal halide combinations involve an initial hydrometallation of the π-system by the transition metal hydride, formed by the reaction of the original metal hydride with the transition metal halide, to form the vinylmetallic intermediate (**99**; equation 38). For the reduction of alkenes, similar alkylmetallic intermediates are implied to be formed. In the case of the reduction of alkenes with $NaBH_4$ in the presence of Co^{II} in alcohol solution, the hydrometallation reaction appears to be reversible as evidenced by the incorporation of an excess of deuterium when $NaBD_4$ was used in the reduction.[56]

$$R-\!\equiv\!-R \;+\; H_nMhal_{m-n} \longrightarrow \text{(99)} \xrightarrow{H_2O} \text{cis-RCH=CHR}$$

(99): R(H)C=C(R)–M–hal_{m-n} with H_{n-1} on M

(99) ↓ cis-vinyl radical ⇌ trans-vinyl radical; each $\xrightarrow{MH}$ cis-alkene and trans-alkene respectively (38)

The mechanisms vary, however, in how the intermediate organotransition metallic intermediates are converted to the reduced products. In most cases, the reduction of alkynes results in the formation of predominantly the *cis*-alkenes; however, significant quantities of the *trans*-alkenes are also formed.[52,53,58] Also, in these cases there is little deuterium incorporation when the reaction mixture is worked up with deuterium oxide. In these cases a homolytic dissociation of the C—M bond is proposed to occur, the resulting vinyl radical then abstracting a hydrogen atom from a metal hydride species.[52] The stereochemistry of this process is consistent with the generally observed stereochemistry in such radical processes involving vinyl radicals. This homolytic process is very reasonable, particularly with those organotransition metal species possessing very weak C—M bonds. However, in the reduction of alkynes using Ti^{III}, quenching the reaction mixture with deuterium oxide results in very high incorporation of deuterium in the product, suggesting that the intermediate vinylmetallic species is stable toward homolytic cleavage under the reaction conditions. The reduction of alkynes with MgH_2 in the presence of Cu^{I} gives the *cis*-alkene as the sole product.[58] In this case, the *cis*-alkene is proposed to be formed by the hydrolysis of the C—Cu bond during the hydrolytic work-up. This is also reasonable in view of the greater stability of C—Cu bonds and their hydrolytic reactivity.

In the case of the reduction of alkenes to alkanes, it does not appear that the stereo-selectivity or -specificity of this process has been studied. It is doubtful that the final steps in the reduction reactions differ from those proposed for the reduction of alkynes, and little stereo-selectivity or -specificity is to be anticipated. The mechanism for the reduction of alkenes with $LiAlH_4$ in the presence zirconium(IV) chloride is proposed to occur *via* a metal exchange reaction followed by hydrolysis (equation 39).[55]

$$LiAlH_4 + ZrCl_4 \rightleftharpoons L_nZr-H \xrightarrow{\text{alkene}} L_nZr\text{–C–C–}H \; \textbf{(100)} \xrightarrow{LiAlH_4}$$

$$L_nZr-H + Al\text{–C–C–}H \; \textbf{(101)} \longrightarrow \text{Product} \quad (39)$$

A possibility not suggested as a possible mechanism for the exclusive formation of the *cis* reduction product of a π-system is the reductive elimination of an organometallic hydride intermediate as illustrated in equation (40).

$$\text{alkenyl–}M^{x}\cdot H \; \textbf{(102)} \longrightarrow \text{alkenyl–}H + M^{x-2} \quad (40)$$

3.3.5.3 Scope and Limitations

The reduction of mono- and di-substituted alkenes with a 1:1 molar ratio of $LiAlH_4$ and Ti^{III}, V^{III}, Cr^{III}, Mn^{II}, Fe^{II}, Fe^{III}, Co^{II} and Ni^{II} chlorides occurs in very high yields, with the relative efficiencies being Co > Ni > Fe^{II} > Fe^{III} > Ti > Cr > V > Mn > Cu > Zn.[52,53] Only Ti, Co and Ni were effective when present in catalytic quantities.[52,53] Ti^{IV} (ref. 54) and Zr^{IV} (ref. 55) chlorides are also effective for the reduction of alkenes to alkanes. Trisubstituted alkenes undergo reduction only slowly, and then only in the presence of Co^{II} and Ni^{II}.[52,53] Tetrasubstituted alkenes apparently do not undergo reduction.

1-Alkynes undergo reduction to 1-alkenes, which in turn are further reduced to the corresponding alkanes. However, when the reduction of phenylacetylene is carried out in the presence of a stoichiometric amount of $FeCl_2$, or a catalytic amount of $NiCl_2$, at –40 °C for a short period of time, styrene is formed in excellent yield (very minor amounts of overreduction also occur).[52] When the reductions are carried out at room temperature for 24 h, however, excellent yields of ethylbenzene are obtained (equations 41 and 42).[52] The diarylacetylene, diphenylacetylene (**106**), is reduced to only *cis*-stilbene by $LiAlH_4$ in the presence of Fe^{II} chloride (equation 43).[52] Dialkylacetylenes, such as (**107**), are cleanly reduced to *cis*-alkenes by $LiAlH_4$ in the presence of Ni^{II} chloride (equation 44).[52] The use of the other transition metal chlorides in the reduction of alkynes results in the formation of small amounts of the *trans*-alkenes in addition to the predominant *cis*-alkene.[52-54]

$$\text{Ph—C≡CH } (\mathbf{103}) \xrightarrow[\text{THF, –40 °C (10 or 60 min)}]{LiAlH_4\text{–}FeCl_2\ (1:1)\text{ or } LiAlH_4\text{–}NiCl_2\ (1:0.1)} \text{PhCH=CH}_2\ (\mathbf{104})\ 92\text{ or }94\% \tag{41}$$

$$(\mathbf{103}) \xrightarrow[\text{THF, r.t., 24 h}]{LiAlH_4\text{–}FeCl_2\ (1:1)\text{ or } LiAlH_4\text{–}NiCl_2\ (1:0.1)} \text{PhCH}_2\text{CH}_3\ (\mathbf{105})\ 94\text{ or }99\% \tag{42}$$

$$\text{Ph—C≡C—Ph } (\mathbf{106}) \xrightarrow[\text{THF}]{LiAlH_4,\ FeCl_2} cis\text{-PhCH=CHPh}\ 86\% \tag{43}$$

$$\text{Pr}^n\text{—C≡C—Me } (\mathbf{107}) \xrightarrow[\text{THF}]{LiAlH_4,\ NiCl_2} cis\text{-Pr}^n\text{CH=CHMe}\ (\mathbf{108})\ 96\text{–}99\% \tag{44}$$

The use of $NaBH_4$ as the hydride source appears to have been employed only with Co^{II} chloride.[56] Mono- and di-substituted alkenes readily undergo reduction, with *cis*-alkenes undergoing reduction faster than *trans*-alkenes. Trisubstituted alkenes react extremely slowly, while tetrasubstituted alkenes apparently do not undergo reduction.

NaH in the presence of *t*-butoxide and Ni^{II} acetate reduces mono- and di-substituted alkenes.[57] Trisubstituted alkenes do not react. 1-Alkynes are reduced to mixtures of *cis*- and *trans*-alkenes, which undergo competitive, further reduction to alkanes. MgH_2, in the presence of Cu^{I} iodide or *t*-butoxide in THF at –78 °C, reduces 1-alkynes to 1-alkenes,[58] which are stable toward further reduction. Disubstituted acetylenes are cleanly reduced to *cis*-alkenes.[58]

The utility of metal hydride–transition metal halide reductions is limited only to alkenes and alkynes which do not contain functional groups which can react with the metal hydride, or with potential low-valent transition metal species. This severely limits the use of this reaction.

3.3.6 REDUCTION OF C=C BY IONIC HYDROGENATION REACTIONS

3.3.6.1 Introduction

'Ionic hydrogenation' is a term used to describe the reduction of a C=X π-system, or the hydrogenolysis of a C—X bond, by the use of a proton source and a hydride source in the same reaction.[59] Protonation of the π-system or the X functional group occurs to form a cationic intermediate, the cationic intermediate then abstracting a hydride from the hydride source. Although many proton sources have been utilized, the most common is trifluoroacetic acid. Similarly, many hydride sources have been investigated, with the most common being an organosilane.

3.3.6.2 Mechanism of the Ionic Hydrogenation Reaction

The rate-determining step in the ionic hydrogenation reaction of carbon–carbon double bonds involves protonation of the C=C to form a carbocation intermediate, followed by the rapid abstraction of hydride from the hydride source (equation 45).[60] There is a very sensitive balance between several factors in order for this reaction to be successful. The proton source must be sufficiently acidic to protonate the C=C to form the intermediate carbocation, yet not so acidic or electrophilic as to react with the hydride source to produce hydrogen. In addition, the carbocation must be sufficiently electrophilic to abstract the hydride from the hydride source, yet not react with any other nucleophile source present, *i.e.* the conjugate anion of the proton source. This balance is accomplished by the use of trifluoroacetic acid as the proton source, and an alkylsilane as the hydride source. The alkene must be capable of undergoing protonation by trifluoroacetic acid, which effectively limits the reaction to those alkenes capable of forming a tertiary or aryl-substituted carbocation. This essentially limits the application of this reaction to the reduction of tri- and tetra-substituted alkenes, and aryl-substituted alkenes.

$$\text{alkene} + F_3CCO_2H \xrightarrow{k_{rd}} \text{carbocation}^+ \xrightarrow[\text{fast}]{-\text{Si-H}} \text{alkane} \quad (45)$$

The ability of an organosilane to transfer a hydride to an intermediate carbocation is, in most cases, quite sensitive to the nature and number of the alkyl or aryl groups bonded to the silicon atom. The general sequence of reactivity decreases in the sequence: $Et_3SiH > (n\text{-}C_8H_{19})_3SiH > Et_2SiH_2 > Ph_2SiH_2 > Ph_3SiH > PhSiH_3$.[61–63]

The stereoselectivity of the ionic hydrogenation reaction appears to be controlled by the steric size of the trialkylsilane in the reduction of $\Delta^{9(10)}$-octalin (**109**; equation 46), which shows a high degree of sensitivity to the steric size of the organosilane hydride donor.[60] However, in the reduction of (**112**; equation 47) there is a considerably smaller change in the range of stereoselectivity of the reduction process.[60] The lower sensitivity in the stereoselectivity of reduction of (**112**) relative to (**109**) has been interpreted

$$(\mathbf{109}) \xrightarrow{CF_3CO_2H} (\mathbf{110}) + (\mathbf{111}) \quad (46)$$

Silane	(110)	(111)
Bu^nSiH_3	22%	78%
Et_2SiH_2	40%	60%
Et_3SiH	42%	58%
Ph_3SiH	58%	42%
$Bu^s{}_3SiH$	72%	28%
$Bu^t{}_2MeSiH$	83%	17%
$Bu^t{}_3SiH$	93%	7%

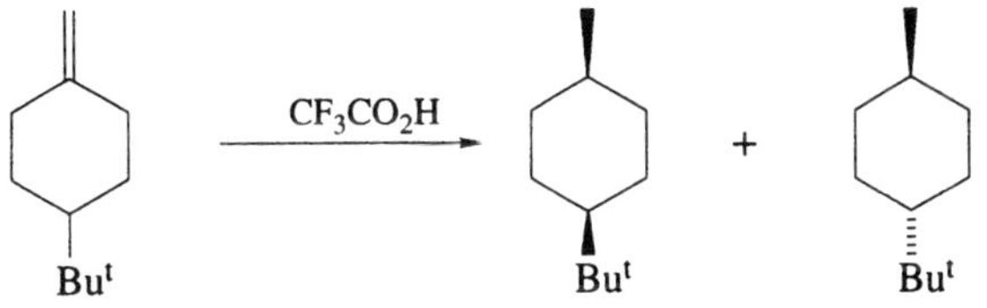

$\xrightarrow{CF_3CO_2H}$ (47)

	(112)	(113)	(114)
Bu^nSiH_3		4%	96%
Et_2SiH_2		8%	92%
Et_3SiH		10%	90%
$Bu^s{}_3H$		16%	84%
$Bu^t{}_2MeSiH$		16%	84%

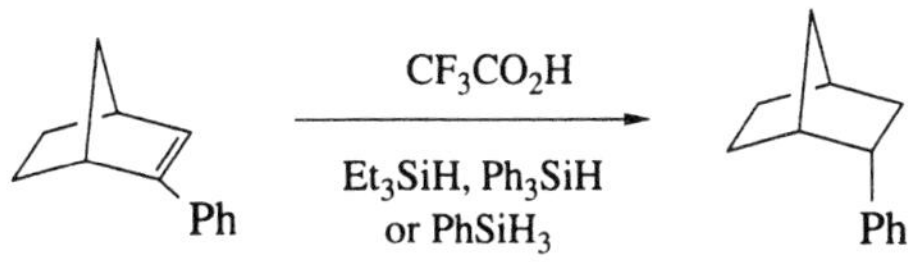

(115) $\xrightarrow[Et_3SiH,\ Ph_3SiH \text{ or } PhSiH_3]{CF_3CO_2H}$ (116) (48)

in terms that steric approach control may not be the sole factor responsible for the stereoselectivities observed in these reactions.[60] However, in the reduction of (**115**; equation 48), the delivery of hydride to the carbocation occurs exclusively from the less sterically hindered *exo* face.[64] It appears that the stereoselectivity of this reduction reaction is sensitive to the structure of the substrate, the structure of the hydride source and perhaps the solvent used for the reaction.

The 'ionic hydrogenation' reaction is not limited to the reduction of appropriately substituted alkenes, but has also been observed to occur with C=O and C=N π-systems.[59] In the case of the reduction of ketones,[67] if protonation of the resulting alcohol followed by loss of a molecule of water produces an aryl-substituted cation, reduction to the corresponding alkane occurs. Similarly alcohols can be induced to undergo 'ionic hydrogenolysis' if a reasonably stabilized carbocation can be formed on dehydration.[63,65,66] Other systems containing good leaving groups such as halides can also be induced to undergo 'ionic hydrogenolysis'.[59]

3.3.6.3 Scope and Limitations

The 'ionic hydrogenation' of carbon–carbon multiple bonds has been demonstrated only for the reduction of carbon–carbon double bonds which on protonation result in the formation of tertiary, or aryl-substituted, carbocations. The ability to use this reduction reaction with functionally substituted alkenes will be limited to those systems that do not contain functional groups that would otherwise react with trifluoroacetic acid or the organosilane.

There has been no report on the success, or failure, of the attempted reduction of C≡C bonds by the ionic hydrogenation type of reaction. The reason for this is not clear. Carbon–carbon triple bonds are capable of being protonated by trifluoroacetic acid to produce what appear to be vinyl cationic species, which should be capable of abstracting a hydride from an organosilane.

3.3.7 REFERENCES

1. J. Hanus, *Chem. Listy*, 1905, **29**, 24.
2. E. J. Corey, W. L. Mock and D. J. Pasto, *Tetrahedron Lett.*, 1961, 347.
3. S. Hünig, R. Müller and W. Thier, *Tetrahedron Lett.*, 1961, 353.
4. E. E. van Tamelen, R. S. Dewey and R. J. Timmons, *J. Am. Chem. Soc.*, 1961, **83**, 3725.
5. F. Aylward and M. Sawistowska, *Chem. Ind. (London)*, 1962, 484.
6. R. Herbert and M. Christl, *Chem. Ber.*, 1979, **112**, 2012.
7. N. J. Cusack, C. B. Reese, A. C. Risius and B. Roozpeikar, *Tetrahedron*, 1976, **32**, 2157.
8. S. F. Foner and R. L. Hudson, *J. Chem. Phys.*, 1958, **28**, 719; N. Wiberg, G. Fischer and H. Bachhuber, *Chem. Ber.*, 1974, **107**, 1456.

9. N. Wiberg, G. Fischer and H. Bachhuber, *Angew. Chem., Int. Ed. Engl.*, 1976, **15**, 385.
10. R. A. Back, *Rev. Chem. Intermed.*, 1984, **5**, 293.
11. A. P. Sylwester and P. B. Dervan, *J. Am. Chem. Soc.*, 1984, **106**, 4648.
12. D. J. Pasto and D. M. Chipman, *J. Am. Chem. Soc.*, 1979, **101**, 2290.
13. D. J. Pasto, *J. Am. Chem. Soc.*, 1979, **101**, 6852.
14. E. W. Garbisch, Jr., S. M. Schildcrout, D. B. Patterson and C. M. Sprecher, *J. Am. Chem. Soc.*, 1965, **87**, 2932.
15. S. Hunig and H. R. Muller, *Angew. Chem., Int. Ed. Engl.*, 1962, **1**, 213.
16. S. Siegel, M. Foreman, R. P. Fisher and S. E. Johnson, *J. Org. Chem.*, 1975, **40**, 3599.
17. W. C. Baird, Jr., B. Franzus and J. H. Surridge, *J. Am. Chem. Soc.*, 1967, **89**, 410.
18. R. Srinivasan and J. N. C. Hsu, *J. Chem. Soc., Chem. Commun.*, 1972, 1213.
19. R. N. McDonald and G. E. Davis, *J. Org. Chem.*, 1969, **34**, 1916.
20. E. E. van Tamelen and R. J. Timmons, *J. Am. Chem. Soc.*, 1962, **84**, 1067.
21. Y. J. Abul-Hajj, *J. Org. Chem.*, 1971, **36**, 2730.
22. S. Siegel, G. M. Foreman and D. Johnson, *J. Org. Chem.*, 1975, **40**, 3589.
23. E. E. van Tamelen, M. Davis and M. F. Deem, *Chem. Commun.*, 1965, 71.
24. S. Uemura, A. Onoe, H. Okazaki, M. Okano and K. Ichikawa, *Bull. Chem. Soc. Jpn.*, 1976, **49**, 1437.
25. E. E. van Tamelen, R. S. Dewey, M. F. Lease and W. H. Pirkle, *J. Am. Chem. Soc.*, 1961, **83**, 4302.
26. J. W. Wilt, G. Gutman, W. J. Ranus, Jr. and A. R. Zigman, *J. Org. Chem.*, 1967, **32**, 893.
27. W. G. Dauben and C. H. Schallhorn, *J. Am. Chem. Soc.*, 1971, **93**, 2254.
28. G. Just and L. Cutrone, *Can. J. Chem.*, 1976, **54**, 867.
29. S. Hünig and M. Schmitt, *Tetrahedron Lett.*, 1984, **25**, 1725.
30. (a) W. Adam and H. J. Eggelte, *J. Org. Chem.*, 1977, **42**, 3987; (b) W. Adam, H. J. Eggelte and A. Rodriguez, *Synthesis*, 1979, 383; (c) W. Adam and I. Erden, *Angew. Chem., Int. Ed. Engl.*, 1978, **17**, 210.
31. G. Nagendrappa, S. N. Moorthy and D. Devaprabhakara, *Indian J. Chem., Sect. B*, 1976, **14**, 81.
32. T. Okuyama, K. Toyoshima and T. Fueno, *J. Org. Chem.*, 1980, **45**, 1604.
33. L. Crombie, P. A. Jenkins and J. Roblin, *J. Chem. Soc., Perkin Trans. 1*, 1975, 1099.
34. D. J. Pasto, *Tetrahedron*, 1984, **40**, 2805.
35. D. J. Pasto and J. K. Borchardt, *Tetrahedron Lett.*, 1973, 2517.
36. A. F. Kluge, K. G. Untch and J. Fried, *J. Am. Chem. Soc.*, 1972, **94**, 9256.
37. D. C. Curry, B. C. Uff and N. D. Ward, *J. Chem. Soc. C*, 1967, 1120.
38. M. O. Abdel-Rahman, M. N. El-Enien, M. Kira and A. H. Zayed, *J. Chem. U. A. R.*, 1966, **9**, 87.
39. K. N. Campbell and L. T. Eby, *J. Am. Chem. Soc.*, 1941, **63**, 216.
40. R. A. Benkeser, G. Schroll and D. M. Sauve, *J. Am. Chem. Soc.*, 1955, **77**, 3378.
41. R. A. Benkeser and F. G. Belmonte, *J. Org. Chem.*, 1984, **49**, 1662.
42. H. O. House and E. F. Kinloch, *J. Org. Chem.*, 1974, **39**, 747.
43. G. M. Whitesides and W. J. Ehmann, *J. Org. Chem.*, 1970, **35**, 3565.
44. B. L. Sondengam, G. Charles and T. M. Akam, *Tetrahedron Lett.*, 1980, **21**, 1069.
45. M. H. P. J. Aerssens and L. Brandsma, *J. Chem. Soc., Chem. Commun.*, 1984, 735.
46. J. D. White and G. L. Larson, *J. Org. Chem.*, 1978, **43**, 4555.
47. (a) G. Stork and J. Tsuji, *J. Am. Chem. Soc.*, 1961, **83**, 2783; (b) G. Stork and S. D. Darling, *J. Am. Chem. Soc.*, 1960, **82**, 1512.
48. P. Angibeaund, M. Larcheveque, H. Normant and B. Tchoubar, *Bull. Soc. Chim. Fr.*, 1968, 595.
49. A. L. Henne and K. W. Greenlee, *J. Am. Chem. Soc.*, 1943, **65**, 2020.
50. C. E. Castro and R. D. Stephens, *J. Am. Chem. Soc.*, 1964, **86**, 4358.
51. C. E. Castro, R. D. Stephens and S. J. Mojé, *J. Am. Chem. Soc.*, 1966, **88**, 4964.
52. E. C. Ashby and J. J. Lin, *J. Org. Chem.*, 1978, **43**, 2567.
53. E. C. Ashby and J. J. Lin, *Tetrahedron Lett.*, 1977, 4481.
54. P. W. Chum and S. E. Wilson, *Tetrahedron Lett.*, 1976, 15.
55. F. Sato, S. Sato and M. Sato, *J. Organomet. Chem.*, 1976, **122**, C25.
56. S.-K. Chung, *J. Org. Chem.*, 1979, **44**, 1014.
57. J.-J. Brunet, L. Mordenti, B. Loubinoux and P. Caubère, *Tetrahedron Lett.*, 1977, 1069.
58. E. C. Ashby, J. J. Lin and A. B. Goel, *J. Org. Chem.*, 1978, **43**, 757.
59. For a review, see D. N. Kursanov, Z. N. Parnes and N. M. Loim, *Synthesis*, 1974, 633.
60. M. P. Doyle and C. C. McOsker, *J. Org. Chem.*, 1978, **43**, 693.
61. D. N. Kursanov, Z. N. Parnes, G. I. Bassova, N. M. Loim and V. I. Zdanovich, *Tetrahedron*, 1967, **23**, 2235.
62. D. N. Kursanov, V. N. Setkina and Y. D. Novikov, *Izv. Akad. Nauk SSSR, Ser. Khim.*, 1964, 1911 (*Chem. Abstr.*, 1965, **62**, 2684).
63. F. A. Carey and H. S. Tremper, *J. Am. Chem. Soc.*, 1968, **90**, 2578.
64. F. A. Carey and H. S. Tremper, *J. Org. Chem.*, 1969, **34**, 4.
65. F. A. Carey and H. S. Tremper, *J. Org. Chem.*, 1971, **36**, 758.
66. F. A. Carey and C.-L. W. Hsu, *J. Organomet. Chem.*, 1969, **19**, 29.
67. C. T. West, S. J. Donnelly, D. A. Kooistra and M. P. Doyle, *J. Org. Chem.*, 1973, **38**, 2675.

3.4
Partial Reduction of Aromatic Rings by Dissolving Metals and by Other Methods

LEWIS N. MANDER
Australian National University, Canberra, Australia

3.4.1 INTRODUCTION

The reduction of aromatic compounds to their dihydro derivatives by dissolving metals is one of the most powerful synthetic procedures available to organic chemists. In one simple reaction it provides a bridge between aromatic and alicyclic molecules, thereby making much of the huge arsenal of aromatic chemistry readily available for the construction of alicyclic structures. It also furnishes the facility to carry through a protracted synthesis a moiety which is inert towards an extensive array of reagents, but which is readily converted by the reductive process into a range of useful functionalities.[1–6]

The most commonly used procedure is that established by Wooster[7] and extensively developed by Birch,[8] *i.e.* the reduction of a solution of the substrate in a mixture of liquid ammonia with an alcohol (usually ethanol or *t*-butyl alcohol) and an inert cosolvent (*e.g.* diethyl ether, tetrahydrofuran) with an alkali metal (lithium, sodium or potassium). Low molecular weight amines have been utilized in place of the ammonia, although the procedure then leads to more extensive reduction. Hexamethylphosphoramide may also serve in place of the ammonia,[9] but there is no apparent advantage to offset its higher cost, toxicity and carcinogenicity. Of rather more interest is the potential of electrochemical and photochemical approaches, which may give complementary outcomes.

The great majority of applications of the 'Birch reduction' have been centered on aromatic ethers, simple alkylbenzenes and bicyclic and polycyclic aromatic hydrocarbons. Over the past decade or so, however, there has been increasing attention given to other derivatives. The reductions of aromatic acids and amides have been placed on a systematic basis, while new procedures have been established for substrates which have until recently been regarded as unsuitable candidates for reduction, *e.g.* aromatic ketones and esters,[6] and even nitriles.[10]

3.4.2 THE BIRCH REDUCTION

Metals of Groups I and II dissolve readily in liquid ammonia. The resulting solutions of solvated electrons are powerful reducing agents that may be used to perform highly selective reactions. Our

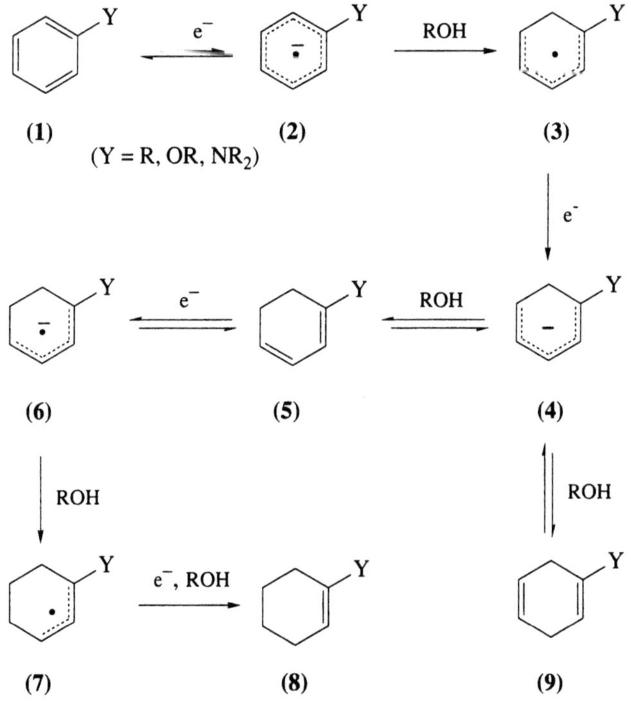

Y
e^-
Y
ROH
Y
(1)
(2)
(3)
(Y = R, OR, NR_2)
e^-
Y
e^-
Y
ROH
Y
(6)
(5)
(4)
ROH
ROH
Y
e^-, ROH
Y
Y
(7)
(8)
(9)

Scheme 1

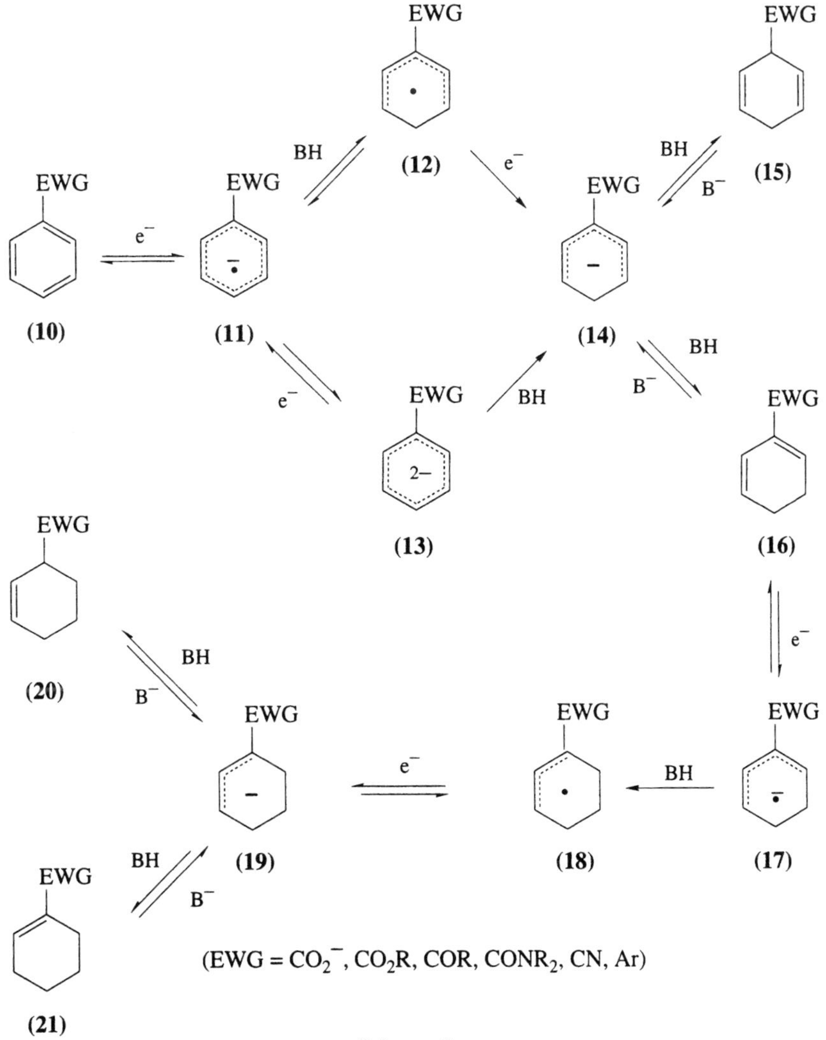

(EWG = CO_2^-, CO_2R, COR, $CONR_2$, CN, Ar)

Scheme 2

understanding of the details of the reduction process as applied to benzenoid substrates (**1**) and (**10**) is summarized in Schemes 1 and 2, respectively.[11] In the first case electrons are added in a reversible step to form a radical anion (**2**), which is protonated and reduced further to a pentadienyl anion (**4**).[12] For the more reactive substrate (**10**), the radical anion (**11**) may be protonated to afford the pentadienyl radical (**12**), or reduced further to a dianion (**13**)[13] which is then protonated to afford the same pentadienyl anion (**14**) that would be formed by reduction of (**12**). Protonation of anions (**4**) and (**14**) occurs predominantly at the central carbon atom[14] to form the unconjugated 1,4-dienes (**9**) and (**15**), respectively, which are resistant to further reduction.

With substrates bearing electron-withdrawing groups (EWG) the radical anions are formed in sufficiently high concentrations that protonation by ammonia proceeds rapidly. However, with hydrocarbons and aromatic ethers *etc.*, it is necessary to displace the initial equilibrium between the substrate and the radical anion through protonation by a stronger acid than ammonia (pK_a *ca.* 35). Alcohols (pK_a 16–20) are normally used for this purpose.

Overreduction may occur as a consequence of initial protonation at the terminal atoms of the cyclopentadienyl anions to form the conjugated 1,3-dienes (**5**) or (**16**), which are then vulnerable to further reduction, affording tetrahydro products, *e.g.* (**8**), (**20**) or (**21**). These products can also be formed when the reaction medium is sufficiently basic to catalyze conjugation of the 1,4-dienes to the 1,3-isomers. The alcohol may then serve an important additional role by functioning as a buffer (*cf.* Section 3.4.3.6, however).

3.4.2.1 Experimental Procedures

Procedures for metal–ammonia reductions have been described in several reviews[1–6] and are outlined for specific substrates in later parts of this section. The following comments are therefore only a summary of the more important factors. Many reductions are relatively insensitive to the choice of conditions, but the majority may be significantly improved by the appropriate choice of metal, proton source, temperature, order of mixing and quenching procedure.

Although it may not always be necessary it is advisable, at least during feasibility studies on a new reaction, that reduction be carried out in ammonia which has been thoroughly dried and then distilled. This is especially important when sodium or potassium are employed since colloidal iron, which invariably contaminates commercial sources of ammonia, catalyzes the formation of amide ion and the reaction of these metals with alcohols; lithium is not significantly affected. There are many examples, however, where reaction mixtures have been deliberately doped by iron salts to suppress overreduction.

Sodium metal is a convenient drying agent for liquid ammonia, and a small amount of an iron salt should be added to catalyze the formation of sodium amide, since sodium itself reacts progressively more slowly with water as the pH of the ammonia solution increases. Oxygen should be rigorously excluded from the reaction, and although ammonia at reflux provides quite good protection, it is advisable to employ an inert atmosphere. Nitrogen is usually satisfactory, but there is a potential for reaction with lithium, so the use of helium or argon has been recommended for reductions with this metal.[3] Temperature is often an important factor and should be monitored with an internal thermometer or sensor. Low temperatures (less than –70 °C) may be crucial for the ring reduction of aromatic esters and ketones, and help to suppress unwanted side reactions with other substrates, *e.g.* hydrogenolysis of methoxy substituents.

The choice of metal can have a profound effect on the outcome of a reduction,[15] and this has been attributed in some cases to the differences in reduction potential (Li = –2.99 V, Na = –2.59 V, K = –2.73 V, Ca = –2.39 V), but variations in outcome are probably due more to the effects of aggregation or coordination of intermediates with the metal cation. Lithium and sodium are usually the most satisfactory, but potassium may be superior, and sometimes essential for specialized applications.[16] Calcium metal has been recommended for reductions in methylamine/ethylenediamine and for the reduction of α-acetoxy ketones, while magnesium appears to offer advantages in some reductions of aromatic amides, esters and polycyclic hydrocarbons.[17] On the whole, however, there appears to be little point to the use of metals other than the common members of Group I. The solubilities of ionic substrates, *e.g.* carboxylate salts, and their products depend very much on the metal counterion, although it is common for insoluble substrates to dissolve as the reduction progresses. Alcohols aid solubility, but a cosolvent is usually necessary, and tetrahydrofuran (which must be dry and oxygen free) is most commonly utilized, although other ethers may be employed.

Tertiary alcohols (butyl and amyl) react very slowly with the metal, but methanol or ethanol protonate the intermediate radical anions more rapidly and can suppress side reactions; propan-2-ol may provide an effective compromise.[5] Water and ammonium salts have also been used for protonation, but have an undesirable effect on many reactions, and should be avoided for quenching purposes while any metal remains. Any excess of metal is best consumed by the addition of a simple diene (isoprene or penta-1,3-diene), since the metal will often react more rapidly with the (protonated) product than with the quenching agent, leading to overreduction. Inverse quenching has been advocated for especially sensitive products.[18]

It has been noted elsewhere[3] that the order of addition of reagents is the most important single variable to influence product distribution. A procedure which is convenient and normally reliable is based on the addition of small pieces of metal to a mixture of substrate, ammonia, solvent and (if appropriate) alcohol at less than –70 °C until the blue colour persists for a reasonable period (approximately 10 min). For less reactive substrates, lithium should be used and the addition of the alcohol delayed so that a higher concentration of metal is maintained during the reduction process. Quenching is best effected by 1,3-pentadiene (or similar substance), followed either by ammonium chloride or another electrophile. In reactions where *in situ* alkylations are performed these are best carried out on lithium derivatives, so in cases where other metals have been used for reduction carefully dried lithium bromide is added before the electrophile.

3.4.2.2 Scope and Limitations

3.4.2.2.1 Substituent effects

Electron-releasing groups direct reduction to unsubstituted 2,5-positions, and while alkyl groups retard reduction (*t*-butyl > isopropyl > ethyl > methyl), alkoxy groups slightly accelerate the rate of reaction.[19] A selection of reaction rates is provided in Table 1. Phenols are rapidly ionized and the resulting salts are resistant to electron addition, although reduction may be effected with high concentrations of lithium.[20] In sharp contrast, β-naphthols are reduced in the hydroxy-bearing ring, furnishing the corresponding β-tetralones.[21]

Table 1 Rates of Reduction of Substituted Benzenes Relative to Benzene[19]

Compound	*Relative rate*	*Compound*	*Relative rate*
p-Di-*t*-butylbenzene	<0.005	*m*-Xylene	0.28
Mesitylene	0.017	*p*-Xylene	0.30
t-Butylbenzene	0.050	Dimethylaniline	0.30
o-Xylene	0.050	Tetralin	0.38
Isopropylbenzene	0.10	Indane	0.94
Aniline	0.10	Benzene	1.0
n-Butylbenzene	0.21	Anisole	3.28
Ethylbenzene	0.25	Sodium benzoate	>200

Groups which allow electron delocalization also accelerate reduction (sodium benzoate reduces >200 times more rapidly than benzene, for example) and afford 1,4-dihydro derivatives, irrespective of alkyl, alkoxy and amino substituents. A major limitation on the choice of electron-withdrawing groups is the ease with which they themselves undergo reduction. Carboxylic acids present no difficulties, but esters, amides, ketones and polycyclic aromatic compounds usually require specially selected conditions. In the absence of an electron-withdrawing group, however, it is often difficult to reduce benzene rings which carry bulky substituents or which are heavily substituted, although there are examples where intramolecular protonation greatly accelerates reaction (see below).

3.4.2.2.2 Secondary reactions

A variety of secondary reactions may reduce the yield of a desired product and in some cases prevent its formation altogether, *e.g.* it is possible for ether groups to be dealkylated, and for both aryl and benzylic heteroatom substituents to undergo hydrogenolysis (see Section 3.4.3.11). Rearomatization may be a problem, usually through the presence of oxygen but in some cases by loss of hydride ion. Tetrahydro products often arise as a consequence of either initial protonation at C-1 or C-5 of the intermediate pentadienyl anions, or base-catalyzed conjugation of 1,4-cyclohexadienes, or as a result of quenching procedures which do not ensure complete consumption of the metal before protonation of the product anions. Ammonia is an excellent solvent for alkylations but may prove to be too nucleophilic for very reactive electrophiles, such as benzyl iodides.

3.4.3 SURVEY OF BIRCH REDUCTIONS

3.4.3.1 Benzenoid Hydrocarbons and Ethers

The use of anisoles (**22**) as synthetic equivalents to cyclohex-2-enones (Scheme 3) has been widespread since the original observations of Birch, and the literature is replete with examples over the past five decades, *e.g.* in the syntheses of steroids,[2] terpenoids[6] and alkaloids.[22,23] The most thorough studies have been carried out within the context of the conversion of estrone derivatives to 19-norandrostane and pregnane derivatives and are instructive for the selection of reagents and reaction parameters for reductions of this general type.[5]

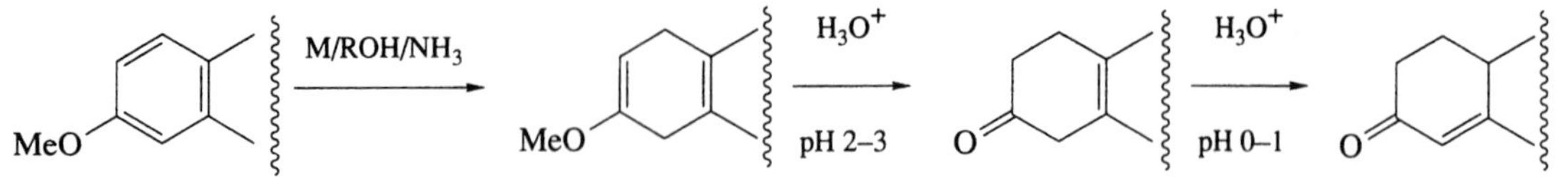

(22) M/ROH/NH$_3$ → H_3O^+, pH 2–3 → H_3O^+, pH 0–1

Scheme 3

The Birch reduction of aromatic hydrocarbons and ethers to the 2,5-dihydro derivatives proceeds most satisfactorily when the substitution pattern allows the addition of hydrogen to two unsubstituted positions in a *para* relationship. If this requirement is satisfied, better yields are obtained from more highly substituted aromatic rings than from (say) anisole itself, which affords a substantial amount (20%) of 1-methoxycyclohexene (*cf.* Scheme 1). Extra substitution presumably hinders protonation at the terminus of the dienyl anion (which would lead to a conjugated diene and overreduction). The utilization of anisole moieties as precursors to cyclohexenones has been of very limited value with many 1,2,3-substitution patterns and more densely substituted derivatives. Compounds (**23**) to (**26**), for example, have only been reduced by employing massive excesses (200–600 equiv.) of lithium metal,[24] while the aromatic ring in (**28**) is completely resistant to reduction.[25]

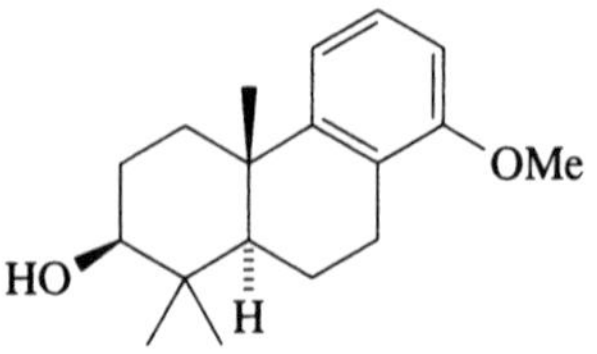

(**23**) 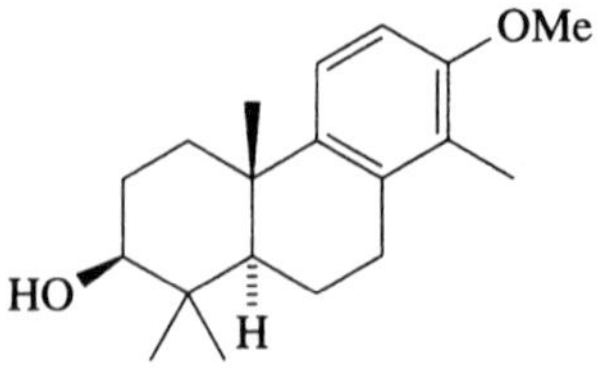 (**24**) 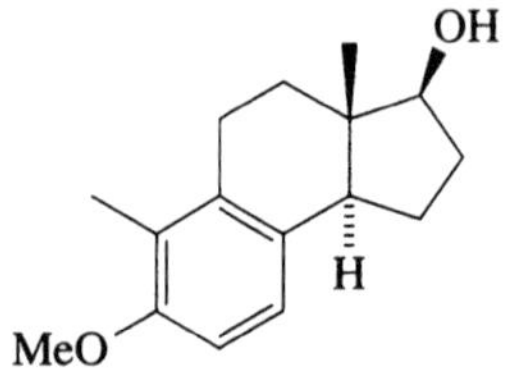 (**25**)

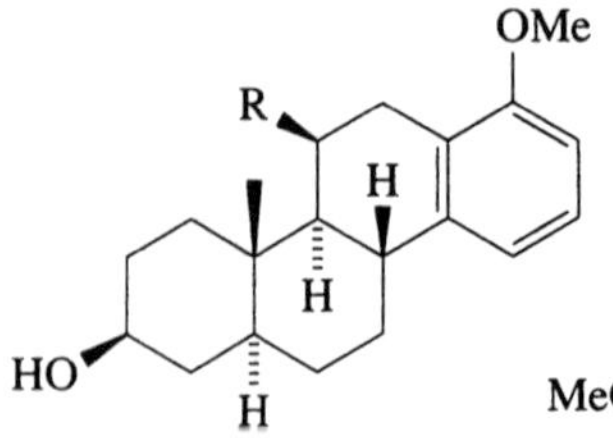

(**26**) R = H
(**27**) R = OH

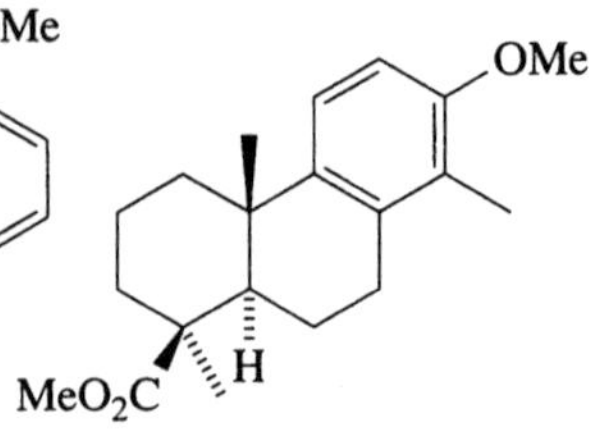

(**28**)

The reduction of silyl aryl ethers in place of the more common alkyl analogs is a useful variation,[26] which would seem to have considerable potential in view of the widespread use of silyl ethers in synthesis. Trimethylsilyl ethers are cleaved during the reaction process (presumably through attack by alkoxide ion), but both *t*-butyldimethylsilyl (**29**) and isopropyldimethylsilyl (**30**) phenyl ethers are stable in the reducing medium and afford the 1,4-dihydroaryl silyl ethers (**31**) and (**32**) in 80–97% yields. The dihydroaromatics (**33**; R = Me, OMe) were used as precursors to the relatively inaccessible β,γ-unsaturated ketones (**34**) through fluoride-mediated hydrolysis of the silyl enol ethers. Alternatively, (**33**; R = Me) could be regiospecifically *C*-acylated to give (**35**).

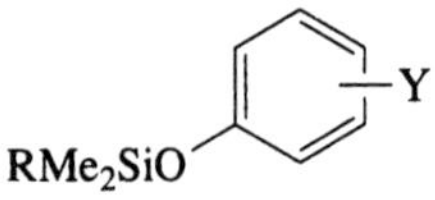

(**29**) R = Bu^t
(**30**) R = Pr^i

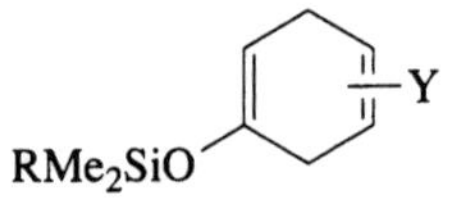

(**31**) R = Bu^t
(**32**) R = Pr^i

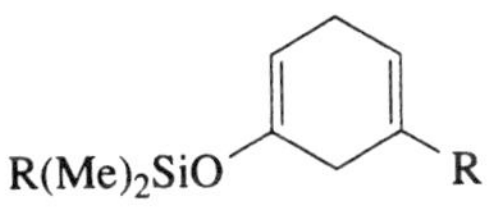

R(Me)$_2$SiO R O R HO O

(33) 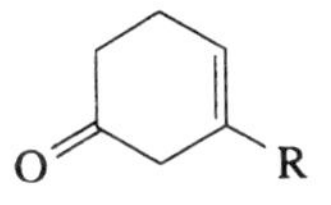 **(34)** 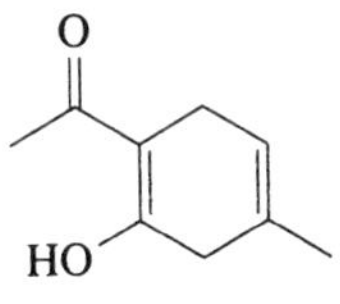 **(35)**

3.4.3.1.1 Intramolecular protonation of Birch intermediates

A number of isolated reports in the literature indicate that neighboring hydroxy functions enhance the rates of reduction of substrates which would otherwise be very slow to react. It has been suggested that the rate accelerations may stem from both complexation with the metal counterion associated with the intermediate radical anion and intramolecular protonation. It was found, for example, that the yield of reduction products from the 11β-hydroxy derivative (**27**) was significantly enhanced (>69%)[27] when compared with the 33% yield obtained from the analog (**26**), although the reaction was still carried out with *ca.* 600 equiv. of lithium metal. Selective reduction of the A-ring in the tetrahydropyranyl ether (**36**) could be achieved satisfactorily, but in the alcohol (**37**) concomitant reduction of the 13-phenyl group took place, indicating participation by the 17α-hydroxy.[28]

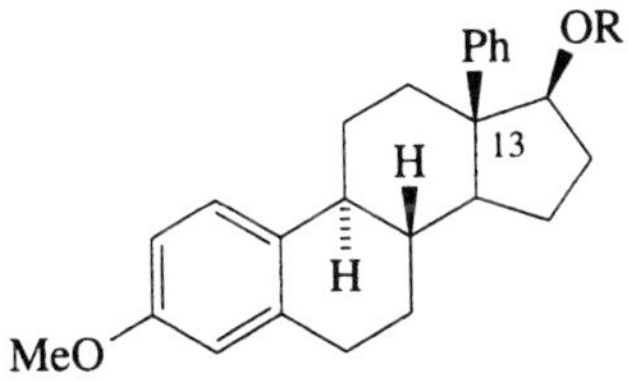

(36) R = THP
(37) R = H

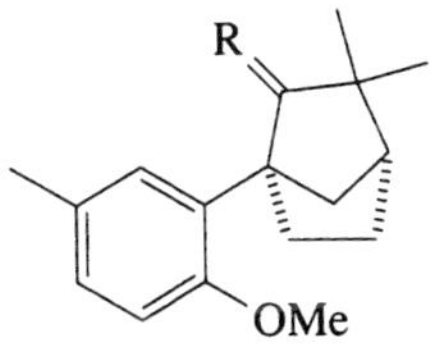

(38) R = O
(39) R = CH_2

Lithium–ammonia reduction of the arylnorbornanone (**38**) proceeded satisfactorily *via* the *exo*-alcohol to give the normal dihydroaromatic derivative, whereas the methylene analog (**39**) gave a 1:1 mixture of dihydro and tetrahydro products.[29] Recent studies on the reduction of dimethanoanthracenes (**40**) to (**42**) have provided some very interesting examples of the intramolecular protonation of radical anions (Scheme 4). Reduction of the parent hydrocarbon (**40**) or the methoxy derivative (**41**) proceeds very slowly to give the expected dihydro compounds (**43**) and (**44**), respectively. Alcohol (**42**), however, is

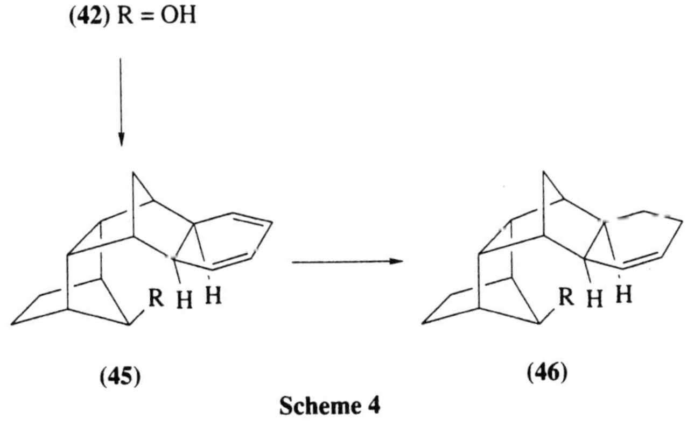

(40) R = H → **(43)** R = H
(41) R = OMe → **(44)** R = OMe
(42) R = OH

(45) → **(46)**

Scheme 4

reduced within minutes to the tetrahydro product (**46**), which arises through protonation at substituted positions on the more hindered face of the aromatic ring and is therefore clearly intramolecular in nature. It may be assumed that diene (**45**) is an intermediate, and it has been found that a sample prepared independently is reduced under the same conditions to (**46**).[30]

During a total synthesis of the diterpenoid enmein, the tricyclic alcohols (**47**) and (**48**) were converted into a range of C-ring ketones (Scheme 5). The degree of reduction and the ease with which it occurs, relative to (**23**), make it apparent that intramolecular protonation is involved here also.[31]

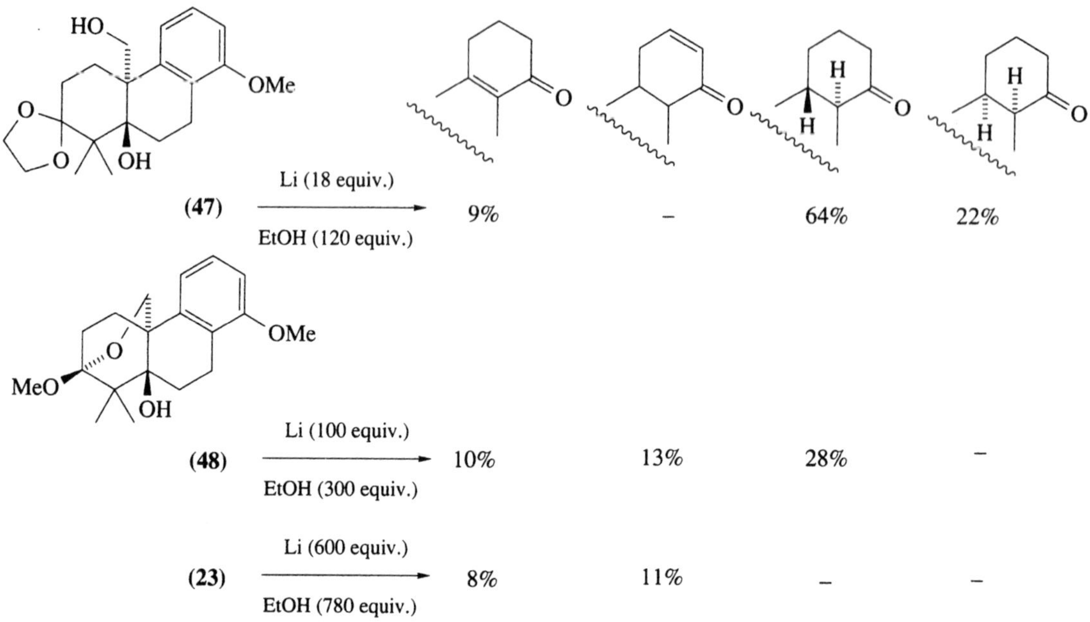

(**47**)	Li (18 equiv.), EtOH (120 equiv.)	9%	–	64%	22%
(**48**)	Li (100 equiv.), EtOH (300 equiv.)	10%	13%	28%	–
(**23**)	Li (600 equiv.), EtOH (780 equiv.)	8%	11%	–	–

Scheme 5

3.4.3.2 Alkyl- and Alkoxy-naphthalenes

Naphthalene (**49**) reacts with sodium in liquid ammonia to form a red complex which is quenched by methanol to form the 1,4-dihydro derivative (**51**), and may therefore be assumed to be the dianion (**50**).[32] If alcohol is present in the reaction mixture, then the 1,4,5,8-tetrahydro derivative (**52**) is formed (Scheme 6). 1-Naphthyl derivatives are reduced in the unsubstituted ring, whereas there is a marked preference for reduction of the substituted ring in 2-methyl- and 2-methoxy-substituted isomers. This preference is not as well defined for bulkier substituents, however, and is reversed in the case of 2-*t*-butylnaphthalene; further details are in a 1970 review.[3] Because of the facile isomerization of the 1,4-dihydro isomers to conjugated derivatives, overreduction occurs readily and ether groups are fairly readily hydrogenolyzed. It is therefore important to select the reaction conditions with care.[33]

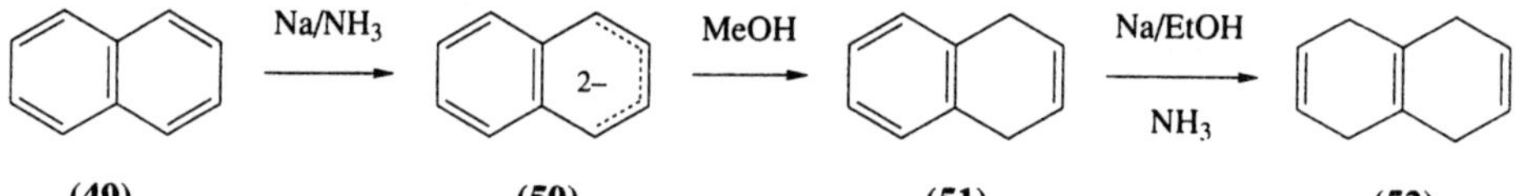

(**49**) —Na/NH_3→ (**50**) (2–) —MeOH→ (**51**) —$Na/EtOH$, NH_3→ (**52**)

Scheme 6

3.4.3.3 Biphenyls, Fluorenes and Fused Polycyclic Hydrocarbons

There has been some confusion over the structure of the dihydro reduction product of biphenyl (**53**), first reported 50 years ago. However, it has been firmly established that it is the 1,4-dihydro derivative (**54**), as would be expected if the second aryl group is viewed as an EWG substituent, and from theoretical considerations.[34] The conflicting analyses have been due to the ease with which the products undergo isomerization. Biphenyls with substituents at C-2 and C-3 tend to undergo reduction in the substituted ring (*e.g.* 80% in the 3-methyl derivative). Reduction of 4-methylbiphenyl is indiscriminate, however, while the 4-methoxy analog gives only deoxygenated products.[35]

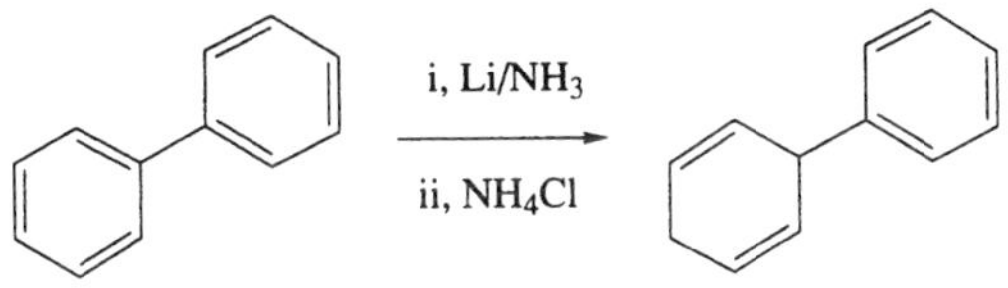

i, Li/NH$_3$

ii, NH$_4$Cl

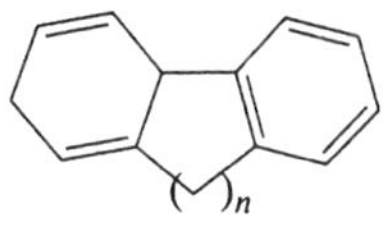

(53) **(54)** **(55)**

Fluorene has been reported to afford the 3,9a-dihydro product, but it is almost certain that this is the 2,4a-dihydro isomer (**55**; $n = 1$) by analogy with biphenyl.[3] 9,10-Dihydrophenanthrene (**56**) is reduced as expected to (**55**; $n = 2$), but spontaneously reverts to the starting material on standing. These systems do not require the presence of alcohol for reduction and it is consequently possible to alkylate the intermediate anions with alkyl halides, as (**56**) gives (**57**).[36] These products are much more stable and structural analysis is simplified accordingly; oxidation of the doubly allylic methylene occurs readily to afford the dienone (**58**; Scheme 7). Dienones of this type have potential as intermediates for the synthesis of natural products.[37] Anthracene and phenanthrene are both readily reduced in the central ring to form the 9,10-dihydro derivatives as might be expected, but to avoid further reduction it is necessary to have an iron salt present. Further examples are reviewed elsewhere.[3]

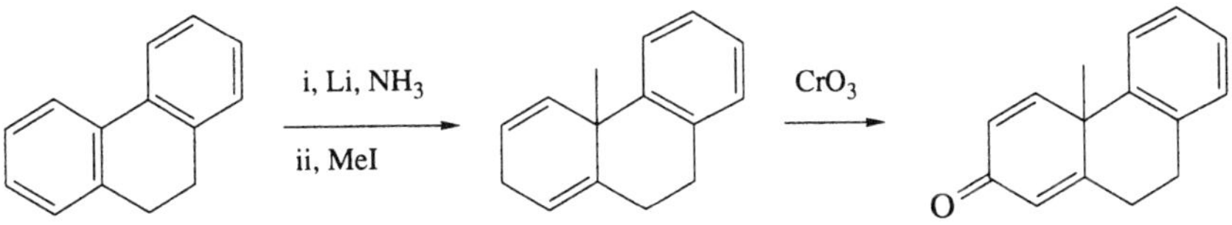

i, Li, NH$_3$

ii, MeI

CrO_3

(56) **(57)** **(58)**

Scheme 7

3.4.3.4 Phenols

Phenols are normally resistant to reduction as a consequence of ionization. This property was put to good effect in the reduction of 7-methoxyfluoren-2-ol (**59**; Scheme 8).[38] Because the phenoxide ring does not serve as an electron sink, the 5,8-dihydro derivative is formed, while the styrenic double bond is protected from reduction. However, phenols have been reduced with concentrated solutions of lithium (Scheme 9),[20] and 4-methoxyphenol (**63**) is reduced to (**64**) under surprisingly mild conditions (Scheme 10).[39]

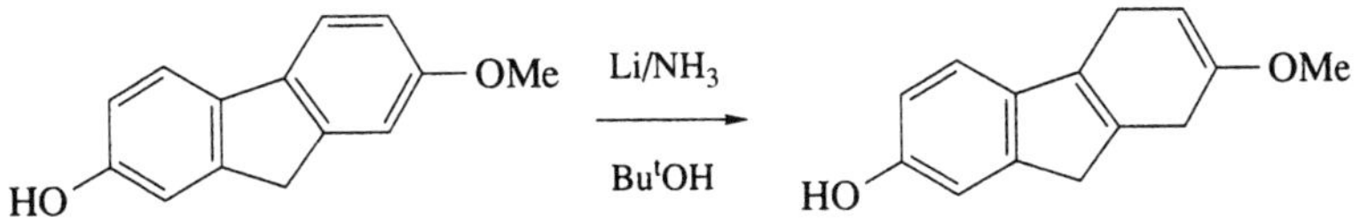

HO — OMe Li/NH$_3$ Bu^{t}OH HO — OMe

(59) **(60)**

Scheme 8

HO Li (4 M)/NH$_3$ Bu^{t}OH HO

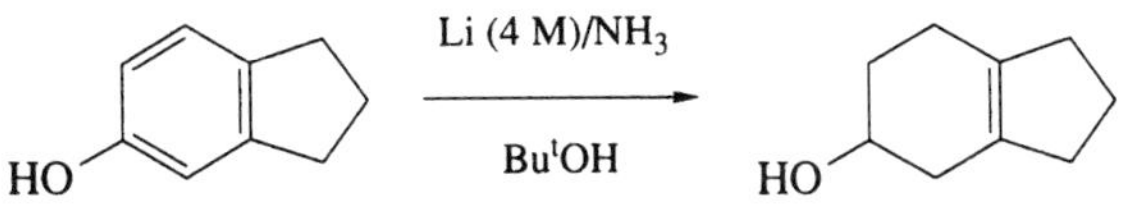

(61) **(62)**

Scheme 9

2-Naphthols are reduced to β-tetralones (which are presumably protected from further reaction as the corresponding enolate anions) even more readily and it is surprising to find that even 6-methoxy-2-naphthol (**65**) is converted into ketone (**66**; Scheme 11).[40] Reduction of the equilenin ketal (**67**) was found to be highly dependent on the choice of metal. Reduction of (**67**; Scheme 12) or its sodium salt by lithium in the absence of an alcohol afforded (after acid hydrolysis) equilin (**68**) as the major product, whereas

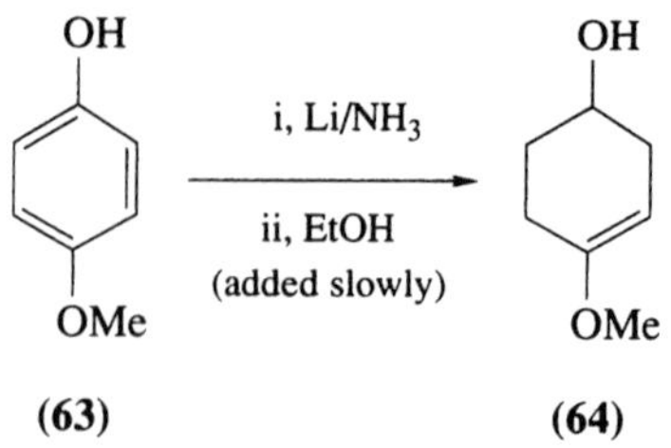

OH OH

i, Li/NH_3

ii, EtOH
(added slowly)

OMe OMe

(63) (64)

Scheme 10

the tetrahydro derivative (**69**) was obtained by reduction with potassium. It was shown that (**68**), the presumed intermediate in the formation of (**69**), was reduced directly and was not isomerized to the conjugated 8-ene beforehand. Similar results were obtained by reduction of 2-naphthol (**71**; Scheme 13) with sodium (again without an added proton source) which afforded a 3:2 mixture of (**72**) and (**73**). Reduction of (**67**) by calcium, however, yielded the A ring reduced derivative (**70**) as a mixture of epimers.[15]

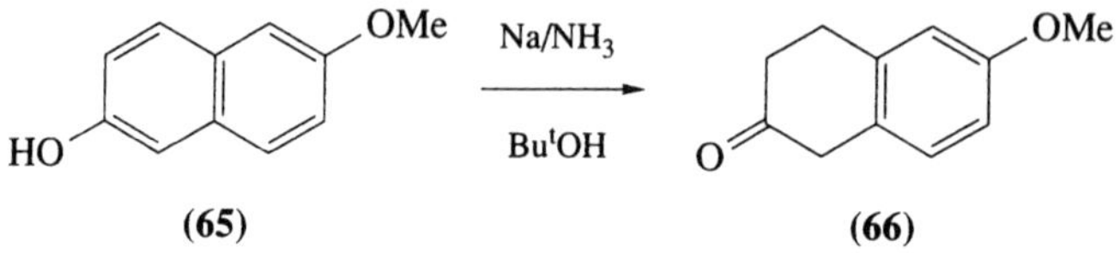

OMe Na/NH_3 OMe

HO Bu^tOH O

(65) (66)

Scheme 11

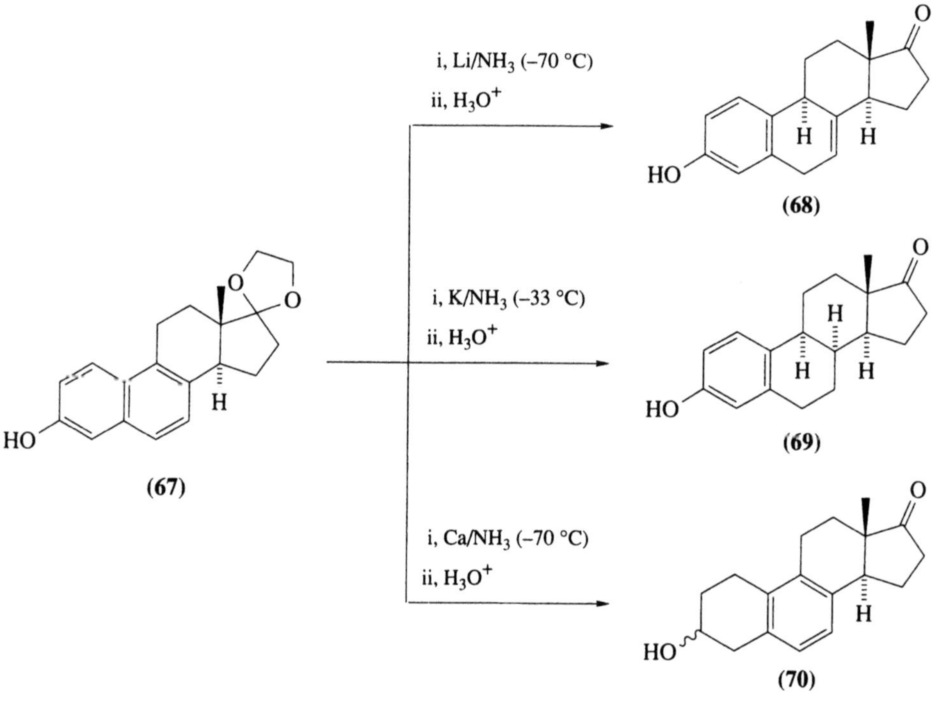

i, Li/NH_3 (−70 °C)
ii, H_3O^+

(68)

i, K/NH_3 (−33 °C)
ii, H_3O^+

(67) (69)

i, Ca/NH_3 (−70 °C)
ii, H_3O^+

(70)

Scheme 12

i,Na/NH_3 (−70 °C)

ii, NH_4Cl

(71) (72) + (73)

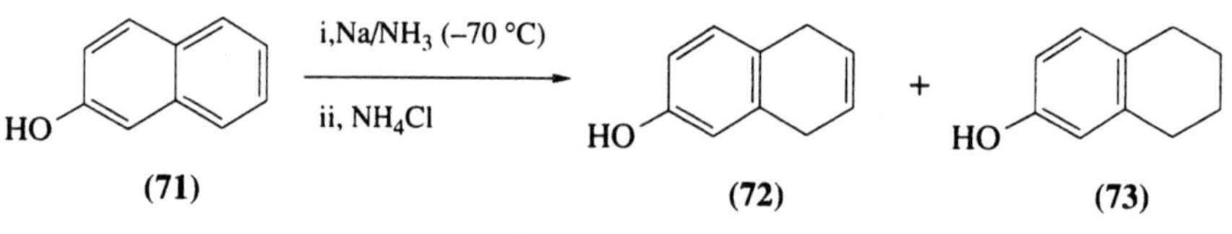

Scheme 13

3.4.3.5 Aromatic Amines

N,N-Dimethylaniline was one of the earliest substrates for reduction examined by Birch. This and subsequent studies have shown that a variety of aniline derivatives are reduced initially to the 2,5-dihydro derivatives (**74**), but also that these are easily isomerized to the 2,3-dihydro isomers (**75**) either *in situ* (leading to tetrahydro derivatives) or during work-up procedures.[41] The dienamines formed in this way may be hydrolyzed to cyclohexenones, treated with electrophiles, or utilized in [4 + 2] cycloadditions. The less basic morpholino dienamines (**76**) are more stable and afford the most reliable outcomes.[42]

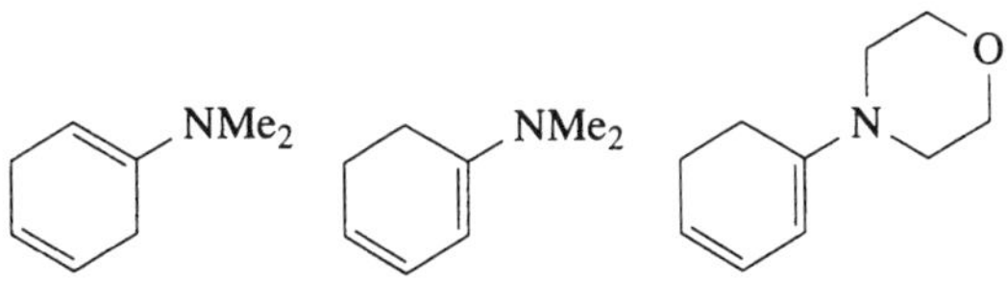

NMe_2 NMe_2 N O

(**74**) (**75**) (**76**)

3.4.3.6 Carboxylic Acids

Aromatic acids are reduced by metal–ammonia solutions very much more readily than simple hydrocarbons and ethers. In contrast to the normal requirements for the latter derivatives, it is often possible to achieve reduction with close to stoichiometric quantities of metal. The addition of aromatic carboxylic acids to liquid ammonia (or *vice versa*) results in the immediate precipitation of the ammonium salt. As the metal is added, however, the precipitate usually dissolves as reduction proceeds, especially if lithium is used. If reduction is carried out in carefully dried, redistilled ammonia, as little as 2.2 mol of lithium are consumed in some cases, thereby demonstrating that the substrate is reduced much more readily than the ammonium ions, which instead react with the intermediates from reduction of the substrate. However, protonation by NH_4^+ is not essential since reduction proceeds equally well on preformed metal carboxylates (although low solubility is then often a problem). The addition of an alcohol is not necessary, but it may serve as a useful buffer and can often improve solubility. The presence of alcohol can nevertheless be deleterious, since it facilitates isomerization of the initially formed 1,4-dihydro isomer to the 3,4-isomer and in this way affords the possibility of further reduction.[43]

3.4.3.6.1 Reductive alkylations

The reduction of benzoic acids affords the dienolate (**77**) which may be alkylated *in situ* by a variety of electrophiles to afford 1-substituted derivatives. Clearly, the addition of alcohol must be avoided or limited to small quantities. It may also be necessary to remove the ammonia before adding the electrophile. The vast majority of applications have been based on reactions with alkyl halides to form (**78**), but additions of (**77**) to formaldehyde,[44] epoxides[45] and α,β-unsaturated esters[46] to form the range of adducts (**79**) to (**81**) have also been reported (Scheme 14).

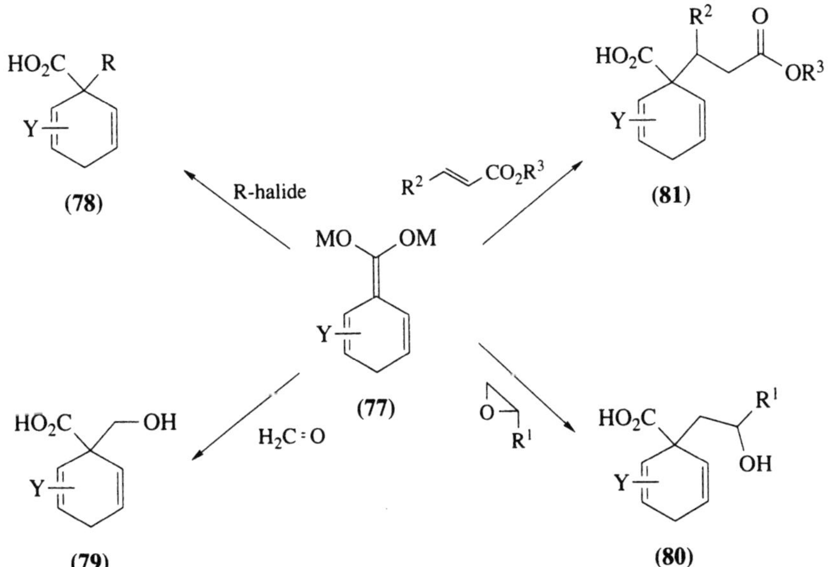

HO_2C R Y (**78**) R-halide R^2 CO_2R^3 R^2 O HO_2C OR^3 Y (**81**)

MO OM Y (**77**)

HO_2C OH Y (**79**) $H_2C{=}O$ O R^1 R^1 HO_2C OH Y (**80**)

Scheme 14

The 1,4-dihydrobenzoic acids derived from reductive alkylation may undergo facile rearomatization with either loss of the carboxylic acid group or the alkyl group. The gibberellin synthesis intermediate (**82**), for example, was found to be especially labile, forming (**83**) simply on exposure to air.[47] Oxidative decarboxylation may be deliberately achieved with lead tetraacetate or electrochemically.[48] Loss of the 1-alkyl group is likely to be a problem when the alkyl moiety can form a reasonably stable free radical, since a chain reaction may then be sustained.[49]

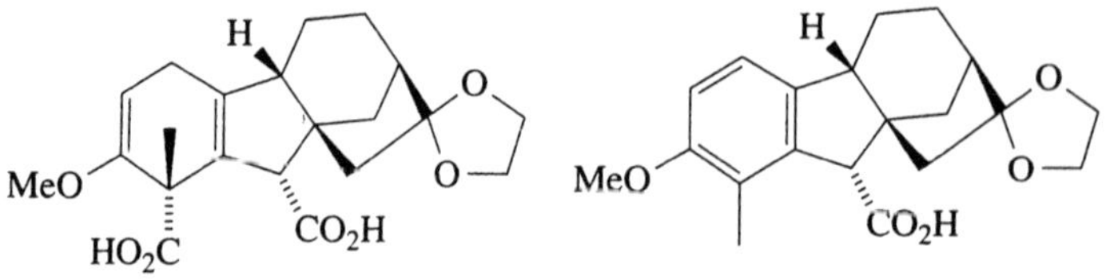

MeO HO$_2$C H O O CO$_2$H

(82)

MeO H O O CO$_2$H

(83)

3.4.3.6.2 Alkylbenzoic acids

Following the early studies of Birch,[8] a comprehensive study of the reduction of alkylbenzoic acids was undertaken by Van Bekkum *et al.*[43] These authors found that 2-alkyl derivatives gave almost quantitative yields of 1,4-dihydro products irrespective of the procedures employed. With other alkyl derivatives, generally excellent yields of 1,4-dihydro compounds were again obtained from combinations of Li–NH_3 (NH_4Cl quench), Li–NH_3–H_2O, or Li–NH_3–EtOH (NH_4Cl quench). With the last combination, however, significant quantities of tetrahydro products from 4-alkyl derivatives were obtained, and if the NH_4Cl quench was omitted, 4,5-dihydro isomers became the major products. If an excess of NH_4Cl was added before the lithium metal, then extensive reduction of the carboxy group took place. *C*-Alkylation experiments were also conducted and invariably gave 1-alkyl-1,4-dihydro products when alkylation occurred. Not surprisingly, alkylation did not occur when water had been added, while variable results were obtained in the presence of ethanol. In the absence of an added proton source the yields were excellent.

3.4.3.6.3 Tetrahydrobenzoic acids

The first sets of experiments carried out on the reduction of benzoic acids gave mixtures of dihydro and tetrahydro derivatives. This is hardly surprising in the light of the Van Bekkum study. The first systematic investigation appears to have been carried out on *p*-isopropylbenzoic acid (**84**; Scheme 15),[50] for which it was found that reduction by lithium–ammonia at –70 °C followed by the addition of ethanol and then ammonium chloride at 5 min intervals furnished a 2:1 mixture of *trans*- and *cis*-1,4-dihydro derivatives (**85**). In the absence of ammonium chloride the 3,4-dihydro isomer (**86**) was obtained and this could be reduced further to a 2.5:1 mixture of the tetrahydro derivatives (**87**) (ammonium chloride quench) and (±)-phellandric acid (**88**).

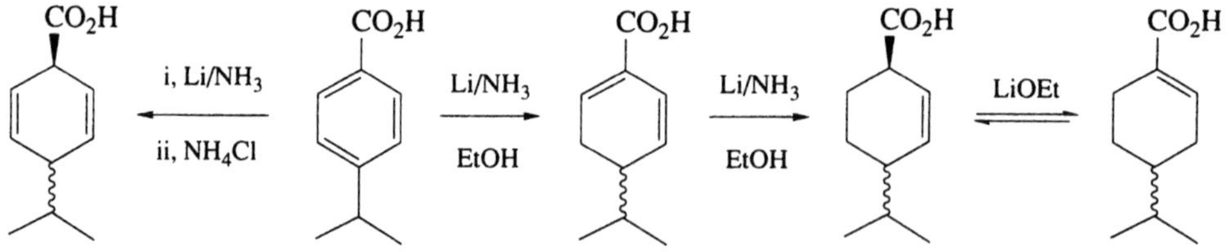

CO$_2$H (85) ← i, Li/NH_3; ii, NH_4Cl — CO$_2$H (84) — Li/NH_3, EtOH → CO$_2$H (86) — Li/NH_3, EtOH → CO$_2$H (87) ⇌ LiOEt CO$_2$H (88)

Scheme 15

Indane-2-carboxylic acid (**89**) has been converted in a similar manner (Scheme 16) to the hydrindene acid (**90**).[51] A crucial step in the conversion of (–)-abietic acid into (+)-kaurene (**93**) and (+)-phyllocladene (**94**) involved the reduction of (**91**) to the mixture of epimers (**92**) combined with other double-bond isomers (Scheme 17).[52]

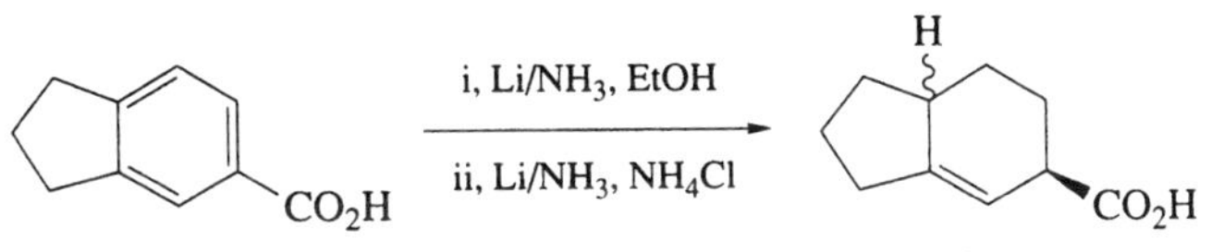

Scheme 16

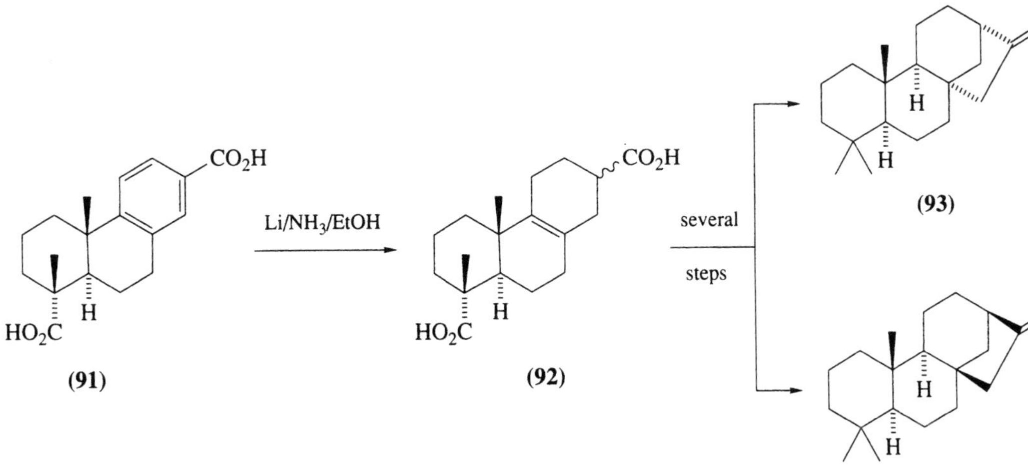

Scheme 17

3.4.3.6.4 Methoxybenzoic acids

One of the synthetically most useful groups of aromatic acids are those which also bear alkoxy functions, in that the reduced products are potential cyclohexenones, while oxidative rearomatization may afford useful aromatic substitution patterns. The electron densities of the intermediate radical anions are dominated by the carboxy group so that 1,4-dihydro products are formed initially, although the problems of subsequent isomerization and overreduction are similar to those encountered with alkylbenzoic acids. The main problem with anisoic acids is the potential for hydrogenolysis of the alkoxy function(s). Alkoxy groups in a *meta* relationship to the acyl group present no difficulties, but *para* substituents are invariably hydrogenolyzed[53] and *ortho* substituents are often vulnerable to reductive cleavage.[54] The extent of *ortho* alkoxy loss depends very much on the substrate and reaction conditions. A simple Li–NH_3 reduction of 2-methoxybenzoic acid under 'aprotic' conditions at –33 °C, for example, results in *ca.* 70% loss of the methoxy group,[55] whereas 2,5-dimethoxybenzoic acid under the same conditions furnishes a quantitative yield of the 1,4-dihydro derivative. Hydrogenolysis can be suppressed through the use of methanol as the proton source,[56] but one-pot reductive alkylation is then impractical.

(i) Reduction of 3-methoxybenzoic acid

Early studies on the reduction of 3-methoxybenzoic acid (**95**) led to a variety of products and a number of uncertainties in structural assignments. A careful study of the reduction was therefore undertaken and revealed that when (**95**) was reduced by sodium–ethanol–ammonia the 1,4-dihydro derivative (**96**) was formed initially, but isomerized rapidly and quantitatively to the 4,5-dihydro compound (**97**) under the reaction conditions. Further reduction could then occur to give the tetrahydro compound (**98**; Scheme 18).[57]

The method of choice for the preparation of the 1,4-dihydro compound (**99**) therefore entails reduction of (**95**) with lithium in ammonia in the absence of alcohol, followed by quenching with ammonium chloride. Acid (**99**) is also susceptible to rearrangement under acidic conditions, however, and the isomer (**100**) is formed at pH 4–5 (conditions which might readily prevail during the isolation procedure). 3-Methoxybenzoic acid (**95**) undergoes reductive alkylation at C-1 without difficulty, and has been used extensively in synthesis.

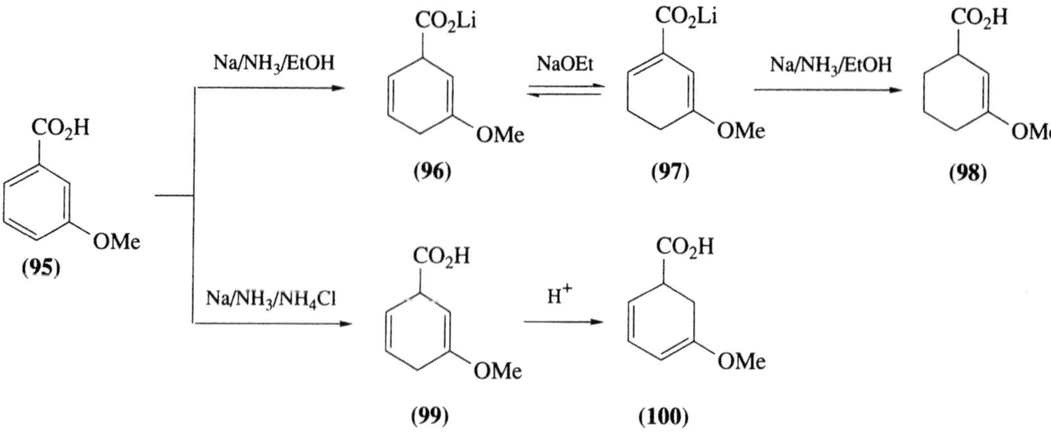

Na/NH$_3$/EtOH
CO_2Li
OMe
(96)
NaOEt
CO_2Li
OMe
(97)
Na/NH$_3$/EtOH
CO_2H
OMe
(98)

CO_2H
OMe
(95)

Na/NH$_3$/NH$_4$Cl
CO_2H
OMe
(99)
H$^+$
CO_2H
OMe
(100)

Scheme 18

(ii) Reduction of 2-methoxybenzoic acid and derivatives

Reduction of 2-methoxybenzoic acid (**101**) with lithium or sodium in ammonia with methanol as the proton source affords the 1,4-dihydrobenzoic acid (**102**; R = H) in good yields,[56] but with more weakly acidic alcohols or in the absence of methanol (conditions which are necessary for reductive alkylation), the methoxy group suffers partial hydrogenolysis; the extent to which this occurs depends on the metal used and the reaction temperature. For example, in the reductive alkylation of (**101**) to (**102**; R = Me), the major product from reduction by lithium in ammonia under reflux (–33 °C) followed by quenching with methyl iodide is (**103**) (corresponding to *ca.* 70% hydrogenolysis of the methoxy group). The use of sodium in place of lithium reduces the loss significantly (<50%),[54] while reduction with potassium completely suppresses the hydrogenolysis.[58]

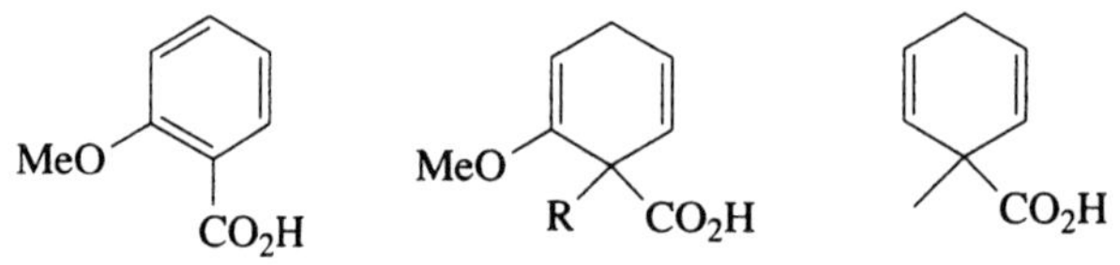

MeO
CO_2H
(101)

MeO
R CO_2H
(102)

CO_2H
(103)

The variable loss of the methoxy substitutent is in fact a feature common to most 2-methoxybenzoic acid derivatives and appears to stem from the ammonium ions which are generated on mixing of the acid with ammonia. Pretreatment with one equivalent of base prevents hydrogenolysis completely in most cases,[58] although the reduced solubility of the carboxylate salt can lead to incomplete reduction. Potassium salts are reasonably soluble, but reduction of the benzoate esters may afford the best solution (see Section 3.4.3.7).

3.4.3.6.5 Naphthoic acids

The metal–ammonia reduction of naphthoic acids occurs initially in the ring bearing the carboxy group, the position of which is an important factor in governing the final outcome of the reduction.

1-Naphthoic acid (**104**) gives dihydro acid (**106**), resulting from 1,4-reduction of the carboxy-bearing ring (Scheme 19).[59] The intermediacy of the enediolate (**105**) is indicated, since it may be trapped by a strong proton source such as ammonium chloride, or an alkyl halide.[60] If the simple 1,4-dihydro compound (**106**; R = H) is required, rapid quenching of the reaction mixture with ammonium chloride gives superior results[61] to carrying out the reduction in the presence of an alcohol, particularly ethanol; the latter conditions can lead to isomerization and/or overreduction.[62]

2-Methoxy-1-naphthoic acids (**107**; R^1 = H, OMe) have been reduced with lithium[63] or sodium[64] in ammonia, with no loss of the *ortho* methoxy substituent being reported. Thus, in the presence of ethanol the 1,4-dihydro acids (**108**; R^2 = H) are obtained, while reduction in the absence of a proton source and quenching with methyl iodide affords the alkylated acids (**108**; R^1 = H, OMe, R^2 = Me).[65] The conditions used for the reductive alkylation in the earlier examples (Na/NH$_3$/–70 °C) would enhance retention of the

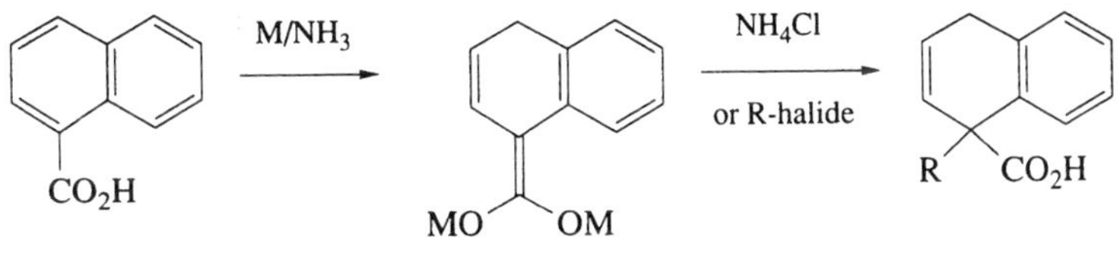

M/NH$_3$; NH$_4$Cl or R-halide

(104) (105) (106)

Scheme 19

methoxy group for reasons mentioned earlier. 4-Methoxy-1-naphthoic acid loses the methoxy group completely during reduction by metal–ammonia solutions, affording (**106**; R = H). Reductive methylation with lithium in ammonia, followed by a methyl iodide quench, gives a mixture of mono- and dialkylated acids (**106**; R = Me) and (**109**).[66] Reduction of the second ring occurs only after reduction of the first ring is complete. Dihydro acids (**108**; R^2 = Me, R^1 = H or 7-OMe) have been reduced further and the products utilized in model studies for the construction of the A-ring region of gibberellins,[65] while acid (**108**; R^1 = 6-OMe, R^2 = H) has been used as a convenient source of the corresponding β-keto ester.[67]

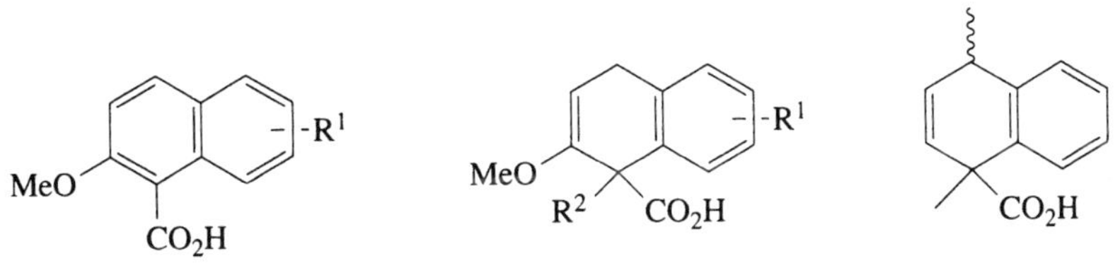

(107) (108) (109)

2-Naphthoic acids similarly undergo preferential reduction in the carboxy-bearing ring. If alcohol and limited quantities of metal (*ca.* 5 equiv.) are used, the 1,2,3,4-tetrahydro derivatives are formed in excellent yield. Reduction in the second ring also proceeds cleanly in the expected sense when additional metal is added, and acids (**110**) and (**111**) have been obtained in this way.[68]

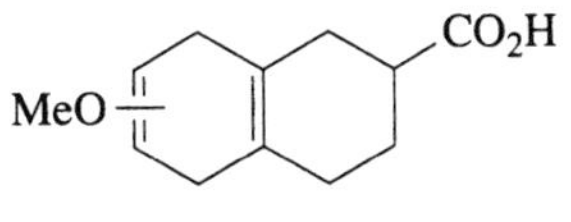

(**110**) 2,6-isomer
(**111**) 2,7-isomer

In the absence of alcohol the reduction may be halted at the dihydro stage, but with difficulty. Stoichiometric control (3 equiv. of metal) gives reasonable results, but it seems to be most effective to use 5 equiv. of lithium with added iron(III) chloride.[69] Thus, from the reduction of (**112**), quenching with ammonium chloride gives either the 1,2- or 1,4-dihydro products (**114**) and (**115**), respectively, while reaction with methyl iodide affords excellent yields of the 2-methyl-1,2-dihydro derivatives (**113**; Scheme 20).

Initial reduction of the C(1)—C(2) π-bond, in contrast to the normal 1,4-pathway encountered with most aromatic carbonyl compounds, is presumably a consequence of the energetic advantage in preserving aromaticity in the second ring. Similar results have been obtained with 2-acetylnaphthalenes and 3,4-dihydrophenanthren-1(2*H*)-ones (Section 3.4.3.9.3). Methoxy substituents in the ring bearing the carboxy group are prone to hydrogenolysis. Methoxy is invariably lost from C-1, C-3 and C-4 when reduction proceeds to the tetrahydro stage, but retention at C-3 and C-4 is possible when reduction is limited to the dihydro stage in the presence of anhydrous iron(III) chloride. Acids (**116**) to (**119**) have been prepared thus in 65–80% yields.[66]

Controlled hydrogenolysis of a methoxy substituent has been used to advantage in an expeditious synthesis of the tetrahydronaphthoic acid (**122**), an important starting material for the total synthesis of gibberellins, *i.e.* regioselective metallation of 1,6-dimethoxynaphthalene (**120**) at C-7 followed by carboxylation to afford (**121**) and then reduction furnished (**122**) in 71% overall yield (Scheme 21).[70]

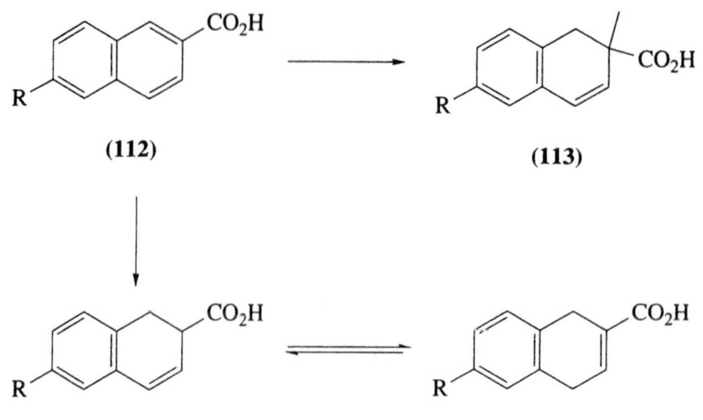

CO_2H
R

(112) (113)

(114) (115)

Scheme 20

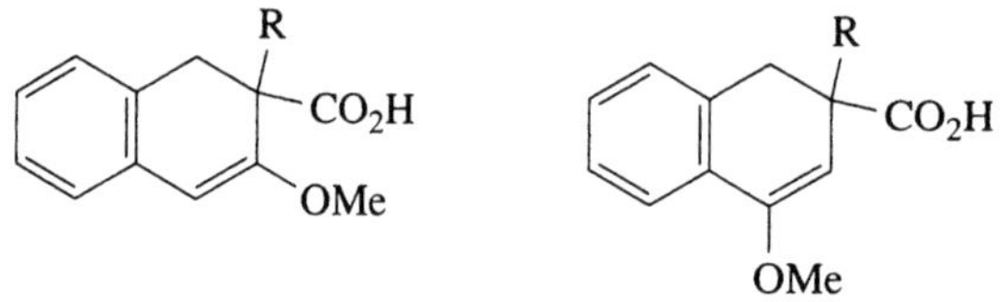

R
CO_2H
OMe

(116) R = H
(117) R = Me

(118) R = H (isolated as keto acid)
(119) R = Me

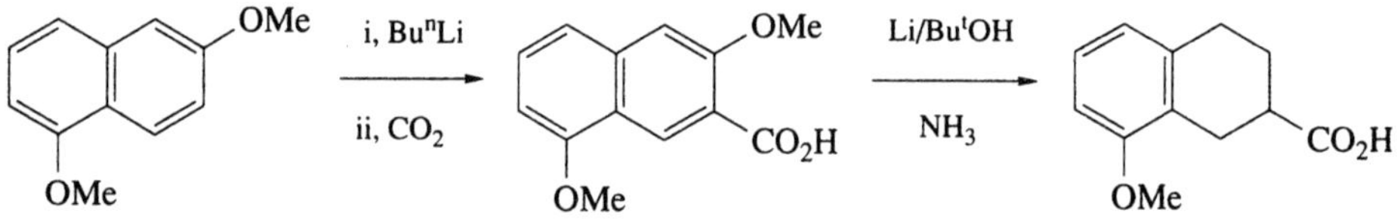

OMe
i, Bu^nLi
ii, CO_2
CO_2H
Li/Bu^tOH
NH_3

(120) (121) (122)

Scheme 21

3.4.3.6.6 *Biphenylcarboxylic acids*

While 2-phenylbenzoic acid is reduced smoothly (Li or Na, NH_3; NH_4Cl) to the dihydro derivative (**123**), the outcome from the reduction of the C-4 isomer is very sensitive to reaction conditions. It was found that in addition to the expected products (**124**) and (**125**) (the latter arising from isomerization of the 1,4-dihydro isomer), there was formed a 25% yield of a mixture of alcohol (**126**) and hydrocarbon (**127**).[71] In a subsequent investigation, in which rapid quenching procedures were employed, a 3:1 mixture of (**124**) and (**125**) was obtained, indicating that the previous overreduction had occurred during the quenching process. Greatly improved results were obtained by substituting the corresponding *t*-butyl ester as a substrate, for which reduction occurs only in the more-substituted ring.[72]

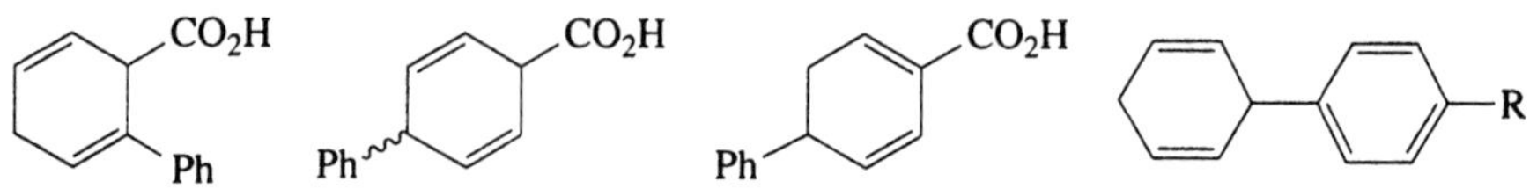

CO_2H
Ph
R

(123) (124) (125)

(126) R = CH_2OH
(127) R = Me
(128) R = CO_2H

3.4.3.6.7 Intramolecular reductive alkylation

Reduction of acid (**129**) by lithium in ammonia furnished the spiro-fused bicyclic acid (**130**) as the major product (Scheme 22). Equally good results were obtained with calcium metal, but lower yields were obtained with sodium and potassium.[73] The bromo and iodo analogs of (**129**), and the isomeric acid (**131**) failed to give any bicyclic products, however.[74]

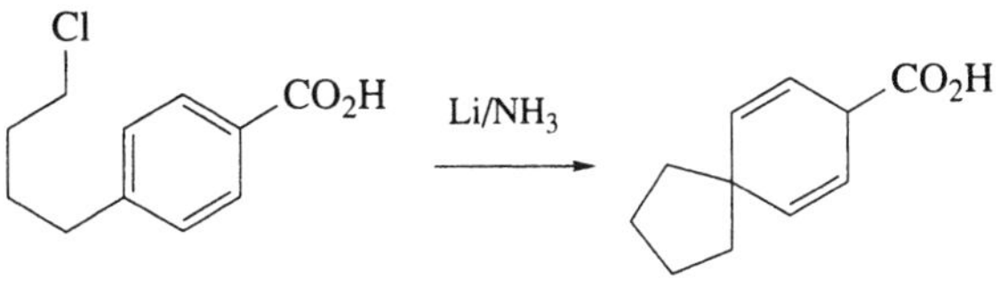

Cl CO_2H Li/NH_3 CO_2H

(129) **(130)**

Scheme 22

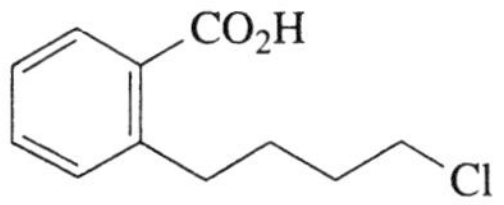

CO_2H Cl

(131)

3.4.3.7 Aromatic Carboxylic Esters

Early studies on the reduction of aromatic esters by metal–ammonia systems afforded little, if any, ring-reduced compounds but rather the products of functional group reduction,[75] and this became the prevailing wisdom for four decades. Nevertheless, a completely general procedure for ring reduction appears to be the low temperature reduction by sodium[76] or potassium[58] in the presence of one equivalent of *t*-butyl alcohol. Even lithium has been shown to give excellent results and may be the preferred metal if the reduction is followed by *in situ* alkylation (see below);[77] lithium fails to give ring reduction with 2,6-disubstituted substrates, however. The role of the proton source has been examined in one systematic study,[78] the results of which are summarized in Table 2. It appears that methoxy substituents promote reduction and, in the case of methyl 2,6-dimethoxybenzoate (**132**), obviate the need for an added proton source.

Rabideau has advocated the use of water as the proton source, added prior to the addition of the metal (sodium), followed by inverse quenching with aqueous ammonium chloride. The procedure is of little generality, however, since it was found to be unsatisfactory for methyl esters and reduction was inhibited by *para* ethyl, isopropyl and *t*-butyl substituents.[79]

Quenching the ester reductions with ammonium chloride affords the 1,4-dihydrobenzoates in 90–100% yields.[80] On the other hand, reaction with alkyl bromides or iodides, *in situ*, generally affords the C-1-alkylated products, again with good to excellent efficiency. These procedures have been successfully applied to more complex substrates, *e.g.* the gibbanes (**133**) and (**134**; Scheme 23).[81] In these cases the alkylations were completely stereoselective, with the approach of the electrophile being controlled by the neighboring B-ring carboxy. However, no diastereoselectivity was observed in the reductive methylation of the (–)-menthyl ester (**135**).[82]

Alkylation of potassium enolates is not always fruitful, and so counterion exchange with lithium bromide prior to addition of the electrophile has been recommended.[58] Reduction of aromatic esters instead of acids provides a number of potential advantages. The esters tend to be more soluble than carboxylate salts, hydrogenolysis of 2-alkoxy substituents does not appear to present the same problem, and the products are more stable. This can be important when enol ether functions are generated, allowing the necessarily acidic work-up procedures for carboxylic acids to be avoided. Indeed, the hydrolysis of enol ether functions may be very slow in aqueous acid and is best achieved through catalysis by mercury(II) nitrate.[81]

It is of considerable interest to find that *t*-butyl esters may be reduced satisfactorily, in contrast to aliphatic *t*-butyl esters, which are cleaved to acids under Birch conditions.[83] Also noteworthy is the regioselective reduction of the ester (**136**; Scheme 24),[72] in marked contrast to the reduction of the corresponding acid. Hydronaphthoate esters (**137**), (**138**) and (**139**) have been prepared (Scheme 25) and are envisaged as useful intermediates for diterpene synthesis.

Table 2 Reduction of Benzoate Esters by Metals in Liquid Ammonia[a]

Substrate	*Metal*	*Proton source*	*Yield of 1,4-dihydrobenzoate (%)*
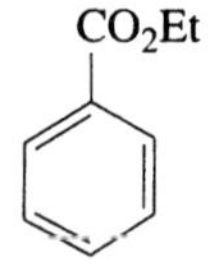 CO_2Et (ethyl benzoate)	Na	—	<5
	Na	Bu^tOH	>95
	K	—	30–50
	K	Bu^tOH	>95
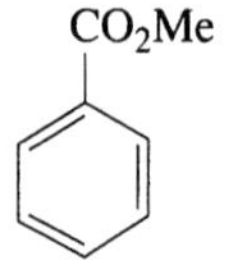 CO_2Me (methyl benzoate)	K	—	20–30
	K	Bu^tOH	>95
	Li	Bu^tOH	>95
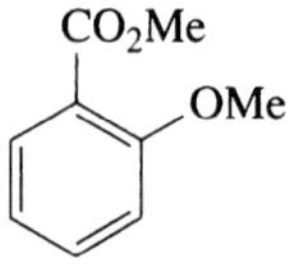 CO_2Me, OMe (methyl 2-methoxybenzoate)	K	—	60–70
	K	Bu^tOH	>95
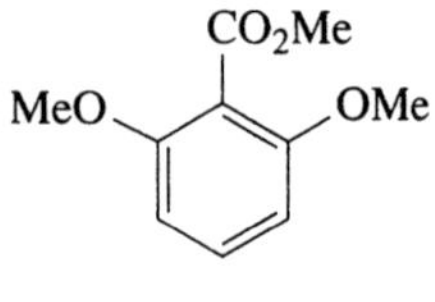 CO_2Me, MeO, OMe **(132)**	K	—	>95
	K	Bu^tOH	>95

[a]Reductions carried out at –78 °C with 2.5 gram equiv. of metal and 1.5 mol equiv. of Bu^tOH (if added).

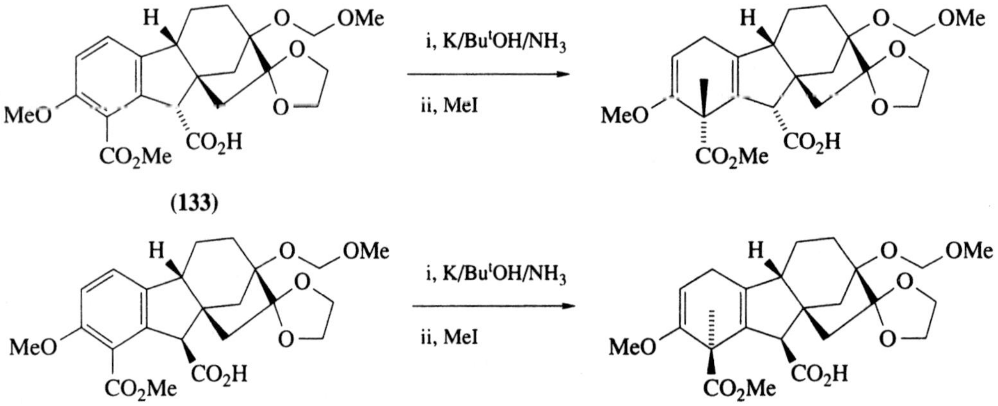

(133) i, $K/Bu^tOH/NH_3$; ii, MeI

(134) i, $K/Bu^tOH/NH_3$; ii, MeI

Scheme 23

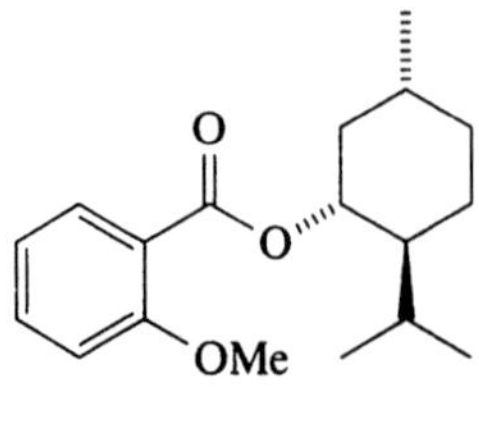

(135)

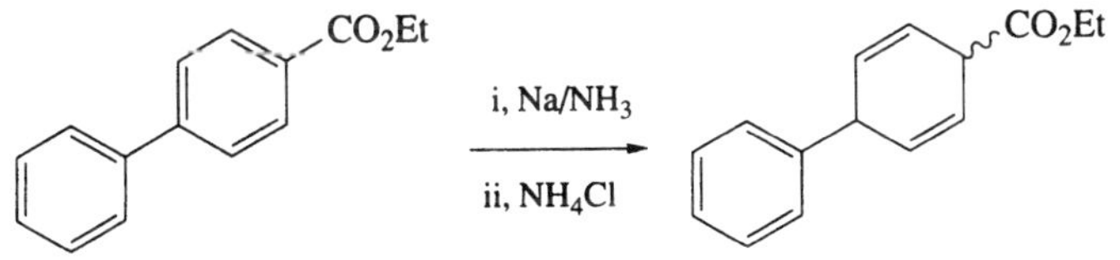

i, Na/NH_3
ii, NH_4Cl

(136)

Scheme 24

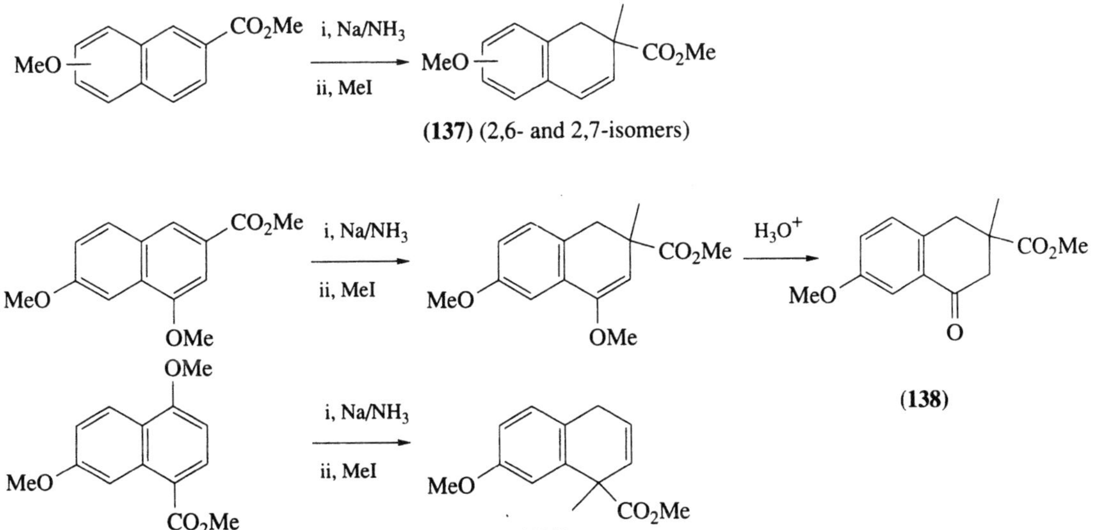

i, Na/NH_3
ii, MeI

(137) (2,6- and 2,7-isomers)

i, Na/NH_3
ii, MeI

H_3O^+

(138)

i, Na/NH_3
ii, MeI

(139)

Scheme 25

3.4.3.8 Aromatic Amides

The ring reduction of aromatic amides, and in particular benzamides, was once thought not to be possible,[84] and in fact sodium–liquid ammonia reductions of secondary amides had been recommended as a route to benzaldehydes.[85] It was found subsequently, however, that under the appropriate conditions 1,4-dihydrobenzamides could be prepared using sodium in ammonia in the presence of an alcohol. Both primary and secondary amides could be reduced, although increasing the steric bulk of the alkyl group decreased the yields of the 1,4-dihydro compound dramatically, while tertiary benzamides gave either reduction of the carbonyl group or a very low yield of the 1,4-dihydro compound.[53,86]

In a more recent study it was found that the conditions (potassium metal, ammonia, –78 °C, 1 equiv. *t*-butyl alcohol) which had proved to be so effective with esters also gave excellent results with the 2-methoxybenzamides (**140**) and (**141**; Scheme 26), although the successful outcome was tentatively attributed to the presence of the *ortho* methoxy. While both (**140**) and (**141**) gave high yields of the *C*-methylated derivatives (**142**) and (**143**), respectively, only the less-hindered (**140**) could be alkylated with 2-bromoethyl acetate, to give (**144**). Amide (**141**) was prepared by C-6 lithiation of (**140**) followed by *C*-methylation, and opens up the exciting prospect of combining two powerful methodologies.[82]

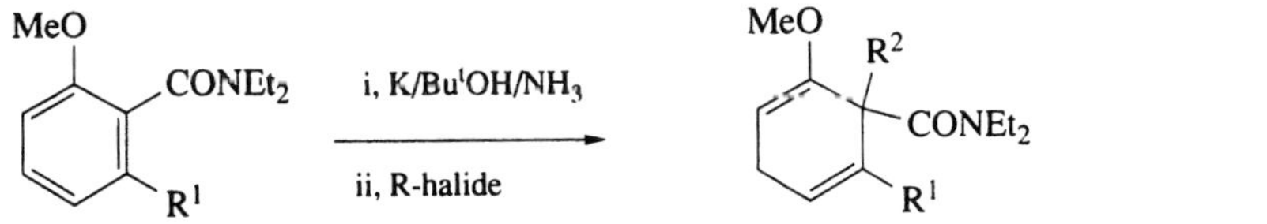

i, $K/Bu^tOH/NH_3$
ii, R-halide

(140) $R^1 = H$
(141) R = Me

(142) $R^1 = H$, $R^2 = Me$
(143) R^1, $R^2 = Me$
(144) $R^1 = H$, $R^2 = CH_2CH_2OAc$

Scheme 26

Although the potential for achieving enantioselective syntheses through the reductive alkylation of esters incorporating a chiral auxiliary into the substrate appears to be poor, the potential for stereocontrol in the corresponding reactions of chiral amides is excellent. The L-prolinol derived oxazepine (**145**) afforded the *C*-methylated product (**146**) as an 85:15 mixture with its *cis* diastereomer, while higher levels of diastereoselectivity were obtained with bulkier halides (*e.g.* EtI, 99:1 and $CH_2{=}CHCH_2Br$, 98:2) (Scheme 27). A range of prolinolamides were also studied, including (**147**; R = H, Me, CH_2OMe), which affords even better stereoselectivity independent of R, *e.g.* (**148**) was formed as a 260:1 mixture with its C-1 epimer (Scheme 28).[87]

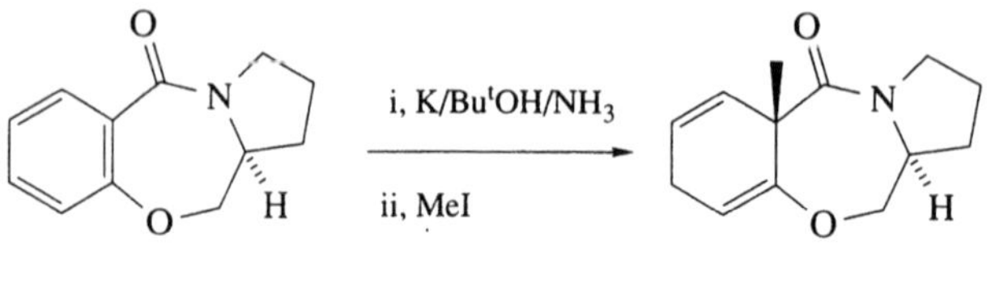

i, K/Bu^tOH/NH$_3$; ii, MeI

(**145**) (**146**)

Scheme 27

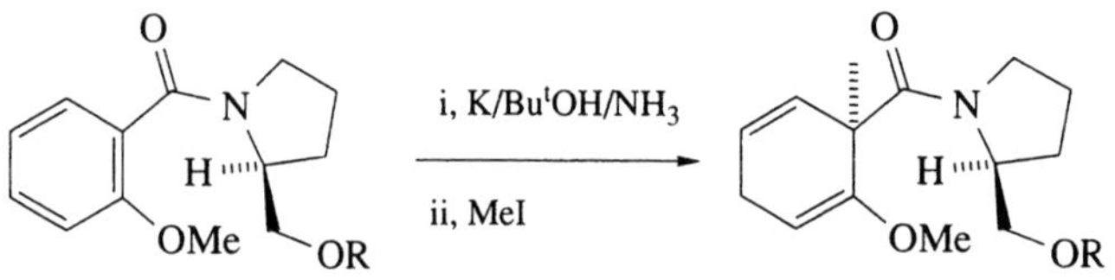

i, K/Bu^tOH/NH$_3$; ii, MeI

(**147**) (**148**)

Scheme 28

3.4.3.9 Aromatic Ketones

As with esters, the prevailing view held for many years was that ring reduction of aromatic ketones was not feasible. The first indication that this was incorrect was provided by the reduction of the hydrophenanthrenone (**149**; Scheme 29).[88] Further isolated examples of ring reduction in naphthyl ketones and molecules containing such moieties have been reported subsequently, but it is clear that these systems are much more easily reduced than simple benzenoid systems. Optimal procedures for the reduction and reductive alkylation of such systems were first established by Watanabe and Narisada for acetophenone and its methoxy derivatives.[16]

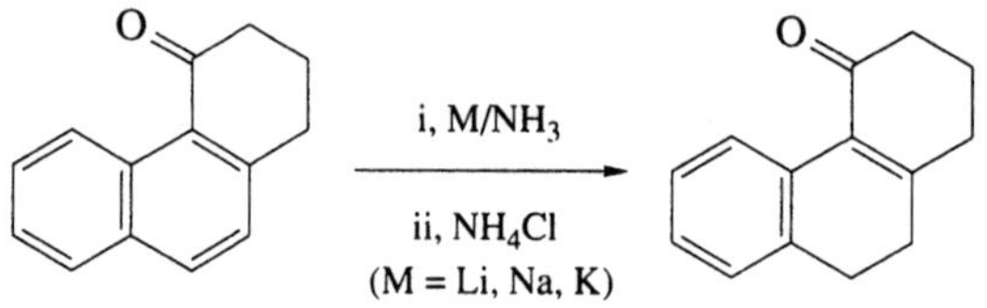

i, M/NH$_3$; ii, NH$_4$Cl (M = Li, Na, K)

(**149**) (**150**)

Scheme 29

3.4.3.9.1 Acetophenones

A systematic study of the reductive alkylation of acetophenones[16] revealed that the desired transformation (Scheme 30) required a careful selection of reagents and conditions. The best results were obtained from reduction by potassium in ammonia at –78 °C, with *t*-butyl alcohol as the proton source. Exchange of the potassium counterion of the enolate (**152**; M = K) for lithium then ensured regioselective alkylation at C-1 to give (**153**) in 80–90% yields (Scheme 30). Metals other than potassium as the reductant led to undesirable side reactions with the carbonyl group, which included simple reduction to the methylcarbinol and ethylbenzene (lithium or sodium), while the absence of a proton source or presence of a strong one (H_2O or acetic acid) during the reduction encouraged formation of the pinacol (**154**). Sodium as a counterion with the enolate (**152**; M = Na) led to overalkylated and *O*-alkylated products. The successful outcomes with potassium as reductant may be due to the fact that the potassium pinacolates are unstable (addition of KNH_2 to benzopinacol, for example has been shown to cause dissociation to benzophenone ketyl).[89]

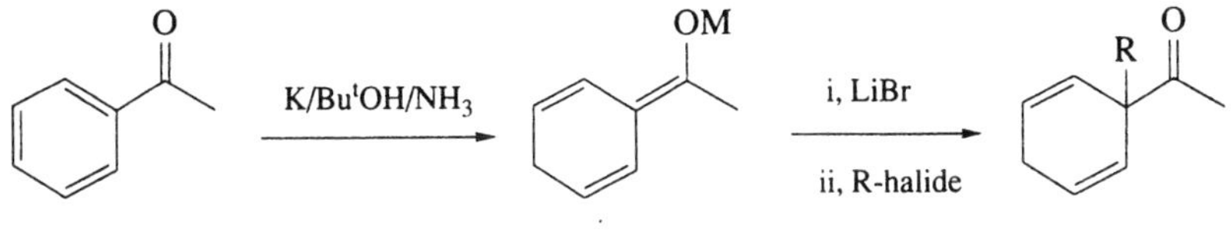

K/Bu^{t}OH/NH$_3$; i, LiBr; ii, R-halide

(151) **(152)** **(153)**

Scheme 30

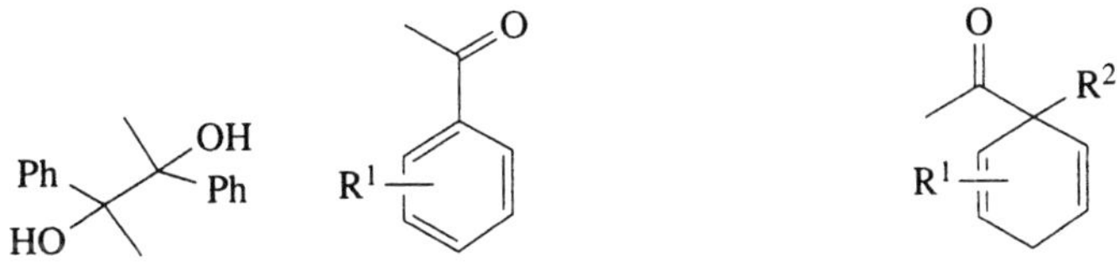

(154)

(155) R^1 = 2-OMe
(156) R^1 = 3-OMe
(157) R^1 = 4-OMe
(158) R^1 = 2,5-$(OMe)_2$

(159) R^1 = H, R^2 = Et
(160) R^1 = H, R^2 = $CH_2CH{=}CH_2$
(161) R^1 = H, R^2 = CH_2CO_2Et
(162) R^1 = 2-OMe, R^2 = Me
(163) R^1 = 3-OMe, R^2 = Me
(164) R^1 = 2,5-$(OMe)_2$, R^2 = CH_2Ph

Acetophenone and several derivatives, (**155**) to (**158**), were subjected to reductive alkylation using the appropriate conditions. As might be expected from results with benzoic acids, the methoxy group was lost from 4-methoxyacetophenone, but the 2- and 3-derivatives (**155**) and (**156**) gave good yields of alkylated 2,5-cyclohexadienes, as did the 2,5-dimethoxy compound (**158**). A range of products with various electrophiles was obtained in good to excellent yield, including (**159**) to (**164**).[16,90]

3.4.3.9.2 1-Tetralones and 1-indanones

Early reports indicated that 1-tetralones are reduced by sodium or lithium in liquid ammonia at –33 °C to the corresponding tetralols or tetralins, although a very recent study has shown that up to seven products (four of them dimeric) may be formed when lithium is utilized.[91] Nevertheless, the reductive alkylation conditions applied successfully to the ring reduction of acetophenone also succeed with 1-tetralones. 1-Tetralone itself was transformed into (**165**) in 60% yield,[16] while modifications to the original procedure allowed the methoxylated derivatives (**166**) to (**169**) to be prepared in excellent yields.[92]

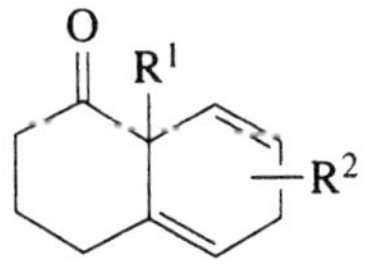

(165) R^1 = Me, R^2 = H
(166) R^1 = Me, R^2 = 5-OMe
(167) R^1 =Me, R^2 = 7-OMe
(168) R^1 = Me, R^2 = 5,7-$(OMe)_2$
(169) R^1 = $CH_2C(Br){=}CH_2$, R^2 = 5-OMe

Keto acid (**170**) has been transformed into the diastereomeric mixture of keto acids (**171**) and (**172**), where the predominant reaction pathway involves approach of the electrophile *anti* to the carboxylate group (Scheme 31). It was found to be essential to neutralize the ammonium ions generated from addition of the substrate acid to the liquid ammonia before adding the potassium metal, however.[92]

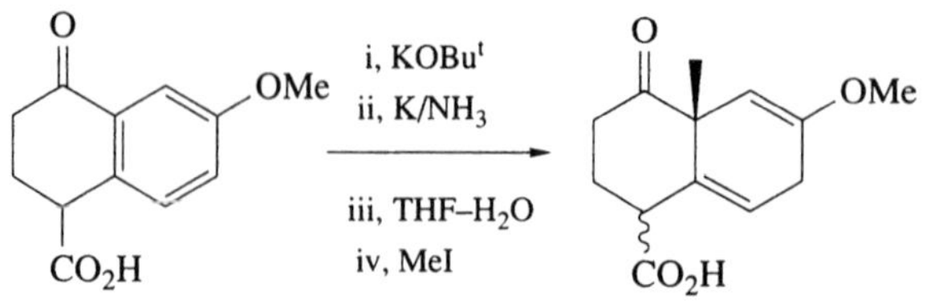

i, $KOBu^t$
ii, K/NH_3
iii, THF–H_2O
iv, MeI

(**170**) CO_2H OMe

(**171**) 4α 64%
(**172**) 4β 19%

Scheme 31

1-Indanones also undergo reductive alkylation, although not as well as 1-tetralones. Good yields were obtained with methyl iodide and 1-indanone as well as its 4-, 6- or 7-methoxy derivatives to give the dihydroindanones (**173**) to (**176**).[93]

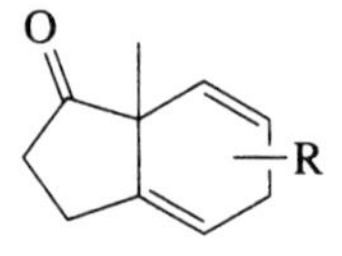

(**173**) R = H
(**174**) R = 4-OMe
(**175**) R = 6-OMe
(**176**) R = 7-OMe

3.4.3.9.3 Acetylnaphthalenes and hydrophenanthrenones

The reduction and reductive alkylation of acetylnaphthalenes follows a similar course to that found for naphthoic acids, in that the ring bearing the acetyl group is preferentially reduced. Simple reduction products are stable enough to be isolated, and regiospecific alkylations of the generated enolates do not appear to be as dependent on the cation as they are with acetophenones, 1-tetralones and 1-indanones. 1-Acetylnaphthalene may be reduced by lithium, sodium or potassium in liquid ammonia at –33 °C to the dienolate (**177**),[94,95] which is alkylated regiospecifically at C-1 by methyl iodide to give (**178**; R = Me) and also protonated at C-1 to give (**178**; R = H) when the reaction mixture is subjected to an inverse quench with a large excess of aqueous ammonium chloride.[96] Traditional quenching procedures afford the 3,4-dihydro isomer (**179**) as a result of isomerization (Scheme 32). The 1-methoxy group of 1,7-dimethoxynaphthalene (**180**) directs electrophilic attack to the *para* position; sodium/ammonia solutions then readily hydrogenolyze this group, thereby giving ready access to 1-alkanoyl-6-methoxynaphthalenes (**182**; Scheme 33).[97]

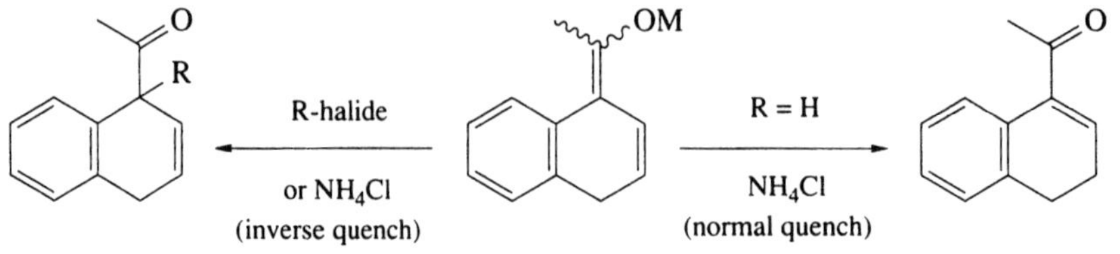

R-halide or NH_4Cl (inverse quench) ← OM → R = H NH_4Cl (normal quench)

(**178**) (**177**) (**179**)

Scheme 32

Early studies on the reduction and reductive alkylation of 2-acetylnaphthalenes (**184**) afforded mixtures of dihydro and tetrahydro derivatives. In one case, the alcohol (**188**) was the major product.[98] A comprehensive examination of the reductive process and quenching methods has been carried out recent-

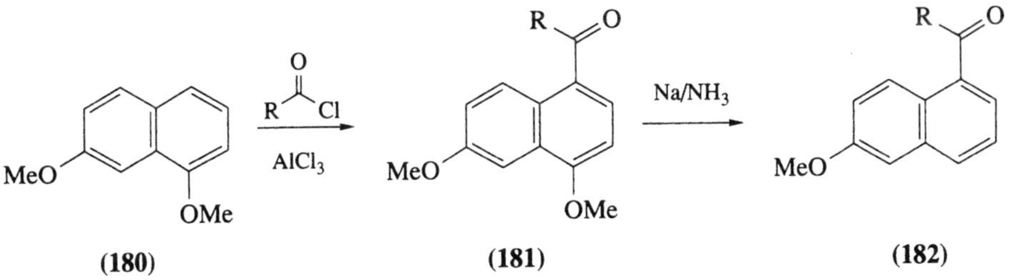

(180) (181) (182)

Scheme 33

ly, however, and the conditions required to optimize the yields of either dihydro derivatives (**186**) or tetrahydro products (**187**) defined, as summarized in Scheme 34.[95]

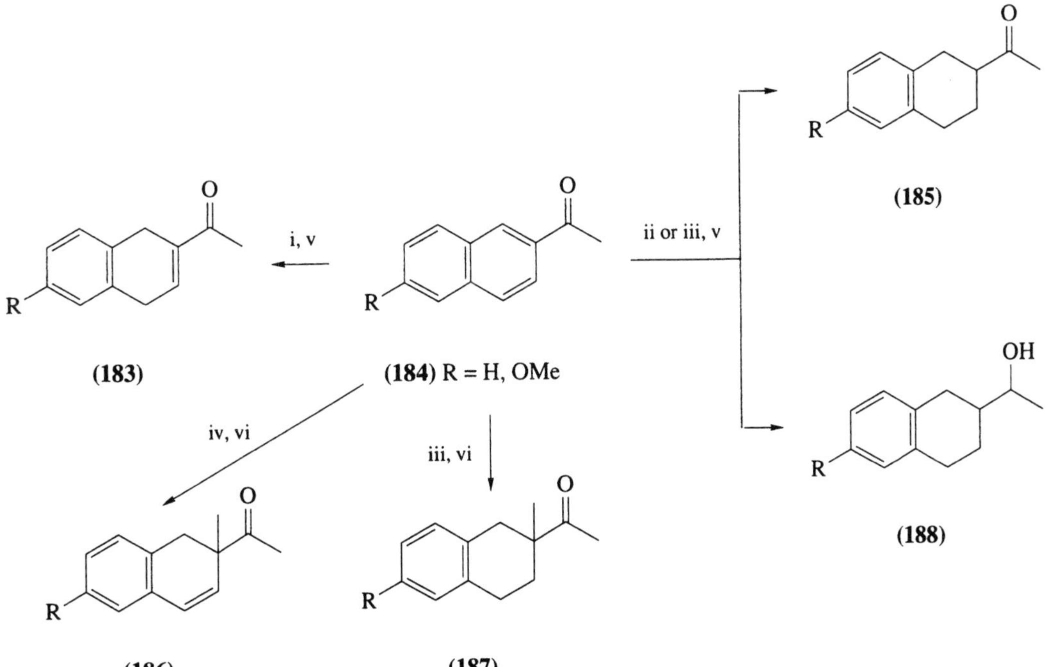

(183) (184) R = H, OMe (185) (186) (187) (188)

i, K/NH_3; ii, Na/NH_3; iii, Li/NH_3; iv, $Li/NH_3/FeCl_3$; v, NH_4Cl; vi, MeI

Scheme 34

Dihydrophenanthren-4(1*H*)-one (**189**) is readily reduced by lithium, sodium or potassium in ammonia at –33 °C, presumably to the dienolate (**190**), since reaction with methyl iodide affords (**191**). Quenching with ammonium chloride, however, leads to the conjugated enone (**192**; Scheme 35).

The isomeric ketone (**193**) reduces equally well with lithium, sodium or potassium in refluxing ammonia, but to the tetrahydro stage, *i.e.* to ketone (**194**); optimal yields were obtained with 4 mol equiv. of metal and catalytic amounts of iron(III) chloride.[99] The presence of the iron salts in this example did not limit the reduction to the dihydro stage, contrary to the findings with 2-acetylnaphthalenes, but gave slightly higher yields than did reductions in their absence. Reductive alkylation of (**193**) affords a variety of products depending on the metal used (Scheme 36). Potassium and sodium are effective reductants, but encourage bisalkylation, yielding mostly (**195**) with a *cis* ring fusion. Alkylation of the lithium enolate is regiospecific, affording mainly (**196**), but less stereoselective.

The B-ring in the tetracyclic compounds (**197**; R = H, OMe) has been reduced with sodium and alcohol in liquid ammonia to afford the α,β-unsaturated ketones (**198**; R = H, OMe; Scheme 37).[100]

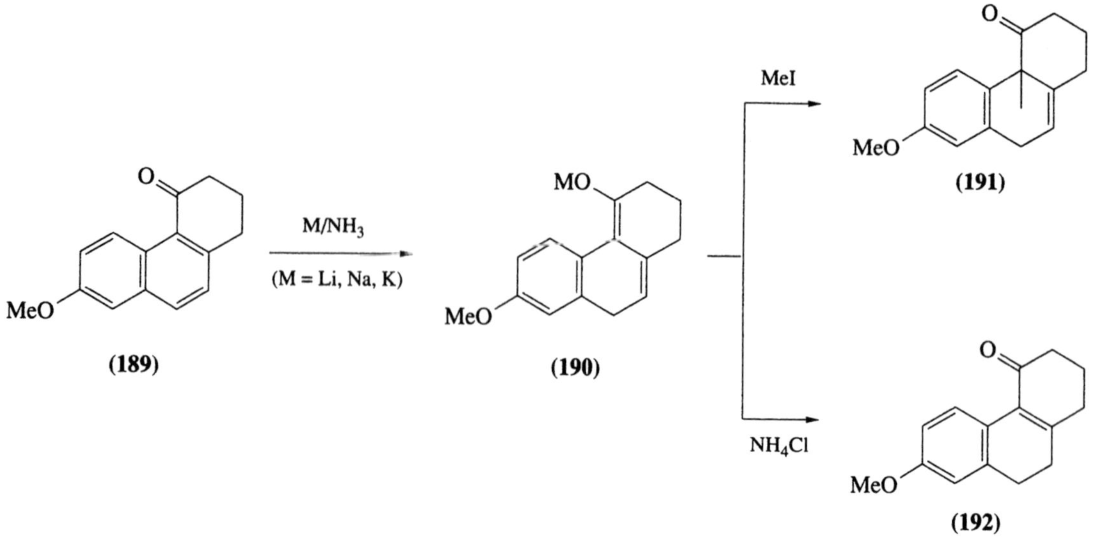

MeI

O

MeO

(191)

O

MO

M/NH_3

(M = Li, Na, K)

MeO

MeO

(189)

(190)

O

NH_4Cl

MeO

(192)

Scheme 35

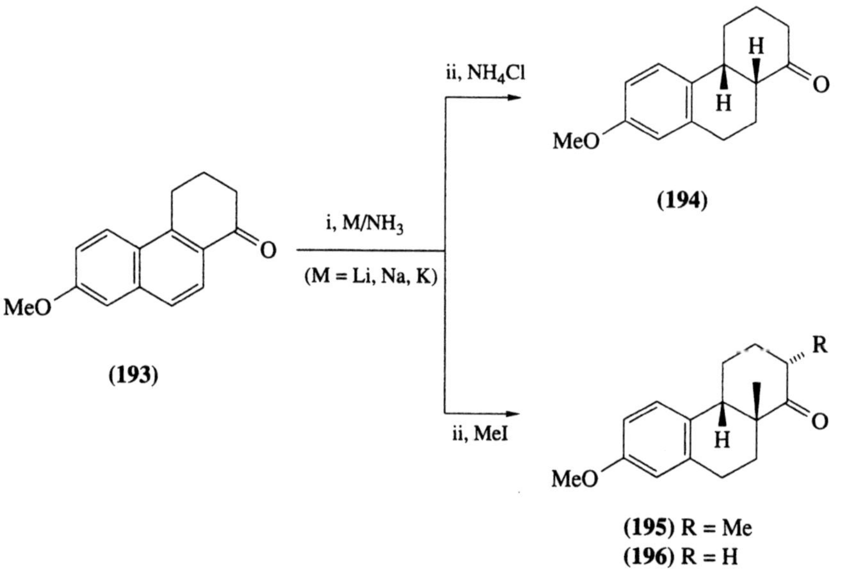

H

ii, NH_4Cl

O

H

MeO

(194)

i, M/NH_3

O

(M = Li, Na, K)

MeO

(193)

R

ii, MeI

O

H

MeO

(195) R = Me
(196) R = H

Scheme 36

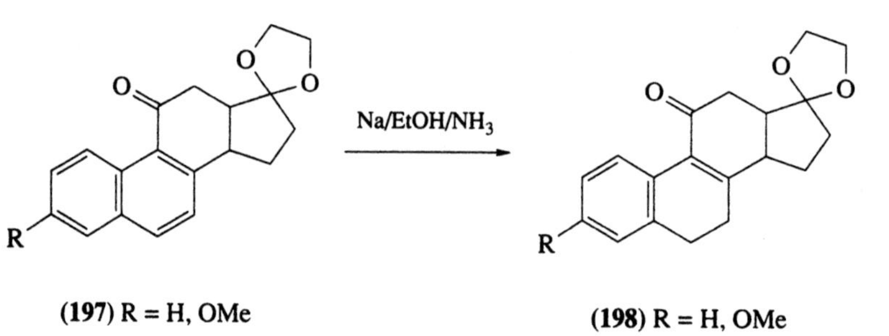

O

O

O

O

O

O

$Na/EtOH/NH_3$

R

R

(197) R = H, OMe

(198) R = H, OMe

Scheme 37

3.4.3.10 Aryl- and Benzyl-silanes

3.4.3.10.1 Arylsilanes

Birch reduction of aryltrialkylsilanes by lithium–liquid ammonia–ethanol at –70 °C has been studied systematically and a typical conversion is indicated in Scheme 38.[101] The major products are usually the 1,4-dihydro derivatives in accord with ESR evidence, which shows that a trimethylsilyl group stabilizes aromatic radical anions.[102] A significant exception is *p*-trimethylsilyltoluene (**199**), which affords the 2,5-dihydro product (**200**) in 95% yield (Scheme 39) and is probably typical of arylsilanes bearing a *para* substituent which is electron releasing. Cleavage of allylic silyl groups in the products by ethoxide ion is a significant problem, especially at higher temperatures (–30 °C). Further reduction to tetrahydro and perhydro derivatives also occurs under these conditions. Electrochemical reduction (in methylamine–lithium chloride) affords better yields of the simple dihydroarylsilanes.[103]

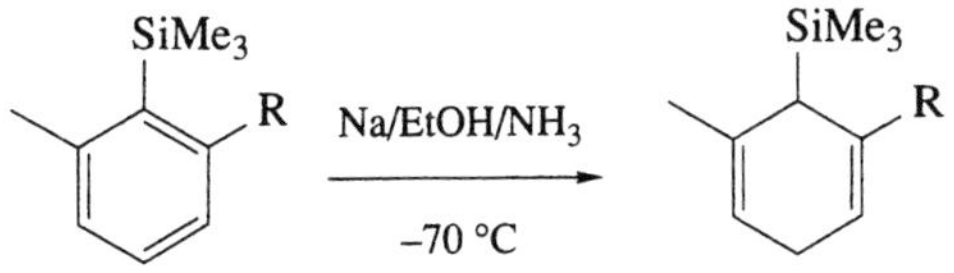

$SiMe_3$ R Na/EtOH/NH_3 –70 °C $SiMe_3$ R

Scheme 38

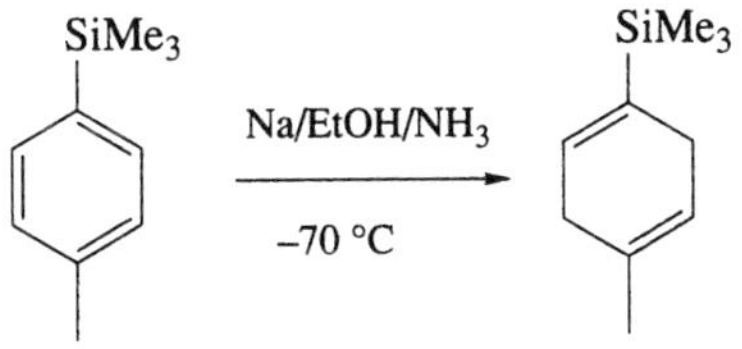

$SiMe_3$ Na/EtOH/NH_3 –70 °C $SiMe_3$

(199) **(200)**

Scheme 39

The 'activating' influence of the trimethylsilyl function is also apparent in the reduction of naphthalene derivatives (Scheme 40). The 1,4-disilyl derivative (**201**) is reduced exclusively in the substituted ring to give (**202**), while 2-trimethylsilylnaphthalene (**203**) affords a 4:1 mixture of (**204**) and (**205**), respectively.[102] As would be expected, the reduction of *C*-silylbenzoic acids and their esters is controlled by the carboxy function.[6]

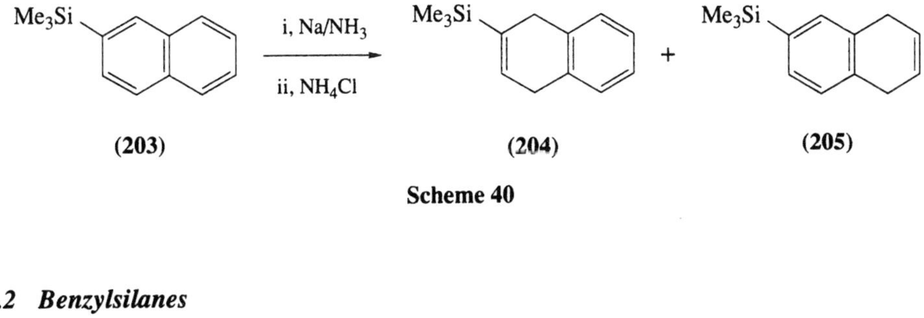

$SiMe_3$ $SiMe_3$ i, Na/NH_3 ii, NH_4Cl $SiMe_3$ $SiMe_3$

(201) **(202)**

Me_3Si i, Na/NH_3 ii, NH_4Cl Me_3Si + Me_3Si

(203) **(204)** **(205)**

Scheme 40

3.4.3.10.2 Benzylsilanes

The Birch reduction of benzylsilanes generally proceeds in high yield and affords the opportunity to prepare alkylidenecyclohexene structures by protodesilylation of the dihydroaromatic products, examples of which are provided in Scheme 41.[104] Reduction of the parent benzyltrimethylsilane proceeds satisfactorily with ethanol as the proton source, but fails with some substrates. The substitution of *t*-butyl alcohol for ethanol, however, avoids the generation of nucleophilic ethoxide, thereby furnishing a more satisfactory procedure. The methodology has also been used in the preparation of γ-hydroxysilanes which were subjected to oxidative fragmentations.[105]

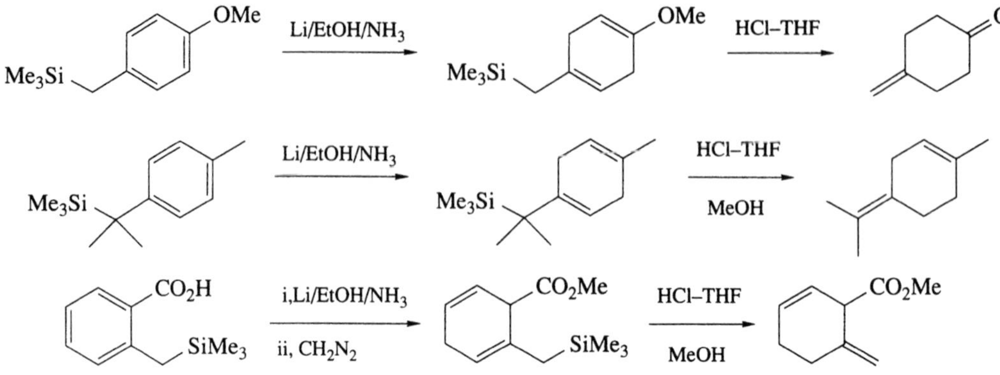

Scheme 41

3.4.3.11 Hydrogenolysis during Birch Reductions

3.4.3.11.1 Hydrogenolysis of nuclear substituents

Hydrogenolysis of halogen and sulfur groups is to be expected during Birch reductions, but if the substrate is sufficiently activated it is possible to preserve a fluorine substituent.[106] The thioether function has also been retained in the reduction of 3-methylthiobenzoic acid to its 1,4-dihydro derivative, but not the *ortho* isomer.[107] Hydrogenolysis of simple aryl alkyl ethers is rare, but there are a number of well-defined situations in which this is likely to occur, *e.g.* ethylenedioxybenzene is reduced to phenol.[108] It has already been noted in earlier sections (3.4.3.6 and 3.4.3.7) that alkoxy groups in a *para* relationship to a carbonyl group are invariably lost during metal–ammonia reductions. The methoxy group of 4-methoxybiphenyl is also lost during reduction,[109] but its loss may be prevented by the presence of a strategically located hydroxy group (converted to phenoxide during the reaction, *cf.* Scheme 8), as in the formation of (**207**; Scheme 42).[110] Biaryl ethers undergo reductive fission, a reaction which has been applied extensively to structure elucidation in the dibenzylisoquinoline alkaloids.[111]

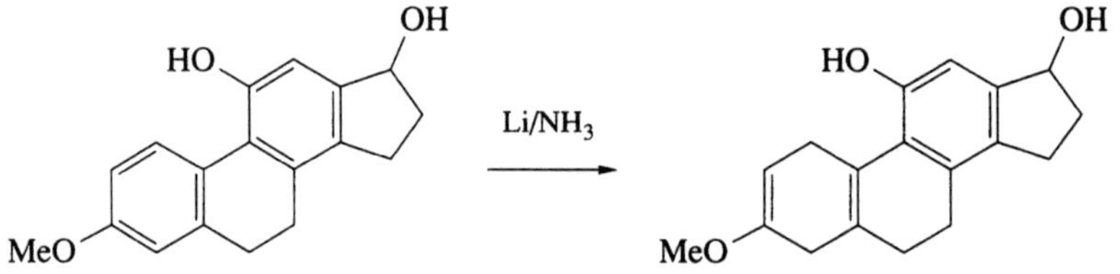

(**206**) (**207**)

Scheme 42

Deoxygenation of phenols may be achieved by reduction of aryl diethyl phosphates with lithium or sodium in liquid ammonia.[112] A recent application of the methodology is outlined in Scheme 43.[113] The reaction works well with a variety of substituted phenols, but not with dihydric phenols or naphthols. The alternative reduction of aryl sulfonates has also been examined, but the limited solubility of these derivatives can present difficulties.

3.4.3.11.2 Protocols for the retention of benzylic substituents

As a general rule, oxygen substituents are lost from benzylic positions during metal–ammonia reductions. An important set of exceptions are those compounds which possess a methoxy group in a *para* relationship to the benzylic substituent. Thus, 4-methoxybenzyl alcohol (**208**) may be reduced in high yield

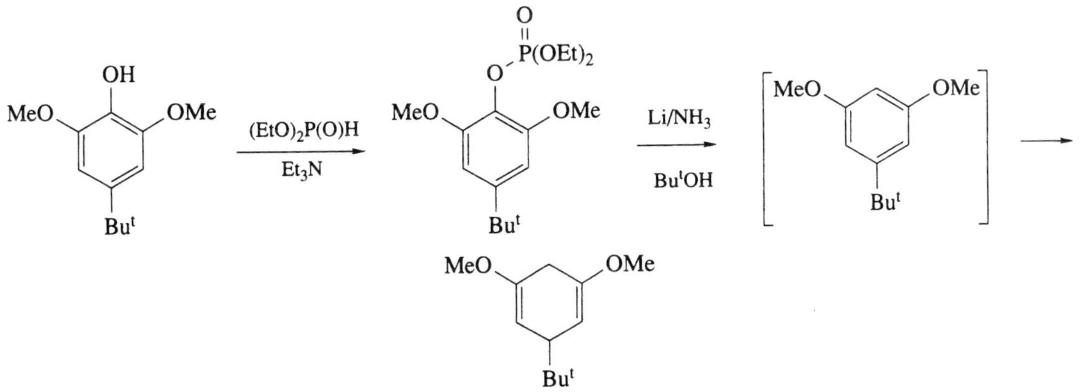

OH MeO OMe Bu^t $(EtO)_2P(O)H$ Et_3N O P(OEt)$_2$ MeO OMe Bu^t Li/NH$_3$ Bu^{t}OH [MeO OMe Bu^t] → MeO OMe Bu^t

Scheme 43

to the 2,5-dihydro derivative (**209**; Scheme 44), whereas benzyl alcohol itself and the 2- and 3-methoxy derivatives are reduced to the dihydrotoluenes.[114]

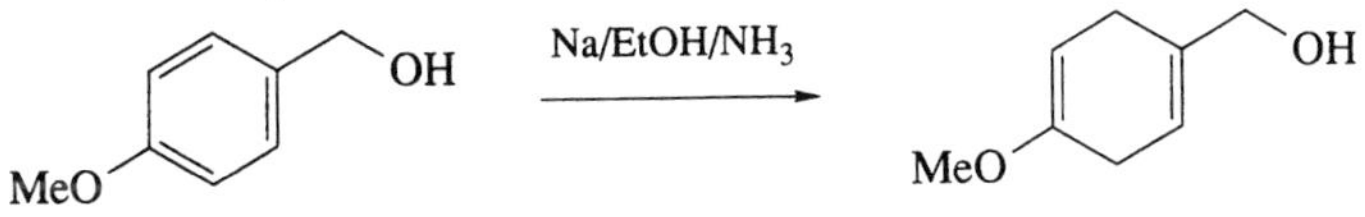

MeO OH Na/EtOH/NH$_3$ MeO OH

(208) **(209)**

Scheme 44

Reduction of 1-(4-methylphenyl)ethanol affords a 4:1 mixture of the 2′,5′-dihydro derivative with 4-ethyltoluene,[115] showing again the influence of an electron-releasing substituent in the *para* position. The result with 4-methoxybenzyl alcohol appears to be completely general. Benzylic hydroxy groups may also be preserved when the proximal aryl ring is hydroxylated, as in the formation of (**207**; Scheme 42) and in the reduction outlined in Scheme 45.[116]

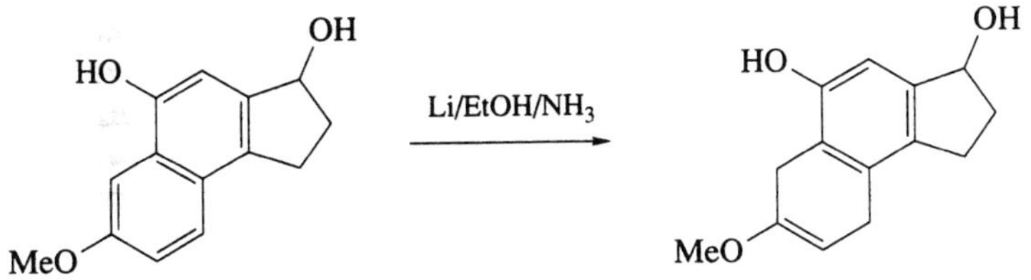

HO OH MeO Li/EtOH/NH$_3$ HO OH MeO

Scheme 45

Metal–ammonia reduction of acetal (**210**) was reported to afford the deoxygenated derivative (**211**; Scheme 46) as the major product,[117] but the choice of a dioxolane protecting group turns out to be unfortunate, since hydrogenolysis is probably a consequence of strain in the five-membered ring. It is possible to retain the acetal function during reduction of the acyclic or 1,3-dioxane analogs and 30–75% yields of the dihydro derivatives (**212**) have been reported.[118]

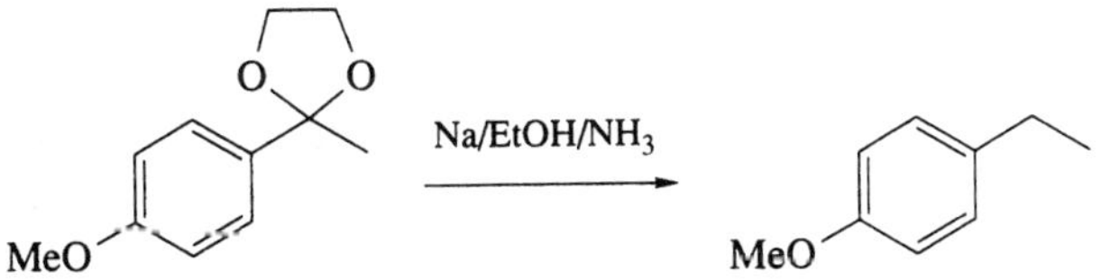

O O MeO Na/EtOH/NH$_3$ MeO

(210) **(211)**

Scheme 46

Benzylic amines are resistant to hydrogenolysis under Birch reduction conditions, unless the amine is quaternary, as in the Emde reaction. A means of preserving an aryl carbonyl group is therefore to protect it as an aminal derivative. An illustration is provided by the preparation of aldehyde (**213**), an important flavor constituent of cumin, as outlined in Scheme 47.[119]

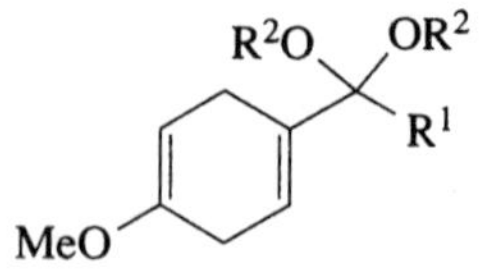

(212)

R^1 = H, Me; R^2 = Me, Et, $-CH_2CH_2-$

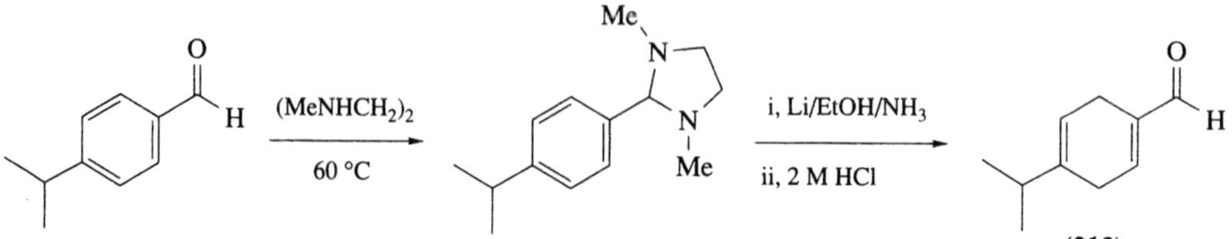

Scheme 47

3.4.4 FURTHER METHODS FOR THE REDUCTION OF AROMATIC RINGS

Extensive investigations have been made into further methods for the reduction of aromatic rings based on the use of dissolving metals in other solvents, especially the lower molecular weight amines (the Benkeser reduction), electrochemical methods (cathodic reductions), photochemical methods and the reaction of radical anions with silylating reagents rather than proton sources. The aim of much of this work has been to produce the normal Birch products more conveniently or cheaply, but very often the outcome has been quite distinct. The alternative method may then provide access to products which are not so easily obtained by the standard metal–liquid ammonia methodology.

3.4.4.1 The Benkeser Reduction

Solutions of Group I metals in the lower molecular weight amines are more potent reductants than those in liquid ammonia, and as a general rule substrates are more extensively reduced than by the Birch method.[120] Naphthalene (**49**; Scheme 48), for example, is reduced by a solution of lithium in ethylamine to a 1:1 mixture of $\Delta^{1(9)}$- and Δ^9-octalins (**214**) and (**215**). If ethylenediamine is employed as the medium, the completely saturated decahydronaphthalene is formed, while the proportion of (**215**) may be increased to 80% by utilizing a (1:1) mixture of ethylamine with dimethylamine.[121] The formation of the more-substituted alkene appears to be a general result for such primary and secondary amine mixtures and has been used to good effect in the reduction of both toluene and cumene to their 3,4,5,6-tetrahydro derivatives (**216**) and (**217**), respectively, in *ca.* 80% yields. A comprehensive review of these kinds of reducing systems, which also draws comparisons with the Birch method, is available,[122] but more recent-

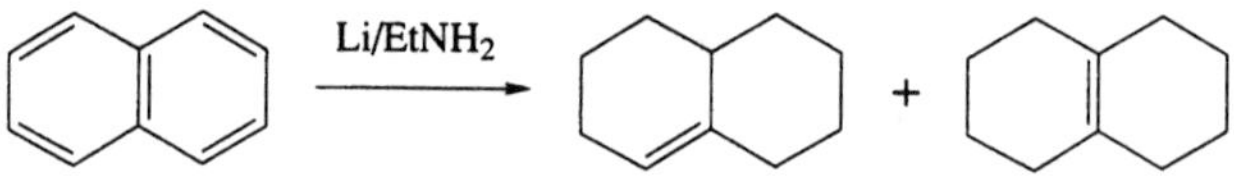

(49) **(214)** **(215)**

Scheme 48

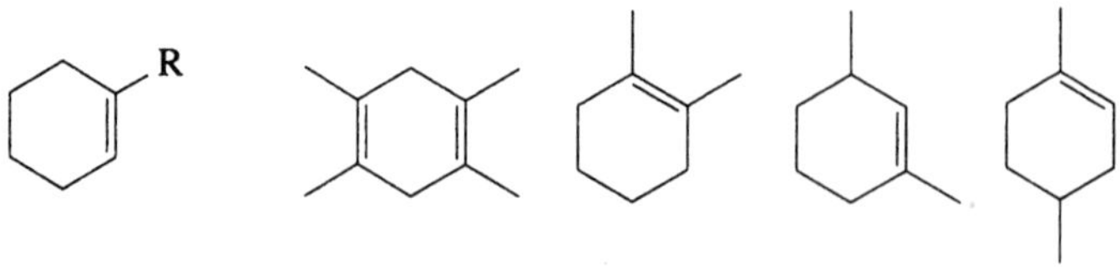

(216) R = Me **(219)** **(220)** **(221)** **(222)**

(217) R = Pr^i

(218) R = Bu^t

ly an improved system has been described. This is based on solutions of calcium in methylamine–ethylenediamine (1:1) and products (**218**) to (**225**), for example, have been formed in generally high yields from the parent aromatic hydrocarbons.[123]

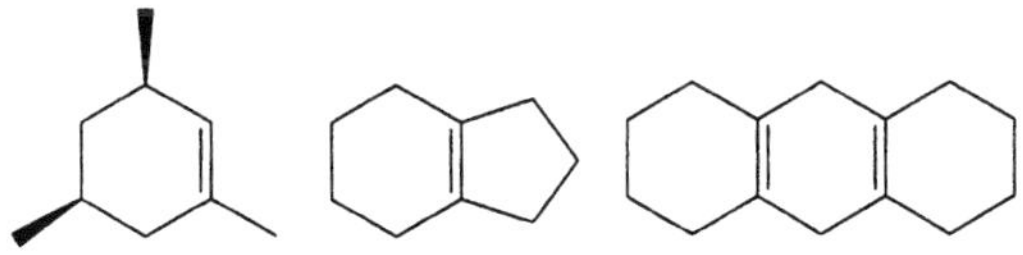

(**223**) (**224**) (**225**)

3.4.4.2 Electrochemical Methods

Electrochemical methods for the reduction of aromatic substrates utilizing ammonia and amines as solvents with lithium salts as electrolytes have been successful. Toluene was reduced to the 2,5-dihydro derivative in 95% yield in methylamine–lithium chloride if an undivided cell was used, while a 53:47 mixture of 3- and 4-methylcyclohexenes was formed in a divided cell.[124]. Of greater interest, however, are attempts to achieve these reductions in aqueous media. In one experiment utilizing a two-phase mixture of substrate in aqueous tetra-*n*-butylammonium hydroxide and a mercury cathode, anisole was reduced on a preparative scale (15 g) to its 2,5-dihydro derivative in 80% yield. The optimal temperature for most reductions appeared to be 60 °C and under these conditions, even suspensions of high molecular weight substrates could be successfully reduced, *e.g.* steroid (**226**) afforded a >90% chemical yield of (**227**). Much higher coulombic yields were obtained when a small amount of THF was added to the mixture, however.[125]

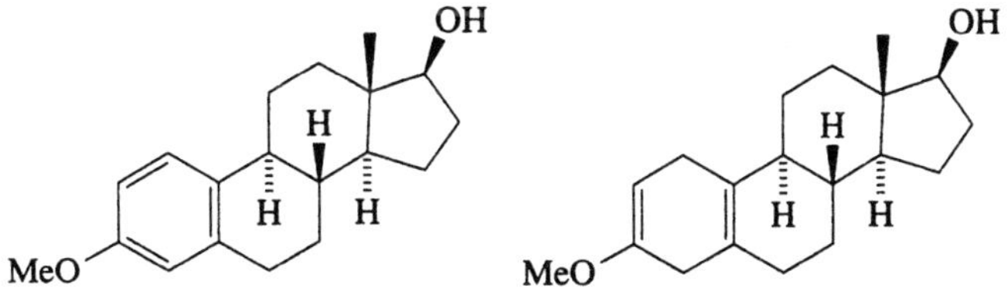

OH H H H MeO (**226**)

OH H H H MeO (**227**)

3.4.4.3 Photochemical Methods

Photochemical electron transfer reactions of electron donor–acceptor pairs in polar solvents provide a convenient and effective method for the generation of radical cations which can be trapped by complex metal hydrides. One of the most effective systems is based on irradiation of a solution of substrate, sodium borohydride and 1,4- or 1,3-dicyanobenzene. A range of bi- and poly-cyclic aromatic hydrocarbons has been converted into the dihydro derivatives in this way.[126] An especially important aspect of this route to dihydroaromatic compounds is that it may give access to products which are regioisomeric with the standard Birch reduction products. Thus, *o*-xylene is converted into the 1,4-dihydro product (**229**) rather than the 'normal' 3,6-dihydro isomer (**228**). The *m*- and *p*-xylenes are similarly reduced to (**230**) and (**231**), respectively.[127]

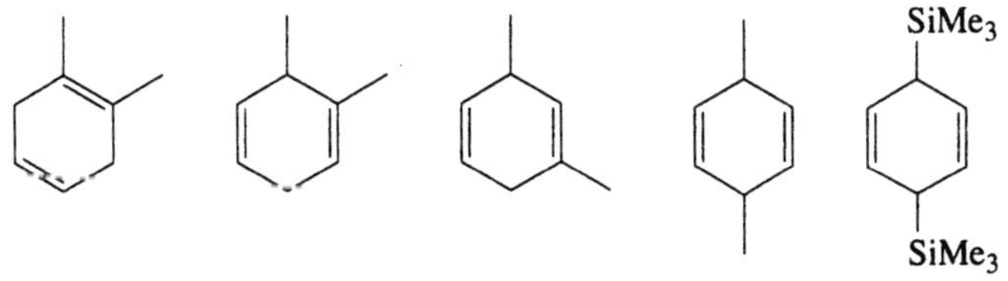

$SiMe_3$ $SiMe_3$

(**228**) (**229**) (**230**) (**231**) (**232**)

3.4.4.4 Reductive Silylations

An interesting alternative to the Birch reduction with useful synthetic potential is provided by reductive silylation. Benzene, on treatment with lithium metal and trimethylsilyl chloride in THF for a prolonged period, for example, affords a high yield of the 1,4-disilyl derivative (**232**).[128] Naphthalene is converted into a 2:1 mixture of the 1,2- and 1,4-adducts (**233**) and (**234**), respectively (Scheme 49).[129]

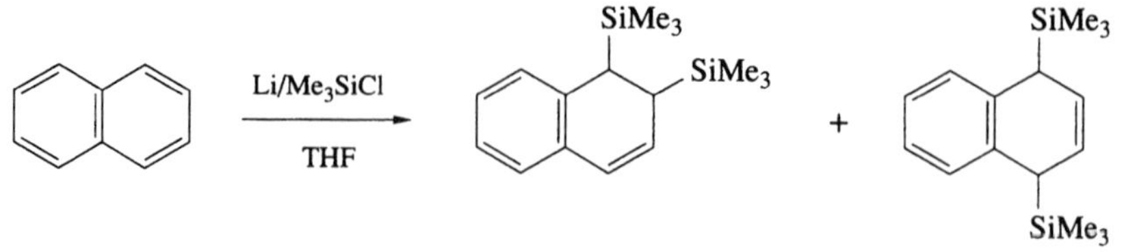

Li/Me_3SiCl, THF; (**49**) → (**233**) + (**234**)

Scheme 49

Estradiol bis(trimethylsilyl) ether (**235**; Scheme 50) has been converted into a diasteromeric mixture of adducts (**236**) which were hydrolyzed by aqueous acetic acid to a mixture of the isomeric enones (**237**; 36% yield) and (**238**; 9% yield).[130] This provides an especially direct route for the introduction of functionality at C-1, which would otherwise require a multistep sequence. Reductive silylation has also been used to convert anisole into enone (**239**)[131] which, following a kinetic resolution, has been used in an enantioselective synthesis of (+)-α-curcumene (**240**; Scheme 51).[132] The silyl substituent was employed to control the stereochemical and regiochemical aspects of subsequent steps.

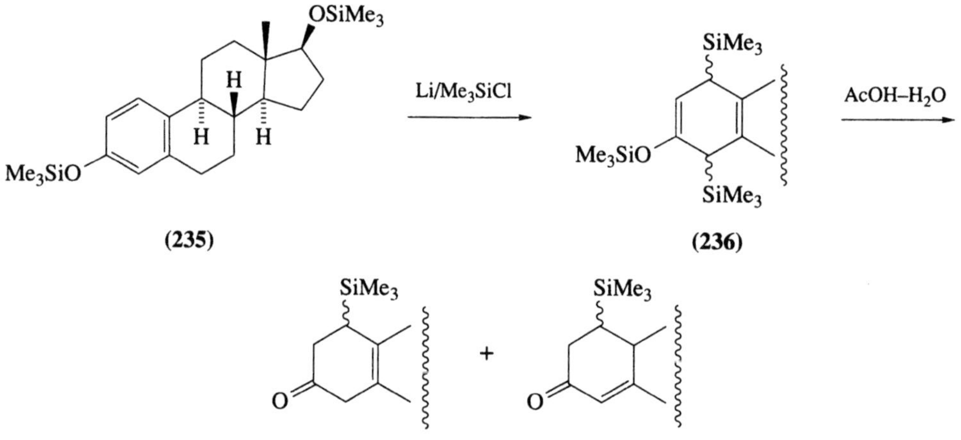

(**235**) → Li/Me_3SiCl → (**236**) → $AcOH-H_2O$ → (**237**) + (**238**)

Scheme 50

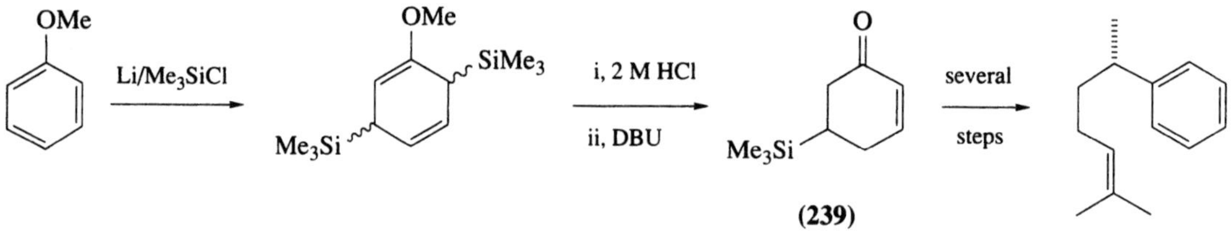

Anisole → Li/Me_3SiCl → i, 2 M HCl; ii, DBU → (**239**) → several steps

Scheme 51

3.4.4.5 Organometallic Procedures

Unlike most η^2-coordinated complexes, those derived from osmium(II), *e.g.* $[Os(NH_3)_5(\eta^2\text{-}C_6H_6)]$ possess a reasonable degree of kinetic stability and are formed regioselectively. For example, on treatment with a mixture of zinc amalgam and $[Os(NH_3)_5(OTf)_3]$, anisole rapidly forms the 2,3-isomer (**241**), although this rearranges to the 3,4-isomer ($k \sim 1\ s^{-1}$ at 20 °C). The complexes are more reactive towards catalytic hydrogenation than the parent aromatic ligand and are readily reduced to η^2-coordinated cycloalkenes which may then be demetallated by oxidation. Hydrogen is added stereoselectively *anti* to the metal centre and (**241**) affords either the tetrahydro or dihydro derivative, depending on the reaction conditions (Scheme 52).[133] The method has the potential to give reduction products in which the alkenic bond is in a complementary location to that expected from the Birch process.

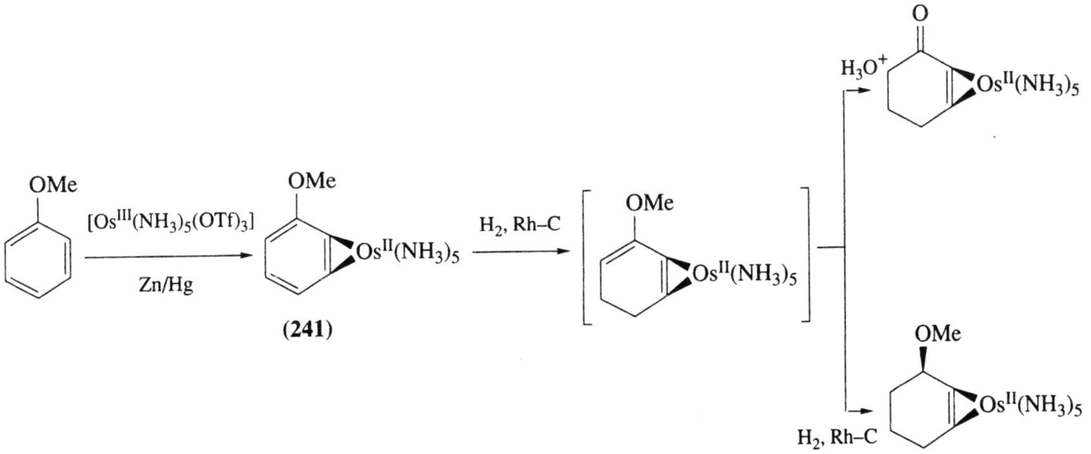

OMe [$Os^{III}(NH_3)_5(OTf)_3$] Zn/Hg → OMe $Os^{II}(NH_3)_5$ (**241**) → H_2, Rh–C → [OMe $Os^{II}(NH_3)_5$] → H_3O^+ → O $Os^{II}(NH_3)_5$; → H_2, Rh–C → OMe $Os^{II}(NH_3)_5$

Scheme 52

3.4.5 REFERENCES

1. H. Smith, 'Organic Reactions in Liquid Ammonia', Wiley-Interscience, New York, 1963, vol. 1, part 2.
2. F. J. Kakis, in 'Steroid Reactions', ed. C. Djerassi, Holden–Day, San Francisco, 1963, p. 267.
3. R. G. Harvey, *Synthesis*, 1970, 161.
4. A. J. Birch and G. S. R. Subba Rao, in 'Advances in Organic Chemistry, Methods and Results', ed. E. C. Taylor, Wiley-Interscience, New York, 1972, p. 1.
5. H. L. Dryden, Jr., in 'Organic Reactions in Steroid Chemistry', ed. J. Fried and J. A. Edwards, Van Nostrand–Reinhold, New York, 1972, vol. 1, p. 1.
6. J. M. Hook and L. N. Mander, *Nat. Prod. Reports*, 1986, **3**, 35.
7. C. B. Wooster and K . L . Godfrey , *J. Am. Chem. Soc.*, 1937, **59**, 596; C. B. Wooster (Du Pont), *US Pat.* 2 182 242 (1939) (*Chem. Abstr.*, 1940, **34**, 1993).
8. A. J. Birch, *Q. Rev., Chem. Soc.*, 1950, **4**, 69; A. J. Birch and H. Smith, *Q. Rev., Chem. Soc.*, 1958, 7, 17.
9. W. Kotlarek, *J. Org. Chem.*, 1975, **40**, 2841.
10. A. G. Schultz and M. Macielag, *J. Org. Chem.*, 1986, **51**, 4983.
11. H. E. Zimmerman, in 'Molecular Rearrangements', ed. P. de Mayo, Wiley-Interscience, New York, 1963, vol.1, p. 347.
12. A. J. Birch, A. L. Hinde and L. Radom, *J. Am. Chem. Soc.*, 1980, **102**, 3370.
13. P. W. Rabideau, N. K. Peters and D. L. Huser, *J. Org. Chem.*, 1981, **46**, 1593.
14. P. W. Rabideau and D. L. Huser, *J. Org. Chem.*, 1983, **48**, 4266.
15. D. J. Marshall and R. Deghenghi, *Can. J. Chem.*, 1969, **47**, 3127.
16. M. Narisada and F. Watanabe, *J. Org. Chem.*, 1973, **38**, 3887.
17. P. Markov and C. Ivanov, *Tetrahedron Lett.*, 1962, 1139.
18. P. W. Rabideau and E. G. Burkholder, *J. Org. Chem.*, 1978, **43**, 4283.
19. A. P. Krapcho and A. A. Bothner-By, *J. Am. Chem. Soc.*, 1959, **81**, 3658.
20. J. Fried, N. A. Abraham and T. S. Santhanakrishnan, *J. Am. Chem. Soc.*, 1967, **89**, 1044.
21. A. J. Birch, *J. Chem. Soc.*, 1944, 430; J. A. Barltrop and J. E. Saxton, *J. Chem. Soc.*, 1952, 1038.
22. W. Nagata, T. Wakabayashi, M. Narisada, Y. Hayase and S. Kamata, *J. Am. Chem. Soc.*, 1971, **93**, 5740.
23. R. Grewe and W. Friedrichsen, *Chem. Ber.*, 1967, **100**, 1550.
24. R. B. Turner, K. H. Gänshirt, P. E. Shaw and J. D. Tauber, *J. Am. Chem. Soc.*, 1966, **88**, 1776; D. K. Banerjee, K. M. Damodaran, P. S. N. Murthy and V. Paul, *Proc. Indian Acad. Sci., Sect. A*, 1978, **87**, 239; W. S. Johnson, B. Bannister and R. Pappo, *J. Am. Chem. Soc.*, 1956, **78**, 6331.
25. F. Fringuelli, V. Mancini and A. Taticchi, *Tetrahedron*, 1969, **25**, 4249.
26. R. E. Donaldson and P. L. Fuchs, *J. Org. Chem.*, 1977, **42**, 2032.
27. W. S. Johnson, R. Pappo and W. F. Johns, *J. Am. Chem. Soc.*, 1956, **78**, 6339.
28. T. B. Windholz, R. D. Brown and A. A. Patchett, *Steroids*, 1965, **6**, 409.
29. D. F. MacSweeney and R. Ramage, *Tetrahedron*, 1971, **27**, 1481.
30. E. Cotsaris and M. N. Paddon-Row, *J. Chem. Soc., Perkin Trans. 2*, 1984, 1487.
31. E. Fujita, M. Shibuya, S. Nakamura, Y. Okada and T. Fujita, *J. Chem. Soc., Perkin Trans. 1*, 1974, 165.
32. W. Hückel and H. Bretschneider, *Justus Liebigs Ann. Chem.*, 1939, **540**, 157.
33. B. Weinstein and A. H. Fenselau, *J. Org. Chem.*, 1964, **29**, 2102; 1965, **30**, 3209.
34. P. J. Grisdale, T. H. Regan, J. C. Doty, J. Figueras, Jr. and J. L. R. Williams, *J. Org. Chem.*, 1968, **33**, 1116.
35. P. W. Rabideau, N. K. Peters and D. L. Huser, *J. Org. Chem.*, 1981, **46**, 1593.
36. P. W. Rabideau and R. G. Harvey, *J. Org. Chem.*, 1970, **35**, 25.
37. A. L. Campbell and J. D. McChesney, *Synth. Commun.*, 1979, **9**, 471.

38. J. Fried and N. A. Abraham, *Tetrahedron Lett.*, 1965, 3505.
39. P. Radlick and H. T. Crawford, *J. Org. Chem.*, 1972, **37**, 1669.
40. N. A. Nelson, R. S. P. Hsi, J. M. Schuck and L. D. Khan, *J. Am. Chem. Soc.*, 1960, **82**, 2573.
41. G. Stork and W. N. White, *J. Am. Chem. Soc.*, 1956, **78**, 4604; B. B. Millward, *J. Chem. Soc.*, 1960, 26.
42. A. J. Birch, E. G. Hutchinson and G. S. R. Subba Rao, *J. Chem. Soc. C*, 1971, 637.
43. H. Van Bekkum, C. B. Van Den Bosch, G. Van Minnen-Pathuis, J. C. De Mos and A. M. Van Wijk, *Recl. Trav. Chim. Pays-Bas*, 1971, **90**, 137.
44. G. W. Holbert and B. Ganem, *J. Am. Chem. Soc.*, 1978, **100**, 352.
45. W. J. Sipio, *Tetrahedron Lett.*, 1985, **26**, 2039.
46. G. S. R. Subba Rao, H. Ramanathan and K. Raj, *J. Chem. Soc., Chem. Commun.*, 1980, 315.
47. A. J. Baker and A. C. Goudie, *J. Chem. Soc., Chem. Commun.*, 1972, 951.
48. A. J. Birch and J. Slobbe, *Tetrahedron Lett.*, 1976, 2079.
49. I. K. Zhurkovich and D. V. Ioffe, *J. Org. Chem. USSR*, 1974, **10**, 216.
50. F. Camps, J. Coll and J. Pascual, *J. Org. Chem.*, 1967, **32**, 2563.
51. J. V. Turner, B. F. Anderson and L. N. Mander, *Aust. J. Chem.*, 1980, **33**, 1061.
52. A. Tahara, M. Shimagaki, S. Ohara, T. Tanaka and T. Nakata, *Chem. Pharm. Bull.*, 1975, **23**, 2329.
53. A. J. Birch, P. Hextall and S. Sternhell, *Aust. J. Chem.*, 1954, **7**, 256; M. E. Kuehne and B. F. Lambert, *J. Am. Chem. Soc.*, 1959, **81**, 4278.
54. H. O. House, R. C. Strickland and E. J. Zaiko, *J. Org. Chem.*, 1976, **41**, 2401.
55. A. J. Birch and J. Slobbe, *Aust. J. Chem.*, 1977, **30**, 1045.
56. O. L. Chapman and P. Fitton, *J. Am. Chem. Soc.*, 1963, **85**, 41.
57. M. E. C. Biffin, A. G. Moritz and D. B. Paul, *Aust. J. Chem.*, 1972, **25**, 1329.
58. J. M. Hook, L. N. Mander and M. Woolias, *Tetrahedron Lett.*, 1982, **23**, 1095.
59. A. J. Birch, *J. Chem. Soc.*, 1944, 430.
60. J. Slobbe, *J. Chem. Soc., Chem. Commun.*, 1977, 82.
61. P. W. Rabideau, E. G. Burkholder and M. J. Yates, *Synth. Commun.*, 1980, **10**, 627.
62. J. L. Marshall and T. K. Folsom, *J. Org. Chem.*, 1971, **36**, 2011.
63. E. L. Eliel and T. E. Hoover, *J. Org. Chem.*, 1959, **24**, 938.
64. P. K. Oommen, *Aust. J. Chem.*, 1975, **28**, 2095.
65. M. D. Bachi, J. W. Epstein, Y. Herzberg-Minzly and H. J. E. Loewenthal, *J. Org. Chem.*, 1969, **34**, 126.
66. A. R. K. Murthy, N. S. Sundar and G. S. R. Subba Rao, *Tetrahedron*, 1982, **38**, 2831.
67. P. K. Oommen, *Aust. J. Chem.*, 1976, **29**, 2087.
68. D. J. Beames, J. A. Halleday and L. N. Mander, *Aust. J. Chem.*, 1972, **25**, 137.
69. G. S. R. Subba Rao, A. R. K. Murthy and N. S. Sundar, *Indian J. Chem., Sect. B*, 1978, **16**, 1027.
70. L. N. Mander and S. G. Pyne, *Aust. J. Chem.*, 1981, **34**, 1899; A. L. Cossey, L. Lombardo and L. N. Mander, *Tetrahedron Lett.*, 1980, **21**, 4383.
71. D. Franks, M. C. Grossel, R. C. Haywood and L. J. S. Knutsen, *J. Chem. Soc., Chem. Commun.*, 1978, 941.
72. P. W. Rabideau, S. J. Nyikos, D. L. Huser and E. G. Burkholder, *J. Chem. Soc., Chem. Commun.*, 1980, 210.
73. M. Julia and B. Malassiné, *Tetrahedron Lett.*, 1972, 2495.
74. S. G. Davies and G. H. Whitham, *J. Chem. Soc., Perkin Trans 1*, 1978, 1479.
75. M. S. Kharasch, E. Sternfeld and F. R. Mayo, *J. Org. Chem.*, 1940, **5**, 362.
76. H. J. E. Loewenthal, 'Guide for the Perplexed Organic Experimentalist', Heyden, London, 1978, p. 133.
77. R. J. Hamilton, L. N. Mander and S. P. Sethi, *Tetrahedron*, 1986, **42**, 2881.
78. R. J. Hamilton, Ph.D. Thesis, Australian National University, 1982: reported in ref. 6.
79. P. W. Rabideau, D. M. Wetzel and D. M. Young, *J. Org. Chem.*, 1984, **49**, 1544.
80. S. Bhattacharyya, B. Basu and D. Mukherjee, *Tetrahedron*, 1983, **39**, 4221.
81. J. M. Hook, L. N. Mander and R. Urech, *J. Org. Chem.*, 1984, **49**, 3250.
82. A. G. Schultz, J. P. Dittami, F. P. Lavieri, C. Salowey, P. Sundararaman and M. B. Szymula, *J. Org. Chem.*, 1984, **49**, 4429.
83. A. G. M. Barrett, P. A. Prokopiou, D. H. R. Barton, R. B. Boar and J. F. McGhie, *J. Chem. Soc., Chem. Commun.*, 1979, 1173.
84. A. J. Birch, *J. R. Inst. Chem.*, 1957, **81**, 100.
85. A. J. Birch, J. Cymerman-Craig and M. Slaytor, *Aust. J. Chem.*, 1955, **8**, 512.
86. L. Dickson, C. A. Matuszak and A. H. Qazi, *J. Org. Chem.*, 1978, **43**, 1007.
87. A. G. Schultz, M. Macielag, P. Sundararaman, A. G. Taveras and M. Welch, *J. Am. Chem. Soc.*, 1988, **110**, 7828; A. G. Schultz, M. Macielag, D. E. Podhorez and J. C. Suhadolnik, *J. Org. Chem.*, 1988, **110**, 2456.
88. S. Mejer, *Bull. Acad. Pol. Sci., Ser. Sci., Chim.*, 1962, **10**, 469 (*Chem. Abstr.*, 1963, **59**, 1553h).
89. C. B. Wooster and D. S. Latham, *J. Am. Chem. Soc.*, 1936, **58**, 76.
90. L. N. Mander and M. Woolias, *Synthesis*, 1979, 185.
91. Z. Marcinow and P. W. Rabideau, *J. Org. Chem.*, 1988, **53**, 2117.
92. J. M. Brown, T. M. Cresp and L. N. Mander, *J. Org. Chem.*, 1977, **42**, 3984; N. N. Marinovic and H. Ramanathan, *Tetrahedron Lett.*, 1983, **24**, 1871.
93. T. L. Gilchrist, C. W. Rees and D. Tuddenham, *J. Chem. Soc., Perkin Trans. 1*, 1981, 3214; Z. Lidert and C. W. Rees, *J. Chem. Soc., Chem. Commun.*, 1982, 499; Z. Lidert and C. W. Rees, *J. Chem. Soc., Chem. Commun.*, 1983, 317.
94. S. Mejer and R. Pacut, *Pol. J. Chem.*, 1978, **52**, 529 (*Chem. Abstr.*, 1978, **89**, 42 871c); S. Mejer and Z. Marcinow, *Bull. Acad. Pol. Sci., Ser. Sci., Chim.*, 1976, **24**, 175 (*Chem. Abstr.*, 1976, **85**, 94 121v).
95. G. S. R. Subba Rao and N. S. Sundar, *J. Chem. Soc., Perkin Trans. 1*, 1982, 875.
96. P. W. Rabideau, C. A. Husted and D. M. Young, *J. Org. Chem.*, 1983, **48**, 4149.
97. A. Chatterjee, S. R. Raychaudhuri and S. K. Chatterjee, *Tetrahedron Lett.*, 1978, 3487.
98. V. M. Kapoor and A. M. Mehta, *Synthesis*, 1975, 471.
99. N. S. Sundar and G. S. R. Subba Rao, *J. Chem. Soc., Perkin Trans. 1*, 1982, 1381.

100. S. Meyer and S. Respondek, *Bull. Acad. Pol. Sci., Ser. Sci., Chim.*, 1966, **14**, 611 (*Chem. Abstr.*, 1967, **66**, 105 088a).
101. C. Eaborn, R. A. Jackson and R. Pearce, *J. Chem. Soc., Perkin Trans. 1*, 1975, 470.
102. H. Alt, E. R. Franke and H. Bock, *Angew. Chem., Int. Ed. Engl.*, 1969, **8**, 525.
103. C. Eaborn, R. A. Jackson and R. Pearce, *J. Chem. Soc., Perkin Trans. 1*, 1974, 2055.
104. D. J. Coughlin and R. G. Salomon, *J. Org. Chem.*, 1979, **44**, 3784.
105. S. R. Wilson, P. A. Zucker, C.-W. Kim and C. A. Villa, *Tetrahedron Lett.*, 1985, **26**, 1969.
106. D. W. Jessup, J. W. Paschal and P. W. Rabideau, *J. Org. Chem.*, 1977, **42**, 2620.
107. C.-P. Chuang, J. C. Gallucci, D. J. Hart and C. Hoffman, *J. Org. Chem.*, 1988, **53**, 3218.
108. A. J. Birch, *J. Chem. Soc.*, 1947, 102.
109. A. J. Birch and G. Nadamuni, *J. Chem. Soc., Perkin Trans. 1*, 1974, 545.
110. A. J. Birch and G. S. R. Subba Rao, *Tetrahedron Lett.*, 1967, 857.
111. M. Tomita, E. Fujita and F. Murai, *Yakugaku Zasshi*, 1951, **226**, 1035.
112. G. W. Kenner and N. R. Williams, *J. Chem. Soc.*, 1955, 522.
113. F. J. Sardina, A. D. Johnston, A. Mouriño and W. H. Okamura, *J. Org. Chem.*, 1982, **47**, 1576.
114. A. J. Birch, *J. Proc. R. Soc. N. S. W.*, 1950, **33**, 245.
115. A. J. Birch and G. S. R. Subba Rao, *Aust. J. Chem.*, 1969, **22**, 2037.
116. S. P. Khanapure, B. G. Hazra and K. G. Das, *J. Chem. Soc., Perkin Trans. 1*, 1981, 1360.
117. A. R. Pinder and H. Smith, *J. Chem. Soc.*, 1954, 113.
118. K. S. J. Stapleford, *Synth. Commun.*, 1982, **12**, 651.
119. F. Camps, J. Coll and J. Pascual, *An. Quim.*, 1968, **64**, 1089 (*Chem. Abstr.*, 1969, **70**, 106 678n); A. J. Birch and K. P. Dastur, *Aust. J. Chem.*, 1973, **26**, 1363.
120. R. A. Benkeser, R. E. Robinson, D. M. Sauve and O. H. Thomas, *J. Am. Chem. Soc.*, 1955, **77**, 3230.
121. E. M. Kaiser and R. A. Benkeser, *Org. Synth.*, 1970, **50**, 88.
122. E. M. Kaiser, *Synthesis*, 1972, 391.
123. R. A. Benkeser, F. G. Belmonte and J. Kang, *J. Org. Chem.*, 1983, **48**, 2796.
124. R. A. Benkeser, E. M. Kaiser and R. F. Lambert, *J. Am. Chem. Soc.*, 1964, **86**, 5272.
125. E. Kariv-Miller, K. E. Swenson and D. Zemach, *J. Org. Chem.*, 1983, **48**, 4210.
126. M. Yasuda, C. Pac and H. Sakurai, *J. Org. Chem.*, 1981, **46**, 788.
127. G. A. Epling and E. Florio, *Tetrahedron Lett.*, 1986, **27**, 1469.
128. D. R. Weyenberg and L. H. Toporcer, *J. Am. Chem. Soc.*, 1962, **84**, 2843; J. Dunoguès, R. Calas and N. Ardoin, *J. Organomet. Chem.*, 1972, **43**, 127.
129. D. R. Weyenberg and L. H. Toporcer, *J. Org. Chem.*, 1965, **30**, 943.
130. O. N. Minailova, T. L. Ivanenko and K. K. Pivnitakii, *Zh. Obshch. Khim.*, 1980, **50**, 2813 (*Chem. Abstr.*, 1981, **94**, 140 015z).
131. M. Laguerre, J. Dunoguès, R. Calas and N. Duffaut, *J. Organomet. Chem.*, 1975, **93**, C17.
132. M. Asaoka, K. Shima and H. Takei, *Tetrahedron Lett.*, 1987, **28**, 5669.
133. W. D. Harman, W. P. Schaefer and H. Taube, *J. Am. Chem. Soc.*, 1990, **112**, 2682.

3.5
Partial Reduction of Enones, Styrenes and Related Systems

EHUD KEINAN
Technion — Israel Institute of Technology, Haifa, Israel

and

NOAM GREENSPOON
The Weizmann Institute of Science, Rehovot, Israel

3.5.1 INTRODUCTION

Although the conjugated systems of polyenes and unsaturated carbonyl compounds are somewhat similar in terms of structure and bonding, their chemical reactivities are significantly different. The number of available methods for partial reduction of unsaturated carbonyls is overwhelmingly greater than the number of known methods for polyene reduction, and this review deals almost exclusively with unsaturated carbonyl compounds, concentrating upon the reduction of the C=C double bond. We subdivide it according to synthetic methodology rather than substrate nature, an arrangement that provides a general view clarifying the scope and limitations of a given method and saving repetition. Dienes and styrenes are discussed at the end of the review.

Methods for reduction of enones may be divided conveniently into four historically based classes. The earliest procedures employed dissolving metals; more recent developments, such as reduction with low-valent transition metal compounds and electrochemical processes, may also be included in this category as they all proceed *via* sequential addition of electrons and protons to the substrate molecule. These methods are discussed in Section 3.5.2.

Catalytic hydrogenation (Section 3.5.3) may be regarded as the second generation of reducing systems. Indeed, both heterogeneous and homogeneous catalytic hydrogenation replaced many of the earlier dissolving metal techniques.

The discovery of metal hydrides and complex metal hydrides, particularly those of boron and aluminum in the early 1940s, revolutionized the reduction of organic functional groups. These reagents may be regarded as the third generation of reducing systems. Extensive studies have led to a broad variety of hydridic reagents whose reducing power and selectivity are controlled by appropriate modification of the ligands in the metal coordination sphere.[1] Applications of these hydridic reagents as well as other main group metal hydrides, such as silicon and tin derivatives, are discussed in Section 3.5.4. A variety of transition metal hydrides, such as iron, copper, chromium and cobalt compounds are employed in stoichiometric quantities, and are presented in Section 3.5.5. Composite reducing systems consisting of a transition metal catalyst and a relatively nonreactive hydride donor represent the fourth generation. Many of the transfer hydrogenation methods may be included within this fourth category as well. Therefore, although in many respects several transfer hydrogenation techniques resemble regular catalytic hydrogenations, they are discussed in Section 3.5.6, which deals with composite reducing systems. Biocatalyzed reduction is a useful approach to enantioselective reduction of enones under very mild conditions. A number of microbiological and enzymatic transformations, as well as biomimetic reactions, are outlined in Section 3.5.7. A few other methods that were not included above are presented as miscellaneous techniques in Section 3.5.8. Finally, partial reductions of dienes and styrenes are discussed in Section 3.5.9.

3.5.2 ELECTRON TRANSFER REDUCTIONS

3.5.2.1 Dissolving Metal Reduction

A variety of organic functional groups are reduced by active metals either in the presence of a proton donor or followed by treatment with a proton donor. This approach is one of the earliest reduction procedures in organic chemistry. Although its importance has decreased with the development of catalytic hydrogenation and metal hydride reduction, there remain a substantial number of dissolving metal reductions that are still in use due to their advantageous selectivity. Dissolving metal reductions of α,β-unsaturated carbonyl compounds have been discussed in several review articles.[2–10]

Metals commonly utilized include the alkali metals, mainly lithium, sodium and potassium, and also calcium, zinc, magnesium, tin and iron. Alkali metals and calcium have been used in liquid ammonia,[10] in low molecular weight aliphatic amines,[11] in hexamethylphosphoramide,[12] in ether or in THF containing crown ethers,[13c] or in very dilute solutions in polyethers such as 1,2-dimethoxyethane (DME).[11a,13] Reactions with metal solutions in liquid ammonia often use a cosolvent, such as ether, THF or DME, to increase solubility of the organic substrate in the reaction mixture. These same metals, as well as zinc and magnesium, have been used as suspensions in various solvents including ether, toluene, xylene, *etc.*

In all procedures a proton source (frequently ethanol, 2-propanol, *t*-butyl alcohol or even water) is provided in the reaction medium, added with the substrate, or added during the work-up procedure.

Sodium amalgam, aluminum amalgam, zinc, zinc amalgam, tin and iron have been added directly to solutions of the substrate in hydroxylic solvents such as ethanol, 2-propanol, *n*-butyl alcohol, *i*-pentyl alcohol, acetic acid, water or aqueous mineral acid. With hydroxylic solvents, and especially with relatively acidic ones, metal amalgams are often used rather than free metals, to minimize the release of hydrogen gas as a side product.

The dissolving metal reductions are better classified as 'internal' electrolytic reductions in which an electron is transferred from the metal surface (or from the metal in solution) to the substrate. Reduction with low-valent metal ions may also be included in this general class (*vide infra*).

The generally accepted mechanism for dissolving metal reduction of enones (Scheme 1)[10] involves reversible addition of an electron to a vacant orbital of the substrate (S), yielding a radical anion ($S^{-}\cdot$). The latter can be protonated to give a neutral radical, which may either dimerize or accept another electron and a proton. Alternatively, stepwise or simultaneous reversible addition of two electrons to S can give a dianion capable of accepting two protons. The sequence and timing of these steps depend upon the substrate, the homogeneity and reduction potential of the medium, and the presence and nature of proton donors in the medium, among other factors.

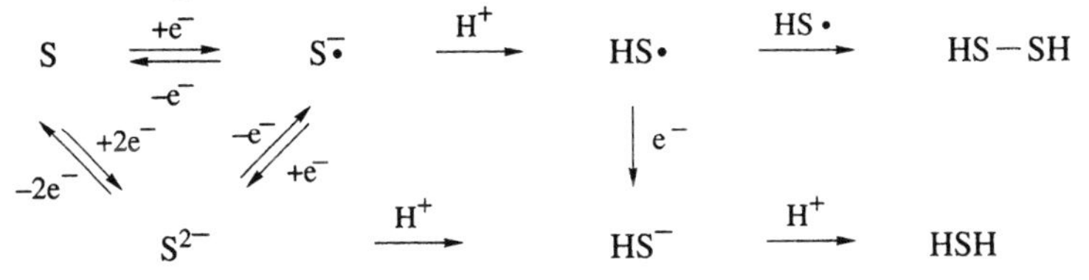

Scheme 1

The stereochemistry of reduction has been extensively studied. Metal–ammonia reduction of steroid and terpenoid enones with a β-carbon at the fusion of two six-membered rings leads, in general, to the thermodynamically more stable isomer at that position.[14] Stork has formulated a more general rule, namely that the product will be the more stable of the two isomers having the newly introduced β-hydrogen axial to the ketone ring.[15] This rule has correctly predicted the stereochemical outcome of many metal–ammonia reductions, with very few exceptions. The rule is rationalized in terms of stereoelectronic effects in the transition state (either the radical anion or the dianion stage). For example, in reduction of octalones of the type shown in Scheme 2, only two (**1** and **2**) of three possible anionic transition states involving a half-chair conformation of the enone-containing ring would be allowed stereochemically.[15]

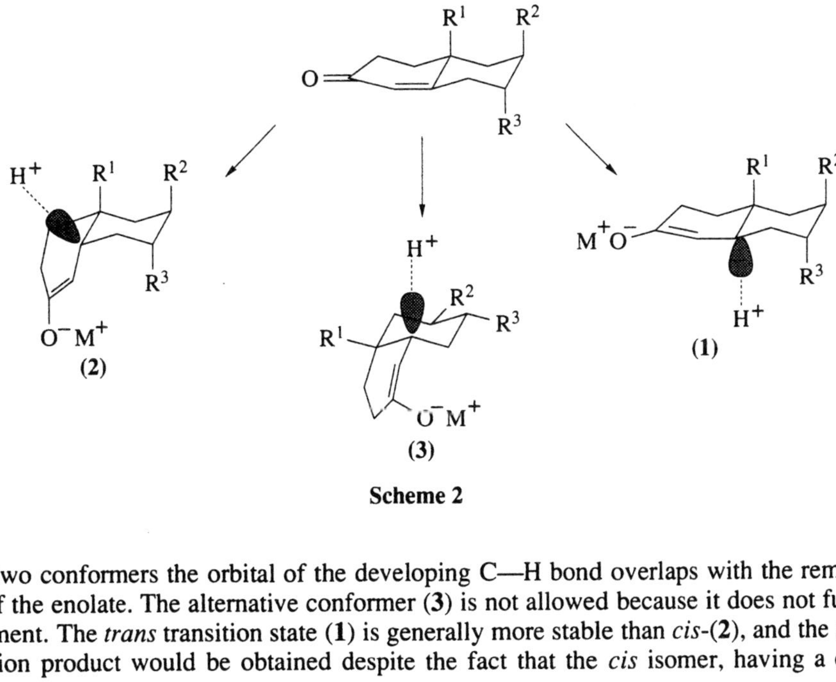

Scheme 2

In these two conformers the orbital of the developing C—H bond overlaps with the remainder of the π-system of the enolate. The alternative conformer (**3**) is not allowed because it does not fulfil the overlap requirement. The *trans* transition state (**1**) is generally more stable than *cis*-(**2**), and the *trans*-2-decalone reduction product would be obtained despite the fact that the *cis* isomer, having a conformation related to (**3**) should be more stable when R^2 and/or R^3 are larger than a hydrogen atom. This rule of 'axial protonation' has been found to be widely applicable to metal–ammonia reductions of octalones, steroids and other fused-ring systems. Representative examples are given in Scheme 3.[15–18]

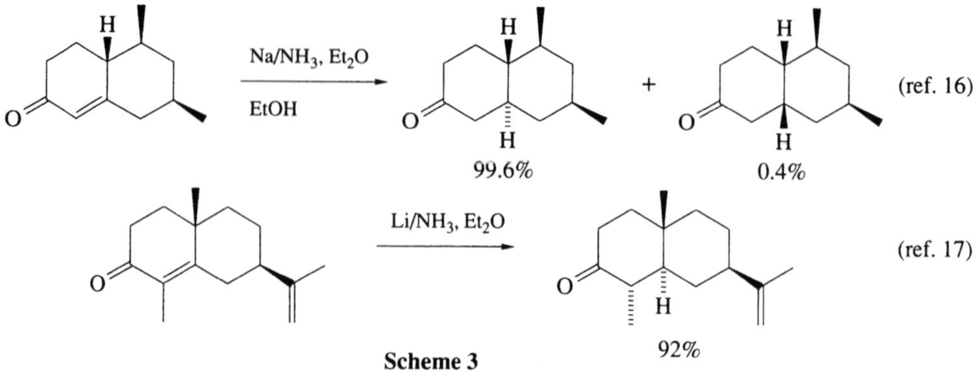

Na/NH$_3$, Et$_2$O; EtOH → 99.6% + 0.4% (ref. 16)

Li/NH$_3$, Et$_2$O → 92% (ref. 17)

Scheme 3

Generally, the conditions employed in the work-up of metal–ammonia reductions leads to products having the more stable configuration at the α-carbon atom, but products having the less stable configuration at this center have been obtained by kinetic protonation of enolate intermediates.[19,20] A more detailed discussion of stereochemistry in metal–ammonia reduction of α,β-unsaturated carbonyl compounds is given in ref. 10.

3.5.2.1.1 Scope and limitations

Before the introduction of metal–ammonia solutions for the reduction of α,β-unsaturated carbonyl compounds,[10] sodium, sodium amalgam, or zinc in protic media were most commonly employed for this purpose. Some early examples of their use include the conversion of carvone to dihydrocarvone with zinc in acid or alkaline medium,[21] and of cholest-4-en-3-one to cholestanone with sodium in alcohol.[22,23] These earlier methods are complicated by a variety of side reactions, such as over-reduction, dimerization, skeletal rearrangements, acid- or base-catalyzed isomerizations and aldol condensations, most of which can be significantly minimized by metal–ammonia reduction.

A wide variety of α,β-unsaturated ketones have been reduced to saturated ketones, usually in good yield, by metal solutions, mainly in liquid ammonia.[10,24–26] The reduction is applicable to compounds with any degree of substitution on the double bond. Although only 2 equiv. of these metals are required for the conversion of an enone to a saturated ketone, it is often convenient to employ the metal in excess. A suspension of lithium bronze (Li_4NH_3)[27] in ether allows the employment of the metal in stoichiometric amounts. Proton donors are often employed to reduce competing side reactions, such as dimerization. The presence of proton donors in the medium may lead to the conversion of an α,β-unsaturated ketone to the saturated alcohol, but at least 4 equiv. of metal must obviously be present for this type of reduction to take place.

Alcohols, such as methanol and ethanol, lead to the sole formation of saturated alcohols from unsaturated ketones when the former are present in excess during the reduction. Mixtures of ketones and alcohols are generally formed when 1 equiv. of these proton donors is employed.[28] These alcohols have an acidity comparable to that of saturated ketones, and when they are present, an equilibrium can be established between the initially formed metal enolate and the saturated ketone. The latter is then reduced to the saturated alcohol. Such reductions generally do not occur to a very significant extent when 1 equiv. of *t*-butyl alcohol[29] or some less acidic proton donor, such as triphenylcarbinol,[28] is employed. The acidity of the ketone involved, as well as the solubility of the metal enolate in the reaction medium, are important in determining whether alcohols are formed.

Even though the reaction conditions may lead to formation of the metal enolate in high yield, further reduction may occur during the quenching step of the reaction. Alcohols such as methanol and ethanol convert metal enolates to saturated ketones much faster than they react with metals in ammonia,[30,31] and quenching of reduction mixtures with these alcohols will usually lead to partial or complete conversion to alcoholic product rather than to the saturated ketone. Rapid addition of excess solid ammonium chloride is the commonly employed quench procedure if ketonic products are desired,[32] but other reagents that destroy solvated electrons before neutralization may be employed, such as sodium benzoate,[33] iron(III) nitrate,[34,35] sodium nitrite,[36] bromobenzene,[37] sodium bromate,[38] 1,2-dibromoethane[4] and acetone.[14]

3.5.2.1.2 Reduction–alkylation

The versatility of metal–ammonia reduction was considerably advanced by the discovery that the lithium enolates of unsymmetrical ketones generated during reduction can undergo *C*-alkylation with alkyl halides and carbonation with carbon dioxide.[39,40] These enolate-trapping reactions allow regiospecific introduction of groups at the carbon atoms of unsymmetrical ketones *via* the appropriate enone precursors. This procedure has been widely employed for ketones of a variety of structural types.[29,39–45] The procedure usually involves generation of a specific lithium enolate of an unsymmetrical ketone by reduction of the corresponding α,β-unsaturated ketone with 2 equiv. of lithium in liquid ammonia that contains no proton donor or just a single equivalent of one. This enolate is then reacted with excess alkylating agent (Scheme 4).

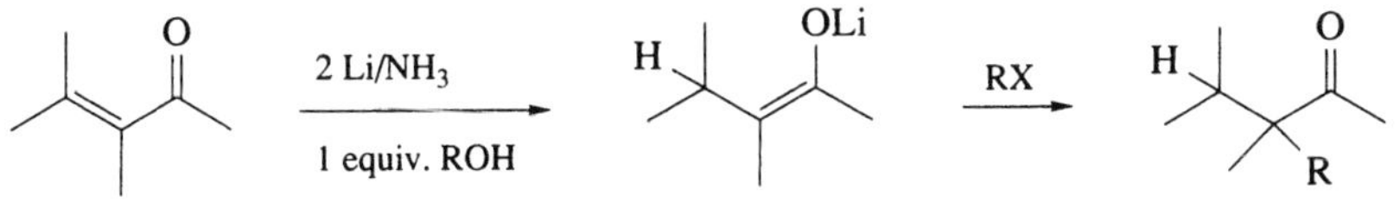

2 Li/NH_3, 1 equiv. ROH; RX

Scheme 4

This reduction–alkylation sequence has been extensively used in the total synthesis of natural products, such as progesterone and lupeol (Scheme 5).[46,47]

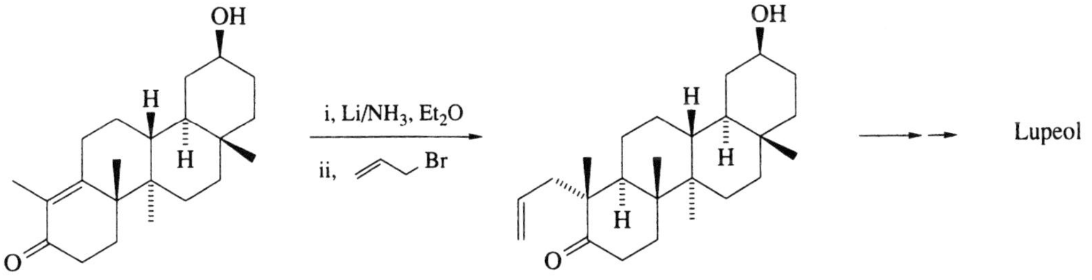

i, Li/NH_3, Et_2O; ii, allyl bromide → → Lupeol

Scheme 5

If the ammonia is removed and replaced by anhydrous ether, the intermediate lithium enolate can be converted to a β-keto ester by carbonation, followed by acidification and treatment with diazomethane, as illustrated in Scheme 6.[48]

i, Li/NH_3; ii, CO_2/Et_2O; iii, H_3O^+; iv, CH_2N_2

Scheme 6

3.5.2.1.3 Dimerization processes

Because of the intermediacy of radical anions and/or hydroxyallyl free radicals in dissolving metal reductions of enones, dimerization may compete with simple reduction. Scheme 7 shows the three types of dimers that may be produced.

The dimerization products shown in Scheme 7 are generally the major ones obtained in electrochemical reductions[49–52] (*vide infra*) or reductions at metal surfaces,[49,53] in which radical anion intermediates must diffuse to a surface before further electron transfer can occur. In metal–ammonia solutions, however, simple reduction is generally favored over dimerization. These solutions provide high concentrations of available electrons, favoring the probability of the radical ion or hydroxyallyl radical accepting a second electron.

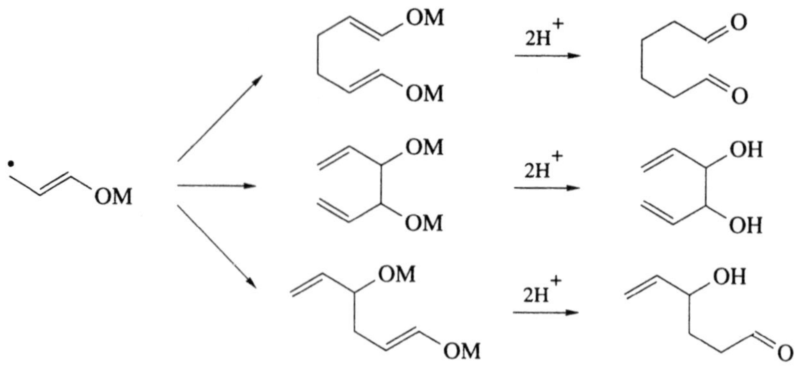

Scheme 7

3.5.2.1.4 Alkene synthesis

Appropriate quenching of a reductively formed lithium enolate with a carboxylic acid anhydride,[54,55] chloride,[56] methyl chloroformate[57] or diethyl phosphorochloridate yields the corresponding enol esters, enol carbonates or enol phosphates. These derivatives may be transformed into specific alkenes *via* reductive cleavage of the vinyl oxygen function,[58] as illustrated by the example in Scheme 8.

50%

i, Li/NH_3, Et_2O; ii, $(EtO)_2POCl$, Et_2O; iii, Li/$EtNH_2$

Scheme 8

3.5.2.1.5 Intramolecular reactions

Dissolving metal reductions of unsaturated ketones involve intermediates with carbanionic character at the β-position. Therefore, intramolecular displacements, additions, and eliminations may occur during the reduction of polyfunctional enones. Many α,β-unsaturated carbonyl compounds have structural features which allow such intramolecular reactions. The examples given in Scheme 9 include intramolecular substitution of a tosylate leaving group,[59] addition to a ketone to form a cyclopropanol,[60] and elimination of an acetate group to give an unconjugated enone.[61]

This approach has been applied in the synthesis of a perhydroindanedione skeleton *via* intramolecular addition to an ester group,[62] in formation of a stable steroidal hemiacetal,[63] and in lithium–ammonia conversion of a bicyclic unsaturated triester into a tricyclic keto diester (Scheme 10).[64]

α,β-Unsaturated ketones with leaving groups at the γ-position normally undergo reductive elimination with metals in ammonia to give metal dienolates as an initial product. Quenching these enolates with ammonium chloride allows the isolation of the β,γ-unsaturated ketone. The latter can isomerize under basic conditions to the conjugated enone. Such processes have been reported with a broad variety of leaving groups, such as hydroxide,[65,66] alkoxide[67] and acetate,[61] as well as during fission of a lactone[68–70] or an epoxide ring.[71]

α,β-Unsaturated carbonyl compounds having a leaving group at the β-position react with dissolving metals to give metal enolates, which may undergo elimination to yield new α,β-unsaturated carbonyl compounds that are susceptible to further reduction.[44,72–78] For example, β-alkoxy-α,β-unsaturated esters[73,74] and acids[79] have been found to undergo double reduction. This procedure was used as a key step in the total synthesis of eremophilane sesquiterpenes (Scheme 11).[73]

Both linear and cross-conjugated dienones are reduced by solutions of metals in liquid ammonia. For example, steroidal 4,6-dien-3-ones (Scheme 12) and related compounds are reduced initially to 3,5-dienolates;[45,80–87] while addition of ammonium chloride to the latter leads to formation of the noncon-

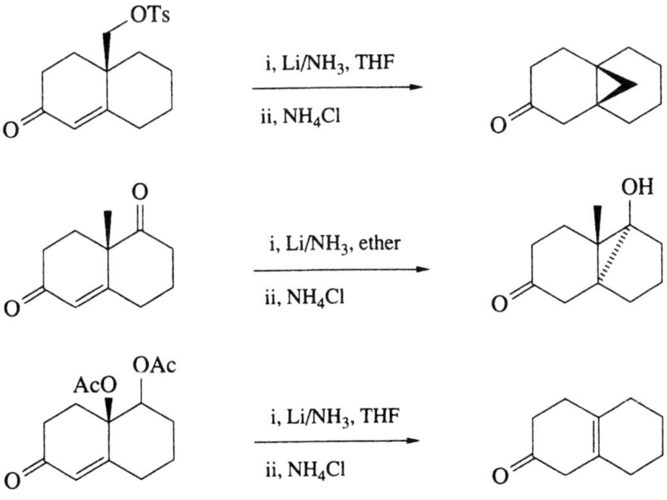

i, Li/NH_3, THF
ii, NH_4Cl

OTs

i, Li/NH_3, ether
ii, NH_4Cl

OH

AcO OAc

i, Li/NH_3, THF
ii, NH_4Cl

Scheme 9

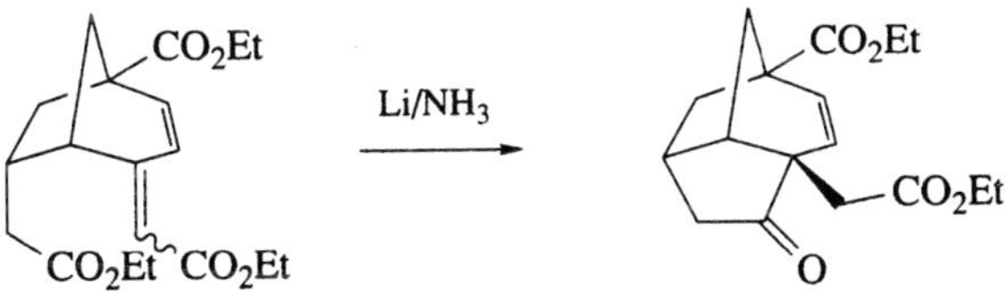

CO_2Et Li/NH_3 CO_2Et CO_2Et CO_2Et CO_2Et

Scheme 10

OMe

EtO_2C i, Li/NH_3, Et_2O −33 °C, 12 min ii, NH_4Cl, −70 °C EtO_2C

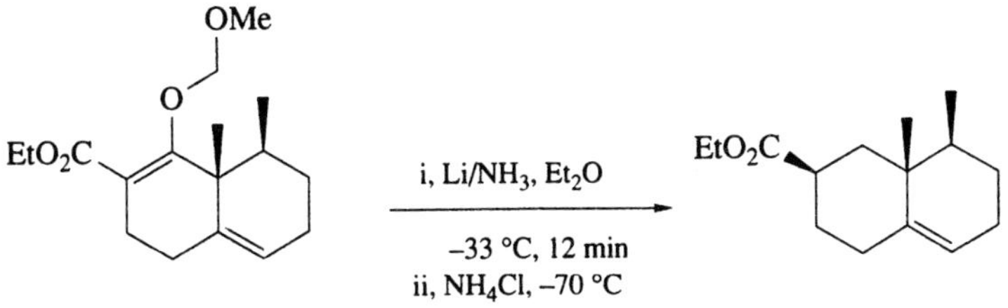

Scheme 11

jugated 5-en-3-one system,[84] addition of proton donors such as ethanol or water initiates isomerization leading to the more stable, conjugated 4-en-3-one skeleton.[81,82] Treatment of the dienolate with excess methyl iodide rather than a proton donor gives the 4,4-dimethyl-5-en-3-one.[45,88]

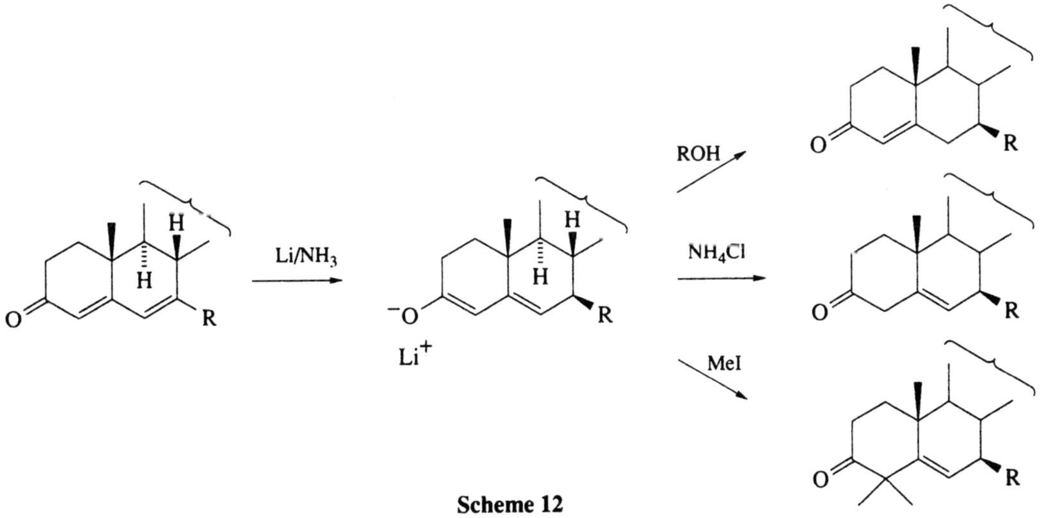

Li/NH_3 ROH NH_4Cl MeI Li^+ R

Scheme 12

Linearly conjugated dienones may be completely reduced to saturated alcohols using excess lithium in liquid ammonia.[89] In variously substituted dienones, the less-substituted double bond is often selectively reduced under these conditions. For example, treatment of a steroidal 14,16-dien-20-one with lithium in liquid ammonia (with or without 1-propanol) leads mainly to reduction of the 16,17 double bond (Scheme 13),[90,91] and the less-substituted double bond of cross-conjugated steroidal dienones,[4,45,92,93] santonin, or related substrates is selectively reduced under these conditions (Scheme 13).[68–70,94]

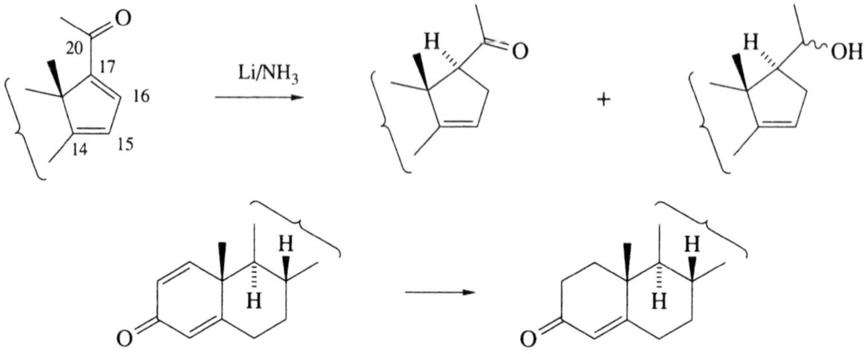

Li/NH_3

Scheme 13

3.5.2.1.6 *Chemoselectivity*

Although a host of organic functionalities are reduced by dissolving metals,[2,3,5–9] it is often possible to reduce double bonds of α,β-unsaturated carbonyl systems without affecting other reducible groups. Internal, isolated alkenes are normally stable to metal–ammonia solutions unless they have very low lying antibonding orbitals[95] or special structural features that stabilize radical anion intermediates.[96] However, terminal alkenes may be reduced by dissolving metals.[97] Mono- and poly-cyclic aromatic compounds undergo reduction with dissolving metals in liquid ammonia (Birch reduction),[2,3,5,8,98,99] but these reactions are generally slow unless proton donors are added. It is therefore possible to reduce α,β-unsaturated ketones selectively in the presence of aromatic rings.[100–103] Selective reduction preserving a reducible indole ring was performed.[104]

Ethynyl carbinols are reduced to allyl alcohols and eventually to alkenes with metal–ammonia solutions containing proton donors.[105] However, by excluding proton donors, selective reduction of conjugated enones has been carried out despite the presence of ethynyl carbinol groups.[35,106–108] Similarly, selective reduction of conjugated enones containing allylic alcohols has also been achieved.[35,106,108] Carbon–halogen bonds of alkyl and vinyl halides are readily cleaved by metals in ammonia.[5,8,9] Yet, as shown in Scheme 14, alkyl fluoride may be retained by limiting reaction times,[93] and a rather sensitive vinyl chloride functionality is preserved by using an inverse addition technique.[109]

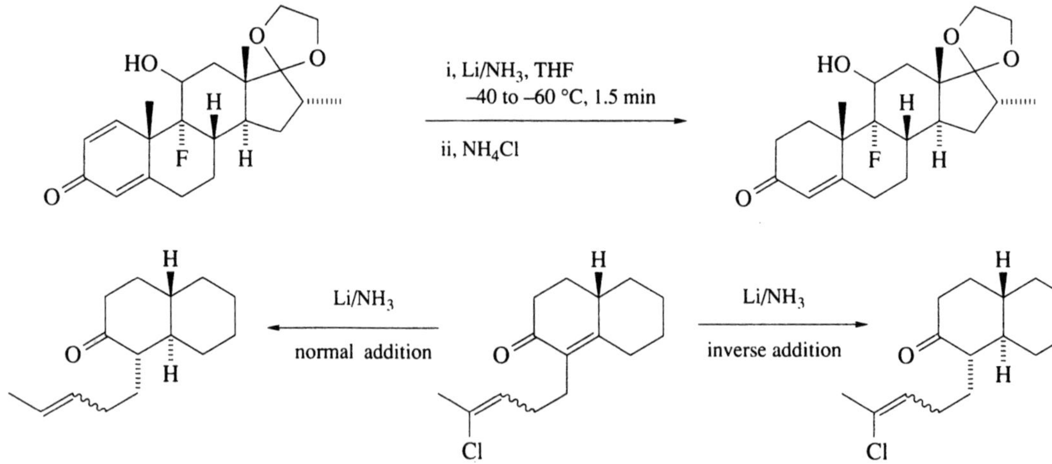

i, Li/NH_3, THF
–40 to –60 °C, 1.5 min
ii, NH_4Cl

Li/NH_3 normal addition

Li/NH_3 inverse addition

Scheme 14

Many other examples of chemoselective enone reduction in the presence of other reducible functionalities have been reported. For instance, the C—S bonds of many sulfides and thioketals are readily cleaved by dissolving metals.[5,8,9,110] Yet, there are examples of conjugate reduction of enones in the presence of a thioalkyl ether group.[110,111] Selective enone reduction in the presence of a reducible nitrile group was illustrated with another steroidal enone.[112] While carboxylic acids, because of salt formation, are not reduced by dissolving metals, esters[113] and amides[2,8] are easily reduced to saturated alcohols and aldehydes or alcohols, respectively. However, metal–ammonia reduction of enones is faster than that of either esters or amides. This allows selective enone reduction in the presence of esters[114] and amides[37,115,116] using short reaction times and limited amounts of lithium in ammonia.

3.5.2.2 Reduction with Low-valent Transition Metals

Low-valent species of early transition metals, such as chromium(II),[117] titanium(II), titanium(III),[118] vanadium, molybdenum and tungsten are useful reducing agents.[119] Electron-deficient alkenes and alkynes are easily reduced by chromium(II) sulfate, (*Z*)-alkenes being more rapidly reduced than the corresponding (*E*)-isomers.[120] Titanium(III) species are weaker reducing agents, exhibiting higher chemoselectivity.[121] Several steroid enediones have been reduced by chromium(II) chloride;[122] interestingly, reduction of cholest-4-ene-3,6-dione yields a different product from that obtained by titanium(III) reduction of the identical substrate (Scheme 15).[121c]

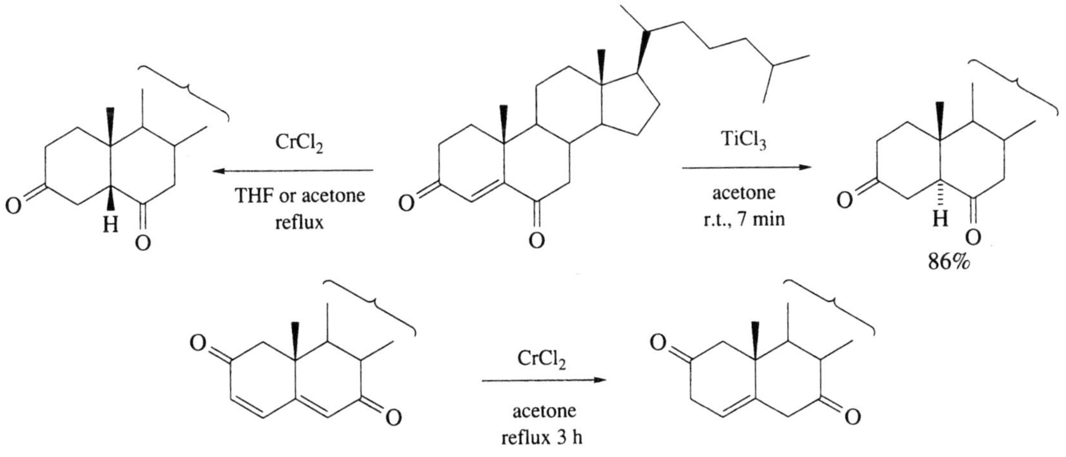

$CrCl_2$, THF or acetone, reflux; $TiCl_3$, acetone, r.t., 7 min; 86%

$CrCl_2$, acetone, reflux 3 h

Scheme 15

Solutions of chromium bis(ethylenediamine)diacetate complex in methanol are capable of reducing simple α,β-unsaturated ketones to the corresponding saturated ketones. Useful yields are obtained provided a proton donor (AcOH) and a good hydrogen donor (BuSH) are present in the reaction mixture (Scheme 16).[123]

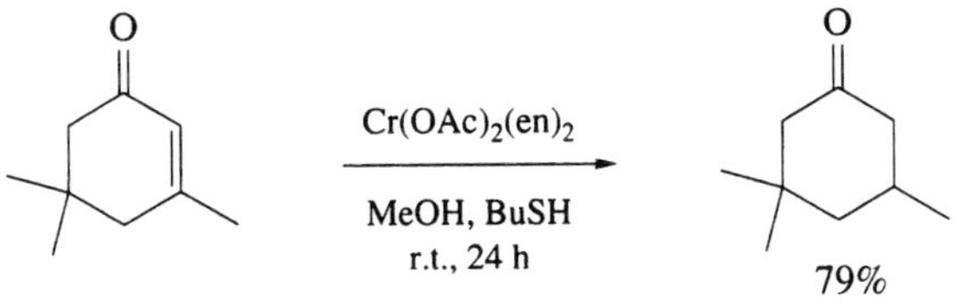

$Cr(OAc)_2(en)_2$, MeOH, BuSH, r.t., 24 h; 79%

Scheme 16

Reductive dimerization of α,β-unsaturated ketones is effected by either Cr^{II} or V^{II} chloride to give 1,6-diketones, but aliphatic α,β-unsaturated aldehydes are dimerized to the allylic glycals.[124] Interestingly, nonconjugated aldehydes are stable towards these reagents. Similar pinacolic couplings of aldehydes and ketones with Ti^{II} reagents were developed by Corey.[125]

Highly reactive metallic titanium, prepared from $TiCl_3$ and potassium, reduces enol phosphates to alkenes, permitting regioselective synthesis of dienes from α,β-unsaturated ketones.[126]

3.5.2.3 Electrochemical Reduction

The electrochemical reduction of α,β-unsaturated ketones and related compounds[5] in aprotic media in the absence of metal cations can, in some cases, lead to relatively stable anion radicals.[12c,127] However, in the presence of proton donors the latter are protonated to form hydroxyallyl radicals, which tend to dimerize more rapidly than they diffuse back to the electrode to undergo further reduction (Scheme 17).[12c] Although these allyl radicals prefer to dimerize by coupling at the β-position, if this position is sterically hindered, as in the case of cholest-4-en-3-one, coupling at the carbonyl carbon may be observed, yielding pinacols.[128]

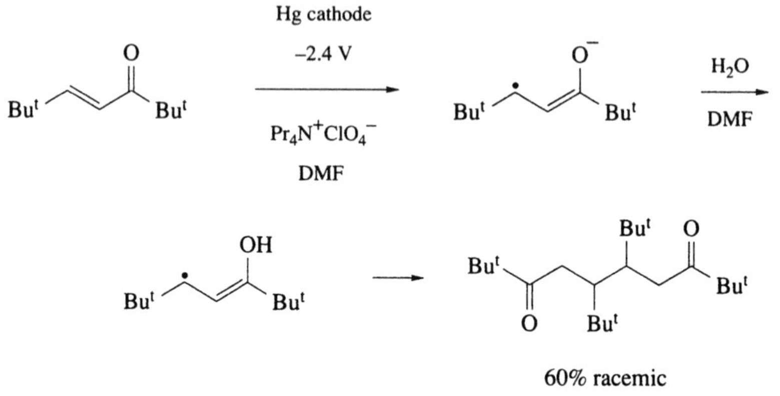

Hg cathode, −2.4 V, $Pr_4N^+ClO_4^-$, DMF; H_2O, DMF

60% racemic

Scheme 17

Reductive dimerizations have been recorded when unsaturated carbonyl compounds are reacted with various metals, such as lithium, sodium, sodium amalgam, potassium, aluminum amalgam, zinc or magnesium.[129,130] Formation of monomeric reduction products is impeded in these reactions because the intermediate allylic radical must diffuse back to the electrode surface or metal particle for further reduction. A possible solution to this problem might be concurrent electrochemical generation of a soluble reducing agent that can intercept radical intermediates before their dimerization. For example, solutions of magnesium in liquid ammonia can be generated electrochemically.[131c] Similarly, tertiary amine salts, such as yohimbine hydrochloride, can participate in the electrochemical reduction of enones (Scheme 18),[131a,b] *via* concurrent reduction of the amine to a radical which transfers a hydrogen atom to the intermediate allyl radical.

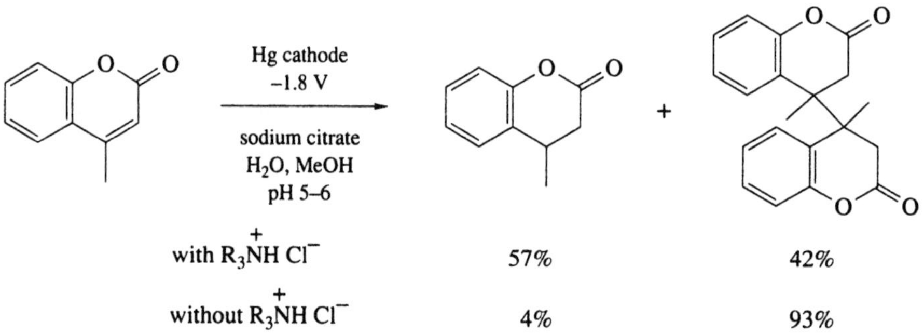

Hg cathode, −1.8 V, sodium citrate, H_2O, MeOH, pH 5–6

with $R_3\overset{+}{N}H\ Cl^-$	57%	42%
without $R_3\overset{+}{N}H\ Cl^-$	4%	93%

Scheme 18

Reductive dimerization of enones to form a new carbon–carbon bond at the β-position, known as hydrodimerization or electrohydrodimerization, has considerable synthetic utility.[132] For example, high yields of cyclic products are achieved when cyclization is kinetically favorable, leading to three- to six-membered rings from the corresponding unsaturated diesters (Scheme 19).[132d]

The product ratio in electrochemical reduction of benzalacetone is significantly altered by surfactants and various cations, which cause micellar and/or ion-pairing effects. Using these additives, it is possible to control the partitioning of the initially formed radical anion between the two main reaction pathways: either dimerization or further reduction to the saturated ketone.[133] Additionally, micellar surfactants allow the use of aqueous media without cosolvents.

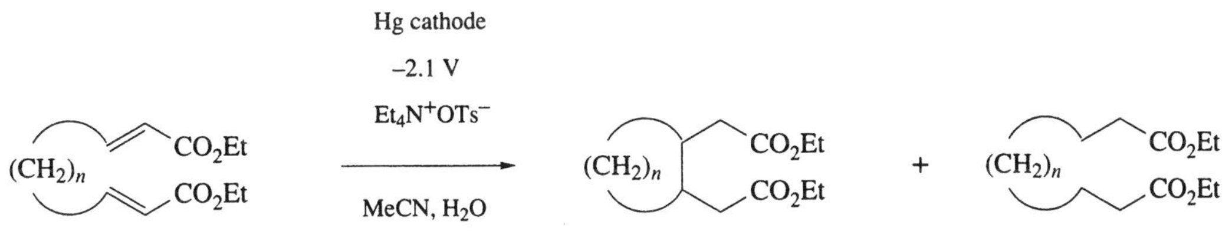

Hg cathode
–2.1 V
$Et_4N^+OTs^-$
MeCN, H_2O

n		
2	41%	48%
3	100%	–
4	90%	–
6	–	43%
8	–	50%

Scheme 19

3.5.3 CATALYTIC HYDROGENATION

Addition of molecular hydrogen to conjugated systems has been extensively reviewed.[5,134–136] Enones can be converted to saturated ketones or to unsaturated or saturated alcohols. Usually, double bonds conjugated to the carbonyl moiety are reduced prior to nonconjugated ones. 1,2-Reduction to allylic alcohols *via* catalytic hydrogenation is quite rare, and this transformation is more conveniently performed with hydridic reducing agents, such as boron and aluminum hydrides (*vide infra*). Nevertheless, there are a number of reported cases where 1,2-reduction is preferred over 1,4-selectivity. Citronellal, for example, is reduced preferentially at the carbonyl function using nickel on silica gel as a catalyst, while hydrogenation catalyzed by $Pd/BaSO_4$ yields the corresponding saturated aldehyde.[137] Reduction to the saturated alcohol is achieved by catalytic hydrogenation over nickel,[138] copper chromite,[139] or nickel–aluminum alloy in NaOH.[140]

Enones are reduced to saturated ketones by catalytic hydrogenation provided the reaction is stopped following the absorption of 1 mol of hydrogen.[141] A number of catalysts were found useful for this, including platinum,[142] platinum oxide,[143,144] Pt/C,[141] Pd/C,[141,145] Rh/C,[141] tris(triphenylphosphine)rhodium chloride,[146,147] nickel–aluminum alloy in 10% aqueous NaOH,[148] and zinc-reduced nickel in an aqueous medium.[149] Mesityl oxide is formed from acetone and reduced in a single pot to methyl isobutyl ketone using a bifunctional catalyst which comprised palladium and zirconium phosphate (Scheme 20).[150]

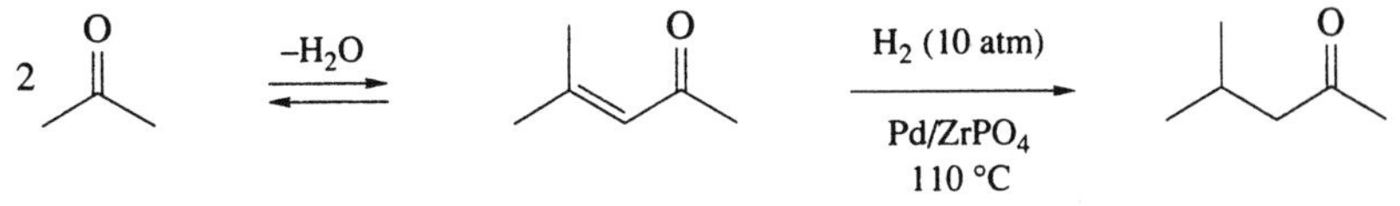

2 (acetone) $\xrightleftharpoons{-H_2O}$ (mesityl oxide) $\xrightarrow[Pd/ZrPO_4,\ 110\ °C]{H_2\ (10\ atm)}$ (methyl isobutyl ketone)

Scheme 20

The ease and the stereochemical course of hydrogenation of α,β-unsaturated ketones are particularly influenced by the nature of the solvent and the acidity or basicity of the reaction mixture. Some efforts have been made to rationalize the effect of the various parameters on the relative proportions of 1,2- to 1,4-addition, as well as on the stereochemistry of reduction.[151] For example, the product distribution in β-octalone hydrogenation in neutral media is related to the polarity of the solvent if the solvents are divided into aprotic and protic groups. The relative amount of *cis*-β-decalone decreases steadily with decreasing dielectric constant in aprotic solvents, and increases with dielectric constant in protic solvents, as exemplified in Scheme 21 (dielectric constants of the solvents are indicated in parentheses).[152] Similar results were observed in the hydrogenation of cholestenone and of testosterone.[153] In polar aprotic solvents 1,4-addition predominates, whereas in a nonpolar aprotic solvent hydrogenation occurs mainly in the 1,2-addition mode.

Acids and bases have a crucial effect on product stereochemistry in hydrogenation of ring-fused enone systems, as illustrated in Scheme 22.[154]

The increased amounts of *trans*-fused product obtained in basic solutions was suggested to arise from hydrogenation of the relatively flat enolate ion which adsorbs irreversibly onto the catalyst surface. Hydrogenation proceeds by hydride ion transfer from the metal catalyst, followed by protonation. Conversely, in an acidic medium protonation occurs initially, followed by irreversible adsorption on the catalyst, and then transfer of a hydride ion.[151] The stereochemistry of reduction is also related to catalyst activity, catalyst concentration, pressure and stirring rate, as they all affect hydrogen availability at the catalyst

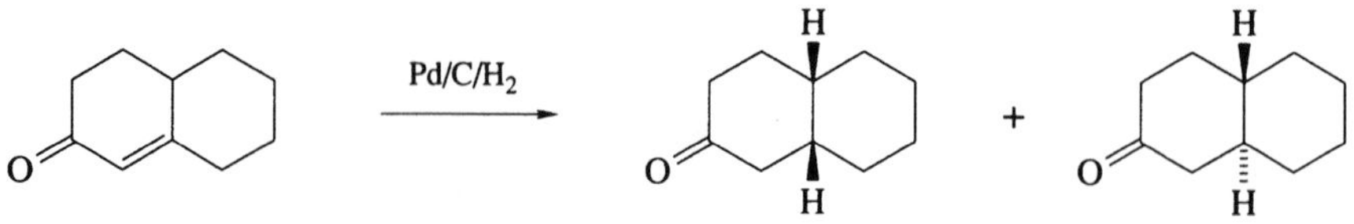

Pd/C/H_2

DMF (38)	79:21
AcOEt (6)	57:43
diethyl ether (4.34)	58:42
hexane (1.89)	48:52
methanol (33.6)	41:59
propanol (21.8)	68:32
t-butyl alcohol (10.9)	91:9

Scheme 21

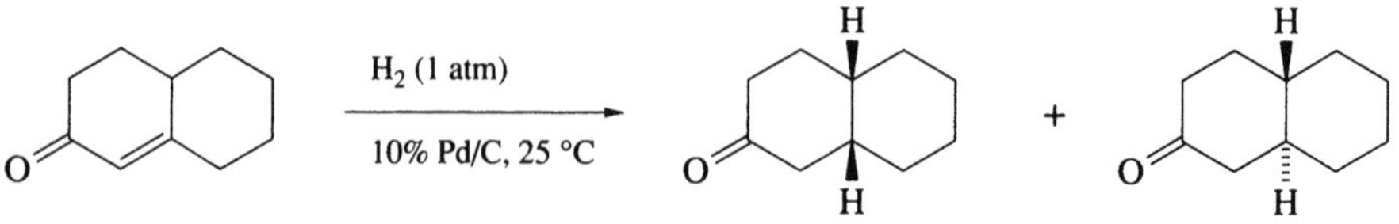

H_2 (1 atm)
10% Pd/C, 25 °C

EtOH	53:47
EtOH, H_2O, HCl	93:7
EtOH, KOH	35–50:65–50

Scheme 22

surface. Under conditions of low hydrogen availability a reversible adsorption is favorable and therefore the product stereochemistry is determined by the relative stability of the *cis* and *trans* adsorbed species. However under conditions of high hydrogen availability, product stereochemistry is determined mainly by the nature of the initial adsorption.[151,152] Platinum catalysts, more than palladium varieties, give products determined by the initial adsorption.

Substrate structure has an important influence on the stereoselectivity of hydrogenation. For example, hydrogenation of hydrindenone having a trisubstituted double bond gives mainly the *cis* product (Scheme 23),[155] whereas similar compounds with a tetrasubstituted double bond tend to give the *trans* isomer. This phenomenon has been rationalized in terms of the preferred conformation of the adsorbed enone, which minimizes steric interactions.[155,156] A high degree of stereocontrol can be achieved in hydrogenation of substrates containing allylic or homoallylic alcohols. Results are rationalized by the coordination of the homogeneous catalyt Ir(COD)Py(PCy_3)PF_6 with the hydroxy groups.[157]

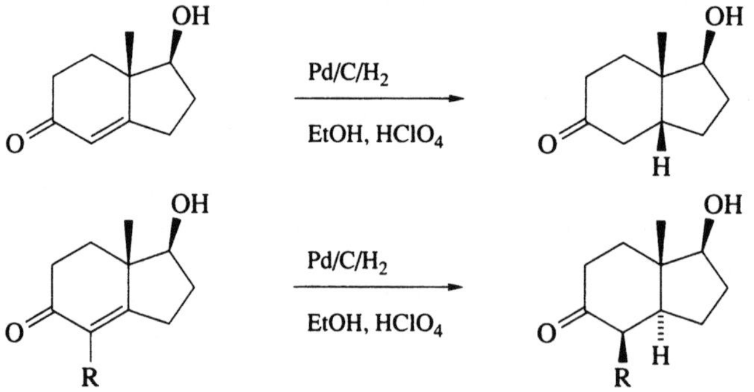

Pd/C/H_2
EtOH, $HClO_4$

Pd/C/H_2
EtOH, $HClO_4$

Scheme 23

The key step in the synthesis of 2-deoxycrustecdysone from the corresponding 20-oxo steroid is the stereoselective catalytic hydrogenation of the α,β-unsaturated lactone to afford a 2:3 mixture of δ- and γ-lactones, respectively.[158] This crude product was converted into the thermodynamically more stable γ-lactone by treatment with aqueous NaOH. α,β-Unsaturated aldehydes and ketones are hydrogenated to produce allylic alcohols in high regioselectivity using hydridoiridium phosphine complexes, such as $[Ir(PEt_2Ph)_4]_4$.[159]

In the case of multiply unsaturated carbonyl compounds, regioselectivity is also sensitive to the nature of the catalyst, to reaction conditions, and to the structure and degree of substitution of the hydrogenated double bonds. For example, hydrogenation of 3,5-heptadien-2-one over nickel on alumina or nickel on zinc oxide occurs mainly at the γ,δ-double bond. But if the catalyst is modified by the addition of lead or cadmium, reduction occurs mainly at the α,β-double bond (Scheme 24).[160]

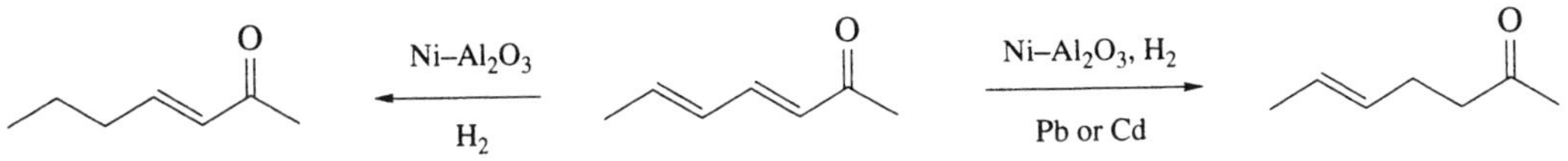

Ni–Al_2O_3 / H_2 ; Ni–Al_2O_3, H_2 / Pb or Cd

Scheme 24

Selective reduction of the γ,δ-double bond of the dienal shown in Scheme 25 was achieved by hydrogenation over palladium on carbon inhibited by quinoline and sulfur. Without inhibition, hydrogenation to the saturated aldehyde was observed.[161]

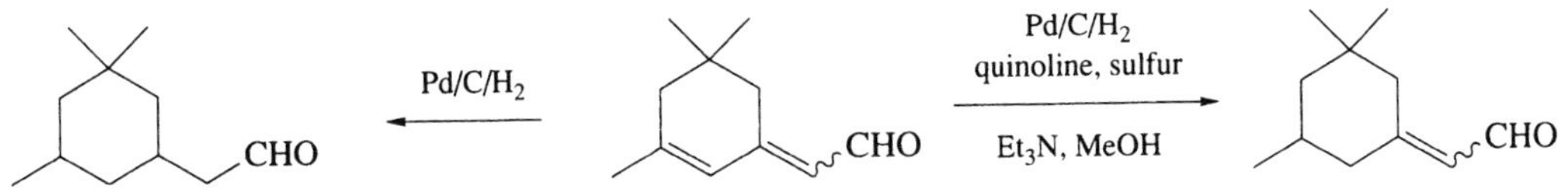

CHO ; Pd/C/H_2 ; CHO ; Pd/C/H_2 quinoline, sulfur / Et_3N, MeOH ; CHO

Scheme 25

Homogeneous catalysts, such as $RhCl(PPh_3)_3$[147] and $RuCl_2(PPh_3)_3$[162] have proved efficient in the selective hydrogenation of enones and dienones. For example, the hydrogenation selectivity of 1,4-androstadiene-3,17-dione to 4-androstene-3,17-dione is increased by elevated pressures, low temperatures and the presence of optimal amount of amines (Scheme 26).[162]

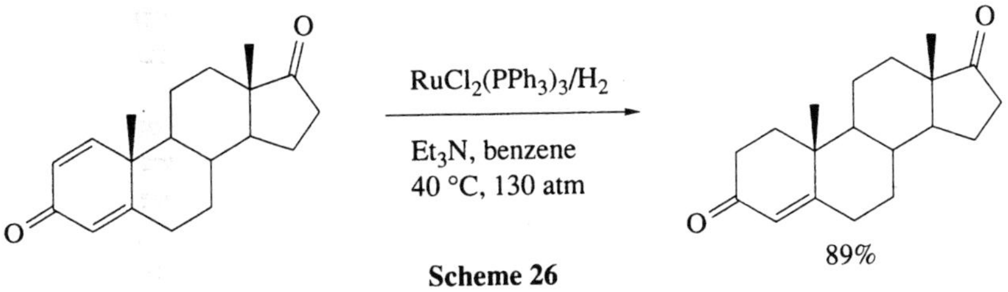

$RuCl_2(PPh_3)_3/H_2$ / Et_3N, benzene 40 °C, 130 atm ; 89%

Scheme 26

The solvated ion pair $[(C_8H_{17})_3NMe]^+[RhCl_4]^-$, formed from aqueous rhodium trichloride and Aliquat-336 in a two-phase liquid system, hydrogenates α,β-unsaturated ketones and esters selectively at the C=C double bond.[163] The reduction of benzylideneacetone follows first-order kinetics in substrate below 0.2 M, and approaches second-order in hydrogen at partial pressures below 0.12 atm (1 atm = 101.3 kPa). The catalysis also depends on the nature of the solvent, the phase transfer catalyst and stirring rates.

The homogeneous water soluble hydrogenation catalyst $K_3(Co(CN)_5H)$ is very active for hydrogenating conjugated dienes and α,β-unsaturated ketones under phase transfer reaction conditions.[164] Thus, conjugated dienes are converted into monoenes, generally with overall 1,4-addition to yield (*E*)-alkenes, and α,β-unsaturated ketones are reduced to saturated ketones in high yields. These conditions are not useful with α,β-unsaturated aldehydes, as they lead to polymerization of the starting material.

Of special interest is the formation of new asymmetric centers by homogeneous catalytic hydrogenation using chiral ligands. This approach is particularly appealing for the synthesis of optically active amino acids *via* hydrogenation of the corresponding α,β-unsaturated *N*-acetylaminocarboxylic acid derivatives. This rather successful methodology has been extensively reviewed;[165] although most of the catalysts employed in these asymmetric hydrogenation reactions are various chiral complexes of rhodium, other metals were also examined, usually affording lower optical yields.[166]

3.5.4 REDUCTION WITH MAIN GROUP METAL HYDRIDES

3.5.4.1 Boron Hydrides

Although $NaBH_4$ does not attack isolated alkenes, C–C double bonds conjugated to strong anion-stabilizing groups may be reduced by this reagent.[167–169] Rationalization of the regioselectivity of borohydride reduction of α,β-unsaturated aldehydes and ketones has been attempted using the 'hard' and 'soft' acid and base concept[170] (*vide infra*, discussion of aluminum hydrides). It is assumed that the relatively soft hydrides add preferentially to the enone system *via* a 1,4-mode while hard reagents attack the carbonyl carbon. Borohydrides are considered softer than the corresponding aluminum hydrides. Replacement of a hydride group on boron by alkoxide makes it a harder reagent. Lithium salts are harder than sodium species. Thus, $LiAlH_4$ gives more 1,2-attack than $LiBH_4$, which, in turn, gives more than $NaBH_4$. $NaBH(OMe)_3$ yields more 1,2-reduction product than $NaBH_4$, and when production of alkoxyborates is prevented, 1,4-reduction predominates. This implies that slow addition of borohydride to a substrate solution should help to build up alkoxyborate species and increase the relative amount of 1,2-reduction. Generally, aldehydes undergo more 1,2-reduction than the corresponding ketones.

The reduction of α,β-unsaturated aldehydes and ketones by $NaBH_4$ leads, in general, to substantial amounts of fully saturated alcohols. In alcoholic solvents, saturated β-alkoxy alcohols can be formed *via* conjugate addition of the solvent.[171] This latter process becomes the main reaction path when reduction is performed in 2-propanol in the presence of sodium isopropoxide. In base, a homoallylic alcohol can become the major product of borohydride reduction of an enone.[171] Analysis of the influence of substrate structure on $NaBH_4$ reduction has shown that increasing steric hindrance on the enone increases 1,2-attack.[171]

$NaBH_4$ reduction of 3-substituted 5,5-dimethylcyclohex-2-enones in alkaline solution of water dioxane occurs exclusively at the 1,2-positions. The rate of reduction is strongly dependent on the 3-substituent. A Hammett-type correlation revealed similar reaction characteristics to those of borohydride reduction of substituted acetophenones.[172]

In order to study the factors determining the regioselectivity of sodium borohydride reduction of α,β-unsaturated ketones, reactions with 3-methylcyclohexenone, carvone and cholestenone were carried out in 2-propanol, diglyme, triglyme or pyridine.[173] Mixtures of 1,2- and 1,4-reduction products were obtained in the alcoholic and ether solvents, whereas pure 1,4-reduction was observed in pyridine. Addition of triethylamine to $NaBH_4$ in diglyme led to formation of triethylamine borine, $Et_3N{\cdot}BH_3$. Similarly, with pyridine, pyridine·borine could be isolated, leading to exclusive 1,4-reductions.

The results were interpreted in terms of steric requirements of the actual reducing species. It was suggested that attack of BH_4^- proceeds exclusively along the 1,4-reduction mode, whereas alkoxyborohydrides (formed as reaction products) prefer the 1,2-reduction mode. The pyridine·borine itself does not reduce enones under the reaction conditions, but it inhibits formation of alkoxyborohydrides.[173] The same trend was observed with aluminum hydride reductions. When $LiAlH_4$ was first reacted with pyridine to form lithium tetrakis(dihydro-*N*-pyridyl)aluminate, 1,4-reduction predominated.[173]

Low regioselectivity is observed in reduction of enones with a 2:1 mixture of sodium cyanoborohydride and zinc chloride in ether at room temperature.[174] A mixture containing 1,2- and 1,4-reduction products is obtained in a ratio that is greatly dependent upon substrate.

From the reduction in methanol of a series of substituted 2-aryl-(*Z*)- and 2-aryl-(*E*)-cinnamates by $NaBH_4$ at room temperature, it was concluded that the facile reduction to give dihydrocinnamates proceeds through an early transition state of considerable polarity.[167,175]

Several organoborohydrides were found to effect the selective 1,4-reduction of enones. For example, lithium and potassium tri-*s*-butylborohydrides (L- and K-selectride) and lithium triethylborohydride were found useful for conjugate reduction of α,β-unsaturated ketones and esters. In general, β-unsubstituted cyclohexenones undergo exclusive 1,4-reduction to the corresponding ketone enolates, which can be protonated or alkylated in high yields. Ketones such as 5-*t*-butylcyclohex-2-en-1-one are cleanly reduced to the saturated ketone using K-selectride at –78 °C in THF (Scheme 27).[176] This regioselectivity, however, is not general, but is a result of steric hindrance of the alkene, as well as the size of the ring. Thus alkyl substitution at the β-position completely suppresses the 1,4-reduction mode. While enones in five- and seven-membered rings are reduced preferentially in a 1,2-manner, six-membered ring enones are reduced in a 1,4-mode. Trapping the intermediate enolate by an alkylating agent (*e.g.* MeI, allyl bromide) results in an efficient reductive alkylation. Accordingly when the reduction of α,β-unsaturated esters is performed in dry ether solvents, the major reaction product arises from carbonyl condensation. However addition of a proton source such as *t*-butyl alcohol results in 1,4-reduction.

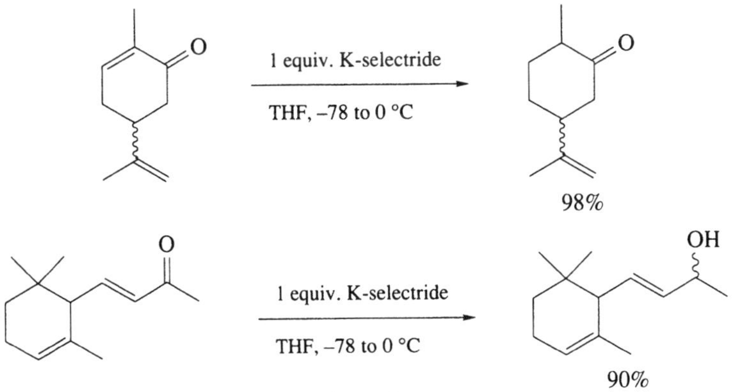

Scheme 27

Reduction of α,β-unsaturated aldehydes and ketones with 9-borabicyclo[3.3.1]nonane (9-BBN-H) proceeds selectively and cleanly to form the corresponding allylic alcohols.[177] The reaction tolerates a large variety of functionalities, such as nitro, carboxylic acid, amide, nitrile, sulfide, disulfide, epoxide, *etc.* Hydroboration of the double bond is a much slower reaction, which does not interfere with carbonyl reduction. For example, 1,2-reduction of cyclohexenone at room temperature with excess of 9-BBN-H in THF is completed within 10 min, while hydroboration of the double bond requires three days. Amine·borane complexes have been used for selective 1,2-reduction of α,β-unsaturated ketones and aldehydes. Thus, cyclohexen-2-one was reduced to hexenol with *t*-butylamine·borane.[178] Borohydride exchange resin, which is obtained from $NaBH_4$ and the quaternary ammonium resin Amberlite 400 IRA, reduces cinnamaldehyde and benzalacetone to the corresponding allylic alcohols.[179] 1-Pyrrolylborane·THF complex reduces α,β-unsaturated ketones and aldehydes to the corresponding allylic alcohols in high yields.[180] Borohydride reduction of α,β-unsaturated carbonyl compounds has been widely applied in natural product chemistry. A number of α,β-unsaturated ketone derivatives of gibberellins are reduced to the corresponding saturated alcohols by $NaBH_4$.[181–184] Sodium borodeuteride reduction of gibberellin A_3 3-ketone affords gibberellin A_1 and its 3-epimer.[181,182] Attack of hydride proceeds stereospecifically from the β-face at C-1. Protonation at C-2 proceeds with limited selectivity. Thus, reduction of the above-mentioned gibberellin with either $NaBH_4$–CuCl in deuterated methanol or $NaBH_4$–LiBr followed by treatment with D_2O gave 2-deuteriogibberellin A_1 methyl ester together with some 3-epi-GA_4 with approximately 2:1 ratio of the 2β:2α deuterides. Using L-selectride for the reduction of a similar gibberellin enone derivative resulted mainly in the 1,2-reduction product, affording the 3α-allylic and saturated alcohols in 47% and 30% yields, respectively.[183]

Substituted gibberellins, such as 1α- and 1β-hydroxy GA_5 and GA_{20} were prepared from a single enone precursor by 1,2-reduction with $NaBH_4$ (Scheme 28). Conversely, catalytic hydrogenation of the same enone with 10% Pd/$CaCO_3$ in pyridine afforded the 1,4-reduction product, 1-oxo-GA_{20}.[184]

The stereoselective 1,2-reduction of the α,β-unsaturated ketone group was one of the key steps in Corey's prostaglandin synthesis (Scheme 29).[185] Various boron and aluminum hydride reagents gave mixtures of the corresponding (15*S*)- and (15*R*)-allylic alcohols in various ratios.[186] Best selectivity was obtained with highly hindered lithium trialkylborohydrides, and even better with R = *p*-PhC_6H_4NHCO This derivative was reduced with thexyl-di-*s*-butylborohydride and tri-*s*-butylborohydride with (15*S*):(15*R*) ratios of 88:12 and 89:11, respectively.[185] 1,2-Reduction of an α,β-unsaturated aldehyde with $NaBH_4$ represents one of the steps in the total synthesis of 6,15-hydroxyperezone.[187]

Stereoselective reduction of an enone lactone was a key step in the construction of the 20-hydroxyecdysone side chain. Totally different mixtures of products were obtained when the reduction was carried out with sodium borohydride or by catalytic hydrogenation (Scheme 30).[158] In all cases, the 1,4-reduction mode is preferred. With borohydride, however, this process is followed by a subsequent reduction of the saturated ketone and base-catalyzed rearrangement of the δ-lactone into a γ-lactone.

Stereoselective conjugate reduction of α,β-unsaturated amides with L-selectride was achieved by using a chiral sultam, temporarily attached to the molecule.[188] Alkynic ketones are enantioselectively reduced to the corresponding propargylic alcohols in up to 92% *ee* using *B*-(3-pinanyl)-9-borabicyclo[3.3.1]nonane.[189] The conjugate reduction of acyclic α,β-unsaturated ketones can provide selectively regio- and stereo-chemically defined enolates that are unattainable by other methods. A knowledge of enone ground-state conformational preferences allows one to predict which enolate geometrical isomer will predominate in these reactions (Scheme 31).[190]

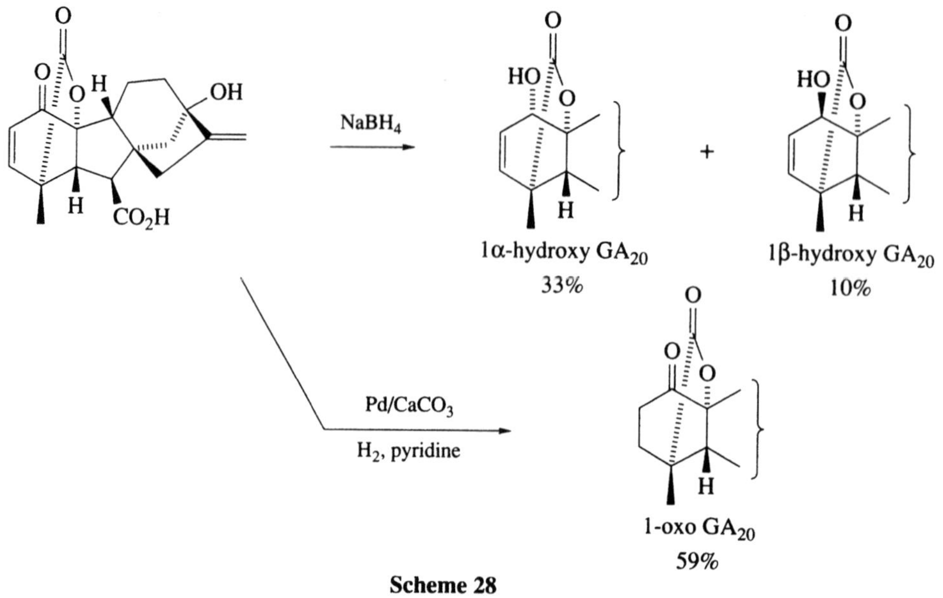

$NaBH_4$

1α-hydroxy GA_{20} 33%

1β-hydroxy GA_{20} 10%

$Pd/CaCO_3$

H_2, pyridine

1-oxo GA_{20} 59%

Scheme 28

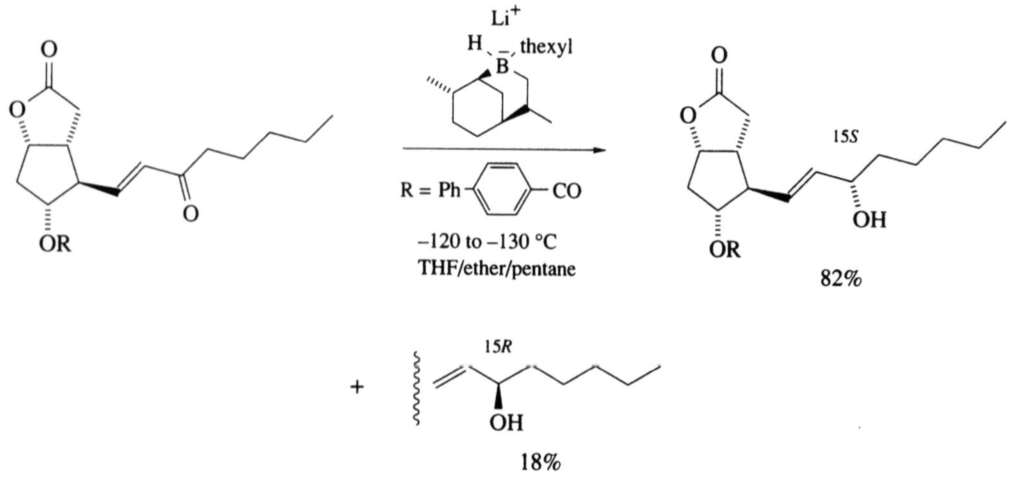

Li^+ thexyl

R = Ph–C_6H_4–CO

–120 to –130 °C

THF/ether/pentane

15*S* 82%

15*R* 18%

Scheme 29

Thus, enones that exist preferentially as *s-trans* conformers will give rise to (*E*)-enolates, whereas conjugate addition of hydride to *s-cis* enones will lead to (*Z*)-enolates. These can be trapped by trimethylsilyl chloride (TMS-Cl) to give the corresponding silyl enol ethers (Scheme 32).[190]

Sodium cyanoborohydride ($NaBH_3CN$) or tetrabutylammonium cyanoborohydride in acidic methanol or acidic HMPA reduces α,β-unsaturated aldehydes and ketones to the corresponding allylic alcohols.[191] This system is limited to enones in which the double bond is not further conjugated, in which case the allylic hydrocarbon is formed in substantial amounts. Thus, reduction of chalcone gives mainly 1,3-diphenylpropene (48%) as well as 26% of the allylic ether. Cyclic enones are also not good substrates, as competing 1,4-addition gives large fractions of saturated alcohols.[191]

Lithium *n*-butylborohydride is prepared by reacting equimolar amounts of *n*-butyl lithium and borane·dimethyl sulfide complex.[192] This reagent effectively reduces enones in toluene–hexane mixtures at –78 °C to give, in most cases, high yields of the corresponding allylic alcohols.[192] Conjugated cyclopentenones, however, give mixtures of 1,2- and 1,4-reduction products. Under identical reaction conditions, saturated ketones are reduced to alcohols. The latter process can take place in the presence of simple esters.

Regioselective 1,2-reduction of enones to the corresponding allylic alcohols is achieved with $NaBH_4$ in the presence of lanthanoid ions, such as La^{3+}, Ce^{3+}, Sm^{3+}, Eu^{3+}, Yb^{3+} and Y^{3+}.[193] This procedure is complementary to those giving predominantly 1,4-selectivity, such as $NaBH_4$ in pyridine.[173] The general

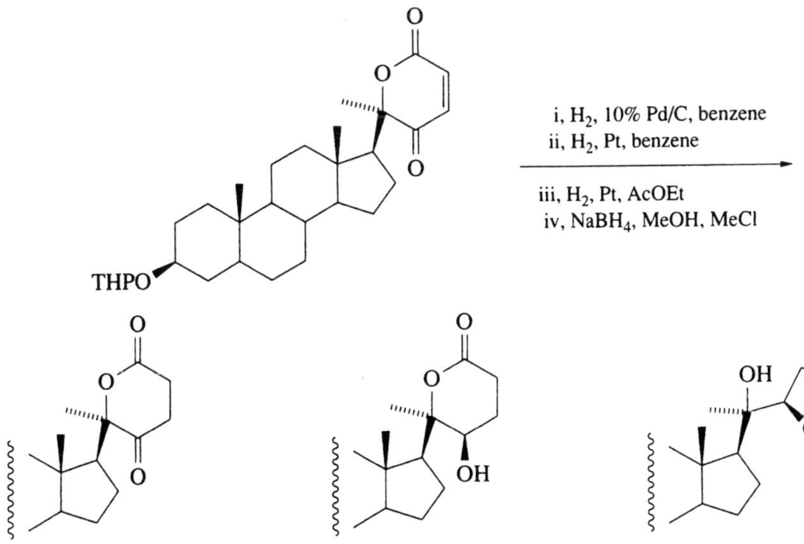

i, H_2, 10% Pd/C, benzene
ii, H_2, Pt, benzene
iii, H_2, Pt, AcOEt
iv, $NaBH_4$, MeOH, MeCl

THPO

i	98%	–	–
ii	17%	–	81%
iii	–	80%	20%
iv	–	5%	78%

Scheme 30

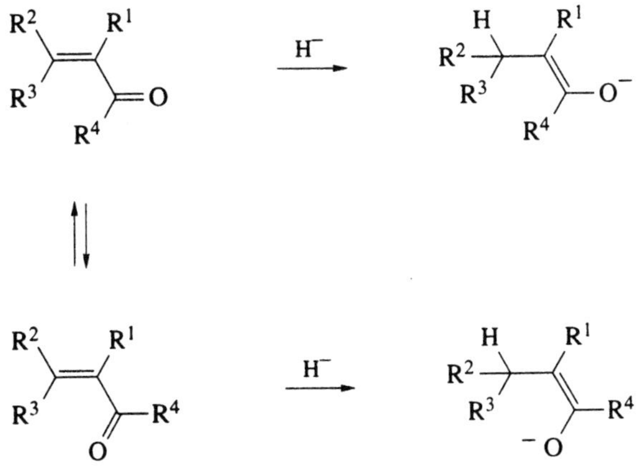

Scheme 31

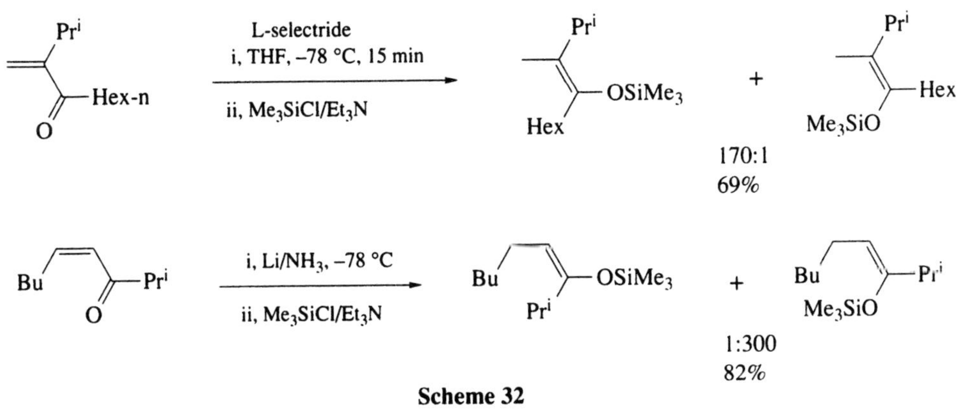

L-selectride
i, THF, –78 °C, 15 min
ii, Me_3SiCl/Et_3N

170:1
69%

i, Li/NH_3, –78 °C
ii, Me_3SiCl/Et_3N

1:300
82%

Scheme 32

utility of $NaBH_4$–$CeCl_3$ selective reduction is illustrated by the conversion of cyclopentenone to cyclopentenol in 97% yield and only 3% of cyclopentanol, although conjugate reduction of cyclopentenone systems by most hydride reagents is usually highly favored (Scheme 33).

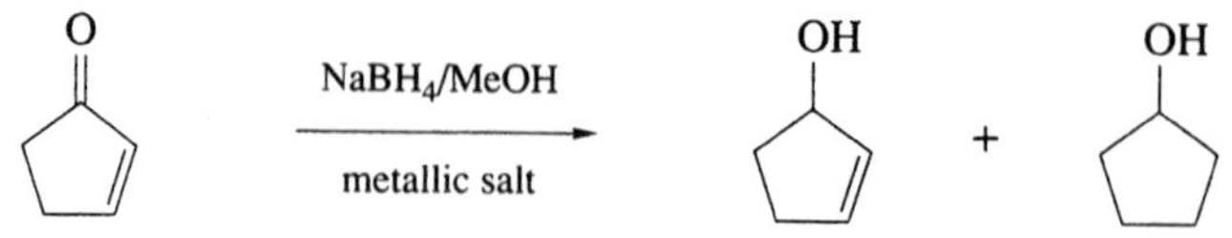

NaBH$_4$/MeOH, metallic salt

Metallic salt	Allylic alcohol	Saturated alcohol
La^{3+}	90	10
Ce^{3+}	97	3
Sm^{3+}	94	6
Eu^{3+}	93	7
Yb^{3+}	89	11
Y^{3+}	86	14
Li^{+}	1	99

Scheme 33

Thus, reaction of equimolar amounts of α,β-unsaturated ketones and either samarium or cerium chloride hexahydrate in methanol with sodium borohydride produced high yields of the corresponding allylic alcohols.[193] This method has gained considerable success and has often been used in regio- and stereo-selective syntheses of natural products.[194,195] Trivalent lanthanoid ions were shown to promote selective 1,2-reduction of conjugated aldehydes in the presence of saturated aldehydes. For example, citral is quantitatively reduced to geraniol with $NaBH_4/ErCl_3$, while citronellal is sluggishly reduced (13%) under the same conditions.[196,197] Conversely, 1,4-reduction of γ-pyrones occurs with tritiated $NaBH_4/Ce(NO_3)_3$.[198]

A mechanistic study of the role of the lanthanide cations suggests that they catalyze decomposition of borohydride by the hydroxylic solvent to afford alkoxyborohydrides, which may be responsible for the observed regioselectivity. The stereoselectivity of the process is also modified by the presence of Ln^{3+} ions, in that axial attack of cyclohexenone systems is enhanced.[193] α,β-Unsaturated aldehydes are regioselectively reduced to allylic alcohols by bis(triphenylphosphine)copper(I) tetrahydroborate in the presence of Lewis acid catalyst.[199]

β-Dialkylamino conjugated enones are reduced to the corresponding γ-amino alcohols with $NaBH_4$ in the presence of $FeCl_3$. These amino alcohols could be converted into conjugated enones by chromic acid oxidation and deamination (Scheme 34).[200] On the other hand, β-acylamino conjugated enones are reduced by $NaBH_4$ to afford β,γ-unsaturated γ-acylamino alcohols, which are regioselectively hydrolyzed to conjugated enones.

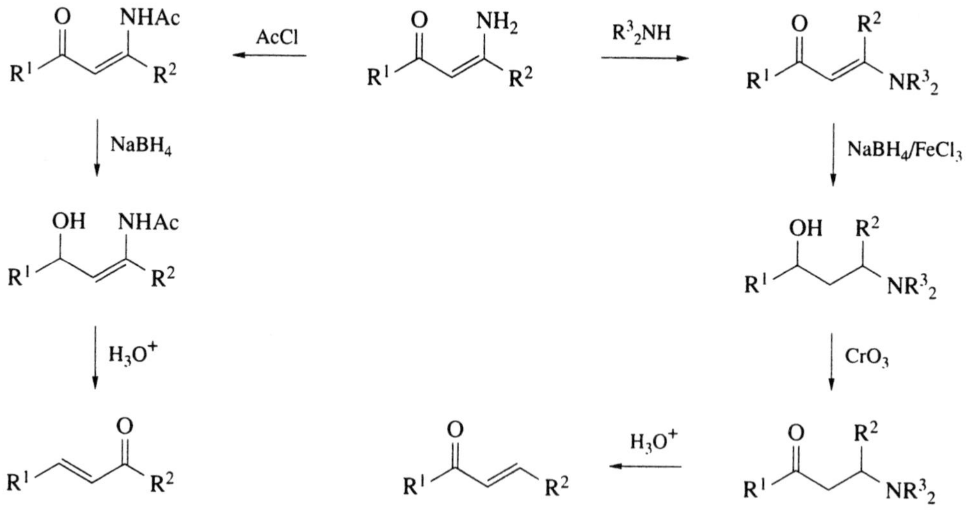

Scheme 34

β,β-Bis(alkylthio)-α,β-unsaturated ketones are selectively reduced in a 1,2-manner to the corresponding allylic alcohols by tritiated $NaBH_4$ in ethanol.[201] Enaminones are similarly reduced by tritiated

$NaBH_4$ to the corresponding β-amino carbonyl compounds, while reduction with $LiAlH_4$ resulted in deaminated products. This is explained on the basis of the HSAB principle, with $NaBH_4$ being considered a soft base and the vinylic β-carbon being a soft acid.[202] Reduction of β-sulfenylated α,β-unsaturated ketones with $NaBH_4$ in the presence of catalytic amounts of $CoCl_2$ or $NiCl_2$ in methanol produces the corresponding desulfenylated, saturated ketones (Scheme 35).[203] These substrates, however, were not affected by combinations of $NaBH_4$ and other metal salts, including $FeCl_2$, $FeCl_3$, CuI and $CuCl_2$. Conjugate reduction of substituted cinnamic acids was achieved with tritiated $NaB[^3H]_4$–$NiCl_2$ in methanol.[204]

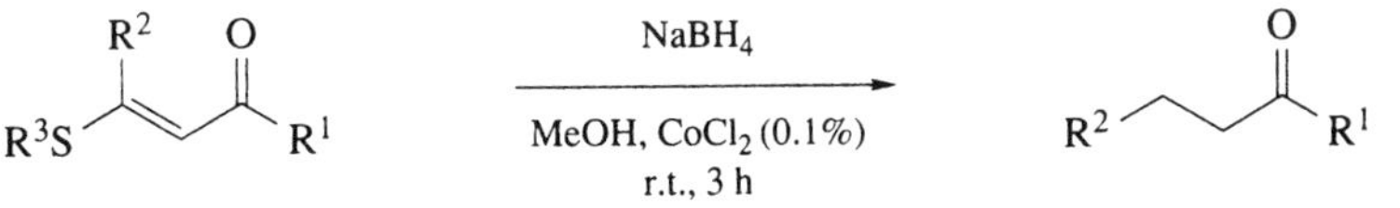

R^3S ... R^2, O, R^1 → $NaBH_4$ / MeOH, $CoCl_2$ (0.1%), r.t., 3 h → R^2 ... O, R^1

Scheme 35

3.5.4.2 Aluminum Hydrides

The properties of complex metal hydrides, particularly those of aluminum, and their use in organic synthesis have been compared in a number of papers, review articles and monographs.[205–209] Useful tables, listing the most appropriate hydride reagents for selective reduction of various polyfunctional compounds, have been published.[1,208–211] Use of chiral metal alkoxyaluminum hydride complexes in asymmetric synthesis has also been reviewed.[212]

The two modes of reduction of α,β-unsaturated aldehydes and ketones, 1,2- and 1,4-addition of metal hydride to the enone system, lead respectively to either an allylic alcohol or a saturated ketone. It has been suggested that the relative importance of these paths depends upon substrate hardness or softness, as defined in terms of the coefficients of the lowest unoccupied molecular orbital (LUMO) (*vide supra*, the discussion of borohydrides). While 1,2-addition is considered to be a mainly charge-controlled process, 1,4-addition is a frontier orbital controlled process.[213] These considerations predict, for example, that the 1,4-addition of a given metal hydride to cyclopentenone should always be faster than a similar addition to cyclohexenone.[214] Moreover, in cases where the enone system is further conjugated to a phenyl ring, as in cinnamaldehyde, increased frontier orbital control should render the enone more prone to 1,4-addition.[215] Obviously, the course of reduction of conjugated carbonyl compounds is also highly influenced by the nature of the metal hydride. According to Pearson's concept of soft and hard acids and bases,[216,217] hard metal hydrides add preferentially to the 2-position and soft metal hydrides to the 4-position of the conjugated enone system.[213–215] As shown in Table 1, these predictions agree well with representative experimental results.[214,218]

Table 1 Ratio of 1,4- to 1,2-Reduction Products

	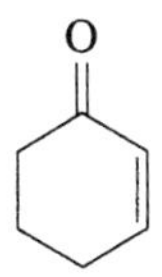	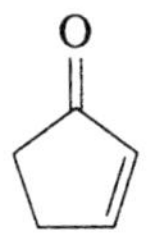	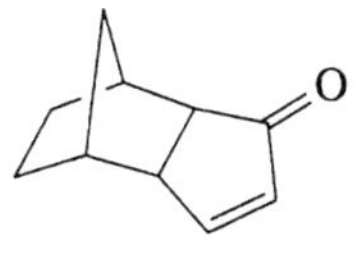
$LiAlH(OMe)_3$	5:95	10:90	24:76
$LiAlH_4$	22:78	86:14	100:0
$LiAlH(SMe)_3$	56:44	95:5	
$LiAlH(OBu^t)_3$	78:22	100:0	100:0
$LiAlH(SBu^t)_3$	95:5	100:0	

Because of their electrophilic nature, Li^+ cations accelerate the reduction of carbonyl compounds by $LiAlH_4$ or $NaBH_4$. Li^+-complexing agents, such as cryptands, crown ethers or polyamines decrease the rate of reduction.[219] In the case of α,β-unsaturated ketones, this slow down is associated with altered regioselectivity. For example, $LiAlH_4$ reduction of cyclohexenones in the absence of the cryptand proceeds predominantly with 1,2-reduction. In the presence of the cryptand, 1,4-attack is favored. This selectivity is more pronounced with $LiAlH_4$ than with $NaBH_4$ (Scheme 36)[219] and is also dependent on solvent. For example, with diethyl ether the 1,2-attack prevails, whereas when the cation is complexed, 1,4-addition predominates.

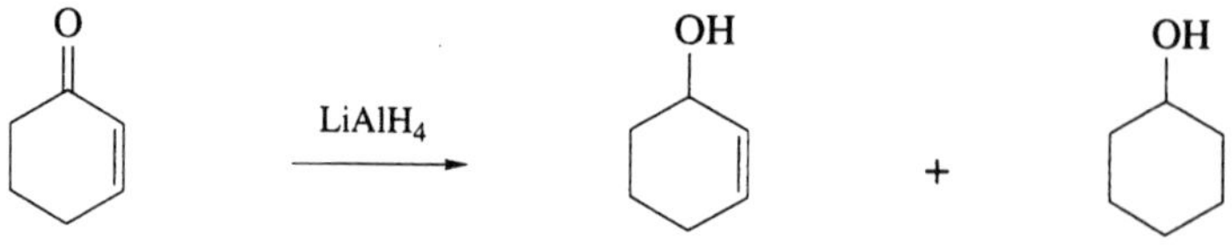

$LiAlH_4$			
THF		86:14	98%
ether		98:2	98%
THF, 1.2 equiv. cryptand		14:86	85%
ether, 1.2 equiv. cryptand		23:77	80%

cryptand =

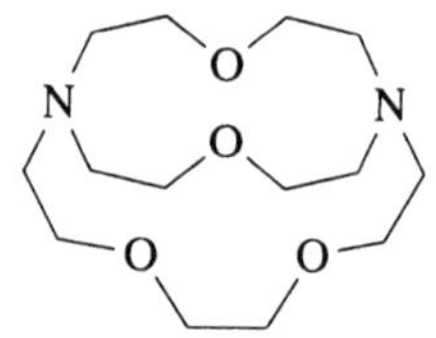

Scheme 36

This effect is explained in terms of frontier molecular orbitals treatment.[219] The regioselectivity of reduction depends upon the relative values of the C-1 and C-3 atomic coefficients in the LUMO. The atom with the larger coefficient corresponds to the predominant site of attack. When Li^+ is complexed by the α-enone, the C-1 coefficient is larger than that of C-3, and C-1 attack is favored. In the absence of such complexation, the C-3 coefficient is larger, leading to 1,4-attack. The strength of carbonyl–Li^+ interaction is strongly dependent upon the solvent, the nature of the complexing agent, and the interaction between the Li^+ ion and the reducing agent. Thus, in strongly coordinating solvents, such as pyridine,[173] 1,4-reduction predominates.

Steric and electronic factors in the enone substrate may also alter selectivity. For example, the high tendency of $LiAlH(OBu^t)_3$ to undergo 1,4-addition with simple enones is modified in the example given in Scheme 37.[220]

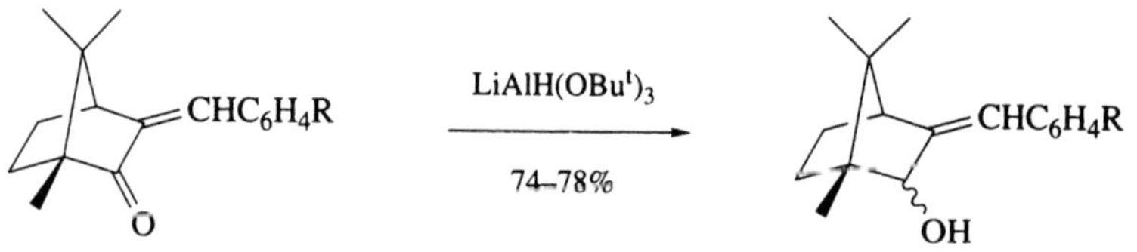

CHC_6H_4R $\xrightarrow[74–78\%]{LiAlH(OBu^t)_3}$ CHC_6H_4R OH

R = H, *m*-F, *p*-Me, *p*-MeO

Scheme 37

β,β-Bis(alkylthio)-α,β-unsaturated ketones undergo two-step reduction with $LiAlH_4$. The first step is a 1,2-addition to the carbonyl, leading to allylic alcohols, which may undergo further reduction of the alkenic bond.[221] The ratio of 1,2- to 1,4-addition of aluminum hydride to an α,β-unsaturated ketone is highly dependent on the enone structure, solvent, relative initial concentrations of reactants, temperature, and softness or hardness of the hydride reagent. These reductions can be controlled to proceed with either 1,2- or 1,4-addition, with high selectivity.[205] Scheme 38[222–225] illustrates the prominent tendency of $LiAlH_4$ and $LiAlH(OMe)_3$ to yield 1,2- rather than 1,4-adducts, as compared to $LiAlH(OBu^t)_3$.

The reagent $NaAlH_2(OCH_2CH_2OMe)_2$ favors 1,2-addition to cyclic enones[223,226–230] with greater selectivity than with either $LiAlH(OMe)_3$[214] or AlH_3.[218] In the reduction of 9-oxoisolongifolene to the allylic 9α- or 9β-alcohols, reversal of stereochemistry occurs when $NaAlH_2(OCH_2CH_2OMe)_2$ is used instead of $LiAlH_4$ or $NaBH_4$.[231] While the latter two reagents lead to formation of the thermodynamically more stable α-alcohol as the major product, increased steric bulk of the former seems to favor the less stable β-isomer. Chiral compounds containing enones are often reduced to the corresponding allylic alcohols with high stereoselectivity.[232] For example (–)-*cis*-carveol is obtained by $LiAlH_4$ reduction of carvone at low temperatures.[233]

Sterically unhindered enones, such as cyclohexenone, are reduced by $LiAlH(OBu^t)_3$ to give predominantly the corresponding saturated ketone,[214] but more sterically congested systems are cleanly reduced *via* the 1,2-mode to give the allylic alcohol, usually with high stereoselectivity.[234–238]

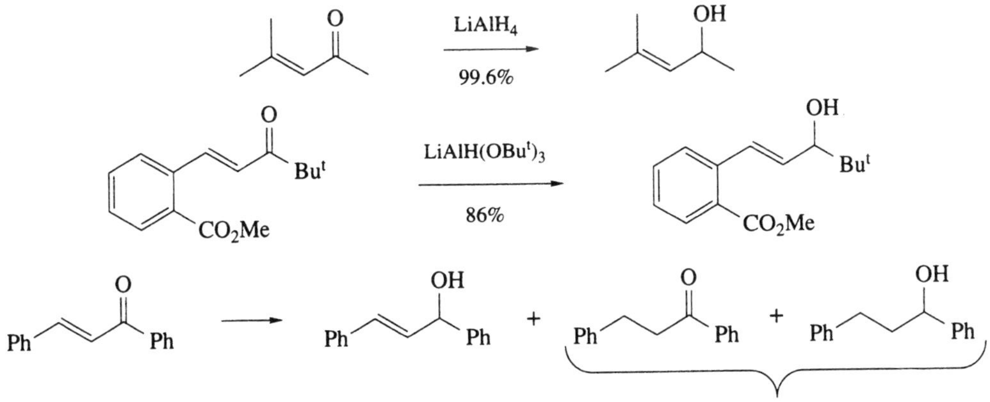

$LiAlH_4$ 99.6%

$LiAlH(OBu^t)_3$ 86%

$LiAlH(OMe)_3$	94%	6%
$LiAlH(OBu^t)_3$	38%	62%

Scheme 38

1,2-Reduction has been reported for other hydride reagents, such as diisobutylaluminum hydride (DIBAL-H),[213,230,239,240] aluminum hydride[218] and 9-BBN-H.[241] α,β-Unsaturated lactones are also reduced by DIBAL-H to the corresponding lactols in high yields.[242]

1,4-Reduction of enones can be effected with high selectivity with $AlH(OBu^t)_2$, $AlH(OPr^i)_2$, $AlH(NPr^i{}_2)_2$ forming saturated ketones in 90–100% yield. $AlH(NPr^i{}_2)_2$ exhibited the lowest selectivity, as no 1,4-reduction of mesityl oxide or isophorone is observed with this reagent. Reacting $AlH(NPr^i{}_2)_2$ with methyl vinyl ketone or cyclohexanone led to a mixture of products. *trans*-Chalcone also undergoes quantitative 1,4-reduction with the above-mentioned hydrides.[240] Similarly, reduction of 9-anthryl styryl ketone or anthracene-9,10-diyl-bis(styryl ketone) with $LiAlH(OBu^t)_3$ affords the saturated ketone as the sole product.[243] Hydrides such as $LiAlH(OBu^t)_3$ and $LiAlH(SBu^t)_3$ favor 1,4-reduction in cyclopentenones,[214,215,218,244–247] as in the example in Scheme 39.[248,249]

$LiAlH(OBu^t)_3$

Scheme 39

Scheme 40 illustrates an interesting two-step selective reduction of an enone system, first with sodium hydride and $NaAlH_2(OCH_2CH_2OMe)_2$ and then with the same reagent in the presence of 1,4-diazabicyclo[2.2.2]octane. Specific reduction, however is not achieved with $NaBH_4$, $LiBH_4$, $LiBHBu^s{}_3$ or 9-BBN-H.[250]

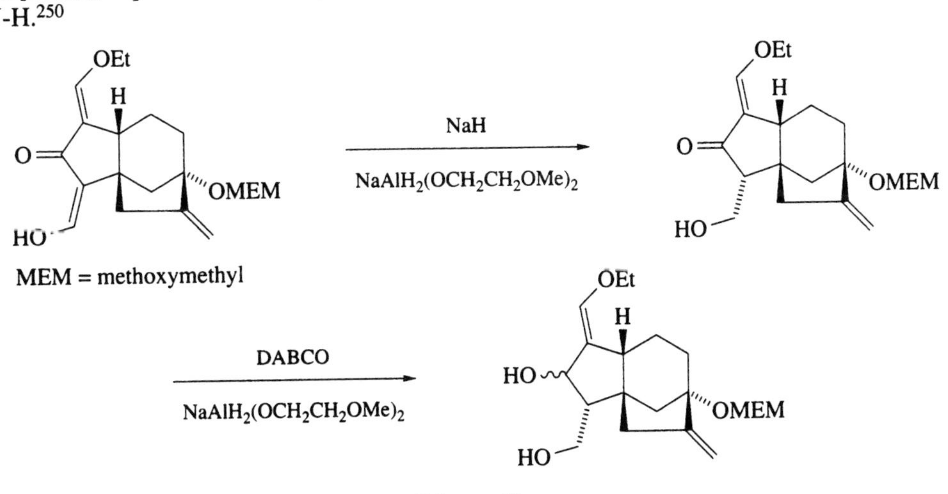

NaH, $NaAlH_2(OCH_2CH_2OMe)_2$

MEM = methoxymethyl

DABCO, $NaAlH_2(OCH_2CH_2OMe)_2$

Scheme 40

Both $LiAlH(OMe)_3$ and $NaAlH_2(OCH_2CH_2OMe)_2$ are convenient reducing agents for low temperature, copper-mediated 1,4-reduction, as shown for the latter by Scheme 41.[223,251] Partial reduction of cyclopentenedione has been achieved with several of these reagents.[252–254]

$NaAlH_2(OCH_2CH_2OMe)_2$, Cu_2Br_2; THF, 2-butanol, –78 to –20 °C, 98%

Scheme 41

There are a number of cases where a less reactive enone group remains intact while a more reactive saturated ketone present in the same substrate is selectively reduced,[255–258] but there are a number of examples of polyfunctional natural products where simultaneous reduction of both saturated and unsaturated ketones or of preferential reduction of the unsaturated one is achieved with $LiAlH(OBu^t)_3$.[259–261]

1,2-Reduction of enol ethers or enol esters of 1,3-diketones, followed by acid-catalyzed allylic rearrangement of the reduction product (see p. 85 in ref. 5) is a useful route to α,β-unsaturated ketones. Aliphatic[262,263] and alicyclic[264] enones have thus been prepared in good yields at low temperatures with $NaAlH_2(OCH_2CH_2OMe)_2$.[265,266]

Reduction of α,β-unsaturated aldehydes can afford either an unsaturated or saturated primary alcohol, or a mixture of both, depending on reaction conditions. For example, while addition of cinnamaldehyde to $NaAlH_2(OCH_2CH_2OMe)_2$ in benzene gives 97% 3-phenylpropanol, inverse addition (of the reducing agent to solution of the substrate) yields 94% cinnamyl alcohol.[267,268] Reduction with $LiAlH_4$ is similarly dependent on the addition sequence. The more sterically hindered hydride $LiAlH(OBu^t)_3$ is highly selective for 1,2-reduction of aldehydes, even under conditions of normal addition. For example, it reduces cinnamaldehyde cleanly to cinnamyl alcohol, without affecting the alkenic bond.[269–271] Similar behavior is exhibited by $NaAlH_2(OCH_2CH_2OMe)_2$, which reduces 2-butenal to 2-butenol in 97% yield.[268] On the other hand, hydrides, such as $LiAlH(OMe)_3$[206,269,270] and $NaAl_2H_4(OCH_2CH_2NMe_2)_3$,[272] usually yield the saturated primary alcohol.[273–275]

Regioselectivity of enone reduction with DIBAL-H is very susceptible to minor structural changes in the substrate. While five-membered exocyclic enones provide the allylic alcohols, which are the normal products for this reagent, reduction of chromones possessing exocyclic six-membered enones yield saturated ketones (Scheme 42).[276] This was explained by the strict coplanarity of the enone function in the five-membered structure, whereas the enones giving rise to saturated ketones are slightly twisted. Reduction of isoflavones with DIBAL-H under these conditions provides the corresponding isoflavan-4-ones in very high selectivity.[276] Reduction of a substituted dihydropyranone with DIBAL-H in benzene affords the corresponding glycal.[277]

DIBAL-H; toluene/THF, –65 °C; 65%

DIBAL-H; toluene/THF, –65 °C; 50%

Scheme 42

The 'ate' complex $LiAlHBu^nBu^i_2$ is prepared from DIBAL-H and *n*-butyllithium in either THF or toluene–hexane. This reagent is more effective for selective 1,2-reduction of enones to the corresponding allylic alcohol than is DIBAL-H alone.[278] The reagent also reduces esters, lactones and acid chlorides to the corresponding alcohols, and epoxides to the respective alcohols. α,β-Unsaturated ketones derived from dehydration of aldol products from 1-(arylthio)cyclopropanecarbaldehydes and ketones were selectively reduced by this 'ate' complex or by DIBAL-H itself, yielding the allylic alcohols with minor amounts of the 1,4-reduction product.[279] Yields were typically higher with this reagent than with DIBAL-H.

Enones may be deoxygenated with $LiAlH_4/AlCl_3$ to give the corresponding alkenic hydrocarbons. The reactive species seem to be $AlHCl_2$ or AlH_2Cl, which act as both Lewis acids and hydride donors. The reaction involves initial 1,2-reduction to form the allylic alcohol, followed by substitution of the allylic hydroxy group by hydride (mainly *via* an S_N2' mechanism) to form the corresponding mixture of alkenes.[280] This technique has been applied to the deoxygenation of natural products. By using mixtures of $LiAlH_4$ and $AlCl_3$, flavanone and chalcones were transformed into flavan and diarylpropenes, respectively.[281] In a few cases, such as with 2-bromocyclopentenone or 2-hydroxycyclohexenone, only 1,2-reduction with $LiAlH_4/AlCl_3$ is observed.[282]

Conjugate reduction is the major pathway of enone reduction with a mixture of $LiAlH_4$ and excess CuI in THF.[283] It has been shown that the active reducing agent in this mixture is an H_2AlI species and not the copper hydride. Enones that have an *s-cis* geometry are reduced much more slowly than similar ketones with an *s-trans* conformation, and no reduction was observed with cyclohexenone and 3,3,5-trimethylcyclohexenone. These results suggest that the mechanism involves coordination of the metal to the carbonyl, forming a six-center transition state.[283]

Enones with two alkyl groups at the β-position are reduced very sluggishly under these conditions. Other metal salts, such as HgI_2, $TiCl_3$ and $HgCl_2$, premixed with $LiAlH_4$ in THF similarly give rise to 1,4-reduction. Yields and selectivities were found to be much lower than with CuI. H_2AlI was found to react in the same manner as $LiAlH_4$/CuI, and the series H_2AlI, $HAlI_2$, H_2AlBr, $HAlBr_2$, H_2AlCl and $HAlCl_2$ was therefore prepared. Of these, the iodo compounds exhibited the highest reactivity. $HAlI_2$ reduces enones at a slower rate than H_2AlI. Diisobutylaluminum-2,6-di-*t*-butyl-4-methylphenoxide was found to be an efficient and stereoselective reducing agent for 1,2-reduction of the enone at position 15 in prostaglandin synthesis,[284] and was used successfully in the synthesis of the lichen macrolide (+)-aspicilin.[285]

Chiral lithium alkoxyaluminum hydride complexes can be used to obtain optically active allylic alcohols.[286–289] These reagents are more selective than the polymer-supported $LiAlH_4$ and $LiAlH_4$–monosaccharide complexes.[290]

α,β-Alkynic ketones are selectively reduced to the corresponding propargylic alcohols with $LiAlH(OMe)_3$. Asymmetric 1,2-reduction of alkynic ketones is an effective method for preparing optically active propargylic alcohols in high yield and high enantioselectivity. Common chiral reductants for this purpose include the Mosher–Yamaguchi reagent,[291–293] the Vigneron–Jacquet complex,[294–296] and $LiAlH_4$/2,2′-dihydroxy-1,1′-binaphthyl/methanol (*R* and *S*) complexes,[297] Landor's chiral $LiAlH_4$–monosaccharide complexes,[298–300] and the $LiAlH_4$–*N*-methylephedrine–*N*-ethylaniline complex.[288]

Asymmetric reduction of d_1-geranial, d_1-neral and related linear terpenic aldehydes can be achieved with $LiAlH_4$–dihydroxybinaphthyl complex with 72–91% *ee* (Scheme 43).[301]

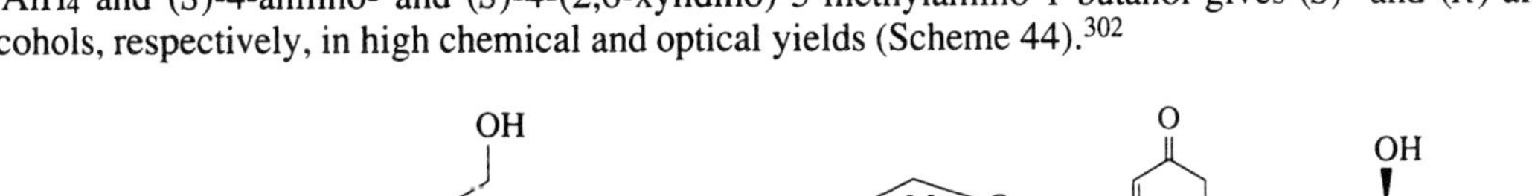

D O → $LiAlH_4$ / (*S*)-(–)-2,2′-dihydroxy-1,1′-binaphthyl, ethanol, THF, –100 °C → H D OH

(*S*)-(+) 91% (91% *ee*)

Scheme 43

Asymmetric reduction of prochiral α,β-unsaturated ketones with chiral hydride reagents derived from $LiAlH_4$ and (*S*)-4-anilino- and (*S*)-4-(2,6-xylidino)-3-methylamino-1-butanol gives (*S*)- and (*R*)-allylic alcohols, respectively, in high chemical and optical yields (Scheme 44).[302]

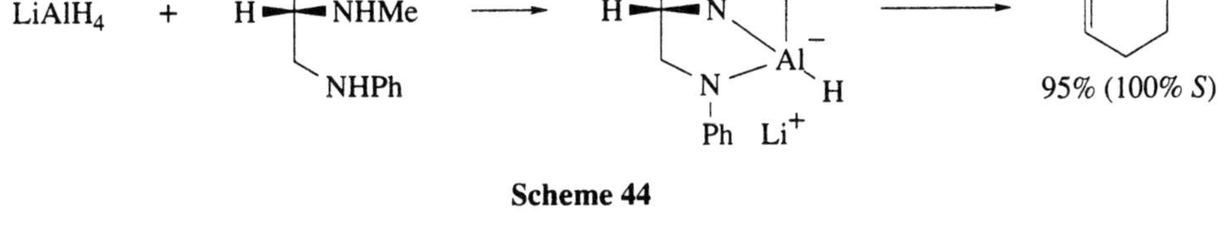

$LiAlH_4$ + H–(OH)(NHMe)(NHPh) → H–(Me N, O, Al, N Ph, H, Li^+) → (O) → OH

95% (100% *S*)

Scheme 44

A modified aluminum hydride prepared by treating $LiAlH_4$ in THF with equimolar amounts of ethanol and optically pure *S*-(–)-2,2′-dihydroxy-1,1′-binaphthyl, gives allylic alcohols of very high optical purity,[186,303] especially attractive in prostaglandin synthesis.[186] Considering both chemical and optical yields, these reductions compete with standard microbiological methods.[304] Asymmetric reduction of α,β-unsaturated ketones is also achieved with $LiAlH_4$, partially decomposed by (–)-*N*-methylephedrine and ethylaniline.[288] This reagent converts open chain enones into the corresponding optically active allylic alcohols in high chemical (92–100%) and optical yields (78–98% *ee*).

3.5.4.3 Silicon Hydrides

The hydrogen in the Si—H bond is slightly hydridic in nature, as would be expected from the relative electronegativities of silicon (1.7) and hydrogen (2.1). Therefore, silanes may function as hydride transfer agents toward highly electrophilic species such as carbonium ions. The hydridic nature of the Si—H bond may be significantly increased upon interaction with strong anionic ligands, such as fluoride and alkoxides (*vide infra*). In addition, the average bond energy of the Si—H and C—H bonds (70 and 99 kcal mol^{-1}, respectively; 1 kcal = 4.18 kJ) suggests that Si—H bonds should be susceptible to hydrogen atom abstraction by carbon radicals. Thus, the dehalogenation of alkyl halides with hydridosilane under homolytic conditions is explained in terms of a radical chain mechanism.[305] Alternatively, silanes readily transfer a hydride ligand to variety of transition metal complexes *via* oxidative addition, allowing for highly selective transition metal catalyzed reduction processes (*vide infra*, Section 3.5.6.2).

A useful reduction method involving hydridosilane in strongly acidic media, 'ionic hydrogenation', is useful for reduction of a number of organic functional groups.[306] The ionic hydrogenation reaction is based on the principle that the carbonium ion formed by protonation of the double bond reacts with an hydride donor to form the hydrogenated product. Reduction conditions generally involve reflux in strongly acidic media in the presence of the silane. Obviously, reduction is possible only when the substrate can produce carbonium ions under the given conditions. A hydrogenation pair most useful for many reduction processes is trifluoroacetic acid and a hydridosilane, which exhibits the following order of reactivity:[306] Et_3SiH > (*n*-octyl)$_3$SiH > Et_2SiH_2 > Ph_2SiH_2 > Ph_3SiH > $PhSiH_3$.

These reducing systems tolerate carboxylic acid derivatives, nitriles, nitro groups, sulfonic esters, aromatic rings and, occasionally, alkenes, alkyl halides, ethers and alcohols as well. Reduction may be chemoselective in compounds containing many functionalities, with the functional groups most easily capable of stabilizing a carbonium ion being reduced most readily. Thus, for example, aliphatic alkenes are reduced only when they are branched at the alkene carbon atom. With α,β-unsaturated ketones, the reduction can be directed almost exclusively to the C—C double bond. Thus, using only 1 equiv. of silane, enones are reduced to saturated ketones (Scheme 45).[307]

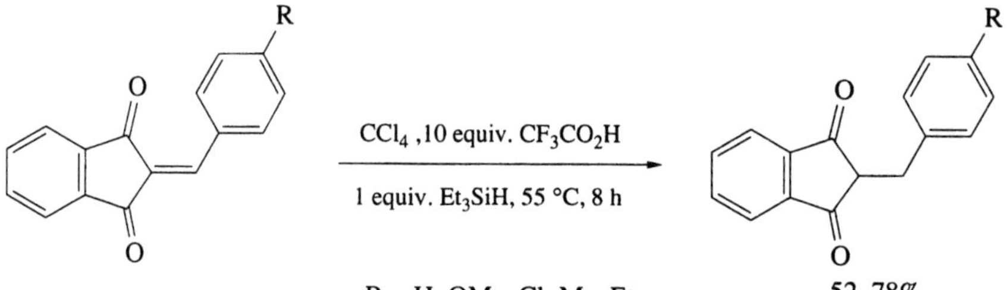

R O O CCl$_4$,10 equiv. CF_3CO_2H 1 equiv. Et_3SiH, 55 °C, 8 h R O O

R = H, OMe, Cl, Me, Et 52–78%

Scheme 45

With excess silane, further reduction of the saturated ketone to the corresponding saturated alcohol occurs in high yield. In the case of chalcones, excess silane may effect complete reduction and deoxygenation to yield the corresponding alkane.[307,308]

The reaction of conjugated enones and dienones with trimethyl- and triethyl-silane in the presence of $TiCl_4$, followed by aqueous work-up produces the corresponding saturated ketones. This Lewis acid catalysis is particularly useful for conjugated reduction of sterically hindered systems (Scheme 46).[309] α,β-Unsaturated esters are not reduced under these conditions.

Anionic activation of Si—H bonds[310] by fluorides, such as KF or CsF, or by potassium phthalate, $KHCO_3$, KSCN, *etc*., yields powerful hydridic reagents that reduce the carbonyl group of aldehydes, ketones and esters,[311] and 1,2-reductions of α,β-unsaturated aldehydes and ketones occur with high selectivity.[311] The analogous activation of hydridosilanes by fluoride ions is also achieved under acidic conditions with boron trifluoride etherate, in which the latter compound is consumed and fluorosilanes are formed.[312]

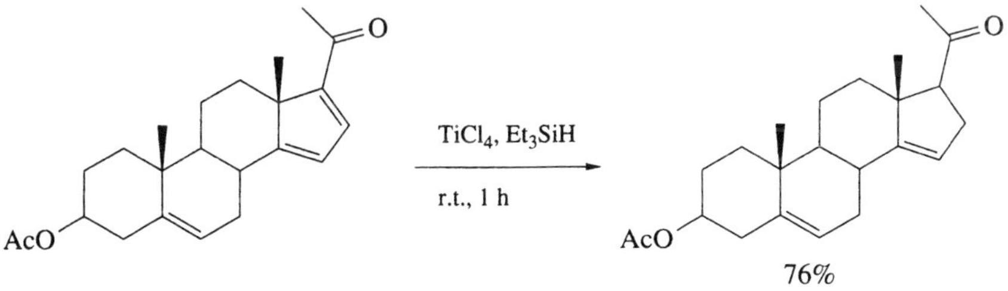

$TiCl_4$, Et_3SiH

r.t., 1 h

AcO AcO 76%

Scheme 46

Effective anionic activation of trichlorosilane can be carried out with either catechol or 2,2′-dihydroxybiphenyl in THF yielding bis(diolato)hydridosilicates.[313] Such reagents exhibit reducing power that is reminiscent of the complex aluminum hydrides. Even tertiary amines are useful activators of trichlorosilane, enhancing its hydridic character.[314]

3.5.4.4 Tin Hydrides

The special characteristics of organotin hydrides as reducing agents are rationalized by the fact that the tin–hydrogen bond is both weaker and less polar than the B—H or Al—H bonds.[315] These characteristics are manifested in reactions that proceed by either a free radical chain or polar mechanism, depending on the substrate, catalyst and reaction conditions.

α,β-Unsaturated aldehydes and ketones are readily reduced by organotin hydrides under rather mild conditions, but the reaction is often obscured by subsequent transformations of the adducts.[316] On heating or under UV irradiation, the organotin monohydrides add mainly at the 1,4-positions of the enone system to form the enol stannane.[315,316] The latter may be hydrolyzed (or cleaved by a second equivalent of tin hydride)[316e] resulting in overall reduction of the double bond (Scheme 47). The mechanistic pathway was demonstrated in reactions carried out with deuterated tin hydrides and/or in deuterated methanol.[317]

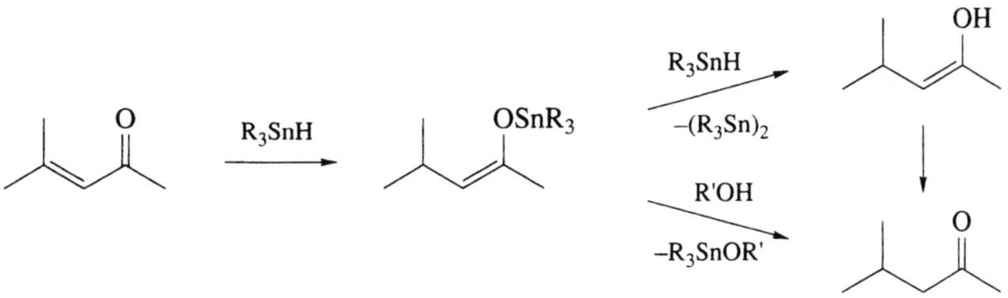

R_3SnH $OSnR_3$ R_3SnH $-(R_3Sn)_2$ OH R'OH $-R_3SnOR'$ O

Scheme 47

Sterically unhindered enones may produce mixtures of products, including carbon stannylated species. For example, methyl vinyl ketone gives rise to significant quantities of the inverted 1,4-adduct, where tin binds at the 4-position, leading to a β-stannyl ketone. In the case of methyl propenyl ketone, addition occurs at positions 3 and 4, producing an α-stannyl ketone (Scheme 48).[316j]

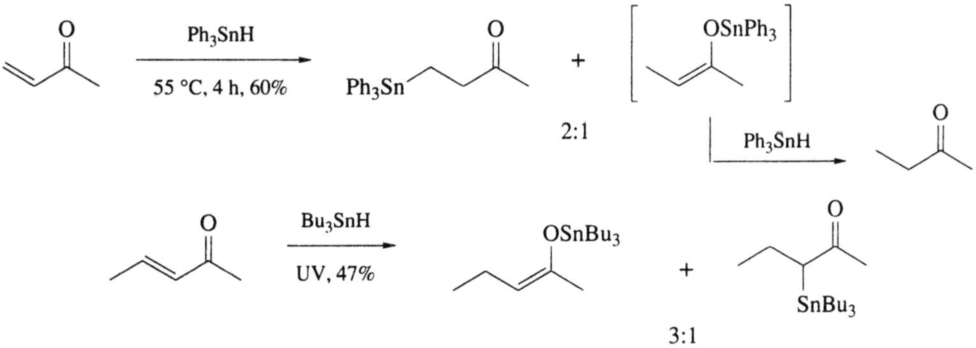

O Ph_3SnH 55 °C, 4 h, 60% Ph_3Sn O + [$OSnPh_3$] 2:1 Ph_3SnH O

O Bu_3SnH UV, 47% $OSnBu_3$ + O $SnBu_3$ 3:1

Scheme 48

In this class of reagents, diphenylstannane exhibited the highest regioselectivity, affording essentially pure 1,4-reduction. Other hydrides, such as Bu_3SnH or Ph_3SnH, give mixtures of 1,2- and 1,4-reduction products and they usually require free radical initiation.[318]

In the case of α,β-unsaturated esters and nitriles, hydrostannation may proceed *via* either a polar or radical mechanism. Compounds containing a terminal multiple bond form the α-stannyl derivative, according to a polar mechanism, while β-adducts are formed according to the radical pathway.[319] Other conditions being equal, triarylstannanes are more active than trialkylstannanes in radical processes. In general, α,β-unsaturated nitriles undergo the polar addition more actively than do the corresponding esters. However, with acrylonitrile, the homolytic mechanism is significant as well.[320] With trialkylstannanes under the action of azobis(isobutyronitrile) or UV irradiation or with triphenylstannane on heating, β-adducts are formed exclusively. Mixtures of α- and β-adducts are produced on thermal addition of trialkylstannanes (Scheme 49).[320] The α/β ratio increases with solvent polarity.

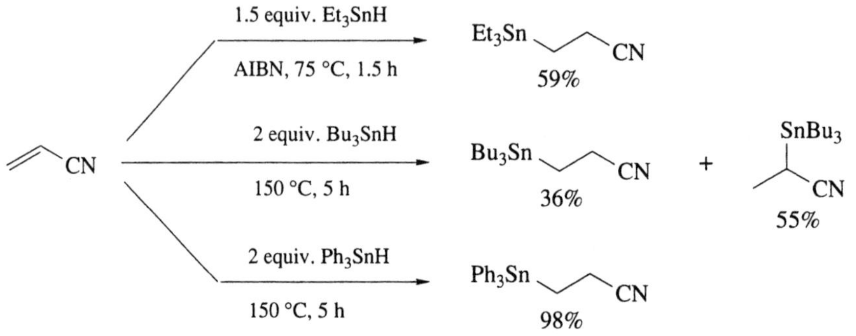

1.5 equiv. Et_3SnH, AIBN, 75 °C, 1.5 h → Et_3Sn–CH2CH2–CN 59%

2 equiv. Bu_3SnH, 150 °C, 5 h → Bu_3Sn–CH2CH2–CN 36% + CH3CH($SnBu_3$)CN 55%

2 equiv. Ph_3SnH, 150 °C, 5 h → Ph_3Sn–CH2CH2–CN 98%

Scheme 49

Hydrostannation of α-alkynic esters generally produces a mixture of products.[315]

3.5.5 REDUCTIONS WITH STOICHIOMETRIC AMOUNTS OF TRANSITION METAL HYDRIDES

3.5.5.1 Copper Hydrides

The known preference of organocopper reagents to engage in 1,4-addition to α,β-unsaturated carbonyl compounds,[321] prompted an extensive search for analogous hydridocopper reagents that would undergo conjugate addition to enones. Indeed, reaction of copper(I) bromide with either 2 equiv. of lithium trimethoxyaluminum hydride or 1 equiv. of sodium bis(2-methoxyethoxy)aluminum dihydride ('Vitride' by Eastman or 'Red-Al' by Aldrich) in THF produces a heterogeneous mixture capable of 1,4-reduction of α,β-unsaturated ketones, esters[322] and nitriles.[323] The exact composition of these reagents is not yet known. Reductions usually take place between –20 and –78 °C to give moderate yields of the saturated carbonyl compound, along with varying amounts of the 1,2-reduction product (Scheme 50). The use of lithium trimethoxyaluminum deuteride with CuBr produces the saturated ketone deuterated at the β-position. Addition of D_2O before the aqueous work-up leads to deuterium incorporation at the α-position. Because these reagents react with other functional groups (saturated ketones and aldehydes and alkyl bromides being reduced almost as rapidly as enones), their chemoselectivity is limited.

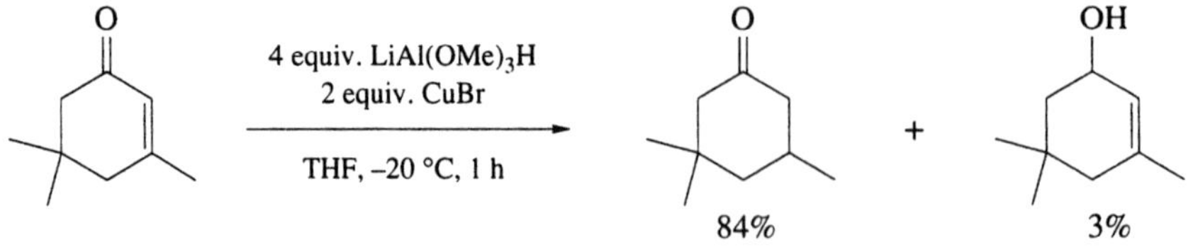

4 equiv. $LiAl(OMe)_3H$, 2 equiv. CuBr, THF, –20 °C, 1 h → saturated ketone 84% + allylic alcohol (OH) 3%

Scheme 50

Combination of $LiAlH_4$ and catalytic amounts of CuI in HMPA/THF (1:4) is useful for 1,4-reduction of α,β-unsaturated ketones, aldehydes and esters.[324] Reactions carried out at –78 °C for 1 h resulted predominantly in the 1,4-reduction product, but traces of the saturated and allylic alcohols were also formed.[324] It was claimed that the ratio between $LiAlH_4$ and CuI (10:1) as well as the presence of HMPA generates a hydridocuprate species which acts as the actual reducing agent. In contrast, in previously reported work using either $LiAlH_4$ or AlH_3 and CuI (in a 4:1 ratio) in THF, it was suggested that the active reductant is H_2AlI[283] (*vide supra*). An improved system based on diisobutylaluminum hydride (DIBAL-H) as the hydride donor and MeCu as the catalyst effects clean conjugate reduction of a variety of α,β-unsaturated carbonyl compounds without 1,2-reduction products. The presence of HMPA, probably acting as a ligand, was found to be of crucial importance for this reducing system, as shown in Scheme 51.[325] Other coordinating solvents including pyridine, DMF and DMSO, did not lead to comparable regioselectivity. Chemoselectivity is demonstrated by the selective 1,6-reduction of methyl sorbate in the presence of a saturated ketone, and the conjugate reduction of the enone of progesterone with only minor reduction of the saturated ketone in this molecule.

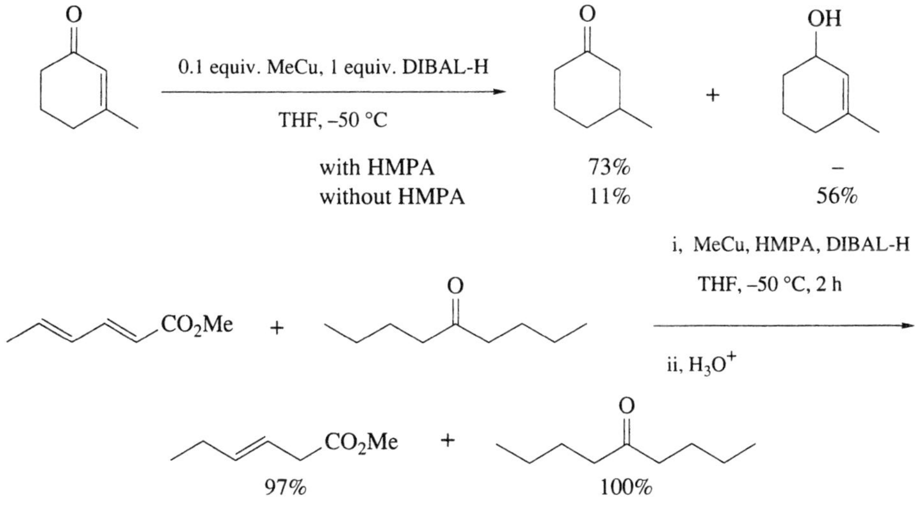

0.1 equiv. MeCu, 1 equiv. DIBAL-H

THF, –50 °C

with HMPA	73%	–
without HMPA	11%	56%

i, MeCu, HMPA, DIBAL-H

THF, –50 °C, 2 h

ii, H_3O^+

CO_2Me + … → CO_2Me 97% + 100%

Scheme 51

A series of heterocuprate complexes Li^+HRCu^-, with R representing a nontransferable ligand such as 1-pentynyl, Bu^tO^- or PhS^-, was generated in toluene from DIBAL-H and CuI by addition of RLi. These reagents were used for clean 1,4-reduction of α,β-unsaturated ketones, esters[326] and nitriles. Yields, however, were quite low in several cases due to the strong basicity of these reagents. Although HMPA was found to facilitate 1,4-reduction in substrates where the β-carbon is highly substituted, enone reduction in multifunctional compounds resulted in low yields (Scheme 52). In a related, independent study, the hydridocuprate complex was prepared by addition of RLi (R = alkyl or alkynyl) to a suspension of CuH in ether or in THF. These reagents were used for clean conjugate reduction of α,β-unsaturated carbonyls,[327] but with poor chemoselectivity, as saturated aldehydes and ketones were reduced under these conditions to the corresponding alcohols, and various tosylates and bromides were reductively cleaved.

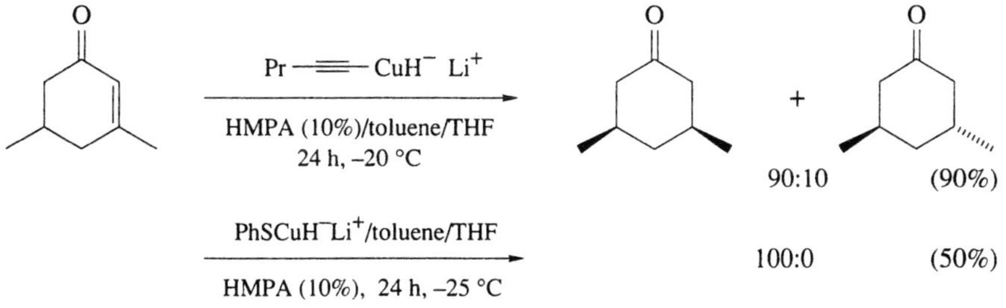

Pr—≡—CuH^- Li^+

HMPA (10%)/toluene/THF
24 h, –20 °C

90:10 (90%)

$PhSCuH^-Li^+$/toluene/THF

HMPA (10%), 24 h, –25 °C

100:0 (50%)

Scheme 52

Polyhydridocopper complexes, such as $LiCuH_2$, Li_2CuH_3, Li_3CuH_4, Li_4CuH_5 and Li_5CuH_6 were prepared[328] by $LiAlH_4$ reduction of Li_nCuMe_{n-1} Reduction of α,β-unsaturated carbonyl compounds with any of these hydrides in ether or in THF produced mixtures of 1,4- and 1,2-reduction products. These reagents also reduce ketones, alkyl halides, alkyl tosylates and aryl halides.

The stable, well-characterized copper(I) hydride cluster $[(PPh_3)CuH]_6$[329] is a useful reagent for conjugate reduction of α,β-unsaturated carbonyl compounds.[330] This hydride donor is chemically compatible with chlorotrimethylsilane, allowing formation of silyl enol ethers *via* a reductive silation process (Scheme 53).

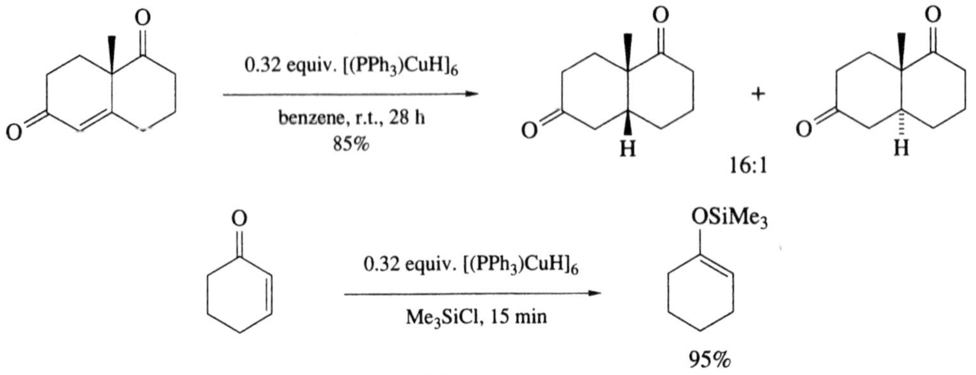

0.32 equiv. $[(PPh_3)CuH]_6$; benzene, r.t., 28 h; 85%; 16:1

0.32 equiv. $[(PPh_3)CuH]_6$; Me_3SiCl, 15 min; $OSiMe_3$; 95%

Scheme 53

3.5.5.2 Iron Hydrides

Iron hydrides were also used for selective 1,4-reduction of enones.[315b] For example, tetracarbonylhydridoferrate, $NaHFe(CO)_4$, prepared in methanol, reduces benzalacetone to benzylacetone. Addition of this reagent to an ethanolic solution containing both an aldehyde and a ketone results in reductive alkylation of the ketone. The reaction probably involves base-catalyzed aldol condensation of the aldehyde and the ketone, followed by elimination of water to give the corresponding α,β-unsaturated ketone. The latter is then reduced by the tetracarbonylhydridoferrate, to afford the saturated ketone.[331] Interestingly, $NaHFe(CO)_4$ in THF reduces α,β-unsaturated carbonyl compounds to the corresponding saturated alcohols with high stereospecificity. For example, (+)- and (–)-carvones are reduced to (–)- and (+)-neodihydrocarveol, respectively.[332]

The binuclear hydride $NaHFe_2(CO)_8$,[333,334] is also useful for clean conjugate reductions. This reagent is capable of selective 1,4-reduction of α,β-unsaturated ketones, aldehydes, esters, nitriles, amides and lactones in good yields (Scheme 54). Reductions are generally performed at –50 °C in a THF solution of $NaHFe_2(CO)_8$ and HOAc. Usually, 2 or more equiv. of the reagent are required for the reduction of 1 equiv. of substrate.

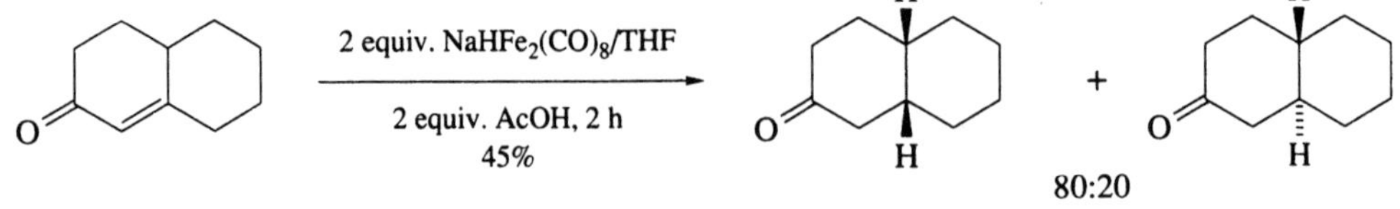

Ph–CH=CH–CHO → 2 equiv. $NaHFe_2(CO)_8$/THF; 2 equiv. AcOH, 0.5 h; 90% → Ph–CH₂CH₂–CHO

Ph–CH=CH–CN → 2 equiv. $NaHFe_2(CO)_8$/THF; 2 equiv. AcOH, 3 h; 96% → Ph–CH₂CH₂–CN

2 equiv. $NaHFe_2(CO)_8$/THF; 2 equiv. AcOH, 2 h; 45%; 80:20

Scheme 54

According to a detailed mechanistic study,[334] the reaction involves reversible, regiospecific addition of $NaHFe_2(CO)_8$ to the C=C double bond of the enone, affording the corresponding binuclear iron enolate. Cleavage of the latter to the mononuclear iron enolate represents the rate-determining step, protonolysis of which provides the saturated ketone.

3.5.5.3 Other Transition Metal Hydrides

The intermetallic hydride $LaNi_5H_6$ is an effective reagent for conjugate reduction of enones. Reduction of the resulting saturated carbonyl occurs very slowly with this reagent, giving high yields of the 1,4-reduction product.[335]

α,β-Unsaturated carbonyl compounds are reduced selectively and in good yields (55–80%) to the corresponding saturated derivatives by the hydridochromium complex $NaHCr_2(CO)_{10}$ in THF at 66 °C. This latter complex is prepared by stirring chromiumhexacarbonyl with potassium graphite (C_8K) in dry THF with subsequent addition of water.[336]

Excess hydridocobaltcarbonyl reduces α,β-unsaturated ketones and aldehydes in moderate yield and good regioselectivity. The reaction involves complexation of the double bond to cobalt, followed by migratory insertion of hydride into the enone, forming an oxaallyl cobalt complex.[337] Poor chemoselectivity is one of the major drawbacks of this reaction, as simple alkenes are rapidly hydroformylated to the corresponding aldehyde under the reaction conditions (25 °C, 1 atm of CO).

α,β-Unsaturated ketones and esters are selectively 1,4-reduced by $Et_4N[\mu\text{-}HMo_2(CO)_{10}]$ and HOAc in refluxing THF.[338] Benzalacetone is quantitatively reduced to benzylacetone under these conditions. However, reduction of cinnamaldehyde gives a mixture of dihydrocinnamaldehyde (3%), cinnamyl alcohol (85%), and phenylpropane (12%).

3.5.6 COMPOSITE REDUCING SYSTEMS

Composite reducing systems comprise at least two components, namely a relatively inactive source of hydride ions and a transfer agent to deliver the hydride selectively from that donor to a target functionality. This family of reducing systems will, therefore, selectively transfer a hydride ion to various electrophilic functional groups, including α,β-unsaturated carbonyl compounds. The acceptor properties of the latter make them excellent ligands for low-valent, electron-rich transition metals and, obviously, good substrates for selective reduction with nonreactive hydride donors.

Such multiple-component reducing systems offer high flexibility because they involve a large number of independent variables that can be tailored to various synthetic tasks, especially in comparison to metal hydride reduction which utilizes a single reagent. Thus, appropriate modification of the hydride donor, judicious selection of a transition metal transfer agent, and in some cases, use of a cocatalyst, provide an opportunity for creating a wide variety of reducing systems that exhibit improved chemoselectivity, as well as regio- and stereo-control.

3.5.6.1 Transfer Hydrogenation using Alcohols as Hydrogen Donors

Catalytic transfer of hydrogen from an organic donor to a variety of unsaturated organic acceptors is widely documented.[339] This approach has also been applied to the reduction of α,β-unsaturated carbonyl compounds, utilizing a catalyst and an organic compound with a low enough oxidation potential to be oxidized under the reaction conditions by the unsaturated carbonyl substrate.[339] With respect to enone reduction, the most commonly used hydrogen donors are primary or secondary alcohols. Temperatures for catalytic transfer hydrogenation are usually in the range 100–200 °C, depending upon the hydride source.

When α,β-unsaturated ketones are heated with a primary or secondary alcohol in the presence of $RuCl_2(PPh_3)_3$ or $RuHCl(PPh_3)_3$ at 200 °C, hydrogen is transferred selectively to the alkenic double bond.[340–342] The competing equilibrium that reduces the saturated ketone back to the alcohol may be suppressed by use of a primary alcohol such as benzyl alcohol or, more conveniently, by the use of boiling ethylene glycol, since saturated ketones are readily separated from insoluble glyoxal polymers.[343] Polyvinyl alcohol can also be used as a convenient hydrogen donor.[344] α,β-Unsaturated ketones give higher yields than the corresponding aldehydes, which undergo self-condensation. α,β-Unsaturated esters undergo transesterification side reactions with the donor alcohol.

Studies on the role of a Ru^{II} catalyst as well as the mechanism of hydrogen transfer in enone reduction with benzyl alcohol at 170–190 °C revealed that $RuCl_2(PPh_3)_3$ is converted by the primary alcohol into $RuH_2(CO)(PPh_3)_3$, which then hydrogenates benzylideneacetone.[345]

Transfer hydrogenation catalyzed by $RuCl_2(PPh_3)_3$ has been applied to the synthesis of cyclododecane-1,2-dione in 53% yield from the corresponding 1,2-diol using benzylideneacetone as the hydrogen acceptor.[346] 5,5-Dimethylcyclohexane-1,3-dione reacts *via* its enol tautomer on heating with ethylene glycol in the presence of $RuCl_2(PPh_3)_3$ to give 3,3-dimethylcyclohexanol, 3,3-dimethylcyclohexanone and its corresponding ketal.[347]

Vinyl ketones, such as methyl vinyl ketone, are not reduced in the presence of $RuCl_2(PPh_3)_3$ on heating with common primary or secondary alcohols, but they are reduced on heating with allylic alcohols, such as hex-1-en-3-ol, using hydrated $RuCl_3$, $RuCl_2(PPh_3)_3$, $RuHCl(PPh_3)_3$, $RuH(OAc)(PPh_3)_3$ or, most efficiently, $Ru_3O(OAc)_7$ (Scheme 55).[348] Surprisingly, other ketones, including acetophenone or benzylideneacetone, are not reduced under these conditions.

OH + O →(Ru complex, 100 °C) O + O

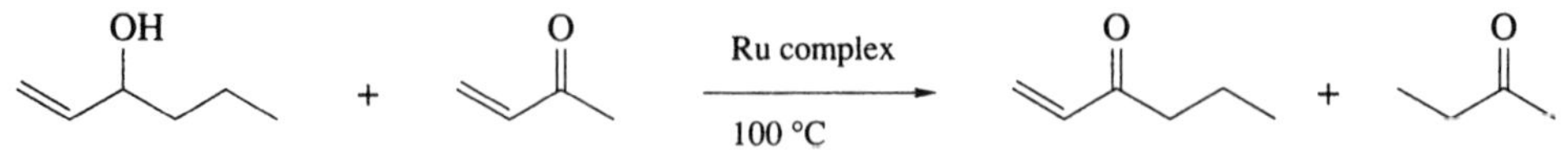

Scheme 55

As in hydrogen transfer between alcohols and saturated ketones, the rate-determining step in the reaction with α,β-unsaturated ketones is hydrogen abstraction from the α-carbon atom. It has been suggested that the hydrogen atom is transferred directly to the β-carbon of the enone, yielding an η^3-oxaallyl complex which following protonation yields the saturated ketone.[340]

Unsaturated esters also undergo transfer hydrogenation under $RuCl_2(PPh_3)_3$ catalysis to the saturated esters, but significant transesterification reaction with the reacting alcohol also occurs.[341] Simple alkenes are reduced, in general, very slowly under the reaction conditions, although $RuCl_2(PPh_3)_3$ is reported to catalyze hydrogen transfer from indoline to cycloheptene in refluxing toluene, to give cycloheptane and indole,[349] and other Ru^{II} complexes catalyze hydrogen transfer from alcohols to diphenylacetylene to yield *cis*-stilbene.[350] α,β-Unsaturated aldehydes are selectively reduced in THF with formic acid and trialkylamine and catalytic amounts of $RuCl_2(PPh_3)_3$ to produce allylic alcohols.[351]

Transfer hydrogenation of a prochiral alkene in the presence of a chiral catalyst may lead to a chiral saturated product. For example, tiglic acid (MeCH═C(Me)CO_2H) is hydrogenated at 120 °C by either 2-propanol in the presence of $Ru_4H_4(CO)_8$(–)-diop$)_2$[352] (diop = 2,3-*O*-isopropylidene-2,3-dihydroxy-1,4-bis(diphenylphosphino)butane) or by benzyl alcohol in the presence of $Ru_2Cl_4(diop)_3$ at 190 °C.[353] The optical purities reported for the resulting saturated acids, however, do not exceed 10–15%, a lower figure than that obtained by catalytic hydrogenation with hydrogen gas.

Prochiral α,β-unsaturated esters can also be asymmetrically hydrogenated by benzyl alcohol or 1-phenylethanol and catalytic $Ru_2Cl_4(diop)_3$,[353] but the optical purities of the resulting esters are even lower than those obtained from hydrogenating the corresponding acids. Enantioselectivity is also observed in transfer hydrogenation of α,β-unsaturated ketones, such as PhCH═CHCOMe, by racemic 1-phenylethanol in the presence of Ru^{II} chloro complexes containing optically active tertiary phosphines, including diop and neomenthyldiphenylphosphine. Thus the optical purity of 1-phenylpropan-1-ol enriched in the (*S*)-(–)-isomer is 11% when reacted under these conditions with benzylideneacetone.[354]

Asymmetric hydrogen transfer from optically active monosaccharides, such as 1,2-α-D-glucofuranose, to prochiral enones is catalyzed by $RuCl_2(PPh_3)_3$ in diphenyl ether at 180 °C or by $RuH_2(PPh_3)_4$ in toluene at 100 °C (Scheme 56).[355] Catalytic hydrogen transfer from sugars with free anomeric hydroxy groups was studied with 2,3;5,6-di-*O*-isopropylidene-D-mannofuranose and $RuH_2(PPh_3)_4$. In an excess of enone acceptor, these sugars were converted in high yields into the corresponding lactones (Scheme 56).[356]

The 1,4-reduction of styryl ketones by 1-phenylethanol using $RuH(PPh_3)_4$ catalyst can be carried out at 50 °C, a relatively low temperature for transfer hydrogenation. An electron-withdrawing group present in the enone system increases the initial rate of reduction, suggesting a transfer of hydrogen to the enone by an intermediate with hydride ion character.[357] Isotope labeling of the alcohol donors shows that hydrogen is regioselectively transferred from the carbinol carbon to the β-carbon of the enone, with the hydroxylic proton being transferred to the α-position. Cleavage of an O—H bond is the rate-determining step in this reaction.[358]

High catalytic activities, with turnovers of up to 900 cycles min^{-1}, is displayed in the transfer hydrogenation of α,β-unsaturated ketones, such as benzylideneacetone and chalcone, using 2-propanol and catalytic amounts of [Ir(3,4,7,8-Me_4-phen)COD]Cl (phen = 1,10-phenanthroline; COD = 1,5-cyclooctadiene) in a weakly alkaline medium.[359] Other Ir-chelated complexes are also active catalysts in this reaction, with over 95% selectivity for the 1,4-reduction mode. Divalent lanthanide derivatives, such as SmI_2 or YbI_2 in stoichiometric quantities, in THF and *t*-butyl alcohol or methanol reduce ethyl cinnamate and cinnamic acid to give the saturated derivatives.[360] Similarly, 3-methylcyclohexenone is reduced to 3-methylcyclohexen-1-ol in 67% yield, but α,β-unsaturated aldehydes are nonselectively reduced with these systems.

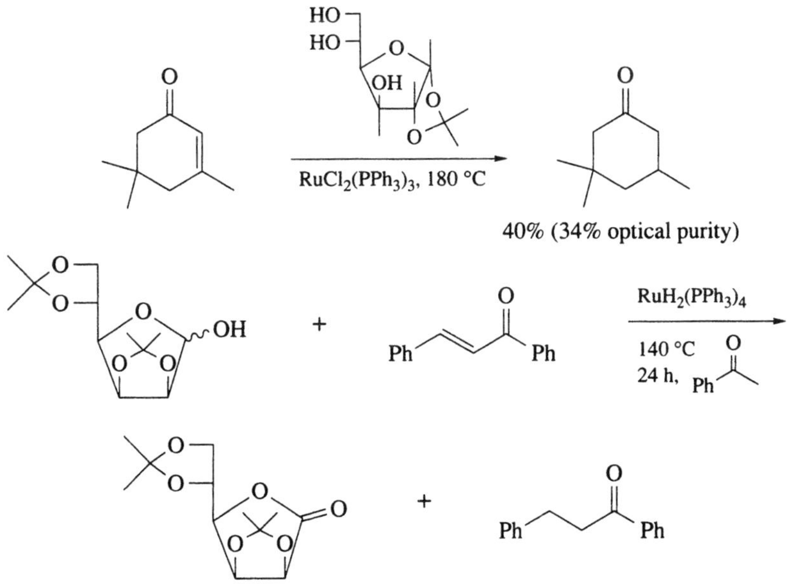

RuCl$_2$(PPh$_3$)$_3$, 180 °C

40% (34% optical purity)

RuH$_2$(PPh$_3$)$_4$

140 °C 24 h

Scheme 56

3.5.6.2 Transition Metal Catalyzed Reductions with Group 14 Metal Hydrides

Group 14 metal hydrides, especially those of silicon and tin, are satisfactory nonreactive hydride donors, as in the absence of a catalyst they are, generally, poor reducing agents. Transition metal complexes are attractive transfer agents because they insert readily into Si—H or Sn—H bonds and they also bind specifically to various functional groups.

Indeed, a combination of tributyltin hydride, Pd^0 catalyst and a weak acid, such as ammonium chloride, forms an effective, yet mild tool for conjugate reduction of α,β-unsaturated aldehydes and ketones.[361] Similar results are obtained with other acidic cocatalysts, such as zinc chloride, acetic acid and tributyltin triflate.[362] With this system, reductions occur with high regioselectivity, providing a useful approach for deuterium incorporation into either the β- or α-position by using either tributyltin deuteride or D_2O, respectively (Scheme 57).[361]

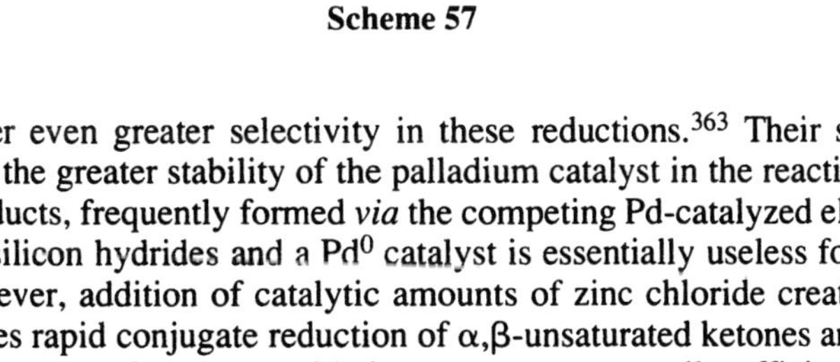

Bu$_3$SnD, PdL$_4$ / H$_2$O

Bu$_3$SnH, PdL$_4$ / D$_2$O

Scheme 57

Silicon hydrides offer even greater selectivity in these reductions.[363] Their superiority over tin hydrides is manifested by the greater stability of the palladium catalyst in the reaction solution, and the absence of diene side products, frequently formed *via* the competing Pd-catalyzed elimination processes.

The combination of silicon hydrides and a Pd^0 catalyst is essentially useless for reduction of electron-deficient alkenes. However, addition of catalytic amounts of zinc chloride creates a new three-component mixture that enables rapid conjugate reduction of α,β-unsaturated ketones and aldehydes.[364] In fact, soluble palladium complexes of various oxidation states were equally efficient catalysts, an obvious practical advantage of this approach. The generality of the method with respect to the substrate, its experimental simplicity, and its easy applicability to large-scale work make it a method of choice for conjugate reduction of unsaturated ketones and aldehydes.

The reaction was found to be both regio- and stereo-selective. In all cases where diphenyldideuteriosilane was used to reduce unsaturated ketones, deuterium was stereoselectively introduced at the less-hindered face of the substrate and regioselectively at the β-position (Scheme 58). Conversely, when reductions were carried out in the presence of traces of D_2O, deuterium incorporation occurred at the α-position just as with the tin hydride reductions of Scheme 57.[364]

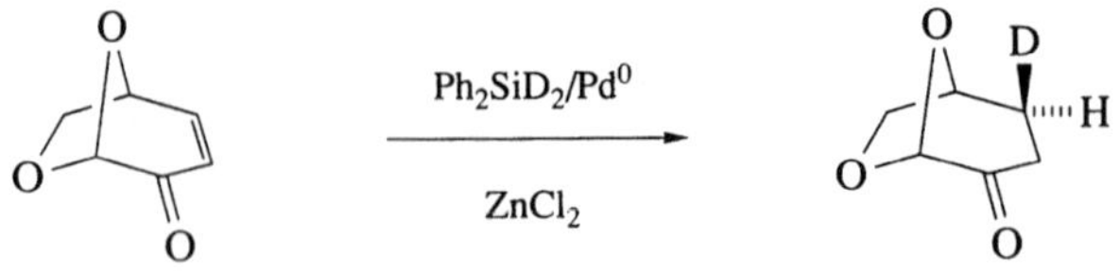

Ph_2SiD_2/Pd^0, $ZnCl_2$

Scheme 58

This method is highly selective for unsaturated ketones and aldehydes, whereas reduction of α,β-unsaturated carboxylic acid derivatives, such as esters, amides and nitriles, is very sluggish. Thus, benzylideneacetone was selectively and cleanly reduced in the presence of methyl cinnamate, cinnamonitrile or cinnamamide.[364]

Combination of silicon hydrides with catalytic amounts of a ruthenium(II) complex in tetrahydrofuran, chloroform or benzene has afforded a new reducing system capable of efficient reduction of α,β-unsaturated carboxylic acids, esters, amides, *etc.*[365] Addition of a weak proton source, such as a sterically hindered phenol significantly increases reaction rates. The ruthenium mixture was found to exhibit the same regioselectivity observed with the above-described palladium systems.

The order of reactivity of this Ru/silane combination to various functional groups differs greatly from that of its Pd/silane/$ZnCl_2$ analog. While the latter is very useful for allylic reductions and essentially useless for unsaturated esters, the Ru-based system exhibits opposite reactivity. This complementary chemoselectivity is illustrated by the reduction of cinnamyl cinnamate (Scheme 59), a substrate containing both an allylic carboxylate and an α,β-unsaturated ester.[365] Each of these can be reduced separately by silicon hydride and the appropriate transition metal catalyst.

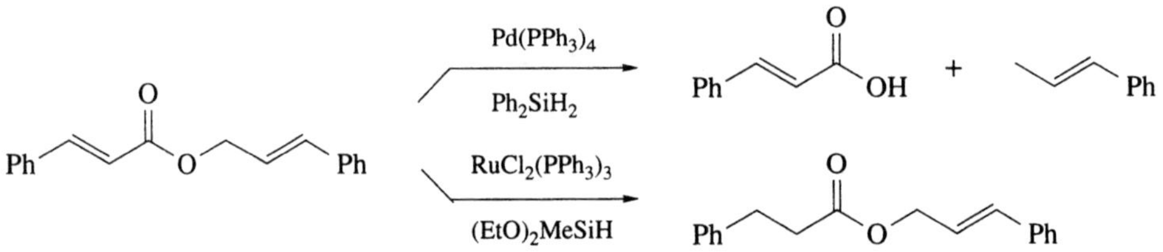

$Pd(PPh_3)_4$, Ph_2SiH_2; $RuCl_2(PPh_3)_3$, $(EtO)_2MeSiH$

Scheme 59

Early transition metal complexes, including those of Group 6, have rarely been used to catalyze transfer hydrogenation[366] and hydrogenation with hydrogen gas,[367] and similarly little is known about hydrosilation with these catalysts. Under mild thermal conditions, catalytic amounts of $Mo(CO)_6$ and phenylsilane engender a powerful reducing system, suitable for conjugate reduction of α,β-unsaturated ketones, carboxylic acids, esters, amides, *etc.* The mixture is especially useful for conjugate reduction of unsaturated nitriles, usually difficult to reduce with other media (Scheme 60).[368] Although the reaction also works with mono- and di-hydridosilanes, the general order of silane reactivity is $PhSiH_3 > Ph_2SiH_2 > Me(EtO)_2SiH >$ PMHS, $PhMe_2SiH$, Et_3SiH.

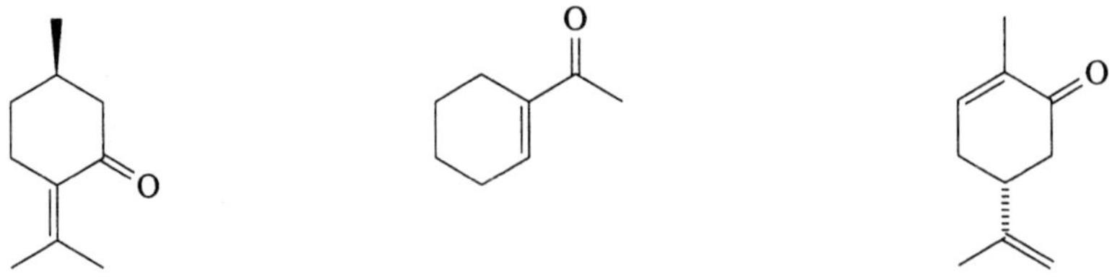

Pulegone 3 h, 100% — Acetylcyclohexene 4.5 h, 100% — Carvone 24 h, no reaction

Scheme 60

Of special interest are the relative rates of reduction of various cyclic enones, such as carvone, acetylcyclohexene and pulegone (Scheme 60). While the enone system in carvone is frozen in its transoid form, in acetylcyclohexene it is flexible and may adopt either transoid or cisoid conformation. Acetylcyclohexene is completely reduced, while essentially no reaction is observed with carvone, demonstrating the clear preference of the cisoid form and indicating that the molybdenum atom interacts simultaneously with both the alkenic bond and the carbonyl of the enone system. Accordingly, pulegone, which is frozen in the cisoid form, is reduced much faster than the other two compounds. A similar phenomenon was observed in enone hydrogenation catalyzed by arene-chromium tricarbonyl complex, where the cisoid conformation is also markedly preferred.[367c] With Pd^0 catalyst, however, enones behave as monodentate ligands and reductions of the above-mentioned substrates proceed at comparable rates.[364] These reactivity characteristics may be utilized for chemoselective differentiation between similar enones. For example, benzalacetone is quantitatively reduced to benzylacetone in the presence of carvone.[368] Allylic heterosubstituents and α-halo carbonyl compounds are also reduced very efficiently under these conditions.[369]

Highly regioselective reduction of α,β-unsaturated ketones and aldehydes to give either the corresponding saturated carbonyls or allylic alcohols as the predominant product is effected by hydrosilation catalyzed by tris(triphenylphosphine)chlororhodium (Wilkinson catalyst), followed by methanolysis of the resulting adducts.[370] Regiospecific deuteration is also achieved by using deuteriosilanes. Product distribution is mainly dependent upon the structure of the hydrosilane employed. In general, monohydridosilanes afford the 1,4-adduct (silyl enol ether), which may be hydrolyzed to the corresponding saturated carbonyl compound. Diaryl- or dialkyl-dihydridosilanes produce mainly silyl ethers (1,2-adduct), which may be hydrolyzed to the corresponding allylic alcohol. Other factors controlling the regioselectivity of this method include the enone structure, the hydridosilane/substrate ratio, the solvent and temperature. Although regioselectivity here is generally satisfactory (Scheme 61),[370] in some cases mixtures of 1,2- and 1,4-reduction products are obtained, even under maximally optimized conditions. The reaction is usually completed within 30–120 min at 0–80 °C in benzene, or in the absence of solvent, using 1.1 equiv. of the hydridosilane and 0.1 mol % of the Rh^I catalyst.

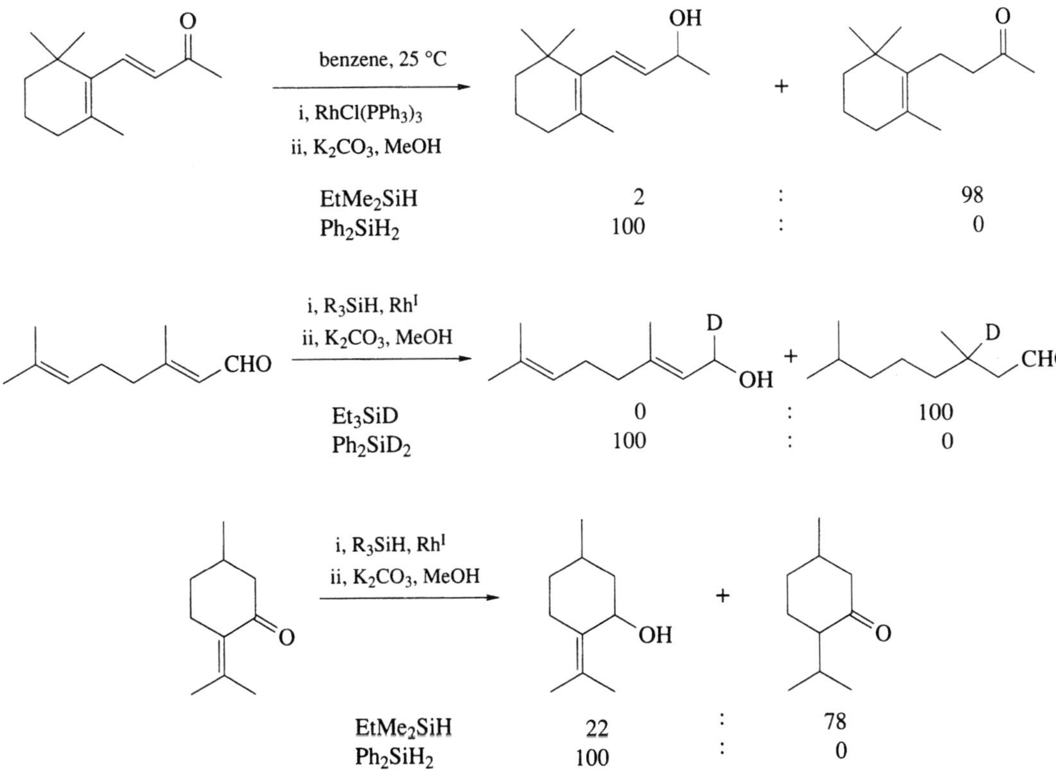

benzene, 25 °C; i, $RhCl(PPh_3)_3$; ii, K_2CO_3, MeOH

	OH (1,2)		O (1,4)
$EtMe_2SiH$	2	:	98
Ph_2SiH_2	100	:	0

i, R_3SiH, Rh^I; ii, K_2CO_3, MeOH

	D, OH		D, CHO
Et_3SiD	0	:	100
Ph_2SiD_2	100	:	0

i, R_3SiH, Rh^I; ii, K_2CO_3, MeOH

	OH		O
$EtMe_2SiH$	22	:	78
Ph_2SiH_2	100	:	0

Scheme 61

Treatment of α,β-unsaturated esters with triethylsilane in benzene in the presence of catalytic amounts of $RhCl(PPh_3)_3$ at room temperature yields the corresponding saturated esters. Conjugated diene esters are reduced to the β,γ- or γ,δ-unsaturated esters, depending upon their substitution pattern (Scheme 62).[371]

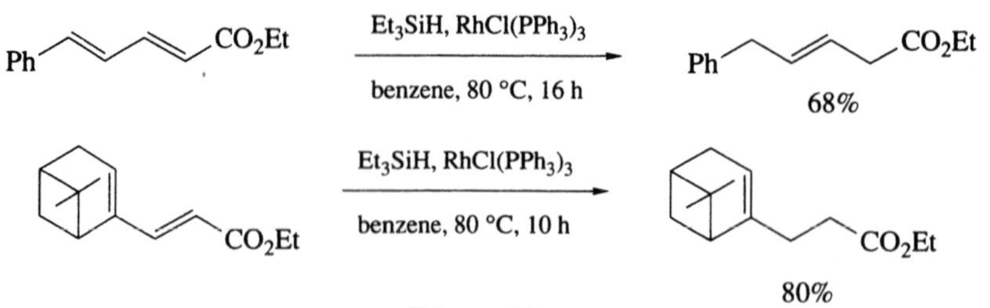

Ph–CH=CH–CH=CH–CO_2Et →(Et_3SiH, $RhCl(PPh_3)_3$; benzene, 80 °C, 16 h) Ph–CH_2–CH=CH–CH_2–CO_2Et 68%

→(Et_3SiH, $RhCl(PPh_3)_3$; benzene, 80 °C, 10 h) 80%

Scheme 62

Other Rh catalysts were also employed for hydrosilation of α,β-unsaturated carbonyl compounds and unsaturated nitriles. Thus, $Rh(acac)_2$ and a tetrakis(μ-acetato)dirhodium cluster were used as catalysts in the hydrosilation[372] of α,β-unsaturated aldehydes. These reactions, however, are not chemoselective, as alkynes, conjugated dienes and alkenes are also hydrosilylated, and allylic heterosubstituents are reductively cleaved.

Optically active, saturated carbonyl compounds and allylic alcohols were prepared *via* 1,4- and 1,2-asymmetric hydrosilation of enones using Rh^I catalysts bearing chiral ligands. For example, 1,4-hydrosilation of α,β-unsaturated ketones afforded the corresponding optically active ketones in 1.4–15.6% *ee* (Scheme 63).[373]

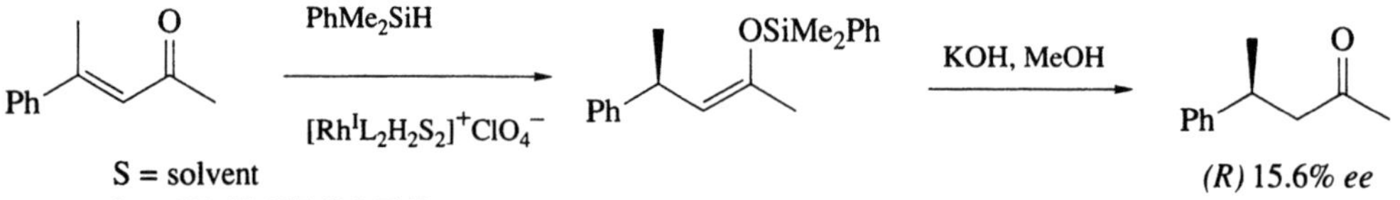

→($PhMe_2SiH$; $[Rh^IL_2H_2S_2]^+ClO_4^-$) $OSiMe_2Ph$ →(KOH, MeOH) *(R)* 15.6% *ee*

S = solvent
L = *(R)*-($PhCH_2$)MePhP

Scheme 63

Asymmetric 1,2-hydrosilation in benzene of α,β-unsaturated ketones with dihydridosilanes and a chiral Rh^I catalyst produced allylic alcohols with up to 69% *ee*.[374] Highly stereoselective hydrosilation of an α,β-unsaturated aldehyde was achieved with triethylsilane and nonchiral Wilkinson catalyst.[375] Dehydrofaranal was thus stereoselectively reduced to the insect pheromone (3*S*,4*R*)-faranal with 85% *de* (Scheme 64).

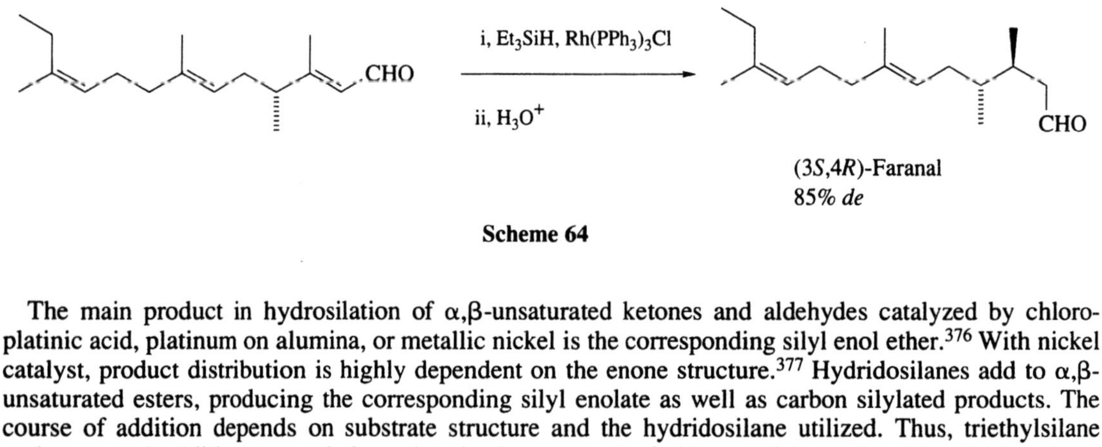

CHO →(i, Et_3SiH, $Rh(PPh_3)_3Cl$; ii, H_3O^+) CHO

(3*S*,4*R*)-Faranal
85% *de*

Scheme 64

The main product in hydrosilation of α,β-unsaturated ketones and aldehydes catalyzed by chloroplatinic acid, platinum on alumina, or metallic nickel is the corresponding silyl enol ether.[376] With nickel catalyst, product distribution is highly dependent on the enone structure.[377] Hydridosilanes add to α,β-unsaturated esters, producing the corresponding silyl enolate as well as carbon silylated products. The course of addition depends on substrate structure and the hydridosilane utilized. Thus, triethylsilane undergoes 1,4-addition to methyl acrylate in the presence of chloroplatinic acid, while trichlorosilane with either chloroplatinic acid or Pt/C gives the β-silyl ester (Scheme 65).[378]

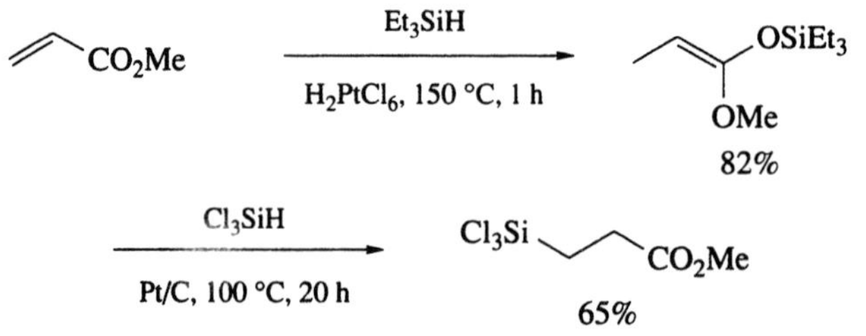

CH_2=CH–CO_2Me →(Et_3SiH; H_2PtCl_6, 150 °C, 1 h) CH_3CH=C(OMe)$OSiEt_3$ 82%

→(Cl_3SiH; Pt/C, 100 °C, 20 h) $Cl_3SiCH_2CH_2CO_2Me$ 65%

Scheme 65

This approach was successfully applied to the total synthesis of (±)-muscone.[379] Treatment of the α,β- and β,γ-enone mixture (Scheme 66) with triethylsilane in refluxing glyme containing catalytic amounts of chloroplatinic acid afforded 1-triethylsilyloxycyclotetradecene. The two isomeric enones rapidly equilibrate under these conditions.

O O + Et_3SiH H_2PtCl_4 $OSiEt_3$

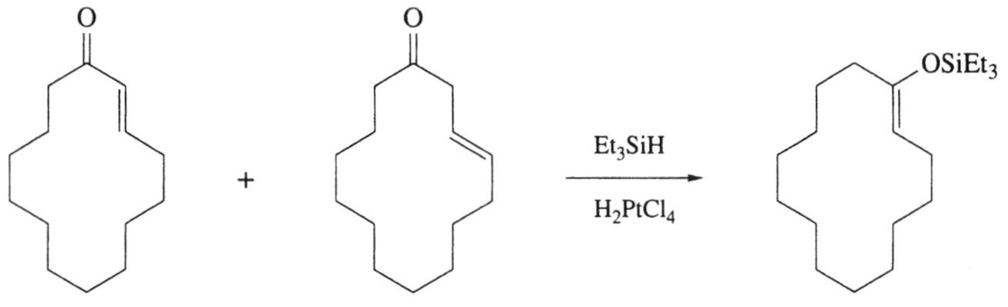

Scheme 66

Selective reduction of pregna-14,16-dien-20-ones to pregn-14-en-20-ones is achieved *via* hydrosilation with tetramethyldisiloxane and catalytic amounts of chloroplatinic acid (Scheme 67).[380] α,β-Unsaturated esters are also reduced to the corresponding saturated esters under these conditions.[381]

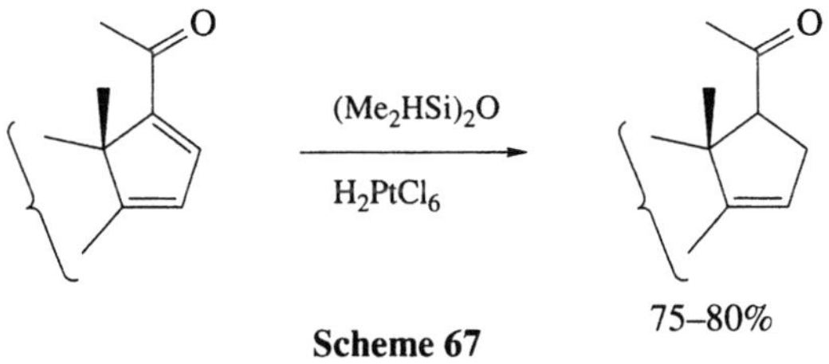

O $(Me_2HSi)_2O$ H_2PtCl_6 O

75–80%

Scheme 67

The platinum dimer $(Pt(\mu\text{-}H)(SiR_3)(PR'_3))_2$ also catalyzes the hydrosilation of α,β-unsaturated aldehydes and ketones. Several aldehydes and ketones were hydrosilated in high yield in the presence of this dimer[382] at 60–100 °C and trialkylsilanes, including $MePh_2SiH$, $EtMe_2SiH$ and Et_3SiH. Triethoxysilane was inert under these reaction conditions. Excellent regioselectivity was generally observed except in cases of highly sterically hindered enones such as tetraphenylcyclopentadienone, where the 1,2-reduction mode was observed. Saturated aldehydes and ketones were not reduced under these reaction conditions, and unsaturated carboxylic acids and esters were only sluggishly reduced. Unfortunately, terminal alkenes and alkynes were efficiently hydrosilated. A suggested mechanism involves cleavage of the platinum dimer to a platinum hydride species, its coordination to the alkene, and subsequent transfer of the R_3Si group to the carbonyl oxygen, affording a π-allyl platinum complex. Hydride migration from Pt to the allylic ligand produces the corresponding silyl enol ether.

3.5.6.3 Transition Metal Catalyzed Reductions with other Hydrogen Donors

Aromatic aldehydes and DMF can serve as hydrogen donors and transfer their formyl hydrogen to α,β-unsaturated ketones in the presence of $RuCl_2(PPh_3)_3$. However, in some cases, decarbonylation of the aldehyde is so severe that no transfer hydrogenation is observed.[383]

A particularly convenient hydrogen donor is formic acid, which not only hydrogenates α,β-unsaturated ketones,[384] but also terminal alkenes in the presence of a variety of ruthenium complexes under mild conditions.[385] Trialkylammonium formate and catalytic amounts of palladium on carbon form a convenient reducing system for reduction of a number of organic functional groups, including α,β-unsaturated aldehydes, ketones and esters.[386] Conjugated dienes reduce to monoenes with 1 equiv. of reagent fairly selectively. Typical reductions are carried out at 100 °C with 10% excess formic acid, 30% excess triethyl- or tributyl-amine, and 1 mol % of palladium in the form of 10% Pd/C. Progress of the reduction is conveniently monitored by measuring the amount of CO_2 evolved. Two examples are given in Scheme 68.[386] The chemoselectivity of this system is somewhat limited, as it affects many other functionalities, such as halo- and nitro-aromatic compounds,[387] allylic heterosubstituents,[388] and terminal alkynes and alkenes.[386]

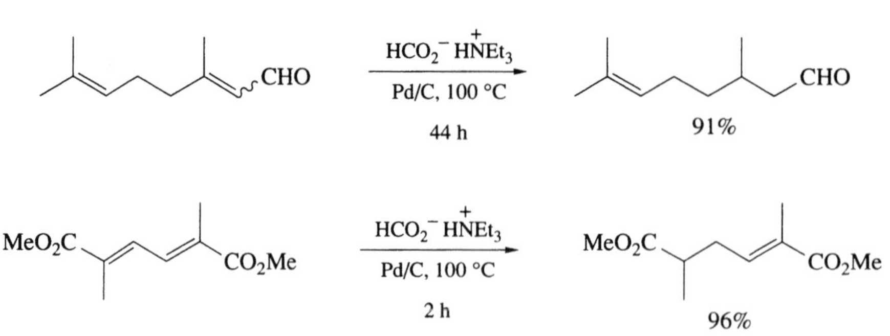

CHO $\xrightarrow[\text{Pd/C, 100 °C}]{HCO_2^- \overset{+}{H}NEt_3}$ CHO

44 h 91%

MeO_2C CO_2Me $\xrightarrow[\text{Pd/C, 100 °C}]{HCO_2^- \overset{+}{H}NEt_3}$ MeO_2C CO_2Me

2 h 96%

Scheme 68

Triisobutylaluminum reduces α,β-unsaturated ketones to allylic alcohols in pentane at room temperature. These products however are accompanied by substantial amounts of tertiary alcohols arising from 1,2-addition of the isobutyl group. The extent of these reactions depends both on the structure of the enone and on the ratio between reagent and substrate. Under similar experimental conditions, bis(*N*-methylsalicylaldimine)nickel catalyzes conjugate reduction of α,β-unsaturated ketones by triisobutylaluminum.[389] In all cases, 1,2-reduction products were also obtained (probably *via* noncatalyzed reduction) and in some cases, side products containing isobutyl group were also formed. The reaction is interpreted in terms of a catalytic cycle involving a hydridonickel intermediate formed by reaction of $Bu^i{}_3Al$ with the nickel complex. Addition of the hydridonickel to the alkene affords a nickel enolate that undergoes transmetallation to give an aluminum enolate. The latter is finally hydrolyzed to the saturated ketone.

A number of composite reducing systems consisting of heterogeneous mixtures of transition metal salts, sodium alkoxides and sodium hydride were developed, which are useful for reduction of various organic functional groups.[390] In organic chemistry, sodium hydride is generally used as a base for proton abstraction. Although some substrates can be reduced by NaH, it is by itself a poor reducing agent. Typical reducing systems (known as complex reducing agents, CRA)[390] are prepared from a transition metal chloride or acetate, sodium *t*-pentoxide and sodium hydride (in 1:1:4 ratio) in either THF or DME. Obviously, neither the exact structure of the actual reducing entity nor their reduction mechanism is fully understood.

The CRA reagents involving nickel salts exhibit reducing properties that are significantly different from those of the corresponding CRA prepared from zinc or magnesium salts. It was demonstrated that the three-component mixture, $NaH/RONa/Ni(OAc)_2$ (NiCRA) reduces carbon–carbon double bonds.[391] Conversely, the mixture $NaH/RONa/ZnCl_2$ (ZnCRA) reduces alkenes poorly but effectively reduces saturated carbonyl functionalities, particularly when mixed with alkaline or alkaline earth metal salts.[392] These observations led to the expected complementary regioselectivity when reducing α,β-unsaturated carbonyl compounds with these reagents. Indeed NiCRA exhibits very high regioselectivity for 1,4-reduction of a number of α,β-unsaturated ketones, while under the same conditions ZnCRA exhibits high regioselectivity for 1,2-reduction.[393] Addition of magnesium bromide enhances the activity of both reagent mixtures. The general applicability of CRA reagents is somewhat limited due to their high basicity as well as to their tendency to undergo side reactions *via* one-electron transfer processes.

3.5.7 BIOCHEMICAL REDUCTIONS

3.5.7.1 Enzymatic Reductions

Much work has been published on the microbiological reduction of α,β-unsaturated ketones. Under anaerobic conditions the reduction of Δ^4-3-keto steroids by *Clostridium paraputrificum* led to the 3-keto 5β-derivatives (Scheme 69).[394] Similar transformations were observed previously with *Bacillus putrificus*,[395] *Penicillium decumbens*,[396] *Rhizopus nigricans*[397] or *Aspergillus niger*.[398] In most cases further reduction led to the corresponding 3α-hydroxy 5β-derivatives.

Highly enantioselective conjugate reductions of substituted cyclopentenones and cyclohexenones were reported by Kergomard using *Beauveria sulfurescens* (ATCC 7159) under anaerobic conditions.[399] The reaction takes place only with substrates containing a small substituent in the α-position and hydrogen in the β-position. The saturated ketones obtained were, in some cases, accompanied by saturated alcohols.

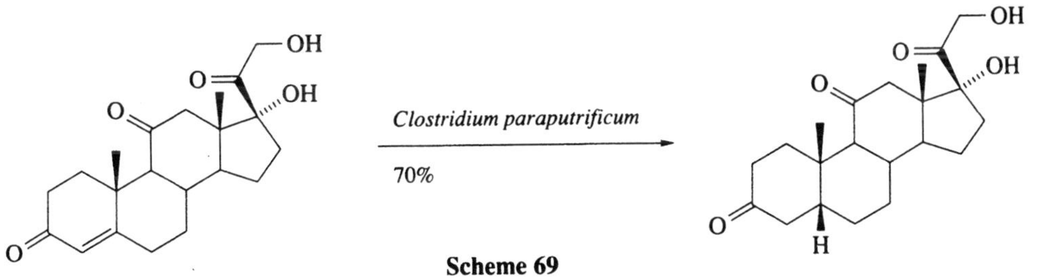

Scheme 69

A number of useful transformations, including enantioselective reductions of acyclic substrates, are illustrated in Scheme 70.

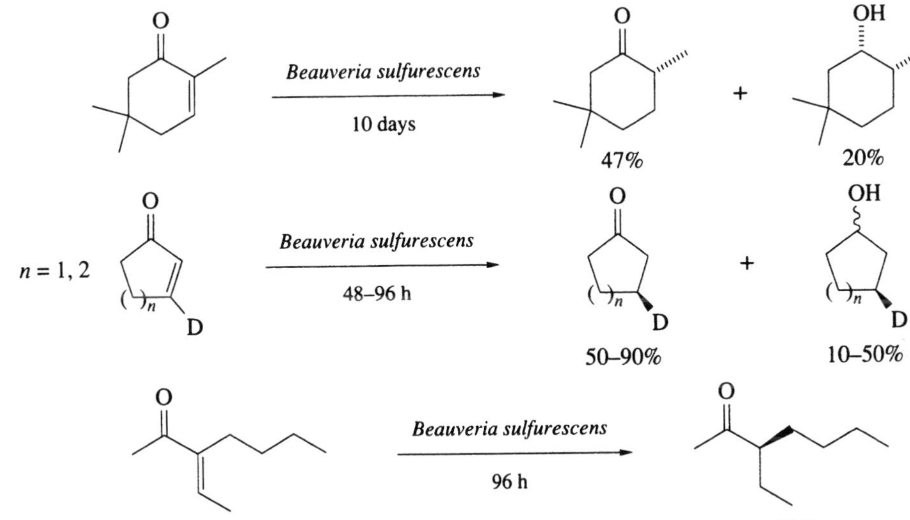

Scheme 70

Both naturally occurring enantiomers of carvone were selectively reduced by *B. sulfurescens* (Scheme 71). (–)-Carvone was reduced to (+)-dihydrocarvone (*trans*) and further to (+)-neodihydrocarveol, whereas (+)-carvone was reduced to (–)-isodihydrocarvone (*cis*), which was then converted to (–)-neoisodihydrocarveol.[400] Similar reductions with identical stereoselectivities were observed earlier with *Pseudomonas ovalis* (strain 6-1) and with a strain of *Aspergillus niger*.[400]

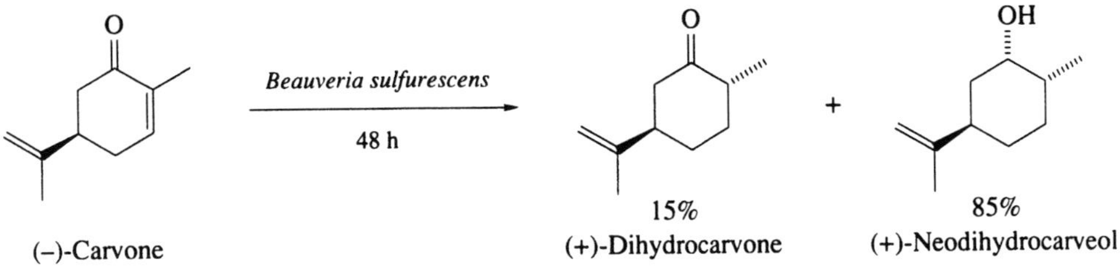

Scheme 71

The reduction of α,β-unsaturated aldehydes by *Beauveria sulfurescens* proceeds along two mechanistic pathways; (i) reversible formation of the corresponding allylic alcohols, and (ii) irreversible formation of thc saturated alcohol (Scheme 72).[401] The latter involves initial, slow 1,4-reduction, followed by fast reduction of the resultant saturated aldehyde. A similar sequence was proposed for the reduction of geranial and geraniol to (*R*)-citronellol with *Saccharomyces cerevisiae*.

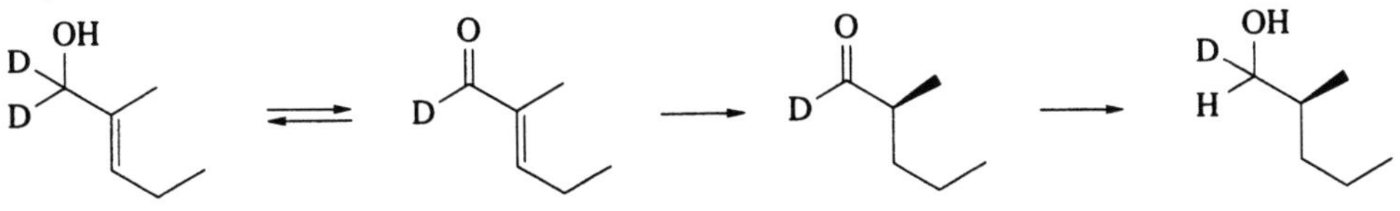

Scheme 72

The above-described reducing characteristics of *B. sulfurescens* were found to be a general phenomenon exhibited by many types of eukaryotic organisms (six fungi) and prokaryotes (more than 20 *Actinomycetes* and *Clostridium* species).[402] For example, in conjugate reduction of cyclohexenone derivatives the addition of two hydrogen atoms across the alkene occurs with *trans* stereochemistry, as shown in Scheme 73 where X represents a small alkyl group and Y, a hydrogen atom. In all cases, the 1,4-reduction mode was completed within 48 h. As these characteristics are shared by many organisms, it was suggested that they all contain very similar reducing enzymes.[402]

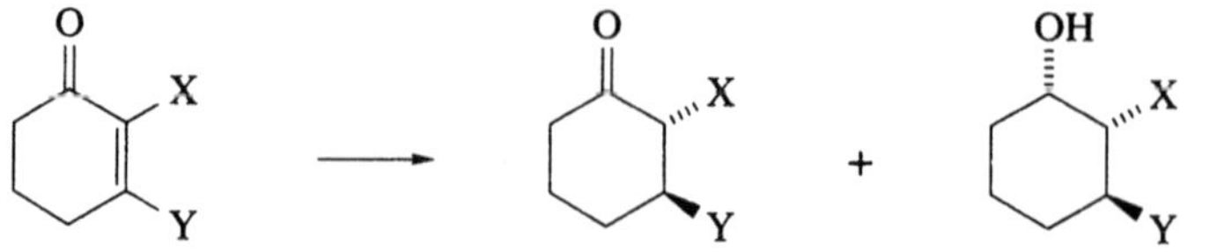

Scheme 73

α,β-Unsaturated ketones bearing perfluoroalkyl groups are reduced by baker's yeast (Scheme 74).[403] Perfluoroalkyl alkenyl ketones give mainly the saturated ketone, along with a small amount of optically active saturated alcohol. Substrates having perfluoroalkyl groups attached to the alkene moiety give mixtures of optically active allylic as well as saturated alcohols, whose relative concentration is time-dependent.

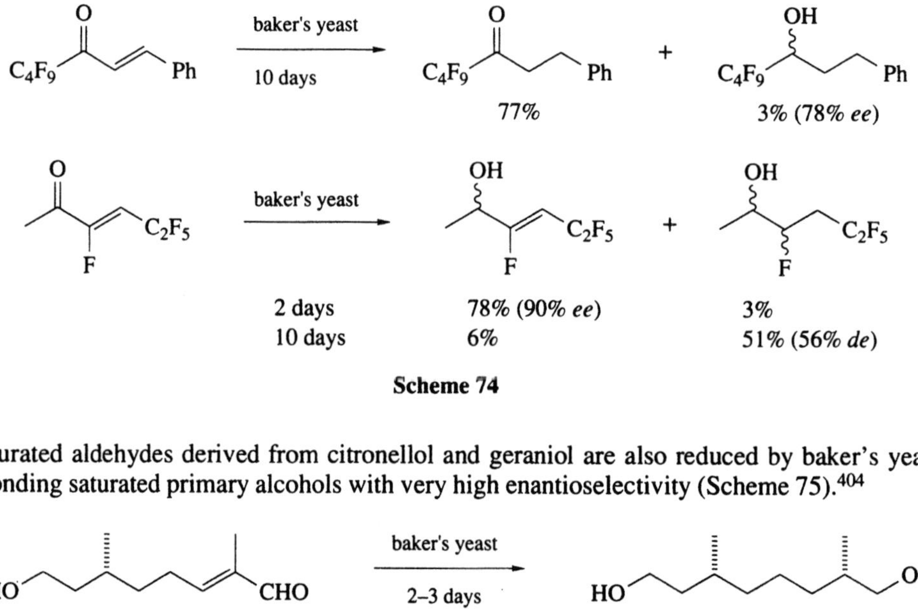

baker's yeast, 10 days: 77% + 3% (78% *ee*)

baker's yeast: 2 days 78% (90% *ee*), 3%; 10 days 6%, 51% (56% *de*)

Scheme 74

Unsaturated aldehydes derived from citronellol and geraniol are also reduced by baker's yeast to the corresponding saturated primary alcohols with very high enantioselectivity (Scheme 75).[404]

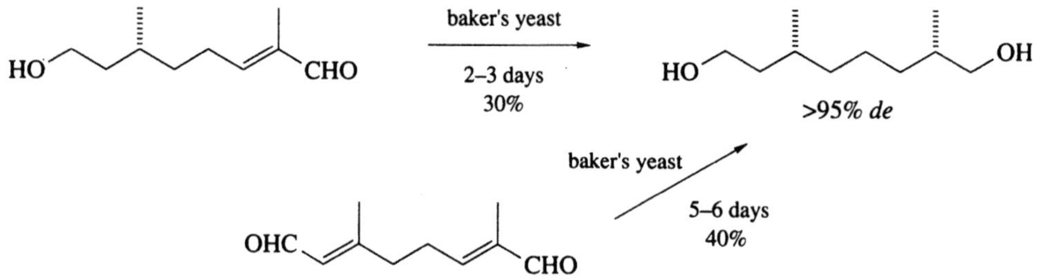

baker's yeast, 2–3 days, 30%; baker's yeast, 5–6 days, 40%; >95% *de*

Scheme 75

Two key chiral building blocks used in the total synthesis of α-tocopherol were prepared *via* microbial reduction of unsaturated carbonyl compounds with baker's yeast and with *Geotrichum candidum*.[405]

Similarly, a key intermediate in the total synthesis of optically active natural carotenoids was prepared by microbial reduction of oxoisophorone with baker's yeast.[406] An alternative approach to the synthesis of α-tocopherol employs a chiral building block that was obtained by baker's yeast reduction of 2-methyl-5-phenylpentadienal.[407]

Microbial reduction of enones has been applied to prostaglandin synthesis. For example, enantioselective reduction of the enone system in $\Delta^{8(12)}$-15-dehydro-PGE$_1$ with *Flavobacterium sp.* (NRRL B-3874) provided optically pure (–)-15-epi-$\Delta^{8(12)}$-PGE_1.[408] NADH-dependent enoate reductase isolated from clostridia is a catalyst for conjugate reduction of α,β-unsaturated carboxylic acids and aldehydes.[409] Enoyl CoA hydratase in D_2O converts crotonyl CoA to (2*R*,3*S*)-2-deuterio-3-hydroxybutyrylCoA.[410] Substituted chalcones are selectively reduced in a 1,4-manner by incubation with *Corynebatesium equi* IFO 3730 to give saturated ketones.[411]

As a general rule in enzymatic reductions, the 1,4-reduction of enones is preferred over the 1,2-reduction mode. However, when an electronegative substituent, such as halogen, is introduced that stabilizes the double bond, enzymatic reduction to allylic alcohols may be achieved.[304]

3.5.7.2 Biomimetic Reductions with NAD(P)H Models

A number of pyridine nucleotide linked dehydrogenases catalyze the reversible hydrogenation–dehydrogenation of the double bond in α,β-unsaturated ketones.[412] Similar biomimetic conjugate reductions of α,β-unsaturated aldehydes and ketones occur with NAD(P)H models, such as 3,5-diethoxycarbonyl-2,6-dimethyl-1,4-dihydropyridine (Hantzsch ester). With highly electron-deficient alkenes, such as maleic acid, maleic anhydride, diethyl maleate, diethyl fumarate, *etc.*, reductions proceed well.[413] Similarly, the alkenic bond of 1-phenyl-4,4,4-trifluorobut-2-en-1-one is reduced by dihydropyridines under mild conditions (Scheme 76).[414] Tracer experiments showed that hydrogen is transferred directly from the 4-position of the pyridine ring to the β-position of the enone system. The reaction thus parallels the enzymatic reduction of androstenedione.[415]

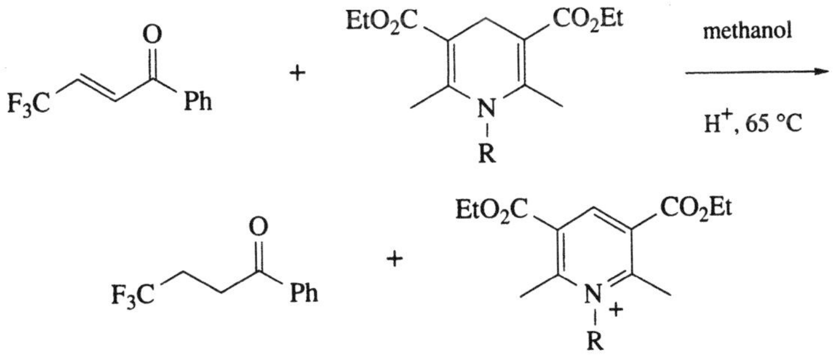

O F3C Ph + EtO2C CO2Et N R methanol H+, 65 °C

O F3C Ph + EtO2C CO2Et N+ R

Scheme 76

However, these reaction conditions (refluxing methanol or photoactivation at room temperature) are useful only for the reduction of highly activated double bonds.[416] Nevertheless, it was found that the reaction is promoted by silica gel,[417] broadening the scope of reducible enone substrates. The method is highly chemoselective as no alcoholic products are observed, and saturated carbonyls, nitro, cyano, sulfinyl and sulfonyl groups remain intact under the reaction conditions (Scheme 77).

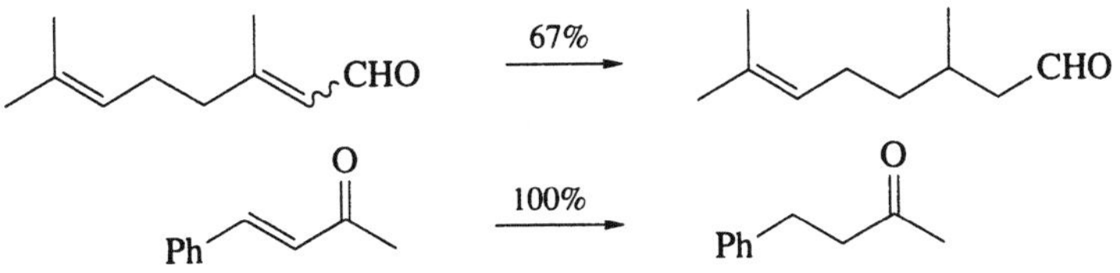

CHO 67% CHO

O Ph 100% O Ph

Scheme 77

Pandit has provided evidence for the Lewis acid catalysis postulated to operate in these reduction reactions.[418] The reduction of various cinnamoylpyridines by 1,4-dihydropyridine derivatives to the corresponding saturated ketones is catalyzed by zinc or magnesium cations. The reduction rate was fastest in the case of 2-cinnamoylpyridine, in which the metal ion can complex simultaneously to both the nitrogen and oxygen sites (Scheme 78). This example is regarded as a model of Lewis acid catalysis of the NADH-dependent enzymatic reduction of Δ^4-3-keto steroids.

In a similar manner, iminium salts derived from α,β-unsaturated aldehydes and ketones are reduced by Hantzsch ester (Scheme 79).[419] The ratio between the 1,4- and 1,2-reduction products depends upon the pK_a of the amine component. α-Keto-β,γ-unsaturated esters are reduced by NAD(P)H models in the presence of $Mg(ClO_4)_2$. At room temperature 1 equiv. of the reducing agent effects 1,4-reduction of the

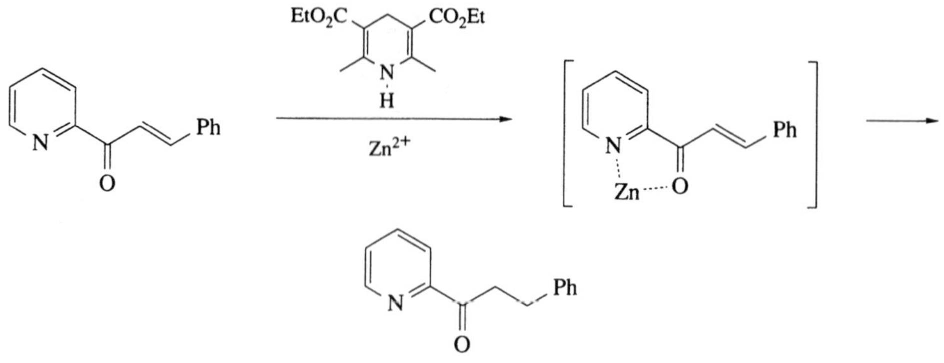

Scheme 78

substrate. At a higher temperature, excess reagent reduces the product to the corresponding α-hydroxy esters.[420] Several α,β-unsaturated esters, ketones and nitriles can be reduced by 1-benzyl-1,4-dihydronicotinamide upon selective photoexcitation in the presence of Ru(2,2′-bipyridine)3.[421]

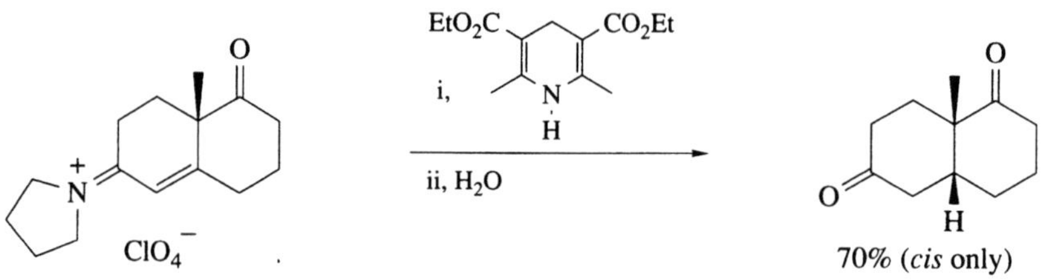

Scheme 79

An autorecycling system for the specific 1,4-reduction of α,β-unsaturated ketones and aldehydes was based on 1,5-dihydro-5-deazaflavin, which can be regarded as an NADH model.[422] The reaction occurs on heating the substrate with catalytic amounts of 5-deazaflavin in 98% formic acid, typically at 120 °C for 24 h (Scheme 80).

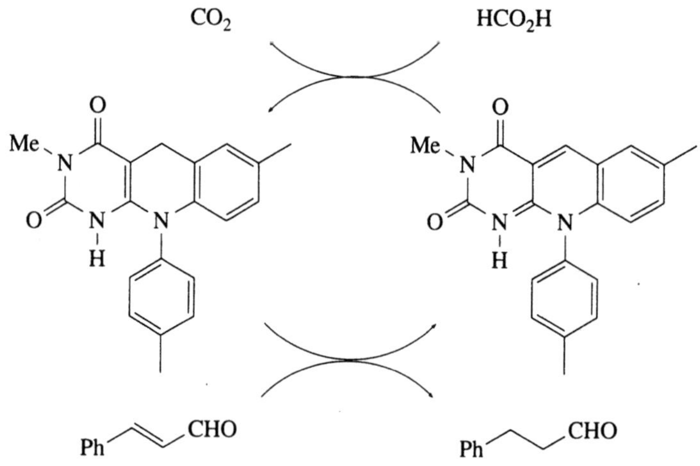

Scheme 80

The iminium salts of 3,3,5-trimethylcyclohex-2-en-1-one were reduced with 1,4-dihydronicotinamide sugar pyranosides to give the corresponding optically active saturated ketone in enantiomeric excess ranging over 3–31%. The product stereochemistry changed sensitively with structural variations in the sugar residues.[423]

The cobalt(I) cobalamin catalyzed reduction of α-methyl-α,β-unsaturated carbonyl compounds produces the corresponding saturated derivatives having an (*S*)-configuration at the α-carbon (Scheme 81).[424] The highest enantiomeric excess (33%) is exhibited by the (*Z*)-configurated methyl ketone. The

(*E*)-configurated enone is reduced by this system to the corresponding (*R*)-product with poor enantiomeric excess.

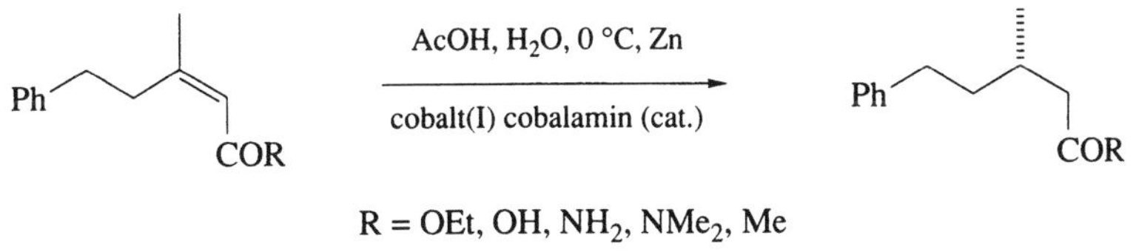

AcOH, H_2O, 0 °C, Zn; cobalt(I) cobalamin (cat.)

R = OEt, OH, NH_2, NMe_2, Me

Scheme 81

3.5.8 MISCELLANEOUS REDUCING AGENTS

Several techniques utilizing miscellaneous reagents, that were not mentioned in the preceding sections, have been reported to effect the 1,4-reduction of α,β-unsaturated aldehydes and ketones.

Sodium dithionite under nitrogen at 80 °C in a water/benzene mixture and in the presence of a phase transfer catalyst reduces dienoic carboxylic acids and esters in a 1,6-mode.[425] Enones such as isophorone, pulegone and carvone are similarly reduced to the saturated ketones. Citral and dimethyl maleate are also reduced in a 1,4-manner in moderate yield.[426]

2-Phenylbenzothiazoline reduced α,β-unsaturated carbonyl compounds in a 1,4-fashion in the presence of stoichiometric amounts of aluminum chloride.[427] No 1,2-reduction products or saturated alcohols were detected. The reagent reduces unsaturated esters and aldehydes much less effectively.

Condensation of an α,β-unsaturated ketone with benzylamine gives the corresponding Schiff base. Treatment with a base such as potassium *t*-butoxide affects rearrangement to a benzaldehyde derivative, as shown in Scheme 82.[428] Hydrolysis of the latter with dilute acetic acid furnishes the corresponding saturated ketone with concomitant formation of benzaldehyde.

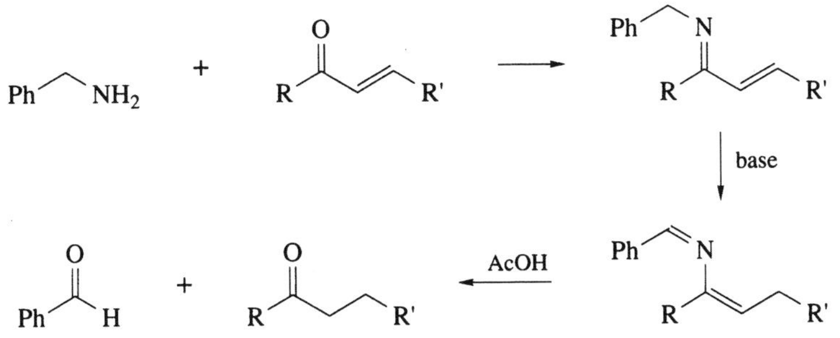

Scheme 82

A reagent prepared from tellurium powder and sodium borohydride in ethanol engenders 1,4-reduction of α,β-unsaturated aldehydes, ketones and esters in high yield and with good regio- and chemo-selectivity (no 1,2-reduction and no reduction of isolated double bonds).[429] Enedicarbonyl compounds are reduced to their saturated analogs by treatment with NaI and HCl in acetone.[430] Activated alkenes of the type $Ph_2C{=}CXY$ where X and Y are CO_2Et, COPh, CO_2Et and CN react in THF at low temperature with lithium amides having hydrogen at the α-position to give the corresponding saturated derivatives.[431] Carvone is reduced in a 1,4-manner by stereospecific addition of thiophenol to the conjugated double bond followed by desulfurization.[432] Pyridazin-3-ones[433] and allene-1,3-dicarboxylic acids[434] are reduced at the C—C double bond with Zn dust in acetic acid. α,β-Unsaturated esters and amides are reduced to the corresponding saturated compounds with magnesium in methanol.[435]

Anthracene hydride (the anion derived from 9,10-dihydroanthracene) reacts rapidly with chalcone to form an anionic Michael adduct along with a chalcone dimerization product (Scheme 83).[436] Prolonged reaction in the presence of anthracene hydride cleaves the Michael adduct into anthracene and the enolate of the saturated ketone. The partial structure RCCCO is essential for this fragmentation, as mesityl oxide, for example, gave only the Michael adduct.

Photolysis of 4a-methyl-4,4a,9,10-tetrahydro-2-(3*H*)-phenanthrone in 2-propanol gave rearranged and 1,4-reduction products, along with traces of 1,2-reduction and small amounts of coupling products.[437]

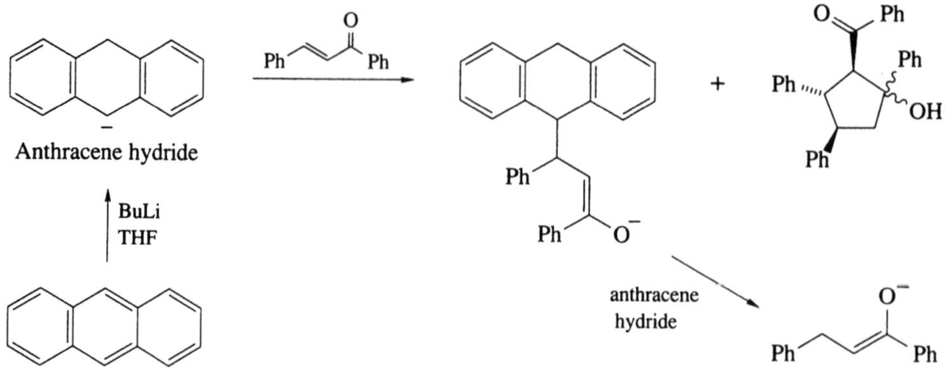

Scheme 83

2-Propanol doped on dehydrated alumina reduces various aldehydes and ketones at room temperature to the corresponding alcohols.[438] α,β-Unsaturated aldehydes are selectively reduced under these conditions to the corresponding allylic alcohols. For example, citral is converted to geraniol in 88% yield.

α,β-Unsaturated nitriles are reduced to saturated nitriles with triethylamine–formic acid azeotrope in DMF.[439]

α,β-Unsaturated ketones are reduced to allylic alcohols with β-branched trialkylaluminum compounds, such as $Bu^i{}_3Al$ and tris-[(*S*)-2-methylbutyl]aluminum. The latter reagent reduces prochiral enones to optically active allylic alcohols with 7%–15% enantiomeric excess.[440,441]

3.5.9 PARTIAL REDUCTION OF CONJUGATED DIENES AND STYRENES

3.5.9.1 Dissolving Metal Reduction

Alkenes conjugated to aromatic rings can be selectively reduced to the corresponding alkylarenes using dissolving metal techniques.[442] These reductions are generally stereoselective, with *trans* addition of a hydrogen molecule across the double bond.[443] As is usually observed in dissolving metal reduction of enones, here also, in the absence of a proton donor, dimerization of the initially formed anion radical is the major reaction product (Scheme 84).[444]

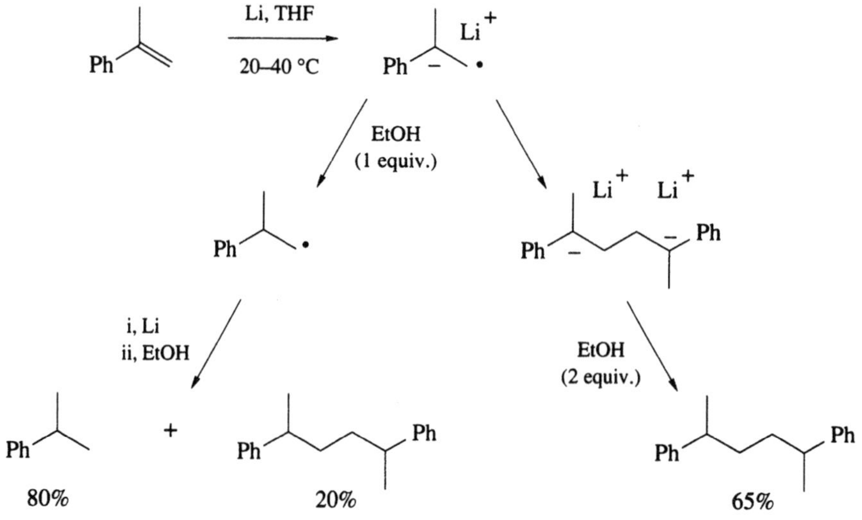

Scheme 84

Carbon–carbon double bonds conjugated to multiple bonds other than aromatic systems may also be reduced with metals. Again the nature of the reduction product is dependent on the availability of a proton donor in the reaction medium. In the absence of an excess of proton donors, dimerization of the initially formed anion radical is observed.[445] Both the reduction of 1,3-dienes[446] and trapping experiments with trimethylsilyl chloride[447] (Scheme 85) have suggested that the initial anion radical formed from these acyclic dienes possesses *cis* configuration, when formed at low temperatures and/or in nonpolar media. This configuration of the tight ion pair intermediate is presumably favored for electrostatic reasons.

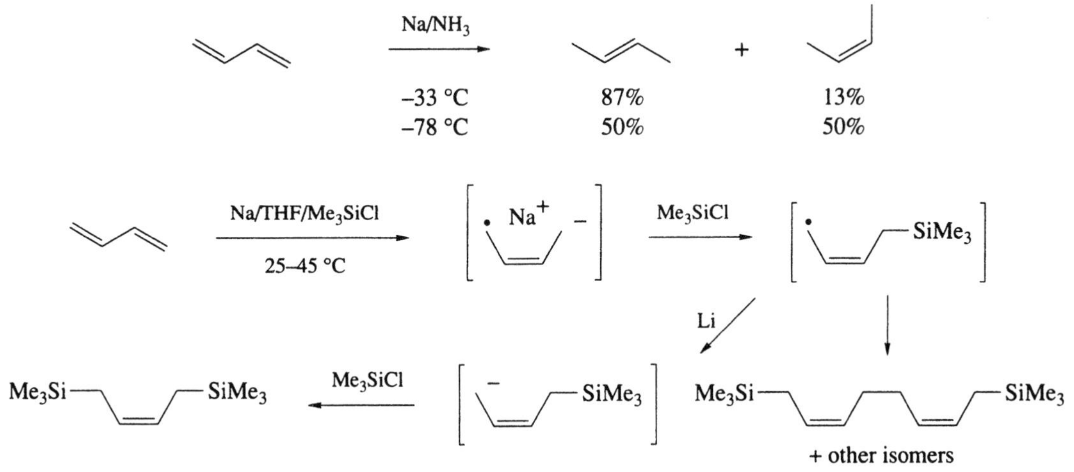

Na/NH$_3$

–33 °C 87% 13%

–78 °C 50% 50%

Na/THF/Me$_3$SiCl

25–45 °C

Na$^+$

Me$_3$SiCl

SiMe$_3$

Li

Me$_3$Si SiMe$_3$ Me$_3$SiCl SiMe$_3$ Me$_3$Si SiMe$_3$

+ other isomers

Scheme 85

Selective reduction of ergosterol and its derivatives has been carried out using electron transfer reduction. Product distribution depends on the hydroxy substituent as well as on the solvent and the metal used (Scheme 86).[448]

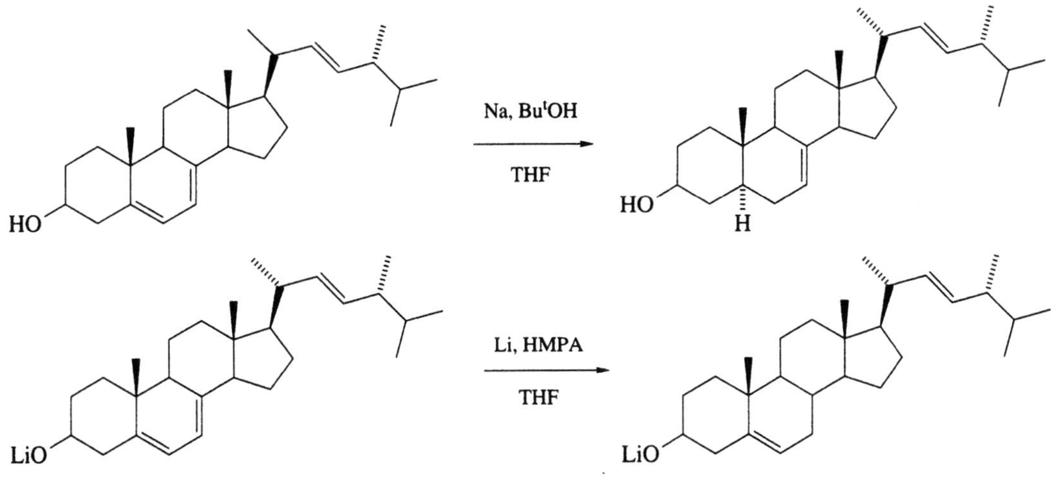

Na, Bu^{t}OH

THF

HO HO H

Li, HMPA

THF

LiO LiO

Scheme 86

Various diphenylethylenes are reduced in high yield by magnesium in methanol.[449] Expectedly, reductions of conjugated double bonds *via* one-electron transfer processes can also be carried out electrochemically. Scheme 87 illustrates both the reduction of a conjugated alkene and partial trapping of the intermediate anion radical to yield a mixed hydrodimerization product.

3.5.9.2 Catalytic Hydrogenation

Partial hydrogenation of conjugated dienes or polyenes is a nontrivial synthetic transformation. This reaction commonly yields a mixture of products arising from 1,2- and 1,4-addition of hydrogen to the butadiene moiety, as well as complete hydrogenation of both double bonds. Unless there is a substantial difference between the two double bonds in terms of substitution pattern and electronic density, selectiv-

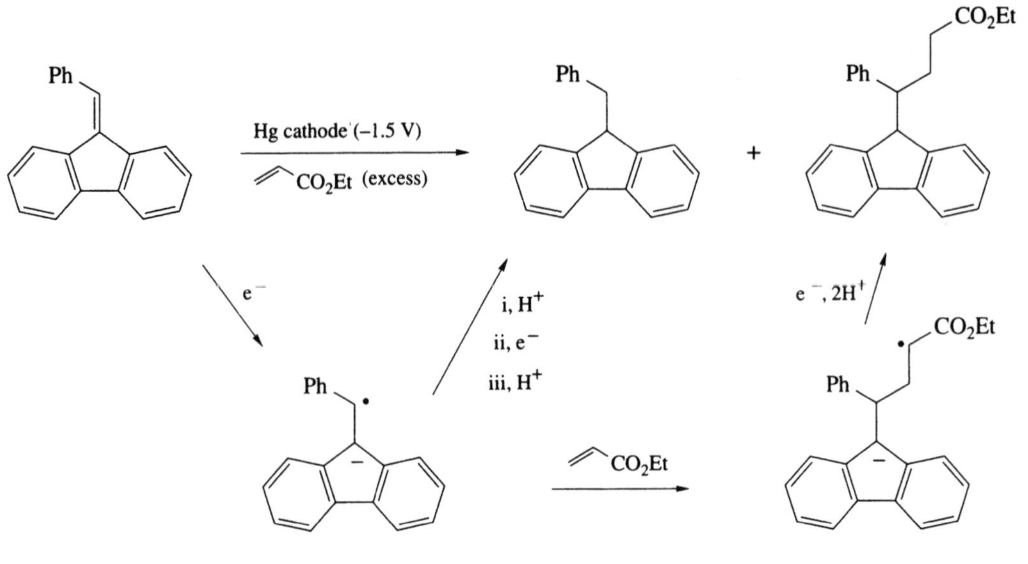
Ph
Hg cathode (–1.5 V)
CO2Et (excess)
Ph
+
CO2Et
Ph
e−
i, H+
ii, e−
iii, H+
e−, 2H+
CO2Et
Ph
Ph
CO2Et

Scheme 87

ity of these reactions is rather poor. Chemists often meet this challenge by a trial and error approach, using heterogeneous hydrogenation catalysts.

Alkene hydrogenation rates usually decrease with increasing degree of substitution on the double bond. This general tendency may also be influenced by the steric bulk and position of the substituents. For instance, in ring systems, exocyclic double bonds are reduced much faster than endocyclic ones.

A number of steroidal dienes have been selectively hydrogenated at the sterically less-hindered position using various catalysts. Raney nickel and PtO_2 are the most commonly used heterogeneous catalysts for this purpose.[450] Several examples of selectivity in heterogeneous catalytic hydrogenation of dienes are illustrated in Scheme 88.[451–457]

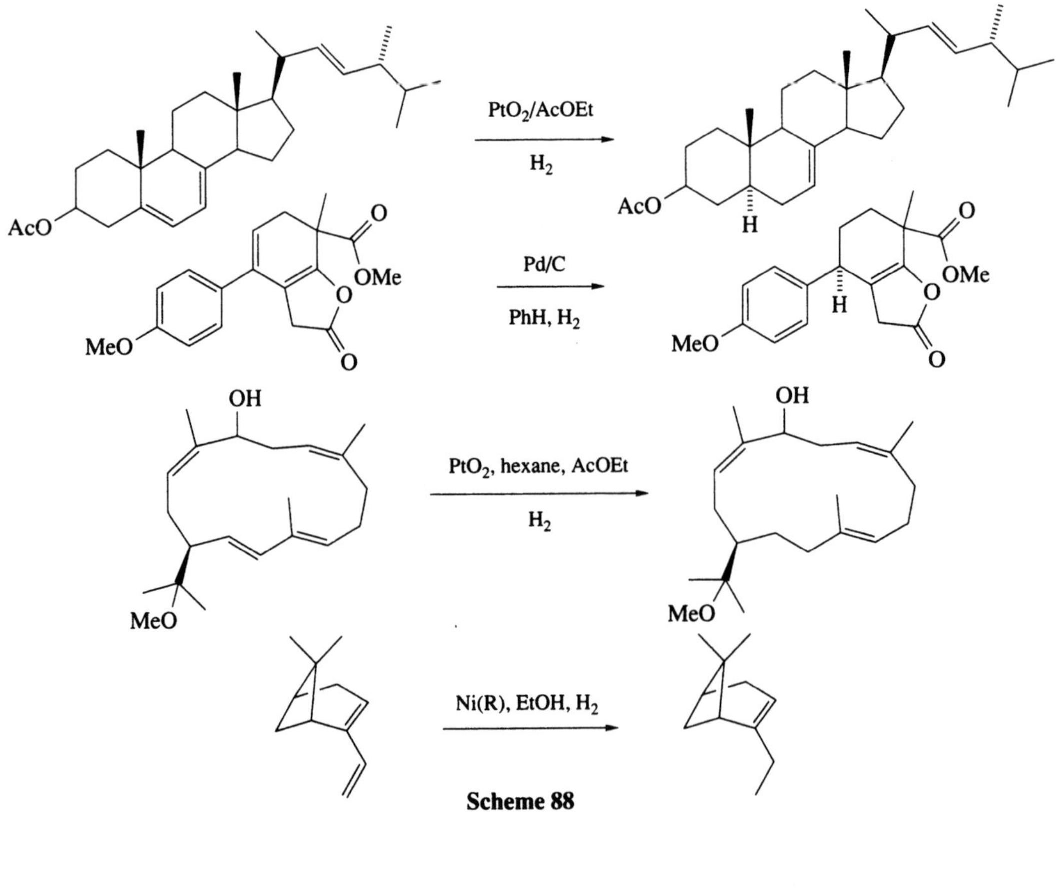
AcO
PtO2/AcOEt
H2
AcO
H
MeO
O
OMe
O
O
Pd/C
PhH, H2
MeO
H
O
OMe
O
O
OH
PtO2, hexane, AcOEt
H2
OH
MeO
MeO
Ni(R), EtOH, H2

Scheme 88

Utilization of homogeneous catalysts for selective hydrogenation of steroidal dienes proved to be beneficial, although the reason for selectivity in these reactions has not been extensively explored. Partial reduction of conjugated dienes may be carried out with soluble chromium complexes. For example, 1,4-hydrogenation of conjugated dienes is selectively achieved with (methylbenzoate)$Cr(CO)_3$[367b] to afford the *cis*-monoalkene. The reaction is highly stereospecific for *cis*-dienes or for cyclic 1,3-dienes held in the required *cisoid* configuration, and affords a product with a *cis* double bond. This catalyst has been successfully employed in partial hydrogenation of several diene intermediates in prostaglandin synthesis. In all cases, the less-substituted double bond was selectively hydrogenated (Scheme 89)[458]

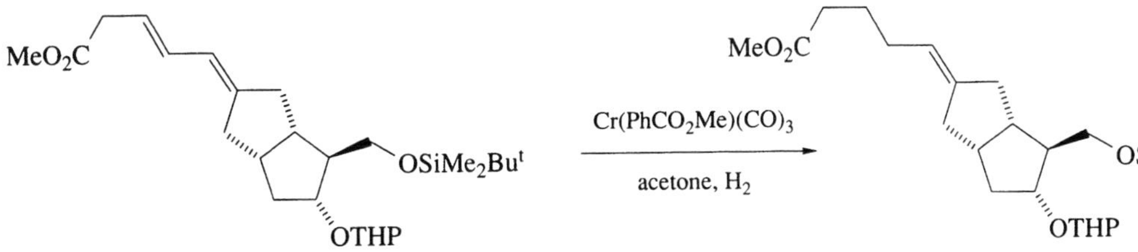

MeO2C OSiMe2Bu^t OTHP → Cr(PhCO2Me)(CO)3, acetone, H2 → MeO2C OSiMe2Bu^t OTHP

Scheme 89

However, the reaction conditions required for catalytic hydrogenation with Cr(arene)$(CO)_3$ are relatively forcing (160 °C/30 atm). A closely related complex $Cr(CO)_3(MeCN)_3$, with relatively labile acetonitrile ligands, operates under far milder conditions (40 °C/1.5 atm).[459] Similarly, the photoactivated complex $Cr(CO)_6$, catalyzes 1,4-hydrogen addition to dienes under 1 atm of hydrogen at room temperature.[460] Nonconjugated 1,4-dienes are also hydrogenated with (arene)$Cr(CO)_3$ at high temperatures, probably *via* isomerization to the conjugated system prior to reduction. No reaction occurs with either 1,5- or 1,6-dienes, owing to the inability of the chromium complex to affect conjugation of the double bonds.

The pentacyano cobalt anion $Co(CN)_5^{3-}$ has found synthetic application in the reduction of conjugated dienes to monoenes.[461] This catalyst is prepared by reaction of cobalt(II) chloride with potassium cyanide under a nitrogen atmosphere in either aqueous (more common) or nonaqueous (*e.g.* methanol) solvents. The preference for aqueous solutions has limited the applications of this reaction due to poor water solubility of most organic substrates. This solution requires preactivation by addition of hydrogen (1 atm) for 1–2 h at 20 °C to generate the active species $(Co(CN)_5H)^{3-}$. This solution is then used immediately, as the actual catalyst tends to loose activity upon standing. A number of simple dienes, including butadiene, isoprene, 1-phenylbutadiene, cyclohexa-1,3-diene, and cyclopentadiene, were successfully hydrogenated under these mild conditions. However, the catalyst is rather sensitive to steric hindrance, which inhibits hydrogenation. For example, 2,5-dimethylhexa-2,4-diene is not reduced by this system. Generally, monoenes are not reduced by $[Co(CN)_5]^{3-}$, but activated monoalkenes, such as styrenes and α,β-unsaturated carbonyl compounds, can be hydrogenated (Scheme 90). Acrylic acid and acrolein do not react under these conditions.

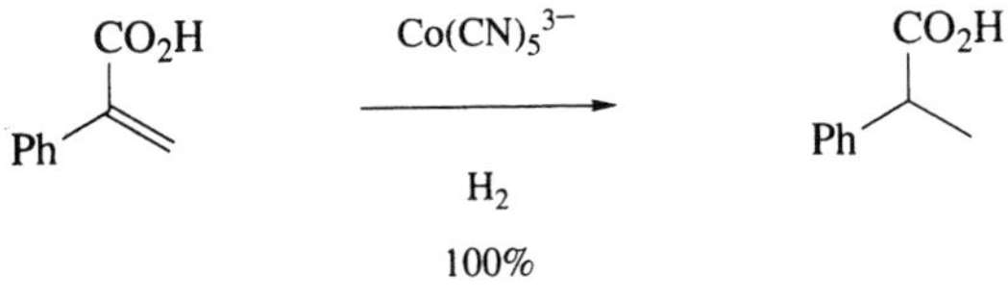

Ph–C(=CH2)–CO2H → $Co(CN)_5^{3-}$, H2, 100% → Ph–CH(CH3)–CO2H

Scheme 90

With terminal 1,3-dienes and either $RhH(PPh_3)_4$ or $(Rh(CO)_2(PPh_3))_2$ in the presence of excess phosphine at 50–100 °C and 15 atm, selective hydrogenation of the terminal double bond is achieved (Scheme 91). Hydrogen uptake must be monitored in these reactions in order to prevent further hydrogenation of the partially reduced product. As is the case for the cobalt cyanide system, here also, substitution at the double bond inhibits the rate of hydrogenation and limits its general applicability.

While most catalysts reduce the least-substituted double bond preferentially, the novel catalyst system $CoBr(PPh_3)_3 \cdot BF_3OEt_2$ can selectively hydrogenate conjugated dienes to monoenes *via* 1,2-hydrogen addition to the more substituted double bond (Scheme 92). Unfortunately, hydrogenation of functionalized alkenes, such as methyl vinyl ketone, methyl acrylate and *n*-butyl vinyl ether, does not occur even under forcing conditions.

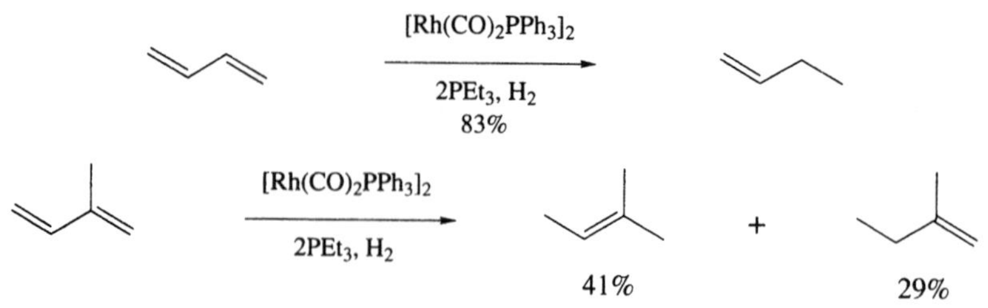

[Rh(CO)$_2$PPh$_3$]$_2$
2PEt$_3$, H$_2$
83%

[Rh(CO)$_2$PPh$_3$]$_2$
2PEt$_3$, H$_2$
41% + 29%

Scheme 91

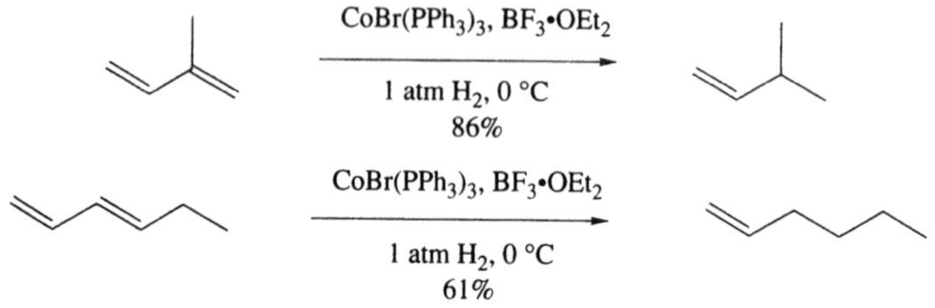

CoBr(PPh$_3$)$_3$, BF$_3$•OEt$_2$
1 atm H$_2$, 0 °C
86%

CoBr(PPh$_3$)$_3$, BF$_3$•OEt$_2$
1 atm H$_2$, 0 °C
61%

Scheme 92

Hydrogenation of penta-1,3-dienes to pent-2-ene is selectively achieved with RuH(PPh$_3$)$_3$Cl.[462] Butadiene and isoprene are hydrogenated in 62% yield using Cp$_2$VCl$_2$·BuLi, affording *trans*-but-2-ene (84%), *cis*-but-2-ene (14%) and but-1-ene (2%).[463] Homogeneous hydrogenation of penta-1,3-diene using LiAlH$_4$ as a catalyst affords *trans*-pent-2-ene (57%), *cis*-pent-2-ene (10%) and pent-1-ene (8%).[464] Tricarbonylcyclopentadienylhydrido-molybdenum and -tungsten were used for hydrogenation of hexa-2,4-diene, 4-methylpentadiene, and octa-2,4,6-triene, producing in all cases a mixture of monoalkene products.[465]

The hydrogenation of alkenes conjugated to aromatic systems is usually much less difficult than partial reduction of polyenes, as the aromatic ring is less susceptible to reduction. Many catalysts of Group 10 metals have been used for this transformation, including Raney nickel,[466] Pd adsorbed on carbon,[467] Pd adsorbed on calcium carbonate,[468] Pd on barium sulfate,[469] metallic palladium,[470] PtO$_2$,[471] chloroplatinic acid[472] or platinum metal.[473] Normally, the stereochemistry of the reduced product arises from *syn* addition of a hydrogen molecule to the less-hindered face of the double bond, as exemplified in the catalytic hydrogenation of a steroidal styrene over metallic palladium (Scheme 93).[474]

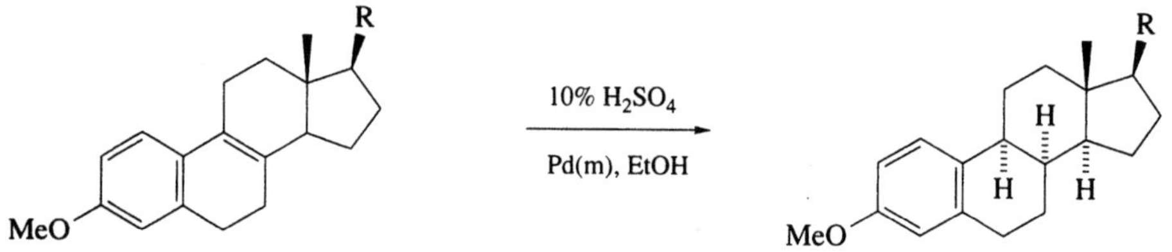

R
MeO
10% H$_2$SO$_4$
Pd(m), EtOH
R
H
H H
MeO

Scheme 93

A number of other soluble transition metal complexes may catalyze hydrogenation of styrene. These include HCo(CO)$_4$,[475] SnCl/H$_2$PtCl$_2$,[476] KNi$_2$(CN)$_2$,[477] RhCl(PPh$_3$)$_3$,[478] RhH(CO)(PPh$_3$)$_3$,[479] K$_3$Co(CN)$_5$,[480] Cr(acac)$_3$, Co(acac)$_3$, Fe(acac)$_3$, Ti(OPr^i)$_4$, VO(OEt)$_3$, which in the presence of trialkylaluminum catalyze hydrogenation of *trans*-stilbene.[481]

3.5.9.3 Miscellaneous Reducing Agents

Reactive styrenes, such as acenaphthylene[482] and compounds in which the double bond is part of an allylic alcohol are reduced with LiAlH$_4$.[483] The double bond of benzalquinaldine is reduced by heating with thiocresol in HCl, affording the corresponding hydrogenated product, as well as the disulfide as a by-product.[484] Benzyl alcohol in KOH may act as a reducing agent for several fluorenyl styrenes.[485] Hydrazine can reduce the double bond of various stilbenes.[486]

3.5.10 REFERENCES

1. H. C. Brown and S. Krishnamurthy, *Tetrahedron*, 1979, **35**, 567.
2. (a) A. J. Birch and H. Smith, *Q. Rev., Chem. Soc.*, 1958, **12**, 17; (b) A. J. Birch and G. S. R. Subba Rao, *Adv. Org. Chem.*, 1972, **8**, 1.
3. C. Djerassi, in 'Steroid Reactions', Holden-Day, San Francisco, 1963, p. 299.
4. H. L. Dryden, Jr., in 'Organic Reactions in Steroid Chemistry', ed. J. Fried and J. A. Edwards, Van Nostrand Reinhold, New York, 1972, vol. 1, p. 1.
5. H. O. House, 'Modern Synthetic Reactions', 2nd edn., Benjamin, Menlo Park, CA, 1972.
6. F. Johnson, *Chem. Rev.*, 1968, **68**, 375.
7. F. J. McQuillin, in 'Technique of Organic Chemistry', ed. A. Weissberger, Interscience, New York, 1963, vol. 11, part 1, chap. 9.
8. H. Smith, 'Organic Reactions in Liquid Ammonia', Wiley, New York, 1963.
9. M. Smith, in 'Reduction', ed. R. L. Augustine, Dekker, New York, 1968, chap. 2.
10. D. Caine, *Org. React. (N.Y.)*, 1976, **23**, 1.
11. (a) M. C. R. Symons, *Q. Rev., Chem. Soc.*, 1959, **13**, 99; (b) U. Schindewolf, *Angew. Chem., Int. Ed. Engl.*, 1968, **7**, 190; (c) J. L. Dye, *Acc. Chem. Res.*, 1968, **1**, 306.
12. (a) H. Normant, *Angew. Chem., Int. Ed. Engl.*, 1967, **6**, 1046; (b) H. Normant, *Bull. Soc. Chim. Fr.*, 1968, 791; (c) K. W. Bowers, R. W. Giese, J. Grimshaw, H. O. House, N. H. Kolodny, K. Kronberger and D. K. Roe, *J. Am. Chem. Soc.*, 1970, **92**, 2783; (d) M. Larcheveque, *Ann. Chim. (Paris)*, 1970, **5**, 129.
13. (a) J. L. Down, J. Lewis, B. Moore and G. Wilkinson, *J. Chem. Soc.*, 1959, 3767; (b) C. Agami, *Bull. Soc. Chim. Fr.*, 1968, 1205; (c) J. L. Dye, M. G. DeBacker and V. A. Nicely, *J. Am. Chem. Soc.*, 1970, **92**, 5226; (d) C. J. Pedersen, *J. Am. Chem. Soc.*, 1967, **89**, 7017; 1970, **92**, 386, 391.
14. D. H. R. Barton and C. H. Robinson, *J. Chem. Soc.*, 1954, 3045.
15. G. Stork and S. D. Darling, *J. Am. Chem. Soc.*, 1960, **82**, 1512; 1964, **86**, 1761.
16. M. J. T. Robinson, *Tetrahedron*, 1965, **21**, 2475.
17. (a) R. Howe and F. J. McQuillin, *J. Chem. Soc.*, 1956, 2670; (b) G. L. Chetty, G. S. Krishna Rao, S. Dev and D. K. Banerjee, *Tetrahedron*, 1966, **22**, 2311.
18. F. J. McQuillin, *J. Chem. Soc.*, 1955, 528.
19. A. J. Birch, H. Smith and R. E. Thornton, *J. Chem. Soc.*, 1957, 1339.
20. H. E. Zimmerman, *J. Am. Chem. Soc.*, 1956, **78**, 1168.
21. O. Wallach, *Justus Liebigs Ann. Chem.*, 1894, **279**, 366.
22. O. Wallach, *Justus Liebigs Ann. Chem.*, 1983, **275**, 103.
23. O. Diels and E. Abderhalden, *Ber. Dtsch. Chem. Ges.*, 1906, **39**, 884.
24. L. H. Knox, E. Blossey, H. Carpio, L. Cervantes, P. Crabbé, E. Velarde and J. ·A. Edwards, *J. Org. Chem.* 1965, **30**, 2198.
25. J. A. Barltrop and A. C. Day, *Tetrahedron*, 1966, **22**, 3181.
26. T. A. Spencer, R. A. J. Smith, D. L. Storm and R. M. Villarica, *J. Am. Chem. Soc.*, 1971, **93**, 4856.
27. R. H. Mueller, and J. G. Gillick, *J. Org. Chem.*, 1978, **43**, 4647.
28. L. E. Hightower, L. R. Glasgow, K. M. Stone, D. A. Albertson and H. Smith, *J. Org. Chem.*, 1970, **35**, 1881.
29. H. Smith, B. J. L. Huff, W. J. Powers and D. Caine, *J. Org. Chem.*, 1967, **32**, 2851.
30. J. F. Eastham and D. R. Larkin, *J. Am. Chem. Soc.*, 1959, **81**, 3652.
31. H. O. House, *Rec. Chem. Prog.*, 1967, **28**, 98.
32. W. L. Jolly and L. Prizant, *Chem. Commun.*, 1968, 1345.
33. A. P. Krapcho and A. A. Bothner-By, *J. Am. Chem. Soc.*, 1959, **81**, 3658.
34. D. C. Burke, J. H. Turnbull and W. Wilson, *J. Chem. Soc.*, 1953, 3237.
35. I. N. Nazarov and I. A. Gurvich, *J. Gen. Chem. USSR (Engl. Transl.)*, 1955, **25**, 1077.
36. A. J. Birch, E. Pride and H. Smith, *J. Chem. Soc.*, 1958, 4688.
37. G. Büchi, S. J. Gould and F. Näf, *J. Am. Chem. Soc.*, 1971, **93**, 2492.
38. M. E. Kuehne, *J. Am. Chem. Soc.*, 1961, **83**, 1492.
39. G. Stork, P. Rosen and N. L. Goldman, *J. Am. Chem. Soc.*, 1961, **83**, 2965.
40. G. Stork, P. Rosen, N. L. Goldman, R. V. Coombs and J. Tsuji, *J. Am. Chem. Soc.*, 1965, **87**, 275.
41. M. J. Weiss, R. E. Schaub, G. R. Allen, Jr., J. F. Poletta, C. Pidacks, R. B. Conrow and C. J. Coscia, *Tetrahedron*, 1964, **20**, 357; M. J. Weiss, R. E. Schaub, J. F. Poletta, G. R. Allen, Jr. and C. J. Coscia, *Chem. Ind. (London)*, 1963, 118.
42. A. Coulombeau, *Bull. Soc. Chim. Fr.*, 1970, 4407.
43. R. Deghenghi, C. Revesz and R. Gaudry, *J. Med. Chem.*, 1963, **6**, 301.
44. R. M. Coates and R. L. Sowerby, *J. Am. Chem. Soc.*, 1971, **93**, 1027.
45. (a) R. Deghenghi and R. Gaudry, *Tetrahedron Lett.*, 1962, 489; (b) R. E. Schaub and M. J. Weiss, *Chem. Ind. (London)*, 1961, 2003.
46. G. Stork and J. E. McMurry, *J. Am. Chem. Soc.*, 1967, **89**, 5464.
47. G. Stork, S. Uyeo, T. Wakamatsu, P. A. Grieco and J. Labovitz, *J. Am. Chem. Soc.*, 1971, **93**, 4945.
48. T. A. Spencer, T. D. Weaver, R. M. Villarica, R. J. Friary, J. Posler and M. A. Schwartz, *J. Org. Chem.*, 1968, **33**, 712.
49. K. W. Bowers, R. W. Giese, J. Grimshaw, H. O. House, N. H. Kolodny, K. Kronberger and D. K. Roe, *J. Am. Chem. Soc.*, 1970, **92**, 2783.
50. C. L. Perrin, *Prog. Phys. Org. Chem.*, 1965, **3**, 165.
51. M. M. Baizer and J. P. Petrosrich, *Prog. Phys. Org. Chem.*, 1970, **7**, 189.
52. D. Miller, L. Mandell and R. A. Day, Jr., *J. Org. Chem.*, 1971, **36**, 1683.
53. J. Wiemann, S. Risse and P.-F. Casals, *Bull. Soc. Chim. Fr.*, 1966, 381.
54. B. J. L. Huff, Ph. D. Thesis, Georgia Institute of Technology, 1969 (*Diss. Abstr. B*, 1969, **29** (12), 4589).
55. G. Stork, M. Nussim and B. August, *Tetrahedron Suppl.*, 1966, **8**, 105.
56. P. Angibeaund, H. Rivière and B. Tchoubar, *Bull. Soc. Chim. Fr.*, 1968, 2937.

57. T. A. Spencer, R. J. Friary, W. W. Schmiegel, J. F. Simeone and D. S. Watt, *J. Org. Chem.*, 1968, **33**, 719.
58. R. E. Ireland and G. Pfister, *Tetrahedron Lett.*, 1969, 2145.
59. G. Stork and J. Tsuji, *J. Am. Chem. Soc.*, 1961, **83**, 2783.
60. P. S. Venkataramani, J. E. Karoglan and W. Reusch, *J. Am. Chem. Soc.*, 1971, **93**, 269.
61. T. A. Spencer, K. K. Schmiegel and W. W. Schmiegel, *J. Org. Chem.*, 1965, **30**, 1626.
62. R. G. Carlson and R. G. Blecke, *J. Chem. Soc., Chem. Commun.*, 1969, 93.
63. M. Tanabe, J. W. Chamberlin and P. Y. Nishiura, *Tetrahedron Lett.*, 1961, 601.
64. B. M. Trost, in 'Abstracts of Papers, Joint Conference CIC–ACS', Toronto, Canada, May 24–29, 1970, Organic Section, paper 42.
65. C. Amendolla, G. Rosenkranz and F. Sondheimer, *J. Chem. Soc.*, 1954, 1226.
66. T. Anthonsen, P. H. McCabe, R. McCrindle, and R. D. H. Murray, *Tetrahedron*, 1969, **25**, 2233.
67. T. Masamune, A. Murai, K. Orito, H. Ono, S. Numata and H. Suginome, *Tetrahedron*, 1969, **25**, 4853.
68. H. Bruderer, D. Arigoni and O. Jeger, *Helv. Chim. Acta*, 1956, **39**, 858.
69. R. Howe, F. J. McQuillin and R. W. Temple, *J. Chem. Soc.*, 1959, 363.
70. K. S. Kulkarni and A. S. Rao, *Tetrahedron*, 1965, **21**, 1167.
71. K. Irmscher, W. Beerstecher, H. Metz, R. Watzel and K.-H. Bork, *Chem. Ber.*, 1964, **97**, 3363.
72. A. Spassky-Pasteur, *Bull. Soc. Chim. Fr.*, 1969, 2900.
73. R. M. Coates and J. E. Shaw, *Tetrahedron Lett.*, 1968, 5405; *J. Org. Chem.*, 1970, **35**, 2597.
74. R. M. Coates and J. E. Shaw, *J. Org. Chem.*, 1970, **35**, 2601.
75. R. E. Ireland and J. A. Marshall, *J. Org. Chem.*, 1962, **27**, 1615.
76. M. Vandewalle and F. Compernolle, *Bull. Soc. Chim. Belg.*, 1966, **75**, 349.
77. M. Vanderwalle and F. Compernolle, *Bull. Soc. Chim. Belg.*, 1967, **76**, 43.
78. D. S. Watt, J. M. McKenna and T. A. Spencer, *J. Org. Chem.*, 1967, **32**, 2674.
79. J. E. Shaw and K. K. Knutson, *J. Org. Chem.*, 1971, **36**, 1151.
80. J. A. Campbell and J. C. Babcock, *J. Am. Chem. Soc.*, 1959, **81**, 4069.
81. A. F. Daglish, J. Green and V. D. Poole, *J. Chem. Soc.*, 1954, 2627.
82. F. Johnson, G. T. Newbold and F. S. Spring, *J. Chem. Soc.*, 1954, 1302
83. J. A. Marshall and H. Roebke, *J. Org. Chem.*, 1968, **33**, 840.
84. M. Nussim, Y. Mazur and F. Sondheimer, *J. Org. Chem.*, 1964, **29**, 1120.
85. H. Van Kamp, P. Westerhof and H. Niewind, *Recl. Trav. Chim. Pays-Bas*, 1964, **83**, 509.
86. E. Wenkert, A. Afonso, J. B. Bredenberg, C. Kaneko and A. Tahara, *J. Am. Chem. Soc.*, 1964, **86**, 2038.
87. P. Westerhof and E. H. Reerink, *Recl. Trav. Chim. Pays-Bas*, 1960, **79**, 771.
88. K. P. Dastur, *Tetrahedron Lett.*, 1973, 4333.
89. A. Zürcher, H. Heusser, O. Jeger and P. Geistlich, *Helv. Chim. Acta*, 1954, **37**, 1562.
90. G. Bach, J. Capitaine and C. R. Engel, *Can. J. Chem.*, 1968, **46**, 733.
91. H. Heusser, M. Roth, O. Rohr and R. Anliker, *Helv. Chim. Acta*, 1955, **38**, 1178.
92. W. F. Johns, *J. Org. Chem.*, 1971, **36**, 711.
93. E. Shapiro, T. Legatt, L. Weber, M. Steinberg and E. P. Oliveto, *Chem. Ind. (London)*, 1962, 300.
94. W. Cocker, B. Donnelly, H. Gobinsingh, T. B. H. McMurry and M. A. Nisbet, *J. Chem. Soc.*, 1963, 1262.
95. B. R. Ortiz de Montellano, B. A. Loving, T. C. Shields and P. D. Gardner, *J. Am. Chem. Soc.*, 1967, **89**, 3365.
96. D. J. Marshall and R. Deghenghi, *Can. J. Chem.*, 1969, **47**, 3127.
97. T. G. Halsall, D. W. Theobald and K. B. Walshaw, *J. Chem. Soc.*, 1964, 1029.
98. A. J. Birch, *Q. Rev., Chem. Soc.*, 1950, **4**, 69.
99. R. G. Harvey, *Synthesis*, 1970, 161.
100. W. Nagata, T. Terasawa, S. Hirai and K. Takeda, *Tetrahedron Lett.*, 1960, 727; *Chem. Pharm. Bull.*, 1961, **9**, 769.
101. W. S. Johnson, E. R. Rogier, J. Szmuszkovicz, H. I. Hadler, J. Ackerman, B. K. Bhattacharyya, B. M. Bloom, L. Stalmann, R. A. Clement, B. Bannister and H. Wynberg, *J. Am. Chem. Soc.*, 1956, **78**, 6289.
102. W. F. Johns, *J. Org. Chem.*, 1963, **28**, 1856.
103. W. S. Johnson, J. M. Cox, D. W. Graham and H. W. Whitlock, Jr., *J. Am. Chem. Soc.*, 1967, **89**, 4524.
104. M. V. R. Koteswara Rao, G. S. Krishna Rao and S. Dev, *Tetrahedron*, 1966, **22**, 1977.
105. F. B. Colton, L. N. Nysted, B. Riegel and A. L. Raymond, *J. Am. Chem. Soc.*, 1957, **79**, 1123.
106. A. Bowers, H. J. Ringold and E. Denot, *J. Am. Chem. Soc.*, 1958, **80**, 6115.
107. I. A. Gurvich, V. F. Kucherov and T. V. Ilyakhina, *J. Gen. Chem. USSR (Engl. Transl.)*, 1961, **31**, 738.
108. P. S. Venkataramani, J. P. John, V. T. Ramakrishnan and S. Swaminathan, *Tetrahedron*, 1966, **22**, 2021.
109. P. T. Lansbury, P. C. Briggs, T. R. Demmin and G. E. DuBois, *J. Am. Chem. Soc.* 1971, **93**, 1311.
110. H. Kaneko, K. Nakamura, Y. Yamoto and M. Kurokawa, *Chem. Pharm. Bull.*, 1969, **17**, 11.
111. R. E. Schaub and M. J. Weiss, *J. Org. Chem.*, 1961, **26**, 3915.
112. P. Beak and T. L. Chaffin, *J. Org. Chem.*, 1970, **35**, 2275.
113. E. Wenkert and B. G. Jackson, *J. Am. Chem. Soc.*, 1958, **80**, 217.
114. G. Stork and F. H. Clarke, Jr., *J. Am. Chem. Soc.*, 1955, **77**, 1072; 1961, **83**, 3114.
115. S. Dubé and P. Deslongchamps, *Tetrahedron Lett.*, 1970, 101.
116. W. G. Dauben, W. W. Epstein, M. Tanabe and B. Weinstein, *J. Org. Chem.*, 1963, **28**, 293.
117. J. R. Hanson, *Synthesis*, 1974, 1.
118. J. E. McMurry, *Acc. Chem. Res.*, 1974, **7**, 281.
119. T.-L. Ho, *Synthesis*, 1979, 1.
120. (a) C. E. Castro, R. D. Stephens and S. J. Mojé *J. Am. Chem. Soc.*, 1966, **88**, 4964; (b) A. Zurqiyah and C. E. Castro, *Org. Synth., Coll. Vol.*, 1973, **5**, 993.
121. (a) E. Knecht, *Ber. Dtsch. Chem. Ges.*, 1903, **36**, 166; (b) P. Karrer, Y. Yen and I. Reichstein, *Helv. Chim. Acta*, 1930, **13**, 1308; (c) L. Blaszczak and J. E. McMurry, *J. Org. Chem. Soc.*, 1974, **39**, 258.
122. (a) J. R. Hanson and E. Premuzic, *J. Chem. Soc. C*, 1969, 1201; (b) J. R. Hanson and E. Premuzic, *Angew. Chem., Int. Ed. Engl.*, 1968, **7**, 247.
123. H. O. House and E. F. Kinloch, *J. Org. Chem.*, 1974, **39**, 1173.

124. J. B. Conant and H. B. Cutter, *J. Am. Chem. Soc.*, 1926, **48**, 1016.
125. E. J. Corey, R. L. Danheiser and S. Chandrasekaran, *J. Org. Chem.*, 1976, **41**, 260.
126. S. C. Welch and M. E. Walters, *J. Org. Chem.* 1978, **43**, 2715.
127. (a) A. Berndt, *Angew. Chem., Int. Ed. Engl.*, 1967, **6**, 251; (b) A. Berndt, *Tetrahedron Lett.*, 1970, 177.
128. (a) P. Bladon, J. W. Cornforth and R. H. Jaeger, *J. Chem. Soc.*, 1958, 863; (b) H. Lund, *Acta Chim. Scand.*, 1957, **11**, 283.
129. R. C. Fuson, *Rec. Chem. Prog.*, 1951, **12**, 1.
130. (a) C. G. Overberger and A. M. Schiller, *J. Org. Chem.*, 1961, **26**, 4230; (b) E. L. Totton, N. C. Camp, III, G. M. Cooper, B. D. Haywood and D. P. Lewis, *J. Org. Chem.*, 1967, **32**, 2033; (c) H. Rosen, Y. Arad, M. Levy and D. Vofsi, *J. Am. Chem. Soc.*, 1969, **91**, 1425; (d) F. Matsuda, *Tetrahedron Lett.*, 1966, 6193; (e) A. Zysman, G. Dana and J. Wiemann, *Bull. Soc. Chim. Fr.*, 1967, 1019; (f) J. Wiemann, M. R. Monot, G. Dana and J. Chuche, *Bull. Soc. Chim. Fr.*, 1967, 3293; (g) E. Touboul, F. Weisbuch and J. Wiemann, *Bull. Soc. Chim. Fr.*, 1967, 4291; (h) C. Glacet, *C. R. Hebd. Seances Acad. Sci.*, 1948, **227**, 480; (i) J. Wiemann and R. Nahum, *C. R. Hebd. Seances Acad. Sci.*, 1954, **238**, 2091.
131. (a) R. N. Gourley, J. Grimshaw and P. G. Miller, *J. Chem. Soc. C*, 1970, 2318; (b) L. Horner and D. H. Skaletz, *Tetrahedron Lett.*, 1970, 3679; (c) A. Spassky-Pasteur, *Bull. Soc. Chim. Fr.*, 1969, 2900.
132. (a) M. M. Baizer, *J. Org. Chem.*, 1964, **29**, 1670; 1966, **31**, 3847; (b) M. M. Baizer and J. D. Anderson, *J. Org. Chem.*, 1965, **30**, 1348, 1351, 1357, 3138; (c) J. D. Anderson, M. M. Baizer and E. J. Prill, *J. Org. Chem.*, 1965, **30**, 1645; (d) J. D. Anderson, M. M. Baizer and J. P. Petrovich, *J. Org. Chem.*, 1966, **31**, 3890, 3897; (e) J. H. Wagenknecht and M. M. Baizer, *J. Org. Chem.*, 1966, **31**, 3885; (f) M. R. Ort and M. M. Baizer, *J. Org. Chem.*, 1966, **31**, 1646; (g) M. M. Baizer and J. D. Anderson, *J. Electrochem. Soc.*, 1964, **111**, 223, 226; M. M. Baizer, *J. Electrochem. Soc.*, 1964, **111**, 215; (h) For reviews, see: M. M. Baizer, J. D. Anderson, J. H. Wagenknecht, M. R. Ort and J. P. Petrovich, *Prog. Electrochem. Acta*, 1967, **12**, 1377; J. D. Anderson, J. P. Petrovich and M. M. Baizer, *Adv. Org. Chem.*, 1969, **6**, 257; M. M. Baizer and J. P. Petrovich, *Prog. Phys. Org. Chem.*, 1970, **7**, 189.
133. D. A. Jaeger, D. Bolikal and B. Nath, *J. Org. Chem.*, 1987, **52**, 276.
134. B. R. James, 'Homogeneous Hydrogenation', Wiley-Interscience, New York, 1973.
135. (a) P. N. Rylander, 'Hydrogenation Methods', Academic Press, London, 1985; (b) P. N. Rylander, 'Catalytic Hydrogenation in Organic Syntheses', Academic Press, New York, 1979.
136. M. Freifelder, 'Catalytic Hydrogenation in Organic Synthesis', Wiley-Interscience, New York, 1978.
137. G. R. Ames and W. Davey, *J. Chem. Soc.*, 1956, 3001.
138. J.-J. Brunet, P. Gallois and P. Caubère, *J. Org. Chem.*, 1980, **45**, 1937, 1946.
139. H. Adkins and R. Connor, *J. Am. Chem. Soc.*, 1931, **53**, 1091.
140. N. F. Hayes, *Synthesis*, 1975, 702.
141. E. Breitner, E. Roginski and P. N. Rylander, *J. Org. Chem.*, 1959, **24**, 1855.
142. A. Skita, *Ber. Dtsch. Chem. Ges.*, 1915, **48**, 1486.
143. C. Weygand and W. Meusel, *Ber. Dtsch. Chem. Ges. A*, 1943, **76**, 498.
144. R. Adams, J. W. Kern and R. L. Shriner, *Org. Synth., Coll. Vol.*, 1932, **1**, 101.
145. R. L. Augustine, *J. Org. Chem.*, 1958, **23**, 1853.
146. R. E. Harmon, J. L. Parsons, D. W. Cooke, S. K. Gupta and J. Schoolenberg, *J. Org. Chem.*, 1969, **34**, 3684.
147. (a) C. Djerassi and J. Gutzwiller, *J. Am. Chem. Soc.*, 1966, **88**, 4537; (b) A. J. Birch and K. A. M. Walker, *J. Chem. Soc. C*, 1966, 1894.
148. P. L. Cook, *J. Org. Chem.*, 1962, **27**, 3873.
149. K. Sakai and K. Watanabe, *Bull. Chem. Soc. Jpn.*, 1967, **40**, 1548.
150. Y. Watanabe, Y. Matsumura, Y. Izumi and Y. Mizutani, *Bull. Chem. Soc. Jpn.*, 1974, **47**, 2922.
151. R. L. Augustine, *Adv. Catal.*, 1976, **25**, 56, and references cited therein.
152. R. L. Augustine, *Ann. N.Y. Acad. Sci.*, 1967, **145**, 19.
153. F. J. McQuillin, W. O. Ord and P. L. Simpson, *J. Chem. Soc.*, 1963, 5996.
154. (a) R. L. Augustine, *J. Org. Chem.*, 1958, **23**, 1853; (b) R. L. Augustine and A. D. Broom, *J. Org. Chem.*, 1960, **25**, 802; (c) R. L. Augustine, D. C. Migliorini, R. E. Foscante, C. S. Sodano and M. J. Sisbarro, *J. Org. Chem.*, 1969, **34**, 1075; (d) S. Nishimura, M. Shimahara and M. Shiota, *J. Org. Chem.*, 1966, **31**, 2394; (e) M. G. Combe, H. B. Henbest, W. R. Jackson, *J. Chem. Soc. C*, 1967, 2467; (f) H. B. Henbest, W. R. Jackson and I. Malunowicz, *J. Chem. Soc. C*, 1967, 2469; (g) I. Jardine, R. W. Howsam and F. J. McQuillin, *J. Chem. Soc. C*, 1969, 260; (h) H. J. E. Loewenthal, *Tetrahedron*, 1959, **6**, 269; (i) L. Velluz, J. Valls and G. Nominé, *Angew. Chem., Int. Ed. Engl.*, 1965, **4**, 181.
155. T. C. McKenzie, *J. Org. Chem.*, 1974, **39**, 629.
156. Z. G. Hajos and D. R. Parrish *J. Org. Chem.*, 1973, **38**, 3239.
157. (a) J. W. Suggs, S. D. Cox, R. H. Crabtree and J. M. Quirk, *Tetrahedron Lett.*, 1981, **22**, 303; (b) G. Stork and D. E. Kahne, *J. Am. Chem. Soc.*, 1983, **105**, 1072.
158. (a) T. Kametani, M. Tsubuki, H. Furuyama and T. Honda, *J. Chem. Soc., Perkin Trans. 1*, 1985, 557; (b) T. Kametani, M. Tsubuki, K. Higurashi and T. Honda, *J. Org. Chem.*, 1986, **51**, 2932.
159. E. Farnetti, M. Pesce, J. Kašpar, R. Spogliarich and M. Graziani, *J. Chem. Soc., Chem. Commun.*, 1986, 746.
160. (a) N. V. Borunova, L. K. Friedlin, L. I. Gvinter, T. Atabekov, V. A. Zamureenko and I. M. Kustanovich, *Izv. Akad. Nauk SSSR, Ser. Khim.*, 1972, **6**, 1299 (*Chem. Abstr.*, 1972, **77**, 87 461n); (b) L. K. Friedlin, L. I. Gvinter, N. V. Borunova, S. F. Dymova and I. M. Kustanovich, *Katal. Reakts. Zhidk. Faze, Tr. Vses. Konf., 3rd*, 1972, 309 (*Chem. Abstr.*, 1973, **79**, 115 066z).
161. P. C. Traas, H. Boelens and H. J. Takken, *Synth. Commun.*, 1976, **6**, 489.
162. (a) S. Nishimura and K. Tsuneda, *Bull. Chem. Soc. Jpn.*, 1969, **42**, 852; (b) S. Nishimura, T. Ichino, A. Akimoto and K. Tsuneda, *Bull. Chem. Soc. Jpn.*, 1973, **46**, 279; (c) S. Nishimura, T. Ichino, A. Akimoto, K. Tsuneda and H. Mori, *Bull. Chem. Soc. Jpn.*, 1975, **48**, 2852.
163. J. Azran, O. Buchman, I. Amer and J. Blum, *J. Mol. Catal.*, 1986, **34**, 229.
164. D. L. Reger, M. M. Habib and D. J. Fauth, *J. Org. Chem.*, 1980, **45**, 3860.

165. (a) S. Masamune, W. Choy, J. S. Petersen and L. R. Sita, *Angew Chem., Int. Ed. Engl.*, 1985, **24**, 1; (b) V. Čaplar, G. Comisso and V. Šunjić, *Synthesis*, 1981, 85; (c) J. W. ApSimon and R. P. Seguin, *Tetrahedron*, 1979, **35**, 2797.
166. (a) J. D. Morrison and W. F. Masler, *J. Org. Chem.*, 1974, **39**, 270; (b) F. Alario, Y. Amrani, Y. Colleuille, T.-P. Dang, J. Jenck, D. Morel and D. Sinou, *J. Chem. Soc., Chem. Commun.*, 1986, 202; (c) U. Nagel and E. Kinzel, *J. Chem. Soc., Chem. Commun.*, 1986, 1098; (d) D. P. Riley and R. E. Shumate, *J. Org. Chem.*, 1980, **45**, 5187; (e) I. Ojima and T. Suzuki, *Tetrahedron Lett.*, 1980, **21**, 1239; (f) I. Ojima, T. Kogure, N. Yoda, T. Suzuki, M. Yatabe and T. Tanaka, *J. Org. Chem.*, 1982, **47**, 1329; (g) B. D. Vineyard, W. S. Knowles, M. J. Sabacky, G. L. Bachman and D. J. Weinkauff, *J. Am. Chem. Soc.*, 1977, **99**, 5946; (h) H. B. Kagan and T.-P. Dang, *J. Am. Chem. Soc.*, 1972, **94**, 6429; (i) K. Achiwa, *J. Am. Chem. Soc.*, 1976, **98**, 8265; (j) K. Kashiwabara, K. Hanaki and J. Fujita, *Bull. Chem. Soc. Jpn.*, 1980, **53**, 2275; (k) J.-C. Poulin and H. B. Kagan, *J. Chem. Soc., Chem. Commun.*, 1982, 1261; (l) K.-i. Onuma, T. Ito and A. Nakamura, *Bull. Chem Soc. Jpn.*, 1980, **53**, 2012; (m) A. Miyashita, A. Yasuda, H. Takaya, K. Toriumi, T. Ito, T. Souchi and R. Noyori, *J. Am. Chem. Soc.*, 1980, **102**, 7932; (n) M. D. Fryzuk and B. Bosnich, *J. Am. Chem. Soc.*, 1977, **99**, 6262; (o) M. D. Fryzuk and B. Bosnich, *J. Am. Chem. Soc.*, 1979, **101**, 3043; (p) U. Nagel, *Angew. Chem., Int. Ed. Engl.*, 1984, **23**, 435; (q) S. Miyano, M. Nawa and H. Hashimoto, *Chem. Lett.*, 1980, 729; (r) U. Schmidt, A. Lieberknecht, H. Griesser and F. Bartkowiak, *Angew. Chem., Int. Ed. Engl.*, 1984, **23**, 318; (s) J. Köttner and G. Greber, *Chem. Ber.*, 1980, **113**, 2323; (t) T.-P. Dang and H. B. Kagan, *J. Chem. Soc., Chem. Commun.*, 1971, 481; (u) T. Minami, Y. Okada, R. Nomura, S. Hirota, Y. Nagahara and K. Fukuyama, *Chem. Lett.*, 1986, 613; (v) J. Bakos, I. Tóth and B. Heil, *Tetrahedron Lett.*, 1984, 4965; (w) J. D. Morrison, R. E. Burnett, A. M. Aguiar, C. J. Morrow and C. Philips, *J. Am. Chem. Soc.*, 1971, **93**, 1301; (x) M. Tamura and K. Harada, *Bull. Chem. Soc. Jpn.*, 1980, **53**, 561; (y) V. Massoneau, P. Le Maux and G. Simonneaux, *Tetrahedron Lett.*, 1986, **27**, 5497; (z) P. Le Maux and G. Simonneaux, *J. Organomet. Chem.*, 1983, **252**, C60.
167. J. H. Schauble, G. J. Walter and J. G. Morin, *J. Org. Chem.*, 1974, **39**, 755.
168. A. Hassner and C. H. Heathcock, *J. Org. Chem.*, 1964, **29**, 1350.
169. E. Schenker, in 'Newer Methods of Preparative Organic Chemistry', ed W. Foerst, Academic Press, New York, 1968, vol. 4, p. 196.
170. J. Bottin, O. Eisenstein, C. Minot and N. T. Anh, *Tetrahedron Lett.*, 1972, 3015.
171. M. R. Johnson and B. Rickborn, *J. Org. Chem.*, 1970, **35**, 1041.
172. S. Geribaldi, M. Decouzon, B. Boyer and C. Moreau, *J. Chem. Soc., Perkin Trans. 2*, 1986, 1327.
173. W. R. Jackson, and A. Zurqiyah, *J. Chem. Soc., Chem. Commun.*, 1965, 5280.
174. S. Kim, C. H. Oh, J. S. Ko, K. H. Ahn and Y. J. Kim, *J. Org. Chem.*, 1985, **50**, 1927, and references cited therein.
175. S. B. Kadin, *J. Org. Chem.*, 1966, **31**, 620.
176. B. Ganem and J. M. Fortunato, *J. Org. Chem.*, 1976, **41**, 2194.
177. S. Krishnamurthy and H. C. Brown, *J. Org. Chem.*, 1975, **40**, 1864.
178. G. C. Andrews and T. C. Crawford, *Tetrahedron Lett.*, 1980, **21**, 693.
179. A. R. Sande, M. H. Jagadale, R. B. Mane and M. M. Salunkhe, *Tetrahedron Lett.*, 1984, **25**, 3501.
180. M. Añez, G. Uribe, L. Mendoza and R. Contreras, *Synthesis*, 1981, 214.
181. Z. J. Duri and J. R. Hanson, *J. Chem. Soc., Perkin Trans.* 1, 1984, 363.
182. J. MacMillan and C. L. Willis, *J. Chem. Soc., Perkin Trans.* 1, 1984, 357.
183. (a) M. H. Beale, *J. Chem. Soc., Perkin Trans. 1*, 1985, 1151; (b) M. H. Beale, J. MacMillan, C. R. Spray, D. A. H. Taylor and B. O. Phinney, *J. Chem. Soc., Perkin Trans. 1*, 1984, 541.
184. B. Voigt and G. Adam, *Tetrahedron*, 1983, **39**, 449.
185. E. J. Corey, K. B. Becker and R. K. Varma, *J. Am. Chem. Soc.*, 1972, **94**, 8616.
186. R. Noyori and M. Suzuki, *Angew. Chem., Int. Ed. Engl.*, 1984, **23**, 847; J. A. Marshall and W. Y. Gung, *Tetrahedron Lett.*, 1988, **29**, 1657.
187. P. Joseph-Nathan, M. E. Garibay, and R. L. Santillan, *J. Org. Chem.*, 1987, **52**, 759.
188. W. Oppolzer and G. Poli, *Tetrahedron Lett.*, 1986, **27**, 4717.
189. (a) M. M. Midland, D. C. McDowell, R. L. Hatch and A. Tramontano, *J. Am. Chem. Soc.*, 1980, **102**, 867; (b) L. E. Overman, K. L. Bell and F. Ito, *J. Am. Chem. Soc.*, 1984, **106**, 4192.
190. A. R. Chamberlin and S. H. Reich, *J. Am. Chem. Soc.*, 1985, **107**, 1440.
191. R. O. Hutchins and D. Kandasamy, *J. Org. Chem.*, 1975, **40**, 2530.
192. S. Kim, Y. C. Moon and K. H. Ahn, *J. Org. Chem.*, 1982, **47**, 3311.
193. (a) J.-L. Luche, *J. Am. Chem. Soc.*, 1978, **100**, 2226; (b) J.-L. Luche and A. L. Gemal, *J. Am. Chem. Soc.*, 1979, **101**, 5848; (c) A. L. Gemal and J.-L. Luche, *J. Am. Chem. Soc.*, 1981, **103**, 5454.
194. C. W. Jefford, T. W. Wallace, N.-T. H. Can and C. G. Rimbault, *J. Org. Chem.*, 1979, **44**, 689.
195. (a) A. P. Kozikowski and P.-u. Park, *J. Org. Chem.*, 1984, **49**, 1674; (b) R. Zamboni and J. Rokach, *Tetrahedron Lett.*, 1982, **23**, 4751; (c) W. C. Still and V. J. Novack, *J. Am. Chem. Soc.*, 1984, **106**, 1148; (d) P. A. Grieco, J. Inanaga and N.-H. Lin, *J. Org. Chem.*, 1983, **48**, 892; (e) E. R. Larson and S. J. Danishefsky, *J. Am. Chem. Soc.*, 1983, **105**, 6715; (f) S. J. Danishefsky and C. Maring, *J. Am. Chem. Soc.*, 1985, **107**, 7762; (g) C.-Y. J. Tu and D. Lednicer, *J. Org. Chem.*, 1987, **52**, 5624; (h) S. J. Danishefsky and R. R. Webb, Jr., *J. Org. Chem.*, 1984, **49**, 1955; (i) P. A. Jacobi, C. S. R. Kaczmarek and U. E. Udodong, *Tetrahedron Lett.*, 1984, **25**, 4859.
196. A. L. Gemal and J.-L. Luche, *Tetrahedron Lett.*, 1981, **41**, 4077.
197. J.-L. Luche and A. L. Gemal, *J. Am. Chem. Soc.*, 1979, **101**, 5848.
198. G. A. Poulton and T. D. Cyr, *Synth. Commun.*, 1980, **10**, 581.
199. G. W. J. Fleet and P. J. C. Harding, *Tetrahedron Lett.*, 1981, **22**, 675.
200. (a) C. Kashima and Y. Yamamoto, *Chem. Lett.*, 1978, 1285; (b) C. Kashima, Y. Yamamoto and Y. Tsuda, *J. Org. Chem.*, 1975, **40**, 526.
201. (a) B. Myrboh, C. V. Asokan, H. Ila and H. Junjappa, *Synthesis*, 1984, 50; (b) C. V. Asokan, H. Ila and H. Junjappa, *Synthesis*, 1985, 163; (c) B. Myrboh, H. Ila and H. Junjappa, *J. Org. Chem.*, 1983, **48**, 5327; (d) R. K. Dieter, Y. J. Lin and J. W. Dieter, *J. Org. Chem.*, 1984, **49**, 3183.

202. (a) S. Carlsson and S.-O. Lawesson, *Tetrahedron*, 1982, **38**, 413; (b) A.-B. A. G. Ghattas, K. A. Jørgensen and S.-O. Lawesson, *Acta Chim. Scand., Ser. B*, 1982, **36**, 505.
203. (a) T. Nishio and Y. Omote, *Chem. Lett.*, 1979, 1223; (b) T. Nishio and Y. Omote, *J. Chem. Soc., Perkin Trans. 1*, 1981, 934.
204. J. Novák and C. A. Salemink, *J. Chem. Soc., Perkin Trans. 1*, 1982, 2403.
205. J. Málek, *Org. React. (N.Y.)*, 1985, **34**, 1.
206. H. C. Brown and P. M. Weissman, *J. Am. Chem. Soc.*, 1965, **87**, 5614.
207. (a) H. C. Brown, S.-C. Kim and S. Krishnamurthy, *J. Org. Chem.*, 1980, **45**, 1; (b) H. C. Brown, P. K. Jadhav and A. K. Mandal, *Tetrahedron*, 1981, **37**, 3547; (c) M. Fieser and L. F. Fieser, 'Reagents for Organic Synthesis', Wiley-Interscience, New York, 1967–1988, vols. 1–13; (d) S.-I. Yamada and K. Koga, in 'Selective Organic Transformations', ed. B. S. Thyagarajan, Wiley-Interscience, New York, vol. 1, 1970; (e) B. D. James, *Rec. Chem. Prog.*, 1970, **31**, 199; (f) J. Vit, *Org. Chem. Bull.*, 1970, **42**, 1 (*Chem. Abstr.*, 1971, **74**, 99 073p); (g) D. M. S. Wheeler and M. M. Wheeler, in 'Organic Reactions in Steroid Chemistry', ed. J. Fried and J. A. Edwards, Van Nostrand-Reinhold, New York, 1972, vol. 1, chap. 2; (h) J. Málek and M. Černý, *Synthesis*, 1972, 217; (i) A. S. Kushner and T. Vaccariello, *J. Chem. Educ.*, 1973, **50**, 154, 157; (j) H. Mishima, *Yuki Gosei Kagaku Kyokai Shi*, 1974, **32**, 1014 (*Chem. Abstr.*, 1975, **82**, 138 613b); (k) C. F. Lane, *Chem. Rev.*, 1976, **76**, 773; (l) C. F. Lane, in 'Aspects of Mechanistic Organometallic Chemistry' (Proceedings of Symposium), ed. J. H. Brewster, Plenum Press, New York, 1978, p. 181; (m) J. R. Boone and E. C. Ashby, *Top. Stereochem.*, 1979, **11**, 53; (n) P. A. Bartlett, *Tetrahedron*, 1980, **36**, 2; (o) S. O. Kim, *Hwakhak Kwa Kongop Ui Chinbo*, 1980, **20**, 293 (*Chem. Abstr.*, 1981, **94**, 102 222g).
208. Ref. 5, chap. 2.
209. E. R. H. Walker, *Chem. Soc. Rev.*, 1976, **5**, 23.
210. A. Hajós, 'Komplexe Hydride', VEB Deutscher Verlag der Wissenschaften, East Berlin, 1966; A. Hajós, 'Complex Hydrides and Related Reducing Agents in Organic Synthesis', Elsevier Amsterdam, 1979.
211. H. C. Brown, in 'Boranes in Organic Chemistry', Cornell University Press, Ithaca, NY, 1972, chaps. 12 and 13.
212. (a) D. R. Boyd and M. A. McKervey, *Q. Rev., Chem. Soc.*, 1968, **22**, 95; (b) J. Mathieu and J. Weill-Raynal, *Bull. Soc. Chim. Fr.*, 1968, 1211; (c) T. D. Inch, *Synthesis*, 1970, 466; (d) J. D. Morrison and H. S. Mosher, 'Asymmetric Organic Reactions', Prentice Hall, Englewood Cliffs, NJ, 1971, pp. 116, 202, 386 (reprint edn., American Chemical Society, Washinton, DC, 1976); (e) H. J. Schneider and R. Haller, *Pharmazie*, 1973, **28**, 417; (f) J. W. Scott and D. Valentine, Jr., *Science (Washington, D.C.)*, 1974, 184, 943; (g) D. Valentine, Jr. and J. W. Scott, *Synthesis*, 1978, 329; (h) J. W. ApSimon and R. P. Seguin, *Tetrahedron*, 1979, **35**, 2797.
213. O. Eisenstein, J. M. Lefour, C. Minot, N. T. Anh and G. Soussan, *C.R. Hebd. Seances Acad. Sci., Ser. C*, 1972, **274**, 1310.
214. J. Durand, N. T. Anh and J. Huet, *Tetrahedron Lett.*, 1974, 2397.
215. J. Bottin, O. Eisenstein, C. Minot and N. T. Anh, *Tetrahedron Lett.*, 1972, 3015.
216. R. G. Pearson, *J. Chem. Educ.*, 1968, **45**, 581.
217. J. Seyden-Penne, *Bull. Soc. Chim. Fr.*, 1968, 3871.
218. H. C. Brown and H. M. Hess, *J. Org. Chem.*, 1969, **34**, 2206.
219. A. Loupy and J. Seyden-Penne, *Tetrahedron*, 1980, **36**, 1937.
220. (a) J.-C. Richer and A. Rossi, *Can. J. Chem.*, 1972, **50**, 438; (b) J. A. Marshall and J. A. Ruth, *J. Org. Chem.*, 1974, **39**, 1971.
221. (a) R. B. Gammill, L. T. Bell and S. A. Nash, *J. Org. Chem.*, 1984, **49**, 3039; (b) R. B. Gammill, D. M. Sobieray and P. M. Gold, *J. Org. Chem.*, 1981, **46**, 3555.
222. M. E. Cain, *J. Chem. Soc.*, 1964, 3532.
223. M. F. Semmelhack, R. D. Stauffer and A. Yamashita, *J. Org. Chem.*, 1977, **42**, 3180.
224. J. E. Baldwin, R. C. Thomas, L. I. Kruse and L. Silberman, *J. Org. Chem.*, 1977, **42**, 3846.
225. (a) P. L. Southwick, N. Latif, B. M. Fitzgerald and N. M. Zaczek, *J. Org. Chem.*, 1966, **31**, 1; (b) J. Durand and J. Huet, *Bull. Soc. Chim. Fr.*, 1978, 428.
226. P. A. Bartlett and W. S. Johnson, *J. Am. Chem. Soc.*, 1973, **95**, 7501.
227. G. D. Prestwich, F. B. Whitfield and G. Stanley, *Tetrahedron*, 1976, **32**, 2945.
228. (a) R. L. Markezich, W. E. Willy, B. E. McCarry and W. S. Johnson, *J. Am. Chem. Soc.*, 1973, **95**, 4414; (b) W. S. Johnson, B. E. McCarry, R. L. Markezich and S. G. Boots, *J. Am. Chem. Soc.*, 1980, **102**, 352.
229. P. C. Traas, H. Boelens and H. J. Takken, *Recl. Trav. Chim. Pays-Bas*, 1976, **95**, 57.
230. K. E. Wilson, R. T. Seidner and S. Masamune, *J. Chem. Soc., Chem. Commun.*, 1970, 213.
231. D. V. Banthorpe, A. J. Curtis and W. D. Fordham, *Tetrahedron Lett.*, 1972, 3865.
232. F. M. Houser, W. P. Ellenberger and T. C. Adams, Jr., *J. Org. Chem.*, 1984, **49**, 1169.
233. R. E. Ireland and P. Maienfisch, *J. Org. Chem.*, 1988, **53**, 640.
234. N. Lander and R. Mechoulam, *J. Chem. Soc., Perkin Trans. 1*, 1976, 484.
235. R. A. Finnegan and P. L. Bachman, *J. Org. Chem.*, 1965, **30**, 4145.
236. D. Caine, P.-C. Chen, A. S. Frobese and J. T. Gupton, *J. Org. Chem.*, 1979, **44**, 4981.
237. R. E. Ireland, M. I. Dawson, S. C. Welch, A. Hagenbach, J. Bordner and B. Trus, *J. Am. Chem. Soc.*, 1973, **95**, 7829.
238. (a) S. J. Danishefsky, J. F. Kerwin, Jr., and S. Kobayashi, *J. Am. Chem. Soc.*, 1982, **104**, 358; (b) S. J. Danishefsky, J. Morris, G. Mullen and R. B. Gammill, *J. Am. Chem. Soc.*, 1980, **102**, 2838.
239. E. Winterfeldt, *Synthesis*, 1975, 617.
240. E. C. Ashby and J. J. Lin, *Tetrahedron Lett.*, 1976, 3865.
241. H. C. Brown, *U.S. NTIS, AD Rep.*, 1976, AD-A026132 (*Chem. Abstr.*, 1976, **85**, 176 353m).
242. P. G. Williard and L. A. Grab, *Tetrahedron Lett.*, 1984, **25**, 5009.
243. H. J. Williams, *J. Chem. Soc., Perkin Trans. 1*, 1973, 1852.
244. W. L. Dilling and R. A. Plepys, *J. Chem. Soc., Chem. Commun.*, 1969, 417.
245. W. L. Dilling and R. A. Plepys, *J. Org. Chem.*, 1970, **35**, 1971.
246. J.-P. Bugel, P. Ducos, O. Gringore and F. Rouessac, *Bull. Soc. Chim. Fr.*, 1972, 4371.

247. P. R. Story and S. R. Fahrenholtz, *J. Am. Chem. Soc.*, 1965, **87**, 1623.
248. J.-B. Wiel and F. Rouessac, *J. Chem. Soc., Chem. Commun.*, 1976, 446.
249. J.-B. Wiel and F. Rouessac, *Bull. Soc. Chim. Fr.*, 1979, 273.
250. E. J. Corey and J. Gorzynski Smith, *J. Am. Chem. Soc.*, 1979, **101**, 1038.
251. M. F. Semmelhack and R. D. Stauffer, *J. Org. Chem.*, 1975, **40**, 3619.
252. M. Vandewalle and E. Madeleyn, *Tetrahedron*, 1970, **26**, 3551.
253. C. J. Sih, R. G. Salomon, P. Price, R. Sood and G. P. Peruzzotti, *J. Am. Chem. Soc.*, 1975, **97**, 857.
254. M. Suzuki, T. Kawagishi, T. Suzuki and R. Noyori, *Tetrahedron Lett.*, 1982, **23**, 4057.
255. P. A. Grieco, N. Fukamiya and M. Miyashita, *J. Chem. Soc., Chem. Commun.*, 1976, 573.
256. (a) Z. G. Hajos, D. R. Parrish and E. P. Oliveto, *Tetrahedron Lett.*, 1966, 6495; (b) Z. G. Hajos, D. R. Parrish and E. P. Oliveto, *Tetrahedron*, 1968, **24**, 2039.
257. G. Saucy, R. Borer and A. Fürst, *Helv. Chim. Acta*, 1971, **54**, 2034.
258. G. Saucy and R. Borer, *Helv. Chim. Acta*, 1971, **54**, 2121.
259. E. Fujita, T. Fujita and Y. Nagao, *Tetrahedron*, 1969, **25**, 3717.
260. R. E. Ireland and D. M. Walba, *Tetrahedron Lett.*, 1976, 1071.
261. K. F. Cohen, R. Kazlauskas and J. T. Pinhey, *J. Chem. Soc., Perkin Trans. 1*, 1973, 2076.
262. G. Stork, G. A. Kraus and G. A. Garcia, *J. Org. Chem.*, 1974, **39**, 3459.
263. G. Stork and G. A. Kraus, *J. Am. Chem. Soc.*, 1976, **98**, 2351.
264. (a) R. Pappo and P. W. Collins, *Tetrahedron Lett.*, 1972, 2627; (b) R. Pappo and C. J. Jung, *Ger. Pat.* 2 321 984 (1973) (*Chem. Abstr.* 1974, **80**, 26 827b); (c) M. M. S. Bruhn and R. Pappo, *Ger. Pat.* 2 415 765 (1974) (*Chem. Abstr.*, 1975, **82**, 86 119y); (d) R. Pappo and C. J. Jung, *US Pat.* 3 969 391 (1976) (*Chem. Abstr.*, 1977, **86**, 55 057e); (e) C. J. Sih, J. B. Heather, G. P. Peruzzotti, P. Price, R. Sood and L.-F. Hsu Lee, *J. Am. Chem. Soc.*, 1973, **95**, 1976; (f) C. J. Sih, J. B. Heather, R. Sood, P. Price, G. P. Peruzzotti, L.-F. Hsu Lee and S. S. Lee, *J. Am. Chem. Soc.*, 1975, **97**, 865; (g) C. J. Sih and J. B. Heather, *US Pat.* 3 968 141 (1976) (*Chem. Abstr.*, 1977, **86**, 29 416b).
265. (a) C. J. Sih, J. B. Heather, G. P. Peruzzotti, P. Price, R. Sood and L. F. Hsu Lee, *J. Am. Chem. Soc.*, 1973, **95**, 1676; (b) C. J. Sih, J. B. Heather, R. Sood, P. Price, G. P. Peruzzotti, L. F. Hsu Lee and S. S. Lee, *J. Am. Chem. Soc.*, 1975, **97**, 865.
266. Y. Asaka, T. Kamikawa and T. Kubota, *Tetrahedron Lett.*, 1972, 1597.
267. V. Bazant, M. Čapka, M. Černý, V. Chvalovský, K. Kochloefl, M. Kraus and J. Málek, *Tetrahedron Lett.*, 1968, 3303.
268. M. Čapka, V. Chvalovský, K. Kochloefl and M. Kraus, *Collect. Czech. Chem. Commun.*, 1969, **34**, 118.
269. H. C. Brown and N. M. Yoon, *J. Am. Chem. Soc.*, 1966, **88**, 1464.
270. H. C. Brown, *U. S. C. F. S. T. I., AD Rep.*, 1966, AD 645 581 (*Chem. Abstr.*, 1967, **67**, 99 306x).
271. H. C. Brown and P. M. Weissman, *Isr. J. Chem.*, 1963, **1**, 430.
272. O. Kriz, J. Machacek and O. Strouf, *Collect. Czech. Chem. Commun.*, 1973, **38**, 2072.
273. J. V. Frosch, I. T. Harrison, B. Lythgoe and A. K. Saksena, *J. Chem. Soc., Perkin Trans. 1*, 1974, 2005.
274. W. Sucrow, *Tetrahedron Lett.*, 1970, 4725.
275. G. Büchi, B. Gubler, R. S. Schneider and J. Wild, *J. Am. Chem. Soc.*, 1967, **89**, 2776.
276. S. Antus, A. Gottsegen and M. Nógrádi, *Synthesis*, 1981, 574.
277. (a) S. J. Danishefsky, J. F. Kerwin, Jr. and S. Kobayashi, *J. Am. Chem. Soc.*, 1982, **104**, 358; (b) S. J. Danishefsky, N. Kato, D. Askin and J. F. Kerwin, Jr., *J. Am. Chem. Soc.*, 1982, **104**, 360.
278. S. Kim and K. H. Ahn, *J. Org. Chem.*, 1984, **49**, 1749.
279. B. M. Trost and L. N. Jungheim, *J. Am. Chem. Soc.*, 1980, **102**, 7910.
280. H. J. Williams, *Tetrahedron Lett.*, 1975, 1271.
281. (a) M. M. Bokadia, B. R. Brown, D. Cobern, A. Roberts and G. A. Somerfield, *J. Chem. Soc.*, 1962, 1658; (b) J. Broome, B. R. Brown, A. Roberts and M. S. A. White, *J. Chem. Soc.*, 1960, 1406.
282. (a) S. R. Wilson and S. M. Haque, *Tetrahedron Lett.*, 1984, **25**, 3151; (b) J. K. Sutherland and G. B. Tometzki, *Tetrahedron Lett.*, 1984, **25**, 881.
283. (a) E. C. Ashby and J. J. Lin, *Tetrahedron Lett.*, 1975, 4453; (b) E. C. Ashby, J. J. Lin and R. Kovar, *J. Org. Chem.*, 1976, **41**, 1939.
284. S. Iguchi, H. Nakai, M. Hayashi, H. Yamamoto and K. Maruoka, *Bull. Chem. Soc. Jpn.*, 1981, **54**, 3033.
285. G. Quinkert, N. Heim, J. Glenneberg, U.-M. Billhardt, V. Autze, J. W. Bats and G. Durner, *Angew. Chem. Int. Ed. Engl.*, 1987, **26**, 362.
286. (a) O. Červinka, O. Křiž and J. Červenka, *Z. Chem.*, 1971, **11**, 109; (b) O. Červinka and O. Kříž, *Collect. Czech. Chem. Commun.*, 1973, **38**, 294.
287. R. Noyori, I. Tomino and M. Nishizawa, *J. Am. Chem. Soc.*, 1979, **101**, 5843.
288. S. Terashima, N. Tanno and K. Koga, *J. Chem. Soc., Chem. Commun.*, 1980, 1026.
289. (a) S. Terashima, N. Tanno and K. Koga, *Tetrahedron Lett.*, 1980, **21**, 2753; (b) S. Terashima, N. Tanno and K. Koga, *Chem. Lett.*, 1980, 981.
290. J. Huton, M. Senior and N. C. A. Wright, *Synth. Commun.*, 1979, **9**, 799.
291. N. Cohen, R. J. Lopresti, C. Neukom and G. Saucy, *J. Org. Chem.*, 1980, **45**, 582.
292. R. S. Brinkmeyer and V. M. Kapoor, *J. Am. Chem. Soc.*, 1977, **99**, 8339.
293. W. S. Johnson, R. S. Brinkmeyer, V. M. Kapoor and T. M. Yarnell, *J. Am. Chem. Soc.*, 1977, **99**, 8341.
294. J.-P. Vigneron and V. Bloy, *Tetrahedron Lett.*, 1979, 2683.
295. J.-P. Vigneron and V. Bloy, *Tetrahedron Lett.*, 1980, **21**, 1735.
296. J.-P. Vigneron and J. M. Blanchard, *Tetrahedron Lett.*, 1980, **21**, 1739.
297. M. Nishizawa, M. Yamada and R. Noyori, *Tetrahedron Lett.*, 1981, **22**, 247.
298. S. R. Landor, B. J. Miller and A. R. Tatchell, *J. Chem. Soc. C*, 1966, 1822.
299. S. R. Landor, B. J. Miller and A. R. Tatchell, *Proc. Chem. Soc., London*, 1964, 227.
300. S. R. Landor, B. J. Miller and A. R. Tatchell, *J. Chem. Soc. C*, 1971, 2339.
301. M. Nishizawa and R. Noyori, *Tetrahedron Lett.*, 1980, **21**, 2821.
302. T. Sato, Y. Gotoh, Y. Wakabayashi and T. Fujisawa, *Tetrahedron Lett.*, 1983, **24**, 4123.

303. (a) R. Noyori, I. Tomino and M. Nishizawa, *J. Am. Chem. Soc.*, 1979, **101**, 3843; (b) R. Noyori, I. Tomino, M. Yamada and M. Nishizawa, *J. Am. Chem. Soc.*, 1984, **106**, 6717; (c) R. Noyori, *Pure Appl. Chem.*, 1981, **53**, 2315.
304. C. J. Sih, J. B. Heather, R. Sood, P. Price, G. P. Peruzzotti, L. F. Hsu Lee and S. S. Lee, *J. Am. Chem. Soc.*, 1975, **97**, 865.
305. Y. Nagai, *Intra-Sci. Chem. Rep.*, 1970, **4**, 115.
306. D. N. Kursanov, Z. N. Parnes and N. M. Loim, *Synthesis*, 1974, 633.
307. Z. N. Parnes, N. M. Loim, V. A. Baranova and D. N. Kursanov, *Zh. Org. Khim.*, 1971, **7**, 2066 (*Chem. Abstr.*, 1972, **76**, 13 495).
308. D. N. Kursanov, N. M. Loim, V. A. Baranova, L. V. Moiseeva, L. P. Zalukaev and Z. N. Parnes, *Synthesis*, 1973, 420.
309. E. Yoshii, T. Koizumi, I. Hayashi and Y. Hiroi, *Chem. Pharm. Bull.*, 1977, **25**, 1468.
310. G. G. Furin, O. A. Vyazankina, B. A. Gostevsky and N. S. Vyazankin, *Tetrahedron*, 1988, **44**, 2675.
311. R. J. P. Corriu, R. Perz and C. Reyé, *Tetrahedron*, 1983, **39**, 999.
312. M. P. Doyle, C. T. West, S. J. Donnelly and C. C. McOsker, *J. Organomet. Chem.*, 1976, **117**, 129.
313. M. Kira, K. Sato and H. Sakurai, *J. Org. Chem.*, 1987, **52**, 948.
314. R. A. Benkeser, *Acc. Chem. Res.*, 1971, **4**, 94.
315. (a) H. G. Kuivila, *Synthesis*, 1970, 499; (b) A. Hajós, 'Complex Hydrides and Related Reducing Agents in Organic Synthesis', Elsevier, Amsterdam, 1979; (c) Y. I. Baukov and I. F. Lutsenko, *Organometal. Chem. Rev., Sect. A*, 1970, **6**, 355.
316. (a) H. G. Kuivila and O. F. Beumel, *J. Am. Chem. Soc.*, 1961, **83**, 1246; (b) H. G. Kuivila and O. F. Beumel, *J. Am. Chem. Soc.*, 1958, **80**, 3798; (c) G. J. M. Van Der Kerk, J. G. A. Luijten and J. G. Noltes, *Chem. Ind. (London)*, 1956, 352; (d) G. J. M. Van Der Kerk, J. G. Noltes and J. G. A. Luijten, *J. Appl. Chem.*, 1957, **7**, 356; (e) G. J. M. Van Der Kerk and J. G. Noltes, *J. Appl. Chem.*, 1959, **9**, 106; (f) J. G. Noltes and G. J. M. Van Der Kerk, *Chem. Ind. (London)*, 1959, 294; (g) I. F. Lutsenko, S. V. Ponomarev and O. P. Petri, *Zh. Obshch. Khim.*, 1962, **32**, 896; (h) M. Pereyre and J. Valade, *C. R. Hebd. Seances Acad. Sci.*, 1964, **258**, 4785; (i) M. Pereyre and J. Valade, *C. R. Hebd. Seances Acad. Sci.*, 1965, **260**, 581; (j) M. Pereyre and J. Valade, *Bull. Soc. Chim. Fr.*, 1967, 1928; (k) M. Pereyre, G. Colin and J. Valade, *Tetrahedron Lett.*, 1967, 4805; (l) A. J. Leusink and J. G. Noltes, *Tetrahedron Lett.*, 1966, 2221.
317. M. Pereyre and J. Valade, *Tetrahedron Lett.*, 1969, 489.
318. H. Laurent, P. Esperling and G. Baude, *Liebigs Ann. Chem.*, 1983, 1996.
319. (a) B. R. Laliberte, W. Davidsohn and M. C. Henry, *J. Organomet. Chem.*, 1966, **5**, 526; (b) A. J. Leusink and J. G. Noltes, *J. Organomet. Chem.*, 1969, **16**, 91; (c) W. P. Neumann, H. Niermann and R. Sommer, *Justus Liebigs Ann. Chem.*, 1962, **659**, 27; (d) M. Pereyre, G. Colin and J. Valade, *Bull. Soc. Chim. Fr.*, 1968, 3358; (e) S. Matsuda, S. Kikawa and I. Omae, *J. Organomet. Chem.*, 1969, **18**, 95.
320. (a) A. J. Leusink and J. G. Noltes, *Tetrahedron Lett.*, 1966, 335; (b) W. P. Neumann and R. Sommer, *Justus Liebigs Ann. Chem.*, 1964, **675**, 10.
321. G. H. Posner, *Org. React. (N. Y.)*, 1972, **19**, 1.
322. (a) M. F. Semmelhack and R. D. Stauffer, *J. Org. Chem.*, 1975, **40**, 3619; (b) M. F. Semmelhack, R. D. Stauffer and A. Yamashita, *J. Org. Chem.*, 1977, **42**, 3180.
323. M. E. Osborn, J. F. Pegues and L. A. Paquette, *J. Org. Chem.*, 1980, **45**, 167.
324. T. Saegusa, K. Kawasaki, T. Fujii and T. Tsuda, *J. Chem. Soc., Chem. Commun.*, 1980, 1013.
325. T. Tsuda, T. Hayashi, H. Satomi, T. Kamamoto and T. Saegusa, *J. Org. Chem.*, 1986, **51**, 537.
326. R. K. Boeckman, Jr. and R. Michalak, *J. Am. Chem. Soc.*, 1974, **96**, 1623.
327. S. Masamune, G. S. Bates and P. E. Georghiou, *J. Am. Chem. Soc.*, 1974, **96**, 3686.
328. E. C. Ashby, J. J. Lin and A. B. Goel, *J. Org. Chem.*, 1978, **43**, 183.
329. T. H. Lemmen, K. Folting, J. C. Huffman and K. G. Caulton, *J. Am. Chem. Soc.*, 1985, **107**, 7774.
330. W. S. Mahoney, D. M. Brestensky and J. M. Stryker, *J. Am. Chem. Soc.*, 1988, **110**, 291.
331. (a) G. F. Cainelli, M. Panunzio and A. Umani-Ronchi, *J. Chem. Soc., Perkin Trans. 1*, 1975, 1273; (b) G. F. Cainelli, M. Panunzio and A. Umani-Ronchi, *Tetrahedron Lett.*, 1973, 2491.
332. M. Yamashita, K. Miyoshi, Y. Okada and R. Suemitsu, *Bull. Chem. Soc. Jpn.*, 1982, **55**, 1329.
333. G. P. Boldrini and A. Umani-Ronchi *J. Organomet. Chem.*, 1979, **171**, 85.
334. (a) J. P. Collman, R. G. Finke, P. L. Matlock, R. Wahren and J. I. Brauman, *J. Am. Chem. Soc.*, 1976, **98**, 4685; (b) J. P. Collman, R. G. Finke, P. L. Matlock, R. Wahren, R. G. Komoto and J. I. Brauman, *J. Am. Chem. Soc.*, 1978, **100**, 1119.
335. T. Imamoto, T. Mita and M. Yokomoto, *J. Chem. Soc., Chem. Commun.*, 1984, 163.
336. G. P. Boldrini and A. Umani-Ronchi, *Synthesis*, 1976, 596.
337. R. W. Goetz and M. Orchin, *J. Am. Chem. Soc.*, 1963, **85**, 2782.
338. D. H. Gibson and Y. S. El-Omrani, *Organometallics*, 1985, **4**, 1473.
339. (a) R. A. W. Johnstone, A. H. Wilby and I. D. Entwistle, *Chem. Rev.*, 1985, **85**, 129; (b) G. Brieger and T. J. Nestrick, *Chem. Rev.*, 1974, **74**, 567; (c) G. W. Parshall, *Catal. Rev.*, 1981, **23**, 107.
340. Y. Sasson and J. Blum, *J. Org. Chem.*, 1975, **40**, 1887.
341. Y. Sasson and J. Blum, *Tetrahedron Lett.*, 1971, 2167.
342. V. Z. Sharf, L. Kh. Freidlin, B. M. Shekoyan and V. N. Krutii, *Izv. Akad. Nauk SSSR, Ser. Khim.*, 1976, 575; 1977, 834 (*Bull. Acad. Sci. USSR, Div. Chem. Sci. (Engl. Transl.)*, 1976, **25**, 557; 1977, **26**, 758).
343. Y. Sasson, M. Cohen and J. Blum, *Synthesis*, 1973, 359.
344. G. Descotes and J. Sabadie, *Bull. Soc. Chim. Fr.*, 1978, 158.
345. G. Speier and L. Markó, *J. Organomet. Chem.*, 1981, **210**, 253.
346. S. L. Regen and G. M. Whitesides, *J. Org. Chem.*, 1972, **37**, 1832.
347. Y. Sasson, J. Blum and E. Dunkelblum, *Tetrahedron Lett.*, 1973, 3199.
348. Y. Sasson and G. L. Rempel, *Can. J. Chem.*, 1974, **52**, 3825.
349. T. Nishiguchi, H. Imai, Y. Hirose and K. Fukuzumi, *J. Catal.*, 1976, **41**, 249.
350. A. Dobson, D. S. Moore and S. D. Robinson, *J. Organomet. Chem.*, 1979, **177**, C8.

351. B. T. Khai and A. Arcelli, *Tetrahedron Lett.*, 1985, **26**, 3365.
352. M. Bianchi, U. Matteoli, G. Menchi, P. Frediani, F. Piacenti and C. Botteghi, *J. Organomet. Chem.*, 1980, **195**, 337.
353. K. Ohkubo, I. Terada and K. Yoshinaga, *Inorg. Nucl. Chem. Lett.*, 1979, **15**, 421.
354. (a) K. Ohkubo, K. Hirata, K. Yoshinaga and M. Okada, *Chem. Lett.*, 1976, 183; (b) K. Ohkubo, K. Hirata and K. Yoshinaga, *Chem. Lett.*, 1976, 577; (c) K. Ohkubo, T. Shoji, I. Terada and K. Yoshinaga, *Inorg. Nucl. Chem. Lett.*, 1977, **13**, 443.
355. G. Descotes and D. Sinou, *Tetrahedron Lett.*, 1976, 4083.
356. G. Descotes, J. P. Praly and D. Sinou, *J. Mol. Catal.*, 1979, **6**, 421.
357. (a) D. Beaupere, P. Bauer and R. Uzan, *Can. J. Chem.*, 1979, **57**, 218; (b) D. Beaupere, L. Nadjo, R. Uzan and P. Bauer, *J. Mol. Catal.*, 1982, **14**, 129; (c) D. Beaupere, P. Bauer, L. Nadjo and R. Uzan, *J. Organomet. Chem.*, 1982, **231**, C49; (d) D. Beaupere, P. Bauer, L. Nadjo and R. Uzan, *J. Mol. Catal.*, 1983, **18**, 73.
358. D. Beaupére, L. Nadjo, R. Uzan and P. Bauer, *J. Mol. Catal.*, 1983, **20**, 185, 195.
359. A. Camus, G. Mestroni and G. Zassinovich, *J. Organomet. Chem.*, 1980, **184**, C10.
360. (a) J. L. Namy, J. Souppe, J. Collin and H. B. Kagan, *J. Org. Chem.*, 1984, **49**, 2045; (b) P. Girrard, J. L. Namy and H. B. Kagan, *J. Am. Chem. Soc.*, 1980, **102**, 2693.
361. E. Keinan and P. A. Gleize, *Tetrahedron Lett.*, 1982, **23**, 477.
362. (a) P. Four and F. Guibe, *Tetrahedron Lett.*, 1982, **23**, 1825; (b) T. X. Yang, P. Four, F. Guibe and G. Balavoine, *Nouv. J. Chim.*, 1984, **8**, 611.
363. (a) E. Keinan and N. Greenspoon, *J. Org. Chem.*, 1983, **48**, 3545; (b) E. Keinan and N. Greenspoon, *Isr. J. Chem.*, 1984, **24**, 82; (c) N. Greenspoon and E. Keinan, *J. Org. Chem.*, 1988, **53**, 3723.
364. (a) E. Keinan and N. Greenspoon, *J. Am. Chem. Soc.*, 1986, **108**, 7314; (b) E. Keinan and N. Greenspoon, *Tetrahedron Lett.*, 1985, **26**, 1353.
365. N. Godinger, M. Sc. Thesis, Weizmann Institute of Science, January 1986.
366. (a) T. Tatsumi, M. Shibasaki and H. Tominaga, *J. Mol. Catal.*, 1981, **13**, 331; (b) T. Tatsumi, K. Hashimoto, H. Tominaga, Y. Mizuta, K. Hata, M. Hidai and Y. Uchida, *J. Organomet. Chem.*, 1983, **252**, 105; (c) Y. Lin and X. Lu, *J. Organomet. Chem.*, 1983, **251**, 321.
367. (a) L. Markó and Z. Nagy-Magos, *J. Organomet. Chem.*, 1985, **285**, 193; (b) E. N. Frankel, *J. Org. Chem.*, 1972, **37**, 1549; (c) M. Sodeoka and M. Shibasaki, *J. Org. Chem.*, 1985, **50**, 1147.
368. E. Keinan and D. Perez, *J. Org. Chem.*, 1987, **52**, 2576.
369. D. Perez, N. Greenspoon and E. Keinan, *J. Org. Chem.*, 1987, **52**, 5570.
370. I. Ojima, T. Kogure and Y. Nagai, *Tetrahedron Lett.*, 1972, 5035; (b) I. Ojima and T. Kogure, *Organometallics*, 1982, **1**, 1390; (c) I. Ojima, M. Nihonyanagi, T. Kogure, M. Kumagai, S. Horiuchi and K. Nakatsugawa, *J. Organomet. Chem.*, 1975, **94**, 449.
371. H.-J. Liu and B. Ramani, *Synth. Commun.*, 1985, **15**, 965.
372. A. J. Cornish, M. F. Lappert, G. L. Filatovos and T. A. Nile, *J. Organomet. Chem.*, 1979, **172**, 153.
373. T. Hayashi, K. Yamamoto and M. Kumada, *Tetrahedron Lett.*, 1975, 3.
374. (a) T. Kogure and I. Ojima, *J. Organomet. Chem.*, 1982, **234**, 249; (b) I. Ojima and T. Kogure, *Chem. Lett.*, 1975, 985.
375. M. Kobayashi, T. Koyama, K. Ogura, S. Seto, F. J. Ritter and I. E. M. Brüggemann-Rotgans, *J. Am. Chem. Soc.*, 1980, **102**, 6602.
376. (a) D. L. Bailey, *US Pat.* 2 917 530 (1959) (*Chem. Abstr.*, 1960, **54**, 6549); *US Pat.* 2 970 150 (1961) (*Chem. Abstr.*, 1961, **55**, 16423); (b) E. Lukevics, *Latv. PSR Zinat. Akad. Vestis*, 1963, 111; (c) A. D. Petrov and S. I. Sadykh-Zade, *Dokl. Akad. Nauk SSSR*, 1959, **121**, 119; (d) A. D. Petrov, V. F. Mironov, V. A. Ponomarenko, S. I. Sadykh-Zade and E. A. Chernyshev, *Izv. Akad. Nauk SSSR*, 1968, 954; (e) S. I. Sadykh-Zade and A. D. Petrov, *Zh. Obshch. Khim.*, 1959, **29**, 3194.
377. (a) E. Frainnet, *Pure Appl. Chem.*, 1969, **19**, 489; (b) E. Frainnet and R. Bourhis, *Bull. Soc. Chim. Fr.*, 1966, 2134; (c) R. Bourhis, E. Frainnet and F. Moulines, *J. Organomet. Chem.*, 1977, **141**, 157.
378. (a) A. D. Pétrov and S. I. Sadikh-Sadé, *Bull. Soc. Chim. Fr.*, 1959, 1932; (b) A. D. Petrov, S. I. Sadikh-Sadé and E. I. Filatova, *Zh. Obshch. Khim.*, 1959, **29**, 2936.
379. G. Stork and T. L. Macdonald, *J. Am. Chem. Soc.*, 1975, **97**, 1264.
380. E. Yoshii, H. Ikeshima and K. Ozaki, *Chem. Pharm. Bull.*, 1972, **20**, 1827.
381. E. Yoshii, Y. Kobayashi, T. Koizumi and T. Oribe, *Chem. Pharm. Bull.*, 1974, **22**, 2767.
382. A. P. Barlow, N. M. Boag and F. G. A. Stone, *J. Organomet. Chem.*, 1980, **191**, 39.
383. J. Blum, Y. Sasson and S. Iflah, *Tetrahedron Lett.*, 1972, 1015.
384. H. Imai, T. Nishiguch and K. Fukuzumi, *Chem. Lett.*, 1976, 655.
385. (a) M. E. Vol'pin, V. P. Kukolev, V. O. Chernyshev and I. S. Kolomnikov, *Tetrahedron Lett.*, 1971, 4435; (b) I. S. Kolomnikov, Y. D. Koreshov, V. P. Kukolev, V. A. Mosin and M. E. Vol'pin, *Izv. Akad. Nauk SSSR, Ser. Khim.*, 1973, 175 (*Bull. Acad. Sci. USSR, Div. Chem. Sci. (Engl. Transl.)*, 1973, **22**, 180).
386. N. A. Cortese and R. F. Heck, *J. Org. Chem.*, 1978, **43**, 3985.
387. N. A. Cortese and R. F. Heck, *J. Org. Chem.*, 1977, **42**, 3491.
388. J. Tsuji and T. Yamakawa, *Tetrahedron Lett.*, 1979, 613.
389. A. M. Caporusso, G. Giacomelli and L. Lardicci, *J. Org. Chem.*, 1982, **47**, 4640.
390. P. Caubère, *Angew. Chem., Int. Ed. Engl.*, 1983, **22**, 599.
391. J.-J. Brunet, L. Mordenti, B. Loubinoux and P. Caubère, *Tetrahedron Lett.*, 1978, 1069.
392. J.-J. Brunet, L. Mordenti and P. Caubère, *J. Org. Chem.*, 1978, **43**, 4804.
393. L. Mordenti, J.-J. Brunet and P. Caubère, *J. Org. Chem.*, 1979, **44**, 2203.
394. A. Fauve and A. Kergomard, *Tetrahedron*, 1981, **37**, 899.
395. L. Mamoli, R. Roch and H. Teschen, *Z. Physiol. Chem.*, 1939, **261**, 287.
396. T. L. Miller and E. J. Hessler, *Biochim. Biophys. Acta*, 1970, **202**, 354.
397. H. C. Murray and D. H. Peterson, *US Pat.* 2 659 741 (1953) (*Chem. Abstr.*, 1954, **48**, 13 737c).
398. H. C. Murray and D. H. Peterson, *US Pat.* 2 649 402 (1953).

399. (a) A. Kergomard, M. F. Renard and H. Veschambre, *J. Org. Chem.*, 1982, **47**, 792; (b) A. Kergomard, M. F. Renard and H. Veschambre, *Tetrahedron Lett.*, 1978, 5197; (c) G. Dauphin, J.-C. Gramain, A. Kergomard, M. F. Renard and H. Veschambre, *Tetrahedron Lett.*, 1980, **21**, 4275; (d) G. Dauphin, J.-C. Gramain, A. Kergomard, M. F. Renard and H. Veschambre, *J. Chem. Soc., Chem. Commun.*, 1980, 318.
400. (a) Y. Noma, S. Nonomura, H. Ueda and C. Tatsumi, *Agric. Biol. Chem.*, 1974, **38**, 735; (b) Y. Noma and S. Nonomura, *Agric. Biol. Chem.*, 1974, **38**, 741.
401. (a) M. Bostmembrun-Desrut, G. Dauphin, A. Kergomard, M. F. Renard and H. Veschambre, *Tetrahedron*, 1985, **41**, 3679; (b) M. Desrut, A. Kergomard, M. F. Renard and H. Veschambre, *Tetrahedron*, 1981, **37**, 3825.
402. (a) A. Kergomard, M. F. Renard and H. Veschambre, *Agric. Biol. Chem.*, 1985, **49**, 1497; (b) M. Desrut, A. Kergomard, M. F. Renard and H. Veschambre, *Biochem. Biophys. Res. Commun.*, 1983, **110**, 908; (c) A. Kergomard, M. F. Renard and H. Veschambre, *Agric. Biol. Chem.*, 1982, **46**, 97; (d) A. Kergomard, M. F. Renard, H. Veschambre, C. A. Groliere and J. Dupy-Blanc, *Agric. Biol. Chem.*, 1986, **50**, 487.
403. T. Kitazume and N. Ishikawa, *Chem. Lett.*, 1984, 587.
404. (a) P. Gramatica, P. Manitto and L. Poli, *J. Org. Chem.*, 1985, **50**, 4625; (b) P. Gramatica, P. Manitto, B. M. Ranzi, A. Delbianco, and M. Francavilla, *Experientia*, 1982, **38**, 775.
405. H. G. W. Leuenberger, W. Boguth, R. Barner, M. Schmid and R. Zell, *Helv. Chim. Acta*, 1979, **62**, 455.
406. H. G. W. Leuenberger, W. Boguth, E. Widmer and R. Zell, *Helv. Chim. Acta*, 1976, **59**, 1832.
407. C. Fuganti and P. Grasselli, *J. Chem. Soc., Chem. Commun.*, 1979, 995.
408. M. Miyano, C. R. Dorn, F. B. Colton and W. J. Marsheck, *J. Chem. Soc., Chem. Commun.*, 1971, 425.
409. (a) H. Simon, H. Günther, J. Bader and W. Tischler, *Angew. Chem., Int. Ed. Engl.*, 1981, **20**, 861; (b) I. Thanos and H. Simon, *Angew. Chem., Int. Ed. Engl.*, 1986, **25**, 462.
410. J. D. Rozzell, *Tetrahedron Lett.*, 1982, **23**, 1767.
411. H. Ohta, J. Konishi and G. Tsuchihashi, *Chem. Lett.*, 1983, 1895.
412. (a) B. Eckstein and A. Nimrod, *Biochim. Biophys. Acta*, 1977, **1**, 499; (b) I. A. Watkinson, D. C. Wilton, A. D. Rahimtula and M. Akhtar, *Eur. J. Biochem.*, 1971, **1**, 23.
413. E. A. Braude, J. Hannah and R. Linstead, *J. Chem. Soc.*, 1960, 3257.
414. B. E. Norcross, P. E. Klinedinst, Jr. and F. H. Westheimer, *J. Am. Chem. Soc.*, 1962, **84**, 797.
415. J. S. McGuire, Jr. and G. M. Tompkins, *Fed. Proc., Fed. Am. Soc. Exp. Biol.*, 1960, **19**, A29.
416. (a) Y. Ohnishi, M. Kagami and A. Ohno, *Chem. Lett.*, 1975, 125; (b) Y. Ohnishi, M. Kagami, T. Numakunai and A. Ohno, *Chem. Lett.*, 1976, 915.
417. K. Nakamura, M. Fujii, A. Ohno and S. Oka, *Tetrahedron Lett.*, 1984, **25**, 3983.
418. R. A. Gase and U. K. Pandit, *J. Am. Chem. Soc.*, 1979, **101**, 7059.
419. (a) M. J. de Nie-Sarink and U. K. Pandit, *Tetrahedron Lett.*, 1979, 2449; (b) U. K. Pandit, F. R. Mas Cabré, R. A. Gase and M. J. de Nie-Sarink, *J. Chem. Soc., Chem. Commun.*, 1974, 627.
420. L. H. P. Meijer and U. K. Pandit, *Tetrahedron*, 1985, **41**, 467.
421. C. Pac, M. Ihama, M. Yasuda, Y. Miyauchi and H. Sakurai, *J. Am. Chem. Soc.*, 1981, **103**, 6495.
422. F. Yoneda, K. Kuroda and K. Tanaka, *J. Chem. Soc., Chem. Commun.*, 1984, 1194.
423. N. Baba, T. Makino, J. Oda and Y. Inouye, *Can. J. Chem.*, 1980, **58**, 387.
424. A. Fischli and D. Süss, *Helv. Chim. Acta*, 1979, **62**, 2361.
425. (a) F. Camps, J. Coll, A. Guerrero, J. Guitart and M. Riba, *Chem. Lett.*, 1982, 715; (b) O. Louis-Andre and G. Gelbard, *Tetrahedron Lett.*, 1985, **26**, 831; F. Camps, J. Coll and J. Guitart, *Tetrahedron*, 1986, **42**, 4603; K. Okada, H. Tanino and K. Hashizume, *Tetrahedron Lett.*, 1984, **25**, 4403.
426. O. Louis-André and G. Gelbard, *Bull. Soc. Chim. Fr.*, 1986, 565.
427. H. Chikashita, M. Miyazaki and K. Itoh, *Synthesis*, 1984, 308.
428. S. K. Malhotra, D. F. Moakley and F. Johnson, *J. Am. Chem. Soc.*, 1967, **89**, 2794.
429. M. Yamashita, Y. Kato and R. Suemitsu, *Chem. Lett.*, 1980, 847.
430. M. D'Auria, G. Piancatelli and A. Scettri, *Synthesis*, 1980, 245.
431. U. Melamed and B.-A. Feit, *J. Chem. Soc., Perkin Trans. 1*, 1980, 1267.
432. G. Solladié and J. Hutt, *Bull. Soc. Chim. Fr.*, 1986, 643.
433. G. R. Brown, A. J. Foubister and B. Wright, *J. Chem. Soc., Chem. Commun.*, 1984, 1373.
434. F. W. Nader, A. Brecht and S. Kreisz, *Chem. Ber.*, 1986, **119**, 1196.
435. (a) I. K. Youn, G. H. Yon and C. S. Pak, *Tetrahedron Lett.*, 1986, **27**, 240; (b) R. Brettle and S. M. Shibib, *Tetrahedron Lett.*, 1980, **21**, 2915; (c) G. Palmisano, B. Daniel, G. Lesma, R. Riva and S. Riva, *J. Org. Chem.*, 1984, **49**, 4138.
436. H. Stamm, A. Sommer, A. Onistschenko and A. Woderer, *J. Org. Chem.*, 1986, **51**, 4979.
437. A. C. Chan and D. I. Schuster, *J. Am. Chem. Soc.*, 1986, **108**, 4561.
438. G. H. Posner and A. W. Runquist, *Tetrahedron Lett.*, 1975, 3601.
439. K. Nanjo, K. Suzuki and M. Sekiya, *Chem. Pharm. Bull.*, 1977, **25**, 2396.
440. G. Giacomelli, A. M. Caporusso and L. Lardicci, *Tetrahedron Lett.*, 1981, **22**, 3663.
441. G. Stork and A. F. Kreft, III, *J. Am. Chem. Soc.*, 1977, **99**, 3850.
442. W. S. Johnson, W. A. Vredenburgh and J. E. Pike, *J. Am. Chem. Soc.*, 1960, **82**, 3409.
443. (a) B. Loev and C. R. Dawson, *J. Am. Chem. Soc.*, 1956, **78**, 1180; (b) B. Loev and C. R. Dawson, *J. Am. Chem. Soc.*, 1956, **78**, 6095; (c) B. Loev and C. R. Dawson, *J. Org. Chem.*, 1959, **24**, 980; (d) C. R. Hauser, H. M. Taylor and T. G. Legford, *J. Am. Chem. Soc.*, 1960, **82**, 1786; B. Heath-Brown, *Chem. Ind. (London)*, 1969, 1595; (e) W. A. Remers, G. J. Gibs, C. Pidacks and M. J. Weiss, *J. Org. Chem.*, 1971, **36**, 279.
444. H. Gilman and J. C. Bailie, *J. Am. Chem. Soc.*, 1943, **65**, 267.
445. D. R. Weyenberg, *J. Org. Chem.*, 1965, **30**, 3236.
446. N. L. Bauld, *J. Am. Chem. Soc.*, 1962, **84**, 4345.
447. D. R. Weyenberg, L. H. Toporcer and L. E. Nelson, *J. Org. Chem.*, 1968, **33**, 1975.
448. (a) D. H. R. Barton, X. Lusinchi, L. Magdzinski and J. S. Ramirez, *J. Chem. Soc., Chem. Commun.*, 1984, 1236; (b) M. Anastasia, P. Allevi, P. Ciuffreda, A. Fiecchi and A. Scala, *J. Org. Chem.*, 1984, **49**, 4297.
449. J. A. Proffit and H. H. Org, *J. Org. Chem.*, 1979, **44**, 3972.

450. (a) G. Rosenkranz, J. Romo, E. Batres and C. Djerassi, *J. Org. Chem.*, 1951, **16**, 298; (b) R. Antonucci, S. Bernstein, D. Giancola and K. J. Sax, *J. Org. Chem.*, 1951, **16**, 1891; (c) R. C. Anderson, R. Stevenson and F. S. Spring, *J. Chem. Soc.*, 1952, 2901; (d) G. D. Laubach and K. J. Brunings, *J. Am. Chem. Soc.*, 1952, **74**, 705; (e) G. D. Laubach, E. C. Schreiber, E. J. Agnello and K. J. Brunings, *J. Am. Chem. Soc.*, 1956, **78**, 4743; (f) S. C. Bernstein, R. Littell and J. H. Williams, *J. Org. Chem.*, 1953, **18**, 1418; (g) W. Tadros and A. L. Boulos, *Helv. Chim. Acta*, 1975, **58**, 668.
451. W. S. Johnson, R. G. Christiansen and R. E. Ireland, *J. Am. Chem. Soc.*, 1957, **79**, 1995.
452. (a) E. L. May and E. M. Fry, *J. Org. Chem.*, 1957, **22**, 1366; (b) N. B. Eddy, J. G. Murphy and E. L. May, *J. Org. Chem.*, 1957, **22**, 1370; (c) B. C. Joshi, C. F. Chignell and E. L. May, *J. Med. Chem.*, 1965, **8**, 694.
453. B. B. Millward, *J. Chem. Soc.*, 1960, 26.
454. P. A. Wender and D. A. Holt, *J. Am. Chem. Soc.*, 1985, **107**, 7771.
455. G. Mehta and K. S. Rao, *J. Org. Chem.*, 1984, **49**, 3848.
456. G. Mchta, A. Srikrishna, A. V. Reddy and M. S. Nair, *Tetrahedron*, 1981, **37**, 4543.
457. G. Ohloff, H. Farnow and G. Schade, *Chem. Ber.*, 1956, 89, 1549; K. M. R. Pillai, W. V. Murray, I. Shooshani, D. L. Williams, D. Gordon, S. Y. Wang and F. Johnson, *J. Med. Chem.*, 1984, **27**, 1131.
458. M. Shibasaki, M. Sodeoka and Y. Ogawa, *J. Org. Chem.*, 1984, **49**, 4096.
459. M. A. Schroeder and M. S. Wrighton, *J. Organomet. Chem.*, 1974, **74**, C29.
460. M. A. Schroeder and M. S. Wrighton, *J. Am. Chem. Soc.*, 1973, **95**, 5764.
461. T. Funabiki, M. Matsumoto and K. Tarama, *Bull. Chem. Soc. Jpn.*, 1972, **45**, 2723.
462. P. S. Hallman, D. Evans, J. A. Osborn and G. Wilkinson, *J. Chem. Soc., Chem. Commun.*, 1967, 305.
463. T. Tajima and E. Kunioka, *J. Org. Chem.*, 1968, **33**, 1689.
464. L. H. Slaugh, *Tetrahedron*, 1966, **22**, 1741.
465. A. Miyake and H. Kondo, *Angew. Chem., Int. Ed Engl.*, 1968, **7**, 880.
466. (a) K. Mayer and E. Eich, *Pharmazie*, 1984, **39**, 537; (b) G. M. Badger, *J. Chem. Soc.*, 1948, 999; (c) C. F. Koelsch, *J. Org. Chem.*, 1961, **26**, 1003; (d) Z.-i. Horii, T. Sakai and Y. Tamura, *Chem. Pharm. Bull.*, 1961, **9**, 446; (e) G. A. Page and D. S. Tarbell, *J. Am. Chem. Soc.*, 1953, **75**, 2053; (f) K. Dimroth, W. Kinzebach and M. Soyka, *Chem. Ber.*, 1966, **99**, 2351; (g) F. Benington, R. D. Morin and L. C. Clark, Jr., *J. Org. Chem.*, 1959, **24**, 917; (h) H. E. Johnson and D. G. Crosby, *J. Org. Chem.*, 1963, **28**, 2794.
467. (a) R. Sala, G. Doria and C. Passarotti, *Tetrahedron Lett.*, 1984, **25**, 4565; (b) G. A. Olah and G. K. Surya Prakash, *Synthesis*, 1978, 397; (c) P. K. Oommen, *Bull. Chem. Soc. Jpn.*, 1976, **49**, 1985; (d) *Synth. Meth.*, 1948, **1**, 669; (e) *Synth. Meth.*, 1973, **27**, 872; (f) H. W. Thompson, *J. Org. Chem.*, 1971, **36**, 2577; (g) J. H. Billman, J. I. Stiles and J. Tonnis, *Synth. Commun.*, 1971, **1**, 127; (h) R. Aneja, S. K. Mukerjee and T. R. Seshadri, *Chem. Ber.*, 1960, **93**, 297; (i) W. J. Horton and E. G. Paul, *J. Org. Chem.*, 1959, **24**, 2000; (j) D. K. Banerjee, S. Chatterjee, C. N. Pillai and M. V. Bhatt, *J. Am. Chem. Soc.*, 1956, **78**, 3769; (k) *Synth. Meth.*, 1957, **11**, 152; (l) R. A. Barnes and R. Miller, *J. Am. Chem. Soc.*, 1960, **82**, 4960; (m) T. Hayashi, Y. Okamoto, K. Kabeta, T. Hagihara and M. Kumada, *J. Org. Chem.*, 1984, **49**, 4224; (n) W. H. Pearlman and M. R. J. Pearlman, *J. Am. Chem. Soc.*, 1950, **72**, 5781; (o) A. T. Blomquist and R. J. Cedergren, *Can. J. Chem.*, 1968, **46**, 1053.
468. (a) L. J. Sargent and L. F. Small, *J. Org. Chem.*, 1951, **16**, 1031; (b) K. M. R. Pillai, W. V. Murray, I. Shooshani, D. L. Williams, D. Gordon, S. Y. Wang and F. Johnson, *J. Med. Chem.*, 1984, **27**, 1131.
469. (a) A. Jung and M. Brini, *Bull. Soc. Chim. Fr.*, 1965, 587; (b) D. Horton and W. Weckerle, *Carbohydr. Res.*, 1975, **44**, 227.
470. (a) T. W. Campbell and R. N. McDonald, *J. Org. Chem.*, 1959, **24**, 1246; (b) T. W. Campbell and R. N. McDonald, *J. Org. Chem.*, 1959, 24, 1246; (c) P. Cañas-Rodriguez, *J. Chem. Soc., Perkin Trans. 1*, 1972, 554; (d) T. W. Russell and D. M. Duncan, *J. Org. Chem.*, 1974, **39**, 3050.
471. (a) G. Wolf and M. A. Seligman, *J. Am. Chem. Soc.*, 1951, **73**, 2082; (b) W. E. McEwen, R. H. Terss and I. W. Elliott, *J. Am. Chem. Soc.*, 1952, **74**, 3605; (c) A. Bertho and H. Kurzmann, *Chem. Ber.*, 1957, **90**, 2319; (d) M. P. Cava and I. Noguchi, *J. Org. Chem.*, 1973, **38**, 60; (e) J. A. Weisbach, C. Burns, E. Macko and B. Douglas, *J. Med. Chem.*, 1963, **6**, 91; (f) A. Smith and J. H. P. Utley, *J. Chem. Soc., Chem. Commun.*, 1965, 427; (g) M. N. Preobrazhenskaya, *Russ. Chem. Rev. (Engl. Transl.)*, 1967, **36**, 753; (h) C. S. Marvel, R. E. Allen and C. G. Overberger, *J. Am. Chem. Soc.*, 1946, **68**, 1088.
472. (a) A. G. Anderson, Jr., A. F. Montana, A. A. MacDonald and G. M. Masada, *J. Org. Chem.*, 1973, **38**, 1445; (b) L. A. Jones and R. Watson, *Can. J. Chem.*, 1973, **51**, 1833,
473. B. E. Maryanoff and H. R. Almond, Jr., *J. Org. Chem.*, 1986, **51**, 3295.
474. K. Junghans, *Chem. Ber.*, 1974, **107**, 3191.
475. Y. Takegami, C. Yokokawa, Y. Watanabe, H. Nasada and Y. Okuda, *J. Org. Chem.*, 1964, **37**, 1190.
476. G. C. Bond and M. Hellier, *J. Catal.*, 1966, **7**, 217.
477. W. H. Dennis, Jr., D. H. Rosenblatt, R. R. Richmond, G. A. Finseth and G. T. Davis, *Tetrahedron Lett.*, 1968, 1821.
478. F. H. Jardine, J. A. Osborn and G. Wilkinson, *J. Chem. Soc. A*, 1967, 1574.
479. C. O'Connor and G. Wilkinson, *J. Chem. Soc. A*, 1968, 2665.
480. J. Kwiatek, I. L. Mador and J. K. Seyler, *J. Am. Chem. Soc.*, 1962, **84**, 304.
481. M. F. Sloan, A. S. Matlack and D. S. Breslow, *J. Am. Chem. Soc.*, 1963, **85**, 4014.
482. (a) I. Goodman, *J. Chem. Soc.*, 1951, 2209; (b) K. Hafner, G. Hafner-Schneider and F. Bauer, *Angew. Chem., Int. Ed. Engl.*, 1968, **7**, 808.
483. F. A. Hochstein and W. G. Brown, *J. Am. Chem. Soc.*, 1948, **70**, 3484.
484. H. Gilman, J. L. Towle and R. K. Ingham, *J. Am. Chem. Soc.*, 1954, **76**, 2920.
485. (a) Y. Sprinzak, *J. Am. Chem. Soc.*, 1956, **78**, 466; (b) M. Avramoff and Y. Spinzak, *J. Am. Chem. Soc.*, 1958, **80**, 493.
486. (a) Huang-Minlon, *J. Am. Chem. Soc.*, 1948, **70**, 2802; (b) T. Kauffmann, C. Kosel and W. Schoeneck, *Chem. Ber.*, 1963, **96**, 999.

3.6
Partial and Complete Reduction of Pyridines and their Benzo Analogs

JAMES G. KEAY
Reilly Industries, Indianapolis, IN, USA

3.6.1 INTRODUCTION

This review deals with the formation of reduced pyridines and their benzo analogs from the parent heteroaromatic bases. Included are acridines, isoquinolines, pyridines and quinolines and their quaternary ammonium salts and *N*-oxides. The formation of the reduced species by other methods, *e.g.* Hantzsch dihydropyridine synthesis, is not addressed.

3.6.2 CHEMICAL REDUCTION

This encompasses all reduction methods other than hydrogenation and includes hydride equivalents and one-electron reductions. Catalytic hydrogen transfer is included under hydrogenation (see Section 3.6.3).

3.6.2.1 Metal Hydride Reductions

Hydride reagents formally add hydrogen to polarized carbon–heteroatom double bonds, the most widely used being based on aluminum (*e.g.* lithium aluminum hydride, LAH) and boron (*e.g.* sodium borohydride, NBH). Complex metal hydrides are generally sufficiently nucleophilic to react with π-electron-deficient heteroaromatics such as pyridines. The modification of these reagents with, for example, alcohols, alkyl groups, amines and Lewis acids greatly affects both their reactivity and selectivity. Solvents have a profound effect on the activity of these reagents, particularly so in the case of borohydrides. LAH and its derivatives are generally used in ethers, whereas borohydride reductions are frequently carried out in alcohols. These reagents can also be used to generate other active species *in situ* (*e.g.* hydrides of Si, Sn and Cu). During the reaction each hydrogen is sequentially replaced by the reduced substrate (Scheme 1). The electronic and steric character of the reductant therefore changes as the reaction proceeds, *e.g.* **(1)** → **(2)** → **(3)**, and changes in regio- and enantio-selectivity may occur. The operation of both direct hydride transfer and single-electron transfer (SET) processes has been proposed as mechanisms.[1] Stable radical intermediates have been detected in the reductions of 1,10-phenanthroline, 2,2-bipyridine and isoquinoline with lithium di- and tri-*t*-butoxyaluminum hydrides. The initial radical pair formed (**4**; Scheme 2) can undergo further reduction to (**5**) or to the dissociated free radical anion (**6**) prior to reaction.

$$\underset{\textbf{(1)}}{LiAlH_4} \xrightarrow{R} \underset{\textbf{(2)}}{LiAl(RH)H_3} \xrightarrow{R} \underset{\textbf{(3)}}{LiAl(RH)_2H_2} \longrightarrow\longrightarrow LiAl(RH)_4 \xrightarrow{H_2O} LiOH + Al(OH)_3 + 4RH_2$$

Scheme 1

$$\underset{\textbf{(4)}}{\overset{+\bullet}{LiAlH_2(OBu^t)_2}\ (\overset{-\bullet}{Ar})}$$

$$\underset{\textbf{(5)}}{ArH^-} + Li^+ + AlH(OBu^t)_2 \qquad \underset{\textbf{(6)}}{Ar^{\bullet -}} + Li^+ + H^\bullet + AlH(OBu^t)_2$$

$$\xrightarrow{H_2O} ArH_2 + LiOH + Al(OH)_3 + Bu^tOH + H_2$$

Scheme 2

Alanes and boranes are initially electrophilic rather than nucleophilic, and coordinate onto free electron pairs in the molecule prior to hydride delivery. The chemistry of metal hydrides has been extensively reviewed,[2] as has the reduction of nitrogen heterocycles with complex metal hydrides.[3] Generally, NBH reductions, carried out in alcohols, produce tetrahydropyridines, whilst LAH generates mainly dihydropyridines.

3.6.2.1.1 Pyridines with borohydrides

Pyridine itself is not reduced by NBH under normal conditions and ring reduction occurs only if electron-withdrawing groups are attached. 3-Cyanopyridine (**7**) when reduced with NBH in ethanol gives 3-cyano-1,4,5,6-tetrahydropyridine (**8**; Scheme 3) as the major product. In aprotic solvents such as pyridine or diglyme, the 1,4-dihydropyridine (**9**) is produced.[4] Reactions of pyridines substituted with electron-withdrawing groups at the 2- and 4-position generally occur at the substituent.[5] Although generally not reactive towards NBH, carboxylic acid esters at the 2- and 6-position of the pyridine ring are reduced to the alcohol.

Sodium cyanoborohydride (**10**) produces mainly 1,4-dihydropyridines (**11**) in the reduction of 3,5-dicyano- and 3,5-diethoxycarbonyl-pyridines, diborane produces more of the 1,2-isomer.[6] With NBH, mixtures of 1,2- and 1,4-dihydro adducts are produced, the latter predominating when carried out in pyridine solution.[6] Nicotinamide (**13**) in ethanol can be reduced to (**8**) in moderate yield; at 140 °C in diglyme the tetrahydropyridine (**8**) was isolated in admixture with the piperidine (**14**), presumably *via* dehydration of the amide.[7] 3-Nitropyridine affords 3-nitropiperidine in moderate yield when reduced in ethanol.[5] The carboxylic acid and halo derivatives of pyridine are generally not reactive toward NBH.

Quinolines, like pyridines, are susceptible to attack at both the 2- and 4-position; however, they are more reactive, quinoline (**15**; R = H) forming a mixture of 1,2-dihydroquinoline (**16**; R = H) and 1,2,3,4-tetrahydroquinoline (**17**; R = H) when reacted in hot ethanol or diglyme.[5] Electron-withdrawing groups

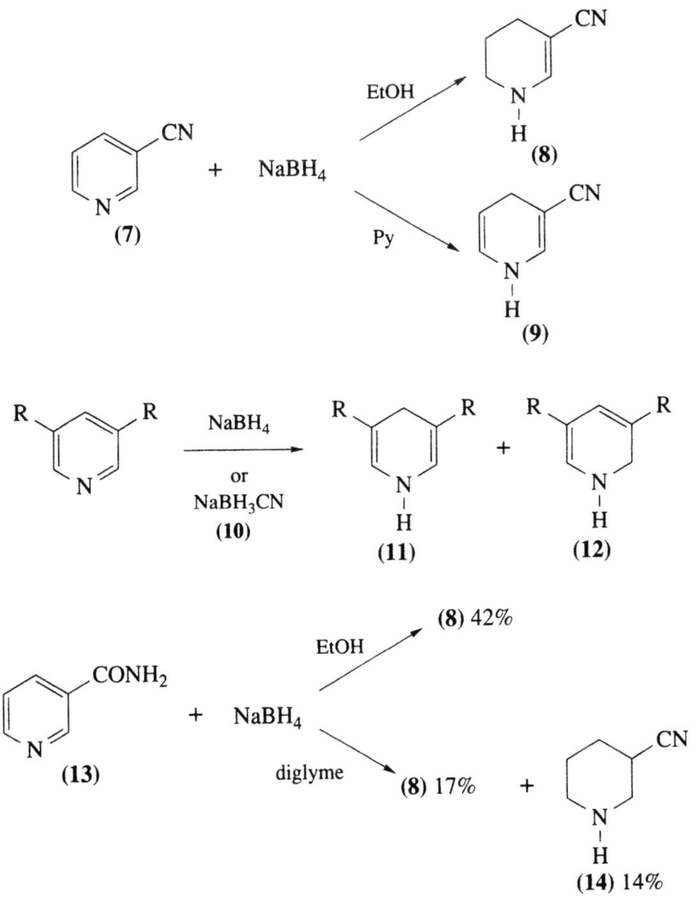

Scheme 3

in the 3-position (**15**; R = CN, $CONH_2$, CO_2Et) gave high yields of 1,4-dihydroquinolines (**18**), whereas halogentated quinolines (**15**; R = Br, F) produced 1,2-dihydroquinolines (**16**) in only moderate yield.[5] 3-Dimethylaminoquinoline (**15**; R = NMe_2) was unreactive under the conditions employed. 2- and 4-cyanoquinolines undergo reduction to give substituent- and/or ring-reduced products, depending upon the solvent employed.[5] Reduction in the presence of carboxylic acids has been used to facilitate reduction with and without alkylation. Quinoline and NBH in carboxylic acids produce *N*-alkyl-1,2,3,4-tetrahydroquinolines (**19**; Scheme 4), whereas sodium cyanoborohydride produces 1,2,3,4-tetrahydroquinolines (**20**).[8] The reduction of quinolines substituted in the benzene ring with nitro groups produces the 1,2-dihydroquinolines (**21**) in high yield when reduced with NBH in acetic acid at low temperature.[9] At higher temperatures the 1,2-dihydroquinoline (**22**) was produced in moderate yield.[8]

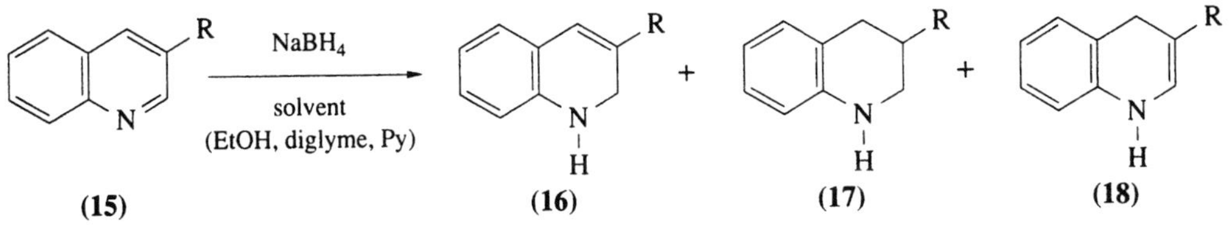

Isoquinoline (**23**; R = H; Scheme 5) undergoes reduction to 1,2,3,4-tetrahydroisoquinoline (**24**; R = H) when treated with NBH in aqueous solvents.[5] Isoquinolines bearing an electron-withdrawing group in the 4-position undergo reduction with NBH to give 1,2-dihydroisoquinolines (**25**) or 1,2,3,4-tetrahydroisoquinolines (**24**), depending upon the solvent.[5] Substituent reduction may also occur. 1,2-Dihydroisoquinolines (**25**) are the expected products in aprotic solvents. With 1- and 3-cyanoisoquinolines both ring and substituent reduction are known to occur. Ring reduction is favored with the former and substituent reduction predominates with the latter producing both amides and amidines.[5] Like quinoline, isoquinoline undergoes reductive alkylation to *N*-alkyl-1,2,3,4-tetrahydroisoquinolines (**26**; Scheme 5) with

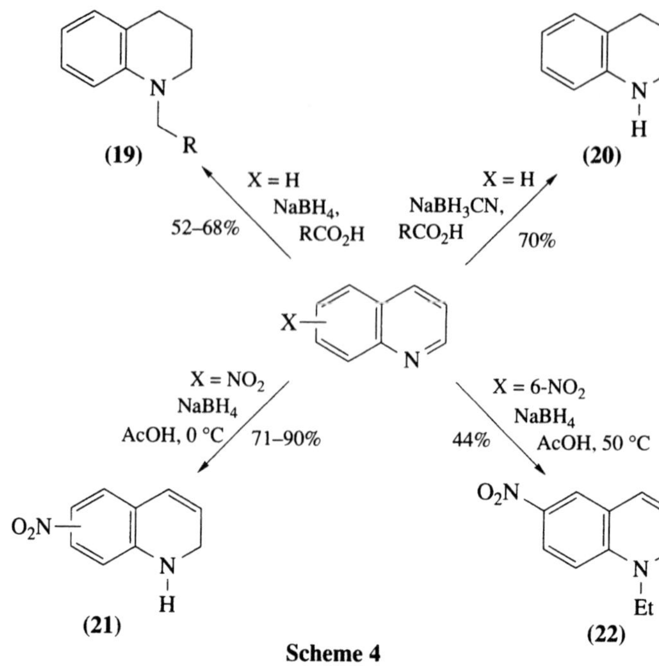

(19) (20)

X = H, $NaBH_4$, RCO_2H, 52–68%

X = H, $NaBH_3CN$, RCO_2H, 70%

X = NO_2, $NaBH_4$, AcOH, 0 °C, 71–90%

X = 6-NO_2, $NaBH_4$, AcOH, 50 °C, 44%

(21) (22)

Scheme 4

NBH, in the presence of carboxylic acids. Reduction with sodium cyanoborohydride gives reduction but no alkylation under similar conditions.[8] The reduction of 5-nitroisoquinoline produced dihydroisoquinoline (**27**) or tetrahydroisoquinoline (**28**) with NBH in acetic acid, depending on the solvent and temperatures employed.

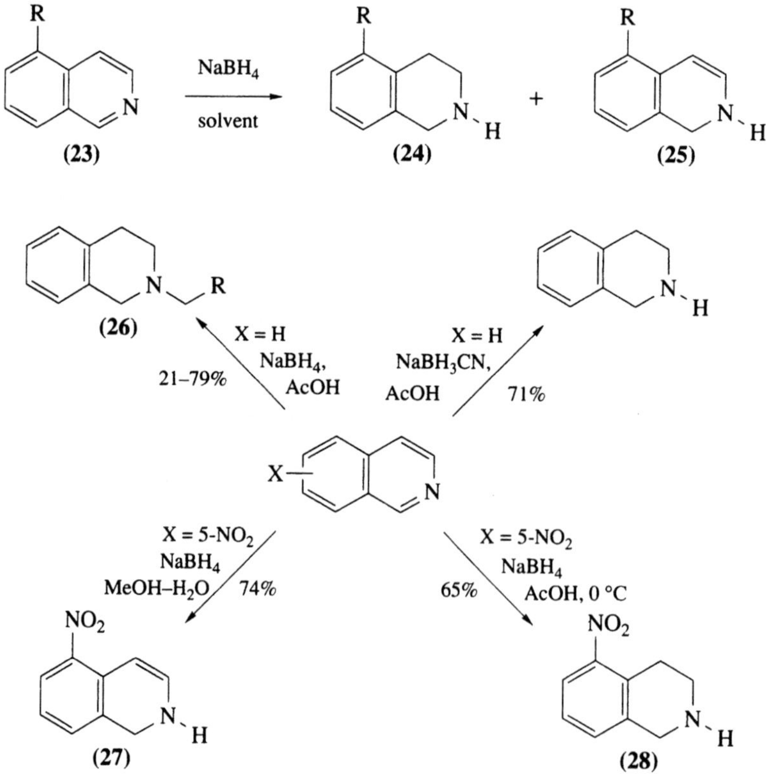

(23) $\xrightarrow[\text{solvent}]{NaBH_4}$ (24) + (25)

(26)

X = H, $NaBH_4$, AcOH, 21–79%

X = H, $NaBH_3CN$, AcOH, 71%

X = 5-NO_2, $NaBH_4$, MeOH–H_2O, 74%

X = 5-NO_2, $NaBH_4$, AcOH, 0 °C, 65%

(27) (28)

Scheme 5

Sodium triethylborohydride (**29**) reacts with isoquinoline to yield the boron-activated enamine (**30**) which undergoes attack by electrophiles to produce 4-substitued isoquinolines (**31**; Scheme 6) in a convenient 'one-pot' synthesis.[10]

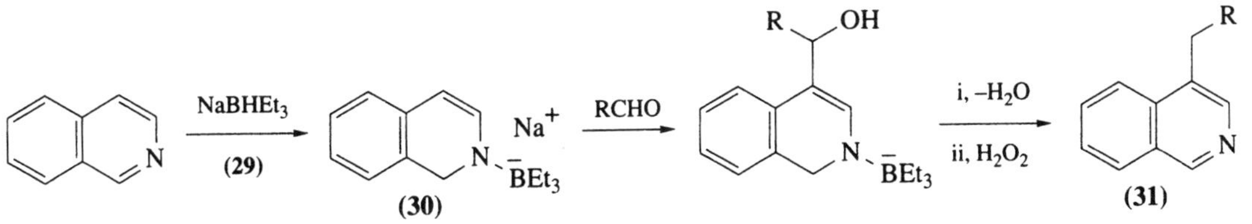

NaBHEt$_3$ (**29**) → (**30**) Na$^+$ BEt$_3$ → RCHO → i, –H$_2$O; ii, H$_2$O$_2$ → (**31**)

R = H, alkyl, aryl; 25–65%

Scheme 6

3.6.2.1.2 Pyridines with aluminum hydrides

Lithium aluminum hydride (LAH) reacts with pyridines and their analogs in aprotic solvents to give dihydro- and tetrahydro-pyridines. In the absence of proton sources dihydropyridines normally predominate, solutions of pyridine and LAH form lithium complexes (**32**; Scheme 7), which likely consist of both 1,2- and 1,4-dihydropyridines.[11] This intermediate has been used as a reducing agent for ketones, and reaction with alkyl halides generates 3-substituted pyridines (**33**) in good yield.[12]

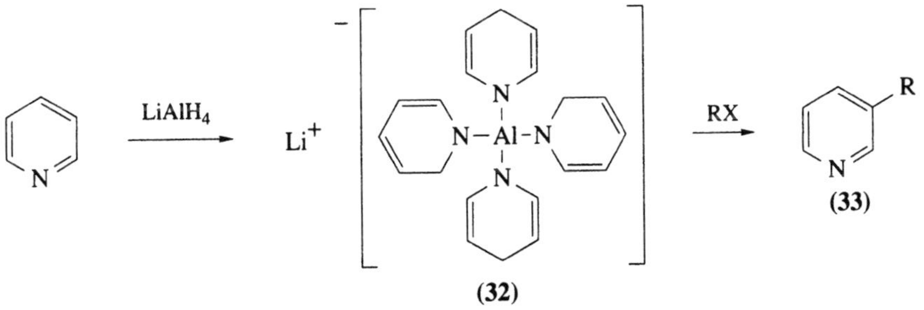

LiAlH$_4$ → Li$^+$ (**32**) → RX → (**33**)

Scheme 7

3,5-Dicyanopyridine produces mixtures of 1,2-dihydropyridines (**35**) and 1,4-dihydropyridines (**36**; Scheme 8)) with LAH. The more hindered alkoxyaluminates favor 1,4-dihydropyridine formation.[13] 1-Alkyl-2-pyridones (**37**) are reduced when aluminum chloride is used in conjunction with LAH. Tetrahydropyridines (**38**) are the major products produced, minor amounts of the piperidines (**39**) are also formed.[14]

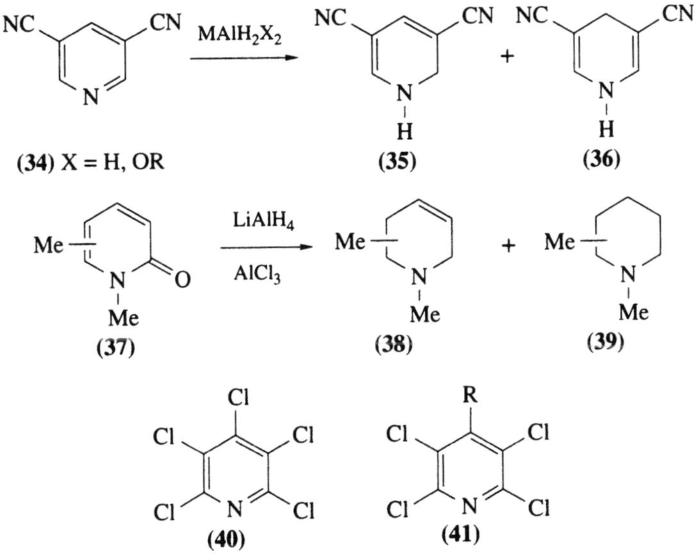

(**34**) X = H, OR → MAlH$_2$X$_2$ → (**35**) + (**36**)

(**37**) → LiAlH$_4$, AlCl$_3$ → (**38**) + (**39**)

(**40**) (**41**)

Scheme 8

Pentachloropyridines (**40**) and 4-substituted tetrachloropyridines (**41**) undergo dehalogenation on treatment with LAH. Minor amounts of ring-reduced products are observed.[15]

Quinolines generally produce 1,2-dihydroquinolines with aluminum hydrides. Diisobutylaluminum hydride and Red-Al have been used to reduce quinolines and quinoline-*N*-boranes (**42**; Scheme 9). The reactions are believed to proceed *via* aminoalanes or boranes and the dihydroquinolines produced are quenched and trapped as the carbamates by the addition of chloroformate esters.[16] Minor amounts of 1,4-dihydro- and 1,2,3,4-tetrahydro-quinolines are also formed. The possession of a 2-substituent hinders reaction. At low temperatures the kinetic product, *i.e.* (**43**), is favored, but at higher temperatures reduction at the 4-position predominates which subsequently produces the tetrahydro derivative (**44**). A similar approach has been employed with isoquinoline, generating 1,2-disubstituted tetrahydroisoquinolines (**45**; Scheme 9).[17]

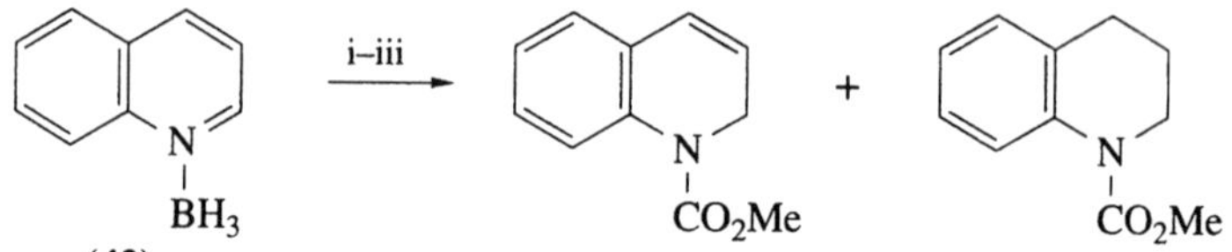

Temperature (°C)	*Time* (h)	*Yield* (%)	(**43**) (%)	(**44**) (%)
–78	0.33	90	100	–
25	18	85	44	56

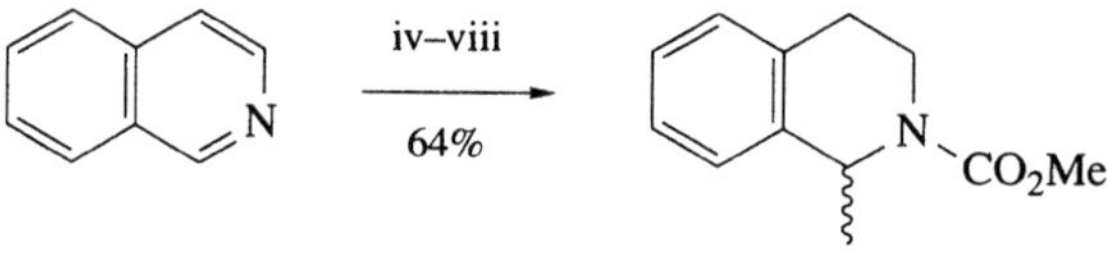

i, 2 equiv. $NaAl(OR)_2H_2$; ii, $ClCO_2Me$; iii, H_3O^+; iv, BH_3•THF, –78 °C; v, MeLi, –78 °C; vi, DIBAL-H, –78 to 25 °C; vii, $ClCO_2Me$, –25 to 0 °C; viii, HCl

Scheme 9

3.6.2.1.3 Pyridinium salts with borohydrides

Coenzyme NAD(P)$^+$ involves a nicotinamide derivative and the transfer of a hydrogen species in a stereoselective manner *via* a 1,4-dihydropyridine (**47**; Scheme 10) and has led to much work on the reduction of NAD and its simpler derivatives (**46**; R = $PhCH_2$, $C_{12}H_{25}$). Most work has been done on the stereoselective reduction of ketones,[18] but heteroaromatic cations have also been reduced (see Section 3.6.2.3).[19] Both single-electron transfer (SET) and direct hydride transfer mechanisms have been proposed.[20]

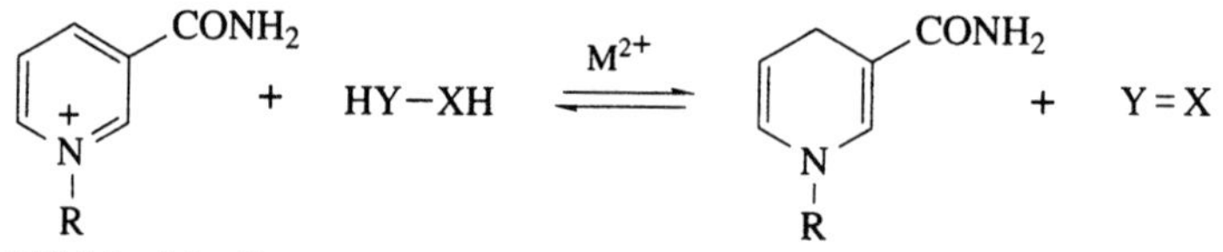

(**46**) M = Mg, Zn (**47**)

Scheme 10

Pyridinium salts (**48**; Scheme 11) with borohydride can give three different products, 1,2-, 1,4- and 1,6-dihydropyridines. 1,4-Dihydropyridines (**49**) are thermodynamically favored, being *ca.* 2 kcal mol^{-1} (1 cal = 4.18 J) more stable than the 1,2(6)-dihydropyridines (**50**). The chemistry of dihydropyridines has been extensively reviewed.[3,21–23] The dihydropyridine systems produced, being enamines, normally undergo protonation in protic solvents to give the iminium salts and hence further reduction. Mixtures of piperidines (**51**) and tetrahydropyridines (**52**) often result, with the former produced by initial hydride attack at the 4-position.[3] Bulky *N*-substituents are known to favor 4-attack.[24] Electron donors in the 3-position direct hydride attack predominantly to the 2-position when reduced as the *N*-alkoxycarbonylpyridinium salts (**53**; Scheme 12) with a variety of boron- and aluminum-based reducing agents.[25] 3-Methoxy and 3-methylthio derivatives gave exclusively 1,2-dihydropyridines (**54**) with NBH,

$K(Pr^{i}O)_3BH$ and $Li(Bu^{t}O)_3AlH$. 3-Trimethylsilyl and 3-trialkylstannyl groups gave very high regiospecificity (80–100%) to the 1,6-dihydropyridines (**55**). 3-Alkylpyridines produced predominantly the 1,2-adducts (**54**) and electron-withdrawing groups in the 3-position produced mixtures of all three isomers. Dihydropyridines containing electron-withdrawing groups at the 3-position are often resistant to further reduction, particularly in the case of the 1,6- and 1,4-dihydro (**56**) derivatives. 1,2-Dihydropyridines are converted to the tetrahydropyridines.[3,26]

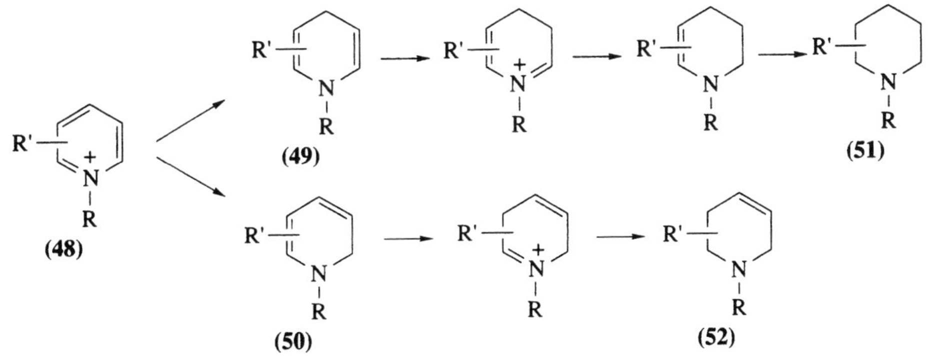

(**48**) (**49**) (**51**) (**50**) (**52**)

Scheme 11

(**54**) R = MeO, MeS — $NaBH_4$ or $K(Pr^{i}O)_3BH$ — (**53**) — $NaBu^{s}_3BH$ or $Li(Bu^{t}O)_3AlH$ — (**55**) R = Me_3Sn, Bu_3Sn, Me_3Si

$NaBH_4$

(**56**) R = CN

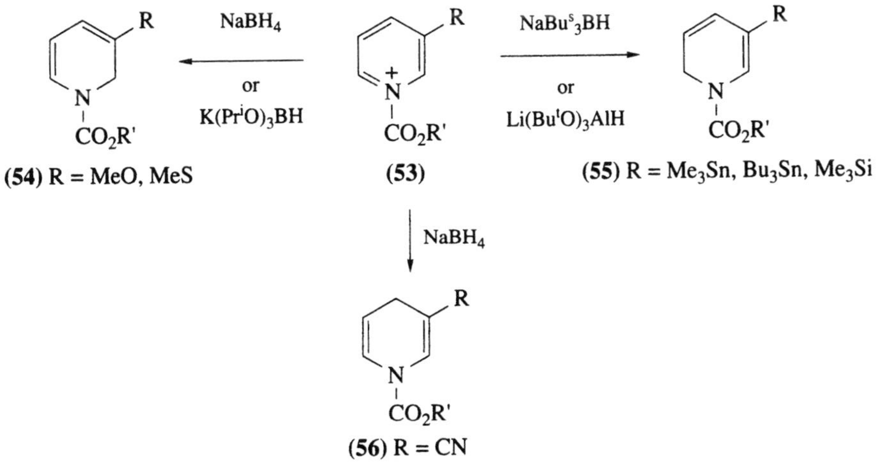

Scheme 12

Dihydropyridines may be prevented from undergoing further reduction by carrying out reductions in aprotic solvents or in protic systems at high pH. Phase transfer methods have also been used to remove the dihydro adducts from the reactive medium prior to further reduction. An elegant solution to the problem of overreduction was developed by Fry; reduction is carried out in the presence of cyanide ion, which traps the dihydropyridine system as 2-cyano-1,2,3,6-tetrahydropyridine (**57**; Scheme 13).[27] The addition of alkoxide re-forms the dihydropyridine (**58**). Separation of 1,4-dihydropyridines from mix-

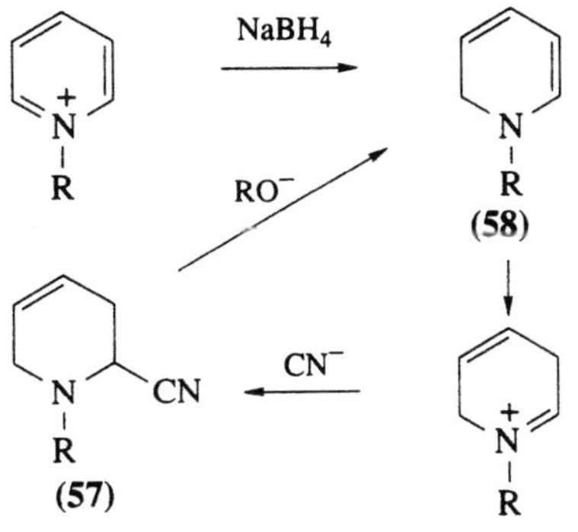

$NaBH_4$ (**58**) RO^- CN^- (**57**)

Scheme 13

tures with the 1,2-dihydropyridines can be achieved because 1,2-dihydropyridines undergo Diels–Alder reactions with maleic anhydride to produce isoquinuclidines (Scheme 13).[28] The regiospecific delivery of hydride to the 4-position has been achieved in high selectivity (**59**) → (**60**) (>90%) by the *in situ* generation of a 'copper hydride' reagent from lithium tri-*t*-butoxyaluminum hydride and copper(I) bromide.[29]

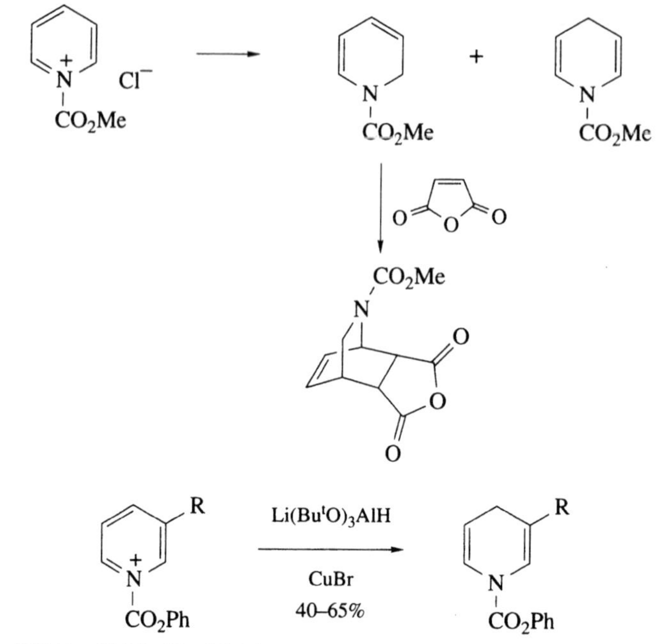

Li(Bu^{t}O)$_3$AlH, CuBr, 40–65%

(**59**) R = H, Me, Et, CO_2Me (**60**)

Scheme 13 ***(continued)***

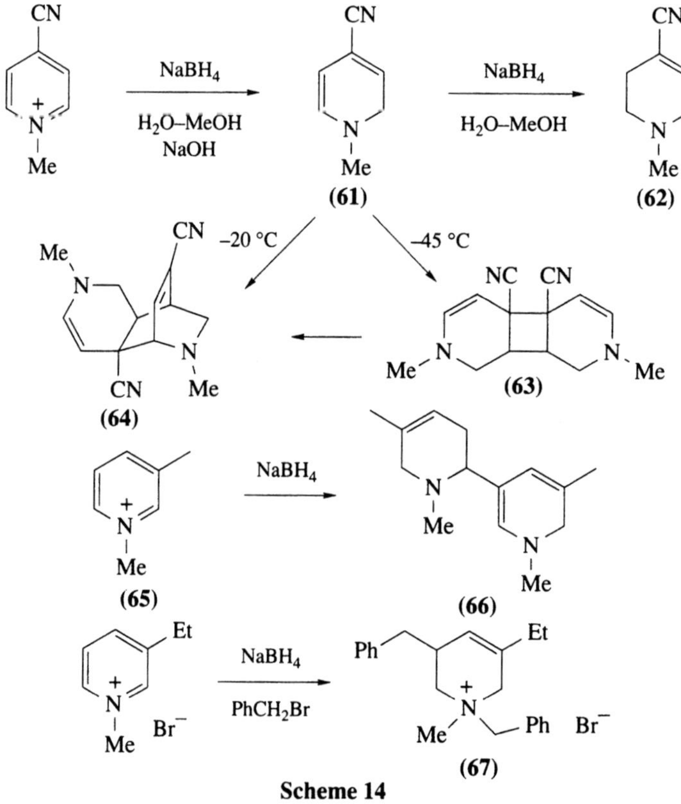

$NaBH_4$, H_2O–MeOH, NaOH (**61**) $NaBH_4$, H_2O–MeOH (**62**)

–20 °C (**64**) –45 °C (**63**)

(**65**) $NaBH_4$ (**66**)

$NaBH_4$, $PhCH_2Br$ (**67**)

Scheme 14

When reduced in aqueous methanol, 1-methyl-4-cyanopyridinium iodide gave the tetrahydropyridine derivative (**62**; Scheme 14). In the presence of sodium hydroxide two different products could be isolated at different temperatures, resulting from [2 + 2] (**63**) and [4 + 2] (**64**) cycloadditions, with the latter formed at higher temperatures or by thermal rearrangements of (**63**).[30] Similar behavior has been observed with other pyridinium salts.[31] Other dimerization processes involving dihydropyridines also occur, as with the C(2)–C(3) dimer (**66**) formed *via* enamine–iminium reactivity during the reduction of 1,3-dimethylpyridinium salts (**65**).[32] Conducting reductions in the presence of external electrophiles enables substituted pyridines, *e.g.* (**67**), to be prepared by reaction of the enamine system produced.[33] Benzomorphans and indole alkaloids can be conveniently prepared from reduced pyridines.[34,35] Pyridine *N*-oxides tend to undergo deoxygenation on reduction. *N*-Aminopyridinium salts undergo ring reduction to di- and tetra-hydropyridines, similar to their carbon analogs.[36]

Quinolinium salts (**68**; Scheme 15) can undergo attack at either the 2- or 4-position. The former normally predominates and the latter leads to 1,2,3,4-tetrahydroquinolines (**69**). 3-Substituents generally produce mixtures of the 1,2- and 1,4-dihydro adducts.[5] Isoquinolinium salts (**70**; Scheme 15) produce both 1,2-dihydroisoquinolines (**71**) and 1,2,3,4-tetrahydroisoquinolines (**72**). Reduction in protic solvents normally produces the tetrahydro adducts, in anhydrous pyridine or dimethylformamide the reduction generally stops at the 1,2-dihydroisoquinoline. Reaction of the enamine system of 1,2-dihydroisoquinolines with electrophiles has been used as a method for generation of 4-substituted isoquinolines.[10,37,38] With some 4-substituents reduction with borohydride does not proceed past the dihydro adduct even in protic solvents.[39] Certain 1-(2-nitrophenylmethyl)isoquinolinium salts, *e.g.* (**73**), have been found to undergo substituent cleavage on reduction with borohydride.[40]

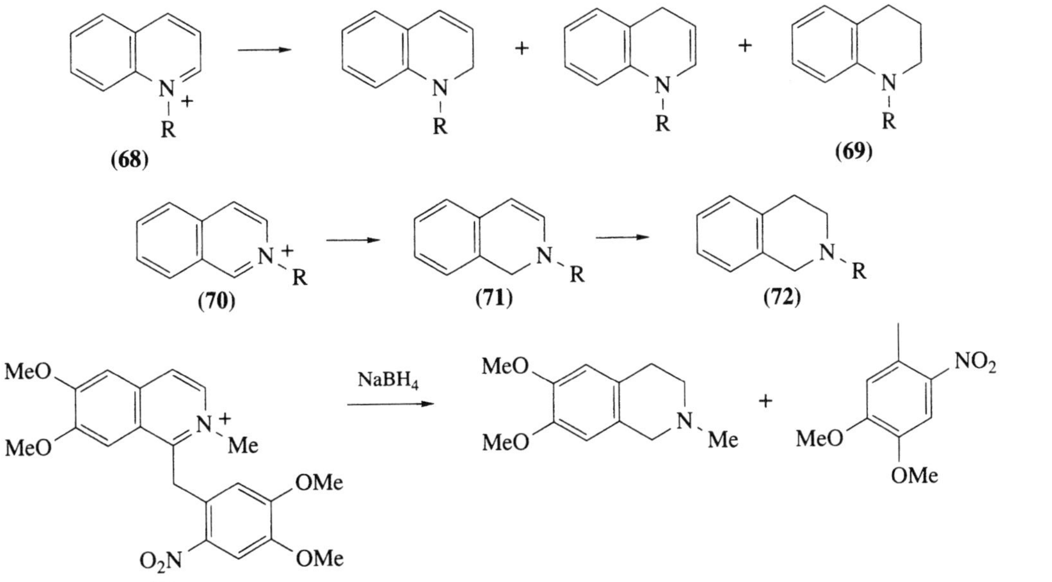

(68) **(69)** **(70)** **(71)** **(72)** **(73)** $NaBH_4$

Scheme 15

3.6.2.1.4 Pyridinium salts with aluminum hydrides

Aluminum hydrides react readily with pyridinium species producing both di- and tetra-hydropyridines. Prolonged heating with an excess of the hydride reagent produces 1,3-pentadiene derivatives (**74**; Scheme 16) from C—N bond cleavage as the major product. The preference is for initial 1,2-dihydropyridine formation, as illustrated by the only minor amounts of piperidines (**75**) produced.[41] *N* Oxides and alkoxypyridinium salts (**76**; Scheme 16) undergo deoxygenation prior to ring reduction, giving tetrahydropyridine (**77**) and piperidine (**78**) along with the parent pyridine. Ring-oxygenated analogs also undergo reduction, the 3-hydroxy-1-phenylpyridinium salt (**79**) produced the tetrahydropyridine (**80**) when treated with LAH in refluxing THF. The 3,2-dimer (**81**) was isolated when the 1-methyl analog was employed.[42]

Quinolinium[43] and isoquinolinium salts form predominantly 1,2-dihydro products with LAH reductions in aprotic solvents. With the latter the enamine system revealed is amenable to functionalization

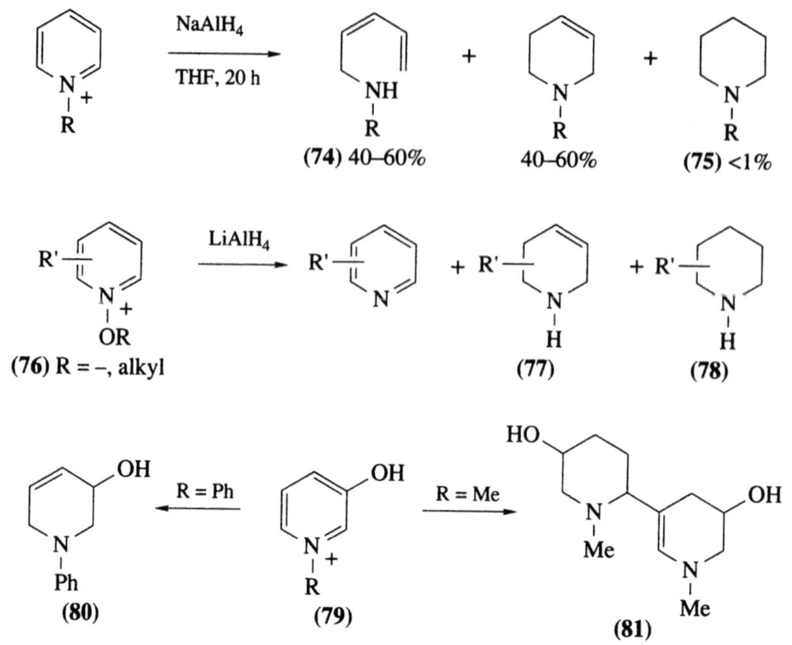

NaAlH$_4$, THF, 20 h

(74) 40–60%

40–60%

(75) <1%

LiAlH$_4$

(76) R = –, alkyl

(77)

(78)

R = Ph

R = Me

(80)

(79)

(81)

Scheme 16

with electrophiles and offers synthetic routes to alkaloids.[38] A variety of useful synthetic transformations have been performed on 1,2-dihydroisoquinolines.[44]

3.6.2.1.5 Other hydrides

Sodium hydride reduction of quinoline in HMPA leads to a 2:3 mixture of 1,2-dihydroquinoline (**82**) and 1,4-dihydroquinoline (**83**) isolated as the *N*-methoxycarbonyl derivatives.[45] *In situ* produced copper hydride reagents react with pyridinium species with high regioselectivity generating 1,4-dihydropyridine

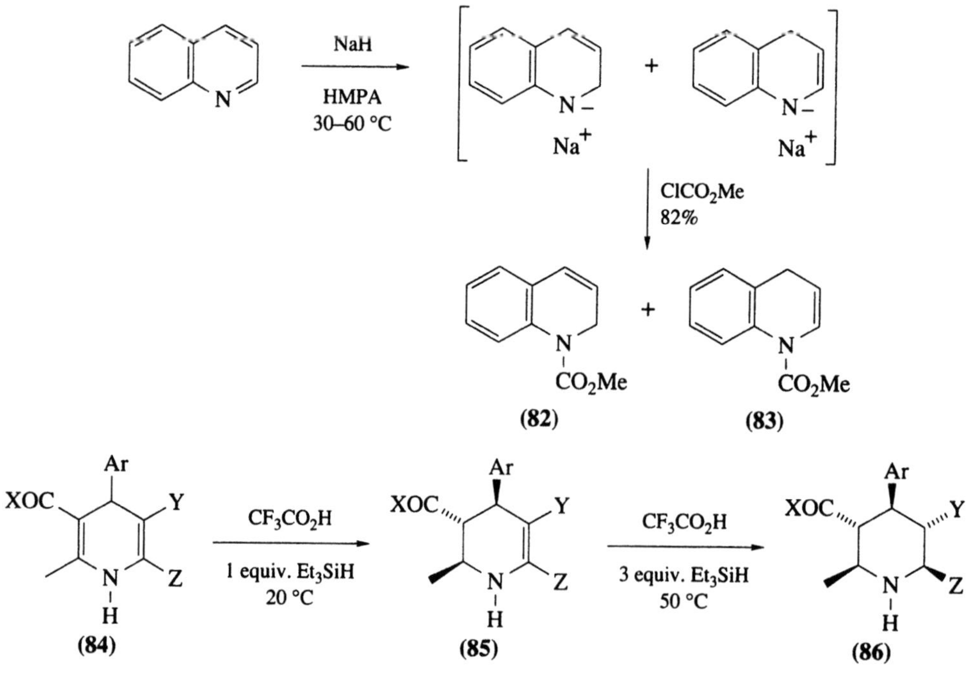

NaH, HMPA, 30–60 °C

$ClCO_2Me$, 82%

(82)

(83)

CF_3CO_2H, 1 equiv. Et_3SiH, 20 °C

CF_3CO_2H, 3 equiv. Et_3SiH, 50 °C

(84)

(85)

(86)

X = Me, OMe; Y = CO_2R, COMe, CN; Z = Me, CO_2Et

Scheme 17

(see **59** → **60**; Scheme 13; Section 3.6.2.1.3).[29] Magnesium and zinc hydrides produce dihydropyridinyl species, which subsequently reduce ketones, nitriles and heterocycles.[46] Triethylsilane in trifluoroacetic acid conveniently reduces Hantzsch type 1,4-dihydropyridines (**84**; Scheme 17) to all-*trans* isomers of 1,2,3,4-tetrahydropyridines (**85**) and piperidines (**86**). Reduction is controlled by temperature and the number of equivalents of silane employed. Good to moderate yields of products are obtained.[47]

3.6.2.2 Reduction of Pyridinium Salts with Dihydropyridines

Heteroaromatic cations undergo reduction when treated with 1,4-dihydronicotinamide.[48] An early study showed that the 10-methylacridinium ion (**87**) was rapidly reduced in a redox reaction to the 9,10-dihydro adduct by 1,4-dihydronicotinamides (**88**; Scheme 18).[49] A variety of systems including pyridines, isoquinolines, quinolines and phenanthridines have been studied using this and related procedures.[48] The selective reduction of pyridinium and quinolinium salts with 1-benzyl-1,2-dihydroisonicotinamide (**89**) has been achieved.[50] The selective conversion to the thermodynamically more stable 1,4-dihydro species (**90**; Scheme 18) is rationalized by the reversibility in the formation of the kinetic products (*i.e.* the 1,2-adducts) in the presence of pyridinium ions. In the pyridinium case 1,6-dihydro adducts were also observed in some cases. Reactivity in such systems is sometimes hindered due to hydration of the dihydropyridine system. This is particularly so in aqueous systems designed to replicate biological activity. Dihydroazines derived from isoquinolines and 3,5-disubstituted pyridines have been reported to overcome some of these difficulties.[51]

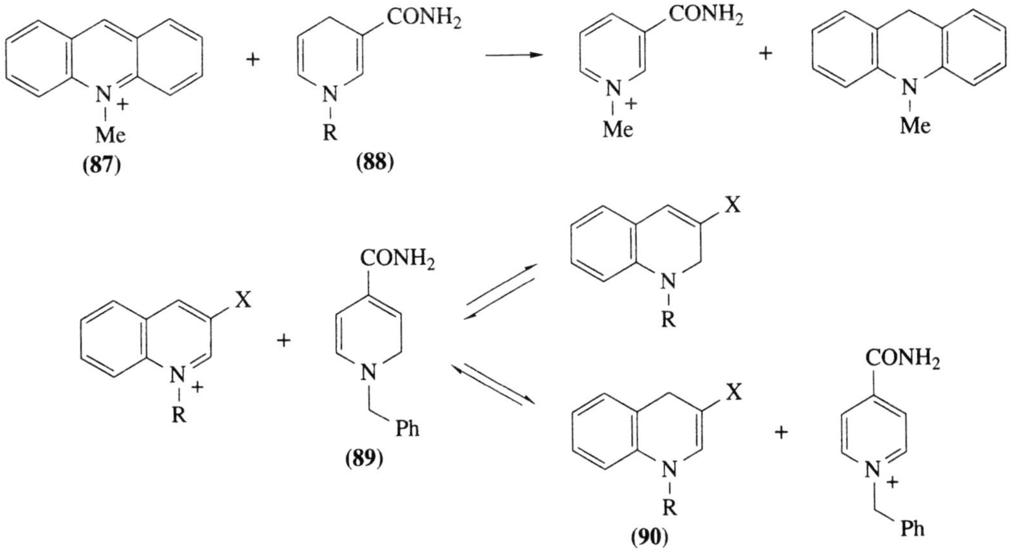

Scheme 18

3.6.2.3 Reduction with Sodium Dithionite

Pyridinium salts can be reduced with the mildly nucleophilic dithionite ion to produce mixtures of dihydropyridines. Much work has been carried out since the early observation that NADPH could be obtained from NADP with this reagent.[52] Reduction produces mainly the 1,4-dihydro isomer with minor amounts of the 1,2- or 1,6-dihydro adducts.[53] The reduction is believed to proceed *via* a stable and sometimes isolable sulfinate intermediate (**91**). In acid solution this decomposes with loss of sulfur dioxide to form the dihydropyridine (**92**; Scheme 19). Various substituted pyridinium species undergo this reaction and the isomer ratio obtained is dependent upon the nature of the solvent, temperature and pH.[22,53] With 1-methyl-4-carbamoylpyridinium bromide (**93**) and dithionite in aqueous sodium carbonate at 0–5 °C for 10 min the only isolated product was the 1,2,5,6-tetrahydropyridine (**94**) obtained in 16% yield, but the 3-carbamoyl salt gave the 1,4-dihydronicotinamide (**95**) in 90% yield, free of the 1,6-isomer.[53] The 3-chloropyridinium ion (**96**) gave a moderate yield of the 1,4-dihydropyridine (**97**).

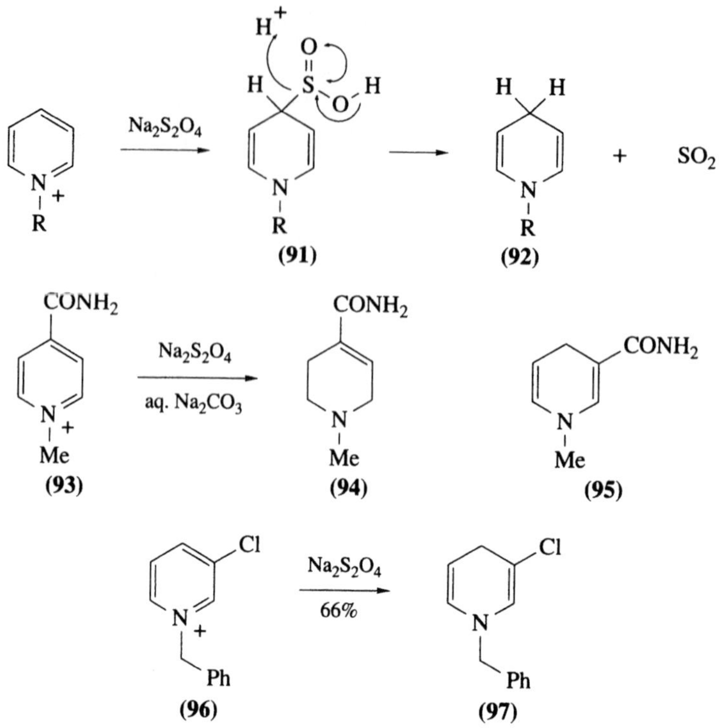

$Na_2S_2O_4$

(91) (92)

+ SO_2

$Na_2S_2O_4$

aq. Na_2CO_3

(93) (94) (95)

$Na_2S_2O_4$

66%

(96) (97)

Scheme 19

3.6.2.4 Reduction with Formic Acid

The reduction of pyridinium species with formic acid in the presence of its salts at elevated temperatures is often referred to as Lukes reduction and has been reviewed.[54] Mixtures of tetrahydropyridines (**98**) and piperidines (**99**; Scheme 20) commonly result. The isolated double bond is resistant to further reduction under these conditions. Reductive cleavage and subsequent cyclization have also been observed, producing ketones as by-products.[55]

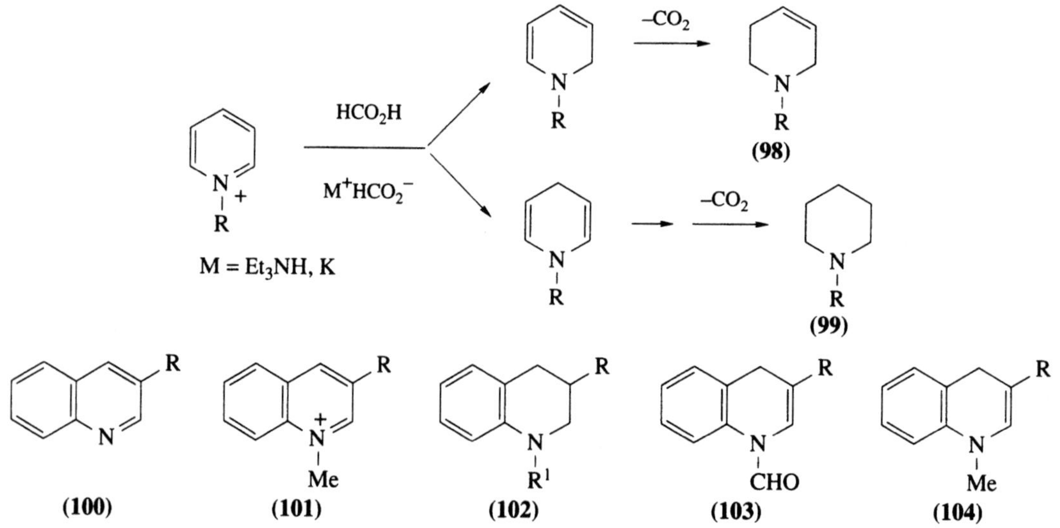

HCO_2H

$M^+HCO_2^-$

M = Et_3NH, K

$-CO_2$

(98)

(99)

(100) (101) (102) (103) (104)

R = OH, OMe, MeCO, $CONH_2$; R^1 = Me, CHO

Scheme 20

On reduction with triethylammonium formate, quinolines (**100**) and their 1-methylquinolinium salts (**101**) possessing an electron-donating group in the 3-position gave the *N*-formyl- and the *N*-methyl-1,2,3,4-tetrahydroquinolines (**102**) as the major products. With electron-withdrawing groups present the quinolines produced mixtures of (**102**; R^1 = CHO) and the 1,4-dihydroquinolines (**103**), (**101**) gave only (**104**) under the same conditions.[56] The 2-, 3- and 4-methylquinolines, hydroxymethylquinolines, carbaldehyde- and cyano-quinolines and their quaternary salts produce a variety of ring-reduced and substituent-modified products on reaction with triethylammonium formate.[57] Formates have also been used to reduce NADP mimics to 1,4-dihydropyridines.[58] Catalytic hydrogen transfer has also been achieved with ruthenium complexes in the presence of formic acid. 2-Methyltetrahydroquinoline is produced on heating 2-methylquinoline and the ruthenium–phosphine complex $RuCl_2(PPh_3)_2$ in the presence of formic acid.[59]

3.6.2.5 One-electron Reduction

Dissolving metal and electrochemical reduction are some of the oldest reactions known. Both occur with initial electron transfer to give, in the case of pyridines (**105**), the radical anion (**106**; Scheme 21), which can then react with a proton or with another radical leading to dimerization (**107**). Dimerization is therefore expected to be the major product in aprotic solvents. Reductions are more successful when labile hydrogens are available, generating dihydropyridines (**108**) and (**109**), tetrahydropyridine (**110**) and piperidine (**111**). Dissolving metal reductions have to a large extent been superseded by metal hydride reductions or catalytic hydrogenation. However, in certain applications, *i.e.* Birch reductions, they are still widely used. Electrochemical reductions proceed by similar mechanisms and are prone to the same dimer formation reactions mentioned above.

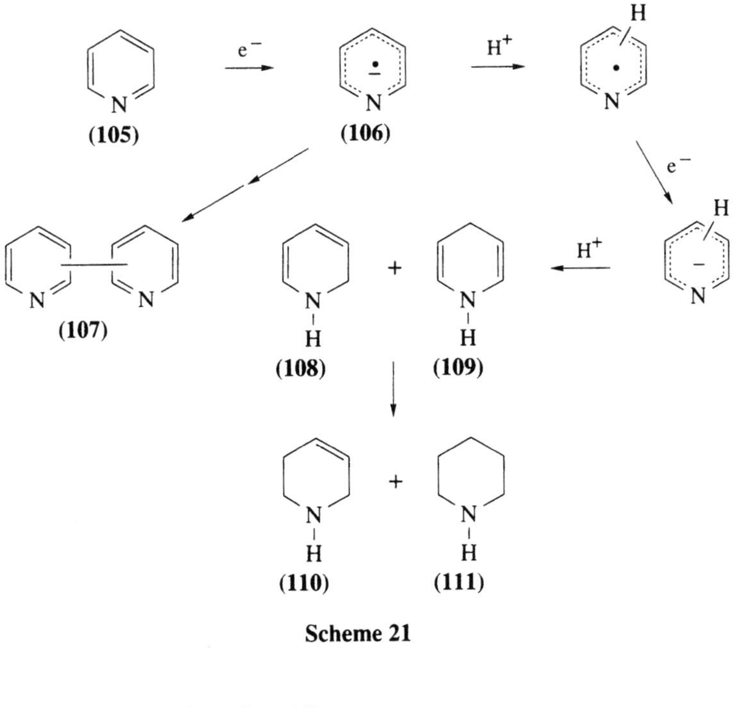

Scheme 21

3.6.2.5.1 Electrochemical reduction of pyridines

The electrochemical behavior of pyridines has been reviewed.[60–62] As usual, electron-withdrawing groups facilitate reduction, which occurs at the cathode and is most often carried out in aqueous sulfuric acid. However, polar aprotic solvents (*e.g.* MeCN, DMF) and liquid ammonia have been used for the polarographic reduction of pyridine.[60] Pyridine, itself, was first reduced to piperidine (**111**) at a lead cathode in 10% sulfuric acid nearly 100 years ago,[63] with dimeric piperidines also formed. Among partially reduced pyridines only 1,2,5,6-tetrahydropyridine (**110**) has been reported, the unpolarizable double bond being incapable of further reduction. Alkylpyridines similarly generate mixtures of alkylpiperidines and alkyl-1,2,5,6-tetrahydropyridines.[61] In some cases the tetrahydropyridine predominates, as do the all-*cis* products. In the presence of aldehydes and ketones the reductive alkylation of both pyridine and alkylpyridines to alkylpiperidines and tetrahydropyridines is achieved. Both carbon and nitrogen alkylation have been reported but 2-piperidinecarbinols (**112**; Scheme 22) and 4-piperidinecarbinols (**113**) predominate, along with minor amounts of the di- and tri-alkylated products. This reaction is related to the Emmert reaction. Dihydropyridines also undergo electroreduction, the 1,4-dihydropyridine (**115**) was further reduced to the tetrahydropyridine (**116**) in moderate yield at a mercury cathode in tetraethylammonium tetrafluoroborate solution.[64]

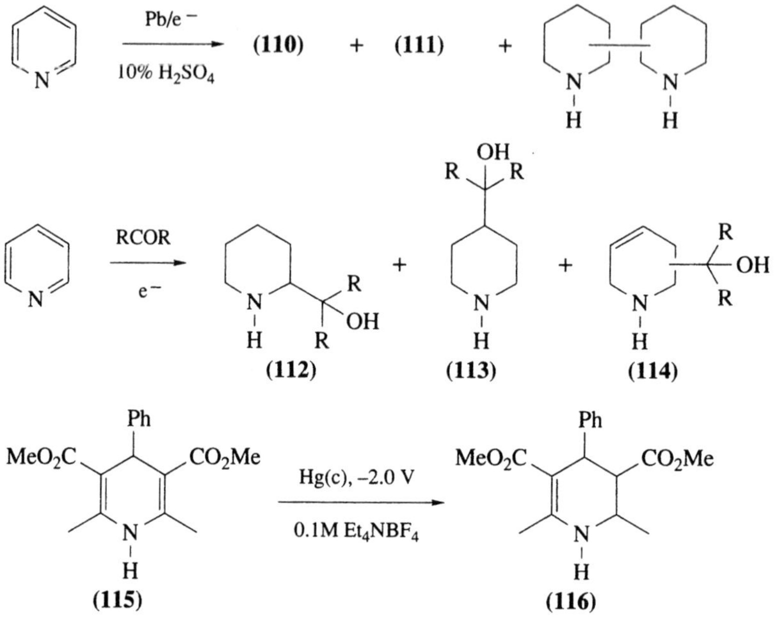

Pb/e⁻, 10% H_2SO_4 → (**110**) + (**111**) +

RCOR, e⁻ → (**112**) + (**113**) + (**114**)

Hg(c), −2.0 V, 0.1M Et_4NBF_4: (**115**) → (**116**)

Scheme 22

Electroreduction of pyridine derivatives bearing carboxylic acid derivatives (**117**; Scheme 23) leads to a variety of products from substituent transformation through ring reduction to ring cleavage.[61]

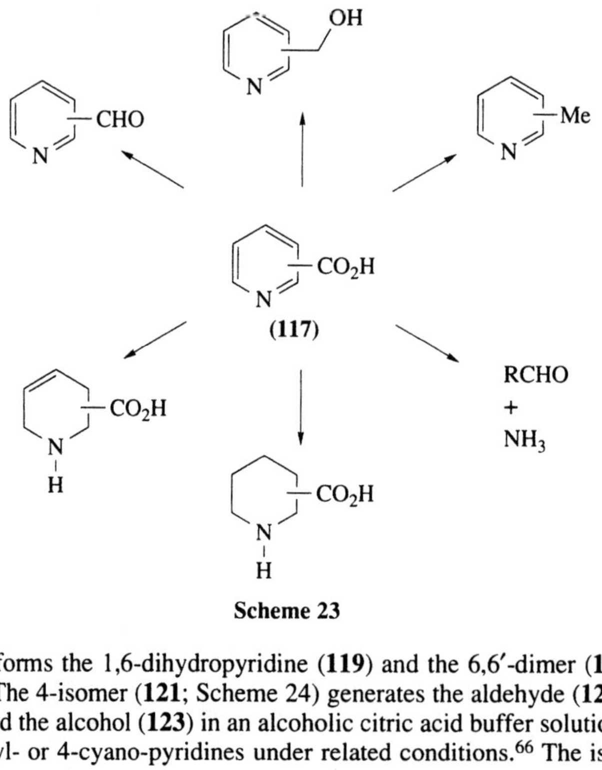

(**117**)

RCHO + NH_3

Scheme 23

Nicotinamide (**118**) forms the 1,6-dihydropyridine (**119**) and the 6,6′-dimer (**120**) upon polarography in aqueous solution.[65] The 4-isomer (**121**; Scheme 24) generates the aldehyde (**122**) when reduced in 0.8 M hydrochloric acid and the alcohol (**123**) in an alcoholic citric acid buffer solution. The thioamides produce the 4-aminomethyl- or 4-cyano-pyridines under related conditions.[66] The isomeric cyanopyridines

(**124**) can undergo a variety of reactions including pyridine formation, dimerization with and without nitrile cleavage as well as reduction of the cyano group to the aminomethylpyridines (**125**).[67]

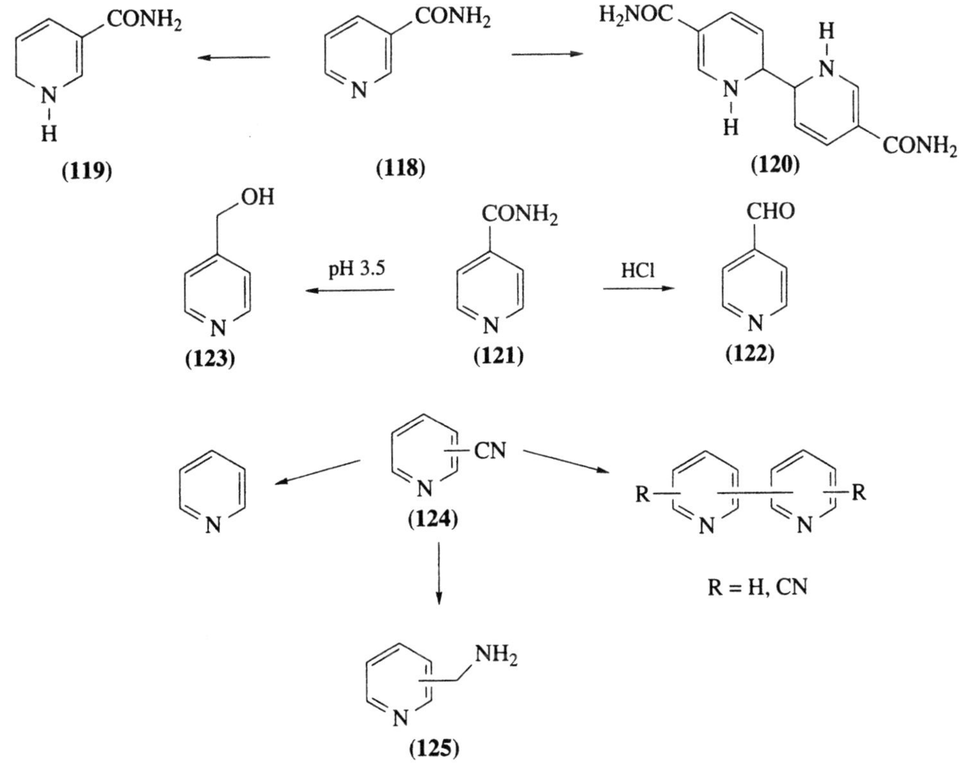

(**119**) (**118**) (**120**)

(**123**) pH 3.5 (**121**) HCl (**122**)

(**124**) R = H, CN

(**125**)

Scheme 24

Nitropyridines exhibit substituent reductions including through-nitrogen (azo/azoxy) coupling. Aminopyridines have been found to be unreducible when using voltammetry.[68] Hydroxypyridines and pyridones generally yield mixtures of piperidines and 1,2,5,6-tetrahydropyridines. Halogenated derivatives

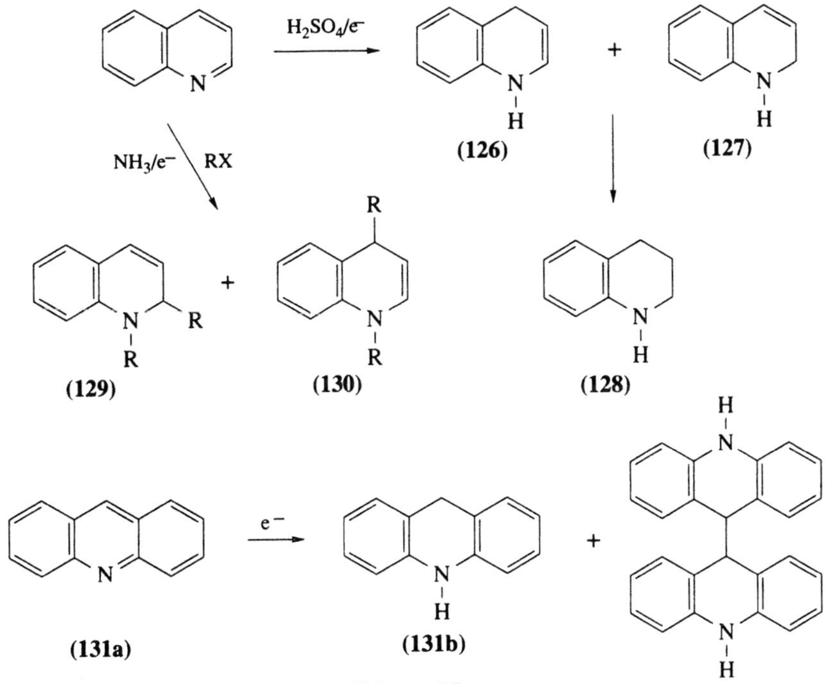

H_2SO_4/e^- (**126**) + (**127**)

NH_3/e^- RX

(**129**) + (**130**) (**128**)

(**131a**) e^- (**131b**) +

Scheme 25

tend to undergo dehalogenation on electroreduction.[69] *N*-Oxides undergo preferential deoxygenation in most cases, some overreduction or substituent reaction may occur.[70]

Quinoline and isoquinoline undergo reduction more readily than does pyridine. In aqueous sulfuric acid quinoline forms both dihydroquinolines (**126** and **127**) and tetrahydroquinolines (**128**). In non-aqueous media (DMF, liquid ammonia) the anion radical may be quenched by alkyl halides to generate both 1,2-dialkyldihydroquinolines (**129**) and 1,4-dialkyldihydroquinolines (**130**).[71] The product distribution is dependent on the nature of the alkylating agent. With isoquinoline preferential alkylation in the benzene ring has been found to occur in some cases.[72] Acridine (**131a**) is reduced to 9,10-dihydro derivative (**131b**) both in acidic and alkaline aqueous solution; 9,9′-dimerization also occurs.

3.6.2.5.2 Electrochemical reduction of pyridinium salts

In deference to their biological significance much work has been done on this class of compound as model studies to analyze or mimic the behavior of pyridine nucleotides and bipyridinium herbicides, pyridinium ions (**132**; Scheme 26) being more susceptible to reduction. Dimerization is a common fate of the neutral radical formed. The most thermodynamically stable product, the 4,4′-bipyridyl species predominate, and other dimers can sometimes rearrange to the 4,4′-bipyridyl isomer.[73] Radicals with mero-

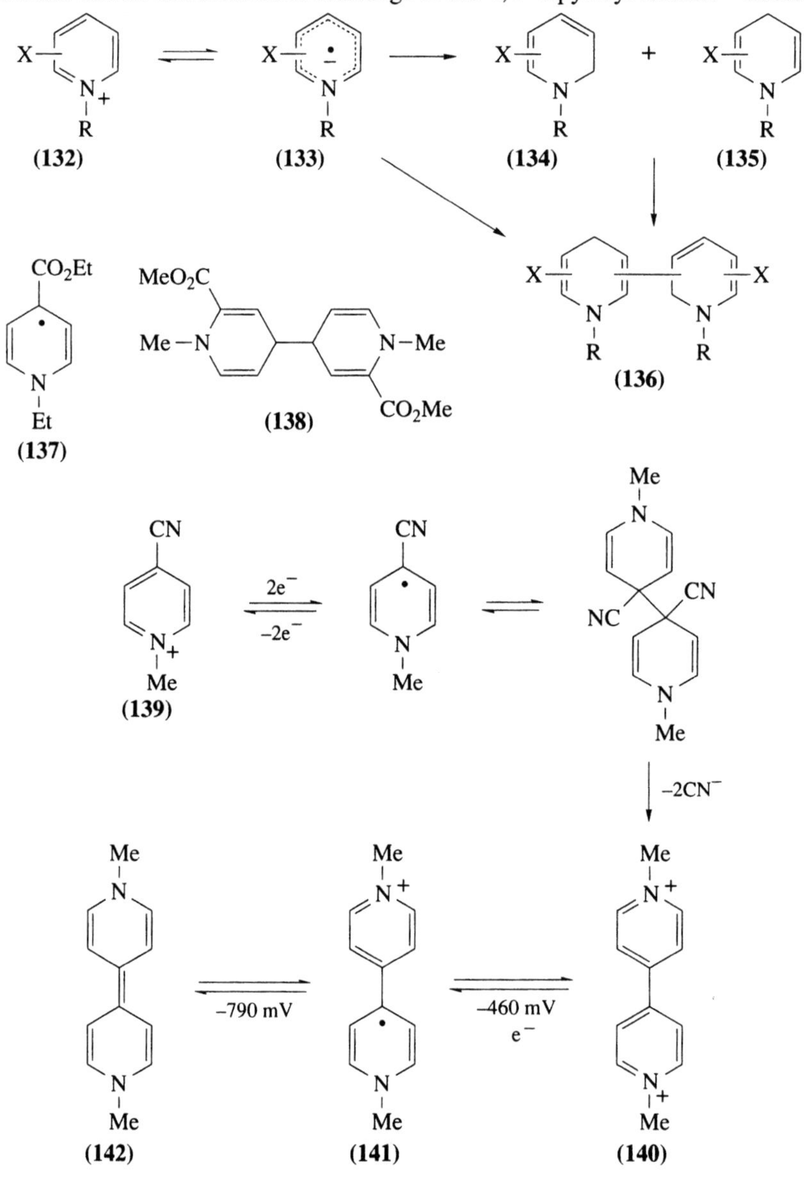

(132) **(133)** **(134)** **(135)** **(136)** **(137)** **(138)** **(139)** **(140)** **(141)** **(142)**

Scheme 26

stabilization by an α- or δ-substituent are surprisingly stable, the radical (**137**) even surviving distillation. Methyl *N*-methyl-2-pyridiniumcarboxylate, methyl *N*-methyl-3-pyridiniumcarboxylate, and methyl *N*-methyl-4-pyridiniumcarboxylate radicals are stabilized in the order 4 > 2 > 3. The 4-ester forms a stable radical, the 3-isomer dimerizes irreversibly, and the 2-isomer dimerizes reversibly to the bipyridine (**138**). Dimerization can occur with both ionic or radical species. Polarographic reduction of 4-cyano-1-methylpyridinium salts (**139**; Scheme 26) suggested that reversible radical formation was followed by irreversible anion generation and dimerization.[74] Loss of cyanide generates the potent herbicidal bipyridinium species (**140**).[75] The ion radical, *e.g.* (**141**), has been proposed as one of the reactive intermediates responsible for the biological effects exhibited by these molecules. The intermediacy of pyridinyl radicals has also been proposed in NADP/NADH(H) interconversion. Pyridinium species have been used as coatings or additives in attempts at the surface modification of electrodes and subsequently to increase selectivity on reduction.[76]

When the pyridinium ring bears reducible substituents reduction in acidic media favors substituent reductions. At high pH, dimers and dihydropyridine are the primary products. Thus, methyl *N*-methyl-2-pyridinecarboxylate (**143**) generates either the alcohol (**144**) or the dimer (**145**), depending on the pH of the electrolyte.[73,75] In the case of 4-nitroarylpyridinium salts (**146**; Scheme 27) preferential ring rather than substituent reduction occurred. Under the conditions employed (graphite or platinum electrodes in MeCN) no ring reduction was observed in the absence of the nitro group. The specific orientation of the nitro group was also found to be important, the wave potential being 50–160 mV more negative when *meta* than for the *ortho* and *para* isomers.[77] Cathodic reduction of *N*-amino- and *N*-acylamino-pyridinium ions (**147**) cleaves the N—N bond, forming the pyridine and the amine or amide species, respectively.[78]

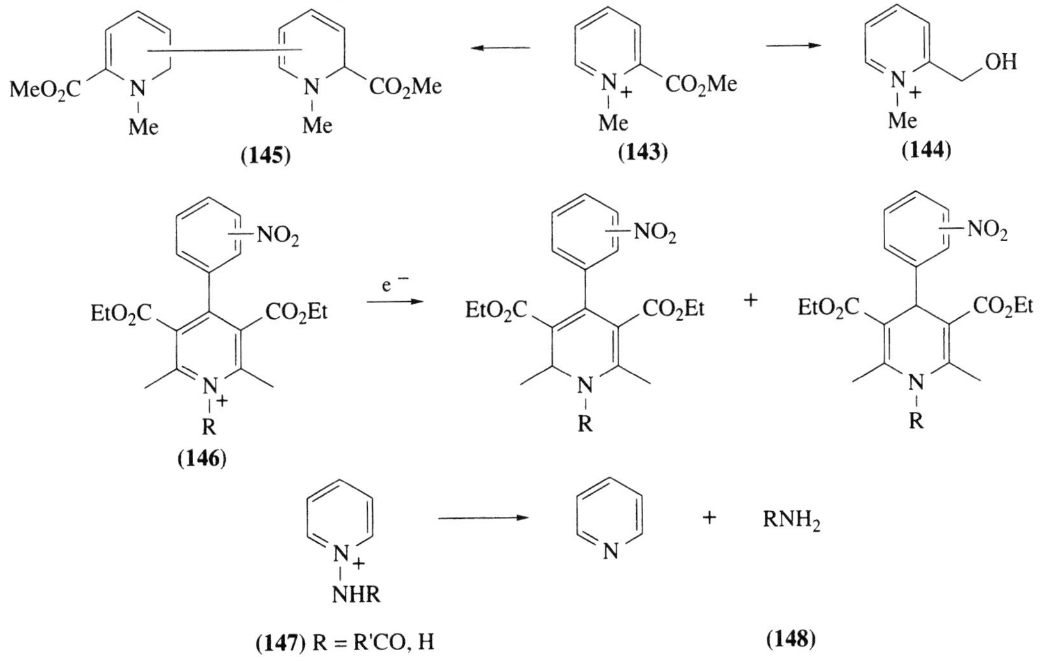

(**145**) (**143**) (**144**)

(**146**)

(**147**) R = R'CO, H (**148**)

Scheme 27

3.6.2.5.3 Dissolving metal reductions

The reduction of organic compounds with metals has been known for a long time. The relative reducing powers of the metals are in accordance with their electrode potentials and the alkali and some alkaline earth metals are capable of reducing a large number of functionalities, including the pyridine ring.[79] When carried out in the presence of a proton source dimerization of the initial radical ions formed is supressed. Thus, sodium in alcohols (Ladenburg reduction)[80] will reduce pyridines to mixtures of tetrahydropyridines (**110**) and piperidines (**111**) as with electroreduction (Section 3.6.2.5.1) and reaction with metal hydrides (Section 3.6.2.1.5). The position of initial attack will dictate which product is finally

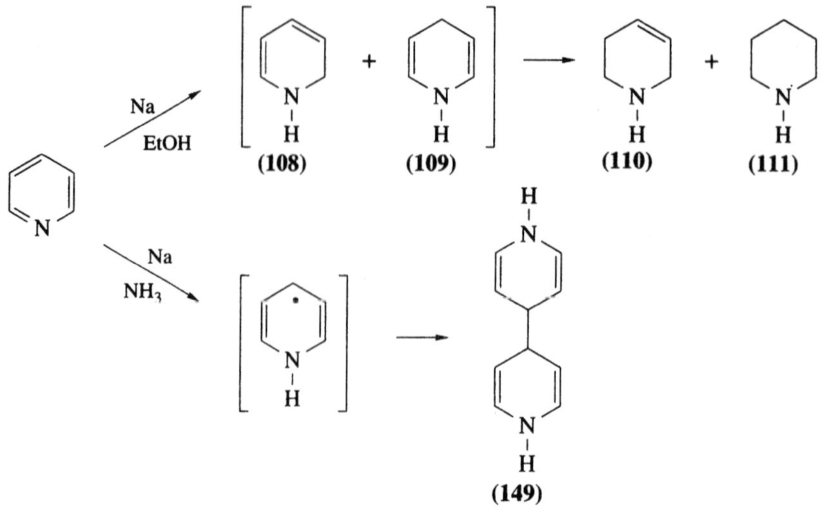

Scheme 28

formed. In aprotic solvents sodium produces tetrahydrobipyridyls (**149**) *via* dimerization of the radical ion formed. Alkylpyridines behave similarly. Ferles has studied reductions of this type with a variety of carbonyl and carboxylic acid derivatives.[54,81] Bipyridines are found as by-products in the Chichibabin reaction.[82]

Sodium in liquid ammonia and ethanol reduces quinoline to 1,2-dihydroquinoline (**150**; Scheme 29)[83] and isoquinoline to tetrahydroisoquinoline (**151**).[84] Zinc and formic acid can also be used. Using higher

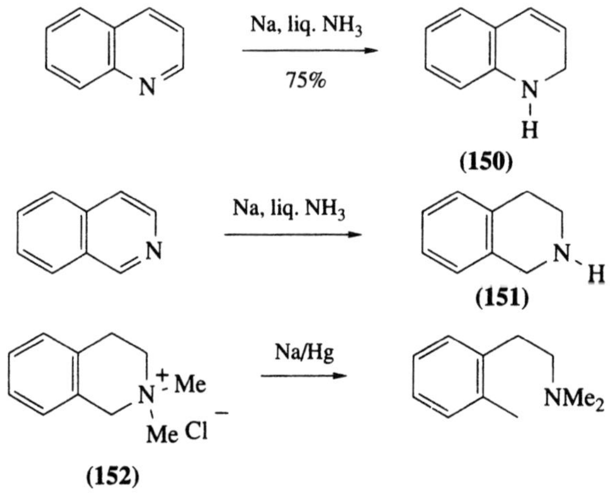

Scheme 29

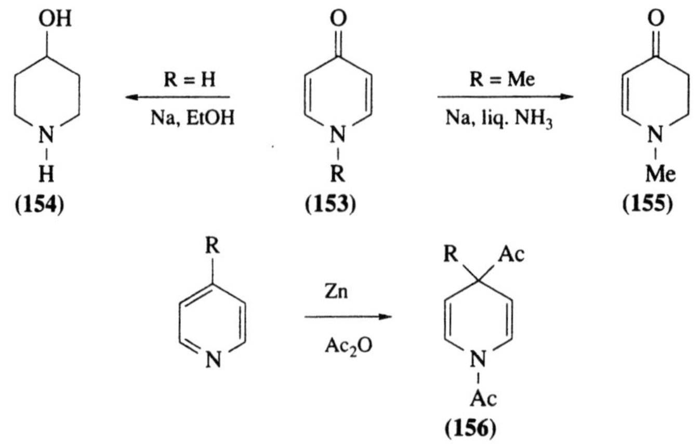

Scheme 30

boiling alcohols (*n*-butyl or *t*-butyl alcohols; Birch reduction) can lead to further reduction. Sodium amalgam reductions generally proceed only on pyridinium ions. However, a variant of the Hofmann degradation, the Emde degradation, will cause reductive cleavage of quaternary nitrogen species (*e.g.* **152**).[85]

4-Pyridone (**153**) is converted to 4-hydroxypiperidine (**154**; Scheme 30) by sodium in ethanol. However, with sodium or lithium in liquid ammonia *N*-methyl-4-pyridone (**153**; R = Me) produced the dihydropyridone (**155**) in moderate yield.[86] Higher conversions were achieved with sodium tetraalkoxyaluminates. Pyridines are generally not reducible by zinc and carboxylic or hydrochloric acids. However, zinc and acetic anhydride converts 4-substituted pyridines to their 1,4-diacyldihydropyridine derivatives (**156**).[87] Tetrahydrobipyridyls have also been found in this reaction.[88] Pyridinium ions are reduced more easily and give mixtures of tetrahydropyridines and piperidine. Substitutents may also be reduced. Zinc and hydrochloric acid along with tin and hydrochloric acid effect complete reduction of the pyridinium ring.

3.6.3 HYDROGENATION

Catalytic hydrogenation is the most frequently employed method of saturating the pyridine ring. Complete reduction to the piperidine normally occurs, the intermediates formed being reactive under the conditions employed. Heterogeneous catalysts continue to be the most popular method for a variety of uses ranging from the synthesis of intermediates to the denitrogenation of fossil fuels. Extensive reviews on the reduction of pyridines have been published.[79,89–93]

For low pressure hydrogenation both supported and unsupported platinium (as its oxide) are effective, however, rhodium is the more effective under mild conditions. Nickel is used at higher pressures and temperatures and *N*-alkylation may occur when alcohols are present. Palladium and ruthenium have likewise been used for a variety of pyridine derivatives. Reductions of pyridine compounds are often carried out in acidic media, as the strongly coordinating amines formed on hydrogenation frequently cause catalyst inhibition. At elevated temperatures hydrogenolysis of the carbon–nitrogen bond can also occur. Generally, the rate of this cleavage is only significant above 120 °C but will obviously depend on the nature of the substituents. Alkylamines, ammonia and alkanes are generally produced. Pyridine and its derivatives may also be reduced by Raney nickel with 2-propanol as the hydrogen source *via* its dehydrogenation to acetone and catalytic hydrogen transfer.[94] Similarly, nickel–aluminium alloy in potassium hydroxide reduces pyridines (*e.g.* **157**; Scheme 31), quinolines and isoquinolines in moderate to high yields,[95] but much higher metal/substrate ratios are required. The accelerating effects of 2-substituents has been observed both in traditional high pressure hydrogenation with nickel or ruthenium and in catalytic transfer hydrogenation.[95] 2-Methylpyridine (**157**) is hydrogenated at a faster rate than pyridine itself. Indeed, 2-methylpyridine can be hydrogenated in the presence of 3- and 4-methylpyridine, this is most likely due to a reduced propensity toward catalyst poisoning due to steric effects, and with 2,6-di-

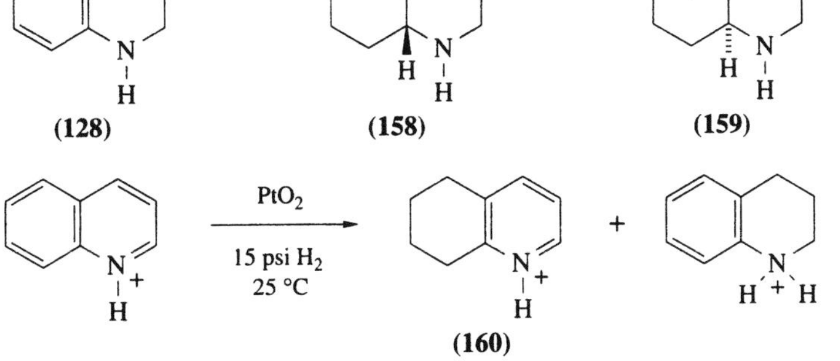

(**157**) —i, ii; 77%→ Cl^-

i, aq. KOH, Ni–Al, 25 °C, 4 h; ii, HCl

(**128**) (**158**) (**159**)

—PtO_2; 15 psi H_2; 25 °C→ (**160**) +

Scheme 31

methylpyridine an additional rate enhancement is observed. However, nonplanar 2-phenylpyridine is hydrogenated more slowly due to the impedance of nickel adsorption to the catalyst surface. In the case of quinoline, hydrogenation with nickel occurs rapidly and preferentially in the pyridine ring, generating 1,2,3,4-Tetrahydroquinoline (**128**). Extended hydrogenation times or more forcing conditions generate *cis*-decahydroquinolines (**158**) and *trans*-decahydroquinolines (**159**). 1,2,3,4-Tetrahydroquinolines can be generated exclusively on hydrogenation with a variety of catalysts.[96] Selective reduction of the benzene ring, however, is more difficult, generating mixtures of tetrahydroquinolines. Some selectivity can be achieved by reduction of quinolines in strong acid over platinum.[97] The major product is the 5,6,7,8-tetrahydroquinoline derivative (**160**; Scheme 31). This effect is more pronounced in the isoquinoline series.

Isoquinoline itself hydrogenates less readily than does quinoline. Studies of the catalytic hydrogenation of isoquinoline over Raney nickel have shown that the reaction proceeds consecutively *via* 1,2,3,4-tetrahydroisoquinoline (**161**), 5,6,7,8-tetrahydroisoquinoline (**162**) and *cis*-decahydroisoquinoline (**163**) to *trans*-decahydroisoquinoline (**164**).[98] Acridine undergoes similar hydrogenative isomerizations with supported Pd.[99] Both *cis*- and *trans*-1,2,3,4,4a,9,9a,10-octahydroacridines were formed at 150 °C and only three (**165, 166** and **167**) of the six possible stereoisomers of perhydroacridine were found, with (**165**) and (**166**) predominant. Alumina was preferred to carbon as a support.[100]

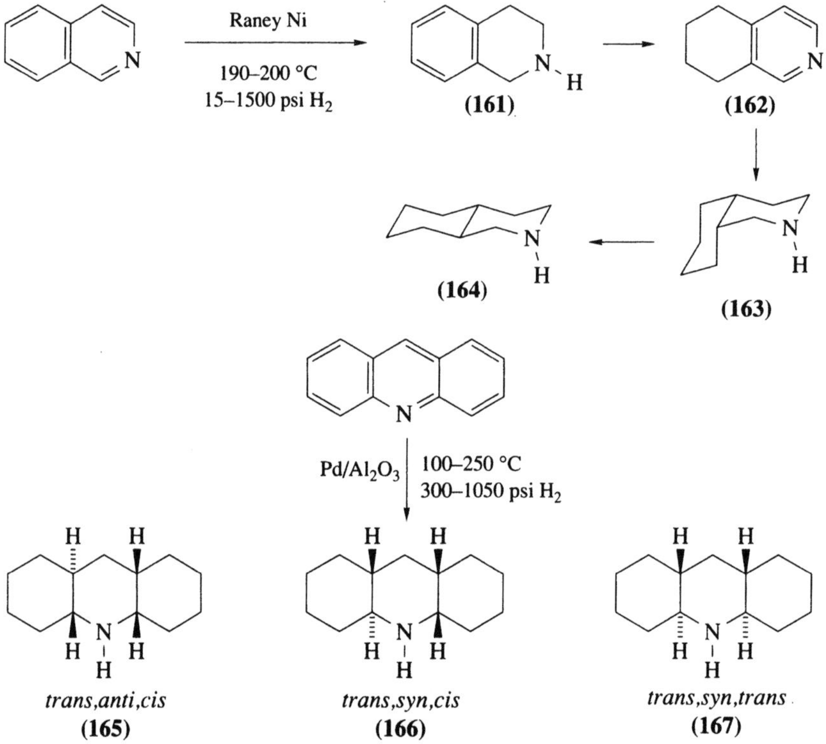

Raney Ni, 190–200 °C, 15–1500 psi H_2 → (**161**) → (**162**) → (**163**) → (**164**)

Pd/Al_2O_3, 100–250 °C, 300–1050 psi H_2

trans,anti,cis (**165**) *trans,syn,cis* (**166**) *trans,syn,trans* (**167**)

Scheme 32

The hydrogenation of amino- and hydroxy-pyridines is controlled to a large extent by the tautomeric forms in which they prefentially exist. 2-Aminopyridines (Scheme 33) are reduced to the amidine (**168**) by Pt, Rh or Ru, which is resistant to further hydrogenation but will undergo hydrogenolysis to piperidine (**169**) and the amine (**170**). 3-Aminopiperidine (**171**) forms readily when the parent heterocycle is hydrogenated over PtO_2 in hydrochloric acid. However, 4-aminopyridine requires much higher pressures for reduction to 4-aminopiperidine (**172**). 2- and 4-pyridones (Scheme 33) generate 2-piperidone (**173**) and 4-hydroxypiperidine (**174**), respectively, when hydrogenated in acetic acid. 2-Pyridone is hydrogenated more readily using Pt.[90] Equal rates of hydrogen uptake were observed, however, when Pd on carbon was used. Varying degrees of hydrogenolysis are observed on reduction of 3-hydroxypyridine to 3-hydroxypiperidine.

Pyridines containing electron-withdrawing groups may undergo only partial ring reduction. Functional groups which readily undergo hydrogenation, *e.g.* nitro and cyano, will be reduced prior to the heteroaromatic ring, whereas those groups strongly resistant to reduction, *e.g.* carboxylic acids and esters, will survive ring reduction unchanged. In addition, when a product can be a vinylogous amide (*e.g.* **175**; Scheme

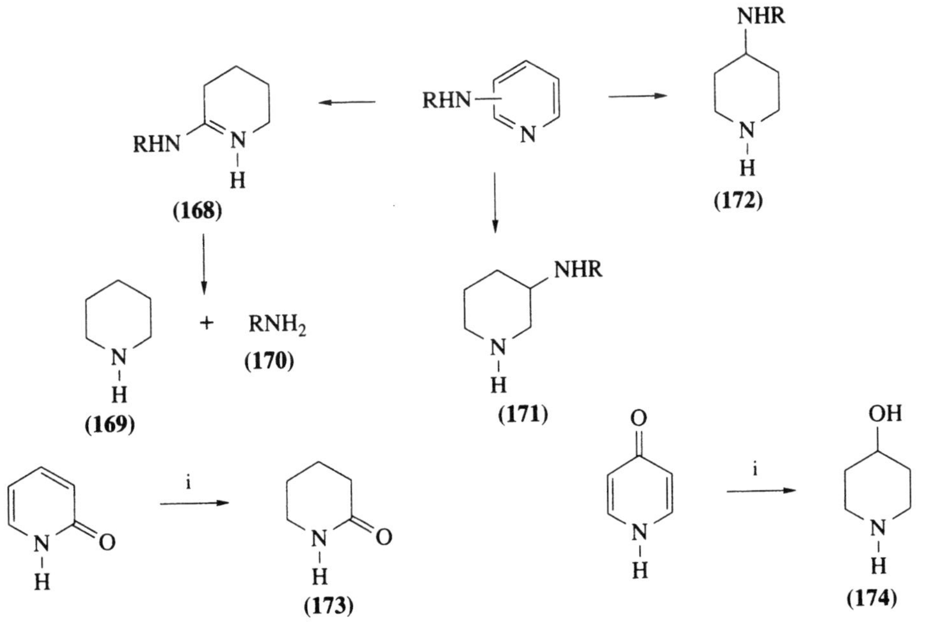

(168) (169) + RNH_2 (170) (171) (172)

(173) (174)

i, PtO_2, 20 °C, 15 psi H_2

Scheme 33

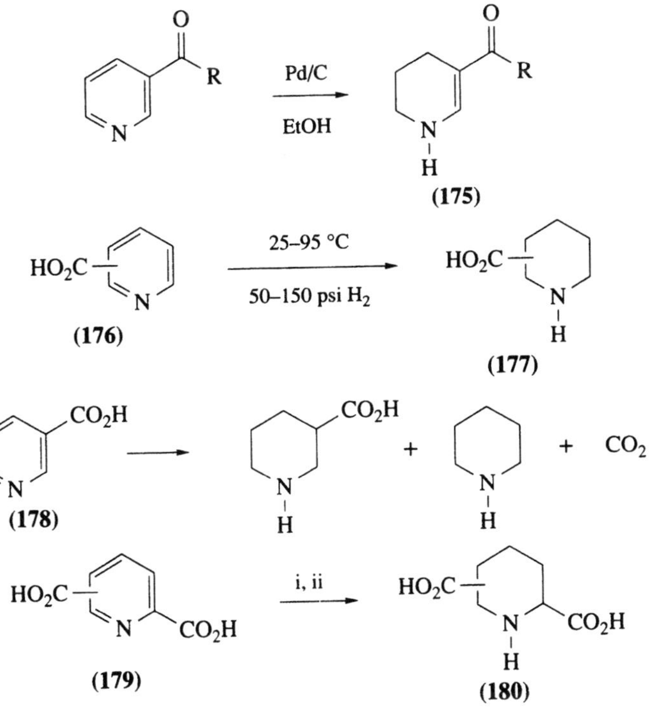

Pd/C, EtOH (175)

(176) 25–95 °C, 50–150 psi H_2 (177)

(178) + + CO_2

(179) i, ii (180)

i, HCl; ii, Pt, 20 °C, 15 psi H_2

Scheme 34

34), it generally resists further hydrogenation. The pyridine monocarboxylic acids (**176**) are reduced readily to their piperidine derivatives (**177**) over Rh, Pd or Ru under increasingly forcing conditions. Platinum may also be used for this conversion. In the case of nicotinic acid (**178**), piperidine, the product of decarboxylation, is also formed. This can be prevented by the addition of an equivalent of inorganic base when using Pt or Rh. Hydrogenation of pyridinedicarboxylic acids (**179**) over platinum in hydrochloric acid gives the *cis*-piperidinedicarboxylic acids (**180**).[101] Pyridinium salts are reduced more readily than pyridines, the tertiary amines formed do not inhibit catalyst activity. Partial hydrogenation as in the free pyridines may also occur. In the case of pyridine *N*-oxides deoxygenation generally occurs first.

Hydrogenation with homogeneous catalysis has received increasing attention over the last few decades.[93] Hexarhodium hexadecacarbonyl (**181**; Scheme 35) under water gas shift conditions forms 1,2,3,4-tetrahydroquinoline (**128**) or *N*-formyl-1,2,3,4-tetrahydroisoquinoline (**182**) with the parent heteroaromatic.[102] When hydrogen is substituted for carbon monoxide, 4-methylquinoline is reduced to 4-methyl-5,6,7,8-tetrahydroquinoline (**183**) exclusively.[102] Isoquinoline behaves similarly generating *N*-formyl-1,2,3,4-tetrahydroisoquinoline (**184**). Similar reductions under water gas shift or synthesis gas conditions using transition metal carbonyls derived from Mn, Co and Fe, have been recorded.[103,104] Promotion by phase transfer agents is observed in some cases.

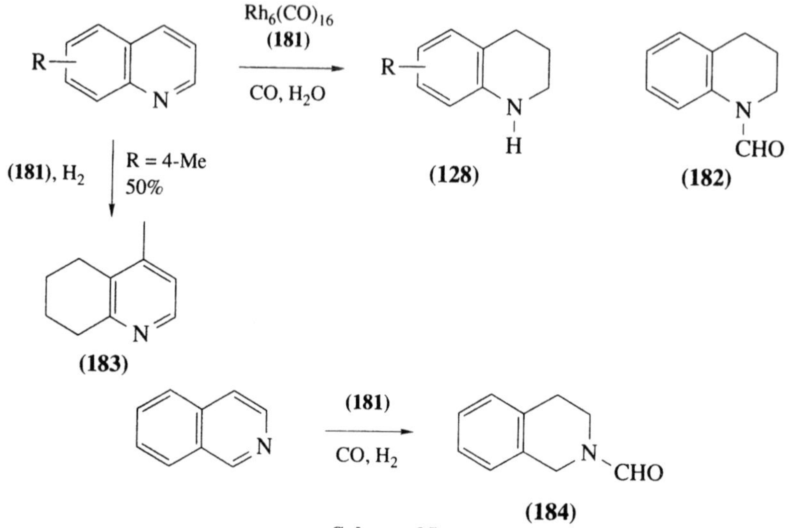

$Rh_6(CO)_{16}$ (**181**); CO, H_2O → (**128**), (**182**)

(**181**), H_2; R = 4-Me; 50% → (**183**)

(**181**); CO, H_2 → (**184**)

Scheme 35

Styrene–divinylbenzene copolymer supported chlorotris(triphenylphosphine)rhodium was also found effective for the regioselective reduction of quinoline and acridines.[105] The supported catalyst was more active than its homogeneous analog. Rhodium supported on soluble organic polymers as macroligands, showed sufficient activity so as to allow the hydrogenation of pyridine to piperidine under mild conditions.[106] The partial reduction of *N*-alkyl-3-carbamoylpyridinium salts (**185**) proceeded selectively when bis(dimethylglyoximate)chloro(pyridine)cobalt (**186**; $CoCl(DMG)_2Py$) was employed as catalyst. When sodium hydrogencarbonate was present no overreduction was observed. An increased preference for 1,4-dihydropyridines (**187**) over 1,2-dihydropyridines (**188**) and 1,6-dihydropyridines (**189**) was observed.[107]

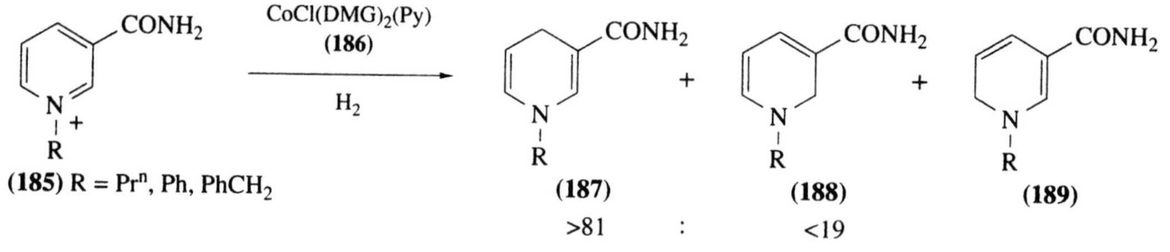

(**185**) R = Pr^n, Ph, $PhCH_2$ —$CoCl(DMG)_2(Py)$ (**186**), H_2→ (**187**) + (**188**) + (**189**)

(**187**) >81 : (**188**) <19

Scheme 36

3.6.4 REFERENCES

1. I. Ya. Postovskii, O. N. Chupakhin and A. I. Matern, *Khim. Geterotsikl. Soedin.*, 1984, 1299 and references cited therein (*Chem. Abstr.*, 1985, **102**, 5201a).
2. A. Hajós, 'Complex Metal Hydrides and Related Reducing Agents in Organic Synthesis', Elsevier, New York, 1979.
3. (a) R. E. Lyle and P. S. Anderson, *Adv. Heterocycl. Chem.*, 1966, **6**, 45; (b) J. G. Keay, *Adv. Heterocycl. Chem.*, 1986, **39**, 1.
4. S.-I. Yamada, M. Kuramoto and Y. Kikugawa, *Tetrahedron Lett.*, 1969, 3101.
5. Y. Kikugawa, M. Kuramoto, I. Saito and S.-I. Yamada, *Chem. Pharm. Bull.*, 1973, **21**, 1914, 1927.
6. E. Booker and U. Eisner, *J. Chem. Soc., Perkin Trans. 1*, 1975, 929.

7. Y. Kikugawa, S. Ikegami and S.-I. Yamada, *Chem. Parm. Bull.*, 1969, **17**, 98.
8. G. W. Gribble and P. W. Heald, *Synthesis*, 1975, 650.
9. K. V. B. Rao and D. Jackman, *J. Heterocycl. Chem.*, 1973, **10**, 213.
10. D. E. Minter and M. A. Re, *J. Org. Chem.*, 1988, **53**, 2653.
11. P. T. Lansbury and J. O. Peterson, *J. Am. Chem. Soc.*, 1962, **84**, 1756.
12. R. F. Francis, C. D. Crews and B. S. Scott, *J. Org. Chem.*, 1978, **43**, 3227.
13. J. Kuthan, J. Procházková and E. Janečková, *Collect. Czech. Chem. Commun.*, 1968, **33**, 3558.
14. M. Holík and M. Ferles, *Collect. Czech. Chem. Commun.*, 1967, **32**, 2288.
15. F. Binns, S. M. Roberts and H. Suschitzky, *J. Chem. Soc. C*, 1970, 1375.
16. D. E. Minter and P. L. Stotter, *J. Org. Chem.*, 1981, **46**, 3965; B. K. Blackburn, J. F. Frysinger and D. E. Minter, *Tetrahedron Lett.*, 1984, **25**, 4913.
17. D. J. Brooks, D. S. Dowell, D. E. Minter and M. C. Villarreal, *J. Org. Chem.*, 1984, **49**, 130.
18. A. I. Meyers and J. D. Brown, *Tetrahedron Lett.*, 1988, **29**, 5617.
19. J. W. Bunting and J. L. Bolton, *Tetrahedron*, 1986, **42**, 1007.
20. L. L. Miller and J. R. Valentine, *J. Am. Chem. Soc.*, 1988, **110**, 3982; D. D. Tanner and A. Kharrat, *J. Am. Chem. Soc.*, 1988, **110**, 2968.
21. U. Eisner and J. Kuthan, *Chem. Rev.*, 1972, **72**, 1.
22. J. Kuthan and A. Kurfürst, *Ind. Eng. Chem., Prod. Res. Dev.*, 1982, **21**, 191.
23. D. M. Stout and A. I. Meyers, *Chem. Rev.*, 1982, **82**, 223.
24. R. E. Lyle and C. B. Boyce, *J. Org. Chem.*, 1974, **39**, 3708.
25. R. J. Sundberg, G. Hamilton and C. Trindle, *J. Org. Chem.*, 1986, **51**, 3672.
26. F. Liberatore, V. Carelli and M. Cardellini, *Tetrahedron Lett.*, 1968, 4735.
27. E. M. Fry, *J. Org. Chem.*, 1964, **29**, 1647.
28. F. W. Fowler, *J. Org. Chem.*, 1972, **37**, 1321.
29. D. L. Comins and A. H. Abdullah, *J. Org. Chem.*, 1984, **49**, 3392.
30. F. Liberatore, A. Casini, V. Carelli, A. Arnone and R. Mondelli, *Tetrahedron Lett.*, 1971, 3829.
31. P. Zarin, E. Liepins, E. S. Lavrinovich and A. K. Arens, *Khim. Geterotsikl. Soedin.*, 1974, 115 (*Chem. Abstr.*, 1974, **80**, 95 804d).
32. A. Casini, B. Di Rienzo, F. M. Moracci, S. Tortorella, F. Liberatore and A. Arnone, *Tetrahedron Lett.*, 1978, 2139.
33. J. P. Kutney, R. Greenhouse and V. E. Ridaura, *J. Am. Chem. Soc.*, 1974, **96**, 7364.
34. J. Bosch, D. Mauleón, F. Boncompte and R. Granados, *J. Heterocycl. Chem.*, 1981, **18**, 263; M. S. Allen, A. J. Gaskell and J. A. Joule, *J. Chem. Soc. C*, 1971, 736; M.-L. Bennasar, R. Lavilla, M. Alvarez and J. Bosch, *Heterocycles*, 1988, **27**, 789.
35. H.-P. Husson, *Bull. Soc. Chim. Belg.*, 1982, **91**, 985.
36. E. E. Knaus and K. Redda, *J. Heterocycl. Chem.*, 1976, **13**, 1237.
37. A. R. Katritzky, M. H. Ibrahim, J.-Y. Valnot and M. P. Sammes, *J. Chem. Res. (S)*, 1981, 70.
38. C. K. Bradsher, *Chem. Heterocycl. Compd.*, 1981, **38** (1), 381.
39. G. Thuillier, B. Marçot and P. Rumpf, *Bull. Soc. Chim. Fr.*, 1969, 2045.
40. J. L. Neumeyer, M. McCarthy, K. K. Weinhardt and P. L. Levins, *J. Org. Chem.*, 1968, **33**, 2890.
41. M. Ferles, O. Kocián and A. Šilhánková, *Collect. Czech. Chem. Commun.*, 1974, **39**, 3532.
42. W. R. Ashcroft and J. A. Joule, *Heterocycles*, 1981, **16**, 1883.
43. G. Jones, *Chem. Heterocycl. Compd.*, 1977, **32** (1), 29.
44. J. Knabe, *Adv. Heterocycl. Chem.*, 1986, **40**, 105.
45. M. Natsume, S. Kumadaki, Y. Kanda and K. Kiuchi, *Tetrahedron Lett.*, 1973, 2335.
46. A. J. de Koning, P. H. M. Budzelaar, B. G. K. van Aarssen, J. Boersma and G. J. M. van der Kerk, *J. Organomet. Chem.*, 1981, **217**, C1.
47. U. Rosentreter, *Synthesis*, 1985, 210.
48. J. W. Bunting and J. L. Bolton, *Tetrahedron*, 1986, **42**, 1007, and references cited therein; A. Ohno and K. Ushio, *Coenzymes Cofactors*, 1987, **2** (B), 275.
49. D. J. Creighton, J. Hajdu, G. Mooser and D. S. Sigman, *J. Am. Chem. Soc.*, 1973, **95**, 6855.
50. A. Nuvole, G. Paglietti, P. Sanna and R. M. Acheson, *J. Chem. Res. (M)*, 1984, 3245.
51. H. Awano and W. Tagaki, *J. Chem. Soc., Chem. Commun.*, 1985, 994.
52. O. Warburg, W. Christian and A. Griese, *Biochem. Z.*, 1935, **282**, 157.
53. G. Paglietti, P. Sanna, A. Nuvole, F. Soccolini and R. M. Acheson, *J. Chem. Res. (M)*, 1983, 2326; R. M. Acheson and G. Paglietti, *J. Chem. Soc., Perkin Trans. 1*, 1976, 45.
54. M. Ferles and J. Pliml, *Adv. Heterocycl. Chem.*, 1970, **12**, 71.
55. R. Lukes and J. Jizba, *Collect. Czech. Chem. Commun.*, 1954, **19**, 941.
56. M. Ferles and O. Kocián, *Collect. Czech. Chem. Commun.*, 1979, **44**, 3141.
57. M. Ferles and O. Kocián, *Collect. Czech. Chem. Commun.*, 1979, **44**, 2238; 1981, **46**, 1518.
58. A. Ohno, S. Ushida and S. Oka, *Bull. Chem. Soc. Jpn.*, 1984, **57**, 506.
59. Y. Watanabe, T. Ohta, Y. Tsuji, T. Hiyoshi and Y. Tsuji, *Bull. Chem. Soc. Jpn.*, 1984, **57**, 2440.
60. H. Lund and I. Tabaković, *Adv. Heterocycl. Chem.*, 1984, **36**, 293; H. Lund, *Adv. Heterocycl. Chem.*, 1970, **12**, 262.
61. J. E. Toomey, Jr., *Adv. Heterocycl. Chem.*, 1984, **37**, 167.
62. R. F. Nelson, *Tech. Chem. (N.Y.)*, 1975, **5** part 2, 269.
63. F. B. Ahrens, *Z. Elekrochem. Angew. Phys. Chem.*, 1896, **2**, 577.
64. E. Wehinger, S . Kazda and A. Knorr (Bayer A.-G.) *Ger. Pat.*, 3 239 273 (1984) (*Chem. Abstr.*, 1984, **101**, 72 619k).
65. C. O. Schmakel, K. S. V. Santhanam and P. J. Elvinga, *J. Electrochem. Soc.*, 1974, **345**, 121.
66. H. Lund, *Acta Chem. Scand.*, 1963, **17**, 2325.
67. J. E. Toomey, Jr., G. A. Chaney and M. Wilcox, *Stud. Org. Chem. (Amsterdam)*, 1987, **30**, 245.
68. A. G. Pozdeeva and E. G. Novikov, *Zh. Prikl. Khim.*, 1969, **42**, 2626 (*Chem. Abstr.*, 1970, **72**, 74 009t).

69. Ref. 61, p. 193.
70. D. Cantin, J. M. Richard, J. Alary and D. Serve, *Electrochim. Acta*, 1988, **33**, 1047.
71. W. H. Smith and A. J. Bard, *J. Am. Chem. Soc.*, 1975, **97**, 6491.
72. C. Degrand and H. Lund, *Acta Chem. Scand., Ser. B*, 1977, **31**, 593.
73. S. Kashti-Kaplan, J. Hermolin and E. Kirowa-Eisner, *J. Electrochem. Soc.*, 1981, **128**, 802.
74. A. Webber, E. Kirowa-Eisner, J. Osteryoung and J. Hermolin, *J. Electrochem. Soc.*, 1982, **129**, 2725.
75. A. Webber and J. Osteryoung, *J. Electrochem. Soc.*, 1982, **129**, 2731.
76. N. Oyama and F. C. Anson, *J. Electrochem. Soc.*, 1980, **127** (3), 640; *Anal. Chem.*, 1980, **52**, 1192.
77. J. Ogle, L. Baumane, J. Stradyns, G. Duburs, V. Kadish, R. Gavars and V. Lusis, *Khim. Geterotsikl. Soedin.*, 1985, 1099 (*Chem. Abstr.*, 1986, **105**, 23 788p).
78. F. Pragst, A. Henrion, W. Abraham and G. Michael, *J. Prakt. Chem.*, 1987, **329**, 1071.
79. M. Hudlicky, 'Reductions in Organic Chemistry', Horwood, Chichester, 1984.
80. A. Ladenburg, *Ber. Dtsch. Chem. Ges.*, 1892, **25**, 2768.
81. M. Ferles and A. Tesařová, *Collect. Czech. Chem. Commun.*, 1967, **32**, 1631.
82. C. K. McGill and A. Rappa, *Adv. Heterocycl. Chem.*, 1988, **44**, 1.
83. W. Hückel and L. Hagedorn, *Chem. Ber.*, 1957, **90**, 752.
84. A. J. Birch and D. Naispuri, *Tetrahedron*, 1959, **6**, 148.
85. A. J. Birch, *Q. Rev., Chem. Soc.*, 1950, **4**, 78.
86. P. Guerry and R. Neier, *Synthesis*, 1984, 485.
87. P. M. Atlani, J.-F. Biellmann and J. Moron, *Tetrahedron*, 1973, **29**, 391.
88. A. T. Nielsen, D. W. Moore, G. M. Muha and K. H. Berry, *J. Org. Chem.*, 1964, **29**, 2175.
89. M. Freifelder, *Adv. Catal.*, 1963, **14**, 203.
90. P. N. Rylander, in 'Catalytic Hydrogenation Over Platinum Metals', Academic Press, New York, 1967, p. 375.
91. M. Freifelder, 'Practical Catalytic Hydrogenation, Techniques and Applications', Wiley, New York, 1971.
92. P. N. Rylander, in 'Catalytic Hydrogenation in Organic Syntheses', Academic Press, New York, 1979, p. 213.
93. B. R. James, 'Homogeneous Hydrogenation', Wiley, New York, 1973.
94. W. J. le Noble, S. Srivastava, J. Minore and C. K. Cheung, *J. Org. Chem.*, 1985, **50**, 394.
95. G. Lunn and E. B. Sansone, *J. Org. Chem.*, 1986, **51**, 513; L. K. Keefer and G. Lunn, *Chem. Rev.*, 1989, **89**, 459.
96. J. E. Shaw and P. R. Stapp, *J. Heterocyl. Chem.*, 1987, **24**, 1477.
97. J. Z. Ginos, *J. Org. Chem.*, 1975, **40**, 1191.
98. H. Okazaki, M. Soeda, Y. Ikefuji and R. Tamura, *Appl. Catal.*, 1988, **43**, 71.
99. K. Sakanishi, M. Ohira, I. Mochida, H. Okazaki and M. Soeda, *J. Chem. Soc., Perkin Trans. 2*, 1988, 1769.
100. I. Mochida, M. Ohira, K. Sakanishi, H. Fujitsu and H. Okazaki, *Nippon Kagaku Kaishi*, 1987, 1033 (*Chem. Abstr.*, 1988, **108**, 112 189y).
101. L. I. Mastafanova, K. F. Turchin, M. I. Evstratova, Y. N. Sheinker and L. N. Yakhontov, *Khim. Geterotsikl. Soedin.*, 1985, 367 (*Chem. Abstr.*, 1985, **103**, 195 965u).
102. S.-I. Murahashi, Y. Imada and Y. Hirai, *Tetrahedron Lett.*, 1987, **28**, 77.
103. T. J. Lynch, M. Banah, H. D. Kaesz and C. R. Porter, *J. Org. Chem.*, 1984, **49**, 1266.
104. R. H. Fish, A. D. Thormodsen and G. A. Cremer, *J. Am. Chem. Soc.*, 1982, **104**, 5234.
105. R. H. Fish and H. Heinemann, *Report*, 1985, LBL-19772; Order No. DE85 016 529; *Energy Res. Abstr.*, 1985, **10**, abstr. 41129 (*Chem. Abstr.*, 1987, **106**, 159 269y).
106. E. A. Karakhanov, A. S. Loktev, V. S. Pshezhetskii, I. P. Stepanova and A. G. Dedov, *Khim. Geterotsikl. Soedin.*, 1984, 1032 (*Chem. Abstr.*, 1984, **101**, 211 014z).
107. T. Okamoto, S. Yamamoto and S. Oka, *J. Mol. Catal.*, 1987, **39**, 219.

3.7
Partial and Complete Reduction of Pyrroles, Furans, Thiophenes and their Benzo Analogs

GORDON W. GRIBBLE
Dartmouth College, Hanover, NH, USA

3.7.1 INTRODUCTION

A fundamental chemical tactic in the synthetic manipulation of heterocycles by organic chemists is the reduction of the heterocyclic ring. This chapter will review advances in the partial and complete reduction of pyrroles, furans, thiophenes and their benzo analogs, indoles, benzo[*b*]furans and benzo[*b*]thiophenes. As appropriate, coverage will include the reduction of those benzo analogs of lesser interest: isoindole, carbazole, benzo[*c*]furan (isobenzofuran), dibenzofuran and dibenzothiophene.

Unlike alkenes and π-deficient heterocycles, such as pyridine, quinoline and isoquinoline, the π-excessive heterocycles that are the subject of the present chapter are inherently more difficult to reduce. Nonetheless, many new and extremely useful reduction technologies for these classes of heterocycles have been developed over the past decade. It is these advances that will be the main focus of this chapter.

3.7.2 REDUCTION OF PYRROLES

3.7.2.1 Catalytic Hydrogenation

The catalytic hydrogenation of pyrroles to pyrrolidines (2,3,4,5-tetrahydropyrroles), although an old transformation (equation 1),[1,2] has not been heavily exploited as a preparative route to pyrrolidines. Nevertheless, sufficient examples exist so as to illustrate the importance of this method. As would be expected, the hydrogenation of 2,5-disubstituted pyrroles proceeds with *cis* stereoselectivity (equation 2).[3,4]

H_2, PtO_2; AcOH, 1 atm; 100 h, r.t.; 63% (1)

H_2, 5% Rh/Al_2O_3; AcOH, 60 psi; 4 h, r.t.; 79% (2)

From the several examples that have been reported (equations 1 and 2, and *vide infra*), it seems evident that rhodium is superior to platinum and palladium as the hydrogenation catalyst for the reduction of pyrroles. For example, the hydrogenation of fused pyrrole (**1**) affords pyrrolizidine (**2**) in excellent yield (equation 3),[5] and, in a synthesis of anatoxin *a*, Bates and Rapoport found that pyrrole ester (**3**) gives (**4**) upon rhodium-catalyzed hydrogenation (equation 4).[6] The latter transformation does not proceed with a platinum catalyst. In a synthesis of 2-acetyl-1-pyrroline, an important aroma component of cooked rice, Buttery and coworkers hydrogenated 2-acetylpyrrole to 2-(1-hydroxyethyl)pyrrolidine using Rh/Al_2O_3.[7] Whereas Labouta *et al.* found that Rh/Al_2O_3 effects the hydrogenation of the pyrrole ring and the propenoic acid unsaturation in (**5**), leading to the γ-aminobutanoic acid (GABA) analog (**6**; equation 5),[8] Paine and Dolphin were able to remove selectively the benzyl group in (**7**) using Pd/C without

(**1**) → (**2**): H_2, 5% Rh/C; EtOH; 91% (3)

(**3**) → (**4**): H_2, 5% Rh/Al_2O_3; MeOH, H_2SO_4; 44 d, 50 psi; 78% (4)

hydrogenating the pyrrole ring (equation 6).[9] The pyrrole ring in carbamate (**8**) can be cleanly hydrogenated using rhodium as the catalyst (equation 7).[8]

CO_2H (**5**) — H_2, 5% Rh/Al_2O_3; aq. HCl, 18 h; 300 kPa, r.t.; 45% → CO_2H (**6**) (5)

BnO_2C, N–H, NC, CO_2Me (**7**) — H_2, 10% Pd/C; THF, Et_3N; 1.5 h, r.t., 1 atm; 96% → HO_2C, N–H, NC, CO_2Me (6)

HO, CO_2Me, N–CO_2Et (**8**) — H_2, 5% Rh/Al_2O_3; EtOH, 300 kPa, 20 h, r.t.; 73% → HO, CO_2Me, N–CO_2Et (7)

The use of rhodium–amino acid complexes in catalytic hydrogenation has been reported by Rajca to be capable of reducing a wide variety of aromatic and heteroaromatic compounds under mild conditions (1 atm, 22 °C, DMF; 1 atm ≈ 101 kPa).[10] Thus, the rhodium–anthranilic acid catalyzed hydrogenation of pyrrole under these mild conditions yields a 2:1 mixture of pyrrolidine and 2,5-dihydropyrrole (53% conversion after 8 h).[10] Recently, Lunn found that the hydrogenation of pyrrole can be carried out with a nickel–aluminum alloy, as generated with aqueous KOH, to give pyrrolidine in 58% yield, albeit relatively slowly (4 d, r.t.).[11]

3.7.2.2 Dissolving Metals and Metals in Acid

Although the Birch reduction (alkali metals in liquid NH_3) of the pyrrole ring is apparently unknown[12] (*cf.* equation 8),[13] the partial reduction of pyrroles to 2,5-dihydropyrroles using Zn/HCl has been of considerable utility.[2,14] For example, pyrrole gives 2,5-dihydropyrrole as the major product upon treatment with Zn dust/20% aq. HCl (equation 9),[2,15] and Lemal and McGregor observed that 2,5-dimethylpyrrole gives a mixture of *trans*- (78%) and *cis*-2,5-dimethyl-2,5-dihydropyrrole (22%) under similar conditions (47% yield).[14,16] In one of these studies, Hudson and Robertson demonstrated that 2,5-dihydropyrrole is not reduced to pyrrolidine under these reaction conditions.[15] Using these same conditions, Schumacher and Hall reported the reduction of 2-benzylpyrrole to the 2,5-dihydro derivative (67%) in a synthesis of the antibiotic anisomycin.[17]

Na, MeOH; NH_3; 91% (8)

Zn, 20% HCl; 0–10 °C, 7 h; 71% → 79% + 21% (9)

3.7.2.3 Miscellaneous Methods

Apart from catalytic hydrogenation and dissolving metal methods, very few other techniques exist for the reduction of the pyrrole ring. However, one example is the improved use of hypophosphorus acid that has been described by Scott and coworkers in a reduction of pyrrole-2-carboxylic acid to the prolyl hydroxylase inhibitor (*S*)-3,4-dehydroproline (after resolution) (equation 10).[18]

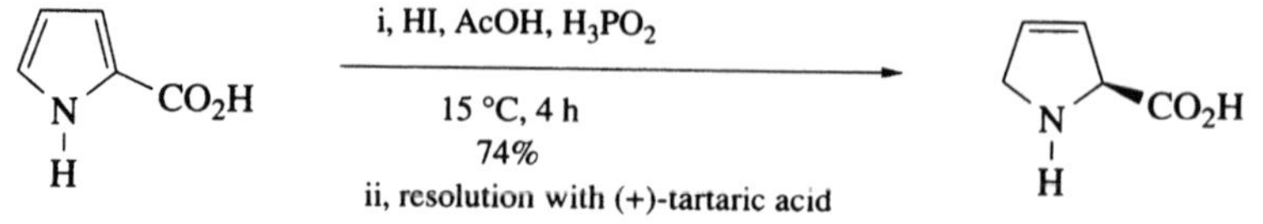

i, HI, AcOH, H_3PO_2; 15 °C, 4 h; 74%; ii, resolution with (+)-tartaric acid (10)

3.7.3 REDUCTION OF FURANS

3.7.3.1 Catalytic Hydrogenation

The catalytic hydrogenation of furans to tetrahydrofurans, which is an important industrial process, has a long history and can be accomplished with a variety of catalysts.[19–24] Nevertheless, new studies of these reductions continue to be reported.[10,25] For example, as shown in Scheme 1, Karakhanov *et al.* have reported that polymer-containing catalysts are particularly efficient in the hydrogenation of furan[25] and le Noble and coworkers have discovered that the combination of isopropyl alcohol and Raney nickel provides for a convenient reduction of furan in a transfer hydrogenation process.[26] The same Russian group has also reported the selective hydrogenation of alkenic double bonds in unsaturated furan ketones using a Raney Pd–Al catalyst (Scheme 1).[27]

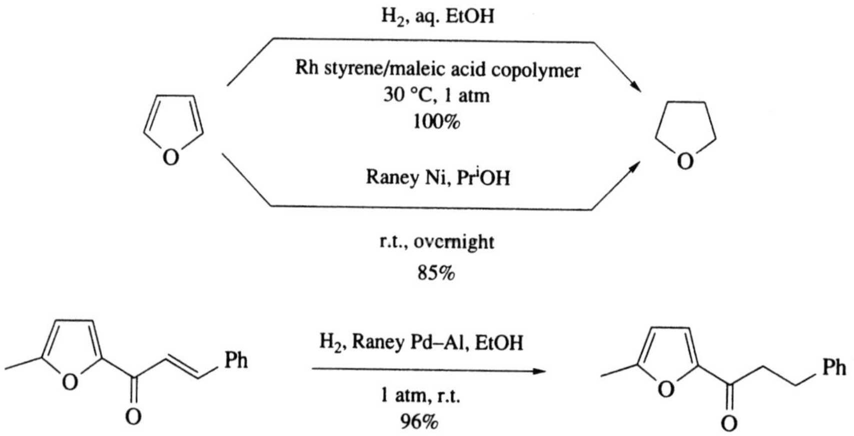

H_2, aq. EtOH; Rh styrene/maleic acid copolymer; 30 °C, 1 atm; 100%

Raney Ni, Pr^iOH; r.t., overnight; 85%

H_2, Raney Pd–Al, EtOH; 1 atm, r.t.; 96%

Scheme 1

A study by Soós has demonstrated that the formation of 5-hydroxy-2-pentanone in the hydrogenation of 2-methylfuran in aqueous alkaline media does not involve 2-methyl-4,5-dihydrofuran as an intermediate,[28] as was earlier believed. Soós proposes that 2-methylfuran is hydrated on the catalyst surface (Pd/C) to give 2-hydroxy-2-methyl-2,5-dihydrofuran, which is hydrogenated to afford eventually 5-hydroxy-2-pentanone (Scheme 2).[28]

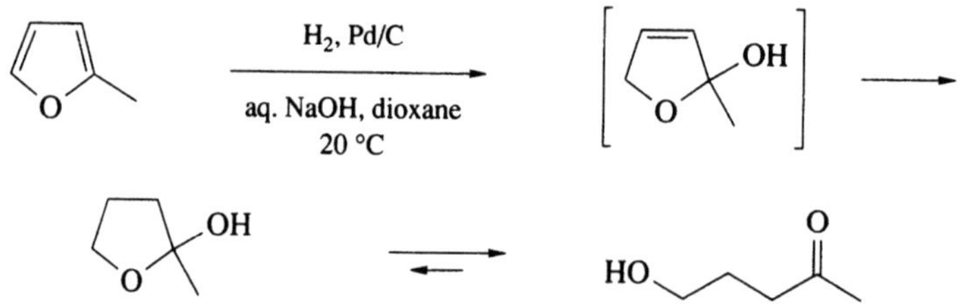

H_2, Pd/C; aq. NaOH, dioxane; 20 °C

Scheme 2

Although normally very difficult,[24] the reduction of furans to 4,5-dihydrofurans can be accomplished by trapping, and later regenerating, the dihydro product (equation 11) as the trifluoroacetate (**9**).[29]

$$\text{2-}R^1\text{-5-}R^2\text{-furan} \xrightarrow[CF_3CO_2H]{H_2,\ Pd/C} \left[\text{3-}OCOCF_3\text{-2-}R^1\text{-5-}R^2\text{-tetrahydrofuran}\ \mathbf{(9)}\right] \xrightarrow[\text{40 min, 48–57\%}]{130\text{–}175\ ^\circ C} \text{2-}R^1\text{-5-}R^2\text{-4,5-dihydrofuran} \quad (11)$$

$R^1 = R^2 = H$; $R^1 = Me$, $R^2 = H$; $R^1 = H$, $R^2 = CO_2Me$

3.7.3.2 Dissolving Metals

Although furan itself is reduced by Li/MeNH2 only under forcing conditions to give a complex mixture,[30] the Birch reduction of furoic acids proceeds easily, is synthetically useful, and has been scrutinized by several groups, as summarized in equations (12)–(14).[31–33] The reduction of 5-alkyl-2-furoic acids is stereorandom, leading to equal amounts of *cis*- and *trans*-dihydro products (equation 13).[31,32] Attempts thus far to effect an asymmetric Birch reduction of furoic acids by chiral protonation have been disappointing (equation 14).[34]

$$\text{2-furoic acid} \xrightarrow[\text{MeOH, 90\%}]{\text{3 equiv. Li/NH}_3} \text{2,5-dihydrofuran-2-}CO_2H \quad (12)$$

$$\text{5-R-2-furoic acid} \xrightarrow[\text{ii, HCl, MeOH; 40–83\%}]{\text{i, Li/NH}_3\text{, MeOH}} cis\text{-product } (50\%) + trans\text{-product } (50\%) \quad (13)$$

R = Me, Et, Pr^i, Pr^n, Bu^t, Bn

$$\text{3-furoic acid} \xrightarrow[\text{or iii–v 3\% } ee]{\text{i, ii 85\%}} \text{methyl 2,3-dihydrofuran-3-carboxylate} \quad (14)$$

i, Na/NH3, Pr^iOH; ii, CH_2N_2; iii, Na/NH3, 1,2:5,6-di-*O*-isopropylidene-α-D-glucofuranose; iv, H^+; v, CH_2N_2

Kinoshita and coworkers have found that, in contrast to the result obtained using Pr^iOH (equation 14), in the presence of nucleophilic alcohols (MeOH, EtOH) the Birch reduction of 3-furoic acid leads to the corresponding acetal (equation 15).[33]

$$\text{3-furoic acid} \xrightarrow[\text{ii, CH}_2\text{N}_2\text{; 92\%}]{\text{i, Na/NH}_3\text{, EtOH}} \text{2-EtO-tetrahydrofuran-4-}CO_2Me \quad (15)$$

The reduction of 5-phenyl-2-furoic acid takes a completely different course, reflecting the benzylic nature of the intermediate dihydrofuran ring (equation 16).[32]

$$\text{5-Ph-2-furoic acid} \xrightarrow[\text{ii, HCl, MeOH; 45\%}]{\text{i, Na/NH}_3\text{, MeOH}} Ph(CH_2)_3CH(OH)CO_2Me\ (89\%) + PhCH_2CH{=}CHCH(OH)CO_2Me\ (11\%) \quad (16)$$

Birch and Slobbe have found that the intermediate anion from the reduction of 2-furoic acid can be alkylated *in situ* (equation 17).[35] These dihydrofuroic acids can be oxidatively decarboxylated with LTA to afford the corresponding 2-alkylfurans[36] in what clearly represents an attractive alternative to the standard two-step metallation–alkylation of the furan ring.

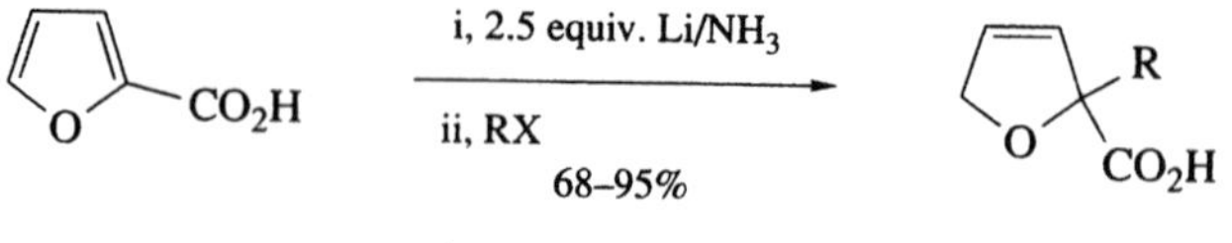

i, 2.5 equiv. Li/NH_3; ii, RX; 68–95% (17)

R = Me, Et, Pr^i, Bn, allyl; X = I, Br, Cl, OMs

3.7.3.3 Ionic Hydrogenation

A very convenient and versatile new reduction method, which has been intensively studied for the reduction of thiophenes (*vide infra*), makes use of the combination of a trialkylsilane and an acid, usually trifluoroacetic acid (TFA). This so-called 'ionic hydrogenation' provides a novel reduction of furans to tetrahydrofurans (Scheme 3).[37] Although Bolestova *et al.* find that the conditions are critical in order to avoid polymerization, the yields of tetrahydrofurans can be quite good. The reduction is accelerated by the presence of boron trifluoride etherate but fails with carbonyl-substituted furans, such as methyl pyromucate.[37]

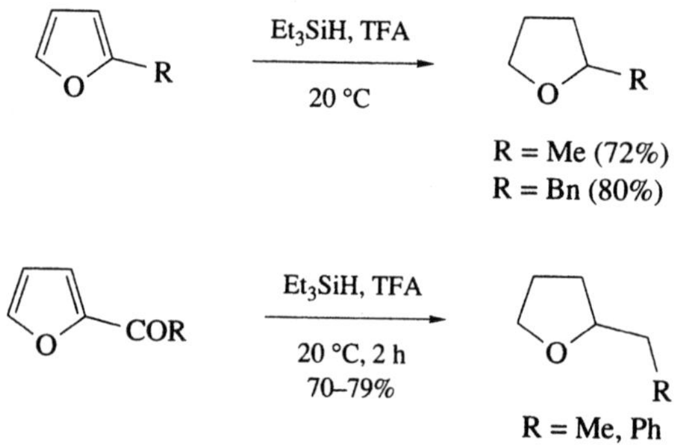

Et_3SiH, TFA; 20 °C

R = Me (72%)
R = Bn (80%)

Et_3SiH, TFA; 20 °C, 2 h; 70–79%

R = Me, Ph

Scheme 3

3.7.4 REDUCTION OF THIOPHENES

Several excellent reviews of the early work on thiophene reduction are available.[38–40] The present discussion will not cover the important industrial topic of catalytic hydrodesulfurization of thiophenes in petroleum.[41–43]

3.7.4.1 Catalytic Hydrogenation

The catalytic hydrogenation of the thiophene ring is inherently difficult because of its high degree of aromaticity and the tendency of the sulfur atom to poison the catalyst. Nevertheless, a number of successful hydrogenation conditions have been developed. For example, Greenfield *et al.* have observed that thiophene and several 2-substituted derivatives can be hydrogenated to the tetrahydro derivatives (equation 18) using synthesis gas and a cobalt catalyst under relatively harsh conditions.[44]

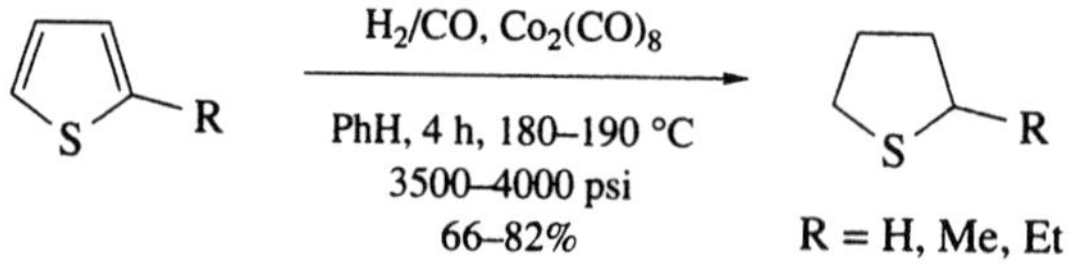

H_2/CO, $Co_2(CO)_8$; PhH, 4 h, 180–190 °C; 3500–4000 psi; 66–82% (18)

R = H, Me, Et

More recently, several milder reducing conditions have been described (equations 19–21).[10,45–48] Noteworthy is the *cis* hydrogenation of thiophene (**10**) performed by Confalone and coworkers in a synthesis of the vitamin biotin (equation 20).[46]

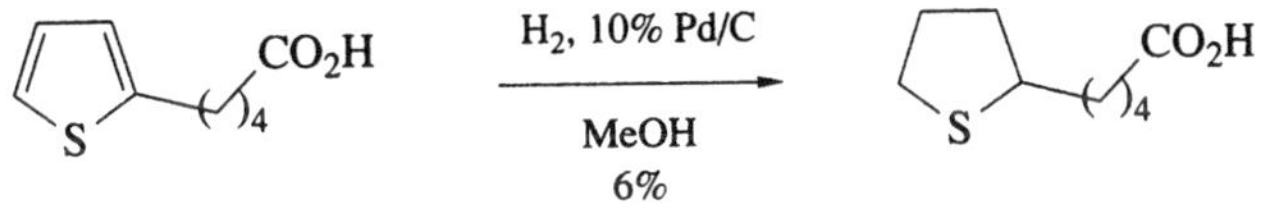

(19)

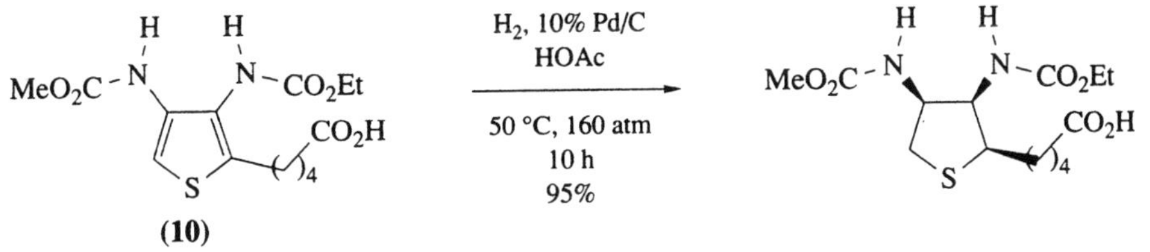

(20)

(10)

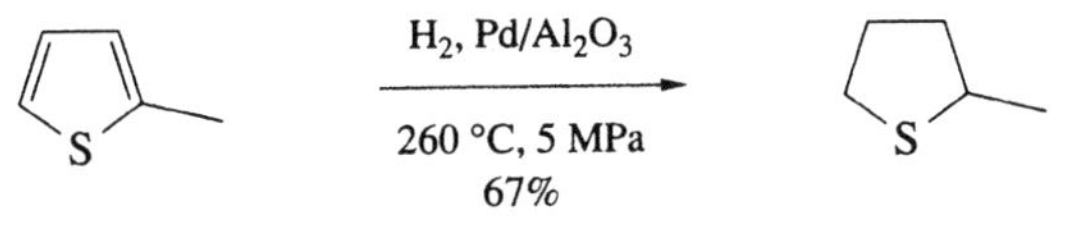

(21)

3.7.4.2 Dissolving Metals

Although the Birch reduction of thiophenes leads to ring cleavage (equation 22),[12,49] the corresponding reaction of thiophene-2-carboxylic acids and 2-acylthiophenes takes a different course. For example, thiophene-2-carboxylic acid gives either the 2,5-dihydro derivative or the cleavage product as the major product, depending on the reaction conditions (equation 23),[50,51] although, in the former case, the reaction mixture is contaminated with the three overreduction products (**11**)–(**13**).[50] However, a modification in the procedure that employs the dry lithium carboxylate prevents overreduction.[52]

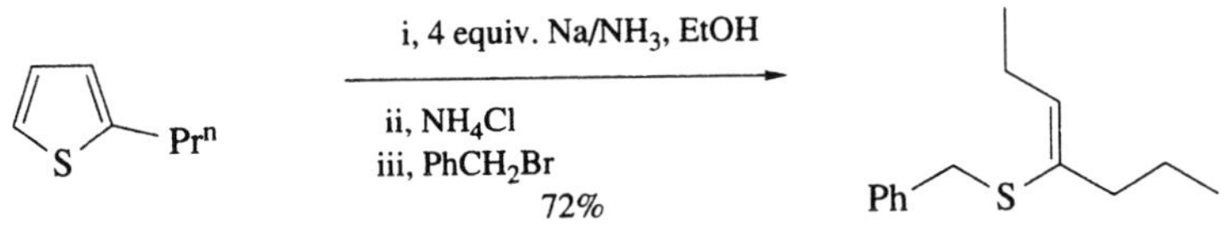

(22)

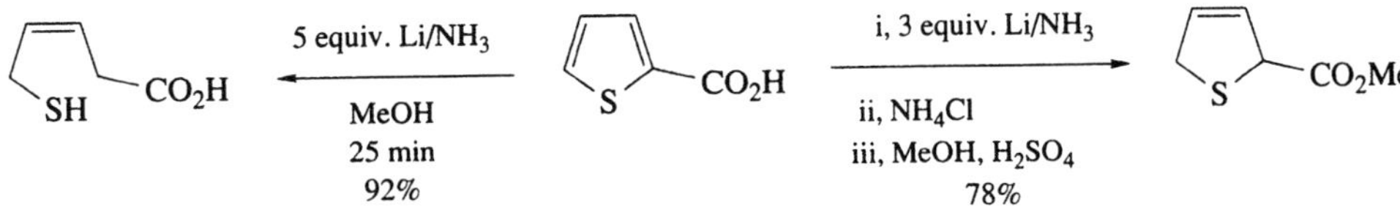

(23)

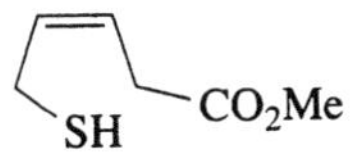

(11) 9.5% 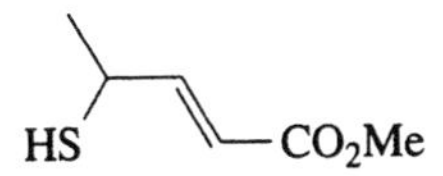 **(12)** 7% 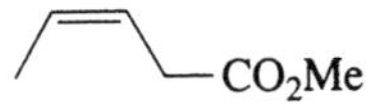 **(13)** 5%

Nishino *et al.* found that 2-acylthiophenes are also reduced to the corresponding 2,5-dihydro derivatives (equation 24)[53] and the intermediate anions can be intercepted with alkylating agents (equation 25).[54] The resulting compounds can be easily converted into the corresponding 1,3-dienyl ketones (equation 25). Dmitrienko and coworkers have found that the Birch reduction of thiophene-3-carboxylic acid leads to the 2,3-dihydro derivative.[55]

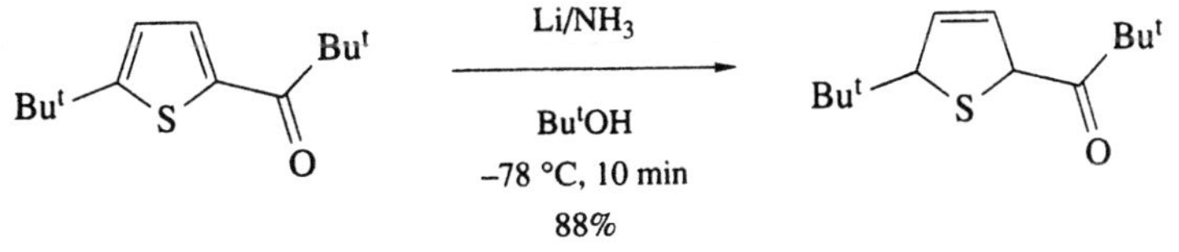

(24)

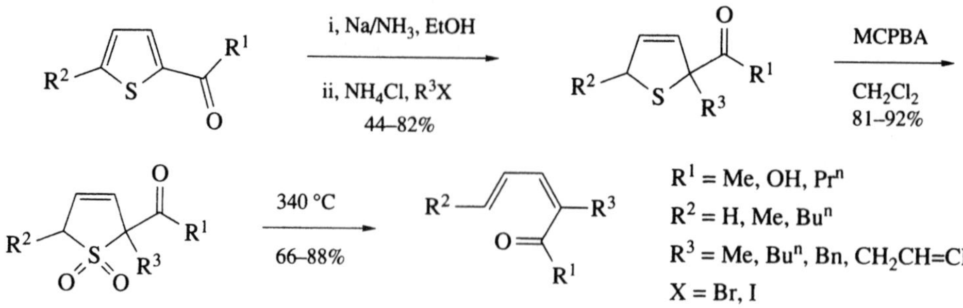

$R^1 = Me, OH, Pr^n$
$R^2 = H, Me, Bu^n$
$R^3 = Me, Bu^n, Bn, CH_2CH{=}CH_2$
X = Br, I

(i, Na/NH_3, EtOH; ii, NH_4Cl, R^3X; 44–82%. MCPBA, CH_2Cl_2, 81–92%. 340 °C, 66–88%) (25)

3.7.4.3 Ionic Hydrogenation

An extremely powerful and efficient new reduction of thiophenes to tetrahydrothiophenes involves ionic hydrogenation with a trialkylsilane and an acid. Typically, a mixture of triethylsilane and trifluoroacetic acid is employed (equations 26 and 27).[56] This reduction is reasonably general, although thiophene itself is poorly reduced (equation 26) and 2,5-diphenylthiophene is inert to these conditions. Not surprisingly, 2-benzoylthiophene is converted to 2-benzyltetrahydrothiophene (75%).[56]

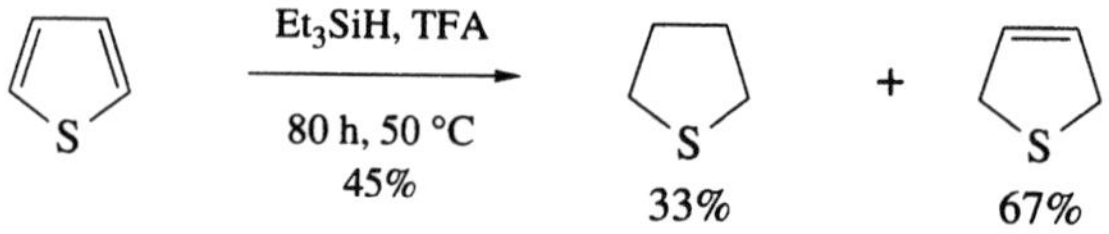

(Et_3SiH, TFA; 80 h, 50 °C; 45%) 33% + 67% (26)

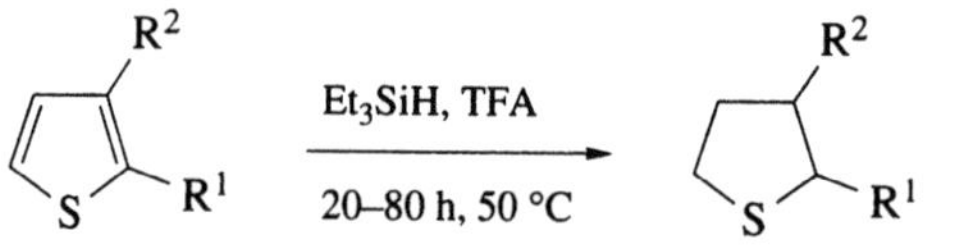

(Et_3SiH, TFA; 20–80 h, 50 °C) (27)

$R^1 = Me, R^2 = H$ (80%)
$R^1 = H, R^2 = Me$ (60%)

For this ionic hydrogenation of thiophenes, Kursanov and coworkers have proposed the pathway shown in Scheme 4, based on the labeling results obtained for the reduction of 2,5-dimethylthiophene.[56]

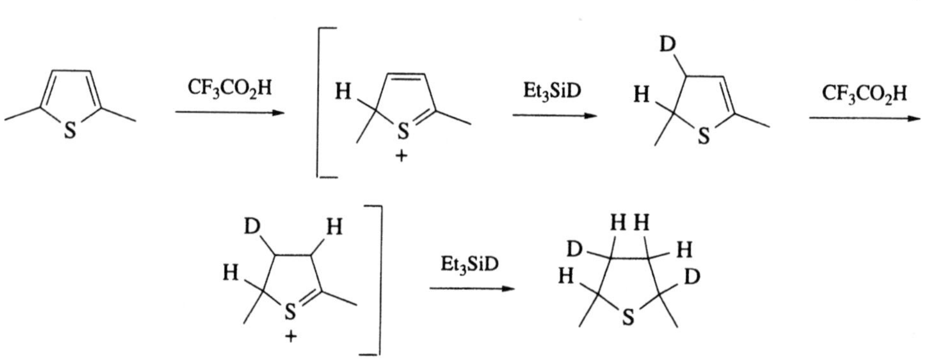

(CF_3CO_2H; Et_3SiD; CF_3CO_2H; Et_3SiD)

Scheme 4

Zav'yalov and Dorofeeva have utilized this reduction to synthesize the biotin analogs shown in Scheme 5.[57]

This ionic hydrogenation can be improved by the addition of either boron trifluoride etherate[58] or certain salts and acids.[59] Furthermore, a more powerful set of conditions involving Et_3SiH and HCl–$AlCl_3$ has been used to reduce thiophenes that are recalcitrant under the usual Et_3SiH/TFA conditions (equations 28 and 29).[60] In fact, the intermediate salt (**14**) can be isolated.

In what might be viewed as a variation on the ionic hydrogenation theme, this same Russian group has found that the combination of zinc/aluminum halide/*p*-toluenesulfonic acid reduces 2-alkylthiophenes mainly to the tetrahydro derivatives (equation 30).[61] The minor product, 2-alkyl-2,5-dihydrothiophene, is

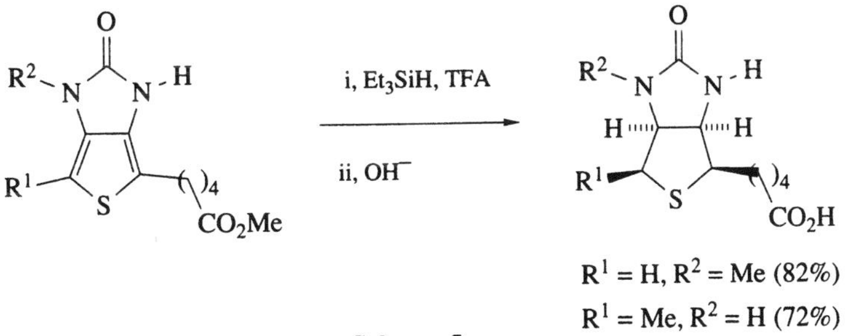

i, Et_3SiH, TFA

ii, OH^-

$R^1 = H$, $R^2 = Me$ (82%)

$R^1 = Me$, $R^2 = H$ (72%)

Scheme 5

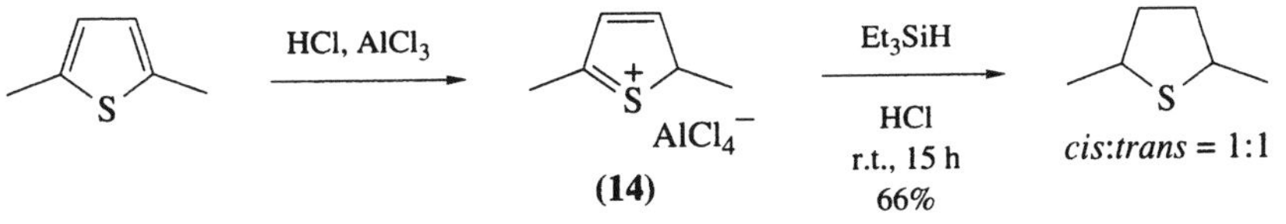

HCl, $AlCl_3$; $AlCl_4^-$ **(14)** ; Et_3SiH / HCl, r.t., 15 h, 66% ; *cis*:*trans* = 1:1 (28)

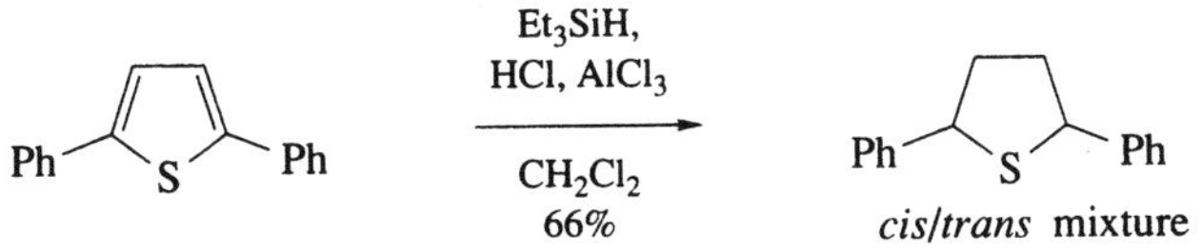

Et_3SiH, HCl, $AlCl_3$ / CH_2Cl_2, 66% ; *cis*/*trans* mixture (29)

not converted to the 2-alkyltetrahydrothiophene under the reaction conditions, suggesting that the reduction proceeds *via* a 4,5-dihydrothiophene intermediate, as shown in Scheme 4.

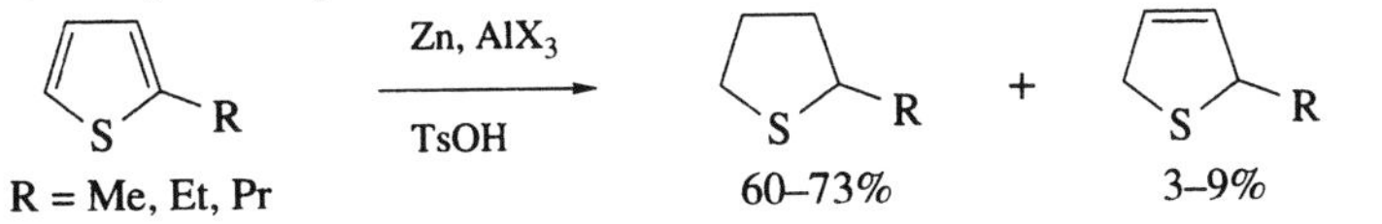

R = Me, Et, Pr ; Zn, AlX_3 / TsOH ; 60–73% + 3–9% (30)

Perhaps surprisingly, Yamanochi and Yamane describe the reduction of 2-thiopheneacetonitrile with $NaBH_4/MeSO_3H$ (DMSO, 70 °C, 2 h) to give 2-(2-thienyl)ethylamine (64% yield) in which the thiophene ring is not reduced.[62]

3.7.4.4 Electrochemical

The electrochemical reduction of thiophene[63] and 5-methyl-2-thiophenecarboxylic acid[64] has been reported but would appear not to be of preparative interest. Thus, thiophene affords a mixture of 2,5-dihydrothiophene (56%) and tetrahydrothiophene (44%) in 55% yield upon electrolysis in aqueous DMF.[63]

3.7.4.5 Miscellaneous Methods

In a process that is related to the ionic hydrogenation method discussed above, Wristers has discovered that the unusual combination of H_2/isopentane/HF/TaF_5 reduces thiophene to tetrahydrothiophene in 80% yield.[65] These conditions also reduce dibenzothiophene to hexahydrodibenzothiophene. It has also been reported that Zn/TFA reduces thiophenes to 2,5-dihydrothiophenes (equations 31 and 32).[66] However, both acetic acid and HCl/benzene are ineffective with zinc in this regard.[66]

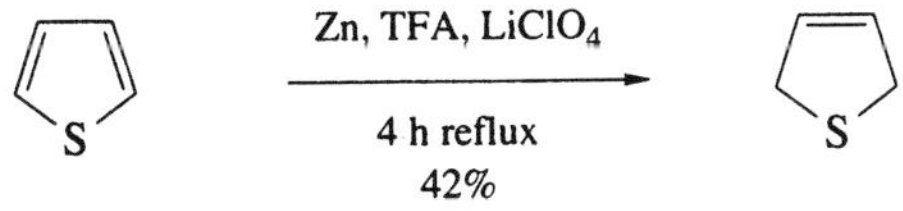

Zn, TFA, $LiClO_4$ / 4 h reflux, 42% (31)

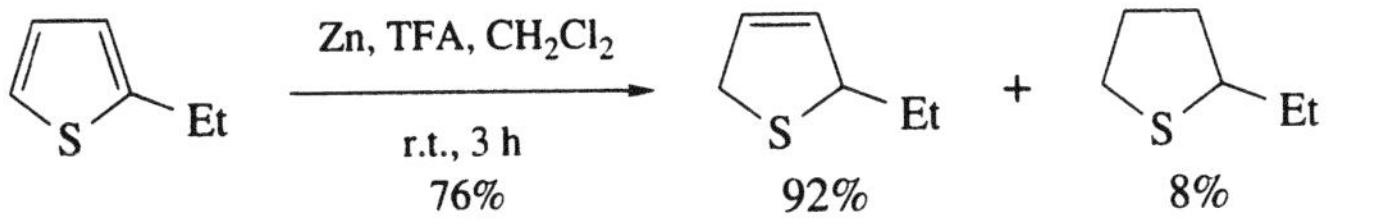

Zn, TFA, CH_2Cl_2 / r.t., 3 h, 76% ; 92% + 8% (32)

3.7.5 REDUCTION OF INDOLES

The reduction of indoles to indolines (2,3-dihydroindoles) has been intensively studied since the turn of the century and several reviews to the early literature are available.[67–71] Nevertheless, over the past decade, a number of new and exciting methods for effecting this important reduction have been discovered.

3.7.5.1 Catalytic Hydrogenation

The catalytic hydrogenation of indoles has a long history and a number of catalysts have been investigated in this context.[67–71] Although indoles are readily hydrogenated, it has not always been possible to control the site and degree of reduction. Moreover, rarely was it possible in these early works to establish a set of conditions that was generally applicable to a variety of different indoles. For example, although Noland and Hammer were able to hydrogenate indole to indoline in 92% yield with Raney nickel (EtOH, 100 atm, 100 °C), the same conditions with 2-methylindole give mostly overreduction (equation 33).[72] Another example is seen in the recent work of Shaw and Stapp, who examined several catalysts in the hydrogenation of indoles.[73] The highest conversions were observed with 5% platinum/alumina, but with erratic results (equation 34). The reduction of carbazole was also accomplished using this same catalyst and the products were 1,2,3,4-tetrahydrocarbazole (36%), *cis*-1,2,3,4,4a,9a-hexahydrocarbazole (5%) and a mixture of dodecahydrocarbazoles (6%).[73]

H_2, Raney Ni; EtOH, 100 atm; 118 °C, 4 h; 72% — 53% + 47% (33)

H_2, 5% Pt/Al_2O_3; n-$C_{16}H_{34}$; 227 °C, 11 h, 2200 psi (equilibrium) (34)

$R^1 = R^2 = H$ (82%)
$R^1 = Me, R^2 = H$ (65%)
$R^1 = R^2 = Me$ (28%)

Interestingly, platinum/silica gel effects the hydrogenolysis of the C—N bond (equation 35).[74]

H_2, Pt/SiO_2; 200 °C; 94% — 49% + 30% + 15% + 6% (35)

In accord with the propensity for indoles to protonate at the C-3 position,[75,76] Smith and Utley found that the catalytic hydrogenation of indoles in the presence of fluoboric acid proceeds smoothly to the indoline stage (equation 36).[77] It seems that these acidic conditions are strong enough to protonate the in-

H_2, PtO_2; HBF_4, EtOH (36)

$R^1 = R^2 = H$ (>85%)
$R^1 = H, R^2 = Me$ (>95%)
$R^1, R^2 = –(CH_2)_4–$ (100%)

dole ring, but do not facilitate the polymerization of indole, which is frequently encountered with mineral acid. These hydrogenation conditions have been used to prepare a series of psychotropic hexahydro-γ-carbolines (**15**; equation 37).[78]

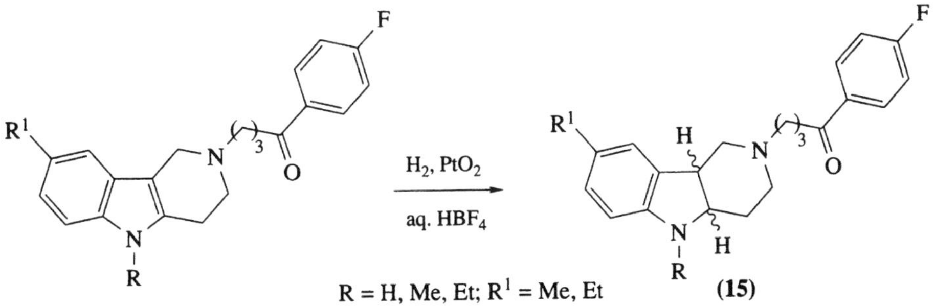

H_2, PtO_2, aq. HBF_4 (37)

R = H, Me, Et; R^1 = Me, Et (**15**)

Ames *et al.* have reported that indoles can be cleanly reduced to indolines under very mild conditions with palladium in the presence of acid (perchloric, acetic, sulfuric, phosphoric) (equation 38).[79]

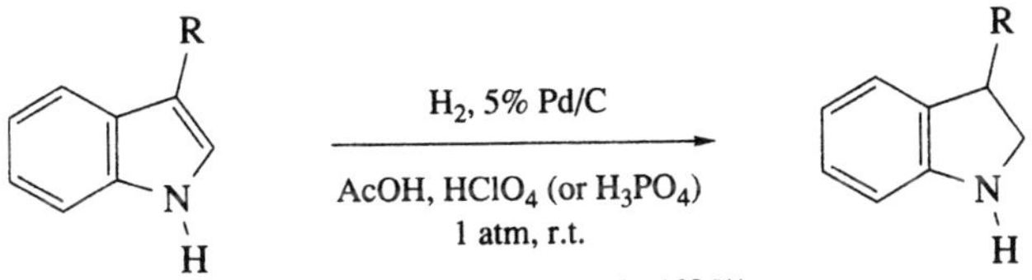

H_2, 5% Pd/C; AcOH, $HClO_4$ (or H_3PO_4), 1 atm, r.t. (38)

R = H (74%); R = Me (69%)

Raney nickel (RaNi) also accomplishes the hydrogenation of indoles to indolines (equations 39 and 40)[70] but the danger of overreduction can be a problem (equation 40). Under slightly different conditions (H_2, RaNi, MeOH/H_2SO_4, 3 h, 60 °C), the yield of indoline from indole is 95%.[80] These latter conditions have been reported to reduce a variety of substituted indoles (*i.e.* 4-OH, 5-OMe, 5-Cl, 5-Bu^t, 2-Ph, 2-Et-5-Pr) to the corresponding indolines.[80] The use of RaNi is particularly valuable in the reduction of *N*-acetylindoles to *N*-acetylindolines (**16**; equation 41).[70] Lunn's conditions (Ni/Al, aq. KOH) also serve to reduce indole to indoline in 73% yield.[11] A ruthenium catalyst ($RuCl_2(PPh_3)_3$, H_2, 310 psi, 85 °C; 1 psi = 6895 Pa) has been used to reduce indole to indoline but the yield was not reported, and this was the only example investigated.[81]

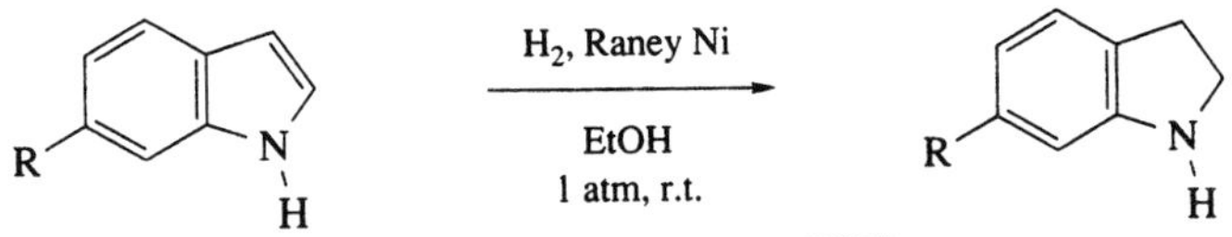

H_2, Raney Ni; EtOH, 1 atm, r.t. (39)

R = H (75%); R = OMe (76%)

H_2, Raney Ni; EtOH, 1 atm, r.t., 84% (40)

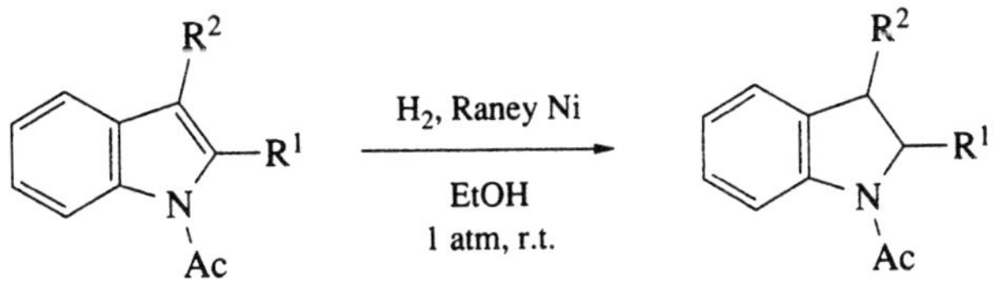

73% + 27%

H_2, Raney Ni; EtOH, 1 atm, r.t. (41)

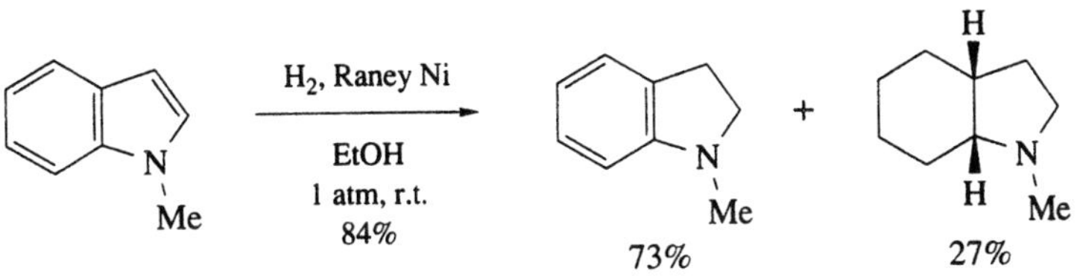

(**16**)

$R^1 = R^2$ = H (99%); R^1 = H, R^2 = Me (98%); $R^1 = R^2$ = Me (97%)

An interesting variation on conventional catalytic hydrogenation is catalytic transfer hydrogenation, in which formic acid or formate serves as the hydrogen source. Thus, indoles are smoothly reduced (and formylated) to the corresponding *N*-formylindoles (equations 42 and 43).[82] Since it is difficult to avoid formylation of the indoline, it is best to employ triethylammonium formate to maximize yields of the reduction product (equation 43).

10% Pd/C, HCO_2H; 0.5 h, 70 °C; 68% — 46% + 54% (CHO) (42)

CO_2Me — 10% Pd/C, HCO_2H; Et_3NHO_2CH; 1.5 h, 110 °C; 83% — CO_2Me, CHO (43)

Watanabe and coworkers have reported that the ruthenium-catalyzed reduction of indoles using formic acid as the hydrogen source leads to indolines (equation 44).[83] It seems likely that acid is required in these reductions (equation 42–44) in order to protonate the indole C-3 position, since, in at least one report using only ammonium formate and palladium, the indole ring is not reduced (equation 45).[84] Admittedly, these conditions are much milder than those that use formic acid.

HCO_2H, $RuCl_2(PPh_3)_3$; 180 °C, 6 h, PhH; R = H (53%); R = Me (50%) (44)

NC — HCO_2NH_4, 10% Pd/C; MeOH, r.t., 24 h; 20% (45)

3.7.5.2 Dissolving Metals and Metals in Acid

In addition to the general reviews on indole reduction that were cited above, several reviews and key papers on the dissolving metal reduction of indoles should be consulted for reference to the early work.[12,85,86]

Whereas *N*-phenylindoles are reduced to the corresponding dihydroindoles with Na/NH_3 (equations 46 and 47)[87–89] and indole-2-carboxylic acid is reduced to the indoline with $Li/NH_3/PhNH_2$ (equation 48),[90] indole itself is reduced to 4,7-dihydroindole (Li/NH_3, MeOH) in 51% yield.[91] These latter conditions also reduce carbazole to 1,4-dihydrocarbazole in 73% yield.[91]

i, Na/NH_3, THF, 10–15 min; ii, NH_4Cl; 88% (46)

In some cases, very high chemoselectivity can be achieved depending on the reaction conditions. Thus, Remers and coworkers demonstrated that 5-methoxy-1-methylindole can be reduced either in the pyrrole ring or in the benzene ring, depending upon the availability of a good proton source (MeOH) (equation 49).[86]

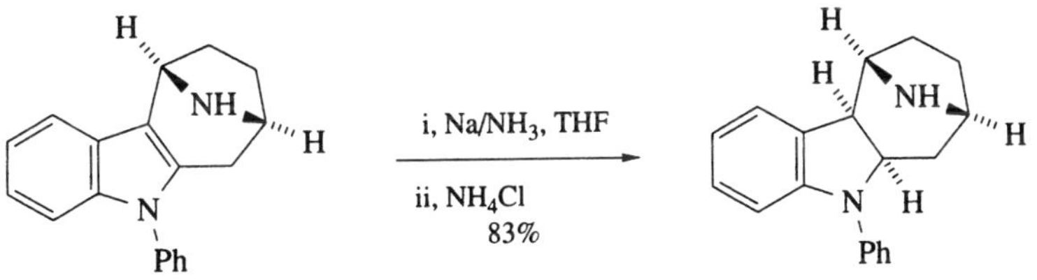

(47)

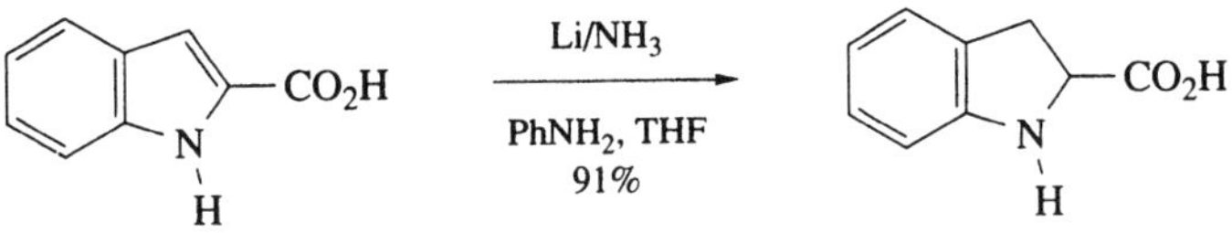

(48)

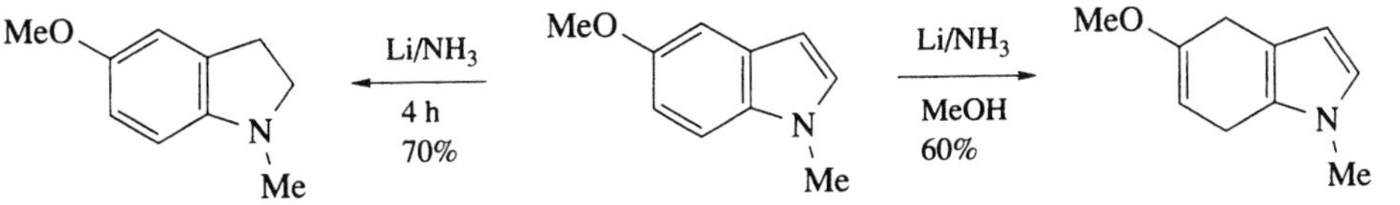

(49)

However, Sauer and coworkers have been able to reduce selectively the styrene double bond in ergoline derivative (**17**) without affecting the indole double bond (equation 50).[92]

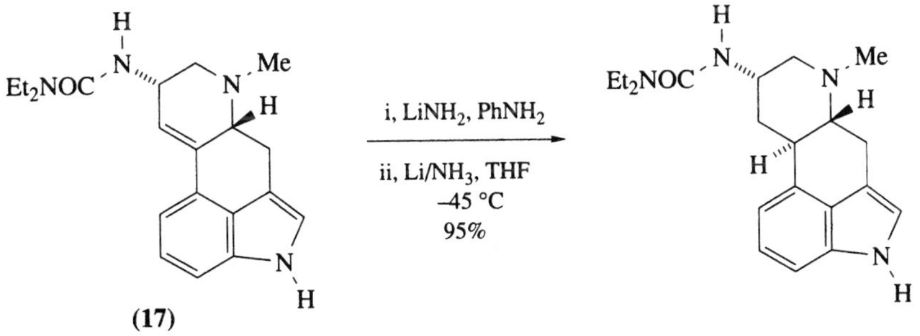

(50)

Independently, Youn and Pak, and Fagan *et al.* found that Mg/MeOH reduces indoles to indolines that have a carbonyl group at the 2-position (equation 51).[93,94] This research supplements a reduction reported by Corey and coworkers that uses Sn/HCl (equation 52).[95] However, ethyl indole-3-carboxylate is not reduced with Mg/MeOH.[93]

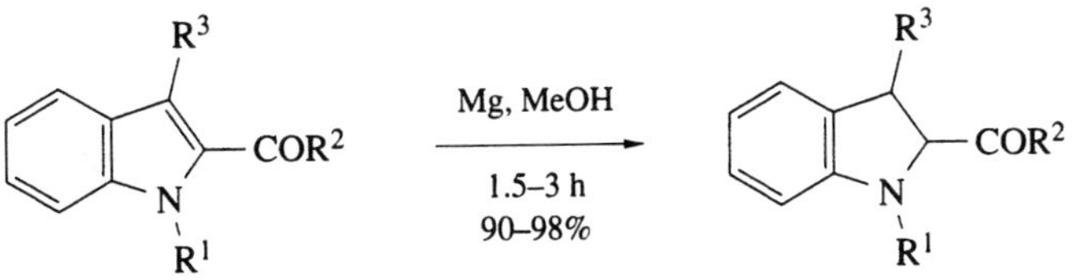

(51)

R^1 = H, Ac; R^2 = OMe, NH_2, NHMe; R^3 = H, Me

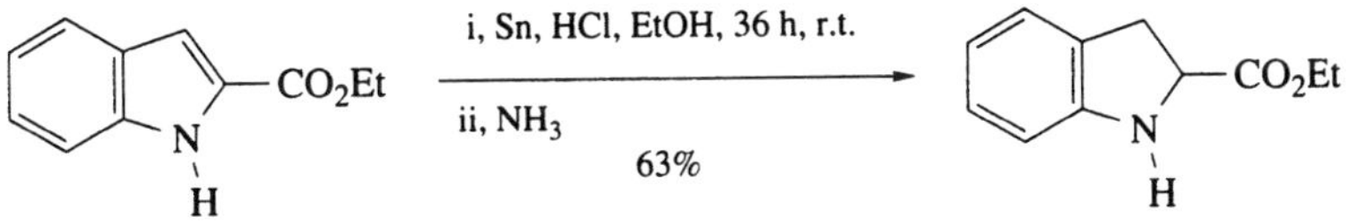

(52)

Although the early attempts to reduce indole to indoline with metals (Sn and Zn) and mineral acid (*e.g.* hydrochloric acid) invariably led to polymerization, Dolby and Gribble found that 85% phosphoric acid is a suitable medium in combination with Zn for a successful indole to indoline reduction (69%).[96] However, the reactions of 2,3-dimethylindole and 1,2,3,4-tetrahydrocarbazole with Zn/H_3PO_4 give much lower yields of reduction product.[96]

A related procedure utilizes zinc dust amalgam and HCl. Thus, several indoles are reduced to indolines in excellent yield (equation 53).[97,98] Somewhat similarly, the facile reduction of the indole double bond in pyridylindole (**18**) with zinc and acetic acid has been described (equation 54).[99] In this example, presumably, protonation of the pyridine nitrogen facilitates the reduction of the indole ring. Interestingly, Birch conditions (Na, EtOH, reflux) with (**18**) result in reduction of only the pyridine ring (93% yield).[100]

Zn(Hg), aq. HCl; 90–100 °C, 120 min (53)

$R^1 = R^2 = R^3 = H$ (97%)
$R^1 = R^3 = H$, R^2 = cyclohexyl (92%)
$R^1 = H$, $R^2 = R^3 = Me$ (90%)
$R^1 = R^3 = H$, $R^2 = Ph$ (95%)
$R^1 = R^2 = Me$, $R^3 = H$ (97%)
$R^1 = R^2 = H$, $R^3 = Et$ (95%)
$R^1 = Me$, $R^2 = R^3 = H$ (100%)
$R^1 = R^3 = H$, $R^2 = Me$ (100%)

(**18**) Zn, AcOH; 80–90 °C, 1.5 h; 62% (54)

3.7.5.3 Boron Hydrides and Related Methods

As will be seen in this section, mixed boron hydrides and complexes of boron probably represent the most general and efficient reagents for the reduction of the indole double bond.

Although a few examples were known previously, Monti and Schmidt, in a careful study, found that diborane is able to reduce *N*-unsubstituted indoles to indolines (equation 55).[101] *N*-Alkylindoles are not reduced under these conditions.

i, BH_3–THF; ii, NaOMe, MeOH; iii, HCl (55)

$R^1 = R^2 = H$ (44%)
$R^1 = Me$, $R^2 = H$ (50%)
$R^1 = H$, $R^2 = Me$ (66%)
$R^1 = R^2 = Me$ (93%)
$R^1, R^2 = -(CH_2)_4-$ (98%)

The mechanism proposed to account for this somewhat remarkable result is shown in Scheme 6, involving as the key intermediate the indolylborane (**19**). Methoxide addition to (**19**) serves to activate the indole C-3 position towards protonation and the resulting indolenium ion (**20**) is reduced by a boron hydride species to give indoline.

Although indoles are not normally reduced by either sodium borohydride or lithium aluminum hydride,[68] Gribble and coworkers have found that $NaBH_4$ in carboxylic acid media reduces indoles to *N*-alkylindolines in excellent yield (equations 56 and 57).[102] This unusual reaction involves C-3 protonation of the indole, borohydride reduction to give an intermediate indoline and subsequent *N*-alkylation under the reaction conditions to give the *N*-alkylindoline. The reduction of *N*-alkylindoles with $NaBH_4$/HOAc gives the corresponding *N*-alkylindolines. Indeed, the combination of $NaBH_4/RCO_2H$ is an excellent one for amine alkylation.[102,103]

Although most common carboxylic acids can be used successfully in this context (equation 57),[102] formic acid gives the indole dimer derived side product (**21**; equation 58)[104] and trifluoroacetic acid gives the condensation product (**22**; equation 59),[105] in addition to the expected products. The particular prob-

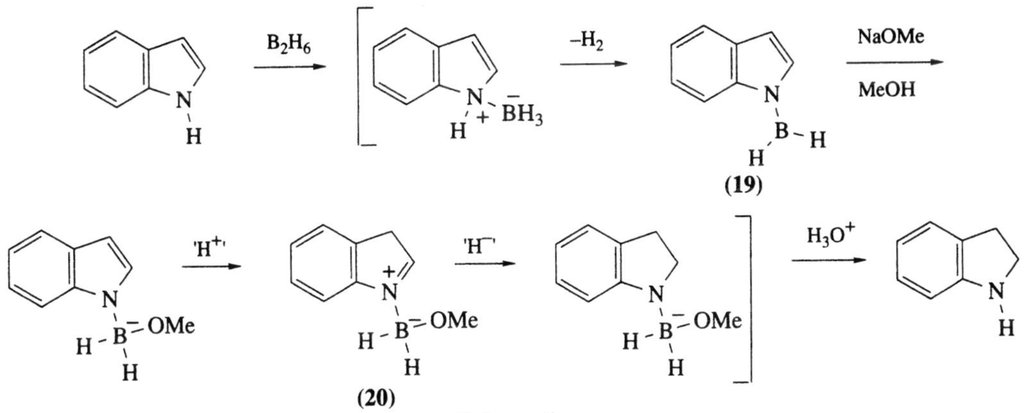

B_2H_6 —H$_2$ NaOMe MeOH

(19)

'H$^+$' 'H$^-$' H_3O^+

(20)

Scheme 6

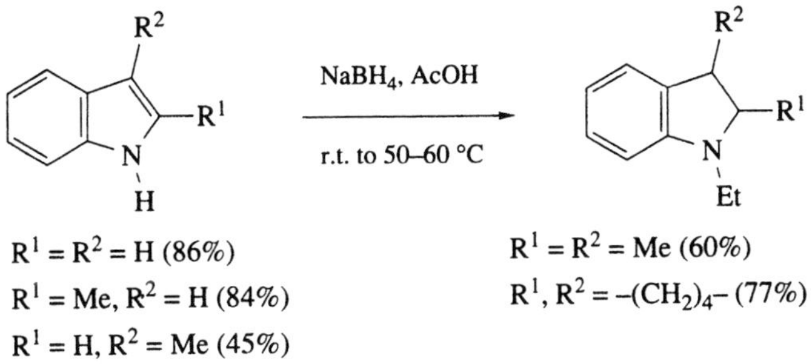

NaBH$_4$, AcOH; r.t. to 50–60 °C (56)

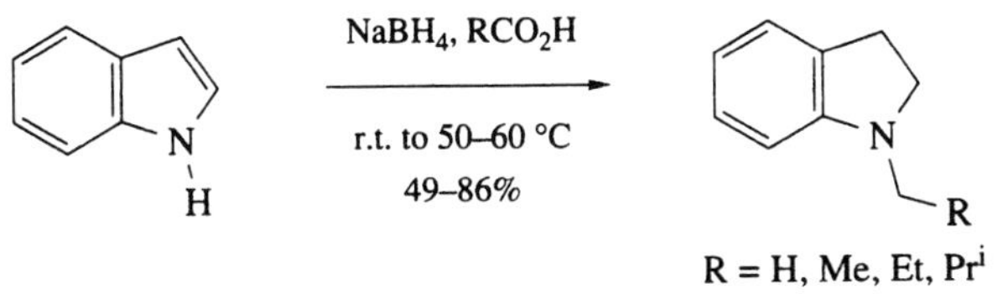

$R^1 = R^2 = H$ (86%)
R^1 = Me, R^2 = H (84%)
R^1 = H, R^2 = Me (45%)

$R^1 = R^2$ = Me (60%)
$R^1, R^2 = -(CH_2)_4-$ (77%)

NaBH$_4$, RCO$_2$H; r.t. to 50–60 °C; 49–86% (57)

R = H, Me, Et, Pr^i

lem encountered with formic acid in this indole reductive *N*-methylation sequence can be circumvented by using a combination of paraformaldehyde/NaBH$_3$CN/HOAc.[106]

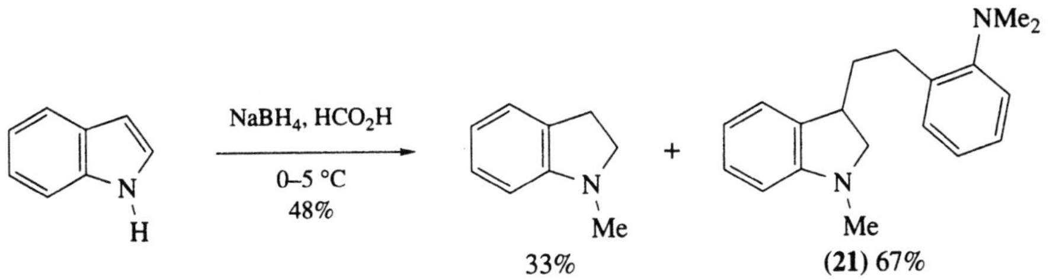

NaBH$_4$, HCO$_2$H; 0–5 °C; 48% (58)

33% **(21)** 67%

NaBH$_4$, TFA; 25 °C to reflux; 6 h; 58% 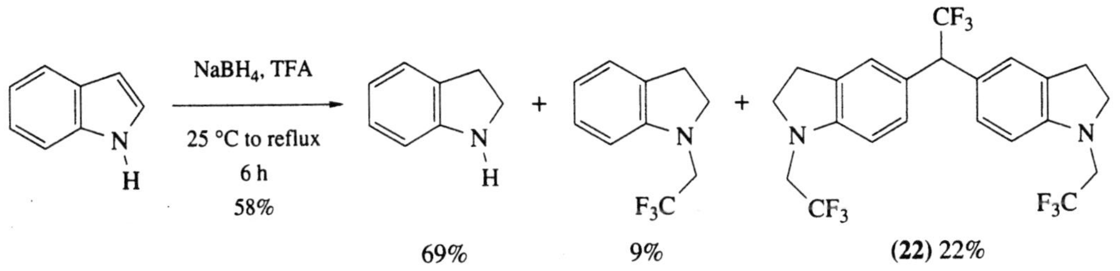 (59)

69% 9% **(22)** 22%

In a companion study, Gribble and coworkers found that the combination NaBH$_3$CN/HOAc is superlative for reducing indoles to indolines rapidly and under mild conditions (equation 60).[102,107] With this reagent combination, *N*-alkylation occurs only at higher temperatures.[108]

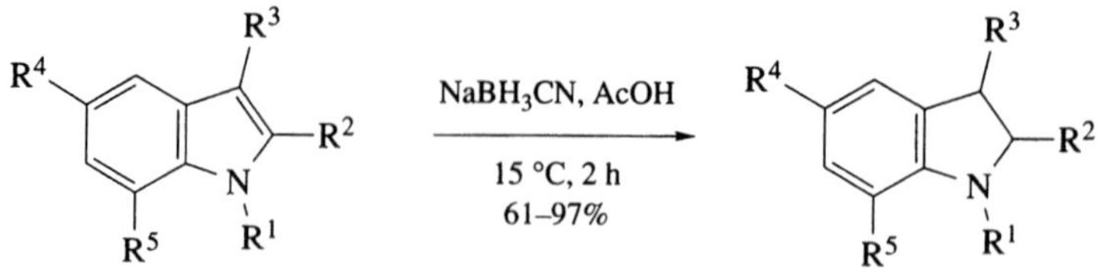

(60)

R^1 = H, Me, Et, Pr^n, Bn, Ph; R^2 = H, Me; R^3 = H, Me; R^2, R^3 = $-(CH_2)_4-$; R^4 = H, Br; R^5 = H, Me

NaBH3CN, AcOH; 15 °C, 2 h; 61–97%

Only those indoles that are sufficiently basic to be protonated by acetic acid are reduced under these conditions. Thus, 5-nitroindole and 2,3-diphenylindole are recovered unchanged,[107] whereas 5,6-dimethoxyindole is cleanly reduced to 5,6-dimethoxyindoline in 86% yield with $NaBH_3CN$/HOAc.[109,110] This differential reactivity has been exploited by Cava and Rawal in their synthesis of CC-1065 analogs in which only the more basic double bond in (**23**) is reduced (equation 61).[111–113] Indeed, this same tactic has since been utilized by Boger,[114] Moody[115,116] and Sundberg,[117] and their coworkers, in their respective synthetic efforts towards CC-1065 and the related phosphodiesterase inhibitors PDE-I and PDE-II. Likewise, Joule and coworkers have utilized $NaBH_3CN$/HOAc in a chemoselective reduction of the more basic indole double bond in benzodipyrrole (**24**; equation 62).[118]

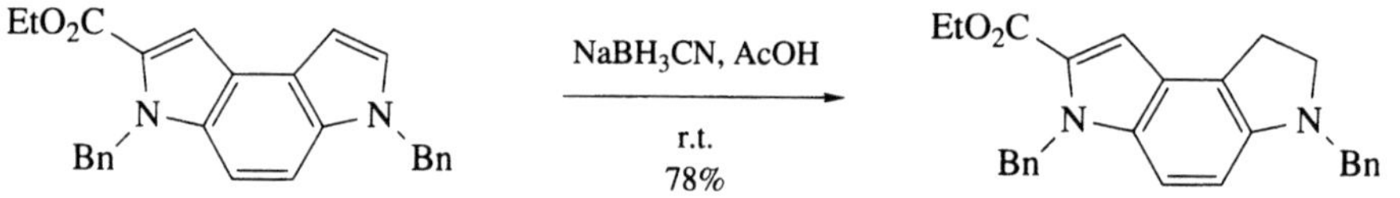

(**23**) NaBH3CN, AcOH; r.t.; 78% (61)

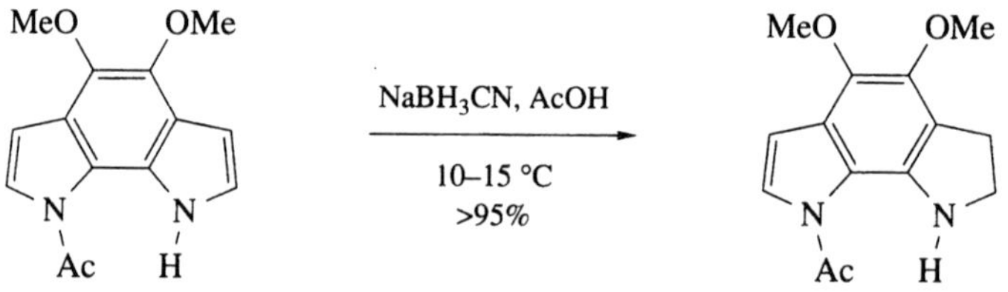

(**24**) NaBH3CN, AcOH; 10–15 °C; >95% (62)

Although the reduction of indoles that contain a basic nitrogen atom fails with $NaBH_4$ (or $NaBH_3CN$) in acetic acid, this reaction is successful using the stronger trifluoroacetic acid, and many examples are known (equations 63–66).[102,119–126] Not surprisingly, the bromine is removed in the reduction of 2-bromoergolines (equation 65).[123] The reduction of β- and γ-carbolines invariably gives the *cis* product (equations 63 and 66),[119,121,124] the result of hydride transfer to the iminium ion from the least hindered face and leading to the more stable *cis* ring fusion. As would be expected, $NaBH_3CN$/TFA is also capable of reducing these basic indoles to indolines (equations 67 and 68).[127–129] This combination has also been used to reduce methyl and ethyl *N*-alkylindole-2-carboxylates to the corresponding indolines in excellent yield.[94]

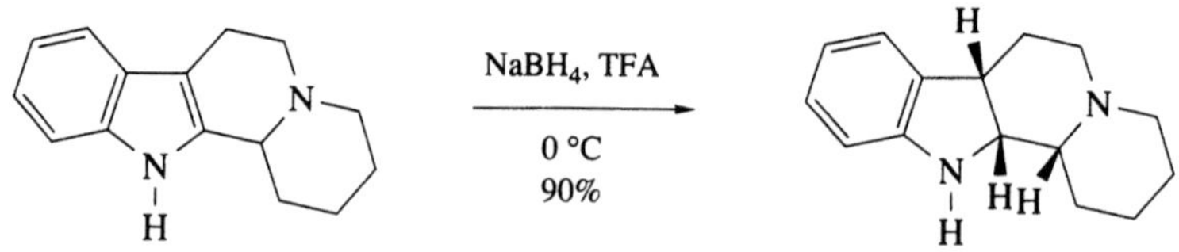

NaBH4, TFA; 0 °C; 90% (63)

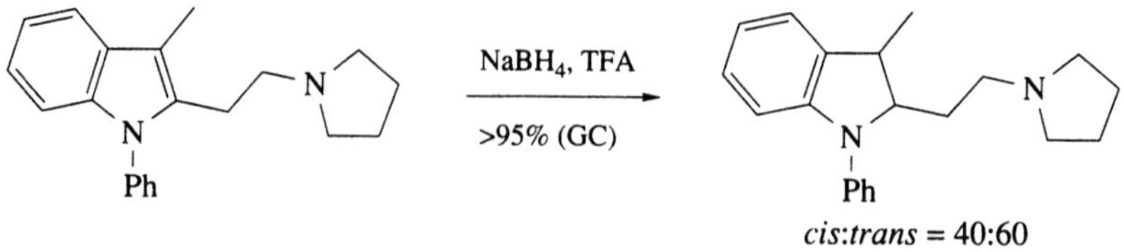

NaBH4, TFA; >95% (GC) (64)

cis:*trans* = 40:60

NaBH$_4$, TFA; 0 °C, 4 h; 62% (65)

NaBH$_4$, TFA; 86% (66)

NaBH$_3$CN; TFA; 84% (67)

cis:*trans* = 44:56

NaBH$_3$CN; TFA; 81% (68)

In some instances, it is essential to use NaBH$_3$CN/TFA rather than NaBH$_4$/TFA, in order to preclude *N*-trifluoroethylation, which can occur using the latter conditions (equations 69 and 70).[105,121,129]

KBH$_4$; TFA; 89% (69)

Whereas indoles that contain a basic nitrogen are smoothly reduced to the corresponding indolines with NaBH$_4$/TFA (*vide supra*), this reagent combination with simple indoles can lead to complications (*i.e.* **22**; equation 59).

The utility of sodium borohydride (and related metal hydrides) in carboxylic acid media in the reduction of indoles has been reviewed recently and several additional examples can be found cited in this review.[103] There is increasing evidence to suggest that acyloxyborohydride species are the actual reducing agents that form under these conditions.[130]

Sodium borohydride in concert with AlCl$_3$–pyridine also reduces indoles to indolines (equation 71),[131] but this reagent combination is less effective than those reported above using carboxylic acids.

Somewhat surprisingly, it has been reported that both tetra-*n*-butylammonium borohydride and zinc borohydride, in the absence of acid, are capable of reducing indoles to indolines (equations 72 and

NaBH$_4$, TFA, 75%: 47% + 13% + 20% + 20%

(70)

NaBH$_4$, AlCl$_3$, pyridine, 0 °C to r.t., 7 h

(71)

$R^1 = R^2 = R^3 = H$ (76%)
$R^1 = R^2 = H, R^3 = Me$ (75%)
$R^1 = R^3 = H, R^2 = Me$ (80%)
$R^1 = Me, R^2 = R^3 = H$ (56%)
$R^1 = R^3 = H, R^2 = Ph$ (15%)
$R^1 = R^2 = H, R^3 = (CH_2)_2NHBz$ (56%)

73),[132,133] although only the latter method would appear to be comparable in terms of yield and versatility to those methods described above. The mechanisms of these reductions are unclear, but it may be that reduction is occurring during work-up, as has been found by Maryanoff and coworkers during their studies,[129] and/or that diborane is involved.

i, $Bu^n{}_4NBH_4$, CH_2Cl_2, reflux 10 h; ii, 10% HCl, r.t., 30 min

(72)

$R^1 = R^2 = R^3 = H$ (48%)
$R^1 = R^3 = H, R^2 = Me$ (66%)
$R^1 = R^2 = Me, R^3 = H$ (44%)
$R^1 = R^2 = H, R^3 = Me$ (6%)
$R^1 = H, R^2, R^3 = -(CH_2)_4-$ (57%)

$Zn(BH_4)_2$, Et_2O, r.t., 2 d

(73)

$R^1 = R^2 = R^3 = H$ (92%)
$R^1 = R^3 = H, R^2 = Me$ (94%)
$R^1 = R^2 = H, R^3 = Me$ (38%)
$R^1 = H, R^2 = R^3 = Me$ (39%)
$R^1 = H, R^2, R^3 = -(CH_2)_4-$ (30%)
$R^1 = Me, R^2 = R^3 = H$ (0%)

Pioneered by Berger, Kikugawa and Maryanoff, and their coworkers, it has been found that borane and amine–boranes in conjunction with acid are capable of reducing indoles to indolines (equations 74–77).[89,128,129,134–138] Depending on the exact system, these methods may or may not be more efficient than the use of the chemically related $NaBH_4(NaBH_3CN)/RCO_2H$ reagent combination. Okamoto and coworkers have employed 2-amino-4-methylpyridine–borane in HOAc to reduce indole to indoline in 86% yield.[139]

i, $Me_3N{\bullet}BH_3$, dioxane; ii, HCl, Δ, 30 min; 80% (74)

$R^1 = R^2 = H$ (80%); $R^1, R^2 = -(CH_2)_4-$ (87%) (*cis*)

$Py{\bullet}BH_3$; 10% HCl–EtOH, 5–20 min, r.t. (75)

$R^1 = R^2 = R^3 = H$ (82%)
$R^1 = R^3 = H, R^2 = Me$ (98%)
$R^1 = R^3 = H, R^2 = Ph$ (56%)
$R^1 = R^2 = H, R^3 = Me$ (82%)
$R^1 = Me, R^2 = R^3 = H$ (54%)

$Py{\bullet}BH_3$, TFA; 0 °C, 2 min; 85–96% (76)

R = H, Ac, Bn, CHO, CO_2Bn

$BH_3{\bullet}THF$, TFA; 0 °C, 2 min; 86% (77)

Berger and coworkers have found that the reaction can be manipulated so as to give the *trans*-fused carboline product (equation 78).[88] This stereoselective reduction was later extended to secondary amines in the γ-carboline series by Elliott and Guzik.[140,141]

i, $BH_3{\bullet}THF$; ii, H_3O^+ (78)

R = Me (65–75%); R = H (>80%)

This stereoselective reduction is thought to involve the pathway shown in Scheme 7, wherein the key step is intramolecular hydride delivery from an amine–borane complex (**25**). This mechanism is supported by the appropriate deuterium labeling experiments.[88]

Zagorevskii *et al.* have observed the same stereoselective *trans* reduction with several tetrahydro-γ-carbolines by generating the diborane from $NaBH_4$ *in situ* (equation 79).[142] As expected, treatment of these carbolines with pyridine–borane and acid gives the *cis*-fused isomers (equation 80).[142] Bosch and coworkers have also utilized an *in situ* generation of diborane to effect the chemoselective reduction of the indole double bond during the synthesis of a new indolomorphan.[143]

Chu and coworkers observed that the indole ester (**26**) can be reduced selectively using pyridine–borane/HCl (equation 81),[144] but Fritz and coworkers found that the same conditions reduce both the indole double bond and the carbonyl group in ketone (**27**; equation 82) but only the indole double bond in amide (**28**; equation 83).[145]

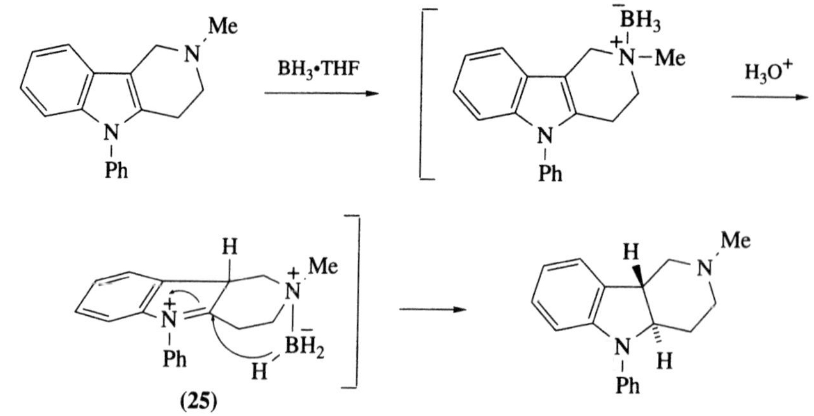

BH_3•THF

H_3O^+

(25)

Scheme 7

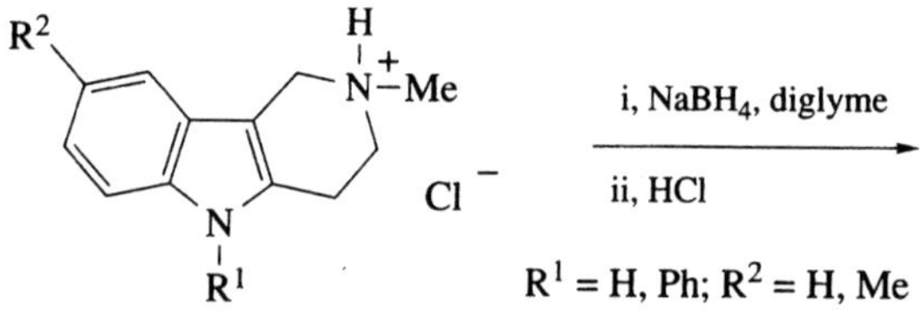

i, $NaBH_4$, diglyme

ii, HCl

R^1 = H, Ph; R^2 = H, Me

(79)

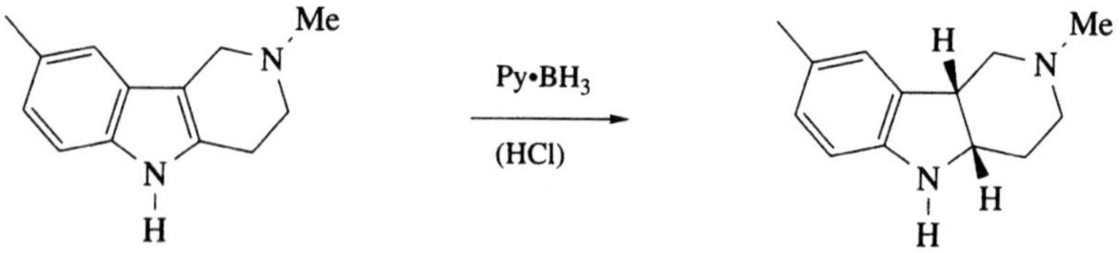

Py•BH_3

(HCl)

(80)

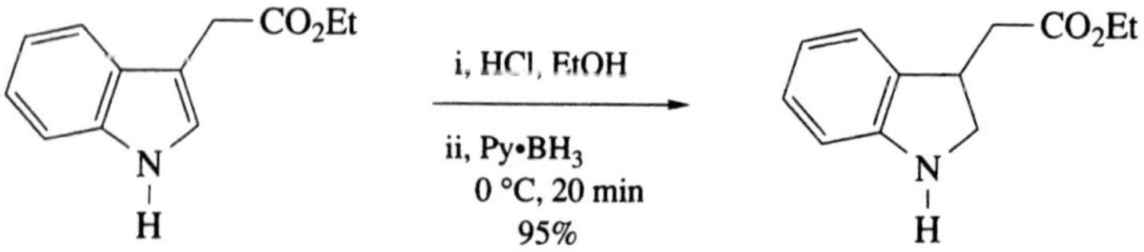

(26)

i, HCl, EtOH

ii, Py•BH_3
0 °C, 20 min
95%

(81)

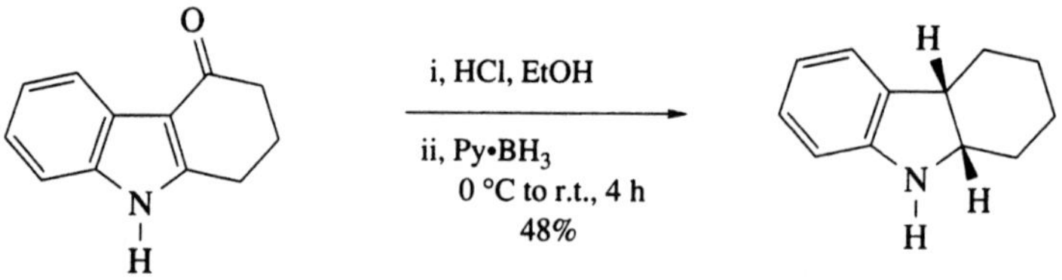

(27)

i, HCl, EtOH

ii, Py•BH_3
0 °C to r.t., 4 h
48%

(82)

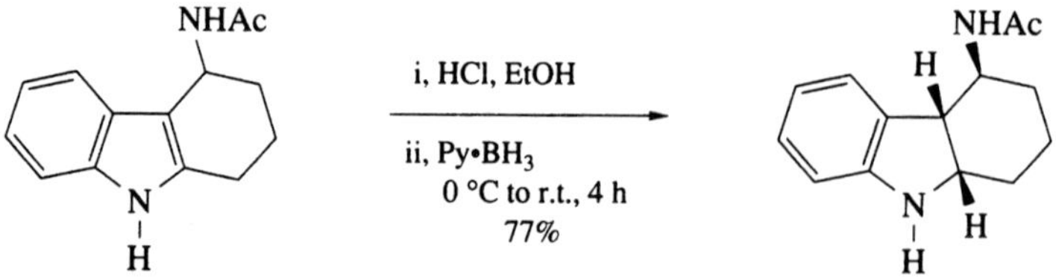

(28)

i, HCl, EtOH

ii, Py•BH_3
0 °C to r.t., 4 h
77%

(83)

Interestingly, whereas $NaBH_3CN$/HOAc fails to reduce the indole ring in the PDE thiophene analog (**29**), this double bond is reduced using borane–TFA (equation 84).[146] These conditions are also reported by Cava and coworkers to reduce a furan analog of (**29**).[113]

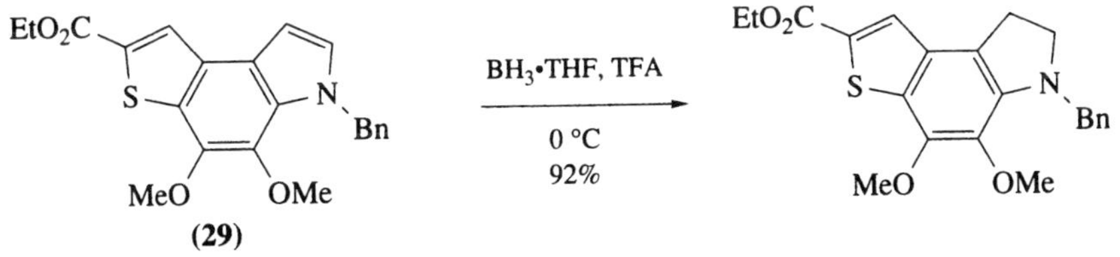

BH_3•THF, TFA; 0 °C; 92% (84)

(**29**)

The ionic hydrogenation of indoles to indolines using Et_3SiH/TFA has been described by Parnes *et al.*[147] and utilized by McKenzie,[127] Magnus,[148,149] Hlasta[150] and Ward,[151] and their coworkers in several different situations (equations 85–89). For the most part, the chemoselectivity and stereoselectivity that are observed with this reagent combination parallel that which was observed with $NaBH_4$/TFA (*vide supra*).

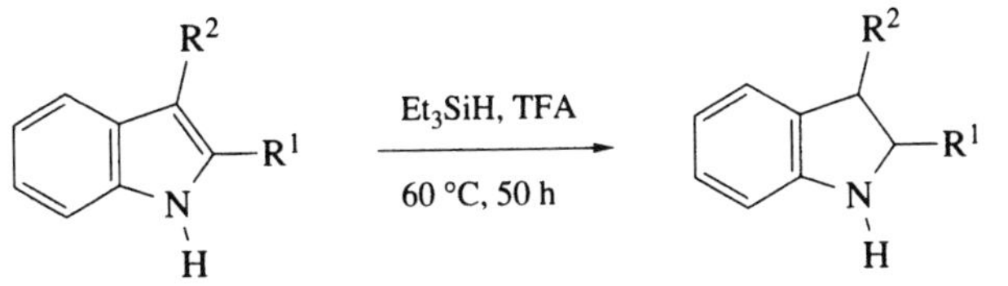

Et_3SiH, TFA; 60 °C, 50 h (85)

R^1 = Me, R^2 = H (83%); R^1 = R^2 = Me (80%) (*cis*:*trans* = 42:58)

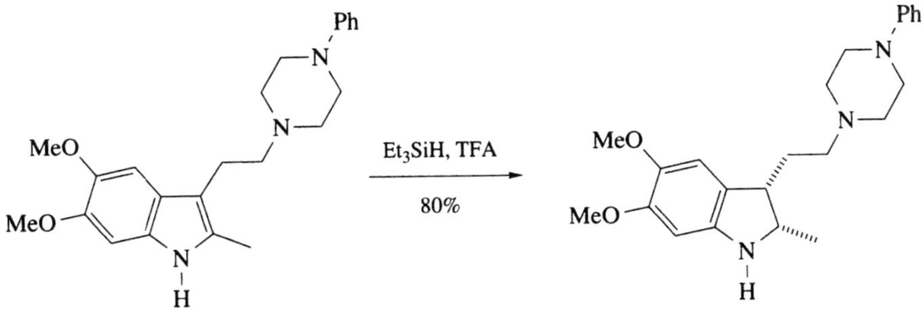

Et_3SiH, TFA; 80% (86)

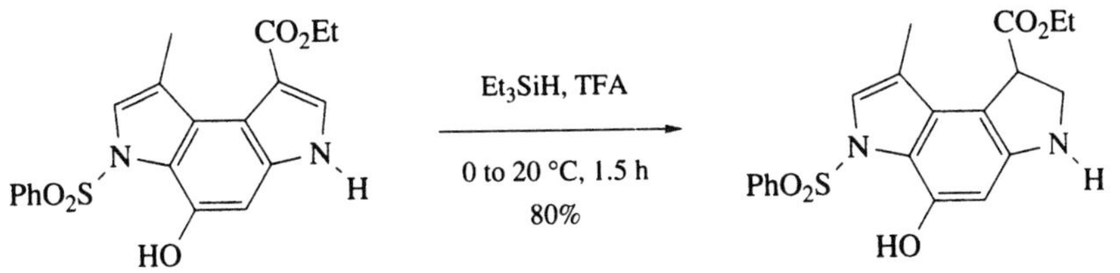

Et_3SiH, TFA; 0 to 20 °C, 1.5 h; 80% (87)

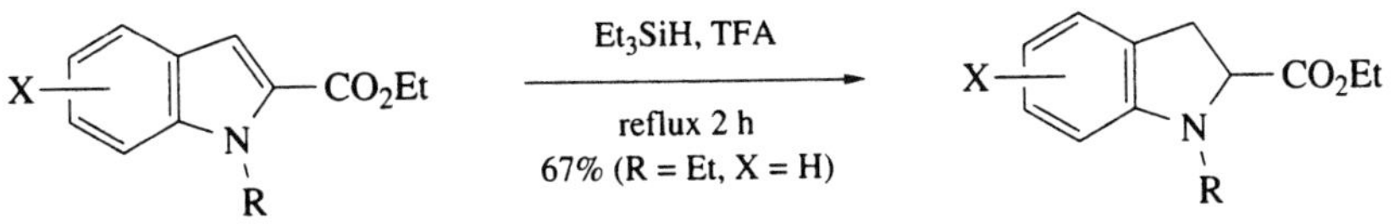

Et_3SiH, TFA; reflux 2 h; 67% (R = Et, X = H) (88)

R = Me, Et, Pr, Bn, allyl; X = H, 5-F, 7-Cl, 5,7-diCl_2

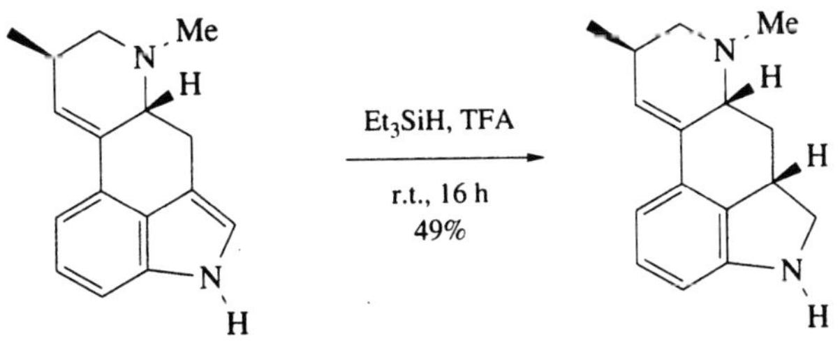

Et_3SiH, TFA; r.t., 16 h; 49% (89)

In summary, myriad boron hydride and silicon hydride reduction methods are available for the reduction of indoles to indolines. It is interesting to compare the various methods for the stereoselective reduction of 2,3-dimethylindole to *cis*- and *trans*-2,3-dimethylindoline (Table 1). While most methods show a preference for the more stable *trans* isomer, only with $Zn(BH_4)_2$ is this preference pronounced.

Table 1 Stereochemistry of the Reduction of 2,3-Dimethylindole

Conditions	*Cis(%)*	*Trans(%)*	*Yield(%)*	*Ref.*
$Zn(BH_4)_2$, r.t.	6	94	39	133
$Zn(BH_4)_2$, 40 °C	18	82	61	133
H_2, Pt/Al_2O_3	18	82	28	73
$NaBH_3CN$, HOAc	25	75	90	107
BH_3–TFA	33	67	82	129
Et_3SiH, TFA	42	58	80	147
B_2H_6, NaOMe, HCl	47	53	93	101
BH_3–Me_3N, HCl	57	43	72	134
Zn, TFA	63	37	—	147
Zn, H_3PO_4	65	35	19	96
Zn, HCl	70	30	—	147

3.7.5.4 Miscellaneous Methods

Two indole reduction methods that have not been studied in detail are the use of phosphonium hydriodide (equation 90)[152] and electrolysis (equations 91–93).[153–155]

PH_4I, HI (90)

electrolysis, MeCN, H_2SO_4, 26 °C, 20 min, 99% (91)

electrolysis, $Bu^n_4N^+OH^-$, aq. THF, 70%; 86% + 14% (92)

electrolysis, 54% H_2SO_4, 90–95% (93)

The isomeric isoindoles are very susceptible to reduction. But because no new methods to accomplish this reduction have been reported in the years since an excellent review[156] was published over 20 years ago, nothing further on this topic will be presented here.

3.7.6 REDUCTION OF BENZO[*b*]FURANS

Unlike the enamine double bond of indoles, the corresponding double bond of benzo[*b*]furans is less basic and, therefore, is much less susceptible to the protonation–reduction tactic exploited so successfully with indoles (*vide supra*). Nevertheless, several reduction methods are possible and reviews of early work are available.[157,158]

3.7.6.1 Catalytic Hydrogenation

Studies of the hydrogenation of benzo[*b*]furans with several Pt, Pd and Rh catalysts (equations 94–97) revealed that Pd gives the highest yields of dihydrobenzo[*b*]furans (equations 94 and 96).[159-163] A Pd/styrene–maleic acid copolymer catalyst reduces benzofuran to dihydrobenzo[*b*]furan in quantitative yield.[25] Platinum gives significant amounts of the cleavage product 2-ethylphenol and rhodium gives overreduction (equation 95).[161] A nickel catalyst results in the deoxygenation of a benzylic acetate (equation 97).[163]

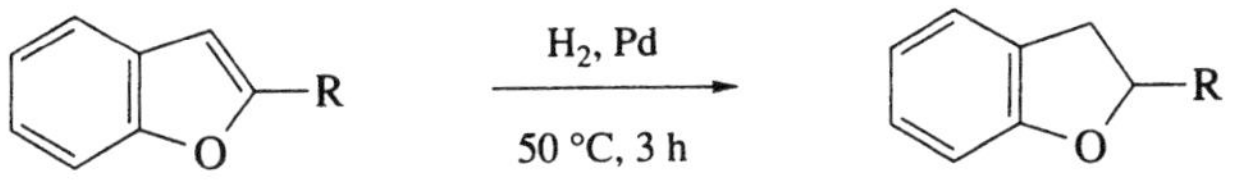

R
O
H_2, Pd
50 °C, 3 h
R
O

(94)

R = H (97%); R = Me (22%); R = Et (33%); R = Pr^n (5%)

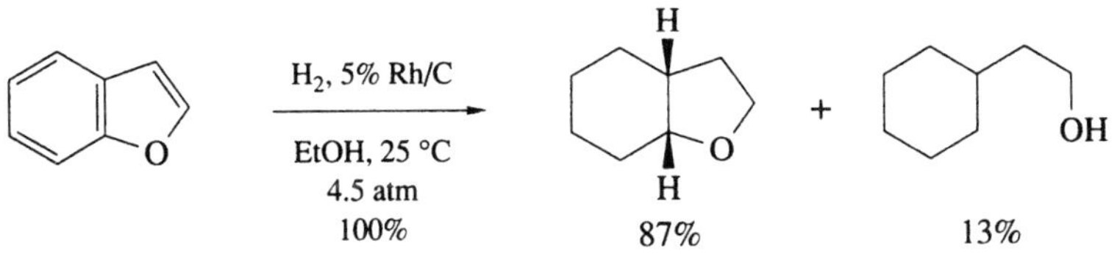

O
H_2, 5% Rh/C
EtOH, 25 °C
4.5 atm
100%
H
O
H
87%
+
OH
13%

(95)

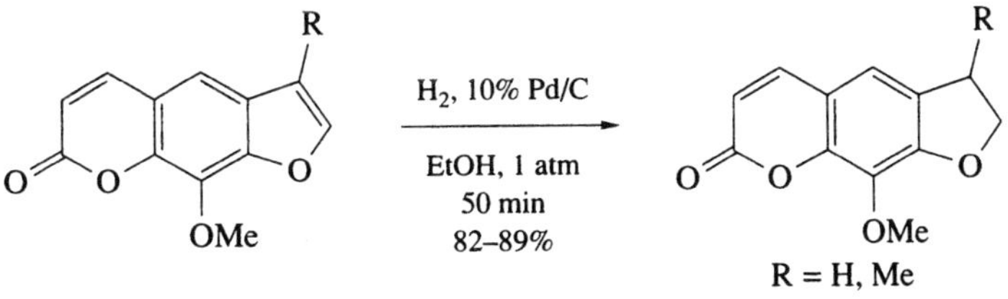

R
O
O
O
OMe
H_2, 10% Pd/C
EtOH, 1 atm
50 min
82–89%
R
O
O
O
OMe
R = H, Me

(96)

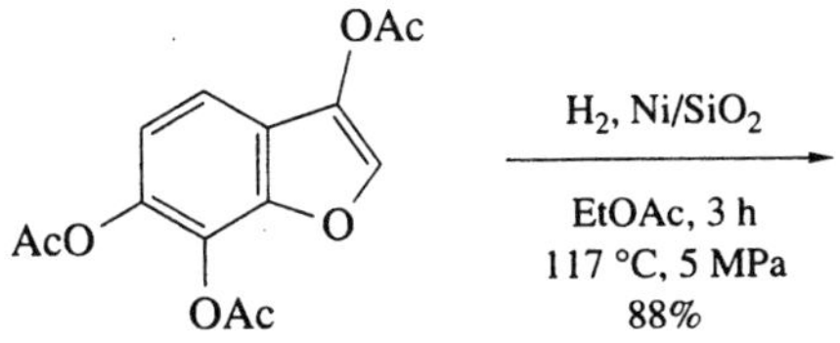

OAc
AcO
O
OAc
H_2, Ni/SiO_2
EtOAc, 3 h
117 °C, 5 MPa
88%
AcO
O
OAc

(97)

A careful study by Matsumoto and coworkers on the hydrogenation of the diterpene precursor (**30**) reveals that the less active catalyst palladium gives a higher percentage of the (*S*)-isomer than does the more active catalyst platinum (equation 98).[164]

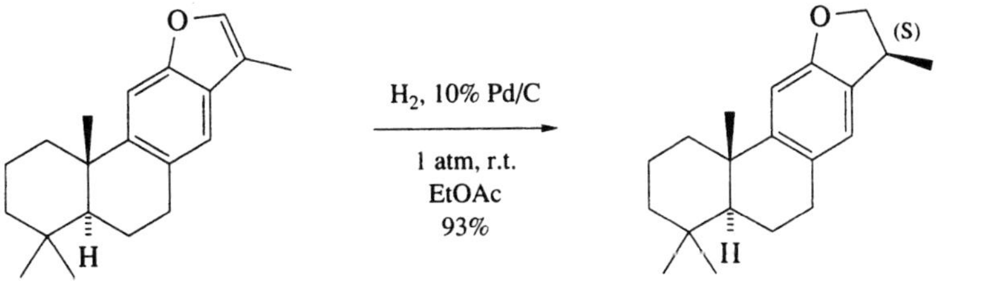

O
H
H_2, 10% Pd/C
1 atm, r.t.
EtOAc
93%
O
(S)
H

(98)

(**30**)

(S):*(R)* = 5:1
(PtO_2/AcOH gives *(S)*:*(R)* = 2:1)

Interestingly, Venugopalan and Balasubramanian found that only the pyran ring is hydrogenated in furopyran (**31**) using Pd/C (equation 99).[165] Somewhat similarly, Horaguchi and coworkers noted that the furan double bond in 7-methoxycyclohepta[*cd*]benzofuran (**32**) is unaffected during the hydrogenation of the cycloheptadiene portion of the molecule (equation 100).[166]

H_2, 10% Pd/C, acetone, 1 atm, r.t., 83% (99)

(31)

H_2, 7% Pd/C, EtOH, 1 atm, r.t., 1.5 h, 93% (100)

(32)

Rapoport and coworkers have used a hydrogenation–dehydrogenation protocol to introduce tritium into psoralen derivatives (equation 101).[167] For the same class of compounds, Heindel and coworkers have discovered a very simple catalytic transfer hydrogenation of the furan double bond (equation 102).[168,169]

i, T_2, 10% Pd/C, AcOH, 4 h, 1 atm, r.t.; ii, 10% Pd/C, Ph_2O reflux, 28 h, 12% (101)

10% Pd/C, cyclohexene, 95% EtOH, reflux, 30 min (102)

R^1 = H, R^2 = Me (64%)
R^1 = Me, R^2 = H (70%)

Although reduction of the isomeric benzo[*c*]furans (isobenzofurans) has been much less studied than the reduction of benzo[*b*]furans, some examples of their reduction are known.[170] Thus, 1,3-diphenylbenzo[*c*]furan can be hydrogenated to afford the two products shown in equation (103).[171]

H_2, 5% Pd/C, EtOAc, 50 psi, 20 °C, 12 h, 72% (103)

46% + 54%

3.7.6.2 Dissolving Metals

The dissolving metal reduction of benzo[*b*]furans and dibenzofurans is a classical technique and, depending on the conditions, can give rise to a variety of products.[157,158] Several examples are summarized in equations (104–108).[172–175] The cleavage of dibenzofurans to afford 2-hydroxybiphenyls (equation 107) is a particularly valuable route to these compounds.[174] The reduction of 3-carbomethoxybenzofurans with Mg/MeOH (equation 108)[175] is reminiscent of the indole reduction reported earlier (equation 51).

2 equiv. Li, Bu^{t}OH; 68% (104)

Li/NH_3, EtOH; 70% (105)

i, Na/NH_3; ii, NH_4Cl; 82% (106)

i, 3 equiv. Li, dioxane, Δ, 3 h; ii, HCl; 100% (107)

Mg, MeOH; r.t. (108)

R = Me (92%); R = Cl (99%)

In contrast to their observations noted earlier, Horaguchi and coworkers were able to reduce (**32**) using Na/EtOH (equation 109).[166]

Na, EtOH; reflux, 40 min; 88% (109)

(**32**) 23% 77% Na EtOH

The Birch-type reduction of 1,3-diphenylbenzo[*c*]furan (**33**) has been studied in detail by Smith and McCall (equation 110).[171]

i, Na, THF; ii, MeOH; 80% (110)

(**33**) 90% 10%

3.7.6.3 Boron Hydrides and Related Methods

Although much less explored than the analogous reductions of indoles (*vide supra*), the reduction of benzofurans with mixed metal hydrides is showing promise as a useful and versatile methodology.

Borisova *et al.* have found that benzofurans are reduced to dihydrobenzofurans using $NaBH_4$ (or KBH_4)/TFA (equation 111).[176] In the case of 2-ethyl-3-methylbenzofuran, the ratio of *cis*:*trans* products is 40:60. In contrast to this result, Cava and coworkers did not observe reduction of the benzofuran double bond in (**34**) using BH_3/TFA (equation 112).[113]

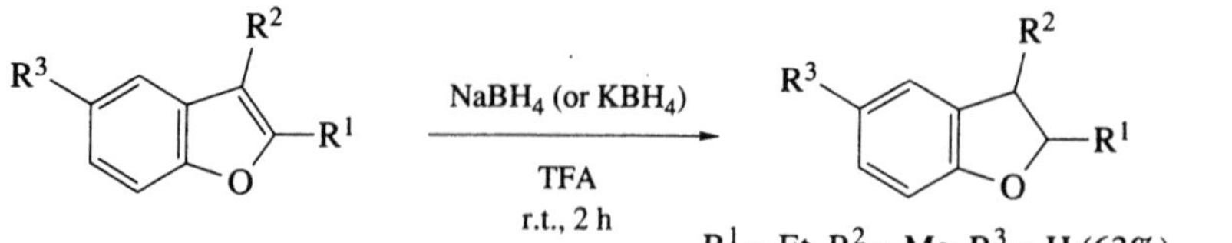

$NaBH_4$ (or KBH_4), TFA, r.t., 2 h (111)

R^1 = Et, R^2 = Me, R^3 = H (63%)
R^1, R^2 = $-(CH_2)_4-$, R^3 = H (67%)
R^1, R^2 = $-(CH_2)_4-$, R^3 = OH (52%)

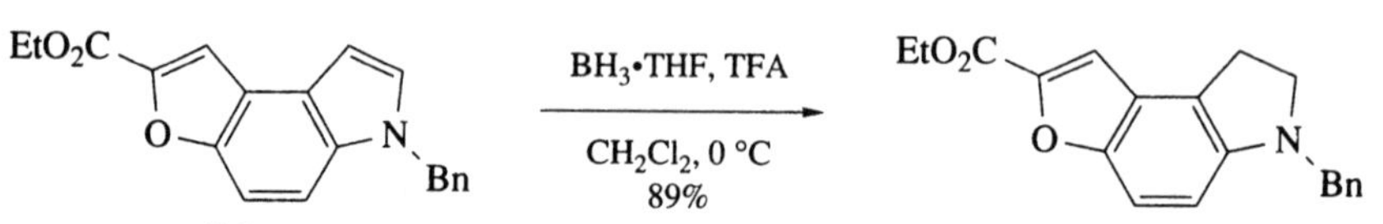

(**34**) BH_3•THF, TFA, CH_2Cl_2, 0 °C, 89% (112)

Karakhanov and coworkers successfully applied the ionic hydrogenation tactic, using Et_3SiH/TFA, to the reduction of several benzofurans (equation 113).[177] In the case of 2,3-dimethylbenzofuran, the ratio of *cis:trans* products is 19:81. It is interesting to note the deuterium-labeling results reported by these Russian scientists. Thus, whereas treatment of benzofuran (or 2-methylbenzofuran) with Et_3SiD/TFA gives rise to the 2-deuterio derivative, indicating the expected C-3 protonation, the same treatment of 3-methylbenzofuran gives rise to the 3-deuterio derivative, indicating C-2 protonation (Scheme 8).

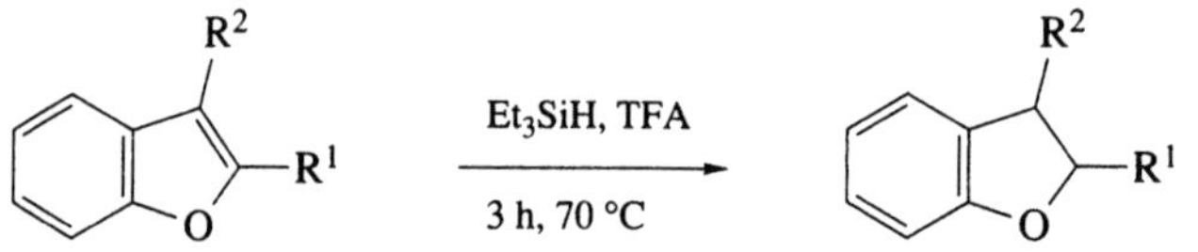

Et_3SiH, TFA, 3 h, 70 °C (113)

R^1 = R^2 = H (81%)
R^1 = H, R^2 = Me (96%)
R^1 = Me, R^2 = H (96%)
R^1 = Et, R^2 = H (96%)
R^1 = R^2 = Me (94%)

CF_3CO_2H → [ ] → Et_3SiD → D

CF_3CO_2H → [ ] → Et_3SiD → D

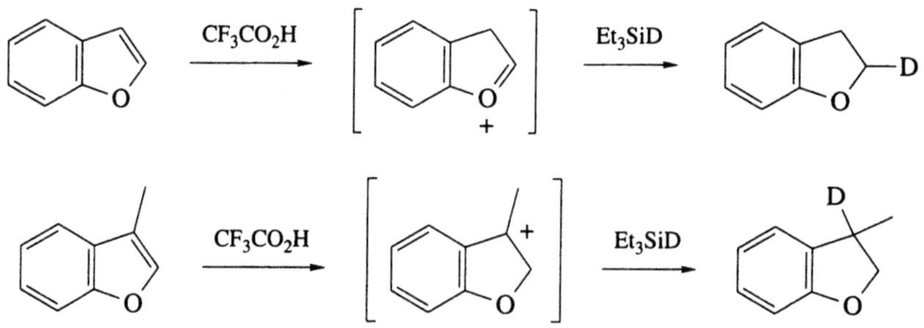

Scheme 8

3.7.6.4 Miscellaneous Methods

A photoreduction of 3-phenylbenzofurans has been reported by Parkanyi and coworkers (equation 114)[178] and benzo[*b*]furans can be electrochemically reduced (equation 115).[154] Further electrolysis of 2,3-dihydrobenzofuran gives 2,3,4,7-tetrahydrobenzofuran. Similarly, electrolysis of 5-methoxybenzofuran gives 69% of the dihydro product. In the absence of water, only 2-ethylphenol is observed.[154]

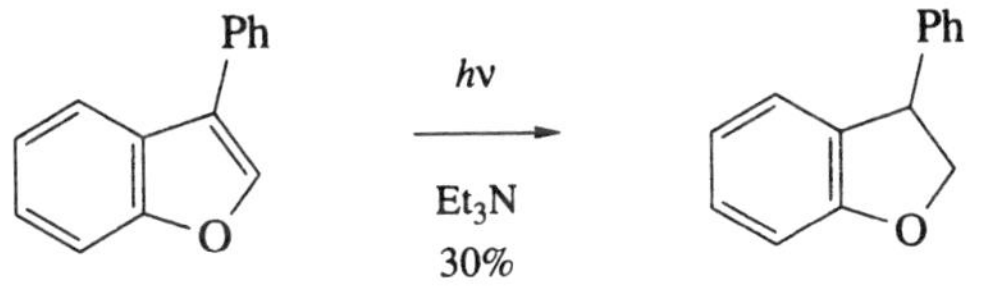

(114)

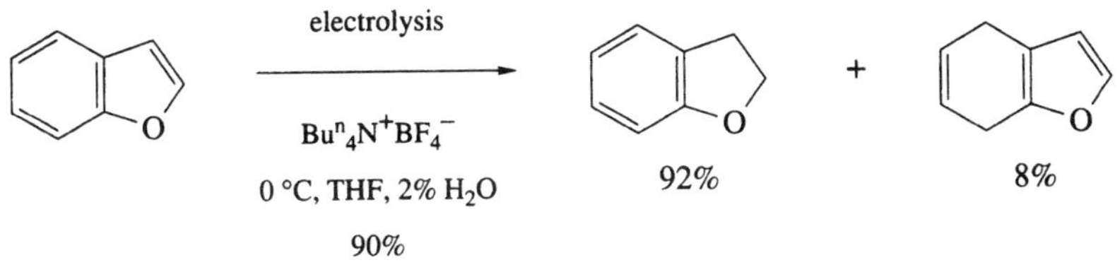

(115)

3.7.7 REDUCTION OF BENZO[*b*]THIOPHENES

The early reductions of benzo[*b*]thiophenes, including hydrosulfurization reactions, have been reviewed.[42,43,179–181]

3.7.7.1 Catalytic Hydrogenation

The catalytic hydrogenation of benzo[*b*]thiophene to 2,3-dihydrobenzo[*b*]thiophene using both a PdS catalyst and a rhodium catalyst have been reported (equation 116).[182–184] Fish and coworkers have also accomplished this reduction using a ruthenium catalyst.[81]

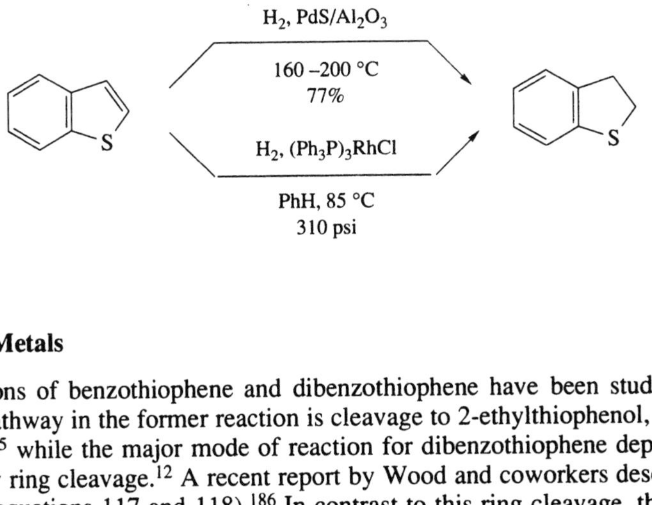

(116)

3.7.7.2 Dissolving Metals

The Birch reductions of benzothiophene and dibenzothiophene have been studied for a number of years.[12] The major pathway in the former reaction is cleavage to 2-ethylthiophenol, under several conditions with Na/NH_3,[185] while the major mode of reaction for dibenzothiophene depends on the order of mixing, but is mainly ring cleavage.[12] A recent report by Wood and coworkers describes the use of calcium in this regard (equations 117 and 118).[186] In contrast to this ring cleavage, the conjugated benzothiophene (**35**) is cleanly reduced to the 2,3-dihydro derivative by the action of Mg/MeOH (equation 119).[175]

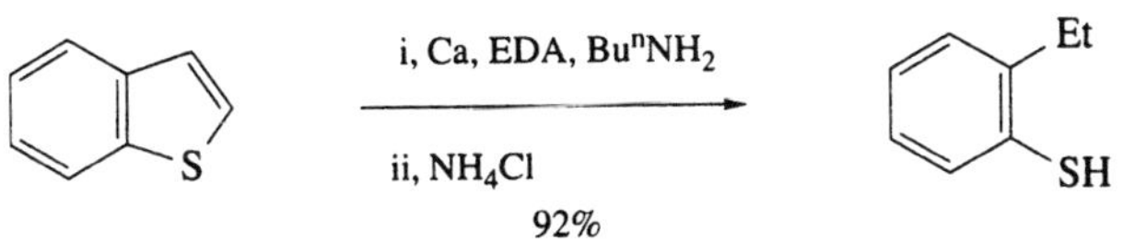

(117)

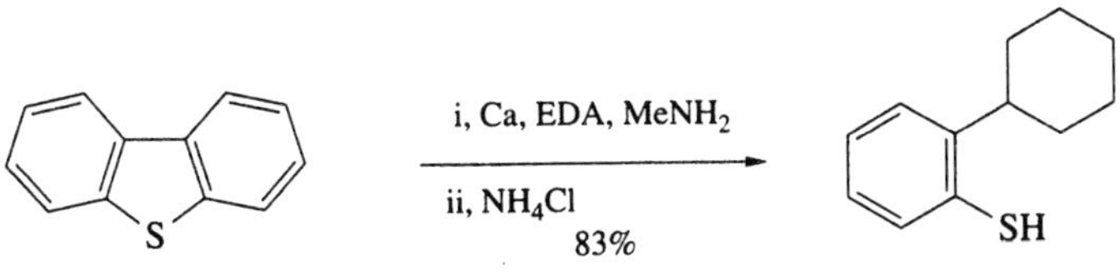

(118)

(35) Mg, MeOH, r.t., 91% (119)

3.7.7.3 Ionic Hydrogenation

The only reduction of benzothiophenes involving metal hydrides is ionic hydrogenation utilizing Et_3SiH/TFA (equation 120).[56,187] The rate of reduction of benzothiophene itself is very sluggish and much slower than that of benzofuran (equation 113).

Et_3SiH, TFA, 50 °C, 20–125 h (120)

R^1 = Me, R^2 = H (80%)
R^1 = H, R^2 = Me (90%)
R^1 = R^2 =H (55%)

The deuterium-labeling studies parallel those which were observed with benzofurans (Scheme 8). Thus, 2-methylbenzothiophene is protonated at C-3 and 3-methylbenzothiophene is protonated at C-2 to give the products (**36**) and (**37**), respectively, after deuteride transfer to the respective intermediate carbocations.[187]

(**36**) (**37**)

3.7.7.4 Electrochemical

The electrolytic reduction of benzo[*b*]thiophene has been described by Pacut and Kariv-Miller (equation 121).[188] Because the two products can be easily separated, this electrochemical Birch reduction is a very good preparation of 2,3-dihydrobenzo[*b*]thiophene.

electrolysis, $Bu^n_4NBF_4$, aq. THF, 0 °C, 87% → 86% + 14% (121)

3.7.8 REFERENCES

1. A. H. Corwin, *Heterocycl. Compds.*, 1950, **1**, 339.
2. L. H. Andrews and S. M. McElvain, *J. Am. Chem. Soc.*, 1929, **51**, 887.
3. M. Tokuda, Y. Yamada, T. Takagi, H. Suginome and A. Furusaki, *Tetrahedron*, 1987, **43**, 281.
4. W. W. Turner, *J. Heterocycl. Chem.*, 1986, **23**, 327.
5. E. E. Schweizer and K. K. Light, *J. Am. Chem. Soc.*, 1964, **86**, 2963.
6. H. A. Bates and H. Rapoport, *J. Am. Chem. Soc.*, 1979, **101**, 1259.
7. R. G. Buttery, L. C. Ling and B. Juliano, *Chem. Ind. (London)*, 1982, 958.
8. I. M. Labouta, P. Jacobsen, P. Thorbek, P. K. Larsen and H. Hjeds, *Acta Chem. Scand., Ser. B*, 1982, **36**, 669.
9. J. B. Paine, III and D. Dolphin, *J. Org. Chem.*, 1988, **53**, 2787.
10. I. Rajca, *Pol. J. Chem.*, 1981, **55**, 775 (*Chem. Abstr.*, 1982, **97**, 91 423g).
11. G. Lunn, *J. Org. Chem.*, 1987, **52**, 1043.
12. A. J. Birch and J. Slobbe, *Heterocycles*, 1976, **5**, 905.

13. A. J. Birch, R. W. Rickards and K. S. J. Stapleford, *Aust. J. Chem.*, 1969, **22**, 1321.
14. G. G. Evans, *J. Am. Chem. Soc.*, 1951, **73**, 5230.
15. C. B. Hudson and A. V. Robertson, *Tetrahedron Lett.*, 1967, 4015.
16. D. M. Lemal and S. D. McGregor, *J. Am. Chem. Soc.*, 1966, **88**, 1335.
17. D. P. Schumacher and S. S. Hall, *J. Am. Chem. Soc.*, 1982, **104**, 6076.
18. J. W. Scott, A. Focella, U. O. Hengartner, D. R. Parrish and D. Valentine, Jr., *Synth. Commun.*, 1980, **10**, 529.
19. R. C. Elderfield and T. N. Dodd, Jr., *Heterocycl. Compds.*, 1950, **1**, 160.
20. A. P. Dunlop and F. N. Peters, in 'The Furans', Reinhold, New York, 1953, chap. 15.
21. P. Bosshard and C. H. Eugster, *Adv. Heterocycl. Chem.*, 1966, **7**, 377.
22. Z. Dudzik and M. Gasiorek, *Przem. Chem.*, 1975, **54**, 637 (*Chem. Abstr.*, 1976, **84**, 89 903b).
23. F. M. Dean, *Adv. Heterocycl. Chem.*, 1982, **30**, 167.
24. V. A. Slavinskaya, R. A. Karakhanov, L. Yu. Brezhnev, I. I. Geiman, L. F. Bulenkova and A. K. Strautinya, *Chem. Heterocycl. Compds. (Engl. Transl.)*, 1982, 997.
25. E. A. Karakhanov, A. S. Loktev, V. S. Pshezhetskii, I. P. Stepanova and A. G. Dedov, *Chem. Heterocycl. Compds. (Engl. Transl.)*, 1984, 835.
26. S. Srivastava, J. Minore, C. K. Cheung and W. J. le Noble, *J. Org. Chem.*, 1985, **50**, 394.
27. R. A. Karakhanov, T. I. Odintsova, V. B. Yakovlev and A. P. Rodin, *React. Kinet. Catal. Lett.*, 1987, **33**, 219 (*Chem. Abstr.*, 1988, **108**, 75 134d).
28. J. Soós, *React. Kinet. Catal. Lett.*, 1987, **34**, 333 (*Chem. Abstr.*, 1988, **109**, 109 600r).
29. S. Morikawa, *Noguchi Kenkyusho Jiho*, 1983, 49 (*Chem. Abstr.*, 1984, **101**, 72 532b).
30. A. O. Bedenbaugh, J. H. Bedenbaugh, J. D. Adkins and W. A. Bergin, *J. Org. Chem.*, 1970, **35**, 543.
31. H. R. Divanford and M. M. Joullié, *Org. Prep. Proced. Int.*, 1978, **10**, 94.
32. T. Masamune, M. Ono and H. Matsue, *Bull. Chem. Soc. Jpn.*, 1975, **48**, 491.
33. T. Kinoshita, K. Miyano and T. Miwa, *Bull. Chem. Soc. Jpn.*, 1975, **48**, 1865.
34. T. Kinoshita and T. Miwa, *J. Chem. Soc., Chem. Commun.*, 1974, 181.
35. A. J. Birch and J. Slobbe, *Tetrahedron Lett.*, 1975, 627.
36. A. J. Birch and J. Slobbe, *Tetrahedron Lett.*, 1976, 2079.
37. G. I. Bolestova, Z. N. Parnes and D. N. Kursanov, *J. Org. Chem. USSR (Engl. Transl.)*, 1979, **15**, 1129.
38. F. F. Blicke, *Heterocycl. Compds.*, 1950, **1**, 249.
39. S. Gronowitz, *Adv. Heterocycl. Chem.*, 1963, **1**, 1.
40. L. I. Belen'kii and Ya. L. Gol'dfarb, *Chem. Heterocycl. Compds.*, 1985, **44**, 457.
41. S. C. Schuman and H. Shalit, *Catal. Rev.*, 1970, **4**, 245.
42. R. J. Angelici, *Acc. Chem. Res.*, 1988, **21**, 387.
43. C. M. Friend and J. T. Roberts, *Acc. Chem. Res.*, 1988, **21**, 394.
44. H. Greenfield, S. Metlin, M. Orchin and I. Wender, *J. Org. Chem.*, 1958, **23**, 1054.
45. G. Claeson and H. G. Jonsson, *Ark. Kemi*, 1969, **31**, 83 (*Chem. Abstr.*, 1969, **71**, 91 191t).
46. P. N. Confalone, G. Pizzolato and M. R. Uskoković, *J. Org. Chem.*, 1977, **42**, 135.
47. T. P. Voskresenskaya, V. A. Semikolenov, V. A. Likholobov, A. P. Shepelin, P. A. Zhdan and A. V. Mashkina, *Kinet. Katal.*, 1982, **23**, 382 (*Chem. Abstr.*, 1982, **97**, 23 058s).
48. A. A. Zirka, R. S. Sagitullin, R. M. Masagutov, A. Kh. Sharipov, N. K. Lyapina and A. V. Mashkina, *React. Kinet. Catal. Lett.*, 1984, **25**, 171 (*Chem. Abstr.*, 1984, **101**, 191 596k).
49. M. Sado, H. Abe, N. Inuzuka, K. Shirai and T. Kumamoto, *Bull. Chem. Soc. Jpn.*, 1983, **56**, 1665.
50. W. G. Blenderman, M. M. Joullié and G. Preti, *J. Org. Chem.*, 1983, **48**, 3206.
51. Ya. L. Gol'dfarb, A. V. Semenovsky, E. P. Zakharov, G. V. Davydova and F. M. Stoyanovich, *Izv. Akad. Nauk SSSR, Ser. Khim.*, 1979, 480 (*Chem. Abstr.*, 1979, **90**, 168 011m).
52. W. G. Blenderman and M. M. Joullié, *Synth. Commun.*, 1981, **11**, 881.
53. K. Nishino, S. Yano, Y. Kohashi, K. Yamamoto and I. Murata, *J. Am. Chem. Soc.*, 1979, **101**, 5059.
54. K. Kosugi, A. V. Anisimov, H. Yamamoto, R. Yamashiro, K. Shirai and T. Kumamoto, *Chem. Lett.*, 1981, 1341.
55. G. Lange, M. E. Savard, T. Viswanatha and G. I. Dmitrienko, *Tetrahedron Lett.*, 1985, **26**, 1791.
56. D. N. Kursanov, Z. N. Parnes, G. I. Bolestova and L. I. Belen'kii, *Tetrahedron*, 1975, **31**, 311.
57. S. I. Zav'yalov and O. V. Dorofeeva, *Izv. Akad. Nauk SSSR, Ser. Khim.*, 1980, 1634 (*Chem. Abstr.*, 1981, **94**, 3963t).
58. Z. N. Parnes, G. I. Bolestova and D. N. Kursanov, *Zh. Org. Khim.*, 1977, **13**, 476 (*Chem. Abstr.*, 1977, **87**, 21 664f).
59. Z. N. Parnes, Yu. I. Lyakhovetsky, S. P. Dolgova, A. S. Pakhomov and D. N. Kursanov, *Izv. Akad. Nauk SSSR, Ser. Khim.*, 1977, 2526 (*Chem. Abstr.*, 1978, **88**, 62 239k).
60. Z. N. Parnes, Yu. I. Lyakhovetsky, M. I. Kalinkin, D. N. Kursanov and L. I. Belen'kii, *Tetrahedron*, 1978, **34**, 1703.
61. F. M. Latypova, Yu. I. Lyakhovetsky, N. K. Lyapina and Z. N. Parnes, *Izv. Akad. Nauk SSSR, Ser. Khim.*, 1986, 2155 (*Chem. Abstr.*, 1987, **107**, 58 768a).
62. T. Yamanochi and H. Yamane (Asahi Chemical Co.), *Jpn. Pat.* 63 107 973 (1988) (*Chem. Abstr.*, 1988, **109**, 210 887q).
63. S. G. Mairanovskii, L. I. Kosychenko and S. Z. Taits, *Elektrokhimiya*, 1977, **13**, 1250 (*Chem. Abstr.*, 1978, **88**, 43 133).
64. V. P. Gul'tyai, I. V. Proskurovskaya, T. Ya. Rubinskaya, A. V. Lozanova, A. M. Moiseenkov and A. V. Semenovsky, *Bull. Acad. Sci. USSR, Div. Chem. Sci. (Engl. Transl.)*, 1979, 1457.
65. J. Wristers, *J. Am. Chem. Soc.*, 1977, **99**, 5051.
66. Yu. I. Lyakhovetsky, M. I. Kalinkin, Z. N. Parnes, F. M. Latypova and D. N. Kursanov, *J. Chem. Soc., Chem. Commun.*, 1980, 766.
67. P. L. Julian, E. W. Meyer and H. C. Printy, *Heterocycl. Compds*, 1952, **3**, 115.
68. B. Robinson, *Chem. Rev.*, 1969, **69**, 785.
69. R. J. Sundberg, 'The Chemistry of Indoles', Academic Press, New York, 1970, p. 129.

70. T. Tóth and A. Gerecs, *Acta Chim. Acad. Sci. Hung.*, 1971, **67**, 229.
71. W. A. Remers, *Chem. Heterocycl. Compds.*, 1972, **25**, 163; R. K. Brown, *Chem. Heterocycl. Compds.*, 1972, **25**, 462.
72. W. E. Noland and C. F. Hammer, *J. Org. Chem.*, 1960, **25**, 1536.
73. J. E. Shaw and P. R. Stapp, *J. Heterocycl. Chem.*, 1987, **24**, 1477.
74. M. J. Guttieri and W. F. Maier, *J. Org. Chem.*, 1984, **49**, 2875.
75. Ref. 69, p. 3.
76. V. A. Budylin, A. N. Kost and E. D. Matveeva, *Chem. Heterocycl. Compds. (Engl. Transl.)*, 1972, 52.
77. A. Smith and J. H. P. Utley, *J. Chem. Soc., Chem. Commun.*, 1965, 427.
78. Y. Nagai, H. Uno , M. Shimiz u an d T . Karasaw a, *Jp n. Pat.* 52 89 699 (1977) (*Chem. Abstr.*, 1978, **88**, 62 374a).
79. D. E. Ames, H. R. Ansari, A. D. G. France, A. C. Lovesey, B. Novitt and R. Simpson, *J. Chem. Soc. C*, 1971, 3088.
80. Nippon Kayaku Co., Ltd., *Jpn. Pat.* 59 65 072 (1984) (*Chem. Abstr.*, 1984, **101**, 171 090n).
81. R. H. Fish, J. L. Tan and A. D. Thormodsen, *Organometallics*, 1985, **4**, 1743.
82. Y. Kikugawa and M. Kashimura, *Synthesis*, 1982, 785.
83. Y. Watanabe, T. Ohta, Y. Tsuji, T. Hiyoshi and Y. Tsuji, *Bull. Chem. Soc. Jpn.*, 1984, **57**, 2440.
84. G. R. Brown and A. J. Foubister, *Synthesis*, 1982, 1036.
85. S. O'Brien and D. C. C. Smith, *J. Chem. Soc.*, 1960, 4609.
86. W. A. Remers, G. J. Gibs, C. Pidacks and M. J. Weiss, *J. Org. Chem.*, 1971, **36**, 279.
87. B. Heath-Brown, *Chem. Ind. (London)*, 1969, 1595.
88. J. G. Berger, S. R. Teller, C. D. Adams and L. J. Guggenberger, *Tetrahedron Lett.*, 1975, 1807.
89. J. G. Berger, P. Tahbaz, A. T. McPhail and K. D. Onan, *Tetrahedron Lett.*, 1983, **24**, 2469.
90. H. G. Schouten, *US Pat.* 4 485 241 (1984) (*Chem. Abstr.*, 1985, **102**, 78 722f).
91. J. W. Ashmore and G. K. Helmkamp, *Org. Prep. Proced. Int.*, 1976, **8**, 223.
92. G. Sauer, G. Haffer and H. Wachtel, *Synthesis*, 1986, 1007.
93. I. K. Youn and C. S. Pak, *Bull. Korean Chem. Soc.*, 1987, **8**, 434 (*Chem. Abstr.*, 1988, **108**, 221 543e).
94. G. P. Fagan, C. B. Chapleo, A. C. Lane, M. Myers, A. G. Roach, C. F. C. Smith, M. R. Stillings and A. P. Welbourn, *J. Med. Chem.*, 1988, **31**, 944.
95. E. J. Corey, R. J. McCaully and H. S. Sachdev, *J. Am. Chem. Soc.*, 1970, **92**, 2476.
96. L. J. Dolby and G. W. Gribble, *J. Heterocycl. Chem.*, 1966, **3**, 124.
97. M. Iizuka and H. Yasuoka (Hodogaya Chemical Co.), *Jpn. Pat.* 69 32 781 (1969) (*Chem. Abstr.*, 1970, **72**, 66 804h).
98. A. N. Kost, L. P. Makhno, A. K. Sheinkman and G. A. Markus, *Metody Poluch. Khim. Reakt. Prep.*, 1967, 39 (*Chem. Abstr.*, 1969, **70**, 114 929k).
99. G. W. Gribble, Ph.D. Dissertation, University of Oregon, 1967.
100. L. J. Dolby and G. W. Gribble, *Tetrahedron*, 1968, **24**, 6377.
101. S. A. Monti and R. R. Schmidt, III, *Tetrahedron*, 1971, **27**, 3331.
102. G. W. Gribble, P. D. Lord, J. Skotnicki, S. E. Dietz, J. T. Eaton and J. L. Johnson, *J. Am. Chem. Soc.*, 1974, **96**, 7812.
103. G. W. Gribble and C. F. Nutaitis, *Org. Prep. Proced. Int.*, 1985, **17**, 317.
104. G. W. Gribble and S. W. Wright, *Heterocycles*, 1982, **19**, 229.
105. G. W. Gribble, C. F. Nutaitis and R. M. Leese, *Heterocycles*, 1984, **22**, 379.
106. G. W. Gribble and C. F. Nutaitis, *Synthesis*, 1987, 709.
107. G. W. Gribble and J. H. Hoffman, *Synthesis*, 1977, 859.
108. Y. Kumar and L. Florvall, *Synth. Commun.*, 1983, **13**, 489.
109. C. G. Chavdarian, D. Karashima, N. Castagnoli, Jr. and H. K. Hundley, *J. Med. Chem.*, 1978, **21**, 548.
110. M. E. Flaugh, D. L. Mullen, R. W. Fuller and N. R. Mason, *J. Med. Chem.*, 1988, **31**, 1746.
111. V. H. Rawal and M. P. Cava, *J. Chem. Soc., Chem. Commun.*, 1984, 1526.
112. V. H. Rawal and M. P. Cava, *J. Am. Chem. Soc.*, 1986, **108**, 2110.
113. V. H. Rawal, R. J. Jones and M. P. Cava, *J. Org. Chem.*, 1987, **52**, 19.
114. D. L. Boger, R. S. Coleman and B. J. Invergo, *J. Org. Chem.*, 1987, **52**, 1521.
115. R. E. Bolton, C. J. Moody, C. W. Rees and G. Tojo, *J. Chem. Soc., Perkin Trans. 1*, 1987, 931.
116. R. E. Bolton, C. J. Moody, C. W. Rees and G. Tojo, *Tetrahedron Lett.*, 1987, **28**, 3163.
117. R. J. Sundberg, G. S. Hamilton and J. P. Laurino, *J. Org. Chem.*, 1988, **53**, 976.
118. P. Meghani, J. D. Street and J. A. Joule, *J. Chem. Soc., Chem. Commun.*, 1987, 1406.
119. G. W. Gribble, J. L. Johnson and M. G. Saulnier, *Heterocycles*, 1981, **16**, 2109.
120. J. Stanton and M. H. Ackerman, *J. Med. Chem.*, 1983, **26**, 986.
121. N. F. Kucherova, N. M. Sipilina, N. N. Novikova, I. D. Silenko, S. G. Rozenberg and V. A. Zagorevskii, *Chem. Heterocycl. Compds. (Engl. Transl.)*, 1980, 1051.
122. O. Repić and D. J. Long, *Tetrahedron Lett.*, 1983, **24**, 1115.
123. G. Sauer and G. Haffer, *Ger. Pat.* 3 411 981 (1985) (*Chem. Abstr.*, 1986, **105**, 43 137d).
124. R. K. Shakhatuni and F. R. Shiroyan, *Arm. Khim. Zh.*, 1983, **36**, 313 (*Chem. Abstr.*, 1983, **99**, 158 213a).
125. J. Le Men, L. Le Men-Oliver, J. Levy, M. C. Levy-Appert-Colin and J. Hannart, *Ger. Pat.* 2 410 651 (1974) (*Chem. Abstr.*, 1975, **82**, 43 640u).
126. D. Herlem and F. K. Huu, *Tetrahedron*, 1979, **35**, 633.
127. A. E. Lanzilotti, R. Littell, W. J. Fanshawe, T. C. McKenzie and F. M. Lovell, *J. Org. Chem.*, 1979, **44**, 4809.
128. B. E. Maryanoff and D. F. McComsey, *J. Org. Chem.*, 1978, **43**, 2733.
129. B. E. Maryanoff, D. F. McComsey and S. O. Nortey, *J. Org. Chem.*, 1981, **46**, 355.
130. D. A. Evans, K. T. Chapman and E. M. Carreira, *J. Am. Chem. Soc.*, 1988, **110**, 3560.
131. Y. Kikugawa, *Chem. Pharm. Bull.*, 1978, **26**, 108.
132. T. Wakamatsu, H. Inaki, A. Ogawa, M. Watanabe and Y. Ban, *Heterocycles*, 1980, **14**, 1441.
133. H. Kotsuki, Y. Ushio and M. Ochi, *Heterocycles*, 1987, **26**, 1771.

134. J. G. Berger, *Synthesis*, 1974, 508.
135. J. G. Berger, F. Davidson and G. E. Langford, *J. Med. Chem.*, 1977, **20**, 600.
136. Y. Kikugawa, *J. Chem. Res. (S)*, 1977, 212.
137. Y. Kikugawa, *J. Chem. Res. (S)*, 1978, 184.
138. Y. Kikugawa, K. Saito and S. Yamada, *Synthesis*, 1978, 447.
139. Y. Okamoto, T. Osawa, Y. Kurasawa, T. Kinoshita and K. Takagi, *J. Heterocycl. Chem.*, 1986, **23**, 1383.
140. A. J. Elliott and H. Guzik, *Tetrahedron Lett.*, 1982, **23**, 1983.
141. A. J. Elliott, H. Guzik, M. S. Puar and A. T. McPhail, *J. Chem. Soc., Perkin Trans. 2*, 1983, 1599.
142. V. A. Zagorevskii, S. G. Rozenberg, N. M. Sipilina, L. U. Bykova and A. P. Rodionov, *Zh. Vses. Khim. Ova.*, 1982, **27**, 102 (*Chem. Abstr.*, 1982, **96**, 217 013y).
143. J. Bonjoch, F. Boncompte, N. Casamitjana and J. Bosch, *Tetrahedron*, 1986, **42**, 6693.
144. C. K. Chu, J. Suh and H. G. Cutler, *J. Heterocycl. Chem.*, 1986, **23**, 1777.
145. T. Vogel, H.-U. Huth and H. Fritz, *Liebigs Ann. Chem.*, 1982, 739.
146. R. J. Jones and M. P. Cava, *J. Chem. Soc., Chem. Commun.*, 1986, 826.
147. Z. N. Parnes, V. A. Budylin, E. Yu. Beilinson and A. N. Kost, *Zh. Org. Khim.*, 1972, **8**, 2564 (*Chem. Abstr.*, 1973, **78**, 84 176h).
148. P. D. Magnus, T. Gallagher, J. Schultz, Y.-S. Or and T. P. Ananthanarayan, *J. Am. Chem. Soc.*, 1987, **109**, 2706.
149. P. Carter, S. Fitzjohn, S. Halazy and P. D. Magnus, *J. Am. Chem. Soc.*, 1987, **109**, 2711.
150. D. J. Hlasta, D. Luttinger, M. H. Perrone, M. J. Silbernagel, S. J. Ward and D. R. Haubrich, *J. Med. Chem.*, 1987, **30**, 1555.
151. J. S. Ward, R. W. Fuller, L. Merritt, H. D. Snoddy, J. W. Paschal, N. R. Mason and J. S. Horng, *J. Med. Chem.*, 1988, **31**, 1512.
152. C. B. Hudson and A. V. Robertson, *Aust. J. Chem.*, 1967, **20**, 1935.
153. H. Hannebaum, H. Nohe and H . R. Mueller (BASF), *Ger. Pat.* 2 403 446 (1975) (*Chem. Abstr.*, 1975, **83**, 210 783x); H. Nohe and H. Hannebaum (BASF), *Ger. Pat.* 2 658 951 (1978) (*Chem. Abstr.*, 1978, **89**, 109 092v.)
154. E. Kariv-Miller, D. F. Dedolph, C. M. Ryan and T. J. Mahachi, *J. Heterocycl. Chem.*, 1985, **22**, 1389.
155. V. N. Krasnosel'skii, E. I. Steblina and L. A. Ponomareva, *Elektrokhimiya*, 1983, **19**, 944 (*Chem. Abstr.*, 1983, **99**, 122 226t).
156. J. D. White and M. E. Mann, *Adv. Heterocycl. Chem.*, 1969, **10**, 113.
157. R. C. Elderfield and V. B. Meyer, *Heterocycl. Compds.*, 1951, **2**, 43.
158. A. Mustafa, 'The Chemistry of Heterocyclic Compounds, Benzofurans', ed. A. Weissberger and E. C. Taylor, Wiley-Interscience, New York, 1974.
159. E. A. Karakhanov, A. G. Dedov and A. S. Loktev, *Chem. Heterocycl. Compds. (Engl. Transl.)*, 1981, 993.
160. R. A. Ellison and F. N. Kotsonis, *J. Org. Chem.*, 1973, **38**, 4192.
161. A. G. Anastassiou and R. P. Cellura, *J. Chem. Soc., Chem. Commun.*, 1969, 1521.
162. J. K. MacLeod, B. R. Worth and R. J. Wells, *Aust. J. Chem.*, 1978, **31**, 1545.
163. E. Tenor, *Ger. Pat.* 147 847 (1981) (*Chem. Abstr.*, 1981, **95**, 187 058b).
164. T. Matsumoto, S. Imai, T. Yoshinari and K. Tsuruta, *Bull. Chem. Soc. Jpn.*, 1987, **60**, 2401.
165. B. Venugopalan and K. Balasubramanian, *Heterocycles*, 1985, **23**, 81; note that the reduction product in this paper is drawn incorrectly, *i.e.* missing the furan double bond.
166. T. Horaguchi, K. Tanemura and T. Suzuki, *J. Heterocycl. Chem.*, 1988, **25**, 39.
167. S. T. Isaacs, H. Rapoport and J. E. Hearst, *J. Labelled Comp. Radiopharm.*, 1982, **19**, 345.
168. N. D. Heindel, N. Foster and M. Choudhuri, *J. Org. Chem.*, 1983, **48**, 3817.
169. N. D. Heindel, N. Foster and T. Varkey, *J. Heterocycl. Chem.*, 1986, **23**, 1579.
170. W. Friedrichsen, *Adv. Heterocycl. Chem.*, 1980, **26**, 135.
171. J. G. Smith and R. B. McCall, *J. Org. Chem.*, 1980, **45**, 3982.
172. S. D. Darling and K. D. Wills, *J. Org. Chem.*, 1967, **32**, 2794.
173. H. Gilman and C. W. Bradley, *J. Am. Chem. Soc.*, 1938, **60**, 2333.
174. T. Keumi, C. Murata, Y. Sasaki and H. Kitajima, *Synthesis*, 1980, 634.
175. E. A. Boyle, F. R. Mangan, R. E. Markwell, S. A. Smith, M. J. Thomson, R. W. Ward and P. A. Wyman, *J. Med. Chem.*, 1986, **29**, 894.
176. L. N. Borisova, S. G. Rozenberg, N. F. Kucherova and V. A. Zagorevskii, *Chem. Heterocycl. Compds. (Engl. Transl.)*, 1981, 869.
177. E. A. Karakhanov, E. A. Dem'yanova, E. G. Shkarin and E. A. Viktorova, *Chem. Heterocycl. Compds. (Engl. Transl.)*, 1975, 1261.
178. C. Parkanyi, A. Lablache-Combier, I. Markó and H. Ofenberg, *J. Org. Chem.*, 1976, **41**, 151.
179. D. K. Fukushima, *Heterocycl. Compds*, 1951, **2**, 149.
180. B. Iddon and R. M. Scrowston, *Adv. Heterocycl. Chem.*, 1970, **11**, 177.
181. R. M. Scrowston, *Adv. Heterocycl. Chem.*, 1981, **29**, 171.
182. A. A. Zirka, G. L. Ryzhova, Yu. G. Slizhov and A. V. Mashkina, *Neftekhimiya*, 1983, **23**, 548 (*Chem. Abstr.*, 1983, **99**, 175 066x).
183. A. A. Zirka, G. L. Ryzhova, Yu. G. Slizhov and A. V. Mashkina, *React. Kinet. Catal. Lett.*, 1983, **23**, 7 (*Chem. Abstr.*, 1983, **100**, 138 415d).
184. R. H. Fish, J. L. Tan and A. D. Thormodsen, *J. Org. Chem.*, 1984, **49**, 4500.
185. W. Hückel and I. Nabih, *Chem. Ber.*, 1956, **89**, 2115.
186. K. V. Wood, R. G. Cooks, J. A. Laugal and R. A. Benkeser, *Anal. Chem.*, 1985, **57**, 692.
187. G. I. Bolestova, A. N. Korepanov, Z. N. Parnes and D. N. Kursanov, *Izv. Akad. Nauk SSSR, Ser. Khim.*, 1974, 2547 (*Chem. Abstr.*, 1975, **82**, 125 212v).
188. R. Pacut and E. Kariv-Miller, *J. Org. Chem.*, 1986, **51**, 3468.

3.8
Partial and Complete Reduction of Heterocycles Containing More Than One Heteroatom

MALCOLM SAINSBURY
University of Bath, UK

3.8.1 INTRODUCTION

To cover the reductive behavior of all types of heterocycles bearing more than one heteroatom within the confines of one brief chapter is impossible, thus the content of this chapter has been edited to reflect those topics and systems in which there is most interest. Many polycyclic and less common heterocyclic molecules are not included. Similarly the weighting of the subject matter is not uniform, and greater emphasis is placed on those areas where the reduction of heterocycles gives access to useful synthons or provides templates which exert stereochemical control in target-orientated syntheses.

3.8.2 REDUCTION OF HETEROCYCLES CONTAINING TWO NITROGEN ATOMS

3.8.2.1 Five-membered Ring Systems

3.8.2.1.1 Pyrazoles, indazoles, pyrazolines and pyrazolidines

The pyrazole ring is resistant to most reducing agents and survives intact when other heterocycles are cleaved. An illustration is provided by the reaction of the pyrazoloimidazolidinone (**1**) with LAH.[1] Here the imidazolidine unit is sensitized to nucleophilic attack by the presence of a lactam carbonyl group, and the product formed is the monocyclic pyrazole (**2**). The stability of pyrazoles to reduction has been exploited in a synthesis of aldehydes from acyl halides. Thus *N*-acylpyrazoles (**3**) on reduction with LAH produce complexes of the type (**4**); these when hydrolyzed yield aldehydes and the parent pyrazoles (**5**).[2]

$LiAlH_4$, THF, 89%

(**1**) (**2**)

$LiAlH_4$, Et_2O ; H_2O/HCl ; + RCHO

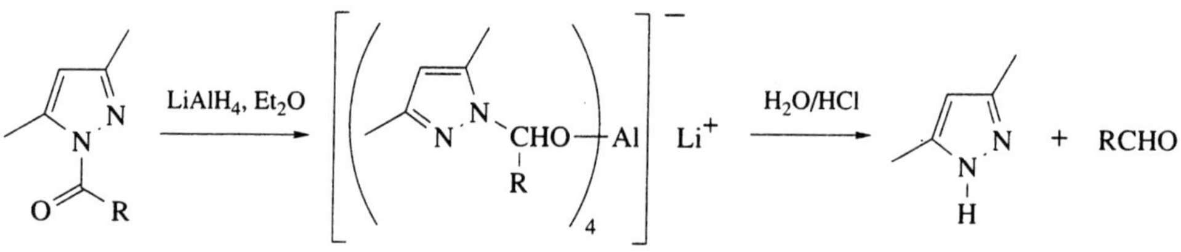

(**3**) (**4**) (**5**)

Unsaturated groups bonded to the carbon atoms of pyrazoles are also reduced, either by chemical reagents such as sodium and ethanol,[3] or through hydrogenation over nickel or palladium catalysts.[4] Nitropyrazoles can be converted into the corresponding amino compounds by exposure to dissolving metals in aqueous acids, but pyrazoles (**6**) are not stable to catalytic hydrogenation in the presence of acids and pyrazolines (**7**) are formed. Now the reaction can proceed further to yield pyrazolidines (**8**), and ultimately acyclic 1,3-diaminopropanes (**9**).[5] Over-reduction is not a problem if neutral media are employed and the hydrogenation of 2-pyrazolidines (**10**) in organic solvents over palladium or Adams' catalyst only serves to reduce the azo group attached to C-5 giving the de-azo derivatives (**11**).[6] LAH reduces both pyrazolines and pyrazolinium salts (**12**) to pyrazolidines (**13**).[7]

(6) → H_2, M, H^+ → (7) → [H] → (8) → [H] → (9)

(10) → H_2, Pd or Pt → (11)

(12) → $LiAlH_4$, Et_2O → (13)

Catalytic hydrogenation of pyrazolin-5-ones (**14**) under mild conditions affords the corresponding pyrazolidinones (**15**), the lactam units of which may be reduced to give the pyrazolidines (**16**) by treatment with LAH.[8] *N*-Sulfonylpyrazolin-4-ones (**17**) can be reduced with L-selectride to afford pyrazolinols (**18**), without affecting the alkenic bond of the heterocycle.[9] The N—N bond of pyrazolidines is cleaved by hydrogenation over Raney nickel as catalyst and, for example, hydrogenolysis of the 1,5-diazabicyclooctane (**19**) yields the 1,5-diazaoctane (**20**).[10] Similar reactions do not occur with *N*-acyl derivatives, although ring opening can now be achieved through the action of sodium in liquid ammonia. In this way the dicarbamate (**21**) is transformed into the 1,3-diaminocyclopentane (**22**).[11]

(14) → H_2, M, atmospheric pressure → (15) → $LiAlH_4$, THF → (16)

(17) → L-selectride → (18)

(19) → Ni, H_2, EtOH, H_2O → (20)

The nucleus of indazole, like that of pyrazole, is stable to reduction, so that treatment of the amido derivative (**23**) with LAH leads only to the amine (**24**).[12] Indazolium salts (**25**) are more reactive and undergo reduction with potassium borohydride to give indazolines (**26**).[13]

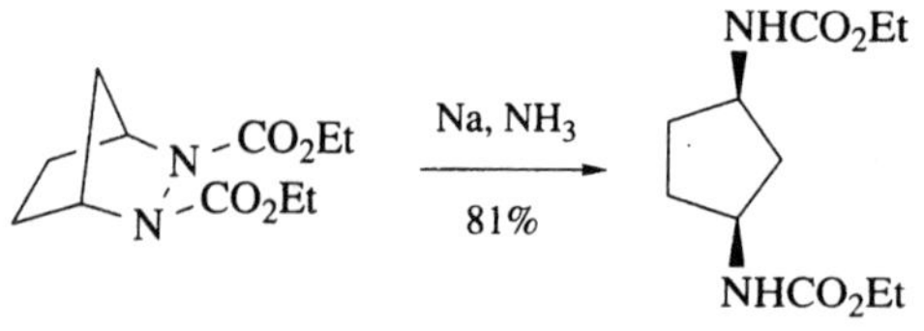

NHCO$_2$Et

N-CO$_2$Et
N-CO$_2$Et

Na, NH$_3$
81%

NHCO$_2$Et

(21) **(22)**

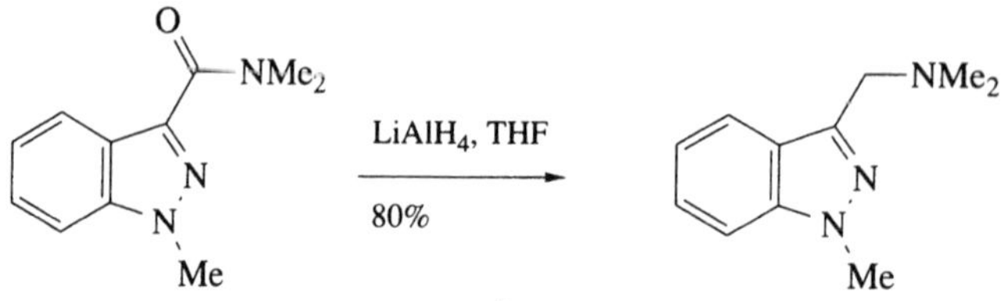

O
NMe$_2$
N
N
Me

LiAlH$_4$, THF
80%

NMe$_2$
N
N
Me

(23) **(24)**

R
+
N−Me
N
Me MeSO$_4^-$

KBH$_4$

R
N−Me
N
Me

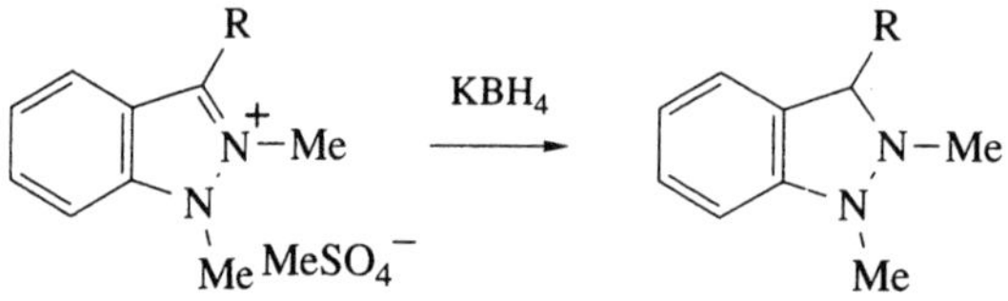

(25) **(26)**

3.8.2.1.2 Imidazoles, benzimidazoles, imidazolines and imidazolium salts

The aromaticity of the imidazole nucleus ensures stability towards reduction, and when benzimidazole (**27**) is hydrogenated over Adams' catalyst in acetic acid the carbocyclic ring is reduced first to give the tetrahydrobenzimidazole (**28**). However, if the solvent is changed to acetic anhydride, *N*-acylation promotes the reduction of the heterocycle and the 1,3-diacetylbenzimidazoline (**29**) is then formed (Scheme 1).[14] Imidazole (**30**) under these conditions gives 1,3-diacetylimidazoline (**31**). Imidazolium salts (**32**) are easily reduced and treatment with excess sodium borohydride in 95% aqueous ethanol culminates in the formation of 1,2-diamines, (**33**) or (**34**). Either N—C bond may cleave, although if the substituent R^1 is benzyl the major products are benzylamines (**33**; R^1 = Bn).[15]

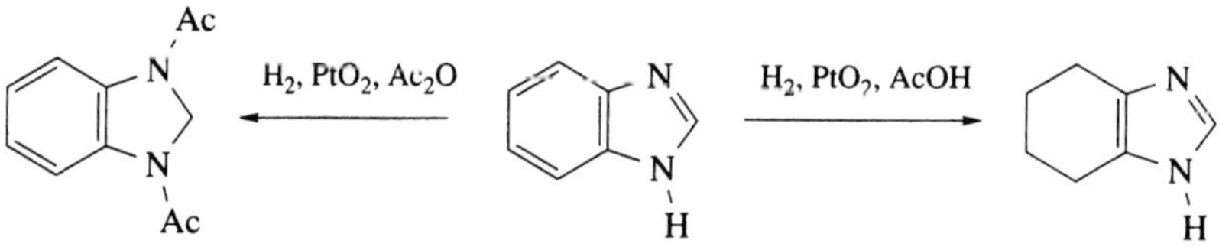

Ac
N
N
Ac

H$_2$, PtO$_2$, Ac$_2$O

N
N
H

H$_2$, PtO$_2$, AcOH

N
N
H

(29) **(27)** **(28)**

Scheme 1

N
N
H

H$_2$, PtO$_2$, Ac$_2$O

N
Ac
N
Ac

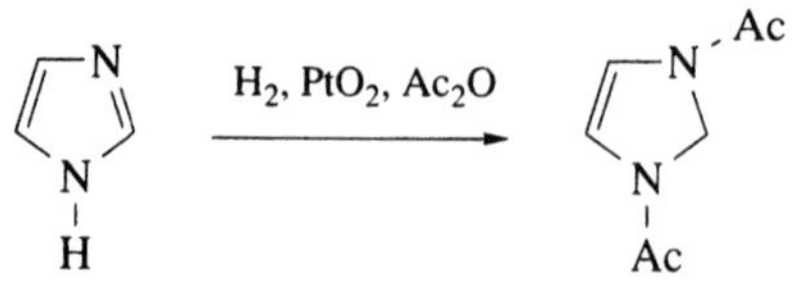

(30) **(31)**

2-Imidazolines (**35**) also undergo reductive decyclization to diamines (**36**) with metal hydrides, but here LAH is a superior reagent to sodium borohydride, and other hydrides such as sodium cyanoborohydride fail to react.[16] In the case of benzimidazolium salts (**37**) LAH in diethyl ether or THF, or sodium borohydride in water, gives benzimidazolines (**38**). Ring scission does not occur.[17]

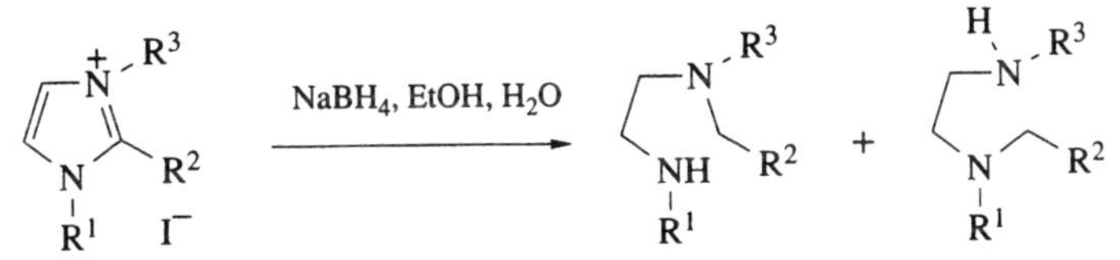

(32) → (33) + (34): $NaBH_4$, EtOH, H_2O

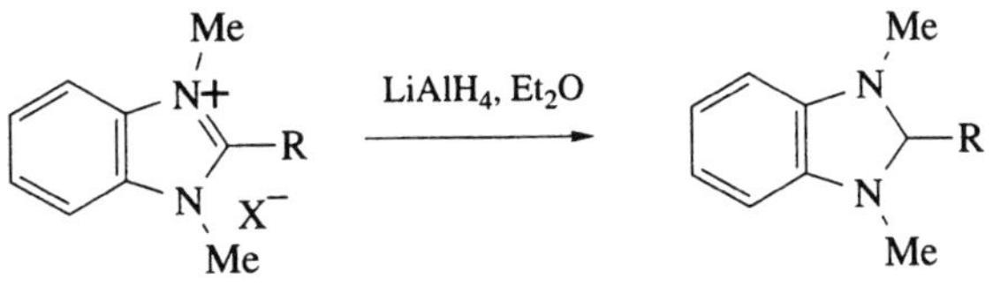

(35) → (36): $LiAlH_4$, THF

(37) → (38): $LiAlH_4$, Et_2O

Imidazolidin-2-ones (**39**) are inert to reduction with sodium borohydride, but LAH converts them into the corresponding imidazolidines (**40**), the rate of the reaction being determined by the size of the *N*-substituents.[18] *N*-Substituted guanidines are similarly unaffected by treatment with sodium borohydride or lithium borohydride, but if an acetoxymethyl substituent is present, as in the derivative (**41**), selective *O*-deacetylation occurs to give the corresponding alcohol (**42**).[19] The carbonyl groups of hydantoins (**43**) survive treatment with sodium amalgam, or Raney nickel in sodium hydroxide, although side chain double bonds are reduced giving dihydro derivatives (**44**).[20] On the other hand, 2,4-dithiohydantoins (**45**) are desulfurized by exposure to Raney nickel in boiling ethanol, and yield the parent imidazoles (**46**).[21]

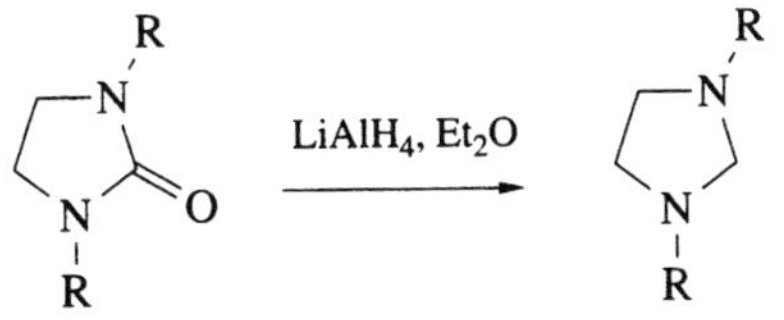

(39) → (40): $LiAlH_4$, Et_2O

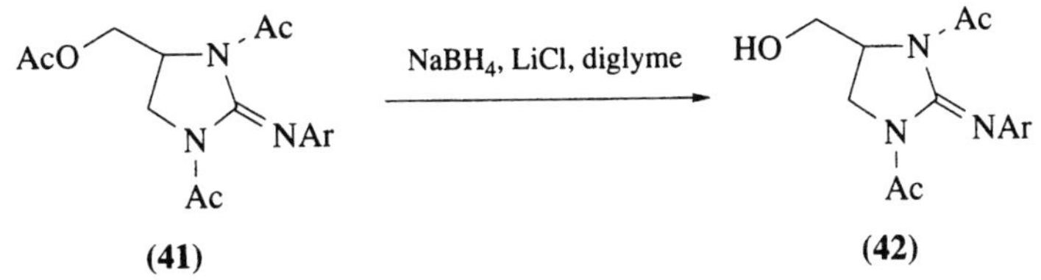

(41) → (42): $NaBH_4$, LiCl, diglyme

(43) → (44): H_2, Ni, NaOH

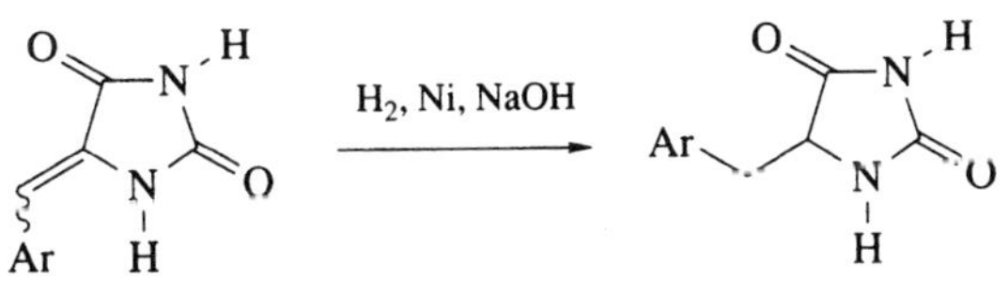

H_2, Ni, EtOH

(45) (46)

3.8.2.2 Six-membered Ring Systems

3.8.2.2.1 *Pyridazines, pyridazinium salts, dihydropyridazines, tetrahydropyridazines, cinnolines and phthalazines*

Pyridazine *N*-oxides (**47**) are partially deoxygenated by hydrogenation at atmospheric pressure over palladium on carbon affording a mixture of the corresponding mono-*N*-oxides (**48**) and (**49**).[22] By contrast, electrochemical methods cause ring opening, provided that the solvent contains acetic anhydride, and under these conditions pyridazine (**50**) itself is converted into 1,4-diacetamidobutane (**51**).[23] A similar ring scission is achieved if *N,N*-diacylhexahydropyridazines are treated with sodium in liquid ammonia; thus the dicarbamate (**52**) affords the lyxose derivative (**53**),[24] and treatment of the bicyclic amide (**54**) with sodium amalgam effects ring opening to give the pyrrolidinone (**55**).[25]

H_2, Pd, C

(47) (48) 55% (49) 42%

Hg cathode, DMF, Ac_2O

(50) (51)

Na, NH_3, –60 °C, 87%

(52) (53)

Na, Hg

(54) (55)

Pyridazines (**56**) are reduced by treatment with a mixture of dimethyl sulfate (which forms the corresponding salts *in situ*) and sodium borohydride to yield 1,6-dihydropyridazines (**57**; R^1 = Me), together with minor amounts of 1,4,5,6-tetrahydropyridazines (**58**; R^1 = Me). Normally 1,6-dihydropyrazines are unstable and decompose in air, but *N*-acylated derivatives are more stable. Thus if methyl chloroformate replaces dimethyl sulfate in the reaction mixture both 1,4-dihydro- (**59**; R^1 = CO_2Me) and 1,6-dihydropyridazines (**57**; R^1 = CO_2Me) can be isolated, the latter predominating if there is an electron-withdrawing group at C-6 in the starting material (Scheme 2).[26]

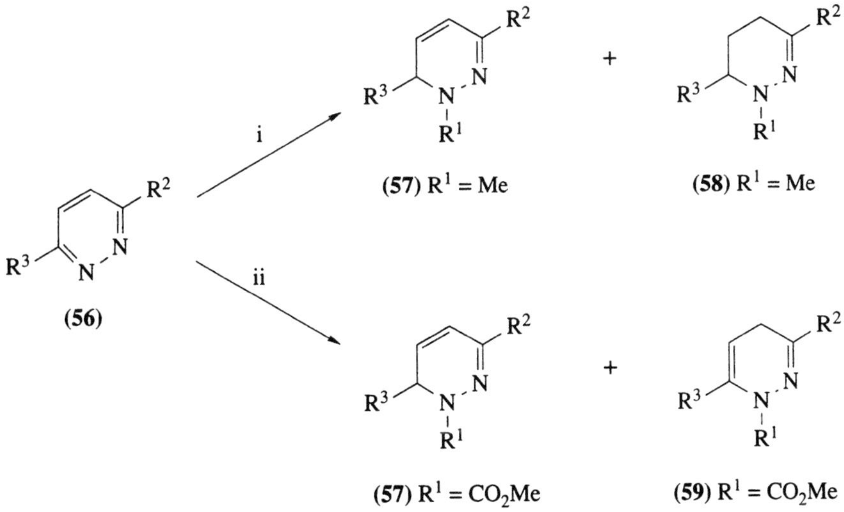

(**56**)

(**57**) R^1 = Me + (**58**) R^1 = Me

(**57**) R^1 = CO_2Me + (**59**) R^1 = CO_2Me

i, Me_2SO_4, $NaBH_4$; ii, $ClCO_2Me$, $NaBH_4$

Scheme 2

LAH converts 1-alkylpyridazin-6-ones (**60**) into the corresponding 1,6-dihydropyridazines (**57**; R^1 = Me, R^3 = H), whereas excess reagent reduces 4,5-dihydropyridazin-6-ones (**61**) to a mixture of 1,4,5,6-tetrahydro- (**62**) and hexahydro-pyrazines (**63**).[27] Cinnolines (**64**) and phthalazines (**66**) both undergo ring opening during cathodic reduction, giving diamines (**65**) and (**67**), respectively.[28]

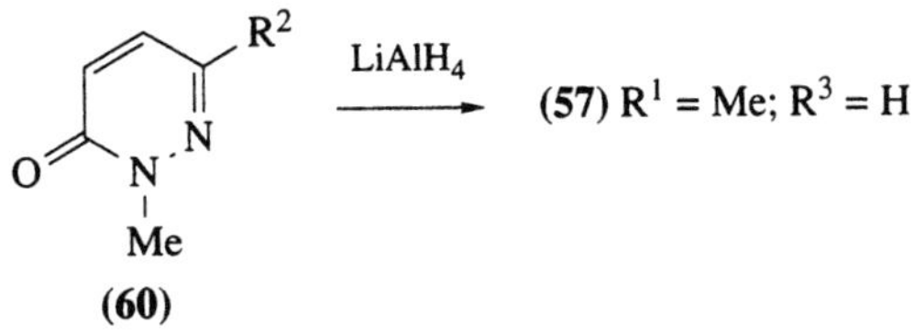

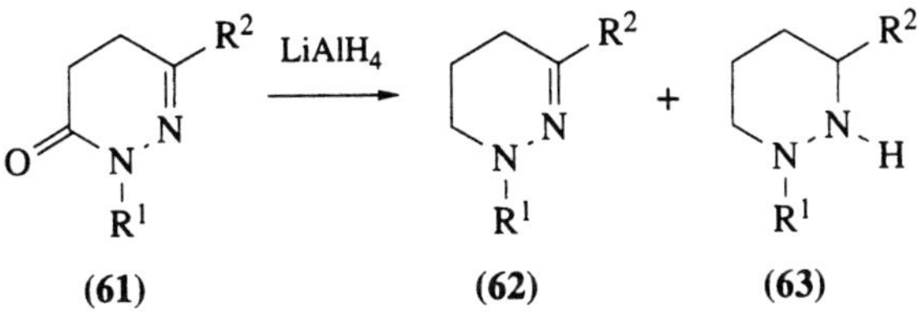

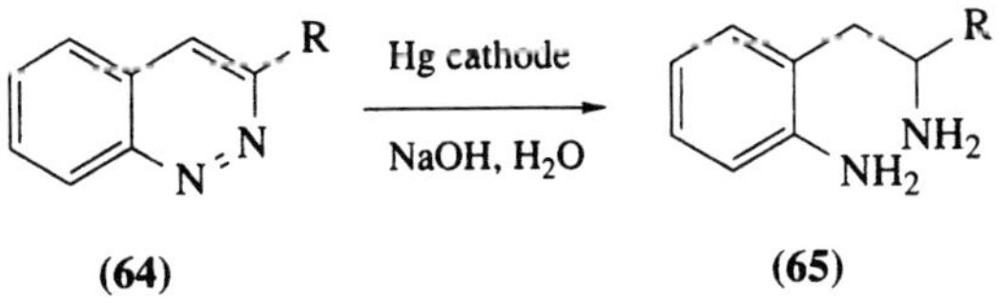

(**60**) $\xrightarrow{LiAlH_4}$ (**57**) R^1 = Me; R^3 = H

(**61**) $\xrightarrow{LiAlH_4}$ (**62**) + (**63**)

(**64**) $\xrightarrow[NaOH,\ H_2O]{Hg\ cathode}$ (**65**)

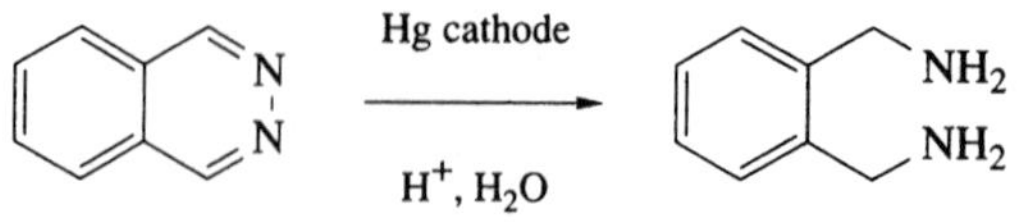

Hg cathode

H^+, H_2O

(66) (67)

3.8.2.2.2 Pyrimidines and quinazolines

Metal hydrides react with pyrimidines (**68**) to give 3,4-dihydropyrimidines (**69**),[29] but electrochemical reduction of the analogs (**70**) gives radical anions (**71**) which may dimerize or, in the presence of a proton source, yield dihydro derivatives (**72**). The latter exist in equilibrium with acyclic tautomers (**73**),[30] and at high potentials further reduction produces anions (**74**) and (**75**) which may then be protonated to give pyrroles (**76**).[31] *N*-Benzyldihydropyrimidines (**77**; R = Bn) are reduced by sodium borohydride in ethanol to give hexahydro derivatives (**78**), but if the *N*-substituent is a phenyl group excess reagent leads to diamines (**79**; Scheme 3).[32] Similar products form when dihydropyrimidinium salts (**80**) are treated with this reagent, although reduction with LAH yields amino imines (**81**). Cathodic reduction of quinazoline (**82**) in aqueous alkali gives first dihydro- (**83**) and then tetrahydro-quinazoline (**84**).[33]

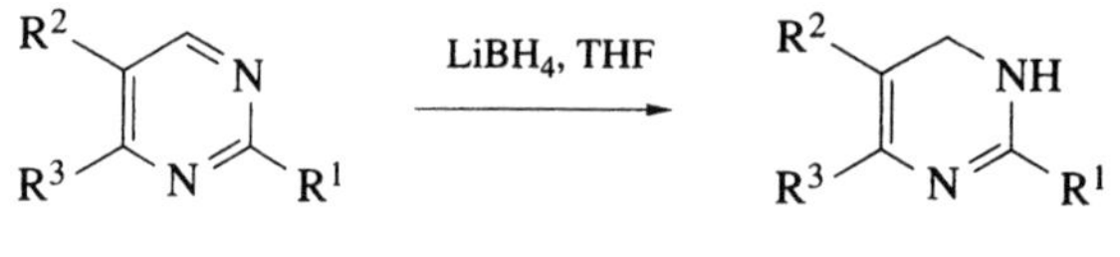

$LiBH_4$, THF

(68) (69)

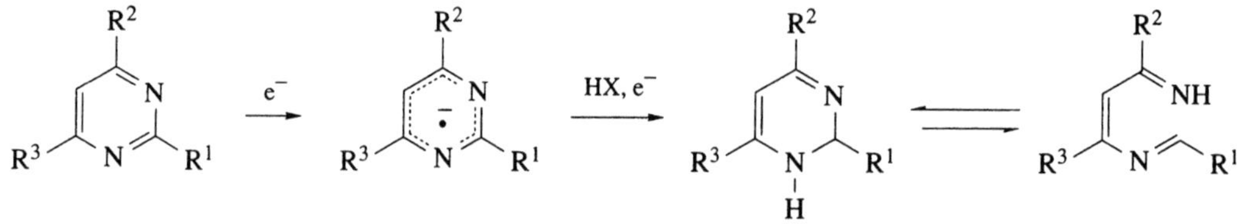

(70) (71) (72) (73)

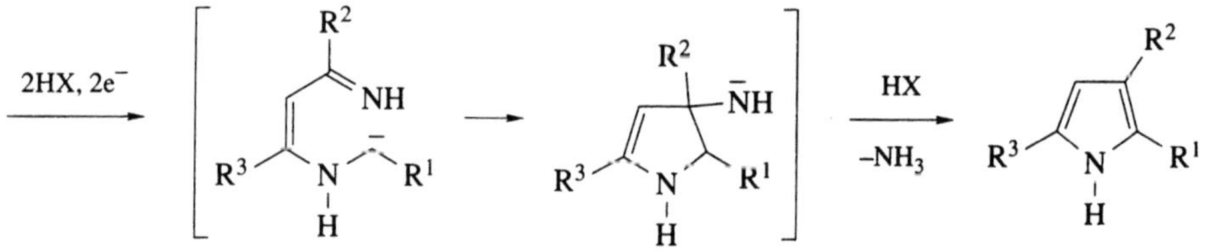

(74) (75) (76)

$NaBH_4$, EtOH R = Bn

$NaBH_4$, EtOH R = Ph

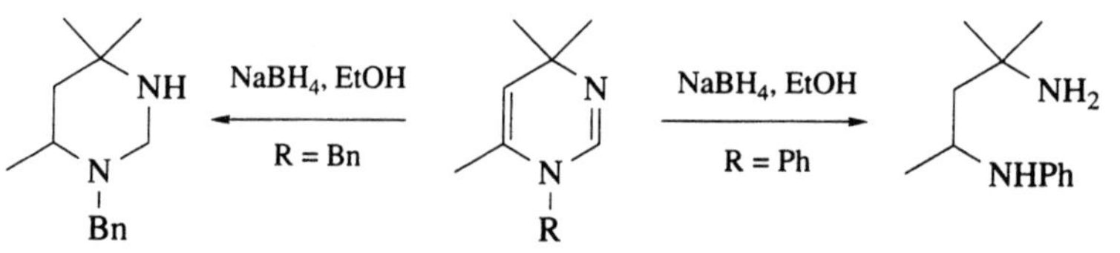

(78) (77) (79)

Scheme 3

Pyrimidin-2(1*H*)-ones (**85**; X = O) and -thiones (**85**; X = S) on reduction with sodium borohydride afford mixtures of 1*H*-3,4- (**86**) and 1*H*-3,6-dihydropyrimidinones (**87**), the product ratios depending on the acidity of the medium, and some further reduction to the fully reduced pyrimidinones may also occur. Similar results are obtained with LAH as the reductant, although 1-arylpyrimidine-2-thiones mainly form the corresponding 3,6-dihydropyridine-2-thiones (**88**), especially if a hydroxy or methoxy group occupies an *ortho* position in the aryl substituent. This suggests that the group may complex with the reagent and facilitate an intramolecular hydride transfer reaction.[34] L-Selectride reduces *N*-benzyl-

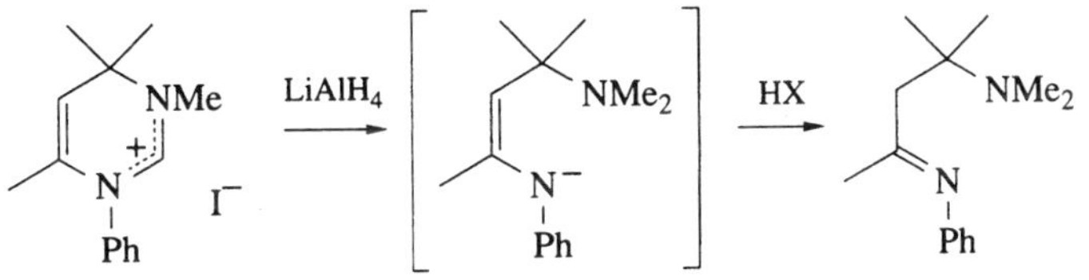

LiAlH$_4$ HX

(80) (81)

Hg cathode, NaOH, H$_2$O

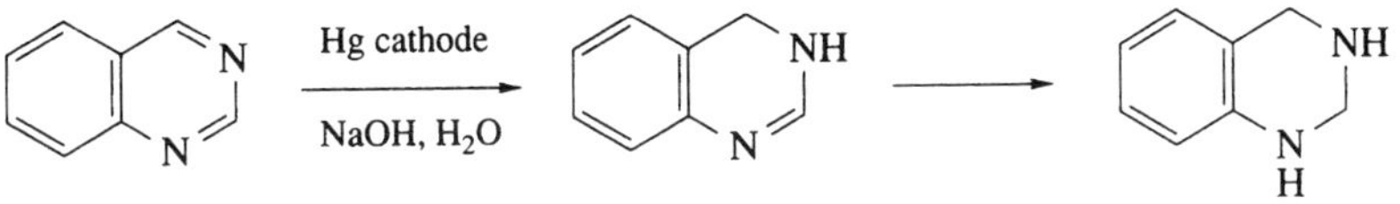

(82) (83) (84)

uracils (**89**) to dihydrouracils (**91**; R = H), and the intermediate anions (**90**) can be quenched by alkyl halides, rather than by water, to give 5-alkyl derivatives (**91**; R = Alkyl).[35]

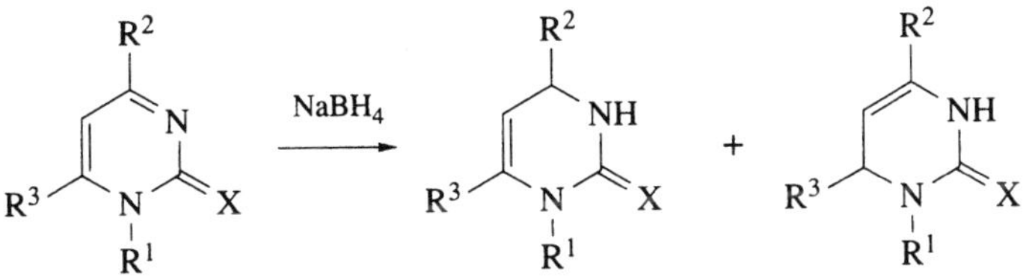

NaBH$_4$

(85) (86) (87)

(85) LiAlH$_4$

R^1 = Ar, X = S

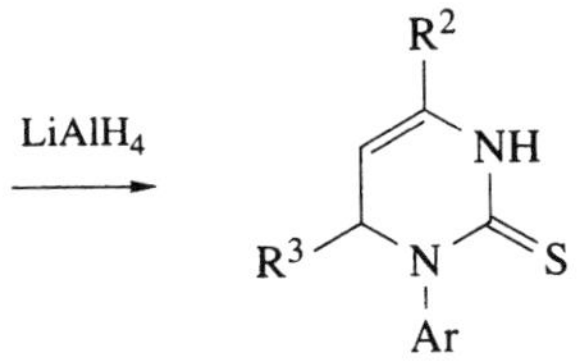

(88)

L-selectride, THF, –78°C RY

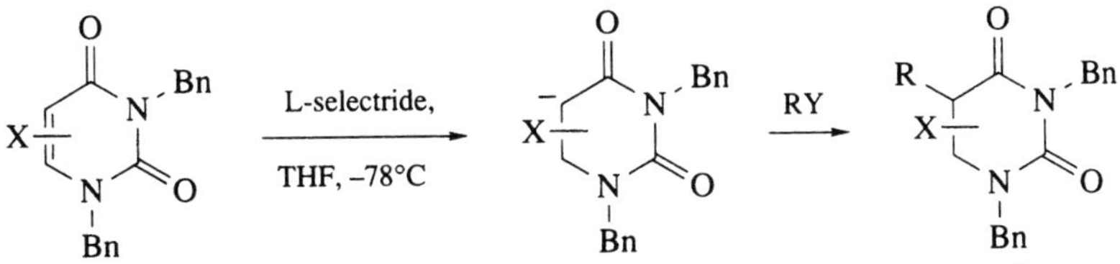

(89) (90) (91)

X = H, 5-F, 5-C≡CSiMe$_3$ or 6-CO$_2$Et

3.8.2.2.3 Pyrazines and quinoxalines

Dissolving metals and amalgams all reduce pyrazines to hexahydropyrazines (piperazines). The same products are formed by the hydrogenation of pyrazines over Adams' catalyst at high pressure, and in the case of 2,5-dibenzylpyrazine (**92**) hydrogenation affords both the *cis*- and *trans*-piperazines (**93**) and (**94**) respectively.[36] For the tetracarbonyl derivative (**95**), reduction is easier and a tetrahydro compound (**96**) is formed by hydrogenation at atmospheric pressure. This product, however, undergoes spontaneous oxidation back to the dihydropyrazine (**97**), which is also produced by direct reduction of the parent pyrazine with sodium dithionate.[37] The stability of the 1,4-tetrapyrazine system is utilized in a synthesis of chiral alcohols (**99**) *via* the reduction of the corresponding bis-diamides (**98**) with lithium triethylborohydride.[38] Benzopyrazine (quinoxaline) (**100**; R = H) on electrochemical reduction yields the 1,4-dihydro derivative (**101**; R = H), but if an alkyl group is present at C-2 the 3,4-dihydro formulation (**102**) is preferred.[33]

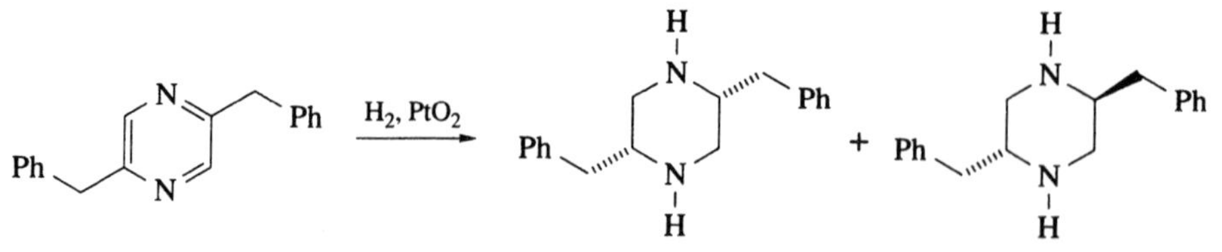

H_2, PtO_2

(92) **(93)** 26% **(94)** 51%

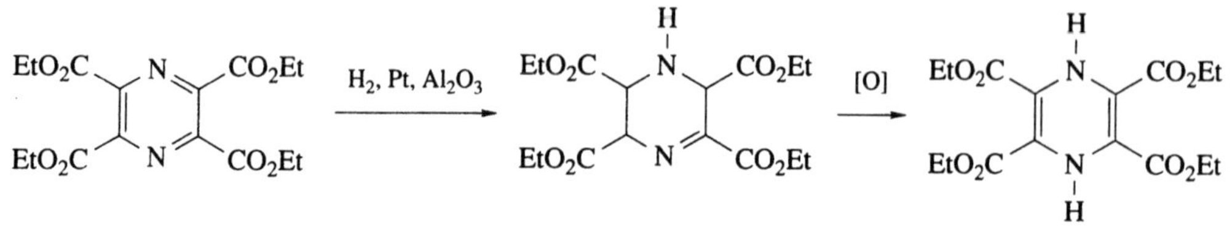

H_2, Pt, Al_2O_3 [O]

(95) **(96)** **(97)**

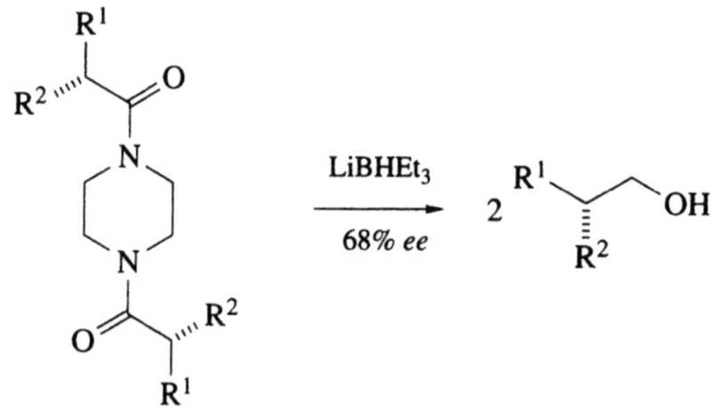

LiBHEt$_3$, 68% *ee*

(98) **(99)**

Hg cathode, H^+, H_2O

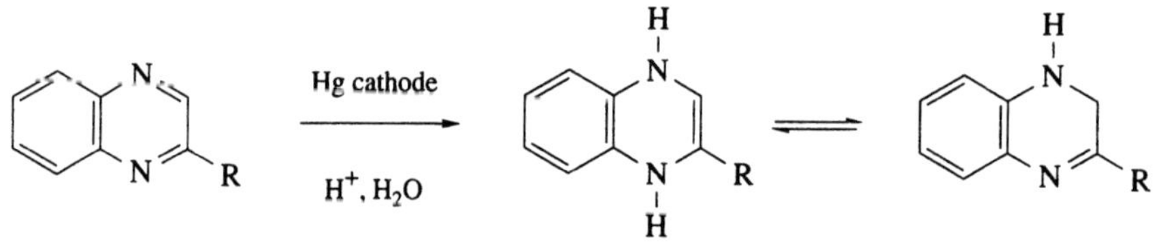

(100) **(101)** **(102)**

3.8.3 REDUCTION OF HETEROCYCLES CONTAINING ONE NITROGEN ATOM AND ONE OXYGEN ATOM

3.8.3.1 Five-membered Ring Systems

3.8.3.1.1 Isoxazoles, isoxazolium salts, isoxazolines and isoxazolidines

(i) Isoxazoles

As early as 1891 Claisen noted that isoxazoles, when treated with sodium and either pentyl alcohol or moist diethyl ether, undergo reductive ring opening to give 1,3-enaminoketones.[39] These products may then be hydrolyzed to 1,3-dicarbonyl compounds. In more recent times the use of readily available isoxazoles as masked 1,3-dicarbonyl compounds has found wide application in synthesis, although dissolving metal reductants are not now in favor.[40] Instead hydrogenation over Raney nickel as catalyst is frequently employed as, for example, in the synthesis of hentriacontane-14,16-dione (**104**), a component of natural waxes, from the isoxazole (**103**).[41] The reductive ring scission of the diisoxazole (**105**) occurs in a similar manner, but here hydrolysis of the bis-1,3-enaminoketone (**106**) in the presence of acids is assumed to afford a 'polyketide intermediate' (**107**) which undergoes spontaneous cyclodehydration to the acylrescorcinol (**108**),[42] thereby mimicing the last stages of the biosynthesis of similar compounds through the 'acetate pathway'.

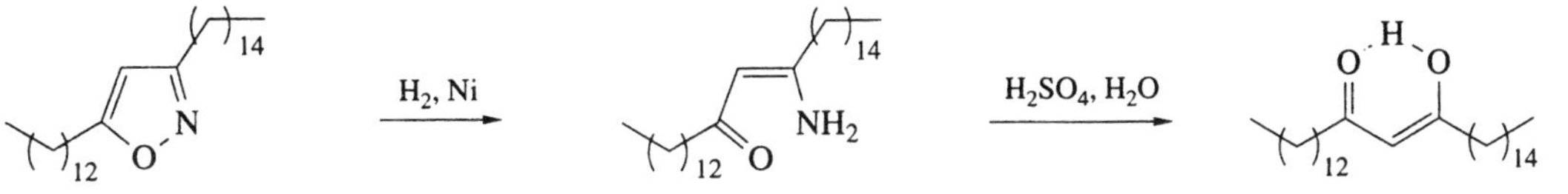

H_2, Ni

H_2SO_4, H_2O

(**103**) (**104**) 75% yield overall

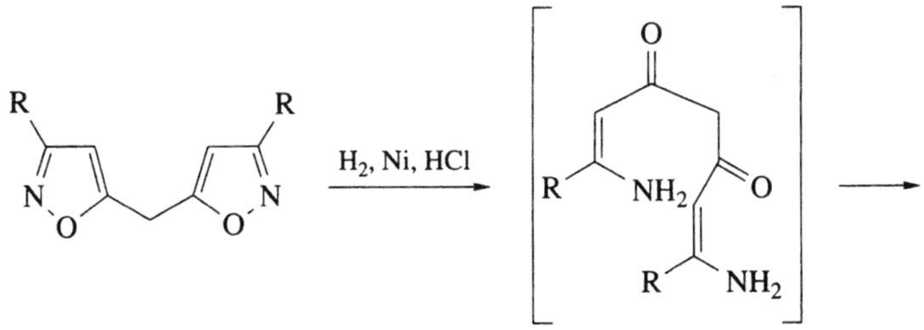

H_2, Ni, HCl

(**105**) (**106**)

$-H_2O$

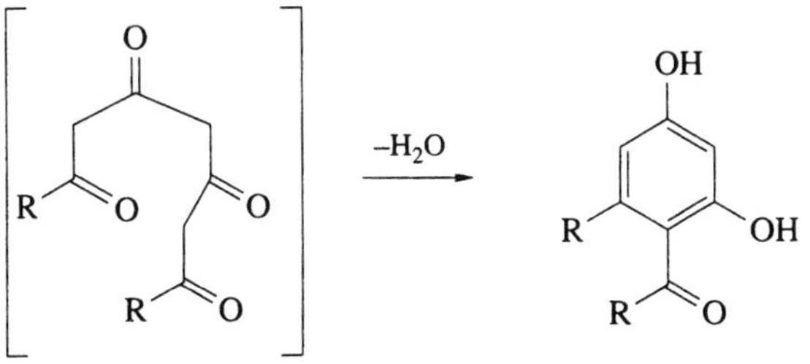

(**107**) (**108**)

Hydrogenation of isoxazolylsulfonamides (**109**) over Raney nickel catalyst also gives enaminoketones, but these products undergo cyclodehydration without the need of acid treatment, affording 1,2,6-thiadiazine-1,1-diones (**110**) in good yields.[43] Stork *et al.* have developed this type of reaction into a general synthesis of pyridines (**112**), using 4-(oxoalkyl)isoxazoles (**111**) as starting materials. In this instance the hydrogenation step is achieved using palladium on carbon as catalyst and the cyclization step is promoted by the addition of triethylamine to the reaction mixture. Intermediate dihydropyridines are formed which are easily oxidized to the corresponding pyridines, often simply by exposure to air.[44]

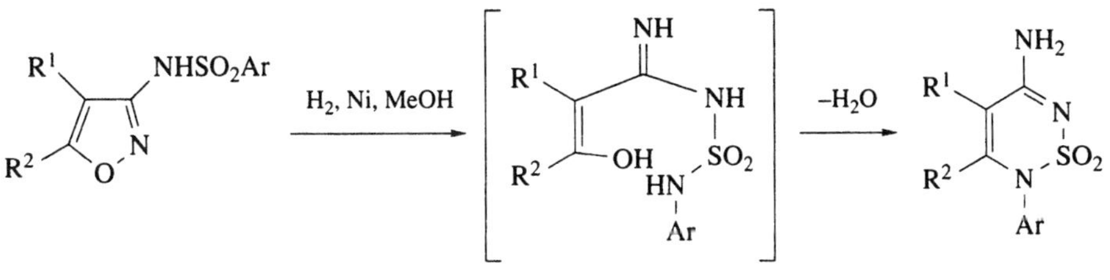

H_2, Ni, MeOH

$-H_2O$

(**109**) (**110**)

Other reagents recommended for the reduction of isoxazoles to 1,3-enaminoketones include samarium diiodide, molybdenum hexacarbonyl in aqueous acetonitrile, and diiron nonacarbonyl in water.[45] Electrochemical methods have also been employed.[46] A particularly mild method involves reaction with the reduced form of the coenzyme lipoamide (LA_2) complexed with iron(II).[47] Birch reduction (sodium and

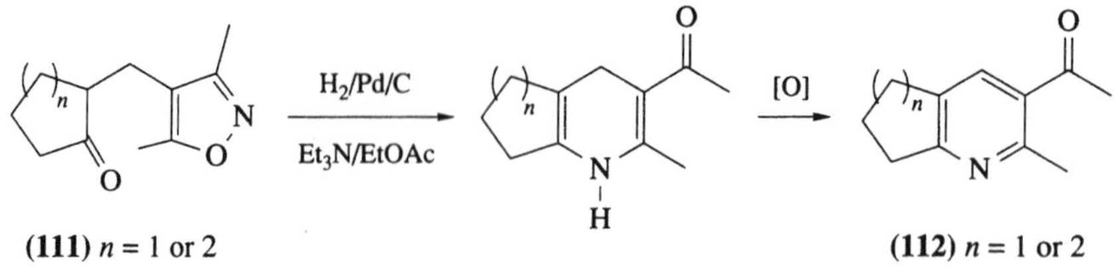

(111) n = 1 or 2

(112) n = 1 or 2

liquid ammonia) of isoxazoles gives 1,3-aminoketones rather than 1,3-enaminoketones, and exposure of these compounds to acids tends to cause elimination of ammonia and the production of enones. Treatment of isoxazoles with metal hydride reductants gives variable results and the products formed seem to depend on the nature and substitution pattern of the groups attached to the heterocycle. Thus, unless an electron-withdrawing group is present at C-4 of the isoxazole ring, sodium borohydride or lithium diisobutylaluminum hydride fail to react at the nucleus, although vulnerable substituents (such as carbonyl groups) are reduced in the expected manner, and 5-chloroisoxazoles may be partially dechlorinated if they are reacted with sodium borohydride in dimethylsulfoxide as solvent.[48] When reduced with lithium aluminum hydride isoxazoles (**113**) afford aziridine alcohols (**114**), together with smaller amounts of aziridines (**115**) and 3-aminoalcohols (**116**).[49] However, for 4-cyanoisoxazoles (**117**) the reduction takes a different course and acyclic intermediates are assumed to form, which on acidification recyclize to give 5-aminoisoxazoles (**118**).[50]

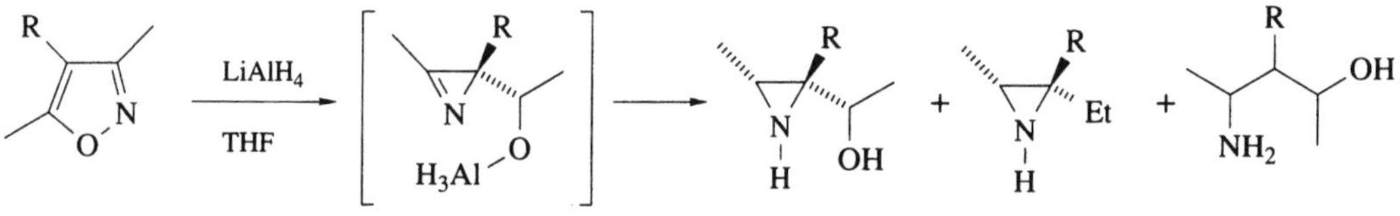

(113) (114) (115) (116)

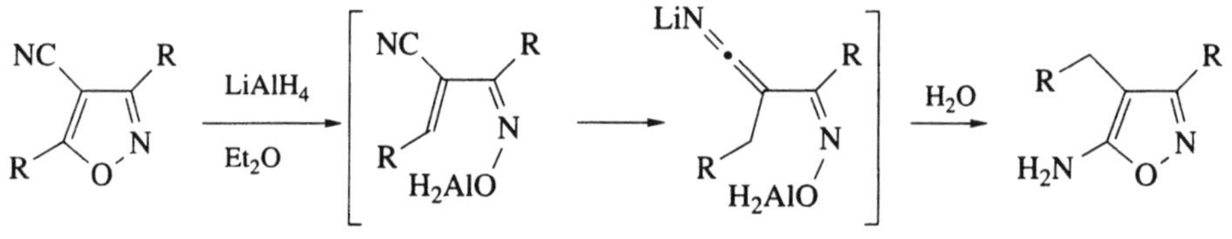

(117) (118)

(ii) Isoxazolium salts

The reduction of isoxazolium salts with metal hydrides is also not entirely predictable, and here the progress of the reaction depends on (i) the reagent, (ii) the solvent, and (iii) the ring substituents.[51] Generally, isoxazolium salts (**119**) on reduction with sodium borohydride in aqueous acetonitrile give 4-isoxazolines (**120**), their borane complexes, 3-isoxazolines (**121**), isoxazolidines (**122**), and minor amounts of 3-hydroxypropylamines (**123**). In some cases, however, particularly if the substituent group at C-3 is aryl and the reaction is carried out in ethanol rather than acetonitrile, the major products are

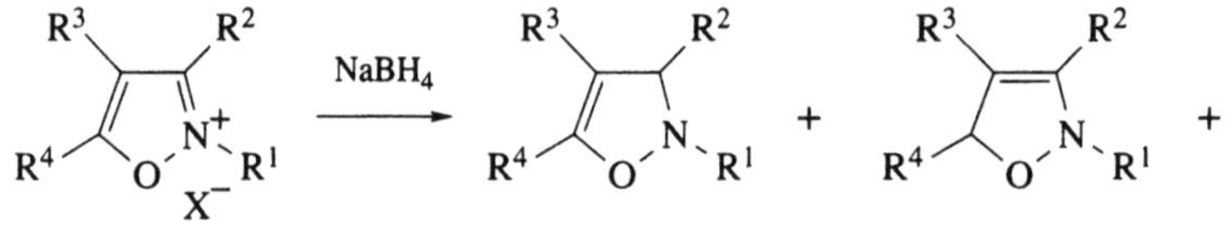

(119) (120) (121)

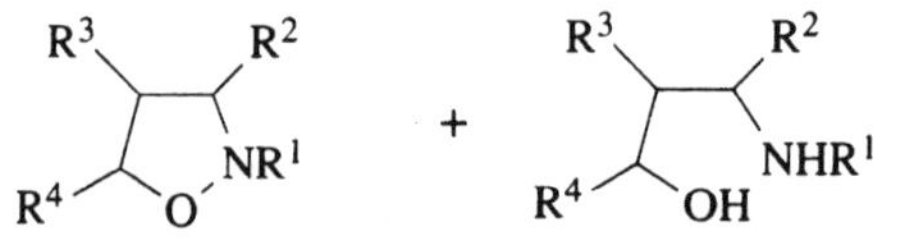

(122) (123)

isoxazolines. Best yields (*ca.* 60%) of isoxazolidines are obtained if LAH in tetrahydrofuran is used as the reductant.

(iii) Isoxazolines

2-Isoxazolines, easily formed by the [3 + 2] cycloaddition of nitrile oxides and alkenes, are extremely versatile synthons for 1,3-aminoalcohols and 1,3-hydroxycarbonyl compounds, and through them to enones and 1,3-dienes. The conditions required to obtain this array of products have been known for many years, but their application to target-orientated syntheses is of much more recent origin.[52]

LAH is commonly used to cleave 2-isoxazolines to 1,3-aminoalcohols, often with a high degree of diastereoselectivity. In the case of 3,5-diphenyl-2-isoxazoline (**124**), for example, the two isomeric alcohols (**125**) and (**126**) are formed in the ratio 19:1.[53] In other cases the nature, position and number of the substituent groups on the ring influence the isomer preferences.[54] Thus, although most groups at C-4 and C-5 are *anti* directing, a hydroxy group at C-4 induces *syn* addition of hydride ion. This stereoselectivity has found application in the syntheses of amino sugars, amino acids and amino polyols. An illustration is the reduction of the two acetals (**127**) and (**129**) which give products conforming to the xylo- (**128**) and arabino- (**130**) series, respectively. In each case the yield of the reduction is high and the diastereoselectivity exceeds 95%.[55]

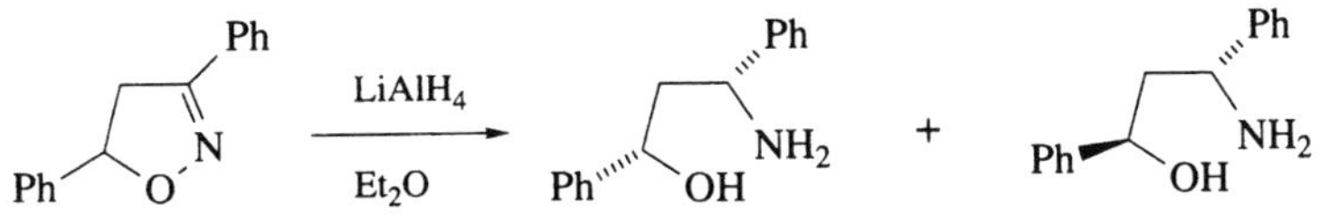

$LiAlH_4$, Et_2O

(**124**) → (**125**) 95% + (**126**) 5%

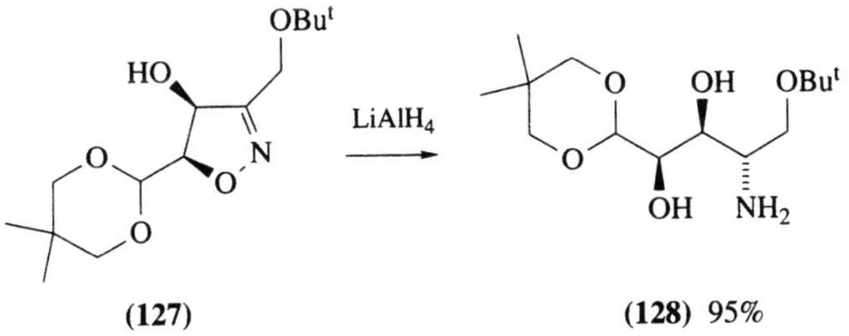

$LiAlH_4$

(**127**) → (**128**) 95%

$LiAlH_4$

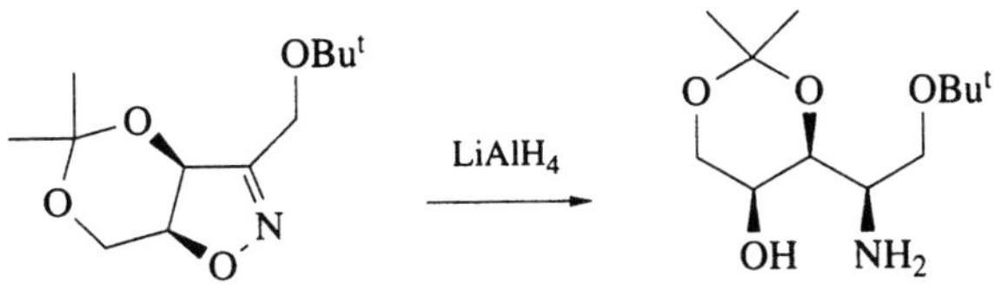

(**129**) → (**130**) 73%

In this last case there are no problems with the choice of the reductant, but where the substrate bears groups themselves liable to reduction then catalytic hydrogenation can sometimes be used. This has been adopted by Kametani *et al.* in the formation of the β-lactams (**132**), related to thienamycin, from isoxazoline-4-carboxylic esters (**131**; Scheme 4).[56] DIBAL has also been employed as reductant, and this reagent was selected by Burri *et al.* in their synthesis of the antibiotic vermiculine (**134**) from the 2-methyl-2-isoxazoline (**133**; Scheme 5).[57]

Hydrogenation of 2-isoxazolines over Raney nickel as catalyst and in the presence of acids often leads first to 1,3-hydroxyimines, and then, through hydrolysis, to 1,3-hydroxycarbonyl compounds. In some cases the stereochemical integrity of the starting material is maintained, but in others hydrolysis of the intermediate imines may cause scrambling of stereochemistry at the α-carbon atom. Rapid protonation of the imine function minimizes this possibility and the use of Lewis acids, such as boron trichloride or aluminum trichloride which release hydrochloric acid on contact with moist methanol, is frequently recommended. Boric acid serves a similar purpose and is effective in, for example, the stereocontrolled reductive ring opening of the 2-isoxazoline (**135**) *en route* to crispatic acid (**136**; Scheme 6).[58]

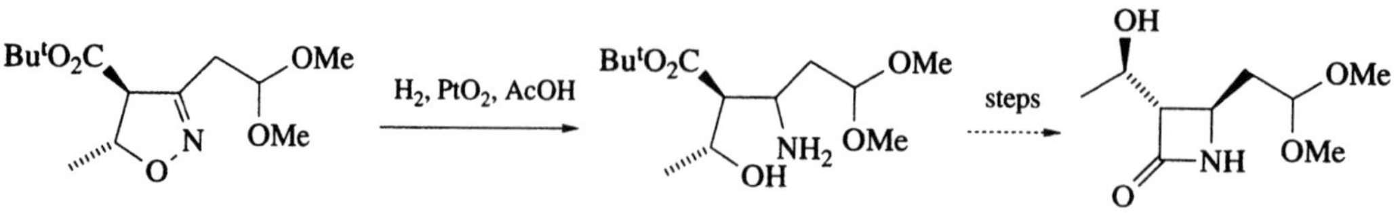

(131) H_2, PtO_2, AcOH → steps → (132)

Scheme 4

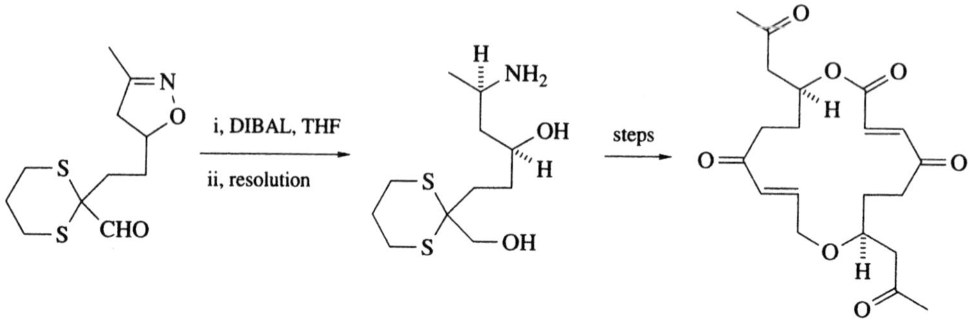

(133) i, DIBAL, THF; ii, resolution → steps → (134)

Scheme 5

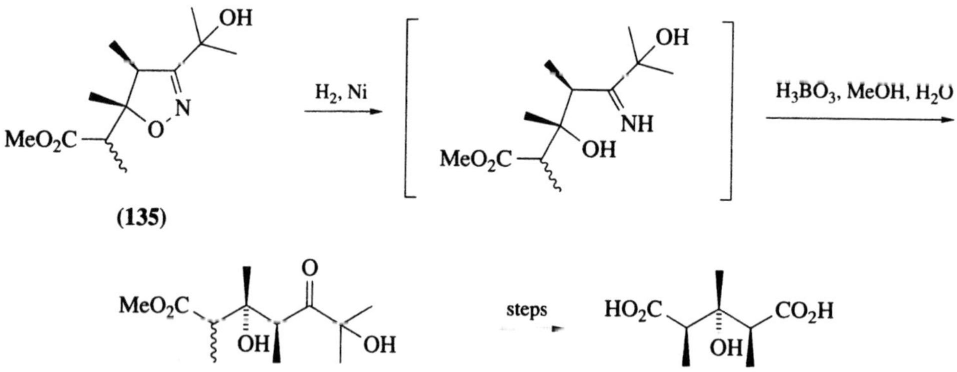

(135) H_2, Ni → [ ] → H_3BO_3, MeOH, H_2O → steps → (136)

Scheme 6

(iv) Isoxazolidines

1,3-Hydroxyamines are formed by the reductive ring opening of isoxazolidines. Thus treatment of the derivative **(137)** with sodium in liquid ammonia, followed by hydrolysis of the carbamate unit, affords *N*-benzyl-*threo*-β-hydroxyornithine (**138**; R = Bn). A wide variety of reductants can be used to accomplish this type of reaction, providing account is taken of the reactions of the substituent groups about the heterocycle. So, for example, if the same substrate **(137)** is hydrogenated over Pearlman's catalyst [$(Pd(OH)_2/C$] not only is the ring cleaved, but the 2-*N*-benzyl group is also removed to afford *threo*-β-hydroxyornithine (**138**; R = H) directly.[59] Perhaps the mildest method for the formation of 1,3-hydroxy-

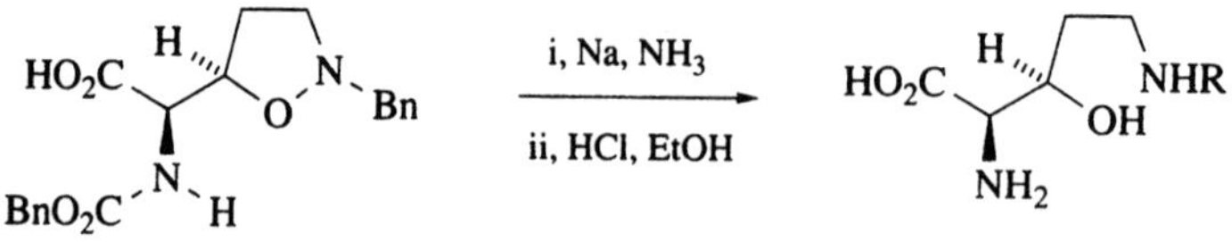

(137) i, Na, NH_3; ii, HCl, EtOH → (138)

amines involves the reaction of isoxazolidines with the reduced form of the lipoamide A_2–iron complex (previously described in Section 3.8.3.1.1i). Hydrogenation of isoxazolin-5-ones over palladium on carbon as catalyst gives the corresponding isoxazolidinones. Despite some confusion in the literature,[60] in the case of benzylidene derivatives (**139**) it is possible to control the reduction so that only the exocyclic double bond is reduced, giving first benzylisoxazolinones (**140**), and then isoxazolidinones (**141**). When the benzylisoxazolinones (**140**) are treated with hydrogen chloride in acetic acid they undergo ring fragmentation with loss of carbon dioxide, possibly as shown, to afford intermediate iminium salts, which on hydrolysis yield α,β-unsaturated ketones (**142**).[60]

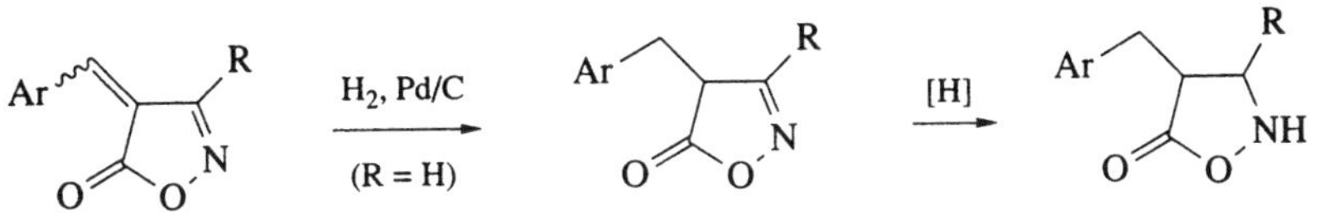

(139) **(140)** **(141)**

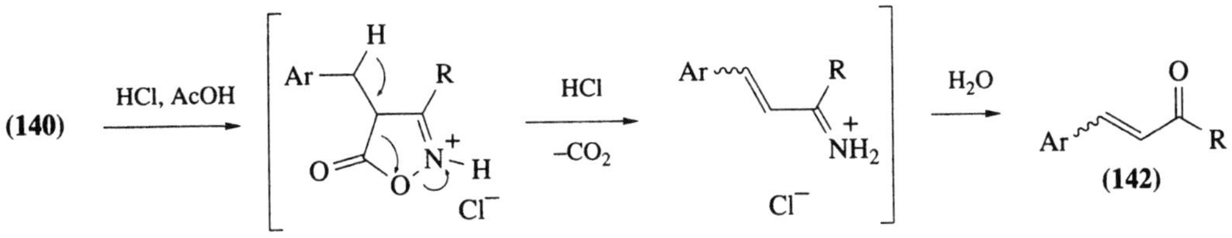

(140) → **(142)**

(v) Benzisoxazoles

1,2-Benzisoxazoles (**143**) behave similarly to monocyclic isoxazoles and are reductively cleaved to 2-hydroxyarylalkylimines (**144**) by catalytic hydrogenation over palladium on barium sulfate, although if the catalyst is Raney nickel deamination occurs, and under severe conditions the aromatic ring may be reduced to yield the appropriate cyclohexanol (**145**). Deamination also results from the chemical reduction of benzisoxazoles with hydrazine, but now the reaction terminates with the formation of phenols (**146**). Sodium and various alcohols normally convert benzisoxazoles into 2-hydroxyarylalkylamines (**147**; Scheme 7).[61] 2,1-Benzisoxazoles (**148**) are rather more stable and are normally prepared by the reductive cyclization of 2-nitroarylcarbonyl compounds. Hydrogenation at high pressure over Raney nickel catalyst will, however, cause degradation to toluidines (**149**).[62]

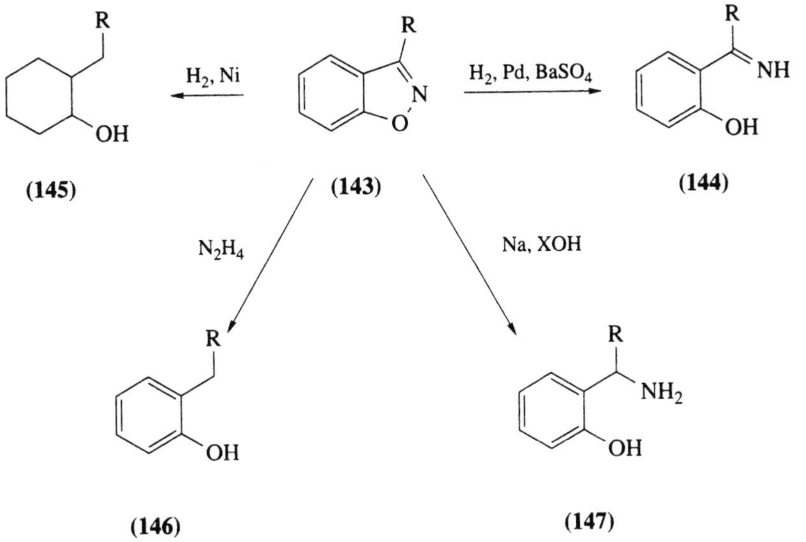

Scheme 7

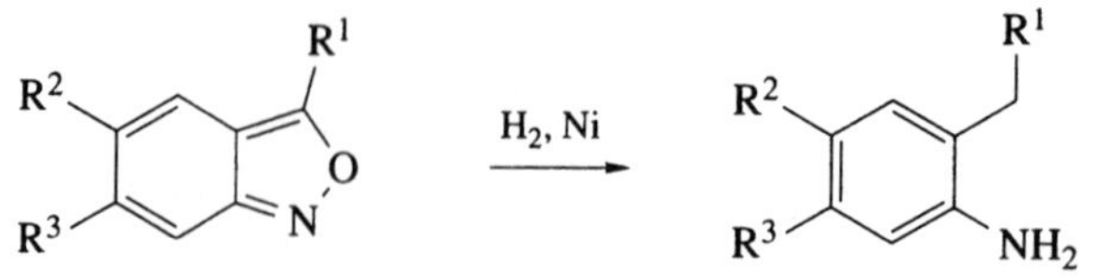

(148) (149)

3.8.3.1.2 Oxazoles, benzoxazoles, oxazolium salts, oxazolines and oxazolinones

Oxazoles (**150**) are reduced to oxazolines (**151**) by treatment with sodium and ethanol, and these products may then undergo further reduction to oxazolidines (**152**) and eventually to amino alcohols (**153**). Reduction with LAH follows the same course,[63] as does electrochemical reduction.[64] In the same way 2,2-bis(oxazoles) (**154**) when reacted with LAH afford diaminodiols (**155**).[65]

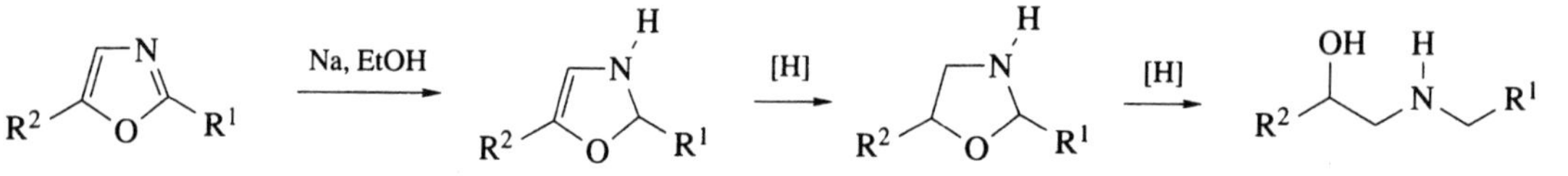

(150) (151) (152) (153)

(154) (155)

The heterocyclic ring of benzoxazoles (**156**) can be cleaved by exposure to diborane giving borazoles (**157**; possibly as shown in Scheme 8). Treatment of these products with hydrochloric acid leads to the formation of 2-aminophenols (**158**).[66] Catalytic hydrogenation of the parent compound (**156**; R = H) over molybdenum sulfide has a similar end result although, since the conditions of the reaction are quite severe, phenol and 2-toluidine can form as by-products.[67]

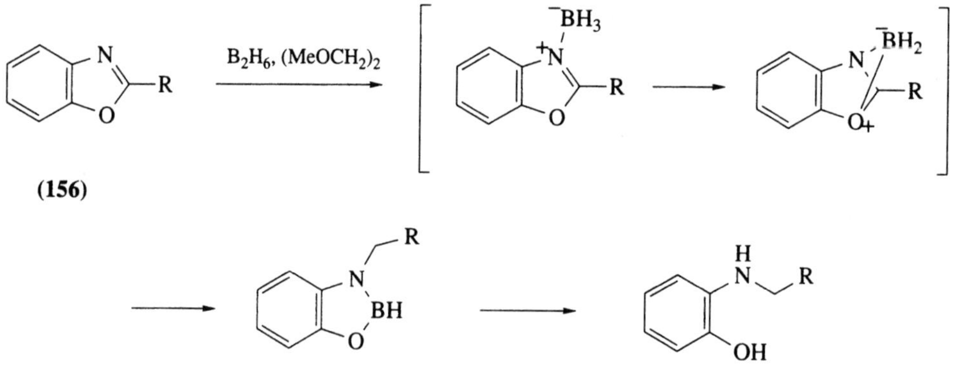

(156) (157) (158)

Scheme 8

The oxazolium salt (**160**), formed from the oxazole (**159**) by reaction with methyl triflate in acetonitrile, may be reductively ring opened by treatment with phenylsilane in the presence of cesium fluoride to give the azomethine ylide (**161**) (presumably this species is in tautomeric equilibrium with the corresponding oxazoline). The azomethine ylide can be trapped as an adduct with a suitable dipolarophile, such as dimethyl acetylenedicarboxylate (DMAD). In the case of this reagent the adduct (**162**) can be oxidized to the corresponding pyrrole (**163**) through reaction with 2,3-dichloro-5,6-dicyano-1,4-benzoquinone (DDQ) (Scheme 9).[68]

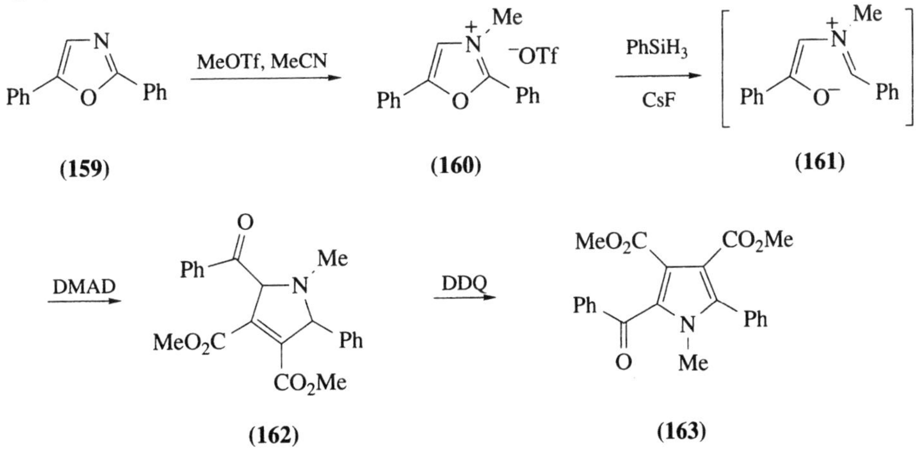

Scheme 9

2-Oxazolines are inert to reduction by LAH, but they are readily hydrolyzed to acids, C-2 providing the carbon atom of the carboxylic acid group. This is a valuable property since carbonyl groups in side chains are reduced normally. An example is shown in Scheme 10, where the 2-oxazolines (**164**) are used as starting materials for the synthesis of ω-hydroxyalkanoic acids (**165**).[69]

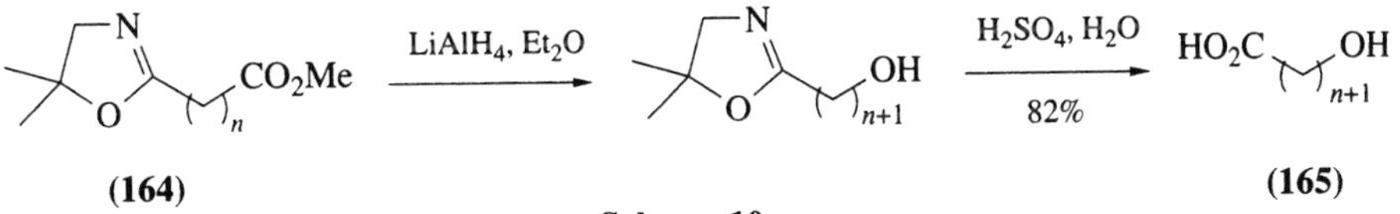

Scheme 10

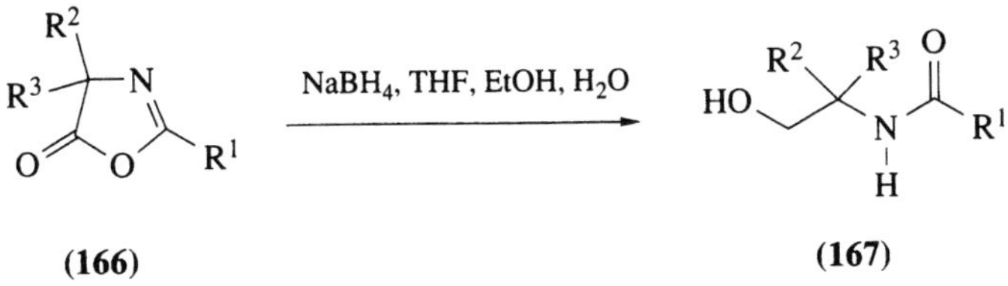

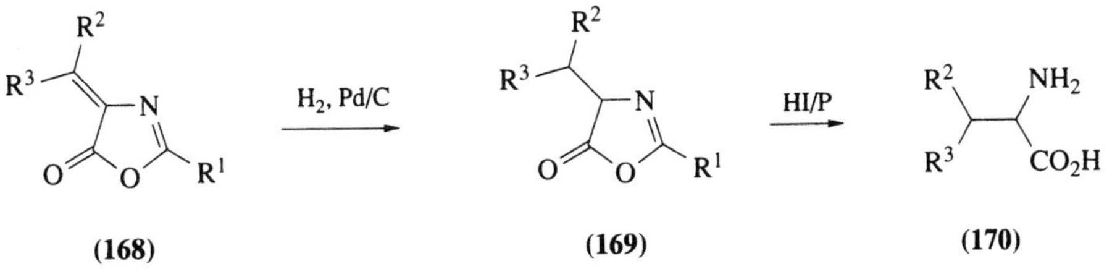

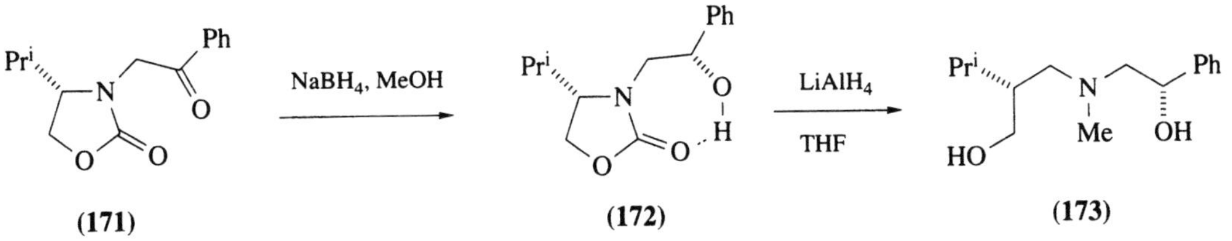

Oxazolinones are not so stable and 2-oxazolin-5-ones (**166**), for example, react with sodium borohydride to yield 2-hydroxyamides (**167**),[70] and their alkylidene derivatives (**168**) are converted into the corresponding alkyloxazolinones (**169**) by hydrogenation over palladium on carbon as catalyst. These compounds may then be cleaved to α-amino acids (**170**) by treatment with hydriodic acid and phosphorus. 3-Oxazolin-2-ones are reduced by LAH firstly to oxazolidin-2-ones[71] and, if *N*-alkylated, then on to 2-hydroxyalcohols.[72] For example, if (*S*)-4-isopropyl-3-phenacyl-1,3-oxazolidin-2-one (**171**) is reduced first with sodium borohydride in methanol, or with lithium butoxyaluminum hydride in THF, the alcohol (**172**) is formed as the major product (71–83%). When this compound is isolated and reduced further with LAH ring scission occurs to produce the acyclic diol (**173**). Should the oxazolidinone (**171**) be reduced directly with LAH there is again a preference for the production of the diol (**173**).[73]

3.8.3.2 Six-membered Ring Systems

3.8.3.2.1 Oxazines, benzoxazines and phenoxazines

(i) 1,2-Oxazines

Dissolving metal reduction of 2*H*-3,6-dihydrooxazines (**174**) effects ring opening and the formation of aminoenols (**175**),[74] whereas hydrogenation of these heterocycles, or of the isomeric 4*H*-5,6-dihydrooxazines (**177**), over a variety of catalysts gives tetrahydrooxazines (**176**).[75] The imine bond of 4*H*-5,6-dihydrooxazines is also reduced by reaction with LAH, but here, as with hydrogenation under high pressure (1500 psi, over Pt or Pd; 1 psi = 6895 Pa),[76] the process continues a further step to afford acyclic 4-aminobutanols (**178**; Scheme 11).[77] Aluminum amalgam in aqueous THF, or hydrogenation over palladium in dry ethanol containing triethyl borate, may be also used to reduce dihydrooxazines to 1,4-amino alcohols. However, in the case of the esters (**179**) the reaction mechanisms for the two types of reduction differ. For the first the final products are pyrroles (**180**) and for the second both pyrroles and pyrrolidines (**181**) are produced (Scheme 12).[78]

Scheme 11

As indicated above, dihydrooxazines, and through them tetrahydrooxazines, can act as masked 4-aminobutanols and advantage is often made of the fact that dihydrooxazines are synthesized by the cycloaddition of nitroso compounds and 1,3-dienes. In this way the relative stereochemistry of the substituents at C-4 and C-6 in the heterocycle is set and this then determines the regioselectivity of addition reactions to the double bond. A good example is provided by the synthesis of the tetraacetylaminoallose (**183**; Scheme 13), which proceeds through a double hydrogenolysis of the *N*-carbamoyltetrahydrooxazine (**182**).[79]

The comparative stability of the 1,2-oxazine ring towards various reductants is well illustrated by the construction of the neurotoxin gephyrotoxin (**189**), in the form of its racemate (Scheme 14).[80] This synthesis begins with the intramolecular cycloaddition of the acylnitroso species (**184**), to give the adduct (**185**). This product may be reduced to the tetrahydrooxazine (**186**), without ring opening, by hydrogenation over 5% palladium on carbon as catalyst in methanol solution. After reaction with propylmagnesium bromide, the derived enamine (**187**) may be further reduced, again without ring opening, to the tetrahydrooxazine (**188**) by treatment with sodium cyanoborohydride in methanol at pH 3.8–5.4. The exclusive formation of this product is possibly due to the fact that only in the least stable of the two possible chair

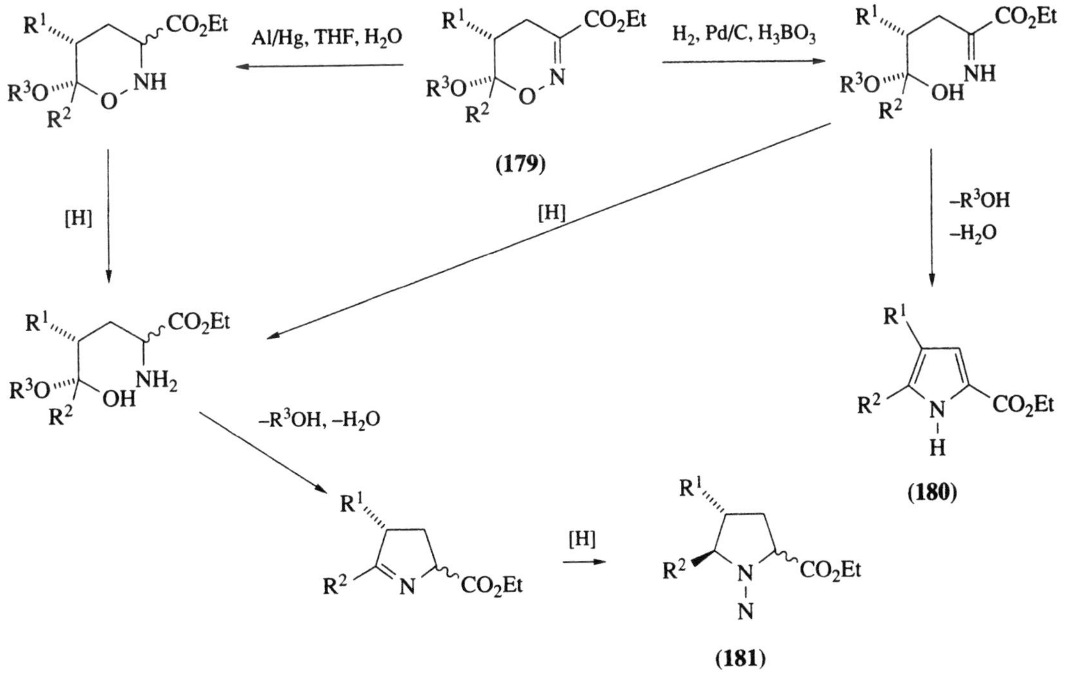

Scheme 12

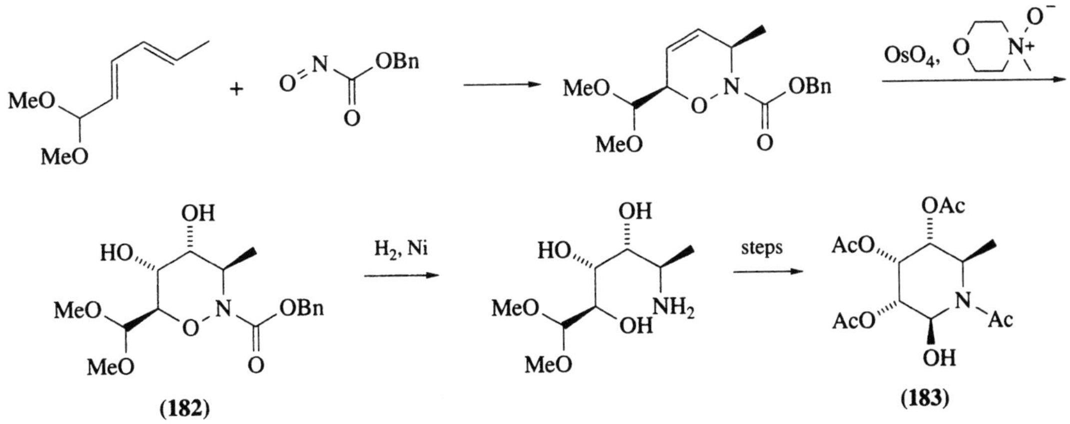

Scheme 13

transition states for the reaction can the incoming reagent avoid severe steric interaction with the bulky butyl substituent of the substrate. Finally the tetrahydrooxazine (**188**) is cleaved by reduction with zinc and acetic acid and the product then undergoes spontaneous recyclization to (±)-gephyrotoxin.

(ii) 1,3-Oxazines

Much interest lies in the use of dihydro-1,3-oxazines (**190**) as enolate equivalents, since, if an alkyl group is carried at C-2, these compounds may be deprotonated and the anions formed reacted with numerous types of electrophiles. Reduction of the imine bond of the products (**191**), is then conveniently effected by treatment with sodium borohydride. The tetrahydrooxazines (**192**) which are formed may then be ring opened by hydrolysis with aqueous acid (Scheme 15). This topic and its utility in synthesis has been well reviewed.[81,82]

(iii) 1,4-Oxazines, 1,4-benzoxazines, phenoxazines and tetrahydro-1,4-oxazines

As with tetrahydro-1,3-oxazines the tetrahydro-1,4-oxazine ring system is stable to reduction by complex metal hydrides, and, for example, LAH can be used to reduce the lactam carbonyl function of mor-

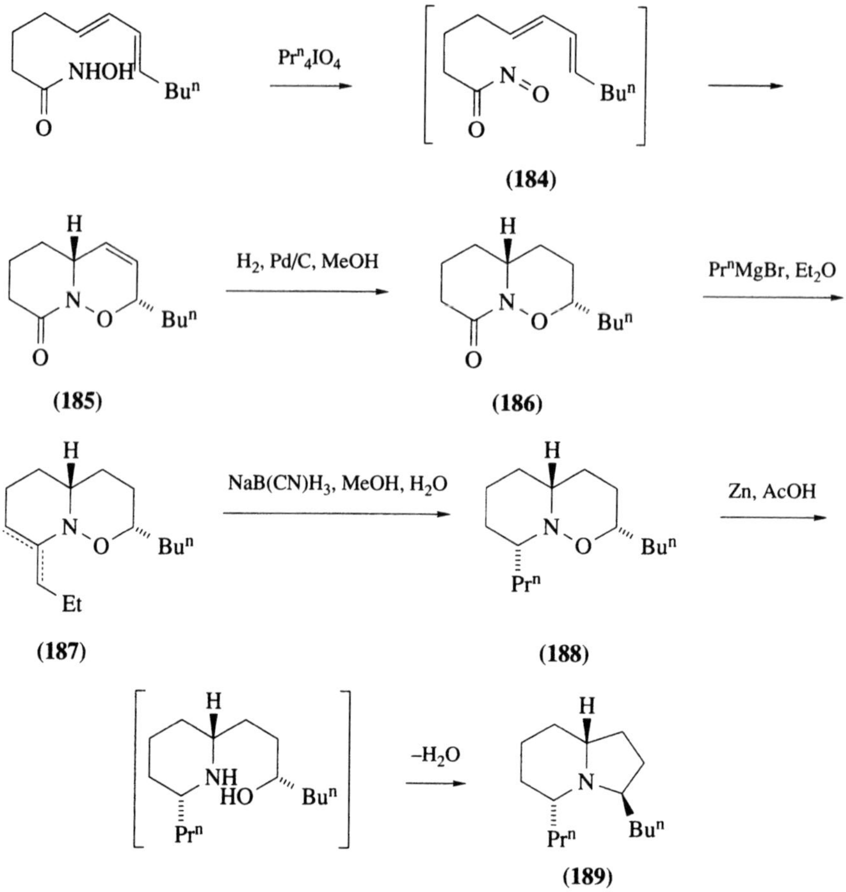

Scheme 14

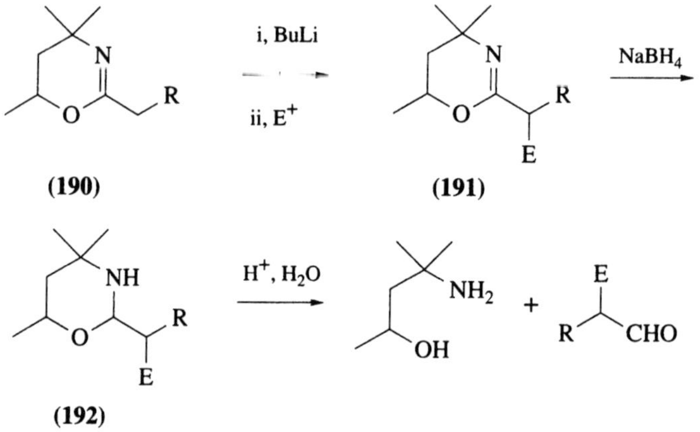

Scheme 15

pholinones (**193**) to give the corresponding morpholines (**194**).[83] Likewise *N*-alkyldihydro-1,4-benzoxazin-3-ones (**195**) react with this reagent to form benzomorpholines (**196**). The same products can be obtained by the reduction of 2*H*-1,4-benzoxazines (**197**), and indeed a synthesis of benzomorpholines requires the hydrogenation of 2-nitrophenyloxymethyl ketones (**198**) over Raney nickel as catalyst (Scheme 16).[84] Since the aromatic nuclei of benzoxazines and phenoxazines are inert to most reductants, substituent groups can be reduced in the expected manner. Similarly the restoration of aromaticity in the products ensures that the reduction of phenoxazin-3(3*H*)-ones (**199**) to 3-hydroxyphenoxazines (**200**) is easily achieved as, for example, by reaction with sodium dithionate in aqueous acetone.[85]

Oxazinones (**201**) and (**204**), which are available in high states of optical purity, can be reacted with *N*-bromosuccinimide and then treated with metal alkyls (RM) to form the corresponding 3-alkyl derivatives

(**202**) and (**205**). In the case of the former, cleavage to the zwitterionic L-amino acids (**203**) can be achieved in *ca.* 98% *ee* by hydrogenolysis in ethanol solution over palladium as catalyst; the chiral auxiliary is lost in the form of bibenzyl. For the second class of compounds ring opening and deprotection to give D-amino acids (**206**) is best carried out through reduction with lithium and liquid ammonia in tetrahydrofuran containing ethanol. The optical and product yields in this second procedure are also excellent and between them these two reaction sequences provide access to a wide range of natural and unnatural amino acids in a predictable manner.[86] (Scheme 17).

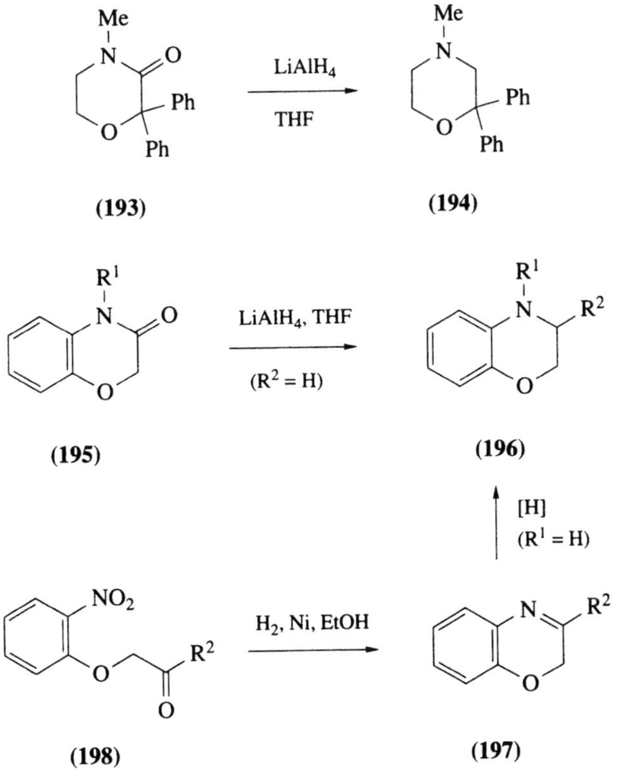

(**193**) —$LiAlH_4$, THF→ (**194**)

(**195**) —$LiAlH_4$, THF (R^2 = H)→ (**196**)

(**198**) —H_2, Ni, EtOH→ (**197**) —[H] (R^1 = H)→ (**196**)

Scheme 16

(**199**) —$Na_2S_2O_4$, acetone, H_2O→ (**200**)

(**201**) —i, NBS, CCl_4; ii, RM→ (**202**) —H_2, Pd, EtOH→ (**203**)

(**204**) —i, NBS, CCl_4; ii, RM→ (**205**) —Li, NH_3, THF, EtOH→ (**206**)

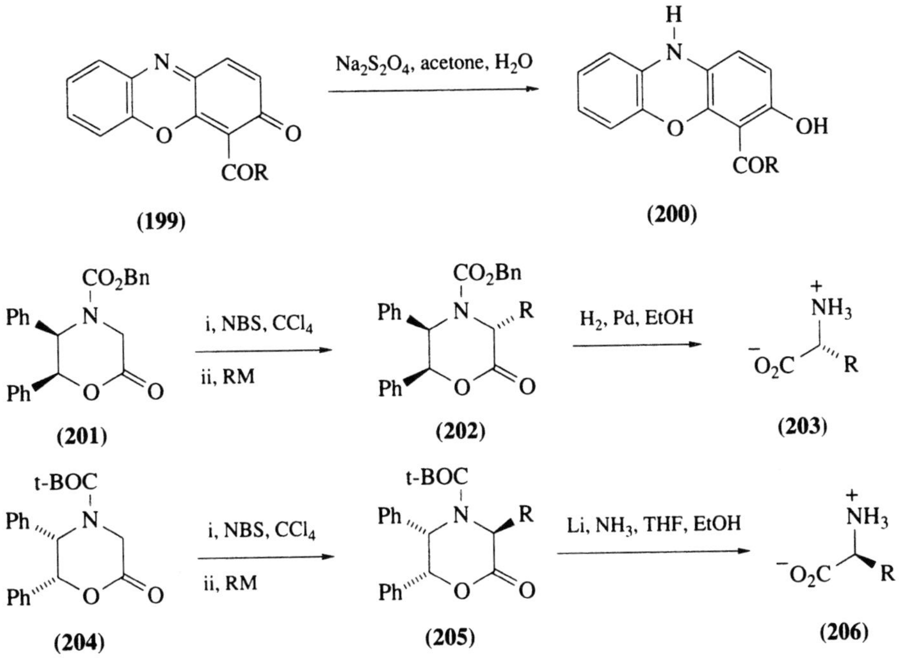

Scheme 17

3.8.4 REDUCTION OF HETEROCYCLES CONTAINING ONE NITROGEN ATOM AND ONE SULFUR OR SELENIUM ATOM

3.8.4.1 Five-membered Ring Systems

3.8.4.1.1 Isothiazoles

The ring system of isothiazoles (**207**) is stable to treatment with sodium borohydride and LAH, and these reagents are employed to reduce ring substituents such as nitro or carbonyl groups to the corresponding amines or alcohols, respectively. Similarly, catalytic hydrogenation of 4-nitroisothiazoles affords the corresponding amines. In this case, however, the reaction should be carried out below 10 °C, otherwise ring scission occurs, probably to form enaminothiones (**208**).[87]

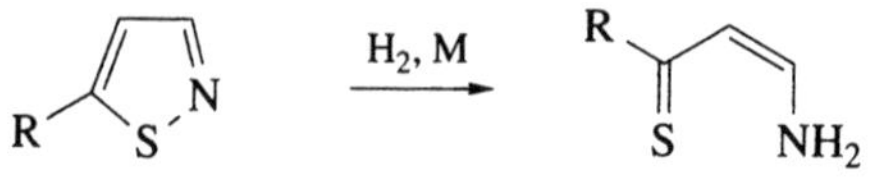

H_2, M

(207) **(208)**

3.8.4.1.2 Thiazoles and thiazolines

Thiazoles are deactivated towards electrophilic substitution, and thus direct reaction with hydride reductants to give thiazolines should be facilitated. There are indeed some examples of this type of reaction,[88] but it is more common to reduce *N*-alkylated thiazolium salts (**209**). These compounds are converted first by reaction with sodium borohydride into 4-thiazolines (**210**), which in protic solvents become protonated and undergo further reduction to yield thiazolidines (**211**).[89] Similarly the isoquinolinium salt (**213**), formed by the acid-promoted cyclization of the isoquinoline (**212**), is converted into the tetrahydroisoquinoline (**214**) (presumably *via* an intermediate 1,2-dihydroisoquinoline) by reaction with sodium borohydride.[90]

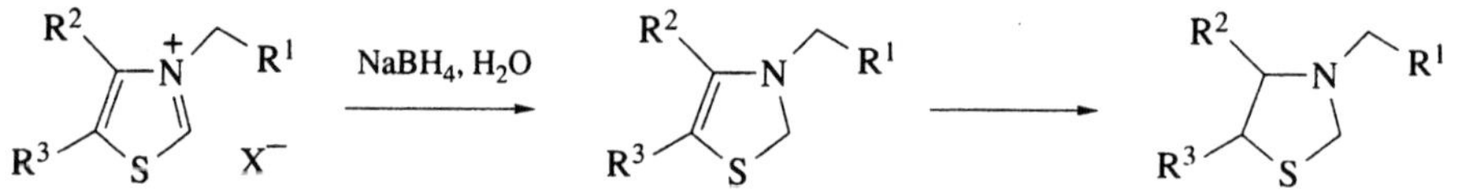

$NaBH_4$, H_2O

(209) **(210)** **(211)**

HX $NaBH_4$, HX

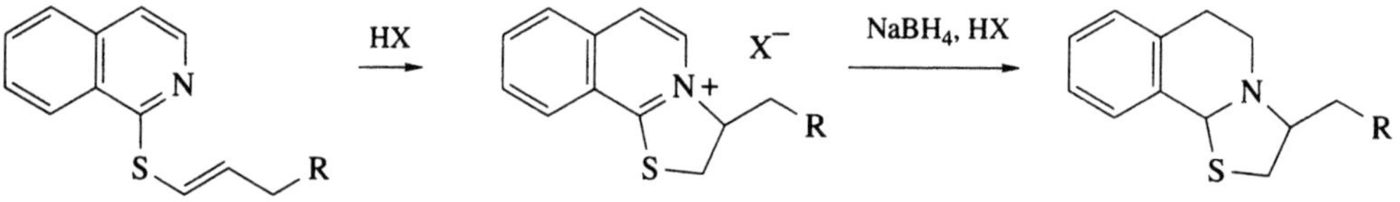

(212) **(213)** **(214)**

4-Alkylidene-1,3-thiazolin-5-ones (**215**) are also reduced by the action of sodium borohydride, but in this case only the exocyclic double bond reacts.[91] Similarly, this reagent may be used to reduce carbonyl groups in the side chains of thiazoles without any effect on the heterocycle itself. The thiazole ring also survives hydrogenation over Raney nickel catalyst at atmospheric pressure,[92] but the fully reduced thiazole ring system is vulnerable to desulfurization. This occurs under quite mild conditions, and thus when

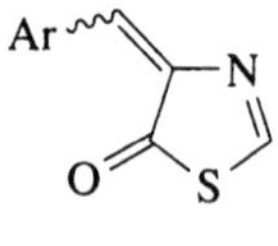

(215)

the tricyclic thiazolidine (**216**) is treated with silver nitrate and the product, without isolation, is reacted with sodium borohydride the pyrrolidine (**217**) is formed.[93] Sodium in liquid ammonia also causes ring opening, and reaction of the thiazoline-2-thiones (**218**) with this reagent system affords azadienyldithiolates (**219**). The products recyclize to thiazoles (**220**) when acidified with hydrochloric acid.[94]

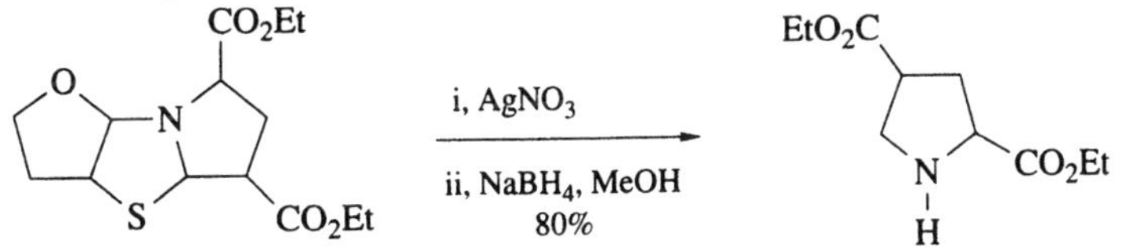

i, $AgNO_3$; ii, $NaBH_4$, MeOH, 80%

(**216**) (**217**)

Na, NH_3; HCl; $2Na^+$

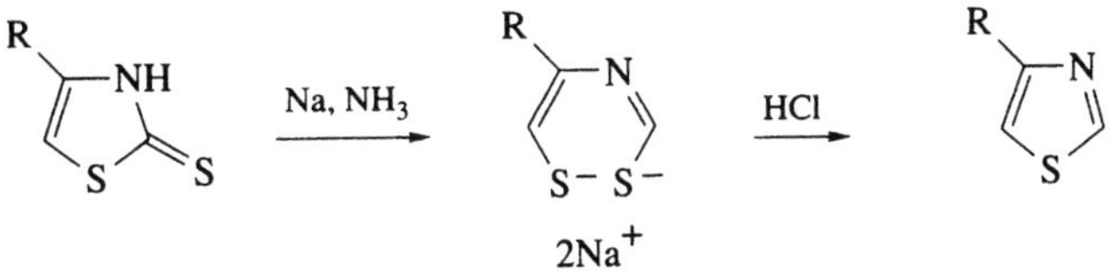

(**218**) (**219**) (**220**)

The carbonyl group and the thiocarbonyl function of the thiazolidine derivative (**221**) are not changed by electrochemical reduction in an acidic medium, but the oximino unit at C-5 is transformed into an amino group and salts (**222**) are produced.[95]

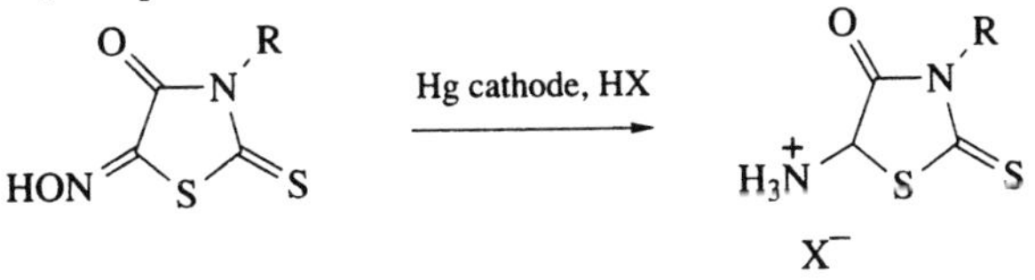

Hg cathode, HX

(**221**) (**222**)

3.8.4.1.3 *Benzothiazoles and benzothiazolium salts*

Benzothiazoles are rather more prone to reductive decyclization than are thiazoles, although it is possible to form benzothiazolines (**224**) through the sodium borohydride reduction of benzothiazolium salts (**223**).[96] On hydrolysis these products yield 2-aminothiophenols (**225**). Prolonged exposure of benzothiazolines to excess sodium borohydride also causes ring opening to aminothiophenols,[97] as does reaction with LAH. In this last case the substrates for reduction may be benzothiazoles (**226**), which presumably initially form benzothiazolines with the reagent (Scheme 18).[98]

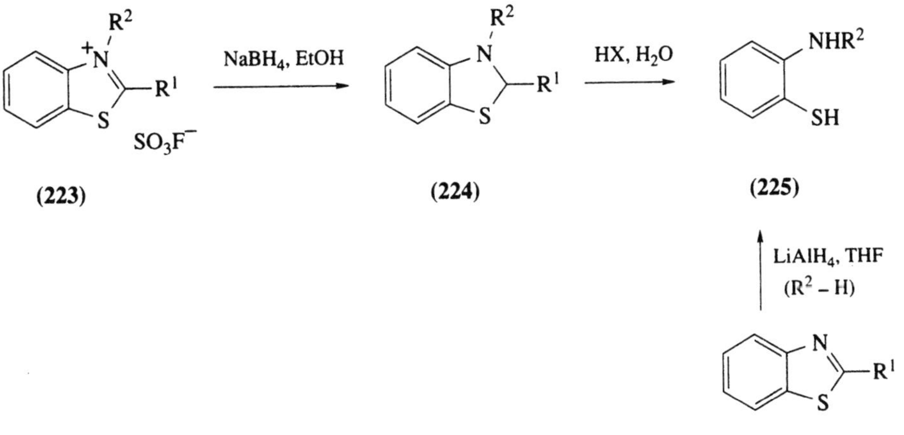

$NaBH_4$, EtOH; HX, H_2O; $LiAlH_4$, THF (R^2 – H)

(**223**) (**224**) (**225**) (**226**)

Scheme 18

3.8.4.1.4 *Isoselenazoles*

Isoselenazoles are very easily cleaved by fragmentation of the C—Se bond. Reagents such as sodium borohydride are effective and, for example, treatment of the pyrimidoisoselenazole (**227**) with this reagent affords the aminopyrimidinone (**228**).[99]

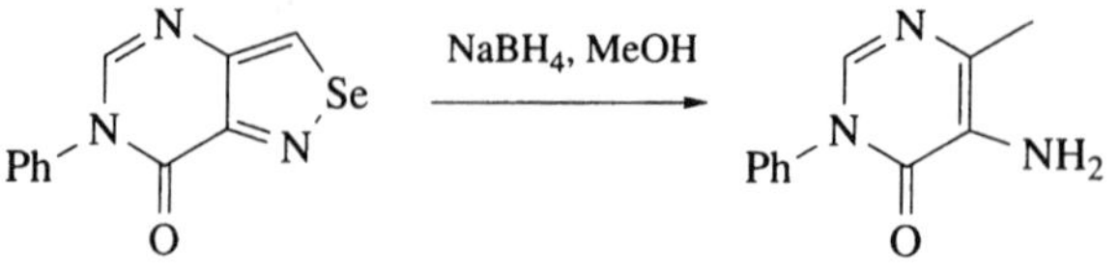

$NaBH_4$, MeOH

(**227**) (**228**)

3.8.4.2 Six-membered Ring Systems

3.8.4.2.1 *Thiazines, benzothiazines and phenothiazines*

(i) 1,3-Thiazines

Reduction of 5-acyl-1,3-thiazines (**229**) with sodium cyanoborohydride in acid media, or with triethylsilane, gives 3,6-dihydro-2*H*-1,3-thiazines (**230**),[100] whereas sodium borohydride has no effect upon the unsaturation of the ring and only reduces the carbonyl group of the side chain. Some other reagents cause ring opening under very mild conditions; thus treatment of 1,3-thiazinium salts (**231**) with hydrogen sulfide is sufficient to yield thioenethioamides (**232**).[101] LAH also cleaves the 1,3-thiazine ring; the 1,3-benzothiazin-4-one (**233**), for example, forms 2-mercaptobenzylamine (**234**), plus a small amount of benzoisothiazole (**235**).[102]

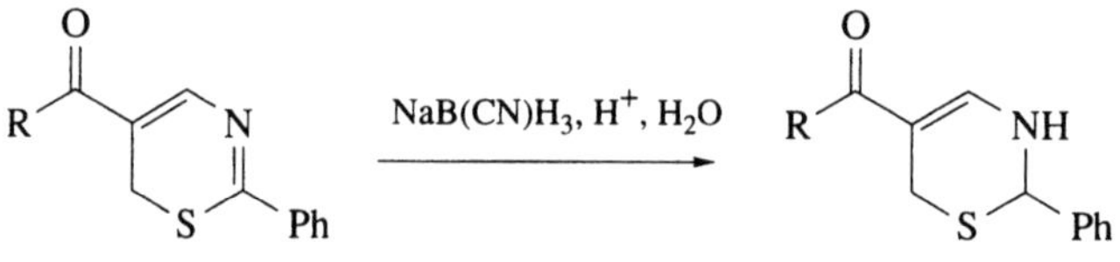

$NaB(CN)H_3$, H^+, H_2O

(**229**) (**230**)

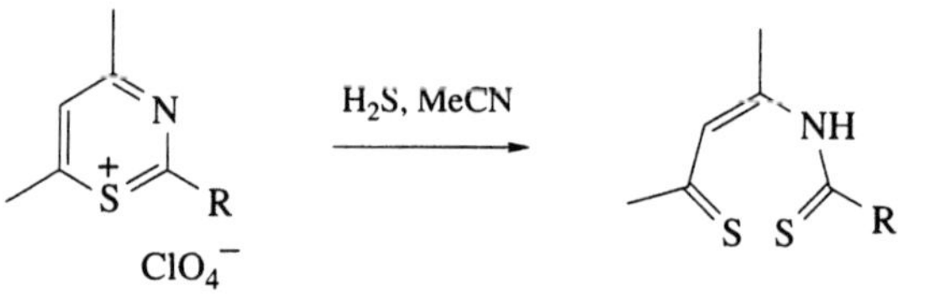

H_2S, MeCN

ClO_4^-

(**231**) (**232**)

$LiAlH_4$ [H]

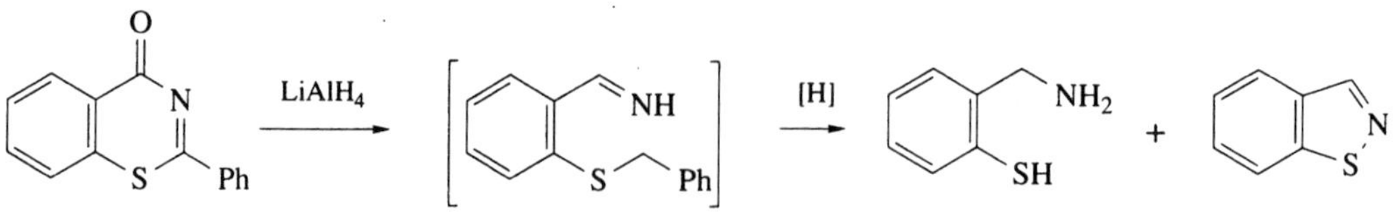

(**233**) (**234**) (**235**)

(ii) Benzo-1,4-thiazines

2*H*-Benzo-1,4-thiazines (**236**) undergo ring opening and ring contraction when hydrogenated over Raney nickel, or palladium on carbon, as catalyst. Here the initial step in the reaction is reductive fragmentation of a C—S bond. Imino intermediates (**237**) may then form, which recyclize to benzothiazoles (**238**).[103]

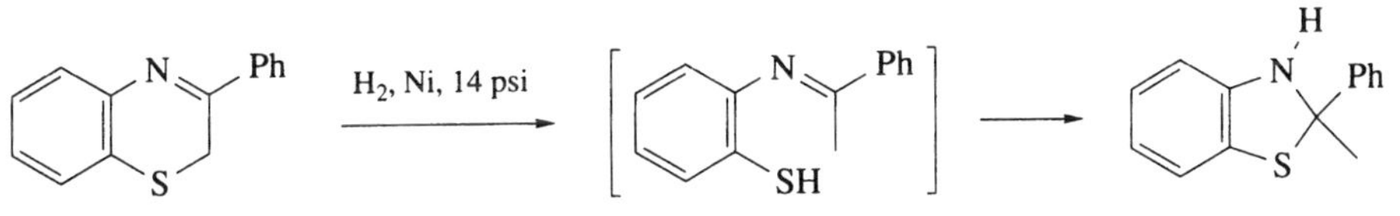

(**236**) (**237**) (**238**)

(iii) Phenothiazines

Phenothiazines, like phenoxazines (Section 3.8.3.2.2i), are relatively stable compounds and unsaturated substituents attached to the carbocycles can be reduced by the usual reagents, leaving the nuclei intact. There is one important difference, however, namely that phenothiazines may be desulfurized. An early example is provided by the degradation of phenothiazines (**239**) to diphenylamines (**240**) by hydrogenolysis over Raney nickel catalyst.[104]

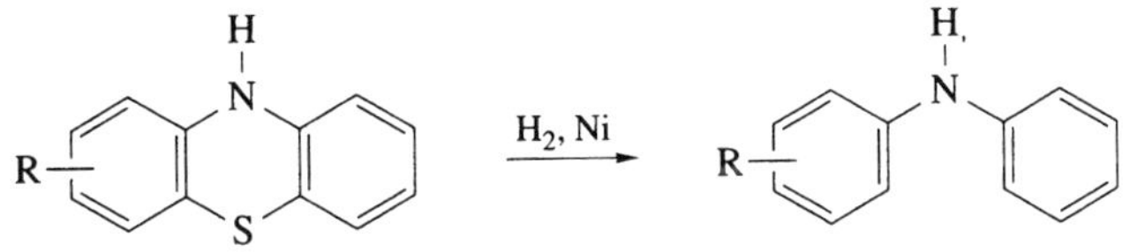

(**239**) (**240**)

3.8.5 REDUCTION OF HETEROCYCLES CONTAINING TWO OXYGEN ATOMS OR TWO SULFUR ATOMS

3.8.5.1 Dioxolanes, Dioxanes, Dithioles and Benzodithioles

3.8.5.1.1 1,3-Dioxolanes and 1,3-dioxanes

Reductive cleavage of a C—O bond of 1,3-dioxolanes may be achieved by a variety of methods; commonly chemical reductants are employed and, although LAH alone is ineffective, the mixed reductants LAH/$AlCl_3$ or LAH/$AlBr_3$ cleave 2-substituted dioxolanes (**241**) to hydroxy ethers (**242**), sometimes with a high degree of diastereotopic selectivity.[105] This is usually explained in terms of the selective complexation of the reagent with one of the oxygen atoms of the ring, which then allows the delivery of hydride ion to one face of the heterocycle, rather than to the other.

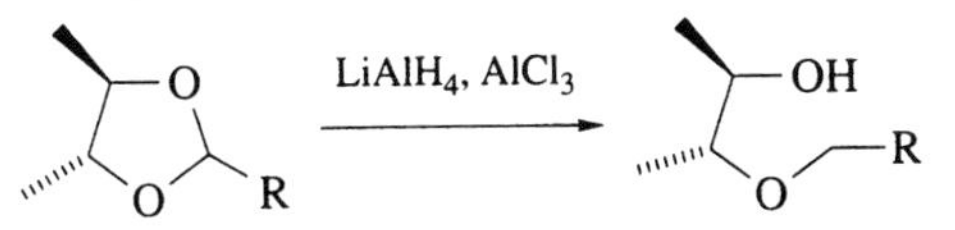

(**241**) (**242**)

The same methodology can be applied to the ring opening of 1,3-dioxanes and this has been used in effecting the chiral reduction of ketones to alcohols. As an illustration, cyclohexyl methyl ketone (**243**) can be converted into the dioxane (**244**) by reaction with (2*R*,4*S*)-pentanediol and cleaved with LAH/$AlBr_3$ to afford (2*R*)-cyclohexylethanol (**245**) in good yield and in high enantiomeric excess. Interestingly, the stereoselectivity of the ring-opening reaction can be reversed if the reagent is changed to triethylsilane and a Lewis acid (preferably $TiCl_4$). Now the 1,3-dioxane (**244**) is opened to give the alternative enantiomer (**246**; Scheme 19).[106]

Stereoselective monodeprotection of one hydroxy group in a polyhydroxylated system is a common requirement in synthesis, and in the case of cyclic orthoesters this can be accomplished through reactions

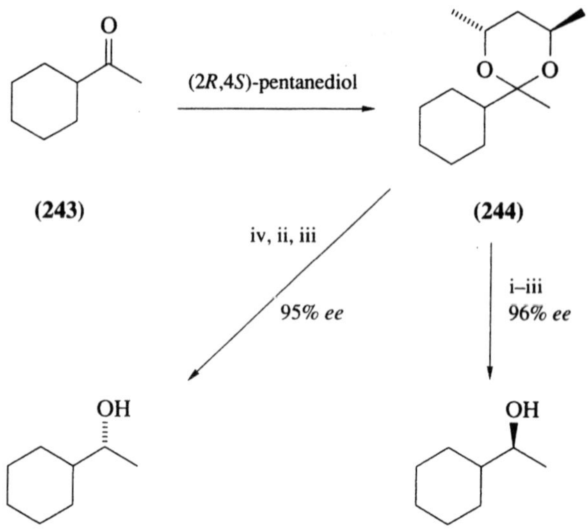
(2R,4S)-pentanediol
iv, ii, iii
95% ee
i–iii
96% ee
OH
OH

(243) (244) (246) (245)

i, AlH_3, $AlBr_3$; ii, $(COCl)_2$, DMSO, Et_3N; iii, Na_2CO_3, H_2O; iv, Et_3SiH, $TiCl_4$

Scheme 19

with DIBAL in nonpolar media. The ring-opening process is highly regioselective, and for the unsymmetrical substrate (**247**) only the hydroxy ether (**248**) is formed (Scheme 20). Ring size is important in this type of reaction, however, and only 1,3-dioxolanes and -dioxanes give satisfactory results.[107]

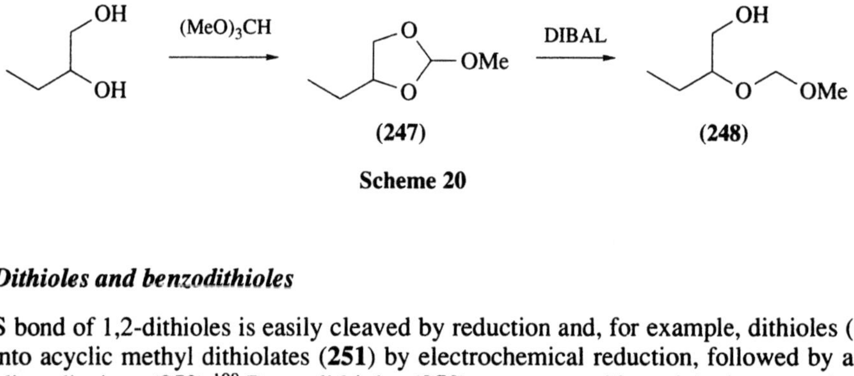
OH
OH
$(MeO)_3CH$
O
O
OMe
DIBAL
OH
O
OMe

(**247**) (**248**)

Scheme 20

3.8.5.1.2 Dithioles and benzodithioles

The S—S bond of 1,2-dithioles is easily cleaved by reduction and, for example, dithioles (**249**) can be converted into acyclic methyl dithiolates (**251**) by electrochemical reduction, followed by alkylation of the intermediate dianions (**250**).[108] Benzodithioles (**252**) are more stable and the heterocycle survives reduction with zinc and trifluoroacetic acid, although the exocyclic double bond is saturated to give 2-alkyldithiolanes (**253**).[109]

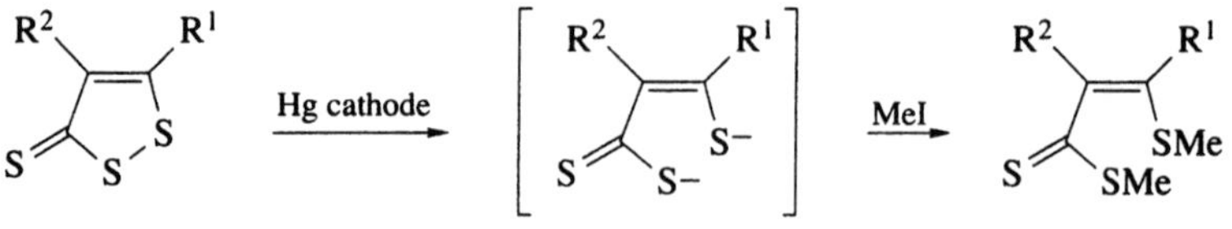
Hg cathode
MeI
SMe
SMe

(**249**) (**250**) (**251**)

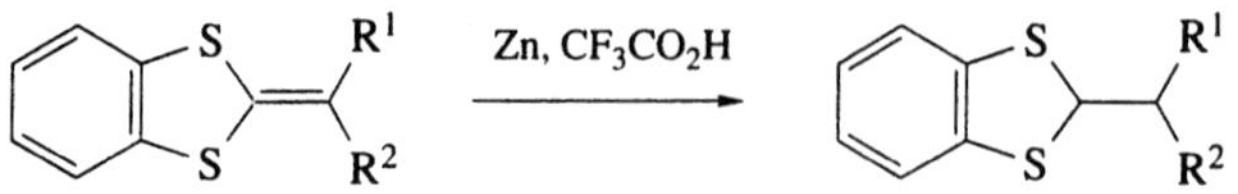
Zn, CF_3CO_2H

(**252**) (**253**)

Benzo-1,3-dithiole-2-thiones (**254**) are cleaved to thiocatechols (**256**) by treatment with DIBAL-H, or diborane·dimethyl sulfide. Sodium borohydride, LAH or Red-Al are much less efficient reagents for this purpose. Under more controlled conditions it is possible to obtain good yields of the intermediate benzodithiolanes (**255**).[110]

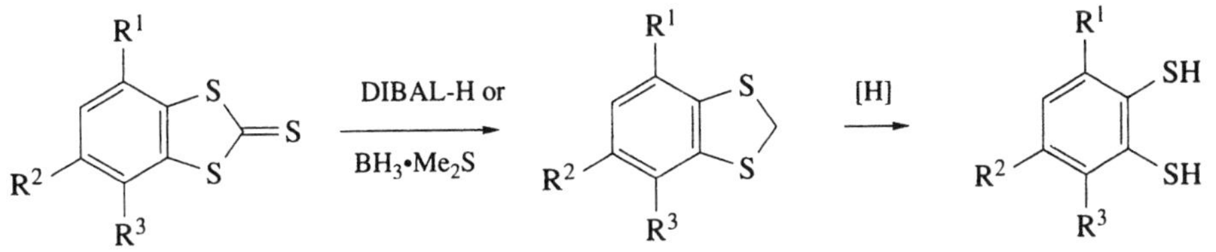

DIBAL-H or $BH_3 \cdot Me_2S$; [H]

(**254**) (**255**) (**256**)

3.8.6 REDUCTION OF HETEROCYCLES CONTAINING THREE NITROGEN ATOMS

3.8.6.1 Triazoles, Triazolium Salts and Benzotriazoles

3.8.6.1.1 1,2,3-Triazoles and 1,2,3-benzotriazoles

The ring system of 1,2,3-triazoles (**257**) is stable to dissolving metals in acids,[111] and 1,2,3-triazolium salts are resistant to reduction unless they are substituted at both positions N-1 and N-2. Thus whereas salts of the type (**258**) are not reduced, analogs of the form (**259**) are converted into triazolines (**260**) by reaction with sodium borohydride.[112]

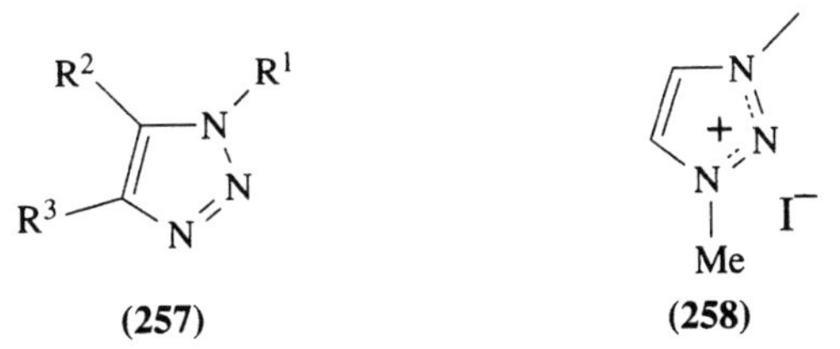

(**257**) (**258**)

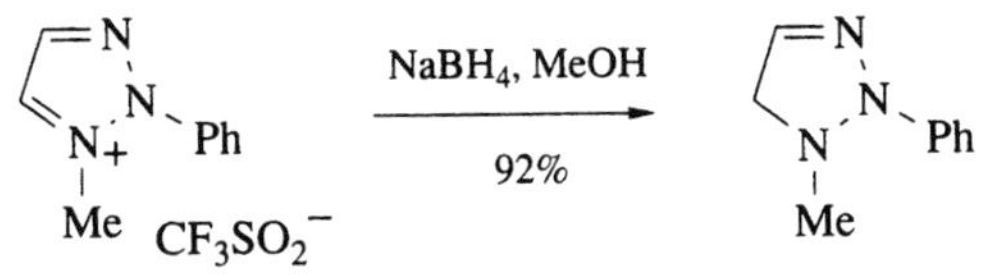

$NaBH_4$, MeOH; 92%

(**259**) (**260**)

Benzotriazoles, like their monocyclic counterparts, are often synthesized by reductive cyclizations and it is only when they are activated as salts that reductions with metal hydrides occur. Potassium borohydride, for example, reacts with aqueous solutions of the salt (**261**) to give 2-methylaminoaniline (**262**). Some *N*-demethylation leading to the triazole (**263**) also takes place, and this may be caused by the prevailing alkaline reaction conditions. LAH in diethyl ether is a more effective reagent for the reductive ring opening of benzotriazoles, and this reagent also cleaves 1,3-dimethylbenzotriazolium methylsulfate (**264**) to 1,2-di(methylamino)benzene (**265**).[113]

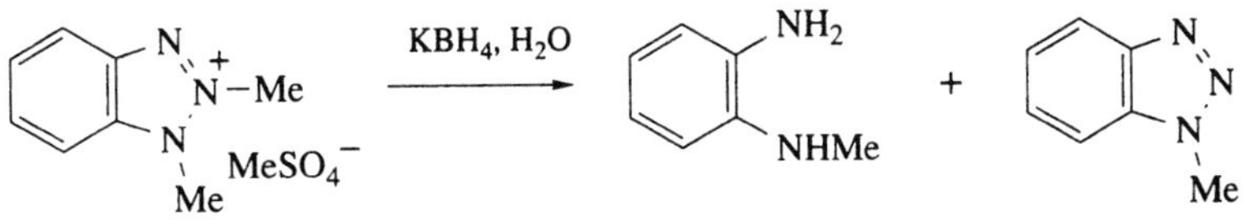

KBH_4, H_2O

(**261**) (**262**) (**263**)

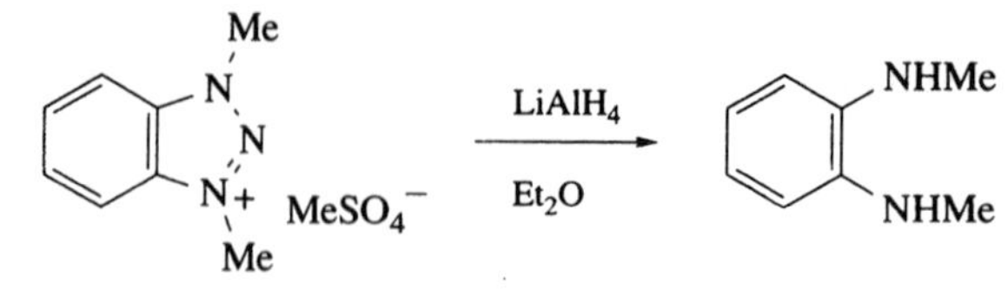

(264) $\xrightarrow[Et_2O]{LiAlH_4}$ (265)

3.8.6.1.2 1,2,4-Triazolium salts

1,2,4-Triazolium salts are readily reduced by treatment with sodium borohydride to give the corresponding triazolines. However, these products are unstable to conditions of acid work-up and fragment to give aldehydes. This property has been exploited to homologate aldehydes using 3-methylthio-1,4-diphenyl-1,2,4-triazolium chloride (**266**) as a masked carboxylate equivalent, as illustrated in Scheme 21.[114]

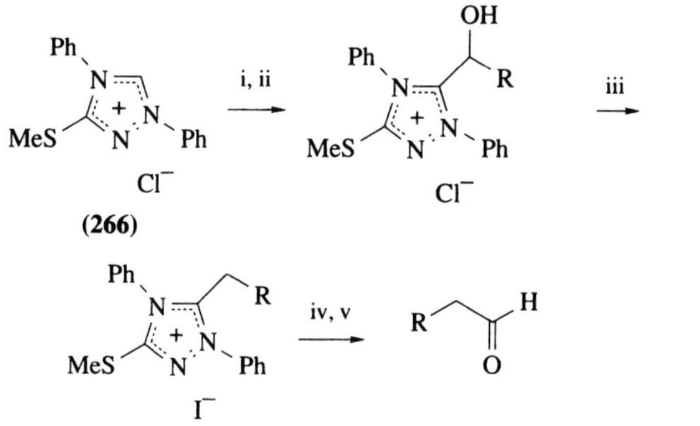

(266)

i, Et_3N; ii, RCHO; iii, $SOCl_2$, KI, $NaHSO_3$; iv, $NaBH_4$; v, HX, HCHO

Scheme 21

LAH reduces 1,2,4-triazol-5-ones (**267**) to both triazoles (**268**) and triazolines (**270**). The course of the reaction depends on the nature of the *N*-substituent; thus when R = H triazoles are formed in modest yield (15–30%), but when R = alkyl triazolium salts (**269**) are produced as intermediates and these are then reduced to triazolines (**270**; Scheme 22).[115]

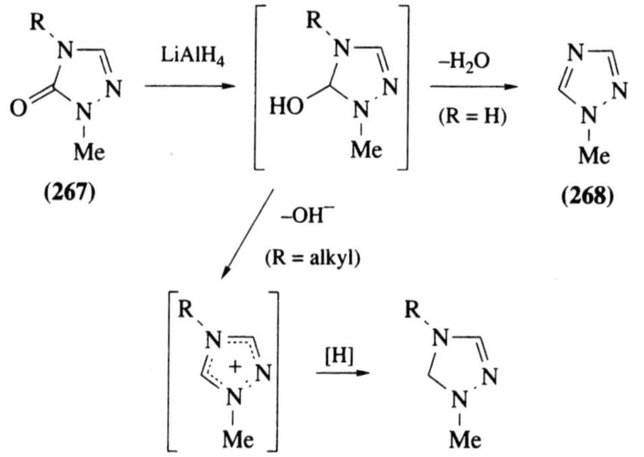

(267) (268) (269) (270)

Scheme 22

3.8.7 REDUCTION OF HETEROCYCLES CONTAINING TWO NITROGEN ATOMS AND ONE OXYGEN ATOM

3.8.7.1 Oxadiazoles and Benzoxadiazoles

3.8.7.1.1 1,2,4-Oxadiazoles

1,2,4-Oxazadiazoles (**271**) undergo ring opening either by catalytic hydrogenation over platinum, palladium or nickel, or by treatment with LAH. In the first case O–N bond cleavage occurs to give iminoamides (**272**), whereas during chemical reduction the products are amino oximes (**273**) (Scheme 23). In some instances the first-formed reduction products may undergo rearrangement; thus hydrogenation of azo derivatives (**274**) eventually affords 1,2,4-triazoles (**275**),[116] and vinylogous amides (**276**) give rise to pyrimidines (**277**).[117] Amino oximes (**279**) are also obtained when oxadiazolin-5-ones (**278**) are reacted with LAH, indicating that oxadiazoles are intermediates in the reaction.[118] 3-Phenyl-1,2,4-oxadiazolines (**280**) yield anions (**281**) when reacted with sodium in liquid ammonia; these products combine with alkyl halides, and where R = Me equal amounts of the two *N*-alkyl isomers (**282**) and (**283**) are formed.[119]

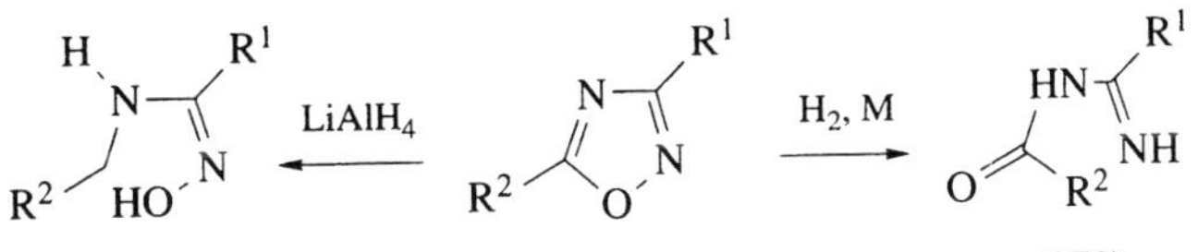

(**273**) (**271**) (**272**)

Scheme 23

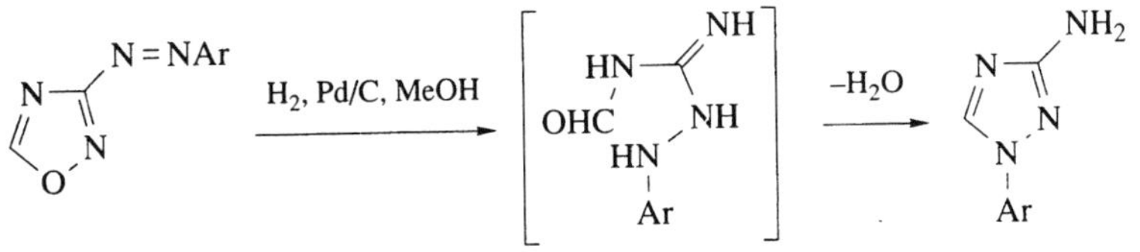

(**274**) (**275**)

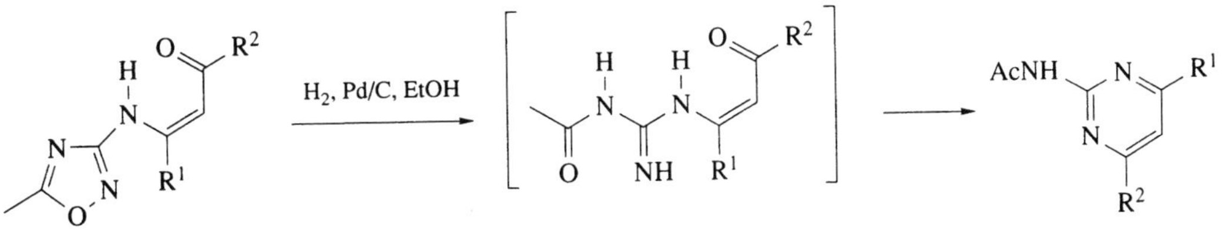

(**276**) (**277**)

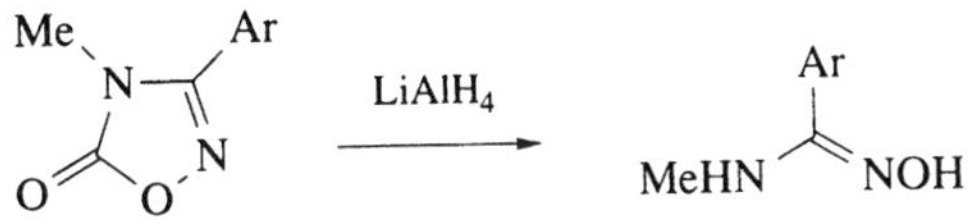

(**278**) (**279**)

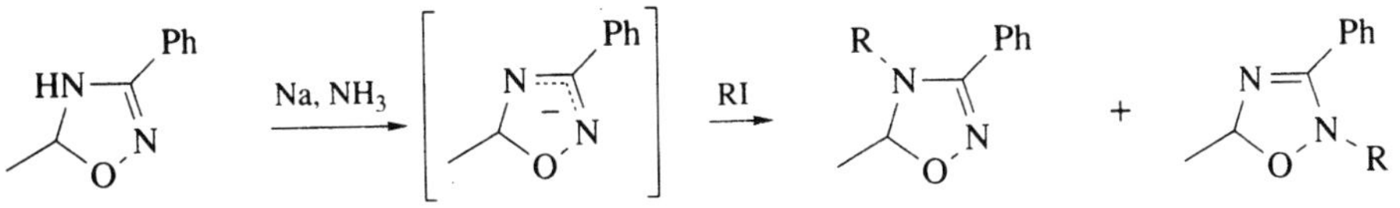

(**280**) (**281**) (**282**) (**283**)

3.8.7.1.2 1,2,5-Oxadiazoles and 1,2,5-benzoxadiazoles

1,2,5-Oxadiazole-2-oxides (furoxans; **284**) undergo ring opening and reduction to 1,2-diketones (**285**) when treated with zinc and acetic acid, but with tin and hydrochloric acid, or with hydriodic acid alone, only deoxygenation to the parent heterocycles (**286**) is observed (Scheme 24).[120] Hydrogenation over palladium on carbon as catalyst converts benz-1,2,5-oxadiazole (**287**) into its tetrahydro derivative (**288**), and causes some ring opening to give 1,2-phenylenediamine (**289**; Scheme 25).[121] If electron-donating groups are present at C-4 the proportion of the corresponding ring-opened product is increased. Chemical reduction with tin and hydrochloric acid only gives 1,2-phenylenediamines.

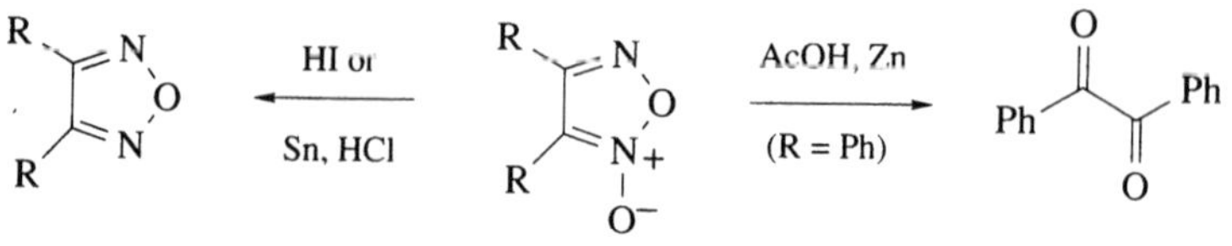

(**286**) ← HI or Sn, HCl — (**284**) — AcOH, Zn (R = Ph) → (**285**)

Scheme 24

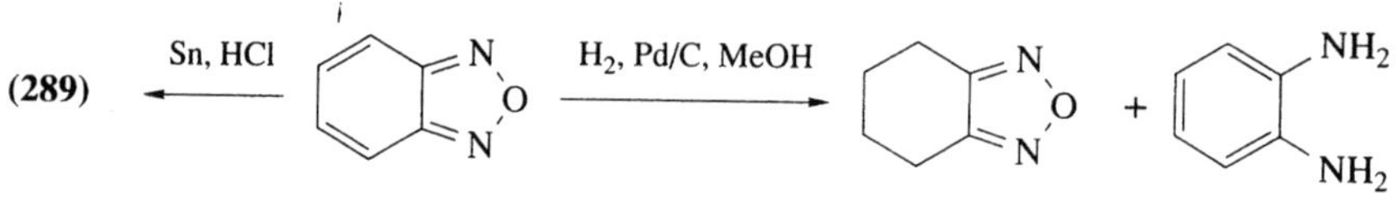

(**289**) ← Sn, HCl — (**287**) — H_2, Pd/C, MeOH → (**288**) 80% + (**289**) 9%

Scheme 25

3.8.7.1.3 1,3,4-Oxadiazoles

1,3,4-Oxazdiazoles are much more stable to reduction and it is possible to reduce substituent groups without affecting the nucleus; however, salts (**290**) are susceptible to reductive decyclization and treatment with sodium sulfide, for example, leads to acylthioacylhydrazines (**291**).[122] LAH reacts in a similar manner, and converts the 1,3,4-oxadiazolidine (**292**) into the *N*-methylaminotetrahydroisoquinoline (**293**).[123]

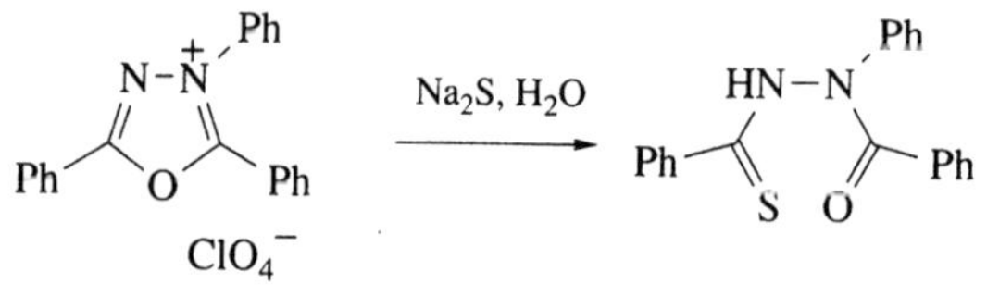

(**290**) ClO_4^- — Na_2S, H_2O → (**291**)

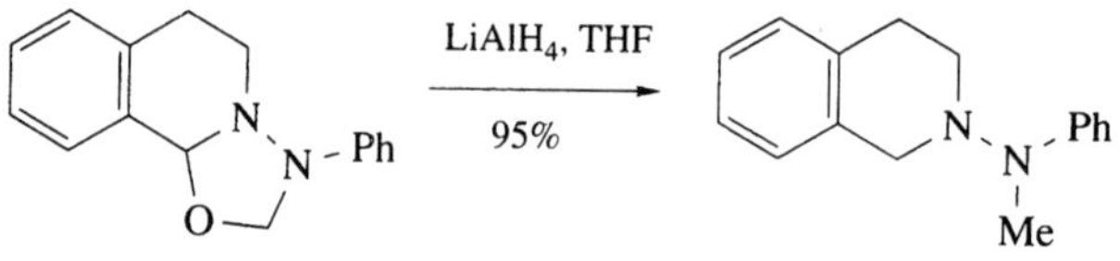

(**292**) — $LiAlH_4$, THF, 95% → (**293**)

3.8.8 REFERENCES

1. E. E. Schweizer and K.-J. Lee, *J. Org. Chem.*, 1984, **49**, 4848.
2. W. Reid, G. Deuschel and A. Kotelko, *Justus Liebigs Ann. Chem.*, 1961, **642**, 121.
3. G. Seybold and H. Eilingsfeld, *Liebigs Ann. Chem.*, 1979, 1271.
4. F. Fabra, E. Fos and J. Vilarrasa, *Tetrahedron Lett.*, 1979, 3179.
5. A. N. Kost, G. A. Golubeva and R. G. Stepanov, *Zh. Obshch. Khim.*, 1962, **32**, 2240.
6. J. Stephanidou-Stephanatou, *Synthesis*, 1985, 296.
7. J. Elguero, R. Jaquier and D. Tizané, *Tetrahedron*, 1971, **27**, 123.
8. M. J. Kornet, *J. Pharm. Sci.*, 1978, **67**, 1471.
9. W. H. Pirkle and D. J. Hoover, *J. Org. Chem.*, 1980, **45**, 3407.

10. C. Verkoyen and P. Rademacher, *Chem. Ber.*, 1984, **117**, 1659.
11. J. M. Mellor and N. M. Smith, *J. Chem. Soc., Perkin Trans. 1*, 1984, 2927.
12. H. R. Snyder, C. B. Thompson and R. L. Hinman, *J. Am. Chem. Soc.*, 1952, **74**, 2009.
13. A. V. El'tsov and M. Z. Girshovich, *Zh. Org. Khim.*, 1967, **3**, 1332.
14. M. Hartmann and L. Panizzon, *Helv. Chim. Acta*, 1938, **21**, 1692.
15. E. F. Godefroi, *J. Org. Chem.*, 1968, **33**, 860.
16. M. W. Anderson, R. C. F. Jones and J. Saunders, *J. Chem. Soc., Chem. Commun.*, 1982, 282.
17. A. V. El'tsov, *Zh. Org. Khim.*, 1965, **1**, 1112.
18. H. A. Bates, N. Condulis and N. L. Stein, *J. Org. Chem.*, 1986, **51**, 2228.
19. D. Huber, G. Leclerc and G. Andermann, *Tetrahedron Lett.*, 1986, **27**, 5731.
20. J. Elks, D. F. Elliott and B. A. Hems, *J. Chem. Soc.*, 1944, 629.
21. A. H. Cook, I. Heilbron and A. L. Levy, *J. Chem. Soc.*, 1947, 1598.
22. S. Sueyoshi and I. Suzuki, *Chem. Pharm. Bull.*, 1975, **23**, 2767.
23. H. Lund and J. Simonet, *C. R. Hebd. Seances Acad. Sci., Ser. C*, 1973, **277**, 1387.
24. R. R. Schmidt, III, *Acc. Chem. Res.*, 1986, **19**, 250.
25. T. L. Gilchrist, D. Hughes and R. Wasson, *Tetrahedron Lett.*, 1987, **28**, 1573.
26. C. Kaneko, T. Tsuchiya and H. Igeta, *Chem. Pharm. Bull.*, 1974, **22**, 2894.
27. E. Testa and L. Fontanella, *Farmaco, Ed. Sci.*, 1971, **26**, 950.
28. H. Lund and I. Tabaković, *Adv. Heterocycl. Chem.*, 1984, **36**, 308.
29. R. S. Shadbolt and T. L. V. Ulbricht, *J. Chem. Soc. C*, 1968, 733.
30. P. J. Elvinga, S. J. Pace and J. E. O'Reilly, *J. Am. Chem. Soc.*, 1973, **95**, 647.
31. P. Martigny and H. Lund, *Acta Chem. Scand., Ser. B*, 1979, **33**, 575.
32. C. Kashima, M. Shimizu, A. Katoh and Y. Omote, *J. Heterocycl. Chem.*, 1984, **21**, 441.
33. H. Lund, *Adv. Heterocycl. Chem.*, 1970, **12**, 274.
34. C. Kashima, K. Harada, A. Katoh, M. Shimizu and Y. Omote, *Nippon Kaguku Kaishi*, 1987, 1194.
35. S. J. Hannon, N. G. Kundu, R. P. Hertzberg, R. S. Blatt and C. Heidelberger, *Tetrahedron Lett.*, 1980, **21**, 1105.
36. R. D. Chambers, W. K. R. Musgrave and P. G. Urben, *J. Chem. Soc., Perkin Trans. 1*, 1974, 2584.
37. H. I. X. Mager and W. Berends, *Recl. Trav. Chim. Pays-Bas*, 1960, **79**, 282.
38. K. Soai, H. Hayashi, A. Shinozaki, H. Umebayashi and Y. Yamada, *Bull. Chem. Soc. Jpn.*, 1987, **60**, 3450.
39. L. Claisen, *Ber. Dtsch. Chem. Ges.*, 1891, **24**, 3900.
40. D. N. McGregor, U. Corbin, J. E. Swigor and L. C. Cheney, *Tetrahedron*, 1969, **25**, 389.
41. G. Bianchi and M. DeAmici, *J. Chem. Soc., Chem. Commun.*, 1978, 962.
42. S. Auricchio, S. Morrocchi and A. Ricca, *Tetrahedron Lett.*, 1974, 2793.
43. H. A. Albrecht, J. F. Blount, F. M. Konzelmann and J. T. Plati, *J. Org. Chem.*, 1979, **44**, 4191.
44. G. Stork, M. Ohashi, H. Kamachi and H. Kakisawa, *J. Org. Chem.*, 1971, **36**, 2784.
45. P. G. Baraldi, A. Barco, S. Benetti, G. P. Pollini and D. Simoni, *Synthesis*, 1987, 857.
46. I. Surov and H. Lund, *Acta Chem. Scand., Ser. B*, 1986, **40**, 831.
47. M. Kijima, Y. Nambu and T. Endo, *J. Org. Chem.*, 1985, **50**, 1140.
48. J. G. Keay, *Adv. Heteocycl. Chem.*, 1986, **39**, 38.
49. A. L. Khurana and A. M. Unrau, *Can. J. Chem.*, 1975, **53**, 3011.
50. A. Alberola, A. M. González, M. A. Laguna and F. J. Pulido, *J. Org. Chem.*, 1984, **49**, 3423.
51. A. Alberola, A. M. González, M. A. Laguna and F. J. Pulido, *Synthesis*, 1984, 510.
52. A. P. Kozikowski, *Acc. Chem. Res.*, 1984, **17**, 410.
53. V. Jäger, W. Schwab and V. Buss, *Angew. Chem., Int. Ed. Engl.*, 1981, **20**, 601.
54. V. Jäger, H. Grund, V. Buss, W. Schwab, I. Müller, R. Schohe, R. Franz and R. Ehrler, *Bull. Soc. Chim. Belg.*, 1983, **92**, 1039.
55. V. Jäger, I. Müller and E. F. Paulus, *Tetrahedron Lett.*, 1985, **26**, 2997.
56. T. Kametani, T. Nagahara, Y. Suzuki, S. Yokohama, S.-P. Huang and M. Ihara, *Heterocycles*, 1980, **14**, 403.
57. K. F. Burri, R. A. Cardone, W. Y. Chen and P. Rosen, *J. Am. Chem. Soc.*, 1978, **100**, 7069.
58. D. P. Curran and C. J. Fenk, *Tetrahedron Lett.*, 1986, **27**, 4865.
59. J. Wityak, S. J. Gould, S. J. Hein and D. A. Keszler, *J. Org. Chem.*, 1987, **52**, 2179.
60. G. Shaw, *J. Chem. Soc.*, 1950, 720; 1951, 1017.
61. H. Böshagen and E. Schaufstätter, *Angew. Chem.*, 1960, **72**, 1000.
62. F. Korte and O. Behner, *Justus Liebigs Ann. Chem.*, 1959, **621**, 51.
63. N. G. Gaylord and D. J. Kay, *J. Am. Chem. Soc.*, 1956, **78**, 2167.
64. B. Woodward and H. Asai, *Yakugaku Zasshi*, 1971, **91**, 425, 436.
65. Pliva Tvornica Farmaceut., *Fr. Pat.* 2 187 761 (1974) (*Chem. Abstr.*, 1974, **81**, 13 101f).
66. K. K. Knapp, P. C. Keller and J. V. Rund, *J. Chem. Soc., Chem. Commun.*, 1978, 971.
67. K. Konuma, H. Hasegawa and K. Itabashi, *Nippon Kagaku Kaishi*, 1986, 1771 (*Chem. Abstr.*, 1987, **107**, 96 634m).
68. E. Vedejs and J. W. Grissom, *J. Am. Chem. Soc.*, 1988, **110**, 3238.
69. D. Haidukewych and A. I. Meyers, *Tetrahedron Lett.*, 1972, 3031.
70. P. Truitt and J. Chakravarty, *J. Org. Chem.*, 1970, **35**, 864.
71. H. Hoffmann, R. Wagner and J. Uhl, *Chem. Ber.*, 1971, **104**, 2134.
72. C. D. Lunsford, R. P. Mays, J. A. Richman, Jr., and R. S. Murphy, *J. Am. Chem. Soc.*, 1960, **82**, 1166.
73. H. Takahashi, N. Yamada, K. Higashiyama and K. Kawai, *Chem. Pharm. Bull.*, 1985, **33**, 84.
74. S. Kojima, *Nippon Kagaku Kaishi*, 1954, **57**, 371.
75. O. Wichterle and M. Kolinský, *Chem. Listy*, 1953, **47**, 1787 (*Chem. Abstr.*, 1955, **49**, 201h).
76. C. Hippeli and H.-U. Reissig, *Synthesis*, 1987, 77.
77. H. E. Winberg, *US Pat.* 2 628 978 (1953) (*Chem. Abstr.*, 1954, **48**, P1429f).
78. E. J. T. Chrystal, T. L. Gilchrist and W. Stretch, *J. Chem. Res. (S)*, 1987, 180.
79. A. Defoin, H. Fritz, G. Geffroy and J. Streith, *Tetrahedron Lett.*, 1986, **27**, 4727.

80. H. Iida, Y. Watanabe and C. Kibayashi, *J. Am. Chem. Soc.*, 1985, **107**, 5534.
81. A. I. Meyers, in 'Heterocycles in Organic Synthesis', Wiley, New York, 1974, p. 201.
82. M. Sainsbury, in 'Comprehensive Heterocyclic Chemistry', ed. A. R. Katritzky and C. W. Rees, Pergamon Press, Oxford, 1984, vol. 3, p. 995.
83. S. O. Wintrop and L. G. Humber, *J. Org. Chem.*, 1961, **26**, 2834.
84. S. P. Gupta, S. S. Chatterjee, P. C. Jain and N. Anand, *Synthesis*, 1974, 660.
85. W. Schäfer and I. Geyer, *Tetrahedron*, 1972, **28**, 5261.
86. P. J. Sinclair, D. Zhai, J. H. Reibenspies and R. M. Williams, *J. Am. Chem. Soc.*, 1986, **108**, 1103; R. M. Williams, D. Zhai and P. J. Sinclair, *J. Org. Chem.*, 1986, **51**, 5021; R. M. Williams, P. J. Sinclair, D. Zhai and D. Chen, *J. Am. Chem. Soc.*, 1988, **110**, 1547; D. Zhai, W. Zhai and R. M. Williams, *J. Am. Chem. Soc.*, 1988, **110**, 2501.
87. K. R. H. Wooldridge, *Adv. Heterocycl. Chem.*, 1972, **14**, 18.
88. D. Suciu, *J. Prakt. Chem.*, 1972, **314**, 961.
89. G. M. Clarke and P. Sykes, *J. Chem. Soc. C*, 1967, 1411.
90. H. Singh and K. Lal, *J. Chem. Soc., Perkin Trans. 1*, 1972, 1799.
91. M. D. Bachi, *J. Chem. Soc., Perkin Trans. 1*, 1972, 310.
92. V. G. Ermolaeva and M. N. Shchukina, *Zh. Obshch. Khim.*, 1963, **33**, 2716.
93. G. A. Kraus and J. O. Nagy, *Tetrahedron Lett.*, 1981, **22**, 2727.
94. S. Hoff and A. P. Blok, *Recl. Trav. Chim. Pays-Bas*, 1974, **93**, 18.
95. V. O. Muzychenko and B. S. Zimenkovskii, *Farm. Zh. (Kiev)*, 1985, 37 (*Chem. Abstr.*, 1986, **105**, 42 700v).
96. E. J. Corey and D. L. Boger, *Tetrahedron Lett.*, 1978, 5.
97. G. Liso, G. Trapani, A. Reho and A. Latrofa, *Synthesis*, 1985, 288.
98. S. Hünig, D. Scheutzow, H. Schlaf and H. Quast, *Justus Liebigs Ann. Chem.*, 1972, **765**, 110.
99. T. Ueda, Y. Shibita, J. Sakakibara, M. Inoue and T. Ishida, *Chem. Pharm. Bull.*, 1982, **30**, 3424.
100. J.-P. Pradère, J.-C. Rozé and G. Duguay, *J. Chem. Res. (S)*, 1982, 72.
101. R. R. Schmidt, III and D. Schwille, *Chem. Ber.*, 1969, **102**, 269.
102. D. Bourgoin-Lagay and R. Boudet, *C. R. Hebd. Seances Acad. Sci., Ser. C*, 1967, **264**, 1304.
103. M. Wilhelm and P. Schmidt, *J. Heterocycl. Chem.*, 1969, **6**, 635.
104. K. H. Shah, B. D. Tilak and K. Venkataraman, *Proc. Indian Acad. Sci., Sect. A*, 1948, **28**, 142.
105. W. J. Richter, *J. Org. Chem.*, 1981, **46**, 5119.
106. A. Mori, K. Ishihara and H. Yamamoto, *Tetrahedron Lett.*, 1986, **27**, 987.
107. M. Takasu, Y. Naruse and H. Yamamoto, *Tetrahedron Lett.*, 1988, **29**, 1947.
108. A. Darchen, P. Berthelot, C. Vaccher, M. N. Viana, M. Debaert and J. L. Burgot, *J. Heterocycl. Chem.*, 1986, **23**, 1603.
109. M. Barbero, S. Cadmuro, M. Ceruti, I. Degani, R. Fochi and V. Regondi, *Gazz. Chim. Ital.*, 1987, **117**, 227.
110. U. Jordis, *J. Chem. Res. (S)*, 1986, 432.
111. H. Lind and H. Kristinsson, *Synthesis*, 1974, 198.
112. T. Isida, T. Akiyama, M. Mihara, S. Kozima and K. Sisido, *Bull. Chem. Soc. Jpn.*, 1973, **46**, 1250.
113. L. I. Rudaya and A. V. El'tsov, *Zh. Org. Khim.*, 1970, **10**, 2142.
114. G. Doleschall, *Tetrahedron Lett.*, 1980, **21**, 4183.
115. J. Daunis, Y. Guindon, R. Jacquier and P. Viallefont, *Bull. Soc. Chim. Fr.*, 1971, 3296.
116. J. A. Maddison, P. W. Seale, E. P. Tiley and W. K. Warburton, *J. Chem. Soc., Perkin Trans. 1*, 1974, 81.
117. M. Ruccio, N. Vivona and G. Cusmano, *J. Heterocycl. Chem.*, 1974, **11**, 829.
118. Y. Royer, M. Sélim and P. Rumpf, *Bull. Soc. Chim. Fr.*, 1973, 1060.
119. C. Malavaud, M.-T. Boisden and J. Barrans, *Bull. Soc. Chim. Fr.*, 1973, 2996.
120. A. Angeli, *Gazz. Chim. Ital.*, 1893, **23** (i), 417.
121. V. Ceré, D. DalMonte and E. Sandri, *Tetrahedron*, 1972, **28**, 3271.
122. G. V. Boyd and A. J. H. Summers, *J. Chem. Soc. C*, 1971, 2311.
123. R. K. Grashey and K. Adelsberger, *Angew. Chem., Int. Ed. Engl.*, 1962, **1**, 267.

3.9
Hydrozirconation of C=C and C≡C and Hydrometallation by Other Metals

JAY A. LABINGER
California Institute of Technology, Pasadena, CA, USA

3.9.1 INTRODUCTION

Hydrometallation, as represented by equation (1), is the addition of a metal–hydrogen bond across an unsaturated group to yield an alkyl metal intermediate or product. From the point of view of organic synthesis, the latter is of interest as a nonstabilized carbanion equivalent, suitable for addition to C=X (see Volume 1, Part 1) or C=C (Volume 4, Chapter 1.6) π-bonds, coupling reactions (Volume 3, Part 2), oxidation (Volume 7, Chapter 4.2) and other transformations; and hydrometallation is just an alternative route to these reagents. Although hydrometallation virtually always participates as a key step in the overall mechanism of catalytic hydrogenation (as well as related catalytic processes, such as hydroformylation), its major applications are not found in the realm of reductions, except for special cases such as stereo- and regio-specific isotopic labelling.

$$\mathrm{M{-}H} + \mathrm{C{=}C} \longrightarrow \underset{\mathrm{M}\quad\;\mathrm{H}}{\mathrm{C{-}C}} \tag{1}$$

There are basically two reasons for making hydrometallation the route of choice: if an alkene or alkyne is the most convenient (accessible, cheap) starting material, or if alternative methods for generating the organometallic reagent are incompatible with other functional groups and/or the key synthetic goals. The classical preparation of lithium or Grignard reagents from organic halides and the corresponding metals will generally not be viable when there are other readily reducible groups present, or when stereochemistry at the reacting center is crucial.

The ideal system for hydrometallation would have the following attributes: a conveniently prepared starting reagent (the metal hydride); favorable kinetics and thermodynamics and high selectivity (in all its aspects) with all alkenes and alkynes that might be used as substrates; and a resulting organometallic reagent that is stable and highly versatile for subsequent transformation, exhibiting the full spectrum of organometallic reactivity. Obviously, there is no single system which is going to satisfy all of these criteria. However, two tactics may overcome the limitations of a single metal: transmetallation and catalyzed hydrometallation. These are represented in Scheme 1. In the first, M_1H readily undergoes hydrometallation, but M_1R cannot be used for the desired transformation. Hence the organic group is transferred to M_2, which can undergo the reaction but which has no suitable direct hydrometallation route. In the second, M_2 has both a suitable hydrometallation precursor and the desired organometallic reagent versatility, but direct hydrometallation does not proceed readily. In such cases addition of a catalytic amount of an M_1 compound may achieve the target organometallic reagent synthesis. Hydroalumination is a prime representative. These cases are discussed in Sections 3.9.3.4.2 and 3.9.4.2, respectively.

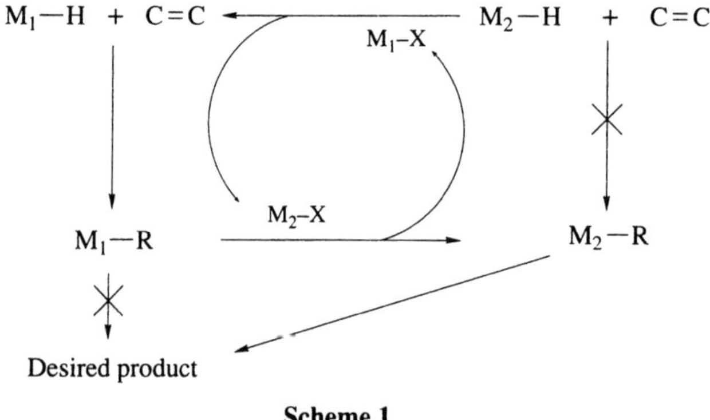

Scheme 1

Until around 15 years ago, hydrometallation was essentially limited to main group examples; for transition metals, the existence of appropriate hydrides, the stability of the organometallic reagents derived therefrom, and their utility for organic transformations were all considered questionable at best. This situation changed completely in the mid 1970s, when Schwartz and coworkers developed hydrozirconation. Here, for the first time, a stable and (moderately) well-characterized transition metal hydride could be added to unsaturated hydrocarbons to produce stable and well-characterized organometallic compounds, which have subsequently been demonstrated to effect a wide variety of synthetic transformations, either directly or following transfer to a second metal center.[1] In fact, many transition metal hydrides are better hydrometallation reagents: M_1 in Scheme 1 is nearly always a transition metal.

Throughout this chapter we will use the abbreviation zr = Cp_2ZrCl—; thus zrR = Cp_2ZrRCl.

3.9.2 MECHANISTIC ASPECTS

Generally the detailed mechanism will include steps such as precoordination of the alkene, dissociation of an auxiliary ligand, *etc.* Bercaw has noted that most commonly the actual hydrometallation step is not rate determining.[2] Since the most important hydrometallation reagent is an insoluble polymer, it is easy to see why mechanistic understanding has not kept pace with applications in this field.

3.9.2.1 Thermodynamic Considerations

Scheme 2 represents the key equilibria in hydrometallation. Note that: (i) association of the alkene (or alkyne) with the metal hydride (**1**) requires a vacant coordination site in the latter, whether (**2**) is an actual intermediate or merely a transition state. Since coordinatively unsaturated species such as (**1**) are rarely stable, a preliminary dissociation of an oligomer, or of an extra ligand, will be required. The latter may well be the limiting step in a given system; (ii) if metal–alkene bonding is strong, (**2**) can be such a stable intermediate that hydrometallation does not take place; on the other hand, if the metal–alkene interaction is very weak (**2**) may never form at all. In either case the organometallic reagent (**3**) will not be obtained. A delicate balance may thus be essential.

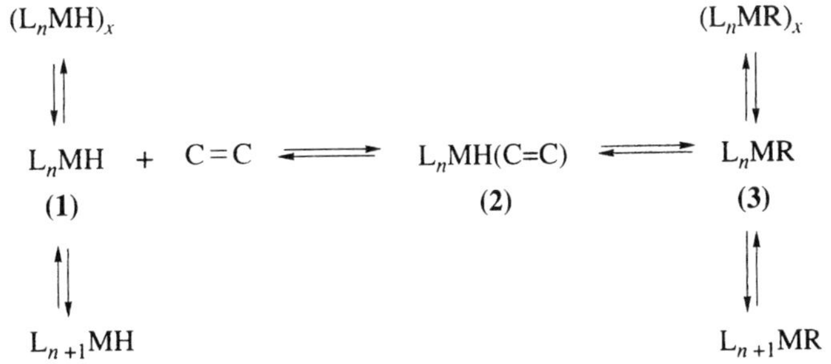

$$
\begin{array}{ccccc}
(L_nMH)_x & & & & (L_nMR)_x \\
\updownarrows & & & & \updownarrows \\
L_nMH + C{=}C & \rightleftharpoons & L_nMH(C{=}C) & \rightleftharpoons & L_nMR \\
(\mathbf{1}) & & (\mathbf{2}) & & (\mathbf{3}) \\
\updownarrows & & & & \updownarrows \\
L_{n+1}MH & & & & L_{n+1}MR
\end{array}
$$

Scheme 2

Schwartz's reasoning for optimizing these thermodynamic considerations led to the development of hydrozirconation.[1] Hydride complexes of the late transition metals do not in general exhibit the hydrometallation reaction, probably because the alkene complexes are too stable. This may be understood from the Dewar–Chatt–Duncanson model for alkene bonding, wherein back donation of metal *d*-electrons to the alkene π^*-orbital is a major contributor. For metal centers with d^0-electron configurations, there should be substantial stabilization of (**3**) with respect to (**2**). Such metals are only found towards the left end of the Periodic Table, particularly Groups IIIA to VA.

Compare equations (2) and (3). Niobium complex (**4**) is a d^2-metal center: back bonding is important, and hydrometallation does not proceed, even when an additional ligand PR_3 is provided. Only when another good π-acceptor ligand (CO) is added does the alkenyl product (**5**) form.[3] In contrast, zirconium complex (**6**) is d^0, and here hydrometallation proceeds directly; no intermediate alkyne complex can be detected.[4] This qualitative picture is reinforced by theoretical studies on the 'bent metallocene' structure.[5] As shown in Figure 1, for a d^0-complex the two highest-filled orbitals are essentially unchanged or stabilized during hydrometallation. However, for a d^2-complex the next orbital, labelled ($1a_1 + \pi^*$), is also occupied, and it is strongly destabilized by hydrometallation; Hoffmann estimates that the thermodynamics are about 50 kcal mol^{-1} worse for d^2 than for d^0 (1 cal = 4.2 J). Note that inclusion of a good π-acceptor ligand such as CO would restabilize the $1a_1$ orbital and cancel most or all of this unfavorable factor.

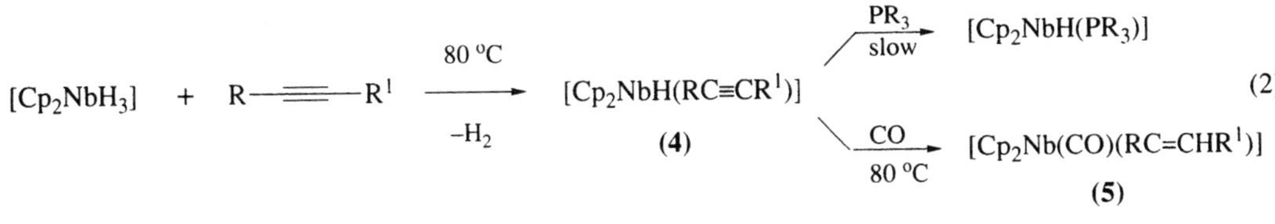

$$
[Cp_2NbH_3] + R{-}{\equiv}{-}R^1 \xrightarrow[-H_2]{80\,^\circ C} \underset{(\mathbf{4})}{[Cp_2NbH(RC{\equiv}CR^1)]} \begin{cases} \xrightarrow[\text{slow}]{PR_3} [Cp_2NbH(PR_3)] \\ \xrightarrow[80\,^\circ C]{CO} \underset{(\mathbf{5})}{[Cp_2Nb(CO)(RC{=}CHR^1)]} \end{cases} \quad (2)
$$

$$
\underset{(\mathbf{6})}{[Cp_2ZrHCl]} + R{-}{\equiv}{-}R^1 \xrightarrow{25\,^\circ C} [Cp_2ZrCl(RC{=}CHR^1)] \quad (3)
$$

Thermodynamic analysis of the overall hydrometallation process is straightforward: using the cycle of Scheme 3 we get the following expression for equation (4): $\Delta H^0 \approx D(\text{M—H}) - D(\text{M—R}) - X$, where $X \approx$ 36–38 kcal mol^{-1}.[6,7] Further estimating ΔS^0 at –27 to –30 eu, it may be concluded that equation (4) is thermodynamically favored as long as the metal–hydride bond is no more than 27–30 kcal mol^{-1} stronger than the metal–alkyl bond.

A survey of available thermochemical data suggests that this requirement is nearly always satisfied, although there are some cases that are close. The difference between $[CpMo(CO)_3H]$ and $[CpMo(CO)_3Et]$ has been estimated to be around 29 kcal mol^{-1};[8] clean hydrometallation with this system has not been

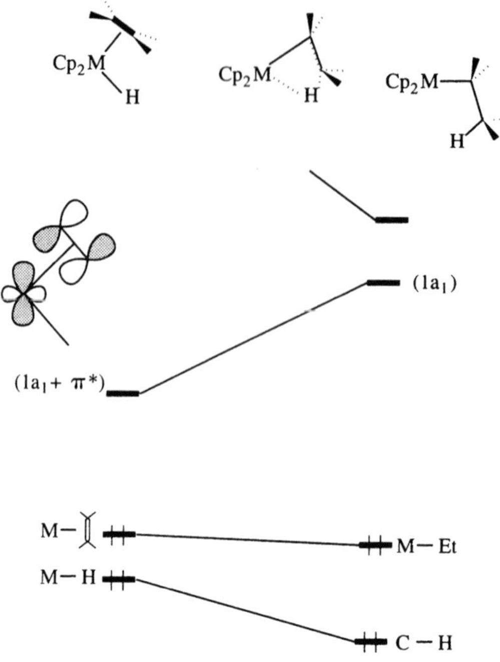

Figure 1 Orbital correlation diagram for hydrometallation in bent metallocene complexes (reproduced from ref. 5 by permission of the American Chemical Society)

$$L_nM{-}H \longrightarrow L_nM\bullet + H\bullet$$
$$H\bullet + C_2H_4 \longrightarrow Et\bullet$$
$$Et\bullet + L_nM\bullet \longrightarrow L_nM{-}Et$$
$$L_nM{-}H + C_2H_4 \longrightarrow L_nM{-}Et \qquad (4)$$

Scheme 3

demonstrated. Actinides[9] and early transition metals[10] exhibit smaller differences between M—H and M—R bond strengths than do the middle and late transition metals. This important trend has been attributed to bond polarity[9,11] or to repulsions between filled metal *d*-orbitals and alkyl group orbitals in the latter group.[10] We thus have a second thermodynamic reason for expecting early transition metals to provide superior hydrometallation reagents.

For hydrometallation of alkynes, similar analysis shows that *D*(M—H) – D(M—vinyl) need only be less than *ca.* 37 kcal mol^{-1}; metal–vinyl bonds are generally stronger than metal–alkyl bonds,[6,9] so this should be easy to achieve. Wayland has concluded that on thermodynamic grounds '...virtually any transition metal should be capable of accomplishing this reaction'.[7]

As for competing formation of (**2**), we calculate that (**2**) will be more stable than (**3**) when the strength of the metal–alkene interaction is greater than the quantity {36 kcal mol^{-1} – [*D*(M—H) – *D*(M—R)]}. Unfortunately, data for absolute metal–alkene bond energies are virtually nonexistent. Some estimates can be made from thermochemical data and (possibly questionable) approximations: $D(L_2Pt{-}C_2H_4) \approx$ 73 kcal mol^{-1},[12] $D(Fe(CO)_4{-}C_2H_4) \approx$ 28 kcal mol^{-1};[13] or from theoretical calculations $D(Cl_3Pt^-{-}C_2H_4)$ ≈ 28 kcal mol^{-1}.[14] In general (especially for later transition metals) *D*(M—H) is considerably greater than *D*(M—C); hence stability of (**2**) will present a problem unless the metal–alkene bond is considerably weaker than in these typical cases. Again we see the important advantage for d^0-metals.

For main group metals, which exhibit relatively small differences between M—H and M—C bond strengths and no strong metal–alkene interactions, the thermodynamics of hydrometallation should be even more favorable than for early transition metals. This does not appear to be the case, especially for hydroalumination (Volume 8, Chapter 3.11). The reason is almost certainly the additional stability of the metal hydride reagent conferred by aggregation: organoaluminum hydrides exist as rather tight dimers, and the bridging bond energy is enough to shift the equilibrium away from the hydrometallation side in many cases.[15]

Similar arguments apply to selectivity issues. Hydrometallation should be preferred for an alkyne over an alkene; for a terminal over an internal double bond; and for anti-Markovnikov addition. Equation (5) is never thermodynamically preferred over hydrometallation of an alkene;[7] however, hydrometallation in the opposite sense to give an alkoxide (equation 6) will be preferred if the metal–oxygen bond is at least 24 kcal mol^{-1} stronger than the metal-carbon bond. The latter condition is probably not generally satisfied for later transition metals,[16] but surely is for the more oxophilic early transition metals[10,17] as well as main group metals (with the latter, even saturated C—O functional groups may not be tolerated). For terminal alkynes, elimination of H_2 to form an alkynyl complex (equation 7) will be favored over hydrometallation if the metal–alkynyl bond is at least 35 kcal mol^{-1} stronger than the metal–vinyl bond, a situation which could arise:[12,18] a difference as large as 38 kcal mol^{-1} was reported for a uranium system.[9] For all these issues it must be kept in mind, of course, that kinetic factors can and usually do have more impact on selectivity than thermodynamic factors. We turn to that aspect next.

$$\text{M–H} + \text{Me}_2\text{C=O} \longrightarrow \text{M–CMe}_2\text{–OH} \quad (5)$$

$$\text{M–H} + \text{Me}_2\text{C=O} \longrightarrow \text{Me}_2\text{CH–O–M} \quad (6)$$

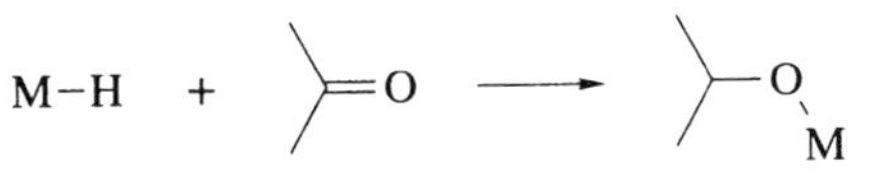

$$\text{M–H} + \text{R–C≡CH} \longrightarrow \text{M–C≡C–R} + \text{H}_2 \quad (7)$$

3.9.2.2 Detailed Mechanism of Hydrometallation

A general picture for the mechanism is shown in Scheme 4, which is based upon a theoretical analysis by Thorn and Hoffmann.[19] Here distinction between (**2**) and (**2a**) reflects the general assumption, supported by calculations, that the insertion step requires the M—H and C═C groups to be *cis* and coplanar, which need not be the case for the first-formed and/or thermodynamically most stable alkene complex (**2**). Thorn and Hoffmann conclude that most or all metal hydrides will have some pathway that leads to hydrometallation without a large kinetic barrier, so long as none of the key intermediates along the way is too stable. The same inference was drawn for the bent metallocene systems discussed earlier (Figure 1): a kinetic barrier to insertion, found only for the d^2-cases, is a consequence of the thermodynamic stabilization of alkene complex (**2**).[5]

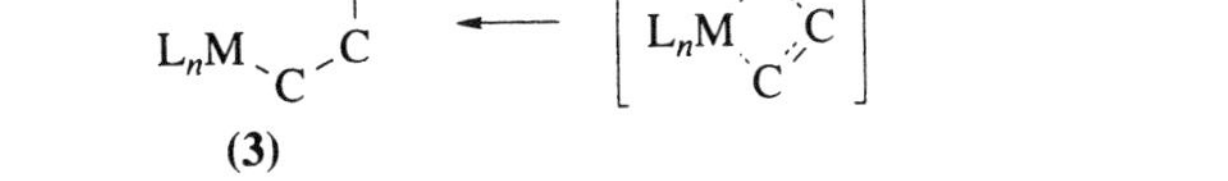

$$\underset{(\mathbf{1})}{\text{L}_n\text{MH}} + \text{C=C} \longrightarrow \underset{(\mathbf{2})}{\text{L}_n\text{MH (C=C)}} \longrightarrow \underset{(\mathbf{2a})}{\text{L}_n\text{M(H)(C=C)}} \longrightarrow \left[\text{L}_n\text{M}\cdots\text{H}\cdots\text{C}\cdots\text{C}\right]^{\ddagger} \longrightarrow \underset{(\mathbf{3})}{\text{L}_n\text{M–C–C–H}}$$

Scheme 4

When the barrier to hydrometallation is solely a consequence of the strength of the metal–alkene interaction, the implication is unfavorable for selectivity. Suppose the kinetic scheme of equation (8) operates; then the rate for formation of (**3**) from a given alkene will be related to the product of the constants k and K for that alkene. However, factors which tend to make K larger (more stable alkene complex) will make k smaller and thus tend to cancel out, resulting in little kinetic differentiation between substrates.

$$\underset{(\mathbf{1})}{\text{L}_n\text{MH}} + \text{C=C} \underset{}{\overset{K}{\rightleftharpoons}} \underset{(\mathbf{2})}{\text{L}_n\text{MH(C=C)}} \overset{k}{\longrightarrow} \underset{(\mathbf{3})}{\text{L}_n\text{MR}} \quad (8)$$

This is in fact the result found for hydrometallation of substituted styrenes by $[RhH_2Cl(PR_3)_3]$ (equation 9).[20] K and k vary inversely as the substituent X is changed, so the overall rates are fairly constant.

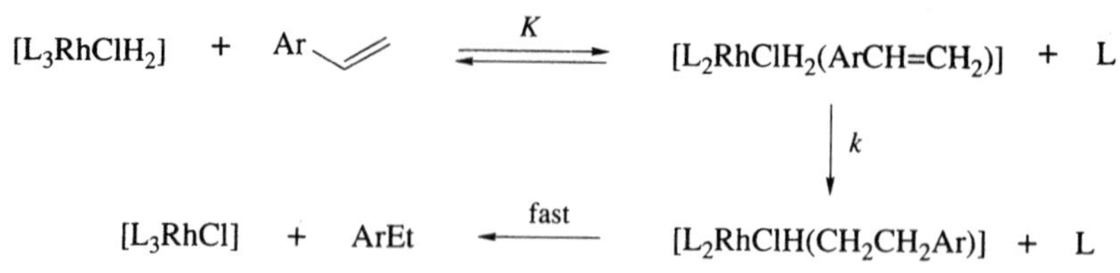

$$[L_3RhClH_2] + ArCH{=}CH_2 \underset{}{\overset{K}{\rightleftharpoons}} [L_2RhClH_2(ArCH{=}CH_2)] + L \quad (9)$$

$$[L_2RhClH_2(ArCH{=}CH_2)] \xrightarrow{k} [L_2RhClH(CH_2CH_2Ar)] + L$$

$$[L_2RhClH(CH_2CH_2Ar)] + L \xrightarrow{\text{fast}} [L_3RhCl] + ArEt$$

A quite different case is presented by the bent metallocene systems, $[Cp_2MH(alkene)]$ where M = Nb or Ta.[2,21] Here the metal–alkene complex is relatively quite stable, so much so that the kinetics of formation of a stable alkyl metal complex cannot be easily studied. Nonetheless it is possible to get a handle on the hydrometallation kinetics by means of dynamic NMR methods (Table 1). The increased rate for propene *versus* ethylene results from both steric destabilization of the ground state and electronic stabilization of the transition state for the former. The first is typical of alkene complexes; the second implies that some partial positive charge develops at the β-carbon during the hydrometallation process, also seen in the trend for substituted styrenes.

Table 1 Kinetic Parameters (at 50 °C) for Hydrometallation in Complexes $[Cp^*{}_2NbH(CH_2CHR)]^2$

R	*k* (s^{-1})	$\Delta G^{\ddagger}$ (kcal mol^{-1})
H	2.62	18.3
Me	890	14.6
Ph	3.18	18.2
p-$Me_2NC_6H_4$	6.80	17.7
p-$MeOC_6H_4$	4.81	17.9
p-MeC_6H_4	3.47	18.1
p-$CF_3C_6H_4$	0.91	19.4

Two conclusions follow from the apparent positive charge build-up. First, anti-Markovnikov addition is favored kinetically as well as thermodynamically, which means that the opposite regiochemistry will be hard to achieve; second, the hydrogen atom behaves more as hydride, H^-, during its migration to the carbon atom. If this accurately represents the chemical nature of the metal–hydrogen bond in the ground state as well, it has clear implications for functional group tolerance. It is not clear to what extent these results can be applied to any other system.

For the ordinary Cp analogs of both Nb and Ta, two isomers of the (substituted alkene) hydride complex are accessible, labelled *endo* and *exo* (Figure 2).[21] These undergo reversible hydrometallation without interconverting, showing that the barrier to alkene rotation is very high and that the hydrogen can move only to the adjacent carbon atom. Such a situation offers potential for regioselectivity modification, as hydrometallation from the *exo* isomer generates the more-substituted metal alkyl (Markovnikov addition). Unfortunately, it has not proven possible in this system to trap that product with added ligand: alkene displacement dominates instead.[21]

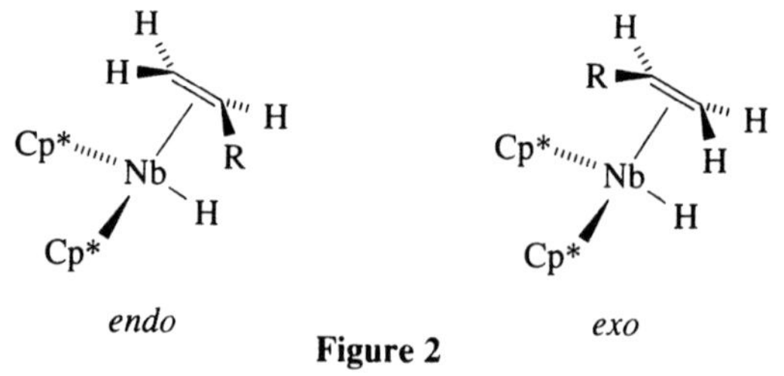

endo *exo*

Figure 2

The other questions raised above—about the relative roles of alkene complexation and hydrometallation, and the possible requirement for structural rearrangement before insertion—cannot be answered in this system, as the alkene binding and geometry are essentially immutable. Roe has shown for a rhodium complex with *trans* alkene and hydride ligands, that rearrangement does not contribute to the hydrometallation barrier: the first step in equation (10) is fast, and k_2 is rate determining.[22] It is notable that for this reaction $\Delta G^{\ddagger} \approx 12$ kcal mol^{-1}, substantially less than that found for the bent niobocene complexes above, and presumably reflecting stronger alkene binding in the latter.

$$trans\text{-}[L_2RhH(C_2H_4)] \underset{}{\overset{K}{\rightleftharpoons}} cis\text{-}[L_2RhH(C_2H_4)] \xrightarrow{k_2} [L_2RhEt] \qquad (10)$$

Summing up to this point: we have concluded that d^0 complexes are preferred for hydrometallation on thermodynamic grounds, and that in the expected absence of strong metal–alkene binding no great kinetic barriers should be anticipated. What mechanistic factors will operate, then, especially as regards selectivity issues?

Unfortunately we have very little information to draw upon. The main hydrometallation reagent, $[Cp_2ZrHCl]$ (**6**), is an insoluble polymer; its reactions are either heterogeneous or limited by the very small extent of dissolution, and detailed kinetics of hydrometallation are essentially unattainable. Several soluble complexes ($Cp^*_2MH_2$, M = Zr,[23] Hf;[24] Cp^*_2ScH[25]) hydrometallate alkenes very rapidly at low temperatures, which at least confirms the prediction that there should be no high barrier. $[Cp^*_2ZrHCl]$ has been prepared[26] but not examined as a hydrometallation reagent. The phenomenology of hydrometallation by $[Cp_2ZrHCl]$—relative reactivities, observed selectivities and the like—is presented in the following sections.

For d^0 metals, does the alkene complex (**2**) have any stable existence as an intermediate? It has commonly been thought that an alkene complex of a d^0 transition metal or main group metal (with no π back bonding component) could not have any lifetime, but would correspond only to a transition state (or at best a negligibly shallow minimum) in the energy surface. For example, gas phase studies on hydroboration were interpreted in terms of a 3-center transition state (structurally equivalent to a boron–alkene complex) as the highest point on the energy surface.[27]

While direct evidence for a somewhat stable intermediate alkene complex has never been obtained, it is possible to infer stability from rearrangement studies. One of the most striking characteristics of hydrozirconation is the facile rearrangement of products obtained from internal alkenes (Scheme 5).[28] The obvious mechanism is a sequence of reversible β-hydrogen additions and eliminations; but does this proceed *via* intermediate alkene complexes (path a) or free alkenes (path b)? Only the starting alkene oct-4-ene and the final *n*-octylzirconium product can be detected,[28] which implies that path (a) is followed: the organic group must always remain associated with the zirconium complex (unless the hydrozirconation rates for oct-3-ene and oct-2-ene are much faster than that for oct-4-ene, and there is nothing that would support this argument). Likewise, ethylene does not displace octene from complexes such as (**7**),[1] which would be anticipated if they are in equilibrium with the free alkene. (Formation of isomerized alkene was found in hydrozirconation of very long chain alkenes, but these reactions required prolonged times and somewhat elevated temperatures.[29])

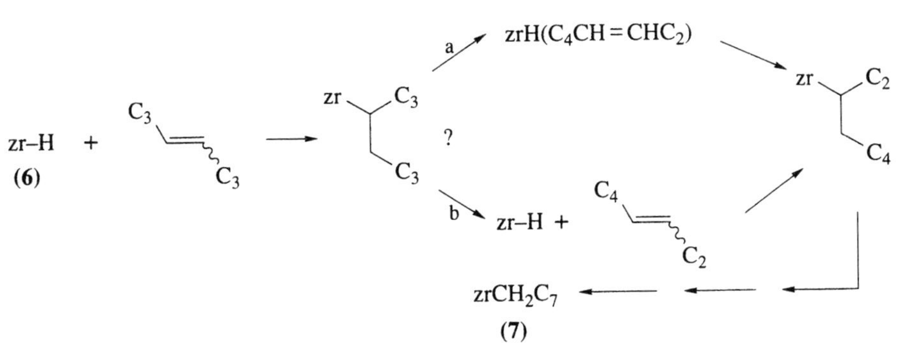

zr–H (**6**) + $C_3CH{=}CHC_3$ → zr–CH(C_3)CH_2C_3 —a→ $zrH(C_4CH{=}CHC_2)$ → zr–CH(C_2)CH_2C_4 → → $zrCH_2C_7$ (**7**)

? —b→ zr–H + $C_4CH{=}CHC_2$

Scheme 5

A related result suggests that even boron–alkene complexes have some stability: the rearrangement shown in equation (11) leads mainly to the *cis* isomer (**8**), whereas hydroboration of 2-methylmethylenecyclohexane under the same conditions gives mostly the *trans* isomer.[30] This shows that the B—H group adds most often to the same face of the double bond from which it is eliminated, and boron thus must remain associated with the methylenecyclohexane. We may conclude that alkene complexes of d^0 metals are in fact viable.

Before leaving this section we briefly examine hydrometallation mechanisms completely different from the 'classic' one depicted in Scheme 2. One class involves metal hydrides which do not have a readily available vacant site and requires conjugated alkenes — 1,3-dienes, styrenes; these proceed *via* a

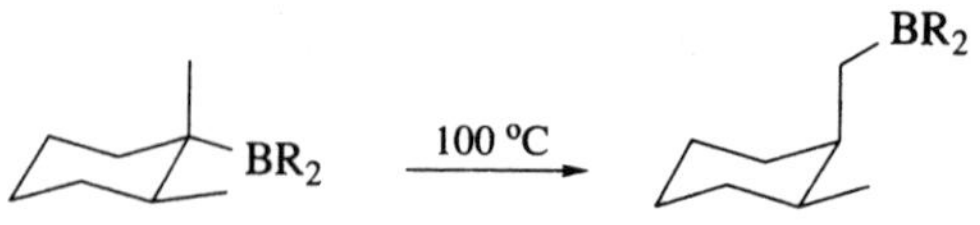

(11)

(8)

free radical pathway, as in Scheme 6.[31,32] More frequently this leads to alkene hydrogenation rather than to a stable hydrometallation product, as indicated by the left hand pathway in Scheme 6; many of the systems do not meet the criteria for favorable hydrometallation thermodynamics discussed earlier. The requirement for conjugation reflects the need to generate a relatively stable alkyl radical; it has been suggested that the regioselectivity of styrene hydrometallation might allow differentiation between the classical (equation 12) and radical (equation 13) mechanisms.[32] The most unequivocal demonstration of the Scheme 6 mechanism is the observation of CIDNP effects in the NMR.[31,32]

Scheme 7 depicts a different radical mechanism; this is a chain path, unlike Scheme 6, that has been

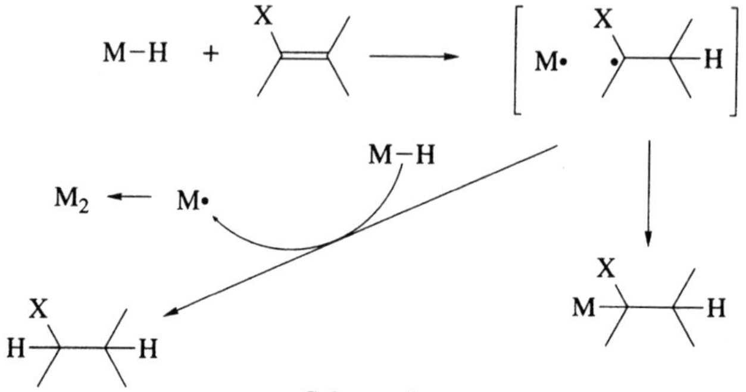

Scheme 6

(12)

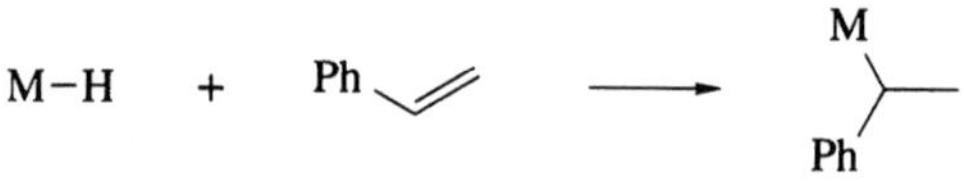

(13)

observed for main group metals such as tin but apparently not for transition metals.[32] Note that it gives the regiochemistry of equation (12), not equation (13). Still another nonclassical possibility is the transfer of hydrogen as a proton instead of a hydrogen atom; this is primarily observed in some cobalt chemistry[33] [for example equation 14, where (Co) = bis(dimethylglyoximato)cobalt and py = pyridine]. Some of these reactions may involve radical pathways as well.[34]

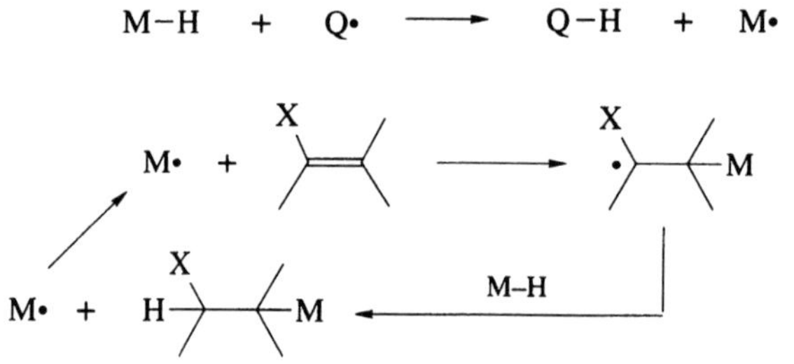

$$M{-}H + Q\bullet \longrightarrow Q{-}H + M\bullet$$

Scheme 7

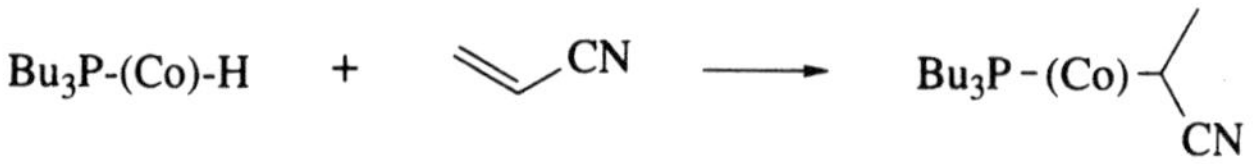

(14)

3.9.3 HYDROZIRCONATION WITH [Cp_2ZrHCl] AND RELATED COMPOUNDS

3.9.3.1 Synthesis, Structure and Properties of Reagents

The synthesis of [Cp_2ZrHCl] (**6**), along with several other zirconocene hydrides, was first reported by Wailes and Weigold in 1970.[35] Treatment of [Cp_2ZrCl_2] with one equivalent (on a hydrogen basis) of either $LiAlH_4$ or $LiAl(OBu^t)_3H$ in THF afforded high (>90%) yields of [Cp_2ZrHCl] as an insoluble white solid. Shortly afterwards, Wailes and coworkers showed that [Cp_2ZrHCl] reacts with alkenes[36] and alkynes,[37] although few of the products were completely characterized, and there was no indication of any particular synthetic value.

In 1974 Schwartz and coworkers, following the rationale sketched out above, demonstrated the potential power of hydrozirconation.[28] Preparation of the [Cp_2ZrHCl] reagent (a commercial preparation is now available) utilized a different reductant—$Na[AlH_2(OCH_2CH_2OMe)_2]$, known commercially as 'Red-Al' or 'Vitride'—from those employed by Wailes. The Red-Al procedure affords the reagent contaminated with small amounts of [Cp_2ZrH_2] and much larger amounts (as much as 30%) of NaCl.[38] Formation of [Cp_2ZrHCl] by direct hydrogenation of the dichloride has been disclosed in a patent.[39]

Purity may be important: attempts to hydrozirconate long chain alkenes failed for certain preparations, especially when $LiAlH_4$ was employed; it was suggested that aluminum-containing impurities might be responsible.[40] An improved preparation uses a filtered ether solution of $LiAlH_4$ as reductant and a wash with CH_2Cl_2 to convert over-reduced [Cp_2ZrH_2] back to [Cp_2ZrHCl]. This gives 77–92% yields and around 95% purity, with the dihydride the only significant contaminant.[41] The absence of salt impurities means that the hydrozirconation reaction mixture will contain no insolubles upon completion, which makes work-up more convenient. One other point is notable: both the absolute and relative rates of hydrozirconation for several unsaturated compounds were found to be quite different for this preparation. Some of the generalizations about mechanism and selectivity in the following sections might not prove to be universally valid, depending instead on the exact nature of the reagent used.

Yet another alternative is *in situ* preparation of [Cp_2ZrHCl], by adding reductant to a solution of [Cp_2ZrCl_2] already containing the unsaturated substrate. The reductant may be a standard hydridic reagent such as Red-Al,[29,42] or a precursor such as a *t*-butyl Grignard reagent, which presumably generates a transient *t*-butylzirconium complex that readily undergoes β-hydrogen elimination.[42] The main advantages are the elimination of the need for inert atmosphere filtration and (possibly) improved reactivity by avoiding insoluble reagents. However, in the latter case hydrozirconation of styrene or butadiene gave some (or mostly) *t*-butylated products as well.[42]

Variants include changing auxiliary ligand X in [Cp_2ZrHX], or substituting cyclopentadienyl rings. The dihydride [Cp_2ZrH_2] hydrozirconates unsaturated molecules, but the products are generally not useful reagents. Instead they undergo secondary reactions, frequently leading to overall hydrogenation. Such reactions are mechanistically complex, involving prior complexation of additional ligand[43–45] or interaction with Cp ring substituents.[46] Often reaction with additional substrate leads to products such as metallocycles; equation (15) shows one such example.[47] Similarly, although a divinylzirconium complex (from double hydrozirconation) was proposed for equation (16),[37] the correct product is (**9**).[48] This problem of instability may effectively preclude the use of [Cp_2ZrH_2] or [Cp_2ZrHR] species as synthetic reagents.

$$[Cp_2ZrH_2] + C_2H_4 \longrightarrow Cp_2Zr(\text{metallacycle}) + C_2H_6 + C_4H_8 + C_4H_{10} \quad (15)$$

$$[Cp_2ZrH_2] + Ph{-}{\equiv}{-}Ph \longrightarrow Cp_2Zr(C_4Ph_4) + PhCH{=}CHPh \quad (16)$$

(**9**)

A number of (substituted Cp) derivatives have been made, but the vast majority are the dihydrides, [Cp'_2ZrH_2]. This alleviates the problem of insolubility: [$(\eta\text{-}C_5H_4Me)_2ZrH_2$][49] and [$CpCp^*ZrH_2$][50] are both dimeric, whereas [$(\eta\text{-}1,2,4\text{-}C_5H_2Me_3)Cp^*ZrH_2$][50] and [$Cp^*_2ZrH_2$][23] are monomers. ([$Cp_2ZrHCl$]

appears to be dimeric in solution;[48] whether its extremely low solubility is due to higher aggregation in the solid state is not known.) Again, dihydrides may not be useful hydrozirconation reagents. Even where clean hydrometallation can be achieved (see Table 7), the presence of excess alkene induces reductive elimination; one would have to worry about reagents used for synthetic transformations doing the same.

Complexes of type [Cp^*_2MHX] (M = Zr, Hf; X = F, Cl, Br,[24,26] OR,[24,51,52] NR_2[52,53]) should exhibit hydrozirconation, and might offer some intriguing advantages such as mechanistic studies uncomplicated by insolubility, and selectivity modulation by steric control. On the other hand, their preparation is likely to be considerably more inconvenient and/or expensive than [Cp_2ZrHCl]. Note that reagents which lead to the precipitation of [Cp_2ZrHCl] ($LiAlH_4$, $LiAlH(OR)_3$) afford instead the dihydrides with substituted Cp.[49] [Cp^*_2ZrHCl] was not made directly, but by comproportionation of the dihydride plus dichloride[26] or the reaction of [$Cp^*_2ZrH_2$] with one equivalent of MeCl.[51] A polymer-supported reagent was obtained by chloromethylating polystyrene, displacing chloride with Cp^-, deprotonating and attaching to Zr and reducing with Red-Al, in hopes of a recyclable hydrozirconation agent. While *n*-alkenes could be hydrozirconated, the presence of uncomplexed Cp groups in the polymer complicated the stoichiometry, and recycling was not achieved.[54]

Reactions of the hafnium analog of (**6**), [Cp_2HfHCl], differ little from those of the zirconium complex.[55] It was suggested that the hafnium reagent is somewhat more sensitive to steric inhibition, as cyclohexene and related cyclic alkenes failed to react; but this might well be due to the specific preparation of the solid (see above), rather than to any inherent molecular differences. There are some changes in the stereochemical outcome of hydrometallation of bicyclic alkenes (see Section 3.9.3.3.3), but it is far from clear that hafnium will provide any features sufficient to overcome the availability and cost advantages of zirconium.

3.9.3.2 Hydrozirconation of Alkenes and Alkynes

Standard conditions for the hydrozirconation of unsaturated hydrocarbons are those originally employed:[28] the hydrocarbon is added to a suspension of a roughly equimolar amount of [Cp_2ZrHCl] in benzene or toluene, and the mixture stirred at ambient or slightly elevated temperatures for 1–3 h. Formation of the hydrozirconation product, [Cp_2ZrRCl], is signalled by dissolution of the white Zr reagent to give a yellow to orange solution. Attempted isolation of the product usually gives an oil, although some crystalline products have been obtained; more frequently characterization is limited to NMR of the solution. Even more frequently the product is not characterized at all, but used directly for subsequent transformation (see Section 3.9.3.4).

Examples of reported hydrozirconations of alkenes and alkynes are summarized in Tables 2–6. Simple alkenes appear always to react as long as they are no more than trisubstituted. The only reported failure is that of a long chain internal alkene, triacont-15-ene,[79] but such alkenes have subsequently been successfully hydrozirconated, albeit at slightly elevated temperature.[40] The order of reactivity, based primarily on qualitative observations, is: terminal alkene > internal alkene (*cis* ≈ *trans*) > exocyclic alkene > cyclic alkene ≈ trisubstituted alkene.[1] Trisubstituted cyclic olefins and tetrasubstituted olefins do not react. Representative examples are shown in Table 2.

Migration of the metal group along the carbon chain is facile; in all cases, the least substituted possible alkylzirconium product (*e.g.* **7**) is the only one obtained. Contrast hydroboration, which takes place under comparably mild conditions; but the migration of boron to the terminal position requires prolonged heating and is seldom a clean transformation (Volume 8, Chapter 3.10). Rearrangement of hydroalumination products is also slow (Volume 8, Chapter 3.11), but for zirconium it is not possible to arrest migration to the terminal position (save for special cases—aromatic or functionalized alkenes—which will be considered in Section 3.9.3.3.2).

Alkynes are universally hydrozirconated as well; the only failure in the literature is that of a perfluorinated compound, $C_7F_{15}C{\equiv}CH$.[69] Alkynes appear to be more reactive than alkenes, both by qualitative comparison and from the results on enynes (Table 6; also see Section 3.9.3.3.1). Furthermore, hydrozirconation of alkynes can compete with reduction of unsaturated functional groups such as nitriles and esters (Table 5), which is generally not true for alkenes. Dienes can be cleanly monohydrozirconated if one of the double bonds is terminal; other cases are considered in Section 3.9.3.3.1.

Hydrozirconation of unsaturated organometallics gives bimetallic compounds, such as substituted analogs of the so-called Tebbe reagent, [$Cp_2Ti(\mu\text{-}CH_2)(\mu\text{-}Cl)AlMe_2$]. (The latter has proven a versatile precursor of a 'methylene equivalent' for Wittig-type reactivity or for alkene metathesis,[80] but is only available as the simple CH_2 version.) Equation (17) gives the desired isomer (**10**) cleanly only for the bulkiest R; analogs were also prepared from alkenyl-boron and -zinc precursors. None of these, however,

Table 2 Hydrozirconation of Simple Alkenes

Alkene	*Product*	*Conditions*[a]	*Characterization*[b]	*Yield (%)*[c]	*Use*	*Comments*	*Ref.*
Oct-1-ene (also *cis*- and *trans*-oct-4-ene)	$zr(CH_2)_7Me$	S	N	96	Transfer to Al		38
Oct-1-ene				86		*In situ* preparation	42
Dec-1-ene	$zr(CH_2)_9Me$	S			Hydrocyanation		56
RCH═CHR ($R = C_nH_{2n+1}$; n = 11, 12, 13, 14, 16, 20)	$zr(CH_2)_{2n+1}Me$	40 °C/THF/50–170 h		(30–80 as RI)		*In situ* preparation	40
Cyclohexene	zrC_6H_{11}	S	N	90	Transfer to Al		38
Norbornene (and other bicyclic alkenes)	 zr	60 °C		(95 as ROH)		See Section 3.9.3.3.3 for stereochemistry	57
H_2C═$CHBu^t$	$zr(CH_2)_2Bu^t$	S	N, A	88	Transfer to Al		38
H_2C═$CHSiMe_3$	$zr(CH_2)_2SiMe_3$	S			Hydrocyanation		56
2-Methyloct-1-ene	$zrCH_2CHMe(CH_2)_5Me$	S				Deuteration	58
H_2C═$CMeCH_2Bu^t$	$zrCH_2CHMeCH_2Bu^t$						59
2-Methylbut-2-ene	$zr(CH_2)_2Pr^i$	S		(100 as RBr)			28
Methylenecyclohexane	$zrCH_2C_6H_{11}$	S	N, A	95	Transfer to Al		38, 60
Styrene	$zr(CH_2)_2Ph$ + $zrCHPhMe$	S	N				59, 61
				(60 as RH)			42
					Oxidation		29
H_2C═$CMePh$ (and other aromatic alkenes)	$zrCH_2CHMePh$			(100 as ROH)		*In situ* preparation	29

[a]Reaction conditions; S = standard conditions (see text). [b]Method of characterization for $[Cp_2ZrRCl]$: N = ^1H NMR; C = ^{13}C NMR; A = elemental analysis; I = infrared; X = X-ray crystallographic structure. [c]Yield of isolated $[Cp_2ZrRCl]$ except for numbers in parentheses which are yields of indicated simple derivative.

Table 3 Hydrozirconation of Alkenes with Oxygen-containing Functional Groups

Alkene	*Product*	*Conditions*[a]	*Characterization*[b]	*Yield (%)*[c]	*Use*	*Comments*	*Ref.*
$CH_2{=}CH(CH_2)_3OH$	$zr(CH_2)_5Ozr$			(56 as RI)		*In situ* preparation	42
$CH_2{=}CH(CH_2)_2O^-K^+$ (and related alkenoxides)	See equation (36)	THF	X	10–40		Dimer with oxygen bridges	62
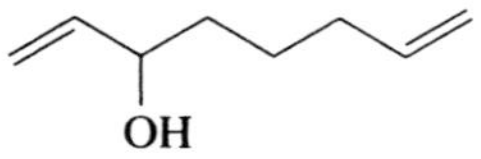 OH (and related dienols)	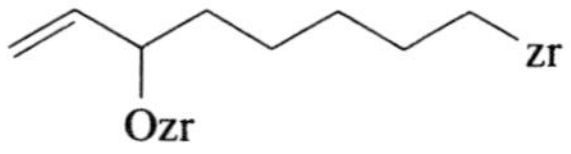 Ozr zr	2 equiv. Zr 10 °C	N	(54–70 as RH)		See Section 3.9.3.3.2 for regiochemistry	63
2-Vinylfuran	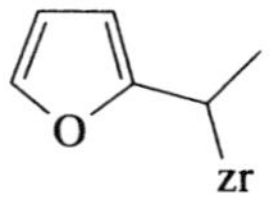 O zr	S			Hydrocyanation		56
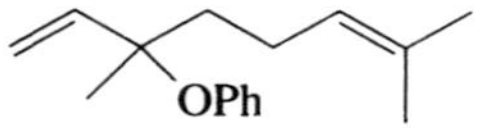 OPh	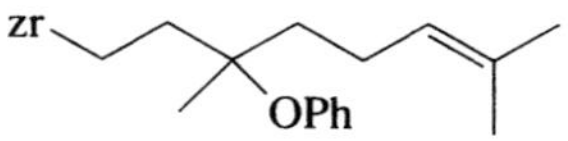 zr OPh	S			Hydrocyanation		56
Oleic acid oxazoline (and other long chain acid oxazolines)	See Scheme 8			(62–82 as RH)		Solvent effect, see Section 3.9.3.3.2 for regiochemistry	58, 64 65

[a–c]See footnotes to Table 2.

Table 4 Hydrozirconation of Simple Alkynes

Alkyne	*Product*	*Conditions*[a]	*Characterization*[b]	*Yield (%)*[c]	*Use*	*Comments*	*Ref.*
C_2H_2	zrCH=CH_2	50 °C	N, A, C	86			61
		S			Pd/coupling		66
Prop-1-yne	zrCH=CHMe	50 °C	N, A, I				61
But-1-yne	zrCH=CHEt	78 °C					37
Hex-1-yne	zrCH=CH$(CH_2)_3$Me	0 °C/CH_2Cl_2		'100'			41
		S			Pd/coupling		66
		S			(η^2-hexyne)Zr		67
Oct-1-yne	zrCH=CH$(CH_2)_5$Me	S	N		Ni/conjugate addition		68
Dec-1-yne	zrCH=CH$(CH_2)_7$Me	S			Hydrocyanation		56
		S			Pd/coupling		66
HC≡CBu^t	zrCH=CHBu^t	S	N, A	89	Transfer to Al		38
					Ni/conjugate addition		68
		50 °C	N, A, I				61
HC≡CPh	zrCH=CHPh	78 °C	N				37
		50 °C	N, A, I				61
		S			Pd/coupling		66
HC≡CC_6H_4-*p*-OMe	zrCH=CHAr	S			Pd/coupling		66
HC≡C$CH_2C_6H_4$-*p*-OMe	zrCH=CHCH_2Ar				Pd/coupling		69
Hex-3-yne	*cis*-zrC(Et)=CHEt	S	N	93	Transfer to Al		38
					Ni/conjugate addition		68
		0 °C/CH_2Cl_2		'100'			41
					Pd/coupling		66
EtC≡CMe		S					4
Pr^{n}C≡CMe	See equation (39)	S					4
Pr^{i}C≡CMe	and	S	N	80		See Section 3.9.3.3.2	4, 38
Bu^{i}C≡CMe	Table 8	S				for regiochemistry	4
Bu^{t}C≡CMe		S					4
PhC≡CPh	*cis*-zrC(Ph)=CHPh	78 °C	N				37

[a–c]See footnotes to Table 2.

Table 5 Hydrozirconation of Alkynes with Functional Groups

Alkyne	*Product*	*Conditions*[a]	*Characterization*[b]	*Yield (%)*[c]	*Use*	*Comments*	*Ref.*
$BuCC{\equiv}CCH_2OH$ (and related)	Direction of addition unknown	2 equiv. Zr		(85 as RH)			70
$HC{\equiv}CCH_2CMe_2OH$	$zrCH{=}CHCH_2CMe_2Ozr$	2 equiv. Zr			Couple to Pd(allyl)		71
$HC{\equiv}COEt$	$zrCH{=}CHOEt$	S			Pd/coupling		66, 72
$HC{\equiv}CCH_2OTHP$	$zrCH{=}CHCH_2OTHP$	S			Ni, Pd/coupling		72
$HC{\equiv}CCH_2OSiMe_2Bu^t$	$zrCH{=}CHCH_2OSiMe_2Bu^t$	S			Hydrocyanation		56
$HC{\equiv}C(CH_2)_4OSiMe_2Bu^t$	$zrCH{=}CH(CH_2)_4OSiMe_2Bu^t$	S	N		Ni/conjugate addition		68
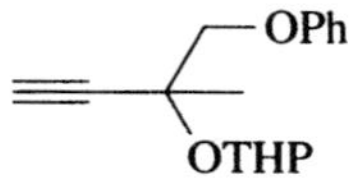 OPh, OTHP (structure)	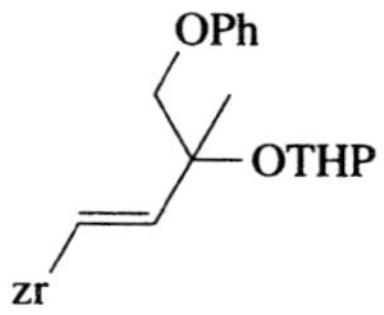 OPh, OTHP, zr (structure)			(10 as RI)			73
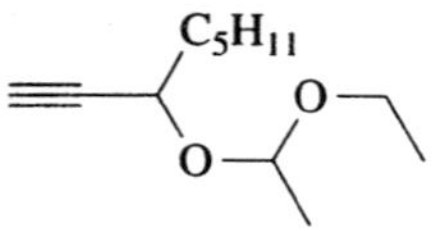 C_5H_{11}, O, O (structure) (and related)	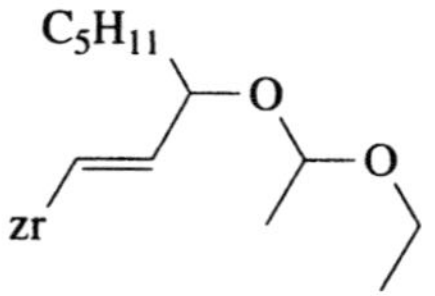 C_5H_{11}, O, O, zr (structure)	S	N		Ni/conjugate addition		68
$HC{\equiv}C(CH_2)_3CO_2Bu^t$	$zrCH{=}CH(CH_2)_3CO_2Bu^t$	S	N	81		Yield 41% with methyl ester	68
$HC{\equiv}C(CH_2)_8CO_2Me$ (and other esters)	$zrCH{=}CH(CH_2)_8CO_2Me$	S			Pd/coupling		66
$HC{\equiv}C(CH_2)_3CN$	$zrCH{=}CH(CH_2)_3CN$	S			Pd/coupling		66
$HC{\equiv}C(CH_2)_3Cl$	$zrCH{=}CH(CH_2)_3Cl$	S			Pd/coupling		66

[a–c]See footnotes to Table 2.

Table 6 Hydrozirconation of Dienes, Diynes and Enynes

Substrate	*Product*	*Conditions*[a]	*Characterization*[b]	*Yield (%)*[c]	*Use*	*Comments*	*Ref.*
1,3-Butadiene (and substituted)	$zrCH_2CH_2CH{=}CH_2$	S		(59–98 as RCHO)			74
1,3-Pentadiene	$zrCH_2CH_2CH{=}CHMe$	S	N	73	Transfer to Al		38
		S			hydrocyanation	See Section 3.9.3.3.2 for regiochemistry	56
1,5-Hexadiene	$zr(CH_2)_4CH{=}CH_2$	S	N	92	Transfer to Al		38
1,7-Octadiene	$zr(CH_2)_6CH{=}CH_2$	S	N	98	Transfer to Al		38
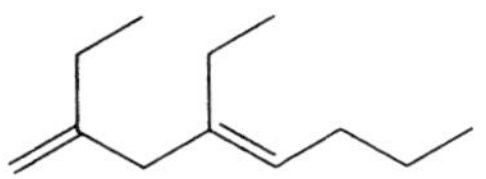	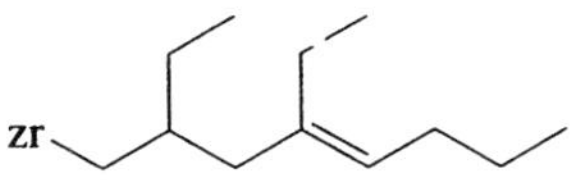 zr			(34 as RCHO)	Bacteriocide		75
4-Vinylcyclohexene	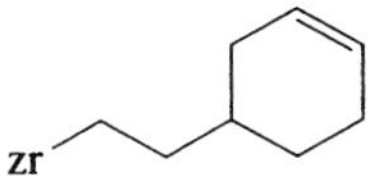 zr			(81 as RI)		*In situ* preparation	42
Vitamin D_3	See equation (41)			(84 as RH)		See Section 3.9.3.4.3	76
$HC{\equiv}C(CH_2)_2C{\equiv}CH$	$zrCH{=}CH(CH_2)_2C{\equiv}CH$	S			Pd/coupling		66
$HC{\equiv}C(CH_2)_nC{\equiv}CTMS$	$zrCH{=}CH(CH_2)_nC{\equiv}CTMS$				Pd/coupling	Protect for mono product	77
$HC{\equiv}CCH{=}CH_2$	$zrCH{=}CHCH{=}CH_2$			75–90			78
$HC{\equiv}CCMe{=}CH_2$	$zrCH{=}CHCMe{=}CH_2$	S				Relatively low yield	66
$HC{\equiv}CCH{=}CHOMe$	$zrCH{=}CHCH{=}CHOMe$					See Section 3.9.3.3.3	78
$HC{\equiv}CCH{=}CH(CH_2)_4Me$	$zrCH{=}CHCH{=}CH(CH_2)_4Me$				Pd/coupling		77
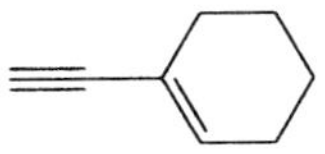	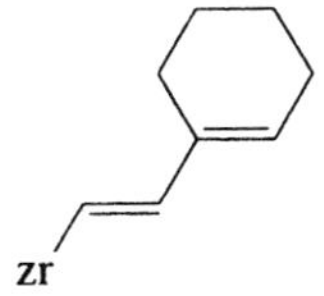 zr						78

[a–c]See footnotes to Table 2.

performed well in alkenation reactions. The same compounds may be alternatively generated by hydroalumination of alkenylzirconium.[81,82]

(6) + R–CH=CH–$AlMe_2$ ⟶ Cp_2Zr(μ-Cl)$AlMe_2$ (10) + Cp_2Zr(μ-Cl)$AlMe_2$ (11) (17)

	(10)	(11)
R = Bu^t	>95%	<5%
R = Bu^n	60%	40%
R = Me	33%	67%

Hydrozirconation of vinylzirconium (**12**), itself obtained by hydrozirconating acetylene, gives the μ-ethylene product (**13**; equation 18) and thus amounts to an overall double hydrozirconation of an alkyne. However, this reaction is not general: attempted hydrozirconation of (substituted vinyl)zirconium does not lead to analogs of (**13**).[61] This may be relevant to the proposed mechanism for isomerization of alkenylzirconium complexes (Section 3.9.3.3.2). Finally, hydrozirconation of an alkynylzirconium species (**14**; equation 19) proceeds 'normally' to give the (μ-alkenediyl)dizirconium complex (**15**).[83] While no attempts to exploit such bimetallic structures synthetically have been reported, it seems quite possible that they might prove useful with sufficient imagination.

(6) + $Cp_2Zr(Cl)CH{=}CH_2$ (12) ⟶ zr–CH_2CH_2–zr (13) (18)

(6) + Cl–Zr$(MeCp)_2$–C≡C–R (14) ⟶ Cp_2Zr(Cl)–C(=CH–$ZrCl(MeCp)_2$) (15) (19)

Hydrometallations with variants of the [Cp_2ZrHCl] reagent are collected in Table 7. As discussed in Section 3.9.3.1, most of these are dihydrides, which cause problems for subsequent synthetic application. Related reactions involving different metals may be found in Section 3.9.4.1.

Table 7 Hydrometallations with Other [Cp_2MHX] Systems

Compound	*Substrate*	*Comments*	*Ref.*
[Cp_2HfHCl]	Hept-1-ene *trans*-Pent-2-ene Isoprene β-Pinene Dodec-1-yne	All very similar to Zr analogs	55
[$Cp*_2ZrH_2$]	Ethylene	Products stable if	84
[$Cp*_2HfH_2$]		no excess ethylene	24
[$Cp*_2ZrH_2$]	Propene	Product unstable	84
[$Cp*_2HfH_2$]	Styrene		24
[$Cp*_2ZrH_2$]	Isobutene	Stable to 75 °C	46
[$Cp*_2HfH_2$]			24
[$Cp*_2HfH_2$]	Bu^tC≡CH	Stable to 125 °C	24
[$Cp*_2(Zr,Hf)H_2$]	MeC≡CH EtC≡CH PhC≡CH MeC≡CMe	Dialkenyl product Dialkenyl Mono/di mixture Mono, unstable	85
[$Cp*_2HfH(CH{=}CHBu^t)$]	Ethylene	Stable to 80 °C	24

3.9.3.3 Selectivity Issues

3.9.3.3.1 Chemoselectivity

Early transition metals are strongly electropositive and form strong M—O and M—N bonds, and their hydrides exhibit strong hydridic or H^- character.[86] Hence reactions of [Cp_2ZrHCl] (**6**) with proton donors such as alcohols and acids, and with polar multiple bonds such as ketones and nitriles, are strongly favored on both thermodynamic and kinetic grounds. Complex (**6**) readily deprotonates or reduces all the above,[87] as well as esters,[87] isonitriles,[88] thioketones,[89] epoxides,[87] CO, CO_2, RNCO and RNCNR;[90] even so weak an acid as CpH.[87] The only hetero multiple bond found not to undergo hydrometallation is the C═N bond in PhCH═NMe. (Imines with neighboring C—H bonds do react, but not by hydrozirconation.[91]) Also several ketones (cyclohexanone, menthone, camphor, *etc.*) reacted readily with (**6**) to give the corresponding alcohol after hydrolysis, but acetophenone was recovered unchanged from the same conditions.[92]

Of course, these findings do not automatically mean that such functional groups cannot be tolerated; that depends upon relative rates. For instance, hydrozirconation of nitriles can be relatively slow: complete reaction of RCN under standard conditions required 72 h, 14 h and 1.5 h, respectively for R = Me, Ph, Bn.[93] Consistent with previous indications that alkynes are more reactive than alkenes towards (**6**), hydrozirconation of an alkyne can be carried out in the presence of a nitrile group (see Table 5),[66] whereas methacrylonitrile gives only C≡N hydrozirconation (equation 20).[78]

(**6**) + CN → N zr (20)

Similarly, alkynic esters have been successfully hydrozirconated (Table 5). HC≡C$(CH_2)_3CO_2Bu^t$ gave the alkenylzirconium product in much higher yield than the corresponding methyl ester, which exhibited carboxy reduction.[68] Clearly the reactivities of the C≡C and ester groups are in close balance here, with steric factors sufficient to tip it towards the former. Hydrozirconation of an alkenic ester would not be expected to work well; no example is reported. In the reaction of an enone with (**6**; equation 21) little if any hydrozirconation of the C═C bond occurs.[94]

(**6**) + Fc(C═O)CH═CH~ —H_3O^+→ Fc(C═O)Pr^n (4%) + Fc~CH═CHCH(OH)Me (96%) (21)

Ether linkages seem to be tolerated well (examples in Tables 3 and 5); this stands in contrast to hydroalumination where the Al—O bond strength generally makes the products unstable. Only for an actual vinyl ether were problems encountered: alkoxide migration appears to be facile in the presumed β-alkoxyalkylzirconium intermediate (**16**; equation 22).[95] The β-alkoxyalkenylzirconium complex obtained from HC≡COEt, though, is stable.[66,72]

(**6**) + ═\O/ → [zr/\OEt] (**16**) → zrOEt (**17**) + C_2H_4 (22)

Functional groups can be protected: carboxylic acids as oxazolines[58,64,65] or silyl esters;[66] and alcohols as THP derivatives[66,72,73] or potassium alkoxides.[62] In fact, it may not be necessary to protect alcohols at all: just use two equivalents of reagent! The first will deprotonate the alcohol to give an alkoxyzirconium complex, and the second can then hydrozirconate the C═C or C≡C bond (equation 23). Several examples are shown in Tables 3 and 5. The presence of such functional groups does frequently have regiochemical consequences, as will be shown in the following section.

═\(\)$_n$OH —(**6**), $-H_2$→ ═\(\)$_n$Ozr —(**6**)→ zr\/\(\)$_n$Ozr (23)

Metal hydrides often give reductive dehalogenation of organic halides, usually via a free radical pathway. [Cp_2ZrHCl] reacts readily with polyhalogen compounds, such as CH_2Cl_2 and $CHCl_3$,[35,41] that are most susceptible to radical attack; less is known about its tolerance for simple alkyl halides. The hydrozirconation of chloroalkenes can lead to some cyclization, as in equation (24); simple reduction of both C=C and C—Cl competes (with the analogous bromide there is much more reduction).[96] Reduction of chlorocyclohexane by (**6**) requires 24 h at 85 °C;[96] hydrozirconation of alkenes containing C—Cl bonds might be achieved cleanly at room temperature. Success with a chloroalkyne has been reported (Table 5), but hydrozirconation of a (*p*-chlorophenyl)alkene is accompanied by some reduction.[29]

(**6**) + [chlorocyclohexene] —(65 °C, 61 h; H_3O^+)→ [bicyclic] 40% + [cyclohexane] 5% + [chlorocyclohexane] 15% + [cyclohexene] 26% + [chlorocyclohexene] 8% (24)

Selecting for C≡C over C=C should be easy, based on comparative reactivities cited earlier. Several examples of hydrozirconation of terminal alkynes in the presence of terminal or internal alkenes were shown in Table 6. Selective hydrozirconation of an internal alkyne in the presence of a terminal alkene might be more difficult; no attempt has been published. A variety of both conjugated and nonconjugated dienes can be cleanly monohydrozirconated; examples may be found in Table 6. Even if both double bonds are in terminal positions, as in 1,3-butadiene, there is no competing double addition or 1,4-addition[74]—a significant and useful contrast to many reactions of main group hydrides. The presence of a double bond in a conjugating position can alter regiochemistry (Section 3.9.3.3.2) or the outcome of subsequent transformation (Section 3.9.4.3.1). In contrast, some α,ω-diynes could not be cleanly monohydrozirconated; it was necessary instead to half protect with EtMgBr/TMSCl to get single addition.[77]

Complications are found for 2,4-dienes, with which three equivalents of [Cp_2ZrHCl] react to give saturated alkylzirconium products; details of these reactions are not understood.[1] It is not known whether similar problems will arise for nonconjugated, diinternal alkenes; no examples have been studied. The hydrozirconation of 1,3-dienes was initially reported to be much slower than that of simple alkenes;[74] but later work suggests this is dependent upon the mode of reagent preparation.[41]

3.9.3.3.2 Regioselectivity

With simple alkenes terminal alkylzirconium compounds are formed, even if the original alkene is not terminal! Presumably this reflects steric disfavoring of a secondary alkylzirconium; the only examples of the latter are found with cyclic alkenes such as cyclohexene or norbornene. The postulated mechanism for the relocation of Zr to a terminal position was shown in Scheme 5, but in truth very little detailed information on this process is available. The intermediate secondary alkyl and internal alkene complexes must be in close energetic balance, so that the rearrangement is rapid on the time scale of the overall hydrozirconation; but is the final stable product still in moderately rapid equilibrium with these less stable intermediates? This question is highly relevant to targets such as regio- or stereo-specifically deuterium-labelled alkanes. Exchange in alkyl(hydrido)zirconium complex (**19**) suggests equilibration (equation 25),[43] but alkyl(chloro)zirconium complexes (such as **18**) may differ. Experiments to probe this issue, such as migration of deuterium labels or magnetization transfer in NMR, would be of interest here.

[methylenecyclohexane] —(**6**)→ $Cp_2Zr(Cl)CH_2$–C(H)(cyclohexyl) (**18**) —$LiAlD(OR)_3$→ $Cp_2Zr(D)CH_2$–C(H)(cyclohexyl) (**19**) ⇌(3h, 25 °C, toluene) $Cp_2Zr(H)CH_2$–C(D)(cyclohexyl) (**19a**) (25)

Tertiary alkylzirconium compounds appear to be generally inaccessible even as intermediates. Tetrasubstituted alkenes are not hydrozirconated at all;[1] also, zirconium is unable to migrate past a tertiary position in a carbon chain. Attempted hydrozirconation of an internal alkene which is 'blocked' at both ends (**20**; equation 26) gives only isomerized alkene (**21**).[97] The only exception, which appears to involve a tertiary intermediate, is found in the hydrozirconation of dienol (**22**) in equation (27).[63]

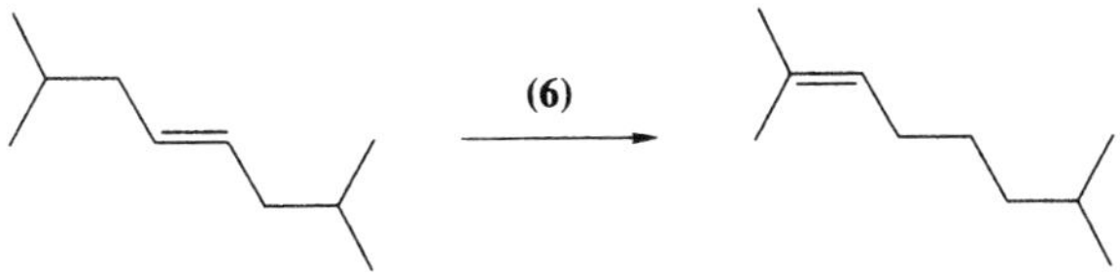

(**20**) —(**6**)→ (**21**) (26)

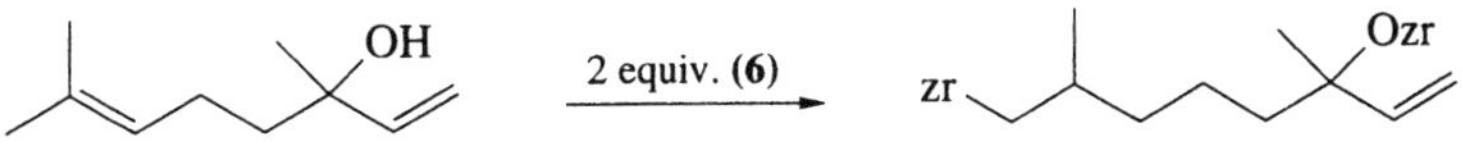

(**22**) —2 equiv. (**6**)→ zr…Ozr (27)

The exclusively terminal regioselectivity can be altered, however, for alkenes containing groups such as aromatic rings, additional double bonds, or oxygen centers. A number of workers have noted that styrene hydrozirconation is not completely regioselective: after subsequent transformation/cleavage steps, the relative amounts of the 'normal' and 'reversed' products, $PhCH_2CH_2X$ and PhCHXMe, respectively, range from 67:33 to 90:10.[29,42,56,59] Unfortunately, such results are not unequivocal: do they represent the static population of organozirconium isomers; competitive reaction of rapidly equilibrating isomers; or even isomerization brought about during the cleavage reaction? Ratios of the isomeric alcohols (equation 28) depend upon oxidant.[29] Only one published study reports NMR of the initial hydrozirconation product (**23**) itself: there is no evidence for the reversed isomer (**24**),[61] but it is not clear how carefully they looked. Recent work suggests a ratio for normal and reversed products of around 85:15 (equation 29). There is no fast equilibrium between (**23**) and (**24**), since complete scrambling of a deuterium label among all alkyl positions was not observed.[98]

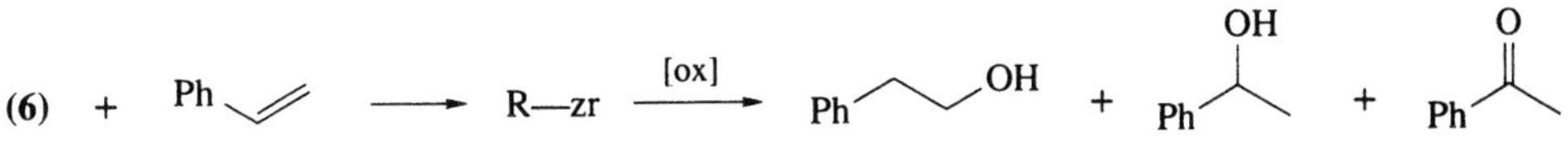

(**6**) + Ph⁀ → R—zr —[ox]→ Ph⁀OH + PhCH(OH)Me + PhC(O)Me (28)

	PhCH₂CH₂OH	PhCH(OH)Me	PhC(O)Me
[ox] = O_2	67%	20%	13%
[ox] = TBHP	93%	7%	–

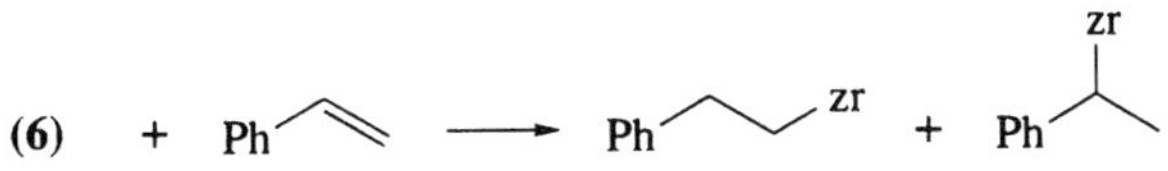

(**6**) + Ph⁀ → Ph⁀zr + PhCH(zr)Me (29)

(**23**) 85% (**24**) 15%

Why is the prohibition against secondary alkylzirconium relaxed for phenyl substituents? Buchwald *et al.* suggest that the flat phenyl group is less sterically demanding than an alkyl,[56] while others have proposed an electronic effect favoring benzylic zirconium compounds.[59] Evidence supports the latter: for internal aromatic alkenes hydrozirconation initially gives mostly the benzylic isomer (**25**) (based on the alcohol products of oxidation), which slowly (48–96 h at 40 °C) converts to the terminal isomer (**26**; equation 30).[29]

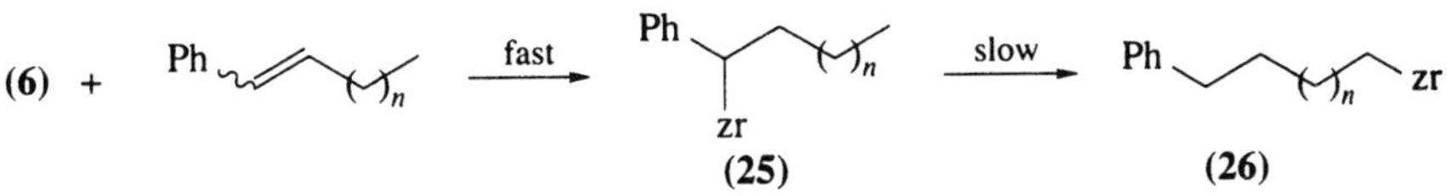

(**6**) + Ph⁀()ₙ —fast→ (**25**) —slow→ (**26**) (30)

Similarly, with 1,3-dienes there is some tendency towards Markovnikov addition, yielding an allylic product, although again there are conflicting results. Buchwald *et al.* conclude that hydrozirconation of 1,3-pentadiene proceeds as in equation (31),[56] whereas Schwartz finds only terminal isomer (**27**), but sees isomerism with a more-substituted diene (equation 32).[1] Again, this may be attributable to electronic stabilization of an allylic zirconium compound. No indication of η^3-allyl formation has ever been observed, even though $[Cp_2ZrRCl]$ is formally coordinatively unsaturated.

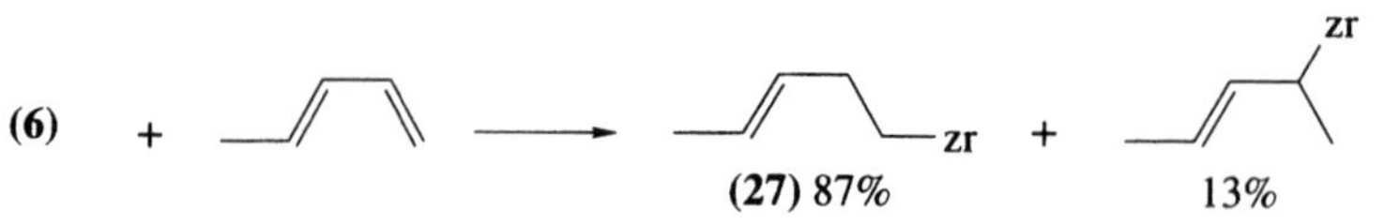

(6) + → (27) 87% zr + 13% zr (31)

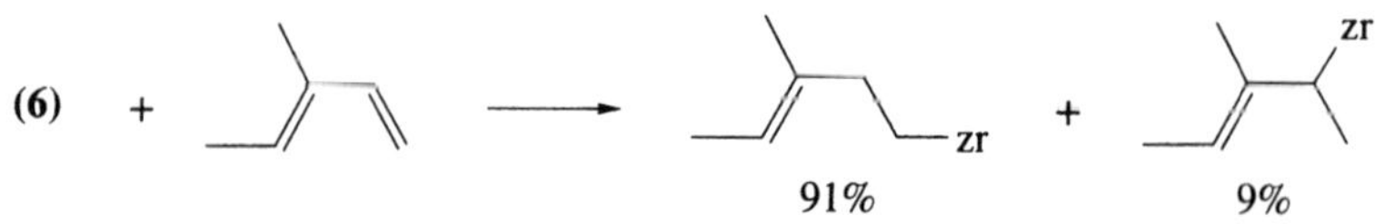

(6) + → 91% zr + 9% zr (32)

Oxygen- and nitrogen-containing functional groups may significantly perturb regiochemistry; here it does appear that additional coordination is involved. Stability of the five-membered chelate ring in intermediate (**28**) was proposed to account for the 3-hydroxystearate obtained in Scheme 8.[64] Hydrozirconation of unsaturated acid oxazolines with the double bond in the 2,3- or 3,4-position, followed by deuterolysis, gives mixtures of 2-, 3- and ω-deuteroacids, depending upon the exact structure, while if the double bond is initially in the terminal position, only ω-deuteration is found.[58,65] These reactions may be accompanied by positional and *cis–trans* isomerization in 'unreacted' alkene; clearly one must proceed with caution in applying hydrozirconation to synthetic transformations of such compounds. Hydrozirconation of 2-vinylfuran gives only the internal product (Table 3),[56] which may be due to *O*-coordination, aromatic stabilization or reduced steric effect of the flat furan ring (see above), or a combination of these.

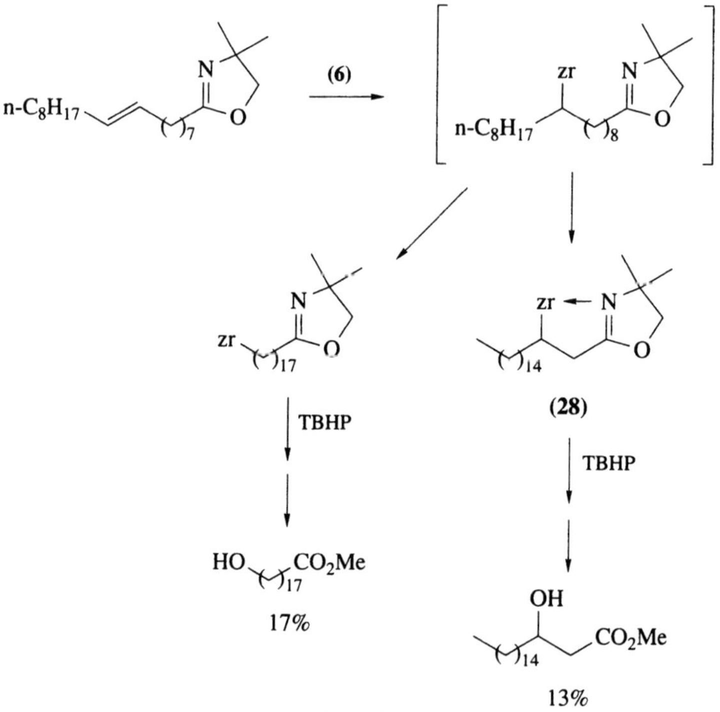

n-C_8H_{17} $(\,)_7$ N O —(6)→ [zr N n-C_8H_{17} $(\,)_8$ O]

zr $(\,)_{17}$ N O —TBHP→ HO $(\,)_{17}$ CO_2Me 17%

zr ← N $(\,)_{14}$ O (**28**) —TBHP→ OH $(\,)_{14}$ CO_2Me 13%

Scheme 8

Definitive evidence for the role of heteroatom coordination may be seen in the hydrozirconation of alkenoxide (**29**; equation 33) where the five-membered ring favors (**30**) (confirmed by the X-ray crystal structure).[62] A related structure was obtained in equation (34); the *O*-coordinated product (**31**) does not isomerize to the terminal alkylzirconium product (**16**) below 140 °C.[95] (At 140 °C it slowly decomposes to **17**, presumably *via* **16**; *cf.* equation 22 above.) In contrast, hydrozirconation of dienols shows preferred reactivity of the double bond remote from the —OH group; an example is shown in Table 3.[63] Probably the first equivalent of (**6**) reacts with the —OH group both to increase steric bulk and prevent the oxygen from coordinating to the second Zr, which does the actual hydrozirconation. After protolysis, oct-1-en-3-ol is the sole product obtained.

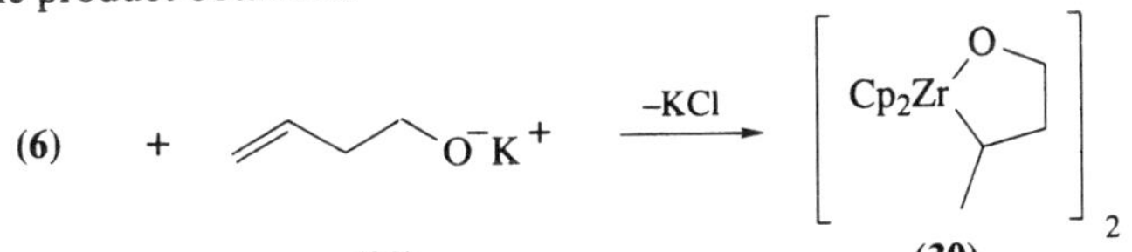

(33)

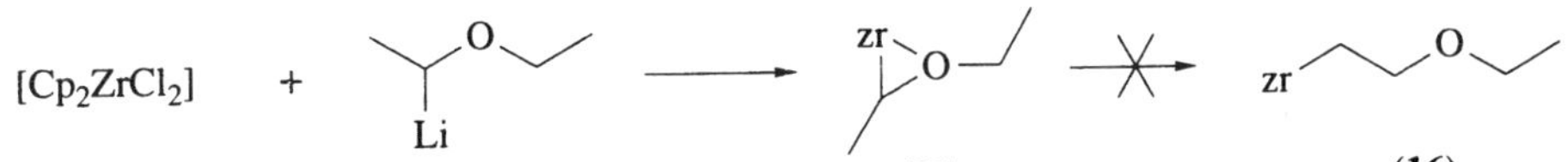

(34)

Terminal alkynes are always hydrozirconated to the terminal alkenylzirconium product,[1] even in the presence of functional groups (Table 5). Internal alkynes give internal alkenyls, without the migrations found for alkenes. With asymmetric internal alkynes RC≡CR^1, the zirconium goes preferentially to the carbon atom bearing the sterically smaller R group, subject to both kinetic and thermodynamic control. For the example shown in equation (35), reaction of (**6**) with a slight excess of (**32**) gives only a slight preference for the favored isomer (**33a**); the composition is stable to prolonged standing at room temperature. Addition of a small amount of (**6**) effects slow isomerization to virtually all (**33a**). Further such results are shown in Table 8.[4] Note that very high regioselectivity can be attained, even where the steric differentiation between the ends is relatively modest (*e.g.* Me *versus* Et). A doubly hydrozirconated intermediate is proposed for the isomerization (equation 36), to account for the requirement for excess [Cp_2ZrHCl].[4] As noted earlier, such a species (**13**) has been isolated for C_2H_2 (equation 18), although not for any substituted alkynes.[61]

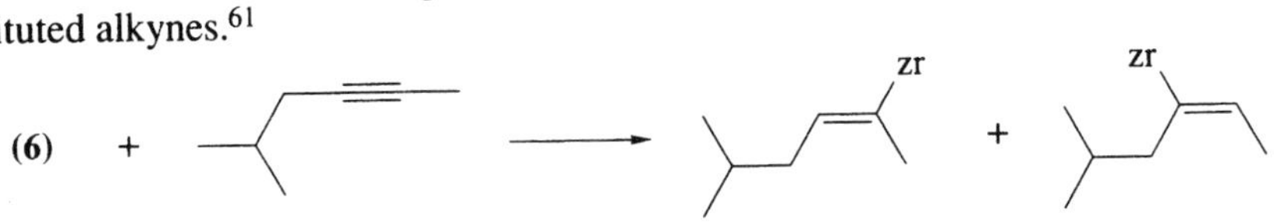

(35)

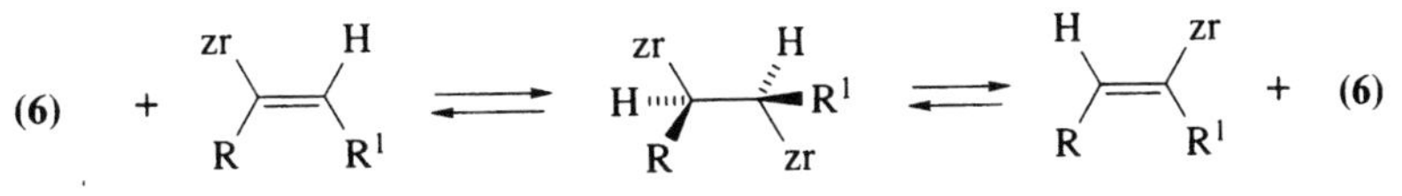

(36)

Table 8 Regiochemistry of Internal Alkyne Hydrozirconation

R	*R^1*	*zrCR═CHR^1:zrCR^1═CHR* Initial	Final
H	Bu^n	>98:2	Same
Me	Et	55:45	89:11
Me	Pr^n	69:31	91:9
Me	Pr^i	84:16	>98:2
Me	Bu^i	55:45	>95:5
Me	Bu^t	>98:2	Same

Since most of the regioselectivity patterns appear to be sterically driven, one would expect that Cp* analogs would show even more pronounced preferences. Terminal alkenes show exclusive selectivity for terminal products, as do terminal alkynes (Table 7). It would be of interest to see whether rearrangement occurs with internal alkenes. A rearrangement has been observed in the alkenylzirconium (**34**), obtained from an internal alkyne (equation 37).[85] This differs from the simple Cp analogs, both in the rearrangement of an alkyne-derived product and the formation of an η^3-allyl complex. The mechanism is not known with certainty, and may depend upon the presence of the hydride ligand in (**34**).

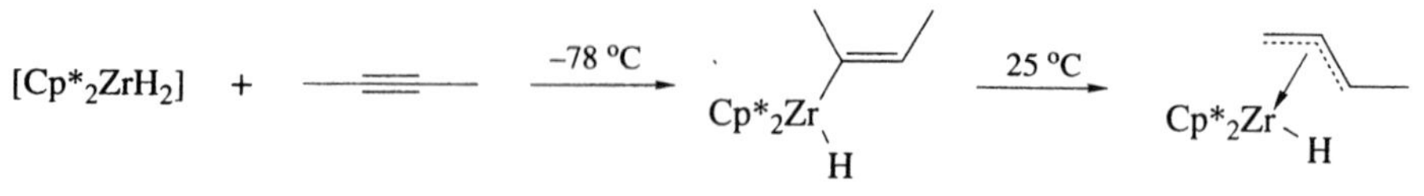

$$[Cp^*_2ZrH_2] + \text{—≡—} \xrightarrow{-78\ ^\circ C} Cp^*_2Zr(H)(\text{alkenyl})\ \textbf{(34)} \xrightarrow{25\ ^\circ C} Cp^*_2Zr(H)(\eta^3\text{-allyl}) \quad (37)$$

3.9.3.3.3 Diastereoselectivity

The mechanism for hydrometallation would appear to require *cis* addition to C=C and C≡C bonds, and this result is essentially always obtained.[1] Apparent exceptions can probably be ascribed to subsequent chemistry. For example, hydrozirconation/oxygenation of 2-methylindene gives a *cis–trans* mixture of alcohols, but the loss of specificity occurs at the oxygenation step, not during hydrozirconation (see Section 3.9.3.4.1).[29] Overall hydrocyanation of terminal alkynes, starting with hydrozirconation, also gives a mixture of isomers; it is not known how this occurs.[56] Reaction of $[Cp_2ZrHCl]$ with diphenylacetylene at 100 °C for prolonged periods gives mostly *trans*-stilbene,[99] but this obviously involves more than simple hydrozirconation.

For acyclic alkenes, the facile rearrangement process obviates most possibilities of stereoselective transformation. The one exception is that of stereospecific isotopic labelling, for which hydrozirconation is a potentially powerful technique: compounds of type Bu^{t}CHDCHDX (**36**), useful for NMR-based mechanistic studies, are conveniently obtained as in Scheme 9.[100] This requires that four separate processes — hydrozirconation of alkyne, cleavage of alkenylzirconium, hydrozirconation of alkene and cleavage of alkylzirconium — all be completely stereospecific.

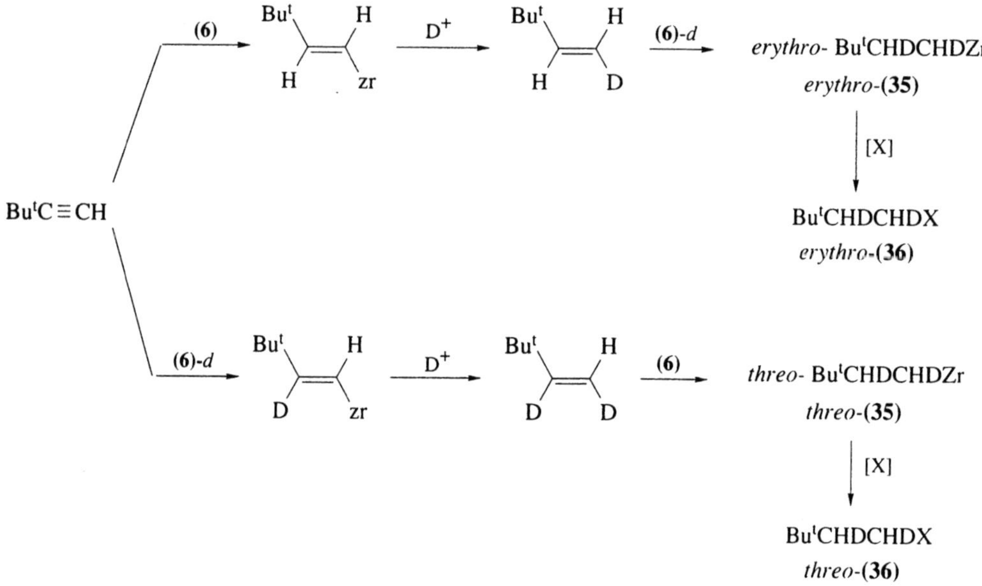

Top sequence: Bu^{t}C≡CH $\xrightarrow{\textbf{(6)}}$ Bu^t(H)C=C(H)(Zr) $\xrightarrow{D^+}$ Bu^t(H)C=C(H)(D) $\xrightarrow{\textbf{(6)}\text{-}d}$ *erythro*-Bu^{t}CHDCHDZr, *erythro*-(**35**) $\xrightarrow{[X]}$ Bu^{t}CHDCHDX, *erythro*-(**36**)

Bottom sequence: Bu^{t}C≡CH $\xrightarrow{\textbf{(6)}\text{-}d}$ Bu^t(D)C=C(H)(Zr) $\xrightarrow{D^+}$ Bu^t(D)C=C(H)(D) $\xrightarrow{\textbf{(6)}}$ *threo*-Bu^{t}CHDCHDZr, *threo*-(**35**) $\xrightarrow{[X]}$ Bu^{t}CHDCHDX, *threo*-(**36**)

Scheme 9

With the possible exception of the third, these can all be easily satisfied. Labinger *et al.* report that either isomer of (**35**) prepared as in Scheme 9 is contaminated with about 10% of the opposite diastereomer; the composition does not change on standing.[100] How does this small amount of net *trans*-hydrozirconation occur? One possibility is that the initial hydrozirconation occasionally takes place with the wrong regiochemistry; this could scramble stereochemistry, as shown in equation (38). Of course, this mechanism also scrambles isotope locations: isomers such as $[Bu^tCH_2CD_2ZrClCp_2]$ should be present as well. None were observed, but it must be noted that the state of NMR equipment at the time of this study was not what it is today; a reinvestigation might be rewarding.

Stereospecific preparation of PhCHDCHDX compounds by this route has not been successful: application of both the top and bottom sequences in Scheme 9 to phenylacetylene results in the same product mixture, although ^{2}H NMR results do not show the complete scrambling required by the mechanism of equation (38).[98] Further study appears warranted to determine whether this approach is limited in scope.

(38)

threo-(**35**) *erythro*-(**35**)

With cyclic alkenes, fairly high diastereoselectivity can be achieved. Hydrozirconation of norbornene, followed by TBHP oxidation, gives the *exo*-alcohol in 95% yield, while the more complex bicyclics camphene (**37**) and β-pinene (**38**) give the mixtures shown in equations (39) and (40), respectively.[57] The hafnium analog shows the opposite preference with β-pinene and does not react with camphene at all, attributed to significantly greater steric demand.[55] Diastereoselectivity in cyclohexanols obtained by hydrozirconation of ketones has also been examined.[92]

(**37**) —(**6**)→ —TBHP→ 97% + 3% (39)

(**38**) —[Cp_2MHCl]→ —TBHP→ M = Zr 87%, M = Hf 7% + 13%, 93% (40)

Hydrozirconation of vitamin D_3 (**39**) gives a mixture of four diastereomers resulting not only from addition to either face of the reacting exocyclic double bond, but also isomerization of the 'nonreacting' 5,6-double bond (equation 41), even though no isomerization of unreacted (**39**) was detected.[76] No explanation was offered. Partial isomerization of a double bond during hydrozirconation of a triple bond has also been observed.[78]

(**39**) (5*Z*) —2 equiv. (**6**)→ —H^+→

(5*Z*,10*S*) 38% + (5*Z*,10*R*) 44% + (5*E*,10*S*) trace + (5*E*,10*R*) 18% (41)

Addition of $[Cp^*_2MH_2]$ (M = Zr, Hf) to alkynes is, not surprisingly, also strictly *cis*; although two isomers are obtained from the Hf compound and but-2-yne, they appear to be conformational and not configurational.[85]

3.9.3.3.4 *Enantioselectivity*

Virtually nothing has been done in this arena. The sole study involves the use of complexes of the type $[Cp'_2ZrH_2]$, where Cp′ bears a chiral substituent, as catalysts for the hydrogenation of prochiral alkenes; a maximum of 2% *ee* was obtained.[101] It is known that chiral zirconium complexes of the so-called *ansa* type, developed by Brintzinger and coworkers,[102] effect very high stereoregularity in propene oligomerization and polymerization.[103] It seems extremely likely that the corresponding —ZrHCl complexes would afford excellent asymmetric induction in the hydrozirconation of prochiral alkenes.

3.9.3.4 Synthetic Utilization of Hydrozirconation

The synthetic utility of organozirconium complexes has been reviewed,[104] and many aspects are considered in greater depth in Volume 1, Chapter 1.5. One general issue which should be mentioned is the stability of organozirconium reagents and effects thereof on yields. Thermal stability appears rarely to be a problem; the most common mode of decomposition for organometallics, β-hydrogen elimination, does not operate here for reasons discussed earlier. Compounds are quite sensitive to proton sources, and reduction (R—Zr to R—H) can compete even without any obvious proton source; examples are presented in the following section. Still, it is usually possible to achieve overall synthetic transformations based upon hydrozirconation in quite high yields; causes of reduced yields are rarely examined.

3.9.3.4.1 *Reactions of alkyl- and alkenyl-zirconium complexes*

Since zirconium is strongly electropositive, these complexes exhibit considerable carbanion character (Zr^+—R^-) and are subject to electrophilic cleavage just as are main group organometallics. Typical reaction patterns are shown in Scheme 10. The simplest such reaction is protonolysis: all hydrozirconation products react readily with aqueous acid. The net result of this overall sequence is hydrogenation. Protodezirconation of both alkyl- and alkenyl-zirconium complexes appears always to proceed with retention of both regio- and stereo-chemistry, so this provides a method for clean *cis* hydrogenation, with high selectivity for triple bonds or terminal double bonds in multiply unsaturated substrates. It is, of course, a stoichiometric hydrogenation, so it will rarely be the method of choice for large scale applications. Catalytic hydrogenation based on hydrozirconation can be accomplished, especially by starting

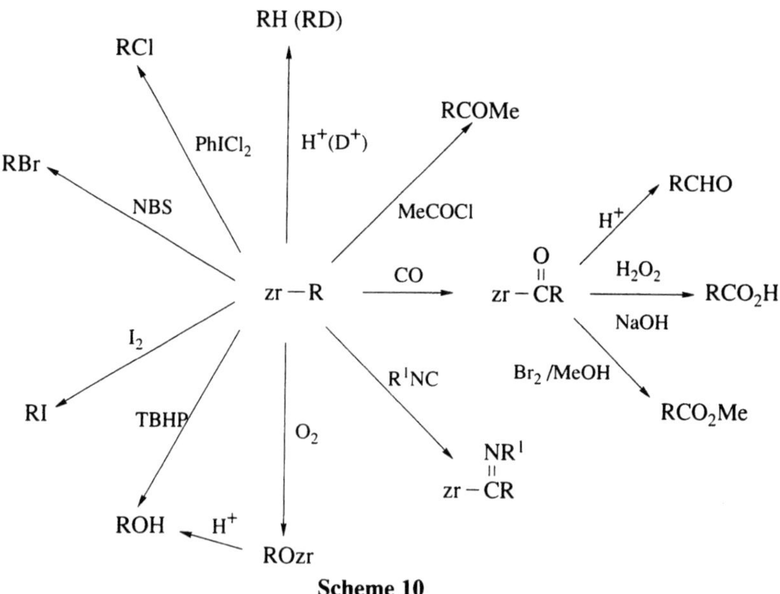

Scheme 10

with [Cp_2ZrH_2] or [Cp_2ZrR_2] complexes, but these do not appear to be very good catalytic systems. Generally high temperatures are required, around 100 °C,[36,99] and clean selective reductions are hard to achieve: hydrogenation of diphenylacetylene gives a mixture of *cis*- and *trans*-stilbenes.[99]

Perhaps the most useful application of the hydrozirconation/protodezirconation sequence is for specific introduction of deuterium label; some examples were presented in the previous section. Although most such work indicates that deuterium incorporation is essentially complete, Svenson and coworkers have found that around 20% of the alkane thus obtained from simple alkenes is undeuterated. The amount of alkane-d_0 depends upon the mode of preparation of (**6**) and increases with longer hydrozirconation reaction times, so transfer of hydrogen to give alkane prior to hydrolysis is probably responsible. This '*in situ*' hydrogenation was found to be even more pronounced for α,β-unsaturated carboxylic acid oxazolines, giving 70% of the undeuterated saturated product.[58]

Protonolysis can also be effected with weaker acids; in some cases this may be essential. For example, hydrozirconation of alkynic alcohols $Me(CH_2)_nC{\equiv}C(CH_2)_mOH$ with 2 equiv. of (**6**), followed by hydrolysis with 2% aqueous NH_4Cl, affords the corresponding *cis*-alkenol in high yield and selectivity, whereas use of aqueous HCl causes both *cis–trans* isomerization and allylic rearrangement.[70] Protonolysis by weak acid may be slow enough for other reactions to compete: TBHP can be used as an oxidant (see below).

Hydrozirconation is not sufficiently reversible to liberate the organic group as an alkene by displacement; but that can be achieved *via* hydride abstraction with trityl ion. This generates a sequence for net alkene isomerization in the counterthermodynamic direction (equation 42).[1]

(**6**) + [alkene] → [alkyl]–zr —(Ph_3C^+, 92%)→ [terminal alkene] (42)

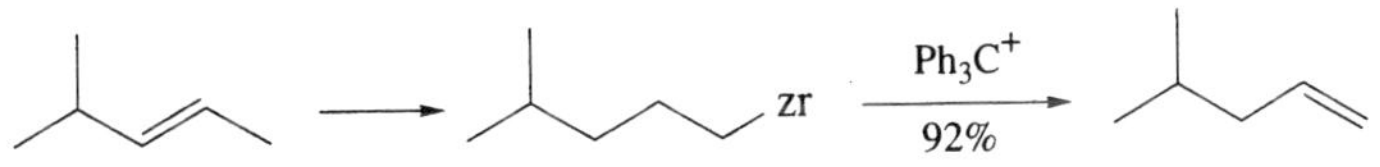

Halogenative cleavage likewise appears completely general for both alkyl- and alkenyl-zirconium complexes (Scheme 10),[1] and proceeds with complete retention of configuration at carbon.[100] For the hydrozirconation products of 3-methyl-1,3-dienes, such as (**40**), cyclization competes (equation 43).[1,74] Analogs without the 3-methyl subsitutent, or with more remote double bonds, give only the uncyclized products in good yields.[74]

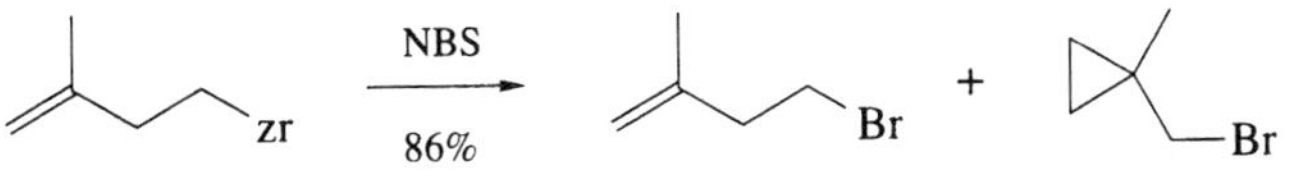

(**40**) zr —(NBS, 86%)→ Br (56%) + cyclopropyl–Br (44%) (43)

Conversion of R—Zr to R—OH can be effected by a variety of oxidants, including (dry) O_2, TBHP, MCPBA, CrO_2Cl_2[1,105] and $MoO_5{\cdot}2HMPA$.[55] The mechanism of autoxidation with O_2 appears to be closely similar to that established for main group organometallics (Scheme 11).[100,105] Most telling is the stereochemical outcome: from either primary alkyl (**35**)[100] or the secondary alkyl obtained by hydrozirconation of 2-methylindene,[29] the mixture of alcohols shows 50% racemization and 50% retention of configuration at the reacting carbon center. This requires that oxidation by ROO— proceeds with retention, which was separately established.[105] Hydrozirconation/autoxidation of hexa-1,5-diene gives 4% of cyclopentylmethanol, attributed to cyclization of the free radical intermediate.[105] Oxidation of alkenylzirconium complexes does not proceed well, with either O_2 or TBHP.[1]

$$R{-}zr + O_2 \longrightarrow [R{\bullet}\ zrOO{\bullet}] \longrightarrow zr{-}OOR$$

$$zr{-}OOR + R{-}zr \longrightarrow 2\ zr{-}OR$$

Scheme 11

CO insertion works well for both alkyl and alkenyl derivatives.[1,106] The resulting acyls may subsequently be converted to aldehydes, acids or esters as shown in Scheme 10; yields (based on zrR) typically range from 50–99%.[106] Two such applications have appeared in the patent literature.[75,107] CO insertion proceeds with stereochemical retention,[100] as in all metal systems studied. The Cp* analogs show quite different CO chemistry. Although insertion to form an acyl (*e.g.* **41**) is the first step, facile rearrangement to alkenoxides (**42**) ensues (equations 44[23] and 45[85]).

$[Cp^*_2ZrHBu^i] \xrightarrow[-50\,^{\circ}C]{CO} Cp^*_2Zr(H)(COBu^i)$ **(41)** $\xrightarrow{-20\,^{\circ}C} Cp^*_2Zr(H)(OCH{=}CHPr^i)$ **(42)** (44)

$Cp^*_2M(\text{diene}) + 2\,CO \xrightarrow{25\,^{\circ}C} Cp^*_2M(O_2C_2\ldots)$ (45)

Insertion of isocyanides R^1NC (R^1 = Me_3C or Me_3Si), which are isoelectronic to CO, appears to be quite similar (Scheme 10); this step has been used to achieve overall hydrocyanation (see Section 3.9.3.4.3).[56] C—S and C—N bonds may also be generated. SO_2 inserts with retention of configuration to yield $[Cp_2ClZrO_2SR]$,[100] while NO leads to *N*-nitrosohydroxylamine derivatives $[Cp_2ClZr\{ON(NO)R\}]$.[1] Neither of the last two has been exploited synthetically.

3.9.3.4.2 Transmetallations

Hydrozirconation products can be used directly as carbanion equivalents, but such reactions are generally not highly successful. Alkylzirconium complexes react with acetyl chloride to give methyl ketones (Scheme 10),[28] but yields are not always good and alkenyls do not react.[38] Addition to ethylene oxide takes place slowly, while carboxylation with CO_2 does not take place at all.[1] This reduced reactivity (compared to main group organometallics) may be largely overcome by first transferring the alkyl or alkenyl group to a more reactive metal center, such as aluminum. (**43**) and (**44**) react with acetyl chloride to give the methyl ketones in near quantitative yields (equation 46).[38] Complications were found only for cyclohexene, which gave virtually none of the desired methyl cyclohexyl ketone (acyclic secondary alkylaluminum reagents are, of course, not accessible by this route, since they cannot be prepared by hydrozirconation), and for certain dienes where cyclization can apparently occur during the transmetallation step (equation 47). Transfer of both alkyl and alkenyl from Zr to Al proceeds with retention of configuration.[38]

(6) + R–CH=CH₂ ⟶ zr–CH₂CH₂R; (6) + R–C≡C–R¹ ⟶ zr–C(R)=CHR¹; $\xrightarrow[-zrCl]{AlCl_3}$ Cl_2Al–CH₂CH₂R **(43)** / Cl_2Al–C(R)=CHR¹ **(44)** $\xrightarrow[-AlCl_3]{R^2COCl}$ R^2CO–CH₂CH₂R / R^2CO–C(R)=CHR¹ (46)

(6) + 1,6-heptadiene ⟶ zr–(CH₂)₅CH=CH₂ $\xrightarrow{AlCl_3}$ cyclopentyl–CH₂–$AlCl_2$ (47)

This hydrozirconation/transmetallation sequence provides a route to net hydroalumination of alkenes and alkynes; the direct reaction is often difficult to achieve. The use of catalytic amounts of Zr will be discussed in Section 3.9.4.2

Even acyl groups may be transferred from Zr to Al, as in equation (48); this generates an acylaluminum or 'acyl anion equivalent' (**45**). Its acyl anion chemistry appears limited, however: only protonolysis to the aldehyde was achieved in good yield, and that could have been done directly with the acylzirconium precursor.[38]

The thermodynamics of the transmetallation reaction should depend largely upon electronegativity considerations: the reaction should proceed to give the more metal–organic product. However, this may

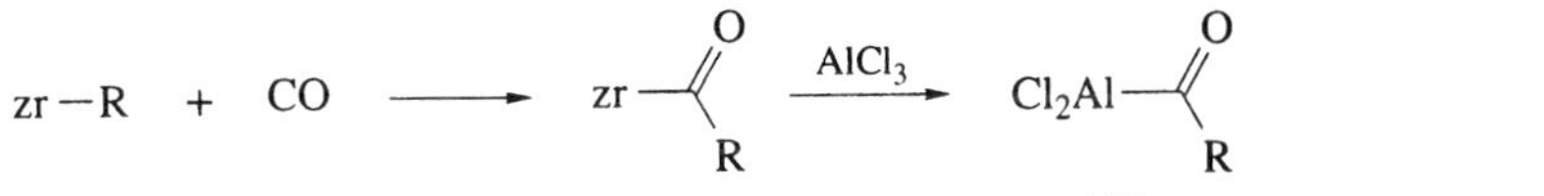

$$\text{zr—R} + \text{CO} \longrightarrow \text{zr—C(=O)R} \xrightarrow{\text{AlCl}_3} \text{Cl}_2\text{Al—C(=O)R} \quad \textbf{(45)} \tag{48}$$

be modulated by other ligands: R_3Al will alkylate $[Cp_2ZrCl_2]$, whereas $[Cp_2ZrRCl]$ alkylates $AlCl_3$. Similarly, $[Cp_2ZrCl(alkenyl)]$ transfers the alkenyl group to R_2AlCl,[38] but the direction of transmetallation can be reversed by addition to BuLi to form an ate complex.[108] Transmetallation from Zr to Li has been reportedly achieved by decomposition of a zirconium ate complex, but no details are available;[109] this would be rather surprising on thermodynamic grounds.

Transfer of alkenyl from Zr to other metals appears more general than alkyl transfer.[38] With CuCl, while the alkenylcopper product was not characterized, it exhibits characteristic decomposition and conjugate addition reactions as shown in equations (49) and (50).[110] In contrast, $[Cp_2ZrCl(octyl)]$**(7)** does not react with CuI.[111] Similarly, transfer of dienyl groups to Bu_3SnCl gives the corresponding dienyltin compounds in 60–90% yields, sought for study in Diels–Alder reactions.[78] Transfer of the octyl group from **(7)** to $SnCl_4$ gave up to 81% yield of (octyl)$SnCl_3$, but further transfers were more difficult.[111]

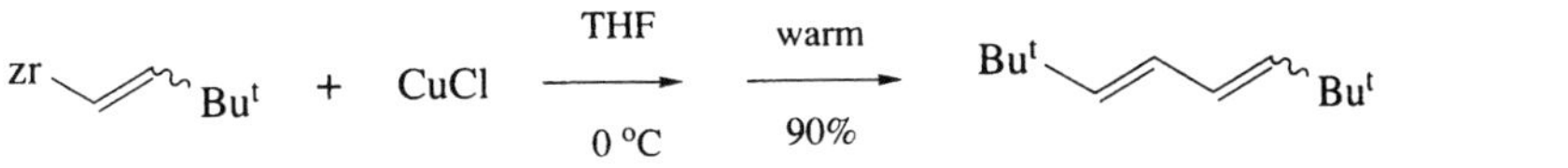

$$\text{zr–CH=CH–Bu}^t + \text{CuCl} \xrightarrow[0\,^\circ\text{C}]{\text{THF}} \xrightarrow[90\%]{\text{warm}} \text{Bu}^t\text{–CH=CH–CH=CH–Bu}^t \tag{49}$$

$$\text{zr–CH=CH–Bu}^t + \text{CuOTf} + \text{CH}_2\text{=CH–C(O)CH}_3 \xrightarrow[-78\,^\circ\text{C}]{\text{LiI}} \xrightarrow[73\%]{\text{H}^+} \text{Bu}^t\text{–CH=CH–CH}_2\text{CH}_2\text{–C(O)CH}_3 \tag{50}$$

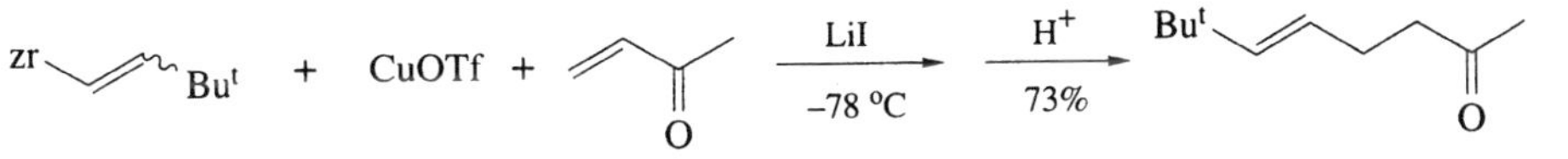

Although the transmetallation sequence of equation (**46**) failed with catalytic amounts of $AlCl_3$, other applications can be done catalytically. For example, (alkenyl)zirconium groups (**46**) may be coupled with organic halides, as in Scheme 12.[72,104,112,113] Catalysts are zerovalent Ni or Pd complexes and yields are typically 70% or better, with sterically crowded alkenyl groups causing problems which may be overcome by an additional cocatalyst such as $ZnCl_2$.[114] The mechanisms of both Ni-[114] and Pd-catalyzed[72] alkenyl–aryl coupling have been investigated, and both appear to proceed *via* reductive elimination from a diorgano-nickel or -palladium intermediate formed by successive transmetallation and oxidative addition, although the oxidation state couples involved for the two metals are different. In a variant, (**46**) can be added to a preformed allylpalladium complex with similar results;[71,115] an application is presented in the following section.

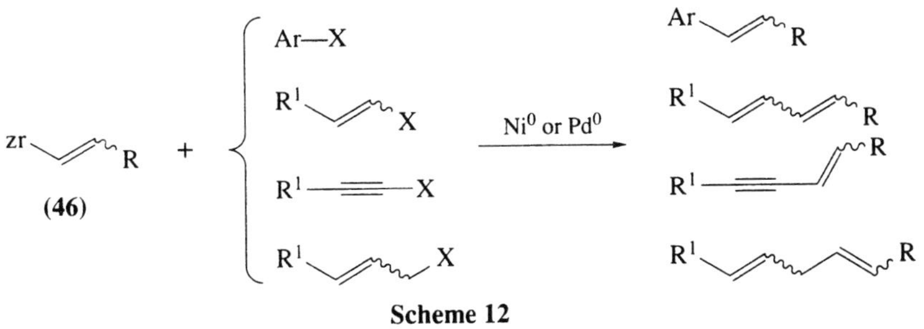

$$\text{zr–CH=CH–R}\ \textbf{(46)} + \left\{\begin{array}{l}\text{Ar–X}\\ \text{R}^1\text{–CH=CH–X}\\ \text{R}^1\text{–C≡C–X}\\ \text{R}^1\text{–CH=CH–CH}_2\text{–X}\end{array}\right. \xrightarrow{\text{Ni}^0 \text{ or Pd}^0} \begin{array}{l}\text{Ar–CH=CH–R}\\ \text{R}^1\text{–CH=CH–CH=CH–R}\\ \text{R}^1\text{–C≡C–CH=CH–R}\\ \text{R}^1\text{–CH=CH–CH}_2\text{–CH=CH–R}\end{array}$$

Scheme 12

Lastly, (**46**) undergoes conjugate addition to enones, catalyzed by Ni complexes (equation 51).[68] Yields vary depending upon structures, but may be significantly improved by reducing the precatalyst $[Ni(acac)_2]$ with DIBAL-H.[68] The reaction appears to be mechanistically similar to conjugate addition with copper reagents, involving electron transfer.[114] This system has potential advantages over copper-based routes, both in reagent preparation and in avoiding the need for an additional ligand in the copper ate reagents. The initial product of conjugate addition is a zirconium enolate;[116] instead of simply hydrolyzing to the γ,δ-enone as in equation (51), it may be used directly for additional transformations such as generation of selenium reagents.[117] A more elaborate example is shown below.

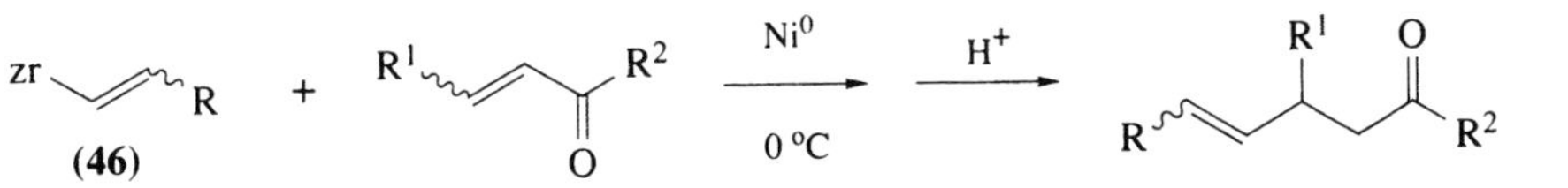

$$\text{zr–CH=CH–R}\ \textbf{(46)} + \text{R}^1\text{–CH=CH–C(O)–R}^2 \xrightarrow[0\,^\circ\text{C}]{\text{Ni}^0} \xrightarrow{\text{H}^+} \text{R–CH=CH–CH(R}^1\text{)–CH}_2\text{–C(O)–R}^2 \tag{51}$$

3.9.3.4.3 *Practical examples*

We will attempt only to give an indication of the scope and flavor of hydrozirconation-based synthetic routes. Terminally functionalized long chain alkanes, which have a variety of important industrial uses, may be obtained from shorter chain terminal alkenes *via* metathesis, hydrozirconation and cleavage (equation 52).[40]

$$2\,C_nH_{2n+1}\text{-CH=CH}_2 \xrightarrow{WCl_6/SnMe_4} C_nH_{2n+1}\text{-CH=CH-}C_nH_{2n+1} \xrightarrow{(6)} zr\text{-}(CH_2)_2(CH_2)_{2n+1} \xrightarrow{[X]} X\text{-}(CH_2)_2(CH_2)_{2n+1} \quad (52)$$

A method for the hydrocyanation of alkenes and alkynes is based upon hydrozirconation followed by isonitrile insertion (Scheme 13); overall yields range from 45–90%.[56] Certain examples do not exhibit the clean regiochemistry represented here, and these were discussed in Section 3.9.3.3.2. The hydrozirconation/protonolysis of equation (53) results in high yield, stereoselective formation of *C*-methyldeoxysugars, an alternative to free radical deoxygenation.[118]

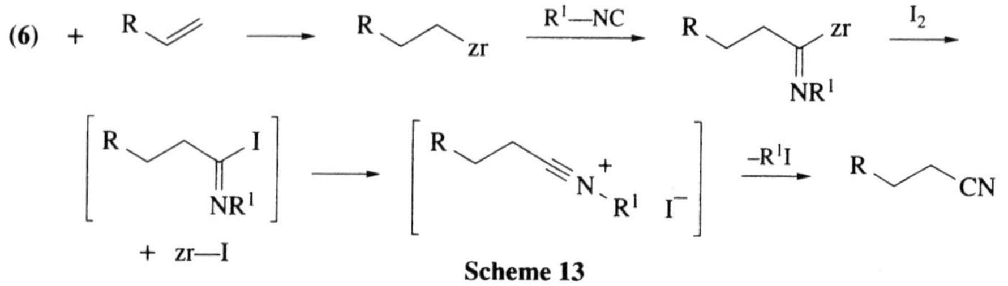

(6) + R–CH=CH₂ → R–CH₂CH₂–zr —(R¹–NC)→ R–CH₂CH₂–C(=NR¹)–zr —(I_2)→ [R–CH₂CH₂–C(=NR¹)–I + zr–I] → [R–CH₂CH₂–C≡N⁺–R¹ I⁻] —($-R^1I$)→ R–CH₂CH₂–CN

Scheme 13

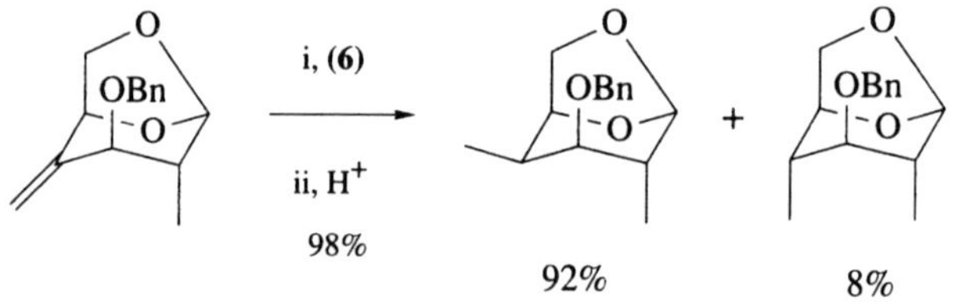

(53): OBn / O bicyclic exo-methylene sugar —(i, (6); ii, H^+; 98%)→ C-methyl product 92% + epimer 8%

ω-Alkynic isobutyl amides such as (**48**), of interest as insecticides, can be made by Pd-catalyzed coupling, as shown in Scheme 14.[77] Similarly, hydrozirconation of a wide variety of alkynes, followed by Pd-catalyzed coupling to an iododeoxyuridine (**49**; dR = 2′-deoxyribose), gives alkenyldeoxyuridines (**50**) which have antitumor and antiviral properties.[66]

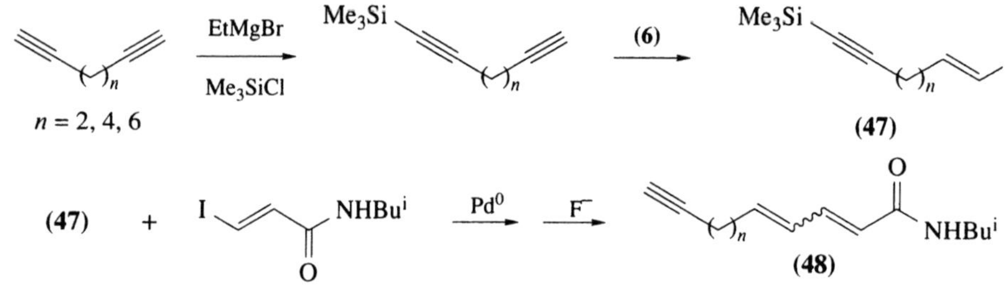

Diyne (n = 2, 4, 6) —(EtMgBr; Me_3SiCl)→ Me_3Si-diyne —(6)→ Me_3Si–C≡C–(CH₂)ₙ–CH=CH–zr (**47**)

(**47**) + I–CH=CH–C(=O)NHBuⁱ —(Pd^0)→ —(F^-)→ (**48**) (O, NHBuⁱ)

Scheme 14

5-Iodouracil nucleoside (H, N, O, I, N, $dR(SiMe_3)_2$) (**49**) + zr–CH=CH₂ —(i, Pd^0; ii, H_2O; 83%)→ 5-vinyluracil nucleoside (H, N, O, N, dR) (**50**)

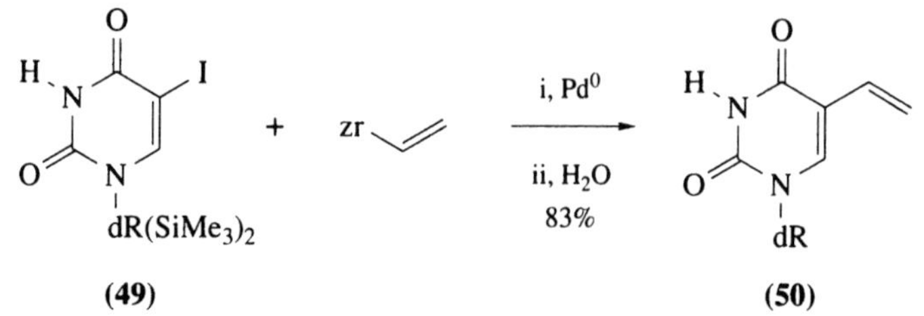

Addition of alkenylzirconium (**51**) to steroid-derived allylpalladium (**52**), followed by hydrogenation and hydrolysis, gives 25-hydroxycholesterol in 66% overall yield from (**52**; Scheme 15).[71] Finally, several approaches to prostaglandins or their precursors have been based upon Ni-catalyzed conjugate addition of alkenylzirconium to enones;[68] one example is depicted in Scheme 16. A prostaglandin synthesis utilizing hydrozirconation has been patented.[119]

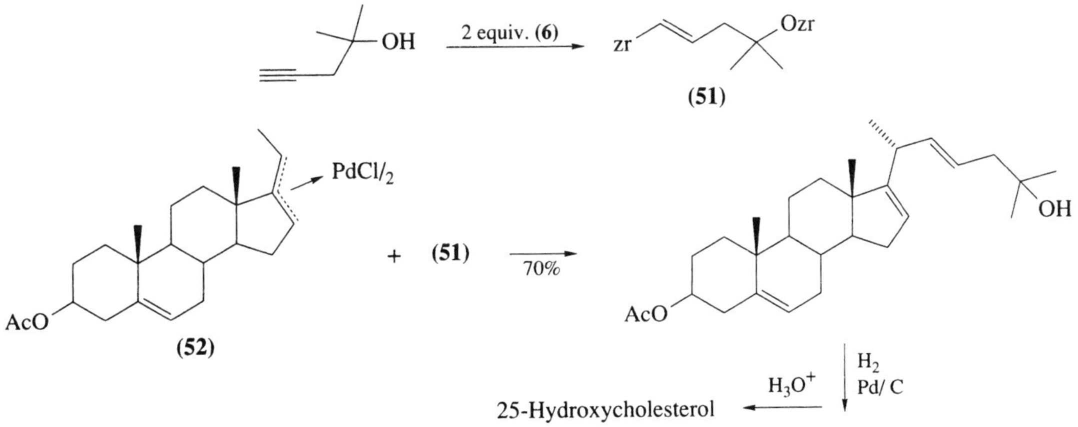

Scheme 15

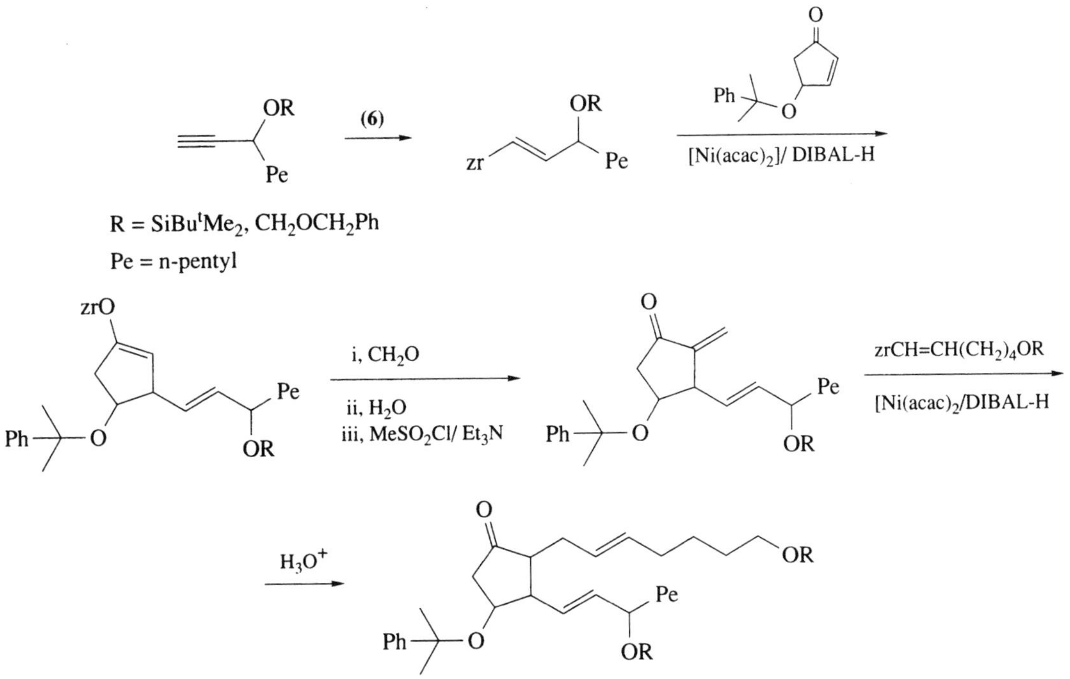

Scheme 16

3.9.4 OTHER HYDROMETALLATIONS

3.9.4.1 Hydrometallation with Other Transition Metals

Although zirconium is only one out of over 50 potentially usable metals in this class (including the lanthanides and actinides), virtually all synthetic applications of hydrometallation with transition metals involve zirconium! Why is this so? The primary reason derives from the near requirement of a d^0-metal center for hydrometallation of a general alkene or alkyne. For later transition metals, hydrometallation to give a stable organometallic product can usually be achieved only for special cases—conjugated dienes, alkenes with electronegative substituents, *etc*. This is due to the relative stability of the η^2-complex, as discussed previously.

One exception to this exclusion of late transition metals may be copper. CuH—MgX_2 complexes, prepared *in situ* from MgH_2/CuX[120] or from NaH/CuX/MgX_2,[121] react with terminal alkynes to give alkenylcopper species. So far, these have only been used as sources for dialkenyl-coupling products, as in equation (54),[121] but there is no obvious reason why other copper-based procedures should not be accessible, as was found for alkenylcopper obtained by transmetallation from Zr (Section 3.9.3.4.2). Alkylcopper cannot be made by hydrometallation (nor by transmetallation), as CuH does not add to alkenes,[120] except for special ones such as enones.[122]

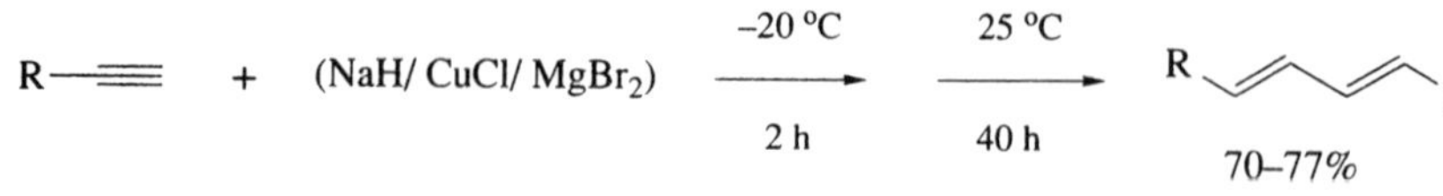

$$\text{R—≡} + (\text{NaH/ CuCl/ MgBr}_2) \xrightarrow[2\text{ h}]{-20\ ^\circ\text{C}} \xrightarrow[40\text{ h}]{25\ ^\circ\text{C}} \text{R–CH=CH–CH=CH–R}\quad 70\text{–}77\% \tag{54}$$

What about other early transition metals? Stable titanium hydrides with d^0-configurations are uncommon, owing to ready reduction to titanium(III) species. Hydrotitanation has been found only for conjugated dienes[123] and trienes[124] where stable η^3-allyls result; such reactivity is found throughout the Periodic Table.[125] With [Cp^*_2ScH] (obtained either as a THF adduct, which is monomeric, or an unsolvated polymer),[126] hydrometallation of alkenes and alkynes does take place rapidly even at –80 °C, but complications set in. Ethylene readily undergoes additional insertion into the Sc—C bond to give a polymer, while substituted alkenes such as propene and isobutene react to form vinyl complexes, such as (**53**), by the so-called σ-bond metathesis mechanism (Scheme 17).[25,126] The predominance of σ-bond metathesis reactions is probably related to the lower coordination here than in the hydrozirconation reagents. It is faster for C—H bonds with higher *s*-character; thus terminal alkynes are not hydrometallated at all, but instead undergo the reaction in equation (55),[126] even though the expected hydrometallation product (**53**) is stable.

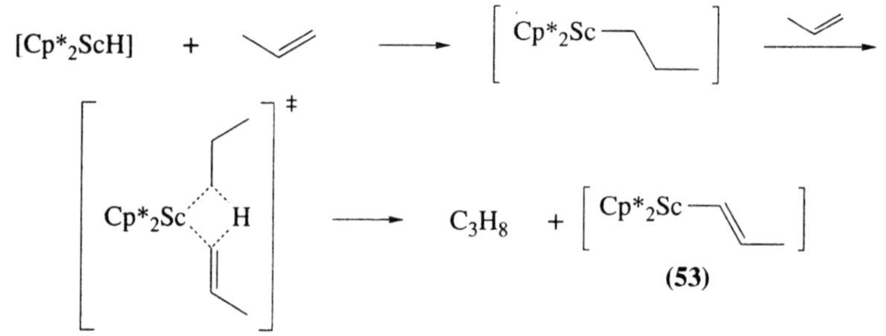

[Cp^*_2ScH] + propene ⟶ [Cp^*_2Sc–propyl] —(propene)⟶ [Cp^*_2Sc···H four-centre transition state]‡ ⟶ C_3H_8 + [Cp^*_2Sc–vinyl] (**53**)

Scheme 17

$$(\mathbf{53}) \not\leftarrow [\text{Cp}^*_2\text{ScH}] + \text{≡—Me} \longrightarrow [\text{Cp}^*_2\text{ScC≡CMe}] + \text{H}_2 \tag{55}$$

The situation for lanthanides and actinides is similar. A number of [$(Cp^*_2LnH)_2$] complexes have been prepared and found to be very active hydrogenation catalysts. The mechanism in Scheme 18 appears to operate, with alkyl complex (**54**) the dominant species present in solution and its hydrogenolysis the rate-determining step.[127] In the absence of added H_2, polymerization and/or σ-bond metathesis sets in (although in contrast to Sc, the result of the latter is an η^3-allyl).[128] No reactions with one equivalent of alkene or alkyne apparently were performed, though, and it is possible that stable alkyllanthanides might be isolable under the right conditions, as with Sc above. The related actinide complexes [{Cp^*_2M(μ-

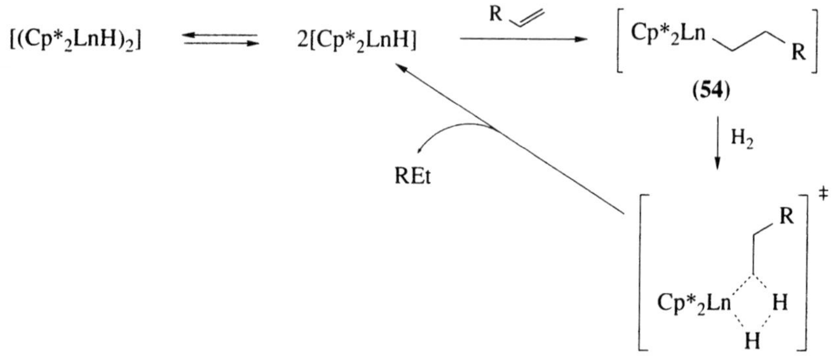

[$(Cp^*_2LnH)_2$] ⇌ 2[Cp^*_2LnH] —(R-alkene)⟶ [Cp^*_2Ln–CH₂CH₂R] (**54**) —(H_2)⟶ [Cp^*_2Ln···H···H four-centre transition state]‡ ⟶ 2[Cp^*_2LnH] + REt

Scheme 18

H)H}2] (M = Th, U) are also hydrogenation catalysts, albeit much less active; for Th the reaction with ethylene afforded thermally stable but light sensitive [$Cp^*_2ThEt_2$].[129]

For all these cases — Sc, lanthanides, actinides — it is probable that conditions for clean hydrometallation could be worked out, but it is questionable whether subsequent transformations would also be clean, and the starting hydrides are all much more expensive and difficult to synthesize than [Cp_2ZrHCl].

Moving the other way to Group VA, the system depicted in equation (2) presents opportunities for highly regioselective additions to alkynes, although subsequent reactions may be complex (equation 56).[3] This, coupled with the difficulty of effecting hydrometallation, makes it unlikely that such systems will find much use in synthesis.

[Cp_2NbH_3] + ≡—Pr^i —(78 °C, $-H_2$)→ —(CO, 75 °C)→ $Cp_2Nb(CO)$ (H, Pr^i alkenyl) >95% —(D^+)→ D, H, Pr^i alkene (56)

MeOSO$_2$F → H, Pr^i alkene + H, Pr^i alkene

3.9.4.2 Catalytic Hydrometallation with Main Group Metals

Although main group organometallics are frequently more powerful synthetic reagents (or at least better understood!) than their transition metal counterparts, their preparation by direct hydrometallation is often difficult or impossible. The most probable cause is the greater tendency for aggregation of the main group hydrides, as other factors should be no less favorable (Section 3.9.2). Both alkali and alkaline earth hydrides are polymeric and do not hydrometallate at all. Group IIIA hydrides are generally dimeric, and hydroalumination works reasonably well for alkynes but not alkenes; hydroboration of course is quite general. As for Group IVB hydrides, these are monomeric, but they have no low-lying orbital available for interaction with the unsaturated substrate and hence are even less reactive.

This problem may be generally solved by catalyzed hydrometallation, which proceeds as shown in Scheme 1. Here the actual hydrometallating species is a transition metal hydride, but only catalytic amounts are needed. The following survey of such methods is brief; more details on the two most important systems, hydroalumination and hydrosilylation, may be found in Volume 8, Chapters 3.11 and 3.12 respectively.

3.9.4.2.1 Groups IA, IIA and IIB

Only one report involving Li has appeared: 'active' LiH, prepared by hydrogenolysis of BuLi, adds to terminal alkenes in the presence of transition metal halides to give (after hydrolysis) the corresponding alkane. VCl_3 was found to give the best alkane yield, but only a small amount of deuterium was incorporated when D_2O was used for cleavage, indicating a stable alkyllithium compound was not obtained. [Cp_2TiCl_2] gave better D incorporation but lower yields.[130] It is not known whether this is a viable general route to alkyllithium reagents for other uses. The selectivity is unusual, though: with VCl_3 as catalyst internal alkenes and alkynes are unreactive. We thus have a method for hydrogenation of C═C in the presence of C≡C, which is otherwise difficult.[130]

Much more has been done on the preparation of Grignard reagents by catalyzed hydromagnesation. This topic has been recently reviewed;[131] only highlights will be summarized here. The basic reaction is essentially a 'transalkenation', the mechanism of which is shown in Scheme 19. The best catalysts are titanium complexes, $TiCl_4$ and [Cp_2TiCl_2], although there is good evidence that the actual catalytic species MY_n is a titanium(III) species, at least for the latter.[132] Only terminal alkenes are hydrometallated,[131] so isomerization can be a major complication.[132] The usual anti-Markovnikov regioselectivity is observed, except for styrene.[131] In some cases carbometallation may be a complication, as in equation (57),[42] although this example utilizes a stoichiometric amount of Zr.

As an alternative, an isolable magnesium hydride can be used directly: successful examples include RMgH;[132,133] Et_2NMgH;[134] and MgH_2. For the last, reaction with terminal alkenes using ZrX_4 as catalyst affords dialkylmagnesium compounds in 80–85% yield.[135] A special preparation of MgH_2 (by direct hy-

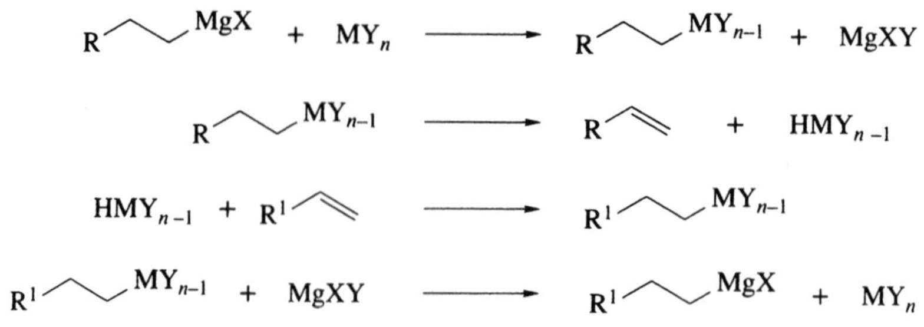

Scheme 19

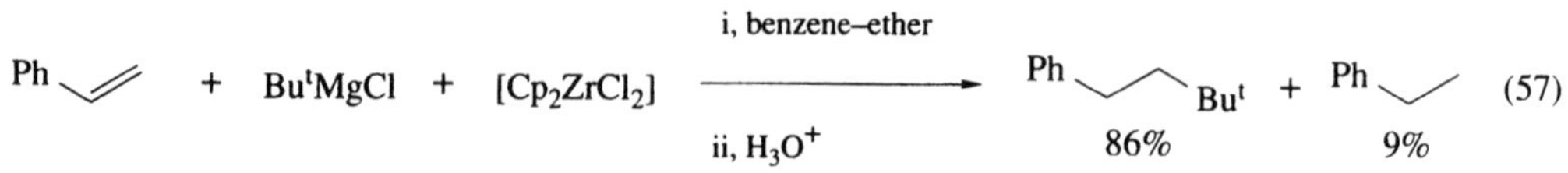

$$\mathrm{Ph{\diagup\!\!\!=} + Bu^tMgCl + [Cp_2ZrCl_2] \xrightarrow[\text{ii, } H_3O^+]{\text{i, benzene–ether}} PhCH_2CH_2Bu^t\ (86\%) + PhCH_2CH_3\ (9\%)} \quad (57)$$

drogenation of Mg metal in the presence of $CrCl_3$) undergoes Ti-catalyzed 1,4-addition to isoprene.[136] This hydromagnesation route to allylic Grignards, which can also be achieved with other starting Mg reagents, avoids the complication of coupling frequently found in the direct preparation from the halide,[131] and has been applied to complex syntheses such as that of polyisoprene polyols.[137] Other transition metal catalysts for hydromagnesation include $[Cp_2ZrCl_2]$[133] and $NiCl_2$.[133,138]

Internal and TMS-protected terminal alkynes are hydromagnesated with high stereoselectivity and (for the latter) excellent regioselectivity, providing a source of vinyl Grignard reagents;[131] Scheme 20 shows a typical use.[139] Note that the Grignard can be thus obtained directly from an alcohol simply by using two equivalents of the Mg reagent, which would not be possible for preparation from a halide and Mg metal; also note the *cis–trans* isomerization which takes place under the reaction conditions.

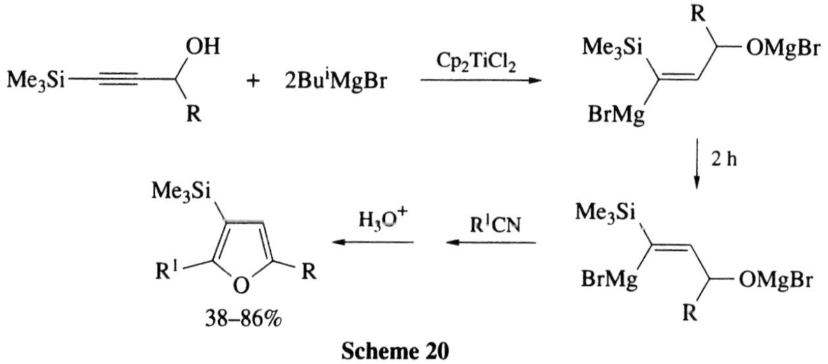

Scheme 20

Organozinc compounds are obtained analogously from ZnH_2 with Zr or Ni catalysts,[133,140] but little has been done to demonstrate the generality or utility of this method.

3.9.4.2.2 Group IIIB

The vast majority of work here is on hydroalumination of alkenes. Hydroboration (Volume 8, Chapter 3.10) generally proceeds perfectly well without any requirement for catalyst. One reason for employing a catalyst might be to allow the use of boron hydrides such as $LiBH_4$[141] or $NaBH_4$,[142] which are somewhat more convenient to handle than the classic hydroboration reagents, and catalyzed hydroboration has been used to control the regio- and stereo-chemistry in 1,3-diol synthesis from allylic alcohol derivatives.[143]

Hydroalumination of alkenes by LAH, catalyzed by Ti and Zr chloride complexes, was demonstrated by two Japanese groups (equation 58).[144,145] Subsequently this reaction has been extended to other sources of Al—H and other catalysts. Further information on catalyzed hydroalumination may be found in Chapter 3.11 of this volume as well as in an earlier review.[146] Hydrogallation by $NaGaH_4/[Cp_2MCl_2]$ (M = Ti, Zr) has been briefly examined and appears quite similar;[147] little if anything is known about the synthetic utility of organogallium reagents.

$$LiAlH_4 \quad + \quad R\diagup\!\!\diagdown\!\!\!\!= \quad \xrightarrow{MX_n} \quad LiAl(CH_2CH_2R)_4 \tag{58}$$

3.9.4.2.3 Group IVB

This group is mechanistically distinct from all those discussed above, in two ways. First, as already noted, no hydride compound with a vacant orbital is readily accessible, precluding coordination of the unsaturated substrate and direct hydrometallation. Uncatalyzed hydrosilylation therefore follows a radical mechanism, such as the one in Scheme 7, and requires high temperatures for 'unactivated' alkenes. Second, the (catalytic metal)–hydride complex is generated by oxidative addition of the (Group IVB metal)–H bond (Scheme 21); hence late transition metal catalysts are required, such as H_2PtCl_6, $[Co_2(CO)_8]$, $[(Ph_3P)_3RhCl]$, *etc.* Hydrosilylation has been extensively studied, and several reviews are available;[148] see also Chapter 3.12 of this volume. More recently hydrogermylation and hydrostannylation have been examined.[149,150]

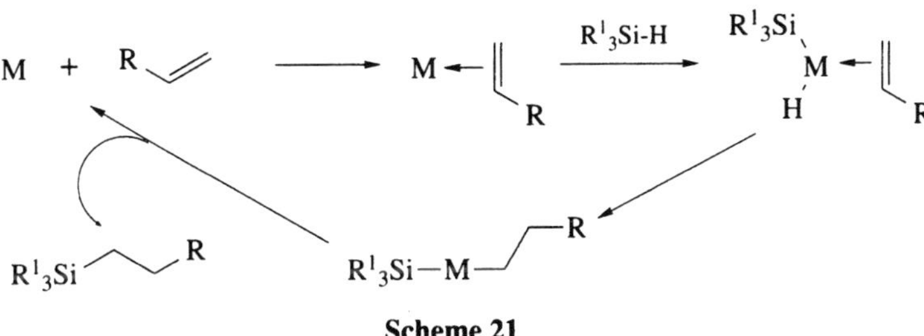

Scheme 21

3.9.5 REFERENCES

1. J. Schwartz and J. A. Labinger, *Angew. Chem., Int. Ed. Engl.*, 1976, **15**, 333.
2. N. M. Doherty and J. E. Bercaw, *J. Am. Chem. Soc.*, 1985, **107**, 2670.
3. J. A. Labinger and J. Schwartz, *J. Am. Chem. Soc.*, 1975, **97**, 1596.
4. D. W. Hart, T. F. Blackburn and J. Schwartz, *J. Am. Chem. Soc.*, 1975, **97**, 679.
5. J. W. Lauher and R. Hoffmann, *J. Am. Chem. Soc.*, 1976, **98**, 1729.
6. P. O. Stoutland, R. G. Bergman, S. P. Nolan and C. D. Hoff, *Polyhedron*, 1988, **7**, 1429.
7. B. B. Wayland, *Polyhedron*, 1988, **7**, 1545.
8. S. P. Nolan, R. Lopez de la Vega, S. L. Mukerjee, A. A. Gonzalez, K. Zhang and C. D. Hoff, *Polyhedron*, 1988, **7**, 1491.
9. L. E. Schock, A. M. Seyam, M. Sabat and T. J. Marks, *Polyhedron*, 1988, **7**, 1517.
10. T. Ziegler, V. Tschinke, L. Versluis, E. J. Baerends and W. Ravenek, *Polyhedron*, 1988, **7**, 1625.
11. J. A. Labinger and J. E. Bercaw, *Organometallics*, 1988, **7**, 926.
12. A. Evans, C. T. Mortimer and R. J. Puddephatt, *J. Organomet. Chem.*, 1975, **85**, 101; 1975, **96**, C58.
13. G. Pilcher and H. A. Skinner, in 'The Chemistry of the Metal–Carbon Bond', ed. F. R. Hartley and S. Patai, Wiley, New York, 1982, vol. 1, p. 43.
14. T. Ziegler and A. Rauk, *Inorg. Chem.*, 1979, **18**, 1558.
15. T. Mole and E. A. Jeffrey, 'Organoaluminum Compounds', Elsevier, New York, 1972, p. 67.
16. H. E. Bryndza, P. J. Domaille, W. Tam, L. K. Fong, R. A. Paciello and J. E. Bercaw, *Polyhedron*, 1988, **7**, 1441.
17. A. R. Dias and J. A. Martinho Simões, *Polyhedron*, 1988, **7**, 1531.
18. A. R. Bulls, J. E. Bercaw, J. M. Manriquez and M. E. Thompson, *Polyhedron*, 1988, **7**, 1409.
19. D. L. Thorn and R. Hoffmann, *J. Am. Chem. Soc.*, 1978, **100**, 2079.
20. J. Halpern and T. Okamoto, *Inorg. Chim. Acta*, 1984, **89**, L53.
21. B. J. Burger, B. D. Santarsiero, M. S. Trimmer and J. E. Bercaw, *J. Am. Chem. Soc.*, 1988, **110**, 3134.
22. D. C. Roe, *J. Am. Chem. Soc.*, 1983, **105**, 7770.
23. J. M. Manriquez, D. R. McAlister, R. D. Sanner and J. E. Bercaw, *J. Am. Chem. Soc.*, 1978, **100**, 2716.
24. D. M. Roddick, M. D. Fryzuk, P. F. Seidler, G. L. Hillhouse and J. E. Bercaw, *Organometallics*, 1985, **4**, 97.
25. G. Parkin, E. Bunel, B. J. Burger, M. S. Trimmer, A. van Asselt and J. E. Bercaw, *J. Mol. Catal.*, 1987, **41**, 21.
26. P. T. Barger and J. E. Bercaw, *Organometallics*, 1984, **3**, 278.
27. T. P. Fehlner, *J. Am. Chem. Soc.*, 1971, **93**, 6366.
28. D. W. Hart and J. Schwartz, *J. Am. Chem. Soc.*, 1974, **96**, 8115.
29. T. Gibson, *Organometallics*, 1987, **6**, 918.
30. B. Rickborn and S. E. Wood, *J. Am. Chem. Soc.*, 1971, **93**, 3940.
31. R. L. Sweany and J. Halpern, *J. Am. Chem. Soc.*, 1977, **99**, 8335.
32. B. Wassink, M. J. Thomas, S. C. Wright, D. J. Gillis and M. C. Baird, *J. Am. Chem. Soc.*, 1987, **109**, 1995 and refs. therein.

33. R. D. W. Kemmitt and D. R. Russell, in 'Comprehensive Organometallic Chemistry', ed. G. Wilkinson, F. G. A. Stone and E. W. Abel, Pergamon Press, Oxford, 1982, vol. 5, p. 95 and refs. therein.
34. S. Derenne, A. Gaudemer and M. D. Johnson, *J. Organomet. Chem.*, 1987, **322**, 229.
35. P. C. Wailes and H. Weigold, *J. Organomet. Chem.*, 1970, **24**, 405.
36. P. C. Wailes, H. Weigold and A. P. Bell, *J. Organomet. Chem.*, 1972, **43**, C32.
37. P. C. Wailes, H. Weigold and A. P. Bell, *J. Organomet. Chem.*, 1971, **27**, 373.
38. D. B. Carr and J. Schwartz, *J. Am. Chem. Soc.*, 1979, **101**, 3521.
39. G. J. Lynch (Monsanto), *US Pat.* 4 147 709 (1979) (*Chem. Abstr.*, 1979, **91**, 20 723).
40. T. Gibson and L. Tulich, *J. Org. Chem.*, 1981, **46**, 1821.
41. S. L. Buchwald, S. J. LaMaire, R. B. Nielsen, B. T. Watson and S. M. King, *Tetrahedron Lett.*, 1987, **28**, 3895.
42. E. Negishi, J. A. Miller and T. Yoshida, *Tetrahedron Lett.*, 1984, **25**, 3407.
43. K. I. Gell and J. Schwartz, *J. Am. Chem. Soc.*, 1978, **100**, 3246.
44. M. Yoshifuji, K. I. Gell and J. Schwartz, *J. Organomet. Chem.*, 1978, **153**, C15.
45. K. I. Gell and J. Schwartz, *J. Organomet. Chem.*, 1978, **162**, C11.
46. D. R. McAlister, D. K. Erwin and J. E. Bercaw, *J. Am. Chem. Soc.*, 1978, **100**, 5966.
47. U. Dorf, K. Engel and G. Erker, *Angew. Chem., Int. Ed. Engl.*, 1982, **21**, 914.
48. D. G. Bickley, N. Hao, P. Bougeard, B. G. Sayer, R. C. Burns and M. J. McGlinchey, *J. Organomet. Chem.*, 1983, **246**, 257.
49. S. B. Jones and J. L. Petersen, *Inorg. Chem.*, 1981, **20**, 2889.
50. P. T. Wolczanski and J. E. Bercaw, *Organometallics*, 1982, **1**, 793.
51. P. T. Wolczanski, R. S. Threlkel and J. E. Bercaw, *J. Am. Chem. Soc.*, 1979, **101**, 218; G. L. Hillhouse and J. E. Bercaw, unpublished results.
52. G. L. Hillhouse and J. E. Bercaw, *J. Am. Chem. Soc.*, 1984, **106**, 5472.
53. P. T. Wolczanski and J. E. Bercaw, *J. Am. Chem. Soc.*, 1979, **101**, 6450.
54. B.-H. Chang, R. H. Grubbs and C. H. Brubaker, Jr., *J. Organomet. Chem.*, 1985, **280**, 365.
55. G. A. Tolstikov, M. S. Miftakhov and F. A. Valeev, *Izv. Akad. Nauk SSSR, Ser. Khim.*, 1979, 2576 (*Bull. Acad. Sci. USSR, Div. Chem. Sci.*, 1979, **28**, 2392).
56. S. L. Buchwald and S. J. LaMaire, *Tetrahedron Lett.*, 1987, **28**, 295.
57. M. S. Miftakhov, N. N. Siderov and G. A. Tolstikov, *Izv. Akad. Nauk SSSR, Ser. Khim.*, 1979, 2748 (*Bull. Acad. Sci. USSR, Div. Chem. Sci.*, 1979, **28**, 2556).
58. J. Alvhall, S. Gronowitz, A. Hallberg and R. Svenson, *J. Am. Oil Chem. Soc.*, 1984, **61**, 430.
59. Y. Liu, Q. Guo and X. Lei, *Youji Huaxue*, 1984, 33 (*Chem. Abstr.*, 1984, **101**, 37 891).
60. K. I. Gell, B. Posin, J. Schwartz and G. M. Williams, *J. Am. Chem. Soc.*, 1982, **104**, 1846.
61. G. Erker, K. Kropp, J. L. Atwood and W. E. Hunter, *Organometallics*, 1983, **2**, 1555.
62. H. Takaya, M. Yamakawa and K. Mashima, *J. Chem. Soc., Chem. Commun.*, 1983, 1283.
63. C. Qian, K. Lu and H. Zhou, in 'Fundamental Research in Organometallic Chemistry, Proceedings of the 1st China–Japan–United States Trilateral Seminar on Organometallic Chemistry', ed. M. Tsutsui, Y. Ishii and Y. Huang, Van Nostrand–Reinhold, New York, 1982, p. 705.
64. R. Svenson and S. Gronowitz, *Chem. Scr.*, 1982, **19**, 149.
65. J. Alvhäll, S. Gronowitz, A. Hallberg and R. Svenson, *Chem. Scr.*, 1985, **24**, 170.
66. P. Vincent, J.-P. Beaucourt and L. Pichat, *Tetrahedron Lett.*, 1982, **23**, 63; P. Vincent, J.-P. Beaucourt, L. Pichat, J. Balzarini and E. De Clercq, *Nucleosides Nucleotides*, 1985, **4**, 447.
67. S. L. Buchwald, B. T. Watson and J. C. Huffman, *J. Am. Chem. Soc.*, 1987, **109**, 2544.
68. J. Schwartz, M. J. Loots and H. Kosugi, *J. Am. Chem. Soc.*, 1980, **102**, 1333.
69. L. Crombie, M. A. Horsham and R. J. Blade, *Tetrahedron Lett.*, 1987, **28**, 4879.
70. C. Qian, D. Deng, H. Zhou and L. Shi, *Huaxue Xuebao*, 1984, **42**, 80 (*Chem. Abstr.*, 1984, **100**, 191 312); C. Qian, D. Deng, H. Zhou and L. Shi, *Youji Huaxue*, 1983, 42 (*Chem. Abstr.*, 1983, **99**, 21 897).
71. M. Riediker and J. Schwartz, *Tetrahedron Lett.*, 1981, **22**, 4655.
72. E. Negishi, T. Takahashi, S. Baba, D. E. Van Horn and N. Okukado, *J. Am. Chem. Soc.*, 1987, **109**, 2393.
73. G. A. Tolstikov, M. S. Miftakhov, N. A. Danilova and F. Z. Galin, *Zh. Org. Khim.*, 1983, **19**, 1857 (*Chem. Abstr.*, 1984, **100**, 513 266).
74. C. A. Bertelo and J. Schwartz, *J. Am. Chem. Soc.*, 1976, **98**, 262.
75. Maruzen Oil Co., *Jpn. Pat.* 82 120 543 (1982) (*Chem. Abstr.*, 1982, **97**, 215 548).
76. A. W. Messing, F. P. Ross, A. W. Norman and W. H. Okamura, *Tetrahedron Lett.*, 1978, 3635.
77. L. Crombie, A. J. W. Hobbs, M. A. Horsham and R. J. Blade, *Tetrahedron Lett.*, 1987, **28**, 4875.
78. M. D. Fryzuk, G. S. Bates and C. Stone, *Tetrahedron Lett.*, 1986, **27**, 1537.
79. K. Maruyama, K. Terada and Y. Yamamoto, *J. Org. Chem.*, 1980, **45**, 737.
80. S. H. Pine, R. Zahler, D. A. Evans and R. H. Grubbs, *J. Am. Chem. Soc.*, 1980, **102**, 3270.
81. F. M. Hartner, Jr. and J. Schwartz, *J. Am. Chem. Soc.*, 1981, **103**, 4979.
82. F. M. Hartner, Jr., S. M. Clift, J. Schwartz and T. H. Tulip, *Organometallics*, 1987, **6**, 1346.
83. G. Erker, W. Frömberg, K. Angermund, R. Schlund and C. Krüger, *J. Chem. Soc., Chem. Commun.*, 1986, 372.
84. F. D. Miller and R. D. Sanner, *Organometallics*, 1988, **7**, 818.
85. C. McDade and J. E. Bercaw, *J. Organomet. Chem.*, 1985, **279**, 281.
86. J. A. Labinger and K. H. Komadina, *J. Organomet. Chem.*, 1978, **155**, C25.
87. P. Etievant, B. Gautheron and G. Tainturier, *Bull. Soc. Chim. Fr., Part 2*, 1978, 292.
88. W. Fromberg and G. Erker, *J. Organomet. Chem.*, 1985, **280**, 355.
89. D. E. Laycock and H. Alper, *J. Org. Chem.*, 1981, **46**, 289.
90. G. Fachinetti, C. Floriani, A. Roselli and S. Pucci, *J. Chem. Soc., Chem. Commun.*, 1978, 269; S. Gambarotta, C. Floriani, A. Chiesi-Villa and C. Guastini, *J. Am. Chem. Soc.*, 1983, **105**, 1690; S. Gambarotta, S. Strologo, C. Floriani, A. Chiesi-Villa and C. Guastini, *J. Am. Chem. Soc.*, 1985, **107**, 6278.
91. K. S. Ng, D. E. Laycock and H. Alper, *J. Org. Chem.*, 1981, **46**, 2899.

92. E. Cesarotti, A. Chiesa, S. Maffi and R. Ugo, *Inorg. Chim. Acta*, 1982, **64**, L207.
93. W. Frömberg and G. Erker, *J. Organomet. Chem.*, 1985, **280**, 343; for the crystal structure of one such product, see: G. Erker, W. Frömberg, J. L. Atwood and W. E. Hunter, *Angew. Chem., Int. Ed. Engl.*, 1984, **23**, 68.
94. P. Etievant, G. Tainturier and B. Gautheron, *C. R. Hebd. Seances Acad. Sci., Ser. C*, 1976, **283**, 233.
95. S. L. Buchwald, R. B. Nielsen and J. C. Dewan, *Organometallics*, 1988, **7**, 2324; G. Erker, R. Schlund and C. Krüger, *J. Chem. Soc., Chem. Commun.*, 1986, 1403.
96. W. Tam and M. F. Rettig, *J. Organomet. Chem.*, 1976, **108**, C1.
97. D. W. Hart and J. Schwartz, unpublished results.
98. J. E. Nelson, J. E. Bercaw and J. A. Labinger, *Organometallics*, 1989, **8**, 2484.
99. P. Meunier, B. Gautheron and S. Couturier, *J. Organomet. Chem.*, 1982, **231**, C1.
100. J. A. Labinger, D. W. Hart, W. E. Seibert, III and J. Schwartz, *J. Am. Chem. Soc.*, 1975, **97**, 3851.
101. S. Couturier and B. Gautheron, *J. Organomet. Chem.*, 1977, **157**, C61; S. Couturier, G. Tainturier and B. Gautheron, *J. Organomet. Chem.*, 1980, **195**, 291.
102. A. Schäfer, E. Karl, L. Zsolnai, G. Huttner and H. H. Brintzinger, *J. Organomet. Chem.*, 1987, **328**, 87 and refs. therein.
103. P. Pino, P. Cioni and J. Wei, *J. Am. Chem. Soc.*, 1987, **109**, 6189.
104. E. Negishi and T. Takahashi, *Synthesis*, 1988, 1; E. Negishi and T. Takahashi, *Aldrichimica Acta*, 1985, **18**, 31.
105. T. F. Blackburn, J. A. Labinger and J. Schwartz, *Tetrahedron Lett.*, 1975, 3041.
106. C. A. Bertelo and J. Schwartz, *J. Am. Chem. Soc.*, 1975, **97**, 228.
107. R. L. Kidwell and G. J. Lynch (Monsanto), *US Pat.* 4 151 186 (1979) (*Chem. Abstr.*, 1979, **91**, 38 939).
108. E. Negishi and L. D. Boardman, *Tetrahedron Lett.*, 1982, **23**, 3327.
109. J. Schwartz, in 'New Applications of Organometallic Reagents in Organic Synthesis', ed. D. Seyferth, Elsevier, New York, 1976, p. 461.
110. M. Yoshifuji, M. J. Loots and J. Schwartz, *Tetrahedron Lett.*, 1977, 1303.
111. V. G. Kumar Das and O. G. Chee, *J. Organomet. Chem.*, 1987, **321**, 335.
112. E. Negishi, N. Okukado, A. O. King, D. E. Van Horn and B. I. Spiegel, *J. Am. Chem. Soc.*, 1978, **100**, 2254.
113. Y. Hayashi, M. Riediker, J. S. Temple and J. Schwartz, *Tetrahedron Lett.*, 1981, **22**, 2629.
114. F. M. Dayrit and J. Schwartz, *J. Am. Chem. Soc.*, 1981, **103**, 4466.
115. J. S. Temple, M. Riediker and J. Schwartz, *J. Am. Chem. Soc.*, 1982, **104**, 1310.
116. M. J. Loots and J. Schwartz, *J. Am. Chem. Soc.*, 1977, **99**, 8045.
117. J. Schwartz and Y. Hayashi, *Tetrahedron Lett.*, 1980, **21**, 1497.
118. N. K. Kochetkov, A. F. Sviridov, D. V. Yashunskii, M. S. Ermolenko and V. S. Borodin, *Izv. Akad. Nauk SSSR, Ser. Khim.*, 1986, 441 (*Bull. Acad. Sci. USSR, Div. Chem. Sci.*, 1986, **35**, 408).
119. D. L. Coffen, P. S. Manchand and L. K. Truesdale (Hoffmann-LaRoche), *Eur. Pat. Appl.* 153 689 (1985) (*Chem. Abstr.*, 1986, **104**, 186 231).
120. E. C. Ashby, J. J. Lin and A. B. Goel, *J. Org. Chem.*, 1978, **43**, 757.
121. S. Acyutha Rao and M. Periasamy, *J. Chem. Soc., Chem. Commun.*, 1987, 495.
122. R. K. Boeckman, Jr. and R. Michalak, *J. Am. Chem. Soc.*, 1974, **96**, 1623; S. Masamune, G. S. Bates and P. E. Georghiou, *J. Am. Chem. Soc.*, 1974, **96**, 3686.
123. H. A. Martin and F. Jellinek, *J. Organomet. Chem.*, 1968, **12**, 149.
124. H. Lehmkuhl and S. Fustero, *Liebigs Ann. Chem.*, 1980, 1353.
125. U. Franke and E. Weiss, *J. Organomet. Chem.*, 1978, **152**, 39 and refs. cited therein.
126. M. E. Thompson, S. M. Baxter, A. R. Bulls, B. J. Burger, M. C. Nolan, B. D. Santarsiero, W. P. Schaefer and J. E. Bercaw, *J. Am. Chem. Soc.*, 1987, **109**, 203.
127. G. Jeske, H. Lauke, H. Mauermann, H. Schumann and T. J. Marks, *J. Am. Chem. Soc.*, 1985, **107**, 8111.
128. G. Jeske, H. Lauke, H. Mauermann, P. N. Swepston, H. Schumann and T. J. Marks, *J. Am. Chem. Soc.*, 1985, **107**, 8091.
129. P. J. Fagan, J. M. Manriquez, E. A. Maatta, A. M. Seyam and T. J. Marks, *J. Am. Chem. Soc.*, 1981, **103**, 6650.
130. E. C. Ashby and S. A. Noding, *J. Org. Chem.*, 1980, **45**, 1041.
131. F. Sato, *J. Organomet. Chem.*, 1985, **285**, 53.
132. E. C. Ashby and R. D. Ainslie, *J. Organomet. Chem.*, 1983, **250**, 1.
133. L. Yu. Gubaidullin, R. M. Sultanov and V. M. Dzhemilev, *Izv. Akad. Nauk SSSR, Ser. Khim.*, 1982, 714 (*Bull. Acad. Sci. USSR, Div. Chem. Sci.*, 1982, **31**, 638).
134. O. S. Vostrikova, R. M. Sultanov and V. M. Dzhemilev, *Izv. Akad. Nauk SSSR, Ser. Khim.*, 1983, 1901 (*Bull. Acad. Sci. USSR, Div. Chem. Sci.*, 1983, **32**, 1724).
135. B. Bogdanović, M. Schwickardi and P. Sikorsky, *Angew. Chem., Int. Ed. Engl.*, 1982, **21**, 199.
136. B. Bogdanović and M. Maruthamuthu, *J. Organomet. Chem.*, 1984, **272**, 115.
137. Y. Kobayashi, F. Sato, T. Miyakoshi, Y. Fujita, M. Shiono, K. Kanehira and S. Suzuki, *Synth. Commun.*, 1986, **16**, 597.
138. T. Amano, T. Ota, K. Yoshikawa, T. Sano, Y. Ohuchi, F. Sato, M. Shiono and Y. Fujita, *Bull. Chem. Soc. Jpn.*, 1986, **59**, 1656.
139. F. Sato and H. Katsuno, *Tetrahedron Lett.*, 1983, **24**, 1809.
140. V. M. Dzhemilev, O. S. Vostrikova and R. M. Sultanov, *Izv. Akad. Nauk SSSR, Ser. Khim.*, 1983, 1195 (*Bull. Acad. Sci. USSR, Div. Chem. Sci.*, 1983, **32**, 1080).
141. K. Isagawa, H. Sano, M. Hattori and Y. Otsuji, *Chem. Lett.*, 1979, 1069.
142. H. S. Lee, K. Isagawa, H. Toyoda and Y. Otsuji, *Chem. Lett.*, 1984, 673.
143. D. A. Evans, G. C. Fu and A. H. Hoveyda, *J. Am. Chem. Soc.*, 1985, **110**, 6917.
144. F. Sato, S. Sato, H. Kodama and M. Sato, *J. Organomet. Chem.*, 1977, **142**, 71.
145. K. Isagawa, K. Tatsumi and Y. Otsuji, *Chem. Lett.*, 1976, 1145.

146. F. Sato, in 'Fundamental Research in Organometallic Chemistry', ed. Y. Ishii and M. Tsutsui, Plenum Press, New York, 1978, vol. 2, p. 81.
147. V. V. Gavrilenko, V. S. Kolesov and L. I. Zakharkin, *Izv. Akad. Nauk SSSR, Ser. Khim.*, 1985, 681 (*Bull. Acad. Sci. USSR, Div. Chem. Sci.*, 1985, **34**, 623).
148. J. F. Harrod and A. J. Chalk, in 'Organic Synthesis via Metal Carbonyls', ed. I. Wender and P. Pino, Wiley, New York, 1977, vol. 2, p. 673; J. L. Speier, *Adv. Organomet. Chem.*, 1979, **17**, 407; I. Fleming, in 'Comprehensive Organic Chemistry', ed. D. H. R. Barton and W. D. Ollis, Pergamon Press, Oxford, 1979, vol. 3, p. 568.
149. Y. Ichinose, H. Oda, K. Oshima and K. Utimoto, *Bull. Chem. Soc. Jpn.*, 1987, **60**, 3468.
150. I. M. Salimgareeva, N. G. Bogatova, A. A. Panasenko, L. M. Khalilov, I. I. Purlei, V. K. Mavrodiev and V. P. Yur'ev, *Izv. Akad. Nauk SSSR, Ser. Khim.*, 1983, 1605 (*Bull. Acad. Sci. USSR, Div. Chem. Sci.*, 1983, **32**, 1456); N. G. Bogatova, I. M. Salimgareeva and V. P. Yur'ev, *Izv. Akad. Nauk SSSR, Ser. Khim.*, 1984, 930 (*Bull. Acad. Sci. USSR, Div. Chem. Sci.*, 1984, **33**, 856); I. M. Salimgareeva, N. G. Bogatova, A. A. Panasenko, Yu. B. Monakov and V. P. Yur'ev, *Izv. Akad. Nauk SSSR, Ser. Khim.*, 1985, 1391 (*Bull. Acad. Sci. USSR, Div. Chem. Sci.*, 1985, **34**, 1272).

3.10
Hydroboration of C=C and C≡C

KEITH SMITH and ANDREW PELTER
University College of Swansea, UK

3.10.1 GENERAL CONSIDERATIONS

3.10.1.1 Introduction

Net hydrogenation of carbon–carbon double and triple bonds is readily achieved by a two-stage process involving initial addition of BH across the multiple bond, and subsequent protonolysis of the B—C bond (Scheme 1).

The method is particularly attractive since the first stage, known as hydroboration, can be applied to almost any nonaromatic carbon–carbon multiple bond and exhibits very useful selectivity features. Hydroboration is also the most attractive means of synthesizing organoboranes, which are arguably the most versatile organometallic reagents available to the organic chemist.[1] The reaction has received extensive

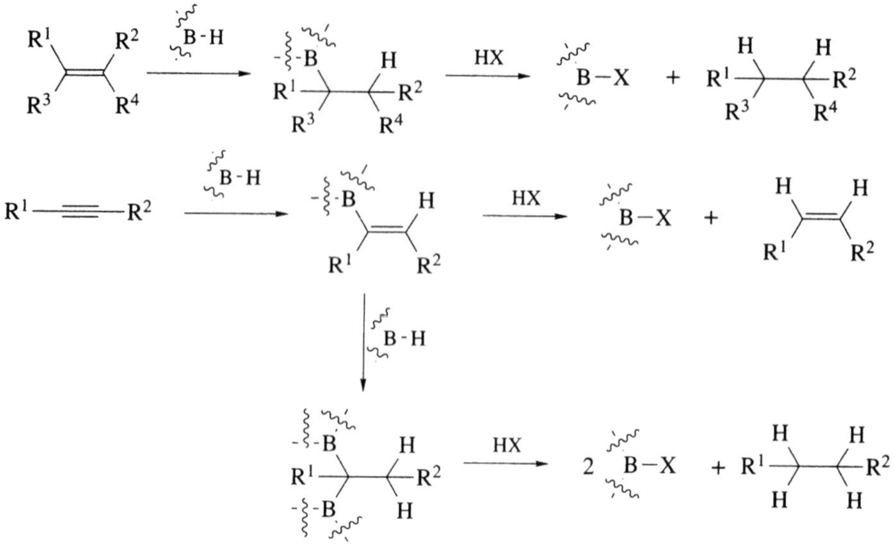

Scheme 1

study and is consequently fairly well understood. Furthermore, a whole range of hydroborating agents has been developed. This provides considerable flexibility with which to control the selectivity of the reaction and the nature of the organoboranes produced. For these reasons the bulk of this review is devoted to the various aspects of the hydroboration reaction. Treatment is necessarily brief, but a recent monograph[1] deals with the topic more fully.

A number of types of organoboranes, including aryl, benzyl, *t*-butyl and methyl derivatives, cannot be obtained *via* hydroboration. Their preparation can be achieved by a number of approaches, the most general of which involves the reaction of a reactive organometallic reagent with a boron halide or alkoxide;[1,2] the organometallic reagent may sometimes be prepared *in situ*.

The reactions of organoboranes are in general not discussed here, but are distributed throughout 'Comprehensive Organic Synthesis' wherever appropriate reaction types are under consideration. However, protonolysis of organoboranes is treated in Section 3.10.7 because of its necessary role in the reduction of carbon–carbon multiple bonds *via* hydroboration (Scheme 1). Reviews of other synthetically useful organoborane reactions appear elsewhere.[1,3–6]

3.10.1.2 Fundamentals of the Hydroboration Reaction

Although different hydroborating agents show considerable differences in ease of reaction and in degree of selectivity (see below), several features are common to all hydroborations.[1,3–6] Thus, the addition across a double bond occurs cleanly in a *cis* fashion. When the two faces of the double bond are sterically differentiated, addition generally occurs preferentially from the less-hindered face, and the regioselectivity is such as to place the boron atom at the less-substituted end of the double bond. Thus, in the hydroboration of α-pinene the organoborane produced consists almost entirely of a single diastereoisomer (equation 1).[1,6] Substantial stereofacial selectivities are also sometimes achieved in acyclic cases.[7]

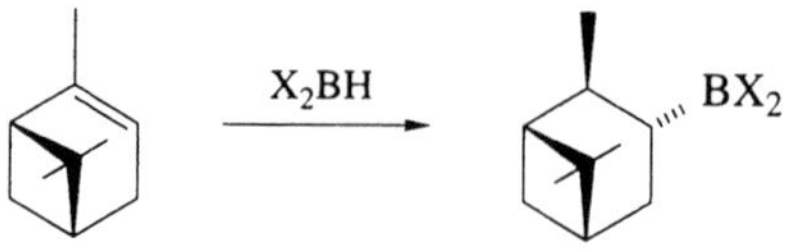

X_2BH BX_2 (1)

Substituents which withdraw electron density from a double bond tend to cause alkenes to react more slowly and to produce greater proportions of products having boron on the side bearing the electron-withdrawing substituent. Electron-donating substituents have exactly the opposite effect. For example, hydroboration of *para*-substituted styrenes with borane–THF produces the 'internal' isomers in proportions which increase as the electron-withdrawing capabilities of the substituents increase (equation 2).[1,8]

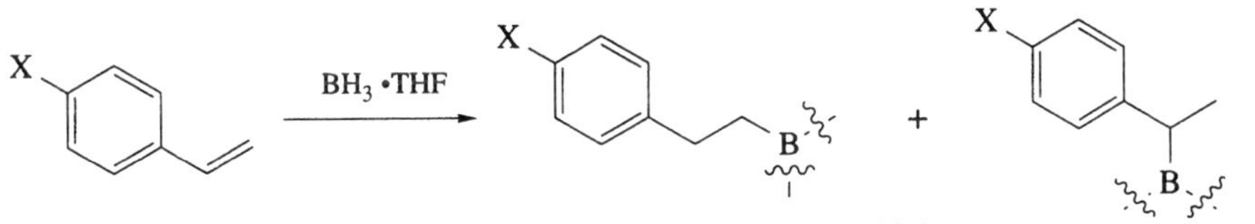

X = OMe 93:7
X = H 81:19
X = CF_3 66:34

(2)

If hydroboration proceeds to place boron at a position proximate to a substituent then complications may arise. For example, β-chloroalkylboranes readily eliminate a chloroborane species to give an alkene. If this occurs during the course of the hydroboration reaction then further hydroboration may result, as is observed in the hydroboration of 3-chlorocyclopentene with borane–THF (Scheme 2).[1,9] The ease of elimination of groups other than halide may be crudely correlated with the leaving group capabilities of the substituents. Thus, β-alkoxyalkylboranes are more stable than β-haloalkyl compounds, though they may still undergo elimination reactions relatively easily under acid or base catalysis.[4] β-Aminoalkylboranes generally require addition of propanoic acid and heat in order to effect clean elimination.[4,10,11] The geometry of the intermediates may also affect their stability, and compounds in which substituents are in a fixed orientation about a ring may be more stable than acyclic analogs.[4]

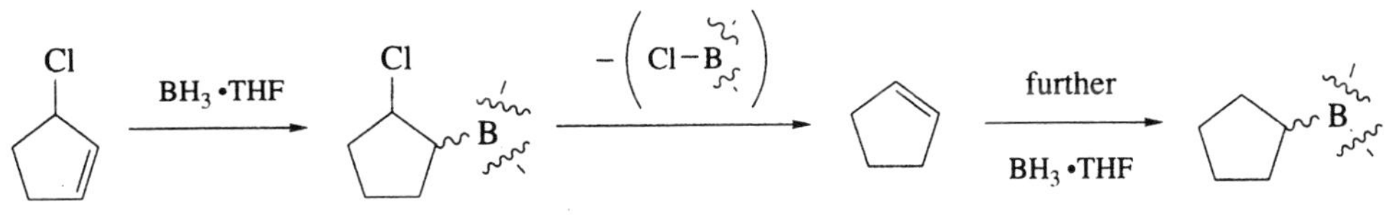

Scheme 2

Haloalkylboron compounds in which the halogen atom is more distant than the β-position are more stable than β-haloalkylboranes,[4] while α-haloalkylboranes undergo ready rearrangement.[12] β,γ-Unsaturated organoboranes may show a tendency to undergo metallotropic rearrangement (Scheme 3).[1,4]

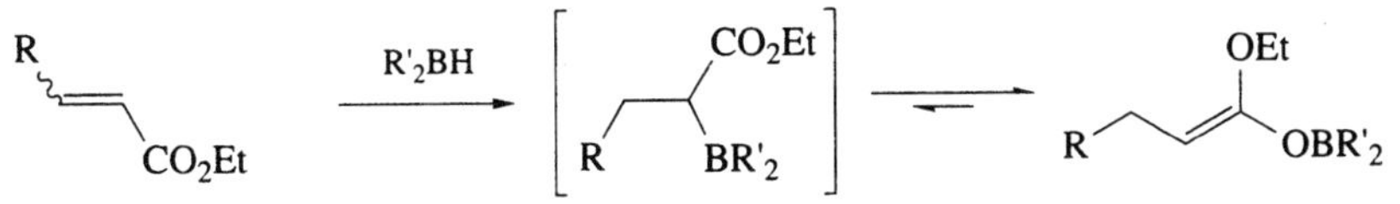

Scheme 3

Dienes and trienes provide opportunities for selective hydroboration of particular double bonds, independent hydroboration of all double bonds, or production of cyclic or oligomeric organoboron structures. However, the products obtained are highly dependent on the nature of the hydroborating agent used and are therefore not discussed further here. Similarly, alkynes give vinylboranes or diborylalkanes in proportions which depend on the reagent and conditions used. As a general rule, however, greater selectivity is available with substituted boranes, such as R_2BH or X_2BH, than with borane itself.[1,4] Because of the differences in selectivity observed with different reagents it is necessary to consider the various types of reagent separately, and this is done in the following subsections. A brief review of trends with different reagents has appeared elsewhere.[13]

3.10.2 HYDROBORATION WITH DIBORANE AND BORANE COMPLEXES

3.10.2.1 Hydroboration with Borane–Tetrahydrofuran

Alkenes react with gaseous diborane only under forcing conditions.[1,3–6,14] However, in the presence of an ether solvent the reaction is much more rapid, and simple alkenes react readily and quantitatively to produce trialkylboranes (equation 3). Diborane has significantly greater solubility in THF than in most other ethers, and forms relatively stable solutions at moderate concentrations. In such solutions the reagent exists substantially in the form of a complex between monomeric borane and THF. This reagent forms the basis against which the properties of other reagents may be compared.

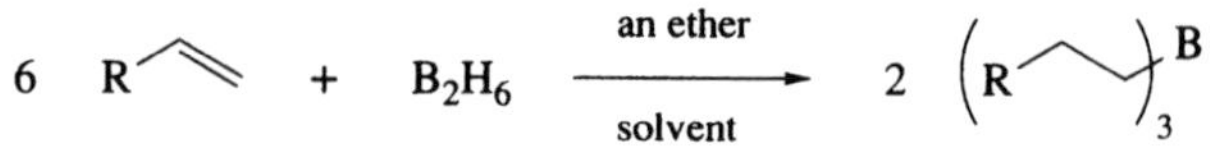

$$6\ \mathrm{R}\diagup\!\!\!\!= \;+\; B_2H_6 \xrightarrow[\text{solvent}]{\text{an ether}} 2\,(\mathrm{R}\diagup\!\diagdown)_3\mathrm{B} \qquad (3)$$

The reactions of alkenes with borane–THF occur sequentially to give first monoalkylboranes, then dialkylboranes, and finally trialkylboranes (Scheme 4).[1] Mono- and di-alkylboranes, like borane itself, exist as hydrogen-bridged dimers in the free state, although they may be complexed as the monomers by suitable Lewis bases. In most cases they will be referred to in this section as though they were monomeric, on the grounds of simplicity and clarity, unless there is a particular reason for consideration of the dimers. It is, however, important for understanding the mechanism of hydroboration that equilibria involving dimers should be recognized (see Section 3.10.6).

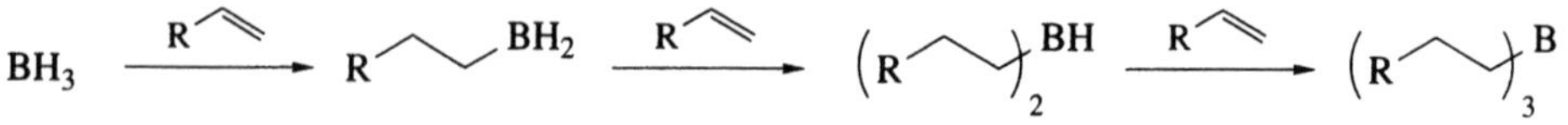

$$BH_3 \xrightarrow{\mathrm{R}\diagup\!=} \mathrm{R}\diagup\!\diagdown BH_2 \xrightarrow{\mathrm{R}\diagup\!=} (\mathrm{R}\diagup\!\diagdown)_2BH \xrightarrow{\mathrm{R}\diagup\!=} (\mathrm{R}\diagup\!\diagdown)_3B$$

Scheme 4

Unhindered alkenes react rapidly as far as the trialkylborane stage, and it is not possible to obtain pure mono- or di-alkylboranes by control of the stoichiometry in such cases. However, alkenes which are somewhat hindered, such as trialkylethylenes, react readily only as far as the dialkylborane stage. Even with some 1,2-dialkylethylenes the rate of the third hydroboration step is sufficiently slower than the first two to allow clean production of dialkylboranes. This is the case with cyclohexene, for example (equation 4).[1]

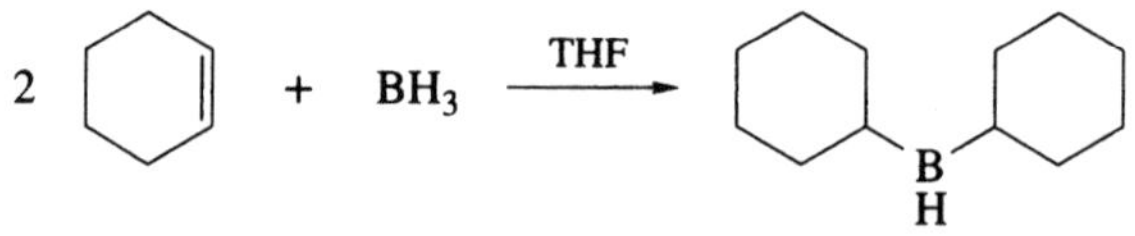

$$2\ \text{cyclohexene} + BH_3 \xrightarrow{\text{THF}} (C_6H_{11})_2BH \qquad (4)$$

Dicyclohexylborane, Chx_2BH

Tetrasubstituted ethylenes and other very hindered alkenes, such as 1,2-di-*t*-butylethylene and 1,3-dimethylcyclohexene, react readily only as far as the monoalkylborane stage. The production of 1,1,2 trimethylpropylborane (thexylborane) in this way is particularly important (equation 5).[1,15]

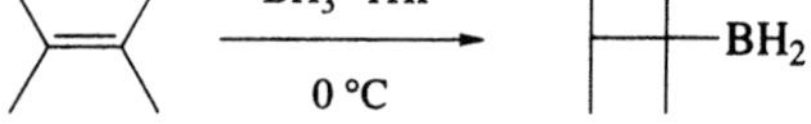

$$Me_2C{=}CMe_2 \xrightarrow[0\,^{\circ}\mathrm{C}]{BH_3\cdot\mathrm{THF}} Me_2CH{-}CMe_2{-}BH_2 \qquad (5)$$

Thexylborane

Although it is general that hydroboration of simple alkenes places boron predominantly at the less-hindered end of the original double bond (see Section 3.10.1.2), the degree of selectivity varies significantly with the nature of the alkene. Figure 1 (see also equation 2) indicates the site of boron placement for a number of representative alkenes.[6,16] It should also be recognized that the regioselectivity increases for the second and particularly for the third hydroboration step (see later). Thus, even though only 6% of the boron atoms are placed at internal positions on hydroboration of unsubstituted 1-alkenes, the trialkylboranes produced contain *ca.* 82% of tri-*n*-alkylborane and *ca.* 18% of di-*n*-alkylmono-*s*-alkylborane.[1,4,5] Less selective hydroborations produce even greater proportions of isomeric organoboranes.

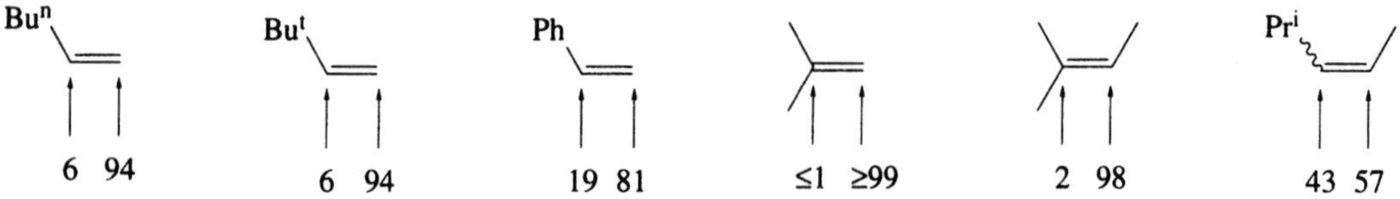

Figure 1 Sites of attack on alkenes by the boron atom of borane–THF (%)

Although considerable selectivity for the less-hindered face is shown in the hydroboration of α-pinene (see equation 1) borane–THF shows only moderate selectivity for the different faces of most simple substituted cyclohexenes and methylenecyclohexanes (*e.g.* Figure 2).[1,17–20]

Figure 2 Facial selectivities for attack on alkenes by the boron atom of borane–THF (%)

Functional groups can substantially modify the regioselectivity (see Section 3.10.1.2, in particular equation 2 and Scheme 2). Some typical regioselectivities for functionally substituted alkenes are shown in Figure 3.[1,4,9,21–23] More extensive accounts of the hydroboration of functional alkenes appear elsewhere.[4,14]

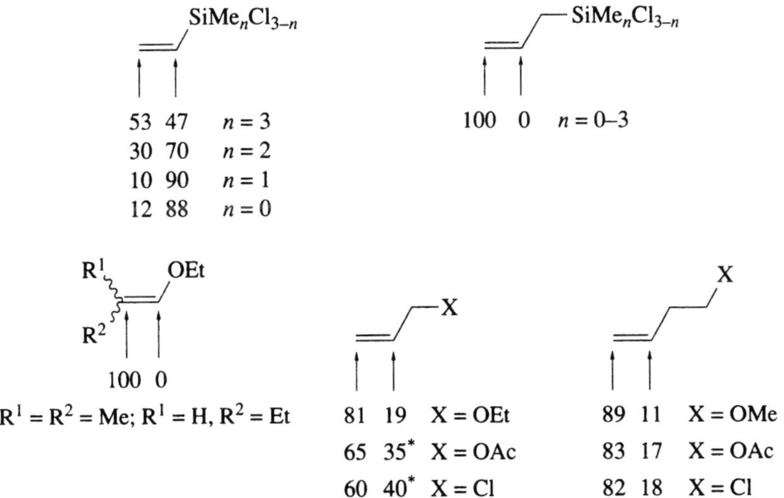

Figure 3 Sites of attack by the boron atom of borane–THF on functionalised alkenes (%). *The products of attack at this point are unstable under the reaction conditions

The hydroboration of simple, open-chain dienes can be somewhat complicated, leading to mixtures of regioisomers, cyclic and open-chain structures, and oligomeric structures of different sizes, all depending on reaction conditions and/or reactant proportions.[1,24] However, by control of stoichiometry, thermal isomerization of product mixtures, control of temperature during the hydroboration stage, subsequent treatment of a trialkylborane with further borane, or similar 'tricks', it is often possible to arrive at an unique product. Several important cyclic dialkylboranes can be made in this way, as shown in equations (6),[25] (7)[26–28] and (8).[29] Similar treatment of trienes can lead to bicyclic systems (*e.g.* equation 9).[1,24,30]

2 [butadiene] + 2 BH_3 —(i, 0 °C; ii, heat)→ [bis(borinane) with B–H–B bridges: H, B B, H] (6)

[diene, R, $(\ )_n$, R] —(i, BH_3•THF; ii, heat)→ [cyclic borane: H, B, R, $(\ )_n$, R] (7)

R = H; n = 1; heat at 170 °C for borinane; for details see Section 3.10.4.3
R = Me; n = 1; heat at 70 °C for 3,5-dimethylborinane
R = Me; n = 2; heat at 70 °C for 3,6-dimethylborepane

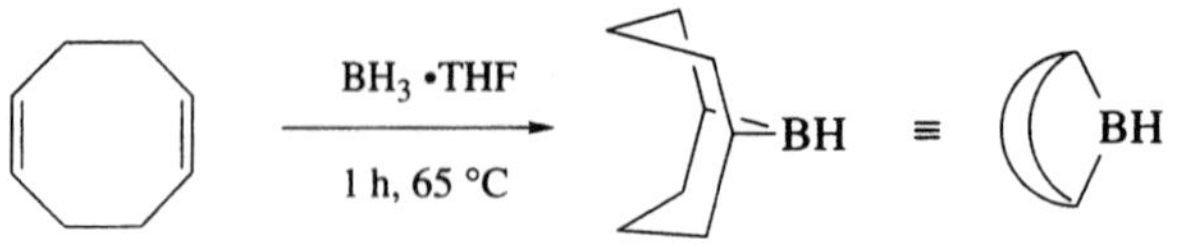

(8)

9-Borabicyclo[3.3.1]nonane, 9-BBN-H

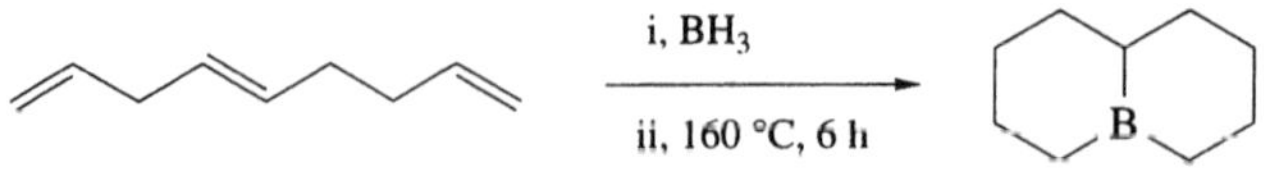

(9)

The reactions of simple, acyclic allenes with borane–THF are usually complicated. However, cyclic allenes give cleaner reactions, which can be controlled to give monohydroboration products, involving predominant attack of boron on the central carbon atom, though the intermediate vinylboranes have not generally been isolated.[1,31,32]

Terminal alkynes also generally give complicated mixtures of products in reactions with borane–THF.[1,33–35] However, internal alkynes react in a more controlled manner and it is possible to obtain reasonable yields of the corresponding (*Z*)-trialkenylboranes from such reactions (equation 10).[33]

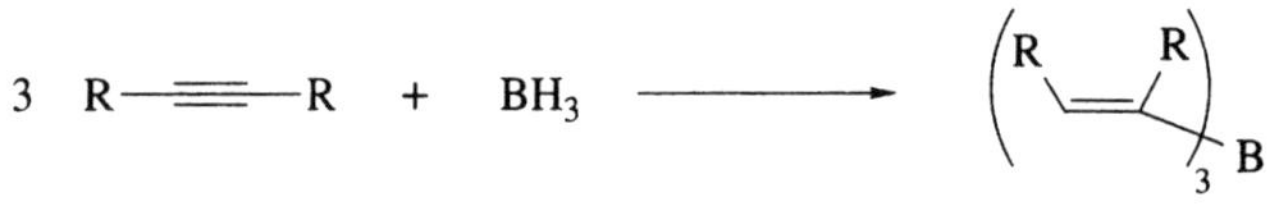

(10)

In the case of unsymmetrical alkynes the regioselectivity is only modest.[34] Use of excess borane–THF results in dihydroboration. Many of the problems encountered in hydroboration reactions of borane–THF can be overcome by the use of more selective reagents such as dialkylboranes or dihaloboranes.

3.10.2.2 Hydroboration with other Sources of Borane

Early reports of hydroboration reactions often recommended *in situ* generation of diborane from sodium tetrahydroborate (borohydride) and trifluoroborane etherate or some similar mixture.[3] Indeed, it is still reasonable to use this method for simple hydroborations, provided that nothing more complicated than oxidation of the resultant organoborane is intended. Otherwise, the approach should be avoided, particularly since borane is now commercially available in the form of several Lewis base complexes.

In principle, all borane–Lewis base complexes are potential hydroborating agents. To a first approximation, however, the stronger the complex, the less reactive it is as a hydroborating agent, because free borane may be required for hydroboration (see Section 3.10.6).[36]

Borane–dimethyl sulfide is a particularly useful reagent.[37,38] The complex is more stable than borane–THF and can be prepared in neat form. It is soluble in a range of organic solvents and retains sufficient 'looseness' to allow it to react readily with alkenes. It carries out most of the reactions of borane–THF, perhaps somewhat more slowly, and gives the same selectivity almost irrespective of solvent. It has been used to hydroborate alkynylsilanes at the α-position.[39] Borane–thioxane is an alternative to borane-dimethyl sulfide.[40]

Similar claims are also made for diphenylamine–borane, which is a solid and relatively stable reagent.[41] Aliphatic amine–borane complexes, however, require elevated temperatures in order to effect hydroboration.[1,3,42] Although this is in general a serious drawback, on occasions it may be advantageous by allowing slow liberation of borane, which leads to cyclization products rather than polymers. An example of the use of triethylamine–borane in this way is shown in equation (11).[30] Alternatively, methyl iodide or a Lewis acid may be added to complex the amine and thus liberate the reactive, free borane[1,42]

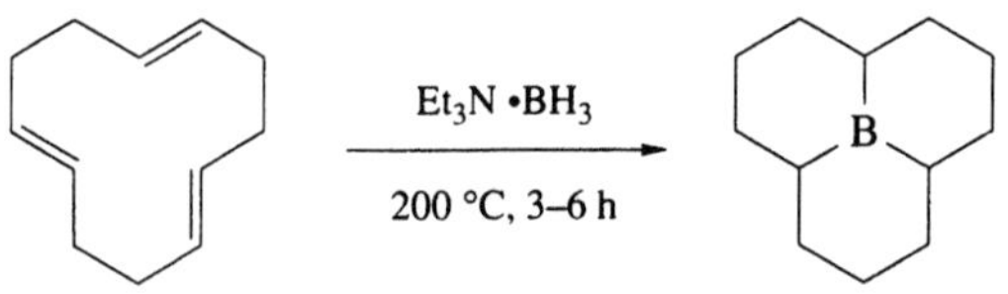

(11)

Hydroborates in the presence of an ester or of methyl iodide have also been used for hydroborations.[43] For example, the system lithium tetrahydroborate–ethyl acetate provides a simple route to dialkylboryl compounds (equation 12).[43a] Alkynes provide alkenylboron compounds.

$$\mathrm{R{\diagup\!\!\!=}} \quad + \quad \mathrm{LiBH_4} \quad + \quad \mathrm{EtOAc} \longrightarrow \mathrm{Li}\left(\mathrm{R}\diagup\!\!\diagdown\right)_2 \mathrm{B(OEt)_2} \qquad (12)$$

Sodium acetoxytrihydroborate is also effective,[44] while certain titanium, rhodium and cobalt complexes catalyze hydroborations with sodium tetrahydroborate.[45] In the latter case the facial selectivities can be quite different from, even opposite to, those observed with borane–THF.[45a]

3.10.3 HYDROBORATION WITH MONOSUBSTITUTED BORANES

3.10.3.1 Hydroboration with Thexylborane and other Alkyl- and Aryl-boranes

Thexylborane is the most easily prepared (see equation 5) and widely studied monoalkylborane.[15] However, it may be expected that the hydroboration characteristics of most other monoalkylboranes will be broadly comparable.[1]

Thexylborane slowly isomerizes to a *primary* alkylborane at ambient temperature, so it is usual to prepare it freshly and to use it, at 0 °C, when it is required. However, addition of an equimolar quantity of *N*,*N*-diethylaniline produces a reagent which appears significantly more stable yet retains the hydroborating and reducing capabilities.[46]

The reactions of hindered alkenes with thexylborane are generally accompanied by a significant amount of dehydroboration to give 2,3-dimethyl-2-butene. Thus, a very large excess of the latter is needed in order to produce (monomeric) dithexylborane.[47] Hydroboration of α-pinene (1:1 ratio) with thexylborane results in *ca.* 75% dehydroboration.[15a] However, alkenes of somewhat lower steric requirements, such as simple 1,2-disubstituted ethylenes, can be converted fairly cleanly into the corresponding thexylmonoalkylboranes (*e.g.* equation 13).[15] These are mixed dialkylboranes and behave like other fairly hindered dialkylboranes (Section 3.10.4.3). They can hydroborate alkenes of lower steric requirements to give totally mixed trialkylboranes.

$$\text{Thexyl–BH}_2 \quad + \quad \text{(norbornene)} \xrightarrow{\text{THF, 20 °C}} \text{Thexyl–B(H)–(norbornyl)} \qquad (13)$$

Unhindered alkenes react with thexylborane to give thexyldialkylboranes. Thus, if mixed thexyldi-*n*-alkylboranes are required, an alternative procedure involving thexylchloroborane must be adopted (Section 3.10.4.5).

Thexylborane is somewhat more regioselective than borane–THF in hydroborations of simple alkenes (Figure 4; compare Figure 1). The differences can be almost wholly attributed to the greater relative importance of the final hydroboration step, which involves a highly regioselective dialkylborane.[1]

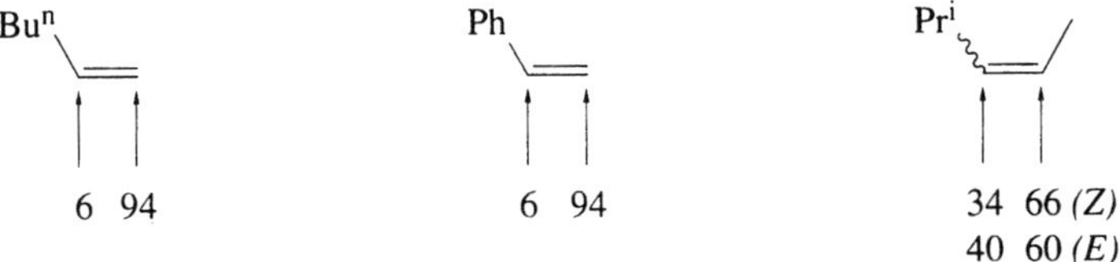

Figure 4 Regioselectivity of attack by the boron atom of thexylborane on alkenes (%)

Thexylborane is more reliable than borane for permitting clean formation of ring compounds from dienes. Under kinetic control there is a preference for formation of five- or seven-membered rather than six-membered rings (equations 14 and 15).[1,15,48] This regioselectivity can be altered by modification of the substitution pattern or by thermal isomerization after the initial hydroboration.

A number of other monoalkylborane derivatives can be prepared by treating an amine adduct of thexylborane with an alkene (*e.g.* equation 16);[49–51] by displacement of 2,3-dimethyl-2-butene or another

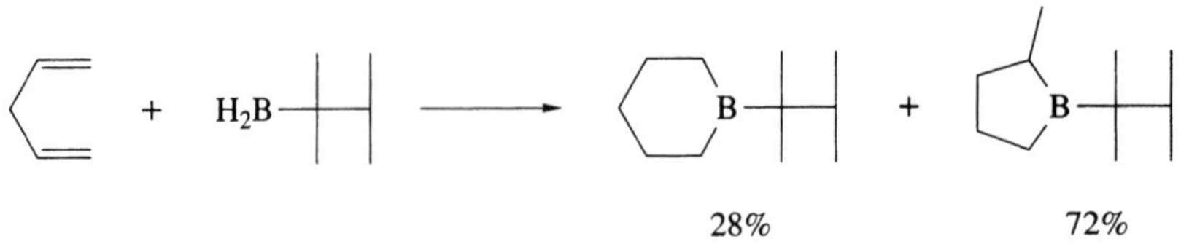

H_2B (14)

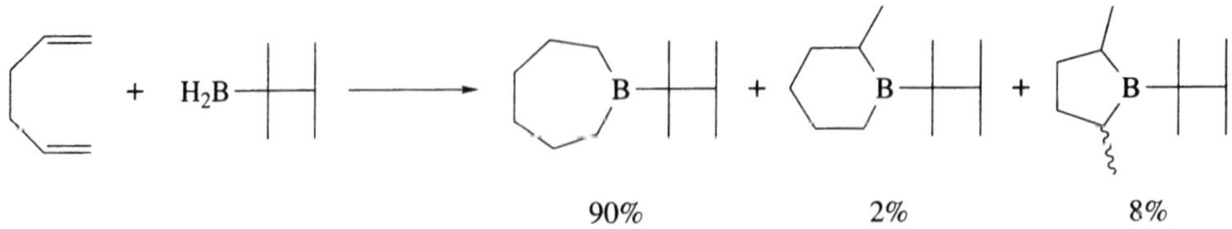

28% 72%

H_2B (15)

90% 2% 8%

hindered alkene from a dialkylborane by addition of TMEDA or triethylamine;[52,53] by redistribution of a dialkylborane with further borane (*e.g.* equation 17);[54] or by reduction of a compound of the general type RBX_2 with a hydride reagent (*e.g.* equation 18).[55] *t*-Butylborane and several arylboranes have been produced by reduction of boroxin derivatives.[56]

$BH_2 \cdot (TMEDA)_{1/2}$ → $BH_2 \cdot (TMEDA)_{1/2}$ + (16)

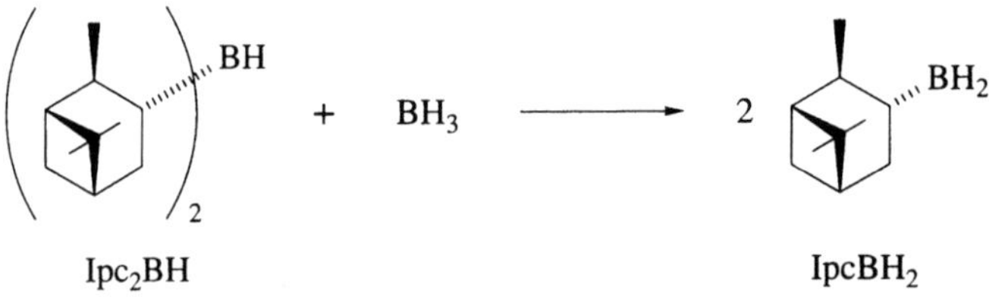

Ipc$_2$BH + BH_3 → 2 IpcBH$_2$ (17)

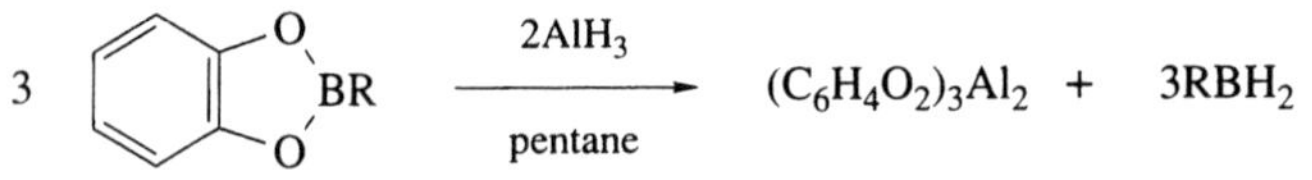

$$3\ \text{C}_6\text{H}_4\text{O}_2\text{BR} \xrightarrow[\text{pentane}]{2AlH_3} (C_6H_4O_2)_3Al_2 + 3RBH_2 \quad (18)$$

Recently, methylborane has been prepared from lithium methyltrihydroborate (equation 19).[57] Other alkylboranes have been obtained by similar methods.[58]

$$LiMeBH_3 + HCl_{(g)} \longrightarrow LiCl + H_2 + MeBH_2 \quad (19)$$

Most monoalkylboranes are not stable for prolonged periods. They tend to redistribute to give mixtures containing dialkylborane, trialkylborane and borane itself. Methylborane appears to be more stable than many,[57] but the others can be stabilized and stored as TMEDA or other trialkylamine complexes.[49,59–61] The free borane is liberated when required by addition of trifluoroborane.

Although systematic comparisons have not been made, it appears that regioselectivities and reactivities of monoalkyl- and monoaryl-boranes are broadly comparable to those of thexylborane. Strangely, however, methylborane appears to be *more* regioselective.[57]

Monoisopinocampheylborane can be prepared in high enantiomeric purity and is an important reagent for asymmetric hydroboration (see Section 3.10.5).

3.10.3.2 Hydroboration with Monohaloboranes

The early literature associated with monochloroborane hydroborations contains errors and it is probably appropriate to ignore publications prior to about 1970.[4] Monochloroborane complexes THF more strongly than does borane and this slows down its reactions with alkenes in this solvent. Thus, other reactions may compete and mixtures may result. The diethyl ether complex is weaker and hydroborations occur readily in this solvent. Dialkylchloroboranes, monoalkylchloroboranes (equation 20), and divinylchloroboranes (equation 21) can be prepared under appropriate conditions.[62]

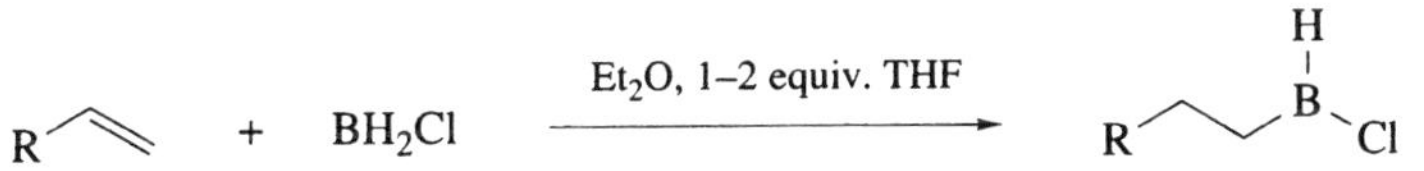

$$\text{R-CH=CH}_2 + BH_2Cl \xrightarrow{Et_2O,\ 1\text{–}2\ \text{equiv. THF}} \text{RCH}_2\text{CH}_2\text{BHCl} \qquad (20)$$

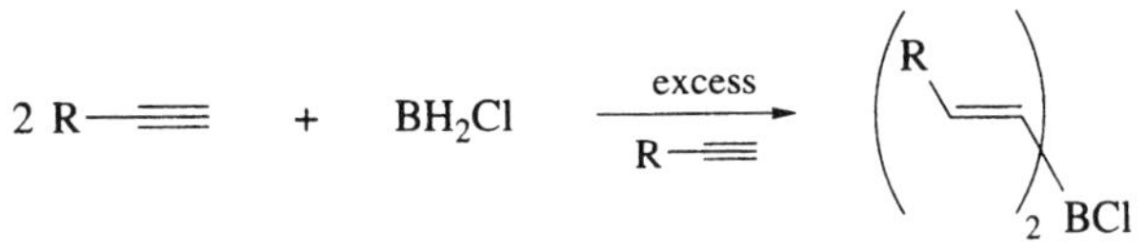

$$2\ \text{R-C}\equiv\text{CH} + BH_2Cl \xrightarrow[\text{R-C}\equiv\text{CH}]{\text{excess}} (\text{RCH=CH})_2\text{BCl} \qquad (21)$$

The reactions of monochloroborane in diethyl ether with α,ω-dienes lead initially to mixtures which are partially polymeric and possibly composed of different ring sizes. However, careful distillation can lead to both depolymerization and isomerization, so that in several cases a single cyclic product can be obtained (*e.g.* Scheme 5).[63]

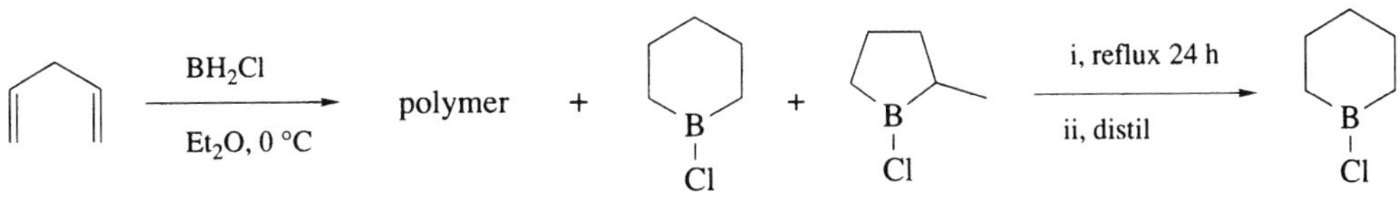

1,4-pentadiene $\xrightarrow[Et_2O,\ 0\ °C]{BH_2Cl}$ polymer + 1-chloroborinane + 1-chloro-2-methylborolane $\xrightarrow[\text{ii, distil}]{\text{i, reflux 24 h}}$ 1-chloroborinane

Scheme 5

The preparation of monochloroborane in ether is capricious and the product is not stable.[62] A more convenient source of monochloroborane is its dimethyl sulfide complex, which is readily prepared by reaction of borane–dimethyl sulfide with either tetrachloromethane (equation 22)[64] or trichloroborane–dimethyl sulfide (equation 23).[65] The latter reaction has also been modified for synthesis of the bromoborane[65] and iodoborane[66] complexes.

$$BH_3{\bullet}SMe_2 + CCl_4 \xrightarrow{\text{heat}} BH_2Cl{\bullet}SMe_2 + CHCl_3 \qquad (22)$$

$$2\ BH_3{\bullet}SMe_2 + BX_3{\bullet}SMe_2 \xrightarrow{\text{heat}} 3\ BH_2X{\bullet}SMe_2 \qquad (23)$$

Monochloborane–dimethyl sulfide coexists with small amounts of the borane and dichloroborane complexes, but the bromoborane–dimethyl sulfide complex appears to be almost pure.[65,66] These complexes react readily with alkenes at 25 °C and can be used for hydroborations in a variety of solvents. Dialkylhaloboranes are obtained in high yield as their dimethyl sulfide complexes (equation 24), but dimethyl sulfide is readily removed under reduced pressure if required.[64,67] The reagents are also useful for cyclic hydroborations of dienes such as cyclooctadiene (equation 25).[68] An alternative approach to dialkylbromoboranes involves the reaction of dialkyl(methylthio)boranes with bromine.[69]

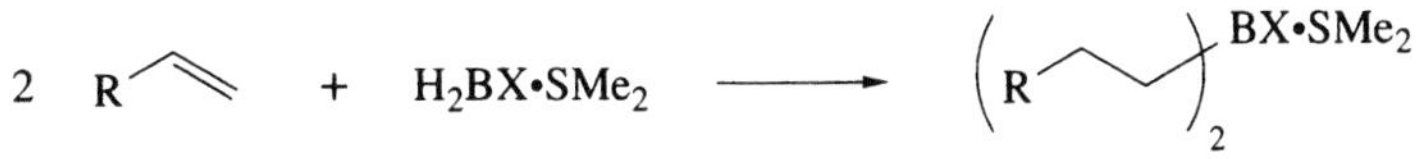

$$2\ \text{R-CH=CH}_2 + H_2BX{\bullet}SMe_2 \longrightarrow (\text{RCH}_2\text{CH}_2)_2BX{\bullet}SMe_2 \qquad (24)$$

$$\text{1,5-cyclooctadiene} \xrightarrow[\text{ii, heat; iii, } BX_3\ (\text{removes } SMe_2)]{\text{i, } H_2BX{\bullet}SMe_2} \text{9-X-9-BBN} \qquad (25)$$

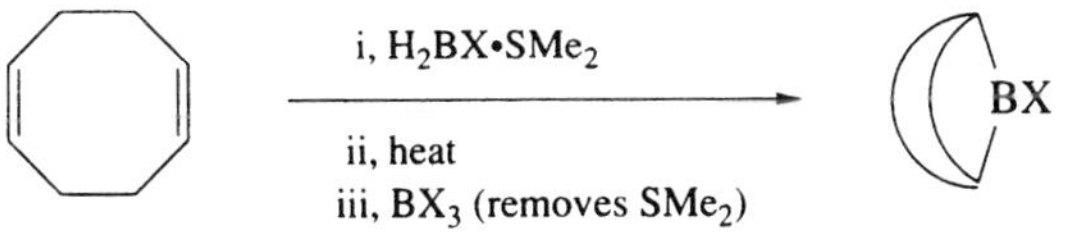

Monohaloboranes generally exhibit significantly greater regioselectivities than borane–THF in hydroboration reactions (Figure 5),[67] though there may be some exceptions to the rule.

	Bu^n (C-1)	Bu^n (C-2)	Ph (C-1)	Ph (C-2)	cis-2-pentene (C-2)	cis-2-pentene (C-3)
BH_3•THF	6	94	19	81	55	45
BH_2Cl•OEt_2	<0.5	>99.5	4	96	58	42
BH_2Cl•SMe_2	0.8	99.2	7	93	–	–
BH_2Br•SMe_2	0.4	99.6	4	96	63	37

Figure 5 Regioselectivities of hydroborations with haloboranes (%)

3.10.4 HYDROBORATION WITH DISUBSTITUTED BORANES

3.10.4.1 General Considerations

The structures of some of the more common examples from a whole range of disubstituted hydroborating agents are given in Figure 6.[1]

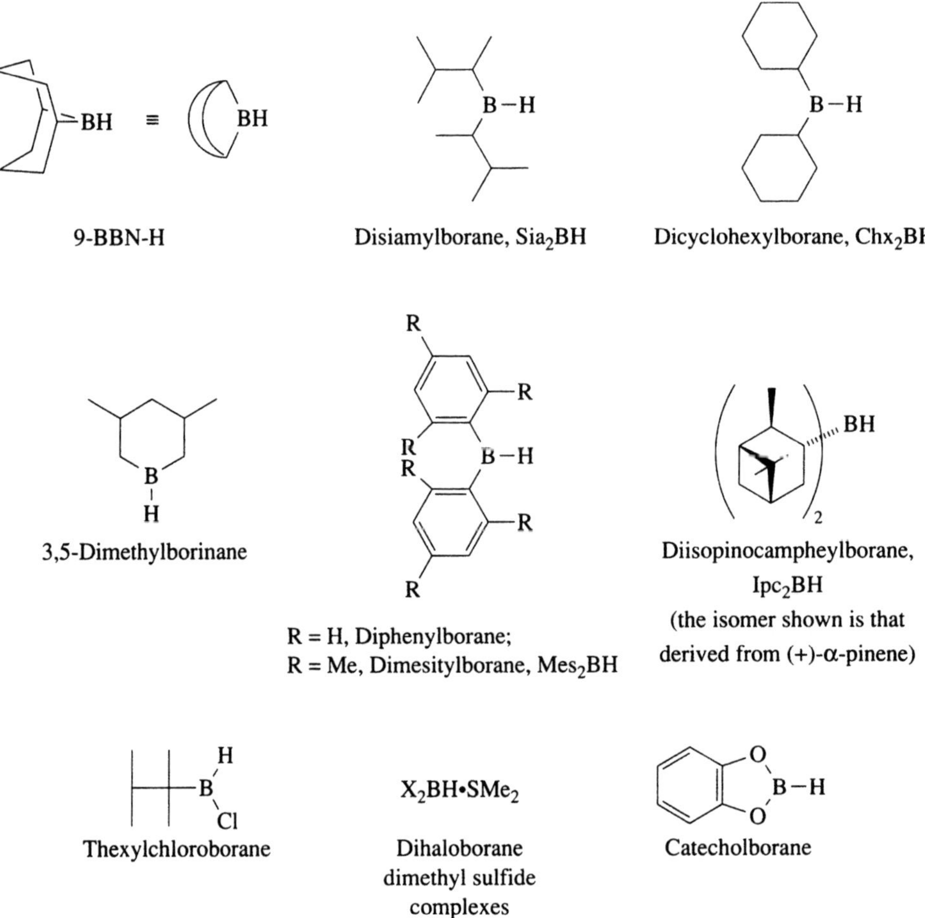

Figure 6 Some common disubstituted borane hydroborating agents

9-Borabicyclo[3.3.1]nonane (9-BBN-H) has received greater study than the others and a separate section is devoted to its reactions, with the other dialkyl- and diaryl-boranes considered as a group. Separate sections are also devoted to dihaloboranes, alkylchloroboranes and catecholborane and its analogs.

3.10.4.2 Hydroboration with 9-BBN-H

9-BBN-H is readily prepared (Section 3.10.2.1, equation 8) and is commercially available. It shows considerable stability, even in air for limited periods,[70] and is therefore a very convenient hydroborating agent.[1,71,72] Unlike di-primary-alkylboranes it is not prone to disproportionation, but it is substantially less hindered than other di-*s*-alkylboranes such as dicyclohexylborane and disiamylborane. Thus, it hydroborates hindered alkenes such as 2,3-dimethyl-2-butene slowly.[72] It is less sensitive to steric factors and more sensitive to electronic factors than disiamylborane. Thus, it shows relatively little ability to discriminate between (*E*)/(*Z*) pairs[73] but readily discriminates between 4-methoxystyrene and 4-(trifluoromethyl)styrene.[72]

All dialkylboranes are highly regioselective, but 9-BBN-H is often the most regioselective of all, frequently giving almost total regioselectivity. Figure 7 shows some typical regioselectivities;[72,74–76] values for borane–THF and for disiamylborane are given for comparison.

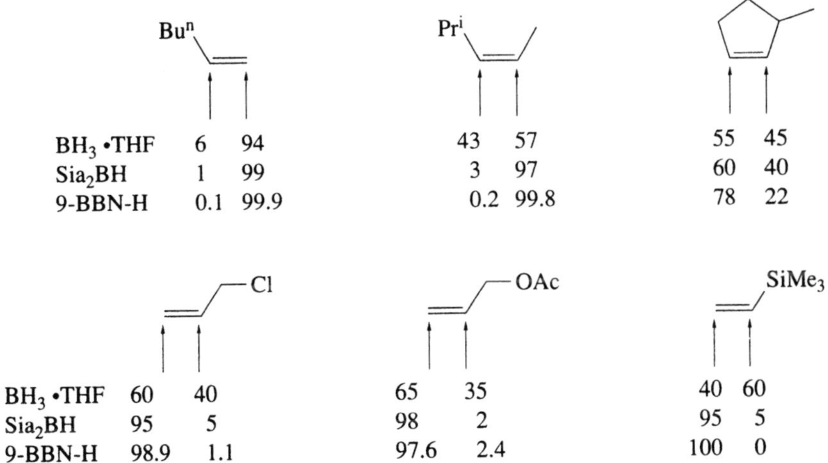

Figure 7 Regioselectivities of attack by the boron atom of dialkylboranes on alkenes (%)

9-BBN-H tolerates many functional groups, and this, coupled with its high regioselectivity, allows the clean synthesis of a number of functionalized organoboranes (*e.g.* equation 26),[74] including many derived from unsaturated heterocyclic compounds.[77] It also shows impressive stereofacial selectivity in the hydroboration of cyclic alkenes (*e.g.* equations 27–29),[75,78] and sometimes in the cases of acyclic alkenes.[7]

BH + Cl ⟶ B Cl (26)

9-BBN-H ⟶ B + B (27)

95% 5%

9-BBN-H ⟶ B + B (28)

97% 3%

Furthermore, the products of reactions with 9-BBN-H seem to be less prone to isomerization than those derived using borane–THF. For example, 1-methylcyclooctene is readily converted into 9-(*trans*-2-methylcyclooctyl)-9-BBN (equation 30),[75] whereas its reaction with borane–THF produces a complex mixture.

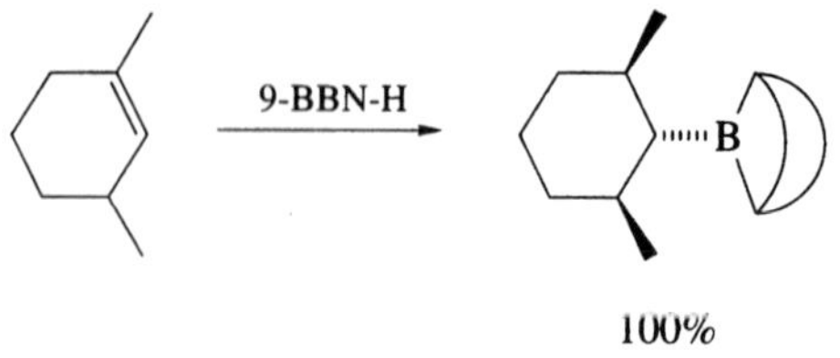

9-BBN-H

(29)

100%

Hydroboration of acyclic, symmetrical, nonconjugated dienes with one equiv. of 9-BBN-H produces almost statistical mixtures of mono- and di-hydroborated species,[74] but cyclic analogs may show substantial deviations from the statistical mixture. For example, monohydroboration of 1,5-cyclooctadiene occurs to the extent of 85% using 1:1 stoichiometry (equation 31),[79,80] whereas disiamylborane gives predominantly dihydroboration product under such conditions.

(30)

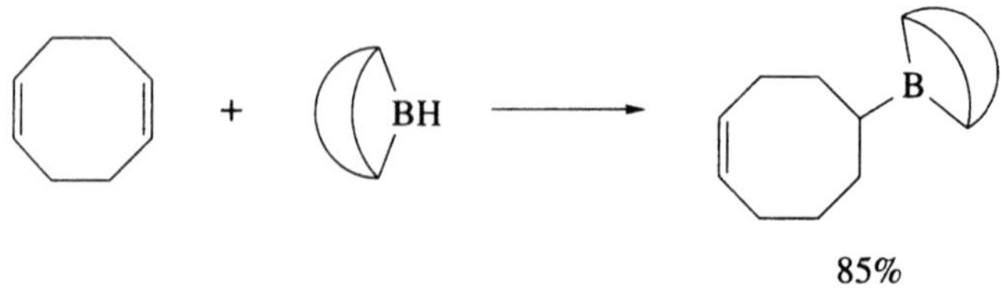

(31)

85%

Unsymmetrical, nonconjugated dienes are generally easier to monohydroborate because of intrinsic differences between the two double bonds. Both 9-BBN-H and disiamylborane favor attachment to a terminal double bond rather than an internal double bond (*e.g.* equation 32).[79] On the other hand, 2-methyl-1,5-hexadiene reacts predominantly at the 1-position with 9-BBN-H and almost exclusively at the 6-position with disiamylborane.[79] The products of dihydroboration of α,ω-dienes with 9-BBN-H can be redistributed with borane–dimethyl sulfide to give boracyclanes.[81] In this way, some of the problems sometimes associated with direct hydroboration of dienes with borane (see Section 3.10.2.1) may be overcome.

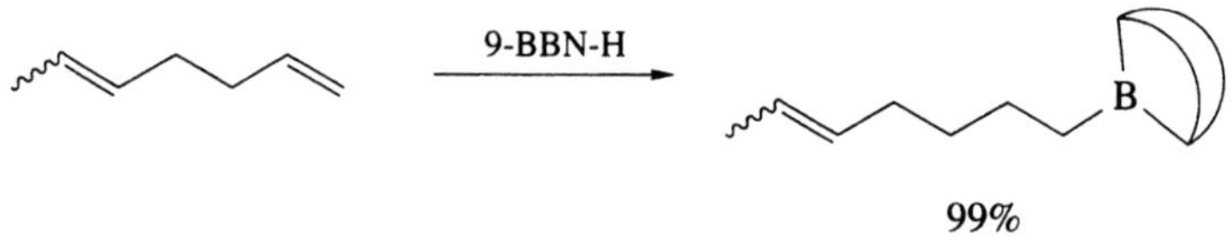

9-BBN-H

(32)

99%

Conjugated dienes are deactivated relative to isolated double bonds towards 9-BBN-H, so monohydroboration is difficult to achieve unless the second hydroboration stage is slowed by intrinsic lack of reactivity of the system or by steric crowding, when it is possible to produce allylic organoboranes (*e.g.* equation 33).[82]

Allylic-9-BBN derivatives are also available by hydroboration of allenes with 9-BBN-H. Unlike disiamylborane, 9-BBN-H shows a marked preference for attachment of boron to one of the termini of the allene system rather than to the central carbon atom. Allene itself is dihydroborated by 9-BBN-H, but substituted allenes give rise predominantly to allylboranes (*e.g.* equation 34).[83]

9-BBN-H (33)

BH + → B (34)

Hydroboration of internal alkynes with 9-BBN-H in stoichiometric amounts is a useful way of synthesizing vinylboranes (Scheme 6).[84,85] Further hydroboration of the product is slow, but can be used to produce *gem*-diboryl compounds (Scheme 6).[84]

R—≡—R → (9-BBN-H) → (9-BBN-H, 25 °C, 24 h)

Scheme 6

The regioselectivity of monohydroboration is governed both by steric and electronic effects, as illustrated in Figure 8. Once again, 9-BBN-H is more susceptible to electronic factors and less to steric factors than is disiamylborane.[84]

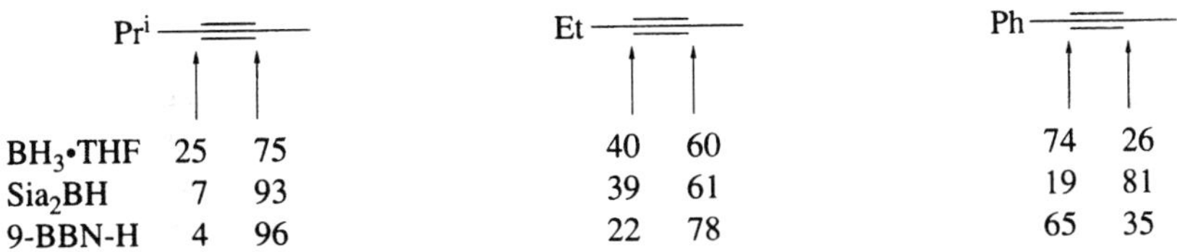

	Pr^i—≡—		Et—≡—		Ph—≡—	
BH_3•THF	25	75	40	60	74	26
Sia_2BH	7	93	39	61	19	81
9-BBN-H	4	96	22	78	65	35

Figure 8 Regioselectivities in hydroborations of alkynes (%)

1-Alkynes give mainly 1,1-diborylalkanes with 9-BBN-H and the reaction becomes essentially quantitative with a 1:2 ratio of reactants.[1,4,42] The alkenylborane can be made to predominate by use of excess alkyne, but more hindered reagents, such as disiamylborane, give the alkenylborane readily with 1:1 stoichiometry.[85a] 9-BBN-H reacts with alkenes more readily than with alkynes, thereby permitting the synthesis of organoboranes containing an alkyne unit (*e.g.* equation 35),[85b] in contrast to the preferential hydroboration of the alkyne group by other dialkylboranes.

Pr^n → (9-BBN-H) → Pr^n (35)

3.10.4.3 Hydroboration with other Dialkyl- and Diaryl-boranes

Few publications[77,85a] have explored the different properties of a range of hydroborating agents with a single substrate. More commonly, the properties of a particular reagent have been compared only with those of borane–THF or 9-BBN-H. Nevertheless, it appears from these reports that there are essentially two other types of diorganylboranes; (i) those which are very hindered, *i.e.* more hindered than 9-BBN-H; and (ii) those which are less hindered than 9-BBN-H, *i.e.* di-primary-alkylboranes.

Possibly the most hindered reagent of all — dimesitylborane, an air-stable white solid with poor solubility in ether solvents — reacts only slowly with alkenes.[86] Reaction with a simple 1-alkene requires 8 h at 25 °C to go to completion, and that with cyclohexene proceeds only to the extent of 20% after 24 h at 65 °C. Other heterogeneous hydroborations have been helped by sonication[87] and perhaps this might help with dimesitylborane. However, even in the absence of such treatment dimesitylborane readily monohydroborates both terminal and internal alkynes,[86] and it seems therefore to be a very discriminating reagent, though no enynes have yet been hydroborated with it.

The regioselectivity of hydroboration of unsymmetrical alkynes is greater with dimesitylborane than with any other known hydroborating agent (Figure 9).[86] The reagent is dominated by steric rather than electronic effects and there are therefore no problems with dihydroboration. It appears likely that the reagent will also tolerate functional groups.[86] Thus, it is a very useful reagent for production of alkenylboranes. Unfortunately, it is not available *via* hydroboration, though it is now commercially available.

	Pr^n—≡—		Ph—≡—	
	↑	↑	↑	↑
BH_3•THF	40	60	74	26
Sia_2BH	39	61	19	81
Catecholborane	40	60	27	73
9-BBN-H	22	78	65	35
Br_2BH•SMe_2	25	75	64	36
Mes_2BH	10	90	2	98

Figure 9 Regioselectivities of disubstituted boranes in hydroborations of alkynes (%)

Highly hindered dialkylboranes, such as diisopinocampheylborane and disiamylborane, are more easily prepared. However, they may undergo a significant amount of retrohydroboration in reactions with relatively hindered alkenes.[1,88] Dicyclohexylborane is less prone to this problem and is also less prone to isomerization than disiamylborane,[89] yet is still highly regioselective. It is a solid which can be stored for considerable periods under dry nitrogen.[4] Its hydroboration properties have been less well studied, for historical reasons, than those of disiamylborane, but the two reagents appear to be very comparable. Its reactions may benefit from sonication.[87] Section 3.10.4.2 provides much comparative information on disiamylborane and 9-BBN-H and coverage here is therefore brief. From what little is known about diphenylborane, it appears comparable to dicyclohexylborane or disiamylborane in its ability to hydroborate alkenes.[90]

The rates of reaction of alkenes with disiamylborane vary over a much wider range than those with borane–THF, and alkynes react more rapidly than comparable alkenes. The great selectivity shown allows ready distinction between different alkenes, or between two different double bonds in a nonconjugated diene (*e.g.* equation 36).[91] Selective monohydroboration of conjugated dienes is more troublesome. However, selective monohydroboration of some conjugated dienes, such as 1,3-cyclohexadiene, can be achieved with disiamylborane,[92] as is also the case with 9-BBN-H.

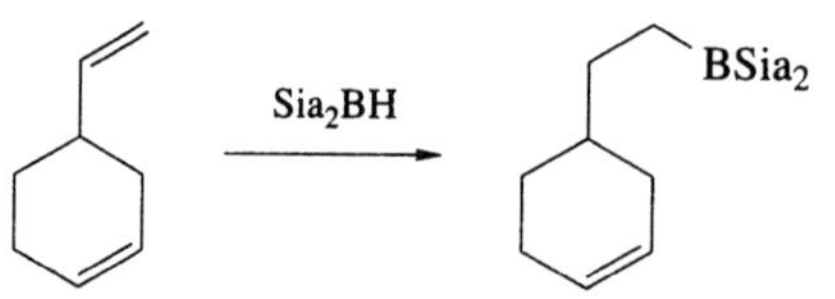

Sia_2BH → $BSia_2$ (36)

In general, the regioselectivities and stereoselectivities of reactions with disiamylborane are more susceptible to steric effects and less to electronic effects, but are otherwise similar to those of 9-BBN-H (Section 3.10.4.2).[1]

Both terminal and internal alkynes react readily to produce vinylboranes with disiamylborane or dicyclohexylborane (*e.g.* equation 37), and dihydroboration is often incomplete even with excess reagent.[33–35,85] This is in contrast to the situation with 9-BBN-H.

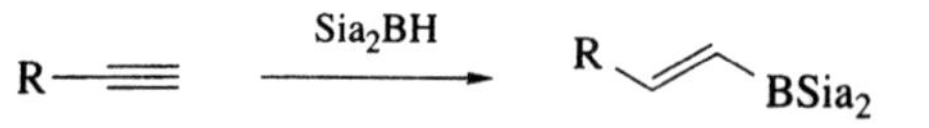

R—≡ —(Sia_2BH)→ R⁄⁼⁄$BSia_2$ (37)

Conjugated diynes may be monohydroborated with disiamylborane, while with dicyclohexylborane dihydroboration is easier (Scheme 7).[93]

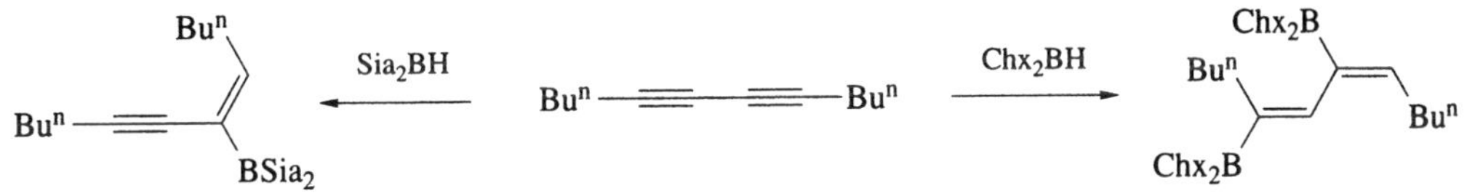

Bu^n—≡—C(BSia$_2$)=CH—Bu^n ← Sia_2BH — Bu^n—≡—≡—Bu^n — Chx_2BH → Bu^nCH=C($BChx_2$)—C($BChx_2$)=CHBu^n

Scheme 7

Disiamylborane and dicyclohexylborane, in contrast to 9-BBN-H, reduce the triple bond of enynes in preference to the double bond.[93,94] Hydroboration of functional alkynes is also possible (*e.g.* equations 38[95] and 39[96]; see also ref. 97).

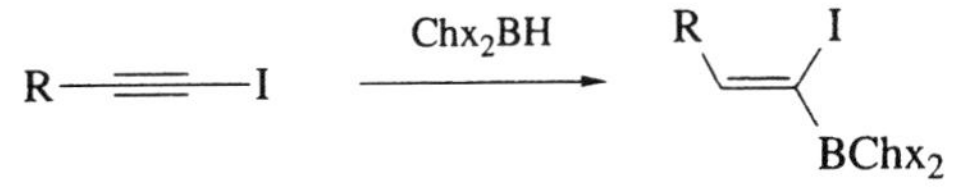

$$R{-}C{\equiv}C{-}I \xrightarrow{Chx_2BH} RCH{=}C(I)BChx_2 \qquad (38)$$

$$R{-}C{\equiv}C{-}CH_2Cl \xrightarrow{Sia_2BH} RCH{=}C(CH_2Cl)BSia_2 \qquad (39)$$

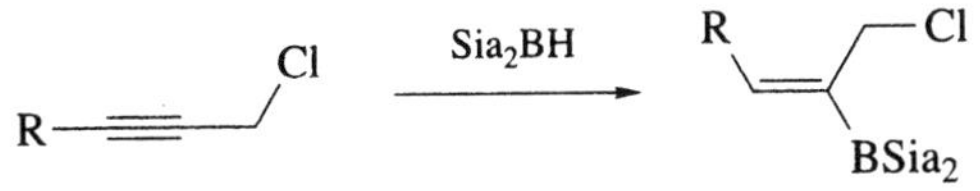

Less hindered dialkylboranes cannot generally be prepared by direct stoichiometric hydroboration of 2 equiv. of an alkene with borane (Section 3.10.2.1), although 3,5-dimethylborinane and 3,6-dimethylborepane are exceptions.[27,28] Borinane itself is obtained by a sequence involving hydroboration (diene:BH_3 = 3:2), thermal isomerization, and then redistribution with further borane–THF,[98,99] but there is as yet no similar direct route to the parent borepane.[100]

The simplest approach to pure acyclic di-primary-alkylboranes involves reduction of a species of the type R_2BX with a hydride reagent such as sodium or potassium hydride (equation 40),[101] aluminum hydride,[102] or lithium aluminum hydride.[103,104]

$$(n\text{-}C_5H_{11})_2B{-}Br \xrightarrow{NaH} (n\text{-}C_5H_{11})_2B{-}H \qquad (40)$$

The dialkylboranes must either be used immediately or converted into more stable products; otherwise redistribution takes place. Such redistribution reactions (*e.g.* equation 41) can be used in reverse to generate dialkylboranes,[105] but this method does not generally give rise to a single component.

$$3\ Et_2B{-}H \rightleftharpoons BH_3 + 2\ Et_3B \qquad (41)$$

Pure dialkylboranes can be stored as amine complexes or as dialkyldihydroborates, in a manner analogous to that used to stabilize monoalkylboranes (Section 3.10.3.1). They are then reliberated by reactions with appropriate electrophiles.[1,4,42] Dimethylborane has recently been obtained from its hydroborate in this way.[57]

Di-primary-alkylboranes react readily at 25 °C with most alkenes. Nevertheless, even mixtures of mono-, di- and tri-alkylboranes with compositions which correspond overall to that of a simple dialkylborane distinguish impressively between alkenes of different steric requirements.[106] Of course, mixtures of different trialkylboranes, $R^1{}_nBR^2{}_{n-3}$ result in such cases. Di-primary-alkylboranes also exhibit regioselectivity which is often only marginally worse than that of disiamylborane.[106] For example, 3,5 dimethylborinane hydroborates 1-hexene to give 99% of the 1-hexyl isomer (equation 42).[28] Dimethylborane is generally somewhat less regioselective than other dialkylboranes.[57]

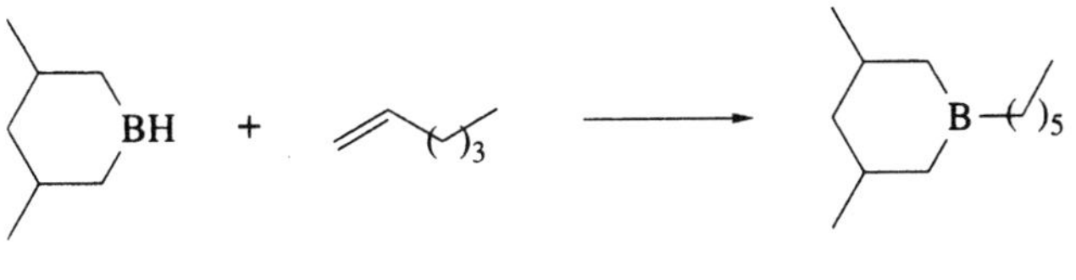

$$\text{3,5-dimethylborinane (BH)} + CH_2{=}CH(CH_2)_3CH_3 \longrightarrow \text{3,5-dimethylborinane–}B(CH_2)_5CH_3 \qquad (42)$$

Di-primary-alkylboranes monohydroborate internal alkynes, but readily dihydroborate 1-alkynes.[85a] Thus, it is difficult by this approach to make 1-alkenyldi-*n*-alkylboranes.

Lithium tri-*n*-alkylhydroborates add B—H across the double bonds of substituted styrenes.[107] The reaction places boron at the α-position, unlike hydroboration with dialkylboranes which places boron at the β-position. It is therefore useful for preparation of unusual organoboranes,[107] but the mechanism is presumably quite different from that occurring in normal hydroborations. The reaction is also very restricted in scope.

3.10.4.4 Hydroboration with Dihaloboranes

The most convenient dihaloborane reagent is the dibromoborane–dimethyl sulfide complex. It is commercially available as a neat liquid or is readily prepared by a redistribution reaction, of a type which is also applicable to chloro and iodo analogs (equation 43).[65,66] Unlike the chloro analog, however, it reacts readily with alkenes[108–110] and alkynes[111] in dichloromethane without need for a decomplexing agent such as trichloroborane.

$$BH_3{\bullet}SMe_2 \quad + \quad 2\ BX_3{\bullet}SMe_2 \quad \rightleftharpoons \quad 3\ BHX_2{\bullet}SMe_2 \qquad (43)$$

The selectivities exhibited by dibromoborane–dimethyl sulfide are also interesting. In contrast to the situation with 9-BBN-H, internal alkynes are hydroborated faster than either terminal alkynes or terminal alkenes. Unlike reactions with disiamylborane, 1,1-disubstituted ethylenes are hydroborated in preference to simple 1-alkenes. These properties can lead to interesting possibilities for selective hydroboration of polyunsaturated molecules (*e.g.* Schemes 8 and 9).[112]

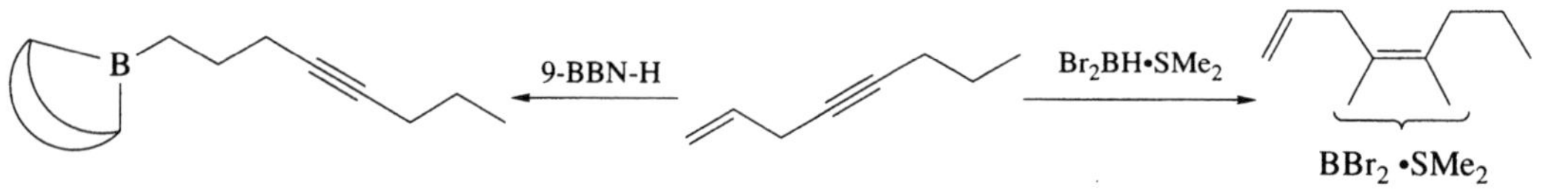

B 9-BBN-H $Br_2BH{\bullet}SMe_2$ $BBr_2{\bullet}SMe_2$

Scheme 8

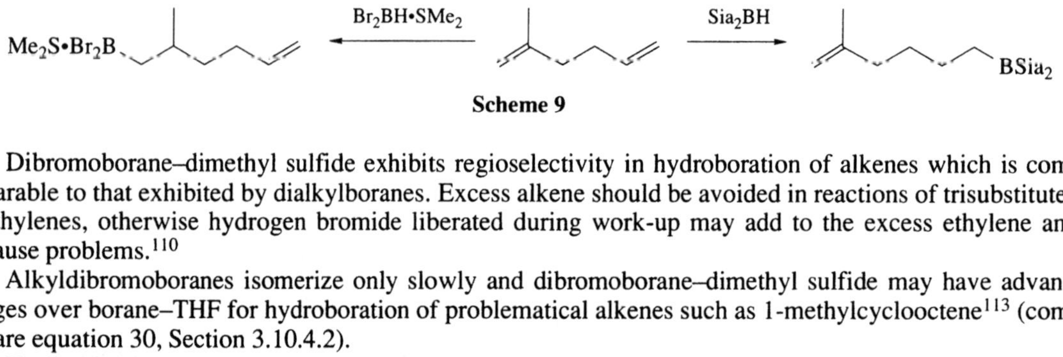

$Me_2S{\bullet}Br_2B$ $Br_2BH{\bullet}SMe_2$ Sia_2BH $BSia_2$

Scheme 9

Dibromoborane–dimethyl sulfide exhibits regioselectivity in hydroboration of alkenes which is comparable to that exhibited by dialkylboranes. Excess alkene should be avoided in reactions of trisubstituted ethylenes, otherwise hydrogen bromide liberated during work-up may add to the excess ethylene and cause problems.[110]

Alkyldibromoboranes isomerize only slowly and dibromoborane–dimethyl sulfide may have advantages over borane–THF for hydroboration of problematical alkenes such as 1-methylcyclooctene[113] (compare equation 30, Section 3.10.4.2).

If alkyldichloroboranes are specifically required, dichloroborane–dimethyl sulfide is the reagent of choice.[114] It is more stable and more convenient than the dichloroborane–diethyl ether complex, but its hydroborating properties are very similar.[115] Dichloroborane complexes ethers even more strongly than monochloroborane, and its reactions with alkenes in this solvent are slow and lead to mixtures. Therefore, it is generally used in pentane and trichloroborane is added to liberate uncomplexed dichloroborane. Under these conditions it readily gives alkyldichloroboranes on reaction with alkenes or alkenyldichloroboranes on reaction with alkynes.[114,115] The latter reaction has been applied to alkynylsilanes (equation 44).[116]

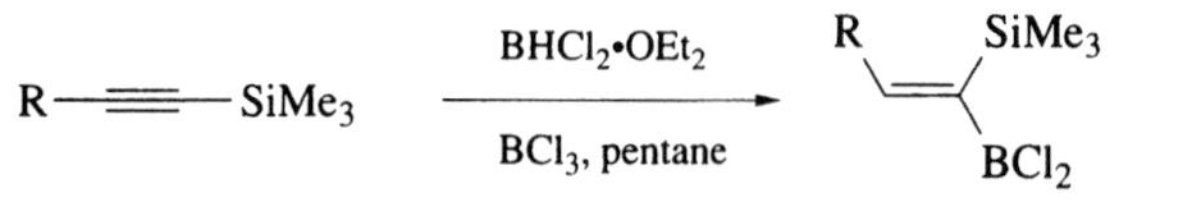

$$R{-}{\equiv}{-}SiMe_3 \xrightarrow[BCl_3,\ \text{pentane}]{BHCl_2{\bullet}OEt_2} \text{R(H)C=C(SiMe}_3\text{)(BCl}_2\text{)} \qquad (44)$$

3.10.4.5 Hydroboration with Alkylhaloboranes

Although several monoalkylchloroboranes have been prepared by the 1:1 reaction of an alkene with monochloroborane–diethyl ether (see Section 3.10.3.2),[62] almost the only one to receive detailed investigation is thexylchloroborane. This reagent is most useful as its dimethyl sulfide complex, prepared according to equation (45)[117,118] or equation (46).[119]

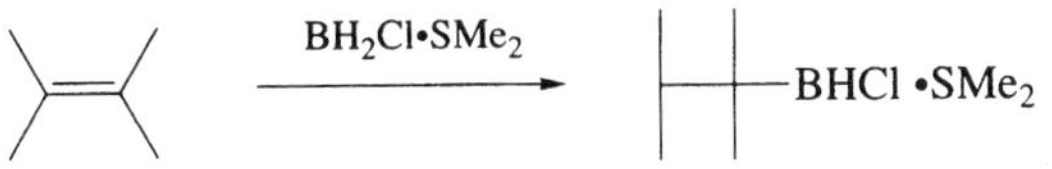

$$\text{Me}_2\text{C=CMe}_2 \xrightarrow{\text{BH}_2\text{Cl}\cdot\text{SMe}_2} \text{ThxBHCl}\cdot\text{SMe}_2 \quad (45)$$

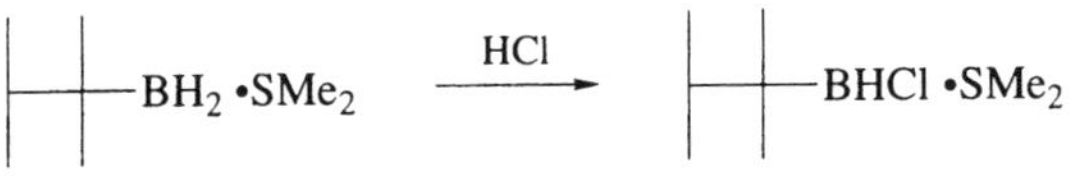

$$\text{ThxBH}_2\cdot\text{SMe}_2 \xrightarrow{\text{HCl}} \text{ThxBHCl}\cdot\text{SMe}_2 \quad (46)$$

Alkylbromoborane–dimethyl sulfide complexes are available by controlled reduction of alkyldibromoborane complexes (equation 47).[120]

$$\text{RBBr}_2\cdot\text{SMe}_2 \xrightarrow[\text{Et}_2\text{O}]{0.25\ \text{LiAlH}_4} \text{RBHBr}\cdot\text{SMe}_2 \quad (47)$$

Thexylchloroborane–dimethyl sulfide is very sensitive to steric factors, reacting, for example, 1000 times faster with 1-hexene than with cyclohexene.[121] In this respect it resembles the dialkylboranes, but it appears to be more sensitive to electronic factors than the latter, and its ability to discriminate between (*E*)- and (*Z*)-isomers is exceptional. For example, it reacts with (*Z*)-3-hexene 100 times more rapidly than with (*E*)-3-hexene.[121]

Reactions of thexylchloroborane with 1-alkenes and unhindered 1,2-disubstituted ethylenes give thexylmonoalkylchloroboranes cleanly (equation 48).[122] Regioselectivities in hydroboration of such alkenes are comparable to those with 9-BBN-H. For the less reactive alkenes, however, the overall yield of thexylmonoalkylchloroborane and the regioselectivity observed in the hydroboration are often lower due to the *in situ* production of thexylborane, formed by slow redistribution of thexylchloroborane during the long reaction times.[117,122]

$$\text{ThxBHCl}\cdot\text{SMe}_2 + \text{RCH=CH}_2 \longrightarrow \text{ThxB(Cl)CH}_2\text{CH}_2\text{R} + \text{SMe}_2 \quad (48)$$

The hydroborating characteristics of alkylbromoborane complexes have not been so extensively investigated, but a range of alkenes appears to be accommodated.[120]

From a synthetic point of view the advantage of alkylhaloborane reagents is that they provide access to mixed dialkylhaloboranes and thereby ultimately to totally mixed trialkylboranes which would be difficult or impossible to obtain by simple stepwise hydroborations. Thus, reactions of simple, unhindered alkenes are of greatest significance, and these work well.

Conversion of mixed dialkylhaloboranes into totally mixed trialkylboranes can be achieved *via* reaction with organometallic reagents[119] or *via* reduction–hydroboration.[120,123] The application of the latter approach for synthesis of totally mixed organoboranes is shown in Scheme 10.

$$\text{R}^1\text{BBr}_2\cdot\text{SMe}_2 \xrightarrow{0.25\ \text{LiAlH}_4} \text{R}^1\text{BHBr}\cdot\text{SMe}_2 \xrightarrow{\text{alkene 2}}$$

$$\text{R}^1\text{R}^2\text{BBr}\cdot\text{SMe}_2 \xrightarrow{\text{MeO}^-,\ \text{MeOH}} \text{R}^1\text{R}^2\text{BOMe} \xrightarrow[\text{alkene 3}]{0.33\ \text{LiAlH}_4} \text{R}^1\text{R}^2\text{R}^3\text{B}$$

Scheme 10

3.10.4.6 Hydroboration with Catecholborane and Related Compounds

Acyclic dialkoxyboranes, $(RO)_2BH$, readily disproportionate,[124] but a number of heterocyclic analogs, such as the trimethyldioxaborinane (**1**), catecholborane (**2**; 1,3,2-benzodioxaborole) and 1,3,2-dithiaborolane (**3**) are more stable.[1] Unfortunately, because of the low Lewis acidity of (**1**) hydroboration occurs only at elevated temperatures. Nevertheless, it has been employed in the hydroboration of alkenes,[125] 1,3-dienes and allenes.[126]

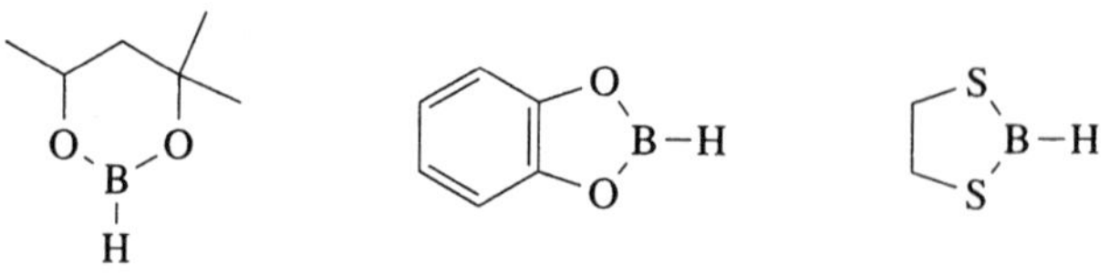

(**1**) (**2**) (**3**)

Catecholborane is somewhat more reactive than (**1**), though still far less reactive than dialkylboranes.[127,128] It is readily prepared by the reaction of catechol with borane–THF, is stable at 0 °C, and hydroborates alkenes slowly and alkynes (equation 49) more rapidly in refluxing THF.[128,129] The rate of reaction can be considerably enhanced by use of Wilkinson's catalyst, *N*,*N*-dimethylaniline–borane or lithium borohydride.[130]

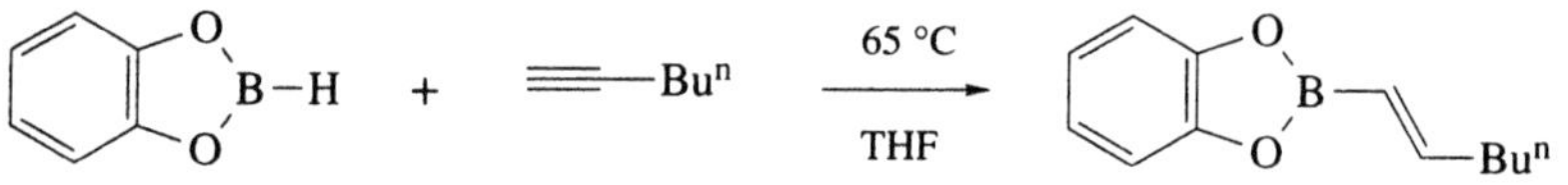

$$\text{(catecholborane) B–H} + \equiv\!\!-\text{Bu}^n \xrightarrow[\text{THF}]{65\ ^\circ\text{C}} \text{(catechol)B–CH=CH–Bu}^n \qquad (49)$$

Regioselectivities appear to approach those achieved with disiamylborane (*e.g.* see Figure 9, Section 3.10.4.3).[129] The *cis* additions to alkynes[131] give alkenylborane derivatives which are useful synthetic intermediates.[132]

1,3,2-Dithiaborolane (**3**) is potentially as useful as catecholborane, but has not been so fully investigated. It is quite stable, and more electrophilic and more reactive towards alkenes than catecholborane.[133]

3.10.5 CHIRAL HYDROBORATION

3.10.5.1 General Considerations

Asymmetric induction during the course of a hydroboration reaction is a well-established phenomenon.[1,134,135] The induction may result from the presence of an asymmetric center in the alkene undergoing hydroboration, or by the incorporation of asymmetric groups into the hydroborating agent. A dramatic example of the former type of induction occurs in the hydroboration of α-pinene; each of the newly generated asymmetric centers is almost exclusively the stereochemistry shown in equation (1) (Section 3.10.1.2). In less rigid systems the degree of control may be less, but even so some impressive inductions have been noted (*e.g.* equation 50).[136] The hydroborating agent chosen may dramatically influence the degree of induction achieved in such asymmetric hydroborations.[137]

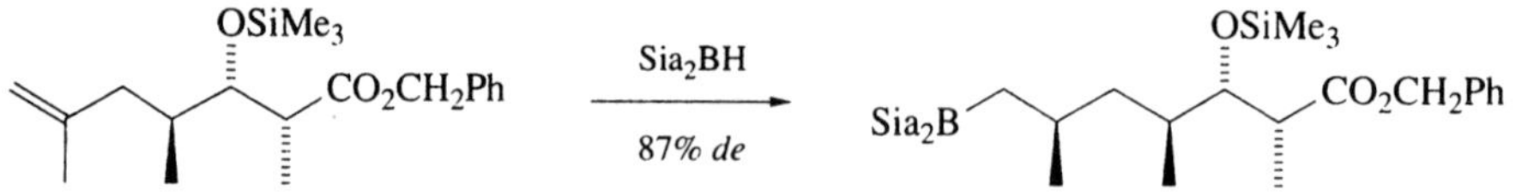

$$\text{(alkene; OSiMe}_3\text{, CO}_2\text{CH}_2\text{Ph)} \xrightarrow[87\%\ de]{\text{Sia}_2\text{BH}} \text{Sia}_2\text{B–(chain; OSiMe}_3\text{, CO}_2\text{CH}_2\text{Ph)} \qquad (50)$$

The alternative approach, in which a chiral hydroborating agent is utilized, is more flexible in that there is no requirement for asymmetry in the alkene. However, if such asymmetry should be present of necessity, then there is the possibility of utilizing the principles of 'double asymmetric induction' by correct choice of chiral reagent.[138] Little work has been done in this field as yet and it will not be discussed further.

Chiral hydroborating agents can be generated by two approaches: (i) by hydroboration of a readily available chiral alkene, the classic example of which is α-pinene; or (ii) by a designed synthesis, followed by resolution of the product. The former method has the advantages that it is quick and easy to generate a reagent, and is inexpensive if the chiral alkene is cheap, as is the case with a number of naturally occurring terpenes. The disadvantage is that the molecule does not necessarily incorporate the appropriate structural design features for the specific purpose required. Nevertheless, a number of terpene-derived hydroborating agents have been prepared and can be very useful for asymmetric hydroboration of certain classes of alkene.[1,134] Most notable among them are diisopinocampheylborane

(Ipc_2BH), monoisopinocampheylborane ($IpcBH_2$) and dilongifolylborane (Lgf_2BH) (Figure 10)[1,134,139] but a number of others, such as dicaranylborane[140] have also been used.

The second approach is more tedious and more expensive, but may provide a reagent with ideal structural features. To date the approach has been used only for generation of chiral *trans*-2,5-dimethylborolane (Figure 10), the synthesis of which involves initial production of the ring system as a mixture of *cis* and *trans* isomers, separation and resolution of the pure *trans* compound, and then manipulation of the boron-bound group to obtain the free borane.[138]

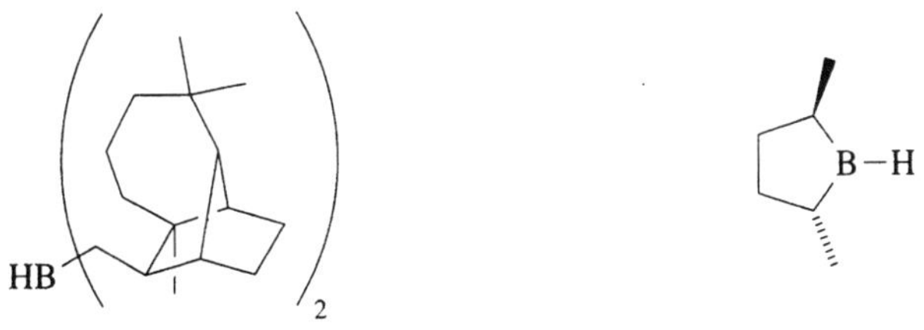

Dilongifolylborane, Lgf_2BH (available cheaply only from (+)-longifolene)

trans-2,5-Dimethylborolane (available in both enantiomeric forms)

Figure 10 Structures of some chiral hydroborating agents; see also equation (17), Section 3.10.3.1, for structures of $IpcBH_2$ and Ipc_2BH, each of which is cheaply available in either enantiomeric form

Dilongifolylborane is simply prepared by admixture of borane–dimethyl sulfide and 2 equiv. (+)-longifolene in THF, whereupon the product crystallizes out of solution as the dimer, and can readily be separated from the solvent. It is used as a suspension for hydroboration reactions.[141]

Diisopinocampheylborane can be prepared in a similar way from (+)- or (–)-α-pinene, and early work was carried out with reagent so prepared. However, α-pinene is often available only in purities up to *ca.* 95%, so that Ipc_2BH produced by direct hydroboration can also be somewhat impure. Fortunately, equilibration of the reagent (dimethyl sulfide must first be removed if borane–dimethyl sulfide is used for the hydroboration) with α-pinene, at 0 °C over several days, results in preferential incorporation of the major enantiomer of α-pinene into the Ipc_2BH. This then becomes available in 98–99% enantiomeric purity.[142]

Monoisopinocampheylborane cannot be directly obtained by controlled hydroboration of α-pinene. However, treatment of Ipc_2BH with one-half equivalent of TMEDA liberates α-pinene and gives a bis-complex of $IpcBH_2$ with TMEDA. Monoisopinocampheylborane of high optical purity is readily obtained from the complex by treatment with trifluoroborane etherate (equation 51).[134] Alternatively, redistribution of Ipc_2BH with further borane can be used to give $IpcBH_2$ (equation 17, Section 3.10.3.1).[53,54,60,143] Again, the reagent is available in essentially 100% enantiomeric purity.[1,134]

$$Ipc_2BH \xrightarrow[\text{ii, } BF_3\bullet OEt_2]{\text{i, 0.5 TMEDA}} IpcBH_2 \quad + \quad \text{(α-pinene)} \qquad (51)$$

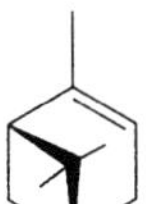

3.10.5.2 Survey of Reactions

Four basic types of simple alkene are susceptible to asymmetric hydroboration, as indicated in Figure 11.

Type I Type II Type III Type IV

Figure 11 General types of alkenes which undergo asymmetric hydroboration

Types I and IV are easily controlled from a *regioselectivity* viewpoint, but types II and III ($R^1 \neq R^2$) generally require highly regioselective reagents in order to provide a single regioisomer. Type I has the disadvantage that the incipient asymmetric center is at a site relatively remote from the chiral groups on boron and so this type of alkene is the most difficult to influence stereochemically.

No single reagent has the appropriate properties to give high asymmetric induction with all four alkene types. Table 1 gives some results for selected examples of the four types with the four major chiral hydroborating agents.[134,138,142,144]

Table 1 Asymmetric Induction During Hydroboration[a]

Alkene	*Type*	Ipc_2BH	$IpcBH_2$	Lgf_2BH	*Dimethylborolane*
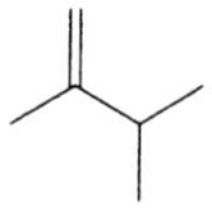	I	30	(1.5)[b]	–	1.4
	II	98.4	24	78	95.2
	III	13	73	–	97.0
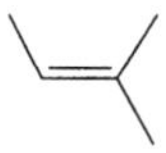	IV	14	53	70	94.2
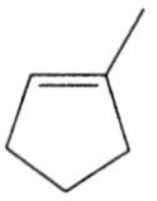	IV	22	66	62	97.0

[a] Figures are enantiomeric excess (%) for the chiral alcohol obtained following hydroboration–oxidation. [b] Figure refers to 2-methyl-1-butene, not 2,3-dimethyl-1-butene.

As can be seen from Table 1, there is no really successful system for asymmetric hydroboration of type I alkenes. The best available appears to be diisopinocampheylborane. Type II alkenes are the most easily hydroborated in a chiral fashion and diisopinocampheylborane is again the reagent of choice. *trans*-2,5-Dimethylborolane gives comparable results, but is much less convenient to obtain. This reagent is, however, significantly better than other reagents for chiral hydroboration of type III or type IV alkenes. Diisopinocampheylborane is of minimal use in these cases because of competitive retrohydroboration.[1] However, Lgf_2BH and $IpcBH_2$ both provide respectable levels of induction for type IV, and $IpcBH_2$ likewise for type III. (Lgf_2BH has yet to be tested for type III.)

The monoalkylborane derived from a modified terpene, 2-ethylapopinene, may give somewhat better results than $IpcBH_2$, but at a heavy price in convenience.[145] The reactions of 1-phenylcycloalkenes with $IpcBH_2$ appear to be particularly favorable, 1-phenylcyclopentene giving 85% *ee* and 1-phenylcyclohexene giving 97% *ee* on hydroboration with this reagent.[146] For less favorable alkenes $IpcBH_2$ can be used in a modified manner which allows the product to be upgraded to essentially 100% optical purity.[147] The process consists of reacting the alkene with $IpcBH_2$ (1:1 ratio) in diethyl ether and then crystallizing the dialkylborane dimer, sometimes after 'aging' the mixture, from the solution. A specific diastereoisomer (that produced in predominance during the hydroboration step) crystallizes in reasonable yield (generally around 70%) and almost 100% optical purity (*e.g.* Scheme 11).[147] After washing, the product is available for further transformation.

One particular transformation of $IpcBHR^*$ involves the reaction with acetaldehyde, which results in displacement of α-pinene and production of a chirally pure alkyldiethoxyborane (equation 52).[147] These derivatives can be subjected to a whole range of organoborane reactions, which result in attachment of the chiral unit to other groups, thereby providing a plethora of synthetic possibilities.[148] They can also be reduced to the corresponding chiral alkylboranes (R^*BH_2), which offer interesting possibilities as alternative chiral hydroborating agents.

Asymmetric hydroborations of several heterocyclic alkenes with diisopinocampheylborane appear to be particularly favorable, resulting in products of almost 100% enantiomeric purity (*e.g.* equation 53).[149] The reagent has also been used in asymmetric syntheses of a number of complex molecules. Discussion is beyond the present scope but a single example is given as an illustration (equation 54).[150]

A number of models have been proposed in order to account for the favored enantiomers produced during asymmetric hydroboration reactions.[4,134] None of the early models were without criticism and were based on scanty theoretical background. Recently, however, much more has become known about the mechanism of the hydroboration reaction (see Section 3.10.6). From these studies it appears that monomeric dialkylboranes (and alkylboranes, *etc.*) are the active intermediates in hydroboration re-

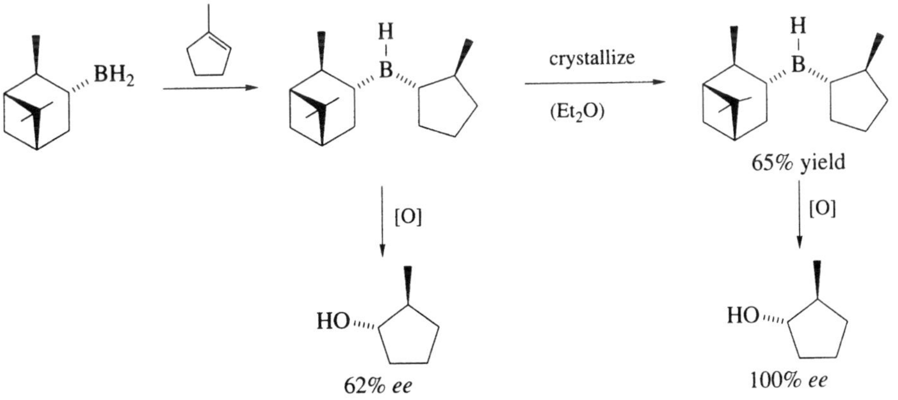

BH_2 → crystallize (Et_2O) → 65% yield

[O] → HO — 62% *ee*

[O] → HO — 100% *ee*

Scheme 11

IpcBHR* + 2 MeCHO → [α-pinene] + $R^*B(OEt)_2$ (52)

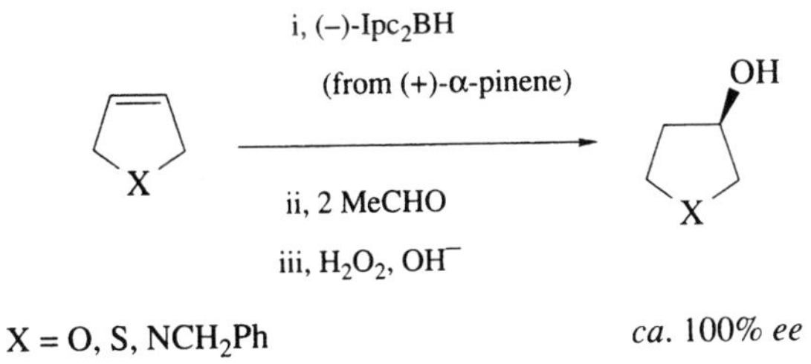

X = O, S, NCH_2Ph — i, (–)-Ipc_2BH (from (+)-α-pinene); ii, 2 MeCHO; iii, H_2O_2, OH^- → OH, X — *ca.* 100% *ee* (53)

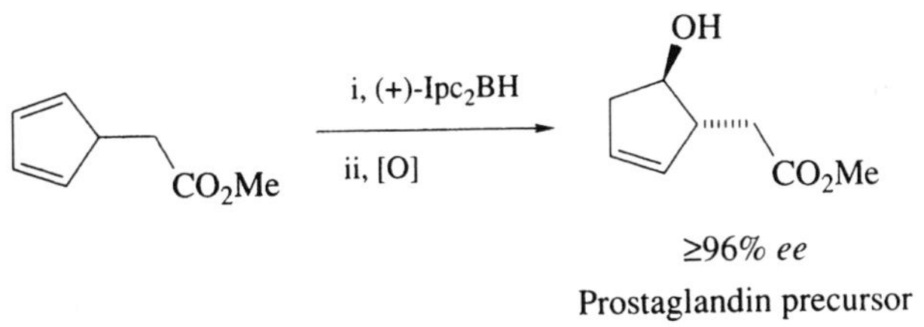

CO_2Me — i, (+)-Ipc_2BH; ii, [O] → OH, CO_2Me — ≥96% *ee* — Prostaglandin precursor (54)

actions. Calculations which assume gas phase monomeric structures are not, therefore, entirely unreasonable. Using modified MM2 calculations, optimal geometries of transition states have been calculated,[151] and these predict the correct stereochemistries for various asymmetric hydroboration reactions. Figure 12 shows the calculated preferred conformations for the reactions of (*E*)-2-butene with $IpcBH_2$ and of (*Z*)-2-butene with Ipc_2BH.

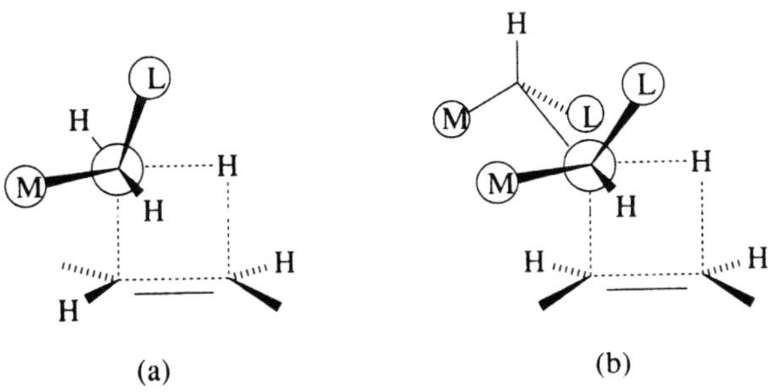

Figure 12 Calculated (ref. 151) transition state geometries for reactions of: (a) $IpcBH_2$ + (*E*)-2-butene; (b) Ipc_2BH + (*Z*)-2-butene; M represents the CH_2 unit and L the CHMe unit adjacent to the boron-bonded carbon atom of each Ipc group

3.10.6 MECHANISM OF THE HYDROBORATION REACTION

The early literature concerning the mechanism of the hydroboration reaction is confused and confusing. Only in relatively recent times, through the careful, systematic and painstaking work of Brown and coworkers, has the mechanism become clear.[1,152] The complexity of hydroboration of simple alkenes with borane–THF requires that the problem be simplified if useful kinetic data are to be obtained. Thus, much of this kind of work has utilized either 2,3-dimethyl-2-butene, which reacts readily only as far as the monoalkylborane stage,[153,154] or a monofunctional hydroborating agent such as 9-BBN-H.[155–162] The information provided by such studies has then led to more appropriate determinations of kinetic data for reactions of other hydroborating agents.[163–167] As a result, it is now clear that *all* hydroborations follow a similar mechanistic course involving the reaction of a free, monomeric borane with the appropriate alkene, although the manifestations in terms of kinetics may vary from case to case.

In view of this conclusion it is intriguing that *ab initio* calculations for the reaction of $BH_3{\cdot}OH_2$ (a model for a borane–ether complex) with ethylene suggest a direct, S_N2-like displacement of water by ethylene, rather than prior dissociation.[168] Previous calculations[169] could be criticized for not taking solvent into account, but the experimental evidence now suggests that the actual hydroborating agent does not involve solvent. This does not imply that solvent or added coordinating agent is unimportant. There is ample evidence that coordinating solvents play a major catalytic role in the reactions.[1,152] However, that role must be in facilitating the *production* of monomer, either by providing a species (the complex) with a lower dissociation energy than the dimer or by reacting directly with the dimer to produce uncomplexed monomer together with complexed monomer.

Most calculations[169a–e] have considered the interaction of unsubstituted borane and ethylene. They suggest the likely formation of a loose three-center complex, which then reorganizes to a four-center transition state[169e] with little or no activation energy. The experimental evidence shows, however, that there may be significant activation energies in reactions of some hindered alkenes with substituted boranes.[155,156]

A few calculations have considered reactions of substituted boranes or ethylenes,[151,169c,f,g,170] and reasonable qualitative accounts of observed selectivities can sometimes be produced.

3.10.7 PROTONOLYSIS OF ORGANOBORANES

3.10.7.1 General Considerations

The overall process of reduction of an alkene to an alkane, or of reduction of an alkyne to either an alkene or an alkane, *via* hydroboration requires that the intermediate organoborane be protonolyzed (Section 3.10.1.1). However, simple trialkylboranes are remarkably resistant to hydrolysis. For example, trimethylborane gives only 69% of hydroxydimethylborane after 7 h at 180 °C with 1 equiv. of water.[171] Similar resistance is shown to alcohols, phenols and amines, and even mineral acids do not completely protonolyze trialkylboranes with any ease.[4] Thus, aqueous or anhydrous hydrogen bromide removes only one alkyl group from tributylborane after reflux for 1 h.[172]

Anhydrous hydrogen fluoride is much more effective and cleaves all three groups at room temperature in an autoclave.[173] Alternatively, trialkylboranes are readily protonolyzed by carboxylic acids.[174] Two alkyl groups are removed from many trialkylboranes by treatment with excess of an anhydrous carboxylic acid at room temperature, and the third is removed on heating. Typically, the trialkylborane is refluxed for 2–3 h with propanoic acid in diglyme to effect complete conversion (equation 55). The method is considered in greater detail in Section 3.10.7.2. Alternatively, 2,2-dimethylpropanoic acid can be used to catalyze the reaction with water or other protic species.[175,176]

$$R_3B \xrightarrow[\text{diglyme, 165 °C}]{\text{excess EtCO}_2\text{H}} 3\,RH \tag{55}$$

All three alkenyl groups of a trialkenylborane can be removed at 0 °C with a carboxylic acid,[33,177] and alkenyldialkylboranes undergo preferential protonolysis of the alkenyl group.[178] Stereochemistry is retained during protonolysis, so the sequence hydroboration–protonolysis, applied to internal alkynes, provides a useful, general method for synthesis of (*Z*)-alkenes (Scheme 12).[178]

If acidic conditions are undesirable, alkenyldialkylboranes can be converted into alkenyltrialkylborates by addition of butyllithium; this allows hydrolysis under alkaline conditions.[179] Alternatively, alkenylbo-

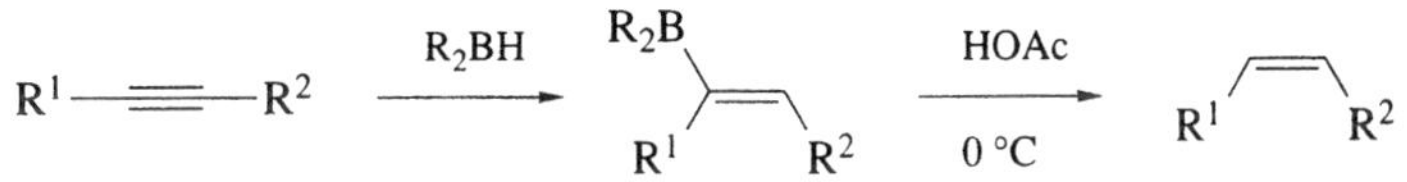

Scheme 12

ron compounds have been cleaved by ammoniacal silver oxide,[180] or under the influence of palladium acetate catalysis.[181] Methanolysis can also be used for cleavage of many alkenylboron compounds.[178]

Arylboron compounds appear to be comparable in reactivity to alkenylboron compounds, but comparative data are not available. Alkynylboranes are fairly readily hydrolyzed,[182] as are benzylboranes, particularly when substituted by electron-withdrawing groups on the benzene ring.[8a] However, none of these groups are available *via* hydroboration and so further discussion here is inappropriate.

Allylboron compounds can sometimes be obtained *via* hydroboration of allenes or 1,3-dienes with dialkylboranes (*e.g.* see Section 3.10.4.2). Allylboron compounds are particularly readily hydrolyzed, and hydrolysis is accompanied by transposition of the double bond.[183] For example, tricrotylborane reacts with water at room temperature to give 2 mol of 1-butene and 1 mol of crotyldihydroxyborane (Scheme 13).

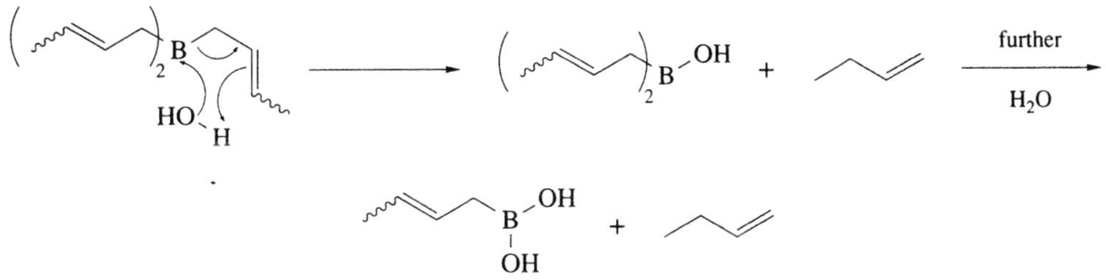

Scheme 13

Allyl groups are selectively cleaved from allyldialkylboranes and all three allyl groups can be readily hydrolyzed from a triallylborane.[183]

A number of other reagents, including alkanethiols and β-dicarbonyl compounds, protonolyze organoboranes. However, they are not generally employed in synthetic procedures and are not discussed here. More exhaustive treatment of protonolyses with other reagents is given elsewhere.[4,14] The remaining sections of this chapter deal with the synthetically useful protonolyses of alkylboron compounds with carboxylic acids and of alkenylboron compounds utilizing carboxylic acids or methanol.

3.10.7.2 Protonolysis of Alkylboranes with Carboxylic Acids: Synthesis of Alkanes

Carboxylic acids readily cleave the first alkyl group from a trialkylborane at 20 °C. The second group is cleaved somewhat more slowly, and the third requires the use of elevated temperatures.[4,14,174] For practical purposes it is usual to employ around 4 equiv. of propanoic or butanoic acid in refluxing diglyme at 165 °C, though use of other solvents is not precluded. The reaction is quite sensitive to steric factors, and tri-*s*-alkylboranes react substantially more slowly than tri-primary-alkylboranes. Nevertheless, some 'mixed' di-*s*-alkylmono-primary-alkylboranes show only a relatively modest preference for cleavage of the primary-alkyl group upon partial protonolysis, though *t*-alkyl groups are substantially less readily cleaved than primary-alkyl groups.[174,184]

Protonolysis with carboxylic acids differs from that with mineral acids in that it appears to involve coordination of the carbonyl oxygen atom of the carboxylic acid to the boron atom of the organoborane, followed by intramolecular proton transfer (Scheme 14).[174,185] This is presumably the reason for the relative ease of protonolysis with carboxylic acids.

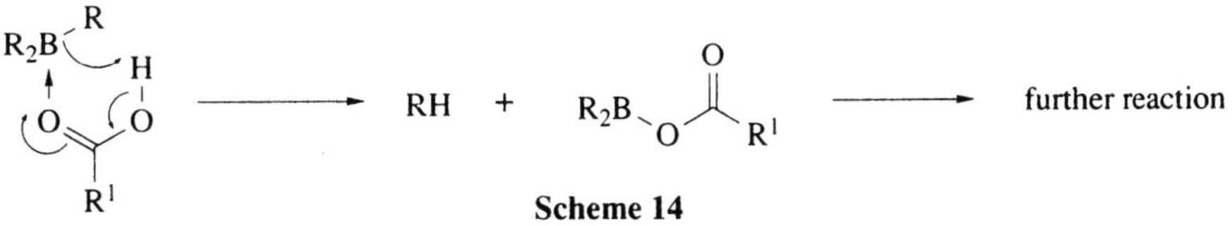

Scheme 14

Replacement of alkyl groups by substituents bearing nonbonding electrons renders the boron atom less electrophilic, and this, together with internal coordination from its own acyloxy groups, explains why the acyloxyboron intermediates become progressively less reactive towards protonolysis.[174]

Protonolysis occurs with retention of stereochemistry at the carbon atom originally attached to boron.[174,186] Thus, by use of a deuteriated hydroborating agent (*e.g.* 9-BBN-D) or of a deuteriated carboxylic acid, or both, deuterium atoms can be introduced at specific and predictable locations in a molecule. An example is the conversion of 2-norbornene into *exo,exo*-2,3-dideuterionorbornane (equation 56).[174]

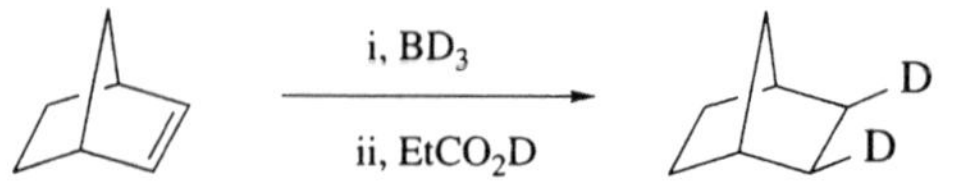

i, BD_3; ii, $EtCO_2D$ (56)

The protonolysis reaction tolerates functionalities such as halide or ether groups in the alkylborane.[174] However, β-dialkylaminoalkylboron compounds undergo elimination to give alkenes under these conditions (see Section 3.10.1.2).[4,10,11] Also, systems which are intrinsically labile to either acid or heat, such as some terpenoids, may give problems. For example, enantiomerically pure limonene produces 1-menthene which is substantially racemized on hydroboration–protonolysis (Scheme 15).[174]

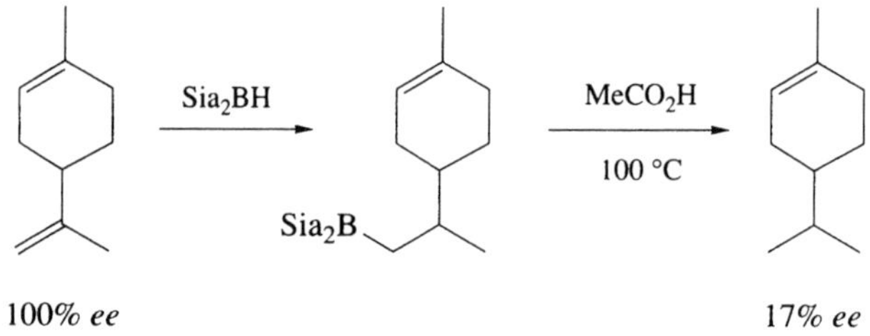

Sia_2BH → Sia_2B ; $MeCO_2H$, 100 °C

100% *ee* 17% *ee*

Scheme 15

3.10.7.3 Protonolysis of Alkenylboranes: Synthesis of Alkenes and Dienes

1-Alkenyldialkylboranes undergo simple protonolysis of the alkenyl group on treatment with hydrochloric acid, but other alkenyldialkylboranes rearrange under these conditions.[187] This problem does not arise on protonolysis with a carboxylic acid, and hydroboration with a dialkylborane followed by protonolysis with a carboxylic acid is therefore of general utility for conversion of alkynes into alkenes.[178] Internal alkynes cleanly give (*Z*)-alkenes (Scheme 12, Section 3.10.7.1).[33,177,178] The process can also be used for stereospecific production of (*E*)-1-deuterio-1-alkenes from terminal alkynes (equation 57).[34,188] Tritiated alkenes have been prepared similarly.[189]

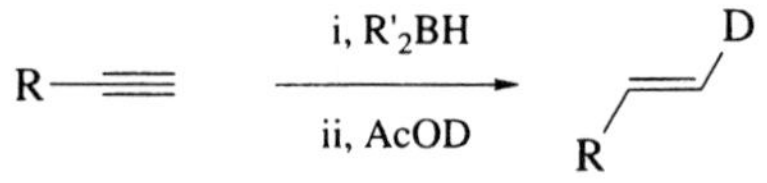

R—≡ ; i, R'_2BH; ii, AcOD → D, R (57)

The combination of deuterioboration and then deuterolysis allows the synthesis of (*Z*)-1,2-dideuterioalkenes and such a process has been used for preparation of dideuteriocycloalkenes (e.g. equation 58).[190–193]

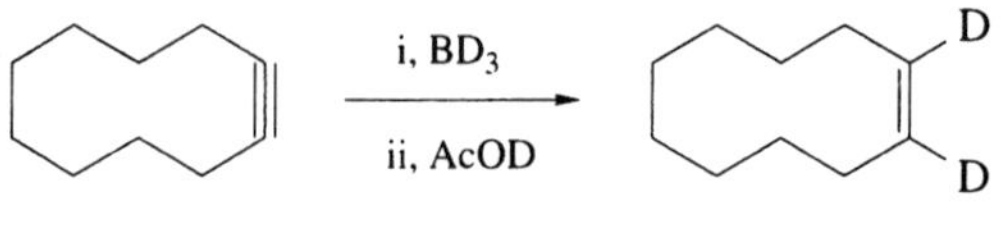

i, BD_3; ii, AcOD → D, D (58)

The protonolysis of alkenylboranes occurs under sufficiently mild conditions that a number of functional groups, including halo, ether and some ester groups, may be tolerated.[14,178b] For example, 1-haloalkynes can be converted into (*Z*)-haloalkenes (*e.g.* equation 59).[95,194]

α,β-Alkynoic esters in most cases undergo the reaction satisfactorily, although a mixture of (*Z*)- and (*E*)-isomers may sometimes occur (*e.g.* equation 60),[195] probably as a result of tautomerism of the intermediate organoborane (see Section 3.10.1.2).

$$\text{Bu}{-}{\equiv}{-}\text{Br} \xrightarrow[\text{ii, AcOH}]{\text{i, Chx}_2\text{BH}} \text{Bu}\diagdown{=}\diagup\text{Br} \qquad (59)$$

$$\text{Bu}^t{-}{\equiv}{-}\text{CO}_2\text{Et} \xrightarrow[\text{ii, AcOH}]{\text{i, Sia}_2\text{BH}} \underset{85\%}{\text{Bu}^t\diagdown{=}\diagup\text{CO}_2\text{Et}} + \underset{15\%}{\text{Bu}^t\diagdown{=}\diagdown\text{CO}_2\text{Et}} \qquad (60)$$

For some functionalized alkenylboranes, cleavage with methanol may be preferable to cleavage with a carboxylic acid. Relatively unhindered alkenyldialkylboranes, such as 9-alkenyl-9-BBNs, are readily cleaved by 1 equiv. of methanol under gentle heating.[178] More hindered derivatives may require catalysis by 2,2-dimethylpropanoic acid, but these mild, almost neutral conditions may be beneficial for acid-labile functionalities. For example, methanolysis has been used for production of enol ethers (*e.g.* equation 61).[196]

$$\left(\text{(EtO)(EtS)C=CH}\right)_3\text{B} \xrightarrow{\text{MeOH}} 3\ \text{(EtO)(EtS)C=CH}_2 \qquad (61)$$

Advantage may be taken of the selective hydroboration of alkynes as compared to alkenes by dicyclohexylborane or disiamylborane (Section 3.10.4.3) to provide syntheses of dienes and their deuteriated derivatives (*e.g.* equation 62).[34,93,197] Dibromoborane–dimethyl sulfide may be used similarly (Section 3.10.4.4).[112] The approach has been used for the stereospecific synthesis of an allylic hydroxydiene (equation 63).[198]

$$\text{1-ethynylcyclohexene} \xrightarrow[\text{ii, AcOD}]{\text{i, R}_2\text{BH}} \text{1-(2-deuteriovinyl)cyclohexene} \qquad (62)$$

$$\text{enyne–CH}_2\text{OAc} \xrightarrow[\text{ii, AcOH; iii, NaOH}]{\text{i, Sia}_2\text{BH}} \text{dienol–OH} \qquad (63)$$

Diynes can lead to enynes or to dienes depending on the dialkylborane chosen as the hydroborating agent, the stoichiometry and the substitution pattern of the diyne (*e.g.* equations 64 and 65).[93,199]

$$\text{Bu}^n{-}{\equiv}{-}{\equiv}{-}\text{Bu}^n \xrightarrow[\text{ii, AcOH}]{\text{i, Sia}_2\text{BH}} \text{Bu}^n\text{CH=CH}{-}{\equiv}{-}\text{Bu}^n \qquad (64)$$

$$\text{Bu}^n{-}{\equiv}{-}{\equiv}{-}\text{Bu}^n \xrightarrow[\text{ii, AcOH}]{\text{i, 2 Chx}_2\text{BH}} \text{Bu}^n\text{CH=CH}{-}\text{CH=CH}\text{Bu}^n \qquad (65)$$

Hydroboration–protonolysis is thus a general, stereospecific and versatile route from alkynes to alkenes, enynes and dienes.

3.10.8 REFERENCES

1. A. Pelter, K. Smith and H. C. Brown, 'Borane Reagents', Academic Press, London, 1988.

2. (a) K. Smith, *Chem. Soc. Rev.*, 1974, **3**, 443; see also refs. 4 and 14; (b) H. C. Brown and U. S. Racherla, *J. Org. Chem.*, 1986, **51**, 427.
3. H. C. Brown, 'Hydroboration', Benjamin, New York, 1962; 2nd printing with Nobel Lecture, Benjamin/Cummings, Reading, MA, 1980; H. C. Brown, 'Boranes in Organic Chemistry', Cornell University Press, Ithaca, NY, 1972.
4. A. Pelter and K. Smith, in 'Comprehensive Organic Chemistry', ed. D. H. R. Barton and W. D. Ollis, Pergamon Press, Oxford, 1979, vol. 3, part 14, p. 689.
5. H. C. Brown and K. Smith, in 'Kirk–Othmer Encyclopaedia of Chemical Technology', 3rd edn., Wiley, New York, 1980, vol. 12, p. 793.
6. H. C. Brown, G. W. Kramer, A. B. Levy and M. M. Midland, 'Organic Synthesis *via* Boranes', Wiley, New York, 1975.
7. W. C. Still and J. C. Barrish, *J. Am. Chem. Soc.*, 1983, **105**, 2487.
8. (a) H. C. Brown and R. L. Sharp, *J. Am. Chem. Soc.*, 1966, **88**, 5851; (b) J. Klein, E. Dunkelblum and M. A. Wolff, *J. Organomet. Chem.*, 1967, **7**, 377.
9. H. C. Brown and E. F. Knights, *Isr. J. Chem.*, 1968, **6**, 691; *J. Am. Chem. Soc.*, 1968, **90**, 4439.
10. D. J. Pasto and R. Snyder, *J. Org. Chem.*, 1966, **31**, 2777; J. W. Lewis and A. A. Pearce, *J. Chem. Soc. B*, 1969, 863; J.-J. Barieux and J. Gore, *Bull. Soc. Chim. Fr.*, 1971, 1649, 3978.
11. C. T. Goralski, B. Singaram and H. C. Brown, *J. Org. Chem.*, 1987, **52**, 4014; B. Singaram, C. T. Goralski, M. V. Rangaishenvi and H. C. Brown, *J. Am. Chem. Soc.*, 1989, **111**, 384.
12. C. F. Lane and H. C. Brown, *J. Am. Chem. Soc.*, 1971, **93**, 1025.
13. H. C. Brown, in 'Current Trends in Organic Synthesis' ed. H. Nozaki, Pergamon Press, Oxford, 1983, p. 247.
14. B. M. Mikhailov and Yu. N. Bubnov, 'Organoboron Compounds in Organic Synthesis', Harwood, London, 1984; see also refs. 4 and 42.
15. (a) E. Negishi and H. C. Brown, *Synthesis*, 1974, 77; (b) H. C. Brown, E. Negishi and J.-J. Katz, *J. Am. Chem. Soc.*, 1975, **97**, 2791; (c) H. C. Brown, J.-J. Katz, C. F. Lane and E. Negishi, *J. Am. Chem. Soc.*, 1975, **97**, 2799.
16. H. C. Brown and G. Zweifel, *J. Am. Chem. Soc.*, 1960, **82**, 4708.
17. D. J. Pasto and F. M. Klein, *J. Org. Chem.*, 1968, **33**, 1468.
18. J. Klein and D. Lichtenberg, *J. Org. Chem.*, 1970, **35**, 2654.
19. H. C. Brown and G. Zweifel, *J. Am. Chem. Soc.*, 1961, **83**, 2544.
20. M. H. Gordon and M. J. T. Robinson, *Tetrahedron Lett.*, 1975, 3867.
21. J. A. Soderquist and H. C. Brown, *J. Org. Chem.*, 1980, **45**, 3571.
22. P. R. Jones and J. K. Myers, *J. Organomet. Chem.*, 1972, **34**, C9; G. L. Larson, D. Hernández and A. Hernández, *J. Organomet. Chem.*, 1974, **76**, 9.
23. H. C. Brown and M. K. Unni, *J. Am. Chem. Soc.*, 1968, **90**, 2902; H. C. Brown and R. M. Gallivan, Jr., *J. Am. Chem. Soc.*, 1968, **90**, 2906; H. C. Brown and R. L. Sharp, *J. Am. Chem. Soc.*, 1968, **90**, 2915.
24. H. C. Brown and E. Negishi, *Tetrahedron*, 1977, **33**, 2331; *Pure Appl. Chem.*, 1972, **29**, 527; E. Negishi, S. U. Kulkarni and H. C. Brown, *Heterocycles*, 1976, **5**, 883.
25. D. E. Young and S. G. Shore, *J. Am. Chem. Soc.*, 1969, **91**, 3497; E. Breuer and H. C. Brown, *J. Am. Chem. Soc.*, 1969, **91**, 4164.
26. H. C. Brown, E. Negishi and P. L. Burke, *J. Am. Chem. Soc.*, 1972, **94**, 3561.
27. H. C. Brown and E. Negishi, *J. Organomet. Chem.*, 1971, **28**, C1.
28. E. Negishi and H. C. Brown, *J. Am. Chem. Soc.*, 1973, **95**, 6757.
29. E. F. Knights and H. C. Brown, *J. Am. Chem. Soc.*, 1968, **90**, 5280.
30. H. C. Brown, E. Negishi and W. C. Dickason, *J. Org. Chem.*, 1985, **50**, 520.
31. D. Devaprabhakara and P. D. Gardner, *J. Am. Chem. Soc.*, 1963, **85**, 1458.
32. I. Mehrotra and D. Devaprabhakara, *Tetrahedron Lett.*, 1972, 4871.
33. H. C. Brown and G. Zweifel, *J. Am. Chem. Soc.*, 1961, **83**, 3834.
34. G. Zweifel, G. M. Clark and N. L. Polston, *J. Am. Chem. Soc.*, 1971, **93**, 3395.
35. G. Zweifel and H. Arzoumanian, *J. Am. Chem. Soc.*, 1967, **89**, 291.
36. H. C. Brown and J. Chandrasekharan, *J. Am. Chem. Soc.*, 1984, **106**, 1863.
37. L. M. Braun, R. A. Braun, H. R. Crissman, M. Opperman and R. M. Adams, *J. Org. Chem.*, 1971, **36**, 2388; C. F. Lane, *J. Org. Chem.*, 1974, **39**, 1437; H. C. Brown, A. K. Mandal and S. U. Kulkarni, *J. Org. Chem.*, 1977, **42**, 1392.
38. K. Smith, *Chem. Ind. (London)*, 1987, 603.
39. P. C. Bulman Page and S. Rosenthal, *Tetrahedron Lett.*, 1986, **27**, 5421.
40. H. C. Brown and A. K. Mandal, *Synthesis*, 1980, 153.
41. C. Camacho, G. Uribe and R. Contreras, *Synthesis*, 1982, 1027.
42. H. C. Brown, M. Zaidlewicz and E. Negishi, in 'Comprehensive Organometallic Chemistry', ed. G. Wilkinson, F. G. A. Stone and E. W. Abel, Pergamon Press, Oxford, 1981, vol. 7, part 45, p. 111.
43. (a) H. C. Brown, V. Somayaji and S. Narasimhan, *J. Org. Chem.*, 1984, **49**, 4822; (b) H. M. Bell, C. W. Vanderslice and A. Spehar, *J. Org. Chem.*, 1969, **34**, 3923.
44. C. Narayana and M. Periasamy, *Tetrahedron Lett.*, 1985, **26**, 1757, 6361.
45. (a) H. S. Lee, K. Isagawa and Y. Otsuji, *Chem. Lett.*, 1984, 363; H. S. Lee, K. Isagawa, H. Toyoda and Y. Otsuji, *Chem. Lett.*, 1984, 673; (b) Y. Aoyama, T. Watanabe, H. Onda and H. Ogoshi, *Tetrahedron Lett.*, 1983, **24**, 1183; (c) N. Satyanarayana and M. Periasamy, *Tetrahedron Lett.*, 1984, **25**, 2501.
46. A. Pelter, D. J. Ryder and J. H. Sheppard, *Tetrahedron Lett.*, 1978, 4715.
47. E. Negishi, J.-J. Katz and H. C. Brown, *J. Am. Chem. Soc.*, 1972, **94**, 4025.
48. H. C. Brown and E. Negishi, *J. Am. Chem. Soc.*, 1972, **94**, 3567.
49. H. C. Brown, J. R. Schwier and B. Singaram, *J. Org. Chem.*, 1979, **44**, 465.
50. H. C. Brown, N. M. Yoon and A. K. Mandal, *J. Organomet. Chem.*, 1977, **135**, C10.
51. H. C. Brown and A. K. Mandal, *Synthesis*, 1978, 146.
52. H. C. Brown, E. Negishi and J.-J. Katz, *J. Am. Chem. Soc.*, 1972, **94**, 5893.

53. H. C. Brown, J. R. Schwier and B. Singaram, *J. Org. Chem.*, 1978, **43**, 4395.
54. A. Pelter, D. J. Ryder, J. H. Sheppard, C. Subrahmanyam, H. C. Brown and A. K. Mandal, *Tetrahedron Lett.*, 1979, 4777.
55. H. C. Brown and S. K. Gupta, *J. Am. Chem. Soc.*, 1971, **93**, 4062.
56. M. F. Hawthorne, *J. Am. Chem. Soc.*, 1959, **81**, 5836.
57. H. C. Brown, T. E. Cole, M. Srebnik and K.-W. Kim, *J. Org. Chem.*, 1986, **51**, 4925.
58. T. E. Cole, R. K. Bakshi, M. Srebnik, B. Singaram and H. C. Brown, *Organometallics*, 1986, **5**, 2303.
59. H. C. Brown and P. K. Jadhav, *J. Org. Chem.*, 1981, **46**, 5047.
60. H. C. Brown and N. M. Yoon, *J. Am. Chem. Soc.*, 1977, **99**, 5514.
61. M. F. Hawthorne, *J. Am. Chem. Soc.*, 1961, **83**, 2541.
62. H. C. Brown and N. Ravindran, *J. Am. Chem. Soc.*, 1976, **98**, 1785.
63. H. C. Brown and M. Zaidlewicz, *J. Am. Chem. Soc.*, 1976, **98**, 4917.
64. W. E. Paget and K. Smith, *J. Chem. Soc., Chem. Commun.*, 1980, 1169.
65. H. C. Brown and N. Ravindran, *Inorg. Chem.*, 1977, **16**, 2938.
66. K. Kinberger and W. Siebert, *Z. Naturforsch., Teil B*, 1975, **30**, 55.
67. H. C. Brown, N. Ravindran and S. U. Kulkarni, *J. Org. Chem.*, 1979, **44**, 2417.
68. H. C. Brown and S. U. Kulkarni, *J. Org. Chem.*, 1979, **44**, 2422.
69. A. Pelter, K. Rowe, D. N. Sharrocks, K. Smith and C. Subrahmanyam, *J. Chem. Soc., Dalton Trans.*, 1976, 2087.
70. J. A. Soderquist and H. C. Brown, *J. Org. Chem.*, 1981, **46**, 4599.
71. C. F. Lane and H. C. Brown, *Heterocycles*, 1977, **7**, 453.
72. H. C. Brown, R. Liotta and C. G. Scouten, *J. Am. Chem. Soc.*, 1976, **98**, 5297.
73. H. C. Brown, D. J. Nelson and C. G. Scouten, *J. Org. Chem.*, 1983, **48**, 641.
74. J. C. Chen, *J. Organomet. Chem.*, 1978, **156**, 213; H. C. Brown and J. C. Chen, *J. Org. Chem.*, 1981, **46**, 3978.
75. H. C. Brown, R. Liotta and L. Brener, *J. Am. Chem. Soc.*, 1977, **99**, 3427.
76. J. A. Soderquist and A. Hassner, *J. Organomet. Chem.*, 1978, **156**, C12.
77. H. C. Brown, J. V. N. Vara Prasad and S.-H. Zee, *J. Org. Chem.*, 1985, **50**, 1582; 1986, **51**, 439; H. C. Brown, P. V. Ramachandran and J. V. N. Vara Prasad, *J. Org. Chem.*, 1985, **50**, 5583.
78. L. Brener and H. C. Brown, *J. Org. Chem.*, 1977, **42**, 2702.
79. R. Liotta and H. C. Brown, *J. Org. Chem.*, 1977, **42**, 2836.
80. H. C. Brown and K. S. Bhat, *J. Org. Chem.*, 1986, **51**, 445.
81. H. C. Brown, G. G. Pai and R. G. Naik, *J. Org. Chem.*, 1984, **49**, 1072.
82. H. C. Brown, R. Liotta and G. W. Kramer, *J. Org. Chem.*, 1978, **43**, 1058.
83. H. C. Brown, R. Liotta and G. W. Kramer, *J. Am. Chem. Soc.*, 1979, **101**, 2966; L. Chevolot, J. Soulié and P. Cadiot, *Tetrahedron Lett.*, 1974, 3435.
84. H. C. Brown, C. G. Scouten and R. Liotta, *J. Am. Chem. Soc.*, 1979, **101**, 96.
85. (a) H. C. Brown, D. Basavaiah and S. U. Kulkarni, *J. Organomet. Chem.*, 1982, **225**, 63; (b) C. A. Brown and R. A. Coleman, *J. Org. Chem.*, 1979, **44**, 2328.
86. A. Pelter, B. Singaram and H. C. Brown, *Tetrahedron Lett.*, 1983, **24**, 1433.
87. H. C. Brown and U. S. Racherla, *Tetrahedron Lett.*, 1985, **26**, 2187.
88. H. C. Brown, N. R. Ayyangar and G. Zweifel, *J. Am. Chem. Soc.*, 1964, **86**, 1071.
89. G. Zweifel and C. C. Whitney, *J. Org. Chem.*, 1966, **31**, 4178.
90. P. Jacob, III, *J. Organomet. Chem.*, 1978, **156**, 101.
91. H. C. Brown and A. W. Moerikofer, *J. Am. Chem. Soc.*, 1961, **83**, 3417; 1963, **85**, 2063; H. C. Brown and G. Zweifel, *J. Am. Chem. Soc.*, 1961, **83**, 1241.
92. G. Zweifel, K. Nagase and H. C. Brown, *J. Am. Chem. Soc.*, 1962, **84**, 190.
93. G. Zweifel and N. L. Polston, *J. Am. Chem. Soc.*, 1970, **92**, 4068.
94. G. M. Clark, K. G. Hancock and G. Zweifel, *J. Am. Chem. Soc.*, 1971, **93**, 1308.
95. G. Zweifel and H. Arzoumanian, *J. Am. Chem. Soc.*, 1967, **89**, 5086.
96. G. Zweifel, A. Horng and J. T. Snow, *J. Am. Chem. Soc.*, 1970, **92**, 1427.
97. G. Zweifel, A. Horng and J. E. Plamondon, *J. Am. Chem. Soc.*, 1974, **96**, 316.
98. H. C. Brown and E. Negishi, *J. Organomet. Chem.*, 1971, **26**, C67.
99. E. Negishi, P. L. Burke and H. C. Brown, *J. Am. Chem. Soc.*, 1972, **94**, 7431.
100. P. L. Burke, E. Negishi and H. C. Brown, *J. Am. Chem. Soc.*, 1973, **95**, 3654.
101. P. J. Maddocks, A. Pelter, K. Rowe, K. Smith and C. Subrahmanyam, *J. Chem. Soc., Perkin Trans. 1*, 1981, 653; J. A. Soderquist and I. Rivera, *Tetrahedron Lett.*, 1988, **29**, 3195.
102. H. C. Brown and S. K. Gupta, *J. Am. Chem. Soc.*, 1971, **93**, 1818.
103. H. C. Brown and S. K. Gupta, *J. Organomet. Chem.*, 1971, **32**, C1.
104. H. C. Brown, E. Negishi and S. K. Gupta, *J. Am. Chem. Soc.*, 1970, **92**, 6648.
105. R. Köster and P. Binger, *Inorg. Synth.*, 1974, **15**, 141.
106. R. Köster, G. Griasnow, W. Larbig and P. Binger, *Justus Liebigs Ann. Chem.*, 1964, **672**, 1.
107. H. C. Brown and S.-C. Kim, *J. Org. Chem.*, 1984, **49**, 1064.
108. H. C. Brown and N. Ravindran, *J. Am. Chem. Soc.*, 1977, **99**, 7097.
109. H. C. Brown, N. Ravindran and S. U. Kulkarni, *J. Org. Chem.*, 1980, **45**, 384.
110. H. C. Brown and U. S. Racherla, *J. Org. Chem.*, 1986, **51**, 895.
111. H. C. Brown and J. B. Campbell, Jr., *J. Org. Chem.*, 1980, **45**, 389.
112. H. C. Brown and J. Chandrasekharan, *J. Org. Chem.*, 1983, **48**, 644.
113. H. C. Brown and U. S. Racherla, *J. Org. Chem.*, 1983, **48**, 1389.
114. H. C. Brown and N. Ravindran, *J. Org. Chem.*, 1977, **42**, 2533.
115. H. C. Brown and N. Ravindran, *J. Am. Chem. Soc.*, 1976, **98**, 1798.
116. A. Hassner and J. A. Soderquist, *J. Organomet. Chem.*, 1977, **131**, C1.
117. H. C. Brown, J. A. Sikorski, S. U. Kulkarni and H. D. Lee, *J. Org. Chem.*, 1980, **45**, 4540.

118. H. C. Brown and J. A. Sikorski, *Organometallics*, 1982, **1**, 28.
119. G. Zweifel and N. R. Pearson, *J. Am. Chem. Soc.*, 1980, **102**, 5919.
120. S. U. Kulkarni, D. Basavaiah, M. Zaidlewicz and H. C. Brown, *Organometallics*, 1982, **1**, 212.
121. J. A. Sikorski and H. C. Brown, *J. Org. Chem.*, 1982, **47**, 872.
122. H. C. Brown, J. A. Sikorski, S. U. Kulkarni and H. D. Lee, *J. Org. Chem.*, 1982, **47**, 863.
123. S. U. Kulkarni, H. D. Lee and H. C. Brown, *J. Org. Chem.*, 1980, **45**, 4542.
124. H. Steinberg, in 'Organoboron Chemistry', Wiley, New York, 1964, vol. 1, p. 105.
125. W. G. Woods and P. L. Strong, *J. Am. Chem. Soc.*, 1966, **88**, 4667.
126. R. H. Fish, *J. Am. Chem. Soc.*, 1968, **90**, 4435.
127. R. H. Fish, *J. Org. Chem.*, 1973, **38**, 158.
128. H. C. Brown and J. Chandrasekharan, *J. Org. Chem.*, 1983, **48**, 5080.
129. H. C. Brown and S. K. Gupta, *J. Am. Chem. Soc.*, 1971, **93**, 1816; 1972, **94**, 4370; 1975, **97**, 5249.
130. D. Männig and H. Nöth, *Angew. Chem., Int. Ed. Engl.*, 1985, **24**, 878; Y. Suseela, A. S. Prasad and M. Periasamy, *J. Chem. Soc., Chem. Commun.*, 1990, 446; A. Arase, Y. Nunokawa, Y. Masuda and M. Hoshi, *J. Chem. Soc., Chem. Commun.*, 1991, 205.
131. D. E. Bergbreiter and D. P. Rainville, *J. Org. Chem.*, 1976, **41**, 3031.
132. C. F. Lane and G. W. Kabalka, *Tetrahedron*, 1976, **32**, 981.
133. S. Thaisrivongs and J. D. Wuest, *J. Org. Chem.*, 1977, **42**, 3243.
134. H. C. Brown, P. K. Jadhav and A. K. Mandal, *Tetrahedron*, 1981, **37**, 3547.
135. J. W. ApSimon and T. L. Collier, *Tetrahedron*, 1986, **42**, 5157.
136. D. A. Evans, J. Bartroli and T. Godel, *Tetrahedron Lett.*, 1982, **23**, 4577.
137. M. M. Midland and Y. C. Kwon, *J. Am. Chem. Soc.*, 1983, **105**, 3725.
138. S. Masamune, B. M. Kim, J. S. Petersen, T. Sato, S. J. Veenstra and T. Imai, *J. Am. Chem. Soc.*, 1985, **107**, 4549.
139. H. C. Brown and P. K. Jadhav, in 'Asymmetric Syntheses', ed. J. D. Morrison, Academic Press, New York, 1983, vol. 2, p. 1.
140. H. C. Brown, J. V. N. Vara Prasad and M. Zaidlewicz, *J. Org. Chem.*, 1988, **53**, 2911.
141. P. K. Jadhav and H. C. Brown, *J. Org. Chem.*, 1981, **46**, 2988.
142. H. C. Brown, M. C. Desai and P. K. Jadhav, *J. Org. Chem.*, 1982, **47**, 5065.
143. H. C. Brown, A. K. Mandal, N. M. Yoon, B. Singaram, J. R. Schwier and P. K. Jadhav, *J. Org. Chem.*, 1982, **47**, 5069.
144. H. C. Brown, P. K. Jadhav and A. K. Mandal, *J. Org. Chem.*, 1982, **47**, 5074.
145. H. C. Brown, R. S. Randad, K. S. Bhat, M. Zaidlewicz, S. A. Weissman, P. K. Jadhav and P. T. Perumal, *J. Org. Chem.*, 1988, **53**, 5513.
146. H. C. Brown, J. V. N. Vara Prasad, A. K. Gupta and R. K. Bakshi, *J. Org. Chem.*, 1987, **52**, 310.
147. H. C. Brown and B. Singaram, *J. Am. Chem. Soc.*, 1984, **106**, 1797; H. C. Brown, P. K. Jadhav and M. C. Desai, *J. Am. Chem. Soc.*, 1982, **104**, 4303.
148. H. C. Brown, R. K. Bakshi and B. Singaram, *J. Am. Chem. Soc.*, 1988, **110**, 1529; *Pure Appl. Chem.*, 1987, **59**, 879.
149. H. C. Brown and J. V. N. Vara Prasad, *J. Am. Chem. Soc.*, 1986, **108**, 2049; H. C. Brown, J. V. N. Vara Prasad and A. K. Gupta, *J. Org. Chem.*, 1986, **51**, 4296.
150. J. J. Partridge, N. K. Chadha and M. R. Uskoković, *J. Am. Chem. Soc.*, 1973, **95**, 7171.
151. K. N. Houk, N. G. Rondan, Y.-D. Wu, J. T. Metz and M. N. Paddon Row, *Tetrahedron*, 1984, **40**, 2257.
152. H. C. Brown, J. Chandrasekharan and K. K. Wang, *Pure Appl. Chem.*, 1983, **55**, 1387.
153. D. J. Pasto, B. Lepeska and T.-C. Cheng, *J. Am. Chem. Soc.*, 1972, **94**, 6083.
154. T. P. Fehlner, *J. Am. Chem. Soc.*, 1971, **93**, 6366.
155. H. C. Brown, C. G. Scouten and K. K. Wang, *J. Org. Chem.*, 1979, **44**, 2589; *Proc. Natl. Acad. Sci. USA*, 1980, **77**, 698.
156. K. K. Wang and H. C. Brown, *J. Org. Chem.*, 1980, **45**, 5303.
157. K. K. Wang, C. G. Scouten and H. C. Brown, *J. Am. Chem. Soc.*, 1982, **104**, 531.
158. D. J. Nelson and H. C. Brown, *J. Am. Chem. Soc.*, 1982, **104**, 4907.
159. D. J. Nelson, C. D. Blue and H. C. Brown, *J. Am. Chem. Soc.*, 1982, **104**, 4913.
160. K. K. Wang and H. C. Brown, *J. Am. Chem. Soc.*, 1982, **104**, 7148.
161. H. C. Brown, K. K. Wang and J. Chandrasekharan, *J. Am. Chem. Soc.*, 1983, **105**, 2340.
162. H. C. Brown, J. Chandrasekharan and K. K. Wang, *J. Org. Chem.*, 1983, **48**, 2901, 3689.
163. J. Chandrasekharan and H. C. Brown, *J. Org. Chem.*, 1985, **50**, 518.
164. H. C. Brown and J. Chandrasekharan, *J. Am. Chem. Soc.*, 1984, **106**, 1863; *Gazz. Chim. Ital.*, 1987, **117**, 377.
165. H. C. Brown and L. T. Murray, *Inorg. Chem.*, 1984, **23**, 2746.
166. H. C. Brown, J. Chandrasekharan and D. J. Nelson, *J. Am. Chem. Soc.*, 1984, **106**, 3768.
167. H. C. Brown and J. Chandrasekharan, *J. Org. Chem.*, 1988, **53**, 4811.
168. T. Clark, D. Wilhelm and P. von R. Schleyer, *J. Chem. Soc., Chem. Commun.*, 1983, 606.
169. (a) S. Dasgupta, M. K. Datta and R. Datta, *Tetrahedron Lett.*, 1978, 1309; (b) T. Clark and P. von R. Schleyer, *J. Organometal. Chem.*, 1978, **156**, 191; (c) M. J. S. Dewar and M. L. McKee, *Inorg. Chem.*, 1978, **17**, 1075; (d) K. R. Sundberg, G. D. Graham and W. N. Lipscomb, *J. Am. Chem. Soc.*, 1979, **101**, 2863; (e) S. Nagase, N. K. Ray and K. Morokuma, *J. Am. Chem. Soc.*, 1980, **102**, 4536; (f) G. D. Graham, S. C. Freilich and W. N. Lipscomb, *J. Am. Chem. Soc.*, 1981, **103**, 2546; (g) M. N. Paddon-Row, N. G. Rondan and K. N. Houk, *J. Am. Chem. Soc.*, 1982, **104**, 7162.
170. D. N. J. White and M. J. Bovill, *J. Chem. Soc., Perkin Trans. 2*, 1983, 225; D. J. Nelson and P. J. Cooper, *Tetrahedron Lett.*, 1986, **27**, 4693.
171. D. Ulmschneider and J. Goubeau, *Chem. Ber.*, 1957, **90**, 2733.
172. J. R. Johnson, H. R. Synder and M. G. van Campen, Jr., *J. Am. Chem. Soc.*, 1938, **60**, 115.
173. G. A. Olah, P. W. Westerman, Y. K. Mo and G. Klopman, *J. Am. Chem. Soc.*, 1972, **94**, 7859.
174. H. C. Brown and K. J. Murray, *Tetrahedron*, 1986, **42**, 5497.

175. R. Köster, K.-L. Amen, H. Bellut and W. Fenzl, *Angew. Chem., Int. Ed. Engl.*, 1971, **10**, 748; R. Köster, H. Bellut and W. Fenzl, *Justus Liebigs Ann. Chem.*, 1974, 54.
176. R. Köster, W. Fenzl and G. Seidel, *Justus Liebigs Ann. Chem.*, 1975, 352; R. Köster, K.-L. Amen and W. V. Dahlhoff, *Justus Liebigs Ann. Chem.*, 1975, 752 and references cited therein.
177. H. C. Brown and G. Zweifel, *J. Am. Chem. Soc.*, 1959, **81**, 1512.
178. (a) H. C. Brown and G. A. Molander, *J. Org. Chem.*, 1986, **51**, 4512; (b) H. C. Brown and K. K. Wang, *J. Org. Chem.*, 1986. **51**, 4514.
179. E. Negishi and K.-W. Chiu, *J. Org. Chem.*, 1976, **41**, 3484.
180. E. J. Corey and T. Ravindranathan, *J. Am. Chem. Soc.*, 1972, **94**, 4013.
181. H. Yatagai, Y. Yamamoto and K. Maruyama, *J. Chem. Soc., Chem. Commun.*, 1978, 702.
182. H. Hartmann and K.-H. Birr, *Z. Anorg. Allg. Chem.*, 1959, **299**, 174.
183. B. M. Mikhailov, *Organomet. Chem. Rev., Sect. A*, 1972, **8**, 1.
184. D. B. Bigley and D. W. Payling, *J. Inorg. Nucl. Chem.*, 1971, **33**, 1157.
185. L. H. Toporcer, R. E. Dessy and S. I. E. Green, *J. Am. Chem. Soc.*, 1965, **87**, 1236; *Inorg. Chem.*, 1965, **4**, 1649.
186. A. G. Davies and B. P. Roberts, *J. Chem. Soc. C*, 1968, 1474; G. W. Kabalka, R. J. Newton, Jr. and J. Jacobus, *J. Org. Chem.*, 1979, **44**, 4185.
187. M. M. Midland and H. C. Brown, *J. Org. Chem.*, 1975, **40**, 2845.
188. R. W. Murray and G. J. Williams, *J. Org. Chem.*, 1969, **34**, 1896.
189. N. H. Nam, A. J. Russo and R. F. Nystrom, *Chem. Ind. (London)*, 1963, 1876.
190. A. C. Cope, G. A. Berchtold, P. E. Peterson and S. H. Sharman, *J. Am. Chem. Soc.*, 1960, **82**, 6370.
191. J. Sicher, M. Svoboda, J. Závada, R. B. Turner and P. Goebel, *Tetrahedron*, 1966, **22**, 659.
192. J. Sicher, M. Svoboda, B. J. Mallon and R. B. Turner, *J. Chem. Soc. B*, 1968, 441.
193. D. S. Sethi, S. Vaidyanathaswamy and D. Devaprabhakara, *Org. Prep. Proced. Int.*, 1970, **2**, 171.
194. C. D. Blue and D. J. Nelson, *J. Org. Chem.*, 1983, **48**, 4538.
195. J. E. Plamondon, J. T. Snow and G. Zweifel, *Organomet. Chem. Synth.*, 1971, **1**, 249.
196. B. M. Mikhailov, T. A. Shchegoleva and E. M. Shashkova, *Zh. Obshch. Khim.*, 1974, **44**, 2193; B. M. Mikhailov, Yu. N. Bubnov and M. Sh. Grigoryan, *Zh. Obshch. Khim.*, 1974, **44**, 2466.
197. V. V. Markova, V. A. Kormer and A. A. Petrov, *Zh. Obshch. Khim.*, 1965, **35**, 1669.
198. E. J. Corey and D. K. Herron, *Tetrahedron Lett.*, 1971, 1641.
199. A. J. Hubert, *J. Chem. Soc.*, 1965, 6669.

3.11
Hydroalumination of C=C and C≡C

JOHN J. EISCH
State University of New York at Binghamton, NY, USA

3.11.1 INTRODUCTION

3.11.1.1 Nature of the Overall Reaction

The addition of aluminum hydrides to alkenic and alkynic carbon–carbon unsaturation, namely the hydroalumination reaction, is the most versatile route to organoaluminum compounds.[1-3] With unsymmetrically substituted C=C and C≡C linkages, the regiochemistry of such hydroaluminations is governed by the interplay of steric and electronic factors, so that these additions can occur in a highly regioselective manner. The addition of dialkylaluminum hydrides to β-substituted styrenes (**2**), for example, can lead regiospecifically to only one of the possible adducts: for steric reasons, a β-*t*-butyl group produces only isomer (**1**); for polar reasons, a β-phenylsulfonyl group yields exclusively isomer (**3**; equation 1).

$$(1) \xleftarrow[E = Bu^t]{R_2AlH} (2) \xrightarrow[E = SO_2Ph]{R_2AlH} (3) \quad (1)$$

In addition to the regiochemical course, the stereochemistry of these additions can be highly selective. Illustrative of this characteristic is the behavior of 1-alkynylsilanes (**5**) toward diisobutylaluminum hydride.[4] In the presence of a Lewis base, such as *N*-methylpyrrolidine, the hydroalumination gives >95% of the *syn* adduct (**4**); in the absence of a donor, the adduct has >95% of the *anti* configuration (**6**; equation 2).

$$(4) \xleftarrow[R_3N]{Bu^i_2AlH} R—≡—SiMe_3\ (5) \xrightarrow{Bu^i_2AlH} (6) \quad (2)$$

Finally, the high locoselectivity[5] or chemoselectivity of hydroalumination toward substrates having both C=C and C≡C bonds is exemplified by the reaction of diisobutylaluminum hydride with 1-octen-7-yne (**7**).[6] The only organoaluminum product detected (**8**) was that of a regioselective *syn* stereospecific addition to the C≡C bond (equation 3).

$$(7) \xrightarrow{Bu^i_2AlH} (8) \quad (3)$$

3.11.1.2 Historical Context and Development

The discovery of the hydroalumination reaction by Ziegler and Gellert in 1949 was the culmination of a series of experiments on the thermal stability of metal alkyls begun by Karl Ziegler in the early 1930s.[7] During an attempt to distill ethyllithium, Ziegler found that the compound decomposed over 100 °C into ethylene and lithium hydride; the ethylene reacted, in steps, with the ethyllithium to give higher *n*-alkyllithiums and these, in turn, eliminated lithium hydride to produce a mixture of higher α-alkenes (Scheme 1).[8]

If lithium hydride would add to ethylene, *in situ*, Ziegler speculated, then the various other steps given in Scheme 1 would become accessible and lithium hydride could thereby catalyze the conversion of ethylene into a mixture of higher alkenes. Indeed, at over 200 °C lithium hydride did begin to effect the oligomerization of ethylene into higher alkenes. However, because of allylic lithiation (**9**) and subsequent elimination of LiH, conjugated dienes (**9a**) were formed and they polymerized to resins that coated the LiH catalyst and thereby retarded the ethylene oligomerization (equation 4).

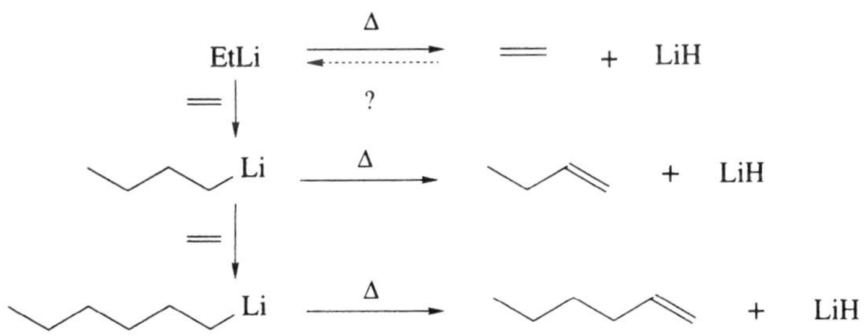

Δ
EtLi ⇌ = + LiH
?
Li —Δ→ + LiH
Li —Δ→ + LiH

Scheme 1

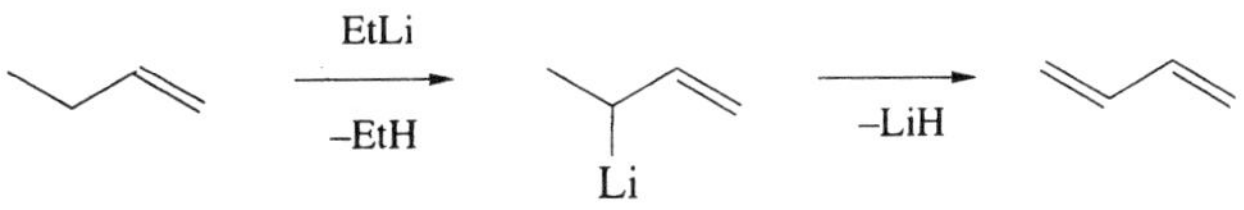

$$\text{(4)}$$

EtLi, –EtH → Li (**9**) → –LiH (**9a**)

With the purpose of employing a more soluble form of complexed lithium hydride in such a process, Ziegler and Gellert heated ethylene in ether under pressure at 120–140 °C with LAH, which had been prepared for the first time by Schlesinger and coworkers.[9] To Ziegler's gratification LAH was smoothly converted into lithium tetraethylaluminate (**10**), which could be isolated in high yield as a crystalline adduct. As proof of structure, the same adduct was also synthesized by the admixture of the lithium and aluminum ethyl derivatives in a 1:1 ratio (Scheme 2). This study represents the first authenticated instance of hydroalumination.

$$4 = \; + \; LiAlH_4 \xrightarrow{120\text{–}140\ °C} LiAlEt_4 \longleftarrow EtLi + Et_3Al$$

(**10**)

Scheme 2

As part of the same comprehensive study Ziegler and Gellert also showed that at 180–200 °C complex (**10**) did catalyze the oligomerization of ethylene into higher α-alkenes. Subsequently, the unstable AlH_3 was found to add even more readily to ethylene and other α-alkenes (**11**) regioselectively to provide quantitative yields of the tri-*n*-alkylaluminum (**12b**). This reaction is thus the prototype for the hydroalumination of unsaturated hydrocarbons (equation 5).

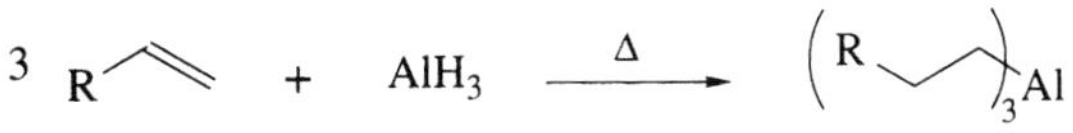

$$3\ R\text{–CH=CH}_2 + AlH_3 \xrightarrow{\Delta} (R\text{–CH}_2\text{CH}_2)_3Al \qquad (5)$$

(**11**) (**12a**) (**12b**)

Since the hydroalumination of an alkene is a readily reversible process, especially at high temperatures, with β,β-disubstituted alkenes or in the presence of nickel,[10] one need not employ the unstable AlH_3 (**12a**) as such but can instead use the commercially available triisobutylaluminum (**13a**).[11] This interesting alkyl can be synthesized directly from activated aluminum metal, hydrogen and isobutylene, and when it is heated at 100–110 °C, it in turn dissociates into the dialkylaluminum hydride and isobutylene. Therefore, AlH_3 can be transferred from (**13a**) to a less volatile α-alkene (**11**) simply by heating (**13a**) and (**11**) and permitting the isobutylene to escape. This hydroalumination by AlH_3 transfer is a widely applicable synthetic method for trialkylaluminums (**12**; Scheme 3).

The hydroalumination of alkynes was first observed by Wilke and Müller in 1955, when they demonstrated that dialkylaluminum hydrides added to disubstituted alkynes stereospecifically in a *syn* fashion (**13b**).[12,13] Subsequently, Zakharkin and coworkers were able to add $NaAlH_4$ to phenylacetylene by employing 5% of Bu^i_2AlH and diglyme as a solvent, but did not ascertain the stereochemistry of reaction.[14] Some 10 years later, in 1966, Slaugh found that LAH in refluxing THF–diglyme mixtures, hydroaluminates such alkynes exclusively in an *anti* manner (**14**; equation 6).[15,16]

During the succeeding decade the regiochemistry and stereochemistry for the hydroalumination of various heteroatom-substituted 1-alkynes was extensively studied and established by the laboratories of Eisch and of Zweifel.[17]

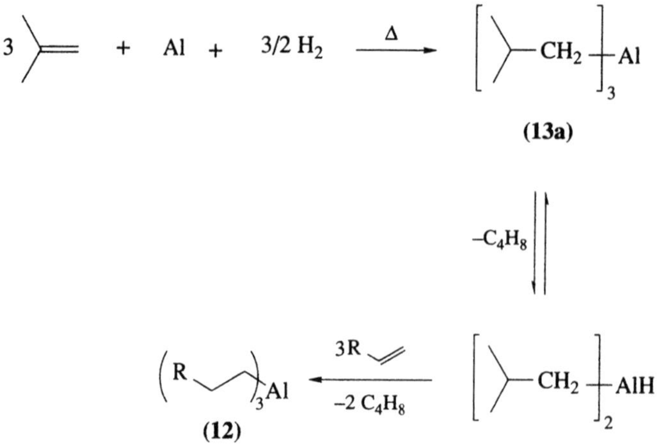

$$3\ \text{(isobutene)} + \text{Al} + 3/2\,\text{H}_2 \xrightarrow{\Delta} [(\text{CH}_3)_2\text{CH}\text{—CH}_2]_3\text{Al}\ \textbf{(13a)}$$

$$\textbf{(13a)} \underset{}{\overset{-\text{C}_4\text{H}_8}{\rightleftharpoons}} [(\text{CH}_3)_2\text{CH}\text{—CH}_2]_2\text{AlH} \xrightarrow[-2\,\text{C}_4\text{H}_8]{3\text{R}\diagup\!\!=} (\text{RCH}_2\text{CH}_2)_3\text{Al}\ \textbf{(12)}$$

Scheme 3

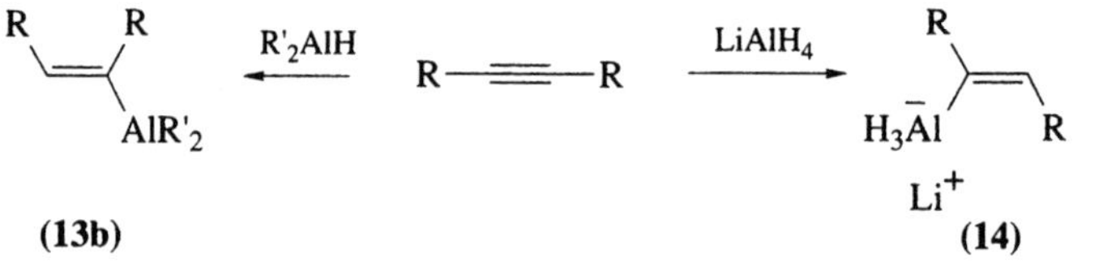

$$\textbf{(13b)}\ \text{R(AlR'}_2\text{)C=CHR} \xleftarrow{\text{R'}_2\text{AlH}} \text{R—≡—R} \xrightarrow{\text{LiAlH}_4} \text{H}_3\bar{\text{Al}}\text{C(R)=CHR}\ \text{Li}^+\ \textbf{(14)} \tag{6}$$

3.11.1.3 Sources of the Aluminum Hydrides

The most commonly employed hydroaluminating agents, DIBAL-H and LAH, are commercially available in a wide variety of ether and hydrocarbon solvents or as the pure reagents.[18] For reactions where a different R_2AlH may be deemed advisable or where the deuterioaluminating agent R_2AlD is required, the corresponding R_2AlCl can be treated with MH or MD (M = Li, Na; equation 7).[19,20]

$$\text{Bu}^i{}_2\text{AlD} \xleftarrow[\text{R = Bu}^i]{\text{LiD/Et}_2\text{O}} \text{R}_2\text{AlCl} \xrightarrow[\text{R = Et}]{\text{NaH}} \text{Et}_2\text{AlH} \tag{7}$$

Aluminum hydride, AlH_3, and dichloroaluminum hydride, Cl_2AlH, can be obtained as etherates by the original method of Schlesinger and coworkers[9] and can be employed for hydroalumination (equation 8).[21]

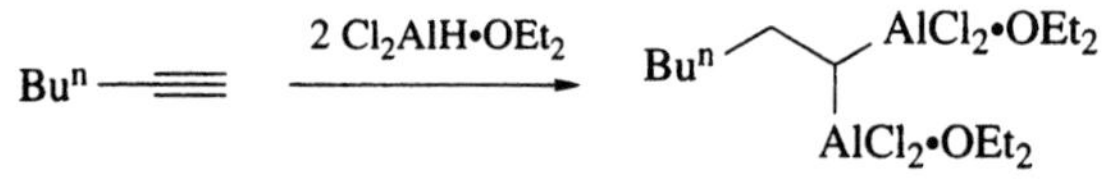

$$\text{Bu}^n\text{—≡} \xrightarrow{2\,\text{Cl}_2\text{AlH•OEt}_2} \text{Bu}^n\text{CH}_2\text{CH(AlCl}_2\text{•OEt}_2)_2 \tag{8}$$

Lithium trialkylaluminum hydrides, such as lithium diisobutyl(methyl)aluminum hydride (**15**), which can be readily prepared by the interaction of diisobutylaluminum hydride with methyllithium in DME, are useful for the *anti* hydroalumination of internal alkynes (equation 9).[22]

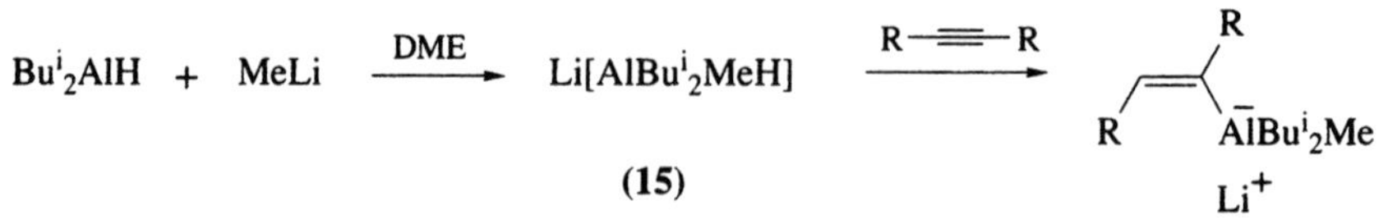

$$\text{Bu}^i{}_2\text{AlH} + \text{MeLi} \xrightarrow{\text{DME}} \text{Li[AlBu}^i{}_2\text{MeH]}\ \textbf{(15)} \xrightarrow{\text{R—≡—R}} \text{RCH=C(R)}\bar{\text{Al}}\text{Bu}^i{}_2\text{Me}\ \text{Li}^+ \tag{9}$$

In addition to simple and complex aluminum hydrides, aluminum alkyls with β-branched chains often serve as hydroaluminating agents themselves, because of their tendency to form Al—H bonds during reaction by the thermal or nickel-catalyzed loss of alkene (*cf.* Scheme 3; equation 10).[23]

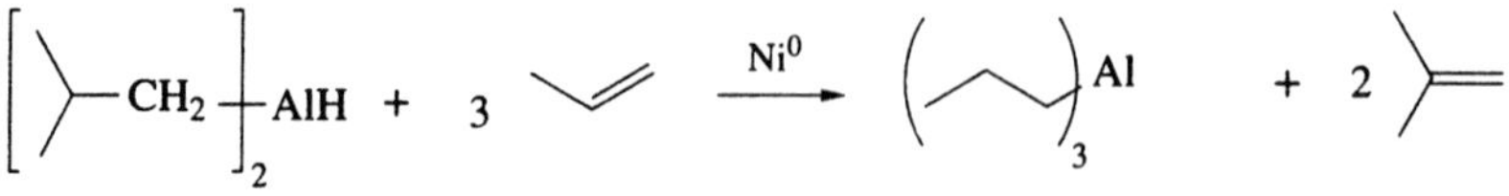

$$[(\text{CH}_3)_2\text{CH}\text{—CH}_2]_2\text{AlH} + 3\ \text{CH}_3\text{CH=CH}_2 \xrightarrow{\text{Ni}^0} (\text{CH}_3\text{CH}_2\text{CH}_2)_3\text{Al} + 2\ \text{(isobutene)} \tag{10}$$

3.11.1.4 Uncatalyzed and Catalyzed Hydroalumination

Typically, hydroaluminations are carried out by heating the substrate with R_2AlH in hydrocarbon media or with $MAlR_3H$ in higher boiling, strongly donating ethers.[17] Metal salts of zirconium,[23] nickel,[24] titanium,[25] and uranium[26] have been found to increase the rate of hydroaluminations. It should be noted, however, that such catalyzed reactions may exhibit a regioselectivity and stereoselectivity different from those of the uncatalyzed hydroalumination.[24] Thus, the hydroalumination of 1-phenylpropyne by Bu^i_2AlH gives an 82:18 ratio of regioisomers (**16**) and (**17**) in the thermal reaction, while the nickel-catalyzed reaction yields a 56:44 ratio of (**16**) and (**17**; equation 11). Furthermore, the uncatalyzed reaction yields only *cis*-1-phenylpropene upon hydrolysis, while the nickel-catalyzed process produces 1-phenylpropene having 6% of the *trans* isomer.

Ph—≡— $\xrightarrow{Bu^i_2AlH}$ (**16**) + (**17**) (11)

	(16)		(17)
thermal:	82	:	18
Ni catalyzed:	56	:	44

It is noteworthy that the transition metals that serve as hydroaluminating catalysts are also active in establishing the equilibrium between aluminum alkyls and their decomposition products, aluminum, hydrogen and alkene (equation 12). Accordingly, these metals, in addition to hafnium, niobium, vanadium, scandium and lanthanum, have found use as activators for the direct synthesis of aluminum alkyls (equation 12, to the left).[2] Probably most of these metal salts will also be capable of accelerating the hydroalumination reaction.

$$(R_2CHCH_2)_3Al \rightleftharpoons 3\ R_2C{=}CH_2 + Al + 3/2\ H_2 \quad (12)$$

3.11.1.5 Potential of Resulting Organoaluminum Reagents in Organic Synthesis

Depending upon the nature of the hydroaluminating agent (R_2AlH, $MAlR_3H$ or LAH), the organoaluminum product will have one to four new carbon–aluminum bonds at which cleavage or insertion reactions can be conducted. Two such reactions are protodealumination (equation 13) and carbonation (equation 14).[27]

(**18**) $\xrightarrow{H_2O}$ (**19**) (13)

(**18**) $\xrightarrow{CO_2}$ (**20**) (14)

Important aspects of such reactions are: (i) the configuration at the original carbon–aluminum bond is usually retained in such cleavages or insertions; (ii) not all three carbon–aluminum bonds undergo reaction with some reagents (CO_2, R_2CO, RX, *etc.*), but vinylic or alkynic carbon–aluminum bonds are generally more reactive than alkyl carbon–aluminum bonds. Thus, in (**18**) carbonation leads to a 96% yield of *cis*-α-phenylcinnamic acid [as salt (**20**)]; any organic products resulting from the R-group in the hydroaluminating reagent itself (*e.g.* isobutyl) should not interfere with isolating the pure desired product. In equations (13) and (14) the isobutane formed by hydrolysis or any isovaleric acid resulting from carbonation can readily be separated from the *cis*-stilbene (**19**) or from the *cis*-α-phenylcinnamic acid [from salt (**20**)] by volatility; and (iii) if the vinyl carbon–aluminum bond of the hydroalumination adduct itself is insufficiently reactive toward certain cleavage agents, it may be activated by first forming the tetrasubstituted aluminate complex with methyllithium (Scheme 4).[28]

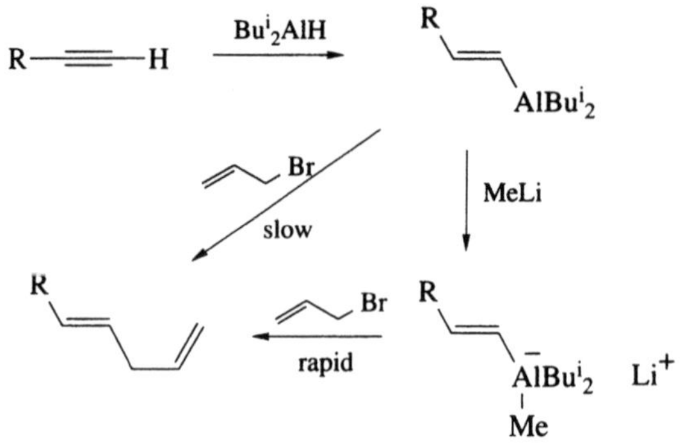

Scheme 4

In light of the foregoing, the utility of hydroalumination in selective organic synthesis will be more applicable to alkynic substrates, where the greater chemical reactivity of the resulting vinyl–aluminum bond will permit the formation of derivatives even with those reagents (CO_2, R_2CO, RX) not cleaving all available carbon–aluminum bonds. Alkenic substrates generally will be less suitable in hydroalumination reactions where the adduct is then required to react with a selective reagent (*e.g.* CO_2) that attacks at most one carbon–aluminum bond. Thus the similar reactivity of two competing alkyl–aluminum bonds will inevitably give a mixture of products and low yields of the desired derivative (Scheme 5). Both (**21**) and (**22**) are unreactive toward further insertion of CO_2 under mild conditions.[29]

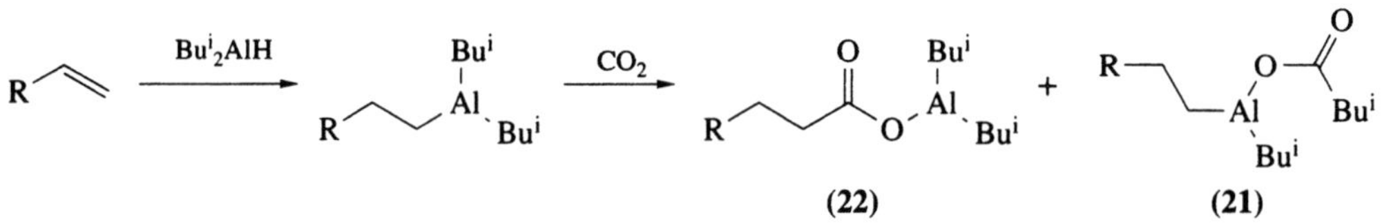

Scheme 5

Only with reagents that attack all available carbon–aluminum bonds (H_2O, D_2O, O_2, X_2, SO_2, *etc.*) will the hydroalumination of alkenic substrates and the derivatization of such adducts prove useful. That two of three C—Al bonds of adduct (**23**) are expended in the formation of side products, which have to be separated, may be an acceptable price to pay for the convenient, high-yielding generation of a desired specifically deuteriated reduction product, (**24**) or (**25**),[30] primary alcohol (**26**),[31] primary halide (**27**)[32] or sulfinic acid.[29] Of course the isobutylated side products should, in each case, be easily separable (Scheme 6).

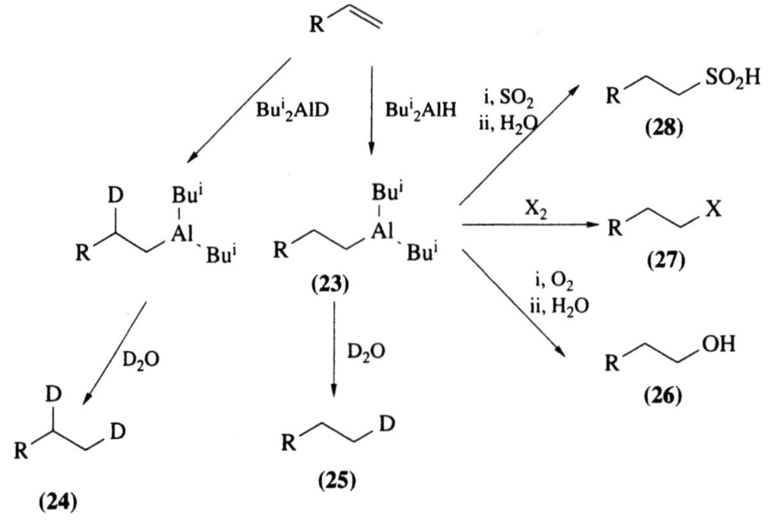

Scheme 6

3.11.2 SCOPE OF HYDROALUMINATION

3.11.2.1 Suitable Alkenic Substrates

The hydroalumination of alkenes, either by direct addition of R_2AlH or by aluminum hydride transfer from triisobutylaluminum (Scheme 3), proceeds readily and without catalysis with terminal alkenes, whether or not there is a chain branch in the β-position (Table 1).

Table 1 Scope of Suitable Alkenic Substrates for Hydroalumination with Dialkylaluminum Hydrides[11,19]

Substrate	*Ease of addition*
$RCH{=}CH_2$ (R = *n*-alkyl)	Very fast
Cycloalkenes (C_4–C_{11})	Fast (C_4, C_8)
	Slow (C_5, C_7, C_9)
	Very slow (C_6, C_{10}, C_{11})
$RR'C{=}CH_2$ (R, R′ = alkyl, aryl)	Fast
$RR'C{=}CH_2$ (R, R′ = alkyl, cycloalkyl)	Less fast
β-Pinene	Less fast
Camphene	Less fast
Limonene	Less fast
$RCH{=}CHR$	Very slow

Thus, all α-alkenes can be readily converted into the tri-*n*-alkylaluminum (>95%) by heating with about 0.33 mol equiv. of $Bu^i{}_2AlH$ or $Bu^i{}_3Al$, until isobutylene ceases to be evolved. With β-pinene hydroalumination can also be achieved in this manner but the result is the formation of R'_2AlH, rather than R'_3Al. At the temperature of reaction, such branched R'_3Al tend to eliminate 1 mol of alkene (equation 15).[33]

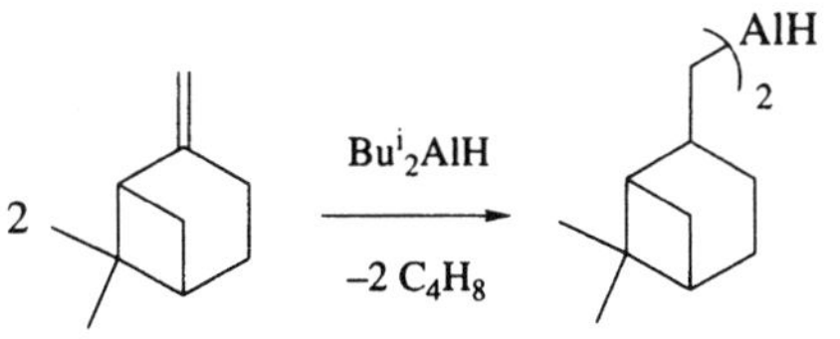

(15)

With alkenes having internal C═C bonds, hydroalumination is disfavored by both kinetic and thermodynamic factors, and the uncatalyzed reaction is generally unfeasible. The hydroalumination of internal alkenes can be catalyzed by the addition of titanium(IV) alkoxides but the same catalysts also promote the isomerization of the secondary aluminum alkyls generated into their primary isomers (equation 16).[34]

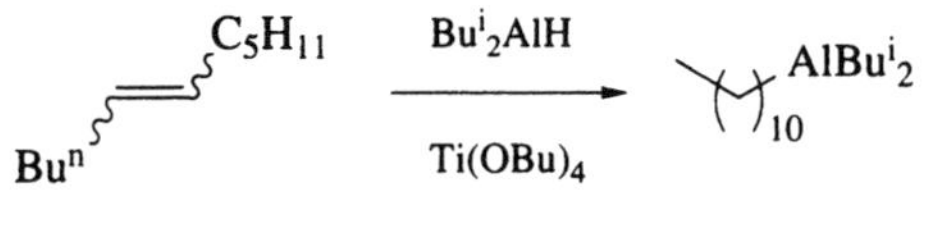

(16)

The hydroalumination of cycloalkenes exhibits two interesting minima in relative rates toward diethylaluminum hydride. The half-life (in min) of a 3:1 ratio of Et_2AlH and cycloalkene at 78 °C varied thus with ring size:[19] C_4, 40; C_5, 200; C_6, 1880; C_7, 110; C_8, 86; C_9, 280; C_{10}, 2500; C_{11}, 1350. The reactivity of cycloundecene is approximately that of an open-chain internal alkene. The low reactivity of cyclohexene can be ascribed to the eclipsing repulsions generated in the transition state by the *syn* attack of Et_2AlH on the C═C bond (equation 17).[35]

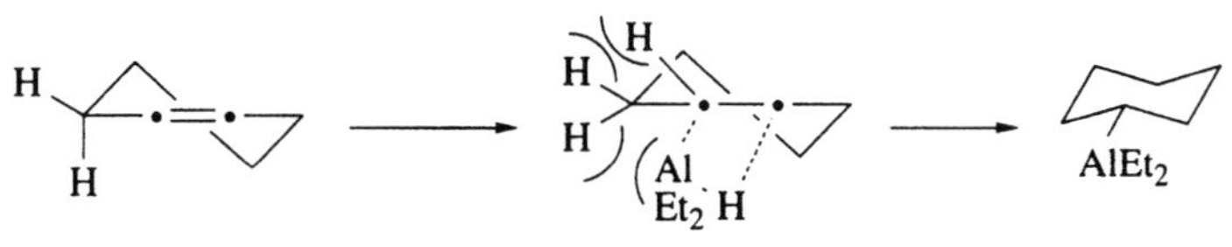

(17)

Such repulsions would be distinctly less with the more planar C_4 and C_5 rings, as well as with the less rigid C_7 and C_8 rings. The minimum in reactivity encountered with the C_{10} ring is indicative that transannular repulsions may be restricting access to one face of the C═C bond, while also generating further transannular repulsions as the vinylic hydrogens bend away from the attacking Et_2AlH molecule (**29**).

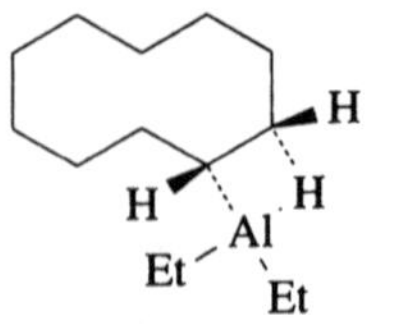

(29)

In contrast with the impracticality of hydroaluminating open-chain internal alkenes, all the unsubstituted cycloalkenes can be usefully hydroaluminated, with titanium catalysis if need be, since isomerization of the C═C bond is of no significance.

3.11.2.2 Suitable Alkynic Substrates

As a class alkynes are much more reactive in hydroalumination than are alkenes. Hence, both terminal and internal alkynes react at feasible rates with both dialkylaluminum hydrides in alkanes and lithium aluminum hydrides ($LiAlR_nH_{4-n}$) in ethers. Selected examples of such additions are presented in Table 2. With alkyl or aryl substituents, it should be noted that R_2AlH adds in a kinetically *syn* manner, (**5**; equation 2) and (**7**; equation 3), and LAH yields the *anti* adduct (**14**; equation 6).

Table 2 Scope of Suitable Alkynic Substrates for Hydroalumination with R'_2AlH or $MAlR'_3H$ Reagents

Substrate	*Reagent*	*Product(s)*	*Ref.*
RC≡CH	R'_2AlH	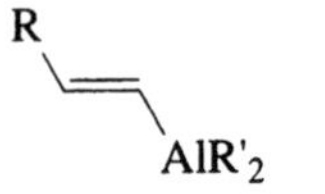 R, AlR'_2	30
RC≡CE (E = $SiMe_3$, $GeMe_3$)	R'_2AlH	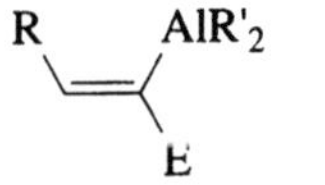 R, AlR'_2, E	4
RC≡CE	R'_2AlH•NR'_3	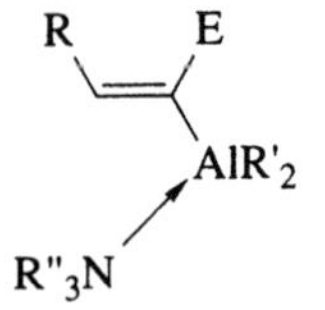 R, E, AlR'_2, R''_3N	4
RC≡CR (R = alkyl, aryl)	R'_2AlH	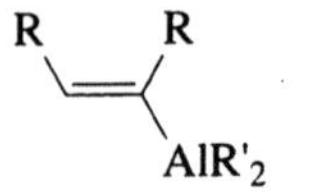 R, R, AlR'_2	12, 13, 30
RC≡CR	$LiAlMeBu^i_2H$	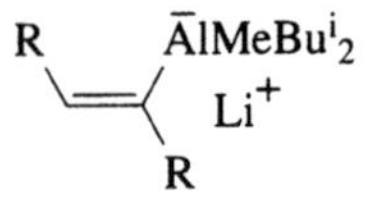 R, $\bar{A}lMeBu^i_2$ Li^+, R	22
RC≡CR''	R'_2AlH	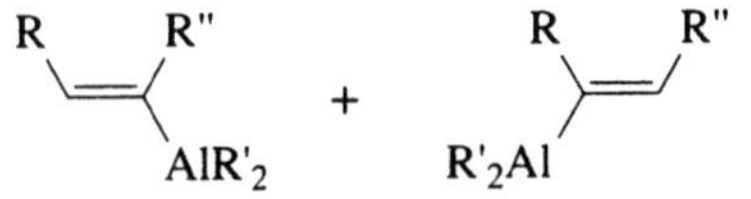 R, R'', AlR'_2 + R, R'', R'_2Al mixture of regioisomers	39, 40
H_2C=$CH(CH_2)_4C$≡CH	R'_2AlH	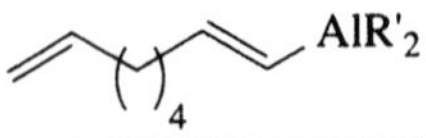 AlR'_2, 4	6

Although reactivity is not a problem with alkynes, the occurrence of side reactions can raise serious difficulties in preparative applications. In addition to the general aspects of reagent selectivity, which are discussed in Section 3.11.2.6, alkynes present the possibility of specific reactions competing with hydroalumination. First, terminal alkynes may undergo alumination (proton–aluminum exchange), which becomes especially prominent with $LiAlR_nH_{4-n}$ (equation 18)[36] and $R_2AlH–R'_3N$ complexes (equation 19).[37]

$$MAlH_4 + 4\,HC\equiv C{-}R \longrightarrow M\left(C\equiv C{-}R\right)_4 + 4\,H_2 \quad (18)$$

M = Li, Na, K; R = alkyl, aryl

$$2\,Et_2AlH{\cdot}NEt_3 + HC\equiv CH \xrightarrow{-2H_2} Et_2Al{-}C\equiv C{-}AlEt_2{\cdot}2Et_3N \quad (19)$$

Secondly, alkynes can add 2 equiv. of R_2AlH with a regioselectivity strongly favoring the geminal dialuminum derivative (equations 8 and 20).[13]

$$n\text{-}C_4H_9C\equiv CH \xrightarrow{2\,Et_2AlH} n\text{-}C_4H_9CH_2CH(AlEt_2)_2 \quad (20)$$

Thirdly, although proton–aluminum exchange may not occur with the starting R_2AlH (equation 19), it may be significant between the hydroalumination adduct and the unconsumed terminal alkyne (equation 21).[38]

$$R_2AlCH{=}CHR' + R'C\equiv CH \longrightarrow R'CH{=}CH_2 + R'C\equiv C{-}AlR_2 \quad (21)$$

Fourthly, the vinylic hydroaluminum adduct (**30**) can either insert a second mole of (**30**) into the vinyl–aluminum bond and thereby dimerize (equation 22),[13] or it can insert a second mole of alkyne (equation 23).[13,39]

$$R''_2Al(R)C{=}CHR' \; (\mathbf{30}) \xrightarrow{R\,=\,H} R''_2AlCH_2CH(AlR''_2)CH(R')CH{=}CHR' \quad (22)$$

$$R''_2Al(R)C{=}CHR' \; (\mathbf{30}) \xrightarrow{R{-}C\equiv C{-}R'} R''_2Al(R)C{=}C(R)C(R'){=}CHR' \quad (23)$$

Finally, in uncatalyzed hydroaluminations with organosubstituted alkynes, the initially formed *syn* adduct can isomerize to the more stable *anti* adduct by way of the dialuminoalkane (equation 24).[40]

$$(Z)\text{-}RCH{=}C(R)AlBu^i_2 \xrightarrow{Bu^i_2AlH} RCH_2C(R)(AlBu^i_2)_2 \xrightarrow{-Bu^i_2AlH} (E)\text{-}RCH{=}C(R)AlBu^i_2 \quad (24)$$

Despite this gamut of competing reactions, reaction conditions can usually be found for preparing either the *syn* or the *anti* adduct of many terminal alkynes with high regio- and stereo-selectivity (Table 2). Failing this the 1-alkynylsilanes can be employed as masked forms of 1-alkynes. In hydroalumination reactions, such silanes have little or no tendency to undergo the side reactions (equations 19–24) to which terminal alkynes are prone. After the silyl-masked alkyne has been hydroaluminated and thereupon derivatized, the silyl group can be removed from the product (Scheme 7).[41]

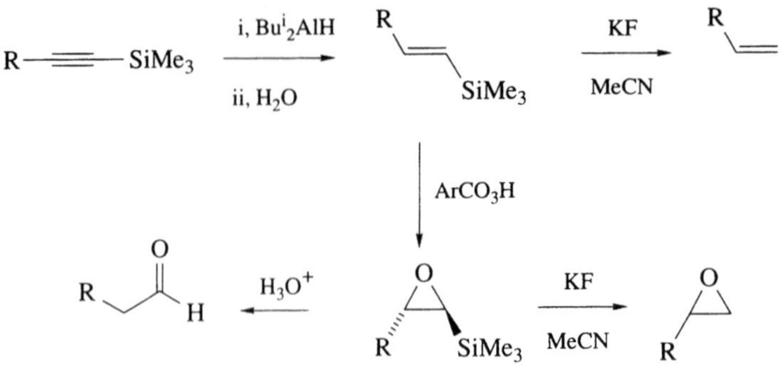

Scheme 7

Although most internal alkynes can be readily hydroaluminated by R_2AlH or LAH (equation 6), the reaction is generally preparatively useful only with symmetrically substituted alkynes. The addition of R_2AlH is stereoselectively *syn* and that of LAH stereoselectively *anti*, but the reaction has a low regioselectivity and thus gives a mixture of regioisomers (equation 11). This problem may be alleviated by enhancing steric factors through the use of a bulky substituent on the C≡C bond (equation 25)[42] but cannot be solved for every internal alkyne.

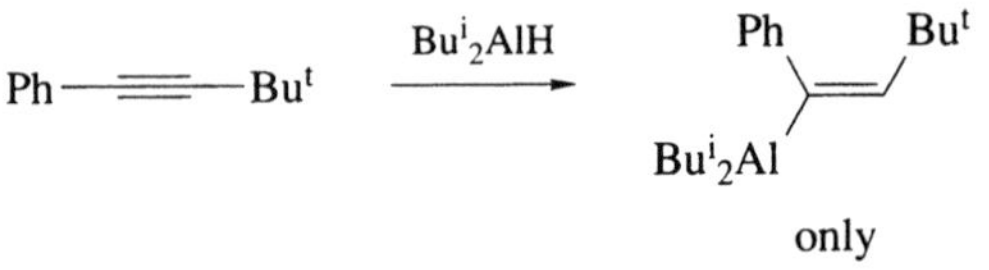

(25)

3.11.2.3 Selective Hydroalumination with Multiunsaturated Substrates

As will become clear in Section 3.11.2.4 many polar and unsaturated functional groups interfere with the hydroalumination of carbon–carbon unsaturation. Hence, the selectivity considered here is the locoselectivity between two different C=C bonds, between a C=C and a C≡C bond or between two C≡C bonds in the same molecule. If such competing groups are not in conjugation, locoselective monohydroalumination of just one such groups can often be easily achieved in the following situations: (i) in dienes, where a terminal C=C bond reacts preferentially;[43] (ii) in alkenynes, where the greater reactivity of the C≡C group leads to selective attack;[6] and (iii) in 1,*n*-alkadiynes, where a terminal C≡C group undergoes addition (Table 3).[44]

With conjugated or cumulated C=C bonds,[45] one bond may respond locoselectively but with the formation of regioisomers as well (**31** and **32**; equation 26). Since one of the regioisomers (**32**) is an allylic aluminum compound, it can generate its allylic isomer (**33**) as well.

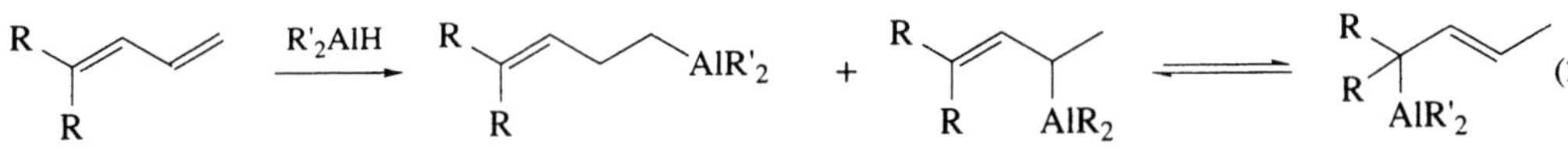

(26)

(**31**) (**32**) (**33**)

3.11.2.4 Interfering Functional Groups

Many functional groups commonly encountered in organic molecules can readily react with the hydroaluminating agent before the carbon–carbon unsaturation does. These groups include: (i) proton sources that cleave the Al—H bond, such as OH, SH, certain NH, CO_2H and even alkynic C—H (*cf.* equation 18); (ii) oxidants that attack the Al—H linkages, such as RO_2R, RS_2R, NO_2, RX, RSO_2X and R_3PO; (iii) electrophiles which add Al—H bonds, such as C=O, C=S, C=N and C≡N derivatives, epoxides, aziridines and episulfides; and (iv) other organometallic compounds, such as RLi[46] or subval-

Table 3 Selective Hydroalumination of Multiunsaturated Substrates with Dialkylaluminum Hydrides

Substrate	*Product(s)*	*Ref.*
$Ph_2C{=}CH_2$	$Ph_2CHCH_2AlR'_2$	45
$Ph_2C{=}C{=}CH_2$	Ph, Ph, R'_2Al + Ph, Ph, AlR'_2	45
$Ph_2C{=}CHCH{=}CH_2$	Ph, Ph, R'_2Al + Ph, Ph, AlR'_2	45
$RCH{=}CH(CH_2)_nCH{=}CH_2$	$RCH{=}CH(CH_2)_nCH_2CH_2AlR'_2$	166
$MeCH{=}CHC{\equiv}CEt$	Et, R'_2Al	163
$RC{\equiv}CCH_2C{\equiv}CSiMe_3$	R, $SiMe_3$, AlR'_2, Et_2O	164
$RC{\equiv}CC{\equiv}CSiMe_3$	R, $\bar{A}lR_3$ [a] Li^+, $SiMe_3$	164

[a] Hydroaluminating agent : $LiAlBu^nBu^i_2H$

ent transition metal complexes (alkene–nickel(0) reagents),[47] which can alter (equation 9) or even destroy (equation 27) the hydroaluminating agent.

$$LiAlH_4 + bipyNiCOD \xrightarrow{THF} LiAlH_2{\bullet}Ni{\bullet}bipy + H_2 \qquad (27)$$

In certain cases, by working with special solvents at low temperatures, one may be able to hydroaluminate a C≡C rather than a competing C≡N group (equation 28)[48] or a C≡C group in addition to a reactive OH group (equation 29).[49]

$$Et{-}C{\equiv}C{-}CN \xrightarrow[-60\ ^\circ C]{LiAlH_4,\ Et_2O} (Et, CN)_2AlH_2Li \qquad (28)$$

$$R{-}C{\equiv}C{-}CH_2OH \xrightarrow[-H_2]{LiAlH_4} R,\ H{-}\bar{A}l(H){-}O \qquad (29)$$

3.11.2.5 Side Reactions

The extent and types of interfering reactions during hydroalumination of alkynes were examined in Section 3.11.2.2 (equations 18–24). To a lesser degree two of these reactions are also encountered in Al—H additions to alkenes and cycloalkenes. Analogous to the reductive dimerization of alkynes (equation 23) is the dimerization of α-alkenes by Bu^i_2AlH, which can be conducted catalytically in hydride to give high yields of dimer (equation 30).[50]

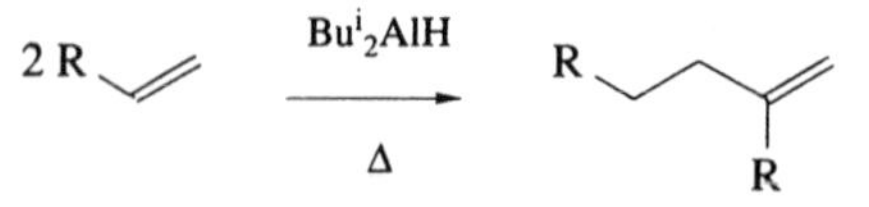

$$2\,R\diagup\!\!\diagdown\!\!= \xrightarrow[\Delta]{Bu^i_2AlH} R\diagdown\!\diagup\!\diagdown\!\!\overset{}{C}(R)\!=\quad (30)$$

The second side reaction with alkenes is the configurational change at the sp^3-hybridized carbon–aluminum bond in the hydroalumination adduct (equation 31).[51]

$$R'R\,C\text{—}AlR''_2 \rightleftharpoons R''_2Al\text{—}C\,R'R \quad (31)$$

Thus, 1,1-dimethylindene (**34**) was subjected to hydroalumination with Bu^i_2AlD in diethyl ether solution. The initially formed adduct was shown to have the *syn* configuration (**35**) by cleaving it with D_2O and demonstrating that only *cis*-dimethylindan-2,3-d_2 (**36**) was formed (Scheme 8). In the absence of donor solvents during the hydroalumination, a 1:1 mixture of (**35**) and its epimer were formed, and deuteriolytic work-up gave both the *cis*- and *trans*-1,1-dimethylindanes-2,3-d_2. This study demonstrates that the hydroalumination of C═C bonds occurs in a kinetically controlled *syn* fashion and sp^3-hybridized carbon centers bonded to tricoordinate aluminum undergo rapid inversion of configuration.[52] If one hopes to attain stereoselective cleavages of such C—Al bonds with retention of configuration, the aluminum center must be kept tetracoordinate. However, even this precaution does not always assure a stereoselective reaction (*cf.* Section 3.11.4).

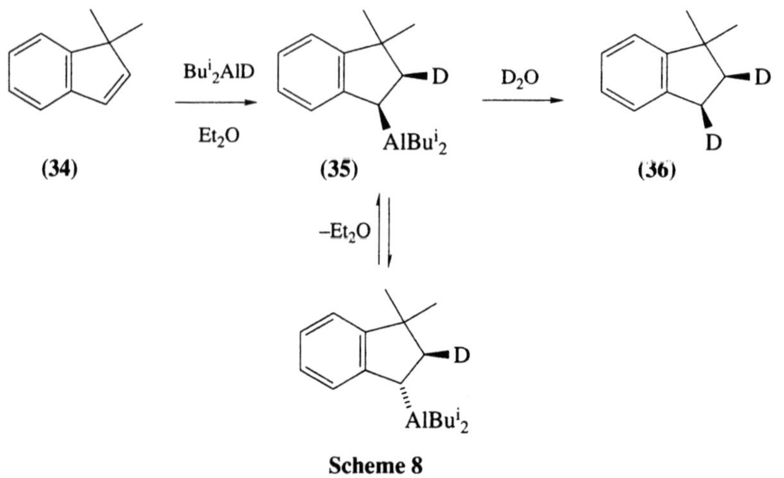

(**34**) $\xrightarrow[Et_2O]{Bu^i_2AlD}$ (**35**) [D, $AlBu^i_2$] $\xrightarrow{D_2O}$ (**36**) [D, D]

(**35**) $\rightleftharpoons$ ($-Et_2O$) epimer [D, $AlBu^i_2$]

Scheme 8

3.11.2.6 Selectivity

3.11.2.6.1 Locoselectivity

The main locoselective features of hydroalumination, which are discussed in some detail in Section 3.11.2.3 are: (i) C≡C bonds are attacked in preference to C═C bonds; (ii) terminal HC≡C and H_2C═CH groups are hydroaluminated much more rapidly than internal C≡C and CH═CH linkages; and (iii) electron-releasing substituents on the carbon–carbon unsaturated group increase the rate of hydroalumination (Table 4).[53] Illustrative are the relative initial rates for PhC≡CE: with E = Ph, Me, H,

SEt, Bu^t, TMS and NMe_2, the rates increase in the same order: 1.0 < 1.1 < 12 < 25 < 28 < 185 < 431 < 19000. By contrast, steric factors appear to have little effect on rate; di-*t*-butylacetylene reacts with Bu^i_2AlH some 20 times faster at 30 °C than does di-*n*-butylacetylene.

Table 4 Relative Reactivity of Alkynes towards Hydroalumination with Diisobutylaluminum Hydride

Alkyne	*Relative reactivity*[a]	*Reactivity ratio*[b]
PhC≡CH	11.8	10.2
PhC≡CMe	1.16	
$n\text{-}C_6H_{13}C{\equiv}CH$	115	18.3
$Pr^nC{\equiv}CPr^n$	6.26	
$n\text{-}C_8H_{17}C{\equiv}CH$	117	16.8
$Bu^nC{\equiv}CBu^n$	6.92	
$Bu^n{\equiv}CBu^n$	6.92	1.11
$Pr^nC{\equiv}CPr^n$	6.26	
$n\text{-}C_8H_{17}C{\equiv}CH$	117	1.01
$n\text{-}C_6H_{13}C{\equiv}CH$	115	
$Bu^tC{\equiv}CBu^t$	151	21.8
$Bu^nC{\equiv}CBu^n$	6.92	
$PhC{\equiv}CBu^t$	27.8	24.0
PhC≡CMe	1.16	
CyC≡CMe	8.22	7.10
PhC≡CMe	1.16	
$p\text{-}MeC_6H_4C{\equiv}CC_6H_4Me\text{-}p$	9.84	1.52
PhC≡CPh	6.47	
PhC≡CPh	1.00 (k_0)	1.00
PhC≡CMe	1.16	1.16
PhC≡CH	11.8	11.8
PhC≡CSEt	24.6	24.6
$PhC{\equiv}CBu^t$	27.8	27.8
$PhC{\equiv}CAlPh_2$	185	185
PhC≡CTMS	431	431
$PhC{\equiv}CNMe_2$	19 000	19 000

[a]Relative reactivities were determined by the method of initial rates and normalized to 35 °C. [b]For pairs of compounds as indicated; for the last eight entries, relative to PhC≡CPh = 1.

3.11.2.6.2 Regioselectivity

The highest regioselectivity in hydroaluminating unsymmetrically substituted C=C and C≡C bonds with R_2AlH occurred in the following structures: (i) with terminal carbon–carbon unsaturation, generally >95% of the adduct has the R_2Al group attached to the terminal carbon (Scheme 3); (ii) with alumino-, sulfonyl-, silyl- or germyl-masked 1-alkynes, R-C≡C-E, where E = Ph_2Al, TMS, Me_3Ge and $PhSO_2$, >98% of the adduct has the R_2Al group attached to the carbon bearing Me_3M (equation 32);[4] (iii) with C≡C bonds bearing amino (R'_2N) or alkoxyl (R′O) groups, the R_2Al group of the R_2AlH becomes attached exclusively to the carbon β to the R'_2N or R′O substituent (equation 33).

$$R{-}{\equiv}{-}E \xrightarrow[E\,=\,Me_3Si\text{ or }Me_3Ge]{R''_2AlH} (R)(H)C{=}C(E)(AlR''_2) \qquad (32)$$

$$R{-}{\equiv}{-}E \xrightarrow[E\,=\,R'_2N\text{ or }R'O]{R''_2AlH} (R)(R''_2Al)C{=}C(H)(E) \qquad (33)$$

Other steric or polar influences on the regiochemistry tend to be less selective:[4,54] (i) a *t*-butyl group on the C≡C bond tends to place the R_2Al group on the distal carbon (equations 25 and 34; 85–100%); and (ii) a R′S or R'_2P group on the C≡C bond tends to place the R_2Al group on the proximal carbon (equation 35; 83–85%).

$$\mathrm{R{-}{\equiv}{-}E} \xrightarrow[\mathrm{E\,=\,Bu^t}]{\mathrm{R''_2AlH}} \mathrm{(R)(R''_2Al)C{=}CH(E)} \quad (34)$$

$$\mathrm{R{-}{\equiv}{-}E} \xrightarrow[\mathrm{E\,=\,R'S\ or\ R'_2P}]{\mathrm{R''_2AlH}} \mathrm{(R)HC{=}C(E)(AlR''_2)} \quad (35)$$

3.11.2.6.3 Stereoselectivity

From all available evidence it is reasonable to conclude that the hydroalumination of both alkenes and alkynes by R_2AlH occurs in a kinetically controlled *syn* manner and that any *anti* adduct formed results from a subsequent isomerization to the more stable configuration. In many situations the *syn* adduct can be isolated or detected prior to isomerization and its conversion to the *anti* adduct monitored.[4,52–54] In Table 5 are summarized the stereoselectivities observed in the hydroaluminations of a spectrum of substituted alkynes under various conditions. It is noteworthy that the thermal isomerization of the *syn* adduct can proceed with remarkable facility (at 0 °C) when a TMS or Me_3Ge group is attached to the C≡C linkage.

Table 5 Stereoselectivities Observed in the Hydroalumination of Heterosubstituted Alkynes by Diisobutylaluminum Hydride

Substrate	*Product(s)*	*Ratio*	*Ref.*
$PhC{\equiv}CBu^t$	*syn*	∞	42
PhC≡CTMS	*syn* > *anti* (R_3N)	96/4	4
	anti > *syn*	96/4	4
$PhC{\equiv}CGeMe_3$	*syn* > *anti* (R_3N)	98/2	4
	anti > *syn*	94/6	4
$PhC{\equiv}CPMe_2$	*syn*	∞	4
$PhC{\equiv}CNMe_2$	*anti*	∞	54
PhC≡CSEt	*syn*	∞	54
$Bu^nC{\equiv}COEt$	*syn* > *anti* (R_3N)	97/3	54
$PhC{\equiv}CSO_2Ph$	*anti*	∞	62
PhC≡CM (M = Li, AlR_2)	*syn*	a	54

[a]The stereochemistry of addition could not be ascertained in these cases, but by analogy to phenylethynylmetallics, PhC≡CE, where E = Si or Ge, such additions most likely are *syn*.

Distinctly less is known about the *anti* hydroalumination adducts obtained from alkynes with $MAlH_4$ or $MAlR_3H$ reagents (equation 9). These adducts could be the direct result of a *trans* addition but, pending further information, they might also arise from initial *syn* addition and subsequent isomerization.

Regardless of mechanistic conclusions, much useful experimental detail has been reported on how various stereochemical outcomes in hydroaluminations can be achieved and maintained.[17,52–54]

3.11.2.7 Modifications in the Hydroalumination Process

3.11.2.7.1 Nature of the aluminum hydride

In general, tricoordinate hydrides such as AlH_3, R_2AlH or Cl_2AlH (even though they may contain some coordinated ether or amine) react toward unsaturated hydrocarbons as electrophilic agents, while alkali aluminum hydrides of the types, $MAlH_4$, $MAlR_2H_2$ or $MAlR_3H$, behave as nucleophilic sources of hydrides. It follows that the electrophilic character of a hydroaluminating agent should be enhanced by adding $AlCl_3$ to R_2AlH or by utilizing Cl_2AlH directly. Such increased electrophilic character may not always be advantageous; sensitive, oxygen-containing masking groups such as acetals and ketals may thereby be more readily attacked (equation 36).[55]

$$\text{(2,2-dialkyl-1,3-dioxolane)} \xrightarrow{R''_2AlH} R_2CH{-}O{-}CH_2CH_2{-}OAlR''_2 \qquad (36)$$

Augmenting the nucleophilic nature of $MAlH_4$ increases the hydroaluminating reactivity towards alkynes and this can be achieved experimentally by replacing the lithium salt (M) with the sodium or potassium salt.[14] However, this nucleophilic activation has its drawbacks, for the AlH_4 thereby becomes a strong Brønsted base and can cause competitive metallations of terminal alkynes (equation 18) and other acidic C—H groups ($CH_2C{\equiv}C$, $ArCH_2$, and $C{=}CCH_2C{=}C$).

3.11.2.7.2 *Solvent effects*

The presence of donor solvents (ethers or amines) or of Lewis bases (R_3N, R_3P, R_2O or R_2S) markedly retards the rate of hydroalumination with neutral R_2AlH reagents because the tricoordinate monomer R_2AlH must function as an electrophile toward the C—C π-bond in the rate-determining step.[3] Despite this retardation donors are often useful adjuvants in hydroaluminations with R_2AlH, because they stabilize the initially formed *syn* configuration (equation 2; Scheme 8).

Donor solvents appear to promote hydroaluminations with $MAlR_3H$ reagents, both through their solubilizing action on such salts and probably by solvating the metal cation and freeing thereby the aluminate anion for attack. Crown ethers would seem to be ideal for this role.

3.11.2.7.3 *Transition metals*

In Section 3.11.1.4 it was pointed out that salts of certain transition metals, lanthanides and actinides promote the hydroalumination reaction. Since such metal salts are introduced into the reaction in their high oxidation states it can be assumed that the metal ions are rapidly reduced to a lower oxidation state and that this state is the active catalyst. For nickel(II) salts, Wilke has shown conclusively that the active agent is a nickel(0)–alkene complex.[56] Analogously, for titanium(IV) salts, such as $TiCl_4$, $Ti(OR)_4$ and Cp_2TiCl_2, it is most likely that a titanium(III) state is involved.[57] The possible role of such metal centers in accelerating hydroalumination will be considered in the next section.

3.11.3 MECHANISMS OF HYDROALUMINATION

3.11.3.1 Reaction Rates and Kinetic Rate Expressions

Careful and extensive kinetic studies have been carried out only for hydroaluminations with dialkylaluminum hydrides. Adequate kinetic information is still lacking on transition metal catalyzed hydroalumination and on the hydroaluminating action of complex metal hydrides, $MAlR_nH_{4-n}$. Preliminary studies on the nickel-catalyzed process have revealed an unstable rate behavior brought about by the deactivation of the catalyst with time.[58]

With $MAlR_nH_{4-n}$ reagents, knowing the kinetic order of the reaction would prove most helpful in discerning whether the Al—H addition occurs initially in an *anti* manner, or whether a *syn* addition is followed by an isomerization to the *anti* adduct. A direct *anti* hydroalumination would likely require the synergistic action of 2 mol of $MAlR_3H$ in the rate-determining step and thus would be second-order in complex hydride (equation 37).

$$R'{-}C{\equiv}C{-}R' \xrightarrow{2\,M^+\,AlR_3H^-} [R'{-}C{\equiv}C{-}R' \cdots H{\cdots}\bar{A}lR_3\ M^+ \ /\ M^+AlR_3H^-] \xrightarrow{-MAlR_3H} (R')(R_3\bar{A}l)C{=}CH(R')\ M^+ \qquad (37)$$

However, there are published reports that can be interpreted as favoring the pathway of *syn* addition and subsequent isomerization.[14] As already noted in equation (18), $NaAlH_4$ in bis(2-methoxyethyl) ether at 120–140 °C converts phenylacetylene into sodium tetrakis(phenylethynyl)aluminate. But if the reaction is conducted in the presence of 5% of Bu^i_2AlH, 95% of the phenylacetylene undergoes hydroalumination instead. This dramatic change in reaction course can be ascribed[14] to a rapid addition of Bu^i_2AlH to phenylacetylene and the aluminum exchange of the vinylaluminum adduct (**37**) with $NaAlH_4$ to regenerate Bu^i_2AlH and continue the catalytic hydroalumination cycle (Scheme 9).

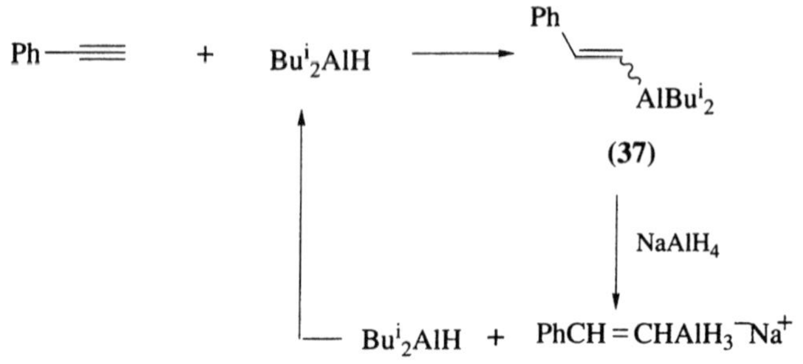

Scheme 9

This study opens up the possibility that hydroaluminations by LAH or $LiAlBu^i_2MeH$ actually occur by the dissociation of such complex hydrides into AlH_3 and R_2AlH at higher temperatures. Steps of conventional *syn* hydroalumination, aluminum exchange and vinylic lithium isomerization could then lead to the *anti* adduct (Scheme 10).

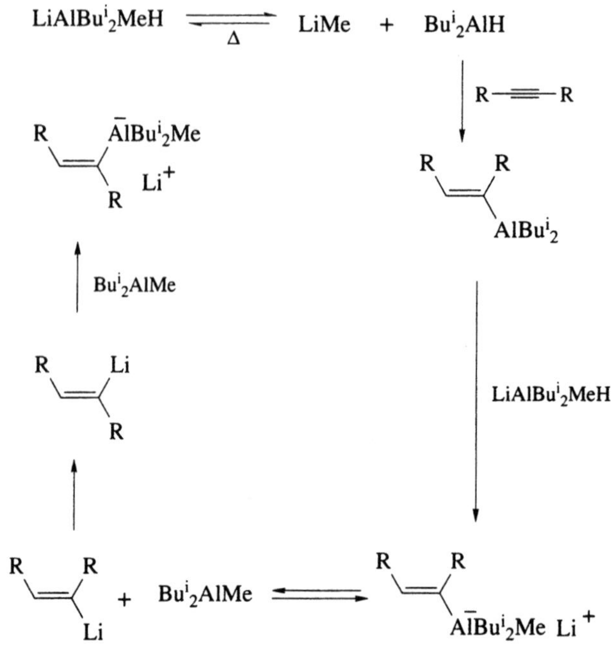

Scheme 10

Kinetic information on the addition of Bu^i_2AlH itself to alkynes has permitted the formulation of a very clear reaction pathway. For additions to masked terminal alkynes, such as the ethynylsilane (**38**)[59] and internal alkynes like 4-octyne (**39**),[60] the reaction is first order in the alkyne and essentially one third order in the hydride (0.37 ± 0.01). The respective Arrhenius relationships for (**38**) and (**39**) were found to be: $k = (1.1 \pm 0.4) \times 10^{10}\, e^{-17\,330 \pm 220/RT}\ l^{1/3}\ mol^{-1/3}\ s^{-1}$; and $k = (4.5 \pm 0.6) \times 10^{10}\, e^{-20\,840 - 50/RT}\ l^{1/3}\ mol^{-1/3}\ s^{-1}$; with activation entropies of −13.4 ± 0.1 eu at −5.2 °C for (**38**) and −11.0 ± 0.3 eu at 30 °C for (**39**). Both reactions exhibit rate retardation as a function of conversion, but such retardation was much less with the alkynylsilane. In addition, the alkynes both underwent *syn* hydroalumination but the alkynylsilane adduct isomerized to the *anti* adduct during the reaction. In fact, equilibration between the *syn* and *anti* adducts of (**38**) proved to be so facile that an equilibrium constant of 120 at –5.2 °C, favoring the *anti* adduct, could be conveniently determined.

Ph—≡—$SiMe_3$ (**38**) Pr^n—≡—Pr^n (**39**)

These kinetic data support a reaction mechanism in which the normally trimeric hydride $(R_2AlH)_3$, must dissociate into its monomer (**40**), which attacks the alkyne in a *syn* fashion in the rate-determining step (Scheme 11). The importance of tricoordinate (**40**) in initiating the electrophilic attack on the alkyne is seen in the effect of donor solvents; the rate of reaction in hexane is 10^3 times faster than that in THF. The latter solvent is known to form a strong 1:1 complex with (**40**).

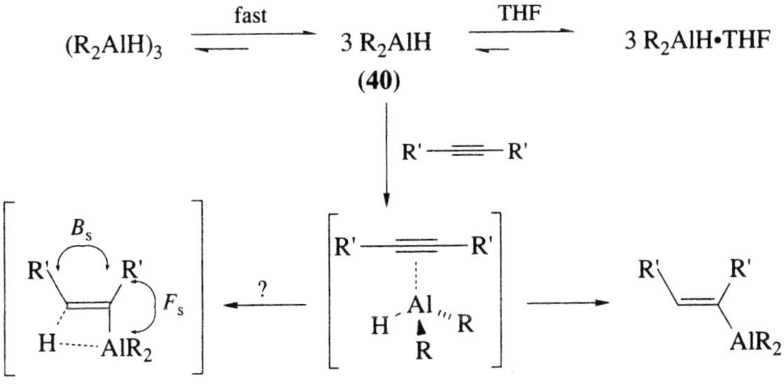

$(R_2AlH)_3$ ⇌ (fast) 3 R_2AlH (**40**) ⇌ (THF) 3 R_2AlH•THF

R'—≡—R'

(**41b**) (**41a**) (**42**)

Scheme 11

The rate retardation observed in hydroaluminating 4-octyne (**39**) was traced back to the formation of a stable 1:1 complex (**44**) between the *syn* adduct (**43**) and the monomeric hydride (**40**) (equation 38). In effect, adduct (**43**) behaves like a donor molecule, such as THF or an amine.

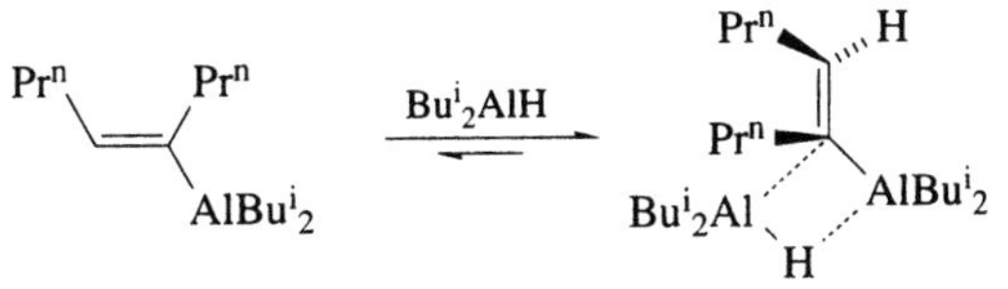

(**43**) ⇌ (Bu^i_2AlH) (**44**) (38)

A decision can also be made on whether the transition state for *syn* addition of the Al—H bond resembles a π-complex (**41a**; Scheme 11), where little geometrical change in the starting alkyne has occurred, or whether a change in hybridization at carbon and some Al—H bond stretching are significant (**41b**). The observed kinetic deuterium isotope effect for Bu^i_2AlH *versus* Bu^i_2AlD is about 1.7. Such a value cannot represent a primary isotope effect and thus would not be consistent with the considerable stretching of the Al—H bond implicit in transition state (**41b**). This isotope effect is more likely connected with the preequilibrium between monomer (**40**) and its trimer.

A further kinetic datum arguing against (**41b**) and for (**41a**) is the effect of bulky substituents. If the transition state resembled (**41b**) large substituents on the alkyne should encounter both back-strain (B_s) and front-strain (F_s) in (**41b**). But, in fact, di-*t*-butylacetylene is some 20 times more reactive than di-*n*-butylacetylene toward Bu^i_2AlH. By passing through an earlier π-complex-like transition state (**41a**), such steric strains could be minimized, and yet the greater electron release of the *t*-butyl group ($\sigma^* = -0.300$) over that of the *n*-butyl groups ($\sigma^* = -0.130$) could still serve to enhance the observed reactivity.[61]

3.11.3.2 Substituent Effects on Relative Rates of Alkynes

The most extensive compilation of initial relative rates for the hydroalumination of alkynes by Bu^i_2AlH is presented in Table 4.[53] Normalized for 35 °C, such initial rates avoid complications setting in at extensive conversions, such as autoretardation through complexation (equation 38), isomerization (equation 24) and bishydroalumination (equation 20). As previously noted, steric properties of the substituents appear not to exert any retarding effect; *t*-butyl-substituted alkynes react faster than their methyl- or *n*-alkyl-substituted counterparts (~20×); and trimethylsilylalkynes are again more reactive than *t*-butylalkynes (~15×).

The principal activating influence seems to be polar, and to be related to the electron-donating character of the substituent as assessed by σ^* values (*cf.* the last eight entries in Table 4).[53] σ^* values that are not directly measurable can generally be calculated from the relation, $\sigma_x^* \approx \sigma^*CH_2X$, for cases where σ^*CH_2X can be observed. The calculated value of σ^* for TMS can thus be estimated as –0.73 and it is thereby evident why this group should activate an alkyne more than the Me_3C group ($\sigma^* = -0.30$). Appropriate σ^* values for SEt, $AlPh_2$ and NMe_2 are neither available nor derivable, but from its very negative Hammett σ_p value (–0.83) the NMe_2 is known to be strongly electron donating. The Allred–Rochow electronegativities of sulfur (2.44) and especially aluminum (1.87) are distinctly less than carbon (2.50), so that both SEt and $AlPh_2$ should release electron density to an adjacent *sp*-hybridized carbon.

Since in every instance the reactivity of an alkyne, RC≡CR, toward Bu^i_2AlH can be related to electron release by R and R′, substituent effects are perfectly consistent with the mechanism depicted in Scheme 11, where R_2AlH attacks the π-cloud of the alkyne electrophilically in the slow step.

3.11.3.3 Substituent Control of Regiochemistry with Alkynes

The variegated regiochemistry with which alkynes, RC≡CR′, respond to Bu^i_2AlH (equations 32–35) can be understood by considering the π-complex-like transition state (**41a**) as unsymmetrical in the sense (**45**) or (**46**), where one carbon is electron deficient.

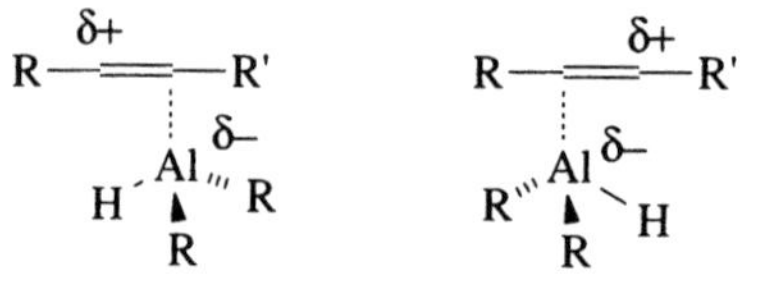

(**45**) (**46**)

The regioisomer derived from (**45**) will be favored when the R-group can sustain partial positive character better than the R′-group, as when R is a π-donor such as R_2N, RO, Me_2P and EtS, since the unshared electrons on the heteroatom can stabilize the positive charge. If R = $PhSO_2$, structure (**45**) would be destabilized, because positive charge would be borne by the carbon attached to a strong electron-withdrawing group, and the regiosisomer derived from (**46**) will be preferred.[62] When R = TMS or Me_3Ge, the alternative transition state (**46**) is favored, even though these groups are σ-electron releasing. The electronic effect operative here would appear to be σ-bond hyperconjugation,[63] wherein such groups stabilize an electron deficiency β to their M—C bond.

3.11.3.4 Stereochemical Effects of Lewis Acids and Bases

Since hydroalumination by neutral aluminum hydrides is an electrophilic attack on a C=C or C≡C linkage, the reaction can be accelerated by Lewis acids such as aluminum halides, and be retarded by Lewis bases like R_3N, R_2O or even unsaturated R_3Al (*cf.* equation 38). Such reagents also exert an effect on the *syn* or *anti* character of the Al—H adduct. Evidence suggests that Lewis acids or bases principally affect the rate of isomerization of the initial *syn* adduct into the generally more stable *anti* adduct; Lewis bases retard such isomerizations, while Lewis acids promote them. The presence of ethers or tertiary amines stabilize the *syn* adducts of alkynyl-silanes and -germanes (**47**) and permit such adducts to be formed in >95% geometrical purity (Scheme 12).[4]

If the donor-free syn adduct (**48**) is generated by adding a Lewis acid, then it can rapidly isomerize to the *anti* adduct (**50**) at 25 °C. Available evidence indicates that the rotational barrier about the C=C bond in (**48**) and (**50**) is very small. A possible explanation is that p_π–p_π bonding with the 3*p*-orbital on aluminum lowers the C=C double bond character. Furthermore, σ-bond hyperconjugation in the transition state for rotation (**49**) reduces its energy and hence the barrier to rotation. That the facile isomeriza-

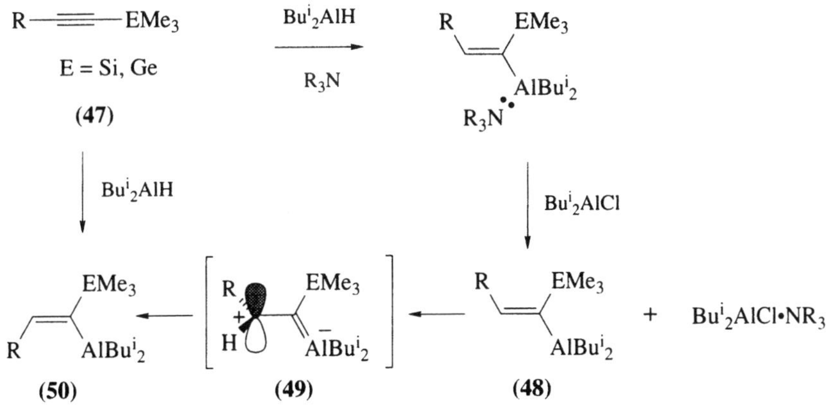

Scheme 12

tion of *syn*-(**48**) into *anti*-(**50**) requires the presence of tricoordinate aluminum speaks for the involvement of the 3*p*-orbital on aluminum in the isomerization process.

An analogous situation arises in the hydroalumination of prochiral alkenes, as already explained in Section 3.11.2.5 and exemplified in the behavior of 1,1-dimethylindene (Scheme 8). The inversion of configuration at the sp^3-hybridized C—Al bond (**51** → **52**) is so facile that the addition of Bu^i_2AlH must be conducted in the presence of a donor (R_2O) to preserve the syn configuration of the initial adduct (**53**; Scheme 13). Again, addition of a stronger Lewis acid liberates (**51**), which, having a tricoordinate Al center, rapidly equilibrates with (**52**). This observation again shows that an available 3*p*-orbital on Al is necessary for configurational inversion.[52]

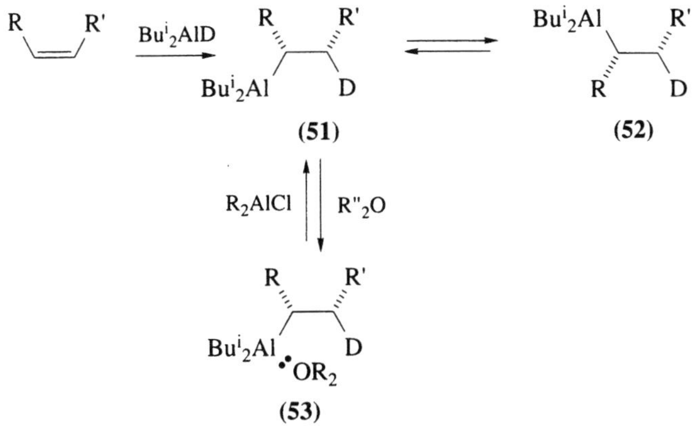

Scheme 13

Syn adducts of purely diorganosubstituted alkynes, such as MeC≡CMe or PhC≡CPh, do not isomerize readily, and the pathway for the formation of *anti* adducts from them involves the addition and elimination of a second hydride (equation 20).

3.11.3.5 Kinetic Effects of Transition Metals

Of the various metal promoters of hydroalumination, only for nickel and possibly for titanium is the mechanistic basis for their catalysis at all clear. In the case of titanocene dichloride, Cp_2TiCl_2, not only does this reagent catalyze hydroalumination but also hydromagnesiation of C=C and C≡C bonds by Grignard reagents.[64] In both instances it is known that R_2AlH or RMgX leads to a reduction and alkylation of the titanium to Cp_2TiR.[65] Such titanium(III) alkyls readily eliminate alkene and form hydrides of the type, Cp_2TiH or $Cp_2TiH{\cdot}HAlR_2$.[57] These hydrides appear to hydrotitanate the unsaturated hydrocarbon and the adduct (**54**) undergoes metal–titanium exchange with R_2AlH or RMgX to regenerate the catalytic agent, Cp_2TiR or Cp_2TiH (Scheme 14).

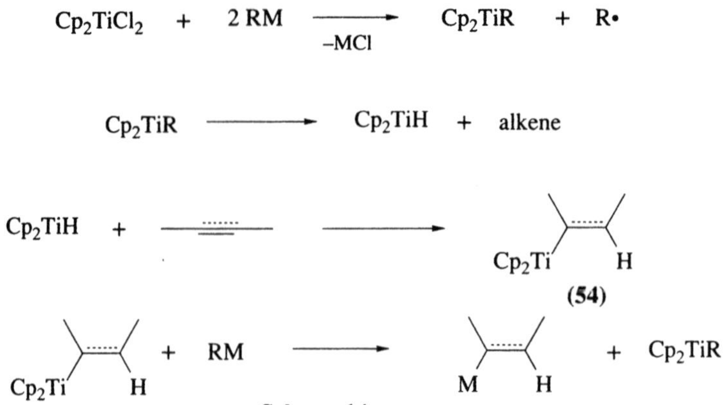

$$Cp_2TiCl_2 + 2\,RM \xrightarrow[-MCl]{} Cp_2TiR + R\bullet$$

$$Cp_2TiR \longrightarrow Cp_2TiH + \text{alkene}$$

$$Cp_2TiH + \text{alkyne} \longrightarrow Cp_2Ti\text{–alkenyl–H}\ \textbf{(54)}$$

$$Cp_2Ti\text{–alkenyl–H} + RM \longrightarrow M\text{–alkenyl–H} + Cp_2TiR$$

Scheme 14

For nickel catalysis by nickel(II) salts, it is now well established that R_3Al or R_2AlH causes reduction to nickel(0), which is held in hydrocarbon solution as a nickel(0)–alkene complex.[56] Wilke's proposal for the nickel-catalyzed, aluminum hydride transfer from R_3Al to another alkene invokes a complex of the R_3Al and the alkene substrate with the nickel(0) center, and the Al—H is transferred in a pericyclic reorganization. Recent evidence in this laboratory, however, assigns a more involved role to nickel(0) in this catalysis.[58] The essence of our proposal is that the actual catalytic carrier is $R_2Al–Ni—H$ (**55**), formed from either R_3Al or R_2AlH, and the catalytic cycle is that depicted in Scheme 15.

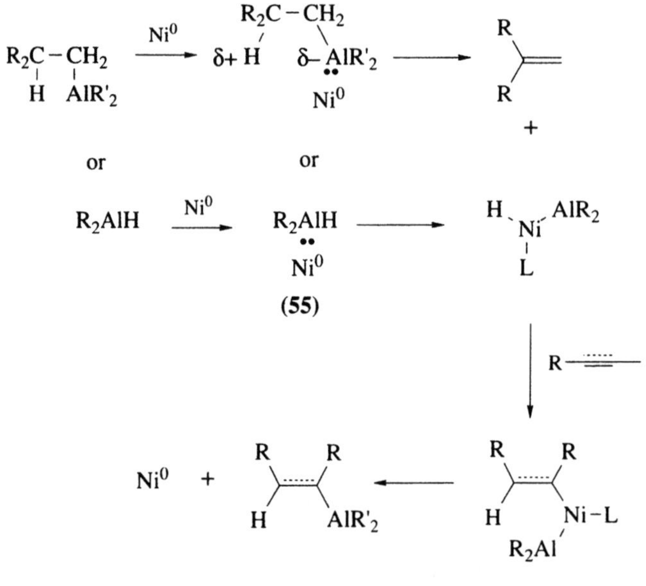

$R_2C–CH_2$ (H, AlR'$_2$) $\xrightarrow{Ni^0}$ $R_2C–CH_2$ (δ+ H, δ– AlR'$_2$ ·· Ni^0) $\longrightarrow$ $R_2C{=}CH_2$ +

or

$R_2AlH \xrightarrow{Ni^0}$ R_2AlH ·· Ni^0 **(55)** $\longrightarrow$ H–Ni(L)–AlR$_2$

$\downarrow$ R–C≡C–H

Ni^0 + (R)(H)C=C(R)(AlR'$_2$) $\longleftarrow$ (R)(H)C=C(R)(Ni–L)(R$_2$Al)

Scheme 15

The experimental observations favoring this proposal are the following: (i) the regioselectivities for nickel-catalyzed hydroaluminations differ significantly from those obtained from the uncatalyzed process (*e.g.* equation 11); (ii) certain hindered alkenes do not undergo uncatalyzed hydroalumination under conditions where the nickel-catalyzed reaction is essentially complete (*e.g.* equation 39); (iii) nickel(0) complexes have been shown to insert into Al—H bonds to yield $R_2Al–Ni—H$ intermediates (equation 40);[66] and (iv) such Al–Ni—H intermediates react with Al—H bonds, with the rate depending upon sub-

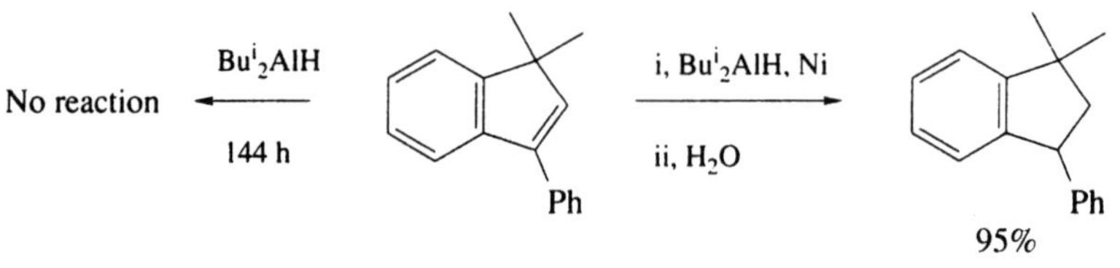

No reaction $\xleftarrow[144\ h]{Bu^i_2AlH}$ 1,1-dimethyl-3-phenylindene $\xrightarrow[ii,\ H_2O]{i,\ Bu^i_2AlH,\ Ni}$ 1,1-dimethyl-3-phenylindane (Ph) 95% (39)

stituents, to evolve dihydrogen and yield aluminum nickelides, causing deactivation of the catalyst (*cf.* equation 27).

$$(Et_3P)_4Ni \xrightarrow[\Delta]{Bu^i_2AlH} (Et_3P)_nNi(H)(AlBu^i_2) \quad (40)$$

3.11.4 CHEMICAL DERIVATIVES OF HYDROALUMINATION ADDUCTS

3.11.4.1 Experimental Conditions: Protolysis and Oxidation

Because organoaluminum compounds react readily with all types of Brønsted acids and with atmospheric oxygen or peroxides, hydroalumination reactions must be conducted anhydrous in an atmosphere of pure nitrogen, and in deoxygenated solvents. Detailed experimental directions for the manipulation and transfer of aluminum alkyls and hydrides are available.[2,3,17,20]

Once the hydroalumination adducts of C═C and C≡C bonds have been cleanly formed, protolysis or oxidation can produce useful derivatives. With H_2O or O_2, generally all available C—Al bonds are destroyed. Protolysis or deuteriolysis of such adducts constitutes an overall reduction or deuteriating reduction of the original C—C unsaturation (equations 13, 39 and 41; Schemes 6–8).[30,52,54,55] Any stereoselectivity or regioselectivity of the original hydroalumination is preserved.

When the adducts are allylic, propargylic or benzylic, varying amounts of rearranged protolytic products can result, depending upon the kind of proton source (equation 42).[67]

$$Ph—≡—Li \xrightarrow[ii,\ D_2O]{i,\ 2\ Bu^i_2AlH} PhCH_2CD_3 \quad (41)$$

$$\text{acenaphthylene} \xrightarrow[ii,\ H\text{-}B]{i,\ Bu^i_2AlH} \text{acenaphthene} + \text{dihydroacenaphthylene} \quad (42)$$

H_2O•Py:	95%	5%
Bu^tOH:	60%	40%

Oxidation of R_3Al compounds with dioxygen proceeds most efficiently with trialkylaluminum compounds (Scheme 6) and, in cases where the resulting isobutyl alcohol can be tolerated, is an efficient route to primary alcohols (equation 43).[31]

$$\xrightarrow[ii,\ O_2;\ iii,\ H_2O]{i,\ R_2AlH} \quad (43)$$

The substitution of aluminum by oxygen can be effected by various peroxide derivatives, such as di-*t*-butyl peroxide,[68] benzoyl peroxide[69,70] and *t*-butyl perbenzoate.[71] The main reaction is accompanied by telltale free-radical side reactions, such as the formation of RR from R_3Al, which become major pathways with aryl and vinyl compounds.[71–74] Oxidation of alkenyl derivative (**56**) with *t*-butyl perbenzoate yields 45% of a mixture of *cis* and *trans* ethers (**57**; equation 44).[71] As of yet there is no generally applicable, highly efficient method for oxidizing vinylaluminum compounds.

$$PhCH{=}C(AlBu^i_2)SiMe_3\ (\mathbf{56}) \xrightarrow{Bu^tOOC(O)Ph} PhCH{=}C(OBu^t)SiMe_3\ (\mathbf{57}) \quad (44)$$

3.11.4.2 Nonmetallodealumination

The cleavage of C—Al bonds can be achieved by nonmetal sources with varying degrees of ease and efficiency. For such reagents not containing carbon (*cf.* Section 3.11.4.4), the following have found some utility: (i) halodealumination by Br_2, BrCN or I_2 with retention of vinylic configuration (equation 45);[22,75–79] (ii) sulfur or selenium insertion to form thiols or selenols in good yields with the consumption of one C—Al bond (equation 46);[80,81] (iii) SO_2 and SO_3 can be inserted into two or even three of the C—Al bonds in R_3Al, and good yields of the sulfinic and sulfonic acids obtained upon hydrolysis (equation 47);[29,82–84] (iv) both thionyl chloride and sulfuryl chloride can react with all three C—Al bonds in R_3Al to give high yields of sulfoxides (equation 48)[85] and sulfonyl chlorides (equation 49)[86] respectively; (v) halides of the elements of Group V (P, As, Sb and Bi) react with aluminum alkyls to yield fully alkylated derivatives, R_3E or partially alkylated halides, R_nEX_{3-n}, depending upon the ratio of reactants and reaction conditions (equation 50).[87–95]

$$\text{RCH=C(R)AlR}_2 \xrightarrow[-R_2AlX]{X_2} \text{RCH=C(R)X} \quad (45)$$

$$R_3Al + S \longrightarrow R_2AlSR \xrightarrow{H_2O} RSH \quad (46)$$

$$R_3Al + SO_n \longrightarrow R_2Al(SO_nR)_2 \xrightarrow{H_2O} RSO_nH \quad (47)$$

$$2\,R_3Al + 3\,SOCl_2 \longrightarrow 3\,R_2SO + 2\,AlCl_3 \quad (48)$$

$$R_3Al + 3\,SO_2Cl_2 \longrightarrow 3\,RSO_2Cl + AlCl_3 \quad (49)$$

$$R_3Al + EX_3 \longrightarrow R_3E + R_{3-n}AlX_n \quad (50)$$

3.11.4.3 Metallodealumination

The conversion of C—Al bonds into carbon–metal bonds by the interaction of R_3Al with the appropriate metal salt is widely applicable to most metals and metalloids (*cf.* Section 3.11.4.2) except those of Groups I and II (equation 51). The utility of this reaction depends upon several factors: (i) the tendency of R_3Al to undergo exchange of one, two or three R-groups with MZ_n (*i.e.* the successive reactivity of R_2AlZ and $RAlZ_2$ toward R_mMZ_{n-m}; (ii) the kinetic stability of the metal alkyl produced (especially of concern with labile transition metal alkyls); and (iii) the ease of separating the desired metal alkyl from the Lewis acidic R_nAl_{3-n} by-products.

$$n\,R_3Al + MX_n \longrightarrow R_nM + n\,R_2AlX \quad (51)$$

The driving force for this aluminum–metal exchange is chiefly the high strength of the resulting Al—Z bond; the same factor underlies the failure of aluminum alkyls to exchange with the very stable salts of the alkali and alkaline earth metals. Instead, these salts form stable complexes (equation 52).[96–100]

$$R_3Al + MZ \longrightarrow M[AlR_3Z] \quad (52)$$

M = L, Na, K, Rb, Cs; Z = X, H, R', OR', NR'_2

Beryllium is an understandable exception, but then its salts are more covalent.[101] However, by the use of less stable salts such as alkoxides, R_3Al can produce aluminates with these metals (equation 53).[102]

$$2\,R_3Al + KOR' \longrightarrow K[AlR_4] + R_2AlOR' \quad (53)$$

The method has been successful for producing the alkyls of beryllium,[103] zinc,[104] cadmium,[105] mercury,[106] boron,[107] gallium,[108] indium,[109,110] germanium,[111] tin,[112] lead[113] and many transition metals.[114]

With transition metals the lability of their carbon–metal bonds often does not permit the isolation of the metal alkyl. Feasible preparation then requires the presence of stabilized ligands on the metal, such as phosphines, 2,2-bipyridyl and cyclopentadienyl. The preparation of dimethyltitanocene from titanocene dichloride and Me_3Al is illustrative.[115] A compilation of typical preparations is provided in Table 6.

Table 6 Preparation of Metal Alkyls by the Aluminum–Metal Exchange Between Aluminum Alkyls and Metal Salts

Metal salt	*Aluminum reagent*	*Product*	*Yield (%)*	*Ref.*
$SnCl_4$	Et_3Al (+KCl)	Et_4Sn	80	112
Et_3SnCl	Et_2AlH	Et_3SnH	89	112
$GeCl_4$	Me_3Al (+NaCl)	Me_4Ge	73	111
$PbCl_2$	$Et_3Al{\bullet}OEt_2$	Et_4Pb	70	113
$ZnCl_2$	Bu^i_3Al	Bu^i_2Zn	95	104
$GaCl_3$	Et_3Al (+KCl)	Et_3Ga	84	108,109
$HgCl_2$	$NaAlEtCl_3$	EtHgCl	91	106

3.11.4.4 Carbodealumination

The most important substitution occurring at the C—Al bond is an aluminum–carbon exchange (equation 54).

$$R_2AlR' + \text{(chiral C)}{-}E \longrightarrow R'{-}\text{(chiral C)} + R_2AlE \qquad (54)$$

Often, only one C—Al bond can be made to react and even then the C—Al bond might require promotion by transition metals like Ni,[116] Pd[117] or Cu[118] or might benefit by activation through aluminate salt formation ($LiAlR_3R'$).[28,119,120] Fortunately, in the vinylic aluminum compounds resulting from the hydroalumination of alkynes (equations 2, 3 and 14) or in the vinylic aluminate complexes arising from Al—H additions to alkynes (equations 6 and 9; Scheme 4), the vinylic carbon–aluminum bond is generally much more reactive than the alkyl C—Al linkages and is thus cleaved preferentially.

Suitable cleaving agents include organic halides such as: primary alkyl bromides, iodides or sulfonates;[28,119,120] allylic,[118,121,122] allenic[123] and propargylic bromides;[120] vinylic or aryl halides with nickel or palladium promoters;[124–127] alkyl chloroformates;[128,129] epoxides;[130,131] and even cyanogen[132] or cyanogen bromide.[133,134] Typical examples of such couplings are depicted in equations (55)–(57).

$$R{-}C{\equiv}CH \xrightarrow[\text{ii, ClCH}_2\text{OEt}]{\text{i, Bu}^i_2\text{AlH}} RCH{=}CHCH_2OEt \qquad (55)$$

$$R{-}C{\equiv}C{-}R \xrightarrow[\text{ii, MeLi; iii, (CN)}_2]{\text{i, Bu}^i_2\text{AlH}} RCH{=}C(R)CN \qquad (56)$$

$$R{-}C{\equiv}CH \xrightarrow[\text{ii, ArI, Ni}]{\text{i, Bu}^i_2\text{AlH}} RCH{=}CHAr \qquad (57)$$

The reactions often proceed in high yield and predominantly with retention of configuration at the vinyl—aluminum bond. Many experimental variations of this substitution reaction remain to be studied systematically and the method appears to have great promise for the synthesis of unsaturated carbon chains. Because only one C—Al bond in R_2AlR' responds to cleavage the method has little appeal for application to the R_2AlH adducts of alkenes (Scheme 3). Little preference for cleavage of the three C—Al bonds would be expected, so conversion to the R′E would be low.

3.11.4.5 Carbalumination

The addition of organoaluminum reagents to unsaturated carbon-containing substrates is probably the most versatile and widely applicable method for forming carbon–carbon bonds from R_3Al compounds. Besides the stepwise insertion of ethylene into the C—Al bonds of triethylaluminum, the so-called Aufbau-Reaktion originally discovered by Ziegler (*cf.* Section 3.11.1.2), most primary alkenes and alkynes are able to undergo a single or several insertions into a carbon–aluminum bond (*cf.* equations 23 and 30). Such reactions can form the basis for the catalytic oligomerizations of alkenes (equation 30)[50] and alkynes (equation 58).[13,39]

$$3\ R{-}C{\equiv}C{-}R \xrightarrow[\text{cat.}]{R'_2AlH} C_6R_6 \qquad (58)$$

But a wide spectrum of other unsaturated carbon centers can insert cleanly in the C—Al bond as well. Possibly the most interesting and most closely akin to Ziegler's ethylene insertion is the preferential methylene insertion brought about by diazomethane into vinylic aluminum bonds (equation 59).[135–139]

$$RCH{=}C(R)AlR'_2 \xrightarrow[-N_2]{CH_2N_2} RCH{=}C(R)CH_2AlR'_2 \qquad (59)$$

For reasons not yet understood, the Simmons–Smith methylenating agent [CH_2Br_2, Zn(Cu)][140] brings about cyclopropanation, instead, to provide cyclopropylaluminum derivatives in good yield (equation 60).[141]

$$RCH{=}C(R)AlR'_2 \xrightarrow[\text{Zn (Cu)}]{CH_2Br_2} \text{cyclopropyl}(R)(R)AlR'_2 \qquad (60)$$

Other addends that generally cleave only one C—Al bond, and that usually being the more reactive vinylic aluminum bond, include: (i) CO_2,[142–144] COS[145] and CS_2;[146] (ii) aldehydes, such as H_2CO[147,148] and MeCHO,[149] and ketones;[150] (iii) esters,[151] lactones,[152] anhydrides[153] and isocyanates;[154] (iv) substrates having C=N,[155] C≡N[156] and C=S[157] linkages; and (v) α,β-unsaturated carbonyl compounds, where 1,4-addition is often attainable.[158,159] Illustrative reactions are given in equations (14), (61) and (62).

$$RCH{=}C(R)AlR_2 \xrightarrow[\text{ii, } H_2O]{\text{i, cyclohexanone}} RCH{=}C(R){-}C_6H_{10}OH \qquad (61)$$

$$RCH{=}C(R)\overset{-}{Al}MeBu^i_2\ Li^+ \xrightarrow[\text{ii, } H_3O^+]{\text{i, } CO_2} RCH{=}C(R)CO_2H \qquad (62)$$

Several influences on the course of inserting C=E units (E = O, N, S) into C—Al bonds merit brief mention: (i) although dialkyl(vinylic)aluminums are often sufficiently reactive to undergo insertion of C=E with retention of configuration, in some cases it may be advantageous to preform the tetraorganoaluminate for better reaction (equation 62);[160] (ii) although saturated R_2AlR' will generally not insert C=E into a given R—Al or R'—Al bond preferentially, such preferential reaction will occur if R' is benzylic or allylic, but then with loss of configuration (equation 63);[161] (iii) for slow insertions of C=E into diisobutyl(vinylic)aluminum adducts the isobutyl groups may cause competitive reduction of C=E linkages by Al—H transfer (Scheme 3).[162] Such Al—H reductions will be especially serious at higher temperatures or in the presence of transition metals.

$$\xrightarrow[\text{ii, } H_3O^+]{\text{i, } R_2CO} \qquad (63)$$

3.11.5 APPLICATIONS IN ORGANIC SYNTHESIS

The stereocontrolled synthesis of alkenes and the use of vinyl- and alkyl-aluminum and -aluminate reagents as nucleophiles has been described above. In addition, the aluminum reagents are also useful in diene synthesis, in reactions with enones and epoxides, and in cyclizations.

3.11.5.1 Dienes, Polyenes and Enynes

A variety of conjugated dienes can be synthesized stereospecifically *via* the hydroalumination of alkynes. For example, the first-formed adduct from an alkyne can insert a second mole of alkyne and upon hydrolysis yield the (*E*,*E*)-1,3-alkadiene (equation 23).[13,39] Alternatively, the first-formed adduct can directly oxidatively dimerize to the (*E*,*E*)-1,3-diene (equation 64).[124]

$$2\ RCH{=}CHAlR'_2 \xrightarrow{\text{CuCl}} RCH{=}CH{-}CH{=}CHR \qquad (64)$$

Besides such homocoupling of alkynes, adduct (**58**) may be heterocoupled, under palladium catalysis, with a vinylic bromide that itself may have been formed by the hydroalumination of a different alkyne (equations 65 and 66).[125]

$$\underset{\textbf{(58)}}{RCH{=}CHAlR'_2} \xrightarrow[\text{catalysis}]{\text{Pd}} RCH{=}CH{-}CH{=}CHR' \qquad (65)$$

+

$$R'CH{=}CHBr \xleftarrow[\text{ii, } Br_2]{\text{i, } R'_2AlH} R'{-}C{\equiv}CH \qquad (66)$$

R = alkyl; R' = H

Conjugated enynes and diynes can be hydroaluminated to produce (*E*,*Z*)- and (*Z*,*Z*)-1,3-alkadienes, respectively (equations 67 and 68).[6,79,163,164]

$$RCH{=}CH{-}C{\equiv}CR' \xrightarrow[\text{ii, } H_2O]{\text{i, } R''_2AlH} RCH{=}CH{-}CH{=}CHR' \qquad (67)$$

$$\text{R—≡—≡—R'} \xrightarrow[\text{ii, } H_2O]{\text{i, } R''_2AlH} \text{R–CH=CH–CH=CH–R'} \qquad (68)$$

Finally, unconjugated dienes and even polyenes can be produced (R = alkyl, aryl; R′ = H) by the transition metal promoted coupling (Cu, Ni, Pd) with an allylic[121,122] or other unsaturated halide (equations 65 and 66; Scheme 4).[124–127]

The preparation of enynes by the locoselective monohydroalumination of diynes can be exemplified by equation (69).[164] In addition, a heterocoupling of adduct (**58**) with 1-halo-1-alkynes can produce conjugated enynes in high yield.[117]

$$\text{(69)} \quad \xrightarrow[\text{ii, } H_2O]{\text{i, } 2\,R''_2AlH}$$

3.11.5.2 Additions to Epoxides and Conjugated Enones

Other promising methods for the stereoselective introduction of the vinyl moiety into functionalized organic substrates are the reactions of such adducts with epoxides to produce homoallylic alcohols (equation 70)[130,131] and with conjugated enones to yield ketones (equation 71).[158,159] Both reactions have major limitations, and their application in a special instance may require a study of the effect of Lewis acids or transition metals as promoters or selectivity agents.

$$\text{RCH=C(R)AlR'}_2 \xrightarrow[\text{ii, } H_3O^+]{\text{i, ethylene oxide}} \text{RCH=C(R)CH}_2\text{CH}_2\text{OH} \qquad (70)$$

$$\text{RCH=C(R)AlR}_2 \xrightarrow[\text{ii, } H_2O]{\text{i, PhCH=CHCOMe}} \text{RCH=C(R)CH(Ph)CH}_2\text{COMe} \qquad (71)$$

3.11.5.3 Carbocyclizations

A scattered but tantalizing array of reactions have been observed, in which hydroalumination brings about the closure of open-chain compounds to carbon rings. Perhaps the most typical of such reactions is the monohydroalumination of 1,5-hexadiene at –40 °C, which is followed by an intramolecular carbalumination (equation 72).[165,166]

$$\text{CH}_2\text{=CH(CH}_2)_2\text{CH=CH}_2 \xrightarrow{R_2AlH} \underset{\textbf{(59)}}{R_2Al(CH_2)_4CH=CH_2} \longrightarrow \text{cyclopentyl–CH}_2AlR_2 \qquad (72)$$

In the presence of diethyl ether the reaction does not proceed beyond the formation of (**59**). Similar treatment of 1,6-heptadiene with Et_2AlH gives the monoadduct, which requires prolonged heating at 100 °C to cyclize to the cyclohexylmethylaluminum derivatives.[167] In contrast, 1,4-pentadiene,[168] 1,7-octadiene, 1,8-nonadiene or 1,9-decadiene, show little tendency to cyclize.[169,170]

Similar cyclizations to substituted cyclopenta rings have also been reported (equations 73 and 74).[43,171] The method holds considerable promise for the synthesis of *cis*-fused C_5 and C_6 ring systems.[172]

$$\xrightarrow[\text{ii, } H_2O]{\text{i, } R_2AlH} \qquad (73)$$

$$\text{(cyclohexenyl)ethyl-}AlR_2 \xrightarrow{180\ °C} \text{bicyclic-}AlR_2 \xrightarrow{-R_2AlH} \text{bicycloalkene} \quad (74)$$

3.11.6 REFERENCES

1. Part 45 of the series, 'Organometallic Compounds of Group III'. Part 44: J. J. Eisch, S. R. Sexsmith and K. C. Fichter, *J. Organomet. Chem.*, 1990, **382**, 273.
2. H. Lehmkuhl, K. Ziegler and H. G. Gellert, *Methoden Org. Chem. (Houben-Weyl)*, ed. E. Müller, 1970, XIII/4, 23.
3. J. J. Eisch, in 'Comprehensive Organometallic Chemistry', ed. G. Wilkinson, Pergamon Press, Oxford, vol. 1, 1982, chap. 6.
4. J. J. Eisch and M. W. Foxton, *J. Org. Chem.*, 1971, **36**, 3520.
5. This term for the site (L., *locus*) selected by a given reagent, out of several intramolecular possibilities, seems to be a superior designation for this course of reaction; J. J. Eisch and J. E. Galle, *J. Org. Chem.*, 1979, **44**, 3279.
6. K. Utimoto, K. Uchida, M. Yamaya and H. Nozaki, *Tetrahedron Lett.*, 1977, 3641.
7. K. Ziegler, *Brennst. Chem.*, 1952, **33**, 193.
8. K. Ziegler and H. G. Gellert, *Justus Liebigs Ann. Chem.*, 1950, **567**, 195.
9. A. E. Finholt, A. C. Bond, Jr. and H. I. Schlesinger, *J. Am. Chem. Soc.*, 1947, **69**, 1199.
10. H. Lehmkuhl, K. Ziegler and H. G. Gellert, *Methoden Org. Chem. (Houben-Weyl)*, 1970, XIII/4, 39.
11. K. Ziegler, H. Martin and F. Krupp, *Justus Liebigs Ann. Chem.*, 1960, **629**, 14.
12. G. Wilke and H. Müller, *Chem. Ber.*, 1956, **89**, 444.
13. G. Wilke and H. Müller, *Justus Liebigs Ann. Chem.*, 1960, **629**, 222.
14. L. L. Ivanov, V. V. Gavrilenko and L. I. Zakharkin, *Izv. Akad. Nauk SSSR, Ser. Khim.*, 1964, 1989 (*Chem. Abstr.*, 1965, **62**, 7660f).
15. L. H. Slaugh, *Tetrahedron*, 1966, **22**, 1741.
16. E. F. Magoon and L. H. Slaugh, *Tetrahedron*, 1967, **23**, 4509.
17. G. Zweifel and J. A. Miller, in 'Organic Reactions,' ed. W. G. Dauben, Wiley, New York, 1984, vol. 32, chap. 2, p. 377.
18. US suppliers include Aldrich, Ethyl, Fluka and Texas Alkyls.
19. K. Ziegler, H. G. Gellert, H. Martin, K. Nagel and J. Schneider, *Justus Liebigs Ann. Chem.*, 1954, **589**, 91.
20. J. J. Eisch, in 'Organometallic Syntheses', ed. J. J. Eisch and R. B. King, Academic Press, New York, 1981, vol. 2, p. 136.
21. A. M. Piotrowski, D. B. Malpass, M. P. Boleslawski and J. J. Eisch, *J. Org. Chem.*, 1988, **53**, 2829.
22. G. Zweifel and R. B. Steele, *J. Am. Chem. Soc.*, 1967, **89**, 5085.
23. W. L. Walsh (Goodrich-Gulf), *Br. Pat.* 878 746 (1961) (*Chem. Abstr.*, 1963, **59**, 6439).
24. J. J. Eisch and M. W. Foxton, *J. Organomet. Chem.*, 1968, **12**, 33.
25. F. Asinger, B. Fell and R. Janssen, *Chem. Ber.*, 1964, **97**, 2515.
26. J. F. Le Marechal, M. Ephritikhine and G. Folcher, *J. Organomet. Chem.*, 1986, **309**, C1.
27. J. J. Eisch and M. W. Foxton, *J. Organomet. Chem.*, 1968, **11**, 7.
28. J. J. Eisch and G. A. Damasevitz, *J. Org. Chem.*, 1976, **41**, 2214.
29. K. Ziegler, F. Krupp, K. Weyer and W. Larbig, *Justus Liebigs Ann. Chem.*, 1960, **629**, 251.
30. G. Wilke and H. Müller, *Justus Liebigs Ann. Chem.*, 1958, **618**, 267.
31. K. Ziegler, F. Kropp and K. Zosel, *Justus Liebigs Ann. Chem.*, 1960, **629**, 241.
32. K. Weyer, Dissertation, Technische Hochschule Aachen, 1956 (cited in ref. 3, p. 213).
33. K. Ziegler, W. R. Kroll, W. Larbig and O. W. Steudle, *Justus Liebigs Ann. Chem.*, 1960, **629**, 53.
34. F. Asinger, B. Fell and F. Theissen, *Chem. Ber.*, 1967, **100**, 937.
35. J. J. Eisch, N. E. Burlinson and M. P. Boleslawski, *J. Organomet. Chem.*, 1976, **111**, 137.
36. L. I. Zakharkin and V. V. Gavrilenko, *Izv. Akad. Nauk SSSR, Otd. Khim. Nauk*, 1963, 1146 (*Chem. Abstr.*, 1963, **59**, 8772e).
37. P. Binger, *Angew. Chem.*, 1963, **75**, 918.
38. V. V. Gavrilenko, B. A. Palei and L. I. Zakharkin, *Izv. Akad. Nauk SSSR, Ser. Khim.*, 1968, 910 (*Chem. Abstr.*, 1968, **69**, 77 310).
39. J. J. Eisch and W. C. Kaska, *J. Am. Chem. Soc.*, 1966, **88**, 2213.
40. J. J. Eisch and W. C. Kaska, *J. Am. Chem. Soc.*, 1963, **85**, 2165.
41. J. J. Eisch and J. E. Galle, *J. Organomet. Chem.*, 1988, **341**, 293.
42. J. J. Eisch and R. Amtmann, *J. Org. Chem.*, 1972, **37**, 3410.
43. J. J. Eisch and G. R. Husk, *J. Org. Chem.*, 1966, **31**, 3419.
44. J. A. Miller and G. Zweifel, *J. Am. Chem. Soc.*, 1983, **105**, 1383.
45. J. J. Eisch and G. R. Husk, *J. Organomet. Chem.*, 1974, **64**, 41.
46. L. I. Zakharkin and V. V. Gavrilenko, *Zh. Obshch. Khim.*, 1962, **32**, 689 (*Chem. Abstr.*, 1962, **57**, 16 109a).
47. J. J. Eisch, L. E. Hallenbeck and K. I. Han, *J. Am. Chem. Soc.*, 1986, **108**, 7763.
48. H. Westmijze, H. Kleijn and P. Vermeer, *Synthesis*, 1979, 430.
49. R. Rossi and A. Carpita, *Synthesis*, 1977, 561.
50. K. Ziegler, H. G. Gellert, K. Zosel, E. Holzkamp, J. Schneider, M. Söll and W. R. Kroll, *Justus Liebigs Ann. Chem.*, 1960, **629**, 121.
51. J. J. Eisch and K. C. Fichter, *J. Am. Chem. Soc.*, 1974, **96**, 6815.
52. J. J. Eisch and K. C. Fichter, *J. Organomet. Chem.*, 1983, **250**, 63.

53. J. J. Eisch and S. G. Rhee, *Justus Liebigs Ann. Chem.*, 1975, 565.
54. J. J. Eisch, H. Gopal and S. G. Rhee, *J. Org. Chem.*, 1975, **40**, 2064.
55. L. I. Zakharkin and I. M. Khorlina, *Izv. Akad. Nauk SSSR, Otd. Khim. Nauk*, 1959, 2255 (*Chem. Abstr.*, 1960, **54**, 10 837h).
56. K. Fischer, K. Jonas, P. Misbach, R. Stabba and G. Wilke, *Angew. Chem., Int. Ed. Engl.*, 1973, **12**, 943.
57. J. J. Eisch and J. E. Galle, *J. Organomet. Chem.*, 1978, **160**, C8.
58. J. J. Eisch, S. R. Sexsmith and K. C. Fichter, *J. Organomet. Chem.*, 1990, **382**, 273.
59. J. J. Eisch and S. G. Rhee, *J. Am. Chem. Soc.*, 1975, **97**, 4673.
60. J. J. Eisch and S. G. Rhee, *J. Am. Chem. Soc.*, 1974, **96**, 7276.
61. R. W. Taft, Jr., in 'Steric Effects in Organic Chemistry', ed. M. S. Newman, Wiley, New York, 1956, chap. 13.
62. J. J. Eisch, M. Behrooz and S. Dua, *J. Organomet. Chem.*, 1985, **285**, 121.
63. W. Hanstein, H. J. Berwin and T. G. Traylor, *J. Am. Chem. Soc.*, 1970, **92**, 829.
64. G. D. Cooper and H. L. Finkbeiner, *J. Org. Chem.*, 1962, **27**, 1493.
65. H. L. Finkbeiner and G. D. Cooper, *J. Org. Chem.*, 1962, **27**, 3395.
66. J. J. Eisch and M. Singh, unpublished studies, 1989.
67. J. J. Eisch and G. R. Husk, unpublished studies, 1963.
68. G. A. Razuvaev, L. P. Stepovik, V. A. Dodonov and G. A. Nesterov, *J. Gen. Chem. USSR (Engl. Transl.)*, 1961, **31**, 2180.
69. G. A. Razuvaev and E. V. Mitrofanova, *J. Gen. Chem. USSR (Engl. Transl.)*, 1968, **38**, 251.
70. G. A. Razuvaev and E. V. Mitrofanova and G. G. Petukhov, *J. Gen. Chem. USSR (Engl. Transl.)*, 1961, **31**, 2180.
71. J. J. Eisch and D. A. Komar, unpublished studies, 1976.
72. A. G. Davies and B. P. Roberts, *J. Chem. Soc. (B)*, 1968, 1074.
73. S. Czernecki, C. Georgulis and E. Michel, *J. Organomet. Chem.*, 1977, **140**, 127.
74. G. A. Razuvaev, E. V. Mitrofanova and G. G. Petukhov, *J. Gen. Chem. USSR (Engl. Transl.)*, 1960, **30**, 1976.
75. E. J. Corey, J. A. Katzenellenbogen and G. H. Posner, *J. Am. Chem. Soc.*, 1967, **89**, 4245.
76. E. J. Corey, H. A. Kirst and J. A. Katzenellenbogen, *J. Am. Chem. Soc.*, 1970, **92**, 6314.
77. G. Zweifel and C. C. Whitney, *J. Am. Chem. Soc.*, 1967, **89**, 2753.
78. B. A. Palei, V. V. Gavrilenko and L. I. Zakharkin, *Izv. Akad. Nauk SSSR, Ser. Khim.*, 1969, 2760 (*Chem. Abstr.*, 1970, **72**, 79 143s).
79. G. Zweifel and W. Lewis, *J. Org. Chem.*, 1978, **43**, 2739.
80. L. I. Zakharkin and V. V. Gavrilenko, *Izv. Akad. Nauk SSSR, Otd. Khim. Nauk*, 1960, 1391 (*Chem. Abstr.*, 1961, **55**, 361d).
81. R. E. Leech and J. E. Knap (Union Carbide), *US Pat.* 2 998 455 (1960) (*Chem. Abstr.*, 1962, **56**, 2 333c).
82. E. B. Baker and H. H. Sisler, *J. Am. Chem. Soc.*, 1953, **75**, 5193.
83. G. C. Bassler and A. F. Stang (Armour), *Ger. Pat.* 1 158 057 (1961) (*Chem. Abstr.*, 1964, **60**, 10 550b).
84. A. J. Rutkowski and A. F. Turbak (Esso), *US Pat.* 3 121 737 (1959) (*Chem. Abstr.*, 1964, **60**, 10 550).
85. P. W. K. Flanagan (Continental Oil), *US Pat.* 2 947 787 (1959) (*Chem. Abstr.*, 1961, **55**, 380b).
86. P. W. K. Flanagan (Continental Oil), *Ger. Pat.* 1 124 033 (1962) (*Chem. Abstr.*, 1962, **57**, 11 023i).
87. O. Y. Okhlobystin and L. I. Zakharkin, *Izv. Akad. Nauk SSSR, Otd. Khim. Nauk*, 1958, 1006 (*Chem. Abstr.*, 1959, **53**, 1 122g).
88. S. B. Becker (Standard Oil), *US Pat.* 3 036 132 (1954) (*Chem. Abstr.*, 1962, **57**, 11 237e).
89. L. Maier (Monsanto), *Ger. Pat.* 1 240 080 (1964).
90. L. Maier, *Helv. Chim. Acta*, 1964, **47**, 2129.
91. C. W. Huffman and M. Hamer (Internat. Minerals & Chem.), *US Pat.* 3 149 137 (1962) (*Chem. Abstr.*, 1965, **62**, 4 053b).
92. L. Maier, *Helv. Chim. Acta*, 1964, **47**, 27.
93. J. J. Menn and K. Szabo, *J. Econ. Entomol.*, 1965, **58**, 734 (*Chem. Abstr.*, 1965, **63**, 7 600f).
94. Y. Takashi and I. Aishima, *J. Organomet. Chem.*, 1967, **8**, 209.
95. W. Stamm and A. Breindel, *Angew. Chem.*, 1964, **76**, 99.
96. H. Lehmkuhl, *Angew. Chem.*, 1963, **75**, 1090.
97. K. Ziegler, R. Köster, H. Lehmkuhl and K. Reinert, *Justus Liebigs Ann. Chem.*, 1960, **629**, 33.
98. K. Ziegler, H. Lehmkuhl and E. Lindner, *Chem. Ber.*, 1959, **92**, 2320.
99. L. I. Zakharkin and V. V. Gavrilenko, *J. Gen. Chem. USSR (Engl. Transl.)*, 1962, 688.
100. G. J. Sleddon, *Chem. Ind. (London)*, 1961, 1492.
101. N. A. Bell and G. E. Coates, *J. Chem. Soc. (A)*, 1968, 628.
102. K. Ziegler, *Bull. Soc. Chim. Fr.*, 1963, 1456.
103. G. Moretti and A. Turchi (Montecatini), *Br. Pat.* 928 716 (1963) (*Chem. Abstr.*, 1963, **59**, 14 022f).
104. E. Hüther (K. Ziegler), *US Pat.* 3 124 604 (1956) (*Chem. Abstr.*, 1964, **60**, 15 909a).
105. H. Jenkner, *Z. Naturforsch., Teil B*, 1957, **12**, 809.
106. L. I. Zakharkin and O. Y. Okhlobystin, *Dokl. Akad. Nauk SSSR*, 1957, **116**, 236 (*Chem. Abstr.*, 1958, **52**, 6 167d).
107. R. Köster, *Justus Liebigs Ann. Chem.*, 1958, **618**, 31.
108. R. Dötzer and F. Engelbrecht (Siemens), *Ger. Pat. Discl.* 115 977 (1960) (*Chem. Abstr.*, 1964, **60**, 6 867h).
109. J. J. Eisch. *J. Am. Chem. Soc.*, 1962, **84**, 3605.
110. H. Jenker (Kali-Chemie), *Ger. Pat.* 972 855 (1953) (*Chem. Abstr.*, 1960, **54**, 24 551d)
111. F. Glockling and J. R. C. Light, *J. Chem. Soc. (A)*, 1967, 623.
112. W. P. Neumann, *Justus Liebigs Ann. Chem.*, 1962, **653**, 157.
113. L. I. Zakharkin and O. Y. Okhlobystin, *Izv. Akad. Nauk SSSR, Otd. Khim. Nauk*, 1959, 1942 (*Chem. Abstr.*, 1960, **54**, 9 737i).
114. P. C. Wailes, R. S. P. Coutts and H. Weigold, 'Organometallic Chemistry of Titanium, Zirconium and Hafnium', Academic Press, New York, 1974, chap. 2.

115. H. Bestian and K. Clauss, *Ger. Pat.* 1 037 446 (1958) (*Chem. Abstr.*, 1960, **54**, 18 546f)
116. E. Negishi and S. Baba, *J. Chem. Soc., Chem. Commun.*, 1976, 596.
117. E. Negishi, N. Okukado, A. O. King, D. E. Van Horn and B. I. Spiegel, *J. Am. Chem. Soc.*, 1978, **100**, 2254.
118. R. A. Lynd and G. Zweifel, *Synthesis*, 1974, 658.
119. K. Uchida, K. Utimoto and H. Nozaki, *J. Org. Chem.*, 1976, **41**, 2215.
120. S. Baba, D. E. Van Horn and E. Negishi, *Tetrahedron Lett.*, 1976, 1927.
121. H. Matsushita and E. Negishi, *J. Am. Chem. Soc.*, 1981, **103**, 2882.
122. E. Negishi, S. Chatterjee and H. Matsushita, *Tetrahedron Lett.*, 1981, **22**, 3737.
123. F. Sato, H. Kodama and M. Sato, *Chem. Lett.* 1978, 789.
124. G. Zweifel and R. L. Miller, *J. Am. Chem. Soc.*, 1970, **92**, 6678.
125. S. Baba and E. Negishi, *J. Am. Chem. Soc.*, 1976, **98**, 6729.
126. M. Sato, K. Takai, K. Oshima and H. Nozaki, *Tetrahedron Lett.*, 1981, **22**, 1609.
127. K. Takai, K. Oshima and H. Nozaki, *Tetrahedron Lett.*, 1980, **21**, 2531.
128. N. Okukado and E. Negishi, *Tetrahedron Lett.*, 1978, 2357.
129. G. Zweifel and R. A. Lynd, *Synthesis*, 1976, 625.
130. S. Warwel, G. Schmitt and B. Ahlfaenger, *Synthesis*, 1975, 632.
131. L. E. Overman and K. L. Bell, *J. Am. Chem. Soc.*, 1981, **103**, 1851.
132. G. Zweifel, J. T. Snow and C. C. Whitney, *J. Am. Chem. Soc.*, 1968, **90**, 7139.
133. G. D. Brindell and D. W. Marshall, *US Pat.* 3 304 317 (1967) (*Chem. Abstr.*, 1967, **66**, 115 788t).
134. O. Scherer and K. Uhl (Hoechst), *Ger. Pat.* 1 179 194 (1964) (*Chem. Abstr.*, 1965, **62**, 453f).
135. H. Hoberg, *Justus Liebigs Ann. Chem.*, 1967, **703**, 1.
136. H. Hoberg, *Justus Liebigs Ann. Chem.*, 1962, **656**, 1.
137. H. Hoberg, *Justus Liebigs Ann. Chem.*, 1966, **695**, 1.
138. L. Almaski, I. Felmeri and A. Gants, *Proc. Acad. Sci. USSR (Engl. Transl.)*, 1958, **118**, 175.
139. H. Hoberg, *Angew. Chem.*, 1966, **78**, 492.
140. H. E. Simmons and R. D. Smith, *J. Am. Chem. Soc.*, 1958, **80**, 5323.
141. G. Zweifel, G. M. Clark and C. C. Whitney, *J. Am. Chem. Soc.*, 1971, **93**, 1305.
142. G. Zweifel and R. B. Steele, *J. Am. Chem. Soc.*, 1967, **89**, 2754.
143. V. M. Bulina, L . L. Ivanov an d Y. B. Pyatnova, *Zh. Org. Khim.*, 1973, **9**, 491 (*Chem. Abstr.*, 1973, **78**, 158 856d).
144. R. L. Danheiser and H. Sard, *J. Org. Chem.*, 1980, **45**, 4810.
145. J. Weidlein, *J. Organomet. Chem.*, 1971, **32**, 181.
146. W. Dahlig, S. Pasynkiewicz and T. Wojnarowski, *Rocz. Chem.*, 1960, **34**, 401.
147. E. Negishi, A. O. King, W. L. Klima, W. Patterson and A. Silveira, Jr., *J. Org. Chem.*, 1980, **45**, 2526.
148. E. Negishi, L. F. Valente and M. Kobayashi, *J. Am. Chem. Soc.*, 1980, **102**, 3298.
149. H. Newman, *Tetrahedron Lett.*, 1971, 4571.
150. R. A. Lynd and G. Zweifel, unpublished results cited in ref. 17.
151. Y. Baba, *Bull. Chem. Soc. Jpn.*, 1968, **41**, 1022.
152. H. Reinheckel and R. Gensike, *J. Organomet. Chem.*, 1967, **10**, 28.
153. H. Reinheckel and K. Haage, *Angew. Chem.*, 1966, **78**, 491.
154. J. R. Horder and M. F. Lappert, *J. Chem. Soc. (A)*, 1968, 2004.
155. H. Lehmkuhl and H. D. Kobs, *Justus Liebigs Ann. Chem.*, 1968, **719**, 11.
156. H. Reinheckel and D. Jahnke, *Chem. Ber.*, 1964, **97**, 2661.
157. H. Reinheckel and D. Jahnke (Deutsche Akad. Wissensch.), *Ger. Pat.* 1 238 913 (1965) (*Chem. Abstr.*, 1967, **67**, 63 804u).
158. J. Hooz and R. B. Layton, *Can. J. Chem.*, 1973, **51**, 2098.
159. P. W. Collins, E. Z. Dajani, M. M. S. Bruhn, C. H. Brown, J. R. Palmer and R. Pappo, *Tetrahedron Lett.*, 1975, 4217.
160. E. Negishi, 'Organometallics in Organic Synthesis', Wiley, New York, 1980.
161. J. J. Eisch and K. C. Fichter, *J. Am. Chem. Soc.*, 1975, **97**, 4772.
162. S. Pasynkiewicz and E. Sliwa, *J. Organomet. Chem.*, 1965, **3**, 121.
163. I. Knox, S. C. Chang and A. H. Andrist, *J. Org. Chem.*, 1977, **42**, 3981.
164. J. A. Miller and G. Zweifel, *J. Am. Chem. Soc.*, 1983, **105**, 1383.
165. K. Ziegler, *Angew. Chem.*, 1956, **68**, 721.
166. R. Reinäcker and G. Goethel, *Angew. Chem.*, 1967, **79**, 862.
167. R. Schimpf, Dissertation, University of Bochum, 1969.
168. G. Hata and A. Miyake, *J. Org. Chem.*, 1963, **28**, 3237.
169. R. Polster, *Justus Liebigs Ann. Chem.*, 1962, **654**, 20.
170. G. Goethel, Dissertation, Technische Hochschule Aachen, 1966.
171. K. H. Müller and P. Borner (Schering), *US Pat.* 3 274 271 (1962) (*Chem. Abstr.*, 1964, **60**, 13 164f)
172. The author acknowledges with gratitude the support of the National Science Foundation through Grant No. CHE-8714911 during the preparation of this chapter.

3.12
Hydrosilylation of C=C and C≡C

TAMEJIRO HIYAMA and TETSUO KUSUMOTO
Sagami Chemical Research Center, Kanagawa, Japan

3.12.1 INTRODUCTION

The addition reaction of hydrosilanes to alkenes and alkynes is called hydrosilylation, a reaction which was discovered when a mixture of 1-octene and trichlorosilane was heated in the presence of diacetyl peroxide. This reaction is particularly useful for the synthesis of organosilicon compounds. Since then, a variety of catalysts and conditions have been reported.[1] Characteristic features of the reaction are discussed in this chapter.

3.12.1.1 Catalysts

Various reagents and conditions which generate silyl radicals from hydrosilanes induce hydrosilylation of alkenes and alkynes. Peroxides like diacetyl peroxide, dibenzoyl peroxide, di-*t*-butyl peroxide and AIBN along with its derivatives are reagents of choice. UV irradiation or γ-irradiation are also applicable. Sometimes, simple heating of a mixture of a substrate and a hydrosilane affords a hydrosilylated product. Since the addition of the silyl radical is reversible, the silyl group attaches itself to the terminal carbon of the substrate to give the thermally more favorable, substituted carbon radical, and hydrosilylation of terminal alkynes gives rise to *trans*-vinylsilanes.[2]

Transition metal complexes, particularly those of Group VIII, are popular catalysts. Of these, Pt and Rh are employed most frequently. A Cr catalyst is a unique catalyst for stereoselective hydrosilylation of 1,3-dienes. Catalysts based on Fe, Os and Ir are rarely used.

Hexachloroplatinic(IV) acid (H_2PtCl_6, **1**) is the catalyst of choice for hydrosilylation with a wide variety of silanes and substrates. This catalyst is commercially available as a hexahydrate; dissolved in isopropyl alcohol, the catalyst is activated.[3] With the platinum catalyst, hydrosilylation proceeds under mild conditions with stereospecificity, namely, *syn* addition to C≡C. The modification of the structure into $[A^+]_2[PtCl_6]^{2-}$ (**2**), wherein A = K, Na, R_4N, R_4P and R_4As, has been studied. In comparison with H_2PtCl_6, the complex (**2**; A = K, Na) is less reactive, but the R_4N analog exhibits similar reactivity.[4a] Platinum–carbon, though inactive by itself, is activated when exposed to air to achieve triple hydrosilylation with $RSiH_3$, giving fully alkylated silanes.[4b] Ultrasonic waves also activate the catalyst.[4c] The platinum phosphine complexes $[PtCl_2(PPh_3)_2]$ and $[Pt(PPh_3)_4]$ also catalyze hydrosilylation, but the reactivity decreases in the order $H_2PtCl_6 > [PtCl_2(PPh_3)_2] > [Pt(PPh_3)_4]$. By choosing an appropriate catalyst, selective hydrosilylation is attainable, as shown in equation (1).[5]

$$HC\equiv CH + HSiR_3 \longrightarrow CH_2{=}CH{-}SiR_3 + R_3Si{-}CH_2CH_2{-}SiR_3 \quad (1)$$

	Catalyst	Yield (%)	
$HSiCl_3$	H_2PtCl_6	90	0
	$[PtCl_2(PPh_3)_2]$	54	15
	$[Pt(PPh_3)_4]$	0	0
$HSiCl_2(n\text{-}C_6H_{13})$	H_2PtCl_6	trace	82
	$[PtCl_2(PPh_3)_2]$	5	53
	$[Pt(PPh_3)_4]$	87	3

Iridium complexes other than $[\{IrCl(cyclooctene)_2\}_2]$[6] are much less reactive and scarcely used for synthetic reactions.

Palladium complexes $[Pd(PPh_3)_4]$, $[PdCl_2(PPh_3)_2]$ and $[PdCl_2(PhCN)_2]$ are not as reactive as H_2PtCl_6 toward hydrosilylation but they are more selective: selective 1,4-addition to 1,3-dienes occurs to give allylsilanes, and for asymmetric hydrosilylation with (*R*)-(*S*)-PPFA-Pd see Section 3.12.7.

Rhodium complexes of various types, *e.g.* $[RhCl(PPh_3)_3]$, $[RhCl(CO)(PPh_3)_2]$, $[\{RhCl(CO)_2\}_2]$, $[RhH(CO)(PPh_3)_3]$, $[\{RhCl(C_8H_{14})_2\}_2]$ and $[\{RhCl(C_2H_4)_2\}_2]$, are available for the hydrosilylation of alkenes and alkynes as well as enones and ketones. Under strictly deoxygenated conditions with the pure rhodium complex, the reaction is extremely slow, and a trace amount of oxygen or peroxide is necessary to activate the catalyst.[7]

Ruthenium complexes are not much employed for hydrosilylation, since they are not in general reactive enough. $[RuCl_2(PPh_3)_3]$ and $[RuHCl(PPh_3)_3]$ are examples,[8] and [(1,4-diaza-1,3-diene)RuHCl(COD)] specifically catalyzes the hydrosilylation of isoprene to give an allylsilane.[9]

Typical examples of nickel catalysts for hydrosilylation are nickel(II) complexes like $[NiCl_2(PPh_3)_2]$, $[NiCl_2(PEt_3)_2]$, $[NiCl_2(dppe)]$ and $[NiCl_2(dmpe)]$ and nickel(0) complexes like $[Ni(PPh_3)_2(CH_2{=}CH_2)]$ and $[Ni(diphos)(CH_2{=}CH_2)]$ as well as the Ziegler-type catalyst $[Ni(acac)_2]{-}AlEt_3$. However, unexpected reaction paths are sometimes observed, such as the disproportionation of chlorohydrosilane (equation 2) and disilylation of alkynes (*cf.* Section 3.12.2.3).[10]

$$2HSiMeCl_2 \xrightarrow{[NiCl_2P_2],\ >120\ °C} MeSiH_2Cl + MeSiCl_3 \quad (2)$$

$P_2 = Me_2PCH_2CH_2PMe_2$, 1,1'-bis(dimethylphosphino)ferrocene

Bis(tetracarbonylcobalt) is the most effective among Co catalysts. The reactive species are $[HCo(CO)_4]$ and $[R_3SiCo(CO)_4]$, both of which are produced by the initial reaction between $[Co_2(CO)_8]$

and $HSiR_3$. Quite a few examples of iron catalysts are recorded, the most useful for organic synthesis being $[Fe(CO)_5]$.

Polynuclear metal carbonyls, for example $[Rh_4(CO)_{12}]$, $[Rh_6(CO)_{16}]$, $[Co_4(CO)_{12}]$, $[Co_6(CO)_{16}]$, $[Co(CO)_4]_4Sn$, $[Ru_3(CO)_{12}]$, $Et_4N[HRu_3(CO)_{11}]$, $[Ir_4(CO)_{12}]$, $[Re_2(CO)_{12}]$, $[Mn_2(CO)_{10}]$ and $[Cr(CO)_6]$, are also active for hydrosilylation.[11]

Aluminum chloride facilitates hydrosilyation of the more-substituted alkenes (see Section 3.12.4.3).[12]

The metal catalysts mentioned above may be supported on polymers to facilitate the recovery of the catalyst from reaction mixtures. Rh and Pt can be complexed with a polymer phosphine such as $[\text{—COC}_6\text{H}_4\text{CONH(CH}_2)_m\text{N(PPh}_2)\text{—}]_n$ or $SiO_2(\equiv SiCH_2CH_2PPh_2)$.[13] Cobalt carbonyl adsorbed on silica gel or alumina is also reported.[14]

3.12.1.2 Hydrosilanes Available

The basic industrial process for hydrosilane synthesis is the reaction of metallic silicon with hydrogen chloride, giving trichlorosilane. The Si—X (X = Cl, Br, OR, H) bond is converted into an Si—C bond by the reaction with organo-lithium or -magnesium reagents, and the Si—Cl bond is also readily transformed to Si—H with lithium aluminum hydride as the reagent of choice. Various kinds of hydrosilanes are available commercially, such as the chlorosilanes $HSiCl_3$, $HSiMeCl_2$, $HSiMe_2Cl$, $HSiEtCl_2$, $HSiPh_2Cl$; the alkoxysilanes $HSi(OMe)_3$, $HSiMe(OMe)_2$, $HSi(OEt)_3$; the hydrosilanes containing only organic groups $HSiMe_3$, $HSiEt_3$, $HSiMe_2Ph$, $HSiEt_2Me$, $HSiPh_3$, H_2SiEt_2, H_2SiPh_2, H_3SiPh, $H_2SiMePh$, $HSiMe_2CH_2CH{=}CH_2$, $HSiMePhCH{=}CH_2$; the polysilyl compounds $HSiMe_2\text{—O—}SiMe_2H$, $HSiMe_2\text{—O—}SiMe_3$, $HSiMe_2\text{—}C_6H_4\text{—}SiMe_2H$, $HSiMe_2\text{—O—}SiMe_2\text{—O—}SiMe_2H$, $HSiMe(OSiMe_3)_2$, $HSi(OSiMe_3)_3$, $(SiHMeO)_5$; the polymethylhydrosiloxane $Me_3Si\text{—O—}(SiHMeO)_n\text{—}SiMe_3$ (n = 3–35); and the copolymers which contain the —(SiHMeO)— unit.

3.12.1.3 Solvents

As most hydrosilanes are oils, hydrosilylation is usually carried out without solvent using an equimolar amount or an excess of the hydrosilane. In cases where a substrate is solid or extremely volatile, solvents are employed. Both nonpolar solvents (hexane, benzene, toluene, xylene) and polar solvents (THF, diglyme, nitrobenzene, acetone) are applicable, although some polar solvents may affect the yield and selectivity of hydrosilylation,[15] as in equation (3).

$$\text{PhCH=CH}_2 + \text{HSiR}_3 \ (1:1) \xrightarrow[100\ ^\circ\text{C}]{\text{H}_2\text{PtCl}_6} \text{PhCH}_2\text{CH}_2\text{SiR}_3 + \text{PhCH(Me)SiR}_3 \quad (3)$$

$HSiCl_3$	No solvent	57 : 43
	THF	100 : 0
$HSiMe_2Cl$	No solvent	85 : 15
	THF	86 : 14

3.12.1.4 Mechanism

The mechanism of the hydrosilylation reaction varies depending on the kind of catalyst and other factors discussed above, and is classified roughly into three types: (i) radical-initiated hydrosilylation; (ii) transition metal catalyzed hydrosilylation; and (iii) ionic hydrosilylation.

In the presence of a radical initiator like peroxide or AIBN, or under UV or γ-ray irradiation, the Si—H bond is cleaved homolytically to afford a silyl radical (**3**; Scheme 1), which adds across a C=C or C≡C bond. The regioselectivity of the addition is governed mainly by the stability of the newly generated radical center. The resulting radical (**4**) picks up hydrogen from another molecule of hydrosilane to complete the catalytic cycle and regenerate the active silyl radical (**3**).[16]

Most of the transition metal catalyzed reactions are explained in terms of the hydrometallation mechanism (Harrod–Chalk mechanism) summarized in Scheme 2.[17] The alkene coordinates to the metal (**M**) to give an alkene complex (**5**), which oxidatively adds to the Si—H bond of the hydrosilane to give (**6**).

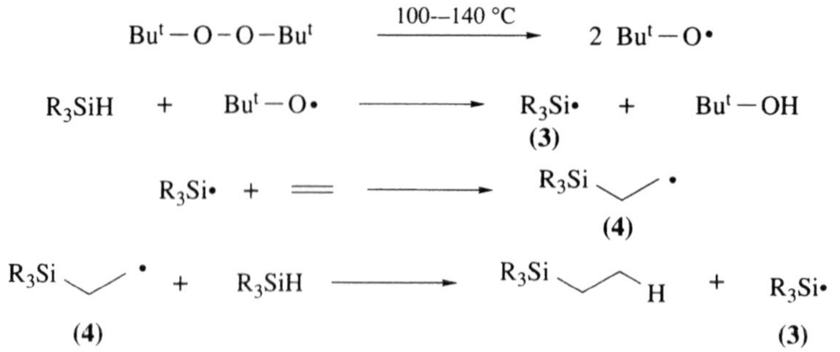

$$\mathrm{Bu^t{-}O{-}O{-}Bu^t \xrightarrow{100\text{--}140\,°C} 2\ Bu^t{-}O\bullet}$$

$$\mathrm{R_3SiH + Bu^t{-}O\bullet \longrightarrow R_3Si\bullet\ (3) + Bu^t{-}OH}$$

$$\mathrm{R_3Si\bullet + {=} \longrightarrow R_3Si\text{CH}_2\text{CH}_2\bullet\ (4)}$$

$$\mathrm{R_3Si\text{CH}_2\text{CH}_2\bullet\ (4) + R_3SiH \longrightarrow R_3Si\text{CH}_2\text{CH}_2\text{H} + R_3Si\bullet\ (3)}$$

Scheme 1

The *cis* ligand alkene then inserts into the M—H bond (hydrometallation) of (**6**) to afford an alkyl metal complex (**7**), which interacts with another molecule of alkene to induce the coupling of the alkyl and silyl ligands and to regenerate the active catalyst (**5**). This last reaction is the rate-determining step of the catalytic cycle.

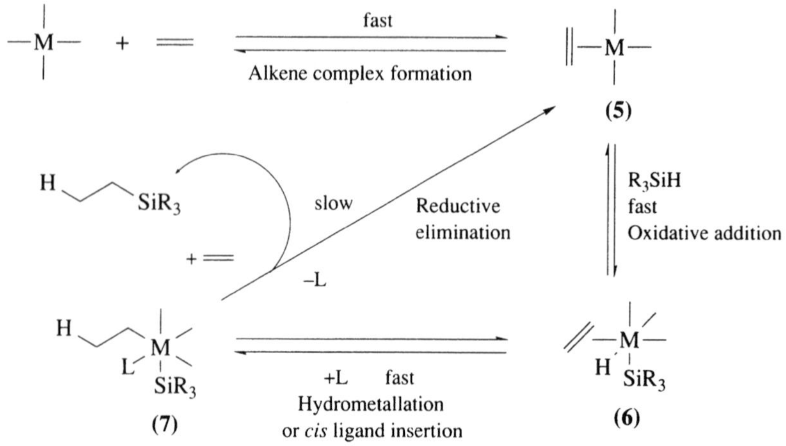

—M— + = ⇌ (fast; Alkene complex formation) (5)

(5) ⇌ (R$_3$SiH, fast, Oxidative addition) (6)

(6) ⇌ (+L, fast, Hydrometallation or *cis* ligand insertion) (7)

(7) + = → (slow, Reductive elimination, –L) H–CH$_2$CH$_2$–SiR$_3$ + (5)

L = neutral ligand or solvent

Scheme 2

As a model of the oxidative addition product, a platinum(0) complex (**8**) has been found to afford an isolable platinum(II) compex (**9**), which has a Si—Pt—H bond.[18]

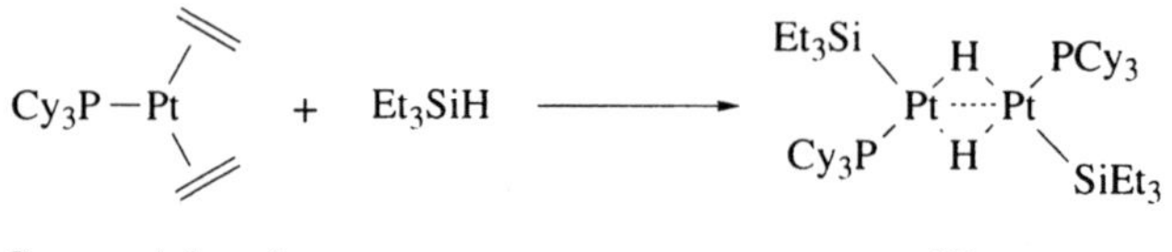

$$\mathrm{Cy_3P{-}Pt(\text{CH}_2{=}\text{CH}_2)_2\ \textbf{(8)} + Et_3SiH \longrightarrow [(Et_3Si)(Cy_3P)Pt(\mu\text{-}H)_2Pt(PCy_3)(SiEt_3)]\ \textbf{(9)}} \qquad (4)$$

(**8**) Cy = cyclohexyl

The stereochemistry of the oxidative addition has been studied using optically active 1-naphthyl(methyl)phenylsilane (**11**) and its deuterium derivative (**10**). The H/D exchange reaction of the hydrosilane proceeds with 100% retention of configuration (equation 5). Addition of the hydrosilane (**11**) to 1-octene gives 1-naphthyl(methyl)octyl(phenyl)silane (**12**) with retention of the configuration of the silicon (equation 6).[19]

Reaction of the Pt complex (**13**) with (**11**) gives a silylplatinum complex (**14**), the silyl group of which has the same configuration as that of (**11**). The oxidative addition is thus assumed to take place from the front side of the Si—H bond (equation 7). The hydrosilane (**11**) reacts with a manganese complex (**15**) to give rise to a complex (**16**; equation 8), which has a structure similar to the intermediate of the oxidative addition step.[20]

Photocatalyzed hydrosilylation of alkenes using [Fe(CO)$_5$], [Ru$_3$(CO)$_{12}$], [(CO)$_4$CoSiR$_3$] and [RhCl(PPh$_3$)$_3$] often produces 1-alkenylsilanes, in addition to normal hydrosilylated products. This is attributed to a different mechanism from that discussed above. Using a low temperature UV irradiation

(5)

(+)-(10) + H_2 —catalyst→ (+)-(11) + HD

(+)-(11)
100% exchange
100% retention

Catalyst

$[IrClCO(PPh_3)_2]$
$[Co_2(CO)_8]$
$[R_3Si^*Co(CO)_4]$
H_2PtCl_6
$[PtCl_2(C_2H_4)_2]$

(6)

(11) + 1-octene —catalyst→ (12)

Catalyst	*Yield (%)*	*Retention (%)*
Pt/C	85	100
H_2PtCl_6	75	100
$PtCl_2(C_2H_4)_2$	56	92

(7)

(11) + Me—Pt(Me)(PMe$_2$Ph)—PMe$_2$Ph (13) —catalyst→ Si(Me)—Pt(H)(PMe$_2$Ph)$_2$ (14)

(8)

(11) + (15) → (16)

technique in a matrix and FTIR spectrometry, hydrosilylation of ethene with a $[(CO)_4CoSiR_3]$[21] or $[Cp^*(CO)_2FeSiR_3]$ catalyst[22] has been studied, and a mechanism, which involves insertion of an alkene into the M—Si bond (silylmetallation), is proposed. The catalytic cycle is demonstrated in Scheme 3 for the cobalt-catalyzed hydrosilylation.[21]

Tertiary amines catalyze the addition of trichlorosilane with electron deficient alkenes, *e.g.* acrylonitrile, and such alkynes as phenylethyne (Scheme 4). The reaction is understood in terms of the nucleophilic addition of a trichlorosilyl anion (**17**), which is generated by proton abstraction from a zwitterion (**18**) by tertiary amines.[23]

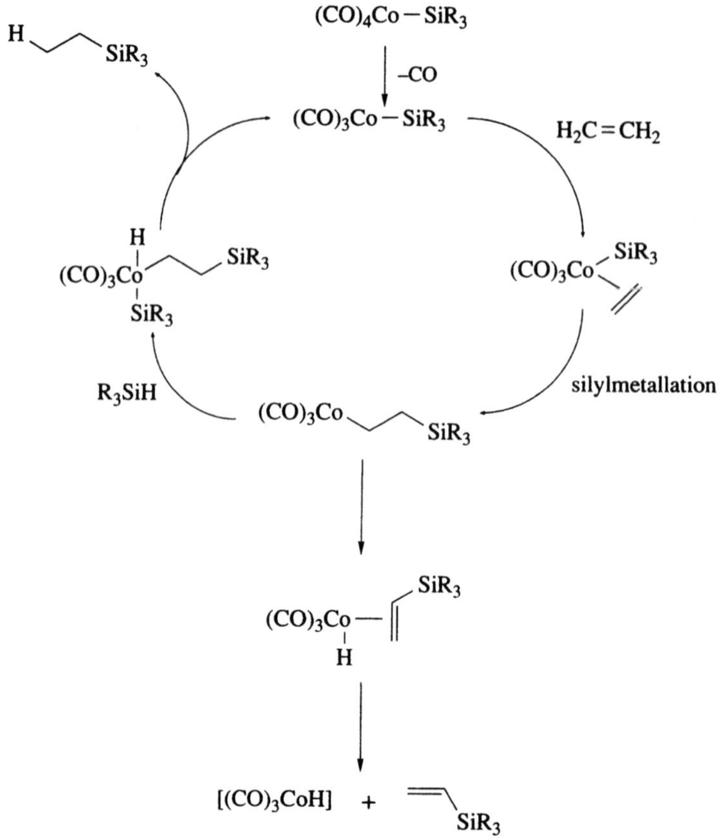

Scheme 3

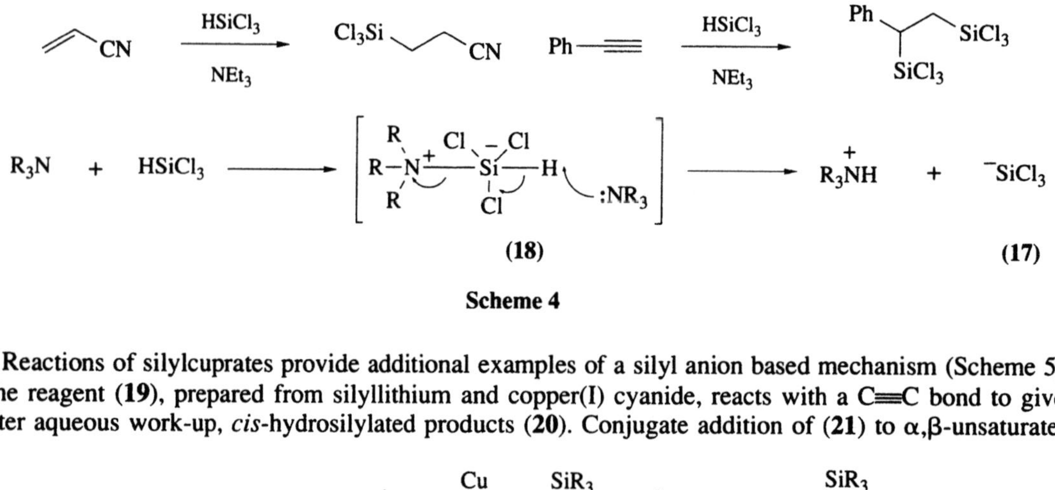

Scheme 4

Reactions of silylcuprates provide additional examples of a silyl anion based mechanism (Scheme 5). The reagent (**19**), prepared from silyllithium and copper(I) cyanide, reacts with a C≡C bond to give, after aqueous work-up, *cis*-hydrosilylated products (**20**). Conjugate addition of (**21**) to α,β-unsaturated

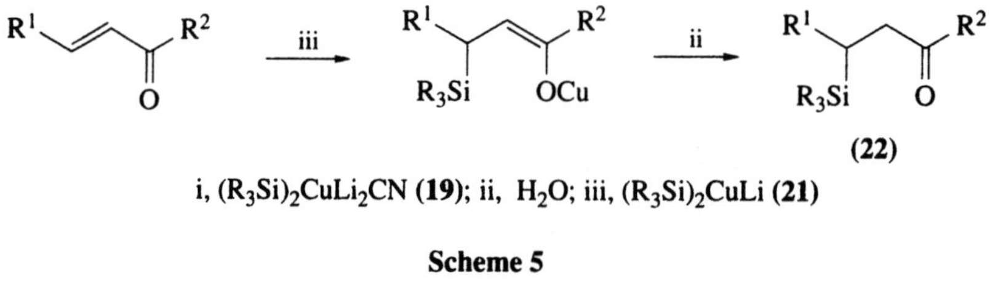

i, $(R_3Si)_2CuLi_2CN$ (**19**); ii, H_2O; iii, $(R_3Si)_2CuLi$ (**21**)

Scheme 5

ketones and esters is an alternative way for hydrosilylating such electron deficient alkenes. Thus, β-silyl ketones or esters (**22**) are easily accessible and are known to be versatile synthetic intermediates.[24]

3.12.2 HYDROSILYLATION OF CARBON–CARBON TRIPLE BONDS

3.12.2.1 Hydrosilylation of Acetylene

Acetylene is converted into vinylsilane (**23**) under appropriate hydrosilylation conditions with a Pt, Rh, Ru or Al catalyst.[25] However, depending on the catalyst and conditions, the initial 1:1 product (**23**) undergoes further reaction to afford 1,2-disilylethane (**24**; equation 9).[1b] Formation of 1,1-disilylethane is scarcely recorded.

$$\equiv \; + \; HSiR_3 \longrightarrow \underset{\mathbf{(23)}}{CH_2{=}CHSiR_3} \longrightarrow \underset{\mathbf{(24)}}{R_3SiCH_2CH_2SiR_3} \qquad (9)$$

To obtain vinylsilanes from alkynes, transition metal complexes of Group VIII combined with a main group metal chloride are particularly effective.[1b] In the presence of a heterogeneous catalyst like Pd/γ-alumina, Rh/carbon and polymer bound Pt, trichlorosilane gives trichlorovinylsilane with atmospheric pressure of acetylene.[1b] Platinum supported on sulfur-containing silica gel is a practical catalyst for 1,2-dihydrosilylation, as exemplified in equation (10).[26]

$$\equiv \xrightarrow[\text{conversion 85\%, yield 93\%}]{Et_3SiH,\ H_2PtCl_6{-}AlCl_3,\ 50\ ^\circ C} CH_2{=}CHSiEt_3$$

$$\equiv \xrightarrow[30\ ^\circ C,\ \text{yield 65\%}]{Cl_2MeSiH,\ SiO_2[{\equiv}Si(CH_2)_3]_2S/Pt} Cl_2MeSiCH_2CH_2SiMeCl_2 \qquad (10)$$

Thermal hydrosilylation of acetylene is feasible, but requires a high reaction temperature. A rather unusual example is shown in equation (11).[27]

$$1\text{-}(HSiCl_2)\text{naphthalene} \; + \; \equiv \xrightarrow[30\%]{590\ ^\circ C} \text{1-sila-phenalene } (SiCl_2) \qquad (11)$$

Hydrosilylation of acetylene with bifunctional hydrosilanes, *e.g.* (**25**; equation 12) and (**27**; equation 13), leads to polymers, *e.g.* (**26**) and (**28**) respectively, which contain silicon in the polymer main chain.[28–30]

$$\underset{\mathbf{(25)}}{Fe(C_5H_4CH_2SiMe_2H)_2} \xrightarrow[H_2PtCl_6]{\equiv} Fe(C_5H_4CH_2SiMe_2CH{=}CH_2)_2 \xrightarrow[\text{catalyst}]{\mathbf{(25)}} \underset{\mathbf{(26)}}{\left[Fe(C_5H_4CH_2SiMe_2CH_2CH_2)_2\right]_n} \qquad (12)$$

$$\underset{\mathbf{(27)}}{HSiMe_2{-}\!\left(O{-}SiMe_2\right)_n\!{-}H} \xrightarrow[Pt/C]{\equiv} \underset{\mathbf{(28)}}{\left[SiMe_2\!\left(O{-}SiMe_2\right)_n\!CH_2CH_2\right]_m} \qquad (13)$$

3.12.2.2 Hydrosilylation of Monosubstituted Alkynes

Hydrosilylation of monosubstituted alkynes (**29**) should give, *a priori*, three isomeric products: (**30**; β-*Z*), (**31**; β-*E*) and (**32**; α), as illustrated in equation (14). Though the product distribution changes depending on the kind of substituent R, the silane, the solvent and the reaction temperature, it is well accepted that (**30**) is derived from radical-initiated hydrosilylation, whereas (**31**) is derived from transition metal catalyzed hydrosilylation. Table 1 summarizes typical examples. For instance, 1-hexyne (**29**; R = Bu^n) is converted into (**30**; R = Bu^n, SiR^1_3 = $SiCl_3$) with $(PhCOO)_2$, but into (**31**; R = Bu^n, SiR^1_3 = $SiCl_3$) with H_2PtCl_6 at low temperature.[2b] When a large excess of $HSiCl_3$ was employed and the reaction time was prolonged, the yield of the 1:1 adducts decreased to 62% and 1:2 adducts were produced in fair amounts ($Cl_3Si(CH_2)_6SiCl_3$, 20% and $Bu^nCH(SiCl_3)CH_2SiCl_3$, 18%). Platinum catalyst (**1**) is more effective in giving (**31**), particularly at low temperature.[31,32] Hydrosilylation of 1-hexyne (**29**; R = Bu^n) with Et_3SiH and the Pt catalyst (**1**) again gave (**31**; R = Bu^n, SiR^1_3 = $SiEt_3$) at ambient temperature,[33] but at 100 °C the formation of (**32**; R = Bu^n, SiR^1_3 = $SiEt_3$) increased.[34] The reactivity of 1-alkynes in hydrosilylation with Et_3SiH and $[\{(Cy_3P)(Me_3Ge)(\mu\text{-}H)Pt\}_2]$ catalyst has been studied,[35] and the following relative order was disclosed: $Pr^nC{\equiv}CH$ (1.0), $Bu^nC{\equiv}CH$ (1.4), n-$C_5H_{11}C{\equiv}CH$ (1.5), $MeEtC(OH)C{\equiv}CH$ (6.8), c-$C_5H_9C{\equiv}CH$ (7.4), $Me_2C(OH)C{\equiv}CH$ (9.5) and $PhCH(OH)C{\equiv}CH$ (12.9). The factors controlling the reactivity and regioselectivity of hydrosilylation are correlated to the ^{13}C NMR chemical shifts of C-1 and C-2 of 1-alkynes. The chemical shifts of C-2 generally appear at lower field than C-1 and show a linear dependence on reactivity: those 1-alkynes with C-2 absorption at low magnetic field are the more reactive.[36]

$$R{-}C{\equiv}CH \xrightarrow[\text{catalyst}]{HSiR^1_3} \underset{(30)\ \beta\text{-}(Z)}{RCH{=}CHSiR^1_3} + \underset{(31)\ \beta\text{-}(E)}{RCH{=}CHSiR^1_3} + \underset{(32)\ \alpha}{R(R^1_3Si)C{=}CH_2} \qquad (14)$$

(**29**)

Table 1 Hydrosilylation of Monosubstituted Ethynes (**29**)

R in (**29**)	$HSiR^1_3$	*Catalyst*	*Yield* (%)	*Products ratio* β-(*Z*) (**30**)	β-(*E*) (**31**)	α (**32**)	*Ref.*
Bu^n	$HSiCl_3$	$(PhCOO)_2$	36	77	23	—	2b
Bu^n	$HSiCl_3$	Pd/C	93	~0	~100	—	2b
Bu^n	$HSiCl_3$	H_2PtCl_6 (71 °C)	92	—	78	22	31
Bu^n	$HSiCl_3$	H_2PtCl_6 (5 °C)	—	—	95	5	32
Bu^n	$HSiEt_3$	H_2PtCl_6 (20 °C)	100	—	82	18	33
Bu^n	$HSiEt_3$	H_2PtCl_6 (100 °C)	78	—	42	58	34
Bu^n	$HSiEt_3$	$[RhCl(PPh_3)_3]$	95	65	35	—	37
Bu^n	$HSiMe_2Ph_3$	$[RhCl(PPh_3)_3]$	98	72	28	—	37
Bu^n	$HSiMe_2Et$	$[RhCl(PPh_3)_3]$	95	81	19	—	37
Ph	$HSiCl_3$	H_2PtCl_6	96	—	69	31	38
Ph	$HSiCl_2Me$	H_2PtCl_6	97	—	82	18	38
Ph	$HSiEt_3$	H_2PtCl_6	93	—	81	19	38
Ph	$HSiCl_3$	$[RhCl(PPh_3)_3]$	60	15	57	28	38
Ph	$HSiCl_2Me$	$[RhCl(PPh_3)_3]$	98	11	71	18	38
Ph	$HSiEt_3$	$[RhCl(PPh_3)_3]$	35	45	36	19	38
$MeCl_2Si$	$HSiClMe_2$	H_2PtCl_6	94	—	only	—	39
$MeCl_2Si$	$HSiClMe_2$	$[RhCl(PPh_3)_3]$	85	—	60	40	39
$MeCl_2Si$	$HSiClMe_2$	$[Pt(PPh_3)_4]$	95	—	4	96	39
Me_3Si	$HSiClMe_2$	$[Pt(PPh_3)_4]$	—	—	80	20	39

In contrast to Pt catalysts, Rh catalysts like $[RhCl(PPh_3)_3]$ afford the β-(*Z*) product (**30**) mainly. The product ratio depends on the donor capability of the phosphine ligand: (**30**) is preferred in the order of $P(o\text{-}MeOC_6H_4)_3 \gg PPh_3 \approx P(o\text{-tolyl})_3 \approx P(p\text{-}MeOC_6H_4)_3 \gg P(OMe)_3 \approx P(OEt)_3 \approx P(OPh)_3$. In other words, acceptor ligands favor *syn* addition leading to the β-(*E*) adduct (**31**).[40] It is also claimed that $[RhCl(PPh_3)_3]$ strictly purified under a completely deoxygenated atmosphere exhibits low activity, but the reactivity is much enhanced by a trace of oxygen or *t*-butyl peroxide.[7] The catalyst prepared *in situ* from $[\{RhCl(C_8H_{14})_2\}_2]$ and 2 mol of PPh_3 is the most reactive rhodium system.[41]

Phenylacetylene (**29**; R = Ph) shows similar reactivity to other 1-alkynes. H_2PtCl_6 catalyst (**1**) affords (**31**; R = Ph) preferentially along with a small amount of (**32**; R = Ph), irrespective of the substituents on the silanes used. The catalyst system, which consists of H_2PtCl_6 and an additive of $AlCl_3$, $SiCl_4$, $GeCl_4$ or $CeCl_3$, increases the yield of the 1:1 adducts and enhances the selectivity in favor of the β-(*E*) product

(**31**).[25] In contrast, the regioselectivity using Rh catalysts such as [RhCl(PPh_3)$_3$], [RhCl(CO)(PPh_3)$_2$] and [Rh(acac)(CO)$_2$] depends on the electronic nature of the hydrosilane: chlorosilanes give (**31**) and (**32**), and trialkylsilanes give (**30**) as the major products, respectively. The ligand effect in the series [RhCl(CO)P_2] has been studied: *syn* addition leading to (**31**) is preferred in the order P = PPh_3 > PPh_2Bu^n > $PPhEt_2$ > PPh_2Pr^i > $PPhBu^n{}_2$ > PEt_2Pr^i > $PPhPr^i{}_2$ > $PBu^n{}_3$.[42]

Hydrosilylation of silylalkynes (**29**; R = $SiR^2{}_3$) gives (**31**; R = $SiR^2{}_3$) as the major product.[1b] The steric repulsion overrides the electronic effect which should encourage the formation of (**32**; R = $SiR^2{}_3$). This trend applies also to the [Pt(PPh_3)$_4$] catalyst, but unusual selectivity is reported in the platinum(0)-catalyzed reaction of $MeCl_2SiC{\equiv}CH$ with Me_2ClSiH, whereby (**32**; R = $SiCl_2Me$) is obtained with exceptionally high selectivity (96%). An additional example of α-adduct (**32**) formation is the H_2PtCl_6-catalyzed hydrosilylation using a trisilacyclohexane, $CH_2(SiMe_2CH_2)_2Si(Me)H$.[43]

The nickel(0) catalyst prepared by reduction of [Ni(acac)$_2$] with Et_3Al induces the reaction of 1-pentyne with $(EtO)_3SiH$ to give a 2:1 adduct (**33**) in 80% yield instead of the normal 1:1 adduct (equation 15).[44]

1-pentyne + $HSi(OEt)_3$ —[Ni(acac)$_2$], $AlEt_3$, 80 %→ $Si(OEt)_3$ (**33**) (15)

Many other examples are recorded of monosubstituted alkynes with substituents $ClCH_2$,[45] $BrCH_2$,[46] CF_3,[47] $HOCR^2{}_2$,[48] $R^2{}_2NCH_2$,[48] R^2OCH_2,[48] 2-thienyl[48] and others.[1b] In all cases the corresponding vinylsilanes are produced in high yields.

An alternative route to the hydrosilylation product of monosubstituted alkynes (**29**) is that which involves silylmetallation of the alkynes (equation 16).[24] A cuprate reagent (**34**) prepared from CuCN and 2 mol of $Me_2PhSiLi$ adds in a *syn* manner to the C≡C bond with the Me_2PhSi group at the terminal carbon. Quenching with water affords a product of type (**31**).

R—≡ (**29**) —i→ (R, (Cu), $SiMe_2Ph$) —ii→ R, $SiMe_2Ph$ (16)

i, $(Me_2PhSi)_2CuLi_2CN$ (**34**); ii, H_2O R = H, 78%; R = Me, 94%; R = Ph, 67%

3.12.2.3 Hydrosilylation of Disubstituted Alkynes

The hydrosilylation of disubstituted alkynes using platinum catalysts is slower than the hydrosilylation of monosubstituted alkynes but exhibits similar selectivities.[35] Examples are shown for 2-butyne and 2-pentyne in equations (17) and (18), respectively.[49] In both cases *syn* addition takes place exclusively. The regioselectivity (65:35) for 2-pentyne is attributed to the general trend that the silyl group prefers the alkyne carbon which has the less bulky substituent (compare the regioselectivity of 1-propyne, where silicon adds to the terminal carbon with a selectivity of 96:4[36]). As a trimethylsilyl group is larger than an alkyl group, hydrosilylation of 1-trimethylsilyl-1-alkynes takes place highly selectively, as shown in equation (19).[50]

2-butyne —i, $HSiMeCl_2$, H_2PtCl_6; ii, MeMgBr; 47%→ $SiMe_3$ (17)

≡—Et —Et_3SiH, [{(Cy_3P)($BnMe_2Si$)(μ-H)Pt}$_2$], 97%→ Et_3Si, Et + Et, $SiEt_3$ (18)

65:35

In contrast to platinum catalysts, rhodium catalysts often effect *anti* addition (equation 20).[51]

n-C_6H_{13}—≡—$SiMe_3$ —(i, $HSiMeCl_2$, H_2PtCl_6; ii, MeMgBr)→ (Z)-n-C_6H_{13}(SiMe$_3$)C=C(SiMe$_3$) 96% + n-C_6H_{13}CH=C($SiMe_3$)$_2$ trace (19)

Ph—≡—Ph —($HSiEt_3$, [$(CH_2NMe)_2C$=RhCl(COD)], air, 20 °C, 24 h, 100%)→ Ph(Et_3Si)C=CHPh (20)

A nickel catalyst induces, in addition to normal hydrosilylation, double silylation (equation 21)[52] and silacyclopentadiene formation (equation 22).[53]

Bu^n—≡—Bu^n —($HSiMeCl_2$, [Ni(bipy)Et_2], r.t.)→ Bu^n(Cl_2MeSi)C=C($SiMeCl_2$)Bu^n 72% + Bu^n(Cl_2MeSi)C=CHBu^n 24% (E:Z = 73:27) (21)

Et—≡—Et + HMe_2Si–$SiMe_2H$ —([$NiCl_2(PEt_3)_2$], 90 °C)→ 2,3,4,5-tetraethyl-1,1-dimethylsilacyclopentadiene 95% + H_2SiMe_2 (22)

3.12.3 HYDROSILYLATION OF CONJUGATED CARBON–CARBON TRIPLE BONDS

Three reaction pathways are possible for vinylalkyne (**34**): 1,2-addition to the C≡C moiety to give (**35**) and/or (**36**); 1,2-addition to the C=C moiety; and 1,4-addition leading to *e.g.* (**37**; equation 23). The reaction favoring the second path remains to be found, but conjugate addition to give allenes (**37**) is sometimes observed as a minor reaction. In most cases, only the alkyne moiety undergoes hydrosilylation, the regioselectivity of which depends on the substituent R connected to the alkynic carbon.[54–60] When R is H or CH_2OMe, (**36**) is produced exclusively, whereas (**35**) predominates when R is bulky, like trialkylsilyl or trisubstituted alkyl. A typical example is illustrated in equation (24).

(**34**) —($HSiR^1{}_3$, H_2PtCl_6)→ (**35**) ($R^1{}_3Si$) + (**36**) ($SiR^1{}_3$) + (**37**) ($R^1{}_3Si$–CH$_2$CH=C=CHR) (23)

(cyclohexenyl)C≡C–C(OH)Et$_2$ —($HSiEt_3$, H_2PtCl_6)→ (cyclohexenyl)($SiEt_3$)C=CH–C(OH)Et$_2$ (24)

In some cases, consecutive 1,2-addition and 1,4-addition take place (equation 25).[61]

CH_2=C(Me)C≡CH —($HSiMeCl_2$, H_2PtCl_6)→ CH_2=C(Me)CH=CH$SiMeCl_2$ → Me_2C=CHCH($SiMeCl_2$)$_2$ (25)

Only a few examples are reported of hydrosilylation of conjugated diynes. 1,4-Bis(trimethylsilyl)butadiyne (**38**) gives at first trisilyl-substituted butenyne (**39**), wherein the hydrosilane-derived silyl group has added to an inner alkyne carbon. The butenyne (**39**) undergoes further hydrosilylation in various ways depending on the electronic nature of the hydrosilanes (Scheme 6). Dichloromethylsilane reacts with the remaining alkynic bond of (**39**) to give (**40**).[62] Trialkylsilane reacts with the intermediate (**41**) in a conjugate manner to give the allene (**42**).[63] This type of reaction is sensitive to the steric bulk of the substituents. The substrate whose C≡C is sterically less crowded undergoes 1,2-addition (equation 26).

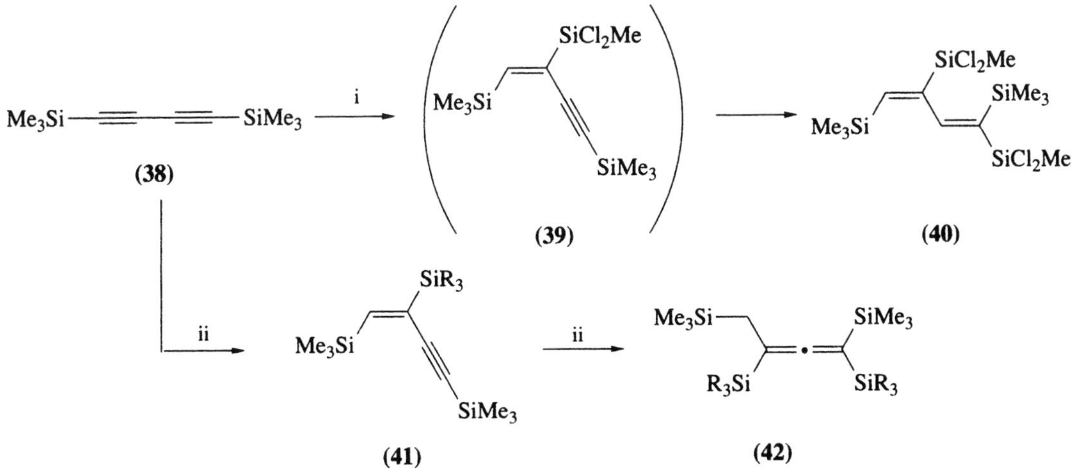

Me3Si—≡—≡—SiMe3 (**38**) —i→ (**39**) → (**40**)

(**38**) —ii→ (**41**) —ii→ (**42**)

i, $HSiCl_2Me$, H_2PtCl_6; ii, $HSiR_3$ (R = Me, Et, Ph), H_2PtCl_6 or $[RhCl(PPh_3)_3]$

Scheme 6

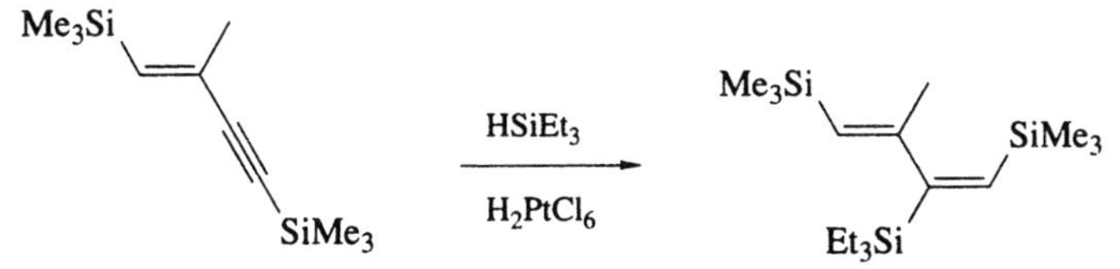

$$\xrightarrow[H_2PtCl_6]{HSiEt_3} \quad (26)$$

3.12.4 HYDROSILYLATION OF CARBON–CARBON DOUBLE BONDS

3.12.4.1 Hydrosilylation of Ethylene

Hydrosilylation of ethylene (equation 27) is the prototype for the addition of hydrosilanes across carbon–carbon double bonds, and is achieved by means of such metal complex catalysts as H_2PtCl_6,[64] $[RhCl(PPh_3)_3]$,[65] $[\{RhCl(CO)_2\}_2]$ (equation 28),[15] $[Co_2(CO)_8]$,[65] $[HCo(CO)_4]$,[66] $[NiCl_2(dmpf)]$,[10,67] $[Pd(PPh_3)_4]$[68] and also with the aid of γ-rays, which initiate radical reactions.[69]

$$H_2C{=}CH_2 \quad + \quad HSiR_3 \xrightarrow{\text{catalyst}} EtSiR_3 \qquad (27)$$

R = Cl, OEt, Me, Et, Ph *etc.*

$$H_2C{=}CH_2 \; + \; HSiCl_3 \xrightarrow[\text{60 atm ethylene, 95\%}]{[\{RhCl(CO_2\}_2],\ 20\ ^\circ C,\ 65\ h} EtSiCl_3 \qquad (28)$$

When trialkylsilanes are employed, vinylsilane is often produced as the main product. This oxidative hydrosilylation can be carried out with $[Me_3SiCo(CO)_4]$,[21] $[Cp^*{}_2Rh_2Cl_4]$ (equation 29),[70] $[\{Ir(OMe)(COD)\}_2]$,[6b] $[Cp^*Fe(CO)_2SiMe_3]$,[22] or $Na[HRu_3(CO)_{11}]$[11d,11e] catalysts. This unusual reaction is understood in terms of silylmetallation rather than the ordinary hydrometallation mechanism.[7]

$$H_2C{=}CH_2 \ + \ HSiEt_3 \xrightarrow{[Cp^*_2Rh_2Cl_4]} SiEt_4 \ + \ {=}\!\!\diagdown_{SiR_3} \qquad (29)$$

	$SiEt_4$	vinyl-SiR_3
1 atm	74%	26%
7 atm	40%	60%

3.12.4.2 Hydrosilylation of Monosubstituted Alkenes

In principle, two regioisomers are produced by the reaction of terminal alkenes and hydrosilanes: one where the terminal carbon forms a bond to the silyl group (normal product) and the other which has the silyl group at the inner carbon (branched product). Platinum and rhodium catalysts yield in general the normal products, whereas palladium catalysts sometimes prefer the reaction pathway leading to the branched products. Ruthenium and rhodium afford additional by-products such as alkenylsilanes and alkanes. A typical example is the hydrosilylation of 1-octene (equation 30).[71]

$$\underset{\text{(excess)}}{R\text{-}CH{=}CH_2} \xrightarrow[\text{catalyst}]{HSiR^1_3} R\text{-}CH_2CH_2\text{-}SiR^1_3 \ + \ R\text{-}CH{=}CH\text{-}SiR^1_3 \ + \ R^*SiR^1_3 \ + \ R{-}Et \qquad (30)$$

$R = n\text{-}C_6H_{13}$, $HSiR^1_3 = HSiMe_2(n\text{-}C_{10}H_{21})$, $R^* = n\text{-}C_5H_{11}CH{=}CHCH_2$

Catalyst				
H_2PtCl_6	96%	–	–	–
$[Ru_3(CO)_{12}]$	11%	47%	43%	87%
$[RhCl(PPh_3)_3]$	60%	32%	8%	38%
$R = Ph$, $HSiR^1_3 = HSiEt_3$				
$[Ru_3(CO)_{12}]$	~0	93%	–	96%
$R = Bu^n$, $HSiR^1_3 = HSiEt_3$, $R^* = Pr^nCH{=}CHCH_2$				
$[Ru_3(CO)_{12}]$	~0	83%	16%	not determined

The product distribution depends on the mole ratio of the substrate alkene to hydrosilane. In the presence of a large excess of alkene, dehydrogenative hydrosilylation is apt to become the major process, as illustrated in equation (31).[70]

$$R\text{-}CH{=}CH_2 \xrightarrow[\text{catalyst}]{HSiR^1_3} R\text{-}CH_2CH_2\text{-}SiR^1_3 \ + \ R\text{-}CH{=}CH\text{-}SiR^1_3 \ + \ R^*SiR^1_3 \qquad (31)$$

$R = n\text{-}C_5H_{11}$, $HSiR^1_3 = HSiEt_3$, catalyst = $[Cp^*_2Rh_2Cl_4]$, $R^* = Bu^nCH{=}CHCH_2$

Substrate:hydrosilane			
1:1	94%	0%	3%
5:1	34%	52%	8%
$R = MeOCO$, $HSiR^1_3 = HSiMeEt_2$, catalyst = $[Co_2(CO)_8]$			
1:3	58%	0%	–
1:1	10%	30%	–
2.5:1	5%	94%	–

Vinylsilanes give 1,1-disilylethane and/or 1,2-disilylethane depending on the substituent of the hydrosilane (equation 32) and the metal catalyst (equation 33). Thus by choosing these variables properly, both types of disilylethane are readily accessible.[72,73]

Under radical conditions, hydrosilanes add across terminal alkenes to give 1-silylalkanes. Di-*t*-butyl peroxide is the catalyst of choice (equation 34). Benzoyl peroxide or AIBN is much less effective.[74]

Intramolecular hydrosilylation affords silacycles. Diphenyl(4-pentenyl)silane (**43**) undergoes ring closure upon treatment with a transition metal catalyst, with or without a solvent, to give the five-membered ring compound (**44**) preferentially, rather than the silacyclohexane (**45**; equation 35).[75] The mode of ring closure has been studied by changing *n* in (**46**), as summarized in equation (36).[76] The silylalkene (**46**) with *n* = 0 or 1 gave polymers by intermolecular hydrosilylation and no monomeric cyclic products.

$$Si\text{-CH=CH}_2 + HSi' \xrightarrow{[Pd(PPh_3)_4]} Si\text{-CH(Si')CH}_3 + Si\text{-CH}_2CH_2\text{-}Si' \quad (32)$$

$Si = Cl_3Si$, $Si' = SiCl_3$	90%	2%
$Si = Cl_3Si$, $Si' = SiMeCl_2$	3%	90%
$Si = Cl_2MeSi$, $Si' = SiCl_3$	1%	89%
$Si = ClMe_2Si$, $Si' = SiCl_3$	0%	80%

$$Cl_3Si\text{-CH=CH}_2 + HSiCl_3 \xrightarrow{i} Cl_3Si\text{-CH(SiCl}_3)CH_3 + Cl_3Si\text{-CH}_2CH_2\text{-}SiCl_3 \quad (33)$$

Catalyst		
H_2PtCl_6–cyclohexanone	0%	92%
$[RhCl(PPh_3)_3]$	6%	90%
$[NiCl_2(PPh_3)_2]$	30%	50%
$[Pd(PPh_3)_4]$	81%	4%

i, vinylsilane:hydrosilane:catalyst = 1:1:2x10^{-4} at 120 °C

$$\text{1-decene} \xrightarrow[64\%]{i} \text{n-}C_{10}H_{21}SiEt_3 \quad (34)$$

i, $HSiEt_3$ (6 mol), Bu^tOOBu^t (0.2 mol), 170 °C, 0.5 h

(43) → (44) + (45) (35)

$H_2PtCl_6 \cdot 6H_2O$, neat, 20 °C	93%	20–22:1
Phenylalanine•Pt•C_2H_4•Cl_2, toluene, 80 °C	89%	100:0
$[RhCl(PPh_3)_4]$, neat, 80 °C	89%	6.6:1

(46) $\xrightarrow[\text{hexane, reflux, 48 h}]{H_2PtCl_6}$ (36)

n	*Yield (%)*	*Ratio*
2	46	0:100
3	58	90:10
4	70	53:47
5	16	54:46
6	2	100:0

Allyl halides (equation 37), vinyl halides (equation 38) and vinyl ethers are good substrates for the preparation of organosilanes having functional groups at the β- or γ-position. Reduction of these substrates sometimes takes place to give the parent alkenes (equation 37), which are derived from β-elimination of the regioisomeric hydrosilylation intermediates.[77]

$$\text{ClCH}_2\text{CH=CH}_2 \xrightarrow[\text{HSiR}_3]{\text{H}_2\text{PtCl}_6} \text{Cl(CH}_2)_3\text{SiR}_3 + \text{CH}_3\text{CH=CH}_2 \tag{37}$$

$HSiEt_3$	5%	70%
$HSiPh_2Me$	70%	10%

$$\text{FCH}_2\text{CH=CH}_2 \xrightarrow[h\nu]{\text{HSiCl}_3} \text{F(CH}_2)_3\text{SiCl}_3 \tag{38}$$

3.12.4.3 Hydrosilylation of Disubstituted and More Highly Substituted Alkenes

With H_2PtCl_6 as the catalyst, highly substituted alkenes undergo double-bond migration before hydrosilylation to give rise to the products which have silyl groups at the least-substituted, commonly terminal, carbons. These substrates generally require longer reaction times.[1c,3,78] Three examples are given in equation (39).

+ $HSiMeCl_2$ —(H_2PtCl_6, 100 °C)→ $MeCl_2Si$ … (82%) + … $SiMeCl_2$ (18%)

+ $HSiCl_3$ —(H_2PtCl_6, 100 °C)→ Cl_3Si … (70%) + … $SiCl_3$ (30%) (39)

+ $HSiCl_3$ —(H_2PtCl_6, 100 °C, 78%)→ $SiCl_3$ (98%) + $SiCl_3$ (2%)

Disubstituted alkenes which have an allylic quaternary carbon center (**47**) undergo skeletal rearrangement as shown in equation (40).[79]

(**47**) + $HSiCl_3$ —($[\{PtCl_2(CH_2{=}CH_2)\}_2]$, reflux, 19 h)→ $SiCl_3$ (32%) + Bu^t, $SiCl_3$ (14%) (40)

Wilkinson's catalyst is less reactive towards 1,1- and 1,2-disubstituted alkenes, and low yields and side reactions are often observed. However, the catalyst (**48**) is reactive enough to afford straightforward products in fair yields without double-bond migration (equation 41).[80]

Fully substituted alkenes are hydrosilylated with aluminum chloride without double-bond migration.[81a] The products have tertiary alkyl groups and thus may be utilized as silyl protecting groups for alcohols and related functional groups.[81b] For example, chlorodimethylthexylsilane (**49**) is readily prepared from 2,3-dimethyl-2-butene (equation 42).

In contrast to the H_2PtCl_6-catalyzed hydrosilylation, the formation of positional isomers is negligible under radical conditions, although the stereochemical results are random. Thus 1-methylcyclohexene

Ph(Me)C=CH₂ ⟶ $PhCH(Me)CH_2SiMeEt_2$ + $PhC(Me)=CHSiMeEt_2$ + $PhC(=CH_2)CH_2SiMeEt_2$

i, 20%	76.3	:	15.1	:	8.6
ii, 61.2%	100	:	0	:	0

PhCH=CHMe ⟶ $PhCH_2CH(SiMeEt_2)Me$ + $Ph(CH_2)_3SiMeEt_2$ (41)

i, 16.2%	64.6	:	26.1
ii, 52.1%	95.2	:	4.8

i, $HSiMeEt_2$, $[RhCl(PPh_3)_3]$, 60 °C, 6 h; ii, $HSiMeEt_2$, $[\{RhCl(CH_2{=}CH_2)_2\}_2]$ (**48**), 60 °C, 6 h

$$Me_2C{=}CMe_2 + HSiMe_2Cl \xrightarrow[\substack{25\ °C\\93\%}]{AlCl_3\ (5\ mol\ \%)} ClSiMe_2CMe_2CHMe_2 \quad \textbf{(49)} \tag{42}$$

The following can be prepared similarly:

Cl—SiMe₂—CEt₂CHEt₂ 80%; Cl—SiMe₂—C(Me)(Pr)CHMe₂ 80%; Cl—SiMe₂—(1-cyclohexylcyclohexyl) 67%

gives a 5.5–9:1 mixture of (**50**) and (**51**).[2a] In contrast, thermal reaction induces some double-bond isomerization, and a third product (**52**) is produced (equation 43).

1-methylcyclohexene ⟶ (**50**) SiCl₃ + (**51**) SiCl₃ + (**52**) cyclohexyl-CH₂SiCl₃ (43)

	(**50**)		(**51**)		(**52**)
i, 100%	85	:	15	:	0
ii, 44%	89	:	11	:	0
iii, 100%	58	:	23	:	19

i, $HSiCl_3$, $(AcO)_2$, reflux (45 °C), 9 h; ii, $HSiCl_3$, UV, reflux, 44 h; iii, $HSiCl_3$, 300 °C, 24 h

Hydrosilylation of (-)-β-pinene (**54**) under radical conditions results in opening of the strained cyclobutane ring to give only (**53**; equation 44).[82a] However, Pt-catalyzed hydrosilylation gives the expected product, which, after conversion of Si—Cl to Si—H, is used for the asymmetric reduction of ketones.[82b]

(**53**) SiCl₃ ⟵(i)— (**54**) —(ii)⟶ $SiMe_nCl_{3-n}$ ⟶ $SiMe_nH_{3-n}$ (44)

$n = 1, 2$

i, $HSiCl_3$, $(AcO)_2$ or Hg lamp; ii, $HSiCl_2Me$ or $HSiClMe_2$, H_2PtCl_6

An enamine (**55**) reacts with trichlorosilane in acetonitrile in the absence of catalyst to give a hydrosilylated product (**56**; equation 45). The regiochemistry is opposite to the standard pattern, the silyl group attaching itself to the more-hindered carbon.[83]

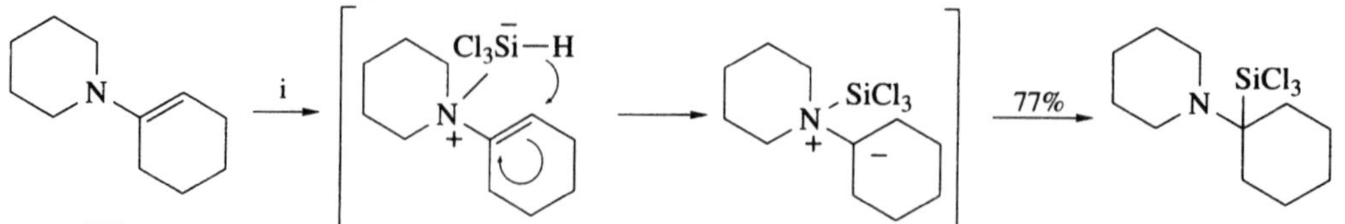

(45)

i, $HSiCl_3$, MeCN, reflux, 2.5 h

Alkenes are reduced with hydrosilanes in an acidic solvent like trifluoroacetic acid. The reaction is called 'ionic hydrogenation' and is useful particularly for reduction of highly substituted alkenes. The reaction is assumed to proceed through a carbocationic intermediate, to which hydride is delivered from the hydrosilane.[84] An example is given in equation (46), but see also Chapters 1.1 and 3.3, this volume.

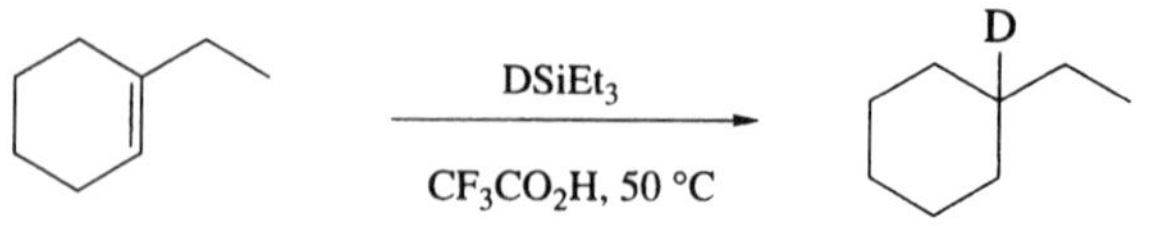

(46)

3.12.5 HYDROSILYLATION OF POLYENES

3.12.5.1 Hydrosilylation of Acyclic Polyenes

Polyenes whose C═C bonds are not conjugated behave as the substituted alkenes described above, and thus the hydrosilylation products are those in which the silyl group attaches itself to the less-hindered alkenic carbons.[85] When dihydrosilanes are employed, both silacarbocycles and polymers are produced (Scheme 7).[86]

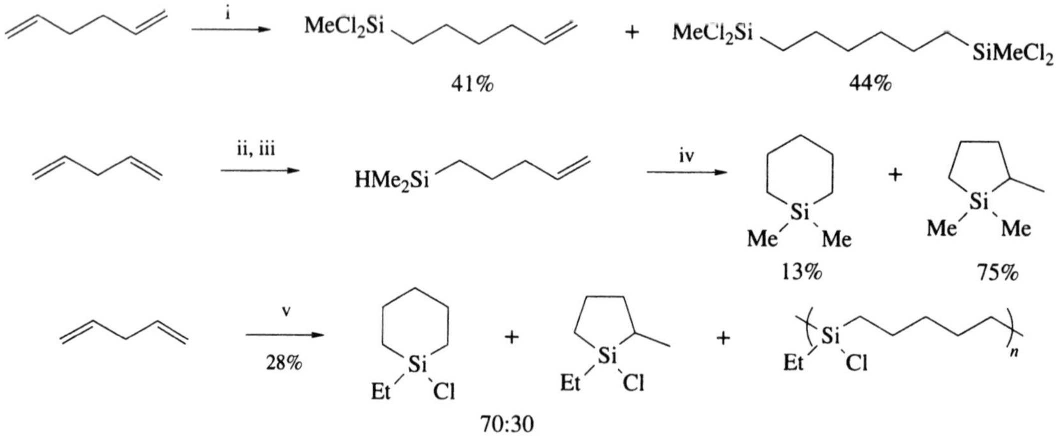

i, $HSiMeCl_2$, H_2PtCl_6, 140–150 °C, 13 h; ii, $HSiMe_2Cl$, H_2PtCl_6; iii, $LiAlH_4$; iv, H_2PtCl_6; v, $H_2SiClEt$, H_2PtCl_6

Scheme 7

Conjugated dienes like 1,3-dienes give 1:1 adducts of various types: usually 1,2- and 1,4-addition products, and the regio- and stereo-isomers of each type, but also 1:2 adducts, and dimerization and reduction products. The product distribution depends on the kinds of dienes, silanes and particularly on the kind of catalyst.[87,88] Of these, chromium hexacarbonyl is specific for the formation of (*Z*)-crotylsilanes from butadienes.[11g] Similar (*Z*)-selective hydrosilylation is observed with $HSiCl_3$ and $[Pd(PPh_3)_4]$ catalyst.[68] The reaction pattern of 1,3-butadiene is summarized in Scheme 8.

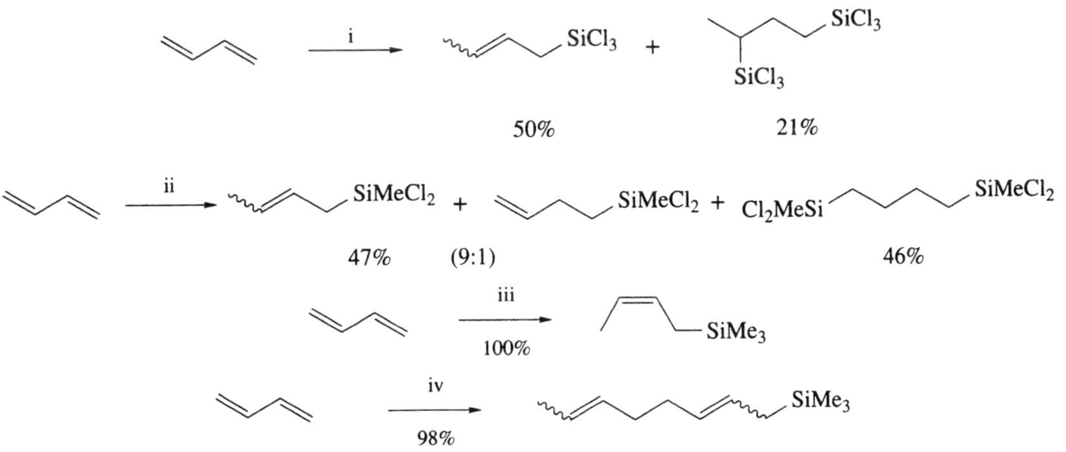

50% 21%

47% (9:1) 46%

100%

98%

i, Pt–C, $HSiCl_3$, 160 °C, 1.3 h; ii, $HSiMeCl_2$, H_2PtCl_6, 200 °C, 4 h; iii, $HSiMe_3$, $[Cr(CO)_6]$, $h\nu$;
iv, $HSiMe_3$, $[Pd(PPh_3)_2]$ (fumaric anhydride), 85 °C, 4.5 h

Scheme 8

Hydrosilylation of isoprene (**57**) has been well studied and the results are summarized in Table 2.[1d,89] Since 1,3-dienes are relatively reactive, metal complexes which do not catalyze hydrosilylation of common alkenes may be employed. Pd, Rh, Ni and Cr catalysts prefer 1,4-addition products with the (*Z*)-configuration. No catalyst is available which prefers 1,2-addition with high selectivity.

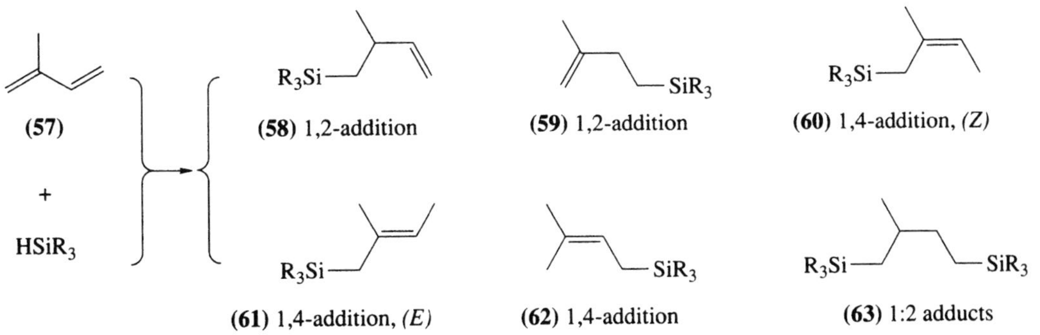

(**57**) + $HSiR_3$ →

(**58**) 1,2-addition (**59**) 1,2-addition (**60**) 1,4-addition, (*Z*)

(**61**) 1,4-addition, (*E*) (**62**) 1,4-addition (**63**) 1:2 adducts

Scheme 9

Table 2 Hydrosilylation of Isoprene (**57**)

Catalyst	*Hydrosilane*	*Conditions*	*Yield (%)* (**58**)	(**59**)	(**60**)	(**61**)	(**62**)	(**63**)
H_2PtCl_6	$HSiCl_3$	165 °C, 17 h					72	6
	$HSiCl_2Me$	165 °C, 17 h		31			57	9
	$HSiMe_3$	165 °C, 17 h		27	20[a]		<2	24
$[Pt(PPh_3)_4]$	$HSiCl_2Me$	48–82 °C, 25 h		38				38
$[PdCl_2(PhCN)_2]/PPh_3$	$HSiCl_2Me$	70 °C, 6 h			95	—		
$[PdCl_2(PPh_3)_2]$	$HSiCl_3$	80 °C, 4 h			100	—		
$[Pd(PPh_3)_4]$	$HSiCl_3$	110 °C, 6 h			83		4	
$[RhCl(PPh_3)_3]$	$HSiClMe_2$	130 °C, 15 h			19	—	41	
	$HSiMe_2Ph$	80 °C, 2 h			27		71	
$[Rh(acac)_3]$	$HSiEt_3$	60 °C, 15 h		5[b]		94[c]		
Ni (vapor)	$HSi(OEt)_3$	−196 to −5 °C			100	—	—	
$[Ni(acac)_3]/AlEt_3$	$HSiEt_3$	20 °C, 4 h			96[d]			
$[Cr(CO)_6]$. $h\nu$	$HSiMe_3$	30 °C			60	—	40	
[HRu(COD){(−)-menthyl—N=CH—}$_2$]Cl	$HSi(OEt)_3$	25 °C, 150 h	38	7	—	—	25	—

[a]Yield for (**60**) and (**61**). [b]Yield for (**58**) and (**59**). [c]Yield for (**60**), (**61**) and (**62**). [d] Yield for (**60**) and (**61**).

Myrcene (**64**) undergoes hydrosilylation preferentially at the terminal vinyl moiety to give a homoallylsilane (**65**), as shown in equation (47).[90]

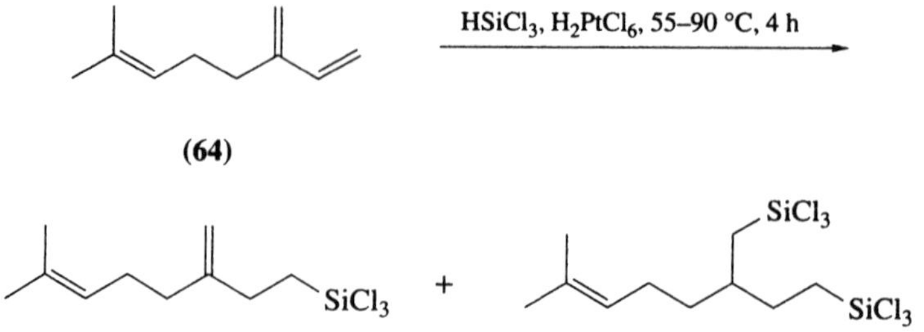

$HSiCl_3$, H_2PtCl_6, 55–90 °C, 4 h

(**64**)

$SiCl_3$ + $SiCl_3$ $SiCl_3$ (47)

(**65**) 41% 19%

The 1,2-dihydrodisilane (**66**) reacts with 2,3-dimethylbutadiene, in the presence of a Ni catalyst, to give both the normal hydrosilylation product and (*Z*)-1,4-disilyl-2-butene (Scheme 10). The nickel catalyst is assumed to be reduced to nickel(0), which oxidatively adds to the Si—Si bond of (**66**) in competition with addition to the Si—H bond.[53]

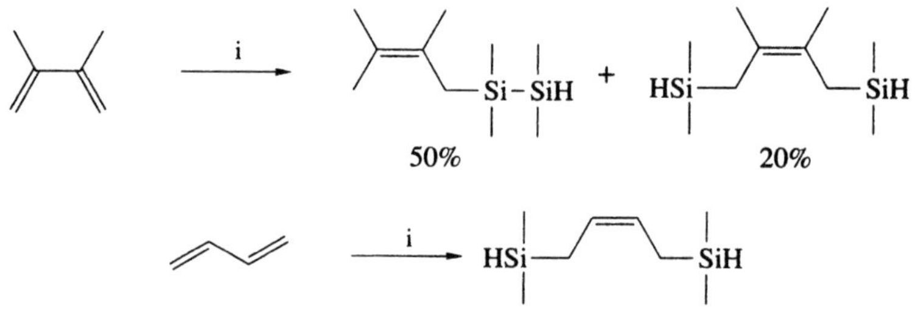

i Si–SiH + HSi SiH

50% 20%

i HSi SiH

69%

i, $HSiMe_2$–$SiMe_2H$ (**66**), $[NiCl_2(PEt_3)_2]$, 90 °C

Scheme 10

3.12.5.2 Hydrosilylation of Cyclic Polyenes

Cyclic conjugated dienes intrinsically have a fixed cisoid diene system and therefore readily undergo 1,4-addition to give allylsilanes regardless of the kind of catalyst and substituents on the hydrosilanes (equations 48 and 49). Nonconjugated dienes like (**67**) and (**68**) give the same allylic silane as the conjugated isomers. The results are ascribed to alkene isomerization into conjugated dienes followed by hydrosilylation (equations 50 and 51).[91–95]

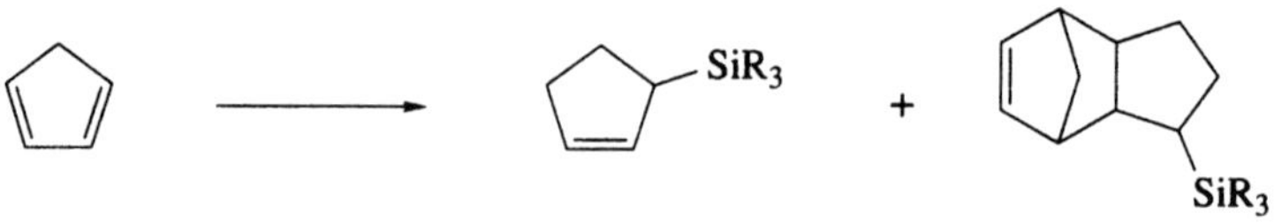

SiR_3 + SiR_3 (48)

i	52%	40%
ii	80–90%	–

i, $HSiCl_3$, H_2PtCl_6, 250 °C; ii, $HSiCl_xMe_{3-x}$, $[PdCl_2(PhCN)_2]/PPh_3$, 80–90 °C

+ $HSi(OEt)_3$ —i or ii→ $Si(OEt)_3$ (49)

i, $[Ni(acac)_2]/Al(OEt)Et_2$, 20 °C, 97%; ii, $[Co_2(CO)_8]$, 20 °C, 88%

In contrast, a large ring polyene undergoes normal hydrosilylation. For example, (*Z,E,E*)-1,5,9-cyclododecatriene (**69**) gives the monoadducts (**70**) and (**71**), in which the two C═C bonds remain intact

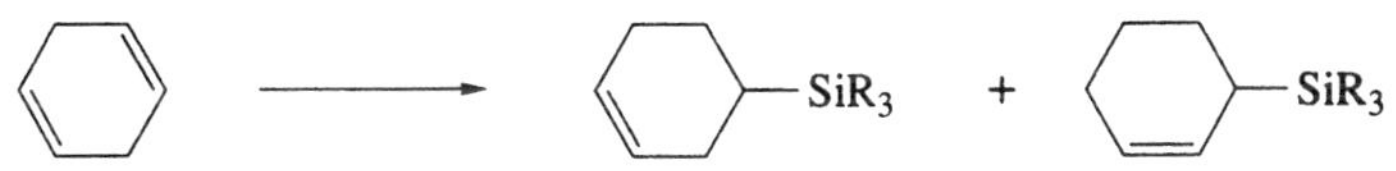

(50)

(67)	i	41%	55%
	ii	–	52%

i, $HSiMeCl_2$, $[NiCl_2(PPh_3)_2]$, 120 °C; ii, $HSi(OEt)_3$, $[Ni(acac)_2]/Al(OEt)Et_2$, 20 °C

(51)

(68)	i	35%	–
	ii	53%	42%
	iii	32%	68%

i, H_2PtCl_6, 45 °C; ii, $[H_2Pt(C_2H_4)(Py)]$; iii, Pt–C, 200 °C

(equation 52). As in many electrophilic reactions of (**69**), the (*E*)-C=C bonds are more reactive than (*Z*)-C=C bonds.[96]

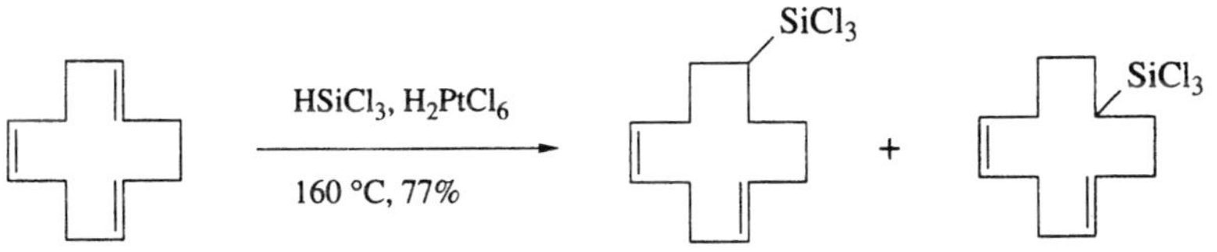

(52)

(69) **(70)** **(71)**

Norbornadiene (**72**) gives a tricyclic product (**73a**) in addition to the normal *endo*- and *exo*-hydrosilylated products (**73b**) and (**73c**) (equation 53). The tricyclic product is preferred under radical conditions.[97]

(53)

(72)			(73a)		(73b)		(73c)
	i	80%	30	:	64	:	6
	ii	77%	94	:	6	:	0
	iii	25%	100	:	0	:	0
	iv	80%	53	:	7	:	28

i, $HSiMeCl_2$, H_2PtCl_6, 50–60 °C; ii, $HSiMeCl_2$, Pt–C, 50–60 °C; iii, $HSiCl_3$, Pt–C, 35–50 °C; iv, $HSiCl_3$, AIBN, 80–100 °C

Dicyclopentadiene (**74**) produces 1:1 adduct(s) (**75**) by reaction with $HSiClMe_2$ and various metal catalysts (Scheme 11). A second addition is achieved only with γ-rays. The resulting 1:2 adducts (**76**) are converted into thermally fairly stable polymers (**77**).[98]

3.12.6 HYDROSILYLATION OF CONJUGATED ENONES

Transition metal catalyzed hydrosilylation of α,β-unsaturated ketones and aldehydes (**78**) proceeds in a 1,4- and/or 1,2-fashion to give enol silyl ethers (**79**) (or the saturated carbonyl compounds **80** after acidic work-up) and/or allyl silyl ethers (**81**) (or allyl alcohols **82**), respectively (equation 54). Monohydrosilanes combined with a Pt or Rh catalyst prefer 1,4-addition. This reaction is an alternative method for the preparation of enol silyl ethers (**79**). Diphenylsilane with $[RhCl(PPh_3)_3]$ catalyst reduces the car-

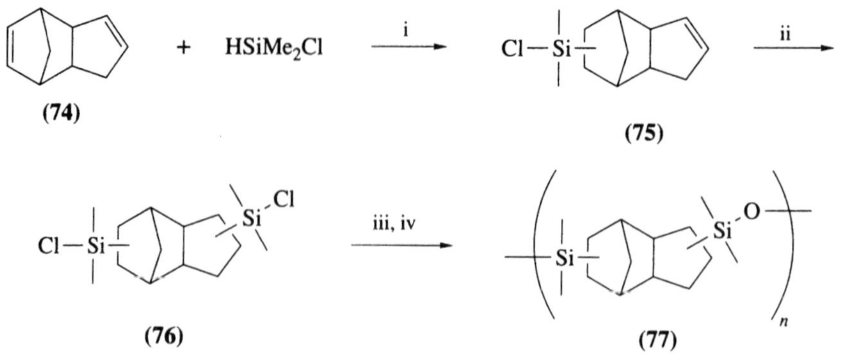

(74) + $HSiMe_2Cl$ —i→ (75) —ii→

(76) —iii, iv→ (77)

i, H_2PtCl_6; ii, $HSiMe_2Cl$, γ-ray; iii, $H_2O/NaHCO_3$; iv, tetramethylguanidine, CF_3CO_2H

Scheme 11

bonyl group preferentially to afford allyl alcohols (**82**) after aqueous work-up, whereas the phenylsilane/[$Mo(CO)_6$] catalyst combination leads to the formation of the saturated ketone (**80**).[99–101]

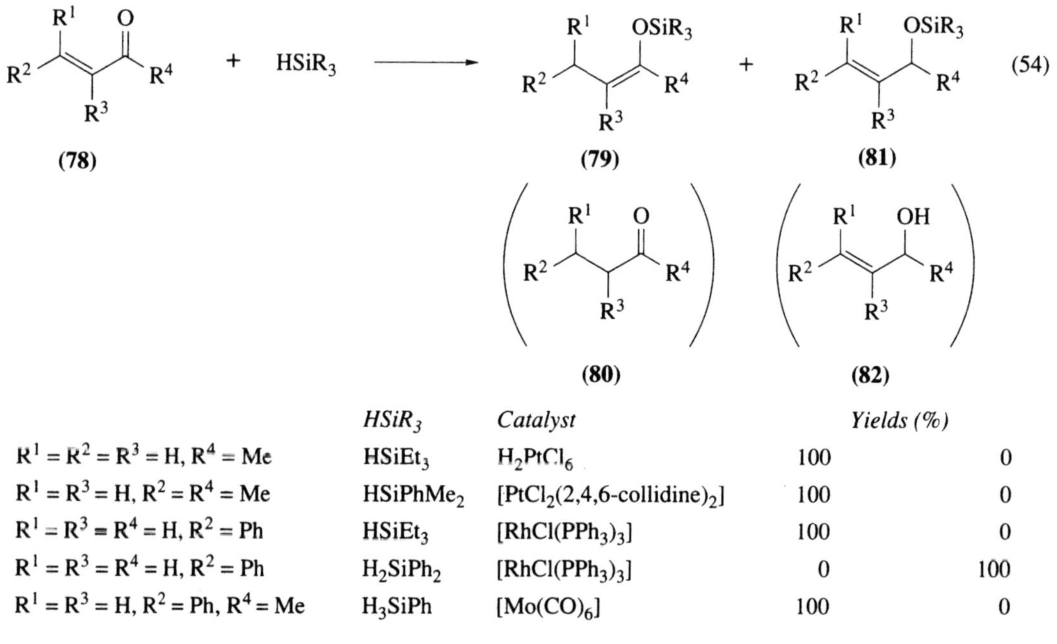

(78) + $HSiR_3$ ⟶ (79) + (81) (54)

(80) (82)

	$HSiR_3$	Catalyst	Yields (%)	
$R^1 = R^2 = R^3 = H, R^4 = Me$	$HSiEt_3$	H_2PtCl_6	100	0
$R^1 = R^3 = H, R^2 = R^4 = Me$	$HSiPhMe_2$	[$PtCl_2$(2,4,6-collidine)$_2$]	100	0
$R^1 = R^3 = R^4 = H, R^2 = Ph$	$HSiEt_3$	[$RhCl(PPh_3)_3$]	100	0
$R^1 = R^3 = R^4 = H, R^2 = Ph$	H_2SiPh_2	[$RhCl(PPh_3)_3$]	0	100
$R^1 = R^3 = H, R^2 = Ph, R^4 = Me$	H_3SiPh	[$Mo(CO)_6$]	100	0

Two modes of reaction are possible for α,β-unsaturated esters (**83**): hydrosilylation of the alkenic moiety to give (**84**) and/or (**85**) and 1,4-reduction to give ketene silyl acetals (**86**) (or saturated esters **87** after aqueous work-up), as illustrated in equation (55). The parent acrylates undergo hydrosilylation, whereas substituted acrylates favor the formation of (**86**) (or **87**).[102,103]

Alkynyl ketones (**88**) behave like substituted alkynes (**29**). Thus, the triple bond reacts first to give vinylsilanes (**89**) and/or (**90**) (equation 56), which are sometimes reduced further to allyl silyl ethers (**91**) and/or (**92**).[104]

As an alternative route to the reduction of α,β-unsaturated ketones accompanied by introduction of a silyl group at the β-position, a silylcuprate reagent is applicable (equation 57).[24b]

3.12.7 ASYMMETRIC HYDROSILYLATION OF CARBON–CARBON DOUBLE BONDS

Hydrosilylation of monosubstituted alkenes with palladium catalysts and trichlorosilane follows a course which favors branched products. By using a chiral phosphine ligand, asymmetric reaction is feasible. Initially, menthyldiphenylphosphine (MDPP, **93**) and neomenthyldiphenylphosphine (NMDPP, **94**) were employed with little success.[105] Later, (*R*)-*N*,*N*-dimethyl-1-[(*S*)-2-diphenylphosphinoferrocenyl]ethylamine [(*R*)-(*S*)-PPFA] (**95**) and its enantiomer were prepared, and these have proved to be the

(55)

(83) + H–Si → (84) + (85) + (86) (87)

(83)	H–Si	Catalyst	Yields (%) (84)	(85)	(86)	(87)
CH_2=$CHCO_2Me$	$HSiMe_3$	H_2PtCl_6	33	4	–	22
CH_2=$C(Me)CO_2Me$	$HSiMe_3$	H_2PtCl_6	–	69	–	–
MeCH=$CHCO_2Me$	$HSiMe_3$	H_2PtCl_6	17	–	–	28(54)*
CH_2=$CHCO_2Et$	$HSiMe_2Ph$	$[RhCl(PPh_3)_3]$	–	85	–	
CH_2=$CHCO_2Et$	$HSiMe_2Cl$	$[RhCl(PPh_3)_3]$	65	–	–	
MeCH=$CHCO_2Me$	$HSiMe_2Ph$	$[RhCl(PPh_3)_3]$	–	–	70	

*$MeCH_2CH(CO_2Me)CHMeCH_2CO_2Me$

(56)

(88) —$HSiEt_3$, H_2PtCl_6→ (89) + (90) ((91), (92))

(88)		(89)	:	(90)	(91)/(92)
HC≡CCOMe	70%	0.9	:	1	(<15%)
Me_3SiC≡CCOMe	39%	5	:	1	(<15%)

(57)

i, $(PhMe_2Si)_2CuLi$; ii, H_2O → 3-methyl-3-($SiMe_2Ph$)cyclohexanone, 99%

best ligands so far for asymmetric hydrosilylation.[106] Thus, styrene is hydrosilylated with trichlorosilane and the palladium catalyst (**96**) to give trichloro-1-phenylethylsilane with moderately good asymmetric induction (Scheme 12). The absolute configuration and the optical purity of the product was determined by conversion into 1-phenylethanol.[107]

The transformation of conjugated dienes to allylsilanes can also be carried out with asymmetric induction. The efficiency of the asymmetric synthesis for acyclic dienes is moderate, but for cyclic dienes it remains to be improved.[108,109] Nevertheless, these reactions provide us with a method for the preparation

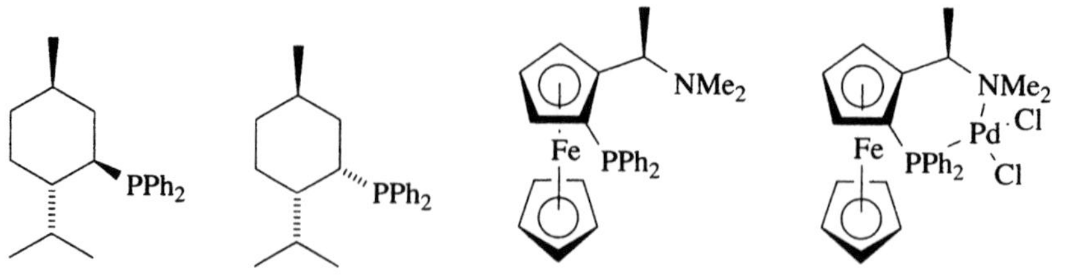

MDPP (**93**) NMDPP (**94**) *(R)-(S)*-PPFA (**95**) *(R)-(S)*-PPFA-Pd (**96**)

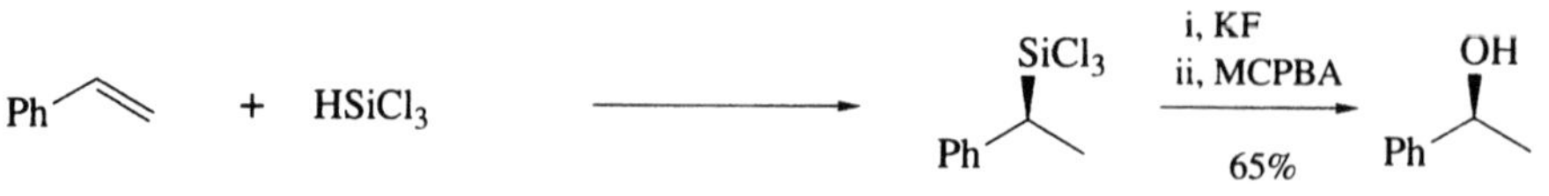

$[PdCl_2(MDPP)_2]$	87%	34% *ee (S)*
$[PdCl_2(NMDPP)_2]$	87%	22% *ee (R)*
(R)-(S)-PPFA-Pd (**96**)	>95%	52% *ee (S)*

Scheme 12

of a chiral allylsilane (**97**; Scheme 13), which has been useful in the study of stereochemical aspects of the reactions of organosilicon compounds.

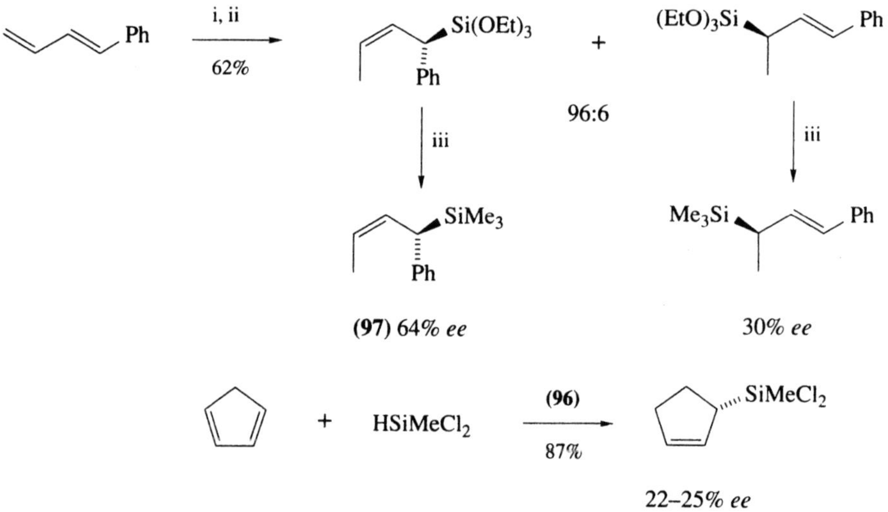

i, (**96**), $HSi(OEt)_3$; ii, NEt_3, EtOH; iii, MeMgBr

Scheme 13

Although asymmetric 1,4-reduction of conjugated enones is recorded,[110] the enantiomeric excess is not high (equation 58).[111]

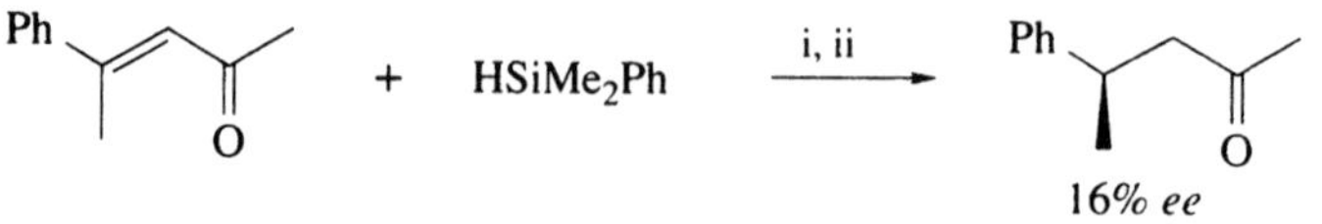

(58)

16% *ee*

i, $[Rh\{(R)\text{-}PMePh(CH_2Ph)\}_2H_2S_2]^+ClO_4^-$; ii, H_2O

Conjugate addition of a silylcuprate reagent (**34**) to the α,β-unsaturated acid amide or ester of a pertinent chiral amine or alcohol, respectively, is an alternative route to β-silyl amides and esters. Conjugate addition of a cuprate reagent to a β-silyl acrylamide also gives a chiral β-silyl amide. The diastereomeric excess of the newly produced chiral center is fairly high, as summarized in Scheme 14.[112]

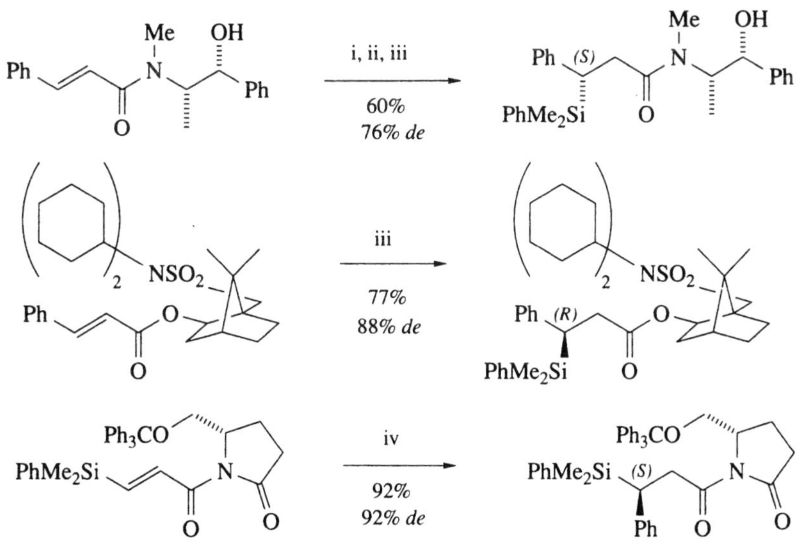

i, Bu^nLi; ii, $TiCl_4$; iii, $(PhMe_2Si)_2CuLi_2CN$ (**34**); iv, PhMgBr, $CuBr{\bullet}Me_2S$

Scheme 14

3.12.8 SYNTHETIC EXTENSION AND APPLICATION

The addition of Si and H across a π-bond is formally a reduction, since Si is more electropositive than carbon. However, unambiguous reduction is only complete when the silicon is replaced by hydrogen. This is easily possible by the protodesilylation of allylsilanes and vinylsilanes.

3.12.8.1 Reaction of Vinylsilanes with Electrophiles

The C—Si bond of vinylsilanes is transformed to C—H by hydrogen iodide with retention of configuration (Scheme 15).[1] The proton attaches itself to the carbon which has the silyl group, because the resulting carbonium ion is stabilized by the silyl group (β-effect). The silyl group is then attacked by a nucleophile to give the alkene. A silicon-bridged carbonium ion (**98**) is sometimes assumed to explain the stereochemical outcome (equation 59).[113] The overall reaction is thus unambiguously reduction of the triple bond to a double bond, but the reaction scheme is applicable to various other electrophiles.[114]

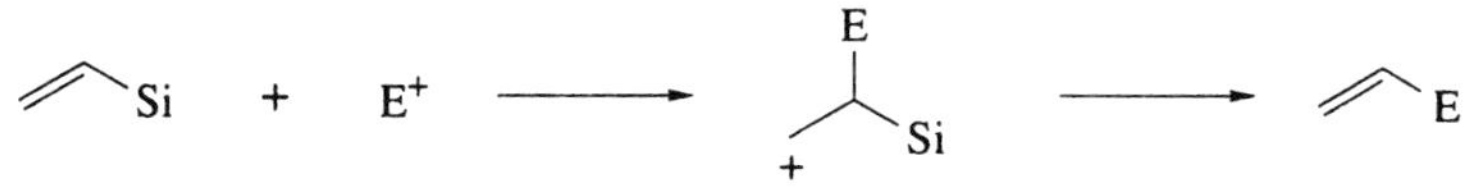

Scheme 15

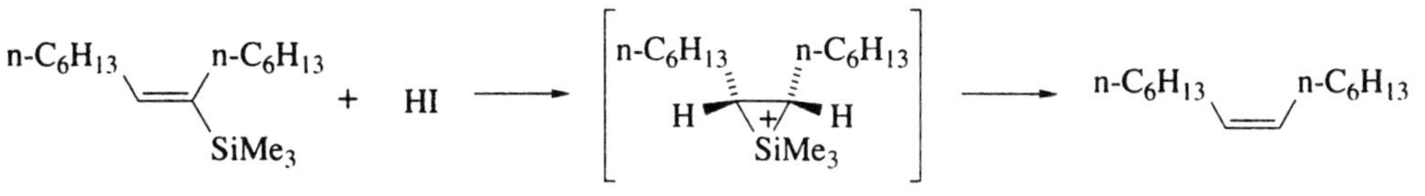

(59)

(98)

Bromine and chlorine react with vinylsilanes to afford vinyl bromide and chloride with net inversion of configuration. Addition of these halogens proceeds with *anti* stereochemistry. Elimination of halosilane in the presence of a nucleophile like F^- or RO^- also is assumed to take place in an *anti* manner (Scheme 16). The same transformation using iodine is applicable only to 1,2-dialkylvinylsilanes.[115] The process is a reliable method for the preparation of vinyl halides of defined configuration. The reactions are not, overall, reductions, and they are included here only to emphasize the usefulness of the products of hydrosilylation.

The reactions of vinylsilanes with carbon electrophiles is discussed in Volumes 2–4.

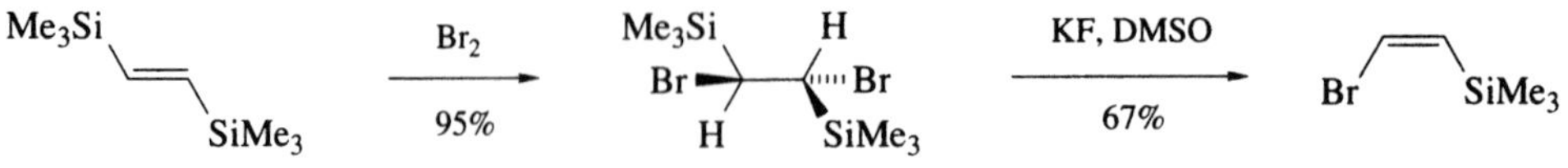

Me3Si, SiMe3 —Br2, 95%→ Me3Si, H, Br, Br, H, SiMe3 —KF, DMSO, 67%→ Br, SiMe3

Scheme 16

3.12.8.2 Coupling Reactions of Vinylsilanes with Organic Halides

In the presence of a fluoride ion generating reagent such as tris(diethylamino)sulfonium difluorotrimethylsilicate (TASF, **99**) and a palladium catalyst, vinyltrimethylsilane couples with vinyl and aryl iodides.[116] For substituted vinylsilanes, the fluorodimethylsilyl group is particularly effective in the coupling reaction, which takes place with retention of configuration of both substrates (Scheme 17). Silicon-based C—C bond formation is as useful as that which employs organoborons[117] for the synthesis of stereo-defined conjugated dienes.

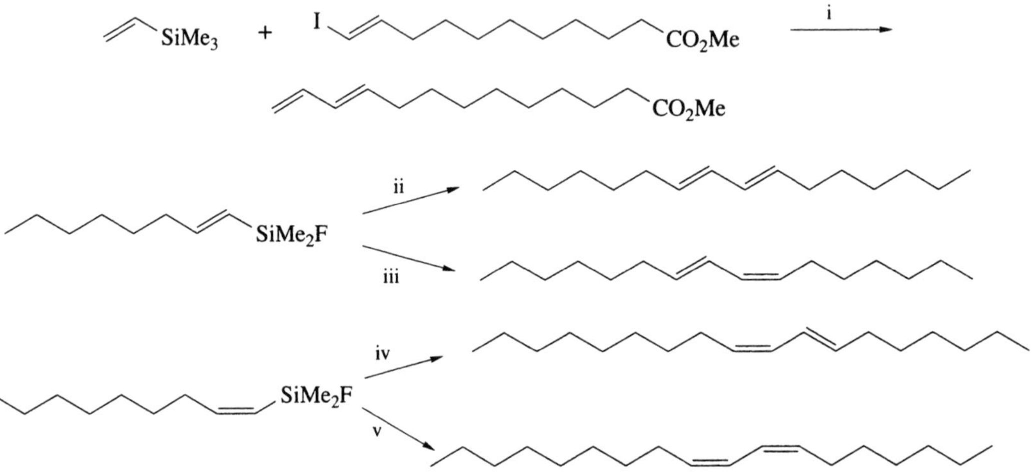

SiMe3 + I ... CO_2Me —i→ CO_2Me

$SiMe_2F$ —ii→ ; —iii→

$SiMe_2F$ —iv→ ; —v→

i, [{(allyl)PdCl}$_2$], P(OEt)$_3$, THF, **(99)**; ii, [{(allyl)PdCl}$_2$], THF, **(99)**, *(E)*-1-iodo-1-octene; iii, [{(allyl)PdCl}$_2$], THF, **(99)**, *(Z)*-1-iodo-1-octene; iv, [Pd(PPh$_3$)$_4$], DMF, **(99)**, *(E)*-1-iodo-1-octene; v, [Pd(PPh$_3$)$_4$], DMF, **(99)**, *(Z)*-1-iodo-1-octene

Scheme 17

3.12.8.3 Reaction of Allylsilanes with Electrophiles

Allylsilanes are easily protodesilylated, completing the reduction process. In the presence of Lewis acids, allylsilanes also react with electrophiles like aldehydes, ketones and acid halides through an S_E2' mechanism involving *anti* stereochemistry. These reactions are extensively discussed in Volumes 2–4.

3.12.8.4 Aldol Reactions

Enol silyl ethers undergo aldol reactions with aldehydes, acetals and their equivalents with the aid of a Lewis acid catalyst. These reactions are discussed in Volumes 2–4. Enol silyl ethers prepared by hydrosilylation of α,β-unsaturated ketones with a rhodium catalyst can be used for aldol reactions with aldehydes or ketones *in situ* under neutral conditions (equation 60).[118]

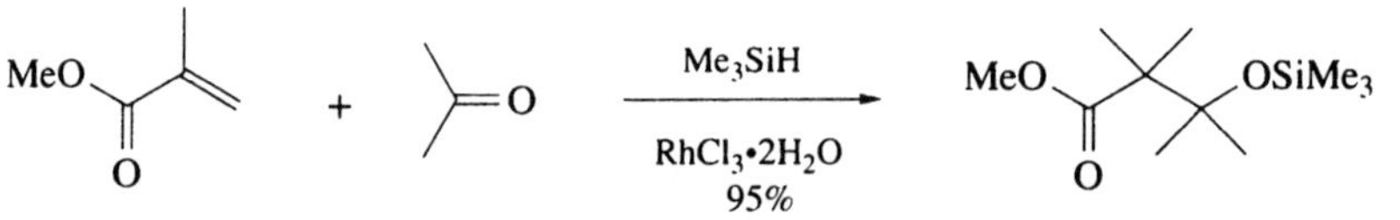

MeO, O + O —Me_3SiH; $RhCl_3{\cdot}2H_2O$; 95%→ MeO, O, $OSiMe_3$ (60)

3.12.8.5 Synthetic Reactions of Pentafluorosilicates

Trichlorosilanes are readily transformed to the pentafluorosilicate compounds (**100**) and (**101**) by reaction with excess potassium fluoride. The C—Si bond of the hexavalent organosilicon species is converted into a C—X (X = halogen or heteroatom like hydroxy and alkoxy) or a C—C bond. Thus, through the hydrosilylation reaction net transformations of alkenes and alkynes to halides, ethers and alcohols, as well as the construction of carbon frameworks, is easily achieved (Scheme 18).[119]

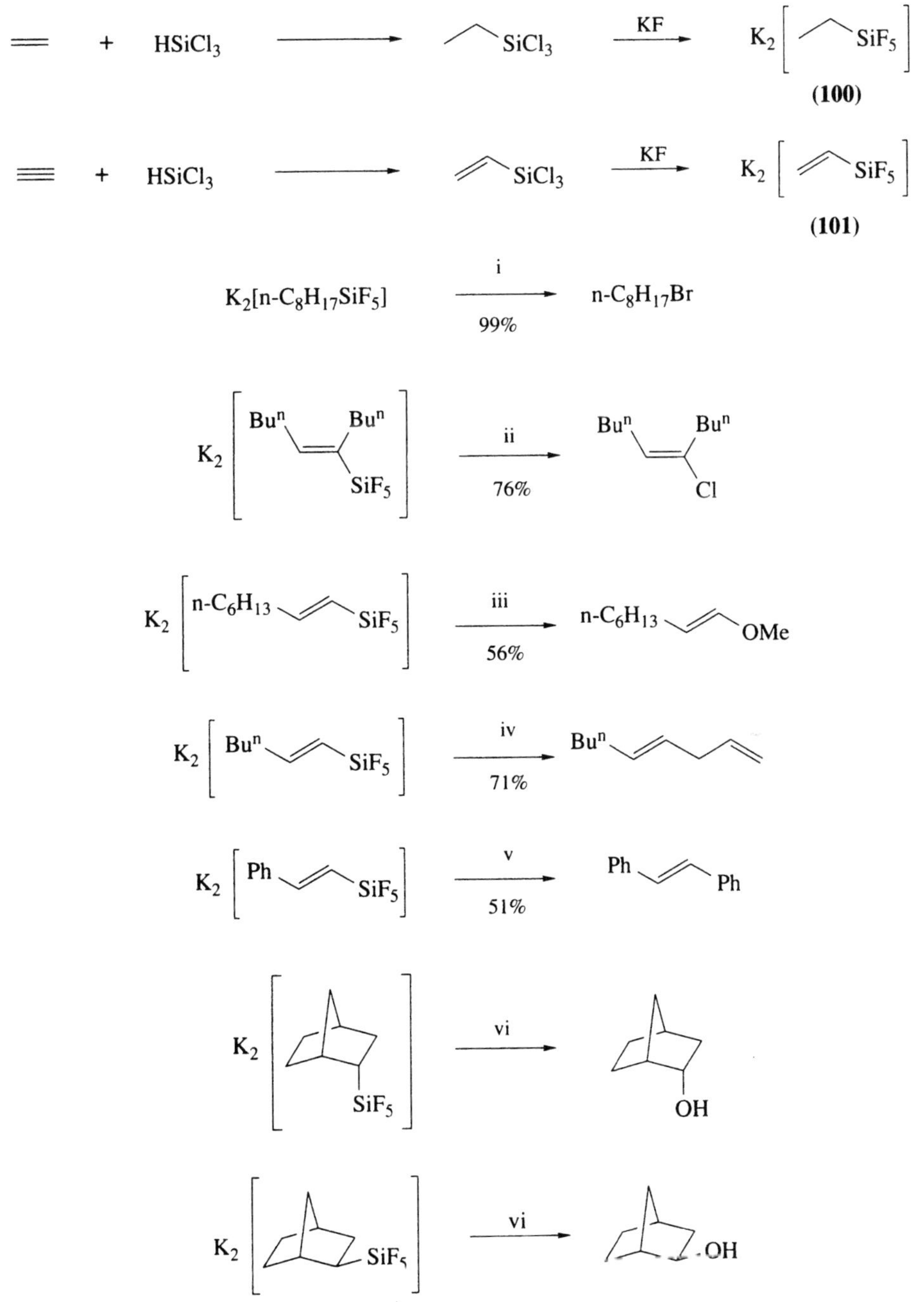

═ + $HSiCl_3$ ⟶ $SiCl_3$ —KF→ K_2[SiF_5] (**100**)

≡ + $HSiCl_3$ ⟶ $SiCl_3$ —KF→ K_2[SiF_5] (**101**)

$K_2[n\text{-}C_8H_{17}SiF_5]$ —i, 99%→ $n\text{-}C_8H_{17}Br$

K_2[Bu^n Bu^n SiF_5] —ii, 76%→ Bu^n Bu^n Cl

K_2[$n\text{-}C_6H_{13}$ SiF_5] —iii, 56%→ $n\text{-}C_6H_{13}$ OMe

K_2[Bu^n SiF_5] —iv, 71%→ Bu^n

K_2[Ph SiF_5] —v, 51%→ Ph Ph

K_2[SiF_5] —vi→ OH

K_2[SiF_5] —vi→ OH

i, NBS; ii, $CuCl_2$; iii, MeOH, $Cu(OAc)_2$; iv, CH_2=$CHCH_2Cl$, $Pd(OAc)_2$;
v, PhI, $Pd(OAc)_2$, PPh_3, NEt_3; vi, MCPBA

Scheme 18

3.12.8.6 Oxidative Transformation of C—Si Bonds

Modification of the last reaction in the previous section has led to a new version of the C—Si to C—O transformation that is highly versatile in organic synthesis. Hydrosilylation of alkenes or alkynes with a hydrosilane of type $HSiR_nX_{3-n}$ (R = organic group, X = H, F, Cl, R^1O, $R^1{}_2N$), followed by oxidation with 30% hydrogen peroxide in the presence of a base such as KF, potassium hydroxide or potassium hydrogencarbonate, leads to alcohols, ketones and acetals.[120] Typical examples are summarized in Scheme 19. This transformation is applicable to intramolecular hydrosilylation, which allows us to introduce contiguous chiral centers in a predictable manner, starting with a properly substituted alkenol, *e.g.* (**102**) and (**103**), as shown in Scheme 20.[121]

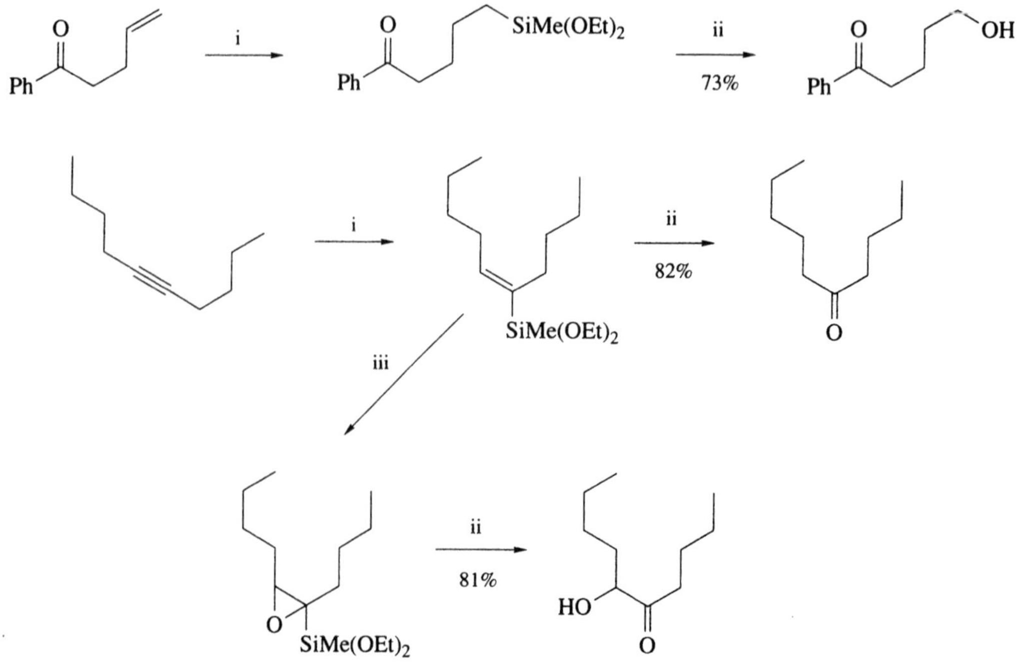

i, $HSiMe(OEt)_2$, H_2PtCl_6; ii, 30% H_2O_2, $KHCO_3$; iii, MCPBA

Scheme 19

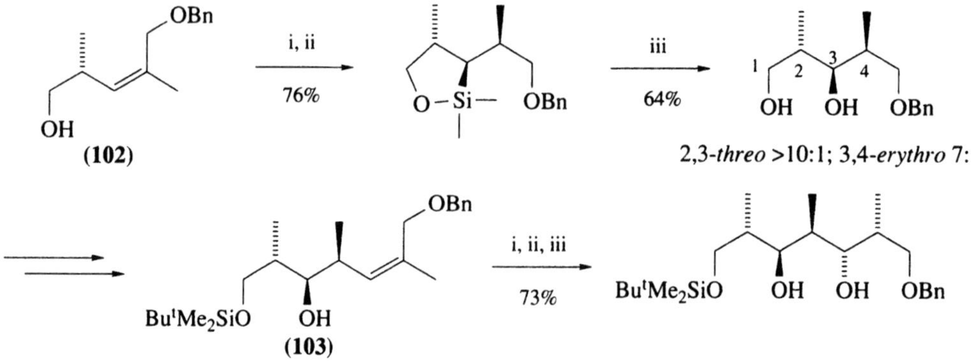

2,3-*threo* >10:1; 3,4-*erythro* 7:1

i, $(HMe_2Si)_2NH$, r.t. to 60 °C; ii, $H_2PtCl_6{\cdot}6H_2O$ (0.1 mol %), 60 °C; iii, H_2O_2, KF, $KHCO_3$, MeOH–THF, r.t.

Scheme 20

3.12.8.7 Hydrosilylation in the Presence of CO

The mechanism of hydrosilylation on a transition metal catalyst involves an intermediate with a carbon–transition metal bond. Carbon monoxide, if present, inserts into the carbon–metal bond to give an acyl metal complex, which can undergo enolization and silylation. A one-carbon extended aldehyde enol silyl ether is produced, useful for further synthetic transformations. Dicobalt octacarbonyl is the best-studied catalyst (Scheme 21).[122] A ruthenium cluster catalyst exhibits similar reactivity (equation 61),[11d] but rhodium carbonyl catalyzes the reaction only of alkynes to afford α,β-unsaturated aldehydes containing a β-silyl group (equation 62).[123]

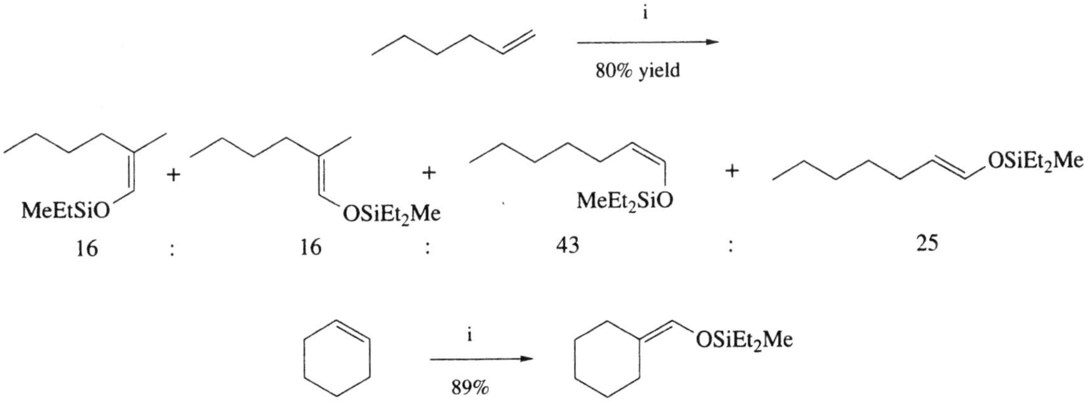

i
80% yield

MeEtSiO + OSiEt$_2$Me + MeEt$_2$SiO + OSiEt$_2$Me

16 : 16 : 43 : 25

i
89%
OSiEt$_2$Me

i, $HSiEt_2Me$, CO, $[Co_2(CO)_8]$, 140 °C

Scheme 21

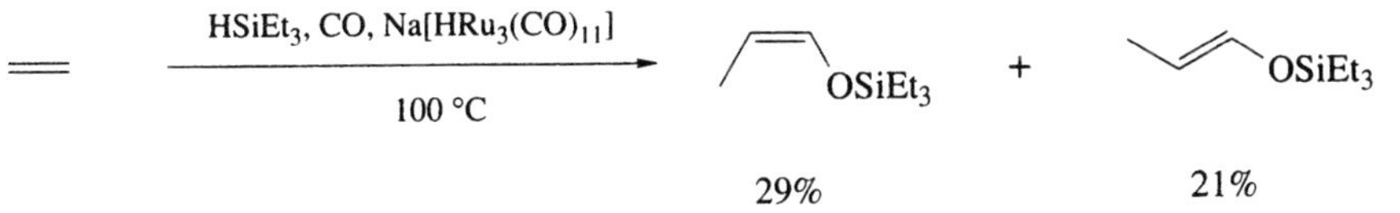

$HSiEt_3$, CO, $Na[HRu_3(CO)_{11}]$
100 °C
OSiEt$_3$ + OSiEt$_3$ (61)
29% 21%

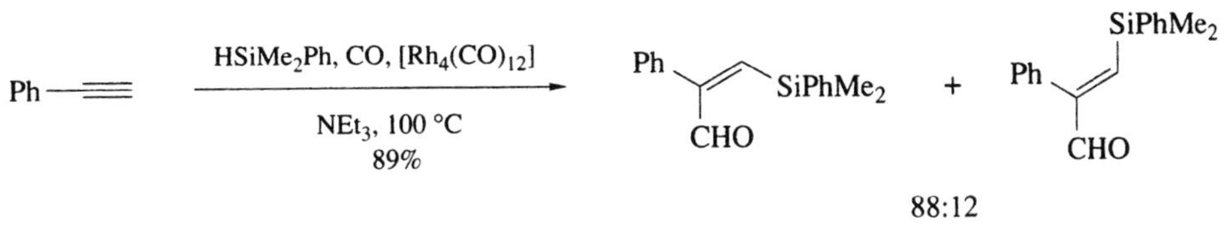

Ph—≡ $HSiMe_2Ph$, CO, $[Rh_4(CO)_{12}]$
NEt_3, 100 °C
89%
Ph SiPhMe$_2$ CHO + Ph SiPhMe$_2$ CHO (62)
88:12

3.12.9 REFERENCES

1. (a) J. F. Harrod and A. J. Chalk, in 'Organic Syntheses via Metal Carbonyls' ed. I. Wender and P. Pino, Wiley, New York, 1977, vol. 2; (b) E. Lukevics, Z. V. Belyakova, M. G. Pomerantseva, M. G. Voronkov, *J. Organomet. Chem. Libr.*, 1977, **5**, 1; (c) J. L. Speier, *Adv. Organomet. Chem.*, 1979, **17**, 407; (d) I. Ojima and T. Kogure, *Rev. Silicon, Germanium, Tin Lead Compd.*, 1981, **5**, 8; (e) C. Eaborn and R. W. Bott, 'Organometallic Compounds of the Group IV Elements', ed. A. G. MacDiarmid, Dekker, New York, 1968, vol. 1, part 1; (f) I. Ojima, in 'The Chemistry of Organic Silicon Compounds', ed. S. Patai and Z. Rappoport, Wiley, 1989, part 2, chap. 25.
2. (a) T. G. Selin and R. West, *J. Am. Chem. Soc.*, 1962, **84**, 1860; (b) R. A. Benkeser, M. L. Burrous, L. E. Nelson and J. V. Swisher, *J. Am. Chem. Soc.*, 1961, **83**, 4385.
3. J. L. Speier, J. A. Webster and G. H. Barnes, *J. Am. Chem. Soc.*, 1957, **79**, 974.
4. (a) I. G. Iovel, Y. Sh. Goldberg, M. V. Shymanska and E. Lukevics, *Organometallics*, 1987, **6**, 1410; (b) A. Onopchenko and E. T. Sabourin, *J. Org. Chem.*, 1987, **52**, 4118; (c) B.-H. Han and P. Boudjouk, *Organometallics*, 1983, **2**, 769.
5. H. Watanabe, M. Asami and Y. Nagai, *J. Organomet. Chem.*, 1980, **195**, 363.
6. (a) D. C. Apple, K. A. Brady, J. M. Chance, N. E. Heard and T. A. Nile, *J. Mol. Catal.*, 1985, **29**, 55; (b) L. A. Oro, M.-J. Fernandez, M. A. Esteruelas and M. S. Jimenez, *J. Mol. Catal.*, 1986, **37**, 151.
7. A. D. Calhoun, K. R. Lung, T. A. Nile, L. L. Stokes and S. C. Smith, *Transition Met. Chem.*, 1983, **8**, 365.
8. (a) B. Marciniec and J. Guli'ński, *J. Mol. Catal.*, 1981, **10**, 123; (b) B. Marciniec and J. Guli'ński, *J. Organomet. Chem.*, 1983, **253**, 349.
9. M. Brockmann, H. tom Dieck and I. Kleinwächter, *J. Organomet. Chem.*, 1986, **309**, 345.
10. Y. Kiso, M. Kumada, K. Tamao and M. Umeno, *J. Organomet. Chem.*, 1973, **50**, 297.
11. (a) G. K. I. Magomedov, G. V. Druzhkova, V. G. Syrkin and O. V. Shkol'nik, *Koord. Khim.*, 1980, **6**, 767 (*Chem. Abstr.*, 1980, **93**, 167 148m); (b) G. K. I. Magomedov, K. A. Andrianov, O. V. Shkol'nik, B. A. Izmailov and V. N. Kalinin, *J. Organomet. Chem.*, 1978, **149**, 29; (c) I. Ojima, T. Fuchikami and M. Yatabe,

J. Organomet. Chem., 1984, **260**, 335; (d) G. Süss-Fink and J. Reiner, *J. Mol. Catal.*, 1982, **16**, 231; (e) G. Süss-Fink, *Angew. Chem., Int. Ed. Engl.*, 1982, **21**, 73; (f) N. A. Kuz'mina, L. V. Il'inskaya, R. G. Gasanov, E. Ts. Chukovskaya and R. Kh. Freidlina, *Izv. Akad. Nauk SSSR, Ser. Khim.*, 1986, 212 (*Chem. Abstr.*, 1987, **106**, 33 155q); (g) M. S. Wrighton and M. A. Schroeder, *J. Am. Chem. Soc.*, 1974, **96**, 6235.

12. M. G. Voronkov, S. N. Adamovich, L. V. Sherstyannikova and V. B. Pukhnarevich, *J. Gen. Chem. USSR (Engl. Transl.)*, 1983, **53**, 706.
13. (a) Z. M. Michalska and B. Ostaszewski, *J. Organomet. Chem.*, 1986, **299**, 259; (b) B. Marciniec, W. Urbaniak and P. Pawlak, *J. Mol. Catal.*, 1982, **14**, 323; (c) V. S. Brovko, N. K. Skvortsov and V. O. Reikhsfel'd, *J. Gen. Chem. USSR (Engl. Transl.)*, 1981, **51**, 335; (d) A. L. Prignano and W. C. Troger, *J. Am. Chem. Soc.*, 1987, **109**, 3586.
14. C. L. Reichel and M. S. Wrighton, *J. Am. Chem. Soc.*, 1981, **103**, 7180.
15. M. C. Musolf and J. L. Speier, *J. Org. Chem.*, 1964, **29**, 2519.
16. K. Y. Choo and P. P. Gaspar, *J. Am. Chem. Soc.*, 1974, **96**, 1284.
17. J. F. Harrod and A. J. Chalk, *J. Am. Chem. Soc.*, 1964, **86**, 1776; A. J. Chalk and J. F. Harrod, *J. Am. Chem. Soc.*, 1965, **87**, 16; J. F. Harrod and A. J. Chalk, *J. Am. Chem. Soc.*, 1965, **87**, 1133; J. F. Harrod and A. J. Chalk, *J. Am. Chem. Soc.*, 1966, **88**, 3491.
18. M. Green, J. A. K. Howard, J. Proud, J. L. Spencer, F. G. A. Stone and C. A. Tsipis, *J. Chem. Soc., Chem. Commun.*, 1976, 671.
19. L. H. Sommer, J. E. Lyons and H. Fujimoto, *J. Am. Chem. Soc.*, 1969, **91**, 7051.
20. C. Eaborn, D. J. Tune and D. R. M. Walton, *J. Chem. Soc., Dalton Trans.*, 1973, 2255; F. Carré, E. Colomer, R. J. P. Corriu and A. Vioux, *Organometallics*, 1984, **3**, 1272.
21. F. Seitz and M. S. Wrighton, *Angew. Chem., Int. Ed. Engl.*, 1988, **27**, 289.
22. C. L. Randolph and M. S. Wrighton, *J. Am. Chem. Soc.*, 1986, **108**, 3366.
23. R. A. Benkeser, K. M. Foley, J. B. Grutzner and W. E. Smith, *J. Am. Chem. Soc.*, 1970, **92**, 697; S. C. Bernstein, *J. Am. Chem. Soc.*, 1970, **92**, 699.
24. (a) I. Fleming, T. W. Newton and F. Roessler, *J. Chem. Soc., Perkin Trans. 1*, 1981, 2527; (b) D. J. Ager, I. Fleming and S. K. Patel, *J. Chem. Soc., Perkin Trans. 1*, 1981, 2520.
25. M. G. Voronkov and S. P. Sushchinskaya, *J. Gen. Chem. USSR (Engl. Transl.)*, 1986, **56**, 555.
26. C.-Y. Hu, X.-M. Han and Y.-Y. Jiang, *J. Mol. Catal.*, 1986, **35**, 329.
27. (a) E. A. Chernyshev and S. A. Shchepinov, *J. Gen. Chem. USSR (Engl. Transl.)*, 1970, **40**, 1732; (b) E. A. Chernyshev, N. G. Komalenkova, V. A. Sharapov, A. I. Gusev and A. V. Kisin, *J. Gen. Chem. USSR (Engl. Transl.)*, 1987, **57**, 1544.
28. G. Greber and M. L. Hallensleben, *Makromol. Chem.*, 1967, **104**, 77; *Angew. Chem.*, 1965, **77**, 511.
29. E. Husemann, *Ger. Pat.* 1 167 031 (*Chem. Abstr.*, 1964, **61**, 10 803b).
30. A. M. Polyakova, M. D. Suchkova, V. V. Korshak and V. M. Vdovin, *Izv. Akad. Nauk SSSR, Ser. Khim.*, 1965, 1267 (*Chem. Abstr.*, 1965, **63**, 13 306b).
31. R. A. Benkeser, R. F. Cunico, S. Dunny, P. R. Jones and P. G. Nerlekar, *J. Org. Chem.*, 1967, **32**, 2634.
32. J. Yoshida, K. Tamao, M. Takahashi and M. Kumada, *Tetrahedron Lett.*, 1978, 2161.
33. V. B. Pukhnarevich, L. I. Kopylova, B. A. Trofimov and M. G. Voronkov, *J. Gen. Chem. USSR (Engl. Transl.)*, 1975, **45**, 2600.
34. M. G. Voronkov, V. B. Pukhnarevich, I. I. Tsykhanskaya, N. I. Ushakova, Yu. L. Gaft and I. A. Zakharova, *Inorg. Chim. Acta*, 1983, **68**, 103.
35. C. A. Tsipis, *J. Organomet. Chem.*, 1980, **188**, 53.
36. C. A. Tsipis, *J. Organomet. Chem.*, 1980, **187**, 427.
37. I. Ojima, M. Kumagai and Y. Nagai, *J. Organomet. Chem.*, 1974, **66**, C14.
38. V. B. Pukhnarevich, L. I. Kopylova, M. Čapka, J. Hetflejš, É. N. Satsuk, M. V. Sigalov, V. Chvalovský and M. G. Voronkov, *J. Gen. Chem. USSR (Engl. Transl.)*, 1980, **50**, 1259.
39. H. Matsumoto, Y. Hoshino and Y. Nagai, *Chem. Lett.*, 1982, 1663.
40. K. A. Brady and T. A. Nile, *J. Organomet. Chem.*, 1981, **206**, 299; J. P. Howe, K. R. Lung and T. A. Nile, *J. Organomet. Chem.*, 1981, **208**, 401.
41. H. M. Dickers, R. N. Haszeldine, L. S. Malkin, A. P. Mather and R. V. Parish, *J. Chem. Soc., Dalton Trans.*, 1980, 308.
42. L. I. Kopylova, V. B. Pukhnarevich, I. I. Tsykhanskaya, É. N. Satsuk, B. V. Timokhin, V. I. Dmitriev, V. Chvalovský, M. Čapka, A. V. Kalabina and M. G. Voronkov, *J. Gen. Chem. USSR (Engl. Transl.)*, 1981, **51**, 1588.
43. G. Fritz, H. Wilhelm and A. Oleownik, *Z. Anorg. Allg. Chem.*, 1981, **478**, 97.
44. M. F. Lappert, T. A. Nile and S. Takahashi, *J. Organomet. Chem.*, 1974, **72**, 425.
45. E. Lukevics, O. A. Pudova, R. Ya. Sturkovich and A. P. Gaukhman, *J. Gen. Chem. USSR (Engl. Transl.)*, 1985, **55**, 1352.
46. S. I. Sadykh-Zade, I. A. Shikhiev and E. M. Kholilova, *Zh. Obshch. Khim.*, 1964, **34**, 1393.
47. N. P. Smetankina and N. I. Miryan, *Zh. Obshch. Khim.*, 1967, **37**, 1383.
48. E. Lukevics, R. Ya. Sturkovich and O. A. Pudova, *J. Organomet. Chem.*, 1985, **292**, 151.
49. J. W. Ryan and J. L. Speier, *J. Org. Chem.*, 1966, **31**, 2698.
50. P. F. Hudrlik, R. H. Schwartz and J. C. Hogan, *J. Org. Chem.*, 1979, **44**, 155.
51. M. F. Lappert and R. K. Maskell, *J. Organomet. Chem.*, 1984, **264**, 217.
52. K. Tamao, N. Miyake, Y. Kiso and M. Kumada, *J. Am. Chem. Soc.*, 1975, **97**, 5603.
53. H. Okinoshima, K. Yamamoto and M. Kumada, *J. Am. Chem. Soc.*, 1972, **94**, 9263.
54. M. D. Standnichuk and A. A. Petrov, *J. Gen. Chem. USSR (Engl. Transl.)*, 1962, **32**, 3449.
55. A. I. Kakhniashvili and D. Sh. Ioramashvili, *J. Gen. Chem. USSR (Engl. Transl.)*, 1970, **40**, 1539.
56. K. M. Rustamov and R. A. Sultanov, *Zh. Vses. Khim. Ova.*, 1985, **30**, 585 (*Chem. Abstr.*, 1986, **105**, 172 560k).
57. T. Sh. Gazarov, G. A. Saryev, R. A. Sultanov and Sh. A. Tarverdiev, *Zh. Vses. Khim. Ova.*, 1984, **29**, 232 (*Chem. Abstr.*, 1984, **101**, 38 532s).

58. D. Sh. Ioramashvili and A. I. Kakhniashvili, *Soobshch. Akad. Nauk. Gruz. SSR*, 1979, **94**, 349 (*Chem. Abstr.*, 1980, **92**, 58 855y).
59. A. I. Kakhniashvili, T. Sh. Gvaliya and D. Sh. Ioramashvili, *Soobshch. Akad. Nauk Gruz. SSR*, 1972, **68**, 609 (*Chem. Abstr.*, 1973, **78**, 97 755w).
60. I. M. Gverdtsiteli and T. G. Talakvadze, *Soobshch. Akad. Nauk Gruz. SSR*, 1975, **79**, 601 (*Chem. Abstr.*, 1976, **84**, 59 690n).
61. M. Licchelli and A. Greco, *Tetrahedron Lett.*, 1987, **28**, 3719.
62. H. Bock and H. Seidl, *J. Am. Chem. Soc.*, 1968, **90**, 5694.
63. T. Kusumoto and T. Hiyama, *Chem. Lett.*, 1985, 1405.
64. M. Čapka, P. Svoboda, V. Bažant and V. Chvalovský, *Collect. Czech. Chem. Commun.*, 1971, **36**, 2785.
65. A. J. Chalk, *J. Organomet. Chem.*, 1970, **21**, 207.
66. Y. L. Baay and A. G. MacDiarmid, *Inorg. Chem.*, 1969, **8**, 986.
67. Y. Kiso, M. Kumada, K. Maeda, K. Sumitani and K. Tamao, *J. Organomet. Chem.*, 1973, **50**, 311.
68. J. Tsuji, M. Hara and K. Ohno, *Tetrahedron*, 1974, **30**, 2143.
69. Yu. M. Lugovoi, A. G. Shostenko, V. E. Myshkin and V. P. Krapchatov, *Kinet. Katal.*, 1981, **22**, 1055.
70. (a) A. Millan, M.-J. Fernandez, P. Bentz and P. M. Maitlis, *J. Mol. Catal.*, 1984, **26**, 89; (b) Y. Seki, K. Takeshita, K. Kawamoto, S. Murai and N. Sonoda, *J. Org. Chem.*, 1986, **51**, 3890.
71. (a) A. Onopchenko, E. T. Sabourin and D. L. Beach, *J. Org. Chem.*, 1984, **49**, 3389; (b) K. Takeshita, Y. Seki, K. Kawamoto, S. Murai and N. Sonoda, *J. Org. Chem.*, 1987, **52**, 4864.
72. B. Marciniec, J. Guli'ński and W. Urbaniak, *Synth. React. Inorg. Metal-Org. Chem.*, 1982, **12**, 139.
73. B. Marciniec, E. Mać'kowska, J. Guli'ński and W. Urabaniak, *Z. Anorg. Allg. Chem.*, 1985, **529**, 222.
74. N. M. K. El-Durini and R. A. Jackson, *J. Organomet. Chem.*, 1982, **232**, 117.
75. H. Sakurai, T. Hirose and A. Hosomi, *J. Organomet. Chem.*, 1975, **86**, 197.
76. J. V. Swisher and H.-H. Chen, *J. Organomet. Chem.*, 1974, **69**, 83.
77. (a) Z. V. Belyakova, M. G. Pomerantseva and Z. V. Belikova, *J. Gen. Chem. USSR (Engl. Transl.)*, 1974, **44**, 2399; (b) Z. V. Belyakova, M. G. Pomerantseva and S. A. Golubtsov, *J. Gen. Chem. USSR (Engl. Transl.)*, 1965, **35**, 1053; (c) D. Cooper, R. N. Haszeldine and M. J. Newland, *J. Chem. Soc. A*, 1967, 2098.
78. (a) J. W. Ryan and J. L. Speier, *J. Am. Chem. Soc.*, 1964, **86**, 895; (b) T. G. Selin and R. West, *J. Am. Chem. Soc.*, 1962, **84**, 1863.
79. R. A. Benkeser and R. F. Cunico, *J. Organomet. Chem.*, 1966, **6**, 441.
80. G. Kuncová and V. Chvalovský, *Collect. Czech. Chem. Commun.*, 1980, **45**, 2085.
81. (a) K. Oertle and H. F. Wetter, *Tetrahedron Lett.*, 1985, **26**, 5511; (b) H. F. Wetter and K. Oertle, *Tetrahedron Lett.*, 1985, **26**, 5515; (c) K. Yamamoto and M. Takemae, *Synth. Lett.*, 1990, 259.
82. (a) E. Frainnet and R. Calas, *Bull. Soc. Chim. Fr.*, 1955, 760; (b) D. Wang and T. H. Chan, *Tetrahedron Lett.*, 1983, **24**, 1573.
83. D. C. Snyder, *J. Organomet. Chem.*, 1986, **301**, 137.
84. (a) D. N. Kursanov, Z. N. Parnes, and N. M. Loim, *Synthesis*, 1974, 633; (b) Z. N. Parnes, G. A. Khotimskaya, Yu. I. Lyakhovetsky and P. V. Petrovskii, *Izv. Akad. Nauk SSSR, Ser. Khim.* 1971, 1562 (*Chem. Abstr.*, 1971, **75**, 109 902v).
85. M. Kumada, K. Naka and Y. Yamamoto, *Bull. Chem. Soc. Jpn.*, 1964, **37**, 871.
86. K. I. Kobrakov, T. I. Chernysheva and N. S. Nametkin, *Dokl. Akad. Nauk SSSR*, 1971, **198**, 1340 (*Chem. Abstr.*, 1971, **75**, 129 908e).
87. D. L. Bailey and A. N. Pines, *Ind. Eng. Chem.*, 1954, **46**, 2363.
88. Z. V. Belyakova, M. G. Pomerantseva, K. K. Popkov, L. A. Efremova and S. A. Golubtsov, *J. Gen. Chem. USSR (Engl. Transl.)*, 1972, **42**, 879.
89. A. J. Cornish, M. F. Lappert, J. J. Macquitty and R. K. Maskell, *J. Organomet. Chem.*, 1979, **177**, 153.
90. L. D. Nasiak and H. W. Post, *J. Organomet. Chem.*, 1970, **23**, 91.
91. V. F. Mironov, N. G. Maksimova and V. V. Nepomnia, *Izv. Akad. Nauk SSSR, Ser. Khim.*, 1967, 329 (*Chem. Abstr.*, 1967, **67**, 32 719r).
92. I. Ojima, M. Kumagai and Y. Miyazawa, *Tetrahedron Lett.*, 1977, 1385.
93. A. J. Cornish, M. F. Lappert and T. A. Nile, *J. Organomet. Chem.*, 1977, **132**, 133.
94. A. J. Cornish, M. F. Lappert and T. A. Nile, *J. Organomet. Chem.*, 1977, **136**, 73.
95. K. Yamamoto and M. Kumada, *J. Organomet. Chem.*, 1968, **13**, 131.
96. H. Takahashi, H. Okita, M. Yamaguchi and I. Shiihara, *J. Org. Chem.*, 1963, **28**, 3353.
97. H. G. Kuivila and C. R. Warner, *J. Org. Chem.*, 1964, **29**, 2845.
98. Y. K. Kim. D. B. Bourrie and O. R. Pierch, *J. Polym. Sci., Polym. Chem. Ed.*, 1978, **16**, 483.
99. S. I. Sadykh-Zade and A. D. Petrov, *J. Gen. Chem. USSR (Engl. Transl.)*, 1959, **29**, 3159.
100. R. Rumin, *J. Organomet. Chem.*, 1983, **247**, 351.
101. (a) I. Ojima and T. Kogure, *Organometallics*, 1982, **1**, 1390; (b) E. Keinan and D. Perez, *J. Org. Chem.*, 1987, **52**, 2576.
102. E. Yoshii, Y. Kobayashi, T. Koizumi and T. Oribe, *Chem. Pharm. Bull.*, 1974, **22**, 2767.
103. I. Ojima, M. Kumagai and Y. Nagai, *J. Organomet. Chem.*, 1976, **111**, 43.
104. V. B. Pukhnarevich, I. I. Tsykhanskaya and M. G. Voronkov, *Izv. Akad. Nauk SSSR, Ser. Khim.*, 1984, 427 (*Chem. Abstr.*, 1984, **100**, 209 970u).
105. K. Yamamoto, Y. Kiso, R. Ito, K. Tamao and M. Kumada, *J. Organomet. Chem.*, 1981, **210**, 9.
106. T. Hayashi, K. Tamao, Y. Katsuro, I. Nakae and M. Kumada, *Tetrahedron Lett.*, 1980, **21**, 1871.
107. (a) Y. Kiso, K. Yamamoto, K. Tamao and M. Kumada, *J. Am. Chem. Soc.*, 1972, **94**, 4373; (b) K. Yamamoto, T. Hayashi, Y. Uramoto, R. Ito and M. Kumada, *J. Organomet. Chem.*, 1976, **118**, 331; (c) W. R. Cullen and N. F. Han, *J. Organomet. Chem.*, 1987, **333**, 269; (d) T. Hayashi, Y. Matsumoto, I. Morikawa and Y. Ito, *Tetrahedron: Asymmetry*, 1990, **1**, 151; (e) T. Hayashi, S. Hengrasmee and Y. Matsumoto, *Chem. Lett.*, 1990, 1377.
108. T. Hayashi and K. Kabeta, *Tetrahedron Lett.*, 1985, **26**, 3023.
109. T. Hayashi, K. Kabeta, T. Yamamoto, K. Tamao and M. Kumada, *Tetrahedron Lett.*, 1983, **24**, 5661.

110. T. Hayashi, K. Yamamoto and M. Kumada, *Tetrahedron Lett.*, 1975, 3.
111. I. Ojima and K. Hirai, *Asymmetric Synth.*, 1985, **5**, 102.
112. (a) W. Oppolzer, R. J. Mills, W. Pachinger and T. Stevenson, *Helv. Chim. Acta*, 1986, **69**, 1542; (b) I. Fleming and N. D. Kindon, *J. Chem. Soc., Chem. Commun.*, 1987, 1177.
113. K. Utimoto, M. Kitai and H. Nozaki, *Tetrahedron Lett.*, 1975, 2825.
114. E. W. Colvin, 'Silicon in Organic Synthesis', Butterworths, London, 1981.
115. J.-P. Pillot, J. Dunoguès and R. Calas, *Synth. Commun.*, 1979, **9**, 395.
116. Y. Hatanaka and T. Hiyama, *J. Org. Chem.*, 1988, **53**, 918; 1989, **54**, 268; *Yuki Gosei Kagaku Kyokai Shi*, 1990, **48**, 834.
117. (a) N. Miyaura and A. Suzuki, *Yuki Gosei Kagaku Kyokai Shi*, 1988, **46**, 848; (b) A. Suzuki, *Pure Appl. Chem.*, 1985, **57**, 1749; 1986, **58**, 629.
118. (a) S. Sato, I. Matsuda and Y. Izumi, *Tetrahedron Lett.*, 1986, **27**, 5517; (b) A. Revis and T. K. Hilty, *Tetrahedron Lett.*, 1987, **28**, 4809.
119. (a) K. Tamao, J. Yoshida, H. Yamamoto, T. Kakui, H. Matsumoto, M. Takahashi, A. Kurita, M. Murata and M. Kumada, *Organometallics*, 1982, **1**, 355; (b) J. Yoshida, K. Tamao, T. Kakui, A. Kurita, M. Murata, K. Yamada and M. Kumada, *Organometallics*, 1982, **1**, 369; (c) J. Yoshida, K. Tamao, H. Yamamoto, T. Kakui, T. Uchida and M. Kumada, *Organometallics*, 1982, **1**, 542; (d) K. Tamao, T. Kakui, M. Akita, T. Iwahara, R. Kanatani, J. Yoshida and M. Kumada, *Tetrahedron*, 1983, **39**, 983.
120. (a) K. Tamao, *Yuki Gosei Kagaku Kyokai Shi*, 1988, **46**, 861; (b) K. Tamao, N. Ishida, T. Tanaka and M. Kumada, *Organometallics*, 1983, **2**, 1694; (c) K. Tamao and K. Maeda, *Tetrahedron Lett.*, 1986, **27**, 65; (d) I. Fleming and P. E. J. Sanderson, *Tetrahedron Lett.*, 1987, **28**, 4229; (e) I. Fleming, R. Henning and H. Plaut, *J. Chem. Soc., Chem. Commun.*, 1984, 29.
121. K. Tamao, T. Nakajima, R. Sumiya, H. Arai, N. Higuchi and Y. Ito, *J. Am. Chem. Soc.*, 1986, **108**, 6090.
122. S. Murai and N. Sonoda, *Angew. Chem., Int. Ed. Engl.*, 1979, **18**, 837.
123. (a) I. Matsuda, A. Ogiso, S. Sato and Y. Izumi, *J. Am. Chem. Soc.*, 1989, **111**, 2332; (b) I. Matsuda, A. Ogiso and S. Sato, *J. Am. Chem. Soc.*, 1990, **112**, 6120.

4.1
Reduction of Saturated Alkyl Halides to Alkanes

TSUNEO IMAMOTO
Chiba University, Japan

4.1.1 INTRODUCTION AND SCOPE OF THE REVIEW

The reduction of saturated alkyl halides to alkanes, as represented in equation (1), is the most fundamental reaction of reductive dehalogenations of organic halides. The importance of these reductions has stimulated considerable investigation, and a number of successful approaches have been reported hitherto. Numerous reducing agents or reagent systems are available and many of them have been applied to practical organic synthesis with notable success.

$$\mathrm{R{-}X} \xrightarrow{\text{reducing agent}} \mathrm{R{-}H} \qquad (1)$$

R = alkyl group; X = F, Cl, Br, I

In 1980, Pinder presented an excellent review dealing with the hydrogenolysis of organic halides.[1] Many other reviews[2–12] are also available, although they are not devoted to reductive dehalogenation but deal with it as part of the broader topic of reduction.

This chapter surveys the reduction of saturated alkyl halides to alkanes. Reductive β-eliminations of vicinal dihalides to alkenes are also described briefly. Reduction of vinyl and aryl halides is covered in this volume, Chapter 4.5; hydrogenolysis of allyl and benzyl halides is covered in this volume, Chapter 4.7, and reduction of α-halo-substituted carbonyl compounds CX—CO to carbonyl compounds CH—CO is covered in this volume, Chapter 4.8.

This subject matter will be treated from the aspect of the type of reagent. Emphasis is made on chemoselective dehalogenation, since the replacement of a halogen atom by hydrogen must often be accomplished in complex molecules containing other functional groups. Reagents and reducing conditions capable of effecting reductive dehalogenations are illustrated by many examples, including practical applications. The mechanisms of the reductions are also discussed briefly in some sections.

4.1.2 GENERAL OBSERVATIONS

The order of ease of reductive dehalogenation of organic halides in the same type of structural environment is: I > Br > Cl >> F. This order is parallel with the dissociation energy of carbon–halogen bonds (H_3C—I: 234 kJ mol^{-1}; H_3C—Br: 293 kJ mol^{-1}; H_3C—Cl: 351 kJ mol^{-1}; H_3C—F: 452 kJ mol^{-1}) and is generally observed in the reduction of alkyl halides.[3] Consequently, selective reduction of di- or polyhalides containing different halogen atoms is possible. Fluorides are often removed only with difficulty and examples of such reductions are comparatively limited.

The reactivities of primary, secondary and tertiary halides are largely dependent on the reagents employed. The ring size of halogenated cycloalkanes and the steric environment also influence the rate of reduction. In general, when the reduction proceeds through an S_N2 type of reaction, primary halides are most easily reduced. On the other hand, when the reduction proceeds through an S_N1 process or involves free radicals as intermediates, the order of reduction falls in the expected sequence: tertiary > secondary > primary.

4.1.3 CATALYTIC HYDROGENOLYSIS

Organic halides are known to be subject to hydrogenolysis in the presence of a catalyst, such as Pd/C or Raney nickel. This catalytic hydrogenolysis is one of the more convenient ways of removing halogens under mild conditions and numerous examples have been reported. Several reviews dealing with this subject have been published.[1,5,6]

Catalytic hydrogenolysis is often inhibited by the release of hydrogen halide produced during the reaction, since both hydrogen ions and halide ions are known to deactivate the catalysts. Hydrogen ions exhibit a particularly strong inhibiting effect, and, therefore, the hydrogenolysis of organic halides is usually carried out in the presence of 1 mol or more of a base. The added base is to neutralize the acid and to maintain the catalytic metal in a low valency or nucleophilic state. Sodium, potassium, calcium and barium hydroxides are generally employed, while sodium acetate, an amine or ammonia are often used for base-labile compounds.

The rate of catalytic hydrogenolysis depends significantly on the structure and halogen involved. Among the types of halides, alkyl halides are much less readily hydrogenolyzed than aryl, vinyl, benzyl and allyl halides (see this volume, Chapters 4.5 and 4.7). Generally, the order of ease of dehalogenation of alkyl halides is I > Br ~ Cl >> F and tertiary > secondary > primary halides. Alkyl fluorides are very difficult to reduce to the corresponding hydrocarbons by conventional catalytic hydrogenation procedures.

Among many catalysts, Pd/C is less susceptible to the poisoning effect of halide ions; hence, this catalyst is preferred by most investigators. Platinum and rhodium are not effective for catalytic dehalogenations. Schemes 1 and 2 illustrate representative Pd/C catalyzed hydrogenolyses of an alkyl iodide and an alkyl bromide.[13,14]

Raney nickel possesses remarkable hydrogenolysis ability, but the catalyst is readily deactivated by the halide ion released. Therefore, massive amounts of the catalyst are usually required to obtain satisfactory results. An example showing the reduction of a primary and a secondary alkyl chloride is shown in Scheme 3.[15]

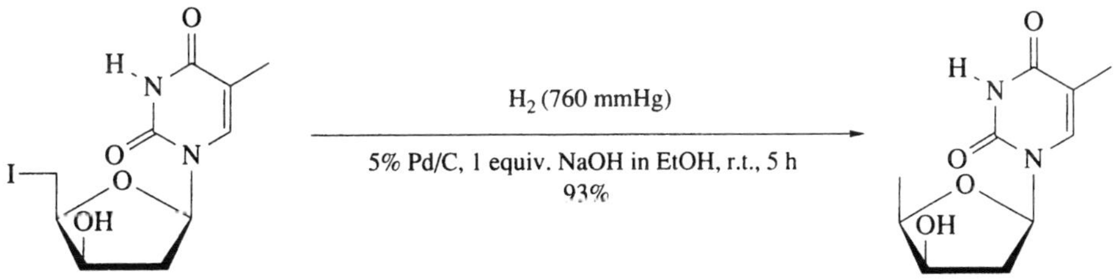
H
N
O
O
N
I
O
OH
H₂ (760 mmHg)
5% Pd/C, 1 equiv. NaOH in EtOH, r.t., 5 h
93%
H
N
O
O
N
O
OH

Scheme 1

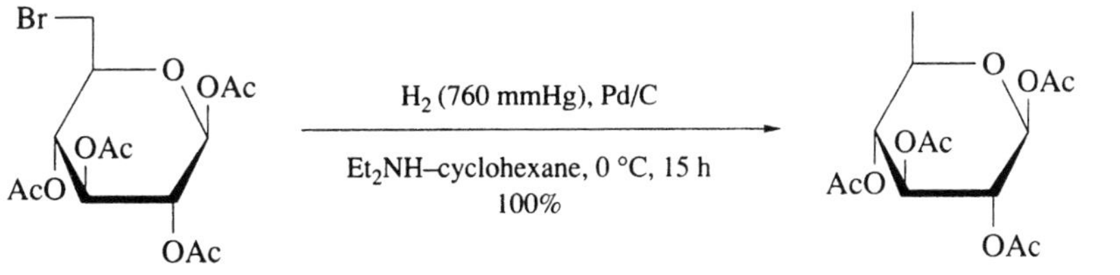
Br
O
OAc
OAc
AcO
OAc
H₂ (760 mmHg), Pd/C
Et₂NH–cyclohexane, 0 °C, 15 h
100%
O
OAc
OAc
AcO
OAc

Scheme 2

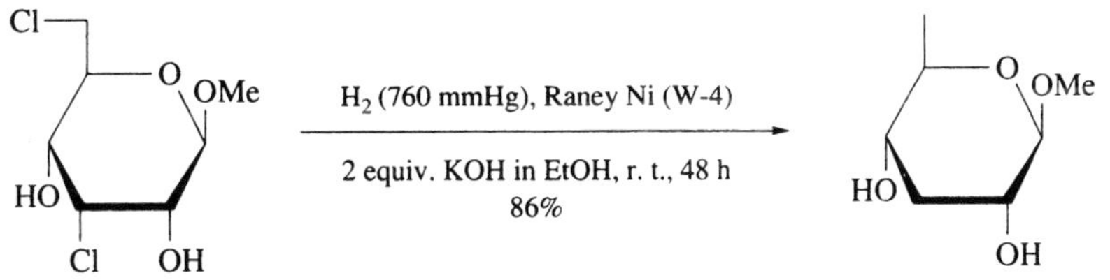
Cl
O
OMe
HO
Cl
OH
H₂ (760 mmHg), Raney Ni (W-4)
2 equiv. KOH in EtOH, r. t., 48 h
86%
O
OMe
HO
OH

Scheme 3

4.1.4 REDUCTION WITH METALS OR LOW-VALENT METAL SALTS

4.1.4.1 Reductions with Metals (Dissolving Metal Method)

Organic halides are reduced by reaction with a metal in the presence or absence of a proton donor. The metals commonly used are the alkali metals (Li, Na, K, Na–K alloy), alkaline earths (Mg, Ca, Ba) and other metals (Zn, Fe, *etc.*).[2]

A classical method using Na– or Li–liquid ammonia (Birch reduction conditions) is effective for reductive dehalogenations of aryl and vinylic halides, but it is not always successfully applied to alkyl halides, although cyclopropyl halides and bridgehead halogens are exceptions.[16,17] Under such conditions, the reactions are often accompanied by side reactions, such as elimination, the Wurtz coupling reaction, cyclization and reduction of carbonyl compounds. An example, a synthesis of pentaprismane (**1**), is shown in Scheme 4.[17]

A currently employed method for dehalogenation of alkyl halides is to use a Li– or Na–alcohol reagent system.[18-20] This method is effective not only for simple alkyl halides but also for the reduction of a halogen atom attached to a bridgehead. Carbon–carbon unsaturated bonds are not affected under the conditions, as shown in Scheme 5.[18]

Alkyl halides are reduced with zinc powder in alcoholic potassium hydroxide[21] or in the presence of a catalytic amount of nickel chloride.[22,23] The former method is employed for the monoreduction of geminal dihalocyclopropyl methyl ketones.[21]

Reduction of alkyl fluorides, which is difficult using conventional dissolving metal methods, can be accomplished using a potassium–crown ether reagent system.[24] Potassium metal dissolves in toluene or diethylene glycol dimethyl ether (diglyme) in the presence of dicyclohexyl-18-crown-6 to form a blue-black solution. The resultant reagent exhibits powerful reducing ability, higher than sodium naphthalene;[25] unactivated primary, secondary and tertiary alkyl fluorides are reduced at room temperature by this reagent to the corresponding alkanes in high yields, as exemplified in Scheme 6.[24]

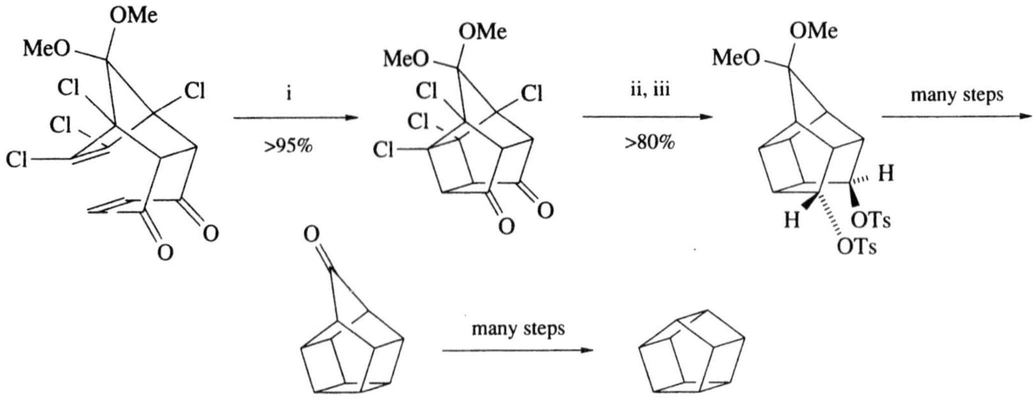

i, $h\nu$ /acetone; ii, Li/liq.NH_3/THF/Bu^tOH–H_2O; iii, TsCl, pyridine

Scheme 4

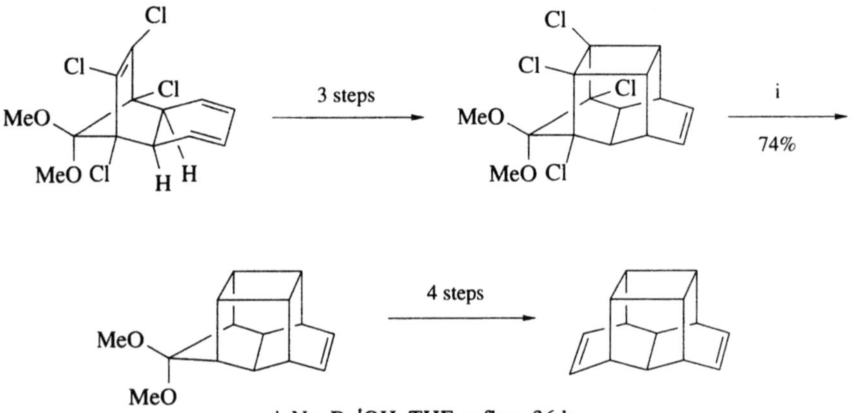

i, Na, Bu^tOH, THF, reflux, 36 h

Scheme 5

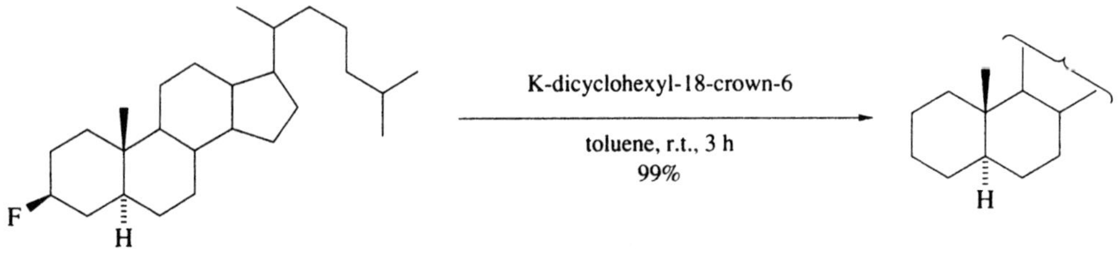

Scheme 6

4.1.4.2 Reduction with Low-valent Metal Salts

4.1.4.2.1 Chromium(II) salts

Chromium(II) salts, such as $Cr(OAc)_2$, $CrCl_2$, $Cr(ClO_4)_2$ and $CrSO_4$ are used for reductive dehalogenation. The reactions are usually carried out in dimethylformamide (DMF) or dimethyl sulfoxide (DMSO). Addition of ethylenediamine or ethanolamine as a chelating agent enhances remarkably the reducing ability.[26,27] Generally, the ease of hydrogenolysis is I > Br > Cl and tertiary > secondary > primary halides. Reduction with Cr^{II} involves a radical process, which in some cases is accompanied by side reactions, such as rearrangement and dimerization. Use of a hydrogen atom donor, such as a thiol, is effective for suppressing such side reactions.[28,29] Chemoselective dehalogenation with Cr^{II} is possible, as exemplified in Scheme 7.[28]

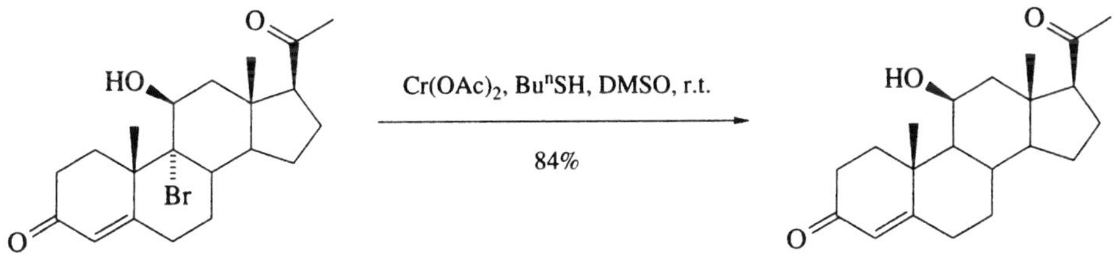

$Cr(OAc)_2$, Bu^nSH, DMSO, r.t.

84%

Scheme 7

An electrochemical reduction procedure using a chromium(II) salt as the catalyst is effective for the dehalogenation of β-hydroxy halides.[30] This procedure is a convenient entry to deoxynucleosides, as shown in Scheme 8.

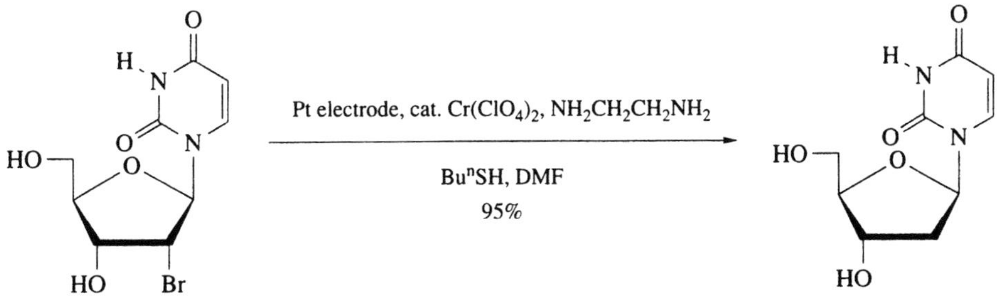

Pt electrode, cat. $Cr(ClO_4)_2$, $NH_2CH_2CH_2NH_2$

Bu^nSH, DMF

95%

Scheme 8

4.1.4.2.2 Samarium(II) iodide

Divalent samarium is known to reduce alkyl halides.[31,32] However, reductions of iodides and bromides in tetrahydrofuran (THF) require a long reaction time and chlorides are not reduced even at refluxing temperature.

Addition of hexamethylphosphoramide (HMPA) to SmI_2 in THF results in the formation of a purple solution of an HMPA complex. The complex exhibits remarkably enhanced reducing power, as shown in equation (2).[33] Thus, bromides and iodides are very rapidly reduced at room temperature to the corresponding alkanes in essentially quantitative yields; ester functions remain intact under these conditions. Chlorides are reduced at refluxing temperature, while fluorides are inert under the conditions.

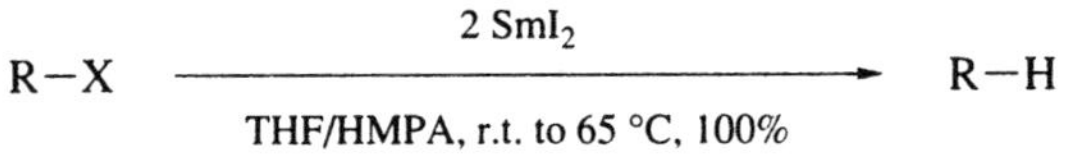

$$\text{R–X} \xrightarrow[\text{THF/HMPA, r.t. to 65 °C, 100\%}]{2\ \text{SmI}_2} \text{R–H} \quad (2)$$

X = Cl, Br, I

4.1.4.2.3 Other low-valent metal salts

Some other low-valent metal species have proved effective in hydrogenolysis of alkyl halides. Nickel tetracarbonyl in THF reductively converts polyhalomethyl groups into di- or mono-halomethyls in good yields.[34] An ionic vanadium carbonyl hydride, $[(\eta^5\text{-}C_5H_5)V(CO)_3H]^-$, is capable of reducing alkyl bromides, but not chlorides.[35] Monoreduction of geminal dibromocyclopropanes proceeds in essentially quantitative yields under these conditions. The titanium(III) chloride/magnesium,[36] titanocene dichloride $[(\eta^5\text{-}C_5H_5)_2TiCl_2]$/magnesium[37] and titanocene dichloride/isopropylmagnesium bromide[38] reagent systems exhibit relatively high activity toward alkyl halides, but the reaction with halocycloalkanes tends to provide significant amounts of alkenes.

Reductions of vicinal dibromides have been extensively studied, using new reagents or reagent systems which involve low-valent metal species, such as $[(\eta^5\text{-}C_5H_5)Cr(NO)_2]_2$,[39] iron–graphite,[40] $TiCl_3$–$LiAlH_4$ or $TiCl_4$–$LiAlH_4$,[41] VCl_3–$LiAlH_4$,[42] Zn–$TiCl_4$,[43] Zn–$[(\eta^5\text{-}C_5H_5)_2TiCl_2]$[44] and $CrCl_3$/$LiAlH_4$.[45] These reagents promote reductive β-elimination to afford the corresponding alkenes in good to excellent yields. In most cases, the reactions proceed through predominant *anti* elimination and the corresponding (*Z*)- and (*E*)-alkenes are obtained with high isomeric purity. A representative example is shown in Scheme 9.[43]

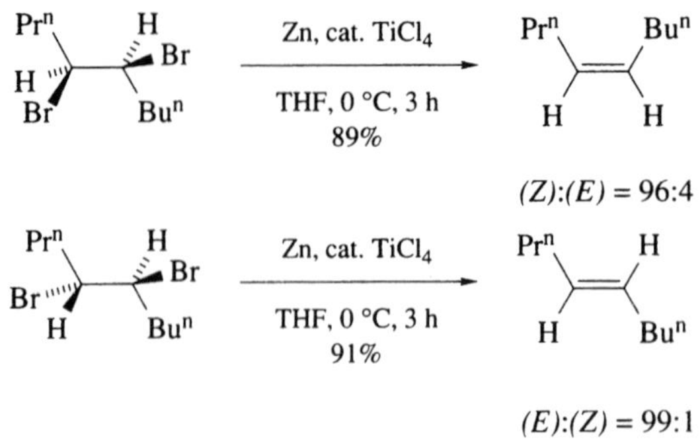

Scheme 9

4.1.5 REDUCTION WITH METAL HYDRIDES

4.1.5.1 Tributyltin Hydride

Tributyltin hydride ($Bu^n{}_3SnH$) is an extremely versatile reagent for replacing halogen atoms in organic molecules by hydrogen atoms.[8,9] Iodides and bromides tend to react spontaneously with this reagent in the absence of a solvent, but usually radical initiators, such as 2,2′-azobisisobutyronitrile (AIBN) or UV irradiation, are used as well as thermal initiation. Chlorides require heating or catalysis by a free radical source. Alkyl fluorides are practically unreactive. Recently, it has been reported that the use of triethylborane as a catalyst significantly accelerates the reduction; alkyl iodides and bromides are reduced even at –78 °C.[46]

As free radicals are involved as intermediates in tin hydride reductions, the nature of the organic group strongly influences the rate of reductions. The order of facility of dehalogenation is tertiary > secondary > primary. Bridgehead halides are also reduced by $Bu^n{}_3SnH$.[47–50] Stepwise reductions of trihalomethyl groups and geminal dihalides, including geminal dihalocyclopropanes, are possible,[51–54] as exemplified in Scheme 10.[51]

$R^1 = PhSO_2, PhCH_2CO_2, EtCO_2, HCO; R^2 =$ Me or Et

i, $Bu^n{}_3SnH$, C_6H_6, 80 °C, 6–8 h; ii, 2 $Bu^n{}_3SnH$, AIBN, C_6H_6, 8–10 h

Scheme 10

Vicinal dichlorides react with 2 equiv. of $Bu^n{}_3SnH$ to give the corresponding saturated hydrocarbons in essentially quantitative yields, whereas vicinal dibromides lead to alkenes in high yields.[55]

A major side reaction encountered in these dehalogenations is intramolecular radical cyclization in systems having unsaturated moieties; cyclization to form five- or six-membered ring systems is frequently observed (see Volume 4, Chapter 4.2).

It is emphasized that the $Bu^n{}_3SnH$ reductions are highly chemoselective. Thus, a variety of other functional groups, such as alcohol, alkene, alkyne, ketone, nitrile, nitro groups, ester, epoxide, and even peroxide[56] and nitrate ester[57] are tolerated. The method using $Bu^n{}_3SnH$ is, therefore, extremely useful for dehalohydrogenolysis in complex molecules having other sensitive functionalities. Several representative examples employed in the synthesis of natural products are illustrated in Schemes 11–15.[57–61]

The method can be applied to the synthesis of deuterium-[62–65] or tritium-labeled[65] compounds by using $Bu^n{}_3SnD$ or $Bu^n{}_3SnT$. The former reagent ($Bu^n{}_3SnD$) is commercially available (Aldrich) and the latter

i, I_2, N_2O_4, $CHCl_3$, 0 °C to r.t., 16 h; ii, $Bu^n{}_3SnH$, C_6H_6, reflux, 48 h; iii, Zn, AcOH, r.t.,16 h

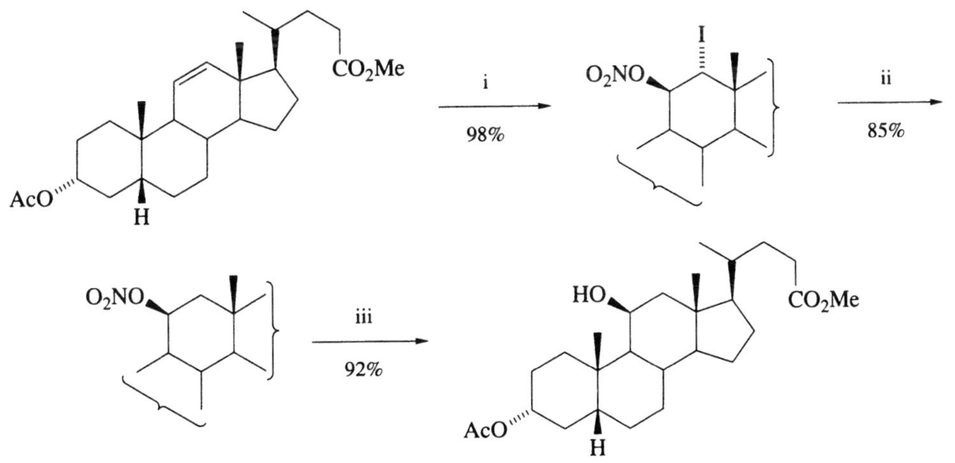

Scheme 11

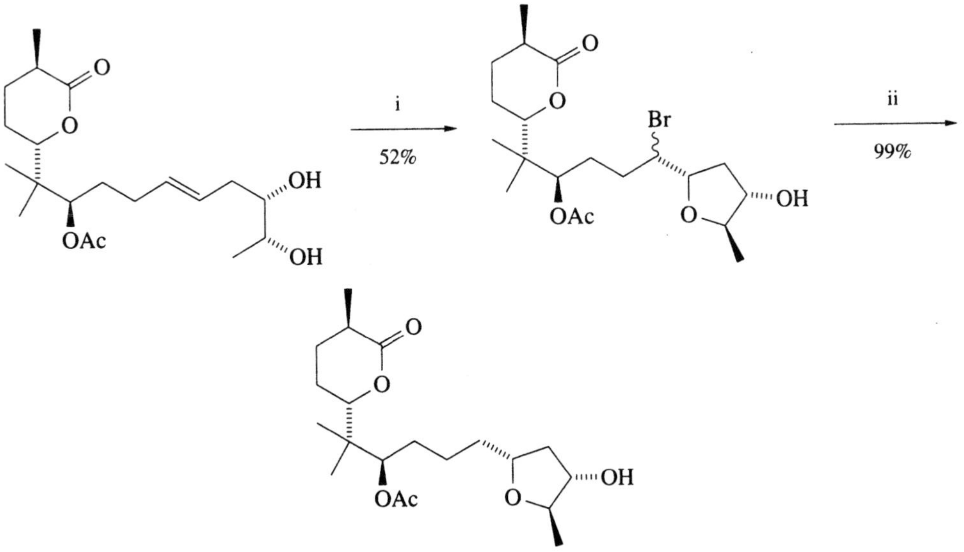

i, NBS, Et_2O–MeCN (5:1), –110 °C, 4 h; ii, $Bu^n{}_3SnH$, AIBN, C_6H_6, reflux, 2 h

Scheme 12

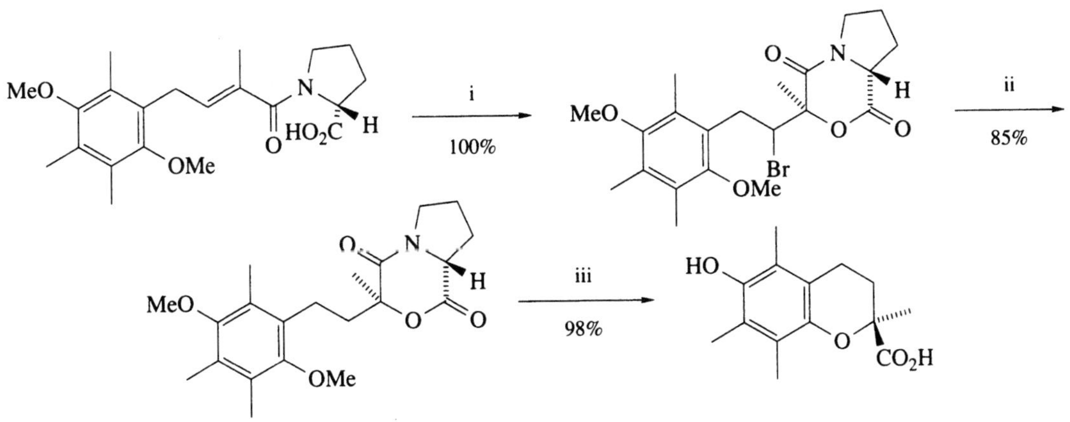

i, NBS, Bu^tOK, –20 °C, 2 h, then r.t., 48 h; ii, $Bu^n{}_3SnH$, C_6H_6, 90 °C, 10 h; iii, conc. HCl, reflux, 10 h

Scheme 13

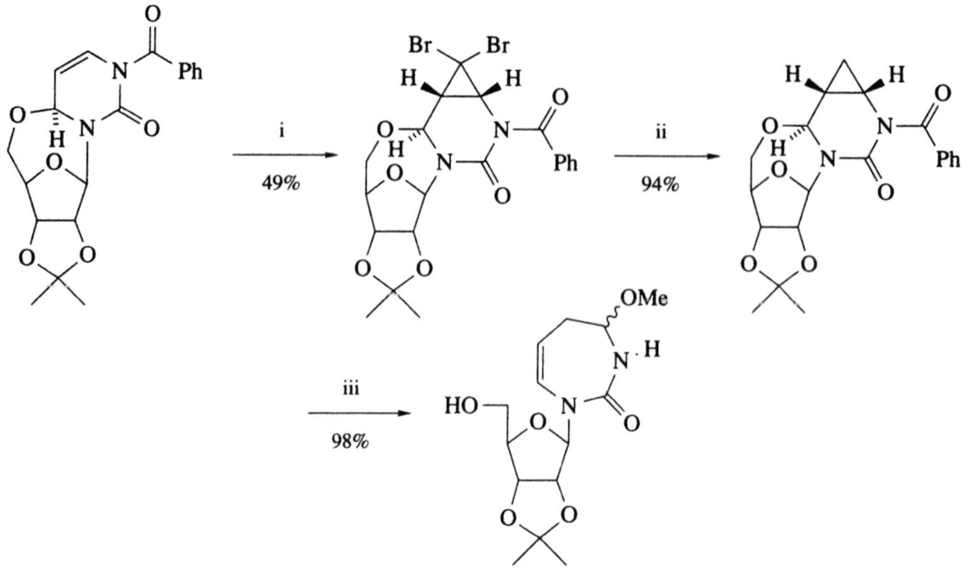

i, $PhHgCBr_3$, C_6H_6, reflux, 52 h; ii, $Bu^n{}_3SnH$, 25 °C, 41 h; iii, MeONa/MeOH, r.t., 1 min

Scheme 14

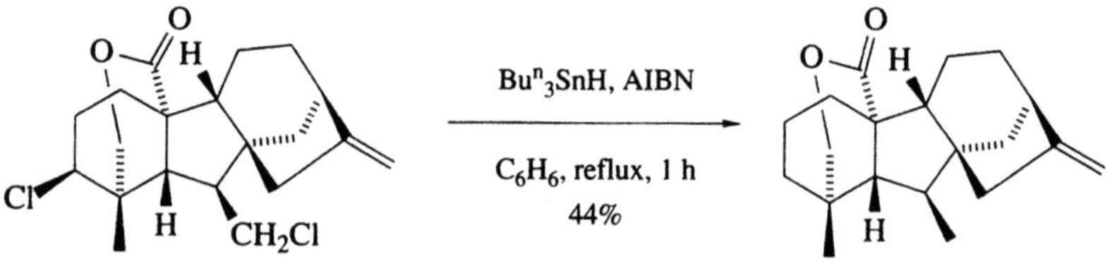

$Bu^n{}_3SnH$, AIBN, C_6H_6, reflux, 1 h, 44%

Scheme 15

($Bu^n{}_3SnT$) is prepared by the reaction of $Bu^n{}_3SnLi$ with tritiated water or by the reaction of $Bu^n{}_3SnCl$ with tritiated sodium borohydride.[65] Representative preparations of labeled compounds with a high degree of stereoselectivity are shown in Schemes 16 and 17.[62,65]

Because organotin hydrides are in general highly toxic, some modifications of the reagents have been devised. The $NaBH_4$–$Bu^n{}_3SnCl$ system, where $Bu^n{}_3SnCl$ is used as the catalyst, is effective,[66,67] and immobilization of organotin hydrides or dihydrides on reticulated polystyrene,[68] alumina or silica gel[69] is another approach. The organic products are not contaminated with organotin residues and the reagents are regenerated with $NaBH_4$ or $LiAlH_4$.

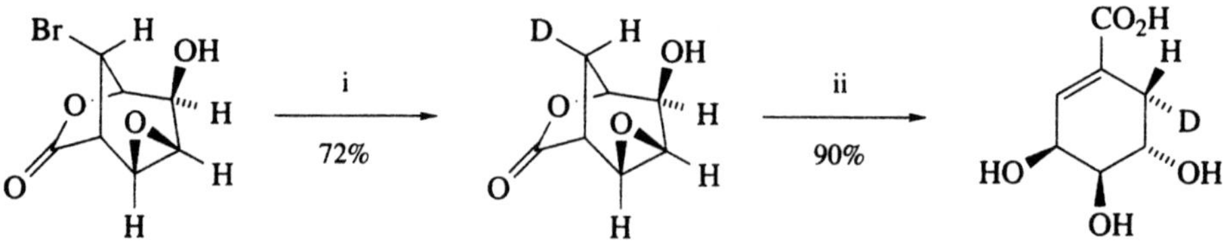

i, $Bu^n{}_3SnH$, AIBN, toluene, reflux, 2 h; ii, KOH, MeOH–H_2O, r.t., 24 h

Scheme 16

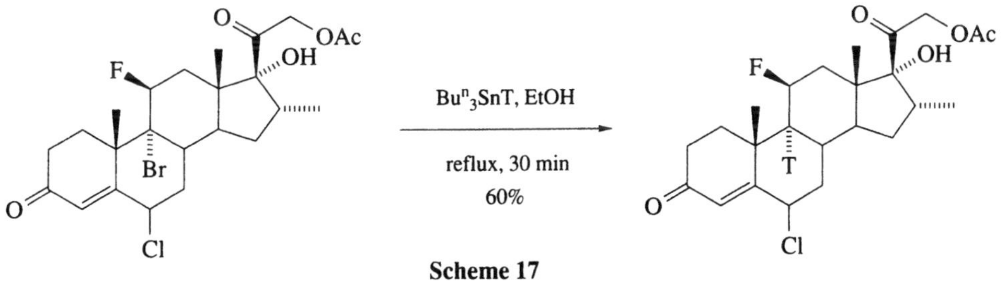

Bu^n_3SnT, EtOH
reflux, 30 min
60%

Scheme 17

4.1.5.2 Copper Hydrides

Various copper hydrides, such as $2LiAlH(OMe)_3{\cdot}CuI$,[70] $LiCuH(n\text{-}C_4H_9)$,[71] $(KCuH_2)_n$,[72] Li_nCuH_{n+1} (n = 1–5)[73] are known to reduce alkyl halides under mild conditions. Among these, the reagent prepared from $LiAlH(OMe)_3$ and CuI reveals unique reactivity. It reduces bromides and chlorides at room temperature in THF in excellent yields. It is interesting to note that the reduction of both *exo*- and *endo*-2-bromonorbornane proceeds with 100% retention, whereas the corresponding phenyl methanesulfonates undergo complete inversion of stereochemistry, as shown in Scheme 18.[70] The results suggest that the copper hydride reagent may attack the bromides from the front side to form copper complexes which, in turn, undergo ligand reorganization and finally release the deuterated products.

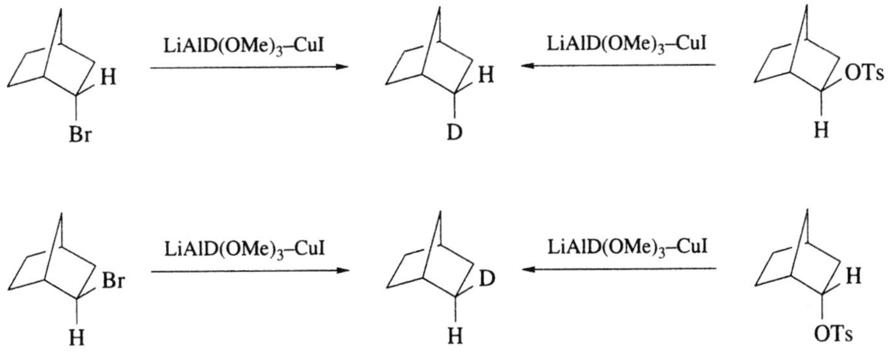

$LiAlD(OMe)_3$–CuI

Scheme 18

4.1.5.3 Other Metal Hydride Reagents

4.1.5.3.1 Organosilane hydrides

Organosilane hydrides are generally inert toward alkyl halides, while they exhibit potential reducing ability when mixed with a catalytic amount (5 mol %) of aluminum chloride.[74] For instance, bromocyclohexane is reduced to cyclohexane in 90% yield by treatment with $Et_3SiH/AlCl_3$ at 0–40 °C; however, bromocycloheptane under similar conditions is converted to a 60:40 mixture of cycloheptane and methylcyclohexane in 65% combined yield.[74]

Recently, Chatgilialoglu *et al.* have reported that tris(trimethylsilyl)silane is like organotin hydrides in being capable of reducing organic halides (equation 3).[75] The data available so far are still limited, but it seems that this reagent rivals tributyltin hydride and is more acceptable from an ecological and toxicological perspective.

$$\text{R–X} \xrightarrow[h\nu \text{ or } (PhCO_2)_2]{(Me_3Si)_3SiH} \text{R–H} \qquad (3)$$

X = Cl, Br, I

4.1.5.3.2 Complex reducing agents (CRA)

Sodium hydride itself exhibits only weak reducing ability, but it becomes highly reactive when coupled with sodium alkoxide and transition metal salts. The reagent system designated by Caubère *et al.* as 'complex reducing agents (CRA)' can be used for reductive dehalogenation of organic halides.[10]

Among various CRAs, NaH–*t*-pentyl-ONa–Ni(OAc)$_2$ (NiCRA) and NaH–*t*-pentyl-ONa–ZnCl$_2$ (ZnCRA) possess particularly notable reducing ability toward alkyl halides.[76] With NiCRA, the general reactivity follows the trend RI > RBr > RCl and primary > secondary > tertiary. ZnCRA reduces only primary halides, following the trend RI > RBr >> RCl. Elimination is not observed in the reduction of β-halo ethers and α-halo acetals.

4.1.6 REDUCTION WITH COMPLEX METAL HYDRIDES

4.1.6.1 Reduction with $LiAlH_4$

Alkyl halides are reduced with $LiAlH_4$ (LAH) to the corresponding saturated hydrocarbons in good to high yields. In view of its ready availability, convenient handling, and ease of removal after completion of reaction, LAH is attractive for the reduction of organic halides. The reduction is usually carried out in ethereal solvents. The classical method using slurries of LAH in ether is not always effective for the reduction, because of its low solubility. Use of THF, dimethoxyethane (DME) or diglyme as the solvent is more effective; for instance, the reduction of *n*-octyl iodide is essentially complete in 0.25 h in these solvents, while the reduction in ether is sluggish, requiring 24 h for 90% completion.[77]

For efficiency and reasons of economy THF is recommended. Representative dehalogenations in THF are shown in Table 1.[78] Most alkyl halides are reduced at 25 °C to alkanes, except for bicyclic halides, such as *exo*-2-bromonorbornane. In most cases, the reaction exhibits the typical characteristics of an S_N2 substitution reaction in the reactivity order I > Br > Cl; primary > secondary > tertiary. In all cases, no alkene, or only traces of it, is formed as a side product as a result of elimination. This method provides a convenient synthetic procedure for hydrodehalogenation of alkyl halides.

Table 1 Reduction of Alkyl Halides with LAH in THF[78]

Alkyl halide[a]	*Temperature* (°C)	*Time* (h)	*Product*	*Yield* (%)
1-Iodooctane	25	0.25	*n*-Octane	94
1-Bromooctane	25	0.5	*n*-Octane	96
1-Chlorooctane	25	24	*n*-Octane	73
1-Chlorooctane	65	4	*n*-Octane	94
2-Bromooctane	25	24	*n*-Octane	70
2-Bromooctane	65	24	*n*-Octane	97
1-Bromo-2-methylpentane	25	6	2-Methylpentane	92
Bromocyclopentane	25	24	Cyclopentane	91
Iodocyclohexane	65	6	Cyclohexane	93
Bromocyclohexane	25	24	Cyclohexane	5
Bromocyclohexane[b]	65	12	Cyclohexane	96
Chlorocyclohexane	25	24	Cyclohexane	Trace
Bromocycloheptane[c]	25	24	Cycloheptane	94
exo-2-Bromonorbornane	25	24	Norbornane	3

[a]Unless otherwise stated, the concentrations of the alkyl halide and LAH are 0.25 M. [b]1.0 M LAH. [c]0.5 M LAH.

The mechanism of LAH reduction of alkyl halides has been investigated by several research groups. Ashby and his coworkers have extensively studied the mechanism using cyclizable probes, hydrogen donor trapping agents and optically active alkyl halides.[79–83] Their exhaustive investigations have led to the following conclusion: alkyl chlorides are reduced entirely by an S_N2 pathway. In the case of bromides, it appears that S_N2 and a single electron transer (SET) are in competition, with the S_N2 process being strongly favored. For alkyl iodides it appears that SET is the major reaction pathway. On the other hand, Newcomb *et al.* have recently claimed that SET is not an important process for LAH reduction of unhindered primary halides, including iodides.[84,85] The mechanism of LAH reduction of cyclopropyl bromides has also been studied. These reductions are believed to proceed predominantly through an SET process.[86,87]

4.1.6.2 Lithium Aluminum Hydride–Metal Salt Systems

Ashby *et al.* have studied reductive dehalogenation by the combined use of LAH and first row transition metal halides (Co^{II}, Ni^{II}, Fe^{II}, Fe^{III}, Mn^{II}, Ti^{III} and V^{III}).[88,89] Table 2 shows that $CoCl_2$, $NiCl_2$ and $TiCl_3$ notably elevate the reducing ability of LAH, and the added salts need only be present in catalytic amounts (10 mol %). Among them, the reagent system LAH–$NiCl_2$ shows the highest reactivity; that is, primary, secondary and cyclic halides (I, Br, Cl) are reduced in essentially quantitative yield at room temperature.

Table 2 Reduction of Alkyl Halides by LAH–Transition Metal Chlorides at Room Temperature in THF[89]

Alkyl halide	*Transition metal chloride*[a]	*Time* (h)	*Product*	*Yield* (%)
1-Chlorodecane	None	24	*n*-Decane	68
	$FeCl_3$		*n*-Decane	100
	$CoCl_2$ (10 mol %)		*n*-Decane	100
	$NiCl_2$ (10 mol %)		*n*-Decane	100
	$TiCl_3$ (10 mol %)		*n*-Decane	100
1-Fluorodecane	None	24	*n*-Decane	0
	$FeCl_2$		*n*-Decane	16
	$CoCl_2$ (10 mol %)		*n*-Decane	10
	$NiCl_2$ (10 mol %)		*n*-Decane	7
	$TiCl_3$ (10 mol %)		*n*-Decane	9
Bromocyclohexane	None	24	Cyclohexane	0
	$FeCl_2$		Cyclohexane	97
	$CoCl_2$		Cyclohexane	99
	$NiCl_2$		Cyclohexane	99
	$TiCl_3$		Cyclohexane	100
Chlorocyclohexane	None	24	Cyclohexane	0
	$CoCl_2$		Cyclohexane	92
	$CoCl_2$ (10 mol %)		Cyclohexane	3
	$NiCl_2$		Cyclohexane	95
	$NiCl_2$ (10 mol %)		Cyclohexane	5
	$TiCl_3$		Cyclohexane	95
	$TiCl_3$ (10 mol %)		Cyclohexane	95

[a]Molar ratio of LAH to transition metal chloride is 1:1, except when noted.

It has been reported that the LAH–$AgClO_4$ (1%) system also exhibits enhanced reactivity toward halides.[90] The reagent system is effective for such substrates as geminal dihalocyclopropanes, which exhibit poor S_N2 reactivity.

The combination of LAH with Lewis acid has a powerful reducing ability. The use of aluminum chloride[91] or cerium chloride[92] under more drastic conditions enables primary alkyl fluorides to be converted to the corresponding saturated hydrocarbons in good to excellent yields.

4.1.6.3 Sodium Borohydride

Sodium borohydride, a representative borohydride reagent, behaves as an effective source of nucleophilic hydride in an aprotic polar solvent, such as DMSO, sulfolane, HMPA, DMF or diglyme, and is used for the reduction of alkyl halides.[93,94] As shown in Table 3, primary and secondary iodides, bromides and chlorides are converted to hydrocarbons at temperatures between 25 and 100 °C using sodium borohydride. Vicinal dihalides, such as 1,2-dibromooctane, are smoothly converted to the corresponding saturated hydrocarbons, in contrast to the reductions using $LiAlH_4$ or low-valent metal salts, which predominantly afford alkenes.

Sodium borohydride reduction offers a significant advantage in synthetic applications. The method allows the reductive removal of halides selectively without affecting other functional groups, such as ester, carboxylic acid, nitrile and sulfone. A typical chemoselective dehalogenation is illustrated in Scheme 19.[95]

Some alkyl halides (primary I, Br, Cl and secondary Br) are reduced by $NaBH_4$ under phase-transfer conditions, as shown in equation (4).[96] This reaction is carried out by the addition of a concentrated aqueous solution of $NaBH_4$ to a stirred solution of substrate and catalyst in a suitable solvent, such as toluene. Yields are nearly quantitative in most cases, although an excess of sodium borohydride is required.

Table 3 Reduction of Alkyl Halides with $NaBH_4$ in Polar Aprotic Solvents[93]

Alkyl halide	*Molar ratio ($NaBH_4$/halide)*	*Solvent*	*Temperature (°C)*	*Time (h)*	*Product*	*Yield (%)*
1-Iododecane	2	DMSO	85	0.25	Decane	93
1-Iododecane	2	Sulfolane	100	0.25	Decane	93
1-Iodododecane	1	HMPA	25	0.025	Dodecane	87
1-Iodooctane	5.3	Diglyme	45	1	Octane	91
1-Bromodecane	2	DMSO	85	1.5	Decane	94
1-Bromododecane	1	HMPA	25	0.022	Docecane	73
1-Bromododecane	2	DMSO	85	1.5	Dodecane	95
1-Bromododecane	2	Sulfolane	100	1.5	Dodecane	96
ω-Bromoundecanoic acid	2	DMSO	25	2.5	Undecanoic acid	98
1-Chlorododecane	2	DMSO	85	4	Dodecane	91
1-Chlorododecane	2	Sulfolane	100	6	Dodecane	85
2-Iodooctane	3	DMSO	85	1	Octane	82
2-Iodooctane	3	Sulfolane	100	1	Octane	81
2-Bromododecane	3	DMSO	85	18	Dodecane	86
2-Bromododecane	3	Sulfolane	100	18	Dodecane	69
2-Chlorodecane	3	DMSO	85	18	Decane	68
Ethyl 5-bromovalerate	4	HMPA	25	0.5	Ethyl valerate	85
Styrene dibromide	4	DMSO	85	1.5	Ethylbenzene	65
Styrene dibromide	4	Sulfolane	100	1.5	Ethylbenzene	64
1,2-Dibromooctane	4	HMPA	70	2	Octane	84

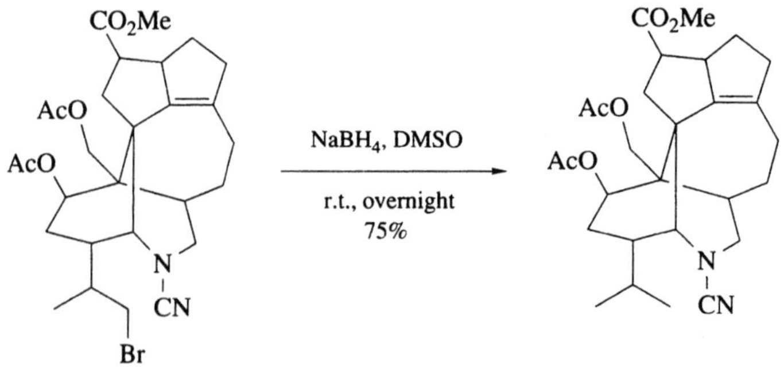

Scheme 19

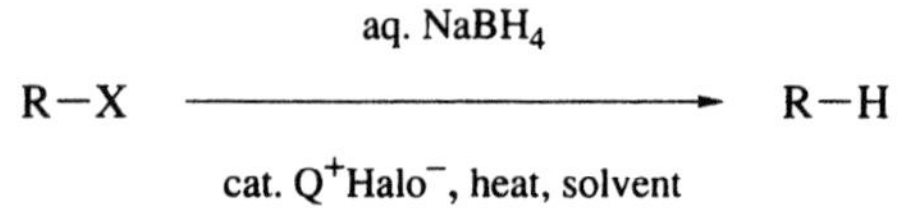

$$R{-}X \xrightarrow[\text{cat. } Q^+\text{Halo}^-\text{, heat, solvent}]{\text{aq. } NaBH_4} R{-}H \quad (4)$$

R = alkyl group; X = Cl, Br, I; Q = $(alkyl)_4N$, $(alkyl)_4P$

4.1.6.4 Lithium Triethylborohydride

Triethylborohydride ($LiEt_3BH$) is the most powerful nucleophilic reducing agent for the reduction of alkyl halides.[97,98] This reagent is far more powerful and cleaner than lithium aluminum hydride and lithium borohydride, as indicated in Figure 1.[98] The replacement of three of the four hydrogen atoms in $LiBH_4$ by ethyl groups enhances the nucleophilicity by a factor of 10 000. This enhanced nucleophilicity arises from the electron-donating effect of the three ethyl groups.

Representative reductions of alkyl halides with $LiBEt_3H$ are shown in Table 4. Most alkyl halides, including chlorides, are reduced under mild conditions in essentially quantitative yields. Even hindered alkyl halides, such as neopentyl bromide and *exo*-norbornyl bromide, undergo facile reduction to the corresponding alkanes in more than 96% yield with this reagent.

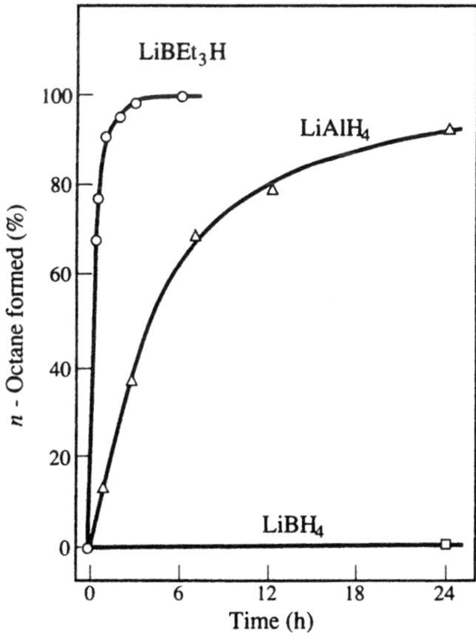

Figure 1 Rates of reduction of n-octyl chloride (0.25 M) with representative complex metal hydrides (0.5 M) in THF at 25 °C (reproduced from ref. 98 by permission from S. Krishnamurthy and H. C. Brown)

Table 4 Reduction of Alkyl Halides with Lithium Triethylborohydride in THF[98]

Alkyl halide[a]	*Temperature* (°C)	*Time* (h)	*Products* (%) *RX*	*RH*	*Other*
1-Iodooctane	25	0.02		100	
1-Bromooctane	25	0.03		100	
1-Chlorooctane	25	1	9	91	
1-Chlorooctane	25	3		100	
1-Bromo-2-methylpentane	25	0.08		98	
Neopentyl bromide	25	24	14	86	
Neopentyl bromide	65	3	5	96	
2-Bromooctane	25	3		98	
2-Bromo-2-methylpentane	25	24	8	<2	84, alkenes
Iodocyclohexane	25	12		92	2, alkene
Bromocyclohexane	25	24	34	64	
Bromocyclohexane	65	6	4	88	2, alkene
Chlorocyclohexane	25	24	99	1	
Chlorocyclohexane	65	48	58	34	
Bromocyclopentane	25	1		99	
Bromocycloheptane	25	3		99	
Bromocyclooctane	25	24		95	
exo-Norbornyl bromide	25	24	84	12	
exo-Norbornyl bromide	65	12	40	59	
exo-Norbornyl bromide	65	72	4	96	

[a] In all cases solutions are 0.25 M in alkyl halide and 0.5 M in $LiEt_3BH$.

This reagent is conveniently prepared by the reaction of triethylborane with lithium hydride in THF.[98] The THF solution (1.0 M) is commercially available from Aldrich under the name of 'Super-Hydride'.

A corresponding deuterated derivative is conveniently prepared from lithium deuteride in quantitative yield; the reagent 'Super-Deuteride' is also commercially available from Aldrich. The ready availability of this reagent offers a simple means of introducing deuterium into the system with stereochemical inversion at the center undergoing substitution. An example is shown in Scheme 20.

Br $\xrightarrow[96\%]{LiEt_3BD,\ THF,\ 65\ °C}$ H, D

Scheme 20

The reduction of alkyl halides by $LiEt_3BH$ exhibits typical S_N2 characteristics, but it has been suggested by Ashby *et al.* that the reductions of secondary alkyl iodides and bromides proceed in part by an SET pathway.[99]

4.1.6.5 Other Borohydrides and Related Reagents

Cyanoborohydride and its modified reagents have been used for reductive dehalogenations. Thus, the combination of sodium or tetrabutylammonium cyanoborohydride, sodium or potassium 9-cyano-9-hydro-9-borabicyclo[3.3.1]nonanate [9-BBNCN] (**2**) or polymeric cyanoborane (**3**) in HMPA furnishes an efficient and mild system for the reduction of alkyl halides.[100] The reagents are selective in that other functional groups, including ester, carboxylic acid, amide, cyano, alkene, nitro, sulfone, ketone, aldehyde and epoxide, are essentially inert under the reduction conditions; thus, the reduction procedure is attractive for synthetic schemes which demand minimum damage to sensitive portions of the molecule.

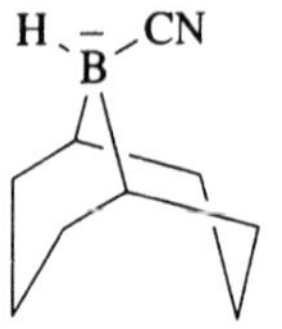

H, CN, B

(**2**)

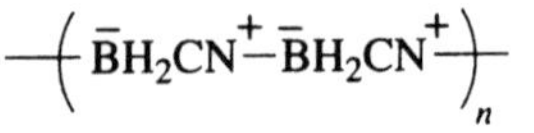

$-\!\!\left(\bar{B}H_2CN^{+}\!-\!\bar{B}H_2CN^{+}\right)_n$

(**3**)

A zinc-modified cyanoborohydride reagent, prepared *in situ* by the reaction of sodium borohydride with zinc chloride in ether, reduces tertiary halides in high yields, whereas primary and secondary halides remain intact.[101] Similar reactivity is observed with lithium 9,9-di-*n*-butyl-9-borabicyclo[3.3.1]-nonanate (**4**), as shown in equation (5).[102]

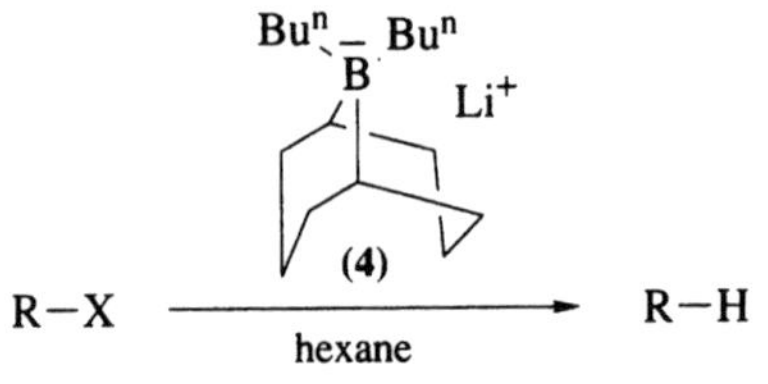

$$R{-}X \xrightarrow[\text{hexane}]{(\mathbf{4})} R{-}H \qquad (5)$$

(**4**): Bu^n, Bu^n, B, Li^+

Yoon and Kim have prepared potassium triphenylborohydride (KPh_3BH) and examined the reducing ability toward alkyl halides.[103] *n*-Octyl iodide is reduced within 1 h at 0 °C, whereas *n*-octyl bromide takes 24 h for complete reduction. Secondary bromides are inert under these conditions.

4.1.7 MISCELLANEOUS REDUCTION PROCEDURES

Some anionic species have been examined for the reduction of organic halides. Among these are $(MeO)_2PK$,[104] Ph_2PK,[105] $K(EtO)_2P(O)H$,[106,107] NaTeH,[108] $Bu^n{}_3SnLi$[109] and $MeSOCH_2Na$,[110] which have been examined for selective reduction of alkyl halides. These reagents possess a strong nucleophilic characteristic and are not always suitable for the reduction of simple alkyl halides, but they are suited to selective monoreduction of geminal dihalocyclopropanes (equation 6). Table 5 summarizes various methods of monoreduction, including the representative ones mentioned in the previous sections.

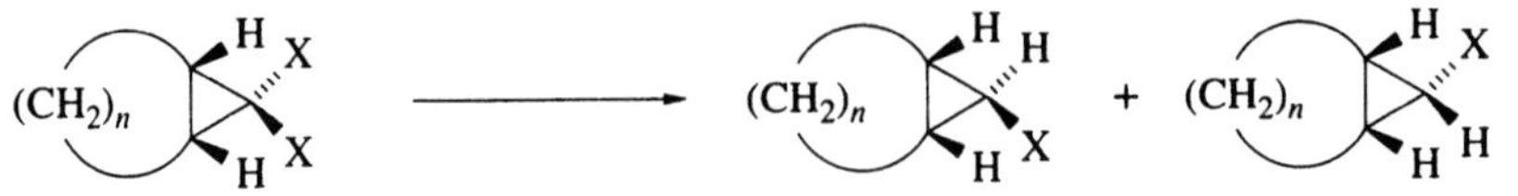

$(CH_2)_n$ [H, X, H, X] → $(CH_2)_n$ [H, H, H, X] + $(CH_2)_n$ [H, X, H, H] (6)

Phosphorus reagents, particularly $(MeO)_2PK$ and Ph_2PK, provide a high yield of monoreduction with excellent stereoselectivity, which is in contrast to other methods, such as LAH reduction.

Reductive elimination of 1,2-dibromoalkanes has been studied by many investigators using sodium disulfide,[121–123] poly(styrylmethylthiol),[124] methyl- or phenyl-selenolate,[125] bis(2-thienyl)ditelluride/$NaBH_4$,[126] sodium telluride/rongalite[127] and trimethylstannylsodium.[128] Among these, sodium sul-

Table 5 Monoreduction of Cyclopropyl Geminal Dihalides

	Reagent or method	*Solvent*	*Yield (%)*	*exo:endo*	*Ref.*
$n = 4$, X = Br	$(MeO)_2PK$	Liquid NH_3	100	100:0	104
	$(MeO)_2PK$	DMSO	97	99:1	104
	$HP(O)(OEt)_2/Et_3N$		83	17:83	106
	NaHTe	EtOH	55	24:76	108
	$MeSOCH_2Na$	DMSO	72	99–90:1–10	110
	MeMgBR	THF	72	27:73	111
	LAH	Et_2O	73	25:75	112
	$LAH/AgClO_4$	Et_2O	67	7:93	90
	$NaAlH_2(OCH_2CH_2OMe)_2$	C_6H_6	65	74:26	113
	$NaBH_4$	DMF	79	36:64	114
	$CrSO_4$	DMF	25	0:100	115
	$Cr(OAc)_2$	DMSO	82	9:91	116
	n-Bu_3SnH		82	29:71	117
	Hg cathode	$MeOH/H_2O$	80–90	16:84	118
$n = 6$, X = Br	$(MeO)_2PK$	Liquid NH_3	82	100:0	104
	$HP(O)(OEt)_2/NEt_3$		74	22:78	106
	NaHTe	EtOH	59	33:67	108
	$MeSOCH_2Na$	DMSO	71	95:5	110
	$LAH/AgClO_4$	THF	75	17:83	90
	$NaBH_4$	DMSO	90	36:64	114
	$Bu^n{}_3SnH$		84	0:100	117
	$Bu^n{}_3SnH/Et_3B$	PhMe	84	18:82	46
	$CrSO_4$	DMF	32	0:100	115
	Zn	AcOH	95	10:90	119
	Hg cathode	MeOH	80–90	0:100	118
$n = 4$, X = Cl	Ph_2PK	DMSO	88	94:6	105
	LAH/$h\nu$	THF	93	28:72	120
	$Bu^n{}_3SnH$		83	36:64	117
	Hg cathode	MeOH/HCl	80–90	0:100	118

fide provides high yields of the corresponding alkenes. The reductions are simply carried out in DMF or phase-transfer conditions. This method compares favorably with the methods using Zn–$TiCl_4$[43] and electrochemical reduction.[129]

4.1.8 REFERENCES

1. A. R. Pinder, *Synthesis*, 1980, 425.
2. H. O. House, 'Modern Synthetic Reactions', 2nd edn., Benjamin, New York, 1972.
3. R. D. Chambers and S. R. James, in 'Comprehensive Organic Chemistry', ed. D. H. R. Barton and W. D. Ollis, Pergamon Press, Oxford, 1979, vol. 1, p. 493.
4. J. Zabicky and S. Ehrlich-Rogozinski, in 'The Chemistry of the Carbon–Halogen Bond', ed. S. Patai, Wiley, London, 1973, p. 143.
5. P. N. Rylander, 'Catalytic Hydrogenation in Organic Syntheses', Academic Press, New York, 1979, p. 235.
6. M. Freifelder, 'Catalytic Hydrogenation in Organic Synthesis', Wiley, New York, 1978, p. 121.
7. T.-L. Ho, *Synthesis*, 1979, 1.
8. M. Pereyre, J.-P. Quintard and A. Rahm, 'Tin in Organic Synthesis', Butterworths, London, 1987, p. 35.
9. W. P. Neumann, *Synthesis*, 1987, 665.
10. P. Caubère, *Angew. Chem., Int. Ed. Engl.*, 1983, **22**, 599.
11. B. Ganem and J. O. Osby, *Chem. Rev.*, 1986, **86**, 763.
12. J. Casanova and L. Eberson, in 'The Chemistry of the Carbon–Halogen Bond', ed. S. Patai, Wiley, London, 1973, p. 979.
13. J. Matulic-Adamic, M. Pavela-Vrancic and V. Skaric, *J. Chem. Soc., Perkin Trans. 1*, 1988, 2681.
14. W. Schüep and E. Hardegger, *Helv. Chim. Acta*, 1970, **53**, 1336.
15. E. H. Williams, W. A. Szarek and J. K. N. Jones, *Can. J. Chem.*, 1971, **49**, 796.
16. A. J. Duggan and S. S. Hall, *J. Org. Chem.*, 1975, **40**, 2238.
17. P. E. Eaton, Y.-S. Or and S. J. Branca, *J. Am. Chem. Soc.*, 1981, **103**, 2134.
18. T.-C. Chou, K.-S. Chung and C. T. Lin, *J. Org. Chem.*, 1988, **53**, 5168.
19. W.-D. Fessner, H. Prinzbach and G. Rihs, *Tetrahedron Lett.*, 1983, **24**, 5857.
20. H.-M. Hutmacher, H.-G. Fritz and H. Musso, *Angew. Chem., Int. Ed. Engl.*, 1975, **14**, 180.
21. R. Barlet, *J. Org. Chem.*, 1978, **43**, 3500.
22. I. Colon, *J. Org. Chem.*, 1982, **47**, 2622.
23. J. Yamashita, Y. Inoue, T. Kondo and H. Hashimoto, *Bull. Chem. Soc. Jpn.*, 1985, **58**, 2709.
24. T. Ohsawa, T. Takagaki, A. Haneda and T. Oishi, *Tetrahedron Lett.*, 1981, **22**, 2583.
25. J. F. Garst and F. E. Barton, II, *Tetrahedron Lett.*, 1969, **10**, 587.
26. J. K. Kochi and P. E. Mocadlo, *J. Am. Chem. Soc.*, 1966, **88**, 4094.
27. J. K. Kochi, D. M. Singleton and L. J. Andrews, *Tetrahedron*, 1968, **24**, 3503.

28. D. H. R. Barton, N. K. Basu, R. H. Hesse, F. S. Morehouse and M. M. Pechet, *J. Am. Chem. Soc.*, 1966, **88**, 3016.
29. M. D. Bachi, J. W. Epstein, Y. Herzberg-Minzly and H. J. E. Loewenthal, *J. Org. Chem.*, 1969, **34**, 126.
30. J. Wellmann and E. Steckhan, *Angew. Chem., Int. Ed. Engl.*, 1980, **19**, 46.
31. P. Girard, J. L. Namy and H. B. Kagan, *J. Am. Chem. Soc.*, 1980, **102**, 2693.
32. H. B. Kagan and J. L. Namy, *Tetrahedron*, 1986, **42**, 6573.
33. J. Inanaga, M. Ishikawa and M. Yamaguchi, *Chem. Lett.*, 1987, 1485.
34. T. Kunieda, T. Tamura and T. Takizawa, *Chem. Pharm. Bull.*, 1977, **25**, 1749.
35. R. J. Kinney, W. D. Jones and R. G. Bergman, *J. Am. Chem. Soc.*, 1978, **100**, 635.
36. S. Tyrlik and I. Wolochowicz, *J. Chem. Soc., Chem. Commun.*, 1975, 781.
37. T. R. Nelson and J. J. Tufariello, *J. Org. Chem.*, 1975, **40**, 3159.
38. E. Colomer and R. J. P. Corriu, *J. Organomet. Chem.*, 1974, **82**, 367.
39. B. W. S. Kolthammer, P. Legzdins and D. T. Martin, *Tetrahedron Lett.*, 1978, **19**, 323.
40. D. Savoia, E. Tagliavini, C. Trombini and A. Umani-Ronchi, *J. Org. Chem.*, 1982, **47**, 876.
41. G. A. Olah and G. K. S. Prakash, *Synthesis*, 1976, 607.
42. T.-L. Ho and G. A. Olah, *Synthesis*, 1977, 170.
43. F. Sato, T. Akiyama, K. Iida and M. Sato, *Synthesis*, 1982, 1025.
44. S. G. Davies and S. E. Thomas, *Synthesis*, 1984, 1027.
45. Y. Okude, T. Hiyama and H. Nozaki, *Tetrahedron Lett.*, 1977, **18**, 3829.
46. K. Miura, Y. Ichinose, K. Nozaki, K. Fugami, K. Oshima and K. Utimoto, *Bull. Chem. Soc. Jpn.*, 1989, **62**, 143.
47. J.-L. Stein, L. Stella and J. M. Surzur, *Tetrahedron Lett.*, 1980, **21**, 287.
48. R. C. Fort, Jr. and J. Hiti, *J. Org. Chem.*, 1977, **42**, 3968.
49. T.-Y. Luh and L. M. Stock, *J. Org. Chem.*, 1977, **42**, 2790.
50. E. W. Della and H. K. Patney, *Synthesis*, 1976, 251.
51. Y. Vo Quang, D. Carniato, L. Vo Quang and F. Le Goffic, *Synthesis*, 1985, 62.
52. W. Boland and L. Jaenicke, *Chem. Ber.*, 1977, **110**, 1823.
53. E. V. Dehmlow and M. Lissel, *Liebigs Ann. Chem.*, 1979, 181.
54. T. Ando, T. Ishihara, E. Ohtani and H. Sawada, *J. Org. Chem.*, 1981, **46**, 4446.
55. R. J. Strunk, P. M. DiGiacomo, K. Aso and H. G. Kuivila, *J. Am. Chem. Soc.*, 1970, **92**, 2849.
56. A. J. Bloodworth and H. J. Eggelte, *J. Chem. Soc., Chem. Commun.*, 1982, 865.
57. M. Do-Trong, W. Kreiser and E. Strube, *J. Steroid Biochem.*, 1983, **19**, 783.
58. J. D. White, M. A. Avery, S. C. Choudhry, O. P. Dhingra, B. D. Gray, M. Kang, S. Kuo and A. J. Whittle, *J. Am. Chem. Soc.*, 1989, **111**, 790.
59. H. Yoda and K. Takabe, *Chem. Lett.*, 1989, 465.
60. V. E. Marquez, K. V. B. Rao, J. V. Silverton and J. A. Kelley, *J. Org. Chem.*, 1984, **49**, 912.
61. B. M. Fraga, A. G. Gonzalez, F. G. Tellado, Z. J. Duri and J. R. Hanson, *J. Chem. Soc., Perkin Trans. 1*, 1984, 1109.
62. K. E. Coblens, V. B. Muralidharan and B. Ganem, *J. Org. Chem.*, 1982, **47**, 5041.
63. E. W. Della and H. K. Patney, *Synthesis*, 1976, 251.
64. T.-Y. Luh and L. M. Stock, *J. Org. Chem.*, 1977, **42**, 2790.
65. H. Parnes and J. Pease, *J. Org. Chem.*, 1979, **44**, 151.
66. E. J. Corey and J. W. Suggs, *J. Org. Chem.*, 1975, **40**, 2554.
67. D. E. Bergbreiter and J. R. Blanton, *J. Org. Chem.*, 1987, **52**, 472.
68. N. M. Weinshenker. G. A. Crosby and J. Y. Wong, *J. Org. Chem.*, 1975, **40**, 1966.
69. H. Schumann and B. Pachaly, *Angew. Chem., Int. Ed. Engl.*, 1981, **20**, 1043.
70. S. Masamune, P. A. Rossy and G. S. Bates, *J. Am. Chem. Soc.*, 1973, **95**, 6452.
71. S. Masamune, G. S. Bates and P. E. Georghiou, *J. Am. Chem. Soc.*, 1974, **96**, 3686.
72. T. Yoshida and E. Negishi, *J. Chem. Soc., Chem. Commun.*, 1974, 762.
73. E. C. Ashby, J. J. Lin and A. B. Goel, *J. Org. Chem.*, 1978, **43**, 183.
74. M. P. Doyle, C. C. McOsker and C. T. West, *J. Org. Chem.*, 1976, **41**, 1393.
75. C. Chatgilialoglu, D. Griller and M. Lesage, *J. Org. Chem.*, 1988, **53**, 3641; M. Lesage, C. Chatgilialoglu and D. Griller, *Tetrahedron Lett.*, 1989, **30**, 2733; M. Ballestri, C. Chatgilialoglu, K. B. Clark, D. Griller, B. Giese and B. Kopping, *J. Org. Chem.*, 1991, **56**, 678.
76. R. Vanderesse, J.-J. Brunet and P. Caubère, *J. Org. Chem.*, 1981, **46**, 1270.
77. S. Krishnamurthy, *J. Org. Chem.*, 1980, **45**, 2550.
78. S. Krishnamurthy and H. C. Brown, *J. Org. Chem.*, 1982, **47**, 276.
79. E. C. Ashby, R. N. DePriest and T. N. Pham, *Tetrahedron Lett.*, 1983, **24**, 2825.
80. E. C. Ashby, R. N. DePriest, A. B. Goel, B. Wenderoth and T. N. Pham, *J. Org. Chem.*, 1984, **49**, 3545.
81. E. C. Ashby and T. N. Pham, *J. Org. Chem.*, 1986, **51**, 3598.
82. E. C. Ashby and T. N. Pham, *Tetrahedron Lett.*, 1987, **28**, 3197.
83. E. C. Ashby, T. N. Pham and A. Madjdabadi, *J. Org. Chem.*, 1988, **53**, 6156; E. C. Ashby, T. N. Pham and A. Amrollah-Madjdabadi, *J. Org. Chem.*,1991, **56**, 1596.
84. S.-U. Park, S.-K. Chung and M. Newcomb, *J. Org. Chem.*, 1987, **52**, 3275.
85. M. Newcomb and D. P. Curran, *Acc. Chem. Res.*, 1988, **21**, 206.
86. M. A. McKinney, S. M. Anderson, M. Keyes and R. Schmidt, *Tetrahedron Lett.*, 1982, **23**, 3443.
87. J. Hatem, J. M. Meslem and B. Waegell, *Tetrahedron Lett.*, 1986, **27**, 3723.
88. E. C. Ashby and J. J. Lin, *Tetrahedron Lett.*, 1977, **18**, 4481.
89. E. C. Ashby and J. J. Lin, *J. Org. Chem.*, 1978, **43**, 1263.
90. N. Shimizu, K. Watanabe and Y. Tsuno, *Chem. Lett.*, 1983, 1877.
91. G. R. Pettit and R. L. Smith, *Can. J. Chem.*, 1964, **42**, 572.
92. T. Imamoto, T. Takeyama and T. Kusumoto, *Chem. Lett.*, 1985, 1491.

93. R. O. Hutchins, D. Kandasamy, F. Dux, III, C. A. Maryanoff, D. Rotstein, B. Goldsmith, W. Burgoyne, F. Cistone, J. Dalessandro and J. Puglis, *J. Org. Chem.*, 1978, **43**, 2259.
94. S. Krishnamurthy and H. C. Brown, *J. Org. Chem.*, 1980, **45**, 849.
95. H. Irikawa, S. Yamamura and Y. Hirata, *Tetrahedron*, 1972, **28**, 3727.
96. F. Rolla, *J. Org. Chem.*, 1981, **46**, 3909.
97. H. C. Brown and S. Krishnamurthy, *J. Am. Chem. Soc.*, 1973, **95**, 1669.
98. S. Krishnamurthy and H. C. Brown, *J. Org. Chem.*, 1983, **48**, 3085.
99. E. C. Ashby, B. Wenderoth, T. N. Pham and W. S. Park, *J. Org. Chem.*, 1984, **49**, 4505.
100. R. O. Hutchins, D. Kandasamy, C. A. Maryanoff, D. Masilamani and B. E. Maryanoff, *J. Org. Chem.*, 1977, **42**, 82.
101. S. Kim, Y. J. Kim and K. H. Ahn, *Tetrahedron Lett.*, 1983, **24**, 3369.
102. H. Toi, Y. Yamamoto, A. Sonoda and S.-I. Murahashi, *Tetrahedron*, 1981, **37**, 2261.
103. N. M. Yoon and K. E. Kim, *J. Org. Chem.*, 1987, **52**, 5564.
104. G. F. Meijs and I. R. Doyle, *J. Org. Chem.*, 1985, **50**, 3713.
105. G. F. Meijs, *J. Org. Chem.*, 1987, **52**, 3923.
106. T. Hirao, T. Masunaga, Y. Ohshiro and T. Agawa, *J. Org. Chem.*, 1981, **46**, 3745.
107. T. Hirao, S. Kohno, Y. Ohshiro and T. Agawa, *Bull. Chem. Soc. Jpn.*, 1983, **56**, 1881.
108. A. Osuka, K. Takechi and H. Suzuki, *Bull. Chem. Soc. Jpn.*, 1984, **57**, 303.
109. W. Reimann, H. G. Kuivila, D. Farah and T. Apoussidis, *Organometallics*, 1987, **6**, 557.
110. C. L. Osborn, T. C. Shields, B. A. Shoulders, C. G. Cardenas and P. D. Gardner, *Chem. Ind. (London)*, 1965, 766.
111. D. Seyferth and B. Prokai, *J. Org. Chem.*, 1966, **31**, 1702.
112. C. W. Jefford, D. Kirkpatrick and F. Delay, *J. Am. Chem. Soc.*, 1972, **94**, 8905.
113. L. Sydnes and L. Skattebøl, *Tetrahedron Lett.*, 1974, **8**, 3703.
114. J. T. Groves and K. W. Ma, *J. Am. Chem. Soc.*, 1974, **96**, 6527.
115. H. Nozaki, T. Aratani and R. Noyori, *Tetrahedron*, 1967, **23**, 3645.
116. T. Shirafuji, K. Oshima, Y. Yamamoto and H. Nozaki, *Bull. Chem. Soc. Jpn.*, 1971, **44**, 3161.
117. D. Seyferth, H. Yamazaki and D. L. Alleston, *J. Org. Chem.*, 1963, **28**, 703.
118. A. J. Fry and R. H. Moore, *J. Org. Chem.*, 1968, **33**, 1283.
119. C. L. Osborn, T. C. Shields, B. A. Shoulders, J. F. Krause, H. V. Cortez and P. D. Gardner, *J. Am. Chem. Soc.*, 1965, **87**, 3158.
120. A. L. J. Beckwith and S. H. Goh, *J. Chem. Soc., Chem. Commun.*, 1983, 907.
121. K. Fukunaga and H. Yamaguchi, *Synthesis*, 1981, 879.
122. J. Nakayama, H. Machida and M. Hoshino, *Tetrahedron Lett.*, 1983, **24**, 3001.
123. D. Landini, L. Milesi, M. L. Quadri and F. Rolla, *J. Org. Chem.*, 1984, **49**, 152.
124. V. Janout and P. Čefelín, *Tetrahedron Lett.*, 1983, **24**, 3913.
125. M. Sevrin, J. N. Denis and A. Krief, *Tetrahedron Lett.*, 1980, **21**, 1877.
126. L. Engman, *Tetrahedron Lett.*, 1982, **23**, 3601.
127. H. Suzuki and M. Inouye, *Chem. Lett.*, 1985, 225.
128. H. G. Kuivila and Y. M. Choi, *J. Org. Chem.*, 1979, **44**, 4774.
129. J. Casanova and H. R. Rogers, *J. Org. Chem.*, 1974, **39**, 2408.

4.2
Reduction of Saturated Alcohols and Amines to Alkanes

STUART W. McCOMBIE
Schering-Plough Corporation, Bloomfield, NJ, USA

4.2.1 INTRODUCTION

This chapter will deal with reductive processes which, either directly or following appropriate derivatization, replace hydroxy and amino groups attached to saturated carbon with hydrogen, as indicated in equation (1). Such processes are usually referred to as deoxygenations and reductive deaminations. In general, we will not deal with functional group removal from allylic or benzylic carbon, except where the behavior of these systems is of particular relevance to the fully saturated case; the same remarks apply to processes which proceed through carbonyl compounds or imines, since the reduction of C=X to CH_2 is covered in detail in Chapters 1.13 and 1.14 of this volume.

$$R^1R^2R^3C{-}OH \quad \text{or} \quad R^1R^2R^3C{-}NH_2 \longrightarrow R^1R^2R^3C{-}H \qquad (1)$$

In the context of total synthesis, the reduction in the level of functionalization implicit in these processes is not always at variance with synthetic objectives, a simple example being the aldol condensation–deoxygenation sequence of Scheme 1, which could replace a (frequently more difficult) direct enolate alkylation. The conversion of readily available, polyfunctionalized materials such as carbohydrates into specifically deoxygenated or deaminated derivatives provides a variety of chiral synthons for the assembly of more complex substances.

The second, major use for deoxygenations and reductive deaminations is the modification of complex natural products, and the need to achieve regio- and chemo-selectivity in such processes has provided significant impetus to the search for novel methods which carry out these transformations under mild

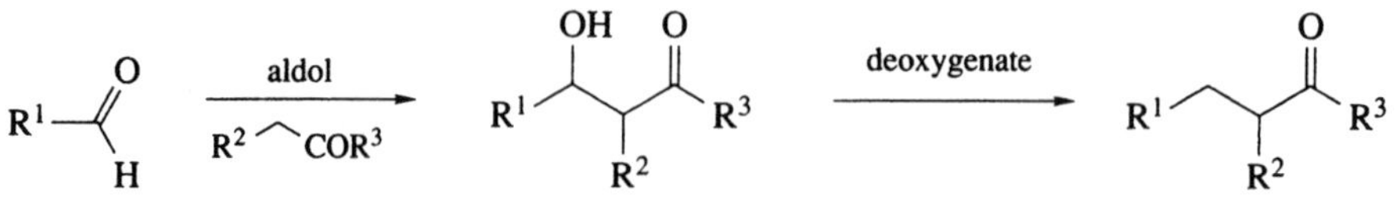

Scheme 1

conditions. Such alteration of natural products is important both for structure elucidations and correlations, and for the exploration of biological structure–activity relationships. For example, therapeutic advantage may be gained by removing a functional group which is not essential to intrinsic activity, but which is a site for metabolic inactivation.

Chemically, methods for deoxygenation and reductive deamination have seen additions in the last two decades which depart from those traditional approaches using hydride reduction of activated alcohols and diazene generation from aliphatic amines. New methods for amine activation, new complex hydrides and, in particular, methods using electron transfer and free radical processes have greatly expanded the range of substrate molecules which can be subject to efficient functional group removal.

In the following sections, the advantages and disadvantages of each method will be illustrated by pertinent examples. The various techniques for removing hydroxy or amino are arranged according to the basic mechanistic process involved in the key step.

4.2.2 DEOXYGENATION OF ALCOHOLS AND THEIR DERIVATIVES

4.2.2.1 Hydride Reduction Methods

The LAH reduction of tosylates[1] is generally useful for primary and some unhindered secondary alcohols. The relatively indiscriminate reactivity of LAH towards many other functional groups means that these must be protected, as exemplified by the synthesis of the 6-deoxygalactose derivative (**2**) shown in equation (2). Primary tosylates are also reduced by $NaBH_4$ in dipolar, aprotic solvents,[2,3] and cyclododecyl tosylate was reduced slowly.[4] Mesylates were also reduced: $NaBH_4$ in DMSO converted (**3**) to (**4**) without significant reduction or cleavage of the secondary sulfonates.[5] The reduction of RCH_2OTs also took place with $NaBH_3CN$ in HMPA; although the reactions were slow, these conditions did permit the survival of nitriles, esters and aryl ketones.[6] More activated sulfonates such as triflates would presumably react more rapidly with $NaBH_3CN$. For primary alcohols, conversion to the bromo or iodo compound should also be considered, since this affords a number of choices (hydride, electron transfer or stannane) for the subsequent reduction step (see Chapter 4.1, this volume).

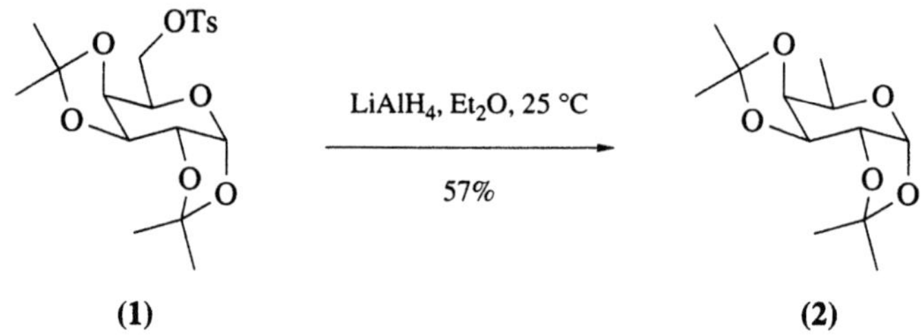

LiAlH$_4$, Et$_2$O, 25 °C; 57% (2)

(**1**) (**2**)

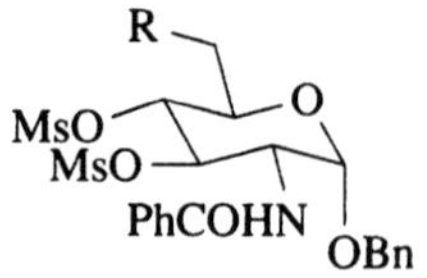

(**3**) R = OMs
(**4**) R = H

When secondary sulfonates are treated with LAH, competing reactions frequently occur. For polyoxygenated, carbohydrate-derived substrates, complexation of the reagent, inductive deactivation and lone pair–nucleophile repulsion all can cause severe retardation of the desired S_N2 process, and S—O scission to the parent alcohol is the usual result.[2,7] In cyclohexanes, elimination may be the dominant process when the tosylate is axial and/or the β-H is activated, an extreme case being that shown in equation (3).

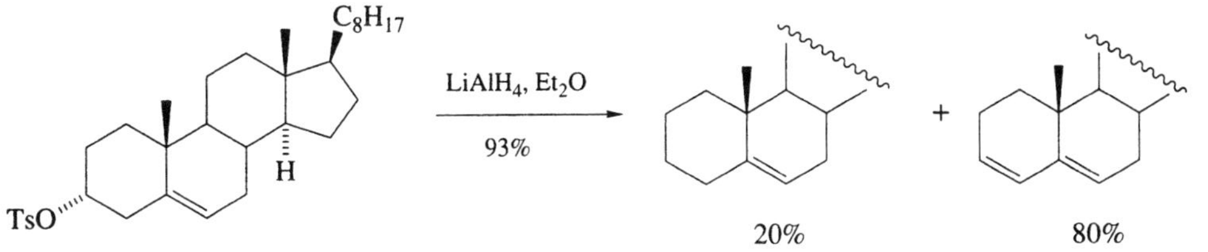

C_8H_{17} TsO H LiAlH$_4$, Et$_2$O 93% 20% + 80% (3)

In a study of the reduction of primary neopentyl and cycloalkyl mesylates and tosylates, two groups[8] showed that $LiBEt_3H$ in THF is superior to LAH and L-selectride in both rate and selectivity for reduction *versus* elimination. While cyclohexyl tosylate still gave 20% alkene, elimination was not seen for the cyclo-pentyl, -heptyl and -octyl esters. The lower rate for neopentyl systems was used to advantage in the selective reduction of (**5**) to (**6**) at 25 °C. In a more complex system,[9] $LiBEt_3H$ reduction of (**7**; R^2 = OMs, OTs) regenerated the alcohol (**7**; R^2 = OH), whereas the more hindered 2-propanesulfonate (**8**) afforded the deoxy compound (**9**). Methods for preparing other hindered sulfonates ($ROSO_2Bu^t$ and $ROSO_2CPh_2CF_3$) have been described,[10] but their behavior towards LAH and other hydrides remains to be systematically explored.

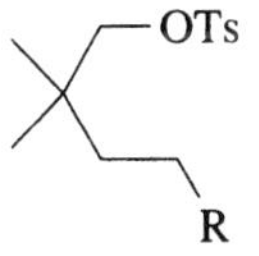

(**5**) R = OTs
(**6**) R = H

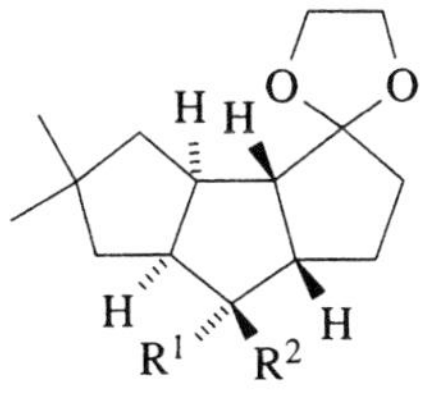

(**7**) R^1 = H, R^2 = OH
(**8**) R^1 = H, R^2 = OSO_2Pr^i
(**9**) R^1 = R^2 = H

Chemoselective reagents of the general type Li(CuHR) were developed by Masamune *et al.*[11] for reducing halides and sulfonates. A reagent prepared from CuI and $2LiAlH(OMe)_3$ in THF reduced several mesylates efficiently in a net S_N2 process, and the related reagent $LiCuHBu^n$ converted mesylate (**10**) to (**11**) without reducing the ester. Under the same conditions, these reagents reduced ketones and substrates susceptible to conjugate addition.

As might be anticipated, attempts to apply sulfonate–hydride processes to tertiary alcohols lead to elimination and rearrangement reactions. However, the relative ease of carbonium ion formation from these systems permits the use of 'ionic hydrogenations'[12] in which an alcohol (or ether, alkene or ester) is reacted with a protic or Lewis acid (usually CF_3CO_2H) in the presence of a mild, acid-stable hydride donor such as Et_3SiH. Although most frequently used for benzylic alcohols, some aliphatic cases have been studied and examples are the efficient conversion of (**12**) to (**13**) and of (**14**) to (**15**) in Et_3SiH–CF_3CO_2H. 2-Octanol was reduced to octane with BF_3–Et_3SiH.[13] The stereochemical course of these reactions is known for some alkene reductions, and would presumably follow a similar course for tertiary alcohol or ether precursors. The methylcholestene (**16**) gave mostly the equatorial product (**17**),[14] but mixtures were obtained from 1,2-dimethylcycloalkenes. These reactions are subject to rearrangement of the carbonium ion prior to hydride capture: 5-cholestene gave an isomer mixture derived by prior back-

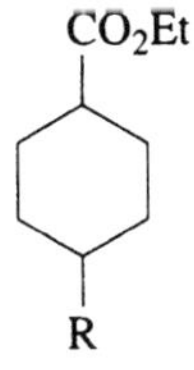

CO_2Et R

(**10**) R = OMs
(**11**) R = H

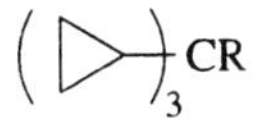

(**12**) R = OH
(**13**) R = H

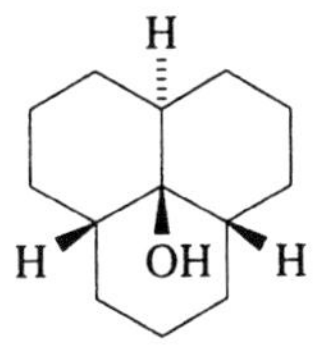

(**14**)

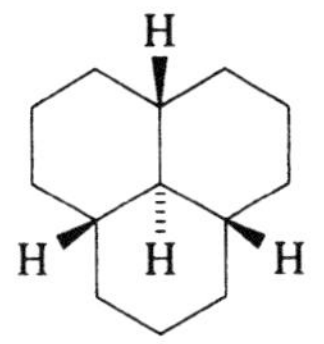

(**15**)

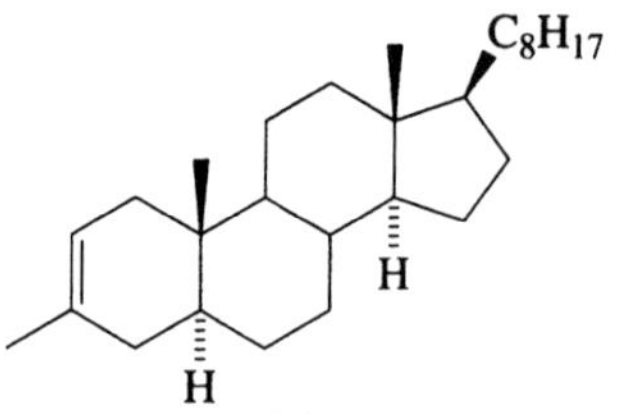

(16) 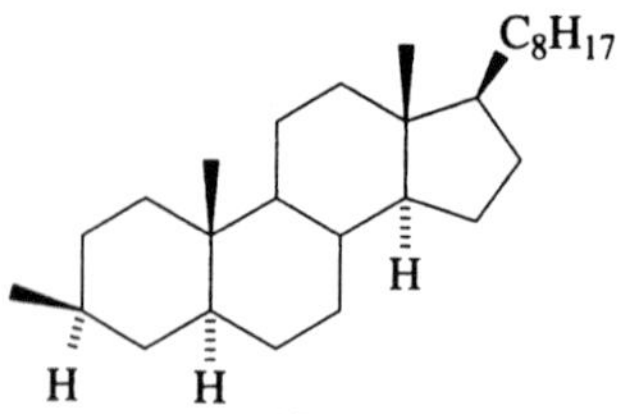 (17) 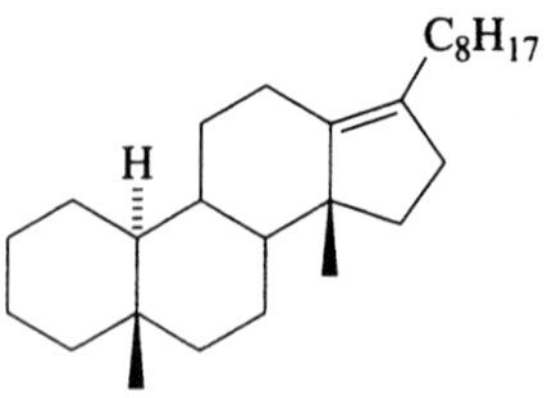 (18)

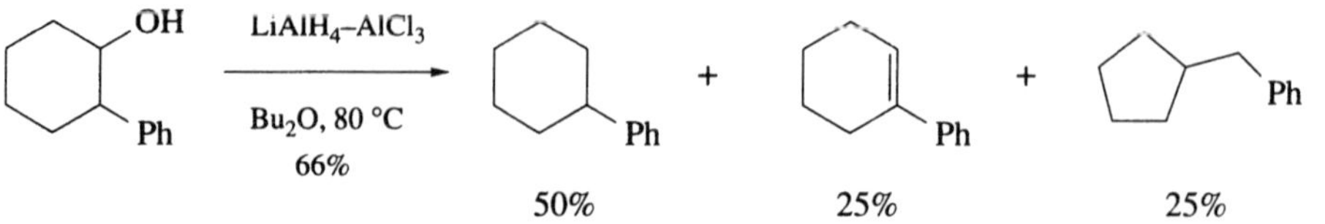

$$\text{2-phenylcyclohexanol} \xrightarrow[\text{Bu}_2\text{O, 80 °C, 66\%}]{\text{LiAlH}_4\text{–AlCl}_3} \text{phenylcyclohexane (50\%)} + \text{phenylcyclohexene (25\%)} + \text{benzylcyclopentane (25\%)} \quad (4)$$

bone rearrangement to (**18**), and the obtention of the all-*trans* product (**15**) from *cis-cis-trans* (**14**) was a consequence of complete 1,2-hydride shift before reduction.[14] For the same reason, reduction of saturated secondary and tertiary alcohols by LAH–$AlCl_3$ mixtures corresponding to 'Cl_2AlH' frequently produced alkenes and rearranged materials in addition to the simple deoxy compound, as shown in equation (4).[15]

The reduction of 1,2-diols and their derivatives by hydride or other methods often constitutes a special case. The following guidelines may be useful, and will be exemplified in later sections. For 1,2-diols with a primary center as the target for reduction, selective activation (a bulky reagent such as mesitylenesulfonyl chloride[16] may be advantageous) can be followed by hydride reduction, which may proceed *via* the epoxide if the second hydroxy is not blocked. In carbohydrates, reduction of bis-secondary 1,2-disulfonates frequently occurs *via* S—O cleavage, epoxide formation and 1,2-diaxial opening (see Chapter 4.4, this volume). Removal of both hydroxy groups is commonly carried out through the corresponding alkene, but can also be done directly with certain photoreduction and free radical methods; the latter provide the methods of choice for monodeoxygenation of diols at the more substituted center.

4.2.2.2 Catalytic Hydrogenolysis

Despite the haptophilic behavior of the hydroxy group towards both heterogeneous and homogeneous noble metal catalyst systems,[17] a C—OH bond unactivated by adjacent alkenic or aromatic groups (see Chapter 4.7) is quite resistant to hydrogenolysis, and forcing conditions are usually required.

In a series of papers,[18] Adkins and coworkers reported hydrogenolysis of alcohols over $CuCrO_2$, BaO–$CuCrO_2$ and Ni catalysts. At the elevated temperatures and pressures needed to induce reduction, cleavage of C—C bonds also took place *via* dehydrogenation–decarbonylation and retroaldol processes. For simple systems, primary alcohols were converted to the next lower hydrocarbon (equation 5), whereas 2-octanol and cyclohexanol were simply deoxygenated. Similar results were obtained with NiO–Al_2O_3 catalysts.[19] Rearrangement of branched compounds was avoided using Co–Al_2O_3, but only partial conversion was seen under very forcing conditions.[20]

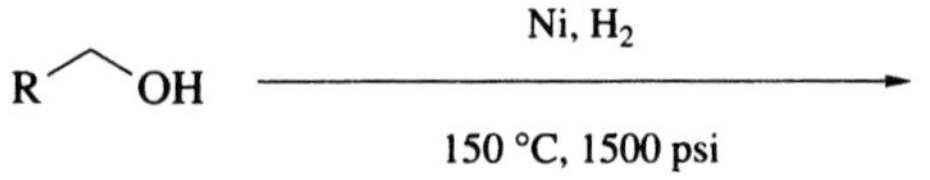

$$\text{RCH}_2\text{CH}_2\text{OH} \xrightarrow[\text{150 °C, 1500 psi}]{\text{Ni, H}_2} \text{R–H} \quad (5)$$

Some selectivity for monodeoxygenation of glycols was obtained over Ni or $CuCrO_2$, and depended upon the separation of the hydroxy groups: under conditions which did not affect the 1,4-diol, cyclohexane-1,3-diol gave cyclohexanol. The examples of equations (6) and (7) show that there was some selectivity for removing primary hydroxy groups.

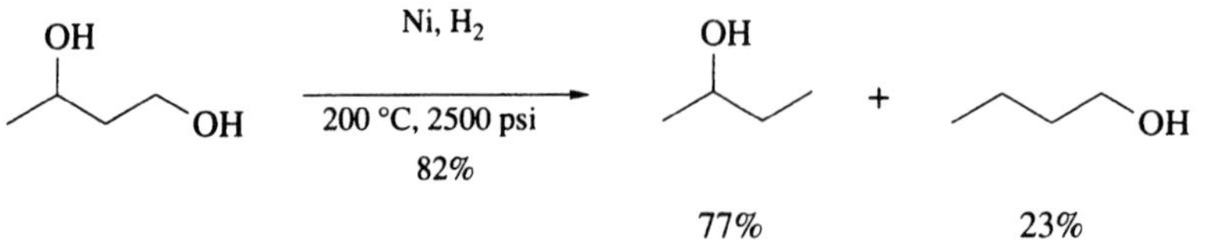

$$\text{butane-1,3-diol} \xrightarrow[\text{200 °C, 2500 psi, 82\%}]{\text{Ni, H}_2} \text{2-butanol (77\%)} + \text{1-butanol (23\%)} \quad (6)$$

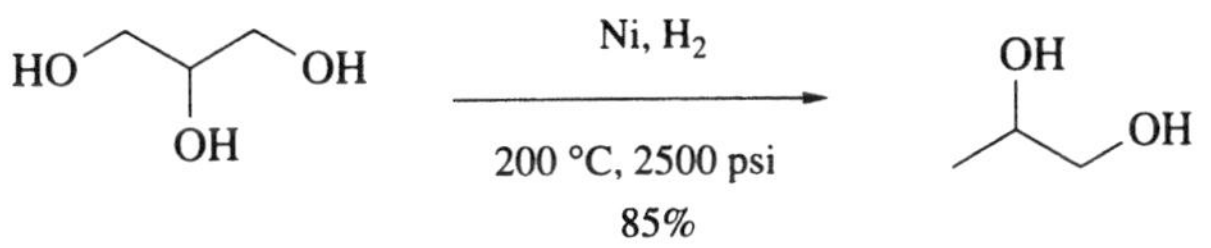

HO–CH₂CH(OH)CH₂–OH → (Ni, H_2; 200 °C, 2500 psi; 85%) → $CH_3CH(OH)CH_2OH$ (7)

The more facile reduction[21] of tertiary alcohols and esters and of secondary tosylates over Pt in HOAc or CF_3CO_2H was ascribed to alkene formation prior to reduction. The deoxygenation of tertiary alcohols with Raney nickel in refluxing toluene[22] also proceeds through the alkenes; these conditions also effected the transformation shown in equation (5).

The two-step sequence shown in Scheme 2 has been used to deoxygenate a variety of monohydric, acyclic alcohols.[23] Reaction of the alcohol with DCC (other carbodiimides gave less satisfactory results in the reductive step) catalyzed by CuCl gave the *O*-alkyl-*N*,*N*′-dicyclohexylisourea, which was hydrogenated over Pd–C in EtOAc to the deoxy compound. Excellent yields were obtained (Table 1) for primary, secondary and tertiary acyclic systems with reduction times of 2–30 h, the tertiary case being the most reactive. However, this method was not useful for most cycloalkanols, the reductive step being extremely slow. The *O*-isobornyl derivative (**19**) was reduced readily, but with rearrangement, giving *endo*-isocamphane (**20**).

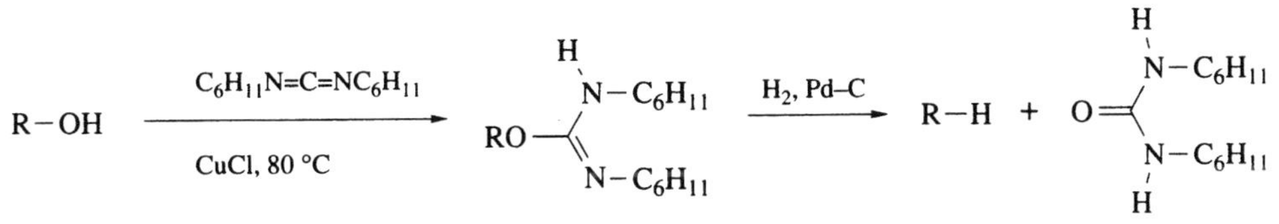

R–OH → ($C_6H_{11}N{=}C{=}NC_6H_{11}$; CuCl, 80 °C) → RO–C(=N–C_6H_{11})–NH–C_6H_{11} → (H_2, Pd–C) → R–H + O=C(NH–C_6H_{11})₂

Scheme 2

Table 1 Formation and Hydrogenation of *O*-Alkyl-*N*,*N*′-dicyclohexylisoureas[23]

Alcohol	*Isourea, yield* (%)[a]	*Hydrocarbon, yield* (%)[b]
Octan-1-ol	93	97 (81)
Octan-2-ol	94	98
Octan-4-ol	93	97
Tetradecanol	93	96 (76)
2-Phenylethanol	94	91
2-Methylheptan-2-ol	c	94

[a]Isolated yield, 0.10 mol scale. [b]Determined by GLC; overall yields in parentheses represent examples in which the intermediate isourea was not isolated. [c]Not cited.

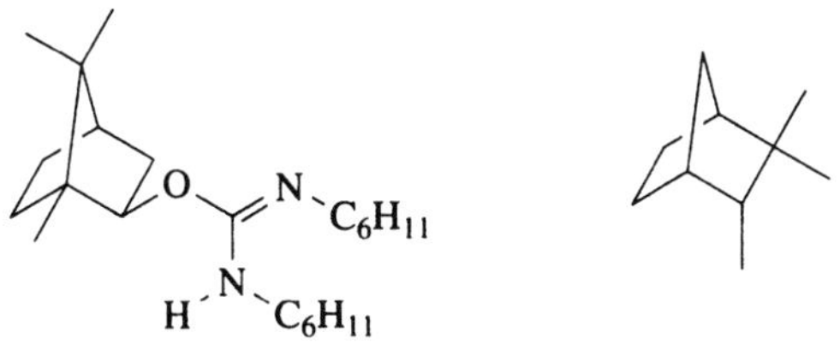

(**19**) (**20**)

4.2.2.3 Electron Transfer Reductions

When a carboxylic, phosphoric or sulfonic ester (or thioester) of an aliphatic alcohol is made to accept electrons, several reaction pathways are available to the resulting radical anion, as shown in Scheme 3.

The desirable fragmentation (A) to give the alkyl radical represents a viable method for overall deoxygenation, providing that alternative processes involving O—X fission (B), or H-abstraction (C) or net two-electron reduction (D; the Bouveault–Blanc reduction of esters, for example) followed by O—X fission, can be minimized. Conditions have been established which cause (A) to be the dominant process for X = C, P and S; this section will compare the merits of these processes for alcohol deoxygenation.

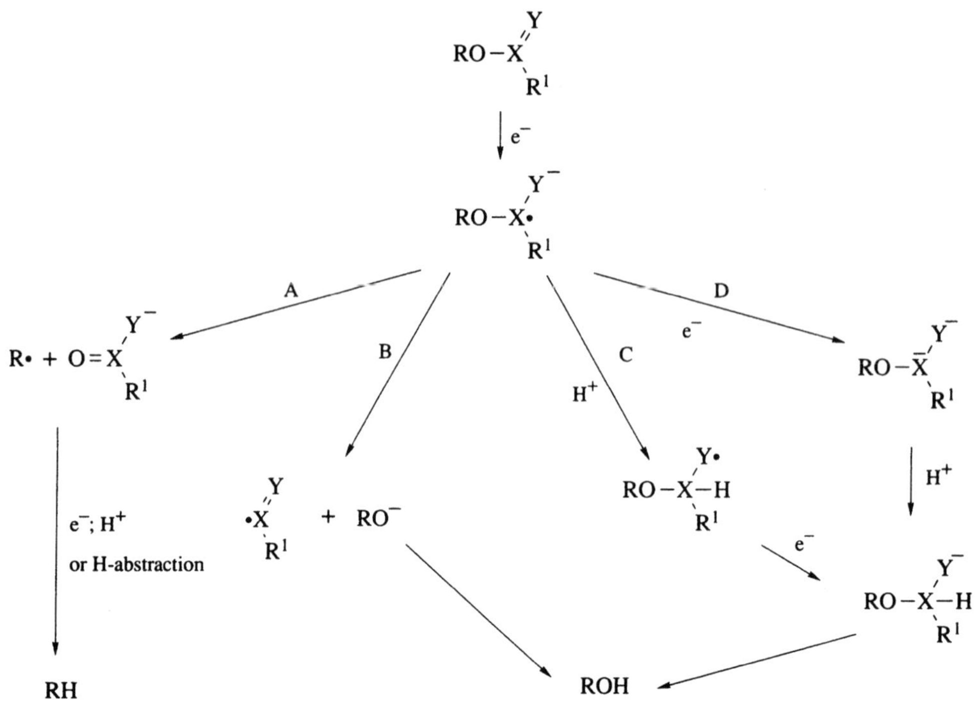

Scheme 3

From the viewpoint of experimental conditions, three methods have been used to supply the solvated electrons required: dissolving metals, electrochemistry and photochemistry. Several groups have explored the reduction of esters by dissolving metals, and this proves to be a general method for deoxygenating tertiary and hindered secondary alcohols. For primary and unhindered secondary systems, substantial amounts of alcohol tend to be regenerated, although this can be minimized by using esters of bulky carboxylic acids. Reductions of the esters of 3β,6β,5α-cholestanediol illustrate salient points.[24] The diacetate (**21**) with Li–$EtNH_2$ or Li–$NH_2CH_2CH_2NH_2$ gave the 6-deoxy compound (**22**) and diol (**23**) in 1.3:1 ratio, improved to 2:1 by using K–Bu^tNH_2 and 18-crown-6. When the pivalate (**24**) was similarly reduced, the RH:ROH ratio was 8.8:1. Compelling evidence for a carbon radical intermediate was provided by the cyclosteroid example shown in equation (8), and is consistent with the very efficient

OAc —(Li–$EtNH_2$, 45%)→ (85%) + (15%) (8)

C_8H_{17}

(21) $R^1 = R^2 =$ OAc
(22) $R^1 =$ OH, $R^2 =$ H
(23) $R^1 = R^2 =$ OH
(24) $R^1 = R^2 = O_2CBu^t$

(25) $R^1 =$ OAc, $R^2 =$ Me
(26) $R^1 =$ Me, $R^2 =$ OAc

deoxygenation[25] of tertiary acetates (**25**) and (**26**) by Na–HMPA–Bu^{t}OH. Similar reductions have been effected with Na–K alloy in THF in the presence of $(MeOCH_2CH_2OCH_2CH_2O)_3N$,[26] avoiding side reactions associated with cleavage of 18-crown-6.

Several methods are available which minimize the O—X cleavage process,[27] of which the most general was developed by Ireland and coworkers.[28] A wide variety of alkyl and cycloalkyl *N,N,N',N'*-tetramethylphosphorodiamidates were cleanly deoxygenated (Scheme 4) with essentially no regeneration of the starting alcohol. Examples are the conversion of (**27**) to (**28**) and of cholesterol to cholest-5-ene, with near-quantitative yields in the reduction step. For tertiary alcohols, the *O,O*-diethylphosphates were satisfactory, as in the conversion of (**29**) to adamantane. Metal–amine reductions of *N,N*-dimethylthiocarbamates,[29] *N,N*-dimethylsulfamates[30] and triflates[31] have also been reported; in some carbohydrate examples, equatorially disposed β-hydroxy groups survived without alkene formation, but this was not generally the case for these reactions proceeding *via* carbanionic intermediates.

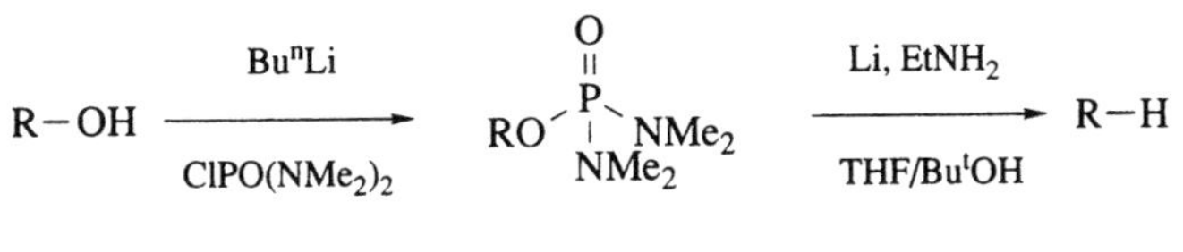

R–OH —(Bu^nLi / $ClPO(NMe_2)_2$)→ $RO-P(=O)(NMe_2)_2$ —(Li, $EtNH_2$ / THF/Bu^{t}OH)→ R–H

Scheme 4

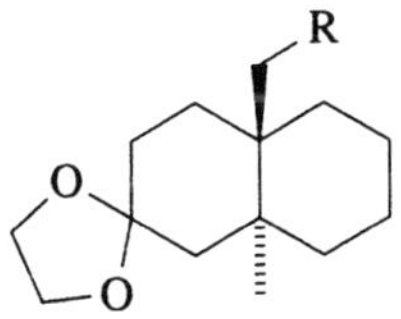

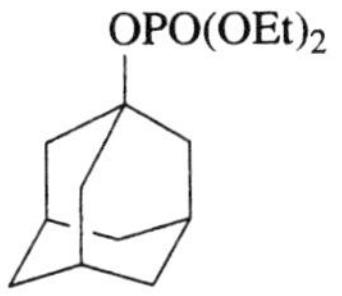

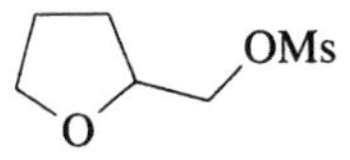

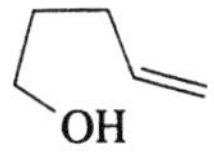

(**27**) R = OH
(**28**) R = H

OPO(OEt)$_2$ (**29**)

OMs (**30**)

OH (**31**)

Mesylates, which tend to undergo O—S scission with dissolving metals, were reduced electrochemically in DMF. Yields of 70–85% were reported[32] under conditions which left esters, nitriles and epoxides unchanged. The conversion of (**30**) to (**31**) again indicated a carbanionic intermediate.

Photochemical methods have been used for deoxygenation under milder conditions. Photolysis of acetates at 254 nm in 95:5 HMPA–H_2O was generally useful for molecules which did not contain chromophoric groups, and removed primary, secondary and tertiary oxygens.[33] Pivalates were generally superior to acetates, as exemplified by the formation of deoxy compound (**32**) in 65% yield from (**33**) and in 75% yield from (**34**). Homolytic scission from the radical anion followed by H-atom abstraction is clearly involved; metal–amine reduction of (**35**) did not give (**32**), but led to complete ring scission *via* elimination from C-2. In appropriate cases, interfering chromophores could be modified by protection: deoxygenation of testosterone acetate was achieved by photoreduction of ketal (**36**) to (**37**) in HMPA–H_2O, followed by acid hydrolysis to (**38**) in 60% overall yield. In all of these direct photoreductions of acetates, pivalates or triflates, the use of HMPA–H_2O was essential for deoxygenation.

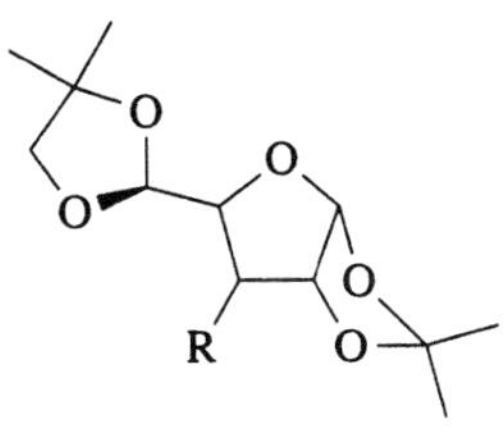

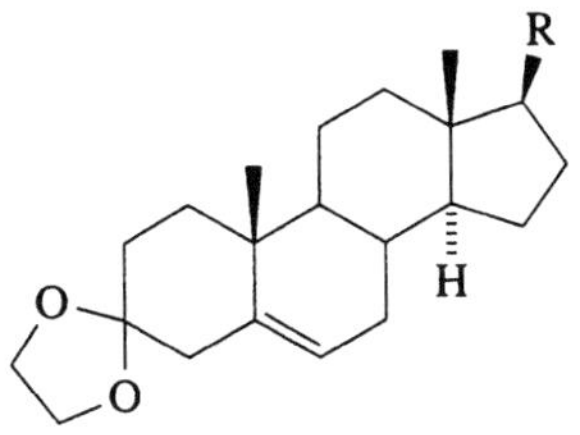

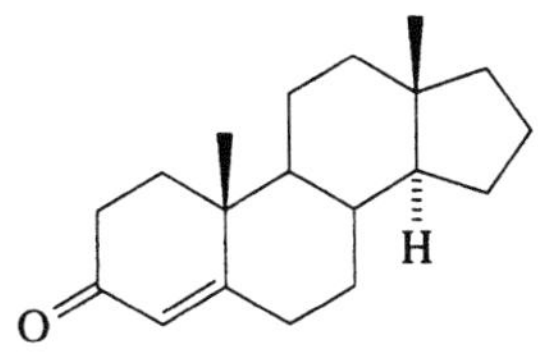

(**32**) R = H
(**33**) R = OAc
(**34**) R = O_2CBu^t
(**35**) R = OPO(NMe$_2$)$_2$

(**36**) R = OAc
(**37**) R = H

(**38**)

For substrates with UV-absorbing functional groups, an interesting photosensitized ester deoxygenation has been described.[34] Irradiation (Pyrex) of an EtOH–H_2O solution of the *m*-trifluorobenzoyl derivative (**39**) containing *N*-methylcarbazole (MCZ) as the sensitizer gave 91% of (**40**) with recovery of MCZ and *m*-$CF_3C_6H_4CO_2H$. This process was specific for secondary alcohols and did not proceed through carbanionic intermediates: the adenosine triester (**41**; A = adenine) was converted cleanly to the 2′,3′-dideoxy compound (**42**) with no 2′,3′-alkene. A radical process was confirmed by incorporation of deuterium from CD_3CD_2OH, but not from D_2O. This method would seem to be particularly useful for vicinal dideoxygenation, the more so since thiocarbonyl–stannane methods are not satisfactory for this transformation. In a further simplification the unsubstituted benzoate (**43**; U = uracil), which was a poor substrate when irradiated with MCZ alone, was efficiently converted to (**44**) in the presence of $Mg(ClO_4)_2$.

(**39**) R = $O_2CC_6H_4$-*m*-CF_3
(**40**) R = H

(**41**) R = O_2CAr
(**42**) R = H
[Ar = C_6H_4-*m*-CF_3]

(**43**) R = O_2CPh
(**44**) R = H

4.2.2.4 Stannane Reduction of *O*-Thiocarbonyl Derivatives

The free radical deoxygenation of secondary alcohols[35] by the reaction of various *O*-thiocarbonyl derivatives with tri-*n*-butylstannane (Bu_3SnH) shown in equation (9) was the prototype for several modifications and related processes which, taken as a whole, permit the deoxygenation of most types of alcohols and the monodeoxygenation of 1,2 and 1,3-diols. In these two-step processes, the initial derivatization involves either basic or near-neutral conditions and the reductive step occurs in neutral, aprotic solvents at 80–160 °C. These processes are limited only by those side-reactions (intramolecular cyclizations or fragmentations) of carbon radicals which can successfully compete with H-abstraction from an efficient donor (R_3SnH), and frequently provide the method of choice for deoxygenation in polyfunctionalized systems bearing groups sensitive to hydride or dissolving metal reagents. These methods have been reviewed in brief[36] and in detail,[27] and are included in general reviews of carbon radical chemistry[37] and synthetic uses of Bu_3SnH.[38]

$$R^1R^2CHO(C{=}S)X \xrightarrow[\text{toluene, reflux}]{Bu_3SnH} R^1R^2CH_2 \qquad (9)$$

Scheme 5 shows the reactions that occur when a thioester or similar substrate (**45**) interacts with Bu_3SnH under free-radical conditions. After initial addition (which may be reversible) to afford the stabilized intermediate radical (**46**), the desired fragmentation (A) will take place with an ease which depends to some extent on the nature of X (in ROCSX), but is more determined by the temperature, and by the nature of R. Steric compression about the C—O region encourages the fragmentation without seriously affecting the addition step, making these methods very useful for hindered systems. After fragmentation, the major fate of the alkyl radical is H-abstraction (C) to give the deoxy compound (**49**) and the chain carrier ($Bu_3Sn\cdot$); a less common occurrence is (D), where capture of R by the substrate rather than Bu_3SnH results in overall conversion to the isomeric *S*-thioester. More frequent (but usually quite minor) side-reactions arise from abstraction prior to fragmentation (B) and ultimately produce the starting alcohol (**52**) or the ether (**53**), depending upon the particular substrate. An alternative possibility involving S_H2 attack on the singly bonded sulfur exists for dithiocarbonates (xanthates) and was proposed on the basis of ESR (electron spin resonance) measurements, which indicated some production of ROCS· from $ROCS_2Me$ at low temperatures.[39] Under typical deoxygenation conditions, however, experiments using dithiocarbonates with *S*-substituents with different steric and electronic characteristics indicated that the addition of $Bu_3Sn\cdot$ to C=S is the important initial step.[40] The presence of Et_3B allowed some deoxygenations to proceed at ambient temperature,[41] but the mechanistic processes involved were not established.

For a given secondary alcohol, the choice of thiocarbonyl derivative is predicated more by structural features (steric hindrance, other functional groups) present in the alcohol than by the actual reductive step. Barton and McCombie[35] reported comparable results for *S*-methyldithiocarbonyl (MDC), thiobenzoyl (TB) and imidazole-1-thiocarbonyl (ITC) derivatives, and typical preparations of these are shown in

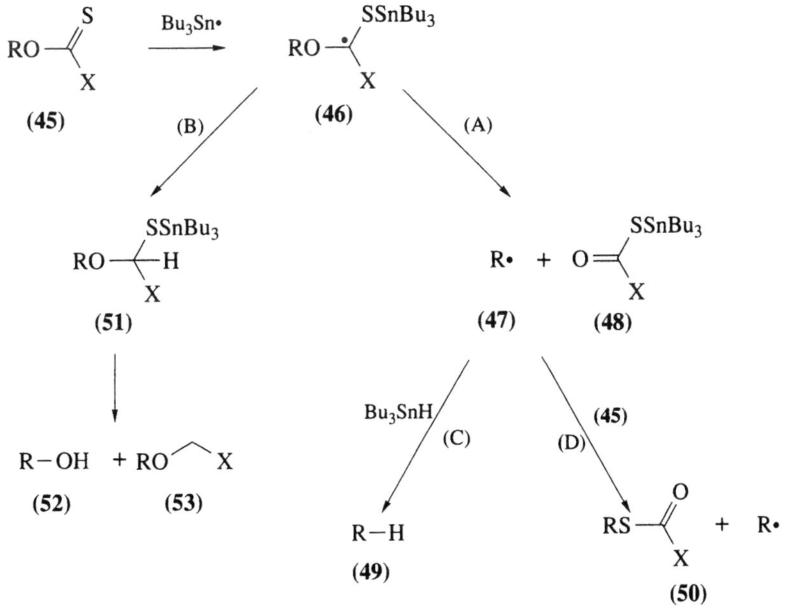

Scheme 5

equations (10)–(12). The ITC derivatives were recommended by Rasmussen *et al.*[42] for carbohydrates, and phenoxythiocarbonyl (PTC) derivatives prepared as shown in equation (13) were introduced by Robins *et al.*[43] for nucleoside deoxygenation.

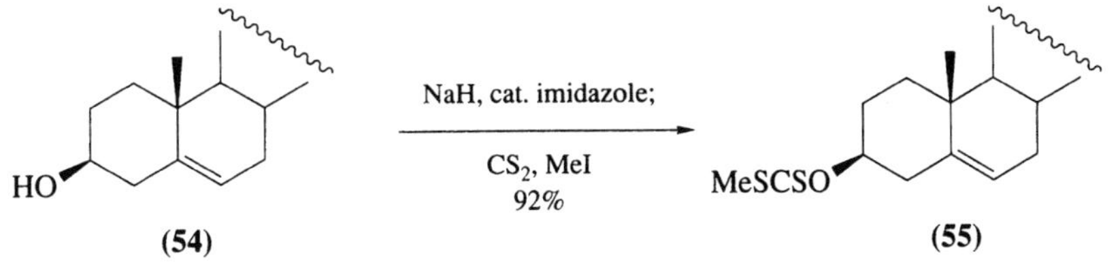

(54) NaH, cat. imidazole; CS$_2$, MeI 92% (55) (10)

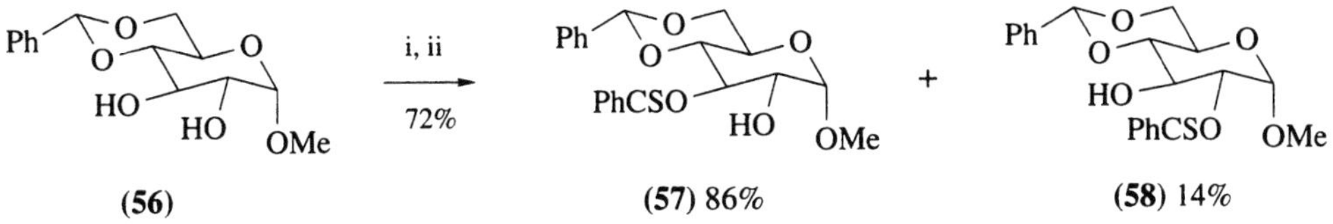

(56) i, ii 72% (57) 86% + (58) 14% (11)

i, PhCCl=NMe$_2$$^+Cl^-$, Py; ii, H$_2$S, Py, 25 °C

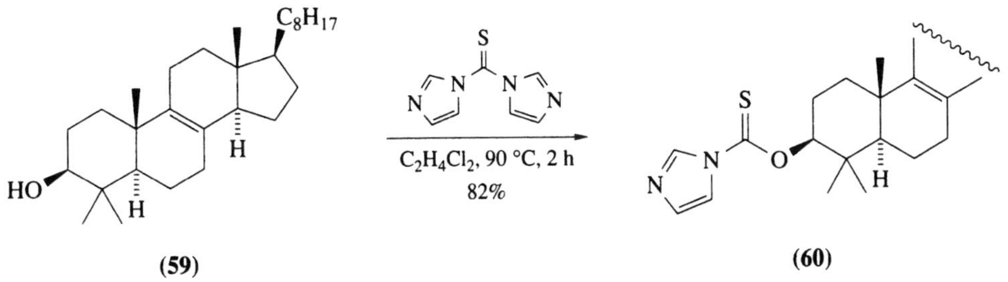

(59) C$_2$H$_4$Cl$_2$, 90 °C, 2 h 82% (60) (12)

Table 2, which compares syntheses of the protected 3-deoxyglucose (**32**) and of cholest-5-ene (**63**), shows that all of the thiocarbonyl derivatives gave good to excellent yields. In general, the MDC and PTC derivatives gave rise to the smallest amounts of by-product, which in the case of the TB and ITC derivatives was mostly the starting alcohol. Although TB derivatives required gradual addition to Bu$_3$SnH solution, and occasionally gave rise to significant amounts of the benzyl ether,[44] the preparative route shown in equation (11) does have one special characteristic: since the prethiolysis intermediate

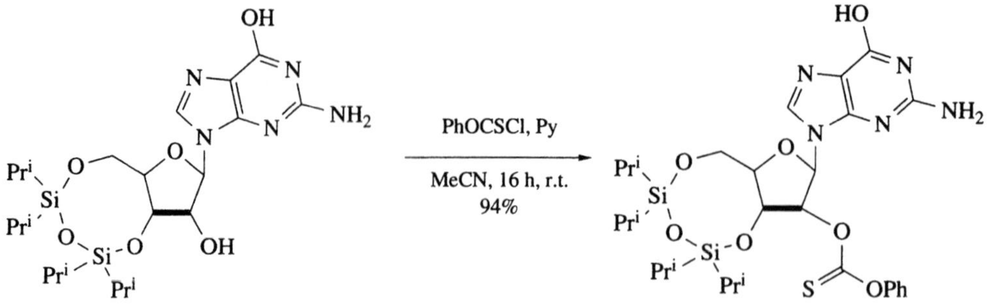

PhOCSCl, Py; MeCN, 16 h, r.t.; 94% (13)

(61) **(62)**

(like the reagent, $PhCCl{=}NMe_2^+Cl^-$) is charged, only a single hydroxy in a polyol is functionalized under mild conditions.[35] For example, (**64**) gave only monoester (**65**); to obtain the bis-TB derivative (**66**), it was necessary to subject isolated (**65**) to a second sequence.

Table 2 Tri-*n*-butylstannane Reduction of *O*-Thiocarbonyl Derivatives of Cholesterol (A) and 1,2,5,6-Di-*O*-ispropylidene D-glucofuranose (B) in Toluene

Alcohol	*Derivative*	*Temperature* (°C)	*Time* (h)	*Yield* (%)[e]	*Ref.*
A	PhCS[a,b]	110	2	73	35
A	MeSCS[b]	110	7	78	35
A	ImCS[b]	110	6	74	35, 42
A	PhOCS[c]	110	7	85	43
B	MeSCS[b]	110	18	85	35
B	ImCS[b]	110	6	74	42
B	PhOCS[c]	80	3	85	43

[a]Thioester added over 0.5–1 h. [b]No initiator added. [c]0.2 equiv. $Bu^t_2O_2$ added. [d]0.2 equiv. AIBN added. [e]Isolated yield.

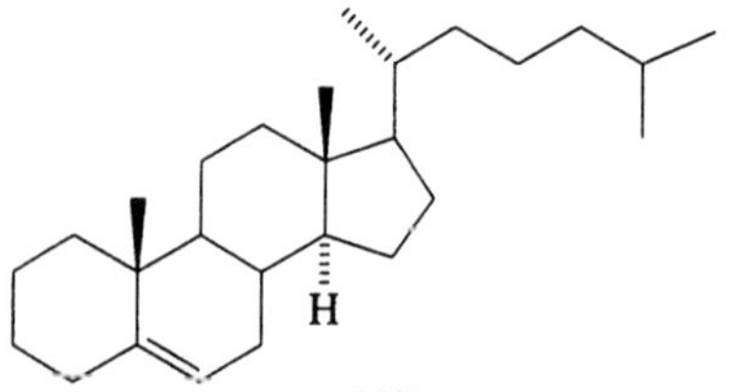

(63)

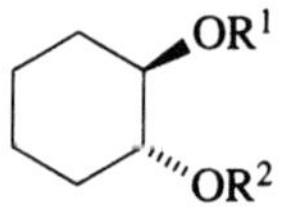

(**64**) $R^1 = R^2 = H$
(**65**) $R^1 = H$, $R^2 = CSPh$
(**66**) $R^1 = R^2 = CSPh$

Secondary hydroxy groups have been removed from a wide range of polyfunctionalized molecules[27] using these processes, which tolerate the presence of esters, ketones, alcohols, C—F bonds, protected amines and alkenes, providing the latter are not situated so as to encourage cyclization of the intermediate radical. The aminoglycoside antibiotics provide some elaborate examples: the protected kanamycin (**67**) was converted to MDC derivative (**68**) and thence to deoxy compound (**69**) in 59% overall yield, and other secondary positions in related systems were deoxygenated *via* *S*-phenyldithiocarbonyl derivatives prepared from the alcohol and PhSCSCl–pyridine.[45] The reductions of lanosterol thioesters[35] and of the hirsutene intermediate (**70**) to (**71**)[46] exemplify deoxygenations at neopentyl centers. As previously noted,[41] inclusion of Et_3B in these reactions permits secondary deoxygenations to be run at ambient temperature, which may be advantageous for thermally labile materials.

Monodeoxygenation of 1,2- and 1,3-diols was achieved[47] *via* their cyclic thiocarbonates, prepared from the diol and *N,N'*-thiocarbonyldiimidazole, by reaction with Bu_3SnH–AIBN followed by alkaline hydrolysis (presumably, F^- would also be effective for the cleavage step). Equation (14) shows this process applied to synthesis of a derivative (**73**) of 5-deoxyglucose. Exclusive secondary deoxygenation is expected on the basis of radical stability; in contrast, the derivative (**72**) was readily converted by an ionic process to an intermediate suitable for 6-deoxygenation, since treatment with KI gave (**74**) quantitatively.

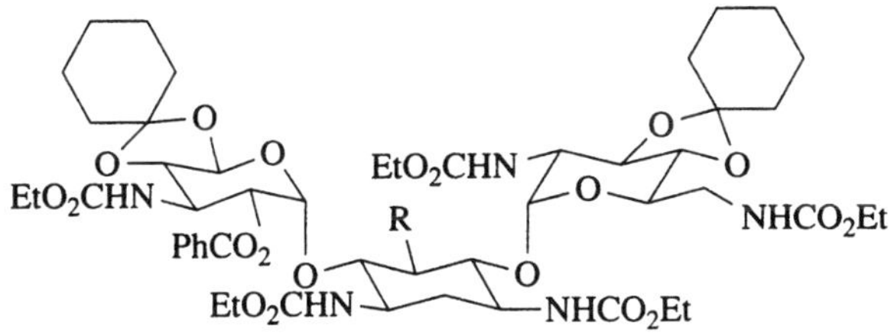

(**67**) R = OH
(**68**) R = OCSSMe
(**69**) R = H

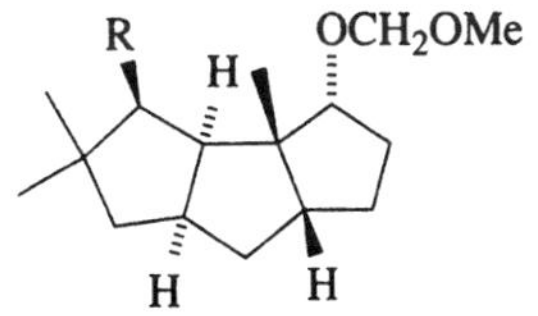

(**70**) R = OCSSMe
(**71**) R = H

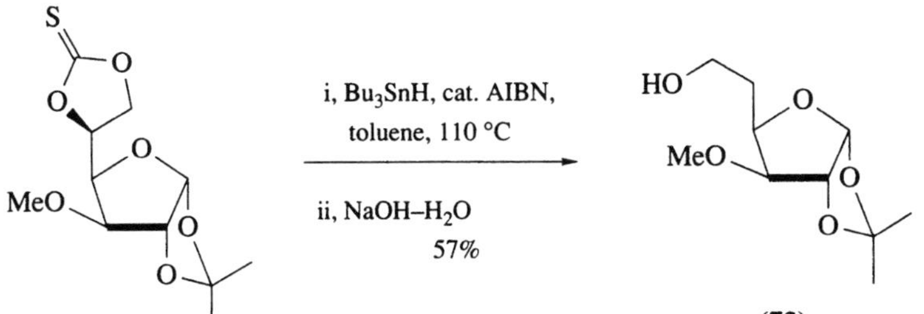

i, Bu_3SnH, cat. AIBN, toluene, 110 °C
ii, NaOH–H_2O
57%

(**72**) (**73**) (14)

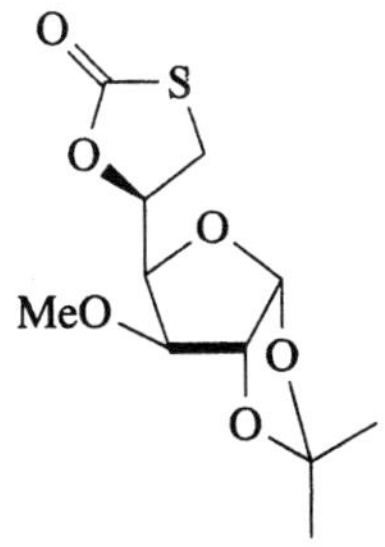

(**74**)

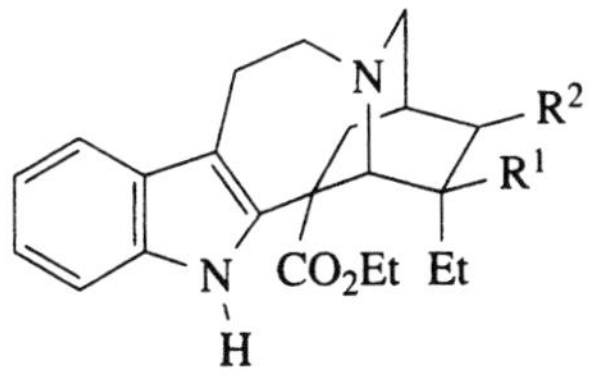

(**75**) R^1, R^2 = –OCSO–
(**76**) R^1 = H, R^2 = OH

Radical stability also accounts for efficient deoxygenation at the tertiary center in the conversion of (**75**) to (**76**) in a synthesis of vinblastine,[48] although a concentration dependent competition between tertiary *versus* secondary reduction was reported for some branched carbohydrate derivatives.[49] When the starting diol contains two secondary centers, mixtures frequently result, as shown in equation (15).

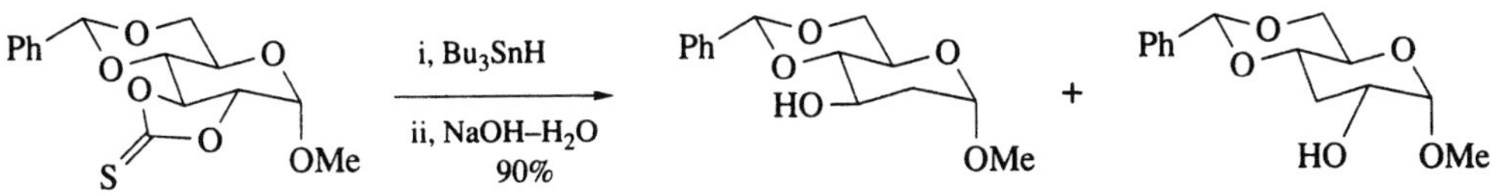

i, Bu_3SnH
ii, NaOH–H_2O
90%

67% 33% (15)

Because the lower degree of stabilization of primary radicals retards the fragmentation (process (A) in Scheme 5), the standard procedure was inefficient for *O*-thiocarbonyl derivatives of primary alcohols, but was rendered acceptable by running the reductive step at 150 °C.[50] Conversion of methylhederagenin (**77**) to monoxanthate (**78**) followed by heating with Bu_3SnH in *p*-cymene gave methyl oleanolate (**79**) in 51% overall yield.

Radical deoxygenation of an isolated tertiary hydroxy, desirable if involvement of adjacent stereocenters (in alkene formation or rearrangements) has to be avoided, presented problems in derivatization. Scheme 6 shows an inventive solution[51] in which the alcohol (**80**) was converted to imidate (**81**). Choice of this electron-rich aromatic system secured smooth thiolysis to (**82**), which was efficiently deoxyge-

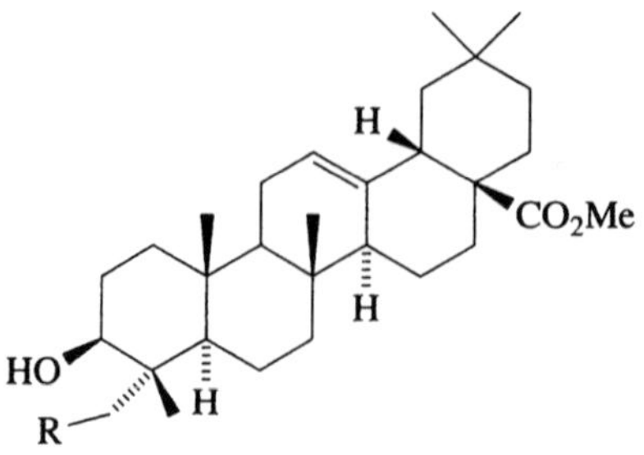

(77) R = OH
(78) R = OCSSMe
(79) R = H

nated by Bu_3SnH (secondary thioformates, although easily prepared, are not satisfactory deoxygenation substrates in most systems).

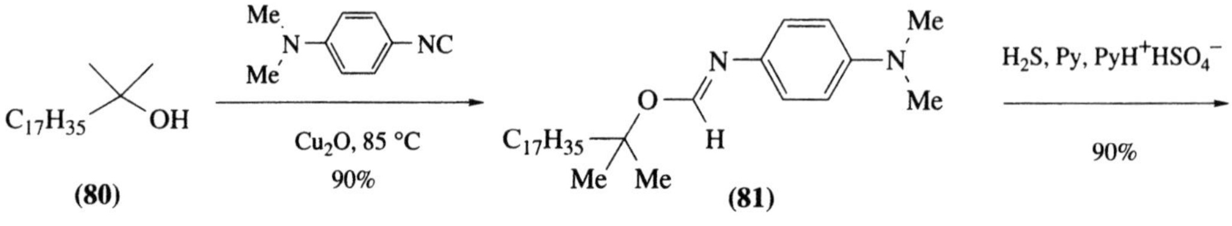

Scheme 6

All of the foregoing reactions, together with those described in Sections 4.2.2.5 and 4.2.3.3, which proceed *via* homolytic C—O or C—N scission, are subject to side reactions (which may become the dominant process) characteristic of carbon-centered free radicals. If the β-carbon bears a group which can yield a relatively stable radical, then alkene formation is the result (Scheme 7).

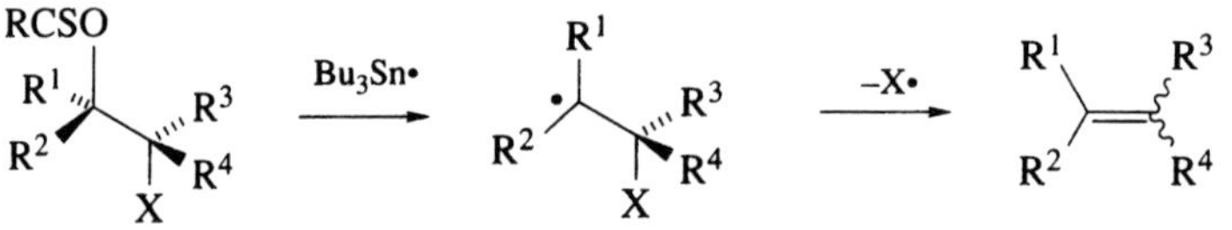

X = NO_2, SPh, SO_2Ph, Cl

Scheme 7

This process occurs for X = O only in two special cases, which have been developed into synthetic methods for alkenes and allylic alcohols. Reaction of Bu_3SnH with 1,2-bis-MDC derivatives gave alkenes from diols[52] in a process which was independent of the geometry of the starting diol. Examples are the conversions of (**84**) to (**85**) and of (**86**) to (**87**).

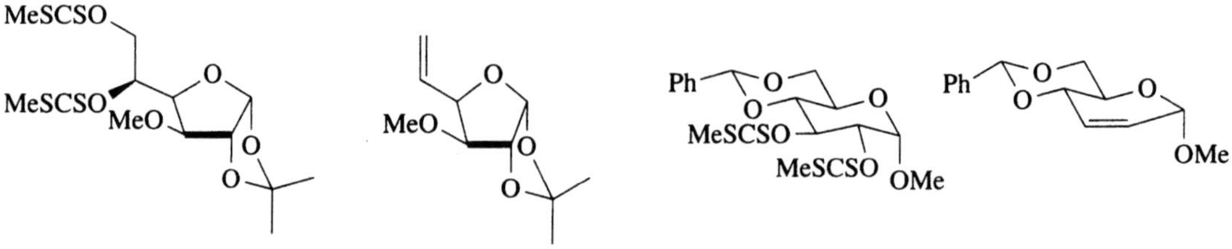

(84) (85) (86) (87)

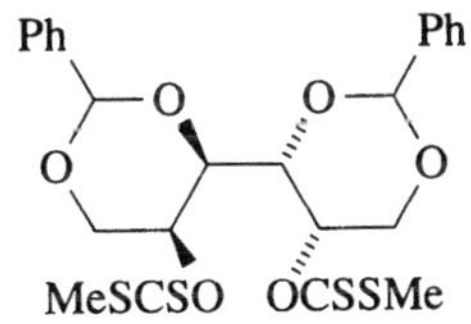

MeSCSO OCSSMe

(88)

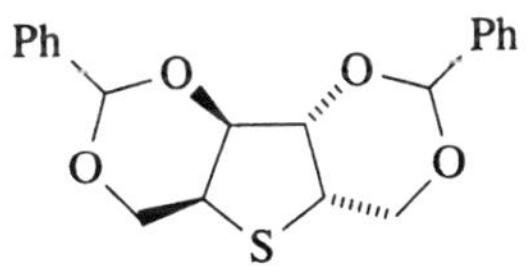

(89)

The formation of alkenes from TB or MDC derivatives of β-hydroxy sulfides and sulfones[53] is also a useful procedure. When the 1,4-bis-MDC derivative (**88**) was reacted with Bu_3SnH, the tetrahydrothiophene (**89**) was produced,[54] presumably by the process shown in Scheme 8. The first intramolecular step is related to the attack on C=S by a carbon radical in the reactions of 1,2-bis-TB derivatives with Bu_3SnH (equation 16).[35]

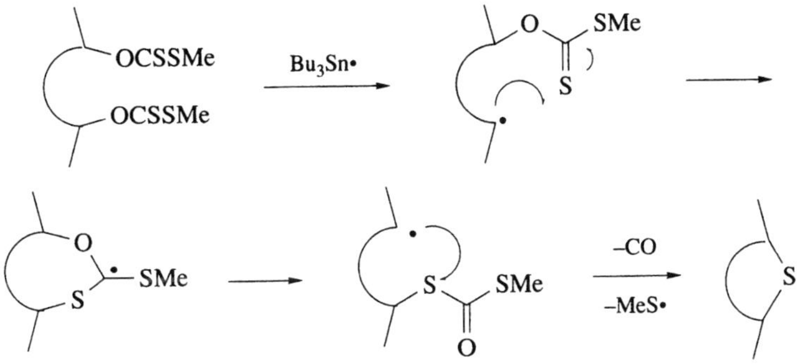

Scheme 8

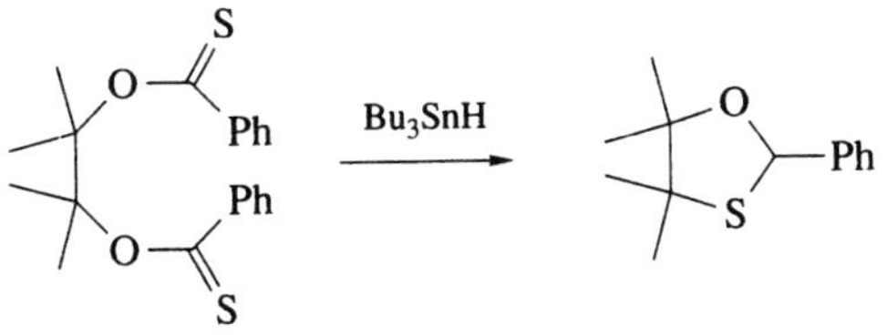

(16)

Alkene formation also took place upon attempted deoxygenation of ITC derivatives of 2,3-epoxy alcohols.[55] This radical-based alternative to the Wharton rearrangement is shown in equation (17), and is analogous to the rearrangement of cyclopropylcarbinyl radicals.

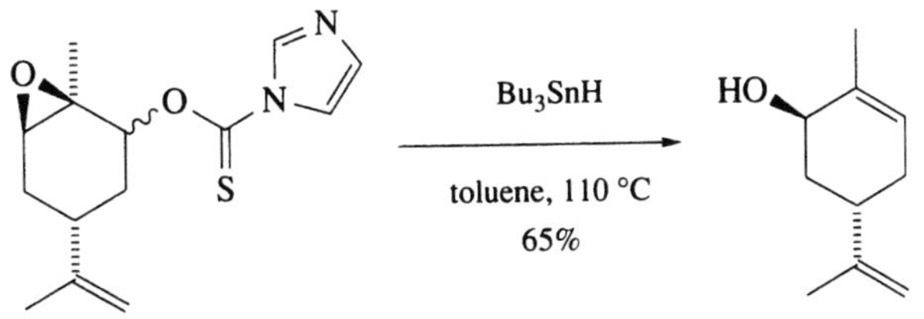

Bu_3SnH; toluene, 110 °C; 65%

(17)

Finally, we note that interception of the carbon radical by a proximate C=C or C≡C system, which has received much attention in recent years as a useful synthetic tool,[37,56] can be a problem when simple deoxygenation is desired at a center five or six atoms removed from the multiple bond. The efficient, planned cyclization[57] of (**90**) to (**91**) also represents an unsuccessful deoxygenation of a secondary alcohol! The initial adduct radicals (to C=S) have also been trapped in intramolecular reactions.[58]

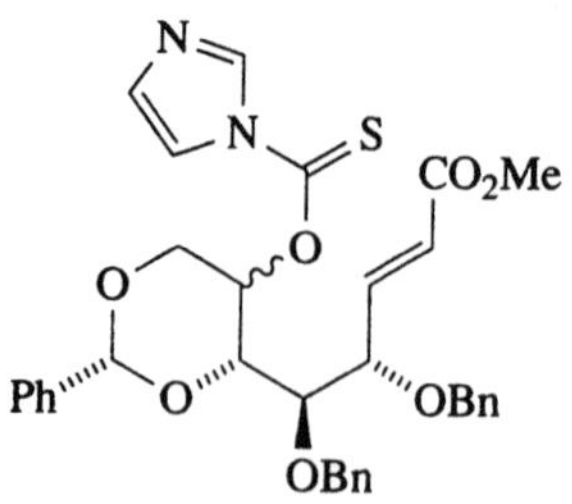

(90)

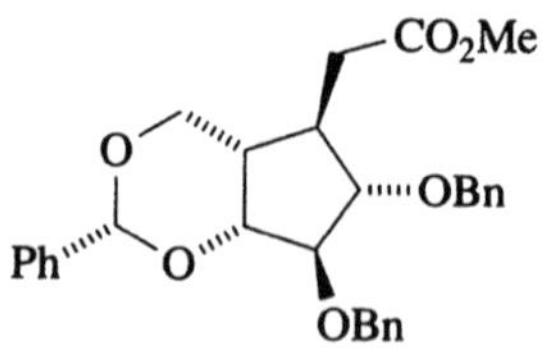

(91)

4.2.2.5 Stannane and Silane Reduction of Esters and Carbonates

From a study of various substrates, Khoo and Lee[59] concluded that direct reduction of *O*-benzoates by Bu_3SnH is satisfactory for benzylic or allylic alcohols, but is not preparatively useful for saturated systems, as shown by the examples of equations (18) and (19). The mechanism probably involves direct attack of $Bu_3Sn\cdot$ on the ethereal oxygen, and is facilitated by electron-withdrawing groups; reduction of $PhCH_2OCOCF_3$ was some four times faster than reduction of $PhCH_2OCOPh$.

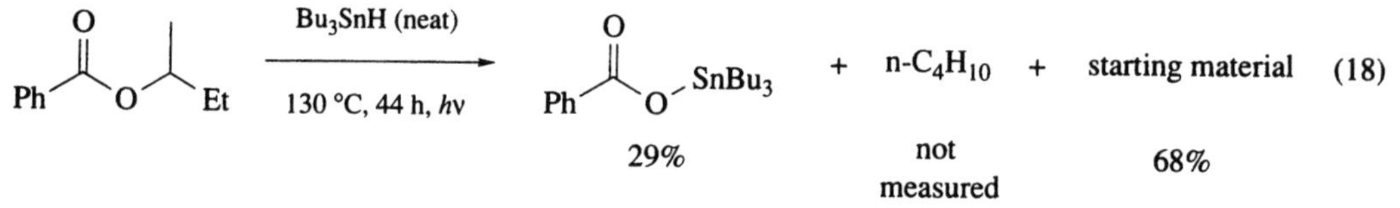

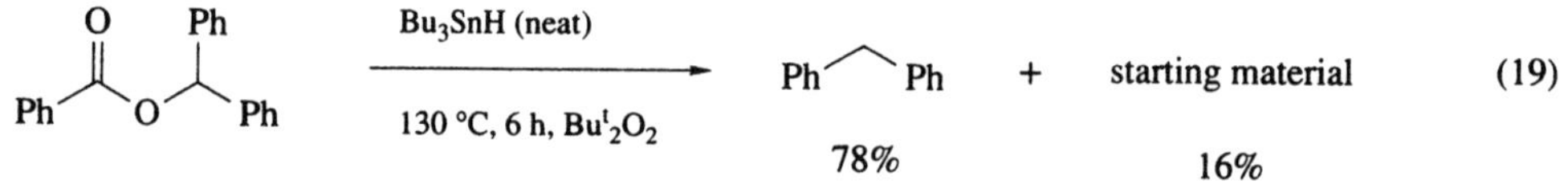

Bu3SnH (neat), 130 °C, 44 h, *hν* → $PhCO_2SnBu_3$ (29%) + n-C_4H_{10} (not measured) + starting material (68%) (18)

Bu3SnH (neat), 130 °C, 6 h, $Bu^t{}_2O_2$ → $PhCH_2Ph$ (78%) + starting material (16%) (19)

A more useful process was found in the reduction of methyloxalates of secondary and tertiary alcohols.[60] One advantage of this method is the high reactivity of the derivatizing reagents [MeOCOCOCl or $(COCl)_2$], which react easily with hindered alcohols. In the gibberellin series, (**92**) could not be converted to a phenoxythiocarbonyl derivative but was deoxygenated in 65% yield to (**93**) *via* (**94**). The final step in the synthesis[61] of bilabolide acetate (**95**) involved a similar reduction of (**96**) at high dilution in toluene, and gave a 2:1 mixture of deoxy compound and starting alcohol.

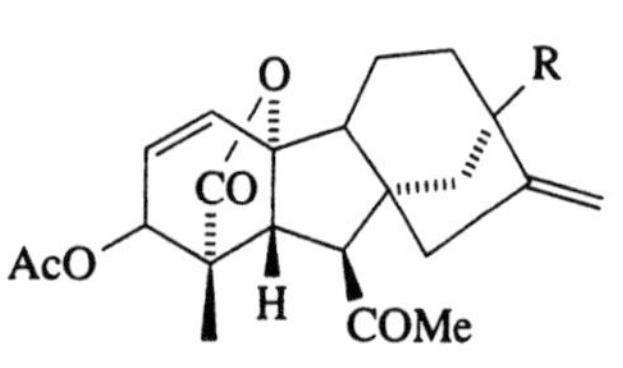

(**92**) R = OH
(**93**) R = H
(**94**) R = O_2CCO_2Me

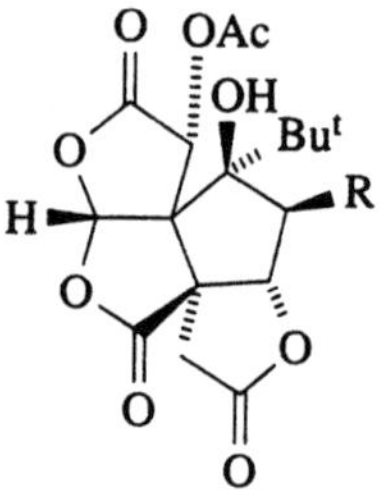

(**95**) R = H
(**96**) R = O_2CCO_2Me

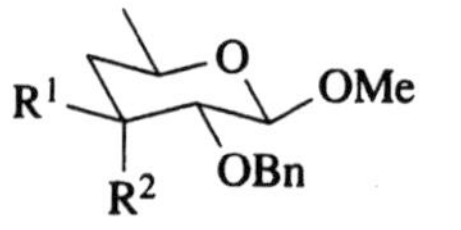

(**97**) R^1 = Ac, R^2 = O_2CPh
(**98**) R^1 = O_2CPh, R^2 = Ac
(**99**) R^1 = Ac, R^2 = H
(**100**) R^1 = H, R^2 = Ac

The reduction of α-ketobenzoates by Bu_3SnH[62] constitutes a special case, probably involving addition of the stannyl radical to the ester carbonyl. Reductions of isomerically pure (**97**) and (**98**) gave the same 4:1 ratio of the epimeric products (**99**) and (**100**).

Simple acetates were reduced at 140 °C with an excess of either Ph_3SiH or *p*-$HPh_2SiC_6H_4SiPh_2H$ and substantial amounts of free radical initiator, necessary to offset dimerization of the chain-carrying silyl radicals. Good yields were obtained (Table 3) for a number of nonfunctionalized primary, secondary and tertiary acetates.[63] Esters lacking any strongly absorbing chromophore, when photolyzed with $HSiCl_3$, gave a mixture of ether and deoxy compound.[64]

Table 3 Reduction of Acetates to Alkanes with Ar_3SiH and $Bu^t{}_2O_2$[63]

Ester	*Method*[a]	*Yield (%)*[b]
c-$C_{12}H_{23}OAc$	A	89
	B	82
$PhCH_2CMe_2OAc$	A	71
n-$C_{12}H_{25}OAc$	A	95
	B	67
Cholesteryl-OAc	A	59

[a]Method A: 1.3 mmol of ester, 2 mmol of p-$C_6H_4(SiPh_2H)_2$ and 1.3 mmol of $Bu^t{}_2O_2$ were reacted for 15 h at 140 °C. Method B: as in A, but 5.2 mmol of Ph_3SiH were used in place of the disilane. [b]Isolated yield of hydrocarbon.

Finally, there are several methods for alcohol deoxygenation which depend upon generating the corresponding alkoxycarbonyl radical, as shown in Scheme 9. Decarboxylation (A) must frequently compete with direct quenching (B), but the decarbonylation (C) is not significant.

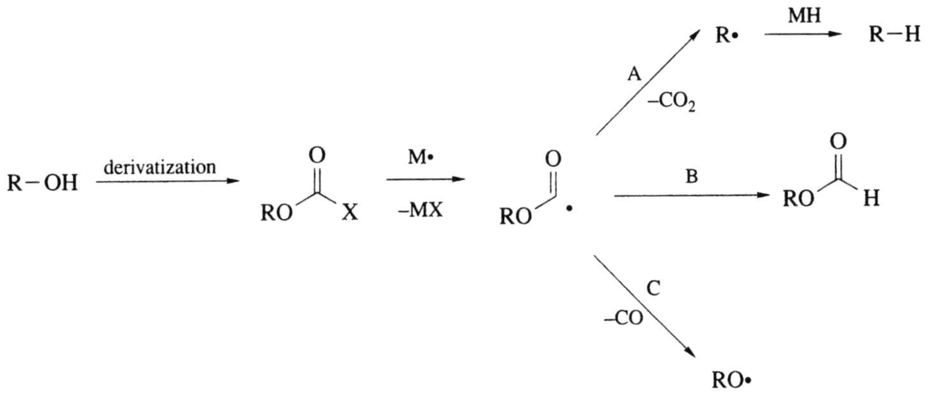

R–OH —derivatization→ ROC(O)X —M•, –MX→ ROC(O)• ; (A) –CO_2 → R• —MH→ R–H ; (B) → ROC(O)H ; (C) –CO → RO•

Scheme 9

Early studies[65] of the ROCOCl–Bu_3SnH reaction at 30–80 °C indicated that the decarboxylation step was dominant only for stabilized educt radicals such as $PhCH_2\cdot$. By using higher temperatures and a less efficient H-donor ($Pr^n{}_3SiH$), good yields of RH were obtained from simple primary and secondary chloroformates, but cholesterol was deoxygenated in only 25% yield.[66] The high affinity of tin-centered radicals for Se^{II} was used to good advantage in reduction of steroidal *Se*-phenylselenocarbonates, as shown in Scheme 10. The corresponding formate and small amounts of the alcohol were side products; as indicated for the reduction of (**101**) shown in equation (20), lower reaction temperatures gave increasing amounts of the formate.[67]

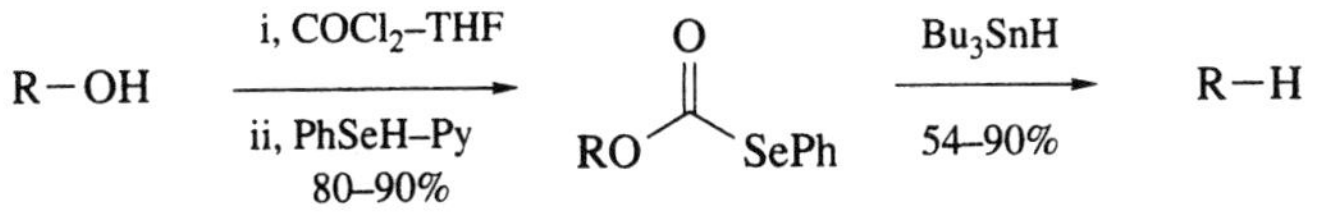

R–OH —i, $COCl_2$–THF; ii, PhSeH–Py; 80–90%→ ROC(O)SePh —Bu_3SnH, 54–90%→ R–H

Scheme 10

Since tertiary chloroformates are difficult to prepare and are unstable above 0 °C, they are not suitable substrates for the foregoing processes. Tertiary alcohols were deoxygenated efficiently[68] *via* the chloroglyoxylates through *in situ* generation of the corresponding thiohydroxamate in the presence of a hindered thiol (Scheme 11). Using Me_3CSH in place of Et_3CSH gave slightly lower yields with corresponding formation of the oxalate thioester, as exemplified by the reduction of (**102**) *via* chloroglyoxylate (**103**): with Me_3CSH, a 2.5:1 mixture of (**104**) and (**105**) was obtained, whereas Et_3CSH gave a high yield of deoxy compound (**104**). To avoid acid-catalyzed skeletal rearrangement when the process was applied to (**106**), the chloroglyoxylate was prepared from the TMS ether (**107**). The product (**108**) was the result of radical quenching from the less hindered side. Applied to a typical secondary alcohol, this reaction gave comparable amounts of deoxy compound and formate even at 178 °C, indicating stepwise loss of CO_2 from the initial radical.

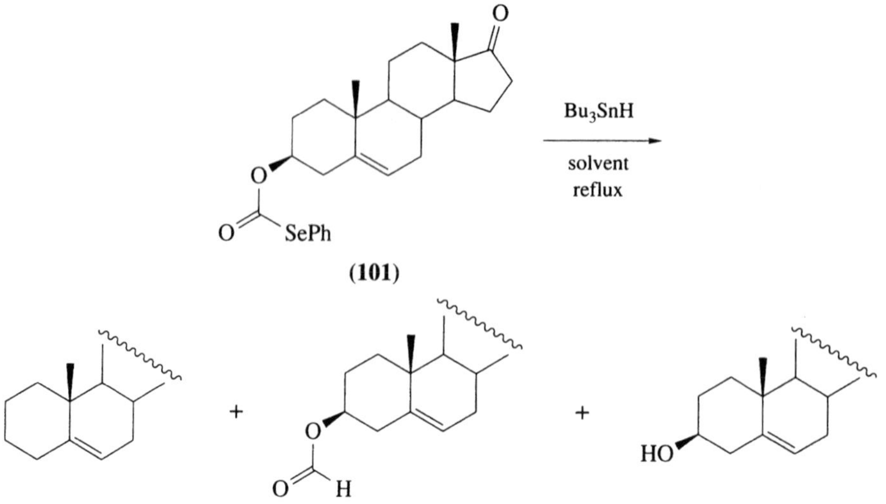

Bu_3SnH, solvent, reflux

(101)

(20)

144 °C (xylene) :	73%	16%	11%
110 °C (toluene) :	39%	42%	14%
80 °C (benzene) :	5%	76%	9%

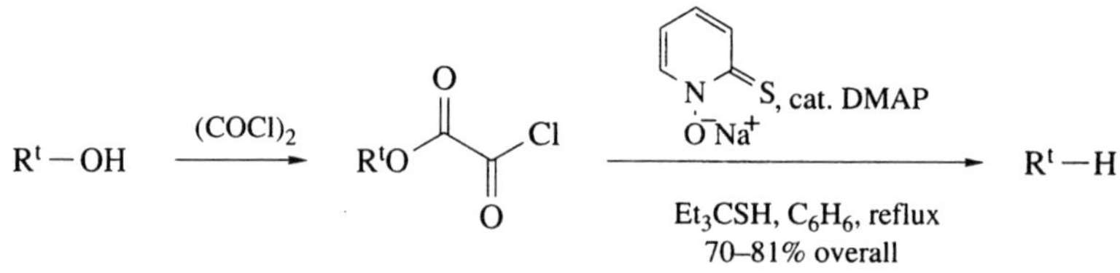

R^t—OH → $(COCl)_2$ → R^tO_2CCOCl → (N-hydroxypyridine-2-thione sodium salt, cat. DMAP; Et_3CSH, C_6H_6, reflux; 70–81% overall) → R^t—H

Scheme 11

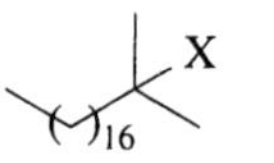

(102) X = OH
(103) X = O_2CCOCl
(104) X = H
(105) X = $O_2CCOSBu^t$

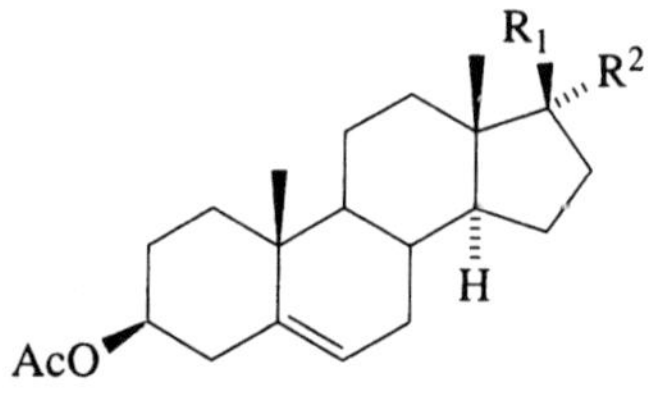

(106) R^1 = OH, R^2 = Me
(107) R^1 = $OSiMe_3$, R^2 = Me
(108) R^1 = Me, R^2 = H

4.2.3 REDUCTIVE DEAMINATION OF AMINES AND THEIR DERIVATIVES

4.2.3.1 Hydride and Hydrogenolysis Methods

Because of the lower electronegativity of nitrogen relative to oxygen, and the lower acidity of, for example, $PhSO_2NH_2$ compared with $PhSO_2OH$, activation of an amine to an extent sufficient to allow an S_N2 displacement which breaks the C—N bond is more difficult than the analogous process for an alcohol. In fact, this process is possible only for primary amines, where double activation or conversion to a heterocyclic quaternary system is possible. Simple quaternary salts are susceptible to S_N2 reactions with strong nucleophiles, but this is not generally a useful deamination method since attack on the least hindered center will take place (equation 21).[69] Systems with additional activation are exceptions: Mannich bases were reduced, in some cases with $NaBH_4$, and more generally by quaternization followed by treatment with $NaBH_3CN$, as shown in equation (22).[70]

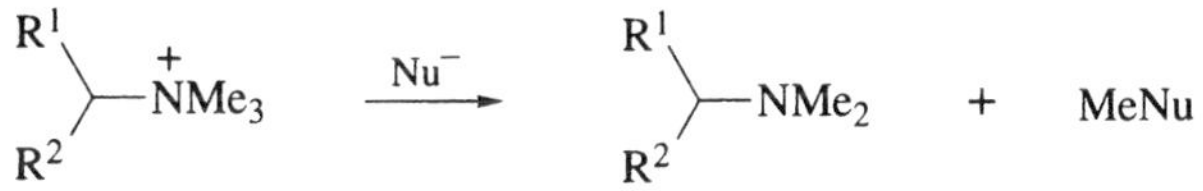

$$R^1R^2CH\overset{+}{N}Me_3 \xrightarrow{Nu^-} R^1R^2CHNMe_2 + MeNu \qquad (21)$$

i, Me_2SO_4; ii, $NaBH_3CN$, HMPA, 2 h, 70 °C; 91% (22)

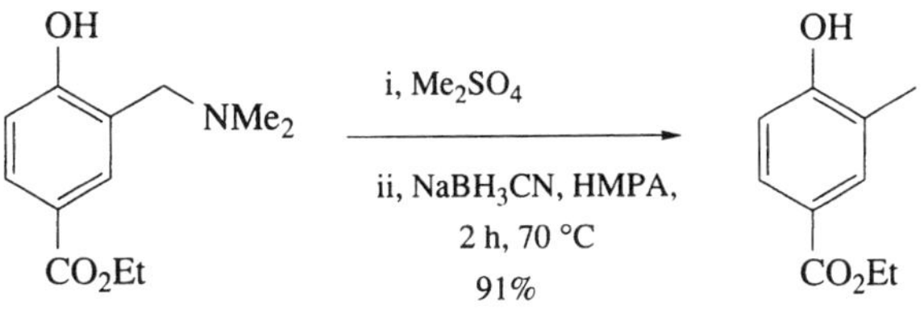

N,N-Disulfonimides, $RN(SO_2R')_2$, have been advanced as amine derivatives suitable for S_N2 processes, at least in those cases where R is a primary radical. *N*-(1-Decyl)-*N,N*-di(*p*-toluene)sulfonimide (**109**) was reduced efficiently to decane by $NaBH_4$ in HMPA at 175 °C but the cyclododecyl analog (**110**) gave desulfonylation to (**111**). *N,N*-Di(trifluoromethane)sulfonimides (triflimides) such as (**112**), prepared from the trifluoromethanesulfonamide with NaH–$(CF_3SO_2)_2O$, are more susceptible to S_N2 reactions with soft nucleophiles, but also eliminate readily: attempted preparation of $Bu^tN(SO_2CF_3)_2$ gave isobutene even at –78 °C. The reactions of triflimides with nucleophiles including $NaBH_4$ have been reviewed.[71]

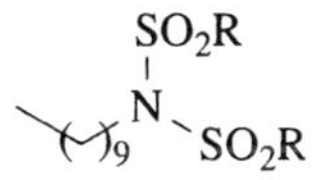

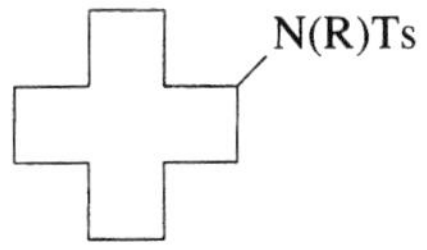

(**109**) R = *p*-MeC_6H_4
(**112**) R = CF_3

(**110**) R = Ts
(**111**) R = H

Katritzky and coworkers have extensively developed the activation of amines by reaction with pyrylium salts to provide *N*-alkyl (or *N*-aryl) pyridinium compounds.[72] When buttressing substituents were present to discourage attack on the pyridine ring, the *N*-alkyl substituent was subject to displacement and elimination processes. In general, primary alkyl substituents reacted with most nucleophiles in a normal S_N2 process as shown in Scheme 12, whereas competition between substitution and elimination took place with the secondary analogs, with elimination dominating the reactions starting from cycloalkylamines.

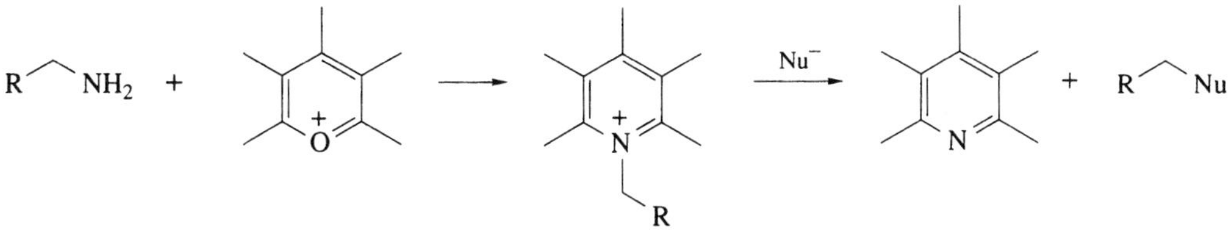

Scheme 12

Direct reductive cleavage of these pyridinium salts was not possible since hydride reagents attacked the ring to give 1,2- or 1,4-dihydropyridines; with suitable substitution patterns, however, these species could be used to achieve overall reductive deamination of RCH_2NH_2 to RCH_3. In the first process,[73] shown in Scheme 13, amines were converted to 2,3,5,6-tetraphenylpyridinium salts (**113**), which were reduced to the 1,4-dihydro compounds (**114**). Pyrolysis of these materials at 180–200 °C then liberated the hydrocarbon: *n*-octane was obtained in 61% overall yield from 1-aminooctane. For aliphatic amines, less satisfactory results were obtained upon heating the 1,2-dihydropyridines derived from 2,4,6-triphenylpyridinium salts, although this was an efficient process for allylic and benzylic amines.[74] The second method also relies upon the formation of 1,4-dihydropyridines such as (**115**), prepared from the amine by treatment with the salt (**116**) followed by $NaBH_4$. On heating at 100–140 °C with DBN, the corresponding alkane was formed in good yield.[75] Cyclohexylamine reacted with (**116**) to give cyclohexene at room temperature.

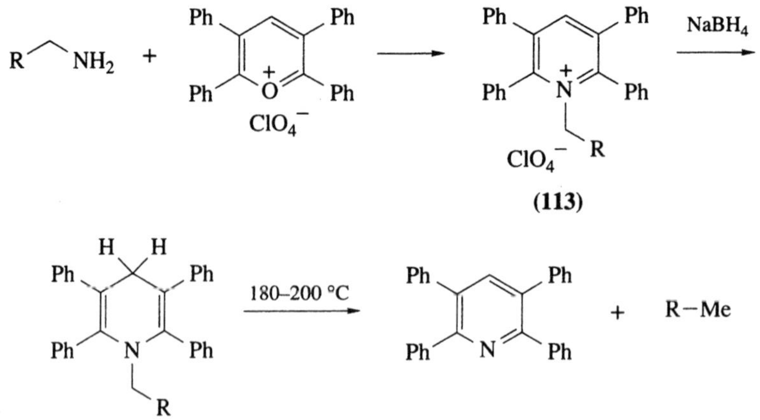

Scheme 13

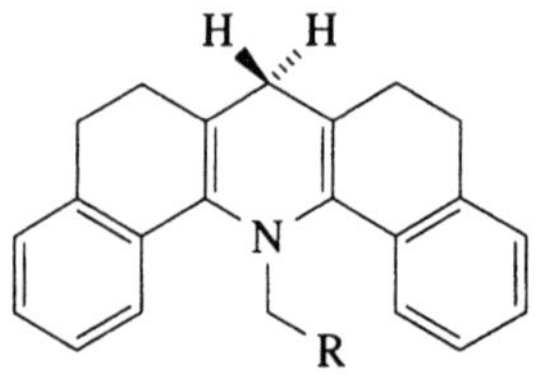

(113)

(114)

(115)

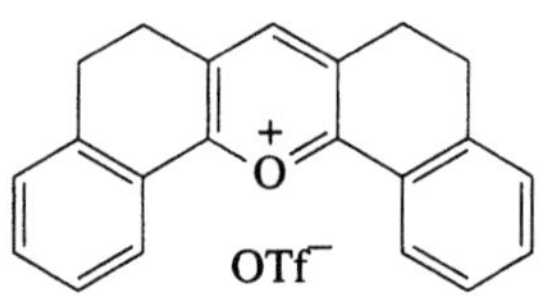

(116)

Completely saturated quaternary ammonium salts, in contrast with those derived from allylic or benzylic amines, are unaffected by catalytic reduction and by mild reducing agents including Na–Hg in protic solvents. More vigorous reduction with Na in hot dioxane or NH_3 (liq.) led to cleavage and elimination to mixtures of alkane and alkene, but little discrimination was seen for cleavage of N—R *versus* N—Me in $RNMe_3^+$, except when the alkyl group was tertiary.[76] Amines of general structure $RNMe_2$ or RNHMe, in which the group R bears a β-hydrogen, are most simply converted to RH *via* the corresponding alkene, prepared by elimination from the quaternary salt or *N*-oxide, as described in Chapters 5.1 and 5.4, Volume 6.

4.2.3.2 Deamination *via* Diazenes and Related Species

Given the high bond energy of molecular N_2, monoalkyl diazenes (diimides) are potentially attractive intermediates for reductive deamination of primary amines. Although RN=NH species are moderately stable in dilute, neutral solutions, they decompose quite readily to give predominantly the alkane by one of two mechanisms, shown in Scheme 14: in neutral solution, a bimolecular process believed to involve triplet pairs and caged radical species occurs, whereas in base the diazene anion loses N_2 to give the corresponding carbanion. These processes have been discussed in some detail,[77] and the stereochemical consequences of having a chiral center at the C—N bond have been examined for a tertiary, benzylic case.[78]

The success of the Wolff–Kishner and related C=O to CH_2 transformations (see Chapter 1.14, this volume) attests to the efficiency of the diazene decomposition route; applied to the reductive deamination of primary amines, this requires methods which transform RNH_2 into RN=NH, which corresponds to an amination–oxidation sequence. Several one- or two-step processes have been described which carry out this transformation. Treatment of amino acids in alkaline solution with excess hydroxylamine-*O*-sulfonic acid (HOS) gave moderate yields of deamination product,[79] as exemplified by the dipeptide case shown in equation (23).

In a related process, HOS in large excess reacted with sulfonamides in hot aqueous alkali to give alkanes,[80] presumably by the process shown in Scheme 15. Yields were generally high when corrected for unchanged sulfonamide, but conversions were quite low even when a large excess of HOS was added in several portions: with 20–25 equiv. of reagent, *N*-cyclohexylbenzenesulfonamide gave 70% cyclohexane at 20% conversion. Although not reported, the use of more recently developed aminating agents such as

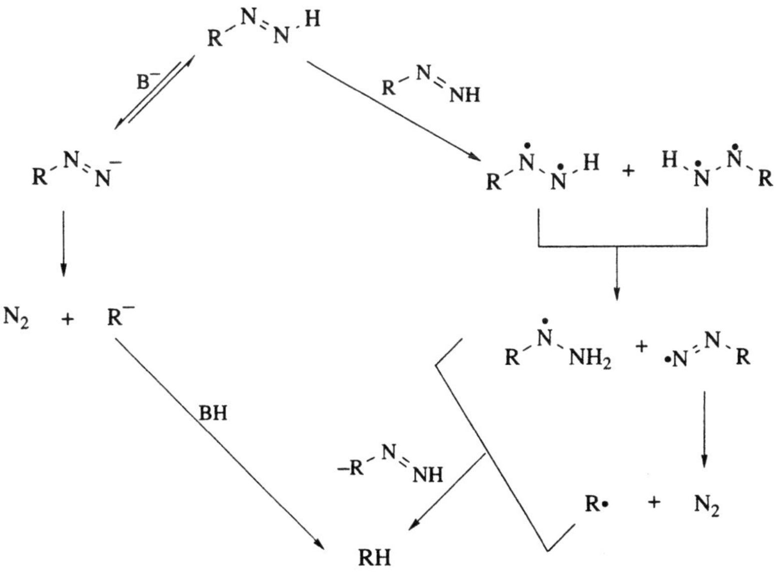

Scheme 14

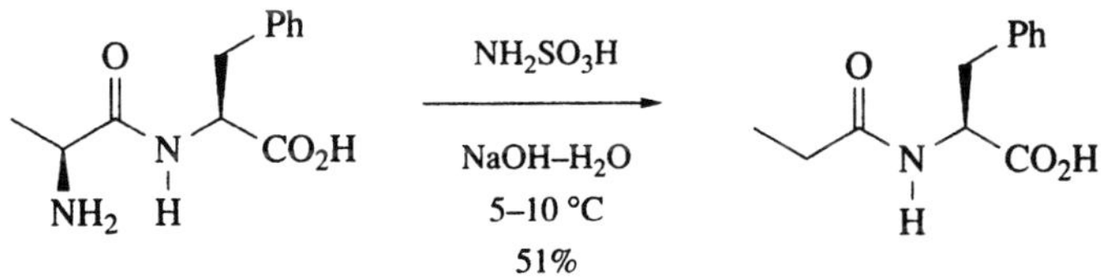

(23)

O-mesitylenesulfonylhydroxylamine[81] or *O*-diphenylphosphinylhydroxylamine[82] might permit these processes to be carried out under nonaqueous conditions.

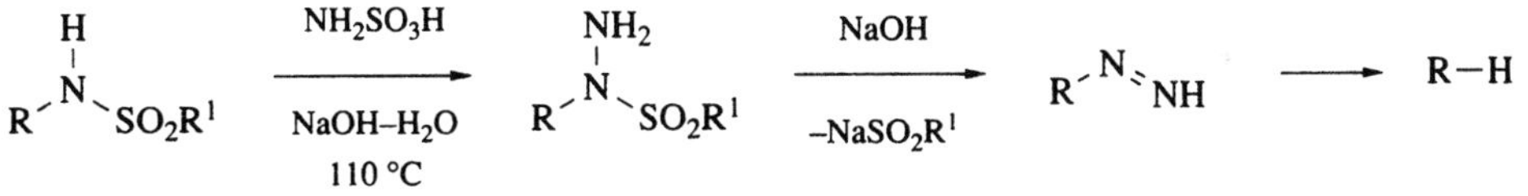

Scheme 15

Difluoramine (HNF_2) has also been used to convert amines to hydrocarbons,[83] according to the process shown in equation (24). No additional base was included, so an excess of the amine was necessary. Yields of 40–77% based on HNF_2 were obtained for simple aliphatic amines. Under these conditions, free radical decomposition of the diazene was probably involved, since a cyclopropylcarbinylamine gave a rearranged product. This ring opening was not detected in the Wolff–Kishner reduction of the corresponding ketone (equations 25 and 26). When an allylic amine was treated with HNF_2, deamination with clean double bond transposition occurred by intramolecular H-transfer in the diazene intermediate (equation 27).

$$3\ RNH_2 + HNF_2 \longrightarrow RH + 2\ RNH_3{}^+F^- \quad (24)$$

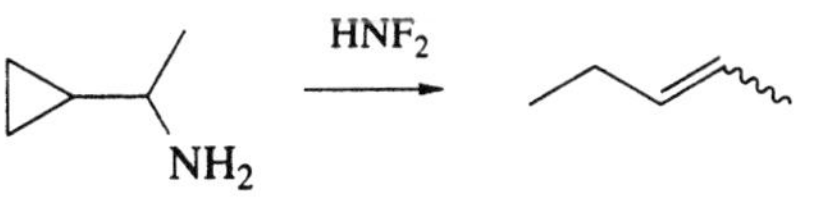

(25)

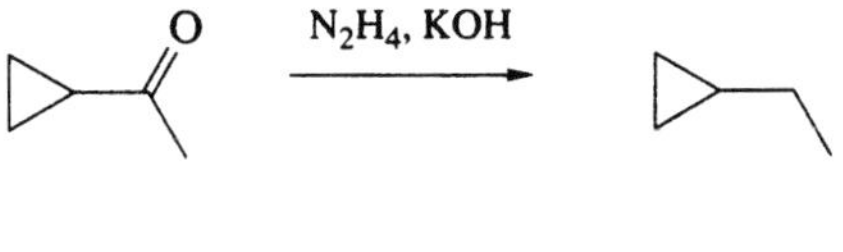

(26)

$$\text{Ph}\diagup\!\!\!\sim\!\!\sim\text{NH}_2 \xrightarrow{\text{NHF}_2} \text{Ph}\diagup\!\!\diagdown\!\!\!= \qquad (27)$$

Finally, we note that deamination could in principle be achieved by generating an aliphatic diazonium ion in the presence of a hydride source; this could lead to an alkane either *via* the diazene, or by S_N2 attack with loss of N_2. Protonating an aliphatic diazo compound, or treating an N^1-alkyl, N^3-aryltriazene with protic or Lewis acid in the presence of a suitable reducing agent ($NaBH_3CN$, $NaBH(OAc)_3$ or R_3SiH) might achieve this transformation.

4.2.3.3 Reduction of Isonitriles and Related Compounds

To an even greater extent than was the case for deoxygenation, the introduction of dissolving metal and free radical reduction processes for suitable amine derivatives has greatly expanded the range of substrates which can be subjected to efficient reactions, and primary amino groups can now be reductively removed from molecules containing a variety of other functional groups. For amines, in contrast to alcohols, the choice of derivative is much more limited since the driving force obtained in going from C—N to C═N (or from N—X to N═X) is not generally sufficient to induce fragmentation from radical or anion radical intermediates; however, highly stable RCN or CN^- fragments are readily ejected upon suitable reduction of isonitriles, which (as such, or generated *in situ*) are now the substrates of choice for reductive deaminations. Isonitriles are readily prepared from primary amines in excellent yield by formylation (with EtOCHO or HCOOAc), followed by dehydration with TsCl–py, $COCl_2$–R_3N or similar reagents.[84]

Table 4 Metal–Ammonia Reduction of Isonitriles to Hydrocarbons[85]

R in RNC	*Metal* [a]	*Yield of RH (%)*
Me_3C	Na	93[b]
c-C_6H_{11}	Li	96[b]
c-C_6H_{11}	Na	93[b]
c-C_6H_{11}	Ca	95[b]
n-$C_{12}H_{25}$	Na	88[c]

[a]2 g equiv. of metal were used. [b]Determined by quantitative IR. [c]Isolated yield.

Ugi and Bodesheim[85] first reported the efficient reduction of isonitriles to hydrocarbons by Na, Li or Ca in liquid ammonia: some results are shown in Table 4. Reduction of c-$C_6H_{11}NC$ was also reported in a study of the metal–amine reduction of nitriles,[86] and the mechanism of isonitrile reduction was examined by Niznik and Walborsky[87] who showed that acyclic, chiral isonitriles gave racemic products. In the reduction of 1,1-diphenyl-2-isocyano-2-methylcyclopropane, those rearrangement and coupling processes which dominated the reduction with Na–NH_3 or metals dispersed in DME were largely suppressed in favor of normal deamination when the reductant was Na–naphthalenide in DME. It seems likely that, depending upon the particular isonitrile and the conditions, these reductions may proceed either by fragmentation of the radical anion (to R· + CN^-), or by addition of a second electron to this species with explusion of the carbanion. Whichever is the case, a carbanion is the ultimate source of RH in the reduction of RNC, and hence fragmentation by β-elimination of a leaving group may occur, as in the related cases of dissolving metal deoxygenation.

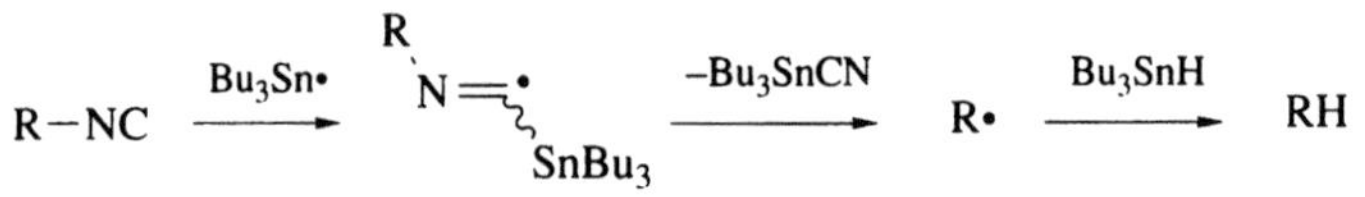

$$\text{R–NC} \xrightarrow{\text{Bu}_3\text{Sn}\bullet} \text{R–N}{=}\dot{\text{C}}\text{–SnBu}_3 \xrightarrow{-\text{Bu}_3\text{SnCN}} \text{R}\bullet \xrightarrow{\text{Bu}_3\text{SnH}} \text{RH}$$

Scheme 16

The Bu_3SnH decyanation of isonitriles (Scheme 16) was first reported by Saegusa,[88] and was developed as a general, mild method for reductive deamination by Barton and coworkers who initially showed that a variety of isonitriles, isothiocyanates and isoselenocyanates reacted with Bu_3SnH under the usual free radical conditions (reflux in benzene, toluene or xylene with catalytic AIBN) to give good yields of the corresponding hydrocarbon.[89] In concert with the normal order of radical stability, primary alkaneisonitriles required higher reaction temperatures, whereas a tertiary example was reduced at 50 °C (Table 5).

Table 5 Reduction of Isonitriles and Isothiocyanates with Bu_3SnH[89]

RNCX	*Solvent* [a]	*Temperature* (°C)	*Yield of RH* (%)
n-$C_{18}H_{37}NC$	Xylene	145	81[c]
$EtO_2CCH_2NHCOCH_2NC$	Benzene	80	71
3-Isocyanocholestane	Toluene	110	56
3-Isothiocyanatocholestane	Toluene	110	83
n-$C_{17}H_{35}CMe_2NC$	Benzene	50	91
meso-PhCH(NC)CH(NC)Ph	Benzene	80	56[d]
(±)-PhCH(NC)CH(NC)Ph	Benzene	80	53[d]

[a]RNCX in the solvent containing AIBN (0.05–0.2 equiv.) was added to Bu_3SnH (1.7–5.0 equiv. in solvent) at the indicated temperature. [b]Isolated yield. [c]Obtained in 73% yield with Bu_3GeH in place of Bu_3SnH. [d]Yield of $PhCH_2CH_2Ph$; stilbene was not produced.

The isonitrile (**117**), isothiocyanate (**118**) and isoselenocyanate (**119**) gave equally high yields of the deamino compound (**120**), and IR monitoring showed that (**117**) was the intermediate in both of the heterocumulene reductions; potential advantages for RNCS over RNC lie in the somewhat better stability (*e.g.* to acids) and the fact that isothiocyanates may be obtained by one-step processes ($CSCl_2$–NaOH, *N,N'*-thiocarbonyldiimidazole or CS_2–DCC) from the amine. The lack of stilbene formation in the reductions of the isomeric 1,2-diisocyano-1,2-diphenylethanes indicates that β-isocyanoalkyl radicals are not prone to fragmentation. Alkene formation was observed from the altrose derivative (**121**), presumably through initial attack at the isonitrile residue. As usual, other β-oxy substituents which did not give stabilized radical fragments were not eliminated: reduction of (**122**) and (**123**) proceeded without alkene formation.

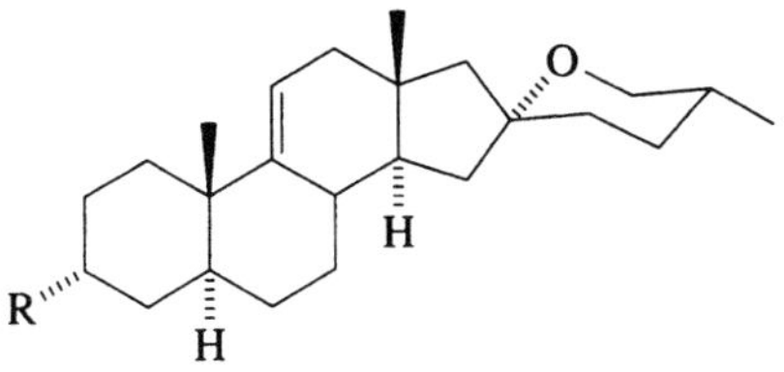

(**117**) R = NC
(**118**) R = NCS
(**119**) R = NCSe
(**120**) R = H

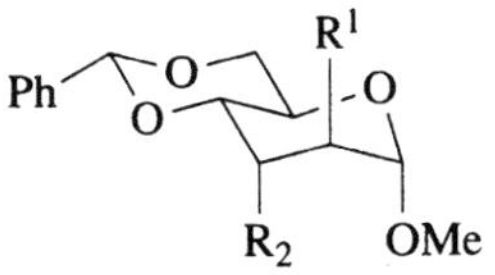

(**121**) R^1 = NC, R^2 = OCSSMe
(**122**) R^1 = NC, R^2 = OH
(**123**) R^1 = NC, R^2 = OMs

In addition to the foregoing steroidal and carbohydrate-derived examples, reductive decyanation has been applied to other areas of natural product chemistry, including the controlled deamination of the aminoglycoside antibiotic fragment neamine (**124**).[90] In addition to the glycylglycine example in Table 5, other amino acids were deaminated: tryptophan ester (**125**) was converted through (**126**) and (**127**) to the indolepropanoate (**128**) in 53% overall yield.[91] The deamination was also used in a synthesis of 6β-substituted penicillanic acid derivatives, as shown in Scheme 17.[92]

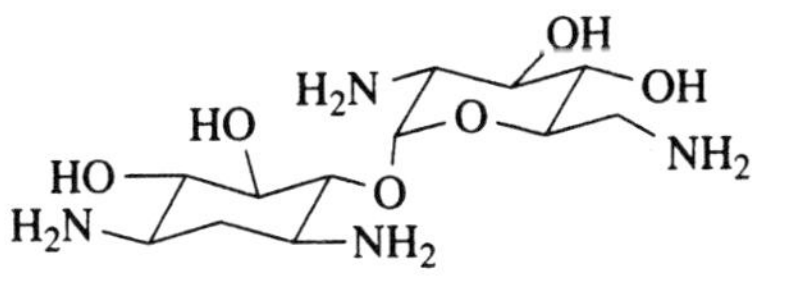

(**124**)

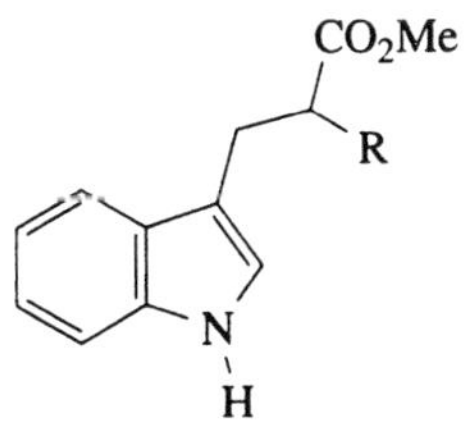

(**125**) R = NH_2
(**126**) R = NHCHO
(**127**) R = NC
(**128**) R = H

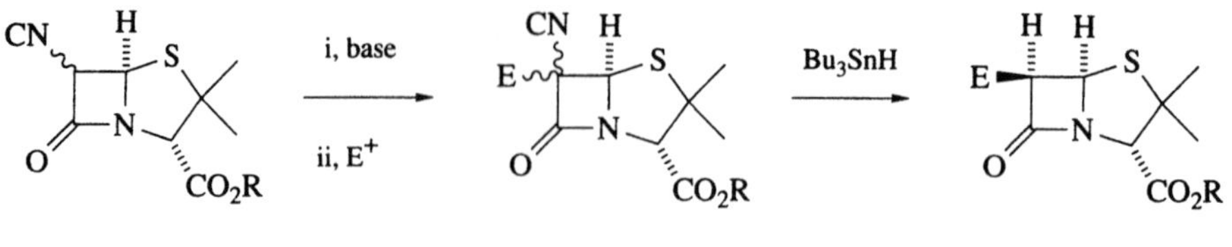

Scheme 17

4.2.4 REFERENCES

1. N. G. Gaylord, in 'Reduction with Complex Metal Hydrides,' Interscience, New York, 1956, p. 855; H. O. House, in 'Modern Synthetic Reactions,' Benjamin, Menlo Park, CA, 1972, p. 45.
2. H. Schmid and P. Karrer, *Helv. Chim. Acta*, 1949, **37**, 1371.
3. H. M. Bell, C. W. Vanderslice and A. Spehar, *J. Org. Chem.*, 1969, **34**, 3923.
4. R. O. Hutchins, D. Hoke, J. Keogh and D. Koharski, *Tetrahedron Lett.*, 1969, 3495.
5. H. Weidmann, N. Wolf and W. Timpe, *Carbohydr. Res.*, 1972, **24**, 184.
6. R. O. Hutchins, B. E. Maryanoff and C. A. Milewski, *J. Chem. Soc., Chem. Commun.*, 1971, 1097.
7. R. S. Tipson, *Adv. Carbohydr. Chem.*, 1953, **8**, 107.
8. S. Krishnamurthy, *J. Organomet. Chem.*, 1978, **156**, 171; R. W. Holder and M. G. Matturro, *J. Org. Chem.*, 1977, **42**, 2166.
9. D. H. Hua, S. Venkataraman, R. A. Ostrander, G.-Z. Sinai, P. J. McCann, J. M. Coulter and M. R. Xu, *J. Org. Chem.*, 1988, **53**, 507.
10. T. Netscher and H. Prinzbach, *Synthesis*, 1987, 683.
11. S. Masamune, P. A. Rossy and G. S. Bates, *J. Am. Chem. Soc.*, 1973, **95**, 6452; S. Masamune, G. S. Bates and P. E. Georghiou, *J. Am. Chem. Soc.*, 1974, **96**, 3686.
12. D. N. Kursanov, Z. N. Parnes and N. M. Loim, *Synthesis*, 1974, 633.
13. M. G. Adlington, M. Orfanopoulos and J. L. Fry, *Tetrahedron Lett.*, 1976, 2955.
14. F. A. Carey and H. S. Tremper, *J. Org. Chem.*, 1971, **36**, 758.
15. J. H. Brewster, S. F. Osman, H. O. Bayer and H. B. Hopps, *J. Org. Chem.*, 1964, **29**, 121.
16. S. E. Creasey and R. D. Guthrie, *J. Chem. Soc., Perkin Trans. 1*, 1974, 1373.
17. H. W. Thompson and R. E. Naipawer, *J. Am. Chem. Soc.*, 1973, **95**, 6379; G. Stork and D. E. Kahne, *J. Am. Chem. Soc.*, 1983, **105**, 1072; D. A. Evans and M. M. Morrissey, *J. Am. Chem. Soc.*, 1984, **106**, 3866.
18. R. Connor and H. Adkins, *J. Am. Chem. Soc.*, 1932, **54**, 4678; B. Wojcik and H. Adkins, *J. Am. Chem. Soc.*, 1933, **55**, 1293; H. Adkins, in 'Reactions of Hydrogen with Organic Compounds over Copper–Chromium Oxide and Nickel Catalysts,' University of Wisconsin Press, Madison, WI, 1937.
19. V. N. Ipatieff, W. W. Thompson and H. Pines, *J. Am. Chem. Soc.*, 1951, **73**, 553.
20. T. A. Ford, H. W. Jacobson and F. C. McGrew, *J. Am. Chem. Soc.*, 1948, **70**, 3793.
21. P. E. Peterson and C. Casey, *J. Org. Chem.*, 1964, **29**, 2325.
22. M. E. Krafft, W. J. Crooks, III, B. Zorc and S. E. Milczanowski, *J. Org. Chem.*, 1988, **53**, 3158.
23. E. Vowinkel and I. Büthe, *Chem. Ber.*, 1974, **107**, 1353.
24. A. G. M. Barrett, P. A. Prokopiou, D. H. R. Barton, R. B. Boar and J. F. McGhie, *J. Chem. Soc., Chem. Commun.*, 1979, 1173; R. B. Boar, L. Joukhadar, J. F. McGhie, S. C. Misra, A. G. M. Barrett, D. H. R. Barton and P. A. Prokopiou, *J. Chem. Soc., Chem. Commun.*, 1978, 68; P. Sengupta, M. Sen and S. Das, *Indian J. Chem., Sect. B*, 1979, **18**, 179.
25. H. Deshayes and J.-P. Pete, *J. Chem. Soc., Chem. Commun.*, 1978, 567.
26. A. K. Bose and P. Mangiaracina, *Tetrahedron Lett.*, 1987, **28**, 2503.
27. These processes have been compared, and examples given in tabular form: W. Hartwig, *Tetrahedron*, 1983, **39**, 2609.
28. R. E. Ireland, D. C. Muchmore and U. Hengartner, *J. Am. Chem. Soc.*, 1972, **94**, 5098.
29. A. G. M. Barrett and P. A. Prokopiou, *J. Chem. Soc., Chem. Commun.*, 1979, 1175.
30. T. Tsuchiya, I. Watanabe, M. Yoshida, F. Nakamura, T. Usui, M. Kitamura and S. Umezawa, *Tetrahedron Lett.*, 1978, 3365.
31. T. Tsuchiya, F. Nakamura and S. Umezawa, *Tetrahedron Lett.*, 1979, 2805.
32. T. Shono, Y. Matsumura, K. Tsubata and Y. Sugihara, *Tetrahedron Lett.*, 1979, 2157.
33. J.-P. Pete and C. Portella, *Synthesis*, 1977, 774; C. Portella, H. Deshayes, J.-P. Pete and D. Scholler, *Tetrahedron*, 1984, **40**, 3635; P. M. Collins and V. R. Z. Munasinghe, *J. Chem. Soc., Chem. Commun.*, 1977, 927.
34. I. Saito, H. Ikehira, R. Kasatani, M. Watanabe and T. Matsuura, *J. Am. Chem. Soc.*, 1986, **108**, 3115.
35. D. H. R. Barton and S. W. McCombie, *J. Chem. Soc., Perkin Trans. 1*, 1975, 1574.
36. D. H. R. Barton and W. B. Motherwell, *Pure Appl. Chem.*, 1981, **53**, 15.
37. M. Ramaiah, *Tetrahedron*, 1987, **43**, 3541.
38. W. P. Neumann, *Synthesis*, 1987, 665.
39. P. J. Barker and A. L. J. Beckwith, *J. Chem. Soc., Chem. Commun.*, 1984, 683.
40. D. H. R. Barton, D. Crich, A. Löbberding and S. Z. Zard, *J. Chem. Soc., Chem. Commun.*, 1985, 646.
41. K. Nozaki, K. Oshima and K. Utimoto, *Tetrahedron Lett.*, 1988, **29**, 6125.
42. J. R. Rasmussen, C. J. Slinger, R. J. Kordish and D. D. Newman-Evans, *J. Org. Chem.*, 1981, **46**, 4843.
43. M. J. Robins, J. S. Wilson and F. Hansske, *J. Am. Chem. Soc.*, 1983, **105**, 4059.
44. E. M. Acton, R. N. Goerner, H. S. Uh, K. J. Ryan and D. W. Henry, *J. Med. Chem.*, 1979, **22**, 518.
45. T. Hayashi, T. Iwaoka, N. Takeda and E. Ohki, *Chem. Pharm. Bull.*, 1978, 1786.

46. K. Tatsuta, K. Akimoto and M. Kinoshita, *J. Am. Chem. Soc.*, 1979, **101**, 6116.
47. D. H. R. Barton and R. Subramanian, *J. Chem. Soc., Perkin Trans. 1*, 1977, 1718.
48. J. P. Kutney, T. Honda, A. V. Joshua, N. G. Lewis and B. R. Worth, *Helv. Chim. Acta*, 1978, **61**, 690.
49. H. Redlich, W. Sudau and H. Paulsen, *Tetrahedron*, 1985, **41**, 4253.
50. D. H. R. Barton, W. B. Motherwell and A. Stange, *Synthesis*, 1981, 743.
51. D. H. R. Barton, W. Hartwig, R. S. Hay Motherwell, W. B. Motherwell and A. Stange, *Tetrahedron Lett.*, 1982, **23**, 2019.
52. A. G. M. Barrett, D. H. R. Barton, R. Bielski and S. W. McCombie, *J. Chem. Soc., Chem. Commun.*, 1977, 866; T. Hayashi, T. Iwaoka, N. Takeda and E. Ohki, *Chem. Pharm. Bull.*, 1978, **26**, 1786.
53. B. Lythgoe and I. Waterhouse, *Tetrahedron Lett.*, 1977, 4223.
54. A. V. Rama Rao, K. A. Reddy, M. K. Gurjar and A. C. Kunwar, *J. Chem. Soc., Chem. Commun.*, 1988, 1273.
55. D. H. R. Barton, R. S. Hay Motherwell and W. B. Motherwell, *J. Chem. Soc., Perkin Trans. 1*, 1981, 2363.
56. D. P. Curran, *Synthesis*, 1988, 439, 489.
57. T. V. RajanBabu, *J. Am. Chem. Soc.*, 1987, **109**, 609.
58. M. D. Bachi and E. Bosch, *J. Chem. Soc., Perkin Trans. 1*, 1988, 1517; K. Nozaki, K. Oshima and K. Utimoto, *Tetrahedron Lett.*, 1988, **29**, 6127.
59. L. E. Khoo and H. H. Lee, *Tetrahedron Lett.*, 1968, 4351.
60. S. C. Dolan and J. MacMillan, *J. Chem. Soc., Chem. Commun.*, 1985, 1588.
61. E. J. Corey and W.-G. Su, *J. Am. Chem. Soc.*, 1987, **109**, 7534.
62. H. Redlich, H.-J. Neumann and H. Paulsen, *Chem. Ber.* 1977, **110**, 2911.
63. H. Sano, M. Ogata and T. Migita, *Chem. Lett.*, 1986, 77; 1988, 119.
64. S. W. Baldwin and S. A. Haut, *J. Org. Chem.*, 1975, **40**, 3885.
65. H. G. Kuivila and E. J. Walsh, Jr., *J. Am. Chem. Soc.*, 1966, **88**, 571, 577.
66. R. A. Jackson and F. Malek, *J. Chem. Soc., Perkin Trans. 1*, 1980, 1207.
67. J. Pfenninger, C. Heuberger and W. Graf, *Helv. Chim. Acta*, 1980, **63**, 2328.
68. D. H. R. Barton and D. Crich, *J. Chem. Soc., Chem. Commun.*, 1984, 774.
69. R. O. Hutchins, D. Kandasamy, F. Dux, III, C. A. Maryanoff, D. Rotstein, B. Goldsmith, W. Burgoyne, F. Cistone, J. Dalessandro and J. Puglis, *J. Org. Chem.*, 1978, **43**, 2259.
70. K. Yamada, N. Itoh and T. Iwakuma, *J. Chem. Soc., Chem. Commun.*, 1978, 1089.
71. R. J. Baumgarten and V. A. Curtis, in 'The Chemistry of Amino, Nitroso and Nitro Compounds and their Derivatives', ed. S. Patai, Wiley, New York, 1982, part 2, suppl. F, p. 929.
72. A. R. Katritzky, *Tetrahedron*, 1980, **36**, 679; A. R. Katritzky and C. M. Marson, *Angew. Chem., Int. Ed. Engl.*, 1984, **23**, 420.
73. A. R. Katritzky, K. Horvath and B. Plau, *J. Chem. Soc., Perkin Trans. 1*, 1980, 2554.
74. A. R. Katritzky, J. Lewis and P.-L. Nie, *J. Chem. Soc., Perkin Trans. 1*, 1979, 442.
75. A. R. Katritzky, S. Bravo-Borja, A. M. El-Mowafy and M. L. Lopez-Rodriguez, *J. Chem. Soc., Perkin Trans. 1*, 1984, 1671.
76. E. Grovenstein, Jr., E. P. Blanchard, Jr., D. A. Gordon and R. W. Stevenson, *J. Am. Chem. Soc.*, 1959, **81**, 4842; E. Grovenstein, Jr. and R. W. Stevenson, *ibid*, 1959, **81**, 4850.
77. E. M. Kosower, *Acc. Chem. Res.*, 1971, **4**, 193.
78. D. J. Cram and J. S. Bradshaw, *J. Am. Chem. Soc.*, 1963, **85**, 1108.
79. G. A. Doldouras and J. Kollonitsch, *J. Am. Chem. Soc.*, 1978, **100**, 341.
80. A. Nickon and A. S. Hill, *J. Am. Chem. Soc.*, 1964, **86**, 1152.
81. Y. Tamura, J. Minamikawa and M. Ikeda, *Synthesis*, 1977, 1.
82. M. J. P. Harger, *J. Chem. Soc., Perkin Trans. 1*, 1981, 3284; W. Klötzer, H. Baldinger, E. M. Karpitschka and J. Knoflach, *Synthesis*, 1982, 592.
83. C. L. Bumgardner, K. J. Martin and J. P. Freeman, *J. Am. Chem. Soc.*, 1963, **85**, 97; C. L. Bumgardner and J. P. Freeman, *ibid*, 1964, **86**, 2233.
84. P. Hoffmann, G. Gokel, D. Marquarding and I. Ugi, in 'Isonitrile Chemistry', ed. I. Ugi, Academic Press, New York, 1971, p. 9.
85. I. Ugi and F. Bodesheim, *Chem. Ber.*, 1961, **94**, 1157.
86. W. Büchner and R. Dufaux, *Helv. Chim. Acta*, 1966, **49**, 1145.
87. G. E. Niznik and H. M. Walborsky, *J. Org. Chem.*, 1978, **43**, 2396.
88. T. Saegusa, S. Kobayashi, Y. Ito and N. Yasuda, *J. Am. Chem. Soc.*, 1968, **90**, 4182.
89. D. H. R. Barton, G. Bringmann, G. Lamotte, R. S. Hay Motherwell and W. B. Motherwell, *Tetrahedron Lett.*, 1979, 2291; D. H. R. Barton, G. Bringmann, G. Lamotte, W. B. Motherwell, R. S. Hay Motherwell and A. E. A. Porter, *J. Chem. Soc., Perkin Trans. 1*, 1980, 2657.
90. D. H. R. Barton, G. Bringmann and W. B. Motherwell, *J. Chem. Soc., Perkin Trans. 1*, 1980, 2665.
91. D. H. R. Barton, G. Bringmann and W. B. Motherwell, *Synthesis*, 1980, 68.
92. D. I. John, E. J. Thomas and N. D. Tyrrell, *J. Chem. Soc., Chem. Commun.*, 1979, 345.

4.3
Reduction of Sulfur–Carbon Bonds and of Other Heteroatoms Bonded to Tetrahedral Carbon

PAUL CAUBÈRE and PHILIPPE COUTROT
Université de Nancy, France

4.3.1 REDUCTION OF SULFUR–CARBON BONDS

4.3.1.1 Introduction

Organic sulfur chemistry is a powerful tool in organic synthesis because of the unique properties of sulfur, derivatives of which are reactive and versatile synthons. However, these useful properties would be of no use if it were not possible to remove sulfur from the final product. Fortunately reductive cleavage of a C—S bond is often easy and constitutes the 'last step' of many syntheses. This formal hydrogenolysis, called desulfurization (but spelled in several different ways), may be performed by numerous reagents and procedures. As usual if different systems are proposed, it means that no one is a panacea, and that they all have their drawbacks. This situation must be accepted. Indeed the diversity of molecules is such that it is impossible to find a unique reagent having all the virtues (inexpensive, catalytic, easily handled, chemo- and stereo-selective...) that an organic chemist wants. So the best we can hope is that reagents will be complementary.

4.3.1.2 Raney Nickel and Nickel(0) Complexes

Desulfurization of organic compounds with Raney nickel is an old reaction[1] with widespread application.[1,2] Raney Ni exists in several forms, called W-1 to W-8,[1b,1c] and differing in the procedures for their preparation. The properties (reactivity, stability, *etc.*) vary from one form to another and with aging. Thus when no information is given about a Raney Ni used, it may be hard to reproduce the reaction. Although sometimes called a catalyst, it is always used in large or very large excess. Raney Ni is sometimes symbolized as Ni(H) or erroneously $Ni(H_2)$. These misleading formulae do not mean that hydrogen is used during the reaction but that hydrogen is included in reagents which have not been degassed. A large number of the properties and drawbacks of Raney Ni have been established for a long time.[1,2] These reagents are well known to desulfurize every substrate containing carbon–sulfur bonds, and yields may vary from low to excellent. The main drawbacks are related to their handling and chemical reactivities. Thus the disadvantages of Raney Ni lie in its preparation, the unpredictable deactivation on aging and the difficulty in determining accurately the weight of Ni used because of the necessity of keeping the pyrophoric solid wet. On the other hand, the main limitations of Raney Ni are the lack of stereoselectivity and the reduction or hydrogenolysis of other functional groups, such as unsaturated C=C groups, nitro and carbonyl groups. Oxidations (*e.g.* alcohol dehydrogenations) as well as rearrangements or condensations may also be observed. A number of side reactions may be diminished and sometimes suppressed by using deactivated Raney Ni. Concerning the mechanism of desulfurization with Raney Ni it appears that the hydrogen atom replacing the sulfur generally comes from hydrogen included in the reagent, although the solvent may also play a part in the reductions. Radical mechanisms intervene during reductive desulfurizations, and racemizations are observed with chiral substrates. However, reductive cleavage of optically active sulfones was found to take place without racemization.

More recent publications confirm or complete the above picture of the scope and limitations. Thus, hindered glycols were obtained by reductive desulfurization of tetrahydrothiophenes (equation 1).[3]

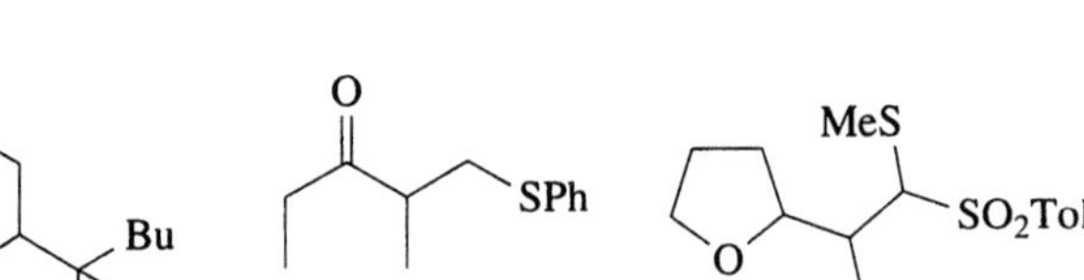

HO OH Bu^t Bu^t S —(Raney Ni; EtOH, reflux; 90%)→ HO OH Bu^t Bu^t (1)

Interesting chemoselectivity of Raney Ni is illustrated by the desulfurization in good yield (with few exceptions) of the functional sulfides (**1**),[4] (**2**),[5] (**3**),[6] (**4**)[7] and (**5**).[8]

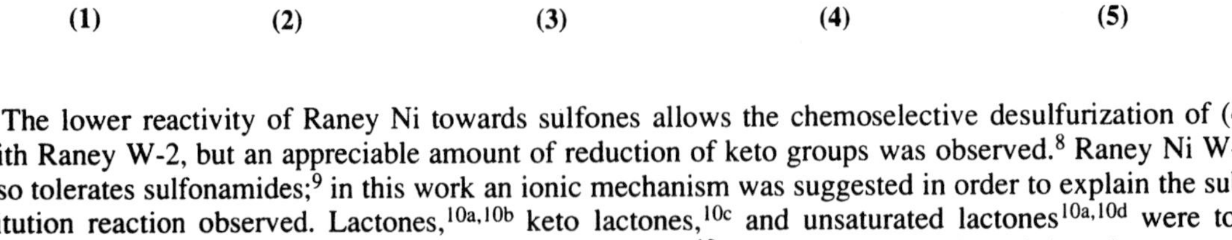

C_5H_{11} Bu^{t}S Y; Y = CO_2Me, CN, COMe **(1)**

O, Bu, SPh, CN **(2)**

O, SPh **(3)**

MeS, SO_2Tol, O, Ph **(4)**

Y, SMe, *n*; Y = H, COMe **(5)**

The lower reactivity of Raney Ni towards sulfones allows the chemoselective desulfurization of (**4**) with Raney W-2, but an appreciable amount of reduction of keto groups was observed.[8] Raney Ni W-2 also tolerates sulfonamides;[9] in this work an ionic mechanism was suggested in order to explain the substitution reaction observed. Lactones,[10a,10b] keto lactones,[10c] and unsaturated lactones[10a,10d] were tolerated during the desulfurization of sulfide or dithioketal[10c] groups, but deactivated (*e.g.* by acetone) Raney Ni was generally used. In contrast, carbon–halide bonds are cleaved where the desulfurization is stereoselective only because the product is the thermodynamically more stable *cis*-lactone (equation 2).[10e,10f]

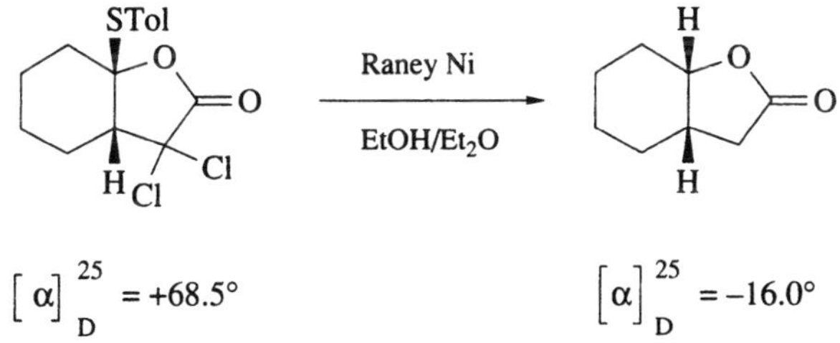

STol O O Raney Ni EtOH/Et_2O H O O (2)

H Cl Cl H

$[\alpha]^{25}_{D} = +68.5°$ $[\alpha]^{25}_{D} = -16.0°$

From comparative studies, it appears that a number of C═C double bonds are tolerated only when Raney Ni has been deactivated.[10a,10d] It has been claimed[11] that desulfurization of thiophenes could lead to acceptable yields of unsaturated products, but this result seems to be due to steric hindrance in the unsaturation formed. Curiously it has been found that Raney Ni desulfurization of unsaturated lithium thiolates resulting from Li–$EtNH_2$ desulfurization (*vide infra*) takes place easily, while direct reduction of the starting heterocycles is unsatisfactory (Scheme 1).[12]

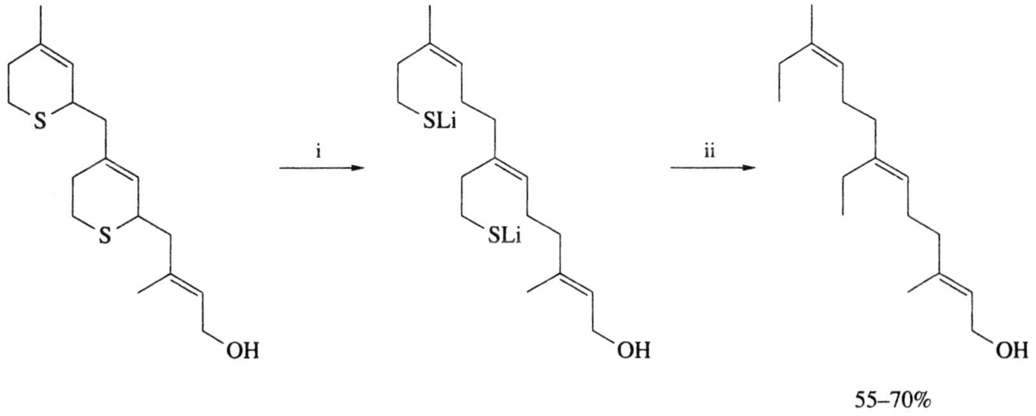

S S OH i SLi SLi OH ii OH

55–70%

i, Li–$EtNH_2$, –78 °C; ii, deactivated Raney Ni, EtOH reflux

Scheme 1

Finally, it has been possible to preserve C═C double bonds during desulfurization of vinyl sulfides (50 to >99%),[13a–13c] and alkenes can be reduced simultaneously with desulfurization, causing the saturation of vinyl sulfides.[14]

Although less commonly used, Raney Ni cleavage of the C—S bond of sulfoxides and sulfones is also of interest. For example, the removal of sulfoxides is one of the steps in the synthesis of spiroketals[15a] and of tetrahydrofuran derivatives (Scheme 2).[15b] Exceptionally, the desulfurization is not accompanied by the hydrogenolysis of a benzyl ether group.

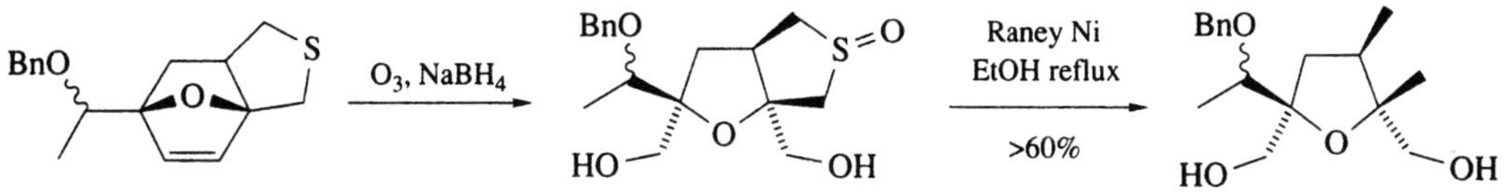

BnO O S O_3, $NaBH_4$ BnO S=O O HO OH Raney Ni EtOH reflux >60% BnO O HO OH

Scheme 2

In contrast, the unexpected hydrogenolysis of a tertiary alcohol was observed during the desulfurization either of 1,3-dithiane derivatives or of their monosulfoxides, as exemplified in Scheme 3.[16] It is intriguing that even deactivated W-2 Raney Ni did not improve the result, but a low temperature did allow convenient desulfurization with a minimum amount of side product.

The removal of sulfoxide[17] or sulfone[18] groups from molecules containing chiral alcohols is not always accompanied by racemization, indicating that oxidation of alcohols by Raney Ni[1] may be avoided.

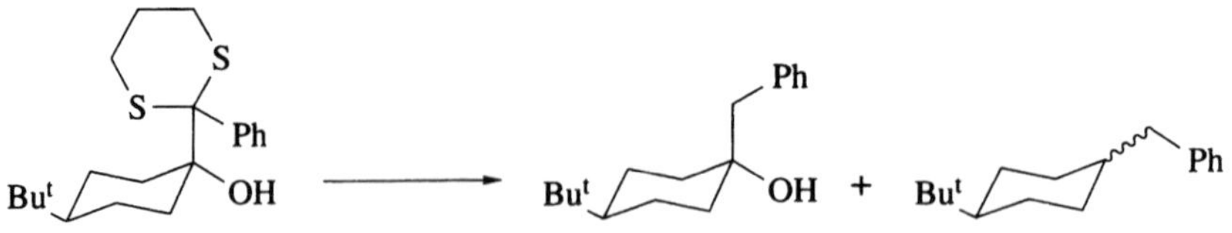

Raney Ni W-2 deactivated or not, 25 °C, EtOH	80%	20%
Raney Ni W-2 freshly prepared, 0 °C, EtOH	96.7%	3.1%

Scheme 3

Desulfurization of thioesters can be used in the synthesis of ethers[19] and particularly of crown ethers,[19b] although yields are rather low (equation 3).

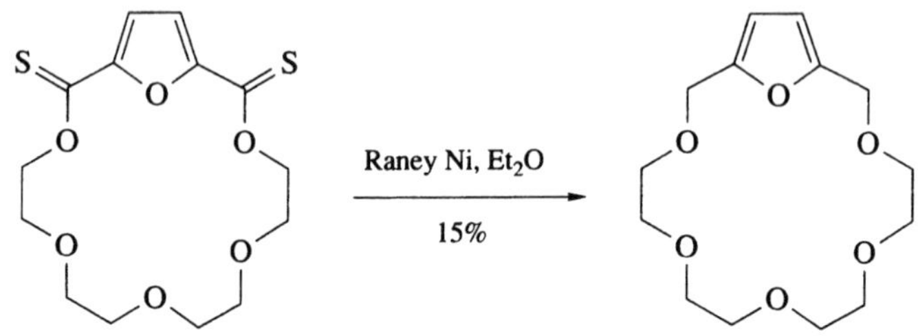

Raney Ni, Et_2O 15% (3)

We can conclude that Raney Ni continues to be useful in synthesis, but it must be kept in mind that, besides the side reactions it gives, the main drawback is the need to use a large excess of a hazardous reagent.

Desulfurization with Raney Ni is always performed under heterogeneous conditions using Ni metal as reagent. It was guessed that a soluble nickel(0) species might be more efficient. Moreover, such studies were expected to give information about the mechanism of desulfurization. That nickel(0) species can insert into C—S bonds by oxidative addition has been established by isolation of the corresponding complexes.[20] On the other hand, [(bipy)Ni(COD)] in small excess (2 mol equiv.) efficiently desulfurizes sulfur-containing heterocycles.[21] The mechanism proposed is given in Scheme 4.

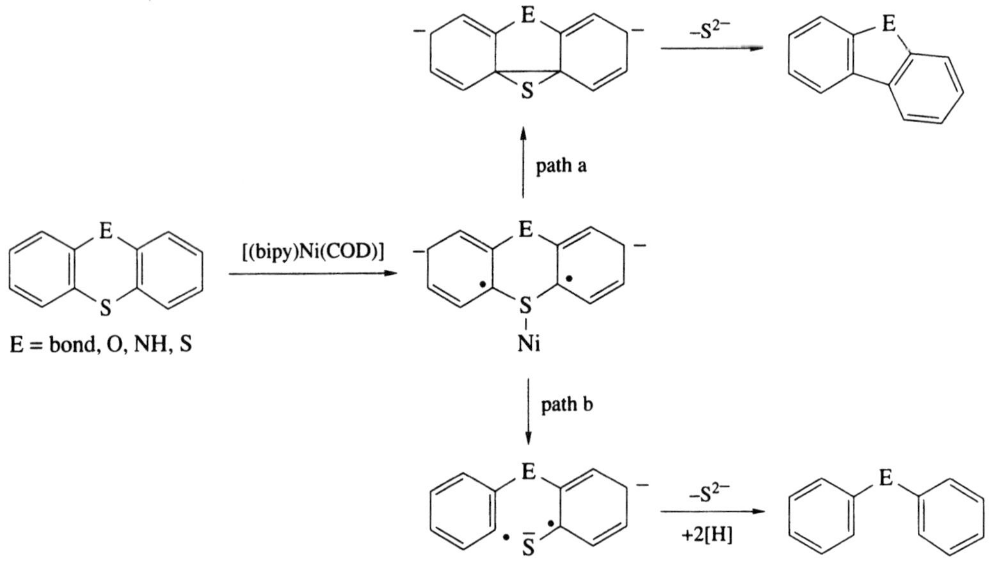

E = bond, O, NH, S

[(bipy)Ni(COD)]

path a

$-S^{2-}$

path b

$-S^{2-}$
+2[H]

Scheme 4

In Scheme 4, path (a) is dominant. In path (b), the hydrogen comes from the solvent (THF). We shall see later that path (b) may be favored by the presence of a complex hydride.[21a,21c] Synthetic application of this desulfurization remains to be demonstrated. It may be that the drawbacks in the preparation of the very sensitive starting complex could be avoided by using one of the better known procedures for the generation of nickel(0) species *in situ*.[22] Finally, nickel(0) species have been proposed as intermediates in the eliminative desulfurization of 1,2-disulfones.[23]

4.3.1.3 Hydrides or Organometallics Combined with Transition Metal Salts or their Complexes

Among the drawbacks of Raney Ni are its highly pyrophoric nature, the loss of activity on storage and the difficulty of knowing the amount of Ni used. One way to avoid these disadvantages would be to generate Ni *in situ* from a known amount of one of its salts. This is possible by reacting a Ni salt with $NaBH_4$; the black reagent thus obtained looks like a metal, often behaves like a metal, but is not a metal: it is a Ni boride.[24] Use of different metal salts has led to the preparation of a number of metal borides,[24] and the reaction has been extended by using lithium aluminum hydride, thus allowing the preparation of metal aluminides.[24] Metal borides and aluminides are well known as desulfurizing reagents.[2c,2d,24]

Among metal borides, Ni boride is the one most often used in desulfurizations. Always used in large (and often very large) excess, it may be prepared before use or preferably generated *in situ* in the presence of the substrate to be desulfurized. The solvent may be water, and EtOH alone, or added to ethers such as THF. In a number of cases $NaBH_4$ is used in even larger excess than the Ni salt. Generation *in situ* allows for more efficient desulfurizations.[11] However, hydrogen is generated in the presence both of the substrate and of Ni boride, which is a hydrogenation catalyst.[24] Finally Ni boride is less reactive than Raney Ni,[13a,25] and does not desulfurize sulfones.[25a] Due to its lower reactivity, Ni boride is more apt to give chemoselective desulfurizations, but is not devoid of side reductions. For example, during the desulfurization of thioalkyl, aryl or heteroaryl ethers, a benzylamino group was not hydrogenolyzed but a carbonyl group was reduced.[26a]

The desulfurizations illustrated in Scheme 5 show that Ni boride may tolerate hindered alkenes,[13a,26b] alcohols[27a] (without oxidoreduction), ketones[26b,28] and esters.[27] However, under certain conditions,

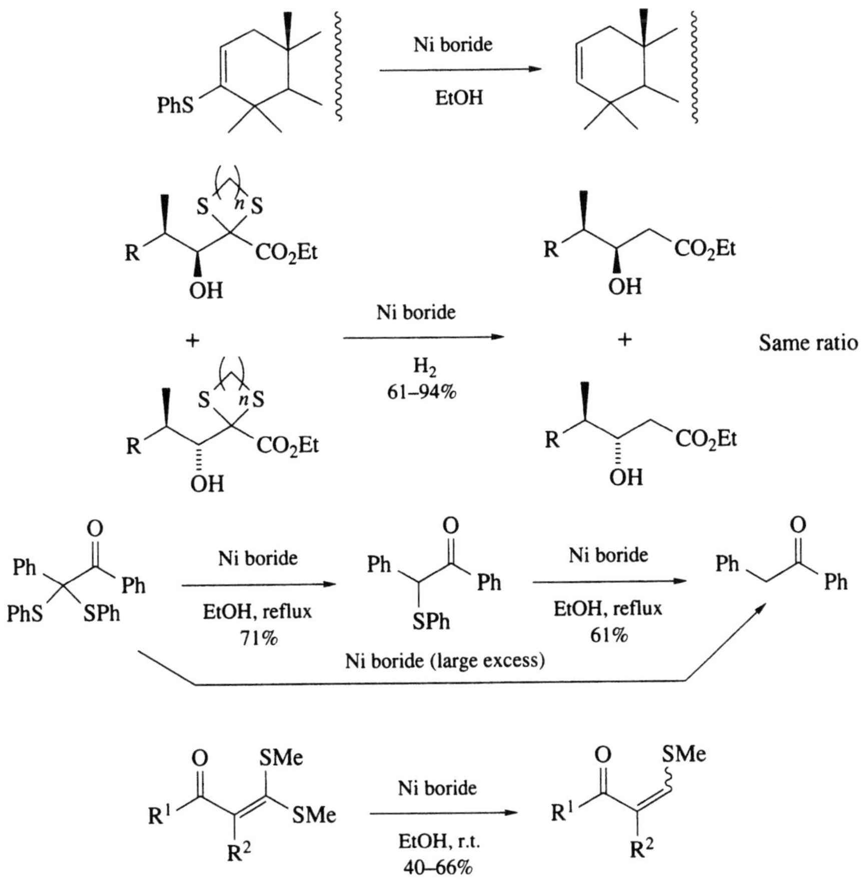

PhS; Ni boride, EtOH

R, S, n, S, CO₂Et, OH; + ; Ni boride, H₂, 61–94%; R, CO₂Et, OH; Same ratio

Ph, O, Ph, PhS, SPh; Ni boride, EtOH, reflux, 71%; Ph, O, Ph, SPh; Ni boride, EtOH, reflux, 61%; Ph, O, Ph; Ni boride (large excess)

O, SMe, R¹, SMe, R²; Ni boride, EtOH, r.t., 40–66%; O, SMe, R¹, R²

Scheme 5

reduction of alkenes may take place along with the desulfurizations.[13a,29] The use of a hydrogen atmosphere improved the procedure.

A number of metal salts or complexes have been combined with $NaBH_4$. Although some results are obtained with Co boride, none of the systems used are as efficient as Ni boride.[26,29] An interesting exception is $[Pd(PPh_3)_4]$ or, better, $[PdCl_2(DPPP)]$ combined with $NaBH_4$, which allows the reductive desulfurization of allyl sulfones.[30] Based on the concept that allyl sulfones undergo substitution by nucleophiles in the presence of palladium(0) complexes this procedure can use the Pd reagents in catalytic amounts, and replacement of $NaBH_4$ by $LiHBEt_3$ seems to improve the procedure (Scheme 6).[30b]

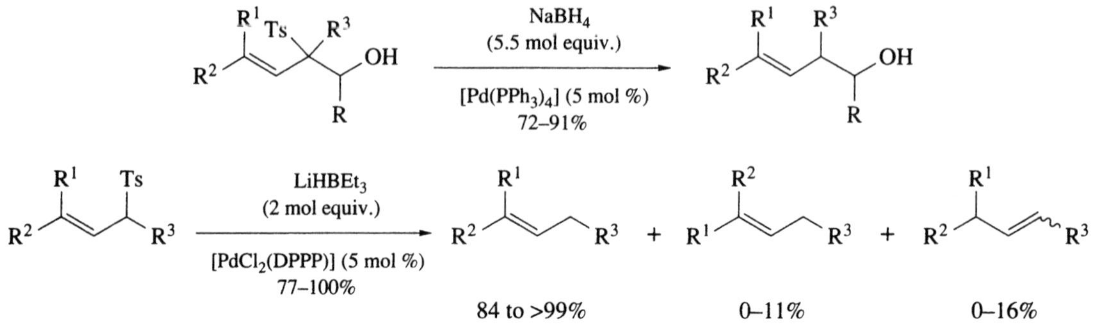

Scheme 6

Note that, while the palladium(0)-containing reagent was of limited application but did not cause isomerization, the palladium(II)-containing reagent, although more versatile, did cause isomerization. $LiHBEt_3$–$[Pd(PPh_3)_4]$ desulfurized allyl thiophenyl ethers, but migrations and (*E*)/(*Z*)-isomerizations were also observed.[31] $LiHBEt_3$ or $NaHBEt_3$ have also been used to generate Fe- or Co-containing reagents which desulfurize thiols.[32]

Reagents of potential interest were also obtained from LAH. The most often used contain Ni or Cu. Thus homogeneous desulfurizations are performed with $[(Cp_2NiAlH_2)^-Li^+]_2$ prepared from $[Cp_2Ni]$ and LAH in THF.[33] The ratio of reagent to substrate varies from 1:1 with thiols to 2:1 with sulfides, sulfoxides or sulfones and to 8:1 with thioketals. Yields, although often moderate, may reach 80%. α-Acyl-substituted, benzyl and aryl C—S bonds are rather easily cleaved, while isolated tetrahedral carbon C—S bonds require more vigorous conditions. It seems that alkenes, methoxy and carbonyl groups as well as esters are tolerated. It is claimed that C–halogen bonds were tolerated, but the examples given are insufficient for a firm conclusion. Nitro groups, conjugated alkenes and terminal alkynes are reduced. A few examples are given in Scheme 7.

The reaction of [(bipy)Ni(COD)] in THF with LAH led to $[(bipy)NiAlH_2]Li{\cdot}n$THF, a rather efficient desulfurization agent.[21a,21c,34] During reactions such as the ones given in Scheme 4, the presence of the hydride favored path b. Some of the other reagents prepared in the same way and containing different metal species[34] deserve to be studied further.

LAH–$CuCl_2$ used in moderate to large excess desulfurizes sulfides and dithioketals in good to very good yields (55–87%),[35] but with allyl sulfides, migration of the double bond occurs. Finally vinyl phenyl sulfones are reductively cleaved in good yields (65%) without reduction of C=C bonds even by an excess of reagent.[36]

Completely different in nature are those curious but efficient reducing agents called 'Metal Complex Reducing Agents' (MCRA),[37] very easily prepared from NaH, a sodium alkoxide and a metal salt. NiCRA–bipy and NiCRA (prepared from NaH, t-$C_5H_{11}OH$ and $Ni(OAc)_2$ with or without bipy, respectively), are not very sensitive and are easily handled. Used in moderate to large excess, they desulfurize alkyl and aryl sulfides, sulfur-containing heterocycles, sulfoxides, sulfones and dithioketals. Under appropriate conditions, dithioketals are selectively transformed into sulfides.[38] So far it has been established that these reagents tolerate ketones, esters, and, to some extent, alkenes. The reactivity of MCRAs is illustrated in Scheme 8.

Desulfurizations of unsaturated sulfur-containing substrates are often accompanied by reductions, migrations or isomerizations, and reagents devoid of these drawbacks are in keen demand. One candidate is a suitable Grignard reagent in the presence of Ni complexes. Thus, Pr^iMgBr (in excess) in the presence of a catalytic amount of $[(Ph_3P)_2NiCl_2]$ stereospecifically desulfurized phenyl thiovinyl substrates in acceptable to good yields,[39] and acetals, ethers, aromatic systems and isolated alkenes appear to be tolerated. The same kind of reductions were performed with aryl thiomethyl and ethyl thiovinyl ethers

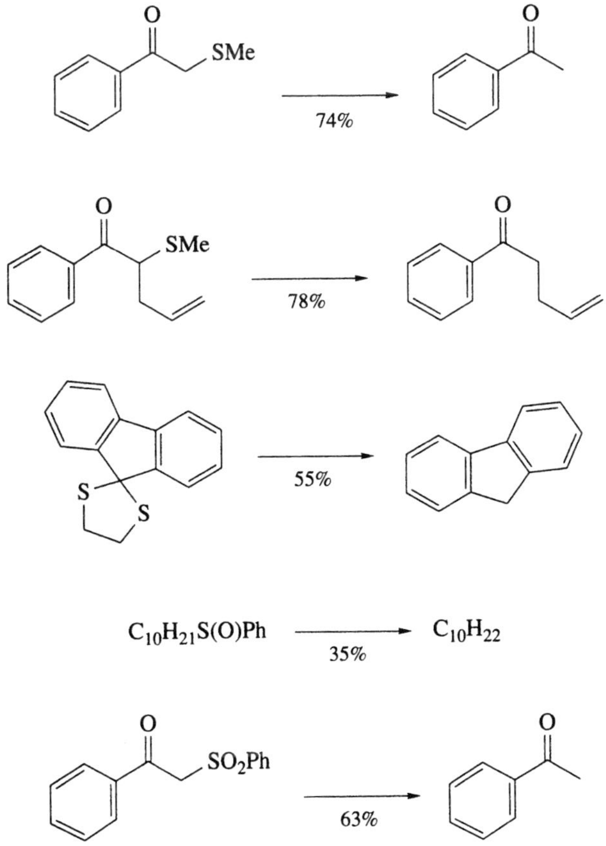

SMe

74%

SMe

78%

S S

55%

$C_{10}H_{21}S(O)Ph$ → $C_{10}H_{22}$
35%

SO_2Ph

63%

Desulfurization with $[(Cp_2NiAlH_2)^{-}Li^{+}]_2$

Scheme 7

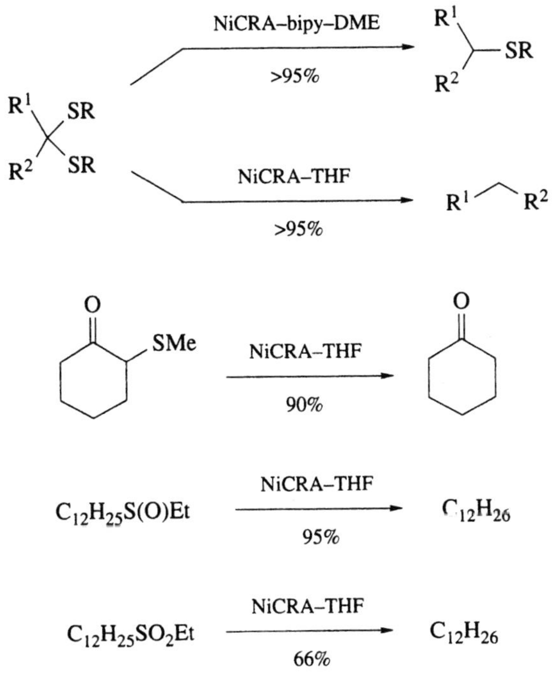

NiCRA–bipy–DME
>95%

R^1 SR
R^2 SR

NiCRA–THF
>95%

R^1 R^2

SMe
NiCRA–THF
90%

$C_{12}H_{25}S(O)Et$ —NiCRA–THF, 95%→ $C_{12}H_{26}$

$C_{12}H_{25}SO_2Et$ —NiCRA–THF, 66%→ $C_{12}H_{26}$

Scheme 8

using Pr^iMgBr or c-$C_6H_{11}MgBr$ and $NiCl_2$–PPh_3 in C_6H_6–Et_2O.[40] However, a change in the nature of the ligand may lead to substitution instead of desulfurization, since substitution of thioalkyl groups by Grignard reagents in the presence of Ni complexes is a well-known synthetic reaction.[8,41] Vinyl sulfones are also desulfurized by BuMgCl in the presence of transition metal complexes either as such or in the presence of DABCO, Bu_3P or Ph_3P.[42a–42d] The most efficient systems contain Ni or Pd and the reagents tolerate isolated alkenes.[42b] The behavior of polyenyl substrates may be more complicated.[42c]

4.3.1.4 Metals and Amalgams

Electropositive metals have been widely used to cleave C—S bonds. Alkali metals are the most commonly used, but Mg, Ca and Zn are also included in a number of procedures. Alkali metals (Li holds the prominent place) are the most widely used to desulfurize sulfides, sulfoxides, sulfones, dithioketals *etc*., and a large amount of information may be obtained from reviews devoted either to reduction by alkali metals[43] or to sulfur chemistry.[2a–2e,44] Before briefly commenting on the scope and limitations of these reductions, it must be emphasized that single electron transfer (SET) takes place during these desulfurizations, which are generally performed in the presence of an excess of a reducing agent. Thus functional groups sensitive to SET may be reduced. However, advantage may be taken of this side reaction, to perform selective desulfurization of α-sulfurated carbonyl substrates (Scheme 9).

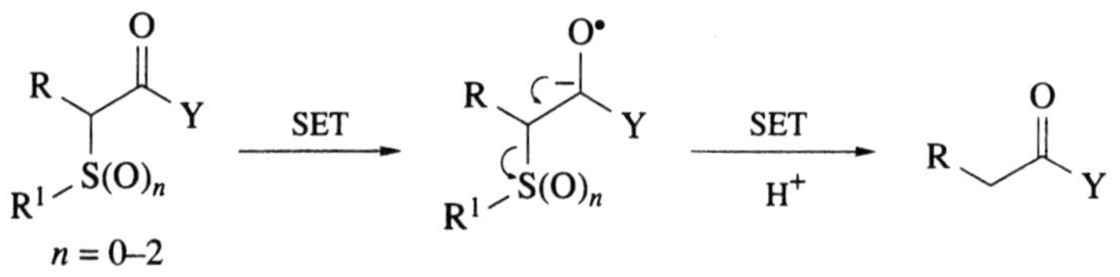

$n = 0$–2

Scheme 9

SET cleavage of C—S bonds is accompanied by eliminations (which are often the only reactions observed) when a good leaving group is present at the vicinal position (Scheme 10).

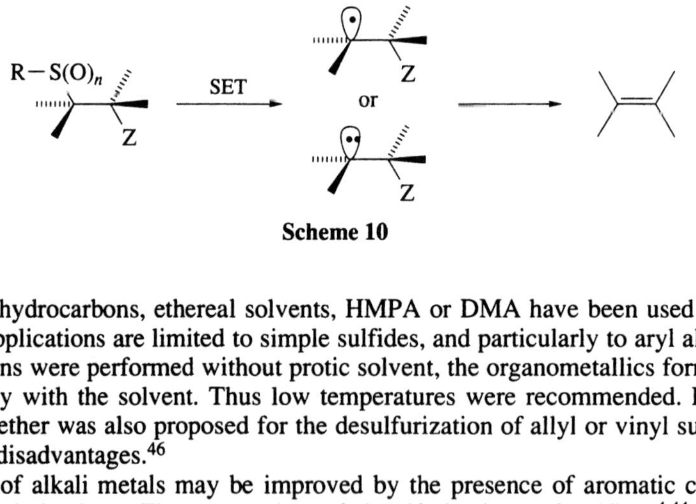

Scheme 10

Li, Na and K in hydrocarbons, ethereal solvents, HMPA or DMA have been used to desulfurize sulfides,[43b,43c,45] but applications are limited to simple sulfides, and particularly to aryl alkyl sulfides. Since these desulfurizations were performed without protic solvent, the organometallics formed may react further and particularly with the solvent. Thus low temperatures were recommended. Potassium graphite (C_8K) in excess in ether was also proposed for the desulfurization of allyl or vinyl sulfones, but this reagent is not free of disadvantages.[46]

The SET ability of alkali metals may be improved by the presence of aromatic compounds such as naphthalene or its derivatives. Thus a number of desulfurizations of sulfides[4,44a,44b,45] or dithioketals[44a,44b,47] have been performed (generally at low temperature in order to diminish side reactions) with Li in the presence of a catalytic amount of naphthalene, or with an excess of lithium naphthalenide or lithium 1-(dimethylamino)naphthalenide (LDMAN; Scheme 11).[47] Note that 1-dimethylaminonaphthalene may be easily removed from the reaction product during the work-up. These desulfurizations tolerate ethers[45] and nitriles[4] but do not respect halides,[45] ketones, esters and a few other structures.[4]

Use of excess alkali metal in liquid ammonia or an amine (generally $EtNH_2$ or ethylenediamine, EDA) is also a widespread procedure used to desulfurize sulfides, dithioketals and sulfones.[2a–2e,43b,43c] From the numerous works published, a number of interesting features emerge. Thus desulfurization of sulfides in liquid NH_3–THF tolerates esters and ketones,[4] while Li–$EtNH_2$ cleaves the C—S bonds of unsaturated substrates without side reactions (Scheme 1).[12] Sodium in liquid ammonia was found to be a better

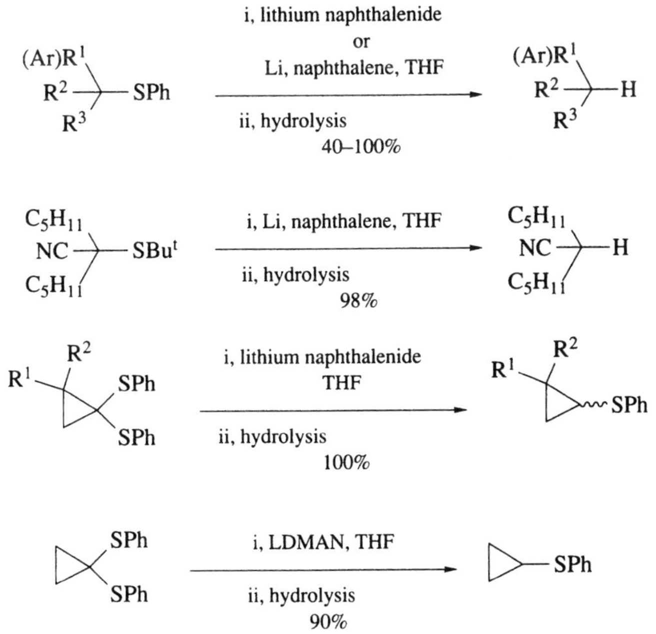
(Ar)R1
R2
SPh
R3
i, lithium naphthalenide
or
Li, naphthalene, THF
ii, hydrolysis
40–100%
(Ar)R1
R2
H
R3
C5H11
NC
SBut
C5H11
i, Li, naphthalene, THF
ii, hydrolysis
98%
C5H11
NC
H
C5H11
R2
R1
SPh
SPh
i, lithium naphthalenide
THF
ii, hydrolysis
100%
R1
R2
SPh
SPh
SPh
i, LDMAN, THF
ii, hydrolysis
90%
SPh

Scheme 11

reagent than Raney Ni, Ni boride or Bu_3SnH (*vide infra*) to desulfurize sulfides in the deoxysugar series.[48] Recently, a careful study of the desulfurization of diaryl sulfides has shown that the reductions depend upon the structure (polycondensed or not) of the substrate.[49]

Unsaturated sulfones were desulfurized by Li–$EtNH_2$ without isomerization (Scheme 12),[50,51] while other procedures were unsuccessful.[50] Isomerization was observed with the same reagent during the desulfurization of allyl and vinyl sulfones.[46]

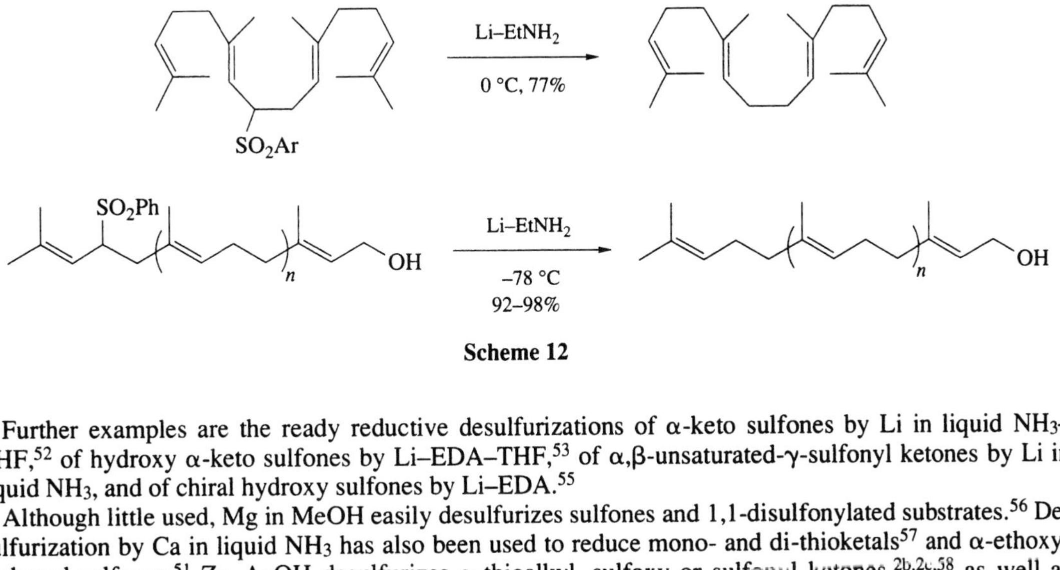
Li–EtNH2
0 °C, 77%
SO2Ar
SO2Ph
n
OH
Li–EtNH2
–78 °C
92–98%
n
OH

Scheme 12

Further examples are the ready reductive desulfurizations of α-keto sulfones by Li in liquid NH_3–THF,[52] of hydroxy α-keto sulfones by Li–EDA–THF,[53] of α,β-unsaturated-γ-sulfonyl ketones by Li in liquid NH_3, and of chiral hydroxy sulfones by Li–EDA.[55]

Although little used, Mg in MeOH easily desulfurizes sulfones and 1,1-disulfonylated substrates.[56] Desulfurization by Ca in liquid NH_3 has also been used to reduce mono- and di-thioketals[57] and α-ethoxycarbonyl sulfones.[51] Zn–AcOH desulfurizes α-thioalkyl, sulfoxy or sulfonyl ketones,[2b,2c,58] as well as γ-sulfonyl-α,β-unsaturated ketones, in good yields,[54] but this procedure does not tolerate acid labile functional groups. This disadvantage has been circumvented by using Zn in excess in the presence of Me_3SiCl,[59a] and still more efficiently using NH_4Cl in THF–H_2O (Scheme 13).[59b,59c]

Zn–NH_4Cl desulfurizes sulfides more slowly than sulfoxides, and α-phenyl sulfides of purely aliphatic esters are hardly desulfurized. Lactones and esters are not hydrolyzed, and α,β-unsaturated ketones are tolerated as well as alkenes, with which no migrations are observed.

Sodium amalgam [Na(Hg)] used in excess in alcohols, is well known to desulfurize sulfides, sulfoxides, and, above all, sulfones,[2a–2c,2e,2f,60,61] but a number of side reactions can occur. It was found[62] that

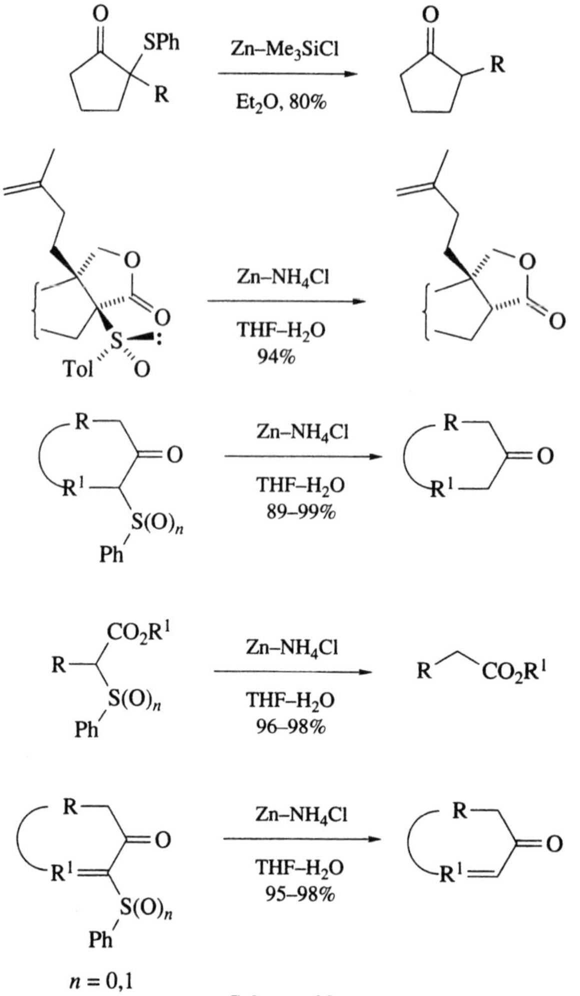

Zn–Me$_3$SiCl, Et$_2$O, 80%

Zn–NH$_4$Cl, THF–H$_2$O, 94%

Zn–NH$_4$Cl, THF–H$_2$O, 89–99%

Zn–NH$_4$Cl, THF–H$_2$O, 96–98%

Zn–NH$_4$Cl, THF–H$_2$O, 95–98%

$n = 0,1$

Scheme 13

6% Na(Hg) used in excess with Na_2HPO_4 in MeOH was particularly efficient, and presented fewer drawbacks. It was possible to perform reductive desulfonylations of sulfones in the presence of alkyl phenyl sulfides.[63] This procedure tolerates ethers, a trimethylsilyl group,[63] esters,[51,63] ketones,[51,64] isolated double[51,63,64a,65] and triple[65] bonds, ketals,[63,66] amides[67] and lactones.[68] A few examples are given in Scheme 14.

However, this procedure also has a number of drawbacks. A large quantity of Hg must be used. Moreover, alkenic bonds are reduced during the desulfurization of conjugated dienyl vinyl sulfones,[42c] and β-eliminations take place in the presence of good vicinal leaving groups.[69]

Since the pioneering work,[70] where Al(Hg) in large excess was shown to desulfurize sulfoxides, sulfones and sulfonamides under mild conditions, this reagent has been widely used.[2a–2c,2e,44b] It is prepared by immersing aluminum foil in 2% aqueous $HgCl_2$, and is often used in THF–H_2O. Efficient in the desulfurization of α-keto[71] and alkoxycarbonyl[72] sulfoxides, it has been used for removing chiral sulfoxides used in asymmetric synthesis. Al(Hg) also desulfurizes α-phenylthio[73] or sulfonyl ketones,[53,74] and allows the monodesulfonylation of 1,1-disulfonyl substrates.[42c] A few examples are given in Scheme 15.

Al(Hg) reduces the carbon–halogen bond[10f] but tolerates a trifluoromethyl group[72a] and isolated alkenes,[72c,73,74] thus allowing the desulfonylation of vinyl sulfones,[36,75] in a reaction which seems to be stereoselective.[36] However, the procedure is not free of drawbacks, since epimerization is observed in the desulfurization of α-thiophenyl ketones.[73] During the desulfurization of functional α-sulfonyl ketones, reduction of nitro groups and, surprisingly, of the keto group, was observed.[76] Desulfurizations of conjugated dienyl vinyl sulfones are accompanied, as with Na(Hg), by the complete reduction of the alkene unit bearing the sulfonyl group.[42c]

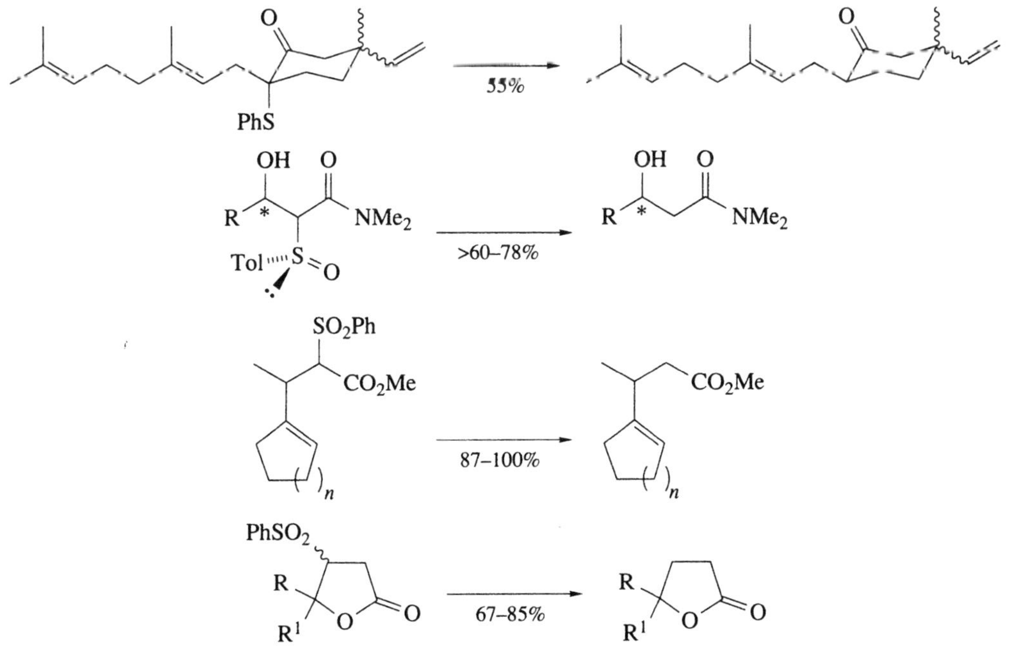

55%

>60–78%

87–100%

67–85%

Desulfurization with 6% $Na(Hg)–MeOH–Na_2HPO_4$ or NaH_2PO_4

Scheme 14

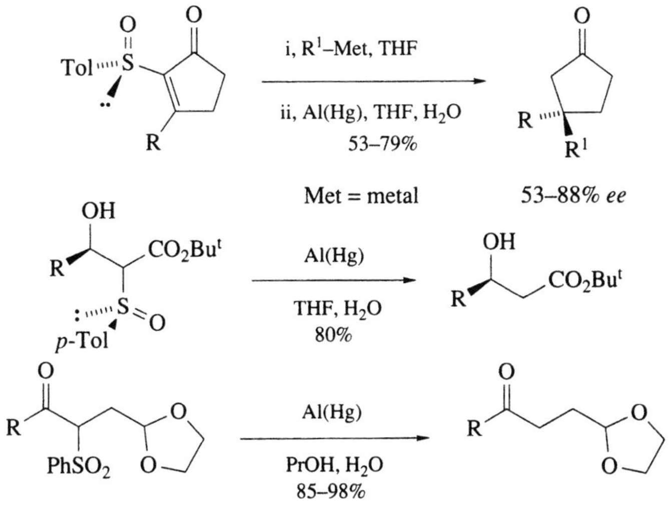

i, R^1–Met, THF

ii, Al(Hg), THF, H_2O
53–79%

Met = metal

53–88% *ee*

Al(Hg)

THF, H_2O
80%

Al(Hg)

PrOH, H_2O
85–98%

R = Pr, hexyl, heptyl, allyl, $C_7H_{15}CO_2Bu^t$

Scheme 15

4.3.1.5 Tin Hydrides

Tin hydrides, generally Bu_3SnH (TBTH) or Ph_3SnH (TPTH) (which may be more efficient), usually used in hot toluene or benzene in the presence of AIBN, cleave C—S bonds by a radical mechanism. Because the C—S bond is not very reactive, reactions performed without AIBN are ineffective. Extensive reviews on tin hydride properties[77] have been published. Information about their use in desulfurizations may also be obtained from other reviews.[44a,44b,78] Evaluation of TBTH in the desulfurization of sulfides has led to a number of important conclusions.[79] Thus, the mechanism of desulfurization is that shown in Scheme 16.

The reaction is of interest only when one of the alkyl groups of the starting sulfide gives stabilized radicals. Moreover, the C—S bond of the thioalkyltin intermediate is much more reactive, and, as a con-

$$Bu_3SnH \xrightarrow{AIBN} Bu_3Sn\bullet$$

$$RSR^1 + Bu_3Sn\bullet \longrightarrow RSSnBu_3 + R^1\bullet$$

$$R^1\bullet + Bu_3SnH \longrightarrow R^1H + Bu_3Sn\bullet$$

$$RSSnBu_3 + Bu_3Sn\bullet \longrightarrow R\bullet + Bu_3SnSSnBu_3$$

$$R\bullet + Bu_3SnH \longrightarrow RH + Bu_3Sn\bullet$$

Scheme 16

sequence, selectivity in desulfurization of unsymmetrical sulfides is not possible. Finally, allyl sulfides may react in an S_H or S_H' sense, leading to reduced or stannylated derivatives. Note that S_H' reductive stannylation of allyl- or propargyl-sulfurated substrates is a well-known reaction.[80]

Desulfurization of sulfides (or thiols) by TBTH–AIBN tolerates nitriles, esters, ketones,[79] ethers,[48,81] lactones,[10f,82] β-lactams[83] and isolated alkenes.[79,83] In contrast, halides are reduced and the stereochemistry of the starting sulfides is not conserved. Yields of these desulfurizations vary considerably, and β-eliminations may occur with vicinally substituted compounds. Some examples of TBTH desulfurizations are given in Scheme 17.

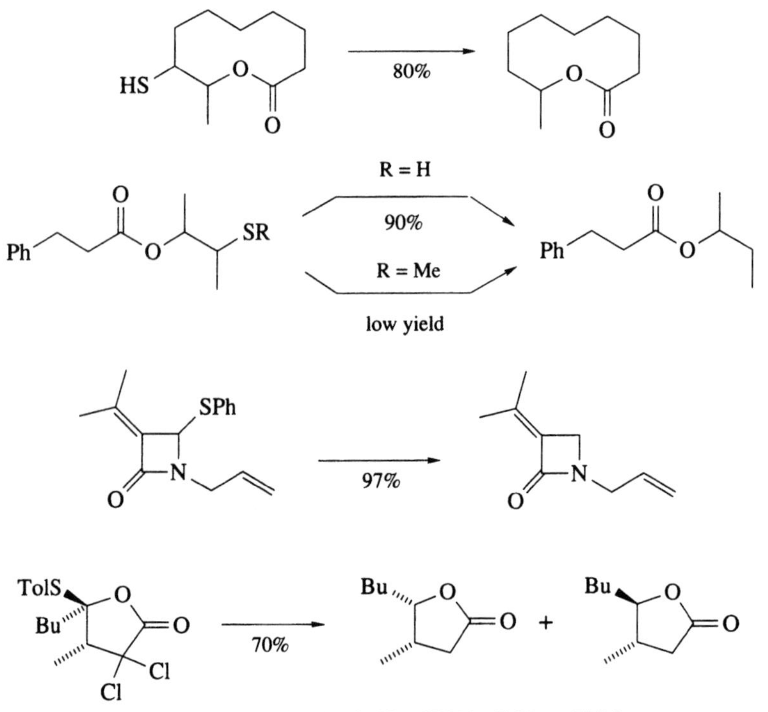

HS ... 80%

SR; R = H, 90%; R = Me, low yield; Ph ... O ... Ph

SPh ... N ... 97%

TolS, Bu, O, Cl, Cl ... 70% ... Bu ... O + Bu ... O

Desulfurization with Bu_3SnH–AIBN in C_6H_6 or PhMe

Scheme 17

Benzothiazolyl alkyl sulfides, prepared from the corresponding alcohols, are easily desulfurized (87–99%) by TBTH–AIBN.[84] This reaction constitutes a good indirect deoxygenation of alcohols. Desulfurization of dithioketals may be controlled. Thus 1,3-dithiolanes with 1 mol equiv. of TBTH lead to β-alkyl (or β-aryl) thiols (64–82%), while 4 mol equiv. of TBTH leads to the completely desulfurized derivative.[85] Desulfurization of heterocyclic thiones is also performed with this reagent.[86] In passing, it should be noted that generation of radicals from C—S bonds and tin hydrides in the presence of radical initiators is of large synthetic interest.[44b,87]

4.3.1.6 Miscellaneous Desulfurizations

The lack of stereoselectivity of the above reagents in a number of reductions of vinyl sulfurated substrates led to a search for new procedures. Thus the problem of stereoselective reduction of vinyl sulfones was solved in many cases by using sodium dithionite.[42c,88] This mild desulfurization proceeds by anionic addition–elimination,[88d] is highly stereoselective, but not devoid of side reactions.[42c,88b]

In the special case of aryl alkyl sulfides, desulfurization may be performed by S_N2 reactions with nucleophiles such as RS^-, RSe^-, and sometimes RO^- or $R^1R^2N^-$.[43b,43c,89] 1,2-Eliminations, where the thioaryl group is the leaving group, have also been used.[43c]

Nucleophilic attack on the sulfur atom of sulfides by RS^- in aprotic or protic solvents may constitute a mild procedure.[2b] BuLi is also a good reagent for attacking one of the sulfur atoms of 2,2-diaryl-1,3-dithiolanes and -dithianes, leading to the corresponding thioalkyl sulfides.[90] In the same way, Bu_3SnLi easily desulfurizes sulfides and dithioketals.[44b,91] Desulfurization of allenyl sulfoxides occurs by attack on sulfur with MeLi.[92] Mild desulfurizations of α-keto sulfoxides and α,β-unsaturated-γ-sulfonyl ketones are obtained with Me_2CuLi.[54,71] Recently SET from SmI_2 was successfully used to desulfurize α-phenylthio and phenylsulfoxy ketones.[93]

Although of little synthetic interest at the present time, mention must be made of desulfurizations by transition metal carbonyls. Thus thiols and/or sulfides are desulfurized by complexes such as $[Mo(CO)_6]$, $[Fe_3(CO)_{12}]$, $[Co_2(CO)_8]$ and $[Os_3(CO)_{12}]$.[94] Coupling desulfurizations of dithioketals have been performed with $[W(CO)_6]$ or $[Mo(CO)_6]$,[95] and a recent review[96] gives further information about the hydrodesulfurization of thiophenes.

4.3.2 REDUCTION OF SELENIUM–CARBON BONDS

4.3.2.1 Introduction

In many respects selenium is like sulfur. Its derivatives are malodorous and they are often considered to be more toxic. The organic chemistry of selenium is younger than the chemistry of sulfur, and reports where reductive cleavage of C—Se is used as a 'last step' always deal with substrates containing nonoxygenated selenium.

With selenium, there are only a few hundred references to reductive deselenation compared with several thousands for the reductive cleavage of the C—S bond. Since the chemistries of selenium and sulfur are so similar, it is not surprising that the main deselenation procedures were borrowed from the desulfurization procedures. However, from the practical point of view, appreciable differences will be encountered due to the weaker C—Se bond and to the greater sensitivity of Se towards nucleophilic attack.

4.3.2.2 Raney Nickel

Reviews devoted to Se chemistry[97] or Raney nickel[1b] show that these are efficient reagents in reductive deselenations. Of course the advantages and drawbacks encountered with desulfurizations (see Section 4.3.1.2) must also be taken into consideration with deselenations. Raney Ni, often freshly prepared, is generally used in excess in solvents such as EtOH, ether, THF or DME, and under conditions milder than those for desulfurizations. From the numerous works published, it appears that the procedure tolerates alkyl and aryl ethers,[98,99,102] ketals,[100a,100b] ketones,[100b,101] esters,[101] lactones[10b,102] and sulfonamides.[9] A few representative examples are given in Scheme 18.

Among the drawbacks of using Raney Ni are the handling of large amounts of a hazardous reagent, and the lack of chemoselectivity against reducible functions such as unsaturation, allylic and benzylic oxygens, 1,3-dithianes, *etc*. Moreover, depending upon the chemical grade of Raney Ni or of the details of its preparation, yields may vary from excellent to very poor. Note, for example, that lactones, which are expected to be tolerated by Raney Ni, may be damaged by basic impurities remaining from the preparation of the reagent.[10b] Finally it appears that deselenations can proceed, at least in part, *via* an alkene, due to the formal elimination of RSeH.[97d] If the carbon atom adjacent to the carbon carrying the Se is a chiral center, the stereochemistry is lost. Note that some of these drawbacks may sometimes be used in synthesis: Raney Ni and H_2 (under pressure) can simultaneously perform deselenation, saturation of an alkene, and debenzylation of a benzyl ester.[98]

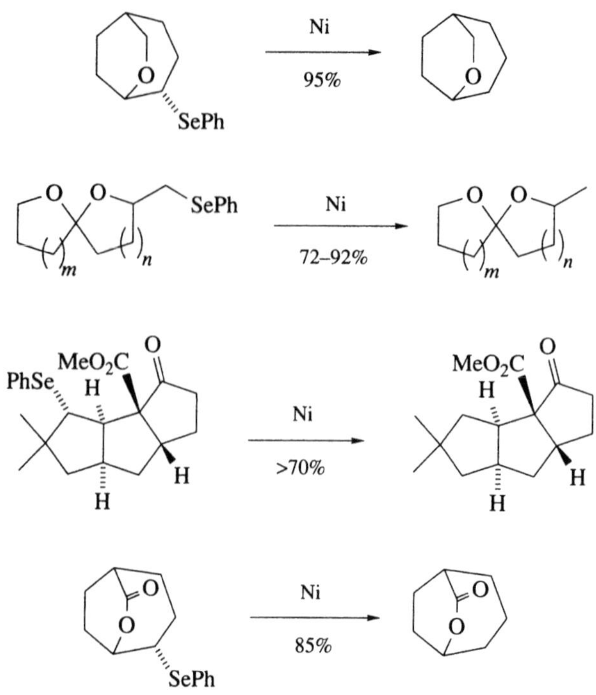

Ni 95%

SePh Ni 72–92%

MeO_2C PhSe H O Ni >70% H H

Ni 85% SePh

Scheme 18

4.3.2.3 Metal Borides

While numerous systems containing hydrides, or organometallics combined with transition metal derivatives, have been used in desulfurizations (see Section 4.4.1.3), deselenations have been performed only with Ni boride.[97f] From a recent careful study,[103] the following main features emerge: (i) the reaction is best performed in THF–MeOH; (ii) $NaBH_4$ must be taken in excess relative to the substrate (the ratio substrate:reagent depends on the structure of the substrate); (iii) to be efficient, $NaBH_4$ must be added to a mixture of the substrate and the Ni salt (no deselenation took place by addition of the substrate to a preformed Ni boride); (iv) hydrogen incorporated into the substrate originated from the MeOH and the $NaBH_4$, but not from THF (this result ruled out a radical mechanism); and (v) the deselenation applies to saturated or vinyl selenides, as well as to diselenoketals, and yields vary, with few exceptions, from good to very good. This procedure tolerates alkyl chlorides (but not alkyl iodides), nitriles, esters,[103] amides,[104] sulfonamides,[9] sulfides and sulfones (an interesting property for selective deselenations) but not sulfinate esters.[103] Moderately hindered alkenes were tolerated,[103,105] but less-hindered systems were reduced. Ketones were more or less tolerated, but the easily reduced ones may lead to the formation of appreciable amounts of alcohols.[103] Moreover, β-selenoketones can give β-elimination, and the usual complications were observed when the selenide group was vicinal to a good leaving group.[103] Allyl selenides were deselenated by $LiHBEt_3$–$[Pd(PPh_3)_4]$ (cat.) in excellent yield (82%) but with migration and (*E*)/(*Z*)-isomerization of the double bonds.[31]

4.3.2.4 Alkali Metals

The two main systems of this kind used in the deselenation of selenides and diselenoketals are Li–$EtNH_2$[97b,97c,97e,97f] and Na in liquid NH_3.[10b,99] Na in HMPA or DMA has also been used with some special aryl alkyl selenides.[43c] This procedure does not deserve special comment since, broadly speaking, the behavior of the reagents towards selenated substrates was very similar to that observed with sulfur-containing substrates (see Section 4.3.1.4). As expected from the difference of bond energies, C—Se bonds are more easily cleaved than C—S bonds. However, few selective deselenations have been performed in the presence of sulfur groups. C—O bonds vicinal to C—Se bonds are, of course, removed with alkene formation, and this property has found interesting applications.[10b,99]

4.3.2.5 Tin Hydrides

The reductive cleavage of the C—Se bonds of selenides and diselenoketals by TBTH or TPTH is a widespread deselenation method mentioned in numerous reviews.[44a,77c,97a,97c–97f,106] A radical mechanism is involved with both hydrides, which are used in refluxing benzene or toluene. TPTH is generally more suitable than TBTH, and freshly distilled TPTH allows shorter reaction times. AIBN is used as a radical initiator with both hydrides, but an interesting point is that deselenation with TPTH can be performed in refluxing toluene *without* AIBN,[97d,107] making this already mild deselenation still more chemoselective, particularly with sulfur-containing substrates. Using appropriate conditions, deselenations may be performed without the reduction or hydrogenolysis of isolated alkenes (even unhindered),[97d,98,102,108,109] alcohols[97d,102] (even allylic[108]), ethers (even benzyl[111]), sulfides,[99,110] dithioketals,[97d] ketones,[113] esters,[97d,98,102,113] lactones,[10b,97d,102,110,112] amides,[104,114] β-lactams[109] and carbamates.[97d,115] A few examples are given in Scheme 19.

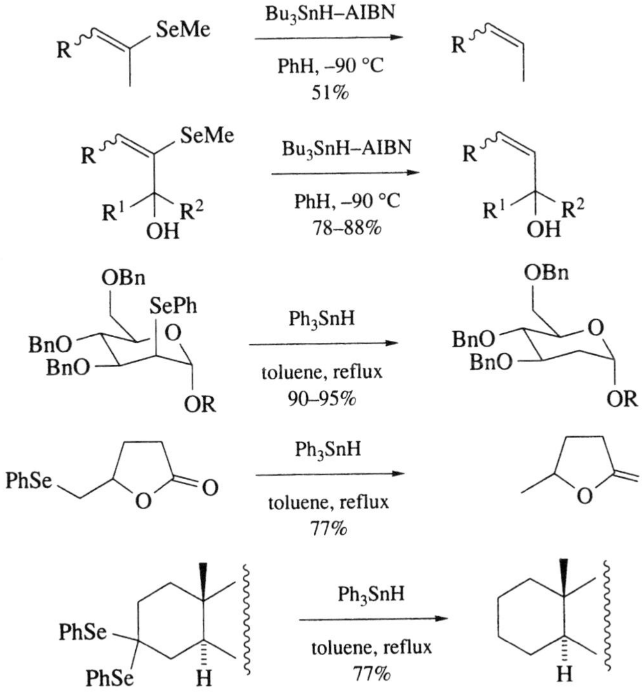

Bu_3SnH–AIBN, PhH, –90 °C, 51%

Bu_3SnH–AIBN, PhH, –90 °C, 78–88%

Ph_3SnH, toluene, reflux, 90–95%

Ph_3SnH, toluene, reflux, 77%

Ph_3SnH, toluene, reflux, 77%

Scheme 19

These deselenations do not tolerate halides, which are generally reduced. Among the drawbacks in addition to side reactions inevitable with radical intermediates, the use of an excess of an expensive reagent and a high reaction temperature may make the procedure less useful, especially with sensitive substrates. Note that the generation of radicals using deselenation by tin hydrides is also very useful in organic synthesis.[87]

4.3.2.6 Miscellaneous Deselenations

Only one kind of deselenation deserves to be mentioned in the present section and deals with the use of nucleophiles. Since selenium is much more sensitive to nucleophilic attack than sulfur, it reacts easily with strong nucleophiles/bases. This property may be a drawback, but is also a useful tool with which to generate organometallics. The main reagents encountered in these deselenations are BuLi[43c,97a,97c,97e,116] and PhSeLi,[43c,97b,117] although others have been used.

Deselenations with organolithium reagents have been successfully performed on mono-, 1,1-di- or 1,1,1-tri-selenides, generally in ethereal solvents such as THF, at more or less low temperature. Of course the deselenated products are obtained after hydrolysis, and yields vary from moderate to very good. One of the main drawbacks with organolithiums is their reactivity towards several functional groups.

Note that the relative insensitivity of sulfur toward nucleophilic attack allows the selective deselenation of thioselenoketals.[118] On the other hand, ketones are tolerated by PhSeLi, a useful property for chemoselective deselenations of α-phenylseleno ketones (Scheme 20).[119]

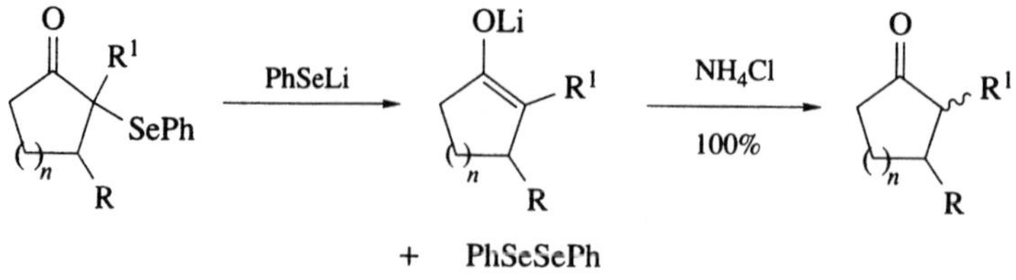

PhSeLi → (OLi enolate) + PhSeSePh; NH_4Cl, 100%

Scheme 20

4.3.3 REDUCTION OF MERCURY–CARBON BONDS

Methods which realize mercury–hydrogen exchange in organomercurials include homolytic decomposition in solvents capable of hydrogen atom transfer, protonolysis by protonic acids and reduction.

The former cannot be considered a good synthetic method because a mixture of reaction products is usually obtained. The protonolysis of organomercury compounds has been extensively studied from a mechanistic standpoint but has found little synthetic utility except for isotopic labelling of organic molecules. Reduction is the most useful method for the demercuration of organomercury compounds, particularly of solvomercurials in the solvomercuration–demercuration reaction, and a lot of reductants are used for this purpose.

The present review considers recent advances in mercury–hydrogen interchange by protonolysis and reducing agents, and especially metal hydride demercuration.

4.3.3.1 Protonolysis

The protic cleavage of the carbon–metal σ-bond ranks among the simplest of all electrophilic substitution processes. As organomercurials are readily prepared in high purity, can be manipulated with ease and are monomeric in solution, most mechanistic studies of the protonolysis of carbon–metal σ-bonds have focused on the protic cleavage of organomercurials. Reviews[120] and a book[121] have been published on this subject.

The cleavage of diphenylmercury in acetic acid with perchloric acid, and the reaction of dialkylmercury with mercury nitrate, may be considered as exhibiting the pure S_E2 mechanism, where an attack of the electrophile at the metal-bonded C-atom is the rate-determining step.[120a]

The coordination of the nucleophilic part Nu of the acid NuH with the mercury atom can occur and may be responsible for a four-center transition state (Scheme 21).

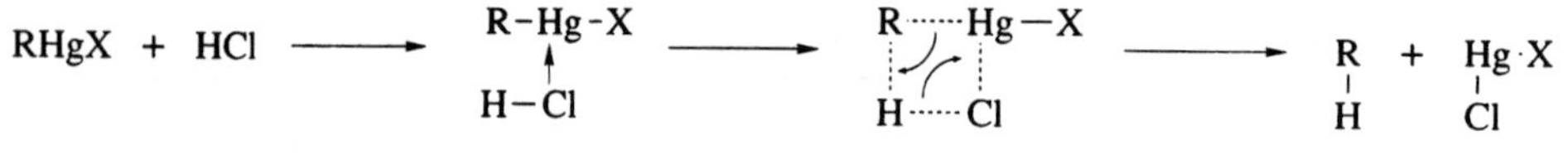

RHgX + HCl → R–Hg–X / H–Cl → R······Hg–X / H······Cl → R–H + Hg–X / Cl

Scheme 21

However, intramolecular nucleophilic participation by the conjugate base during protonolysis of a C—Hg bond is questionable. A study of the acidolysis of the carbon–mercury bond in unsymmetrical dialkylmercurials rather suggests that the reaction proceeds *via* a three-center transition state.[122] In any case, substantial kinetic and stereochemical evidence has led to the idea that reaction occurs by a concerted, front side attack with a transition state that involves a pentacoordinate carbon center.[121] In some cases unimolecular mechanisms, S_E1, also have been observed.[120a,122]

The importance of organomercurials in organic synthesis is due in large part to the valuable solvomercuration–demercuration reaction. As the reaction of proton acids with solvomercurated products derived from alkenes leads back to the alkene,[123] protonolysis has in this procedure little synthetic utility.

However, the protonolysis of alkenylmercurials derived from the solvomercuration of alkynes is known as a method for preparing stereoisomerically pure alkenes. Acidic deuterolysis provides also a convenient route to labelled organic substrates. For instance, in a study on the regio- and stereo-selectivity of the acetoxymercuration of alkynes, protodemercuration with acetic acid of a 3:1 mixture of the vinylmercury compounds (**6**) and (**7**) occurred almost quantitatively to give a 3:1 mixture of the corresponding esters (**8**) and (**9**) (equation 4),[124] and other examples of protodemercuration occurring with retention of configuration are known.[125]

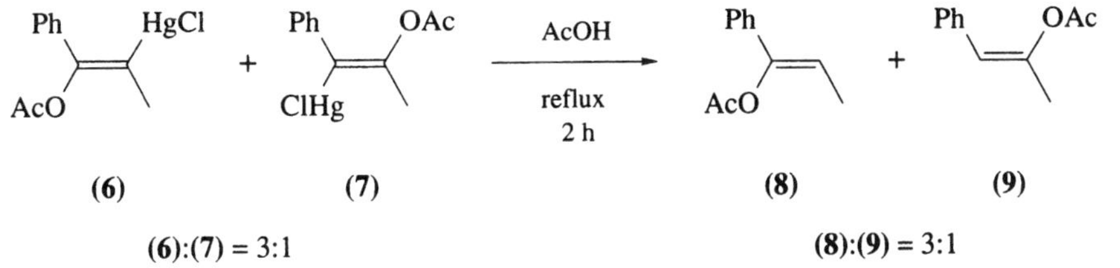

(**6**) + (**7**) —AcOH, reflux, 2 h→ (**8**) + (**9**) (4)

(**6**):(**7**) = 3:1 (**8**):(**9**) = 3:1

However, the alkenylmercury acetate (**10**) undergoes mercury–hydrogen exchange with HCl with inversion. A mercurinium ion intermediate has been proposed, where the vicinal phenyl groups provide the driving force for isomerization (Scheme 22).[126]

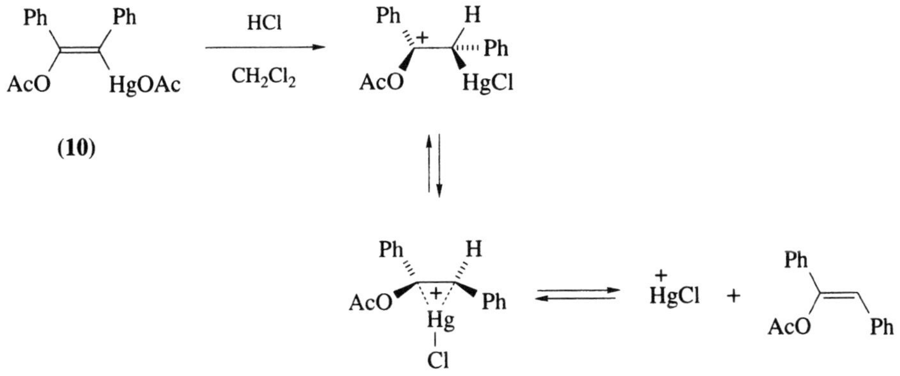

(**10**) —HCl, CH_2Cl_2→

Scheme 22

Acidic deuterolysis and protonolysis have also been used for clarification of the placement of the mercury atoms[127] after the mercuration step, in particular for porphyrin derivatives.[128]

Protodemercuration as the second step of intramolecular *C*-vinylation induced by mercury(II) salts has been recently described: examples are in equations (5)[129] and (6).[130] A large enhancement of the hydrolysis rate was observed when NaI was added to the acid solution.[129b] This effect can be attributed to halide-catalyzed protodemercuration, an established method for cleavage of arylmercurials.[131]

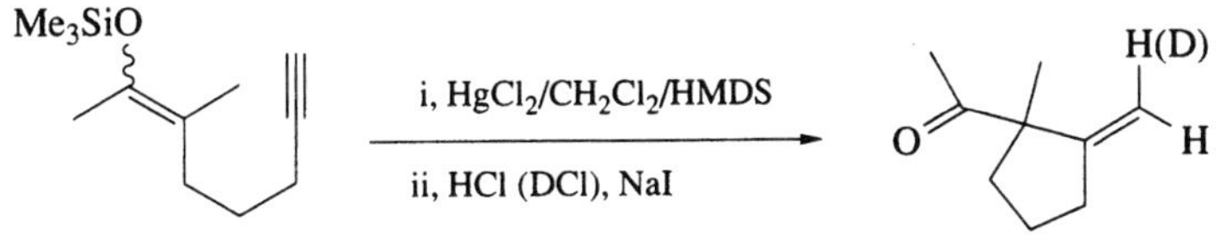

—i, $HgCl_2/CH_2Cl_2$/HMDS; ii, HCl (DCl), NaI→ (5)

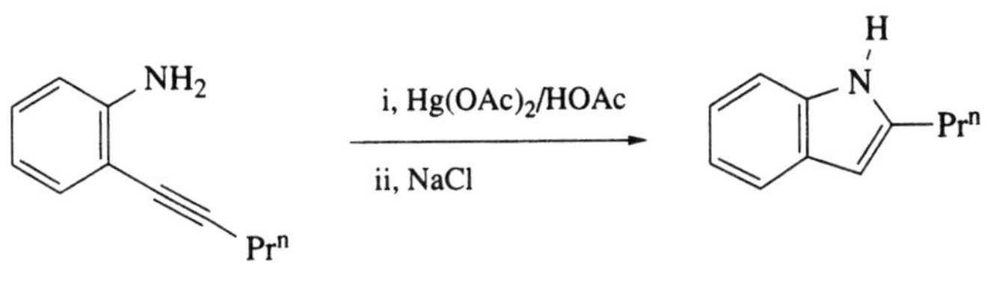

—i, $Hg(OAc)_2$/HOAc; ii, NaCl→ (6)

4.3.3.2 Metal Hydride Demercuration

The reduction of organomercurials by sodium borohydride has been widely studied and the noncage free radical chain mechanism indicated in Scheme 23 is generally accepted.[132,133]

However, recent findings of Singh and Khanna on the LAH reduction of organomercury(II) halides suggested an electron transfer mechanism of the $S_{RN}1$ type, involving attack of R· on the metal hydride (Scheme 24).[134] In other respects, LAH has been relatively little used as a reducing reagent for demercuration in the last decade.[135]

$$RHgX \xrightarrow{NaBH_4} RHgH + NaBH_3X$$

$$R\bullet + RHgH \longrightarrow RH + RHg\bullet$$

$$RHg\bullet \longrightarrow R\bullet + Hg$$

Scheme 23

$$R\bullet + AlH_4^- \longrightarrow RH + AlH_3\bullet^-$$

$$AlH_3\bullet^- + RHgX \longrightarrow AlH_3 + R\bullet + Hg + X^-$$

Scheme 24

A similar process was questioned by Russell and Guo for the reduction of organomercury(II) halides with $NaBH_4$. Although alkyl radicals can abstract a hydrogen atom from $NaBH_4$, hydrogen abstraction from BH_4^- occurs more slowly than abstraction from MeOH/NaOH, or from RHgH, and much less readily than attack upon AlH_4^-. These results confirm that the reduction of alkylmercury(II) halides with $NaBH_4$ in MeOH/NaOH proceeds *via* hydrogen abstraction by the alkyl radical from RHgH and not from $NaBH_4$.[136]

Rearrangements of the initially formed radicals frequently occur. For instance, the reduction of 5-hexenylmercury(II) bromide with $NaBH_4$ or $NaBD_4$ affords 1-hexene and methylcyclopentane, the two expected products of the 5-hexenyl radical.[133a] The intermediate radicals derived from 2,2,2-triphenylethylmercury(II) chloride[137] and 2,2-dimethyl-2-phenylethylmercury(II) chloride have been trapped by oxygen to give the expected products, accompanied by side products arising from radical rearrangements.[138] Trapping of the radicals generated in the metal hydride reduction has also been achieved with alkenes to give coupled products. Some examples of this strategy are given in Scheme 25.[139]

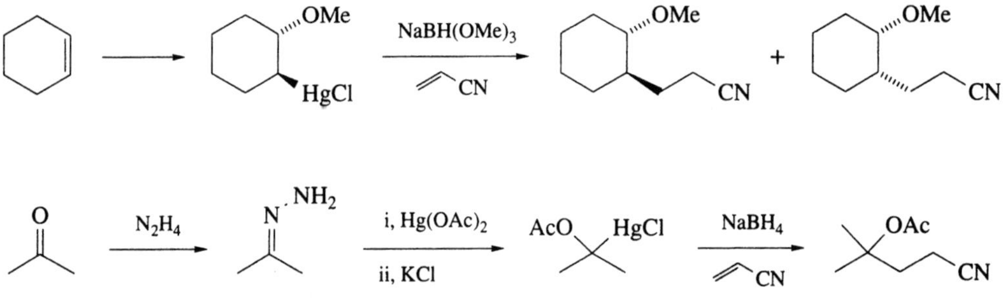

OMe HgCl NaBH(OMe)$_3$ CN OMe CN + OMe CN

O N_2H_4 NH$_2$ N i, Hg(OAc)$_2$ ii, KCl AcO HgCl $NaBH_4$ CN OAc CN

Scheme 25

Stereochemical studies have brought an important contribution to the radical mechanism of the borohydride reduction of organomercurials.[133b] A recent ^2H NMR study of the reduction of norbornyl-type mercurials with $NaBD_4$ confirmed the previous finding of Gray and Jackson[140] about the composition of the product mixture. The major components were clearly nortricyclanol and *anti*-7-norborneol. ^2H NMR data provided accurate measures of the preferred directions of deuterium abstraction by the radical intermediate.[141]

Reduction with $NaBD_4$ of *exo*- and *endo*-norbornylmercury(II) bromide provides mostly *exo*-[2-$^2H_1$]norbornane.[133b] The loss of stereochemistry could also be observed in the sodium borohydride reduction of *cis*- and *trans*-4-methylcyclohexylmercury(II) bromide, which yielded essentially the same mixture of *cis*- and *trans*-4-methyl[1-$^2H_1$]cyclohexane (Scheme 26).[141] These reductions are thus stereoselective, but not stereospecific.

The stereoselectivities in the cleavage of the C—Hg bond with $NaBH_4$ depend upon the nature of the radical intermediates and the reaction conditions. Recent attempts to find optimal conditions for this stereoselective cleavage have been published.[142] The principle is given in Scheme 27 for the oxymercuration–demercuration of α,β-unsaturated esters, generating two asymmetric centers in the 3-alkoxy-2-alkyl ester.

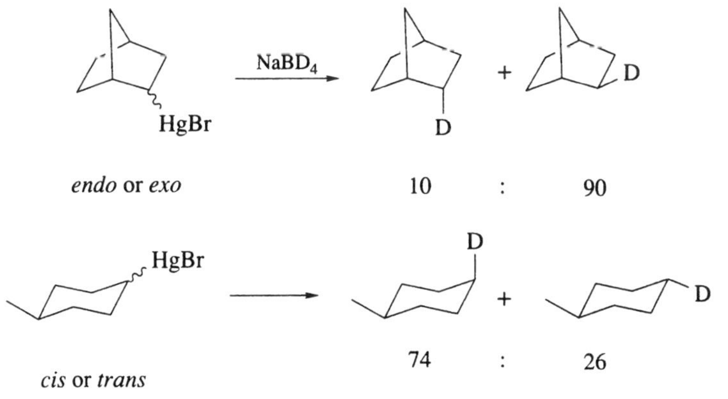

HgBr $\xrightarrow{NaBD_4}$ D + D

endo or *exo* 10 : 90

HgBr ⟶ D + D

cis or *trans* 74 : 26

Scheme 26

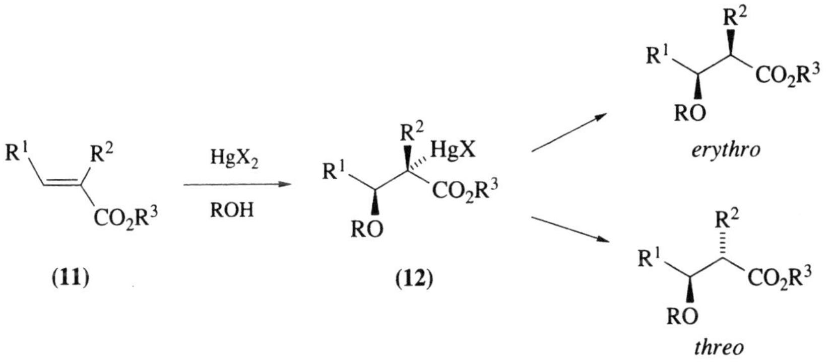

R^1, R^2, CO_2R^3 **(11)** $\xrightarrow[ROH]{HgX_2}$ R^2, HgX, R^1, CO_2R^3, RO **(12)** ⟶ R^1, R^2, CO_2R^3, RO *erythro*; R^1, R^2, CO_2R^3, RO *threo*

Scheme 27

It has been suggested that radical intermediates such as (**14**) are involved in the $NaBH_4$ cleavage, and that they produce the *threo* isomers selectively from the cyclic ethers (**13**), whereas no stereoselectivity occurs with the radicals derived from the tiglic acid adducts (**12**).[143]

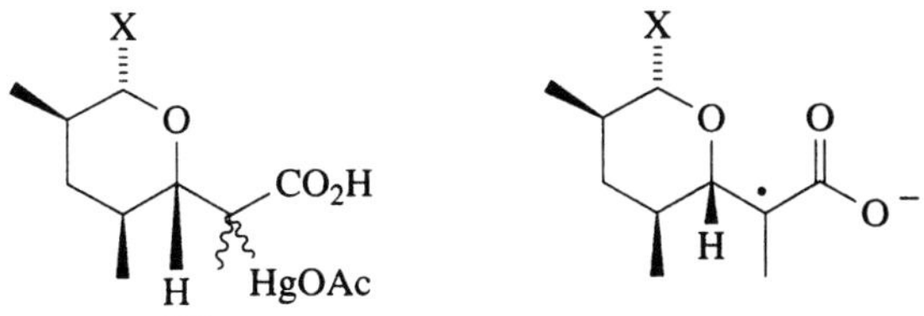

X, O, CO_2H, H, HgOAc **(13)** X, O, O, O^-, H **(14)**

Other factors affecting the diastereoselectivity in the metal hydride demercuration of α-mercury(II) carbonyl compounds have also been identified, which include the nature of the solvent, the amount of hydride used, the mode of addition, the nature of the hydride source and the ligand on mercury. The rationalization of these results is difficult.[142b]

A frequent problem associated with oxymercuration–demercuration is reverse deoxymercuration, which can occur during the reduction step.[144] Brown's method of using a basic homogeneous medium is often used to minimize deoxymercuration.[145] The structure of the organomercurial is also important and affects the choice of hydride, sodium trimethoxyborohydride being preferred to $NaBH_4$ for β-alkoxy-substituted organomercury(II) compounds.[146]

Another example is mercurilactonization, which proceeds with good stereocontrol in the formation of both γ- and δ-lactones, but suffers from elimination during the reductive demercuration. An exploration of a number of reagents and conditions, including $NaBH_4$ and various pHs, hydrogen over Wilkinson's catalyst, Bu_3SnH and Na_2S, indicates that reductive elimination is minimized in alkali borohydride, although it cannot be avoided entirely (Scheme 28).[147]

In contrast, $NaBH_4$/NaOH must be avoided in the demercuration of organomercurials obtained by alkoxymercuration of vinyl ethers bearing conjugated electron-withdrawing groups, because of substantial

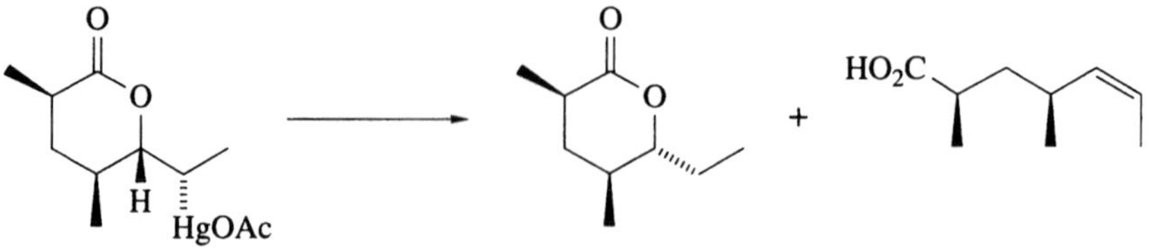

O O HO$_2$C

O O +

H
HgOAc

(15) **(16)** **(17)**

Reagent/Conditions	*Yield* (%)	
	(16)	**(17)**
$NaBH_4$, MeOH, –78 °C	90	10
H_2/[$(Ph_3P)_3RhCl$]/MeOH	0	100
Bu_3SnH	66	30
Na_2S	0	100

Scheme 28

reductive elimination. However, demercuration in simple vinyl ethers can be accomplished with this reagent (Scheme 29).[148]

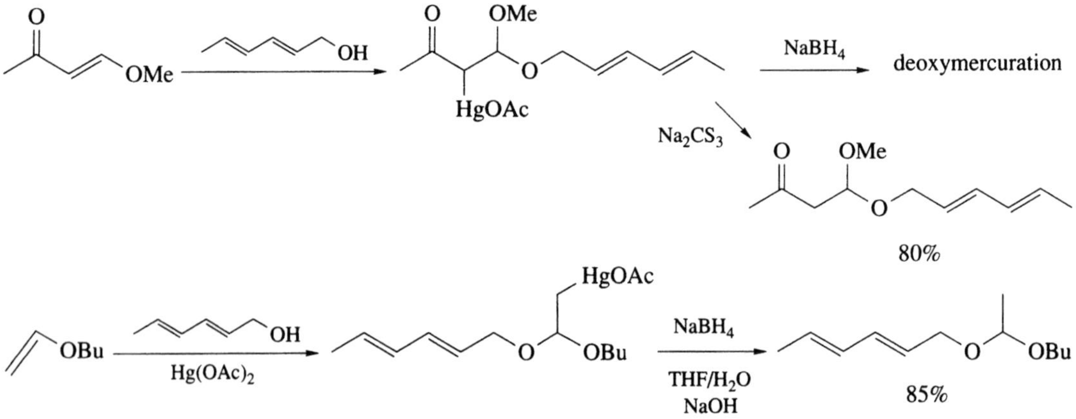

O O OMe NaBH$_4$ deoxymercuration
OMe OH O
HgOAc
Na$_2$CS$_3$ O OMe
O
80%

HgOAc
OBu OH NaBH$_4$
Hg(OAc)$_2$ O OBu THF/H$_2$O O OBu
NaOH 85%

Scheme 29

The products obtained from the oxymercuration–demercuration of alkenylacetates under standard conditions ($NaBH_4$/NaOH 3 M) are diols. However, the yields are significantly lower than with methoxy- and hydroxy-alkenes because of competitive deoxymercuration.[149] Increasing the amount of base results in major increases in the yield of hydrated products. A less basic procedure has been developed, which allows for the survival of the acetate group.

Phase transfer reagents sometimes avoid deoxymercuration during the reduction step, as with the products (**18**) to (**20**) from hydroxymercuration,[150] amidomercuration[151] and from spiroacetals.[152]

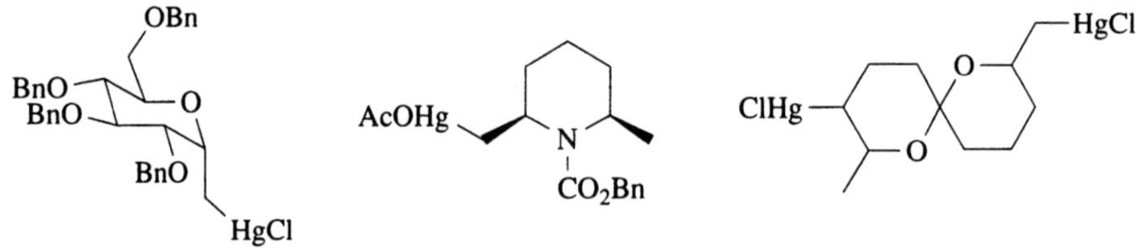

OBn
BnO O
BnO
BnO
HgCl

AcOHg N
CO$_2$Bn

HgCl
O
ClHg
O

(**18**) 73% (**19**) 93% (**20**) >90%

Phase transfer catalysis often avoids other side reactions. For instance, the reaction of organomercurials (**21**) with $NaBH_4$ in the standard conditions leads to symmetrical organomercurials (**22**) whereas Lattes's phase transfer conditions give 2-alkenyl-pyrrolidines (**23**) and -piperidines (**24**) in 35–90% yields.[153,154]

In the reaction between peroxymercurials and sodium borohydride,[155] epoxide formation and deoxymercuration compete with hydrogenodemercuration (Scheme 30). Except for the synthesis of *t*-butyl exo-2-norbornyl peroxide and the *t*-butyl peroxy derivative of styrene, the peroxymercuration and borohydride reduction of nonterminal alkenes is not an attractive method.[155]

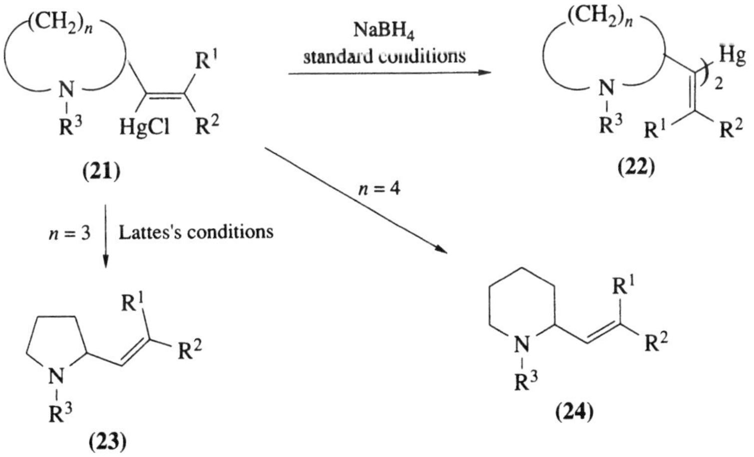

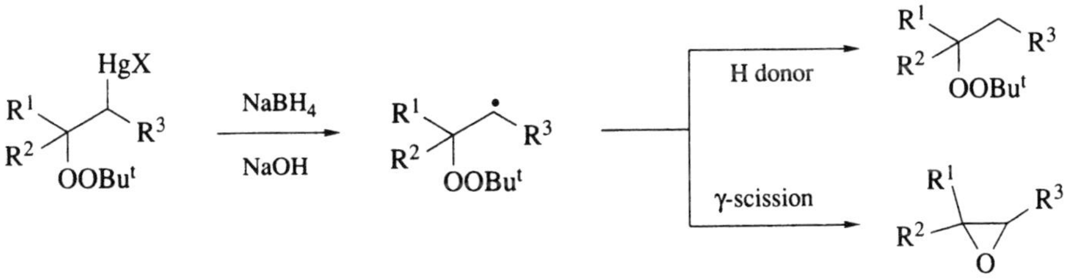

Scheme 30

It has also been reported that cycloperoxymercurials[156–160] give by-products when submitted to hydrogenodemercuration with $NaBH_4$. The amount of by-product is dependent on the structure of the peroxymercurials. Cycloperoxymercurials containing endocyclic mercurio substituents have a strong tendency to deoxymercuriate and poor yields of peroxides are obtained, the major products being unsaturated alcohols (Scheme 31).[156]

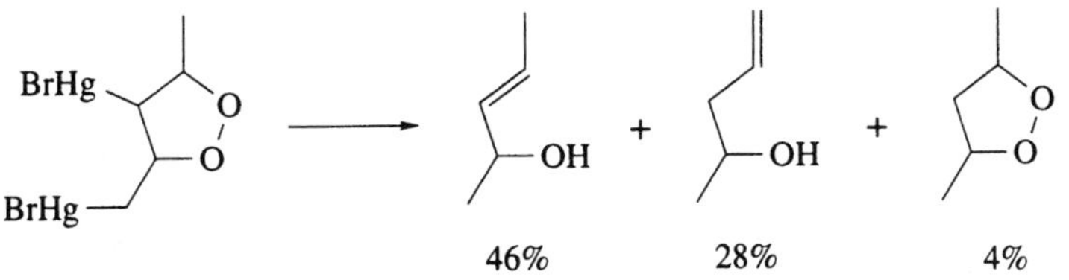

46% 28% 4%

Scheme 31

On the other hand, cycloperoxymercurials which contain exocyclic mercurio substituents are reduced, *via* an exocyclic β-peroxy radical, and give a mixture of the cyclic peroxide resulting from H-atom abstraction and the epoxy alcohol resulting from $S_H{}^i$ radical attack on the peroxide bond (Scheme 32).[157]

The relative amounts of peroxide and epoxy alcohol are dependent on the structure of the intermediate radical. It has been suggested that the critical steric condition for the $S_H{}^i$ reaction is the dihedral angle about the O—C bond between the attacking radical and the leaving oxygen. When the peroxide bond and the radical center are colinear, which can be realized for a six-membered ring, maximum $S_H{}^i$ reaction occurs. For the analogous radical derived from a five-membered ring the O—C dihedral angle has a maximum of approximately 165° in the most favourable conformation for $S_H{}^i$ attack, and the amount of epoxy alcohol ($S_H{}^i$ product) is substantially less than for the six-membered radical. No $S_H{}^i$ reaction is possible from endocyclic radicals because they cannot assume the required conformation to give internal radical attack. These conclusions are consistent with the observation that reduction of peroxymercurials derived from methyl oleate gives no epoxy alcohols, but peroxymercurials derived from methyl linoleate give mixtures of unsaturated epoxides and epoxy alcohols as major reaction products.[158]

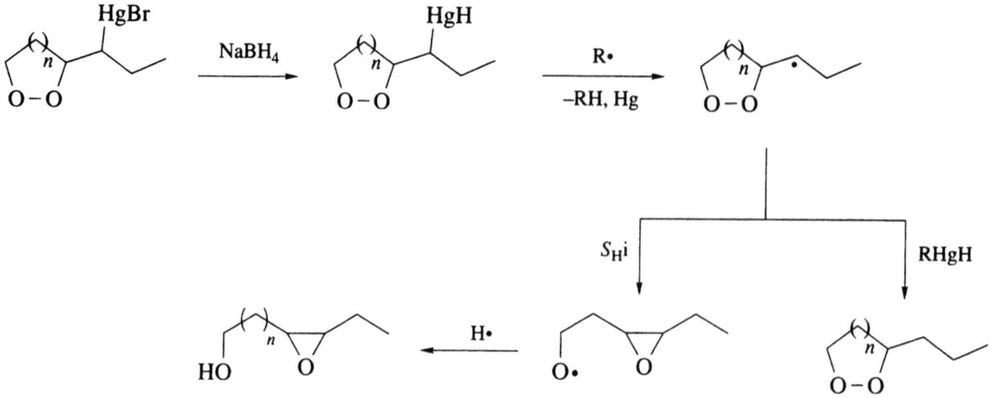

Yield (%)

n	Peroxide	Epoxy alcohol
1	75	25
2	<10	90
3	100	0

Scheme 32

A number of demercurations of RHgX or RHgOR′ use tributyltin,[147,135b] or triphenyltin hydride,[161] but complete removal of tin residues can sometimes be difficult; $NaBH_4$ reduction is then preferred.[152] Deoxymercuration has also been observed during demercuration.[147] The presence of anhydrous sodium acetate avoided this side reaction with Ph_3SnH,[161] and the use of tributyltin hydride instead of $NaBH_4$/NaOH in the demercuration of peroxymercurials led to much improved yields of peroxides (Scheme 33).[155]

Peroxymercurial (X = HgBr)				
Peroxide yields (X = H) (%)				
Bu_3SnH	63	24	59	61
$NaBH_4$/NaOH	0	0	2	13

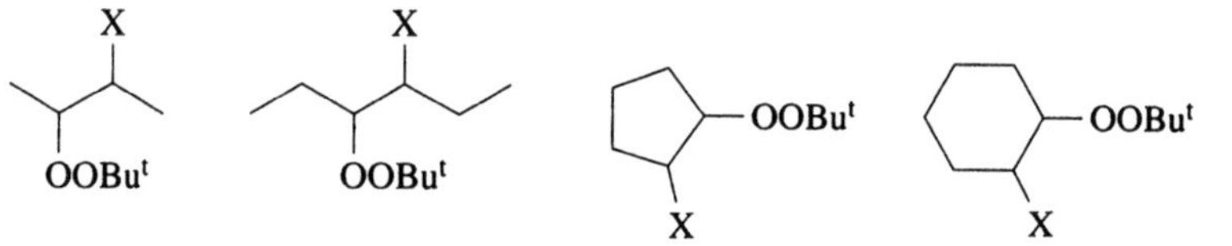

Scheme 33

That epoxides are formed as by-products supports the assumption of β-peroxyalkyl radicals as intermediates, and tributyltin hydride itself will be the principal hydrogen donor. A mechanism has been suggested[155] which involves a hitherto unreported S_H2 displacement of alkylmercury by tributyltin, but another radical mechanism has also been proposed.[162]

Although β-elimination of the mercury group and formation of side products can occur during the reduction of organomercurials, sodium borohydride remains the most important reducing reagent for this purpose. Numerous papers over the past 10 years mention slight modifications in the experimental sections. These include: (i) the hydroxymercuration–demercuration, without internal cyclization, of iridoid glucosides,[163] alkyl-substituted cycloalkenes,[164] methylenecycloalkanes,[165] bicyclooctene, trimethylenenorbornene and related alkenes,[166] cyclopropane[167] and vinylcyclopropanes,[168] alkenylsilanes,[169] and, with internal cyclization, of alkenylphenols[170] and alkenyl alcohols;[171] (ii) the alkoxymercuration–demercuration, without internal cyclization, of 1-octene in aqueous sodium dodecylsulfate,[172] of the dihydropyran group in a total synthesis of (–)-specionin,[173] and, with internal cyclization, of a cyclobutene double bond in a synthesis of (±)-lineatin;[174] (iii) the aminomercuration–demercuration, without internal cyclization, of alkenes,[175] α,β-unsaturated esters,[176] terminal alkynes,[177] and enynes,[178] and, with internal cyclization, of allylamines, dialkylamines, 1,4- and 1,5-hexadiene,[179] 1,5-cyclooctadiene,[180] a butenyl-substituted β-lactam,[181] and a butenyl-substituted aminodioxolane;[182] (iv) the amidomercuration–demercuration, without internal cyclization, of alkenes,[183] and, with internal cyclization, of alkenylcarbamates,[184] (hydroxyalkenyl)carbamates,[185] and substituted acrylanilides;[186] (v) the sulfonamidomercuration of alkenes or dienes,[187] and the phosphoramidomercuration of alkenes;[188] and (vi) the mercury(II)-induced cyclization–demercuration as a chemical mimic of terpenoid biosynthesis of polyenes,[189] for transannular cyclization of cycladienes[190] or cyclatrienes,[191] and for [3,3] sigmatropic rearrangements of dienes.[192]

4.3.3.3 Miscellaneous Methods

In the past 10 years other reductants of the C—Hg bond have been used, mainly thiols and sodium amalgam. Reduction with hydrogen sulfide,[193] sodium dithionite,[194] metals,[143,195] alcohols,[196] alkaline aqueous bases,[197] aromatic amines,[198] Wilkinson's catalyst[143] and electrochemical reductions[199] have also been described. Organomercurials react with thiols by free radical substitutions with an S_H2 mechanism (Scheme 34).[200] The reaction between PhSH and Δ^5-hexenylmercury chloride initiated by light or AIBN gives a mixture of 1-hexene and methylcyclopentane.[201]

$$PhS\bullet + RHgCl \longrightarrow R\dot{H}g(PhS)Cl$$

$$R\dot{H}g(PhS)Cl \longrightarrow R\bullet + PhSHgCl$$

$$R\bullet + PhSH \longrightarrow RH + PhS\bullet$$

Scheme 34

Hydrogenolytic cleavages using H_2S–pyridine, Na_2S–aqueous NaOH, Na_2S–pH3 buffer, Na_2S–MeCN, Na_2CS_3, sodium amalgam (and $NaBH_4$)[143] have also been studied with the mercurial precursors (**13**) of the Prelog–Djerassi lactone and with the mercurials (**12**) derived from α,β-unsaturated acids.[142b] Excellent yields are obtained in reduced products but with variable stereoselectivities. On the basis of the stereochemical results the same radical intermediate (**14**) is suggested in the hydrogen sulfide–pyridine cleavage as in the $NaBH_4$ reduction of cyclic mercurial ethers which produce the *threo* isomer selectively.

In the case of mercurated tiglic acid adducts (**12**), the *erythro* product is obtained with aqueous Na_2S, and an ionic mechanism has been proposed with an intermediate carboxylate enolate (**25**).[143] Sodium trithiocarbonate[143] and, especially, 1,3-propanedithiol[142b] exhibit a preference for retention of configuration leading to the *erythro* isomer. Such sulfur-containing reagents are probably capable of intramolecular hydrogen atom delivery as illustrated schematically in (**26**), and they also promote demercuration by an ionic mechanism.

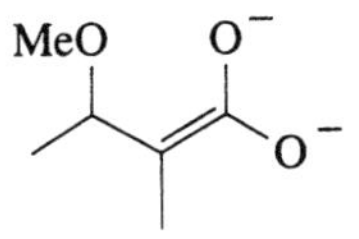

MeO O⁻ O⁻ L$_n$Hg S S H

(25) 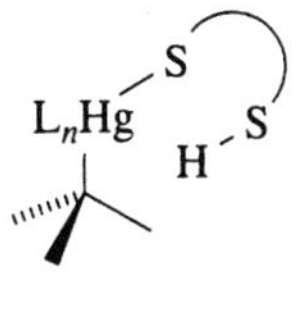 **(26)**

However, a limitation in the use of certain thiol reagents, as for most reagents used for demercuration, can be their propensity to undergo elimination with reversion to the starting material. Sodium sulfide, for example, cannot be used in the reduction step of the mercurilactonization (Scheme 28),[147] and the presence of a base ($NaHCO_3$ or Et_3N) is necessary in the 1,3-propanedithiol reduction of mercurials.[142b] In other cases, thiol reagents (Na_2CS_3) succeed where $NaBH_4$ induces substantial elimination (Scheme 29).[148]

The sodium amalgam cleavage of alkylmercurials involves an ionic mechanism.[202] The stereochemical course of the replacement of mercury by deuterium in a range of organomercury halides or acetates employing 1–2% sodium amalgam/D_2O/NaOH has been investigated by ^2H NMR spectroscopy. The results confirm that sodium amalgam reductions are completely stereospecific with retention at carbon. No rearrangement was observed in the rearrangement prone nortricyclyl–norbornenyl pair (equation 7),[141] although inversion of configuration in the demercuration of the vinylmercurial derived from the acetoxymercuration of diphenylacetylene has been described.[126] Sodium amalgam has also been used for reducing the amidomercurial derived from methyl undecylenate.[203]

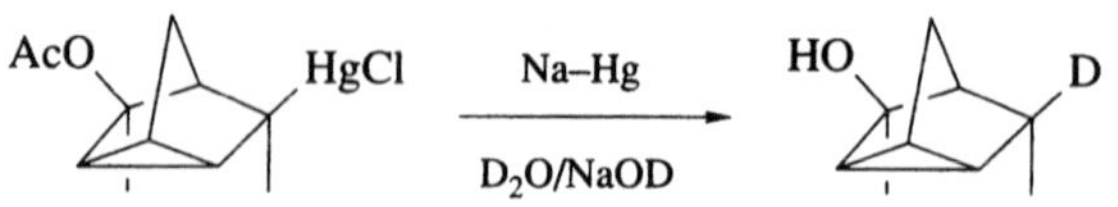

AcO ... HgCl —(Na–Hg, D_2O/NaOD)→ HO ... D (7)

The overall replacement of mercury by hydrogen can be achieved by metal/mercury exchange, *e.g.* Li/Hg in an aprotic medium[195] and then hydrolysis, or Zn/Hg in acetic acid.[126] The dilithioalkene (**30**) prepared by Li/Hg exchange[195] can be obtained free of lithium amalgam if the bromomercury compound (**29**) is used, and not allowed to react with lithium powder but with *t*-butyllithium in cyclopentane (Scheme 35).

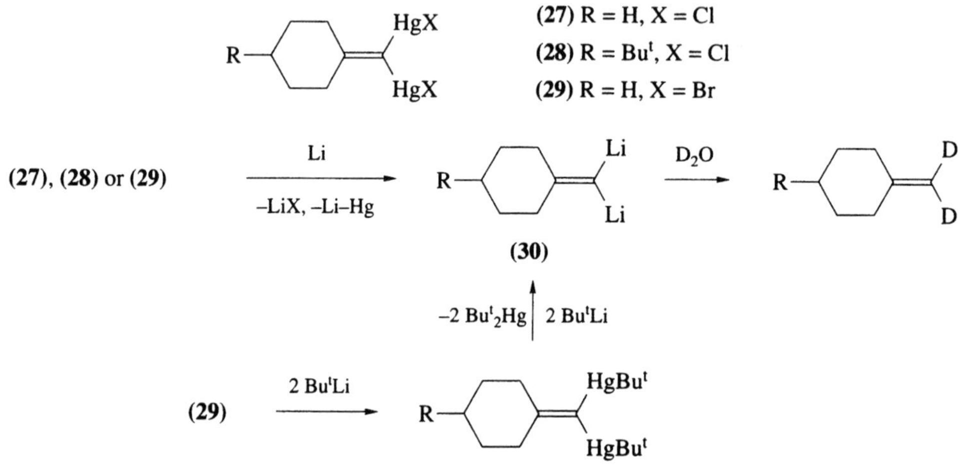

(**27**) R = H, X = Cl
(**28**) R = Bu^t, X = Cl
(**29**) R = H, X = Br

(**27**), (**28**) or (**29**) —(Li; –LiX, –Li–Hg)→ (**30**) —(D_2O)→

(**29**) —(2 Bu^{t}Li)→ HgBu^t/HgBu^t —(2 Bu^{t}Li; –2 Bu$^t_2$Hg)→ (**30**)

Scheme 35

A novel electron transfer free radical mechanism has been elucidated for sodium naphthalenide induced demercuration.[204] A new reductant for the cleavage of the C—Hg bond, *N*-benzyl-1,4-dihydronicotinamide (BNAH), has also been proposed; it reduces alkylmercury(II) acetates *via* an electron transfer chain substitution mechanism.[205]

4.3.4 REDUCTION OF PHOSPHORUS–CARBON BONDS

Reduction of the phosphorus–carbon bond is generally achieved with the aim of preparing new phosphorus compounds, and not with the object of preparing reduced carbon compounds. It is a striking difference from mercury–carbon bond cleavage, which is usually carried out in order to prepare carbon compounds.

4.3.4.1 Reduction of Tervalent Organophosphorus Compounds

The cleavage of the C—P^{III} bond can be achieved by alkali metals (Li, Na, K), transition metal complexes, organometallics (RLi, Na/naphthalene) and by photolysis.[206]

For instance, it is well established that the P—C bond of tertiary phosphines, having at least one aromatic group, can be cleaved by an alkali metal (Li/THF or NH_3, Na or K/dioxan, NH_3 or diglyme) to the corresponding phosphide anion, which can then be alkylated with electrophiles to give substituted tertiary phosphines. It can also give the secondary phosphine upon treatment with water or an alcohol.[207] On the basis of reaction rates of various aromatic tertiary phosphines with potassium in dioxan the following cleavage series has been proposed:[208] α-naphthyl > phenyl > *p*-tolyl > 2,5-dimethylphenyl > ethyl > cyclohexyl.

The selective reductive cleavage of the P—C bond between phosphorus and a phenyl group is a slow reaction especially for large scale preparations.[209] Recent findings report that ultrasonic irradiation accelerates the reduction of phenylphosphines with metallic lithium, and provides a clean source of phosphide anions.[210]

Whereas diphosphines $RPhP(CH_2)_nPPhR$ (n = 2–6) have been prepared from the reduction of the corresponding tertiary diphosphines $Ph_2P(CH_2)_nPPh_2$, CH_2—P cleavage is observed for n = 1.[211] Sodium naphthalenide has also been used as a selective reagent for the reductive cleavage of one aryl–phosphorus bond in these tertiary diphosphines.[212] Similarly, the synthesis of compounds derived from phospholyl anions relies upon a phosphorus-phenylphosphole bond cleavage by alkali metals or sodium naphthalenide (Scheme 36).[213]

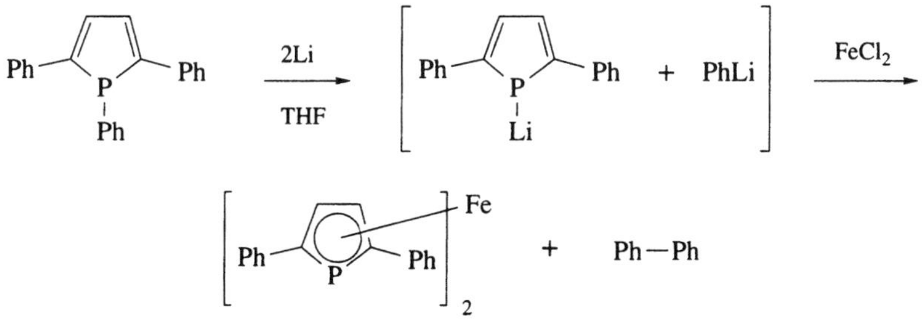

Scheme 36

Recently it has become clear that tertiary phosphine–metal complexes are reactive and liable to undergo carbon–phosphorus bond scission. The reaction between the C—P^{III} bond and the transition metal to which the tertiary phosphine is bound has profound implications on homogeneous catalysis, particularly on the mode of homogeneous catalyst deactivation in hydroformylation (Rh- and Co-catalyzed) and various other hydrogenation/dehydrogenation reactions, including asymmetric hydrogenation.[214]

An interesting example, observed by Nishiguchi and coworkers, is the extensive hydrogenolysis of PPh_3 coordinated to various transition metals. When $[RuH_2(PPh_3)_4]$, $[RhH(PPh_3)_4]$, $[RuCl_2(PPh_3)_3]$ and $[RhCl(PPh_3)_3]$ were heated at 140 °C for 1 h in pyrrolidine, a solvent which shows the highest hydrogen-donating ability in this reaction, the triphenylphosphine ligands were hydrogenolyzed almost completely to benzene. Radical inhibitors did not retard the reaction and radical initiators did not speed up the reaction, indicating that the hydrogenolysis did not proceed by a radical process.[215]

A plausible mechanism was reported for the catalytic formation of benzene resulting from phosphorus–carbon bond cleavage which occurs during propylene hydroformylation catalyzed by triphenylphosphine-substituted rhodium carbonyls under higher H_2 partial pressures (Scheme 37).[216]

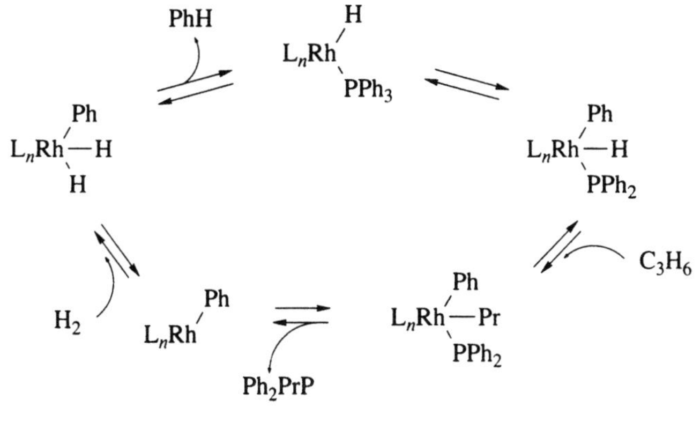

Scheme 37

Alkyllithiums also induce C—P^{III} cleavage. In the reaction of (**31**) with MeLi, P—C cleavage occurs to give Ph_2PMe and the corresponding allenyllithium, which is converted into tris(diphenylphosphine)-allene (**32**) upon hydrolysis.[217]

Reactions of the paramagnetic pentacoordinate $[TaBr_3(PhPMe_2)_2]$ adduct with methyl- or butyllithium (or with the proton sponge 1,8-bis(dimethylamino)naphthalene) lead to methyltantalum(IV) derivatives as a result of P—Me bond cleavage. The cleavage of the P—Me bond rather than the phosphorus–aryl bond is unprecedented.[218]

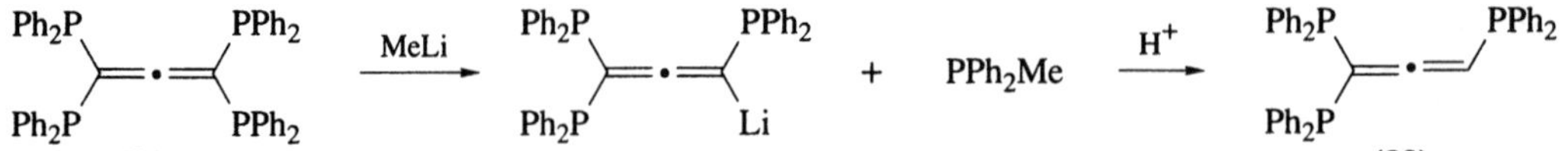

(31) —MeLi→ + PPh_2Me —H^+→ (32)

4.3.4.2 Reduction of Tetravalent Organophosphorus Compounds

The procedures to break the C—P^{IV} bond are rather limited. The main methods are alkaline hydrolysis and LAH reduction of phosphonium salts. Electrochemical reductions can also be used for this purpose.

The alkaline cleavage of quaternary phosphonium salts is a well-documented reaction, which represents one of the most important methods for preparing phosphine oxides. Hydrocarbons are the other products of this reaction (equation 8).[219]

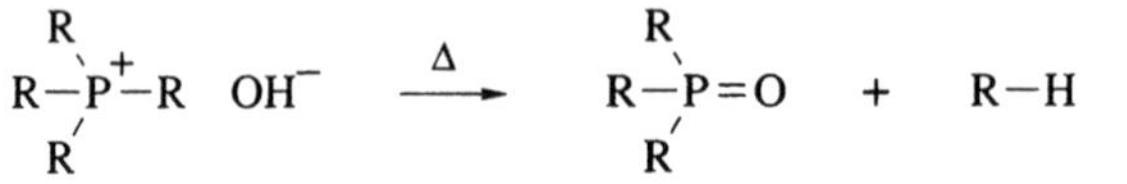

$$R_4P^+ \; OH^- \xrightarrow{\Delta} R_3P{=}O + R{-}H \quad (8)$$

Numerous data[219,220] illustrate a third order rate dependence for this type of reaction, first order with respect to the phosphonium ion concentration and second order with respect to hydroxide ion concentration. The influence of the solvent has been studied.[221] Optically active benzylethylmethylphenylphosphonium hydroxide was shown to decompose stereospecifically with inversion to produce optically active ethylmethylphenylphosphine oxide.[222] On the basis of these findings, the mechanism generally admitted is indicated in Scheme 38.

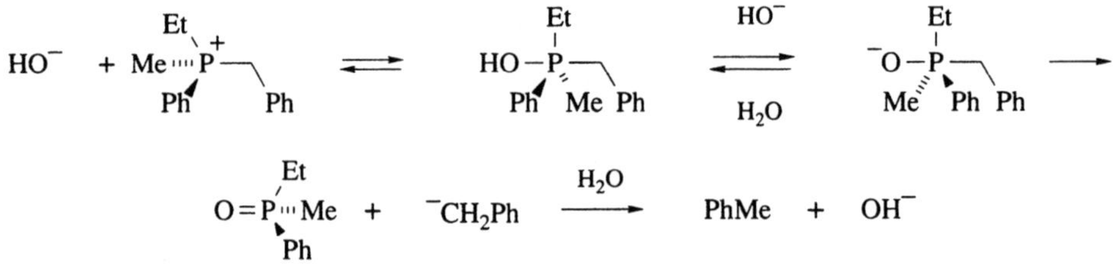

HO^- + Et(Me)(Ph)P^+CH_2Ph ⇌ HO–P(Et)(Ph)(Me)CH_2Ph ⇌ (HO^- / H_2O) ^-O–P(Et)(Me)(Ph)CH_2Ph →

O=P(Et)(Me)(Ph) + $^-CH_2Ph$ —H_2O→ PhMe + OH^-

Scheme 38

The order of ease of displacement of groups from the phosphonium salts is allyl, benzyl > phenyl > methyl > 2-phenylethyl > ethyl, higher alkyls.[219] The usefulness of this method for the preparation of various alkyldiphenylphosphine oxides from the available triphenylphosphine is particularly noteworthy. In contrast, the preparative value of this method is diminished in cases where the carbanions corresponding to the leaving groups on the phosphorus have approximately the same stability. In some cases the reaction is used to prepare the carbon compound (equations 9 to 14).[223]

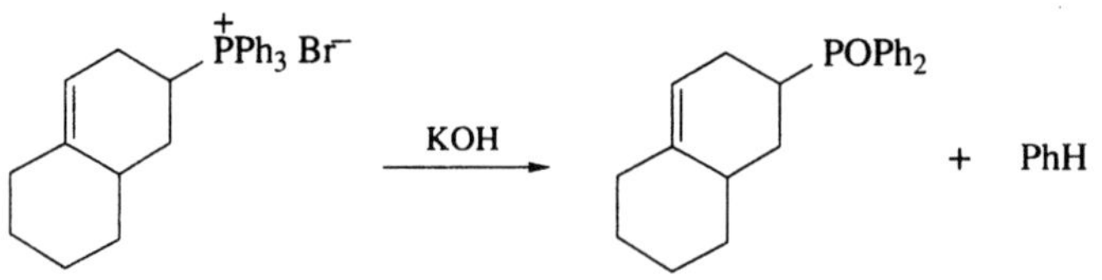

$\overset{+}{P}Ph_3\ Br^-$ —KOH→ $POPh_2$ + PhH (ref. 223a) (9)

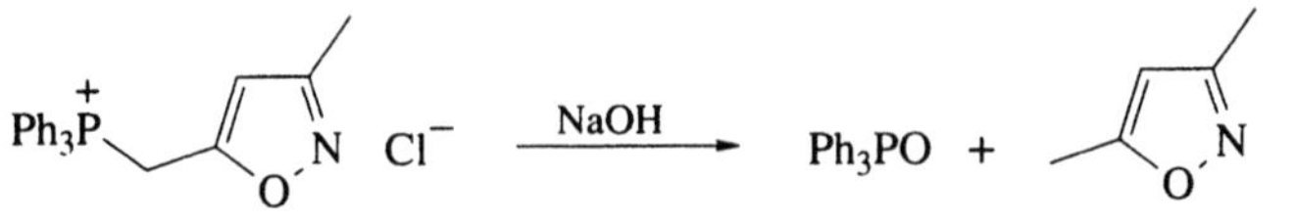

$Ph_3\overset{+}{P}$ … N Cl^- —NaOH→ Ph_3PO + … N (ref. 223b) (10)

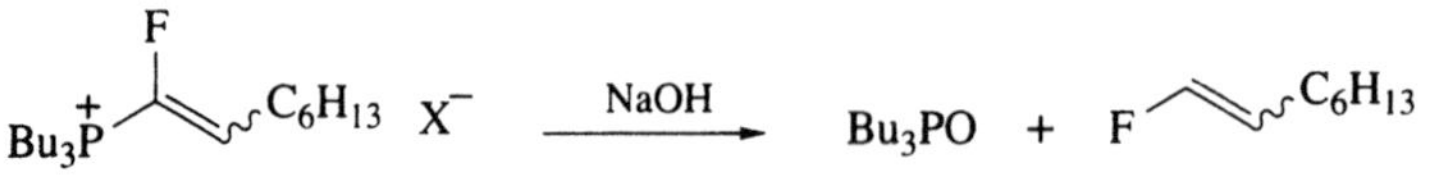

$Bu_3\overset{+}{P}$(F)… C_6H_{13} X^- —NaOH→ Bu_3PO + F… C_6H_{13} (ref. 223c) (11)

$(Z):(E) = 6:94$ $(Z):(E) = 100:1$

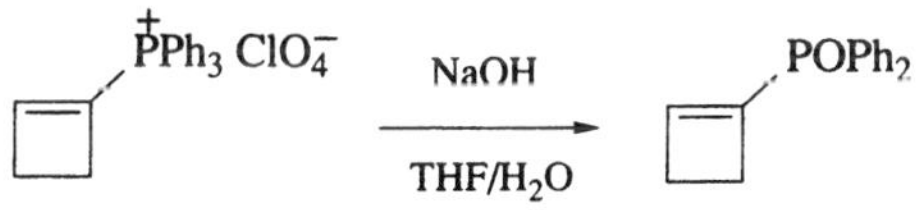

(ref. 223d) (12)

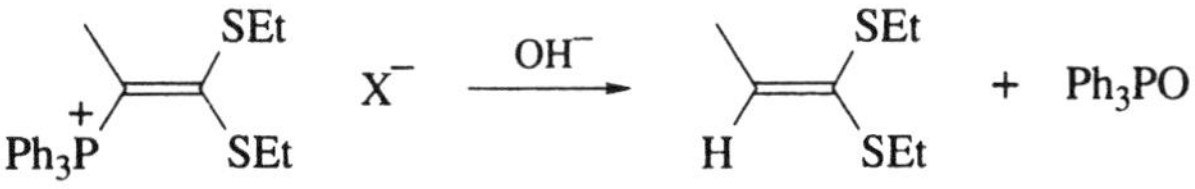

(ref. 223e) (13)

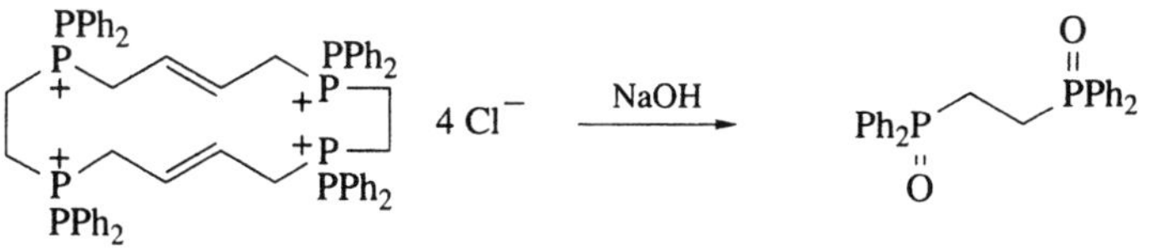

(ref. 223f) (14)

For cyclic derivatives, ring size is an important factor in reaction rate and stereospecificity.[224] Alkali hydroxide induced decompositions of chiral $[R^1R^2R^3R^4P]^+OH^-$ have also been found to occur with retention, inversion or racemization, depending both on the nature of R^n and on the reaction conditions.[225]

Interesting effects on the rate and the regioselectivity of the alkaline cleavage of quaternary phosphonium salts are observed when an *o*- or *p*-methoxy- or *o*- or *p*-dimethylamino-phenyl group is present in the molecule. For example, phosphonium bromide (**33**) undergoes alkaline cleavage in 1:1 dioxane–water to give *N,N*-dimethylaniline (96.5 %), benzene (3.5 %) and benzyldiphenylphosphine oxide (96 %) 10^3 times more rapidly at 38 °C than phosphonium bromide (**34**), which gives only toluene as the hydrocarbon product (Scheme 39).[226]

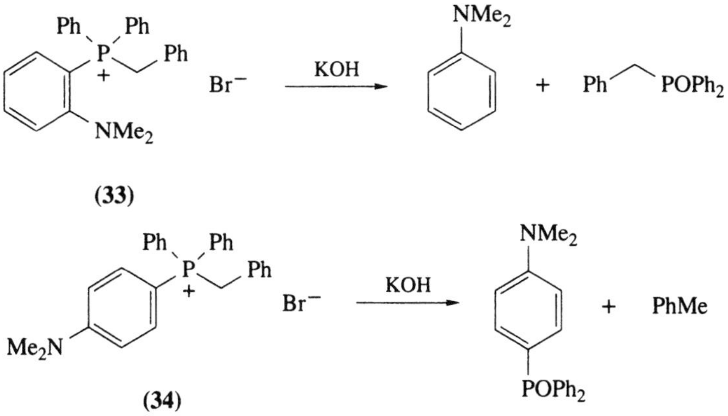

(33)

(34)

Scheme 39

In the base hydrolysis of the unsymmetrical fluoromethylene bisphosphonium salt (**35**) the regioselectivity of C—P bond cleavage can be accounted for by the susceptibility of the phosphonium center to nucleophilic attack by hydroxide ion and not by the relative stability of the newly formed fluoromethylene ylide (Scheme 40).[227]

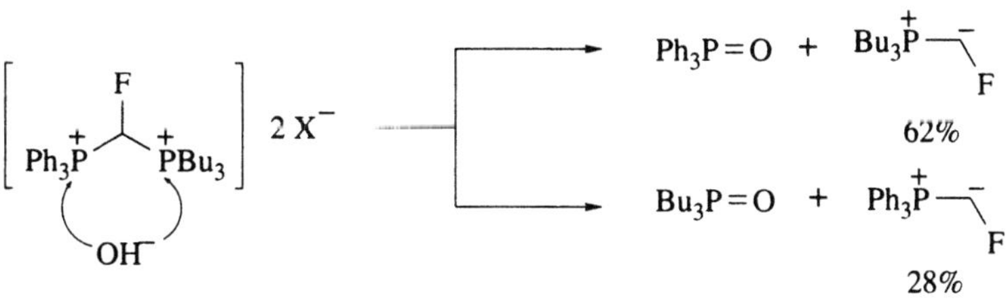

(35)

Scheme 40

In some cases, decomposition of phosphonium salts which have a phenyl group or a negative substituent Z in a β-position to the phosphorus atom can occur *via* elimination reactions. According to the nature of Z, tertiary phosphines (Z = Ph, CN; equation 15)[219] or vinylphosphonium salts (Z = OPh, OAc, OH, Br; equation 16) are obtained. In the latter case, the vinylphosphonium salts cannot be isolated from the aqueous medium because they add water very easily, but trapping with methanol reveals their presence.[228]

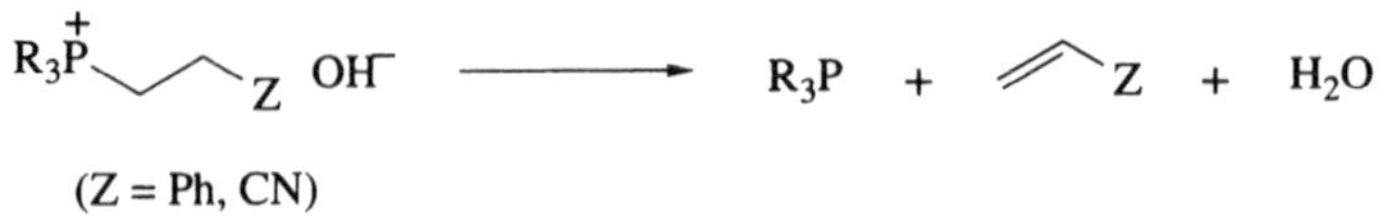

$$R_3\overset{+}{P}CH_2CH_2Z\ OH^- \longrightarrow R_3P + CH_2{=}CHZ + H_2O \qquad (15)$$

(Z = Ph, CN)

$$Ph_3\overset{+}{P}CH_2CH_2Z\ Br^- \xrightarrow[MeOH]{OH^-} \left[Ph_3\overset{+}{P}CH{=}CH_2\ Br^-\right] \longrightarrow Ph_3\overset{+}{P}CH_2CH_2OMe\ Br^- \qquad (16)$$

(Z = OAc, OPh, OH, Br)

Ureidomethylphosphonium salts, which contain a hydrogen substituent on N-1, follow a β-elimination pathway similar to the one indicated in equation (15) giving the tertiary phosphine, but if the N-1 position is blocked by a methyl group the reaction follows the S_{NP} pathway giving the tertiary phosphine oxide (Scheme 41).[229]

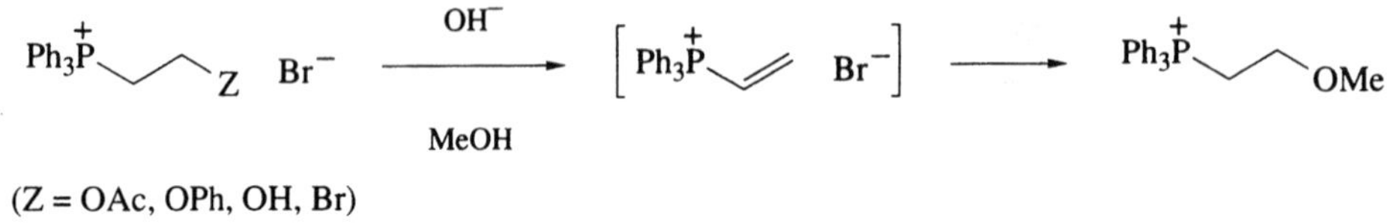

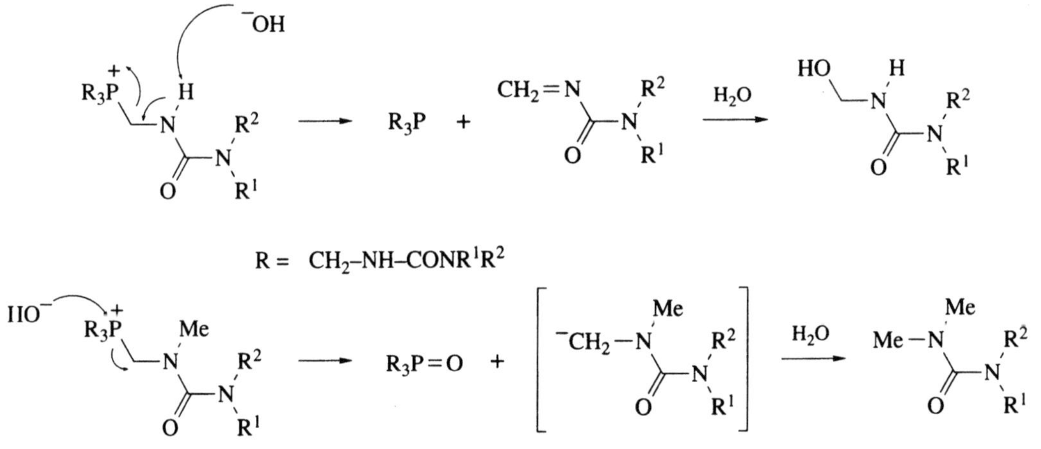

$$R_3\overset{+}{P}CH_2N(H)CONR^1R^2 \xrightarrow{^-OH} R_3P + CH_2{=}NCONR^1R^2 \xrightarrow{H_2O} HOCH_2N(H)CONR^1R^2$$

R = CH_2–NH–CONR[1]R[2]

$$HO^- + R_3\overset{+}{P}CH_2N(Me)CONR^1R^2 \longrightarrow R_3P{=}O + \left[{}^-CH_2{-}N(Me)CONR^1R^2\right] \xrightarrow{H_2O} Me{-}N(Me)CONR^1R^2$$

Scheme 41

Certain vinylphosphonium salts substituted by an electron-withdrawing group (Z = CH═CH2, Ph, PhCO, RCO) react with aqueous bases to give phosphine oxides resulting from an anionotropic migration of a phenyl group from phosphorus to the α-carbon (S_{NP}mig) (Scheme 42).[230]

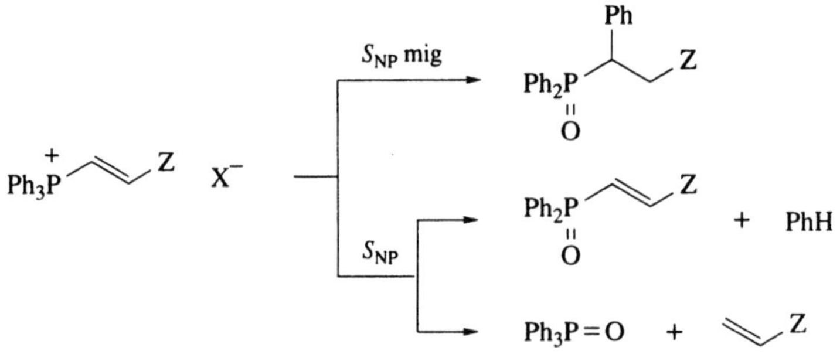

$$Ph_3\overset{+}{P}CH{=}CHZ\ X^- \xrightarrow{S_{NP}\,mig} Ph_2P(O)CH(Ph)CH_2Z$$

$$Ph_3\overset{+}{P}CH{=}CHZ\ X^- \xrightarrow{S_{NP}} Ph_2P(O)CH{=}CHZ + PhH$$

$$Ph_3\overset{+}{P}CH{=}CHZ\ X^- \xrightarrow{S_{NP}} Ph_3P{=}O + CH_2{=}CHZ$$

Scheme 42

Basic hydrolysis of phosphonium salts has been studied in phase transfer conditions. The results indicate that a heterogeneous medium (H_2O/CH_2Cl_2) speeds up the hydrolysis (up to 83 % yield after 15 h at 20 °C in CH_2Cl_2/H_2O, 2% yield without CH_2Cl_2). As a consequence the degradation of phosphonium salts in basic heterogeneous media has to be taken into account when using these reagents as phase transfer catalysts.[231] Micellar effects on competitive hydrolysis and hydration of vinylphosphonium salts also have been observed.[232]

Lithium aluminum hydride is the second method which is used to cleave the C—P^{IV} bond. The mechanism of attack at phosphorus by hydride is similar to the attack of hydroxide ion, with subsequent or simultaneous expulsion of the group which gives the most stable anion. Thus the preferential removal of benzyl groups, one at a time, is observed. 1,2-Vinylene- and 1,4-butadienylene-bis(phosphonium) salts react with LAH to give selective cleavage of the unsaturated bridge between the two phosphorus atoms.[233] Cyclic phosphonium salts with a benzyl group attached to the phosphorus are cleaved in good yield to the corresponding cyclic phosphines by use of LAH.[234] Cleavage with total S_N2' attack occurs with allylic phosphonium salts. Of particular interest is the fact that a triphenylphosphonium salt with a γ-position extremely crowded gives exclusively the γ-hydrogenolysis product (Scheme 43).[235]

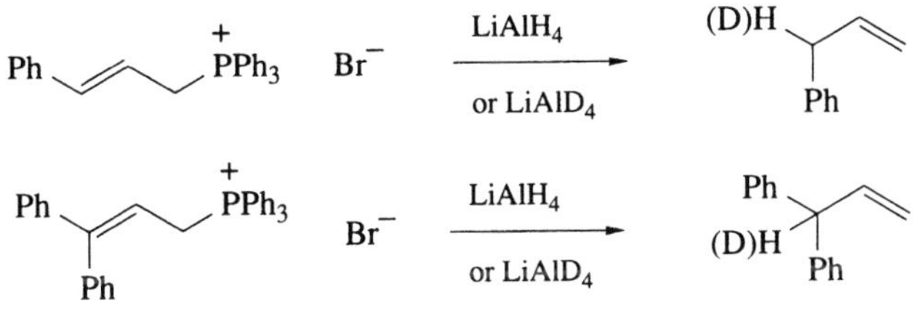

$$\text{PhCH=CHCH}_2\overset{+}{\text{P}}\text{Ph}_3\ \text{Br}^- \xrightarrow[\text{or LiAlD}_4]{\text{LiAlH}_4} \text{(D)H–CH(Ph)–CH=CH}_2$$

$$\text{Ph}_2\text{C=CHCH}_2\overset{+}{\text{P}}\text{Ph}_3\ \text{Br}^- \xrightarrow[\text{or LiAlD}_4]{\text{LiAlH}_4} \text{Ph(D)H–C(Ph)–CH=CH}_2$$

Scheme 43

Metals such as Na or alkali metal amalgams can also be used in the cleavage of the C—P^{IV} bond.[236] In the latter case, reductive cleavage of achiral and optically active quaternary phosphonium salts succeeds in high yields with retention of configuration.[237]

4.3.4.3 Reduction of Pentavalent Organophosphorus Compounds

This short review is limited to the reduction of pentavalent organophosphorus compounds with coordination number four.

Various methods or reagents have been used in the last decade to reduce a C—P^V bond into a C—H bond and the following methods are reported: alkaline hydrolysis, nucleophile-induced cleavage by alcoholates, metal hydrides and metal naphthalenides, metals, thermolysis and photolysis.

Alkaline hydrolysis of ylides is a general method which achieves the C—P/C—H exchange. Protonation of ylides affords a phosphonium hydroxide, which can react with a second OH^- ion and then decompose into hydrocarbon and phosphine oxide. Among the four ligands of the ylide, the ligand that is the most electronegative or the best stabilized as an anion is the best leaving group (equation 17).[238]

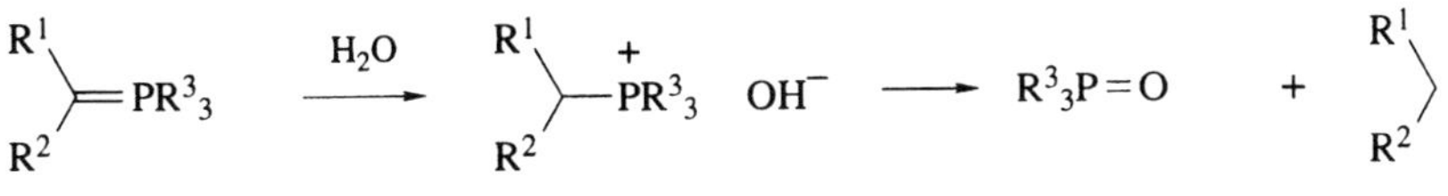

$$R^1R^2C{=}PR^3_3 \xrightarrow{H_2O} R^1R^2CH{-}\overset{+}{P}R^3_3\ OH^- \longrightarrow R^3_3P{=}O + R^1R^2CH_2 \qquad (17)$$

Reductions of certain (acylalkylidene)phosphoranes with Zn–HOAc and conversion of (acylalkoxycarbonylmethylene)phosphoranes to β-keto esters by a thermolysis route have been reported, but gave poor results with (acylethoxycarbonylmethylene)phosphoranes. The desired reductive removal of triphenylphosphine may be achieved in this case with excess Al–Hg in wet THF, with periodic addition of acids (equation 18).[239]

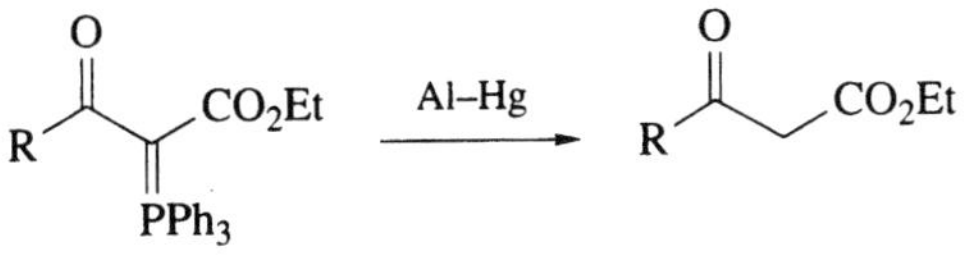

$$RC(O)C(=PPh_3)CO_2Et \xrightarrow{\text{Al–Hg}} RC(O)CH_2CO_2Et \qquad (18)$$

Hydrolysis of alkyl-substituted phosphine oxides is generally difficult, heating with solid sodium or potassium hydroxide is needed in order to break the P—C bond. Hydrolysis of phosphonates and phosphinates occurs with P—O cleavage in preference to P—C cleavage. Exceptions have been reported with several aromatic phosphine oxides containing *p*- and *o*-nitrobenzyl groups and other stabilized systems.[240] Perfluorinated phosphine oxides can also be hydrolyzed to the phosphinic and phosphonic acids.[241]

In a dibenzophosphepin, the degree of exocyclic *versus* endocyclic C—P bond rupture of (**36**) can be correlated with the relative apicophilicity of the nucleophile. Assuming a trigonal bipyramidal intermediate or transition state in these reactions, relatively highly apicophilic nucleophiles (such as H^-) give ring cleavage, whereas poorly apicophilic nucleophiles (such as OH^-) lead to exocyclic C—P bond rupture (Scheme 44).[242]

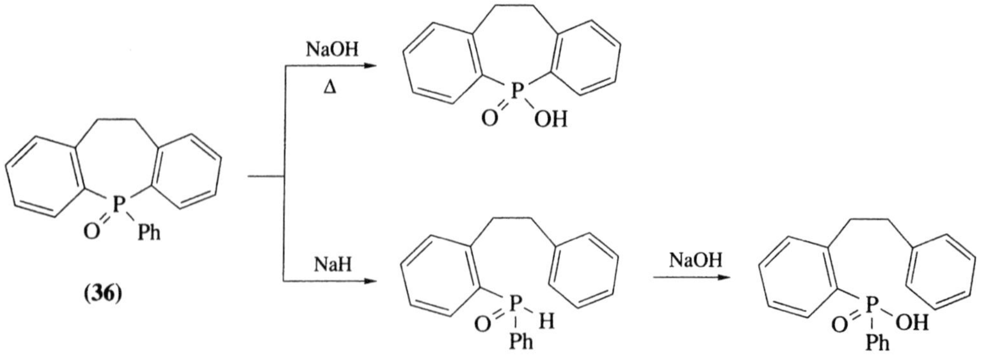

Scheme 44

Rearrangement can occur during this type of reaction: the *N*-*t*-butyl-α-chlorophosphonoamidates (**37a**) and (**37b**) react with methanolic benzyltrimethylammonium methoxide to give, respectively, the products (**38**; R = Me) and (**39**; R = Ph) (Scheme 45).[243]

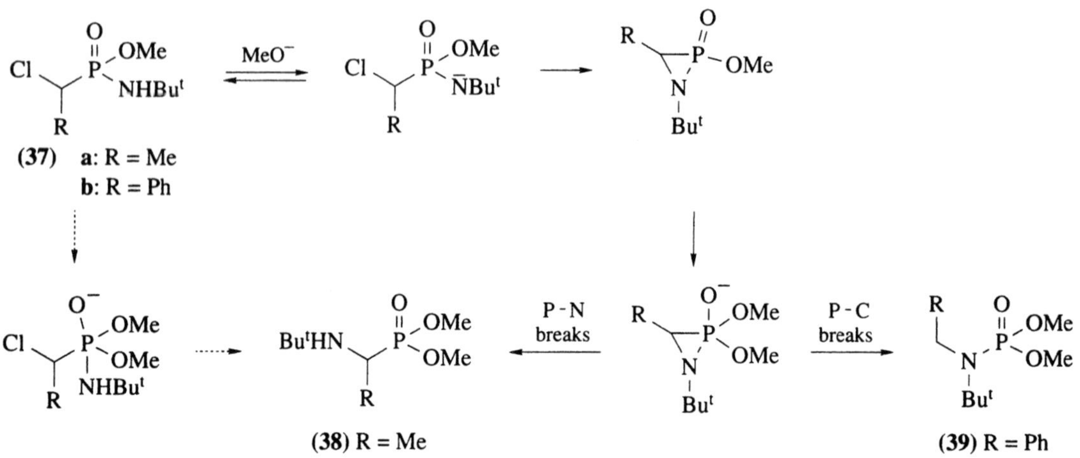

(**37**) **a**: R = Me
b: R = Ph

(**38**) R = Me

(**39**) R = Ph

Scheme 45

Allylphosphine oxides may be used to make allyl alcohols by the reductive removal of the Ph_2PO group with LAH. The reaction involves transposition of the double bond (equation 19).[235,244] $Li(Bu^tO)_3AlH$ breaks selectively the C—P phosphonate bond, and this reagent is chemoselective and can

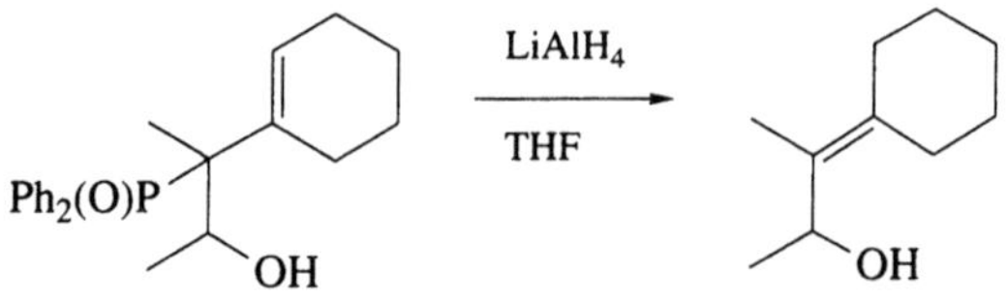

(19)

be used in the presence of keto or ester carbonyl groups. Reduction of the same types of compounds with $NaBH_4/EtOH$ gives the β-hydroxyphosphonates and does not cleave the C—P bond (equation 20).[245]

$$\text{RC(O)CH(CO}_2\text{Et)P(O)(OEt)}_2 \xrightarrow[\text{THF}]{\text{Li(OBu}^t)_3\text{AlH}} \text{RC(O)CH}_2\text{CO}_2\text{Et} \tag{20}$$

The use of the 1-naphthylmethyl moiety as a P—H protecting group in the synthesis of a phosphino macrocycle has been described; the protecting group is removed from the phosphine sulfide by treatment with excess potassium naphthalenide (equation 21).[246]

$$\text{S=P}-\text{CH}_2\text{Nph (macrocycle)} \xrightarrow[\text{THF}]{\text{potassium naphthalenide}} \text{S=P}-\text{H} \tag{21}$$

Nph = naphthyl

Thermolysis is a method that can sometimes be used to cleave the C—P^V bond, particularly in phosphole chemistry. For instance, an optimized phosphole–phosphorin conversion procedure has been described for preparing the first known 2-(2′-pyridyl)phosphorin from 1-phenyl-3,4-dimethylphosphole.[247] The nickel is not only used as a desulfurizing reagent but also as a catalyst for the cleavage of the phosphorus–phenyl bond.[248] Photolysis has also been reported for C—P^V bond cleavage. Thus *p*-nitrophenylmethylphosphonic acid undergoes easy photochemical P—C bond cleavage in alkaline ethanol solution resulting in *p*-nitrotoluene, orthophosphate and ethyl phosphate.[249]

4.3.5 REFERENCES

1. (a) G. R. Pettit, *Org. React.* (*N.Y.*), 1966, **12**, 356; (b) H. Hauptmann and W. F. Walter, *Chem. Rev.*, 1962, **62**, 347; (c) W. A. Bonner and R. A. Grimm, *Chem. Org. Sulfur Compd.*, 1966, **2**, 35.
2. (a) P. D. Magnus, *Tetrahedron*, 1977, **33**, 2019; (b) B. M. Trost, *Chem. Rev.*, 1978, **78**, 363; (c) L. Field, *Synthesis*, 1978, 713; (d) R. K. Dieter, *Tetrahedron*, 1986, **42**, 3029; (e) O. De Lucchi and L. Pasquato, *Tetrahedron*, 1988, **44**, 6755; (f) P. Bravo and G. Resnati, in 'Perspectives in the Organic Chemistry of Sulfur, Twelfth International Symposium on the Organic Chemistry of Sulfur, Nijmegen, The Netherlands', 1986, ed. B. Zwanenburg and A. J. H. Klunder, Elsevier, Amsterdam, 1987, p. 89.
3. J. Nakayama, S. Yamaoka and M. Hoshino, *Tetrahedron Lett.*, 1988, **29**, 1161.
4. S. Kamata, S. Uyeo, N. Haga and W. Nagata, *Synth. Commun.*, 1973, **3**, 265.
5. D. Morgans, Jr. and G. B. Feigelson, *J. Org. Chem.*, 1982, **47**, 1131.
6. I. Paterson and I. Fleming, *Tetrahedron Lett.*, 1979, 995.
7. K. Ogura, A. Yanagisawa, T. Fujino and K. Takahashi, *Tetrahedron Lett.*, 1988, **29**, 5387.
8. R. K. Dieter and Y. J. Lin, *Tetrahedron Lett.*, 1985, **26**, 39.
9. D. H. R. Barton, M. R. Britten-Kelly and D. Ferreira, *J. Chem. Soc., Perkin Trans. 1*, 1978, 1090.
10. (a) T. Takahashi, S. Hashiguchi, K. Kasuga and T. Tsuji, *J. Am. Chem. Soc.*, 1978, **100**, 7424; (b) K. C. Nicolaou, S. P. Seitz, W. J. Sipio and J. F. Blount, *J. Am. Chem. Soc.*, 1979, **101**, 3884; (c) M. Kato, H. Saito and A. Yoshikoshi, *Chem. Lett.*, 1984, 213; (d) M. Kato, A. Ouchi and A. Yoshikoshi, *Chem. Lett.*, 1983, 1511; (e) J. P. Marino and A. D. Perez, *J. Am. Chem. Soc.*, 1984, **106**, 7643; (f) J. P. Marino and R. Fernandez de la Pradilla, *Tetrahedron Lett.*, 1985, **26**, 5381.
11. J. Schut, J. B. F. N. Engberts and H. Wynberg, *Synth. Commun.*, 1972, **2**, 415.
12. P. L. Stotter and R. E. Hornish, *J. Am. Chem. Soc.*, 1973, **95**, 4444.
13. (a) R. B. Boar, D. W. Hawkins, J. F. McGhie and D. H. R. Barton, *J. Chem. Soc., Perkin Trans. 1*, 1973, 654; (b) E. J. Corey and G. Schmidt, *Tetrahedron Lett.*, 1979, 2317; (c) Y. Tominaga, A. Ushirogochi and Y. Matsuda, *J. Heterocycl. Chem.*, 1987, **24**, 1557.
14. Y. Tamura, H. Maeda, S. Akai and H. Ishibashi, *Tetrahedron Lett.*, 1982, **23**, 2209.
15. (a) C. Iwata, K. Hattori, S. Uchida and T. Imanishi, *Tetrahedron Lett.*, 1984, **25**, 2995; (b) L. L. Klein and M. S. Shanklin, *J. Org. Chem.*, 1988, **53**, 5202.
16. E. Juaristi, J. S. Cuiz-Sánchez and F. R. Ramos-Morales, *J. Org. Chem.*, 1984, **49**, 4912.
17. G. Solladié, C. Greck, G. Demailly and A. Solladié-Cavallo, *Tetrahedron Lett.*, 1982, **23**, 5047.
18. T. Sato, Y. Okumura, J. Itai and T. Fujisawa, *Chem. Lett.*, 1988, 1537.
19. (a) J. S. Bradshaw, B. A. Jones and J. S. Gebhard, *J. Org. Chem.*, 1983, **48**, 1127; (b) S. L. Baxter and J. S. Bradshaw, *J. Org. Chem.*, 1981, **46**, 831.

20. K. Osakada, M. Maeda, Y. Nakamura, T. Yamamoto and A. Yamamoto, *J. Chem. Soc., Chem. Commun.*, 1986, 442; E. Wenkert, M. E. Shepard and A. T. McPhail, *J. Chem. Soc., Chem. Commun.*, 1986, 1390.
21. (a) J. J. Eisch and K. R. Im, *J. Organomet. Chem.*, 1977, **139**, C51; (b) J. J. Eisch, L. E. Hallenbeck and K. I. Han, *J. Org. Chem.*, 1983, **48**, 2963; (c) J. J. Eisch, L. E. Hallenbeck and K. I. Han, *J. Am. Chem. Soc.*, 1986, **108**, 7763.
22. A. S. Kende, L. S. Liebeskind and D. M. Braitsch, *Tetrahedron Lett.*, 1975, 3375; I. Colon and D. R. Kelsey, *J. Org. Chem.*, 1986, **51**, 2627 and refs. therein.
23. M. Julia and J. N. Verpeaux, *Tetrahedron Lett.*, 1982, **23**, 2457.
24. B. Ganem and J. O. Osby, *Chem. Rev.*, 1986, **86**, 763.
25. (a) W. E. Truce and F. M. Perry, *J. Org. Chem.*, 1965, **30**, 1316; (b) J. Clark, R. K. Grantham and J. Lydiate, *J. Chem. Soc. C*, 1968, 1122.
26. (a) M. R. Everby and R. D. Waigh, *Synth. Commun.*, 1986, **16**, 779; (b) B. Myrboh, L. W. Singh, H. Ila and H. Junjappa, *Synthesis*, 1982, 307.
27. (a) L. A. Flippin and M. A. Dombroski, *Tetrahedron Lett.*, 1985, **26**, 2977; (b) P. N. Confalone and R. A. Earl, *Tetrahedron Lett.*, 1986, **27**, 2695.
28. W. E. Truce and F. E. Roberts, *J. Org. Chem.*, 1963, **28**, 961.
29. T. Nishio and Y. Omote, *Chem. Lett.*, 1979, 365, 1223; T. Nishio and Y. Omote, *J. Chem. Soc., Perkin Trans. 1*, 1981, 934.
30. (a) H. Kotake, T. Yamamoto and H. Kinoshita, *Chem. Lett.*, 1982, 1331; (b) M. Mohri, H. Kinoshita, K. Inomata and H. Kotake, *Chem. Lett.*, 1985, 451.
31. R. O. Hutchins and K. Learn, *J. Org. Chem.*, 1982, **47**, 4380.
32. H. Alper and T. L. Prince, *Angew. Chem., Int. Ed. Engl.*, 1980, **19**, 315; H. Alper, S. Ripley and T. L. Prince, *J. Org. Chem.*, 1983, **48**, 250.
33. M.-C. Chan, K.-M. Cheng, M. K. Li and T.-Y. Luh, *J. Chem. Soc., Chem. Commun.*, 1985, 1610; M.-C. Chan, K.-M. Cheng, K. M. Ho, C. T. Ng, T. M. Yam, B. S. L. Wang and T.-Y. Luh, *J. Org. Chem.*, 1988, **53**, 4466.
34. J. J. Eisch, L. E. Hallenbeck and M. A. Lucarelli, *Fuel*, 1985, **64**, 440.
35. T. Mukaiyama, *Int. J. Sulfur. Chem., Part B*, 1972, **7**, 173; K. Navasaka, M. Hayashi and T. Mukaiyama, *Chem. Lett.*, 1972, 259.
36. V. Pascali and A. Umani-Ronchi, *J. Chem. Soc., Chem. Commun.*, 1973, 351.
37. P. Caubère, *Angew. Chem., Int. Ed. Engl.*, 1983, **22**, 599.
38. S. Becker, Y. Fort, R. Vanderesse and P. Caubère, *J. Org. Chem.*, 1989, **54**, 4848; S. Becker, Y. Fort and P. Caubère, *J. Org. Chem.*, 1990, **55**, 6194.
39. B. M. Trost and P. L. Ornstein, *Tetrahedron Lett.*, 1981, **22**, 3463.
40. E. Wenkert and T. W. Ferreira, *J. Chem. Soc., Chem. Commun.*, 1982, 840.
41. H. Okamura, M. Miura and H. Takei, *Tetrahedron Lett.*, 1979, 43 and refs. cited therein; E. Wenkert, T. W. Ferreira and E. L. Michelotti, *J. Chem. Soc., Chem. Commun.*, 1979, 637; Z.-J. Ni and T.-Y. Luh, *J. Chem. Soc., Chem. Commun.*, 1988, 1011.
42. (a) J. L. Fabre, M. Julia and J. N. Verpeaux, *Tetrahedron Lett.*, 1982, **23**, 2469; (b) J. L. Fabre and M. Julia, *Tetrahedron Lett.*, 1983, **24**, 4311; (c) T. Cuvigny, C. H. du Penhoat and M. Julia, *Tetrahedron*, 1987, **43**, 859; (d) M. Capet, T. Cuvigny, C. H. du Penhoat, M. Julia and G. Loomis, *Tetrahedron Lett.*, 1987, **28**, 6273.
43. (a) H. Smith, in 'Chemistry in Nonaqueous Ionizing Solvents', ed. G. Jander, H. S. Spandau and C. C. Addison, Interscience, New York, 1963, part 2, p. 191; (b) A. Maercker, *Angew. Chem., Int. Ed. Engl.*, 1987, **26**, 972; (c) M. Tiecco, *Synthesis*, 1988, 749.
44. (a) M. Ramaiah, *Tetrahedron*, 1987, **43**, 3541; (b) E. Block and M. Aslam, *Tetrahedron*, 1988, **44**, 281.
45. C. G. Screttas and M. Micha-Screttas, *J. Org. Chem.*, 1978, **43**, 1064; C. G. Screttas and M. Micha-Screttas, *J. Org. Chem.*, 1979, **44**, 713.
46. D. Savoia, C. Trombini and A. Umani-Ronchi, *J. Chem. Soc., Perkin Trans. 1*, 1977, 123.
47. T. Cohen, W. M. Daniewski and R. B. Weisenfeld, *Tetrahedron Lett.*, 1978, 4665; T. Cohen and J. R. Matz, *Synth. Commun.*, 1980, **10**, 311; T. Cohen, J. P. Sherbine, J. R. Matz, R. R. Hutchins, B. M. McHenry and P. R. Willey, *J. Am. Chem. Soc.*, 1984, **106**, 3245; T. Cohen, J. P. Sherbine, S. A. Mendelson and M. Myers, *Tetrahedron Lett.*, 1985, **26**, 2965 and refs. cited in these refs.
48. T. H. Hashill, P. W. K. Woo and D. R. Watson, *J. Org. Chem.*, 1977, **42**, 1302.
49. M. A. Francisco, A. Kurs, A. R. Katritzky and D. Rasala, *J. Org. Chem.*, 1988, **53**, 596, 4821.
50. P. A. Grieco and Y. Masaki, *J. Org. Chem.*, 1974, **39**, 2135.
51. B. M. Trost, L. Weber, P. Streg, T. J. Fullerton and T. J. Dietsche, *J. Am. Chem. Soc.*, 1978, **100**, 3426 and refs. cited therein.
52. M. J. Kurth and M. J. O'Brien, *J. Org. Chem.*, 1985, **50**, 3846.
53. C. Fehr, *Helv. Chim. Acta*, 1983, **66**, 2512.
54. P. T. Lansbury, R. W. Erwin and D. A. Jeffrey, *J. Am. Chem. Soc.*, 1980, **102**, 1602.
55. C. Bödeker, E. R. de Waard and H. O. Huisman, *Tetrahedron*, 1981, **37**, 1233.
56. A. C. Brown and L. A. Carpino, *J. Org. Chem.*, 1985, **50**, 1749; E. P. Künding and A. F. Cunningham, Jr., *Tetrahedron*, 1988, **44**, 6855.
57. E. L. Eliel and T. W. Doyle, *J. Org. Chem.*, 1970, **35**, 2716; B. C. Newman and E. L. Eliel, *J. Org. Chem.*, 1970, **35**, 3641.
58. G. A. Russell and G. J. Mikol, *J. Am. Chem. Soc.*, 1966, **88**, 5498.
59. (a) S. Kurozumi, T. Toru, M. Kobayashi and S. Ishimoto, *Synth. Commun.*, 1977, **7**, 427; (b) R. A. Holton, R. M. Kennedy, H.-B. Kim and M. E. Krafft, *J. Am. Chem. Soc.*, 1987, **109**, 1597; (c) R. A. Holton, D. J. Crouse, A. D. Williams and R. M. Kennedy, *J. Org. Chem.*, 1987, **52**, 2317.
60. B. M. Trost, *Acc. Chem. Res.*, 1978, **11**, 453.
61. P. J. Kocienski, *Chem. Ind. (London)*, 1981, 548.
62. B. M. Trost, H. C. Arndt, P. E. Strege and T. R. Verhoeven, *Tetrahedron Lett.*, 1976, 3477.
63. R. V. C. Carr and L. A. Paquette, *J. Am. Chem. Soc.*, 1980, **102**, 853.

64. (a) B. M. Trost, W. C. Vladuchick and A. J. Bridges, *J. Am. Chem. Soc.*, 1980, **102**, 3554; (b) B. M. Trost and J. E. Vincent, *J. Am. Chem. Soc.*, 1980, **102**, 5680.
65. J. A. Marshall and J. A. Markwalder, *Tetrahedron Lett.*, 1988, **29**, 4811.
66. M. A. Brimble, D. L. Officer and G. M. Williams, *Tetrahedron Lett.*, 1988, **29**, 3609.
67. R. Annunziata, M. Cinquini, F. Cozzi, F. Montanari and A. Restelli, *J. Chem. Soc., Chem. Commun.*, 1983, 1138.
68. B. M. Trost and T. R. Verhoeven, *J. Am. Chem. Soc.*, 1980, **102**, 4743; V. Bhat and R. C. Cookson, *J. Chem. Soc., Chem. Commun.*, 1981, 1123; J. C. Carretero, S. De Lombaert and L. Ghosez, *Tetrahedron Lett.*, 1987, **23**, 2135.
69. C.-N. Hsiao and H. Shechter, *Tetrahedron Lett.*, 1982, **23**, 1963.
70. E. J. Corey and M. Chaykovsky, *J. Am. Chem. Soc.*, 1964, **86**, 1639; 1965, **87**, 1345.
71. G. H. Posner, J. P. Mallano, H. Hulce and L. L. Frye, *J. Am. Chem. Soc.*, 1982, **104**, 4180; G. H. Posner, T. P. Kogan and M. Hulce, *Tetrahedron Lett.*, 1984, **25**, 383; G. H. Posner and M. Hulce, *Tetrahedron Lett.*, 1984, **25**, 379.
72. (a) C. Mioskowski and G. Solladié, *Tetrahedron*, 1980, **36**, 227; (b) G. Solladié, F. Matloubi-Moghadam, C. Luttmann and C. Mioskowski, *Helv. Chim. Acta*, 1982, **65**, 1602; (c) E. J. Corey, L. O. Weigel, A. R. Chamberlin, H. Cho and D. H. Hua, *J. Am. Chem. Soc.*, 1980, **102**, 6613; (d) G. Solladié and F. Matloubi-Moghadam, *J. Org. Chem.*, 1982, **47**, 91.
73. S. Knapp, R. Lis and P. Michna, *J. Org. Chem.*, 1981, **46**, 624.
74. K. Kondo and D. Tunemoto, *Tetrahedron Lett.*, 1975, 1397.
75. T. Ohnuma, N. Hata, H. Fujiwara and Y. Ban, *J. Org. Chem.*, 1982, **47**, 4713.
76. M. A. Wuonola and R. B. Woodward, *J. Am. Chem. Soc.*, 1973, **95**, 284.
77. (a) H. G. Kuivila, *Synthesis*, 1970, 499; (b) M. Pereyre, J.-P. Quintard and A. Rahm, 'Tin in Organic Synthesis', Butterworth, London, 1987; (c) W. P. Neumann, *Synthesis*, 1987, 665.
78. E. Vedejs, *Acc. Chem. Res.*, 1984, **17**, 358.
79. C. G. Gutierrez and L. R. Summerhays, *J. Org. Chem.*, 1984, **49**, 5206.
80. Y. Ueno, S. Aoki and M. Okawara, *J. Am. Chem. Soc.*, 1979, **101**, 5414; Y. Ueno and M. Okawara, *J. Am. Chem. Soc.*, 1979, **101**, 1893; Y. Ueno, S. Aoki and M. Okawara, *J. Chem. Soc., Chem. Commun.*, 1980, 683; D. P. Curran, *Synthesis*, 1988, 489.
81. G. A. Krafft and P. T. Meinke, *Tetrahedron Lett.*, 1985, **26**, 135; K. C. Nicolaou, D. G. McGarry, P. K. Somers, C. A. Veale and G. T. Furst, *J. Am. Chem. Soc.*, 1987, **109**, 2504.
82. E. Vedejs and D. W. Powell, *J. Am. Chem. Soc.*, 1982, **104**, 2046.
83. J. D. Buynak, M. N. Rao, H. Pajanhesh, R. Y. Chandragekaran, K. Finn, P. de Meester and S. C. Chu, *J. Org. Chem.*, 1985, **50**, 4245.
84. Y. Watanabe, T. Araki, Y. Ueno and T. Endo, *Tetrahedron Lett.*, 1986, **27**, 5385.
85. C. G. Gutierrez, R. A. Stringham, T. Nitasaka and K. G. Glasscock, *J. Org. Chem.*, 1980, **45**, 3393.
86. D. R. Williams and J. L. Moore, *Tetrahedron Lett.*, 1983, **24**, 339.
87. B. Giese, 'Radicals in Organic Synthesis: Formation of Carbon–Carbon Bonds', Pergamon Press, Oxford, 1986; D. P. Curran, *Synthesis*, 1988, 417.
88. (a) J. Bremner, M. Julia, M. Launay and J.-P. Stacino, *Tetrahedron Lett.*, 1982, **23**, 3265; (b) M. Julia and J.-P. Stacino, *Bull. Soc. Chim. Fr.*, 1985, 831; (c) M. Julia and J.-P. Stacino, *Tetrahedron*, 1986, **42**, 2469; (d) M. Julia, H. Lauron, J.-P. Stacino, J. N. Verpeaux, Y. Jeannin and Y. Dromzee, *Tetrahedron*, 1986, **42**, 2475.
89. S. Cabiddu, S. Melis, P. P. Piras and F. Sotgiu, *Synthesis*, 1982, 583.
90. H. Ikehira, S. Tanimoto and T. Oida, *J. Chem. Soc., Perkin Trans. 1*, 1984, 1223.
91. T. Takeda, K. Ando, A. Mamada and T. Fujiwara, *Chem. Lett.*, 1985, 1149.
92. G. Neef, U. Eder and A. Seeger, *Tetrahedron Lett.*, 1980, **21**, 903.
93. G. A. Molander and G. Hahn, *J. Org. Chem.*, 1986, **51**, 1135.
94. H. Alper and C. Blais, *J. Chem. Soc., Chem. Commun.*, 1980, 169; H. Alper, *Pure Appl. Chem.*, 1980, **52**, 607; R. D. Adams, D. A. Katahiro and L.-W. Yang, *Organometallics*, 1982, **1**, 235; H. Alper, F. Sibtain and J. Heveling, *Tetrahedron Lett.*, 1983, **24**, 5329; S. C. Shim, S. Antebi and H. Alper, *Tetrahedron Lett.*, 1985, **26**, 1935; T.-Y. Luh and C. S. Wong, *J. Org. Chem.*, 1985, **50**, 5413 and refs. cited in these references.
95. C. S. Wong, W. S. Leung, L. L. Yeung and T.-Y. Luh, *J. Organomet. Chem.*, 1986, **307**, C49; L. L. Yeung, Y. C. Yip and T.-Y. Luh, *J. Chem. Soc., Chem. Commun.*, 1987, 981; D. K. P. Ng and T.-Y. Luh, *Tetrahedron Lett.*, 1988, **29**, 5131 and refs. cited therein.
96. R. J. Angelici, *Acc. Chem. Res.*, 1988, **21**, 387.
97. (a) D. L. J. Clive, *Aldrichimica Acta*, 1978, **11**, 43; (b) D. L. J. Clive, *Tetrahedron*, 1978, **34**, 1049; (c) A. Krief, in 'Proceedings of the Third International Symposium on Organic Selenium and Tellurium Compounds, Metz, 1979', ed. D. Cagniant and G. Kirsch, Université de Metz, 1981, p. 13; (d) D. L. J. Clive, G. J. Chittattu, V. Farina, W. A. Kiel, S. M. Menchen, C. G. Russell, A. Singh, C. K. Wong and N. J. Curtis, in ref. 97c, p. 43; (e) A. Krief, W. Dumont, A. Cravador, J. N. Denis, S. Halazy, L. Hevesi, D. Labar, J. Luccheti, J. Rémion, M. Sevrin and D. van Ende, *Bull. Soc. Chim. Fr., Part 2*, 1980, 519; (f) C. Paulmier, in 'Selenium Reagents and Intermediates in Organic Synthesis', Pergamon Press, Oxford, 1986.
98. S. V. Ley, B. Lygo, H. Molines and J. A. Morton, *J. Chem. Soc., Chem. Commun.*, 1982, 1251.
99. K. C. Nicolaou, R. L. Magolda, W. J. Sipio, W. E. Barnette, Z. Lysenko and M. M. Joullié, *J. Am. Chem. Soc.*, 1980, **102**, 3784.
100. (a) S. V. Ley and B. Lygo, *Tetrahedron Lett.*, 1982, **23**, 4625; (b) A. M. Doherty, S. V. Ley, B. Lygo and D. J. Williams, *J. Chem. Soc., Perkin Trans. 1*, 1984, 1371.
101. S. V. Ley and P. J. Murray, *J. Chem. Soc., Chem. Commun.*, 1982, 1252.
102. K. C. Nicolaou, *Tetrahedron*, 1981, **37**, 4097.
103. T. G. Back, *J. Chem. Soc., Chem. Commun.*, 1984, 1417; T. G. Back, V. I. Birss, M. Edwards and M. V. Krishna, *J. Org. Chem.*, 1988, **53**, 3815 and refs. cited therein.
104. A. Toshimitsu, K. Terao and S. Uemura, *J. Org. Chem.*, 1986, **51**, 1724.
105. D. H. R. Barton, X. Lusinchi and P. Milliet, *Tetrahedron Lett.*, 1982, **23**, 4949.

106. N. Petragnani, H. M. C. Ferraz and G. V. J. Silva, *Synthesis*, 1986, 157.
107. D. L. J. Clive, G. J. Chittattu, V. Farina, W. A. Kiel, S. M. Menchen, C. G. Russell, A. Singh, C. K. Wong and N. J. Curtis, *J. Am. Chem. Soc.*, 1980, **102**, 4438.
108. J. N. Denis and A. Krief, *Tetrahedron Lett.*, 1982, **23**, 3411.
109. P. J. Giddings, D. I. John and E. J. Thomas, *Tetrahedron Lett.*, 1980, **21**, 399.
110. D. L. J. Clive, G. J. Chittattu and C. K. Wong, *J. Chem. Soc., Chem. Commun.*, 1978, 41.
111. G. Jaurand, J.-M. Beau and P. Sinaÿ, *J. Chem. Soc., Chem. Commun.*, 1981, 572.
112. D. L. J. Clive, C. G. Russell, G. J. Chittattu and A. Singh, *Tetrahedron*, 1980, **36**, 1399.
113. W. P. Jackson, S. V. Ley and A. J. Whittle, *J. Chem. Soc., Chem. Commun.*, 1980, 1173.
114. A. Toshimitsu, T. Aoai, H. Owada, S. Uemura and M. Okano, *J. Org. Chem.*, 1981, **46**, 4727.
115. D. L. J. Clive, V. Farina, A. Singh, C. K. Wong, W. A. Kiel and S. M. Menchen, *J. Org. Chem.*, 1980, **45**, 2120.
116. H. J. Reich, *Acc. Chem. Res.*, 1979, **12**, 22; H. J. Reich, in 'Proceedings of the Fourth International Conference on the Organic Chemistry of Selenium and Tellurium', ed. F. J. Berry and W. R. McWhinnie, Aston University, Birmingham, 1983, p. 258; J. V. Comasseto, *J. Organomet. Chem.*, 1983, **253**, 131.
117. D. Liotta, *Acc. Chem. Res.*, 1984, **17**, 28.
118. L. Wartski, M. El Bouz, J. Seyden-Penne, W. Dumont and A. Krief, *Tetrahedron Lett.*, 1979, 1543.
119. G. Zima, C. Barnum and D. Liotta, *J. Org. Chem.*, 1980, **45**, 2736; D. Liotta, M. Saindane, C. Barnum and G. Zima, *Tetrahedron*, 1985, **41**, 4881; M. Solomon, W. Hoekstra, G. Zima and D. Liotta, *J. Org. Chem.*, 1988, **53**, 5058.
120. (a) O. A. Reutov, *Pure Appl. Chem.*, 1968, **17**, 79; *Tetrahedron*, 1978, **34**, 2827 and refs. cited therein; (b) W. Kitching, *Rev. Pure Appl. Chem.*, 1969, **19**, 1; (c) D. S. Matteson, *Organomet. Chem. Rev., Sect. A*, 1969, 263; (d) L. G. Makarova, in 'Organometallic Reactions', ed. E. I. Becker and M. Tsutsui, Wiley, New York, 1972, vol. 1, p. 325.
121. (a) F. R. Jensen, 'Electrophilic Substitution of Organomercurials', McGraw-Hill, New York, 1968; R. Taylor, in 'Comprehensive Chemical Kinetics', ed. C. H. Bamford and C. F. H. Tipper, Elsevier, Amsterdam, 1972, vol. 13, p. 278; (b) W. A. Nugent and J. K. Kochi, *J. Am. Chem. Soc.*, 1976, **98**, 5979 and refs. cited therein; (c) D. S. Matteson, 'Organometallic Reaction Mechanisms', Academic Press, New York, 1974; (d) L. F. McGahey and F. R. Jensen, *J. Am. Chem. Soc.*, 1979, **109**, 4397 and refs. cited therein; (e) G. A. Olah, *J. Am. Chem. Soc.*, 1972, **94**, 808; (f) D. J. Bencivengo, M. L. Brownawell, M. Y. Li and J. San Filippo, Jr., *J. Am. Chem. Soc.*, 1984, **106**, 3703; (g) D. J. Bencivengo and J. San Filippo, Jr., *Organometallics*, 1983, **2**, 1907.
122. D. Dodd and M. D. Johnson, *J. Chem. Soc. B*, 1969, 1071; J. R. Coad and M. D. Johnson, *J. Chem. Soc. B*, 1967, 633.
123. T. Iwayanagi, M. Matsuo and Y. Saito, *J. Organomet. Chem.*, 1977, **135**, 1.
124. S. Uemura, H. Miyoshi and M. Okano, *J. Chem. Soc., Perkin Trans. 1*, 1980, 1098.
125. M. Giffard and J. Cousseau, *J. Organomet. Chem.*, 1980, **201**, C1.
126. R. D. Bach, R. A. Woodward, T. J. Anderson and M. D. Glick, *J. Org. Chem.*, 1982, **47**, 3707.
127. F. De Sarlo, A. Guarna, A. Goti and A. Brandi, *J. Organomet. Chem.*, 1984, **269**, 115.
128. K. M. Smith and K. C. Langry, *J. Org. Chem.*, 1983, **48**, 500.
129. (a) M. A. Boaventura, J. Drouin and J. M. Conia, *Synthesis*, 1983, 801; *J. Am. Chem. Soc.*, 1985, **107**, 1726.
130. R. C. Larock and L. W. Harrison, *J. Am. Chem. Soc.*, 1984, **106**, 4218.
131. G. B. Deacon, G. N. Stretton and M. J. O'Connor, *Synth. Commun.*, 1983, **13**, 1041.
132. (a) R. C. Larock, 'Organomercury Compounds in Organic Synthesis', Springer-Verlag, Heidelberg, 1985; (b) J. Barluenga and M. Yus, *Chem. Rev.*, 1988, **88**, 487; (c) J. L. Wardell, in 'Comprehensive Organometallic Chemistry', ed. G. Wilkinson, Pergamon Press, Oxford, 1982, vol. 2, p. 863; (d) L. G. Makarova and A. N. Nesmeyanov, 'Methods of Elementoorganic Chemistry', ed. A. N. Nesmeyanov and K. A. Kocheshkov, North-Holland, Amsterdam, vol. 4, 1967; (e) H. Straub, K. P. Zeller and H. Leditsche, *Methoden Org. Chem. (Houben-Weyl)*, 1974, **XIII/2b**, 1.
133. (a) R. P. Quirk and R. E. Lea, *Tetrahedron Lett.*, 1974, 1925; *J. Am. Chem. Soc.*, 1976, **98**, 5973; (b) G. M. Whitesides and J. San Filippo, Jr., *J. Am. Chem. Soc.*, 1970, **92**, 6611.
134. P. R. Singh and R. K. Khanna, *Tetrahedron Lett.*, 1983, **24**, 1411.
135. (a) D. B. Collum, W. C. Still and F. Mohamadi, *J. Am. Chem. Soc.*, 1986, **108**, 2094; (b) D. B. Collum, F. Mohamadi and J. S. Hallock, *J. Am. Chem. Soc.*, 1983, **105**, 6882; (c) W. Adam and K. Sakanishi, *J. Am. Chem. Soc.*, 1978, **100**, 3935; (d) S. N. Suryawanshi and U. R. Nayak, *Tetrahedron Lett.*, 1978, 465.
136. G. A. Russell and D. Guo, *Tetrahedron Lett.*, 1984, **25**, 5239.
137. R. P. Quirk, *J. Org. Chem.*, 1972, **37**, 3554.
138. C. L. Hill and G. M. Whitesides, *J. Am. Chem. Soc.*, 1974, **96**, 870.
139. (a) B. Giese, K. Heuck and U. Lüning, *Tetrahedron Lett.*, 1981, **22**, 2155; (b) B. Giese and U. Erfort, *Chem. Ber.*, 1983, **116**, 1240; (c) T. R. Hoye, A. J. Caruso and M. J. Kurth, *J. Org. Chem.* 1981, **46**, 3550.
140. (a) G. A. Gray and W. R. Jackson, *J. Am. Chem. Soc.*, 1969, **91**, 6205; (b) V. M. A. Chambers, W. R. Jackson and G. W. Young, *J. Chem. Soc., Chem. Commun.*, 1970, 1275.
141. W. Kitching, A. R. Atkins, G. Wickham and V. Alberts, *J. Org. Chem.*, 1981, **46**, 563.
142. (a) S. Thaisrivongs and D. Seebach, *J. Am. Chem. Soc.*, 1983, **105**, 7407 and references cited therein; (b) F. H. Gouzoules and R. A. Whitney, *J. Org. Chem.*, 1986, **51**, 2024; *Tetrahedron Lett.*, 1985, **26**, 3441.
143. P. A. Bartlett and J. L. Adams, *J. Am. Chem. Soc.*, 1980, **102**, 337.
144. (a) J. Oda, T. Nakagawa and Y. Inouye, *Bull. Chem. Soc. Jpn.*, 1967, **40**, 373; (b) R. N. Butler, *Synth. Reagents*, 1981, **4**, 1; (c) F. G. Bordwell and M. L. Douglass, *J. Am. Chem. Soc.*, 1966, **88**, 993; (d) B. Giese, S. Gantert and A. Schulz, *Tetrahedron Lett.*, 1974, 3583.
145. H. C. Brown and P. J. Geoghegan, Jr., *J. Org. Chem.*, 1970, **35**, 1844.
146. (a) B. Giese and K. Heuck, *Tetrahedron Lett.*, 1980, **21**, 1829; (b) B. Giese and K. Heuck, *Chem. Ber.*, 1979, **112**, 3759.
147. P. A. Bartlett, D. P. Richardson and J. Myerson, *Tetrahedron*, 1984, **40**, 2317.
148. R. K. Boeckman, Jr. and C. J. Flann, *Tetrahedron Lett.*, 1983, **24**, 4923.

149. H. C. Brown and G. J. Lynch, *J. Org. Chem.*, 1981, **46**, 531.
150. (a) J. R. Pougny, M. A. M. Nassr and P. Sinaÿ, *J. Chem. Soc., Chem. Commun.*, 1981, 375; (b) J. Barluenga, J. López-Prado, P. J. Campos and G. Asensio, *Tetrahedron*, 1983, **39**, 2863.
151. K. E. Harding and T. H. Marman, *J. Org. Chem.*, 1984, **49**, 2838.
152. W. Kitching, J. A. Lewis, M. T. Fletcher, J. J. De Voss, R. A. I. Drew and C. J. Moore, *J. Chem. Soc., Chem. Commun.*, 1986, 855.
153. S. Arseniyadis and J. Gore, *Tetrahedron Lett.*, 1983, **24**, 3997.
154. M. C. Benhamou, G. Etemad-Moghadam, V. Speziale and A. Lattes, *Synthesis*, 1979, 891.
155. A. J. Bloodworth and J. L. Courtneidge, *J. Chem. Soc., Perkin Trans. 1*, 1982, 1797.
156. A. J. Bloodworth and J. A. Khan, *J. Chem. Soc., Perkin Trans. 1*, 1980, 2450.
157. J. R. Nixon, M. A. Cudd and N. A. Porter, *J. Org. Chem.*, 1978, **43**, 4048.
158. E. Bascetta and F. D. Gunstone, *J. Chem. Soc., Perkin Trans. 1*, 1984, 2207.
159. A. J. Bloodworth, J. A. Khan and M. E. Loveitt, *J. Chem. Soc., Perkin Trans. 1*, 1981, 621.
160. N. A. Porter and P. Zuraw, *J. Chem. Soc., Chem. Commun.*, 1985, 1472.
161. F. Paquet and P. Sinaÿ, *J. Am. Chem. Soc.*, 1984, **106**, 8313; *Tetrahedron Lett.*, 1984, **25**, 3071.
162. (a) M. L. Bullpitt and W. Kitching, *J. Organomet. Chem.*, 1972, **46**, 21; (b) V. M. A. Chambers, W. R. Jackson and G. W. Young, *J. Chem. Soc., Chem. Commun.*, 1970, 1275; *J. Chem. Soc. C*, 1971, 2075.
163. G. Carnevale, E. Davini, C. Iavarone and C. Trogolo, *J. Org. Chem.*, 1988, **53**, 5343.
164. H. C. Brown, G. J. Lynch, W. J. Hammar and L. C. Liu, *J. Org. Chem.*, 1979, **44**, 1910.
165. Y. Senda, S. Kamiyama and S. Imaizumi, *J. Chem. Soc., Perkin Trans. 1*, 1978, 530.
166. H. C. Brown and W. J. Hammar, *Tetrahedron*, 1978, **34**, 3405.
167. (a) C. Battistini, P. Crotti, B. Macchia, F. Macchia and C. H. DePuy, *J. Org. Chem.*, 1978, **43**, 1400; (b) G. Langbein, H. J. Siemann, I. Gruner and C. Müller, *Tetrahedron*, 1986, **42**, 937.
168. S. Nishida, T. Fujioka and N. Shimizu, *J. Organomet. Chem.*, 1978, **156**, 37.
169. J. A. Soderquist and K. L. Thompson, *J. Organomet. Chem.*, 1978, **159**, 237.
170. (a) P. Bravo, C. De Vita, C. Ticozzi, F. Viani and G. Cavicchio, *Gazz. Chim. Ital.*, 1986, **116**, 441; (b) T. Hosokawa, S. Miyagi, S. Murahashi, A. Sonoda, Y. Matsuura, S. Tanimoto and M. Kakudo, *J. Org. Chem.*, 1978, **43**, 719.
171. (a) P. Bravo, C. Ticozzi, F. Viani and A. Arnone, *Gazz. Chim. Ital.*, 1985, **115**, 677; (b) S. J. Danishefsky, W. H. Pearson and D. F. Harvey, *J. Am. Chem. Soc.*, 1984, **106**, 2455; (c) Y. Matsuki, M. Kodama and S. Ito, *Tetrahedron Lett.*, 1979, 2901.
172. M. Livneh, J. K. Sutter and C. N. Sukenik, *J. Org. Chem.*, 1987, **52**, 5039.
173. D. P. Curran, P. B. Jacobs, R. L. Elliott and B. H. Kim, *J. Am. Chem. Soc.*, 1987, **109**, 5280.
174. J. D. White, M. A. Avery and J. P. Carter, *J. Am. Chem. Soc.*, 1982, **104**, 5486.
175. M. B. Gasc, J. Perie and A. Lattes, *Tetrahedron*, 1978, **34**, 1943.
176. J. Barluenga, J. Villamaña and M. Yus, *Synthesis*, 1981, 375.
177. J. Barluenga, F. Aznar, R. Liz and R. Rodes, *J. Chem. Soc., Perkin Trans. 1*, 1983, 1087; 1980, 2732.
178. J. Barluenga, F. Aznar, R. Liz and M.-P. Cabal, *Synthesis*, 1986, 960.
179. J. Barluenga, C. Najera and M. Yus, *J. Heterocycl. Chem.*, 1979, **16**, 1017; 1980, **17**, 917; 1981, **18**, 1297.
180. J. Barluenga, J. Pérez-Prieto, A. M. Bayón and G. Asensio, *Tetrahedron*, 1984, **40**, 1199.
181. T. Aida, R. Legault, D. Dugat and T. Durst, *Tetrahedron Lett.*, 1979, 4993.
182. Y. Saitoh, Y. Moriyama, T. Takahashi and Q. Khuong-Huu, *Tetrahedron Lett.*, 1980, **21**, 75.
183. (a) J. Barluenga, C. Jiménez, C. Nájera and M. Yus, *J. Chem. Soc., Perkin Trans. 1*, 1983, 591; *J. Chem. Soc., Chem. Commun.*, 1981, 670; (b) A. J. Fry and J. A. Simon, *J. Org. Chem.*, 1982, **47**, 5032.
184. W. Carruthers, M. J. Williams and M. T. Cox, *J. Chem. Soc., Chem. Commun.*, 1984, 1235.
185. Y. Tamaru, M. Hojo and Z. Yoshida, *J. Org. Chem.*, 1988, **53**, 5731.
186. S. J. Danishefsky and E. Taniyama, *Tetrahedron Lett.*, 1983, **24**, 15.
187. J. Barluenga, C. Jiménez, C. Nájera and M. Yus, *J. Chem. Soc., Perkin Trans. 1*, 1984, 721; *J. Chem. Soc., Chem. Commun.*, 1981, 1178.
188. A. Koziara, B. Olejniczak, K. Osowska and A. Zwierzak, *Synthesis*, 1982, 918.
189. (a) M. Nishizawa, H. Takenaka and Y. Hayashi, *J. Org. Chem.*, 1986, **51**, 806; (b) M. Nishizawa, H. Yamada and Y. Hayashi, *Tetrahedron Lett.*, 1986, **27**, 187; (c) M. Nishizawa, H. Takenaka, K. Hirotsu, T. Higuchi and Y. Hayashi, *J. Am. Chem. Soc.*, 1984, **106**, 4290; (d) M. Nishizawa, H. Takenaka, H. Nishide and Y. Hayashi, *Tetrahedron Lett.*, 1983, **24**, 2581.
190. E. Tsankova, I. Ognyanov and T. Norin, *Tetrahedron*, 1980, **36**, 669.
191. S. Misumi, T. Ohtsuka, H. Hashimoto, Y. Ohfune, H. Shirahama and T. Matsumoto, *Tetrahedron Lett.*, 1979, 35.
192. (a) N. Bluthe, M. Malacria and J. Gore, *Tetrahedron Lett.*, 1982, **23**, 4263; (b) L. E. Overman, *Angew. Chem., Int. Ed. Engl.*, 1984, **23**, 579.
193. R. J. Ferrier and S. R. Haines, *J. Chem. Soc., Perkin Trans. 1*, 1984, 1689.
194. L. M. Sayre and F. R. Jensen, *J. Org. Chem.*, 1979, **44**, 228.
195. A. Maercker and R. Dujardin, *Angew. Chem., Int. Ed. Engl.*, 1984, **23**, 224.
196. (a) J. Barluenga, F. Aznar and M. Bayod, *Synthesis*, 1988, 144; (b) M. Yamamoto, M. Yoshitake and K. Yamada, *J. Chem. Soc., Chem. Commun.*, 1983, 991.
197. W. Sas, *J. Chem. Soc., Chem. Commun.*, 1984, 862.
198. J. Barluenga, P. J. Lampos, M. A. Roy and G. Asensio, *J. Chem. Soc., Perkin Trans. 1*, 1980, 1420.
199. (a) M. Devaud, *J. Organomet. Chem.*, 1981, **220**, C27; (b) M. S. Mubarak and D. G. Peters, *J. Org. Chem.*, 1985, **50**, 673.
200. K. U. Ingold and B. P. Roberts, 'Free Radical Substitution Reactions', Wiley, London, 1971, p. 34.
201. G. A. Russell and H. Tashtoush, *J. Am. Chem. Soc.*, 1983, **105**, 1398.
202. F. R. Jensen, J. J. Miller, S. J. Cristol and R. S. Beckley, *J. Org. Chem.*, 1972, **37**, 4341.
203. J. Perthuis and P. Poisson, *Bull. Soc. Chim. Fr., Part 2*, 1985, 75.
204. P. R. Singh and R. K. Khanna, *Tetrahedron Lett.*, 1983, **24**, 973.

205. H. Kurosawa, H. Okada and T. Hattori, *Tetrahedron Lett.*, 1981, **22**, 4495.
206. M. Dankowski, K. Praefcke, S. C. Nyburg and W. Wong, *Phosphorus Sulfur*, 1979, **7**, 275.
207. L. Maier, in 'Organic Phosphorus Compounds', ed. G. M. Kosolapoff and L. Maier, Wiley-Interscience, New York, 1972, vol. 1, p. 23.
208. K. Issleib and H. Völker, *Chem. Ber.*, 1961, **94**, 392.
209. V. D. Bianco and S. Doronzo, *Inorg. Synth.*, 1976, **16**, 155.
210. T.-S. Chou, J.-J. Yuan and C.-H. Tsao, *J. Chem. Res. (S)*, 1985, 18.
211. (a) T. Chou, C.-H. Tsao and S. C.-Hung, *J. Org. Chem.*, 1985, **50**, 4329; (b) P. Brooks, M. J. Gallagher and A. Sarroff, *Phosphorus Sulfur*, 1987, **30**, 389.
212. T.-S. Chou, C.-H. Tsao and S. C. Hung, *J. Organomet. Chem.*, 1986, **312**, 53.
213. (a) G. de Lauzon, B. Deschamps, J. Fisher, F. Mathey and A. Mitschler, *J. Am. Chem. Soc.*, 1980, **102**, 994; (b) F. Mathey, *Nouv. J. Chim.*, 1987, **11**, 585.
214. P. E. Garrou, *Chem. Rev.*, 1985, **85**, 171.
215. T. Nishiguchi, K. Tanaka and K. Fukuzumi, *J. Organomet. Chem.*, 1980, **193**, 37.
216. A. G. Abatjoglou, E. Billig and D. R. Bryant, *Organometallics*, 1984, **3**, 923.
217. H. Schmidbaur, Th. Pollok, G. Reber and G. Müller, *Chem. Ber.*, 1987, **120**, 1403.
218. N. Hovnanian and L. G. Hubert-Pfalzgraf, *J. Organomet. Chem.*, 1986, **299**, C29.
219. H. R. Hays and D. J. Peterson, in 'Organic Phosphorus Chemistry', ed. G. M. Kosolapoff and L. Maier, Wiley-Interscience, New York, 1972, vol. 3, p. 349.
220. N. S. Isaacs and O. H. Abed, *Tetrahedron Lett.*, 1986, **27**, 1209.
221. (a) F. Y. Khalil, M. T. Hanna, F. M. Abdel-Halim and M. El-Batouti, *Rev. Roum. Chim.*, 1985, **30**, 571; (b) J. G. Dawber, J. C. Tebby and A. C. Waite, *J. Chem. Soc., Perkin Trans. 2*, 1983, 1923; *Phosphorus Sulfur*, 1984, **19**, 99; (c) F. Y. Khalil and G. Aksnes, *Can. J. Chem.*, 1978, **58**, 1953.
222. W. E. McEwen, K. F. Kumli, A. Blade-Font, M. Zanger and C. A. Van der Werf, *J. Am. Chem. Soc.*, 1964, **86**, 2378.
223. (a) G. H. Posner and S. Lu, *J. Am. Chem. Soc.*, 1985, **107**, 1424; (b) E. Öhler and E. Zbiral, *Chem. Ber.*, 1980, **113**, 2852; (c) D. G. Cox, N. Gurusamy and D. J. Burton, *J. Am. Chem. Soc.*, 1985, **107**, 2811; (d) T. Minami, H. Sako, T. Ikehira, T. Hanamoto and I. Hirao, *J. Org. Chem.*, 1983, **48**, 2569; (e) H. J. Bestmann, R. Engler, H. Hartung and K. Roth, *Chem. Ber.*, 1979, **112**, 28; (f) M. Vincens, J. T. Grimaldo Moron, R. Pasqualini and M. Vidal, *Tetrahedron Lett.*, 1987, **28**, 1259.
224. (a) S. E. Cremer, B. C. Trivedi and F. L. Weitl, *J. Org. Chem.*, 1971, **36**, 3226; (b) J. C. Gallucci and R. H. Holmes, *J. Am. Chem. Soc.*, 1980, **102**, 4379; (c) S. Samaan, *Phosphorus Sulfur*, 1979, **7**, 89.
225. (a) R. Luckenbach, *Chem. Ber.*, 1975, **108**, 803 and references cited therein; (b) D. Valentine, Jr., J. F. Blount and K. Toth, *J. Org. Chem.*, 1980, **45**, 3691.
226. (a) S. M. Cairns and W. E. McEwen, *J. Org. Chem.*, 1987, **52**, 4829; (b) G. L. Keldsen and W. E. McEwen, *J. Am. Chem. Soc.*, 1978, **100**, 7312.
227. D. G. Cox and D. J. Burton, *J. Org. Chem.*, 1988, **53**, 366.
228. H. J. Cristau, H. Christol and M. Soleiman, *Phosphorus Sulfur*, 1978, **4**, 287.
229. A. W. Frank, *Phosphorus Sulfur*, 1978, **5**, 19.
230. (a) H. J. Cristau, F. Plenat and F. Guida-Pietra Santa, *Phosphorus Sulfur*, 1987, **34**, 75; (b) H. J. Cristau, L. Chiche, F. Fallouh, P. Hullot, G. Renard and H. Christol, *Nouv. J. Chim.*, 1984, **8**, 191.
231. H. J. Cristau, A. Long and H. Christol, *Tetrahedron Lett.*, 1979, 349.
232. D. A. Jaeger and D. Bolikal, *J. Org. Chem.*, 1986, **51**, 1350.
233. H. J. Cristau, L. Labaudiniere and H. Christol, *Phosphorus Sulfur*, 1983, **15**, 359.
234. L. Horner, P. Walach and H. Kunz, *Phosphorus Sulfur*, 1978, **5**, 171.
235. T. Hirabe, M. Nojima and S. Kusabayashi, *J. Org. Chem.*, 1984, **49**, 4084.
236. H. J. Meeuwissen, Th. A. Van der Knaap and F. Bickelhaupt, *Tetrahedron*, 1983, **39**, 4225.
237. L. Horner and K. Dickerhof, *Phosphorus Sulfur*, 1983, **15**, 213.
238. H. J. Bestmann and R. Zimmermann, in 'Carbon–Carbon Bond Formation', ed. R. L. Augustine, Dekker, New York, 1979, vol. 1, chap. 3.
239. M. P. Cooke, Jr., *J. Org. Chem.*, 1982, **47**, 4963.
240. G. Aksnes, R. Gierstae and E. A. Wulvik, *Phosphorus Sulfur*, 1988, **39**, 141.
241. T. Mahmood, J. Bao, R. L. Kirchmeier and J. M. Shreeve, *Inorg. Chem.*, 1988, **27**, 2913.
242. Y. Segall, E. Shirin and I. Granoth, *Phosphorus Sulfur*, 1980, **8**, 243.
243. M. J. P. Harger and A. Williams, *Tetrahedron Lett.*, 1986, **27**, 2313.
244. R. R. Arndt and S. Warren, *Tetrahedron Lett.*, 1978, 4089.
245. (a) G. Durrant and J. K. Sutherland, *J. Chem. Soc., Perkin Trans. 1*, 1972, 2582; (b) T. Bottin-Strzalko and J. Seyden-Penne, *Bull. Soc. Chim. Fr., Part 2*, 1984, 161.
246. E. P. Kyba and S.-T. Liu, *Inorg. Chem.*, 1985, **24**, 1613.
247. (a) J. M. Alcaraz, B. Deschamps and F. Mathey, *Phosphorus Sulfur*, 1984, **19**, 45 and refs. cited therein; (b) J. M. Alcaraz, A. Breque and F. Mathey, *Tetrahedron Lett.*, 1982, **23**, 1565.
248. G. Märkl, *Phosphorus Sulfur*, 1977, **3**, 77.
249. N. Iwamoto, Y. Okamoto and S. Takamuku, *Bull. Chem. Soc. Jpn.*, 1986, **59**, 1505.

4.4
Reduction of Epoxides

SINJI MURAI
Osaka University, Japan

and

TOSHIAKI MURAI and SHINZI KATO
Gifu University, Japan

4.4.1 INTRODUCTION

Two types of reductions of epoxides, reduction giving alcohols and deoxygenation to alkenes, are dealt with in this chapter, because of their wide applicability in organic synthesis.[1] These reactions provide access to alcohols and to alkenes when combined with appropriate methods for the synthesis of the starting epoxides, but the two-step conversion of alkenes to alcohols *via* epoxides is the more important.

4.4.2 REDUCTION OF EPOXIDES TO ALCOHOLS

Three major types of reagents, metal hydrides, dissolving metals and hydrogen, have been used to reduce epoxides to alcohols. To be useful as a synthetic transformation the reductive ring opening of epoxides should be regioselective and compatible with other functional groups existing in the same molecule. In this section the emphasis will be on illustrating such selectivity.

4.4.2.1 Metal Hydrides

For more than 40 years LAH (lithium aluminum hydride, $LiAlH_4$)[2,3] has been the most widely employed reagent for reducing epoxides to the corresponding alcohols. Although only 0.25 equiv. of LAH is stoichiometrically necessary, a slight excess of LAH is generally used. Both the reactivity and regioselectivity are highly dependent on the substitution pattern of the epoxides. The reduction of the more-substituted epoxides[4] and of cycloalkene oxides in medium-sized rings (from eight to eleven membered)[5] requires higher temperatures and longer reaction times. The yields and product ratio of the reduction of some representative alkyl- and aryl-substituted epoxides with LAH are shown in Table 1. The opening of the epoxy ring generally takes place at the less-substituted carbon.[4] For β-methylstyrene oxide, only the *trans* isomer is reduced primarily by benzylic attack.[7] The *cis* isomer and other aryl-substituted epoxides undergo ring opening predominantly at the carbon away from the phenyl group. The vinyl group in a 1-methyl-1-vinyl-substituted epoxide[8] does not affect the normal regioselectivity that gives the more-substituted alcohol. However, a steroidal diene monoepoxide undergoes conjugate reduction with lithium aluminum deuteride (equation 1).[9]

Table 1 Reduction of Epoxides with $LiAlH_4$

Epoxide	*Yield (%)*	*Products ratio*	*Ref.*
R (monosubstituted epoxide, O)		R–CH(OH)–CH₃ + R–CH₂CH₂–OH	
R = Me	60	100:0	6
R = $PhCH_2$	84	100:0	4
R = Ph	82	95:5	6
R, R' (1,1-disubstituted epoxide, O)		R, R' tertiary OH + R, R' primary OH	
R = R' = Me	26	95:5	6
R = R' = Ph	97	96:4	4
R = CH_2=CH, R' = Me	48	100:0	8
Ph, R, R' (epoxide, O)		Ph–CH(OH)–CH₂CH₃ + Ph–CH₂–CH(OH)–CH₃	
R = H, R' = Me		10:90	7
R = Me, R' = H		96:4	7
R, R' (trisubstituted epoxide, O)		R, R', OH + R, R', OH	
R = R' = Me	25	100:0	4
R = Me, R' = Bu^t	21	0:100	10
R = Ph, R' = Me		99:1	7
Ph (epoxide, O)		Ph, OH + Ph, OH	
		100:0	7

$$\text{BzO–steroid diene epoxide (R)} \xrightarrow{LiAlD_4} \text{BzO–steroid (R, OH, D)} \quad (1)$$

The reduction of cyclohexene oxides having substituents at various positions has been studied both because of synthetic and mechanistic interest.[11–18] As for alkyl-substituted epoxides, the regioselectivity of ring opening of cyclohexene oxides depends on the substitution pattern. In some cases[9–12] inversion of configuration of the hydroxy group appears to have taken place in the minor products. Some results with rather good regioselectivity are shown in equations (2)–(4).

$LiAlH_4$, Et_2O — OH + OH + OH (2)

99.4:0.57:0.06

$LiAlH_4$, Et_2O — OH + OH + OH (3)

89:9:2

Bu^t — O — $LiAlH_4$, Et_2O — Bu^t OH + Bu^t OH (4)

90:10

In contrast to the low selectivity for a simple β-hydroxycyclohexene oxide (equation 5),[15] high regioselectivities are observed for similar steroidal epoxides (equations 6 and 7).[16,17]

OH — O — $LiAlH_4$ — OH OH + OH OH (5)

60:40

HO — O — $LiAlH_4$, THF, 0 °C, 99% — HO OH (6)

HO — O — $LiAlH_4$ — HO OH (7)

A dramatic reverse in selectivity was observed for 2-substituted cyclohexene oxide by simply changing the *t*-butyl group to a TMS or trimethylgermyl group,[14] as depicted in equation (8). Regiocontrol by the TMS group is further demonstrated in the corresponding cyclopentene and cycloheptene oxides.[19]

O, R — $LiAlH_4$ — OH, R + OH, R (8)

R = CMe_3	0:100
R = $SiMe_3$	99:1
R = $GeMe_3$	96:4

However, this substituent effect does not apply to 3-substituted cyclohexene oxides,[15] where a mixture of isomers is obtained both for the *t*-butyl- and TMS-substituted epoxides (equation 9).[20]

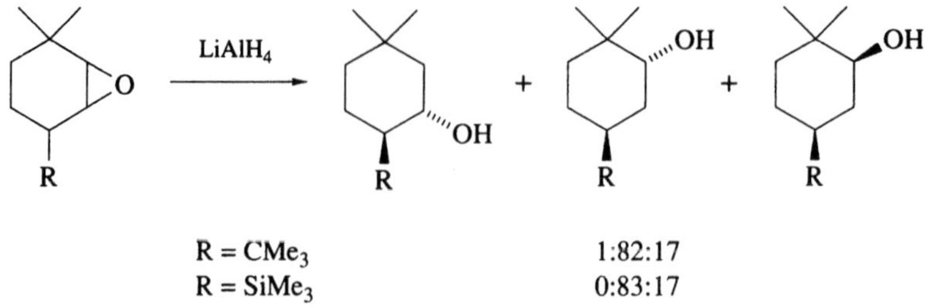

LiAlH$_4$ (9)

R = CMe$_3$ 1:82:17
R = SiMe$_3$ 0:83:17

For epoxy-protected ketones, highly regioselective ring opening leading to β-keto alcohols was attained (equation 10).[21] In the reduction of a diepoxide, 1,4-diol derivatives were predominantly obtained (equation 11).[22]

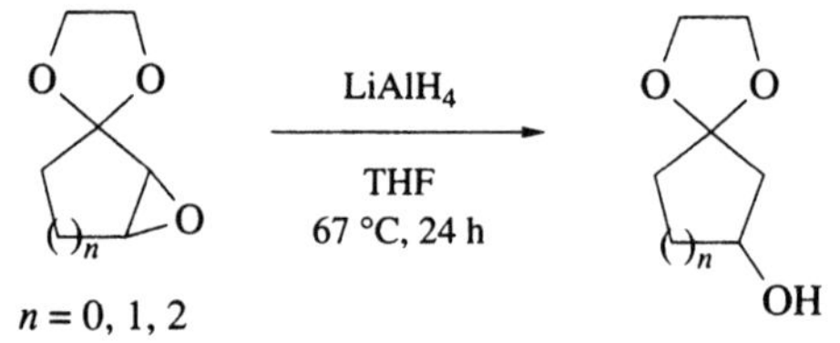

LiAlH$_4$, THF, 67 °C, 24 h (10)

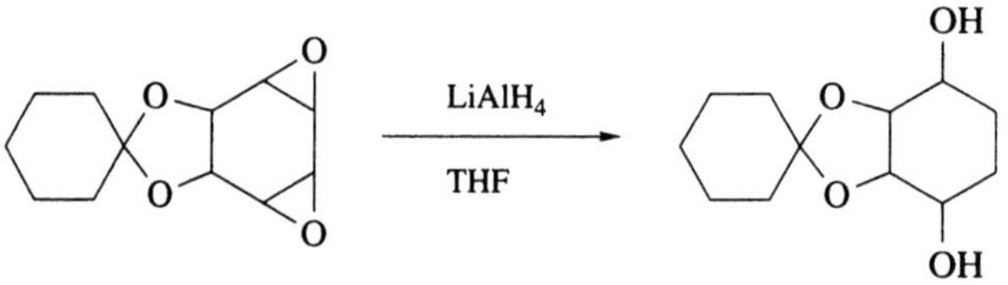

n = 0, 1, 2

LiAlH$_4$, THF (11)

Reductions with LAH are frequently applied to natural product synthesis to yield the more-substituted alcohols. Some recent examples are given in equations (12) and (13).[23,24] In the nucleoside synthesis,[23] the adenosine group can tolerate the reduction conditions.

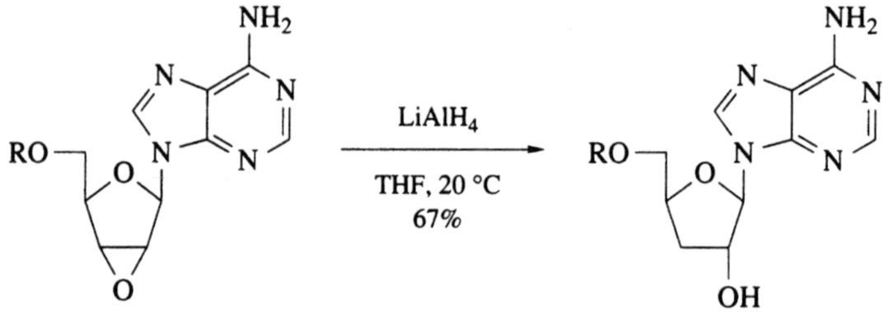

LiAlH$_4$, THF, 20 °C, 67% (12)

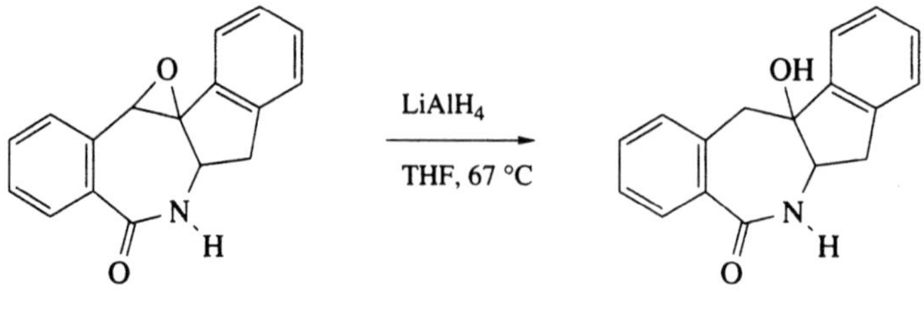

LiAlH$_4$, THF, 67 °C (13)

Sodium borohydride ($NaBH_4$), a relatively mild reducing agent, reduces epoxides only sluggishly[25,26] except for nitro epoxides.[27] In the mixed solvent *t*-butyl alcohol and methanol, sodium borohydride can reduce aryl-substituted epoxides, terminal epoxides and cyclohexene oxide, to the corresponding alcohols.[25,28] The regiochemistry of this reaction is nearly the same as that with LAH. For example the reduction of styrene oxide with potassium borohydride gives only the α-phenethyl alcohol, although the yield is low (equation 14).

$$\text{Ph-epoxide} \xrightarrow[\text{MeOH/Bu}^t\text{OH, 18\%}]{\text{KBH}_4} \text{PhCH(OH)CH}_3 \quad (14)$$

Because of the low reactivity of sodium borohydride, functional groups such as carbamonyl, carbonyl, nitro and cyano groups, which are generally reduced with LAH, can tolerate the reaction conditions.[28] For example, it has been used for the selective reduction of an epoxy group in an ω-cyano epoxide (equation 15).[29]

$$\xrightarrow[\text{MeOH/Bu}^t\text{OH}]{\text{NaBH}_4} \quad (15)$$

2-Phenylpropene oxide is selectively reduced in the presence of *trans*-stilbene oxide even with an excess amount of sodium borohydride (equation 16),[28] and a similar system using lithium borohydride ($LiBH_4$) also shows selective reduction of a styrene oxide in the presence of a benzamide.[30]

$$\xrightarrow[\text{MeOH/Bu}^t\text{OH}]{\text{NaBH}_4} \quad (16)$$

The effect of additives and of modified aluminum and borohydrides has been extensively examined in efforts to enhance reactivity and improve selectivity. The system of LAH and aluminum chloride was early applied to achieving opposite regioselectivity.[5] For example, the reduction of styrene oxide with this system takes place at the benzylic position to give β-phenethyl alcohol as the major product,[5] and the same effect is observed with 1,4-dialkylcyclohexene oxide (equation 17).[14]

$$\xrightarrow{\text{LiAlH}_4,\ \text{AlCl}_3} \quad (17)$$

R = Me	6.5:93.5
R = Bu^t	2.3:97.7

In addition to reversing the regiocontrol, the addition of aluminum chloride also enhances the reactivity of LAH towards tetrasubstituted epoxides, as shown in Table 2. The regioselectivity in the case of unsymmetrically substituted epoxides is not always high.[31,32] Moreover, this enhanced reactivity enables cycloalkene oxides in medium-sized rings to be reduced to the corresponding alcohols smoothly,[33] although a 3:1 ratio of LAH to aluminum chloride is essential for this reaction. Higher proportions of the latter give rise to by-products, whereas lower proportions decrease the reactivity. This system appears to be economically advantageous compared with the reductions using lithium triethylborohydride ($LiBHEt_3$) to be described later.

Another modified metal hydride, lithium triethylborohydride,[34] the so-called 'superhydride', has been introduced as a powerful reducing agent especially suitable for trisubstituted, tetrasubstituted and bicyclic epoxides (Table 3). With trisubstituted epoxides the regiochemistry is completely controlled to give only tertiary alcohols. No skeletal rearrangement is observed for benzonorbornadiene oxide.

Furthermore, 'superhydride' is used as an alternative to LAH in the reduction of carbohydrate epoxides (equation 18).[35] In this example, the C—O bond away from the hydroxy group is selectively cleaved, but a tosyloxy group is also reduced in epoxides having a hydroxy group protected with tosyl. A kinetic study of reductions with 'superhydride' has been reported.[36]

Although diborane $(BH_3)_2$ reduces epoxides, it usually gives a mixture of products in addition to the usual alcohols.[38,39] The reduction of styrene oxide derivatives with diborane has been studied in order to develop a new route to 1,3-diols from epoxides (equation 19).[40] α,β-Unsaturated epoxides[41] undergo

Table 2 Reduction of Tetrasubstituted Epoxides with $LiAlH_4/AlCl_3$[a]

Substrate	*Reaction time* (h)	*Yield* (%)	*Product(s) ratio*
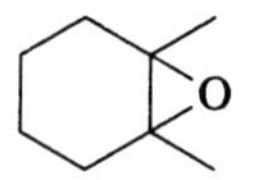	4	94	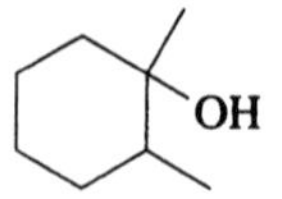
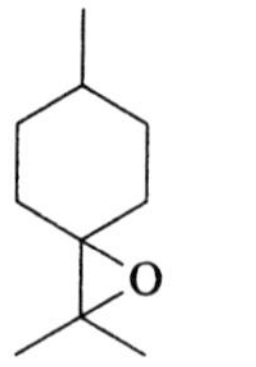	8	93	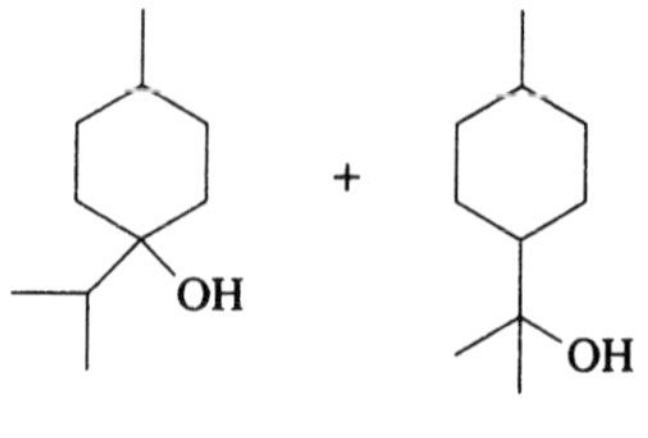 75:25
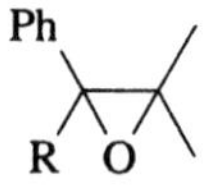 R = Me	4	97	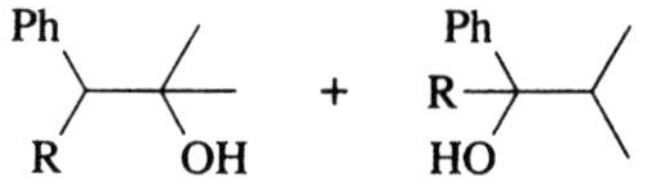 8:92
R = Ph	6	96	17:83

[a] Ref. 32.

Table 3 Reduction of Epoxides with $LiEt_3BH$[a]

Epoxide	*Time*	*Temperature* (°C)	*Yield* (%)	*Product*
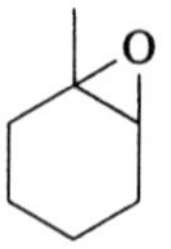	5 min	25	99	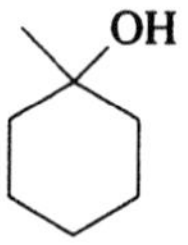
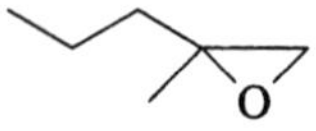	5 min	25	100	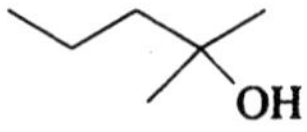
	3–6 h	65	100	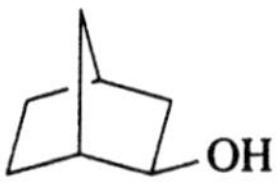
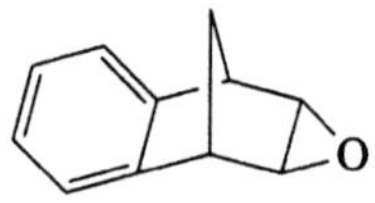	24 h	65	100	

[a] Ref. 34.

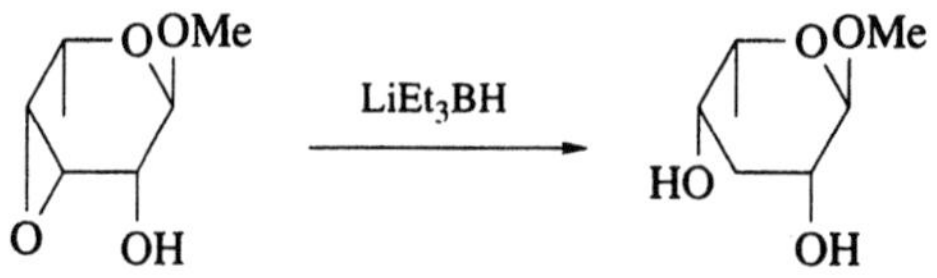

$LiEt_3BH$ (18)

conjugate reduction with diborane to give (*Z*)-allylic alcohols selectively (equation 20), and the selectivity of this reaction does not appear to depend on the substitution pattern of the epoxides.

The reverse selectivity has been described with diborane/boron trifluoride,[38] sodium borohydride/diborane,[42] sodium borohydride/trichlorotrispyridine–rhodium,[43] lithium 9,9-di-*n*-butyl-9-borabicyclo-[3.3.1]nonanate (for structure, see Table 4),[44] sodium cyanoborohydride/boron trifluoride,[45] and

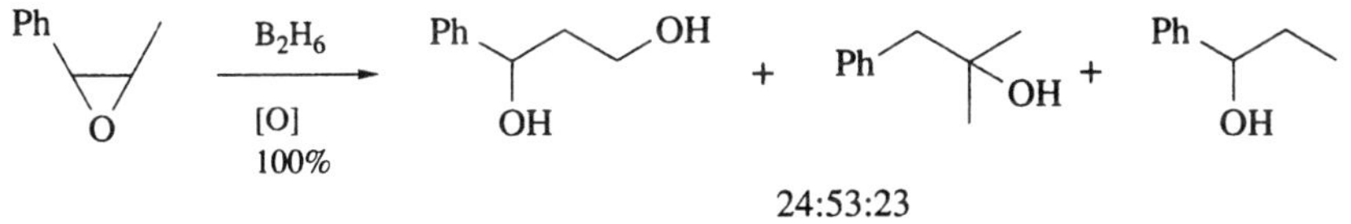

(19)

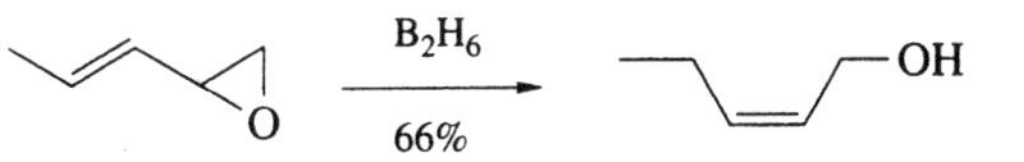

(20)

diborane/morpholine.[46] Representative results are summarized in Table 4. Although the regioselective ring opening of aryl-substituted epoxides leading to the less-substituted alcohol can be achieved by some reagents, only the sodium cyanoborohydride/boron trifluoride system is highly efficient for the conversion of monosubstituted epoxides to primary alcohols.

Table 4 Regiochemistry of Ring Opening of Epoxides with Borohydride Derivatives

Epoxide	*Reagent* [a]	*Yield (%)*	*Product ratio*	*Ref.*
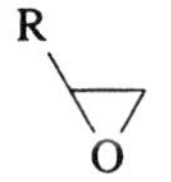			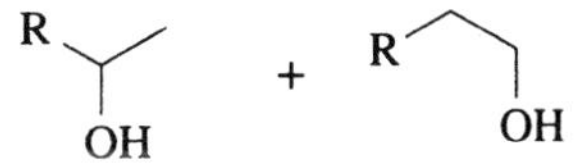	
R = Et	A	95	100:0	42
R = n-C_8H_{17}	B	100	93:7	44
R = $C_{10}H_{21}$	C	83	11:89	45
R = Ph	D	98	100:0	37
	B	100	1:99	44
	C	79	3:97	45
	E	100	100:0	46
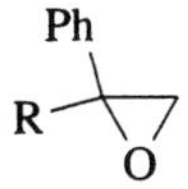			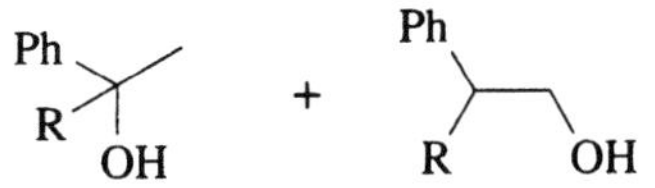	
R = Me	D	100	0:100	37
	B	86	2:98	44
	C	94	1:99	45
R = Ph	D	95	0:100	37
	B	94	17:83	44
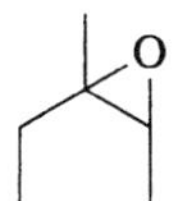			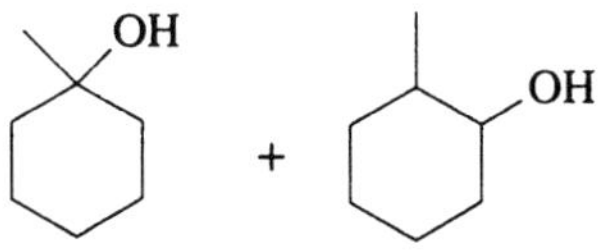	
	A	100	26:74	42
	C	87	3:97	45
	E	94	11:89	46

[a] A, $NaBH_4/B_2H_6$; B, lithium 9,9-di-*n*-butyl-9-nonabicyclo[3.3.1]nonane, see below; C, $NaBH_3CN/BF_3$; D, $NaBH_4/BF_3$; E, $NaBH_4$/morpholine, (BuNO)/BF_3.

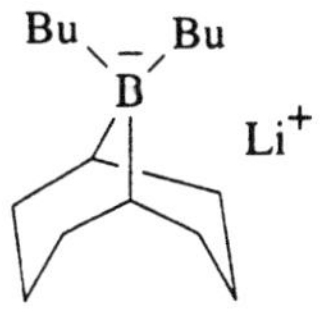

(B)

The reduction of carbohydrate epoxides has also been examined with the sodium borohydride/diborane system.[47] As shown in Table 5, the corresponding alcohols are obtained in good yields, although the regiochemistry of ring opening appears to be much affected by steric and polar factors. It should be noted that arylsulfonyloxy and ketal groups remain intact, whereas they are reduced by LAH.

Table 5 Reduction of Carbohydrate Epoxides with $NaBH_4/BH_3$[a]

Epoxide	*Yield (%)*	*Product ratio*
	80[b]	94:6
	40	
	90	

[a] Ref. 47. [b] p-$ClC_6H_4SO_2$, p-$MeC_6H_4SO_2$.

The system of sodium hydride, sodium alkoxide and metal salts developed by Caubère and his coworkers can selectively reduce epoxides to alcohols (Table 6).[48] The regiochemistry of ring opening is con-

Table 6 Reduction of Epoxides with $NaH/C_5H_{11}ONa/MX_n$[a]

Epoxide	*M in MX_n*	*Yield (%)*	*Product ratio*
	Zn	87	100:0
	Ni	95	10:90
$R = C_6H_{13}$, R' = H	Zn	99	100:0
	Ni	99	30:70
R = Ph, R' = H	Zn	96	100:0
	Ni	96	0:100
R = Ph, R' = Me	Zn	95	100:0
	Ni	93	0:100
	Zn	95	100:0
	Ni	95	35:65

[a] Ref. 48.

trolled using either the Zn or the Ni salt. Thus, with zinc chloride, styrene oxide is converted to α-phenethyl alcohol, whereas the use of nickel chloride results in the selective formation of β-phenethyl alcohol.

The selective conversion of epoxy alcohols to 1,3-diols has been studied using Red-Al (sodium bis(2-methoxyethoxy)aluminum hydride, $[(MeOCH_2CH_2O)_2AlH_2]Na$) because of the recent dramatic progress in the asymmetric synthesis of epoxy alcohols. Although Red-Al has been used as a reducing agent for epoxides since 1971,[49] attention had not been paid until three groups independently reported the reduction of epoxy alcohols to 1,3-diol derivatives in 1982 (Table 7).[50–52] According to Kishi and his coworker,[52] the selectivity of the reduction with Red-Al was improved by lowering the temperature to –20 °C. In some cases the use of THF as a solvent is also critical in enhancing the regioselectivity.[53] For phenyl-substituted epoxy alcohols, DME is superior to THF.[54] As an exceptional case a 1,2-dialkyl-1-hydroxymethyl-substituted epoxide undergoes ring opening at the carbon away from the hydroxy group to give a 1,2-diol (equation 21).

Table 7 Reduction of Epoxy Alcohols

Epoxide	*Reagent* [a]	*Yield (%)*	*Ratio of 1,3- and 1,2-diols*	*Ref.*
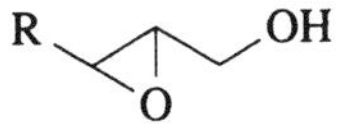 R, OH, O (epoxy alcohol)				
R = Bu^n	A	97	1:150	49
R = *n*-heptyl	B	90	100:1	53
	C	94	4:1	53
R = *n*-decyl	A	93	1:145	54
	B	95	100:1	53
	C	77	5:1	53
R = $BnOCH_2CH_2$	B		150:1	52
	D		1:13	52
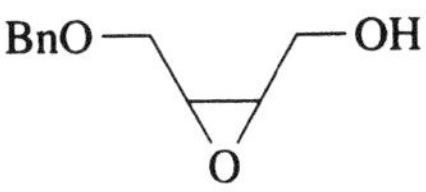 BnO, OH, O	C	95	100:1	51
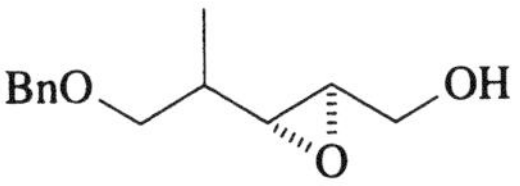 BnO, OH, O	C		>20:1	52
	D		1:8	52
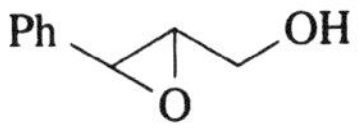 Ph, OH, O	A	76	1:54	53
 Ph, OH, O	C		22:1	54
OH, O	B	95	100:1	53
	C	96	1:6.8	55
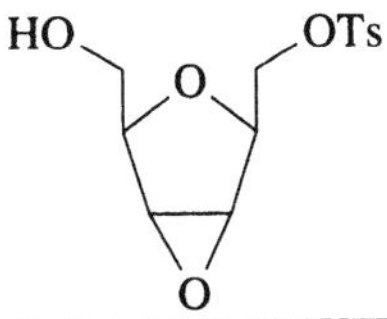 HO, OTs, O, O	B		12:1	50

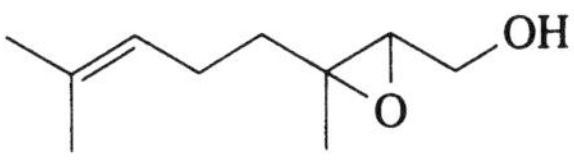

[a] A, $Ti(OPr^i)_4/LiBH_4$; B, Red Al; C, $LiAlH_4$; D, DIBAL-H.

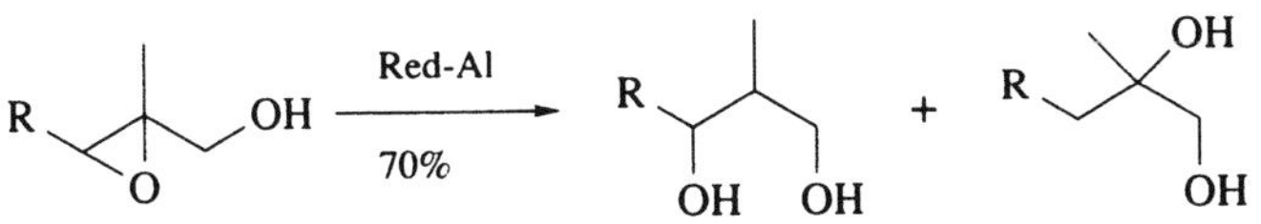

R, OH, O —(Red-Al, 70%)→ R, OH, OH + R, OH, OH (21)

R = C_7H_{15} 1:100

The reverse regiocontrol, giving 1,2-diols, is observed with DIBAL-H (diisobutylaluminum hydride).[51] The remarkable effect of titanium tetraisopropoxide as an additive to lithium borohydride has also been reported.[55] In this reaction benzene is a better solvent than THF, probably because a Ti complex using both oxygens in epoxy alcohols is formed in benzene before the hydride attack. Other metal hydrides used include sodium hydrogen telluride (NaHTe)[56] and an ate complex derived from DIBAL-H and butyllithium,[57] both of which reduce epoxides to alcohols, although they have been tested with only a small number of examples.[56] In the former case the reaction may proceed *via* a 2-hydroxyalkyltellurol intermediate.

4.4.2.2 Dissolving Metals

Lithium and sodium in amine solvents are well known as dissolving metals capable of reducing epoxides to alcohols, but they have not been used as frequently as metal hydrides. In contrast to LAH, lithium in ethylenediamine,[58] and sodium in liquid ammonia[59] reduce bicyclic epoxides within 2 h to give alcohols without rearrangement, as indicated in Table 8. However, the reaction conditions must be rigorously controlled, otherwise the reaction is accompanied by rearrangement, as observed for tricyclic epoxides.[60,61] The reduction of norbornene oxide using lithium and lithium chloride in THF followed by a hydrolytic work-up gives the *exo*-alcohol.[62] Functional groups such as acetyl,[63] carbonyl,[64,65] ethoxycarbonyl[59] and vinyl[63] remain intact under these conditions. The ring opening of epoxides predominantly takes place at the less-substituted carbon[57,58] to give the more-substituted alcohols, except for aryl-substituted epoxides.[58] The epoxy ring in an alkynic epoxide can be reduced selectively upon successive treatment with methyllithium and lithium in ammonia (equation 22).[66]

Table 8 Reduction of Epoxides with a Dissolving Metal[a]

Epoxide	*Metal*	*Yield (%)*	*Product*	*Ref.*
 (norbornene oxide)	Li Na Ca	86 82 99	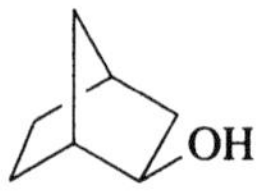 (exo-norbornanol, OH)	58 59 71
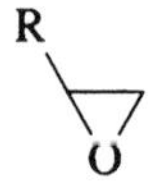 R-epoxide, R = Et, R = Bu	Li Ca	82 100	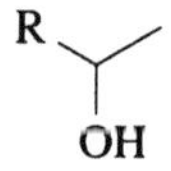 R, OH	58 71
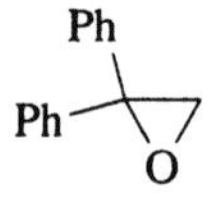 Ph, Ph, O	Na	89	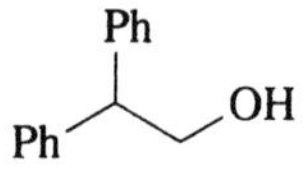 Ph, Ph, OH	59
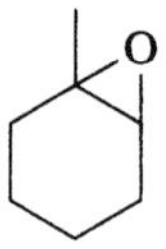 O	Li Ca	93 97	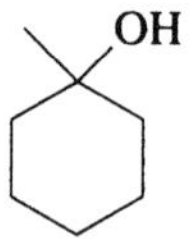 OH	58 71
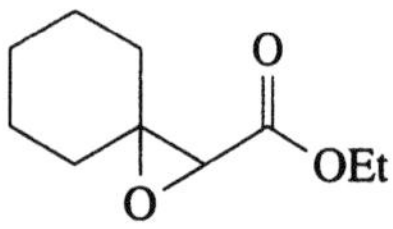 O, O, OEt	Na	46	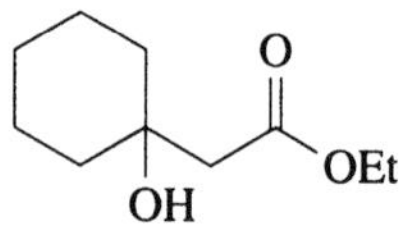 O, OEt, OH	59
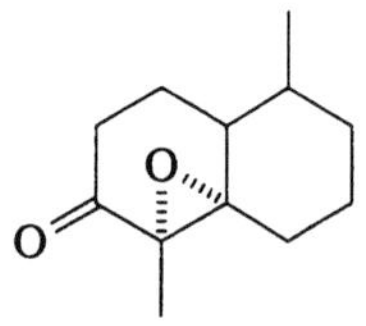 O, O	Li	53	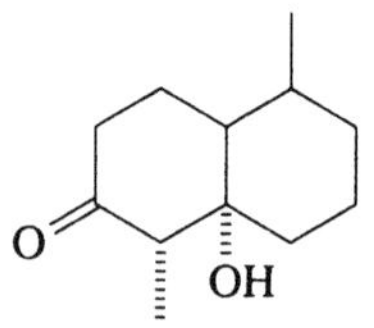 O, OH	64

[a] Ethylenediamine or liquid ammonia was used as a solvent.

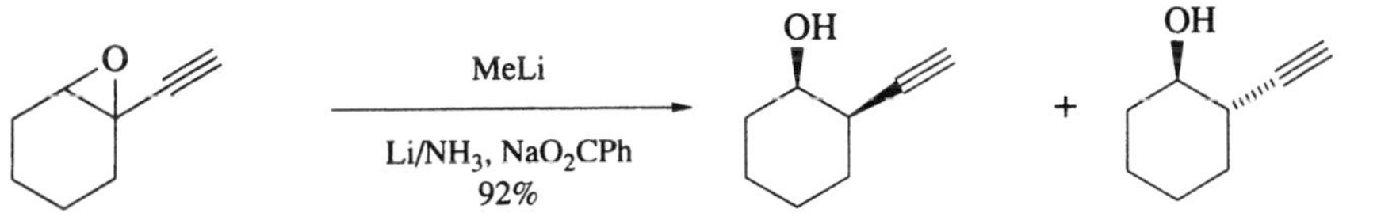

MeLi

Li/NH_3, NaO_2CPh
92%

70:30

(22)

As dissolving metals, aluminum,[67–70] calcium,[71] and zinc[21,72,73] have also been used. In the case of aluminum, water can be used as a solvent.[68,70] Calcium[71] can reduce epoxides in ethylenediamine in the presence of white sand to give similar results to those for lithium. This reagent is particularly suitable from the economical and safety point of view for large-scale reduction. Noteworthy is the regioselectivity using the zinc/trimethylchlorosilane system followed by a hydrolytic work-up,[21,73] as shown in Table 9. Alkyl-substituted epoxides[73] undergo ring opening predominantly at the more-substituted carbon to yield primary alcohols.

Table 9 Reduction of Epoxides with Zn/Me_3SiCl[a]

Epoxide	*Yield (%)*	*Product ratio*	*Ref.*
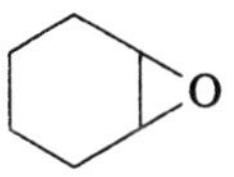	98	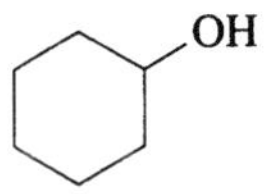	73
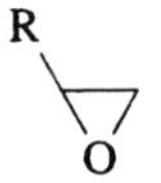 R = n-C_9H_{19}	96	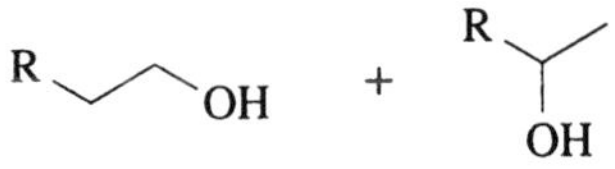 69:31	73
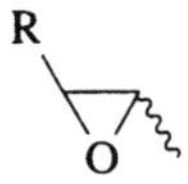 R = C_5H_{11}	97	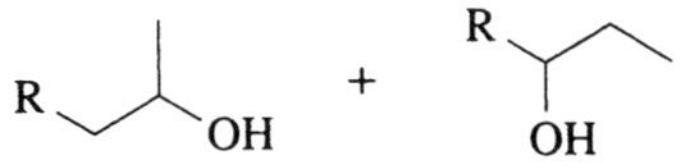 66:34	73
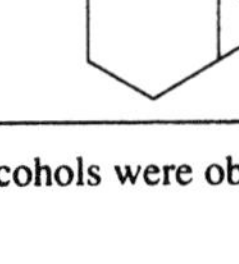	97	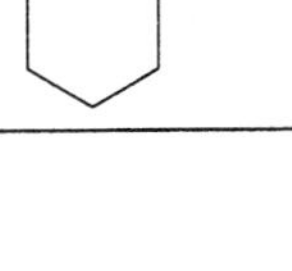	21
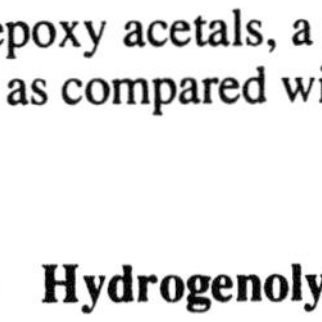	93	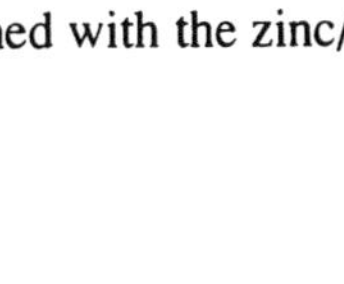	21

[a] The alcohols were obtained after the hydrolytic work-up.

For epoxy acetals, a complete reverse in regioselectivity is attained with the zinc/trimethylchlorosilane system as compared with LAH (equation 10).[21]

4.4.2.3 Hydrogenolysis

Hydrogenolysis of epoxides to yield alcohols has been much reported in the patent literature, because of its importance as an industrial process, but studies on reactivity and selectivity have not been done systematically. The selectivity is highly dependent on the substituents, as in the case of reduction using metal hydrides. As a metal catalyst, Raney Ni was intensively examined in the early stage.[74–83] It usually requires high pressures (*ca.* 100 atm) and temperatures (100 °C), as shown in Table 10. Alcohols, benzene, THF and even water have been used as solvents. Accordingly, a hydroxy group in the epoxides remains intact, and hydrocarbons are formed only as by-products. In some cases by-product formation can be suppressed by adding sodium hydroxide or triethylamine. Ring opening of epoxides takes place exclusively at the more-substituted carbon atoms. Other nickel catalysts include nickel borate[80] and nickel–phosphine,[81] which convert terminal epoxides into primary and secondary alcohols in a ratio of more than 85 to 15 (equation 23).

Table 10 Raney-Nickel Catalyzed Hydrogenolysis of Epoxides

Epoxide	*Conditions*	*Yield (%)*	*Product ratio*	*Ref.*
[structure]	80 °C, 120 atm	100	[structures] + + ; 16:81:3	72
Ph [structure]	25 °C, 1 atm	100	Ph [structure] OH + Ph [structure] OH + Ph [structure]; 30:11:59	73
	25 °C, 1 atm, NaOH	100	5:93:2	73
HO [structure] OH, O	150 °C, 200 atm	88	HO [structure] OH, OH	76
BzO [structure] O			BzO [structure] OH + BzO [structure]	
$R = C_8H_{17}$	24 atm	85	46:54	78
	24 atm, Et_3N	89	100:0	78

$$\text{(propylene oxide)} \xrightarrow{H_2} \text{1-propanol (OH)} \qquad (23)$$

Ni borate, 150 °C, 2 h 98%
Ni–P, 150–190 °C, 3 h 91%

Catalysts based on palladium, such as 10% palladium on charcoal,[84,85] have also been used frequently. In these cases ketones derived from isomerization of the epoxides are often obtained as by-products. As shown in Table 11, both product- and regio-selectivities are highly dependent on the configuration of the starting epoxides. Geneste *et al.*[84] suggested that the selectivity may be controlled at the stage of the approach of epoxides to catalyst.

It is noteworthy that the epoxy ring is selectively reduced to a hydroxy group in the nitrosugar epoxides, with regioselectivity controlled by the nitro group (equation 24).[86]

Palladium on barium sulfate has been used in the hydrogenolysis of epoxides in natural product synthesis.[87] The reaction in equation (25) illustrates its high level of functional group compatibility.

A platinum-based catalyst, such as Adams' catalyst, is an alternative to nickel and palladium, and is more active.[88] For example, when the reaction in equation (24) is carried out in the presence of a platinium catalyst, the nitro group is also reduced.[86] The hydrogenolysis of 2-butene oxide and propylene oxide proceeds at a slightly lower temperature than when using nickel,[84] and for propylene oxide the

Table 11 Pd/C-Catalyzed Hydrogenolysis of Epoxides[a]

Epoxide	*Yield (%)*	*Product ratio*
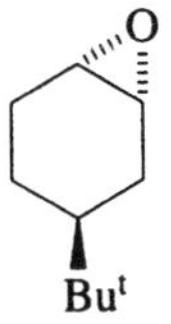	100	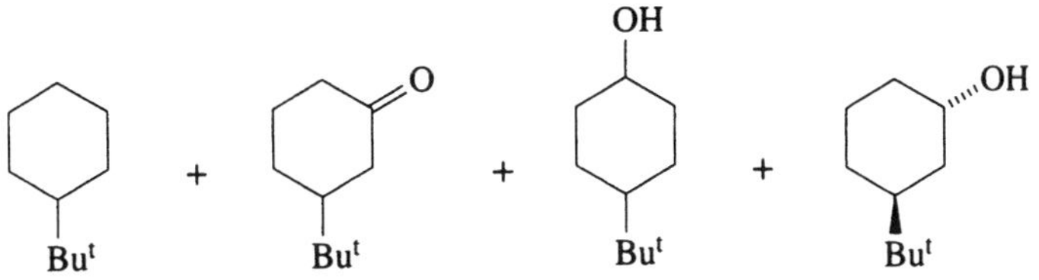 2:20:2.5:75.5
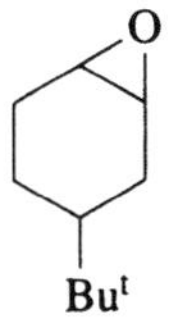	98	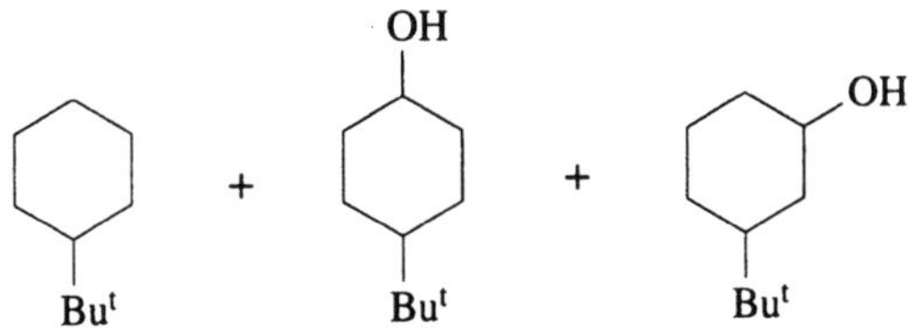 18:6:76
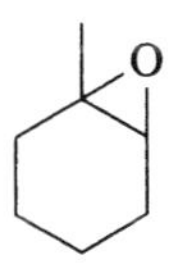	100	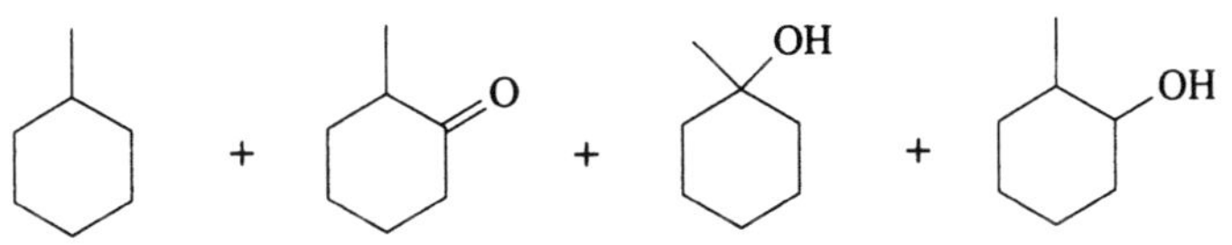 5:32.5:19:43.5

[a] Ref. 84.

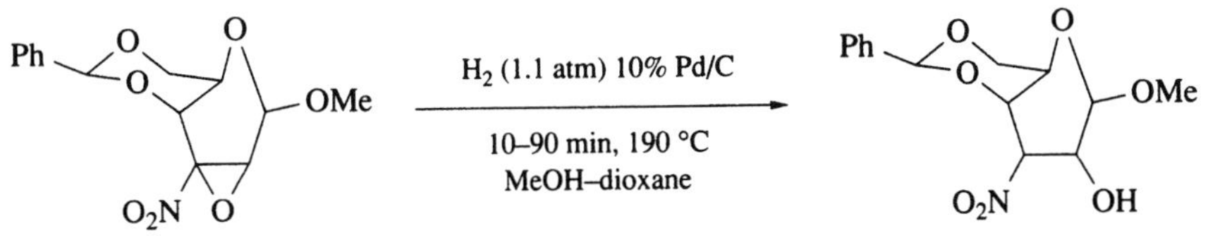

H_2 (1.1 atm) 10% Pd/C; 10–90 min, 190 °C; MeOH–dioxane (24)

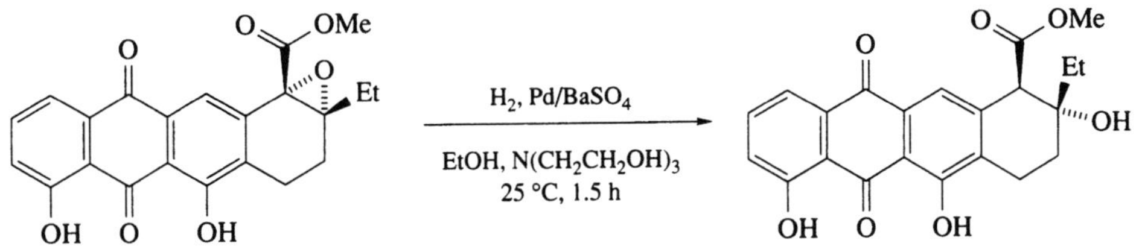

H_2, Pd/$BaSO_4$; EtOH, $N(CH_2CH_2OH)_3$; 25 °C, 1.5 h (25)

regioselectivity is reversed with platinium, giving 2-propanol predominantly.[89] Only a few studies on the mechanism of hydrogenolysis have been reported.[90]

4.4.2.4 Miscellaneous

Low-valent metal salts, such as chromium(II) acetate[91,92] and samarium iodide,[93–95] have been reported to be useful. With chromium(II) acetate, a steroid epoxy ketone is efficiently converted to a β-hydroxy ketone (equation 26).

Samarium iodide is a reagent capable of highly selective reduction of epoxy ketones and esters to the corresponding alcohols (Table 12). Diene monoepoxides are converted with high regio- and stereo-selectivity to the (*E*)-allylic alcohols, and the reaction can be carried out at –90 °C under neutral conditions. As a result functional groups such as alkoxycarbonyl, carbonyl and cyano survive.

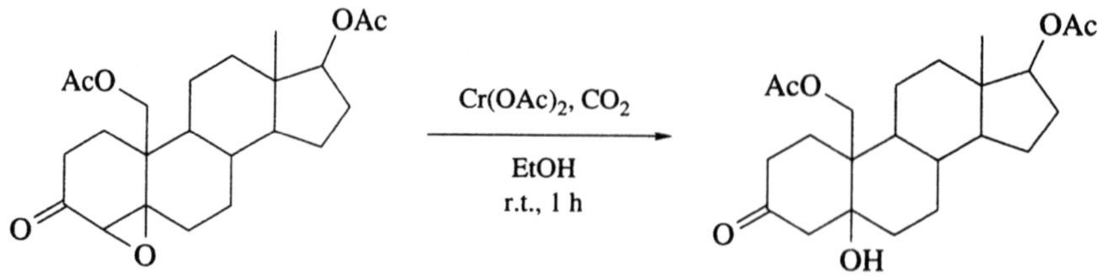

Cr(OAc)$_2$, CO$_2$ / EtOH / r.t., 1 h (26)

Table 12 Reduction of Epoxy Ketones, Esters and Vinyl Epoxides with SmI$_2$

Epoxide	Yield (%)	Product ratio	Ref.
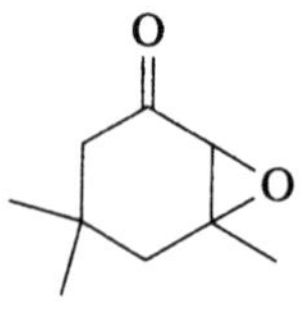	82	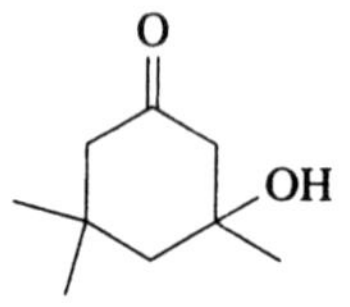	93
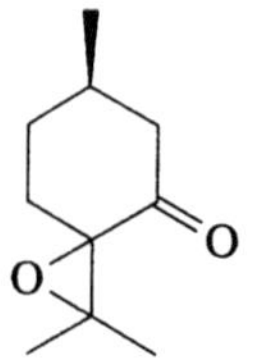	79	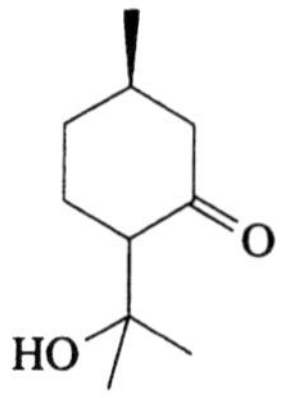	93
Y = CO$_2$Et, R = H	77		94
Y = CO$_2$Et, R = Me	75		
Y = CO$_2$Et	68	>200:1	95
Y = COSEt	80		94
Y = SO$_2$Ph	82		
Y = PO(OEt)$_2$	84		
Y = H	69		
Y = Me	42		
Y = SPh	54		

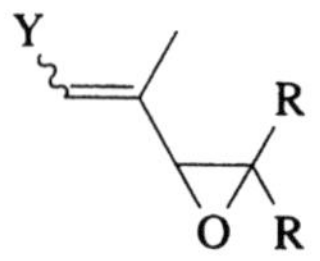

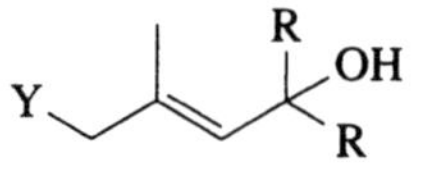

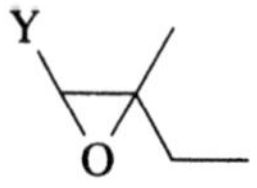

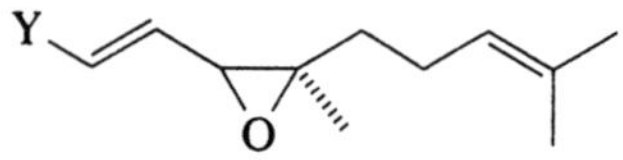

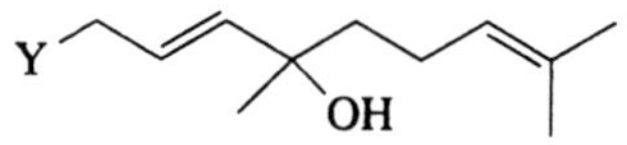

Although electrochemical reduction[96] and enzymic reduction[97] of epoxides to alcohols have been reported, little information is available about their characteristic features.

4.4.3 DEOXYGENATION OF EPOXIDES

Deoxygenation of epoxides to alkenes[1,98,99] is occasionally an important process in the synthesis of complex molecules, and a number of reagents for this transformation have been developed. Conventionally, multistep deoxygenation using hydrogen bromide and zinc in acetic acid has been used.[100,101] However, this system is not highly selective. As a result much effort has been devoted to improving the yield and the stereoselectivity of the reaction. In this section various kinds of reagents capable of one-step deoxygenation will be dealt with in the following order: (i) phosphorus reagents; (ii) silicon reagents; (iii) selenium and tellurium reagents; (iv) low-valent metals; (v) metal carbonyls; and (vi) miscellaneous. Two- or multi-step deoxygenation of epoxides to alkenes[98] will be mentioned only briefly in this section since the one-step process is generally superior. Deoxygenation giving a saturated carbon–carbon bond will not be included, because the process is synthetically less important.

4.4.3.1 Phosphorus Reagents

In 1955 Wittig *et al.* found that triphenylphosphine could induce the deoxygenation of epoxides at 200 °C.[102] Mechanistically, this process probably involves *anti* opening of the epoxide followed by *syn* elimination of triphenylphosphine oxide from a betaine intermediate. Accordingly, the reaction proceeds with inversion of stereochemistry, which means that *trans*-epoxides give *cis*-alkenes. The systems of bis(dimethylamino)phosphorous acid/butyllithium,[103] and lithium diphenylphosphide[104] have been examined

Table 13 Deoxygenation of Epoxides with Phosphorus Reagents

Epoxide	*Reagent* [a]	*Yield (%)*	*Product ratio*	*Ref.*
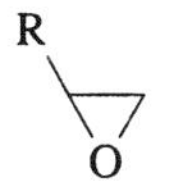 (monosubstituted epoxide R) → R–CH=CH₂			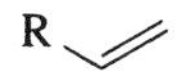	
R = n-$C_{10}H_{21}$	A	60		105
R = n-C_7H_{15}	B	95		106
R = n-$C_{18}H_{37}$	C	95		108
R = PhCH=CH	C	42		106
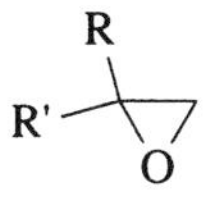 (1,1-disubstituted epoxide R, R') → RR'C=CH₂			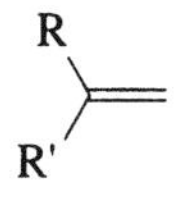	
R = Ph, R' = cyclopropyl	B	54		106
 (*trans*-epoxide R, R') → (*E*)-RCH=CHR' + (*Z*)-RCH=CHR'				
R = MeOCO(CH_2)$_7$, R' = n-C_8H_{17}	A	99	100:0	105
R = MeCO(CH_2)$_3$, R' = n-C_8H_{17}	C	94	>98:2	108
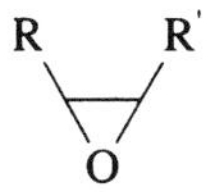 (*cis*-epoxide R, R') → (*E*)-RCH=CHR' + (*Z*)-RCH=CHR'				
R = MeOCO(CH_2)$_7$, R' = n-C_8H_{17}	A	95	0:100	105
R = HCO_2(CH_2)$_8$, R' = Bu^n	C	93	0:100	108
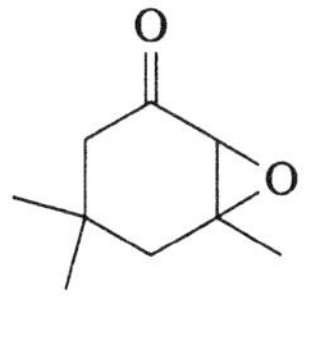 (isophorone oxide) → (3,5,5-trimethylcyclohex-2-enone)	B	95	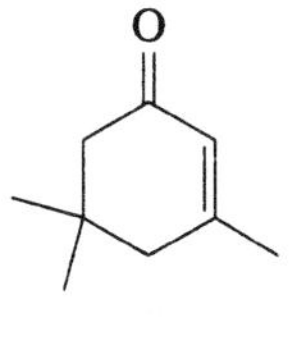	106
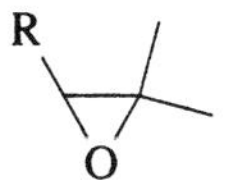 (trisubstituted epoxide R, Me, Me) → RCH=CMe₂			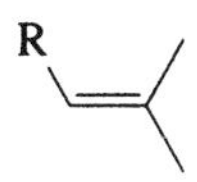	
R = $CH_2CH_2CH(Me)CH_2CH_2OMe$	A	99		105
R = $CH_2CH_2CH(Me)CH_2CH_2OCOH$	C	98		108

[a] A, $(PhO)_3P^+Me\ I^-$; B, P_2I_4; C, Ph_3PHI/Ph_3PI_2.

at lower temperature. The reaction with both systems proceeds with more than 96% inversion at room temperature, although the yield of the product with $Me_2NP(O)H$/butyllithium is not so high. With Ph_2PLi the alkenes were obtained quantitatively from the corresponding epoxides. However, the reaction is highly affected by the steric circumstance around the epoxides. Furthermore, because Ph_2PLi is a strong base, functional groups sensitive to base have to be protected.

For deoxygenation of epoxides with retention of stereochemistry (*trans*-epoxides giving *trans*-alkenes) methyltriphenoxyphosphonium iodide/boron trifluoride etherate,[106] diphosphorous tetraiodide,[106,107] and triphenylphosphine hydriodide/triphenylphosphine diiodide[108] have been introduced. The reaction with these systems generally proceeds below room temperature in high yield (Table 13). For terminal epoxides a slightly higher temperature (gentle reflux of dichloromethane) is needed. Because of the mild reaction conditions functional groups, such as alkoxycarbonyl and carbonyl groups, remain intact. In all cases the reaction begins with *anti* opening of the epoxides, followed by *anti* elimination, leading overall to retention of configuration. Both *trans*- and *cis*-alkenes have been made from a *trans*-epoxide by using either Ph_2PLi or Ph_3PI_2 (equation 27).[109] Trisubstituted steroidal epoxides are converted to the corresponding alkenes with the triphenylphosphine/iodine system.[110]

(27)

	Yield	trans : cis
P_2I_4	80%	>99:<1
Ph_3PI/MeI	77%	<1:>99

4.4.3.2 Silicon and Tin Reagents

Trialkylsilyl metals, such as trimethylsilylpotassium,[111] dimethylphenylsilyllithium,[112] and dimethylphenylsilyldiethylaluminum,[113] have been known to induce deoxygenation of epoxides. The reaction with TMSK and $PhMe_2SiLi$ proceeds with inversion of stereochemistry (Table 14) whereas $PhMe_2SiAlEt_2$ shows complete retention in the case of *trans*-stilbene oxide and cinnamyl alcohol epoxide.

Table 14 Deoxygenation of Epoxides with Silicon and Tin Reagents

Epoxide	*Reagent* [a]	*Yield (%)*	*Product ratio* (trans : cis)	*Ref.*
trans-epoxide				
R = R' = Pr^i	A	93	<1:99	111
R = R' = Ph	B	83	<1:99	112
R = R' = Bu^n	C	80	99:1	113
R = Ph, R' = CH_2OPh	C	88	99:1	113
cis-epoxide				
R = R' = Pr^i	A	75	>92:8	111
R = R' = Ph	B	83	>99:1	112
R = R' = Bu^n	C	67	1:99	113
R = n-C_6H_{13}, R' = CH_2OH	C	65	1:99	113

[a] A, Me_3SiK; B, $Me_2PhSiLi$; C, $(Bu^n{}_3SnAlMe_3)^-$ Li^+

It should be noted that the ate complex $(PhMe_2SiAlMe_3)Li$ shows selectivity opposite to that of $PhMe_2SiAlEt_2$, deoxygenation proceeding with inversion, as in equation (28). The reaction with a similar ate complex $(Bu^n{}_3SnAlMe_3)Li$, however, proceeds with retention of stereochemistry (Table 14). It is noteworthy that the hydroxy group in allylic alcohol epoxides does not affect the reaction using $(Bu^n{}_3SnAlMe_3)Li$, which probably involves a β-oxytin intermediate.[113]

$$\text{Ph-epoxide-CH}_2\text{OBn} + (\text{Me}_2\text{PhSiAlMe}_3)^-\text{Li}^+ \xrightarrow[72\%]{25\ °\text{C, 18 h}} (E)\text{-PhCH=CHCH}_2\text{OBn} + (Z)\text{-PhCH=CHCH}_2\text{OBn} \quad 0:100 \tag{28}$$

4.4.3.3 Selenium and Tellurium Reagents

Triphenylphosphine selenide,[114] potassium selenocyanate,[115] 3-methyl-2-selenoxobenzothiazole (**1**),[116] sodium *O,O*-diethyl phosphorotelluronate,[117] *O,O*-dialkylphosphoroselenoic acid salts[118] and benzeneselenocarboxamide,[119] have all been reported as reagents capable of one-step deoxygenation of epoxides.

(1)

The reaction with (**1**) and $(EtO)_2P(O)TeNa$ proceeds at below room temperatures under acidic conditions, whereas the reaction with Ph_3PSe is carried out using a slightly alkaline solution. For cycloalkene oxides, except for cyclohexene oxide, the reaction is sluggish. In all cases the stereochemistry of the product is retained. The reactions proceed *via* episelenium or epitellurium intermediates followed by extrusion of selenium or tellurium, although the reaction pathways to these intermediates are not the same. For example, the reaction with KSeCN may form a cyclic intermediate after nucleophilic *anti* opening of the epoxide (equation 29), followed by *anti* attack of selenium, forming an episelenium intermediate (equation 30).

(29)

(30)

With sodium *O,O*-dialkyl phosphorotelluronate, terminal epoxides are converted to alkenes very quickly, and *cis*-epoxides are reduced more easily than *trans*-epoxides (equation 31). The corresponding selenoate reagent has been applied to sugar epoxide deoxygenation (equation 32).

R = Pr^n; $(EtO)_2P(O)TeLi$, 58.6%; products 80:9:11

(31)

β-Hydroxy selenides prepared by the reaction of epoxides with selenophenol and methanol are converted to alkenes upon treatment with thionyl chloride/triethylamine in methylene chloride (equations 33 and 34).[120] The overall transformation is formally a deoxygenation of an epoxide. This method is particularly suitable for the formation of tri- and tetra-substituted alkenes.

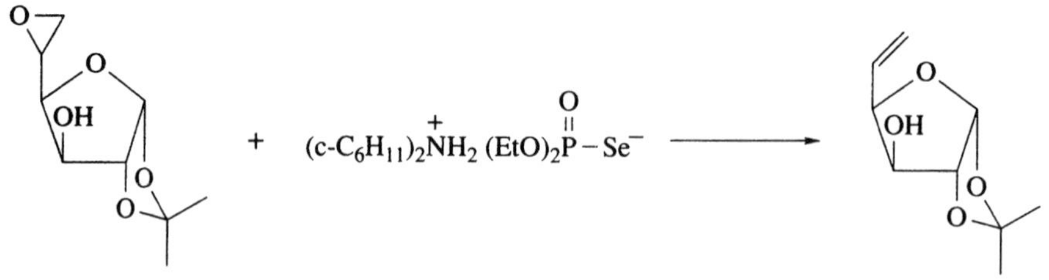

(32)

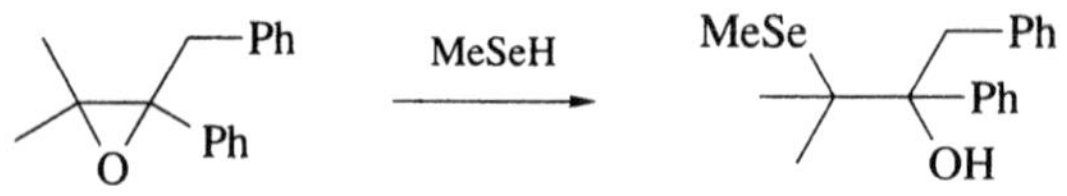

(33)

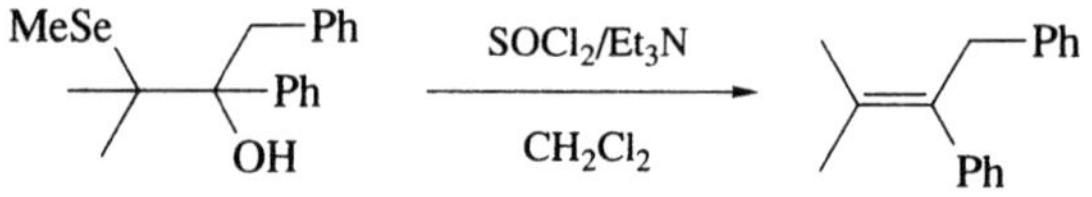

(34)

4.4.3.4 Low-valent Metals

The study of lower valent tungsten halides by Sharpless *et al.* in 1972[121] was the first systematic study of low-valent metals as deoxygenating reagents, although some studies on low-valent chromium[122] and the zinc–copper couple[123] had appeared in the earlier literature. The tungsten reagents were prepared from tungsten hexachloride in three ways: (i) with butyllithium; (ii) with a lithium dispersion; and (iii) with lithium iodide (equations 35–37). In each case the ratio of WCl_6 to lithium reagent is important.

$$WCl_6 + Bu^nLi \xrightarrow{\text{THF}} WCl_6/Bu^nLi \quad (35)$$

$$WCl_6 + \text{Li dispersion} \xrightarrow{\text{THF}} WCl_6/Li \quad (36)$$

$$WCl_6 + LiI \xrightarrow[\text{in vacuo, no solvent}]{130\ ^\circ\text{C}} WCl_6/LiI \quad (37)$$

For isomeric cyclododecene oxides[124] all these tungsten reagents show high reactivity and selectivity (greater than 95% yield, with more than 95% retention), but only the reagent prepared by reaction of WCl_6 with three equimolar amounts of lithium iodide could convert *cis*-4-octene oxide to the *cis*-alkene with high stereoselectivity (equation 38). The reagent of equation (35) has been used for deoxygenation in natural product synthesis (equation 39).[125]

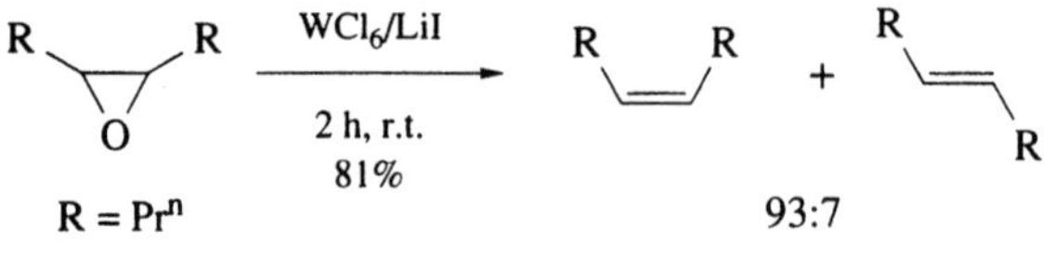

(38)

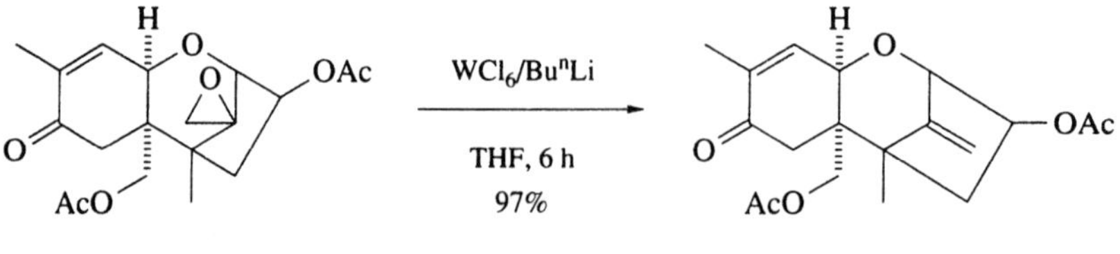

(39)

Iron trichloride/tungsten hexachloride,[126] titanium trichloride/LAH[127] and WCl_6/LAH[128] have also been examined, and show similar results to the Sharpless systems mentioned above. In these cases the stereochemical course is not determined by the kind of reagent; instead, *trans*-alkenes are predominantly formed from both *cis*- and *trans*-epoxides (equations 40 and 41).

(40)

$FeCl_3/Bu^nLi$	81%	94:6
$WCl_6/LiAlH_4$	84%	99:1

(41)

WCl_6/Bu^nLi	86%	89:11
$WCl_6/LiAlH_4$	84%	96:4

Metallocene dichlorides catalyze the deoxygenation of epoxides with sodium amalgam,[129] but the stereoselectivity is not high. However, when magnesium is employed as a reductant in combination with titanocene dichloride,[130] even trisubstituted epoxides are efficiently converted to alkenes (equation 42). With this reagent, *trans*-epoxides are deoxygenated to *trans*-alkenes with 100% selectivity, whereas *cis*-epoxides give *cis*-alkenes with *ca.* 80% selectivity. Furthermore, functional groups such as carbonyl groups, alkenes, ethers and acetals remain intact under these conditions.

$CpTiCl_2$ + Mg + (epoxide) —94%→ (alkene) (42)

Epoxides undergo deoxygenation upon treatment with vanadium or molybdenum acetylacetonate complexes, although their stereoselectivity is low.[131] When the molybdenum(VI) oxo complex $MoO(S_2CNEt_2)_2$ is employed, the deoxygenation proceeds with retention of configuration. For example, *cis*-2-butene oxide is converted to *cis*-2-butene at 130 °C in 83% yield.[132]

An alkylmanganese complex[133] has been examined; reduction of aryl-substituted epoxides and cycloalkene oxides gives the corresponding alkenes in good yield with retention of stereochemistry (equation 43).

+ Bu_3MnLi —(−30 °C, 1 h; 60 °C, 2 h; 59%)→ (43)

10:90

Samarium iodide has also been used for reducing epoxides to alkenes. Compared with other low-valent metal systems this reagent requires prolonged reaction time at room temperature, but the stereoselectivity, with *cis*-epoxides giving *cis*-alkenes, is high.[134]

Metal atoms such as lithium,[135] and first row transition metal atoms,[136] have been studied for deoxygenation of epoxides. Limonene oxide is converted in high yield to the corresponding alkene upon treatment with lithium and biphenyl in DME (equation 44).[137]

—(Li; biphenyl, DME, 153 °C, 4 h)→ (44)

4.4.3.5 Metal Carbonyls

Sodium (cyclopentadienyl)dicarbonylferrate,[138] dicobaltoctacarbonyl,[139] and iron pentacarbonyl[140] have been reported to deoxygenate epoxides efficiently to the corresponding alkenes. The reaction with (cyclopentadienyl)dicarbonylferrate proceeds with inversion of stereochemistry, whereas $Fe(CO)_5$ shows low stereoselectivity. Both $Co_2(CO)_8$ and $Fe(CO)_5$ are applicable to epoxides having carbonyl groups (equations 45 and 46).

$Co_2(CO)_8$, r.t., 18 h, 100%; R = CO_2Me; 99:1 (45)

$Fe(CO)_5$, tetramethylurea, 2 h, 82%; AcO; AcO (46)

The reaction with (cyclopentadienyl)dicarbonylferrate can be directed to give either *trans*- or *cis*-alkenes simply by changing the reaction conditions. Thus, reaction with *cis*-stilbene oxide followed by treatment with sodium iodide in acetone gives *cis*-stilbene, whereas only *trans*-stilbene is obtained when the reaction is prepared at room temperature and is heated under reflux in THF (equation 47). Although deoxygenation of terminal epoxides proceeds very rapidly, internal epoxides require longer times. Consequently, as in equation (48), selective reduction of diepoxides can be carried out.

Ph, Ph; $NaFeCp(CO)_2$; Ph, Ph + Ph, Ph (47)

NaI, acetone; 82% — 100:0
THF, 67 °C; 96% — 0:100

i, $NaFeCp(CO)_2$; ii, NaI, acetone (48)

4.4.3.6 Miscellaneous

Vapor-deposited carbon atoms have been known to reduce epoxides to alkenes but with low stereoselectivity,[141] whereas chemically generated carbon atoms, by thermal decomposition of 5-tetrazolediazonium chloride (**2**), deoxygenate *cis*- and *trans*-2-butene oxide with a high degree of retention of stereochemistry.[142]

N_2^+; N, N–H, N=N; Cl^-

(**2**)

Carbenes or carbenoids have also been reported to deoxygenate epoxides. To generate these species, dimethyl diazomalonate/rhodium(II) acetate[143] and 9-diazofluorene[144] have been employed; both reagents show high stereoselectivity. For example, *cis*-2-butene oxide is converted to *cis*-alkene with more than 93% retention under irradiation (350 nm) using 9-diazofluorene, and functional groups, such as carbonyl, can survive using diazomalonate and rhodium(II) acetate (equation 49).

Trifluoroacetyl iodide, generated *in situ* from trifluoroacetic anhydride and sodium iodide, deoxygenates epoxides.[145] The reaction begins by *anti* opening of the epoxide, followed by the formation of an

$$\text{cyclohexenone oxide} + N_2{=}C(R)_2 \xrightarrow[C_6H_6,\ 80\ ^\circ C]{Rh_2(OAc)_4} \text{cyclohexenone} \quad (49)$$

$R = CO_2Me$

iodonium ion intermediate and subsequent deiodonation to give alkenes. Because of the mild reaction conditions, this reagent has been used for the reduction of compounds having various functional groups (equation 50).[146]

$$\xrightarrow[15\ \text{min, r.t.}\ 60\%]{(CF_3CO)_2O,\ NaI,\ THF/MeCN} \quad (50)$$

Deoxygenation *via* iodohydrin intermediates using chlorosilane/sodium iodide[147] and *p*-toluenesulfonic acid/sodium iodide[148] has also been reported. In a natural product synthesis,[149] an epoxide has been deoxygenated with a large excess of triethylsilane at 300 °C for 30 h (equation 51).

$$\xrightarrow[300\ ^\circ C,\ 36\ h]{HSiEt_3} \quad (51)$$

4.4.4 REFERENCES

1. J. G. Smith, *Synthesis*, 1984, 629.
2. (a) W. G. Brown, *Org. React. (N.Y.)*, 1951, **6**, 469; (b) R. E. Parker and N. S. Isaacs, *Chem. Rev.*, 1959, **59**, 737.
3. J. S. Pizey, in 'Synthetic Reagents', Horwood, Chichester, 1974, vol. 1, p. 115.
4. E. L. Eliel and M. N. Rerick, *J. Am. Chem. Soc.*, 1960, **82**, 1362.
5. M. L. Mihailović, V. Andrejević, J. Milovanović and J. Janković, *Helv. Chim. Acta*, 1976, **59**, 2305.
6. E. L. Eliel and D. W. Delmonte, *J. Am. Chem. Soc.*, 1956, **78**, 3226.
7. R. Guyon and P. Villa, *Bull. Soc. Chim. Fr.*, 1975, 2584 (*Chem. Abstr.*, 1976, **85**, 93 482b).
8. R. S. Lenox and J. A. Katzenellenbogen, *J. Am. Chem. Soc.*, 1973, **95**, 957.
9. E. J. Parish and G. J. Schroepfer, Jr., *Chem. Phys. Lipids*, 1977, **18**, 258.
10. B. Cooke, E. C. Ashby and J. Lott, *J. Org. Chem.*, 1968, **33**, 1132.
11. B. Rickborn and J. Quartucci, *J. Org. Chem.*, 1964, **29**, 3185.
12. N. A. LeBel and G. G. Ecke, *J. Org. Chem.*, 1965, **30**, 4316.
13. B. Rickborn and W. E. Lamke, II, *J. Org. Chem.*, 1967, **32**, 537.
14. D. K. Murphy, R. L. Alumbaugh and B. Rickborn, *J. Am. Chem. Soc.*, 1969, **91**, 2649.
15. B. C. Hartman and B. Rickborn, *J. Org. Chem.*, 1972, **37**, 4246.
16. D. G. Loughhead, *J. Org. Chem.*, 1985, **50**, 3931.
17. M. L. Mihailović, Lj. Lorenć, V. Pavlović and J. Kalvoda, *Tetrahedron*, 1977, **33**, 441.
18. J.-C. Richer, M.-A. Poirier, Y. Maroni and G. Manuel, *Can. J. Chem.*, 1978, **56**, 2049.
19. H. Beisswenger and M. Hanack, *Tetrahedron Lett.*, 1982, **23**, 403.
20. J.-C. Richer, M.-A. Poirier, Y. Maroni and G. Manuel, *Can. J. Chem.*, 1980, **58**, 39.
21. Y. D. Vankar, N. C. Chaudhuri and C. T. Rao, *Tetrahedron Lett.*, 1987, **28**, 551.
22. T. Suami, S. Ogawa and Y. Funaki, *Bull. Chem. Soc. Jpn.*, 1975, **48**, 1545.
23. H. B. J. Chattopadhyaya, *Synthesis*, 1985, 1108.
24. R. Alonso, L. Castedo and D. Dominguez, *Tetrahedron Lett.*, 1986, **27**, 3539.
25. K. Soai, A. Ookawa, H. Oyamada and M. Takase, *Heterocycles*, 1982, **19**, 1371.
26. K. Soai, *Yuki Gosei Kagaku Kyokai Shi*, 1987, **45**, 1148.
27. H. H. Baer and C. B. Madumelu, *Carbohydr. Res.*, 1975, **39**, C8.
28. A. Ookawa, H. Hiratsuka and K. Soai, *Bull. Chem. Soc. Jpn.*, 1987, **60**, 1813.
29. G. D. Maio, E. Vecchi and E. Zeuli, *Gazz. Chim. Ital.*, 1983, **113**, 823.
30. K. Soai, A. Ookawa and H. Hayashi, *J. Chem. Soc., Chem. Commun.*, 1983, 668.

31. K. N. Gurudutt and B. Ravindranath, *Synthesis*, 1983, 888.
32. G. K. Nagarajarao and R. B. Hula, *Indian IN* 158 096 (1986) (*Chem. Abstr.*, 1987, **107**, 7417c).
33. V. Andrejević, M. Bjelaković, M. M. Mihailović and M. L. Mihailović, *Helv. Chim. Acta*, 1985, **68**, 2030.
34. S. Krishnamurthy, R. M. Schubert and H. C. Brown, *J. Am. Chem. Soc.*, 1973, **95**, 8486.
35. H. H. Baer and D. J. Astles, *Carbohydr. Res.*, 1984, **126**, 343.
36. H. C. Brown, S. Narasimhan and V. Somayaji, *J. Org. Chem.*, 1983, **48**, 3091.
37. C. F. Lane, in 'Synthetic Reagents', ed. J. S. Pizey, Horwood, Chichester, 1977, vol. 3, p. 25.
38. H. C. Brown and N. M. Yoon, *Chem. Commun.*, 1968, 1549.
39. H. C. Brown, P. Heim and N. M. Yoon, *J. Am. Chem. Soc.*, 1970, **92**, 1637.
40. P. A. Marshall and R. H. Prager, *Aust. J. Chem.*, 1977, **30**, 141, 151.
41. M. Zaidlewicz, A. Uzarewicz and R. Sarnowski, *Synthesis*, 1979, 62.
42. H. C. Brown and N. M. Yoon, *J. Am. Chem. Soc.*, 1968, **90**, 2686.
43. C. J. Love and F. J. McQuillin, *J. Chem. Soc., Perkin Trans. 1*, 1973, 2509.
44. Y. Yamamoto, H. Toi, A. Sonoda and S. Murahashi, *J. Chem. Soc., Chem. Commun.*, 1976, 672.
45. R. O. Hutchins, I. M. Taffer and W. Burgoyne, *J. Org. Chem.*, 1981, **46**, 5214.
46. W. B. Smith, *J. Org. Chem.*, 1984, **49**, 3219.
47. Y.-L. Fu and M. Bobek, *J. Org. Chem.*, 1980, **45**, 3836.
48. Y. Fort, R. Vanderesse and P. Caubère, *Tetrahedron Lett.*, 1985, **26**, 3111.
49. T. K. Jones and J. H. J. Peet, *Chem. Ind. (London)*, 1971, 995.
50. A. M. Mubarak and D. M. Brown, *J. Chem. Soc., Perkin Trans. 1*, 1982, 809.
51. P. Ma, V. S. Martin, S. Masamune, K. B. Sharpless and S. M. Viti, *J. Org. Chem.*, 1982, **47**, 1378.
52. J. M. Finan and Y. Kishi, *Tetrahedron Lett.*, 1982, **23**, 2719.
53. S. M. Viti, *Tetrahedron Lett.*, 1982, **23**, 4541.
54. Y. Gao and K. B. Sharpless, *J. Org. Chem.*, 1988, **53**, 4081.
55. L. Dai, B. Lou, Y. Zhang and G. Guo, *Tetrahedron Lett.*, 1986, **27**, 4343.
56. D. H. R. Barton, A. Fekih and X. Lusinchi, *Tetrahedron Lett.*, 1985, **26**, 6197.
57. S. Kim and K. H. Ahn, *J. Org. Chem.*, 1984, **49**, 1717.
58. H. C. Brown, S. Ikegami and J. H. Kawakami, *J. Org. Chem.*, 1970, **35**, 3243.
59. E. M. Kaiser, C. G. Edmonds, S. D. Grubb, J. W. Smith and D. Tramp, *J. Org. Chem.*, 1971, **36**, 330.
60. L. A. Paquette and K. H. Fuhr, *J. Am. Chem. Soc.*, 1972, **94**, 9221.
61. L. A. Paquette, K. H. Fuhr, S. Porter and J. Clardy, *J. Org. Chem.*, 1974, **39**, 467.
62. I. Rothberg, L. Schneider, S. Kirsch and R. O'Fee, *J. Org. Chem.*, 1982, **47**, 2675.
63. P. Ceccherelli, M. Curini, R. Pellicciari, R. Coccia and E. Wenkert, *J. Org. Chem.*, 1982, **47**, 3587.
64. W. A. Ayer, L. M. Browne and S. Fung, *Can. J. Chem.*, 1976, **54**, 3276.
65. P. Sengupta, K. Das and S. Das, *Indian J. Chem., Sect. B*, 1985, **24**, 1175.
66. R. G. Carlson and W. W. Cox, *J. Org. Chem.*, 1977, **42**, 2382.
67. A. M. M. Hossain, D. N. Kirk and G. Mitra, *Steroids*, 1976, **27**, 603.
68. E. J. Corey, E. J. Trybulski, L. S. Melvin, Jr., K. C. Nicolaou, J. A. Secrist, R. Lett, P. W. Sheldrake, J. R. Falck, D. J. Brunelle, M. F. Haslanger, S. Kim and S. Yoo, *J. Am. Chem. Soc.*, 1978, **100**, 4618.
69. A. E. Greene, M. A. Teixeira, E. Barreiro, A. Cruz and P. Crabbé, *J. Org. Chem.*, 1982, **47**, 2553.
70. M. Fétizon, I. Hanna and R. Zeghdoudi, *Synth. Commun.*, 1986, **16**, 1.
71. R. A. Benkeser, A. Rappa and L. A. Wolsieffer, *J. Org. Chem.*, 1986, **51**, 3391.
72. D. Bittler, H. Hofmeister, H. Laurent, K. Nickisch, R. Nickolson, K. Petzoldt and R. Wiechert, *Angew. Chem., Int. Ed. Engl.*, 1982, **21**, 696.
73. Y. D. Vankar, P. S. Arya and C. T. Rao, *Synth. Commun.*, 1983, **13**, 869.
74. A. Suzuki, M. Miki and M. Itoh, *Tetrahedron*, 1967, **23**, 3621.
75. A. Kergomard and H. Veschambre, *C. R. Hebd. Seances Acad. Sci., Ser. C*, 1974, **279**, 155 (*Chem. Abstr.*, 1974, **81**, 136 305h).
76. S. Mitsui, Y. Sugi, M. Fujimoto and K. Yokoo, *Tetrahedron*, 1974, **30**, 31.
77. Y. Nagahisa, Y. Sugi and S. Mitsui, *Chem. Ind. (London)*, 1975, 38.
78. S. Suzuki, *US Pat.* 3 975 449 (1976) (*Chem. Abstr.*, 1977, **86**, 54 967q).
79. H. Mueller, K. H. Ross and W. Schroeder, *Ger. Pat.* 2 643 400 (1978) (*Chem. Abstr.*, 1978, **89**, 42 410b).
80. A. Nagakura, M. Moroe, H. Tsuruta and T. Yoshida, *Jpn. Pat.* 78 92 750 (1978) (*Chem. Abstr.*, 1979, **90**, 5991s).
81. E. J. Parish and G. J. Schroepfer, Jr., *Chem. Phys. Lipids*, 1980, **26**, 141.
82. M. Amagasa and T. Aoki, *Jpn Pat.* 73 31 083 (1973) (*Chem. Abstr.*, 1974, **80**, 26 745y).
83. K. Isogai, T. Ogino, T. Hiiro, K. Endo and N. Yokokawa, *Yuki Gosei Kagaku Kyokai Shi*, 1976, **34**, 492 (*Chem. Abstr.*, 1977, **86**, 4881m).
84. G. C. Accrombessi, P. Geneste and J.-L. Olivé, *J. Org. Chem.*, 1980, **45**, 4139.
85. G. C. Accrombessi, P. Geneste, J.-L. Olivé and A. A. Pavia, *Tetrahedron*, 1981, **37**, 3135.
86. H. H. Baer, C. B. Madumelu, Z. S. Hanna and P. G. Potvin, *Carbohydr. Res.*, 1979, **76**, 141.
87. P. N. Confalone and G. Pizzolato, *J. Am. Chem. Soc.*, 1981, **103**, 4251.
88. H. Velgová and A. Trka, *Collect. Czech. Chem. Commun.*, 1983, **48**, 937.
89. M. Bartók and F. Notheisz, *J. Chem. Soc., Chem. Commun.*, 1980, 667.
90. F. Notheisz, A. G. Zsigmond, M. Bartók and G. V. Smith, *J. Chem. Soc., Faraday Trans. 1*, 1987, **83**, 2359.
91. C. H. Robinson and R. Henderson, *J. Org. Chem.*, 1972, **37**, 565.
92. M. Tohoma, R. Mahara, H. Takeshita, T. Kurosawa, S. Ikegawa and H. Nittono, *Chem. Pharm. Bull.*, 1985, **33**, 3071.
93. G. A. Molander and G. Hahn, *J. Org. Chem.*, 1986, **51**, 2596.
94. G. A. Molander, B. E. La Belle and G. Hahn, *J. Org. Chem.*, 1986, **51**, 5259.
95. K. Otsubo, J. Inanaga and M. Yamaguchi, *Tetrahedron Lett.*, 1987, **28**, 4437.
96. K. Bouglel and J. Simonet, *Electrochim. Acta*, 1979, **24**, 481.
97. S. R. Steckbeck, J. A. Nelson and T. A. Spencer, *J. Am. Chem. Soc.*, 1982, **104**, 893.

98. H. N. C. Wong, C. C. M. Fok and T. Wong, *Heterocycles*, 1987, **26**, 1345.
99. P. E. Sonnet, *Tetrahedron*, 1980, **36**, 557.
100. D. R. James, R. W. Rees and C. W. Shoppee, *J. Chem. Soc.*, 1955, 1370.
101. R. R. King and R. Greenhalgh, *Can. J. Chem.*, 1985, **63**, 1089.
102. G. Wittig and W. Haag, *Chem. Ber.*, 1955, **88**, 1654.
103. E. J. Corey and D. E. Cane, *J. Org. Chem.*, 1969, **34**, 3053.
104. (a) E. Vedejs and P. L. Fuchs, *J. Am. Chem. Soc.*, 1971, **93**, 4070; (b) E. Vedejs and P. L. Fuchs, *J. Am. Chem. Soc.*, 1973, **95**, 822; (c) E. Vedejs, K. A. J. Snoble and D. L. Fuchs, *J. Org. Chem.*, 1973, **38**, 1178.
105. K. Yamada, S. Goto, H. Nagase, Y. Kyotani and Y. Hirata, *J. Org. Chem.*, 1978, **43**, 2076.
106. H. Suzuki, T. Fuchita, A. Iwasa and T. Mishina, *Synthesis*, 1978, 905.
107. J. N. Denis, R. Magnane, M. V. Eenoo and A. Krief, *Nouv. J. Chim.*, 1979, **3**, 705.
108. P. E. Sonnet, *Synthesis*, 1980, 828.
109. A. G. Martínez and M. O. Ruiz, *Synthesis*, 1983, 663.
110. Z. Paryzek and R. Wydra, *Tetrahedron Lett.*, 1984, **25**, 2601.
111. P. B. Dervan and M. A. Shippey, *J. Am. Chem. Soc.*, 1976, **98**, 1265.
112. M. T. Reetz and M. Plachky, *Synthesis*, 1976, 199.
113. S. Matsubara, T. Nonaka, Y. Okuda, S. Kanemoto, K. Oshima and H. Nozaki, *Bull. Chem. Soc. Jpn.*, 1985, **58**, 1480.
114. D. L. J. Clive and C. V. Denyer, *J. Chem. Soc., Chem. Commun.*, 1973, 253.
115. J. M. Behan, R. A. W. Johnstone and M. J. Wright, *J. Chem. Soc., Perkin Trans. 1*, 1975, 1216.
116. V. Calò, L. Lopez, A. Mincuzzi and G. Pesce, *Synthesis*, 1976, 200.
117. (a) D. L. J. Clive and S. M. Menchen, *J. Chem. Soc., Chem. Commun.*, 1977, 658; (b) D. L. J. Clive and S. M. Menchen, *J. Org. Chem.*, 1980, **45**, 2347.
118. W. Kudelska and M. Michalska, *Tetrahedron*, 1981, **37**, 2989.
119. A. Ogawa, J. Miyake, S. Murai and N. Sonoda, *Tetrahedron Lett.*, 1985, **26**, 669.
120. J. Rémion and A. Krief, *Tetrahedron Lett.*, 1976, 3743.
121. K. B. Sharpless, M. A. Umbreit, M. T. Nieh and T. C. Flood, *J. Am. Chem. Soc.*, 1972, **94**, 6538.
122. J. K. Kochi, D. M. Singleton and L. J. Andrews, *Tetrahedron*, 1968, **24**, 3503.
123. S. M. Kupchan and M. Maruyama, *J. Org. Chem.*, 1971, **36**, 1187.
124. M. A. Umbreit and K. B. Sharpless, *Org. Synth.*, 1980, **60**, 29.
125. (a) E. W. Colvin and S. Cameron, *J. Chem. Soc., Chem. Commun.*, 1986, 1084; (b) E. W. Colvin and S. Cameron, *Heterocycles*, 1987, **25**, 133.
126. T. Fujisawa, K. Sugimoto and H. Ohta, *Chem. Lett.*, 1974, 883.
127. (a) J. E. McMurry and M. P. Fleming, *J. Org. Chem.*, 1975, **40**, 2555; (b) J. E. McMurry, M. G. Silvestri, M. P. Fleming, T. Hoz and M. W. Grayston, *J. Org. Chem.*, 1978, **43**, 3249.
128. Y. Fujiwara, R. Ishikawa, F. Akiyama and S. Teranishi, *J. Org. Chem.*, 1978, **43**, 2477.
129. M. Berry, S. G. Davies and M. L. H. Green, *J. Chem. Soc., Chem. Commun.*, 1978, 99.
130. R. Schobert, *Angew. Chem., Int. Ed. Engl.*, 1988, **27**, 855.
131. Y. Hayashi and J. Schwartz, *Inorg. Chem.*, 1981, **20**, 3473.
132. K. G. Moloy, *Inorg. Chem.*, 1988, **27**, 677.
133. T. Kauffmann and M. Bisling, *Tetrahedron Lett.*, 1984, **25**, 293.
134. P. Girard, J. L. Namy and H. B. Kagan, *J. Am. Chem. Soc.*, 1980, **102**, 2693.
135. K. N. Gurudutt and B. Ravindranath, *Tetrahedron Lett.*, 1980, **21**, 1173.
136. J. A. Gladysz, J. G. Fulcher and S. Togashi, *J. Org. Chem.*, 1976, **41**, 3647.
137. M. A. Pasha and B. Ravindranath, *Indian J. Chem., Sect. B*, 1983, **22**, 1149.
138. (a) W. P. Giering, M. Rosenblum and J. Tancrede, *J. Am. Chem. Soc.*, 1972, **94**, 7170; (b) M. Rosenblum, M. R. Saidi and M. Madhavarao, *Tetrahedron Lett.*, 1975, 4009.
139. P. Dowd and K. Kang, *J. Chem. Soc., Chem. Commun.*, 1974, 384.
140. H. Alper and D. D. Roches, *Tetrahedron Lett.*, 1977, 4155.
141. (a) J. H. Plonka and P. S. Skell, *J. Chem. Soc. D*, 1970, 1108; (b) P. S. Skell, K. J. Klabunde, J. H. Plonka, J. S. Roberts and D. L. Williams-Smith, *J. Am. Chem. Soc.*, 1973, **95**, 1547.
142. R. H. Parker and D. B. Shevlin, *Tetrahedron Lett.*, 1975, 2167.
143. M. G. Martin and B. Ganem, *Tetrahedron Lett.*, 1984, **25**, 251.
144. C. J. Shields and G. B. Schuster, *Tetrahedron Lett.*, 1987, **28**, 853.
145. P. E. Sonnet, *J. Org. Chem.*, 1978, **43**, 1841.
146. D. N. Sarma and R. P. Sharma, *Chem. Ind. (London)*, 1984, 712.
147. R. Caputo, L. Mangoni, O. Neri and G. Palumbo, *Tetrahedron Lett.*, 1981, **22**, 3551.
148. R. N. Baruah, R. P. Sharma and J. N. Baruah, *Chem. Ind. (London)*, 1983, 524.
149. E. J. Corey and W. Su, *J. Am. Chem. Soc.*, 1987, **109**, 7534.

4.5
Reduction of Vinyl Halides to Alkenes, and of Aryl Halides and Related Compounds to Arenes

MILOŠ HUDLICKÝ
Virginia Polytechnic Institute and State University, Blacksburg, VA, USA

4.5.1 HYDROGENOLYSIS OF THE CARBON–HALOGEN BOND IN VINYL HALIDES

Unlike most reductions of organic compounds, hydrogenolysis of the bond between trigonal carbon and a halogen does not possess many synthetic applications. Who would be interested in converting, for example, laboriously prepared iodobenzene to benzene by replacing iodine by hydrogen? However, the knowledge of reaction conditions for hydrogenolysis of halogen compounds is frequently useful for preventing such a reaction during reductions of other functions in an organic molecule.

Compared with alkyl halides, of which the fluorides practically do not undergo hydrogenolysis,[1] vinyl halides are hydrogenolyzed more readily, sometimes even in preference to reduction of double bonds[2,3,4] and aldehyde groups.[5] The ease of replacement of halogens by hydrogen parallels the strength of the carbon–halogen bonds, although halogens with a higher bond dissociation energy may, on occasion, be hydrogenolyzed preferentially.[6,7]

4.5.1.1 Hydrogenolysis of Vinylic Fluorides

Hydrogenolysis of vinylic fluorides is hardly ever done with the aim of removing fluorine from a molecule. Rather, it occurs as a usually undesirable side reaction, accompanying reductions of other functions in fluorinated compounds.[8–11] It frequently takes place during catalytic hydrogenation of fluorinated alkenes. A clear-cut saturation of the double bonds in polyfluorinated alkenes is rare and usually requires mild reaction conditions,[12–14] and hydrogenolysis of fluorine without reduction of the double bond in catalytic hydrogenation is unknown. Numerous examples are quoted in the review literature,[10,11] and a few samples are shown in Table 1.

Table 1 Hydrogenation and Hydrogenolysis of Polyfluoroalkenes

Starting fluoroalkene	*Reaction conditions*	*Product of hydrogenation (yield %)*		*Product of hydrogenolysis (yield %)*		*Ref.*
$CF_2{=}CF_2$	H_2/Pd, 20 °C	CHF_2CHF_2	(96)	CHF_2CH_2F	(—)	12,13
	H_2/Ni, 90–100 °C		(66)		(14)	
$(CF_3)_2C{=}CF_2$	H_2/Pd, 20 °C	$(CF_3)_2CHCHF_2$	(95)	$(CF_3)_2CHMe$	(—)	12,13
	H_2/Ni, 100°C		(10)		(75)	
F, F, F, F, F, F (cyclobutene)	$H_2/Pd/Al_2O_3$, 40 °C, 1 atm	F, F, F, F, F, F (cyclobutane)	(100)	F, F, F, F, F (cyclobutane)	(—)	15
	H_2/Pd (pumice), 45 °C, 1 atm		(70)		(26)[a]	
	H_2/Ni (pumice), 25 °C, 1 atm		(67)		(22)	
	Ni (Raney), 60 °C, autoclave		(21)		(41)[b]	

[a]The balance (4%) is *trans*-1,2,3,3,4,4-hexafluorocyclobutane. [b]The balance (36%) is 1,1,2,2-tetrafluorocyclobutane.

Catalytic hydrogenation of fluoro- and difluoro-*cis*- and *trans*-butenedioic acids and their esters frequently affords products only of hydrogenolysis, especially when platinum oxide (Adams' catalyst) is used. Hydrogenolysis is favored in polar solvents, and is found to be more effective when rhodium or nickel rather than palladium are used as the catalysts (Table 2). Hydrogenolysis takes place prior to the saturation of the double bonds: fluoro- and 2,3-difluoro-succinic acids do not suffer any loss of fluorine under the same reaction conditions.[8]

Table 2 Catalytic Hydrogenation of Difluoromaleic Acid

Catalyst	*Solvent*	*Temperature (°C)*	*Yield of $HO_2CCHFCHFCO_2H$ (%)*	*Yield of $HO_2CCHFCH_2CO_2H$ (%)*	*Yield of $HO_2CCH_2CH_2CO_2H$ (%)*
10% Pd (C)	Et_2O	–70	37	41	3[a]
10% Pd (C)	Et_2O	25	19	70	11
10% Pd (C)	H_2O	25		67	33
5% Rh (C)	H_2O	25		11	89

[a]19% remained unreacted.

Complete hydrogenolysis of fluorine is observed in unsaturated fluorinated amino acids[9] and their derivatives.[16] γ-Amino-α-fluorocrotonic acid and 5-amino-2-fluoro-2-hexenedioic acid afford, on hydrogenation over platinum oxide, γ-aminobutyric and α-aminoadipic acid, respectively. The same compounds are obtained on treatment of the above fluoroamino acids with hydriodic acid and red phosphorus.[9] Fluorine is also replaced by hydrogen in *trans*-α-fluorostilbene, which on catalytic hydrogenation over palladium gives bibenzyl (equation 1),[17] and in difluorostilbene, which on heating with sodium affords the same product.[18]

$$\text{PhC(X)=C(F)Ph} \xrightarrow[\text{or ii, X = F, 81\%}]{\text{i, X = H, 48\%}} \text{PhCH}_2\text{CH}_2\text{Ph} \qquad (1)$$

i, H_2/10% Pd(C), $CaCO_3$, EtOH, r.t., 2.5 atm, 5 h; ii, Na, MeOH, heat, 5 min

Vinylic fluorine in heavily fluorinated alkenes is readily replaced by hydrogen in reductions with lithium aluminum hydride and sodium borohydride. The isomeric 1-phenylpentafluoropropenes are reduced by lithium aluminum hydride regio- and stereo-specifically to tri- and tetra-fluoropropenes. (equations 2 and 3).[19]

LiAlH$_4$, glyme; −15 to −10 °C, 94%; 0–30 °C, 95% (2)

LiAlH$_4$, glyme, −20 to −15 °C; 84.2% (3)

Hexafluorocyclobutene is reduced with lithium aluminum hydride in ether at −78 °C to 3,3,4,4-tetrafluorocyclobutene in 37% yield.[20] 1,3,3,4,4-Pentafluorocyclobutene and 1,3,3,4,4,5,5-heptafluorocyclopentene on treatment with lithium aluminum hydride or sodium borohydride afford 3,3,4,4-tetrafluorocyclobutene and 3,3,4,4,5,5-hexafluorocyclopentene, respectively (equation 4).[7]

i, 88–92%; or ii, 64–68% (4)

i, LiAlH$_4$, Et$_2$O, 0 °C; ii, NaBH$_4$, diglyme, 0 °C

In 1-chloroperfluorocyclopentene, fluorine is replaced by hydrogen in preference to chlorine in reductions with complex hydrides (equation 5).[6]

NaBH$_4$, diglyme, 0 °C; 88% (5)

In 1-chloro-2-hydroperfluorocycloalkenes, a direct nucleophilic attack by hydrides substitutes hydrogen for chlorine giving (**1**), whereas allylic displacement (S_N2') replaces fluorine and forms (**2**) (equation 6 and Table 3).[7]

LiAlH$_4$; direct → (**1**); allylic → (**2**) (6)

4.5.1.2 Hydrogenolysis of Vinylic Chlorides

Like vinyl fluorides, vinyl chlorides suffer hydrogenolysis during catalytic hydrogenation. Rhodium is claimed to effect addition of hydrogen rather than hydrogenolysis of vinylic chlorides. However, hy-

Table 3 Reduction of 1-Chloro-2-hydroperfluorocycloalkenes with Complex Hydrides (equation 6)[7]

	Products in equation (6)	*Yield of* (**1**) (%)	*Yield of* (**2**) (%)
$n = 2$	$LiAlH_4$, Et_2O, 0 °C	24	40
	$NaBH_4$, diglyme, 0 °C	60	12
$n = 3$	$LiAlH_4$, Et_2O, 0 °C	8	88
	$NaBH_4$, diglyme, 0 °C	8	40

drogenation of both *cis*- and *trans*-1,3-dichloropropene over 5% rhodium on alumina in cyclohexane at 100 °C and 48–49 atm (1 atm = 101 325 Pa) gives only 43–47% of 1,3-dichloropropane, and 31–34% of 1-chloropropane, resulting from replacement of vinylic chlorine by hydrogen prior to the saturation of the double bond.[21] Hydrogenation of 2-chloroheptene over Raney nickel in methanol in the presence of 1 equiv. of potassium hydroxide at room temperature and atmospheric pressure affords heptane in 85% yield.[22] Under similar conditions, but in the presence of 3 equiv. of potassium hydroxide, 1,1,5-trichloro-1-pentene yields 72% of pentane, resulting from hydrogenolysis of both vinylic chlorine and that bonded to tetrahedral carbon.[22] Alkynic chlorine is hydrogenolyzed in the presence of 1 equiv. of potassium hydroxide, 1-chloro-1-hexyne giving hexane in 85% yield.[22]

Selective replacement of the vinylic chlorines takes place in the hydrogenation of chlorinated norbornenes (equations 7 and 8), which suffer saturation of the double bonds as well as hydrogenolysis of the chlorines.[23–25]

X X Cl Cl Cl Cl — i, X = H, 82% or ii, X = OMe, 93% → X X Cl Cl (7)

i, H_2/Raney Ni, KOH, r.t.; ii, H_2/5% Pd(C), EtOH, Et_3N

Cl Cl Cl Cl Cl Cl OAc — H_2/Pd(C), r.t., 1 atm; 33% (82% based on starting material consumed) → Cl Cl Cl Cl OAc (8)

In hexachlorocyclopentadiene, all the chlorines are hydrogenolyzed and both double bonds saturated in vapor phase hydrogenation over palladium on silica at 330 °C, giving cyclopentane in 82% yield.[1]

Vinylic chlorine is replaced by hydrogen in 1-chloro-3,4-dihydro-2-naphthaldehyde during catalytic hydrogenation, which saturates the double bond, but leaves the aldehyde group intact (equation 9).[5]

Cl CHO — H_2/5% Pd(C); NaOH, H_2O, MeOH, 50–60 °C, 1.4–3.5 atm, 3.5 h; 84% → CHO (9)

Selective hydrogenolysis of chlorine in preference to fluorine occurs during catalytic hydrogenation of chlorotrifluoroethylene. In addition to trifluoroethylene, trifluoroethane is formed by reduction of the double bond (equation 10).[3,13]

F F F Cl → F F F + F F F (10)

H_2/5% Pd(C), 130–135 °C, 10 s	85%	15%
H_2/Pd, 20 °C, 1 atm	60%	25%

Replacement by hydrogen of vinylic chlorine without reduction of the double bond can be accomplished by complex hydrides. The efficacy of lithium aluminum hydride is increased by titanium tetrachloride. Thus α-chloro-*trans*-stilbene (**3**) gives 98% of *trans*-stilbene (**4**) with only 1% of bibenzyl (equation 11).[26]

(**3**) —[2 $LiAlH_4$, $TiCl_4$; THF, reflux, 3 h]→ (**4**) 98% + 1% (11)

In 1-chloro-2-hydroperfluorocyclobutene and cyclopentene, direct reaction of hydride substitutes hydrogen for chlorine (**1**), while allylic reduction leads to replacement of fluorine (**2**) (equation 6, Table 3).[7]

Reduction with metals such as sodium, lithium and zinc leaves the double bonds intact, with rare exceptions.[27] In (**5**), both allylic and vinylic chlorine atoms are replaced by hydrogen using sodium or lithium in *t*-butyl alcohol.[27] While sodium does not reduce the double bond, lithium affords a mixture of (**6**) and (**7**) (equation 12).[27]

(**5**) —[Na, Bu^tOH, THF, reflux, 8 h; 62.8%]→ (**6**)

(**5**) —[Li, Bu^tOH, THF, reflux]→ (**6**) 42% + (**7**) 22% (12)

In contrast, zinc in acetic acid replaces only one of the bridge chlorines in (**8**) leaving vinylic and allylic chlorines as well as the double bond intact.[23,25] Similar results are obtained using chromium(II) acetate (equation 13).[25]

(**8**)				(13)
	Zn, AcOH, reflux, 4.5 h	70%	9%	
	$Cr(OAc)_2$, AcOH, reflux, 16 h	78%	18%	

Zinc in acetic acid hydrogenolyzes vinylic chlorine in phenyl (*Z*)-β-chlorostyryl sulfone (equation 14).[28]

—[Zn, AcOH, reflux, 16 h; 86%]→ (14)

4.5.1.3 Hydrogenolysis of Vinylic Bromides

Vinylic bromine can be replaced fairly easily without reduction of the double bond. This is a characteristic difference from hydrogenolysis of vinylic chlorine and especially fluorine. Catalytic hydrogenation of ethyl α-methyl-β-bromocrotonate resulted in replacement of bromine and formation of ethyl α-methylcrotonate (equation 15).[2] In 14-bromocodeinone (**9**), replacement of bromine occurs with a double bond shift to give neopinone (**10**; equation 16).[4]

—[H_2/Pd($BaSO_4$), Et_3N, EtOH, r.t., 1 atm; 71%]→ (15)

(9) $\xrightarrow[81\%]{H_2/10\%\ Pd(C),\ CHCl_3,\ MeOH,\ 25\ °C,\ 1\ atm}$ (10) (16)

Homogeneous catalytic hydrogenation over tetrakis(triphenylphosphine)palladium using hydrogen or sodium formate as hydrogen donors converts β-bromostyrene to styrene in 37% and 35% yields respectively (the other products being polymers).[29]

Electrolytic replacement of vinylic bromine by hydrogen in (*E*)-bromostilbene is complicated by the formation of diphenylacetylene in addition to *trans*- and a small amount of *cis*-stilbene.[30]

An interesting hydrogenolysis of bromine is observed when 2-bromocycloalken-2-ones are treated with diethyl phosphite and triethylamine. The predominant products are 3-cycloalkenones with only minor amounts of 2-cycloalkenones (equation 17).[31]

$\xrightarrow{HP(O)(OEt)_2,\ Et_3N,\ r.t.,\ 15\ h}$ 63% + 3% (17)

4.5.1.4 Hydrogenolysis of Vinylic Iodides

Very scanty documentation is available on replacement of vinylic iodine by hydrogen. Treatment of iodotrifluoroethylene with sodium borohydride in diethylene glycol and water or alcohol affords trifluoroethylene in 73% yield. In the case of chlorotrifluoroethylene and bromotrifluoroethylene no hydrogenolysis, but saturation of the double bond results under identical conditions (equation 18).[32]

$\xrightarrow{NaBH_4,\ H_2O,\ diglyme,\ 0–5\ °C}$ (18)

X = Cl	86%	6%
X = Br	94%	3%
X = I	–	73%

1-Chloro-2-iodotetrafluorocyclobutene and 1-chloro-2-iodohexafluorocyclopentene treated with sodium aluminum hydride in diglyme by addition of the compounds to the reagent give, after decomposition of the intermediate complex with deuterium oxide, 1-chloro-2-deuteriotetrafluorocyclobutene in 38% yield, and 1-chloro-2-deuteriohexafluorocyclopentene in 43% yield, respectively (equation 19).[33] The iodine in iodoalkene (**11**) is replaced by hydrogen on refluxing with zinc and hydrochloric acid to give iodine-free alkene (**12**; equation 20).[34]

$\xrightarrow[43\%]{i,\ NaAlH_4,\ diglyme,\ <10\ °C;\ ii,\ D_2O}$ (19)

(**11**) $\xrightarrow[63\%]{Zn,\ dil.\ HCl,\ reflux,\ 3\ h}$ (**12**) (20)

4.5.2 HYDROGENOLYSIS OF THE TRIGONAL CARBON–HALOGEN BOND IN ARYL HALIDES

Replacement of aromatic halogens by hydrogen can be accomplished by catalytic hydrogenation and by reduction with complex hydrides, with sodium, with zinc, and with inorganic and organic reducing agents.

4.5.2.1 General

Hydrogenolysis of aromatic halides using catalytic hydrogenation takes place easily with rates of replacement of halogen decreasing in the order I > Br > Cl > F. For example, replacement of chlorine in preference to fluorine takes place in chlorofluoropyridines (equation 21).[35]

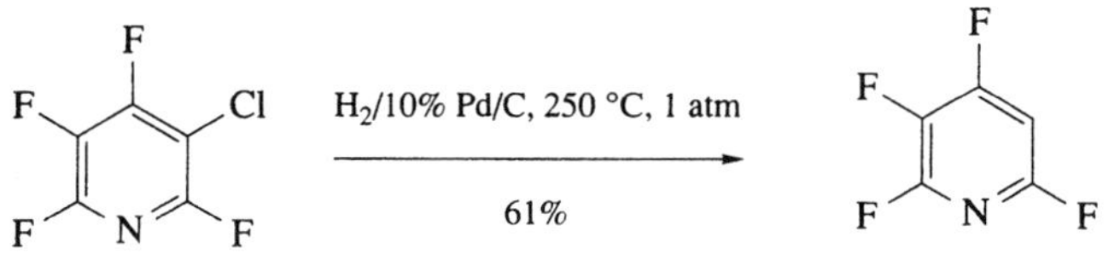
F
F
Cl
F
N
F
H_2/10% Pd/C, 250 °C, 1 atm
61%
F
F
F
N
F

(21)

The same order of ease of replacement applies to reduction of aryl halides with lithium aluminum hydride[36,37] and with zinc (equations 22 and 23).[38]

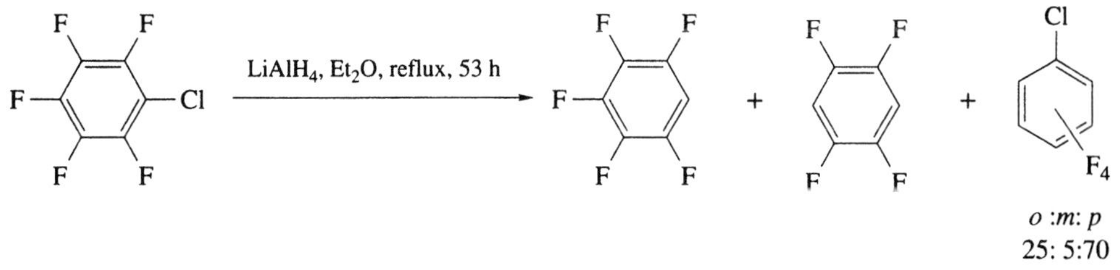
F
F
F
Cl
F
F
$LiAlH_4$, Et_2O, reflux, 53 h
F
F
F
F
F
+
F
F
F
F
+
Cl
F_4
o :m: p
25: 5:70

(22)

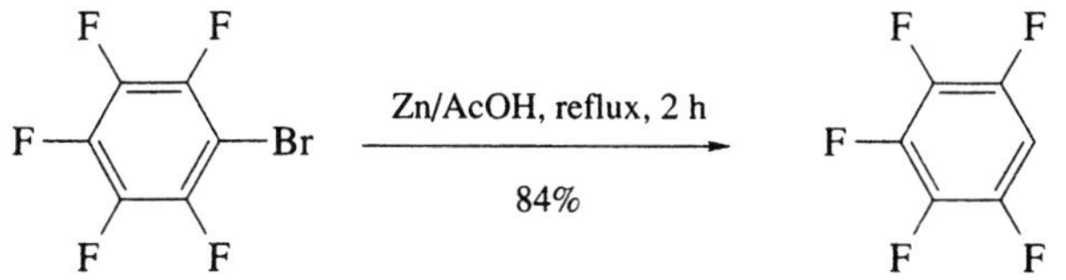
F
F
F
Br
F
F
Zn/AcOH, reflux, 2 h
84%
F
F
F
F
F

(23)

Replacement of fluorine in preference to chlorine is observed in chlorofluoropyridines, where lithium aluminum hydride replaces the fluorines in the α- and γ-positions of lower electron density faster than the chlorine in the β-position (equation 24).[39]

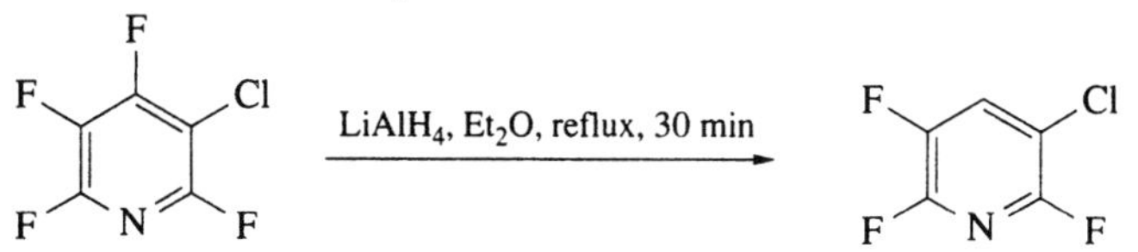
F
F
Cl
F
N
F
$LiAlH_4$, Et_2O, reflux, 30 min
F
Cl
F
N
F

(24)

Another regioselective hydrogenolysis of fluorine with hydrides takes place in perfluorotoluene,[40] and electrolytically in pentafluorobenzoic acid (equations 25 and 26).[41]

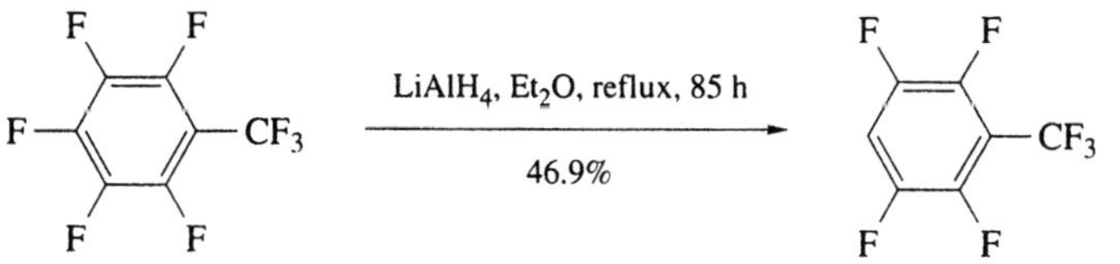
F
F
F
CF_3
F
F
$LiAlH_4$, Et_2O, reflux, 85 h
46.9%
F
F
CF_3
F
F

(25)

Because of the different rates of hydrogenolysis, aryl halides containing different halogens can be reduced selectively. 1-Bromo-2-chlorobenzene is reduced to chlorobenzene in 96% yield on treatment for 2 h at room temperature with 'a complex reducing agent' prepared from nickel(II) chloride, sodium hydride and *t*-pentyl alcohol.[42] 1-Bromo-4-chlorobenzene is converted to chlorobenzene on treatment with

electrolysis, 20% H_2SO_4, –1.2 V, Hg cathode; 75% (26)

triphenyltin hydride,[43] and 4-bromo-2-chlorophenol to 2-chlorophenol on treatment with ethanethiol and aluminum chloride (equation 27).[44]

EtSH, $AlCl_3$, CH_2Cl_2, r.t., 48 h; 61% (27)

The reverse order of replacement of halogens, *i.e.* preferential substitution of chlorine rather than of bromine by hydrogen in bromochloro compounds, is possible if chlorine is activated by its position in the molecule. For example, 5-bromo-6-chloro-2,4-dimethylpyrimidine, in which chlorine is rendered labile by its α-position to one of the ring nitrogens, is reduced to 5-bromo-2,4-dimethylpyrimidine by refluxing with *p*-toluenesulfonylhydrazine in chloroform (equation 28).[45]

$TsNHNH_2$, $CHCl_3$, reflux, 12 h; 84% (28)

Substitution of halogen by hydrogen in aryl halides can take place without, or with only partial, reduction of carbon–carbon double bonds,[46] aldehydes,[46] nitriles,[46] and even readily reducible nitro groups[46,47] present in the molecule (equations 29–32).

HCO_2H, Et_3N, Pd(C), 100 °C (29)

X	Time	CO_2Me (alkene)	CO_2Me (saturated)
X = Cl	1.5 h	22%	55%
X = Br	7.0 h	93%	–

(30)

X	Conditions	CHO	OH	CH_3
4-Br	H_2/Pd(PPh_3)$_4$, AcOK, DMF, 100 °C, 3 h	76%	–	–
2-Br	HCO_2H, Et_3N/5% Pd(C), 50 °C, 2 h	44%	8%	–
2-Br	HCO_2H, Et_3N/Pd(OAc)$_2$, 2P(*o*-Tol)$_3$, 50 °C, 2 h	43%	10%	10%

HCO_2H, Et_3N, 5% Pd(C), 100 °C, 23.5 h; 80%

HCO_2H, Et_3N, 5% Pd/$BaSO_4$, 50 °C, 20 h; 53% (31)

$X-C_6H_4-NO_2 \longrightarrow C_6H_5-NO_2 + C_6H_5-NH_2$ (32)

X			
2-Cl	$H_2/Pd(PPh_3)_4$, AcOK, DMF, 100 °C, 6 h	67%	–
2-Br	$H_2/Pd(PPh_3)_4$, AcOK, DMF, 100 °C, 6 h	80%	–
2-Br	HCO_2H, Et_3N, 5% Pd(C), 50 °C, 48 h	78%	8%
4-Cl	HCO_2H, Et_3N, 5% Pd(C), 100 °C, 1 h	90%	–

4.5.2.2 Hydrogenolysis of Aryl Fluorides

Substitution of hydrogen for fluorine in compounds containing single fluorine atoms linked to aromatic rings occurs during catalytic hydrogenation of *p*-fluorobenzoic acid, which gives benzoic acid and cyclohexanecarboxylic acid on treatment with hydrogen over platinum black,[48] and in hydrogenation of *p*-fluorophenylacetic acid, which affords ethyl cyclohexylacetate on heating with hydrogen, ethanol and Raney nickel (yields of these experiments were not quoted) (equation 33).[49]

p-F–C_6H_4–CH_2CO_2H (Raney Ni, EtOH, 200 °C) → 165 atm, 2 h: p-F–C_6H_4–CH_2CO_2Et; 160 atm, 3.5 h: C_6H_{11}–CH_2CO_2Et (33)

Hydrogenation of (**13**) over palladium black yields mainly (**14**) with only partial hydrogenolysis of fluorine to give (**15**; equation 34).[50]

F–$C_6H_3(NO_2)$–COMe (**13**) $\xrightarrow{H_2/\text{Pd black, AcOH, } H_2SO_4}$ F–$C_6H_3(NH_2)$–Et (**14**) 80% + H_2N–C_6H_4–Et (**15**) 20% (34)

Reduction of *p*-fluoroacetophenone with Raney alloys in 5% sodium hydroxide gives 1-phenylethanol in high yields (equation 35).[51]

F–C_6H_4–COMe $\xrightarrow{\text{Raney alloy, 5\% NaOH, 50 °C, 2 h}}$ C_6H_5–CH(OH)Me (35)

Ni–Al 86%
Cu–Al 93%

Catalytic hydrogenation of pentafluoropyridine (**16**) gives a mixture of the tetrafluoropyridines (**17**)–(**19**), with reduction taking place most easily at the 4-position and to only a trace extent at the 3-position (equation 36).[35]

(**16**) $\xrightarrow{H_2/\text{Pd(C), 320 °C}}$ (**17**) 30% + (**18**) 5% + (**19**) trace (36)

Single fluorine atoms in aromatic fluoro compounds such as *p*-fluorotoluene and *p*-fluoroanisole are replaced by hydrogen using lithium aluminum hydride in the presence of di-*t*-butyl peroxide under irradiation (equation 37).[52]

$$\text{F}-\text{C}_6\text{H}_4-\text{R} \xrightarrow[\substack{\text{R = Me 90\%} \\ \text{R = OMe 93\%}}]{\text{LiAlH}_4,\ \text{THF},\ (\text{Bu}^t\text{O})_2,\ h\nu,\ 1.9\ \text{h}} \text{C}_6\text{H}_5-\text{R} \quad (37)$$

Hydrogenolysis of polyfluoro and perfluoro aromatics by lithium aluminum hydride has already been shown in equations (24) and (25). Additional examples are replacement of fluorine by hydrogen in pentafluorobenzene, where the predominant product is 1,2,4,5-tetrafluorobenzene (**20**), resulting from the reduction by lithium aluminum hydride *para* to the hydrogen (equation 38),[53] and replacement of fluorine in the β-position in 1,2,3,4-tetrafluoronaphthalene (equation 39).[54]

$$\text{C}_6\text{HF}_5 \xrightarrow[\text{reflux, 53\%}]{\text{LiAlH}_4,\ \text{Et}_2\text{O}} \underset{(\textbf{20})\ 72\%}{\text{1,2,4,5-C}_6\text{H}_2\text{F}_4} + \underset{6\%}{\text{C}_6\text{H}_2\text{F}_4} + \underset{2\%}{\text{C}_6\text{H}_2\text{F}_4} \quad (38)$$

$$\text{1,2,3,4-tetrafluoronaphthalene} \xrightarrow[88\%]{\text{LiAlH}_4,\ \text{THF, reflux, 144 h}} \text{1,2,4-trifluoronaphthalene} \quad (39)$$

Removal of fluorine from 1-fluoronaphthalene is accomplished by treatment with ethanethiol and aluminum chloride at room temperature (equation 40).[55]

$$\text{1-fluoronaphthalene} \xrightarrow[87\%]{\text{EtSH, AlCl}_3,\ \text{CH}_2\text{Cl}_2,\ \text{r.t., 7 h}} \text{naphthalene} \quad (40)$$

Further examples of hydrogenolysis of aryl fluorides can be found in the review literature.[10,11]

4.5.2.3 Hydrogenolysis of Aryl Chlorides

A general method for replacement of aromatic chlorine by hydrogen is catalytic hydrogenation over palladium on carbon,[56] calcium carbonate[57a] or silica gel,[57b] over platinum oxide (Adams' catalyst),[56] and over Raney nickel.[58,59] Platinized charcoal is inhibited by some halogen compounds.[56] Reductions, especially those using Raney nickel, have to be carried out in the presence of alcoholic potassium hydroxide to neutralize hydrogen chloride resulting from the hydrogenolysis.[58,59] In dichlorides with equivalent chlorine atoms, partial or total hydrogenolysis can be accomplished, depending upon whether 1 or 2 equiv. of the base are used.[59]

In catalytic hydrogenation, chlorine is replaced by hydrogen in chlorinated aromatic hydrocarbons (equations 41 and 42),[57] phenols (equation 43),[60] amines (equation 44),[56] carboxylic acids (equation 45),[56,59] and nitro compounds (equation 46).[46,47] Hydrogenolysis of chlorine in chloronitro compounds takes precedence over reduction of nitro groups,[46,47] provided that contact with the halogen-free product is not too long.[46] The reaction is achieved using palladium on carbon[46] or tetrakis(triphenylphos-

$$\text{Cl}-\text{C}_6\text{H}_5 \xrightarrow[32\%]{\text{Raney Ni, KOH, MeOH, r.t., 30 min}} \text{C}_6\text{H}_6; \quad \text{Cl}-\text{C}_6\text{H}_5 \xrightarrow[90\%]{\text{NaH}_2\text{PO}_4,\ \text{Pd(C)},\ \text{H}_2\text{O, THF}} \text{C}_6\text{H}_6$$

$$\text{C}_6\text{H}_6 \xleftarrow[50\%]{\text{Raney Ni, 2 KOH, MeOH, r.t., 15 min}} \text{Cl}-\text{C}_6\text{H}_4-\text{Cl} \quad (41)$$

phine)palladium.[47] Hydrogen donors such as sodium hypophosphite,[61] ethanol[62] or ammonium formate[60] can be employed in catalytic hydrogen transfer.

i, 90%, ii, 63% or iii, 30% (42)

i, $H_2/Pd(SiO_2)$, 330 °C, 6 h; ii, $H_2/Ni(Al_2O_3)$; iii, $H_2/Fe(SiO_2)$

$HCO_2NH_4/Pd(C)$, MeOH, r.t., 10 min, 100% (43)

i, 90% or ii (44)

i, Raney Ni, KOH, MeOH, r.t., 70 min; ii, $H_2/Pd(C)$, KOH, MeOH, r.t., 11 min

i, 85% or ii (45)

i, Raney Ni, 2 KOH, MeOH, r.t., 5 h; ii, $H_2/Pd(C)$, 95% EtOH

$H_2/(Ph_3P)_4Pd$, DMF, AcOK, 100 °C, 16 h, 67% (46)

In the reduction of *p*-chlorocinnamic acid with an excess of sodium hypophosphite and palladium (equation 47), the double bond is saturated and the hydrogenolysis of chlorine occurs. With one equivalent of the hypophosphite, only saturation of the double bond takes place, and *p*-chlorohydrocinnamic acid is obtained after 1.5 h.[61]

Pd(C), NaH_2PO_2 (3 equiv.), H_2O, THF, Na_2CO_3, 50 °C, 24 h, 70% (47)

In aromatic heterocycles, chlorine is replaced by hydrogen in catalytic vapor phase hydrogenation over palladium on charcoal at 280 °C. 3,5-Dichlorotrifluoropyridine affords 2,4,6-trifluoropyridine in 75% yield.[35]

H_2/5% Pd(C), AcOH, AcONa, 55–70 °C, 81–87% (48)

(21) **(22)**

Chlorine in 2-chlorolepidine (**21**) is very reactive, since the compound can be considered an imide chloride. It is readily replaced by hydrogen to form lepidine (**22**; equation 48).[63] Similarly (**23**) is converted to (**24**) on catalytic hydrogenation (equation 49).[62] The above reductions may be applied to many six-membered aromatic heterocycles containing chlorine on the carbon next to nitrogen; thus pyridones, quinolones and related compounds can be converted to their parent heterocycles by way of their α-chloro compounds.

(23) $\xrightarrow{\text{H}_2\text{/5\% Pd(C), KOH, EtOH, r.t., 1 atm}}$ 53% (24) (49)

High yields of 1-phenylethanol are obtained on heating chlorinated acetophenones at 50 °C with Raney nickel or Raney copper alloys in 5% sodium hydroxide (equation 50). Copper alloy is somewhat less efficient, as in some cases the hydrogenolysis of chlorine is incomplete.[51]

Cl–C₆H₄–COMe $\xrightarrow{\text{5\% NaOH, 50 °C}}$ PhCH(OH)Me + Cl–C₆H₄–CH(OH)Me (50)

Ni–Al alloy	88%	–
Cu–Al alloy	44%	56%

Examples of hydrogenolysis of aryl chlorides by lithium aluminum hydride are rare. Chlorobenzene is reduced to benzene only in a low yield,[36] but chloronaphthalene gives a good yield of naphthalene with lithium aluminum hydride in the presence of either di-*t*-butyl peroxide[52] or titanium tetrachloride (equation 51).[26]

1-chloronaphthalene $\xrightarrow{\text{i, 96\% or ii, 83\%}}$ naphthalene (51)

i, $LiAlH_4$, $(Bu^tO)_2$, *hν*, r.t., 2 h; ii, 2 $LiAlH_4$, $TiCl_4$, THF, reflux, 3 h

Zinc dust in the presence of nickel chloride, triphenylphosphine and sodium iodide as a catalyst reduces chlorobenzene in methanol after 20 h at 60 °C to benzene in 99% yield, and *p*-chloroanisole to anisole in dimethylformamide/water in 88–97% yields.[64] The yields are considerably lower without sodium iodide (equation 52).[64]

Cl–C₆H₄–OMe $\xrightarrow[\text{97\%}]{\text{Zn, NiCl}_2\text{, Ph}_3\text{P, NaI; DMF/H}_2\text{O (25:1), 70 °C, 2 h}}$ PhOMe (52)

A similar effect on aryl halides is produced by using a suspension of zinc dust, nickel chloride and sodium iodide in moist hexamethylphosphoramide at 60 °C under ultrasound sonication.[65] Chlorobenzene is thus converted to benzene in 64–93% yields.[65]

Chlorine in position 2 in 2-chloro-4,6-dimethylpyrimidine is replaced by hydrogen on heating for 10 min at 100–110 °C with azeotropic (57%) hydriodic acid, giving a 78% yield of 4,6-dimethylpyrimidine.[66]

Hydrogenolysis of aryl chlorides can also be achieved by treatment with triethylsilane in the presence of palladium,[67] and by reduction with isopropyl alcohol under UV irradiation (equation 53).[68]

PhCl $\xrightarrow[\text{ii, Et}_3\text{SiH, 10\% Pd(C), } h\nu\text{, 7 min; 78\%}]{\text{i, Pr}^i\text{OH, } h\nu\text{, 12 h; 72\%}}$ C₆H₆ (53)

4.5.2.4 Hydrogenolysis of Aryl Bromides

Substitution of hydrogen for bromine is easier and faster than for chlorine. In catalytic hydrogenation, 10% palladium on carbon is a more active catalyst than platinum oxide, and this in turn is better than

10% platinum on carbon.[56,69] A palladium complex with triphenylphosphine and Raney nickel in the presence of potassium hydroxide is equally suitable for the hydrogenolysis (Table 4).[59,70] Instead of hydrogen, hydrogen donors such as formic acid (equation 54)[71] and triethylammonium formate[46] are employed in the presence of 10% palladium on carbon[46,71] or palladium diacetate complexed with phosphines.[46]

Table 4 Examples of Reduction of Bromobenzene to Benzene

Reaction conditions	*Yield (%)*	*Ref.*
H_2/Pd ($CaCO_3$), 10% KOH, EtOH, r.t.	99	38
H_2/10% Pd (C), MeOH, r.t., 5 h		56
HCO_2H/10% Pd (C), DMF, reflux, 3 h	85–90	71
HCO_2H, Et_3N/$Pd(OAc)_2$ + $[P(OC_7H_7)_3]_2$, 50 °C, 4.5 h	92	46
$LiAlH_4$, THF, 25 °C, 24 h	95	36
$LiAlH_4$, THF, 65 °C, 6 h	100	36
$NaAlH_2(OCH_2CH_2OMe)_2$, xylene, 100–115 °C, 1 h	53	72
Zn, $NiCl_2$, NaI, H_2O, HMPA, 60 °C, ultrasound	81	65
$ZnBr_2$, K, THF, reflux 4 h; diglyme, reflux 8 h	83	73
Ph_3SnH, reflux	60–75	43
Et_3SiH/10% Pd (C), 4 min	79	67
Me_2CHOH, *hν*, r.t.	72	68

Br

HCO_2H, 10% Pd(C), DMF, reflux, 3 h

81% (54)

Bromoanilines are reduced to anilines by hydrogenation over Raney nickel (equation 55),[59] and *p*-bromoacetophenone is converted to 1-phenylethanol on heating at 50 °C with Raney alloys in 5% sodium hydroxide. The best yields are obtained using copper–aluminum alloy (89–91%).[51]

Br —NH_2 H_2/Raney Ni, KOH, MeOH, r.t., 1 atm —NH_2 (55)

	Time (min)	*Yield* (%)
o-Br	45	90
m-Br	50	91
p-Br	60	85

Reduction of aryl bromides with lithium aluminum hydride takes place in tetrahydrofuran solutions at room temperature.[36] It has also been performed with better results in the presence of di-*t*-butyl peroxide under UV irradiation[52] or in the presence of titanium tetrachloride (equations 56 and 57).[26] Reduction of bromobenzene with sodium bis(methoxyethoxy)aluminum hydride (Red-Al, Vitride) at 100–115 °C gives benzene in 53% yield (Table 4).[72]

Br i, 21% or ii, 96% (56)

i, $LiAlH_4$, THF, 25 °C, 24 h; ii, $LiAlH_4$, THF, $(Bu^tO)_2$, *hν*, r.t.

Br i, 95% or ii, 98% (57)

i, $LiAlH_4$, $(Bu^tO)_2$, THF, *hν*, r.t., 2.3 h; ii, $LiAlH_4$, $TiCl_4$, THF, reflux, 3 h

Replacement of bromine in aryl bromides by hydrogen using zinc takes place at higher temperatures. Bromobenzene gives benzene by refluxing in diglyme with zinc, generated *in situ* from anhydrous zinc bromide and potassium in tetrahydrofuran (Table 4).[73] Zinc powder in boiling acetic acid reduces pentafluorobromobenzene to pentafluorobenzene in 84% yield after refluxing for 2 h. Iron filings and aluminum flake do not react.[38]

Under rather complex conditions, bromobenzene is reduced to benzene by heating with zinc powder and nickel chloride in dimethylformamide or hexamethylphosphoramide containing small amounts of water (Table 4).[65]

A regioselective substitution by hydrogen of the more reactive bromine atoms in the α-positions on the thiophene ring converts 2,3,5-tribromothiophene to 3-bromothiophene by refluxing with zinc powder in acetic acid (equation 58).[74]

Zn, AcOH, reflux, 4 h; 89–90% (58)

Metal salts such as chromium(II) chloride ($CrCl_2$) complexed with ethylenediamine reduce α-bromonaphthalene in dimethylformamide at room temperature in 93–98% yield.[75]

A selective reducing agent for replacement of halogens by hydrogen, triphenylstannane (triphenyltin hydride), reduces aryl bromides to arenes in high yields (Table 4).[43] Triethylsilane (triethylsilicon hydride) converts bromobenzene to benzene in the presence of 10% palladium on carbon in 79% yield (Table 4).[67]

A strongly reducing mixture is prepared by treatment of nickel(II) chloride and *t*-pentyl alcohol with sodium hydride. In dimethoxyethane, *o*-bromochlorobenzene is selectively reduced to chlorobenzene in 96% yield after 2 h at 20 °C.[42]

Isopropyl alcohol under UV irradiation converts bromobenzene to benzene in 72% yield (Table 4).[68] Similar replacement of bromine by hydrogen is accomplished by treatment of aryl bromides dissolved in dichloromethane with a mixture of ethanethiol and anhydrous aluminum chloride.[44,55] This hard acid–soft base combination reacts with polycyclic aromatic halides[55] and halogenated phenols[44] by an addition–elimination mechanism, leading to an aryl ethyl sulfide through a radical anion intermediate. This is converted by another molecule of ethanethiol to the debrominated arene and diethyl disulfide.[55] 1-Bromonaphthalene is thus transformed into naphthalene (equation 59), 2,4,6-tribromophenol into phenol (equation 60), and bromochlorophenols into chlorophenols in 61–91% yields.[44]

$AlCl_3$, EtSH; Br HSEt; H $\bar{A}lCl_3$; –HBr, –$AlCl_3$ (59)

SEt; EtSH, r.t., 8 h; 67%; + EtSSEt

Br; Br; OH; Br; EtSH, $AlCl_3$, CH_2Cl_2, r.t., 49 h; 73%; OH (60)

In 5-bromouracil having no substituents on N-1, bromine is replaced by hydrogen on heating at 180 °C with 1-benzyl-1,4-dihydronicotinamide. Thus (**25**) yields (**26**), (**27**) and (**28**) (equation 61).[76]

4.5.2.5 Hydrogenolysis of Aryl Iodides

As might be expected, substitution of iodine by hydrogen in aryl iodides is easier than in aryl bromides. This is evident in selective reduction of chloroiodo aromatics (equation 62),[36,52] and especially bromoiodo aromatics (equation 63),[77] and also in the selective reduction of iodonitro aromatics to nitro aromatics (equations 64 and 65).[78]

(25) → (26) 54% + (27) 10% + (28) 19% (61)

Bn-N pyridine-CONH$_2$, 180 °C, 3 h

i, 95% or ii, 96% (62)

i, LiAlH$_4$, THF, 25 °C; ii, LiAlH$_4$, (Bu^{t}O)$_2$, THF, $h\nu$, 42 min

NaBH$_4$/Cp$_2$TiCl$_2$, DMF, 70 °C, 24 h (63)

85% 12%

NaBH$_4$, DMSO, D$_2$O, 18–27 °C, 15 min (64)

95% (88% pure D)

Bn-N pyridine-CONH$_2$, Rh(Ph$_3$P)$_3$Cl, MeCN, r.t., 3 h (65)

100%

Reduction of aryl iodides has been accomplished by catalytic hydrogenation using hydrogen or hydrogen donors over palladium[71] or Raney nickel catalysts (equation 66),[59] by reduction with complex hydrides (equations 62–64),[36,52,77,78] with metals[65] and their compounds,[79] with hydrogen iodide,[66] and with organic compounds (equations 65 and 67).[44,79,80] Table 5 shows reaction conditions for reduction of iodobenzene to benzene.

H$_2$/Raney Ni, KOH, MeOH, r.t., 13 h (66)

95%

EtSH, AlCl$_3$, CH$_2$Cl$_2$, r.t. (67)

	Time (min)	*Yield* (%)
o-	9	87
p-	10	82

Table 5 Examples of Reduction of Iodobenzene to Benzene

Reaction conditions	*Yield (%)*	*Ref.*
H_2/Pd ($CaCO_3$), KOH, EtOH, r.t.	100	57
HCO_2H/10% Pd (C), DMF, reflux 3 h	90–95	71
$LiAlH_4$, THF, 25 °C, 24 h	83	36
$NaBH_4/Cp_2TiCl_2$, DMF, 70 °C, 9 h	98	77
Zn, $NiCl_2$, NaI, H_2O, HMPA, 60 °C, ultrasound, 4 h	92	65
$Fe(CO)_5$, K_2CO_3, CO, MeOH, 60 °C, 48 h	88	79

4.5.3 HYDROGENOLYSIS OF TRIGONAL CARBON–OXYGEN BONDS

Hydrogenolysis of the carbon–oxygen bond in alcohols, enols and phenols is relatively difficult, regardless of type.

4.5.3.1 Hydrogenolysis of Enols

In derivatives of enols, cleavage of the trigonal carbon to oxygen bond has been accomplished by catalytic hydrogenation and by sodium. Examples are substitution of the trifluoromethanesulfonyloxy (triflyloxy) group by hydrogen in steroidal enol triflates by catalytic hydrogenation (equation 68),[81] and by catalytic transfer of hydrogen using tributylammonium formate and palladium (equation 69),[82] and formation of cyclohexene in the Birch reduction of 1-methoxy-1,4-cyclohexadiene with sodium in liquid ammonia.[83]

OTf; TfO; H_2/PtO_2, EtOH, r.t., 1 atm; 70% (68)

C_8H_{17}; TfO; $HCO_2H/(Ph_3P)_2Pd(OAc)_2$, DMF, Bu_3N, 60 °C, 1 h; 77%; C_8H_{17} (69)

A more comprehensive discussion of the reduction of enol derivatives can be found in Chapter 4.6, this volume.

4.5.3.2 Hydrogenolysis of Phenols and Their Derivatives

Replacement of a free hydroxy group in phenols by hydrogen is impractical. It takes place only under very energetic conditions, for example on distillation with zinc dust, or by heating with dry lithium aluminum hydride at 350 °C.[84] Such drastic procedures are certainly not suited for compounds carrying sensitive functional groups. Substitution of hydroxy by hydrogen in polycyclic aromatic hydroxy compounds is achieved by extended refluxing with azeotropic (57%) hydriodic acid in acetic acid (equation 70).[85] The same limitation to polycyclic phenols is valid for a very gentle method of hydrogenolysis using an excess of a mixture of ethanethiol with anhydrous aluminum chloride (equation 70).[55]

Though direct hydrogenolysis of free phenols seems hardly feasible, this does not apply to derivatives of phenols such as their ethers, esters, sulfonyl esters, arylisoureas and aryloxytetrazoles. Some phenol ethers are hydrogenolyzed by heating with sodium in ethanol. Pyrogallol trimethyl ether (1,2,3-trimethoxybenzene) loses the central methoxy group, yielding the dimethyl ether of resorcinol (equation 71).[86]

The two reducing agents used for hydrogenolysis of polycyclic phenols, *i.e.* hydriodic acid and ethanethiol with aluminum chloride, are also applicable to polycyclic phenol ethers. Thus 7-methoxybenz[*a*]an-i, 52% or ii, 76% (70)

i, 57% HI, P, AcOH, reflux, 15 h; ii, 2.5 equiv. $AlCl_3$, 5 equiv. EtSH, CH_2Cl_2, r.t., 5 h

OMe, OMe, OMe — Na, EtOH, heat → OMe, OMe (71)

thracene is transformed in 95% yield into benz[*a*]anthracene on refluxing for 15 h with 57% hydriodic acid and acetic acid,[85] and 9-methoxy- and 9-ethoxy-phenanthrene are converted to phenanthrene on treatment with ethanethiol and aluminum chloride (equation 72).[55]

OMe — i, 2 h, 99.7% → ← i, 2.5 h, 96% — OEt (72)

i, 2.5 equiv. $AlCl_3$, 5 equiv. EtSH, CH_2Cl_2, r.t.

Hydriodic acid reduces some esters of polycyclic phenols. 7-Acetoxybenz[*a*]anthracene and 7,12-diacetoxybenz[*a*]anthracene are converted to benz[*a*]anthracene on refluxing with hydriodic acid and acetic acid (equation 73).[85] On the other hand, 9-acetoxyphenanthrene is not reduced, but hydrolyzed to 9-hydroxyphenanthrene.[85]

OAc — i, 99% → ← i, 100% — OAc, OAc (73)

i, 57% HI, AcOH, reflux, 15 h

A more general method for hydrogenolysis of aryl–oxygen bonds is reduction of aryl trifluoromethanesulfonates (triflates). The cleavage can be effected by catalytic hydrogen transfer using triethylammonium formate[87,88] or sodium borohydride[88] as hydrogen donors, and palladium complexes as catalysts. This method is applicable not only to triflates of condensed aromatic phenols, but also to phenyl triflates. The hydrogenolysis of the triflate group occurs in preference to reduction of reducible groups present in the molecule such as nitro[87] and cyano (equations 74–76).[88]

OTf — i, 98%, ii, 81% or iii, 91% → (74)

i, (TfO-α) HCO_2H, Et_3N, DMF, $Pd(OAc)_2$, Ph_3P, 65 °C, 12 h; ii, (TfO-α) $NaBH_4$, $(Ph_3P)_4Pd$, Ph_3P, THF, 62 °C, 15 h; iii, (TfO-β) HCO_2H, Et_3N, DMF, $Pd(OAc)_2$, Ph_3P, 60 °C, 1 h

Other aryl fluoroalkanesulfonates besides trifluoromethanesulfonyl esters of phenols can be reduced to arenes. Moreover, the preparation of such esters and their reduction may be carried out 'in one pot'. A mixture of *m*-methoxyphenol, perfluorooctanesulfonyl fluoride, triethylamine, dimethylformamide, for-

$$\text{TfO–C}_6\text{H}_4\text{–CN} \xrightarrow{\text{i, 69\% or ii, 81\%}} \text{C}_6\text{H}_5\text{–CN} \qquad (75)$$

i, HCO_2H, Et_3N, DMF, $Pd(OAc)_2$, Ph_3P, 65 °C, 12 h; ii, $NaBH_4$, $(Ph_3P)_4Pd$, Ph_3P, THF, 62 °C, 2 h

$$m\text{-}O_2N\text{–C}_6\text{H}_4\text{–OTf} \xrightarrow[\text{79\%}]{\text{HCO}_2\text{H, Pd(OAc)}_2\text{, Ph}_3\text{P, Et}_3\text{N, THF, 75 °C, 30 min}} \text{C}_6\text{H}_5\text{–NO}_2 \qquad (76)$$

mic acid and bis(triphenylphosphine)palladium dichloride heated at 80 °C for 8 h affords a 70% yield of anisole.[89]

Several derivatives of phenols have been found to be especially suited for hydrogenolysis by catalytic hydrogenation. Phenol ethers prepared by the reaction of phenols with 1-phenyl-5-chlorotetrazole or with 2-chlorobenzoxazole are hydrogenolyzed over 5% palladium on carbon or over platinum oxide, but not over Raney nickel. The hydrogenations are run at 35 °C in benzene, ethanol or tetrahydrofuran, and give 35–89% yields of the corresponding hydrocarbons.[90] The reaction sequence is exemplified by conversion of phenylphenols to biphenyls (equation 77).

(77)

Ph–C₆H₄–OH → (i; *o*-Ph; 86%, *p*-Ph; 98%) → Ph–C₆H₄–O–(2-benzoxazolyl)

Ph–C₆H₄–OH → (iii; *o*-Ph; 98%, *p*-Ph; 93%) → Ph–C₆H₄–O–(1-phenyltetrazol-5-yl)

Ph–C₆H₄–O–(2-benzoxazolyl) → (ii; *o*-Ph, 7 h; 82%, *p*-Ph, 16 h; 75%) → C_6H_5–Ph

Ph–C₆H₄–O–(1-phenyltetrazol-5-yl) → (iv; *o*-Ph, 8 h; 82%, *p*-Ph, 7 h;78%) → C_6H_5–Ph

i, Me_2CO, K_2CO_3, 2-chlorobenzoxazole, reflux, overnight; ii, H_2/5% Pd(C), C_6H_6, 35 °C;, iii, Me_2CO, K_2CO_3, 5-chloro-1-phenyltetrazole, reflux, overnight; iv, H_2/5% Pd(C), C_6H_6, 35 °C

In this way, guaiacol affords an 86% yield of anisole, thymol a 72% yield of *p*-cymene, 1- and 2-naphthol 50% and 65% yields of naphthalene respectively, and *p*-chlorophenol a 70% yield of benzene with simultaneous hydrogenolysis of chlorine.[90]

Other derivatives of phenols liable to hydrogenolysis are alkylated arylisoureas: *O*-aryl-*N,N*-dialkylisoureas[91] and *O*-aryl-*N,N'*-dialkylisoureas.[92] Both derivatives suffer easy hydrogenolysis by catalytic hydrogenation under very mild reaction conditions (equation 78).[91,92]

Conversion of phenols to aryl cyanates is carried out by adding equivalent amounts of the phenol and triethylamine dissolved in ether to a stirred solution of cyanogen bromide in the same solvent at –10 °C, filtering triethylamine hydrobromide after 50 min, and evaporating the solvent *in vacuo* at room temperature.[91] Aryl cyanates are transformed to *O*-aryl-*N,N*-diethylisoureas by adding to their suspensions in dry ether with stirring at 0–20 °C a solution of diethylamine in ether, and by evaporating the ether after 15–60 min at room temperature *in vacuo*. Hydrogenolysis is then best effected in ethanol over 5% palladium on carbon at 20 or 45 °C at atmospheric pressure. Yields of hydrogenations are usually 95–100% (78–97% overall based on phenols). *ortho*-Substituted derivatives are hydrogenolyzed more slowly than *para* and *meta* analogs. *p*-Chlorophenol and *p*-nitrophenol are in this way converted into benzene and aniline in 88% and 52% overall yields, respectively. Under the conditions used, chlorine is hydrogenolyzed, and the nitro group is reduced to the amino group.[91]

Ph–OH →(i, 79%) Ph–O–C(=NC$_6$H$_{11}$)NHC$_6$H$_{11}$ →(ii, 95%) C$_6$H$_6$

Ph–OH →(iii) Ph–O–CN →(iv, 88%) Ph–O–C(=NH)NEt$_2$ →(v, 98%) C$_6$H$_6$ (78)

i, $C_6H_{11}N{=}C{=}NC_6H_{11}$, 70 °C, 24 h; ii, H_2/5% Pd(C), EtOAc, 20 °C, 1 h; iii, BrCN, Et_3N, Et_2O, –10 °C, 1 h; iv, Et_2NH; v, H_2/5% Pd(C), EtOH, 20 °C, 8 h

O-Aryl-*N,N'*-dicyclohexylisoureas are synthesized by stirring mixtures of phenols with 2–3 mol equiv. of dicyclohexylcarbodiimide at 45–100 °C for 1–3 d. Hydrogenations are accomplished in isopropyl alcohol or ethyl acetate at 20 °C and atmospheric pressure using 200–300 mg of 5% palladium on carbon or on calcium carbonate per 0.02 mol of the phenol. The hydrogenolysis can also be effected with the crude mixture of the *O*-arylisoureas and the unreacted carbodiimide using as much hydrogen as needed for hydrogenation of the excess carbodiimide.

Table 6 shows results of hydrogenolysis of phenols by all four methods described above.

Table 6 Comparison of Results of Hydrogenolysis of Phenols after their Conversion to 5-Aryloxy-1-phenyltetrazoles (A), 2-Aryloxybenzoxazoles (B), *O*-Aryl-*N,N*-'diethylisoureas (C), and *O*-Aryl-*N,N'*-dicyclohexylisoureas (D) [90–92]

Phenol	*Product*	*Method A*[a] *yield (%)*	*Method B*[b] *yield (%)*	*Method C*[c] *yield (%)*	*Method D*[d] *yield (%)*
Phenol (C$_6$H$_5$–OH)	Benzene			86	85
4-Phenylphenol	Biphenyl	73	75	96	78
1-Naphthol	Naphthalene	44		92	44
MeO–C$_6$H$_4$–OH	MeO–C$_6$H$_5$	81		90	70
Cl–C$_6$H$_4$–OH	Benzene	64		88	76

[a] H_2/cat./tetrazole. [b] H_2/cat./benzoxazole. [c] H_2/cat./diethylisourea. [d] H_2/cat./dicyclohexylisourea.

4.5.4 HYDROGENOLYSIS OF TRIGONAL CARBON–SULFUR BONDS

Hydrogenolysis of vinyl sulfides,[26] vinyl sulfoxides[93] and vinyl sulfones[94] is rare, and may be accompanied by saturation of the double bond.[26] Thus vinylic sulfide (**29**) gives only 5% of 1,5-diphenyl-2-

pentene, and 80% of 1,5-diphenylpentane on reduction with lithium aluminum hydride in the presence of titanium tetrachloride (equation 79).[26]

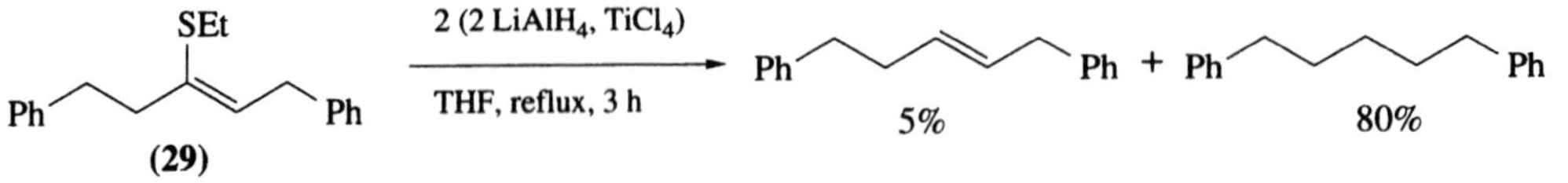

SEt; Ph, Ph **(29)** —2 (2 $LiAlH_4$, $TiCl_4$) / THF, reflux, 3 h→ Ph–Ph 5% + Ph–Ph 80% (79)

Vinylic sulfoxide (**30**) is desulfurized to (**31**) with *t*-butyllithium (equation 80),[93] and sulfone (**32**) affords triphenylethylene on reduction with aluminum amalgam or lithium aluminum hydride (equation 81).[94]

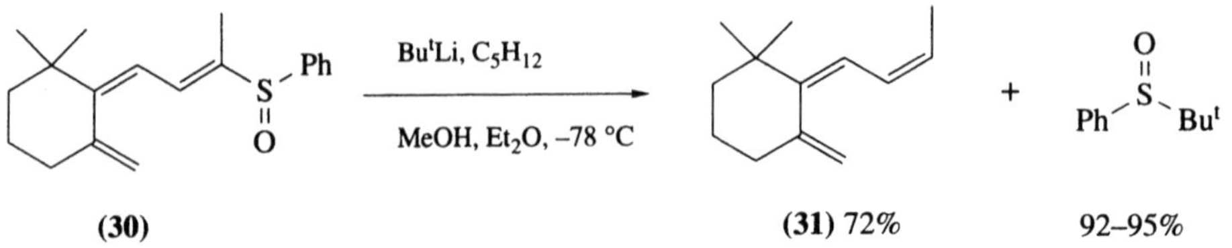

(30) —Bu^tLi, C_5H_{12} / MeOH, Et_2O, –78 °C→ **(31)** 72% + Ph–S(=O)–Bu^t 92–95% (80)

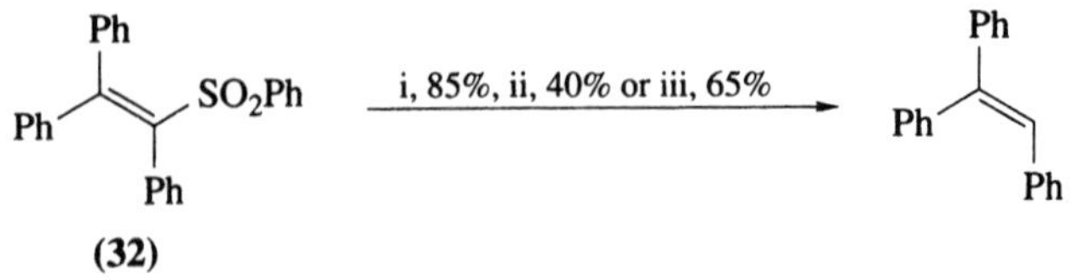

(32) —i, 85%, ii, 40% or iii, 65%→ Ph, Ph, Ph (81)

i, $AlHg_x$, 10% H_2O in THF, reflux, 3 h; ii, $LiAlH_4$, Et_2O, reflux; iii, $LiAlH_4$–$CuCl_2$ (1:2)

Hydrogenolysis of compounds with sulfur atoms attached to aromatic rings such as benzenethiols, and aryl sulfides, disulfides, sulfoxides and sulfones takes place on refluxing with Raney nickel or nickel boride. Sulfur combines with nickel, and hydrogen replaces the sulfur-containing group.

Raney nickel is a stronger desulfurizing agent than nickel boride. On the other hand nickel boride, prepared from nickel(II) chloride and sodium borohydride, is more selective since it does not desulfurize aromatic sulfones. An example of such selectivity is desulfurization of *p*-phenylsulfonylphenyl *p*-tolyl sulfide (**33**), which affords diphenyl sulfone in 91% yield and toluene in 84% yield. In contrast, Raney nickel desulfurizes both the sulfide and the sulfone giving a mixture of benzene and toluene (equation 82).[95]

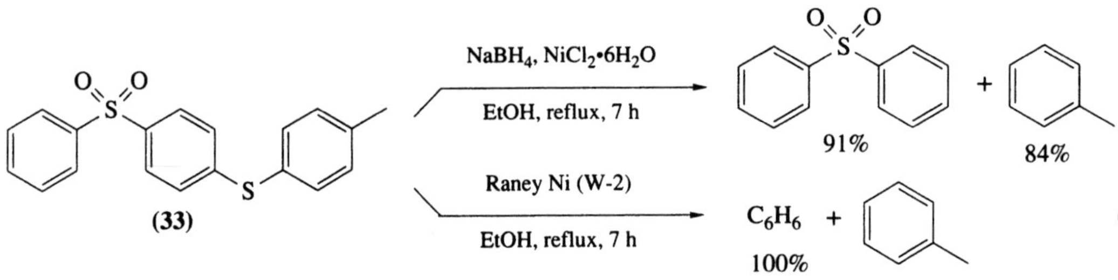

(33) —$NaBH_4$, $NiCl_2$•$6H_2O$ / EtOH, reflux, 7 h→ 91% + 84%

(33) —Raney Ni (W-2) / EtOH, reflux, 7 h→ C_6H_6 + 100% (82)

As is evident from Table 7, nickel boride does not give as high yields as Raney nickel. However, it offers certain advantages over the latter. In addition to higher selectivity, its preparation is faster and more simple than that of Raney nickel, it allows for better dosage, and is not pyrophoric.[95]

1-Naphthyl phenyl sulfide is desulfurized to naphthalene on refluxing with lithium aluminum hydride and titanium tetrachloride,[26] and 1-naphthyl ethyl sulfide and 1-naphthyl isopropyl sulfide are converted to naphthalene on treatment with ethanethiol and anhydrous aluminum chloride (equation 83).[55]

Table 7 Desulfurization of Aromatic Sulfur Compounds by Nickel Boride and Raney Nickel[95]

Compound	*% Yield of toluene with* Nickel boride	Raney nickel
p-MeC_6H_4SH	60	91
$(p$-$MeC_6H_4S)_2$	37	85
$(p$-$MeC_6H_4)_2S$	72	86
$(p$-$MeC_6H_4)_2SO$	80	100
$(p$-$MeC_6H_4)_2SO_2$	0	88

1-(SR)naphthalene —(5 EtSH, 2.5 $AlCl_3$, CH_2Cl_2, r.t.)→ naphthalene (83)

	Time (h)	*Yield* (%)
R = Et	5.5	89
R = Pr	10	91
R = Ph (8 $LiAlH_4$, 4 $TiCl_4$, THF, reflux)	3	56

4.5.5 HYDROGENOLYSIS OF TRIGONAL CARBON–NITROGEN BONDS

4.5.5.1 Hydrogenolysis of Enamines

Substitution of hydrogen for amino or alkylated amino groups is very rare. One example is hydrogenolysis of enamines. Reduction of enamines with complex hydrides, particularly with sodium cyanoborohydride, results in saturation of the double bond and conversion to amines. On the other hand, treatment of enamines with chloroalanes and especially alanes results in the cleavage of the carbon–nitrogen bond and formation of the corresponding alkene and amine; some examples are shown in Table 8.[96] The cleavage proper takes place by an addition–elimination mechanism in which the first step is addition of the alane across the double bond, yielding an aminoalkylalane. Whereas hydrolysis of the addition product with water would give the saturated amine, heating it in refluxing solvents results in elimination of aminoalane to afford an alkene (equation 84).[96]

1-pyrrolidinocyclohexene —(AlH_3 ($LiAlH_4$ + $AlCl_3$), Et_2O, reflux, 45 h)→ 2-(H_2Al)-1-pyrrolidinocyclohexane —(heat, 80%)→ cyclohexene + H_2Al—N(pyrrolidine) (84)

Table 8 Examples of Reduction of Pyrrolidine Enamines[96]

Enamine	*Alane*	*Alkene*	*Yield (%)*	*Amine*	*Yield (%)*
Et	$AlHCl_2$		27	Et, Et	66
	AlH_2Cl	MeCH=CHEt	35		56
	AlH_3		81		13
Pr^n	$AlHCl_2$		19	Pr^n, Et	78
	AlH_2Cl	$EtCH=CHPr^n$	30		56
	AlH_3		85		9
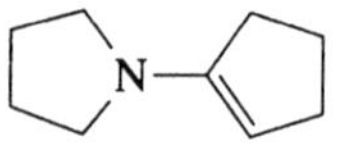	AlH_3	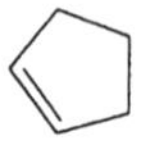	92	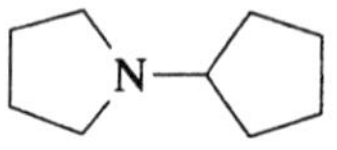	5
	$AlHCl_2$		75		22
	AlH_2Cl		75		15
	AlH_3		80		13
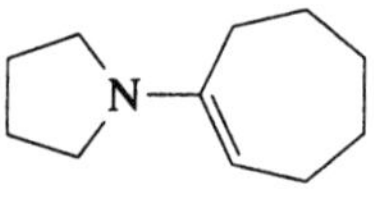	AlH_3		85	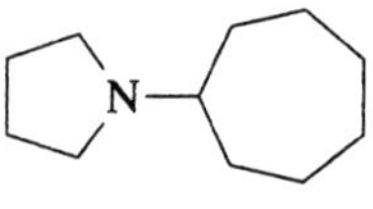	14
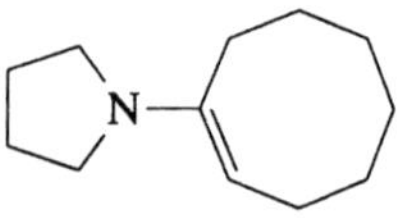	AlH_3		85	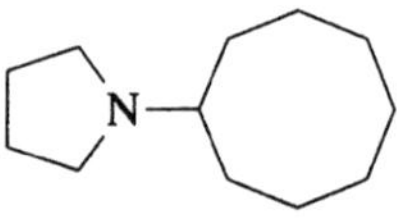	10

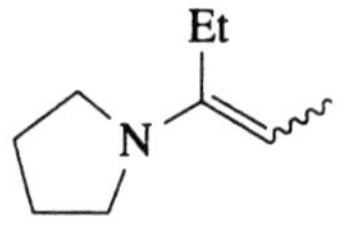

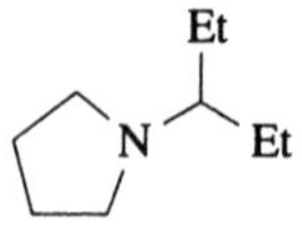

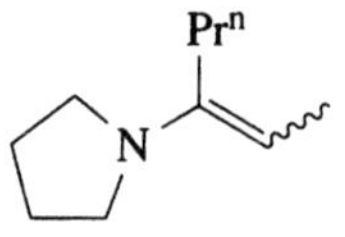

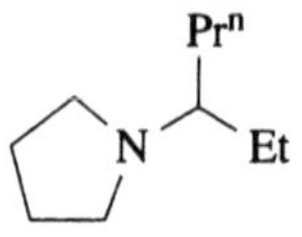

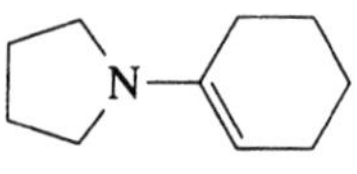

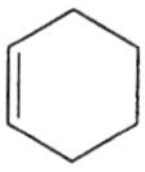

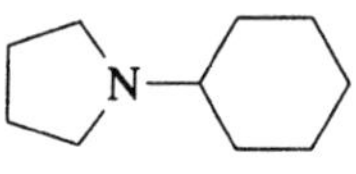

4.5.5.2 Hydrogenolysis of Aromatic Amines

One-step replacement of an aromatic amino group by hydrogen is impractical. Exceptionally, hydrogenolysis took place when *p*-toluidine was treated with solid lithium aluminum hydride at approximately 350 °C. Toluene was obtained in 17% yield.[84]

Equally unusual is hydrogenolysis of a bond between an aromatic ring and a quaternary amino group. Phenyltrialkylammonium iodides are reductively cleaved to benzene and trialkylamines by electrolysis (equation 85).[97]

$$\mathrm{Ph{-}\overset{+}{N}Me_3\ \overset{-}{I}} \xrightarrow{\mathrm{Na_2CO_3,\ H_2O,\ Pb\ cathode,\ 30\ A,\ 2.5\ h}} \underset{75\%}{\mathrm{C_6H_6}} + \underset{77\%}{\mathrm{Me_3N{\cdot}HCl}} \qquad (85)$$

4.5.5.3 Hydrogenolysis of Arenediazonium Salts

A general method of replacement of the primary aromatic amino group by hydrogen is reduction of aromatic diazonium salts, prepared by diazotization, typically by treatment of the primary amine or its salts with sodium or potassium nitrite in strongly acidic aqueous media. Conversion to the diazonium salts may also be accomplished in nonaqueous media using acetic acid,[98] or using alkyl nitrites (equation 86).[99] Most diazonium salts decompose at temperatures above 0–5 °C. Only certain salts such as naphthalene-1,5-sulfonates,[100] tetrafluoroborates[101–103] and hexafluorophosphates[104,105] are stable at room temperature.

Reduction of diazonium salts resulting in replacement of the diazonium group by hydrogen has been known for 125 years, and has been carried out using alcohols,[106–109] sodium stannite,[110] hypophosphorous acid,[111,112] formaldehyde,[113] zinc,[100] and other reagents.[114] In addition to these classical reagents, reviewed in an article in *Organic Reactions*,[114] several new reducing agents have been developed:

$$\mathrm{Ar{-}NH_2} \xrightarrow[\mathrm{NaNO_2,\ H_2SO_4,\ AcOH}]{\mathrm{NaNO_2,\ 3\ HCl,\ H_2O;\ C_5H_{11}ONO,\ THF}} \mathrm{Ar{-}\overset{+}{N}{\equiv}N\ X^-} \quad X = Cl;\ X = C_5H_{11}O;\ X = HSO_4 \tag{86}$$

potassium ferrocyanide,[115] tin in hydrochloric acid,[116,117] sodium borohydride,[102] triethylsilane,[103] tributylstannane (tributyltin hydride),[103] and a roster of organic compounds which provide hydrogen for combinations with the aryl cations[104] or radicals[99,103,108,118] formed by decomposition of the diazonium salts.

The mechanism of displacement of the diazonium group by hydrogen depends on the chemical nature of the diazonium compound, the reducing agents and the reaction conditions. Electron-withdrawing groups in the aromatic ring generally facilitate the replacement.[115]

In appropriately polar solvents, diazonium salts decompose to nitrogen and aryl cations, and formation of arenes may take place by transfer of hydrogen from the reducing agent or from the solvent. Depending on reaction conditions, an ionic mechanism (equation 87),[119] or a radical mechanism (equation 88)[107] may be operating, and experimental evidence is claimed for both. However, the radical mechanism seems to have more support.

$$\mathrm{Ar\overset{+}{N}{\equiv}N\ BF_4^-} \longrightarrow \mathrm{Ar^+ + N_2 + BF_4^-} \xrightarrow{\mathrm{NaBF_4}} \mathrm{ArH + BH_3BF_3 + NaF}$$

$$\mathrm{Ar\overset{+}{N}{\equiv}N\ BF_4^-} \longrightarrow \mathrm{Ar^+ + N_2 + BF_4^-} \xrightarrow{\text{2-phenyl-1,3-dioxolane}} \mathrm{ArH} + \text{2-phenyl-1,3-dioxolan-2-ylium}\ \mathrm{BF_4^-} \tag{87}$$

$$\mathrm{Ar\overset{+}{N}{\equiv}N\ X^-} \xrightarrow{\mathrm{Cu^+}} \mathrm{Ar\bullet + X^- + N_2 + Cu^{2+}}$$

$$\mathrm{Ar\bullet} \xrightarrow{\mathrm{EtOH}} \mathrm{ArH} + \mathrm{CH_3\dot{C}HOH} \longrightarrow \mathrm{CH_3CHO} \tag{88}$$

In reduction of arenediazonium fluoroborate with tributylstannane, formation of the arene takes place not only by hydrogen atom abstraction from the stannane, but also by reaction of the radical intermediate with the solvent. Reduction of 4-methoxybenzenediazonium fluoroborate with tributyltin deuteride in tetrahydrofuran or acetonitrile affords not only deuterated but also deuterium-free arenes (equation 89). There are also small amounts of biphenyl by-products.[103]

$$\mathrm{Ar\overset{+}{N}{\equiv}N\ BF_4^- + Bu_3SnD} \longrightarrow \mathrm{ArN{=}ND + BF_3 + Bu_3SnF} \tag{89}$$

$$\mathrm{Ar\overset{+}{N}{\equiv}N\ BF_4^-} \longrightarrow \mathrm{Ar\bullet} \xrightarrow[\text{40–62\%}]{\mathrm{Bu_3SnD,\ MeCN}} \mathrm{ArH} \qquad \mathrm{ArN{=}ND} \xrightarrow{\text{38–60\%}} \mathrm{ArD + N_2}$$

Finally, hydrogen transfer can occur intramolecularly, as in the reduction of arenediazonium fluoroborates with formamide. The intramolecular transfer has been proven by reaction with *C*-deuterated formamide, which introduced deuterium in place of the diazonium group (equation 90).[101]

Sodium stannite converts (**34**) to (**35**). Both the diazonium salt and the sodium stannite are prepared *in situ* (equation 91).[110]

Tin metal in the presence of aqueous hydrochloric acid in a mixture of dichloromethane and polyethylene glycol replaces the diazonium group with hydrogen in 38–76% isolated yields (equation 92).[116,117]

$$p\text{-}O_2NC_6H_4N_2^+ \xrightarrow{DCONH_2} p\text{-}O_2NC_6H_4N{=}N{-}NH{-}C(O)D \longrightarrow p\text{-}O_2NC_6H_4D \quad (90)$$

$$\textbf{(34)} \xrightarrow[73\%]{SnCl_2,\ NaOH,\ ice} \textbf{(35)} \quad (91)$$

$$[p\text{-}ClC_6H_4N_2^+\ Cl^-] \xrightarrow[HO(CH_2CH_2O)_nH,\ 72\%]{Sn,\ 3\ M\ HCl\ (3\ equiv.),\ 0\ °C\ to\ r.t.,\ 3\ h} ClC_6H_5 \quad (92)$$

The most specific reagent for replacement of a diazonium group by hydrogen is hypophosphorous acid.[111,112] It is commercially available, it works at room temperature, and it is especially selective: it does not reduce any of the common functional groups present in diazonium salts. The reductive removal of the amino group can be carried out in one pot using a large excess (5–15 mol) of aqueous hypophosphorous acid after diazotization (equation 93).[114] Both diazotization and reduction may be carried out in hypophosphorous acid. Under such conditions, an amino group linked to the aromatic ring is replaced by hydrogen, while aliphatic amino groups are left intact (equation 94).[111]

$$p\text{-}MeC_6H_4NH_2 \xrightarrow[67\%]{i,\ NaNO_2,\ HCl,\ H_2O,\ cooling;\ ii,\ H_3PO_2,\ 0\ °C,\ few\ days} MeC_6H_5 \quad (93)$$

$$H_2N(CH_2)_nC_6H_4NH_2 \xrightarrow[ii,\ r.t.,\ 4\ h]{i,\ 50\%\ H_3PO_2\ (15\ mol),\ NaNO_2,\ 5\ °C,\ 45\ min} H_2N(CH_2)_nC_6H_5 \quad (94)$$

	Yield(%)
m-, *n* = 1	87
m-, *n* = 2	70
p-, *n* = 1	84
p-, *n* = 2	71

Historically, the first replacement of a diazonium group by hydrogen was accomplished with alcohols.[109] Of these, ethanol is the best, while methanol tends to convert diazonium compounds to aryl methyl ethers.[114] The hydrogen atom replacing the diazonium group comes from the alcohol, which is converted to the corresponding aldehyde. The reaction is catalyzed by UV irradiation[108] and by copper(I) oxide.[106,107] Copper(I) oxide alone reduces the diazonium compounds, but in poor yields.[107] The reaction can be carried out in aqueous media[108] or in organic solvents.[106,107]

Hydrogen donors other than alcohols have been used for replacement of the diazonium group by hydrogen; these include acetone,[108] tetrahydrofuran,[99,120] dioxane,[120] dioxolane,[120] ethylene glycol dimethyl ether,[120] hexamethylphosphoramide (HMPA)[121] and other compounds.[120]

Very often, the replacement of a diazonium group by hydrogen is carried out using stable crystalline diazonium salts, such as the 1,5-naphthalenedisulfonates,[100] 2-naphthol-1-sulfonates,[100] tetrafluoroborates,[101–103,115,118] hexafluorophosphates[104] and mercury bromides.[72]

Diazonium tetrafluoroborates are reduced with potassium ferrocyanide at rates depending strongly on substituents in the benzene rings. Relative rates of reduction of *p*-nitrobenzenediazonium, benzenediazonium and *p*-methoxybenzenediazonium fluoroborates have been found to be 3062:1:0.0031, showing a span of six orders of magnitude, electron-withdrawing substituents increasing, and electron-releasing groups decreasing the rates.[115]

A useful addition to the roster of reducing agents capable of replacement of diazonium groups by hydrogen in good yields is sodium borohydride. It is applied to dry diazonium fluoroborates in suspension in methanol or in solution in dimethylformamide (equation 95).[102]

$$HO_2C\text{—}C_6H_4\text{—}N_2^+\,BF_4^- \xrightarrow[68\%]{NaBH_4,\ MeOH,\ cooling} HO_2C\text{—}C_6H_5 \quad (95)$$

Another hydride used successfully for the same purpose is tributylstannane (tributyltin hydride) in ether, tetrahydrofuran or acetonitrile.[103] Triethylsilane acts similarly, but more slowly and more selectively: it does not reduce nitro groups (equation 96).[103]

$$MeO\text{—}C_6H_4\text{—}N_2^+\,BF_4^- \xrightarrow{i,\ 94\%,\ ii,\ 71\%\ or\ iii,\ 90\%} MeO\text{—}C_6H_5 \quad (96)$$

i, Bu_3SnH, THF, r.t., 50 min; ii, Bu_3SnH, MeCN, r.t., 1.5 h; iii, Et_3SiH, MeCN, r.t., overnight

Other hydrogen donors which have been applied to the reduction of diazonium fluoroborates are formamide[101] and *N*-benzyl-1,4-dihydronicotinamide (equation 97).[118]

$$O_2N\text{—}C_6H_4\text{—}N_2^+\,BF_4^- \xrightarrow[82\%]{DMF,\ Et_3N,\ r.t.,\ 45\ min} O_2N\text{—}C_6H_5 \quad (97)$$

(alternatively: $PhCH_2$–N(1,4-dihydropyridine)–$CONH_2$, MeOH)

Like diazonium fluoroborates, diazonium fluorophosphates decompose under anhydrous conditions in tetramethylurea to give the parent compounds. The reaction is carried out by adding dry diazonium salts to tetramethylurea at room temperature. The reaction is exothermic, but in some cases warming to 65 °C is desirable (equation 98).[104]

$$O_2N\text{—}C_6H_4\text{—}N_2^+\,PF_6^- \xrightarrow[62\%]{(Me_2N)_2CO,\ r.t.,\ 1.5\ h} O_2N\text{—}C_6H_5 \quad (98)$$

An unexpected displacement of a diazonium group in diazotized aminonaphthalenes by hydrogen was discovered when diazonium mercury(II) bromides, prepared by treatment of diazonium bromides with mercury(II) bromide, were stirred at room temperature for 10–20 min in hexamethylphosphoramide (HMPA) (equation 99).[121]

$$(1\text{-}C_{10}H_7N_2^+)_2\,HgBr_4^{2-} \xrightarrow[70\%]{i} C_{10}H_8 \xleftarrow[94\%]{i} (2\text{-}C_{10}H_7N_2^+)_2\,HgBr_4^{2-} \quad (99)$$

i, HMPA, r.t., 10–20 min

Examples of replacement of diazonium groups by hydrogen by means of various reducing agents are shown in Table 9.

Table 9 Comparison of Various Methods of Replacement of Diazonium Group by Hydrogen

	% Yield of product with								
	Sn/HCl	H_3PO_2	*EtOH/hv*	$EtOH/Cu_2O$	$NaBH_4$	Et_3SiH	Bu_3Sn	$HCONH_2$	$(Me_2N)_2CO$
$CH_3-C_6H_4-\overset{+}{N}_2$		77–83[104]		45[106]		25[103]	87[103]		25–30[104]
$MeO-C_6H_4-\overset{+}{N}_2$	74–87[117] (48–76)[a]					90[103]	94[103]		
$Cl-C_6H_4-\overset{+}{N}_2$	64–100[116] (50–72)[a]					70[103]	89[103]		70[104]
$O_2N-C_6H_4-\overset{+}{N}_2$		55–60[104]		97[106]		72[103]	52[103b]	82[101]	62[104]
$2-CO_2H-C_6H_4-\overset{+}{N}_2$			92.5[108]		77[102]				

[a] Yield of pure compound. [b] Partial reduction to aniline lowered the yield.

4.5.6 REFERENCES

1. J. R. Lacher, A. Kianpour and J. D. Park, *J. Phys. Chem.*, 1956, **60**, 1454.
2. S. M. Kupchan and A. Afonso, *J. Org. Chem.*, 1960, **25**, 2217.
3. C. B. Miller, J. H. Pearson and L. B. Smith, *US Pat.* 2 802 881 (1957) (*Chem. Abstr.*, 1958, **52**, 1198e).
4. H. Conroy, *J. Am. Chem. Soc.*, 1955, **77**, 5960.
5. J. A. Virgilio and E. Heilweil, *Org. Prep. Proced. Int.*, 1982, **14**, 9.
6. D. J. Burton and R. L. Johnson, *J. Am. Chem. Soc.*, 1964, **86**, 5361.
7. D. J. Burton and R. L. Johnson, *Tetrahedron Lett.*, 1966, 2681.
8. M. Hudlicky, *J. Fluorine Chem.*, 1979, **14**, 189; 1983, **23**, 241.
9. V. Tolman and K. Vereš, *Tetrahedron Lett.*, 1964, 1967.
10. M. Hudlicky, 'Chemistry of Organic Fluorine Compounds', 2nd edn., Horwood, Chichester, 1976, p. 174.
11. F. J. Mettile and D. J. Burton, *Fluorine Chem. Rev.*, 1968, **1**, 315.
12. I. L. Knunyants, M. P. Krasuskaya and E. I. Mysov, *Izv. Akad. Nauk SSSR, Otd. Khim. Nauk*, 1960, 1412 (*Chem. Abstr.*, 1961, **55**, 349c).
13. I. L. Knunyants, E. I. Mysov and M. P. Krasuskaya, *Izv. Akad. Nauk SSSR, Otd. Khim. Nauk*, 1958, 906 (*Chem. Abstr.*, 1959, **53**, 1102b).
14. I. L. Knunyants, M. P. Krasuskaya, E. I. Mysov and I. A. Mukhtarov, *Izv. Akad. Nauk SSSR, Otd. Khim. Nauk*, 1962, 2141 (*Chem. Abstr.*, 1963, **58**, 13 802d).
15. L. Hilaire and F. G. Gault, *Bull. Soc. Chim. Fr.*, 1969, 842.
16. M. Hudlicky, *J. Fluorine Chem.*, 1988, **40**, 99.
17. J. Bornstein, M. R. Borden, F. Nunes and H. I. Tarlin, *J. Am. Chem. Soc.*, 1963, **85**, 1609.
18. R. Meier and F. Böhler, *Chem. Ber.*, 1957, **90**, 2344.
19. W. Dmowski, *J. Fluorine Chem.*, 1985, **29**, 273.
20. G. Fuller and J. C. Tatlow, *J. Chem. Soc.*, 1961, 3198.
21. G. E. Ham and W. P. Coker, *J. Org. Chem.*, 1964, **29**, 194.
22. L. Horner, L. Schläfer and H. Kämmerer, *Chem. Ber.*, 1959, **92**, 1700.
23. C. F. Wilcox, Jr. and J. G. Zajacek, *J. Org. Chem.*, 1964, **29**, 2209.
24. K. V. Scherer, Jr., *Tetrahedron Lett.*, 1966, 5685.
25. K. L. Williamson, Y. F. L. Hsu and E. I. Young, *Tetrahedron*, 1968, **24**, 6007.
26. T. Mukaiyama, M. Hayashi and K. Narasaka, *Chem. Lett.*, 1973, 291.
27. P. G. Gassman and P. G. Pape, *J. Org. Chem.*, 1964, **29**, 160.
28. I. Yamamoto, T. Sakai, S. Yamamoto, K. Ohta and K. Matsuzaki, *Synthesis*, 1985, 676.
29. I. Pri-Bar and O. Buchman, *J. Org. Chem.*, 1986, **51**, 734.
30. N. Gatti, W. Jugelt and H. Lund, *Acta Chem. Scand., Ser. B*, 1987, **41**, 646.
31. T. Hirao, T. Masunaga, K. Hayashi, Y. Ohshiro and T. Agawa, *Tetrahedron Lett.*, 1983, **24**, 399.
32. A. L. Anderson, R. T. Bogan and D. J. Burton, *J. Fluorine Chem.*, 1971, **1**, 121.
33. D. J. Burton and F. J. Mettile, *J. Fluorine Chem.*, 1982, **20**, 157.
34. R. N. Haszeldine, *J. Chem. Soc.*, 1953, 922.
35. R. D. Chambers, F. G. Drakesmith and W. K. R. Musgrave, *J. Chem. Soc.*, 1965, 5045.
36. H. C. Brown and S. Krishnamurthy, *J. Org. Chem.*, 1969, **34**, 3918.
37. G. M. Brooke, R. D. Chambers, J. Heyes and W. K. R. Musgrave, *Proc. Chem. Soc.*, 1963, 213.
38. J. F. Tilney-Bassett, *Chem. Ind. (London)*, 1965, 693.
39. R. E. Banks, R. N. Haszeldine, J. V. Latham and I. M. Young, *J. Chem. Soc.*, 1965, 594.
40. D. J. Alsop, J. Burdon and J. C. Tatlow, *J. Chem. Soc.*, 1962, 1801.
41. P. Carrahar and F. G. Drakesmith, *Chem. Commun.*, 1968, 1562.
42. P. Caubère, *Angew. Chem., Int. Ed. Engl.*, 1983, **22**, 599.
43. L. A. Rothman and E. I. Becker, *J. Org. Chem.*, 1960, **25**, 2203.
44. M. Node, T. Kawabata, K. Ohta, M. Fujimoto, E. Fujita and K. Fuji, *J. Org. Chem.*, 1984, **49**, 3641.
45. T. Sakamoto, Y. Kondo and H. Yamanaka, *Synthesis*, 1984, 252.
46. N. A. Cortese and R. F. Heck, *J. Org. Chem.*, 1977, **42**, 3491.
47. Y. Akita, A. Inoue, K. Ishida, K. Terui and A. Ohta, *Synth. Commun.*, 1986, **16**, 1067.
48. F. Swarts, *Bull. Acad. R. Belg.*, 1920, 399 (*Chem. Abstr.* 1922, **16**, 2316).
49. F. L. M. Pattison and B. C. Saunders, *J. Chem. Soc.*, 1949, 2745.
50. H. Oelschläger and P. Schmersahl, *Arch. Pharm.*, 1963, **296**, 324.
51. M. Tashiro, H. Nakamura and K. Nakayama, *Org. Prep. Proced. Int.*, 1987, **19**, 442.
52. A. L. J. Beckwith and S. H. Goh, *J. Chem. Soc., Chem. Commun.*, 1983, 907.
53. G. M. Brooke, J. Burdon and J. C. Tatlow, *J. Chem. Soc.*, 1962, 3253.
54. P. L. Coe, G. M. Pearl and J. C. Tatlow, *J. Chem. Soc. C*, 1971, 604.
55. M. Node, K. Nishide, K. Ohta and E. Fujita, *Tetrahedron Lett.*, 1982, **23**, 689.
56. R. Baltzly and A. P. Phillips, *J. Am. Chem. Soc.*, 1946, **68**, 261.
57. (a) M. Busch and H. Stöve, *Chem. Ber.*, 1916, **49**, 1063; (b) W. Roth and P. von R. Schleyer, *Z. Naturforsch., Teil B*, 1983, **38**, 1697.
58. R. E. Lutz, G. Ashburn and R. J. Rowlett, Jr., *J. Am. Chem. Soc.*, 1946, **68**, 1322.
59. H. Kämmerer, L. Horner and H. Beck, *Chem. Ber.*, 1958, **91**, 1376.
60. M. K. Anwer and A. F. Spatola, *Tetrahedron Lett.*, 1985, **26**, 1381.
61. S. K. Boyer, J. Bach, J. McKenna and E. Jagdmann, Jr., *J. Org. Chem.*, 1985, **50**, 3408.
62. W. D. Guither, D. G. Clark and R. N. Castle, *J. Heterocycl. Chem.*, 1965, **2**, 67.
63. F. W. Neumann, N. B. Sommer, C. E. Kaslow and R. L. Shriner, *Org. Synth., Coll. Vol.*, 1955, **3**, 519.
64. I. Colon, *J. Org. Chem.*, 1982, **47**, 2622.
65. J. Yamashita, Y. Inoue, T. Kondo and H. Hashimoto, *Bull. Chem. Soc. Jpn.*, 1985, **58**, 2709.
66. D. J. Brown and P. Waring, *Aust. J. Chem.*, 1973, **26**, 443.
67. J. D. Citron, J. E. Lyons and L. H. Sommer, *J. Org. Chem.*, 1969, **34**, 638.

68. J. T. Pinhey and R. D. G. Rigby, *Tetrahedron Lett.*, 1969, 1267.
69. B. A. Fox and T. L. Threlfall, *Org. Synth., Coll. Vol.*, 1973, **5**, 346.
70. H. Erlenmeyer and W. Grubenmann, *Helv. Chim. Acta*, 1947, **30**, 297.
71. P. N. Pandey and M. L. Purkayastha, *Synthesis*, 1982, 876.
72. V. Bažant, M. Čapka, M. Černý, V. Chvalovský, K. Kochloefl, M. Kraus and J. Málek, *Tetrahedron Lett.*, 1968, 3303.
73. R. D. Rieke, S. J. Uhm and P. M. Hudnall, *J. Chem. Soc., Chem. Commun.*, 1973, 269.
74. S. Gronowitz and T. Raznikiewicz, *Org. Synth., Coll. Vol.*, 1973, **5**, 149.
75. R. S. Wade and C. E. Castro, *Org. Synth., Coll. Vol.*, 1988, **6**, 821.
76. M. Sako, K. Hirota and Y. Maki, *Tetrahedron*, 1983, **39**, 3919.
77. B. Meunier, *J. Organomet. Chem.*, 1981, **204**, 345.
78. H. M. Bell, C. W. Vanderslice and A. Spehar, *J. Org. Chem.*, 1969, **34**, 3923.
79. J.-J. Brunet and M. Taillefer, *J. Organomet. Chem.*, 1988, **348**, C5.
80. S. Yasui, K. Nakamura and A. Ohno, *Chem. Lett.*, 1984, 377.
81. V. B. Jigajinni and R. H. Wightman, *Tetrahedron Lett.*, 1982, **23**, 117.
82. S. Cacchi, E. Morera and G. Ortar, *Tetrahedron Lett.*, 1984, **25**, 4821.
83. A. J. Birch and G. S. R. Subba Rao, *Aust. J. Chem.*, 1970, **23**, 1641.
84. T. Severin and I. Ipach, *Synthesis*, 1973, 796.
85. M. Konieczny and R. G. Harvey, *J. Org. Chem.*, 1979, **44**, 4813.
86. H. Thoms and W. Siebeling, *Chem. Ber.*, 1911, **44**, 2134.
87. S. Cacchi, P. G. Ciattini, E. Morera and G. Ortar, *Tetrahedron Lett.*, 1986, **27**, 5541.
88. G. A. Peterson, F.-A. Kunng, J. S. McCallum and W. D. Wulff, *Tetrahedron Lett.*, 1987, **28**, 1381.
89. Q.-Y. Chen and Y.-B. He, *Synthesis*, 1988, 896.
90. W. J. Musliner and J. W. Gates, Jr., *J. Am. Chem. Soc.*, 1966, **88**, 4271.
91. E. Vowinkel and H.-J. Baese, *Chem. Ber.*, 1974, **107**, 1213.
92. E. Vowinkel and C. Wolff, *Chem. Ber.*, 1974, **107**, 907.
93. P. G. Theobald and W. H. Okamura, *Tetrahedron Lett.*, 1987, **28**, 6565.
94. V. Pascali and A. Umani-Ronchi, *J. Chem. Soc., Chem. Commun.*, 1973, 351.
95. W. E. Truce and F. M. Perry, *J. Org. Chem.*, 1965, **30**, 1316.
96. J. M. Coulter, J. W. Lewis and P. P. Lynch, *Tetrahedron*, 1968, **24**, 4489.
97. B. Emmert, *Chem. Ber.*, 1909, **42**, 1507.
98. H. H. Hodgson and J. Walker, *J. Chem. Soc.*, 1933, 1620.
99. J. I. G. Cadogan and G. A. Molina, *J. Chem. Soc., Perkin Trans. 1*, 1973, 541.
100. H. H. Hodgson and E. Marsden, *J. Chem. Soc.*, 1940, 207.
101. M. D. Threadgill and A. P. Gledhill, *J. Chem. Soc., Perkin Trans. 1*, 1986, 873.
102. J. B. Hendrickson, *J. Am. Chem. Soc.*, 1961, **83**, 1251.
103. J. Nakayama, M. Yoshida and O. Simamura, *Tetrahedron*, 1970, **26**, 4609.
104. K. G. Rutherford and W. A. Redmond, *J. Org. Chem.*, 1963, **28**, 568.
105. K. G. Rutherford, W. A. Redmond and J. Rigamonti, *J. Org. Chem.*, 1961, **26**, 5149.
106. H. H. Hodgson and H. S. Turner, *J. Chem. Soc.*, 1942, 748.
107. H. H. Hodgson and H. S. Turner, *J. Chem. Soc.*, 1943, 86.
108. M. Kato, T. Tamano and T. Miwa, *Bull. Chem. Soc. Jpn.*, 1975, **48**, 291.
109. P. Griess, *Philos. Trans. R. Soc. London, Ser. A*, 1864, **154**, 683.
110. L. F. Fieser and H. Heymann, *J. Am. Chem. Soc.*, 1942, **64**, 376.
111. N. Kornblum and D. C. Iffland, *J. Am. Chem. Soc.*, 1949, **71**, 2137.
112. J. Mai, *Chem. Ber.*, 1902, **35**, 162.
113. R. Q. Brewster and J. A. Poje, *J. Am. Chem. Soc.*, 1939, **61**, 2418.
114. N. Kornblum, *Org. React. (N.Y.)*, 1944, **2**, 262.
115. M. P. Doyle, J. K. Guy, K. C. Brown, S. N. Mahapatro, C. M. VanZyl and J. R. Pladziewicz, *J. Am. Chem. Soc.*, 1987, **109**, 1536.
116. N. Suzuki, T. Azuma, T. Nomoto and Y. Izawa, *Chem. Ind. (London)*, 1985, 698.
117. N. Suzuki, T. Azuma, K. Watanabe, T. Nomoto, Y. Izawa and H. Tomioka, *J. Chem. Soc., Perkin Trans. 1*, 1987, 1951.
118. S. Yasui, K. Nakamura and A. Ohno, *Tetrahedron Lett.*, 1983, **24**, 3331.
119. D. F. DeTar and T. Kosuge, *J. Am. Chem. Soc.*, 1958, **80**, 6072.
120. H. Meerwein, H. Allendörfer, P. Beckmann, F. Kunert, H. Morschel, F. Pawellek and K. Wunderlich, *Angew. Chem.*, 1958, **70**, 211.
121. M. S. Newman and W. M. Hung, *J. Org. Chem.*, 1974, **39**, 1317.

4.6
Reduction of Ketones to Alkenes

A. RICHARD CHAMBERLIN and DANIEL J. SALL
University of California at Irvine, CA, USA

4.6.1 INTRODUCTION

The carbon–carbon π-bond plays a central role in synthesis, not only in its own right as a functional group present in many target molecules, but also as a synthetic intermediate through which an extraordinary number of important transformations are accomplished. In fact, nearly two complete volumes of 'Comprehensive Organic Synthesis' are devoted to alkene reactions, many of which require stereo- or regio-isomerically homogeneous starting materials if they are to be useful. This chapter examines one of the more important general routes to this fundamentally important functional group, the transformation of ketones into alkenes.

Although there are a few methods for carrying out this reductive process in a single step, it is rarely conducted by such direct methods. In the vast majority of cases, two-step procedures are employed. The ketone can be first converted into a vinyl–heteroatom derivative (*e.g.* a vinyl phosphate) or a substituted hydrazone, followed by reductive cleavage or elimination, respectively. Alternatively, the transformation can be accomplished by reduction of the ketone followed by elimination; however, since both ketone reductions and elimination reactions are covered elsewhere in this series, this chapter will focus on the former methods.

4.6.2 DIRECT METHODS

There are few reports of the direct conversion of ketones into alkenes. Phiel has reported the conversion of pentan-2-one into pent-1-ene in the vapor phase over a supported MnO_2 catalyst (Scheme 1).[1] No other examples were reported, so that that the generality of the selectivity in favor of the less substituted double bond is uncertain.

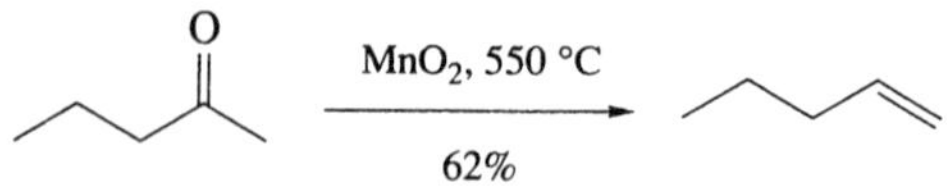

O

MnO_2, 550 °C

62%

Scheme 1

2-Phenylacetophenone has been reduced to (*E*)-stilbene using elemental sulfur in HMPA at elevated temperatures (Scheme 2).[2] Since (*Z*)-stilbene is rapidly converted into the (*E*)-isomer under the reaction conditions, it is unclear whether (*Z*)-stilbene forms first and then isomerizes to the more stable (*E*)-isomer or whether (*E*)-stilbene forms directly. These conditions were also found to convert 2-hydroxy-2-phenylacetophenone into (*E*)-stilbene.

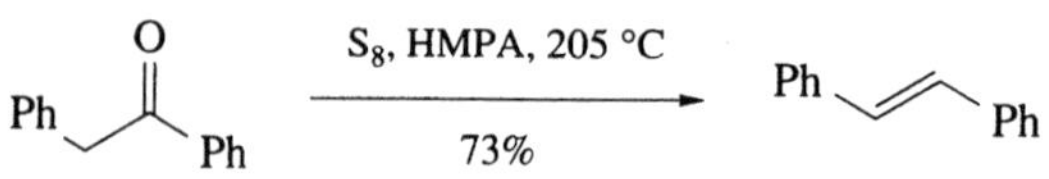

O

Ph Ph

S_8, HMPA, 205 °C

73%

Ph Ph

Scheme 2

Finally, functionalized cyclohexanone derivatives have been converted into the corresponding cyclohexene products by zinc and trimethylsilyl chloride (Scheme 3).[3] The reaction, which proceeds through an ylide-stabilized organozinc carbenoid intermediate rather than a silyl enol ether, shows a slight preference for the formation of the more substituted double bond, and is compatible with acetoxy and halogen functional groups. Similar conditions have been employed to convert 3-keto steroids into the corresponding Δ^2-derivatives regioselectively, as well as selectively reducing the 3-keto group of 3,6-, 3,7-, 3,17-, 3,20-diketo steroids to the Δ^2-products, leaving the other ketone group intact.[4]

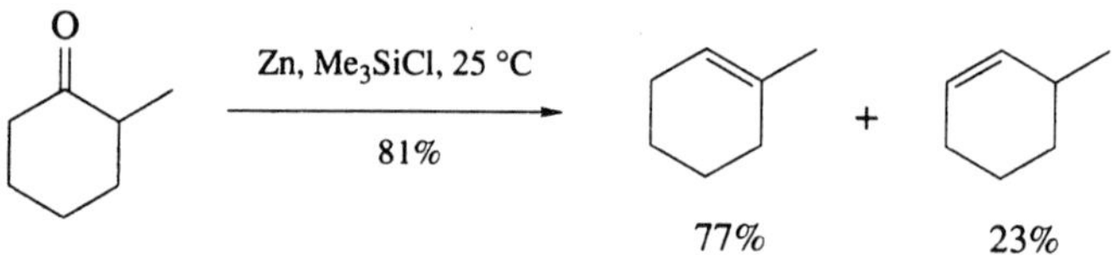

O

Zn, Me_3SiCl, 25 °C

81%

+

77% 23%

Scheme 3

4.6.3 REDUCTION AND SUBSEQUENT DEHYDRATION

One of the most general reaction sequences for the transformation of ketones into alkenes is reduction of the ketone to the corresponding alcohol followed by dehydration. While this method has been widely used, it often suffers from a lack of both stereo- and regio-chemical control in the formation of the double bond. Since the reduction of ketones and the subsequent dehydration of the resultant alcohols are covered in depth in other sections (this volume, Chapter 1.1 and Volume 6, Chapter 5.1), we present here only a few representative examples and divert the reader to these other sections for a detailed analysis of this area. In the total synthesis of (+)-occidentalol (Scheme 4), 1,2-reduction of the enone moiety gave the corresponding allylic alcohol, which was selectively dehydrated in the presence of a tertiary alcohol to give the diene system.[5] Reduction of an enone and subsequent dehydration of the resulting allylic alcohol was also employed in preparing the unsaturated decalin fragment of compactin (Scheme 4).[6] In this instance, the resulting 1,3-diene arose from carbon–carbon double bond migration during the dehydration. Finally, Murae *et al.* have reduced and dehydrated a highly functionalized pyrone to the corresponding dihydropyran in the course of the formal total synthesis of bruceantin.[7] These representative examples illustrate that this methodology can be useful in complex molecules and that it is compatible with an array of functional groups, including amines, aryl and silyl ethers, tertiary alcohols, dithianes, esters and lactones.

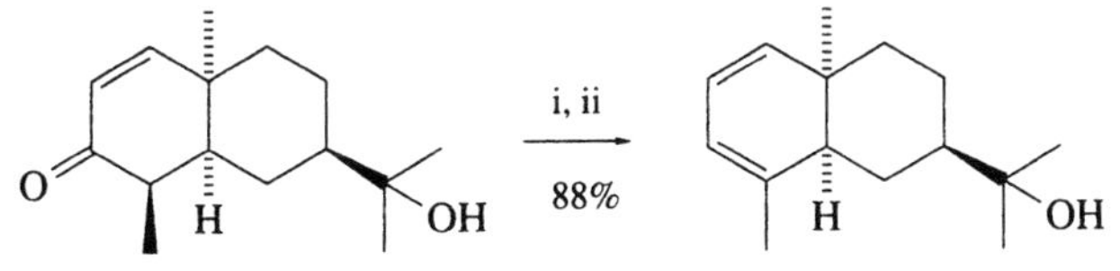

i, ii

88%

(+)-Occidentalol

i, LAH, Et_2O, 25 °C; ii, TsOH, PhH, 25 °C

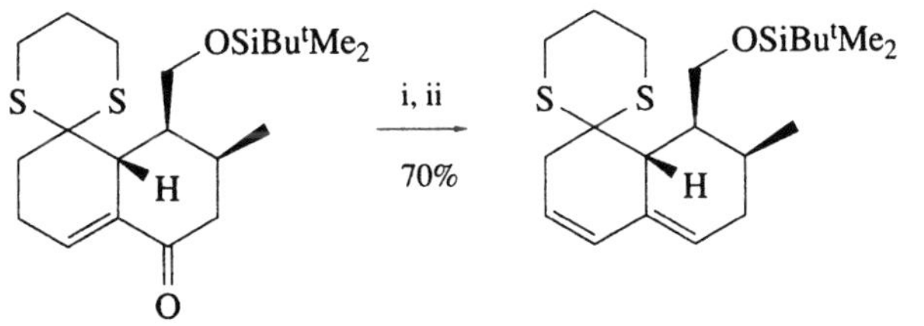

$OSiBu^tMe_2$

i, ii

70%

i, $NaBH_4$, $CeCl_3$; ii, pyridinium *p*-toluenesulfonate, $ClCH_2CH_2Cl$, reflux

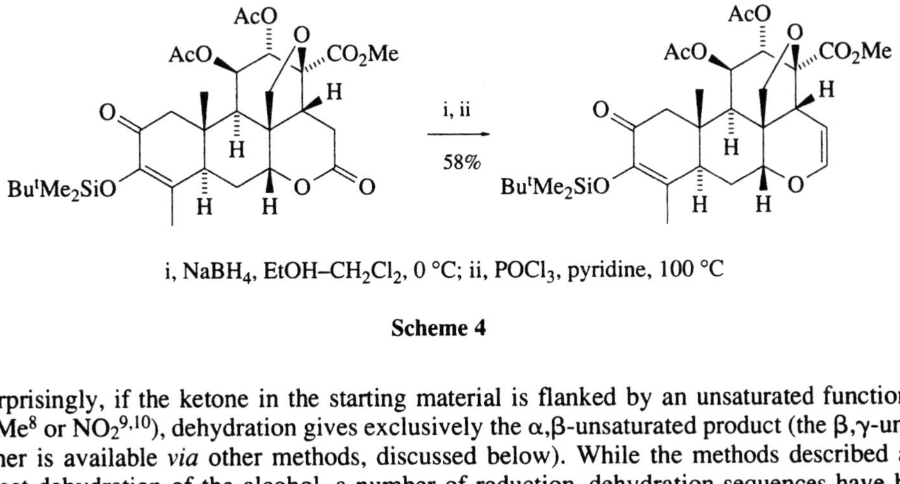

AcO, CO_2Me, Bu^tMe_2SiO

i, ii

58%

i, $NaBH_4$, $EtOH–CH_2Cl_2$, 0 °C; ii, $POCl_3$, pyridine, 100 °C

Scheme 4

Not surprisingly, if the ketone in the starting material is flanked by an unsaturated functional group (*i.e.* CO_2Me[8] or NO_2[9,10]), dehydration gives exclusively the α,β-unsaturated product (the β,γ-unsaturated regioisomer is available *via* other methods, discussed below). While the methods described above involve direct dehydration of the alcohol, a number of reduction–dehydration sequences have been conducted by initial reduction of the ketone and subsequent derivatization of the resultant alcohol (as the phenyl methanesulfonate, methyl *p*-toluenesulfonate, xanthate, acetate, methyl ether, *etc.*). This type of reduction–dehydration sequence is discussed in detail in Volume 6, Chapter 5.1.

4.6.4 REDUCTIVE ELIMINATION OF α-SUBSTITUTED KETONES

A variety of α-substituted ketones (α-halo, -hydroxy, -methoxy and -amino) have been converted into alkenes by reductive elimination. Obviously, this approach, depends not only on the feasibility of α-functionalization but also on the ability to introduce the α-substituent regioselectively. α-Halo ketones, which are readily prepared from the starting ketone,[11] have been subjected to reductive elimination conditions to afford the desired alkenic products. For example, in the preparation of (±)-eriolanin (Scheme 5),[12] the protected dihydroxy keto ester is treated sequentially with LDA and Br_2 to afford the α-bromo ketone. Selective ketone reduction to the halohydrin followed by zinc-induced elimination of HOBr affords the partially deprotected alkenic ester. Similar conditions have been employed in steroidal systems (Scheme 6).[13]

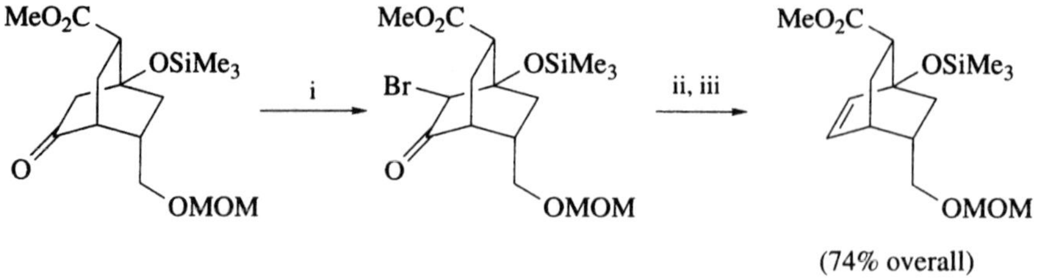

(74% overall)

i, LDA, Br_2, $NaHCO_3$; ii, $NaBH_4$, EtOH; iii, Zn, EtOH

Scheme 5

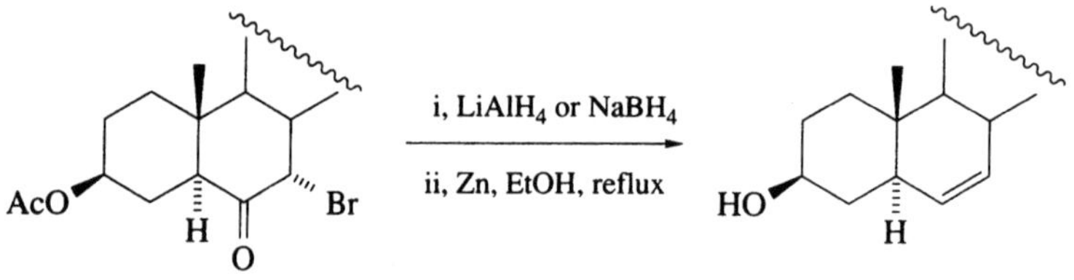

Scheme 6

In 1913, Kishner observed in one instance that under standard Wolff–Kishner reduction conditions, 2-hydroxy-2,6-dimethyloctan-3-one underwent eliminative reduction upon treatment with hydrazine hydrate and base at elevated temperatures to afford 2,6-dimethyloctan-2-ene (Scheme 7).[14] This same reaction was later found to occur in the case of α-methoxy ketones[15] and has since been referred to as the Kishner eliminative reduction. The reaction entails initial formation of the hydrazone and elimination of the α-substituent to afford the intermediate alkenyldiazene, which subsequently collapses to the desired alkene. Given the facile transformation of ketones into α-halo ketones,[11] these conditions have been used to introduce alkenes regioselectively in the 2α-halocholestan-3-one series[16] as shown in Scheme 8. Yields of 2-cholestene parallel the resistance of the α-halogen to undergo competitive elimination reactions.

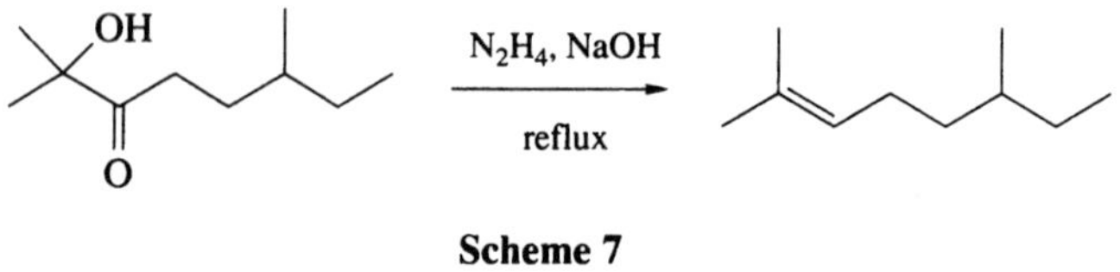

Scheme 7

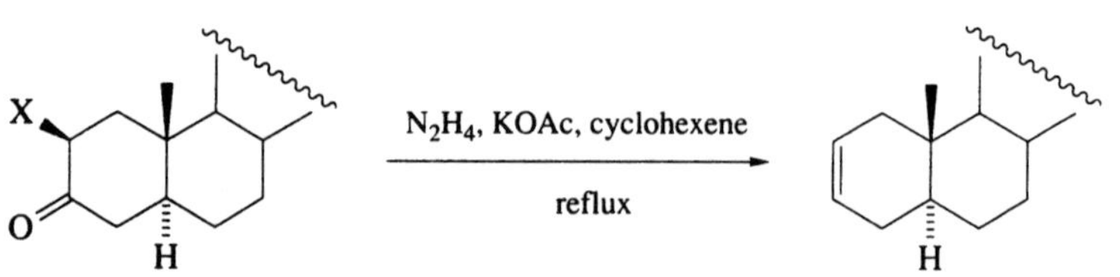

X = F	71%	X = Br	62%
X = Cl	68%	X = I	54%

Scheme 8

Similar conditions have also been employed in the reduction of α-tosyl ketones, but only low yields were realized.[17] On the other hand, a wide variety of other α-substituents have been shown to undergo reductive elimination to afford good yields of the alkenic product (Scheme 9).[17] The temperature at which the reaction occurs in these examples depends on the nature of the substituent and its ease of elimination. Kishner eliminative reduction conditions have also been applied to 2-azacycloalkanones, for which ring size determines the course of reaction.[18] While 1-azacyclohexan-3-ones give only the Wolff–

Kishner reduction product (azacyclohexanes), the seven-membered ring analogs afford the acyclic eliminative reduction products (Scheme 10). This difference is presumably due to the greater ease with which the leaving group in the larger ring can follow an energetically favorable trajectory (*i.e.* perpendicular to the incipient π-bond) during elimination. A wide variety of acyclic α-amino ketones have also been studied as substrates for the eliminative reduction pathway. The yield of alkene (*versus* the normal Wolff–Kishner reduction product) parallels the size of the α-amino substituent and hence the steric bulk around the C—N bond.[18]

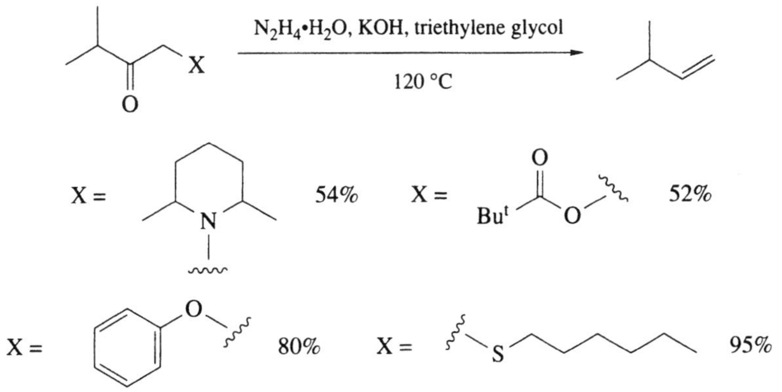

$N_2H_4{\bullet}H_2O$, KOH, triethylene glycol, 120 °C

X = 2,6-dimethylpiperidin-1-yl 54%; X = Bu^tCO_2 52%; X = PhO 80%; X = S-hexyl 95%

Scheme 9

i, 78%; i, 83%

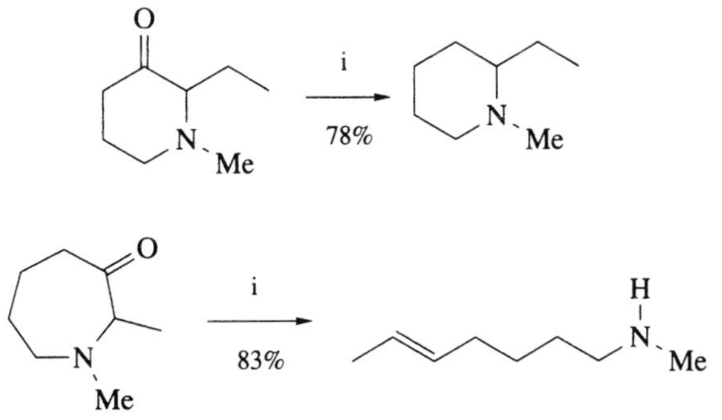

i, $N_2H_4{\bullet}H_2O$, KOH, triethylene glycol, 120 °C

Scheme 10

The most widely applied extension of the Kishner eliminative reduction methodology is the Wharton reaction,[19] in which the hydrazone derivative of an α,β-epoxy ketone undergoes elimination to generate an allylic alcohol (Scheme 11). Since the starting epoxy ketone is derived directly from the corresponding enone, this reaction sequence represents an overall transformation of enone into allylic alcohol in which oxygen undergoes a 1,3-transposition within a concomitant migration of the carbon–carbon double bond.

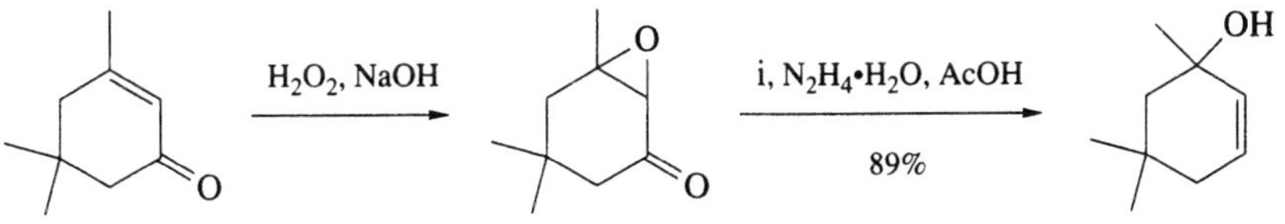

H_2O_2, NaOH; i, $N_2H_4{\bullet}H_2O$, AcOH, 89%

Scheme 11

This transformation has found extensive use in converting cyclopentenone and cyclohexenone ring systems to the rearranged allylic alcohols during the course of the total syntheses of natural products. In the preparation of (±)-quadrone (Scheme 12),[20] the tricyclic enone was epoxidized and the resulting α,β-epoxy ketone treated with hydrazine to afford the allylic alcohol. The cyclopropane-directed epoxidation shown in Scheme 13 gives an allylic alcohol that is taken on to (–)- and (+)-carenones.[21] In the total synthesis of (±)-α-costal (Scheme 14),[22] Liu and Wynn selectively epoxidized an enone carbon–carbon double bond in the presence of an isolated alkene in order to obtain a diene alcohol. Finally, the Wharton reaction has been employed, although with only modest success, in the preparation of 2,6-epoxy-1(2*H*)-benzoxocin sugars (Scheme 15).[23]

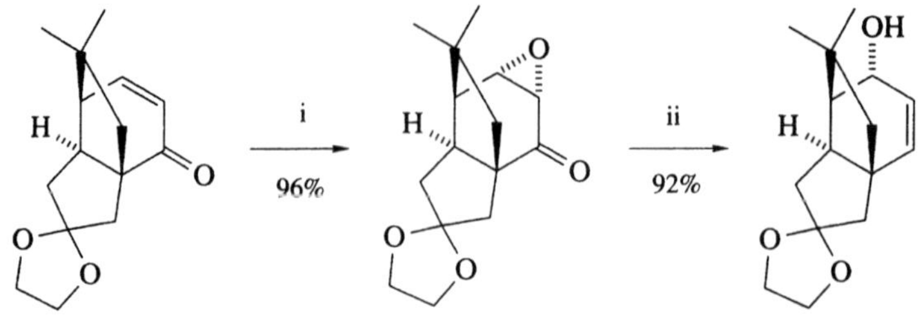

i, Bu^tOOH, NaOH, MeOH, 25 °C; ii, $N_2H_4{\bullet}H_2O$, MeOH, AcOH

Scheme 12

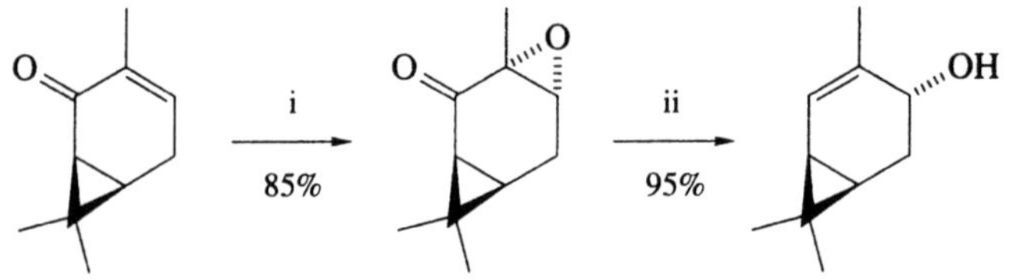

(–)-Carvone

i, 30% H_2O_2, NaOH; ii, Me_3SiCl, $N_2H_4{\bullet}H_2O$, DMF

Scheme 13

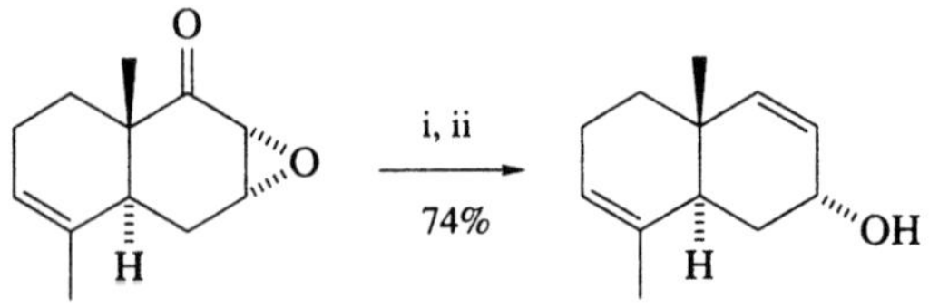

i, N_2H_4, AcOH; ii, $CuCl_2$, THF, H_2O, pH 7

Scheme 14

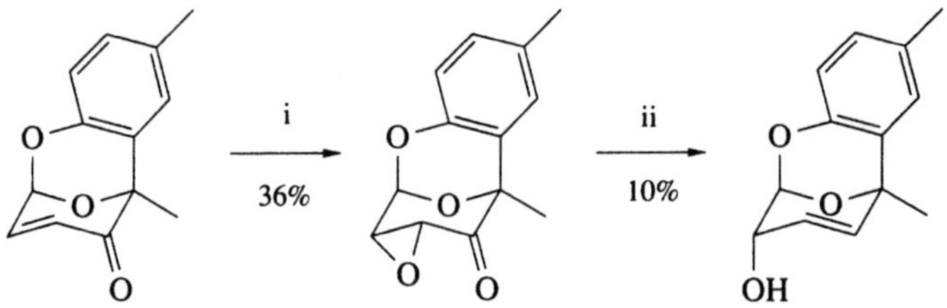

i, NaOCl, dioxane; ii, N_2H_4, AcOH

Scheme 15

While it is not clear from these applications whether the Wharton reaction necessarily proceeds *via trans* diaxial epoxide opening, Ziegler *et al.*[24] in synthetic studies towards quassinoids, have found that there is no such stereoelectronic requirement in a rigid *trans*-decalin ring system (Scheme 16).

The Wharton reaction has also been applied to the conversion of exocyclic α,β-epoxy ketones to the corresponding exocyclic hydroxy methylene products (Scheme 17).[25] Similar conditions transform exocyclic methyl ketones into the exocyclic ethylidene derivative (Scheme 18).[26] Not unexpectedly, the (*Z*)-alkene is isolated in only slightly higher yields than the (*E*)-isomer.

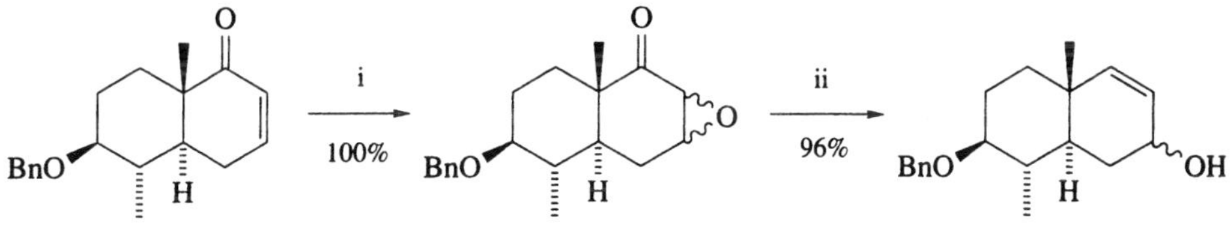

i, 30% H_2O_2; 6 M NaOH, 15 °C; ii, 95% N_2H_4, AcOH, MeOH

Scheme 16

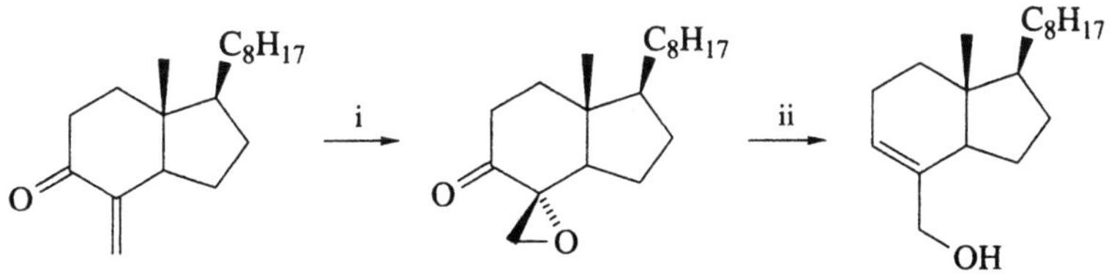

i, 30% H_2O_2, 20% NaOH, MeOH; ii, N_2H_4, MeOH, 0 °C

Scheme 17

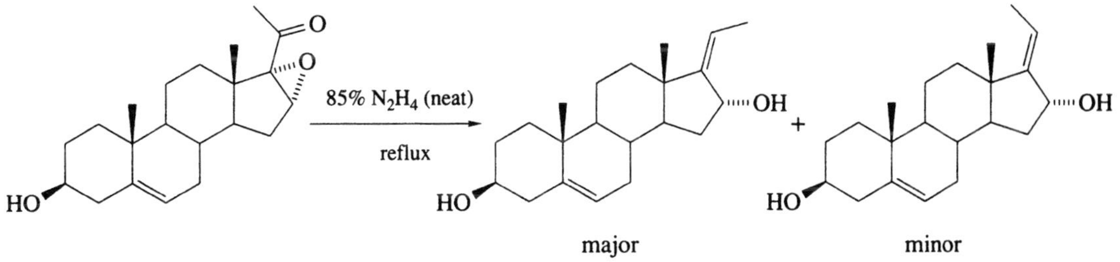

85% N_2H_4 (neat), reflux

major minor

Scheme 18

There are a few applications of the Wharton reaction in acyclic systems. For example, treating an acyclic α,β-epoxy ketone (Scheme 19) with hydrazine hydrate at 0 °C affords the expected product, but as a 1:1-(*E*):(*Z*) mixture.[27] However, the major product (60%) in this particular case is the bicyclic alcohol that arises *via* cyclization of the intermediate vinyl radical (or anion) on the endocyclic carbon–carbon double bond.

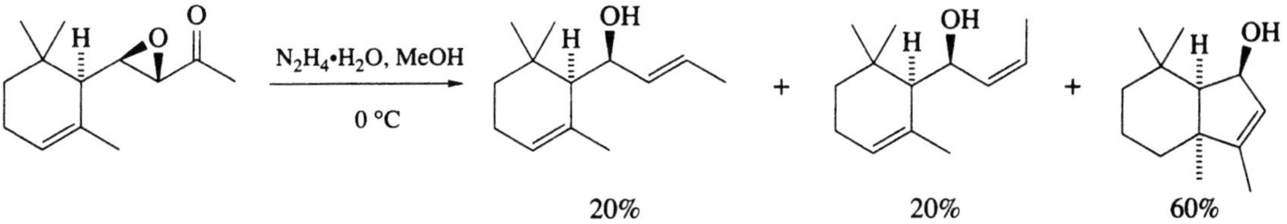

$N_2H_4 \cdot H_2O$, MeOH, 0 °C

20% 20% 60%

Scheme 19

In isolated cases in which the Wharton reaction has failed, alternative strategies have been developed. For example, in a total synthesis of confertin,[28] the starting enone (Scheme 20) was reduced to an allylic alcohol and subsequently epoxidized to afford the epoxy alcohol. Conversion to the phenyl methanesulfonate followed by reduction gave the sought-after rearranged allylic alcohol. An attempted Wharton reaction in the total synthesis of β-2,7,11-cembratriene-4,6-diol (β-CBT; Scheme 21)[29] also failed to give the desired allylic alcohol. In this case, initial epoxide formation was followed by reduction of the ketone (rather than treatment with hydrazine) and the resultant secondary alcohol was converted into the phenyl methanesulfonate. Reduction of the epoxy phenyl methanesulfonate then afforded the formal Wharton product in good yields. This adaptation also gives high yields in acyclic systems.[30]

A final method of alkene formation from α-derivatized ketones entails formation of unsaturated tosylhydrazones from α,β-unsaturated ketones followed by conjugate reduction with sodium cyanoborohydride in acid media.[31] This transformation, which proceeds with regioselective migration of the original double bond, has been applied to the synthesis of quassinoids by Ganem *et al.* (Scheme 22).[32] A similar

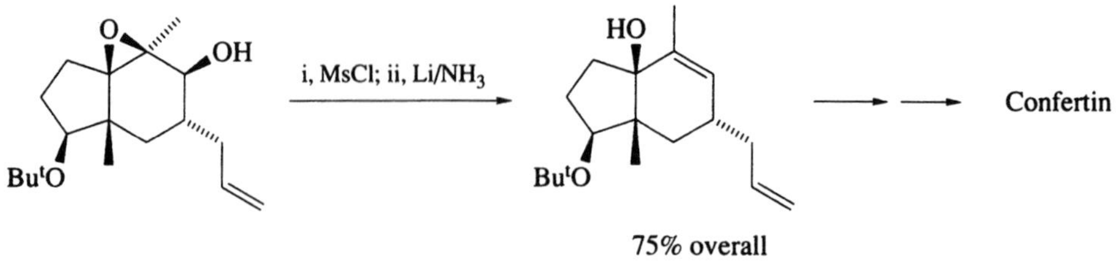

Scheme 20

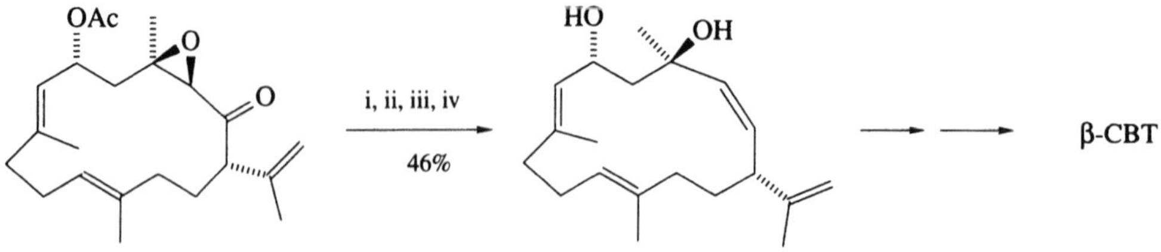

i, $NaBH_4$, 0 °C; ii, MsCl, Et_3N, 0 °C; iii, K_2CO_3, 0 °C; iv, Na/NH_3, –40 °C

Scheme 21

reaction, with diborane as the reducing agent, has been employed in the conversion of Δ^4-cholesten-3-one into 5α-cholest-3-ene and of cholestenone into cholesta-3,5-diene.[33]

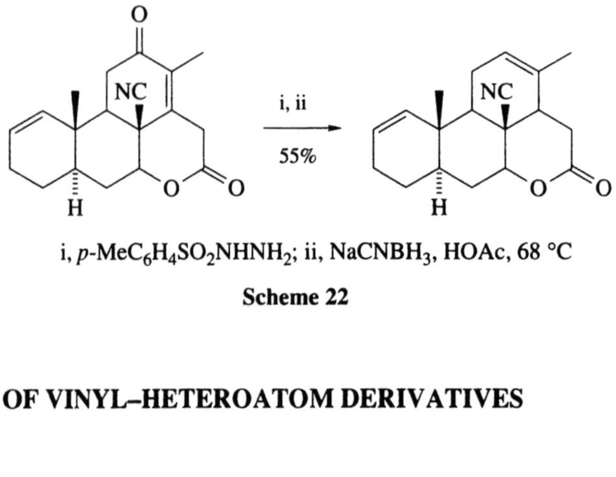

i, p-$MeC_6H_4SO_2NHNH_2$; ii, $NaCNBH_3$, HOAc, 68 °C

Scheme 22

4.6.5 REDUCTION OF VINYL–HETEROATOM DERIVATIVES

4.6.5.1 Vinyl Esters

One of the most useful methods for the transformation of ketones into alkenes is conversion of the ketone, *via* enolate formation, into a vinyl–OR derivative and subsequent reductive cleavage of the sp^2-oxygen bond. Because there are a number of ways to generate either the kinetic or thermodynamic enolate selectively, this route offers one of the most important ways of introducing the alkenic double bond with excellent control of regioselectivity.

4.6.5.1.1 Vinyl phosphates

Within the general category of vinyl esters, the most widely used method for ketone to alkene transformation involves the sequence ketone → vinyl phosphate → alkene, which was originally developed independently by Fetizon[34] and by Ireland.[35] Fetizon's procedure requires α-halogenation of the starting ketone and subsequent vinyl phosphate formation *via* a Perkow reaction with triethyl phosphite (Scheme 23). In contrast, Ireland's method (Scheme 24)[35] offers a direct route to vinyl phosphates *via* controlled enolate formation and trapping. More recently, Ireland has also employed vinyl *N,N,N′,N′*-tetramethylphosphorodiamidate derivatives,[36] rather than the original diethyl phosphates (Scheme 25), although it is not clear if this modification is superior to the original procedure. Besides reduction with Li or Na in an amine solvent, vinyl phosphates have also been reduced to the alkene under aprotic conditions with activated Ti.[37] These reduction conditions afford better yields of the alkene, and enones can be reduced to the corresponding diene without overreduction of the diene products.

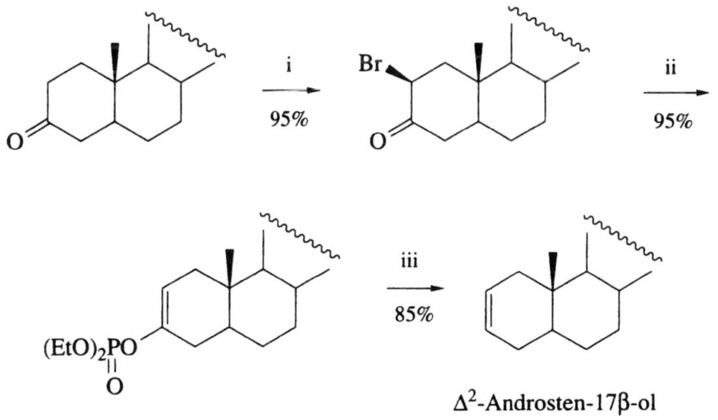

Δ^2-Androsten-17β-ol

i, $Me_3PhN^+Br^-$; ii, $(EtO)_3P$ (neat), Δ; iii, Li/NH_3, Bu^{t}OH

Scheme 23

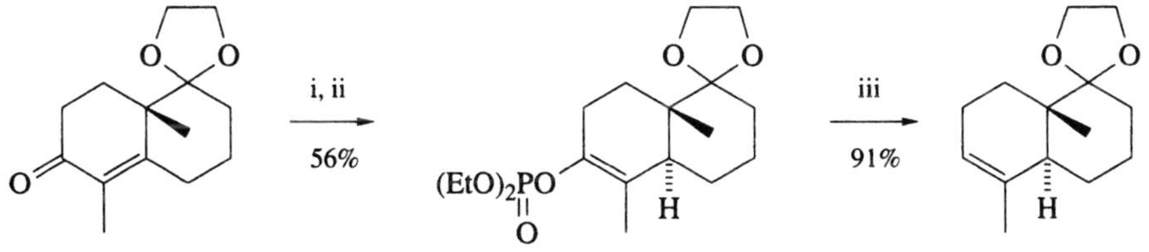

i, Li/NH_3; ii, $(EtO)_2POCl$; iii, Li, $EtNH_2$, Bu^{t}OH

Scheme 24

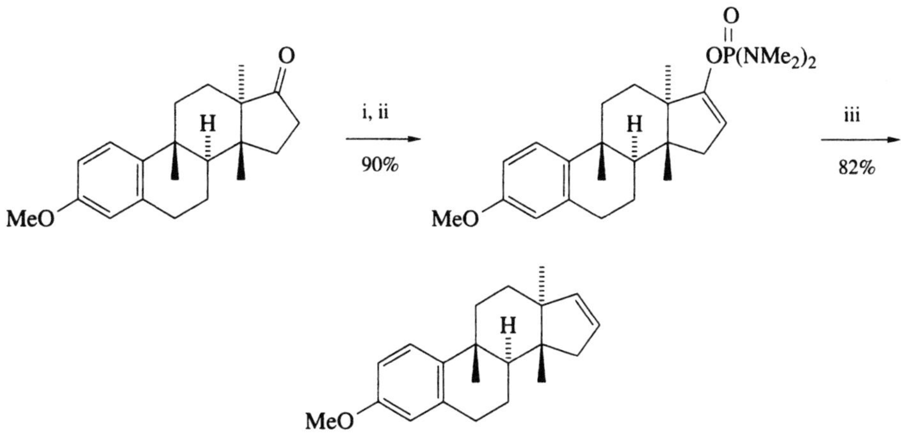

i, LDA; ii, $(Me_2N)_2POCl$, TMEDA; iii, Li/NH_3

Scheme 25

Despite these modifications, Ireland's original method of both vinyl phosphate formation and reduction remains the most popular. In studies on the cationic rearrangements of [4.3.2]propellanes, Smith *et al.*[38] converted a tricyclic ketone, *via* the diethyl vinylphosphate derivative, into the corresponding alkene (Scheme 26). In a similar manner, Kamata *et al.*[39] prepared Δ^{11}- and Δ^6-steroids (Scheme 27) with excellent control of regiochemistry. As shown in this example, esters are susceptible to cleavage under the standard reduction conditions (Li, NH_3, Bu^{t}OH, –35 °C), while acetals survive.

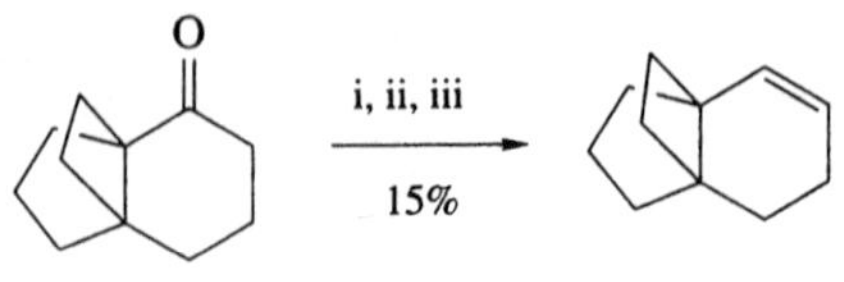

i, LDA; ii, $(EtO)_2POCl$; iii, Li, $EtNH_2$

Scheme 26

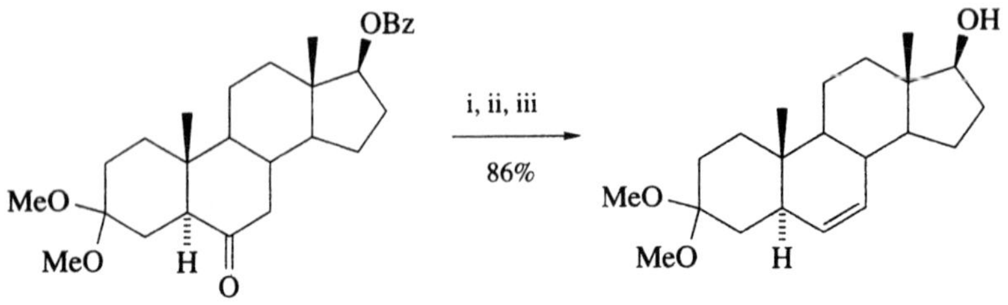

i, LDA, –78 °C; ii, $(EtO)_2POCl$, –78 °C; iii, Li/NH_3, Bu^tOH, –35 °C

Scheme 27

Vinyl *N,N,N′,N′*-tetramethylphosphorodiamidates have also been of value in the total syntheses of natural products, such as (±)-castelanolide (Scheme 28)[40] and (±)-albene (Scheme 29).[41] Acetals survive the reduction, but benzyl ethers are cleaved.

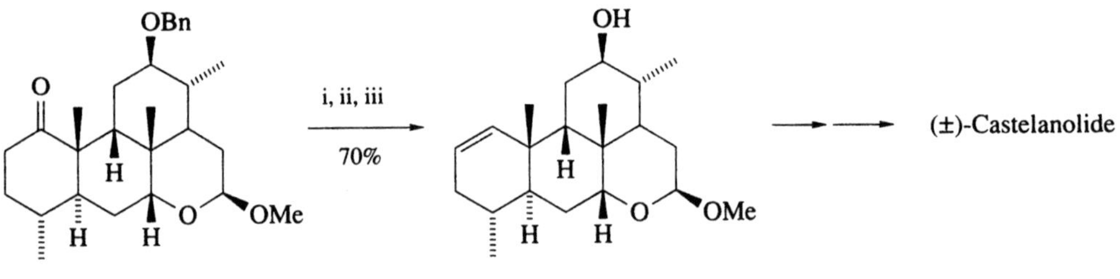

i, $LiNPr^i_2$, TMEDA, –78 °C; ii, $(Me_2N)_2POCl$, 0 °C; iii, $EtNH_2$, Bu^tOH

Scheme 28

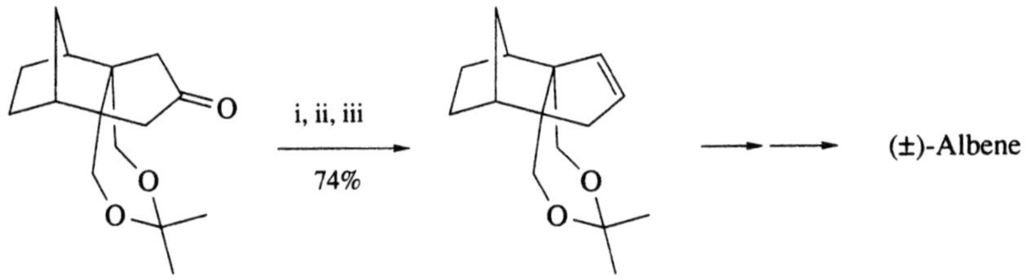

i, $LiNPr^i_2$, –78 °C; ii, $(Me_2N)_2POCl$, HMPA, 0 °C; iii, Li, $EtNH_2$, Bu^tOH, 5 °C

Scheme 29

This methodology has also been applied to the conversion of alkyl esters into vinyl ethers[42] with high stereoselectivity favoring the (*Z*)-isomer (Scheme 30). While standard Li/amine reduction conditions were not applicable, the enol phosphates could be reduced using triethylaluminum and tetrakis(triphenylphosphine)palladium.

In a related reaction, aryl phosphates (prepared from the corresponding phenols) have been reduced to the parent hydrocarbon with Li or Na in NH_3.[43] Yields are increased if electron-donating groups are present on the aromatic ring. While reduction of the aromatic ring itself can be avoided under these conditions, addition of an alcohol to the reducing medium results in isolation of the fully saturated cyclohexane analog. Ireland *et al.*[44] have reported a similar reduction of a phenol to the arene with Li/NH_3 and ethanol.

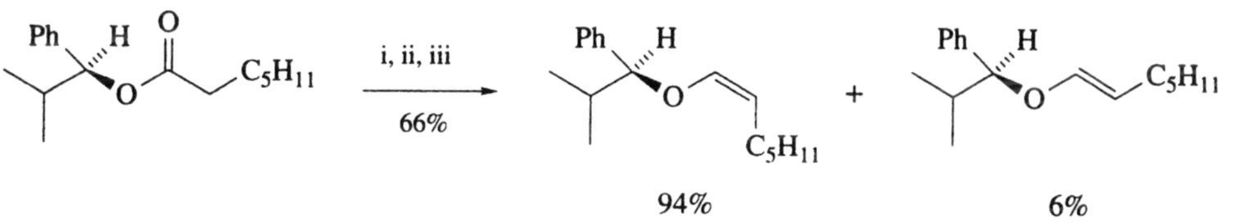

94% 6%

i, $LiNPr^i_2$; ii, $(EtO)_2POCl$, HMPA; iii, Et_3Al, $Pd(PPh_3)_4$

Scheme 30

4.6.5.1.2 Vinyl triflates

While not receiving as much attention as vinyl phosphates, vinyl triflates have been successfully employed as intermediates in the transformation of ketones into alkenes (Scheme 31). As in the case of vinyl phosphates, regioselective alkene formation ultimately results from kinetic or thermodynamic enolate generation (Scheme 32).[45]

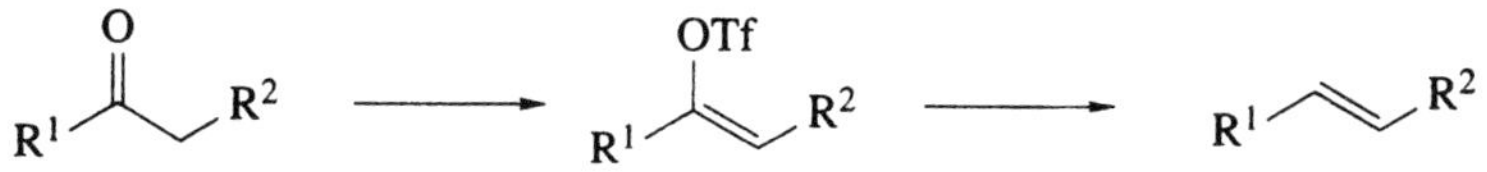

Scheme 31

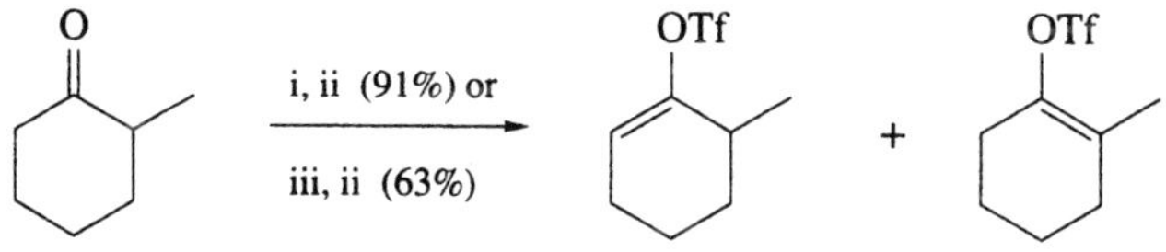

reagents i, ii	95: 5
reagents iii, ii	3:97

i, LDA; ii, Tf_2NPh; iii, Pr^i_2NMgBr

Scheme 32

Vinyl triflates have been prepared from the starting ketones using 2,6-di-*t*-butyl-4-methylpyridine/triflic anhydride,[46] sodium bicarbonate/triflic anhydride[47] in the case of activated systems and lithium hexamethyldisilazide/HMPA/*N*-phenyltriflimide.[45,48] Once formed, vinyl triflates are readily reduced to the corresponding alkene with tributyltin hydride/tetrakis(triphenylphosphine)palladium(0) in the presence[47] or absence of lithium chloride,[45] triethylsilicon hydride/tetrakis(triphenylphosphine)palladium(0)/lithium chloride,[47] or tri-*n*-butylamine/bis(triphenylphosphine)palladium acetate/formic acid.[49] The latter conditions, in which the reducing species is proposed to be tri-*n*-butylammonium formate, are compatible with a wide range of functional groups, including ketones, esters, ethers and alcohols. Dolle *et al.* have utilized this methodology under conditions of thermodynamic control of enolization to convert an enone into a steroidal diene in the preparation of ergosterol β-isomers (Scheme 33).[50]

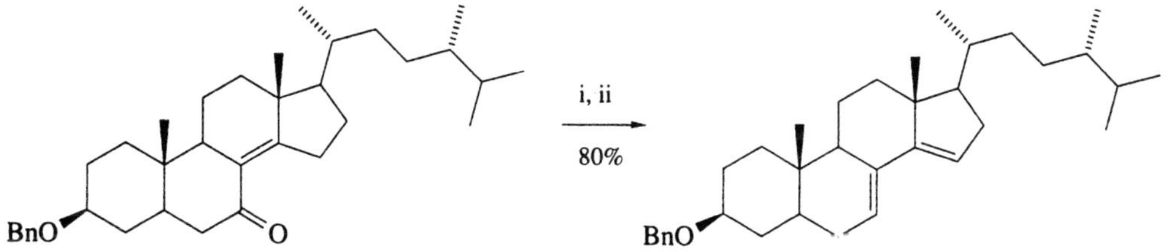

i, Tf_2O, 2,6-di-*t*-butyl-4-methylpyridine; ii, Bu_3N, HCO_2H, cat. $Pd(OAc)_2(Ph_3P)_2$, DMF

Scheme 33

Similar to the reduction of aryl phosphates, phenols can be reduced to arenes *via* the corresponding aryl triflates.[51,52] The reducing systems employed are compatible with a wide array of functional groups, including ketones, ethers, sulfides, aryl ethers, nitro compounds, esters, amides, lactones and cyano groups. Finally, Subramanian *et al.*[53] have employed the nonafluoro-*n*-butyl triflate derivative to effect the conversion of phenols into arenes.

4.6.5.1.3 Vinyl acetates

While not extensively explored, vinyl acetates have also been reduced to the corresponding alkenes with iron pentacarbonyl at elevated temperatures (Scheme 34).[54] In addition, vinyl acetates have been subjected to hydroboration,[55] but elimination of the boronate ester in the intermediate and subsequent hydroboration of the resulting alkene are so rapid that this method is not synthetically useful for ketone to alkene conversion.

OAc

MeO

$[Fe(CO)_5]$, Bu_2O
reflux

35%

MeO

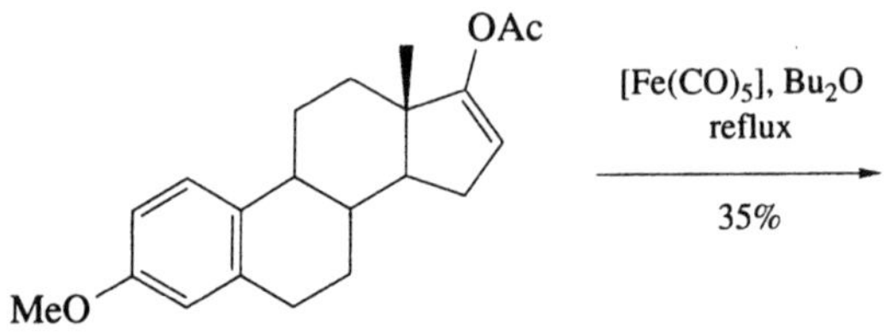

Scheme 34

4.6.5.2 Vinyl Ethers

Although not as extensively employed as vinyl esters, vinyl ethers find occasional use as intermediates in the conversion of ketones into alkenes. Once again, vinyl ethers can be generated regioselectively by kinetic or thermodynamic enolate generation and, therefore, offer another route by which ketones can be regioselectively replaced by alkenic double bonds.

4.6.5.2.1 Vinyl sulfides, selenides and sulfoxides

Within the general class of vinyl ethers, vinyl sulfides are the most frequent choice for carrying out the transformation of ketones into alkenes, according to the synthetic sequence shown in Scheme 35. Vinyl sulfides are generally prepared from the corresponding ketones by either of two strategies. The first involves introduction of an α-alkylthiol group followed by formal dehydration (route A; Scheme 35). As indicated, regioselective introduction of the α-alkylthiol group is accomplished by kinetic or thermodynamic enolate generation and, therefore, allows for regioselective alkene formation. The second process (route B; Scheme 35) entails direct conversion of the ketone into a vinyl sulfide, in which the carbonyl oxygen is replaced by sulfur.

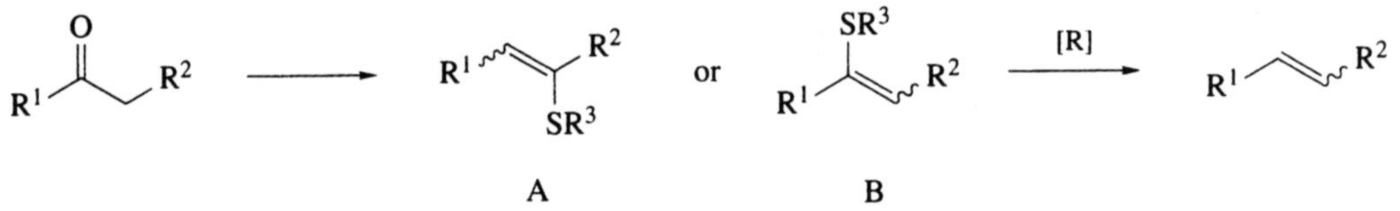

O

R^1 R^2

R^1 R^2

SR^3

A

or

SR^3

R^1 R^2

B

[R]

R^1 R^2

Scheme 35

As an example of the former method (route A; Scheme 35), enolate formation in spirocyclobutanone systems followed by treatment with diphenyl disulfide affords the α-thiophenyl ketone; subsequent hydride reduction provides the α-hydroxythiane, which upon mesylation and base-induced elimination gives the vinyl sulfide in good yields.[56] Alternatively, α-alkylthiol ketones, such as 2-*S*-methyl-1-tetralone can be converted into the corresponding tosylhydrazones, which undergo the Shapiro reaction (Section 4.6.5.4) to afford the target vinyl sulfide.[57] In a related procedure, 4-methylcyclohexanone has been converted into 1-*S*-methyl-5-methylcyclohexene in a one-pot sequence by initial tosylhydrazone formation and subsequent sulfenylation of the dianion generated with Bu^nLi. *In situ* Shapiro reaction of this intermediate then affords the desired vinyl sulfide.[58] Although this latter procedure provides an interesting method for 1,2-carbonyl transposition, it is obviously of little value for the ketone to alkene conversion because the same alkene would be formed in the direct Shapiro reaction of the starting ketone.

Employing the second general method (route B; Scheme 35), vinyl sulfide formation in acyclic systems has been accomplished from ketones by initial thioacetal formation under acidic conditions with benzenethiol, followed by thiol elimination employing copper(I) triflate.[59] If acid sensitive products are generated, the elimination step can be conducted in the presence of 2,6-lutidine or Hünig's base. Finally, vinyl sulfides have been prepared from the starting ketone in a single step by treatment with benzenethiol and *p*-toluenesulfonic acid[60] or with benzenethiol and P_2O_5 in methylene chloride.[60]

Vinyl sulfides have been reduced to the corresponding alkene with isopropylmagnesium bromide/nickel chloride/tri-*n*-octylphosphine[61] or isopropylmagnesium bromide/tris(triphenylphosphine)-nickel chloride at elevated temperatures,[62] in which the reducing agent is postulated as 'NiH'. Raney nickel[63,64] and $NiCl_2/NaBH_4$ ('NiB')[63] have also been found to reduce vinyl sulfides to alkenes; however, 'NiH' and 'NiB' are usually the method of choice because of their functional group compatibility and their unreactivity towards isolated carbon–carbon double bonds. As an example of this methodology, Boar *et al.*[63] have employed a vinyl sulfide intermediate in the conversion of 5α-lanost-8-en-3-one to 5α-lanosta-2,8-diene (Scheme 36). Since it is necessary to maintain an acidic reaction medium in order to avoid overreduction of the isolated alkenic double bond, boric acid is added.

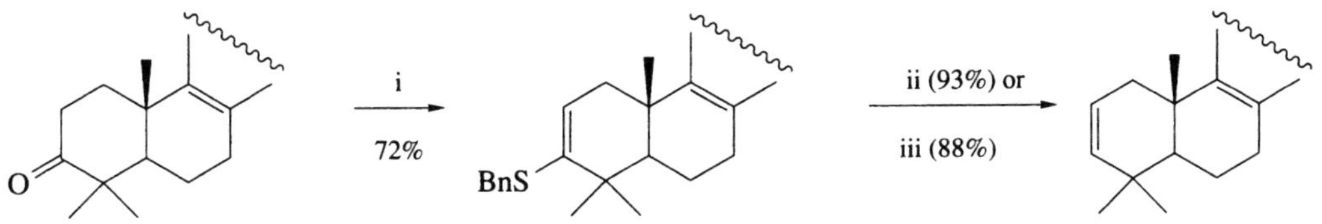

i, BnSH, $BF_3•Et_2O$, AcOH; ii, $NiCl_2$, $NaBH_4$, H_3BO_3, EtOH; iii, Raney Ni

Scheme 36

Thioacetals, prepared from the corresponding ketones, have also been reduced to alkenes using 'moderately active' Raney nickel as the desulfurization agent (Scheme 37).[65] This form of Raney nickel does not reduce the newly formed alkene or react with other isolated carbon–carbon double bonds. Note that under these conditions the saturated hydrocarbon often observed in Raney nickel desulfurizations[66] is not formed. Similar conditions have been used for the conversion of cyclodecanone to cyclodecene.[67] While thioacetals have also served as intermediates in the analogous conversion of enones into dienes, a mixture of diene products is obtained. For instance, dithiane generation from 17β-hydroxy-Δ^4-androsten-3-one, followed by desulfurization affords a 4:1 mixture of 17β-hydroxy-$\Delta^{2,4}$-androstadiene and 17β-hydroxy-$\Delta^{3,5}$-androstadiene, respectively.[65]

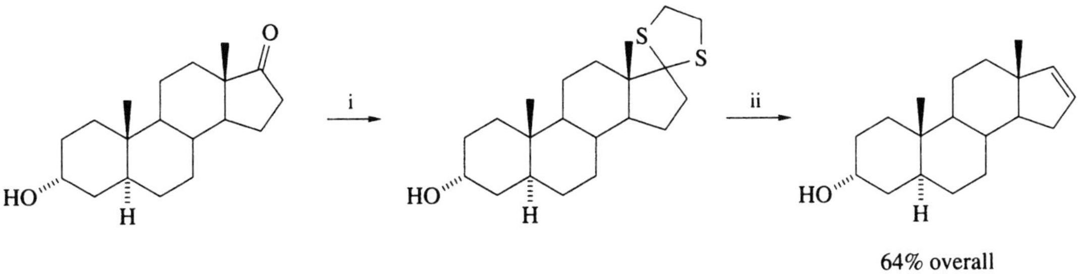

64% overall

i, $HSCH_2CH_2SH$, $BF_3•Et_2O$; ii, Raney Ni (moderately active)

Scheme 37

α-Acetoxy ketones have also been converted to alkenes through the intermediacy of a 1,2-dithiane derivative. In the transformation of pulchellin to pulchellon (Scheme 38), Inayama *et al.*[64] treated the starting α-acetoxy ketone with 1,2-ethanedithiol to generate the diacetoxydithiane intermediate, which upon desulfurization with deactivated Raney nickel afforded the desired alkenic product. The lactam moiety of indanone undergoes a related reductive process, *via* the amide ketene thioacetal, to give indole (Scheme 39).[68]

Vinyl selenides, which behave similarly to vinyl sulfides, have also been prepared from ketones and reduced to the alkenic product. By this route, 3β-hydroxypregn-5-en-20-one was converted into the exocyclic 17-ethylidene derivative (Scheme 40).[69] Vinyl selenides have also been reduced with 'NiB' during

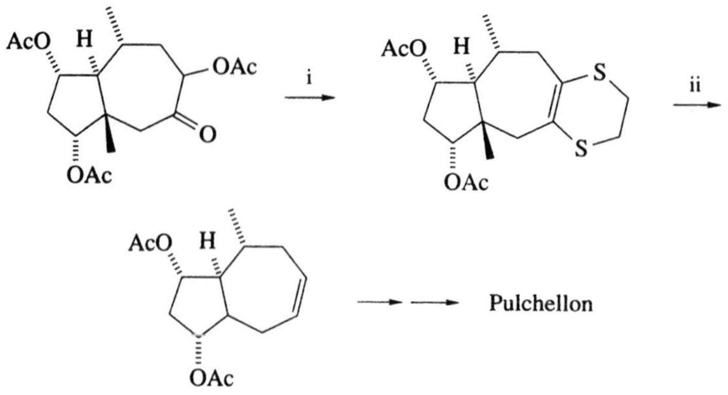

→ → Pulchellon

i, $HSCH_2CH_2SH$, $BF_3 \cdot Et_2O$; ii, Raney Ni (deactivated), acetone, reflux

Scheme 38

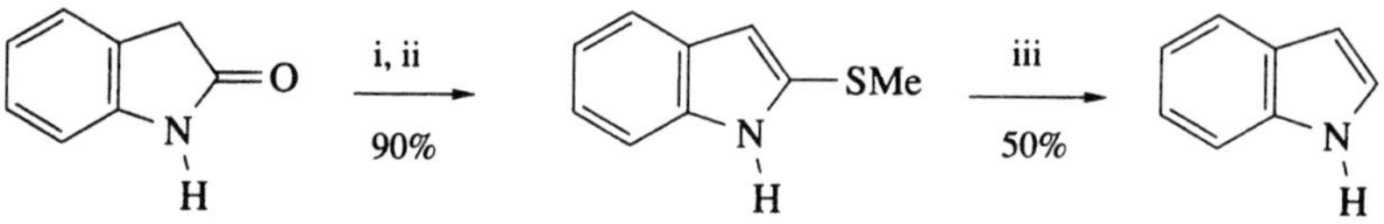

i, Lawesson's reagent; ii, MeI, K_2CO_3; iii, Pr^iMgBr, $[(Ph_3P)_2NiCl_2]$

Scheme 39

the course of synthetic studies on the ergot alkaloids[70] and in acyclic systems.[71] In the latter case, the reduction proved to be nonstereospecific (Scheme 41).

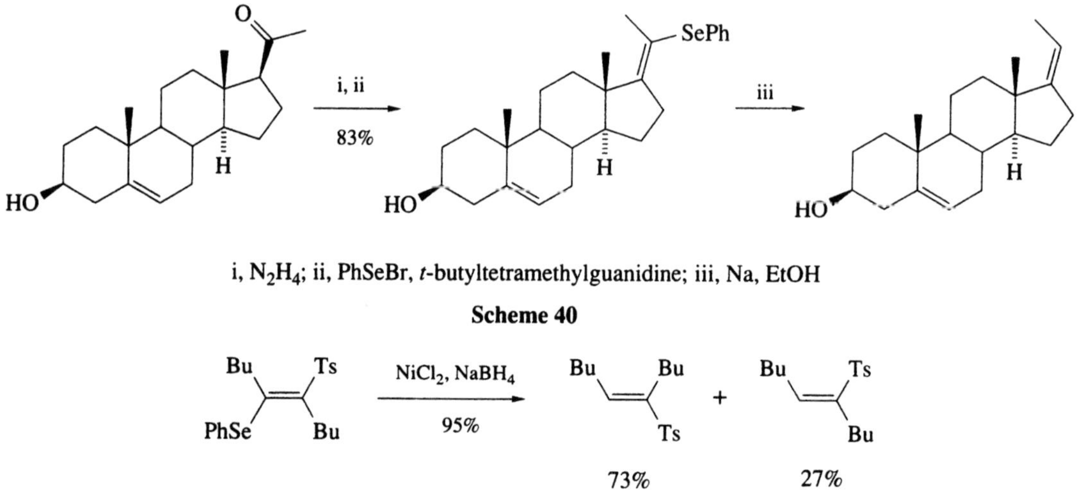

i, N_2H_4; ii, PhSeBr, *t*-butyltetramethylguanidine; iii, Na, EtOH

Scheme 40

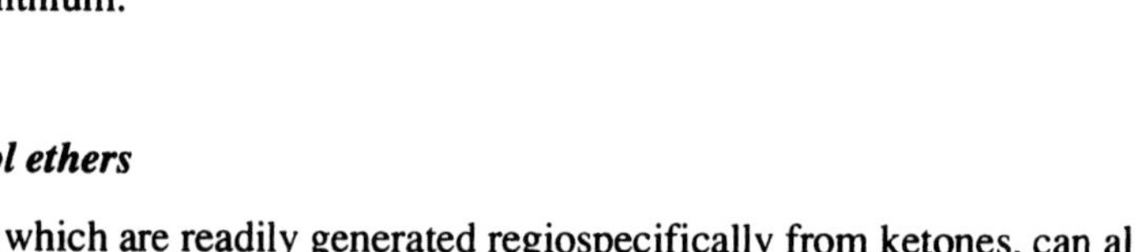

Scheme 41

In addition to vinyl sulfides and selenides, vinyl sulfoxides have been reduced to the corresponding alkenes using *t*-butyllithium.[72]

4.6.5.2.2 Silyl enol ethers

Silyl enol ethers, which are readily generated regiospecifically from ketones, can also be reduced to alkenes, particularly by hydroboration.[73–76] Hydroboration of silyl enol ethers results in the addition of boron to the β-carbon of the double bond to afford *trans*-β-trimethylsilyloxy organoboranes, which in cyclic systems undergo *anti* elimination in the presence of acid to give the alkenic product (Scheme 42).[74] A number of acids have been tested successfully, including carboxylic acids, $BF_3 \cdot Et_2O$ and aqueous HCl. Since enolate formation can be conducted under kinetic or thermodynamic control, either alkene regioisomer is obtainable. In acyclic systems, the initially formed *trans*-β-trimethylsilyloxyorganoborane undergoes rapid *syn* elimination, even in the absence of acid, to afford the alkene product.[76] As a result, the (*E*):(*Z*) ratio of the product follows directly from the stereochemistry of the starting enol ether (Scheme 43).

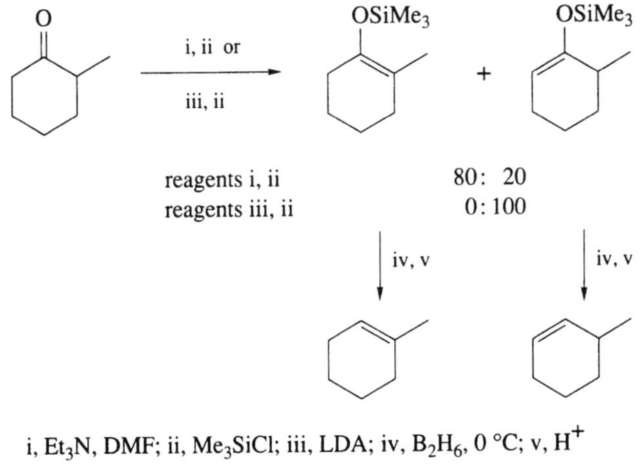

i, ii or iii, ii

reagents i, ii 80 : 20
reagents iii, ii 0 : 100

iv, v

i, Et_3N, DMF; ii, Me_3SiCl; iii, LDA; iv, B_2H_6, 0 °C; v, H^+

Scheme 42

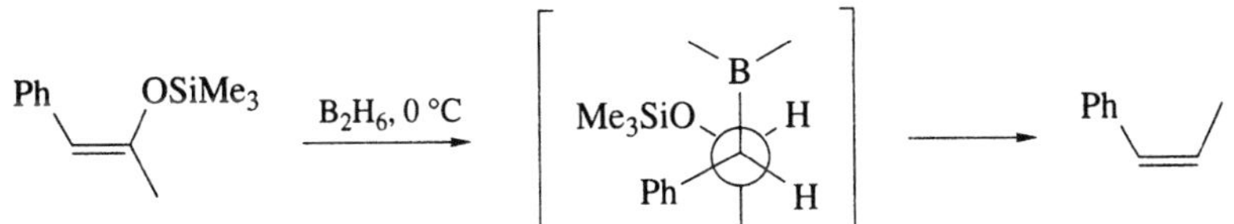

B_2H_6, 0 °C

Scheme 43

4.6.5.2.3 *Alkyl enol ethers*

Only a few cases of the reduction of alkyl enol ethers to the corresponding alkene have been reported. α-Ethoxystyrene and related derivatives are reduced to alkenes by treatment with Grignard reagents.[77] In addition, enol ethers of cyclohexanone derivatives have been cleaved to the cyclohexene products using either DIBAL at elevated temperatures[78] or diborane/boron trifluoride etherate.[79] By analogy to the hydroboration of silyl enol ethers, this latter method involves formation of an intermediate β-ethoxy organoborane which undergoes acid-catalyzed elimination to afford the alkene.

4.6.5.3 Vinyl Halides

Vinyl halides represent yet another important class of intermediates in the conversion of ketones to alkenes. The most widely applied conditions for the conversion of ketones into vinyl halides are those developed by Barton *et al.*[80] for the conversion of 3β-acetoxyandrost-5-ene-17-one into 3β-hydroxyandrosta-5,16-diene (Scheme 44). These conditions of vinyl halide formation and subsequent reduction have been useful in a number of steroid systems for the introduction of a Δ^{16}-carbon–carbon double bond[81,82] and have been shown to be compatible with such functional groups as alcohols, isolated double bonds and acetals. The scope of vinyl iodide formation from hydrazones has been studied by Pross and Sternhell,[83] and recently the original reaction conditions were improved by using sterically hindered guanidine bases rather than triethylamine.[84,85] Haloalkenes have also been prepared from the corresponding ketones by treatment with iodoform and chromium chloride[86] or with phosphorous pentahalides.[87]

The reduction of vinyl halides most often is carried out *via* a metallation–protonation reaction. For instance, sodium in ethanol was employed by Barton *et al.*[80] for the reduction of a steroidal vinyl iodide (Scheme 44). A number of conditions for vinyl halide reduction to alkenes have been reported, representative examples of which include: vinyl chloride and bromide reduction in acyclic systems with

i, 64% N_2H_4, Et_3N, Δ; ii, I_2, Et_3N; iii, Na, EtOH

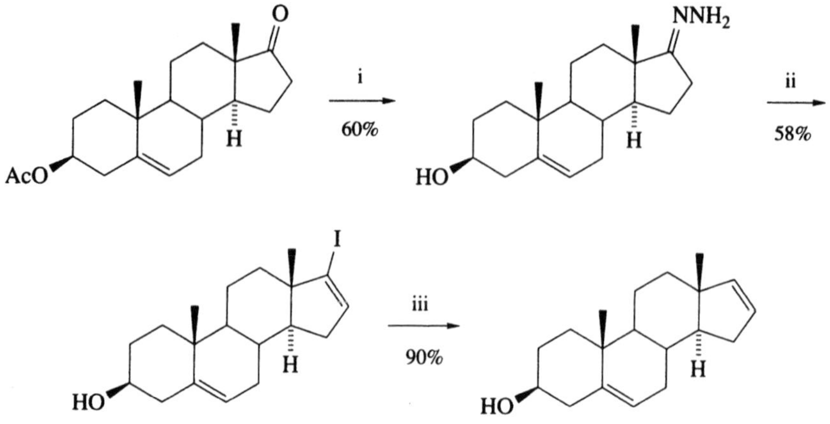

Scheme 44

DIBAL-H/*n*-butyllithium,[88] in cyclic and acyclic systems with iron pentacarbonyl,[54] in cyclodecanes with lithium dihydrodimethoxyaluminate(III)/copper(I) iodide,[89] and in cyclohexane and cyclopentane systems with NaH/sodium *t*-butylpentyl/Ni(OAc)$_2$.[90] The monoreduction of 1,3-diketones can be carried out under similar conditions, as illustrated by the reaction of a substituted cyclohexane-1,3-dione with oxalyl chloride[91] to give the corresponding 1-chlorocyclohexenone, which was subsequently reduced to the enone with zinc–silver couple (Scheme 45).[92] Kropp *et al.*[93] have reported the photolytic reduction of vinyl iodides in acyclic systems; however, when an α-hydrogen is present, formation of the diene product is a limiting side reaction (Scheme 46). For a more extensive discussion of vinyl halide reductions, see the preceding chapter in this volume.

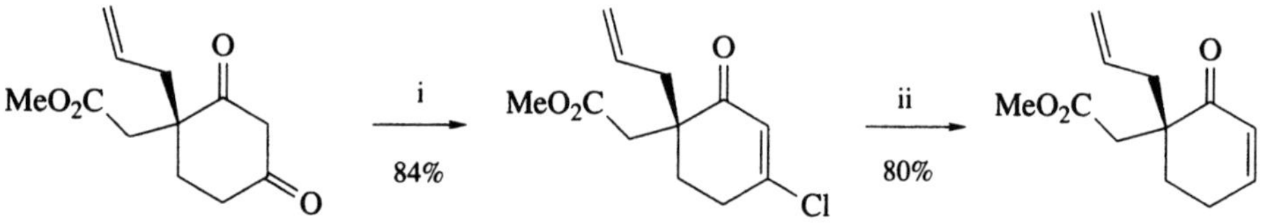

i, $(COCl)_2$, $CHCl_3$, Δ; ii, Zn–Ag, MeOH, 25 °C

Scheme 45

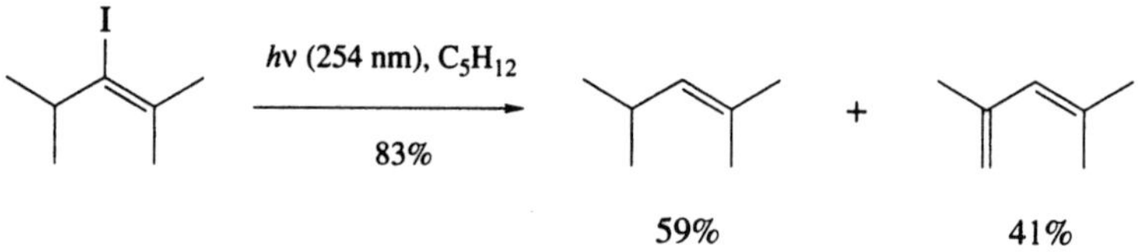

Scheme 46

4.6.5.4 Enamines

Enamines represent a final class of vinyl–heteroatom derivatives that can be prepared from ketones and subsequently reduced to alkenes (Scheme 47). Since enamine formation is covered in depth elsewhere, this section will deal only with the reduction of enamines to alkenes.

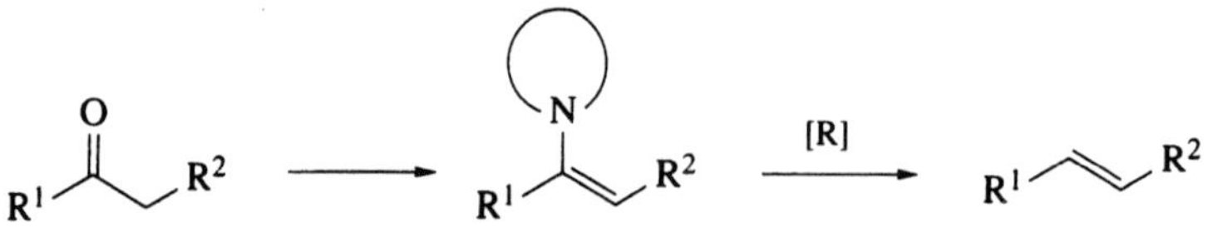

Scheme 47

Cyclic enamines have been reduced to alkenes with lithium aluminum hydride/aluminum chloride.[94] The reducing efficiency of simple aluminum hydrides as well as chlorinated aluminum hydrides has been studied in detail, and AlH_3 has been found to be the most effective for the conversion of cyclohexenamines into cyclohexene (Scheme 48).[95]

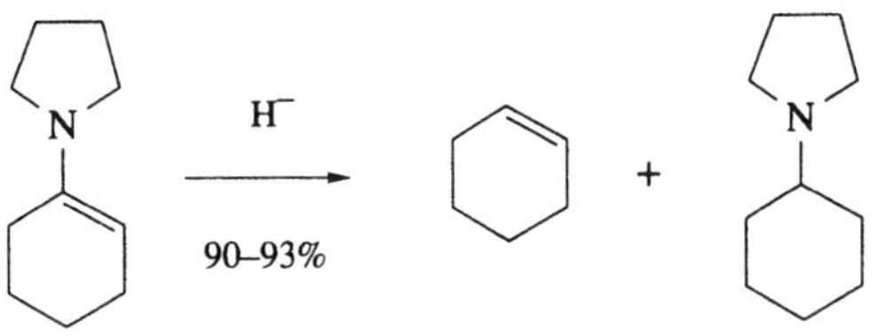

$AlHCl_2$	77%	23%
AlH_2Cl	83%	17%
AlH_3	84%	16%

Scheme 48

As with silyl and alkyl enol ethers, enamines react with boron hydrides to afford the corresponding simple alkenic products. In the case of cyclohexyl enamine, hydroboration gives the β-amino organoborane, which upon acid-induced *anti* elimination affords cyclohexene as the product.[96] Similar conditions were employed by Friary *et al.*[97] in the regioselective preparation of unsaturated proline derivatives (Scheme 49). Brown *et al.*[98] have shown that a single acyclic enamine geometric isomer can be reduced to either the (*E*)- or (*Z*)-alkene depending on the reducing species (Scheme 50): treatment of (*E*)-1-morpholino-1-phenyl-1-propene with borane methyl sulfide followed by methanolysis gives the β-aminoboronate ester, which upon peroxide oxidation undergoes *syn* elimination to afford (*E*)-1-phenyl-1-propene, while hydroboration with 9-BBN gives the trialkyl boronate, which on methanolysis undergoes *anti* elimination to give (*Z*)-1-phenyl-1-propene.

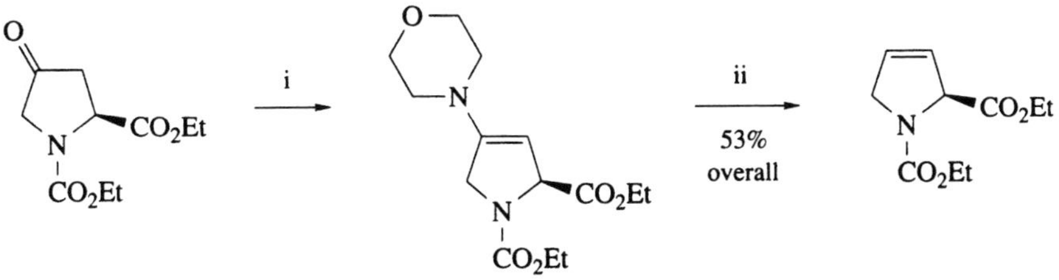

i, morpholine, TsOH; ii, B_2H_6, THF, reflux

Scheme 49

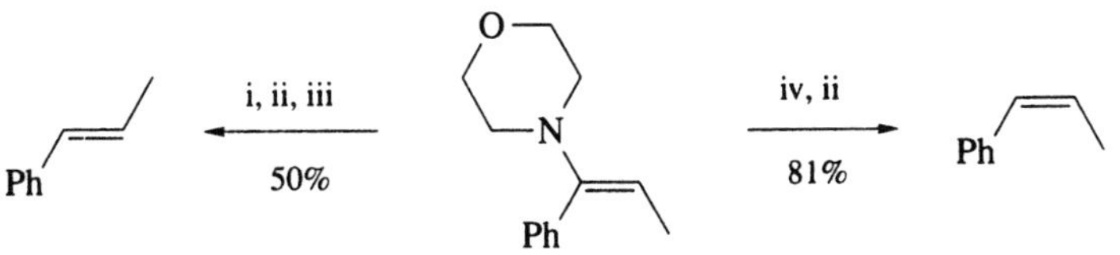

i, $BH_3{\bullet}Me_2S$; ii, MeOH; iii, H_2O_2, NaOH; iv, 9-BBN

Scheme 50

4.6.6 REDUCTIVE ELIMINATION OF HYDRAZONE DERIVATIVES

In the early 1950s Bamford and Stevens reported a thermally induced reaction of ketone tosylhydrazone monosodium salts that gives products derived either from carbene intermediates (aprotic conditions) or carbocations (protic conditions). Both reactions, which are illustrated in Scheme 51, can give alkenes as the major product, although the protic reaction is of marginal synthetic value because mixtures of alkene regioisomers and rearrangement products are normally obtained. On the other hand, the aprotic reaction does on occasion offer excellent regiochemical control in the preparation of alkenes, although it has found more widespread use in the synthesis of multicyclic, strained hydrocarbons. Both reactions have been covered in previous reviews.[99] Treating tosylhydrazones with alkyllithium bases results in a mechanistically different reaction that also provides alkenes, but *via* an alkenyllithium intermediate (Scheme 51). This process, commonly known as the Shapiro reaction, has been applied to the preparation of a large number of structurally diverse alkenes and is also the subject of earlier reviews.[99,100] The Shapiro reaction has itself been modified to allow trapping of the intermediate vinyllithium with electrophiles; however, this aspect of arenesulfonylhydrazone chemistry, which is the subject of a recent comprehensive survey,[101] is not within the scope of this section and is therefore excluded. Related eliminations of α-heteroatom hydrazones are discussed above in Section 4.6.4.

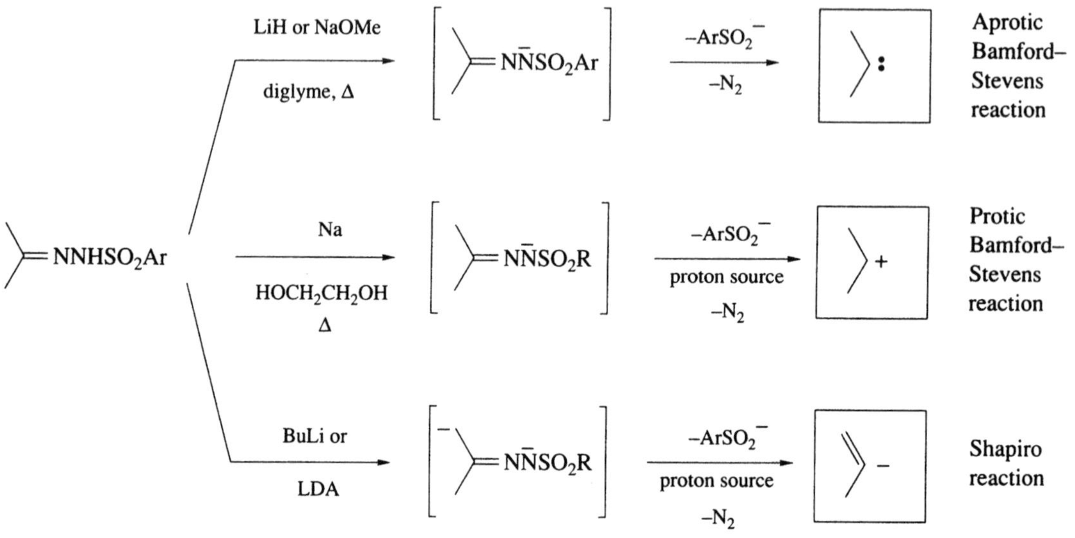

Scheme 51

4.6.6.1 Formation of Arenesulfonylhydrazones

By far the most common sulfonylhydrazone derivatives reported to undergo the Bamford–Stevens or Shapiro reaction have been ketone tosyl- or trisyl-hydrazones (trisyl = triisopropylbenzenesulfonyl), although there are a few reports of the use of benzenesulfonylhydrazones or *N*-aminoaziridine-derived hydrazones. In terms of availability and cost of precursors, tosylhydrazones, which are prepared from ketones and commercially available tosylhydrazine, are preferred. Equimolar amounts of the reactants are generally dissolved in methanol, ethanol or acetic acid, usually in the presence of an acidic catalyst, such as concentrated hydrochloric acid, and allowed to react at room temperature or somewhat above, and then stored in the cold, after which the crystalline tosylhydrazone can be isolated in good yield by filtration. Even hindered ketones, such as camphor, react smoothly at somewhat higher temperatures. Diethyl ether can also serve as the reaction solvent and reportedly is superior to alcohols.[102] Although trisylhydrazones are prepared in an analogous manner (commonly in methanol, ethanol, acetonitrile or diethyl ether), the starting trisylhydrazine is not currently commercially available; it must be prepared from the sulfonyl chloride and hydrazine hydrate.[103] Even hindered ketones, such as camphor and diisopropyl ketone, undergo the reaction in good yield,[103] although extraordinarily crowded ketones can be totally resistant to trisylhydrazone formation (Scheme 52).[104] Arenesulfonylhydrazones of aldehydes are prepared by the same general method, with minor modifications.[15] The formation of β-keto ester tosylhydrazones has also been reported,[105] as has a procedure for converting orthoesters into ester tosylhydrazones[106] and acyl halides into amide trisylhydrazones ('trisylamidrazones').[107]

Arenesulfonylhydrazones exist as mixtures of (*E*)- and (*Z*)-isomers, which usually cannot be physically separated at ambient temperatures[108,109] because the inversion barrier of the azomethine bond is too low, although spectroscopic methods can distinguish between the two isomers.[110] As expected, the (*E*):(*Z*) ratio for any given hydrazone is dependent upon the sizes of the groups attached to the azomethine carbon, with the less sterically crowded (*E*)-isomer usually predominating.

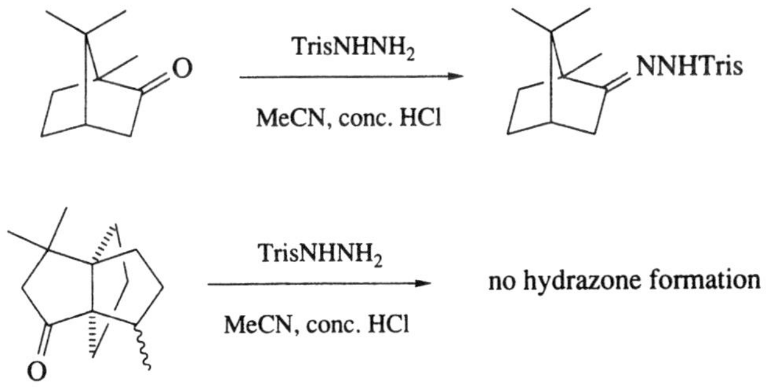

TrisNHNH$_2$

MeCN, conc. HCl

NNHTris

TrisNHNH$_2$

MeCN, conc. HCl

no hydrazone formation

Scheme 52

4.6.6.2 The Aprotic Bamford–Stevens Reaction

The aprotic reaction is generally conducted on a preformed lithium, potassium or sodium salt. Since the N—H proton of arenesulfonylhydrazones is relatively acidic, deprotonation is readily achieved with relatively weak bases, such as LiH, KH or sodium methoxide. The salt is then heated (and sometimes irradiated, presumably encouraging more rapid breakdown of the diazoalkane intermediate to the carbene) to give the product. For aldehyde hydrazone salts, mixtures of alkenes and β-insertion products are usually observed (Scheme 53).[111] On the other hand, the salts of most five- and six-membered ring hydrazones give predominantly alkene products under these conditions (Scheme 54),[112] while medium ring derivatives usually form bicyclic products *via* transannular carbene insertion.[113] Likewise, bridged bicyclic hydrazone salts give alkene products[114] only if an appropriately positioned C—H bond is not present across the ring in the rigidly held boat (Scheme 55).[103]

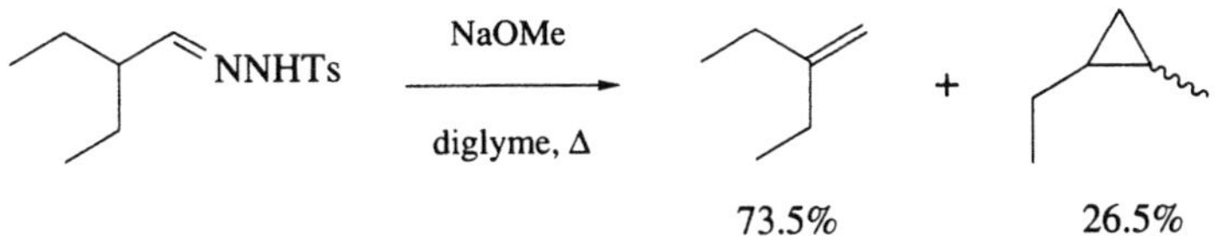

NNHTs

NaOMe

diglyme, Δ

73.5% 26.5%

Scheme 53

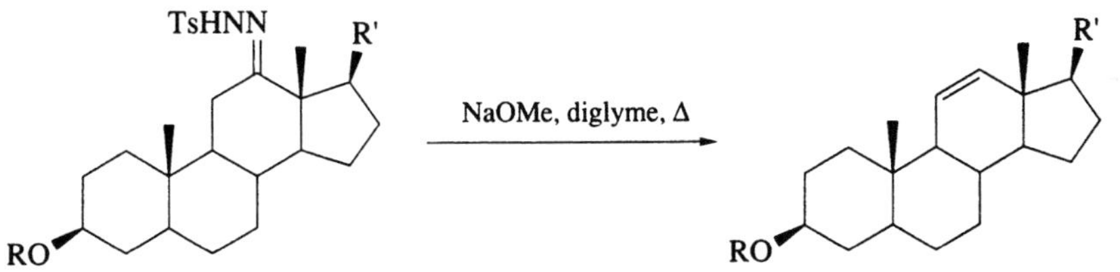

TsHNN R'

NaOMe, diglyme, Δ

RO R' RO

Scheme 54

Br O i, ii Br

NNHTs iii no

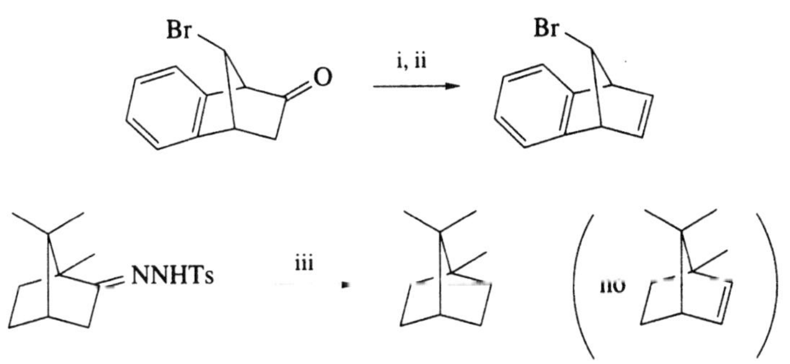

i, TsNHNH$_2$; ii, *h*ν on Li salt; iii, NaOMe, diglyme

Scheme 55

Within the constraints imposed by the intermediacy of a carbene, the aprotic Bamford–Stevens reaction can provide good yields of alkene with occasional examples of excellent regiochemical control in specific cases. In general, the more highly substituted regioisomer predominates. For instance, thermal decomposition of the sodium salt of 2-methylcyclohexanone tosylhydrazone affords a 2:1 mixture of methylcyclohexenes, favoring the 1-methyl isomer (Scheme 56).[115] A slight preference (2:1) for the formation of the less substituted alkene regioisomer was observed in Paquette's synthesis of precapnelladiene,[116] an example that also illustrates that transannular insertion is not always the preferred pathway in medium rings. Although these marginal selectivities are typical, there are exceptions. Some steroidal tosylhydrazones, for example, have been shown to afford exclusively the Δ^2 double bond isomer if the A/B ring juncture is *trans*, while the Δ^3 isomer is formed from the A/B *cis* starting material (Scheme 57).[117] Sugar tosylhydrazone salts have been converted into unsaturated sugars in high yield (Scheme 58),[118] with no competitive formation of the alternative bicyclic vinyl ether regioisomer.

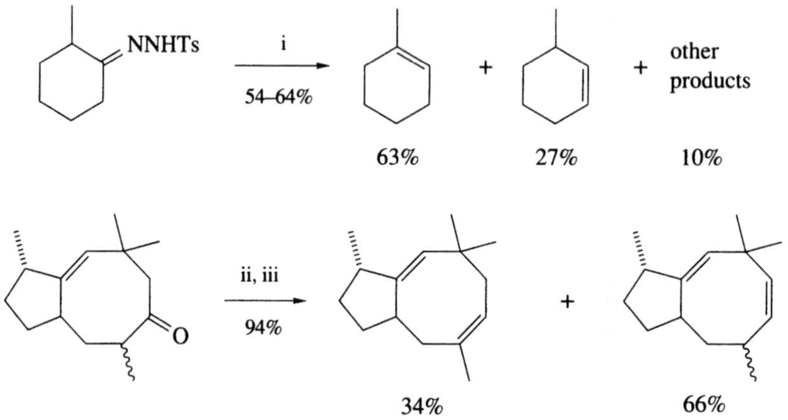

NNHTs; i; 54–64%; other products; 63%; 27%; 10%

ii, iii; 94%; 34%; 66%

i, NaOMe, NMP, Δ; ii, $TsNHNH_2$; iii, Bu^nLi, diglyme, reflux

Scheme 56

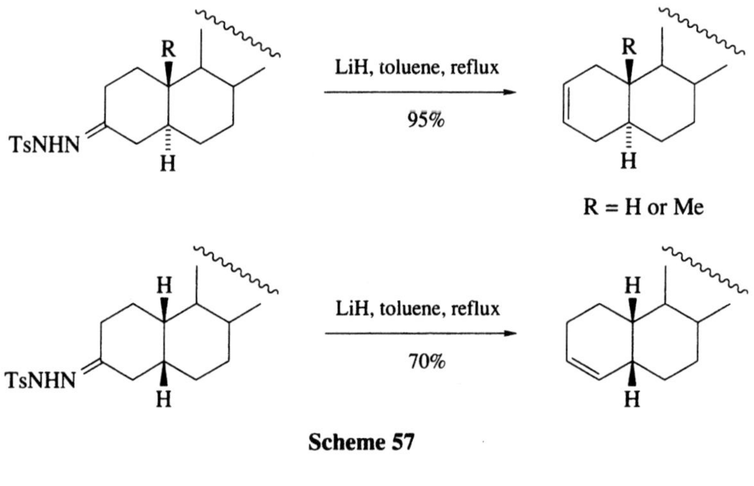

TsNHN; R; H; LiH, toluene, reflux; 95%; R = H or Me

TsNHN; H; H; LiH, toluene, reflux; 70%

Scheme 57

RO; O; i, ii; 96%

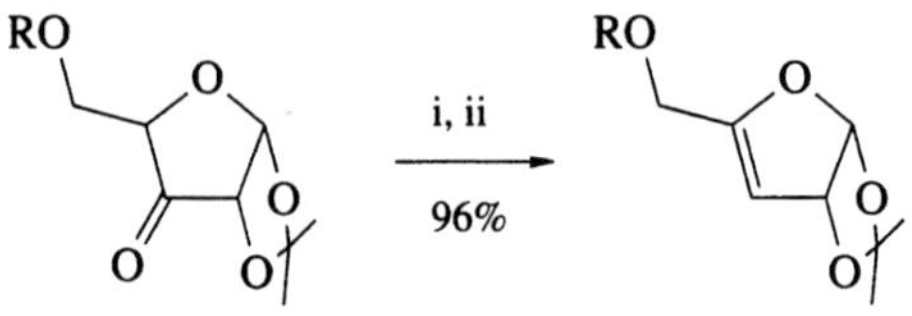

i, $TsNHNH_2$, EtOH; ii, NaOAc, DMF, 100 °C

Scheme 58

The issue of stereoselectivity in acyclic systems has not been studied systematically, but a high degree of control is not expected. There is one report of the formation of an ethylideneazacyclohexane with excellent control of stereoselectivity (Scheme 59).[119] No explanation for this dramatic selectivity was offered, nor is one readily apparent. Note also the strong preference for the formation of the more highly substituted regioisomer.

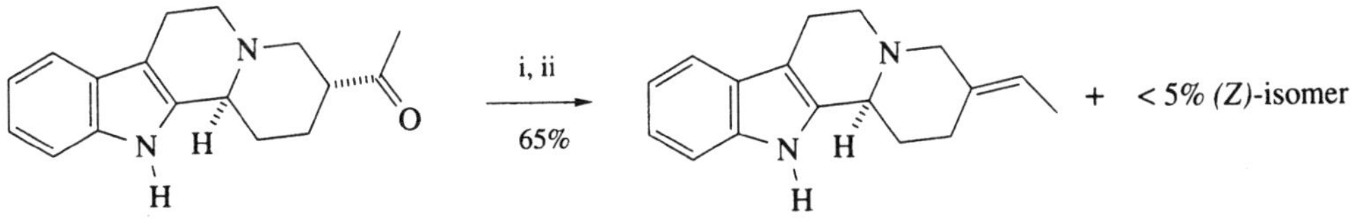

i, ii

65%

+ < 5% (*Z*)-isomer

(±)-Deplancheine

i, $TsNHNH_2$, AcOH; ii, NaOMe, diglyme, 150 °C

Scheme 59

4.6.6.3 The Protic Bamford–Stevens Reaction

The protic reaction on occasion is a useful method of alkene formation, but is far from general because the cation intermediate tends to undergo rearrangements.[120] Further, even for cases in which elimination to an alkene is the predominant pathway, the regioselectivity of the process is often mediocre. A key step in the synthesis of (+)-α-eudesmol and (–)-α-selinene exemplifies this point (Scheme 60).[121] There are, however, isolated examples of excellent selectivity, such as the reaction of a 3-ketotetrahydrofuran tosylhydrazone salt to give the corresponding cyclic enol ether as the major product (Scheme 61),[122] the intro-

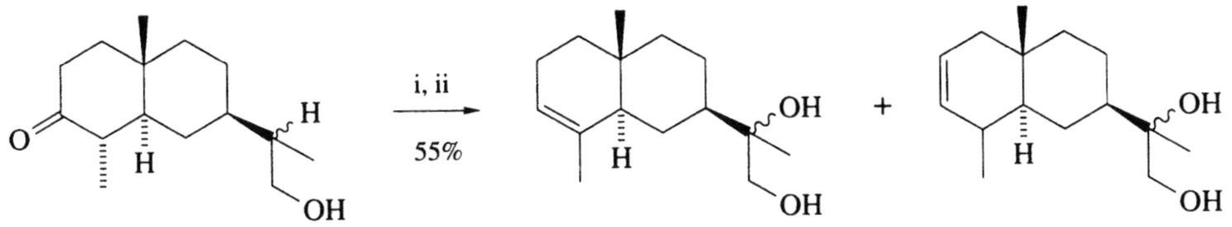

i, ii

55%

31% 69%

i, $TsNHNH_2$, BF_3•Et_2O; ii, Na, $HO(CH_2)_2OH$, Δ

Scheme 60

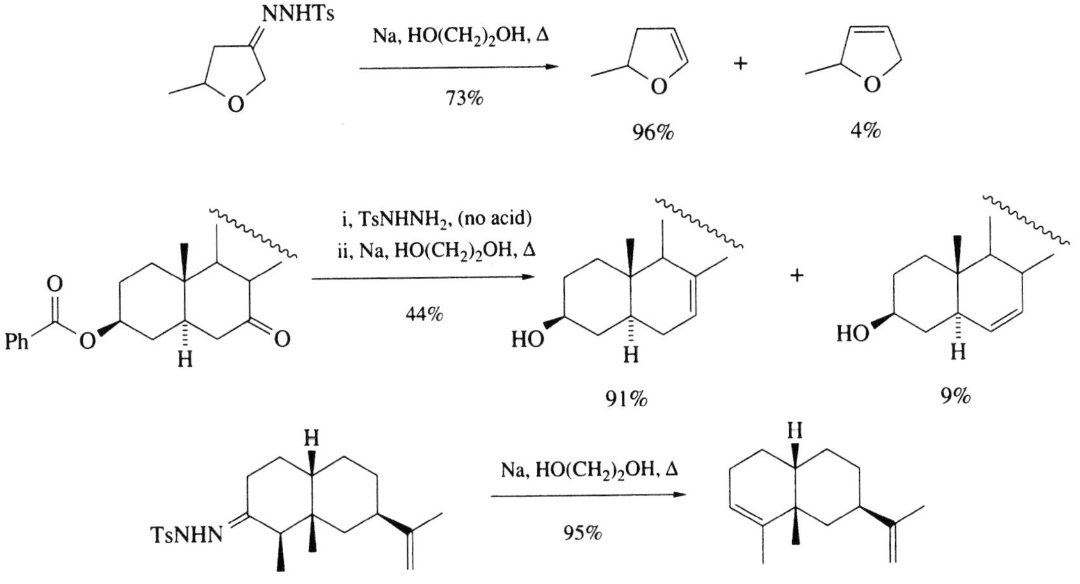

NNHTs

Na, $HO(CH_2)_2OH$, Δ

73%

96% 4%

i, $TsNHNH_2$, (no acid)
ii, Na, $HO(CH_2)_2OH$, Δ

44%

91% 9%

TsNHN

Na, $HO(CH_2)_2OH$, Δ

95%

Scheme 61

duction of a Δ^7 double bond in a steroid skeleton,[123] and the exclusive formation of the more highly substituted alkene regioisomer in a terpenoid decalin ring system (Scheme 61).[124] In some cases, it is the less substituted regioisomer that predominates, as illustrated by one step in the synthesis of azaprostaglandin analogs (Scheme 62).[125] These examples clearly illustrate that predicting regioselectivities in the aprotic Bamford–Stevens reaction without extremely close precedent is treacherous.

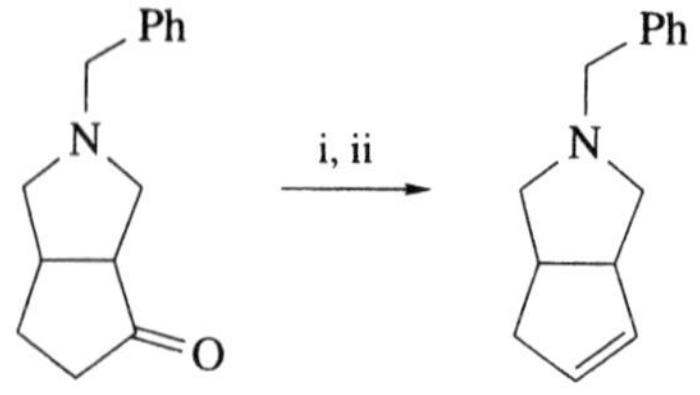

i, $TsNHNH_2$, TFA; ii, $NaO(CH_2)_2ONa$, $HO(CH_2)_2OH$, HMPA

Scheme 62

4.6.6.4 The Shapiro Reaction

The Shapiro reaction is the most generally applicable of the three variations discussed in this section. In contrast to the Bamford–Stevens reaction, the Shapiro modification reliably affords alkenes without skeletal rearrangements or competing insertion reactions. Scheme 63 shows several representative examples of typical Shapiro reactions that would give different results under Bamford–Stevens conditions. Most ketone tosyl- or trisyl-hydrazones with at least one α-proton undergo deprotonation with strong bases to give a dianion that decomposes *in situ* ultimately to the alkene product. A major advantage of the reaction of arenesulfonylhydrazones derived from unsymmetrical ketones is that the regiochemistry of the resultant alkene is highly predictable, almost always favoring the less substituted isomer, although arenesulfonylhydrazones of α,β-unsaturated ketones are an occasional exception to this generalization.

98% 2%

100%

100%

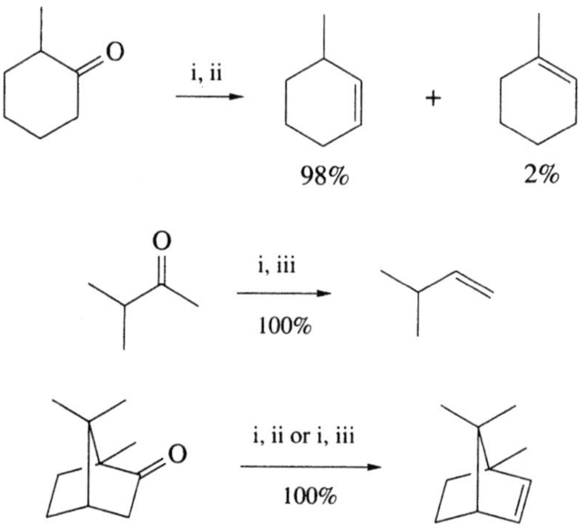

i, $TsNHNH_2$; ii, MeLi, Et_2O; iii, Bu^nLi, hexane

Scheme 63

4.6.6.4.1 Regioselectivity

One of the most valuable attributes of arenesulfonylhydrazones as a source of vinyllithiums is that the reaction exhibits a strong preference for the formation of one of the two possible vinyllithium regioisomers. In general, one may be confident that for unsymmetrically substituted ketone tosyl- or trisyl-hydrazones, deprotonation of the monoanion will occur predominantly at the less substituted α-position, *i.e.* Me > CH_2 > CH, to give, after elimination, the corresponding less substituted alkene product, which is generally favored over the more highly substituted regioisomer by ratios of 50:1 or more (Scheme 63). There are very few exceptions to this generalization. In one bicyclic derivative the bridgehead methine undergoes significant deprotonation to give the more highly substituted lithioalkene as a major by-product.[126] Similarly, the trisylhydrazone of a β-keto acetal undergoes deprotonation at the more highly substituted α-position, which in this case results in elimination of ethoxide followed by conjugate addition–elimination of the resultant monoanion to give a 1,2-pyrazole heterocycle.[127] Other than such rare exceptions, the Shapiro reaction consistently gives good yields of the less substituted alkene regioisomer and has been applied extensively in the synthesis of natural products, including approaches to aconitium alkaloids,[128] luciduline,[129] β-elemol,[130] costunolide[131] and compactin,[132] as well as a dodecahedrane intermediate (Scheme 64).[133]

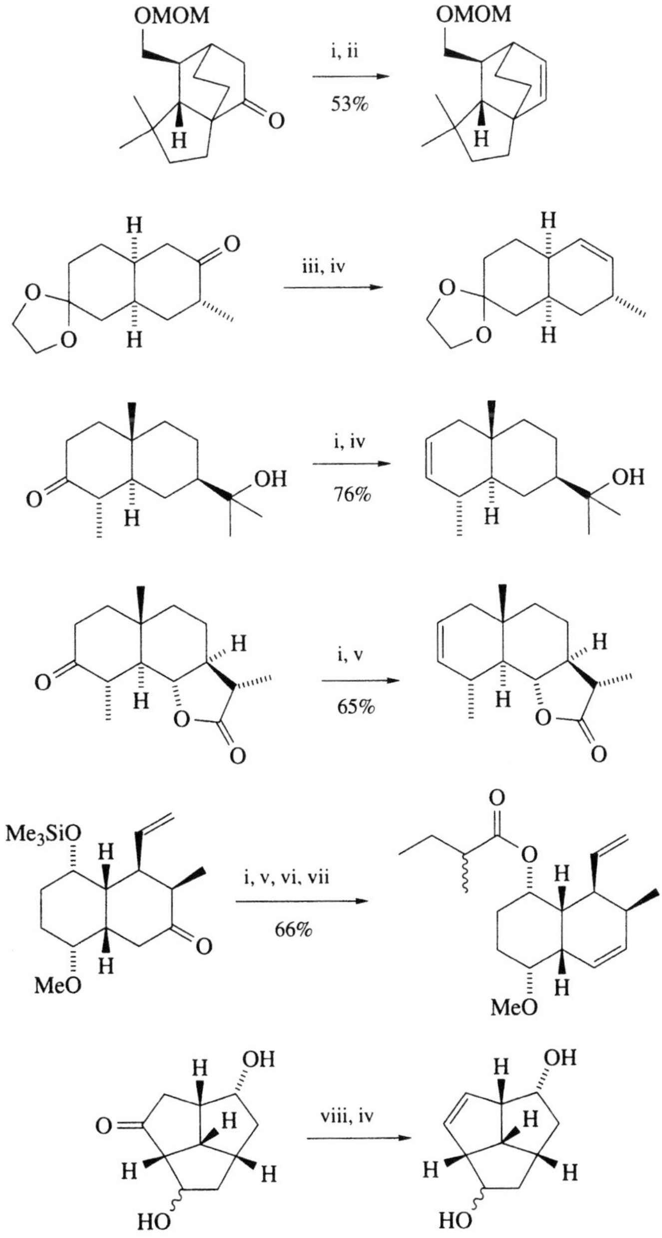

i, $TsNHNH_2$, $BF_3{\bullet}Et_2O$; ii, Bu^nLi, TMEDA; iii, $TsNHNH_2$, 25 °C; iv, MeLi; v, LDA, –78 to 0 °C; vi, 2% aq. HCl, 25 °C; vii, $(BuCO)_2O$, pyridine, DMAP; viii, $TrisNHNH_2$, $MgSO_4$

Scheme 64

Generalizing about regioselectivity in α,β-unsaturated systems is much less straightforward. There can be as many as three acidic sites that could undergo deprotonation and predicting *a priori* which might be preferred is difficult. There are instances of deprotonation at the least highly substituted of several positions and just as many counter examples.[103,134] Nonetheless, high levels of selectivity are often observed. For example, (+)-pulegone trisylhydrazone gives an excellent yield of the diene derived from deprotonation of the endocyclic α-methylene group in preference to either γ-methyl substituent (Scheme 65).[135] Heathcock's compactin synthesis (Scheme 65)[136] and van Tamelen's aphidicolin approach (Scheme 65)[137] both illustrate applications of this reaction in the terpene area.

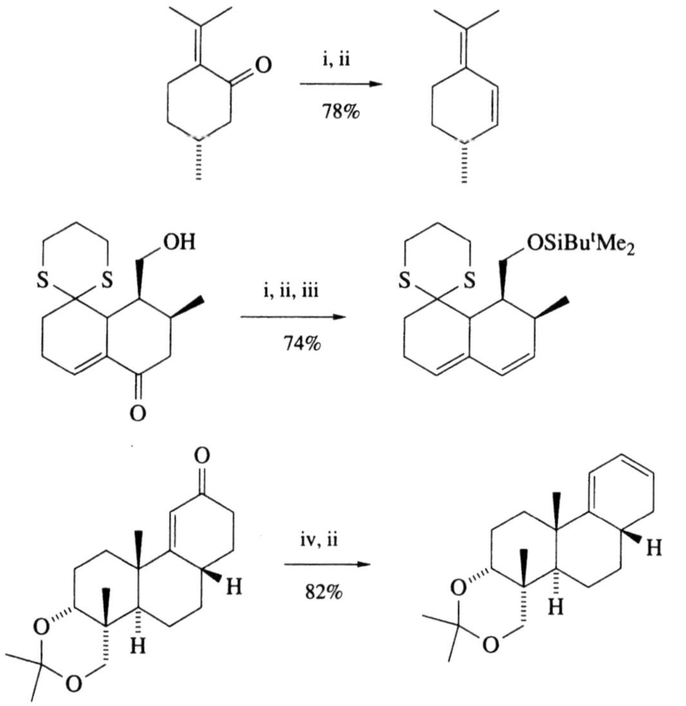

i, ii 78%

OH S S i, ii, iii 74% O OSiBu^{t}Me$_2$ S S

O iv, ii 82% H O H O H O H O

i, TrisNHNH$_2$, HCl; ii, Bu^{n}Li, TMEDA; iii, Bu^{t}Me$_2$SiCl; iv, TsNHNH$_2$, TsOH

Scheme 65

In contrast to the high regioselectivity observed for most saturated ketone arenesulfonylhydrazones, differential substitution adjacent to equivalently substituted α-positions does not appear to afford good regiochemical control. For instance, 3-alkyl-,[103] 3,4-dialkyl-[138] and 3,5-dialkyl-cyclohexanone derivatives[139] exhibit virtually no regioselectivity. One 3-ketosteroid reportedly gives the Δ^2 regioisomer,[140] but the yield is only 50% and no mention is made of the Δ^3 isomer. Finally, only meagre selectivity is observed in the reaction of 2-*n*-butyl-6-methylcyclohexanone tosylhydrazone.[141] These examples are summarized in Scheme 66.

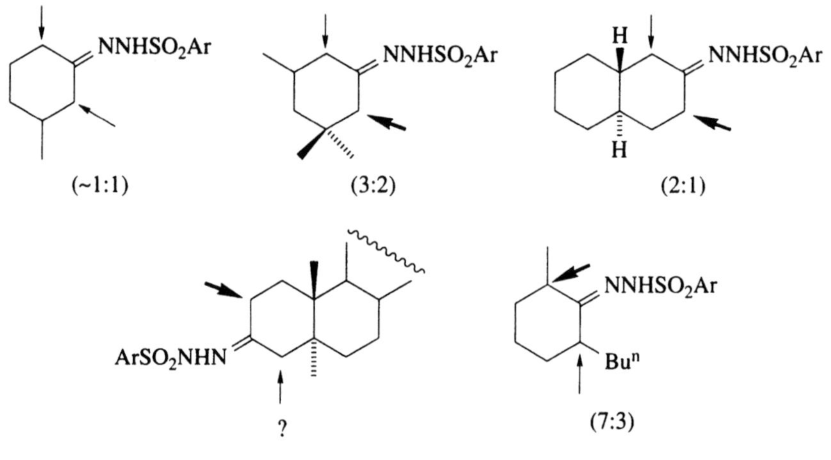

NNHSO$_2$Ar (~1:1) NNHSO$_2$Ar (3:2) H NNHSO$_2$Ar H (2:1)

ArSO$_2$NHN ? NNHSO$_2$Ar Bu^n (7:3)

Scheme 66

Although the observed regiochemical preferences in dianion formation bear a strong resemblance to those of kinetic enolate formation, there are a few examples that dramatize the fact that the regioselectivity cannot be understood exclusively on the basis of kinetically controlled deprotonation in the less substituted position. Specifically, the stereochemistry of the azomethine bond can come in to play as a '*syn*-directing effect', known to occur in the deprotonation of other imine derivatives.[142] Of course, these two factors themselves are not independent, since the (*E*)-hydrazone isomer is usually present in large excess under most conditions (see examples above), so that deprotonation in the less sterically encumbered position is predicted by either rationale, rendering the point immaterial in most cases. However, there are circumstances under which substantial deprotonation occurs at the more substituted position or, in α,β-unsaturated ketone arenesulfonylhydrazones, at a position otherwise not consistent with simple kinetically controlled deprotonation. For example, deprotonation regioselectivity for pulegone tosylhydrazone is a function of the hydrazone stereochemistry, but this effect is observed only in TMEDA and not in benzene.[109]

Similarly, for alkenes derived from saturated methyl ketones the regioselectivity is determined by starting hydrazone (*E*):(*Z*) ratios in some solvents but not in others. Thus 2-octanone trisylhydrazone, which is an inseparable 85:15 mixture of (*E*)- and (*Z*)-isomers, gives an 85:15 ratio of 1-octene:2-octene if vinyllithium formation is carried out in THF, but a 98:2 ratio of the same products when 10% TMEDA–hexane is the solvent.[103] The implication of this observation is that in THF the regioselectivity is determined by azomethine stereochemistry but that in TMEDA–hexane it is not. Note, however, that in this case a *syn*-directing effect does not occur in TMEDA, whereas in the previous example it does. Thus more than 10 years after it was asserted that a 'detailed explanation of the observed solvent dependencies...await further studies owing to the complexities of the reaction system....',[109] little headway has been made.

In the case of β-dicarbonyl compounds, either regioisomer can be obtained, depending upon reaction conditions. Monotosylhydrazones of cyclic β-diketones give the α,β-unsaturated ketone by treatment with potassium carbonate,[143] while the reaction of β-keto esters with stronger base (LDA) gives β,γ-unsaturated esters (Scheme 67).[144] In the former example, only the more acidic α-proton is removed (in ad-

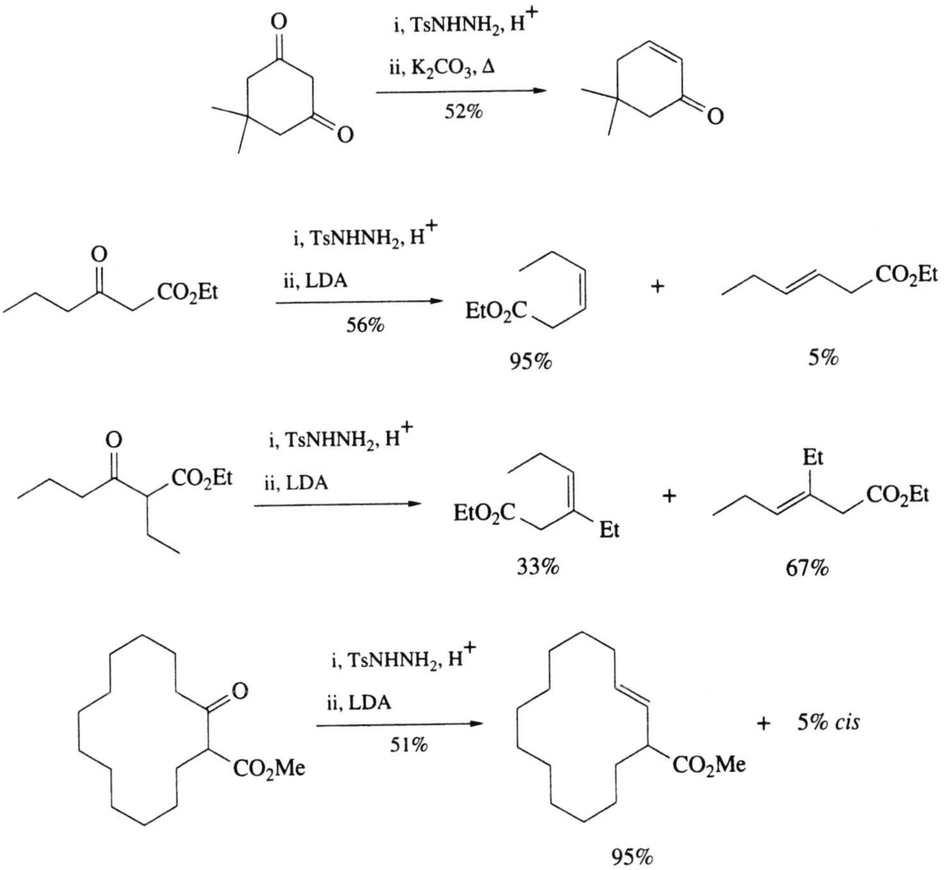

i, $TsNHNH_2$, H^+
ii, K_2CO_3, Δ
52%

i, $TsNHNH_2$, H^+
ii, LDA
56%
EtO_2C 95% + CO_2Et 5%

i, $TsNHNH_2$, H^+
ii, LDA
EtO_2C Et 33% + Et CO_2Et 67%

i, $TsNHNH_2$, H^+
ii, LDA
51%
CO_2Me 95% + 5% *cis*

Scheme 67

dition to the N—H), while LDA forms a trianion which preferentially eliminates in the opposite direction. A similar result was reported by Evans *et al.*, in a synthesis of juvabione (Scheme 68),[145] which made use of an *N*-aminoaziridine-derived hydrazone in place of the usual arenesulfonylhydrazone. The authors noted that such *N*-aminoaziridine derivatives are generally applicable to the Shapiro reaction, but no published study on the subject has appeared.

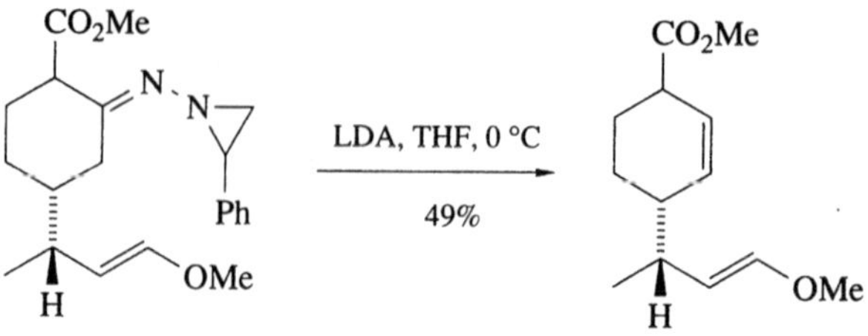

LDA, THF, 0 °C
49%

Scheme 68

4.6.6.4.2 *Stereoselectivity*

The issue of stereoselectivity of alkene formation from acyclic tosyl- or trisyl-hydrazones has been addressed in only a few cases. For straight chain ketones, there is a preference for *cis*-alkene formation (see Scheme 67). A few other examples appear to give lower ratios, which is surprising in view of several reports of highly *cis* selective reactions in which the vinyllithium intermediate was trapped with electrophiles.[101] Hydrocarbon branching in the α′-position degrades the stereoselectivity considerably (Scheme 67).

4.6.6.4.3 *Limitations*

There are some notable limitations of the Shapiro reaction. Most prominently, aldehyde arenesulfonylhydrazones do not form alkenes, but instead suffer addition of the alkyllithium 'base' to the azomethine bond.[146] Several other specific types of hydrazone derivatives also fail to undergo alkene formation, proceeding instead through well-documented alternative pathways. For example, arenesulfonylhydrazones with a leaving group in the α-position can suffer elimination at the monoanion stage, giving a tosylazoalkene as the major product.[147] When the elimination step is the opening of an α,β-oxirane ring, this process initiates the Eschenmoser fragmentation.[148,149] A related elimination has been observed in attempts to form a cyclic enol ether from the tosylhydrazone of a 3-ketotetrahydropyran (Scheme 69),[150] and some cyclic ether derivatives give allenes *via* a related elimination pathway.[151] On the other hand, a polycyclic α-amino ketone was found not to undergo elimination of the amino group (presumably because the conjugate base of the amine is a poorer leaving group), giving instead the expected allylic amine (Scheme 70).[152]

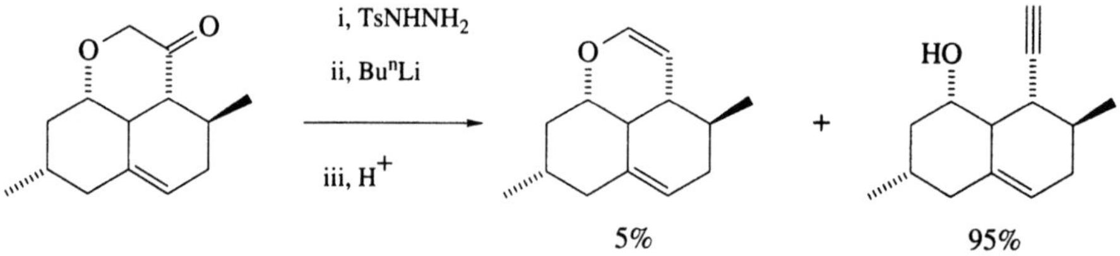

i, $TsNHNH_2$
ii, Bu^nLi
iii, H^+

5% + 95%

Scheme 69

i, $TsNHNH_2$, MeOH
ii, Bu^nLi, –78 to 20 °C
89%

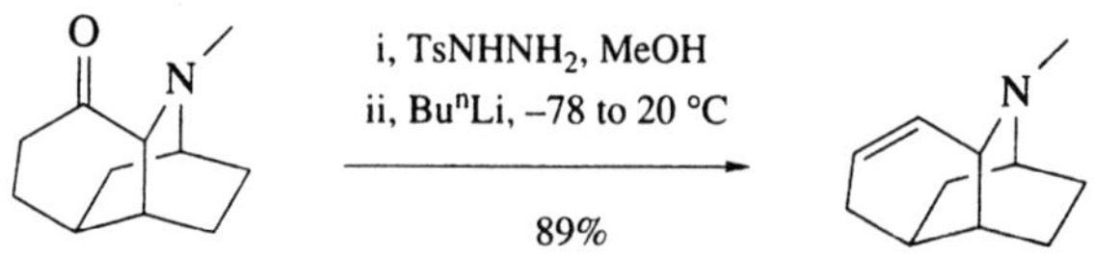

Scheme 70

In contrast to these results, it has been demonstrated that α-(methylthio)tosylhydrazone monoanions, formed by reaction of an initially formed dianion with dimethyl disulfide, undergo *in situ* dianion formation rather than elimination, to give vinyl sulfides (Scheme 71).[58] Since vinyl sulfides can be hydrolyzed to ketones, this sequence constitutes a very effective 1,2-carbonyl transposition.

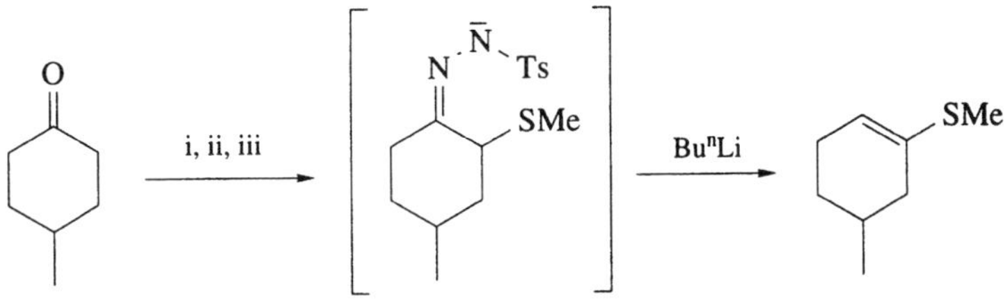

i, $TsNHNH_2$; ii, Bu^nLi, TMEDA, –50 °C; iii, Me_2S

Scheme 71

Finally, hindered arenesulfonylhydrazones can undergo addition of the alkyllithium rather than deprotonation, and, in fact, the normal Shapiro conditions generally do not give any alkene if an α-methine proton must be removed in the initial deprotonation step. Shapiro reported that the use of LDA overcomes the problem,[153] but there are later examples in which this modification is ineffective. In one such case, lithium *t*-butylamide proved to be superior (Scheme 72)[154] and in another it proved necessary to resort to the aforementioned *N*-aminoaziridine derivative in an aprotic Bamford–Stevens reaction, when the normal Shapiro and Bamford–Stevens conditions failed (Scheme 73).[155]

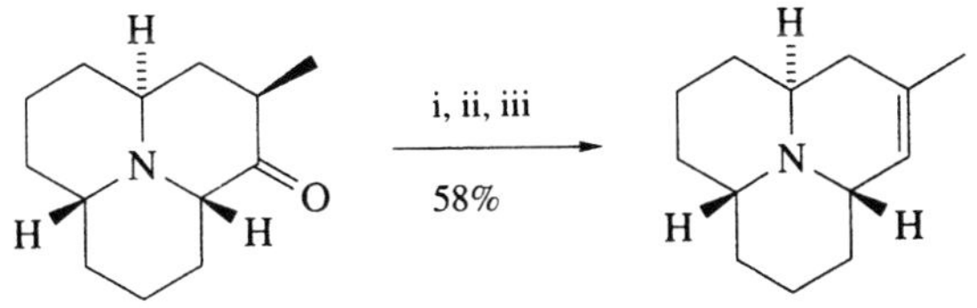

i, N_2H_4, EtOH, Δ; ii, TsCl; iii, Bu^tNHLi

Scheme 72

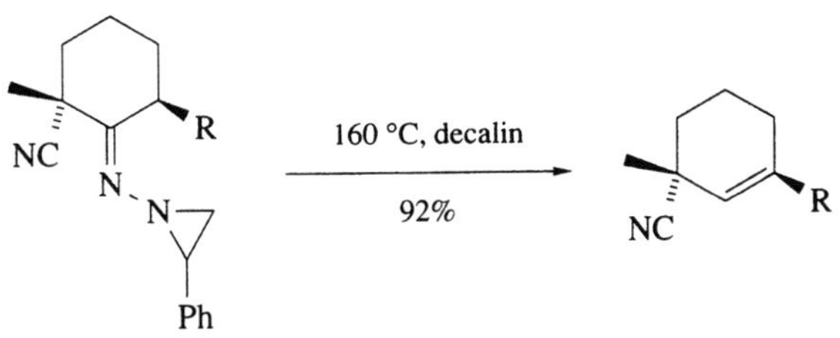

Scheme 73

4.6.7 MISCELLANEOUS RELATED TRANSFORMATIONS

4.6.7.1 Quinones to Arenes

A variation of the ketone to alkene transformation is conversion of quinones into arenes, for which a number of reagent systems have been employed. Over a century ago, the reduction of polycyclic quinones to aromatic systems with phosphorous and hydroiodic acid at elevated temperatures was reported.[156] These conditions, however, resulted in complex reaction mixtures consisting of phenols and polyhydrogenated products. More recently, Konieczny and Harvey[157] have reported the reduction of quinones in the absence of phosphorous using hydroiodic acid in acetic acid, which alleviates the need for high temperatures and results in cleaner reactions and higher yields of the arene products (Scheme 74).

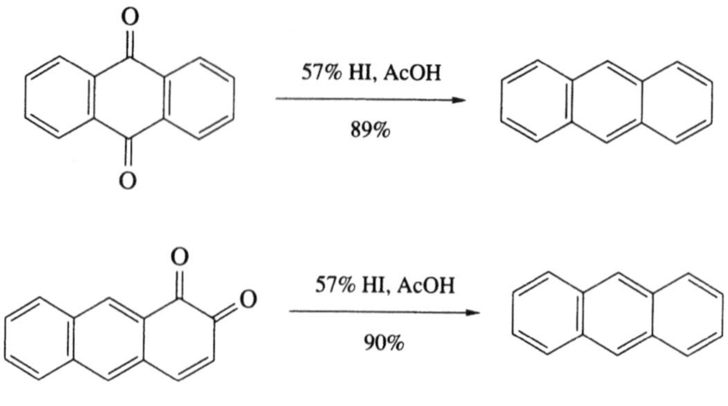

57% HI, AcOH

89%

57% HI, AcOH

90%

Scheme 74

Other methods to reduce polycyclic quinones to arenes that have met with some success include $SnCl_4/HCl$,[158] Zn/NH_3,[159] Zn/pyridine/$MeCO_2H$,[160] $NaBH_4/BF_3{\cdot}Et_2O$,[161] and $NaBH_4$ in MeOH.[162] All of these methods are compatible with aryl halides and the latter two are also compatible with aryl ethers, anilines and phenols. 10-Arylmethylene-9-anthrones have also been reduced to anthracenes with diborane (Scheme 75).[163]

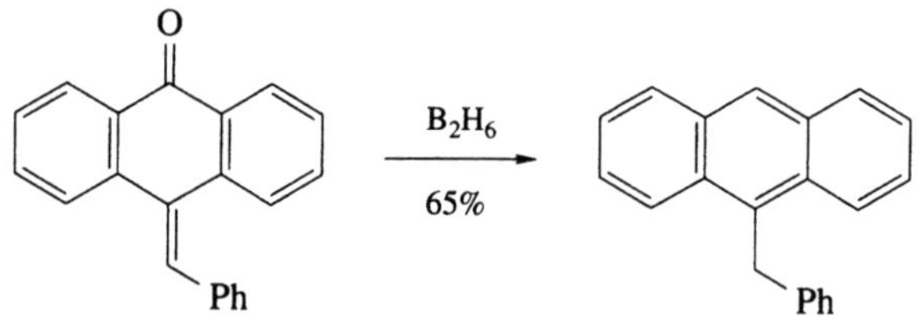

B_2H_6

65%

Scheme 75

The conversion of quinones to phenols and subsequently to arenes is an indirect alternative to this process and is covered in this volume, Chapter 4.5.

4.6.7.2 Ketones and 1,2-Diketones to Alkynes

Ketones and 1,2-diketones are readily converted into alkynes, which can subsequently be reduced to alkenes. Since there are a number of methods to reduce alkynes stereoselectively, this route offers a method of controlling the stereochemistry of acyclic alkenes derived from ketone starting materials. The most common approach for the ketone to alkyne transformation has been conversion of the starting ketone into a vicinal dichloride and subsequent bis elimination induced by strong base.[164]

Acetophenone derivatives have been converted into the corresponding alkynes employing $NaNH_2$ in NH_3[165] or sodium *t*-butoxide[166] as the base in the elimination step. Ferrocenylalkynes[167] have been prepared in a similar manner: treating acetylferrocene with $POCl_3$ in DMF[168] results in the formation of the β-chloro aldehyde, which upon treatment with KOH in dioxane affords the ferrocenylalkyne in high yield. Alkynes have also been prepared from benzyl alkyl ketones by initial hydrazone formation and subsequent oxidation with mercury(I) trifluoroacetate.[169] Ketones have also been converted into the corresponding vinyl halide[170] or vinyl phosphate[171] derivatives, according to the methods described in Section 4.6.5 of this chapter, and then subjected to elimination conditions to afford the target alkynes.

Cyclooctanone has been converted into cyclooctyne by treatment with semicarbazide acetate, resulting in the formation of the semicarbazone, which was oxidized to the 1,2,3-selenadiazole, and then pyrolyzed over glass powder to give cyclooctyne.[172]

1,2-Diketones have also been converted into alkynes, most commonly through the bishydrazone derivatives. Originally, cyclodeca-1,2-dione was converted into cyclodecyne by initial bishydrazone formation and subsequent oxidation with mercury(II) oxide (Scheme 76).[173] More recently, cyclic

1,2-bishydrazones have been converted to the corresponding alkynes with $Pb(OAc)_4$[174] or by irradiating an aqueous methanolic NaOH solution of the bishydrazone.[175] In acyclic systems 1,2-bishydrazones have been oxidized with Cu_2Cl_2 in pyridine or $Cu(OAc)_2$ in the presence of O_2.[176] Finally, substituted diphenyl diketones (benzils) have been converted into 1,2-diphenylalkynes (tolanes) upon heating in triethyl phosphite.[177]

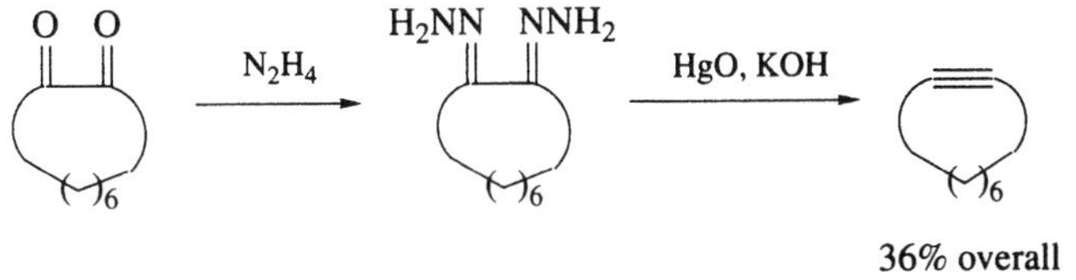

O O — N_2H_4 → H_2NN NNH_2 — HgO, KOH → (cycloalkyne, $(\ )_6$)

36% overall

Scheme 76

4.6.8 REFERENCES

1. D. H. Piehl, *J. Chem. Soc., Chem. Commun.*, 1969, 461.
2. J. Perregaard, I. Thomsen and S.-O. Lawesson, *Acta Chem. Scand., Ser. B*, 1975, **29**, 599.
3. W. B. Motherwell, *J. Chem. Soc., Chem. Commun.*, 1973, 935.
4. P. Hodge and M. N. Khan, *J. Chem. Soc., Perkin Trans. 1*, 1975, 809.
5. M. Sergent, M. Mongrain and P. Deslongchamps, *Can. J. Chem.*, 1972, **50**, 336.
6. C. H. Heathcock, M. J. Taschner, T. Rosen, J. A. Thomas, C. R. Hadley and G. Popjak, *Tetrahedron Lett.*, 1982, **23**, 4747.
7. M. Sasaki and T. Murae, *Tetrahedron Lett.*, 1989, **30**, 355.
8. J. Ficini, J. d'Angelo and J. Noiré, *J. Am. Chem. Soc.*, 1974, **96**, 1213.
9. J. C. Sowden and H. O. L. Fischer, *J. Am. Chem. Soc.*, 1947, **69**, 1048.
10. P. Dampawan and W. W. Zajac, Jr., *Tetrahedron Lett.*, 1982, **23**, 135.
11. A. E. Greene, J.-C. Muller and G. Ourisson, *J. Org. Chem.*, 1974, **39**, 186; R. T. Arnold and S. T. Kulenovic, *J. Org. Chem.*, 1978, **43**, 3687; R. H. Reuss and A. Hassner, *J. Org. Chem.*, 1974, **39**, 1785; R. R. Gallucci and R. Going, *J. Org. Chem.*, 1981, **46**, 2532.
12. M. R. Roberts and R. H. Schlessinger, *J. Am. Chem. Soc.*, 1981, **103**, 724.
13. D. R. James, R. W. Rees and C. W. Shoppee, *J. Chem. Soc.*, 1955, 1370.
14. N. Kishner, *J. Russ. Phys.-Chem. Soc.*, 1913, **45**, 973.
15. D. E. Ames and R. E. Bowman, *J. Chem. Soc.*, 1951, 2752.
16. P. S. Wharton, S. Dunny and L. S. Krebs, *J. Org. Chem.*, 1964, **29**, 958.
17. N. J. Leonard and S. Gelfand, *J. Am. Chem. Soc.*, 1955, **77**, 3272.
18. N. J. Leonard and S. Gelfand, *J. Am. Chem. Soc.*, 1955, **77**, 3269.
19. P. S. Wharton and D. H. Bohlen, *J. Org. Chem.*, 1961, **26**, 3615.
20. S. D. Burke, C. W. Murtiashaw, J. O. Saunders, J. A. Oplinger and M. S. Dike, *J. Am. Chem. Soc.*, 1984, **106**, 4558.
21. D. D. Maas, M. Blagg and D. F. Wiemer, *J. Org. Chem.*, 1984, **49**, 853.
22. H.-J. Liu and H. Wynn, *Tetrahedron Lett.*, 1985, **26**, 4843.
23. F. M. Hauser, W. P. Ellenberger and T. C. Adams, Jr., *J. Org. Chem.*, 1984, **49**, 1169.
24. F. E. Ziegler, K.-J. Hwang, J. F. Kadow, S. I. Klein, U. K. Pati and T.-F. Wang, *J. Org. Chem.*, 1986, **51**, 4573.
25. J. W. Morzycki, J. Jurek and W. J. Rodewald, *Can. J. Chem.*, 1986, **64**, 1540.
26. W. R. Benn and R. M. Dodson, *J. Org. Chem.*, 1964, **29**, 1142.
27. G. Ohloff and G. Uhde, *Helv. Chim. Acta*, 1970, **53**, 531.
28. J. A. Marshall and R. H. Ellison, *J. Am. Chem. Soc.*, 1976, **98**, 4312.
29. J. A. Marshall, E. D. Robinson and R. D. Adams, *Tetrahedron Lett.*, 1988, **29**, 4913.
30. A. Yasuda, H. Yamamoto and H. Nozaki, *Tetrahedron Lett.*, 1976, 2621.
31. R. O. Hutchins, C. A. Milewski and B. E. Maryanoff, *J. Am. Chem. Soc.*, 1973, **95**, 3662.
32. D. G. Batt, N. Takamura and B. Ganem, *J. Am. Chem. Soc.*, 1984, **106**, 3353.
33. C. Djerassi and I. Midgley, *Tetrahedron Lett.*, 1972, **23**, 4673.
34. M. Fetizon, M. Jurion and N. T. Anh, *Chem. Commun.*, 1969, 112.
35. R. E. Ireland and G. Pfister, *Tetrahedron Lett.*, 1969, **20**, 2145.
36. R. E. Ireland, D. C. Muchmore and U. Hengartner, *J. Am. Chem. Soc.*, 1972, **94**, 5098.
37. S. C. Welch and M. E. Walters, *J. Org. Chem.*, 1978, **43**, 2715.
38. A. B. Smith, III, B. A. Wexler, C.-Y. J. Tu and J. P. Konopelski, *J. Am. Chem. Soc.*, 1985, **107**, 1308.
39. S. Kamata, N. Haga, T. Matsui and W. Nagata, *Synthesis*, 1986, 588.
40. P. A. Grieco, R. Lis, S. Ferriño and J. Y. Jaw, *J. Org. Chem.*, 1984, **49**, 2342; G. Vidari, S. Ferriño and P. A. Grieco, *J. Am. Chem. Soc.*, 1984, **106**, 3539; P. A. Grieco, R. Lis, S. Ferrino and J. Y. Jaw, *J. Org. Chem.*, 1982, **47**, 601.
41. B. M. Trost and P. Renaut, *J. Am. Chem. Soc.*, 1982, **104**, 6668.
42. F. Charbonnier, A. Moyano and A. E. Greene, *J. Org. Chem.*, 1987, **52**, 2303.
43. G. W. Kenner and N. R. Williams, *J. Chem. Soc.*, 1955, 522.
44. R. E. Ireland, P. Bey, K.-F. Cheng, R. J. Czarny, J.-F. Moser and R. I. Trust, *J. Org. Chem.*, 1975, **40**, 1000.

45. W. J. Scott, G. T. Crisp and J. K. Stille, *J. Am. Chem. Soc.*, 1984, **106**, 4630.
46. P. J. Stang and W. Treptow, *Synthesis*, 1980, 283.
47. J. R. Hassdenteufel and M. Hanack, *Tetrahedron Lett.*, 1980, **21**, 503.
48. S. L. Schreiber and R. C. Hawley, *Tetrahedron Lett.*, 1985, **26**, 5971.
49. S. Cacchi, E. Morera and G. Ortar, *Tetrahedron Lett.*, 1984, **25**, 4821.
50. R. E. Dolle, S. J. Schmidt and L. I. Kruse, *Tetrahedron Lett.*, 1988, **29**, 1581.
51. S. Cacchi, P. G. Ciattini, E. Morera and G. Ortar, *Tetrahedron Lett.*, 1986, **27**, 5541.
52. G. A. Peterson, F.-A. Kunng, J. S. McCallum and W. D. Wulff, *Tetrahedron Lett.*, 1987, **28**, 1381.
53. L. R. Subramanian, A. G. Martinez, A. H. Fernandez and R. M. Alvarez, *Synthesis*, 1984, 481.
54. S. J. Nelson, G. Detre and M. Tanabe, *Tetrahedron Lett.*, 1973, **14**, 447.
55. H. C. Brown and R. M. Gallivan, Jr., *J. Am. Chem. Soc.*, 1968, **90**, 2906.
56. B. M. Trost, K. Hiroi and S. Kurozumi, *J. Am. Chem. Soc.*, 1975, **97**, 438; K. Hiroi, M. Nishida, A. Nakayama, K. Nakazawa, E. Fujii and S. Sato, *Chem. Lett.*, 1979, 969.
57. S. Kano, T. Yokomatsu, T. Ono, S. Hibino and S. Shibuya, *J. Chem. Soc., Chem. Commun.*, 1978, 414.
58. T. Nakai and T. Mimura, *Tetrahedron Lett.*, 1979, **20**, 531.
59. T. Cohen, G. Herman, J. R. Falck and A. J. Mura, Jr., *J. Org. Chem.*, 1975, **40**, 812.
60. M. Kakimoto, T. Yamamoto and M. Okawara, *Tetrahedron Lett.*, 1979, **20**, 623.
61. E. Wenkert and T. W. Ferreira, *J. Chem. Soc., Chem. Commun.*, 1982, 840.
62. B. M. Trost and P. L. Ornstein, *Tetrahedron Lett.*, 1981, **22**, 3463.
63. R. B. Boar, D. W. Hawkins, J. F. McGhie and D. H. R. Barton, *J. Chem. Soc., Perkin Trans. 1*, 1973, 654.
64. S. Inayama, T. Kawamata and T. Ohkura, *Tetrahedron Lett.*, 1978, **19**, 1557.
65. J. Fishman, M. Torigoe and H. Guzik, *J. Org. Chem.*, 1963, **28**, 1443.
66. G. R. Pettit and E. E. van Tamelen, *Org. React. (N.Y.)*, 1962, **12**, 356.
67. D. A. Ben-Efraim and F. Sondheimer, *Tetrahedron*, 1969, **25**, 2823.
68. E. Wenkert, J. M. Hanna, Jr., M. H. Leftin, E. L. Michelotti, K. T. Potts and D. Usifer, *J. Org. Chem.*, 1985, **50**, 1125.
69. D. H. R. Barton, G. Bashiardes and J.-L. Fourrey, *Tetrahedron Lett.*, 1984, **25**, 1287.
70. I. Ninomiya, C. Hashimoto, T. Kiguchi, D. H. R. Barton, X. Lusinchi and P. Milliet, *Tetrahedron Lett.*, 1985, **26**, 4187.
71. T. G. Back, *J. Chem. Soc., Chem. Commun.*, 1984, 1417.
72. P. G. Theobald and W. H. Okamura, *Tetrahedron Lett.*, 1987, **28**, 6565.
73. J. Klein, R. Levene and E. Dunkelblum, *Tetrahedron Lett.*, 1972, **23**, 2845.
74. G. L. Larson, E. Hernández, C. Alonso and I. Nieves, *Tetrahedron Lett.*, 1975, **16**, 4005.
75. G. L. Larson, D. Hernández and A. Hernández, *J. Organomet. Chem.*, 1974, **76**, 9.
76. G. L. Larson and A. Hernández, *J. Organomet. Chem.*, 1975, **102**, 123.
77. C. M. Hill, R. A. Walker and M. E. Hill, *J. Am. Chem. Soc.*, 1951, **73**, 1663.
78. P. Pino and G. P. Lorenzi, *J. Org. Chem.*, 1966, **31**, 329.
79. D. J. Pasto and C. C. Cumbo, *J. Am. Chem. Soc.*, 1964, **86**, 4343.
80. D. H. R. Barton, R. E. O'Brien and S. Sternhell, *J. Chem. Soc.*, 1962, 470.
81. A. B. Turner and P. T. van Leersum, *Tetrahedron Lett.*, 1983, **24**, 4589.
82. A. B. Turner and P. T. van Leersum, *J. Chem. Soc., Perkin Trans. 1*, 1988, 1653.
83. A. Pross and S. Sternhell, *Aust. J. Chem.*, 1970, **23**, 989.
84. D. H. R. Barton, G. Bashiardes and J.-L. Fourrey, *Tetrahedron Lett.*, 1983, **24**, 1605.
85. D. H. R. Barton, G. Bashiardes and J.-L. Fourrey, *Tetrahedron*, 1988, **44**, 147.
86. K. Takai, K. Nitta and K. Utimoto, *J. Am. Chem. Soc.*, 1986, **108**, 7408.
87. P. F. Hudrlik and A. K. Kulkarni, *Tetrahedron*, 1985, **41**, 1179.
88. S. Kim and K. H. Ahn, *J. Org. Chem.*, 1984, **49**, 1717.
89. S. Masamune, P. A. Rossy and G. S. Bates, *J. Am. Chem. Soc.*, 1973, **95**, 6452.
90. P. Caubère, *Angew. Chem., Int. Ed. Engl.*, 1983, **22**, 599.
91. R. D. Clark and C. H. Heathcock, *J. Org. Chem.*, 1976, **41**, 636.
92. P. A. Zoretic, C. Bhakta and R. H. Khan, *Tetrahedron Lett.*, 1983, **24**, 1125.
93. P. J. Kropp, S. A. McNeely and R. D. Davis, *J. Am. Chem. Soc.*, 1983, **105**, 6907.
94. J. W. Lewis and P. P. Lynch, *Proc. Chem. Soc., London*, 1963, 19.
95. J. M. Coulter, J. W. Lewis and P. P. Lynch, *Tetrahedron*, 1968, **24**, 4489.
96. J. W. Lewis and A. A. Pearce, *Tetrahedron Lett.*, 1964, **5**, 2039.
97. R. J. Friary, Sr., J. M. Gilligan, R. P. Szajewski, K. J. Falci and R. W. Franck, *J. Org. Chem.*, 1973, **38**, 3487.
98. B. Singaram, C. T. Goralski, M. V. Rangaishenvi and H. C. Brown, *J. Am. Chem. Soc.*, 1989, **111**, 384.
99. R. H. Shapiro, *Org. React. (N.Y.)*, 1976, **23**, 405.
100. R. M. Adlington and A. G. M. Barrett, *Acc. Chem. Res.*, 1983, **16**, 55.
101. A. R. Chamberlin and D. J. Sall, *Org. React. (N.Y.)*, 1990, **39**, 1.
102. S. H. Bertz and G. Dabbagh, *J. Org. Chem.*, 1983, **48**, 116.
103. A. R. Chamberlin and F. T. Bond, *J. Org. Chem.*, 1978, **43**, 154.
104. A. B. Smith, III and P. J. Jerris, *J. Org. Chem.*, 1982, **47**, 1845.
105. C. A. Bunnell and P. L. Fuchs, *J. Am. Chem. Soc.*, 1977, **99**, 5184.
106. P. G. Williard, L. A. Grab, and J. P. Springer, *Tetrahedron Lett.*, 1981, **22**, 2731.
107. J. E. Baldwin and J. C. Bottaro, *J. Chem. Soc., Chem. Commun.*, 1981, 1121.
108. R. M. Adlington and A. G. M. Barrett, *J. Chem. Soc., Perkin Trans. 1*, 1981, 2848 and ref. 109.
109. W. G. Dauben, G. T. Rivers and W. T. Zimmerman, *J. Am. Chem. Soc.*, 1977, **99**, 3414.
110. C. A. Bunnell and P. L. Fuchs, *J. Org. Chem.*, 1977, **42**, 2614.
111. L. Friedman and H. Shechter, *J. Am. Chem. Soc.*, 1959, **81**, 5512.
112. J. M. Coxon, M. P. Hartshorn, D. N. Kirk and M. A. Wilson, *Tetrahedron*, 1969, **25**, 3107; for many other examples, see ref. 99.
113. M. A. Luyten and R. Keese, *Tetrahedron*, 1986, **42**, 1687.

114. J. W. Wilt and P. J. Chenier, *J. Am. Chem. Soc.*, 1968, **90**, 7366.
115. J. W. Wilt and W. J. Wagner, *J. Org. Chem.*, 1964, **29**, 2788.
116. W. A. Kinney, M. J. Coghlan and L. A. Paquette, *J. Am. Chem. Soc.*, 1984, **106**, 6868.
117. L. Caglioti, P. Grasselli and A. Selva, *Gazz. Chim. Ital.*, 1964, **94**, 537.
118. V. Nair and A. K. Sinhababu, *J. Org. Chem.*, 1978, **43**, 5013.
119. S. B. Mandal and S. C. Pakrashi, *Heterocycles*, 1987, **26**, 1557.
120. R. H. Shapiro and M. J. Heath, *J. Am. Chem. Soc.*, 1967, **89**, 5734; G. Kaufman, F. Cook, H. Shechter, J. Bayless and L. Friedman, *J. Am. Chem. Soc.*, 1967, **89**, 5736; M. Fetizon and P. Jaudon, *Tetrahedron*, 1977, **33**, 2079; N. Satyanarayana and U. R. Nayak, *Synth. Commun.*, 1985, **15**, 1107.
121. J. P. Kutney and A. K. Singh, *Can. J. Chem.*, 1984, **62**, 1407.
122. M. A. Gianturco, P. Friedel and V. Flanagan, *Tetrahedron Lett.*, 1965, **6**, 1847.
123. R. C. Cambie, P. A. Craw, R. J. Hughes, P. S. Rutledge and P. D. Woodgate, *Aust. J. Chem.*, 1982, **35**, 2111.
124. E. Piers and R. J. Keziere, *Can. J. Chem.*, 1969, **47**, 137.
125. K. Miyajima, M. Takemoto and K. Achiwa, *Heterocycles*, 1988, **27**, 643.
126. B. M. Trost and T. N. Nanninga, *J. Am. Chem. Soc.*, 1985, **107**, 1293.
127. R. Baudouy, J. Sartoretti and F. Choplin, *Tetrahedron*, 1983, **39**, 3293.
128. M. Ihara, Y. Ishida, M. Abe, M. Toyota, K. Fukumoto and T. Kametani, *Chem. Lett.*, 1985, 1127.
129. W. L. Scott and D. A. Evans, *J. Am. Chem. Soc.*, 1972, **94**, 4779.
130. J. P. Kutney and A. K. Singh, *Can. J. Chem.*, 1983, **61**, 1111.
131. P. A. Grieco and M. Nishizawa, *J. Org. Chem.*, 1977, **42**, 1717.
132. N. N. Girotra, R. A. Reamer and N. L. Wendler, *Tetrahedron Lett.*, 1984, **25**, 5371.
133. W. P. Roberts and G. Shoham, *Tetrahedron Lett.*, 1981, **22**, 4895.
134. P.-A. Bütikofer and C. H. Eugster, *Helv. Chim. Acta*, 1983, **66**, 1148; R. T. Taylor, C. R. Degenhardt, W. P. Melega and L. A. Paquette, *Tetrahedron Lett.*, 1977, **18**, 159; L. A. Paquette, R. G. Daniels and R. Gleiter, *Organometallics*, 1984, **3**, 560; J. C. Caille, M. Farnier and R. Guilard, *Can. J. Chem.*, 1986, **64**, 824.
135. M. G. Silvestri, *J. Org. Chem.*, 1983, **48**, 2419.
136. T. Rosen, M. J. Taschner, J. A. Thomas and C. H. Heathcock, *J. Org. Chem.*, 1985, **50**, 1190.
137. E. E. van Tamelen, S. Zawacky, R. K. Russell and J. G. Carlson, *J. Am. Chem. Soc.*, 1983, **105**, 142.
138. T. K. Jones and S. E. Denmark, *Helv. Chim. Acta*, 1983, **66**, 2377.
139. P. C. Traas, H. Boelens and H. J. Takken, *Tetrahedron Lett.*, 1976, **17**, 2287.
140. W. E. Fristad, T. R. Bailey and L. A. Paquette, *J. Org. Chem.*, 1978, **43**, 1620.
141. F. T. Bond and R. A. DiPietro, *J. Org. Chem.*, 1981, **46**, 1315.
142. D. E. Bergbreiter and M. Newcomb, *Asymm. Synth.*, 1983, **2**, 1.
143. G. A. Hiegel and P. Burk, *J. Org. Chem.*, 1973, **38**, 3637.
144. C. A. Bunnell and P. L. Fuchs, *J. Am. Chem. Soc.*, 1977, **99**, 5184.
145. D. A. Evans and J. V. Nelson, *J. Am. Chem. Soc.*, 1980, **102**, 774.
146. E. Vedejs and W. T. Stolle, *Tetrahedron Lett.*, 1977, **18**, 135.
147. C. B. Reese and H. P. Sanders, *J. Chem. Soc., Perkin Trans. 1*, 1982, 2719; R. McCague, C. J. Moody and C. W. Rees, *J. Chem. Soc., Chem. Commun.*, 1982, 497.
148. A. Eschenmoser, D. Felix and G. Ohloff, *Helv. Chim. Acta*, 1967, **50**, 708.
149. M. Tanabe, D. F. Crowe, R. L. Dehn and G. Detre, *Tetrahedron Lett.*, 1967, **8**, 3739.
150. S. J. Hecker and C. H. Heathcock, *J. Am. Chem. Soc.*, 1986, **108**, 4586.
151. A. M. Foster and W. C. Agosta, *J. Org. Chem.*, 1972, **37**, 61.
152. D. A. Evans, A. M. Golob, N. S. Mandel and G. S. Mandel, *J. Am. Chem. Soc.*, 1978, **100**, 8170.
153. K. J. Kolonko and R. H. Shapiro, *J. Org. Chem.*, 1978, **43**, 1404.
154. R. H. Mueller, M. E. Thompson and R. M. DiPardo, *J. Org. Chem.*, 1984, **49**, 2217; J. W. Huffman, G. Shanmugasundaram, R. Sawdaye, P. C. Raveendranath and R. C. Desai, *J. Org. Chem.*, 1985, **50**, 1460.
155. F. Mohamadi and D. B. Collum, *Tetrahedron Lett.*, 1984, **25**, 271.
156. C. Graebe and C. Liebermann, *Liebigs Ann. Chem. Suppl.*, 1870, **7**, 287.
157. M. Konieczny and R. G. Harvey, *J. Org. Chem.*, 1979, **44**, 4813.
158. W. Davies and Q. N. Porter, *J. Chem. Soc.*, 1957, 4967.
159. G. M. Badger and A. R. M. Gibb, *J. Chem. Soc.*, 1949, 799.
160. J. T. Traxler, *Synth. Commun.*, 1977, **7**, 161.
161. D. S. Bapat, B. C. Subba Rao, M. K. Unni and K. Venkataraman, *Tetrahedron Lett.*, 1960, **1**, 15.
162. T. R. Criswell and B. H. Klanderman, *J. Org. Chem.*, 1974, **39**, 770.
163. M. Rabinovitz and G. Salemnik, *J. Org. Chem.*, 1968, **33**, 3935.
164. N. A. Domnin, *J. Gen. Chem. (USSR) (Engl. Transl.)*, 1938, **8**, 851.
165. D. A. H. Taylor, *J. Chem. Soc.*, 1963, 1553.
166. M. S. Newman and D. E. Reid, *J. Org. Chem.*, 1958, **23**, 665.
167. M. Rosenblum, N. Brawn, J. Papenmeier and M. Applebaum, *J. Organomet. Chem.*, 1966, **6**, 173.
168. Z. Arnold and J. Žemlička, *Proc. Chem. Soc., (London)*, 1958, 227.
169. R. J. Theis and R. E. Dessy, *J. Org. Chem.*, 1966, **31**, 624.
170. A. M. Krubiner, N. Gottfried and E. P. Oliveto, *J. Org. Chem.*, 1969, **34**, 3502.
171. E. Negishi, A. O. King and W. L. Klima, *J. Org. Chem.*, 1980, **45**, 2526.
172. H. Meier and I. Menzel, *Chem. Commun.*, 1971, 1059.
173. A. T. Blomquist, R. E. Burge, Jr. and A. C. Sucsy, *J. Am. Chem. Soc.*, 1952, **74**, 3636.
174. A. Krebs, *Tetrahedron Lett.*, 1968, **9**, 4511.
175. H. Meier and I. Menzel, *Synthesis*, 1971, 215.
176. J. Tsuji, H. Takayanagi and Y. Toshida, *Chem. Lett.*, 1976, 147.
177. T. Mukaiyama, H. Nambu and T. Kumamoto, *J. Org. Chem.*, 1964, **29**, 2243.

4.7
Hydrogenolysis of Allyl and Benzyl Halides and Related Compounds

IAN D. ENTWISTLE and WILLIAM W. WOOD
Shell Research Ltd, Sittingbourne, UK

4.7.1 INTRODUCTION

Hydrogenolysis to effect replacement of a group —XR attached to a benzylic or allylic center by its heteroatom (X), is widely used in organic synthesis, especially in protecting group strategies, most frequently involving replacement of one group XR, where X is O, N or S (Scheme 1). In reviewing this area of chemistry we have dealt with the material in a functional, rather than an historical, approach, covering first the catalytic methods, then the hydride reducing reagents, followed by dissolving metal reductions. These three sections form the bulk of the review, with other methods, which are less widely employed, examined in less detail at the end.

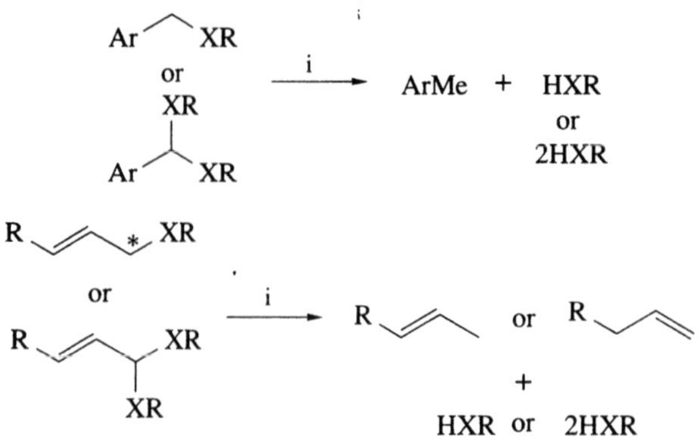

Ar⌒XR or XR/Ar⌒XR —i→ ArMe + HXR or 2HXR

R⌒⌒*XR or R⌒⌒XR(XR) —i→ R⌒⌒ or R⌒⌒ + HXR or 2HXR

i, hydrogenolysis

Scheme 1

4.7.2 HYDROGENOLYSIS WITH PALLADIUM AND ITS DERIVATIVES AND WITH OTHER TRANSITION METAL CATALYSTS

4.7.2.1 Hydrogenolysis over Palladium with Molecular Hydrogen

Debenzylation by reductive cleavage over palladium metal catalysts with molecular hydrogen has been widely utilized for many decades. A comprehensive review[1] of the early literature on hydrogenolysis of benzyl groups emphasized the main applications and described a number of preparative procedures which are still frequently used, along with a wider range of newer chemical and catalytic methods.

The main interest in hydrogenolysis of benzyl derivatives stems from the use of the benzyl group as a protecting group for hydroxy, amino and carboxylic acids.[1] Recovery of the transformed substrate rather than the toluene formed is usually more important. The disadvantages of the general method using supported palladium metal and molecular hydrogen are mainly lack of regioselectivity, and overreductions, effects more frequently experienced with cleavage of allyl derivatives. For many conversions these deficiencies have been resolved by variations in the use of the catalyst (see later sections) and source of hydrogen donor for both allyl and benzyl derivatives. Where regioselectivity, or stereospecificity, is not an important requirement, hydrogenolysis over palladium on charcoal (Pd/C, H_2) remains a widely useful method.

Primary,[2] secondary[3] and tertiary[4] benzyl alcohols (equation 1) can be reduced to the corresponding toluene with Pd/H_2 on a wide range of supports such as charcoal, $BaSO_4$, $SrSO_4$ and $CaCO_3$. Similar cleavages[5,6] of C—OH bonds in aromatic methanols are commonly achieved (equation 2). Concomitant reduction of ring substituents such as halogen, and nitrile can often be a problem. Allylic alcohols[7] are more susceptible to saturation of the double bond. Indeed allylic unsaturation can be reduced without hydrogenolysis in some cases. In many hydrogenolysis reactions the presence of an acid (<1%) enhances the reaction rate.[12]

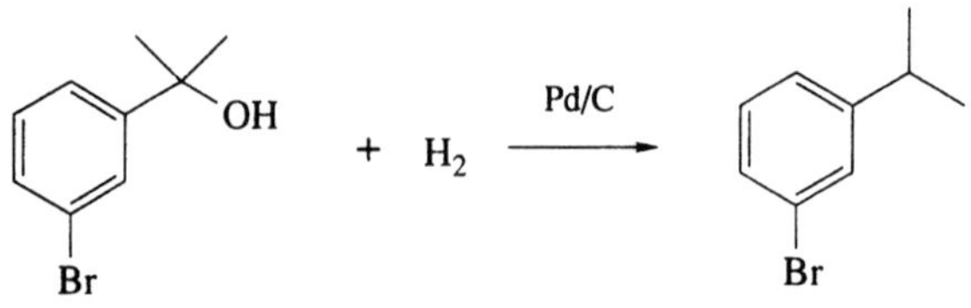

$$\text{(3-Br-C}_6\text{H}_4\text{)C(CH}_3)_2\text{OH} + H_2 \xrightarrow{\text{Pd/C}} \text{(3-Br-C}_6\text{H}_4\text{)CH(CH}_3)_2 \quad (1)$$

$$\text{Ar(Py)C(CH}_3\text{)OH} + H_2 \xrightarrow{\text{10\% Pd/C}} \text{Ar(Py)CHCH}_3 \quad (2)$$

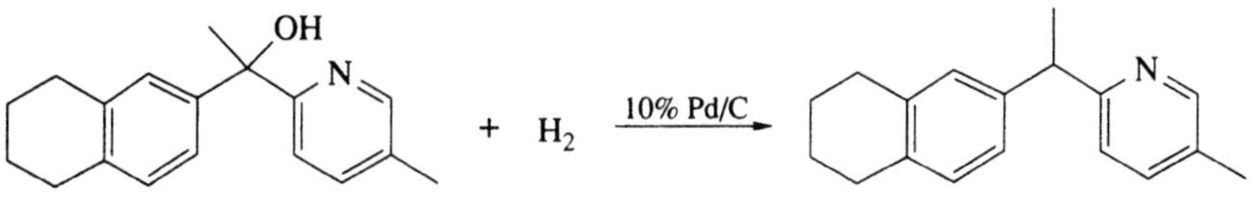

The ether and ester derivatives of benzyl alcohols are widely utilized as protecting groups.[1,8] In the case of protected peptides hydrogenolysis often has advantages over hydrolysis. This conversion is useful especially for ring opening of oxygen heterocycles (equation 3)[9] where inversion of configuration can be achieved, but deprotection of benzyl ethers is the principal application. Other arylalkyl ethers are cleaved (equation 4), and acetals[1] derived from benzaldehyde are debenzylated.

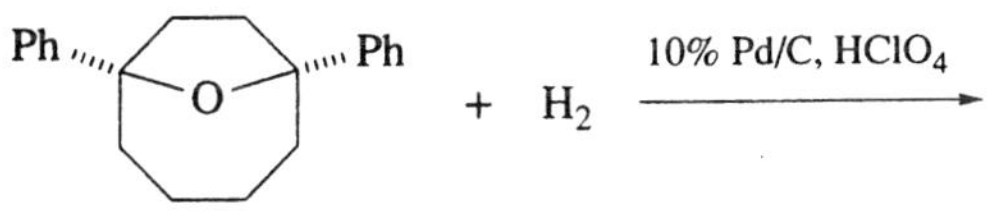

(3)

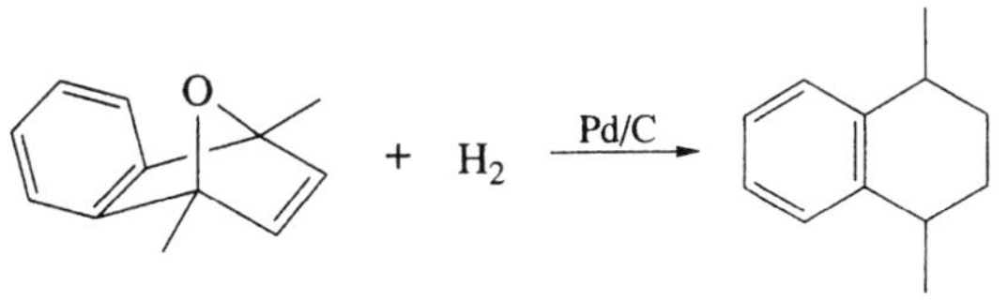

(4)

Many practical procedures have been described.[1] Primarily these are of use in peptide chemistry when deprotection of benzyloxycarbonyl (Z) derivatives is readily achieved with Pd/C and molecular hydrogen (equation 5). A great variety of other types of transformation has been reported, *e.g.* equations (6)–(8).[10–12]

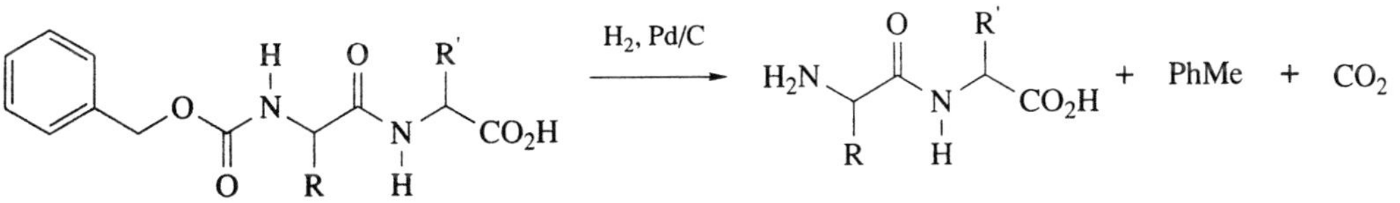

(5)

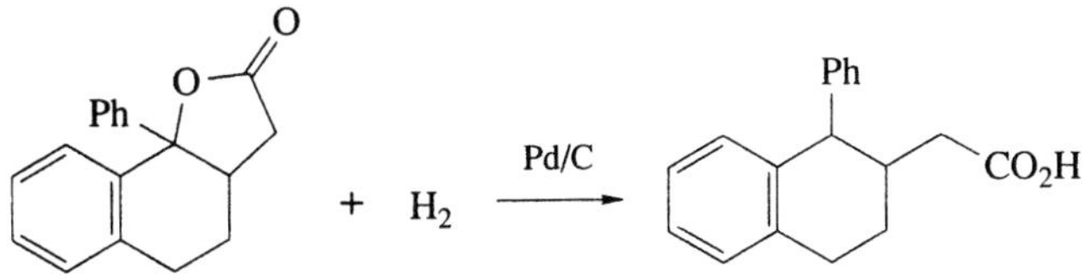

(6)

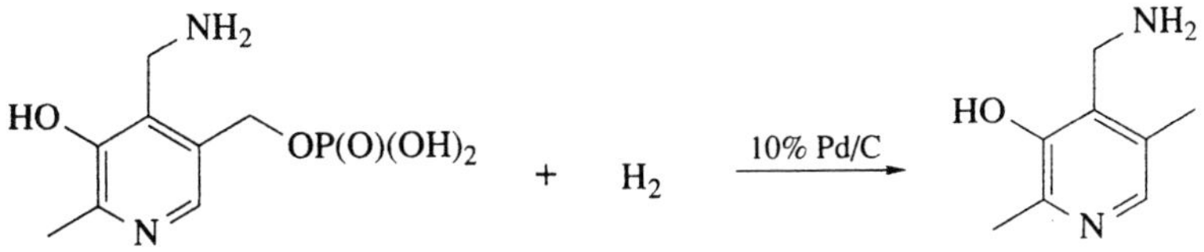

(7)

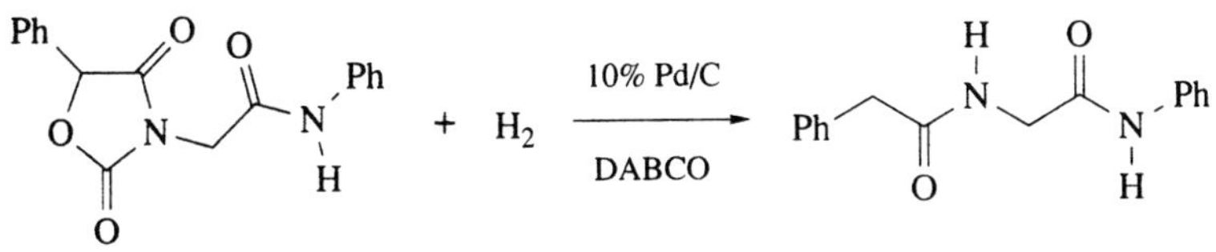

(8)

Simple primary and secondary benzylamines are generally resistant to hydrogenolysis over Pd/C, but helpful substituents (equation 9) or a different aromatic system (equation 10) do undergo cleavage.[13,14] Related dibenzylamine cleavages have been studied,[15] and where one of the aryl groups is substituted cleavage occurs readily. Tertiary benzylamines are cleaved in high yield. A useful *C*-methylation of aromatic rings *via* amino methylation exemplifies this usage (Scheme 2).[16] Cleavage of allylamines under similar conditions has not found much usage.

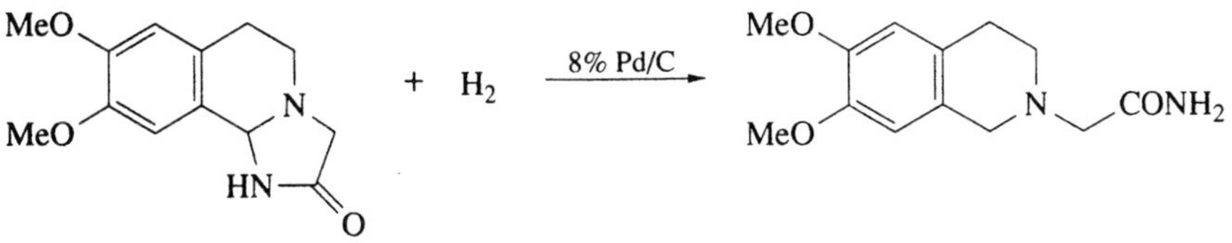

(9)

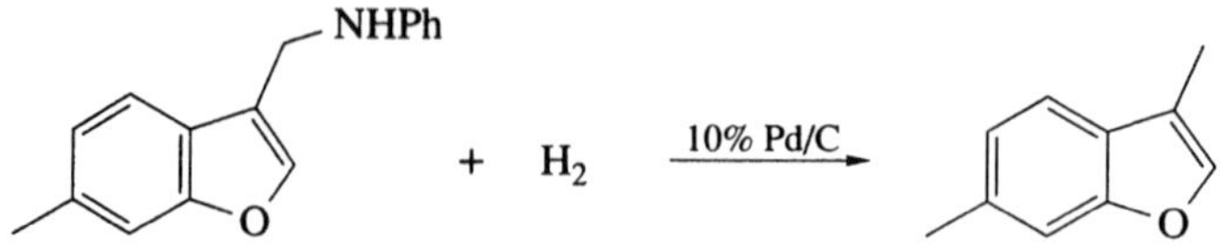

NHPh

+ H_2 —10% Pd/C→ (10)

HO — i → HO N — ii → HO

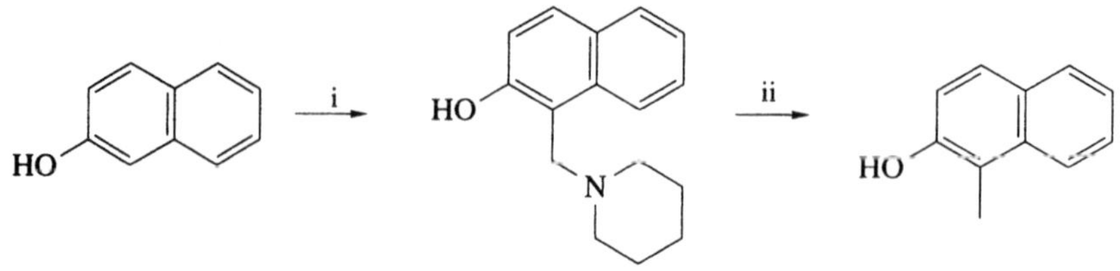

i, HCHO, piperidine; ii, H_2, 1% Pd/$SrCO_3$

Scheme 2

Arylalkyl quaternary ammonium groups are generally cleaved more readily by chemical methods. Reduction over Pd/C is used to advantage in obtaining a sample of desacetimido colchiceine (equation 11) from the natural product.[17]

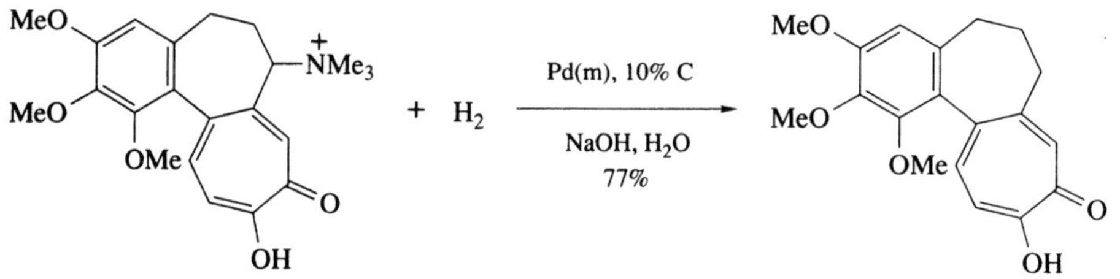

MeO MeO OMe $\overset{+}{N}Me_3$ O OH + H_2 —Pd(m), 10% C; NaOH, H_2O; 77%→ MeO MeO OMe O OH (11)

Catalyst poisoning by sulfur has discouraged development of procedures for cleavage of sulfides. Desulfurization of highly activated allylic methyl sulfides can be effected by addition of lead to the palladium(0) catalyst.[18] In the presence of various tertiary amines satisfactory cleavage of several related allyl sulfides has been utilized (equation 12) using a modified Pd/$CaCO_3$ catalyst.

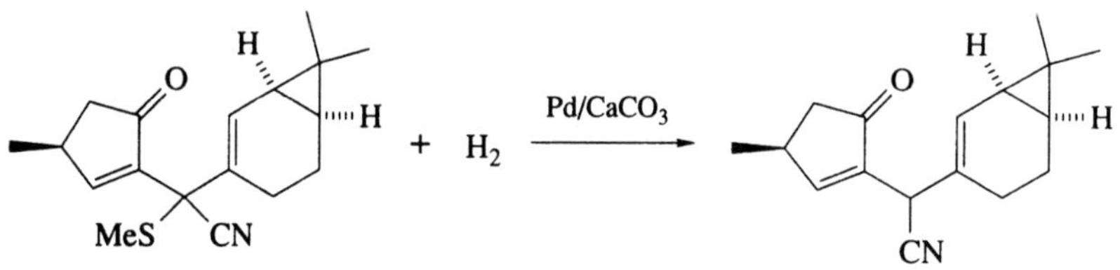

O H H MeS CN + H_2 —Pd/$CaCO_3$→ O H H CN (12)

Both benzyl and allyl halides are rapidly hydrogenolyzed. Cleavage can be achieved selectively in the presence of aryl and alkyl halides. Double bond isomerization can often prevent any stereoselectivity in allylic cleavages.

4.7.2.2 Hydrogenolysis over Palladium with Other Hydride Sources

Many of the difficulties experienced with heterogeneous Pd-catalyzed hydrogenolysis of benzyl groups have been overcome by deploying such alternatives to molecular hydrogen as cyclohexene and formate anion. A recent review of catalytic transfer hydrogenation describes the application to hydrogenolysis of benzyl and allyl derivatives.[19]

In particular the use of both cyclohexene and formates has gained wide acceptance for cleavage of benzyl, benzyloxycarbonyl (Z) or 4-methoxybenzyloxycarbonyl (PMZ) protected amino acids and peptides (Scheme 3).[20] An extensive table listing conditions for debenzylation of over 70 protected amino acids or peptides has been prepared.[19] Generally nonracemized products are obtained since transfer hydrogenolysis with cyclohexene occurs usually under mild conditions of refluxing methanol or ethanol.

Cleavage is effected at lower temperatures using cyclohexadienes,[21] or more usually formic acid or formates when reaction times are often as low as 0.15 h.[22]

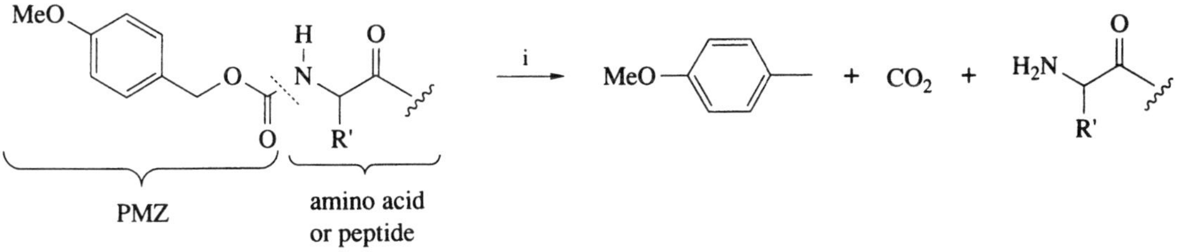

PMZ — amino acid or peptide

i, cyclohexene, ROH, 10% Pd/C, Δ

Scheme 3

Cleavage of *N*-(benzyloxycarbonyl) groups from methionine has been achieved. Using liquid ammonia solution also allowed deprotection of (*S*)-benzylcysteine and other methionine-containing peptides.[23] Simultaneous removal of *t*-butoxycarbonyl protection occurs to advantage when formic acid as donor is used to remove Z-groups.

Similar utilization for cleavage of benzyl alcohols, ethers and esters has made this method the preferred one where regioselectivity is required, but for allylic cleavage double bond isomerization becomes a problem.[24] A number of benzylic ethers and alcohols (equation 13), and benzhydric ethers and acetals (equation 14) have been cleaved in relatively high yield using cyclohexene in the presence of $AlCl_3$.[25] Mild conditions of cleavage are frequently required for debenzylation of carbohydrates and a number of protected carbohydrates have been cleaved with formic acid as donor (equation 15).[26]

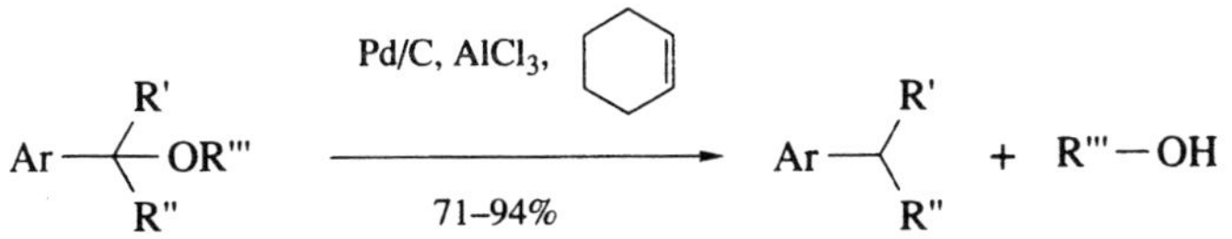

Pd/C, $AlCl_3$, cyclohexene; 71–94% (13)

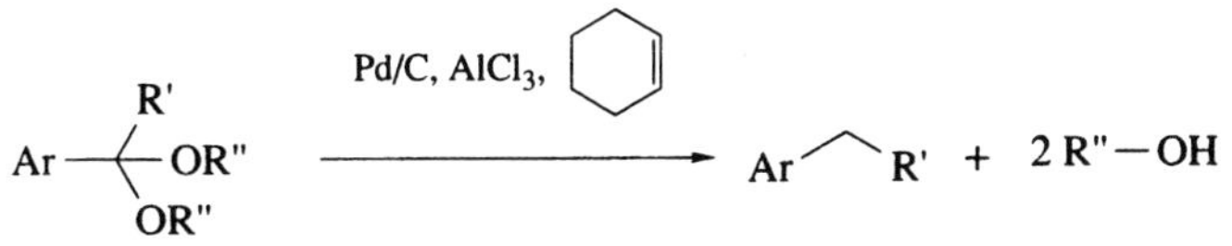

Pd/C, $AlCl_3$, cyclohexene (14)

$$\text{2,3,4,6-Tetra-}O\text{-benzyl-}\alpha\text{-D-glucopyranose} \xrightarrow[\text{Pd/C}]{HCO_2H} \text{D-Glucose} \qquad (15)$$

Alcohols or phenols[27] protected by benzyloxycarbonyl groups are very smoothly reduced over Pd/C using phosphinic acid or sodium phosphinate as hydrogen donor. Effective cleavage of an allylic *t*-amine has been demonstrated (equation 16) utilizing ammonium formate as donor.[28] The lack of regiospecificity could possibly be overcome by selective use of catalysts. For example, double bond migration can be influenced by Pd/Hg or Pd/Pb catalysts.[29]

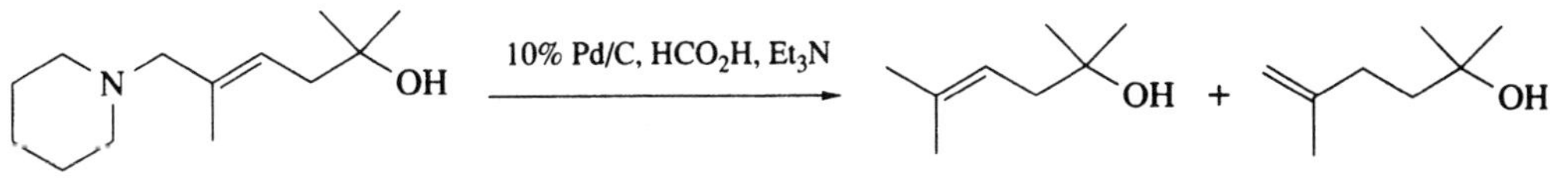

10% Pd/C, HCO_2H, Et_3N (16)

The successful application of catalytic hydrogen transfer reduction to benzylic reductions has not been mirrored for allylic cleavage. Use of homogeneous or ligated palladium catalysts has however extended hydrogenolysis by transfer hydrogenation to allyl transformations. The extensive use of the electrophilic properties of π-allyl complexes in many synthetic strategies is widely documented.[30] Reduction of such (π-allyl)palladium intermediates with hydride from decarboxylated formate was shown to yield 1-alkenes (equation 17), where the hydride attack was preferentially on the more-substituted side of the π-allylic system. The conversion (equation 18) indicates the regioselectivity and use of allyl as a protecting group. Isomeric allyl acetates or phenyl ethers were also selectively reduced (equation 17). Later work refined this method and showed that greater selectivity for 1-alkenes could be achieved using PBu^n_3 as ligand.[31]

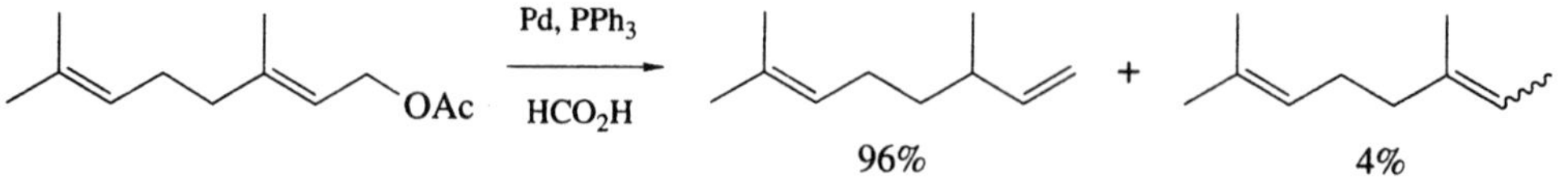

Pd, PPh_3 / HCO_2H; OAc; 96%; 4% (17)

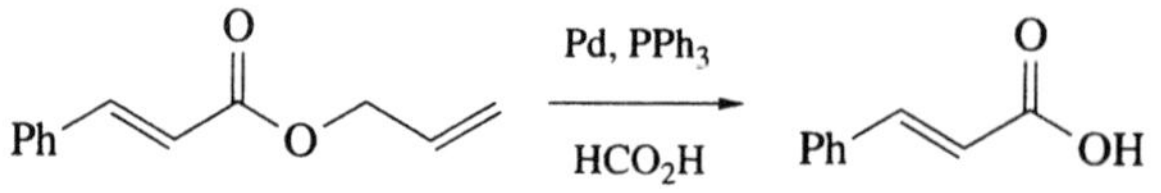

Ph; O; O; Pd, PPh_3 / HCO_2H; Ph; O; OH (18)

A greater selectivity toward formation of (*E*)-2-alkenes from allylic acetates is achieved using 2-propanol as a hydrogen donor and Pd^0/SmI_2 as catalyst generated from $PdCl_2/PPh_3$ (Scheme 4). In the formation of (±)-limonene from (–)-carvyl acetate, optical activity is lost by this method.[32] Although very high yields are attainable, control of the regiochemistry and stereochemistry is variable.[32] Conversion of allyl propionate, phenyl ether and other substituted allyl acetates to 1-alkenes with $Pd(PPh_3)_4$ and formic acid, occurs in only modest yield. Triallylamine yielded a mixture of products containing diallylamine and allylamine.[33]

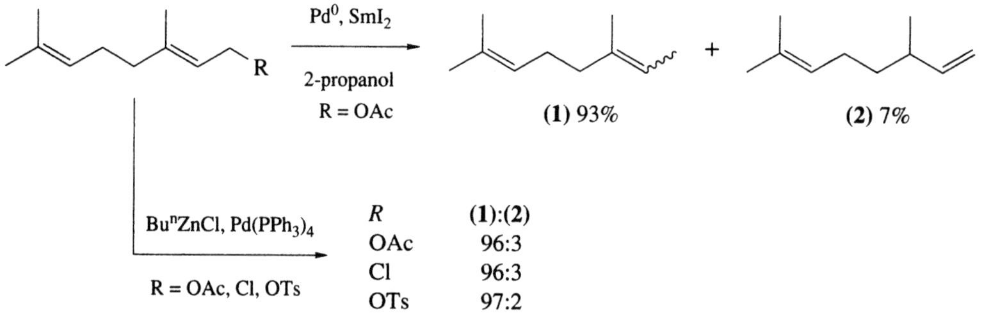

R; Pd^0, SmI_2 / 2-propanol / R = OAc; **(1)** 93%; **(2)** 7%

Bu^nZnCl, $Pd(PPh_3)_4$ / R = OAc, Cl, OTs

R	**(1)**:**(2)**
OAc	96:3
Cl	96:3
OTs	97:2

Scheme 4

A study of the cross coupling of allylic electrophiles with organometallics provides a useful regio- and stereo-selective hydrogenolysis of allylic acetates, halides and tosylates.[34] The reaction of geranyl acetate with *n*-butylzinc chloride (Scheme 4) in the presence of $Pd(PPh_3)_4$ yielded a mixture of (*E*)/(*Z*)-isomers of 2,6-dimethyl-2,6-octadiene (96% yield). The presence of β-hydrogens on the alkylzinc and the size of the alkyl group, appear to be critical for regio- and stereo-selectivity.

As a useful alternative source of H^- various metal hydrides have been studied. Allylic acetates such as geranyl and cinnamyl are efficiently hydrogenolyzed with $Pd(PPh_3)_4$ and $NaBH_3CN$.[35] For many other examples of allylic acetate cleavage better regioselectivity was observed when $NaBH_4$ was used, but very variable *cis*/*trans* isomer ratios were found. Conjugation of the allylic unsaturation to aromatic rings produces conjugated alkenes exclusively. Overall control of the 1-ene/2-ene ratio was variable.

Greater regioselectivity and control over (*E*)/(*Z*) ratios of the alkene products can be obtained by utilizing less labile leaving groups such as —OPh, —SPh, —SO_2Ph, —SePh, —Cl and —$OSiMe_2Bu^t$, in conjunction with more powerful hydride donors. Of these the most effective appears to be the hindered hydride $LiBHEt_3$.[36] This achieves selectivity for 2-enes and retention of (*E*)-geometry, and appears to attack rapidly at the less-hindered side of the π-allyl complex. $NaBH_3CN$ and ammonium formate control the product geometry less effectively. A useful alternative catalyst [$PdCl_2$(DPPP)] for regiospecific cleavage of allylic *p*-tolyl sulfones with $LiBHEt_3$ (equation 19) illustrates the importance of the π-allyl complex in controlling the specificity.[37]

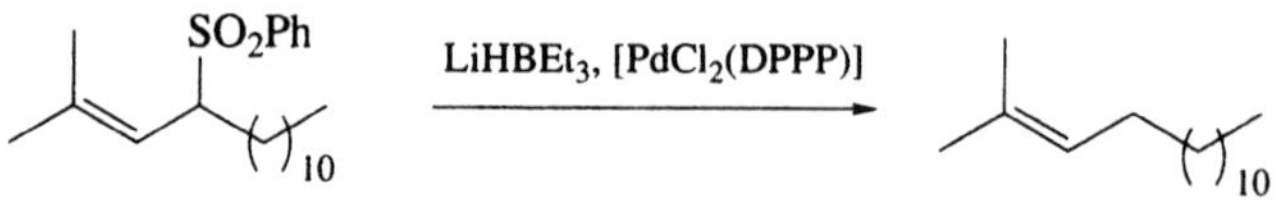

SO_2Ph; 10; $LiHBEt_3$, [$PdCl_2$(DPPP)]; 10 (19)

Steric control of the mechanism has been clearly demonstrated by the reduction of 3-tosyl-1-cyclohexane derivatives (Scheme 5). This control has been deployed in the synthesis of coenzyme Q_{10}. The wider use of these reductions has been studied and very high stereoselectivity is observed for hydrogenolysis of farnesyl derivatives, —Cl, —OPh, —$OSiMe_2Bu^t$, —SPh, —SOMe, —SO_2Me and $^+N({=}CH_2)Me$.

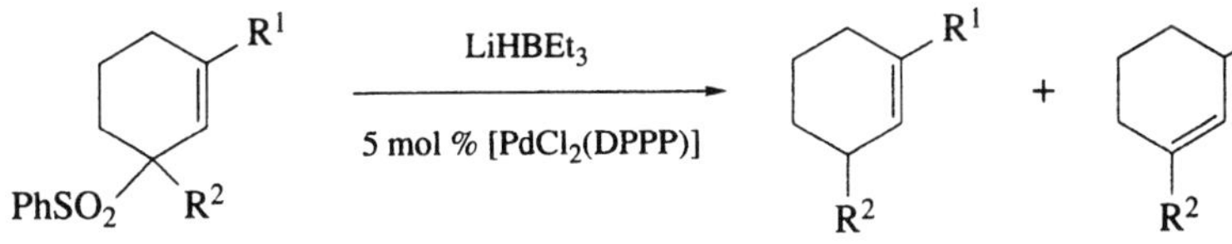

LiHBEt$_3$, 5 mol % [PdCl$_2$(DPPP)]

R^1 = H, R^2 = Bn	0	>99
R^1 = Bn, R^2 = H	>99	0

Scheme 5

Increased chemoselectivity has been achieved by using Pd^0 and Pd^{II} activation of the allylic centers, but many reagents such as the strong hydride donors are often incompatible with electrophilic groups such as ketones and aldehydes. Stereospecific reduction of the model compounds (**3**) and (**4**) in Scheme 6 suggested that the hydride could be delivered by an internal transfer and thus need not have high nucleophilicity. Hydrogenolysis with such a hydride source as tributyltin hydride with free radical scavenging fulfils this requirement. Its chemoselectivity is effectively demonstrated by the high yield transformations (equations 20 and 21). Some allylic amines can also be cleaved in moderate yields (equation 22).[38] Substrates which possess a relatively acidic hydrogen α to the allylic unit are incompatible with the low basicity of Bu_3SnH and undergo β-hydride elimination to yield the corresponding diene.[39] Reduction of these allylic systems can be effected with polymethylhydrosiloxane (PMHS) as donor (equation 23),[40] which has advantages such as stability, nontoxicity and lower cost compared to metal hydrides. In cleavages where the formation of the π-allyl intermediate is slow, for example when the allyl unit is sterically crowded and competing reactions cause yield loss, PMHS is a more effective donor (Scheme 7). Both Pd^0 and Pd^{II} catalyst complexes can be used with PMHS. Michael acceptors are readily reduced by Bu_3SnH but are unaffected by PMHS. The greater chemoselectivity of PMHS as a hydride donor is also demonstrated by the failure to reduce allylic acid chlorides which are sensitive to Bu_3SnH. While greater chemoselectivity is an advantage of the use of either Bu_3SnH or PMHS, less regioselectivity is observed than for the other hydrides described.

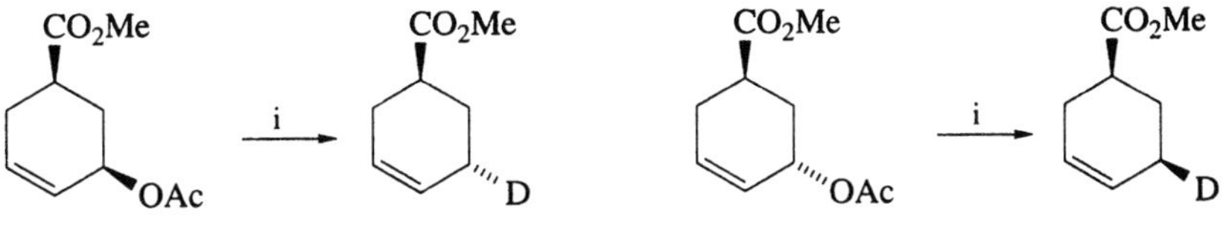

(**3**) (**4**)

i, NaBD$_4$, Pd(PPh$_3$)$_4$

Scheme 6

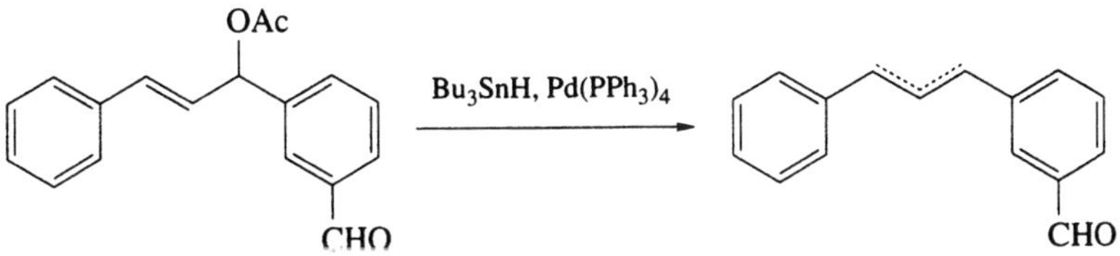

Bu$_3$SnH, Pd(PPh$_3$)$_4$ (20)

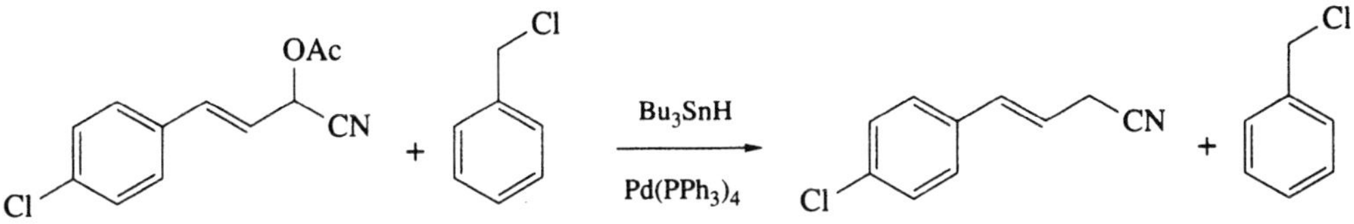

Bu$_3$SnH, Pd(PPh$_3$)$_4$ 99% 99% (21)

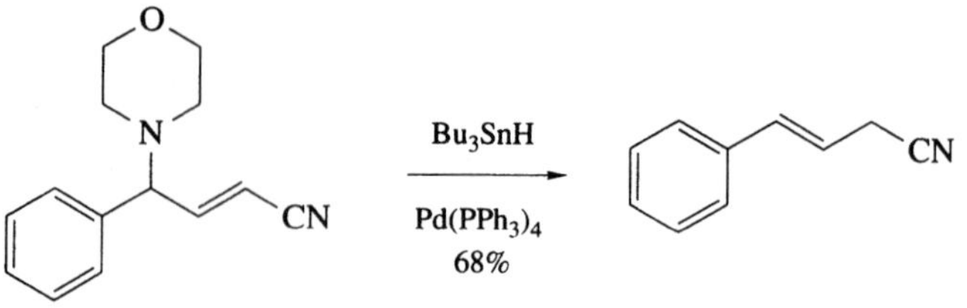

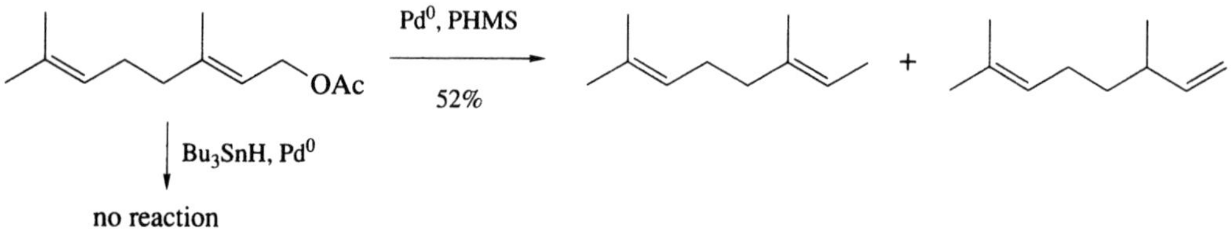

Scheme 7

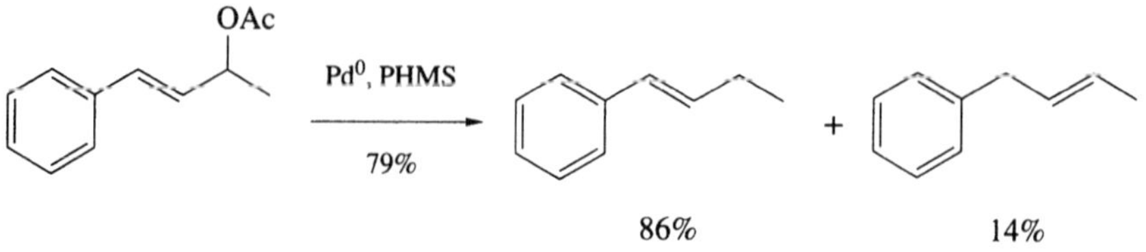

Cleavage of the allylic C—N bond over Pd^0 with hydrides has been reported only rarely. The unusual hydrogenolysis of allylic nitro compounds[41] by Pd^0-catalyzed hydride transfer (Scheme 8) could have more extensive use than the reported synthesis of homoallylic alcohols. Regioselectivity is dependent on the nature of the π-allyl unit and on substitution as well as the nature of the hydride (equations 24–26). Good nucleophilic hydride sources such as $NaBH_4$ with small ligands are utilized to obtain 2-alkenes regioselectively. High regioselectivity is also evident with α,β-unsaturated ester reductions.

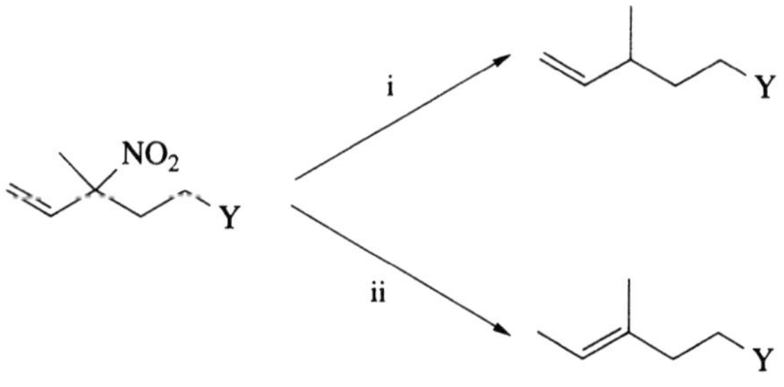

Y = CO_2Me, CN, SO_2Ph

i, HCO_2NH_4, Pd^0; ii, $NaBH_4$, Pd^0

Scheme 8

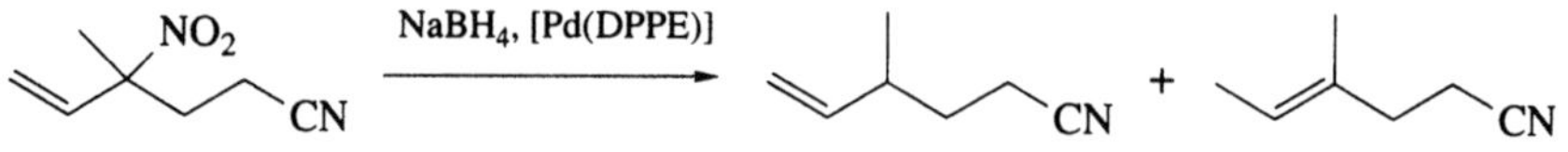

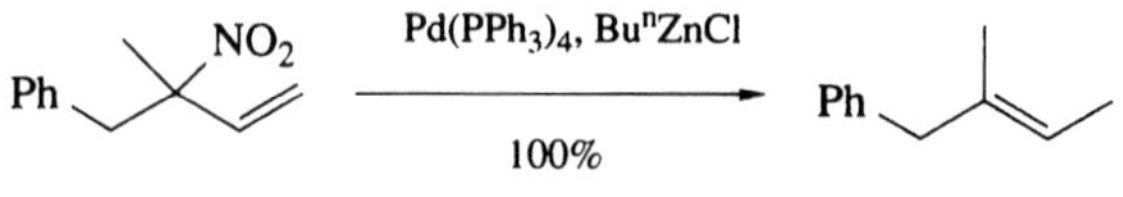

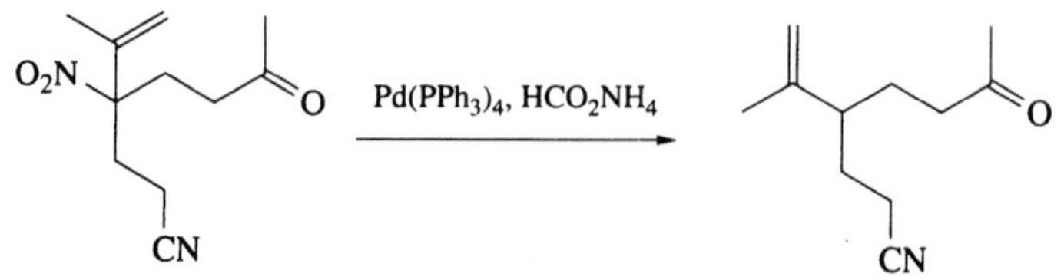

4.7.2.3 Hydrogenolysis over Other Transition Metal Catalysts

Both palladium and platinum metals are widely used as hydrogenation catalysts. Although only palladium has found significant usage for benzylic and allylic cleavage of C–heteroatom bonds, catalyst poisoning has discouraged its use for the cleavage of C—S bonds. Where deactivated catalysts have been required, both rhodium and rhenium metals or their ligated complexes have, however, been utilized as well as doped palladium. Desulfurization of benzylic and allylic compounds is readily achieved with Rh^0 complexes.[42] Regioselectivity is affected markedly by alkyl substituents on the alkenic bond of the substrate (equations 27 and 28). Terminal alkenes are often isomerized to inner alkenes and the stereochemistry is uncontrolled. Benzylic thiols, sulfides and disulfides are cleaved in high yield to the hydrocarbon with molecular hydrogen over $Rh(PPh_3)_3Cl$ or $Rh(PPh_3)_2(CO)Cl$.[43] Chemoselectivity is achieved more readily with these catalysts.

SPh

[RhH(PPh$_3$)$_4$]

85%

+

(27)

75%

25% *(E)*:*(Z)* = 100:0

[RhH(PPh$_3$)$_4$]

SPh

74%

+

(28)

16%

84%

Copper chromite ($CuCr_2O_4$) has historically been widely used as a hydrogenation catalyst. Generally because of its low catalytic activity its chemoselectivity is useful, although it does require high temperature and autoclave pressure conditions. It is effective for cleavage of benzylic alcohols,[44] primary and secondary benzylic esters[45] and ethers.[46] Efficient cleavage of benzylamines has also been utilized (equation 29).[47] Other copper salts and copper alloys have found infrequent use.

OH

Me$_2$N

NMe$_2$

OMe

$CuCrO_4$, H_2

OH

OMe

(29)

The wider utility of palladium metal catalysts and hydride cleavage of allylic systems has more or less replaced interest in using platinum-based catalysts such as Adams' catalyst. Where this has been used for conjugation-stabilized allylic centers good yields have been achieved.[47] The use of platinum for debenzylation at low H_2 pressure is effective.[48]

4.7.2.4 Hydrogenolysis over Raney Nickel

Reduction of benzylic oxygen functional groups with Raney nickel has occasionally been reported, but there are very few examples in the recent literature. Hydrogenolysis over W-2 Raney nickel has been used to selectively remove benzyl ether protection in the presence of a 4-methoxybenzyl protecting group (equation 30).[49]

OPr^i

O

O

OBn

MeO

W-2 Raney Ni, EtOH, H$_2$

100%

OPr^i

O

O

OH

MeO

(30)

Benzylic halides have also been reduced to hydrocarbons by hydrogenolysis over Raney nickel (equation 31).[50] Benzylic amines have been cleaved by hydrogenolysis over Raney nickel (equation 32)[51] or by treatment of a methiodide salt with basic Raney nickel (equation 33).[52]

Raney Ni, EtOH, NaOH, H_2, 90% (31)

Raney Ni, H_2, 80% (32)

Raney Ni, NaOH, 100% (33)

Raney nickel is most frequently used to reduce organic sulfur compounds, usually to hydrocarbons.[53] Many different types of organosulfur derivatives can be reduced in this way using various grades of Raney nickel, usually in alcoholic solvents. Thiols in general are readily reduced, but benzylic thiols with α-hydroxy groups form products from alternative reactions (equation 34).[54] Benzylic alkyl and aryl sulfides are readily reduced to alkanes with Raney nickel in the presence of a variety of other functional groups (equations 35–39).[55–59] When a sulfide is removed from an asymmetric center with Raney nickel stereochemical integrity is lost. However, in this instance, oxidation to the sulfone, followed by desulfurization, retained optical activity (equations 40 and 41).[60]

Ni, Me_2CO, 71%; 45% + 31% + 24% (34)

Raney Ni, 82% (35)

Raney Ni, EtOH, 100% (36)

Raney Ni, EtOH (37)

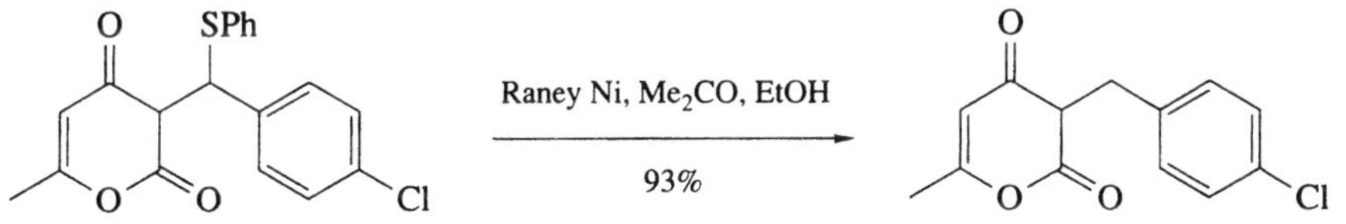
O
SPh
Cl
Raney Ni, Me2CO, EtOH
93%

(38)

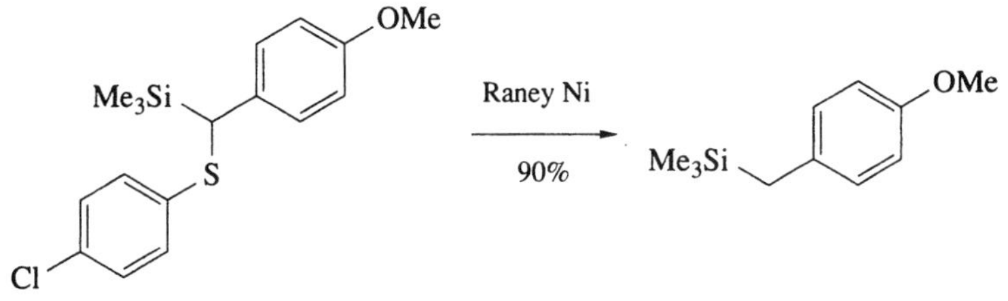
OMe
Me3Si
S
Cl
Raney Ni
90%

(39)

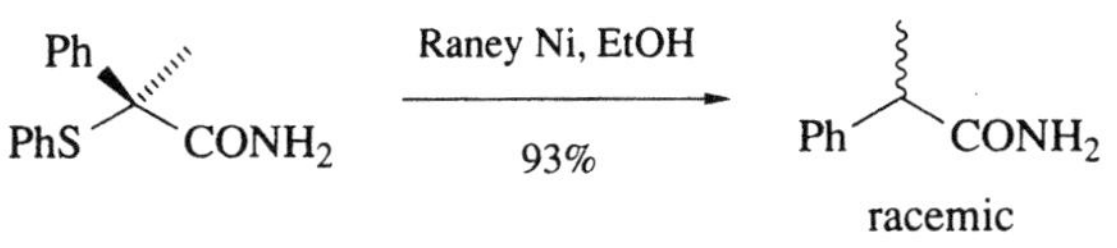
Ph
PhS
CONH2
Raney Ni, EtOH
93%
racemic

(40)

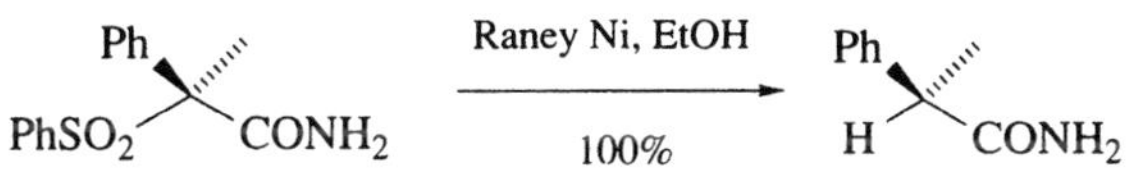
Ph
PhSO2
CONH2
Raney Ni, EtOH
100%
H

(41)

4.7.3 HYDROGENOLYSIS WITH HYDRIDE REDUCING AGENTS

4.7.3.1 Lithium Aluminum Hydride

On treatment with LAH, benzylic halides and related compounds are reduced to the corresponding hydrocarbon (equations 42 and 43),[61,62] although in some cases coupled, dimeric products can also be obtained. Selectivity between halides is possible,[63] and in some examples LiH has been added to the reaction mixture.[64]

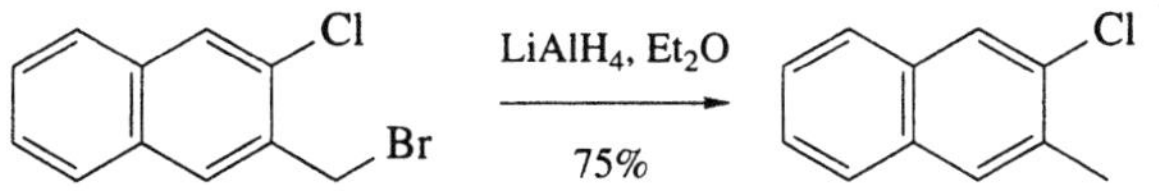
Cl
Br
LiAlH4, Et2O
75%

(42)

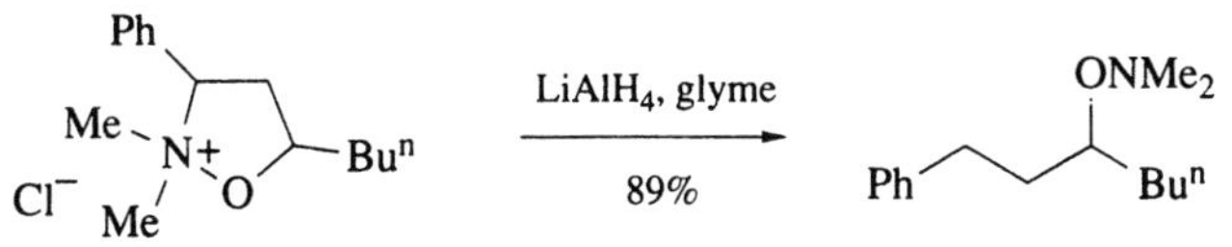
Ph
Me
N+
Cl−
Me
O
Bun
LiAlH4, glyme
89%
ONMe2

(43)

Reduction of allylically placed functional groups with LAH normally occurs *via* one of two mechanisms: S_N2 or S_N2'.[65] In acyclic systems the mechanism followed depends on the nature of the leaving group. It has been shown (Table 1) that good leaving groups generally give unrearranged products (S_N2), while poor leaving groups give rearranged products (S_N2').[66] The evidence suggests that reduction of triphenylphosphonium salts is a suitable method for the specific formation of S_N2' products, even when the structural features of the substrate favor an S_N2 mechanism.

The regio- and stereo-chemical outcome of reductions of cyclic allylic systems is controlled by more complex factors.[65] Reduction of the allyl bromides (**6**) and (**7**) occurs exclusively by *syn* γ-attack (Scheme 9),[67] but other structural features influence the reaction. For example, where the allylic double bond is in a thermodynamically stable position, rearrangement does not normally occur.[68] The reaction has also been studied in complex carbohydrate alkenes.[69]

Reduction of allylic carbonyl derivatives usually leads to the corresponding allylic alcohol; however, in some cases, intermediate oxidation states can be obtained. Reduction of amide derivatives such as *N*-acylpyrazoles results in the formation of an aldehyde in good yield.[70]

Table 1 Reduction of Allylic Functional Groups with LAH

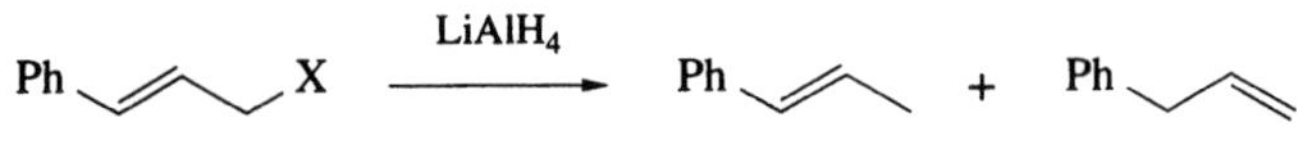
LiAlH4
Ph
X
Ph
+
Ph

(5) **(5a)** **(5b)**

X	*Reaction time (h)*	*Yield (%)*	*Ratio (5a):(5b)*
Br	0.5	100	100:0
Cl	0.5	78	100:0
Ts	1.0	50	100:0
$^{+}SMe_2Br^{-}$	0.5	90	90:10
$^{+}NEt_3Br^{-}$	0.5	62	76:24
$^{+}NBu_3Br^{-}$	6.0	2	76:24
$P(O)(OEt)_2$	0.5	46	2:98
$^{+}PBu_3Br^{-}$	0.5	90	7:93
$^{+}PPh_3Br^{-}$	0.5	95	0:100

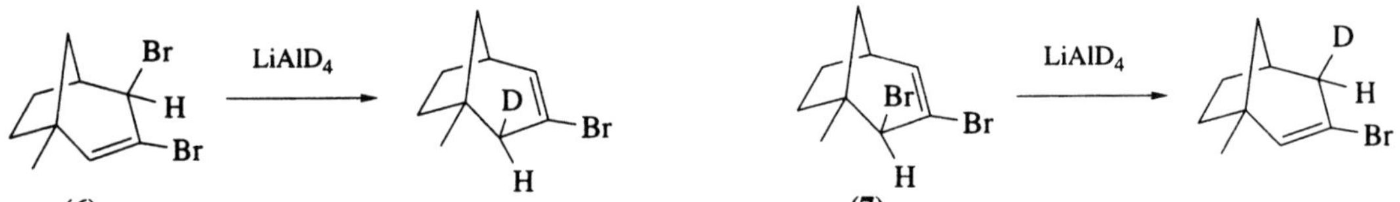
Br
H
Br
LiAlD4
D
Br
H
Br
Br
H
LiAlD4
D
H
Br

(6) **(7)**

Scheme 9

Certain benzylic alcohols, ethers, acetals and thiols, which are normally resistant to reduction by LAH, can be reduced to the hydrocarbon when the substrate is activated with a Lewis acid. $AlCl_3$ has been most widely used in this role. Various acetals can also be reduced to ethers using the same reagent (equation 44).[71] A related procedure has been used to convert aldehydes into sulfides (equation 45).[72]

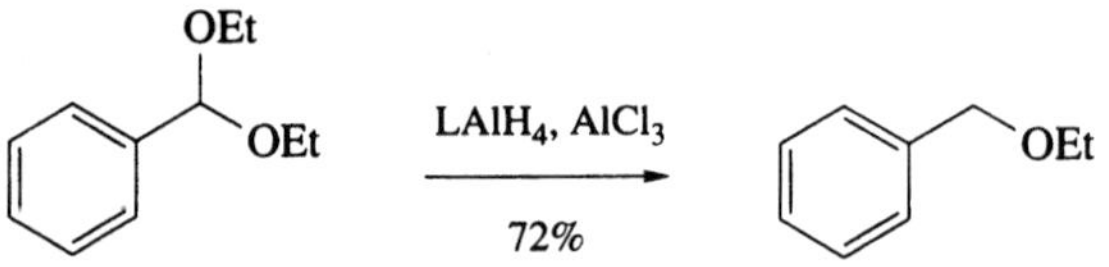
OEt
OEt
LAlH4, AlCl3
72%
OEt

(44)

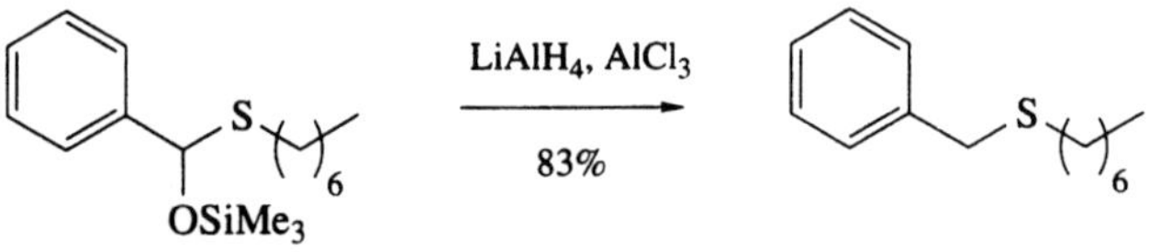
S
6
OSiMe3
LiAlH4, AlCl3
83%
S
6

(45)

A variety of other Lewis acids have been used in conjunction with LAH, including nickel, copper and titanium salts, the resulting reagents being used to reduce sulfides and benzyl ethers to hydrocarbons (equations 46–48).[73–75] Addition of nickelocene to LAH produces a specific desulfurization reagent which does not reduce ketones and esters.[76]

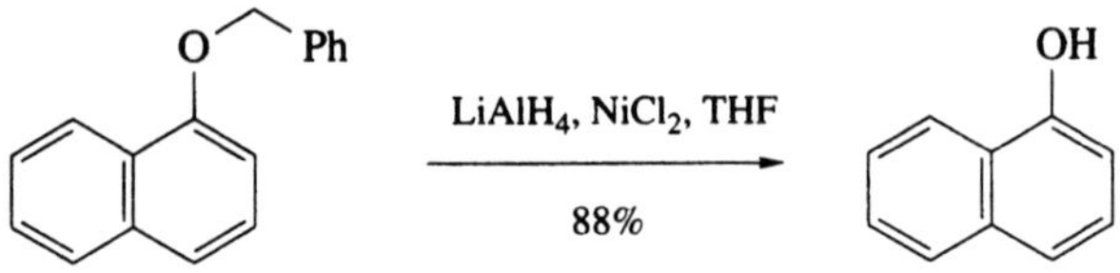
O
Ph
LiAlH4, NiCl2, THF
88%
OH

(46)

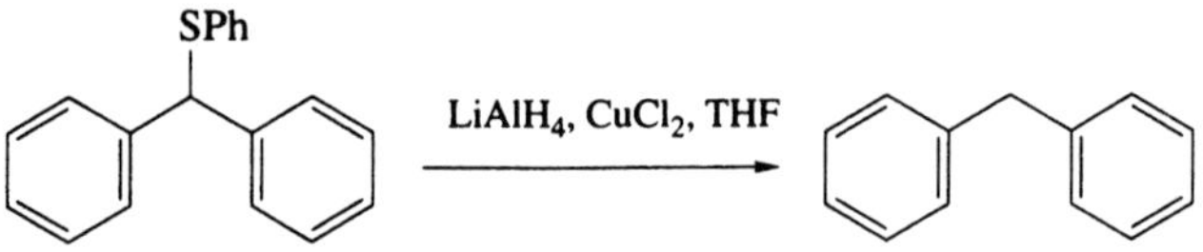
SPh
LiAlH4, CuCl2, THF

(47)

$LiAlH_4$, $TiCl_4$, THF; 84% (48)

The enhanced reductive power of Lewis acid/LAH mixtures observed with benzyl substituents is reflected in allyl compounds. Thus allyl methyl ethers are readily demethoxylated with LAH/$TiCl_4$ (equation 49),[77] with complete rearrangement of the double bond and some (*Z*)-selectivity. Allyl alcohols are reduced to alkenes with the same reagent or with LAH/$AlCl_3$ (equation 50).[78] With both these reagents double bond migration was observed to varying degrees as well as diene formation. Primary allyl alcohols have been reduced to alkenes with titanocene and LAH (equation 51).[79]

OMe Ph; $LiAlH_4$, $TiCl_4$; 93%; Ph; (*Z*):(*E*) = 78:22 (49)

HO; $LiAlH_4$/$AlCl_3$, 80%; $LiAlH_4$/$TiCl_4$, 65% (50)

OH; $LiAlH_4$, $TiCl_2Cp_2$; 85% (51)

4.7.3.2 Derivatives of Lithium Aluminum Hydride

Reduction of organic compounds in general with alkoxyaluminum hydrides has been reviewed.[80] Benzylic aldehydes, ketones and carboxylic acids are reduced to benzylic alcohols with $NaAlH_2(OC_2H_4OMe)_2$. However this reagent can also be used to reduce the benzylic alcohol so formed to a hydrocarbon, particularly when there are electron-donating groups on the aromatic ring (equation 52). It should be noted however that aromatic halides are unstable towards reduction under these conditions.[81] DIBAL-H has also been used to reductively detosylate at the benzylic position (equation 53).[82]

HO, OH; $Na(OCH_2CH_2OMe)_2AlH_2$, PhH; 95%; HO (52)

S, OTs; DIBAL-H, PhMe; 70%; S (53)

4.7.3.3 Sodium Borohydride

Some unusual benzylic functional groups can be reduced to hydrocarbons using $NaBH_4$ alone in alcohols (equation 54).[83] Choice of solvent can be used to enhance (or reduce) the reductive power of $NaBH_4$. Thus in DMSO (or sulfolane), $NaBH_4$ effectively reduces primary, secondary and tertiary benzylic halides to alkanes, leaving nitro, ester and carboxylic acids untouched (equation 55).[84] There has also been one report of the reductive cleavage of a benzylic trichloromethyl group with $NaBH_4$ in another rate-enhancing aprotic solvent, DMF (equation 56).[85]

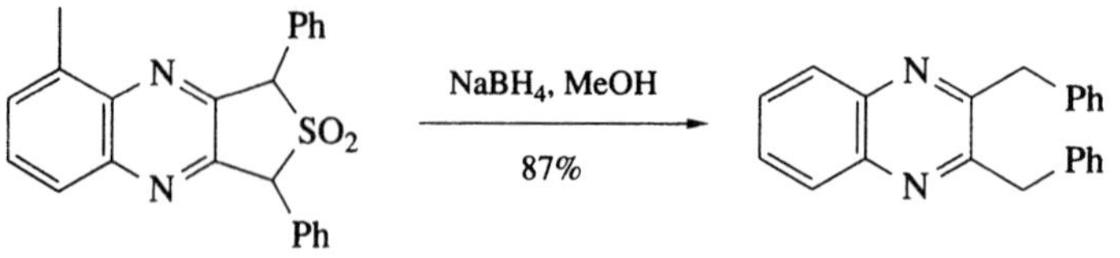

NaBH$_4$, MeOH; 87% (54)

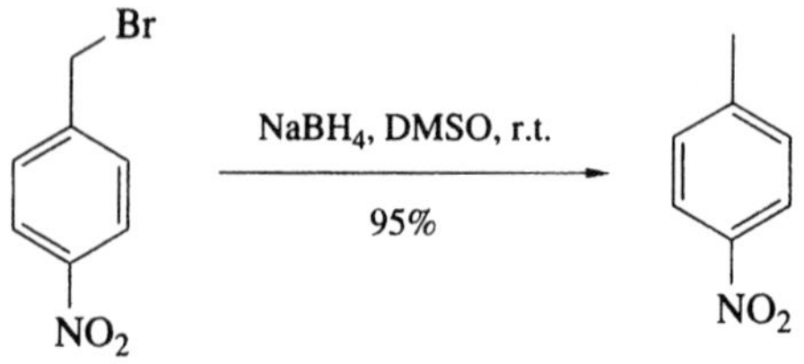

NaBH$_4$, DMSO, r.t.; 95% (55)

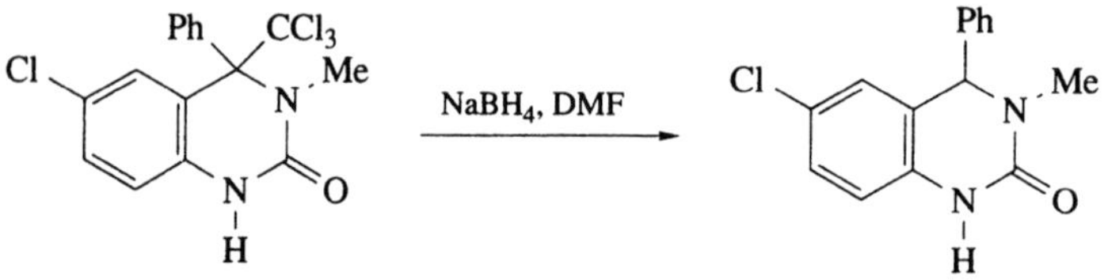

NaBH$_4$, DMF (56)

Benzylic alcohols can be reduced with $NaBH_4$ under acidic conditions using either mineral acid (equation 57)[86] or trifluoroacetic acid.[87] These reactions involve formation and reduction of a benzylic carbonium ion. Similarly the Lewis acid $AlCl_3$ can be used in place of a Brønsted acid, reducing diaryl and aryl alkyl ketones and carbinols to hydrocarbons in excellent yield.[88] Benzylic carbonium ions can also be trapped with $NaBH_4$ after generation under solvolytic conditions. Thus a variety of aryl-, diaryl- and triaryl-methyl chlorides can be reduced in aqueous diglyme containing sodium hydroxide (equation 58).[89]

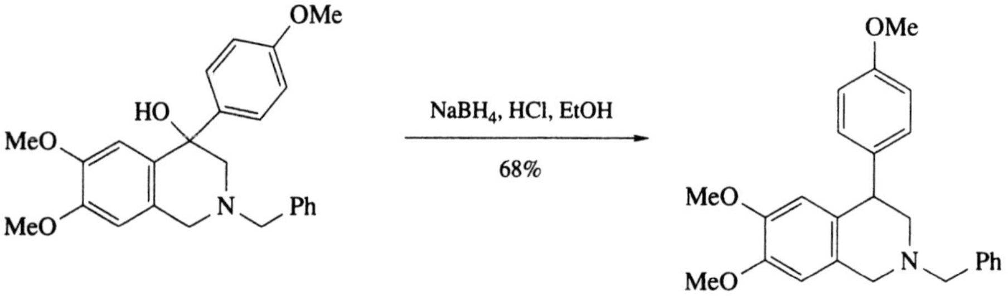

NaBH$_4$, HCl, EtOH; 68% (57)

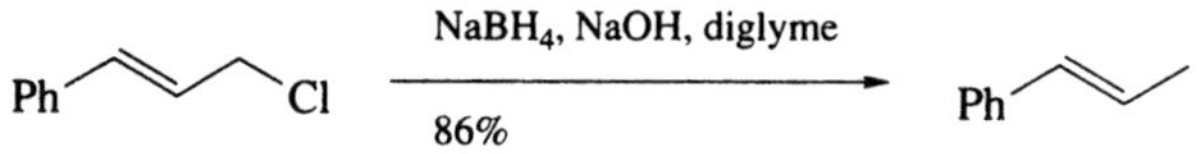

NaBH$_4$, NaOH, diglyme; 86% (58)

Nickel boride has been used to reduce benzylic dithioacetals to sulfides or hydrocarbons (equation 59).[90] More recently the reagent has been applied to the reduction of allylic functional groups. The reagent effectively reduces allyl alcohols, acetates, trifluoroacetates and benzoates, but leaves methyl and benzyl ethers untouched and appears to lead to the thermodynamically more stable alkene (equation 60).[91,92]

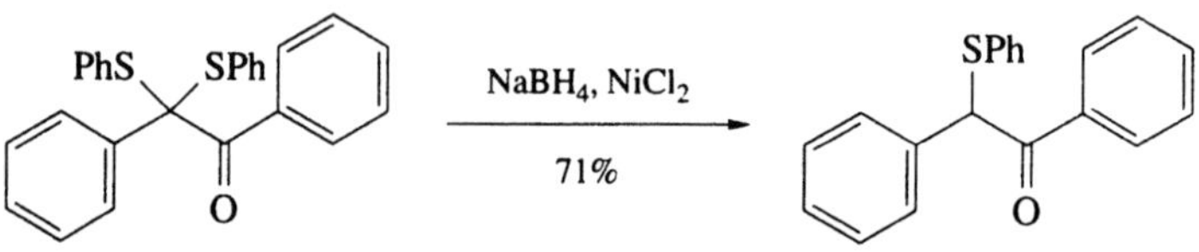

NaBH$_4$, NiCl$_2$; 71% (59)

Me3SiO, Me3SiO — NaBH4, NiCl2 — 60% — Me3SiO (60)

4.7.3.4 Derivatives of Sodium Borohydride

The reductive power of $NaCNBH_3$ can be modified by the addition of Lewis acids in a similar manner to $NaBH_4$ itself. Thus addition of $ZnCl_2$ to $NaCNBH_3$ produces a reagent which is stable in aqueous media and is capable of reducing a variety of benzylic and allylic functional groups to the corresponding hydrocarbon (equations 61–63).[93]

Br — $NaCNBH_3$, $ZnCl_2$ — 99% (61)

Br — $NaCNBH_3$, $ZnCl_2$ — 90% (62)

Br — $NaCNBH_3$, $ZnCl_2$ — 90% — 67% + 33% (63)

Zinc iodide can be used in a similar manner to the chloride in this type of reaction, but the reagent so formed has been examined in more detail and has been shown to reduce benzylic and allylic aldehydes, ketones and alcohols to hydrocarbons (equation 64).[94] The reagent is much more reactive than that formed between ZnI_2 and $NaBH_4$. The ZnI_2/$NaCNBH_3$ reagent is, however, relatively unselective, attacking esters and nitro groups to some extent. Unactivated carbonyls are reduced to alcohols.

Br, O, Et, MeS — $NaCNBH_3$, ZnI_2, CH_2Cl_2 — 70% — Br, MeS (64)

$NaCNBH_3$ can also be used in an important protecting group strategy to reductively cleave 4-methoxybenzylidene acetals. As shown in Scheme 10, either 6-*O*- or 4-*O*-4-methoxy benzyl ethers can be obtained, depending on additive and solvent.[95,96]

4.7.3.5 Other Hydride Reducing Agents

The reducing system produced from an organosilane and BF_3 has been applied to both benzylic and allylic systems. Among the many different types of alcohol reduced to hydrocarbons by the reagent, the reduction of benzylic alcohols can be effected with various organosilanes as hydride donors (equation 65).[97]

Radical reduction of allylic and benzylic functional groups with Bu_3SnH and AIBN has been applied to benzylic nitro groups and benzylic and allylic sulfides, the former technique providing a high yielding chemoselective method of reducing secondary and tertiary benzylic nitro groups (equation 66).[98] Unsym-

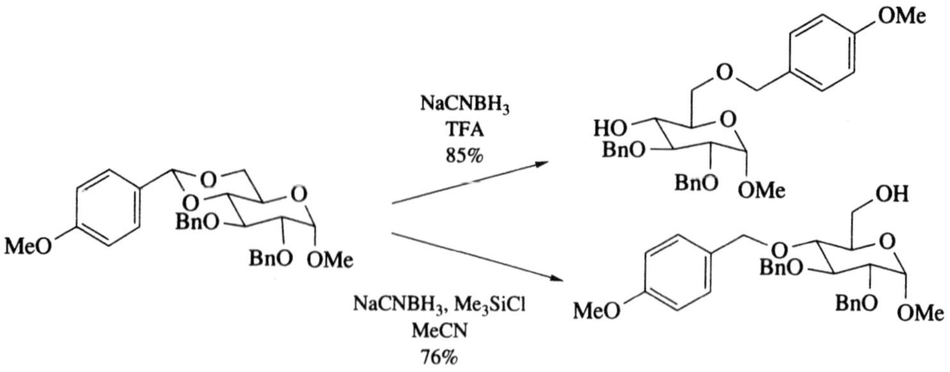

NaCNBH$_3$
TFA
85%

NaCNBH$_3$, Me$_3$SiCl
MeCN
76%

Scheme 10

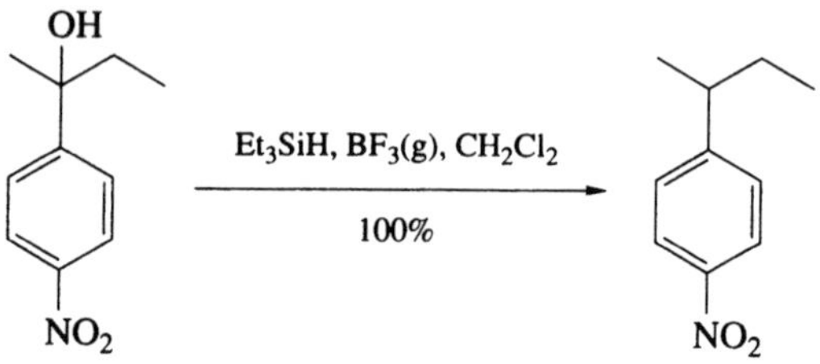

Et$_3$SiH, BF$_3$(g), CH$_2$Cl$_2$
100%

(65)

metrical benzyl and allyl sulfides have also been reduced with Bu$_3$SnH.[99] In the former case the enhanced stability of benzylic radicals results in the formation of toluenes and tin sulfides (equation 67). In the latter case cleavage of the allyl—S bond predominated over other reactions (equation 68) (with the exception of methylallyl sulfides). Some preference for reduction without rearrangement of the allyl system was detected. An allylic analog of the Barton–McCombie reaction has also been reported (equation 69).[100]

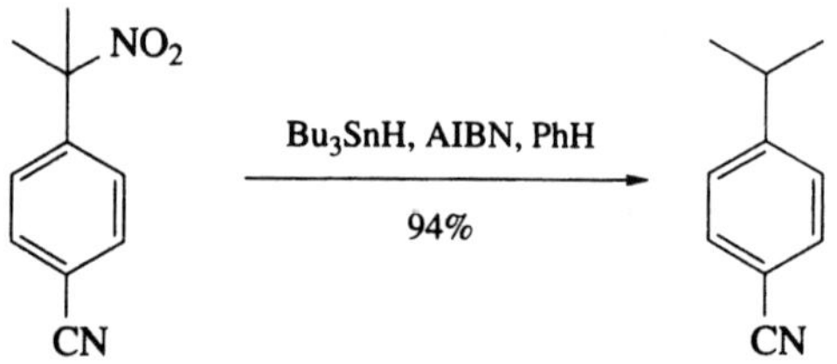

Bu$_3$SnH, AIBN, PhH
94%

(66)

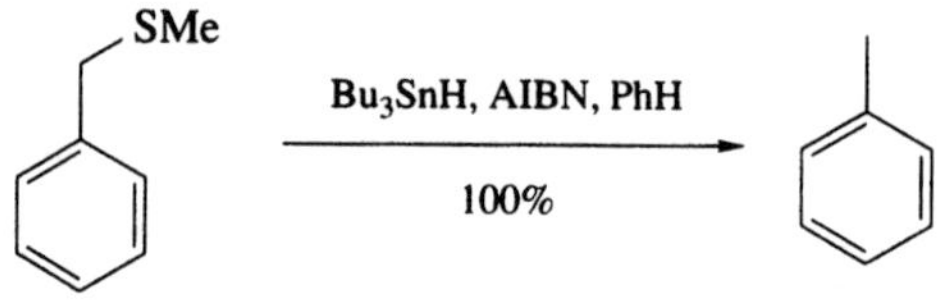

Bu$_3$SnH, AIBN, PhH
100%

(67)

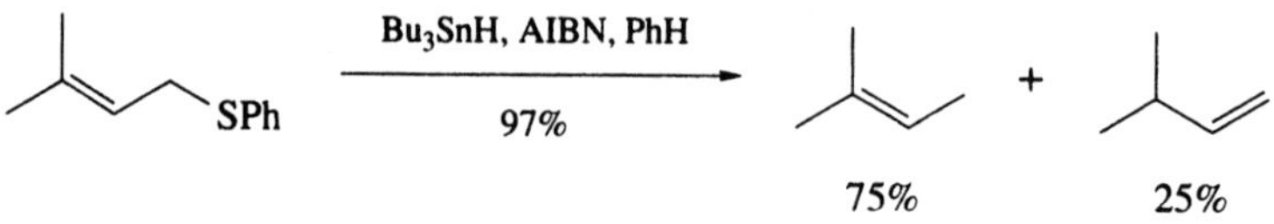

Bu$_3$SnH, AIBN, PhH
97%
75% + 25%

(68)

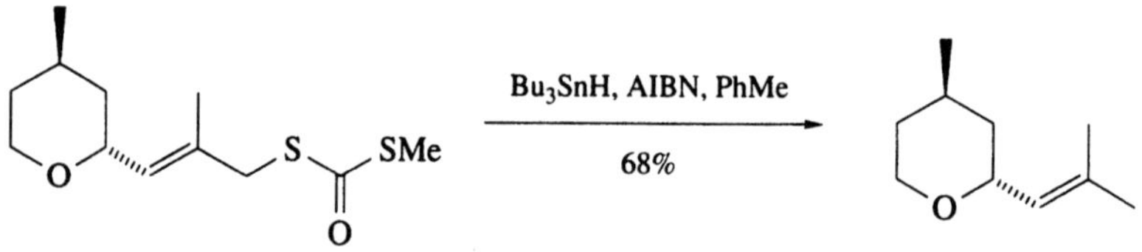

Bu$_3$SnH, AIBN, PhMe
68%

(69)

4.7.4 DISSOLVING METAL REDUCTIONS

4.7.4.1 Lithium and Sodium in Liquid Ammonia

Lithium and sodium dissolved in liquid ammonia have been extensively employed in the reduction of benzylic hydroxyls, carbonyl compounds and acetals. These reductions have been reviewed.[101] Over-reduction of benzyl alcohols by Li/NH_3 to 1,4-dihydro derivatives can be avoided using an ammonium chloride quench.[102] Aryl halides are unstable under these conditions.

Sodium in liquid ammonia, often with alcohols as cosolvents, provides a less vigorous reducing system for benzylic alcohols. Terminal and disubstituted alkenes are reduced by the Li/NH_3 system, while only terminal alkenes usually succumb to Na/NH_3, allowing benzylic alcohols containing unsaturation to be effectively reduced with the latter reagent.[101] Benzylic ketals are readily reduced to hydrocarbons without overreduction using Na/NH_3/alcohol (equation 70).[103] Lithium in ammonia can lead to over-reduction.

MeO — Na, NH_3, MeOH, 74% — MeO (70)

Allylic alcohols are reduced with lithium or sodium in ammonia, or low molecular weight amines either with or without alcohols.[101] The thermodynamically more stable product is often formed, leading to rearrangement in some cases (equation 71).[104] Methyl and cyclic ethers are similarly reduced (equations 72 and 73),[105,106] as are allylic acetates, halides and epoxides (equation 74 and 75).[107,108]

Benzylic and allylic sulfides and sulfones are readily reduced to hydrocarbons using lithium or sodium in alcoholic solvents or in amines.[109,110] Allylic sulfones are reduced in a similar manner (Scheme 11),[111] either with or without migration of the double bond, depending on the reaction conditions used.

Benzylic and allylic tertiary amines may be reduced to hydrocarbons using one of two methods. Oxidation to an amine oxide followed by reduction with lithium in liquid ammonia provides one of these

HO — Na, NH_3, MeOH, 78% (71)

OMe, THPO — Li, $EtNH_2$, 50% — THPO (72)

Li, $EtNH_2$, NH_4Cl, 90% — OH (73)

OH, OAc — Li, NH_3, NH_4Cl — OH (74)

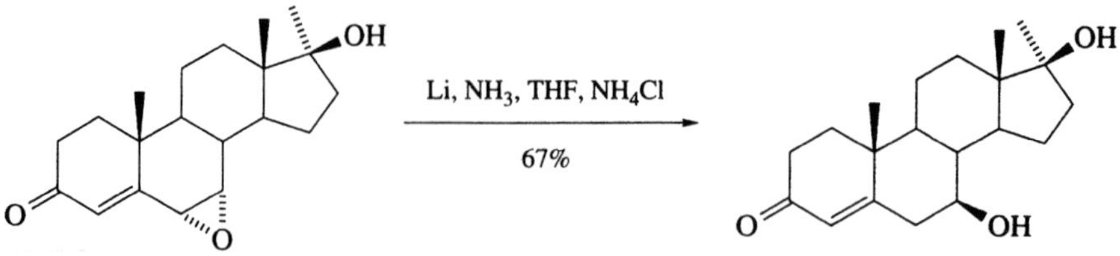

Li, NH_3, THF, NH_4Cl

67%

(75)

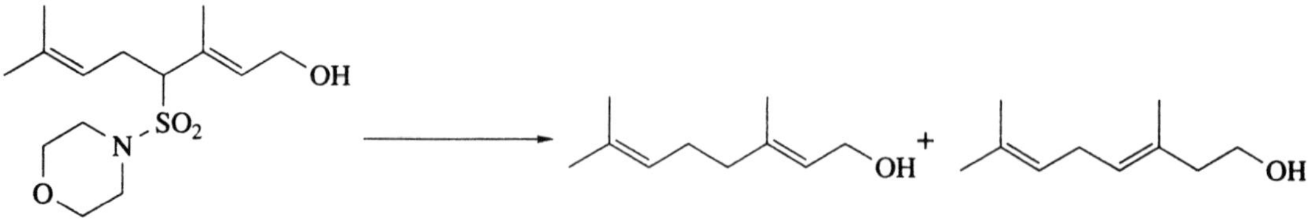

Li/NH_3, Bu^tOH; 70% 7% 93%

Na/NH_3, dibenzo-18-crown-6; 75% 95% 5%

Scheme 11

(equation 76).[112] Alternatively, formation of a labile quaternary salt, followed by reduction and hydrolysis, leads to secondary amine products (Scheme 12).[113]

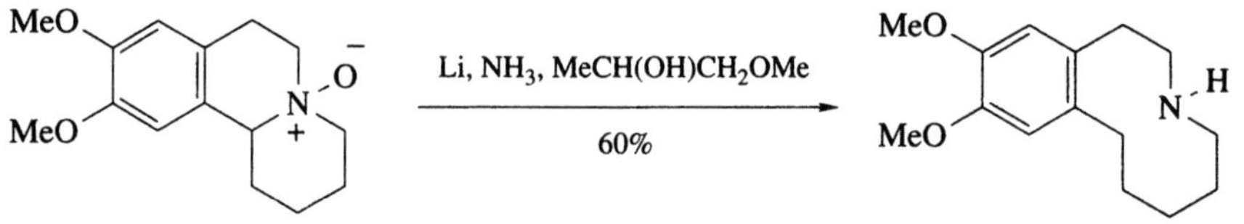

Li, NH_3, $MeCH(OH)CH_2OMe$

60%

(76)

$NH_2NH_2 \cdot H_2SO_4$

Li, NH_3

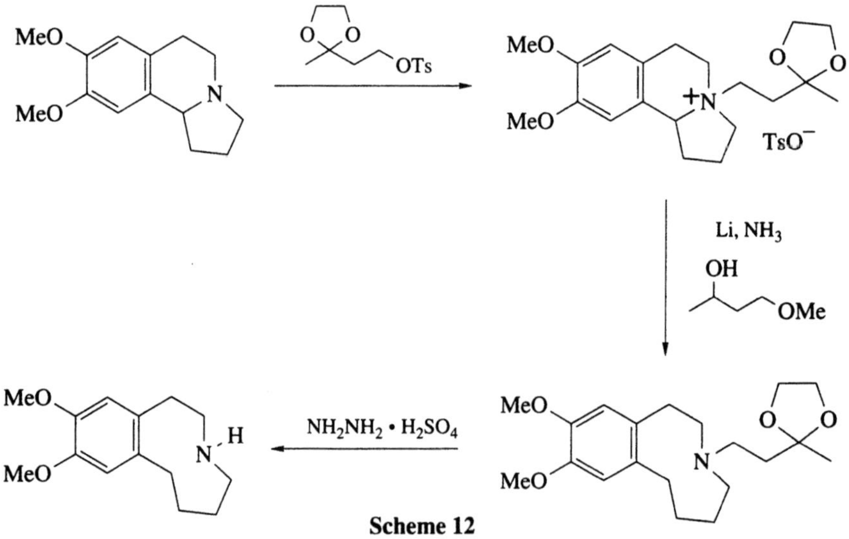

Scheme 12

4.7.4.2 Other Metals

Zinc metal and zinc amalgam have been widely used for the reduction of benzylic and allylic functional groups. Primary, secondary and tertiary benzylic alcohols are readily reduced by zinc amalgam or zinc dust under acidic conditions (equation 77).[114] Other acid-stable functional groups are unaffected (*e.g.* aryl—Cl, aryl—OH, CO_2H). The reaction can also be carried out under basic conditions (equation 78).[115] Esters of benzyl alcohols are similarly cleaved under acidic or basic conditions without reduction of the resulting carboxylic acid (equation 79).[116]

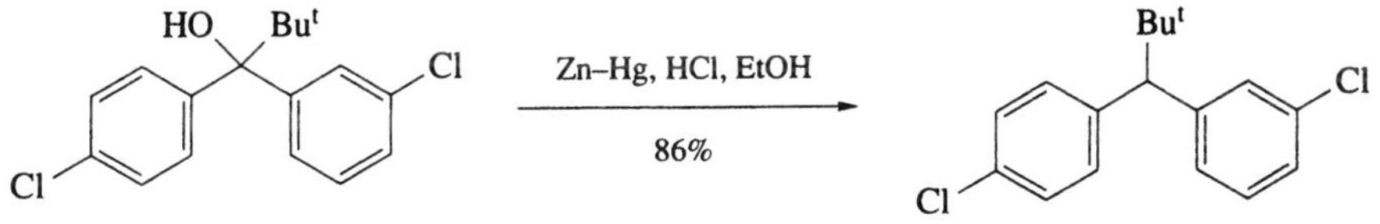

Zn–Hg, HCl, EtOH; 86% (77)

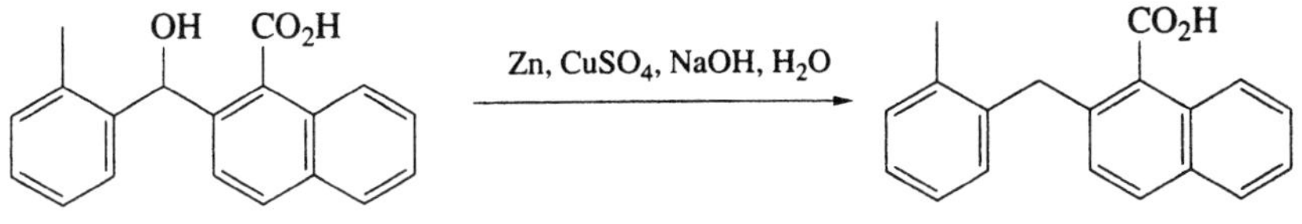

Zn, $CuSO_4$, NaOH, H_2O (78)

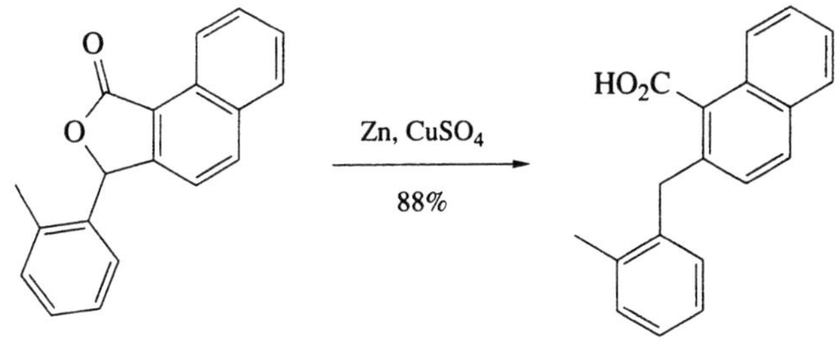

Zn, $CuSO_4$; 88% (79)

Primary, secondary and tertiary benzylic halides are cleaved with zinc dust in acetic acid or under basic conditions (equations 80 and 81).[117,118] Benzylic C—N bonds can also be cleaved (equation 82),[119] with particular ease when quaternary ammonium salts are reduced (equation 83).[120] There has also been a recent report of cyclopropane cleavage with zinc metal and zinc chloride (equation 84).[121]

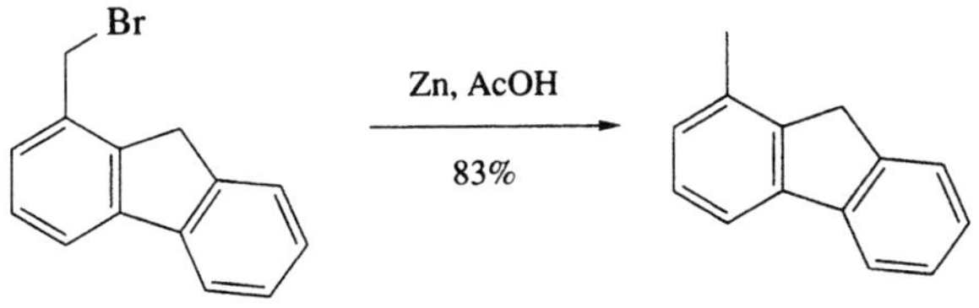

Zn, AcOH; 83% (80)

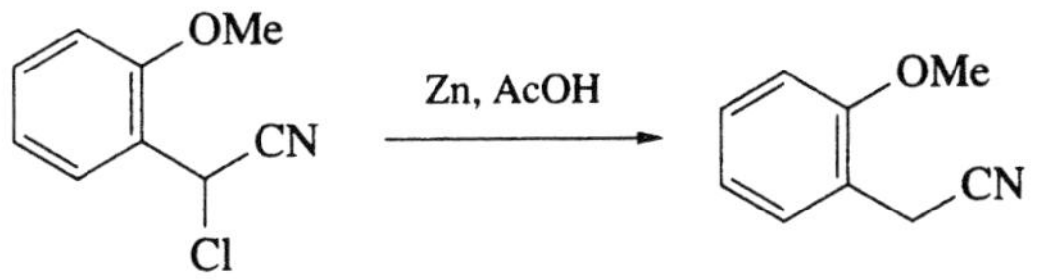

Zn, AcOH (81)

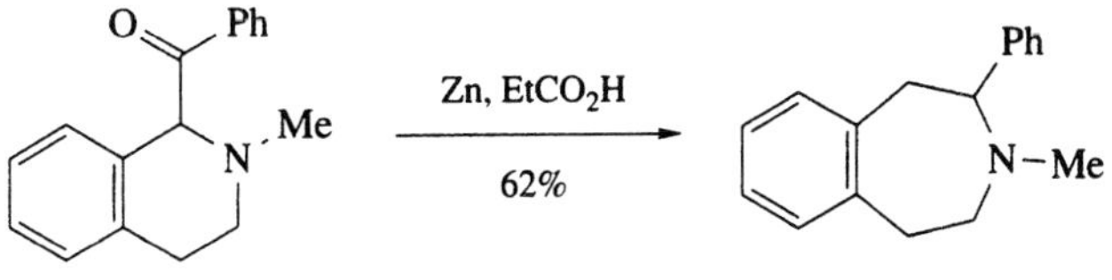

Zn, $EtCO_2H$; 62% (82)

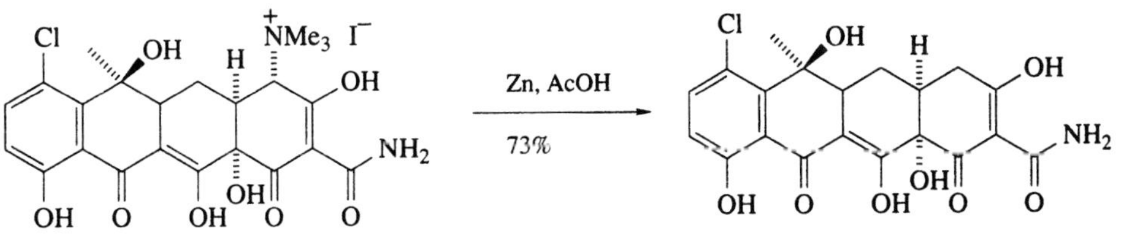

Zn, AcOH; 73% (83)

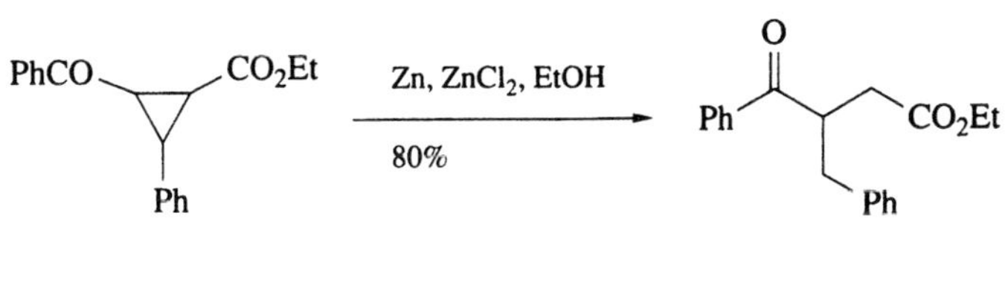

Zn, $ZnCl_2$, EtOH; 80% (84)

Allylic functional groups can be reduced with zinc metal (*cf.* equation 83). Alcohols, esters and strained rings are all cleaved under the usual conditions in good yield and with similar chemoselectivity to the benzylic situation (equations 85–87).[122–124] Allylic sulfonate esters are similarly reduced (equation 88).[125]

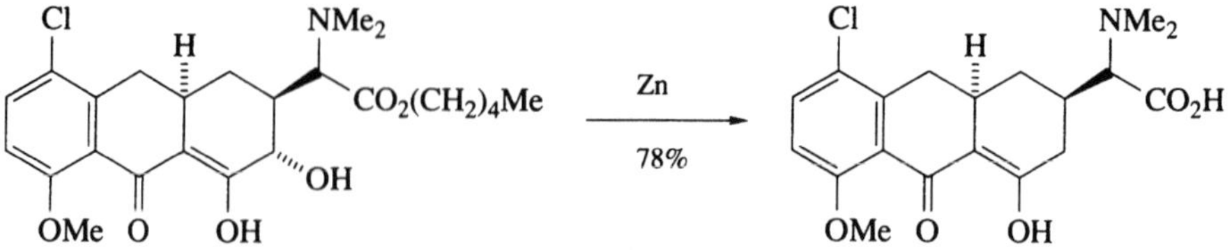

Zn, 78% (85)

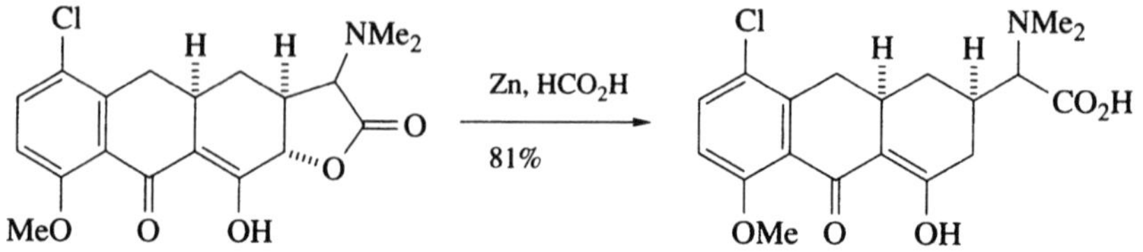

Zn, HCO_2H, 81% (86)

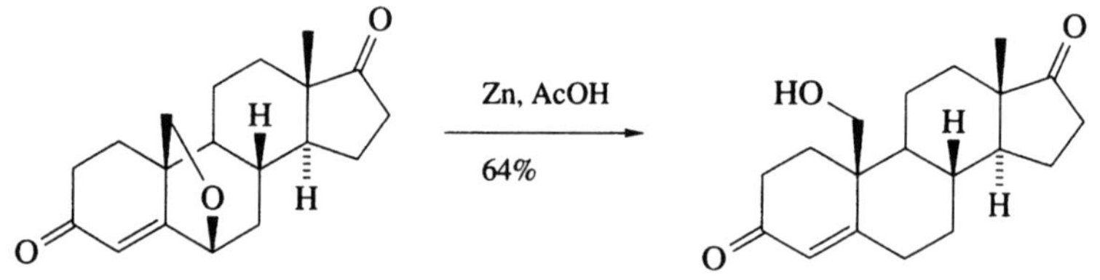

Zn, AcOH, 64% (87)

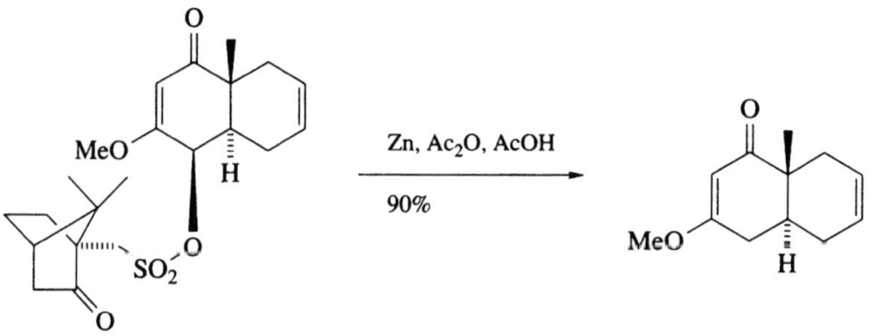

Zn, Ac_2O, AcOH, 90% (88)

4.7.5 ELECTROLYTIC REDUCTION PROCESSES

Electrochemical hydrogenolysis of allylic and benzylic centers can be achieved, but application of this reduction has not been either widely used or acclaimed. It is less convenient experimentally than the comparable method using dissolving metals and control of selectivity is more difficult to achieve than with other methods. Only limited applications are found. Generally electroreduction is carried out using cathodes of metals of high overvoltage such as copper, cadmium, lead and mercury, with anodes of platinum.

Selective reduction of aromatic acids and their derivatives to aldehydes can be achieved electrochemically.[126] Reduction of the carbinolamines is observed, as in the formation of anilinomethylpyridine (Scheme 13) and 4-methylpyridine.[127] Reductive cleavage of benzyl methyl ethers can be achieved leading to hydrocarbons when the ring substituents are electron attracting.[128]

A more useful application of the electron-attracting effect of aryl subtituents is found in the cleavage of 4-picolyl derivatives of protected cysteine and tyrosine in peptide synthesis (Scheme 14).[129] Catalytic cleavage of these derivatives is usually unattractive for S-containing acids. Protecting groups removable also include 3- and 4-picolyl esters.[130]

Esters of secondary benzyl alcohols can also be electrohydrogenolyzed (equation 89).[131] Electrolysis of resolved methyl *O*-benzoylatrolactate gave racemized 2-phenylpropionic acid. This result is in sharp contrast with the 77–92% inversion of configuration claimed for the electrochemical dechlorination in equation (90).[132]

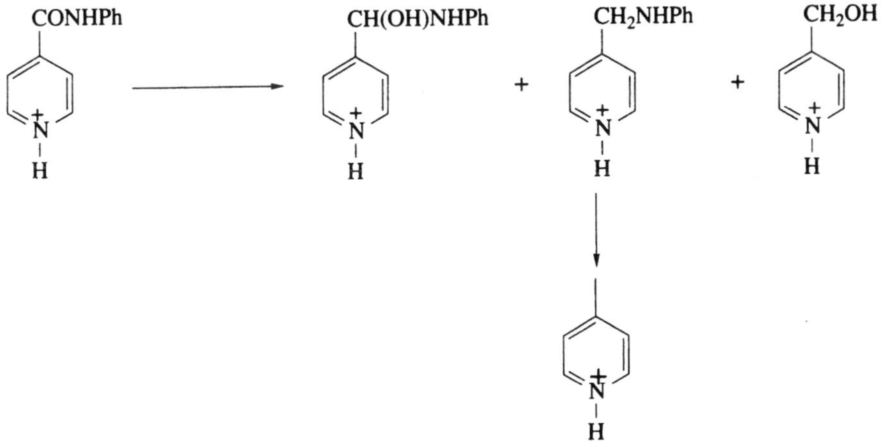

CONNHPh CH(OH)NHPh CH_2NHPh CH_2OH

Scheme 13

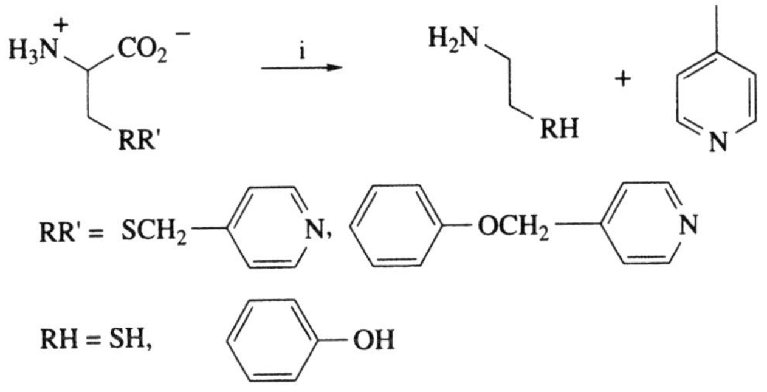

i, Hg, cathode, 0.25 M H_2SO_4

Scheme 14

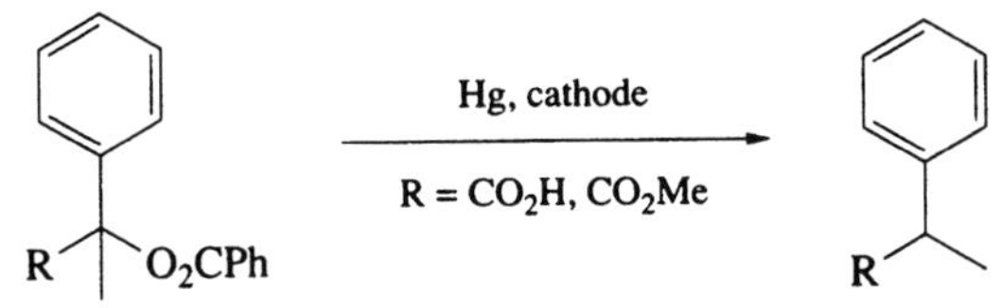

Hg, cathode

R = CO_2H, CO_2Me (89)

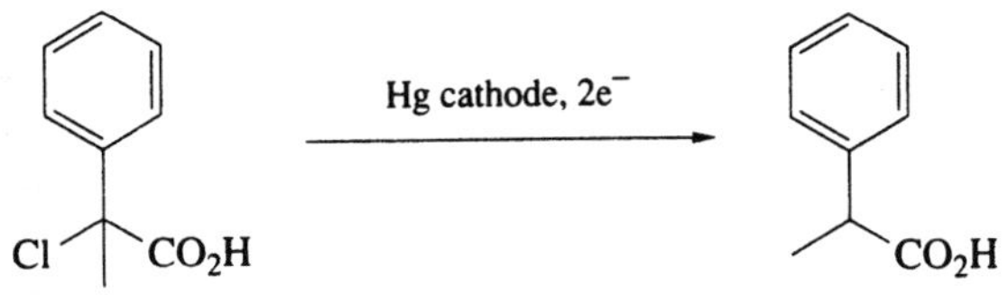

Hg cathode, $2e^-$ (90)

In contrast with other methods only infrequent use of electrohydrogenolysis of benzylic or allylic esters has been made in synthetic strategies. Control of the electrode potentials and electrolyte pH have been shown to be important and essential for chemoselectivity, and would need to be studied in many instances before a particular application was made. The instability of the product towards further electrolysis often causes problems (however see equation (91) for a successful application).[133]

Greater control over the product selectivity is observed when more precise control of the cell conditions and process is used. Since in general the current distribution at the working electrode is not uniform and the effective potential is therefore not the same over the whole electrode, it is difficult to reduce selectively one of two electroactive groups when potentials differ by less than 0.15 V.

One useful approach involves indirect electrochemical reduction using an electron transfer catalyst which is selected for inertness to any anion radicals or dianions present.[134] Thus allylic sulfones (Scheme

electrohydrogenolysis → chemical manipulation → 64% (91)

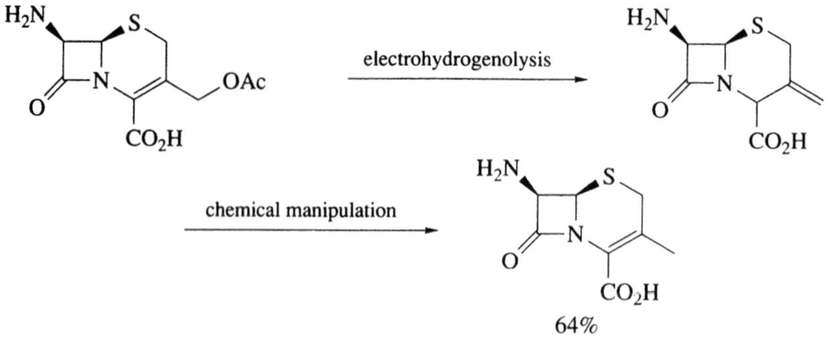

15) are preparatively converted to the hydrocarbon without loss of the alkylsulfonyl group. Both sulfonyl groups are lost by direct electroreduction.

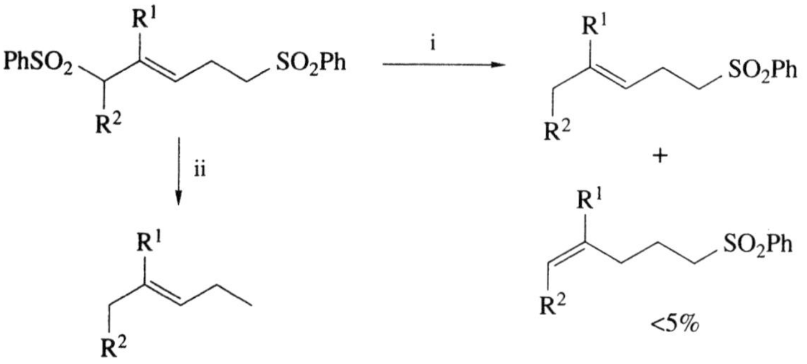

<5%

$R^1 = Me, R^2 = H$

$R^1 = Me, R^2 = CH_2CH{=}CMe_2$

i, electrolysis with anthracene; ii, electrolysis without anthracene

Scheme 15

An alternative approach to the commonly utilized cleavage of allylic isoprenoid compounds (Scheme 16) using attack of $+2e^-$ in place of H^- at a Pd^0 π-allyl complex also requires a catalyst addition.[135] Electrolysis of allylic acetates with Pb cathode and Pt anode in the presence of $Pd(PPh_3)_4$ in MeCN, gave moderate yields of the inner alkenes from both terminal and inner acetates (equations 92 and 93). Control of the regioselectivity was clearly dictated by the attack of H^+ from the less-hindered side of the allylic carbanions.

Electrolytic desulfurization of *o*-aminobenzyl sulfide, sulfoxide or sulfone illustrates carbon–sulfur bond cleavage. Careful control of the reaction conditions is important to achieve chemoselectivity (Scheme 17).[136] Trifluoromethyl substituents are not reduced under the conditions used (Cd sheet ca-

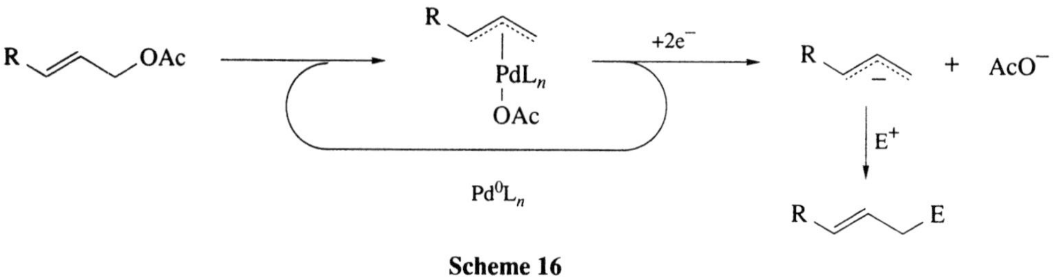

$+2e^-$; E^+ ; Pd^0L_n

Scheme 16

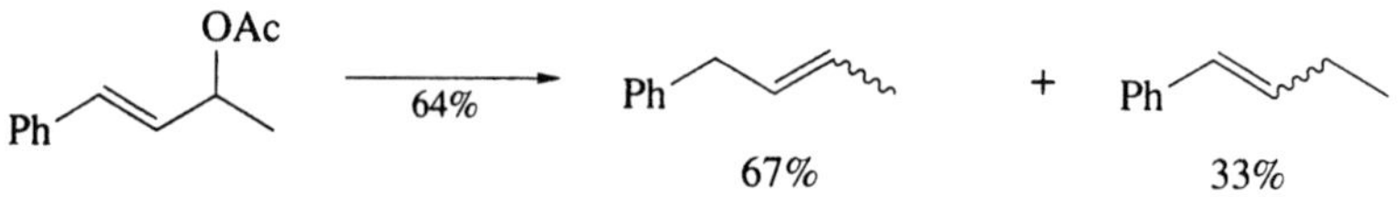

64% → 67% + 33% (92)

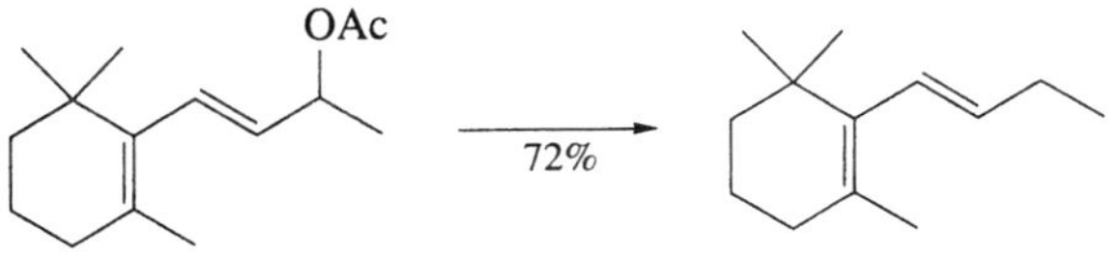

OAc

72% (93)

thode and Pt-coated Ti anode), and surprisingly the 2-amino substituent did not interfere with cathodic discharge. An alternative cleavage of the chloroacetanilide of the sulfalimine salt obtained *via* Scheme 17 by electrolysis over a Hg cathode in phosphate-buffered electrolyte, gave good yields of the corresponding toluene.[137] Some dehalogenation of the chloroacetyl group does occur, illustrating the difficulty of controlling chemoselectivity.

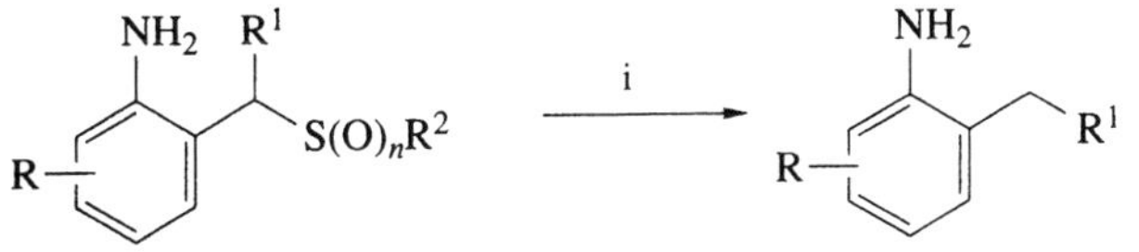

NH_2 R^1 $S(O)_nR^2$ R i NH_2 R^1 R

R^1 = alkyl; R^2 = alkyl, aryl; R = CF_3; n = 0–2

i, electrolysis, Cd cathode, Pt anode

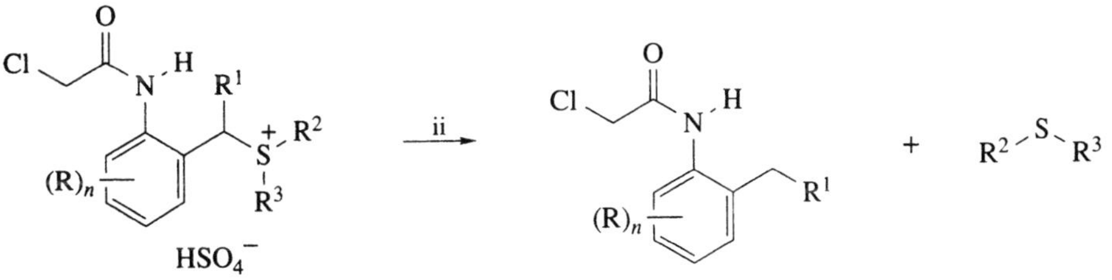

Cl O N H R^1 S^+ R^2 R^3 $(R)_n$ HSO_4^- ii Cl O N H R^1 $(R)_n$ + R^2 S R^3

ii, Hg cathode, Pt anode

Scheme 17

4.7.6 BIOMIMETIC REDUCING AGENTS

Attempts to mimic enzymatic reduction processes have been reported.[138] Of major interest has been the involvement of NAD(P)H in redox reactions.[139] The most investigated model compounds employed as donors are analogs or derivatives of 1,4-dihydronicotinamide. In the presence of a catalyst the hydrogenolysis of cinnamyl acetate (Scheme 18) is readily effected with *N*-propyl-1,4-dihydronicotinamide. $RhCl(PPh_3)_2$ appears to be more effective than $PdCl_2(PPh_3)_2$ or $RuCl_2(PPh_3)_3$.[140] This reduction appears to have a complex mechanism as it is very sensitive to added salts such as $LiClO_4$, which is essential for a high conversion. $Mg(ClO_4)_2$ effectively prevents cleavage. The high yields are also dependent on the use of acetonitrile in the solvent. Although the chemoselectivity is high, ketones and esters being inert, the type of allylic structure that can be reduced appears to be restricted to compounds with allylic unsaturation conjugated to aryl rings. This requirement assists the formation of mainly 2-alkenes from acetates, chlorides and phenylsulfonyl derivatives. *N*-Phenyl-1,4-dihydronicotinamide (BNAH) has also been utilized in a visible light catalyzed reduction of sulfonate esters of methyl mandelate (Scheme 19).[141] A number of functional groups, including keto, nitro and ester groups, were unaffected. These conversions are accelerated by addition of photosensitizers such as tris(bipy)ruthenium(II) dichloride or tetraphenylporphine.

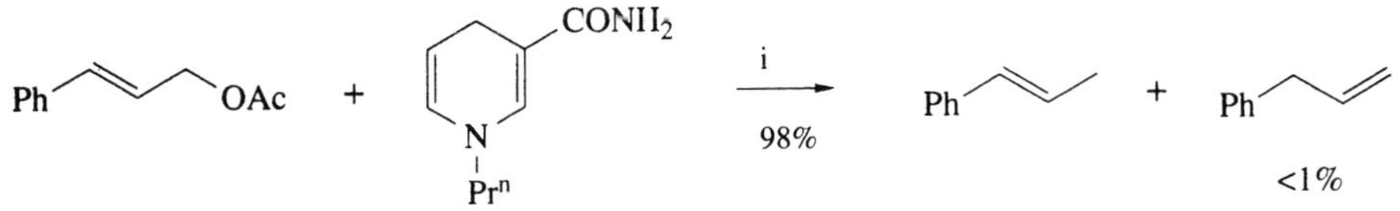

Ph OAc + $CONH_2$ N Pr^n i 98% Ph + Ph <1%

i, $RhCl(PPh_3)_3$, $LiClO_4$, MeCN, dark

Scheme 18

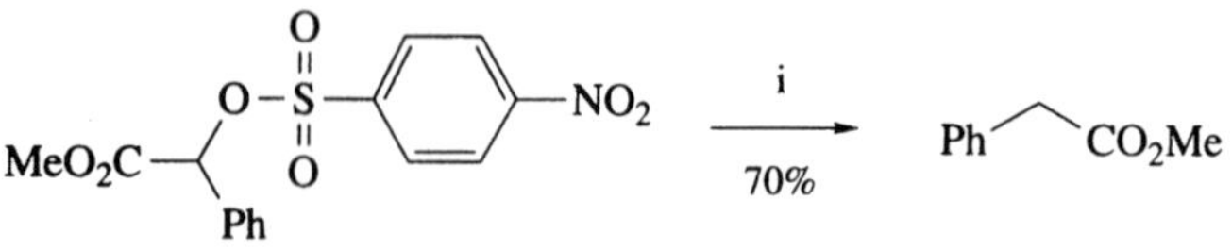

i, $h\nu$, BNAH, Py, tris(bipy)ruthenium(II) dichloride

Scheme 19

4.7.7 PHOTOLYTIC REDUCTION PROCESSES

Photochemically aided hydrogenolysis of allylic C–heteroatom bonds has attracted little interest. A few methods for debenzylation of ethers, amines and thiols, catalyzed by irradiation, have been utilized (see previous section also). An unusual desulfurization of benzyl thiol with triethyl phosphite (equation 94) in high yield, catalyzed by UV irradiation, is a useful alternative to other methods described in other parts of this chapter.[142]

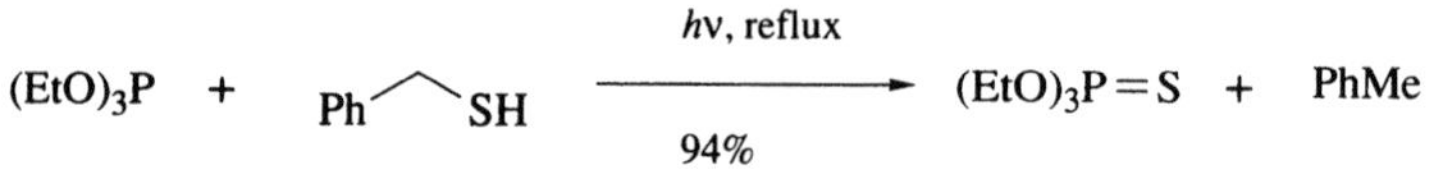

$$(EtO)_3P + PhCH_2SH \xrightarrow[94\%]{h\nu,\ reflux} (EtO)_3P{=}S + PhMe \quad (94)$$

4.7.8 MISCELLANEOUS REDUCING AGENTS

Historically, hydrogen derived from dissociation of hydrogen iodide has been utilized widely for reduction. Acceleration of the hydrogenolysis rate is achieved by addition of phosphorus to remove the iodine formed.[143] In general reduction with HI is efficient for acid-tolerant substrates and is a useful alternative to catalytic methods, especially to achieve chemoselectivity (equations 95 and 96).[144,145] Application of HI to cleavage of allylic centers has not been useful as allylic and other unsaturated centers are more susceptible to HI addition and elimination effects.

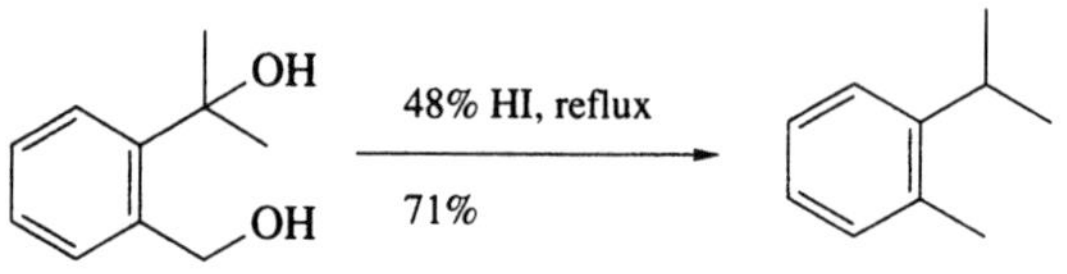

48% HI, reflux; 71% (95)

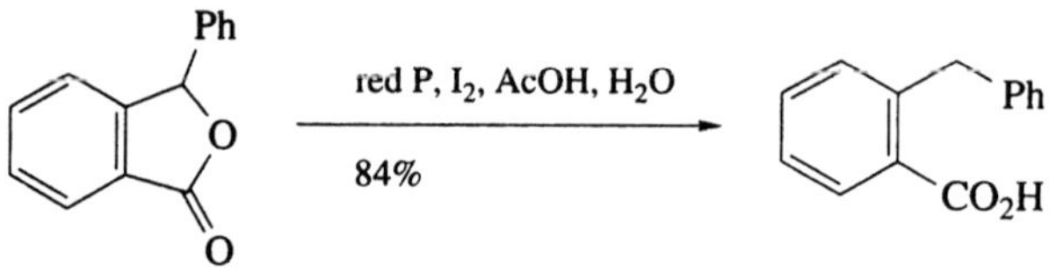

red P, I_2, AcOH, H_2O; 84% (96)

The use under neutral conditions of this reagent has more recently been effective for the cleavage of primary, secondary and tertiary benzyl alcohols in refluxing benzene.[146] The hydrocarbons from a range of variously substituted alcohols were obtained in 60–98% yields (Scheme 20).

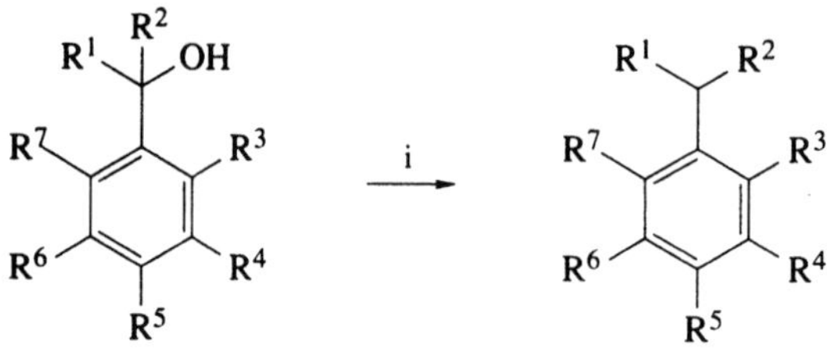

i, P_2I_4, PhH, N_2, Δ

R^1, R^2 = H, Me, Ph, Bu^t

R^3–R^7 = H, Me, Ph, NO_2, CO_2Et, CN, Br

Scheme 20

As an alternative mild application of HI, diiododimethylsilane (Me_2SiI_2) has merit for reduction of secondary and tertiary benzylic alcohols to hydrocarbons at room temperature in near quantitative yields.[147] Side chain carbonyl groups were unaffected, but benzyl alcohol itself was converted to the iodide. TMS-Cl–NaI–MeCN is recognized as an equivalent to TMS-I and is a less costly alternative to Me_2SiI_2. Applications of these milder reagents have not been reported for allylic cleavages.

As an alternative to the use of tin metal, tin(II) chloride has been widely utilized. Less acidic conditions can be employed than is normal for tin metal dissolution and examples of allylic systems (equation 97)[148] have been cleaved as well as secondary and tertiary benzylic centers. Primary benzyl centers can be cleaved *via* the iodide (equation 98).[149]

$SnCl_2$, aq. HCl (97)

$SnCl_2$, HCl, reflux, 85% (98)

4.7.9 REFERENCES

1. W. H. Hartung and R. Simonoff, *Org. React. (N.Y.)*, 1953, **7**, 263.
2. R. Baltzly and J. S. Buck, *J. Am. Chem. Soc.*, 1943, **65**, 1984.
3. W. H. Hartung and F. S. Crossley, *J. Am. Chem. Soc.*, 1934, **56**, 158.
4. M. Stiles and A. J. Sisti, *J. Org. Chem.*, 1960, **25**, 1691.
5. N. I. Shujkin and I. F. Bel'skij, *Chem. Ber.*, 1958, **91**, 948.
6. F. C. Uhle, J. E. Kreuger and A. E. Rogers, *J. Am. Chem. Soc.*, 1956, **78**, 1932.
7. M. A. Miropo'skaya, N. I. Fedorova, A. Ya. Veinberg, M. Ts. Yanotovski and G. I. Samokhvalov, *Zh. Obshch. Khim.*, 1962, **32**, 2214 (*Chem. Abstr.*, 1962, **58**, 8892d); T. W. Russell and R. C. Hoy, *J. Org. Chem.*, 1971, **36**, 2018.
8. T. W. Greene, 'Protective Groups in Organic Synthesis', Wiley, New York, 1981.
9. A. C. Cope, M. A. McKervey and N. M. Weinshenker, *J. Am. Chem. Soc.*, 1967, **89**, 2932.
10. H. Dannenberg and S. Laufer, *Chem. Ber.*, 1954, **87**, 733.
11. W. C. Cunningham and J. W. Thanassi, *Experientia*, 1979, **35**, 451.
12. J. C. Sheehan and G. D. Laubach, *J. Am. Chem. Soc.*, 1951, **73**, 4752.
13. S. Kano, T. Ebata and S. Shibuya, *Heterocycles*, 1980, **14**, 43.
14. D. Beke and L. Tökés, *Chem. Ber.*, 1962, **95**, 2122.
15. H. Dahn, U. Solms and P. Zoller, *Helv. Chim. Acta.* 1952, **35**, 2117.
16. J. W. Cornforth, O. Kauder, J. E. Pike and R. Robinson, *J. Chem. Soc.*, 1955, 3348.
17. E. E. van Tamelen, T. A. Spencer, D. S. Allen, Jr. and R. L. Orvis, *Tetrahedron*, 1961, **14**, 8.
18. T. F. Braish and P. L. Fuchs, *Synth. Commun.*, 1986, **16**, 785.
19. I. D. Entwistle, R. A. W. Johnstone and A. H. Wilby, *Chem. Rev.*, 1985, **85**, 129.
20. A. E. Jackson and R. A. W. Johnstone, *Synthesis*, 1976, 685.
21. A. M. Felix, E. P. Heimer, T. J. Lambros, C. Tzougraki and J. Meienhofer, *J. Org. Chem.*, 1978, **43**, 4194.
22. B. ElAmin, G. M. Anantharamaiah, G. P. Royer and G. E. Means, *J. Org. Chem.*, 1979, **44**, 3442; K. M. Sivanandaiah and S. Gurusiddappa, *J. Chem. Res. (S)*, 1979, 108.
23. K. Kuromizu and J. Meienhofer, *J. Am. Chem. Soc.*, 1974, **96**, 4978.
24. J. Tsuji and T. Yamakawa, *Tetrahedron Lett.*, 1979, 613.
25. G. A. Olah and G. K. S. Prakash, *Synthesis*, 1978, 397; G. A. Olah, G. K. S. Prakash and S. C. Narang, *Synthesis*, 1978, 825.
26. V. S. Rao and A. S. Perlin, *Carbohydr. Res.*, 1980, **83**, 175.
27. I. D. Entwistle, *Tetrahedron Lett.*, 1979, 555.
28. T. R. Weir, B. A. Patel and R. F. Heck, *J. Org. Chem.*, 1980, **45**, 4926.
29. R. A. W. Johnstone and A. H. Wilby, *Tetrahedron*, 1981, **37**, 3667.
30. J. Tsuji, 'Organic Synthesis with Palladium Compounds', 1980, Springer-Verlag, Berlin.
31. J. Tsuji, I. Minami and I. Shimizu, *Synthesis*, 1986, 623.
32. T. Tabuchi, J. Inanaga and M. Yamaguchi, *Tetrahedron Lett.*, 1986, **27**, 601.
33. H. Hey and H.-J. Arpe, *Angew. Chem., Int. Ed. Engl.*, 1973, **12**, 928.
34. H. Matsushita and E. Negishi, *J. Org. Chem.*, 1982, **47**, 4161.
35. R. O. Hutchins, K. Learn and R. P. Fulton, *Tetrahedron Lett.*, 1980, **21**, 27.

36. R. O. Hutchins and K. Learn, *J. Org. Chem.*, 1982, **47**, 4381.
37. M. Mohri, H. Kinoshita, K. Inomata and H. Kotake, *Chem. Lett.*, 1985, 451.
38. E. Keinan and N. Greenspoon, *Tetrahedron Lett.*, 1982, **23**, 241.
39. J. Tsuji, T. Yamakawa, M. Kaito and T. Mandai, *Tetrahedron Lett.*, 1978, 2075.
40. E. Keinan and N. Greenspoon, *J. Org. Chem.*, 1983, **48**, 3545.
41. N. Ono, I. Hamamoto, A. Kamimura and A. Kaji, *J. Org. Chem.*, 1986, **51**, 3734.
42. K. Osakada, K. Matsumoto, T. Yamamoto and A. Yamamoto, *Chem. Ind. (London)*, 1984, 634.
43. A. S. Berenblyum, L. K. Ronzhin, M. V. Emolaev, I. V. Kalechits and M. L. Khidekel, *Izv. Akad. Nauk SSSR, Ser. Khim.*, 1973, **11**, 2650; *Chem. Abstr.*, 1974, **80**, 95 402w.
44. E. J. Skerrett and D. Woodcock, *J. Chem. Soc.*, 1952, 2804.
45. B. B. Elsner, H. E. Strauss and E. J. Forbes, *J. Chem. Soc.*, 1957, 578.
46. W. S. Emerson *et al.*, *J. Am. Chem. Soc.*, 1950, **72**, 5314.
47. K. H. Scheit, *Chem. Ber.*, 1966, **99**, 3884.
48. D. B. Clayson, *J. Chem. Soc.*, 1949, 2016.
49. Y. Oikawa, T. Tanaka, K. Horita and O. Yonemitsu, *Tetrahedron Lett.*, 1984, **25**, 5397.
50. P. S. Bailey, S. S. Bath, W. F. Thomsen, H. H. Nelson and E. E. Kawas, *J. Org. Chem.*, 1956, **21**, 297.
51. A. McCoubrey, *J. Chem. Soc.*, 1951, 2931.
52. S. Sugasawa and S. Ushioda, *Tetrahedron*, 1959, **5**, 48.
53. H. Hauptmann and W. F. Walter, *Chem. Rev.*, 1962, **62**, 347.
54. C. Djerassi, M. Gorman and J. A. Henry, *J. Am. Chem. Soc.*, 1955, **77**, 4647.
55. M. Sakamoto, Y. Omote and H. Aoyama, *J. Org. Chem.*, 1984, **49**, 396.
56. Y. Tamura, H. D. Choi, H. Shindo and H. Ishibashi, *Chem. Pharm. Bull.*, 1982, **30**, 915.
57. F. C. Copp, K. W. Franzmann, J. Gilmore and W. B. Whalley, *J. Chem. Soc., Perkin Trans. 1*, 1983, 909.
58. M. Moreno-Mañas and R. Pleixats, *Synthesis*, 1984, 430.
59. H. Ishibashi, H. Nakatani, Y. Umei and M. Ikeda, *Tetrahedron Lett.*, 1985, **26**, 4373.
60. W. A. Bonner, *J. Am. Chem. Soc.*, 1952, **74**, 1034.
61. W. E. Parham and C. D. Wright, *J. Org. Chem.*, 1957, **22**, 1473.
62. A. Liguori, G. Sindona and N. Uccella, *Tetrahedron*, 1983, **39**, 683.
63. L. D. Moore, *J. Chem. Eng. Data*, 1964, **9**, 251.
64. C. S. Marvel and B. D. Wilson, *J. Org. Chem.*, 1958, **23**, 1483.
65. R. M. Magid, *Tetrahedron*, 1980, **36**, 1901.
66. T. Hirabe, M. Nojima and S. Kusabayashi, *J. Org. Chem.*, 1984, **49**, 4084.
67. C. W. Jefford, S. N. Mahajan and J. Gunsher, *Tetrahedron*, 1968, **24**, 2921.
68. G. Ohloff, H. Farrow and G. Schade, *Chem. Ber.*, 1956, **89**, 1549.
69. B. Fraser-Reid, S. Y.-K. Tam and B. Radatus, *Can. J. Chem.*, 1975, **53**, 2005.
70. W. Ried and F. J. Königstein, *Angew. Chem.*, 1958, **70**, 165.
71. E. L. Eliel and M. N. Rerick, *J. Org. Chem.*, 1958, **23**, 1088.
72. R. S. Glass, *Synth. Commun.*, 1976, **6**, 47.
73. V. L. Tweedie and M. Cuscurida, *J. Am. Chem. Soc.*, 1957, **79**, 5463.
74. T. Mukaiyama, K. Narasaka, K. Maekawa and M. Furusato, *Bull. Chem. Soc. Jpn.*, 1971, **44**, 2285.
75. T. Mukaiyama, M. Hayashi and K. Narasaka, *Chem. Lett.*, 1973, 291.
76. M.-C. Chan, K.-M. Cheng, M. K. Li and T.-Y. Luh, *J. Chem. Soc., Chem. Commun.*, 1985, 1610.
77. H. Ishikawa and T. Mukaiyama, *Chem. Lett.*, 1976, 737.
78. Y. Fujimoto and N. Ikekawa, *Chem. Pharm. Bull.*, 1976, **24**, 825.
79. F. Sato, Y. Tomuro, H. Ishikawa, T. Oikawa and M. Sato, *Chem. Lett.*, 1980, 103.
80. J. Málek and M. Černý, *Synthesis*, 1972, 217.
81. M. Černý and J. Málek, *Collect. Czech. Chem. Commun.*, 1970, **35**, 2030.
82. C. G. M. Janssen, P. M. van Lier, P. Schipper, L. H. J. G. Simons and E. F. Godefroi, *J. Org. Chem.*, 1980, **45**, 3159.
83. E. J. Moriconi, T. E. Brady and R. E. Misner, *J. Org. Chem.*, 1971, **36**, 479.
84. R. O. Hutchins, D. Hoke, J. Keogh and D. Koharski, *Tetrahedron Lett.*, 1969, 3495.
85. M. Yamamoto and H. Yamamoto, *Chem. Pharm. Bull.*, 1981, **29**, 2135.
86. A. Brossi, G. Grethe, S. Teitel, W. C. Wildman and D. T. Bailey, *J. Org. Chem.*, 1970, **35**, 1100.
87. G. W. Gribble, R. M. Leese and B. E. Evans, *Synthesis*, 1977, 172.
88. A. Ono, N. Suzuki and J. Kamimura, *Synthesis*, 1987, 736.
89. H. M. Bell and H. C. Brown, *J. Am. Chem. Soc.*, 1966, **88**, 1473.
90. W. E. Truce and F. E. Roberts, *J. Org. Chem.*, 1963, **28**, 961.
91. D. N. Sarma and R. P. Sharma, *Tetrahedron Lett.*, 1985, **26**, 371.
92. D. N. Sarma and R. P. Sharma, *Tetrahedron Lett.*, 1985, **26**, 2581.
93. S. Kim, Y. J. Kim and K. H. Ahn, *Tetrahedron Lett.*, 1983, **24**, 3369.
94. C. K. Lau, C. Dufresne, P. C. Bélanger, S. Piétré and J. Scheigetz, *J. Org. Chem.*, 1986, **51**, 3038.
95. R. Johansson and B. Samuelsson, *J. Chem. Soc., Chem. Commun.*, 1984, 201.
96. J. Garegg, H. Hultberg and S. Wallin, *Carbohydr. Res.*, 1982, **108**, 97.
97. M. G. Adlington, M. Orfanopoulos and J. L. Fry, *Tetrahedron Lett.*, 1976, 2955.
98. N. Ono, H. Miyake, R. Tamura and A. Kaji, *Tetrahedron Lett.*, 1981, **22**, 1705.
99. C. G. Gutierrez and L. R. Summerhays, *J. Org. Chem.*, 1984, **49**, 5206.
100. T. Cohen and M.-T. Lin, *J. Am. Chem. Soc.*, 1984, **106**, 1130.
101. E. M. Kaiser, *Synthesis*, 1972, 391.
102. G. H. Small, A. E. Minnella and S. S. Hall, *J. Org. Chem.*, 1975, **40**, 3151.
103. A. R. Pinder and H. Smith, *J. Chem. Soc.*, 1954, 113.
104. A. J. Birch, *J. Chem. Soc.*, 1945, 809.
105. S. K. Dasgupta, D. R. Crump and M. Gut, *J. Org. Chem.*, 1974, **39**, 1658.
106. T. Kobayashi and H. Tsuruta, *Synthesis*, 1980, 492.

107. J. A. Marshall and M. T. Pike, *J. Org. Chem.*, 1968, **33**, 435.
108. K. Irmscher, W. Beerstecher, H. Metz, R. Watzel and K.-H. Bork, *Chem. Ber.*, 1964, **97**, 3363.
109. M. Oki, W. Funakoshi and A. Nakamura, *Bull. Chem. Soc. Jpn.*, 1971, **44**, 828.
110. R. E. Dabby, J. Kenyon and R. F. Mason, *J. Chem. Soc.*, 1952, 4881.
111. A. M. Moiseenkov, E. V. Polunin and A. V. Semenovsky, *Angew. Chem., Int. Ed. Engl.*, 1981, **20**, 1057.
112. J. P. Yardley, *Synthesis*, 1973, 543.
113. J. P. Yardley, R. W. Rees and H. Smith, *J. Med. Chem.*, 1967, **10**, 1088.
114. E. J. Skerrett and D. Woodcock, *J. Chem. Soc.*, 1950, 2718.
115. E. Clar and D. G. Stewart, *J. Chem. Soc.*, 1951, 687.
116. M. S. Newman and R. Gaertner, *J. Am. Chem. Soc.*, 1950, **72**, 264.
117. L. A. Pinck and G. E. Hilbert, *J. Am. Chem. Soc.*, 1946, **68**, 751.
118. A. R. Pinder and R. Robinson, *J. Chem. Soc.*, 1952, 1224.
119. C. Reby and N. J. Gardent, *Bull. Soc. Chim. Fr.*, 1972, 1574.
120. J. H. Boothe, G. E. Bonvicino, C. W. Waller, J. P. Petisi, R. W. Wilkinson and R. B. Broschard, *J. Am. Chem. Soc.*, 1958, **80**, 1654.
121. W. S. Murphy and S. Wattanasin, *J. Chem. Soc., Perkin Trans. 1*, 1986, 1445.
122. J. J. Korst, J. D. Johnston, K. Butler, E. J. Bianco, L. H. Conover and R. B. Woodward, *J. Am. Chem. Soc.*, 1968, **90**, 439.
123. A. Dobrev and C. Ivanov, *Synthesis*, 1977, 562.
124. H. Ueberwasser, K. Heusler, J. Kalvoda, Ch. Meystre, P. Wieland, G. Anner and A. Wettstein, *Helv. Chim. Acta*, 1963, **46**, 344.
125. A. J. Speziale, J. A. Stephens and Q. E. Thompson, *J. Am. Chem. Soc.*, 1954, **76**, 5011.
126. J. H. Wagenknecht, *J. Org. Chem.*, 1972, **37**, 1513.
127. H. Lund, *Acta Chem. Scand.*, 1963, **17**, 2325.
128. R. F. Garwood, Naser-ud-din and B. C. L. Weedon, *J. Chem. Soc., Chem. Commun.*, 1968, 923.
129. A. Gosden, D. Stevenson and G. T. Young, *J. Chem. Soc., Chem. Commun.*, 1972, 1123.
130. A. Gosden, R. Macrae and G. T. Young, *J. Chem. Res. (S)*, 1977, 22.
131. R. E. Erickson and C. M. Fischer, *J. Org. Chem.*, 1970, **35**, 1604.
132. B. Czochralska, *Chem. Phys. Lett.*, 1967, **1**, 239.
133. M. Ochiai, O. Aki, A. Morimoto, T. Okada, K. Shinozaki and Y. Asahi, *Tetrahedron Lett.*, 1972, **23**, 2341.
134. H. Lund and J. Simonet, *Acta Chem. Scand., Ser. B*, 1977, **31**, 909.
135. S. Torii, H. Tanaka, T. Katoh and K. Morisaka, *Tetrahedron Lett.*, 1984, **25**, 3207.
136. Monsanto, *US Pat.* 4 404 069 (1982) (*Chem. Abstr.*, 1984, **100**, 34 241c).
137. Monsanto, *US Pat.* 4 493 755 (1983) (*Chem. Abstr.*, 1985, **102**, 122 075y).
138. B. S. Green, Y. Ashani and D. M. Chipman, (eds.), 'Chemical Approaches to Understanding Enzyme Catalysis; Biomimetic and Transition State Analogs', Elsevier, Amsterdam, 1982.
139. R. J. Kill and D. A. Widdowson, 'The Redox Chemistry of 1,4-Dihydronicotinic acid Derivatives', in 'Bioorganic Chemistry', ed. E. E. van Tamelen, Academic Press, New York, 1978, vol. 4.
140. K. Nakamura, A. Ohno and S. Oka, *Tetrahedron Lett.*, 1983, **24**, 3335.
141. K. Nakamura, S. Yasui, A. Ohno and S. Oka, *Tetrahedron Lett.*, 1983, **24**, 2001.
142. F. W. Hoffmann, R. J. Ess, T. C. Simmons and R. S. Hanzel, *J. Am. Chem. Soc.*, 1956, **78**, 6414.
143. R. N. Renaud and J. C. Stephens, *Can. J. Chem.*, 1974, **52**, 1229.
144. W. E. Parham and Y. A. Sayed, *Synthesis*, 1976, 116.
145. M. G. Sarngadharan and T. R. Seshadri, *Indian J. Chem.*, 1966, **4**, 503.
146. H. Suzuki, H. Tani, H. Kubota, N. Sato, J. Tsuji and A. Osuka, *Chem. Lett.*, 1983, 247.
147. W. Ando and M. Ikeno, *Tetrahedron Lett.*, 1979, 4941.
148. I. Ichimoto and C. Tatsumi, *Agric. Biol. Chem.*, 1970, **34**, 961.
149. J. W. Flesher, S. Soedigdo and D. R. Kelley, *J. Med. Chem.*, 1967, **10**, 932.

4.8
Reduction of α-Substituted Carbonyl Compounds —CX—CO— to Carbonyl Compounds —CH—CO—

ALBERT J. FRY
Wesleyan University, Middletown, CT, USA

4.8.1 GENERAL CONSIDERATIONS

4.8.1.1 Scope of the Review

This review addresses in particular the reductive cleavage of groups X from α-substituted carbonyl compounds (equation 1). Since such studies have almost always involved ketones, the carbonyl group is understood herein to be that of a ketone unless stated otherwise. X is usually a heteroatom (most often halogen, but also oxygen, sulfur or nitrogen, and occasionally other atoms such as selenium or tellurium), but can be carbon in special cases. Since many reductive cleavages of compounds of type (**1**) involve production of enol and enolate ion intermediates, alternative reaction paths involving other components of the medium are sometimes observed. Where these involve carbon–carbon bond formation, they are more properly treated elsewhere in this series. A few such reactions are however referred to herein where appropriate from a mechanistic point of view, and the review also covers reductive conversions of (**1**) to enol derivatives.

O X ⟶ O H (1)

(**1**)

4.8.1.2 Mechanisms of Bond Cleavage

Reductive cleavages of the type with which we are concerned here exhibit some common features. The strength of the bond being cleaved is an important determinant of the ease of cleavage. Thus, the ease of reduction of α-halocarbonyl compounds follows the order: I > Br > Cl > F. By the same token, cleavage of α-ketols (**1**; X = OH) is more facile if the hydroxy group is first converted to a sulfonate or other ester.[1] In principle the dependence upon bond strength can be taken advantage of in order to remove one of several α-substituents selectively. Reports of this type of selective reduction are uncommon, but a few examples will be noted herein. Enolate ions and/or enols are frequently generated in the cleavage process, and ease of reduction is therefore also usually correlated with the stability of the incipient enolate or enol. For example, 2-substituted 1,3-dicarbonyl compounds are reduced more readily than α-substituted ketones, which in turn are more easily reduced than α-substituted esters and amides.

Many reagents have been introduced for the reductive cleavage of compounds (**1**), especially where X is a halogen atom.[2–4] The various methods fall into several quite different mechanistic categories. Conversions corresponding to equation (1) can be effected by: (i) electron transfer, (ii) strong nucleophiles, (iii) metal hydrides, (iv) metal carbonyls, and (v) heterogeneous hydrogenation catalysts. We describe here the general features of each type of reduction; specific reducing agents and α-substituents are taken up in subsequent sections. It is sometimes unclear which route is involved in a given case; for example, it may be difficult to distinguish between mechanisms (i) and (ii) with some combinations of reagents. In such cases the assignment of a particular procedure to a specific mechanistic category may be arbitrary.

Electron transfer agents were among the first to be used to effect the conversion shown in equation (1). The cleavage may be effected with active metals, by low-valent transition metal ions, or electrochemically, at a cathode. The nature of the electron transfer step in such reactions is not well understood. It is tempting to postulate that the initial step is injection of an electron into the LUMO of the carbonyl compound to afford intermediate (**2**), followed by loss of X^- from (**2**) to afford a radical (**3**), which is then reduced further (Scheme 1). However, the strong dependence of the reduction rate on the strength of the C—X bond[5] and the fact that axial α-halocyclohexanones are much easier to reduce by electron transfer than their equatorial epimers[6] imply that the LUMO energy is lowered by overlap of the C—X bond with the C=O group.

Another common method for cleavage of substituents α to a carbonyl group involves reaction of the substituted carbonyl compound with an appropriate nucleophile. Such reactions are believed to occur by nucleophilic attack upon the α-substituent to afford an enolate (equation 2). Nucleophiles which can accomplish this conversion encompass a wide variety of structural types. Some of the best methods for reduction of halocarbonyl compounds fall into this category.

Most methods in the literature for reductive removal of α-substituents involve either electron transfer or nucleophilic cleavage, but other mechanistic pathways have been encountered with certain reagents or

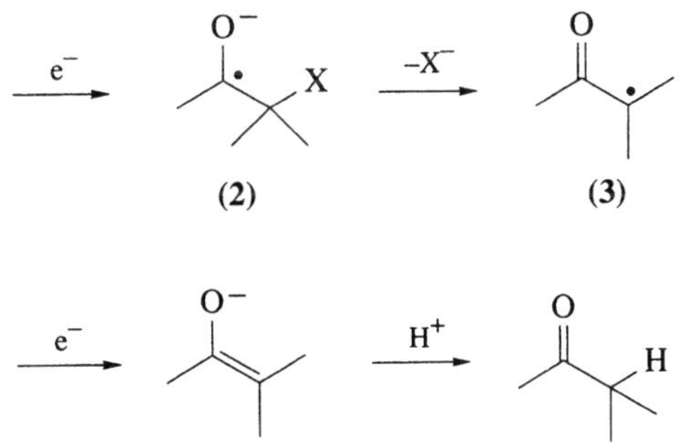

(1) $\xrightarrow{e^-}$ (2) $\xrightarrow{-X^-}$ (3)

(3) $\xrightarrow{e^-}$ (4) $\xrightarrow{H^+}$

Scheme 1

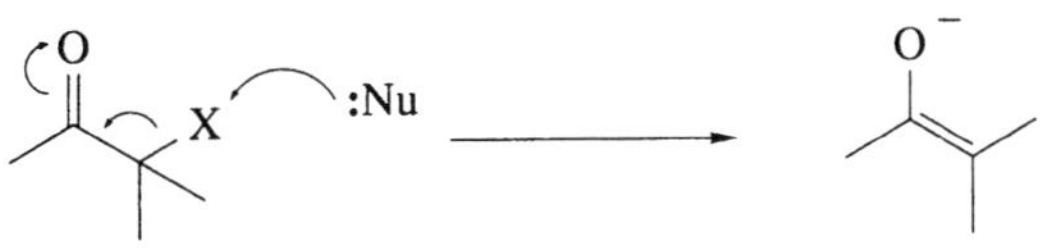

(1) + :Nu ⟶ (4) + Nu–X (2)

combinations of reagents. Metal hydrides have been used for the reductive cleavage process; the most frequently employed hydrides have been those of tin. In general the reactions of tin hydrides with α-substituted carbonyl compounds appear to be of free radical type, generating (**3**) in the key cleavage step (equation 3). Transition metal carbonyls have also been used to effect equation (1), especially with α-halo ketones. Such reactions appear to proceed either by metal enolates or by π-complexes (**5**). Most activity in this area has involved attempts to build up carbocyclic systems through cycloaddition of such complexes with added dienes, and is not relevant in the present context. Some studies, however, have afforded reduced ketones, and will be discussed here. Finally, hydrogenolysis of α-substituents can frequently be effected at noble metal hydrogenation catalysts.

(1) + R• ⟶ (3) + R–X (3)

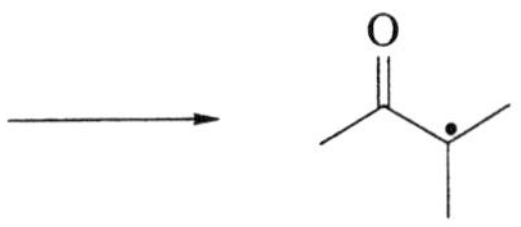

—M(CO)$_n$

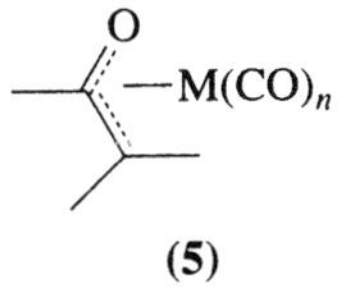

(5)

4.8.2 THE CARBON–HALOGEN BOND

4.8.2.1 Reduction By Electron Transfer

α-Halocarbonyl compounds are very easily reduced by electron transfer agents. The remarkable ease of reduction of such compounds can be seen from inspection of a few representative electrochemical reduction potentials: –0.34 V for bromoacetone and –1.15 V for chloroacetone, compared with –2.23 and <–2.6 V for typical alkyl bromides and chlorides, respectively, or even –1.27 V for benzyl bromide (potentials reported relative to the saturated calomel electrode, or SCE).[7] As these data demonstrate, α-halo- (especially bromo- and iodo-) carbonyl compounds are among the easiest of all organic compounds to reduce by electron transfer. The reductive dehalogenation of such compounds can be carried out by a wide variety of metals and low-valent metal ions. In particular, the great ease with which bromo ketones are reduced has led to a profusion of reagents for reductive cleavage of such compounds.

4.8.2.1.1 Reductive cleavages by metals

Reduction of chloro, bromo and iodo ketones can be effected readily by active metals. Zinc in acetic acid has generally been used for this purpose,[8] but many other metals would presumably serve as well, in view of the extreme ease with which this type of compound is reduced. Fry and Herr showed that the reduction of bromo ketones can be carried out even by a metal as low in the electromotive series as metallic mercury, and they provided an approximate thermodynamic analysis of the factors to be considered in addition to the reduction potential of the metal itself.[9] The initial intermediate in the reduction of halo ketones by zinc is presumably a zinc enolate (**6**), but this is immediately cleaved by the protic solvent to afford the corresponding enol, which then tautomerizes to afford the corresponding carbonyl compound. Reduction in the presence of a deuterated or tritiated solvent should therefore permit introduction of a deuterium (or tritium) atom into the α-position (equation 4). Indeed, reduction of mono-, di- and tri-halo esters and amides in D_2O does afford products containing one, two and three α-deuteriums, respectively, in synthetically useful yields and isotopic abundances.[10] However, reduction of halo ketones in this solvent is accompanied by base-promoted loss (exchange) of the isotope from the product α-deuterio ketone. Reduction in DOAc (or HOAc containing TOAc when the tritiated ketone is desired), where exchange is slower, eliminates this problem, however. The direction of approach of the proton donor upon the intermediate enol is governed by stereoelectronic considerations,[8] and reduction in this solvent can therefore be used to introduce the isotope stereoselectively, as formation of (**7**–**9**; equations 5–7)[11–12] illustrates. Reduction of α-chloro esters and ketones is facilitated by addition of sodium iodide to the medium, probably by formation of the iodocarbonyl compound.[13]

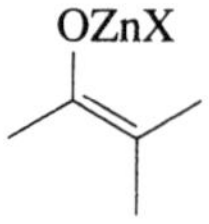

OZnX

(**6**)

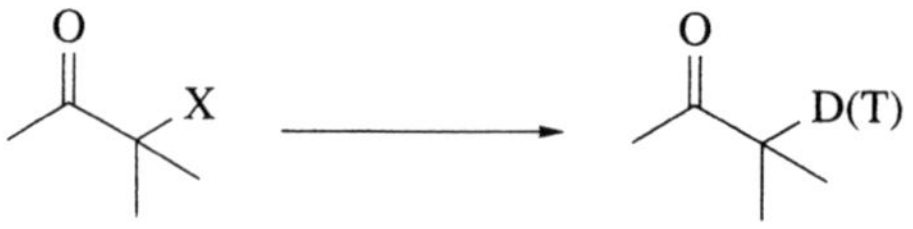

(**1**) → D(T) (4)

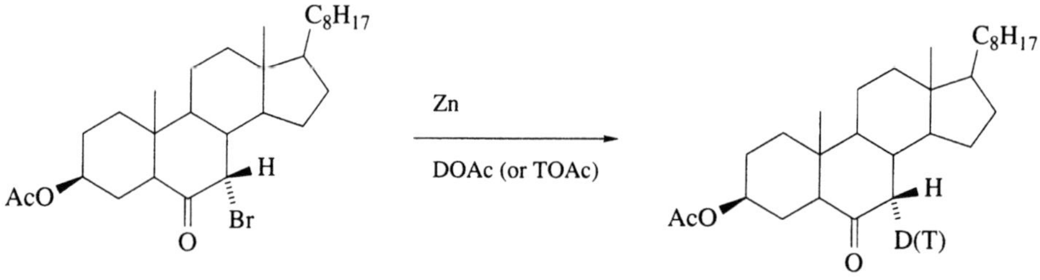

C8H17, AcO, H, Br, O —Zn, DOAc (or TOAc)→ C8H17, AcO, H, D(T), O (**7**) (5)

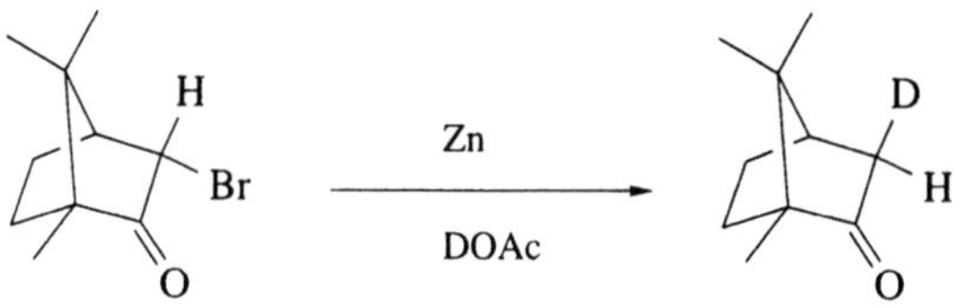

H, Br, O —Zn, DOAc→ D, H, O (**8**) (6)

If the reaction between an α-halo ketone and zinc is carried out in an aprotic solvent in the presence of an electrophilic reagent, the zinc enolate (**6**) can be trapped. Products corresponding to reaction at carbon are observed with carbon electrophiles (alkyl halides or aldehydes; equation 8),[14] but reaction occurs at oxygen with halosilanes and acid anhydrides (equation 9).[15–17]

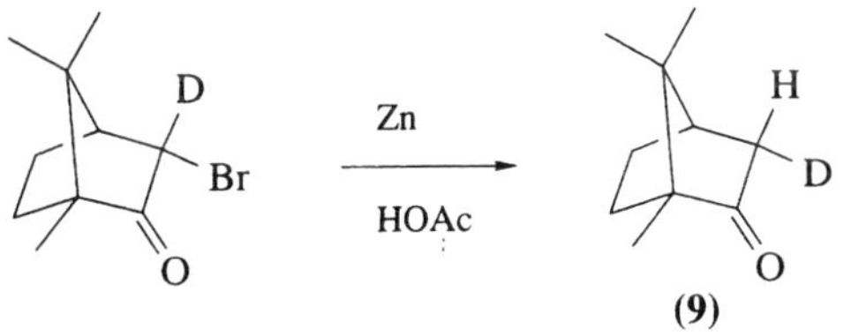

Zn, HOAc → (9) (7)

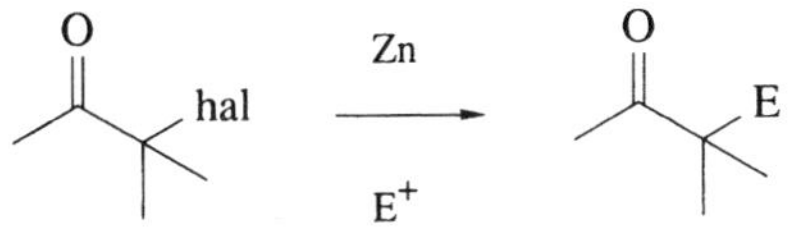

Zn, E^+ (8)

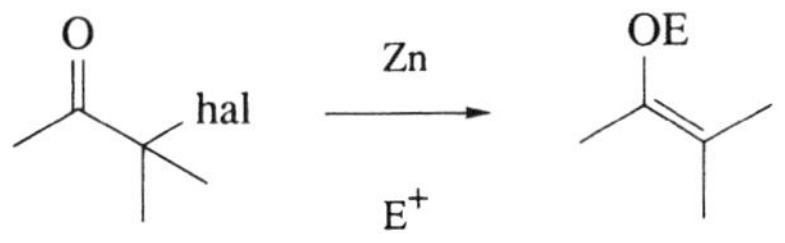

Zn, E^+ (9)

4.8.2.1.2 Reductive cleavages by low-valent metal ions

A number of low-valent metal ions have been shown to reduce α-halocarbonyl compounds.[18] The most commonly used species for this purpose have been chromium(II)[19] and low-valent titanium[20] salts, although vanadium(II), samarium(II), iron(II) and tin(II) salts have also been used.[17,21,22] Chloro, bromo and iodo ketones can all be reduced by chromium(II) and titanium(III) salts. Selective reductions are possible: axial halides are reduced in preference to equatorial,[23] and α,α-dihalo ketones can be selectively reduced to the corresponding monohalides (equation 10).[17] The use of samarium(II) iodide has recently been advocated for such α-cleavages.[22] α-Halo esters and ketones are reduced instantaneously at –78 °C in excellent yields. α-Acetoxy esters are stable to this reagent.

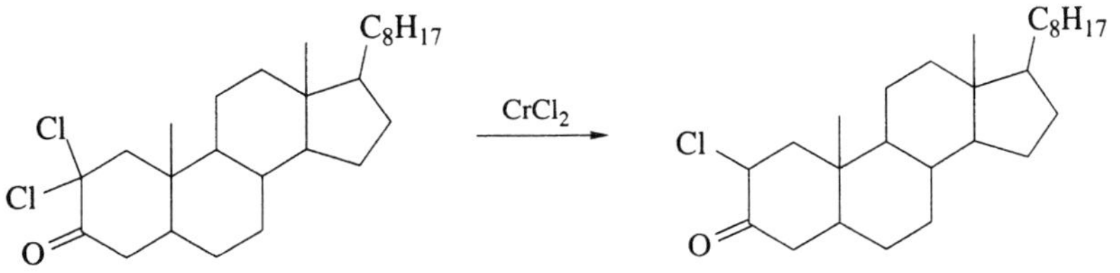

$CrCl_2$ (10)

4.8.2.1.3 Electrochemical reductive cleavages

α-Halocarbonyl compounds are readily reduced electrochemically.[5] In principle, the ease with which this can be effected should readily permit selective removal of α-halogen atoms in the presence of most other organic functional groups, but this point seems not to have been investigated. Reduction is best carried out under conditions which ensure that the initially formed enolate is quickly protonated, thus preventing undesired side reactions. Typically, this means conducting the electrolysis in an acidic protic solvent such as acetic acid or an aprotic solvent containing a proton donor such as phenol or malonic ester.[24] Curious behavior is observed during the electrochemical reduction of α,α,α-trifluoroacetophenone (**10**). When the solution is monitored during electrolysis, it is possible to observe the build-up and eventual decay of the corresponding monofluoride (**12**), but not of the difluoride (**11**).[5] It appears that the ease of reduction of these substances follows the order (**11**) > (**10**) > (**12**) > (**13**). In other words, (**11**) appears to fall out of its expected place in the sequence. It is not obvious why this should be so.

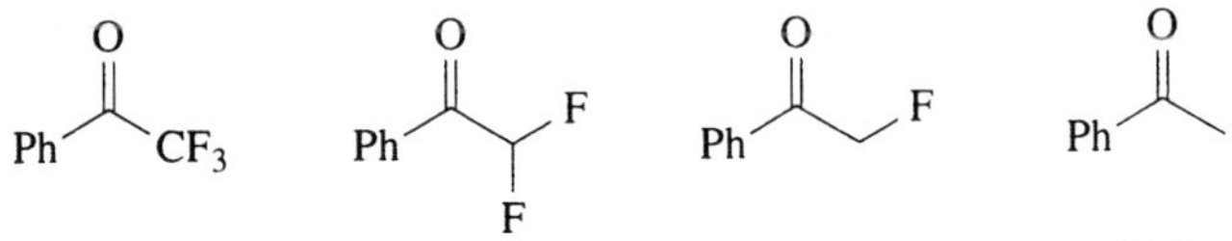

(10) **(11)** **(12)** **(13)**

In order to inquire into the stereochemical course of electrochemical reduction of α-halocarbonyl compounds, it is first necessary to appreciate that the problem should not even be posed in these terms. Two separate questions have to be addressed: (i) is there a preferred geometry for reduction of the halo ketone to an enolate? and (ii) is there a preferred geometry for ketonization of the intermediate enolate or enol? The answer to the first of these questions is a most definite yes. Equatorial α-halocyclohexanones and *exo*-α-halo ketones in the camphor series are reduced at potentials which are considerably more negative than those required for their axial and *endo* epimers, respectively.[5,6] These facts are consistent with a transition state for electron transfer in which the halogen atom of the C—X bond is nearer the electrode surface at the point of electron transfer. Electron transfer then generates the enolate, whose ketonization path is no different than when it is produced by other processes.[8]

A possibly special stereochemical situation is presented by α-halocarbonyl compounds in the cyclopropane series.[25] Electrochemical reduction of the α-halo acid (**14**) proceeds predominantly with stereochemical inversion, whereas reduction of the corresponding carboxylate proceeds predominantly with retention of configuration.

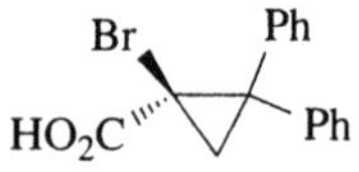

Br Ph
HO_2C Ph

(14)

4.8.2.2 Reduction By Strong Nucleophiles

As mentioned in Section 4.8.1.2, reductive dehalogenation of halocarbonyl compounds can be carried out by a large number of nucleophilic reagents according to equation (2). Such dehalogenations occur more cleanly and rapidly, and under milder conditions, if carried out in such a way that nucleophilic attack on the halogen atom is assisted by electrophilic attack upon the carbonyl group. In particular, considerations based upon hard and soft acid–base (HSAB) theory[26] suggest that the combination of a hard acid (to coordinate with the carbonyl oxygen) and a soft base to attack the α-substituent ought to be especially efficient at such cleavages (equation 11). This principle works remarkably well, and a number of dehalogenation methods have been based upon it.

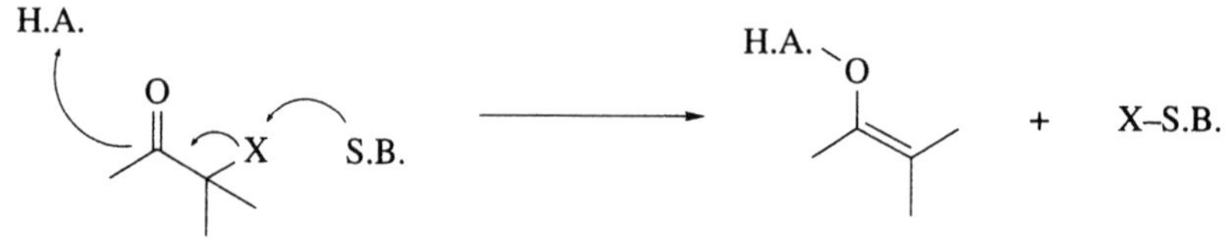

H.A. O X S.B. → H.A. O + X–S.B. (11)

Although many nucleophiles dehalogenate α-halocarbonyl compounds, the most generally used have been halides, especially iodide ion, thiols and other nucleophilic sulfur species, and phosphines. All of these are effective by themselves, but all benefit by addition of an electrophile to the medium to assist cleavage by coordination to the carbonyl (equation 11).

4.8.2.2.1 Reductive cleavages by halide salts

Historically, sodium and potassium iodides were the first halide salts found to dehalogenate halo ketones. The reductions are carried out in acidic media, in order to protonate the ketone.[27] The acyclic bromine atom in (**15**) could be removed in this manner without affecting either the acetoxy group at the same site or the sterically hindered bromine atom at C-12.[28] By using metal ions[29] or other Lewis acids[30] to assist the dehalogenation process, even bromide and chloride salts can be used to effect dehalogenation. Among the best reagents of this type is a combination of $SnCl_2$ and NaCl in aqueous THF.[29] Rather than using a mixture of two salts, the two functions can be combined into one by using an iodide salt of which the cation is a good Lewis acid. Cerium(III) iodide is particularly effective in this respect because of the high 'oxaphilicity' of lanthanide ions. This substance readily dehalogenates many halo ketones at room temperature.[31] Cerium enolates (**16**) are produced initially; these can be trapped by carbonyl compounds in the medium to form aldols or by D_2O to afford the deuterated ketone.[32]

Trimethylsilyl iodide (TMS—I), as such or generated *in situ*, is especially effective as a mild dehalogenation agent for halo ketones.[33] The reaction appears to generate the enol silyl ether initially, and this is converted to the carbonyl compound during work-up (equation 12). Both cyclic and acyclic α-bromo and α-chloro ketones are dehalogenated, generally in excellent yields. Methyltrichlorosilane, on the other

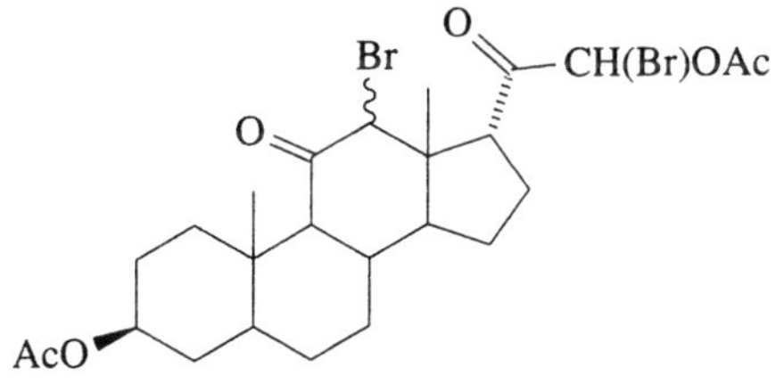

Br O CH(Br)OAc O AcO

(15) either diastereomer

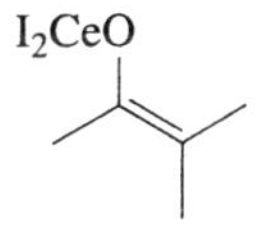

I_2CeO

(16)

hand, reacts only slowly with chloro ketones and cyclic bromo ketones,[34] and might be useful for selective dehalogenation of an acyclic α-bromo ketone in the presence of one of the latter functionalities.

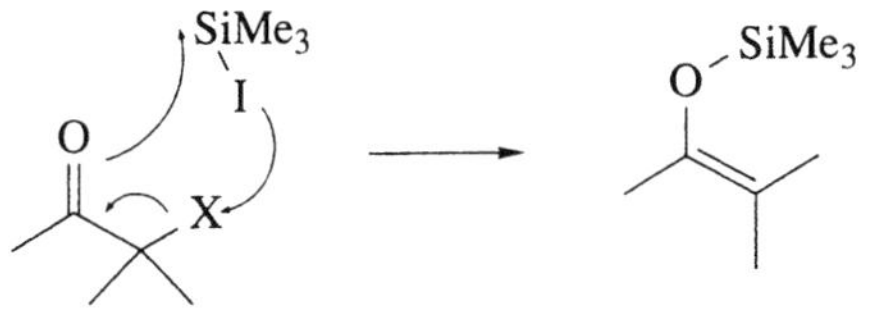

SiMe₃ I O X → O SiMe₃ (12)

4.8.2.2.2 *Reductive cleavages by sulfur species*

Thiols efficiently dehalogenate α-halo and α,α-dihalo ketones.[35] Reduction is typically carried out in ethanol containing a weak base. Although this system will also reduce α-thio ketones, it is believed that the halo ketone is reduced, and that reaction proceeds as in equation (11), with a proton from the solvent serving as Lewis acid to assist dehalogenation by thiol (or thiolate). The reactions are of course accelerated by inclusion of a metal salt in the medium.[36]

An especially interesting application of HSAB theory was reported by Fuji and coworkers.[37] They found that chloro and fluoro ketones (**17**) are converted to the reduced dithioacetals (**20**) by the combination of ethanethiol (EtSH) and $AlCl_3$ (Scheme 2). Diethyl disulfide (EtSSEt) is ineffective at promoting this transformation, whereas either EtSH or EtSSEt could be used in concert with $AlCl_3$ to convert a bromo or iodo ketone (**18**) to (**20**). It was suggested that ketones (**17**) are first converted to the corresponding halodithioacetals (**19**), which are then dehalogenated in a sense (equation 13) which is inverse to that shown in equation (11). Apparently chlorine and fluorine are too hard to permit attack by the thiol, even when assisted by $AlCl_3$. Substrates (**18**) presumably react as in equation (11), with either EtSH or EtSSEt serving as soft base, to afford the dehalogenated ketone (**21**), which is then converted to (**20**).

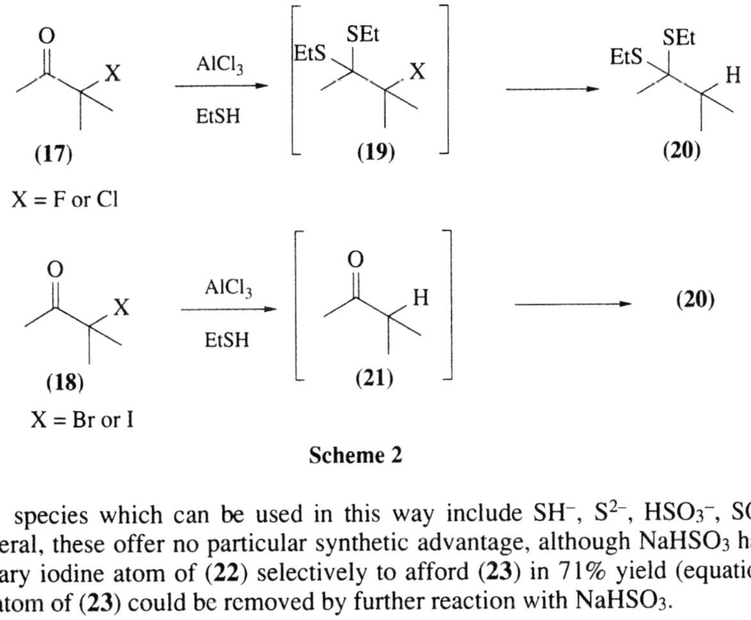

O X —$AlCl_3$/EtSH→ [EtS SEt X] → EtS SEt H

(17) X = F or Cl **(19)** **(20)**

O X —$AlCl_3$/EtSH→ [O H] → **(20)**

(18) X = Br or I **(21)**

Scheme 2

Other sulfur species which can be used in this way include SH^-, S^{2-}, HSO_3^-, SO_3^{2-}, $S_2O_3^{2-}$ and $S_2O_4^{2-}$.[38] In general, these offer no particular synthetic advantage, although $NaHSO_3$ has been found to remove the tertiary iodine atom of (**22**) selectively to afford (**23**) in 71% yield (equation 14).[39] The remaining iodine atom of (**23**) could be removed by further reaction with $NaHSO_3$.

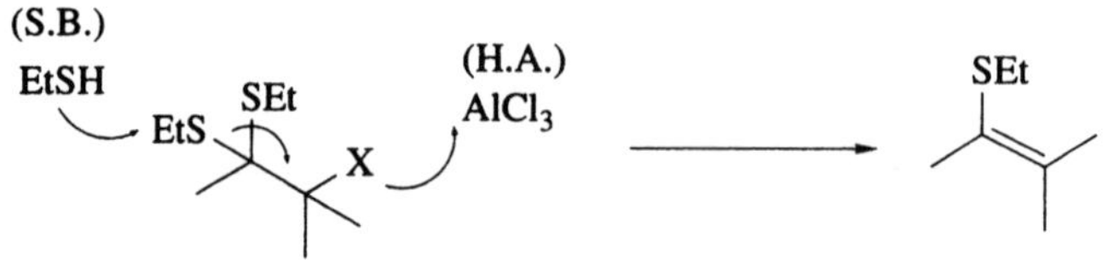

(S.B.) EtSH EtS SEt X (H.A.) $AlCl_3$ → SEt (13)

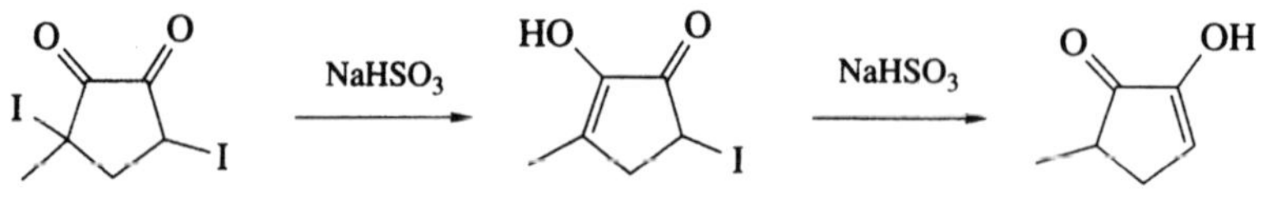

NaHSO$_3$ NaHSO$_3$ (14)

(22) (23)

4.8.2.2.3 *Reductive cleavages by other nucleophiles*

While a number of other nucleophilic agents besides those already mentioned have been found to effect reductive dehalogenation of α-halocarbonyl compounds, the only ones of synthetic interest are organic phosphines (R_3P) and organotelluride ions (RTe^-). Tertiary phosphines react with halo ketones to form enol phosphonium salts, which can be either isolated or immediately hydrolyzed to afford the carbonyl compound (equation 15).[40] A competing reaction involves direct displacement of halogen by phosphorus to form an α-ketophosphonium salt; this is especially a problem with primary halo ketones. The α,α-dichloro ketone (**24**) could be converted *via* the enol phosphonium salt (**25**) to the corresponding monohalide (**26**) or to the mixed bromo chloro ketone (**27**; Scheme 3).[40] PI_3 and P_2I_4[41] and diphenylphosphine[42] are also useful dehalogenation agents. The latter is believed to occur according to the general path described in equation (11), with phosphorus acting as a soft base and hydrogen as a hard acid.

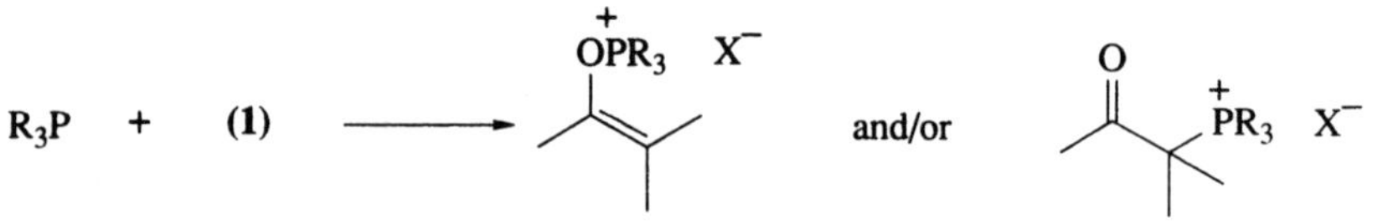

R_3P + (**1**) → $\overset{+}{O}PR_3$ X^- and/or $\overset{+}{P}R_3$ X^- (15)

(24) (25) (26) (27)

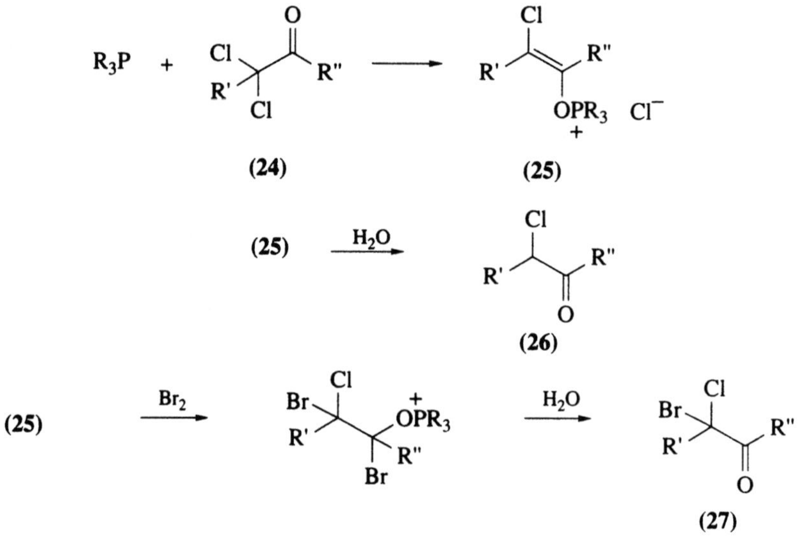

Scheme 3

The organotelluride salt (**28**) readily dehalogenates a variety of α-halocarbonyl compounds, including ketones, acids and esters.[43] It is not clear whether the first step is displacement of halide to afford an α-tellurocarbonyl compound, followed by attack upon the latter by (**28**), or whether (**28**) attacks the α-substituent directly. Reduction can be effected by $NaBH_4$ containing a catalytic quantity of (**28**).[43]

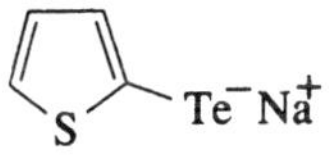

(28)

4.8.2.3 Other Reduction Methods

Transition metal carbonyls, including $Fe(CO)_5$, $Fe_2(CO)_9$, $[HFe(CO)_4]^-$, $Co_2(CO)_8$ and $Mo(CO)_6$, together with related reagents such as nickel and palladium salts, all effect dehalogenation of α-halocarbonyl compounds.[3,4,18] Yields in general are not high, and a number of competing reactions, including dimerization of the halocarbonyl compound, are sometimes observed. The reactions have sometimes been used for obtaining α-deuteriocarbonyl compounds by inclusion of D_2O in the medium.[44]

A number of tin hydrides have been shown to dehalogenate α-halocarbonyl compounds. Species which have been used include trialkyltin hydrides (R_3SnH) and dialkyltin hydrides (R_2SnH_2).[45] Dehalogenation of α-bromo ketones can be carried out under very mild conditions (–78 °C, 10 min) by a mixture of $SnCl_2$ and DIBAL-H in the presence of TMEDA.[46] The reaction probably involves SnH_2 as an intermediate. It has been reported that reduction of (**29**) by $NaBH_4$ affords (**30**), except in the presence of a trace of $Pb(OAc)_2$, $Ni(OAc)_2$ or $Hg(OAc)_2$, in which case the α-bromine is removed selectively to afford (**31**).[47] The latter reductions may involve intermediate metal hydrides.

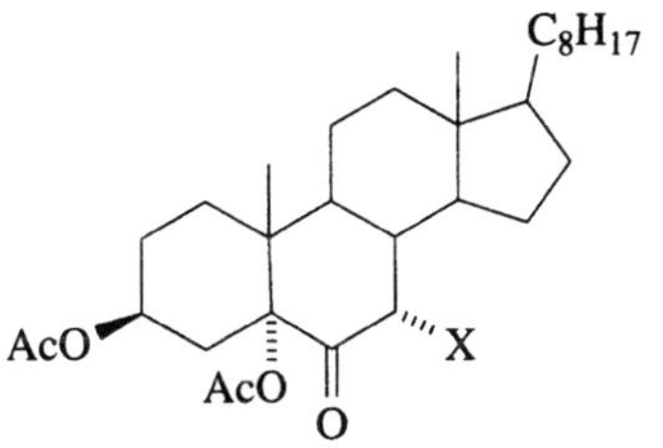

(**29**) X = Br
(**31**) X = H

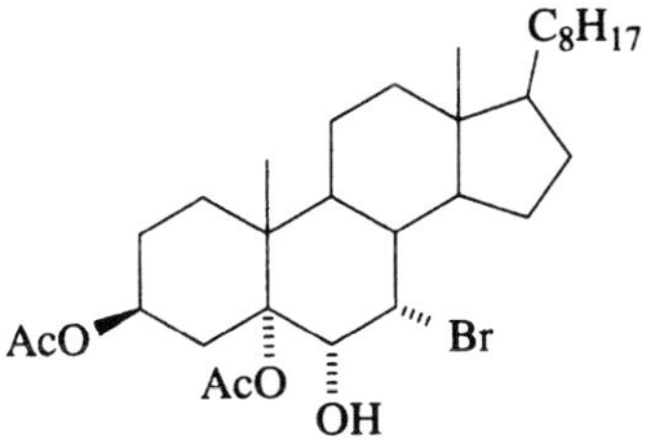

(**30**)

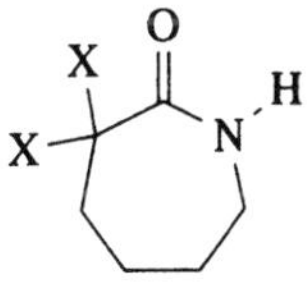

(**32**) X = Br or Cl

Finally, hydrogenolysis of the halogen atom of α-halocarbonyl compounds can be carried out at noble metal catalysts.[48] Palladium is frequently used for this purpose. Selective removal of just one of the halogen atoms in the dihalides (**32**) can be effected in NaOAc/HOAc over a palladium catalyst.[49]

4.8.3 THE CARBON–OXYGEN BOND

Carbon–oxygen bonds are generally stronger and hence harder to break than structurally related carbon–halogen bonds. Since bond cleavage involves ejection of the α-oxygen as an anion, the ease of bond cleavage is correlated with the leaving group ability of the oxygenated function; thus, the ease of reducibility decreases in the sequence: sulfonate > acyloxy > hydroxy or alkoxy.[5] Reductive cleavage is also faster in acidic media, where the neutral leaving group can be ejected.

4.8.3.1 Reduction By Electron Transfer

4.8.3.1.1 Reductive cleavages by metals

Active metals readily cleave α-ketols and α-ketol acetates. Reductions can be carried out with the aid of lithium, barium or calcium in ammonia,[50] or with zinc or tin, which are usually used in acidic media.[51]

Zinc is a relatively mild reducing agent and is therefore somewhat selective. Axial steroidal ketol acetates are reduced more readily than equatorial (equations 16 and 17).[51] On the other hand, metal–ammonia systems are powerful reductants; thus, calcium and barium reduce axial and equatorial isomeric ketol acetates with equal ease.[50] Lithium is a more powerful reductant and frequently overreduces α-ketols

(Section 4.8.3.1.3).[5] The phenacyl group has been suggested as a protecting group for acids and phenols.[52] It is readily removed by zinc in acetic acid at room temperature.

Zn, HOAc, 64% (16)

no reaction (17)

4.8.3.1.2 *Reductive cleavages by low-valent metal ions*

The more negative potentials characteristic of α-oxygenated substrates compared to their halogenated counterparts call for relatively powerful reductants such as chromium(II), titanium(II) and samarium(II) to effect α-cleavage, although reduction can be facilitated in a series of related substances by increasing the leaving group ability of the oxygenated function (Section 4.8.3).

α-Ketols are not reduced efficiently by low-valent metal ions. One of the few reports of this conversion was made by McMurry and coworkers, who found that 1-acetyl-1-cyclohexanols are converted to the corresponding ketones by a mixture of $TiCl_4$ and zinc (in which the active reagent is probably Ti^{II}).[53] Unfortunately, the reaction is not general; other α-ketols afford rearranged products under the same conditions. Deoxygenation of ketols generally requires that the hydroxy group be converted to a derivative such as an ester or sulfonate before reduction. Even so, chromium(II) requires relatively vigorous conditions to effect reduction.[54] The major exception to this generalization is the reduction of α,β-epoxy ketones, which can be carried out to produce the corresponding β-ketols (aldols; equation 18) by chromium(II) under mild conditions, driven in this case by relief of strain in the three-membered ring.[50] Reduction affords the corresponding α,β-unsaturated carbonyl compound if reduction is carried out under more highly acidic conditions. Samarium(II) reduces a variety of α-oxygenated ketones very readily at –78 °C in THF/MeOH in excellent yields, again except for α-ketols.[22a] These, however, could be reduced cleanly by adding a solution of the ketol and Ac_2O to SmI_2, in order to form and reduce the ketol acetate *in situ*. α-Oxygenated esters, including even α-hydroxy esters, can be reduced by SmI_2 in THF/HMPA at room temperature.[22b]

Cr^{II}; H^+, H_2O (18)

4.8.3.1.3 *Electrochemical reductive cleavages*

The electrochemical reduction of α-oxygenated carbonyl compounds has been reviewed.[5] In a series of phenacyl derivatives, $PhCOCH_2OR$, where R is acetyl, benzoyl, phenyl and hydrogen, the first two are about 0.1 V (or in energy terms a few kcal mol^{-1}; 1 cal = 4.18 J) easier to reduce than the latter two. Ring strain (**33**), use of an acidic solvent and buttressing by a second activating group (**34**) all favor the cleavage process. Electrochemical reduction of α-ketols and ketol acetates (R = H or Ac) cannot be stopped at the ketone stage (**35**), because the latter are reduced at about the same potential as the starting ketol. An overall four-electron process takes place, and the saturated alcohol (**36**) is formed (equation 19). Reduction is stereoselective: reduction of a series of steroidal ketol acetates afforded the equatorial alcohol in 90–100% yield.[5] Vinylogous α-cleavage with bond migration has been observed with cephalosporanic acid derivatives (**37**; equation 20).[55]

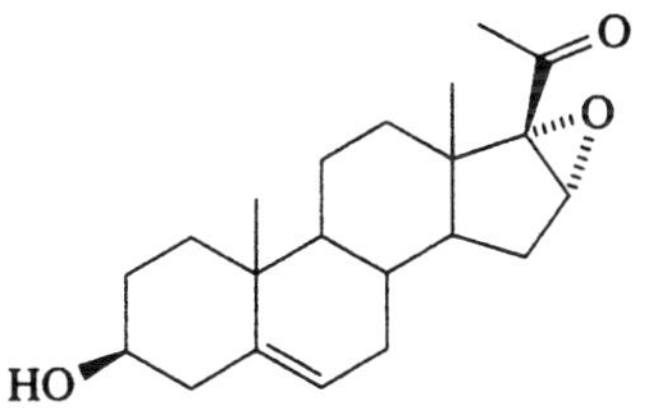

(33) 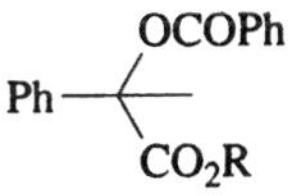 (34)

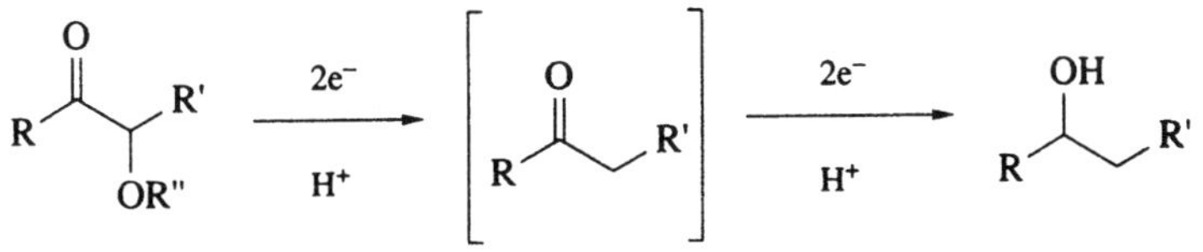

(19)

(35) (36)

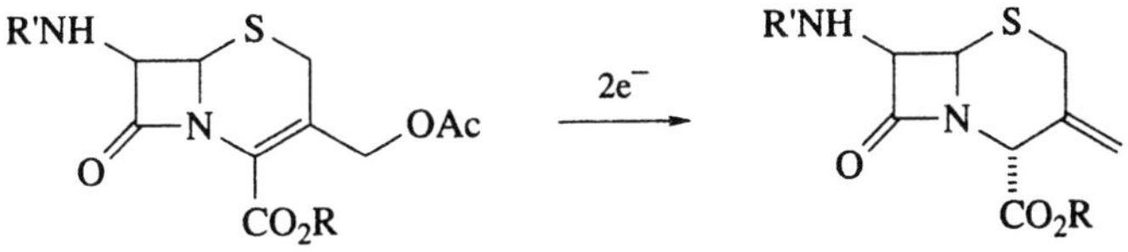

(20)

(37)

4.8.3.2 Other Reduction Methods

Although very common with α-halocarbonyl compounds, cleavage of α-oxygenated substances by nucleophiles is almost unknown. A mixture of phenyl trimethylsilyl selenide (**38**) and $MgBr_2$ dehalogenates substituted benzoin acetates under mild conditions.[56] The reaction probably proceeds by the HSAB path (equation 11); the precise function of $MgBr_2$ is unclear.

$PhSeSiMe_3$

(**38**)

Reductive cleavage of α-ketol acetates can also be effected in moderate yield by $Fe(CO)_5$.[57] Ketones, esters and alkenes are unreactive toward this reagent.

4.8.4 THE CARBON–SULFUR BOND

The following discussion includes removal of a variety of sulfur substituents, including not just alkyl- and aryl-thio (RS), but also sulfinyl (RSO), and sulfonyl (RSO_2) groups α to carbonyl groups.

4.8.4.1 Reduction By Electron Transfer

4.8.4.1.1 Reductive cleavages by metals

Alkylthio groups can be removed (equation 21) by reaction of the α-alkylthio ketone with either zinc in wet ether in the presence of TMS–Cl,[58] lithium in amine solvents or in THF containing an electron mediator such as naphthalene or trimesitylborane,[59] or $Na(Hg)_x$ in methanol buffered with Na_2HPO_4.[60] The enolate formed by Li reduction in NH_3 can be alkylated (equation 22).[61]

(21)

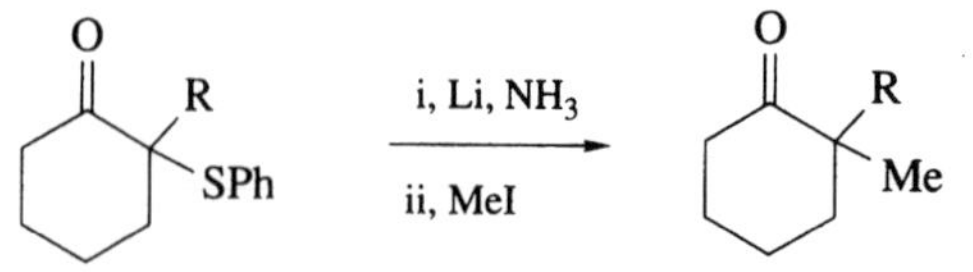

(22)

Sulfinyl and sulfonyl groups are easier to reduce than α-(alkylthio)carbonyl compounds. Aluminum amalgam has frequently been used for removal of sulfinyl and sulfonyl groups.[62] This reagent is typically used in refluxing (wet) THF; reduction of β-keto sulfones can be carried out at room temperature by $Na(Hg)_x$, if the solution is buffered with Na_2HPO_4.[60]

Phenacylsulfonamides (**39**) are readily reduced by zinc in acetic acid. The phenacylsulfonamido group has therefore been suggested as a protecting group for primary and secondary amines (equation 23), as well as a temporary acidifying group to facilitate conversion of primary to secondary amines.[63]

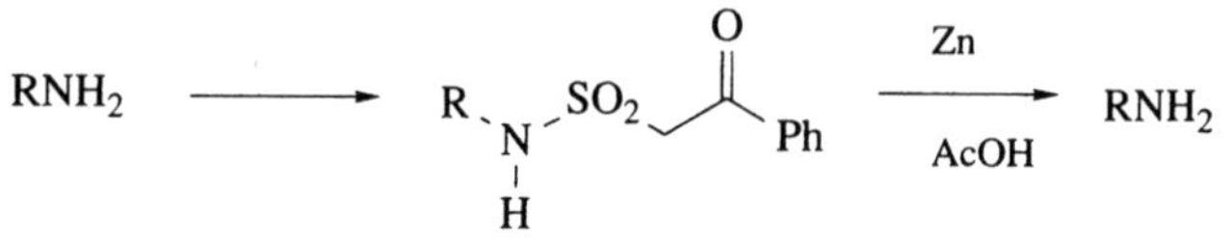

(23)

(39)

4.8.4.1.2 Reductive cleavages by low-valent metal ions

The potent reductant SmI_2 readily reduces α-alkylthio, α-sulfinyl and α-sulfonyl ketones at –78 °C.[22] A mixture of iron(II) polyphthalocyanine and thiophenol has been used to reduce α-halo, α-alkylthio, and α,α-bis(alkylthio) ketones.[64] The iron compound apparently reacts as an electron transfer mediator; the actual source of electrons is the thiophenol, which is converted to diphenyl disulfide in the course of the reaction.

4.8.4.1.3 Electrochemical reductive cleavages

α-Alkylthio groups are most readily removed electrochemically when sulfur carries a positive charge, as in phenacylsulfonium ions (**40**), or when a second activating group is present (**41**).[5] Torii reported a synthetic route to 2,2-disubstituted acetates (**42**) using the 2-benzothiazolylthio moiety as a blocking group to prevent overalkylation. The group is easily removed electrochemically after alkylation (equation 24).[65] A similar sequence has been employed using the phenylsulfonyl group as prosthetic group (equation 25).[66]

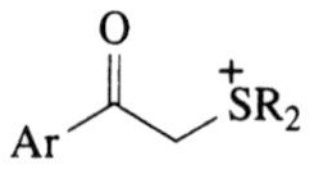

(40) 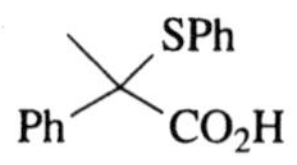 **(41)**

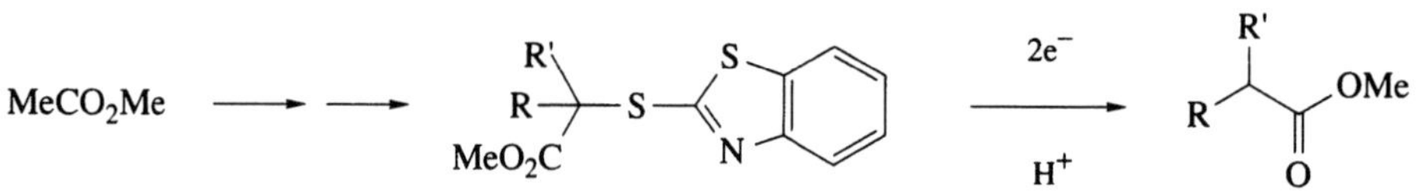

(24)

(42)

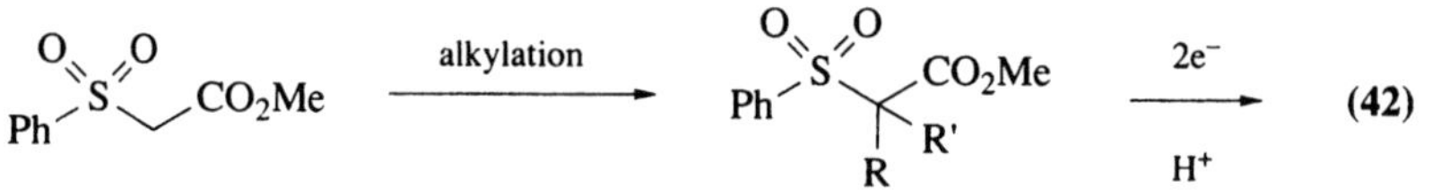

(25)

4.8.4.2 Other Reduction Methods

Thiolate ions react with α-(alkylthio)carbonyl compounds to afford disulfides and the corresponding reduced ketone (equation 26).[67] The reaction apparently involves direct nucleophilic attack by thiolate on the sulfur atom of the alkylthio group. Other soft bases, such as cyanide ion, thiourea and tertiary phosphines, also effect this conversion.[67] Raney nickel of course readily desulfurizes α-alkylthiocarbonyl compounds. The reaction is quite selective; for example, the ester, ketone and alkenic moieties of (**43**) are unaffected by the Raney nickel treatment (equation 27).[68] Raney nickel reduction of (**44**) is reported to proceed with retention of configuration in ethanol and with inversion in acetone.[69] Telluride salts also desulfurize α-alkylthio ketones.[43]

O SR + R'S⁻ ⟶ O H + RSSR' (26)

O S-*p*-Tol Et CO_2Me (**43**) —Ni⟶ O Et CO_2Me (27)

Ph—C(CONH$_2$)(SO$_2$Ph)

(**44**)

4.8.5 THE CARBON–NITROGEN BOND

Carbon–nitrogen bonds are harder to cleave than carbon–oxygen bonds. Cleavage generally requires a powerful reductant. The nitrogen atom should carry a positive charge, or the leaving group should be highly stabilized, preferably both of these.[5] Reduction of quaternary ammonium ions, the Emde reduction for example, is generally most efficient where one of the four groups is benzylic.

4.8.5.1 Electrochemical Reduction

Electrochemical reductive cleavage of α-amino ketones becomes easier as the acidity of the medium is increased, indicating that they are reduced as their conjugate acids.[5] As with α-ketols (Section 4.8.3.1.3), reductive cleavage of the carbon–heteroatom bond is frequently accompanied by reduction of the carbonyl group (equation 28).[5]

O N —4e⁻, Pb⟶ OH N H (28)

4.8.6 THE CARBON–CARBON BOND

Indanedione dehydro dimers (**45**) undergo facile carbon–carbon bond cleavage (equation 29), either electrochemically,[5] or by a variety of chemical reductants such as Pt/H_2, Na/NH_3 or Zn/HCl.[70] Presumably the process is markedly assisted here by the production of two β-diketonate anions. By the same token, acyclic 1,1,2,2-tetracarboxylates (**46**) and the related cyclic derivatives (**47**) are readily reduced to the corresponding malonates and linear tetraesters in 95–100% yield (equations 30–31).[5]

(**45**) → 2 [indanedione-Ar anion] (29)

$$\text{(46)} \xrightarrow[2H^+]{2e^-} 2\ CH_2X_2 \quad (30)$$

(**47**) $X = CO_2Me$ → (n = 1 or 2) [$2e^-$, $2H^+$] (31)

4.8.7 THE CARBON–SELENIUM BOND

Cleavage of carbon–selenium bonds α to carbonyl groups follows similar principles to those obeyed by the corresponding carbon–sulfur bonds, except that milder conditions are required, as befits the lower bond strength of the carbon–selenium bond. Thus cleavage is very facile if selenium carries a positive charge (**48** → **49**; equation 32)[71] or if the acidity of the leaving group is enhanced, as in the aryl selenide (**50**).[72] Nickel boride, used in the latter reaction, has been shown to reductively remove selenium in the presence of sulfur.[72]

(**48**) —Mg^0→ (**49**) (32)

(**50**)

4.8.8 REFERENCES

1. H. Lund, *Acta Chem. Scand.*, 1960, **14**, 1927.
2. A. J. Waring, in 'Comprehensive Organic Chemistry', ed. D. H. R. Barton and W. D. Ollis, Pergamon Press, Oxford, 1979, vol. 1, p. 1017.
3. A. R. Pinder, *Synthesis*, 1980, 425.
4. D. Perez, N. Greenspoon and E. Keinan, *J. Org. Chem.*, 1987, **52**, 5570.
5. A. J. Fry, 'Synthetic Organic Electrochemistry', 2nd edn., Wiley, New York, 1989, chap. 5.
6. A. J. Fry, *Top. Curr. Chem.*, 1972, **34**, 1.

7. Ref. 5, p. 95.
8. H. O. House, 'Modern Synthetic Reactions', 2nd edn., Benjamin, Menlo Park, 1972, chap. 3.
9. A. J. Fry and D. Herr, *Tetrahedron Lett.*, 1978, 1721.
10. L. S. Trzupek, E. R. Stedronsky and G. M. Whitesides, *J. Org. Chem.*, 1972, **37**, 3300.
11. E. J. Corey and G. A. Gregoriou, *J. Am. Chem. Soc.*, 1959, **81**, 3127.
12. R. R. Sauers and C. K. Hu, *J. Org. Chem.*, 1971, **36**, 1153.
13. G. Stork and F. H. Clarke, Jr., *J. Am. Chem. Soc.*, 1961, **83**, 3114.
14. T. A. Spencer, R. W. Britton and D. S. Watt, *J. Am. Chem. Soc.*, 1967, **89**, 5727.
15. G. M. Rubottom, R. C. Mott and D. S. Krueger, *Synth. Commun.*, 1977, **7**, 327.
16. S. Hashimoto, A. Itoh, Y. Kitagawa, H. Yamamoto and H. Nozaki, *J. Am. Chem. Soc.*, 1977, **99**, 4192.
17. B. Ellis and V. Petrow, *J. Chem. Soc.*, 1953, 3869.
18. R. Noyori and Y. Hayakawa, *Org. React. (N.Y.)*, 1983, **29**, 163.
19. J. R. Hanson and E. Premuzic, *Angew. Chem., Int. Ed. Engl.*, 1968, **7**, 247.
20. T.-L. Ho and C. M. Wong, *Synth. Commun.*, 1973, **3**, 237.
21. T.-L. Ho and G. A. Olah, *Synthesis*, 1976, 807.
22. (a) G. A. Molander and G. Hahn, *J. Org. Chem.*, 1986, **51**, 1135; (b) K. Kusuda, J. Inanaga and M. Yamaguchi, *Tetrahedron Lett.*, 1989, **30**, 2945.
23. R. M. Evans, J. C. Hamlet, J. S. Hunt, P. G. Jones, A. G. Long, J. F. Oughton, L. Stephenson, T. Walker and B. M. Wilson, *J. Chem. Soc.*, 1956, 4356.
24. A. J. Fry and M. Susla, *J. Am. Chem. Soc.*, 1989, **111**, 3225.
25. R. Annino, R. E. Erickson, J. Michalovic and B. McKay, *J. Am. Chem. Soc.*, 1966, **88**, 4424.
26. R. G. Pearson, 'Hard and Soft Acids and Bases', Dowden, Hutchinson and Ross, Stroudsburg, PA, 1973.
27. A. L. Gemal and J.-L. Luche, *Tetrahedron Lett.*, 1980, **21**, 3195.
28. G. A. Fleischer and E. C. Kendall, *J. Org. Chem.*, 1951, **16**, 573.
29. A. Ono, J. Kamimura and N. Suzuki, *Synthesis*, 1987, 406.
30. J. M. Townsend and T. A. Spencer, *Tetrahedron Lett.*, 1971, 137.
31. T.-L. Ho, *Synth. Commun.*, 1979, **9**, 241.
32. S. Fukuzawa, T. Tsuruta, T. Fujinami and S. Sakai, *J. Chem. Soc., Perkin Trans. 1*, 1987, 1473.
33. G. A. Olah, M. Arvanaghi, and Y. D. Vankar, *J. Org. Chem.*, 1980, **45**, 3531.
34. G. A. Olah, A. Husain, B. P. Singh, and A. K. Mehrotra, *J. Org. Chem.*, 1983, **48**, 3667.
35. R. Seshadri, W. J. Pegg and M. Israel, *J. Org. Chem.*, 1981, **46**, 2596.
36. A. Ono, E. Fujimoto and M. Ueno, and *Synth. Commun.*, 1986, **16**, 653.
37. K. Fuji, M. Node, T. Kawabata and M. Fujimoto, *Chem. Lett.*, 1984, 1153.
38. A. Ono, T. Maruyama and J. Kamimura, *Synthesis*, 1987, 1093.
39. G. Hesse and K. Mix, *Chem. Ber.*, 1959, **92**, 2427.
40. H. Hoffmann and H. J. Diehr, *Tetrahedron Lett.*, 1962, 583.
41. J. N. Denis and A. Krief, *Tetrahedron Lett.*, 1981, **22**, 1431.
42. I. J. Borowitz, K. C. Kirby, Jr., P. E. Rusek and E. Lord, *J. Org. Chem.*, 1969, **34**, 2687.
43. L. Engman and M. P. Cava, *J. Org. Chem.*, 1982, **47**, 3946.
44. R. Noyori, Y. Hayakawa, M. Funakura, H. Takaya, S. Murai, R. Kobayashi and S. Tsutsumi, *J. Am. Chem. Soc.*, 1972, **94**, 7202.
45. H. G. Kuivila and L. W. Menapace, *J. Org. Chem.*, 1963, **28**, 2165.
46. T. Oriyama and T. Mukaiyama, *Chem. Lett.*, 1984, 2069.
47. T. Goto and Y. Kishi, *Tetrahedron Lett.*, 1961, 513.
48. R. L. Augustine, in 'Catalytic Hydrogenation', Dekker, New York, 1965.
49. R. J. Wineman, E. T. Hsu and C. E. Anagnostopoulos, *J. Am. Chem. Soc.*, 1958, **80**, 6233.
50. C. Djerassi, 'Steroid Reactions', Holden–Day, San Francisco, 1963.
51. R. S. Rosenfeld and T. F. Gallagher, *J. Am. Chem. Soc.*, 1955, **77**, 4367.
52. J. B. Hendrickson and C. Kandall, *Tetrahedron Lett.*, 1970, 343.
53. J. E. McMurry, M. G. Silvestri, M. P. Fleming, T. Hoz and M. W. Grayston, *J. Org. Chem.*, 1978, **43**, 3249.
54. B. M. Trost, S. A. Godleski and J. Ippen, *J. Org. Chem.*, 1978, **43**, 4559.
55. M. Ochiai, O. Aki, A. Morimoto, T. Okada, K. Shinozaki and Y. Asaki, *J. Chem. Soc., Perkin Trans. 1*, 1974, 258.
56. S. I. Pennanen, *Synth. Commun.*, 1988, **18**, 1097.
57. S. J. Nelson, G. Detre and M. Tanabe, *Tetrahedron Lett.*, 1973, 447.
58. S. Kurozumi, T. Toru, M. Kobayashi and S. Ishimoto, *Synth. Commun.*, 1977, **7**, 427.
59. S. Kamata, S. Uyeo, N. Haga and W. Nagata, *Synth. Commun.*, 1973, **3**, 265.
60. B. M. Trost, H. C. Arndt, P. E. Strege and T. R. Verhoeven, *Tetrahedron Lett.*, 1976, 3477.
61. R. M. Coates, H. D. Pigott and J. Ollinger, *Tetrahedron Lett.*, 1974, 3955.
62. P. G. Gassman and G. D. Richmond, *J. Org. Chem.*, 1966, **31**, 2355.
63. J. B. Hendrickson and R. J. Bergeron, *Tetrahedron Lett.*, 1970, 345.
64. H. Inoue, T. Nagata, H. Hata and E. Imoto, *Bull. Chem. Soc. Jpn.*, 1979, **52**, 469.
65. S. Torii, H. Okumoto and H. Tanaka, *J. Org. Chem.*, 1980, **45**, 1330.
66. B. Lamm and K. Ankner, *Acta Chem. Scand., Ser. B*, 1978, **32**, 193.
67. M. Oki, W. Funakoshi and A. Nakamura, *Bull. Chem. Soc. Jpn.*, 1971, **44**, 828; M. Oki and W. Funakoshi, *Bull. Chem. Soc. Jpn.*, 1971, **44**, 832.
68. G. H. Posner and E. Asirvatham, *J. Org. Chem.*, 1985, **50**, 2589.
69. R. A. Grimm and W. A. Bonner, *J. Org. Chem.*, 1967, **32**, 3470.
70. F. M. Beringer, S. A. Galton and S. J. Huang, *Tetrahedron*, 1963, **19**, 809.
71. M. Hori, T. Kataoka, H. Shimizu and K. Tsutsumi, *Tetrahedron Lett.*, 1989, **30**, 981.
72. T. G. Back, V. I. Birss, M. Edwards and M. V. Krishna, *J. Org. Chem.*, 1988, **53**, 3815.

Author Index

This Author Index comprises an alphabetical listing of the names of over 7000 authors cited in the references listed in the bibliographies which appear at the end of each chapter in this volume.

Each entry consists of the author's name, followed by a list of numbers, each of which is associated with a superscript number. For example

Abbott, D. E., 6[12,12c], 10[40], 573[53,54]

The numbers indicate the text pages on which references by the author in question are cited; the superscript numbers refer to the reference number in the chapter bibliography. Citations occurring in the text, tables and chemical schemes and equations have all been included.

Although much effort has gone into eliminating inaccuracies resulting from the use of different combinations of initials by the same author, the use by some journals of only one initial, and different spellings of the same name as a result of transliteration processes, the accuracy of some entries may have been affected by these factors.

Subject Index